clinically oriented Anatomy

FOURTH EDITION

Introduction
1 / Thorax
2 / Abdomen
3 / Pelvis and Perineum
4 / Back
5 / Lower Limb
6 / Upper Limb
7 / Head
8 / Neck
9 / Cranial Nerves
Index

clinically oriented Anatomy

FOURTH EDITION

Keith L. Moore, Ph.D., F.I.A.C., F.R.S.M.
Professor Emeritus of Anatomy and Cell Biology
Faculty of Medicine, University of Toronto
Toronto, Ontario, Canada
Professor of Anatomy and Cell Science
Department of Anatomy, Faculty of Medicine
University of Manitoba
Winnipeg, Manitoba,Canada

Arthur F. Dalley II, Ph.D.
Professor, Department of Cell Biology
Vanderbilt University School of Medicine
Nashville, Tennessee
Formerly Professor, Department of Biomedical Science
Creighton University School of Medicine
Omaha, Nebraska

with the developmental assistance and dedication of
Lisa S. Donohoe, B.A.
Marion E. Moore, B.A.

LIPPINCOTT WILLIAMS & WILKINS
A **Wolters Kluwer** Company
Philadelphia • Baltimore • New York • London
Buenos Aires • Hong Kong • Sydney • Tokyo

Editor: Paul J. Kelly
Managing Editor: Crystal Taylor
Marketing Manager: Christine Kushner
Development Editor: Lisa S. Donohoe
Illustrator: J/B Woolsey Associates
Production Editor: Karen Ruppert

351 West Camden Street
Baltimore, Maryland 21201–2436 USA

227 East Washington Square
Philadelphia, Pennsylvania 19106

Printed in Canada

First Edition, 1980
Second Edition, 1985
Third Edition, 1992

Library of Congress Cataloging-in-Publication Data

Moore, Keith L.
Clinically oriented anatomy / Keith L. Moore, Arthur F. Dalley II; with the developmental assistance and dedication of Lisa S. Donohoe, Marion E. Moore. — 4th ed.
p. cm.
Includes bibliographical references and index.
ISBN 0-683-06141-0 (alk. paper)
1. Human anatomy. I. Dalley, Arthur F. II. Title.
[DNLM: 1. Anatomy. QS 4 M822c 1999]
QM23.2.M67 1999
611—dc21
DNLM/DLC
for Library of Congress 98-48704
CIP

To purchase additional copies of this book, call our customer service department at **(800) 638–3030** or fax orders to **(301) 824–7390.** International customers should call **(301) 714–2324.**

***Visit Lippincott Williams & Wilkins on the Internet*:** **http://www.lww.com.** Lippincott Williams & Wilkins customer service representatives are available from 8:30 am to 6:00 pm, EST, Monday through Friday, for telephone access.

03
6 7 8 9 10

Dedicated to the memory of

John Charles Boileau Grant

(1886–1973)

MC, MB, CHB, Hon DSC (Man.), FRCS (Edin.)

Professor J.C. Boileau Grant graduated from Edinburgh University in 1908 and became one of the most renowned teachers of human anatomy in the world. His books *Grant's Method of Anatomy, Grant's Atlas of Anatomy*, and *Grant's Dissector* are still widely used. Some of the classic drawings from his *Atlas* appear in this book. Professor Grant was Professor and Head of Anatomy at the University of Manitoba from 1919 to 1930, after which he went to the University of Toronto. He retired as Professor and Chairman of Anatomy in Toronto in 1959 and for several years was Visiting Professor of Anatomy at the University of California, Los Angeles School of Medicine. The above painting of Professor Grant hangs in the anatomy lobby of the Medical Sciences Building of the University of Toronto Faculty of Medicine, near the *J.C.B. Grant Museum of Anatomy,* which contains many specimens prepared under his careful scrutiny. His books and the museum serve as memorials to this outstanding teacher of anatomy.

To Marion

my best friend, wife, and colleague

for her love, unconditional support, and understanding • (KLM)

To Muriel

my bride, best friend, counselor, and mother to

Tristan, Denver, and Skyler, our sons,

with love and appreciation for their support and understanding,

good humor, and—most of all—patience • (AFD)

To Students

You will remember some of what you hear,

much of what you read,

more of what you see,

and almost all of what you experience and understand fully.

Preface

Two decades have passed since the manuscript for the first edition of *Clinically Oriented Anatomy* was submitted; seven years have passed since the third edition was published. This long interval resulted from a decision to make this book even more student friendly by preparing a *major revision* with significant new changes in the text–art program. This emphasis is also reflected in the modern design of the book.

Clinical Emphasis

Clinically Oriented Anatomy has been widely acclaimed for the relevance of its clinical correlations. As in previous editions, the fourth edition places clinical emphasis on anatomy that is important in general practice, diagnostic radiology, emergency medicine, and general surgery. Special attention has been directed toward assisting students in learning the anatomy they will need to know in the 21st century, and to this end new features have been added and existing features updated.

Medical Imaging and Case Studies. Each chapter, except the last, ends with sections on medical imaging and case studies accompanied by clinicoanatomical problems. The medical imaging displays, which pull together various combinations of radiographs, MRIs, CTs, correlative line art, and explanatory text, are new to this edition and should help prepare future professionals who need to be familiar with diagnostic images.

Surface Anatomy. Also a new feature, these special sections, identified by screened text, have been created because of anatomy's relationship to physical examination and diagnosis.

Clinical Correlations. Popularly known as the "blue boxes," the clinical information has grown, and much of it is now supported by photographs and/or dynamic color illustrations to help with understanding the practical value of anatomy. Each clinical blue box now appears with titles, and there is a complete list of titles in the book's front matter to help users locate information.

Extensive New Art Program. The art program in this book has undergone a major revision: many of the clinical conditions are now supported by photographs and/or color illustrations; multipart illustrations often combine dissections, line art, and medical images; most tables appear in color and are illustrated to aid the student's understanding of the structures described; and the text and illustrations have been developed to work together for optimum pedagogical effect, aiding the learning process and reducing the amount of searching required to find structures.

Terminology. The English terminology herein adheres to the new *Terminologia Anatomica* (1998), approved by the International Federation of Associations of Anatomists (IFAA), which includes terms such as *deep popliteal nodes* currently used in English-speaking countries and *nodi profundi poplitei* used in Europe, Asia, and other parts of the world. Eponyms, although not endorsed by the IFAA, appear in parentheses in this book—sternal angle (angle of Louis), for example—to assist students who will hear eponymous terms during their clinical studies.

Introduction. Students from many countries have written to express their views of this book—gratifyingly, mostly congratulatory. A common student plea was to strengthen the introductory chapter of the book, especially the description of the nervous system. The nervous system, often overwhelming for a beginning student, is one of the most clinically relevant aspects of first-year courses. In response, the introduction has been completely rewritten and the number of illustrations more than tripled. This chapter presents information and concepts that will prepare students to study the regional anatomy in the chapters that follow.

Summary of Key Features

Most of these features have already been discussed, but they are listed here so that they may be seen at a glance.

- More than 500 new full-color illustrations
- Multipart illustrations combining dissections, line art, and medical images such as MRIs
- Illustrated and multicolored tables
- Expanded clinical correlations ("blue boxes")
- Screened text highlighting surface anatomy
- Displays at the end of every chapter focusing on diagnostic imaging
- Boldface type to highlight key terms

Addition to Authorship

The fourth edition welcomes **Dr. Arthur F. Dalley II** as a co-author. Dr. Dalley is Professor of Cell Biology at Vanderbilt University School of Medicine in Nashville, Tennessee (formerly Professor of Anatomy and Director of Gross Anatomy at Creighton University School of Medicine in Omaha, Nebraska). Dr. Dalley was awarded the *Dedicated Teacher Award* by Creighton University in 1991 and has received the American Medical Student Association *Golden Apple Award* ten times. In 1998 he received the Creighton University School of Medicine's *Distinguished Continuing Medical Educator Award*—a first for a basic scientist at Creighton. Along with his teaching responsibilities, Dr. Dalley currently holds the following positions: President of the American Association of Clinical Anatomists; consulting editor of *Netter's Atlas of Human Anatomy*; consultant in gross anatomy for *Stedman's Medical Dictionary*; and associate editor of *Clinical Anatomy*.

Commitment to Educating Students

This book is written for health science students, keeping in mind those who may not have had a previous acquaintance with anatomy. Dr. Dalley and I have tried to present the material in an interesting way so that it can be easily integrated with what will be taught in more detail in other disciplines such as Physical Diagnosis, Medical Rehabilitation, and Surgery. We hope that this text will serve two purposes: to educate and to excite. If students develop enthusiasm for clinical anatomy, the goals of this book will have been fulfilled.

Keith L. Moore
University of Toronto
Faculty of Medicine

Acknowledgments

We wish to thank the following colleagues who were invited by the publisher to assist with the development of this fourth edition through their critical analysis and review of an initial draft of the manuscript.

- *Erle K. Adrian Jr*, PhD, Professor and Deputy Chairman, Department of Cellular & Structural Biology, University of Texas Health Sciences Center at San Antonio
- *Edward T. Bersu*, PhD, Associate Professor, Department of Anatomy, University of Wisconsin Medical School, Madison
- *William D. Davenport Jr*, PhD, Associate Professor of Oral Pathology, Associate Professor of Anatomy, Director, Research Histology Laboratory, Coordinator of Support Technologies, Louisiana State University School of Dentistry
- *David Dean*, PhD, Assistant Professor, Department of Neurological Surgery, Case Western Reservc University
- *Richard L. Drake*, PhD, Professor and Vice Chairman, Department of Cell Biology, Neurobiology, and Anatomy, University of Cincinnati Medical Center
- *Andrew Evan*, PhD, Professor of Anatomy, Indiana University School of Medicine
- *Virginia L. Naples,* PhD, Associate Professor of Biological Sciences, Northern Illinois University
- *Sharon C. Oberg*, PhD, Associate Professor, Department of Cell Biology and Anatomy, Medical University of South Carolina
- *Bruce A. Richardson*, PhD, Professor, Division of Basic Medical Sciences, California College of Podiatric Medicine, Touro University College of Osteopathic Medicine
- *William J. Swartz*, PhD, Professor, Department of Cell Biology and Anatomy, Louisiana State University School of Medicine

Several students—who have since graduated—were also invited by the publisher to review an initial draft of the manuscript:

- *Anna Bloxham*, MD, Yale University School of Medicine
- *Sharon Liu*, DO, Philadelphia College of Osteopathic Medicine
- *Anna Monias*, MD, Mt. Sinai Medical School
- *Stacie B. Peddy*, MD, University of Maryland School of Medicine
- *Kara M. Villareal*, MD, University of Arizona College of Medicine

In addition to reviewers, many people—some of them unknowingly—helped us by perusing parts of the manuscript and/or providing constructive criticism of the text and illustrations in the third edition:

- *Dr. Peter Abrahams*, Consultant Clinical Anatomist, University of Cambridge and examiner to the Royal College of Surgeons of Edinburgh
- *Dr. Robert D. Acland*, Professor of Surgery/Microsurgery, Division of Plastic and Reconstructive Surgery, University of Louisville

- *Dr. Anne Agur*, Associate Professor of Anatomy and Cell Biology, University of Toronto Faculty of Medicine
- *Dr. Anna Marie Arenson*, Assistant Professor of Medical Imaging, University of Toronto Faculty of Medicine
- *Dr. Julian J. Baumel*, Professor Emeritus of Biomedical Sciences, Creighton University School of Medicine
- *Dr. Edna Becker*, Associate Professor of Medical Imaging, University of Toronto Faculty of Medicine
- *Dr. Helen L. Block*, Attending Physician, Emergency Department, Long Island Jewish Medical Center
- *Dr. Donald R. Cahill*, Professor of Anatomy (formerly Chair), Mayo Medical School, Editor-In-Chief of *Clinical Anatomy*
- *Dr. Joan Campbell*, Assistant Professor of Medical Imaging, University of Toronto Faculty of Medicine
- *Dr. Carmine D. Clemente*, Professor of Anatomy and Orthopedic Surgery, University of California, Los Angeles School of Medicine
- *Dr. James D. Collins*, Professor of Radiological Sciences, University of California, Los Angeles School of Medicine/Center for Health Sciences
- *Dr. Raymond F. Gasser*, Professor of Anatomy, Louisiana State University School of Medicine
- *Dr. Ralph Ger*, Professor of Anatomy and Structural Biology, Albert Einstein College of Medicine; Professor of Surgery, State University of New York at Stony Brook; Associate Chairman, Department of Surgery, Nassau County Medical Center
- *Dr. Masoom Haider*, Assistant Professor of Medical Imaging, University of Toronto Faculty of Medicine
- *Dr. Duane E. Haines*, Professor and Chairman, Department of Anatomy, University of Mississippi Medical Center
- *Dr. Walter Kuchareczyk*, Professor and Chair of Medical Imaging, University of Toronto Faculty of Medicine; Clinical Director of Tri-Hospital Magnetic Resonance Centre
- *Dr. E.L. Lansdown*, Professor Emeritus of Medical Imaging, University of Toronto Faculty of Medicine
- *Dr. Michael von Lüdinghausen*, University Professor, Anatomy Institute, University of Würtzburg
- *Dr. Shirley McCarthy*, Director of MRI, Department of Diagnostic Radiology, Yale University School of Medicine
- *Dr. Marita L. Nelson*, Professor of Pathology, John A. Burns School of Medicine, University of Hawaii
- *Dr. Todd R. Olson*, Professor of Anatomy and Structural Biology, Albert Einstein College of Medicine
- *Dr. David Peck*, Associate Professor of Anatomy and Neurobiology, University of Kentucky School of Medicine
- *Dr. T.V.N. Persaud*, Professor of Human Anatomy and Cell Science Faculties of Medicine & Dentistry, University of Manitoba
- *Dr. Thomas H. Quinn*, Professor of Biomedical Sciences, Creighton University School of Medicine
- *Dr. George E. Salter*, Professor of Anatomy, Department of Cell Biology, University of Alabama at Birmingham

- *Dr. Tamiko Sato*, Associate Professor of Cell Biology and Anatomy, New York Medical College
- *Dr. Tatsuo Sato*, Professor and Head, Second Department of Anatomy, Tokyo Medical and Dental University Faculty of Medicine
- *Professor Colin P. Wendell-Smith*, Department of Anatomy and Physiology, University of Tasmania
- *Dr. Eugene J. Wenk*, Professor of Cell Biology and Anatomy, New York Medical College
- *Dr. David G. Whitlock*, Professor of Anatomy, University of Colorado Medical School

Our grateful appreciation is extended to **Marion Moore**, the senior author's wife (of 50 years), for her friendly help and patience with the computer preparation of the manuscript. She prepared many drafts of each chapter and provided tactful, constructive criticism as the "polishing" process progressed. We also wish to acknowledge the excellent work of **Lisa Donohoe**, the Developmental Editor. In addition to bringing a new perspective to the organization of the text and conceptualization of many illustrations, she developed a student-friendly text–art coordination. Art plays a major role in facilitating learning. Although the number of illustrations from *Grant's Atlas of Anatomy* has been greatly reduced and replaced by new art, we gratefully acknowledge the excellent art done by the following: *Dorothy Foster Chubb, Elizabeth Blackstock, Nancy Joy, Nina Kilpatrick, David Mazierski, Stephen Mader, Bart Vallecoccia, Sari O'Sullivan,* and *Kam Yu.*

The superb new art was prepared by *J/B Woolsey Associates*, Elkins Park, Pennsylvania. The following people worked with the authors and editors: *Art Development*—John Woolsey, Craig Durant, and Todd Smith; *Art Director*—Craig Durant assisted by Allison Cantley, Laura Colangelo, Mark Desman, Joel Dubin, Robert Fedirko Jr, Jin Ho Park, and Regina Santoro.

Many thanks also to those at Lippincott Williams & Wilkins who participated in the development of this edition: *Paul Kelly*—Acquisitions Editor; *Nancy Evans*—Editorial Director; *Crystal Taylor*—Managing Editor; *Karen Ruppert*—Production Editor; *Mike Standen*—Editorial Assistant; and *Danielle Jablonski*—Development Assistant.

Keith L. Moore
Arthur F. Dalley II

Figure Credits

Introduction

Numbered Figures

Agur AMR. Grant's Atlas of Anatomy. 9th ed. Baltimore: Williams & Wilkins, 1991. I.36, I.38

Wicke L. (Taylor AN, editor and translator) Atlas of Radiologic Anatomy. 6th English ed. Baltimore: Williams & Wilkins, 1998. A translation of: Wicke L. Roentgen-Anatomie Normalbefunde. 5th ed. Munich: Urban & Schwarzenberg, 1995. I.39 (CT), I.40 (ultrasound), I.41, I.42

Unnumbered Figures Appearing in Boxes

Agur AMR. Grant's Atlas of Anatomy. 9th ed. Baltimore: Williams & Wilkins, 1991. Newborn Skull on page 25

Roche Lexikon Medizin. 4th ed. Munich: Urban & Schwarzenberg, 1998. Varicose veins on page 35

Willis MC. Medical Terminology: The Language of Health Care. Baltimore: Williams & Wilkins, 1996. Atheromatous plaque and thrombus on page 34

Chapter 1

Numbered Figures

Agur AMR. Grant's Atlas of Anatomy. 9th ed. Baltimore: Williams & Wilkins, 1991. Thorax orientation figure used throughout chapter, 1.1, 1.2A, 1.4 (modified), 1.5, 1.7*A&B*, 1.9, 1.13, 1.14 (modified), 1.16, 1.17, 1.19, 1.26 (modified), 1.27, 1.28, 1.29, 1.30, 1.40, 1.44, 1.46, 1.47 (*B* modified), 1.49 (*B* modified), 1.53 (modified), 1.53 (modified), 1.54, 1.56, 1.58*A*, 1.59, 1.60, 1.62, 1.63, 1.64, 1.66*B&D*, 1.68*A* (photograph), 1.68*B*, 1.69, 1.71

Cahill DR, Orland MJ, Reading CC. Atlas of Human Cross-Sectional Anatomy with CT and MR Images. 2nd ed. New York: Wiley-Liss, 1990. 1.70

Clemente CD. Gray's Anatomy of the Human Body. 30th American ed. Baltimore: Williams & Wilkins, 1985. (Redrawn from Tandler, 1912.) 1.43

Unnumbered Figures Appearing in Boxes

Agur AMR. Grant's Atlas of Anatomy. 9th ed. Baltimore: Williams & Wilkins, 1991. Aortic and pulmonary valves on page 132; Aorta on page 148; Double aorta on page 149

Lippert H. Lehrbuch Anatomie. 4th ed. Munich: Urban & Schwarzenberg, 1996. Bronchoscopic view inserts on page 105

Roche Lexikon Medizin. 4th ed. Munich: Urban & Schwarzenberg, 1998. Inflamed carcinoma of the breast on page 78; Bronchial asthma on page 105; Bronchiolar carcinoma on page 112; Pericardiocentesis on page 119

Chapter 2

Numbered Figures

Agur AMR. Grant's Atlas of Anatomy. 9th ed. Baltimore: Williams & Wilkins, 1991. 2.1*B*, 2.5*A*, 2.6, 2.7, 2.12, 2.19, 2.25*A* (modified), 2.26, 2.27, 2.31*B&C*, 2.32, 2.35*B&C*, 2.36*B&C*, 2.38*C–F*, 2.43*A&B*, 2.46 (spleen), 2.47C, 2.50*B–D*, 2.51, 2.53, 2.55*B* (modified), 2.58, 2.59, 2.60, 2.61*A&B*, 2.63, 2.64, 2.66, 2.68*A*, 2.71, 2.72*B*, 2.73, 2.74*A*, 2.75, 2.76, 2.78, 2.79, 2.80, 2.81, 2.82*B&C*, 2.83, 2.84, 2.85, 2.86, 2.87

Anderson JE. Grant's Atlas of Anatomy. 8th ed. Baltimore: Williams & Wilkins, 1983. 2.38

Gartner LP, Hiatt JL. Color Atlas of Histology. 2nd ed. Baltimore: Williams & Wilkins, 1994. 2.47*D*

Haines DE. Neuroanatomy: An Atlas of Structures, Sections, and Systems. Baltimore: Williams & Wilkins, 1994. 2.41 (photographic insert)

Unnumbered Figures Appearing in Boxes

Agur AMR. Grant's Atlas of Anatomy. 9th ed. Baltimore: Williams & Wilkins, 1991. Testis on page 205; Kidneys and ureters on page 288

Roche Lexikon Medizin. 4th ed. Munich: Urban & Schwarzenberg, 1998. Posterior view of right anterior abdominal wall on page 206; Cirrhosis of the liver on page 271; Aneurysm on page 306

Willis MC. Medical Terminology: The Language of Health Care. Baltimore: Williams & Wilkins, 1996. Radiograph of kidney stones on page 322

Chapter 3

Numbered Figures

Agur AMR. Grant's Atlas of Anatomy. 9th ed. Baltimore: Williams & Wilkins, 1991. 3.3*B&C*, 3.4*A&B*, 3.5*B–E*, 3.8*A*, 3.10, 3.11, 3.13*B*, 3.14*A*, 3.15, 3.17, 3.18, 3.20, 3.23 *top*, 3.26*A*, 3.27, 3.29, 3.31, 3.32, 3.34, 3.36*A*, 3.40, 3.41*D*, 3.42, 3.43, 3.44, 3.47, 3.48, 3.49, 3.50, 3.51, 3.52*C*, 3.53, 3.55, 3.56

Figures Appearing in Tables

Moore KL, Agur AMR. Essential Clinical Anatomy. Baltimore: Williams & Wilkins, 1995. 3.1

Chapter 4

Numbered Figures

Agur AMR: Grant's Atlas of Anatomy. 9th ed. Baltimore: Williams & Wilkins, 1991. 4.1*B*, 4.5, 4.6*A*, 4.7, 4.8, 4.12, 4.15, 4.16*A*, 4.19, 4.23*A*, 4.24*A*, 4.26, 4.27, 4.30, 4.32, 4.33, 4.37, 4.38, 4.39

Wicke L. (Taylor AN, editor and translator) Atlas of Radiologic Anatomy. 6th English ed. Baltimore: Williams & Wilkins, 1998.

A translation of: Wicke L. Roentgen-Anatomie Normalbefunde. 5th ed. Munich: Urban & Schwarzenberg, 1995. 4.40, 4.41, 4.42

Roland LP. Merritt's Textbook of Neurology. 9th ed. Baltimore: Williams & Wilkins, 1995. 4.43

Unnumbered Figures Appearing in Boxes

Agur AMR: Grant's Atlas of Anatomy. 9th ed. Baltimore: Williams & Wilkins, 1991. Surface anatomy of the back on page 474

Roland LP. Merritt's Textbook of Neurology. 9th ed. Baltimore: Williams & Wilkins, 1995. Lumbar myelogram and CT scan on page 447

Chapter 5

Numbered Figures

Agur AMR. Grant's Atlas of Anatomy. 9th ed. Baltimore: Williams & Wilkins, 1991. 5.3*B*, 5.5, 5.8, 5.9, 5.13*A*, 5.14*C*, 5.15, 5.16*B*&*C*, 5.17, 5.21, 5.23, 5.26, 5.28*D*, 5.29, 5.30, 5.32, 5.33*A*&*B*, 5.34, 5.35, 5.36, 5.37*A*, 5.38, 5.39, 5.42*B*, 5.44, 5.45, 5.50, 5.51, 5.54, 5.55*A*&*C*, 5.56, 5.57, 5.58, 5.59, 5.60*B–D*, 5.64, 5.65, 5.66, 5.67, 5.73*C*, 5.74, 5.51, 5.77, 5.78

Wicke L. (Taylor AN, editor and translator). Atlas of Radiologic Anatomy. 6th English ed. Baltimore: Williams & Wilkins, 1998. A translation of: Wicke L. Roentgen-Anatomie Normalbefunde. 5th ed. Munich: Urban & Schwarzenberg, 1995. 5.3*A*, 5.70, 5.71*B*, 5.72, 5.73*A*&*B*, Orientation for 5.73*C*

Unnumbered Figures Appearing in Boxes

Adams JC, Hamblen DL. Outline of Orthopaedics. 11th ed. Edinburgh: Churchill-Livingston, 1990. Baker's cyst on page 629

Agur AMR. Grant's Atlas of Anatomy. 9th ed. Baltimore: Williams & Wilkins, 1991. Surface anatomy photographs on page 567; Illustration and photograph of lateral aspect of the leg on page 568; Surface anatomy photographs on pages 569–570; Radiograph on page 587; Surface anatomy photographs on pages 587 and 593; Illustrations and photographs of the foot on pages 643 and 644

Roche Lexikon Medizin. 4th ed. Munich: Urban & Schwarzenberg, 1998. Varicose veins on page 527; Arthrosis on page 536

Willis MC. Medical Terminology: The Language of Health Care. Baltimore: Williams & Wilkins, 1996. Hips on page 615

Yochum TR, Rowe LJ. Essentials of Skeletal Radiology. 2nd ed. Baltimore: Williams & Wilkins, 1995. March fracture MRI on page 513

Figures Appearing in Tables

Agur AMR. Grant's Atlas of Anatomy. 9th ed. Baltimore: Williams & Wilkins, 1991. 5.10, 5.11, 5.12 *Left*, 5.16 *Right*

Chapter 6

Numbered Figures

Agur AMR. Grant's Atlas of Anatomy. 9th ed. Baltimore: Williams & Wilkins, 1991. 6.2, 6.5, 6.14, 6.23*D*, 6.24*B*, 6.25, 6.26, 6.28, 6.29, 6.31, 6.32, 6.33, 6.34, 6.35, 6.36 (modified), 6.39*A–C*, 6.39*E*, 6.41, 6.42, 6.43, 6.44*A–C*, 6.47, 6.48, 6.51 (modified), 6.53*B*, 6.55, 6.57*A*&*B*, 6.58, 6.59, 6.60, 6.63, 6.65C, 6.66*B*&*D*, 6.67, 6.73, 6.74*B–D*, 6.75, 6.77*B*, 6.78*A*, 6.79, 6.80, 6.81, 6.82, 6.83

Salter RB. Textbook of Disorders and Injuries of the Musculoskeletal System. 3rd ed. Baltimore: Williams & Wilkins, 1998. 6.84, 6.85

Wicke L. (Taylor AN, editor and translator) Atlas of Radiologic Anatomy. 6th English ed. Baltimore: Williams & Wilkins, 1998. A translation of: Wicke L. Roentgen-Anatomie Normalbefunde. 5th ed. Munich: Urban & Schwarzenberg, 1995. 6.78*B*

Unnumbered Figures Appearing in Boxes

Agur AMR. Grant's Atlas of Anatomy. 9th ed. Baltimore: Williams & Wilkins, 1991. Axilla on page 694; Surface anatomy photographs (anterior and posterior views) on pages 718–719; Medial aspect of arm on page 732; Elbow, arm, and axilla on page 733; Vulnerable position of the ulnar nerve on pages 761 and 775; Surface anatomy photographs on page 779; Shoulder joint on page 793

Anderson MK, Hall SJ. Sports Injury Management. Baltimore: Williams & Wilkins, 1995. Biceps tendon rupture on pages 724 and 727; Skier's (gamekeeper's) thumb on page 810

Backhouse KM, Hutchings RT. Color Atlas of Surface Anatomy. Baltimore: Williams & Wilkins, 1986. Surface anatomy photograph on page 780

Moore KL, Agur AMR. Essential Clinical Anatomy. Baltimore: Williams & Wilkins, 1995. Posterior aspect of arm on page 732

Roland LP. Merritt's Textbook of Neurology. 9th ed. Baltimore: Williams & Wilkins, 1995. Winged scapula on page 689

Salter RB. Textbook of Disorders and Injuries of the Musculoskeletal System. 3rd ed. Baltimore: Williams & Wilkins, 1998. Separation of humeral epiphysis on pages 724 and 766; Radiograph on page 799

Figures Appearing in Tables

Agur AMR. Grant's Atlas of Anatomy. 9th ed. Baltimore: Williams & Wilkins, 1991. 6.1*E*, 6.2, 6.3, 6.4 *Bottom*, 6.6, 6.7, 6.8 *Left*, 6.9, 6.11*A*, 6.12

Chapter 7

Numbered Figures

Agur AMR. Grant's Atlas of Anatomy. 9th ed. Baltimore: Williams & Wilkins, 1991. 7.1*A*, 7.1*C*, 7.2, 7.3, 7.5*A–C*, 7.7, 7.8, 7.10, 7.11*C*, 7.15, 7.17, 7.18, 7.19, 7.20, 7.22, 7.27, 7.29, 7.30, 7.31, 7.32, 7.33, 7.35, 7.39, 7.41, 7.44, 7.45, 7.49, 7.50, 7.52, 7.53*B*, 7.54, 7.55, 7.57, 7.61, 7.63, 7.67, 7.68, 7.70, 7.71, 7.74, 7.75, 7.76, 7.77, 7.78, 7.79, 7.83, 7.84, 7.85, 7.91

Unnumbered Figures Appearing in Boxes

Agur AMR. Grant's Atlas of Anatomy. 9th ed. Baltimore: Williams & Wilkins, 1991. Mental foramen on page 837; Skulls on page 847; CT of child's head on pages 849 and 906; Photograph on page 916

Ger R, Abrahams P, Olson T. Essentials of Clinical Anatomy, 3rd ed. New York: Parthenon Publishing Group, 1996. Photograph on page 874

Leung AKC, Wong AL, Robson WLLM. Ectopic thyroid gland simulating a thyroglossal duct cyst. Can J Surg 1995;38:87. Aberrant thyroid gland on page 947

Moore KL, Persaud TVN. The Developing Human: Clinically Oriented Embryology. 6th ed. Philadelphia: W.B. Saunders Company, 1998. Cleft lip on page 929; Cleft palate on page 940

Moore KL, Persaud TVN. The Developing Human: Clinically Oriented Embryology. 6th ed. Philadelphia: W.B. Saunders Company, 1998. (Courtesy of Dr. Gerald S. Smyser, Altru Health System, Grand Forks, ND.) Sagittal MRI on page 892

Roche Lexikon Medizin, 3. Auflage, Munich, Germany: Urban & Schwarzenberg, 1998. Hyphema on page 908; Mastoiditis on page 969

Sadler TW. Langman's Medical Embryology. 7th ed. Baltimore: Williams & Wilkins, 1995. Photograph of hydrocephalus on page 892; Thyroglossal cyst on page 947

Skin Cancer Foundation, New York, NY. Squamous cell carcinoma on page 869

Smith Kline Corporation. Essentials of Neurological Examination, 1978. Testing sensory function of CN V on page 862

Welch Allen, Inc. Skaneateles Falls, NY. Retinal detachment on page 906; Otoscopic examination and tympanic membrane on page 966; Otitis media on page 969

Figures Appearing in Tables

Agur AMR. Grant's Atlas of Anatomy. 9th ed. Baltimore: Williams & Wilkins, 1991. *7.7A&B*, 7.8*A*, 7.9*A&B*, 7.12

Chapter 8

Numbered Figures

Agur AMR. Grant's Atlas of Anatomy. 9th ed. Baltimore: Williams & Wilkins, 1991. 8.2, 8.5, 8.6, 8.7, 8.8, 8.9, 8.11, 8.12, 8.13, 8.17, 8.19, 8.20*B*, 8.22, 8.23, 8.24, 8.25*A*, 8.26, 8.27, 8.28. 8.29, 8.30, 8.32, 8.34, 8.35, 8.36, 8.37 *Bottom*, 8.38*B*, 8.40, 8.41*A&B*, 8.43, 8.48, 8.49, 8.50, 8.51

Sadler TW. Langman's Medical Embryology. 7th ed. Baltimore: Williams & Wilkins, 1995. 8.53

Liebgott B. The Anatomical Basis of Dentistry. Philadelphia: BC Decker Inc., 1986. 8.38*A*

Roche Lexikon Medizin. 4th ed. Munich: Urban & Schwarzenberg, 1998. 8.39

Wicke L. (Taylor AN, editor and translator) Atlas of Radiologic Anatomy. 6th English ed. Baltimore: Williams & Wilkins, 1998. A translation of: Wicke L. Roentgen-Anatomie Normalbefunde. 5th ed. Munich: Urban & Schwarzenberg, 1995. 8.44, 8.45, 8.46, 8.47

Willis MC. Medical Terminology: The Language of Health Care. Baltimore: Williams & Wilkins, 1996. 8.52 (Doppler color flow courtesy of Hoag Memorial Hospital Presbyterian, Newport Beach, CA)

Unnumbered Figures Appearing in Boxes

Agur AMR. Grant's Atlas of Anatomy. 9th ed. Baltimore: Williams & Wilkins, 1991. Contraction of platysma on page 1022; Pyramidal lobe of the thyroid gland on page 1034; Recurrent laryngeal nerves on page 1035; Structures in the anterior neck on page 1064

Moore KL, Persaud TVN. The Developing Human: Clinically Oriented Embryology. 6th ed. Philadelphia: W.B. Saunders Company, 1998. (Courtesy of Dr. DA Kernahan, The Children's Memorial Hospital, Chicago) Radiograph on page 1060

Roche Lexikon Medizin. 4th ed. Munich: Urban & Schwarzenberg, 1998. Scintigram on page 1035

Roland LP. Merritt's Textbook of Neurology. 9th ed. Baltimore: Williams & Wilkins, 1995. Photograph on page 1003

Willis MC. Medical Terminology: The Language of Health Care. Baltimore: Williams & Wilkins, 1996. Occlusion of the carotid artery on page 1019 (color flow Doppler courtesy of Acuson Corp., Mt. View, CA)

Willms JL, Schneiderman H, Algranati PS. Physical Diagnosis: Bedside Evaluation of Diagnosis and Function. Baltimore: Williams & Wilkins, 1994. Palpation of the submandibular nodes on page 1064

Figures in Tables

Agur AMR. Grant's Atlas of Anatomy. 9th ed. Baltimore: Williams & Wilkins, 1991. 8.1*C*, 8.3*B*, 8.3*C&D* (modified), 8.5*A–F*, 8.6*A*

Chapter 9

Numbered Figures

Agur AMR. Grant's Atlas of Anatomy. 9th ed. Baltimore: Williams & Wilkins, 1991. Brain orientation figures used throughout chapter, 9.1, 9.3*A*, 9.5, 9.6*B&C*, 9.7*B*, 9.9*A*, 9.10, 9.13

Unnumbered Figures Appearing in Boxes

Roche Lexikon Medizin. 4th ed. Munich: Urban & Schwarzenberg, 1998. Oculomotor paralysis on page 1093

List of Clinical Blue Boxes

Introduction to Clinically Oriented Anatomy

Chapter 1 Thorax

Chapter 2 Abdomen

Chapter 3 Pelvis and Perineum

Chapter 4 Back

Chapter 5 Lower Limb

Chapter 6 Upper Limb

Chapter 7 Head

Chapter 8 Neck

Chapter 9 Summary of Cranial Nerves

Contents

Orange text indicates Surface Anatomy sections; green text highlights Medical Imaging and Case Studies.

3 Pelvis and Perineum

4 Back

5 Lower Limb

Orange text indicates Surface Anatomy sections; green text highlights Medical Imaging and Case Studies.

Upper Limb

Orange text indicates Surface Anatomy sections; green text highlights Medical Imaging and Case Studies.

7 Head

8 Neck

9 Summary of Cranial Nerves

Orange text indicates Surface Anatomy sections; green text highlights Medical Imaging and Case Studies.

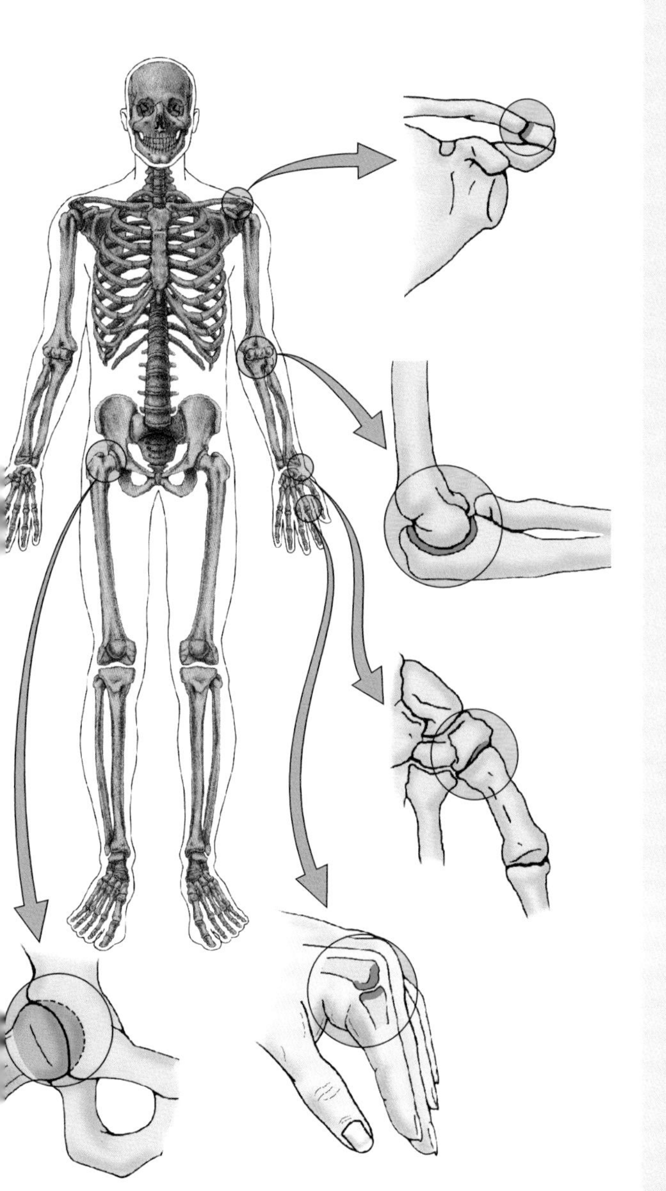

Introduction to Clinically Oriented Anatomy

Anatomy—the study of the structure and function of the body—is one of the oldest basic medical sciences; it was first studied formally in Egypt (approximately 500 BC). The earliest descriptions of anatomy were written on papyruses (paper reed) between 3000 and 2500 BC (Persaud, 1984). Much later, human anatomy was taught in Greece by *Hippocrates* (460–377 BC), who is regarded as the Father of Medicine and a founder of the science of anatomy. In addition to the *Hippocratic Oath*, Hippocrates wrote several books on anatomy. In one he stated, "The nature of the body is the beginning of medical science." *Aristotle* (384–322 BC) was the first person to use the term *anatome*, a Greek word meaning "cutting up or taking apart." The Latin word *dissecare* has a similar meaning.

Vesalius' masterpiece *De Humani Corporis Fabrica*, published in 1543, marked a new era in the history of medicine. The study of anatomy suddenly became an objective discipline based on direct observations as well as scientific principles. *Hieronymus Fabricius* (1537–1619) was responsible for the construction in 1594 of the famous anatomical theatre at Padua. He was one of the teachers of *William Harvey*, and it is believed that Fabricius' discovery of the valves in the veins led Harvey to the discovery of the circulation of blood. The publication in 1628 of Harvey's book *Exercitatio Anatomica De Motu Cordis et Sanguinis in Animalibus*, on the movements of the heart and the circulation of blood in animals, represents a milestone in the history of medicine (Persaud, 1997). By the 17th century, human dissections became an important feature in European medical schools, and anatomical museums were established in many cities.

During the 18th and 19th centuries, anatomists published impressive treatises and lavish atlases with illustrations that introduced new standards for depicting the human body. The shortage of cadavers for dissection and anatomical demonstrations led to illegal means of obtaining human bodies. Professional grave robbers supplied anatomy schools with corpses, in some cases by murdering their victims. Medical students and their teachers had also been involved in body snatching (Persaud, 1997). In Britain, the *Anatomy Act* was passed by Parliament in 1832. It made legal provisions for medical schools to receive unclaimed and donated bodies for anatomical studies. This paved the way for similar legislations in other countries.

Approaches to Studying Anatomy

This book deals mainly with *human gross anatomy*—the examination of body structures that can be seen without a microscope. The three main approaches to studying anatomy are regional, systemic, and clinical.

- **Regional anatomy** (topographical anatomy) is the method of studying the body by regions, such as the thorax and abdomen. Surface anatomy is an essential part of the study of regional anatomy. The *surface anatomy boxes* in this book provide visible knowledge of what lies under the skin and what structures are perceptible to touch (palpable)—organs such as the liver.
- **Systemic anatomy** is the method of studying the body by systems, for example, the circulatory and reproductive systems.
- **Clinical anatomy** emphasizes structure and function as they relate to the practice of medicine and other health sciences. The *clinical correlation boxes* in this book describe the practical applications of anatomy (p. 13).

These approaches to studying anatomy give a three-dimensional view of the body's structures.

Regional Anatomy

Regional anatomy is the study of the regions of the body (Fig. I.1). This approach deals with structural relationships of the parts of the body in the region under study. Most laboratory courses are based on *regional dissections* (e.g., of the thorax [chest]). The computer is a useful adjunct in teaching regional anatomy (Cahill and Leonard, 1997) because it facilitates certain aspects of instruction, such as the display of computed tomography (CT), magnetic resonance imaging (MRI), manipulation of three-dimensional anatomical renderings, and layer separations of tissues in dissections. *Prosections*—carefully prepared dissections for the demonstration of anatomical structures—are also useful; however, learning is most efficient and retention is highest when didactic study is combined with the experience of dissection (Mutyala and Cahill, 1996).

During dissection you observe, palpate, and move parts of the body. In 1770, *Dr. William Hunter*, a distinguished Scottish anatomist and obstetrician, stated:

> *Dissection alone teaches us where we may cut or inspect the living body with freedom and dispatch.*

Surface anatomy is a method for studying the anatomy of the living body at rest and in action. By observing the surface of the body and the structures under it, much can be learned. The aim of this method is to visualize—shape into distinct mental images—structures that lie beneath the skin and are palpable. In people with stab wounds, for example, a physician must be able to visualize the deep structures that might be injured. A knowledge of surface anatomy can also save inefficient memorization of facts because the body is always available to observe and palpate.

The *physical examination* of a person is the clinical application of surface anatomy. Palpation is a clinical technique for examining living anatomy. *Palpation of arterial pulses*, for instance, is part of every physical examination. You will learn to use instruments to observe parts of the body, such as an *ophthalmoscope* to observe the eyes, and to listen to functioning parts of the body (e.g., a *stethoscope* to listen to the heart and

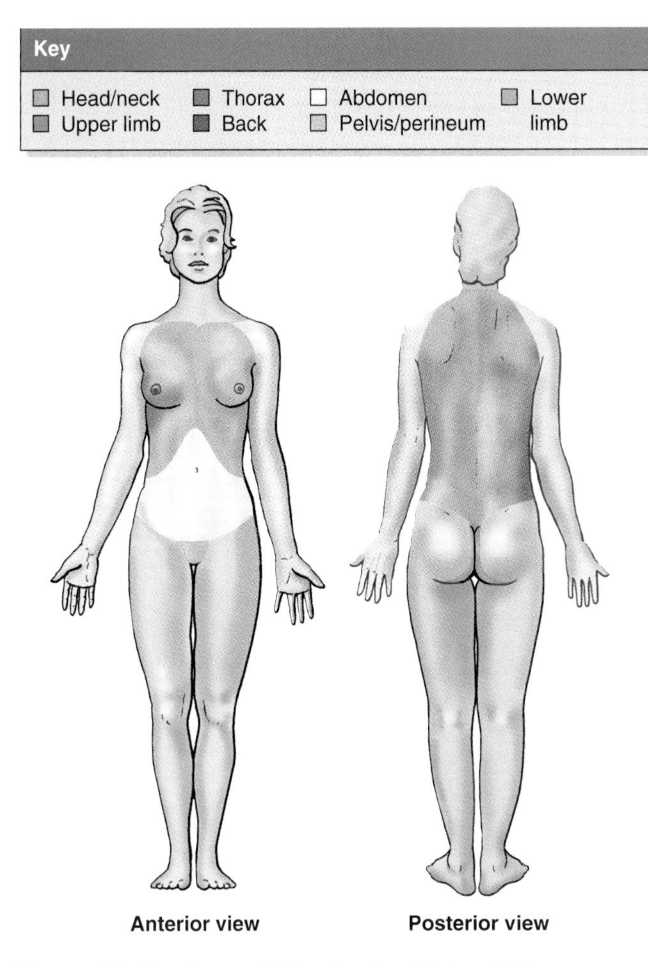

Figure I.1. Regions of the body. All descriptions are expressed in relation to the anatomical position illustrated here.

lungs). You will also use a *reflex hammer* for examining the functional state of nerves and muscles. When reading the surface anatomy boxes in this text, make an effort to associate living anatomy with the anatomy that you learn in lectures and demonstrations.

Systemic Anatomy

Systemic anatomy is the study of the body systems. The systems and their branches of study (in parentheses) are:

- The *integumentary system* (*dermatology*) consists of the skin (L. integumentum, a covering) and its appendages—hair, nails, and sweat glands, for example. The skin, an extensive sensory organ, forms a protective covering and container for the body.
- The *skeletal system* (*osteology*) consists of bones and cartilage; it provides support for the body and is what the muscular system acts on to produce movement. It also protects vital organs such as the heart, lungs, and pelvic organs.
- The *articular system* (*arthrology*) consists of joints and their associated ligaments, connecting the bony parts of the skeletal system and providing the sites at which movements occur. Thus, much of the skeletal, articular, and muscular systems constitute the *locomotor system* and work together to produce locomotion of the body. The structures responsible for locomotion are the muscles, bones, joints, and ligaments of the limbs, as well as the arteries, veins, and nerves that supply oxygen and nutrients to them, remove waste from them, and stimulate them to act.
- The *muscular system* (*myology*) consists of muscles that contract to move parts of the body (e.g., the bones that articulate at joints).
- The *nervous system* (*neurology*) consists of the *central nervous system* (brain and spinal cord) and the *peripheral nervous system* (cranial and spinal nerves), together with their motor and sensory endings. The nervous system controls and coordinates the functions of organs such as the heart and other structures (e.g., muscles) and relates the body to the environment.
- The *circulatory system* (*angiology*) consists of the cardiovascular and lymphatic systems, which function in parallel. The *cardiovascular system* consists of the heart and blood vessels that propel and conduct blood through the body. The *lymphatic system* is a network of lymphatic vessels that withdraws excess tissue fluid (lymph) from the body's interstitial (intercellular) fluid compartment, filters it through lymph nodes, and returns it to the bloodstream.
- The *digestive* or *alimentary system* (*gastroenterology*) consists of the organs associated with ingestion, mastication (chewing), deglutition (swallowing), digestion, and absorption of food, and the elimination of feces (solid waste) remaining after the nutrients have been absorbed.
- The *respiratory system* (*pulmonology*) consists of the air passages and lungs that supply oxygen to the body and eliminate carbon dioxide.
- The *urinary system* (*urology*) consists of the kidneys, ureters, urinary bladder, and urethra, which filter blood and subsequently produce, transport, store, and intermittently excrete urine (liquid waste), respectively.
- The *reproductive* or *genital system* (*gynecology* in females and *andrology* in males) consists of the genital organs (e.g., ovaries, testes, and external genitalia) that are involved in reproduction.
- The *endocrine system* (*endocrinology*) consists of ductless glands, such as the thyroid gland, that produce hormones that are carried by the circulatory system to all parts of the body. These glands influence metabolism and other processes, such as the menstrual cycle.

Clinical Anatomy

Clinical anatomy emphasizes aspects of structure and function of the body that are important in the practice of medicine, dentistry, and the allied health sciences. It incorporates the regional and systemic approaches to studying anatomy and stresses clinical applications. *Endoscopic and imaging techniques* (e.g., examination of the interior of the stomach) also demonstrate living anatomy. Clinical anatomy is exciting to learn because of its emphasis on clinical problems. *Case studies* and *clinically oriented questions*—features of this book—are integral parts of the clinical anatomical approach to studying anatomy.

Anatomicomedical Terminology

Anatomy and Medicine have an international vocabulary. Although you are familiar with common terms for parts and regions of the body, you must learn the correct nomenclature (axilla instead of armpit and clavicle instead of collar bone, for example) that enables precise communication among health care professionals worldwide, as well as among scholars in basic and applied health sciences. Nevertheless, you must also know what the common terms refer to so that you can understand the words patients use when they describe their complaints. You must also use terms that they can understand when you explain their medical problems to them.

The terminology in this book conforms with the new *Terminologia Anatomica: International Anatomical Terminology* (Federative Committee on Anatomical Terminology, 1998)—the reference guide on anatomical language. Anatomical terms are expressed in Latin, but English equivalents are now given for most terms (e.g., the common shoulder muscle—musculus deltoideus—is the deltoid muscle in English). Unfortunately, the terminology commonly used in hospitals may differ from the official terminology. Because this discrepancy may be a source of confusion, this text clarifies commonly confused terms by placing the unofficial designations in parentheses when the terms are first used, for example, *pharyngotympanic tube* (auditory tube, eustachian tube) and *internal thoracic artery* (internal mammary artery). *Eponyms*—terms incorporating the names of people—are not used in the new terminology because they give no clue about the type or location of the structures involved. In addition, some eponyms are historically inaccurate; e.g., Poupart was not the first anatomist to describe the inguinal ligament (Poupart's ligament). Notwithstanding, commonly used eponyms appear in parentheses throughout the book when terms are first used to avoid ambiguity (e.g., *sternal angle* [angle of Louis]). Note that the eponymous term does not indicate that the angle is in the sternum.

Anatomical terminology introduces a large part of medical terminology. To be understood, you must express yourself clearly, using the proper terms in the correct way. Because most terms are derived from Latin and Greek, medical language may be difficult at first; however, as you learn the origin of terms, the words make sense (Squires, 1986; Willis, 1995). For example, the term *decidua,* used to describe the lining of the pregnant uterus, is derived from Latin and means "a falling off." This term is appropriate because the endometrium "falls off" or is expelled after the baby is born, just as the leaves of deciduous trees fall at the end of summer.

To describe the body clearly and to indicate the position of its parts and organs relative to each other, anatomists around the world have agreed to use the same descriptive terms of position and direction. Because clinicians also use these terms, it is important to learn them well. Practice using them so that your meaning is clear when you describe parts of the body in patient histories or during discussions of patients with clinicians.

Anatomical Position

All anatomical descriptions are expressed in relation to the anatomical position to ensure that descriptions are not ambiguous (Fig. I.1). The anatomical position refers to persons—regardless of the actual position they may be in—as if they were standing erect, with their:

- Head, eyes, and toes directed anteriorly (forward)
- Upper limbs by the sides with the palms facing anteriorly
- Lower limbs together with the feet directed anteriorly.

This anatomical position is adopted worldwide for giving anatomicomedical descriptions. By using this position, you can relate any part of the body to any other part. Although gravity causes a downward shift of internal organs in the upright position, it is often necessary to describe the position of organs in a recumbent or supine position because this is the posture in which people are usually examined. Consequently, visualize the anatomical position in your mind's eye when describing patients (or cadavers) lying on their sides, supine (recumbent, lying down, face upward), or prone (face downward).

Anatomical Planes

Anatomical descriptions are based on four imaginary planes (median, sagittal, coronal, and horizontal) that pass through the body in the anatomical position (Fig. I.2):

- **Median plane**—the vertical plane passing longitudinally through the body—dividing it into right and left halves. The term *midsagittal plane* is a superfluous term for the median plane (O'Rahilly, 1997). Parasagittal, used by neuroanatomists and neurologists, is also unnecessary because any plane parallel to the median plane is sagittal by definition. A plane near the median plane is a *paramedian plane.*
- **Sagittal planes** are vertical planes passing through the body *parallel to the median plane.* It is helpful to give a point of reference by naming a structure intersected by the plane you are referring to, such as a sagittal plane through the midpoint of the clavicle.
- **Coronal planes** are vertical planes passing through the body *at right angles to the median plane*, dividing it into anterior (front) and posterior (back) portions.
- **Horizontal (transverse) planes** are planes passing through the body *at right angles to the median and coronal planes.* A horizontal plane divides the body into superior (upper)

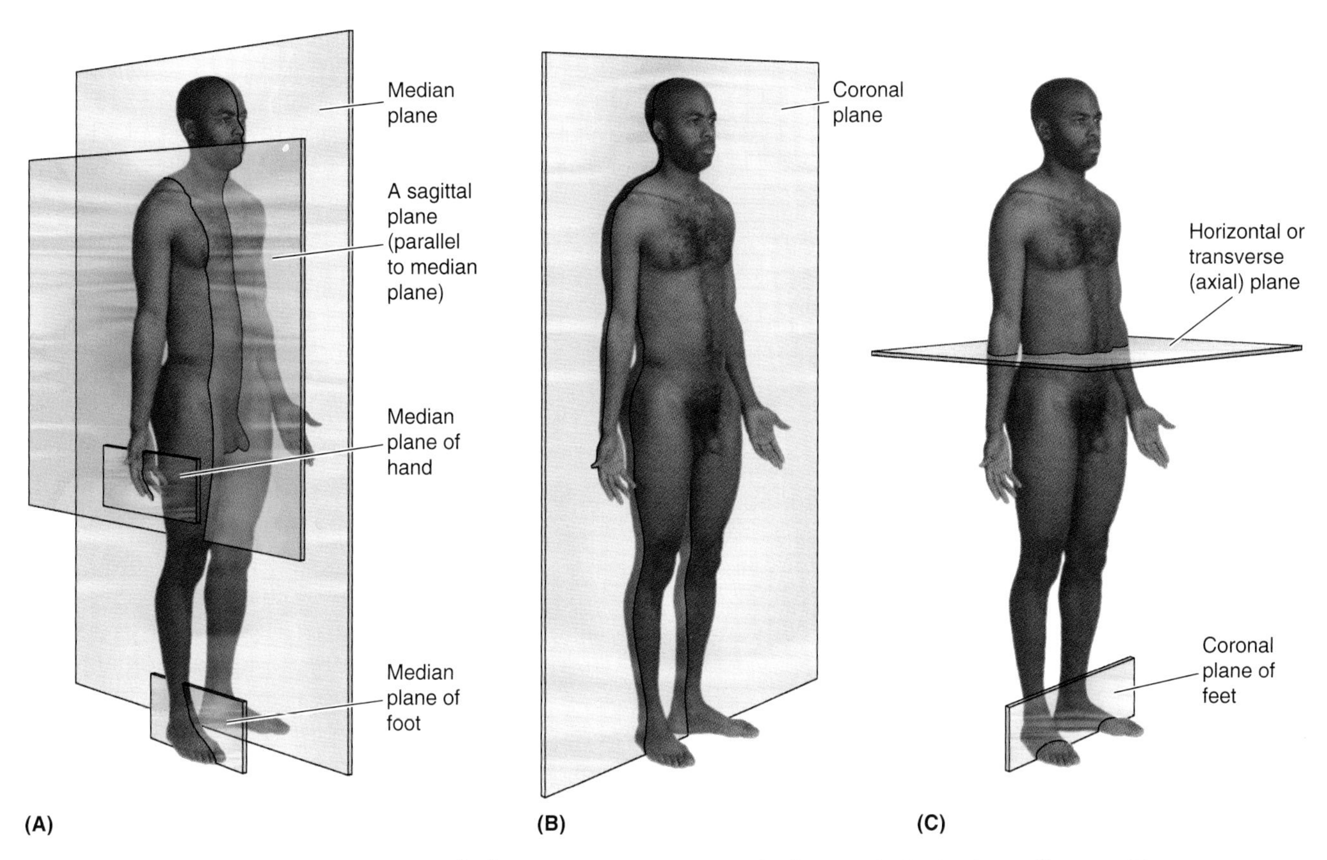

Figure I.2. Anatomical planes. The main planes of reference in the body are illustrated here.

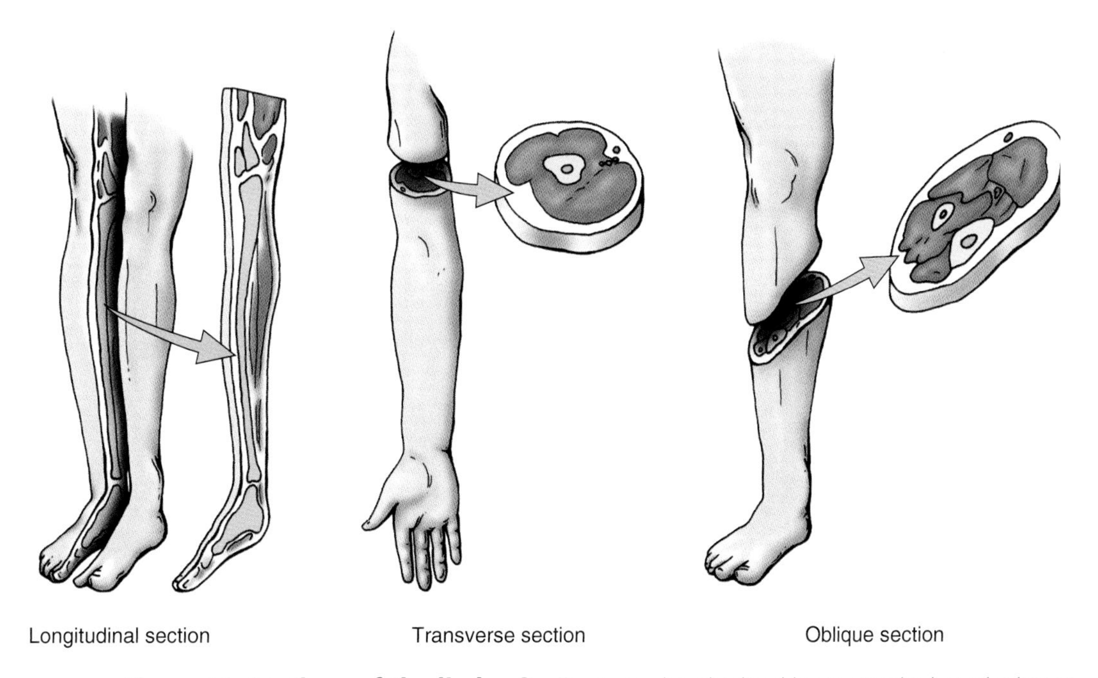

Figure I.3. Sections of the limbs. Sections may be obtained by anatomical sectioning or medical imaging techniques.

and inferior (lower) parts. It is helpful to give a reference point to identify the level of the plane, such as a "horizontal plane through the umbilicus." Radiologists refer to horizontal planes as transaxial or simply *axial planes* that are perpendicular to the long axis of the body and limbs.

Commonly, sections in coronal and horizontal planes are symmetrical, passing through both the right and left members of paired structures, allowing some comparison. The number of sagittal, coronal, and horizontal planes is unlimited. The main use of anatomical planes is to describe sections.

Anatomists create sections of the body and its parts anatomically (Fig. I.3) and clinicians create them by planar imaging, such as CT, to describe and display internal structures. Sections provide views of the body as if cut or sectioned along particular planes.

- **Longitudinal sections** run lengthwise in the long axis of the body or any of its parts, and the term applies regardless of the position of the body.
- **Transverse sections**, or cross sections, are slices of the body or its parts that are cut at right angles to the longitudinal axis of the body or any of its parts; a transverse section through the foot lies in the coronal plane (Fig. I.2*C*).
- **Oblique sections** are slices of the body or any of its parts that are not cut along one of the previously mentioned anatomical planes. In practice, many radiographical images and anatomical sections do not lie precisely in the sagittal, coronal, or horizontal planes; often they are slightly oblique.

Terms of Relationship and Comparison

Various adjectives, arranged as pairs of opposites, describe the relationship of parts of the body in the anatomical position and compare the relative position of two structures with each other (Fig. I.4).

Superficial, intermediate, and **deep** are terms used in dissections to describe the position of one structure, such as a muscle, with respect to other structures such as skin and bone.

Medial is a term that is used to indicate that (in the anatomical position) a structure, such as the 5th digit of the hand (L. manus), or little finger, is near or nearer to the median plane of the body. Conversely, **lateral** stipulates that a structure, the 1st digit of the hand, or thumb, for example, is farther away from the median plane. The terms lateral and medial are not synonymous with *external* (outer) and *internal* (inner). External and internal mean farther from and nearer to the center of an organ or cavity, respectively.

Posterior denotes the back surface of the body or nearer to the back. Because people stand erect, the term **dorsal** (used to describe the backs of quadrupeds) is interchangeable with posterior and is favored by embryologists and neuroanatomists (e.g., the dorsal horns of gray matter in the spinal cord). **Anterior** denotes the front surface of the body and **ventral** is equivalent to anterior. Ventral is favored by neuroanatomists because it is equally applicable to humans and animals used in neuroanatomical research. *Rostral* is often used instead of anterior when describing parts of the brain; it means toward the rostrum (L. beak or nose); however, in humans it denotes nearer the anterior part of the head (e.g., the frontal lobe of the brain is rostral to the cerebellum).

Inferior refers to a structure that is situated nearer the soles of the feet. **Caudal** pertains to the tail (L. cauda) and is a useful directional term when referring to the tail region or the trunk—represented by the coccyx, the small bone at the inferior (caudal) end of the vertebral column. The term caudal is used in embryology because the embryo has a tail until the middle of the 8th week (Moore and Persaud, 1998). **Superior** refers to a structure that is nearer the vertex, the topmost point of the skull. *Cranial* relates to the cranium (Mediev. L. skull) and is a useful directional term when referring to the head region.

Combined terms describe intermediate positional arrangements. For example:

- *Inferomedial* means nearer to the feet and median plane; for example, the anterior parts of the ribs run inferomedially
- *Superolateral* means nearer to the head and farther from the median plane.

Proximal and **distal** are directional terms that are used when contrasting positions nearer the attachment (proximal) or origin of a limb or structure, and away from its attachment (distal) or origin.

Dorsum refers to the superior or dorsal surface (back) of any part that protrudes anteriorly from the body, such as the *dorsum of the tongue, penis, or foot.* It is easier to understand why these surfaces are considered dorsal if one thinks of a plantigrade animal that walks on its soles, such as the bear. The **sole** indicates the inferior aspect or bottom of the foot, much of which is in contact with the ground when standing barefooted. The **palm** refers to the flat of the hand, exclusive of the thumb and fingers, and is the opposite of the dorsum of the hand.

Terms of Laterality

Paired structures having right and left members (e.g., the kidneys) are *bilateral,* whereas those occurring on one side only (e.g., the spleen) are *unilateral.*

Ipsilateral means occurring on the same side of the body; the right thumb and right great (big) toe are ipsilateral, for example.

Contralateral means occurring on the opposite side of the body; the right hand is contralateral to the left.

Terms of Movement

Various terms describe movements of the limbs and other parts of the body (Fig. I.5). Movements take place at joints where two or more bones or cartilages articulate with one another.

Flexion indicates bending or *decreasing the angle* between the bones or parts of the body. Flexion of the upper limb at the elbow joint is an anterior bending; flexion of the lower limb at the knee joint is a posterior bending. **Dorsiflexion** describes flexion at the ankle joint, as occurs when walking uphill or lifting the toes off the ground. **Plantarflexion** turns the foot or toes toward the plantar surface (e.g., when standing on your toes).

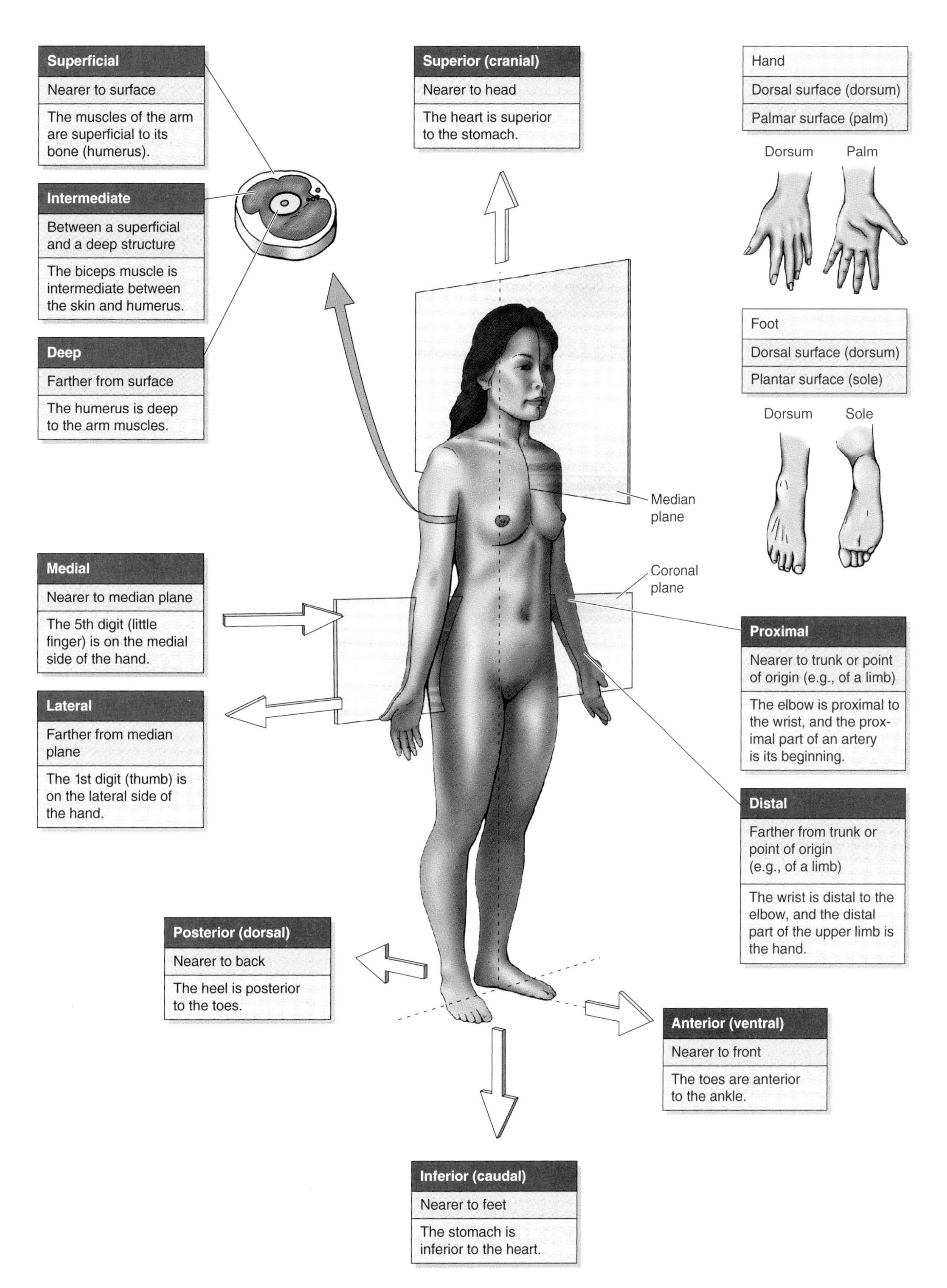

Figure I.4. Terms of relationship and comparison. These terms describe the position of one structure with respect to another.

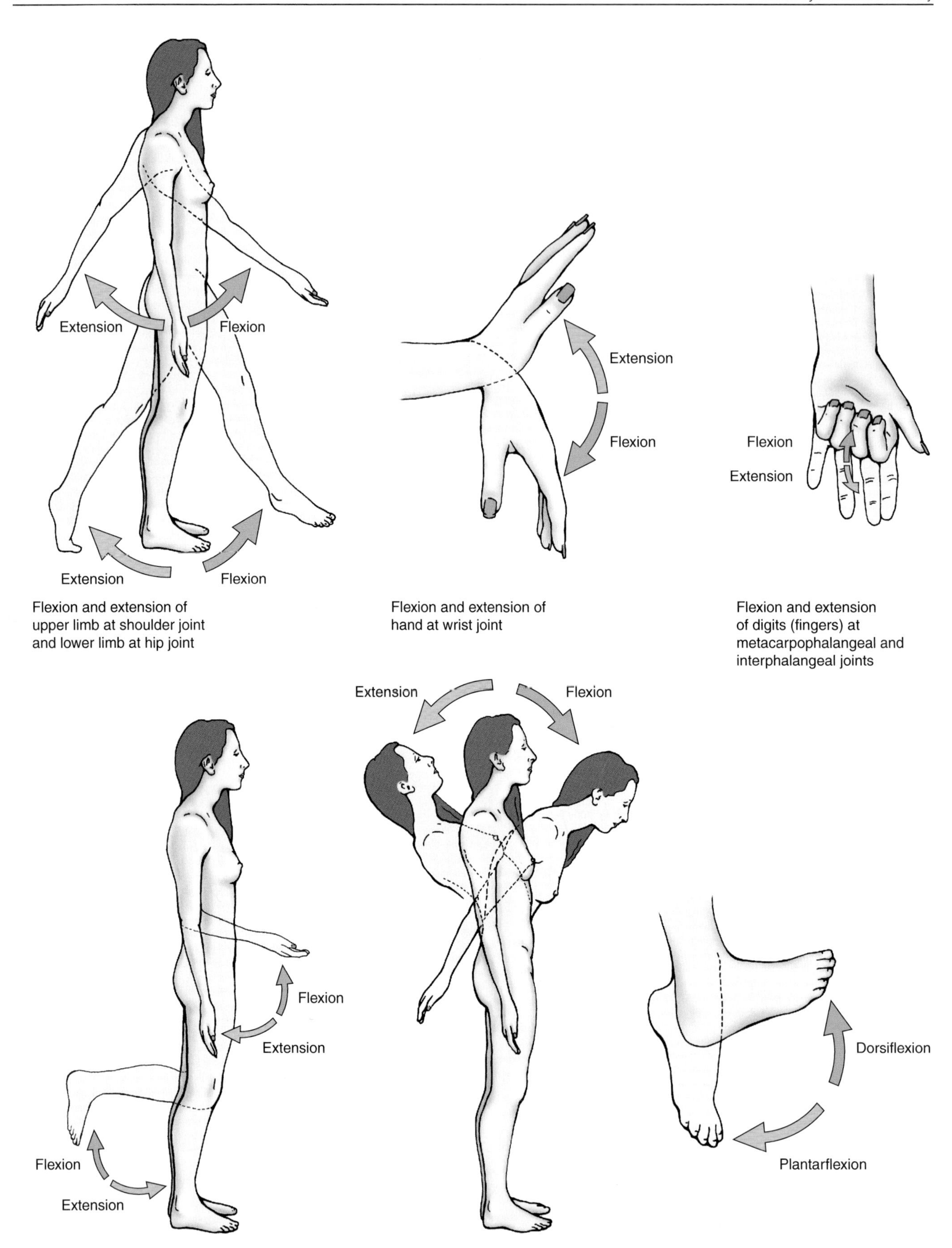

Figure I.5. Terms of movement. These terms describe movements of the limbs and other parts of the body; the movements take place at joints where two or more bones articulate with one another.

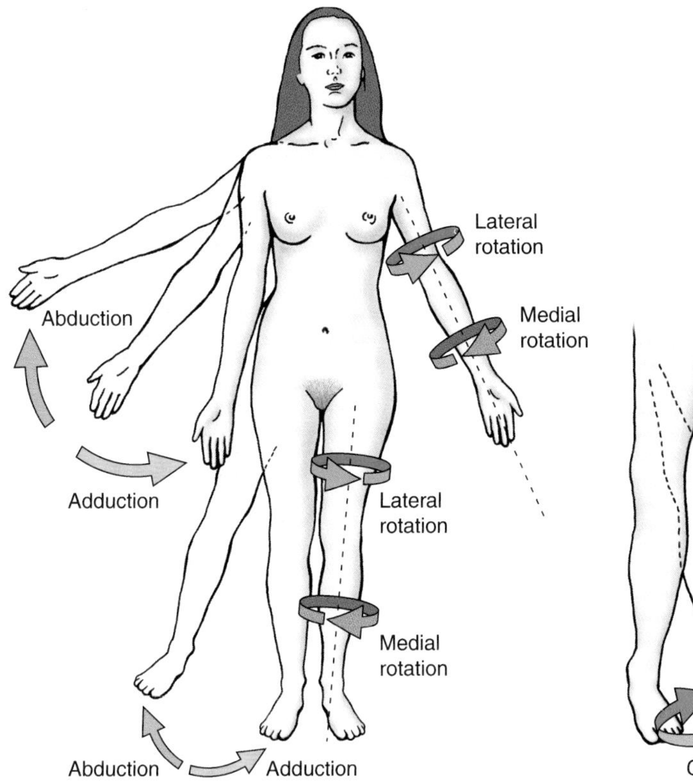

Circumduction

Opposition

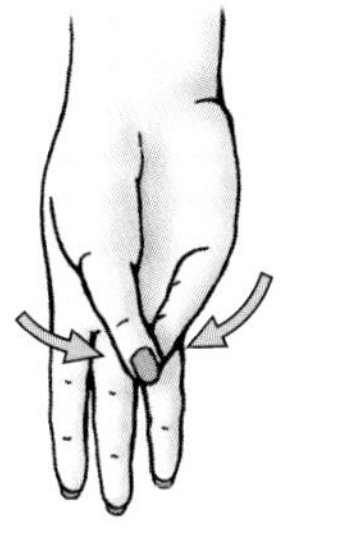

Reposition

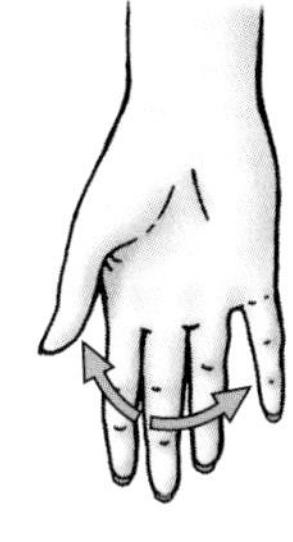

Abduction and adduction of right limbs and rotation of left limbs at the shoulder and hips joints, respectively

Circumduction (circular movement) of lower limb at hip joint

Opposition and reposition of the thumb and little finger

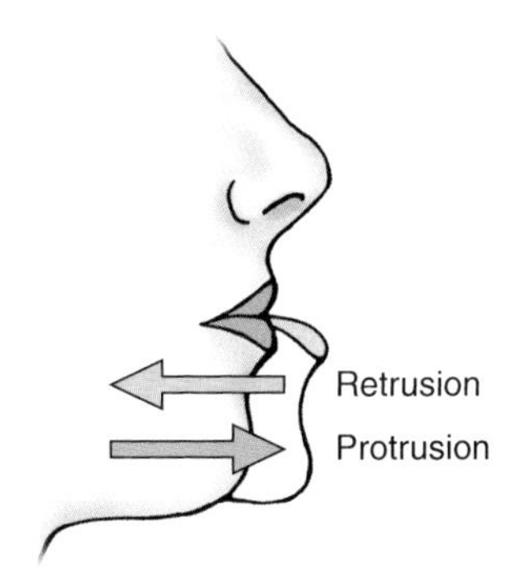

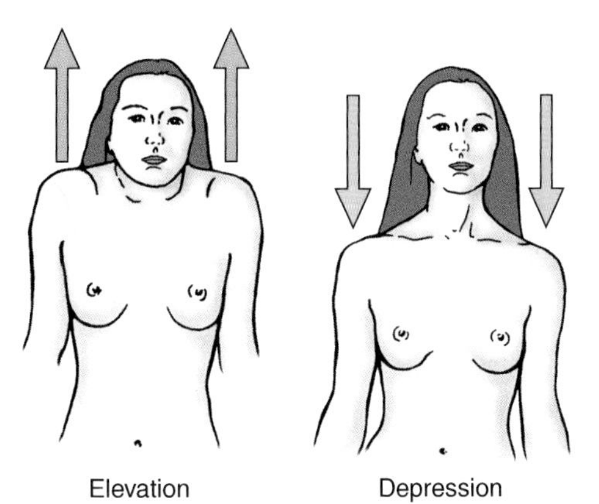

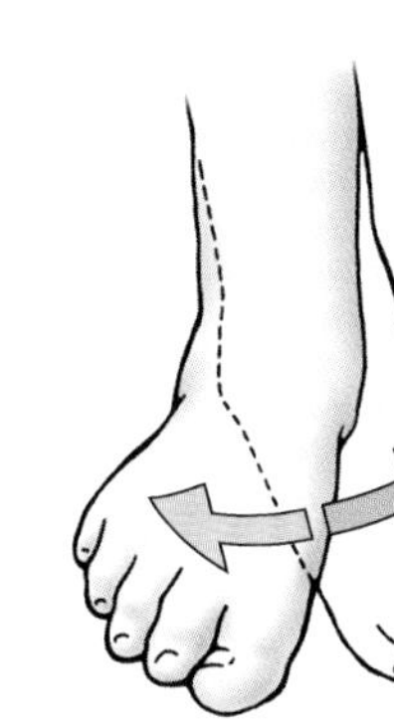

Protrusion and retrusion of jaw at temporomandibular joints

Elevation and depression of shoulders

Inversion and eversion of foot at subtalar and transverse tarsal joints

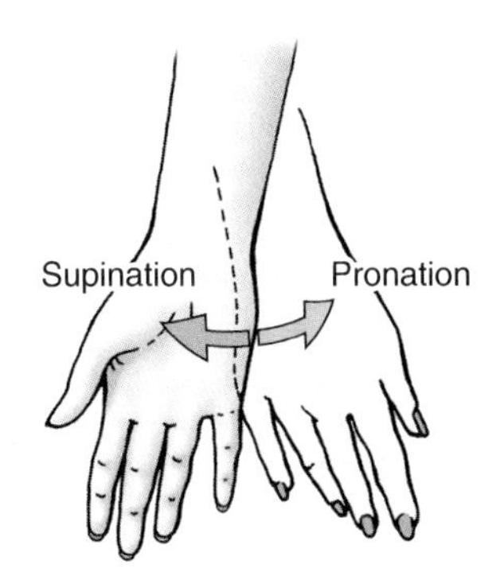

Extension indicates straightening or *increasing the angle* between the bones or parts of the body. Extension usually occurs in a posterior direction, but extension of the lower limb at the knee joint is in an anterior direction. Extension of a limb or part beyond the normal limit—**hyperextension** (overextension)—can cause injury, such as "whiplash" (e.g., hyperextension of the neck during a rear-end automobile collision). An important exception applies at the ankle joint; when your foot is extended, it is plantar*flexed* (e.g., when standing on your toes).

Abduction means moving away from the median plane in the coronal plane (e.g., when moving an upper limb away from the side of the body). In *abduction of the digits* (fingers or toes), the term means spreading them apart—moving the other fingers away from the 3rd or middle finger, or moving the other toes away from the 2nd toe.

Adduction means moving toward the median plane in a coronal plane (e.g., when moving an upper limb toward the side of the body). In *adduction of the digits*, the term means moving them toward the median plane of the hand—moving the other fingers toward the 3rd digit (middle finger).

As you can see by noticing the way the thumbnail faces (laterally instead of posteriorly in the anatomical position), the thumb is rotated 90° relative to the other digits. Therefore, the thumb:

- Flexes and extends in the coronal plane
- Abducts and adducts in the sagittal plane.

Rotation involves turning or revolving a part of the body around its longitudinal axis, such as turning one's head to the side. **Medial rotation** (internal rotation) brings the anterior surface of a limb closer to the median plane, whereas **lateral rotation** (external rotation) takes the anterior surface away from the median plane.

Circumduction is a circular movement that is a combination of flexion, extension, abduction, and adduction occurring in such a way that the distal end of the part moves in a circle. Circumduction can occur at any joint at which all the above-mentioned movements are possible (e.g., the hip joint).

Opposition is the movement by which the pad of the 1st digit (thumb) is brought to another digit pad. We use this movement to pinch, button a shirt, and lift a teacup by the handle. Reposition describes the movement of the 1st digit from the position of opposition back to its anatomical position.

Protrusion is a movement anteriorly (forward) as occurs in protruding the mandible (sticking the chin out). **Retrusion** is a movement posteriorly (backward) as occurs in retruding the mandible (tucking the chin in). The similar terms **protraction** and **retraction** are used most commonly for anterior and posterior movements of the shoulder.

Elevation raises or moves a part superiorly, as in elevating the shoulders when shrugging. **Depression** lowers or moves a part inferiorly, as in depressing the shoulders when standing at ease.

Eversion moves the sole of the foot away from the median plane (turning the sole laterally). When the foot is fully everted it is also dorsiflexed. **Inversion** moves the sole of the foot toward the median plane (facing the sole medially). When the foot is fully inverted it is also plantarflexed.

Pronation is the movement of the forearm and hand that rotates the radius medially around its longitudinal axis so that the palm of the hand faces posteriorly and its dorsum faces anteriorly. When the elbow joint is flexed, pronation moves the hand so that the palm faces inferiorly (e.g., placing the palms flat on a table). When applied to the foot, pronation refers to a combination of eversion and abduction resulting in lowering of the medial margin of the foot. **Supination** is the movement of the forearm and hand that rotates the radius laterally around its longitudinal axis so that the dorsum of the hand faces posteriorly and the palm faces anteriorly (i.e., moving them into the anatomical position). When the elbow joint is flexed, supination moves the hand so that the palm faces superiorly. When applied to the foot, supination generally implies movements resulting in raising the medial margin of the foot.

Structure of Terms

Anatomy is a descriptive science and necessarily requires names for the many structures and processes of the body. Students beginning their studies in anatomy often feel overwhelmed by the many new anatomicomedical terms. Fortunately, there are books to help you learn these terms (Squires, 1986; Willis, 1995). Many terms indicate the shape, size, location, function, or resemblance of one structure to another.

Some muscles have descriptive names to indicate their main characteristics; the *deltoid muscle* that caps the point of the shoulder is triangular for example, like the symbol for *delta*, the 4th letter of the Greek alphabet. The suffix *oid* means "like" something; therefore, deltoid means "like delta." *Biceps* means two-headed and *triceps* means three-headed. Some muscles are named according to their shape—the *piriformis muscle*, for example, is "pearlike." Other muscles are named according to their location. The *temporal muscle* is in the temporal region (temple) of the skull. In some cases, actions are used to describe muscles: for example, the *levator scapulae* elevates the scapula (L. shoulder blade). Thus, there are logical reasons for the names of muscles and other parts of the body, and if you learn their meanings and think about them as you read and dissect, you should have no difficulty remembering their names.

Abbreviations of Terms

Abbreviations of terms are used for brevity in medical histories and in this and other books, such as in tables of muscles, arteries, and nerves, and even in speaking. Clinical abbreviations are used in discussions and descriptions of signs and symptoms. Learning to use them also speeds note-taking. The following are common anatomical and clinical abbreviations (modified from: A clinical anatomy curriculum for the medical student of the 21st century: Gross anatomy. *Clin Anat* 9:71–99, 1996).

a., aa.	artery, arteries
ANS	autonomic nervous system
ant.	anterior
Ao.	aorta
AP	anteroposterior
asc.	ascending
AV	atrioventricular or arteriovenous—depending on the context
b.	bone
br.	branch
C1–C7(8)	cervical vertebrae (C1–C7)/spinal cord segments and spinal nerves (C1–C8)
CA	cancer, carcinoma, and cardiac arrest—depending on the context
CAD	coronary artery disease
CAT	computed axial tomography
CN	cranial nerve (e.g., CN VII, facial nerve)
Co.	coccyx/coccygeal spinal cord segment; coccygeal spinal nerve
com.	common
CNS	central nervous system
CSF	cerebrospinal fluid
CT	computed tomography
desc.	descending
DIP	distal interphalangeal (e.g., the DIP joint of the finger)
ECG/EKG	electrocardiogram; electrocardiography
EEG	electroencephalogram; electroencephalography
e.g.	for example
EMG	electromyogram; electromyography
ext.	extensor/external
FDP	flexor digitorum profundus
FDS	flexor digitorum superficialis
fl.	flexor
G.	Greek (e.g., G. zyon, yoke)
GI	gastrointestinal (e.g., GI tract)
IMA or IMV	inferior mesenteric artery or vein
IML	intermediolateral cell column or nucleus
inf.	inferior
int.	internal
IP	interphalangeal (e.g., IP joint of the finger); this abbreviation is also used when giving a drug intraperitoneally (e.g., an IP injection)
IV	interventricular, intervertebral—depending on the context
I.V.	intravenous or intravenously
IVC	inferior vena cava
IVF	in vitro fertilization
jt.	joint
L, L.	left or Latin (e.g., L. cauda, a tail)
LA	left atrium of the heart
lat.	lateral
LICS	Left intercostal space—the apical impulse or heartbeat is usually heard and felt in the 4th or 5th LICS
lig., ligg.	ligament, ligaments
LLQ	left lower quadrant of the abdomen
LP	lumbar puncture into the subarachnoid space of the lumbar cistern to obtain a sample of CSF
LUQ	left upper quadrant of the abdomen
LV	left ventricle of the heart
m., mm.	muscle, muscles (when muscles are obvious in an illustration or a medical image, the abbreviation is usually omitted)
MAL	midaxillary line
MCL	midclavicular line
MCP	metacarpophalangeal (e.g., MCP joint of the thumb)
med.	medial
MI	myocardial infarction—death of part of the myocardium (heart muscle)
MRA	magnetic resonance angiography
MRI	magnetic resonance imaging
MSL	midsternal line
MTP	metatarsophalangeal (e.g., MTP joint of the foot)
MV	mitral valve (the left AV valve)
n., nn.	nerve, nerves
PA	posteroanterior
PIP	proximal interphalangeal (e.g., PIP joint of the finger)
PNS	peripheral nervous system
post.	posterior
R	right (e.g., on a radiograph)
r., rr.	ramus, rami (branch, branches)
RA	right atrium of the heart
RICS	right intercostal space
RLQ	right lower quadrant of the abdomen
RUQ	right upper quadrant of the abdomen
RV	right ventricle of the heart
S1–S5	sacral vertebrae/spinal cord segments/spinal nerves

SA	sinuatrial (sinoatrial); referring to the sinus venosus of the primordial heart
SCM	sternocleidomastoid (a neck muscle)
SMV	superior mesenteric vein
SNS	somatic nervous system
sup.	superior
supf.	superficial
SVC	superior vena cava
T1–T12	thoracic vertebrae/spinal cord segments/spinal nerves
T & A	tonsillectomy and adenoidectomy
TIA	transient ischemic attack—sudden loss of neurological function caused by a brief period of inadequate perfusion of blood to the brain
TMJ	temporomandibular joint (jaw joint)
v., vv.	vein, veins

Anatomical Variations

Anatomy books for beginning students describe the structure of the body observed in most people (approximately 70%). Students are often frustrated because the bodies they are examining or dissecting do not conform to the atlas or text they are using (Bergman et al., 1988). Often students ignore the variations or inadvertently damage them by attempting to produce conformity. Consequently, *expect anatomical variations when you dissect* or inspect prosected specimens. In a random group of people, individuals differ from each other in physical appearance. The bones of the skeleton vary not only in their basic shape but also in lesser details of surface structure. A wide variation is found in the size, shape, and form of the attachment of muscles. Similarly, considerable variation exists in the division of veins, arteries, and nerves. Veins vary the most and nerves the least. Individual variation must be considered in physical diagnosis and treatment.

Most descriptions in this text assume a normal range of variation. However, the frequency of variation often differs among various human groups, and variations collected in one population may not apply to members of another population. Some variations, such as those occurring in the origin and course of the cystic artery, are clinically important (see Chapter 2) and any surgeon operating without knowledge of them is certain to have problems. In this text, *clinically significant variations appear in the clinical correlation boxes.* Apart from racial and sexual differences, humans exhibit considerable genetic variation, such as polydactyly (extra digits). Approximately 3% of newborn infants show one or more significant congenital anomalies (Moore and Persaud, 1998). Other defects (e.g., atresia [blockage] of the intestine) are not detected until symptoms occur. Students often find variations and congenital anomalies in the cadavers; developing an awareness of the occurrence of variations and a sense of their frequency is one of the many values of dissection.

Skin and Fascia

Because the skin is readily accessible, it is important in physical examinations. The skin is one of the best indicators of general health (Swartz, 1994). The skin provides:

- *Protection* of the body from the environment, abrasions, fluid loss (e.g., in minor burns), harmful substances, and invading microorganisms
- *Heat regulation* through the sweat glands and blood vessels
- *Sensation* (e.g., pain) by way of superficial nerves and their sensory endings.

The skin forms a container for the body's structures (e.g., tissues and organs) and vital substances (especially fluids).

The skin—the body's largest organ—consists of (Fig. I.6):

- The **epidermis**, a superficial cellular layer
- The **dermis**, a deep connective tissue layer.

The deep layers of the dermis contain **hair follicles** with their associated smooth arrector pili muscles and sebaceous glands. Contraction of the **arrector muscles of hairs** (arrector pili muscles) erects the hairs, causing "goose bumps." Hair follicles are generally slanted to one side, and several **sebaceous glands** lie on the side pointed to by the hair (as the hair appears to "enter" the skin [of course, it is emerging from rather than entering the skin]). Consequently, contraction of the arrector muscles causes the hairs to stand up straighter and thus the sebaceous glands are compressed, helping to express their oily secretion onto the skin surface. The skin also contains a large number of **sweat glands**. The evaporation of water (sweat) from the skin by these glands provides a *thermoregulatory mechanism* for heat loss.

The *avascular epidermis* is nourished by the underlying *vascularized dermis.* The epidermis has no blood vessels or lymphatics but it and the dermis are supplied by arteries that extend up to the border between the dermis and the **subcutaneous tissue** (superficial fascia), where they form a deep plexus of anastomosing arteries. The skin is also supplied by **afferent nerve endings**—sensitive to touch, irritation (pain), and temperature.

The deep layer of the dermis is formed by a dense layer of interlacing **collagen** and **elastic fibers**. The bundles of collagen fibers in the dermis are mostly arranged in parallel rows. These fibers provide skin tone and account for the strength and toughness of skin. The collagen fibers produce characteristic tension and wrinkle lines in the skin. The **tension lines** (cleavage or Langer's lines) tend to run longitudinally in the limbs and circumferentially in the neck and trunk (Fig. I.7). Tension lines at the elbows, knees, ankles, and wrists are parallel to the transverse creases that appear when the limbs are flexed; flex your wrist and you will see several of them.

The **subcutaneous tissue** is composed of loose, fatty connective tissue. This highly variable subcutaneous tissue is a

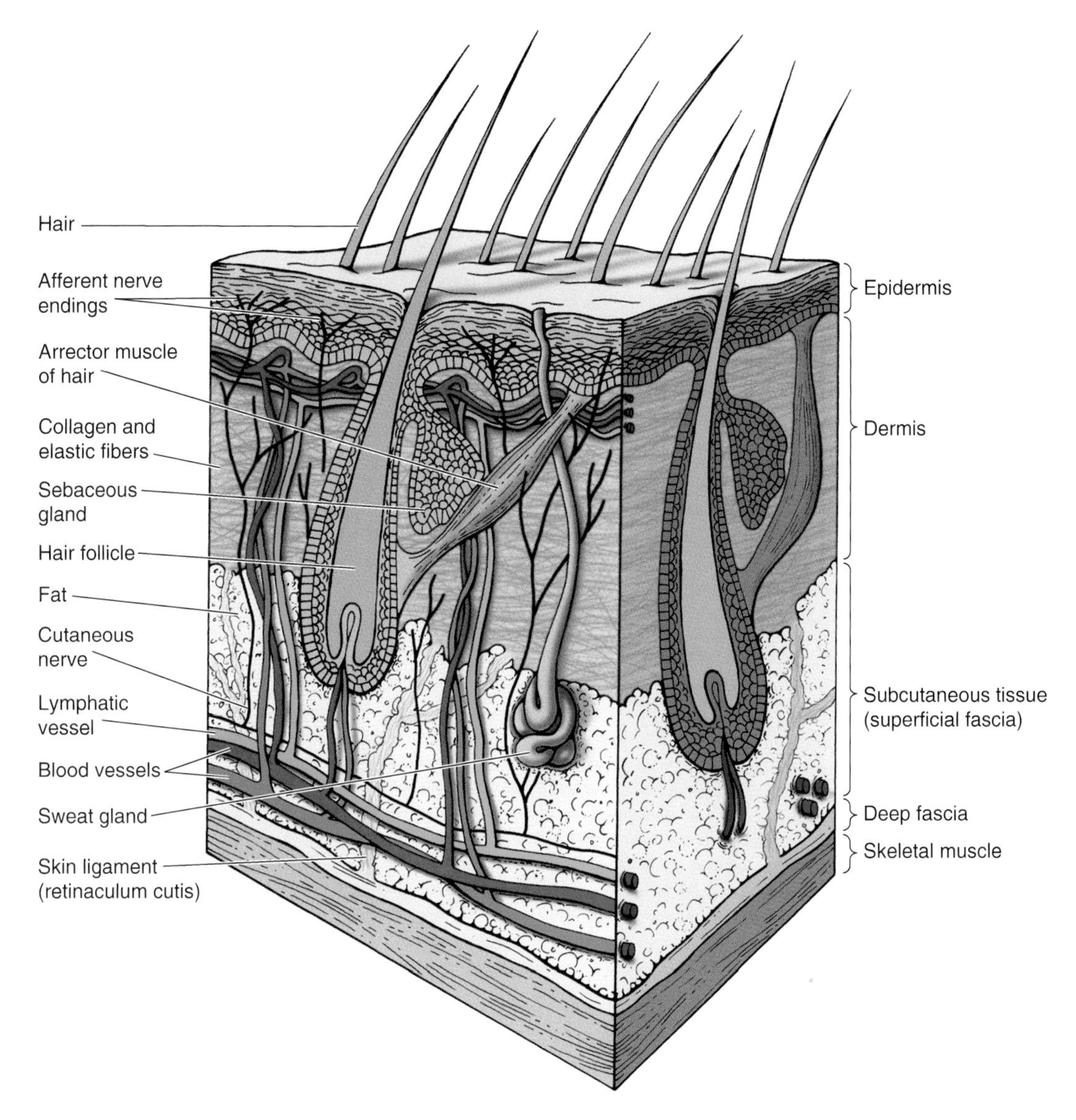

Figure I.6. Schematic drawing of the skin and its appendages. The layered arrangement of the body's covering and the hairs and glands embedded within the skin and subcutaneous tissue are demonstrated here.

Skin Incisions and Wounds

Karl Langer, an Austrian anatomist, studied the tension (cleavage) lines in the skin of cadavers and observed that the skin is always under tension, and when the collagen fibers in the dermis are disturbed by an incision, the wound gapes. Several years later, surgeons were advised that surgical incisions should be made parallel with the tension lines. Skin incisions along these lines usually heal well with little scarring because the lines of force pull the cut surfaces together. An incision across a tension line disrupts and disturbs the collagen fibers and may produce excessive (keloid) scarring. Stab wounds in the skin by an ice pick, for example, are usually slitlike rather than rounded because the pick splits the collagen fibers in the dermis and allows the wound to gape. The direction of the skin slit indicates the predominate direction of the fibers deep to the tension lines.

Stretch Marks in Skin

The collagen and elastic fibers in the dermis form a tough, flexible meshwork of tissue. The skin can distend considerably when the abdomen enlarges during pregnancy, for example; however, it can be stretched too far, damaging the collagen fibers in the dermis. Bands of thin wrinkled skin, initially red but becoming purple and white stretch marks (L. striae gravidarum), appear on the abdomen, buttocks, thighs, and breasts. Stretch marks (L. striae distensae) also form in obese individuals and result from loosening of the fascia and reduced cohesion between the collagen fibers as the skin stretches. Stretch marks generally fade after pregnancy and weight loss; they never disappear completely. ○

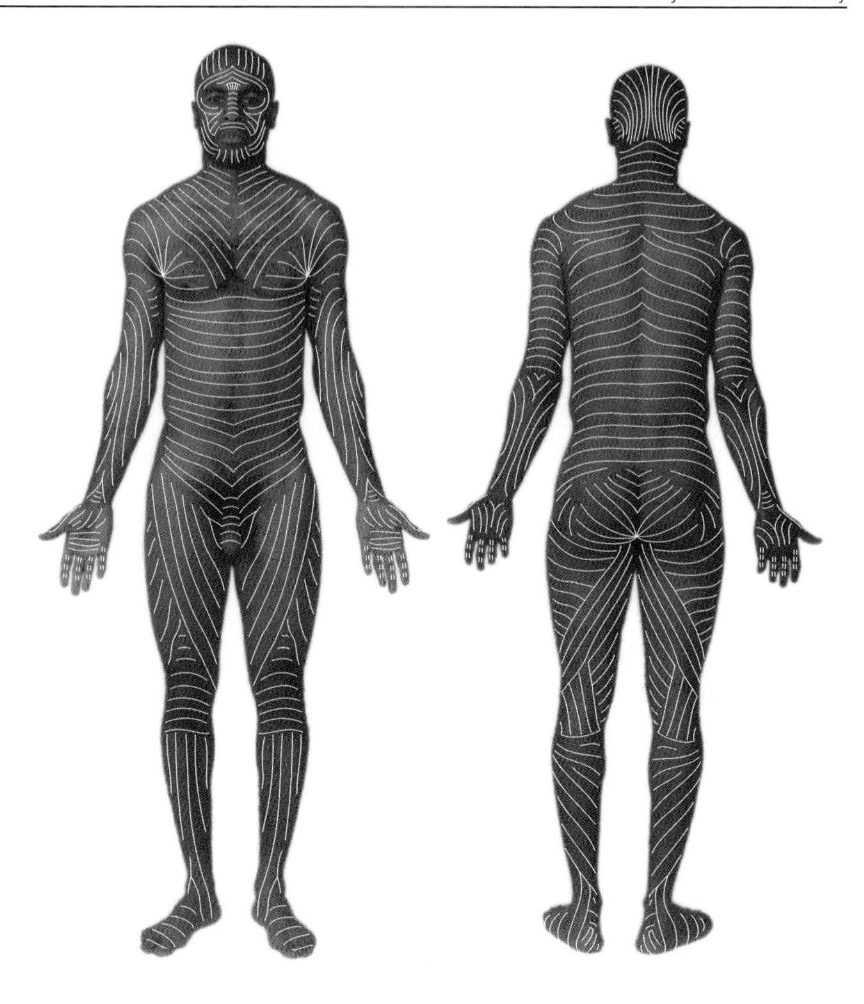

Figure I.7. Schematic drawing of the cleavage (Langer's) lines in the skin. Surgical incisions made in the direction of these lines, which run parallel to the predominant direction of the collagen fiber bundles in the dermis, have less tendency to gape.

thermal regulator, and it provides protection for the skin from bony prominences in the buttocks, for example. Located between the dermis and underlying deep fascia, the subcutaneous tissue contains sweat glands, blood vessels, lymphatics, and **cutaneous nerves.** The distribution of subcutaneous tissue varies considerably in different sites in the same individual. Compare, for example, the subcutaneous tissue and its fat content at the waist and thighs with the anteromedial part of the leg (the shin—the anterior border of the tibia). Also consider the different distribution of subcutaneous tissue and fat between the sexes and in different nutritional states.

Skin ligaments (L. retinacula cutis)—numerous small fibrous bands—extend through the subcutaneous tissue and attach the deep surface of the dermis to the underlying deep fascia. These ligaments determine the mobility of the skin over deep structures. Skin ligaments are particularly well developed in the breast, where they form *suspensory ligaments.* Skin ligaments are also well developed, although short, in the palms and soles.

The **deep fascia** is a dense, organized connective tissue layer that invests deep structures such as the muscles (Gartner and Hiatt, 1997). Groups of muscles in the limbs with similar functions are located in compartments formed by the deep fascia. These *fascial compartments* may contain or direct the spread of an infection or a tumor. Contracting skeletal muscles in the limbs compress veins and function with venous valves to move blood toward the heart (see Fig. I.21). The role of the deep fascia in this *musculovenous pump* is to limit outward expansion of the muscles as they contract. The contraction of the muscles within the fascia surrounding them compress the intramuscular veins, which push the blood out of them toward the heart. Backflow is prevented by the valves.

Skeletal System

The skeleton is composed of bones and cartilages (Fig. I.8). **Bone**—a living tissue—is a highly specialized, hard form of connective tissue that forms most of the skeleton and is the chief supporting tissue of the body. Bones provide:

- Protection for vital structures
- Support for the body
- The mechanical basis for movement
- Storage for salts (e.g., calcium)
- A continuous supply of new blood cells.

Bones take many years to grow and mature. The humerus (arm bone), for example, begins to ossify at the end of the embryonic period (8 weeks); however, ossification is not complete until the person is 20 years old.

Cartilage—a resilient, semirigid form of connective tissue—forms parts of the skeleton where motion occurs (e.g., the **costal cartilages** that attach the ribs to the sternum). Cartilage has no capillary blood supply of its own; consequently, its cells obtain oxygen and nutrients by long-range diffusion. The articulating surfaces of bones participating in a synovial joint are capped with **articular cartilages** that provide gliding surfaces for free movement of the articulating bones (see Fig. I.13).

The proportion of bone and cartilage in the skeleton changes as the body grows; the younger a person is, the greater the contribution of cartilage. The bones of a newborn are soft and flexible because they are mostly composed of cartilage. The skeletal system consists of two main parts (Fig. I.8):

- The **axial skeleton** consists of the bones of the head (*skull*), neck (*hyoid bone* and *cervical vertebrae*), and trunk (*ribs, sternum, vertebrae,* and *sacrum*).
- The **appendicular skeleton** consists of the *bones of the limbs*, including those forming the pectoral (shoulder) and pelvic girdles.

Bones

The differences between the two types of bone, **compact** and **spongy** or cancellous (Fig. I.9), depend on the relative amount of solid matter and on the number and size of the spaces they contain. All bones have a superficial thin layer of compact bone around a central mass of spongy bone, except where the latter is replaced by a **medullary (marrow) cavity**. Within this cavity of adult bones, and between the spicules of spongy bone, blood cells and blood platelets are formed (Ross et al., 1994). The compact bone of the **body**, or **shaft**, that surrounds the medullary cavity is *cortical bone*. The architecture of spongy and compact bone varies according to function. Compact bone provides strength for weightbearing. In long bones designed for rigidity and attachment of muscles and ligaments, the amount of compact bone is greatest near the middle of the body where it is liable to buckle. In addition, long bones have elevations (e.g., ridges, crests, and tubercles) that serve as buttresses (supports) where heavy muscles attach. Living bones have some elasticity (flexibility) and great rigidity (hardness).

Classification of Bones

Bones are classified according to their shape.

- *Long bones* are tubular (e.g., the humerus in the arm).
- *Short bones* are cuboidal and are found only in the ankle (tarsus) and wrist (carpus).
- *Flat bones* usually serve protective functions (e.g., the flat bones of the skull protect the brain).
- *Irregular bones* (e.g., in the face) have various shapes other than long, short, or flat.
- *Sesamoid bones* (e.g., the patella [knee cap]) develop in certain tendons and are found where tendons cross the ends of long bones in the limbs; they protect the tendons from excessive wear and often change the angle of the tendons as they pass to their attachments.

Accessory Bones

Accessory (supernumerary) bones develop when additional ossification centers appear and form extra bones. Many bones develop from several centers of ossification, and the separate parts normally fuse. Sometimes one of these centers fails to fuse with the main bone, giving the appearance of an extra bone; however, careful study shows that the apparent extra bone is a missing part of the main bone. Circumscribed areas of bone are often seen along the sutures of the skull where the flat bones come together, particularly those related to the parietal bone (see lateral view of skull in Chapter 7). These small, irregular, wormlike bones are *sutural bones* (wormian bones). It is important to know that accessory bones are common in the foot so as not to mistake them for bone chips in radiographs and other medical images.

Heterotopic Bones

Bones sometimes form in soft tissues where they are not normally present (e.g., in scars). Horse riders often develop heterotopic bones in their thighs (*rider's bones*), probably because of hemorrhagic (bloody) areas that undergo calcification and eventual ossification. The heterotopic bone formation results from straining the muscles that adduct the thighs. ⊙

Bone Markings and Formations

Bone markings appear wherever tendons, ligaments, and fascia are attached, or where arteries lie adjacent to or enter bones. Other formations occur in relation to the passage of a tendon (often to direct the tendon or improve its leverage) or to control the type of movement occurring at a joint. The various markings and features of bones are (Fig. I.10):

- **Condyle:** rounded articular area (e.g., the lateral femoral condyle)
- **Crest:** ridge of bone (e.g., the iliac crest)
- **Epicondyle:** eminence superior to a condyle (e.g., the lateral epicondyle of the humerus)
- **Facet:** smooth, flat area, usually covered with cartilage, where a bone articulates with another bone (e.g., the *superior costal facet* on the body of a vertebra for articulation with a rib)
- **Foramen:** passage through a bone (e.g., the obturator foramen)

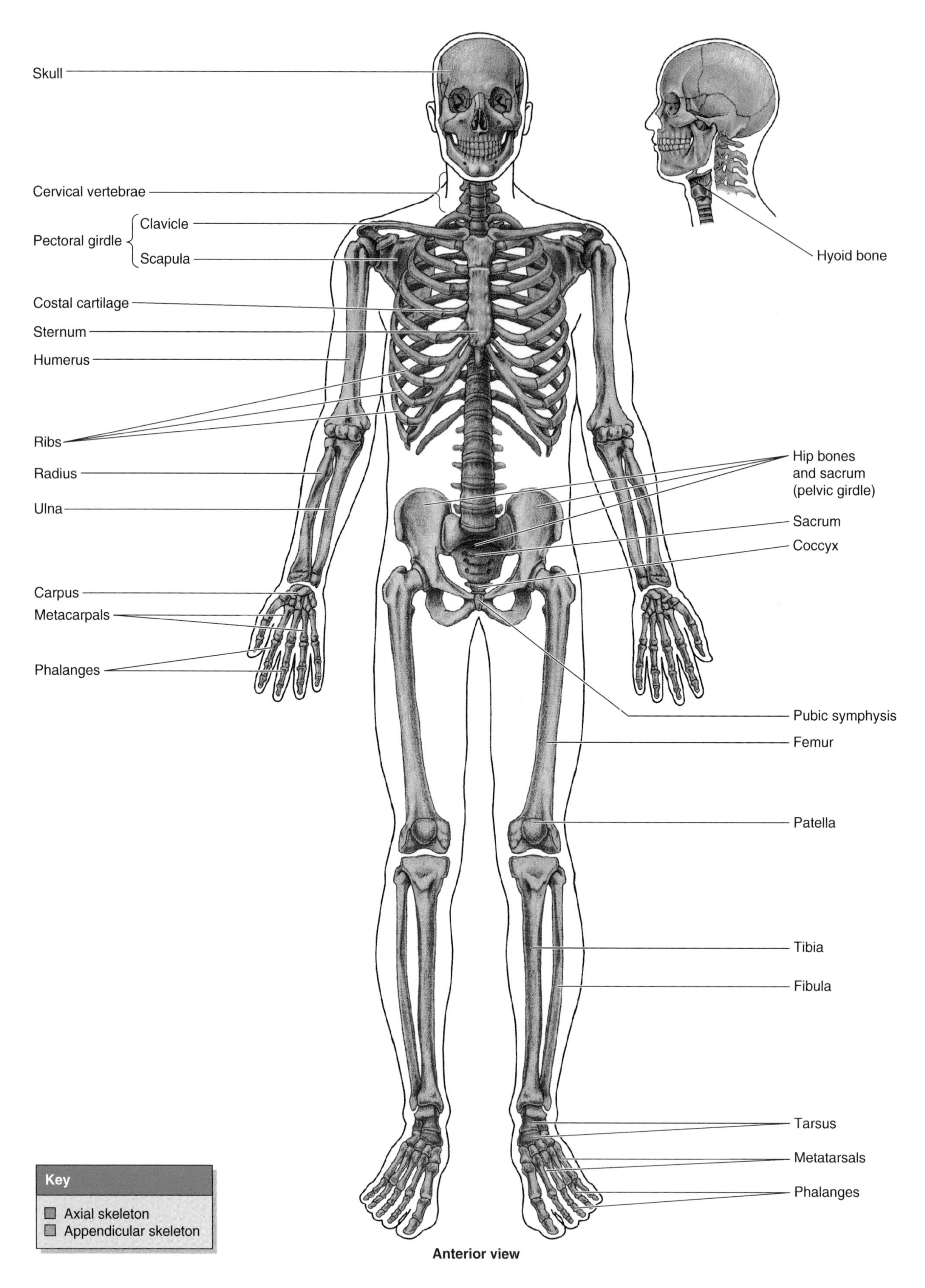

Figure I.8. Skeletal system. The skeleton of the head, neck, and trunk forms the axial skeleton; the skeleton of the limbs forms the appendicular skeleton.

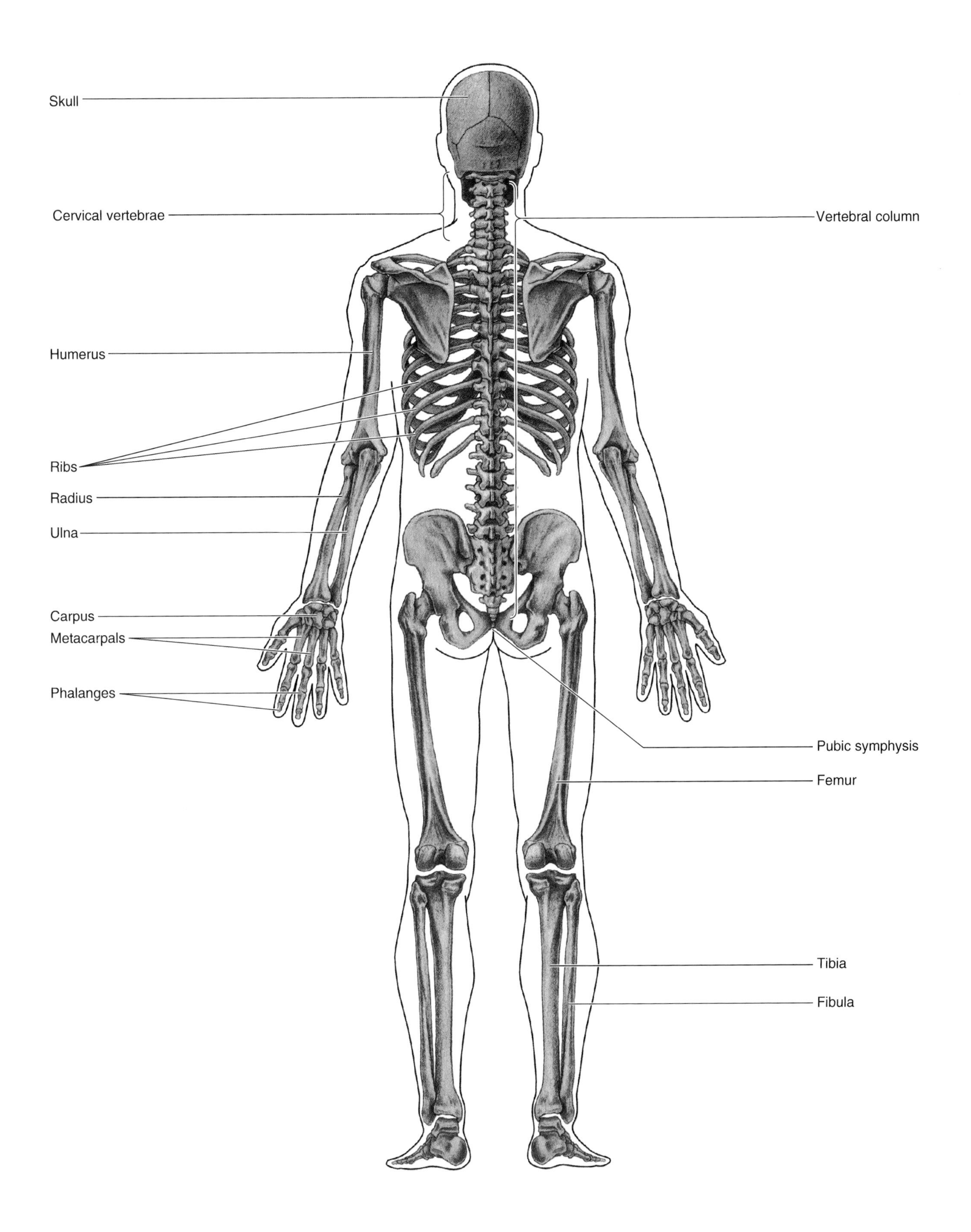

Posterior view

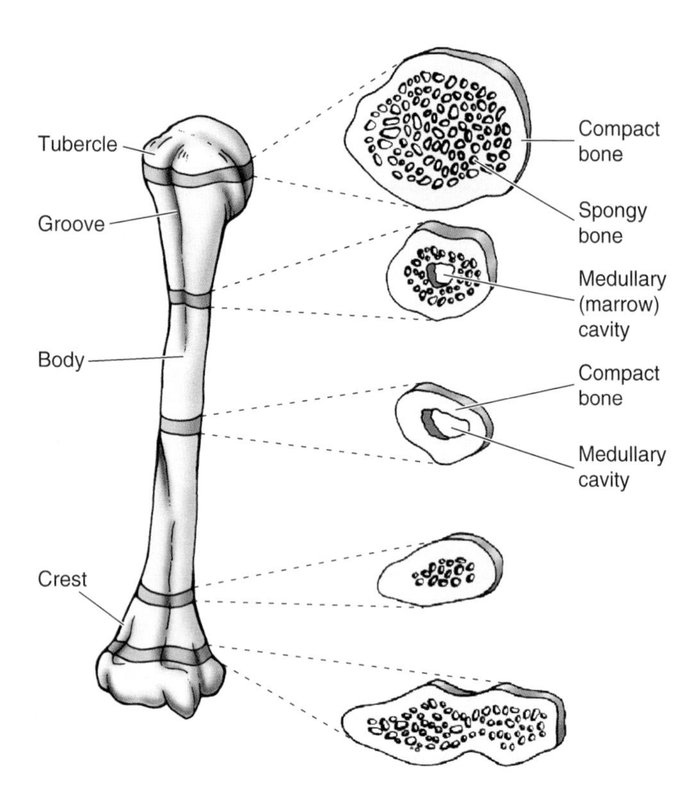

Figure I.9. Transverse sections of the humerus (arm bone). The body, or shaft, of a living bone is a tube of compact bone, the medullary (marrow) cavity of which contains red or yellow marrow, or a combination of both.

- **Fossa:** hollow or depressed area (e.g., the infraspinous fossa of the scapula)
- **Groove:** elongated depression or furrow (e.g., the arterial grooves in the *calvaria*—the domelike superior part of the cranium)
- **Line:** linear elevation (e.g., the soleal line of tibia)
- **Malleolus:** rounded process (e.g., the lateral malleolus of the fibula)
- **Notch:** indentation at the edge of a bone (e.g., the greater sciatic notch)
- **Protuberance:** projection of bone (e.g., the external occipital protuberance)
- **Spine:** thornlike process (e.g., the spine of the scapula)
- **Spinous process:** projecting spinelike part (e.g., the spinous process of a vertebra)
- **Trochanter:** large blunt elevation (e.g., the greater trochanter of the femur)
- **Tubercle:** small raised eminence (e.g., the greater tubercle of the humerus)
- **Tuberosity:** large rounded elevation (e.g., the ischial tuberosity)

Trauma to Bone and Bone Changes

Bones are living organs that hurt when injured, bleed when fractured, remodel in relationship to stresses placed on them, and change with age. Like other organs, bones have blood vessels, lymphatic vessels, and nerves, and they may become diseased. Unused bones, such as in a paralyzed limb, *atrophy* (decrease in size). Bone may be absorbed, which occurs in the mandible when the teeth are extracted. Bones *hypertrophy* (enlarge) when they must support increased weight for a long period.

Trauma to a bone (e.g., during an accident) may break it. For the fracture to heal properly, the broken ends must be brought together approximating their normal position. This is called *reduction of a fracture.* During bone healing, the surrounding fibroblasts (connective tissue cells) proliferate and secrete collagen that forms a *collar of callus* to hold the bones together. Remodeling of bone occurs in the fracture area and the callus calcifies. Eventually, the callus is resorbed and replaced by bone. After several months, little evidence of the fracture remains, especially in young people. Fractures are more common in children than in adults because of the combination of their slender growing bones and carefree activities. Fortunately, many of these breaks are *greenstick fractures* (incomplete breaks caused by bending the bones). Fractures in growing bones heal faster than those in adult bones.

Osteoporosis

During old age, both the organic and inorganic components of bone decrease, producing *osteoporosis*—a reduction in the quantity of bone (atrophy of skeletal tissue). Hence, the bones become brittle, lose their elasticity, and fracture easily.

Sternal Puncture

Examination of bone marrow provides valuable information for evaluating hematological diseases. Because it lies just beneath the skin (i.e., is subcutaneous in location) and is easily accessible, the sternum (breast bone) is a commonly used site for harvesting bone marrow. During a sternal puncture, a wide-bore (large diameter) needle is inserted through the thin cortical bone of the sternum into the spongy bone, and a *sample of red marrow* (marrow that produces blood cells) is aspirated with a syringe for laboratory examination. *Bone marrow transplantation* is sometimes performed in the treatment of leukemia. ⊙

Bone Development

All bones derive from *mesenchyme* (embryonic connective tissue) by two different processes: intramembranous ossification

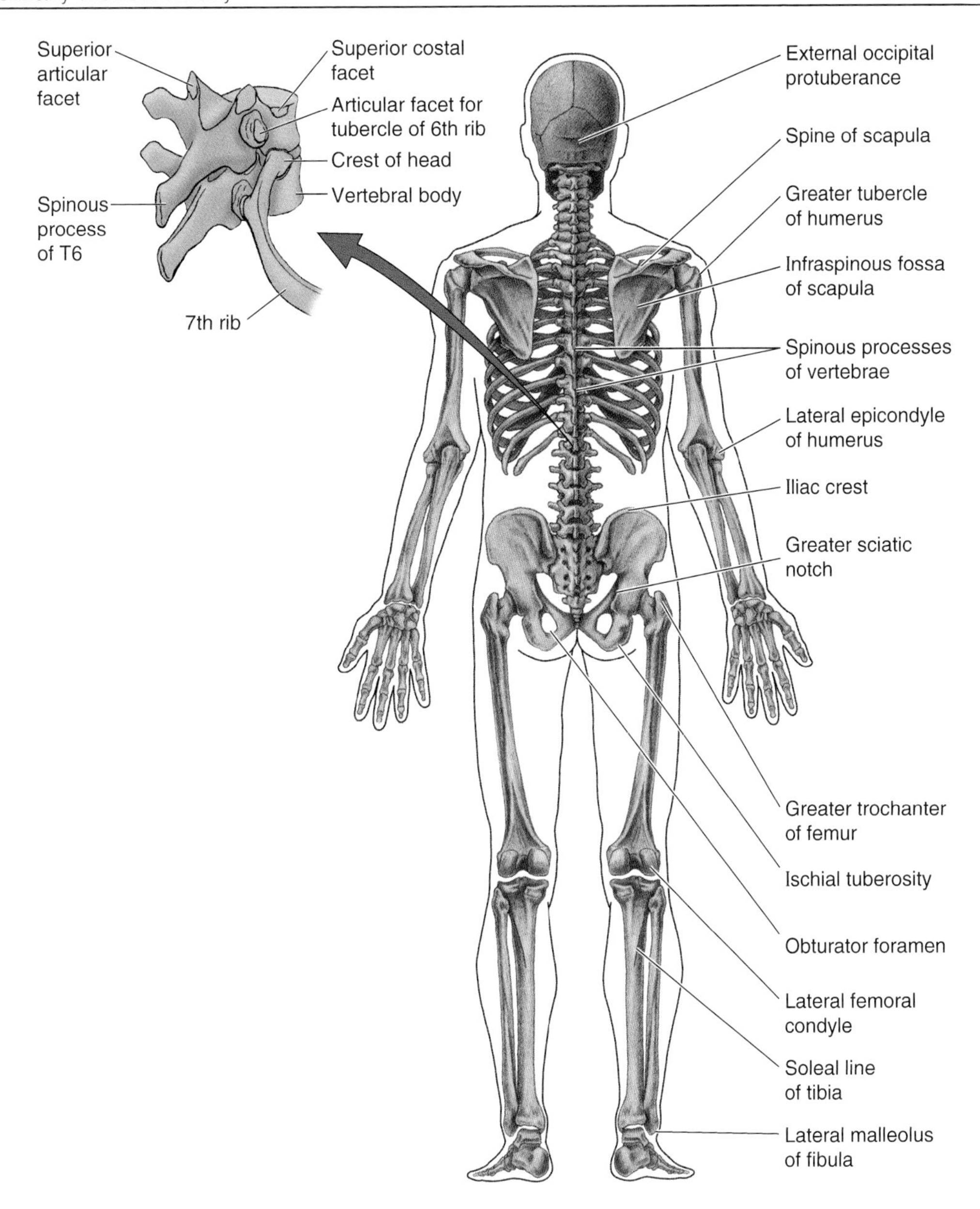

Figure I.10. Bone markings and formations. Markings appear on bones wherever tendons, ligaments, and fascia attach. Other formations relate to joints, the passage of tendons, and the provision of increased leverage.

(directly from mesenchyme) and endochondral ossification (from cartilage derived from mesenchyme). The histology of a bone is the same either way (Cormack, 1993; Gartner and Hiatt, 1997; Moore and Persaud, 1998).

- In *intramembranous ossification* (membranous bone formation), mesenchymal models of bones form during the embryonic period, and direct ossification of the mesenchyme begins in the fetal period.
- In *endochondral ossification* (or cartilaginous bone formation), cartilage models of the bones form from mesenchyme during the fetal period, and subsequently bone replaces most of the cartilage.

A brief description of endochondral ossification helps to explain how long bones grow (Fig. I.11).

The mesenchymal cells condense and differentiate into *chondroblasts*—dividing cells in growing cartilage tissue—that form a cartilaginous bone model. In the midregion of the model, the cartilage calcifies (becomes impregnated with calcium salts) and periosteal capillaries (capillaries from the fibrous sheath surrounding the model) grow into the calcified cartilage of the bone model and supply its interior. These blood vessels, together with associated *osteogenic (bone-forming) cells*, form a **periosteal bud** (Fig. I.11*A*). The capillaries initiate the **primary ossification center**, so named because the bone tissue it forms replaces most of the cartilage in the main

Figure I.11. Development and growth of a long bone. A. Formation of primary and secondary ossification centers. **B.** Growth in length occurs on both sides of the cartilaginous epiphyseal plates (*arrows*). For growth to continue, the bone formed from the primary center in the diaphysis does not fuse with that formed from the secondary centers in the epiphyses until the bone reaches its adult size.

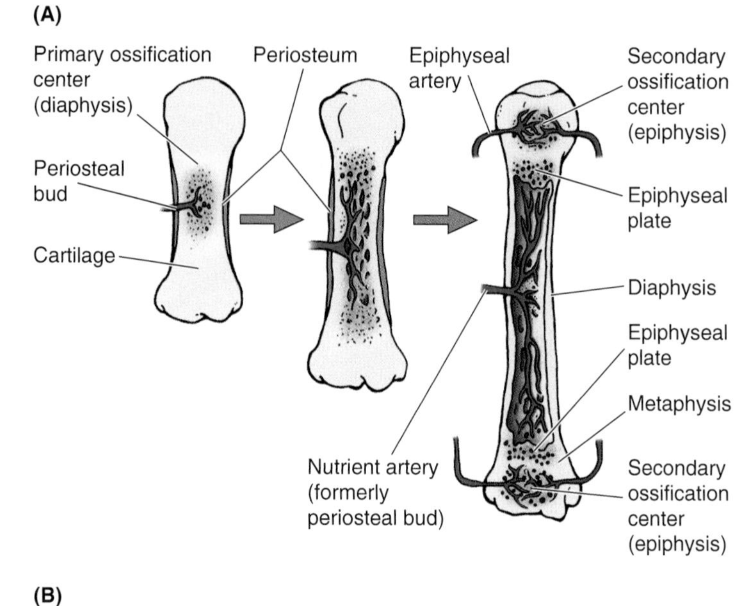

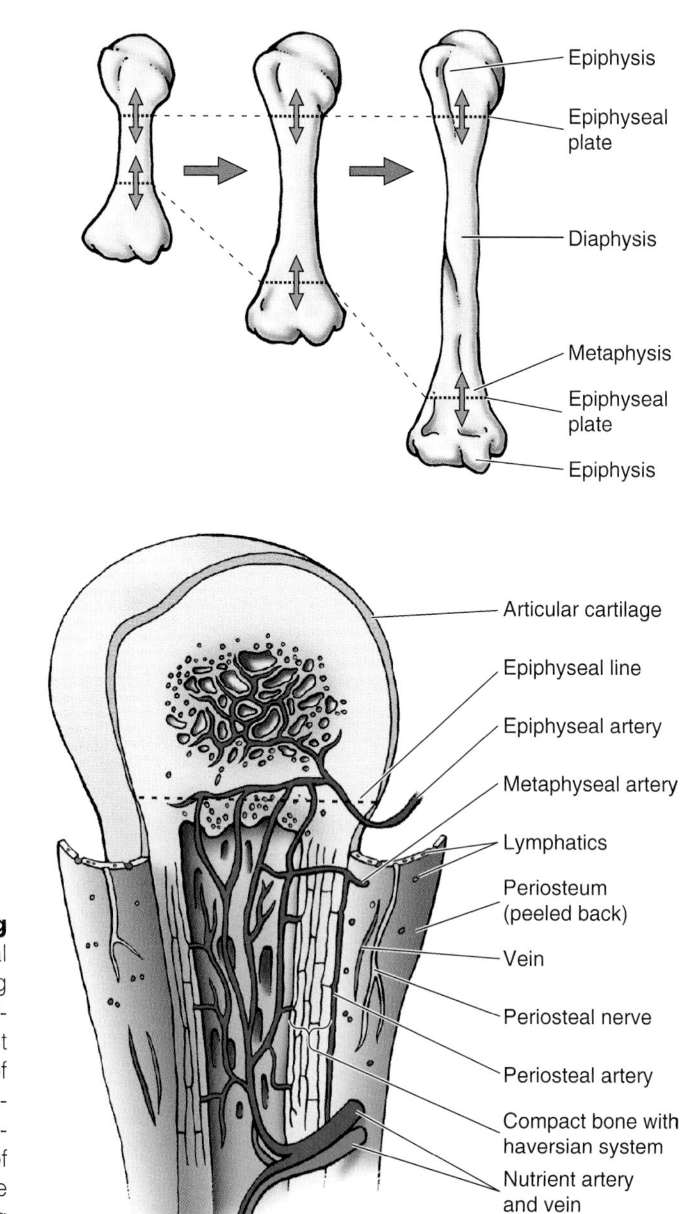

Figure I.12. Vasculature and innervation of a long bone. The epiphysis is supplied with blood by the epiphyseal artery. The diaphysis, metaphyses, and bone marrow of a long bone are supplied mainly by the large nutrient artery (or arteries). Metaphyseal and epiphyseal arteries pierce the compact bone and supply the spongy bone and marrow of the ends of the bone. Branches of the periosteal arteries supply the periosteum. The periosteum is rich in sensory nerves—the periosteal nerves. The bulk of compact bone is composed of haversian canal systems (osteons). The haversian canal in the system houses one or two small blood vessels for nourishing the osteocytes (bone cells).

body of the bone model. The body of a bone ossified from the primary ossification center is the **diaphysis**, which grows as the bone develops.

Most **secondary ossification centers** appear in other parts of the developing bone after birth; the parts of a bone ossified from these centers are **epiphyses**. The chondrocytes in the middle of the epiphysis hypertrophy, and the *bone matrix* (intercellular substance) between them calcifies and begins to break down. **Epiphyseal arteries** grow into the developing cavities with associated osteogenic cells. The flared part of the diaphysis nearest the epiphysis is the **metaphysis**. For growth to continue, the bone formed from the primary center in the diaphysis does not fuse with that from the secondary centers in the epiphyses until the bone reaches its adult size. Thus, during growth of a long bone, cartilaginous **epiphyseal plates** intervene between the diaphysis and epiphyses (Fig. I.11*B*). These growth plates are eventually replaced by bone at each of its two sides, diaphyseal and epiphyseal. When this occurs, bone growth ceases, and the diaphysis fuses with the epiphyses. The seam formed during this fusion process (*synostosis*) is particularly dense and is recognizable in radiographs as an **epiphyseal line**, which marks the zone of fusion between the epiphysis and diaphysis occurring when growth in length has ceased (Fig. I.12). The epiphyseal fusion of bones occurs progressively from puberty to maturity. *Ossification* of short bones is similar to that of the primary ossification center of long bones, and only one short bone, the calcaneus (heel bone), develops a secondary ossification center (Williams et al., 1995).

Vasculature and Innervation of Bones

Arteries enter bones from the **periosteum**—the fibrous connective tissue membrane investing bones (Fig. I.12). **Periosteal arteries** enter at numerous points and supply the bone; these arteries are responsible for nourishment of the compact bone. Consequently, a bone from which the periosteum has been removed dies. Near the center of the body of a bone, a **nutrient artery** passes obliquely through the compact bone and supplies the spongy bone and bone marrow. **Metaphyseal** and **epiphyseal arteries** supply the ends of the bones.

Veins accompany arteries through the nutrient foramina. Many large veins also leave through foramina near the articular ends of the bones. Bones containing red bone marrow have numerous large veins. *Lymphatics* (lymphatic vessels) are abundant in the periosteum.

Nerves accompany blood vessels supplying bones. The periosteum is richly supplied with sensory nerves—**periosteal nerves**—that carry pain fibers. The periosteum is especially sensitive to tearing or tension, which explains the acute pain from bone fractures. Bone itself is relatively sparsely supplied with sensory endings. Within bones, *vasomotor nerves* cause constriction or dilation of blood vessels, regulating bloodflow through the bone marrow.

Joints

A joint is an **articulation**—the place of union or junction between two or more bones or parts of bones of the skeleton. Joints exhibit a variety of form and function. Some joints have no movement; others allow only slight movement, and some are freely movable, such as the shoulder joint.

Classification of Joints

The three types of joint are classified according to the manner or type of material by which the articulating bones are united.

- **Synovial joints are united by an articular capsule** spanning and enclosing a joint cavity. **Articular cartilage** covers the bearing surfaces of the bones. A synovial joint, such as the knee joint, is characterized by a **joint cavity** enclosed by a joint capsule containing **synovial fluid**. The bones in Figure I.13*A* have been pulled apart for demonstration, and the articular capsule has been inflated. Consequently the joint cavity is exaggerated. Normally this cavity is a potential space that contains a small amount of synovial fluid. The **periosteum** invests the bones and blends with

Bone Growth and Assessment of Bone Age

Knowledge of the sites where ossification centers occur, the times of their appearance, the rates at which they grow, and the times of fusion of the sites (times when synostosis occurs) is important in clinical medicine, forensic science, and anthropology. A general index of growth during infancy, childhood, and adolescence is indicated by *bone age*, as determined from radiographs (negative images on X-ray films). The age of a person can be determined by studying the ossification centers in bones. The main criteria are:

- Appearance of calcified material in the diaphysis and/or epiphyses
- Disappearance of the dark line representing the epiphyseal plate (absence of this line indicates that epiphyseal fusion has occurred; fusion occurs at specific times for each epiphysis).

Fusion of epiphyses with the diaphysis occurs 1 to 2 years earlier in girls than in boys.

Determination of bone age can be helpful in predicting adult height in early- or late-maturing adolescents (Behrman et al., 1996). Assessment of bone age also helps to establish the approximate age of human skeletal remains in medicolegal cases. ⭘

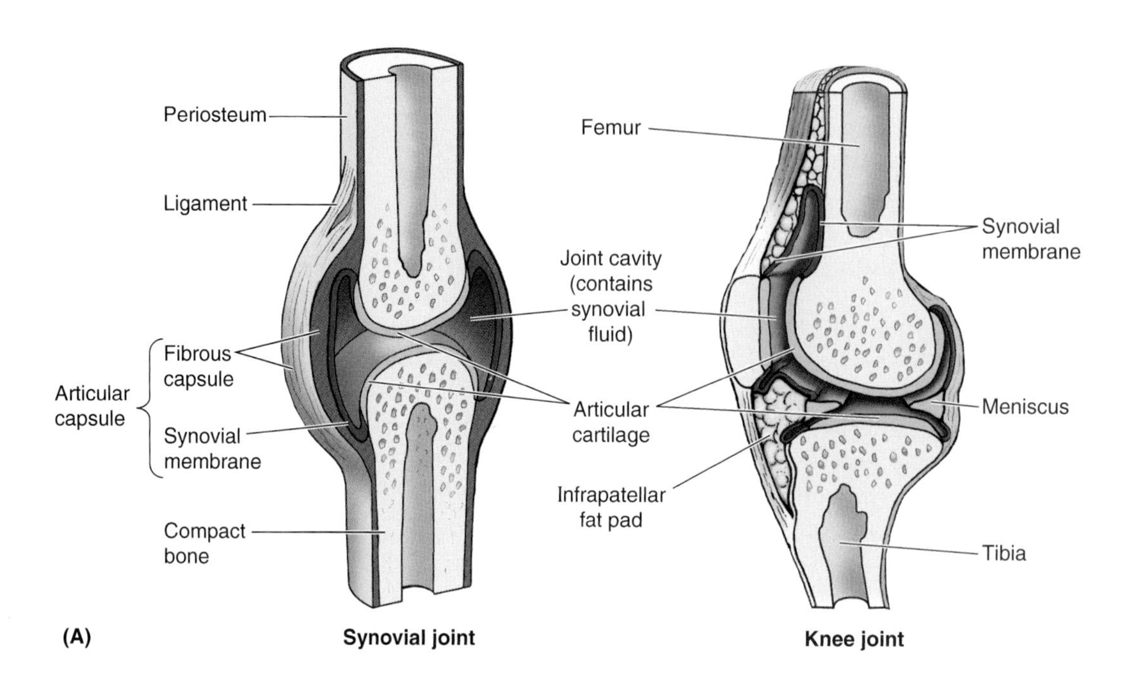

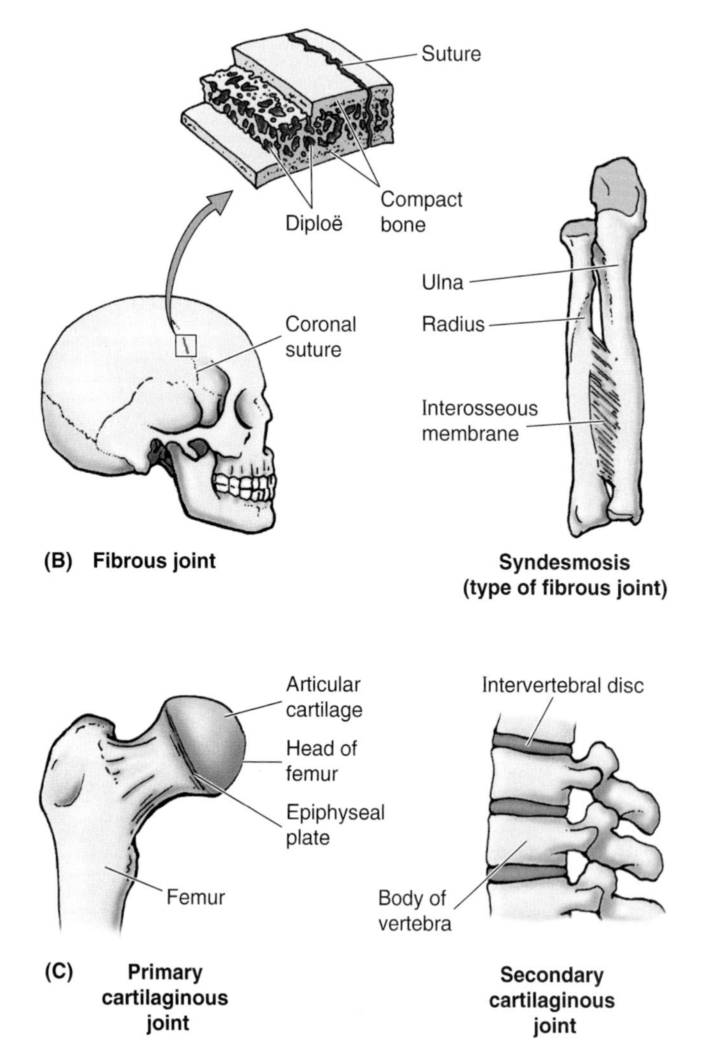

Figure I.13. Various types of joint. A. Synovial joint. **B.** Fibrous joints. **C.** Primary and secondary cartilaginous joints.

Avascular Necrosis

Loss of the arterial supply to an epiphysis or other parts of a bone results in death of bone tissue—*avascular necrosis* (G. nekrosis, deadness). After every fracture, small areas of adjacent bone undergo necrosis. In some fractures, avascular necrosis of a large fragment of bone may occur. A number of clinical disorders of epiphyses in children result from avascular necrosis of unknown etiology (cause). These disorders are referred to as *osteochondroses* (Salter, 1998).

Effects of Disease and Diet on Bone Growth

Some diseases produce early epiphyseal fusion (ossification time), compared with what is normal for the chronological age of the individual; other diseases cause fusion to be delayed. The growing skeleton is sensitive to relatively slight and transient illnesses and to periods of malnutrition. Proliferation of cartilage at the metaphysis slows down during starvation and illness, but degeneration of cartilage cells in the columns continues, producing a dense line of provisional calcification. These lines later become bone with thickened trabeculae, or *lines of arrested growth.*

Displacement and Separation of Epiphyses

Without knowledge of bone growth and the appearance of bones in radiographic and other diagnostic images at various ages, a *displaced epiphyseal plate could be mistaken for a fracture,* and separation of an epiphysis could be interpreted as a displaced piece of a fractured bone. Knowing the patient's age and the location of epiphyses can prevent these anatomical errors. The edges of the diaphysis and epiphysis are smoothly curved in the region of the epiphyseal plate. Bone fractures always leave a sharp, often uneven edge of bone. An injury that causes a fracture in an adult usually causes displacement of an epiphysis in a child. ○

the fibrous capsule of the joints in which the bones participate. All internal structures in a synovial joint that are not covered with articular cartilage are covered by synovial membrane. The bones are separated by the joint cavity but are joined by an **articular capsule**—*a fibrous capsule lined with synovial membrane.*

- **Fibrous joints are united by fibrous tissue.** The amount of movement occurring at a fibrous joint depends in most cases on the length of fibers uniting the articulating bones. The **sutures** of the skull are examples of fibrous joints (Fig. I.13*B*). These bones are close together, either interlocking along a wavy line or overlapping. A **syndesmosis** type of fibrous joint unites the bones with a sheet of fibrous tissue, either a ligament or fibrous membrane. Consequently, this type of joint is partially movable. The **interosseous membrane** in the forearm is a sheet of fibrous tissue that joins the radius and ulna in a syndesmosis. A **gomphosis** or dentoalveolar syndesmosis is a type of fibrous joint in which a peglike process fits into a socket articulation between the root of the tooth and the alveolar process (socket). Mobility of this joint (a loose tooth) indicates a pathological state affecting the supporting tissues of the tooth. However, microscopic movements here give us information (via the sense of proprioception) about how hard we are biting or clenching our teeth, and whether we have a particle stuck between our teeth.

- **Cartilaginous joints are united by hyaline cartilage or fibrocartilage.** In primary cartilaginous joints or synchondroses, the bones are united by hyaline cartilage, which permits slight bending during early life. **Primary cartilaginous joints** are usually temporary unions, such as those present during the development of a long bone (Figs. I.11 and I.13*C*), where the bony epiphysis and the body are joined by an *epiphyseal plate.* Primary cartilaginous joints permit growth in the length of a bone. When full growth is achieved, the epiphyseal plate converts to bone and the epiphyses fuse with the diaphysis. **Secondary cartilaginous joints** or symphyses are strong, slightly movable joints united by fibrocartilage. The **fibrocartilaginous intervertebral discs** between the vertebrae consist of binding connective tissue that joins the vertebrae together. Cumulatively, these joints provide strength and shock absorption as well as considerable flexibility to the vertebral column (spine).

Synovial joints—the most common type of joint—provide free movement between the bones they join and are typical of nearly all limb joints. Their name comes from the lubricating substance (synovial fluid) that is in the **joint cavity** or synovial cavity, which is lined with a synovial membrane or articular cartilage (Fig. I.13*A*). The **synovial membrane** consists of vascular connective tissue that produces synovial fluid. The three distinguishing features of a synovial joint are:

- A **joint cavity**
- Bone ends covered with **articular cartilage**
- Articulating surfaces and joint cavity enclosed by an **articular capsule** (fibrous capsule lined with synovial membrane).

Synovial joints are usually reinforced by *accessory ligaments* that are either separate (*extrinsic*) or are a thickening of a portion of the articular capsule (*intrinsic*).

Some synovial joints have other distinguishing features such as fibrocartilaginous *articular discs,* which are present when the articulating surfaces of the bones are incongruous. The **six major types of synovial joint** are classified according to the shape of the articulating surfaces and/or the type of movement they permit (Fig. I.14).

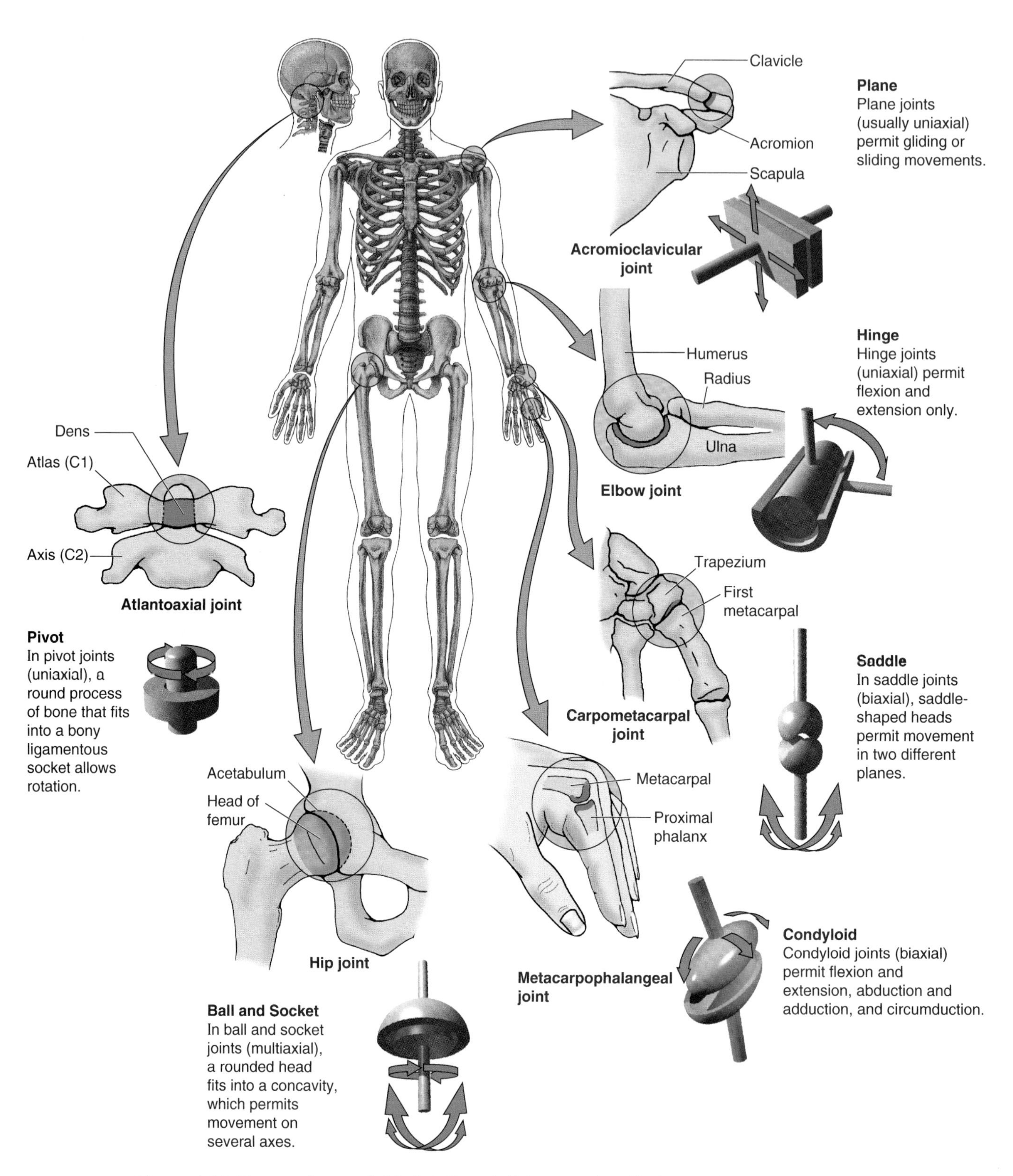

Figure I.14. Types of synovial joint. Synovial joints are classified according to the shape of the articulating surfaces and/or the type of movement they permit. In this type of joint, the articulating bones move freely on one another.

Joints of the Newborn Skull

The bones of the *calvaria* (skull cap) of a newborn infant do not make full contact with each other. At these sites the sutures form wide areas of fibrous tissue—*fontanelles*. The **anterior fontanelle** is the most prominent; lay people call it the baby's "*soft spot*." The fontanelles in a newborn are often felt as ridges because of the overlapping of the cranial bones by molding of the calvaria as it passes through the birth canal. Normally, the anterior fontanelle is flat. A bulging fontanelle may indicate increased intracranial pressure; however, the fontanelle normally bulges during crying. Pulsations of the fontanelle reflect the pulse. A depressed fontanelle may be observed when the baby is dehydrated (Swartz, 1994).

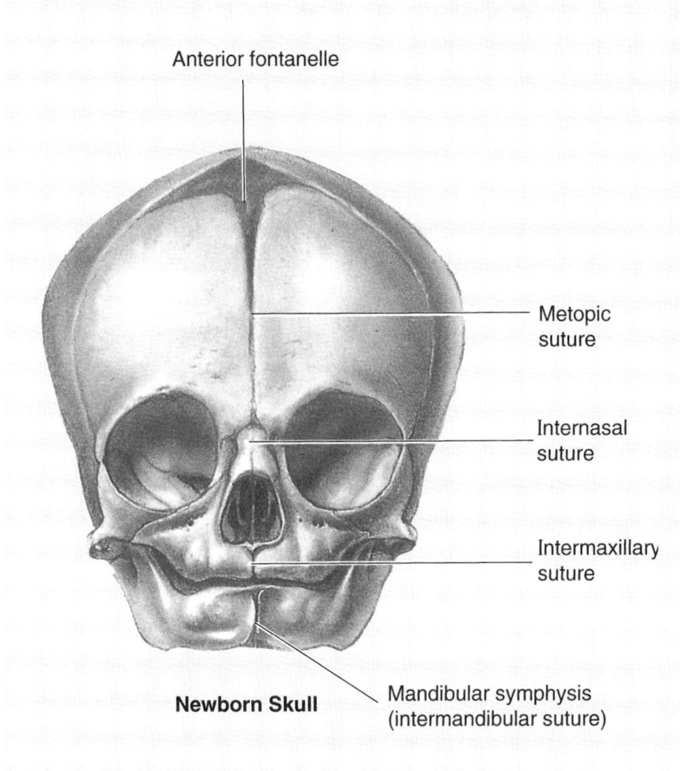

Degenerative Joint Disease

Synovial joints are well designed to withstand wear but heavy use over several years can cause degenerative changes. Some destruction is inevitable during normal activities such as jogging, which wears away the articular cartilages and sometimes erodes the underlying articulating surfaces of the bones. The normal aging of articular cartilage begins early in adult life and progresses slowly thereafter, occurring on the ends of the articulating bones, particularly those of the hip, knee, vertebral column, and hands (Salter, 1998). These *irreversible degenerative changes in joints* result in the articular cartilage becoming less effective as a shock absorber and a lubricated surface. As a result, the articulation becomes increasingly vulnerable to the repeated friction that occurs during joint movements. In some people these changes do not produce significant symptoms; in others they cause considerable pain.

Degenerative joint disease—degenerative arthritis, osteoarthritis, or osteoarthrosis—is often accompanied by stiffness, discomfort, and pain. *Osteoarthritis* is common in older people and usually affects joints that support the weight of their bodies (e.g., the hips and knees). Most substances in the bloodstream, normal or pathological, easily enter the joint cavity. Similarly, traumatic infection of a joint may be followed by *arthritis*—inflammation of a joint—and *septicemia*—blood poisoning.

Arthroscopy

The cavity of a synovial joint can be examined by inserting a cannula and a small telescope (*arthroscope*) into it. This surgical procedure—arthroscopy—enables an orthopaedic surgeon to examine joints for abnormalities such as torn articular discs. Some surgical procedures can also be performed during arthroscopy (e.g., by inserting instruments through small puncture incisions). Because the opening in the articular capsule for inserting the arthroscope is small, healing is more rapid after this procedure than after traditional joint surgery. ○

- **Plane joints** (e.g., the acromioclavicular joint between the acromion of the scapula and the clavicle) are numerous and are nearly always small. *They permit gliding or sliding movements.* The opposed surfaces of the bones are flat or almost flat. Most plane joints allow movement in only one plane (axis); hence, they are *uniaxial joints*. Movement of plane joints is limited by their tight articular capsules.
- **Hinge joints** move in one plane (sagittal) around only one axis (uniaxial) that runs transversely between the bones involved (e.g., the elbow joint). *Hinge joints permit flexion and extension only.* The articular capsule of these joints is thin and lax anteriorly and posteriorly where movement occurs; however, the bones are joined by strong, laterally placed collateral ligaments.
- **Saddle joints** are biaxial with opposing surfaces shaped like a saddle (i.e., they are concave and convex where they articulate with each other). The **carpometacarpal joint** at the base of the 1st digit (thumb) is a saddle joint.
- **Condyloid joints** are also biaxial and allow movement in two planes, sagittal and coronal (e.g., the **metacarpophalangeal [knuckle] joints**); however, movement in one axis

(sagittal) is usually greater (freer) than in the other. Their two axes lie at right angles to each other. Condyloid (knucklelike) joints permit flexion and extension, abduction and adduction, and circumduction.

- **Ball and socket joints** are *multiaxial*; they move in multiple axes and in multiple planes. In these highly movable joints (e.g., the **hip joint**), the spheroidal surface of one bone moves within the socket of another (e.g., the **head of the femur** in the **acetabulum** of the hip bone). Flexion and extension, abduction and adduction, medial and lateral rotation, and circumduction can occur at ball and socket joints.
- **Pivot joints** are uniaxial and allow rotation. In these joints, a rounded process of bone rotates within a sleeve or ring. Examples include the rotation of the radius during pronation and supination (Fig. I.5), and rotation of the atlas (C1 vertebra) around a fingerlike process—the dens (odontoid process)—of the axis (C2 vertebra) during rotation of the head at the **atlantoaxial joint** (Fig. I.14).

Vasculature and Innervation of Joints

Joints receive blood from articular arteries that arise from the vessels around the joint. The arteries often anastomose (communicate) to form networks ("arterial articular networks") to ensure a blood supply to and across the joint in the various positions assumed by the joint. *Articular veins* are communicating veins (L. venae comitantes) that accompany arteries and, like them, are located in the articular capsule, mostly in the synovial membrane.

Joints have a rich nerve supply; the nerve endings are in the articular capsule. In the distal parts of the limbs (hands and feet), the *articular nerves* are branches of the cutaneous nerves supplying the overlying skin. However, most articular nerves are branches of nerves that supply the muscles that cross and therefore move the joint. *Hilton's Law* states that the nerves supplying a joint also supply the muscles moving the joint or the skin covering their attachments.

Joints transmit a sensation, *proprioception*—information that provides an awareness of movement and position of the parts of the body. The synovial membrane is relatively insensitive. Pain fibers are numerous in the fibrous capsule and associated ligaments, causing considerable pain when the joint is injured. The sensory nerve endings respond to the twisting and stretching that occurs during sports activities such as basketball. People with arthritis can confirm that joints are well supplied with pain endings.

Muscular System

Muscle cells—often called *muscle fibers* because they are long and narrow when relaxed—produce contractions that move body parts, including internal organs. The associated connective tissue conveys nerve fibers and capillaries to the muscle fibers as it binds them into bundles or fascicles. Muscles also give form to the body and provide heat. There are three types of muscle (Table I.1):

- *Skeletal muscle*, which moves bones and other structures (e.g., the eyes)
- *Cardiac muscle*, which forms most of the walls of the heart and adjacent parts of the great vessels, such as the aorta
- *Smooth muscle*, which forms part of the walls of most vessels and hollow organs, moves substances through viscera such as the intestine, and controls movement through blood vessels.

Skeletal Muscle

Most skeletal muscles are attached directly or indirectly through tendons to bones, cartilages, ligaments, or fascia, or to some combination of these structures. Some skeletal muscles are attached to organs (the eyeball, for example), to skin (such as facial muscles), and to mucous membrane (intrinsic tongue muscles).

When a muscle contracts and shortens, one of its attachments usually remains fixed and the other one moves. Attachments of muscles are commonly described as the origin and insertion; the *origin* is usually the proximal end of the muscle that remains fixed during muscular contraction, and the *insertion* is usually the distal end of the muscle that is movable. However, some muscles can act in both directions under different circumstances. Therefore, this book usually uses the terms *proximal* and *distal* or *medial* and *lateral* when describing most muscle attachments.

Skeletal muscles produce movements of the skeleton and other parts. Figure I.15 shows the major skeletal muscles; they are often called *voluntary muscles* because individuals can control many of them at will; however, some of their actions are automatic. For example, the diaphragm contracts automatically; a person controls it voluntarily, however, when taking a deep breath. Skeletal muscles are also referred to as "striated" or "striped" muscle because of the striped appearance of their cells (fibers) under microscopy.

Skeletal muscles produce movement by shortening; they pull and never push; however, certain phenomena—such as "popping of the ears" to equalize air pressure, and the musculovenous pump—take advantage of all the expansion of muscle bellies during contraction.

The architecture and shape of skeletal muscles vary. The fleshy part is the **muscle belly** (Fig. I.16*A*). Some muscles are fleshy throughout but most have **tendons** that attach to bones. When referring to the length of a muscle, both the belly and tendons are included—i.e., a muscle's length is the distance between its bony attachments. Some tendons form flat sheets, or **aponeuroses** (Fig. I.16*B*), that anchor one muscle to another, such as the oblique muscles of the anterolateral abdominal wall (Fig. I.15, anterior view). Most muscles are named on the basis of their function or the bones to which

Table I.1. Types of Muscle

Muscle Type	Location	Appearance	Type of Activity	Stimulation
Skeletal ("striated" or "voluntary") muscle 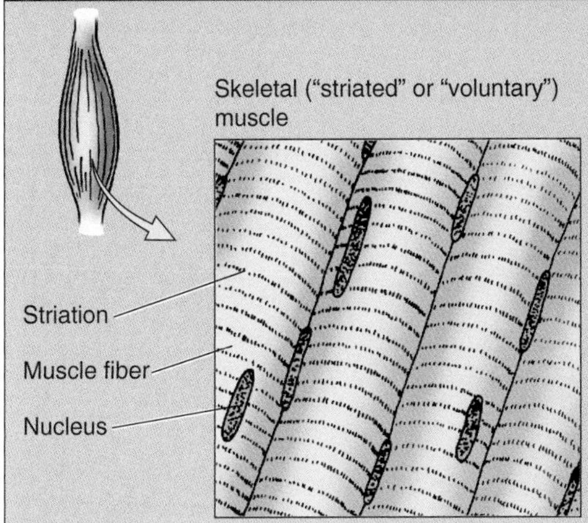	Named muscle (e.g., the biceps of the arm) attached to the skeleton and fascia of limbs, body wall, and head/neck	Large, very long, unbranched, cylindrical fibers with transverse striations (stripes) arranged in parallel bundles; multiple, peripherally located nuclei	Strong, quick intermittent (phasic) contraction above a baseline tonus; acts primarily to produce movement or resist gravity	Voluntary (or reflexive) by the somatic nervous system
Cardiac muscle 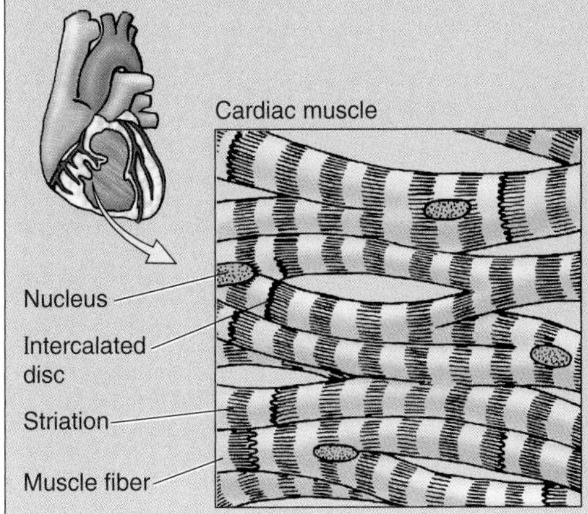	Muscle of heart (myocardium) and adjacent portions of the great vessels (aorta, vena cava)	Branching and anastomosing shorter fibers with transverse striations (stripes) running parallel and connected end-to-end by complex junctions (intercalated discs); single, central nucleus	Strong, quick, continuous rhythmic contraction; acts to pump blood from heart	Involuntary; intrinsically (myogenically) stimulated and propagated; rate and strength of contraction modified by autonomic nervous system
Smooth ("unstriated" or "involuntary") muscle 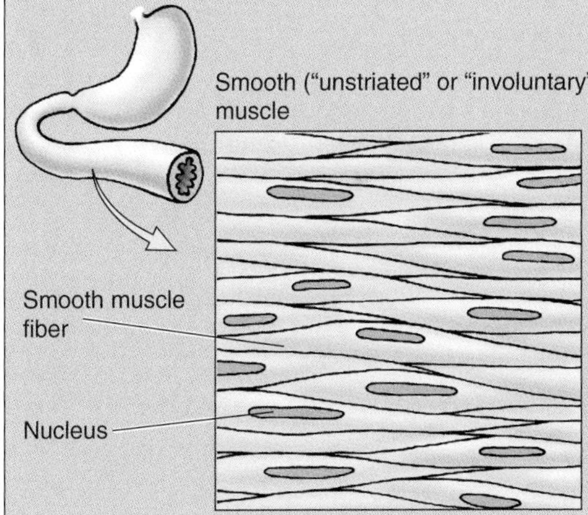	Walls of hollow viscera and blood vessels, iris, and ciliary body of eye; attached to hair follicles of skin (arrector muscle of hair)	Single or agglomerated small, spindle-shaped fibers without striations; single, central nucleus	Weak, slow, rhythmic, or sustained tonic contraction; acts mainly to propel substances (peristalsis) and to restrict flow (vasoconstriction and sphincteric activity)	Involuntary by autonomic nervous system

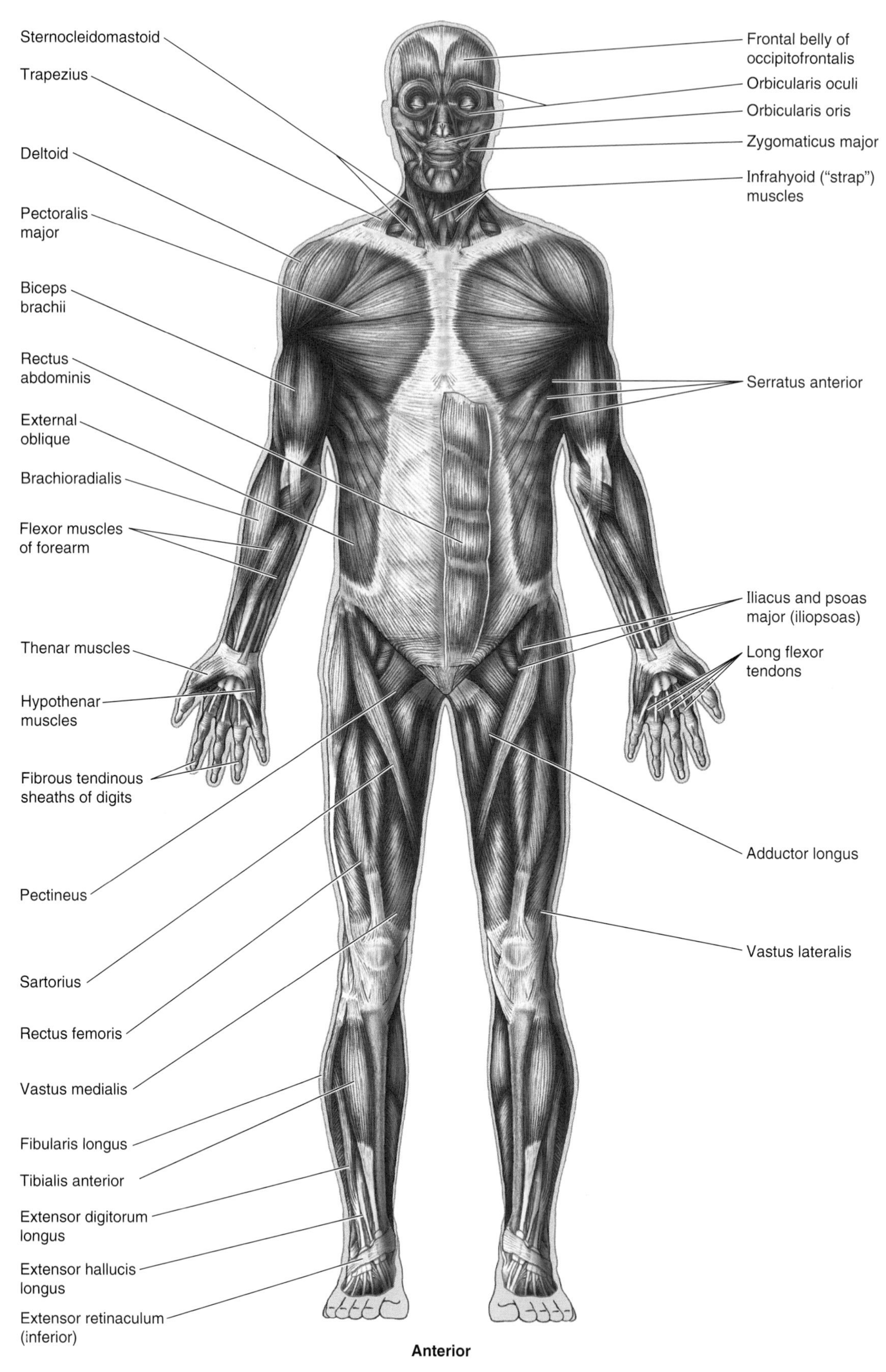

Figure I.15. Skeletal muscles. Most of these muscles produce movements of the skeleton; however, some muscles move other parts (e.g., the eyes, mouth, and scalp). The *orbicularis oris* encircles the mouth and plays an important role in articulation and chewing. The tongue has no bones or joints beyond the hyoid bone, which serves as its base, yet it is mobile.

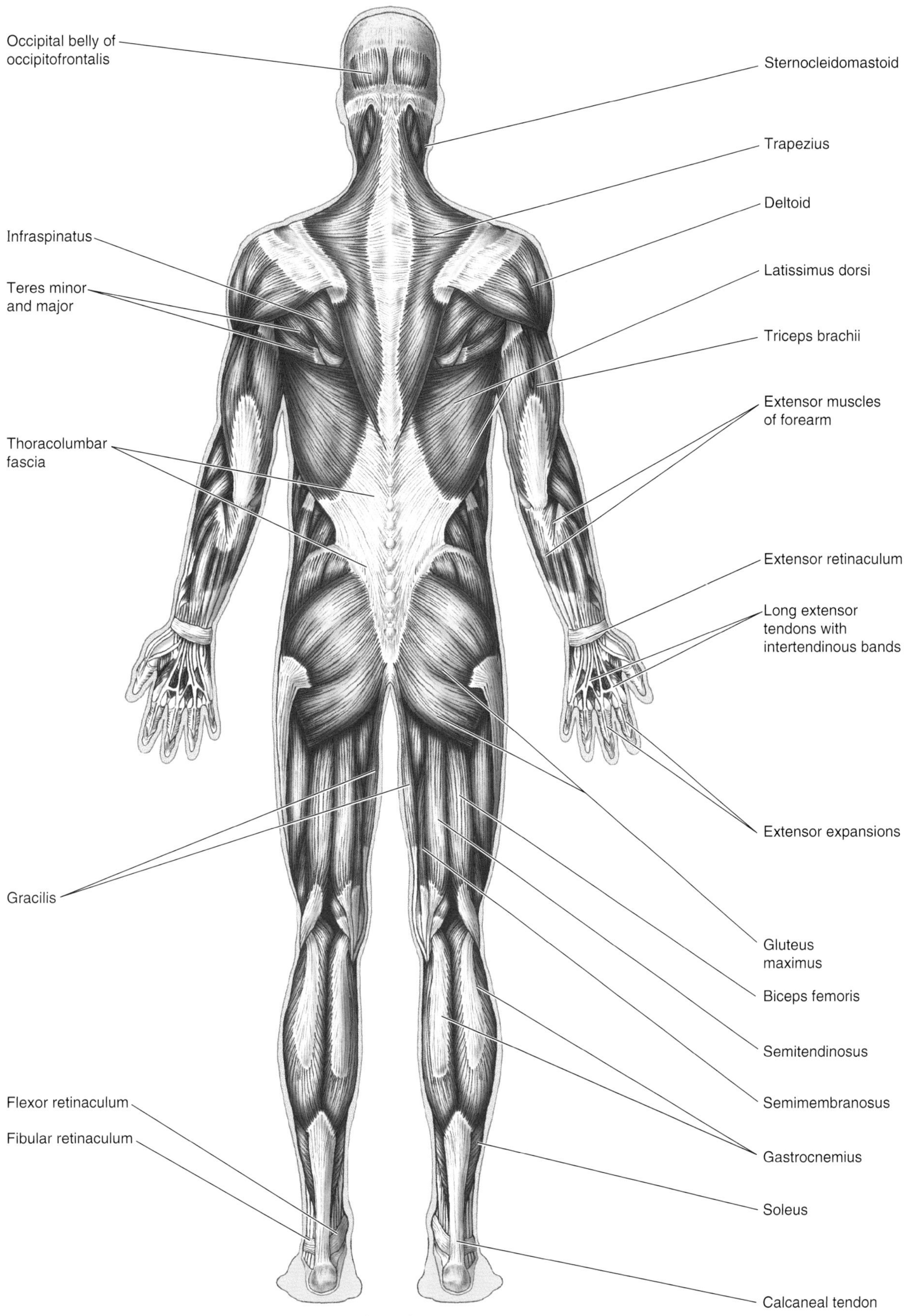

Occipital belly of occipitofrontalis
Sternocleidomastoid
Trapezius
Deltoid
Infraspinatus
Latissimus dorsi
Teres minor and major
Triceps brachii
Extensor muscles of forearm
Thoracolumbar fascia
Extensor retinaculum
Long extensor tendons with intertendinous bands
Extensor expansions
Gracilis
Gluteus maximus
Biceps femoris
Semitendinosus
Semimembranosus
Flexor retinaculum
Fibular retinaculum
Gastrocnemius
Soleus
Calcaneal tendon

Posterior

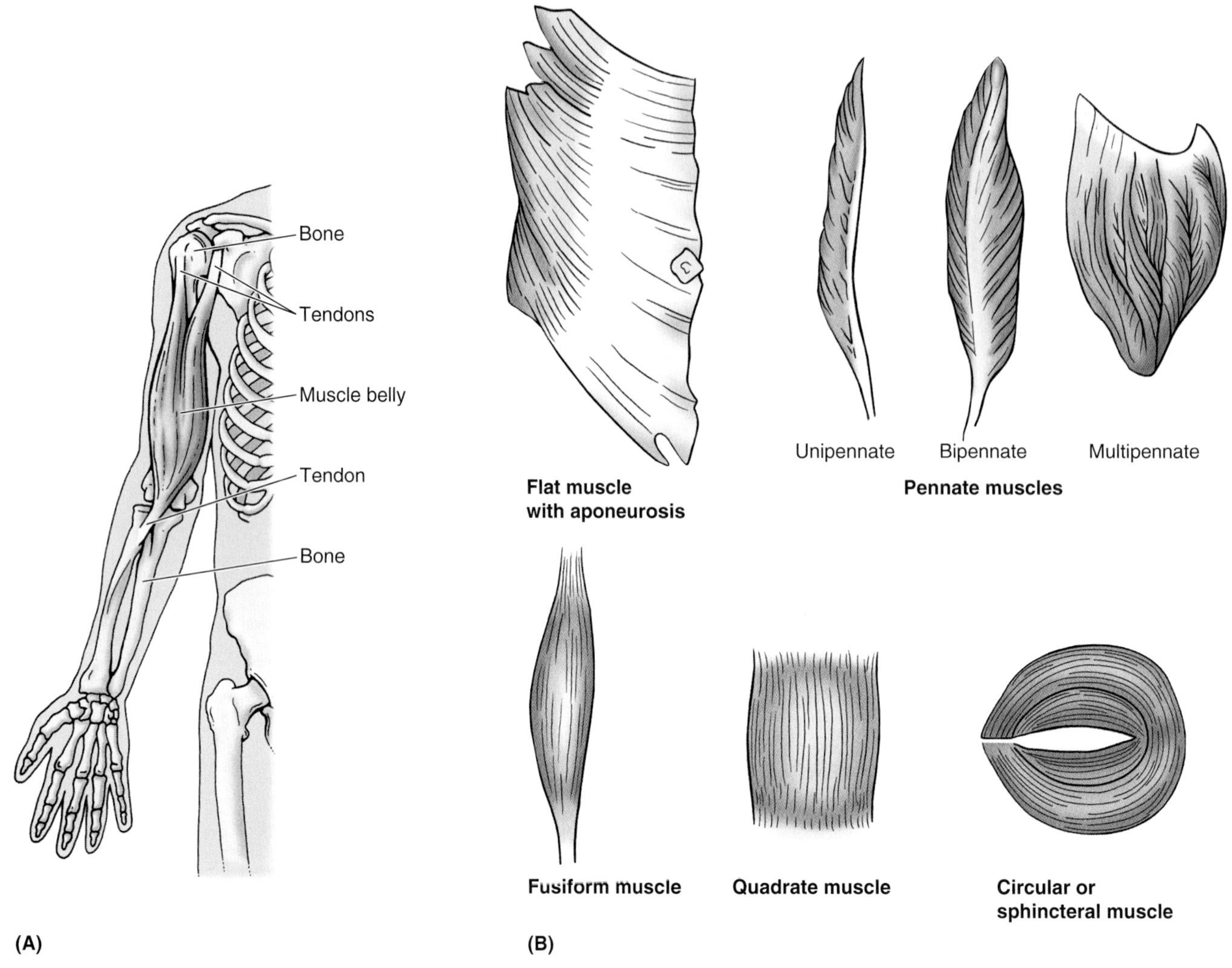

Figure I.16. Architecture and shape of skeletal muscles. A. The fleshy part of a muscle is its belly. In the example shown, the proximal and distal ends of the muscle are attached to bones by tendons. The length of the muscle is the distance between its bony attachments. **B.** Various types of muscle whose shapes are dependent on the arrangement of their fibers.

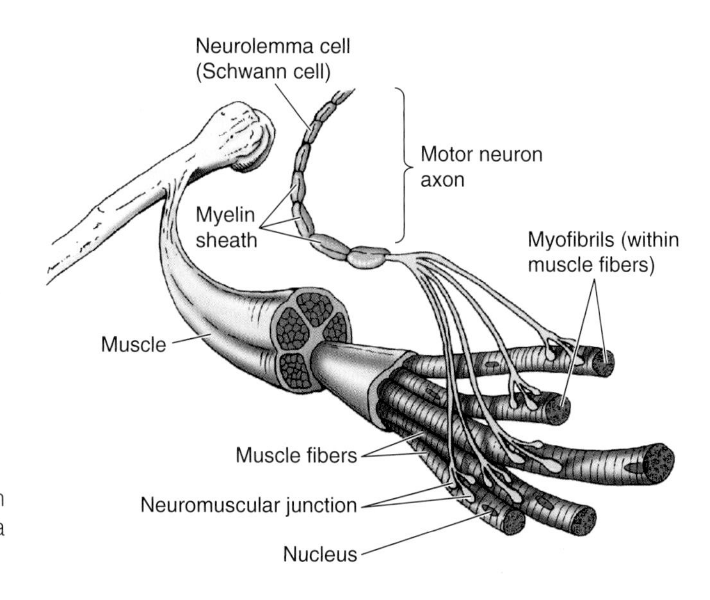

Figure I.17. Motor unit. The aggregate of a motor neuron axon and all the muscle fibers innervated by it constitutes a motor unit.

they are attached (Squires, 1986). The abductor digiti minimi muscle, for example, abducts the little finger. The sternocleidomastoid muscle (cleido—clavicle) attaches inferiorly to the sternum and clavicle and superiorly to the mastoid process of the temporal bone of the skull. Other muscles are named on the basis of their position (medial, lateral, anterior, or posterior) or length (brevis, short; longus, long). Muscles may be described or classified according to their shape (Figs. I.15 and I.16). For example:

- **Flat muscles** with parallel fibers often having an aponeurosis—e.g., the external oblique muscle
- **Pennate muscles**, which are featherlike (L. pennatus, feather) in the arrangement of their fascicles, and may be uni-, bi-, or multipennate—like the deltoid muscle
- A **fusiform muscle**, which is spindle-shaped (round, thick belly, and tapered ends)—e.g., the biceps brachii muscle
- A **quadrate muscle**, which has four equal sides (L. quadratus, square)—e.g., pronator quadratus muscle
- **Circular** or **sphincteral muscle**, which surrounds a body opening or orifice constricting it when contracted—e.g., the orbicularis oculi closes the eye.

The structural unit of a muscle is a muscle fiber. A **motor unit** is the functional unit consisting of a motor neuron and the muscle fibers it controls (Fig. I.17). When a nerve impulse reaches a motor neuron in the spinal cord, another impulse is initiated, which causes all the muscle fibers supplied by that motor unit to contract simultaneously. The number of muscle fibers in a motor unit varies from one to several hundred. The number of fibers varies according to the size and function of the muscle. Large motor units, where one neuron supplies several hundred muscle fibers, are in the large trunk and thigh muscles. In the small eye and hand muscles, where precision movements are required, the motor units include only a few muscle fibers.

Movements result from activation of an increasing number of motor units. During movements of the body, the main muscles are called into action.

- *Prime movers* or *agonists* are the main muscles that activate a specific movement of the body; they contract actively to produce the desired movement.
- *Antagonists* are muscles that oppose the action of prime movers; as a prime mover contracts, the antagonist progressively relaxes, producing a smooth movement.
- *Synergists* prevent movement of the intervening joint when a prime mover passes over more than one joint; they complement the action of the prime movers.
- *Fixators* steady proximal parts of a limb while movements are occurring in distal parts.

The same muscle may act as a prime mover, antagonist, synergist, or fixator under different conditions.

Growth and Regeneration of Skeletal Muscle

Skeletal muscle fibers cannot divide but they can be replaced individually by new muscle fibers derived from *satellite cells of skeletal muscle.* These cells are inside the basement membranes of the muscle fibers and represent a potential source of *myoblasts*—precursors of muscle cells—which are capable of fusing with each other to form new skeletal muscle fibers if required (Cormack, 1993; Gartner and Hiatt, 1997). The number of new fibers that can be produced is insufficient to compensate for major muscle degeneration or trauma. Instead of becoming regenerated effectively, the new skeletal muscle is composed of a disorganized mixture of muscle fibers and fibrous scar tissue. Skeletal muscles are able to grow larger in response to frequent strenuous exercise, such as occurs during body building. This growth results from *hypertrophy* of existing fibers, not from the addition of new muscle fibers. Hypertrophy lengthens and increases the *myofibrils* within the muscle fibers (Fig. I.17), thereby increasing the amount of work the muscle can perform.

Muscle Testing

Muscle testing helps an examiner diagnose nerve injuries. There are two common testing methods:

- The person performs movements that resist those of the examiner. For example, the person keeps the forearm flexed while the examiner attempts to extend it. This technique enables the examiner to gauge the power of the person's movements.
- The examiner performs movements that resist those of the person. When testing flexion of the forearm, the examiner asks the person to flex his or her forearm while the examiner resists the efforts. Usually muscles are tested in bilateral pairs for comparison.

Electromyography (EMG), the electrical stimulation of muscles, is another method for testing muscle action (Basmajian and DeLuca, 1985). The examiner places surface electrodes over a muscle and asks the person to perform certain movements, and then amplifies and records the differences in electrical action potentials of the muscles. A normal resting muscle shows only a baseline activity (tonus), which disappears only during deep sleep, paralysis, and when under anesthesia. Contracting muscles demonstrate variable peaks of phasic activity. EMG makes it is possible to analyze the activity of an individual muscle during different movements. EMG may also be part of the treatment program for restoring the action of muscles. ○

A muscle whose pull usually operates along the line of the bones to which it is attached (e.g., the brachioradialis, Fig. I.15, anterior view) exerts most of its force to maintain contact between the articular surfaces of the joint it crosses (i.e., it resists dislocating forces); this type of muscle is a **shunt muscle.** If the muscle is oriented more transverse to the bone it moves (e.g., the biceps brachii muscle), it is capable of rapid and effective movement; this type of muscle is a **spurt muscle.**

Cardiac Muscle

Cardiac muscle forms the muscular wall of the heart (i.e., it forms the *myocardium*). Some cardiac muscle is also present in the walls of the aorta, pulmonary vein, and superior vena cava (SVC).

Cardiac muscle contractions are not under voluntary control. Heart rate is regulated intrinsically by a *pacemaker* composed of special cardiac muscle fibers that are influenced by the autonomic nervous system (ANS) (p. 45). Cardiac muscle fibers are chains of cardiac muscle cells joined end to end by cell junctions (Cormack, 1993; Ross et al., 1994; Gartner and Hiatt, 1997). Cardiac muscle is also a type of "striated" muscle because of its striped appearance under microscopy (Table I.1). Cardiac muscle fibers have the same general pattern of striations as skeletal muscle fibers; however, they are traversed at intervals by **intercalated discs** (disks)—specialized end-to-end junctions—that are unique to cardiac muscle fibers. Most often the terms "striated" or "striped" refer to voluntary skeletal muscle. Most cardiac muscle cells have a single nucleus, but some contain two.

Smooth Muscle

Smooth muscle (Table I.1)—so-called because of the lack of striations in the appearance of the muscle fibers under microscopy—forms a large part of the middle coat or layer (tunica media) of the walls of most blood vessels (see Fig. I.19) and the muscular part of the wall of the digestive tract. Smooth muscle is also found in skin—arrector muscle of hair associated with hair follicles (Fig. I.6)—and in the eyeball, where it controls lens thickness and pupil size. Like cardiac muscle, smooth muscle is innervated by the ANS; hence, it is *involuntary muscle* that can undergo partial contraction for long periods. This is important in regulating the size of the lumina of tubular structures. In the walls of the digestive tract, uterine tubes, and ureters, the smooth muscle cells undergo rhythmic contractions (peristaltic waves). This process—*peristalsis*—propels the contents along these tubular structures.

Hypertrophy of the Myocardium and Myocardial Infarction

In *compensatory hypertrophy*, the myocardium responds to increased demands by increasing the size of its fibers. When cardiac muscle fibers are damaged by loss of their blood supply during a heart attack, the tissue becomes necrotic (dies) and the fibrous scar tissue that develops forms a **myocardial infarct** (MI)—an area of *myocardial necrosis* (pathologic death of cardiac tissue). Muscle cells that degenerate are not replaced because cardiac muscle cells do not divide. Furthermore, there is no equivalent to the satellite cells of skeletal muscle that can produce new cardiac muscle fibers.

Hypertrophy and Hyperplasia of Smooth Muscle

Smooth muscle cells undergo compensatory hypertrophy in response to increased demands. Smooth muscle cells in the uterine wall during pregnancy increase not only in size but also in number (*hyperplasia*) because these cells retain the capacity for cell division. In addition, new smooth muscle cells can develop from incompletely differentiated cells—**pericytes**—that are located along small blood vessels (Cormack, 1993; Gartner and Hiatt, 1997). ⊙

Cardiovascular System

The heart and blood vessels make up a blood transportation network—the *cardiovascular system* (Fig. I.18). Through this system, the heart pumps blood through the body's vast system of vessels. The blood carries nutrients, oxygen, and waste products to and from the cells. There are three types of blood vessel: arteries, veins, and capillaries. Blood under high pressure leaves the heart and is distributed to the body by a branching system of thick-walled **arteries** (Fig. I.19). The final distributing vessels—**arterioles**—deliver oxygenated blood to **capillaries**, which form the **capillary bed** where the interchange of oxygen, nutrients, waste products, and other substances with the extracellular fluid occurs. Blood from the capillary bed passes into thin-walled **venules**, which resemble wide capillaries. Venules drain into small veins that open into larger **veins**. The largest veins, the superior and inferior venae cavae, return poorly oxygenated blood to the heart.

Arteries

Arteries carry blood from the heart and distribute it to the body. The blood passes through arteries of ever decreasing caliber. Their walls have three coats or tunics:

- **Tunica adventitia**
- **Tunica media**
- **Tunica intima.**

The different types of artery are distinguished from each other on the basis of the thickness and differences in the

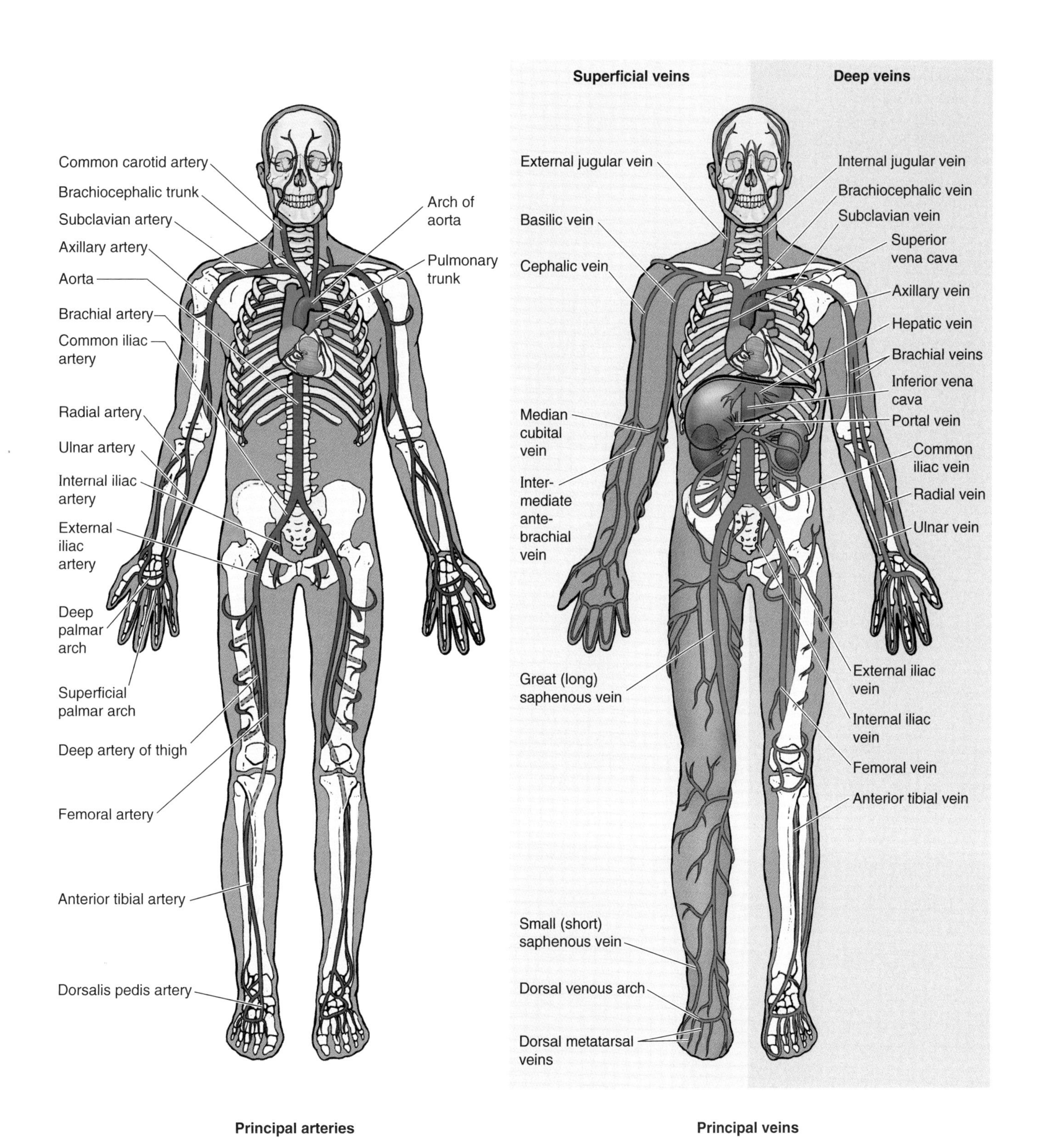

Figure I.18. Cardiovascular system. The heart and blood vessels are designed to distribute blood throughout the body. Most arteries carry oxygenated blood away from the heart and most veins carry deoxygenated blood to the heart; however, the pulmonary arteries arising from the pulmonary trunk carry deoxygenated blood to the lungs, and the pulmonary veins carry oxygenated blood from the lungs to the heart.

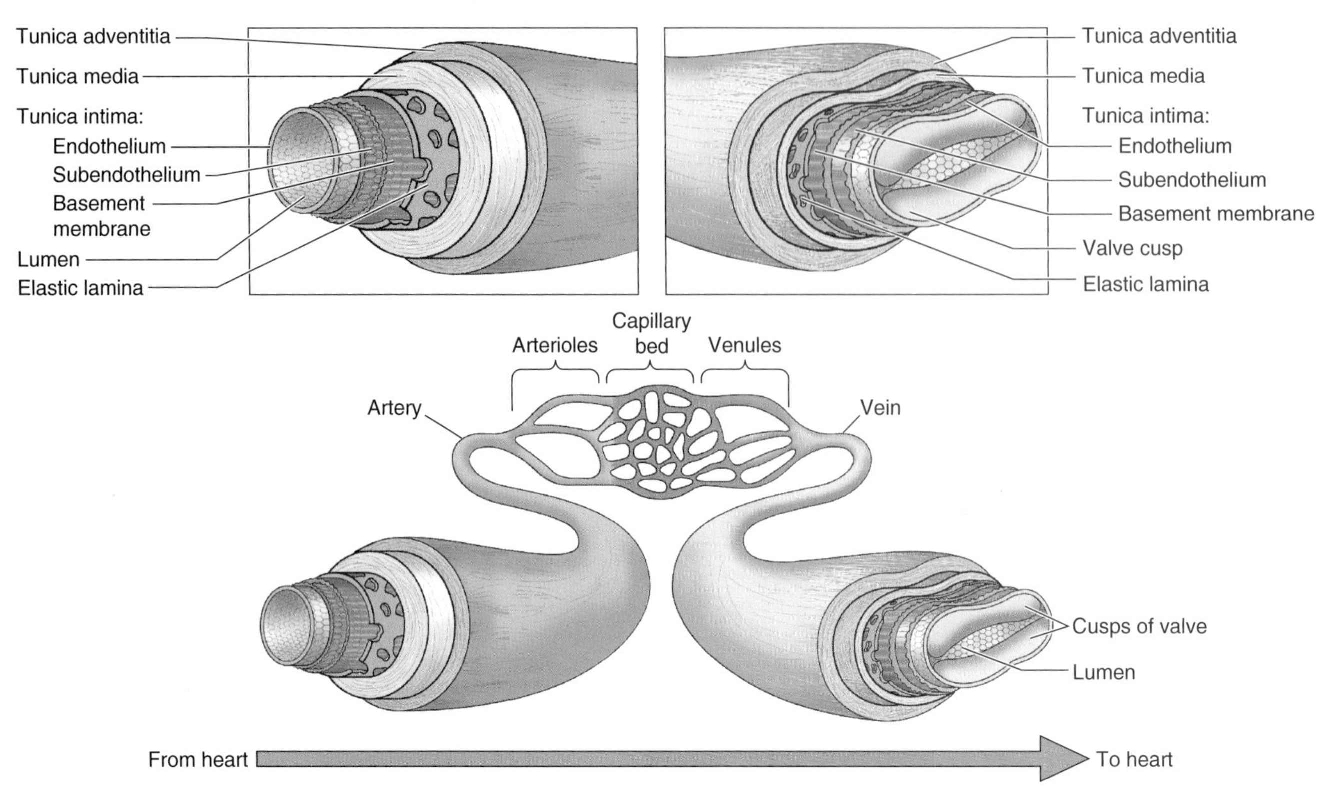

Figure I.19. Structure of blood vessels. The walls of blood vessels are constructed of three concentric coats (L. tunicae). With less muscle, veins are thinner walled than their companion arteries and have wide lumens (L. lumina) that usually appear flattened in tissue sections. Most medium-sized veins that return blood to the heart against the influence of gravity have valves, which break up the columns of blood and, if healthy, allow bloodflow only toward the heart.

Arteriosclerosis and Ischemic Heart Disease

The most common acquired disease of arteries is **arteriosclerosis** (hardening of the arteries)—a group of diseases characterized by thickening and loss of elasticity of the arterial walls. *Atherosclerosis*—a common form of arteriosclerosis—is associated with the buildup of fat (mainly cholesterol) in the arterial walls. Calcium deposits then form atheromatous plaques (*atheromas*)—well demarcated yellow areas or swellings on the intimal surfaces of arteries produced by lipid deposits. Expansion of an atherosclerotic lesion into the tunica intima of elastic and muscular arteries may result in the formation of a *thrombus* (blood clot), which may occlude the artery. The consequences of atherosclerosis include *ischemic heart disease*, resulting from inadequate blood supply, and *myocardial infarction* (necrosis of heart muscle and heart attack), stroke, and gangrene (e.g., in the distal parts of the limbs). ⊕

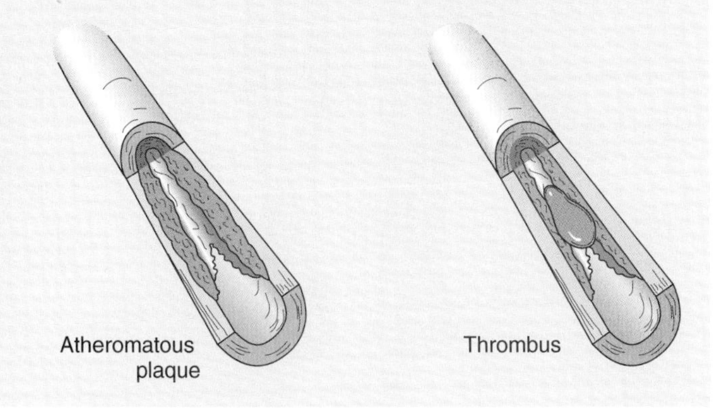

makeup of the coats, especially the tunica media. Artery size is a continuum; that is, there is a gradual change in morphological characteristics from one type to another (Gartner and Hiatt, 1977). There are three types of artery:

- **Elastic arteries** (conducting arteries) are the largest type; the aorta and branches originating from the arch of the aorta are good examples. The body is able to maintain blood pressure in the arterial system between contractions of the heart because of the elasticity of these arteries. This quality allows them to expand when the heart contracts and to return to normal between cardiac contractions.
- **Muscular arteries** (distributing arteries) such as the femoral artery distribute blood to various parts of the body. Their walls chiefly consist of circularly disposed

smooth muscle fibers, which constrict their lumina—the spaces in the interior of the arteries—when they contract. *Muscular arteries regulate the flow of blood* to different parts as required by the body.

- *Arterioles* are the smallest type; they have relatively narrow lumina and thick muscular walls. The degree of arterial pressure within the vascular system is regulated mainly by the degree of tonus (firmness) in the smooth muscle in the arteriolar walls. If the tonus is above normal, *hypertension* (high blood pressure) results.

Veins

Veins return blood from the capillary beds to the heart. The *large pulmonary veins are atypical in that they carry well-oxygenated blood* (commonly referred to as "arterial blood") from the lungs to the heart. Because of the lower blood pressure in the venous system, the walls of veins are thinner than those of their companion arteries (Fig. I.19). There are three sizes of veins: small, medium, and large. The smallest veins are *venules*. These tributaries unite to form larger veins that commonly join to form *venous plexuses*, such as the **dorsal venous arch** of the foot (Fig. I.18). *Medium-sized veins* in the limbs and in some other locations—where the flow of blood is opposed by the pull of gravity—have flap *valves* that permit blood to flow toward the heart but not in the reverse direction. A *large vein*, such as the SVC, is characterized by wide bundles of longitudinal smooth muscle and a well-developed tunica adventitia.

Veins tend to be double or multiple. Those that accompany deep arteries—**accompanying veins** (L. venae comitantes)—surround them in an irregular branching network (Fig. I.20). The accompanying veins occupy a relatively unyielding *vascular sheath* with the artery they accompany. As a result, they are stretched and flattened as the artery expands during contraction of the heart, which aids in driving venous blood toward the heart. Systemic veins are more variable than arteries, and *anastomoses*—natural communications, direct or indirect, between two veins—occur more often between them. The outward expansion of the bellies of contracting skeletal muscles in the limbs, limited by the deep fascia, compresses the veins, "milking" the blood superiorly toward the heart—the **musculovenous pump**—(Fig. I.21). The valves of the veins break up the columns of blood, thus relieving the more dependent parts of excessive pressure, and allow venous blood to flow only toward the heart. The venous congestion that hot and tired feet experience at the end of a busy day is relieved by resting the feet on a footstool that is higher than the trunk. This position of the feet also helps the veins to return blood to the heart.

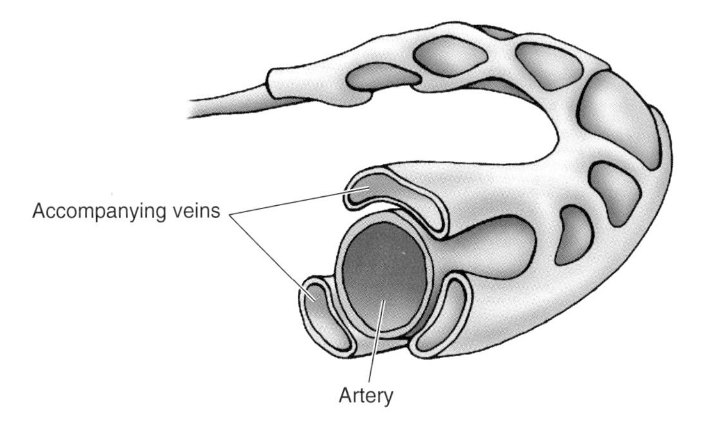

Figure I.20. Accompanying or companion veins (L. venae comitantes). Although most veins of the trunk occur as large single vessels, the veins in the limbs occur as two or more smaller vessels (companion veins), which accompany an artery in a common vascular sheath. Pulsations of the artery within the sheath aid venous return by compressing the smaller, multiple veins.

Varicose Veins

When the walls of veins lose their elasticity, they are weak. A weakened vein dilates under the pressure of supporting a column of blood against gravity. This results in *varicose veins*—abnormally swollen, twisted veins—most often seen in the legs. Varicose veins have a caliber greater than normal, and their valve cusps do not meet or have been destroyed by inflammation. These veins have *incompetent valves*; thus, the column of blood ascending toward the heart is unbroken, placing increased pressure on the weakened walls, further exacerbating the varicosity problem. Varicose veins also occur in the presence of degenerated deep fascia. Such incompetent fascia is incapable of containing the expansion of contracting muscles; thus, the musculovenous pump is ineffective (Fig. I.21). ○

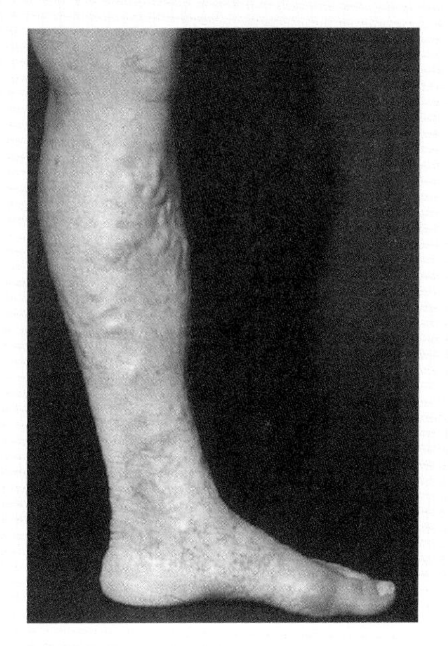

Varicose Veins

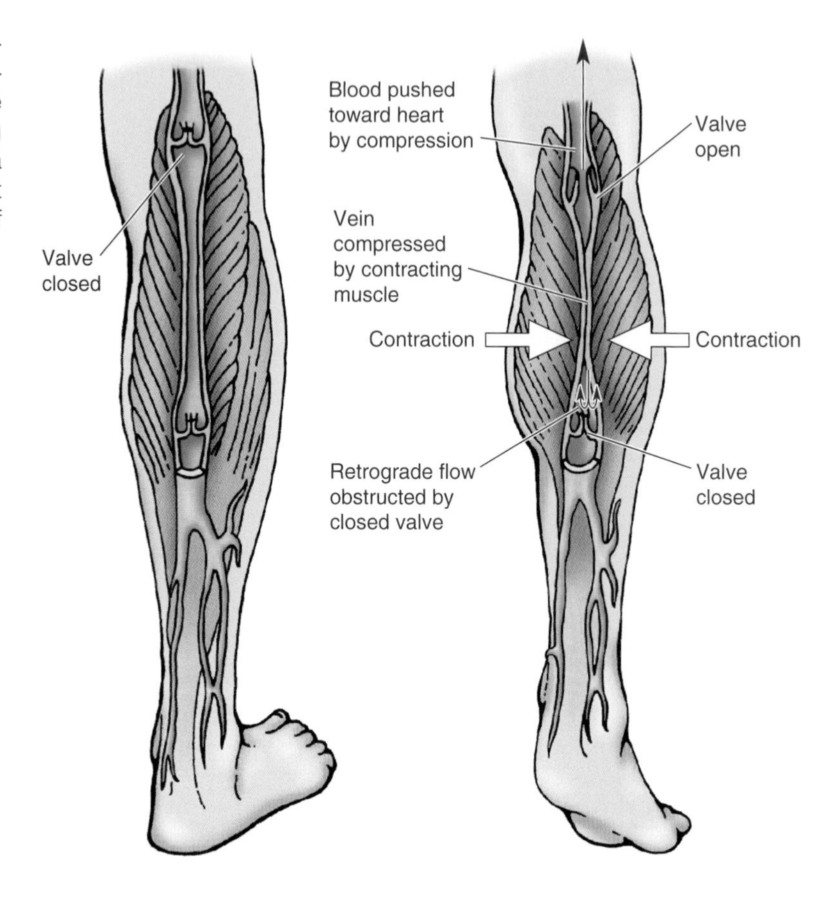

Figure I.21. The musculovenous pump. Muscular contractions in the limbs function with the venous valves to move blood toward the heart. The outward expansion of the bellies of contracting muscles is limited by deep fascia and becomes a compressive force, propelling the blood against gravity. The *black arrows* indicate the direction of bloodflow.

Capillaries

Capillaries are simple endothelial tubes connecting the arterial and venous sides of the circulation. They are generally arranged in networks—**capillary beds**—between the arterioles and venules (Fig. I.19). The blood flowing through the capillary beds is brought to them by **arterioles** and carried away from them by **venules.** As the hydrostatic pressure in the arterioles forces blood through the capillary bed, oxygen, nutrients, and other cellular materials are exchanged with the surrounding tissue. In some regions, such as in the fingers, there are direct connections between the small arteries and veins proximal to the capillary beds they supply and drain. The sites of such communications—*arteriovenous anastomoses (AV shunts)*—permit blood to pass directly from the arterial to the venous side of the circulation without passing through capillaries. *AV shunts are numerous in the skin,* where they have an important role in conserving body heat.

Lymphatic System

The lymphatic system is part of the circulatory system; the other part is the cardiovascular system. The lymphatic system is a vast network of **lymphatics** (lymphatic vessels) that are connected with **lymph nodes**—small masses of lymphatic tissue (Fig. I.22). The lymphatic system, which collects surplus tissue fluid as lymph, also includes **lymphatic organs** such as the spleen. Tissue fluid that enters and is conveyed by a lymphatic vessel is **lymph**, which is usually clear and watery and has the same constituents as blood plasma. The lymphatic system consists of:

- *Lymphatic plexuses*, networks of very small lymphatic vessels—*lymphatic capillaries*—that originate in the intercellular spaces of most tissues
- *Lymphatics*, a bodywide network of lymphatic vessels, originating from *lymphatic plexuses*, along which lymph nodes are located
- *Lymph nodes* through which lymph passes on its way to the venous system
- *Aggregations of lymphoid tissue* in the walls of the alimentary canal and in the spleen and thymus
- *Circulating lymphocytes* formed in *lymphoid tissue* such as lymph nodes and the spleen, and in *myeloid tissue* in red bone marrow.

After traversing one or more lymph nodes, lymph enters larger lymphatic vessels—*lymphatic trunks*—that unite to form either the **thoracic duct** or the **right lymphatic duct** (Fig. I.22). The thoracic duct begins in the abdomen as a sac—the **cisterna chyli** (chyle cistern)—and ascends through the thorax and enters the junction of the **left internal jugular** and **left subclavian veins**—the venous left angle.

- The right lymphatic duct drains lymph from the body's

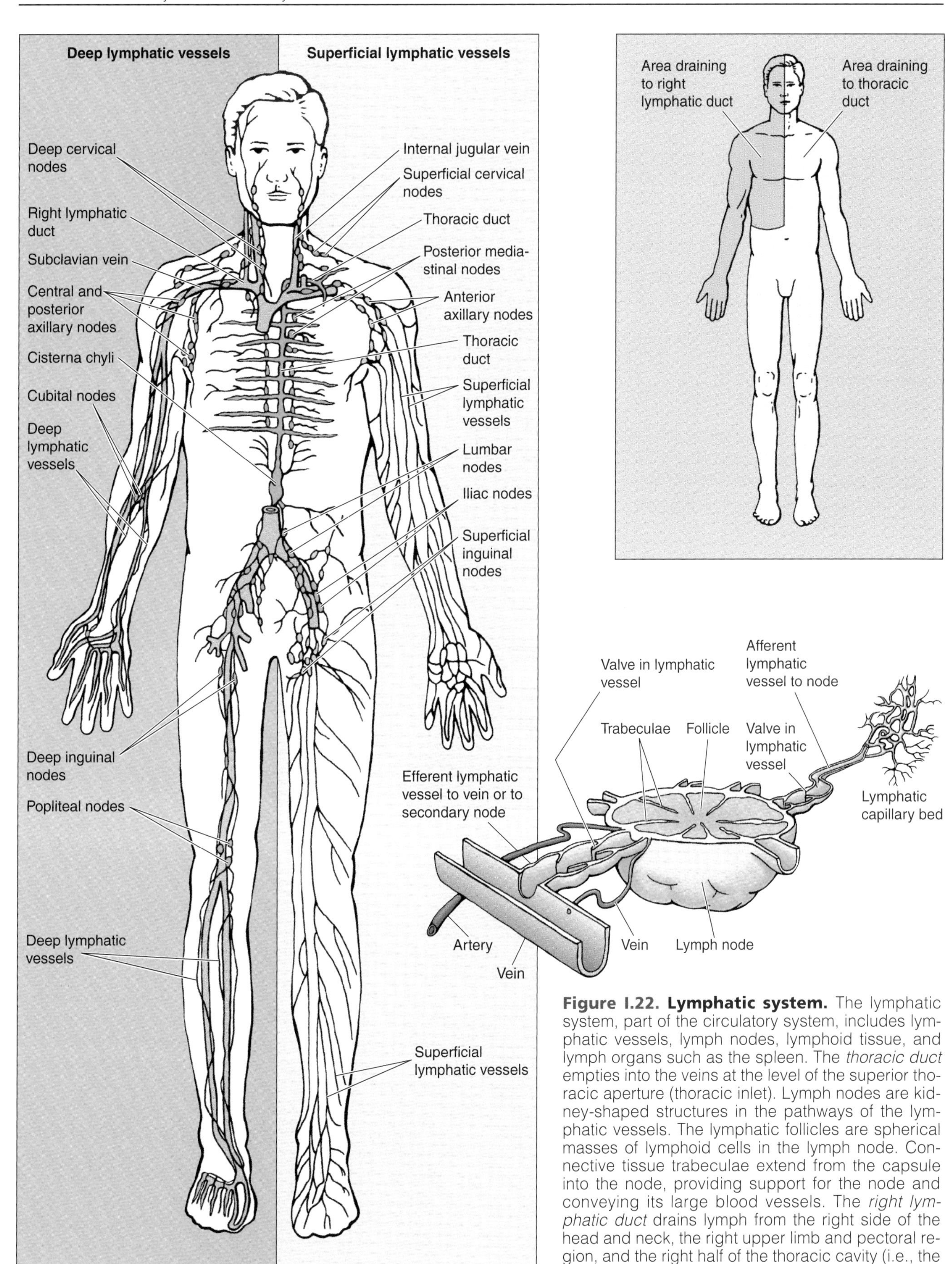

Figure I.22. Lymphatic system. The lymphatic system, part of the circulatory system, includes lymphatic vessels, lymph nodes, lymphoid tissue, and lymph organs such as the spleen. The *thoracic duct* empties into the veins at the level of the superior thoracic aperture (thoracic inlet). Lymph nodes are kidney-shaped structures in the pathways of the lymphatic vessels. The lymphatic follicles are spherical masses of lymphoid cells in the lymph node. Connective tissue trabeculae extend from the capsule into the node, providing support for the node and conveying its large blood vessels. The *right lymphatic duct* drains lymph from the right side of the head and neck, the right upper limb and pectoral region, and the right half of the thoracic cavity (i.e., the upper right quadrant). The *thoracic duct* drains lymph from the remainder of the body.

right upper quadrant (right side of the head and neck, the right upper limb, and the right half of the thoracic cavity).

- The thoracic duct drains lymph from the remainder of the body.

Superficial lymphatic vessels are in the skin and subcutaneous tissue (Fig. I.6). These vessels eventually drain into **deep lymphatic vessels** in the deep fascia between the muscles and the subcutaneous tissue, which accompany the major blood vessels. **The functions of lymphatics include**:

- *Drainage of tissue fluid, collection of lymph plasma* from the tissue spaces, and *transport of lymph* to the venous system
- *Absorption and transport of fat,* in which special lymphatic capillaries (*lacteals*) receive all absorbed fat from the intestine and convey the *chyle* through the thoracic duct to the venous system
- *Formation of a defense mechanism for the body*; when foreign protein drains from an infected area, antibodies specific to the protein are produced by immunologically competent cells and/or lymphocytes and dispatched to the infected area.

Lymphangitis, Lymphadenitis, and Lymphedema

Lymphangitis and lymphadenitis are the inflammation of lymphatic vessels and lymph nodes, respectively. These conditions may occur when the lymphatic system is involved in the spread (metastasis) of cancer cells. *Lymphedema*—the accumulation of interstitial fluid—occurs when lymph does not drain from an area of the body. For instance, if cancerous lymph nodes are surgically removed from the axilla (armpit), lymphedema of the limb may occur. Solid cell growths may permeate lymphatic vessels and form minute *cellular emboli* (plugs), which may break free and pass to regional lymph nodes. In this way, lymphogenous cancer cells spread to other tissues and organs. ⊙

Nervous System

The nervous system enables the body to react to continuous changes in its internal and external environments. It also controls and integrates the various activities of the body, such as circulation and respiration. For descriptive purposes, the nervous system is divided

- Structurally into the *central nervous system* (CNS) and *peripheral nervous system* (PNS)
- Functionally into the *somatic nervous system* (SNS) and *autonomic nervous system* (ANS).

Nervous tissue consists of two main types of cell: **neurons** (nerve cells) and **neuroglia** (glia cells) that support the neurons.

The neuron is the structural and functional unit of the nervous system that is specialized for rapid communication (Fig. I.23). It is composed of a **cell body** and processes—**dendrites** and an **axon**—that carry impulses to and away from the cell body, respectively. **Myelin**—layers of lipid and protein substances—form a **myelin sheath** around some axons, greatly increasing the velocity of impulse conduction. Neurons communicate with each other at **synapses**—points of contact between neurons. The communication occurs by means of *neurotransmitters*—chemical agents released or secreted by one neuron, which may excite or inhibit another neuron, continuing or terminating the relay of impulses or the response to them.

Neuroglia, which are approximately five times as abundant as neurons, are nonneuronal, nonexcitable cells that form a major component (scaffolding) of nervous tissue—supporting, insulating, and nourishing the neurons. In the CNS, neuroglia include the oligodendroglia cells, astrocytes, ependymal cells, and microglia—small neuroglial cells. In the PNS, neuroglia include the satellite cells around the neurons in the spinal ganglia (dorsal root ganglia) and the neurolemma (Schwann) cells that form the myelin and neurolemmal sheaths around peripheral nerve fibers (see Fig. I.26).

Central Nervous System

The CNS consists of the **brain** and **spinal cord** (Fig. I.24). The principal roles of the CNS are to:

- Integrate and coordinate incoming and outgoing neural signals
- Carry out higher mental functions such as thinking and learning.

A collection of nerve cell bodies in the CNS is a **nucleus.** A bundle of nerve fibers (axons) connecting neighboring or distant nuclei of the CNS is a **tract.**

Sections through the brain and spinal cord reveal that they are composed of gray matter and white matter. The nerve cell bodies lie within and constitute the **gray matter**; the interconnecting fiber tract systems form the **white matter** (Fig. I.25). In transverse sections of the spinal cord, the gray matter appears roughly as an H-shaped area embedded in a matrix of white matter. The struts (supports) of the H are **horns**; hence, there are right and left *dorsal (posterior) and ventral (anterior) gray horns.* Three membranous layers—pia mater, arachnoid mater, and dura mater, collectively constituting the **meninges**—and the *cerebrospinal fluid* (**CSF**) surround and protect the CNS. The brain and spinal cord are intimately covered on their outer surface by the innermost layer, a delicate, transparent covering—the **pia mater**. The CSF is located between the **pia mater** and the **arachnoid mater**; fine weblike strands joining the pia and arachnoid extend through

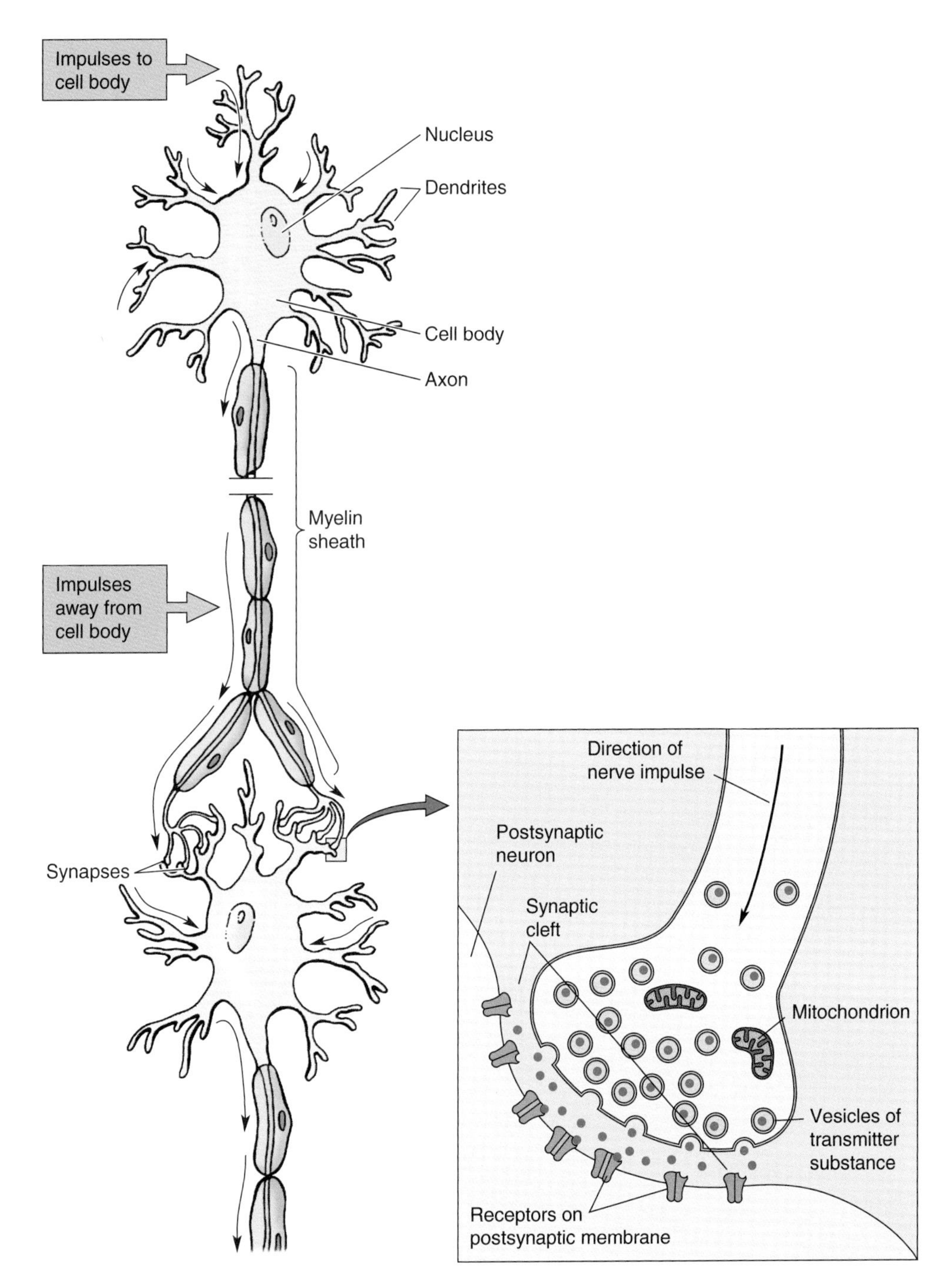

Figure I.23. Structure of a motor neuron. A neuron or nerve cell is composed of a cell body having a nucleus and two types of processes—dendrites and axons. Impulses pass to the cell body through the dendrites (*upper black arrows*) and the axon carries impulses away from the cell body (*lower black arrows*). A neuron influences other neurons at junctional points or synapses. The detailed structure of an *axodendritic synapse* is illustrated (*inset*); neurotransmitter substances diffuse across the narrow space (synaptic cleft) between the two cells and become bound to receptors.

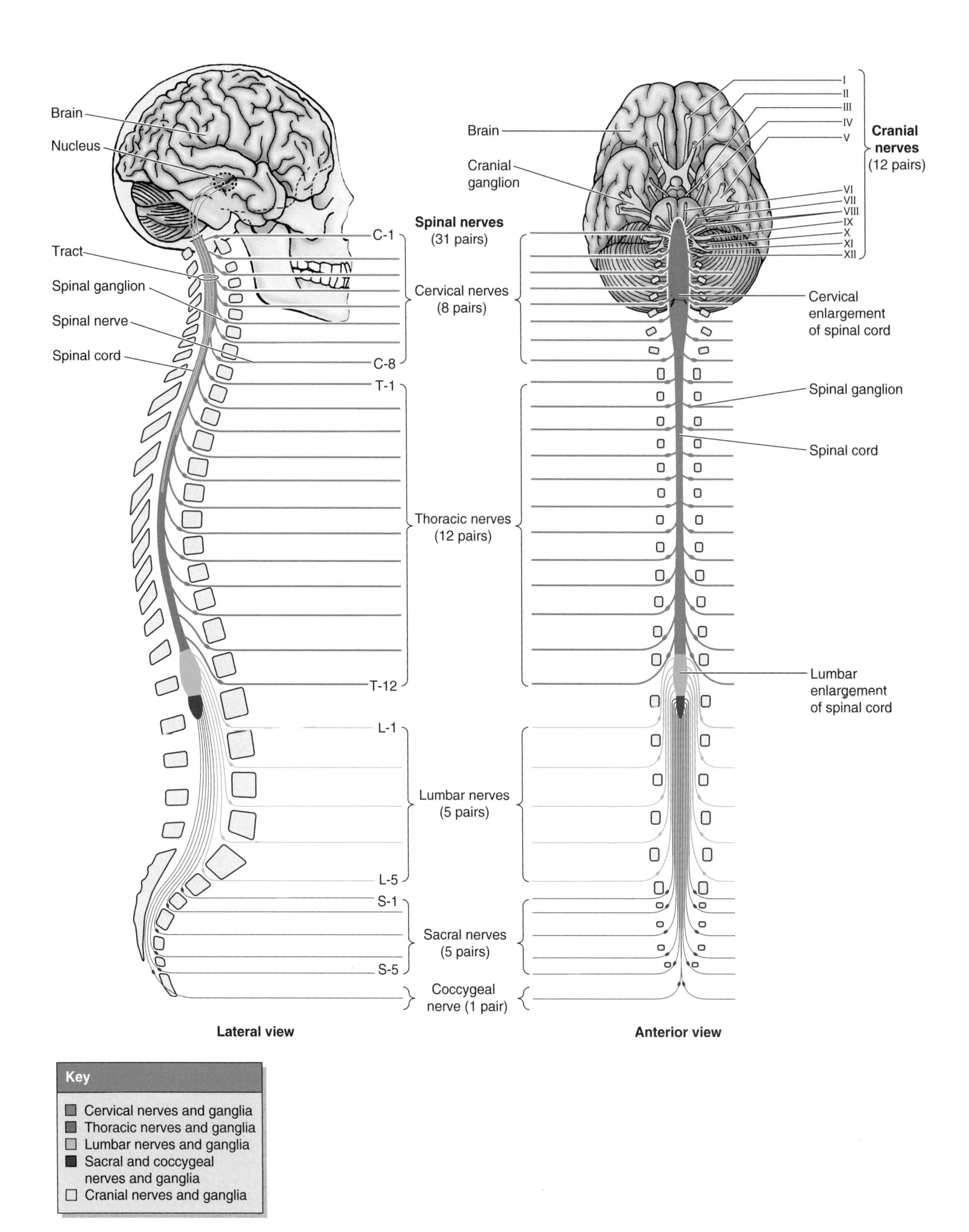

Figure I.24. Basic organization of the nervous system. The brain and spinal cord constitute the CNS. A collection of nerve cell bodies in the CNS is a *nucleus*, and a bundle of nerve fibers connecting neighboring or distant nuclei in the CNS is a *tract*. The PNS consists of nerve fibers and cell bodies outside the CNS. Peripheral nerves are either cranial or spinal nerves. A collection of nerve cell bodies outside the CNS is a ganglion (e.g., a cranial or spinal ganglion).

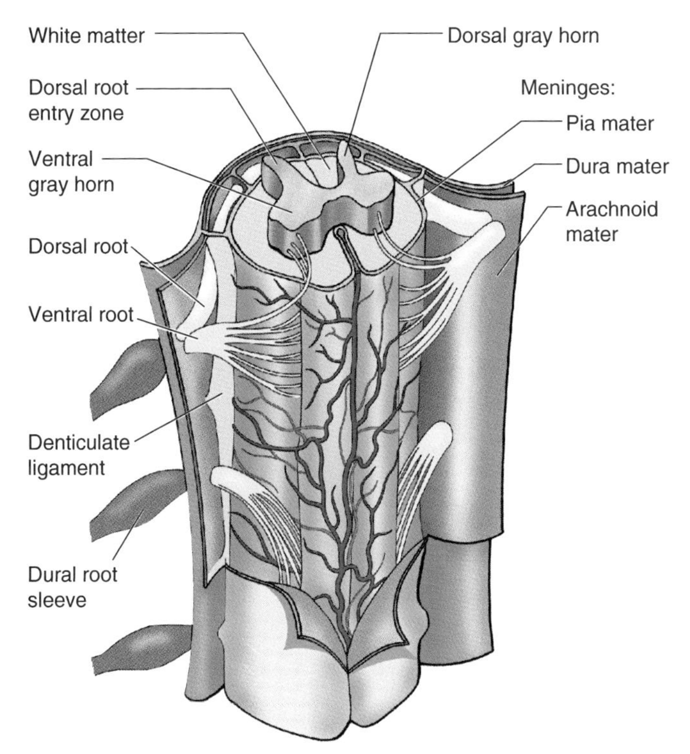

Figure I.25. Ventrolateral view of the spinal cord and spinal meninges—the membranes covering the spinal cord. The dura mater and arachnoid mater are incised and reflected to show the dorsal and ventral roots and the denticulate ligament. The spinal cord is sectioned to show its horns of gray matter. These membranes or coverings extend along the nerve roots to the point where the dorsal and ventral roots join, forming the dural root sleeves that enclose the sensory (dorsal root) ganglia. Distally, the coverings continue as the outer layer (epineurium) of the spinal nerve (see Fig. I.27).

the CSF. External to the pia and arachnoid is the thick, tough **dura mater**, intimately related to the bone of the internal aspect of the neurocranium (braincase).

Peripheral Nervous System

The PNS consists of nerve fibers and cell bodies outside the CNS that conduct impulses to or away from the CNS (Fig. I.24). The PNS is made up of nerves that connect the CNS with peripheral structures. A bundle of nerve fibers (axons) in the PNS, held together by a connective tissue sheath, is a **peripheral nerve,** a strong, whitish cord in living persons. A collection of nerve cell bodies outside the CNS is a ganglion—a **spinal ganglion**, for example.

Peripheral nerves are either cranial or spinal nerves. Eleven pairs of **cranial nerves** arise from the brain; the 12th pair arises mostly from the superior part of the spinal cord. All cranial nerves exit the cranial cavity through foramina (openings) in the cranium (G. kranion, skull). The 31 pairs of **spinal nerves** (cervical, thoracic, lumbar, sacral, and

Damage to the Brain and/or Spinal Cord

When the brain or spinal cord is damaged, in most circumstances the injured axons do not recover. Their proximal stumps begin to regenerate, sending sprouts into the area of the lesion; however, this growth stops in approximately 2 weeks. As a result, permanent disability follows destruction of a tract in the CNS. For discussions of axonal degeneration and regeneration in the CNS, see Barr and Kiernan (1993) and Hutchins et al. (1997). ○

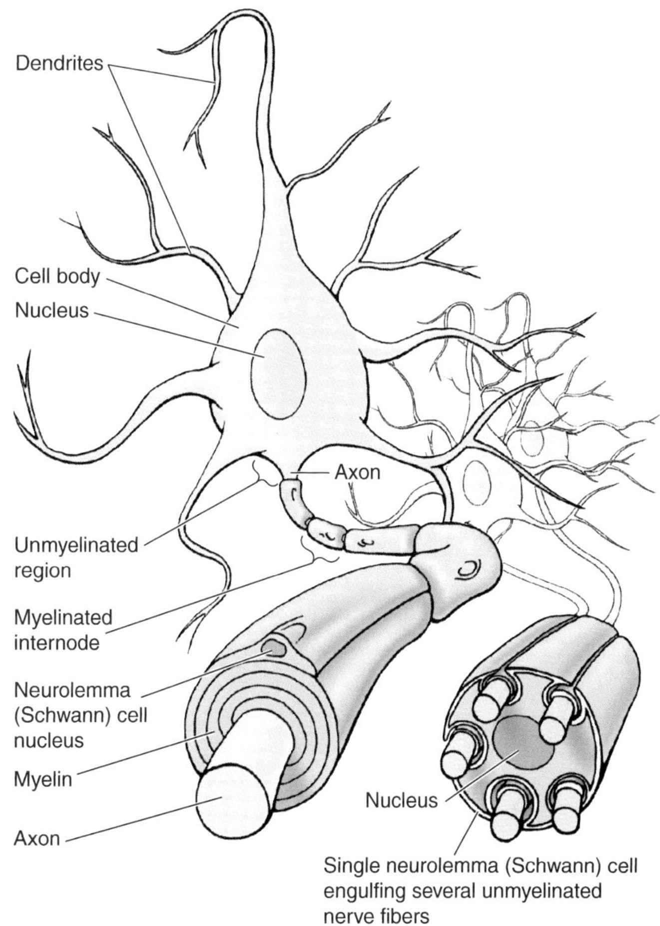

Figure I.26. Myelinated and unmyelinated peripheral nerve fibers. Myelinated nerve fibers have a neurolemmal sheath composed of a continuous series of neurolemma (Schwann) cells that surround the axon and form segments of myelin. The interruptions between segments are nodes of Ranvier. The segments of myelin between adjacent nodes of Ranvier are internodal segments, or internodes. Unmyelinated nerve fibers are engulfed in groups by a single neurolemma cell that does not produce myelin.

coccygeal) arise from the spinal cord and exit through intervertebral foramina in the vertebral column (spine).

The PNS is anatomically and operationally continuous with the CNS. Its *afferent*, or *sensory, fibers* convey neural impulses to the CNS from the sense organs (e.g., the eyes) and from sensory receptors in various parts of the body (e.g., in the skin). Its *efferent*, or *motor, fibers* convey neural impulses from the CNS to the effector organs (muscles and glands).

A **peripheral nerve fiber** (Figs. I.26 and I.27) consists of:

- An axon
- A neurolemmal sheath
- An endoneurial connective tissue sheath.

The **neurolemmal sheath** may take two forms, creating two classes of nerve fiber.

- *Myelinated nerve fibers* have a neurolemmal sheath consisting of a continuous series of Schwann cells that surround an individual axon and form myelin.
- *Unmyelinated nerve fibers* are engulfed in groups by a single neurolemma cell that does not produce myelin; most fibers in cutaneous nerves are unmyelinated.

Peripheral nerves are fairly strong and resilient because the nerve fibers are supported and protected by **three connective tissue coverings** (Fig. I.27):

- *Endoneurium*, a delicate connective tissue sheath that surrounds the neurolemma cells and axons
- *Perineurium*, which encloses a bundle (fascicle) of peripheral nerve fibers, providing an effective barrier against penetration of the nerve fibers by foreign substances
- *Epineurium*, a thick sheath of loose connective tissue that surrounds and encloses the nerve bundles, forming the outermost covering of the nerve; it includes fatty tissues, blood vessels, and lymphatics.

A peripheral nerve is much like a telephone cable: the axons being the individual wires insulated by the neurolemmal sheath (Schwann cells) and endoneurium, the insulated wires bundled by the perineurium, and the bundles surrounded in turn by the epineurium forming the outer wrapping of the "cable."

A **typical spinal nerve** arises from the spinal cord by rootlets, which converge to form two nerve roots (Fig. I.28).

- The *ventral (anterior) root* contains motor fibers from nerve cell bodies in the ventral horn of the spinal cord.
- The *dorsal (posterior) root* carries sensory fibers to the dorsal horn of the spinal cord.

The dorsal and ventral nerve roots unite to form a *mixed spinal nerve* that immediately divides into two rami

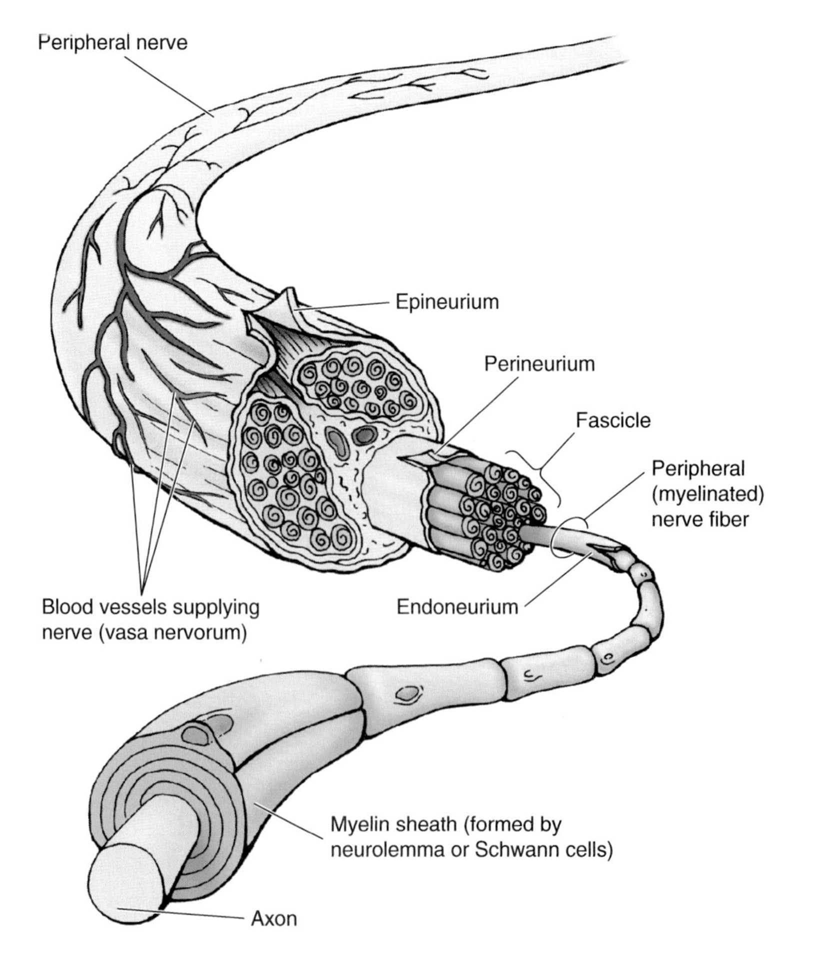

Figure I.27. Arrangement and ensheathment of peripheral, myelinated nerve fibers. All but the smallest peripheral nerves are arranged in bundles (fascicles), and the entire nerve is surrounded by the *epineurium*, a connective tissue sheath. Each small bundle of nerve fibers is also enclosed by a sheath—the *perineurium*. Individual nerve fibers have a delicate connective tissue covering—the endoneurium. The myelin sheath is formed by neurolemma (Schwann) cells.

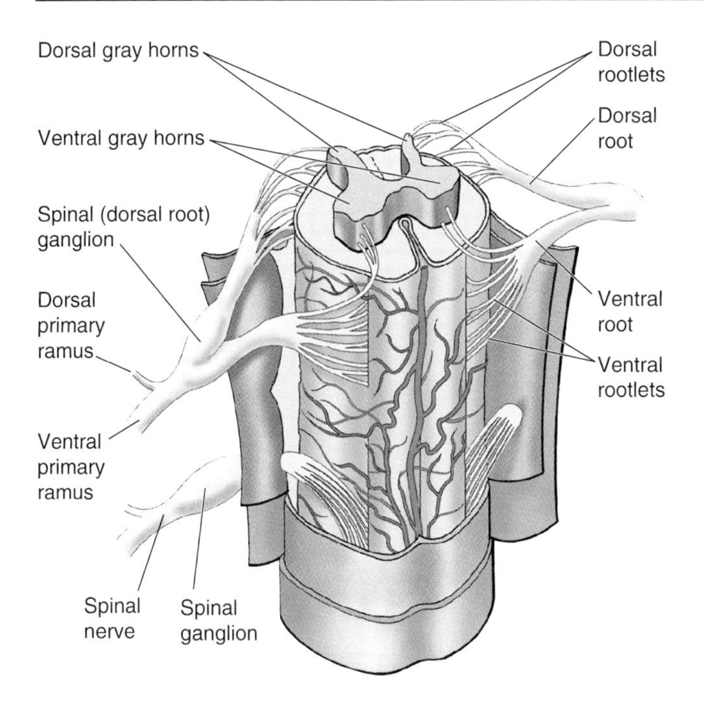

Figure I.28. Ventrolateral view of the spinal cord. The meninges are incised and reflected to show the H-shaped gray matter in the spinal cord, and the dorsal and ventral rootlets and roots of two spinal nerves. The dorsal and ventral rootlets enter and leave the dorsal and ventral gray horns, respectively. The dorsal and ventral nerve roots unite distal to the spinal (dorsal root) ganglion to form a *mixed spinal nerve*, which immediately divides into dorsal and ventral primary rami.

Rhizotomy

The dorsal and ventral roots are the only site where the motor and sensory fibers of a spinal nerve are segregated; therefore, only at this location can the surgeon selectively section either functional element for the relief of intractable pain or spastic paralysis.

Peripheral Nerve Degeneration and Ischemia of Nerves

Neurons do not proliferate in the adult nervous system except those related to the sense of smell in the olfactory epithelium. Therefore, neurons destroyed through disease or trauma are not replaced (Hutchins et al., 1997). When peripheral nerves are crushed or severed, their axons degenerate mainly distal to the lesion because they are dependent on their cell bodies for survival. If the axons are damaged but the cell bodies are intact, regeneration and return of function may occur in certain circumstances. The chance of survival is best when a peripheral nerve is compressed. Pressure on a nerve commonly causes **paresthesia** (pins and needles sensation) in normal people who cross their legs, for example.

A *crushing nerve injury* damages or kills the axons distal to the injury site; however, the neuronal cell bodies usually survive and the connective tissue coverings of the nerve remain intact. No surgical repair is needed for this type of nerve injury because the intact connective tissue sheaths guide the growing axons to their destinations. Regeneration is less likely to occur to a severed peripheral nerve. Sprouting occurs at the proximal ends of the axons; however, the growing axons may not reach their distal targets.

A *cutting nerve injury* requires surgical intervention because regeneration of the axon requires apposition of the cut ends by sutures through the epineurium. The individual bundles of nerve fibers are realigned as accurately as possible. *Anterograde (wallerian) degeneration* is the degeneration of axons detached from their cell bodies (Hutchins et al., 1997). The degenerative process involves the axon and its myelin sheath, even though this sheath is not part of the injured neuron.

Compromising a nerve's blood supply for a long period by compression of the vasa nervorum (Fig. I.27) can also cause nerve degeneration. Prolonged ischemia of a nerve may result in damage no less severe than that produced by crushing or even cutting the nerve. The *Saturday night syndrome*, named after an intoxicated individual who "passes out" with a limb dangling across the arm of a chair or the edge of a bed, is an example of a more serious, often permanent, paresthesia. This condition can also occur from sustained use of a tourniquet during a surgical procedure. If the *ischemia* (inadequate blood supply) is not too prolonged, temporary numbness or paresthesia results. *Transient paresthesias* are familiar to anyone who has had an injection of anesthetic for dental repairs. Pressure on a nerve also causes transient paresthesias (e.g., when people sit cross-legged on the floor or sit for a long time in one position, as on a toilet seat).

(branches): a *dorsal (posterior) primary ramus* and a *ventral (anterior) primary ramus*. As branches of the mixed spinal nerve, the dorsal and ventral rami carry both motor and sensory nerves, as do all their subsequent branches.

- The dorsal rami supply nerve fibers to the synovial joints of the vertebral column, deep muscles of the back, and the overlying skin.
- The ventral rami supply nerve fibers to the much larger remaining area, consisting of the anterior and lateral regions of the trunk and the upper and lower limbs arising from them.

The *components of a typical spinal nerve* (Fig. I.29) include:

- **Somatic fibers**
  —*General sensory (general somatic afferent)* fibers transmit sensations from the body to the spinal cord; they may be exteroceptive sensations (pain, temperature, touch, and pressure) from the skin, or pain and proprioceptive sensations from muscles, tendons, and joints. Propriocep-

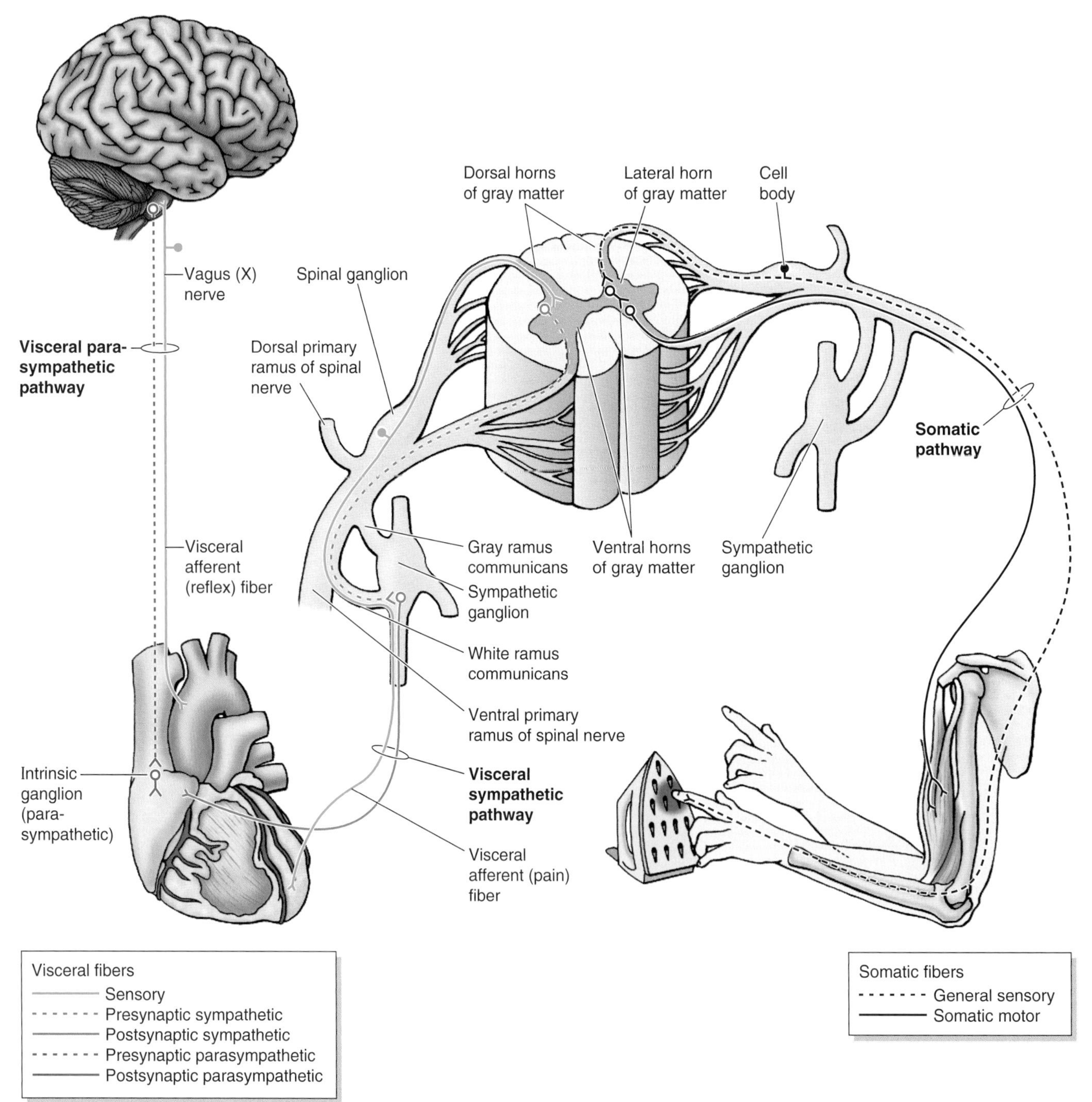

Figure I.29. Components of somatic (spinal) and visceral nerves. Somatic, visceral sympathetic, and visceral parasympathetic pathways are illustrated. The somatic motor system permits voluntary and reflexive movement by causing contraction of skeletal muscles, such as occurs when one touches a hot iron.

tive sensations are unconscious sensations that convey information on joint position and the tension of tendon and muscles, providing information on how the body and limbs are oriented in space.
—*Somatic motor (general somatic efferent)* fibers transmit impulses to skeletal (voluntary) muscles.

- **Visceral sensory and motor fibers**
  —*Visceral sensory (general visceral afferent)* fibers transmit reflex or pain sensations from mucous membranes, glands, and blood vessels.
  —*Visceral motor (general visceral efferent)* fibers transmit impulses to smooth (involuntary) muscle and glandular tissues. Two varieties of fibers—presynaptic and postsynaptic—team together to conduct impulses from the CNS to smooth muscle or glands. This is explained with the ANS (below).
- **Connective tissue coverings** illustrated in Figure I.27.
- **Vasa nervorum**, the blood vessels supplying the nerves.

Both types of sensory fiber—visceral sensory and general sensory—have their cell bodies in spinal ganglia or sensory ganglia of cranial nerves (Fig. I.24).

Somatic Nervous System

The SNS, composed of somatic parts of the CNS and PNS, provides sensory and motor innervation to all parts of the body (G. soma), except the viscera in the body cavities, smooth muscle, and glands (Fig. I.29). The *somatic sensory system* transmits sensations of touch, pain, temperature, and position from sensory receptors. The somatic motor system permits voluntary and reflexive movement by causing contraction of skeletal muscles, such as occurs when one touches a hot iron.

Autonomic Nervous System

The ANS, classically described as the *visceral motor system* (Fig. I.29), consists of fibers that innervate involuntary (smooth) muscle, modified cardiac muscle (the intrinsic stimulating and conducting tissue of the heart), and glands. However, the visceral efferent fibers of the ANS are accompanied by visceral afferent fibers. As the afferent component of autonomic reflexes and in conducting visceral pain impulses, these fibers also play a role in the regulation of visceral function. Thus, some authors consider the visceral afferent fibers to be part of the ANS. In any case, these fibers should be considered along with this system.

The efferent nerve fibers and ganglia of the ANS are organized into two systems or divisions:

- *Sympathetic* (thoracolumbar) division
- *Parasympathetic* (craniosacral) division.

Conduction of impulses from the CNS to the effector organ involves a series of two neurons in both systems. The cell body of the 1st *presynaptic*, or *preganglionic, neuron* is located in the gray matter of the CNS. Its fiber (axon) synapses only on the cell bodies of *postsynaptic*, or *postganglionic, neurons*, the 2nd neurons in the series. The cell bodies of the 2nd neurons are located in autonomic ganglia outside the CNS, with fibers terminating on the effector organ (smooth muscle, modified cardiac muscle, or glands).

The anatomical distinction between the two divisions of the ANS is based primarily on the location of the presynaptic cell bodies. A functional distinction pharmacologically important in medical practice is that the postsynaptic neurons of the two systems generally liberate different neurotransmitter substances: *norepinephrine* by the sympathetic division (except in the case of sweat glands) and *acetylcholine* by the parasympathetic division.

Sympathetic (Thoracolumbar) Division of the ANS

The cell bodies of the presynaptic neurons of the sympathetic division of the ANS are located in the **intermediolateral cell columns** or nucleus (IMLs) of the spinal cord (Fig. I.30). The paired (right and left) IMLs are a part of the gray matter extending between the 1st thoracic (T1) and the 2nd or 3rd lumbar (L2 or L3) segments of the spinal cord. In horizontal sections of this part of the spinal cord, the IMLs appear as small **lateral horns** of the H-shaped gray matter, looking somewhat like an extension of the cross-bar of the H between the dorsal and ventral horns.

The cell bodies of postsynaptic neurons of the sympathetic nervous system occur in two locations, the paravertebral and prevertebral ganglia (Fig. I.31).

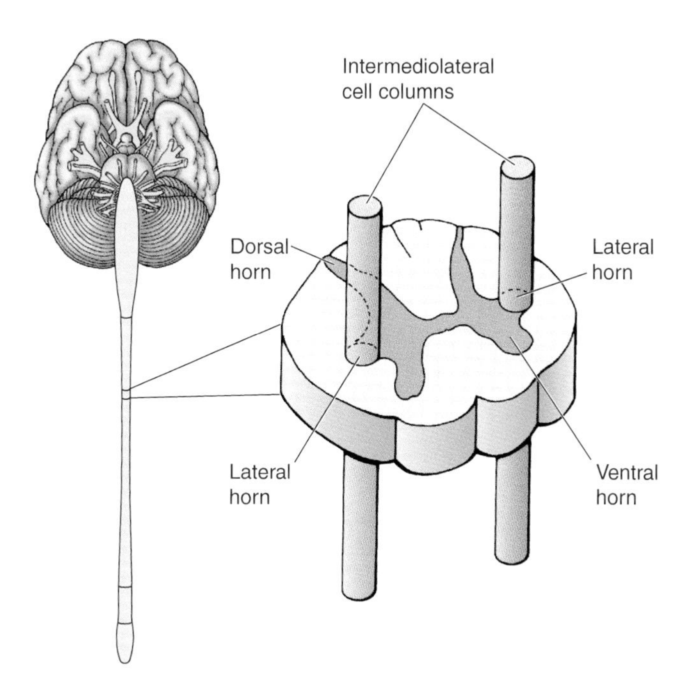

Figure I.30. Intermediolateral cell columns (IMLs). Each IML column or nucleus constitutes the lateral horn of gray matter of spinal cord segments T1 through L2 or L3, and consists of the cell bodies of the presynaptic neurons of the sympathetic nervous system.

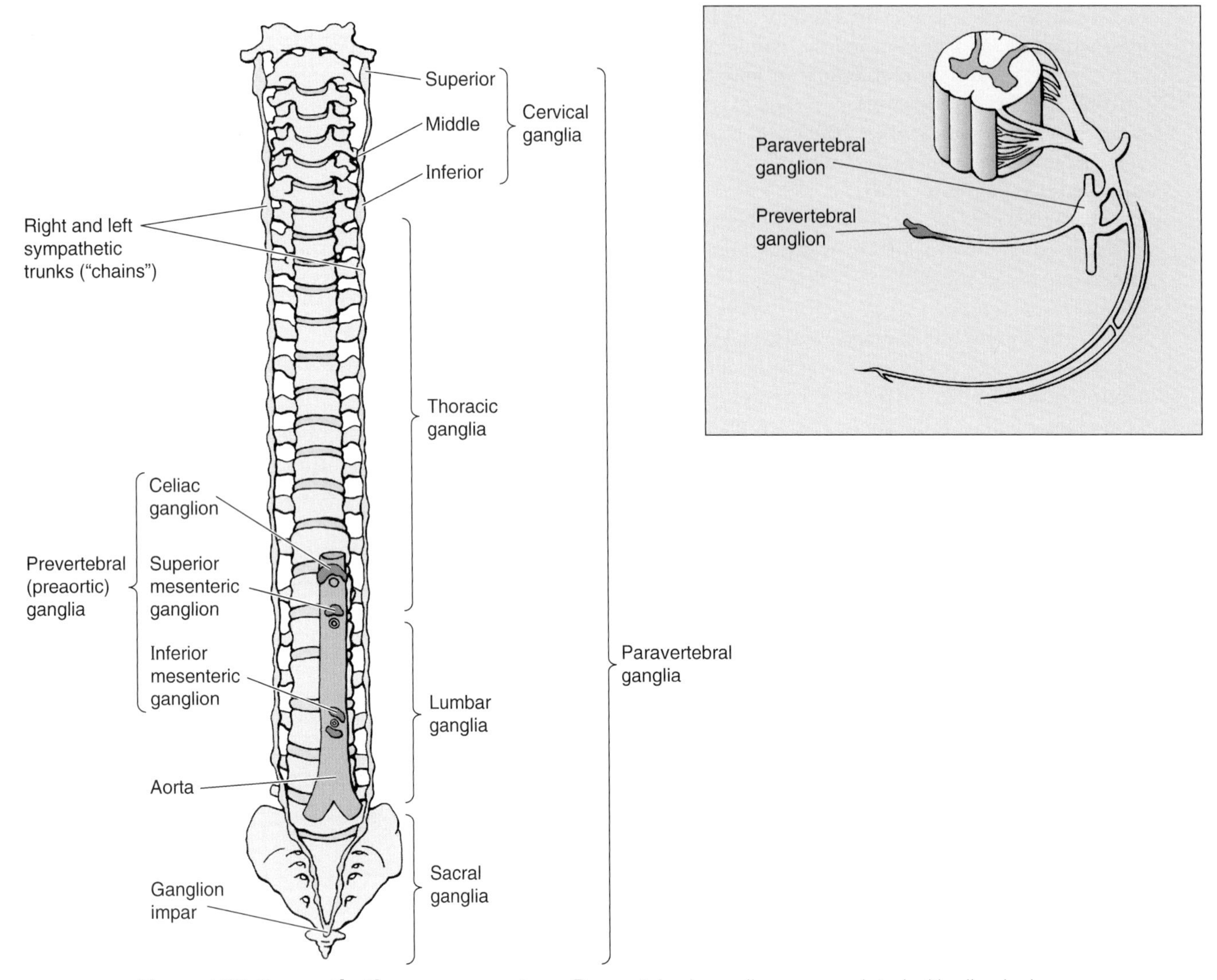

Figure I.31. Sympathetic nervous system. Paravertebral ganglia are associated with all spinal nerves, although at cervical levels eight spinal nerves share three ganglia—superior, middle, and inferior. The paravertebral ganglia are linked to form right and left sympathetic trunks (chains) on each side of the vertebral column. Prevertebral (preaortic) ganglia occur in the plexuses that surround the origins of the main branches of the abdominal aorta, such as the celiac artery, and are specifically involved in the innervation of abdominopelvic viscera—internal organs such as the stomach and the intestine.

- **Paravertebral ganglia** are linked to form right and left *sympathetic trunks (chains)* on each side of the vertebral column and extend essentially the length of this column. The superior paravertebral ganglion—the **superior cervical ganglion** of each sympathetic trunk—lies at the base of the skull. The **ganglion impar** forms inferiorly where the two trunks unite at the level of the coccyx.
- **Prevertebral ganglia** are in the plexuses that surround the origins of the main branches of the abdominal aorta, such as the two large *celiac ganglia*, which surround the origin of the *celiac trunk*, an artery arising from the aorta.

Because they are motor fibers, the axons of presynaptic neurons leave the spinal cord through ventral roots and enter the ventral rami of spinal nerves T1 through L2 or L3 (Fig. I.32). Almost immediately after entering the ventral rami, all the presynaptic sympathetic fibers leave the ventral primary rami of these spinal nerves and pass to the sympathetic trunks through white communicating branches (L. white rami communicantes). Within the sympathetic trunks, presynaptic fibers follow one of three possible courses:

- Enter and synapse immediately with a postsynaptic neuron of the paravertebral ganglion at that level
- Ascend or descend in the sympathetic trunk to synapse with a postsynaptic neuron of a higher or lower paravertebral ganglion
- Pass through the sympathetic trunk without synapsing, continuing on through an abdominopelvic splanchnic nerve to reach the prevertebral ganglia.

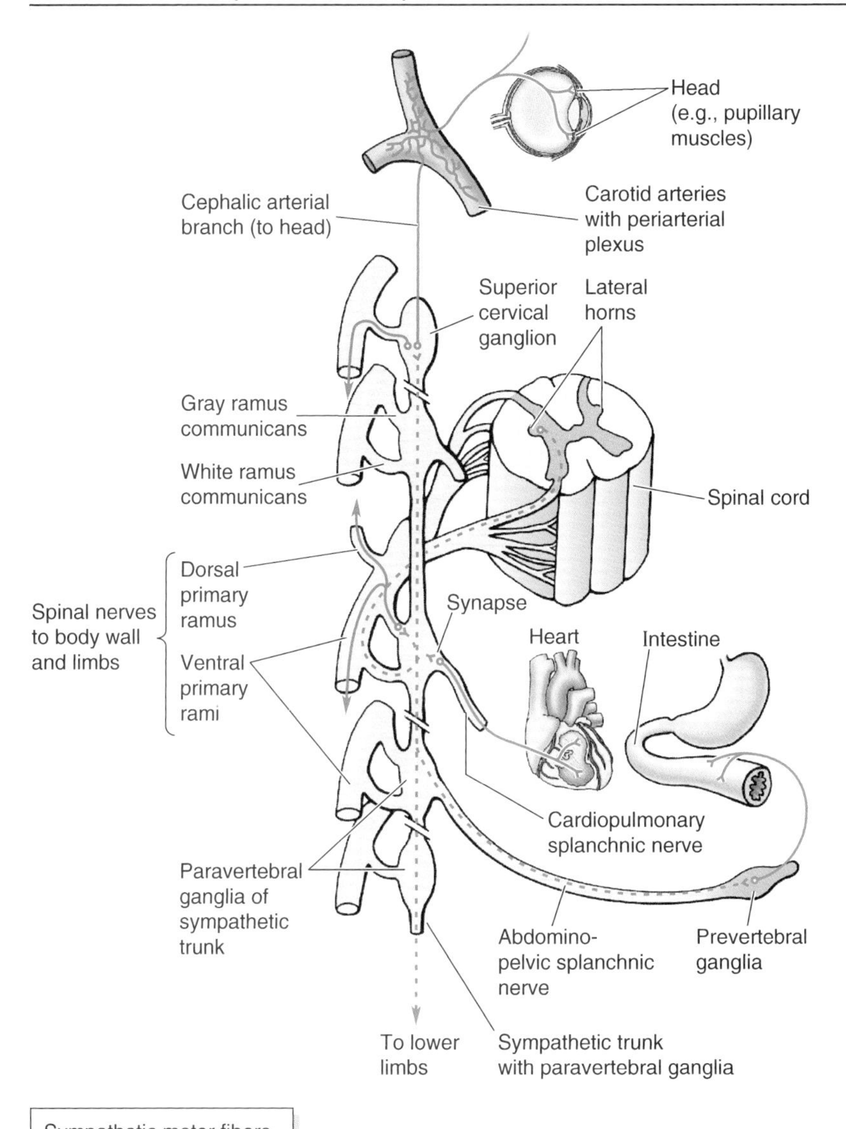

Figure I.32. Courses taken by sympathetic motor fibers. Presynaptic fibers in the sympathetic trunks follow one of three possible courses: (1) enter and synapse immediately with a postsynaptic neuron of the paravertebral ganglion at that level; (2) ascend or descend in the sympathetic trunk to synapse with a postsynaptic neuron of a higher or lower paravertebral ganglion; or (3) pass through the sympathetic trunk without synapsing, continuing on by means of an abdominopelvic splanchnic nerve to reach a prevertebral ganglion.

Presynaptic sympathetic fibers that provide autonomic innervation within the head, neck, body wall, limbs, and thoracic cavity will follow one of the first two courses, synapsing within the paravertebral ganglia. Presynaptic sympathetic fibers innervating viscera within the abdominopelvic cavity follow the 3rd course.

Postsynaptic sympathetic fibers destined for distribution within the neck, body wall, and limbs pass from the paravertebral ganglia of the sympathetic trunks to adjacent ventral rami of spinal nerves (Fig. I.33) through *gray communicating branches or gray rami communicantes*. They enter all branches of the spinal nerve, including the dorsal primary rami, to stimulate contraction of blood vessels (vasomotion) and arrector pili muscles associated with hairs (pilomotion resulting in "goose bumps"), and to cause sweating (sudomotion). Postsynaptic sympathetic fibers that perform these functions in the head (plus innervation of the dilator muscle of the iris) all have their cell bodies in the **superior cervical ganglion** at the superior end of the sympathetic trunk. They pass by means of a *cephalic arterial branch* (L. ramus) to form a **periarterial plexus** of nerves, which follows the branches of the carotid arteries to reach their destination.

Splanchnic nerves convey visceral efferent (autonomic) and afferent fibers to the viscera of the body cavities. Postsynaptic sympathetic fibers destined for the viscera of the thoracic cavity (e.g., the heart, lungs, and esophagus) pass through *cardiopulmonary splanchnic nerves* to enter the cardiac, pulmonary, and esophageal plexuses (Fig. I.32). The presynaptic sympathetic fibers involved in the innervation of viscera of the abdominopelvic cavity (e.g., the stomach and intestines) pass to the prevertebral ganglia through abdominopelvic splanchnic nerves (comprising the greater, lesser, least, and lumbar splanchnic nerves) (Figs. I.32–I.34). All presynaptic sympathetic fibers of the *abdominopelvic splanchnic nerves*, except those involved in innervating the suprarenal (adrenal) glands, synapse here. The postsynaptic fibers from the prevertebral

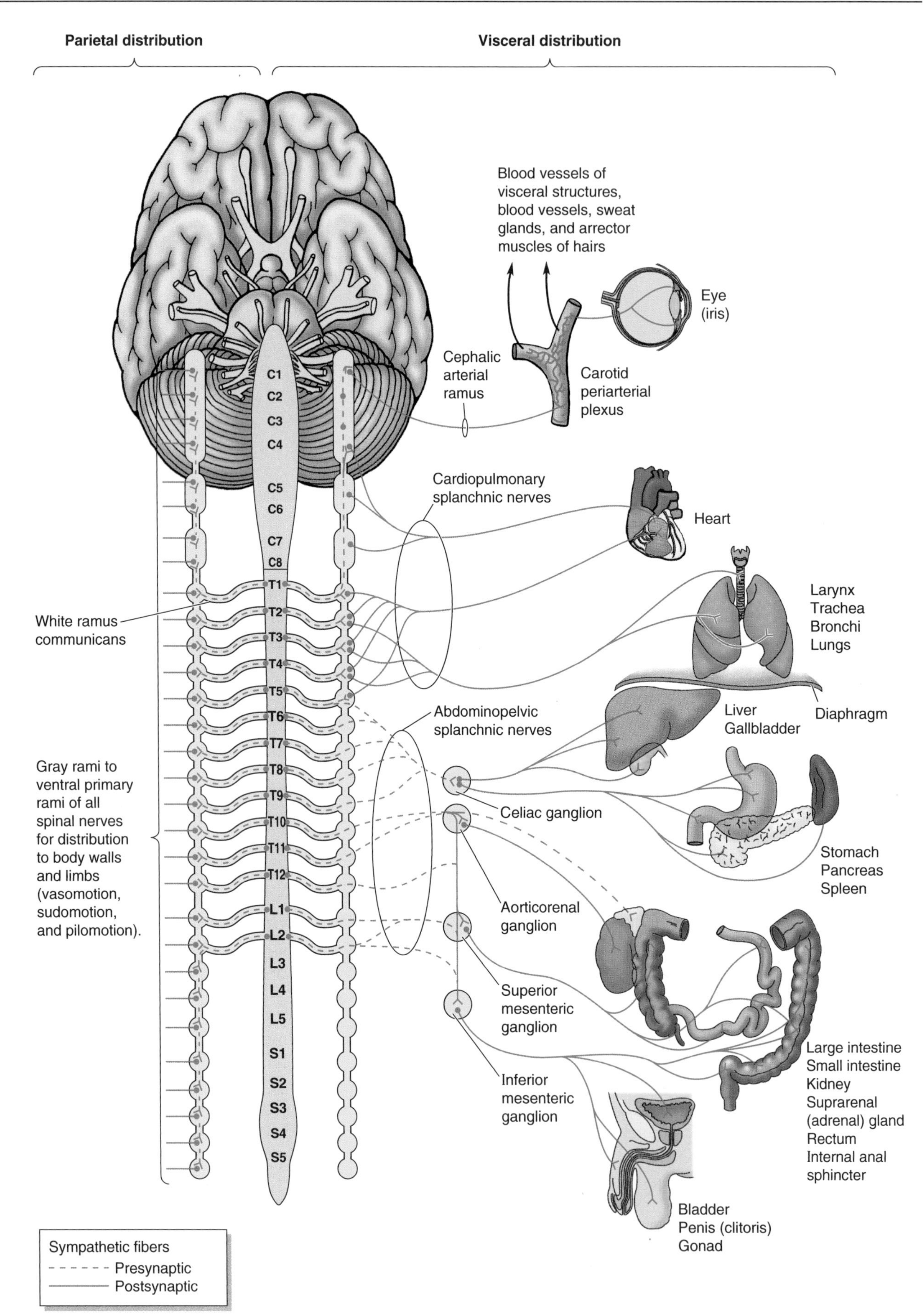

Parietal distribution
Visceral distribution
Blood vessels of visceral structures, blood vessels, sweat glands, and arrector muscles of hairs
Eye (iris)
Cephalic arterial ramus
Carotid periarterial plexus
C1
C2
C3
C4
C5
C6
C7
C8
T1
T2
T3
T4
T5
T6
T7
T8
T9
T10
T11
T12
L1
L2
L3
L4
L5
S1
S2
S3
S4
S5
Cardiopulmonary splanchnic nerves
Heart
Larynx
Trachea
Bronchi
Lungs
White ramus communicans
Diaphragm
Liver
Gallbladder
Abdominopelvic splanchnic nerves
Gray rami to ventral primary rami of all spinal nerves for distribution to body walls and limbs (vasomotion, sudomotion, and pilomotion).
Celiac ganglion
Stomach
Pancreas
Spleen
Aorticorenal ganglion
Superior mesenteric ganglion
Inferior mesenteric ganglion
Large intestine
Small intestine
Kidney
Suprarenal (adrenal) gland
Rectum
Internal anal sphincter
Bladder
Penis (clitoris)
Gonad
Sympathetic fibers
Presynaptic
Postsynaptic

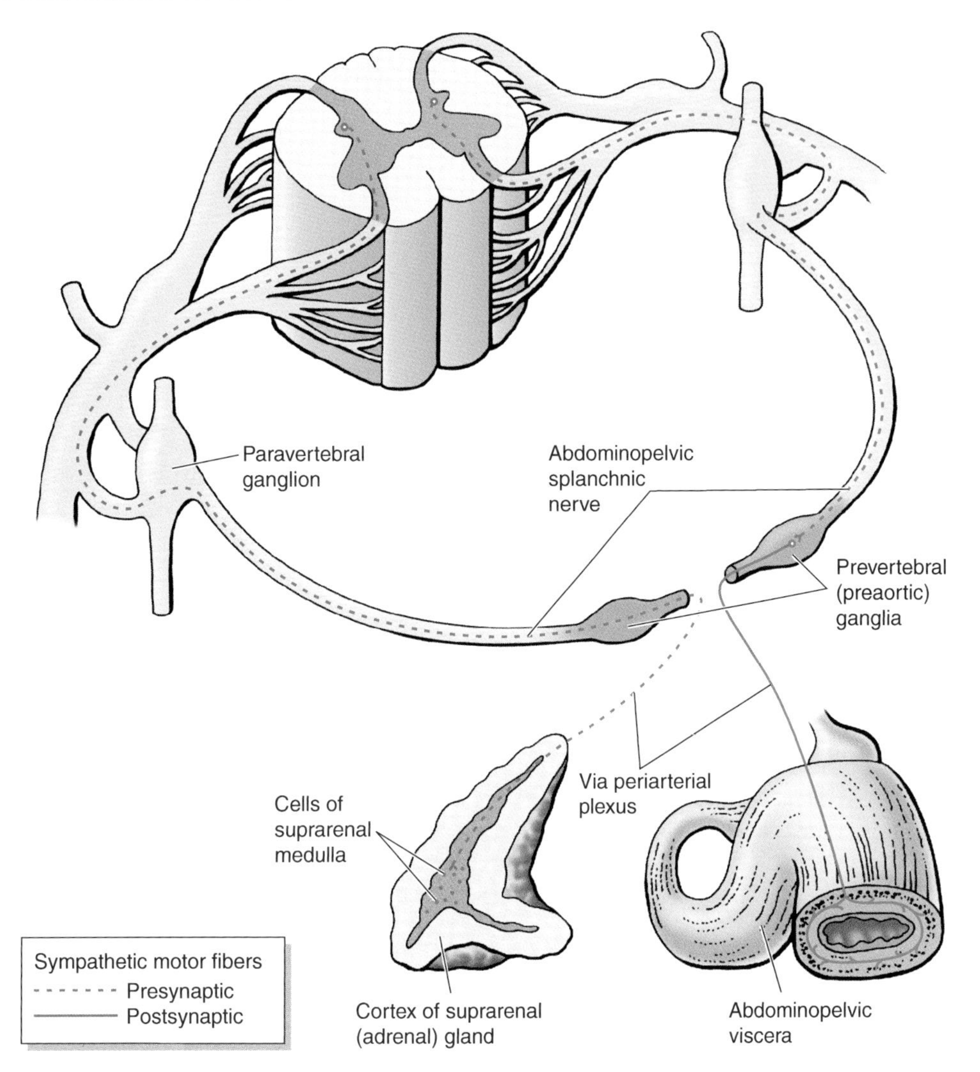

Figure I.34. Sympathetic supply to the medulla of the suprarenal gland. The sympathetic supply to this gland is exceptional. The secretory cells of the medulla are postsynaptic sympathetic neurons that lack axons or dendrites. Consequently, the suprarenal medulla is supplied directly by presynaptic sympathetic neurons. The neurotransmitters produced by medullary cells are released into the bloodstream to produce a widespread sympathetic response.

ganglia form periarterial plexuses, which follow branches of the abdominal aorta to reach their destination.

Presynaptic sympathetic fibers pass from the prevertebral (celiac) ganglia to terminate on cells in the medulla of the suprarenal gland (Fig. I.34). The suprarenal medullary cells function as a special type of postsynaptic neuron that instead of releasing their neurotransmitter substance onto the cells of a specific effector organ, release it into the bloodstream to circulate throughout the body, producing a widespread sympathetic response.

Figure I.33. Distribution of postsynaptic sympathetic nerve fibers. Fibers destined for distribution within the neck, body wall, and limbs pass from the paravertebral ganglia of the sympathetic trunks to adjacent ventral rami of all spinal nerves through gray *rami communicantes*. All postsynaptic sympathetic fibers distributed to the head have their cell bodies in the superior cervical ganglion of the sympathetic trunk. They pass by means of a cephalic arterial ramus to form a carotid periarterial plexus. Splanchnic nerves convey visceral efferent (autonomic) and afferent fibers to the viscera of the body cavities. Postsynaptic sympathetic fibers destined for the viscera of the thoracic cavity (e.g., the heart) pass through *cardiopulmonary splanchnic nerves*. The presynaptic sympathetic fibers involved in the innervation of viscera of the abdominopelvic cavity (e.g., the stomach) pass to the prevertebral ganglia through *abdominopelvic splanchnic nerves*. The postsynaptic fibers from the prevertebral ganglia form periarterial plexuses, which follow branches of the abdominal aorta to reach their destination.

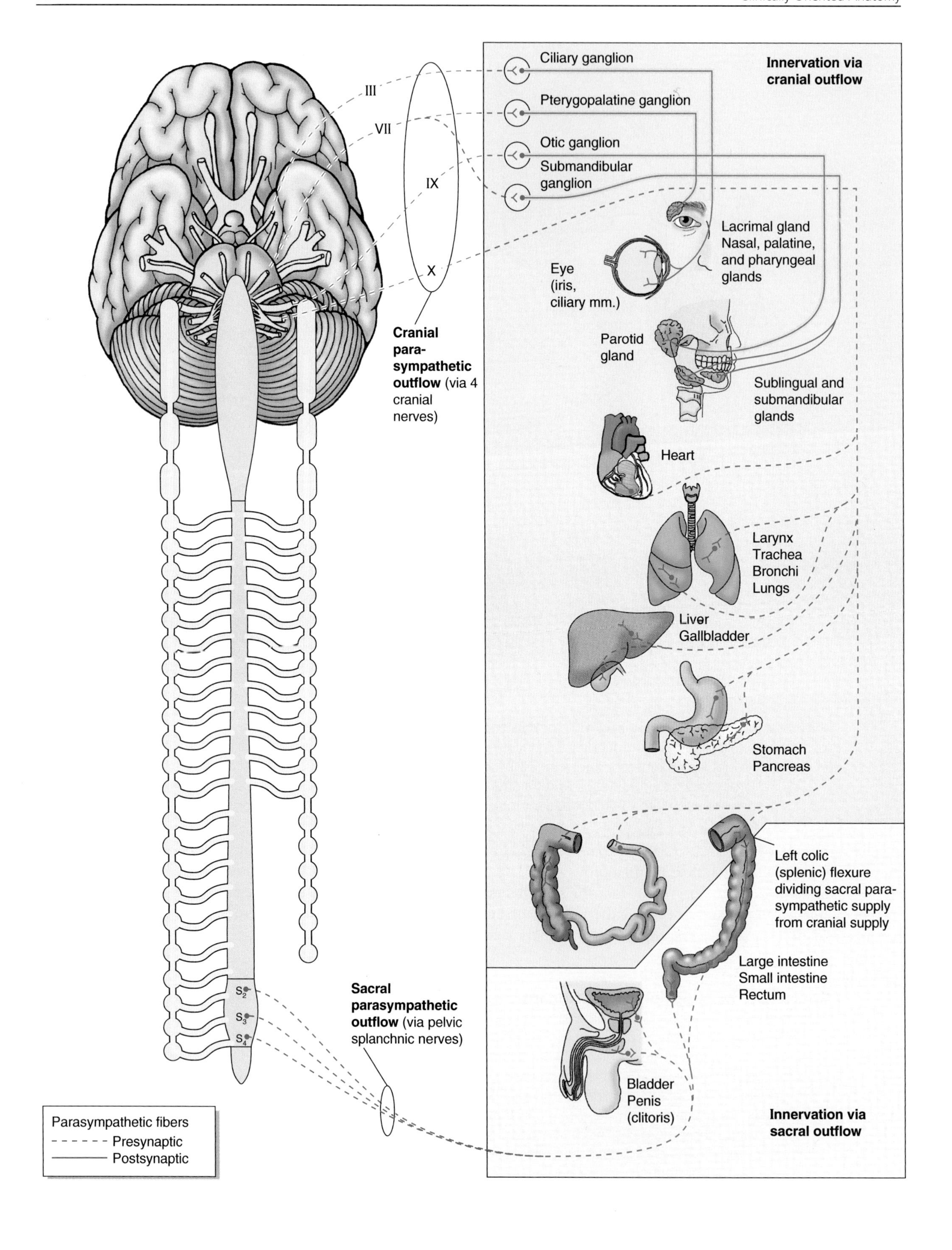

III
VII
IX
X
Cranial parasympathetic outflow (via 4 cranial nerves)
Ciliary ganglion
Pterygopalatine ganglion
Otic ganglion
Submandibular ganglion
Innervation via cranial outflow
Lacrimal gland
Nasal, palatine, and pharyngeal glands
Eye (iris, ciliary mm.)
Parotid gland
Sublingual and submandibular glands
Heart
Larynx
Trachea
Bronchi
Lungs
Liver
Gallbladder
Stomach
Pancreas
Left colic (splenic) flexure dividing sacral parasympathetic supply from cranial supply
Large intestine
Small intestine
Rectum
Bladder
Penis (clitoris)
Innervation via sacral outflow
S2
S3
S4
Sacral parasympathetic outflow (via pelvic splanchnic nerves)
Parasympathetic fibers
Presynaptic
Postsynaptic

Table I.2. Functions of the Autonomic Nervous System (ANS)

Organ, Tract, or System		Effect of Sympathetic Stimulation[a]	Effect of Parasympathetic Stimulation[b]
Eyes	Pupil	Dilates pupil (admits more light for increased acuity at a distance)	Constricts pupil (protects pupil from excessively bright light)
	Ciliary body		Contracts ciliary muscle, allowing lens to thicken for near vision (accommodation)
Skin	Arrector muscle of hair	Causes hairs to "stand on end" ("gooseflesh or goose bumps")	No effect (does not reach)[c]
	Peripheral blood vessels	Vasoconstricts (blanching of skin, lips, and turning fingertips blue)	No effect (does not reach)[c]
	Sweat glands	Promotes sweating[d]	No effect (does not reach)[c]
Other glands	Lacrimal glands	Slightly decreases secretion[e]	Promotes secretion
	Salivary glands	Secretion decreases, becomes thicker, more viscous[e]	Promotes abundant, watery secretion
Heart		Increases the rate and strength of contraction; inhibits the effect of parasympathetic system on coronary vessels, allowing them to dilate[e]	Decreases the rate and strength of contraction (conserving energy); constricts coronary vessels in relation to reduced demand
Lungs		Inhibits effect of parasympathetic system, resulting in bronchodilation and reduced secretion, allowing for maximum air exchange	Constricts bronchi (conserving energy) and promotes bronchial secretion
Digestive tract		Inhibits peristalsis, and constricts blood vessels to digestive tract so that blood is available to skeletal muscle; contracts internal anal sphincter to aid fecal continence	Stimulates peristalsis and secretion of digestive juices Contracts rectum, inhibits internal anal sphincter to cause defecation
Liver and gallbladder		Promotes breakdown of glycogen to glucose (for increased energy)	Promotes building/conservation of glycogen; increases secretion of bile
Urinary tract		Vasoconstriction of renal vessels slows urine formation; internal sphincter of bladder contracted to maintain urinary continence	Inhibits contraction of internal sphincter of bladder, contracts detrusor muscle of the bladder wall causing urination
Genital system		Causes ejaculation and vasoconstriction resulting in remission of erection	Produces engorgement (erection) of erectile tissues of the external genitals
Suprarenal medulla		Release of adrenaline into blood	No effect (does not innervate)

Underlying general principles:

[a] In general, the effects of sympathetic stimulation are catabolic—preparing body to "flee or fight."

[b] In general, the effects of parasympathetic stimulation are anabolic—promoting normal function and conserving energy.

[c] The parasympathetic system is restricted in its distribution to the head, neck, and body cavities (except for erectile tissues of genitalia); otherwise, parasympathetic fibers are never found in the body wall and limbs. Sympathetic fibers, by comparison, are distributed to all vascularized portions of the body.

[d] With the exception of the sweat glands, glandular secretion is parasympathetically stimulated.

[e] With the exception of the coronary arteries, vasoconstriction is sympathetically stimulated; the effects of sympathetic stimulation on glands (other than sweat glands) are the indirect effects of vasoconstriction.

As described earlier, postsynaptic sympathetic fibers are components of virtually all branches of all spinal nerves. By this and other means, they extend to and innervate all the body's blood vessels, sweat glands, and many other structures. Thus, the sympathetic nervous system reaches virtually all parts of the body with the rare exception of avascular tissues such as cartilage and nails.

Parasympathetic (Craniosacral) Division of the ANS

Presynaptic parasympathetic neuron cell bodies are located in

Figure I.35. Parasympathetic (craniosacral) division of the ANS. The presynaptic parasympathetic nerve cell bodies are located in the CNS. Their fibers exit by two roots: (1) In the gray matter of the brainstem, with fibers exiting the CNS within cranial nerves II, VII, IX, and X; these fibers constitute the *cranial parasympathetic outflow*; (2) In the gray matter of the sacral (S2–S4) segments of the spinal cord, with fibers exiting the CNS via the ventral roots of spinal nerves S2 through S4 and the pelvic splanchnic nerves that arise from their ventral rami; these fibers constitute the sacral parasympathetic outflow. The cranial outflow provides parasympathetic innervation of the head, neck, and most of the trunk; the sacral outflow provides the parasympathetic innervation of the pelvic viscera.

two sites within the CNS, their fibers exiting by two routes (Fig. I.35):

- In the gray matter of the brainstem, the fibers exit the CNS within cranial nerves III, VII, IX, and X; these fibers constitute the *cranial parasympathetic outflow*.
- In the gray matter of the sacral segments of the spinal cord (S2 through S4), the fibers exit the CNS through the ventral roots of spinal nerves S2 through S4 and the pelvic splanchnic nerves that arise from their ventral rami; these fibers constitute the *sacral parasympathetic outflow*.

Not surprisingly, the cranial outflow provides parasympathetic innervation of the head, and the sacral outflow provides the parasympathetic innervation of the pelvic viscera. However, in terms of the innervation of thoracic and abdominal viscera, the cranial outflow through the vagus nerve (CN X) is dominant. It provides innervation to all the thoracic viscera and most of the gastrointestinal (GI) tract from the esophagus through most of the large bowel (to its left colic flexure). The sacral outflow supplies only the descending and sigmoid colon and rectum.

Regardless of the extensive influence of its cranial outflow, the parasympathetic system is much more restricted than the sympathetic system in its distribution. The parasympathetic system distributes only to the head, visceral cavities of the trunk, and erectile tissues of the external genitalia. With the exception of the latter, it does not reach the body wall or limbs, and except for the initial parts of the ventral rami of spinal nerves S2 through S4, its fibers are not components of spinal nerves or their branches.

Four discrete pairs of parasympathetic ganglia occur in the head (details are in Chapters 7 and 9). Elsewhere, presynaptic parasympathetic fibers synapse with postsynaptic cell bodies that occur singly in or on the wall of the target organ (*intrinsic* or *enteric ganglia*).

Functions of the Divisions of the ANS

Although both sympathetic and parasympathetic systems innervate involuntarily and often affect the same structures, they have different (usually contrasting) but coordinated effects (Figs. I.33 and I.34, Table I.2). In general, the sympathetic system is a catabolic (energy-expending) system that enables the body to deal with stresses, such as when preparing the body to "flee or fight." The parasympathetic system is primarily a homeostatic or anabolic (energy-conserving) system, promoting the quiet and orderly processes of the body, such as those that allow the body to "feed and assimilate." Table I.2 summarizes the specific functions of the ANS and its divisions.

Visceral Sensation

Visceral afferent fibers have important relationships to the ANS, both anatomically and functionally. We are usually unaware of the sensory input of these fibers, which provides information about the condition of the body's internal environment. This information is integrated in the CNS, often triggering visceral or somatic reflexes, or both. Visceral reflexes regulate blood pressure and chemistry by altering such functions as heart and respiratory rates and vascular resistance. Visceral sensation, which reaches a conscious level, is generally categorized as pain that is usually poorly localized and may be perceived as hunger or nausea. Surgeons operating on patients who are under local anesthesia may handle, cut, clamp, or even burn (cauterize) visceral organs without evoking conscious sensation. However, adequate stimulation such as the following may elicit true pain:

- Sudden distension
- Spasms or strong contractions
- Chemical irritants
- Mechanical stimulation, especially when the organ is active
- Pathological conditions (especially ischemia) that lower the normal thresholds of stimulation.

Normal activity usually produces no sensation but may do so when the blood supply is inadequate (ischemia). Most visceral reflex (unconscious) sensation and some pain travel in visceral afferent fibers that accompany the parasympathetic fibers retrograde. Most visceral pain impulses (from the heart and most organs of the peritoneal cavity) travel centrally along visceral afferent fibers accompanying sympathetic fibers.

Medical Imaging Techniques

Familiarity with medical imaging techniques commonly used in clinical settings enables one to recognize congenital anomalies, tumors, and fractures. The most commonly used medical imaging techniques are:

- Conventional radiography (plain films)
- CT
- Ultrasonography (sonography)
- MRI
- Nuclear medicine imaging.

You must learn to interpret medical images, which, together with surface anatomy, are the only views of a person's anatomy that are commonly available after the 1st year of study. Medical imaging techniques permit the observation of anatomical structures in living persons and the ▶

▶ study of their movements in normal and abnormal activities (e.g., the heart and stomach).

Radiography

Wilhelm Roentgen, a German physicist and Nobel laureate, discovered X-rays in 1895. His discovery allowed us to observe the living structure of the human skeletal system. Bones and joints are readily visualized on radiographs (roentgenograms). **Radiological anatomy** is the study of the structure and function of the body using radiographical techniques. It is an important part of anatomy and is the anatomical basis of **radiology**, the branch of medical science dealing with the use of radiant energy in the diagnosis and treatment of disease. Being able to identify normal structures on radiographs makes it easier to recognize the changes caused by disease and injury. A conventional radiograph (Fig. I.36) is often referred to clinically as a *plain film*. The essence of a radiological examination is that a highly penetrating beam of X-rays transilluminates the patient, showing tissues of differing densities of mass within the body as images of differing densities of light and dark on the X-ray film. A tissue or organ that is relatively dense in mass (e.g., compact bone) absorbs more X-rays than does a less dense tissue such as spongy bone (Table I.3). Consequently, a dense tissue or organ produces a relatively transparent area on the X-ray film because relatively fewer X-rays reach the silver salt/gelatin emulsion in the film. Therefore, relatively fewer grains of silver are developed at this area when the film is processed. A very dense substance is *radiopaque*, whereas a substance of less density is *radiolucent*.

In basic radiological nomenclature, a posteroanterior (PA) radiograph is the standard view of the thorax, or chest. It is one in which the X-rays traverse the patient from posterior (P) to anterior (A); the X-ray tube is posterior to the patient and the X-ray film is anterior (Fig. I.37*A*). An anteroposterior (AP) radiograph is one in which the X-rays traverse ▶

Table I.3. Basic Principles of X-Ray Image Formation

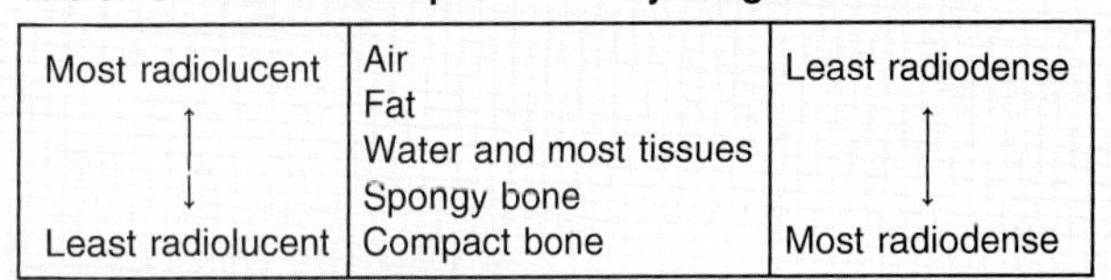

Most radiolucent	Air	Least radiodense
↑	Fat	↑
	Water and most tissues	
↓	Spongy bone	↓
Least radiolucent	Compact bone	Most radiodense

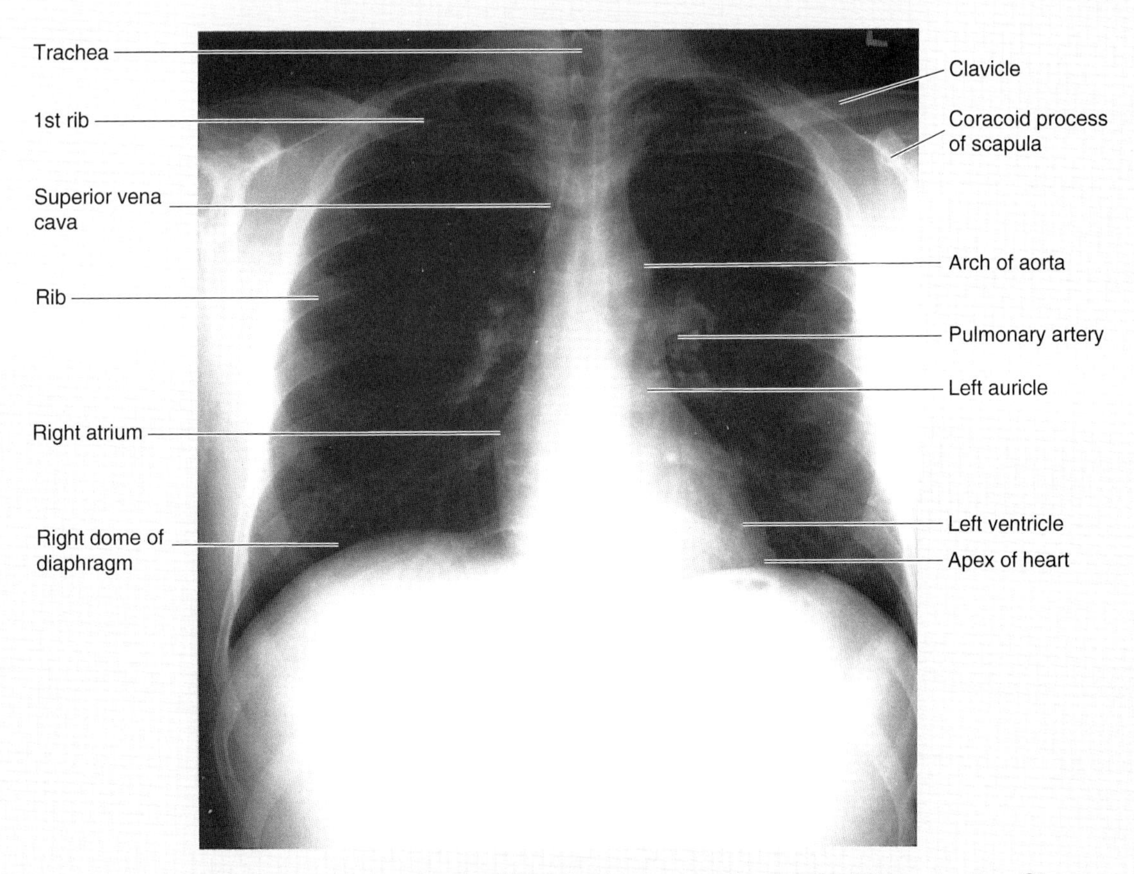

Figure I.36. Radiograph of the thorax (chest). This is a posteroanterior (PA) projection. Observe the arch of the aorta, parts of the heart, and domes of the diaphragm; note that the dome of the diaphragm is higher on the right side. (Courtesy of Dr. E.L. Lansdown, Professor of Medical Imaging, University of Toronto, Toronto, Ontario, Canada)

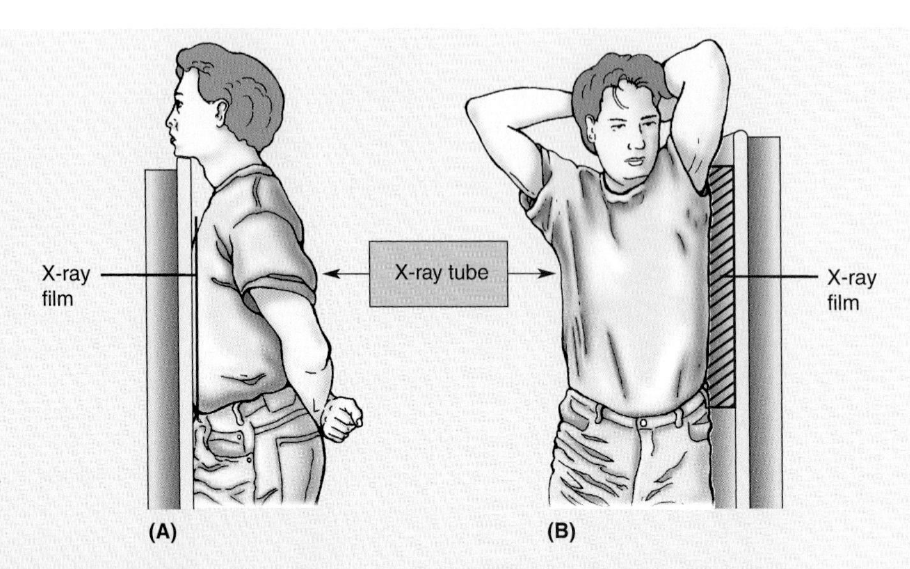

Figure I.37. Orientation of a patient's thorax (chest) during radiography. A. When taking a posteroanterior (PA) projection, the X-rays from the X-ray tube pass through the thorax from the back to reach the X-ray film anterior to the person. **B.** When taking a lateral projection, the X-rays pass through the thorax from the side to reach the X-ray film adjacent to the person's side.

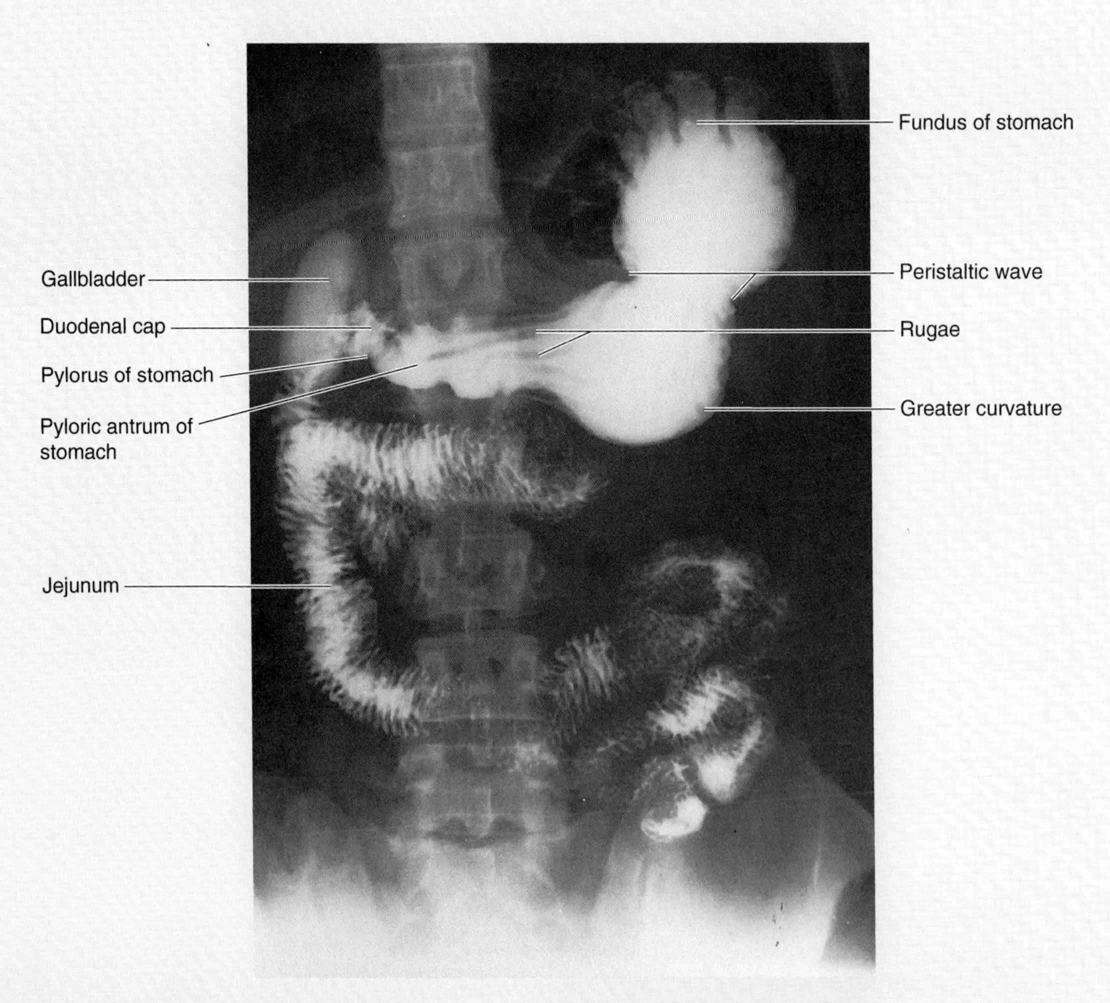

Figure I.38. Radiograph of the stomach, small intestine, and gallbladder. Observe the gastric folds, or rugae—longitudinal folds of the mucous membrane. Also observe the peristaltic wave that is moving the gastric contents toward the duodenum, which is closely related to the gallbladder. (Courtesy of Dr. J. Heslin, Toronto, Ontario, Canada).

▶ the patient from anterior (A) to posterior (P), indicating that the X-ray tube is anterior to the patient and the X-ray film is posterior. A lateral radiograph is made with the part of the patient's body being studied close to the X-ray film (Fig. I.37*B*). The introduction of contrast media (radiopaque substances) allowed the study of form and function of various organs and cavities, such as the stomach and small intestine after swallowing barium (Fig. I.38). Most radiological examinations are performed in at least two projections at right angles to each other. Because each radiograph presents a composite view of the tissues penetrated by the X-ray beam, structures overlap each other. Because of this and the absence of depth, more than one view is usually necessary to detect and localize an abnormality accurately.

Computerized Tomography

CT scans show radiographic images of the body that resemble transverse anatomical sections (Fig. I.39). During this process, a beam of X-rays passes through the body as the X-ray tube moves in an arc or a circle around the body. The amount of radiation absorbed by each different volume element of the chosen body plane varies with the amount of fat, water density tissue, and bone in each element. A multitude of linear energy absorptions is measured and put into a computer. It matches the many linear energy absorptions to each point within the section or plane that is scanned and displays the CT image on a printout or a cathode-ray tube (monitor). The CT image ▶

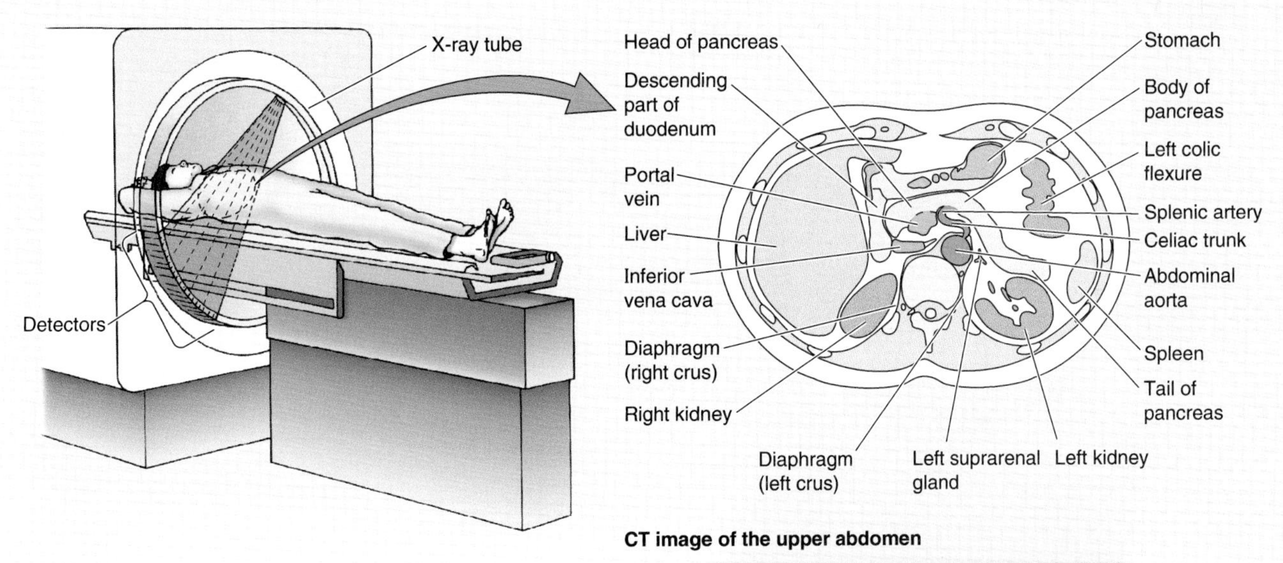

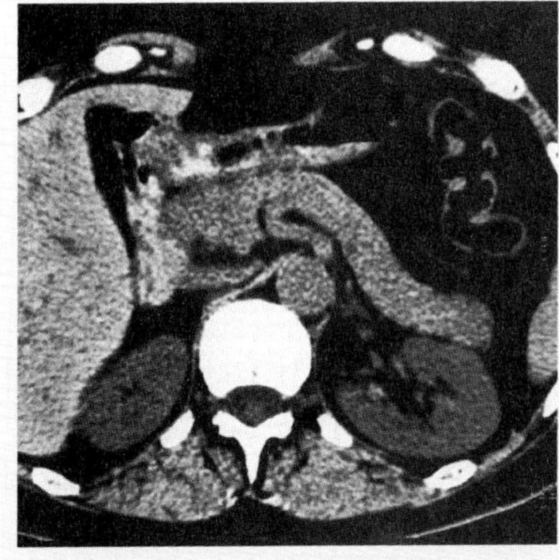

Figure I.39. Technique for producing an abdominal CT scan. The X-ray tube rotates around the person in the CT scanner and sends a fan-shaped beam of X-rays through the person's upper abdomen from a variety of angles. X-ray detectors on the opposite side of the person's body measure the amount of radiation that passes through a horizontal section of the person. A computer reconstructs the CT images from several scans and an abdominal CT scan is produced. The scan is oriented so it appears the way an examiner would view it when standing at the foot of the bed and looking toward a supine person's head.

▶ relates well to radiographs, in that areas of great absorption (e.g., vertebrae) are relatively transparent and those with little absorption are black (Table I.3). CT scans are always displayed as if the viewer were standing at a supine patient's feet.

Ultrasonography

Ultrasonography (sonography) visualizes superficial or deep structures in the body by recording pulses of ultrasonic waves reflecting off the tissues (Fig. I.40). Ultrasonography has the advantage of a lower cost than CT and MRIs, and the machine is portable; it can be performed in the physician's office or at the bedside. A **transducer** in contact with the skin generates high frequency sound waves that pass through the body and reflect off tissue interfaces that lie between tissues of differing characteristics, such as soft tissue and bone. Echoes from the body reflect into the transducer and convert to electrical energy. The electrical signals are recorded and displayed ▶

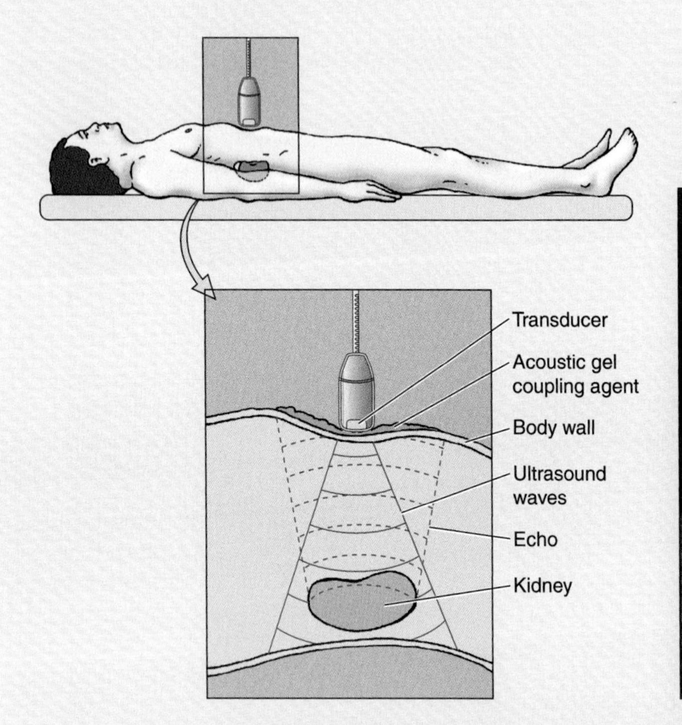

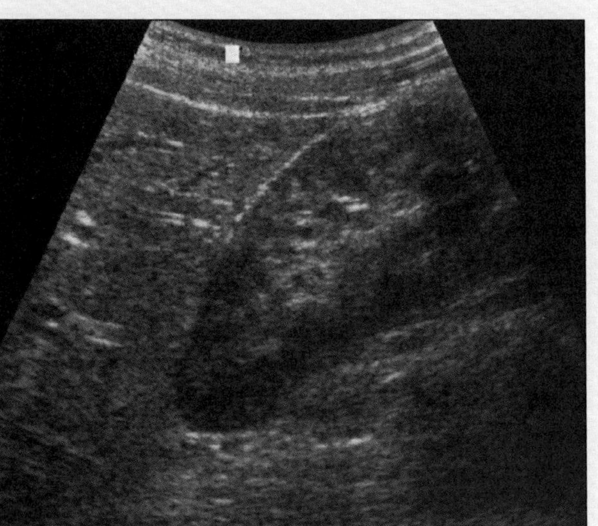

Figure I.40. Technique for producing an abdominal ultrasound scan of the upper abdomen. The image results from the echo of ultrasound waves from abdominal structures of different densities. The ultrasound image of the right kidney is displayed on a monitor.

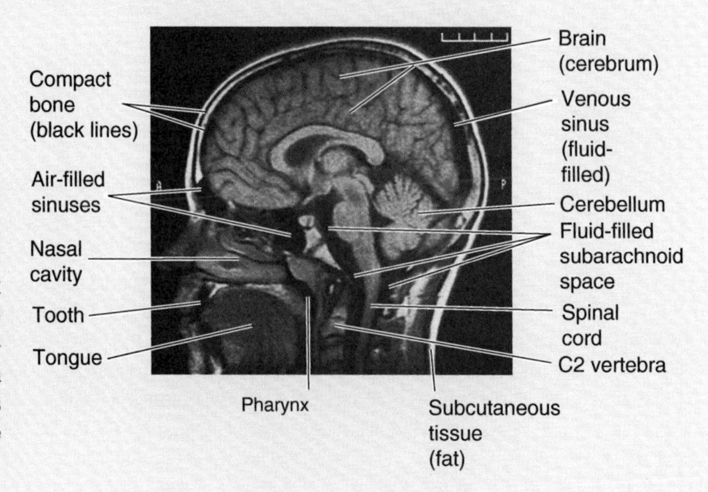

Figure I.41. Median MRI of the head. Many details of the CNS are visible. Structures in the nasal and oral cavities and upper neck are also shown. The black low signal areas superior to the anterior and posterior aspects of the nasal cavity are the air-filled frontal and sphenoidal sinuses. The patient is placed in a strong magnetic field that aligns the body's free protons. The aligned protons are "flipped" by radiowaves, and emit radiowaves as they flip back. The latter radiowaves are detected by an MRI system and processed by a computer, which produces the MRI scan. With the data collected, the MRI system can construct other images (e.g., transverse, sagittal, and coronal MRIs of the head).

▶ on a TV monitor as a cross-sectional image, which can be viewed in real time and recorded as a single image or on videotape.

In *Doppler ultrasonography*, the shifts in frequency between emitted ultrasonic waves and their echoes are used to measure the velocities of moving objects. This technique is based on the principle of the Doppler effect. The *Doppler principle* is used to view moving blood within blood vessels and display bloodflow in color, su-perimposed on the two-dimensional cross-sectional image. A major advantage of ultrasonography is its ability to produce real-time images, demonstrating motion, and its ability to incorporate sound information from vessels (Doppler).

Scanning of the pelvic viscera from the surface of the abdomen requires a fully distended bladder to displace the air-filled intestinal loops out of the pelvis and to create an acoustical window through which to visualize the pelvic organs. *Transvaginal sonography* permits the positioning of the transducer closer to the organ of interest (e.g., the ovary) and avoids fat and gas, which absorb or reflect sound waves. Bone reflects nearly all ultrasound waves, whereas air conducts them poorly. Consequently, the CNS and lungs of adults cannot be examined by ultrasonography.

The appeal of ultrasonography in obstetrics is that it is a noninvasive procedure that can yield useful information about the pregnancy, such as determining whether the pregnancy is intrauterine or extrauterine (ectopic) and whether the embryo is living (Callen, 1994). It has also become a standard method of evaluating growth and development of the embryo and fetus. Measurement of the head circumference is an important measurement of head growth in infants and children and for determining fetal head size (Hadlock, 1994).

Magnetic Resonance Imaging

MRI shows images of the body similar to that by CT scans, but MRI is better for tissue differentiation. MRIs resemble anatomical sections closely, especially in the brain (Fig. I.41). The person is put in a scanner with a strong magnetic field and is pulsed with radiowaves. The signals emitted from the patient are stored in a computer and reconstructed into various images of the body. The appearance of tissues on the generated images can be varied by controlling how radiofrequency pulses are sent and received.

Magnetically aligned free protons in the tissues are excited (flipped) with a radiowave pulse, which then give off minute but measurable energy signals as they flip back. For example, regions giving signals come from tissues that are high in proton density, such as fat or water. The tissue signal is based primarily on three properties of protons in a particular region of the body. These are referred to as T1 relaxation, T2 relaxation, and proton density. Although liquids have a high density of free protons, the excited free protons in moving fluids such as blood tend to move out of the field before they flip and give off their signal and are replaced by unexcited protons. Consequently, moving ▶

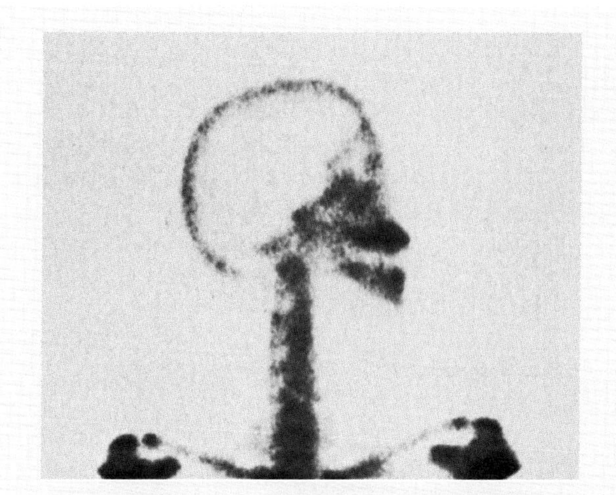

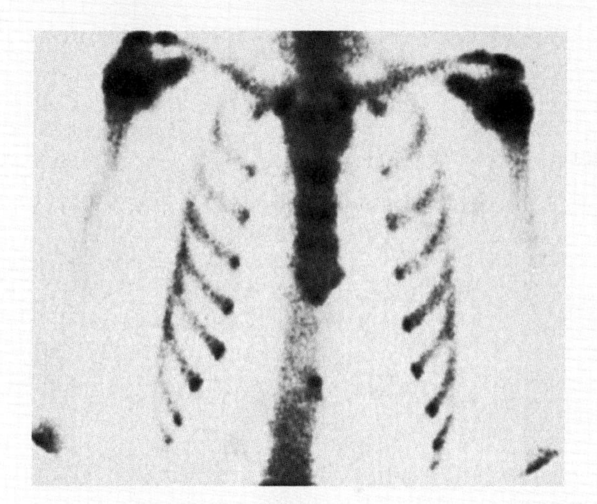

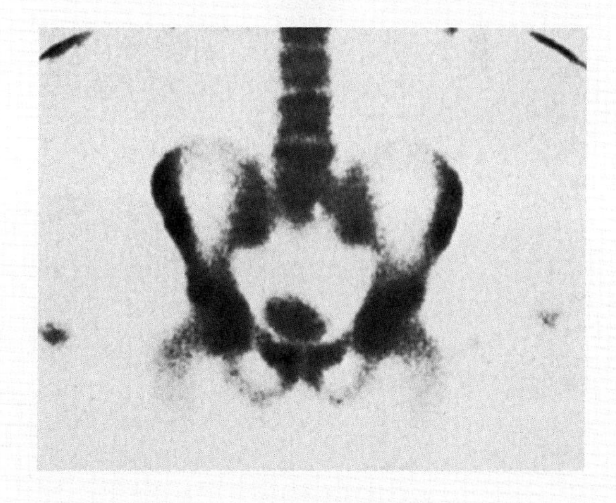

Figure I.42. Bone scans of the head and neck, thorax, and pelvis. The images can be viewed as a whole or in cross section.

▶ fluids appear black. MRI has the capacity to depict anatomy in any plane: transverse, median, sagittal, coronal, and even arbitrary oblique planes. MR scanners produce good images of the head, thorax, abdomen, and limbs. Motion has been a problem but the fast scanners now used can even visualize moving structures, such as the heart and bloodflow, in real time.

Nuclear Medicine Imaging

Nuclear medicine imaging techniques provide information about the distribution of trace amounts of radioactive substances introduced into the body. Nuclear medicine scans show images of specific organs following intravenous (I.V.) injection of a small dose of radioactive material. The radionuclide is tagged to a compound that is selectively taken up by an organ, e.g., technetium-99m methylene diphosphonate (MDP). In general, diphosphonates are used, and MDP is one of the commonest agents (Fig. I.42). Images can be viewed as the whole organ or in cross sections, which are referred to as SPECT (single photon emission computed tomography). ⊙

References and Suggested Readings

Amadio PC: Reaffirming the importance of dissection. *Clin Anat* 9:136, 1996.

Barr ML, Kiernan JA: *The Human Nervous System: An Anatomical Viewpoint,* 6th ed. Philadelphia, JB Lippincott, 1993.

Basmajian JV, DeLuca CJ: *Muscles Alive: Their Functions Revealed by Electromyography,* 5th ed. Baltimore, Williams & Wilkins, 1985.

Behrman RE, Kliegman RM, Arvin AM (eds): *Nelson Textbook of Pediatrics,* 15th ed. Philadelphia, WB Saunders, 1996.

Bergman RA, Thompson SA, Afifi AK, Saadeh FA: *Compendium of Human Anatomic Variation: Text, Atlas, and World Literature.* Baltimore, Urban & Schwarzenberg, 1988.

Cahill DR, Leonard RJ: The role of computers and dissection in teaching anatomy: A comment. *Clin Anat* 10:140, 1997.

Callen PW: *Ultrasonography in Obstetrics and Gynecology,* 3rd ed. Philadelphia, WB Saunders, 1994.

Cormack DH: *Essential Histology.* Philadelphia, JB Lippincott, 1993.

Federative Committee on Anatomical Terminology: *Terminologia Anatomica: International Anatomical Nomenclature.* Stuttgart, Thieme, 1998.

Fitzgerald MJT: *Neuroanatomy. Basic and Clinical,* 2nd ed. London, Baillière Tindall, 1992.

Gartner LP, Hiatt JL: *Color Textbook of Histology.* Philadelphia, WB Saunders, 1997.

Gross AE: Orthopedic surgery: Adult. *In* Gross A, Gross P, Langer B (eds): *A Complete Guide for Patients and Their Families.* Toronto, Harper & Collins, 1989.

Hadlock FP: Ultrasound determination of menstrual age. *In* Callen PW (ed): *Ultrasonography in Obstetrics and Gynecology,* 3rd ed. Philadelphia, WB Saunders, 1994.

Haines DE: *Neuroanatomy, An Atlas of Structures, Sections, and Systems,* 4th ed. Baltimore, Williams & Wilkins, 1995.

Haines DE (ed): *Fundamental Neuroscience.* New York, Churchill Livingstone, 1997.

Hutchins JB, Naftel JP, Ard MD: The cell biology of neurons and glia. *In* Haines DE (ed): *Fundamental Neuroscience.* New York, Churchill Livingstone, 1997.

Jones DJ: Reassessing the importance of dissection: A critique and elaboration. *Clin Anat* 10:123, 1997.

Levi CS, Lyons EA, Schollenberg J, Bristowe JRB: The value of post void scans in the diagnosis of ruptured ectopic pregnancy. *J Ultrasound Med* 1:253, 1982.

Moore KL: Anatomical terminology/clinical terminology. *Clin Anat* 1: 7, 1988.

Moore KL: Meaning of "Normal." *Clin Anat* 2:235, 1989.

Moore KL, Persaud TVN: *The Developing Human: Clinically Oriented Embryology,* 6th ed. Philadelphia, WB Saunders, 1998.

Mutyala S, Cahill DR: Catching up. *Clin Anat* 9:53, 1996.

O'Rahilly R: Making planes plain. *Clin Anat* 10:129, 1997.

Persaud TVN: *Early History of Human Anatomy From Antiquity to the Beginning of the Modern Era.* Springfield, Charles C Thomas, 1984.

Persaud TVN: *A History of Anatomy. The Post-Vesalian Era.* Springfield, Charles C Thomas, 1997.

Ross MH, Romrell LJ, Kaye G: *Histology. A Text and Atlas,* 3rd ed. Baltimore, Williams & Wilkins, 1994.

Salter RB: *Textbook of Disorders and Injuries of the Musculoskeletal System,* 3rd ed. Baltimore, Williams & Wilkins, 1998.

Squires B: *Basic Terms of Anatomy and Physiology,* 2nd ed. Toronto, WB Saunders, 1986.

Stedman's Medical Dictionary, 26th ed. Baltimore, Williams & Wilkins, 1995.

Swartz MH: *Textbook of Physical Diagnosis, History and Examination,* 2nd ed. Philadelphia, WB Saunders, 1994.

Williams PL, Bannister LH, Berry MM, Collins P, Dussek JE, Ferguson MWJ (eds): *Gray's Anatomy. The Anatomical Basis of Medicine and Surgery,* 38th ed. New York, Churchill Livingstone, 1995.

Willis MC: *Medical Terminology: The Language of Health Care.* Baltimore, Williams & Wilkins, 1995.

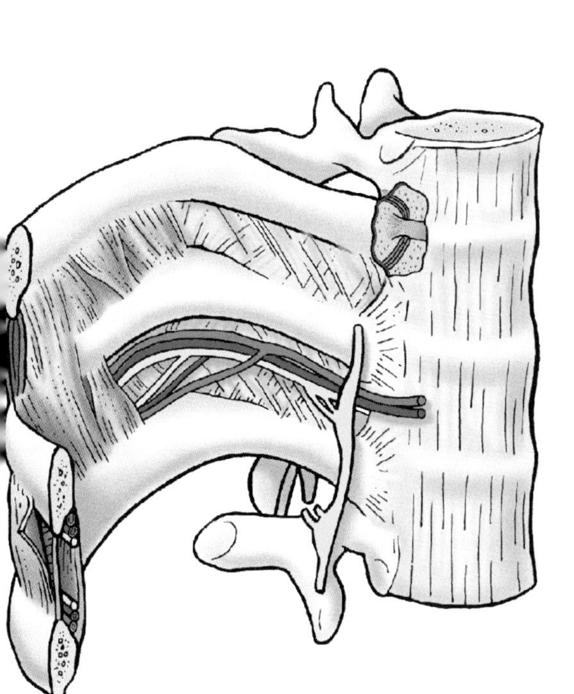

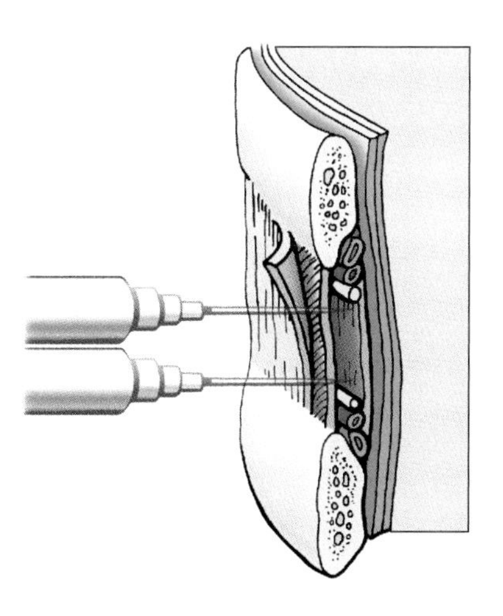

chapter

1 Thorax

The thorax (chest) is the superior part of the trunk between the neck and the abdomen. It is formed by the 12 pairs of ribs, sternum (breast bone), costal cartilages, and 12 thoracic vertebrae (Fig. 1.1). These bony and cartilaginous structures form the **thoracic cage** (rib cage), which surrounds the **thoracic cavity** and supports the *pectoral (shoulder) girdle*. Along with the skin and associated fascia and muscles, the thoracic cage forms the **thoracic (chest) wall**, which lodges and protects the contents of the thoracic cavity—the heart and lungs, for example—as well as some abdominal organs such as the liver and spleen. The thoracic cage provides attachments for muscles of the neck, thorax, upper limbs, abdomen, and back. The muscles of the thorax itself elevate and depress the thoracic cage during breathing. Because the most important structures in the thorax—the heart, trachea, lungs, great vessels, and the thoracic wall itself—are constantly moving, the thorax is one of the most dynamic regions of the body.

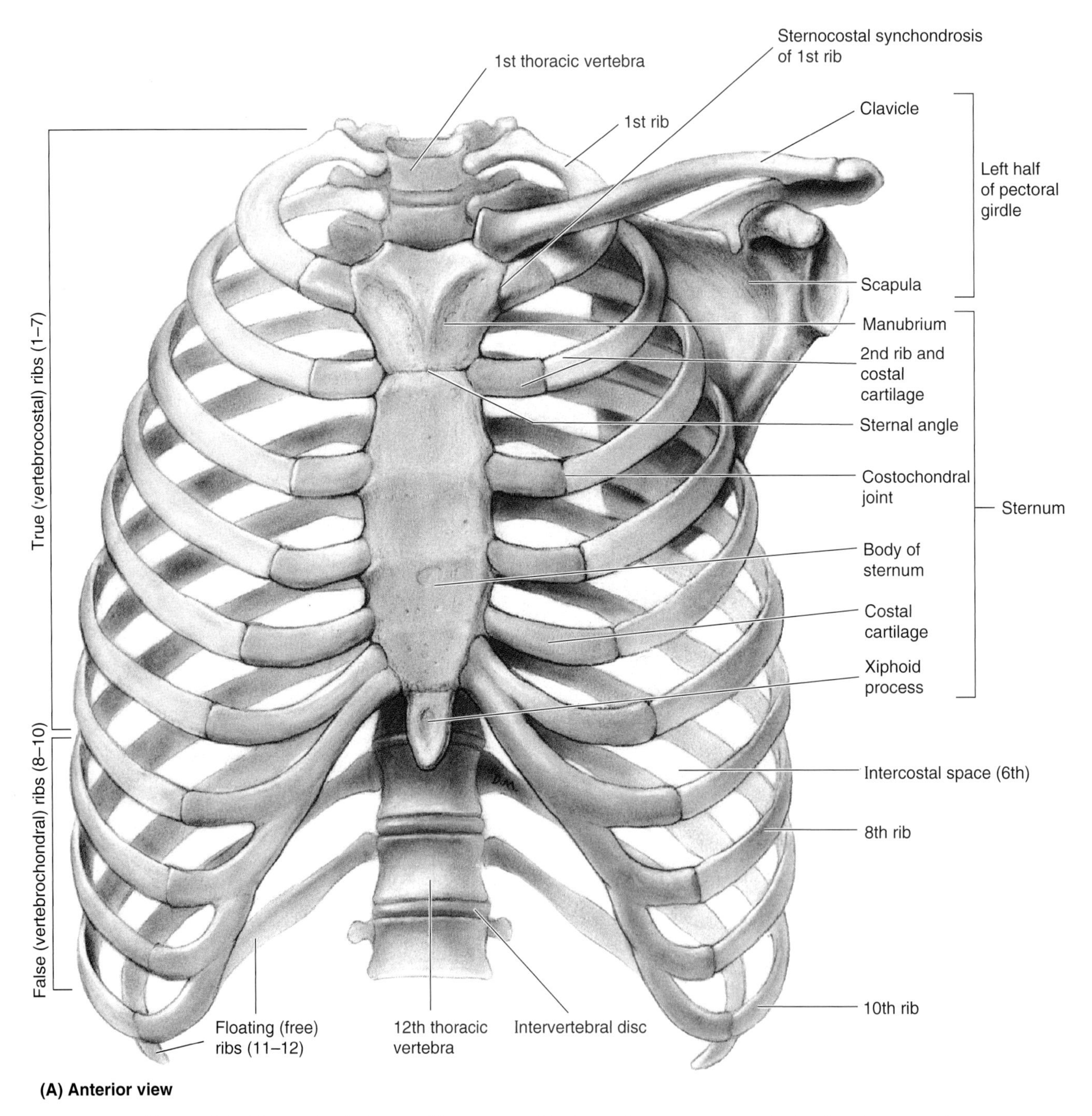

Figure 1.1. Thoracic skeleton. The osteocartilaginous thoracic cage includes the sternum, 12 pairs of ribs and costal cartilages, and 12 thoracic vertebrae and intervertebral (IV) discs. The clavicles and scapulae form the pectoral (shoulder) girdle.

Chest Pain

The significance of pain in the thorax varies from negligible to very serious. Although chest pain can result from pulmonary disease, it is probably the most important symptom of cardiac disease (Swartz, 1994). However, chest pain may also result from intestinal, gallbladder, and musculoskeletal disorders. When evaluating a patient with chest pain, the examination is largely concerned with discriminating between serious conditions and the many minor causes of pain. People who have had a *heart attack* usually describe a "crushing" substernal pain (deep to the sternum) that does not disappear with rest. ○

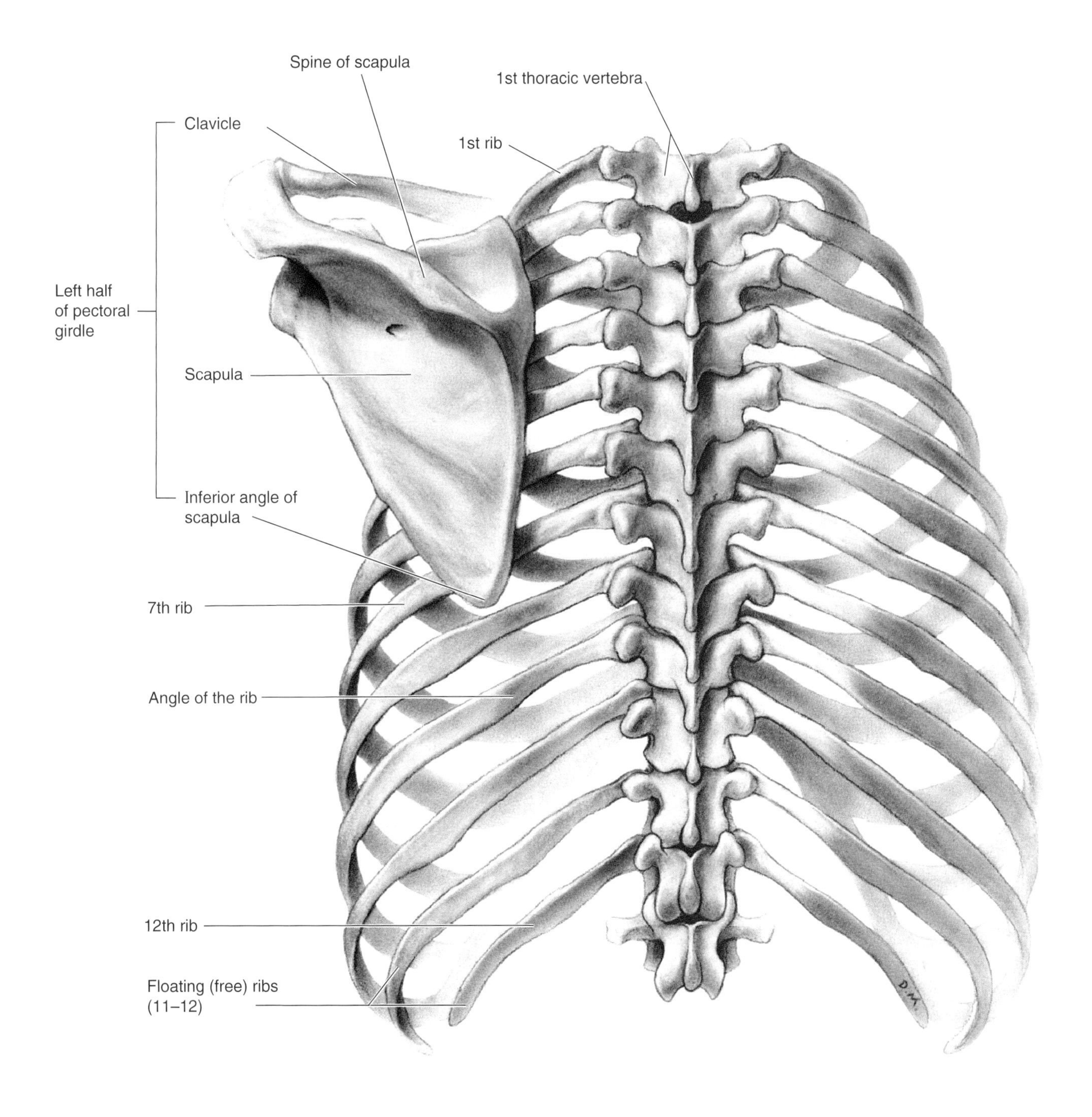

Figure 1.1. *(Continued)*

Thoracic Wall

The thoracic cage is covered by skin, fascia, and muscles, including those attaching the pectoral girdle to the upper limb and trunk. The mammary glands of the breasts (L. mammae) are in the subcutaneous tissue. The function of the thoracic wall is to not only protect the contents of the thoracic cavity but also provide the mechanical function of breathing. With each breath, the muscles of the thoracic wall—working in concert with the diaphragm and muscles of the abdominal wall—vary the volume of the thoracic cavity, first by expanding the capacity of the cavity, thereby allowing the lungs to expand, and then, mostly through their relaxation, decreasing the volume of the cavity, causing the lungs to expel air.

Fascia of the Thoracic Wall

The *subcutaneous tissue* (*superficial fascia*, hypodermis) is a layer composed of loose, irregular connective tissue immediately beneath the skin that is closely attached to the skin by coarse bands—the skin ligaments (*retinacula cutis*). The subcutaneous tissue of the thorax contains variable amounts of fat, sweat glands, blood and lymphatic vessels, cutaneous nerves, and, in the breasts of mature females, the mammary glands. The *deep fascia* (investing fascia) is a thin fibrous membrane, devoid of fat, that is usually dense and loosely attached to the subcutaneous tissue and overlying skin. It closely invests the underlying muscles forming the *epimysium*, a connective tissue envelope. The *deep fascia* invests the muscles and associated tendons up to their attachment to bone, such as the ribs, and is itself attached to the periosteum of the bones. Parts of it are named for the muscle being invested—the pectoralis fascia, for example. The deep fascia helps to hold the parts of the thorax together and presents a barrier to infection.

Skeleton of the Thoracic Wall

The thoracic skeleton forms the **osteocartilaginous thoracic cage** (Fig. 1.1), which protects the thoracic viscera and some abdominal organs. The thoracic skeleton includes:

- 12 pairs of ribs and costal cartilages
- 12 thoracic vertebrae and intervertebral (IV) discs
- The sternum.

The ribs and costal cartilages form the largest part of the thoracic cage.

Ribs and Costal Cartilages

Ribs (L. costae) are curved, flat bones that form most of the thoracic cage (Figs. 1.1 and 1.2). They are remarkably light in weight yet highly resilient. Each rib has a spongy interior containing **bone marrow** (hematopoietic tissue) that forms blood cells. *There are three types of ribs*:

- **True (vertebrocostal) ribs** (the first seven ribs)—so-called because they attach directly to the sternum through their own costal cartilages.
- **False (vertebrochondral) ribs** (the 8th to 10th ribs). Their cartilages are joined to that of the rib immediately superior to them; thus, their connection with the sternum is indirect.
- **Floating (vertebral, free) ribs** (the 11th and 12th ribs). The rudimentary cartilages of these ribs do not connect even indirectly with the sternum; instead, they end in the posterior abdominal musculature.

Typical ribs (3rd to 9th) have a:

- **Head** that is wedge-shaped and has two facets, separated by the **crest of the head** (Fig. 1.2): one facet for articula-

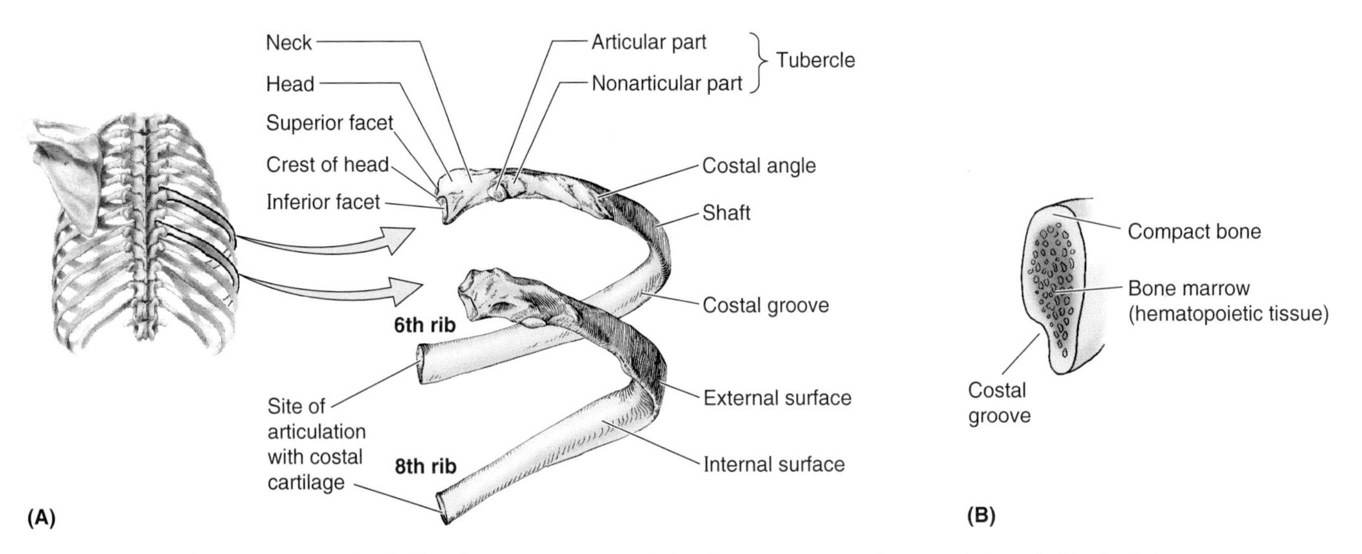

Figure 1.2. Typical ribs. The 3rd through 9th ribs have common characteristics. **A.** Each rib has a head, neck, tubercle, and shaft (body). **B.** Cross section of a rib.

tion with the numerically corresponding vertebra and one facet for the vertebra superior to it.

- **Neck** that connects the head with the body (shaft) at the level of the tubercle.
- **Tubercle** occurring at the junction of the neck and shaft. The tubercle has a smooth *articular part* for articulating with the corresponding transverse process of the vertebra, and a rough *nonarticular part* for attachment of the costotransverse ligament.
- **Shaft** that is thin, flat, and curved—most markedly at the **costal angle** where the rib turns anterolaterally; the concave internal surface has a **costal groove** that protects the intercostal nerve and vessels.

Atypical ribs (1st, 2nd, and 10th to 12th) are dissimilar (Fig. 1.3):

- The **1st rib** is the broadest (i.e., its shaft is widest and is nearly horizontal), shortest, and most sharply curved of the seven true ribs; it has a single facet on its head for articulation with T1 vertebra and two transversely directed grooves crossing its superior surface for the subclavian vessels, which are separated by a **scalene tubercle** and ridge.
- The **2nd rib** is thinner (its shaft is more typical), less curved, and substantially longer than the 1st rib; it has two facets on its head for articulation with the bodies of T1 and T2 vertebrae, and a **tubercle** for muscle attachment.
- The **10th to 12th ribs**, like the 1st rib, have only one facet on their heads.
- The **11th and 12th ribs** are short and have no necks or tubercles.

Costal cartilages prolong the ribs anteriorly and contribute to the elasticity of the thoracic wall. The cartilages increase in length through the first seven and then gradually decrease. The first seven cartilages (and sometimes the 8th; see Fig. 1.16) join the sternum; the 8th, 9th, and 10th articulate with the cartilages just superior to them. In some people the 10th pair of ribs may be floating (free). The 11th and 12th cartilages form caps on the anterior ends of these ribs. **Intercostal spaces** separate the ribs and their costal cartilages from one another. These spaces are occupied by intercostal muscles, vessels, and nerves.

Rib Fractures and Associated Injuries

The short, broad, 1st rib, posteroinferior to the clavicle (collar bone), is rarely fractured because of its protected position (note that it cannot be palpated). When it is broken, however, injury to the *brachial plexus of nerves* and *subclavian vessels* may occur. The 1st rib is clinically important because so many structures cross and attach to it. It has a prominent **scalene tubercle** on its superior surface for attachment of the scalenus anterior muscle. This surface also has two ▶

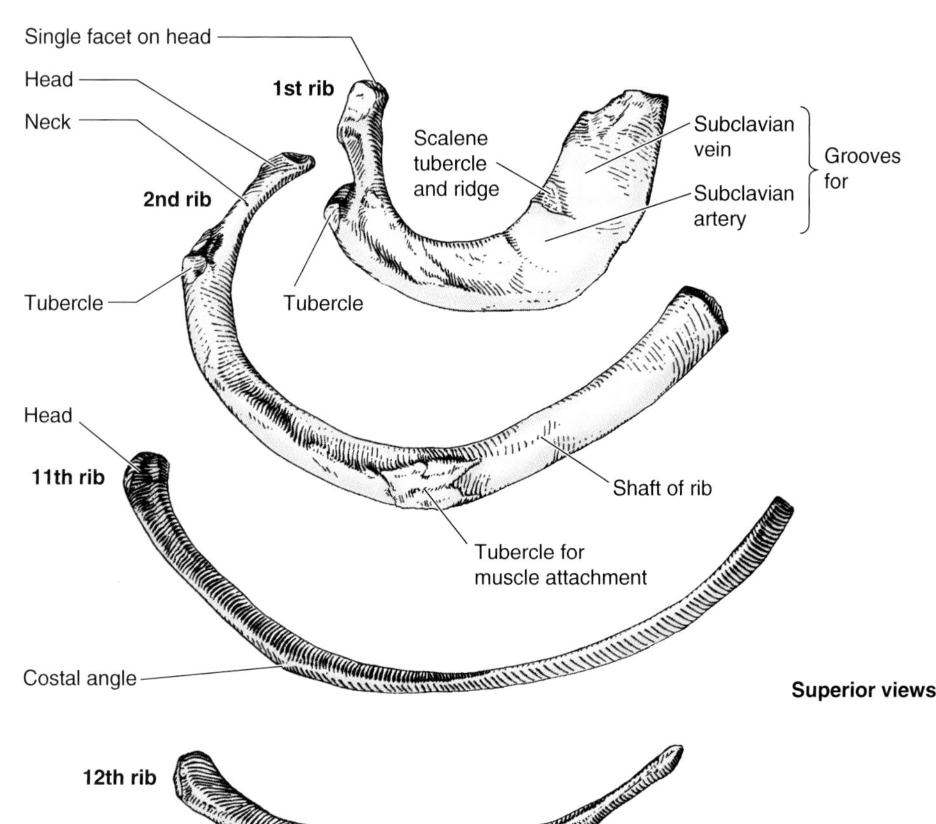

Figure 1.3. Atypical ribs. These ribs differ from typical ribs; the 1st rib is short and flattened, for example, and the tubercle merges with the angle. The 11th and 12th ribs lack necks and tubercles, and the 12th rib is shorter than most ribs.

▶ transversely directed shallow grooves, anterior and posterior to the tubercle, for the subclavian vein, and for the subclavian artery and inferior trunk of the brachial plexus, respectively.

The middle ribs are most commonly fractured. Rib fractures usually result directly from blows or indirectly from crushing injuries. The weakest part of a rib is just anterior to its angle; however, direct violence may fracture a rib anywhere, and its broken end may injure internal organs such as the lung and/or the spleen. *Lower rib fractures* may tear the diaphragm and result in a *diaphragmatic hernia* (see Chapter 2). Rib fractures are painful because the broken parts move during respiration, coughing, laughing, and sneezing. Rib pain may also result from metastasis (spread) of cancer from the breast or prostate. Chest radiographs demonstrate these metastases.

Flail chest ("stove-in chest") occurs when a sizable segment of the anterior and/or lateral thoracic wall moves freely because of *multiple rib fractures*. This condition allows the loose segment of the wall to move paradoxically (inward on inspiration and outward on expiration). Flail chest is an extremely painful injury and impairs ventilation, thereby affecting oxygenation of the blood. During treatment, the loose segment is often fixed by hooks and/or wires so that it cannot move.

Thoracotomy and Bone Grafting

The surgical creation of an anterior opening into the thoracic wall is an *anterior thoracotomy*. H-shaped cuts through the perichondrium of the cartilages are used to shell out segments of costal cartilage to gain entrance to the thoracic cavity. Sometimes surgeons use a piece of rib for autogenous bone grafting in procedures such as reconstruction of the mandible (lower jaw) following tumor excision.

Surgeons also cut through the periosteum and remove pieces of ribs posteriorly—*posterior thoracotomy*—to enter the thoracic cavity and remove a lung tumor, for example. Following the operation, the missing pieces of ribs regenerate from the intact periosteum, but the ribs rarely return to their original form.

Supernumerary Ribs

People usually have 12 ribs on each side, but the number is increased by the presence of cervical and/or lumbar ribs, or decreased by failure of the 12th pair to form. Supernumerary or extra ribs result from the retention and development of the costal processes of cervical and lumbar vertebrae (Moore and Persaud, 1998). The distal part of the transverse processes of these vertebrae develops from the costal processes. In some people, these processes grow unduly large, forming extra ribs.

Cervical ribs (incidence 0.5–1%) articulate with C7 vertebra but rarely attach to the sternum. Cervical ribs may be free, articulate, or fuse with the 1st rib, or attach to the 1st rib by a fibrous band. Cervical ribs are commonly found in asymptomatic people; however, they are clinically significant because they may compress fibers of the inferior trunk of the brachial plexus and cause pain or numbness (paresthesia) in the shoulder and upper limb (*cervical rib syndrome*—one of several *thoracic outlet syndromes* [TOS]). The hand pain is most severe in the 4th and 5th fingers. Often it is the fibrous band extending from the cervical rib to the 1st rib that compresses C8 and T1 nerves, or the inferior trunk of the brachial plexus. A cervical rib may also compress the subclavian artery, resulting in *ischemic muscle pain* in the upper limb (pain resulting from poor blood ▶

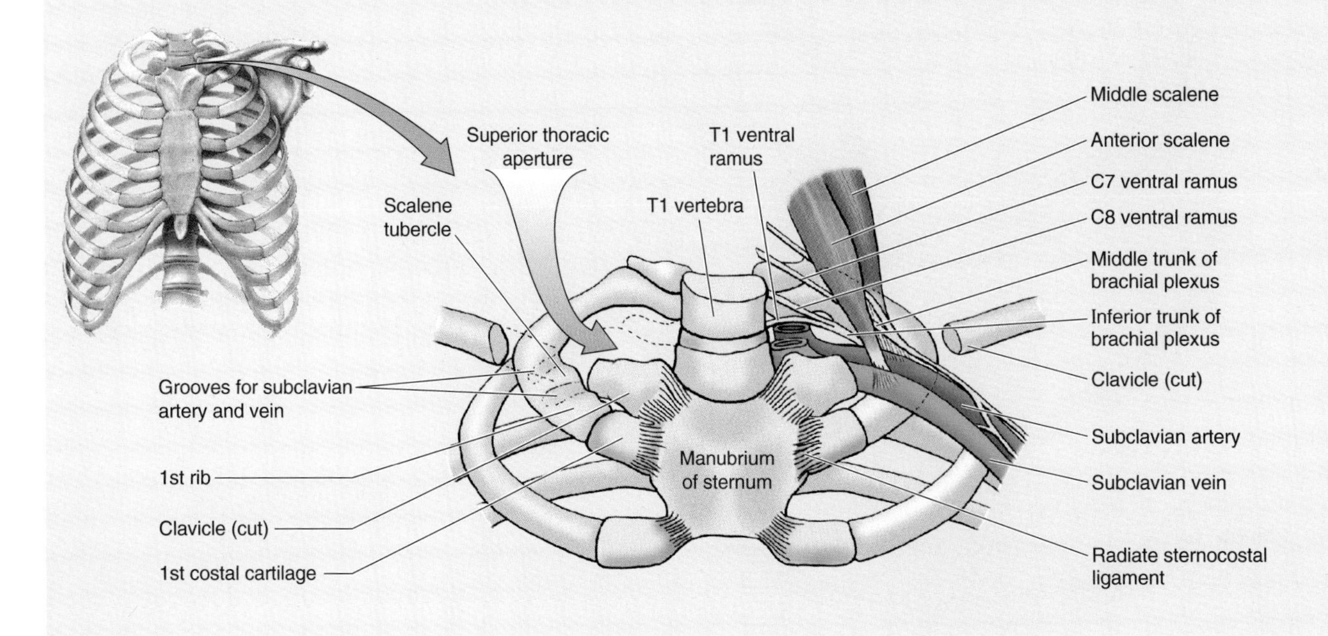

▶ supply to the limb muscles). If the pain is severe, it may be necessary to remove all or part of the cervical rib. Compression of the subclavian artery by a cervical rib or other cause at the superior thoracic aperture may cause the pulse to diminish when the upper limb is not at one's side and, especially, if the angle between the neck and shoulder is increased. The pulse increases when the limb is elevated (Holsen maneuver).

Lumbar ribs are less common than cervical ribs (see Chapter 4) but have clinical significance in that they may confuse the identification of vertebral levels in radiographs and other diagnostic images. In addition, a fractured lumbar rib may be erroneously interpreted as a fractured transverse process of L1 vertebra. If a lumbar transverse process appears unusually long, a lumbar rib may be present.

Protective Function and Variation of Costal Cartilages

Costal cartilages provide resilience to the thoracic cage, preventing many blows from fracturing the sternum and/or ribs. Because of the remarkable elasticity of the ribs and costal cartilages in children, chest compression may produce injury within the thorax even in the absence of a rib fracture. In elderly people, the costal cartilages lose some of their elasticity and become brittle; they may undergo calcification, making them radiopaque (e.g., in radiographs). ○

Thoracic Vertebrae

Thoracic vertebrae (Figs. 1.4 and 1.5) are typical in that they have vertebral arches (neural arches) and seven processes for muscular and articular connections. *Special features of thoracic vertebrae* include:

- Costal facets or demifacets on their bodies for articulation with the heads of ribs
- Costal facets on their transverse processes for articulation with the tubercles of ribs, except for the inferior two or three thoracic vertebrae
- Long spinous processes.

Two *demifacets* or *costal facets*—small plane surfaces on each side of the IVdisc between two adjacent vertebrae that articulate with a rib—are located laterally on the bodies of T2 through T9 vertebrae. The **superior demifacet** at the superior posterolateral margin of the vertebral body articulates with the head of its own rib (rib of the same number); the **inferior demifacet**, located posterolaterally and inferiorly on the vertebral body, articulates with the head of the rib inferior to it (rib numbered one number higher than the vertebra).

The costal facets of other vertebrae vary somewhat.

- T1 has a single facet for the head of the 1st rib and a demifacet for the cranial part of the 2nd rib.

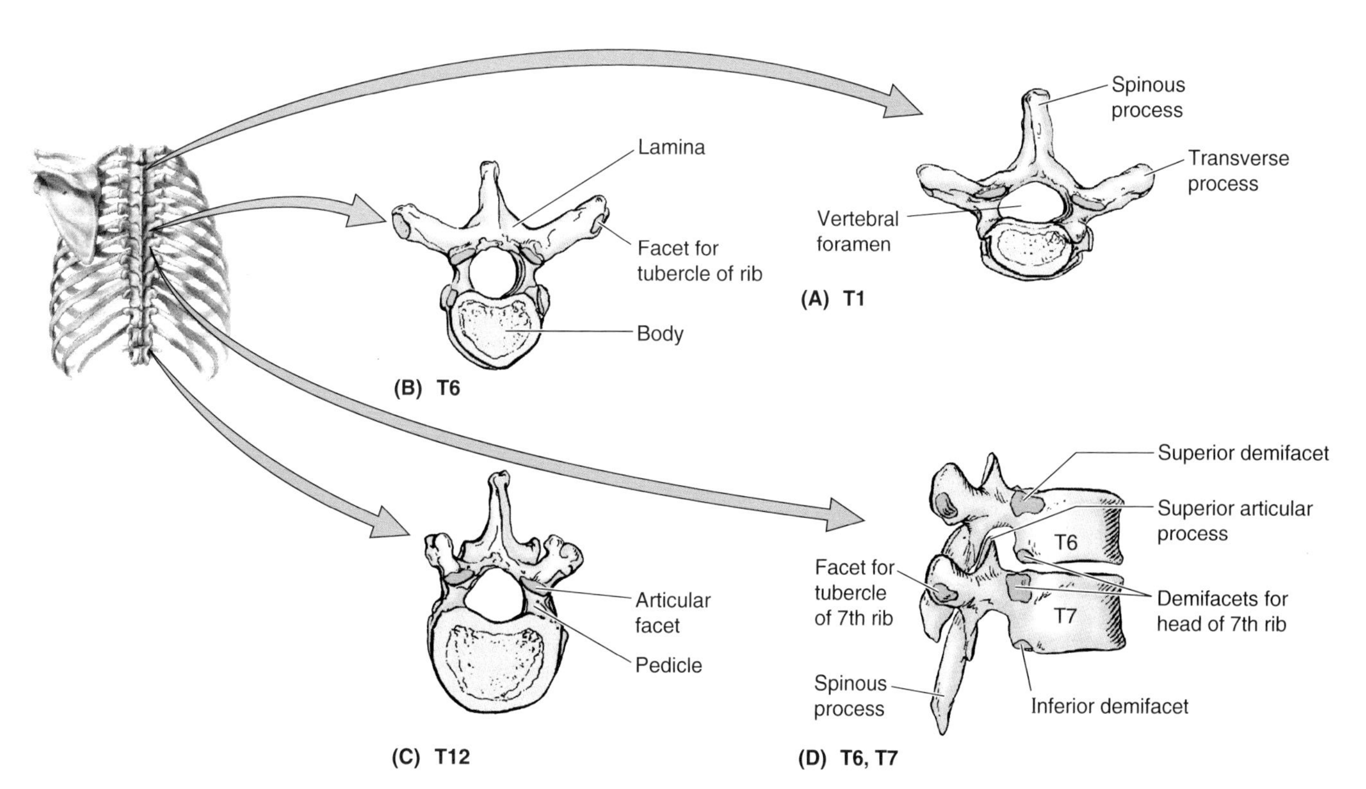

Figure 1.4. Typical thoracic vertebrae. T5 through T9 vertebrae have typical characteristics of thoracic vertebrae. **A–C.** Superior views. **D.** Lateral view of T6 and T7 vertebrae.

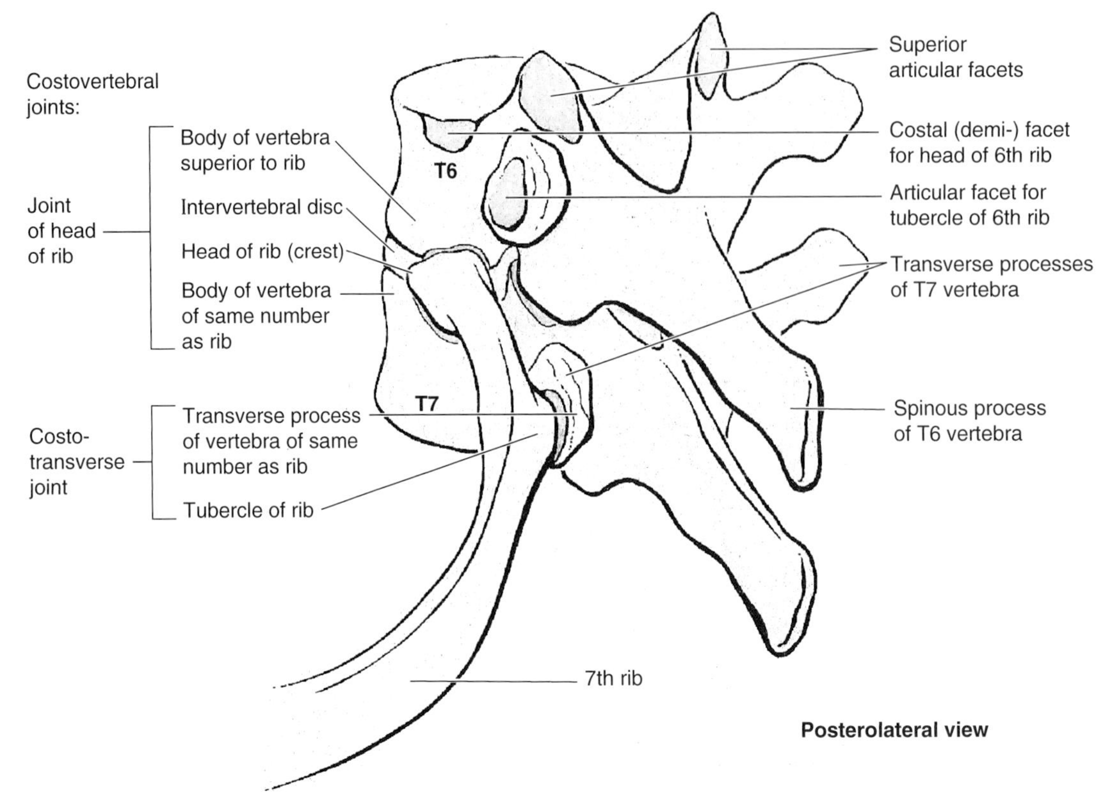

Figure 1.5. Costovertebral articulations of a typical rib. The costovertebral joints include the *joint of the head of the rib*, in which the head of the rib articulates with two adjacent vertebral bodies and the intervertebral (IV) disc between them, and the *costotransverse joint*, in which the tubercle of the rib articulates with the transverse process of a vertebra.

- T10 has only one costal facet that is partly on its body and partly on its pedicle.
- T11 and T12 have only a single costal facet, which is on their pedicles.

The **spinous processes** projecting from the vertebral arches of typical thoracic vertebrae are long and slope inferiorly (Fig. 1.4*D*). They cover the intervals between the **laminae** of adjacent vertebrae, thereby preventing sharp objects such as a knife from entering the **vertebral canal** (spinal canal) and injuring the spinal cord.

Effect of an Aortic Aneurysm on Vertebrae

Movements between adjacent vertebrae are relatively small in the thoracic region, largely because of the associated thoracic cage. This limited movement provides a relative rigidity that protects the heart and lungs. The bodies of T5 through T8 are related to the thoracic aorta, which often flattens their left sides. When the aorta develops an *aneurysm* (localized dilation), the bodies of these vertebrae may be partly eroded by pressure from the aneurysm. These bony changes may be visible on radiographs. ⊙

Sternum

The sternum (G. sternon, chest) is the flat, elongated bone that forms the middle of the anterior part of the thoracic cage (Fig. 1.6). The sternum consists of three parts: manubrium, body, and xiphoid process.

The **manubrium** (L. handle, like the handle of a sword, the sternal body forming the blade) is a roughly triangular bone that lies at the level of the bodies of T3 and T4 vertebrae. The manubrium is the widest and thickest of the three parts of the sternum. The easily palpated concave center of the superior border of the manubrium is the **jugular notch** (suprasternal notch). This notch is deepened in an articulated skeleton (and in life) by the medial (sternal) ends of the clavicles, which are too large for the relatively small *clavicular notches* in the manubrium that receive them, forming the sternoclavicular (SC) joints. Inferolateral to the clavicular notch, the costal cartilage of the 1st rib fuses with the lateral border of the manubrium—the *sternocostal synchondrosis* of the 1st rib (Fig. 1.1*A*). Superior and inferior to the **manubriosternal joint**, the manubrium and body lie in slightly different planes; hence, their junction forms a projecting **sternal angle** (of Louis). This palpable clinical landmark is *located opposite the 2nd pair of costal cartilages* at the level of the IVdisc between T4 and T5 vertebrae. Because the 1st rib is not palpable, rib counting in physical examinations starts with the 2nd rib adjacent to the subcutaneous and easily palpated sternal angle (Fig. 1.6*A*).

The **body of the sternum**—longer, narrower, and thin-

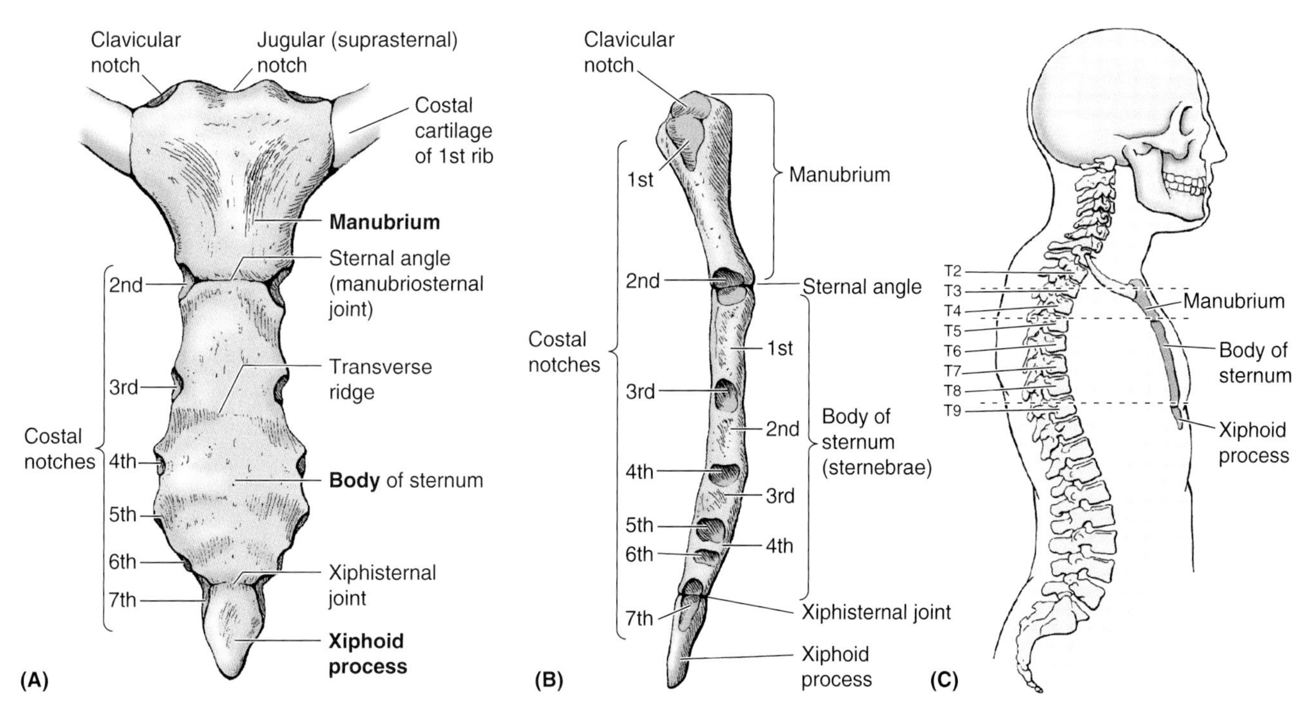

Figure 1.6. Sternum. A. Anterior view. **B.** Lateral view. Observe the thickness of the superior third of the manubrium between the clavicular notches. **C.** Lateral view. The relationship of the sternum to the vertebral column is shown.

ner than the manubrium—is located at the level of T5 through T9 vertebrae (Fig. 1.6, *A–C*). Its width varies because of the scalloping of its lateral borders by the **costal notches**. In young people, four *sternebrae*—primordial segments of the sternum—are obvious. The sternebrae articulate with each other at primary cartilaginous joints (sternal synchondroses). These joints begin to fuse from the inferior end between puberty (sexual maturity) and 25 years of age. The nearly flat anterior surface of the sternum is marked in adults by three variable **transverse ridges** (Fig. 1.6*A*) that represent the lines of fusion (synostosis) of its four originally separate sternebrae.

The **xiphoid process**—the smallest and most variable part of the sternum—is thin and elongated. It lies at the level of T10 vertebra. Although often pointed, the xiphoid process may be blunt, bifid, curved, or deflected to one side or anteriorly. It is cartilaginous in young people but more or less ossified in adults older than 40 years. In elderly people, the xiphoid process may fuse with the sternal body.

The xiphoid process is an important landmark in the median plane because:

- Its junction with the body of the sternum at the **xiphisternal joint** indicates the inferior limit of the central part of the thoracic cavity projected onto the anterior body wall; this joint is also the site of the *infrasternal (subcostal) angle* of the inferior thoracic aperture.
- It is a midline marker for the upper limit of the liver, the central tendon of the diaphragm, and the inferior border of the heart.

Bony Xiphoid Processes

Not uncommonly, people in their early forties suddenly detect their partly ossified xiphoid processes and consult their physicians about the hard lumps in the "pits of their stomachs" (*epigastric fossae*). Never having been aware of their xiphoid processes before, they fear they have developed a tumor or "stomach cancer."

Sternal Fractures

Fractures of the sternum are not common despite its subcutaneous location. Crush injuries can occur after traumatic compression of the thoracic wall in automobile accidents when the driver's chest is forced into the steering column, for example. The installation of air bags in vehicles has reduced the number of sternal fractures. A fracture of the body of the sternum is usually a *comminuted fracture* (the sternum is broken into pieces). Displacement of the bone fragments is uncommon because the sternum is invested by deep fascia and the sternal attachment of the pectoralis major muscles. The most common site of sternal fracture is at the sternal angle, resulting in *dislocation of the manubriosternal joint.* ▶

Median Sternotomy

To gain access to the thoracic cavity for surgical operations in the mediastinum, for *coronary artery bypass grafting*, for example, the sternum is divided ("split") in the median plane and retracted. It is the flexibility of ribs and costal cartilages that enables spreading of the halves of the sternum. *Sternal splitting* also gives good exposure for removal of tumors in the superior lobes of the lungs. After surgery, the halves of the sternum are joined with wire sutures.

Sternal Biopsies

The sternal body is often used for *bone marrow needle biopsy* because of its breadth and subcutaneous position. The needle pierces the thin cortical bone and enters the vascular spongy bone. Sternal biopsy is commonly used to obtain specimens of bone marrow for transplantation and detection of metastatic cancer and *blood dyscrasias* (abnormalities).

Sternal Anomalies

The unfused halves of the developing sternum of the fetus (sternal bars) may fail to unite because of defective ossification. *Complete sternal cleft* is uncommon; such a severe cleft is usually associated with *ectopia cordis*, a congenital condition in which the heart is exposed on the thoracic wall because of maldevelopment of the sternum and pericardium (Moore and Persaud, 1998). Prosthetic material is usually required to close the sternal defect. Sternal clefts involving the manubrium and superior half of the body are V- or U-shaped and can be repaired during infancy by direct apposition and fixation of the cartilaginous sternal halves (Sabiston and Lyerly, 1994).

Sometimes there is a perforation (*sternal foramen*) in the sternal body because of faulty ossification. It is not clinically significant; however, one should be aware of its possible presence so that it will not be misinterpreted on a chest radiograph as a bullet wound. Although the xiphoid process is commonly perforated in elderly persons because of incomplete ossification, this perforation is not clinically significant. In infants the tip of the xiphoid process may protrude anteriorly beneath the skin. This congenital abnormality may persist but it usually does not need surgical correction.

Sex Differences in the Sternum

The body of the sternum is usually shorter and thinner in females than in males. These sex differences may be useful in determining the gender of human skeletal remains (e.g., in medicolegal cases and anthropological studies). ⊙

Joints of the Thoracic Wall

Although movements of the joints of the thoracic wall are frequent, in association with respiration, for example, the range of movement at the individual joints is small. Any disturbance that reduces the mobility of these joints, however, interferes with respiration. During deep breathing, the excursions of the thoracic cage (anteriorly, superiorly, or laterally) are considerable. Straightening the back further increases the anteroposterior (AP) diameter of the thorax. *Joints of the thoracic wall* (Table 1.1) occur between the:

- Vertebrae (intervertebral [IV] joints)
- Ribs and vertebrae (costovertebral joints: joints of the heads of ribs and costotransverse joints)
- Ribs and costal cartilages (costochondral joints)
- Costal cartilages (interchondral joints)
- Sternum and costal cartilages (sternocostal joints)
- Sternum and clavicle (SC joints)
- Parts of the sternum (manubriosternal and xiphisternal joints) in young people—the former and sometimes the latter usually fuses in very old people.

The *IV joints* between the bodies of adjacent vertebrae are joined together by longitudinal ligaments and IV discs. These joints are discussed with the back (see Chapter 4).

Costovertebral Joints

A typical rib articulates with the vertebral column at two joints (Fig. 1.7):

- Joints of heads of ribs
- Costotransverse joints.

Joints of Heads of Ribs. The head of each typical rib articulates with demifacets or costal facets of two adjacent thoracic vertebrae (Fig. 1.4) and the *IV disc* between them. The head articulates with the superior part of the corresponding (same-numbered) vertebra, the inferior part of the vertebra superior to it, and the adjacent disc uniting the two vertebrae. For example, the head of the 6th rib articulates with the superior part of the body of T6 vertebra, the inferior part of T5, and the disc between these vertebrae (Fig. 1.7). The **crest of the head** of the rib attaches to the IV disc by an **intra-articular ligament** within the joint, dividing the enclosed space into two synovial cavities. Exceptions to this general arrangement of articulation occur with the heads of the 1st, sometimes the 10th, and usually the 11th and 12th ribs; they articulate only with their own vertebral bodies (bodies of the same number as the rib). In these cases no intra-articular ligaments exist and the joint cavities are not divided.

An **articular capsule** surrounds each joint and connects the head of the rib with the circumference of the joint cavity. The fibrous capsule is strongest anteriorly where it forms a **radiate ligament** that fans out from the anterior

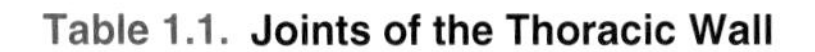

Table 1.1. Joints of the Thoracic Wall

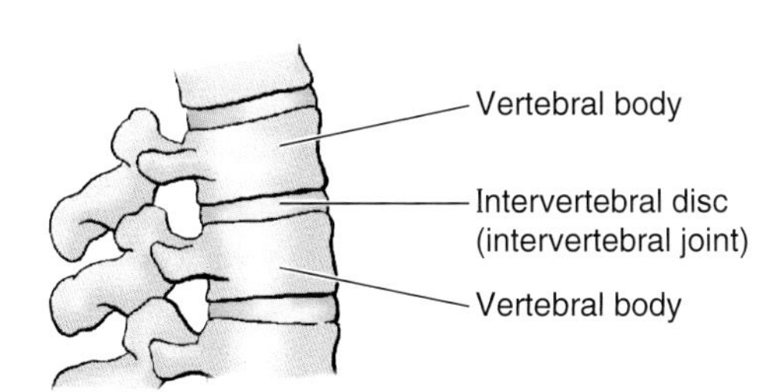

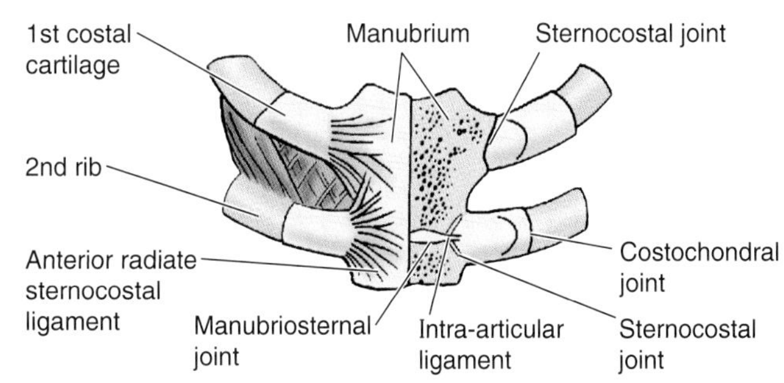

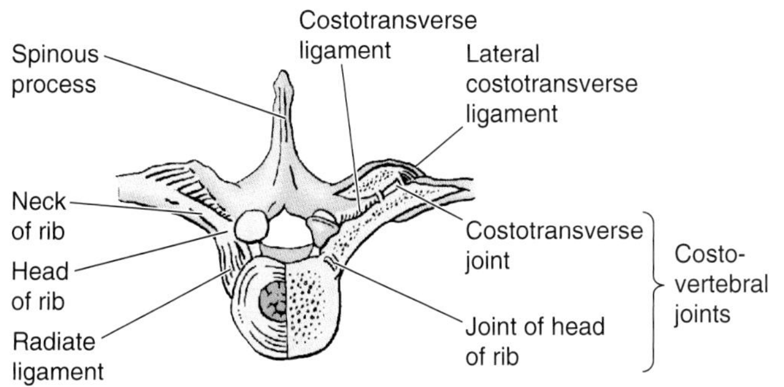

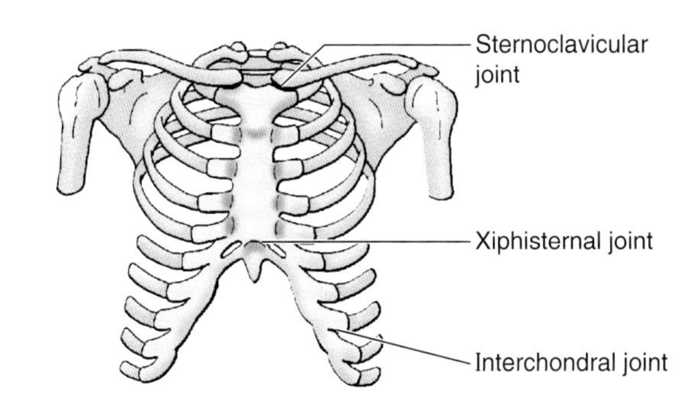

Joint	Type	Articulations	Ligaments	Comments
Intervertebral (IV)	Symphysis (secondary cartilaginous joint)	Adjacent vertebral bodies bound together by disc	Anterior and posterior longitudinal	
Costovertebral Joints of head of rib	Synovial plane joint	Head of each rib with superior demifacet or costal facet of corresponding vertebral body and inferior demifacet or costal facet of vertebral body superior to it	Radiate and intra-articular ligaments of head of rib	Heads of 1st, 11th, and 12th ribs (sometimes 10th) articulate only with corresponding vertebral body.
Costotransverse		Articulation of tubercle of rib with transverse process of corresponding vertebra	Lateral and superior costotransverse	11th and 12th ribs do not articulate with transverse process of corresponding vertebrae.
Costochondral	Primary cartilaginous joint	Articulation of lateral end of costal cartilage with sternal end of rib	Cartilage and bone bound together by periosteum	No movement normally occurs at this joint.
Interchondral	Synovial plane joint	Articulation between costal cartilages of 6th–7th, 7th–8th, and 8th–9th ribs	Interchondral ligaments	Articulation between costal cartilages of 9th and 10th ribs is fibrous.
Sternocostal	1st: primary cartilaginous joint (synchondrosis)	Articulation of first costal cartilages with manubrium of sternum		
	2nd to 7th: synovial plane joints	Articulation of the 2nd to 7th pairs of costal cartilages with sternum	Anterior and posterior radiate sternocostal	
Sternoclavicular	Saddle type of synovial joint	Sternal end of clavicle with manubrium of sternum and first costal cartilage	Anterior and posterior sternoclavicular ligaments; costoclavicular ligament	This joint is divided into two compartments by an articular disc.
Manubriosternal	Secondary cartilaginous joint (symphysis)	Articulation between manubrium and body of sternum		This joint often fuses and becomes a synostosis in older persons.
Xiphisternal	Primary cartilaginous joint (synchondrosis)	Articulation between xiphoid process and body of sternum		

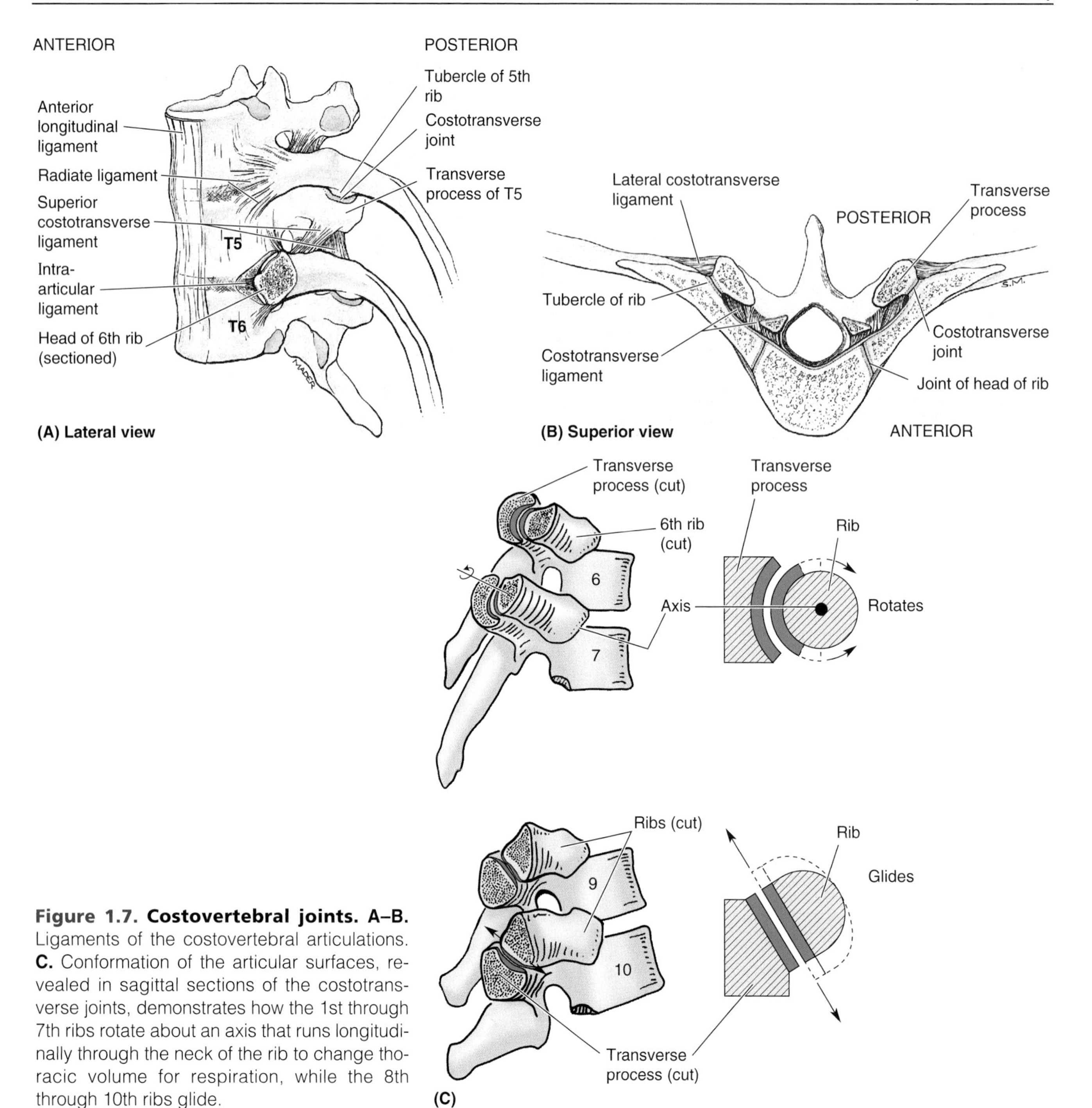

Figure 1.7. Costovertebral joints. A–B. Ligaments of the costovertebral articulations. **C.** Conformation of the articular surfaces, revealed in sagittal sections of the costotransverse joints, demonstrates how the 1st through 7th ribs rotate about an axis that runs longitudinally through the neck of the rib to change thoracic volume for respiration, while the 8th through 10th ribs glide.

margin of the head of the rib to the sides of the bodies of two vertebrae and the IV disc between them. The heads of the ribs connect so closely to the vertebral bodies that only slight gliding movements occur at the joints of the heads of ribs; however, even slight movement here may produce a relatively large excursion of the distal (sternal or anterior) end of a rib.

Costotransverse Joints. The tubercle of a typical rib articulates with the transverse costal facet on the anterior surface of the end of the transverse process of its own vertebra (Fig. 1.7). These small synovial joints are surrounded by thin articular capsules that attach to the edges of the articular facets. A *costotransverse ligament* passing from the neck of the rib to the transverse process, and a **lateral costotransverse ligament** passing from the tubercle of the rib to the tip of the transverse process, strengthen the anterior and posterior aspects of the joint, respectively. A **superior costotransverse ligament** is a broad band that joins the crest of the neck of the rib to the transverse process superior to it. The aperture between this ligament and the vertebra permits passage of the spinal nerve and the dorsal branch of the intercostal artery. The superior costotransverse ligament may be divided into a strong *anterior costotransverse ligament* and a weak *posterior costotransverse* ligament. The strong costotransverse ligaments binding these joints limit their movements to slight gliding. However, the articular surfaces on the tubercles of the superior six ribs are convex and fit into concavities on the transverse processes (Fig. 1.7*C*). As a result, some elevation and depression movements of the distal (sternal) ends of the ribs and sternum in the sagittal plane (Fig. 1.8*B*) are associated with rotation of the ribs. Flat articular surfaces of tubercles and transverse processes of the 7th to 10th ribs (Fig. 1.7*C*) allow both gliding and pivoting here, resulting in elevation and depression of the ribs in the transverse plane (Fig. 1.8*A*).

The floating 11th and 12th ribs do not articulate with transverse processes and have even freer movements as a result.

Costochondral Joints

The costochondral articulations are *hyaline cartilaginous joints*. Each rib has a cup-shaped depression in its sternal end into which the costal cartilage fits (Table 1.1). The rib and its cartilage are firmly bound together by the continuity of the periosteum of the rib with the perichondrium of the cartilage. No movement normally occurs at these joints.

Interchondral Joints

These articulations between the adjacent borders of the 6th and 7th, 7th and 8th, and 8th and 9th costal cartilages are *plane synovial joints* (Table 1.1). Each of these joints usually has a synovial cavity that is enclosed by an **articular capsule**. The joints are strengthened by interchondral ligaments. The articulation between the 9th and 10th costal cartilages is a fibrous joint.

Sternocostal Joints

The 1st to 7th ribs articulate through their costal cartilages with the lateral borders of the sternum as follows (Table 1.1):

- The 1st pair of cartilages articulates with the manubrium only
- The 2nd pair of cartilages articulates with the manubrium and 1st sternal segment (sternebra)
- The 3rd to 5th pairs of cartilages articulate with the 2nd and 3rd sternal segments
- The 6th pair of cartilages articulates with the 4th sternal segment only
- The 7th pair of cartilages articulates with the 4th sternal segment and the xiphoid process.

The 1st pair of costal cartilages articulates with the manubrium by means of **primary cartilaginous joints** or synchondroses. The cartilages unite directly to the hyaline cartilage in the depressions at the superolateral margins of the manubrium. The 2nd to 7th pairs of costal cartilages articu-

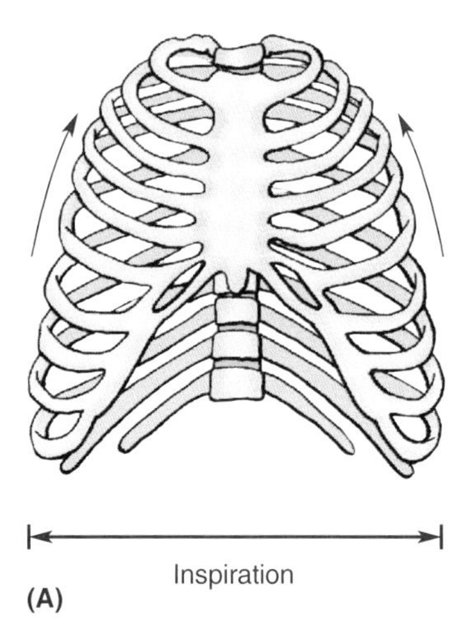

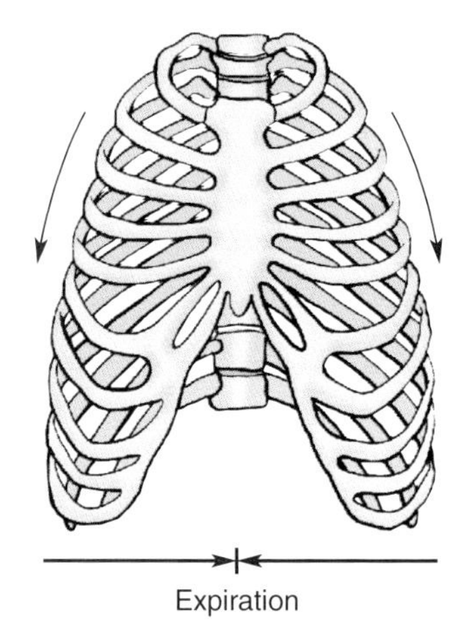

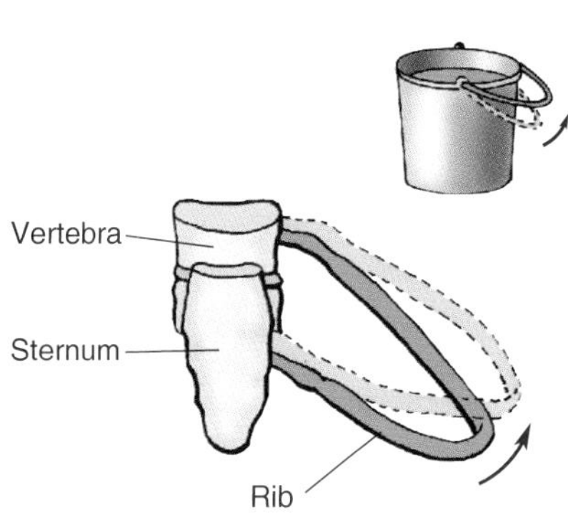

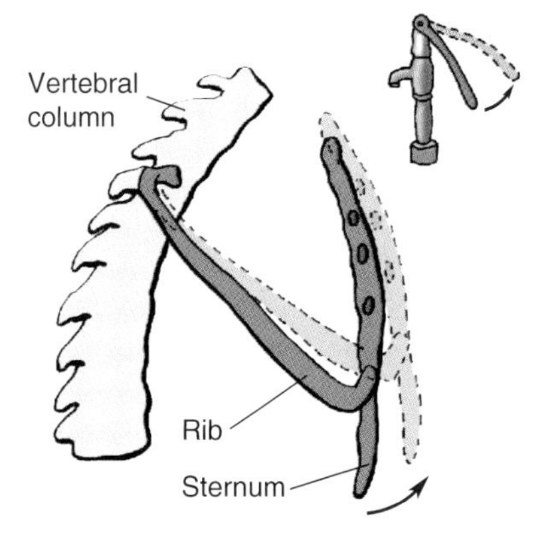

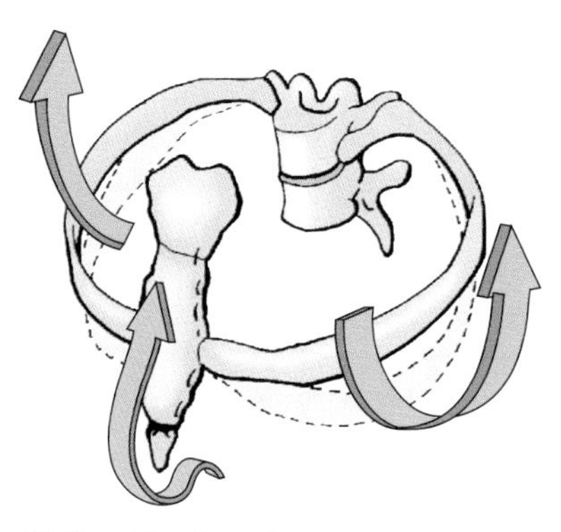

Figure 1.8. Movements of the thoracic wall. A. During inspiration and expiration, observe how the thorax widens during inspiration as the ribs are elevated. **B.** The middle parts of the lower ribs move laterally when they are elevated ("bucket-handle" movement). When the upper ribs are elevated, the anteroposterior (AP) diameter of the thorax is increased ("pump-handle" movement) with a greater excursion (increase) occurring inferiorly. **C.** The combination of movements that occur during inspiration increase the AP and transverse diameters of the thoracic cage.

late with the sternum at **synovial joints** and move during respiration. The weak articular capsules of these joints are reinforced (thickened) anteriorly and posteriorly to form **radiate sternocostal ligaments**. These continue as thin, broad membranous bands passing from the costal cartilages to the anterior and posterior surfaces of the sternum, forming a feltlike covering for this plate of bone.

Dislocation of Ribs

A rib dislocation (*slipping rib syndrome*) is the displacement of a costal cartilage from the sternum—*dislocation of a sternocostal joint* or the displacement of the interchondral joints. Rib dislocations are common in body contact sports; possible complications are pressure on or damage to nearby nerves, vessels, and muscles (Birrer, 1994). *Displacement of interchondral joints* usually occurs unilaterally and involves ribs 8, 9, and 10. Trauma sufficient to displace these joints often injures underlying structures such as the diaphragm and/or liver, causing severe pain, particularly during deep inspiratory movements. The injury produces a lumplike deformity at the displacement site.

Separation of Ribs

A rib separation refers to *dislocation of a costochondral junction* between the rib and its costal cartilage. In separations of the 3rd through 10th ribs, tearing of the perichondrium and periosteum usually occurs. As a result, the rib may move superiorly, overriding the rib above and causing pain. ⊙

Movements of the Thoracic Wall

Movements of the thoracic wall and diaphragm during inspiration produce increases in the intrathoracic volume and diameters of the thorax (Fig. 1.8*A*). Consequent pressure changes result in air being alternately drawn into the lungs (inspiration) through the nose, mouth, larynx, and trachea and expelled from the lungs (expiration) through the same passages. During passive expiration, the diaphragm, intercostal muscles, and other muscles relax, decreasing intrathoracic volume and increasing the *intrathoracic pressure*. The stretched elastic tissue of the lungs recoils, expelling most of the air. Concurrently, *intra-abdominal pressure* decreases.

The *vertical diameter* (height) of the central part of the thoracic cavity increases during inspiration as the diaphragm descends, compressing the abdominal viscera below it. During expiration, the diameter returns to normal as the elastic recoil of the lungs produces subatmospheric pressure in the pleural cavities, between the lungs and the thoracic wall. As a result of this and the absence of resistance to the previously compressed viscera, the domes of the diaphragm ascend, diminishing the vertical diameter. The *transverse diameter* of the thorax increases slightly when the intercostal muscles contract, raising the middle (lateral-most parts) of the ribs—the *bucket handle movement* (Fig. 1.8*B*). The *AP diameter* of the thorax also increases considerably when these muscles contract: movement of the ribs (primarily 2nd through 6th) at the costovertebral joints about an axis passing through the necks of the ribs causes the sternal ends of the ribs to rise—the *pump handle movement*. Because the ribs slope inferiorly, their elevation also results in anterior-posterior movement of the sternum, especially its inferior end, with slight movement occurring at the manubriosternal joint in young people, in whom it has not yet synostosed. The combination of all these movements moves the thoracic cage anteriorly, superiorly, and laterally (Fig. 1.8*C*).

Paralysis of the Diaphragm

Paralysis of half of the diaphragm (hemidiaphragm) because of injury to its motor supply from the *phrenic nerve* does not affect the other half because each dome has a separate nerve supply. One can detect paralysis of the diaphragm radiographically by noting its paradoxical movement. Instead of descending on inspiration, the paralyzed dome is pushed superiorly by the abdominal viscera that are being compressed by the active side; it falls during expiration in response to the positive pressure in the lungs. ⊙

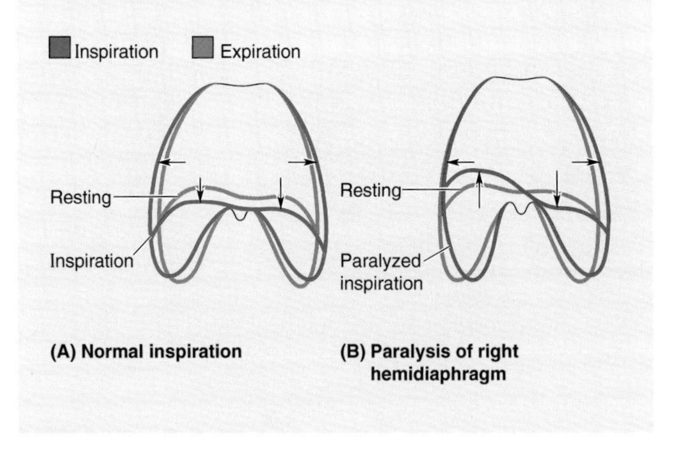

(A) Normal inspiration **(B) Paralysis of right hemidiaphragm**

Breasts

Both men and women have breasts; normally they are well-developed only in women (Fig. 1.9). The **mammary glands** in the breasts are accessory to reproduction in women but are rudimentary and functionless in men, consisting of only a few small ducts. Usually, little fat is present in the male breast, and the glandular system does not normally develop. The breasts are the most prominent superficial structures in the anterior thoracic wall, especially in women. The mammary glands are

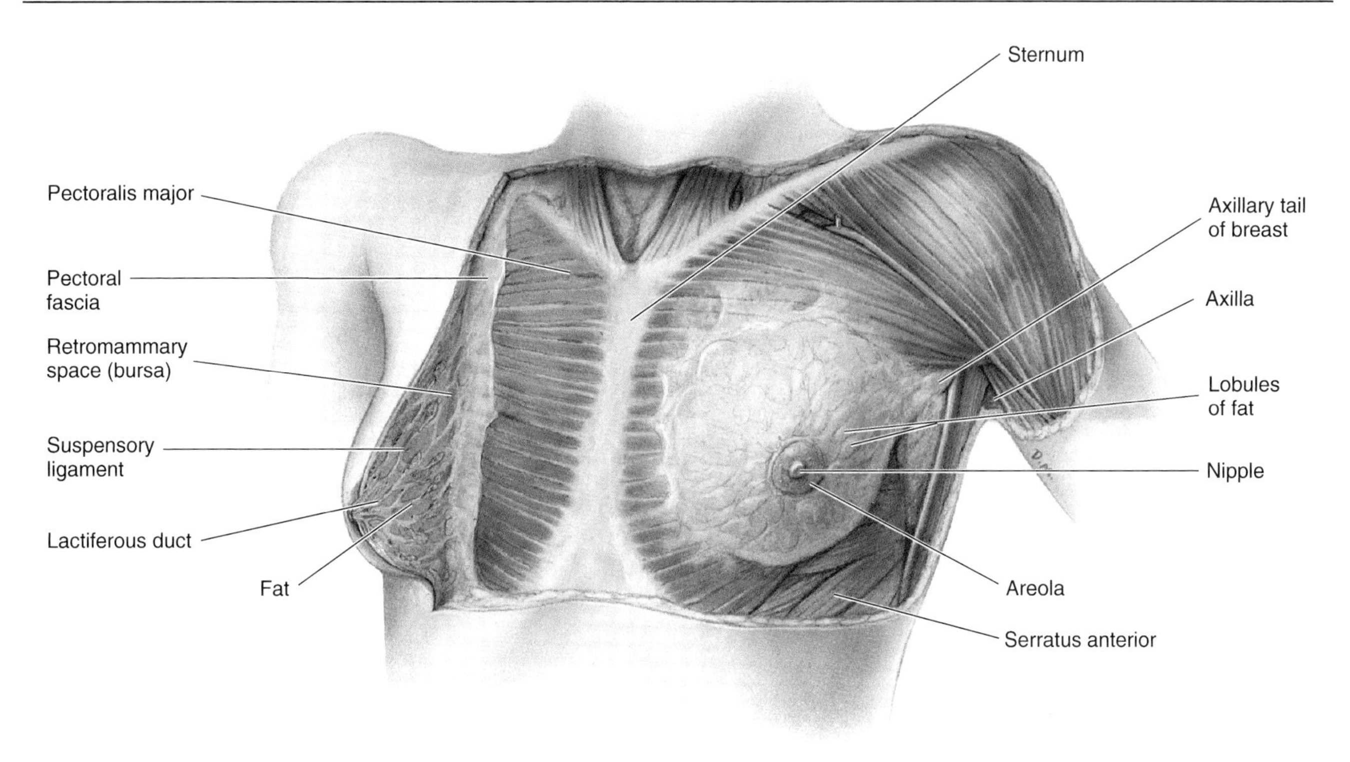

Figure 1.9. Superficial dissection of the pectoral region of the female. On the left side, observe the breast (skin removed) extending from the 2nd through the 6th ribs. Observe also the axillary tail extending into the axilla. The nonlactating breast consists primarily of fat. On the right side, observe the deep pectoral fascia covering the pectoralis major. Observe that the mammary gland lies in the subcutaneous connective tissue, or superficial fascia, between the skin and deep fascia.

in the subcutaneous tissue overlying the pectoral muscles (pectoralis major and minor). The amount of fat surrounding the glandular tissue determines the size of the breasts. At the greatest prominence of the breast is the **nipple**, surrounded by a circular pigmented area of skin—the **areola** (L. small area).

Female Breasts

The roughly circular base (bed) of the female breast (Figs. 1.9 and 1.10) extends:

- Transversely from the lateral border of the sternum to the midaxillary line (MAL) (p. 91)
- Vertically from the 2nd through 6th ribs.

A small part of the mammary gland may extend along the inferolateral edge of the pectoralis major toward the axilla (armpit), forming an **axillary tail** (of Spence). Some women discover this—especially when it may enlarge during a menstrual cycle—and become concerned that it may be a "lump" or enlarged lymph nodes.

Two-thirds of the breast rests on the **deep pectoral fascia** overlying the pectoralis major; the other 3rd rests on the fascia covering the serratus anterior. Between the breast and deep pectoral fascia is a loose connective tissue plane or potential space—the **retromammary space** (bursa). This plane, containing a small amount of fat, allows the breast some degree of movement on the pectoral fascia. The mammary gland is firmly attached to the dermis of the overlying skin by skin ligaments (retinacula cutis)—the **suspensory ligaments** (of Cooper) of the breast. These fibrous condensations of the connective tissue stroma, particularly well-developed in the superior part of the gland, help support the lobules of the gland.

During puberty (8 to 15 years of age), the breasts normally grow because of glandular development and increased fat deposition. The areolae and nipples also enlarge. Breast size and shape result from genetic, racial, and dietary factors. The lactiferous ducts give rise to buds that form 15 to 20 lobules of glandular tissue, which constitute the gland. Each lobule of the breast is drained by a **lactiferous duct** that usually opens independently on the nipple. The ducts converge toward the nipple like the spokes of a bicycle wheel. Deep to the areola, each duct has a dilated portion, the **lactiferous sinus**, in which a small droplet of milk accumulates or remains in the nursing mother. As the infant begins to suckle, compression of the areola (and the lactiferous sinus beneath it) expresses the accumulated droplets and encourages the infant to continue nursing as the hormonally mediated "let down reflex" ensues and the mother's milk is secreted into—not sucked from the gland by—the baby's mouth.

The areolae contain numerous **sebaceous glands** that enlarge during pregnancy and secrete an oily substance that

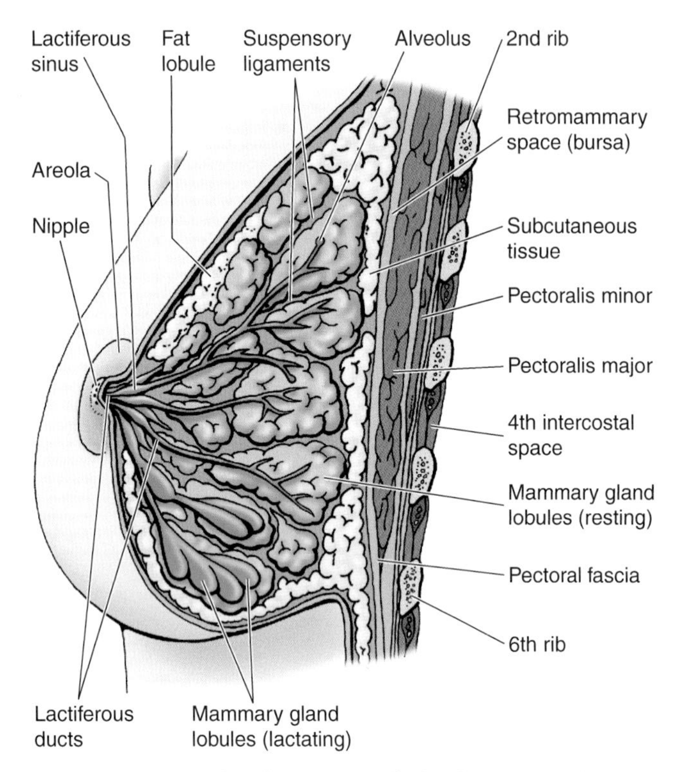

Figure 1.10. Sagittal section of the breast and anterior thoracic wall. The breast consists of glandular tissue and fibrous and adipose tissue between the lobes and lobules of glandular tissue, together with blood vessels, lymphatic vessels, and nerves. The superior two-thirds show schematically the suspensory ligaments and alveoli of the breast with resting mammary gland lobules. The inferior part shows lactating mammary gland lobules.

provides a protective lubricant for the areola and nipple, which are particularly subject to chaffing and irritation as mother and baby begin the nursing experience. The **nipples** are conical or cylindrical prominences in the centers of the areolae. The nipples have no fat, hair, or sweat glands. In young *nulliparous women*—those who have never borne a viable child—the nipples are usually at the level of the *4th intercostal spaces*. However, the position of nipples varies considerably in women, especially in *multiparous women*—those who have given birth to viable children at least twice. Consequently, the nipples are not a reliable guide to the 4th intercostal spaces in adult females. The tips of the nipples are fissured with the lactiferous ducts opening into them. The nipples are composed mostly of circularly arranged smooth muscle fibers that compress the lactiferous ducts during lactation and erect the nipples in response to stimulation, as when a baby begins to suckle.

The **mammary glands** are modified sweat glands; therefore, they have no special capsule or sheath. The rounded contour and most of the volume of the breasts is produced by the fat lobules, except during pregnancy when the mammary glands enlarge and new glandular tissue forms. The milk-secreting **alveoli** are arranged in grapelike clusters. In most women the breasts enlarge slightly during their menstrual cycles because of the increase in gonadotropic hormones (follicle-stimulating hormone [FSH] and luteinizing hormone [LH]).

Changes in the Breasts

Changes, such as branching of the lactiferous ducts, occur in the breast tissues during the menstrual cycles and pregnancy (Ferguson et al., 1992). Although mammary glands are prepared for secretion by midpregnancy, they do not produce milk until shortly after the baby is born. *Colostrum*, a creamy white to yellowish premilk fluid, may secrete from the nipples during the last trimester of pregnancy and during initial episodes of nursing. Colostrum is believed to be especially rich in protein, immune agents, and a growth factor affecting the infant's intestines.

In *multiparous women* the breasts often become large and pendulous. The breasts in elderly women are small and wrinkled because of the decrease in fat and atrophy of glandular tissue.

Breast Quadrants

For the anatomical location and description of tumors, the surface of the breast is divided into four quadrants. For example, a physician's record might state:

A hard irregular mass was felt in the upper inner quadrant of the breast at the two o'clock position, approximately 2.5 cm from the margin of the areola. ⊙

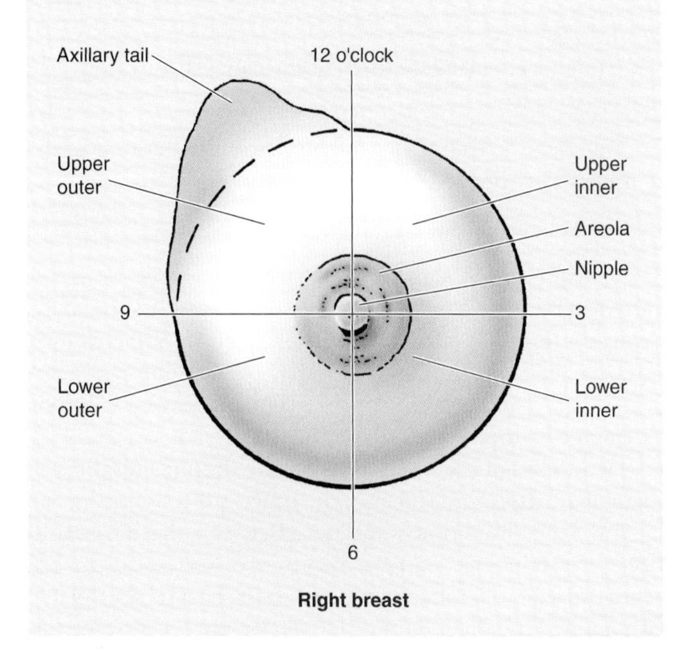

Vasculature of the Breast

The arterial supply of the breast (Fig. 1.11) is derived from:

- Medial mammary branches of perforating branches and anterior intercostal branches of the **internal thoracic artery**, originating from the subclavian artery
- **Lateral thoracic** and **thoracoacromial arteries**, branches of the axillary artery
- **Posterior intercostal arteries**, branches of the thoracic aorta in the 2nd, 3rd, and 4th intercostal spaces.

The venous drainage of the breast is mainly to the **axillary vein**, but there is some drainage to the *internal thoracic vein.*

The lymphatic drainage of the breast is *important* because of its role in the metastasis of cancer cells. Lymph passes from the nipple, areola, and lobules of the gland to the **subareolar lymphatic plexus** (Fig. 1.12), and from it:

- Most lymph (more than 75%), especially from the lateral quadrants of the breast, drains to the **axillary lymph nodes**, initially to the *pectoral (anterior) nodes* for the most part; however, some lymph may drain directly to the other

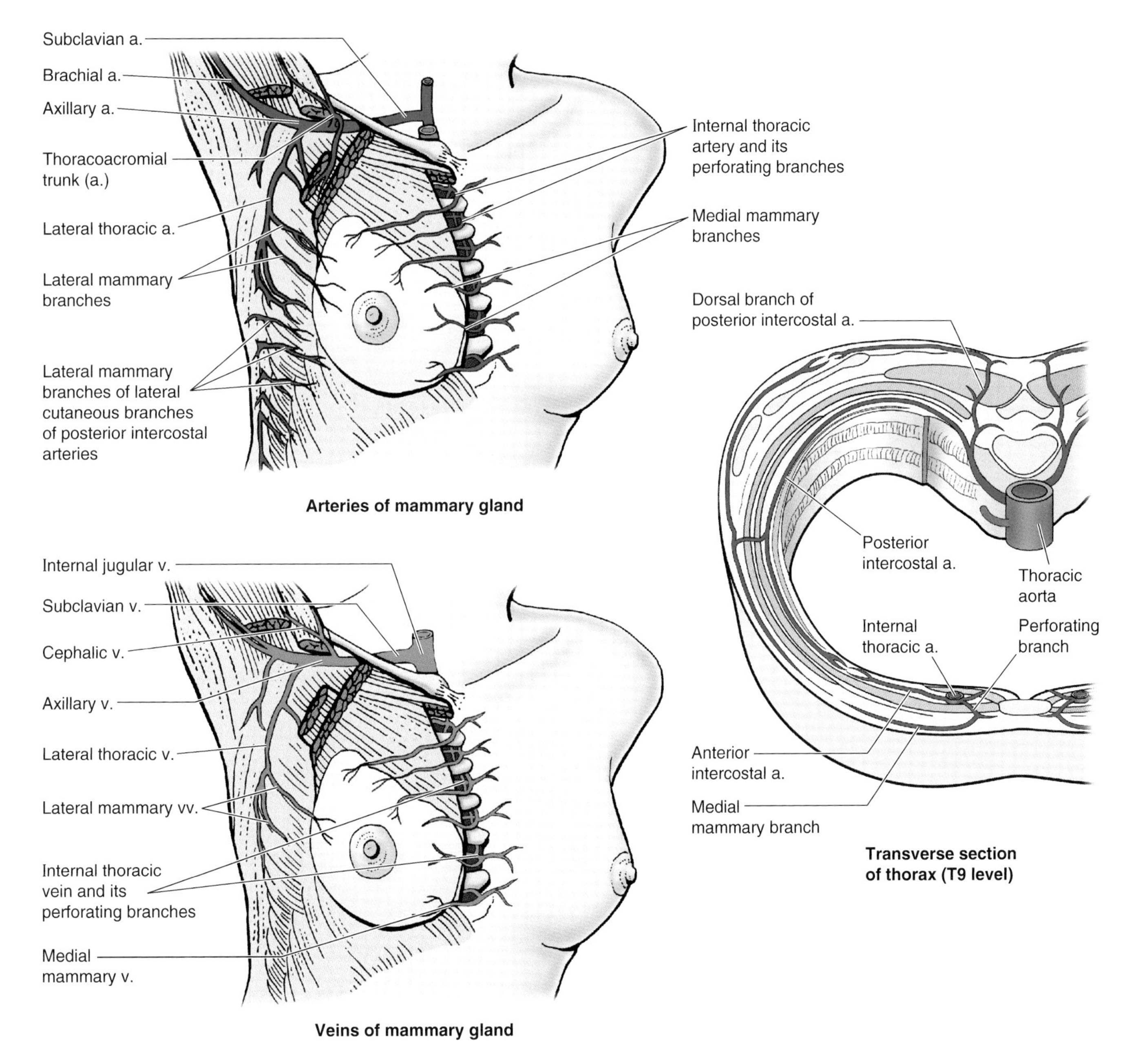

Figure 1.11. Vasculature of the breast. The mammary gland is extremely vascular and is supplied mainly by perforating branches of the internal thoracic artery, by several branches of the axillary artery—mainly the lateral thoracic artery—and by branches arising from the intercostal arteries as they pass deep to the breast. Venous drainage is to axillary (mainly) and internal thoracic veins.

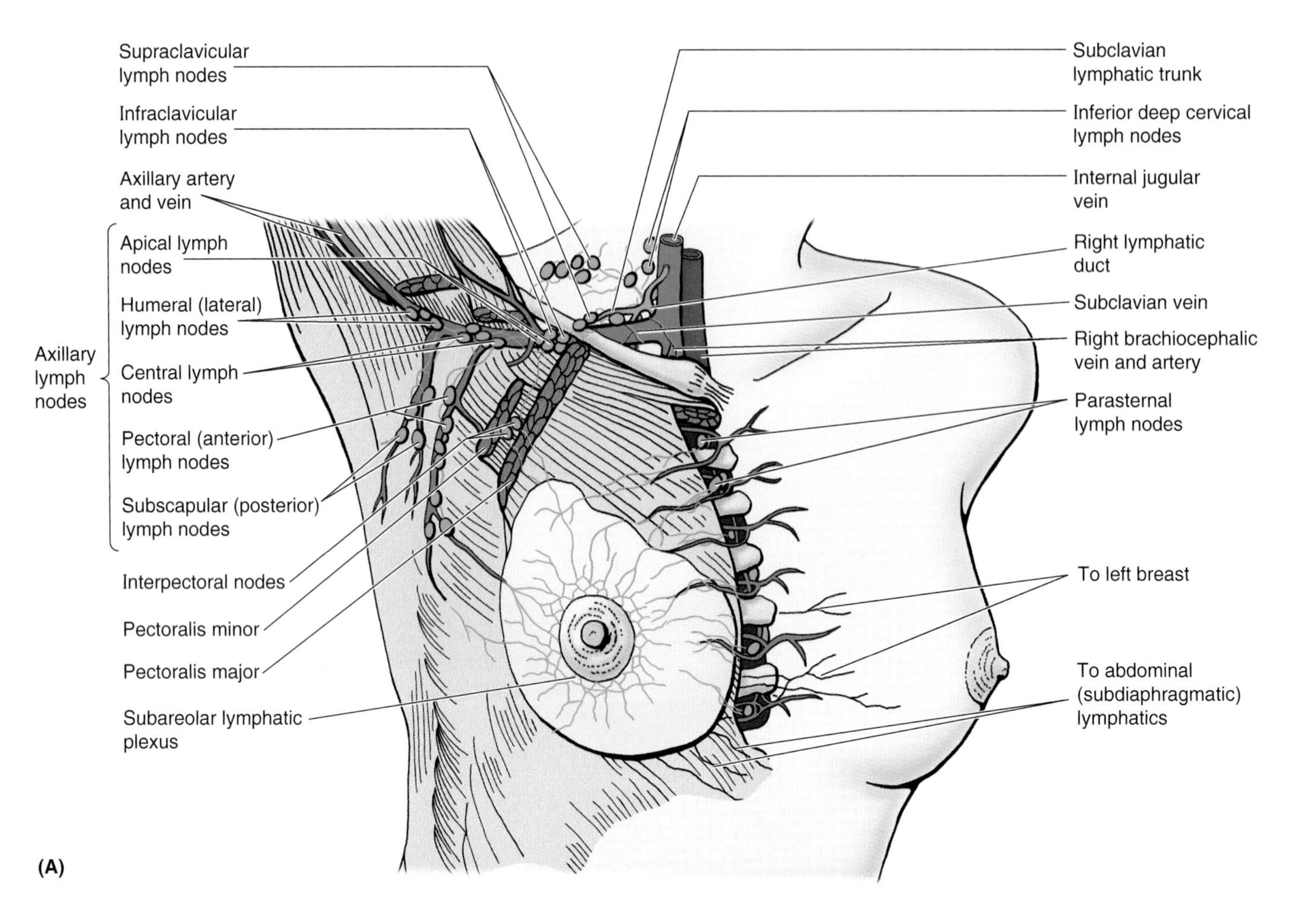

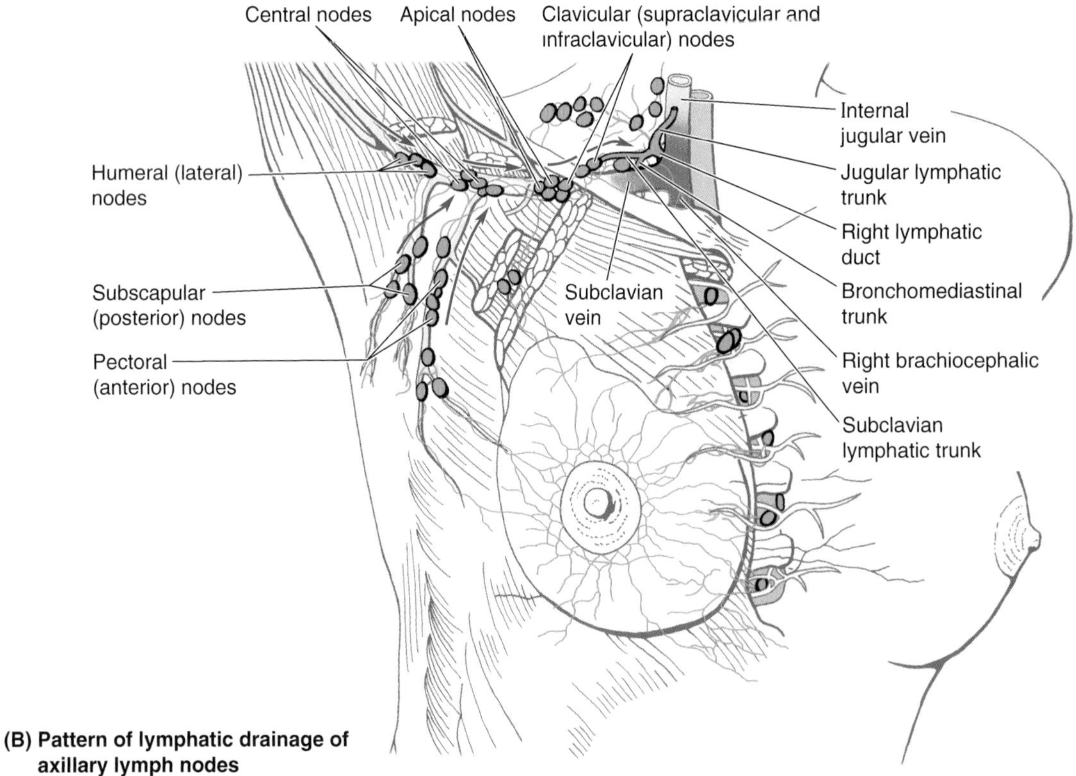

(B) Pattern of lymphatic drainage of axillary lymph nodes

Figure 1.12. Lymphatic drainage of the breast. A. Most lymph drains to the axillary lymph nodes. **B.** The *red arrows* indicate the direction of lymph flow from the axillary lymph nodes to the right lymphatic duct.

axillary nodes or even to interpectoral, deltopectoral, supraclavicular, or inferior deep cervical nodes.

- Most of the remaining lymph, particularly from the medial quadrants, drains to the *parasternal nodes* or to the opposite breast, while lymph from the lower quadrants passes deeply to the inferior phrenic (abdominal) nodes.

Lymphatic vessels in the skin of the breast, except the nipple and areola, drain into the axillary, inferior deep cervical, and infraclavicular nodes, and also into the parasternal nodes of both sides.

Lymph from the axillary nodes drains into infraclavicular and supraclavicular nodes and from them into the **subclavian lymphatic trunk**, which also drains lymph from the upper limb. Lymph from the parasternal nodes enters the **bronchomediastinal trunk**, which drains lymph from the thoracic viscera. The termination of these lymphatic trunks varies. Traditionally, these trunks are described as merging with each other and the **jugular lymphatic trunk**, draining the head and neck, to form a very short, **right lymphatic duct** on the right side, or entering the termination at the thoracic duct on the left side. However, in many, perhaps most, cases the trunks open independently into the *junction of the internal jugular and subclavian veins* to form the brachiocephalic veins. In some cases they open into both of these veins.

Nerves of the Breast

The nerves of the breast derive from anterior and lateral cutaneous branches of the **4th through 6th intercostal nerves** (Fig. 1.13). The ventral primary rami of T1 through T11 are called intercostal nerves because they run within the intercostal spaces. Communicating branches (rami communicantes) connect each ventral ramus to a sympathetic trunk. The branches of the intercostal nerves pass through the deep fascia covering the pectoralis major to reach the skin, including the breast in the subcutaneous tissue overlying this muscle. The branches of the intercostal nerves thus convey sensory fibers to the skin of the breast and sympathetic fibers to the blood vessels in the breasts and smooth muscle in the overlying skin and nipple.

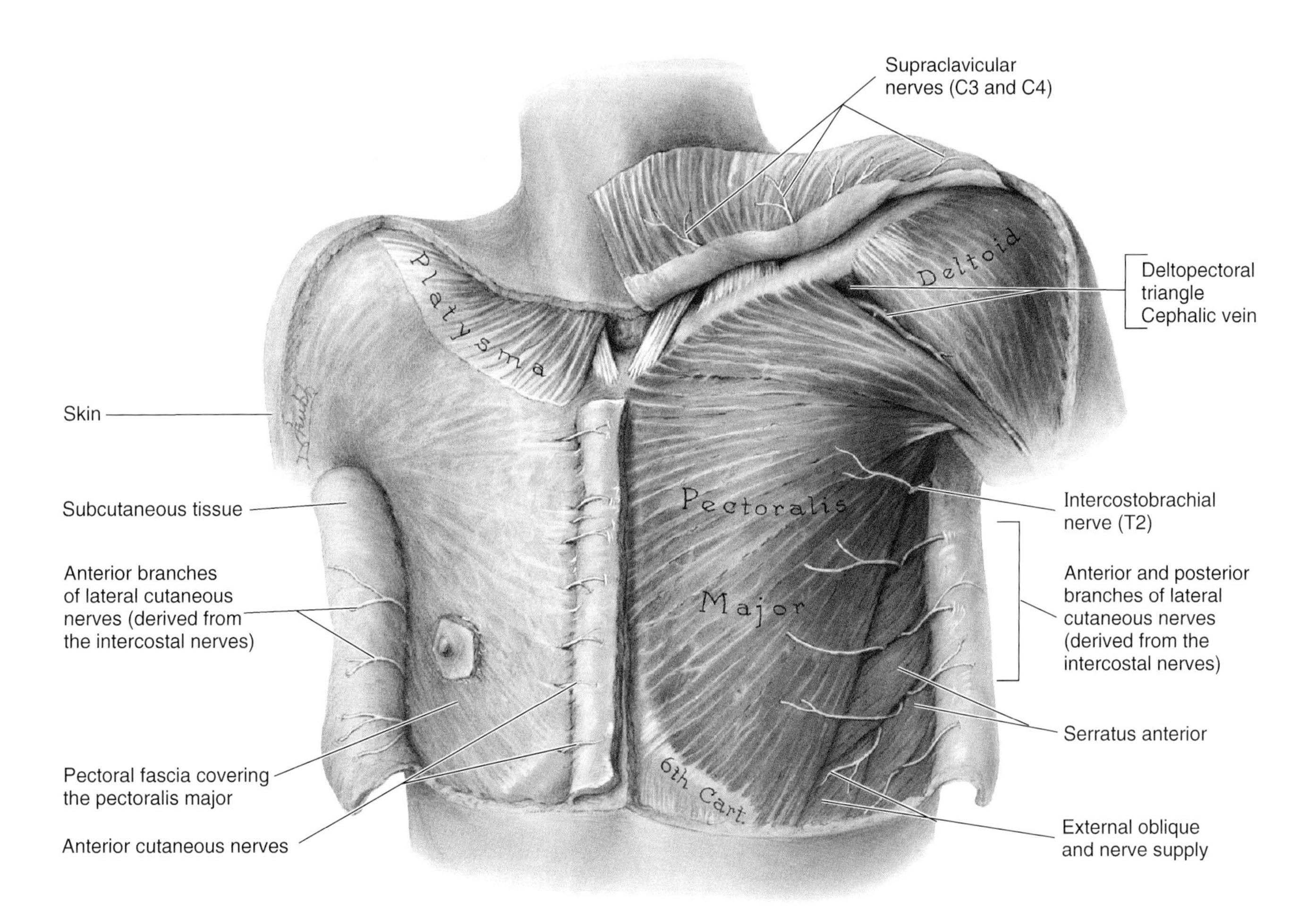

Figure 1.13. Superficial dissection of the pectoral region of a male. The platysma (G. flat plate) is cut short on the right side and is reflected on the left side, together with the underlying supraclavicular nerves. Observe the filmy, deep fascia covering the right pectoralis major. It has been removed on the left side to show the cutaneous nerves.

Carcinoma of the Breast

Understanding the lymphatic drainage of the breasts is of practical importance in predicting the metastasis of *carcinoma of the breast*—**breast cancer**. Carcinomas of the breast are almost all adenocarcinomas derived from the glandular epithelium of the terminal ducts in the mammary gland lobules (Rubin and Farber, 1993). Cancer cells that enter a lymphatic vessel usually pass through two or three groups of lymph nodes before entering the venous system.

Interference with the lymphatic drainage of the breast by cancer may cause deviation of the nipple and produce a leatherlike, thickened appearance of the skin. The skin is thickened or "puffy," with prominent pores that give it an orange peel appearance (*peau d'orange sign*) because of the edema (excess fluid in the subcutaneous tissue) resulting from the blocked lymphatic drainage. The larger dimples result from cancer invasion of the glandular tissue and fibrosis (fibrous degeneration) that cause shortening of the suspensory ligaments. *Subareolar breast cancer* may cause inversion of the nipple by the same mechanism.

The posterior intercostal veins drain into the *azygos/hemiazygos system of veins* alongside the bodies of the vertebrae (see Fig. 1.23), which empties into the superior vena cava (SVC). Through this route, cancer cells can spread from the breast to the vertebrae and from there to the skull and brain. *When breast cancer cells invade the retromammary space* (Fig. 1.10), attach to or invade the deep pectoral fascia overlying the pectoralis major, or metastasize to the interpectoral nodes, the breast elevates when the muscle contracts. This movement is a clinical sign of advanced cancer of the breast. To observe this upward movement, the physician has the patient place her hands on her hips and press to tense her pectoral muscles.

Lymphatic vessels carry cancer cells from the breast to lymph nodes, chiefly those in the axilla. The cells lodge in the nodes, producing nests of tumor cells (*metastases*). Abundant communications between lymphatic pathways and between the axillary, cervical, and parasternal nodes may cause metastases from the breast to develop in the supraclavicular lymph nodes, the opposite breast, or the abdomen. Because the axillary lymph nodes are the most common site of metastases from a breast cancer, enlargement of these palpable nodes in a woman suggests the possibility of breast cancer and may be key to early detection. However, the absence of enlarged axillary lymph nodes is no guarantee that metastasis from a breast cancer has not occurred because the malignant cells may have passed to other nodes, such as the infraclavicular and supraclavicular lymph nodes.

Mastectomy (excision of a breast) is not as common as it once was as a treatment for breast cancer. In *simple mastectomy*, the breast is removed down to the retromammary space. *Radical mastectomy*, a more extensive surgical procedure, involves removal of the breast, pectoral muscles, fat, fascia, and all lymph nodes in the axilla and pectoral region. In current practice, often only the tumor and surrounding tissues are removed; this is a *lumpectomy*, or a wide local excision. ▶

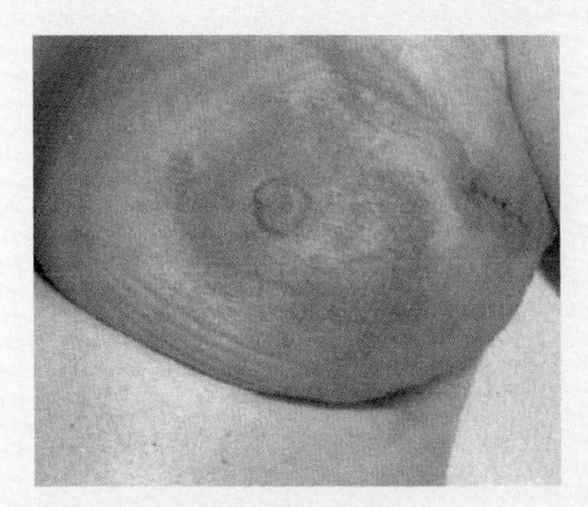

Inflamed carcinoma of the breast

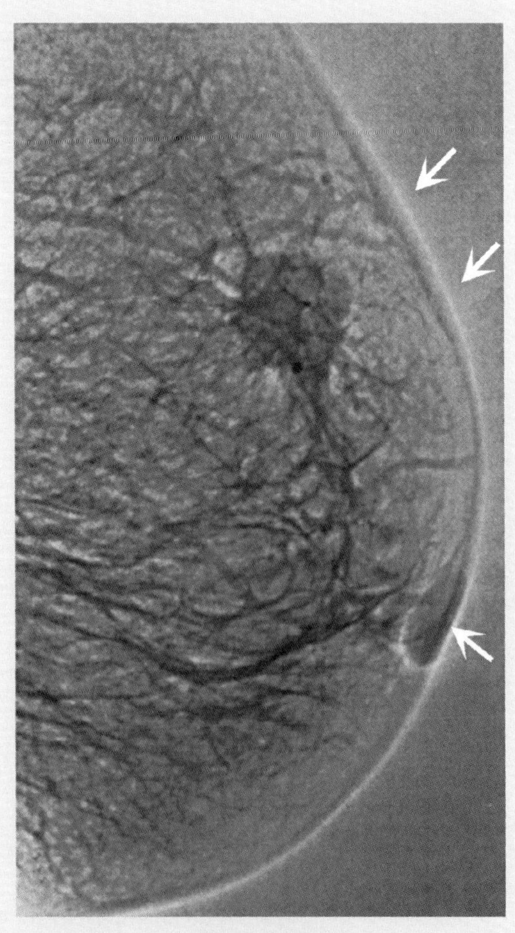

Mammogram

▶ **Mammography** (radiographic examination of breasts) is one of the techniques used to detect breast masses. The carcinoma appears as a large, jagged density in the **mammogram** (paired *white arrows*). Note thickening of the overlying skin. The *lower arrow* indicates the nipple. Mammography is also used by surgeons to guide them when removing breast tumors, cysts, and abscesses.

Polymastia, Polythelia, and Amastia

Supernumerary breasts (exceeding the normal number)—*polymastia*—or nipples (*polythelia*) may occur superior or inferior to the normal breasts, occasionally developing in the axilla or anterior abdominal wall. Usually supernumerary breasts consist only of a rudimentary nipple and areola, which may be mistaken for a mole (nevus) until they change pigmentation with the normal nipples during pregnancy. However, glandular tissue may also occur and further develop with lactation. Extra breasts may appear anywhere along a line extending from the axilla to the groin, the location of the embryonic mammary ridge ("milk line") from which the breasts develop (Moore and Persaud, 1998) and along which breasts develop in animals with multiple breasts. In either sex, there may be no breast development (amastia) or there may be a nipple but no glandular tissue.

Breast Cancer in Men

Approximately 1.5% of breast cancers occur in men. As in women, breast cancer in men usually metastasizes to lymph nodes, bone, pleura, lung, liver, and skin. Carcinoma of the breast affects approximately 1000 men per year in the United States (Swartz, 1994). A visible and/or palpable subareolar mass or secretion from a nipple may indicate a *malignant tumor*. Breast cancer in males tends to infiltrate the deep pectoral fascia, pectoralis major, and the apical group of axillary lymph nodes. Although breast cancer is uncommon in men, the consequences are serious because they are frequently not detected until extensive metastases have occurred, as in bones, for example.

Gynecomastia

Enlargement of the breasts in males—gynecomastia—commonly occurs at puberty but may also accompany aging or be drug related (e.g., after treatment with diethylstilbestrol for prostate cancer). Gynecomastia may also result from a change in the metabolism of sex hormones by the liver (Swartz, 1994). Approximately 40% of postpubertal males with Klinefelter syndrome (XXY trisomy) have gynecomastia (Moore and Persaud, 1998). ⊙

Thoracic Apertures

The thoracic cavity communicates with the neck through the superior thoracic aperture, or thoracic inlet, and with the abdominal cavity through the inferior thoracic aperture, or thoracic outlet (Fig. 1.14). The inferior thoracic aperture is closed by the diaphragm, which arches superiorly into the thorax.

Superior Thoracic Aperture

Structures entering or leaving the thoracic cavity through the oblique, kidney-shaped superior thoracic aperture include the trachea, esophagus, nerves, and vessels that supply and drain the head, neck, and upper limbs. The adult superior thoracic aperture measures approximately 6.5 cm anteroposteriorly and 12.5 cm transversely. (To help you visualize this opening, note that it is just slightly larger than necessary to allow passage of a 2" × 4" piece of lumber.) Because of the obliquity of the 1st pair of ribs, the aperture slopes anteroinferiorly. *The superior thoracic aperture is bounded by the*:

- 1st thoracic (T1) vertebra (posterior landmark)
- 1st pair of ribs and their costal cartilages
- Superior border of the manubrium (anterior landmark).

Thoracic Outlet Syndrome

When clinicians refer to the superior thoracic aperture as the thoracic "outlet," they are emphasizing the important nerves and arteries that pass from the thorax through this aperture into the lower neck and upper limb. Hence, various types of thoracic outlet syndrome exist (Rowland, 1995). The *costoclavicular syndrome* (pallor and coldness of the skin of the upper limb and diminished radial pulse) results from compression of the subclavian artery between the clavicle and 1st rib. The *cervical rib syndrome* results from compression of C8 and T1 nerve roots and the inferior trunk of the brachial plexus. The thoracic outlet syndrome is also discussed in Chapter 6. ⊙

Inferior Thoracic Aperture

The inferior thoracic aperture—the anatomical thoracic outlet—is much more spacious than the superior thoracic aperture. The inferior thoracic aperture is large and irregular in outline (Fig. 1.14). It is also oblique because the posterior thoracic wall is much longer than the anterior wall. The inferior thoracic aperture is closed by the musculotendinous diaphragm separating the thoracic and abdominal cavities. Structures passing from or to the thorax from the abdomen pass through openings that traverse the diaphragm, such as the esophagus and inferior vena cava (IVC), or pass posterior to it (e.g., aorta). *The inferior thoracic aperture is bounded by the:*

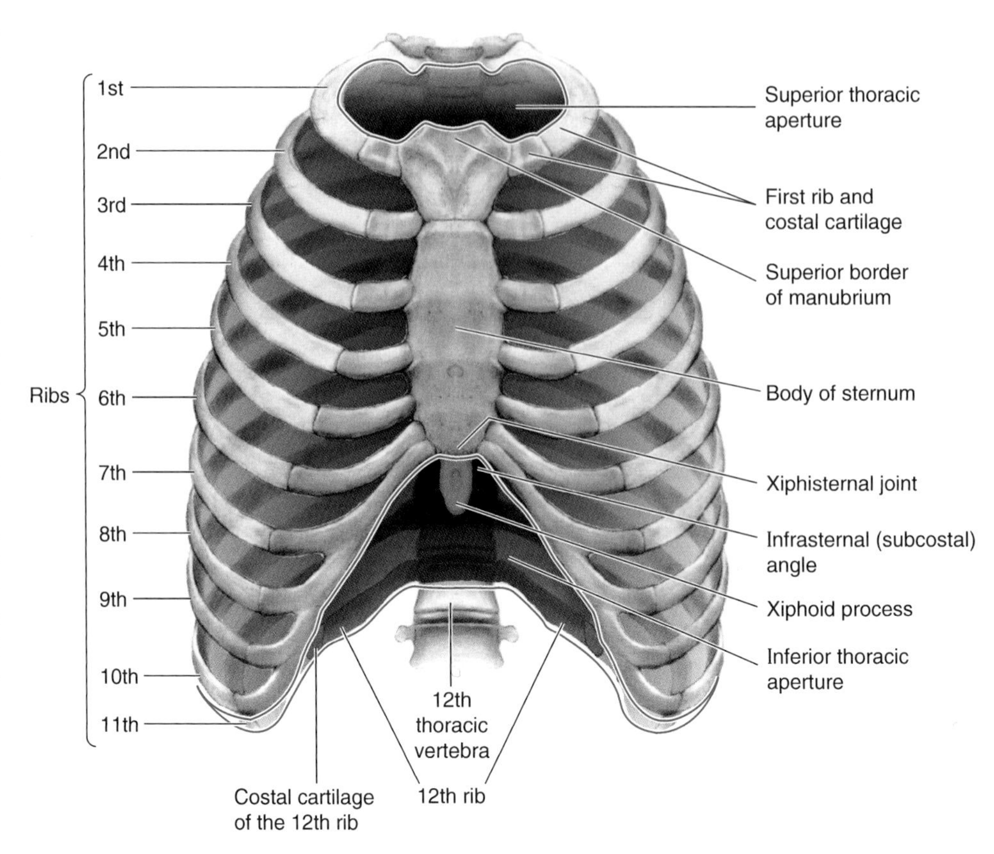

Figure 1.14. Thoracic apertures. The thoracic cavity communicates with the anterior (visceral) compartment of the neck by the superior thoracic aperture (thoracic inlet—clinically, thoracic outlet). The superior thoracic aperture is the "doorway" between the thoracic cavity and the neck region. The inferior thoracic aperture (thoracic outlet) provides attachment for the diaphragm, which separates the thoracic and abdominal cavities; however, the diaphragm protrudes upward so that upper abdominal viscera (e.g., the liver) reside within and receive protection from the thoracic cage. Thoracic structures must perforate the diaphragm to communicate with the abdomen.

- 12th thoracic vertebra (posterior landmark)
- 11th and 12th pairs of ribs
- Costal cartilages of ribs 7 through 10
- Xiphisternal joint (anterior landmark).

Muscles of the Thoracic Wall

Several upper limb muscles attach to the ribs (Figs. 1.15–1.19)—such as the pectoralis major, pectoralis minor, subclavius, and serratus anterior—as do the anterolateral abdominal muscles and some back and neck muscles. The **pectoral muscles** covering the anterolateral thoracic wall usually act on the upper limbs (see Chapter 6); however, the **pectoralis major** and other muscles may also function as accessory muscles of respiration, helping to expand the thoracic cavity when inspiration is deep and forceful (e.g., after a 100-meter dash). The **serratus anterior** overlying the lateral surface of the thorax, which rotates the scapula (L. shoulder blade) and holds it against the thoracic wall, may also function as an accessory muscle of respiration because it elevates the ribs. The **scalene muscles** (scaleni), passing from the neck to the 1st and 2nd ribs, also serve as accessory respiratory muscles by elevating these ribs during forced inspiration.

The serratus posterior, levator muscles of the ribs (levatores costarum), intercostal, subcostal, and transversus thoracis are muscles of the thoracic wall (Table 1.2).

Dyspnea—Difficult Breathing

When patients with respiratory problems such as *asthma* or with *heart failure* struggle to breathe, they use their accessory respiratory muscles to assist the expansion of their thoracic cavities. They lean on a table to fix their pectoral girdles (clavicles and scapulae) so these muscles are able to act on their rib attachments and expand the thorax. ○

The **serratus posterior muscles**, extending from the vertebrae to the ribs, are both inspiratory muscles. The *serratus posterior superior* lies at the junction of the neck and back. It arises from the inferior part of the nuchal ligament (neck ligament, L. ligamentum nuchae) in the neck and the spinous processes of C7 and T1 through T3 vertebrae. This muscle runs inferolaterally and attaches to the superior borders of the

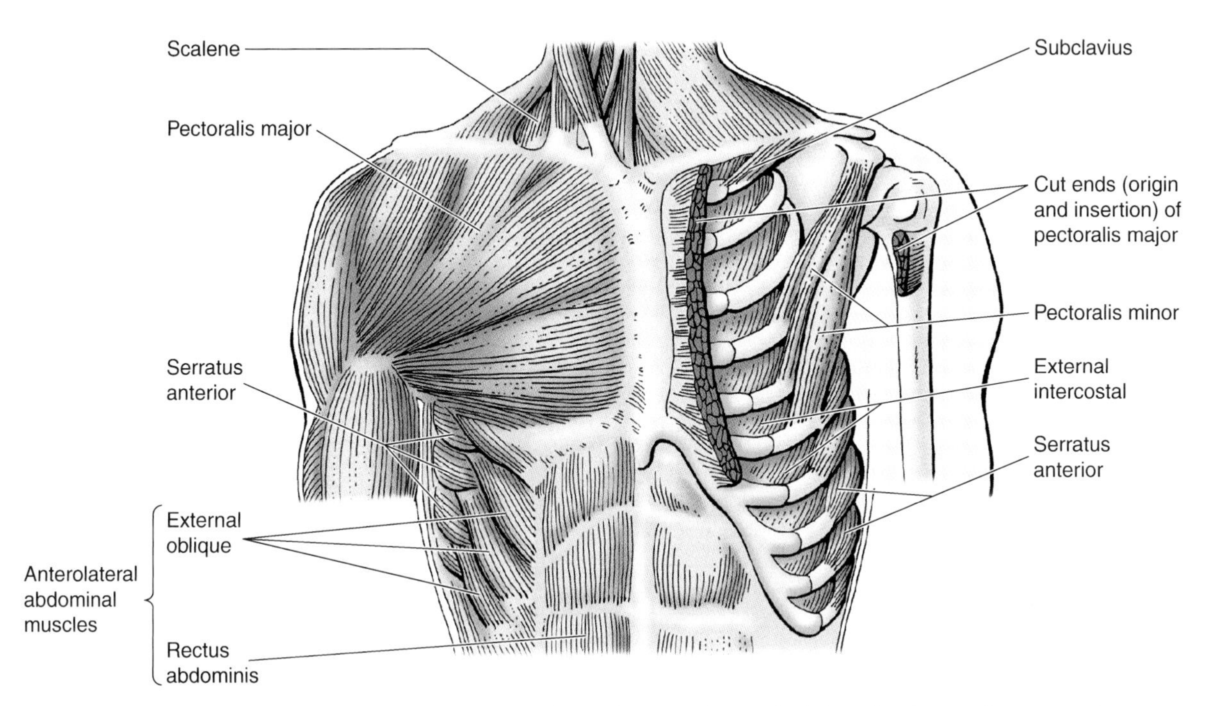

Figure 1.15. Pectoral, external intercostal, and anterolateral abdominal muscles. A portion of the scalene muscles, which extend from the superior two ribs to the cervical transverse processes, is also shown. The pectoralis major has been removed on the left side to expose the pectoralis minor, subclavius, and external intercostal muscles.

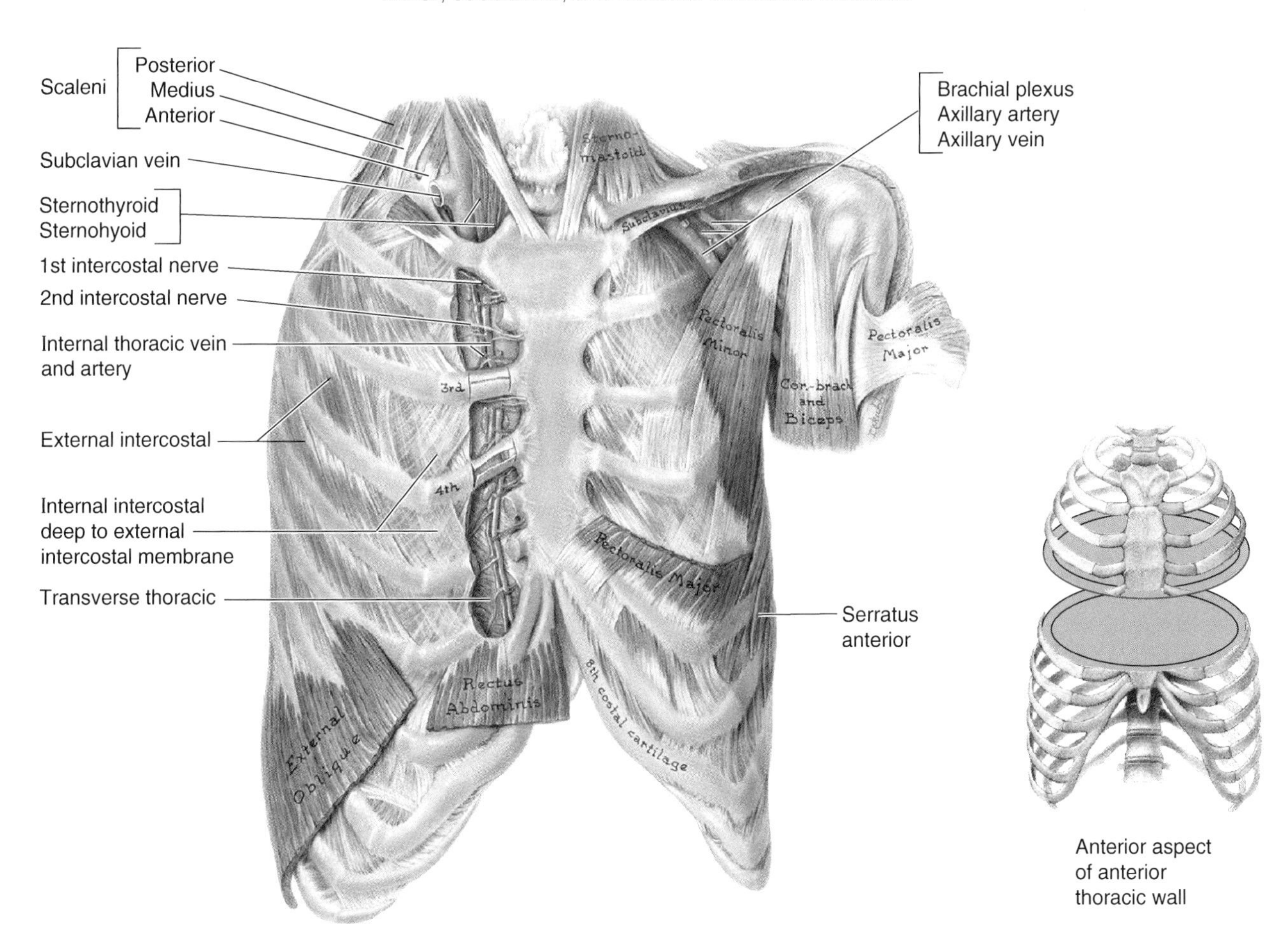

Figure 1.16. Dissection of the anterior aspect of the anterior thoracic wall. Observe the internal thoracic vessels running approximately 1 cm lateral to the edge of the sternum and the parasternal lymph nodes (*green*). Lymph from the medial part of the breast drains into these nodes and is the route by which cancer cells from the breast may spread to the lungs and mediastinum—the mass of tissues and organs separating the lungs. It is not uncommon for the 8th rib to attach to the sternum, as in this specimen. The H-shaped cuts through the perichondrium of the 3rd and 4th costal cartilages were used to shell out pieces of cartilage, illustrating how one type of thoracotomy is performed.

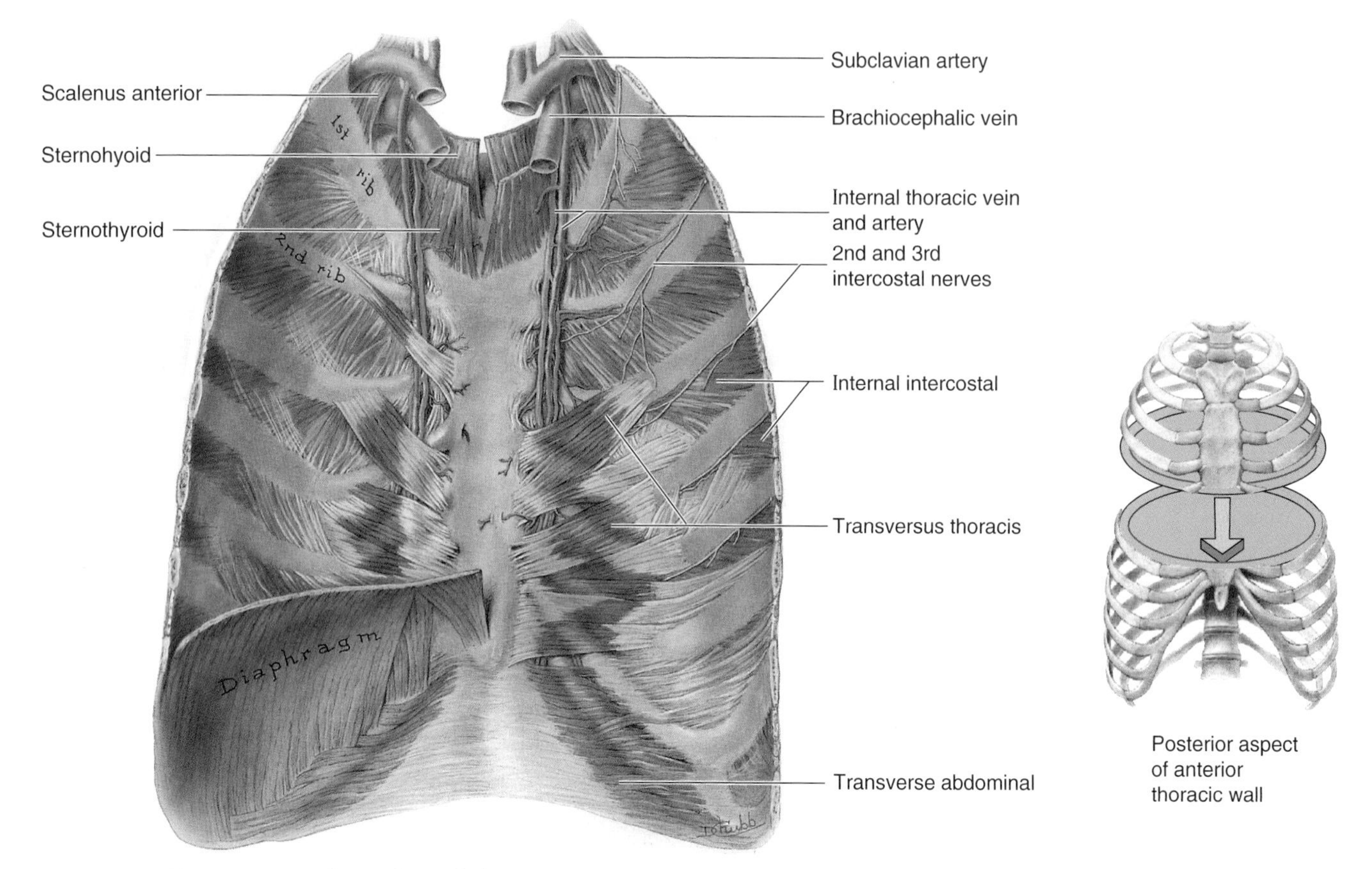

Figure 1.17. Dissection of the posterior aspect of the anterior thoracic wall. Observe the internal thoracic arteries arising from the subclavian arteries and their accompanying veins (L. venae comitantes). Observe that superior to the 2nd costal cartilage, there is only one internal thoracic vein on each side, which drains into the brachiocephalic vein. Note the continuity of the transverse thoracic (L. transversus thoracis) with the transverse abdominal (L. transversus abdominis) inferior to the level of the diaphragm.

2nd through 4th (or 5th) ribs. The serratus posterior superior elevates the superior four ribs, increasing the AP diameter of the thorax and raising the sternum. The *serratus posterior inferior* lies at the junction of the thoracic and lumbar regions (see Chapter 4). It arises from the spinous processes of the last two thoracic spinous processes and the first two lumbar spinous processes. It runs superolaterally and attaches to the inferior borders of the inferior three or four ribs near their angles. The serratus posterior inferior depresses the inferior ribs, preventing them from being pulled superiorly by the diaphragm.

The **levatores costarum muscles** (elevator muscles of the ribs) are attached to the transverse processes of C7 and T1 through T11 vertebrae (Fig. 1.19) and pass inferolaterally to attach to the ribs, close to their tubercles. As their name indicates, these 12 fan-shaped muscles elevate the ribs but have a relatively unimportant inspiratory function.

The **intercostal muscles** occupy the intercostal spaces (see Figs. 1.15–1.17; Table 1.2). The superficial layer is formed by the external intercostals, the middle layer by the internal intercostals, and the deepest layer by the innermost intercostals.

- The **external intercostal muscles** (11 pairs) occupy the intercostal spaces from the tubercles of the ribs posteriorly to the costochondral junctions anteriorly (Figs. 1.15, 1.16, and 1.18). Anteriorly, the muscle fibers are replaced by the **external intercostal membranes** (Fig. 1.18*A*). These muscles run inferoanteriorly from the rib above to the rib below. Each muscle attaches superiorly to the inferior border of the rib above and inferiorly to the superior border of the rib below. These muscles are continuous inferiorly with the *external oblique muscles* in the anterolateral abdominal wall. The external intercostals—*muscles of inspiration*—elevate the ribs.
- The **internal intercostal muscles** (11 pairs) run deep to and at right angles to the external intercostals (Figs. 1.17 and 1.18). Their fibers run inferoposteriorly from the floors of the costal grooves to the superior borders of the ribs inferior to them. The internal intercostals attach to the shafts of the ribs and their costal cartilages as far anteriorly as the sternum and as far posteriorly as the angles of the ribs. Between the ribs posteriorly, medial to the angles, the internal intercostals are replaced by the **internal intercostal membranes** (Fig. 1.18*A*). The inferior internal intercostals are continuous with the *internal oblique muscles* in the anterolateral abdominal wall. The internal intercostals are muscles of expiration.

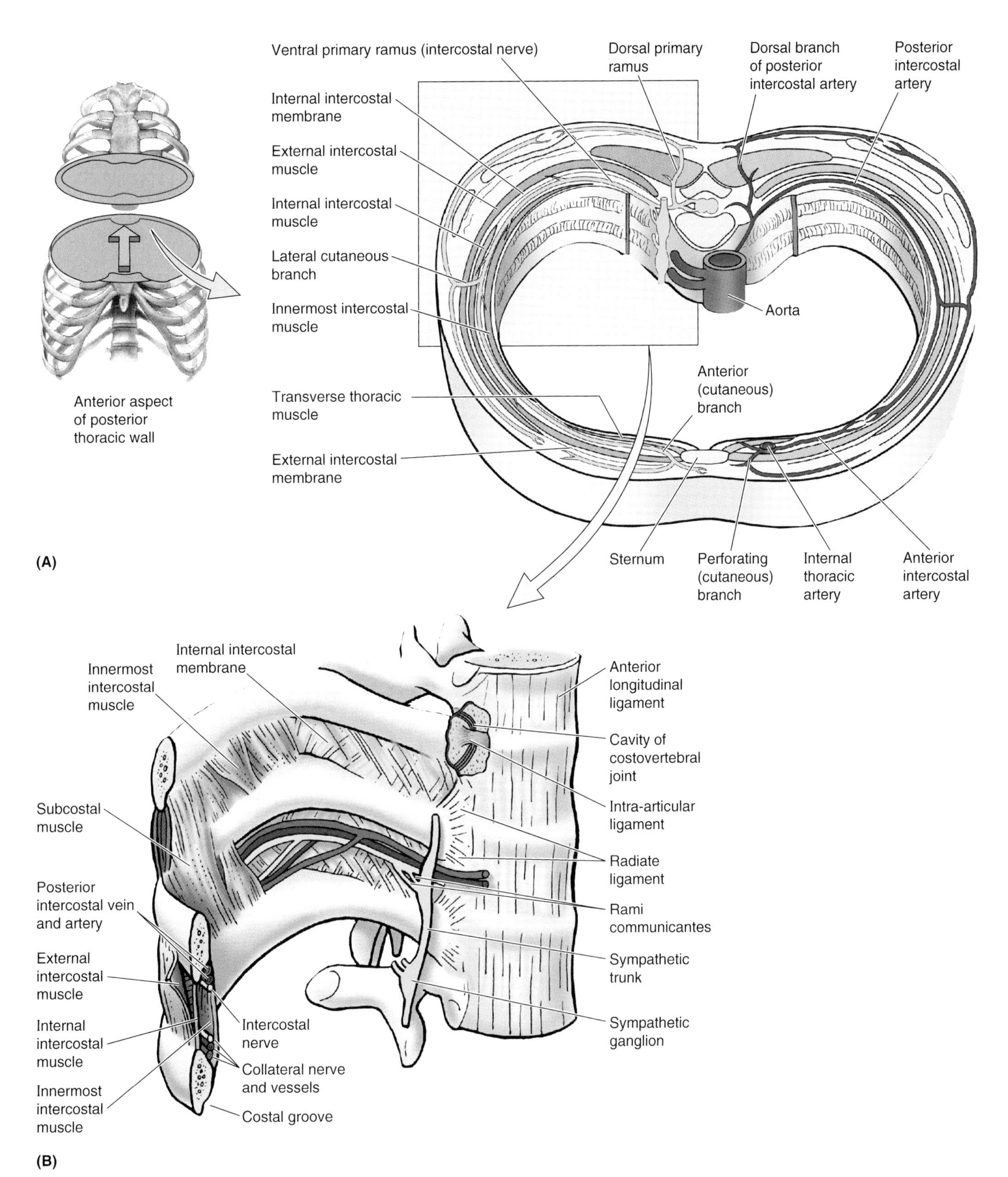

Figure 1.18. Contents of an intercostal space. A. This diagrammatic transverse section is simplified by showing nerves on the right and arteries on the left. Observe the intercostal muscular layers and the intercostal nerves and vessels in relation to them. **B.** This drawing shows the contents of the posterior part of an intercostal space and the cavities of a costovertebral joint. In the most superior intercostal space, observe the innermost intercostal muscle bridging an intercostal space. Also observe the subcostal muscle bridging two intercostal spaces. Note the order of the structures in the dissected intercostal space: **VAN** (**V**ein, **A**rtery, and **N**erve). Also note the collateral nerve and vessels and the attachment of the intercostal nerves to the sympathetic trunk by communicating branches (rami communicantes).

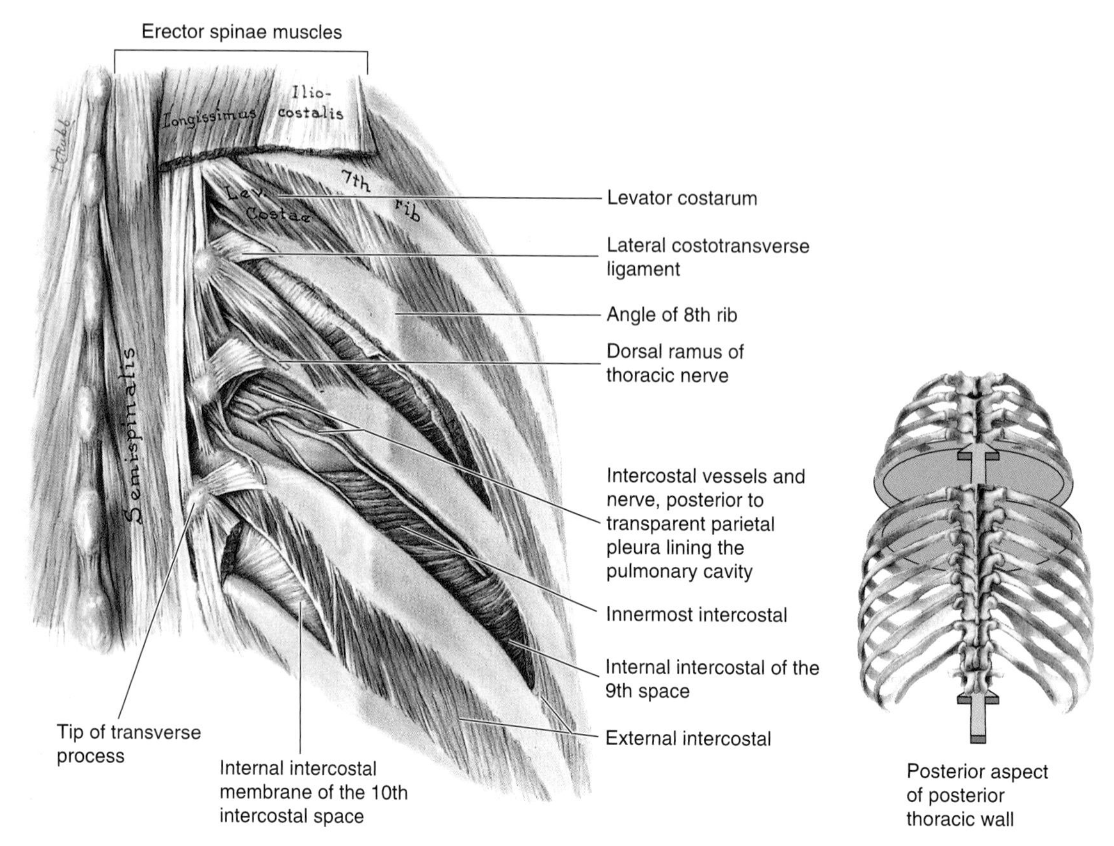

Figure 1.19. Dissection of the posterior part of the thoracic wall. The iliocostalis and longissimus muscles (two of the three components of the erector spinae muscles of the back) have been removed to expose the levatores costarum muscles. In the 8th and 10th intercostal spaces, varying parts of the external intercostal muscle have been removed to expose the underlying internal intercostal membrane, which is continuous with the internal intercostal muscle. In the 9th intercostal space, the levator costarum (elevator of the rib) has been removed to expose the intercostal vessels and nerve.

Table 1.2. Muscles of the Thoracic Wall

Muscle	Superior Attachment	Inferior Attachment	Innervation	Main Action[a]
Serratus posterior superior	Ligamentum nuchae, spinous processes of C7 to T3 vertebrae	Superior borders of 2nd to 4th ribs	2nd to 5th intercostal nerves	Elevate ribs
Serratus posterior inferior	Spinous processes of T11 to L2 vertebrae	Inferior borders of 8th to 12th ribs near their angles	Ventral rami of 9th to 12th thoracic spinal nerves	Depress ribs
Levator costarum	Transverse processes of T7—11	Subjacent ribs between tubercle and angle	Dorsal primary rami of C8—T11 nerves	Elevate ribs
External intercostal				Elevate ribs
Internal intercostal	Inferior border of ribs	Superior border of ribs below		Depress ribs
Innermost intercostal			Intercostal nerve	Probably depress ribs
Subcostal	Internal surface of lower ribs near their angles	Superior borders of 2nd or 3rd ribs below		Elevate ribs
Transversus thoracis	Posterior surface of lower sternum	Internal surface of costal cartilages 2—6		Depress ribs

[a]All intercostal muscles keep intercostal spaces rigid, thereby preventing them from bulging out during expiration and from being drawn in during inspiration. The role of individual intercostal muscles and accessory muscles of respiration in moving the ribs is difficult to interpret despite many electromyographic studies

- The **innermost intercostal muscles** are similar to the internal intercostals and are really deep parts of them. The innermost intercostals are separated from the internal intercostals by the intercostal nerves and vessels (Fig. 1.18). These muscles pass between the internal surfaces of adjacent ribs and occupy the middle parts of the intercostal spaces. Whether their action differs from that of the internal intercostal muscles is unlikely but undetermined.

The **subcostal muscles** are variable in size and shape. These thin muscular slips extend from the internal surface of the angle of one rib to the internal surface of the rib inferior to it. Crossing one or two intercostal spaces, the subcostals run in the same direction as the internal intercostals and lie internal to them (Fig. 1.18*B*). The subcostals probably also elevate the ribs.

The **transversus thoracis, or transverse thoracic muscles,** consist of four or five slips that attach posteriorly to the xiphoid process, the inferior part of the body of the sternum, and the adjacent costal cartilages (Figs. 1.16 and 1.17). They pass superolaterally and attach to the 2nd through 6th costal cartilages. The transverse thoracic is continuous inferiorly with the transverse abdominal in the anterolateral body wall. These muscles appear to have an unimportant expiratory function.

Nerves of the Thoracic Wall

The thoracic wall has 12 pairs of thoracic spinal nerves. As soon as they leave the IV foramina, they divide into ventral and dorsal primary rami (Fig. 1.18*A*). *The ventral rami of T1 through T11 nerves form the intercostal nerves* that run along the extent of the intercostal spaces. The ventral rami of T12 nerves, inferior to the 12th ribs, form the *subcostal nerves* (see Chapter 2). The dorsal rami of thoracic spinal nerves pass posteriorly, immediately lateral to the articular processes of the vertebrae (Fig. 1.19), to supply the bones, joints, muscles, and skin of the back in the thoracic region.

Typical intercostal nerves (3rd through 6th) run along the intercostal spaces posteriorly, between the parietal pleura (serous lining of the thoracic cavity) and the internal intercostal membrane (Figs. 1.18 and 1.19). At first they run across the internal surface of the internal intercostal membrane and muscle near the middle of the intercostal space. Near the angles of the ribs, the nerves pass between the internal intercostal and innermost intercostal muscles. Here the intercostal nerves enter and are sheltered by the **costal grooves** (Fig. 1.18*B*), where they lie just inferior to the intercostal arteries. Collateral branches of these nerves arise near the angles of the ribs and supply the intercostal muscles. The nerves continue anteriorly between the internal and innermost intercostal muscles, giving branches to these and other muscles and giving rise to lateral cutaneous branches in approximately the MAL. Anteriorly, the nerves appear on the internal surface of the internal intercostal muscle. Near the sternum the intercostal nerves turn anteriorly, passing between the costal cartilages as anterior cutaneous branches.

Through the dorsal ramus and the lateral and anterior cutaneous branches of the ventral ramus, each spinal nerve supplies a well-defined, striplike area of skin extending from the posterior median line to the anterior median line. These bandlike skin areas—**dermatomes** (Fig. 1.20)—are each supplied by sensory fibers of a single dorsal root through the dorsal and ventral rami of its spinal nerve. The dermatomes are arranged in a segmental fashion because the thoracoabdominal nerves arise from segments of the spinal cord (Fig. 1.20, *B* and *C*). Closely related dermatomes such as T4, T5, and T6 overlap considerably. In fact, a lesion of a single spinal nerve may not produce a noticeable sensory deficit because of the overlap in the distribution of adjacent nerves. Physicians require an understanding of the segmental, or dermatomal, innervation of the skin so they can determine with a pin whether a particular segment of the spinal cord is functioning normally. The group of muscles supplied by a pair of intercostal nerves is a **myotome** (Fig. 1.20*A*). *Muscular branches of typical intercostal nerves* also supply the subcostal, transversus thoracis, levator costarum, and serratus posterior muscles. *Branches of a typical intercostal nerve include*:

- **Rami communicantes** (communicating branches) that connect each intercostal nerve to the ipsilateral sympathetic trunk (Fig. 1.21). Presynaptic fibers leave each nerve by means of a white ramus and pass to a ganglion of the **sympathetic trunk.** Postsynaptic fibers distributed to the body wall and limbs leave all the ganglia of the sympathetic trunk via gray rami to join the ventral ramus of the nearest spinal nerve, including all intercostal nerves. Sympathetic nerve fibers are distributed through all branches of the intercostal nerves to blood vessels, sweat glands, and smooth muscle.

- **Collateral branches** that arise near the angles of the ribs and help to supply intercostal muscles.

- **Lateral cutaneous branches** that arise beyond the angles of the ribs and pierce the internal and external intercostal muscles approximately halfway around the thorax. The lateral cutaneous branches divide in turn into anterior and posterior branches that supply the skin of the thoracic and abdominal walls.

- **Anterior cutaneous branches** that supply the skin on the anterior aspect of the thorax and abdomen. After penetrating the muscles and membranes of the intercostal space in the parasternal line, the anterior cutaneous branches divide into medial and lateral branches.

- **Muscular branches** that supply the intercostal, subcostal, transversus thoracis, levatores costarum, and serratus posterior muscles.

The 1st and 2nd intercostal nerves are atypical (Fig. 1.16). In the first part of their course, they pass on the internal surfaces of the 1st and 2nd ribs.

- The **1st intercostal nerve** has no anterior cutaneous branch and usually no lateral cutaneous branch. The ventral ramus of the 1st thoracic (T1) spinal nerve divides

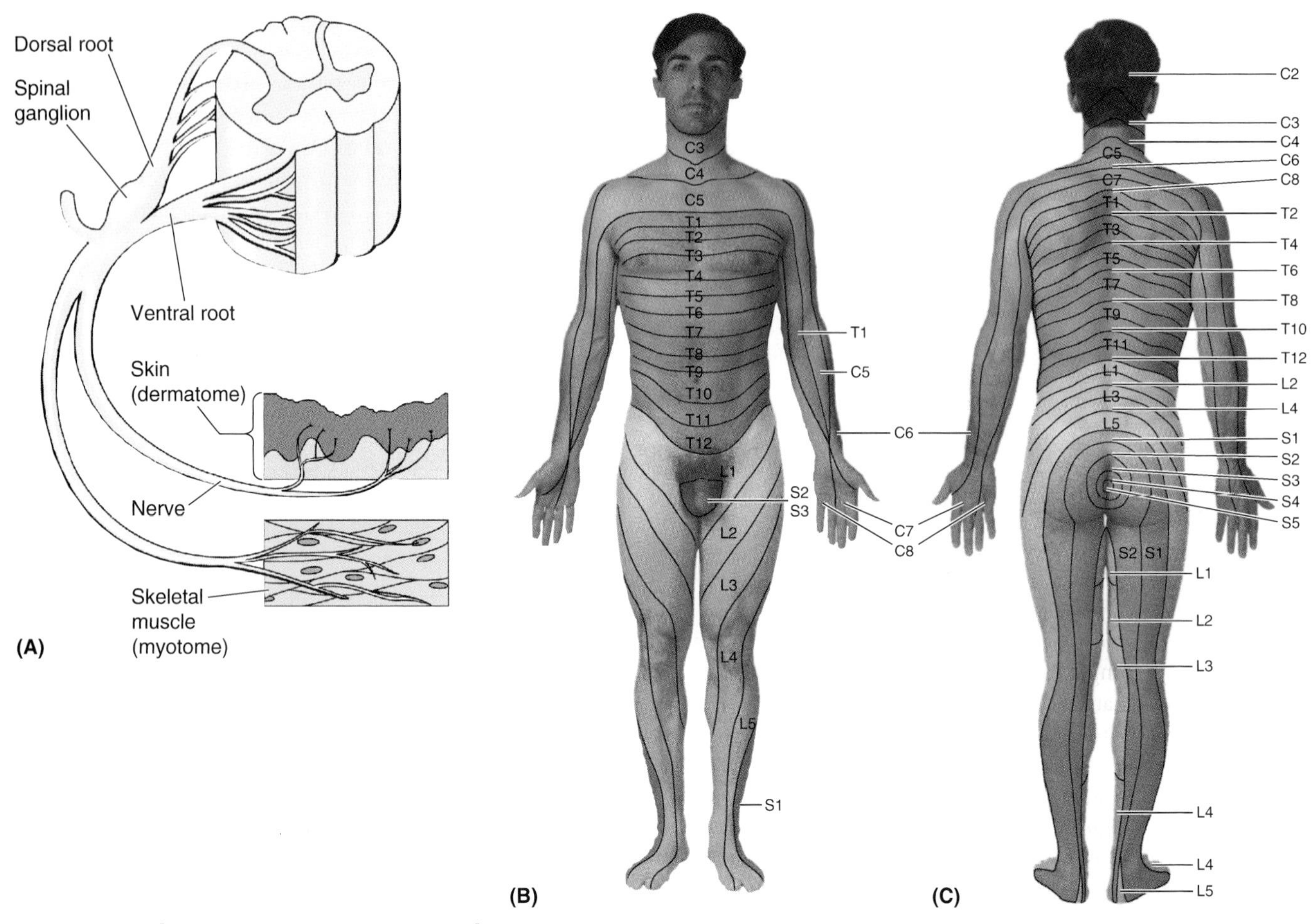

Figure 1.20. Dermatomes and myotomes. Dermatomes are areas of skin innervated by a single (pair of) spinal nerve(s), or one segment of the spinal cord. Except for the limbs, dermatomes are arranged fairly segmentally. C1 nerve lacks a significant afferent component and does not supply the skin. Myotomes are groups of muscles innervated by a single (pair of) spinal nerve(s), or one segment of the spinal cord. Muscles are supplied by one or more segments of the spinal cord.

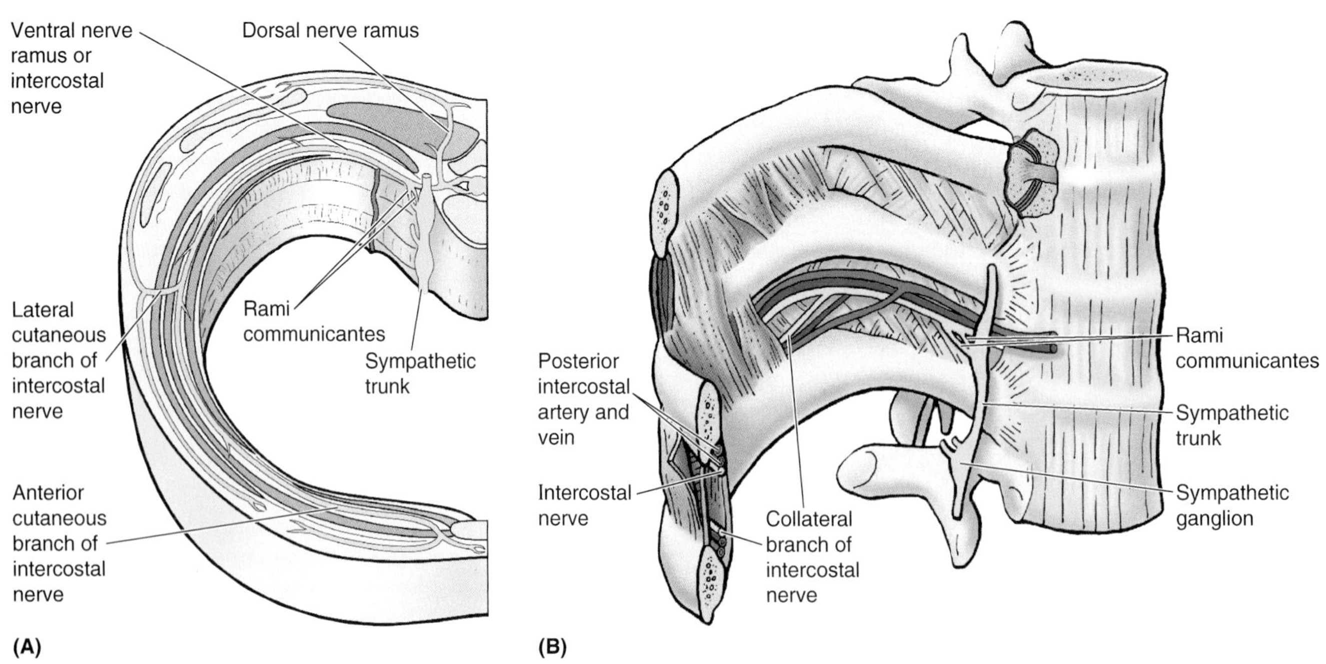

Figure 1.21. Intercostal nerves. A. The dorsal nerve ramus or dorsal primary ramus innervates the deep back muscles and the skin adjacent to the vertebral column. The intercostal nerves are the ventral nerve rami or ventral primary rami of spinal nerves T1 through T11. The intercostal nerves attach to the sympathetic trunk by rami communicantes. **B.** Observe the relationship of the intercostal nerve to the intercostal vessels.

into a large superior and a small inferior part. The superior part joins the *brachial plexus,* the nerve plexus supplying the upper limb, and the inferior part becomes the 1st intercostal nerve.

- The **2nd intercostal nerve** is usually the larger of two branches of the ventral ramus of the 2nd thoracic (T2) spinal nerve; the smaller branch also joins the brachial plexus. The lateral cutaneous branch of the 2nd intercostal nerve—the *intercostobrachial nerve*—emerges from the 2nd intercostal space at the MAL, penetrates the serratus anterior, and enters the axilla and arm (L. brachium). The intercostobrachial nerve usually supplies the floor—skin and subcutaneous tissue—of the axilla and then communicates with the *medial brachial cutaneous nerve* to supply the medial and posterior surfaces of the arm.

Herpes Zoster Infection

A *herpes zoster infection* causes a classic, dermatomally distributed skin lesion—*shingles*—an agonizingly painful condition. *Herpes zoster is a viral disease of the spinal ganglia.* After invading a ganglion, the virus produces a sharp burning pain in the dermatome supplied by the involved nerve (Fig. 1.20). The affected skin area becomes red and vesicular eruptions appear. The pain may precede or follow the skin eruptions. Although primarily a *sensory neuropathy*—pathological change in the nerve—weakness from motor involvement occurs in 0.5 to 5.0% of people, usually in elderly cancer patients (Rowland, 1995). Muscular weakness usually occurs in the same myotomal distribution, as do the dermatomal pain and vesicular eruptions.

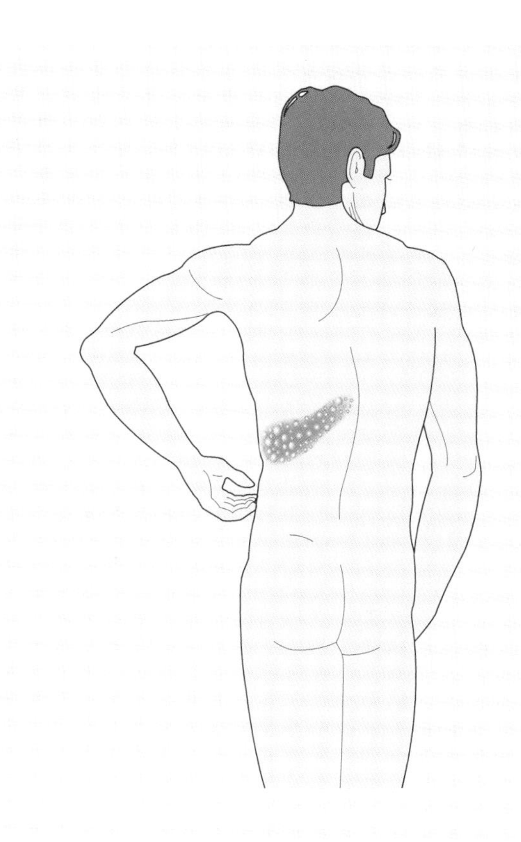

Thoracocentesis

Sometimes it is necessary to insert a hypodermic needle through an intercostal space into the pleural cavity—*thoracocentesis*—to obtain a sample of fluid or to remove blood or pus. To avoid damage to the intercostal nerve and vessels, the needle is inserted superior to the rib, high ▶

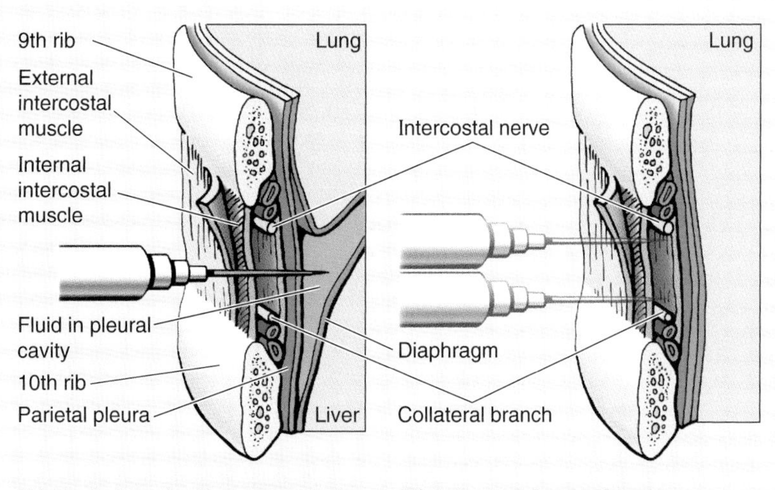

(A) Technique for thoracocentesis (in midaxillary line) **(B) Intercostal nerve block**

▶ enough to avoid the collateral branches. The needle passes through the intercostal muscles and parietal pleura (serous membrane lining the wall of the pleural cavity) into the pleural cavity.

Intercostal Nerve Block

Local anesthesia of an intercostal space is produced by injecting a local anesthetic agent around the intercostal nerves between the paravertebral line—a vertical line corresponding to the tips of the transverse processes of the vertebrae—and the area of required anesthesia. This procedure, *an intercostal nerve block*, involves infiltration of the anesthetic around the intercostal nerve trunk and its collateral branches (Fig. 1.21*B* and p. 87). The term "block" indicates that the nerve endings in the skin and the transmission of impulses through the sensory nerves carrying information about pain are interrupted ("blocked") before the impulses reach the spinal cord and brain. Because any particular area of skin usually receives innervation from two adjacent nerves, considerable overlapping of contiguous dermatomes occurs. Therefore, complete loss of sensation usually does not occur unless two or more intercostal nerves are anesthetized. ○

Vasculature of the Thoracic Wall

Arteries of the Thoracic Wall

The arterial supply to the thoracic wall (Fig. 1.22, Table 1.3) derives from the:

- **Thoracic aorta** through posterior intercostal and subcostal arteries
- **Subclavian artery** through internal thoracic and supreme (superior) intercostal arteries
- **Axillary artery** through superior and lateral thoracic arteries.

The intercostal arteries course through the thoracic wall between the ribs. *Each intercostal space is supplied by three arteries*, a large posterior intercostal artery (and its collateral branch) and a small pair of anterior intercostal arteries.

- The first two posterior intercostal arteries arise from the **supreme (superior) intercostal artery**, a branch of the costocervical trunk of the subclavian artery. Usually nine pairs of posterior intercostal arteries and one pair of subcostal arteries arise posteriorly from the thoracic aorta (Fig. 1.22). Because the aorta is slightly to the left of the vertebral column, the right intercostal arteries have a longer course than those on the left side. The right arteries cross the vertebrae and pass posterior to the esophagus, thoracic duct, azygos vein (see Fig. 1.27), and the right lung and pleura. Each posterior intercostal artery gives off a dorsal branch that accompanies the dorsal ramus of the spinal nerve to supply the spinal cord, vertebral column, back muscles, and skin. Each artery also gives off a small collateral branch that crosses the intercostal space and runs along the superior border of the rib. The terminal and collateral branches of each posterior intercostal artery anasto-

Table 1.3. Arterial Supply to the Thoracic Wall

Artery	Origin	Course	Distribution
Posterior intercostals	Superior intercostal artery (intercostal spaces 1 and 2) and thoracic aorta (remaining intercostal spaces)	Pass between internal and innermost intercostal muscles	Intercostal muscles and overlying skin, parietal pleura
Anterior intercostals	Internal thoracic (intercostal spaces 1–6) and musculophrenic arteries (intercostal spaces 7–9)		
Internal thoracic	Subclavian artery	Passes inferiorly and lateral to sternum between costal cartilages and internal intercostal muscles to divide into superior epigastric and musculophrenic arteries	By way of anterior intercostal arteries to intercostal spaces 1–6
Subcostal	Thoracic aorta	Courses along inferior border of 12th rib	Muscles of anterolateral abdominal wall

Figure 1.22. Arteries of the thoracic wall. The arterial supply to the wall derives from: (*a*) the *thoracic aorta* through the posterior intercostal and subcostal arteries; (*b*) the *axillary artery*, and (*c*) the *subclavian artery* through the internal thoracic and supreme (superior) intercostal arteries.

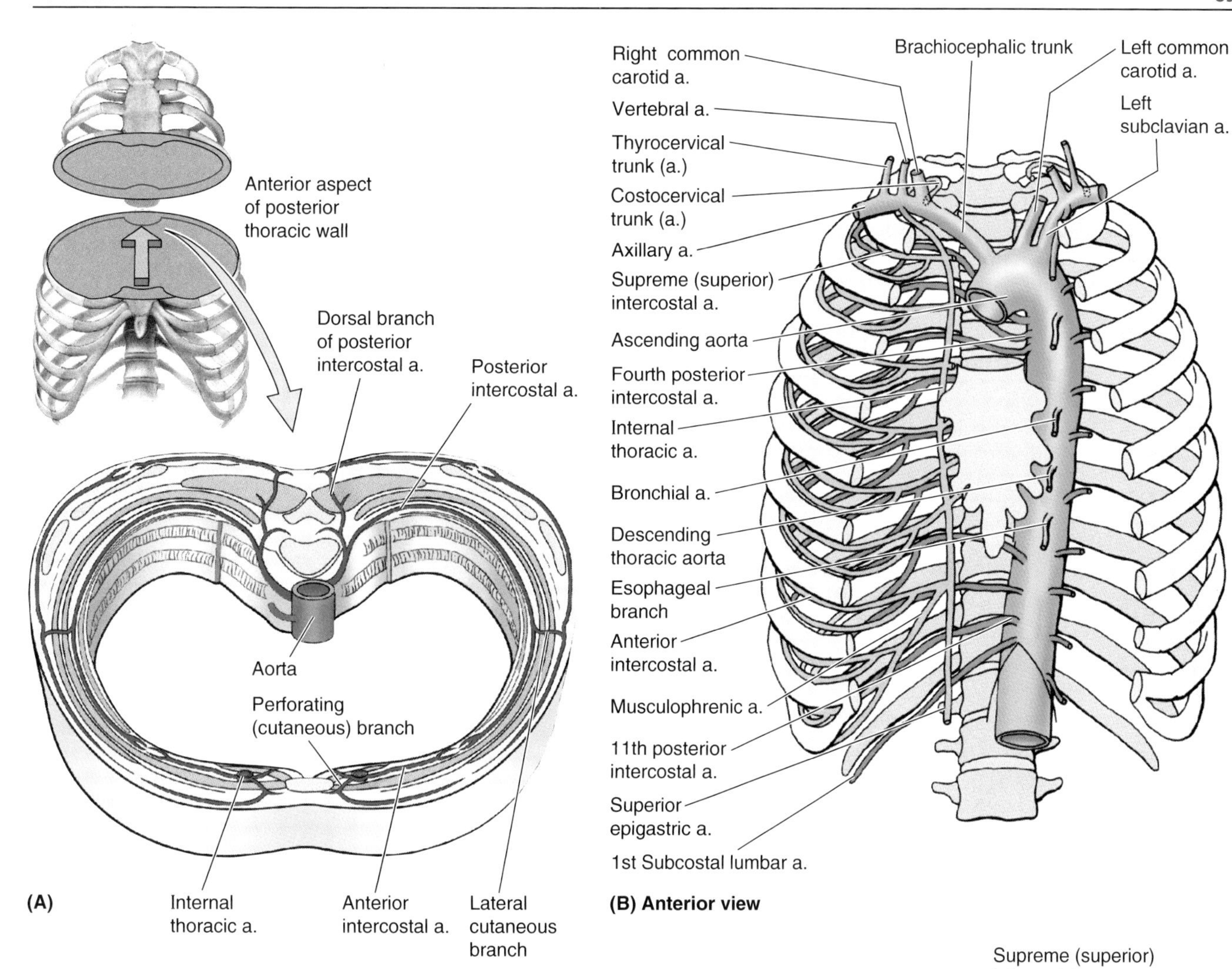

(A)

(B) Anterior view

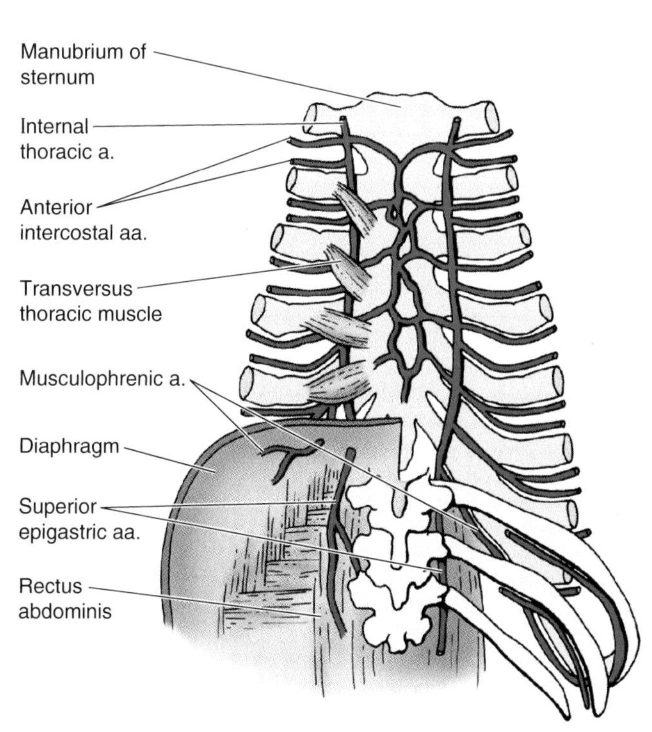

(C) Posterior view

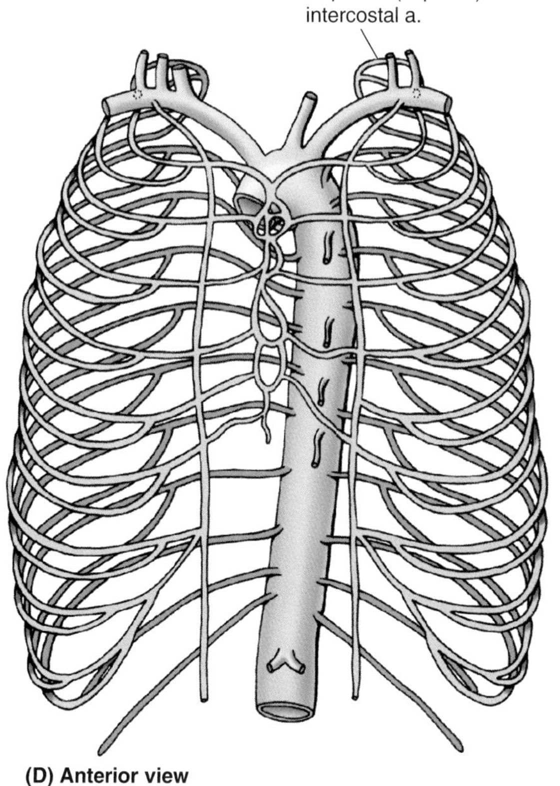

(D) Anterior view

mose anteriorly with anterior intercostal arteries (Fig. 1.22*A*). The posterior intercostal artery accompanies the intercostal nerve through the intercostal space. Close to the angle of the rib, the artery enters the costal groove, where it lies between the intercostal vein and nerve. At first the artery runs between the parietal pleura and the internal intercostal membrane (Fig. 1.19); it then runs between the innermost intercostal and internal intercostal muscles.

- The **internal thoracic arteries** (historically, internal mammary arteries) arise in the root of the neck from the inferior surfaces of the first parts of the **subclavian arteries.** Each internal thoracic artery descends into the thorax posterior to the clavicle and 1st costal cartilage (Figs. 1.16, 1.17, and 1.22). It runs on the internal surface of the thorax slightly lateral to the sternum and lies on the pleura posteriorly. Near its origin, the internal thoracic artery is

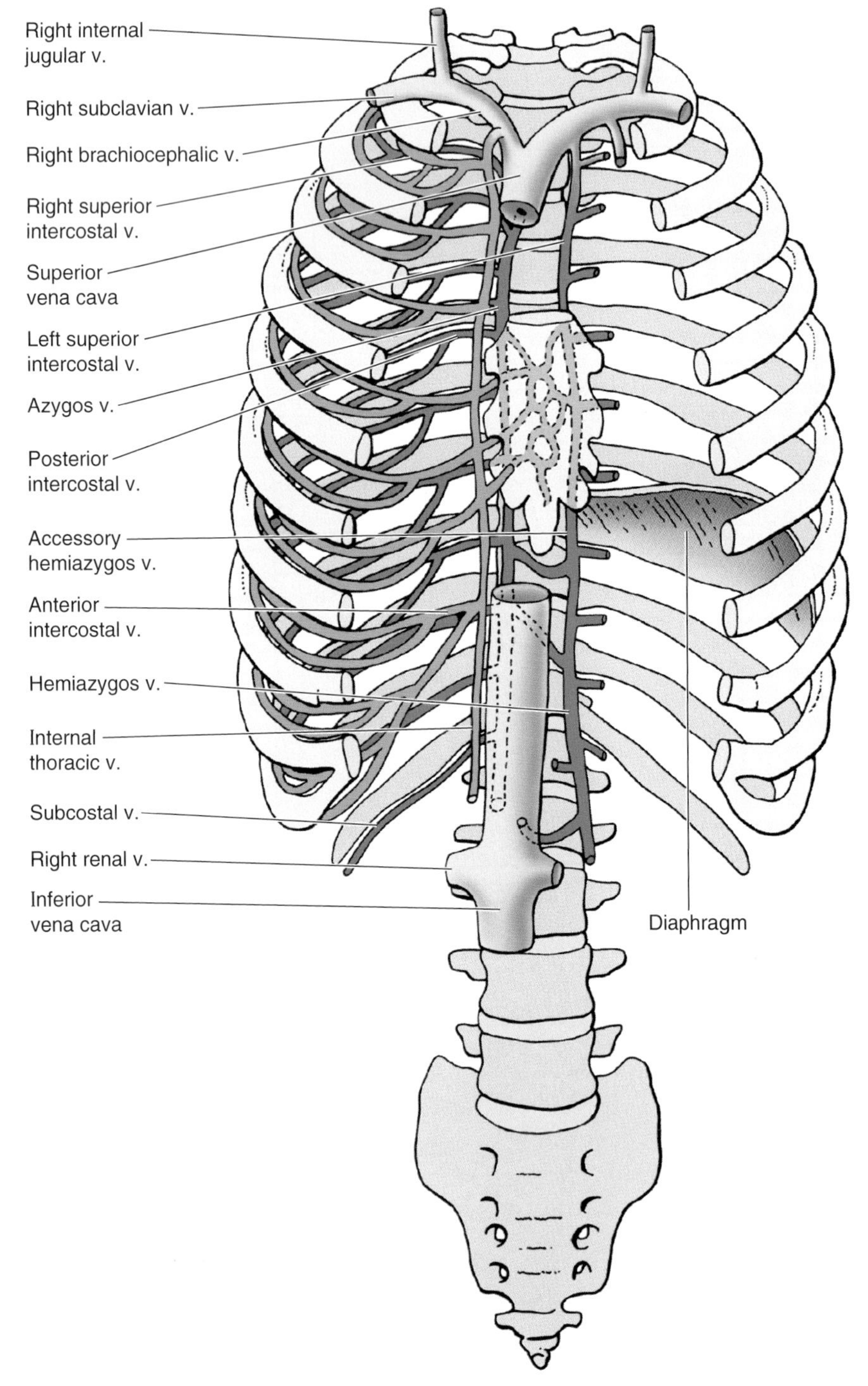

Figure 1.23. Veins of the thoracic wall. Intercostal veins accompany the intercostal arteries and lie deep in the costal grooves. Eleven posterior intercostal veins and one subcostal vein occur on each side. The posterior intercostal veins anastomose with the anterior intercostal veins—tributaries of the internal thoracic veins. Most posterior intercostal veins drain into the azygos system of veins that conveys venous blood to the superior vena cava.

crossed by the phrenic nerve, the nerve to the diaphragm. It then runs inferiorly in the thorax posterior to the superior six costal cartilages and intervening internal intercostal muscles. After descending past the 2nd costal cartilage, the internal thoracic artery runs anterior to the transversus thoracis muscle (Figs. 1.18*A* and 1.22*C*). It ends in the 6th intercostal space, where it divides into the *superior epigastric* and *musculophrenic arteries.*

- A pair of **anterior intercostal arteries** supplies the anterior parts of the upper nine intercostal spaces. These arteries pass laterally, one near the inferior margin of the superior rib and the other near the superior margin of the inferior rib. The anterior intercostal arteries supplying the superior six intercostal spaces derive from the **internal thoracic arteries.** At their origins, the arteries supplying the first two intercostal spaces lie between the parietal pleura and the internal intercostal muscles, whereas those supplying the next four intercostal spaces are separated from the pleura by slips of the transverse thoracic muscle. The arteries supplying the 7th through 9th intercostal spaces derive from the **musculophrenic arteries**, also branches of the internal thoracic arteries. The anterior intercostal arteries supply the intercostal muscles and send branches through them to supply the pectoral muscles, breasts, and skin. *There are no anterior intercostal arteries in the inferior two intercostal spaces*; these spaces are supplied by the posterior intercostal arteries and their collateral branches.

Veins of the Thoracic Wall

The **intercostal veins** accompany the intercostal arteries and nerves and lie deepest (most superior) in the costal grooves (Figs. 1.21 and 1.23). Eleven **posterior intercostal veins** and one **subcostal vein** are on each side. The posterior intercostal veins anastomose with the *anterior intercostal veins* (tributaries of the internal thoracic veins). Most posterior intercostal veins end in the **azygos venous system** of veins that conveys venous blood to the SVC (pp. 90 and 155). In the region of the 1st through 3rd intercostal spaces, these posterior intercostal veins unite to form a trunk (superior intercostal vein) that usually empties into the corresponding *brachiocephalic vein*; however, it may empty into the SVC. The **internal thoracic veins** are the accompanying veins (L. venae comitantes) of the internal thoracic arteries.

Surface Anatomy of the Thoracic Wall

Several imaginary lines facilitate anatomical descriptions, identification of thoracic areas, and the location of lesions such as a bullet wound:

- **Anterior median (midsternal) line** indicates the intersection of the median plane with the anterior chest wall (A).
- **Midclavicular lines** (MCLs) pass through the midpoints of the clavicles, parallel to the anterior median line.
- **Anterior axillary line** runs vertically along the anterior axillary fold (B) that is formed by the border of the pectoralis major as it spans from the thorax to the humerus (arm bone).
- **Midaxillary line** runs from the apex (deepest part) of the axilla, parallel to the anterior axillary line.
- **Posterior axillary line**, also parallel to the anterior axillary line, is drawn vertically along the posterior axillary fold formed by the latissimus dorsi and teres major muscles as they span from the back to the humerus.
- **Posterior median (midvertebral, midspinal) line** is a vertical line described by the tips of the spinous processes of the vertebrae.
- **Scapular lines** are parallel to the posterior median line and cross the inferior angles of the scapulae. ▶

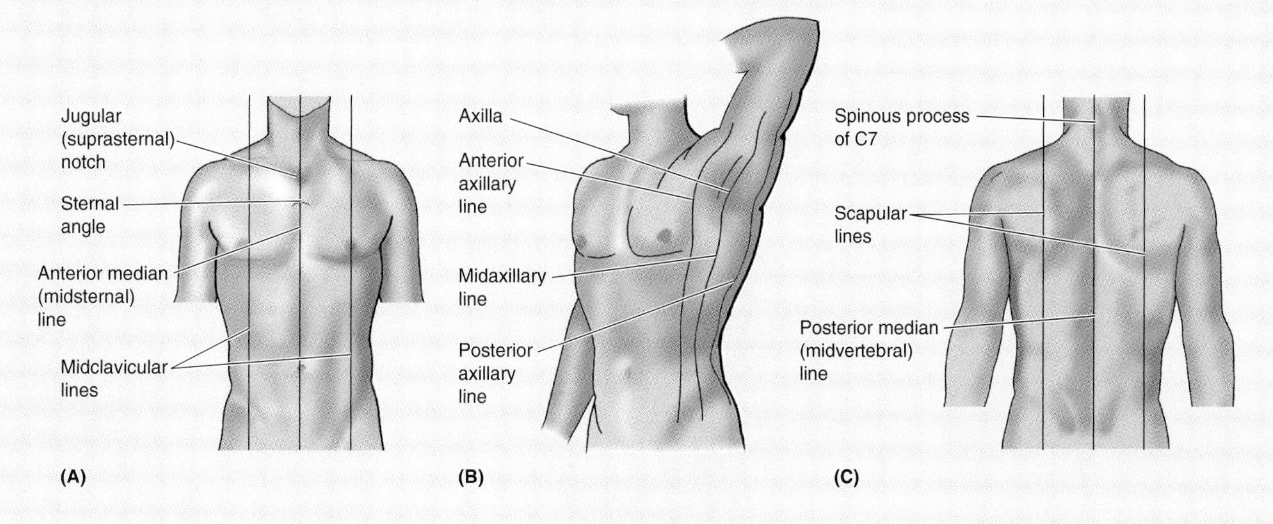

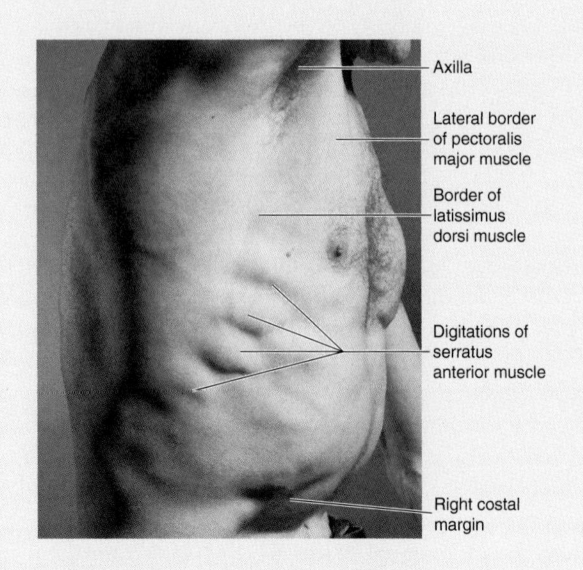

The **clavicles** lie subcutaneously, forming bony ridges at the junction of the thorax and neck. They can be palpated easily throughout their length, especially where their medial ends articulate with the manubrium of the sternum. ▶

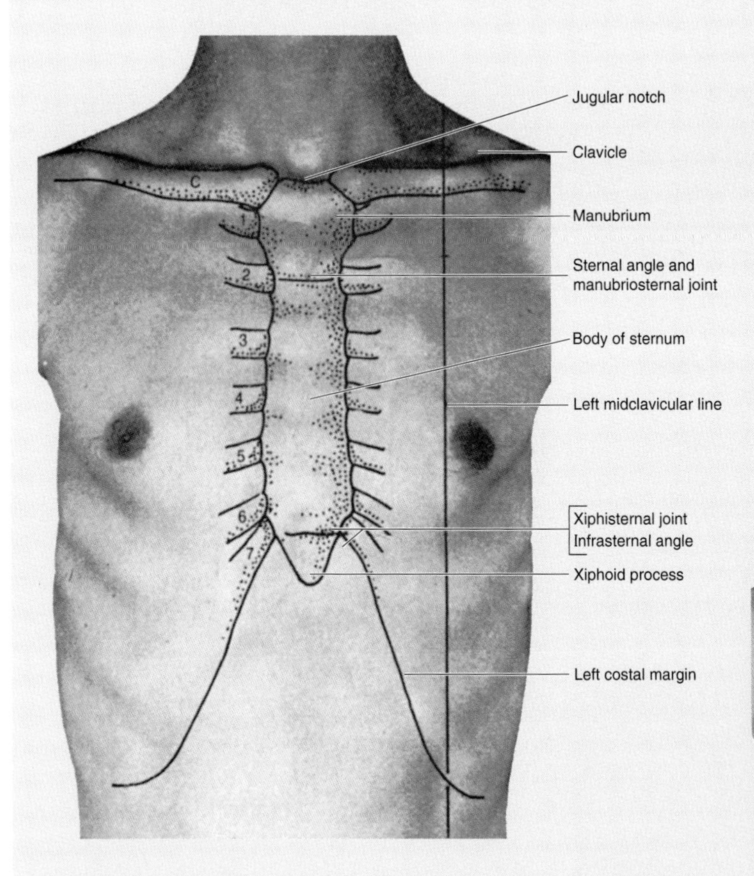

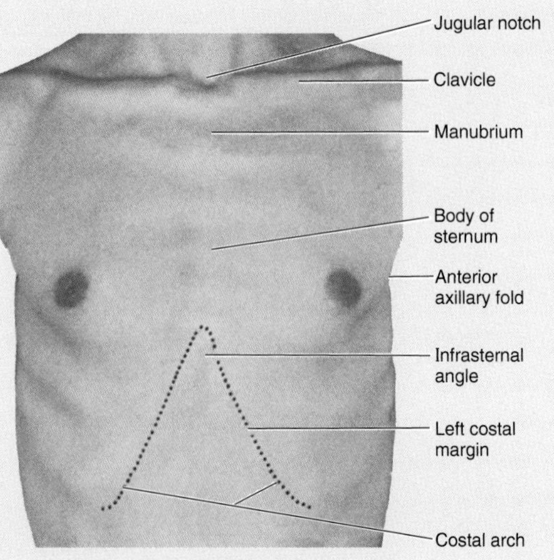

▶ The **sternum** lies subcutaneously in the anterior median line of the thorax and is palpable throughout its length. The **jugular notch** in the manubrium can be palpated between the prominent medial ends of the clavicles. The jugular notch lies at the level of the inferior border of the body of T2 vertebra and the space between the 1st and 2nd thoracic spinous processes.

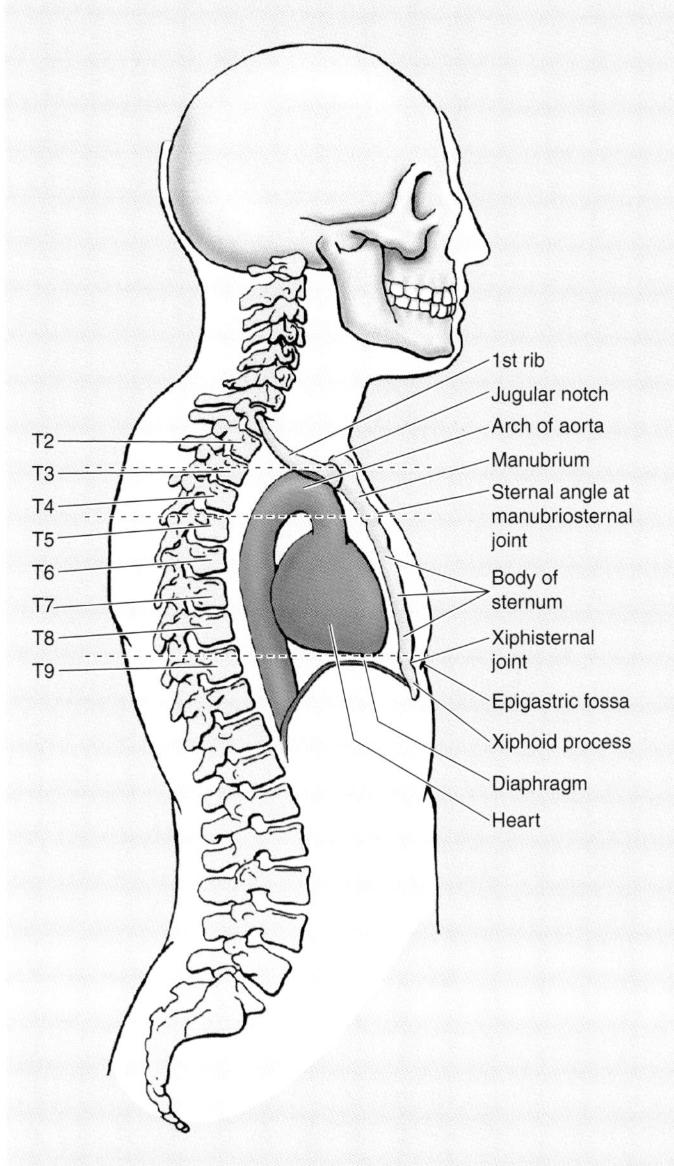

The **manubrium**, approximately 4 cm long, lies at the level of the bodies of T3 and T4 vertebrae and is anterior to the arch of the aorta. The **sternal angle** is palpable and often visible because the *manubriosternal joint* between the manubrium and sternal body moves slightly during forced respiration. The sternal angle lies at the level of the T4/T5 IV disc and the space between the 3rd and 4th thoracic spinous processes. *The sternal angle is flanked by the 2nd pair of costal cartilages.*

The manubrium directly overlies the merging of the brachiocephalic veins to form the SVC. Because it is common practice clinically to insert catheters into the SVC for feeding extremely ill patients and other purposes (Ger et al., 1996), it is essential to know the surface anatomy of this large vein. The SVC passes inferiorly deep to the manubrium and manubriosternal junction, but projecting as much as a fingerbreadth to the right of the margin of the bony structures. The IVC enters the right atrium of the heart opposite the right 3rd costal cartilage.

To count the ribs and intercostal spaces anteriorly, slide the fingers laterally from the sternal angle onto the 2nd costal cartilage and count the ribs and spaces by moving the digits inferolaterally. The 1st intercostal space is inferior to the 1st rib; likewise, the other spaces are inferior to similarly numbered ribs. Posteriorly, the medial end of the spine of the scapula overlies the 4th rib. Palpate the scapular spine and run your fingers medially to palpate the 4th rib. Count the ribs inferiorly until you reach the 8th rib. Move your fingers laterally and palpate the inferior angle of the scapula. These surface guidelines enable you to count the ribs posteriorly.

The **body of the sternum**, approximately 10 cm long, lies anterior to the right border of the heart and vertebrae T5 through T9. The **intermammary cleft** (midline depression, or cleavage, between the mature female breasts) overlies the sternal body. The **xiphoid process** lies in a slight depression—the *epigastric fossa*—where the converging costal margins form the **infrasternal angle**. This angle is used in cardiopulmonary resuscitation (CPR) for locating the proper hand position on the inferior part of the sternal body. You can feel the *xiphisternal joint*, often seen as a ridge, at the level of the inferior border of T9 vertebra.

The **costal margins**, formed by the medial borders of the 7th through 10th costal cartilages, are easily palpable because they extend inferolaterally from the xiphisternal joint. The superior part of the costal margin is formed by the 7th costal cartilage, and its inferior part is formed by the 8th through 10th costal cartilages. The costal margins form the sides of the infrasternal angle. ▶

▶ **Breasts** are the most prominent surface features of the anterior thoracic wall, especially in women. Variation in the size, shape, and symmetry of female breasts—even between the two breasts of one person—is the rule and not the exception. Their flattened superior surfaces show no sharp demarcation from the anterior surface of the wall; however, laterally and inferiorly their borders are well defined. The median area between the breasts is the intermammary cleft. A venous pattern over the breasts is often visible, especially during pregnancy. The **nipple** is surrounded by the slightly raised and circular pigmented **areola**, the color of which varies with a woman's complexion. The areola darkens during pregnancy and retains darkened pigmentation thereafter. The areola is normally dotted with the papular openings of the *areolar glands* (sebaceous glands in the skin forming the areola). On occasion, one or both nipples are inverted; this minor congenital anomaly may make breast feeding difficult. The nipple lies anterior to the 4th intercostal space in men, approximately 10 cm from the anterior median line (AML). The position of the nipple in women is so inconsistent that it is not reliable as a surface landmark for the 4th intercostal space. ⊙

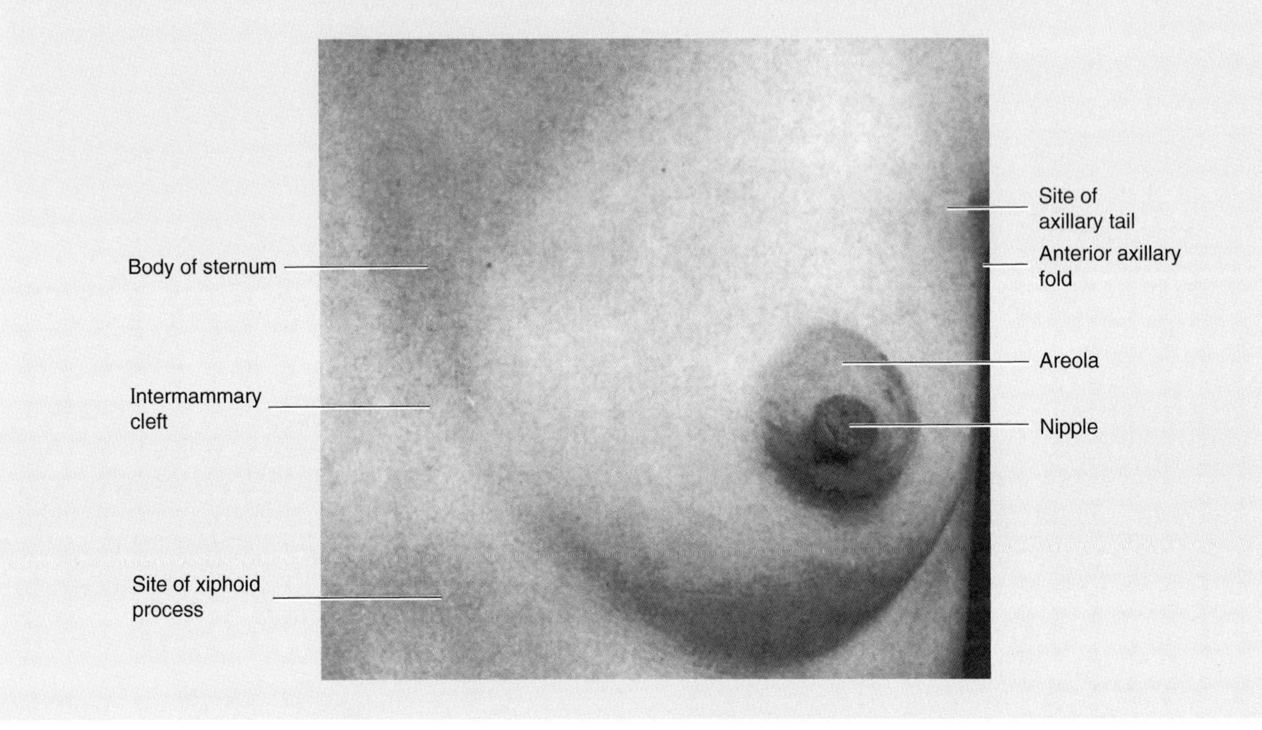

Thoracic Cavity and Viscera

The thoracic cavity has three divisions or compartments:

- Two lateral compartments—the **pulmonary cavities**—that contain the lungs and pleurae (lining membranes)
- A central compartment—the **mediastinum**—that contains all other thoracic structures: the heart, thoracic parts of the great vessels, thoracic part of the trachea, esophagus, thymus, and other structures (e.g., lymph nodes).

The pulmonary cavities are completely separate from each other and, with the lungs and pleurae, occupy the majority of the thoracic cavity. The mediastinum extends from the superior thoracic aperture to the diaphragm (Fig. 1.24).

Pleurae and Lungs

To visualize the relationship of the pleurae and lungs, push your fist into an underinflated balloon (Fig. 1.24). The part of the balloon wall adjacent to the skin of your fist (representing the lung) is comparable to the visceral pleura; the remainder of the balloon represents the parietal pleura. The cavity between the layers of the balloon is analogous to the pleural cavity. At your wrist (root of lung), the inner and outer walls of the balloon are continuous, as are the visceral and parietal layers of pleura, together forming a pleural sac. Note that the lung is outside of but surrounded by the pleural sac, just as your fist was surrounded by but outside of the balloon.

The illustration in the lower left of Figure 1.24 is also helpful in understanding the development of the lungs and pleurae. During the embryonic period, the developing lungs invaginate (grow into) the *pericardioperitoneal canals*, the primordia (beginnings) of the pleural cavities (Moore and Persaud, 1998). The invaginated coelomic epithelium covers the primordia of the lungs and becomes the *visceral pleura* in the same way that the balloon covers your fist. The epithelium lining the walls of the pericardioperitoneal canals forms the *parietal pleura*. During embryogenesis, the pleural cavities become separated from the pericardial and peritoneal cavities;

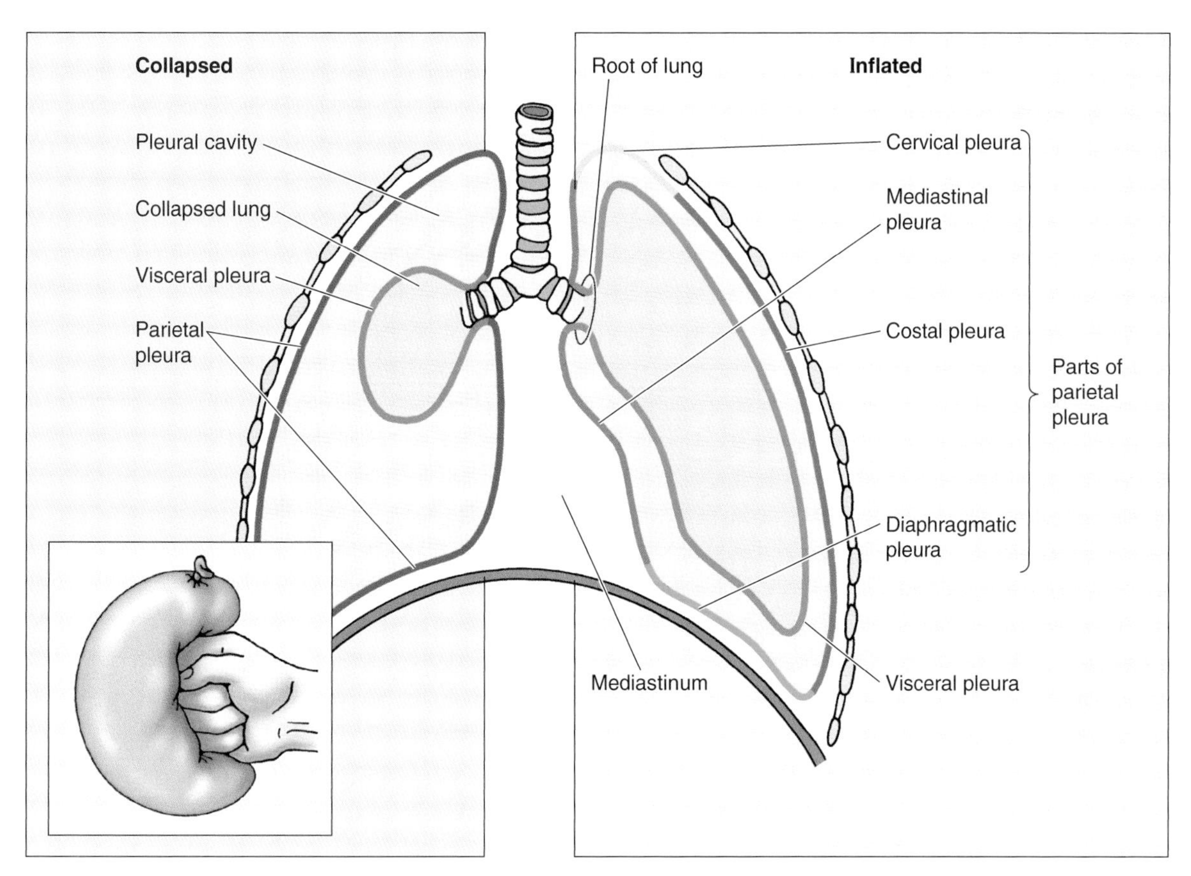

Figure 1.24. Pleurae and lungs. Each lung is invested by a serous membrane arranged like a closed sac—the pleura—which is invaginated by the lung. The *inset* drawing showing a fist invaginating an underinflated balloon demonstrates the relationship of the lung (represented by the fist) to walls of the pleural sac (parietal and visceral layers of pleura). The cavity of the pleural sac (pleural cavity) is comparable to the cavity of the balloon. When a lung is inflated, the parietal and visceral layers of pleura are separated by a lubricated potential serous space—the pleural cavity. If there is an injury to the thoracic wall that allows air to enter the pleural cavity, the elasticity of the lung causes it to collapse so that the entire pleural cavity may become filled with air (pneumothorax). The root of the lung is comprised of the pulmonary vessels and bronchi.

however, a congenital diaphragmatic defect results in a *diaphragmatic hernia* connecting the peritoneal cavity with one of the pleural cavities (usually the left one), and the herniation of abdominal viscera into the thorax (Moore et al., 1994).

Pleurae

Each lung is invested by and enclosed in a serous pleural sac that consists of two continuous membranes, the pleurae (Fig. 1.24):

- The **visceral pleura** (pulmonary pleura) invests the lungs, including the surfaces within the horizontal and oblique fissures; it cannot be dissected from the lungs.
- The **parietal pleura** lines the pulmonary cavities.
- The **pleural cavity**—the potential space between the layers of pleura—contains a capillary layer of *serous pleural fluid*, which lubricates the pleural surfaces and allows the layers of pleura to slide smoothly over each other during respiration. Its surface tension also provides the cohesion that keeps the lung surface in contact with the thoracic wall; consequently, the lung expands and fills with air when the chest expands while still allowing sliding to occur, much like a layer of water between two glass plates.

The **visceral pleura** closely covers the lung and is adherent to all its surfaces (Figs. 1.24 and 1.25). It provides the lung with a smooth, slippery surface, enabling it to move freely on the parietal pleura. The visceral pleura dips into the lung fissures so that the lobes of the lung are also covered with it. The visceral pleura is continuous with the parietal pleura at the **hilum of the lung**, where structures comprising the root of the lung (e.g., the bronchus and pulmonary vessels) enter and leave the lung.

The **parietal pleura** lines the pulmonary cavities and so adheres to the thoracic wall, the mediastinum, and the diaphragm. *The parietal pleura consists of four parts* (Figs. 1.24 and 1.25):

- **Costal pleura** covers the internal surfaces of the thoracic wall

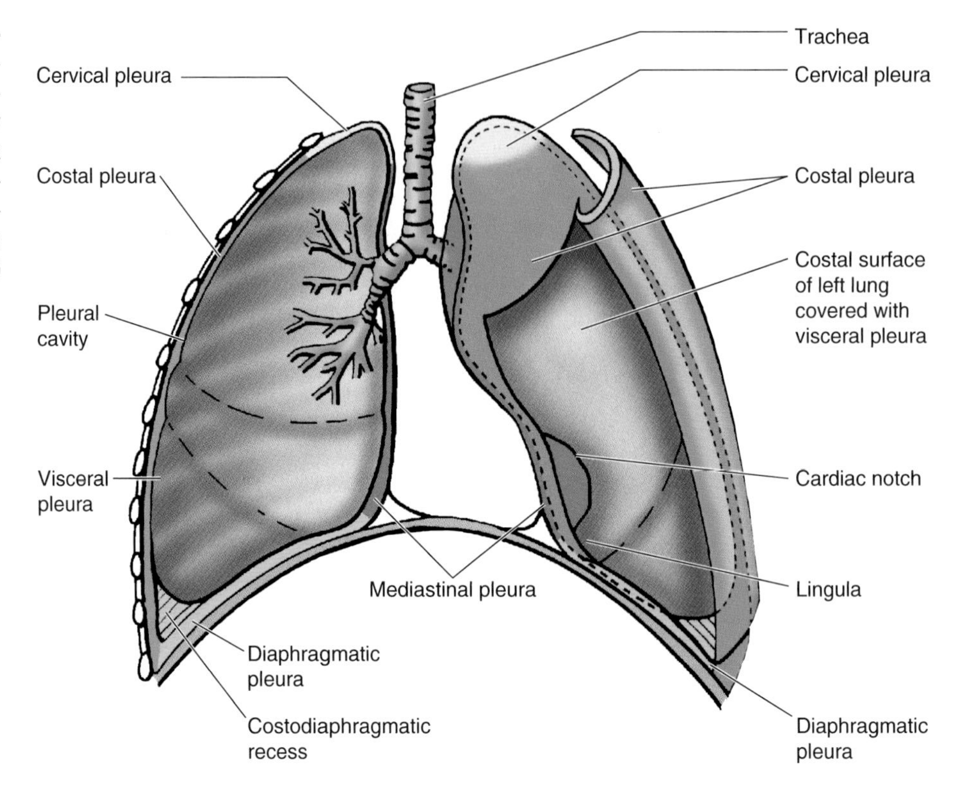

Figure 1.25. Parts of the parietal pleura. The parts of the parietal pleura are named after the portion of the wall of the pulmonary cavity that they cover: costal pleura lines the ribs (thoracic wall); mediastinal pleura, the mediastinum; diaphragmatic pleura, the diaphragm; and cervical pleura (pleural cupula) extends into the neck.

- **Mediastinal pleura** covers the lateral aspects of the mediastinum—the mass of tissues and organs separating the pulmonary cavities and their pleural sacs
- **Diaphragmatic pleura** covers the superior or thoracic surface of the diaphragm on each side of the mediastinum
- **Cervical pleura** (pleural cupula, dome of pleura) extends through the superior thoracic aperture into the root of the neck, forming a cup-shaped pleural dome over the apex of the lung (the part extending above the 1st rib).

The **costal pleura** is separated from the internal surface of the thoracic wall (sternum, ribs, and costal cartilages, intercostal muscles and membranes, and sides of thoracic vertebrae) by **endothoracic fascia**. This thin, extrapleural layer of loose connective tissue forms a natural cleavage plane for the surgical separation of the costal pleura from the thoracic wall, allowing the thoracic surgeon to move and place instruments inside the thoracic wall yet remain outside—thereby preventing potential infection from entering the pleural cavities or sacs. The endothoracic fascia also forms a thin layer of connective tissue between the diaphragm and the diaphragmatic pleura.

The **mediastinal pleura** covers the mediastinum, the space between the pulmonary cavities. It is continuous with the costal pleura anteriorly and posteriorly, the diaphragmatic pleura inferiorly, and the cervical pleura superiorly at the lines of pleural reflection (Fig. 1.25). Superior to the *root of the lung*, the mediastinal pleura is a continuous sheet between the sternum and vertebral column. At the **hilum of the lung**, the mediastinal pleura passes laterally, where it encloses the structures comprising the root (the bronchus and pulmonary vessels, for example) and becomes continuous with the visceral pleura. Inferior to the root of the lung, the mediastinal pleura passes laterally as a double layer from immediately anterior to the esophagus to the lung, where it is continuous with the visceral pleura as the *pulmonary ligament* (see Figs. 1.28 and 1.29). When the root of the lung is severed and the lung is removed, this double layer of pleura "hangs" from the root like the large sleeve of a Chinese robe hangs from the forearm.

The **diaphragmatic pleura** is the part of the parietal pleura that covers the superior surface of the diaphragm, except along its costal attachments and where it is covered with the pericardium, the fibroserous membrane surrounding the heart. A thin layer of endothoracic fascia, the *phrenicopleural fascia*, connects the diaphragmatic pleura with the muscular fibers of the diaphragm.

The **cervical pleura** is the dome-shaped cap of the pleural sac and is the continuation of the costal and mediastinal layers of pleura. The cervical pleura covers the apex of the lung that extends superiorly through the superior thoracic aperture into the root of the neck (Fig. 1.26*A*). The summit of the cervical pleura is 2 to 3 cm superior to the level of the medial third of the clavicle at the level of the neck of the 1st rib (Fig. 1.26*B*). The cervical pleura is strengthened by an extension of the endothoracic fascia—the *suprapleural membrane* (Sibson's fascia)—that attaches to the internal border of the 1st rib and the transverse process of C7 vertebra.

The relatively abrupt lines along which the parietal pleura changes direction from one wall of the pleural cavity to another are **the lines of pleural reflection** (Figs. 1.26 and 1.27).

- The *sternal line of pleural reflection* is sharp or abrupt and occurs where the costal pleura becomes continuous with the mediastinal pleura anteriorly.

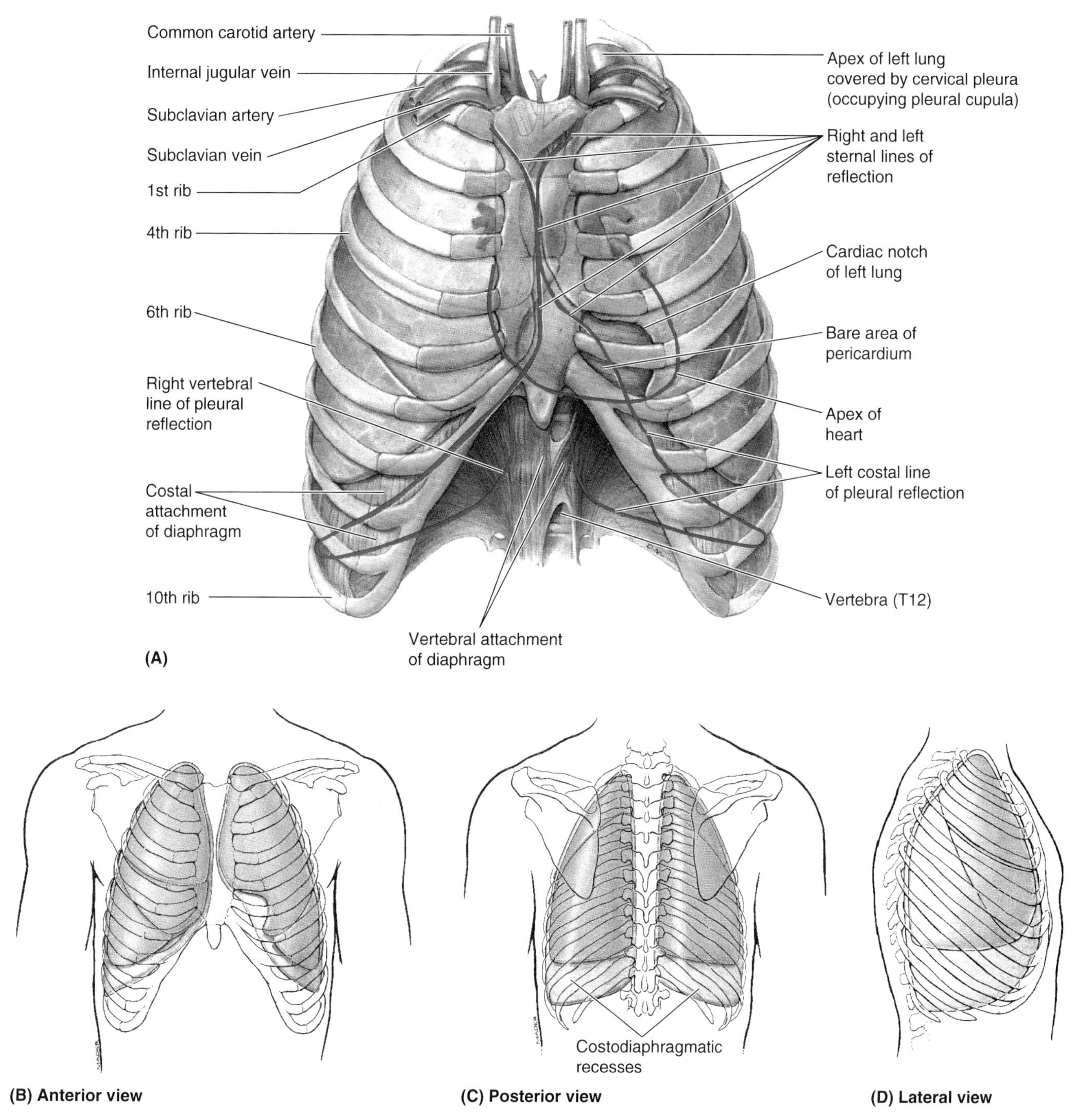

Figure 1.26. Contents of the thorax and outline of the pleura and lungs. A. Observe the apices of the lungs and cervical pleura extending into the neck. Note the cardiac notch in the left lung. Observe that the parietal pleura also deviates away from the median plane toward the left side in the region of the lung's cardiac notch, leaving a "bare area" where the pericardial sac is accessible for needle puncture without traversing the lung or pleural cavity. In **B–D,** observe the outline of the parietal pleura and lungs during quiet respiration and note the costodiaphragmatic recesses not occupied by lung. This is where pleural exudate accumulates when the body is erect. In **A** and **D,** follow the outlines of the oblique and horizontal fissures in the right lung, noting especially how the horizontal fissure parallels the 4th rib.

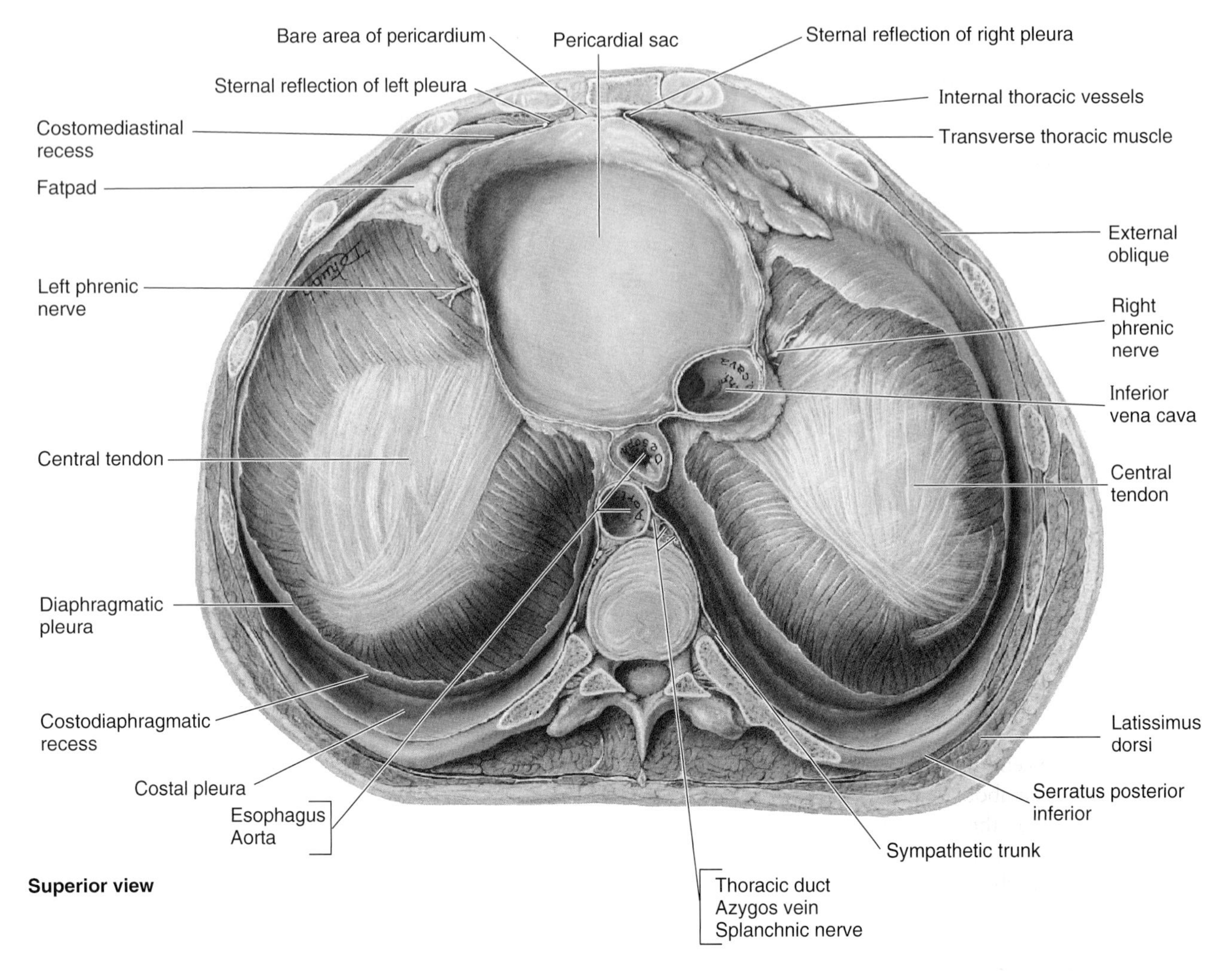

Figure 1.27. Diaphragm and pericardial sac. Most of the diaphragmatic pleura is removed. Observe the pericardial sac situated on the anterior, central part of the diaphragm, mostly on the left side. Also observe the costodiaphragmatic and costomediastinal recesses, and the "bare area" of the pericardium.

- The *costal line of pleural reflection* is also sharp and occurs where the costal pleura becomes continuous with diaphragmatic pleura inferiorly.
- The vertebral line of pleural reflection is a much rounder, gradual reflection where the costal pleura becomes continuous with the mediastinal pleura posteriorly.

The right and left sternal reflections of pleura are indicated by lines that pass inferomedially from the SC joints to the anterior median line at the level of the sternal angle. Here the pleural sacs come in contact and may slightly overlap each other.

The sternal line of pleural reflection on the right side passes inferiorly in the median plane to the posterior aspect of the xiphoid process, where it turns laterally (Fig. 1.24). The sternal line of reflection on the left side passes inferiorly in the median plane to the level of the 4th costal cartilage. Here it passes to the left margin of the sternum and then continues inferiorly to the 6th costal cartilage, creating a "notch" shallower than the cardiac notch of the lung, which allows a part of the pericardium (heart sac) to be in direct contact with the anterior thoracic wall. This is important for pericardiocentesis (p. 119).

The *costal line of pleural reflection* passes obliquely across the *8th rib* in the MCL, the *10th rib* in the MAL, and the *12th rib* at its neck or inferior to it. The *vertebral line of pleural reflection* parallels the vertebral column, running in the paravertebral plane from vertebral level T1 through T12.

The lungs do not occupy the pulmonary cavities completely during expiration; thus the peripheral diaphragmatic pleura is in contact with the lowermost parts of the costal pleura. The potential pleural spaces here are the **costodiaphragmatic recesses**—the pleura-lined "gutters"—that surround the upward convexity of the diaphragm inside the thoracic wall (Figs. 1.25 and 1.27). Similar but smaller pleural recesses are located posterior to the sternum where the costal pleura is in contact with the mediastinal pleura. The potential pleural spaces here are the **costomediastinal recesses**; the left recess is potentially larger (less occupied) because of the cardiac notch in the left lung. The inferior borders of the lungs move further into the pleural recesses during deep inspiration and retreat from them during expiration.

Injuries to the Cervical Pleura and Apex of Lung

Because of the inferior slope of the 1st pair of ribs and the superior thoracic aperture they form, the cervical pleura and apex of the lung project through the "hole" into the neck, posterior to the inferior attachments of the sternocleidomastoid muscles. Consequently, the lungs and pleural sacs may be injured in wounds to the neck. Air enters the pleural cavity producing *pneumothorax*—the presence of air (G. pneuma) in the pleural cavity. The cervical pleura reaches a relatively higher level in infants and young children because of the shortness of their necks. Consequently, the pleura is vulnerable to injury during the first few years after birth.

Injury to Other Parts of the Pleurae

The pleurae descend inferior to the costal margin in three regions, where an abdominal incision might inadvertently enter a pleural sac: the right part of the infrasternal angle, and right and left costovertebral angles. The small areas of pleura exposed in the costovertebral angles inferomedial to the 12th ribs are posterior to the superior poles of the kidneys. The pleura is in danger here from an incision in the posterior abdominal wall when surgical procedures expose a kidney, for example.

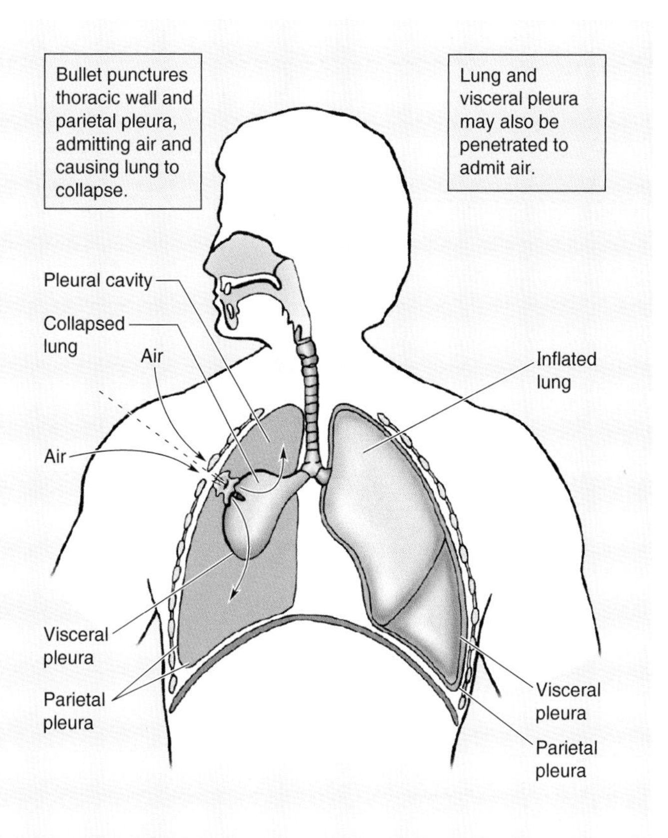

Pulmonary Collapse

If a sufficient amount of air enters the pleural cavity, the surface tension adhering visceral to parietal pleura (lung to thoracic wall) is broken and the lung collapses because of its inherent elasticity (elastic recoil). When a lung collapses, the pleural cavity (normally a potential space) becomes a real space. One lung may be collapsed after surgery, for example, without collapsing the other lung because the pleural sacs are separate. *Laceration of a lung and its visceral pleura* or penetration of the thoracic wall and the parietal pleura result in hemorrhage and the escape of air into the pleural cavity. The amount of blood and air that accumulates determines the extent of pulmonary collapse.

Pneumothorax, Hydrothorax, and Hemothorax

Entry of air into the pleural cavity (*pneumothorax*), resulting from a penetrating wound of the parietal pleura or rupture of a lung from a bullet, for example, results in partial collapse of the lung. Fractured ribs may also tear the parietal pleura and produce pneumothorax. The accumulation of a significant amount of fluid in the pleural cavity (*hydrothorax*) may result from *pleural effusion* (escape of fluid into the pleural cavity). With a chest wound, blood may also enter the pleural cavity (*hemothorax*). Hemothorax results more often from injury to a major intercostal vessel than from laceration of a lung.

Pleurectomy

Obliteration of a pleural cavity by disease such as *pleuritis* (inflammation of the pleura) or during surgery (*pleurectomy*, or excision of a part of the pleura, for example) does not cause appreciable functional consequences; however, it may produce pain during exertion. In other procedures (*pleural poudrage*), adherence of the parietal and visceral layers of pleura is induced by covering the apposing layers of pleura with a slightly irritating powder. Pleurectomy and pleural poudrage are performed to prevent recurring *spontaneous pneumothorax* resulting from lung disease. ▶

Pleuritis (Pleurisy)

During inspiration and expiration, the normally smooth, moist pleurae make no detectable sound during *auscultation of the lungs* (listening to breath sounds); however, inflammation of the pleura—pleuritis (pleurisy)—makes the lung surfaces rough. The resulting friction (*pleural rub*) is detectable with a stethoscope. The inflamed surfaces of pleura may also cause the parietal and visceral layers of pleura to adhere (*pleural adhesion*). Acute pleuritis is marked by sharp, stabbing pain, especially on exertion, such as climbing stairs, when the rate and depth of respiration may be increased even slightly. ⊙

Lungs

The lungs are the vital organs of respiration. Their main function is to oxygenate the blood by bringing inspired air into close relation with the venous blood in the pulmonary capillaries. Although cadaveric lungs may be shrunken, hard to the touch, and discolored in appearance, healthy lungs in living people are normally light, soft, and spongy. They are also elastic and recoil to approximately one-third their size when the thoracic cavity is opened (Fig. 1.24). The lungs are separated from each other by the heart, viscera, and great vessels of the mediastinum. The lungs attach to the heart and trachea by structures that comprise the roots of the lungs. The **root of the lung** is formed by structures entering and emerging from the lung at its hilum (hilus)—the bronchus and pulmonary vessels (Figs. 1.28 and 1.29). The root is enclosed within the area of continuity between the parietal and visceral layers of

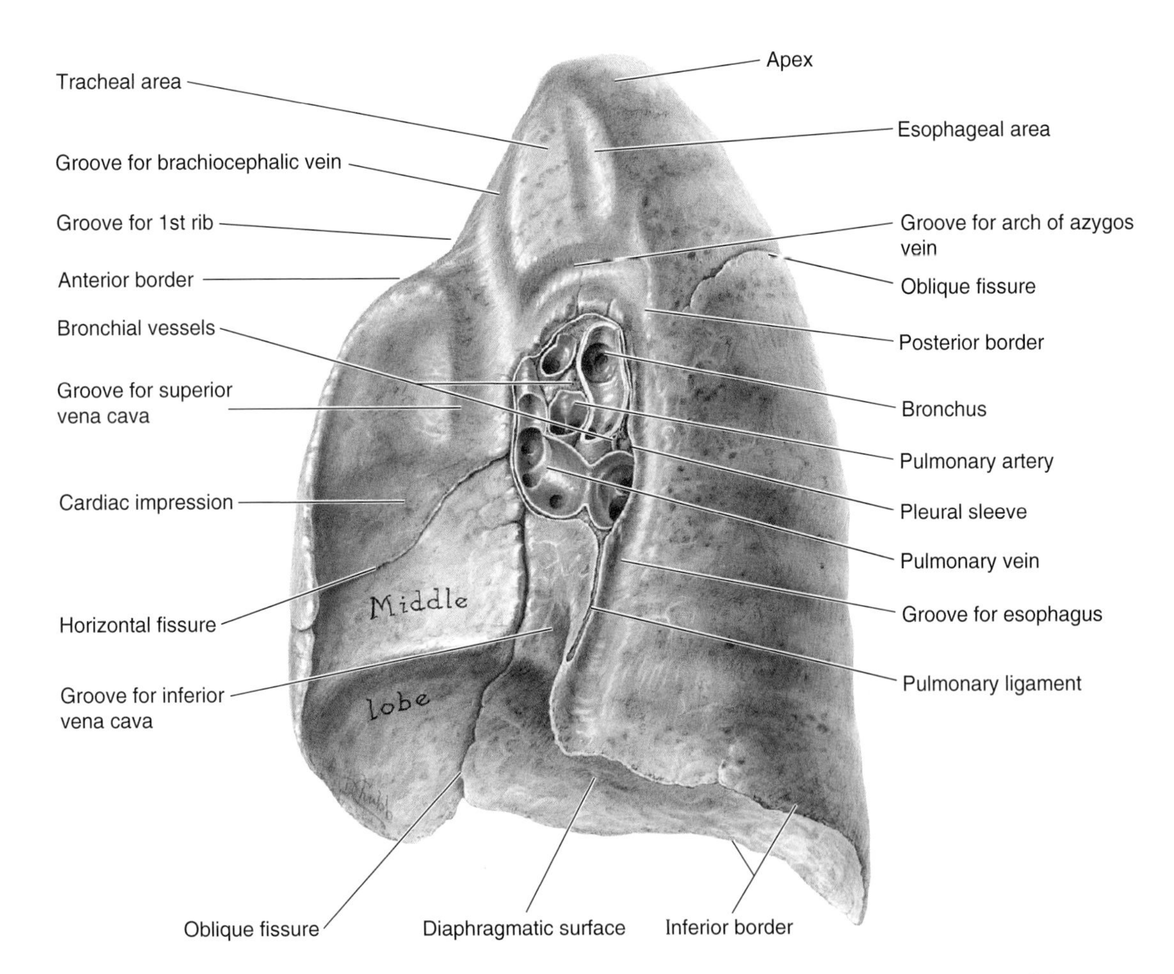

Figure 1.28. Mediastinal surface of the right lung. Observe the somewhat pear-shaped depression, the hilum (doorway) of the lung near the center of this surface, containing the pulmonary vessels and bronchi that constitute the root of the lung, through which these structures (cut here) enter the lung. At the hilum, note that the pulmonary veins lie most anteriorly and inferiorly and that the bronchus is central and posteriorly placed. In the right lung, the superior lobar (eparterial) bronchus may occur superior to the pulmonary artery.

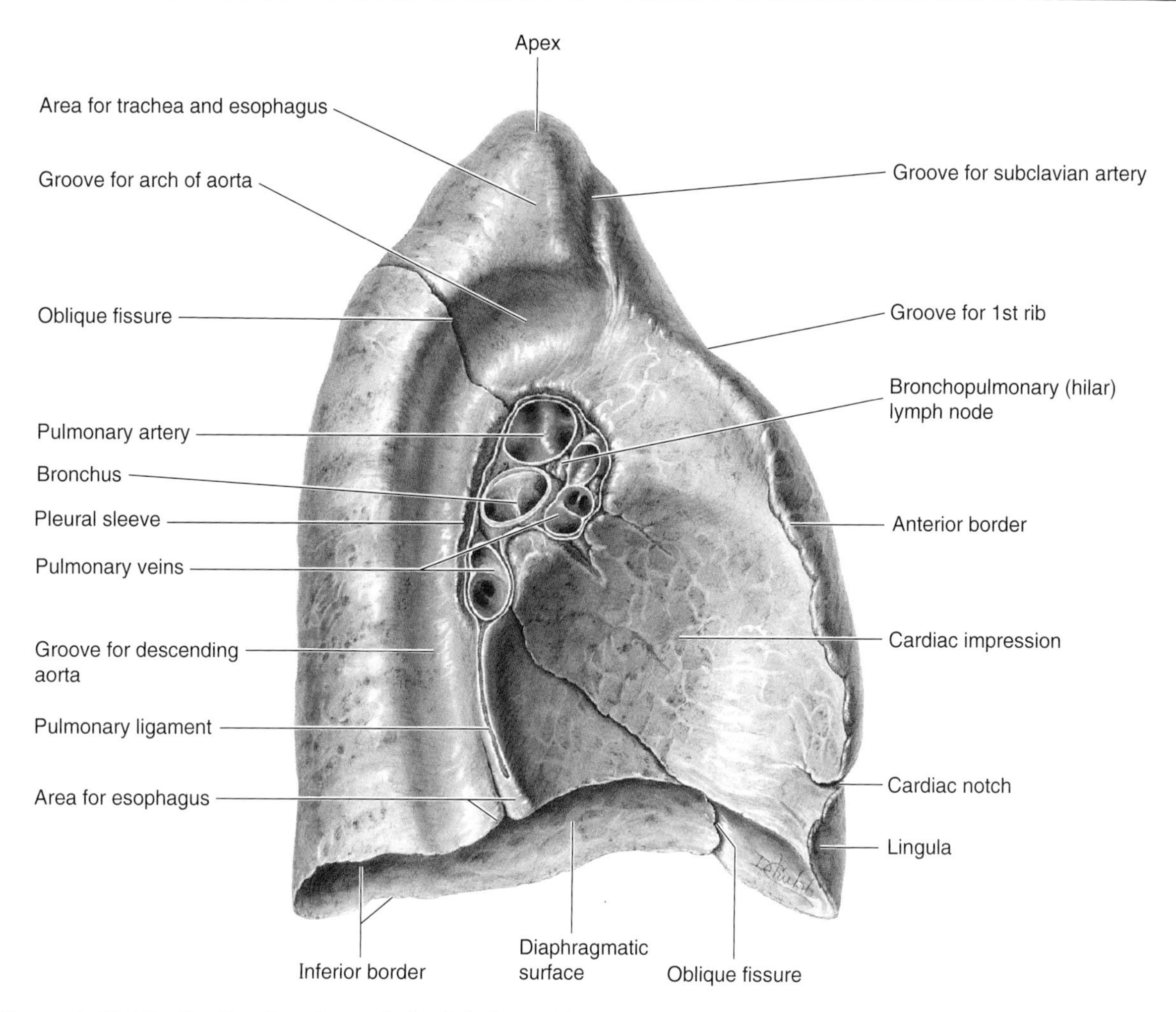

Figure 1.29. Mediastinal surface of the left lung. Near the center, observe the hilum of the lung and its contents comprising the root (cut). Note that at the hilum of the left lung, the pulmonary veins lie most anteriorly and inferiorly, and the bronchus is central and posterior, as in the right lung. However, in the hilum of the left lung, the pulmonary artery is clearly the most superior component of the root of the lung. Also observe the continuation (reflection) of parietal pleura with visceral pleura surrounding the root and forming the pulmonary ligament descending from the root. Examine the area of contact of the trachea and esophagus and the grooves for the arch of the aorta and descending aorta. Observe the large cardiac impression. The impressions of these structures are not maintained when fresh lungs are removed from the thorax.

pleura—the **pleural sleeve** or mesopneumonium (mesentery of the lung). The root of the lung connects the lung with the heart and trachea. The **hilum of the lung** is the area on the medial surface of each lung, the point at which the structures forming the root—the main bronchus, pulmonary vessels, bronchial vessels, lymphatic vessels, and nerves—enter and leave the lung. The hilum can be likened to the area of earth where a plant's roots enter the ground.

The horizontal and oblique fissures divide the lungs into lobes (Fig. 1.30). *The right lung has three lobes, the left lung has two.* The right lung is larger and heavier than the left, but it is shorter and wider because the right dome of the diaphragm is higher and the heart and pericardium bulge more to the left. The anterior margin of the right lung is relatively straight, whereas this margin of the left lung has a deep **cardiac notch.** The cardiac notch primarily indents the anteroinferior aspect of the superior lobe of the left lung. This often creates a thin, tonguelike process of the superior lobe—the *lingula* (L. dim. of lingua, tongue)—which extends below the cardiac notch and slides in and out of the costomediastinal recess during inspiration and expiration. Each lung has:

- *An apex*, the blunt superior end of the lung ascending above the level of the 1st rib into the root of the neck that is covered by cervical pleura
- *Three surfaces* (costal, mediastinal, and diaphragmatic)
- *Three borders* (anterior, inferior, and posterior).

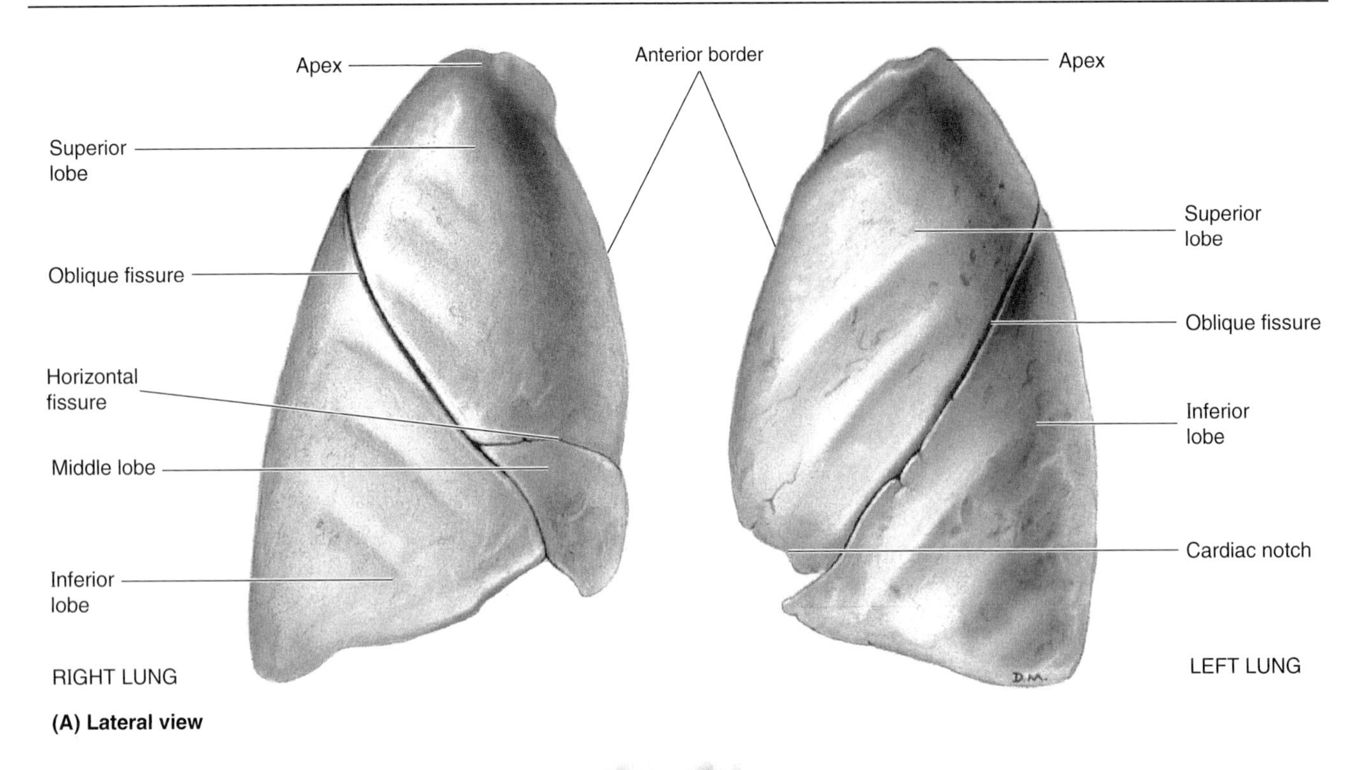

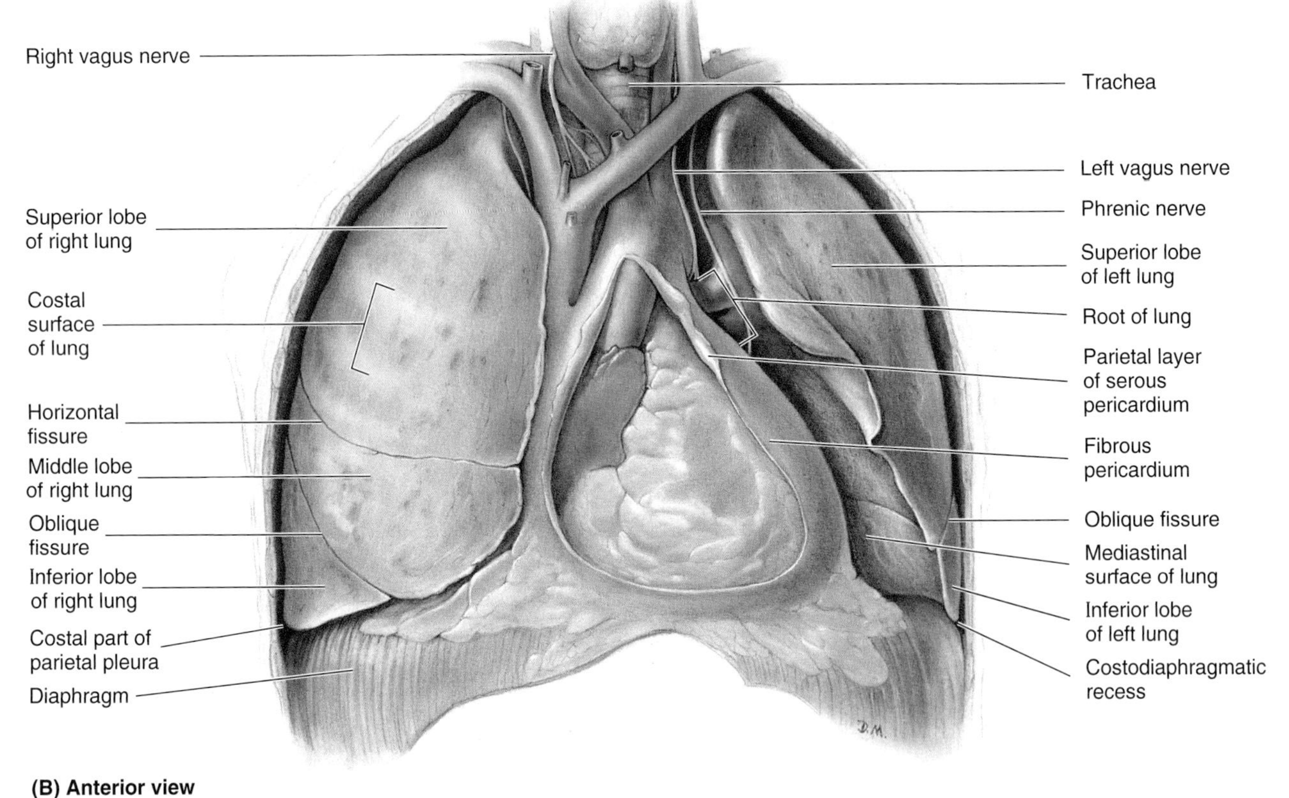

Figure 1.30. Lungs and heart. A. The lungs after removal from the pulmonary cavities. Observe the three lobes of the right lung and the two lobes of the left lung. Although clearly defined in these specimens, the oblique and horizontal fissures may be incomplete or absent in some specimens. **B.** The heart and lungs in situ. The fibrous pericardium is removed anteriorly to expose the heart and great vessels. Observe that the fibrous pericardium is lined by the parietal layer of serous pericardium. Also note the phrenic nerve passing anterior to the root of the lung, while the vagus nerve (CN X) passes posterior to the root.

The lungs in an embalmed cadaver—usually firm to the touch—have contact impressions that are formed by the structures adjacent to them, such as the ribs (Figs. 1.28–1.30). These markings provide clues to the relationships of the lungs; however, they are not visible during surgery or in fresh cadaveric or postmortem specimens. There is a *groove for the esophagus* and a *cardiac impression for the heart* on the mediastinal surface of the right lung. The cardiac impression on the mediastinal surface of the left lung is much larger, and there is a prominent, continuous groove for the *arch of the aorta* and the *descending (thoracic) aorta*, as well as a smaller groove for the esophagus. **Each lung has three surfaces** (Figs. 1.28–1.30):

- *Costal surface*—adjacent to the sternum, costal cartilages, and ribs
- *Mediastinal surface*—including the hilum of the lung and related medially to the mediastinum and posteriorly to the sides of the vertebrae
- *Diaphragmatic surface*—resting on the convex dome of the diaphragm.

The **costal surface of the lung** is large, smooth, and convex. It is related to the costal pleura that separates it from the ribs, costal cartilages, and the innermost intercostal muscles. The posterior part of this surface is related to the bodies of the thoracic vertebrae and is sometimes referred to as the vertebral part of the costal surface.

The **mediastinal surface of the lung** is concave because it is related to the middle mediastinum containing the pericardium and heart. Because two-thirds of the heart is to the left, the pericardial concavity is understandably deeper in the left lung. The mediastinal surface includes the hilum and thus receives the root of the lung, around which the pleura forms a covering, or pleural sleeve. The **pulmonary ligament** hangs inferiorly from the pleural sleeve around the lung root (Figs. 1.28 and 1.29). Some people have difficulty visualizing the root of the lung, the pleural sleeve, and the pulmonary ligament. Put on an extra large lab coat and abduct your upper limb. The root of the lung is comparable to your forearm, and your coat sleeve represents the pleural sleeve. The pulmonary ligament is comparable to the slack of your sleeve that hangs from your wrist, and your wrist, hand, and abducted fingers represent the branching structures of the root—the bronchi and pulmonary vessels.

The **diaphragmatic surface of the lung**—also concave (Figs. 1.28–1.30)—forms the *base of the lung* that rests on the dome of the diaphragm. The concavity is deeper in the right lung because of the higher position of the right diaphragmatic dome, which overlies the large liver. Laterally and posteriorly, the diaphragmatic surface is bounded by a thin, sharp margin (inferior border) that projects into the **costodiaphragmatic recess** of the pleura (Fig. 1.30*B*). **Each lung has three borders:**

- *Anterior border*—where the costal and mediastinal surfaces meet anteriorly and overlap the heart; the *cardiac notch* indents this border of the left lung
- *Inferior border*—which circumscribes the diaphragmatic surface of the lung and separates this surface from the costal and mediastinal surfaces
- *Posterior border*—where the costal and mediastinal surfaces meet posteriorly; it is broad and rounded and lies in the cavity at the side of the thoracic region of the vertebral column.

Variation in the Lobes of the Lung

Occasionally an extra fissure divides a lung, or a fissure is absent. For example, the left lung sometimes has three lobes and the right lung only two. The most common "accessory" lobe is the *azygos lobe*, which appears in the right lung in approximately 1% of people. In these cases, the azygos vein arches over the apex of the right lung and not over the right hilum, isolating the medial part of the apex as an azygos lobe.

Appearance of the Lungs

The lungs are light pink in healthy children and young people who are nonsmokers and live in a clean environment (e.g., the sherpas in the Himalayan mountains). The lungs are often dark and mottled in most adults living in either urban or agricultural areas, especially those who smoke, because of the accumulation of carbon and dust particles in the air and irritants inhaled in tobacco. However, the lungs are capable of handling a considerable amount of carbon without being adversely affected. Special "dust cells" (phagocytes) remove carbon from the gas-exchanging surfaces and deposit it in the "nonactive" connective tissue, which supports the lung, or in lymph nodes receiving lymph from the lungs.

Auscultation and Percussion of the Lungs

Auscultation of the lungs (listening to their sounds with a stethoscope) and percussion of the lungs (tapping the chest over the lungs with the fingers to detect sounds in the apices of the lungs) should always include the root of the neck. When clinicians refer to auscultating the base of the lung, they are not usually referring to its diaphragmatic surface or anatomical base. They are usually referring to the inferoposterior part of the inferior lobe. To auscultate this area, the clinician applies a stethoscope to the posterior thoracic wall at the level of the 10th thoracic vertebra. ▶

Flotation of the Lungs

Fresh healthy lungs always contain some air; consequently, pulmonary tissue removed from them will float in water. Diseased lungs filled with fluid, fetal lungs, and lungs from a stillborn infant that have never expanded sink when placed in water. The lungs of a liveborn infant who dies shortly after birth floats. These observations are of medicolegal significance in determining whether a dead infant was stillborn (born dead) or whether it was born alive and started to breathe.

Apical Lung Cancers

Lung cancer involving a phrenic nerve may result in paralysis of the hemidiaphragm. Because of the intimate relationship of the *recurrent laryngeal nerve* to the apex of the lung (Fig. 1.30*B*), this nerve may be involved in *apical lung cancers*. This involvement usually results in hoarseness owing to paralysis of a vocal fold (cord) because the recurrent laryngeal nerve supplies all but one of the laryngeal muscles (see Chapter 8). ⊙

Trachea and Bronchi

The **main bronchi** (primary bronchi), one to each lung, pass inferolaterally from the bifurcation of the **trachea** at the level of the sternal angle to the hila (pleural of hilum) of the lungs (Fig. 1.31*B*). The walls of the trachea and bronchi are supported by horseshoe or C-shaped rings of hyaline cartilage.

- The **right main bronchus** is wider, shorter, and runs more vertically than the left main bronchus as it passes directly to the hilum of the lung.
- The **left main bronchus** passes inferolaterally, inferior to the arch of the aorta and anterior to the esophagus and thoracic aorta, to reach the hilum of the lung.

The main bronchi enter the hila of the lungs and branch in a constant fashion within the lungs to form the **bronchial tree.** Each main bronchus divides into **lobar bronchi** (secondary bronchi), two on the left and three on the right, each of which supplies a lobe of the lung. Each lobar bronchus divides into several **segmental bronchi** (tertiary bronchi) that supply the bronchopulmonary segments (Figs. 1.31 and 1.32). A **bronchopulmonary segment**:

- Is a pyramidal-shaped segment of the lung, with its apex facing the lung root and its base at the pleural surface
- Is the largest subdivision of a lobe
- Is separated from adjacent segments by connective tissue septa
- Is supplied independently by a segmental (tertiary) bronchus and a tertiary branch of the pulmonary artery
- Is named according to the segmental bronchus supplying it
- Is drained by the intersegmental parts of the pulmonary veins that lie in the connective tissue between and drain adjacent segments
- Is surgically resectable.

Beyond the direct branches of the lobar bronchi—i.e., beyond the segmental bronchi (Fig. 1.31*B*)—are from twenty to twenty-five generations of branches that eventually end in **terminal bronchioles** (Fig. 1.32). Each terminal bronchiole gives rise to several generations of **respiratory bronchioles** and each respiratory bronchiole provides 2 to 11 **alveolar ducts**, each of which gives rise to 5 or 6 **alveolar sacs** lined by alveoli. The **alveolus** (L. small hollow space) is the basic structural unit of gas exchange in the lung (Cormack, 1993; Gartner and Hiatt, 1997). New alveoli continue to develop until the age of approximately 8 years, by which time there are approximately 300 million alveoli (Moore and Persaud, 1998).

Aspiration of Foreign Bodies

Because the right bronchus is wider and shorter and runs more vertically than the left bronchus, foreign bodies are more likely to enter and lodge in it or one of its branches. A potential hazard encountered by dentists is an aspirated foreign body, such as a piece of tooth, filling material, or a small instrument. Such objects are also most likely to enter the right main bronchus. To create a sterile environment and avoid aspiration of foreign objects, dentists may insert a thin rubber dam into the oral cavity—when performing a root canal procedure, for example.

Bronchoscopy

When examining the bronchi with a *bronchoscope*, one observes a keel-like ridge—the **carina** (L. keel of a boat)—between the orifices of the main bronchi. The carina is a cartilaginous projection of the last tracheal ring. Normally the carina lies in a sagittal plane and has a fairly definite edge. If the tracheobronchial lymph nodes in the angle between the main bronchi are enlarged because cancer cells have metastasized from a *bronchogenic carcinoma*, for example, the carina is distorted, widened posteriorly, and immobile. Hence, morphological changes in the carina ▶

▶ are important diagnostic signs to bronchoscopists in assisting with the differential diagnosis of respiratory disease.

The mucous membrane covering the carina is one of the most sensitive areas of the tracheobronchial tree and is associated with the *cough reflex*. For example, when children aspirate a peanut, they choke and cough. Once the peanut passes the carina, coughing usually stops. The resulting chemical *arachidic bronchitis* (inflammation of bronchus) caused by substances released from the peanut (L. arachis*)* and the collapse of the lung (*atelectasis*) distal to the foreign body causes difficult breathing (*dyspnea*). In addition, during postural drainage of the lungs (inverting the patient to "join forces" with gravity), lung secretions pass to the carina and cause coughing, which helps to expel them.

Lung Resections

Knowledge of the anatomy of the bronchopulmonary segments is essential for precise interpretations of radiographs and other diagnostic images of the lungs. Knowledge of these segments is also essential for surgical resection of diseased segments. Bronchial and pulmonary disorders such as tumors or abscesses (collections of pus) often localize in a bronchopulmonary segment, which may be surgically resected. During treatment of *lung cancer*, the surgeon may remove a whole lung (*pneumonectomy*), a lobe (*lobectomy*), or a bronchopulmonary segment (*segmentectomy*).

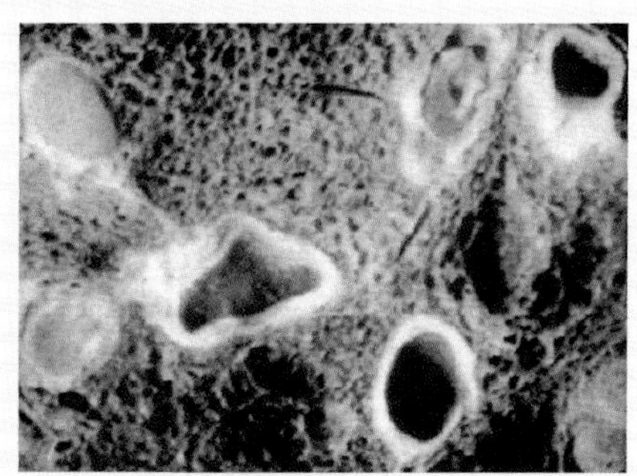

Bronchial asthma (cross-section of lung)

Bronchial Asthma

Bronchial asthma is an increasingly common condition of the lungs in which widespread narrowing of the airways is present, varying over short periods of time with recovery occurring either spontaneously or resulting from treatment. Asthma is caused in varying degrees by contraction (spasm) of smooth muscle, edema of the mucosa, and mucus in the lumen of the bronchi and bronchioles. These changes are caused by the local release of *spasmogens*— ▶

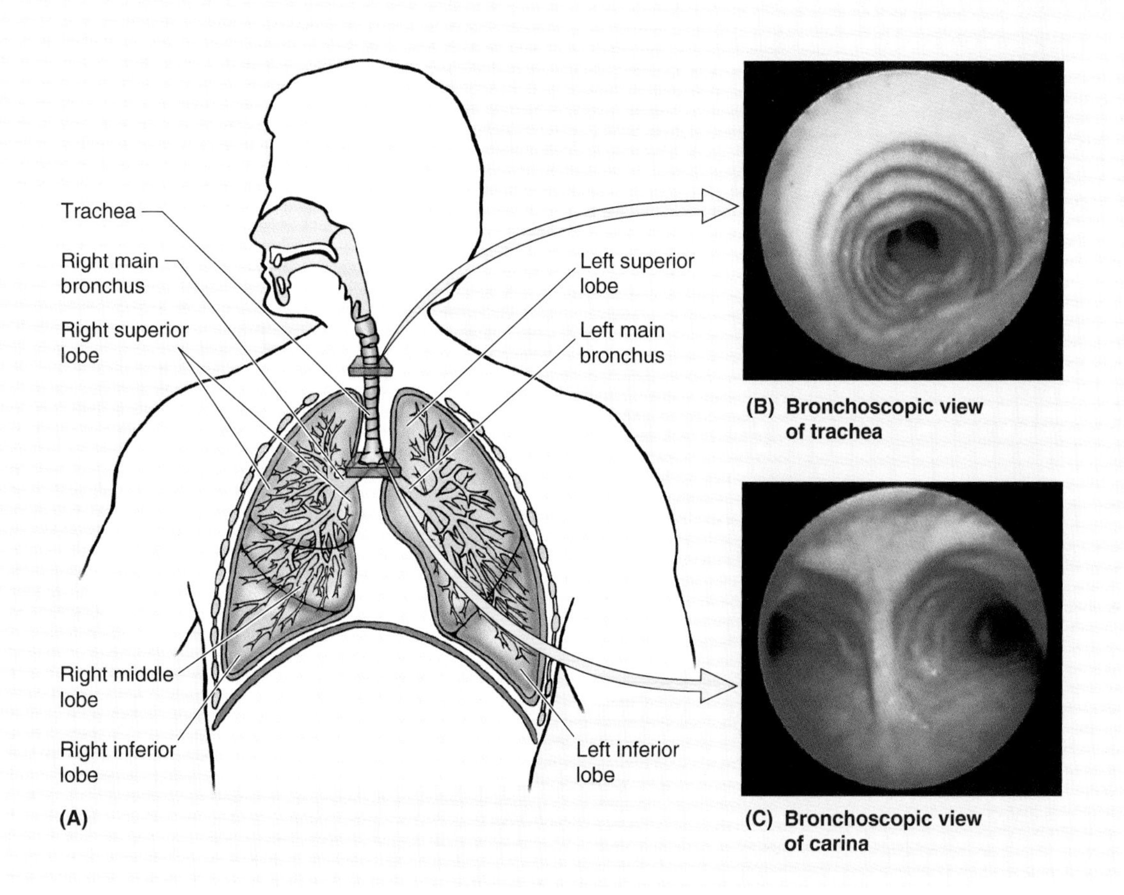

(B) Bronchoscopic view of trachea

(C) Bronchoscopic view of carina

▶ substances causing contraction of smooth muscle—and vasoactive substances influencing the tone and caliber of blood vessels (e.g., histamine or prostaglandins) in the course of an allergic process. The absence of cartilages in the walls of bronchioles is a potential hazard because it allows these airways to constrict and almost close down when the tonus in their smooth muscle cells becomes excessive. In this asthmatic condition, the problem has more to do with expiration than inspiration because the bronchioles drawn open during inspiration also have to remain open during expiration if they are to permit a rapid outflow of air owing to the elastic recoil of the lung tissues. Consequently, in the course of an asthmatic attack, more wheezing noises and breathing difficulties may be experienced during air expulsion than during inspiration (Cormack, 1993). ⊙

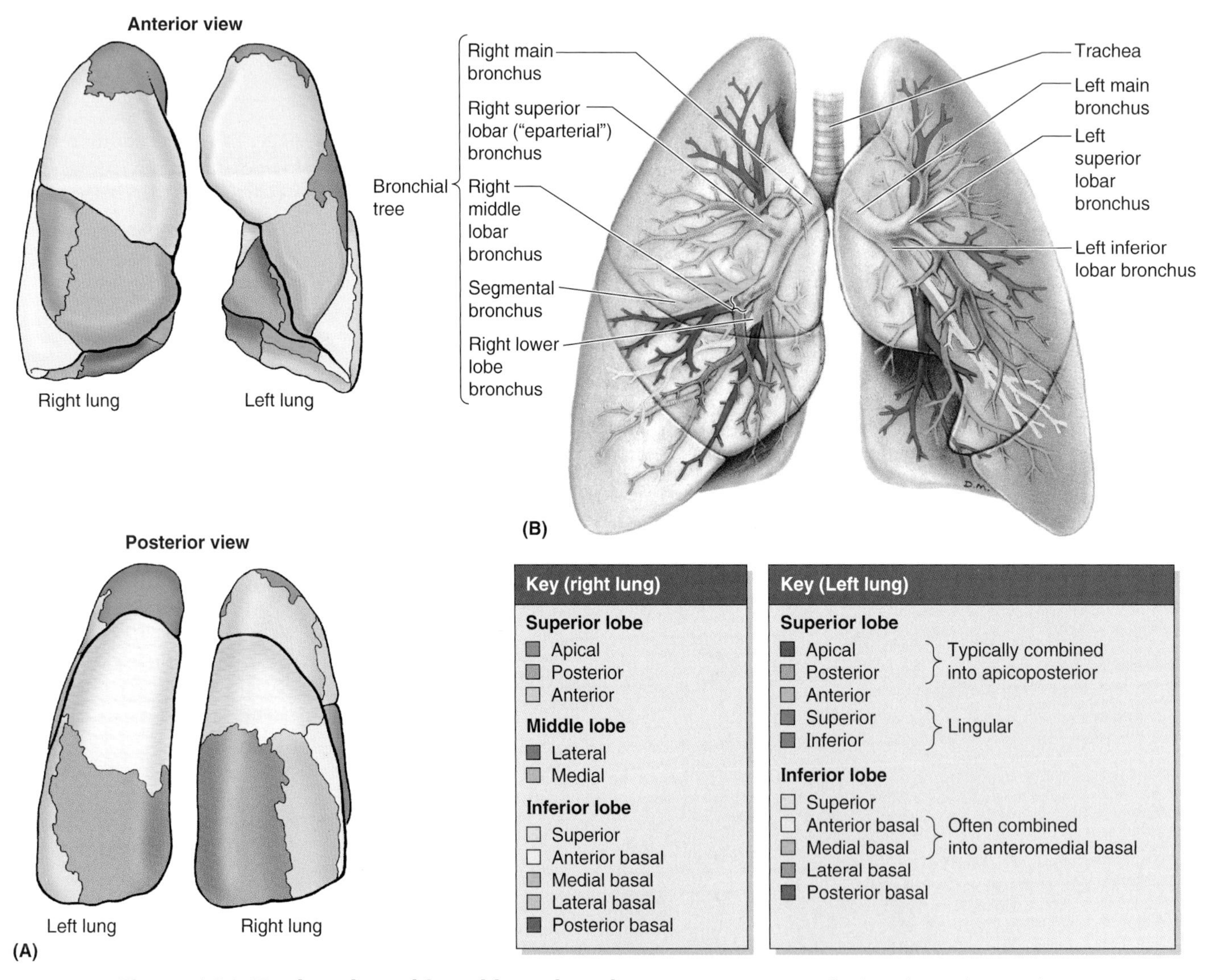

Figure 1.31. Trachea, bronchi, and bronchopulmonary segments. A. Anterior and posterior views of the lungs following injection of a different color of latex into each segmental (tertiary) bronchus, thus demonstrating the bronchopulmonary segments. **B.** Observe that the right main bronchus is more vertical and shorter than the left main bronchus. Although it lies posterior to the anterior margin of the lung and thus is not apparent in this view, the right main bronchus gives off the right superior lobar (lobe) bronchus before entering the hilum of the lung. The lobar bronchi are divided into segmental bronchi (colored), each of which serves a bronchopulmonary segment of the same name.

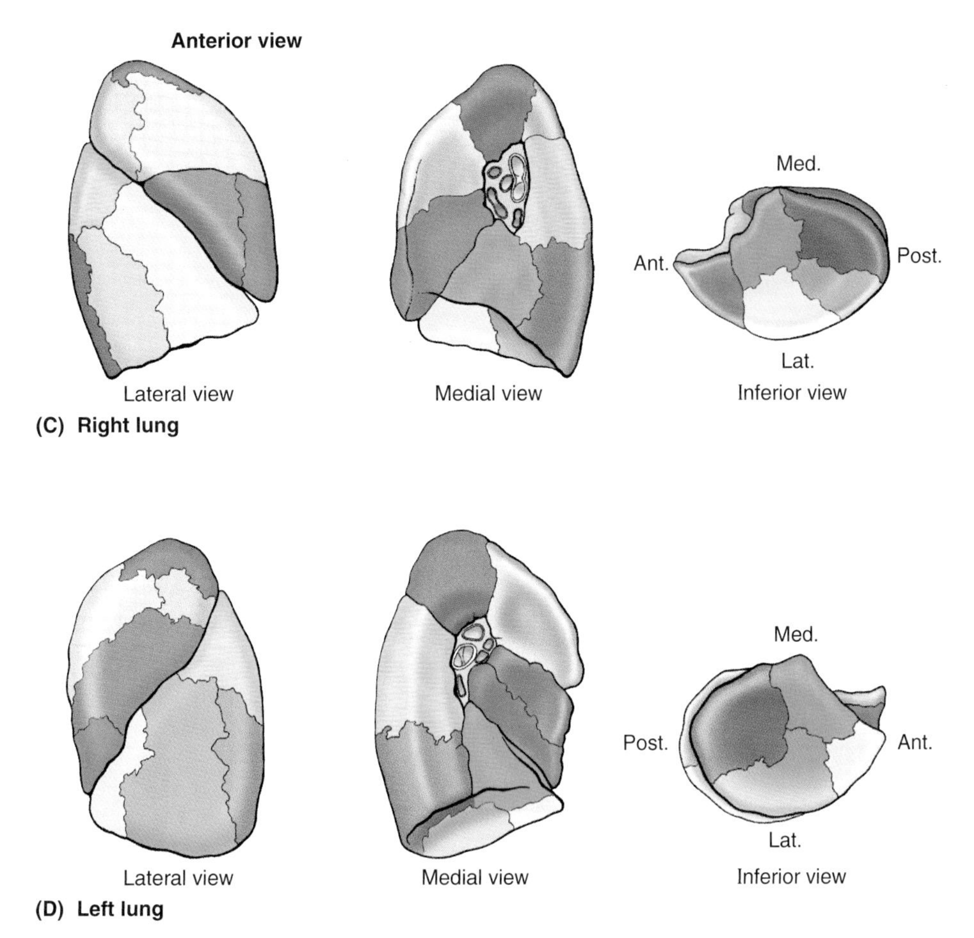

Figure 1.31. *(Continued)* **C–D.** Bronchopulmonary segments of the right and left lungs in lateral, medial, and inferior views.

Vasculature and Nerves of Lungs and Pleurae

Each lung has a large **pulmonary artery** supplying blood to it and two **pulmonary veins** draining blood from it (Fig. 1.33). The right and left pulmonary arteries arise from the **pulmonary trunk** at the level of the sternal angle and carry poorly oxygenated ("venous") blood to the lungs for oxygenation. (Hence they are usually colored blue like veins in anatomical illustrations.) Each pulmonary artery becomes part of the root of the corresponding lung and gives off its 1st branch to the superior lobe before entering the hilum. Within the lung each artery descends posterolateral to the main bronchus and divides into **lobar** and **segmental arteries.** Consequently, an arterial branch goes to each lobe and bronchopulmonary segment of the lung, usually on the anterior aspect of the corresponding bronchus. The arteries and bronchi are paired in the lung, branching simultaneously and running parallel courses.

The **pulmonary veins**, two on each side, carry well-oxygenated ("arterial") blood from the lungs to the left atrium of the heart. (Hence they are usually colored red or purple like arteries in anatomical illustrations.) Beginning in the pulmonary capillaries, the veins unite into larger and larger vessels. Intrasegmental parts of pulmonary veins drain blood from adjacent bronchopulmonary segments into the intersegmental parts of the pulmonary veins in the septa, which separate the segments. The pulmonary veins run independent courses from the arteries and bronchi as they run toward the hilum. The veins from the visceral pleura drain into the pulmonary veins, and the veins from the parietal pleura join the systemic veins in adjacent parts of the thoracic wall.

The **bronchial arteries** supply blood for nutrition of the structures comprising the root of the lungs, the supporting tissues of the lungs, and the visceral pleura (Fig. 1.34*A*). The **left bronchial arteries** arise from the thoracic aorta; however, the single **right bronchial artery** may arise from

- A superior posterior intercostal artery
- A common trunk from the thoracic aorta with the right 3rd posterior intercostal artery
- A left superior bronchial artery.

The small bronchial arteries provide branches to the upper esophagus and then pass along the posterior aspects of the main bronchi, supplying them and their branches as far distally as the respiratory bronchioles. The distal-most branches

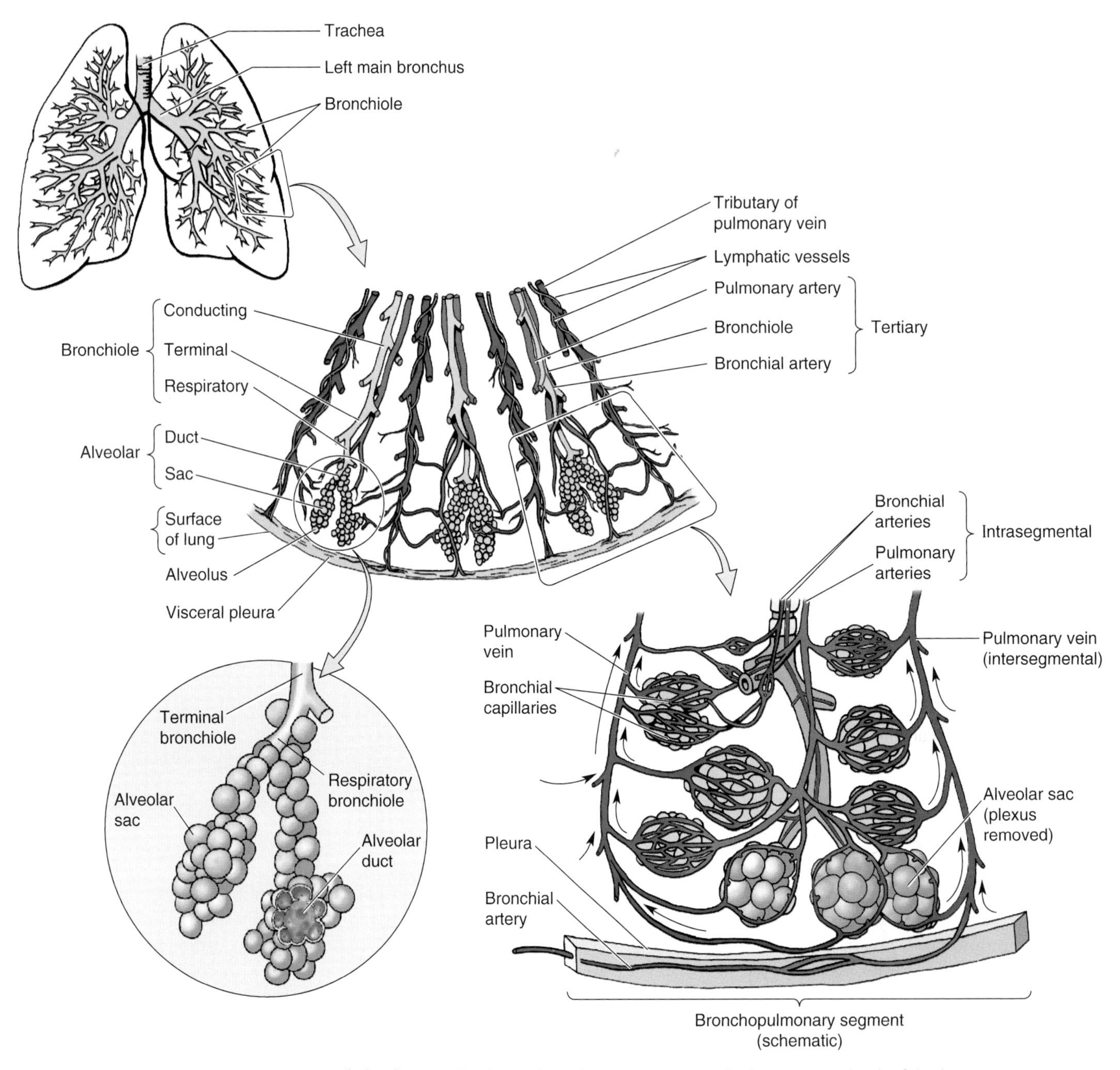

Figure 1.32. Structure of the lungs. The bronchopulmonary segment is the structural unit of the lung. Some 15 or more generations after the segmental bronchus, each terminal bronchiole gives rise to several generations of respiratory bronchioles, and each respiratory bronchiole gives rise to 5 or 6 alveolar sacs lined by alveoli, which are the basic structures for gas exchange. Each intrasegmental pulmonary artery, carrying poorly oxygenated blood, ends in a capillary plexus in the walls of the alveolar sacs and alveoli, where O_2 and CO_2 are exchanged. The pulmonary veins arise from the pulmonary capillaries draining toward and coursing in the septa between adjacent segments to carry well-oxygenated blood to the heart.

of the bronchial arteries anastomose with branches of the pulmonary arteries in the walls of the bronchioles and in the visceral pleura. The parietal pleura is supplied by the arteries that supply the thoracic wall (p. 88).

The **bronchial veins** (Fig. 1.34*B*) drain only part of the blood supplied to the lungs by the bronchial arteries; some blood is drained by the pulmonary veins, especially from the more peripheral parts and the distal root of the lung. The right bronchial vein drains into the **azygos vein** and the left bronchial vein drains into the **accessory hemiazygos vein** or the left superior intercostal vein. Bronchial veins also receive some blood from esophageal veins.

Pulmonary Thromboembolism

Obstruction of a pulmonary artery by a thrombus (blood clot) is a common cause of morbidity (sickness) and mortality (death). An *embolus* (plug) in a pulmonary artery forms when a thrombus, fat globule, or air bubble travels in the blood to the lungs from a leg vein, for example, after it is traumatized during a tibial fracture. The thrombus passes through the right side of the heart to a lung through a pulmonary artery. The thrombus may block a pulmonary artery—pulmonary thromboembolism (PTE)—or one of its branches. The immediate result is partial or complete obstruction of bloodflow to the lung. The blockage results in a sector of lung that is ventilated but not perfused with blood. When a large embolus occludes a pulmonary artery, the patient suffers *acute respiratory distress* because of a major decrease in the oxygenation of blood and may die in a few minutes. A medium-sized embolus may block an artery supplying a bronchopulmonary segment, producing a *thrombotic infarct*—an area of necrotic (dead) tissue.

In physically active people, a collateral circulation—an indirect, accessory blood supply—often exists and develops further when there is a PTE so that infarction is not as likely to occur. Anastomoses with branches of the bronchial arteries abound in the region of the terminal bronchioles. In ill people with impaired circulation in the lung such as *chronic congestion*, PTE commonly results in lung infarction. When an area of visceral pleura is also deprived of blood, it becomes inflamed (*pleuritis*) and irritates or becomes fused to the sensitive parietal pleura, resulting in pain. The pain from parietal pleura is referred to the cutaneous distribution of the intercostal nerves to the thoracic wall, or, in the case of inferior nerves, to the anterior abdominal wall. ○

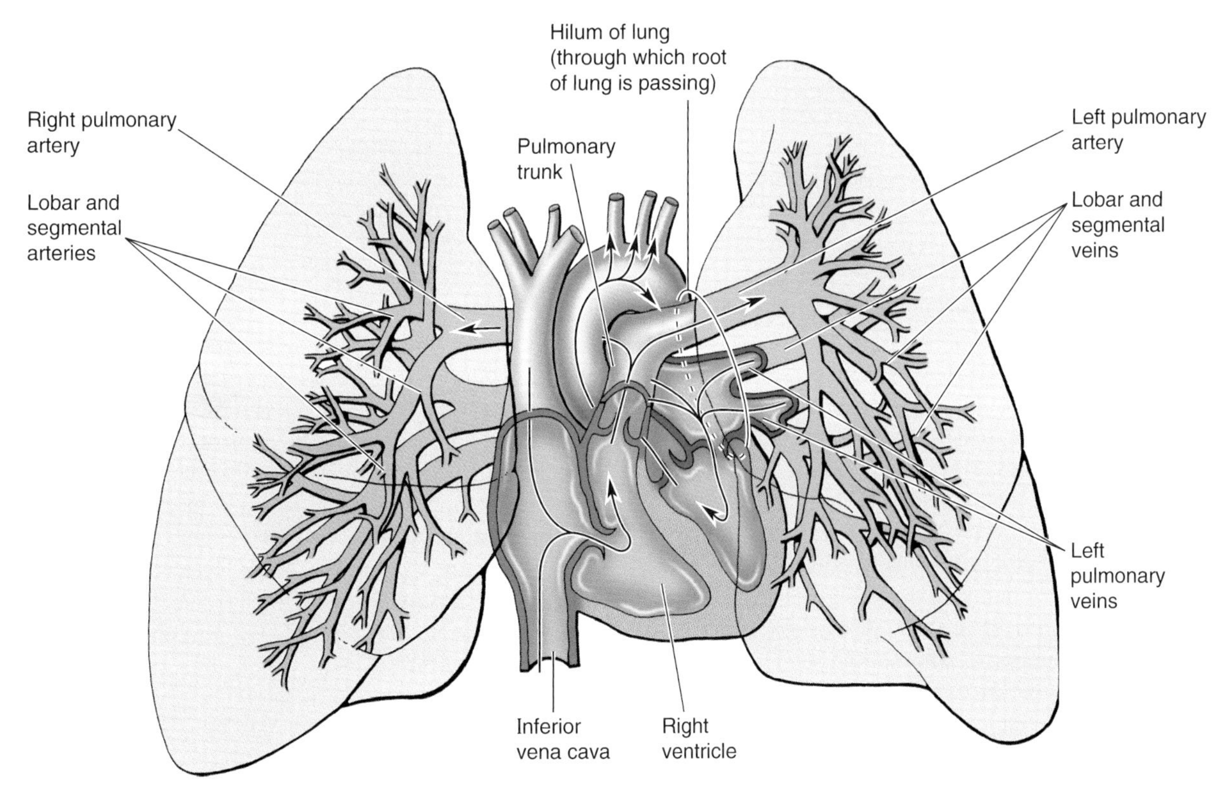

Figure 1.33. Pulmonary circulation. The pulmonary trunk from the right ventricle of the heart divides into right and left pulmonary arteries, which transport poorly oxygenated blood to the lungs. The pulmonary arteries subdivide into lobar and segmental arteries within the lungs. After passing through the pulmonary capillaries where gas exchange (O_2 and CO_2) occurs, the blood drains into the pulmonary veins, which return the well-oxygenated blood to the left atrium of the heart. These vessels of the root of the lung enter and leave the lung at the hilum. Note that the right pulmonary artery passes under the arch of the aorta to reach the right lung; the left pulmonary artery lies completely to the left of the arch.

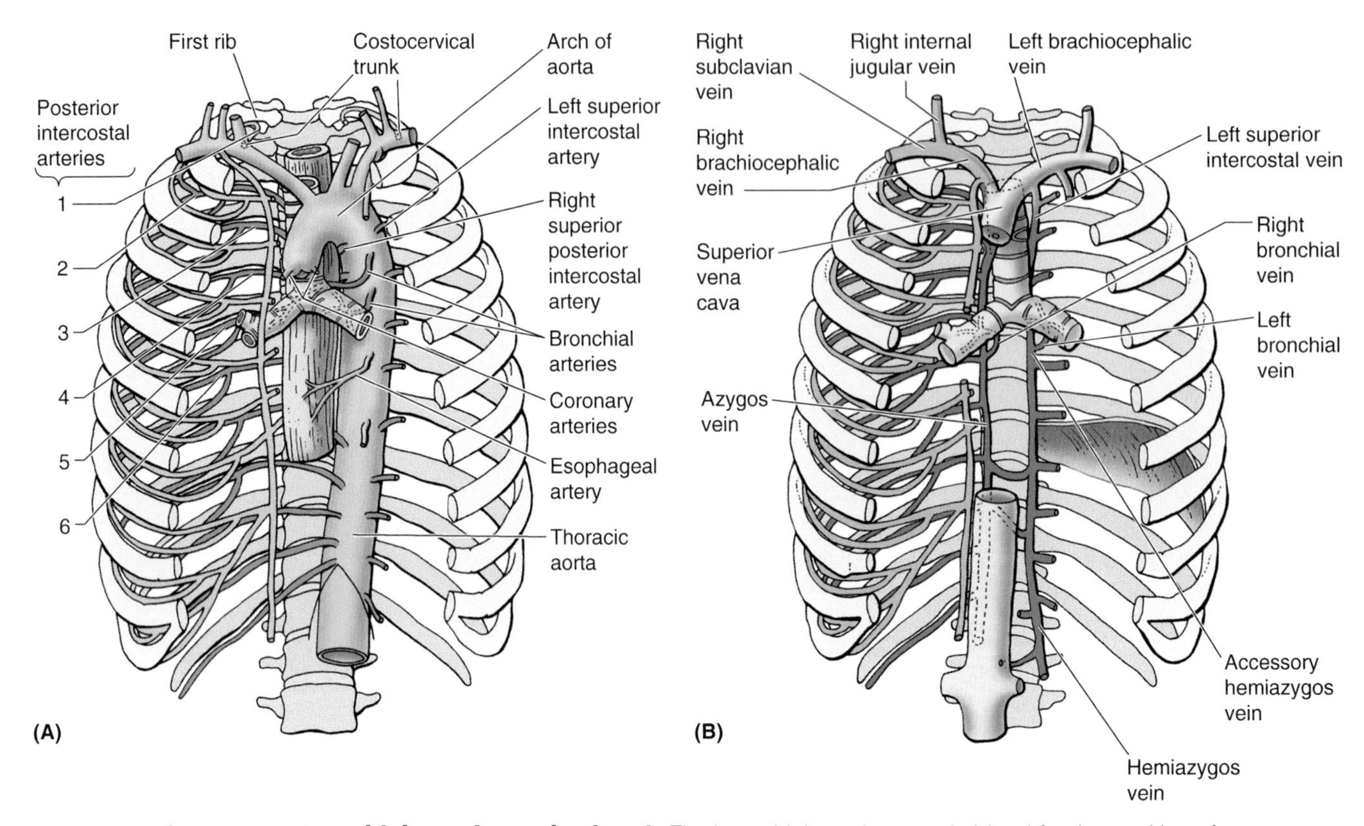

Figure 1.34. Bronchial arteries and veins. A. The bronchial arteries supply blood for the nutrition of the supporting tissues of the lungs and visceral pleura. These arteries arise from the thoracic aorta but the origin of the right bronchial artery is variable. It may arise from (*a*) a superior posterior intercostal artery; (*b*) a common trunk from the thoracic aorta with the right third posterior intercostal artery; or (*c*) the left superior bronchial artery. **B.** The bronchial veins drain some of the blood supplied to the lungs by the bronchial arteries; the rest is drained by the pulmonary veins. The right bronchial vein drains into the azygos vein and the left bronchial vein drains into the accessory hemiazygos vein or the left superior intercostal vein.

The **lymphatic plexuses in the lungs** communicate freely (Fig. 1.35).

The **superficial (subpleural) lymphatic plexus** lies deep to the visceral pleura and drains the lung parenchyma (tissue) and visceral pleura. Lymphatic vessels from the plexus drain into the **bronchopulmonary lymph nodes** (hilar lymph nodes) in the hilum of the lung.

The **deep lymphatic plexus** is located in the submucosa of the bronchi and in the peribronchial connective tissue. It is largely concerned with draining the structures that form the root of the lung. Lymphatic vessels from this plexus drain initially into the **pulmonary lymph nodes** located along the lobar bronchi. Lymphatic vessels from these nodes continue to follow the bronchi and pulmonary vessels to the hilum of the lung, where they also drain into the **bronchopulmonary lymph nodes.** From them, lymph from both the superficial and deep lymphatic plexuses drains to the superior and inferior **tracheobronchial lymph nodes** superior and inferior to the bifurcation of the trachea and main bronchi, respectively. The right lung drains primarily through the respective sets of nodes on the right side, and the superior lobe of the left lung drains primarily through respective nodes of the left side. Many, but not all, of the lymphatics from the lower lobe of the left lung, however, drain to the right superior tracheobronchial nodes; the lymph then continues to follow the right-side pathway.

Lymph from the tracheobronchial lymph nodes passes to the right and left **bronchomediastinal lymph trunks.** These trunks usually terminate on each side at the junction of the subclavian and internal jugular veins; however, the right bronchomediastinal trunk may first merge with other lymphatic trunks converging here to form the very short **right lymphatic duct.** The left bronchomediastinal trunk may terminate in the **thoracic duct.** Lymph from the parietal pleura drains into the lymph nodes of the thoracic wall (intercostal, parasternal, mediastinal, and phrenic). A few lymphatic vessels from the *cervical parietal pleura* drain into the axillary lymph nodes.

The **nerves of the lungs and visceral pleura** are derived from the pulmonary plexuses anterior and (mainly) posterior to the roots of the lungs (Fig. 1.36). These nerve networks contain parasympathetic fibers from the **vagus nerves** (CN X) and sympathetic fibers from the sympathetic trunks. The *parasympathetic ganglion cells*—cell bodies of postsynaptic parasympathetic neurons—are in the **pulmonary plexuses** and along the branches of the bronchial tree. The *sympathetic ganglion cells*—cell bodies of postsynaptic sympathetic neurons—

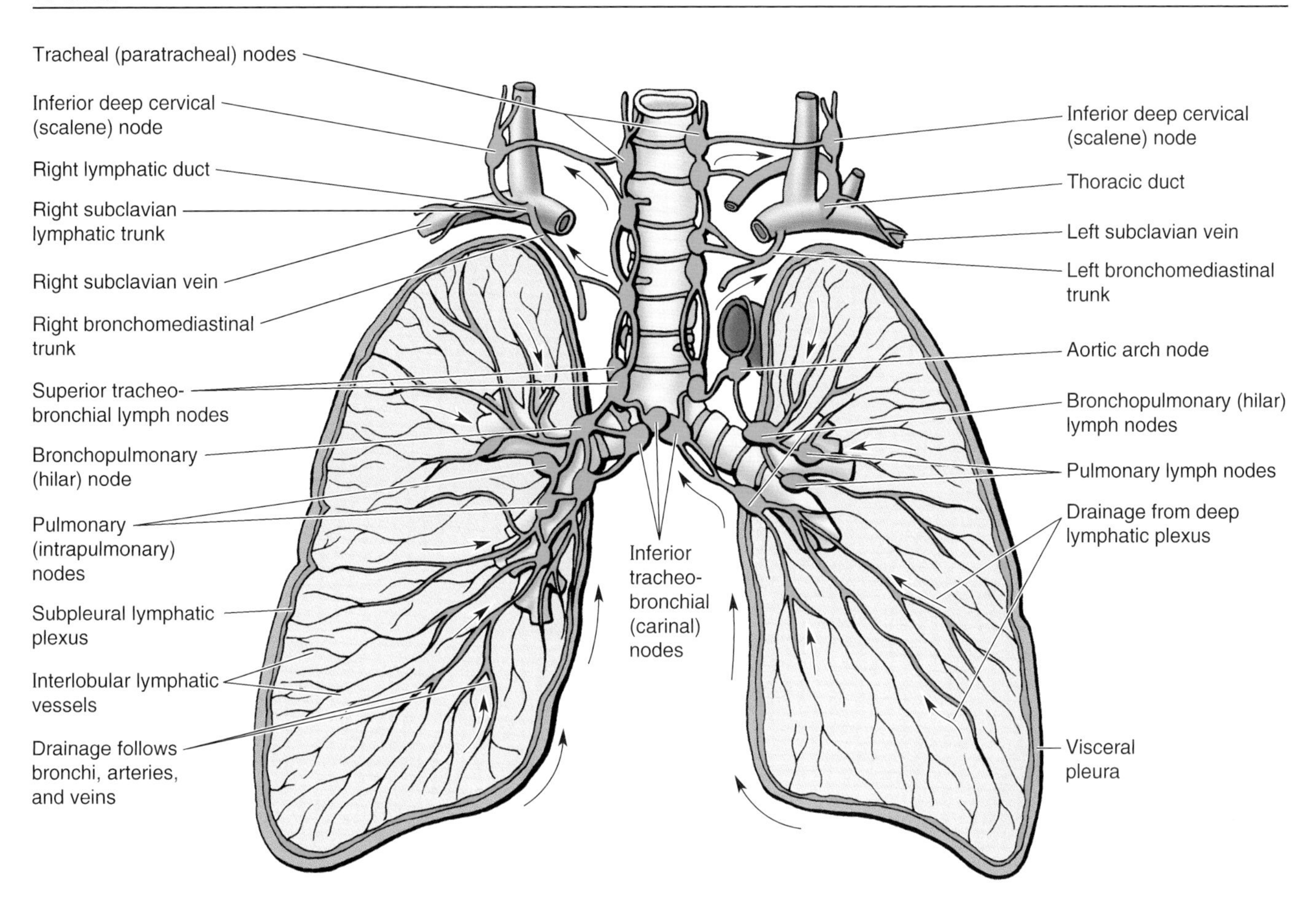

Figure 1.35. Lymphatic drainage of the lungs. The lymphatic vessels originate in superficial subpleural and deep lymphatic plexuses. The subpleural plexus lies deep to the visceral pleura and drains lymph from the surface of the lung to the hilum of the lung, where it enters the bronchopulmonary (hilar) lymph nodes. The deep plexus is in the lung and follows the bronchi and pulmonary vessels to the root of the lung. The lymph passes through pulmonary lymph nodes and enters the bronchopulmonary lymph nodes. All lymph from the lung leaves along the root of the lung and enters the tracheobronchial (carinal) lymph nodes and superior tracheobronchial lymph nodes. From here the lymph traverses a variable number of tracheal nodes and enters the bronchomediastinal trunks.

Inhalation of Carbon Particles and Irritants

Lymph from the lungs carries phagocytes that contain ingested carbon particles from inspired air. In many people, especially cigarette smokers and/or dwellers of urban and agricultural areas, these particles color the surface of the lungs and the lymph nodes a mottled gray to black. *Smoker's cough* results from inhalation of irritants in tobacco.

Pleural Adhesion

If the parietal and visceral layers of pleura adhere (*pleural adhesion*), the lymphatic vessels in the lung and visceral pleura may drain into the axillary lymph nodes. The presence of carbon particles in these nodes is presumptive evidence of a pleural adhesion. ▶

are in the *paravertebral sympathetic ganglia* of the sympathetic trunks. The parasympathetic fibers from the vagus nerves are motor to the smooth muscle of the bronchial tree (*bronchoconstrictor*), inhibitor to the pulmonary vessels (*vasodilator*), and secretor to the glands of the bronchial tree (*secretomotor*). The visceral afferent fibers of CN X are distributed to the:

- Bronchial mucosa and are probably concerned with cough reflexes
- Bronchial muscles and are involved in stretch reception
- Interalveolar connective tissue and are involved in Hering-Breuer reflexes, the mechanism that tends to limit respiratory excursions
- Pulmonary arteries as pressor receptors and pulmonary veins as chemoreceptors.

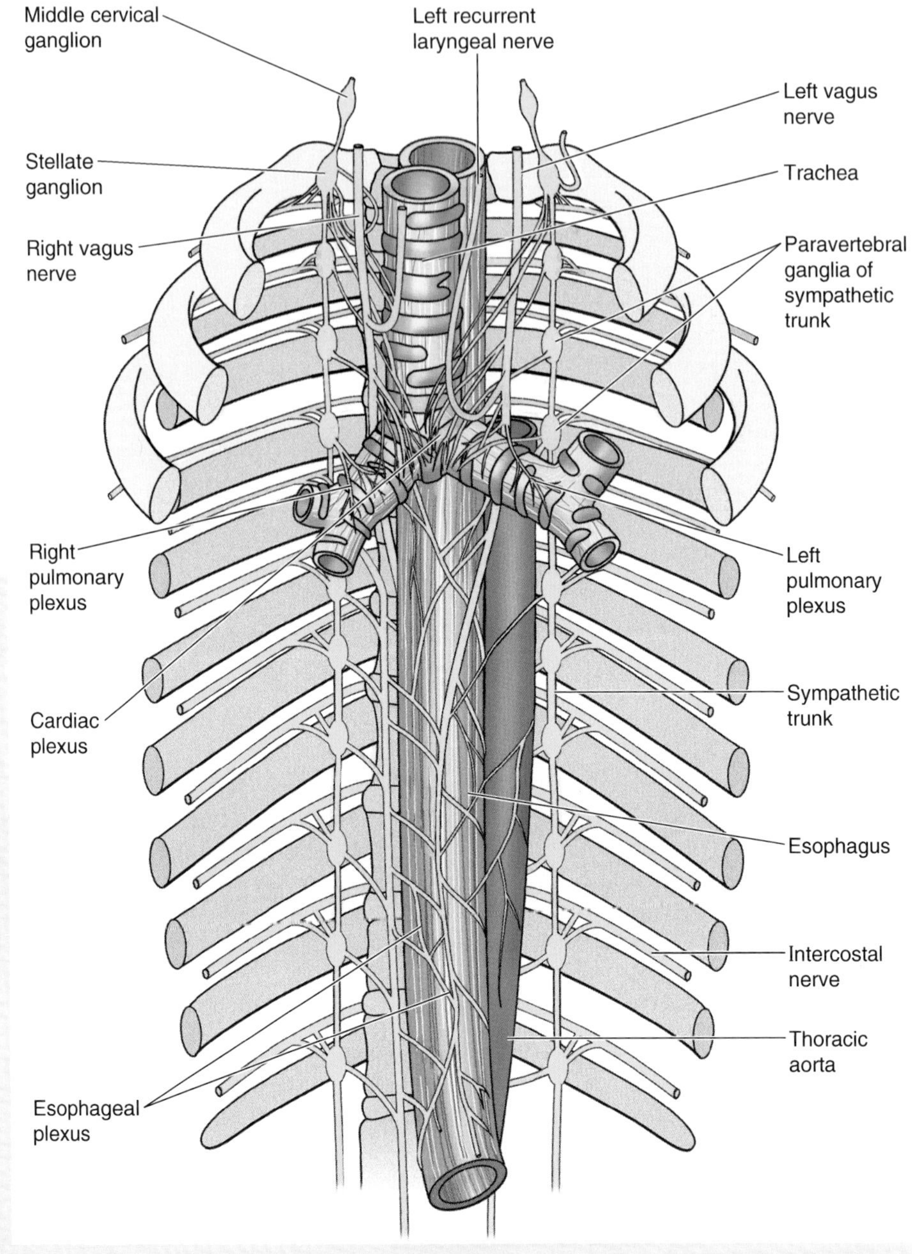

Figure 1.36. Nerves of the lungs and visceral pleura. Observe the right and left pulmonary plexuses, anterior and posterior to the roots of the lungs, receiving sympathetic contributions from the right and left sympathetic trunks and parasympathetic contributions from the right and left vagus nerves (CN X). The vagus nerves continue inferiorly after contributing to the posterior pulmonary plexus and become part of the esophageal plexus, often losing their identity and reforming as anterior and posterior vagal trunks. Branches of the pulmonary plexuses run along the bronchi to the lungs.

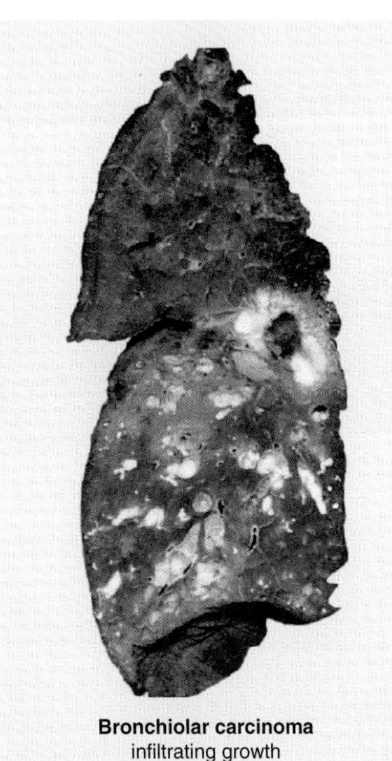

Bronchiolar carcinoma
infiltrating growth

Bronchiolar Carcinoma

Bronchiolar carcinoma (CA) is a common type of lung cancer that arises from the epithelium of the bronchial tree. **Lung cancer** is mainly caused by cigarette smoking. Bronchiolar CA usually metastasizes widely because of the arrangement of the lymphatics. Tumor cells may enter the systemic circulation by invading the wall of a sinusoid or venule in the lung.

Bronchogenic Carcinoma

This squamous cell or oat cell CA arises in the mucosa of the large bronchi and produces a persistent, productive cough or *hemoptysis* (spitting of blood). Malignant cells (cancer cells) metastasize early to the thoracic lymph nodes. Common sites of *hematogenous metastases* (spreading through the blood) of cancer cells from a bronchogenic carcinoma are the brain, bones, lungs, and suprarenal (adrenal) glands. The tumor cells probably enter the systemic circulation by invading the wall of a sinusoid or venule in a lung, and are transported through the pulmonary veins, left heart, and aorta to these structures. Often the lymph nodes superior to the clavicle (supraclavicular lymph nodes) are enlarged with bronchogenic CA owing to metastases of cancer cells from the tumor. Consequently, the *supraclavicular lymph nodes* are considered to be **sentinel lymph nodes** because their enlargement alerts the physician to the possibility of malignant disease in the thoracic and/or abdominal organs. ⊙

Afferent fibers from the visceral pleura and bronchi may accompany sympathetic fibers, mediating nociceptive responses to painful or injurious stimuli. The *sympathetic fibers* are inhibitor to the bronchial muscle (*bronchodilator*), motor to the pulmonary vessels (*vasoconstrictor*), and inhibitor to the alveolar glands of the bronchial tree—type II secretory epithelial cells of the alveoli (Fig. 1.32).

The **nerves of the parietal pleura** derive from the intercostal and phrenic nerves. The *costal pleura* and the peripheral part of the *diaphragmatic pleura* are supplied by the **intercostal nerves.** They mediate the sensations of touch and pain. The central part of the diaphragmatic pleura and the *mediastinal pleura* is supplied by the **phrenic nerves** (Fig. 1.27).

Injury to the Pleurae

The visceral pleura is insensitive to pain because it receives no nerves of general sensation. The parietal pleura, particularly the costal pleura, is extremely sensitive to pain because it is richly supplied by branches of the intercostal and phrenic nerves. Irritation of the parietal pleura produces local pain and referred pain to the areas supplied by the same segments of the spinal cord. Irritation of the costal and peripheral parts of the diaphragmatic pleura results in local pain and referred pain along the intercostal nerves to the thoracic and abdominal walls. Irritation of the mediastinal and central diaphragmatic areas of parietal pleura results in referred pain to the root of the neck and over the shoulder (C3 through C5 dermatomes). ⊕

Surface Anatomy of the Pleurae and Lungs

The cervical pleurae and apices of the lungs pass through the superior thoracic aperture into the **supraclavicular fossae**, which are located superior and posterior to the clavicles and lateral to the tendons of the sternocleidomastoid muscles (Fig. 1.27). The anterior borders of the lungs lie adjacent to the anterior line of reflection of the parietal pleura as far inferiorly as the 4th costal cartilages. Here the margin of the left pleural reflection moves laterally and then inferiorly at the cardiac notch to reach the 6th costal cartilage. The anterior border of the left lung is more deeply indented by its cardiac notch. On the right side, the pleural reflection continues inferiorly from the 4th to the 6th costal cartilage, paralleled closely by the anterior border of the right lung. Both pleural reflections and anterior lung borders pass laterally at the 6th costal cartilages. The pleural reflections reach the *MCL* at the level of the 8th costal cartilage, the 10th rib at the *MAL*, and the 12th rib at the *scapular line*; however, the inferior margins of the lungs reach the MCL at the level of the 6th rib, the MAL at the 8th rib, and the scapular line at the 10th rib, proceeding toward the spinous process of T10 vertebra. They then proceed toward the spinous process of T12 vertebra. Thus, the parietal pleura generally extends approximately two ribs inferior to the ▶

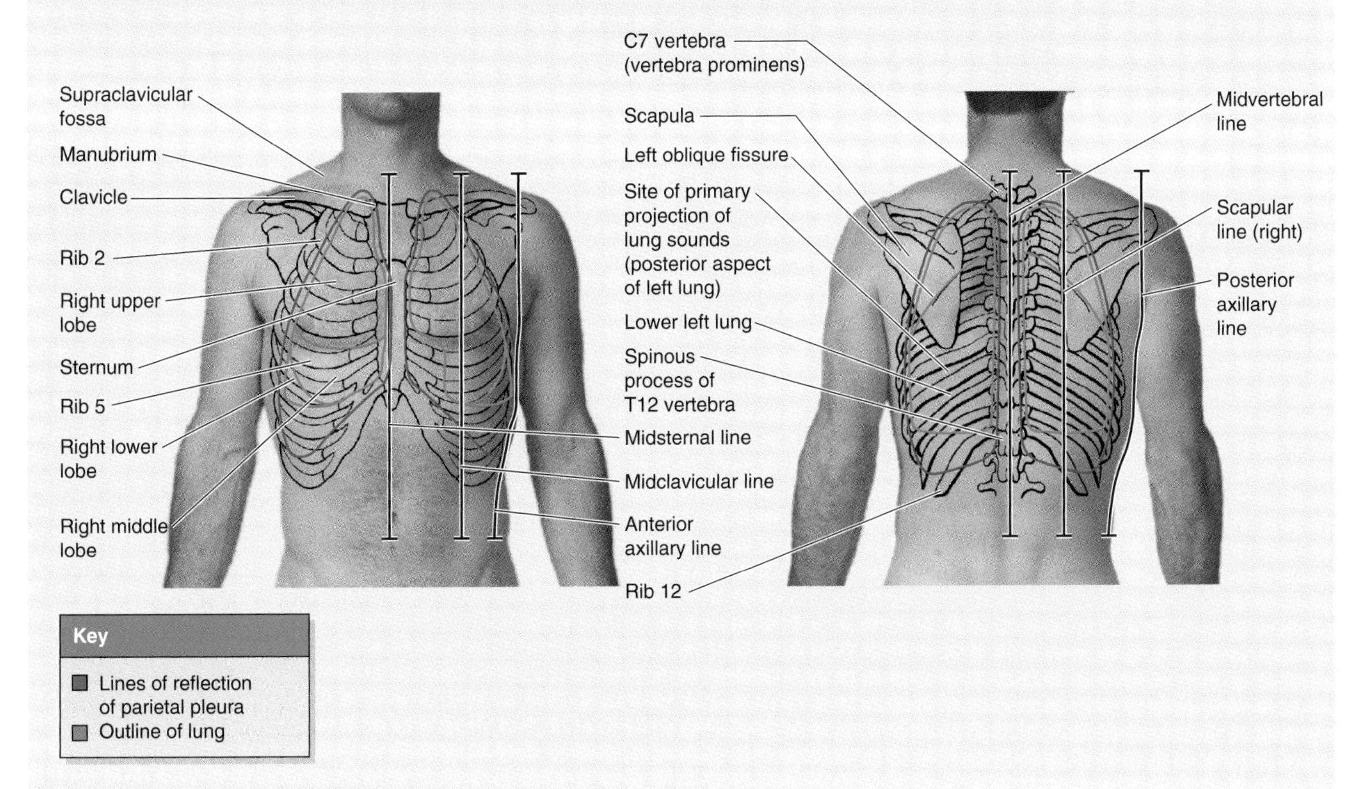

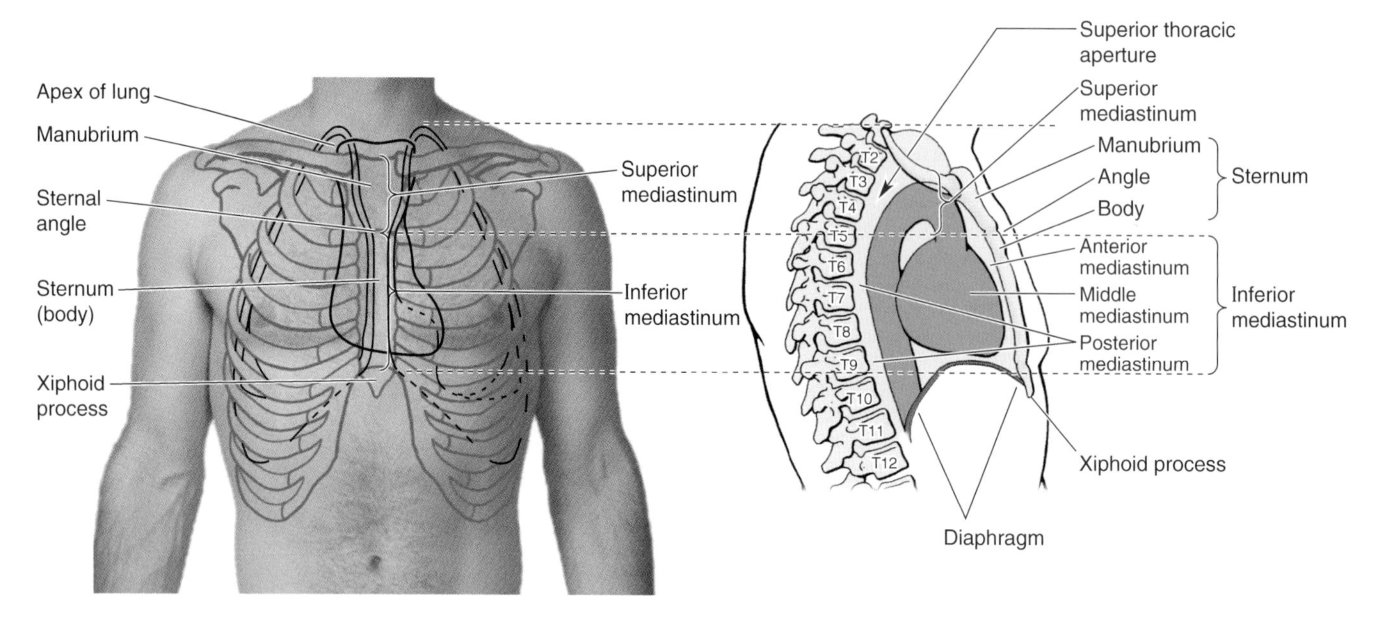

Figure 1.37. Heart and middle mediastinum. The middle mediastinum consists of the fibrous pericardium and its contents (heart and great vessels entering and leaving it). The subdivisions of the mediastinum are illustrated in the drawing on the left. The level of the viscera relative to the subdivisions depends on a person's position. Those illustrated pertain to a person who is lying in the supine position. When standing erect, the division between the superior and inferior mediastina, which passes through the sternal angle and the intervertebral (IV) disc of vertebrae T4 and T5, transects the ascending aorta and tracheal bifurcation. When erect, the division transects the arch of the aorta and the tracheal bifurcation lies below the division. This change occurs because the pericardium and its contents sag inferiorly when one stands.

▶ lung. The *oblique fissure* of the lungs extends from the level of the spinous process of T2 vertebra posteriorly to the 6th costal cartilage anteriorly, which coincides approximately with the vertebral border of the scapula when the upper limb is elevated above the head (causing the inferior angle to be rotated laterally). The *horizontal fissure* of the right lung extends from the oblique fissure along the 4th rib and costal cartilage anteriorly. ⊙

Mediastinum

The mediastinum (Mod. L. middle septum)—occupied by the mass of tissue between the two pulmonary cavities—is *the central compartment of the thoracic cavity* (Fig. 1.37). It is covered on each side by mediastinal pleura and contains all the thoracic viscera and structures except the lungs. The mediastinum extends from the superior thoracic aperture to the diaphragm inferiorly and from the sternum and costal cartilages anteriorly to the bodies of the thoracic vertebrae posteriorly. The mediastinum in living persons is a highly mobile region because it consists primarily of hollow (liquid or air-filled) visceral structures united only by loose connective tissue, often infiltrated with fat. The major structures in the mediastinum are also surrounded by blood and lymphatic vessels, lymph nodes, nerves, and fat. The looseness of the connective tissue and the elasticity of the lungs and parietal pleura on each side of the mediastinum enable it to accommodate movement, volume, and pressure changes in the thoracic cavity—such as those resulting from movements of the diaphragm, thoracic wall, and tracheobronchial tree during respiration, pulsations of the great arteries, and movement of the lungs and heart. This connective tissue becomes more fibrous and rigid with age; hence, the mediastinal structures become less mobile.

Divisions of the Mediastinum

The mediastinum is artificially divided into superior and inferior parts for purposes of description. The **superior mediastinum** extends inferiorly from the superior thoracic aperture to the horizontal plane (often referred to as the transverse thoracic plane), which includes the **sternal angle** anteriorly and passes approximately through the junction (IV disc) of T4 and T5 vertebrae posteriorly (Fig. 1.37). The **inferior mediastinum** between this plane and the diaphragm is further subdivided by the pericardium into the anterior, middle, and posterior parts. *The middle mediastinum contains the heart and great vessels.* Some structures—the esophagus, for example—pass vertically through the mediastinum and therefore lie in more than one mediastinal compartment.

Levels of the Viscera Relative to the Mediastinal Divisions

The division between the superior and inferior mediastinum—the *transverse thoracic plane*—is defined in terms of bony body wall structures. The level of the viscera relative to the subdivisions of the mediastinum depend on the position of the person. When a person is lying supine—or when one is dissecting a cadaver—the level of the viscera relative to the subdivisions of the mediastinum are as shown in Figure 1.37. Anatomical descriptions traditionally describe the level of the viscera as if the person were in this position, that is, lying in bed or on the operating or dissection table. In this position, the abdominal viscera push the mediastinal structures superiorly. However, when standing or sitting erect, the levels of the viscera are as shown in the drawing below (*B*). This occurs because the soft structures in the mediastinum (especially the pericardium and its contents), the heart and great vessels, and the abdominal viscera supporting them sag inferiorly under the influence of gravity. *In the supine position,*

- The arch of the aorta lies superior to the transverse thoracic plane
- The bifurcation of the trachea is transected by the plane
- The central tendon of the diaphragm (or the diaphragmatic surface or inferior extent of the heart) lies at the level of the xiphisternal junction and vertebra T9.

When standing or sitting erect,

- The arch of the aorta is transected by the transverse thoracic plane
- The tracheal bifurcation lies inferior to the plane
- The central tendon of the diaphragm may fall to the level of the middle of the xiphoid process and the T9/T10 IV disc.

This movement of mediastinal structures must be considered during physical and radiological examinations.

Mediastinoscopy and Mediastinal Biopsies

Using a *mediastinoscope,* surgeons can see much of the mediastinum and conduct minor surgical procedures. They insert this tubular, lighted instrument through a small incision at the root of the neck, just superior to the manubrium near the jugular notch. During mediastinoscopy, surgeons can view or biopsy mediastinal lymph nodes to determine if cancer cells have metastasized to them from a bronchogenic carcinoma, for example. The mediastinum can also be explored and biopsies taken by removing part of a costal cartilage.

Widening of the Mediastinum

Radiologists and emergency physicians sometimes observe widening of the mediastinum when viewing chest radiographs (Sauerland, 1994). Any structure in the mediastinum may contribute to pathological widening. It is often observed after trauma resulting from a head-on collision, for example, which produces hemorrhage into the mediastinum from lacerated great vessels such as the aorta or SVC. Frequently, *malignant lymphoma* (cancer of lymphatic tissue) produces massive enlargement of mediastinal nodes and widening of the mediastinum. Enlargement (hypertrophy) of the heart (occurring with congestive heart failure) is a common cause of widening of the inferior mediastinum. ⊙

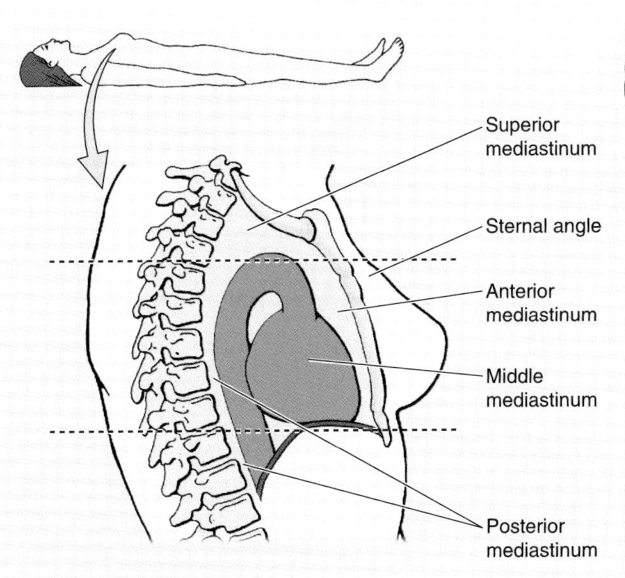

(A) Supine position

(B) Standing position

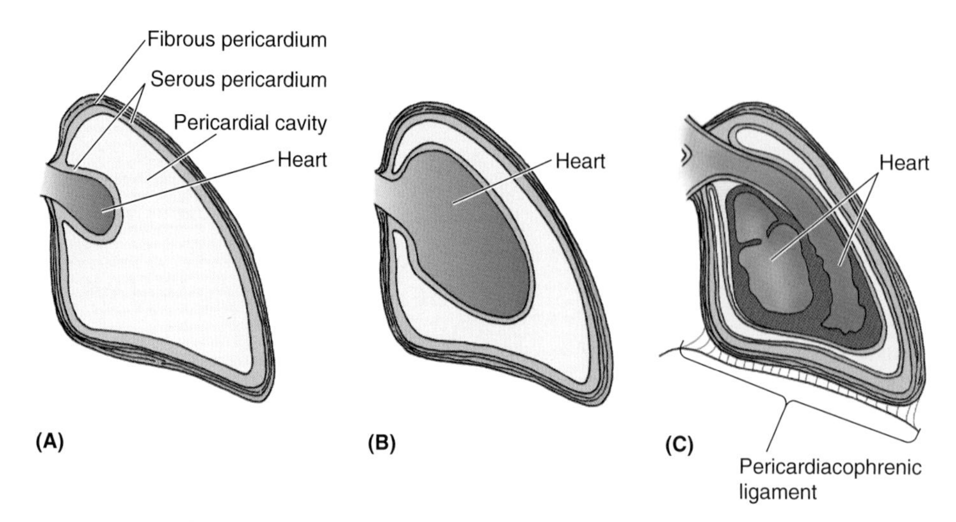

Figure 1.38. Pericardium and heart. A. The heart occupies the middle mediastinum and is enclosed by the pericardium, composed of two parts. The tough, outer fibrous pericardium stabilizes the heart and helps to prevent it from overdilating. Within the fibrous pericardium is a double-layered sac, the serous pericardium. The developing heart invaginates the wall of the serous sac (**B**) and practically obliterates the pericardial cavity (**C**), leaving only a potential space. The pericardium is bound to the central tendon of the diaphragm by the pericardiacophrenic ligament.

Pericardium

The middle mediastinum contains the pericardium, heart, and roots of the great vessels (Fig. 1.37)—ascending aorta, pulmonary trunk, and SVC—passing to and from the heart. The pericardium is a *double-walled fibroserous sac* that encloses the heart and the roots of its great vessels (Figs. 1.30 and 1.38, Table 1.4). The tough external fibrous layer of the sac—the **fibrous pericardium**—is bound to the central tendon of the diaphragm by the **pericardiacophrenic ligament.** Anteriorly, the fibrous pericardium is attached to the sternum by the *sternopericardial ligaments,* which are highly variable in their development. Posteriorly, the fibrous pericardium is bound by loose connective tissue to structures in the posterior mediastinum. Thus the heart is relatively tethered in place inside this fibrous sac. The internal surface of the fibrous pericardium is lined with a glistening serous membrane, the **parietal layer of serous pericardium.** This layer is reflected onto the heart at the great vessels (aorta, pulmonary trunk and veins, and venae cavae) as the **visceral layer of serous pericardium.** The serous pericardium is composed mainly of mesothelium, a single layer of flattened cells forming an epithelium that lines both the internal surface of the fibrous pericardium and the external surface of the heart.

The pericardium is influenced by movements of the heart and great vessels, the sternum, and diaphragm because the fibrous pericardium is

- Fused with the tunica adventitia of the great vessels entering and leaving the heart
- Attached to the posterior surface of the sternum by sternopericardial ligaments
- Fused with the central tendon of the diaphragm.

The fibrous pericardium protects the heart against sudden overfilling because it is so unyielding and closely related to the great vessels that pierce it superiorly. The ascending aorta carries the pericardium superiorly beyond the heart to the level of the sternal angle.

The **pericardial cavity** is the potential space between opposing layers of the parietal and visceral layers of serous pericardium. It normally contains a thin film of fluid that enables the heart to move and beat in a frictionless environment. The **parietal layer of serous pericardium** fuses to the internal surface of the fibrous pericardium. The **visceral layer of serous pericardium** forms the *epicardium*—the external layer of the heart wall—and reflects from the heart and great vessels to become continuous with the parietal layer of serous pericardium, where

- The aorta and pulmonary trunk leave the heart; a digit can be inserted into the *transverse pericardial sinus* located posterior to these large vessels and anterior to the SVC
- The SVC, IVC, and pulmonary veins enter the heart; these vessels are partly covered by serous pericardium that forms the *oblique pericardial sinus.*

These sinuses form during development of the heart as a consequence of the folding of the primordial heart tube. As the heart tube folds, its venous end moves posterosuperiorly (Fig. 1.39) so that the venous end of the tube lies adjacent to the arterial end, separated only by the **transverse pericardial sinus**—a transversely running passage in the pericardial sac

Table 1.4. **Layers of Pericardium and Heart**

Pericardium
External sac called fibrous pericardium
Internal sac called serous pericardium: Parietal layer—lines fibrous pericardium Visceral layer (becomes outermost layer of wall of heart, the epicardium*) Parietal and visceral layers are continuous around roots of great vessels.
The film of fluid in pericardial cavity between visceral and parietal layers of serous pericardium allows heart to move freely within pericardial sac.
Heart
Wall of heart is composed of three layers, from superficial to deep: *Epicardium Myocardium Endocardium

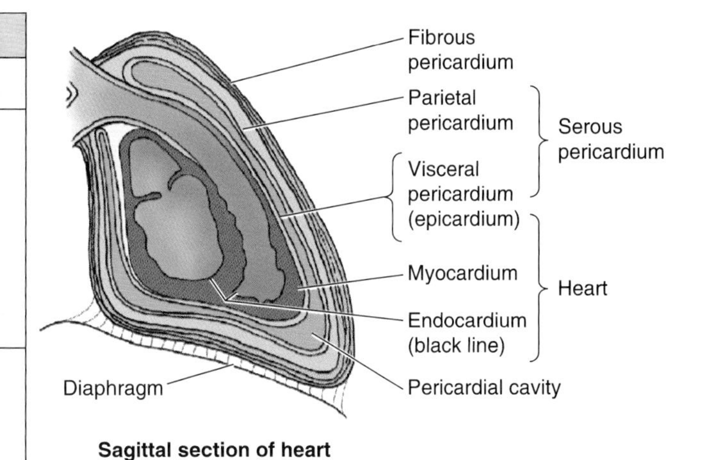

Sagittal section of heart

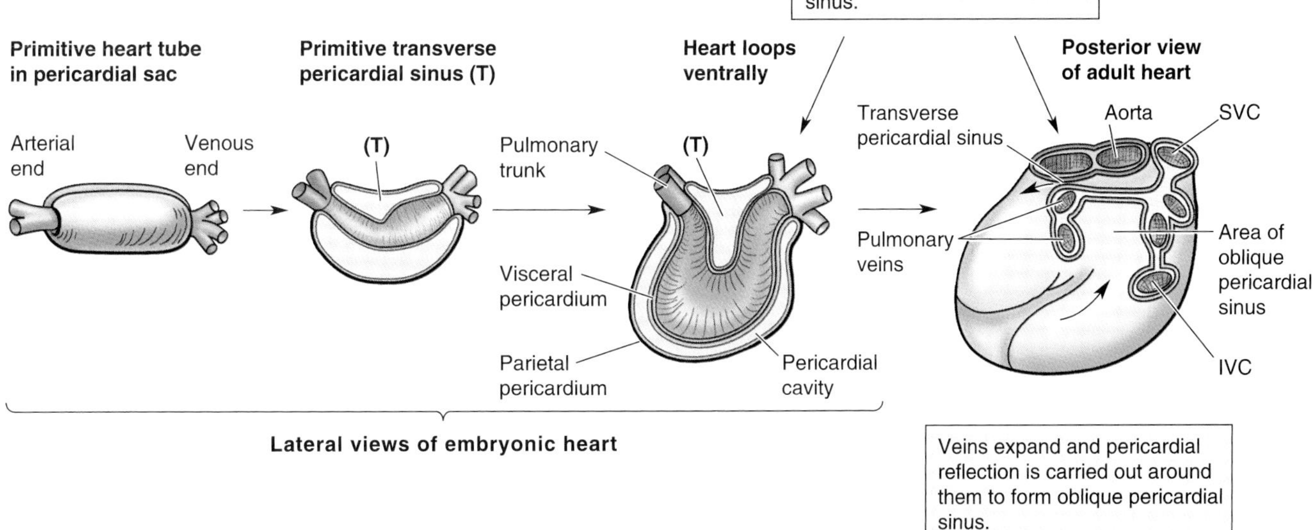

Figure 1.39. Development of the heart and pericardium. The embryonic heart tube invaginates the double-layered pericardial sac (somewhat like placing a wiener in a hotdog bun). The primordial heart then loops ventrally, bringing the primitive arterial and venous ends of the heart together and creating the transverse pericardial sinus (*T*) between them. The veins expand and move apart. The pericardium is reflected around them to form the boundaries of the oblique pericardial sinus.

between the origins of the great vessels (Fig. 1.40). The transverse sinus is posterior to the intrapericardial parts of the pulmonary trunk and ascending aorta and anterior to the SVC and superior to the atria of the heart. As the veins of the heart develop and expand, a pericardial reflection surrounding them forms the **oblique pericardial sinus**, a wide pocketlike recess in the pericardial cavity posterior to the base (posterior aspect) of the heart. The oblique sinus is bounded laterally by the pericardial reflections surrounding the pulmonary veins and IVC and posteriorly by the pericardium overlying the anterior aspect of the esophagus. The oblique sinus can be entered inferiorly and will admit several digits; however, the digits cannot pass around any of these structures because the sinus is a blind sac (cul-de-sac).

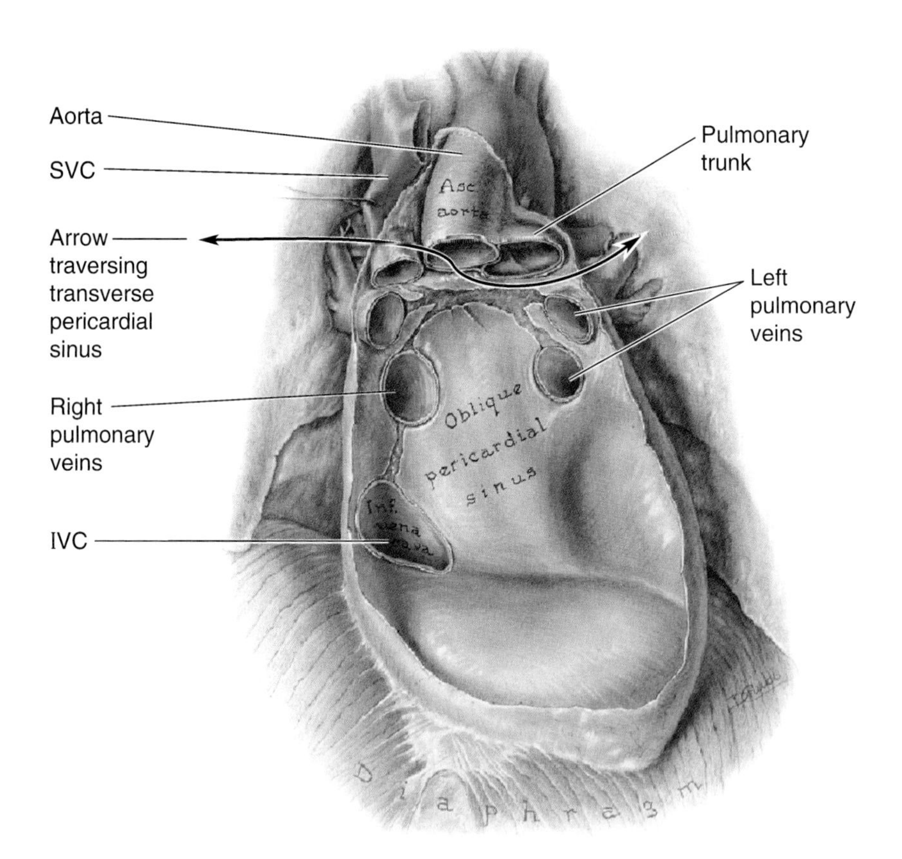

Figure 1.40. Interior of the pericardial sac. To remove the heart from the sac, the eight vessels piercing it were severed. Observe that the oblique pericardial sinus is circumscribed by five veins and that the superior vena cava (SVC) is partly inside and mostly outside the pericardium. Also observe that the peak of the pericardial sac is near the junction of the ascending aorta and the arch of the aorta. Note that the transverse pericardial sinus is bounded anteriorly by the serous pericardium covering the posterior aspect of the pulmonary trunk and ascending aorta, posteriorly by that covering the SVC, and inferiorly by the visceral pericardium covering the atria of the heart.

Surgical Significance of the Transverse Pericardial Sinus

The transverse pericardial sinus is especially important to cardiac surgeons. After the pericardial sac is opened anteriorly, a digit can be passed through the transverse pericardial sinus posterior to the aorta and pulmonary trunk. By passing a surgical clamp or placing a ligature around these vessels, inserting the tubes of a coronary bypass machine, and then tightening the ligature, surgeons can stop or divert the circulation of blood of these large arteries while performing cardiac surgery, such as *coronary artery bypass grafting*. Cardiac surgery is performed while the patient is on cardiopulmonary bypass (Sabiston and Lyerly, 1994).

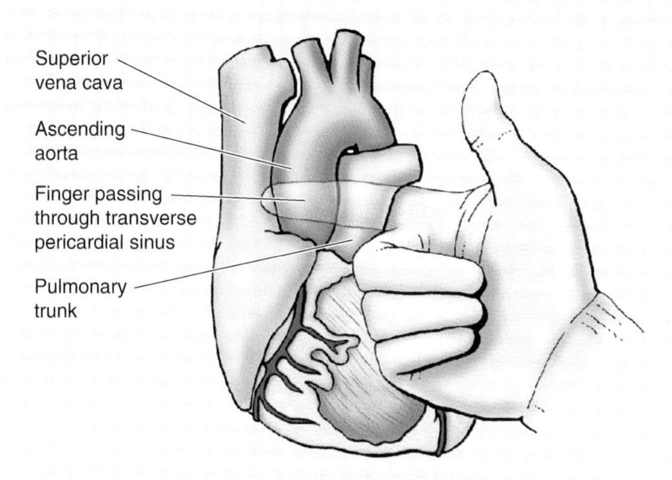

Exposure of the IVC and SVC

After passing through the diaphragm, the entire thoracic part of the IVC (approximately 2 cm) is within the pericardium. Consequently, the pericardial sac must be opened to expose the superior part of the IVC. The same is true for the terminal part of the SVC, which is partly inside and partly outside the pericardial sac.

Pericarditis, Pericardial Effusion, and Cardiac Tamponade

The pericardium may be involved in several disease processes. Inflammation of the pericardium (*pericarditis*) usually causes chest pain, and certain inflammatory diseases may produce *pericardial effusion* (passage of fluid from pericardial capillaries into the pericardial cavity). As a result, the heart becomes compressed (unable to expand and fill fully) and ineffective. A chronically inflamed and thickened pericardium may actually calcify and seriously hamper cardiac efficiency (Sauerland, 1994). Noninflammatory pericardial effusions often occur with *congestive heart failure*—inadequacy of a heart that fails to pump out blood at the same rate that it receives it (i.e., maintain blood circulation).

Usually the opposing layers of serous pericardium make no detectable sound during auscultation. However, pericarditis makes the surfaces rough and the resulting friction (*pericardial friction rub*) sounds like the rustle of silk when listening with a stethoscope. If extensive pericardial effusion exists, the excess pericardial fluid does not allow the heart to expand fully, thereby limiting the inflow of blood to the ventricles. This phenomenon—*cardiac tamponade* (heart compression)—is a potentially lethal condition because the fibrous pericardium is tough and inelastic. Consequently, heart volume is increasingly compromised by fluid outside the heart but inside the pericardial cavity.

Stab wounds that pierce the heart cause blood to enter the pericardial cavity—*hemopericardium*—likewise producing cardiac tamponade. Hemopericardium may also result from perforation of a weakened area of heart muscle following a heart attack (*myocardial infarction [MI]*). Cardiac tamponade may also result from bleeding into the pericardial cavity following cardiac operations. When the fluid in the pericardial cavity accumulates rapidly, the heart is compressed and circulation fails. The veins of the face and neck become engorged because of the backup of blood beginning at the SVC where it enters the pericardium.

Pericardiocentesis (drainage of fluid from the pericardial cavity) is usually necessary to relieve cardiac tamponade. To remove the excess fluid, a wide-bore needle may be inserted through the left 5th or 6th intercostal space near the sternum. This approach to the pericardial sac is possible because the cardiac notch in the left lung and the shallower notch in the left pleural sac leaves part of the pericardial sac exposed—the "bare area" of the pericardium (Figs. 1.26*A* and 1.27). The pericardial sac may also be reached by entering the infrasternal angle and passing the needle superoposteriorly. At this site, the needle avoids the lung and pleurae and enters the pericardial cavity; however, *care must be taken not to puncture the internal thoracic artery*. In patients with *pneumothorax*—air or gas in the pleural cavity—the air may dissect along connective tissue planes and enter the pericardial sac, producing a *pneumopericardium* that can be demonstrated radiographically (Sauerland, 1994). ⭘

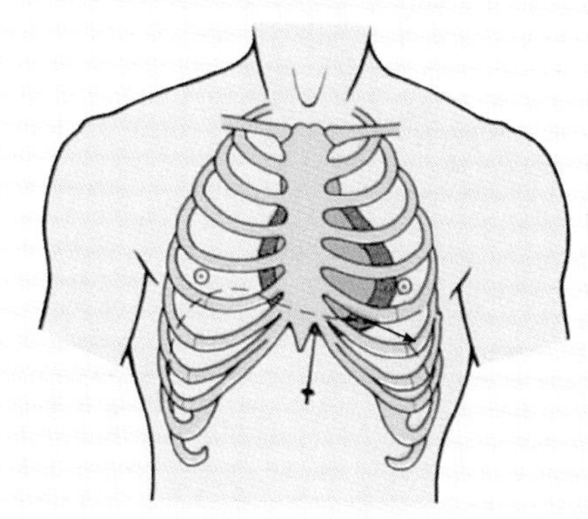

Pericardiocentesis

The **arterial supply of the pericardium** (Fig. 1.41) is mainly from a slender branch of the internal thoracic artery, the **pericardiacophrenic artery**, that often accompanies or at least parallels the phrenic nerve to the diaphragm. Smaller contributions of blood come from the

- *Musculophrenic artery*, a terminal branch of the internal thoracic artery
- *Bronchial, esophageal*, and *superior phrenic* arteries—branches of the thoracic aorta
- *Coronary arteries* (visceral layer of serous pericardium only).

The **venous drainage of the pericardium** (Fig. 1.41) is from the

- *Pericardiacophrenic veins*, tributaries of the brachiocephalic (or internal thoracic) veins
- Variable tributaries of the *azygos venous system.*

The **nerve supply of the pericardium** (Fig. 1.41) is from the

- *Phrenic nerves* (C3 through C5)—primary source of sensory fibers. Pain sensations conveyed by the phrenic nerves are commonly referred to the skin of the ipsilateral supraclavicular region (top of shoulder on the same side).
- *Vagus nerves* —function uncertain.
- *Sympathetic trunks*—vasomotor.

The Heart and Great Vessels

The heart, slightly larger than a clenched fist, is a double self-adjusting muscular pump, the parts of which work in unison to propel blood to all parts of the body. The right side of the heart receives poorly oxygenated ("venous") blood from the body through the SVC and IVC and pumps it through the pulmonary trunk to the lungs for oxygenation. The left side receives well-oxygenated ("arterial") blood from the lungs through the pulmonary veins and pumps it into the aorta for distribution to the body (Fig. 1.42*A*). The heart has four chambers: **right** and **left atria** and **right** and **left ventricles.** The atria are receiving chambers that pump blood into the ventricles—the discharging chambers.

The synchronous pumping actions of the heart's two atrioventricular (AV) pumps (right and left chambers) constitute the **cardiac cycle** (Fig. 1.42, *B–F*). The cycle begins with a period of ventricular relaxation (**diastole**) and ends with a period of ventricular contraction (**systole**). Two heart sounds are heard with a stethoscope: a "lub" sound as the atria transfer blood to the ventricles and a "dub" sound as the ventricles contract and propel blood from the heart. The heart sounds are produced by the snapping shut of the one-way valves that normally keep blood from flowing backward during contractions of the heart.

The wall of each heart chamber consists of three layers (Table 1.4):

- **Endocardium,** a thin internal layer (endothelium and subendothelial connective tissue) or lining membrane of the heart that also covers its valves

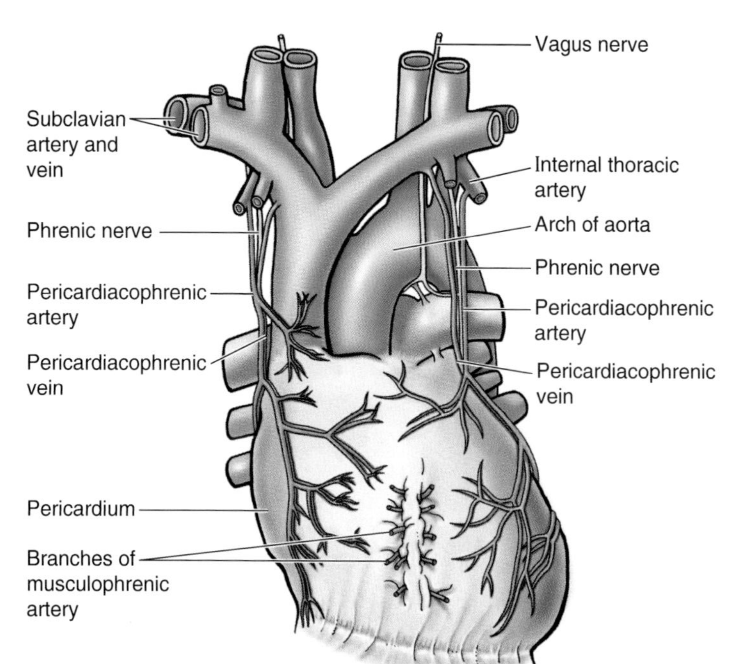

Figure 1.41. Arterial supply and venous drainage of the pericardium. The arteries of the pericardium derive from the internal thoracic arteries and their musculophrenic branches and from the thoracic aorta. The pericardiacophrenic arteries, slender branches of the internal thoracic arteries, accompany or parallel the phrenic nerves to the diaphragm. The veins are tributaries of the brachiocephalic veins. The vagus (CN X) and phrenic nerves are also shown.

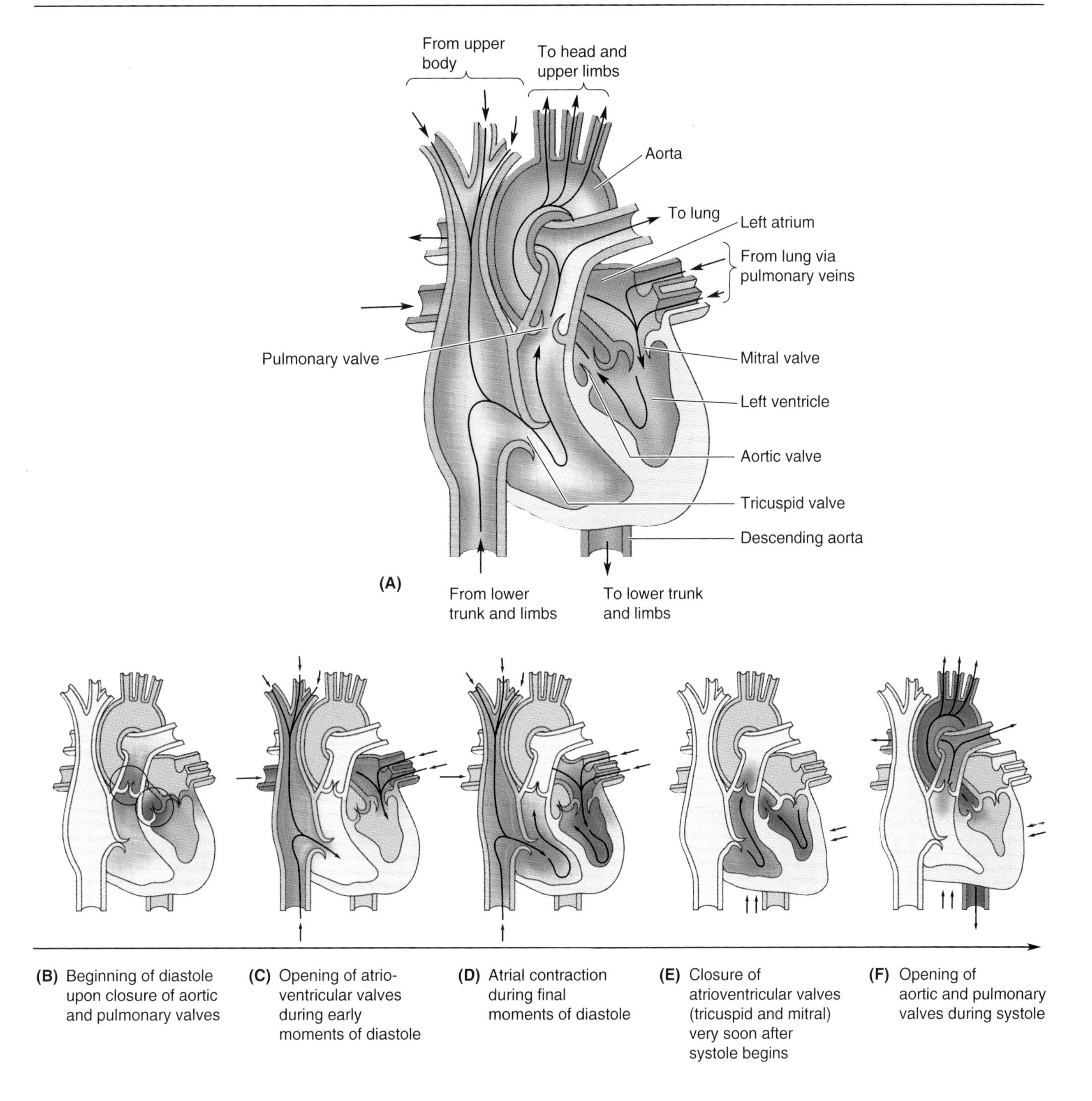

Figure 1.42. Cardiac cycle. The cardiac cycle describes the complete movement of the heart or heartbeat and includes the period from the beginning of one heartbeat to the beginning of the next one. The cycle consists of diastole (ventricular filling) and systole (ventricular emptying).

- **Myocardium,** a thick middle layer composed of cardiac muscle
- **Epicardium,** a thin external layer (mesothelium) formed by the visceral layer of serous pericardium.

The walls of the heart consist mostly of thick myocardium, especially in the ventricles.

When the ventricles contract, they produce a wringing motion because of the spiral orientation of the cardiac muscle fibers. This motion propels the blood from the heart. The muscle fibers are anchored to the *fibrous skeleton of the heart.* This is a complex framework of dense collagen forming four **fibrous rings** (L. annuli fibrosi), which surround the orifices of the valves, a right and left **fibrous trigone**, formed by con-

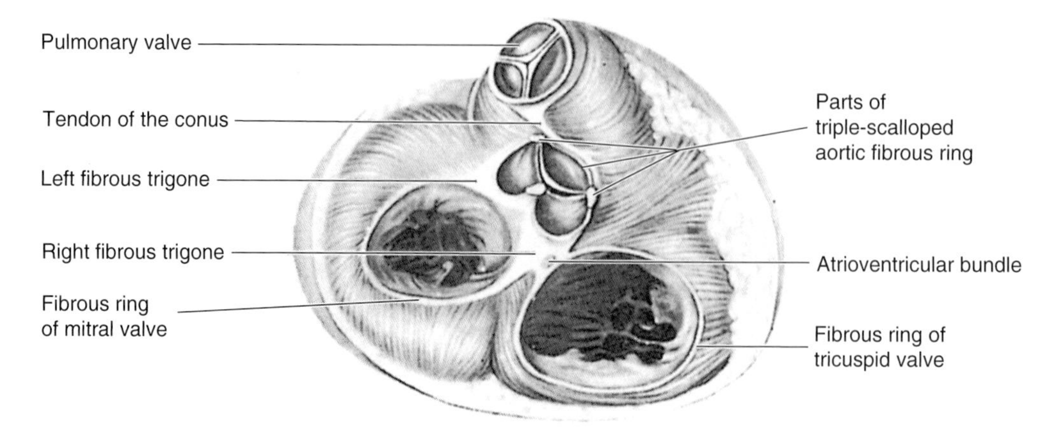

Figure 1.43. Fibrous skeleton of the heart. Posterior view. The atria have been removed at the atrioventricular groove, leaving the ventricles and valves. The fibrous skeleton (white) is composed of four fibrous rings—each encircling a valve, two trigones, and the membranous portions of the interatrial and interventricular septa (the latter are not shown).

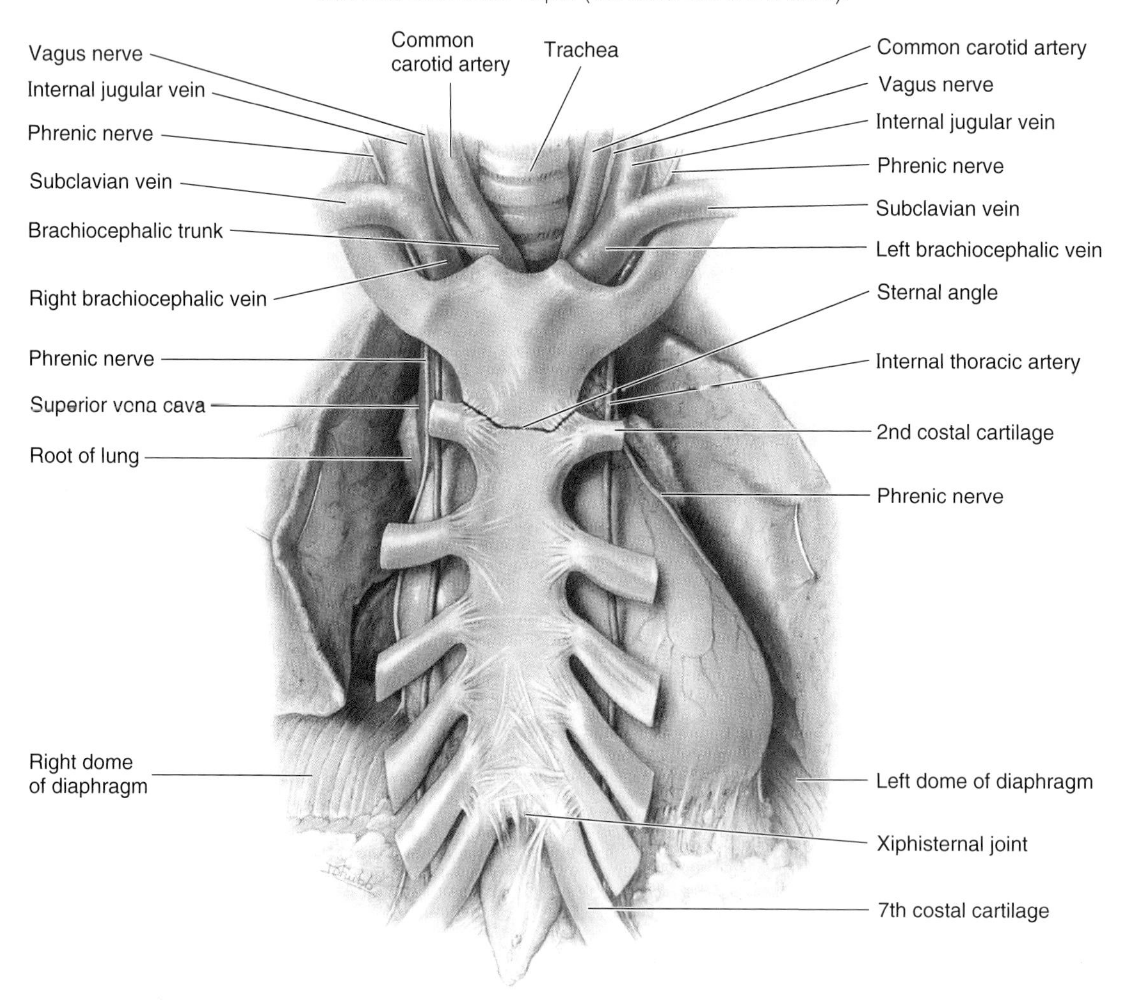

Figure 1.44. Pericardial sac in relation to the sternum. This dissection exposes the pericardial sac posterior to the body of the sternum from just superior to the sternal angle to the level of the xiphisternal joint. Observe that the pericardial sac lies approximately one-third to the right of the midsternal line and two-thirds to the left.

necting the rings, and the membranous parts of the interatrial and interventricular septa. The **fibrous skeleton of the heart** (Fig. 1.43):

- Keeps the orifices of the AV and semilunar valves patent and from being overly distended by the volume of blood pumping through them
- Provides attachments for the leaflets and cusps of the valves

- Provides attachment (origin and insertion) for the myocardium
- Forms an electrical "insulator," by separating the myenterically conducted impulses of the atria and ventricles so that they contract independently, and by surrounding and providing passage for the initial part of the AV bundle.

The heart and roots of the great vessels within the pericardial sac are related anteriorly to the sternum, costal cartilages, and the medial ends of the 3rd through 5th ribs on the left side (Fig. 1.44). The heart and pericardial sac are situated obliquely, approximately two-thirds to the left and one-third to the right of the median plane. The heart is shaped like a tipped-over, three-sided pyramid with an *apex*, now "on the ground," directed anteriorly and to the left; a *base*, opposite the apex, now facing mostly posteriorly; and three sides: the *diaphragmatic surface* on which the pyramid is now resting, the anterior *sternocostal surface*, and the *pulmonary surface*, now facing left. The **base of the heart** (Fig. 1.45, *A* and *B*):

- Is the heart's posterior aspect (opposite the apex) as it lies in the thorax

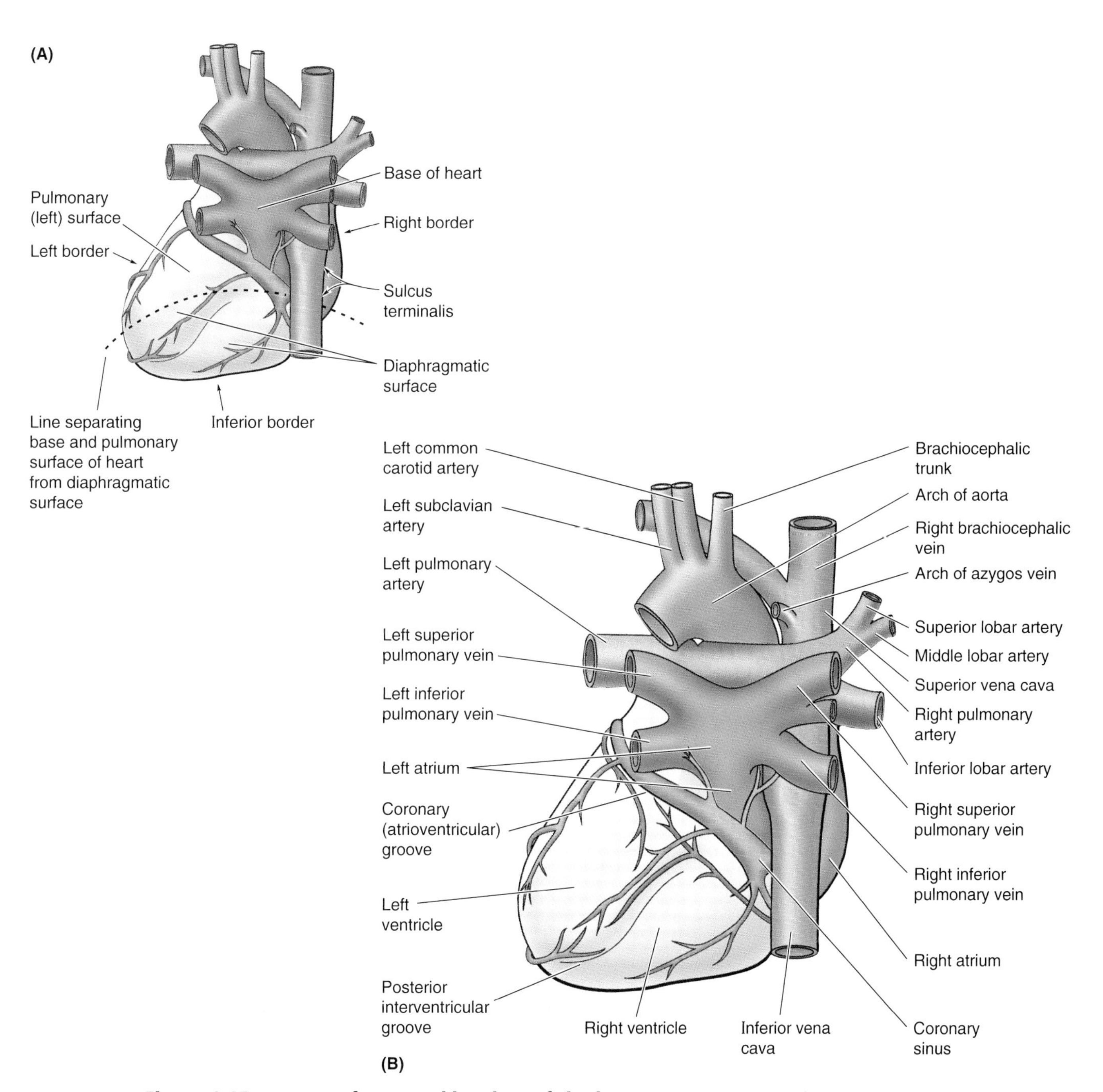

Figure 1.45. Parts, surfaces, and borders of the heart. The relationship of the great vessels of the heart are also shown. **A–B.** Pulmonary (left) surface of the heart and great vessels, including the base and diaphragmatic surface of the heart.

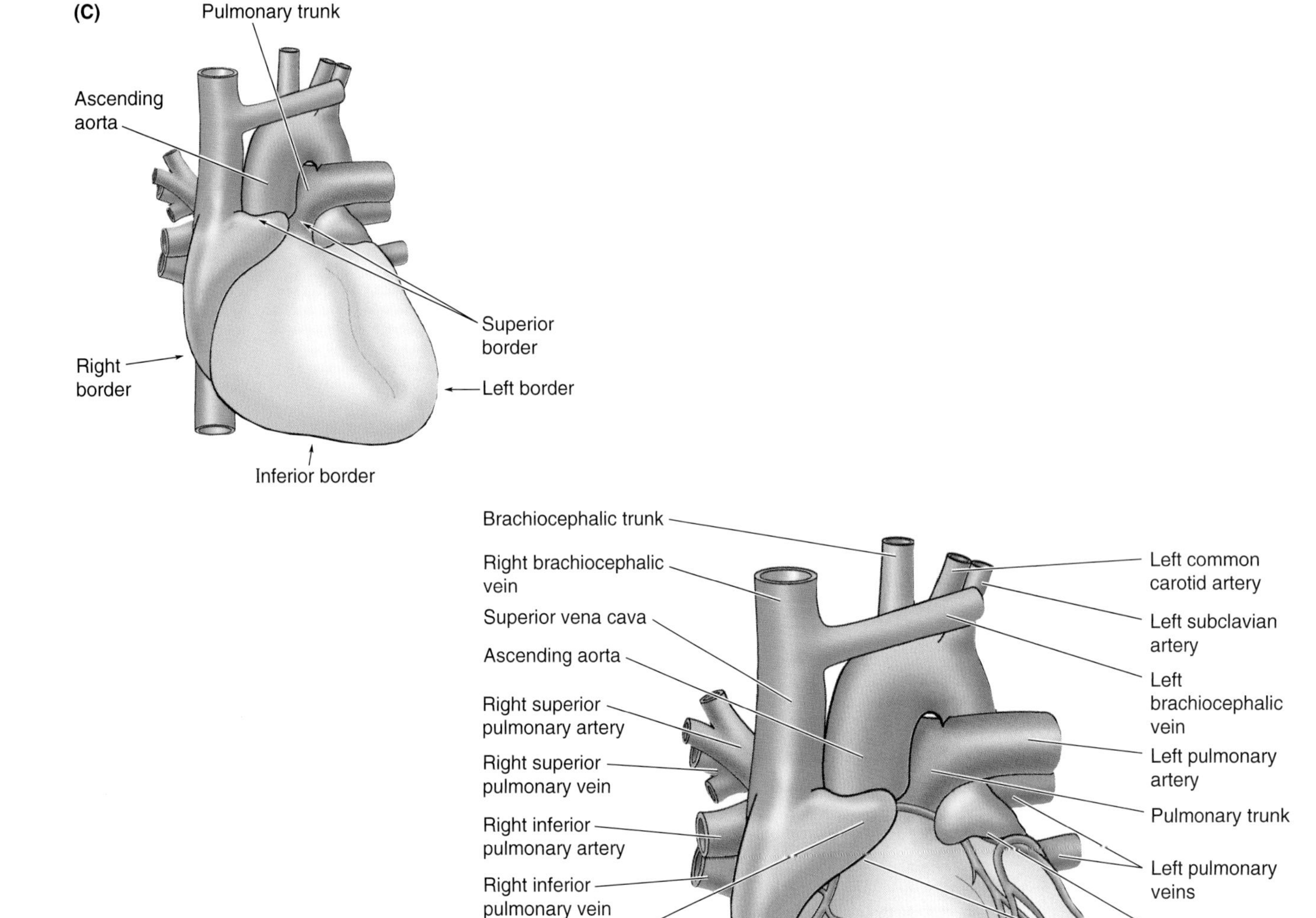

Figure 1.45. *(Continued)* **C–D.** Sternocostal surface of the heart and great vessels. The entire right auricle and much of the right atrium are visible, but only a small part of the left auricle is shown.

- Is formed mainly by the left atrium, with a lesser contribution by the right atrium
- Faces posteriorly toward the bodies of vertebrae T6 through T9, and is separated from them by the pericardium, oblique pericardial sinus, esophagus, and aorta
- Extends superiorly to the bifurcation of the pulmonary trunk and inferiorly to the coronary or AV groove (sulcus)
- Receives the pulmonary veins on the right and left sides of its left atrial portion and the superior and inferior venae cavae at the upper and lower ends of its right atrial portion.

The **apex of the heart** (Fig. 1.45*D*):

- Is formed by the inferolateral part of the left ventricle
- Lies posterior to the left 5th intercostal space in adults, usually 9 cm (or approximately the breadth of the person's hand) from the median plane

- Is where maximal pulsation of the heart (**apex beat**) occurs (i.e., underlies the site where the "heartbeat" may be observed or palpated on the thoracic [chest] wall).

The **three surfaces of the heart** (Fig. 1.45, *A–D*) are the:

- *Anterior (sternocostal) surface*, formed mainly by the right ventricle
- *Diaphragmatic (inferior) surface*, formed mainly by the left ventricle and partly by the right ventricle; it is related mainly to the central tendon of the diaphragm
- *Pulmonary (left) surface,* formed mainly by the left ventricle; it occupies the cardiac impression of the left lung.

The **four borders of the heart** are visible in both anterior and posterior views (Fig. 1.45, *A* and *C*):

- *Right border* (slightly convex), formed by the right atrium and extending between the SVC and IVC
- *Inferior border* (oblique, nearly vertical), formed mainly by the right ventricle and slightly by the left ventricle
- *Left border* (nearly horizontal), formed mainly by the left ventricle and slightly by the left auricle
- *Superior border*, formed by the right and left atria and auricles in an anterior view; the ascending aorta and pulmonary trunk emerge from the superior border, and the SVC enters its right side. Posterior to the aorta and pulmonary trunk and anterior to the SVC, the superior border forms the inferior boundary of the transverse pericardial sinus.

The **pulmonary trunk**, approximately 5 cm long and 3 cm wide, is the arterial continuation of the right ventricle and divides into right and left pulmonary arteries. The pulmonary trunk and arteries conduct poorly oxygenated blood to the lungs for oxygenation (Fig. 1.42*A*).

Right Atrium. This chamber forms the right border of the heart and receives venous blood from the SVC, IVC, and coronary sinus (Fig. 1.45, *B* and *D*). The earlike *right auricle* is a conical muscular pouch that projects from the right atrium like an "add-on" room, increasing the capacity of the atrium as it overlaps the ascending aorta. The **interior of the right atrium** (Fig. 1.46, *A* and *B*) has

- A smooth, thin-walled posterior part—the **sinus venarum**—on which the venae cavae (SVC and IVC) and coronary sinus open, bringing poorly oxygenated blood into the heart
- A rough, muscular anterior wall composed of pectinate muscles (L. musculi pectinati)
- A right AV orifice through which the right atrium discharges the poorly oxygenated blood it has received into the right ventricle.

The smooth and rough parts of the atrial wall are separated externally by a shallow vertical groove, the **sulcus terminalis** or terminal groove (Fig. 1.45*A*), and internally by a vertical ridge, the **crista terminalis** or terminal crest (Fig. 1.46*A*).

The SVC opens into the superior part of the right atrium at the level of the right 3rd costal cartilage. The IVC opens into the inferior part of the right atrium almost in line with the SVC at approximately the level of the 5th costal cartilage. The **opening of the coronary sinus**—a short venous trunk receiving most of the cardiac veins—is between the right AV orifice and the IVC orifice. The **interatrial septum** separating the atria has an oval, thumbprint-sized depression, the **oval fossa** (L. fossa ovalis), which is a remnant of the **oval foramen** (L. foramen ovale) and its valve in the fetus (Moore and Persaud, 1998).

Positional Abnormalities of the Heart

Because of abnormal folding of the embryonic heart, the position of the heart may be completely reversed so that the apex is directed to the right instead of the left—**dextrocardia** (Moore and Persaud, 1998). This congenital anomaly is the most common positional abnormality of the heart, but it is still relatively uncommon. Dextrocardia is associated with mirror image positioning of the great vessels and the arch of the aorta. This anomaly may be part of a general transposition of the thoracic and abdominal viscera (*situs inversus*), or the transposition may affect only the heart (*isolated dextrocardia*). In dextrocardia with situs inversus, the incidence of accompanying cardiac defects is low and the heart usually functions normally; however, in isolated dextrocardia the congenital anomaly is complicated by severe cardiac anomalies such as transposition of the great arteries. For a discussion of the prognosis and treatment of dextrocardia, see Behrman et al. (1996).

Percussion of the Heart

Percussion defines the density and size of the heart. The classical percussion technique is to create vibration by tapping the chest with a finger while listening and feeling for differences in sound wave conduction. Percussion is performed at the 3rd, 4th, and 5th intercostal spaces from the left anterior axillary line to the right anterior axillary line (Swartz, 1994). Normally the percussion note changes from resonance to dullness (because of the presence of the heart) approximately 6 cm lateral to the left border of the sternum. ⊙

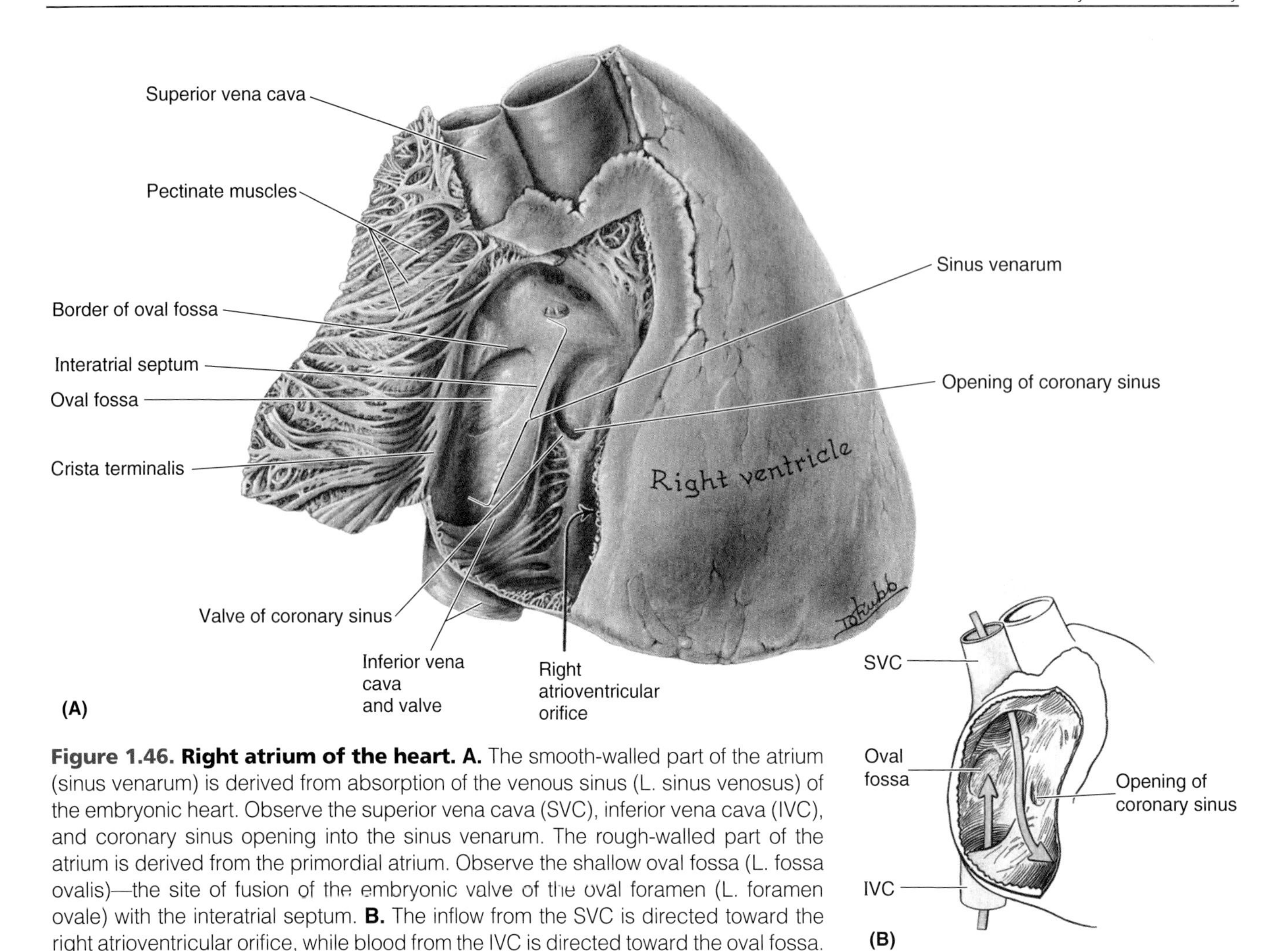

Figure 1.46. Right atrium of the heart. A. The smooth-walled part of the atrium (sinus venarum) is derived from absorption of the venous sinus (L. sinus venosus) of the embryonic heart. Observe the superior vena cava (SVC), inferior vena cava (IVC), and coronary sinus opening into the sinus venarum. The rough-walled part of the atrium is derived from the primordial atrium. Observe the shallow oval fossa (L. fossa ovalis)—the site of fusion of the embryonic valve of the oval foramen (L. foramen ovale) with the interatrial septum. **B.** The inflow from the SVC is directed toward the right atrioventricular orifice, while blood from the IVC is directed toward the oval fossa.

Embryology of the Right Atrium

Understanding the development of the right atrium makes its adult anatomy easier to comprehend and remember (Moore and Persaud, 1998). The primordial atrium is represented in the adult by the *right auricle*. The definitive atrium is enlarged by incorporation of most of the embryonic *venous sinus* (L. sinus venosus). The *coronary sinus* is also a derivative of this venous sinus. The part of the venous sinus incorporated into the primordial atrium becomes the smooth-walled **sinus venarum** of the adult right atrium (Fig. 1.46*A*). The separation between the primordial atrium—the adult auricle—and the sinus venarum—the derivative of the venous sinus—is indicated externally by the **sulcus terminalis** and internally by the **crista terminalis.**

Before birth, the valve of the IVC directs most of the oxygenated blood returning from the placenta by way of the umbilical vein and IVC toward the **oval foramen** in the interatrial septum, through which it passes into the left atrium. The oval foramen has a flaplike valve that permits a right-to-left shunt of blood but prevents a left-to-right shunt. After birth, the oval foramen normally closes as its valve fuses with the interatrial septum. The closed oval foramen is represented in the interatrial septum by the depressed **oval fossa.** The ***border of the oval fossa*** (L. limbus fossae ovalis) surrounds the fossa. The floor of the fossa is formed by the valve of the oval foramen. The rudimentary **IVC valve**, a semilunar crescent of tissue, has no function after birth; it varies considerably in size and is occasionally absent.

Atrial Septal Defects

Congenital malformations of the interatrial septum—usually in the form of incomplete closure of the oval foramen—are referred to as atrial septal defects (ASDs). A probe-sized patency appears in the superior part of the oval fossa in 15 to 25% of adults (Moore and Persaud, 1998). These small openings, by themselves, cause no hemodynamic abnormalities and are therefore of no clinical significance and should not be considered as forms of ASDs. Clinically significant ASDs vary widely in size and location ▶

▶ and may occur as part of more complex congenital heart disease (Sabiston and Lyerly, 1994). Large ASDs allow oxygenated blood from the lungs to be shunted from the left atrium through the ASD into the right atrium, causing enlargement of the right atrium and ventricle and dilation of the pulmonary trunk. This left-to-right shunt of blood overloads the pulmonary vascular system, resulting in enlargement (hypertrophy) of the right atrium and ventricle and pulmonary arteries. ○

Right Ventricle. The right ventricle forms the largest part of the anterior surface of the heart, a small part of the diaphragmatic surface, and almost the entire inferior border of the heart (Fig. 1.46). Superiorly it tapers into an arterial cone, the **conus arteriosus** (infundibulum), that leads into the pulmonary trunk (Fig. 1.47). The interior of the right ventricle has irregular muscular elevations called **trabeculae carneae**. A thick muscular ridge, the *supraventricular crest*, separates the ridged muscular wall of the inflow part of the chamber from the smooth wall of the **conus arteriosus** or outflow part. The inflow part of the ventricle receives blood from the right atrium through the **right AV orifice** (Fig. 1.46*A*), located posterior to the body of the sternum at the level of the 4th and 5th intercostal spaces. The right AV orifice is surrounded by a fibrous ring that is part of the fibrous skeleton of the heart. The fibrous ring around the orifice resists the dilation that might otherwise result from blood being forced through it.

The **tricuspid valve** (Fig. 1.48) guards the right AV orifice. The bases of the valve cusps are attached to the fibrous ring around the orifice. **Tendinous cords** (L. chordae tendineae) attach to the free edges and ventricular surfaces of the anterior, posterior, and septal cusps—much like the cords attaching to a parachute. Because the cords are attached to adjacent sides of two cusps, they prevent separation of the cusps as well as their inversion when tension is applied to the tendinous cords throughout ventricular contraction (systole); that is, the cusps of the tricuspid valve are prevented from prolapsing (being driven into the right atrium) as ventricular pressure rises. Thus, regurgitation of blood (backward flow of blood) from the right ventricle back into the right atrium is blocked by the valve cusps.

The **papillary muscles** (Fig. 1.47) form conical projections with their bases attached to the ventricular wall and tendinous cords arising from their apices. There are usually three papillary muscles (anterior, posterior, and septal) in the right ventricle that correspond in name to the cusps of the tricuspid valve.

- The **anterior papillary muscle**, the largest and most prominent of the three, arises from the anterior wall of the right ventricle. Its tendinous cords attach to the anterior and posterior cusps of the tricuspid valve.
- The **posterior papillary muscle**, smaller than the anterior muscle, may consist of several parts. The posterior papillary muscle arises from the inferior wall of the right ventricle and its tendinous cords attach to the posterior and septal cusps of the tricuspid valve.
- The **septal papillary muscle** arises from the interventricular septum and its tendinous cords attach to the anterior and septal cusps of the tricuspid valve.

The papillary muscles begin to contract before contraction of the right ventricle, tightening the tendinous cords and drawing the cusps together. Contraction is maintained throughout systole. This prevents ventricular blood from passing back (regurgitating) into the right atrium.

The **interventricular septum**—*composed of membranous and muscular parts*—is a strong, obliquely placed partition between the right and left ventricles (Fig. 1.47), forming part of the walls of each. The superoposterior part of the septum is thin and membranous and is continuous with the fibrous skeleton of the heart (Fig. 1.43), which forms the great majority of the septum. The muscular part is thick and bulges into the cavity of the right ventricle because of the higher blood pressure in the left ventricle. The **septomarginal trabecula** (moderator band) is a curved muscular bundle that runs from the inferior part of the interventricular septum to the base of the anterior papillary muscle. This trabecula is important because it carries part of the **right bundle branch of the AV bundle**, a part of the conducting system of the heart to the anterior papillary muscle (see Fig. 1.53).

The **right AV orifice** is large enough to admit the tips of three fingers (Fig. 1.46). When the right atrium contracts, blood is forced through this orifice into the right ventricle, pushing the cusps of the tricuspid valve aside like curtains. The inflow of blood into the right ventricle (*inflow tract*) enters posteriorly, and the outflow of blood into the pulmonary trunk (*outflow tract*) leaves superiorly and to the left. Consequently, the blood takes a U-shaped path through the right ventricle. The inflow (AV) orifice and outflow (pulmonary) orifice are approximately only 2 cm apart.

The **pulmonary valve** (Fig. 1.48) at the apex of the conus arteriosus is at the level of the left 3rd costal cartilage. Each of three semilunar **cusps of the pulmonary valve** (anterior, right, and left) is concave when viewed superiorly. The cusps project into the artery but lie close to its walls as blood leaves the right ventricle. Following relaxation of the ventricle (**diastole**), the elastic recoil of the wall of the pulmonary trunk forces the blood back toward the heart. However, the cusps open up like pockets as they catch the reversed bloodflow and completely close the pulmonary orifice, preventing any significant amount of blood from returning to the right ventricle. Immediately superior to each valve, the wall of the pulmonary trunk is slightly dilated to form a pulmonary sinus. The *pulmonary sinuses* are the spaces at the origin of the pulmonary trunk between the dilated wall of the vessel and each cusp of the pulmonary valve. The blood in the pulmonary sinuses prevents the cusps from sticking to the wall of the pulmonary trunk and failing to close.

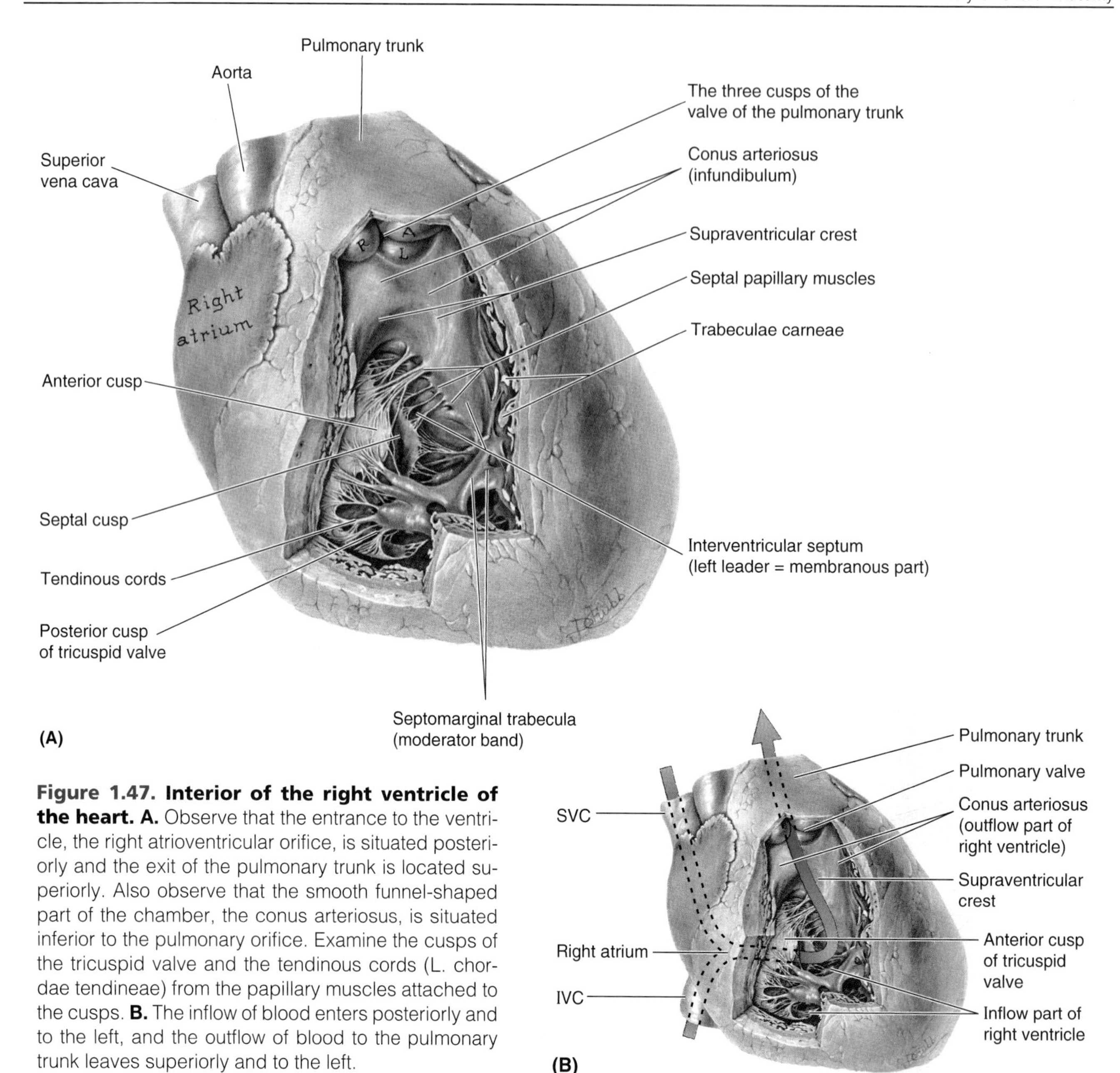

Figure 1.47. Interior of the right ventricle of the heart. A. Observe that the entrance to the ventricle, the right atrioventricular orifice, is situated posteriorly and the exit of the pulmonary trunk is located superiorly. Also observe that the smooth funnel-shaped part of the chamber, the conus arteriosus, is situated inferior to the pulmonary orifice. Examine the cusps of the tricuspid valve and the tendinous cords (L. chordae tendineae) from the papillary muscles attached to the cusps. **B.** The inflow of blood enters posteriorly and to the left, and the outflow of blood to the pulmonary trunk leaves superiorly and to the left.

Ventricular Septal Defects

The membranous part of the interventricular septum develops separately from the muscular part and has a complex embryological origin (Moore and Persaud, 1998). Consequently, this part is the common site of ventricular septal defects (VSDs). This congenital heart anomaly ranks first on all lists of cardiac defects. Isolated VSD accounts for approximately 25% of all forms of congenital heart disease (Moore and Persaud, 1998). The size of the defect varies from 1 to 25 mm. A VSD causes a left-to-right shunt of blood through the defect. A large shunt increases pulmonary bloodflow, which causes severe pulmonary disease (hypertension—increased blood pressure) and may cause *cardiac failure*. The much less common VSDs in the muscular part of the septum frequently close spontaneously during childhood (Creasy and Resnik, 1994).

Pulmonary Valve Stenosis

With *pulmonary valve stenosis* (narrowing), the valve cusps are fused, forming a dome with a narrow central opening. In *infundibular pulmonary stenosis*, the conus arteriosus is underdeveloped. Both types of pulmonary stenosis produce a restriction of right ventricular outflow and may occur together. The degree of hypertrophy of the right ventricle is variable. ▶

Pulmonary Valve Incompetence

The free margins of the cusps of the pulmonary valve are normally thin. If these thicken and become inflexible or are damaged by disease, the valve will not close completely. An *incompetent pulmonary valve* results in a backrush of blood under high pressure into the right ventricle during diastole. Pulmonic regurgitation may be heard through a stethoscope as a **heart murmur**—an abnormal sound from the heart—produced in this case by damage to the cusps of the pulmonary valve. ⊙

Left Atrium. This heart chamber forms most of the base of the heart (Fig. 1.48). The valveless pairs of right and left *pulmonary veins* enter the smooth-walled atrium. The tubular, muscular *left auricle* forms the superior part of the left border of the heart (Fig. 1.45*D*) and overlaps the root of the pulmonary trunk. A semilunar depression in the *interatrial septum* indicates the floor of the oval fossa. The **interior of the left atrium** has

- A larger smooth-walled part and a smaller muscular auricle containing pectinate muscles
- Four pulmonary veins (two superior and two inferior) entering its posterior wall
- A slightly thicker wall than that of the right atrium

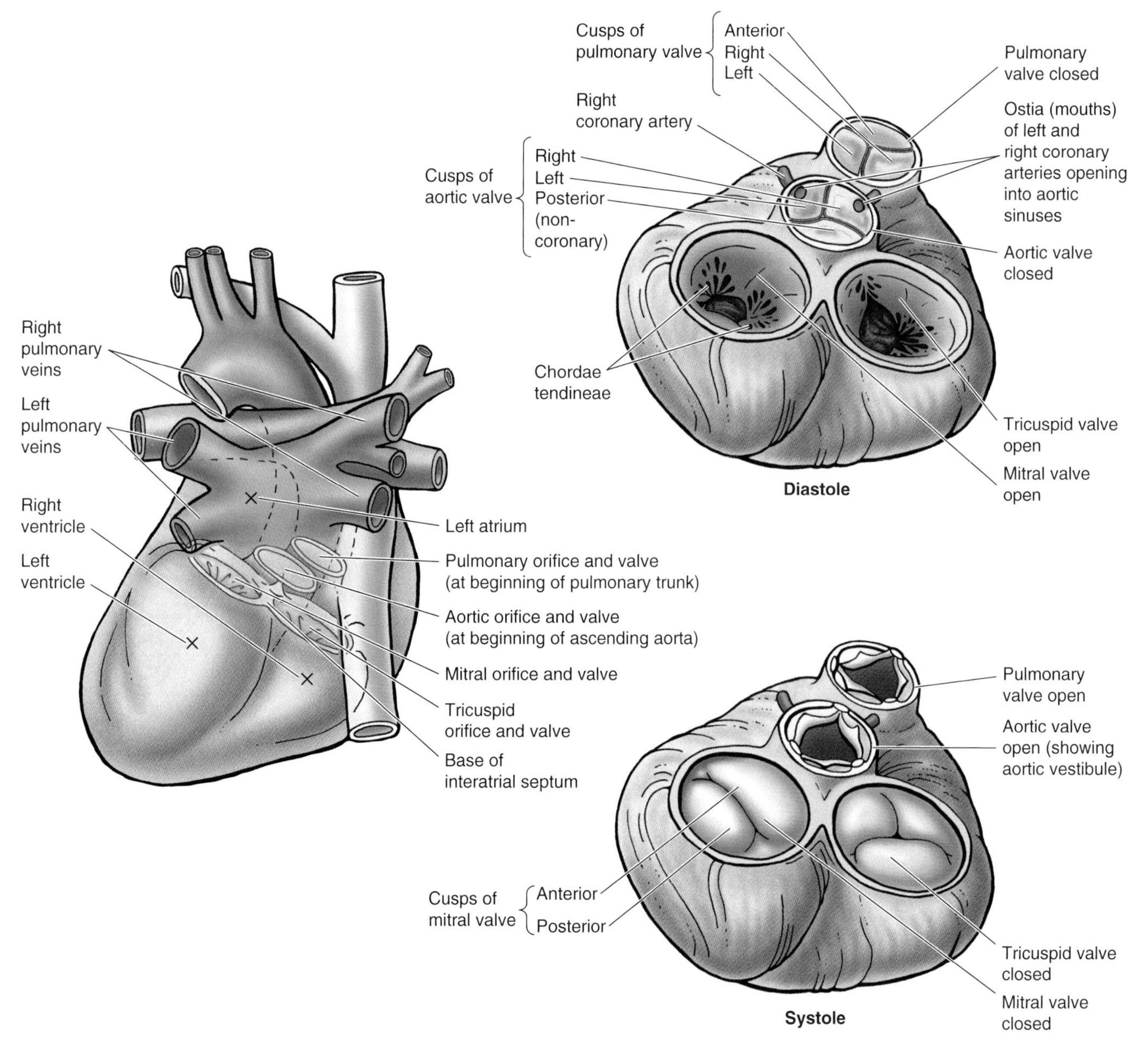

Figure 1.48. Valves of the heart and great vessels. At the beginning of diastole (ventricular filling), the aortic and pulmonary valves are closed; shortly thereafter, the tricuspid and mitral valves open (Fig. 1.42). Shortly after systole (ventricular emptying) begins, the tricuspid and mitral valves close and the aortic and pulmonary valves open.

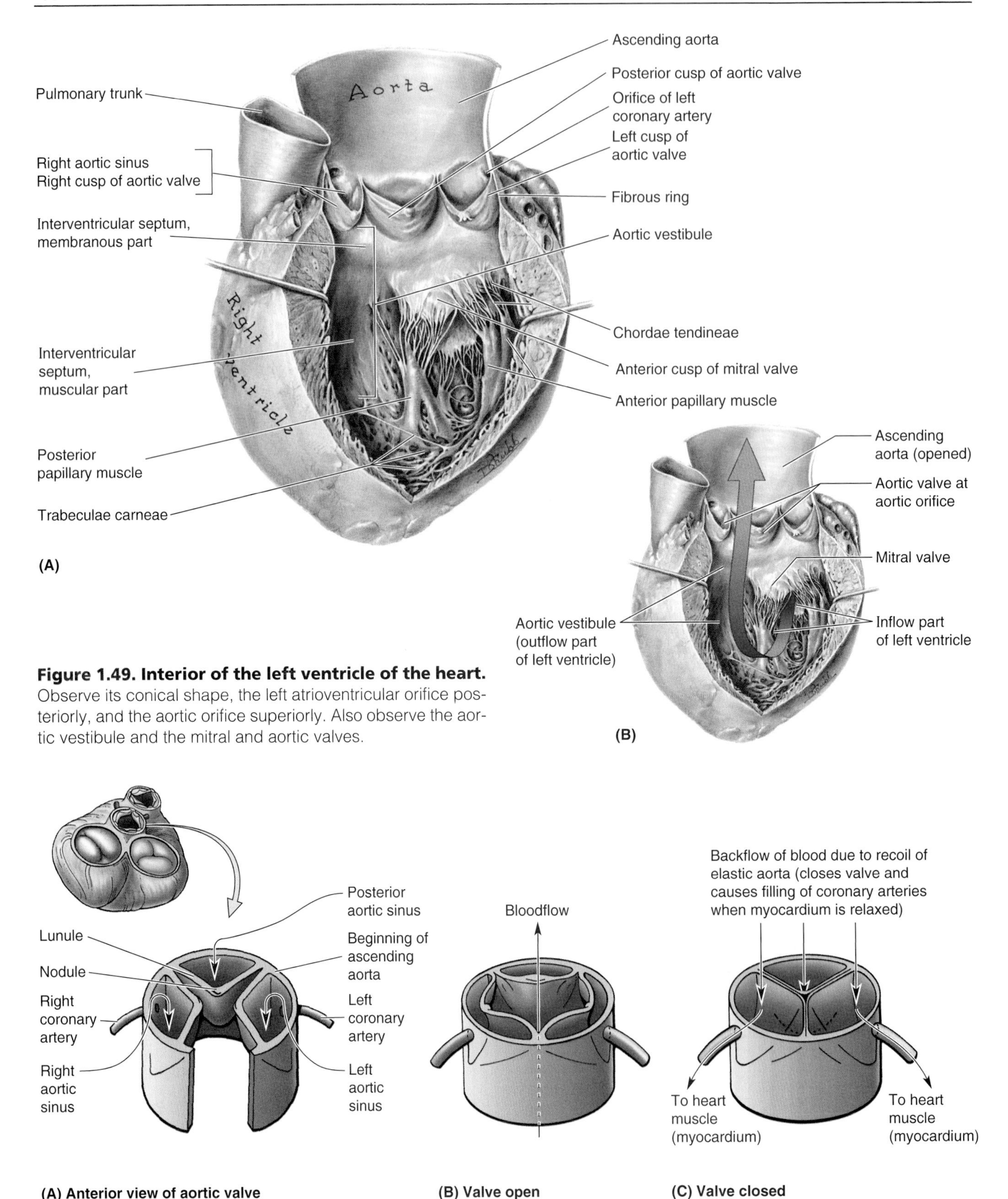

Figure 1.49. Interior of the left ventricle of the heart. Observe its conical shape, the left atrioventricular orifice posteriorly, and the aortic orifice superiorly. Also observe the aortic vestibule and the mitral and aortic valves.

Figure 1.50. Aortic valve, aortic sinuses, and coronary arteries. Like the pulmonary valve, the aortic valve has three semilunar cusps: right, posterior, and left. Each cusp has a fibrous nodule at the midpoint of its free edge and a thin connective tissue area, the lunule, to each side of the nodule. When the valve closes, the nodules and lunules meet in the center. Observe the coronary arteries arising from the aortic sinuses, the dilations between the aortic wall and the concave superior aspect of each of the semilunar cusps of the aortic valve.

- An interatrial septum that slopes posteriorly and to the right
- A left AV orifice through which the left atrium discharges the oxygenated blood it receives into the left ventricle.

The smooth-walled part of the left atrium is formed by absorption of parts of the embryonic pulmonary veins, whereas the rough-walled part, mainly in the auricle, represents the remains of the left part of the primordial atrium (Moore and Persaud, 1998).

Left Ventricle. This chamber forms the apex of the heart, nearly all its left (pulmonary) surface and border, and most of the diaphragmatic surface (Figs. 1.45 and 1.49). Because arterial pressure is much higher in the systemic than in the pulmonary circulation, the left ventricle performs more work than the right ventricle. The **interior of the left ventricle** (Figs. 1.48 and 1.49) has

- A double-leaflet **mitral valve** that guards the left AV orifice
- Walls that are twice as thick as that of the right ventricle
- A conical cavity that is longer than that of the right ventricle
- Walls that are mostly covered with a mesh of **trabeculae carneae** that is finer and more numerous than that of the right ventricle
- Anterior and posterior **papillary muscles** that are larger than those in the right ventricle because this ventricle works harder
- A superoanterior outflow part formed by the smooth-walled *aortic vestibule* leading to the aortic orifice
- An **aortic orifice** (Fig. 1.48) that lies in its right posterosuperior part and is surrounded by a fibrous ring to which the right, posterior, and left *cusps* of the *aortic valve* are attached.

The **ascending aorta**, approximately 2.5 cm in diameter, begins at the aortic orifice.

The **mitral valve** has two cusps, anterior and posterior. The adjective *mitral* derives from the valve's resemblance to a bishop's miter (headdress). The mitral valve is located posterior to the sternum at the level of the 4th costal cartilage. Each of its cusps receives chordae tendineae from more than one papillary muscle. These muscles and their tendinous cords support the mitral valve, allowing the cusps to resist the pressure developed during contractions (pumping) of the left ventricle. The chordae tendineae become taut, preventing the cusps from being forced into the left atrium.

The **aortic valve** (Figs. 1.49 and 1.50), obliquely placed, is located posterior to the left side of the sternum at the level of the 3rd intercostal space. Superior to each valve, dilations of the aortic wall form **aortic sinuses**. The mouth of the right coronary artery is in the right aortic sinus; the mouth of the left coronary artery is in the left aortic sinus; and no artery arises from the posterior aortic (noncoronary) sinus.

Strokes or Cerebrovascular Accidents

Thrombi (clots) form on the walls of the left atrium in certain types of heart disease. If these *thrombi* become detached, or pieces break off, they pass into the systemic circulation and occlude peripheral arteries. Arterial occlusion of an artery in the brain results in a *stroke* or cerebrovascular accident (CVA) that paralyzes the parts of the body previously controlled by the now-damaged (ischemic) area of the brain.

Valvular Insufficiency and Heart Murmurs

The mitral valve is the most frequently diseased of the heart valves. Nodules form on the valve cusps causing irregular (turbulent) bloodflow. Later, the diseased cusps undergo scarring and shortening, resulting in **mitral insufficiency**—defective functioning of the mitral valve. As a result, blood regurgitates into the left atrium when the left ventricle contracts, producing a characteristic *heart murmur*. Turbulent energy in the chambers of the heart and blood vessels produces murmurs (Swartz, 1994). **Aortic insufficiency**—defective functioning of the aortic valve—results in aortic regurgitation (backrush of blood into the left ventricle), producing a heart murmur and a *collapsing pulse* (forcible impulse that rapidly diminishes). Obstruction of bloodflow or passage of blood from a narrow to a larger vessel produces turbulence. Turbulence sets up *eddies* (small whirlpools) that produce vibrations that are audible as a murmur. *Thrills* are the superficial vibratory sensations felt on the skin over an area of turbulence.

Mitral Stenosis

Advanced *valvular stenosis* (narrowing of the mitral orifice) is characterized by a loud murmur that is loudest at the apex of the heart or somewhat medial to this point (Willms et al., 1994). Because mitral stenosis is a disease of mechanical obstruction, it can be corrected by replacement with a *valve prosthesis* (Sabiston and Lyerly, 1994).

Congenital Aortic Stenosis

Congenital aortic stenosis refers to a group of anomalies that cause obstruction of bloodflow from the left ventricle to the aorta. Although the stenosis usually occurs at the aortic valve, the lesion may be above or below the valve. In aortic stenosis, the edges of the aortic valve are usually fused to form a dome with a small opening. Aortic stenosis causes extra work for the heart, resulting in *left ventricular hypertrophy*. ▶

Aneurysm of the Ascending Aorta

The distal part of the ascending aorta receives a strong thrust of blood when the left ventricle contracts. Because its wall is not reinforced by fibrous pericardium, an aneurysm (localized dilation) may develop. Patients usually complain of chest pain that radiates to the back.

Basis for Naming the Aortic and Pulmonary Valves

The following account explains the embryological basis for naming the pulmonary and aortic valves. The truncus arteriosus, the common arterial trunk from both ventricles of the embryonic heart, has four cusps (**A**). The truncus arteriosus divides into two vessels, each with its own valve (pulmonary and aortic), that has three cusps (**B**). The heart undergoes partial rotation so that its apex becomes directed to the left, resulting in the arrangement of cusps as shown in **C**. Consequently, the cusps are named according to their embryological origin, not their postnatal anatomical position. Thus, the pulmonary valve has right, left, and anterior cusps, and the aortic valve has right, left, and posterior cusps. Similarly, the aortic sinuses are named right, left, and posterior. This terminology also agrees with the coronary arteries: note that the right coronary artery arises from the right aortic sinus, superior to the right aortic valve, and that the left coronary has a similar relation to the left aortic valve and sinus. The posterior cusp and sinus does not give rise to a coronary artery; thus, it is also referred to as a "noncoronary" cusp and sinus. ○

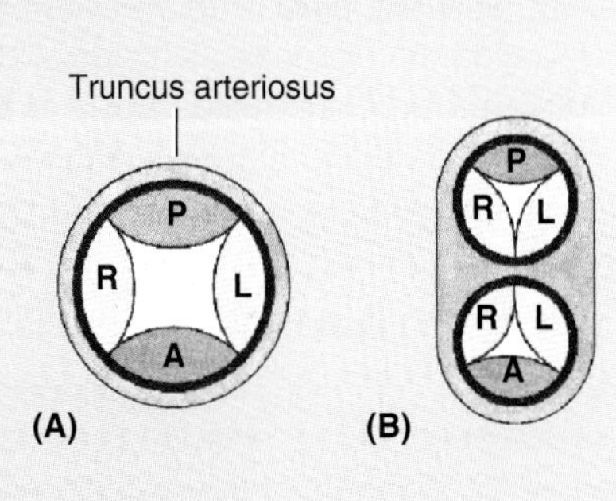

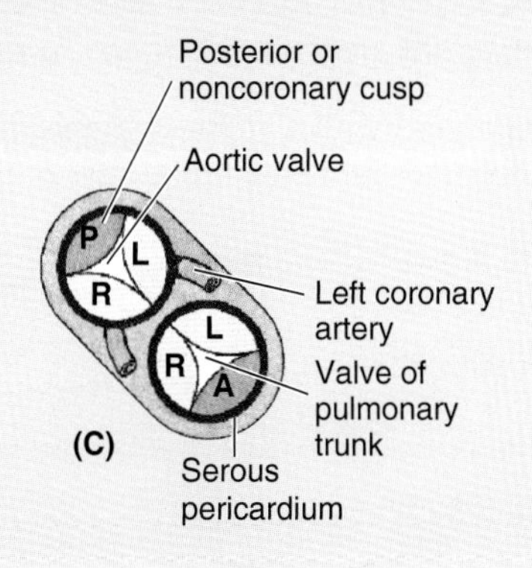

Vasculature and Innervation of the Heart. The blood vessels of the heart comprise the coronary arteries and cardiac veins, which carry blood to and from most of the myocardium (Figs. 1.51 and 1.52). The endocardium and some subendocardial tissue located immediately external to the endocardium receive oxygen and nutrients by diffusion or microvasculature directly from the chambers of the heart. The blood vessels of the heart, normally embedded in fat, for the main part course across the surface of the heart just deep to the epicardium. Occasionally parts of the vessels become embedded within the myocardium. The blood vessels of the heart receive both sympathetic and parasympathetic innervation.

Arterial Supply of the Heart. The **coronary arteries**—the 1st branches of the aorta—supply the myocardium and epicardium of the heart. The right and left coronary arteries arise from the corresponding aortic sinuses at the proximal part of the **ascending aorta**, just superior to the aortic valve (Figs. 1.50 and 1.51, Table 1.5). The coronary arteries supply both the atria and ventricles; however, the atrial branches are usually small and not readily apparent in the cadaveric heart.

No sharp demarcation line exists between the ventricular distribution of the coronary arteries.

Dominance of the coronary arterial system is defined by which artery gives rise to the posterior interventricular artery (posterior descending artery). Dominance of the right coronary artery is typical (Fig. 1.51*A*), with the right coronary artery supplying most of the diaphragmatic surface. The left coronary artery is dominant in approximately 10% of people (Fig. 1.51*B*), and there is codominance in 15% of people. (See the clinical box on *Variations of the Coronary Arteries*, p. 135).

The **right coronary artery** (RCA) arises from the right aortic sinus of the ascending aorta (Fig. 1.50) and runs in the coronary or AV groove (L. sulcus). Near its origin, the RCA usually gives off an ascending **sinuatrial (sinoatrial) (SA) nodal branch** that supplies the *SA node* (Fig. 1.51*A*). The RCA then descends in the coronary groove and gives off the **right marginal branch** that supplies the right border of the heart as it runs toward (but does not reach) the apex of the heart. After giving off this branch, the RCA turns to the left and continues in the coronary groove to the posterior aspect of the heart. At the **crux** (L. cross) of the heart—the junction of the septa and walls of the four heart chambers—the RCA gives rise to the **AV nodal branch**, which supplies the *AV node*. The RCA then gives off the large posterior inter-

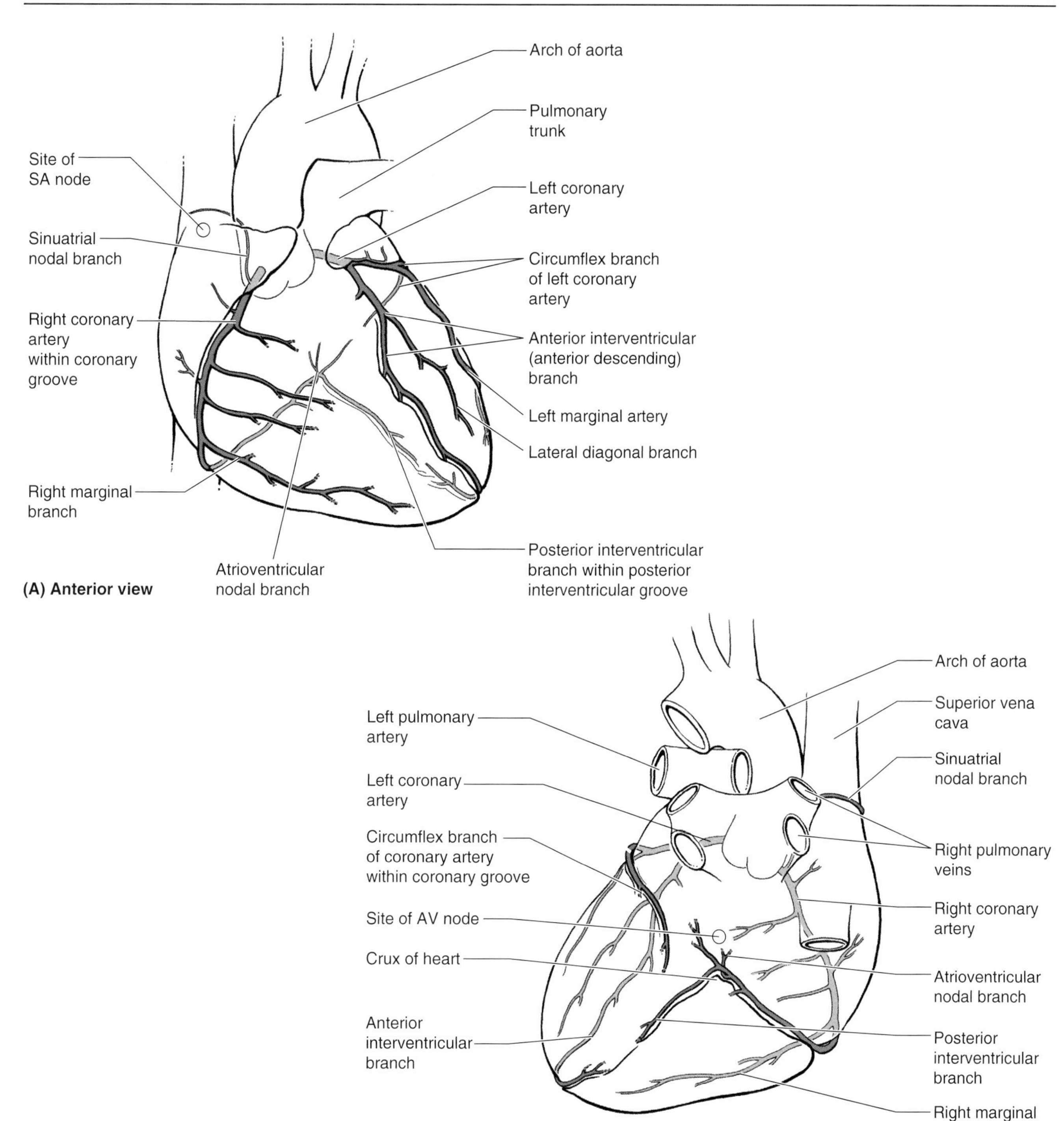

Figure 1.51. Coronary arteries. A–B. Observe that the right coronary artery (RCA) courses in the atrioventricular (coronary) groove to reach the posterior surface of the heart, where it anastomoses with the circumflex branch of the left coronary artery (LCA) (anastomoses are not demonstrated here). Early in its course, it gives off the sinuatrial (SA) nodal artery. **B.** Observe also that the LCA divides into a circumflex branch that passes posteriorly to anastomose with the RCA on the posterior aspect of the heart, and a left anterior descending (LAD) branch, the anterior interventricular artery, which follows the anterior interventricular groove to the apex of the heart and hooks onto the posterior surface of the heart.

Table 1.5. Arterial Supply of the Heart

Artery/Branch	Origin	Course	Distribution	Anastomoses
Right coronary	Right aortic sinus	Follows coronary (AV) groove between the atria and ventricles	Right atrium, SA and AV nodes, and posterior part of IV septum	Circumflex and anterior IV branches of left coronary artery
SA nodal	Right coronary artery near its origin (in 60%)	Ascends to SA node	Pulmonary trunk and SA node	
Right marginal	Right coronary artery	Passes to inferior margin of heart and apex	Right ventricle and apex of heart	IV branches
Posterior IV	Right coronary artery	Runs from posterior IV groove to apex of heart	Right and left ventricles and IV septum	Circumflex and anterior IV branches of left coronary artery
AV nodal	Right coronary artery near origin of posterior IV artery	Passes to AV node	AV node	
Left coronary	Left aortic sinus	Runs in AV groove and gives off anterior interventricular and circumflex branches	Most of left atrium and ventricle, IV septum, and AV bundles; may supply AV node	Right coronary artery
SA nodal	Circumflex branch (in 40%)	Ascends on posterior surface of left atrium to SA node	Left atrium and SA node	
Anterior interventricular	Left coronary artery	Passes along anterior IV groove to apex of heart	Right and left ventricles and IV septum	Posterior IV branch of right coronary artery
Circumflex	Left coronary artery	Passes to left in AV groove and runs to posterior surface of heart	Left atrium and left ventricle	Right coronary artery
Left marginal	Circumflex branch	Follows left border of heart	Left ventricle	IV branches

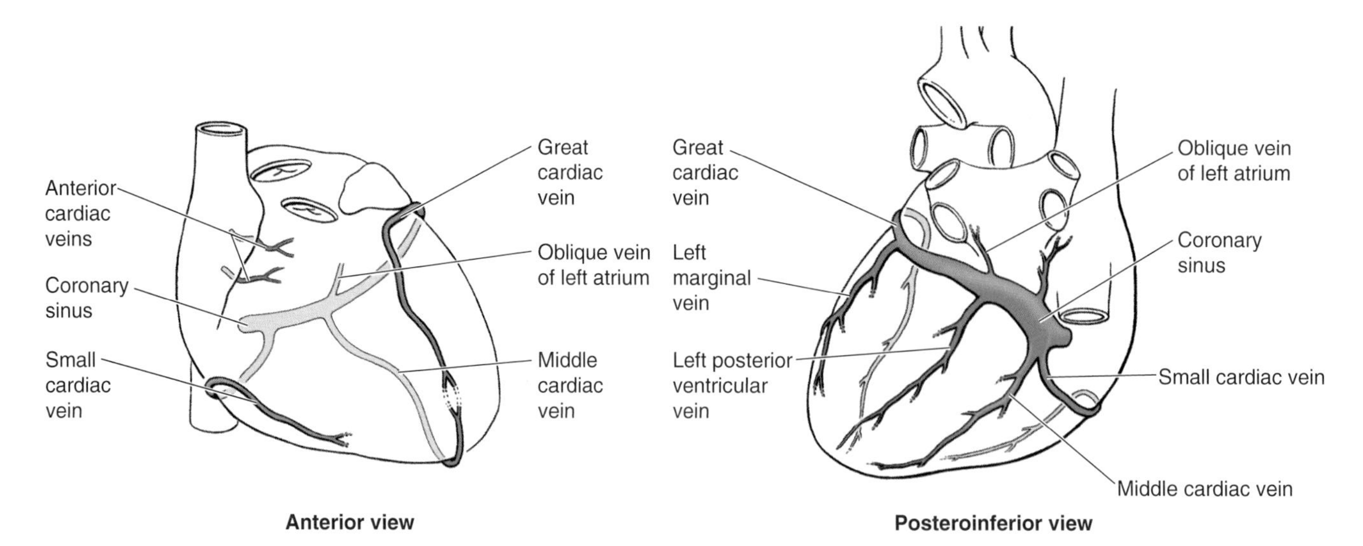

Figure 1.52. Cardiac veins. The coronary sinus, the main drainage vessel of the heart, empties into the right atrium. The great, middle, and small cardiac veins, the oblique vein of the left atrium, and the left posterior ventricular vein are the main vessels draining into the coronary sinus. The anterior cardiac veins drain directly into the right atrium.

ventricular artery that descends in the posterior interventricular groove toward the apex of the heart. The **posterior interventricular branch** supplies both ventricles and sends perforating *interventricular septal branches* to the IV septum. Near the apex of the heart, the RCA anastomoses with the circumflex and anterior interventricular branches of the left coronary artery. **Typically, the RCA supplies**

- The right atrium
- Most of right ventricle
- Part of the left ventricle (the diaphragmatic surface)
- Part (usually the posterior third) of the AV septum
- The SA node (in approximately 60% of people)
- The AV node (in approximately 80% of people).

The **left coronary artery** (LCA) arises from the *left aortic sinus* of the ascending aorta (Fig. 1.50) and passes between the left auricle and pulmonary trunk in the coronary groove. In approximately 40% of people, the **SA nodal branch** arises from the circumflex branch of the LCA and ascends on the posterior surface of the left atrium to the SA node. At the left end of the coronary groove (Fig. 1.51), the LCA divides into two branches, an **anterior interventricular branch** (left anterior descending branch, LAD branch) and a **circumflex branch**. The anterior interventricular branch passes along the interventricular groove to the apex of the heart. Here it turns around the inferior border of the heart and anastomoses with the posterior interventricular branch of the right coronary artery. The anterior interventricular branch supplies both ventricles and the interventricular septum. In many people, the anterior interventricular branch gives rise to a **lateral (diagonal) branch**, which descends on the anterior surface of the heart. The smaller circumflex branch of the LCA follows the coronary groove around the left border of the heart to the posterior surface of the heart. The **left marginal artery**, a branch of the circumflex branch, follows the left margin of the heart and supplies the left ventricle. The circumflex branch of the LCA terminates on the posterior aspect of the heart and often anastomoses with the posterior interventricular branch of the RCA. **Typically, the LCA supplies**

- The left atrium
- Most of the left ventricle
- Part of the right ventricle
- Most of the IV septum (usually its anterior two-thirds), including the AV bundle of conducting tissue, through its perforating *IV septal branches*
- The SA node (in approximately 40% of people).

Variations of the Coronary Arteries

Variations in the branching patterns of the coronary arteries are common. In most people the RCA and LCA share approximately equally in the blood supply to the heart (**A**). In approximately 15% of hearts the LCA is dominant in that the posterior interventricular branch is a branch of the circumflex artery (**B**). A few people have only a single coronary artery (**C**). In other people the circumflex branch arises from the right aortic sinus (**D**). Approximately 4% of people have an accessory coronary artery.

The branches of coronary arteries are considered to be end arteries—ones that supply regions of the myocardium without overlap or anastomoses from other large branches. However, anastomoses do exist between branches of the coronary arteries, subepicardial or myocardial, and between these arteries and extracardiac vessels such as thoracic vessels (Williams et al., 1995). Clinical studies show that anastomoses cannot provide collateral routes quickly enough to prevent the effects of sudden coronary artery occlusion. The functional value of these anastomoses appears to be more effective in slowly progressive coronary artery disease. ▶

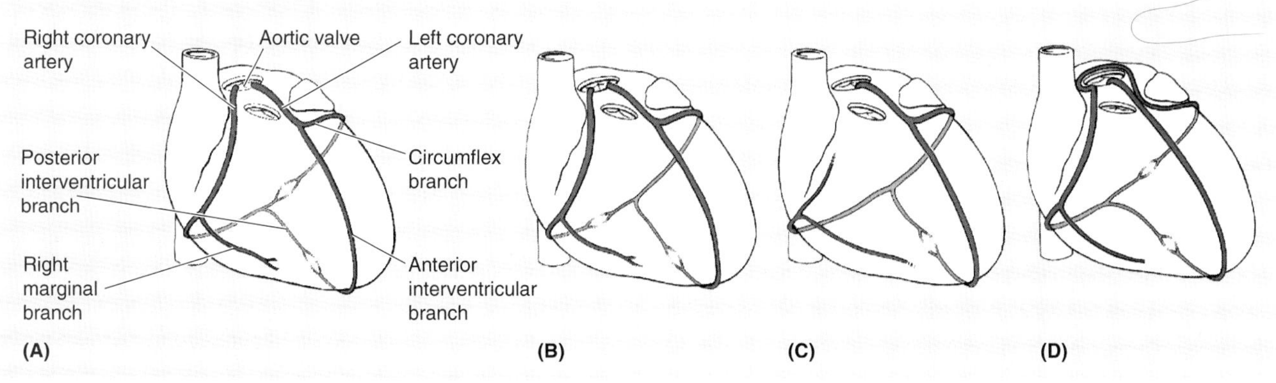

Coronary Atherosclerosis and Myocardial Infarction

With sudden occlusion of a major artery by an embolus (G. embolos, plug), the region of myocardium supplied by the occluded vessel becomes infarcted (rendered virtually bloodless) and soon degenerates (i.e., the tissue dies, or becomes necrotic). An area of myocardium that has undergone necrosis is an MI. The most common cause of **ischemic heart disease** (ischemia—lacking adequate blood supply) is coronary insufficiency, resulting from atherosclerosis of the coronary arteries.

Coronary atherosclerosis begins during early adulthood and slowly results in stenosis of the lumina of the coronary arteries. The atherosclerotic process results in lipid accumulations on the internal walls of the coronary arteries. As coronary atherosclerosis progresses, the collateral channels connecting one coronary artery with the other expand, permitting adequate perfusion of the heart to continue. Despite this compensatory mechanism, the myocardium may not receive enough oxygen when the heart needs to perform increased amounts of work. Strenuous exercise, for example, increases the heart's activity and its need for oxygen. The insufficiency of blood supply to the heart (*myocardial ischemia*) may result in an area of necrosis in the myocardium—MI.

Angina Pectoris

Patients with angina pectoris (chest pain) commonly describe the severe constricting pain as a tightness in the chest. This type of central chest discomfort is usually shortened clinically to **angina** when referring to a severe, often constricting chest pain. Stress produces arterial constriction and is a common cause of angina. Equally common causes are strenuous exercise after a heavy meal and sudden exposure to cold. When food enters the stomach, bloodflow to it and other parts of the digestive tract is increased. As a result, some blood is diverted from other organs, including the heart. *Anginal pain is often relieved by one or two minutes of rest.* Sublingual nitroglycerin (medication placed or sprayed under the tongue for absorption through the oral mucosa) may be administered because it dilates the coronary arteries, increases bloodflow to the heart, and usually relieves the angina.

The pain resulting from myocardial ischemia and myocardial infarction is usually more severe than with angina pectoris, and *the pain resulting from MI does not disappear after one or two minutes of rest.* MI may also follow excessive exertion by a person with stenotic (narrowed) coronary arteries. The straining heart muscle demands more oxygen than the stenotic arteries can provide; as a result, the ischemic area of myocardium undergoes infarction and a *heart attack* occurs. Occlusion of any but the smallest branches of a coronary artery usually results in death of the cardiac muscle fibers it supplies. The damaged muscle is replaced by fibrous tissue and a scar forms.

Coronary Bypass Graft

Some patients with obstruction of the coronary circulation and severe angina undergo a coronary bypass graft operation (for details, see Goldman, 1989). A segment of a vein is connected to the ascending aorta or to the proximal part of a coronary artery and then to the coronary artery distal to the stenosis. The *great saphenous vein* is commonly harvested for coronary bypass surgery because it has approximately the same diameter as the coronary arteries, can be easily dissected from the lower limb, and offers relatively lengthy portions with a minimum occurrence of valves or branching. A coronary bypass graft shunts blood from the aorta to a stenotic coronary artery to increase the flow distal to the obstruction. Simply stated, it provides a detour around the stenotic area (arterial stenosis) or blockage (arterial atresia). Revascularization of the myocardium may also be achieved by surgically anastomosing an internal thoracic artery with a coronary artery.

Coronary Angioplasty

In selected patients, surgeons use *percutaneous transluminal coronary angioplasty* in which they pass a catheter with a small inflatable balloon attached to its tip into the obstructed coronary artery. When the catheter reaches the obstruction, the balloon is inflated and the vessel is stretched to increase the size of the lumen. In other cases, *thrombokinase* is injected through the catheter; this enzyme dissolves the blood clot. ○

Venous Drainage of the Heart. The heart is drained mainly by veins that empty into the coronary sinus and partly by small veins that empty into the right atrium (Fig. 1.52). The **coronary sinus**, the main vein of the heart, is a wide venous channel that runs from left to right in the posterior part of the coronary groove. The coronary sinus receives the anterior interventricular vein or *great cardiac vein* at its left end, and the posterior interventricular or *middle cardiac vein* and *small cardiac veins* at its right end. The *left posterior ventricular vein* and *left marginal vein* also open into the coronary sinus.

The **great cardiac vein** is the main tributary of the coronary sinus. It begins near the apex of the heart and ascends with the anterior interventricular branch of the LCA. At the coronary groove it turns left and runs around the left side of the heart with the circumflex branch to reach the coronary sinus. The great cardiac vein drains the areas of the heart sup-

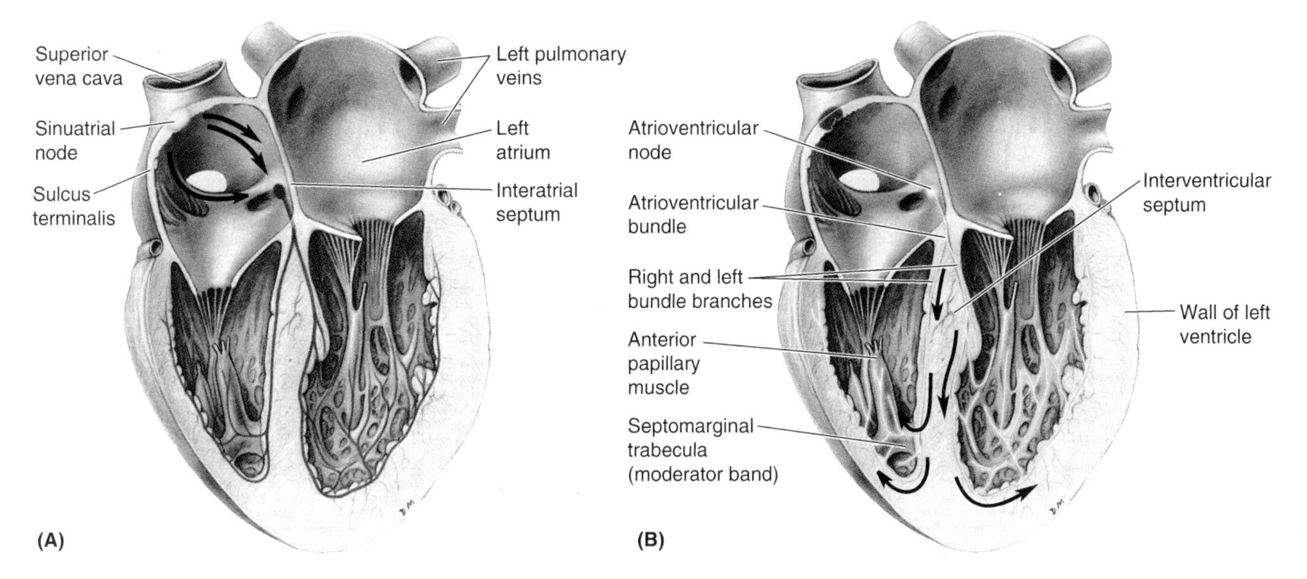

Figure 1.53. Conducting system of the heart. Observe the sinuatrial (SA) node at the superior end of the sulcus (internally, crista) terminalis and the atrioventricular (AV) node in the inferior part of the interatrial septum. The AV bundle begins at the AV node and divides into right and left bundles at the junction of the membranous and muscular parts of the interventricular septum (IVS).

plied by the left coronary artery. The *middle* and *small cardiac veins* drain most of the areas supplied by the right coronary artery.

The **oblique vein of the left atrium** is a small, relatively unimportant vessel postnatally that runs over the posterior wall of the left atrium and merges with the great cardiac vein to form coronary sinus. The oblique vein is the remnant of the embryonic left SVC, which occasionally persists in adults, replacing or augmenting the left SVC.

Several small **anterior cardiac veins** begin over the anterior surface of the right ventricle, cross over the coronary groove, and usually end directly in the right atrium; sometimes they enter the small cardiac vein.

The **smallest cardiac veins** (L. venae cordis minimae) are minute vessels that begin in the capillary beds of the myocardium and open directly into the chambers of the heart, chiefly the atria. Although called veins, they are valveless communications with the capillary beds of the myocardium and may carry blood from the heart chambers to the myocardium. They may also provide a collateral circulation for parts of the heart musculature.

Lymphatic Drainage of the Heart. Lymphatic vessels in the myocardium and subendocardial connective tissue pass to the *subepicardial lymphatic plexus.* Vessels from this plexus pass to the coronary groove and follow the coronary arteries. A single lymphatic vessel, formed by the union of various vessels from the heart, ascends between the pulmonary trunk and left atrium and ends in the inferior *tracheobronchial lymph nodes,* usually on the right side.

Conducting System of the Heart. In the ordinary sequence of events in the cardiac cycle, the atrium and ventricle work together as one pump on each side of the heart. The **impulse-conducting system of the heart** (Fig. 1.53)—which *coordinates the cardiac cycle* (Fig. 1.42)—consists of cardiac muscle cells and highly specialized conducting fibers for initiating impulses and conducting them rapidly through the heart. Nodal tissue initiates the heartbeat and coordinates contractions of the four heart chambers.

The **SA node** is located anterolaterally just deep to the epicardium at the junction of the SVC and right atrium, near the superior end of the sulcus terminalis. The SA node—a small collection of nodal tissue, specialized cardiac muscle fibers, and associated fibroelastic connective tissue—is the **pacemaker of the heart.** The SA node initiates and regulates the impulses for contraction, giving off an impulse approximately 70 times per minute in most people. The contraction signal from the SA node spreads through the musculature of both atria. The SA node is supplied by the SA nodal artery, which is usually a branch of the right coronary artery, but it may arise from the left coronary artery. The SA node is supplied by both divisions of the autonomic nervous system through the *cardiac plexus* (Fig. 1.54). The rate at which the node produces impulses can be altered by nervous stimulation.

- Sympathetic stimulation accelerates the heart rate.
- Parasympathetic stimulation slows down the heart rate.

The **AV node** is a smaller collection of nodal tissue located in the posteroinferior region of the interatrial septum near the opening of the coronary sinus. The signal generated by the SA node passes through the walls of the right atrium, propagated by the cardiac muscle (*myogenic conduction*), which transmits the signal rapidly from the SA node to the AV node. The AV node then distributes the signal to the ventricles through the **AV bundle.** Sympathetic stimulation speeds up conduction and parasympathetic stimulation slows it down. The AV bundle, the only bridge between the atrial and ventricular myocardium, passes from the AV node through the insulating fibrous skeleton of the heart and along the membranous part of the IV septum. At the junction of the membranous and muscular parts of the septum, the AV bundle divides into **right** and **left bundles**. The bundles proceed on each side of the muscular IV septum deep to the endocardium and then ramify into *subendocardial branches* (Purkinje fibers), which extend into the walls of the respective ventricles. The subendocardial branches of the right bundle stimulate the muscle of the IV septum, the anterior papillary muscle through the septomarginal trabecula (moderator band), and the wall of the right ventricle. The left bundle divides near its origin into approximately six smaller tracts, which give rise to subendocardial branches that stimulate the IV septum, the anterior and posterior papillary muscles, and the wall of the left ventricle.

In approximately 80% of people, the AV node is supplied by the *AV nodal artery*, the 1st IV septal branch of the right coronary artery. In other people, it is supplied by an AV nodal artery from the left coronary artery. Usually the AV bundle and left and right bundles are supplied by the IV septal branches of the anterior interventricular branch of the left coronary artery, but the posterior limb of the left bundle may receive blood from both coronary arteries.

Summary of the conducting system of the heart:

The SA node initiates an impulse that is rapidly conducted to cardiac muscle fibers in the atria, causing them to contract.

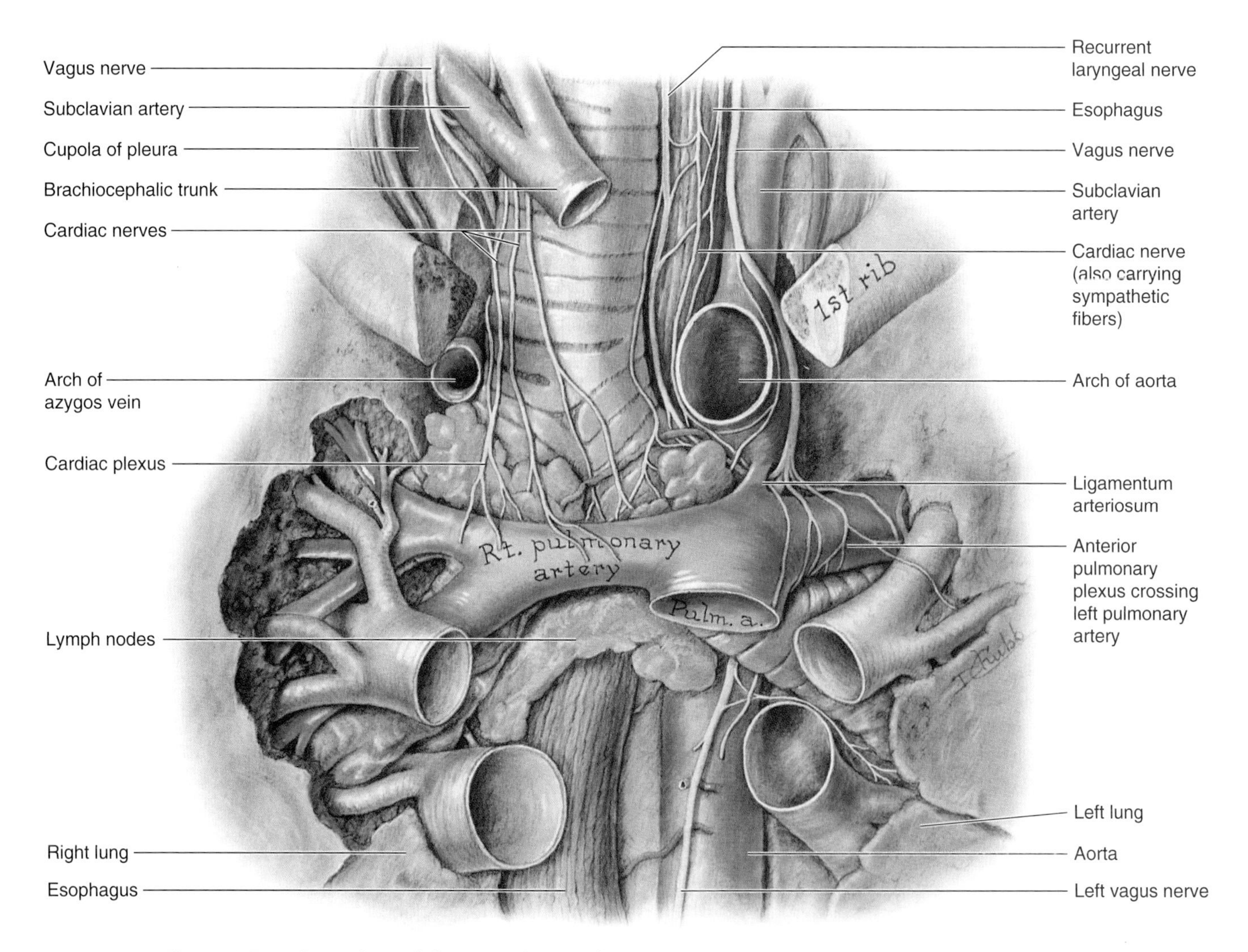

Figure 1.54. Dissection of the superior mediastinum. Observe the cardiac branches of the vagus (CN X) and sympathetic nerves running down the sides of the trachea and forming the cardiac plexus. Although shown lying on the trachea, the primary relationship of the cardiac plexus is to the ascending aorta and pulmonary trunks, which have been removed to expose the plexus.

The impulse spreads by myogenic conduction that rapidly transmits the impulse from the SA node to the AV node.

The signal is distributed from the AV node through the AV bundle and its branches, the right and left bundles, which pass on each side of the IV septum to supply subendocardial branches to the papillary muscles and the walls of the ventricles.

Innervation of the Heart. The heart is supplied by autonomic nerve fibers from superficial and deep **cardiac plexuses** (Fig. 1.54; see also Fig. 1.58*B*). These nerve networks lie anterior to the bifurcation of the trachea, posterior to the ascending aorta, and superior to the bifurcation of the pulmonary trunk.

The **sympathetic supply** is from presynaptic fibers with cell bodies in the lateral horn of the superior five or six thoracic segments of the spinal cord, and postsynaptic sympathetic fibers with cell bodies in the cervical and superior thoracic paravertebral ganglia of the sympathetic trunks. The postsynaptic fibers end in the SA and AV nodes and in relation to the terminations of parasympathetic fibers on the coronary arteries. *Sympathetic stimulation of the nodal tissue increases the heart's rate and the force of its contractions.* Sympathetic stimulation (indirectly) produces dilation of the coronary arteries by inhibiting their constriction. This supplies more oxygen and nutrients to the myocardium during periods of increased activity.

The **parasympathetic supply** is from presynaptic fibers of the vagus nerves. The postsynaptic parasympathetic fibers also end in the SA and AV nodes and directly on the coronary arteries. The cell bodies of the postsynaptic fibers constitute intrinsic ganglia in the vicinity of these structures. *Stimulation of parasympathetic nerves slows the heart rate, reduces the force of the heartbeat, and constricts the coronary arteries,* saving energy between periods of increased demand.

Injury to the AV Node and AV Bundle

If parts of the conducting system of the heart (Fig. 1.53) are affected by the blockage—e.g., stenosis of the anterior interventricular branch (LAD), which gives rise to the septal branches supplying the AV bundle in most people—a **heart block** may occur. In this case, the ventricles may contract independently at their own rate. A *pacemaker* (artificial heart regulator) may have to be implanted in some cases. **Common sites of coronary artery occlusion** are in the

- Anterior IV (LAD) branch of the LCA
- Circumflex branch of the LCA
- RCA and its posterior IV branch.

Damage to the conducting system of the heart, often resulting from ischemia caused by *coronary artery disease,* produces disturbances of cardiac muscle contraction. Damage to the AV node results in a heart block because the atrial excitation does not reach the ventricles. As a result, the ventricles begin to contract independently at their own rate, which is slower than that of the atria. Damage to one of the bundle branches results in a *bundle branch block,* in which excitation passes along the unaffected branch and causes systole of that ventricle. The impulse then spreads to the other ventricle, producing a late asynchronous contraction. With an ASD, the AV bundle usually lies in the margin of the defect. Obviously, this vital part of the conducting system must be preserved during surgical repair of the defect. Destruction of the AV bundle would cut the only physiological link between the atrial and ventricular musculature.

Electrocardiography

The passage of impulses over the heart from the SA node can be amplified and recorded as an *electrocardiogram* (ECG or EKG). Functional testing of the heart includes exercise tolerance tests (treadmill stress tests). This testing is done mainly to check the consequences of possible coronary artery disease. *Exercise tolerance tests* are of considerable importance in detecting the cause of heartbeat irregularities. Heart rate, ECG, and blood pressure readings are monitored as the patient does increasingly demanding exercise on a treadmill. The results show the maximum effort a patient's heart can safely tolerate.

Artificial Cardiac Pacemaker

In some people with a "heart block," a cardiac pacemaker (approximately the size of a pocket watch) is inserted subcutaneously. The artificial pacemaker consists of a pulse generator or battery pack, a wire (lead), and an electrode. Cardiac pacemakers produce electrical impulses that initiate ventricular contractions at a predetermined rate. An electrode with a catheter connected to it is inserted into a vein and its progression through the venous pathway is followed with a fluoroscope—a device for examining deep structures in "real time" (as motion occurs) by means of radiographs. The terminal of the electrode of ▶

▶ the pacemaker is passed through the SVC to the right atrium and through the tricuspid valve into the right ventricle. Here the electrode is firmly fixed to the trabeculae carneae in the ventricular wall and placed in contact with the endocardium.

Restarting the Heart

In most cases of cardiac arrest, first-aid workers perform **cardiopulmonary resuscitation** (CPR) to restore cardiac output and pulmonary ventilation. By applying firm pressure to the chest over the inferior part of the sternal body (external or closed chest cardiac massage), the sternum moves posteriorly 4 to 5 cm. The increased intrathoracic pressure forces blood out of the heart into the great arteries. When the external pressure is released and the intrathoracic pressure falls, the heart again fills with blood. If the heart stops beating (**cardiac arrest**) during heart surgery, the surgeon attempts to restart it using internal or open chest (direct) heart massage.

Fibrillation of the Heart

Fibrillation is multiple, rapid, circuitous contractions or twitchings of muscular fibers, including cardiac muscle. In *atrial fibrillation*, the normal regular rhythmical contractions of the atria are replaced by rapid irregular and uncoordinated twitchings of different parts of the atrial walls. The ventricles respond at irregular intervals to the dysrhythmic impulses received from the atria, but usually circulation remains satisfactory. In *ventricular fibrillation*, the normal ventricular contractions are replaced by rapid, irregular twitching movements that do not pump (i.e., they do not maintain the systemic circulation, including the coronary circulation). The damaged conducting system of the heart does not function normally. As a result, an irregular pattern of uncoordinated contractions occurs in the ventricles, except in those areas that are infarcted. Ventricular fibrillation is the most disorganized of all *dysrhythmias*, and in its presence no effective cardiac output occurs. The condition is fatal if allowed to persist.

Defibrillation of the Heart

An electric shock may be given to the heart through the thoracic wall via large electrodes (paddles). This shock causes cessation of all cardiac movements and a few minutes later the heart may begin to beat more normally. As coordinated contractions and hence pumping of the heart is re-established, some degree of systemic (including coronary) circulation results.

Cardiac Referred Pain

The heart is insensitive to touch, cutting, cold, and heat; however, ischemia and the accumulation of metabolic products stimulate pain endings in the myocardium. The afferent pain fibers run centrally in the middle and inferior cervical branches and especially in the thoracic cardiac branches of the sympathetic trunk. The axons of these primary sensory neurons enter spinal cord segments T1 through T4 or T5, especially on the left side.

Cardiac referred pain is a phenomenon whereby ▶

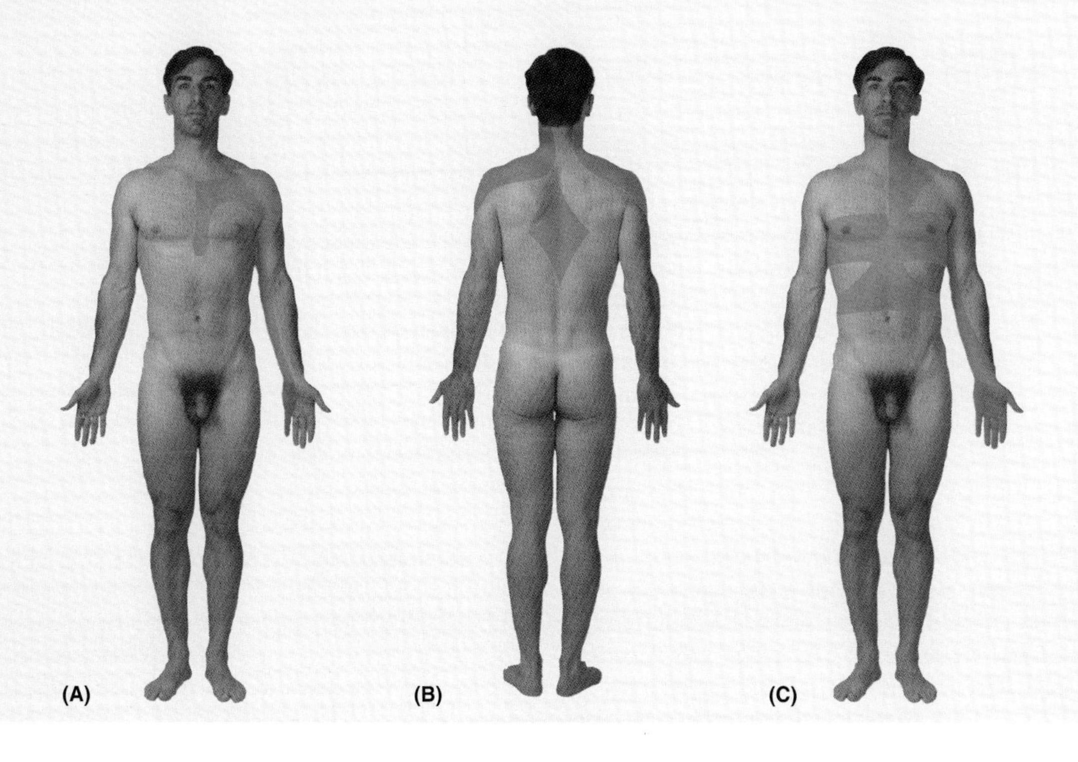

▶ noxious stimuli originating in the heart are perceived by the patient as pain arising from a superficial part of the body—the skin on the left upper limb, for example. Visceral pain is transmitted by visceral afferent fibers accompanying sympathetic fibers and is typically referred to somatic structures or areas such as the upper limb having afferent fibers with cell bodies in the same spinal ganglion, and central processes that enter the spinal cord through the same dorsal roots (Hardy and Naftel, 1997a).

Anginal pain is commonly felt as radiating from the substernal and left pectoral regions to the left shoulder and the medial aspect of the left upper limb. This part of the limb is supplied by the medial cutaneous nerve of the arm. Often the lateral cutaneous branches of the 2nd and 3rd intercostal nerves join the medial cutaneous nerve of the arm. Consequently, cardiac pain is referred to the upper limb because the spinal cord segments of these cutaneous nerves (T1, T2, and T3) are also common to the visceral afferent terminations for the coronary arteries.

Synaptic contacts may also be made with commissural (connector) neurons that conduct impulses to neurons on the right side of comparable areas of the spinal cord. This occurrence explains why pain of cardiac origin, although usually referred to the left side, may be referred to the right side, both sides, or the back. ○

Surface Anatomy of the Heart

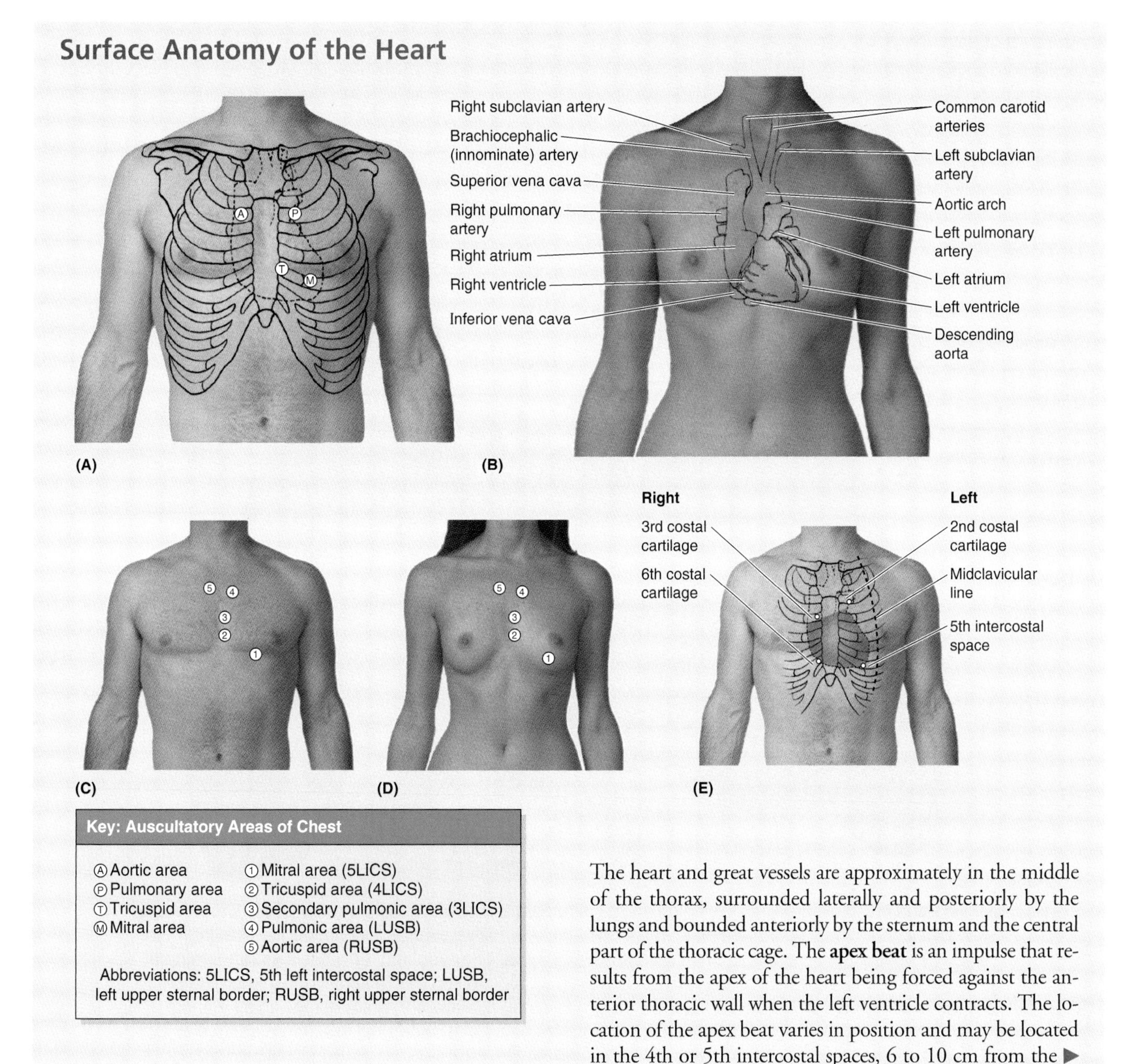

The heart and great vessels are approximately in the middle of the thorax, surrounded laterally and posteriorly by the lungs and bounded anteriorly by the sternum and the central part of the thoracic cage. The **apex beat** is an impulse that results from the apex of the heart being forced against the anterior thoracic wall when the left ventricle contracts. The location of the apex beat varies in position and may be located in the 4th or 5th intercostal spaces, 6 to 10 cm from the ▶

▶ midline. The outline of the heart can be traced on the anterior surface of the thorax by using these guidelines (*E*):

- The superior border corresponds to a line connecting the inferior border of the 2nd left costal cartilage to the superior border of the 3rd right costal cartilage.
- The right border corresponds to a line drawn from the 3rd right costal cartilage to the 6th right costal cartilage; this border is slightly convex to the right.
- The inferior border corresponds to a line drawn from the inferior end of the right border to a point in the 5th intercostal space close to the left MCL; the left end of this line corresponds to the location of the apex and apex beat.
- The left border corresponds to a line connecting the left ends of the lines representing the superior and inferior borders.
- The pulmonary, aortic, mitral, and tricuspid valves are located posterior to the sternum; however, the sounds produced by them are best heard at the auscultatory areas illustrated.

Auscultatory Areas

Clinicians' interest in the surface anatomy of the heart and cardiac valves results from their need to listen to valve sounds. The areas (*A*, *C*, and *D*) are as wide apart as possible so that the sounds produced at any given valve may be clearly distinguished from those produced at other valves. Blood tends to carry the sound in the direction of its flow; consequently, each area is situated superficial to the chamber or vessel into which the blood has passed and in a direct line with the valve orifice. ○

Superior Mediastinum

The superior mediastinum is superior to the transverse thoracic plane passing through the sternal angle and the junction (IV disc) of vertebrae T4 and T5 (Fig. 1.55). From anterior to posterior, the main contents of the superior mediastinum are (Fig. 1.56, *A* and *B*):

- Thymus, a lymphoid organ
- Great vessels related to the heart and pericardium:
  - brachiocephalic veins
  - SVC
  - arch of the aorta, and roots of its major branches:
    - brachiocephalic trunk
    - left common carotid
    - left subclavian artery

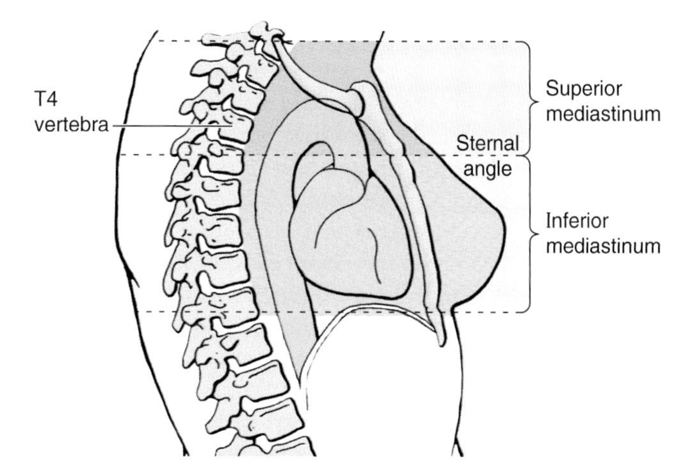

Figure 1.55. Superior mediastinum. The superior mediastinum extends inferiorly from the superior thoracic aperture to the transverse thoracic plane, passing through the sternal angle and the intervertebral (IV) disc of vertebrae T4 and T5.

- Vagus and phrenic nerves
- Cardiac plexus of nerves
- Left recurrent laryngeal nerve
- Trachea
- Esophagus
- Thoracic duct
- Prevertebral muscles.

Thymus. The thymus, a primary lymphoid organ, is located in the lower part of the neck and the anterior part of the superior mediastinum. It lies posterior to the manubrium and extends into the anterior mediastinum, anterior to the pericardium. After puberty, the thymus undergoes gradual involution and is largely replaced by fat. A rich arterial supply to the thymus is derived mainly from the anterior intercostal and the anterior mediastinal branches of the **internal thoracic arteries**. The *veins of the thymus* end in the left brachiocephalic, internal thoracic, and inferior thyroid veins. The *lymphatic vessels of the thymus* end in the parasternal, brachiocephalic, and tracheobronchial lymph nodes.

Age Changes in the Thymus

The flat, flask-shaped lobes of the thymus are a prominent feature of the superior mediastinum during infancy and childhood. In some newborn infants the thymus may also extend superiorly through the superior thoracic aperture into the neck and compress the trachea. The thymus plays an important role in the development and maintenance of the immune system. As puberty is reached, the thymus begins to diminish in relative size. By adulthood it has been largely replaced by adipose tissue and is often scarcely recognizable; however, it continues to produce T-lymphocytes. ○

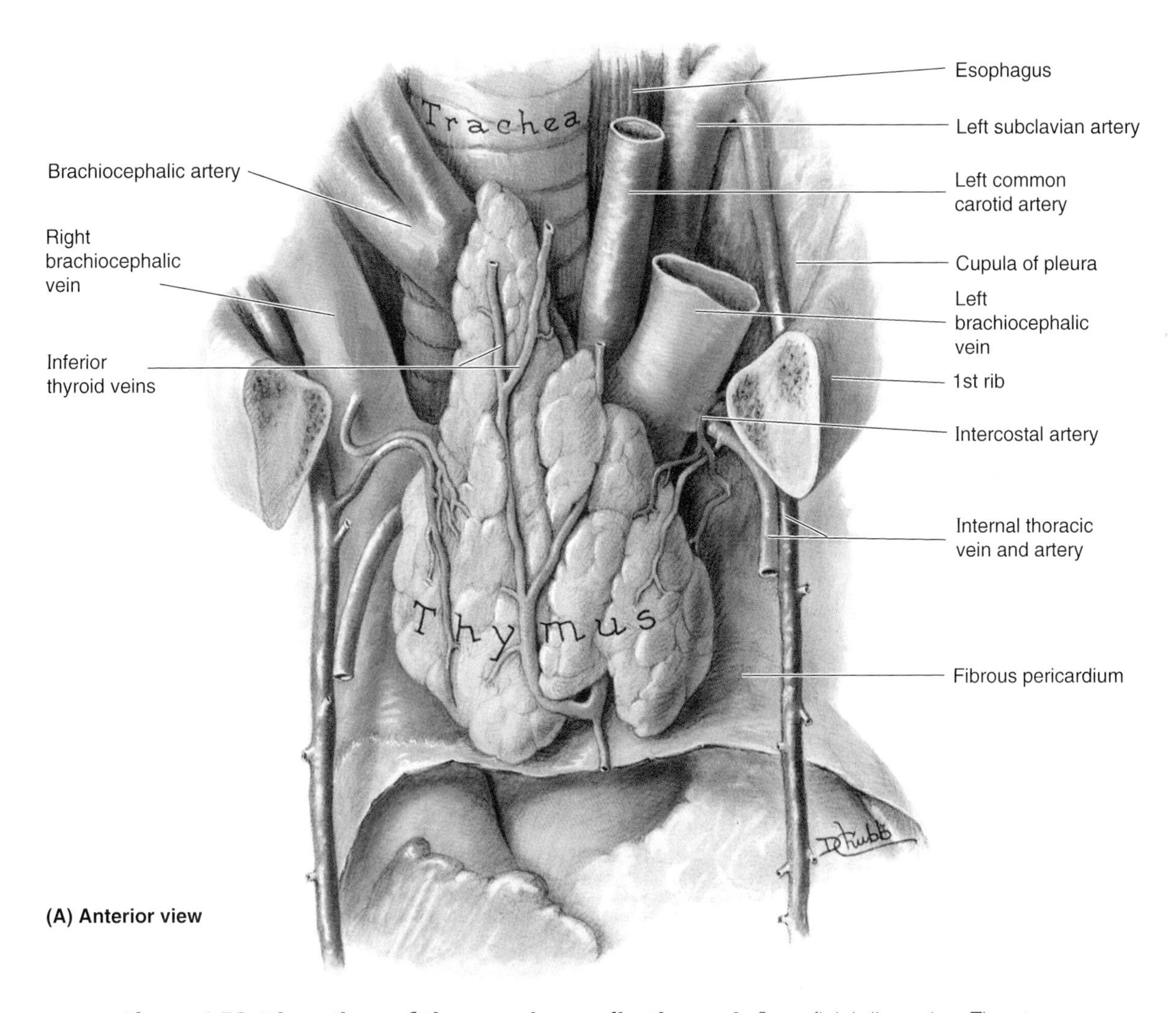

Figure 1.56. Dissections of the superior mediastinum. A. Superficial dissection. The sternum and ribs have been excised and the pleurae removed. It is unusual to see such a distinct thymus in an adult.

Great Vessels. The brachiocephalic veins are formed posterior to the SC joints by the union of the internal jugular and subclavian veins (Fig. 1.56). At the level of the inferior border of the 1st right costal cartilage, the brachiocephalic veins unite to form the SVC. The **left brachiocephalic vein** is over twice as long as the right vein because it passes from the left to the right side, passing across the anterior aspects of the roots of the three major branches of the arch of the aorta, and shunting blood from the head, neck, and left upper limb to the right atrium. The origin of the **right brachiocephalic vein** (i.e., by the union of right internal jugular and subclavian veins—the right "venous angle") receives lymph from the right lymphatic duct, and the origin of the left brachiocephalic vein (left "venous angle") receives lymph from the thoracic duct (Fig. 1.35).

The **SVC** returns blood from all structures superior to the diaphragm, except the lungs and heart. It passes inferiorly and ends at the level of the 3rd costal cartilage, where it enters the right atrium. The SVC lies in the right side of the superior mediastinum, anterolateral to the trachea and posterolateral to the ascending aorta. The **right phrenic nerve** lies between the SVC and the mediastinal pleura (Fig. 1.56*B*). The terminal half of the SVC is in the middle mediastinum, where it lies beside the ascending aorta and forms the posterior boundary of the transverse pericardial sinus.

The **arch of the aorta** (aortic arch), the curved continuation of the ascending aorta (Fig. 1.57, Table 1.6), begins posterior to the 2nd right sternocostal joint at the level of the sternal angle and arches superoposteriorly and to the left. The arch of the aorta ascends anterior to the right pulmonary artery and the bifurcation of the trachea to reach its apex at the left side of the trachea and esophagus, as it passes over the root of the left lung. The arch descends on the left side of the body of T4 vertebra.

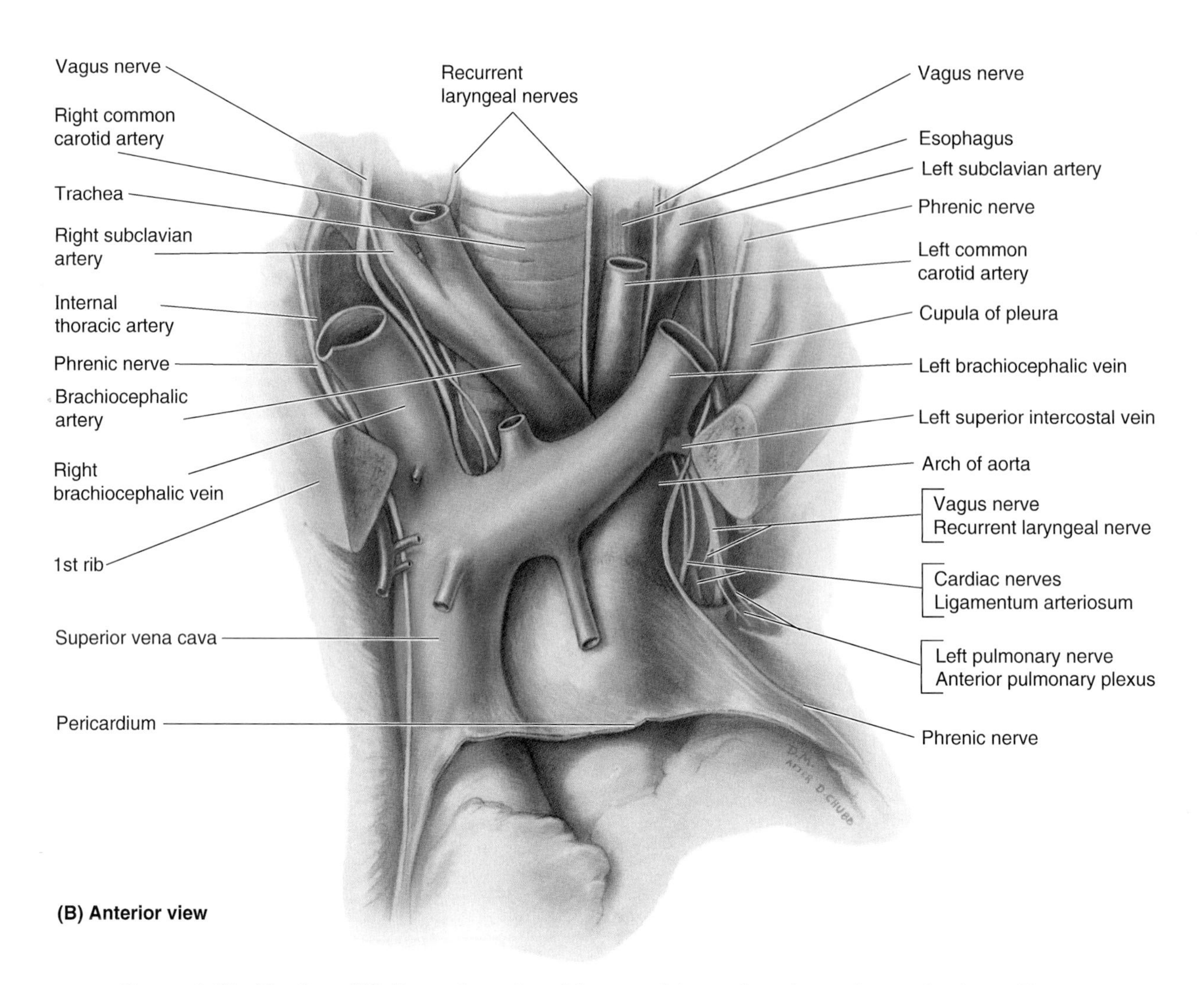

Figure 1.56. *(Continued)* **B.** Deep dissection of the root of the neck and superior mediastinum. The thymus has been removed. Observe the right vagus nerve (CN X) crossing anterior to the right subclavian artery and giving off the right recurrent laryngeal nerve, which passes medially to reach the trachea and esophagus. Note that the left recurrent laryngeal nerve passes inferior and then posterior to the arch of the aorta and ascends between the trachea and esophagus to the larynx.

Figure 1.57. Common pattern of branches of the arch of the aorta (present in approximately 65% of people). The largest branch (*BT*) arises from the beginning of the arch and divides into two branches (*RS* and *RC*). The next artery (*LC*) arises from the superior part of the arch. The 3rd branch (*LS*) arises from the arch approximately 1 cm distal to the left common carotid.

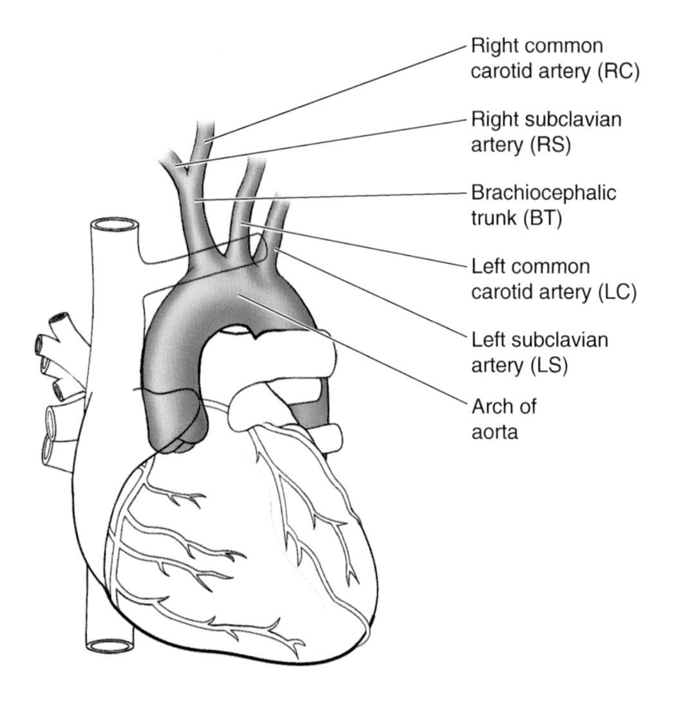

Table 1.6. **Aorta and Its Branches in the Thorax**

Artery	Origin	Course	Branches
Ascending aorta	Aortic orifice of left ventricle	Ascends approximately 5 cm to sternal angle where it becomes arch of aorta	Right and left coronary arteries
Arch of aorta	Continuation of ascending aorta	Arches posteriorly on left side of trachea and esophagus and superior to left main bronchus	Brachiocephalic, left common carotid, left subclavian
Thoracic aorta	Continuation of arch of aorta	Descends in posterior mediastinum to left of vertebral column; gradually shifts to right to lie in median plane at aortic hiatus	Posterior intercostal arteries, subcostal, some phrenic arteries and visceral branches (e.g., esophageal)
Posterior intercostal	Posterior aspect of thoracic aorta	Pass laterally, and then anteriorly parallel to ribs	Lateral and anterior cutaneous branches
Bronchial (1–2 branches)	Anterior aspect of aorta or posterior intercostal artery	Run with the tracheobronchial tree	Bronchial and peribronchial tissue, visceral pleura
Esophageal (4–5 branches)	Anterior aspect of thoracic aorta	Run anteriorly to esophagus	To esophagus
Superior phrenic (vary in number)	Anterior aspects of thoracic aorta	Arise at aortic hiatus and pass to superior aspect of diaphragm	To diaphragm

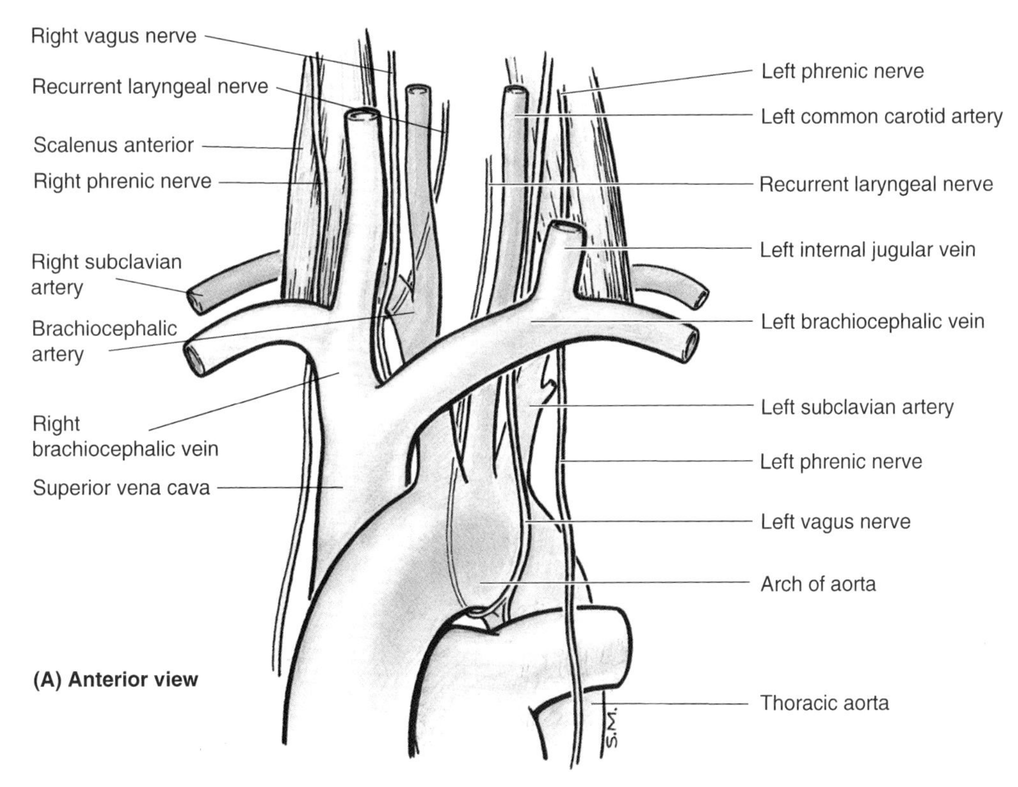

Figure 1.58. Great vessels and nerves. A. Relationships of the great vessels and nerves in the lower neck and superior mediastinum.

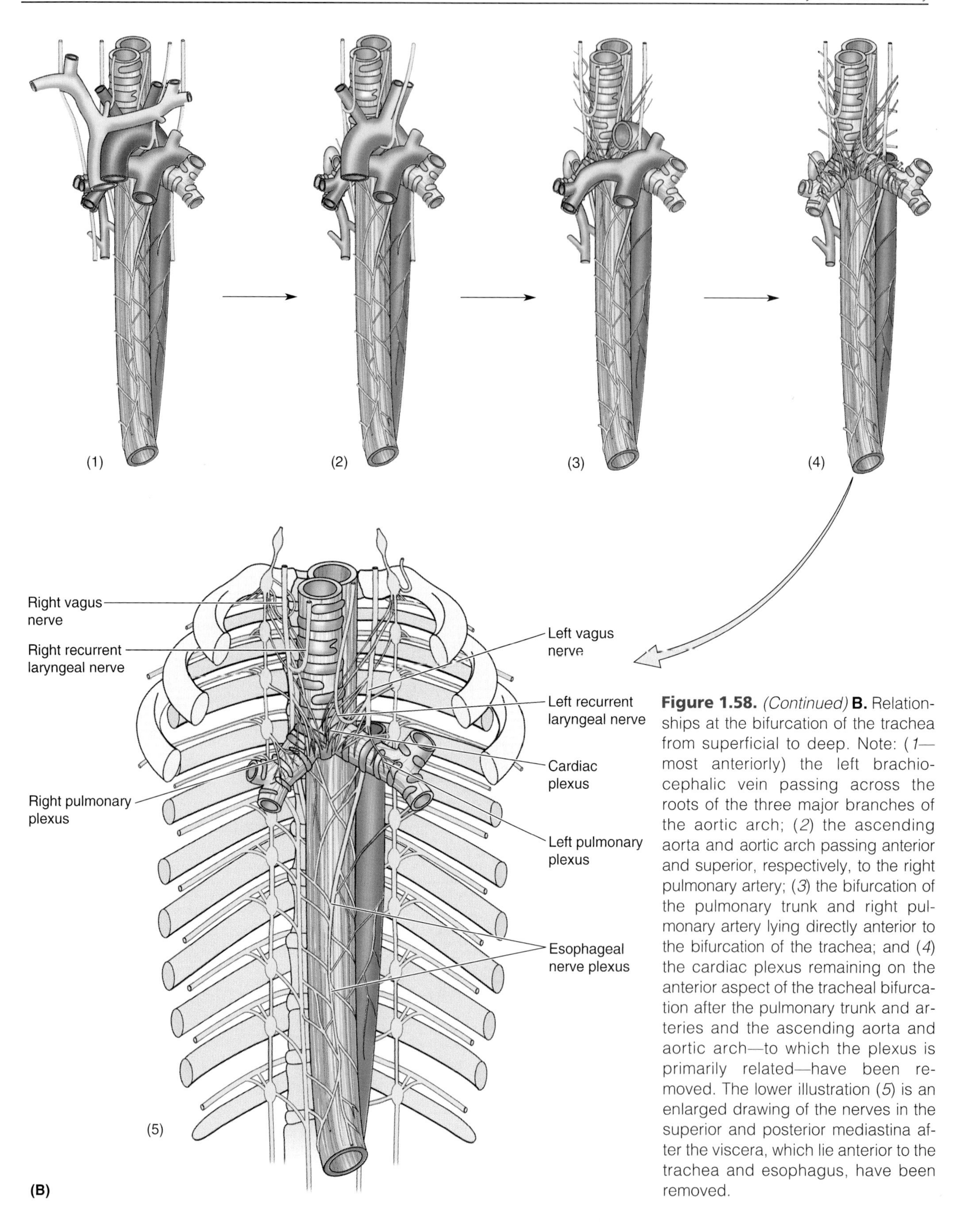

Figure 1.58. *(Continued)* **B.** Relationships at the bifurcation of the trachea from superficial to deep. Note: (*1*—most anteriorly) the left brachiocephalic vein passing across the roots of the three major branches of the aortic arch; (*2*) the ascending aorta and aortic arch passing anterior and superior, respectively, to the right pulmonary artery; (*3*) the bifurcation of the pulmonary trunk and right pulmonary artery lying directly anterior to the bifurcation of the trachea; and (*4*) the cardiac plexus remaining on the anterior aspect of the tracheal bifurcation after the pulmonary trunk and arteries and the ascending aorta and aortic arch—to which the plexus is primarily related—have been removed. The lower illustration (*5*) is an enlarged drawing of the nerves in the superior and posterior mediastina after the viscera, which lie anterior to the trachea and esophagus, have been removed.

The arch of the aorta ends by becoming the thoracic aorta posterior to the 2nd left sternocostal joint. Note that the arch of the azygos vein occupies a corresponding position on the right side of the trachea over the root of the right lung, although its contents are flowing in the opposite direction (Fig. 1.54).

The **ligamentum arteriosum**, the remnant of fetal ductus arteriosus (Moore and Persaud, 1998), passes from the root of the left pulmonary artery to the inferior surface of the arch of the aorta (Figs. 1.54 and 1.56*B*). The **left recurrent laryngeal nerve** hooks beneath the arch of the aorta adjacent to the ligamentum arteriosum and ascends between the trachea and esophagus (Fig. 1.58*A*). *The usual branches of the arch of the aorta* (Figs. 1.57 and 1.58*A*) are the

- Brachiocephalic trunk
- Left common carotid artery
- Left subclavian artery.

The **brachiocephalic trunk**, the first and largest branch of the aortic arch, arises posterior to the manubrium, where it is anterior to the trachea and posterior to the left brachiocephalic vein. It ascends superolaterally to reach the right side of the trachea and the right SC joint, where it divides into the right common carotid and right subclavian arteries.

The **left common carotid artery**, the 2nd branch of the arch, arises posterior to the manubrium, slightly posterior and to the left of the brachiocephalic trunk. It ascends anterior to the left subclavian artery and is at first anterior to the trachea and then to its left. It enters the neck by passing posterior to the left SC joint.

The **left subclavian artery**, the 3rd branch of the arch, arises from the posterior part of the arch, just posterior to the left common carotid artery. It ascends lateral to the trachea and the left common carotid artery through the superior mediastinum; it has no branches in the mediastinum. As it leaves the thorax and enters the root of the neck, it passes posterior to the left SC joint.

Location of the Left Brachiocephalic Vein in Children

The brachiocephalic vein is formed and runs initially in the lower neck rather than in the superior mediastinum in some children. This occurs because children's necks are relatively short. The possible high location of this vein must be kept in mind when doing a *tracheostomy* (creating an opening in the trachea [see Chapter 8] to relieve upper airway obstruction and facilitate ventilation).

Variations in Branches of the Arch of the Aorta

The usual pattern of branches of the arch of the aorta, shown in Figure 1.57, is present in approximately 65% of people. Variations in the origin of the branches of the aortic arch are fairly common. In approximately 27% of people, the left common carotid originates from the brachiocephalic trunk. A brachiocephalic trunk fails to form in approximately 2.5% of people, in these cases each of the four arteries (right and left common carotid and subclavian arteries) originating independently from the arch of the aorta. The left vertebral artery originates from the arch of the aorta in approximately 5% of people. Both right and left brachiocephalic trunks originate from the arch in approximately 1.2% of people. For a description of other variations in the origins of the branches of the arch of the aorta, see Bergman et al. (1988).

A **retroesophageal right subclavian artery** sometimes arises as the last (most left-sided) branch of the arch of the aorta. The artery crosses posterior to the esophagus to reach the right upper limb and may compress the esophagus, causing difficulty in swallowing (*dysphagia*). Less commonly, an accessory artery to the thyroid gland, the *thyroid ima artery* (L. thyroidea ima), arises from the arch of the aorta or the brachiocephalic artery. Because it ascends anterior to the trachea, surgeons are careful when making a median approach to the trachea.

Anomalies of the Arch of the Aorta

The most superior part of the aortic arch is usually approximately 2.5 cm inferior to the superior border of the manubrium, but it may be more superior or inferior. Sometimes the aortic arch curves over the root of the right lung and passes inferiorly on the right side, forming a **right aortic arch**. In some cases the abnormal arch, after passing over the root of the right lung, passes posterior to the esophagus to reach its usual position on the left side. Less frequently, a **double aortic arch** forms a vascular ring around the esophagus and trachea. A trachea that is compressed enough to affect breathing may require surgical division of the vascular ring. An *aneurysm (localized dilation) of the aortic arch* may also exert pressure on the trachea and esophagus, causing difficulty in breathing and swallowing.

Coarctation of the Aorta

The descending aorta in this congenital anomaly has an abnormal narrowing (stenosis) that diminishes the caliber of the aortic lumen, producing an obstruction to bloodflow to the inferior part of the body. The most common site for an isolated coarctation is near the site of insertion of the ductus or ligamentum arteriosum (Moore and Persaud, 1998; Sabiston and Lyerly, 1994). When the coarctation is inferior to the site of the ductus or ligamentum arteriosum (postductal coarctation), a good collateral circulation usu- ▶

▶ ally develops between the proximal and distal parts of the aorta through the intercostal and internal thoracic arteries. This type of coarctation is compatible with many years of life because the collateral circulation carries blood to the descending aorta inferior to the stenosis. The collateral vessels may become so large that they cause notable pulsation in the intercostal spaces and erode the adjacent surfaces of the ribs, which is visible in chest radiographs. ⊙

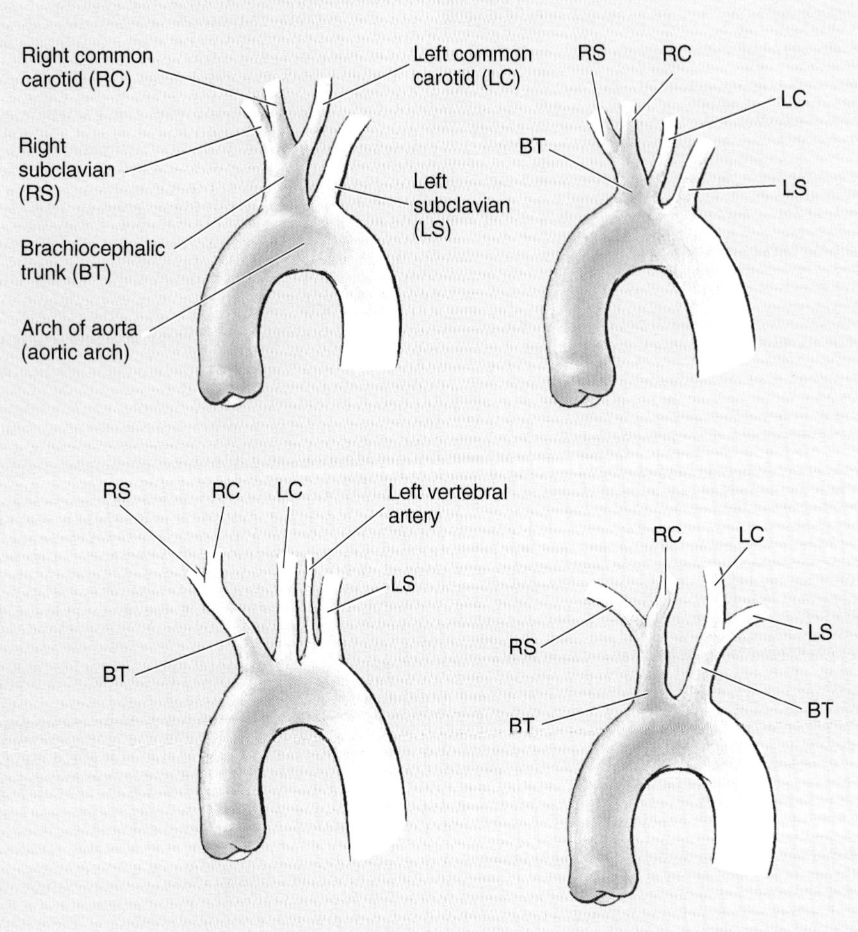

Variations in the origins of the branches of the arch of the aorta

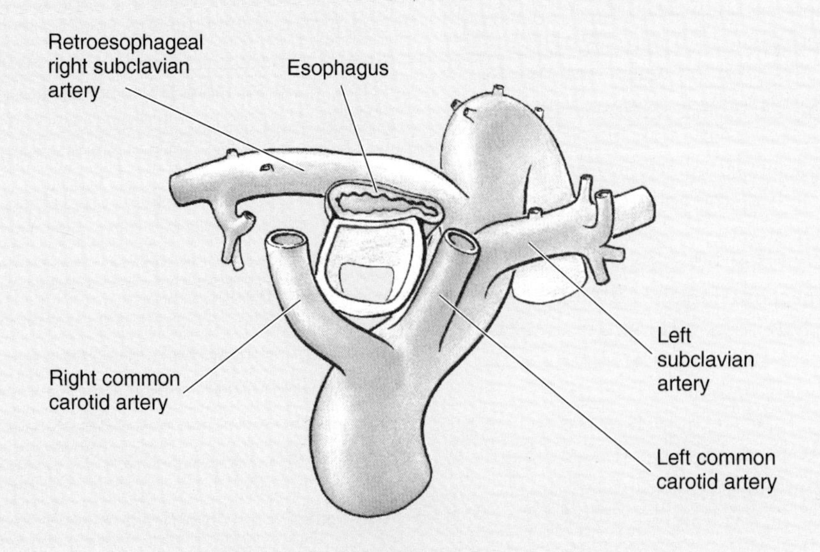

Retroesophageal right subclavian artery

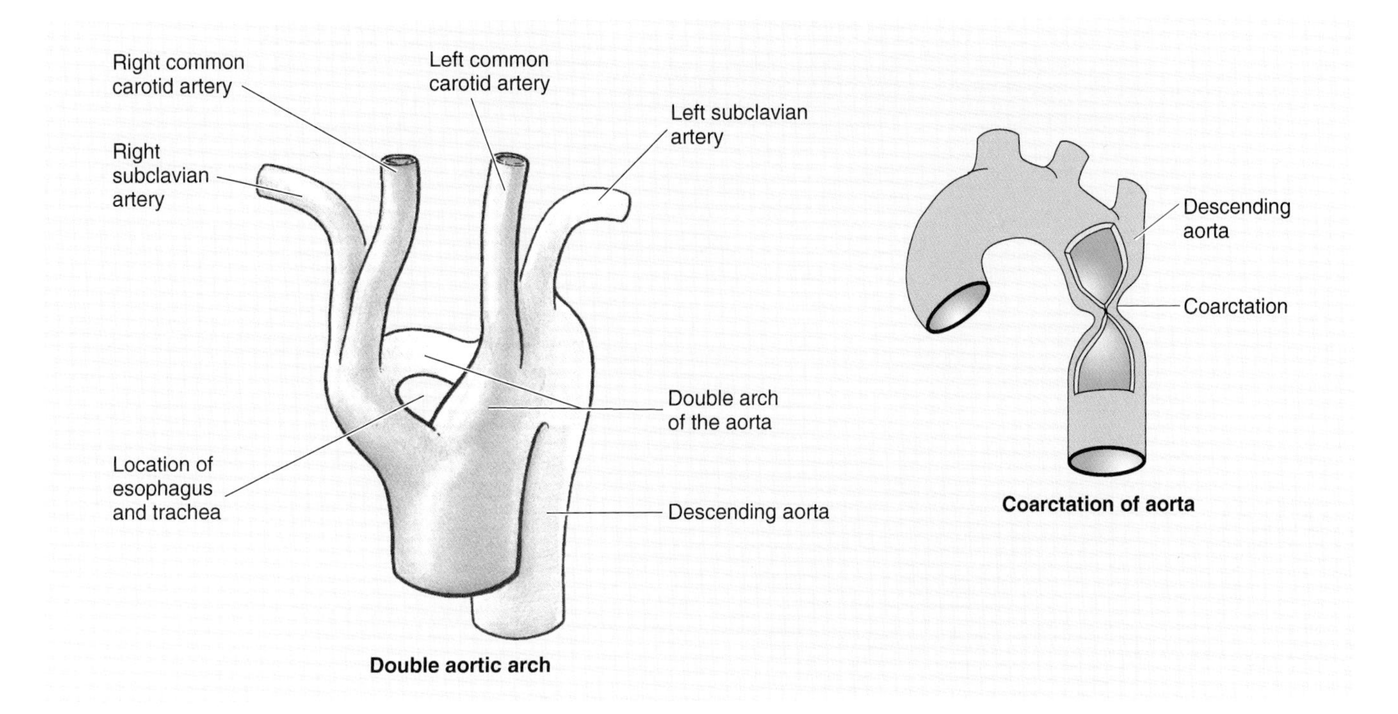

Nerves in the Superior Mediastinum. The *vagus nerves* arise bilaterally from the medulla of the brain, exit the cranium (L. skull), and descend through the neck posterolateral to the common carotid arteries (Fig. 1.58*A*, Table 1.7). Each nerve enters the superior mediastinum posterior to the respective SC joint and brachiocephalic vein.

The **right vagus nerve** enters the thorax anterior to the right subclavian artery, where it gives rise to the **right recurrent laryngeal nerve**. This nerve hooks around the right subclavian artery and ascends between the trachea and esophagus to supply the larynx. The right vagus nerve runs posteroinferiorly through the superior mediastinum on the right side of the trachea. It then passes posterior to the right brachiocephalic vein, SVC, and root of the right lung. Here it divides into many branches that contribute to the **pulmonary plexus** (Fig. 1.58*B*). Usually the right vagus nerve leaves this plexus as a single nerve and passes to the esophagus, where it again breaks up and contributes fibers to the **esophageal nerve plexus**. The right vagus nerve also gives rise to nerves that contribute to the **cardiac plexus**.

The **left vagus nerve** descends in the neck posterior to the left common carotid artery (Fig. 1.58*A*). It enters the mediastinum between the left common carotid artery and left subclavian artery. When it reaches the left side of the aortic arch, the left vagus nerve diverges posteriorly from the left phrenic nerve. It is separated laterally from the phrenic nerve by the *left superior intercostal vein*. As the left vagus nerve curves medially at the inferior border of the aortic arch, it gives off the **left recurrent laryngeal nerve**. This nerve passes inferior to the aortic arch just lateral to the ligamentum arteriosum and ascends to the larynx in the groove between the trachea and esophagus. The left vagus nerve passes posterior to the root of the left lung, where it breaks up into many branches that contribute to the **left pulmonary plexus** (Fig. 1.58*B*). The nerve leaves this plexus as a single trunk and passes to the esophagus, where it joins fibers from the right vagus in the **esophageal nerve plexus**.

Injury to the Recurrent Laryngeal Nerves

The recurrent laryngeal nerves supply all intrinsic muscles of the larynx, except one. Consequently, any investigative procedure (e.g., *mediastinotomy*) or disease process in the superior mediastinum may injure these nerves and affect the voice. Because the left recurrent laryngeal nerve winds around the arch of the aorta and ascends between the trachea and esophagus, it may be involved in a bronchogenic or esophageal carcinoma, enlargement of mediastinal lymph nodes, or an *aneurysm of the arch of the aorta*. In the latter condition, the nerve may be stretched by the dilated arch. ⊙

The **phrenic nerves** (Fig. 1.58*A*) supply the diaphragm; approximately one-third of their fibers are sensory to the diaphragm. Each nerve enters the superior mediastinum between the subclavian artery and the origin of the brachiocephalic vein (Table 1.7). The fact that the phrenic nerves pass anterior to the roots of the lungs provides an important means of distinguishing them from the vagus nerves, which pass posterior to the roots.

The **right phrenic nerve** passes along the right side of the right brachiocephalic vein, SVC, and pericardium over the right atrium. It also passes anterior to the root of the right

Table 1.7. **Nerves of the Thorax**

Nerve	Origin	Course	Distribution
Vagus (CN X)	8 to 10 rootlets from medulla of brainstem	Enters superior mediastinum posterior to sternoclavicular joint and brachiocephalic vein; gives rise to recurrent laryngeal nerve; continues into abdomen	Pulmonary plexus, esophageal plexus, and cardiac plexus
Phrenic	Ventral rami of C3–C5 nerves	Passes through superior thoracic aperture and runs between mediastinal pleura and pericardium	Central portion of diaphragm
Intercostals	Ventral rami of T1 to T11 nerves	Run in intercostal spaces between internal and innermost layers of intercostal muscles	Muscles in and skin over intercostal space; lower nerves supply muscles and skin of anterolateral abdominal wall
Subcostal	Ventral ramus of T12 nerve	Follows inferior border of 12th rib and passes into abdominal wall	Abdominal wall and skin of gluteal region
Recurrent laryngeal	Vagus nerve	Loops around subclavian on right; on left runs around arch of aorta and ascends in tracheoesophageal groove	Intrinsic muscles of larynx (except cricothyroid); sensory inferior to level of vocal folds
Cardiac plexus	Cervical and cardiac branches of vagus nerve and sympathetic trunk	From arch of aorta and posterior surface of heart, fibers extend along coronary arteries and to SA node	Impulses pass to SA node; parasympathetic fibers slow rate, reduce force of heartbeat, and constrict coronary arteries; sympathetic fibers have opposite effect
Pulmonary plexus	Vagus nerve and sympathetic trunk	Forms on root of lung and extends along bronchial subdivisions	Parasympathetic fibers constrict bronchioles; sympathetic fibers dilate them
Esophageal plexus	Vagus nerve, sympathetic ganglia, greater splanchnic nerve	Distal to tracheal bifurcation, the vagus and sympathetic nerves form a plexus around the esophagus	Vagal and sympathetic fibers to smooth muscle and glands of inferior two-thirds of esophagus

lung and descends on the right side of the IVC to the diaphragm, which it pierces near the vena caval foramen.

The **left phrenic nerve** descends between the left subclavian and left common carotid arteries. It crosses the left surface of the arch of the aorta anterior to the left vagus nerve and passes over the left superior intercostal vein. It then descends anterior to the root of the left lung and runs along the pericardium, superficial to the left atrium and ventricle of the heart, where it pierces the diaphragm to the left of the pericardium.

Trachea. The trachea descends anterior to the esophagus and enters the superior mediastinum, inclining a little to the right of the median plane (Fig. 1.59). The posterior surface of the trachea is flat where it is applied to the esophagus. The trachea ends at the level of the sternal angle by dividing into the right and left main bronchi. The trachea terminates superior to the level of the heart and is not a component of the posterior mediastinum.

Esophagus. The esophagus is a fibromuscular tube that extends from the pharynx to the stomach. It is usually flattened anteroposteriorly (Figs. 1.59 and 1.60). It enters the superior mediastinum between the trachea and vertebral column, where it lies anterior to the bodies of the vertebrae T1 through T4. Initially, it inclines to the left but is moved by the arch of the aorta to the median plane opposite the root of the left lung. In the superior mediastinum, the **thoracic duct** usually lies on the left side of the esophagus, deep (medial) to the arch of the aorta (Fig. 1.60*B*). Inferior to the arch, the esophagus again inclines to the left as it approaches and passes through the **esophageal hiatus** in the diaphragm.

Posterior Mediastinum

The posterior mediastinum is located anterior to T5 through T12 vertebrae, posterior to the pericardium and diaphragm, and between the parietal pleura of the two lungs. *The posterior mediastinum contains the*

- Thoracic aorta
- Thoracic duct
- Posterior mediastinal lymph nodes (e.g., tracheobronchial nodes)

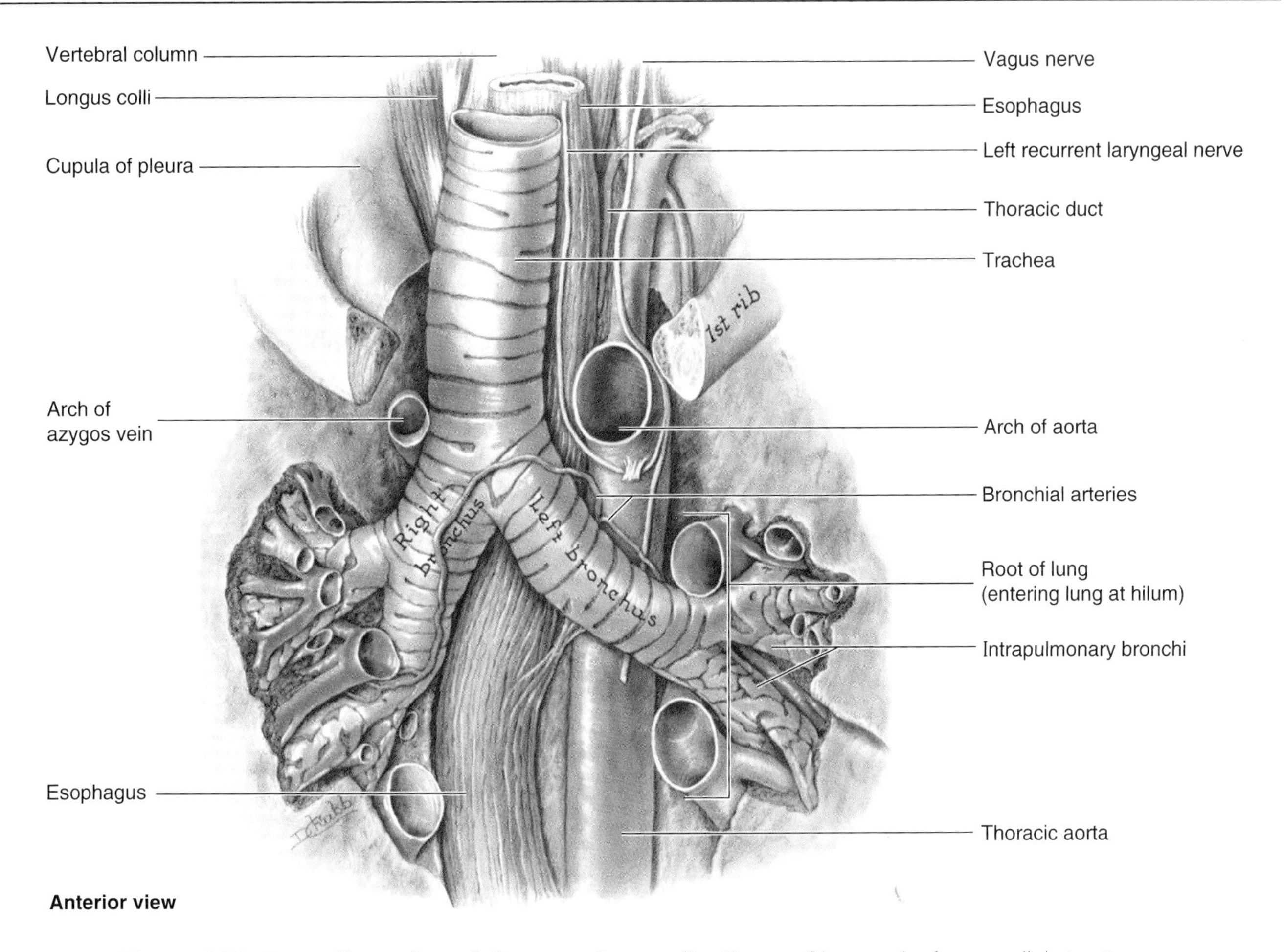

Figure 1.59. Deep dissection of the superior mediastinum. Observe the four parallel structures: trachea, esophagus, left recurrent laryngeal nerve, and thoracic duct. Note that the right bronchus is more vertical, shorter, and wider than the left bronchus. The course of the right bronchial artery is abnormal; usually it passes posterior to the bronchus.

- Azygos and hemiazygos veins
- Esophagus
- Esophageal plexus
- Thoracic sympathetic trunks
- Thoracic splanchnic nerves.

Thoracic Aorta. The thoracic aorta—the thoracic part of the descending aorta—is the continuation of the arch of the aorta (Figs. 1.60*B* and 1.61, Table 1.6). It begins on the left side of the inferior border of the body of T4 vertebra and descends in the posterior mediastinum on the left sides of T5 through T12 vertebrae. As it descends, it approaches the median plane and displaces the esophagus to the right. The **thoracic aortic plexus** (Fig. 1.60*B*), an autonomic nerve network, surrounds it. The thoracic aorta lies posterior to the root of the left lung (Fig. 1.59), the pericardium, and esophagus. The thoracic aorta terminates—changes its name to abdominal aorta—anterior to the inferior border of T12 vertebra and enters the abdomen through the **aortic hiatus** (opening) in the diaphragm. The thoracic duct and azygos vein ascend on its right side after accompanying the aorta through this hiatus (Fig. 1.61). The **branches of (arteries arising from) the thoracic aorta** (Fig. 1.62, Table 1.6) are:

- Bronchial
- Pericardial
- Posterior intercostal
- Superior phrenic
- Esophageal
- Mediastinal
- Subcostal.

The **bronchial arteries** consist of one right and two small left vessels. The bronchial arteries supply the trachea, bronchi, lung tissue, and lymph nodes. The **pericardial arteries** send twigs to the pericardium. The **posterior intercostal arteries** (nine pairs) pass into the 3rd through 11th intercostal

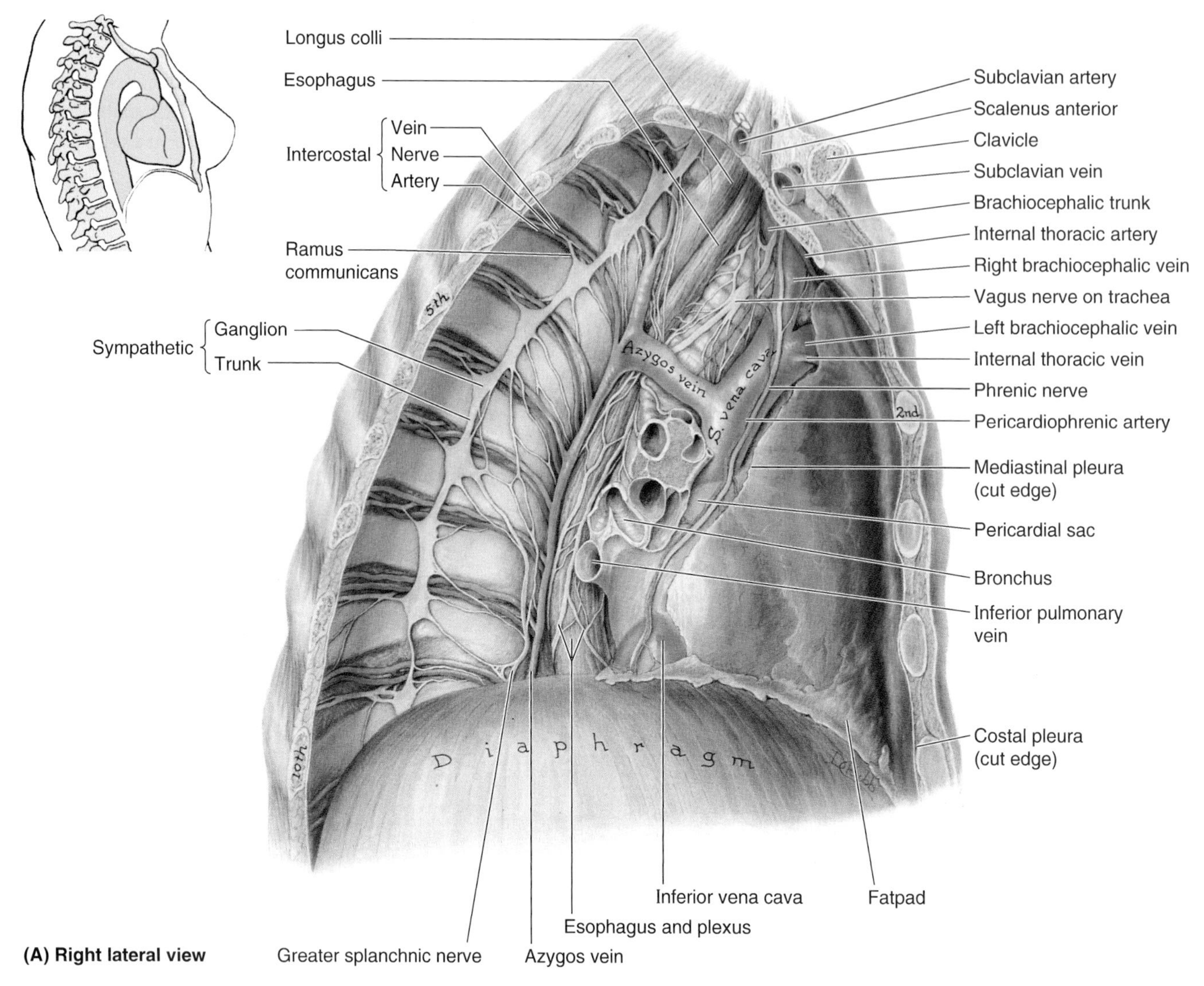

Figure 1.60. Dissections of the mediastinum. A. Right side. Most of the costal and mediastinal pleura has been removed to expose the underlying structures. Observe that this side of the mediastinum, the *"blue side,"* is dominated by the arch of the azygos vein, the superior vena cava (SVC), and the right atrium.

spaces. The **superior phrenic arteries** pass to the posterior surface of the diaphragm, where they anastomose with the musculophrenic and pericardiacophrenic branches of the internal thoracic artery. Usually the two **esophageal arteries** supply the middle third of the esophagus. The **mediastinal arteries** are small and supply the lymph nodes and other tissues of the posterior mediastinum. The **subcostal arteries** that enter the abdomen are in series with the intercostal arteries.

Esophagus. The esophagus descends into the posterior mediastinum from the superior mediastinum, passing posterior and to the right of the arch of the aorta (Fig. 1.61) and posterior to the pericardium and left atrium. The esophagus constitutes the primary posterior relationship of the base of the heart. It then deviates to the left and passes through the **esophageal hiatus** in the diaphragm at the level of T10 vertebra, anterior to the aorta. The esophagus may have three impressions, or "constrictions," in its thoracic part. These may be observed as narrowings of the lumen in oblique chest radiographs that are taken as barium is swallowed. The esophagus is compressed by three structures:

- The aortic arch
- The left main bronchus
- The diaphragm.

No constrictions are visible in the empty esophagus; however, as it expands during filling, the above structures compress its walls.

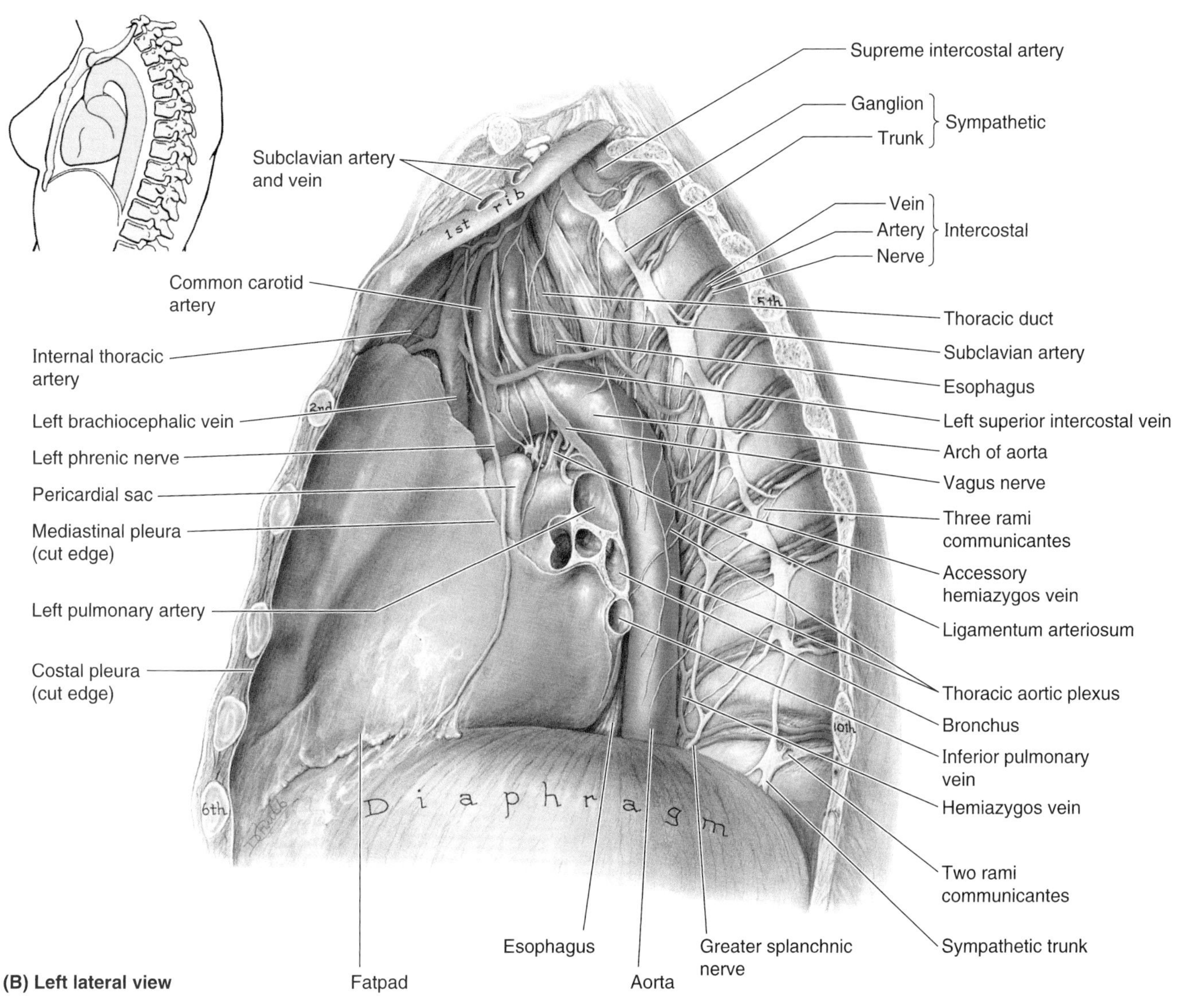

Figure 1.60. *(Continued)* **B.** Left side. Observe that this side of the mediastinum, the *"red side,"* is dominated by the arch of the aorta, the thoracic aorta, the left common carotid and subclavian arteries. Observe the sympathetic trunk attached to intercostal nerves by communicating branches (L. rami communicantes), and the left superior intercostal vein, draining the upper 2 to 3 intercostal spaces and passing anteriorly to enter the left brachiocephalic vein.

Blockage of the Esophagus

The impressions produced in the esophagus by adjacent structures are of clinical interest because of the slower passage of substances at these sites. The impressions indicate where swallowed foreign objects are most likely to lodge and where a stricture may develop following the accidental drinking of a caustic liquid such as lye. ⊙

Thoracic Duct. In the posterior mediastinum, the thoracic duct lies on the anterior aspect of the bodies of the inferior seven thoracic vertebrae (Fig. 1.63). It is the largest lymphatic channel in the body. The thoracic duct conveys most lymph of the body to the venous system (that from the lower limbs, pelvic cavity, abdominal cavity, left side of the thorax, left side of the head and neck, and the left upper limb—i.e., all but the right upper quadrant—see p. 37). The thoracic duct originates from the **chyle cistern** (L. cisterna chyli) in the abdomen and ascends through the aortic hiatus in the diaphragm (Fig. 1.61). The thoracic duct is usually thin-walled and dull white; often it is beaded because of its numerous valves. It ascends in the posterior mediastinum between the thoracic aorta on its left, the azygos vein on its right, the esophagus anteriorly, and the vertebral bodies posteriorly. At the level of T4, T5, or T6 vertebrae, the thoracic duct crosses

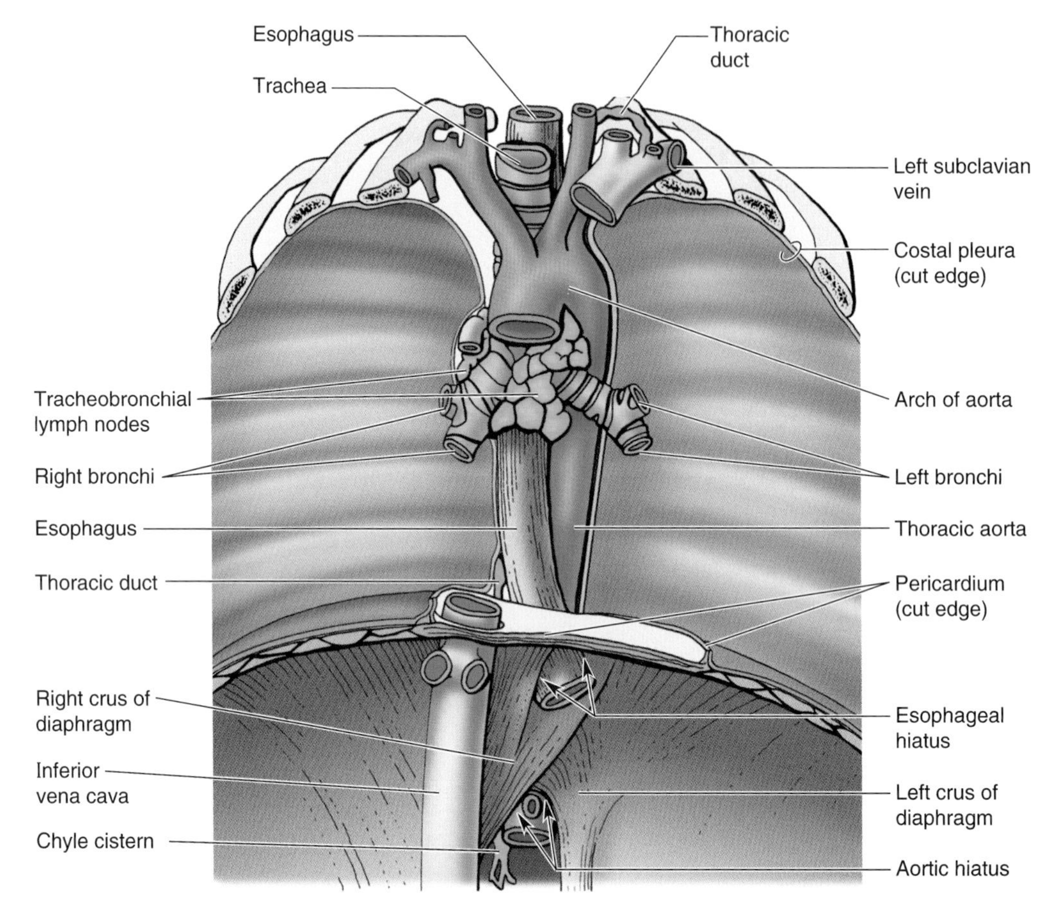

Figure 1.61. Anterior view of esophagus, trachea, bronchi, and aorta. Observe that the arch of the aorta curves posteriorly on the left side of the trachea and esophagus. Observe also that the right crus of the diaphragm is larger and longer than the left one, and that it splits to enclose the esophagus. Note that the thoracic duct in this specimen enters the left subclavian vein.

to the left, posterior to the esophagus, and ascends into the superior mediastinum. The thoracic duct receives branches from the middle and upper intercostal spaces of both sides through several collecting trunks. It also receives branches from posterior mediastinal structures. Near its termination it often receives the jugular, subclavian, and bronchomediastinal lymphatic trunks (although any or all these vessels may terminate independently). The thoracic duct usually empties into the venous system near the union of the left internal jugular and subclavian veins—i.e., the left "venous angle" or origin of the left brachiocephalic vein (Fig. 1.63)—but it may open into the left subclavian vein as illustrated in Figure 1.61.

Laceration of the Thoracic Duct

Because the thoracic duct is thin-walled and may be colorless, it may be difficult to identify. Consequently, it is vulnerable to inadvertent injury during investigative and/or surgical procedures in the posterior mediastinum. Laceration of the thoracic duct during an accident or lung surgery results in lymph escaping into the thoracic cavity at rates ranging from 75 to 200 cc per hour (Woodburne and Burkel, 1994). Lymph (chyle) may also enter the pleural cavity, producing *chylothorax.* This fluid may be removed by a needle tap or by thoracocentesis (p. 87), but in some cases it may be necessary to ligate the thoracic duct. The lymph then returns to the venous system by other lymphatic channels that join the thoracic duct superior to the ligature.

Variations of the Thoracic Duct

Variations of the thoracic duct are common because the superior part of the duct represents the original left member of a pair of vessels in the embryo (Moore and Persaud, 1998). Sometimes two thoracic ducts are present for a short distance. ○

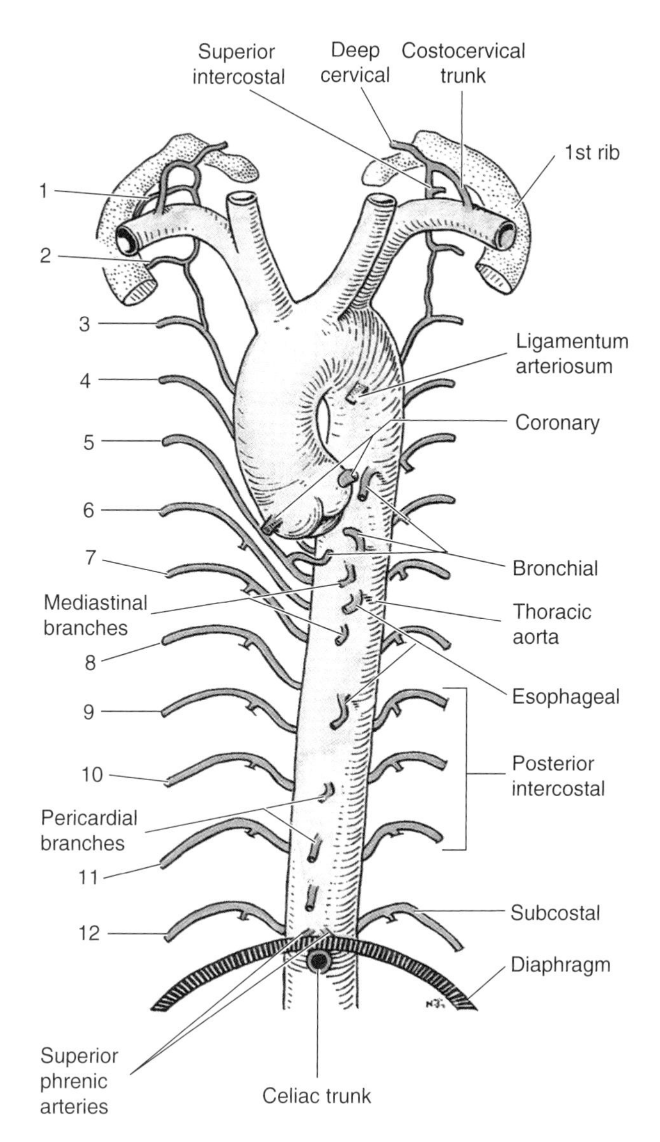

Figure 1.62. Branches of the thoracic aorta. The superior phrenic arteries arising from the inferior part of the thoracic aorta supply the diaphragm. The numbers 1 to 12 indicate posterior intercostal arteries.

Vessels and Lymph Nodes of the Posterior Mediastinum. The thoracic aorta and its branches were discussed previously. *Posterior mediastinal lymph nodes* (Fig. 1.63) lie posterior to the pericardium, where they are related to the esophagus and thoracic aorta. There are several nodes posterior to the inferior part of the esophagus and more (up to eight) anterior and lateral to it. The posterior mediastinal lymph nodes receive lymph from the esophagus, the posterior aspect of the pericardium and diaphragm, and the middle posterior intercostal spaces.

The **azygos system of veins**, on each side of the vertebral column, drains the back and thoracoabdominal walls (Fig. 1.64) as well as the mediastinal viscera. The azygos system exhibits much variation, not only in its origin but also in its course, tributaries, anastomoses, and termination. The *azygos vein* (azygos means paired) and its main tributary, the *hemiazygos vein*, usually arise from "roots" or anastomoses with the posterior aspect of the IVC and/or renal vein, respectively, which merge with the ascending lumbar veins.

The **azygos vein** forms a collateral pathway between the SVC and IVC and drains blood from the posterior walls of the thorax and abdomen. It ascends in the posterior mediastinum, passing close to the right sides of the bodies of the inferior eight thoracic vertebrae. It arches over the superior aspect of the root of the right lung to join the SVC, similar to the way the arch of the aorta passes over the root of the left lung. In addition to the **posterior intercostal veins**, the azygos vein communicates with the vertebral venous plexuses that drain the back, vertebrae, and structures in the vertebral canal (see Chapter 4). The azygos vein also receives the mediastinal, esophageal, and bronchial veins.

The **hemiazygos vein** arises on the left side by the junction of the left subcostal and ascending lumbar veins. It ascends on the left side of the vertebral column, posterior to the thoracic aorta as far as T9 vertebra. Here it crosses to the right, posterior to the aorta, thoracic duct, and esophagus, and joins the azygos vein. The hemiazygos vein receives the inferior three posterior intercostal veins, the inferior esophageal veins, and several small mediastinal veins.

The **accessory hemiazygos vein** begins at the medial end of the 4th or 5th intercostal space and descends on the left side of the vertebral column from T5 through T8. It receives tributaries from veins in the 4th through 8th intercostal spaces and sometimes from the left bronchial veins. It crosses over T7 or T8 vertebrae, posterior to the thoracic aorta and thoracic duct, where it joins the azygos vein. Sometimes the accessory hemiazygos vein joins the hemiazygos vein and opens with it into the azygos vein. The accessory hemiazygos is frequently connected to the left superior intercostal vein, as shown in Figure 1.64. The left superior intercostal vein, which drains the 1st through 3rd intercostal spaces, may communicate with the accessory hemiazygos vein; however, it drains primarily into the left brachiocephalic vein.

Alternate Venous Routes to the Heart

The azygos, hemiazygos, and accessory hemiazygos veins offer alternate means of venous drainage from the thoracic, abdominal, and back regions when *obstruction of the IVC* occurs. In some people, an accessory azygos vein parallels the main azygos vein on the right side. Other people have no hemiazygos system of veins. A clinically important variation, although uncommon, is when the azygos system receives all the blood from the IVC except that from the liver. In these people, the azygos system drains nearly all the blood inferior to the diaphragm, except from the digestive tract. If the SVC is obstructed superior to the entrance of the azygos vein, blood can drain inferiorly into the veins of the abdominal wall and return to the right atrium through the IVC and azygos venous system. ○

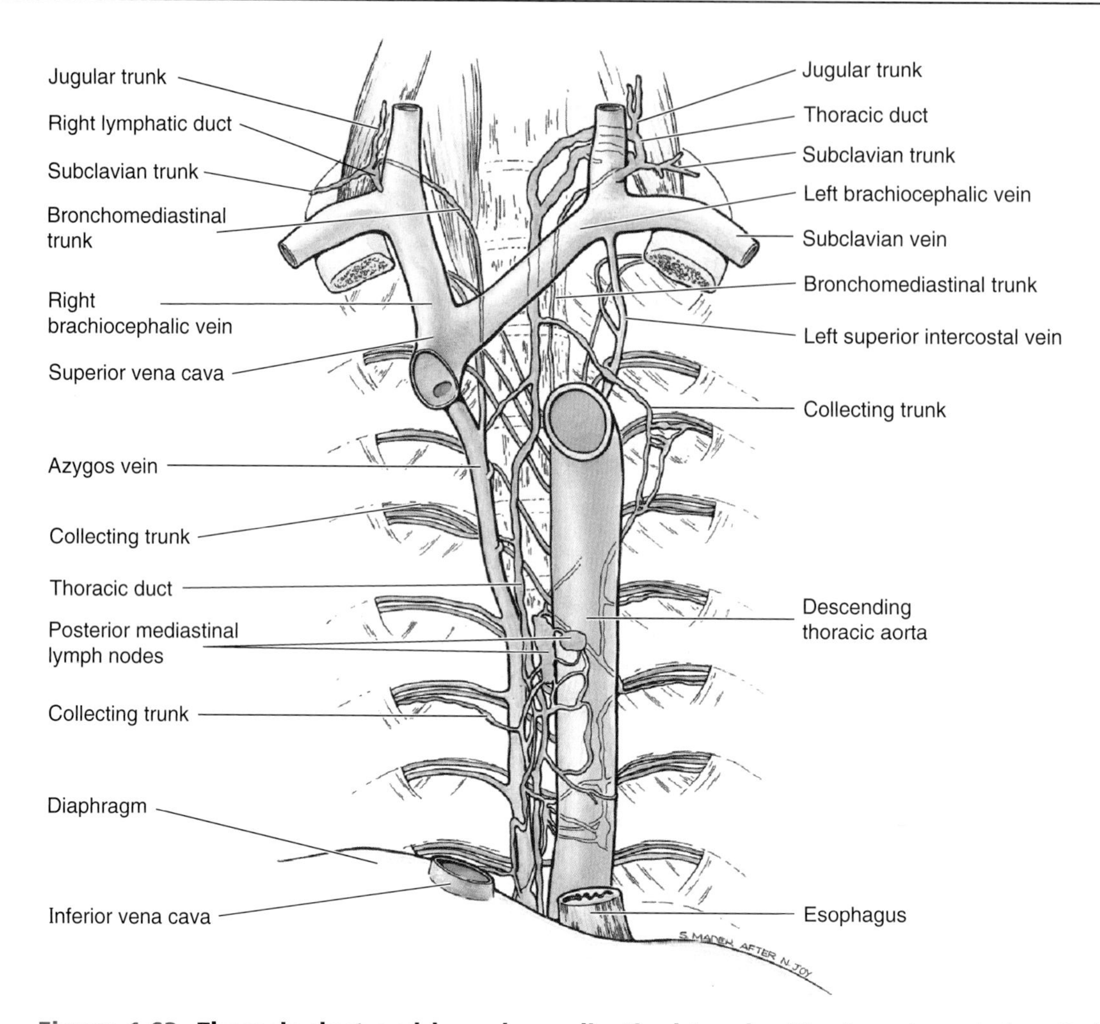

Figure 1.63. Thoracic duct and bronchomediastinal trunks. The thoracic aorta is pulled slightly to the left and the azygos vein slightly to the right to expose the thoracic duct, which ascends on the vertebral column between the azygos vein and the descending thoracic aorta. At approximately the junction of the posterior and superior mediastina (IV disc T4/T5), the thoracic duct passes to the left and continues its ascent to the neck where it arches laterally to open near, or at, the angle of union of the internal jugular and subclavian veins (left "venous angle"). Also observe that the thoracic duct receives branches from the intercostal spaces of both sides through several collecting trunks and branches from posterior mediastinal structures. Near its termination, the thoracic duct receives the jugular, subclavian, and bronchomediastinal trunks. Observe that the right lymphatic duct is formed by the union of the right jugular, subclavian, and bronchomediastinal trunks.

Nerves of the Posterior Mediastinum. The sympathetic trunks and their associated ganglia form a major portion of the autonomic nervous system (Fig. 1.65, Table 1.7). The **thoracic sympathetic trunks** are in continuity with the cervical and lumbar sympathetic trunks. The thoracic trunks lie against the heads of the ribs in the superior part of the thorax, the costovertebral joints in the midthoracic level, and the sides of the vertebral bodies in the inferior part of the thorax. The **lower thoracic splanchnic nerves**—also known as greater, lesser, and least splanchnic nerves—are part of the abdominopelvic splanchnic nerves because they supply viscera inferior to the diaphragm. They consist of presynaptic fibers from the 5th through 12th sympathetic ganglia, which pass through the diaphragm and synapse in prevertebral ganglia in the abdomen. They supply sympathetic innervation for most of the abdominal viscera. These splanchnic nerves are discussed further with the abdomen (see Chapter 2).

Anterior Mediastinum

The anterior mediastinum, the smallest subdivision of the mediastinum, lies between the body of the sternum and the transversus thoracis muscles anteriorly and the pericardium posteriorly. It is continuous with the superior mediastinum at the sternal angle and is limited inferiorly by the diaphragm. The anterior mediastinum consists of loose connective tissue (*sternopericardial ligaments*), fat, lymphatic vessels, a few lymph nodes, and branches of the internal thoracic vessels. In infants and children, the anterior mediastinum contains the inferior part of the thymus. In unusual cases, this gland may extend to the level of the 4th costal cartilages.

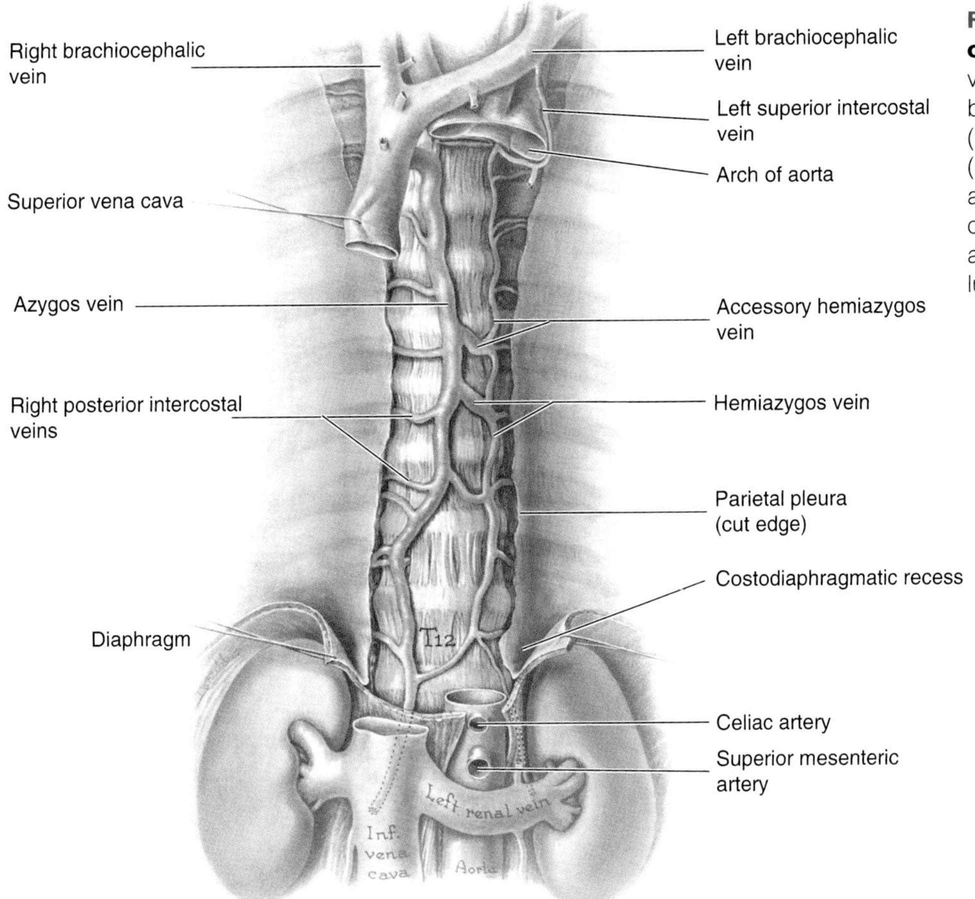

Figure 1.64. Azygos system of veins. Note that the azygos vein forms a direct connection between the inferior vena cava (IVC) and superior vena cava (SVC). The azygos and hemiazygos veins are also continuous inferiorly (below the diaphragm) with the ascending lumbar veins (not shown).

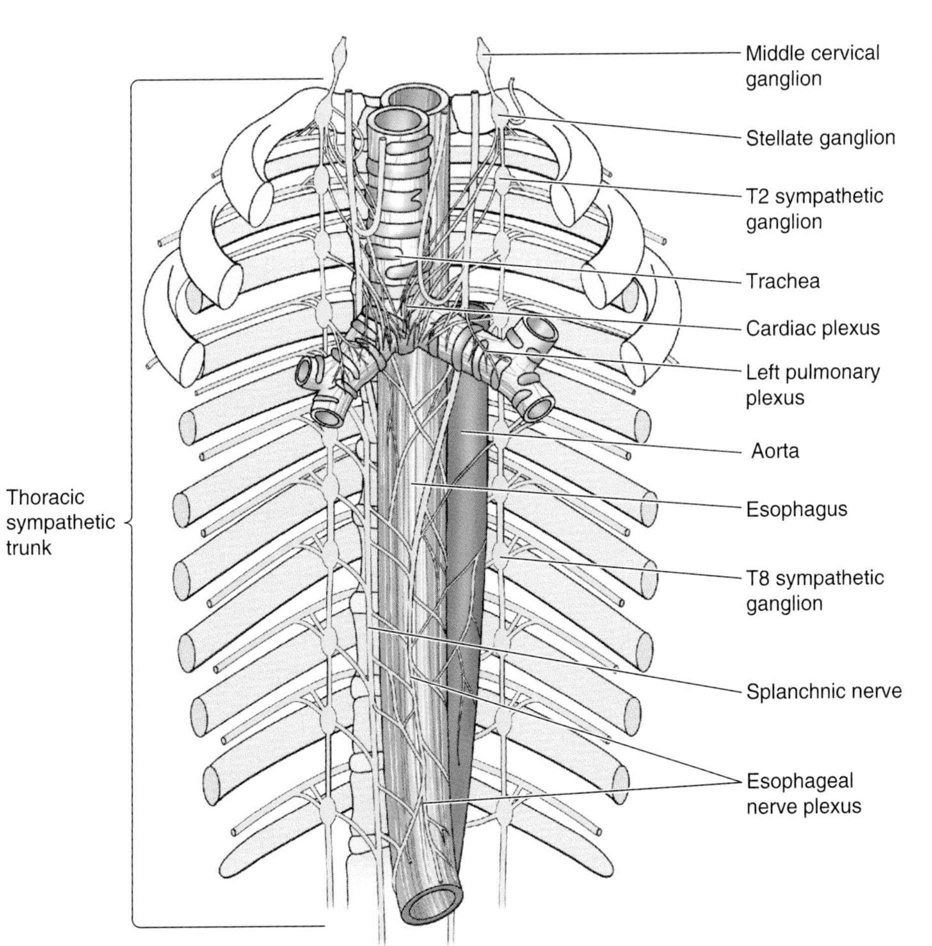

Figure 1.65. Autonomic nerves of the superior and posterior mediastina. Observe the thoracic sympathetic trunks and the stellate or cervicothoracic ganglion, which is formed by the fusion of the inferior cervical ganglion and the 1st thoracic ganglion. The 2nd ganglion is often larger than the adjacent ganglia.

Medical Imaging of the Thorax

The main imaging methods for examination of the thorax are radiography, echocardiography, computerized tomography, and magnetic resonance imaging (MRI).

Radiography

The most frequently used radiograph of the thorax is a **posteroanterior (PA) projection** (Fig. 1.66). The radiologist or technician places the anterior aspect of the patient's thorax against the X-ray film cassette and rotates the shoulders anteriorly to move the scapulae away from the superior parts of the lungs (Fig. 1.66*A*). The patient takes a deep breath and holds it. The deep inspiration causes the diaphragmatic domes to descend, filling the lungs with air (increasing their radiolucency), and moving the costodiaphragmatic parts of the lungs into the costodiaphragmatic recesses. In PA projections, most ribs stand out clearly against the background of the relatively lucent lungs. The inferior ribs tend to be obscured by the diaphragm and the superior contents of the abdomen (e.g., liver), depending on the phase of respiration when the ▶

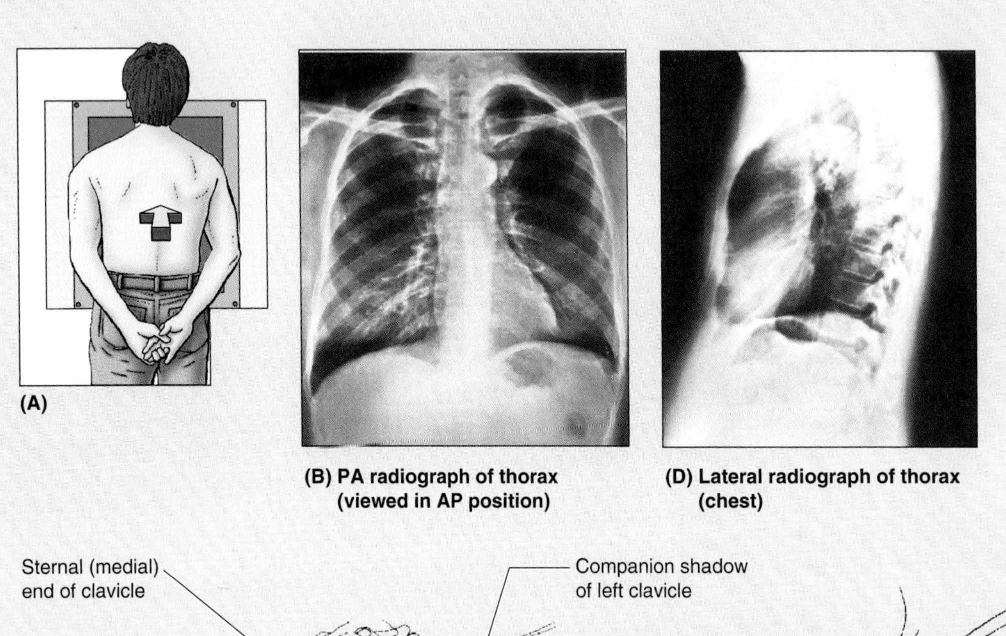

(B) PA radiograph of thorax (viewed in AP position)

(D) Lateral radiograph of thorax (chest)

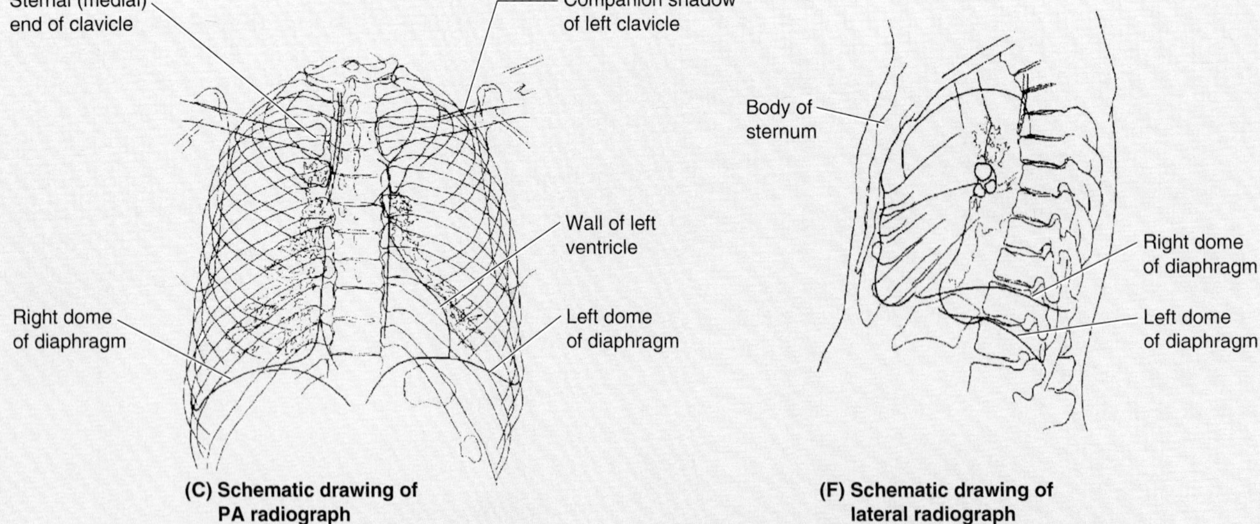

(C) Schematic drawing of PA radiograph

(F) Schematic drawing of lateral radiograph

Figure 1.66. Radiographs of the thorax. A. Orientation of the person's thorax relative to the film cassette and X-ray beam (*arrow*) for a posteroanterior (PA) radiograph. **B.** PA radiograph of the thorax. **C.** Schematic drawing identifying the main structures visible in the PA radiograph. The right dome is usually higher than the left one. **D.** Lateral radiograph of the thorax. **E.** Orientation of the person's thorax relative to the film cassette and X-ray beam (*arrow*) for a lateral radiograph. **F.** Schematic drawing identifying the structures visible in the lateral radiograph.

▶ radiograph was taken. Usually only lateral margins of the manubrium are visible in PA projections. Uncommonly, cervical ribs, missing ribs, forked ribs, and fused ribs are visible. Occasionally, the costal cartilages are calcified in older people and the calcification tends to increase in the inferior cartilages.

Lateral radiographs are made with the side of the thorax against the film cassette and the upper limbs elevated over the head. Lateral radiographs allow better viewing of a lesion or anomaly confined to one side of the chest.

A **PA radiograph** is a composite of the images cast by the soft tissues and bones of the thoracic wall. Soft tissues, including those of the breasts, cast shadows of varying density depending on their composition and thickness. Paralleling the superior margins of the clavicles are shadows cast by the skin and subcutaneous tissues covering these bones. The clavicles, ribs, and inferior cervical and superior thoracic vertebrae are visible. The other thoracic vertebrae are more or less obscured by the sternum and mediastinum. In lateral chest radiographs (Fig. 1.66*D*), the middle and inferior thoracic vertebrae are visible, although they are partially obscured by the ribs. The three parts of the sternum are also visible. In PA projections, the **right and left domes of the diaphragm** are separated by the central tendon, which is obscured by the heart. The right dome of the diaphragm, formed by the underlying liver, is usually approximately one-half an intercostal space higher than the left dome. In a lateral projection, both domes are often visible as they arch superiorly from the sternum. The **lungs**, because of their low density, are relatively lucent compared with surrounding structures. The lungs exhibit a radiodensity similar to that of air and therefore produce paired radiolucent areas. In PA projections, the lungs are obscured inferior to the domes of the diaphragm and anterior and posterior to the mediastinum. The areas obscured in PA projections are usually visible in lateral projections. The *pulmonary arteries* are visible in the hilum of each lung. Intrapulmonary vessels are slightly larger in caliber in the inferior lobes. Transverse sections of the air-filled bronchi have lucent centers and thin walls.

The **heart** casts most of the central radiopaque shadow in PA projections (Fig. 1.66*B*), but the separate chambers of the heart are not distinguishable. Knowledge of the structures forming the **cardiovascular shadow** or silhouette (Fig. 1.67*A*) is important because changes in the shadow may indicate anomalies or functional disease. In AP projections, the borders of the cardiovascular shadow are the

- **Right border:** right brachiocephalic vein, SVC, right atrium, and IVC
- **Left border:** terminal part of the arch of aorta (aortic knob), pulmonary trunk, left auricle, and left ventricle.

The left inferior part of the cardiovascular shadow presents the region of the apex. The typical anatomical apex, if present, is often inferior to the shadow of the diaphragm. The three main *types of cardiovascular shadows* are (Fig. 1.67*B*)

- *Transverse type*, observed in obese persons, pregnant women, and infants
- *Oblique type*, characteristic of most people
- *Vertical type*, present in people with narrow chests.

Fluoroscopy

The movements of different organs (e.g., heart, lungs, and diaphragm) may be studied by means of fluoroscopy. In this system, the X-rays impinge on the input screen of an image-intensifier tube that is coupled to a television monitor. This image-intensifier television system gives superior resolution along with reduced radiation exposure. ▶

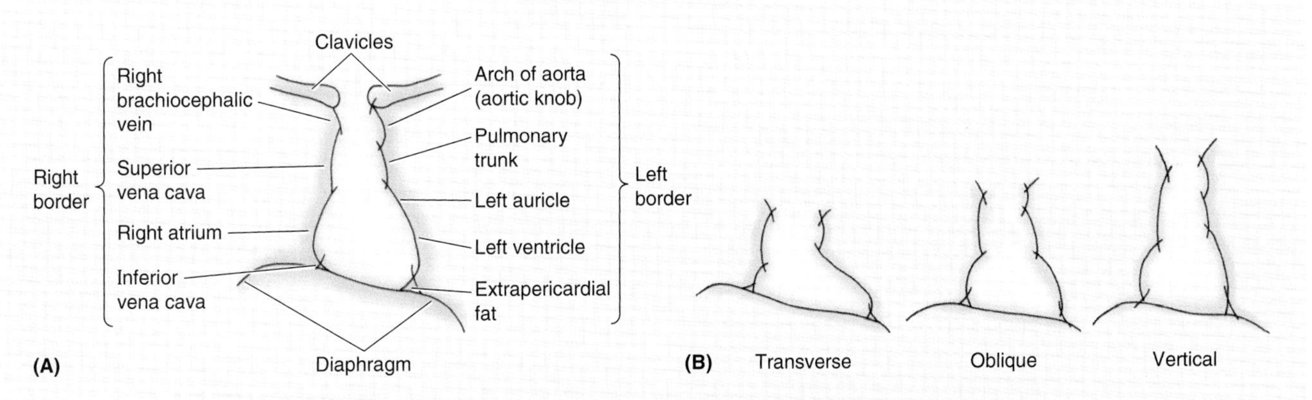

Figure 1.67. Cardiovascular shadows (mediastinal silhouettes). A. Composition of margins of cardiovascular shadow. **B.** Types of cardiovascular shadows.

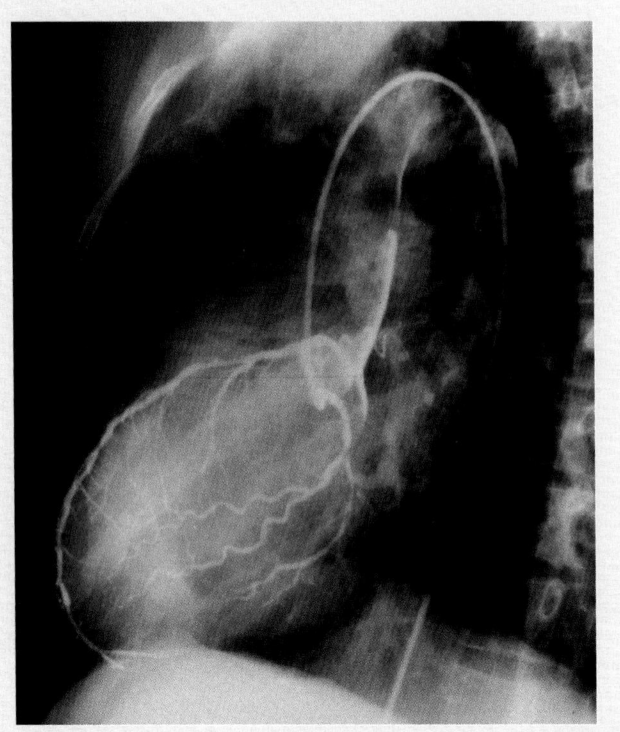

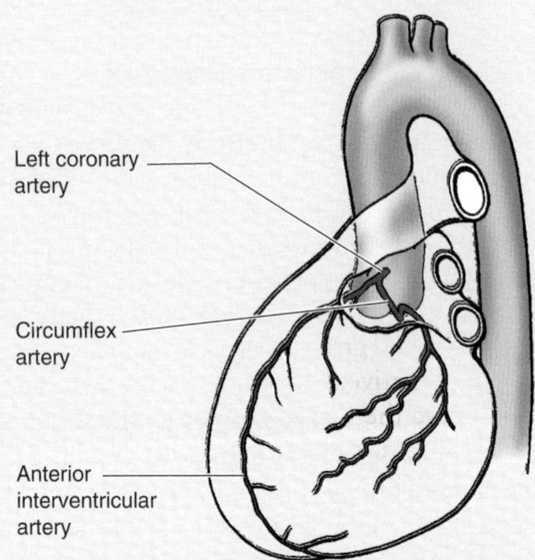

(A) Left lateral view

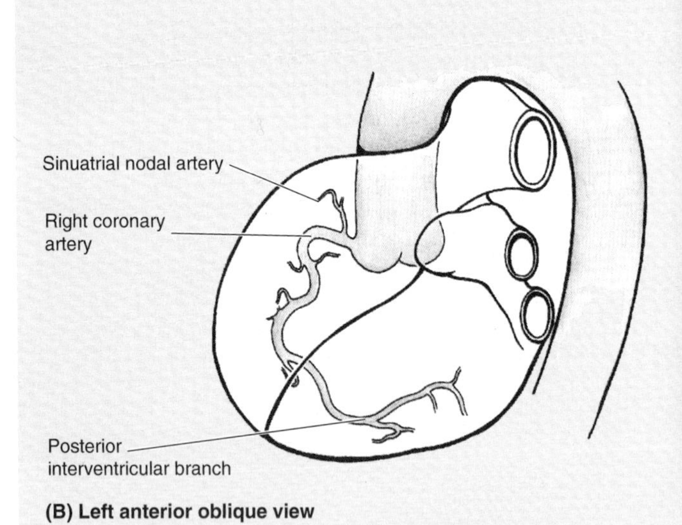

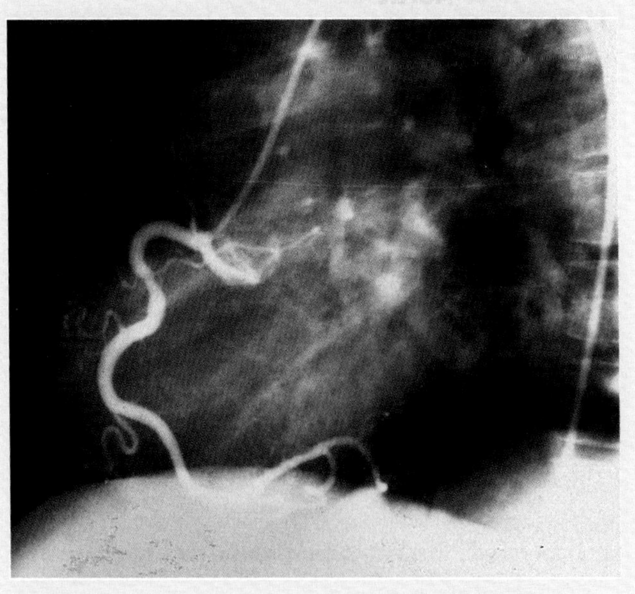

(B) Left anterior oblique view

Figure 1.68. Coronary arteriograms. The orientation drawing shows the course of the left coronary artery shown in the arteriogram (**A**). Observe the catheter in the ascending aorta and the arch of the aorta. **B.** Observe the sinuatrial (SA) nodal artery arising from the right coronary artery. (Courtesy of Dr. I. Morrow, Department of Radiology, Health Sciences Centre, University of Manitoba, Winnipeg, Manitoba, Canada)

Bronchography

Bronchography—radiographic examination of the tracheobronchial tree following introduction of an oily radiopaque material (by means of aerosol or drip)—is rarely performed now. It has been superseded by high-resolution computed tomography (CT) and bronchoscopy.

Cardiac Catheterization

In cardiac catheterization, a radiopaque catheter is inserted into a peripheral vein and passed under fluoroscopic control into the right atrium, right ventricle, pulmonary trunk, and pulmonary arteries, respectively. Using this technique, intracardiac pressures can be recorded and blood samples may be removed. If a radiopaque contrast medium is injected, it can be followed through the heart and great vessels using serially exposed X-ray films. Alternatively, **cineradiography** can be performed to observe the flow of dye in "real time." Both techniques permit study of the circulation through the functioning heart and are helpful in the study of congenital cardiac defects.

Coronary Angiography

Using coronary angiography, the coronary arteries can be visualized with coronary arteriograms (Fig. 1.68). A long, narrow catheter is passed into the ascending aorta via the femoral (thigh) or brachial (arm) arteries. Under fluoroscopic control, the tip of the catheter is placed just inside the opening of a coronary artery. A small injection of radiopaque contrast material is made and cineradiographs are taken to show the lumen of the artery and its branches, as well as any stenotic areas that may be present. In a similar manner, an **aortic angiogram** can be made by injecting radiopaque contrast material into the aorta and into openings of the arteries arising from the arch of the aorta (Fig. 1.69). ▶

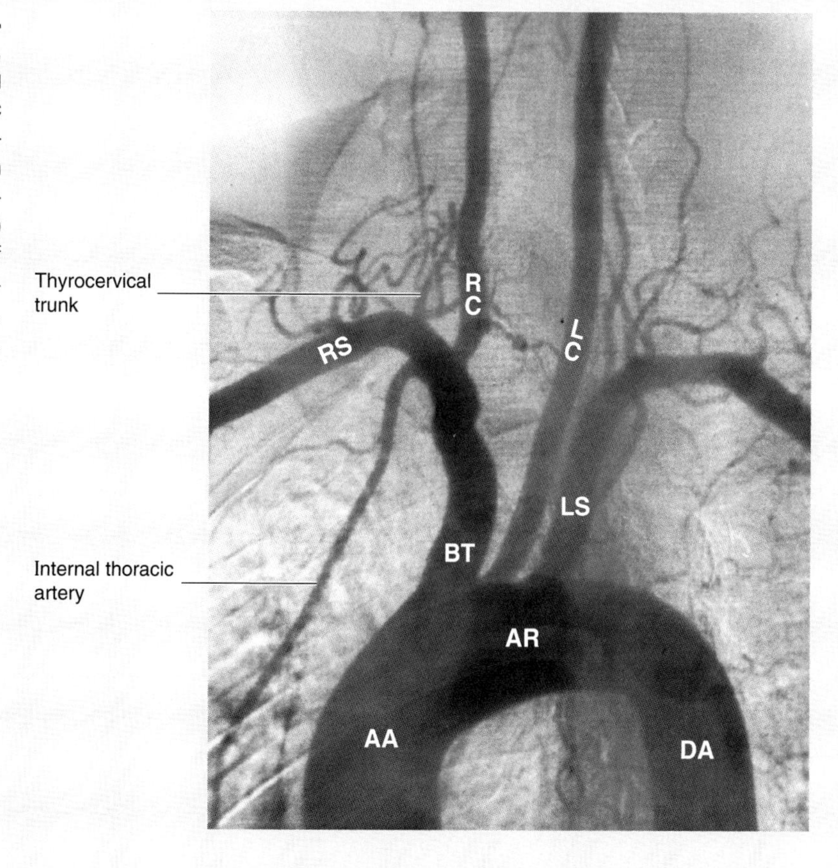

Figure 1.69. Aortic angiogram. Observe the ascending aorta (*AA*), arch of the aorta (*AR*), descending thoracic aorta (*DA*), brachiocephalic trunk (*BT*) branching into the right subclavian (*RS*), and right common carotid (*RC*) arteries. The left subclavian (*LS*) and left common carotid (*LC*) arteries arise directly from the arch of the aorta. (Courtesy of Dr. E.L. Lansdown, Professor of Medical Imaging, University of Toronto, Toronto, Ontario, Canada)

Echocardiography

Echocardiography (ultrasonic cardiography) is a method of graphically recording the position and motion of the heart by the echo obtained from beams of ultrasonic waves directed through the thoracic wall. Echocardiography delineates valvular stenosis and regurgitation, especially on the left side of the heart. This technique may detect as little as 20 mL of fluid in the pericardial cavity, such as that resulting from pericardial effusion. *Doppler echocardiography* is a technique that records the flow of blood through the heart and great vessels by Doppler ultrasonography.

Computed Tomography and Magnetic Resonance Imaging

CT and MRI are commonly used to examine the thorax (Figs. 1.70 and 1.71). CT is sometimes combined with mammography for detection of breast cancer. Before CT scans are taken, an iodide-contrast material is given intravenously. As breast cancer cells have an unusual affinity for iodide, they become recognizable. CT is quicker and less expensive, but MRI is better for detecting and delineating lesions. *CT scans* are always oriented to show how a horizontal section of a patient's body lying on an examination table would appear to the physician who is at the patient's feet. Therefore, the left lateral edge of the image represents the right lateral surface of the patient's body. MRI scans can construct various images of transverse, sagittal, oblique, or coronal sections of the body. CT and MRI scans are mainly used as adjuncts to conventional radiography when high-contrast sensitivity is desired. For reasons of cost, speed, and availability, CT is more widely used than MRI. ⊙

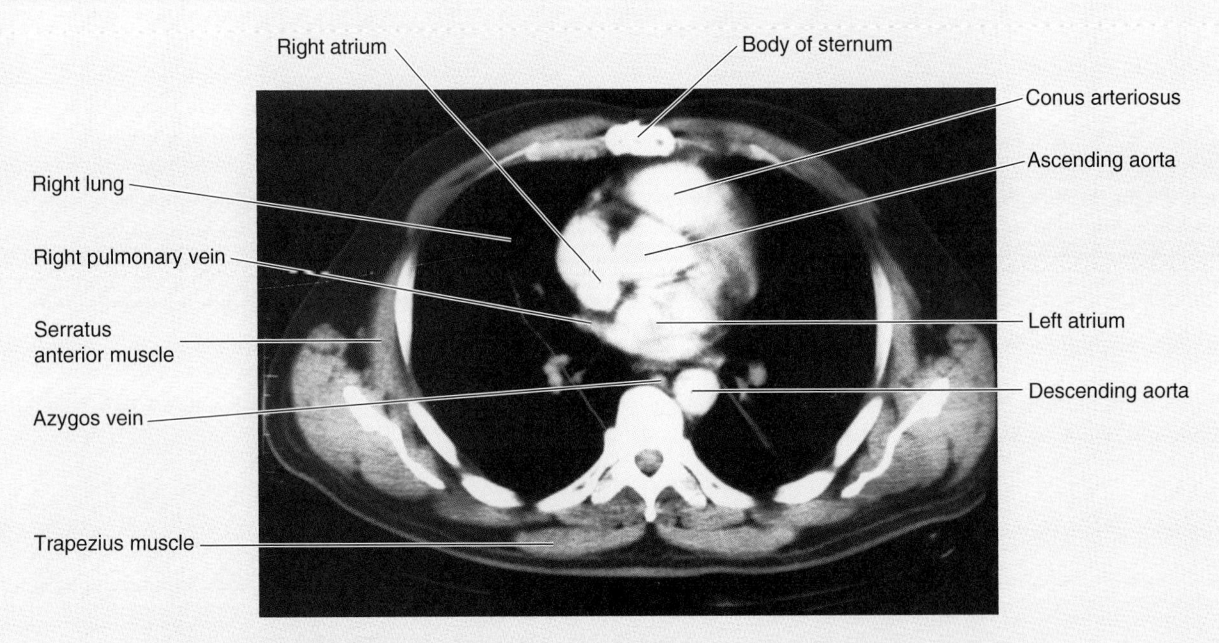

Figure 1.70. Transverse CT scan of the thorax. The scan at the level of T8 vertebra shows the thoracic viscera and wall. (Courtesy of Donald R. Cahill, Professor of Anatomy, Mayo Medical School, Rochester, Minnesota)

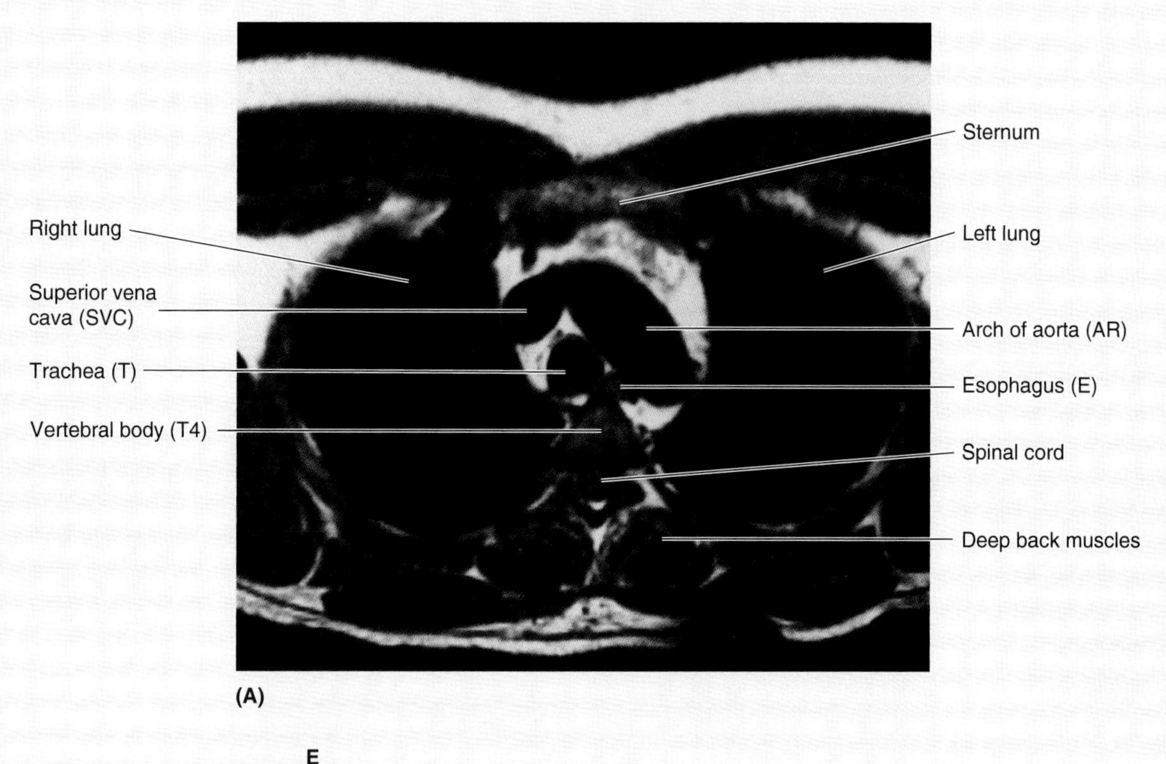

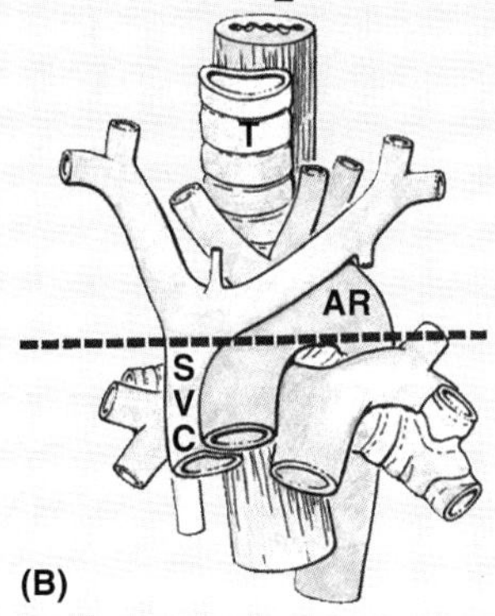

Figure 1.71. Transverse MRI of the thorax. A. Transverse MRI of the superior mediastinum at the level of the arch of the aorta. (Courtesy of Dr. W. Kucharczyk, Chair of Medical Imaging, University of Toronto, and Clinical Director of Tri-Hospital Magnetic Resonance Centre, Toronto, Ontario, Canada). **B.** Diagram showing the level of the MRI.

CASE STUDIES

Case 1.1

A young man who was stabbed in the chest was rushed to a hospital. The stab wound was in the 3rd left intercostal space, just lateral to the sternum. The emergency physician noted that the veins of his face and neck were engorged.

Clinicoanatomical Problems

- What vital structures may have been injured?
- What probably caused the engorgement of his cervical and facial veins?
- What emergency clinical procedure would likely be performed before he was taken to the operating room?

The problems are discussed on page 168.

Case 1.2

During the physical examination of a 12-year-old girl, a young physician was unable to detect a heartbeat. The radial pulse, however, was normal. After considerable thought, the physician was able to detect a normal heartbeat.

Clinicoanatomical Problems

- Where would the physician normally attempt to listen to the apex beat of the heart?
- What congenital anomaly of the heart could account for failure to detect a heartbeat on the left side of the chest?
- Where else would you attempt to detect a heartbeat?

The problems are discussed on page 168.

Case 1.3

A 42-year-old heavy smoker consulted her physician about an alteration in her voice, severe loss of weight, a persistent cough, and blood-stained sputum. Bronchoscopy and radiographs of the thorax were requested. A *distorted tracheal carina* was observed during bronchoscopy. The radiographs of her chest and a subsequent biopsy revealed a bronchogenic carcinoma in the superior (upper) lobe of her left lung.

Clinicoanatomical Problems

- In view of the signs and symptoms, where did cancer cells from the tumor metastasize?
- What superficial nodes would also be likely to be enlarged and palpable?
- What would likely be the cause of the alteration in her voice?
- What caused distortion of the carina of the trachea?

The problems are discussed on page 168.

Case 1.4

A 46-year-old woman consulted her physician about a firm, painless lump in her left breast. During the physical examination, the physician felt a lump in the upper outer quadrant of the breast. He also observed dimpling and thickening of the skin in this quadrant and noticed that her left nipple was noticeably higher than the right one. Palpation of the axilla revealed enlarged, firm lymph nodes. A diagnosis of carcinoma of the breast was made.

Clinicoanatomical Problems

- Where would most lymph from the left upper outer quadrant of the breast transport most of the cancer cells?
- To what other lymph nodes may the lymph carry cancer cells?
- What caused the thickening and dimpling of the skin and elevation of the nipple?

The problems are discussed on page 168.

Case 1.5

A 45-year-old woman was playing tennis and suddenly fell, complaining of a severe pain in her chest and down her left arm. Her playing partner rushed her to the hospital.

Clinicoanatomical Problems

- What likely caused the pain in the woman's chest and arm?
- Why did the woman feel pain along the medial side of the left arm?
- Is visceral pain from the chest usually referred to the left arm?

The problems are discussed on page 168.

Case 1.6

A woman was stabbed in the right side of her lower neck. The stab wound was approximately 2.5 cm superior to the medial third of the clavicle. Shortly after the bleeding was controlled, the woman began breathing rapidly and was given oxygen by the paramedics. Physical examination revealed a significant shift of the apex beat of the heart to the left side of the thorax, and poor breath sounds were heard on the right side of the chest.

Clinicoanatomical Problems

- What structures were likely injured?
- What injuries could cause the shift of the apex beat of the heart to the left side of the thorax?
- What procedure would likely be performed to rectify the abnormal position of the heart?
- What structures are vulnerable during this procedure?

The problems are discussed on page 168.

Case 1.7

A 62-year-old man consulted his physician about his difficulty in breathing. During the physical examination, the physician palpated the man's trachea in the jugular notch. During cardiac systole, he felt the trachea move abnormally. Radiographic studies revealed an *aneurysm of the arch of the aorta*.

Clinicoanatomical Problems

- What is an aneurysm of the arch of the aorta?
- Why is this abnormality common in older people?
- What structures may be compressed by the aneurysm?
- Why does the trachea move abnormally during cardiac systole?

The problems are discussed on page 169.

Case 1.8

While having a heated discussion with a client, a 48-year-old business woman experienced a sudden, crushing substernal pain in her chest that radiated along the medial aspect of her left arm. The client helped her to the couch where the woman attempted to relieve the pain by squirming, stretching, and belching. When her secretary noted that she was pale, perspiring, and writhing in pain, she called a physician and an ambulance. The ambulance attendants administered oxygen and rushed her to the hospital, where she was admitted to the intensive care unit (ICU). She was placed under observation with ECG monitoring for detection of potential fatal arrhythmias. Her blood pressure was low (a sign of *shock*).

On questioning, the resident learned that the patient had had previous attacks of substernal discomfort during stress that she was reluctant to describe as pain. She said that this discomfort always passed when she rested. When the resident asked the patient to describe her present chest pain, she said that it was the worst pain she had ever felt and clenched her fist to demonstrate its viselike nature. She said that when the pain struck, she had a feeling of weakness and nausea. On auscultation, the resident detected an occasional arrhythmia. The ECG was also abnormal.

Diagnosis *Acute myocardial infarction* (MI) caused by coronary atherosclerosis that resulted in ischemia of the myocardium.

Clinicoanatomical Problems

- Define acute MI and coronary atherosclerosis.
- Explain the anatomical basis of the referred pain from the patient's heart to the left side of her chest, shoulder, and medial aspect of her arm.

The problems are discussed on page 169.

Case 1.9

A 58-year-old man who had lived in an industrial area all his life consulted his physician because he was coughing up blood (*hemoptysis*) and was experiencing shortness of breath (*dyspnea*) during exertion. The physician learned that he had been a heavy cigarette smoker for more than 40 years and had had a "smoker's cough" for several years. He stated that his shortness of breath and cough had been getting worse for the last few months. He first noticed that his sputum was blood-streaked approximately 3 weeks earlier and stated that he had experienced chest pain (*angina*) on the left side at that time. Physical examination revealed that his left medial supraclavicular lymph nodes were slightly enlarged and more firm than usual. His breath sounds and resonance on the left side were more diminished than on the right side. The physician requested chest radiographs.

Radiology Report There is obscuration (indistinctness) of the hilum of the left lung by a mass. Normal left mediastinal contours superior to the hilum cannot be recognized, and there is slight radiolucency of the remainder of the left lung. The mediastinum is shifted slightly to the left. These changes are most likely caused by a malignant tumor in the left superior lobe bronchus with metastases to the left hilar lymph nodes.

Endoscopy On examination of the interior of the main bronchi under local anesthesia with a bronchoscope, the otolaryngologist observed a growth that was partly obstructing the origin of the left superior lobe bronchus. He obtained a biopsy of the tumor through the bronchoscope. The enlarged supraclavicular lymph nodes were also biopsied for microscopic examination.

Mediastinoscopy Examination of the mediastinum with a mediastinoscope inserted through a suprasternal incision under anesthesia revealed enlarged tracheobronchial lymph nodes. Through the mediastinoscope, the surgeon biopsied these nodes.

Pathology Report Bronchogenic carcinoma was detected in the bronchial biopsy. The supraclavicular lymph nodes did not show definite tumor involvement, but the mediastinal lymph nodes showed many malignant (cancerous) cells.

Diagnosis *Bronchogenic carcinoma* with metastases to mediastinal lymph nodes.

Clinicoanatomical Problems

- Using your knowledge of the anatomical relations of the bronchi and lungs, state which structures are likely to be involved by direct extension of a malignant tumor of the bronchus.
- Where would you expect tumor cells to metastasize via the lymph and blood?
- What is unusual about the lymph drainage of the inferior lobe of the left lung?
- Explain the probable anatomical basis for metastasis of cancer cells from a bronchogenic carcinoma to the brain.

The problems are discussed on page 169.

Case 1.10

During a violent argument with his wife, a 44-year-old inebriated man was stabbed with a paring knife, the blade of which was 9 cm long. The knife penetrated the 4th intercostal space along the left sternal border. By the time he was taken to the hospital emergency, the patient was semiconscious, in shock, and gasping for breath. In a few moments he became unconscious and died. The coroner, or medical examiner, performed an autopsy.

Autopsy Report Death was caused by excessive loss of blood and cardiac tamponade resulting from a stab wound.

Clinicoanatomical Problems

- Using your knowledge of the surface anatomy of the thorax, what organ(s) would you expect to be punctured by the knife?
- Where would the blood likely accumulate?
- Discuss cardiac tamponade and how this condition probably caused the man's death.

The problems are discussed on page 170.

Case 1.11

A short, thin man with spindly limbs (*gracile habitus*), 42 years of age, complained about recent difficulties in breathing during exercise (*exertional dyspnea*) and extreme fatigue. He stated that other than being physically underdeveloped, he had been well most of his life, until the last year or so when he had had several respiratory infections. Physical examination revealed a prominent right ventricular cardiac impulse. A moderately loud *midsystolic murmur* was heard over the 2nd and 3rd intercostal spaces along the inferior left sternal border. An ECG showed changes suggestive of right ventricular hypertrophy. PA and lateral radiographs of the thorax and angiocardiographic studies were requested.

Radiology Report Radiographs revealed enlargement of the right side of the heart, especially of the right outflow tract, a small aortic knob, dilation of the pulmonary artery and its major branches, and increased pulmonary vascular markings. During right cardiac catheterization, the catheter easily passed from the right atrium into the left atrium. Serial samples of blood for determination of oxygen saturation were taken as the catheter was withdrawn from the left atrium into the right atrium and then through the IVC. These studies revealed increased oxygen saturation of the right atrial blood compared with blood in the IVC. Serial determinations of pressures showed unequal pressures in the atria (slightly higher in the left atrium).

Diagnosis *Atrial septal defect* (ASD) classified as a secundum type with a left-to-right atrial shunt of blood.

Clinicoanatomical Problems

- Was this man likely born with the ASD?
- Where else may defects occur in the interatrial septum?
- What additional complications do you think might occur in this patient in view of the left-to-right shunt of blood?

The problems are discussed on page 170.

Case 1.12

A 16-month-old boy was helping his mother clean up the morning after a cocktail party when he suddenly started to choke. Thinking he must have something in his throat, she straddled him over her forearm and gave him several back blows. Although he seemed to be somewhat better after this, it was not long before he began to cough again. When she observed that he was having difficulty breathing she called her pediatrician who arranged to meet her at the hospital. When asked what the child had been eating when he began to choke, the mother replied, "Nothing! But he could have picked up something that dropped on the floor last night, such as a peanut."

Physical Examination Physical examination revealed that the child was in *respiratory distress* characterized by coughing and dyspnea. On subsequent examination, the pediatrician noted limited movement of the right side of the chest. Auscultation disclosed *reduced breath sounds* over the right lung anteriorly and posteriorly. On percussion, the physician thought there was slight *hyperresonance over the right lung.* She requested a fluoroscopic examination of the thorax and inspiration and expiration chest films.

Radiology Report There is hyperinflation of the middle and inferior lobes of the right lung, with a shift of the heart and other mediastinal structures to the left that decreases on inspiration. It appears that a foreign body is probably lodged in the right middle lobe bronchus, just inferior to the origin of the superior lobe bronchus.

Bronchoscopy Under general anesthesia, the interior of the bronchial tree was examined with a bronchoscope. A foreign object was observed in the right middle lobe bronchus at the site suggested by the radiologist. The bronchoscopist removed the object with some difficulty, using forceps passed through the bronchoscope. The foreign object was a large peanut.

Diagnosis *Bronchial obstruction* resulting from aspiration of a peanut.

Clinicoanatomical Problems

- What is the embryological and anatomical basis for foreign bodies entering the right main bronchus and involving the middle and inferior lobes of the right lung?
- If the peanut had not been removed, the right middle and inferior lobes of the infant's lung might have collapsed. Explain why a lung collapses.
- What would be the appearance of the atelectatic lobes in a radiograph?

- What effect would atelectasis have on the position of the heart, other mediastinal structures, and the diaphragm?

The problems are discussed on page 170.

Case 1.13

A 10-year-old girl, wrapped in a blanket, was carried into the outpatient department. The nurse immediately took her into an examining room and called a physician. As the nurse prepared the child for physical examination, he observed that the child was shivering (chills) and was holding the right side of her chest. He also noted that her respirations were rapid (*tachypnea*) but shallow. The girl had a hacking cough and brought up some sputum containing some blood-tinged mucous material. Her temperature was 41.5°C and her pulse rate was 115. On percussion of the thorax, the physician noted dullness over the right inferoposterior region of the child's chest. On auscultation, he noted suppression of breath sounds on the right side and a pleural rub. When asked to describe the pain, the child said it was a sharp, stabbing pain that became worse when she breathed deeply, coughed, or sneezed. When asked where she first felt pain, she placed her hand over the inferior part of her right chest. When asked where else she experienced pain, she pointed to her umbilical area and right shoulder. The physician requested a complete blood count, a sputum culture, and chest films in both prone and upright positions.

Laboratory Report The white blood cell count is elevated (*leukocytosis*) and many pneumococci were seen in the sputum.

Radiology Report There is an area of consolidation (airless lung) in the posterior part of the base, or diaphragmatic surface, of the right lung. There is also a slight shift of the heart and other mediastinal structures to the right side.

Diagnosis Pleuritis caused by *pneumococcal pneumonitis*.

Clinicoanatomical Problems

- What is the function of the pleurae?
- Discuss pleuritis and pleural effusion.
- How is pus removed from the pleural cavity?
- Using your knowledge of the nerve supply to the pleurae, explain the referral of pain to the right side of the chest, periumbilical area, and right shoulder.
- Explain anatomically why a slight shift of the heart and other mediastinal structures may occur with pneumonitis.

The problems are discussed on page 171.

Case 1.14

During a lengthy trip in the car, a 38-year-old woman experienced *substernal discomfort*, pain in the right side of her thorax, and breathlessness. She said that she felt sick to her stomach (*nausea*) and that she was going to faint (*syncope*). Believing she may have been experiencing a heart attack, her husband drove her to a hospital.

Physical Examination The physician observed evidence of shock and rapid breathing (*tachypnea*). He also noted swollen, tender veins (*varicose veins*), particularly in her right thigh and calf (signs and symptoms of *thrombophlebitis*). On questioning, he learned that she had had painful varicose veins in her lower limbs for some time and that they became extremely painful during her recent long car ride. He also learned that she had been taking birth control pills for approximately 9 years. Examination of her lungs revealed a few small, moist *atelectatic rales* (transitory, light crackling sounds) in the right side of her chest. Auscultation also revealed a pleural rub on the right side. Cardiac examination detected *tachycardia* (rapid beating of the heart) and *arrhythmia* (irregularity of the heartbeat). An ECG was performed that suggested some right heart strain. Radiographs of her thorax, pulmonary angiograms, photoscans, and fluoroscopy were requested.

Radiology Report The radiographs show some increase in radiolucency of the right lung. Fluoroscopy of her lungs revealed poor or absent pulsations in the descending branch of the right pulmonary artery and relative anemia of the right lung that is consistent with *pulmonary thromboembolism (PTE)*. The photoscans (scintigrams) obtained after intravenous injection of radioactive iodinated (^{131}I) human albumin microparticles showed practically no pulmonary bloodflow to the right lung.

Diagnosis *PTE* resulting from the release of a thrombus from a varicose vein in the lower limb.

Clinicoanatomical Problems

- How do you think the radiologist injected the contrast material into the patient's right ventricle during pulmonary angiography?
- What are the main factors involved in thrombogenesis and pulmonary embolism?
- What probably caused the patient's severe substernal discomfort and shoulder pain?

The problems are discussed on page 171.

Case 1.15

During the physical examination of a 15-year-old girl for summer camp, a "machinelike" murmur was heard during auscultation in the 2nd intercostal space near the left sternal edge. On palpation, the physician felt a continuous thrill (vibration) at the same location. Other physical findings were normal. The young woman said she had always been well, although she feels that she gets "out of breath" faster than other girls. Following consultation with her parents and a cardiologist, the family physician decided to conduct further investigations. He ordered PA and lateral chest radiographs and angiocardiography.

Radiology Report The radiographs of the chest reveal slight left ventricular enlargement and slight prominence of the pulmonary artery and aortic knob. The ECG indicates a moderate degree of left ventricular hypertrophy. Angiocardiography was then performed. The catheter was passed to the heart through the femoral vein and IVC into her right atrium, right ventricle, and pulmonary trunk. A small injection of contrast showed the tip of the catheter to be in the thoracic aorta. The catheter was drawn back to the right atrium and a right angiocardiogram was performed that showed an essentially normal right heart. Another catheter was passed via the femoral artery to the ascending aorta and contrast medium injected into it (aortography). The ascending aorta and aortic arch appear normal, but the left and right pulmonary arteries, as well as the thoracic aorta, are opacified. These studies demonstrate the presence of a *patent ductus arteriosus (PDA).*

Diagnosis Left-to-right shunting of blood through a PDA.

Clinicoanatomical Problems

- Discuss the location of the ductus arteriosus and its embryological origin, prenatal function, and postnatal closure.
- What caused the characteristic "machinelike" murmur and left ventricular enlargement?
- How do you think this left-to-right shunting of blood could be prevented?
- What clinical condition do you think might cause right-to-left shunting of blood through the ductus arteriosus?

The problems are discussed on page 172.

DISCUSSION OF CASES

Case 1.1

The knife would puncture the *pericardial sac* and the right ventricle of the heart. Stab wounds that pierce the heart cause blood to enter the pericardial cavity producing *hemopericardium* and *cardiac tamponade.* As blood accumulates in the pericardial sac, the heart's ability to expand and fill with blood following each contraction becomes increasingly compromised and circulation is impaired. The veins of the face become engorged because of the compression of and accumulation (backup) of blood in the SVC, which, in turn, impedes the return of blood from the head and neck. *Pericardiocentesis* was likely performed to remove blood from the pericardial cavity and relieve the cardiac tamponade, allowing the heart to expand more fully to receive blood. A wide-bore needle was probably inserted through the left 5th or 6th intercostal space to the left of the sternum to aspirate the blood.

Case 1.2

The physician would normally attempt to listen to the heartbeat with a stethoscope in the 5th left intercostal space, medial to the interclavicular line. In other words, a little inferomedial to the nipple in a 12-year-old girl. *Dextrocardia*, a congenital anomaly in heart position, would account for failure to detect a heartbeat on the left side. The strongest heartbeat in a person with this abnormality would be heard in the 5th right intercostal space to the right of the inferior end of the sternum.

Case 1.3

The bronchogenic carcinoma (CA—cancer arising in the mucosa of the bronchi) had obviously metastasized to the left *bronchomediastinal lymph nodes.* The *supraclavicular lymph nodes* are usually enlarged also and become palpable when there is carcinoma of a bronchus. Because of this, these lymph nodes are called *sentinel nodes.* Enlargement of the bronchomediastinal nodes can exert pressure on the left *recurrent laryngeal nerve.* In this case, the nerve compression caused paralysis of her left vocal fold (see Chapter 8). This abnormality resulted in the alteration of her voice. The *distortion of the tracheal carina* observed during bronchoscopy was caused by enlargement of the tracheobronchial lymph nodes in the angle between the main bronchi.

Case 1.4

Most cancer cells from the left upper quadrant of the breast are carried by the lymph to the axillary lymph nodes, mainly to the pectoral nodes. The lymph may also carry some cancer cells to the supraclavicular and infraclavicular lymph nodes. Interference with the lymphatic drainage of the breast by invasion of cancer cells can produce *edema* (accumulation of fluid in tissues), which gives the skin a thickened, finely dimpled appearance and has been compared to the skin of an orange ("peau d'orange") or to pigskin leather. Localized depressions of the skin (large dimples, fingertip size or greater) and/or retraction of the nipple results when cancer invades the suspensory ligaments, glandular tissue, or lactiferous ducts. Elevation of the entire breast, making the affected nipple higher than its contralateral partner, results from entry of cancer cells into the retromammary space, the deep pectoral fascia, and interpectoral nodes.

Case 1.5

The severe pain in her chest and arm likely resulted from sudden occlusion of a coronary artery. The strenuous exercise during the tennis game increased her heart's activity and need for oxygen. The insufficient blood supply to the heart (*myocardial ischemia*) resulted in an *MI.* Visceral pain from the heart is transmitted by visceral afferent fibers that accompany sympathetic fibers and is typically referred to somatic structures, such as the left upper limb (referred pain). The visceral afferent fibers enter the spinal cord through the same dorsal roots as the somatic sensory fibers. Although usually referred to the left arm, the pain may be referred to the right arm, both arms, the neck and chin, or the back.

Case 1.6

The structures that may have been injured are the right subclavian artery and vein, the suprapleural membrane, cervical pleura, and the apex of the lung. Injury to the subclavian vessels and parietal pleura could result in the accumulation of air and blood in the pleural cavity (*pneumothorax* and *hemothorax*). This condition may in turn result in partial or total collapse of the right lung (*atelectasis*). Accumulation of air or blood sufficient to increase the volume of the right pulmonary cavity will cause the soft mediastinum (including the heart) to shift to the left (mediastinal shift). A *thoracocentesis* followed by insertion of a valved chest tube (drain) would likely be performed to remove blood from the pleural cavity and allow the lung to be inflated. The intercostal nerve and vessels are vulnerable to injury during thoracocentesis.

Case 1.7

An aneurysm of the arch of the aorta is a dilation of the wall of the arch. This abnormality is common in older people with arterial disease and in certain congenital disorders. The arch of the aorta lies posterior to the manubrium and runs superoposteriorly and to the left, anterior to the trachea. The arch then passes inferiorly to the left of the trachea. Consequently, pressure may be exerted on the trachea and esophagus, causing difficulty in breathing (dyspnea) and swallowing. During ventricular systole (contraction and emptying), blood is forced into the ascending aorta and arch of the aorta, which enlarges the aneurysm and increases the compression of the trachea and esophagus. Abnormal movement ("tugging") of the trachea during systole can be palpated in the jugular notch.

Case 1.8

Acute MI is a disease of the myocardium, characterized by necrosis of ventricular muscle that results from sudden occlusion of a part of the coronary circulation. Blockage of a coronary artery results in dysfunction of the heart as a pump. If a large branch of a coronary artery is involved, the infarcted area may be so extensive that cardiac function is severely disrupted and death occurs. MI may also result from excessive exertion (e.g., running to catch a bus) by a person with stenotic coronary arteries. An *atheroma* (or atheromatous plaque) is a lipid deposit that produces a swelling on the endothelial surface of the blood vessel. Ulceration of the atheroma results in the release of atheromatous debris that is carried along the coronary artery until it reaches the stenotic part. Because it blocks the vessel, no blood can pass to the myocardium, and MI occurs unless a good collateral circulation has developed previously.

Anastomoses exist between the terminations of the right and left coronary arteries in the coronary groove and between the interventricular branches around the apex in approximately 10% of apparently normal hearts. The potential for the development of this collateral circulation probably exists in most if not all hearts. In slow occlusion of a coronary artery, the collateral circulation has time to increase so that adequate perfusion of the myocardium can occur when a potentially ischemic event occurs. Consequently, infarction may not result. On sudden blockage of a large coronary branch, some infarction is probably inevitable, but the extent of the area damaged depends on the degree of development of collateral anastomotic channels.

If large branches of both coronary arteries are partially obstructed, an extracardiac collateral circulation may be utilized to supply blood to the heart. These collaterals connect the coronary arteries with the vasa vasorum in the tunica adventitia of the aorta and pulmonary arteries and with branches of the internal thoracic, bronchial, and phrenic arteries. Reversal of flow in the anterior and smallest cardiac veins (venae cordis minimae) may bring luminal blood to the capillary beds of the myocardium in some regions, providing some additional collateral circulation. However, unless these collaterals have dilated in response to pre-existing *ischemic heart disease*, they are unlikely to be able to supply sufficient blood to the heart during an acute event and prevent MI.

The dominant symptom of MI is deep visceral pain. Afferent pain fibers from the heart run centrally through the middle and inferior cervical branches and the upper thoracic branches of the sympathetic trunks of the neck and thorax. Central processes (axons) of primary sensory neurons enter spinal cord segments T1 through T4 or T5 on the left side. Pain of cardiac origin is often referred to the left side of the chest and along the medial aspect of the arm and upper forearm. These are the areas of the body that send sensory impulses to the same spinal ganglia and segments of the spinal cord that receive cardiac sensation. *Visceral referred pain is the perception of visceral pain* as occurring in remote cutaneous areas.

Case 1.9

Because of the anatomical relations of the bronchi and lungs, some bronchogenic cancers may extend into the thoracic wall, diaphragm, and mediastinum. Involvement of a phrenic nerve may result in paralysis of half of the diaphragm. Direct infiltration of the pleurae produces *pleural effusion* into the pleural cavity. This pleural exudate may be bloody (sanguineous) and may contain exfoliated malignant cells. Because of the close relationship of the recurrent laryngeal nerves to the apices of the lungs, they may be involved in cancer of this region of a lung. An apical tumor often produces hoarseness because of paralysis of the vocal folds.

A tumor of the apex of the lung that invades locally may involve the superior thoracic spinal nerves, the thoracic sympathetic chain, and the cervicothoracic (stellate) ganglion, resulting in the *Horner syndrome* (see Chapter 8). In such a case, pain is likely in the shoulder and axilla. Involvement of the hilar and mediastinal lymph nodes occurs by lymphogenous dissemination of cancer cells. The lymphatic vessels of the lungs originate in superficial and deep plexuses accompanying small blood vessels. Lymph then drains into bronchopulmonary lymph nodes in the hilum. As these nodes enlarge, they increase the size of the hilum of the lung, giving it a lumpy appearance. Enlargement of metastatic tumors here may compress the bronchi, interfering with ventilation of the lung. The bronchopulmonary lymph nodes drain into inferior and superior groups of tracheobronchial nodes that lie in the angles between the trachea and bronchi. They form part of the mediastinal group of lymph nodes that are scattered throughout the mediastinum.

Clinically, the inferior group of tracheobronchial nodes are commonly referred to as carinal lymph nodes because of their relationship to the carina of the trachea. Splaying and fixation of the carina may be associated with bronchogenic carcinoma when it has metastasized to the carinal nodes. These abnormalities can be seen bronchoscopically and radiologically. Enlarged mediastinal lymph nodes may indent the esophagus, which can be observed radiologically as the patient swallows a barium sulfate emulsion. As lymph from vessels in the costal parietal pleura reaches the parasternal lymph nodes via intercostal lymphatic vessels, *lymphogenous spread of bronchogenic carcinoma* may also involve these nodes. Lymph from the entire right lung drains into tracheobronchial nodes on the right side, and most lymph from the left lung drains into nodes on the left side, but some lymph from the inferior lobe of the left lung also drains into nodes on the right side. Thus, tumor cells in tracheobronchial lymph nodes on the right side may spread by lymphogenous dissemination from the inferior lobe of the left lung.

The right and left bronchomediastinal trunks drain lymph from the thoracic viscera and lymph nodes. The right bronchomediastinal trunk may join the right lymphatic duct, and the left trunk may join the thoracic duct, but more commonly they open independently into the junction of the internal jugular and subclavian veins ("venous angles") of their own sides. Thus, lymph from the lungs and pleurae containing tumor cells enters the venous system and heart. After passing through the pulmonary circulation, the blood returns to the heart for distribution to the body. Common sites of hematogenous metastases from bronchogenic carcinoma are lymph nodes, lungs, brain, bones, and suprarenal glands. Often, the medial supraclavicular lymph nodes, particularly on the left side, are enlarged and hard because of the presence of tumor cells.

The anatomical basis for involvement of the supraclavicular (sentinel) lymph nodes is that lymph passes cranially from the thoracic and abdominal viscera via the bronchomediastinal trunks and the thoracic duct to reach the venous system. Backflow of lymph from the thoracic duct can pass into the deep supraclavicular nodes, posterior to the sternocleidomastoid muscles. This occurrence is probably the reason that involvement of the nodes on the left side is most common. The brain is a common site for hematogenous spread of bronchogenic carcinoma. The tumor cells probably enter the blood through the wall of a capillary or venule in the lung and are transported to the brain via the internal carotid artery and vertebral artery systems. Once in the cranium, the tumor cells probably pass between the endothelial cells of the capillaries and enter the brain.

Although most cancer cells from the lung are probably transported to the brain via the arterial system, others may be carried by the venous system. It has been suggested that constant coughing and enlarged mediastinal lymph nodes compress the superior and inferior venae cavae, hampering venous return and causing the blood draining the bronchi to reverse its flow and pass through the bronchial veins into the azygos venous system. From here, blood and tumor cells could pass to the extradural or epidural external vertebral venous plexuses around the spinal dura mater. As this plexus communicates with the cranial dural venous sinuses, tumor cells can be transported to the brain. The passage of tumor cells to the veins of the vertebral column also explains the frequency of metastases to vertebrae.

Case 1.10

The knife, entering the 4th intercostal space at the left sternal border, did not penetrate the left lung because of the cardiac notch in its anterior border. The knife probably nicked the parietal layer of pleura of the left lung and then passed through the conus arteriosus of the right ventricle and the aortic vestibule of the left ventricle, immediately inferior to the aortic orifice. Blood passed from the wounds in both ventricles into the pericardial sac. As blood accumulated rapidly in the pericardial cavity, severe compression of the heart and obstruction to the inflow of blood to ventricles (*cardiac tamponade*) occurred. This pressure increased until it exceeded the pressure in these large veins, preventing normal venous return to the heart and outflow of blood from the heart to the lungs. This explanation accounts for the patient's shock and gasping for breath before his death.

Case 1.11

ASD is a congenital malformation because it is an imperfection in the interatrial septum that develops during embryonic formation of the heart. The common form of ASD is the *secundum type of ASD*, so classified because it results from abnormal development of the oval foramen and secondary septum (L. septum secundum) (Moore and Persaud, 1998). The primary septum (L. septum primum) and secondary septa normally fuse in such a way that no opening remains between the right and left atrium. The site of the prenatal opening, and where most defects occur, is represented by the oval fossa in the adult right atrium.

The left-to-right shunt of blood occurs because left atrial pressure exceeds that in the right atrium during the major part of the cardiac cycle. Some of the patient's blood therefore makes two circuits through the lungs. As a result of this shunting, the work load of the right ventricle increases, and its muscular wall hypertrophies (increases in size). The cavities of the right atrium, right ventricle, and pulmonary artery dilate to accommodate the excess amount of blood in them.

During their first 40 years, a majority of ASD patients have a moderate to good exercise tolerance, despite the fact that the opening in the interatrial septum is often 2 to 4 cm in diameter. Usually patients with ASDs do not exhibit *cyanosis* (bluish coloration of the skin). *Pulmonary vascular disease* (arteriosclerosis) is likely to develop with increased pulmonary artery pressure, particularly if recurrent respiratory infections occur. Severe pulmonary hypertension may eventually result in higher pressure in the right atrium than in the left, reversing the shunt and causing cyanosis, severe disability, and heart failure.

Case 1.12

Because the right main bronchus is wider, shorter, and more vertical than the left main bronchus, foreign bodies more frequently pass into the right than into the left bronchus. Foreign bodies that are commonly found are nuts, hardware, pins, crayons, and dental material (e.g., part of a restored tooth). The right middle and inferior lobes of the right lung are usually involved because the right inferior lobe bronchus is in line with the right main bronchus and because the foreign body often lodges in the proximal inferior lobe ("intermediate") bronchus, superior to the origin of the middle lobe bronchus.

When complete obstruction of a main bronchus occurs, the

chapter

2 Abdomen

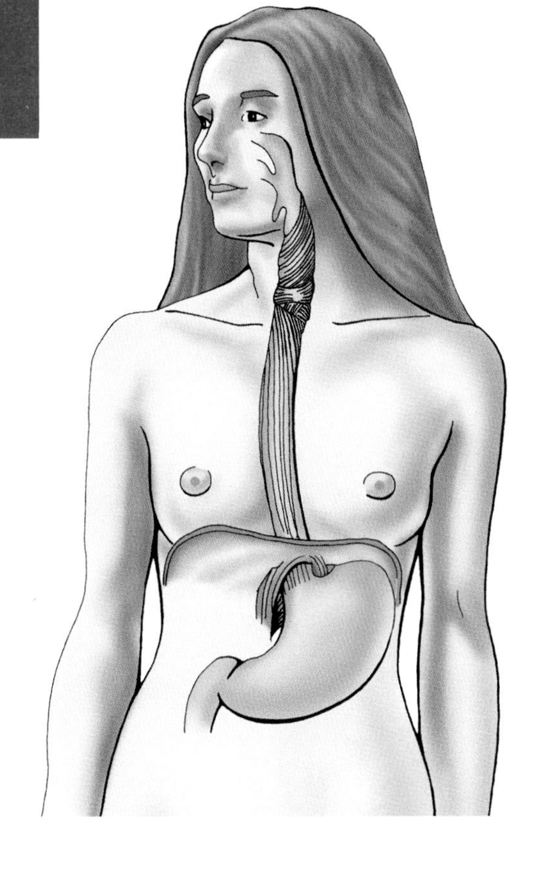

Harrison's Principles of Internal Medicine, 10th ed. New York, McGraw-Hill, 1983.

Ross MH, Romrell LJ, Kaye G: *Histology. A Text and Atlas*, 3rd ed. Baltimore, Williams & Wilkins, 1994.

Rowland LP (ed): *Merritt's Textbook of Neurology*, 9th ed. Baltimore, Williams & Wilkins, 1995.

Rubin E, Farber JL (eds): *Pathology*, 2nd ed. Philadelphia, JB Lippincott, 1993.

Sabiston DC Jr, Lyerly HK: *Sabiston Essentials of Surgery*, 2nd ed. Philadelphia, WB Saunders, 1994.

Sauerland EK: *Grant's Dissector*, 11th ed. Baltimore, Williams & Wilkins, 1994.

Swartz MH: *Textbook of Physical Diagnosis. History and Examination*, 2nd ed. Philadelphia, WB Saunders, 1994.

Swartz MA, Moore ME: *Medical Emergency Manual Differential Diagnosis and Treatment*, 3rd ed. Baltimore, Williams & Wilkins, 1983.

Tandler J: The development of the heart. *In* Keibel F, Mall FP (eds): *Manual of Human Embryology*. Philadelphia, JB Lippincott, 1912.

Williams PL, Bannister LH, Berry MM, Collins P, Dyson M, Dussek JE, Fergusson MWJ(eds): *Gray's Anatomy*, 38th ed. Edinburgh, Churchill Livingstone, 1995.

Willms JL, Schneiderman H, Algranati PS: *Physical Diagnosis Bedside Evaluation of Diagnosis and Function*. Baltimore, Williams & Wilkins, 1994.

Woodburne RT, Burkel WE: *Essentials of Human Anatomy*, 9th ed. New York, Oxford University Press, 1994.

The abdomen—the part of the trunk between the thorax and pelvis—has musculotendinous walls, except posteriorly where the wall includes the lumbar vertebrae and intervertebral (IV) discs (see Fig. 2.3). The abdominal wall encloses the abdominal cavity, containing the peritoneal cavity and abdominal viscera (L. soft parts, internal organs) such as the stomach, intestine, and liver, as well as blood vessels, lymphatics, and nerves.

Abdominal Cavity

The abdominal cavity forms the superior and major part of the **abdominopelvic cavity** (Fig. 2.1*A*). It lies between the **thoracic diaphragm** and the superior pelvic aperture, or **pelvic inlet**. The diaphragm forms the roof of the abdominal cavity; it has no floor of its own because it is continuous with the **pelvic cavity**. The abdominal cavity extends superiorly into the osseocartilaginous **thoracic cage** (rib cage) to the 4th intercostal space (Fig. 2.1*B*). Consequently, some abdominal organs—the spleen, liver, part of the kidneys, and stomach—are protected by the thoracic cage. The greater pelvis (expanded part of the pelvis superior to the pelvic inlet) supports and partly protects the lower abdominal viscera (part of the ileum, cecum, and sigmoid colon). The abdominal cavity contains peritoneum—a serous membrane lining the abdominopelvic cavity and covering most organs, variable amounts of fat, most of the digestive organs (stomach, intestine, liver, gallbladder, and pancreas), and part of the urogenital system (kidneys and ureters), as well as the spleen. In summary, *the abdominal cavity is*:

- The major part of the abdominopelvic cavity
- Located between the diaphragm and the pelvic inlet
- Separated from the thoracic cavity by the thoracic diaphragm
- Continuous inferiorly with the pelvic cavity
- Under cover of the thoracic cage superiorly
- Supported and partially protected inferiorly by the greater pelvis
- The space surrounded by the multilayered abdominal walls
- The location of most digestive organs, the spleen, kidneys, and most of the ureters.

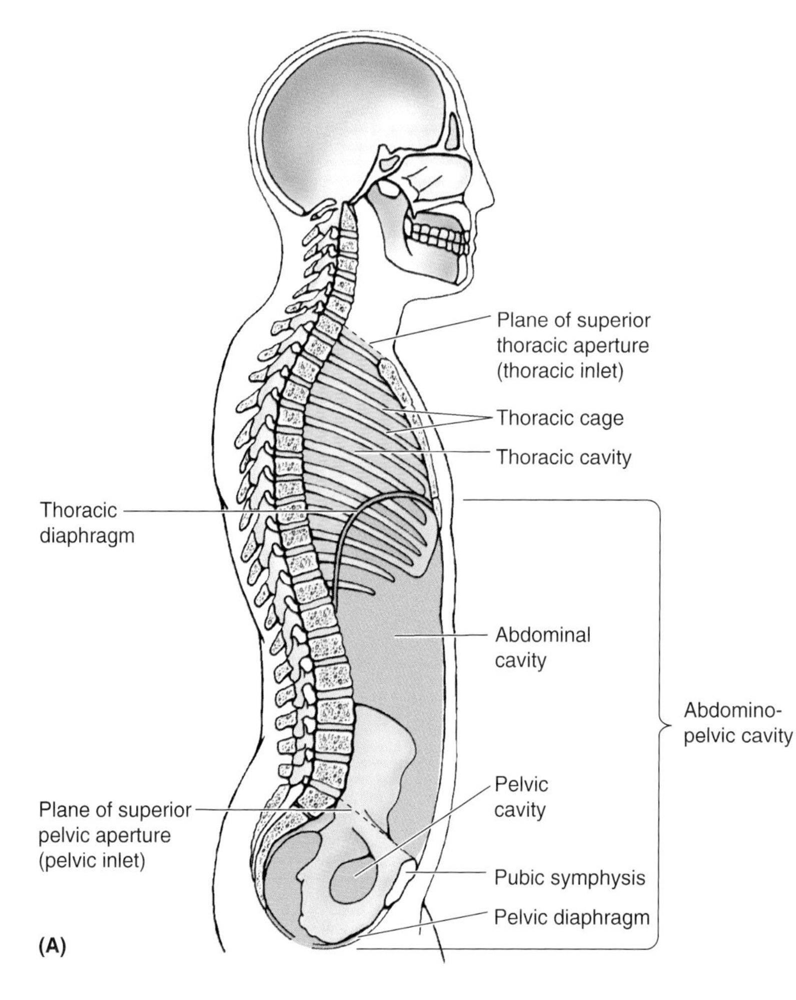

Figure 2.1. Abdominopelvic cavity. A. Median section of the body showing the abdominal and pelvic cavities as subdivisions of the continuous abdominopelvic cavity.

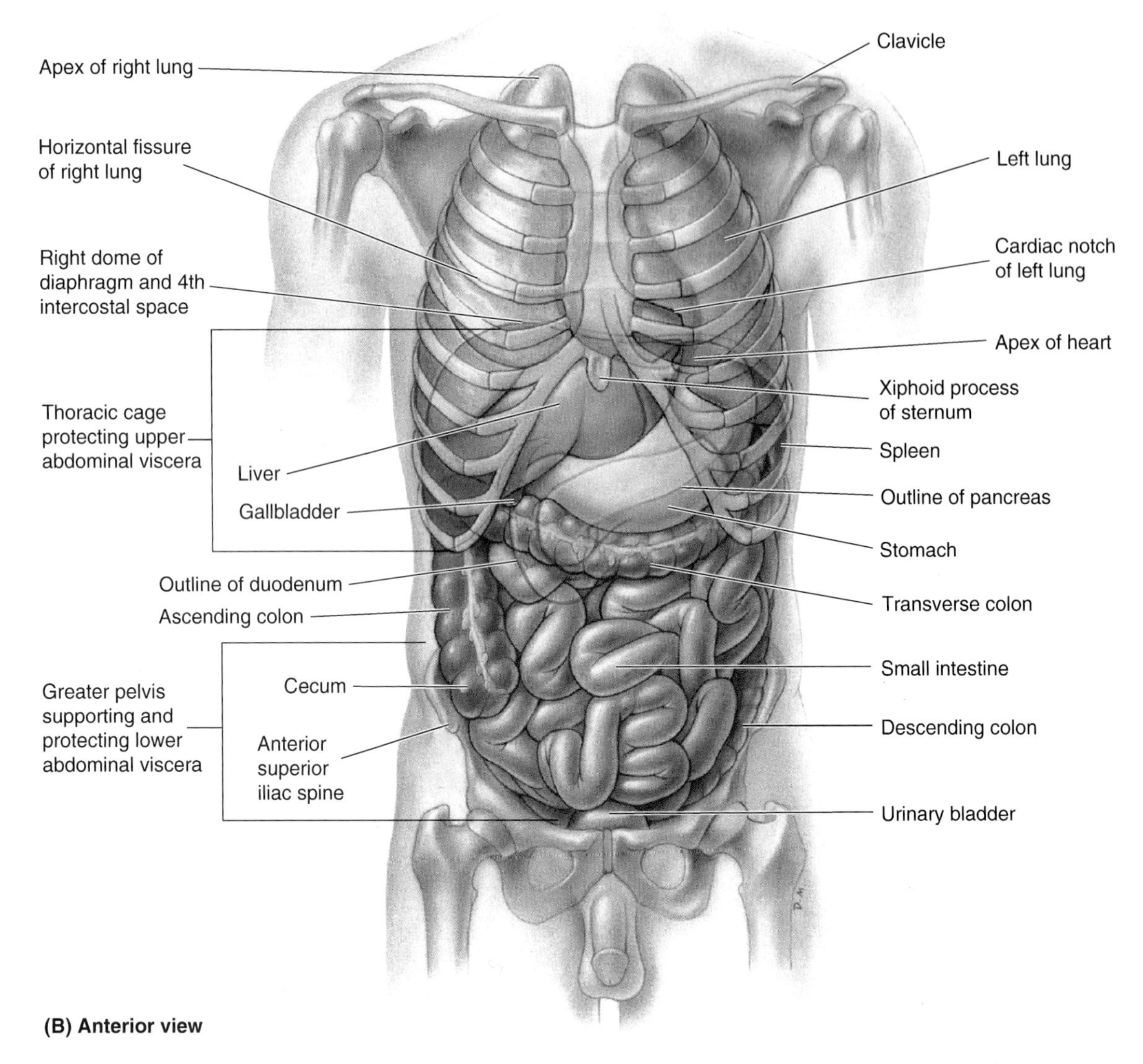

Figure 2.1. *(Continued)* **B.** Overview of the viscera of the thorax and abdomen in situ.

Clinicians use *nine regions of the abdominal cavity to describe the location of abdominal organs or pains* (Fig. 2.2*A*, Table 2.1). The nine regions are delineated by four planes (Fig. 2.2*B*):

- Two horizontal (subcostal and transtubercular planes)
- Two vertical (midclavicular planes).

For general clinical descriptions, four quadrants of the abdominal cavity are defined by two planes (Fig. 2.2*C*):

- One horizontal (transumbilical plane)
- One vertical (median plane).

To elaborate, *the horizontal planes are the*:

- **Subcostal plane** passing through the inferior border of the 10th costal cartilage on each side
- **Transtubercular plane** passing through the iliac tubercles and the body of L5 vertebra; these tubercles—approximately 5 cm posterior to the anterior superior iliac spines—are usually palpable
- **Transumbilical plane** passing through the umbilicus and the IV disc between L3 and L4 vertebrae.

The vertical planes are the:

- **Midclavicular planes** passing from the midpoint of the clavicles (approximately 9 cm from the midline) to the *midinguinal points*—midpoints of the lines joining the anterior superior iliac spines and the superior edge of the pubic symphysis (L. symphysis pubis)
- **Median plane** passing longitudinally through the body, dividing it into right and left halves.

Figure 2.2. Regions of the abdominal wall. A. Division of the wall into nine regions. **B.** The vertical and horizontal abdominal reference planes used to divide the wall into nine regions. **C.** Division of the wall into four quadrants, the simplest and most commonly used subdivisions.

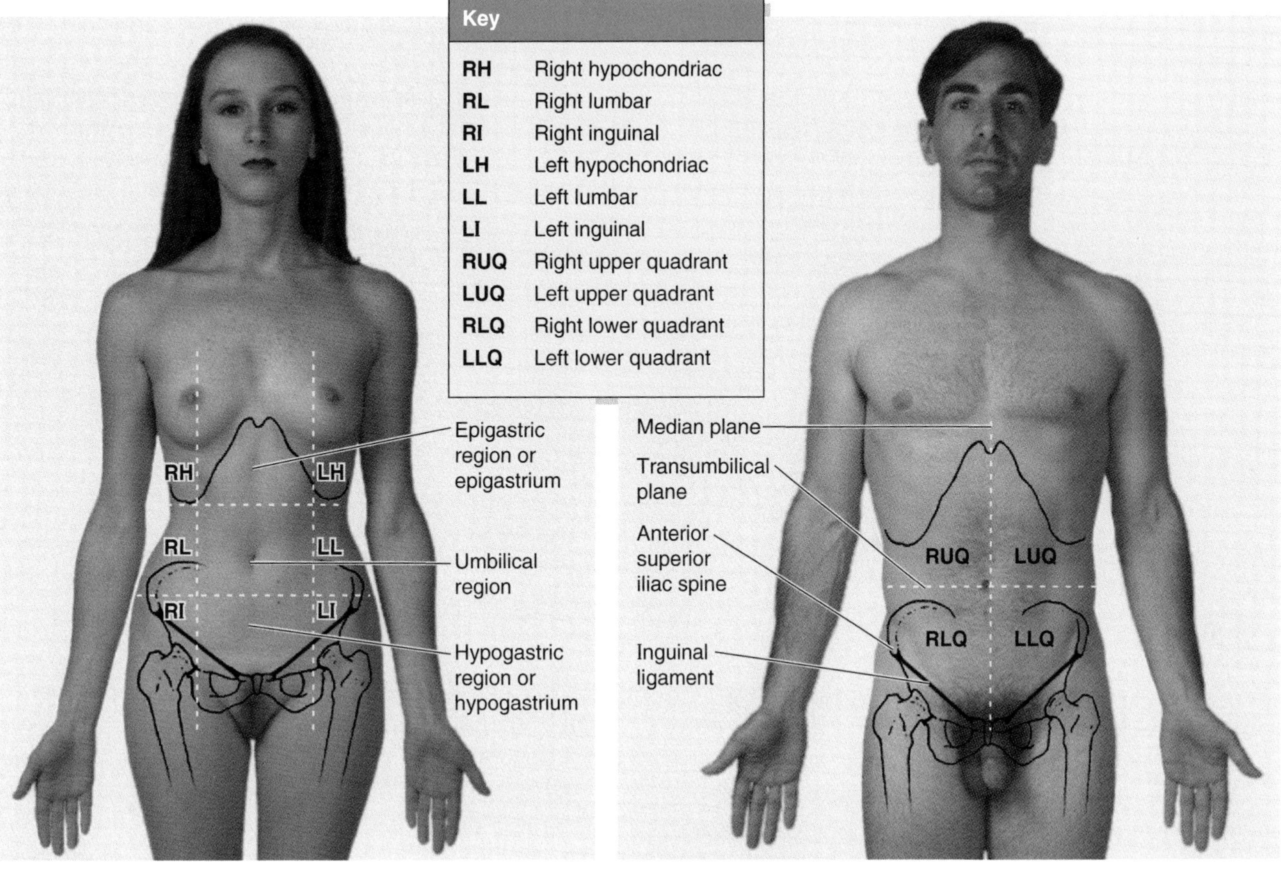

(A) Abdominal regions

(C) Abdominal quadrants

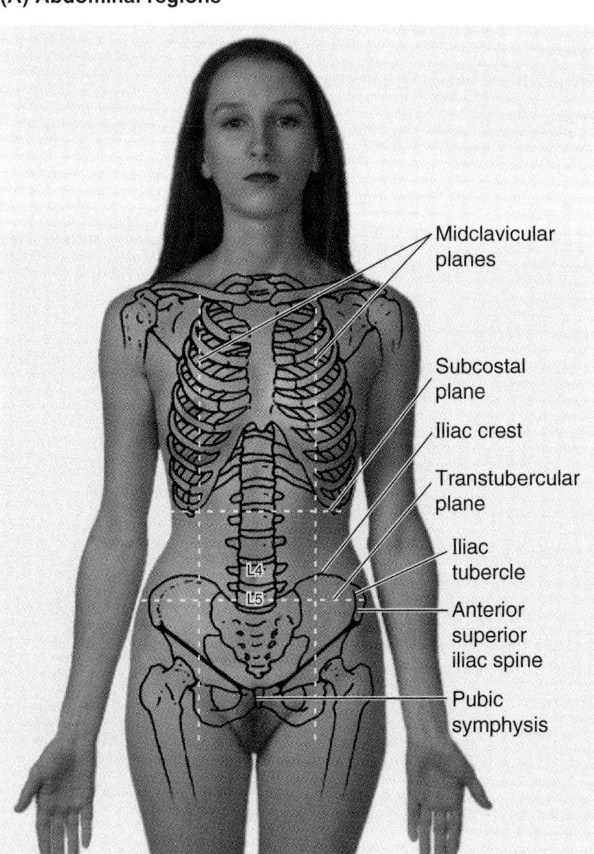

(B) Abdominal reference planes

Right upper quadrant (RUQ)	Left upper quadrant (LUQ)
Liver: right lobe	Liver: left lobe
Gallbladder	Spleen
Stomach: pylorus	Stomach
Duodenum: parts 1-3	Jejunum and proximal ileum
Pancreas: head	Pancreas: body and tail
Right suprarenal gland	Left kidney
Right kidney	Left suprarenal gland
Right colic (hepatic) flexure	Left colic (splenic) flexure
Ascending colon: superior part	Transverse colon: left half
Transverse colon: right half	Descending colon: superior part

Right lower quadrant (RLQ)	Left lower quadrant (LLQ)
Cecum	Sigmoid colon
Vermiform appendix	Descending colon: inferior part
Most of ileum	Left ovary
Ascending colon: inferior part	Left uterine tube
Right ovary	Left ureter: abdominal part
Right uterine tube	Left spermatic cord: abdominal part
Right ureter: abdominal part	Uterus (if enlarged)
Right spermatic cord: abdominal part	Urinary bladder (if very full)
Uterus (if enlarged)	
Urinary bladder (if very full)	

Table 2.1. **Location of Abdominal Structures by Quadrants**

Abdominal Diseases and Abdominal Pain

Diseases of the abdomen are common and often serious. Pain is a common symptom of abdominal disease; for example, pain from *appendicitis* (inflammation of the appendix) begins as a generalized pain that becomes prominent in the epigastric region. The pain then moves toward the umbilicus and finally localizes in the right lower quadrant. Appendicitis ranks high on the list of diseases leading to hospitalization and is a major cause of abdominal pain. *Appendectomy*—surgical removal of the appendix—is the most common reason for emergency abdominal surgery in children and adolescents; however, any abdominal organ is subject to disease or injury. If a hollow gastrointestinal (GI) organ ruptures, its irritating contents enter the peritoneal cavity, producing *peritonitis* (inflammation of the peritoneum) and severe pain.

Location of Structures by Abdominal Quadrants

It is important to know what organs are located in each abdominal quadrant so that one knows where to auscultate, percuss, and palpate them (Table 2.1). Knowledge of the location of organs is also essential for recording findings during a physical examination. For example, part of a typical clinical write-up might be:

> A well-healed paramedian scar with a small, easily reducible central herniation is present in the RUQ. A firm mass is also felt in this region that raises the possibility of cancer of the pylorus of the stomach, pancreatic head, or transverse colon. ○

Anterolateral Abdominal Wall

Although the abdominal wall is continuous, it is subdivided into the *anterior wall, right and left lateral walls* (flanks), and *posterior wall* for descriptive purposes (Fig. 2.3). The anterolateral abdominal wall extends from the thoracic cage to the pelvis. The major part of the wall is musculotendinous. The boundary between the anterior and lateral walls is indefinite.

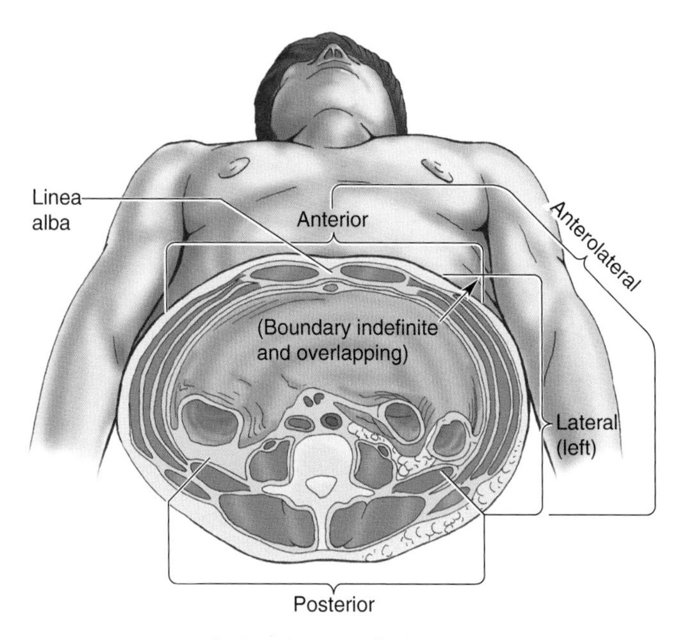

Figure 2.3. Subdivisions of the abdominal wall. Schematic transverse section of the abdomen demonstrating various aspects of the wall and its components. Some abdominal contents, such as the intestine and great vessels, are shown.

Consequently, the combined term *anterolateral abdominal wall* is often used because some structures, such as the muscles and cutaneous nerves, are in both the anterior and lateral walls.

During a physical examination, the anterolateral wall is inspected, palpated, percussed, and auscultated. Surgeons usually incise this wall during abdominal surgery. *The anterolateral abdominal wall is bounded* (Fig. 2.4)

- Superiorly by the cartilages of the 7th through 10th ribs and the xiphoid process of the sternum (breast bone)
- Inferiorly by the inguinal ligament and the pelvic bones.

The wall consists of skin and subcutaneous tissue (superficial fascia) composed mainly of fat, muscles, deep fascia, endoabdominal fascia/fat, and parietal peritoneum. The skin attaches loosely to the subcutaneous tissue except at the umbilicus, where it adheres firmly.

Most of the anterolateral wall consists of three musculotendinous layers, each of which has fibers that run in a different direction. This three-ply structure is similar to that of the intercostal spaces in the thorax (see Chapter 1).

Fascia of the Anterolateral Abdominal Wall

The subcutaneous tissue over most of the wall consists of a layer of connective tissue that contains a variable amount of fat (Fig. 2.4). In morbid obesity the fat is many inches thick, often forming one or more sagging folds (L. panniculi; singular, panniculus, "apron"). *In the inferior part of the wall, the subcutaneous tissue is composed of two layers*:

- A **fatty** (superficial) **layer** (Camper's fascia)
- A **membranous** (deep) **layer** (Scarpa's fascia).

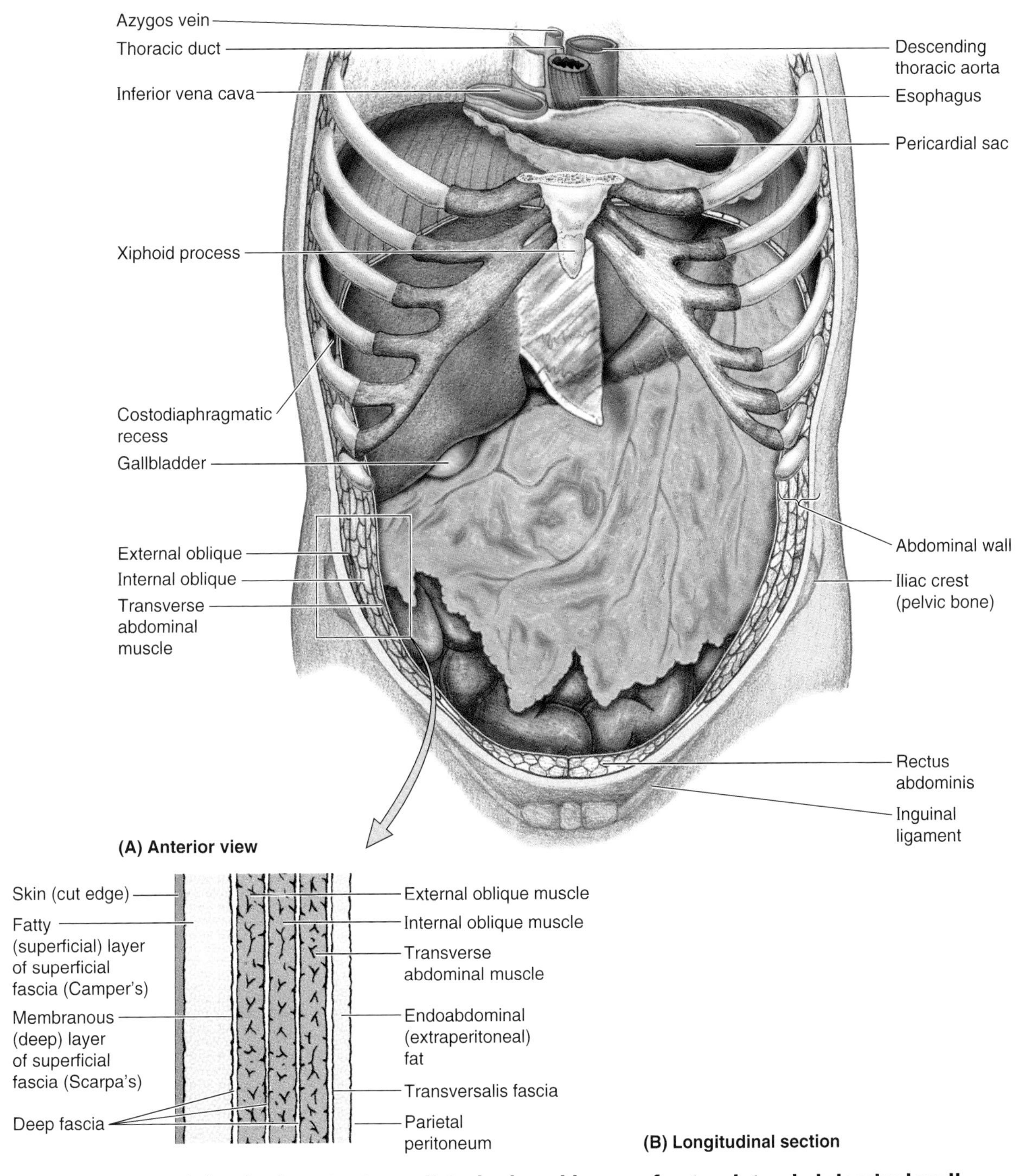

Figure 2.4. Abdominal contents, undisturbed, and layers of anterolateral abdominal wall. **A.** The anterior thoracic and abdominal walls are cut away. Most of the intestine is covered by the apronlike greater omentum, a peritoneal fold hanging from the stomach. Most of the liver and stomach lies under cover of the thoracic cage (rib cage). **B.** The layers of the abdominal wall are illustrated.

A layer of deep fascia invests the external oblique and cannot be separated easily from this muscle. The **deep fascia** in the abdomen is extremely thin, being represented only by the epimysium (fibrous sheath) of the most superficial muscles. A relatively firm, membranous sheet—the **transversalis fascia**—lines most of the abdominal wall. This fascial layer covers the deep surface of the transverse abdominal muscle (L. transversus abdominis) and its aponeurosis; the right and left sides of the fascia are continuous deep to the *linea alba* (Fig. 2.3). The **parietal peritoneum** is internal to the transversalis fascia and is separated from it by a variable amount of **endoabdominal (extraperitoneal) fat** (Fig. 2.4).

Closing Abdominal Skin Incisions

When closing abdominal skin incisions, surgeons include the membranous layer of subcutaneous tissue when suturing because of its strength. Between this layer and the deep fascia covering the rectus abdominis and external oblique muscles is a potential space where fluid may accumulate (e.g., urine from a ruptured urethra). Although there are no barriers (other than gravity) to prevent fluid from spreading superiorly from this space, it cannot spread inferiorly into the thigh because the membranous layer of subcutaneous tissue fuses with the deep fascia of the thigh (fascia lata) along a line approximately 2.5 cm inferior and parallel to the inguinal ligament. A potential space between the transversalis fascia and the parietal peritoneum—the *space of Bogros*—is used for placing prostheses when repairing inguinal hernias, for example (Skandalakis et al., 1995). ⊙

Muscles of the Anterolateral Abdominal Wall

There are five muscles in the anterolateral abdominal wall (Figs. 2.5–2.7): three flat muscles and two vertical muscles. Their attachments, nerve supply, and main actions are listed in Table 2.2. *The three flat muscles of the anterolateral abdominal wall are the:*

- **External oblique**, the superficial muscle arising from the middle and lower ribs by muscular slips that interdigitate with those of the serratus anterior, a pectoral muscle; the fibers of the external oblique pass inferomedially
- **Internal oblique**, the intermediate muscle whose fibers run horizontally at the level of the anterior superior iliac spine; the fibers run obliquely upward superior to this level and obliquely downward inferior to it
- **Transverse abdominal**, the innermost muscle whose fibers, except for the most inferior ones, run more or less transversomedially.

All three flat muscles end anteriorly in a strong, sheetlike aponeurosis. The aponeuroses of these muscles interlace at the **linea alba** (L. white line) with their fellows of the opposite side to form the tough, aponeurotic, tendinous sheath of the rectus muscle—the **rectus sheath** (Figs. 2.5 and 2.8). The decussation and interweaving of the aponeuroses in the rectus sheath is not only from side to side but also from superficial to deep.

The two vertical muscles of the anterolateral abdominal wall are within the rectus sheath:

- **Rectus abdominis**
- **Pyramidalis.**

The other contents of the rectus sheath are the superior and inferior **epigastric arteries** and veins, lymphatic vessels, and **ventral primary rami** of T7 through T12 nerves.

The pyramidalis is a small triangular muscle that is absent in approximately 20% of people. It lies anterior to the inferior part of the rectus abdominis and attaches to the anterior surface of the pubis and the anterior pubic ligament. It ends in the *linea alba*, a fibrous band running a variable distance superior to the pubic symphysis. The pyramidalis tenses the linea alba; when present, surgeons use the attachment of the pyramidalis to the linea alba as a landmark for an accurate median abdominal incision (Skandalakis et al., 1995).

External Oblique Muscle

The external oblique is the largest and most superficial of the three flat anterolateral abdominal muscles (Fig. 2.5). Its muscular part contributes to the anterolateral part of the abdominal wall and its aponeurosis contributes to the anterior part. Most fibers of the external oblique run inferomedially—in the same direction as the fingers do when the hands are in one's side pockets. The attachments, nerve supply, and main actions of the external oblique are presented in Table 2.2. As the muscle fibers pass inferomedially, they become aponeurotic approximately at the midclavicular line (MCL) and form a sheet of tendinous fibers that decussate at the linea alba. Medial to the pubic tubercle, the external oblique aponeurosis attaches to the **pubic crest**. Inferiorly, the inferior margin of the external oblique aponeurosis thickens and folds back on itself to form the **inguinal ligament**, a fibrous band extending between the **anterior superior iliac spine** and the **pubic tubercle** (Fig. 2.5*B*). The inguinal ligament is therefore not a free-standing structure. As it folds back on itself it becomes in-curved to form a trough, which forms the floor of the inguinal canal. Some fibers of the inguinal ligament pass upward to cross the linea alba and blend with the lower fibers of the contralateral aponeurosis. These fibers form the *reflected inguinal ligament* (Fig. 2.6).

Internal Oblique Muscle

The intermediate of the three flat abdominal muscles—the internal oblique—is a thin muscular sheet that fans out anteromedially (Figs. 2.6–2.8). Its fibers also become aponeurotic in roughly the same (midclavicular) line as the external oblique and participate in the formation of the rectus sheath. The inferior aponeurotic fibers of the internal oblique arch over the **spermatic cord** as it passes through the *inguinal canal.* The fibers then descend posterior to the **superficial inguinal ring** (Fig. 2.6) to attach to the pubic crest and pecten pubis (Fig. 2.5*B*), a sharp ridge that is the continuation of the superior pubic ramus. The most inferior, medial tendinous fibers of the internal oblique join with aponeurotic fibers of the more deeply placed transverse abdominal muscle to form the **conjoint tendon**, which turns inferiorly and attaches to the pubic crest and pecten pubis. The attachments, nerve supply, and main actions of the internal oblique are listed in Table 2.2.

Between the internal oblique and transverse abdominal muscles is a **neurovascular plane** that corresponds with a similar plane in the intercostal spaces. In both regions the plane

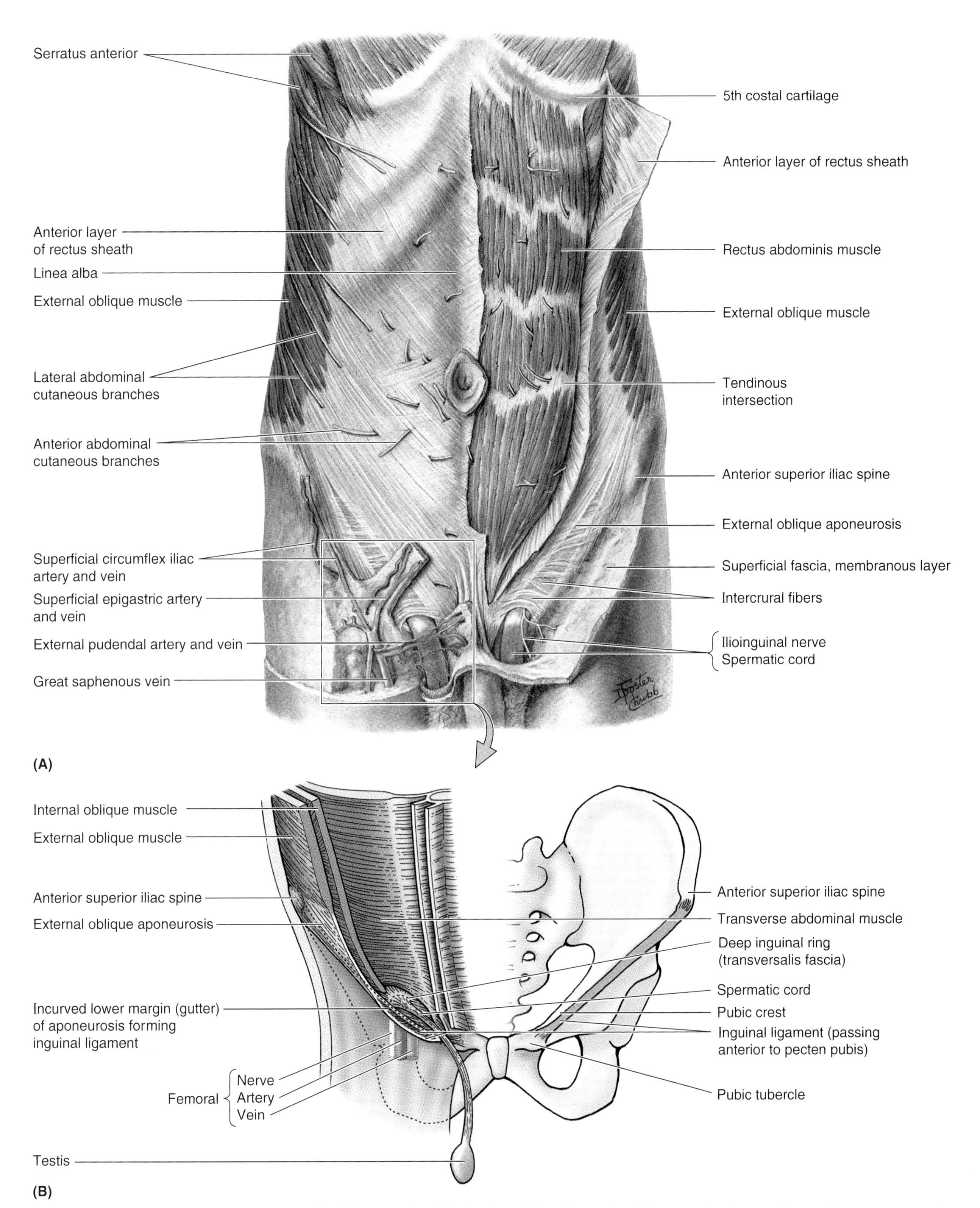

Figure 2.5. Anterolateral abdominal wall. A. Superficial dissection. The anterior layer of the rectus sheath is reflected on the left side. Observe the anterior cutaneous nerves (T7 through T12) piercing the rectus abdominis and the anterior layer of the rectus sheath. **B.** Schematic drawing of the wall showing the three flat abdominal muscles and formation of the inguinal ligament.

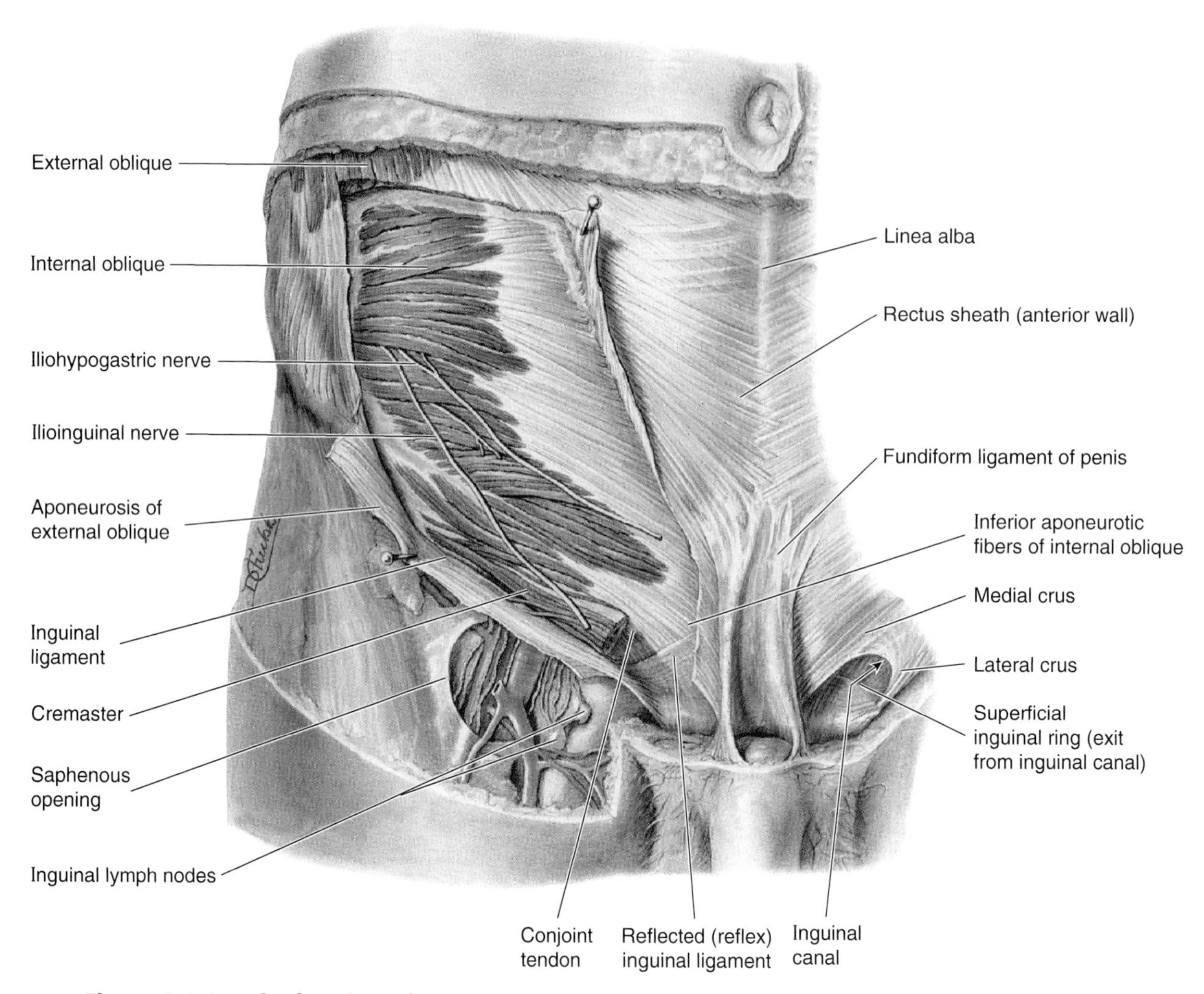

Figure 2.6. Inguinal region of a male. The aponeurosis of the external oblique is partly cut away and the spermatic cord has been cut and removed from the inguinal canal. The reflected (reflex) inguinal ligament is formed by aponeurotic fibers of the external oblique. Observe the iliohypogastric and ilioinguinal nerves (branches of the first lumbar nerve) passing between the external and internal oblique muscles. The ilioinguinal nerve is vulnerable during repair of an inguinal hernia.

lies between the middle and deepest layers of muscle (Figs. 2.4–2.7). *The neurovascular plane of the anterolateral abdominal wall contains the:*

- Thoracoabdominal nerves—anterior abdominal (cutaneous) branches of ventral primary rami (intercostal nerves) of T7 through T11
- Subcostal nerve—the large ventral ramus of T12 nerve
- Iliohypogastric and ilioinguinal nerves—branches of the ventral ramus of L1 nerve
- Inferior intercostal, subcostal, and lumbar arteries
- Deep circumflex iliac artery, a branch of the external iliac artery.

In the anterior part of the abdominal wall, the nerves and vessels leave the neurovascular plane and lie in a more superficial plane.

Transverse Abdominal Muscle

The fibers of the transverse abdominal muscle—the innermost of the three flat abdominal muscles (Fig. 2.7)—run more or less transversomedially, except for the inferior ones that run parallel to those of the internal oblique. The fibers of the transverse abdominal muscle end in an aponeurosis, which contributes to the formation of the rectus sheath (Fig. 2.8). The attachments, nerve supply, and main actions of the transverse abdominal muscle are listed in Table 2.2.

Rectus Abdominis Muscle

A long, broad straplike muscle, the rectus abdominis is the principal vertical muscle of the anterior abdominal wall (Fig. 2.5*A*). Its attachments, nerve supply, and main actions are listed in Table 2.2. The paired rectus muscles—separated by the linea alba—lie close together inferiorly. The rectus abdominis is three times as wide superiorly as inferiorly; it is broad and thin superiorly and narrow and thick inferiorly.

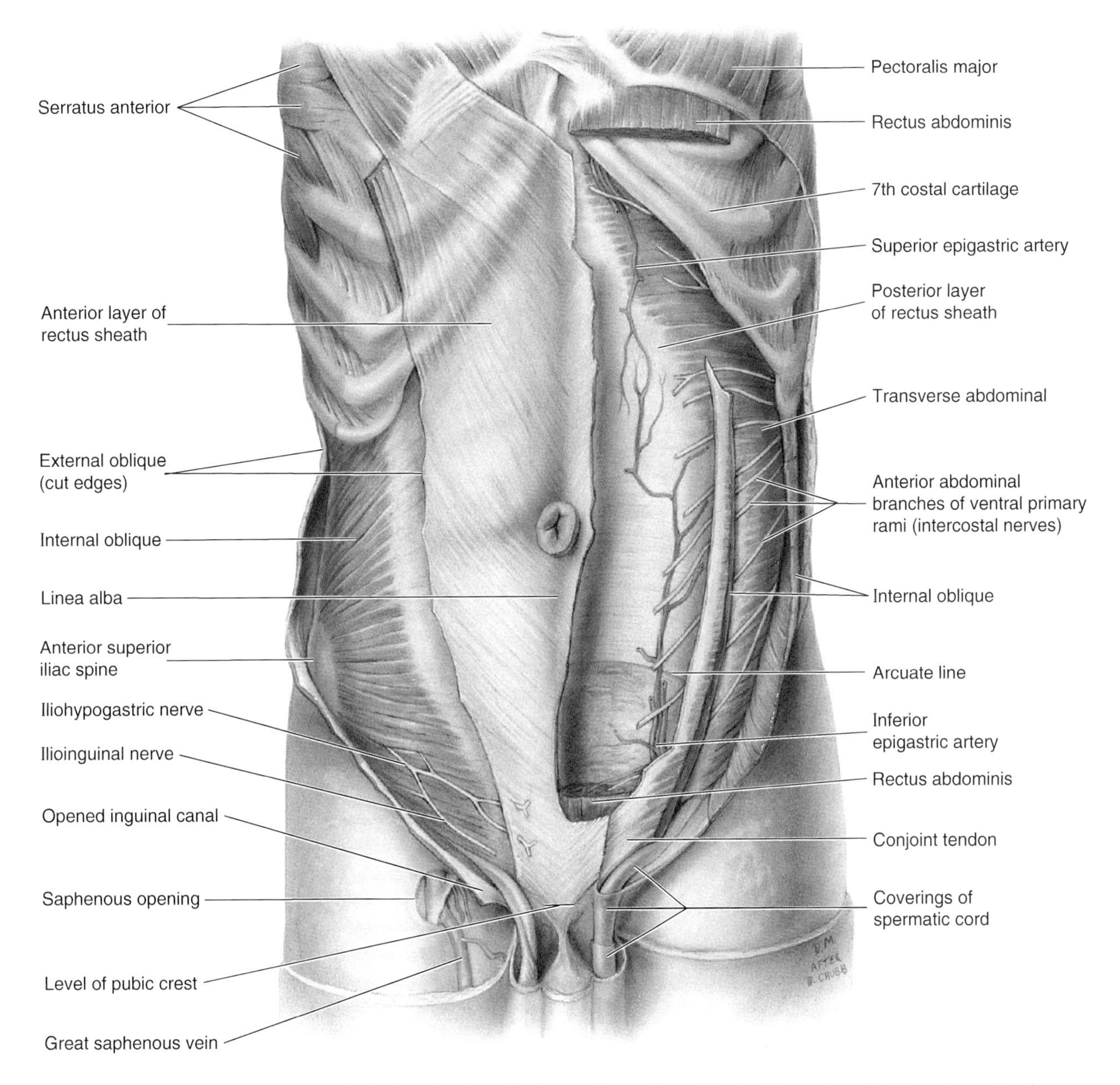

Figure 2.7. Anterolateral abdominal wall, deep dissection. Most of the external oblique is excised on the right side. The rectus abdominis is excised and the internal oblique is divided on the left side. Observe the anastomosis between the superior and inferior epigastric arteries that indirectly unites the arteries of the upper and lower limbs.

Most of the rectus abdominis is enclosed in the rectus sheath (Figs. 2.7 and 2.8*B*). The rectus muscle is anchored transversely by attachment to the anterior layer of the rectus sheath at three or more **tendinous intersections** (Fig. 2.5*A*). When tensed in muscular persons, the stretches of muscle between the tendinous intersections bulge outward. The intersections—indicated by grooves in the skin between the muscular bulges—usually occur at the level of the xiphoid process, umbilicus, and halfway between these structures.

The **rectus sheath** (Figs. 2.6–2.8) is the strong, incomplete fibrous compartment of the rectus abdominis and pyramidalis muscles. It is formed by the decussation and interweaving of the aponeuroses of the flat abdominal muscles. The superior two-thirds of the internal oblique aponeurosis splits into two layers, or laminae, at the lateral border of the rectus abdominis; one lamina passing anterior to the muscle and the other passing posterior to it. The anterior lamina joins the aponeurosis of the external oblique to form the anterior layer of the rectus sheath. The posterior lamina joins the aponeurosis of the transverse abdominal muscle to form the posterior layer of the rectus sheath. The fibers of the anterior and posterior layers of the sheath interlace in the anterior median line to form the complex *linea alba*.

Table 2.2. **Muscles of the Anterolateral Abdominal Wall**

Muscle	Origin	Insertion	Innervation	Main Action
External oblique	External surfaces of 5th–12th ribs	Linea alba, pubic tubercle, and anterior half of iliac crest	Thoracoabdominal nerves (inferior 6 thoracic nerves) and subcostal nerve	Compress and support abdominal viscera,[a] flex and rotate trunk
Internal oblique	Thoracolumbar fascia, anterior two-thirds of iliac crest, and lateral half of inguinal ligament	Inferior borders of 10th–12th ribs, linea alba, and pecten pubis via conjoint tendon	Thoracoabdominal (ventral rami of inferior 6 thoracic) and first lumbar nerves	
Transverse abdominal	Internal surfaces of 7th–12th costal cartilages, thoracolumbar fascia, iliac crest, and lateral third of inguinal ligament	Linea alba with aponeurosis of internal oblique, pubic crest, and pecten pubis via conjoint tendon		Compresses and supports abdominal viscera[a]
Rectus abdominis	Pubic symphysis and pubic crest	Xiphoid process and 5th–7th costal cartilages	Thoracoabdominal nerves (ventral rami of inferior six thoracic nerves)	Flexes trunk (lumbar vertebrae) and compresses abdominal viscera[a]

Approximately 80% of people have an insignificant muscle, the *pyramidalis,* which is located in the rectus sheath anterior to the most inferior part of the rectus abdominis. It extends from the pubic crest of the hip bone to the linea alba. This small muscle draws down on the linea alba.
[a]In so doing, act as antagonists of diaphragm to produce expiration.

The **rectus sheath** has

- An *anterior layer* consisting of the interlaced aponeurosis of the external oblique and the anterior lamina of the internal oblique aponeurosis
- A *posterior layer* consisting of the fused posterior lamina of the internal oblique aponeurosis and the transverse abdominal aponeurosis; the inferior one-fourth of this layer is deficient because the entire aponeuroses of all three flat muscles pass anterior to the rectus abdominis, leaving the posterior surface of the muscle in contact with the transversalis fascia
- A *crescentic line*—the **arcuate line**—that demarcates the transition between the aponeurotic posterior wall of the sheath covering the superior three-fourths of the muscle, and the transversalis fascia covering the inferior one-fourth.

The posterior layer of the rectus sheath is deficient superior to the costal margin because the transverse abdominal muscle passes internal to the costal cartilages and the internal oblique attaches to the costal margin. Hence, superior to the costal margin, the rectus abdominis lies directly on the thoracic wall (Fig. 2.5*A*). The inferior one-fourth of the rectus sheath is also deficient because the internal oblique aponeurosis does not split here to enclose the rectus abdominis. The inferior limit of the posterior layer of the rectus sheath is marked by the arcuate line, which defines the point at which the posterior lamina of the internal oblique and the aponeurosis of the transverse abdominal become part of the anterior rectus sheath, leaving only the relatively thin transversalis fascia to cover the rectus abdominis posteriorly (Fig. 2.8*B*). Inferior to the arcuate line, which usually occurs approximately one-third of the distance from the umbilicus to the pubic crest, the aponeuroses of the three flat muscles pass anterior to the rectus abdominis to form the anterior layer of the rectus sheath.

The **linea alba** (Figs. 2.7 and 2.8)—the fibrous band running vertically the entire length of the anterior abdominal wall—receives the attachments of the oblique and transverse abdominal muscles. This tendinous raphe (G. rhaphe, suture, seam) is narrow inferior to the umbilicus and wide superior to it. The linea alba transmits small vessels and nerves to the skin. In thin muscular persons, a groove is visible in the skin overlying the linea alba. At its middle, underlying the umbilicus, the linea alba contains the **umbilical ring**, a defect in the linea alba through which the fetal umbilical vessels pass to and from the umbilical cord and placenta. All layers of the anterolateral abdominal wall fuse at the umbilicus. As fat accumulates in the subcutaneous tissue postnatally, the skin becomes raised around the umbilical ring and the umbilicus becomes depressed. This occurs 7 to 14 days after birth, when the atrophic umbilical cord "falls off."

Functions and Actions of the Anterolateral Abdominal Muscles

The muscles of the anterolateral abdominal wall:

- Form a strong expandable support for the anterolateral abdominal wall
- Protect the abdominal viscera from injury, such as a low blow in boxing
- Compress the abdominal contents
- Help to maintain or increase the intra-abdominal pressure and, in so doing, oppose the diaphragm and produce expiration
- Move the trunk and help to maintain posture.

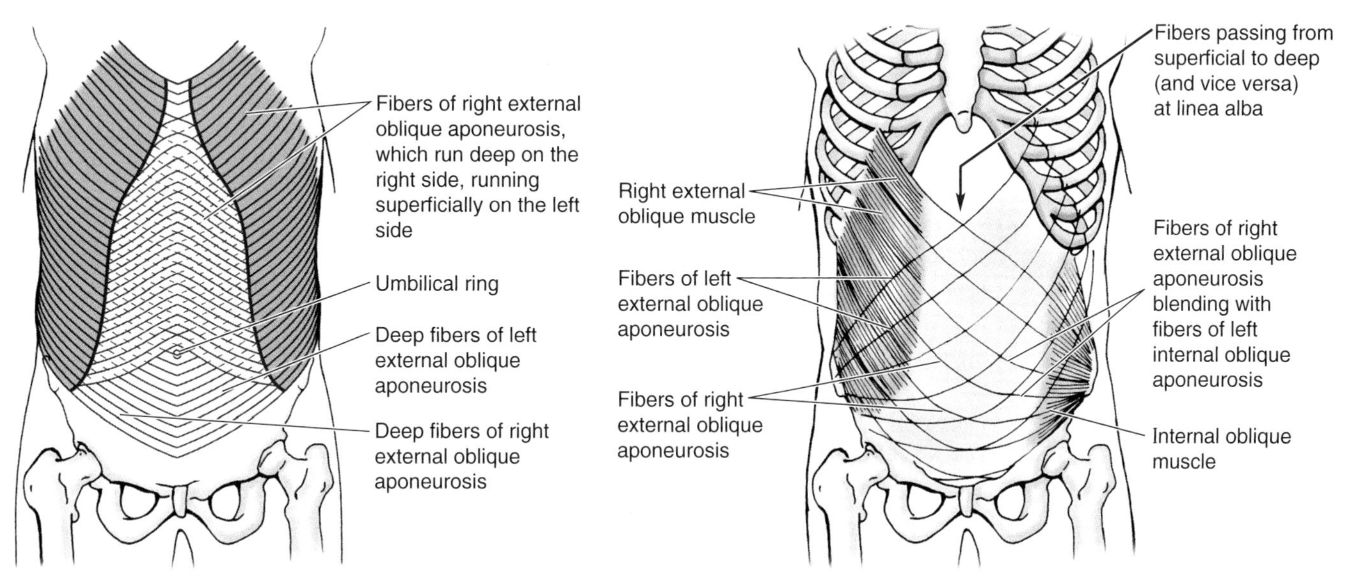

(A) Intramuscular exchange of superficial and deep fibers within aponeuroses of contralateral external oblique muscles.

Intermuscular exchange of fibers between aponeuroses of contralateral external and internal oblique muscles.

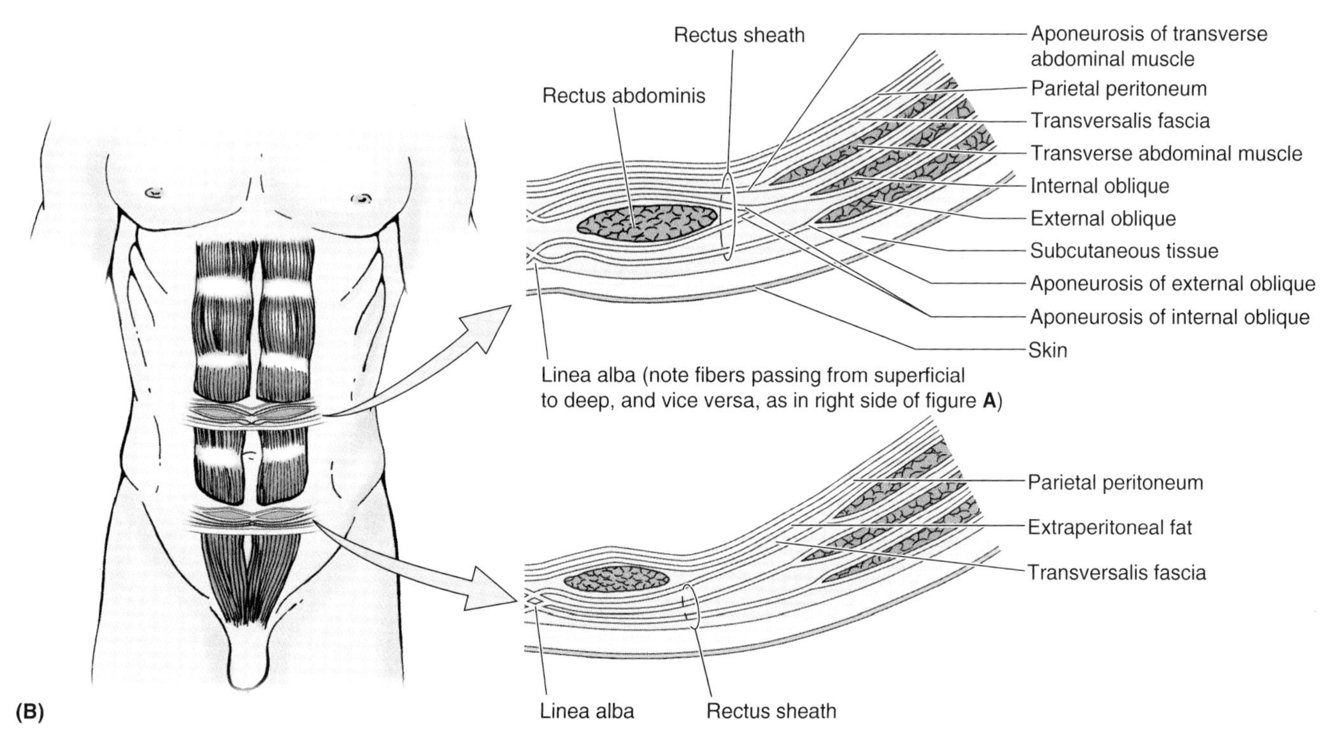

Figure 2.8. Structure of anterolateral abdominal wall. A. Bilaminar aponeuroses of the external and internal oblique muscles. **B.** Transverse sections of the wall superior and inferior to the umbilicus showing the makeup of the rectus sheath.

The oblique and transverse muscles, acting together bilaterally, form a muscular girdle that exerts firm pressure on the abdominal viscera. The rectus abdominis participates little, if at all, in this action.

Acting together, the abdominal muscles compress the abdominal viscera and elevate the diaphragm during respiration. When the diaphragm contracts, the anterolateral abdominal wall expands as its muscles relax to make room for the organs, such as the liver, that are pushed inferiorly. When the diaphragm relaxes, the wall sinks in as the muscles contract. The combined actions of the anterolateral muscles assist in expelling air during expiration and produce the force required for defecation (evacuation of fecal material from the rectum), micturition (urination), and parturition (childbirth). The anterolateral abdominal muscles are also involved in movements of the trunk at the lumbar vertebrae and in controlling the tilt of the pelvis when standing for maintenance of posture. Consequently, strengthening the anterolateral abdominal wall musculature improves standing and sitting posture. The rectus abdominis is a powerful flexor of the thoracic region and especially of the lumbar region of the vertebral column, pulling the anterior costal margin and pubic crest toward each other. The oblique abdominal muscles also assist in movements of the trunk, especially lateral flexion and rotation of the lumbar and lower thoracic vertebral column. The transverse abdominal muscle probably has no appreciable effect on the vertebral column (Williams et al., 1995).

Nerves of the Anterolateral Abdominal Wall

The skin and muscles of the anterolateral abdominal wall (Fig. 2.9, Table 2.3) are supplied mainly by the:

- Thoracoabdominal (former inferior intercostal) nerves—the anterior abdominal (cutaneous) branches of ventral primary rami of the inferior six thoracic nerves (T7 through T11)
- Subcostal nerves (T12)
- Iliohypogastric and ilioinguinal nerves (L1).

The **thoracoabdominal nerves** pass inferoanteriorly from the intercostal spaces and run in the neurovascular plane be-

Protuberance of the Abdomen

A prominent abdomen is normal in infants and young children because their GI tracts contain considerable amounts of air. In addition, their anterolateral abdominal cavities are enlarging and their abdominal muscles are gaining strength. An infant's or young child's relatively large liver also accounts for some bulging. The well-conditioned adult of normal weight has a flat or scaphoid (boat-shaped) abdomen when in the supine position. During pregnancy, protrusion of the abdomen occurs because of the growing fetus and is especially prominent when the baby is "carried low." The abdomen also expands in both sexes following the deposition of excessive fat or the accumulation of feces, fluid such as *ascites* (accumulation of serous fluid in the peritoneal cavity), or flatus (gas in the GI tract). The five common causes of abdominal protrusion all begin with "*f*" (fat, feces, fetus, flatus, and fluid).

An everted umbilicus may be a sign of increased intra-abdominal pressure, usually resulting from ascites or a large mass (e.g., a tumor, a fetus, or an enlarged organ such as the liver). *Liposuction* is a surgical method for removing unwanted subcutaneous fat using a percutaneously placed suction tube and high vacuum pressure. The tubes are inserted subdermally through small skin incisions.

During old age, anterolateral abdominal muscle laxity contributes to abdominal protuberance. Tumors and organomegaly (organ enlargement such as splenomegaly, or enlargement of the spleen) also produce abdominal enlargement. As the abdomen enlarges, the anterolateral abdominal muscles thin out (atrophy), the skin grows, and the nerves and blood vessels lengthen. Reddish, elongated lines or streaks often appear in the abdominal skin of pregnant women (*striae gravidarum*), obese people, and patients with chronic ascites. These "stretch marks" gradually change into thin, scarlike lines—*lineae albicantes*—that are white, pale tan, or slightly bluish bands of discoloration.

When the anterior abdominal muscles are underdeveloped, become atrophic, or have insufficient tonus to resist the increased weight of a protuberant abdomen on the anterior pelvis, the pelvis tilts anteriorly at the hip joints when standing (the pubis descends and the sacrum ascends) producing *lordosis*, an excessive convex curvature of the lumbar region of the vertebral column. (See Chapter 4, *Back*, for further details.)

Physical Examination of the Abdominal Wall

To enable adequate relaxation of the patient's anterolateral abdominal wall, physicians perform this part of the physical examination with the patient in the supine position with thighs and knees semiflexed. Otherwise, the deep fascia of the thighs pulls on the membranous layer of abdominal subcutaneous tissue, tensing the abdominal wall.

Abdominal Hernias

The anterolateral abdominal wall may be the site of hernias. Most hernias occur in the inguinal, umbilical, and epigastric regions. (Inguinal hernias are discussed on pages 205–207.) *Umbilical hernias* are common in newborns because the anterior abdominal wall is relatively weak in the umbilical ring, especially in low–birth-weight infants. Umbilical hernias are usually small and result from increased intra-abdominal pressure in the presence of weakness and incomplete closure of the anterior abdominal wall after ligation of the umbilical cord at birth. Herniation occurs through the *umbilical ring*—the opening in the linea alba. *Acquired umbilical hernias* occur most commonly in women and obese people. Extraperitoneal fat and/or peritoneum protrude into the hernial sac. An *epigastric hernia*—a hernia in the epigastric region through the linea alba—occurs in the midline between the xiphoid process and the umbilicus. Epigastric hernias tend to occur in people older than 40 years and are usually associated with obesity. The hernial sac, composed of peritoneum, is covered only with skin and fatty subcutaneous tissue. ⊙

tween the internal oblique and transverse abdominal muscles to supply the abdominal skin and muscles. The anterior abdominal cutaneous branches pierce the rectus sheath a short distance from the median plane. *Anterior abdominal cutaneous branches of thoracoabdominal nerve(s)*

- T7 through T9 supply the skin superior to the umbilicus
- T10 innervates the skin around the umbilicus
- T11, plus the cutaneous branches of the subcostal (T12), iliohypogastric, and ilioinguinal (L1), supply the skin inferior to the umbilicus.

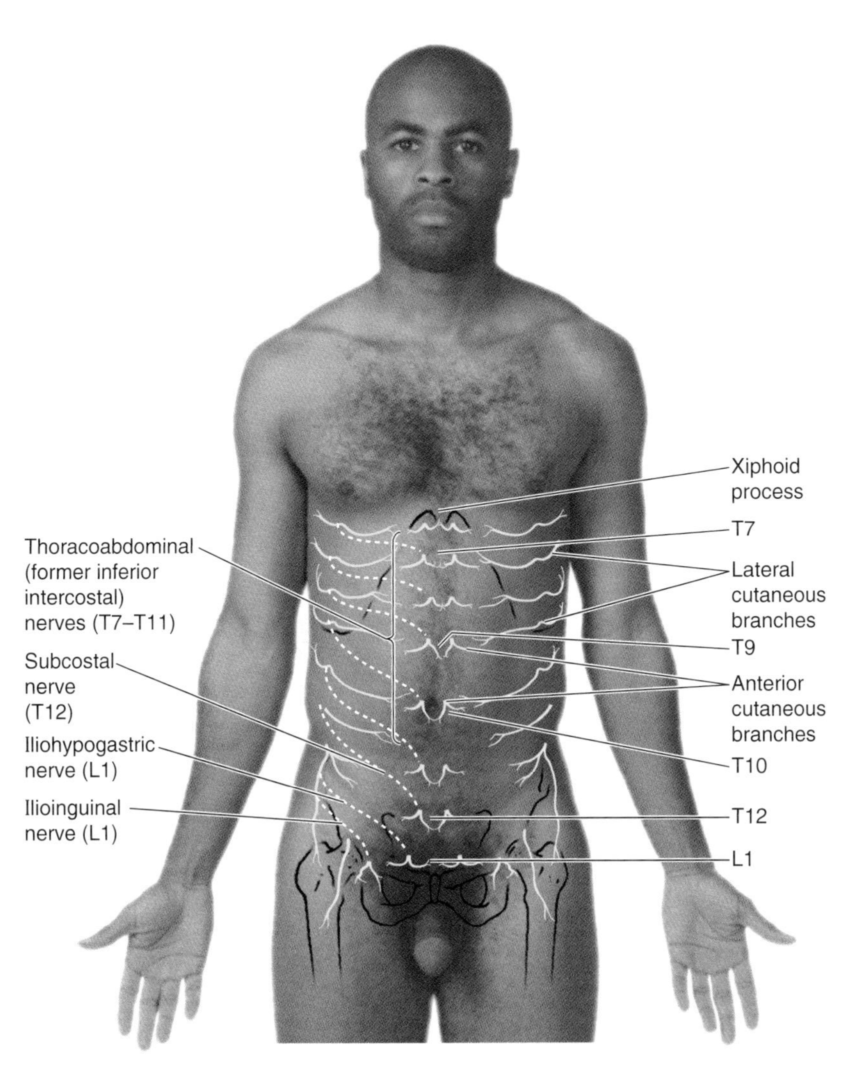

Figure 2.9. Schematic illustration of the distribution of the thoracoabdominal nerves. These nerves are the ventral rami of the thoracic spinal nerves 7 through 11; they supply the anterolateral muscles and overlying skin. These nerves are supplemented by the anterior terminal branches of the subcostal nerve and by a branch of the iliohypogastric nerve of the lumbar plexus. After passing through the muscles, they are sensory nerves of the skin.

Table 2.3. Nerves of the Anterolateral Abdominal Wall

Nerve	Origin	Course	Distribution
Thoracoabdominal (T7–T11)	Continuation of lower intercostal nerves	Run between 2nd and 3rd layers of abdominal muscles	Anterior abdominal muscles and overlying skin; periphery of diaphragm
Subcostal (T12)	Ventral ramus of 12th thoracic nerve	Runs along inferior border of 12th rib	Lowest slip of external oblique muscle and skin over anterior superior iliac spine and hip
Iliohypogastric (L1)	Chiefly from ventral ramus of 1st lumbar nerve	Pierces transverse abdominal muscle; branches pierce external oblique aponeurosis	Skin of hypogastric region and over iliac crest; internal oblique and transverse abdominal
Ilioinguinal (L1)	Ventral ramus of 1st lumbar nerve	Passes between 2nd and 3rd layers of abdominal muscles and passes through inguinal canal	Skin of scrotum or labium majus, mons pubis, and adjacent medial aspect of thigh; internal oblique and transverse abdominal

During their course through the wall, the thoracoabdominal, subcostal, and iliohypogastric nerves communicate with each other.

Vessels of the Anterolateral Abdominal Wall

The blood vessels of the anterolateral abdominal wall (Fig. 2.10) *are:*

- *Superior epigastrics* from the internal thoracic vessels
- *Inferior epigastrics* and *deep circumflex iliacs* from the external iliac vessels
- *Superficial circumflex iliacs and superficial epigastrics* from the femoral artery and greater saphenous vein
- *Anterior and collateral branches of the posterior intercostal vessels* in the 10th and 11th intercostal spaces, and from anterior branches of *subcostal vessels*
- *Branches of the musculophrenic vessels* from the internal thoracic vessels.

The arterial supply to the anterolateral abdominal wall is summarized in Table 2.4.

The **superior epigastric artery** is the direct continuation of the internal thoracic artery. It enters the rectus sheath superiorly through its posterior layer and supplies the upper part of the rectus abdominis and anastomoses with the inferior epigastric artery.

The **inferior epigastric artery** arises from the external iliac artery just superior to the inguinal ligament. It runs superiorly in the transversalis fascia to enter the rectus sheath below the

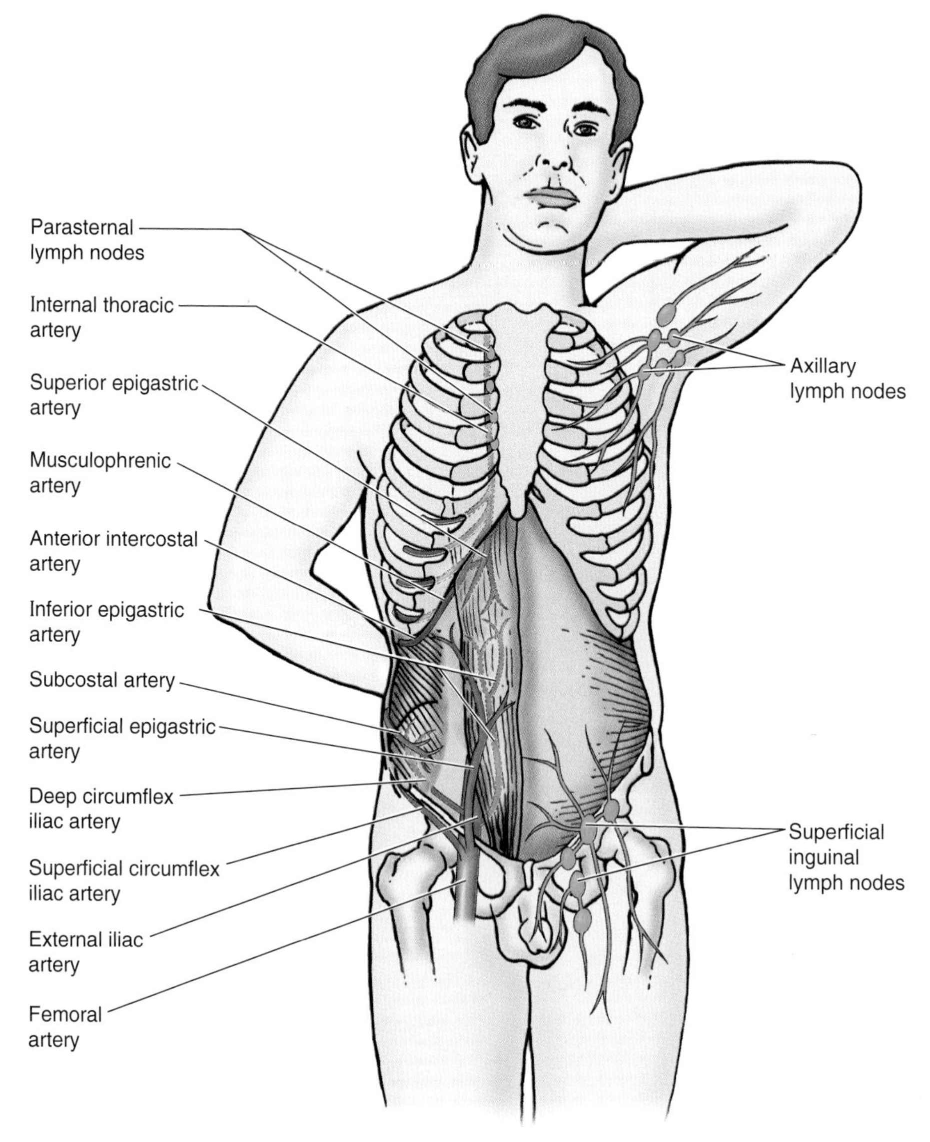

Figure 2.10. Arteries and lymphatics of the anterolateral abdominal wall. Lateral cutaneous branches of the intercostal and subcostal arteries accompany the cutaneous nerves. Anterior cutaneous branches also derive from superior and inferior epigastric arteries. Most superficial lymphatic vessels superior to the umbilicus drain to the axillary lymph nodes; a few drain to the parasternal lymph nodes. Superficial lymphatic vessels inferior to the umbilicus drain to the superficial inguinal lymph nodes.

arcuate line. Its branches enter the lower rectus abdominis and anastomose with those of the superior epigastric artery approximately in the umbilical region.

In lymphatic drainage of the anterolateral abdominal wall (Fig. 2.10):

- Superficial lymphatic vessels accompany the subcutaneous veins; those superior to the umbilicus drain mainly to the *axillary lymph nodes*; however, a few drain to the parasternal lymph nodes. Superficial lymphatic vessels inferior to the umbilicus drain to the *superficial inguinal lymph nodes.*
- Deep lymphatic vessels accompany the deep veins and drain to the external iliac, common iliac, and lumbar (lateral aortic) lymph nodes.

Table 2.4. Arteries of the Anterolateral Abdominal Wall

Artery	Origin	Course	Distribution
Superior epigastric	Internal thoracic artery	Descends in rectus sheath deep to rectus abdominis	Rectus abdominis and superior part of anterolateral abdominal wall
Inferior epigastric	External iliac artery	Runs superiorly and enters rectus sheath; runs deep to rectus abdominis	Rectus abdominis and medial part of anterolateral abdominal wall
Deep circumflex iliac		Runs on deep aspect of anterior abdominal wall, parallel to inguinal ligament	Iliacus muscle and inferior part of anterolateral abdominal wall
Superficial circumflex iliac	Femoral artery	Runs in superficial fascia along inguinal ligament	Subcutaneous tissue and skin over inferior part of anterolateral abdominal wall
Superficial epigastric		Runs in superficial fascia toward umbilicus	Subcutaneous tissue and skin over suprapubic region

Palpation of the Anterolateral Abdominal Wall

Warm hands are especially important when palpating the abdominal wall because cold hands make the anterolateral abdominal muscles tense, producing involuntary spasms (*guarding*) of the muscles. Intense guarding occurs during palpation when an organ (such as the appendix) is inflamed and in itself constitutes a clinically significant sign. The involuntary muscular spasms attempt to protect the viscera from pressure, which is painful when an abdominal infection is present. The common nerve supply of the skin and muscles of the wall explains why these spasms occur. Some patients tend to place their hands behind their heads when lying supine, which also tightens the muscles and makes the examination difficult. Placing the upper limbs at the sides and putting a pillow under the person's knees tends to relax the anterolateral abdominal muscles.

Superficial Abdominal Reflexes

Physicians and surgeons examine the reflexes of the abdominal wall to determine if there is abdominal disease such as appendicitis. The abdominal wall is the only protection most of the abdominal organs have. Consequently, it will react if an organ is diseased or injured. With the person supine and the muscles relaxed, the superficial abdominal reflex is elicited by quickly stroking horizontally, laterally to medially toward the umbilicus. Usually, contraction of the abdominal muscles is felt; this reflex may not be observed in obese people. Similarly, any injury to the abdominal skin results in a rapid reflex contraction of the abdominal muscles.

Abdominal Surgical Incisions

Surgeons use various incisions to gain access to the abdominal cavity. When possible, the incisions follow the cleavage lines (Langer's lines) in the skin. (Refer to the *Introduction* for a description and illustration of these lines.) The incision that allows adequate exposure and, secondarily, the best possible cosmetic effect, is chosen. The location of the incision also depends on the type of operation, the location of the organ(s) the surgeon wants to reach, bony or cartilaginous boundaries, avoidance of (especially motor) nerves, maintenance of blood supply, and minimizing injury to muscles and fascia of the wall while aiming for favorable healing. Thus, before making an incision, the surgeon considers the direction of the muscle fibers and the location of the aponeuroses and nerves. Consequently, a variety of incisions are routinely used, each having specific advantages and limitations.

Instead of transecting muscles, causing irreversible necrosis (death) of muscle fibers, the surgeon splits them in the direction of (between) their fibers. The rectus ▶

▶ abdominis is an exception; it can be transected because its muscle fibers run short distances between tendinous intersections, and its segmental innervation enters at the lateral part of the rectus sheath. Therefore, the nerves can be easily located and preserved. The surgeon chooses the part of the anterolateral abdominal wall that gives the freest access to the targeted organ with the least disturbance to the nerve supply to the muscles. Muscles and viscera are retracted toward, not away from, their neurovascular supply. Cutting a motor nerve paralyzes the muscle fibers supplied by it, thereby weakening the anterolateral abdominal wall. However, because of overlapping areas of innervation between nerves in the anterolateral abdominal wall, one or two small branches of

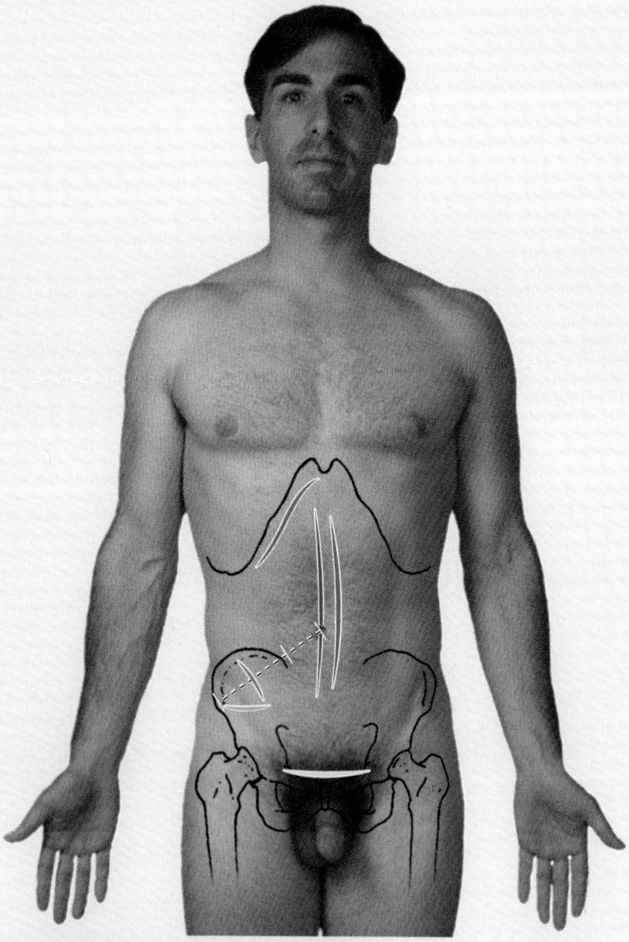

Key

- Median or midline incision
- Left paramedian incision
- Gridiron (muscle-splitting) incision
- Pfannenstiel (suprapubic) incision
- Transverse (abdominal) incision
- Subcostal incision

nerves may usually be cut without a noticeable loss of motor supply to the muscles or loss of sensation to the skin. Little if any communication occurs between nerves from the lateral border of the rectus abdominis to the anterior midline. The following are the most common surgical incisions:

Median or midline incisions can be made rapidly without cutting muscle, major blood vessels, or nerves. They cut through the fibrous tissue of the linea alba, superior and/or inferior to the umbilicus. Because the linea alba transmits only small vessels and nerves to the skin, a midline incision is relatively bloodless and avoids major nerves; however, incisions in some people may reveal abundant and well-vascularized fat. Conversely, because of its relatively poor blood supply, the linea alba may undergo necrosis and subsequent degeneration after incision if its edges are not aligned properly during closure. Median incisions can be made along any part or the length of the linea alba from xiphoid to pubis. Thus, they are good for exploratory procedures. Lower median incisions (below the umbilicus) are frequently used for reaching female pelvic viscera.

Paramedian incisions (lateral, left or right, to the median plane) are made in the sagittal plane, and may extend from the costal margin to the pubic hairline. The incision goes through the anterior layer of the rectus sheath, and the muscle is freed and retracted laterally to prevent tension and injury to the vessels and nerves. The posterior layer of the rectus sheath and the peritoneum is then incised to enter the peritoneal cavity.

Gridiron (muscle-splitting) incisions are often used for an appendectomy. The oblique *McBurney incision* is approximately 2.5 cm superomedial to the anterior superior iliac spine (*McBurney's point*). This incision is less popular now than an almost transverse incision in the line of a skin crease. In either case, the external oblique aponeurosis is incised inferomedially in the direction of its fibers and retracted. The musculoaponeurotic fibers of the internal oblique and transverse abdominal are then split in the line of their fibers and retracted. The *iliohypogastric nerve*, running deep to the internal oblique, is identified and preserved. Carefully made, the entire exposure cuts no musculoaponeurotic fibers; therefore, when the incision is closed, the muscle fibers move together and the abdominal wall is as strong after the operation as it was before. When kept relatively small and done carefully, the gridiron incision provides good access and avoids cutting, tearing, and stretching of nerves.

Pfannenstiel (suprapubic) incisions are made at the pubic hairline. These incisions—horizontal with a slight convexity—are used for most gynecological and obstetrical operations (e.g., for cesarean section and removal of a tubal pregnancy). The linea alba and the anterior layers of the rectus sheaths are transected and resected superiorly ▶

▶ and the rectus muscles are retracted laterally or divided through their tendinous parts allowing reattachment without muscle fiber injury. The iliohypogastric and ilioinguinal nerves are identified and preserved.

Transverse incisions through the anterior layer of the rectus sheath and rectus abdominis provide good access and cause the least possible damage to the nerve supply of the rectus abdominis. This muscle may be divided transversely without serious damage because a new transverse band forms that, when the muscle segments are rejoined, is similar to a tendinous intersection. Transverse incisions are not made through the tendinous intersections because cutaneous nerves and branches of the superior epigastric vessels pierce these fibrous regions of the muscle. Transverse incisions are most useful above the level of the umbilicus. They can be increased laterally as needed to increase exposure but are not good for exploratory procedures because superior and inferior extension is difficult.

Subcostal incisions provide access to the gallbladder and biliary tract on the right side and the spleen on the left. The incision is made parallel but at least 2.5 cm inferior to the costal margin to avoid the 7th and 8th thoracic spinal nerves (Fig. 2.9).

High-risk incisions include pararectus and inguinal incisions. *Pararectus incisions* along the lateral border of the rectus sheath are undesirable because they are likely to cut the nerve supply to the rectus abdominis. Blood supply from the inferior epigastric artery also may be compromised. *Inguinal incisions* for repairing hernias may injure the ilioinguinal nerve directly or it may be inadvertently included in the suture during closure of the incision. In such cases, patients may feel pain in the L1 dermatome region, which includes the scrotum (or the labium majus).

Many abdominopelvic surgical procedures are now performed by endoscopic (minimally invasive) surgery, in which tiny perforations of the abdominal wall allow entry of remotely operated instruments, replacing the larger conventional incisions. Thus, the potential for contamination through the open wound and the time required for healing are minimized.

Incisional Hernia

An incisional hernia is a protrusion of omentum (a fold of peritoneum) or an organ through a surgical incision. Surgeons who make incisions based on a thorough knowledge of anterolateral abdominal wall anatomy will only occasionally have to deal with this problem. However, if the muscular and aponeurotic layers of the abdomen do not heal properly, an incisional hernia can result. Infection, bowel obstruction, and obesity are predisposing factors to incisional hernias.

Injury to Nerves of the Anterolateral Abdominal Wall

Because the inferior thoracic spinal nerves (T11 and T12) and the iliohypogastric and ilioinguinal nerves (L1) supply the abdominal musculature in the inguinal region, injury to them during surgery or an abdominal injury may result in weakening of muscles in the inguinal region, predisposing development of a direct inguinal hernia (pp. 205–207). The *ilioinguinal nerve* supplies motor branches to the fibers of the internal oblique that are inserted into the lateral border of the conjoint tendon. Division of this nerve paralyzes these fibers and weakens the conjoint tendon, which may also result in a direct inguinal hernia.

Peritonitis

Inflammation of the peritoneum, which lines the abdominal cavity and covers viscera such as the stomach, causes pain in the overlying skin and an increase in the tone of the anterolateral abdominal muscles. Rhythmic movements of the anterolateral abdominal wall normally accompany respirations. If the abdomen is drawn in as the chest expands (*paradoxical abdominothoracic rhythm*) and muscle rigidity is present, either peritonitis or pneumonitis (inflammation of the lungs) may be present. A particular feature of a physical examination of a person with an *acute abdomen* (intense abdominal pain) is spasm of the anterolateral *abdominal* muscles—*guarding*—a boardlike muscular rigidity that cannot be willfully suppressed. Refer to page 210 for further discussion of peritonitis. ○

Internal Surface of the Anterolateral Abdominal Wall

The internal (posterior) surface of the anterolateral abdominal wall is covered with **parietal peritoneum** (Fig. 2.11). The infraumbilical part of this surface exhibits several peritoneal folds, some of which contain remnants of vessels that carried blood to and from the fetus (Moore and Persaud, 1998). **Five umbilical peritoneal folds**—two on each side and one in the median plane—pass toward the umbilicus:

- The **median umbilical fold** extends from the apex of the urinary bladder to the umbilicus and covers the **median umbilical ligament**, the remnant of the urachus (reduced allantoic stalk), that joined the apex of the fetal bladder to the umbilicus (Moore and Persaud, 1998).
- Two **medial umbilical folds**, lateral to the median umbilical fold, cover the **medial umbilical ligaments**, the remnants of the occluded fetal umbilical arteries.
- Two **lateral umbilical folds**, lateral to the medial umbilical folds, cover the **inferior epigastric vessels**, which bleed if cut.

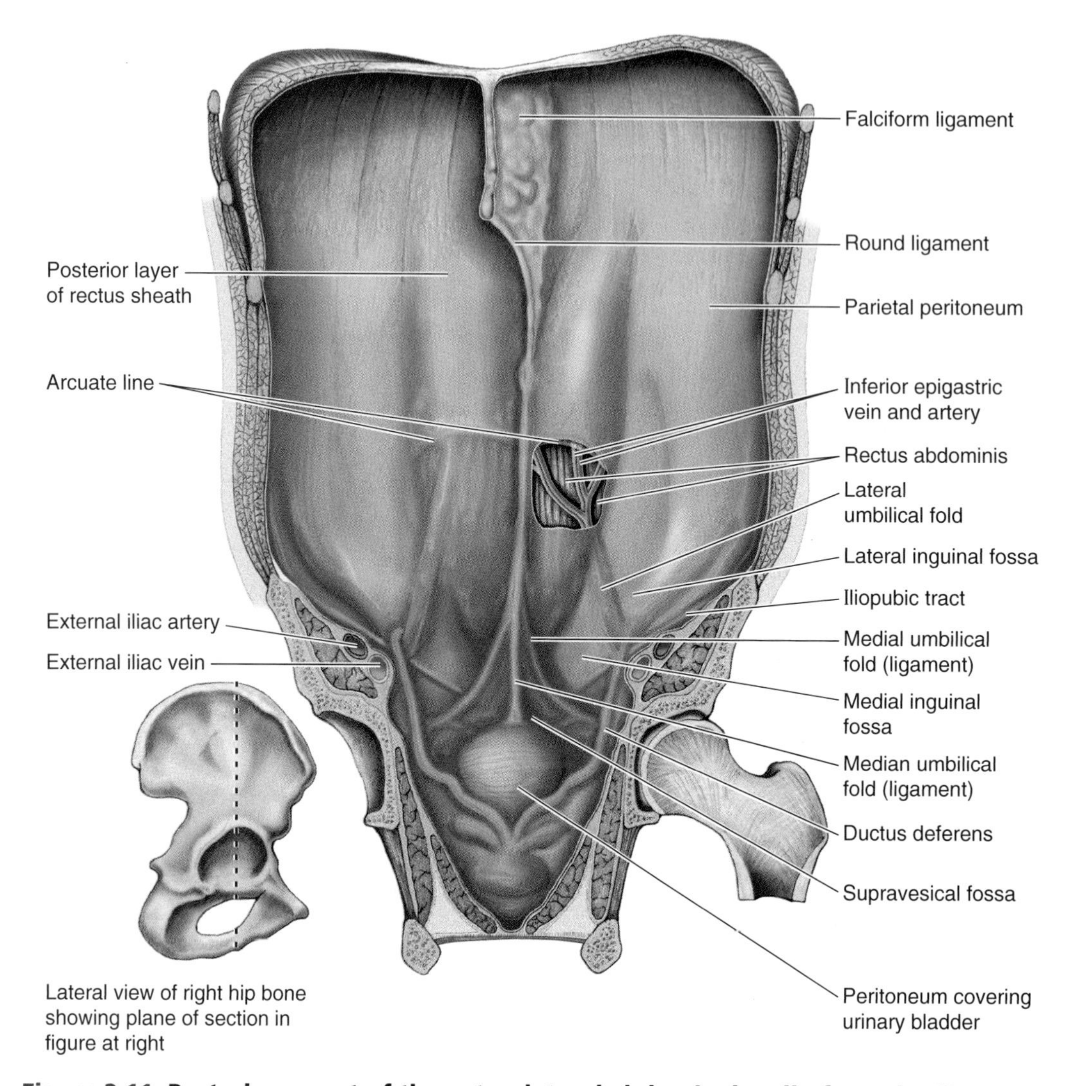

Figure 2.11. Posterior aspect of the anterolateral abdominal wall of a male. Observe the ligaments, folds, and fossae. Note that the lateral umbilical folds cover the inferior epigastric blood vessels; obviously they will bleed if these folds are cut.

The depressions lateral to the umbilical folds are **peritoneal fossae**, each of which is a potential site for a hernia. The location of a hernia in one of these fossae determines how the hernia is classified. *The shallow fossae between the umbilical folds are the:*

- **Supravesical fossae** between the median and medial umbilical folds, formed as the peritoneum reflects from the anterior abdominal wall onto the bladder. These are potential sites for rare external supravesical hernias. The level of the supravesical fossae rises and falls with filling and emptying of the bladder.
- **Medial inguinal fossae** between the medial and lateral umbilical folds (areas also commonly called *inguinal triangles*), which are potential sites for the less common direct inguinal hernias.
- **Lateral inguinal fossae**, lateral to the lateral umbilical folds, include the deep inguinal rings and are potential sites for the most common type of hernia in the lower abdominal wall—*indirect inguinal hernia.* (Refer to pages 205–207 for a discussion of inguinal hernias.)

The supraumbilical part of the internal surface of the anterior abdominal wall has a sagittally oriented peritoneal reflection—the **falciform ligament**—that extends between the upper anterior abdominal wall and the liver and encloses the round ligament of the liver (L. ligamentum teres) in its inferior free edge. The **round ligament** is a fibrous remnant of the umbilical vein, which extended from the umbilicus to the liver prenatally (Moore and Persaud, 1998).

External Supravesical Hernia

An external supravesical hernia leaves the peritoneal cavity through the supravesical fossa. The site of this hernia is medial to that of a direct inguinal hernia (pp. 205–207). The iliohypogastric nerve is in danger of injury during the repair of this type of hernia.

Postnatal Patency of the Umbilical Vein

Before the birth of a fetus, the umbilical vein carries well-oxygenated, nutrient-rich blood from the placenta to the fetus (Moore and Persaud, 1998). Although reference is often made to the "occluded" umbilical vein forming the round ligament of the liver, this vein is patent for some time after birth and is used for *umbilical vein catheterization* for exchange transfusion during early infancy—e.g., in infants with *erythroblastosis fetalis* or *hemolytic disease* of the fetus (Behrman et al., 1996). ⊙

Inguinal Region

The inguinal region (groin) is an important area surgically because it is the site of inguinal hernias. These hernias occur in both sexes but are more common in males. The inguinal region is an area of weaknesses in the inferior part of the anterolateral abdominal wall, especially in males, because of the passage of the spermatic cord through the inguinal canal.

Inguinal Canal

The inguinal canal in adults is an oblique, inferomedially directed passage (approximately 4 cm long) through the inferior part of the anterolateral abdominal wall. It lies parallel and just superior (2–4 cm) to the medial half of the **inguinal ligament** (Figs. 2.12 and 2.13). The main occupant of the inguinal canal is the **spermatic cord** in males and the **round ligament of the uterus** in females. These are functionally and developmentally distinct structures that happen to occur in the same location. The inguinal canal also contains blood and lymphatic vessels and the **ilioinguinal nerve** in both sexes. *The inguinal canal has an opening at each end.*

- The **deep (internal) inguinal ring** (entrance to the inguinal canal) is the site of an *outpouching of the transversalis fascia* approximately 1.25 cm superior to the middle of the inguinal ligament and *lateral to the inferior epigastric artery.* The deep inguinal ring is the beginning of an evagination in the transversalis fascia, forming an opening (like the entrance to a cave) through which the ductus deferens (vas deferens), or round ligament of the uterus in the female, and gonadal vessels pass to enter the inguinal canal. The transversalis fascia continues into the canal, forming the innermost covering (internal fascia) of the structures traversing the canal.
- The **superficial (external) inguinal ring** (exit from the inguinal canal) is a slitlike opening between the diagonal fibers of the aponeurosis of the external oblique, *superolateral to the pubic tubercle,* through which the spermatic cord, or round ligament in the female, emerge from the inguinal canal. The lateral and medial margins of the superficial ring formed by the split in the aponeurosis are called *crura* (L. leglike parts). The **lateral crus** attaches to the pubic tubercle and the **medial crus** to the pubic crest. Fibers arising from the inguinal ligament lateral to the superficial ring arch superolaterally to the superficial ring; these **intercrural fibers** help to prevent the crura from spreading apart (i.e., they keep the "split" in the aponeurosis from expanding).

The inguinal canal has two walls (anterior and posterior), a roof, and a floor (Figs. 2.12 and 2.13):

- **Anterior wall**: formed mainly by the *aponeurosis of the external oblique* with the lateral part of the wall being reinforced by fibers of the internal oblique
- **Posterior wall**: formed mainly by *transversalis fascia* with the medial part of the wall being reinforced by formation of the **conjoint tendon** (inguinal falx)—the merging of the pubic attachments of the internal oblique and transverse abdominal aponeuroses into a common tendon
- **Roof**: formed by the arching *fibers of the internal oblique* and *transverse abdominal muscles*
- **Floor**: formed by the superior surface of the in-curving inguinal ligament, which forms a shallow trough (Fig. 2.5*B*); it is reinforced in its most medial part by the *lacunar ligament,* a reflected part or extension from the deep aspect of the inguinal ligament to the pectineal line of the pecten pubis (Fig. 2.12).

The **iliopubic tract** (deep crural arch) is the thickened inferior margin of the transversalis fascia that appears as a fibrous band running parallel and posterior (deep) to the inguinal ligament. The iliopubic tract—seen only when the inguinal region is viewed from its internal aspect—contributes to the posterior wall of the inguinal canal as it bridges the external iliofemoral vessels from the iliopectineal arch to the superior pubic ramus (Fig. 2.11).

The iliopubic tract demarcates the inferior edge of the deep inguinal ring and the superomedial margin of the femoral canal (containing the femoral vessels). It is a useful landmark during laparoscopic inguinal hernia repair (Skandalakis et al., 1995).

Development of the Inguinal Canal. The testes develop in the extraperitoneal connective tissue in the superior lumbar region of the posterior abdominal wall (Fig. 2.14*A*). The **gubernaculum** is a fibrous cord connecting the primordial **testis** to the anterolateral abdominal wall at the site of the fu-

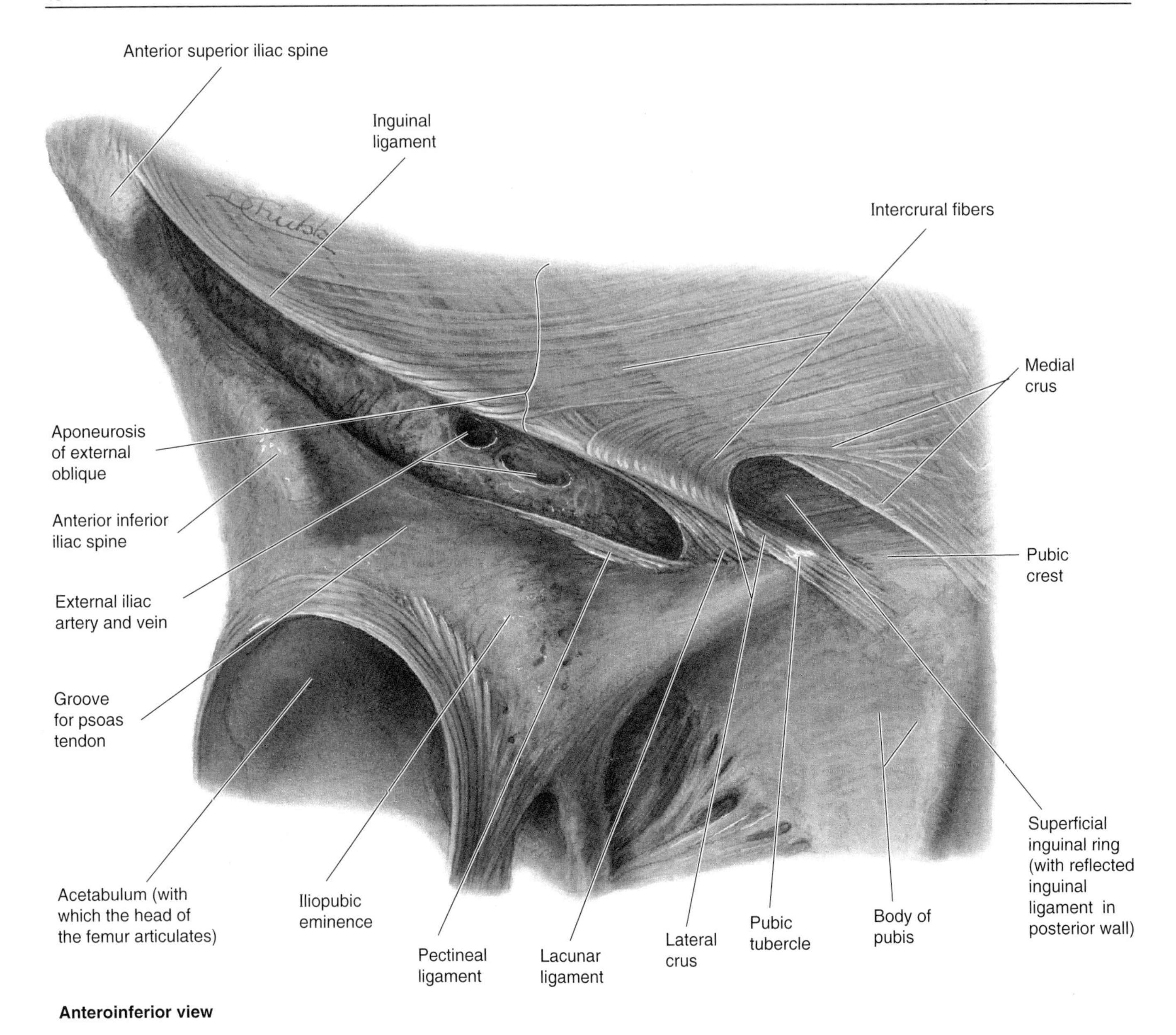

Figure 2.12. Inguinal ligament. Observe that this ligament is formed by the inferior, underturned fibers of the aponeurosis of the external oblique muscle. Observe the superficial inguinal ring, the slitlike opening in the aponeurosis of the external oblique through which the spermatic cord in the male and the round ligament in the female emerge from the inguinal canal.

ture deep ring of the inguinal canal (Moore and Persaud, 1998). A peritoneal diverticulum—the **processus vaginalis**—traverses the developing inguinal canal, carrying muscular and fascial layers of the anterolateral abdominal wall before it as it enters the **primordial scrotum.** By the 12th week of development, the testis has migrated to the pelvis, and by 28 weeks (7th month) it lies close to the developing deep inguinal ring (Fig. 2.14*B*). The testis begins to pass through the inguinal canal—formed by the processus vaginalis—during the 28th week and takes approximately 3 days to traverse it. Approximately 4 weeks later, the testis enters the scrotum (Fig. 2.14*C*). As the testis, its duct (the **ductus deferens**), and its vessels and nerves descend, they are ensheathed by musculofascial extensions of the anterolateral abdominal wall, which account for the presence of their derivatives in the adult scrotum—the internal and external **spermatic fasciae** and **cremaster muscle** (Fig. 2.13). The stalk of the processus vaginalis normally degenerates; however, its distal saccular part forms the **tunica vaginalis testis**—the serous sheath of the testis and epididymis. The gubernaculum is represented postnatally by the **scrotal ligament**, which extends from the testis to the skin of the scrotum.

The **ovaries** also develop in the superior lumbar region of the posterior abdominal wall and migrate to the lateral wall of

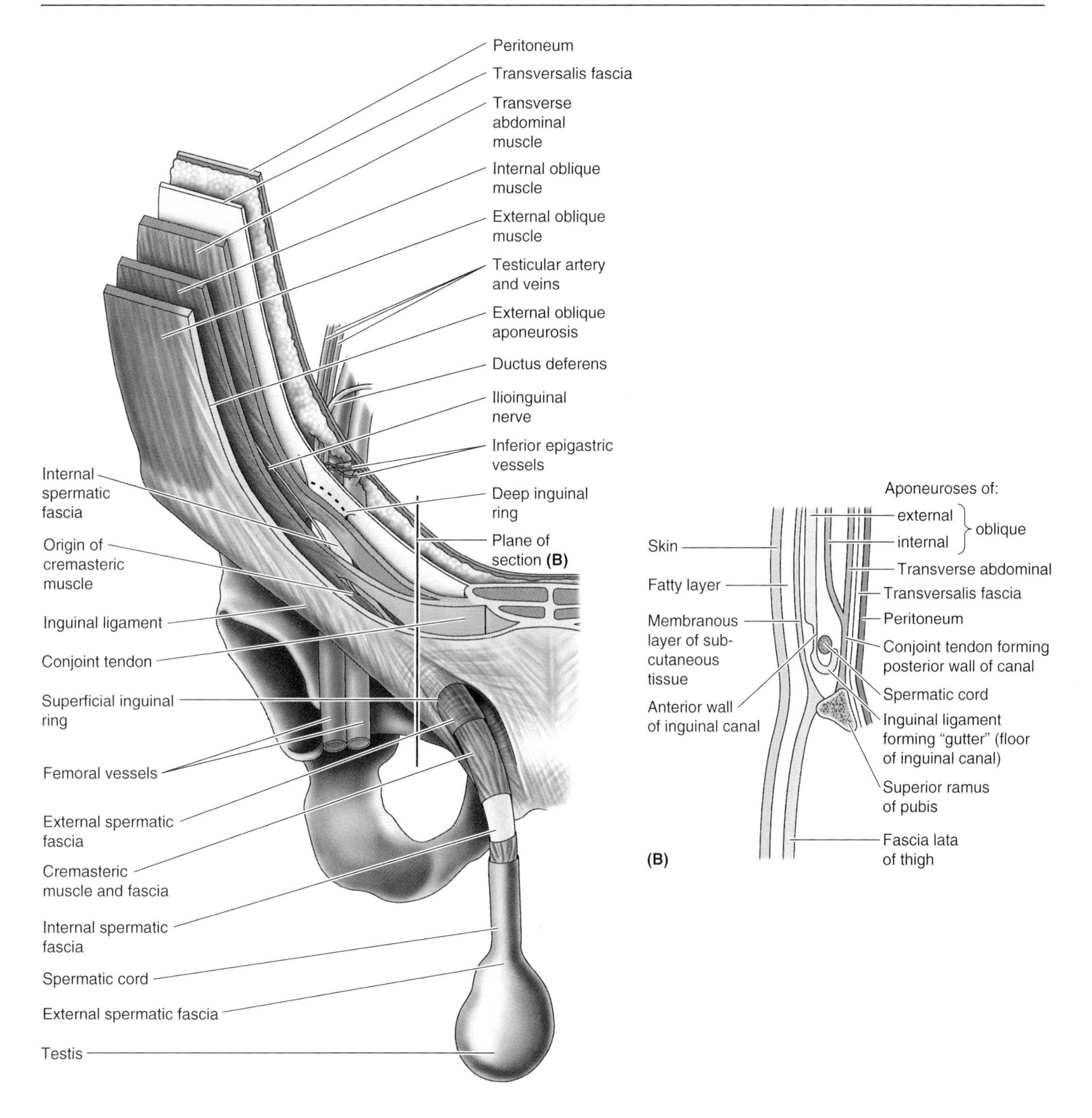

Figure 2.13. Inguinal canal, spermatic cord, and testis. A. Observe the layers of the abdominal wall and the derivation of the coverings of the spermatic cord and testis from them. Observe also the passage of the spermatic cord and ilioinguinal nerve through the inguinal canal. Examine the conjoint tendon—the common tendon of the transverse and internal oblique muscles; it is frequently muscular rather than aponeurotic as shown here and it may be poorly developed. It forms the posterior wall of the medial part of the inguinal canal. **B.** Sagittal section of the anterior abdominal wall and inguinal canal at the plane shown in (**A**).

Figure 2.14. Schematic drawings illustrating formation of the inguinal canals and descent of the testes. **A.** Seven-week embryo showing the testis before its descent from the dorsal abdominal wall. **B.** Fetus at 28 weeks showing the processus vaginalis and the testis passing through the inguinal canal. The processus vaginalis carries fascial layers of the abdominal wall before it. Observe that the testis passes posterior to the processus vaginalis, not through it. **C.** Newborn infant after obliteration of the stalk of the processus vaginalis. The remains of the processus vaginalis have formed the tunica vaginalis of the testis and the remnant of the gubernaculum has formed the scrotal ligament.

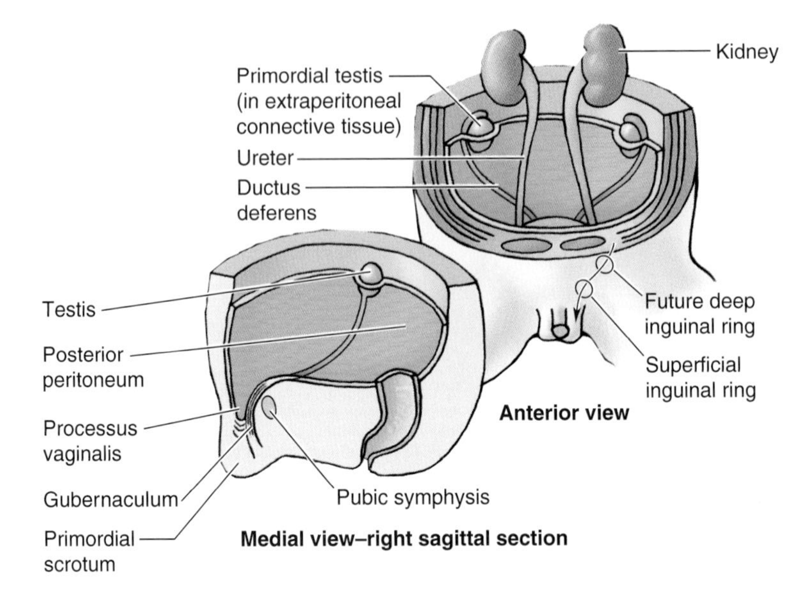

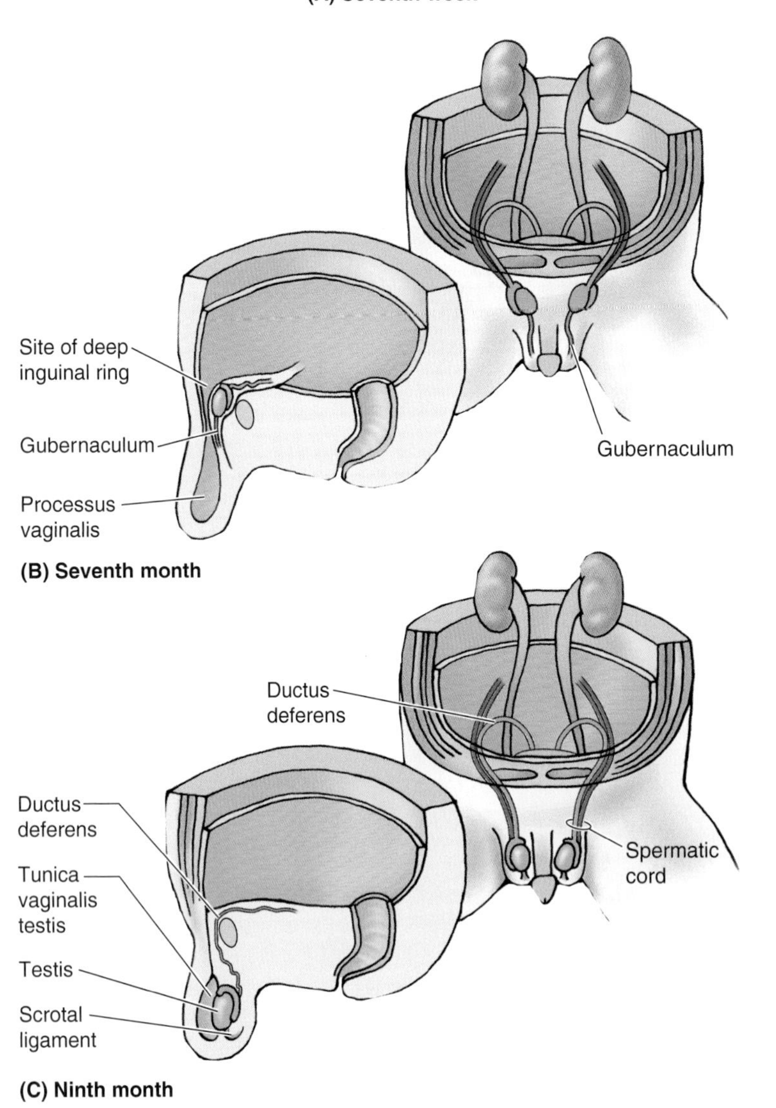

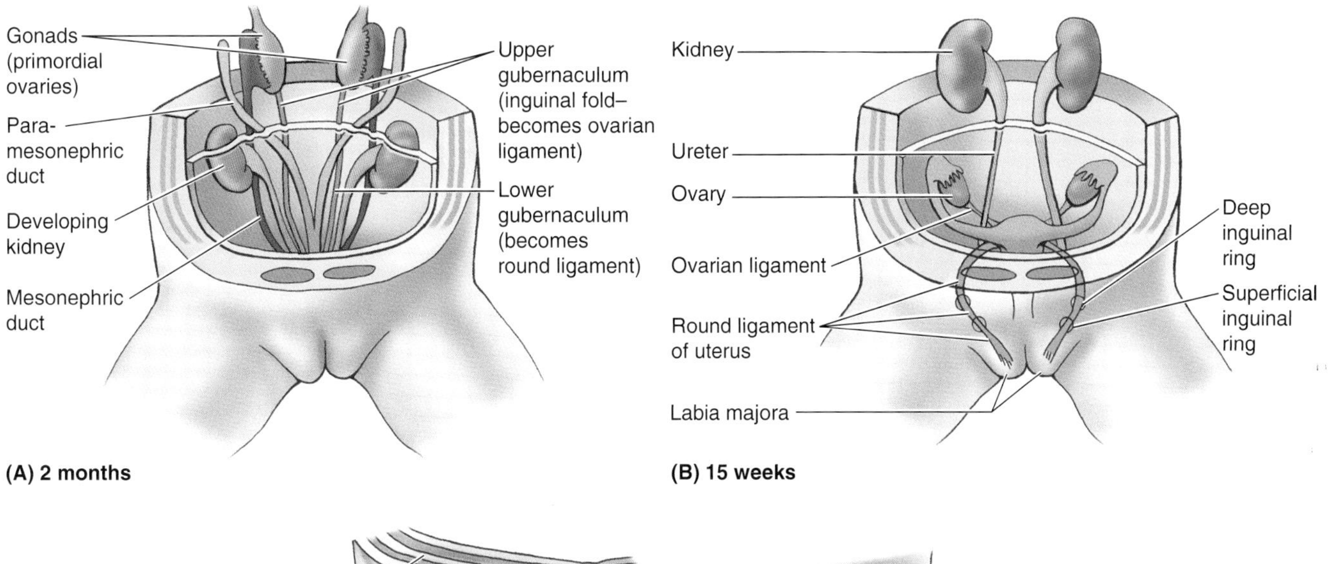

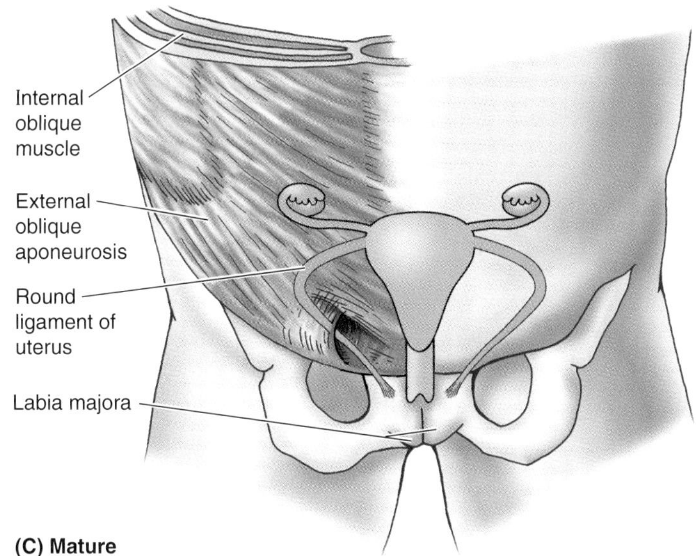

Figure 2.15. Schematic drawings illustrating formation of the inguinal canals in females. **A.** The undifferentiated gonads (primordial ovaries) at 2 months are located on the dorsal abdominal wall. **B.** The ovaries at 15 weeks have descended into the greater pelvis. The gubernaculum has become the ovarian ligament and round ligament of the uterus. The processus vaginalis (not illustrated) formed the inguinal canal on each side as in the male fetus, and the round ligament passes through it and attaches to the subcutaneous tissue of the labium majus. **C.** In the mature female, the processus vaginalis has degenerated but the round ligament and ilioinguinal nerve (not shown) pass through the inguinal canal.

the pelvis (Fig. 2.15, *A–C*). The processus vaginalis of the peritoneum traverses the transversalis fascia at the site of the deep inguinal ring forming the inguinal canal as in the male, and protrudes into the developing labium majus. The **gubernaculum**, a fibrous cord connecting the ovary and primordial uterus to the developing labium majus, has the following adult derivatives:

- The **ovarian ligament** between the ovary and the uterus
- The **round ligament** between the uterus and the labium majus.

Because of the attachment of the ovarian ligaments to the uterus, the ovaries do not descend to the inguinal region; however, the round ligament passes through the inguinal canal and attaches to the subcutaneous tissue of the labium majus (Fig. 2.15, *B–C*).

The processus vaginalis usually disappears by the 6th month of fetal development; for a review of the embryology of the female genital tract, see Moore and Persaud (1998). The inguinal canals in females are narrower than those in males, and the canals in infants of both sexes are shorter and much less oblique than in adults. The superficial inguinal rings in infants lie almost directly anterior to the deep inguinal rings.

Increased Intra-Abdominal Pressure. The deep and superficial inguinal rings in the adult do not overlap because of the oblique path of the inguinal canal. Consequently, increases in intra-abdominal pressure act on the inguinal canal, forcing the posterior wall of the canal against the anterior wall and strengthening this wall, thereby decreasing the likelihood of herniation until the pressures overcome the resistant effect of this mechanism. Furthermore, contraction of the external oblique approximates the anterior wall of the canal to the posterior wall. Contraction of the internal oblique and transverse abdominal muscles makes the roof of the canal descend, constricting the canal.

Palpation of the Inguinal Rings in Adult Males

The superficial inguinal ring is palpable superolateral to the pubic tubercle by invaginating the skin of the upper scrotum with the index finger. The examiner's finger follows the spermatic cord superolaterally to the superficial inguinal ring. If the ring is dilated, it may admit the finger without causing pain. Should a hernia be present, a sudden impulse is felt against either the tip or the pad of the examining finger when the patient is asked to cough (Swartz, 1994). The characteristics of inguinal hernia are discussed on pages 205–207.

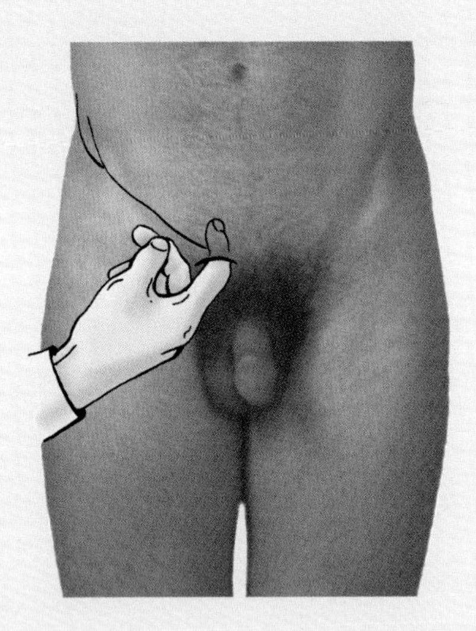

With the palmar surface of the finger against the anterior abdominal wall, the *deep inguinal ring* may be felt as a skin depression superior to the inguinal ligament, 2 to 4 cm superolateral to the pubic tubercle (Willms et al., 1994). ○

Spermatic Cord

The spermatic cord suspends the testis in the scrotum and contains structures running to and from the testis (Fig. 2.13). *The spermatic cord*:

- Begins at the deep inguinal ring lateral to the inferior epigastric vessels
- Passes through the inguinal canal
- Exits at the superficial inguinal ring
- Ends in the scrotum at the posterior border of the testis.

The spermatic cord is surrounded by fascial coverings derived from the anterolateral abdominal wall during prenatal development.

The coverings of the spermatic cord (Figs. 2.13 and 2.16, Table 2.5) *include*:

- **Internal spermatic fascia** derived from the transversalis fascia
- **Cremasteric fascia** derived from the fascia of both the superficial and deep surfaces of the internal oblique muscle
- **External spermatic fascia** derived from the external oblique aponeurosis.

The cremasteric fascia contains loops of the **cremaster muscle**, which is formed by the lowermost fascicles of the internal oblique muscle arising from the inguinal ligament. *The cremaster muscle reflexly draws the testis superiorly in the scrotum*, particularly when it is cold. In a warm environment such as a hot bath, the cremaster relaxes and the testis descends deeply in the scrotum. Both responses occur in an attempt to regulate the temperature of the testis for *spermatogenesis* (formation of sperms), which requires a constant temperature approximately one degree cooler than core temperature. The cremaster is innervated by the genital branch of the genitofemoral nerve (L1, L2), a derivative of the lumbar plexus.

The *constituents of the spermatic cord* (Figs. 2.16 and 2.17) are the:

- **Ductus deferens,** a muscular tube approximately 45 cm long that conveys sperms from the **epididymis** to the ejaculatory duct
- **Testicular artery** arising from the aorta and supplying the testis and epididymis
- **Artery of the ductus deferens** arising from the inferior vesical artery
- **Cremasteric artery** arising from the **inferior epigastric artery**
- **Pampiniform plexus,** a venous network formed by up to 12 veins, draining into the right or left testicular veins
- **Sympathetic nerve fibers** on arteries and sympathetic and parasympathetic nerve fibers on the ductus deferens
- **Genital branch of the genitofemoral nerve** supplying the cremaster muscle
- **Lymphatic vessels** draining the testis and closely associated structures and passing to the lumbar lymph nodes.

Scrotum

The scrotum is a cutaneous sac consisting of two layers (Fig. 2.16, Table 2.5): heavily pigmented skin and the closely re-

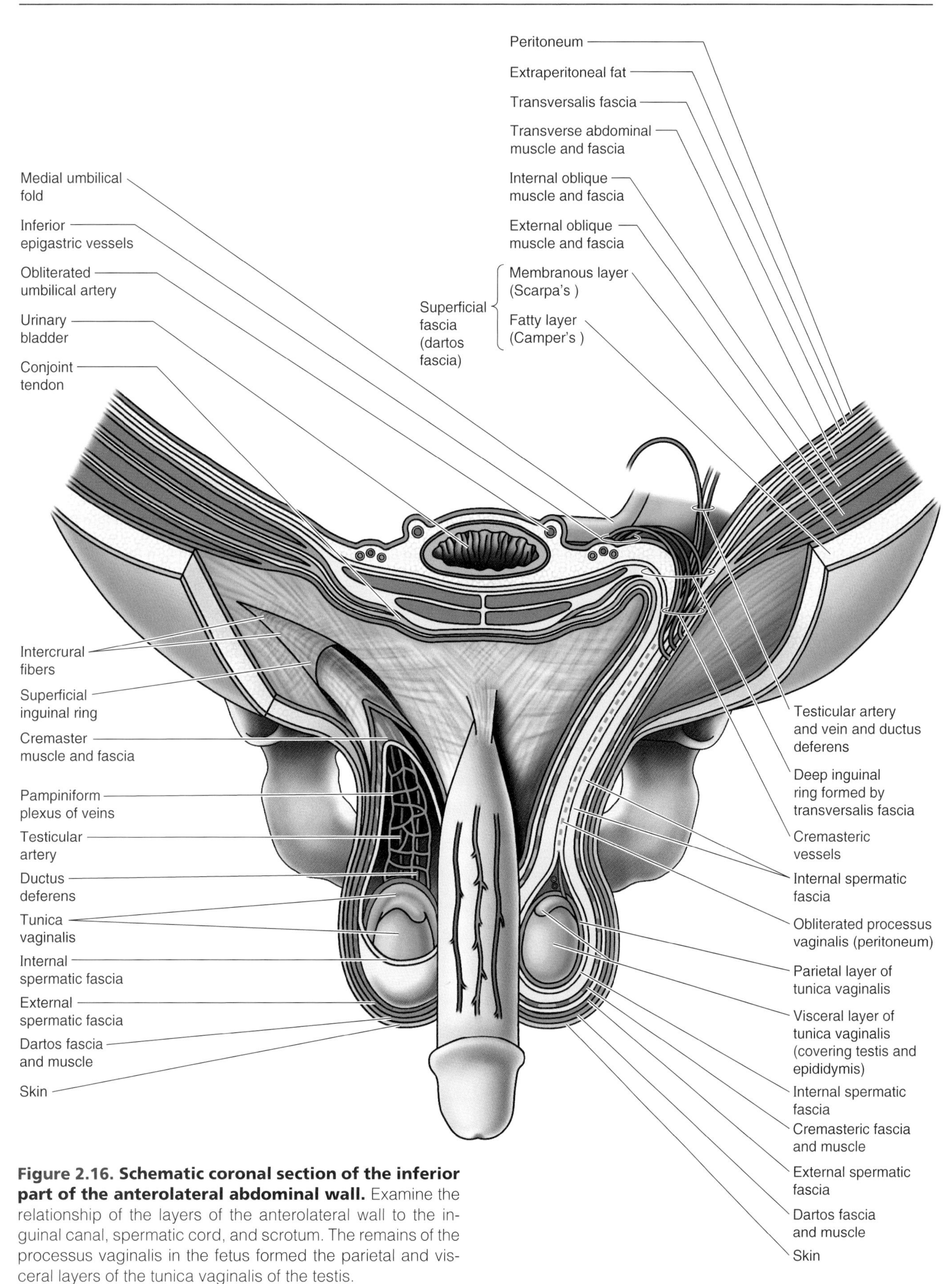

Figure 2.16. Schematic coronal section of the inferior part of the anterolateral abdominal wall. Examine the relationship of the layers of the anterolateral wall to the inguinal canal, spermatic cord, and scrotum. The remains of the processus vaginalis in the fetus formed the parietal and visceral layers of the tunica vaginalis of the testis.

Table 2.5. **Corresponding Layers of the Anterior Abdominal Wall, Spermatic Cord, and Scrotum**

Layers of Anterior Abdominal Wall	Scrotum and Coverings of Testis	Coverings of Spermatic Cord
Skin	Skin	Scrotum (and scrotal septum)
Subcutaneous tissue or superficial fascia	Superficial (dartos) fascia and dartos muscle	
External oblique aponeurosis	External spermatic fascia	External spermatic fascia
Internal oblique muscle	Cremaster muscle	Cremaster muscle
Fascia of both superficial and deep surfaces of the internal oblique muscle	Cremasteric fascia	Cremasteric fascia
Transverse abdominal muscle		
Transversalis fascia	Internal spermatic fascia	Internal spermatic fascia
Extraperitoneal fat		
Peritoneum	Tunica vaginalis	Obliterated processus vaginalis

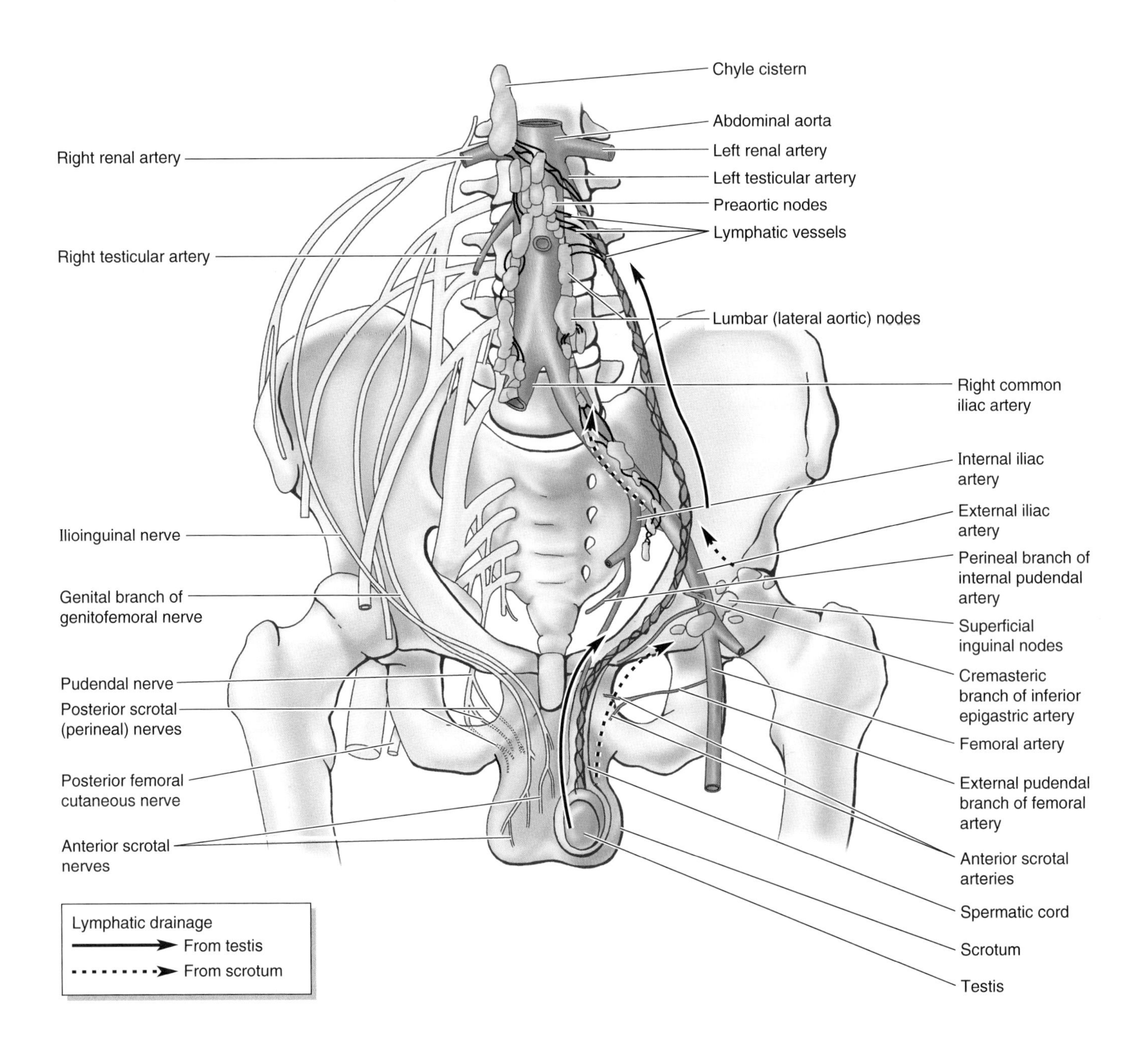

lated **dartos fascia**, a layer of smooth muscle fibers responsible for the rugose (wrinkled) appearance of the scrotum. Because the **dartos muscle** attaches to the skin, its contraction causes the scrotum to wrinkle when cold, which helps to regulate the loss of heat through its skin.

The scrotum is divided internally by the *septum of the scrotum* into right and left compartments and externally by the *scrotal raphe* (see Chapter 3), a cutaneous ridge marking the line of fusion of the embryonic *labioscrotal swellings* (Moore and Persaud, 1998). The superficial dartos fascia of the scrotum is devoid of fat and is continuous anteriorly with the **membranous layer of subcutaneous tissue** (Scarpa's fascia) of the anterolateral abdominal wall and posteriorly with the *subcutaneous tissue of the perineum* (Colles' fascia—the continuation of the membranous layer of subcutaneous tissue into the perineum).

The *development of the scrotum* is closely related to the formation of the inguinal canals. The scrotum develops from the *labioscrotal swellings*—two cutaneous outpouchings of the anterior abdominal wall that fuse to form a pendulous cutaneous pouch, the scrotum (Moore and Persaud, 1998). Later in the fetal period, the testes and spermatic cords enter the scrotum.

The *arterial supply of the scrotum* (Fig. 2.17) is from the:

- Perineal branch of the **internal pudendal artery** that forms **posterior scrotal arteries**
- External pudendal branches of the **femoral artery** that form the **anterior scrotal arteries**
- Cremasteric branch of the **inferior epigastric artery**.

The *scrotal veins accompany the arteries*. The lymphatic vessels of the scrotum drain into the **superficial inguinal lymph nodes**.

The nerves of the scrotum are the:

- Genital branch of the **genitofemoral nerve** (L1, L2), supplying the anterolateral surface of the scrotum
- **Anterior scrotal nerves**—branches of the *ilioinguinal nerve* (L1)—supplying the anterior surface of the scrotum
- **Posterior scrotal nerves**—branches of the perineal branch of the *pudendal nerve* (S2 through S4)—supplying the posterior surface of the scrotum
- Perineal branches of the **posterior femoral cutaneous nerve** (S2, S3) supplying the inferior surface of the scrotum.

Anesthetizing the Scrotum

The anterior third of the scrotum is supplied by the L1 segment of the spinal cord through the ilioinguinal nerve and the posterior two-thirds of the scrotum are supplied mainly by the S3 spinal segment through the perineal and posterior femoral cutaneous nerves. Therefore, to anesthetize the anterior surface of the scrotum, a spinal anesthetic agent has to be injected more superiorly than when anesthetizing the posterior surface of the scrotum. ⊙

Epididymis

The epididymis is formed by minute convolutions of the *duct of the epididymis*, so tightly compacted that they appear solid. The **epididymis** lies on the posterior surface of the testis (Fig. 2.18), which is covered by the tunica vaginalis except at its posterior margin. The convoluted duct of the epididymis becomes progressively smaller as it passes from the head of the epididymis on the superior part of the testis to its tail. The **ductus deferens** or deferent duct begins at the tail of the epididymis as the continuation of the epididymal duct, or duct of the epididymis (Fig. 2.18*B*). The **efferent ductules** transport the sperms from the **rete testis** (L. rete, a net) to the epididymis where they are stored. The rete testis is a network of canals at the termination of the straight (seminiferous) tubules. *The epididymis consists of:*

- A **head**, the superior expanded part that is composed of lobules formed by the coiled ends of 12 to 14 *efferent ductules*
- A **body** that consists of the convoluted *duct of the epididymis*
- A **tail** that is continuous with the ductus deferens, the duct that transports sperms from the epididymis to the ejaculatory duct for expulsion into the prostatic part of the urethra (see Chapter 3).

Testis

The testis (G. orchis) is an ovoid organ that is suspended in the scrotum by the spermatic cord. Usually the left testis (testicle) is suspended (hangs) more inferiorly than the right testis. The testes are covered with a tough fibrous coat—the **tunica albuginea** (Fig. 2.18*B*). The testes produce **sperms** (spermatozoa; male germ cells) and hormones, principally

Figure 2.17. Arterial supply and lymphatic drainage of the testis and scrotum. The pudendal nerve, a branch of the sacral plexus, arises from the ventral primary rami of S2, S3, and S4. This nerve is part of the somatic nervous system and is shown innervating the scrotum. Observe the long, slender testicular artery arising from the abdominal aorta inferior to the renal arteries. The long course of the artery through the abdomen and inguinal canal results from the descent of the fetal testis from the dorsal abdominal wall into the scrotum. Observe that the lymphatic drainage from the testis and scrotum differs, which is clinically significant (p. 204).

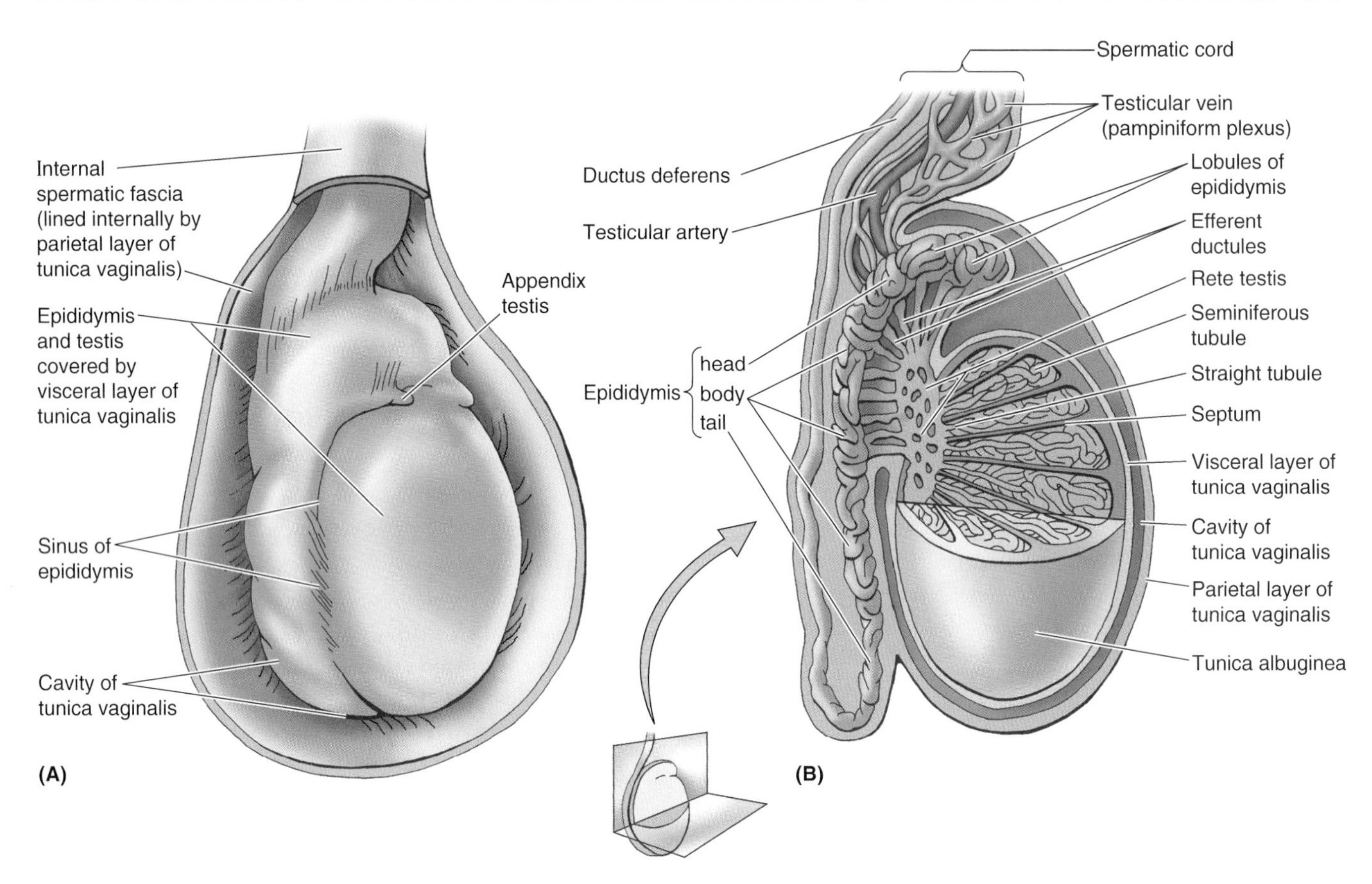

Figure 2.18. Spermatic cord, epididymis, and testis. A. Lateral view showing the distal part of the spermatic cord, the epididymis, and most of the testis covered by the two layers of the tunica vaginalis: the visceral layer, which provides an intimate covering, and the outer parietal layer, which lines the internal spermatic fascia (cut open). The two layers are separated by the cavity (shown in (**B**)) of the tunica vaginalis. **B.** Schematic vertical section of the spermatic cord, epididymis, and testis. Observe that the epididymis is essentially the irregular convoluted duct of the epididymis that continues as the ductus deferens (vas deferens). Observe also the other components of the spermatic cord.

testosterone. The sperms are formed in the long, convoluted **seminiferous tubules** that are joined by straight tubules to the **rete testis.**

The surface of each testis is covered by the **visceral layer of the tunica vaginalis**, except where the testis attaches to the epididymis and spermatic cord. The tunica vaginalis is a closed peritoneal sac partially surrounding the testis, which represents the closed-off distal part of the embryonic processus vaginalis (Fig. 2.14). The visceral layer of the tunica vaginalis is closely applied to the testis, epididymis, and the inferior part of the ductus deferens. The slitlike recess of the tunica vaginalis—the **sinus of the epididymis**—is between the body of the epididymis and the posterolateral surface of the testis (Fig. 2.18*A*). The **parietal layer of the tunica vaginalis**, adjacent to the internal spermatic fascia, is more extensive than the visceral layer and extends superiorly for a short distance into the distal part of the spermatic cord. The small amount of fluid in the cavity of the tunica vaginalis separates the visceral and parietal layers, allowing the testis to move freely in the scrotum.

The **testicular arteries**, long and slender (Figs. 2.17 and 2.18*B*), arise from the anterolateral aspect of the *abdominal aorta* just inferior to the *renal arteries.* They pass retroperitoneally (external or posterior to the peritoneum) in an oblique direction, crossing over the ureters and the inferior parts of the *external iliac arteries* to reach the *deep inguinal rings*. They enter the inguinal canals through the deep rings, pass through the canals, exit them through the superficial inguinal rings, and enter the spermatic cords to supply the testes. The testicular artery or one of its branches anastomoses with the artery of the ductus deferens.

The **testicular veins** emerge from the testis and epididymis and join to form a venous network, the **pampiniform plexus**, consisting of 8 to 12 veins lying anterior to the ductus deferens and surrounding the testicular artery in the spermatic cord. *The pampiniform plexus is part of the thermoregulatory system of the testis*, helping to keep this gland at a constant temperature. The left testicular vein originates in the pampiniform plexus and empties into the left renal vein; the right testicular vein has a similar origin and course but enters the inferior vena cava (IVC).

The *lymphatic drainage of the testis* is to the **lumbar** (**lateral aortic**) and **preaortic lymph nodes** (Fig. 2.17). The *autonomic nerves of the testis* arise as the *testicular plexus of nerves* on the testicular artery, which contains vagal parasympathetic fibers and sympathetic fibers from the T7 segment of the spinal cord.

Cryptorchidism (Undescended Testes)

The words *orchis* (G. testis) and *kryptos* (L. fr. G. hidden) explain the term *cryptorchidism*. The testes are undescended in approximately 3% of full-term and 30% of premature infants (Moore and Persaud, 1998). If a testis is not descended or retractable (capable of being drawn down), the condition is *cryptorchidism*. The undescended testis lies somewhere along the normal path of its prenatal descent, usually in the inguinal canal. The importance of cryptorchidism is its potential for developing malignancy in the testis.

Hydrocele

A hydrocele is the presence of excess fluid in a *persistent processus vaginalis*. This congenital anomaly may be associated with an indirect inguinal hernia (p. 205). The fluid accumulation results from secretion of an abnormal amount of serous fluid from the visceral layer of the tunica vaginalis. The size of the hydrocele depends on how much of the processus vaginalis persists. A **hydrocele of the testis** (*A*) is confined to the scrotum and distends the tunica vaginalis. A **hydrocele of the cord** (*B*) is confined to the spermatic cord and distends the persistent part of the stalk of the processus vaginalis. A congenital hydrocele of the cord and testis may communicate with the peritoneal cavity. Detection of a hydrocele requires *transillumination*, a procedure during which a bright light is applied to the side of the scrotal enlargement in a darkened room. The transmission of light as a red glow indicates excess serous fluid in the scrotum. Newborn infants often have residual peritoneal fluid in their tunica vaginalis; however, this fluid is usually absorbed during the first year. Certain pathological conditions, such as injury and/or inflammation of the epididymis, may also produce a hydrocele in adults—a collection of serous fluid in the tunica vaginalis of the testis.

Hematocele

A *hematocele of the testis* (*C*) is a collection of blood *in the tunica vaginalis* (e.g., resulting from rupture of branches of the testicular artery caused by trauma to the testis). Trauma may produce a scrotal and/or testicular *hematoma* (accumulation of blood, usually clotted, in any extravascular location). Blood does not transilluminate; therefore, transillumination can differentiate a hematocele or hematoma from a hydrocele. A hematocele of the testis may be associated with a *scrotal hematocele* resulting from effusion of blood into the scrotal tissues.

Cysts and Hernias of the Canal of Nuck

Remnants of the processus vaginalis in female infants can enlarge and form cysts in the inguinal canal. A persistent processus vaginalis in the female that traverses the inguinal canal—a *canal of Nuck*—may produce a bulge in the anterior part of the labium majus that may develop into an indirect inguinal hernia.

Epididymitis and Orchitis

Epididymitis, or inflammation of the epididymis, is the most common cause of tender scrotal swelling. Epididymitis and *orchitis* (inflammation of the testis) may be a complication of mumps, an acute communicable disease. The testes and epididymis are swollen and acutely painful.

Spermatocele and Epididymal Cyst

A *spermatocele* is a retention cyst (collection of fluid) in the epididymis (*A*), usually near its head. Spermatoceles contain a milky fluid and are generally asymptomatic. An *epididymal cyst* is a collection of fluid anywhere in the epididymis (*B*). ▶

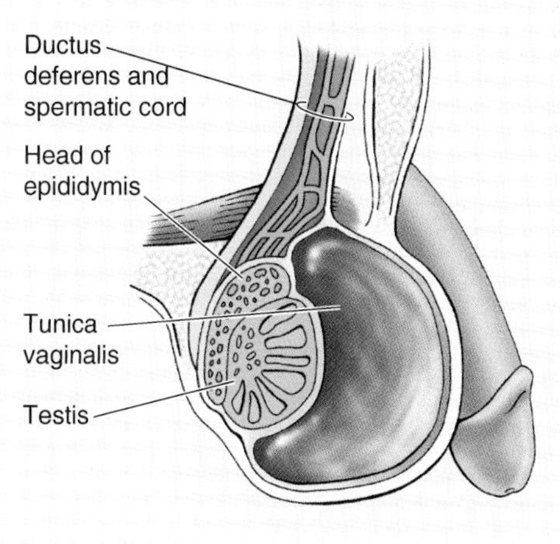

(A) Hydrocele of testis

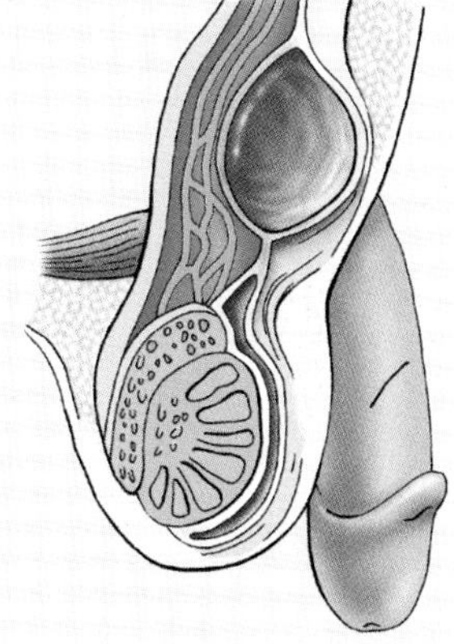

(B) Hydrocele of cord

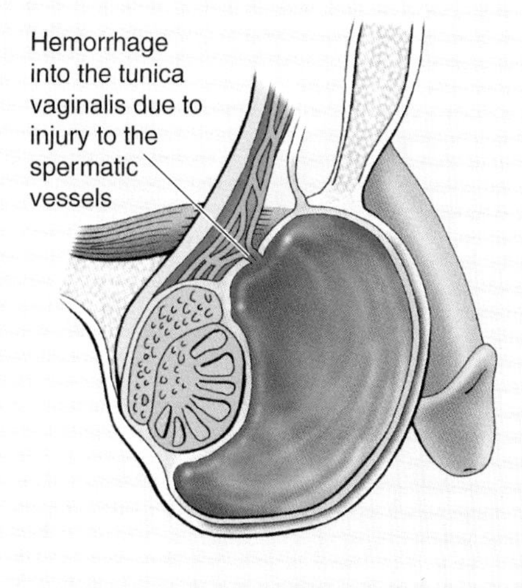

(C) Hematocele of testis

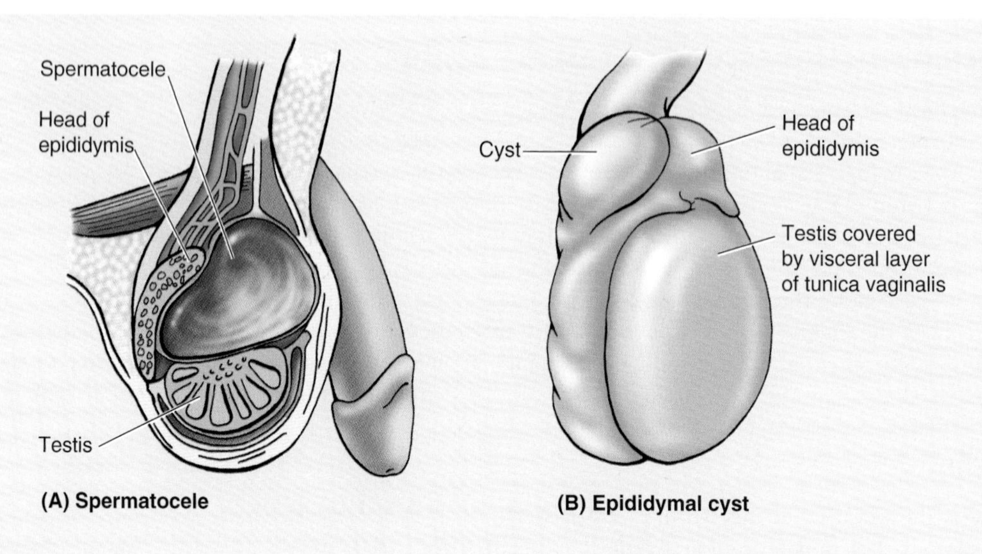

(A) Spermatocele **(B) Epididymal cyst**

Varicocele

The vinelike pampiniform plexus of veins may become dilated (varicose) and tortuous, producing a varicocele that is usually visible only when the man is standing or straining; the enlargement usually disappears when the person lies down. Varicoceles often result from defective valves in the testicular vein. Palpating a varicocele can be likened to feeling a bag of worms. Kidney or renal vein problems can result in distension of the pampiniform veins, especially on the left. Consequently, it is necessary to rule out kidney or other abdominal causes of varicocele, especially if the enlargement is asymmetric.

Cancer of the Testis and Scrotum

Because the testes descend from the posterior abdominal wall to the scrotum during fetal development, their lymphatic drainage differs from that of the scrotum, which is an outpouching of anterolateral abdominal skin (Fig. 2.17). Consequently,

- *Cancer of the testis* metastasizes to the *lumbar lymph nodes*, which lie just inferior to the renal veins
- *Cancer of the scrotum* metastasizes to the *superficial inguinal lymph nodes*, which lie in the subcutaneous tissue inferior to the inguinal ligament and along the terminal part of the great saphenous vein.

Cancer of the Uterus and Labium Majus

Lymphogenous metastasis of cancer most commonly occurs along lymphatic pathways that parallel the venous drainage of the organ that is the site of the primary tumor. This is also true of the uterus; however, some lymphatic vessels follow the course of the round ligament through the inguinal canal, and thus, while occurring less often, metastatic uterine cancer cells (especially from tumors adjacent to the proximal attachment of the round ligament) can spread from the uterus to the labium majus (site of distal attachment of the round ligament, the developmental homologue of the scrotum) and from there to the superficial inguinal nodes, which receive lymph from the skin of the perineum (including the labia).

Cremasteric Reflex

Contraction of the cremaster muscle is elicited by lightly stroking the skin on the medial aspect of the superior part of the thigh with an applicator stick or tongue depressor. This area of skin is supplied by the ilioinguinal nerve. The rapid elevation of the testis on the same side is the *cremasteric reflex*. This reflex is extremely active in children; consequently, hyperactive cremasteric reflexes may simulate undescended testes. A hyperactive reflex can be abolished by having the child sit in a cross-legged, squatting position. If the testes are descended, they can be palpated in the scrotum.

Vestigial Remnants of Embryonic Genital Ducts

When the tunica vaginalis is opened, rudimentary structures may be observed at the superior extremities of the testes and epididymis. These structures are small remnants of genital ducts in the embryo (Moore and Persaud, 1998). They are rarely observed unless pathological changes occur. The *appendix of the testis* is a vesicular remnant of the cranial end of the paramesonephric duct—the embryonic female genital duct that forms half of the uterus. It is attached to the superior pole of the testis. The *appendices of the epididymis* are remnants of the cranial end of the mesonephric duct—the embryonic male genital ▶

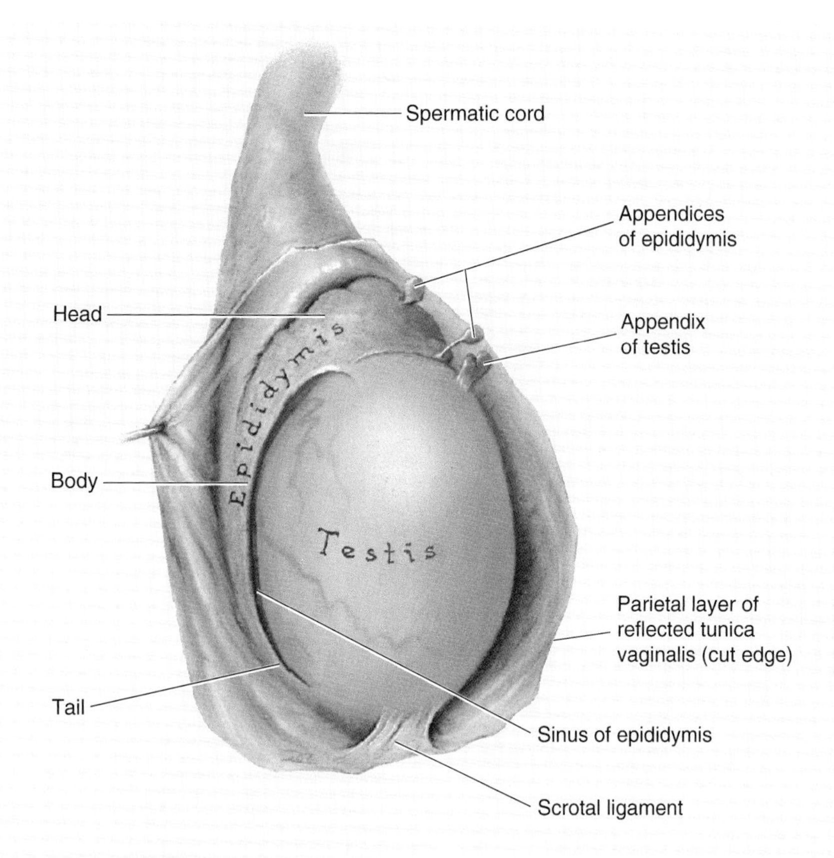

▶ duct that forms part of the ductus deferens—which is attached to the head of the epididymis. This part of the duct, together with the mesonephric tubules associated with it, normally forms the efferent ductules and epididymis.

Vasectomy

A bilateral excision of a segment of the *ductus deferens*—**vasectomy**—is performed to produce sterility in males. To perform a vasectomy, also called a **deferentectomy**, the duct is identified by its firm consistency and isolated on each side by incising the superoanterior scrotal wall. Double ligatures are placed on each duct and the duct is then sectioned between the ligatures; thus, both cut ends are ligated. Sperm can no longer pass to the urethra and therefore degenerate in the epididymis and ductus deferens. However, secretions of the auxiliary genital glands (seminal vesicles, bulbourethral glands, and prostate) can still be ejaculated.

Inguinal Hernias

An inguinal hernia ("rupture") is a protrusion of parietal peritoneum and viscera such as the small intestine, or part of them, through a normal or abnormal opening from the cavity in which they belong. Most hernias are reducible, meaning that they can be returned to their normal place in the peritoneal cavity by appropriate manipulation. *Approximately 90% of abdominal hernias are in the inguinal region*; the two main types are indirect and direct inguinal hernias. Approximately 75% are indirect hernias.

An indirect (congenital) inguinal hernia:

- Is the most common of all abdominal hernias
- Leaves the abdominal cavity **lateral to the inferior epigastric vessels** and enters the deep inguinal ring (*A*)
- Has a **hernial sac** formed by a persistent processus vaginalis and all three fascial coverings of the spermatic cord

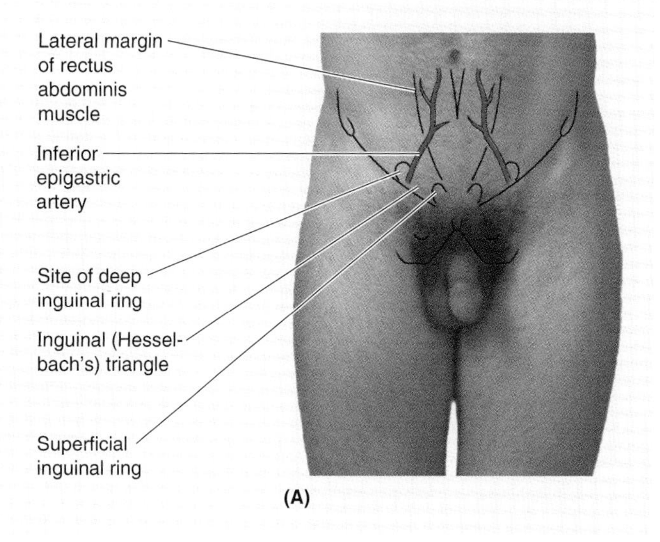

(A)

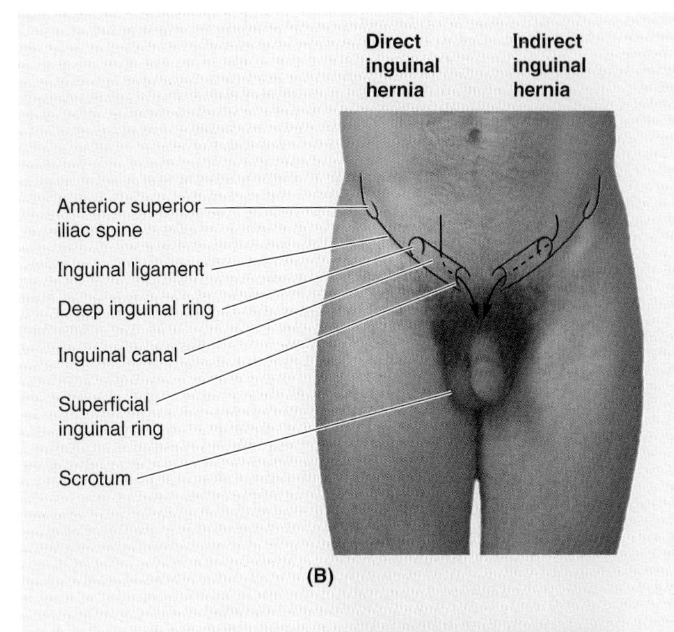

(B)

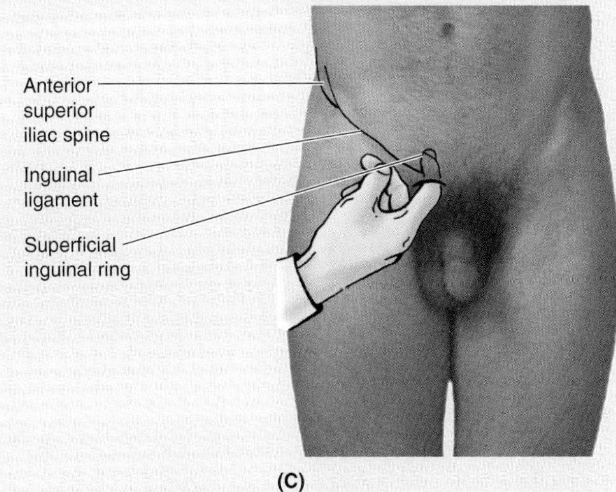

(C)

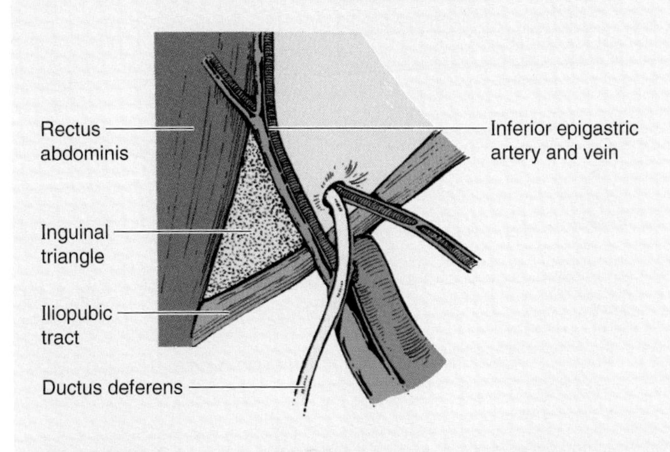

(D) Posterior view of right anterior abdominal wall

- Traverses the entire inguinal canal (*B*)
- Exits through the superficial inguinal ring
- Commonly enters the scrotum.

Palpation of an indirect inguinal hernia is performed using the same technique as that described for palpating the inguinal rings (p. 198).

Normally, most of the processus vaginalis disappears before birth, except for the distal part that forms the tunica vaginalis of the testis (Fig. 2.16). The peritoneal part of the hernial sac of an indirect inguinal hernia is formed by the persisting processus vaginalis. If the entire stalk of the processus vaginalis persists, the hernia extends into the scrotum superior to the testis, forming a complete indirect inguinal hernia (*E*).

Indirect inguinal hernias can occur in women; however, they are approximately 20 times more common in men. If the processes vaginalis persists in females, it forms a small peritoneal pouch—the *canal of Nuck*—in the inguinal canal that may enter the labium majus. Part of the small intestine may herniate into this pouch and through the inguinal canal, forming an indirect inguinal hernia and a bulge in the labium majus.

A direct (acquired) inguinal hernia:

- Leaves the abdominal cavity **medial to the inferior epigastric artery** (*A* and *E*)
- Protrudes through an area of relative weakness in the posterior wall of the inguinal canal
- Has a hernial sac formed by transversalis fascia
- Lies outside the processes vaginalis, which is usually obliterated, parallel to the spermatic cord, and outside the inner one or two fascial coverings of the cord
- Does not traverse the entire inguinal canal—usually only its most medial part (lower end) adjacent to the superficial inguinal ring
- Protrudes through the **inguinal (Hesselbach) triangle** that lies between the inferior epigastric artery superolaterally, the rectus abdominis medially, and the inguinal ligament inferiorly (*D*)
- Emerges through or around the conjoint tendon to reach the superficial inguinal ring, gaining an outer covering of external spermatic fascia, inside or parallel to that on the cord itself (*E*)
- Almost never enters the scrotum; however, when it does it passes lateral to the spermatic cord, deep to the skin and dartos fascia.

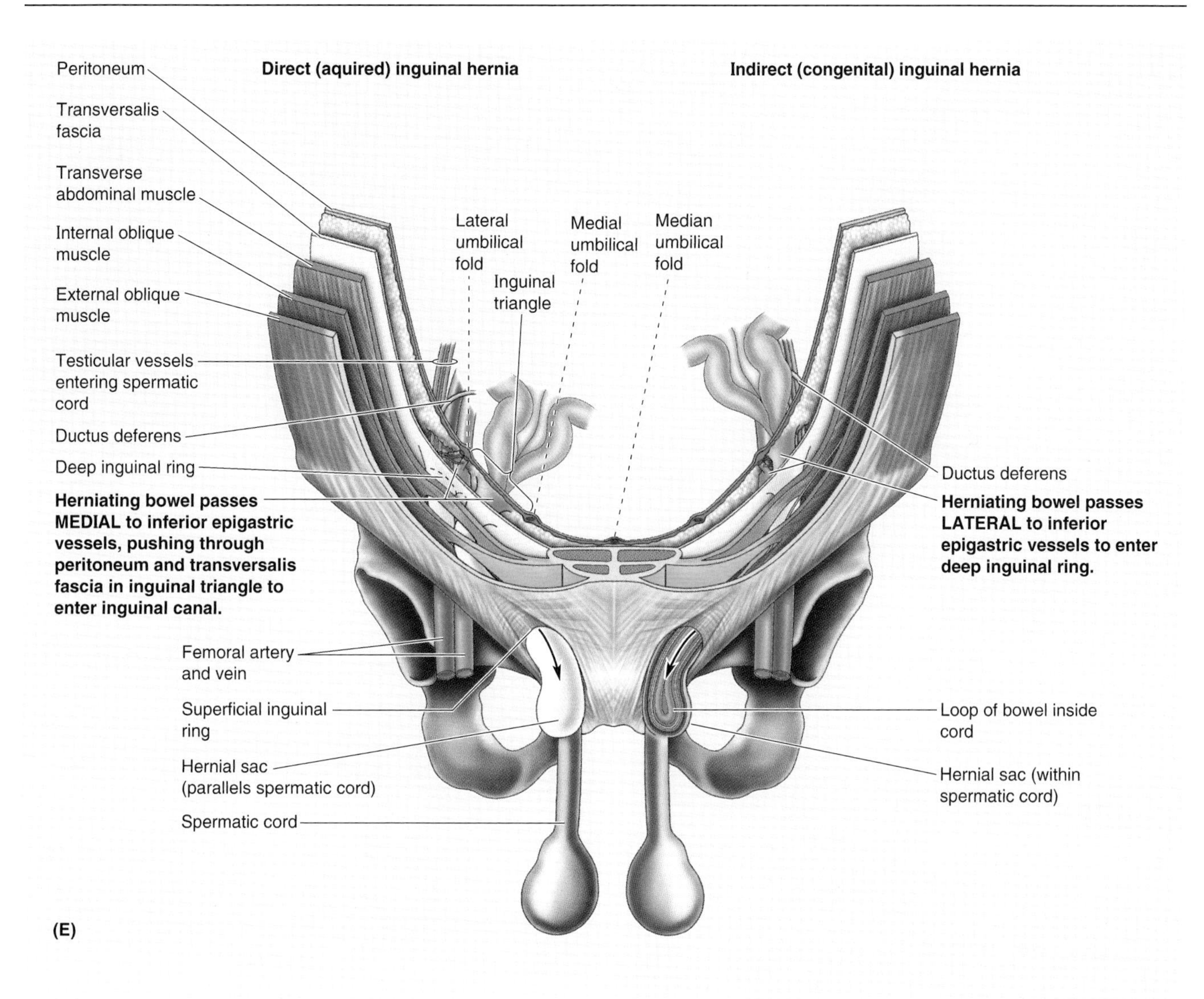

▶ Palpation of a direct inguinal hernia is performed by placing the palmar surface of the index and/or middle finger over the inguinal triangle and asking the person to cough or bear down (strain). If a hernia is present, a forceful impulse is felt against the pad of the finger. The finger can also be placed in the superficial inguinal ring; if a direct hernia is present, a sudden impulse is felt at the side of the finger when the person coughs or bears down. ○

Surface Anatomy of the Anterolateral Abdominal Wall

The **umbilicus** is an obvious feature of the anterolateral abdominal wall and is the reference point for the transumbilical plane. This puckered indentation of skin in the center of the anterior abdominal wall is typically at the level of the IV disc between L3 and L4 vertebrae; however, its position varies with the amount of subcutaneous fat present. The **epigastric fossa** (pit of the stomach) is a slight depression in the epigastric region, just inferior to the **xiphoid process.** This fossa—the site of substantial pain caused by pyrosis (heartburn), for example—an esophageal symptom—is particularly noticeable when a person is in the supine position because the abdominal organs spread out, drawing the anterolateral abdominal wall posteriorly in this region. The 7th through 10th costal cartilages unite on each side, and their medial borders form the **costal margin.** When a person is in the supine position, observe the rise and fall of the abdominal wall with respiration—superiorly ▶

▶ with inspiration and inferiorly with expiration. The rectus abdominis muscles can be palpated and observed when a supine person is asked to raise the head and shoulders against resistance.

The location of the **linea alba** is visible because of the vertical skin groove superficial to this raphe. The groove is usually obvious because the linea alba is approximately 1 cm wide between the two parts of the **rectus abdominis** superior to the umbilicus. Inferior to the umbilicus, the linea alba is linear and not indicated by a groove. Some pregnant women, especially those with dark hair and a dark complexion, have a heavily pigmented line—the *linea nigra*—in the midline skin external to the linea alba. After pregnancy, the color of this line fades. The **pubic symphysis** is a cartilaginous joint that can be felt as a firm resistance in the median plane distal to the linea alba. The bony **iliac crest** at the level of L4 vertebra can be easily palpated as it extends posteriorly from the **anterior superior iliac spine**.

The *semilunar lines* (L. lineae semilunares) are slightly curved, linear impressions in the skin that extend from the inferior costal margin near the 9th costal cartilages to the **pubic tubercles**. These semilunar skin grooves (5–8 cm from the midline) are clinically important because they are parallel with the lateral edges of the rectus sheath. Generally, incisions are not made along the semilunar lines because doing so would interrupt the multiple nerve supply of the rectus muscle. Skin grooves also overlie the *tendinous intersections* of the rectus abdominis, which are clearly visible in persons with well developed rectus muscles. The interdigitating bellies of the **serratus anterior** and **external oblique** muscles are also visible. The site of the inguinal ligament is indicated by the **inguinal groove**, a skin crease that is parallel and just inferior to the inguinal ligament, which is readily visualized by having the person drop one leg to the floor while lying supine on an examining table. The inguinal groove marks the division between the anterolateral abdominal wall and the thigh. ○

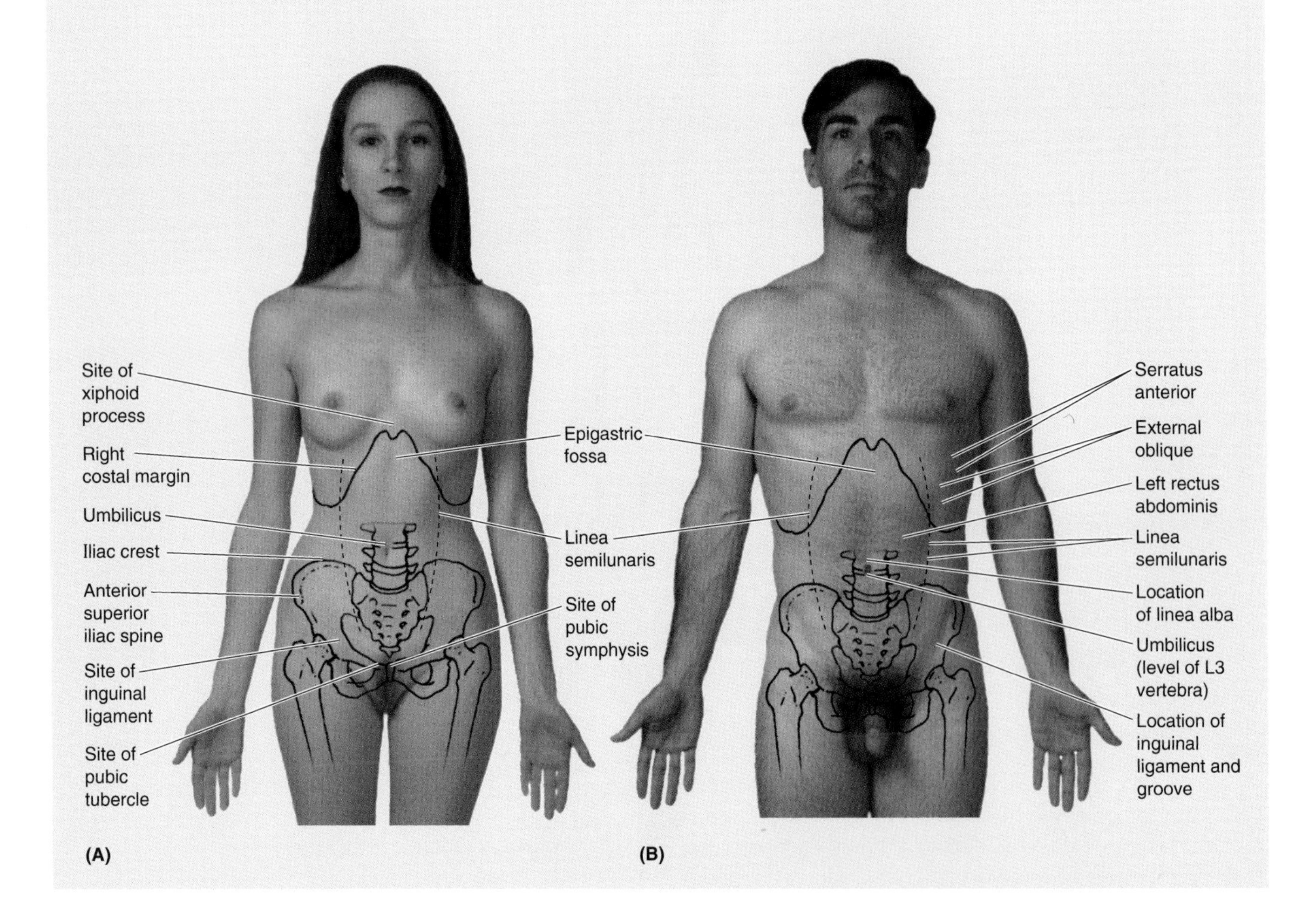

Peritoneum and Peritoneal Cavity

The **peritoneum**—a continuous, glistening, transparent serous membrane—lines the abdominopelvic cavity and invests the viscera (organs). *The peritoneum consists of two continuous layers* (Fig. 2.19):

- **Parietal peritoneum** lining the internal surface of the abdominopelvic wall
- **Visceral peritoneum** investing the viscera such as the stomach and intestines.

Both layers of peritoneum consist of *mesothelium*, a layer of simple squamous epithelial cells. The parietal peritoneum receives the same blood and nerve supply as does the region of the wall it lines. The visceral peritoneum and the organs it covers receive the same blood and nerve supply.

The peritoneum and viscera are in the abdominal cavity. There are no organs in the peritoneal cavity, which is normally empty except for a thin layer of fluid that keeps the peritoneal surfaces moist. *The relationship of the viscera to the peritoneum is as follows*:

- *Intraperitoneal organs* are almost completely covered with visceral peritoneum (e.g., the stomach and spleen). "In-

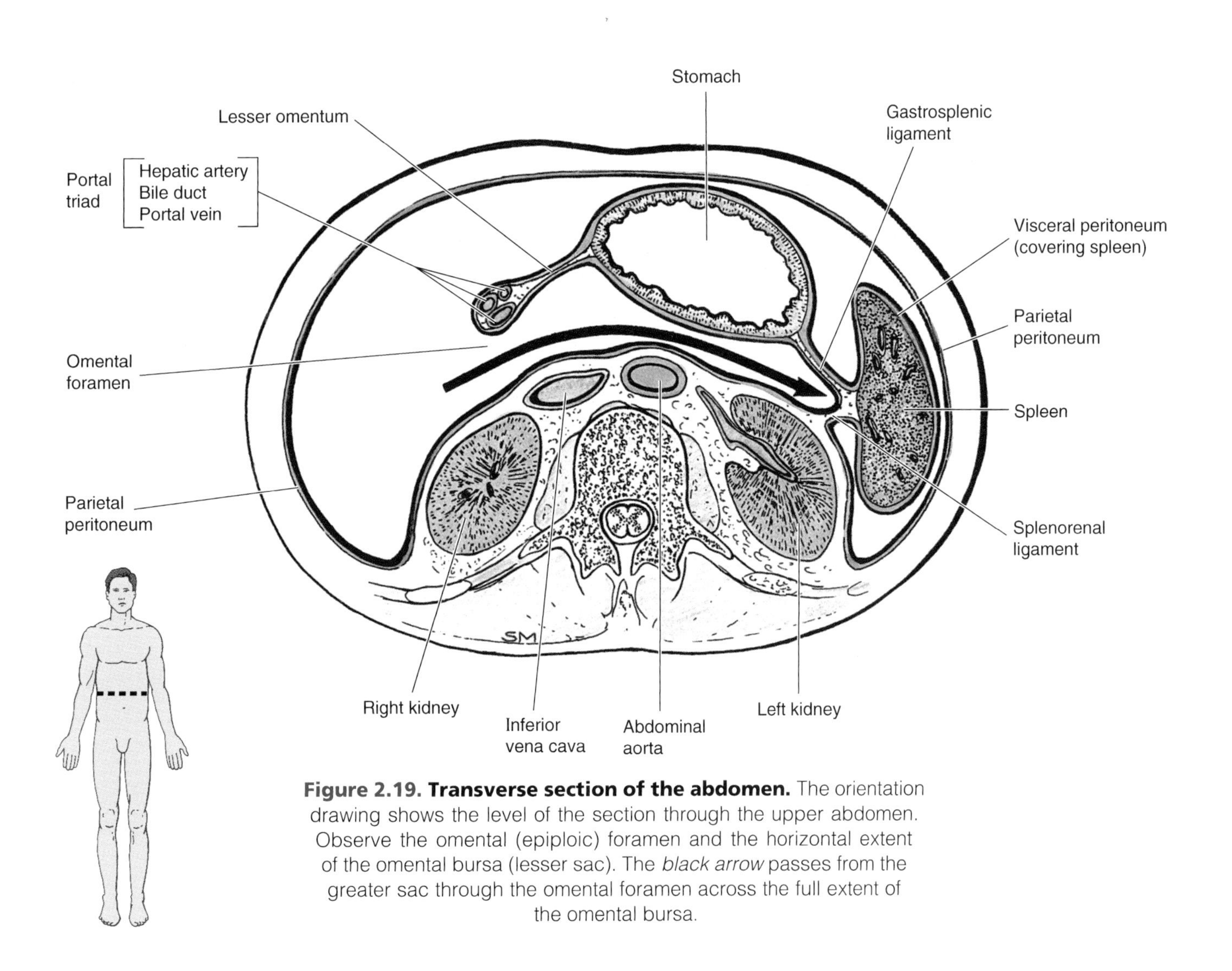

Figure 2.19. Transverse section of the abdomen. The orientation drawing shows the level of the section through the upper abdomen. Observe the omental (epiploic) foramen and the horizontal extent of the omental bursa (lesser sac). The *black arrow* passes from the greater sac through the omental foramen across the full extent of the omental bursa.

traperitoneal" in this case does *not* mean inside the peritoneal cavity (although the term is used clinically for substances injected into the cavity). Intraperitoneal organs have conceptually, if not literally, invaginated into the closed sac, like pressing your fist into an inflated balloon (see the discussion of pleura in Chapter 1).

- *Extraperitoneal* or *retroperitoneal organs* are also outside the peritoneal cavity—external or posterior to the parietal peritoneum—and are only partially covered with peritoneum (usually on just one surface); organs such as the kidneys are between the parietal peritoneum and the posterior abdominal wall and have parietal peritoneum only on their anterior surfaces (unless fat intervenes).

The **peritoneal cavity** is within the abdominal cavity, which is continuous with the pelvic cavity. The *peritoneal cavity is a potential space of capillary thinness between the parietal and visceral layers of peritoneum.* It contains no organs but contains a thin film of **peritoneal fluid** that lubricates the peritoneal surfaces, enabling the viscera to move over each other without friction and allowing the movements of digestion. In addition to lubricating the surfaces of the viscera, the peritoneal fluid contains leukocytes and antibodies that resist infection. The peritoneal fluid is absorbed by lymphatic vessels on the inferior surface of the diaphragm. *The peritoneal cavity is completely closed in males*; however, there is a communication pathway in females to the exterior of the body through the uterine tubes, uterine cavity, and vagina. This communication constitutes a potential pathway of infection from the exterior.

Embryology of the Peritoneal Cavity

Early in its development, the embryonic body cavity (intraembryonic coelom) is lined with *mesoderm*, the primordium of the peritoneum (Moore and Persaud, 1998). At a slightly later stage, the primordial abdominal cavity is lined with *parietal peritoneum* derived from mesoderm, which forms a closed sac. The lumen of the peritoneal sac is the *peritoneal cavity*. As the organs develop they invaginate (protrude) to varying degrees into the peritoneal sac, acquiring a peritoneal covering—the *visceral peritoneum*. A viscus (organ) such as the kidney protrudes only partially into the peritoneal cavity; hence, it is primarily retroperitoneal—always remaining external to the peritoneal cavity and posterior to the peritoneum lining the abdominal cavity. Other viscera, such as the stomach and spleen, protrude completely into the peritoneal sac and are almost completely invested by *visceral peritoneum*. These viscera are connected to the abdominal wall by a *mesentery* of variable length, which is composed of two layers of peritoneum with a thin layer of loose connective tissue between them. Viscera with a mesentery, such as most of the small intestine, are mobile, the degree of which varies with the length of the mesentery.

As organs protrude into the peritoneal sac, their vessels, nerves, and lymphatics remain connected to their extraperitoneal (usually retroperitoneal) sources or destinations so that these connecting structures lie between the layers of the peritoneum forming their mesenteries. Initially, the entire primordial gut was suspended in the center of the peritoneal cavity by a dorsal mesentery attached to the midline of the posterior body wall. As the organs grow, they gradually reduce the size of the peritoneal cavity until it is only a potential space between the parietal and visceral layers of peritoneum. As a consequence, several parts of the gut come to lie against the posterior abdominal wall, and their dorsal mesenteries become gradually reduced because of pressure from overlying organs (Fig. 2.20, *A–D*). For example, during development, the growing coiled mass of small intestine pushes the part of the gut that will become the descending colon to the left side and presses its mesentery against the posterior abdominal wall. The mesentery is held there until the layer of peritoneum that formed the left side of the mesentery and the part of the visceral peritoneum of the colon lying against the body wall fuse with the parietal peritoneum of the body wall. As a result, the colon becomes fixed to the posterior abdomi-

Blockage of the Uterine Tubes

In a technique for testing the patency of the uterine tubes (*hysterosalpingography*), air or radiopaque dye is injected into the uterine cavity, from which it normally flows through the uterine tubes and into the peritoneal cavity (see Fig. 3.52*A*). Failure of the air or dye to enter the peritoneal cavity indicates blockage of the uterine tubes (resulting from inflammatory disease, for example).

Peritonitis, Ascites, Paracentesis, and Intraperitoneal Injection

Because of the well-innervated peritoneum, patients undergoing abdominal surgery experience more pain with large, invasive incisions than they do with small, laparoscopic incisions or vaginal operations. With an abdominal injury resulting from a stab wound, for example, or an infection resulting from perforation of an appendix, the peritoneum becomes inflamed—**peritonitis**. The peritoneum exudes fluid and cells in response to injury or infection. The excess fluid in the peritoneal cavity—**ascites**—can be removed by **paracentesis** (surgical puncture of the peritoneal cavity for the aspiration of fluid).

The surface area of the peritoneum is extensive; therefore, fluid injected into it is absorbed rapidly. For this reason, certain anesthetic agents such as solutions of barbiturate compounds may be injected into the peritoneal cavity by *intraperitoneal (IP) injection*. ⊙

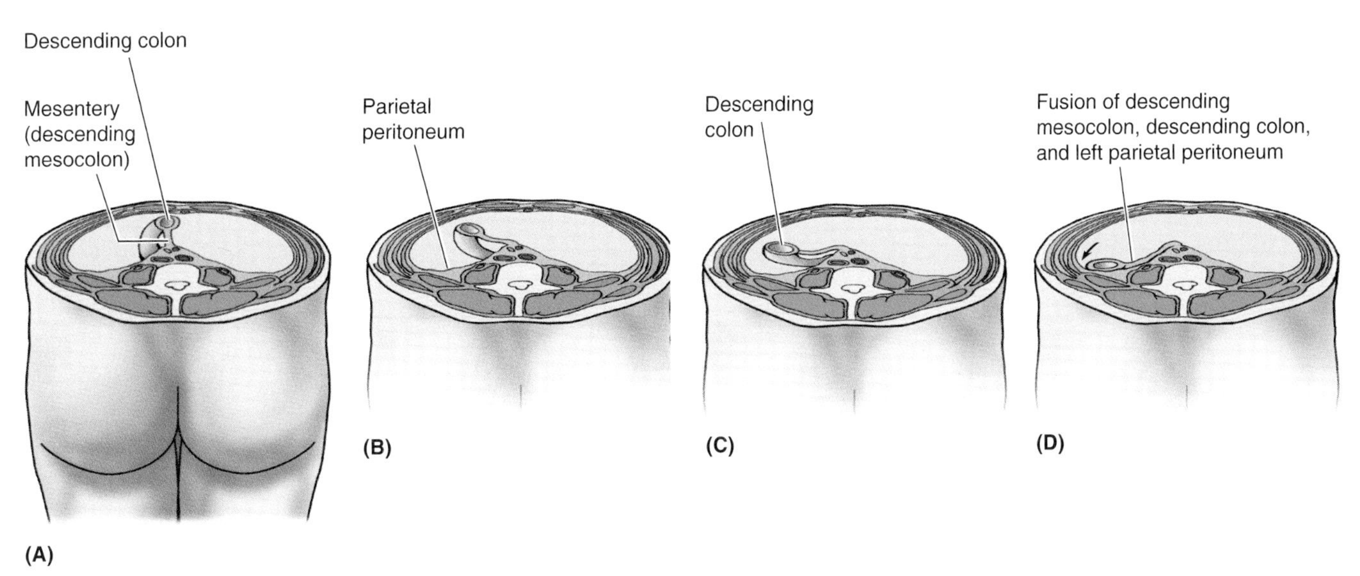

Figure 2.20. Stages of absorption of the descending mesocolon. A–D. The mesocolon gradually fuses with the parietal peritoneum and the descending colon becomes retroperitoneal—external or posterior to the peritoneum—as shown in (**D**). The *arrow* indicates the paracolic gutter and the site where an incision is made during mobilization of the colon during surgery. Sometimes the descending colon retains a short mesentery, similar to the stage shown in (**C**), especially where the colon is in the iliac fossa.

nal wall on the left side with peritoneum covering only its anterior aspect. The descending colon (as well as the ascending colon on the right side) has thus become *secondarily retroperitoneal*, having once been intraperitoneal.

The layers of peritoneum that fused now form a *fusion fascia*—a connective tissue plane in which the nerves and vessels of the descending colon continue to lie. Thus, the descending colon of the adult can be freed from the posterior body wall (surgically mobilized) by incising the peritoneum along the lateral border of the descending colon and then bluntly dissecting along the plane of the fusion fascia, elevating the neurovascular structures from the posterior body wall until the midline is reached. The ascending colon can be similarly mobilized on the right side.

While several parts of the digestive tract and associated organs become secondarily retroperitoneal (e.g., most of the duodenum and pancreas, as well as the ascending and descending parts of the colon), covered with peritoneum only on their anterior surface, other parts and some organs (e.g., the sigmoid colon and the spleen) retain a short mesentery. However, the roots of the short mesenteries do not remain attached to the midline but shift to the left or right by a fusion process like that described for the descending colon.

Rupture of the Intestine

When a penetrating wound ruptures the intestine, gas and other intestinal contents enter the peritoneal cavity, and peritonitis develops. This painful condition is accompanied by exudation of serum, fibrin, cells, and pus into the peritoneal cavity. In addition to severe abdominal pain, tenderness, nausea and/or vomiting, fever, and constipation are present. Because the intense pain worsens with movement, people with peritonitis commonly lie with their knees flexed to relax their anterolateral abdominal muscles and reduce the intra-abdominal pressure and pain.

Flow of Inflammatory Exudate

Infected peritoneal fluid associated with the rupture of organs and peritonitis flows inferiorly along the paracolic gutters (see Fig. 2.23*B*) into the pelvic cavity where absorption of toxins is slow. To facilitate the flow of exudate, patients with peritonitis are often placed in the sitting position at a 45° angle.

Peritoneal Adhesions and Adhesiotomy

If the peritoneum is damaged by a stab wound, for example, parts of the inflamed parietal and visceral layers of peritoneum may adhere because of inflammatory processes. *Adhesions* (scar tissue) may also form after an abdominal operation (e.g., a ruptured appendix) and cause complications such as intestinal obstruction in approximately 2% of patients. *Adhesiotomy* refers to the surgical separation of adhesions. Adhesions are often found during dissection of cadavers (see Fig. 2.32*B*). ▶

Ascites

Under certain pathological conditions such as peritonitis and portal venous congestion, effusion and accumulation of serous fluid occurs. In these cases, the peritoneal cavity may be distended with several liters of abnormal fluid (*ascites*). Widespread metastases (spread) of cancer cells to the abdominal viscera cause exudation of fluid and cells from the venules; the ascitic fluid in this case contains cancer cells and is often blood-stained.

Abdominal Paracentesis

Excess fluid in the peritoneal cavity may have to be removed by *paracentesis*. After injection of a local anesthetic agent, a needle or trocar and a cannula are inserted through the anterolateral abdominal wall into the peritoneal cavity through the linea alba, for example. The needle is inserted superior to the empty urinary bladder and in a location that avoids the inferior epigastric artery. ⊙

Descriptive Terms for Parts of the Peritoneum

Various terms are used to describe the parts of the peritoneum that connect organs with other organs or to the abdominal wall.

A **mesentery** (Fig. 2.21*A*) is a double layer of peritoneum that occurs as a result of the invagination of the peritoneum by an organ and constitutes a continuity of the visceral and parietal peritoneum that provides a means for neurovascular communication between the organ and the body wall. The mesentery connects the organ to the posterior abdominal wall (e.g., the mesentery of the small intestine). The mesentery of the large intestine is the **mesocolon** (Fig. 2.21*B*). Mesenteries have a core of connective tissue containing blood and lymphatic vessels, nerves, lymph nodes, and fat.

An **omentum** is a double-layered extension or fold of peritoneum that passes from the stomach and proximal part of the duodenum to adjacent organs in the abdominal cavity or to the abdominal wall (Fig. 2.21, *A–E*).

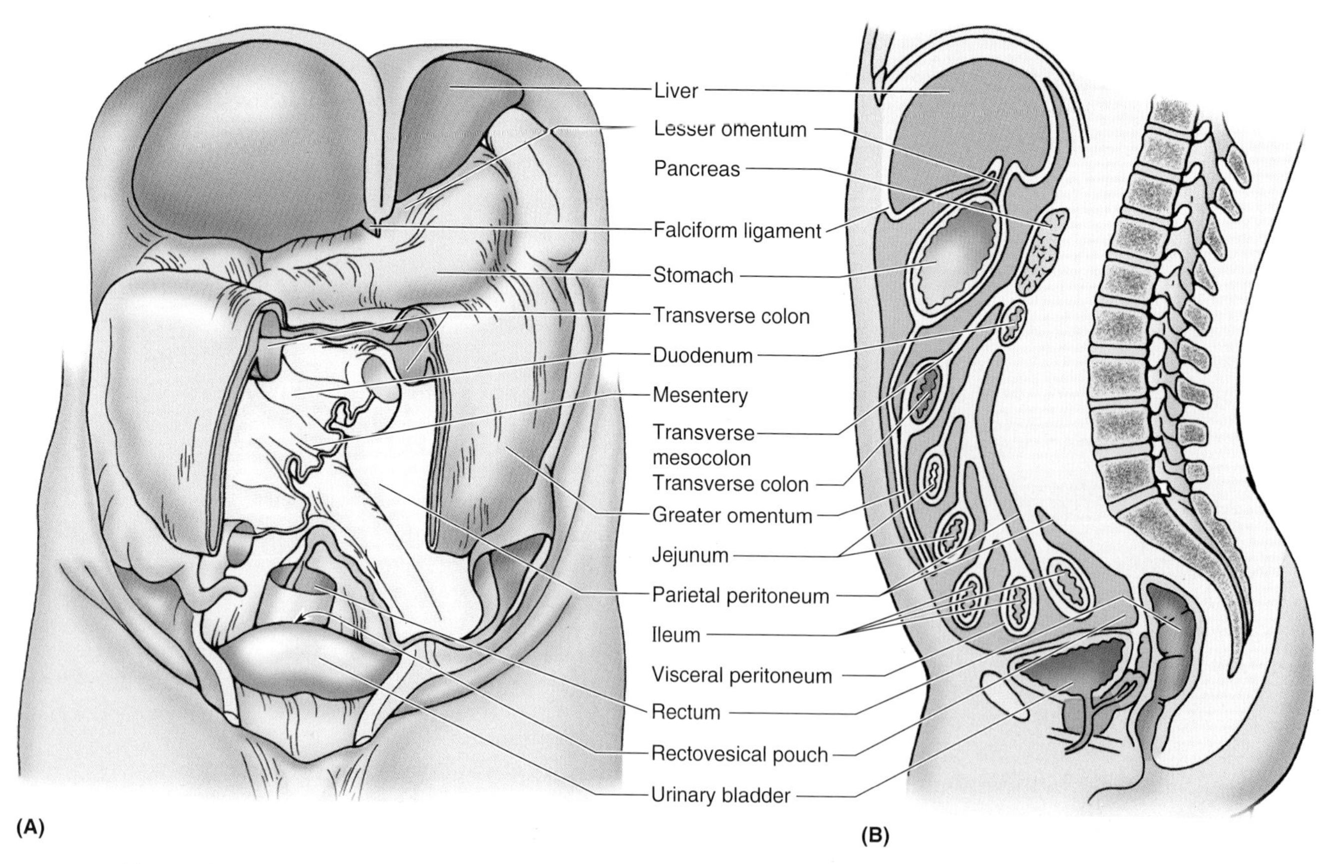

Figure 2.21. Principal parts of the peritoneum. A. Anterior view of the opened peritoneal cavity. Parts of the greater omentum, transverse colon, and small intestine have been cut away to reveal deep structures and the layers of the mesenteric structures. The mesentery of the jejunum and ileum (small intestine) and sigmoid mesocolon have been cut close to their parietal attachments. **B.** Sagittal section of the abdominopelvic cavity of a male, showing the relationships of the peritoneal attachments.

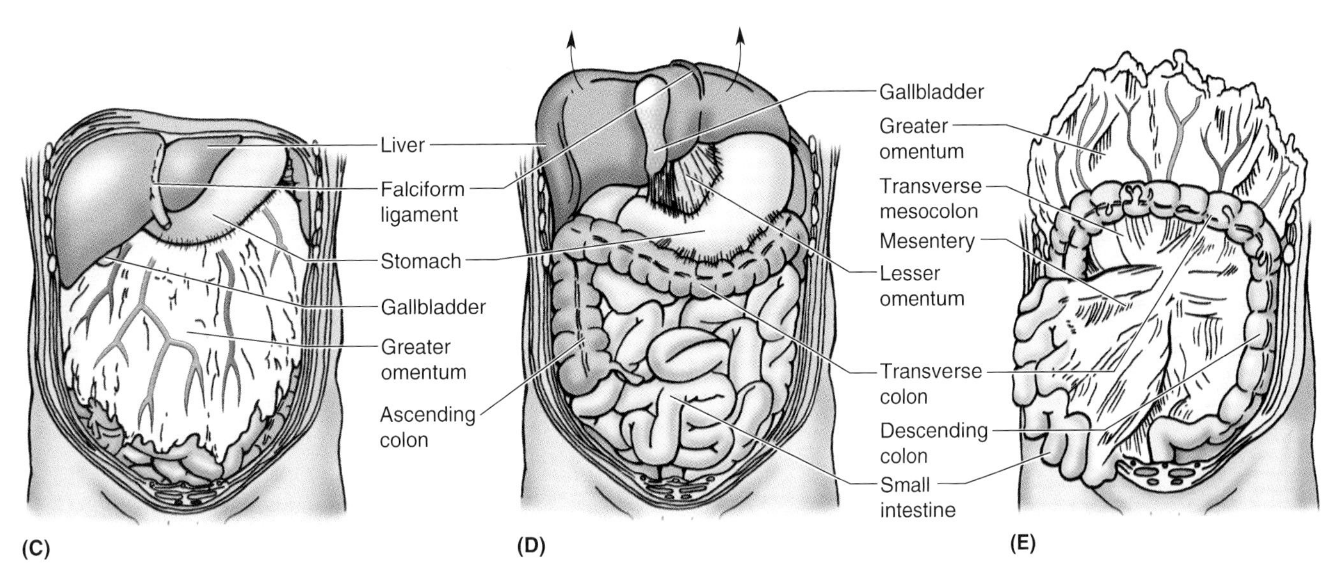

Figure 2.21. *(Continued)* **C.** The greater omentum is shown in its "normal" position covering most of the abdominal viscera. **D.** The lesser omentum, attaching the liver to the lesser curvature of the stomach, is shown by reflecting the liver and gallbladder superiorly. The greater omentum has been removed from the greater curvature of the stomach to reveal the intestines. **E.** The greater omentum has been reflected superiorly and the small intestine has been retracted to the right side to reveal the mesentery of the small intestine. Observe also the mesentery of the transverse colon (transverse mesocolon).

- The **lesser omentum** connects the lesser curvature of the stomach and the proximal part of the duodenum to the liver
- The **greater omentum** is a prominent peritoneal fold that hangs down like an apron from the greater curvature of the stomach and the proximal part of the duodenum. After descending, it folds back and attaches to the anterior surface of the transverse colon and its mesentery.

A **peritoneal ligament** consists of a double layer of peritoneum that connects an organ with another organ or to the abdominal wall. *The liver is connected to the:*

- Anterior abdominal wall by the **falciform ligament** (Fig. 2.22)
- Stomach by the **gastrohepatic ligament** (the membranous portion of the lesser omentum)
- Duodenum by the **hepatoduodenal ligament** (the thickened free edge of the lesser omentum that conducts the portal triad: portal vein, hepatic artery, and bile duct).

The gastrohepatic and hepatoduodenal ligaments are continuous parts of the lesser omentum and are separated only for descriptive convenience.

The stomach is connected to the:

- Inferior surface of the diaphragm by the **gastrophrenic ligament**
- Spleen by the **gastrosplenic ligament** (gastrolienal ligament)) that reflects to the hilum of the spleen
- Transverse colon by the **gastrocolic ligament** (the apron-like part of the greater omentum).

The greater omentum and the gastrosplenic ligament are continuous and are separated only for descriptive purposes.

A **peritoneal fold** is a reflection of peritoneum that is raised from the body wall by underlying blood vessels, ducts, and obliterated fetal vessels (e.g., the medial and lateral *umbilical folds* on the internal surface of the anterolateral abdominal wall). Some peritoneal folds contain blood vessels and bleed if cut, such as the lateral umbilical folds, which contain the inferior epigastric arteries (p. 191).

A **peritoneal recess**, or fossa, is a pouch of peritoneum that is formed by a peritoneal fold (e.g., the inferior recess of the omental bursa between the layers of the greater omentum and the supravesical and umbilical fossae between the umbilical folds).

Functions of the Greater Omentum

The greater omentum, large and fat-laden, prevents the visceral peritoneum from adhering to the parietal peritoneum lining the anterolateral abdominal wall. It has considerable mobility and moves around the peritoneal cavity with peristaltic movements of the viscera. It wraps itself around an inflamed organ such as the appendix, walling it off and thereby protecting other viscera from it. For this reason, the greater omentum is often referred to as the "abdominal policeman." Thus, it is common when entering the abdominal cavity, in either dissection or surgery, to find the omentum markedly displaced from the "normal" position in which it is almost always depicted in anatomical ▶

▶ illustrations. The greater omentum also cushions the abdominal organs against injury and forms insulation against loss of body heat.

Abscess Formation

Perforation of a duodenal ulcer, rupture of the gallbladder, or perforation of the appendix may lead to the formation of a *circumscribed collection of purulent exudate* in the subphrenic recess. The abscess may be walled inferiorly by adhesions (see "Subphrenic Abscesses", p. 264).

Spread of Pathological Fluids

Peritoneal recesses are of clinical importance in connection with the spread of pathological fluids such as pus, a product of inflammation. The recesses determine the extent and direction of the spread of fluids that may enter the peritoneal cavity when an organ is diseased or injured (see also pp. 211 and 217). ⊙

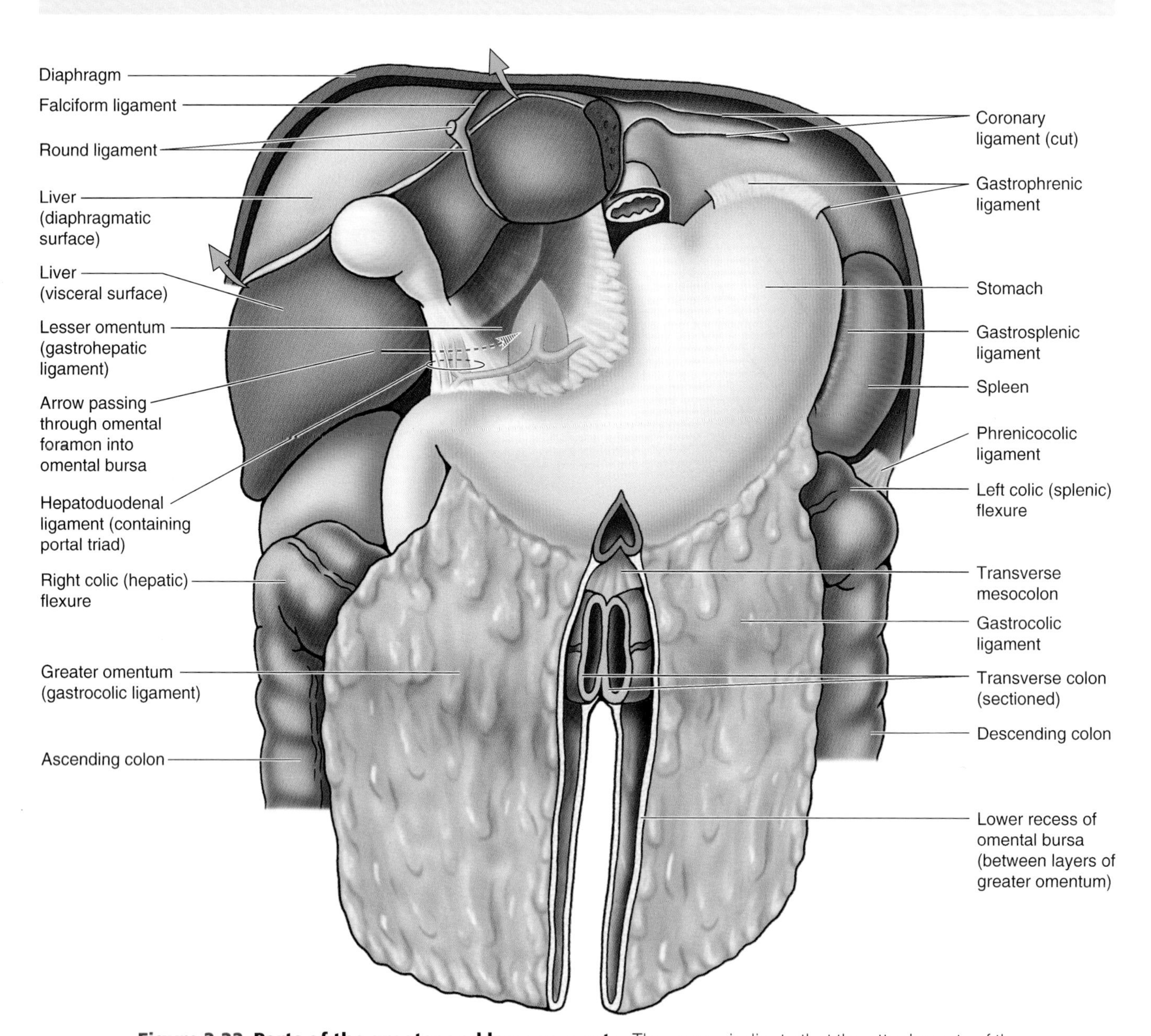

Figure 2.22. Parts of the greater and lesser omenta. The *arrows* indicate that the attachments of the liver are severed and the liver and gallbladder have been reflected superiorly. The central part of the greater omentum has been cut out to show its relation to the transverse colon and mesocolon. The greater omentum consists of the gastrophrenic ligament, gastrosplenic and gastrocolic ligaments, all which arise from the greater curvature of the stomach. Observe the free edge of the hepatoduodenal ligament (part of lesser omentum) containing the portal triad: hepatic artery, bile duct, and portal vein.

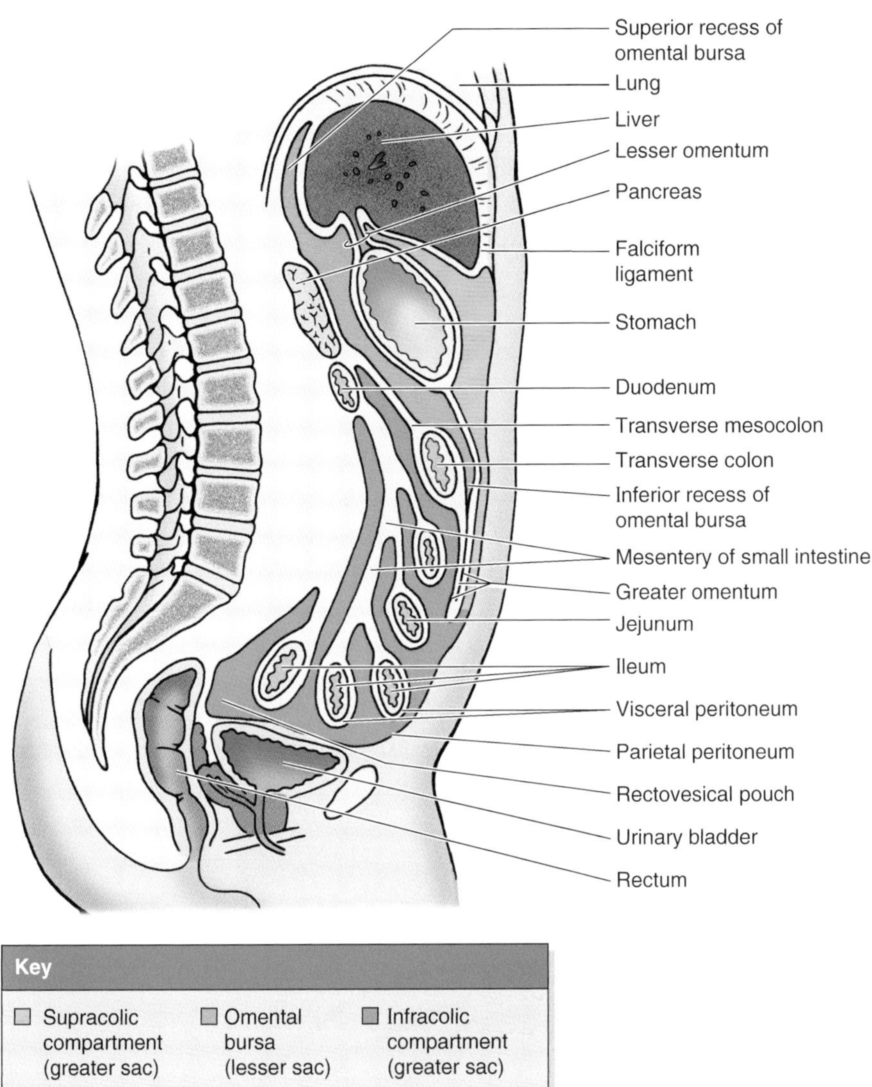

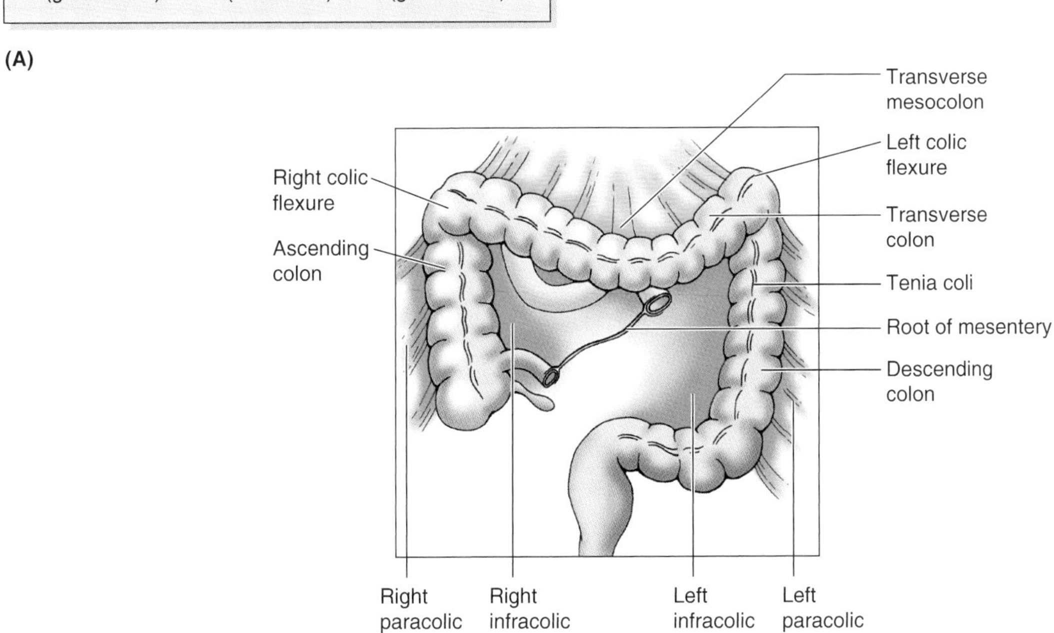

Figure 2.23. Omental (epiploic) bursa and supracolic and infracolic compartments of the greater sac. A. Sagittal section of the abdominopelvic cavity showing the viscera and the arrangement of the peritoneum and mesenteries. **B.** Supracolic and infracolic compartments of the greater sac after removal of the greater omentum. Observe the infracolic spaces and paracolic gutters.

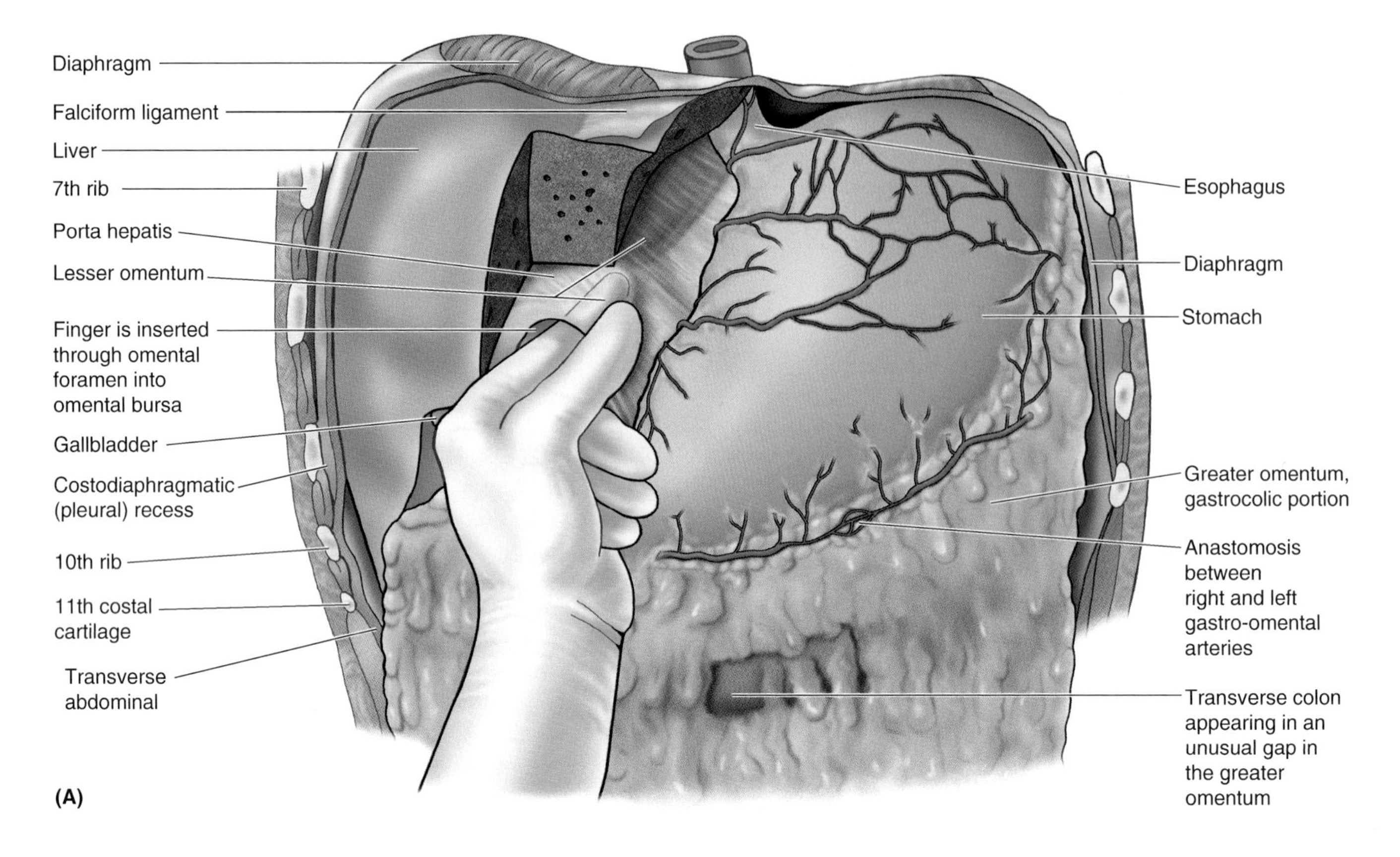

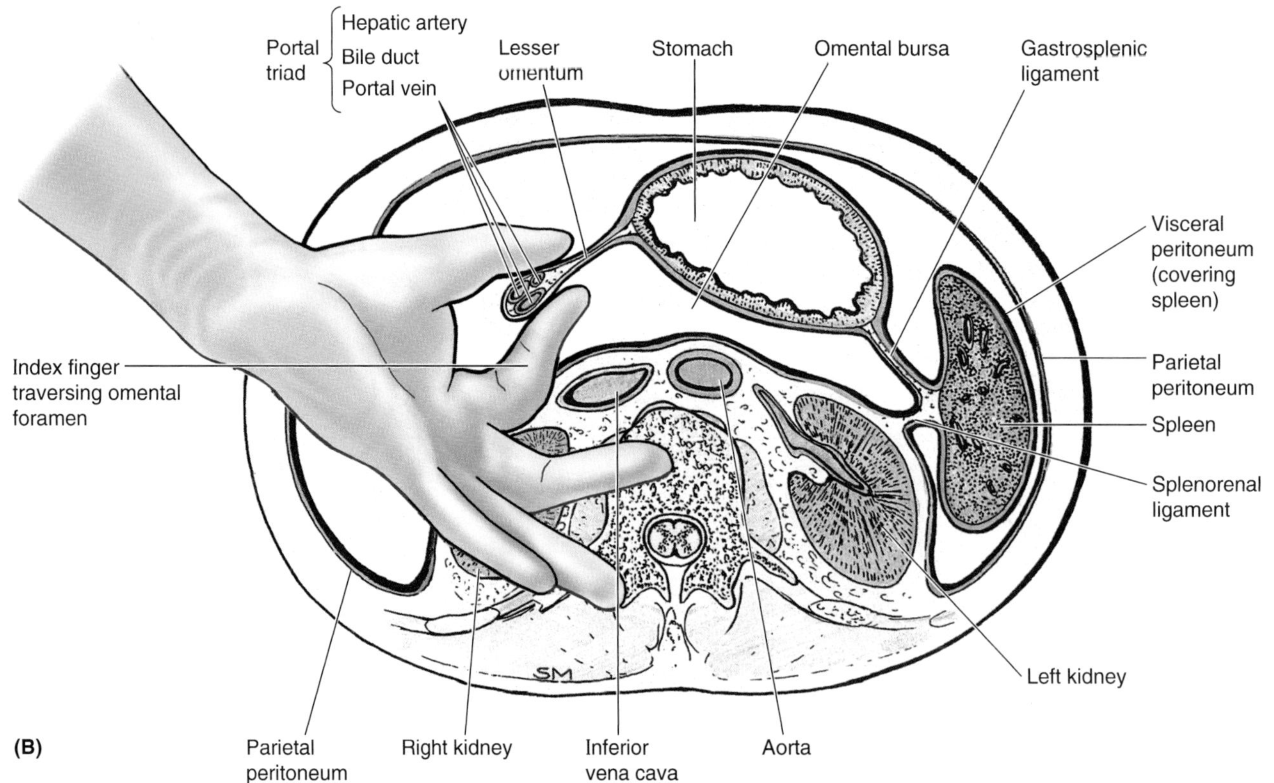

Figure 2.24. Omental (epiploic) foramen and omental bursa. A. The index finger passes from the greater sac through the omental foramen into the omental bursa (lesser sac). **B.** The hepatic artery is compressed between the index finger in the omental foramen and the thumb on the anterior wall of the foramen. Compression of the hepatic artery may be necessary when there is bleeding from the cystic artery during a cholecystectomy (removal of the gallbladder).

Subdivisions of the Peritoneal Cavity

As the fetal organs assume their final positions, the peritoneal cavity is divided into the **greater and lesser peritoneal sacs** (Fig. 2.23*A*).

- The **greater sac** is the main and larger part of the peritoneal cavity
- The **lesser sac** or **omental bursa** lies posterior to the stomach and adjoining structures.

A surgical incision through the anterolateral abdominal wall enters the greater sac.

The **transverse mesocolon**—mesentery of the transverse colon—divides the abdominal cavity into a:

- **Supracolic compartment** containing the stomach, liver, and spleen
- **Infracolic compartment** containing the small intestine and ascending and descending colon.

The infracolic compartment lies posterior to the greater omentum and is divided into right and left infracolic spaces by the **mesentery of the small intestine** (Fig. 2.23*B*). Free communication occurs between the supracolic and infracolic compartments through the **paracolic gutters** (sulci, recesses, fossae)—the grooves between the lateral aspect of the ascending or descending colon and the posterolateral abdominal wall.

Flow of Ascitic Fluid and Pus

The *paracolic gutters are of considerable clinical importance* because they provide pathways for the flow of ascitic fluid and the spread of intraperitoneal infections. Purulent material (consisting of or containing pus) in the abdomen can be transported along the paracolic gutters into the pelvis, especially when the person is upright (erect position); conversely, infections in the pelvis may extend superiorly to a subphrenic recess situated under the diaphragm (p. 264), especially when the person is supine. Similarly, the paracolic gutters provide pathways for the spread of tumor cells that have sloughed from the ulcerated surface of a tumor and entered the peritoneal cavity. ⊙

The **omental bursa** is an extensive saclike cavity that lies posterior to the stomach and adjacent structures and the lesser omentum (Figs. 2.23*A*, 2.24, and 2.25). *The omental bursa has:*

- A **superior recess** that is limited superiorly by the diaphragm and the posterior layers of the coronary ligament of the liver
- An **inferior recess** between the superior part of the layers of the greater omentum.

The omental bursa permits free movement of the stomach on the structures posterior and inferior to it because the anterior and posterior walls of the omental bursa slide smoothly over each other. Most of the inferior recess of the bursa is a potential space sealed off from the main part of the omental bursa posterior to the stomach following adhesion of the anterior and posterior layers of the greater omentum (Fig. 2.25*B*). The omental bursa communicates with the greater peritoneal sac through the **omental foramen** (epiploic foramen, or foramen of Winslow), an opening situated posterior to the free edge of the lesser omentum (hepatoduodenal ligament). The omental foramen can be located by running a finger along the gallbladder to the free edge of the lesser omentum (Fig. 2.24*A*). The omental foramen usually admits two fingers. *The boundaries of the omental foramen are* (Fig. 2.24*B*):

- *Anteriorly*—portal vein, hepatic artery, and bile duct contained in the hepatoduodenal ligament (free edge of the lesser omentum)
- *Posteriorly*—IVC and right crus of diaphragm, covered with parietal peritoneum (they are retroperitoneal)
- *Superiorly*—caudate lobe of the liver, covered with visceral peritoneum
- *Inferiorly*—superior or first part of the duodenum, portal vein, hepatic artery, and bile duct.

Fluid in the Omental Bursa

Perforation of the posterior wall of the stomach results in the passage of its fluid contents into the omental bursa. An inflamed or injured pancreas can also result in the passage of pancreatic fluid into the bursa, forming a *pancreatic pseudocyst.*

Intestine in the Omental Bursa

Although uncommon, a loop of small intestine may pass through the omental foramen into the omental bursa and be strangulated by the edges of the foramen. As none of the boundaries of the foramen can be incised because they contain blood vessels, the swollen intestine must be decompressed using a needle so it can be returned to the greater peritoneal sac through the omental foramen.

Severance of the Cystic Artery

The cystic artery may be accidentally severed during *cholecystectomy*—removal of the gallbladder. The surgeon can control the hemorrhage by compressing the hepatic artery between the index finger in the omental foramen and the thumb on its anterior wall (Fig. 2.24). This procedure allows the surgeon to identify the bleeding artery and clamp it. ⊙

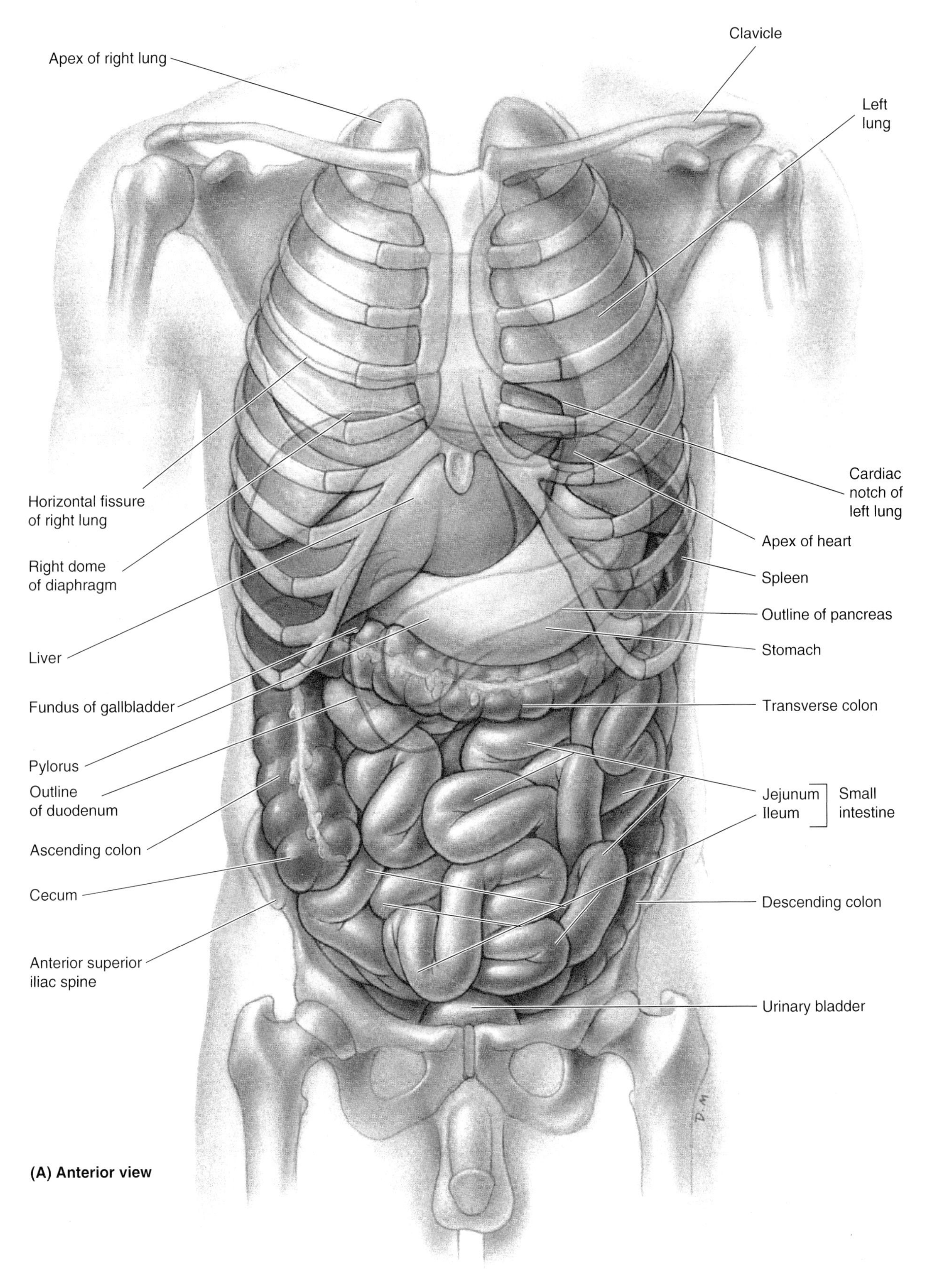

Figure 2.26. Overview of thoracic and abdominal viscera. A–B. Observe that some abdominal organs extend superiorly into the thoracic cage (rib cage) and are protected by it. Observe also that a large part of the small intestine is in the pelvis and that the right kidney is lower than the left kidney; this results from the mass of the liver on the right side.

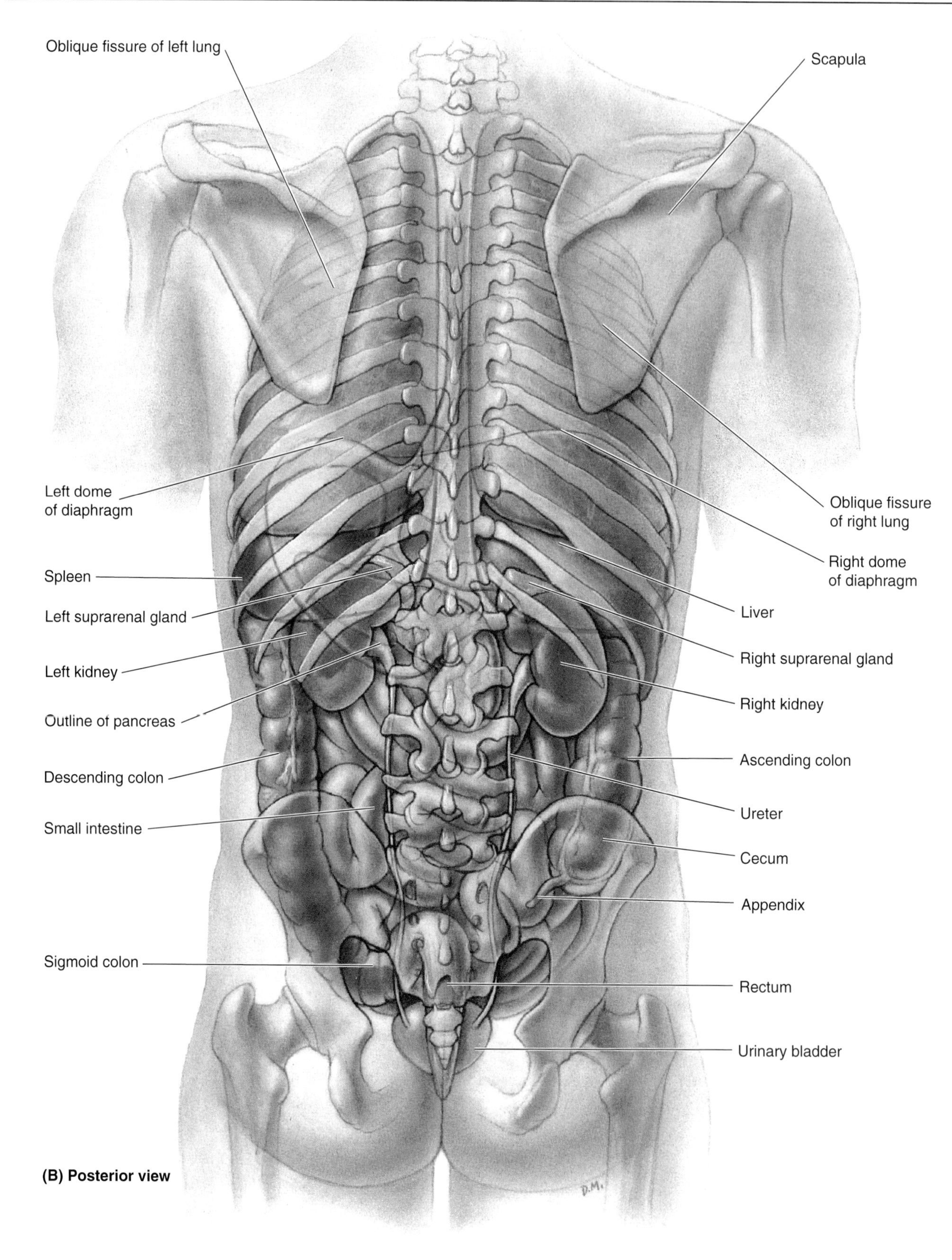

Figure 2.26. *(Continued)* Observe that the appendix projects posteriorly and inferomedially toward the pelvic brim; consequently, it is visible only in the posterior view.

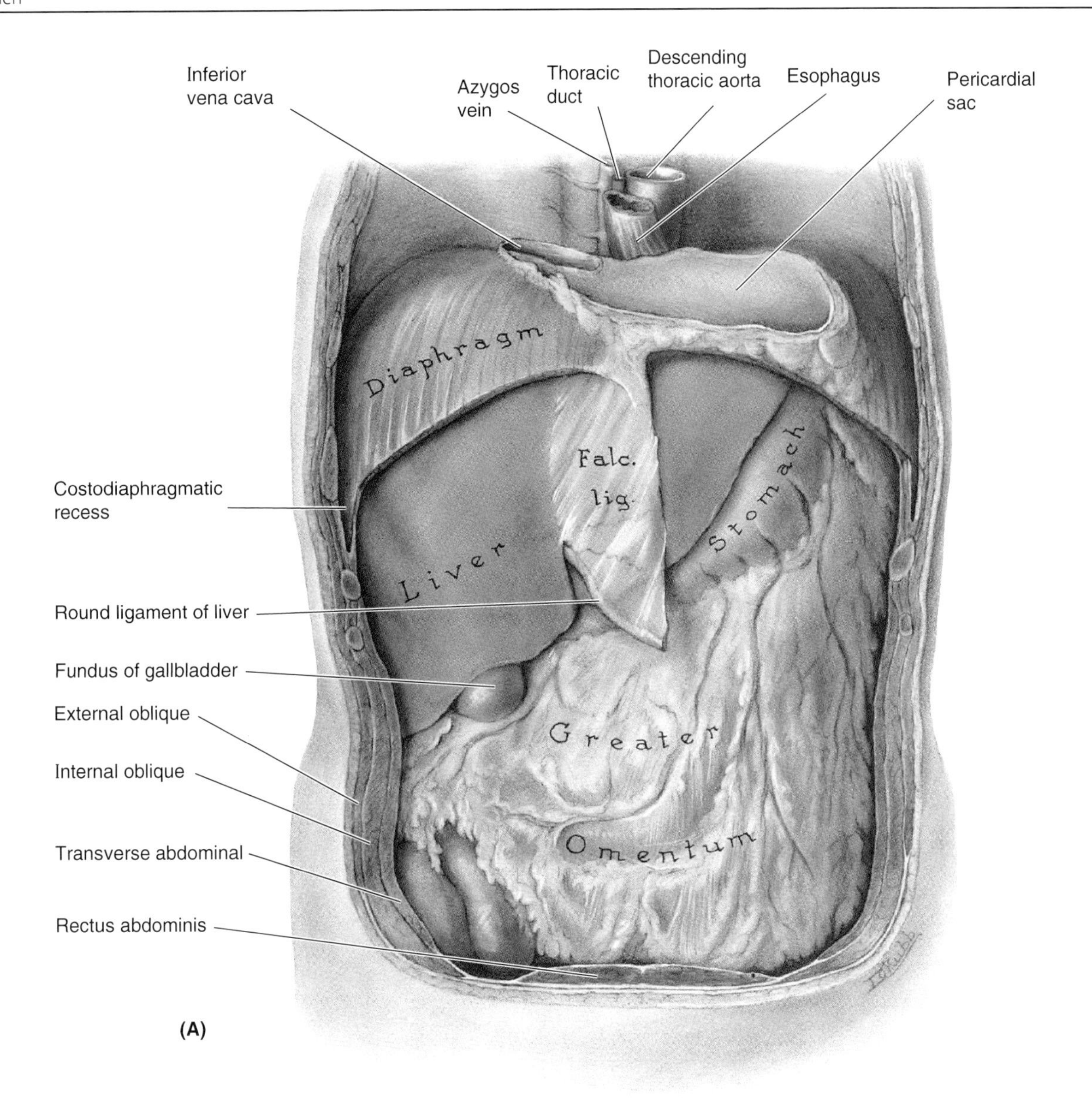

Figure 2.27. Abdominal contents, undisturbed. A. The anterior abdominal and thoracic walls are cut away. Observe the fundus of the gallbladder projecting inferior to the sharp, inferior border of the liver. The falciform ligament (*Falc. lig.*) is severed at its attachment to the anterior abdominal wall. This ligament resists displacement of the liver to the right.

the **abdominal aorta**. The three major branches of the aorta supplying the gut are the **celiac trunk** and the superior and inferior **mesenteric arteries.**

The **portal vein** (Fig. 2.28*B*)—formed by the union of the superior mesenteric and splenic veins—is the main channel of the **portal system of veins**, which collects blood from the abdominal part of the GI tract, pancreas, spleen, and most of the gallbladder and carries it to the liver.

Esophagus

The esophagus is a muscular tube (approximately 25 cm long) with an average diameter of 2 cm that *extends from the pharynx to the stomach* (Fig. 2.29, *A–C*). As seen under fluoroscopy following a barium swallow, the esophagus normally has four constrictions where adjacent structures produce impressions (Williams et al., 1995):

- At its beginning, approximately 15 cm from the incisor teeth and caused by the cricopharyngeus muscle (see Chapter 8), which is referred to clinically as the *upper esophageal sphincter*
- Where it is crossed by the arch of the aorta, 22.5 cm from the incisor teeth
- Where it is crossed by the left main bronchus, 27.5 cm from the incisor teeth

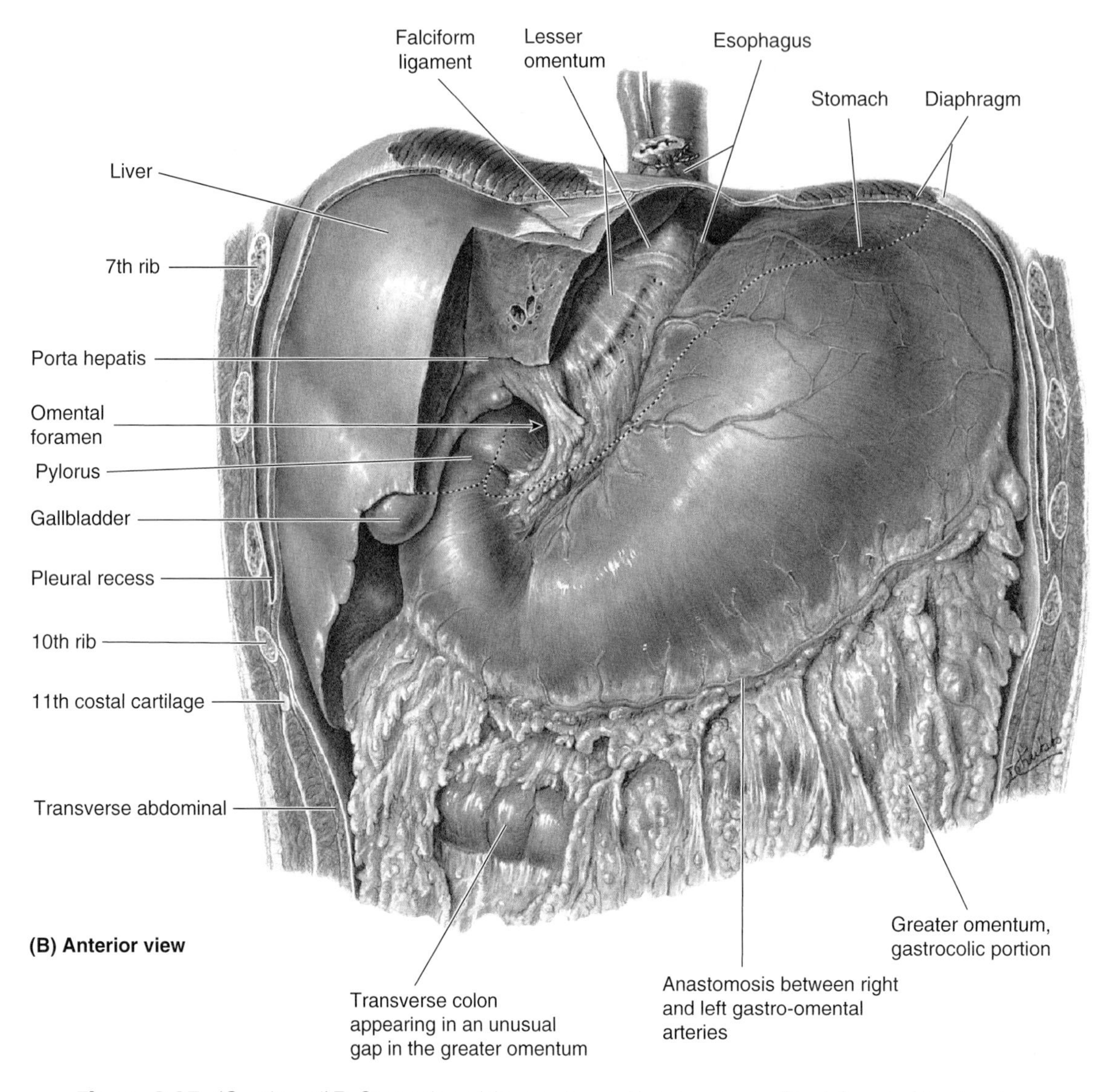

Figure 2.27. *(Continued)* **B.** Stomach and the greater and lesser omenta. The left part of the liver is cut away (position of intact liver is indicated by the *dotted line*) and the stomach is inflated with air. Observe also the omental (epiploic) foramen, the entrance from the greater sac into the omental bursa.

- Where it passes through the diaphragm, approximately 40 cm from the incisor teeth (referred to clinically as the *lower esophageal sphincter*).

These data are clinically important when passing instruments through the esophagus and into the stomach (Williams et al., 1995). The short abdominal part of the esophagus extends from the diaphragm to the cardial (cardiac) orifice of the stomach (Fig. 2.29*C*). The function of the esophagus is to convey food from the pharynx to the stomach. **The esophagus**:

- Follows the curve of the vertebral column as it descends through the neck and mediastinum—the median partition of the thoracic cavity (see Chapter 1)
- Passes through the elliptical **esophageal hiatus** in the muscular right crus of the diaphragm, just to the left of the median plane at the level of T10 vertebra

Figure 2.28. Arterial supply and venous drainage of the gastrointestinal (GI) tract. **A.** Arterial supply. **B.** Venous drainage. The portal vein drains poorly oxygenated, nutrient-rich blood from the GI tract, spleen, pancreas, and gallbladder to the liver. The *black arrow* indicates the communication of the esophageal vein with the azygos venous system.

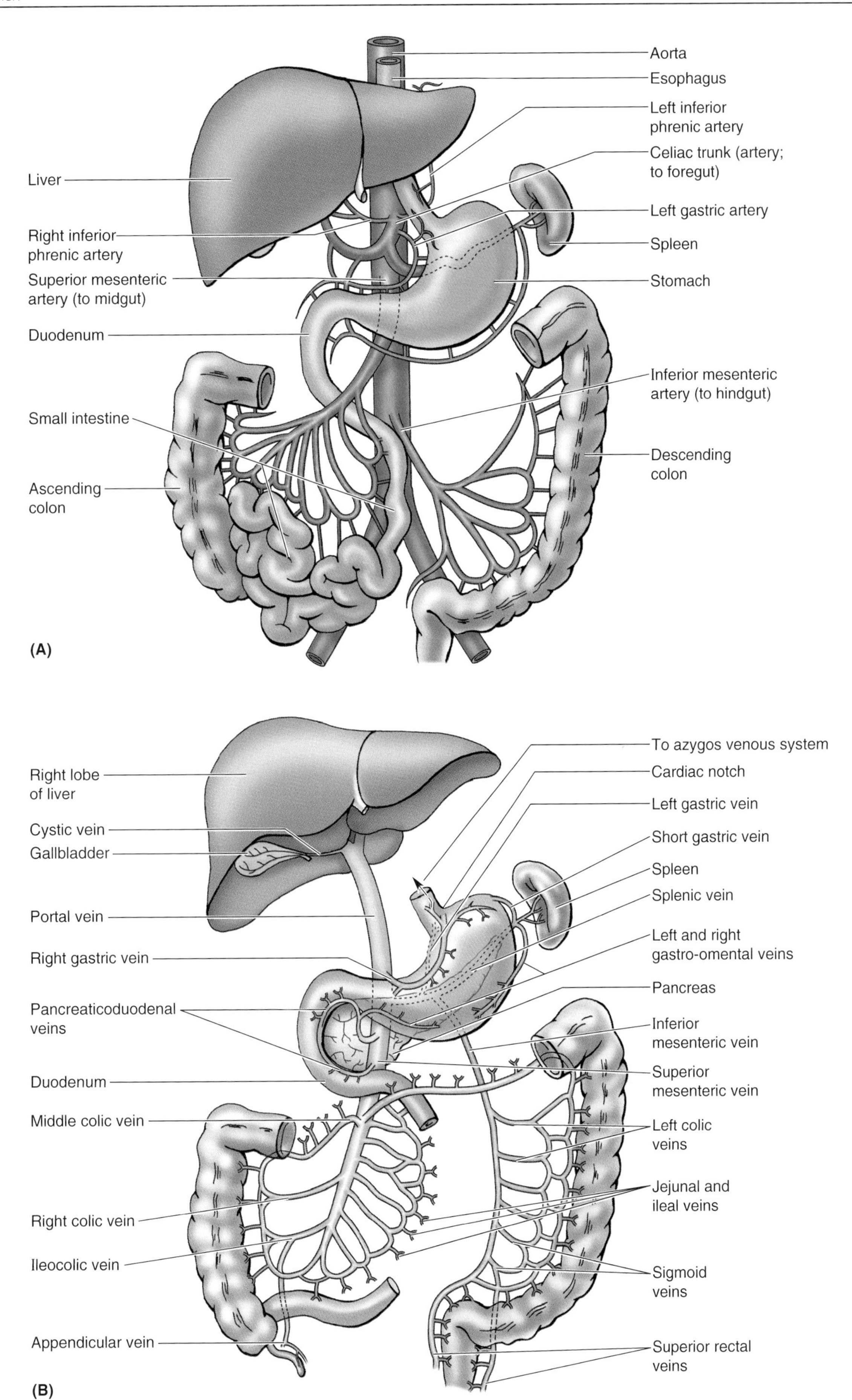
Aorta
Esophagus
Left inferior phrenic artery
Celiac trunk (artery; to foregut)
Liver
Left gastric artery
Right inferior phrenic artery
Spleen
Superior mesenteric artery (to midgut)
Stomach
Duodenum
Inferior mesenteric artery (to hindgut)
Small intestine
Descending colon
Ascending colon
(A)
To azygos venous system
Cardiac notch
Right lobe of liver
Left gastric vein
Short gastric vein
Cystic vein
Gallbladder
Spleen
Splenic vein
Portal vein
Left and right gastro-omental veins
Right gastric vein
Pancreas
Pancreaticoduodenal veins
Inferior mesenteric vein
Superior mesenteric vein
Duodenum
Middle colic vein
Left colic veins
Jejunal and ileal veins
Right colic vein
Ileocolic vein
Sigmoid veins
Appendicular vein
Superior rectal veins
(B)

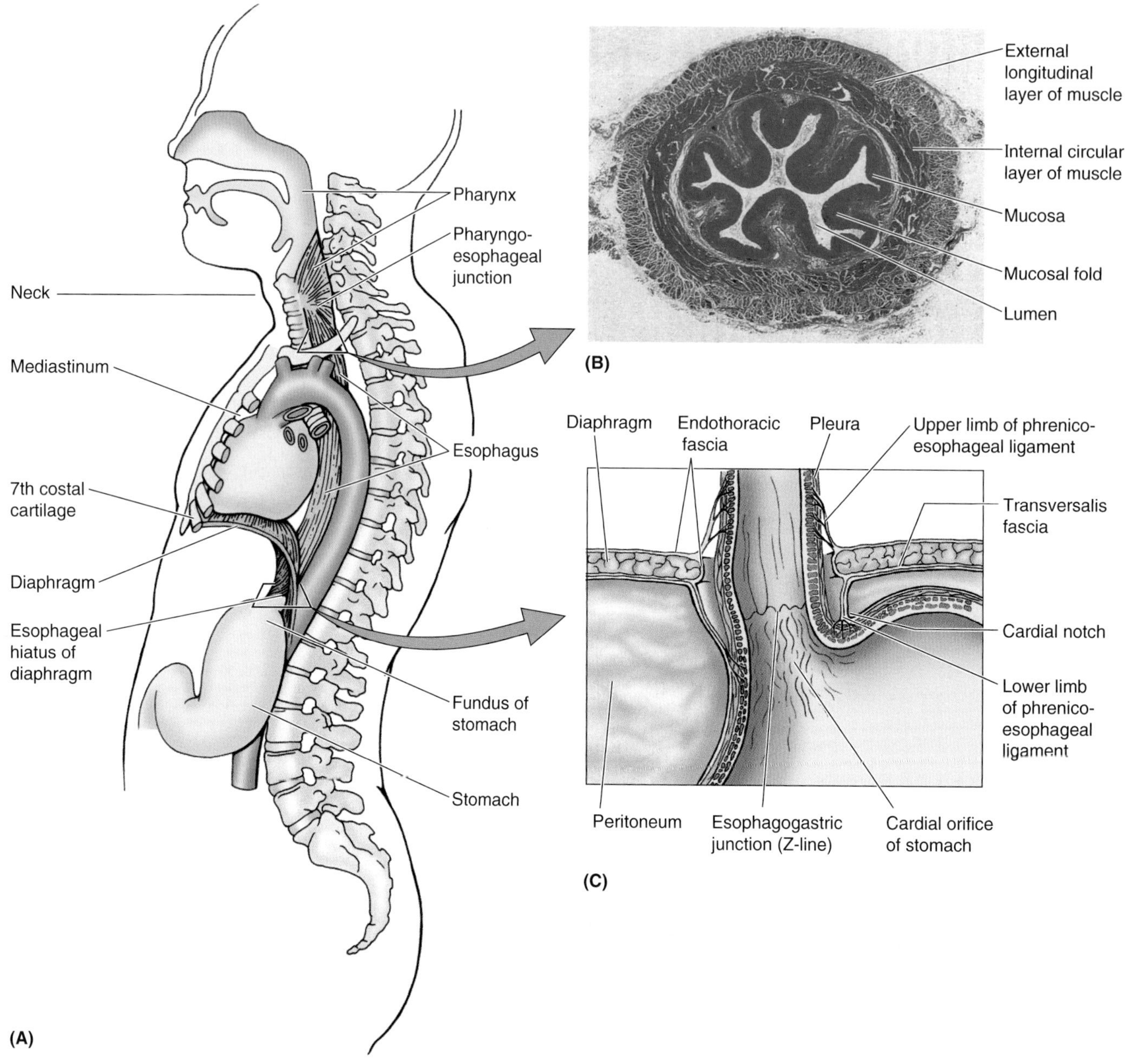

Figure 2.29. Esophagus and associated structures. A. Schematic drawing of a lateral view of the head, neck, and trunk showing the esophagus and the structures associated with it. The esophagus descends posterior to the trachea and leaves the thorax through the esophageal hiatus in the diaphragm. **B.** Transverse section of the esophagus showing the muscular layers and microscopic structure of its wall. **C.** Coronal section of the esophagus, diaphragm, and stomach (superior part). Observe the phrenicoesophageal ligament that connects the esophagus flexibly to the diaphragm; it limits upward movement of the esophagus while permitting some movement during swallowing and respiration.

- Terminates by entering the stomach at the **cardial orifice** of the stomach (Fig. 2.29*C*) to the left of the midline at the level of the 7th left costal cartilage and T11 vertebra
- Is encircled by the **esophageal nerve plexus** distally (Fig. 2.30)
- Is retroperitoneal but is covered anteriorly and laterally by peritoneum.

The esophagus has internal **circular** and external **longitudinal layers of muscle** (Fig. 2.29*B*). In its superior third, the external layer consists of **skeletal muscle**; the inferior third is composed of **smooth muscle**, and the middle third is made up of both types of muscle.

Food passes through the esophagus rapidly because of the peristaltic action of its musculature. The esophagus is attached to the margins of the esophageal hiatus in the di-

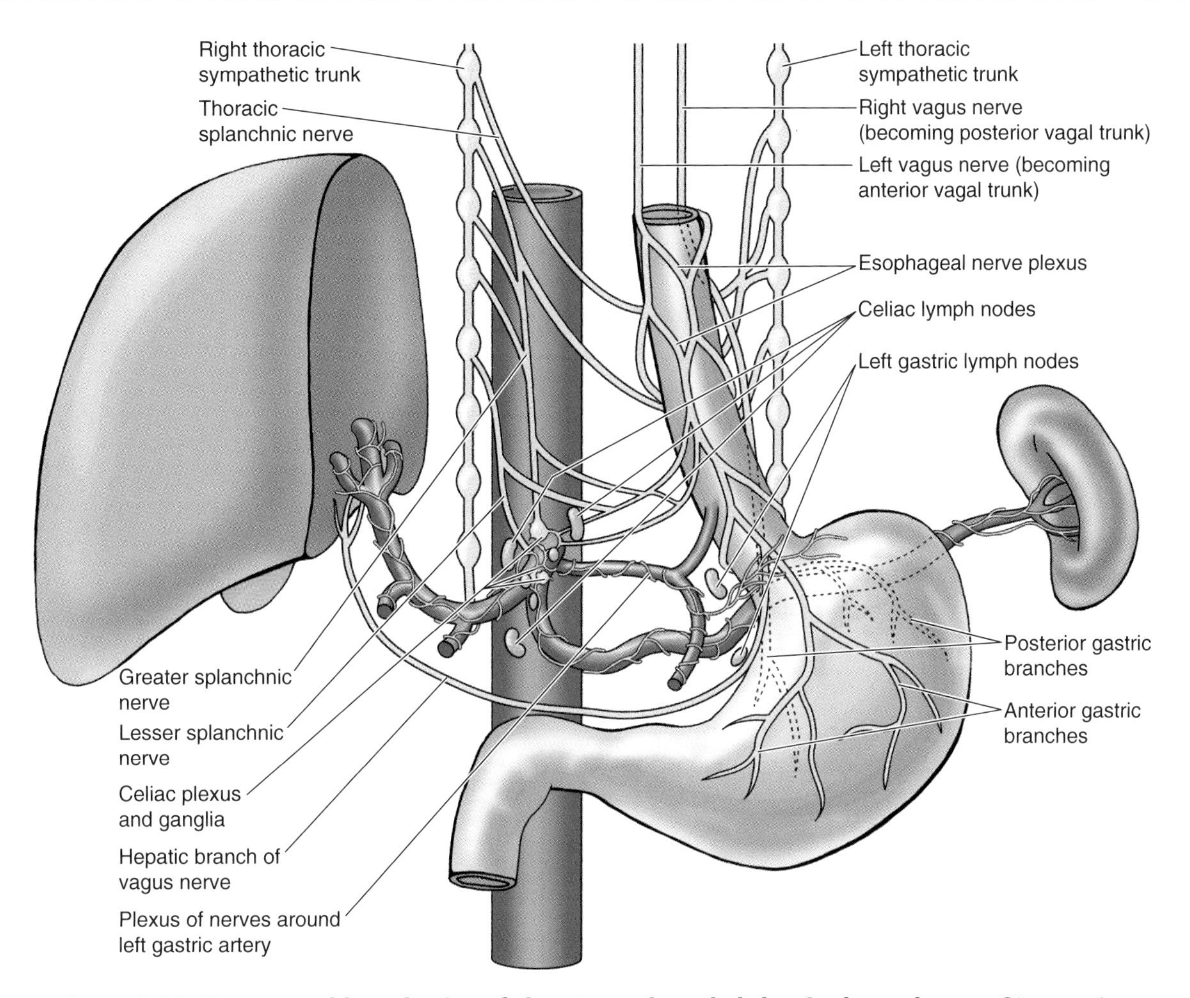

Figure 2.30. Nerves and lymphatics of the stomach and abdominal esophagus. Observe the branches of the anterior and posterior vagal trunks to these organs. As they reach the distal end of the esophagus, each vagus nerve (CN X) divides into several branches and there is an interchange of branches between the two sides to form the esophageal plexus. Observe also the sympathetic nerves supplying these organs through perivascular plexuses from the celiac plexus. The lymphatic vessels of the stomach have a pattern similar to that of the arteries (although the flow is in the opposite direction). Lymph from the stomach and abdominal part of the esophagus drains to the celiac and gastric lymph nodes.

aphragm by the **phrenicoesophageal ligament** (Fig. 2.29*C*), an extension of inferior diaphragmatic fascia. This ligament permits independent movement of the diaphragm and esophagus during respiration and swallowing.

The short, trumpet-shaped abdominal part of the esophagus—approximately 1.25 cm long—passes from the esophageal hiatus in the right crus of the diaphragm to the stomach. The right border of the esophagus is continuous with the lesser curvature of the stomach; however, its left border is separated from the fundus of the stomach by the **cardial notch** (Fig. 2.29*C*). The **esophagogastric junction** lies to the left of T11 vertebra on the horizontal plane that passes through the tip of the xiphoid process. Surgeons and endoscopists designate the Z-line—a jagged line where the mucosa abruptly changes from esophageal to gastric mucosa—as the **esophagogastric junction.** At this junction, the diaphragmatic musculature forming the esophageal hiatus functions as a physiological **esophageal sphincter** that contracts and relaxes. Radiological studies show that food stops here momentarily and that the sphincter mechanism is normally efficient in preventing reflux of gastric contents into the esophagus. When one is not eating, the lumen is normally collapsed above this level to prevent food or stomach juices from regurgitating into the esophagus.

The *arterial supply* of the abdominal part of the esophagus is from the **left gastric artery**, a branch of the celiac trunk, and the **left inferior phrenic artery** (Fig. 2.28*A*). The *venous drainage* is to the *portal venous system* through the **left gastric vein** and into the systemic venous system through esophageal veins entering the **azygos vein** (Fig. 2.28*B*).

The *lymphatic drainage* of the abdominal part of the esophagus is into the **left gastric lymph nodes** (Fig. 2.30); efferent lymphatic vessels from these nodes drain mainly to **celiac lymph nodes.** The *innervation* is from the **vagal trunks** (becoming anterior and posterior gastric nerves), the **thoracic sympathetic trunks**, the greater and lesser **splanchnic nerves**, and the **esophageal nerve plexus** around the left gastric and inferior phrenic arteries (Fig. 2.30).

Esophageal Varices

In *portal hypertension*—an abnormally increased blood pressure in the portal venous system—an abnormally large volume of blood bypassing the liver causes the esophageal veins to enlarge and form *esophageal varices*, collateral channels that may rupture and cause severe hemorrhage. Esophageal varices commonly develop in alcoholics who have developed *cirrhosis of the liver* (pp. 271, 279).

Esophageal Cancer

The incidence of esophageal cancer is low in North America; however, the onset of difficult swallowing (*dysphagia*) in anyone older than 45 years (especially males) raises suspicion of esophageal cancer. Using an *esophagoscope*, a relatively thin, movable fiberoptic tube, the physician can perform *esophagoscopy* and may observe and biopsy a tumor. Cancer cells from a tumor of the abdominal part of the esophagus usually metastasize to the *left gastric lymph nodes* (Fig. 2.30); however, some cancer cells enter the thoracic duct and pass into the venous system.

Pyrosis

Pyrosis (*heartburn*) is the most common type of esophageal discomfort or substernal pain. This burning sensation in the abdominal part of the esophagus is often accompanied by regurgitation of small amounts of food or gastric fluid into the esophagus. Pyrosis (G. a burning) may be associated with *esophageal hiatal or hiatus hernia* (pp. 227–228). As indicated by its common name, heartburn, pyrosis is commonly perceived as a "chest (vs. abdominal) sensation." As mentioned in Chapter 1, determining the nature and source of chest pains can be one of the more challenging aspects of physical diagnosis. ○

Stomach

The stomach is the expanded part of the digestive tract between the esophagus and small intestine. In most people the shape of the stomach resembles the letter J; however, the shape and position of the stomach varies in different persons and even in the same individual because of diaphragmatic movements during respiration, the stomach's contents, and the position of the person (e.g., whether lying down or standing).

The stomach acts as a food blender and reservoir; its chief function is enzymatic digestion. The *gastric juice* gradually converts a mass of food into a liquid mixture—*chyme*—that

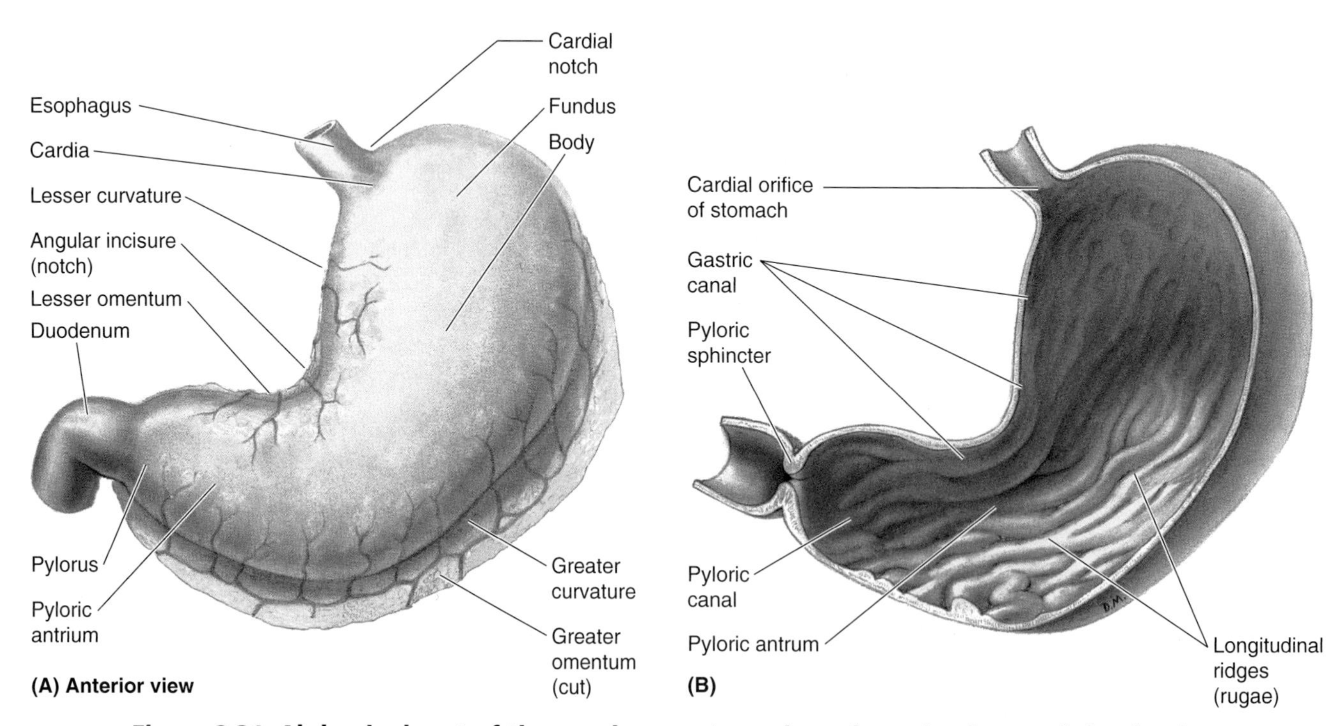

Figure 2.31. Abdominal part of the esophagus, stomach, and proximal part of the duodenum. A. Lateral view of the external surface. Observe the arteries in the omenta that supply the stomach. **B.** Internal surface (mucous membrane). Observe the longitudinal gastric folds, or rugae. Along the lesser curvature, observe several longitudinal mucosal folds extending from the esophagus to the pylorus and forming the gastric canal through which liquids pass.

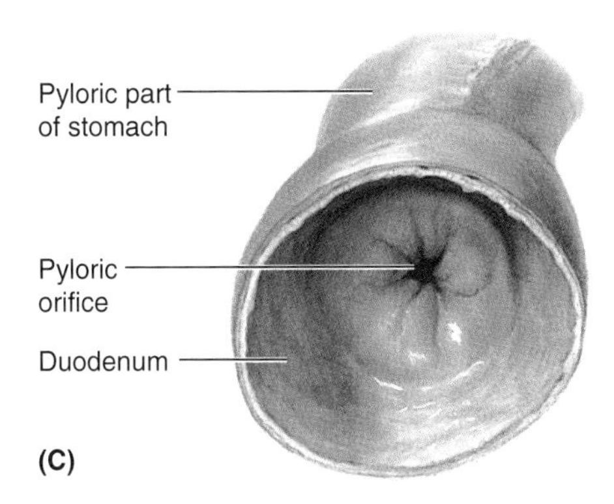

Figure 2.31. *(Continued)* **C.** Pylorus. This is the markedly constricted terminal part of the stomach. The pyloric orifice is the distal opening of the pyloric canal into the duodenum.

passes fairly quickly into the duodenum. An empty stomach is only of slightly larger caliber than the large intestine; however, it is capable of considerable expansion and can hold 2 to 3 liters of food. A newborn infant's stomach, approximately the size of a lemon, can expand to hold up to 30 mL of milk.

Parts of the Stomach

The stomach (Fig. 2.31, *A–C*) has four parts (cardia, fundus, body, and pyloric part) and two curvatures.

- **Cardia** is the part surrounding the cardial orifice.
- **Fundus** is the dilated superior part that is related to the left dome of the diaphragm and is limited inferiorly by the horizontal plane of the **cardial orifice**. The superior part of the fundus usually reaches the level of the left 5th intercostal space. The **cardial notch** is between the esophagus and fundus. The fundus may be dilated by gas, fluid, food, or any combination of these.
- **Body** lies between the fundus and the pyloric antrum.
- **Pyloric part** is the funnel-shaped region of the stomach; its wide part, the **pyloric antrum**, leads into the **pyloric canal**, its narrow part. The **pylorus** (the distal sphincteric region of the pyloric part) is thickened to form the **pyloric sphincter**, which controls discharge of the stomach contents through the *pyloric orifice* into the duodenum.
- **Lesser curvature** forms the shorter concave border of the stomach; the **angular incisure** (notch) is the sharp indentation approximately two-thirds of the distance along the lesser curvature that indicates the junction of the body and the pyloric part of the stomach.
- **Greater curvature** forms the longer convex border of the stomach.

Intermittent emptying of the stomach occurs when intragastric pressure overcomes the resistance of the pyloric sphincter.

The **pylorus** (G. gatekeeper) guards the pyloric orifice; its wall is thicker because it contains more circular smooth muscle. The middle layer of the muscularis externa is greatly thickened to form the **pyloric sphincter** (Fig. 2.31*B*). The pylorus is normally in tonic contraction; it is closed except when emitting *chyme* (G. juice), the semifluid contents of the stomach. At irregular intervals, *gastric peristalsis* passes the chyme through the pyloric canal and pyloric orifice into the small intestine for further mixing, digestion, and absorption.

Interior of the Stomach

The smooth surface of the gastric mucosa—mucous layer of the stomach—is reddish-brown during life, except in the pyloric part where it is pink. When contracted, the gastric mucosa is thrown into longitudinal ridges—**gastric folds**, or **rugae** (Fig. 2.31*B*); they are most marked toward the pyloric part and along the greater curvature. A **gastric canal** (furrow) forms temporarily during swallowing between the longitudinal gastric folds of the mucosa along the lesser curvature. It can be observed radiographically and endoscopically. The gastric canal forms because of the firm attachment of the gastric mucosa to the muscular layer, which does not have an oblique layer at this site. Saliva and small quantities of masticated food and other fluids pass through the gastric canal to the pyloric canal.

Relations of the Stomach

The stomach is covered by peritoneum, except where blood vessels run along its curvatures and in a small area posterior to the cardial orifice. The two layers of the lesser omentum extend around the stomach and leave its greater curvature as the greater omentum (Figs. 2.23, 2.25, and 2.32).

Displacement of the Stomach

Pancreatic pseudocysts and abscesses in the omental bursa may push the stomach anteriorly. This displacement is usually visible in lateral radiographs of the stomach and other diagnostic images such as computed tomographies (CTs). Following *pancreatitis* (inflammation of the pancreas), the posterior wall of the stomach may adhere to the part of the posterior wall of the omental bursa that covers the pancreas. This adhesion occurs because of the close relationship of the posterior wall of the stomach to the pancreas.

Hiatal or Hiatus Hernia

A hiatal or hiatus hernia is a protrusion of a part of the stomach into the mediastinum through the esophageal hiatus of the diaphragm. The hernias occur most often in people after middle age, possibly because of weakening of the muscular part of the diaphragm and widening of the esophageal hiatus. The hernias are often distressful and cause pain. Although clinically there are several types of hiatal hernia (Skandalakis et al., 1995), the two main types are *sliding hiatus hernia* and *paraesophageal hiatus hernia.* ▶

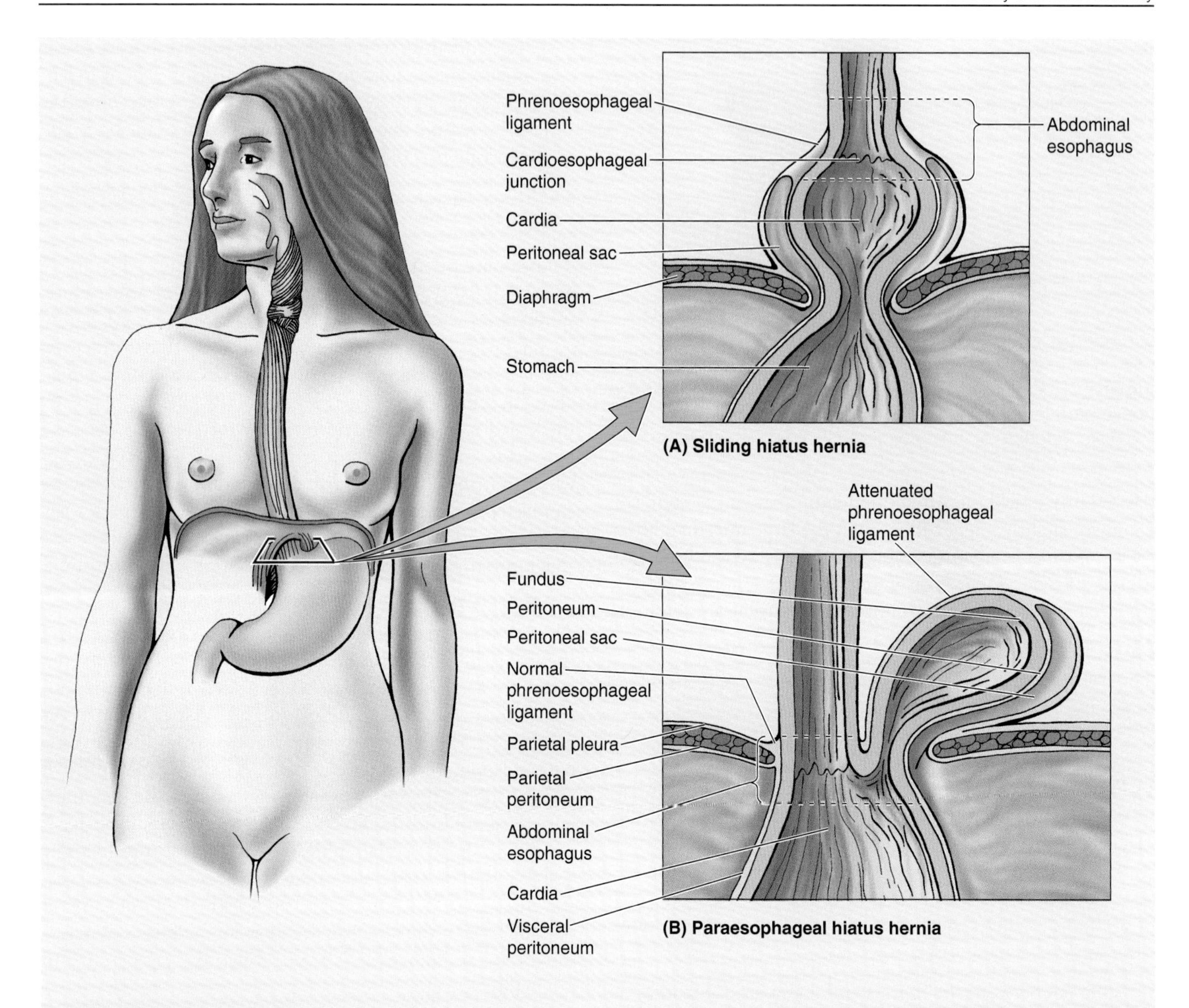

▶ In **sliding hiatus hernia** (*A*), the abdominal part of the esophagus, the cardia, and parts of the fundus of the stomach slide superiorly through the esophageal hiatus into the thorax, especially when the person lies down or bends over. Some regurgitation of stomach contents into the esophagus is possible because the clamping action of the right crus of the diaphragm on the inferior end of the esophagus is weak.

In the less common **paraesophageal hiatus hernia** (*B*), the cardia remains in its normal position; however, a pouch of peritoneum, often containing part of the fundus, extends through the esophageal hiatus anterior to the esophagus. In these cases, usually no regurgitation of gastric contents occurs because the cardial orifice is in its normal position.

Congenital Diaphragmatic Hernia

In congenital diaphragmatic hernia (CDH), part of the stomach and intestine herniate through a large *posterolateral defect in the diaphragm*. This type of hernia occurs in approximately one of every 2200 newborn infants. CDH results from the complex development of the diaphragm (Moore and Persaud, 1998). Because of lung hypoplasia, the mortality rate in these infants is high (approximately 76%). ⊙

- Anteriorly, the stomach is related to the **diaphragm**, the left lobe of liver, and the anterior abdominal wall.
- Posteriorly, the stomach is related to the **omental bursa** and the pancreas; the posterior surface of the stomach forms most of the anterior wall of the omental bursa.

The **bed of the stomach** on which the stomach rests in the supine position is formed by the structures forming the posterior wall of the omental bursa. *From superior to inferior* (Fig. 2.32), *the stomach bed is formed by the:*

- Left dome of the diaphragm
- Spleen
- Left kidney and suprarenal gland

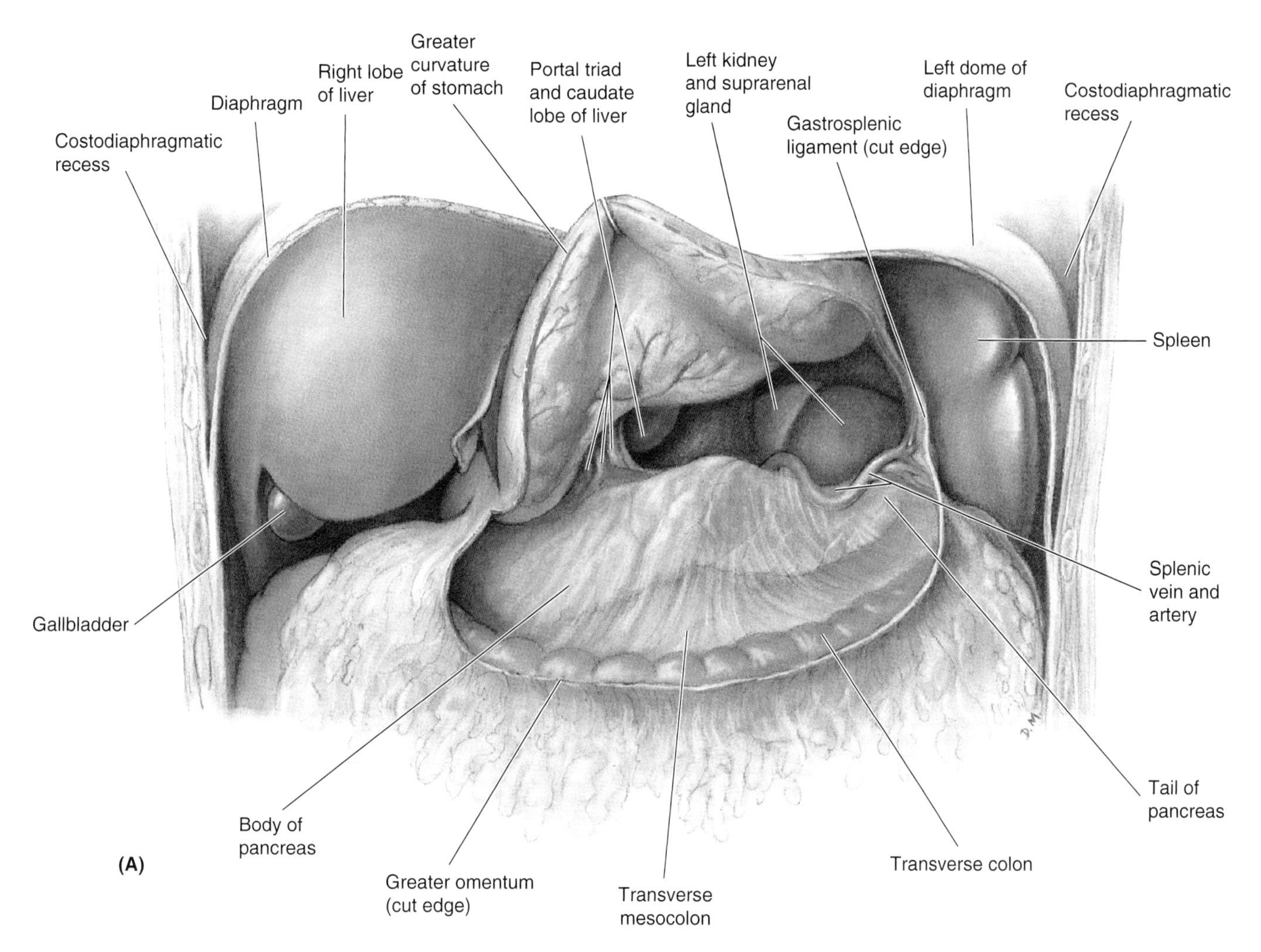

Figure 2.32. Omental (epiploic) bursa and stomach bed. A. Relationships of the omental bursa. The greater omentum and gastrosplenic ligament have been cut along the greater curvature of the stomach, and the stomach has been reflected superiorly. Observe the portal triad in the inferior root of the hepatoduodenal ligament (free edge of lesser omentum).

- Splenic artery
- Pancreas
- Transverse mesocolon and colon.

Vessels and Nerves of the Stomach

The stomach has a rich arterial supply (Fig. 2.33*A*, Table 2.6). The gastric arteries arise from the celiac trunk and its branches.

- **Left gastric artery**: arises directly from the **celiac trunk** and runs in the lesser omentum to the cardia and then turns abruptly to course along the lesser curvature of the stomach and anastomose with the right gastric artery
- **Right gastric artery**: usually arises from the **hepatic artery** and runs to the left along the lesser curvature to anastomose with the left gastric artery
- **Right gastro-omental artery** (gastroepiploic artery): arises as one of two terminal branches of the **gastroduodenal artery**, runs to the left along the greater curvature, and anastomoses with the left gastro-omental artery
- **Left gastro-omental artery**: arises from the **splenic artery** and courses along the greater curvature to anastomose with the right gastro-omental artery
- **Short gastric arteries** (four to five): arise from the distal end of the splenic artery or its splenic branches and pass to the fundus of the stomach.

The **gastric veins** parallel the arteries in position and course (Fig. 2.33*B*). The left and right gastric veins drain into the **portal vein** and the *short gastric veins* and *left gastro-omental vein* drain into the **splenic vein**, which joins the superior mesenteric vein (SMV) to form the portal vein. The *right gastro-omental vein* empties in the SMV. A **prepyloric vein** ascends over the pylorus to the right gastric vein. Because this vein is obvious in living persons, surgeons use it for identifying the pylorus.

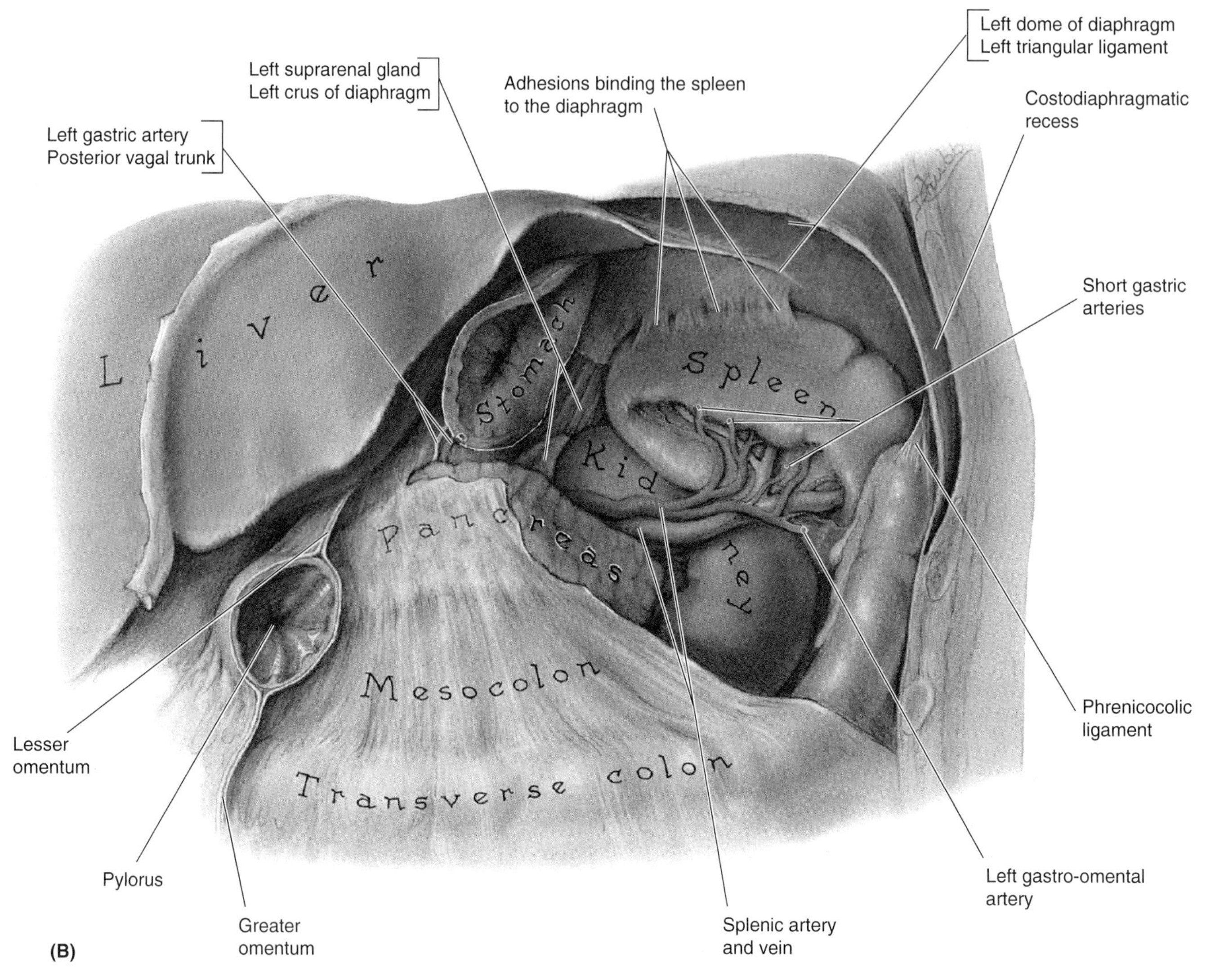

Figure 2.32. *(Continued)* **B.** Stomach bed. The stomach and most of the lesser omentum have been excised, and the peritoneum of the posterior wall of the omental bursa covering the stomach bed is largely removed to reveal the organs in the bed.

The **gastric lymphatic vessels** (Fig. 2.34*A*) accompany the arteries along the greater and lesser curvatures of the stomach. They drain lymph from its anterior and posterior surfaces toward its curvatures, where the **gastric** and **gastro-omental lymph nodes** are located. The efferent vessels from these nodes accompany the large arteries to the **celiac lymph nodes**. *The following is a summary of the lymphatic drainage of the stomach*:

- Lymph from the superior two-thirds of the stomach drains along right and left gastric vessels to the **gastric nodes**; lymph from the fundus and superior part of the body of the stomach also drains along the short gastric arteries and left gastro-omental vessels to the **pancreaticosplenic nodes**.
- Lymph from the right two-thirds of the inferior third of the stomach drains along the right gastro-omental vessels to the **pyloric nodes**.
- Lymph from the left one-third of the greater curvature drains along the short gastric and splenic vessels to the **pancreaticoduodenal nodes**.

The parasympathetic nerve supply of the stomach (Fig. 2.34*B*) is from the *anterior and posterior vagal trunks* and their branches, which enter the abdomen through the esophageal hiatus. The **anterior vagal trunk**, derived mainly from the left vagus nerve (CN X), usually enters the abdomen as a single branch that lies on the anterior surface of the esophagus. It runs toward the lesser curvature of the stomach, where it gives off hepatic and duodenal branches that leave the stomach in the hepatoduodenal ligament. The rest of the anterior vagal trunk continues along the lesser curvature, giving rise to anterior gastric branches. The larger **posterior vagal trunk**, derived mainly from the right vagus nerve, enters the abdomen on the posterior surface of the esophagus and passes toward the lesser cur-

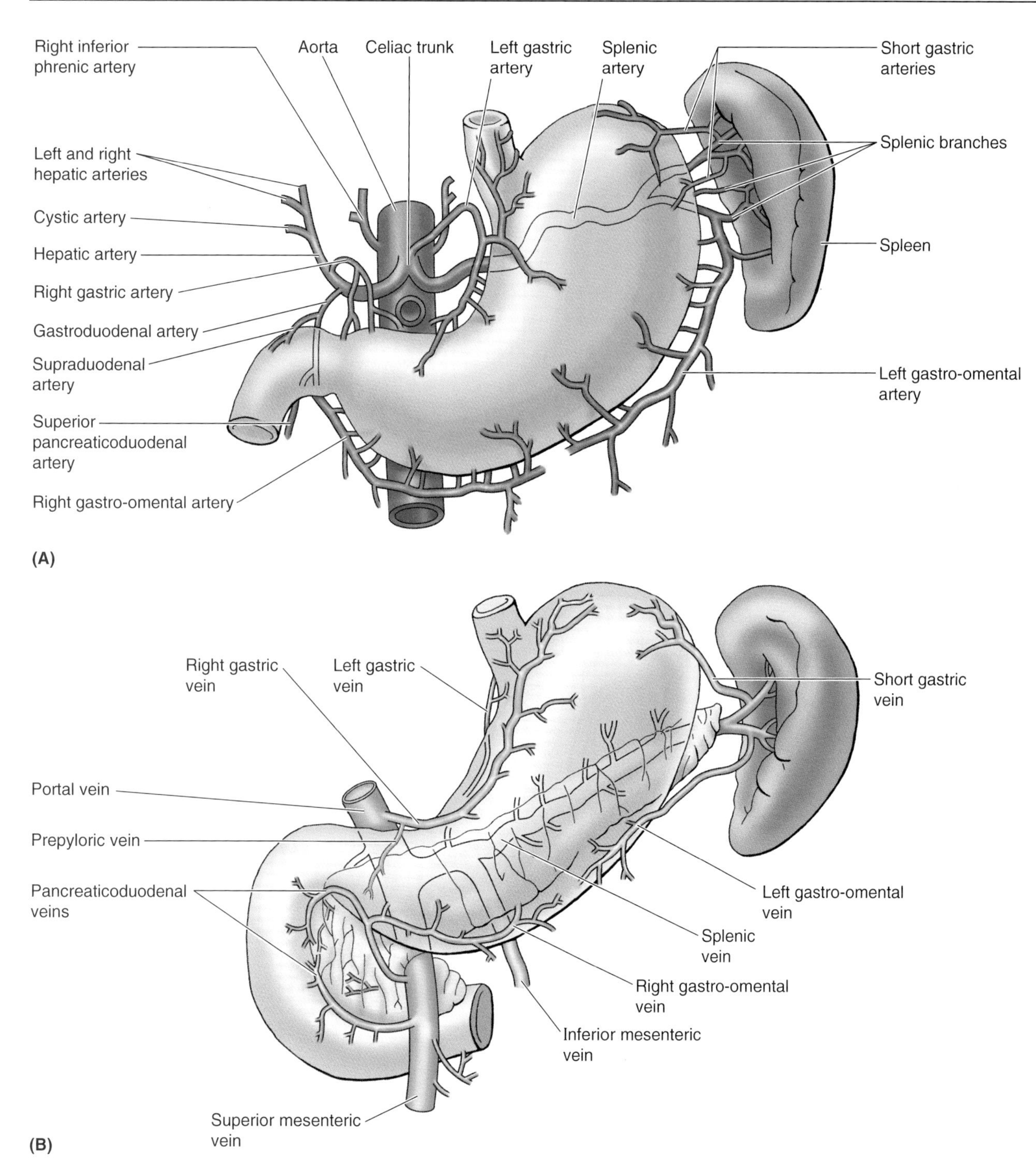

Figure 2.33. Arteries and veins of the stomach and spleen. A. Arterial supply. Observe that the stomach receives its main blood supply from branches of the celiac trunk. The fundus of the stomach is supplied by short gastric arteries arising from the splenic artery. The spleen is supplied by the splenic artery, the largest branch of the celiac trunk, which runs a tortuous course to the hilum of the spleen and breaks up into its terminal (splenic) branches. **B.** Venous drainage. The drainage of the stomach is directly or indirectly into the portal vein. The splenic vein usually receives the inferior mesenteric vein (IMV) and then unites with the superior mesenteric vein (SMV) to form the portal vein as shown here.

vature of the stomach. The posterior vagal trunk supplies branches to the anterior and posterior surfaces of the stomach. It gives off a celiac branch that runs to the *celiac plexus* and then continues along the lesser curvature, giving rise to posterior gastric branches.

The sympathetic nerve supply of the stomach from T6 through T9 segments of the spinal cord passes to the **celiac plexus** through the **greater splanchnic nerve** and is distributed through the plexuses around the gastric and gastro-omental arteries.

Surface Anatomy of the Stomach

The surface markings of the stomach vary because its size and position change under various circumstances (e.g., after a heavy meal). In the supine position, the stomach commonly lies in the right and left upper quadrants, or epigastric, umbilical, and left hypochondriac and lumbar regions as shown in the photograph. In the erect position, the stomach moves inferiorly. In asthenic (thin, weak) persons, the body of the stomach may extend into the pelvis. The stomach is usually partially overlain by the transverse colon near the left colic flexure. *The surface markings of the stomach include*:

- The *cardial orifice*, which usually lies posterior to the 7th left costal cartilage, 2 to 4 cm from the median plane at the level of T11 vertebra
- The *fundus*, which usually lies posterior to the left 5th rib in the midclavicular plane
- The *greater curvature*, which passes inferiorly to the left as far as the 10th left cartilage before turning medially to reach the pyloric antrum
- The *lesser curvature*, which passes from the right side of the cardia to the pyloric antrum; the most inferior part of the curvature is marked by the *angular incisure* (notch), which lies just to the left of the midline
- In the supine position, the *pyloric part* of the stomach usually lies at the level of the 9th costal cartilages at the level of L1 vertebra; the pyloric orifice is approximately 1.25 cm left of the midline
- In the erect position, the *pylorus* usually lies on the right side; its location varies from L2 through L4 vertebrae.

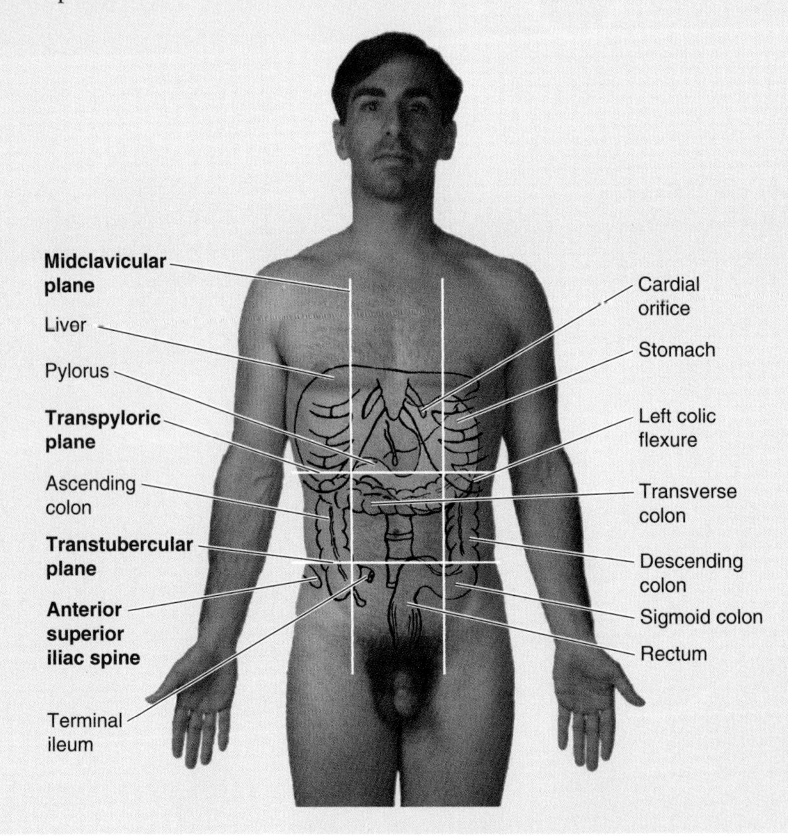

Pylorospasm

Spasmodic contraction of the pylorus sometimes occurs in infants, usually between 2 to 12 weeks of age. Pylorospasm is characterized by failure of the smooth muscle fibers encircling the pyloric canal to relax normally. As a result, food does not pass easily from the stomach into the duodenum and the stomach becomes overly full, usually resulting in vomiting.

Congenital Hypertrophic Pyloric Stenosis

A marked thickening of the smooth muscle in the pylorus affects approximately 1 of every 150 male infants and 1 of every 750 female infants (Moore and Persaud, 1998). The elongated, overgrown pylorus is hard, and severe stenosis (narrowing) of the pyloric canal is present. The proximal part of the stomach is secondarily dilated because of the ▶

▶ pyloric obstruction. Although the cause of *congenital hypertrophic pyloric stenosis* is unknown, genetic factors appear to be involved because of this condition's high incidence in infants of monozygotic twins.

Carcinoma of the Stomach

When the body or pyloric part of the stomach contains a malignant tumor, the mass may be palpable. The incidence of stomach cancer is higher in certain countries (e.g., Scandinavia) than in others (e.g., North America). This type of cancer is also more common in men than in women. The etiological factors are unknown, but this malignancy appears to be related to diet. Since the development of *flexible* fiber endoscopes, gastroscopy has become common. Using **gastroscopy**, physicians can inspect the mucosa of the air-inflated stomach, enabling them to observe gastric lesions and take biopsies. The extensive lymphatic drainage of the stomach and the impossibility of removing all the lymph nodes create a surgical problem. The nodes along the splenic vessels can be excised by removing the spleen, gastrosplenic and splenorenal ligaments, and the body and tail of the pancreas. Involved nodes along the gastro-omental vessels can be removed by resecting the greater omentum; however, removal of the aortic and celiac nodes and those around the head of the pancreas is difficult.

Gastrectomy

Total gastrectomy—removal of the entire stomach—is uncommon. *Partial gastrectomy*—removal of part of the stomach—may be performed to remove a region of the stomach involved by a carcinoma, for example, or to excise the pyloric antrum in some cases of *peptic ulcer disease*. Because the anastomoses of the arteries supplying the stomach provide good collateral circulation, one or more arteries may be ligated during this procedure without seriously affecting the blood supply to the part of the stomach remaining in place. When removing the pyloric antrum, for example, the greater omentum is incised parallel and inferior to the right gastro-omental artery, requiring ligation of all the omental branches of this artery. The omentum does not degenerate, however, because of anastomoses with other arteries such as the omental branches of the left gastro-omental artery, which are still intact. Partial gastrectomy to remove a carcinoma usually also requires removal of all involved regional lymph nodes. Because cancer frequently occurs in the pyloric region, removal of the **pyloric lymph nodes**—and the right **gastro-omental nodes** also receiving lymph drainage from this region—is especially important. As stomach cancer becomes more advanced, the lymphogenous dissemination of malignant cells involves the **celiac lymph nodes**, to which all gastric nodes drain.

Gastric Ulcers

The secretion of acid by parietal cells of the stomach is largely controlled by the vagus nerves; hence, **vagotomy**—*section of the vagal trunks* at the esophageal hiatus—is performed in some persons with **peptic ulcers** (lesions of the mucosa of the stomach commonly associated with the presence of *Helicobacter pylori*) to reduce the production of acid. Research shows that 9 of 10 gastric ulcers are caused by infection with *H. pylori*, which can be treated with antibiotics. Usually mucus covers the mucosa, forming a barrier between the acid and the cells of the mucosa. Sometimes this protection is inadequate and the gastric juices erode the mucosa, forming an ulcer. If the ulcer erodes into the gastric arteries, it can cause life-threatening bleeding. *Vagotomy* may be performed in conjunction with resection of the ulcerated area (*antrectomy*, or resection of the pyloric antrum). In **selective vagotomy**, the stomach is denervated but the vagal branches to the pylorus, biliary tract, intestines, and celiac plexus are preserved. A *parietal cell vagotomy* attempts to denervate even more specifically the area in which the parietal cells are located (Sabiston and Lyerly, 1994; Skandalakis et al., 1995), hoping to affect the acid-producing cells while sparing other abdominal structures supplied by the vagus nerve.

A posterior gastric ulcer may erode through the stomach wall into the pancreas, resulting in referred pain to the back. In such cases, *erosion of the splenic artery* results in severe hemorrhage into the peritoneal cavity. Pain impulses from the stomach are carried by visceral afferent fibers that accompany sympathetic nerves. This fact is evident because the pain of a recurrent peptic ulcer may persist after complete vagotomy, whereas patients who have had a bilateral sympathectomy may have a perforated peptic ulcer and experience no pain.

Visceral Referred Pain

Pain is an unpleasant sensation associated with actual or potential tissue damage and mediated by specific nerve fibers to the brain, where its conscious appreciation may be modified. *Organic pain* arising from an organ such as the stomach varies from dull to severe; however, the pain is poorly localized. It radiates to the dermatome level, which receives visceral afferent fibers from the organ concerned. *Visceral referred pain* from a gastric ulcer, for example, is referred to the epigastric region because the stomach is supplied by pain afferents that reach the T7 and ▶

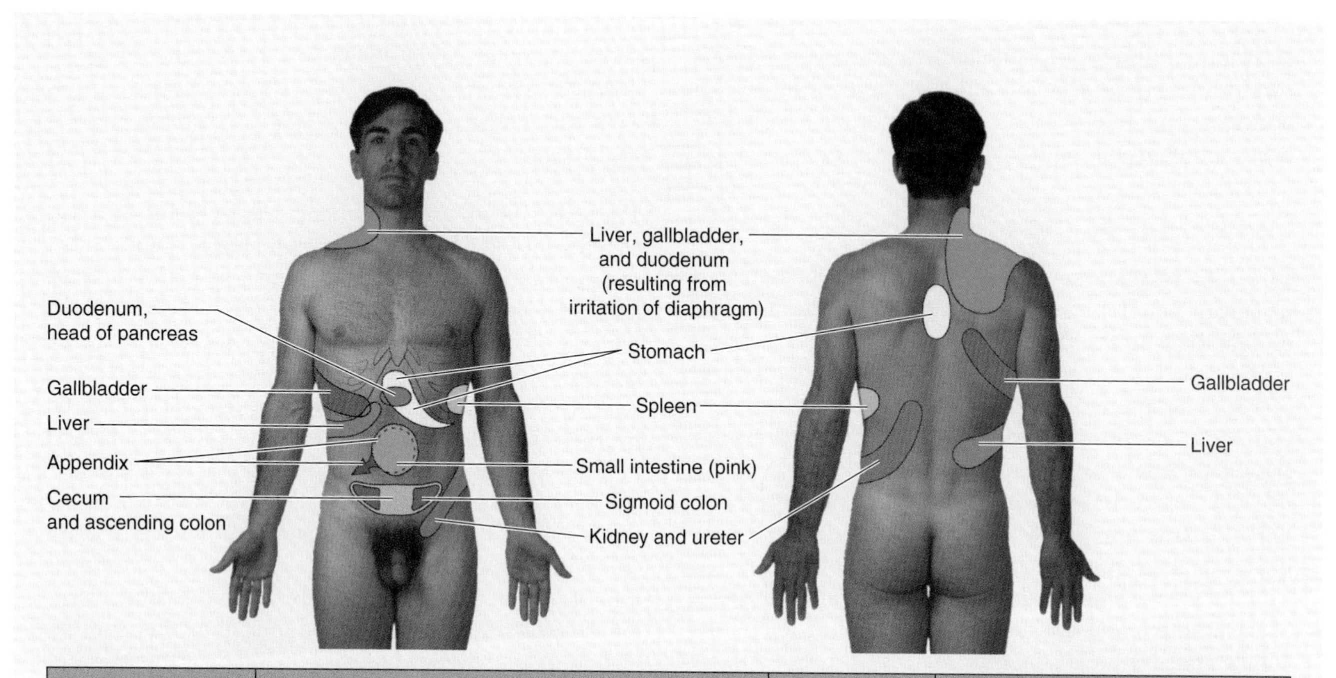

Organ	Nerve Supply	Spinal Cord	Referred Site and Clinical Example
Stomach	Anterior and posterior vagal trunks. Presynaptic sympathetic fibers reach celiac and other ganglia through greater splanchnic nerves.	T6–T9 or T10	Epigastric and left hypochondriac regions (e.g., gastric peptic ulcer)
Duodenum	Vagus nerves. Presynaptic sympathetic fibers reach celiac and superior mesenteric ganglia through greater splanchnic nerves.	T5–T9 or T10	Epigastric region (e.g.,duodenal peptic ulcer) Right shoulder if ulcer perforates
Pancreatic head	Vagus and thoracic splanchnic nerves.	T8–T9	Inferior part of epigastric region (e.g., pancreatitis)
Small intestine (jejunum and ileum)	Posterior vagal trunks. Presynaptic sympathetic fibers reach celiac ganglion through greater splanchnic nerves.	T5–T9	Periumbilical region (e.g., acute intestinal obstruction)
Colon	Vagus nerves. Presynaptic sympathetic fibers reach celiac, superior mesenteric, and inferior mesenteric ganglia through greater splanchnic nerves. Parasympathetic supply to distal colon is derived from pelvic splanchnic nerves through hypogastric nerves and inferior hypogastric plexus.	T10–T12 (proximal colon) L1–L3 (distal colon)	Hypogastric region (e.g., ulcerative colitis) Left lower quadrant (e.g., sigmoiditis)
Spleen	Celiac plexus, especially from greater splanchnic nerve.	T6–T8	Left hypochondriac region (e.g., splenic infarct)
Appendix	Sympathetic and parasympathetic nerves from superior mesenteric plexus. Afferent nerve fibers accompany sympathetic nerves to T10 segment of spinal cord.	T10	Periumbilical region and later to right lower quadrant (e.g., appendicitis)
Gallbladder and liver	Nerves are derived from celiac plexus (sympathetic), vagus nerve (parasympathetic), and right phrenic nerve (sensory).	T6–T9	Epigastric region and later to right hypochondriac region; may cause pain on posterior thoracic wall or right shoulder owing to diaphragmatic irritation
Kidneys/ureters	Nerves arise from the renal plexus and consist of sympathetic, parasympathetic, and visceral afferent fibers from thoracic and lumbar splanchnics and the vagus nerve.	T11–T12	Small of back, flank (lumbar quadrant), extending to groin (inguinal region) and genitals (e.g., renal or ureteric calculi)

Pain is perceived as originating in areas supplied by the somatic nerves entering the spinal cord at the same segment as the sensory nerves from the organ producing the pain. Although the areas of pain are not always as shown, they provide clues for the clinician when determining which organ may be affected.

▶ T8 spinal sensory ganglia and spinal cord segments through the greater splanchnic nerve. The pain is interpreted by the brain as though the irritation occurred in the skin of the epigastric region, which is also supplied by the same sensory ganglia and spinal cord segments.

Pain arising from the parietal peritoneum is of the somatic type and is usually severe. The site of its origin can be localized. The anatomical basis for this localization of pain is that the parietal peritoneum is supplied by somatic sensory fibers through thoracic nerves, whereas a viscus such as the appendix is supplied by visceral afferent fibers in the lesser splanchnic nerve. *Inflamed parietal peritoneum is extremely sensitive to stretching.* When digital pressure is applied to the anterolateral abdominal wall over the site of inflammation, the parietal peritoneum is stretched. When the fingers are suddenly removed, extreme localized pain is usually felt—*rebound tenderness.* ⊕

Table 2.6. **Arterial Supply to Esophagus, Stomach, Duodenum, Liver, Gallbladder, Pancreas, and Spleen**

Artery	Origin	Course	Distribution
Celiac	Abdominal aorta just distal to aortic hiatus of diaphragm	Soon divides into left gastric, splenic, and common hepatic arteries	Supplies esophagus, stomach, duodenum (proximal to bile duct), liver and biliary apparatus, and pancreas
Left gastric	Celiac trunk	Ascends retroperitoneally to esophageal hiatus, where it passes between layers of hepatogastric ligament	Distal portion of esophagus and lesser curvature of stomach
Splenic	Celiac trunk	Runs retroperitoneally along superior border of pancreas; it then passes between layers of splenorenal ligament to hilum of spleen	Body of pancreas, spleen, and greater curvature of stomach
Left gastro-omental (gastroepiploic)	Splenic artery in hilum of spleen	Passes between layers of gastrosplenic ligament to greater curvature of stomach	Left portion of greater curvature of stomach
Short gastric (n = 4–5)	Splenic artery in hilum of spleen	Passes between layers of gastrosplenic ligament to fundus of stomach	Fundus of stomach
Hepatic[a]	Celiac trunk	Passes retroperitoneally to reach hepatoduodenal ligament and passes between its layers to porta hepatis; divides into right and left hepatic arteries	Liver, gallbladder, stomach, pancreas, duodenum, and respective lobes of liver
Cystic	Right hepatic artery	Arises within hepatoduodenal ligament	Gallbladder and cystic duct
Right gastric	Hepatic artery	Runs between layers of hepatogastric ligament	Right portion of lesser curvature of stomach
Gastroduodenal	Hepatic artery	Descends retroperitoneally, posterior to gastroduodenal junction	Stomach, pancreas, first part of duodenum, and distal part of bile duct
Right gastro-omental (gastroepiploic)	Gastroduodenal artery	Passes between layers of greater omentum to greater curvature of stomach	Right portion of greater curvature of stomach
Anterior and posterior superior pancreatico-duodenal	Gastroduodenal artery	Descends on head of pancreas	Proximal portion of duodenum and head of pancreas
Anterior and posterior inferior pancreatico-duodenal	Superior mesenteric artery	Ascends retroperitoneally on head of pancreas	Distal portion of duodenum and head of pancreas

[a]For descriptive purposes, the hepatic artery is often divided into common hepatic artery from its origin to the origin of the gastroduodenal artery, and the remainder of the vessel is called the hepatic artery proper.

Figure 2.34. Lymphatic drainage and innervation of the stomach and small intestine. A. Lymphatic drainage. The *arrows* indicate the direction of lymph flow to the lymph nodes. **B.** Innervation. The nerves are both parasympathetic through the vagus (CN X) nerves and sympathetic from the celiac plexus. The stomach is supplied by the celiac plexus through the periarterial plexuses along the arteries to the stomach. Presynaptic sympathetic fibers reach the celiac and other ganglia through the splanchnic nerves. Postsynaptic fibers are then distributed to the blood vessels and musculature of the stomach. The small intestine is supplied by autonomic and afferent fibers from the celiac and superior mesenteric plexuses.

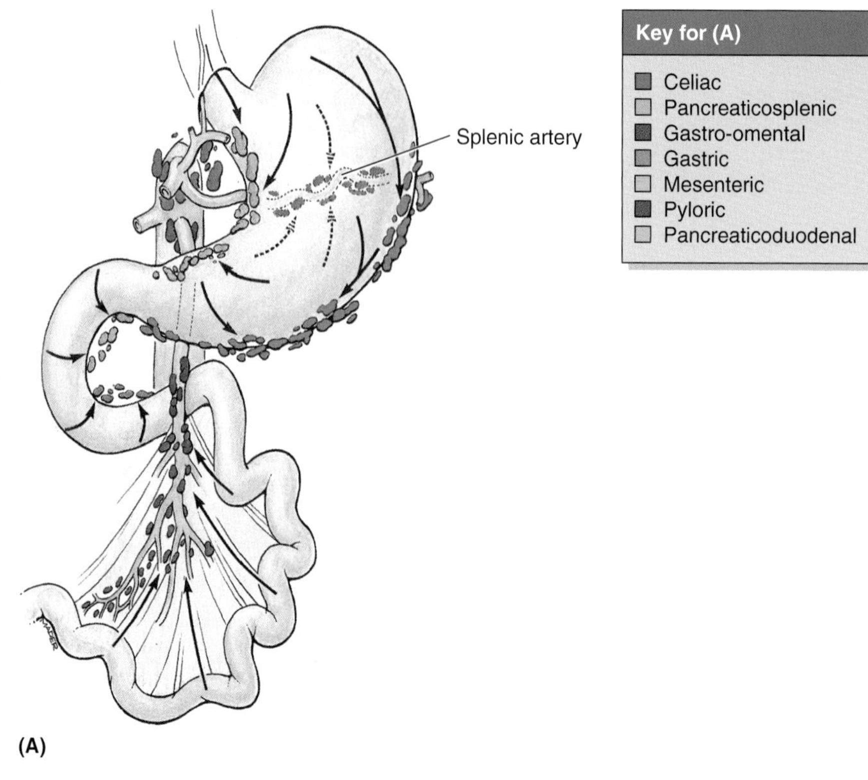

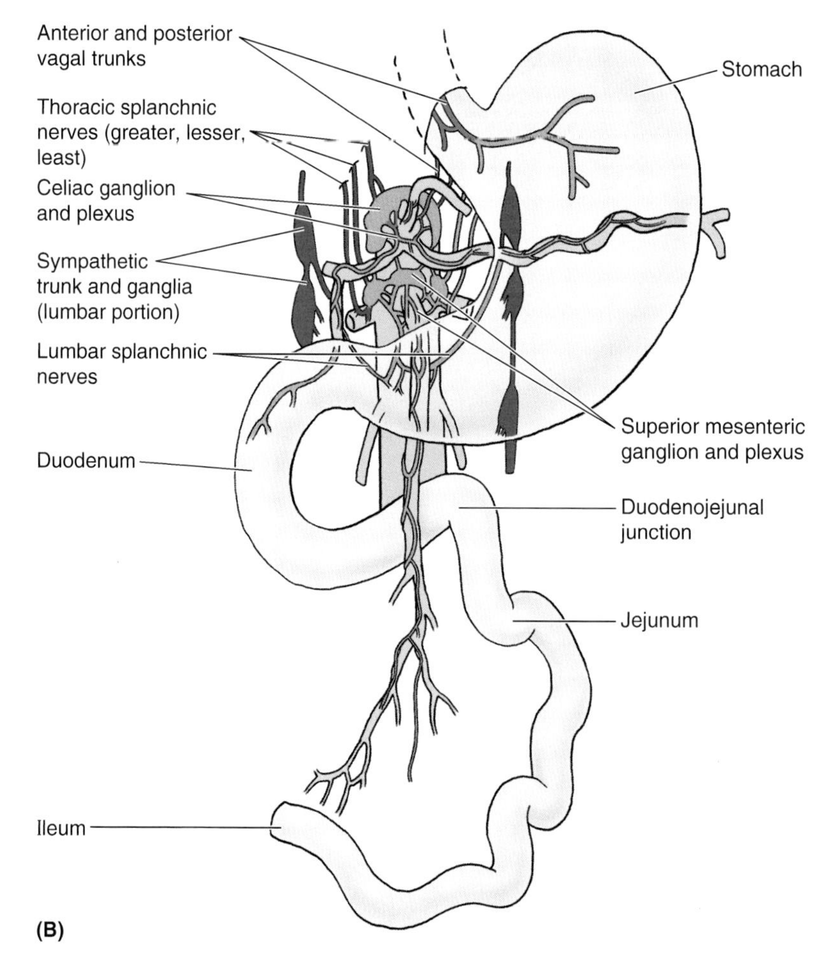

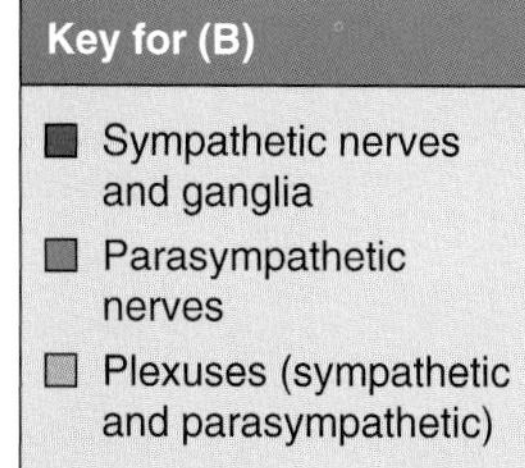

Small Intestine

The small intestine, consisting of the duodenum, jejunum, and ileum (Fig. 2.35), *extends from the pylorus to the ileocecal junction* where the ileum joins the cecum, the first part of the large intestine. The pylorus empties the contents of the stomach into the duodenum, the first of the three parts of the small intestine.

Duodenum

The duodenum—*the first and shortest part of the small intestine*—is also the widest and most fixed part. The duodenum pursues a C-shaped course around the head of the pancreas (Fig. 2.36). The duodenum begins at the pylorus on the right side and ends at the **duodenojejunal junction** on the left side. This junction occurs approximately at the level of the L2 vertebra, 2 to 3 cm to the left of the midline. The junction usually takes the form of an acute angle, the **duodenojejunal flexure**. Most of the duodenum is fixed by peritoneum to structures on the posterior abdominal wall and is considered partially retroperitoneal. *The duodenum is divisible into four parts* (Table 2.7):

- **Superior (1st) part** is short (approximately 5 cm) and lies anterolateral to the body of L1 vertebra
- **Descending (2nd) part** is longer (7–10 cm) and descends along the right sides of L1 through L3 vertebrae
- **Horizontal (3rd) part** is 6 to 8 cm long and crosses L3 vertebra
- **Ascending (4th) part** is short (5 cm) and begins at the left of L3 vertebra and rises superiorly as far as the superior border of L2 vertebra.

The first 2 cm of the superior (first) part of the duodenum—immediately distal to the pylorus—has a mesentery and is mobile. This free part is the *ampulla* (duodenal cap), which has an appearance distinct from the remainder of the duodenum when observed radiographically using contrast medium.

The distal 3 cm of the superior part and the other three parts of the duodenum have no mesentery and are immobile because they are retroperitoneal. The principal relationships of the duodenum are outlined in Table 2.7 and illustrated in Figure 2.36.

The **superior (first) part of the duodenum** ascends from the pylorus and is overlapped by the liver and gallbladder. Peritoneum covers its anterior aspect but it is bare of peritoneum posteriorly, except for the ampulla—the first 2 cm that joins the pylorus. The proximal part has the *hepatoduodenal ligament* (part of the lesser omentum) attached superiorly and the greater omentum attached inferiorly (Fig. 2.22). Posterior to the superior part are the portal vein, bile duct, gastroduodenal artery, and IVC.

The **descending (second) part of the duodenum** runs inferiorly, curving around the head of the pancreas. Initially it lies to the right and parallel to the IVC. The **bile and pancreatic ducts** enter its posteromedial wall. These ducts usually unite to form the **hepatopancreatic ampulla**, which opens on the summit of an eminence located posteromedially in the descending duodenum—the **major duodenal papilla**. The descending part of the duodenum is entirely retroperitoneal. The anterior surface of its proximal and distal thirds is intimately covered with peritoneum; however, the peritoneum reflects from its middle third to form the double-layered mesentery of the transverse colon, the transverse mesocolon.

The **inferior or horizontal (third) part of the duodenum** runs transversely to the left, passing over the IVC, aorta, and L3 vertebra. *It is crossed by the superior mesenteric artery (SMA) and vein and the root of the mesentery* of the jejunum and ileum (Fig. 2.36*B*). Superior to it is the head of the pancreas and its uncinate process. The anterior surface of the horizontal part is covered with peritoneum, except where it is crossed by the superior mesenteric vessels and the root of the mesentery. Posteriorly it is separated from the vertebral column by the right psoas major, IVC, aorta, and the right testicular or ovarian vessels.

The **ascending part (fourth) of the duodenum** runs superiorly and along the left side of the aorta to reach the inferior border of the body of the pancreas. Here it curves anteriorly to join the jejunum at the **duodenojejunal junction** that takes the form of an acute angle—the *duodenojejunal flexure*—which is supported by the attachment of a *suspensory muscle of*

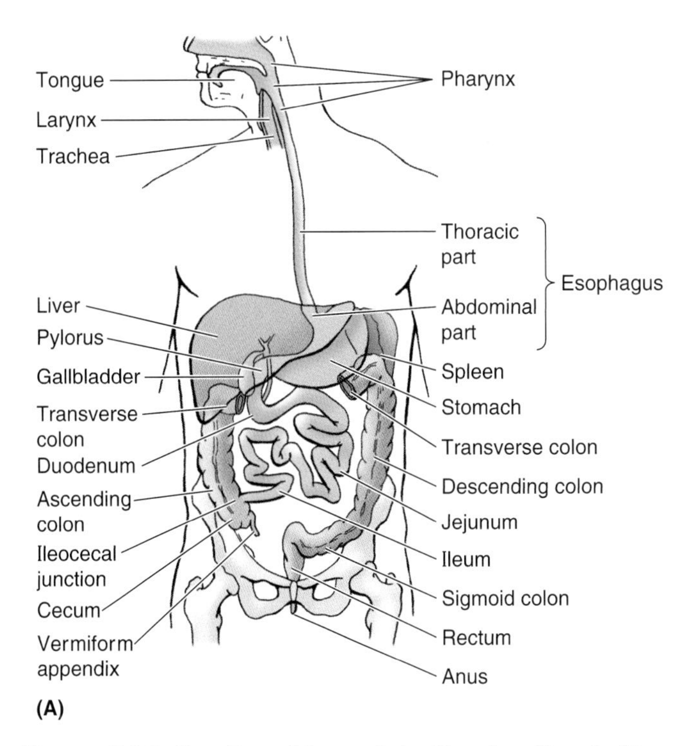

Figure 2.35. Small and large intestine in situ. A. Diagrammatic orientation drawing of the digestive system, extending from the lips to the anus. **B.** Small and large intestine. **C.** Ileocecal region showing its blood supply (see p. 238).

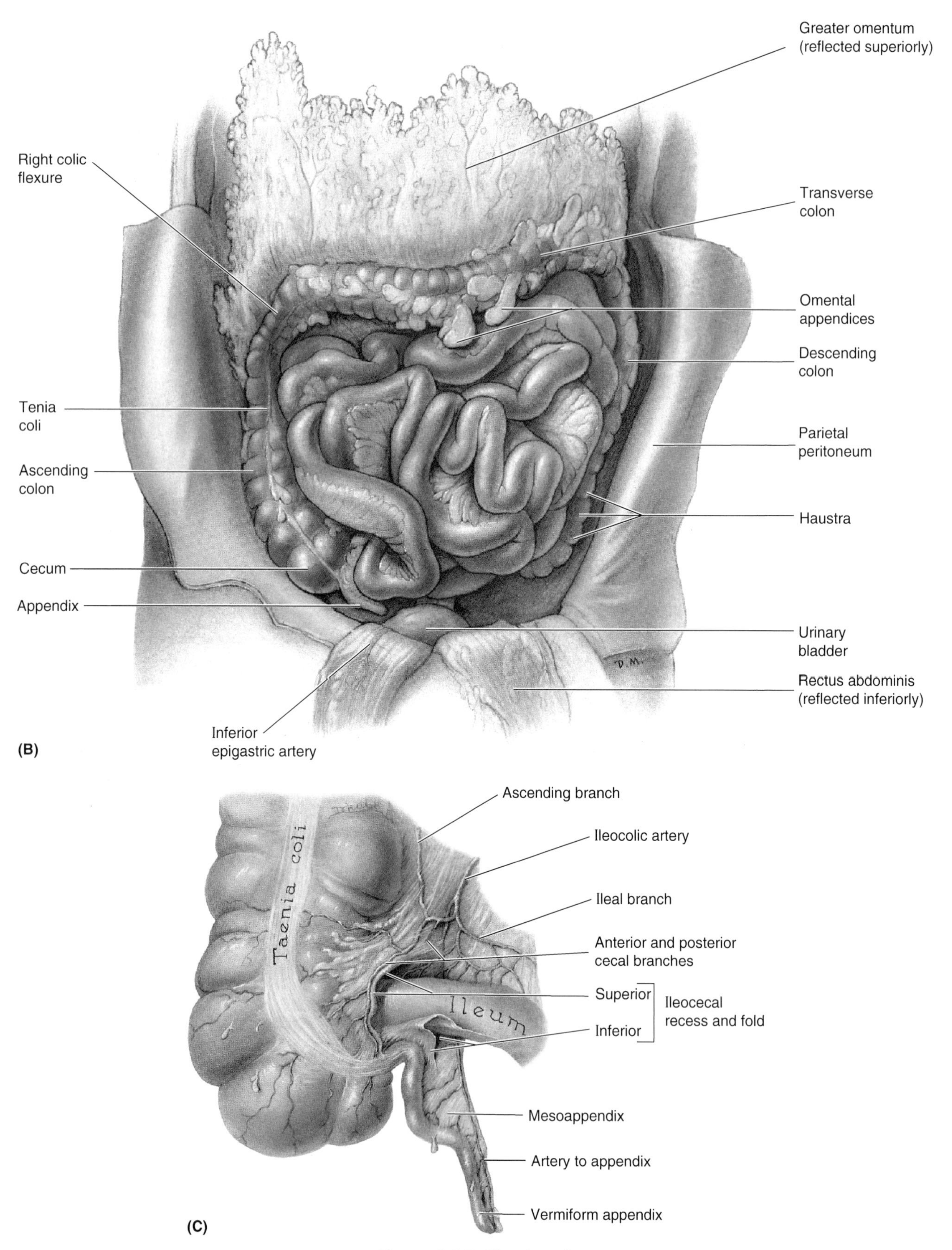

Figure 2.35. *(Continued)*

Table 2.7. **Relationships of the Duodenum**

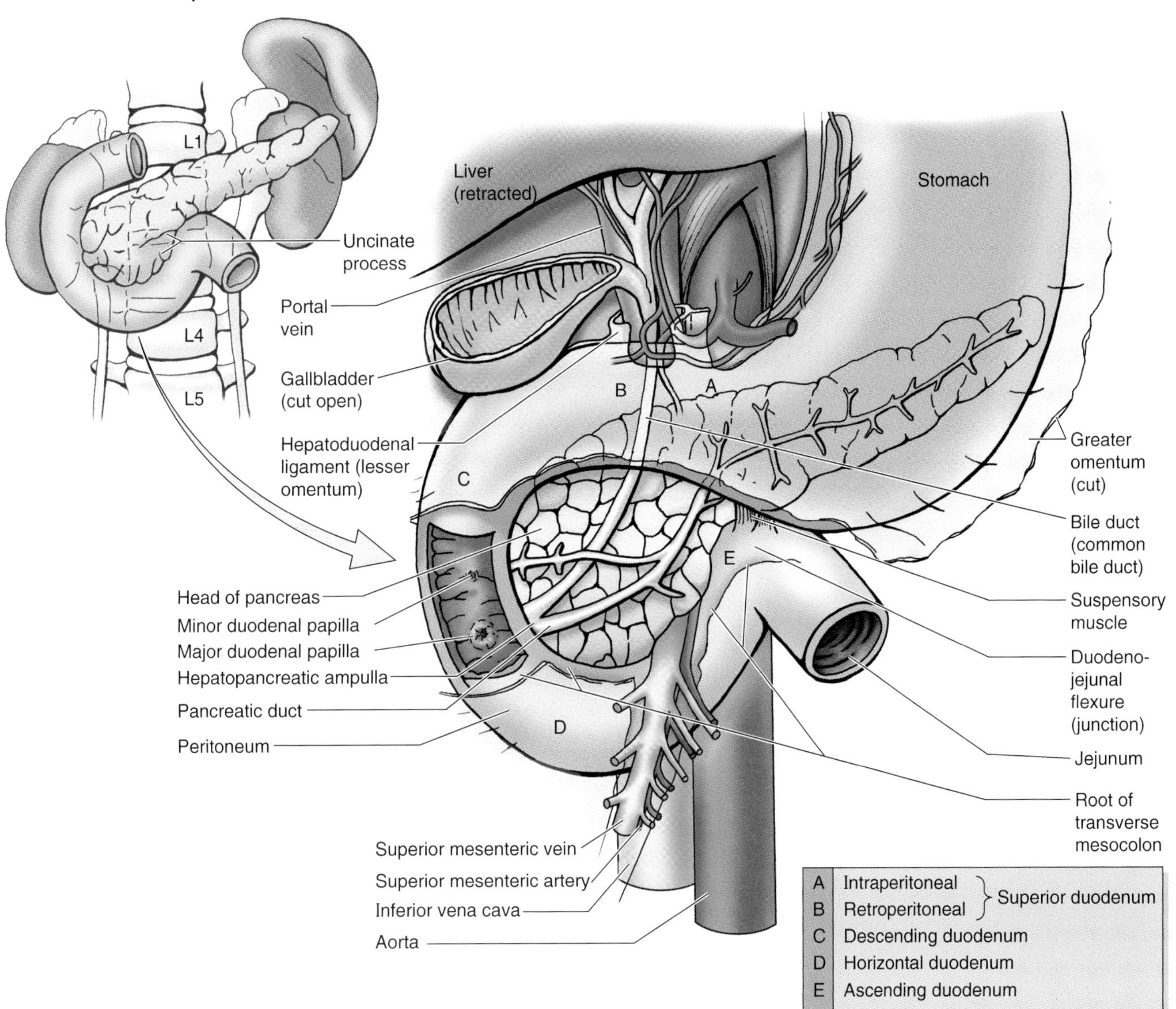

Part of Duodenum	Anterior	Posterior	Medial	Superior	Inferior	Vertebral Level
Superior (1st part)	Peritoneum Gallbladder Quadrate lobe of liver	Bile duct Gastroduodenal artery Portal vein IVC		Neck of gallbladder	Neck of pancreas	Anterolateral to L1 vertebra
Descending (2nd part)	Transverse colon Transverse mesocolon Coils of small intestine	Hilum of right kidney Renal vessels Ureter Psoas major	Head of pancreas Pancreatic duct Bile duct			Right of L2–L3 vertebrae
Horizontal (3rd part)	SMA SMV Coils of small intestine	Right psoas major IVC Aorta Right ureter		Head and uncinate process of pancreas Superior mesenteric vessels		Anterior to L3 vertebra
Ascending (4th part)	Beginning of root of mesentery Coils of jejunum	Left psoas major Left margin of aorta	Head of pancreas	Body of pancreas		Left of L3 vertebra

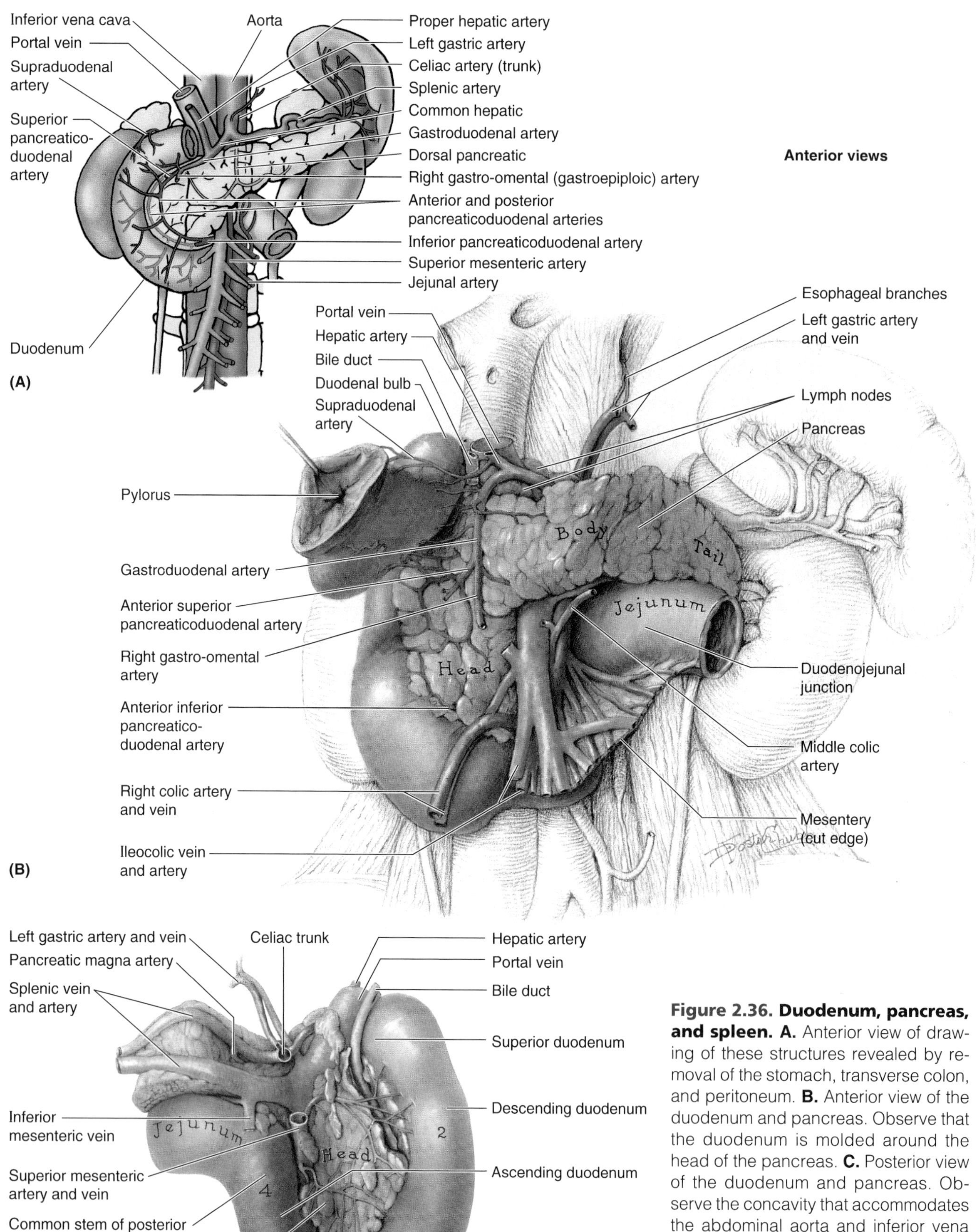

Figure 2.36. Duodenum, pancreas, and spleen. A. Anterior view of drawing of these structures revealed by removal of the stomach, transverse colon, and peritoneum. **B.** Anterior view of the duodenum and pancreas. Observe that the duodenum is molded around the head of the pancreas. **C.** Posterior view of the duodenum and pancreas. Observe the concavity that accommodates the abdominal aorta and inferior vena cava (IVC), which have been removed. Observe also that the bile duct is descending in a fissure (opened up) in the posterior part of the head of the pancreas.

the duodenum (ligament of Treitz). This muscle is composed of a slip of skeletal muscle from the diaphragm and a fibromuscular band of smooth muscle from the third and fourth parts of the duodenum. Contraction of this muscle widens the angle of the flexure, facilitating movement of the intestinal contents. The suspensory muscle passes posterior to the pancreas and splenic vein and anterior to the left renal vein.

Paraduodenal Hernias

Two or three inconstant folds and fossae (recesses) are around the duodenojejunal junction. The *paraduodenal fold and fossa* are large and lie to the left of the ascending (fourth) part of the duodenum. If a loop of intestine enters this fossa, it may strangulate. During repair of a *paraduodenal hernia*, care must be taken not to injure the branches of the inferior mesenteric artery and vein, or the ascending branches of the left colic artery, that are related to the paraduodenal fold and fossa. ⊙

The **duodenal arteries** arise from the celiac trunk and the SMA (Fig. 2.36). The celiac trunk, via the **gastroduodenal artery** and its branch—the **superior pancreaticoduodenal artery**—supplies the duodenum proximal to the entry of the bile duct into the descending (second) part of the duodenum. The SMA, through its branch—the **inferior pancreaticoduodenal artery**—supplies the duodenum distal to the entry of the bile duct. The pancreaticoduodenal arteries lie in the curve between the duodenum and the head of the pancreas and supply both structures. The anastomosis of the superior and inferior pancreaticoduodenal arteries, which occurs approximately at the level of entry of the bile duct—or, according to some authors, at the junction of the descending and horizontal parts of the duodenum—is an anastomosis between the celiac and superior mesenteric arteries. *An important transition in the blood supply of the digestive tract occurs*:

- Proximally (extending from the abdominal part of the esophagus), the blood is supplied by the celiac trunk
- Distally (extending to the left colic flexure), the blood is supplied by the SMA.

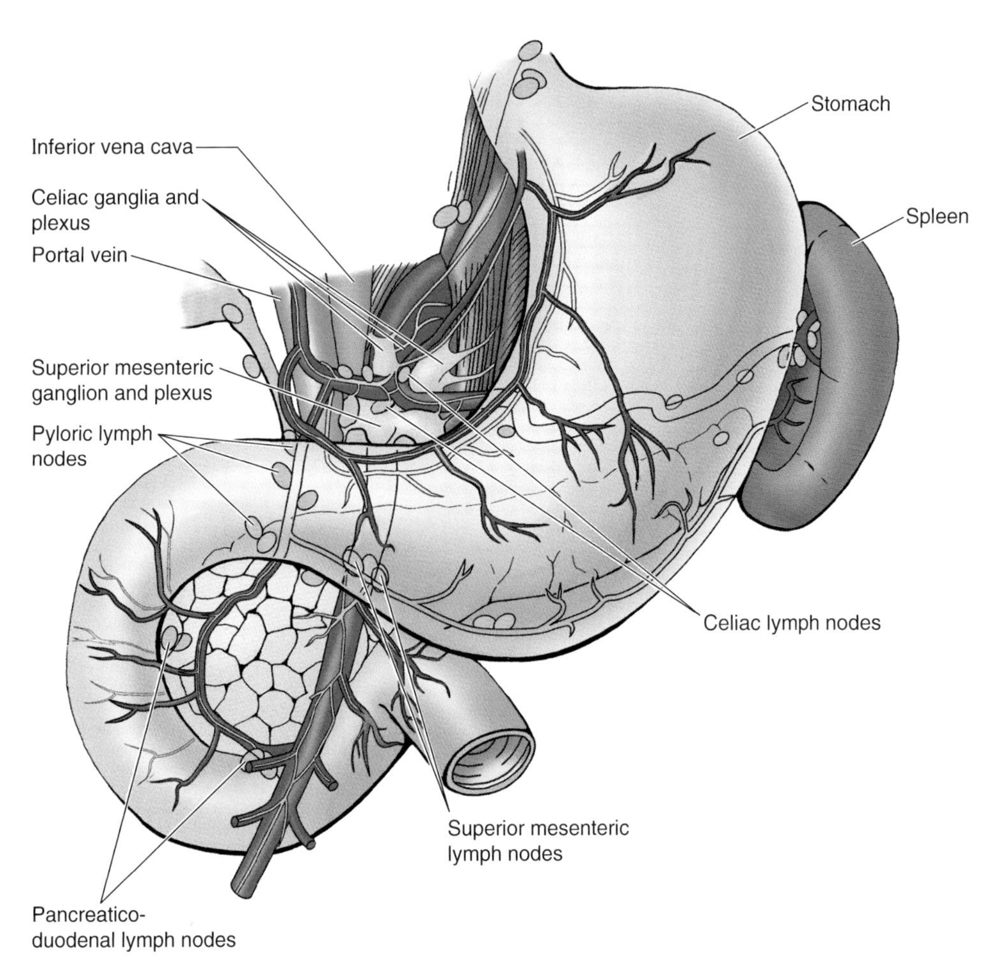

Figure 2.37. Lymphatic drainage of the duodenum, pancreas, and spleen. The close positional relationship of these organs results in their blood vessels and lymphatic vessels being the same in whole or in part. The lymph nodes closely associated with the duodenum and pancreas are the pancreaticoduodenal, pyloric, and superior mesenteric lymph nodes. The lymph from the spleen drains to the pancreaticosplenic and superior mesenteric lymph nodes.

The basis of this transition in blood supply is embryological (Moore and Persaud, 1998).

The **duodenal veins** follow the arteries and drain into the **portal vein**—some directly and others indirectly—through the superior mesenteric and splenic veins.

The **lymphatic vessels of the duodenum** follow the arteries. The *anterior lymphatic vessels of the duodenum* drain into the **pancreaticoduodenal lymph nodes** located along the superior and inferior pancreaticoduodenal arteries, and into the **pyloric lymph nodes** that lie along the gastroduodenal artery (Fig. 2.37). The *posterior lymphatic vessels* pass posterior to the head of the pancreas and drain into the **superior mesenteric lymph nodes**. Efferent lymphatic vessels from the duodenal lymph nodes drain into the **celiac lymph nodes**.

The *nerves of the duodenum* (Fig. 2.37) derive from the **vagus** and **sympathetic nerves** through the celiac and superior mesenteric plexuses on the pancreaticoduodenal arteries.

Duodenal Ulcers

Most (95%) inflammatory erosions of the duodenal wall—*duodenal ulcers*—are in the posterior wall of the superior (first) part of the duodenum. Occasionally an ulcer perforates the duodenal wall, permitting the contents to enter the peritoneal cavity and produce *peritonitis*. Because the superior part of the duodenum closely relates to the liver and gallbladder, either of them may adhere to and be ulcerated by a duodenal ulcer. *Erosion of the gastroduodenal artery* by a duodenal ulcer, a posterior relation of the superior part of the duodenum, results in severe hemorrhage into the peritoneal cavity and peritonitis.

Gallstones in the Duodenum

The proximity of the superior part of the duodenum to the gallbladder explains how a gallstone may ulcerate from the eroded fundus of the gallbladder and enter a perforated duodenum. Because of the intimate relationship of the pancreas to the duodenum, this gland may also be invaded by a posterior duodenal ulcer.

Developmental Changes in the Mesoduodenum

During the early fetal period, the entire duodenum has a mesentery (Moore and Persaud, 1998); however, most of it fuses with the posterior abdominal wall because of pressure from the overlying transverse colon. Because the attachment of the mesoduodenum to the wall is secondary (has occurred through formation of a *fusion fascia*—see p. 211), the duodenum and the closely associated pancreas can be separated (surgically mobilized) from the underlying retroperitoneal viscera during surgical operations involving the duodenum, without endangering the blood supply to the kidney or the ureter. ⭘

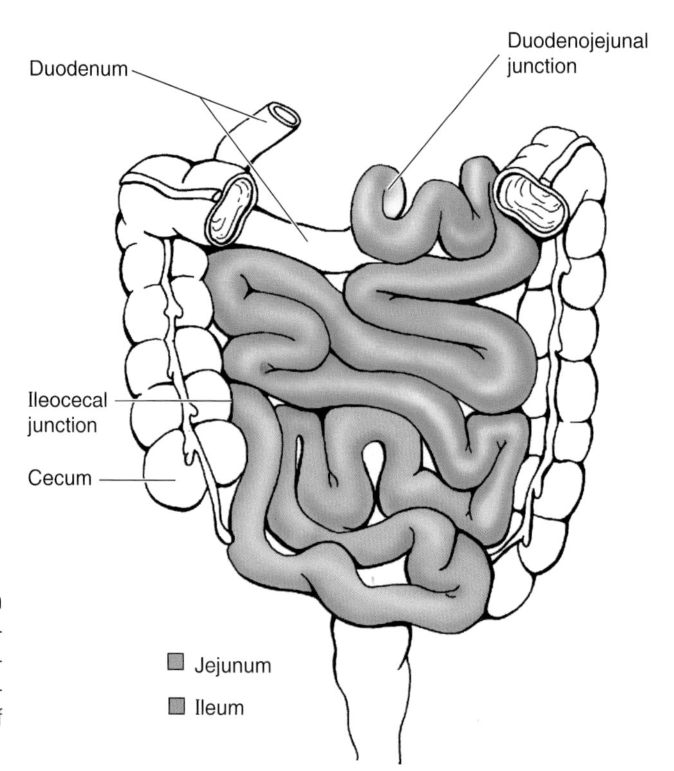

Figure 2.38. Jejunum and ileum. The orientation drawing shows that the jejunum begins at the duodenojejunal flexure and the ileum ends at the cecum. The combined term "jejunoileum" is sometimes used as an expression of the fact that there is no clear external line of demarcation between the jejunum and ileum.

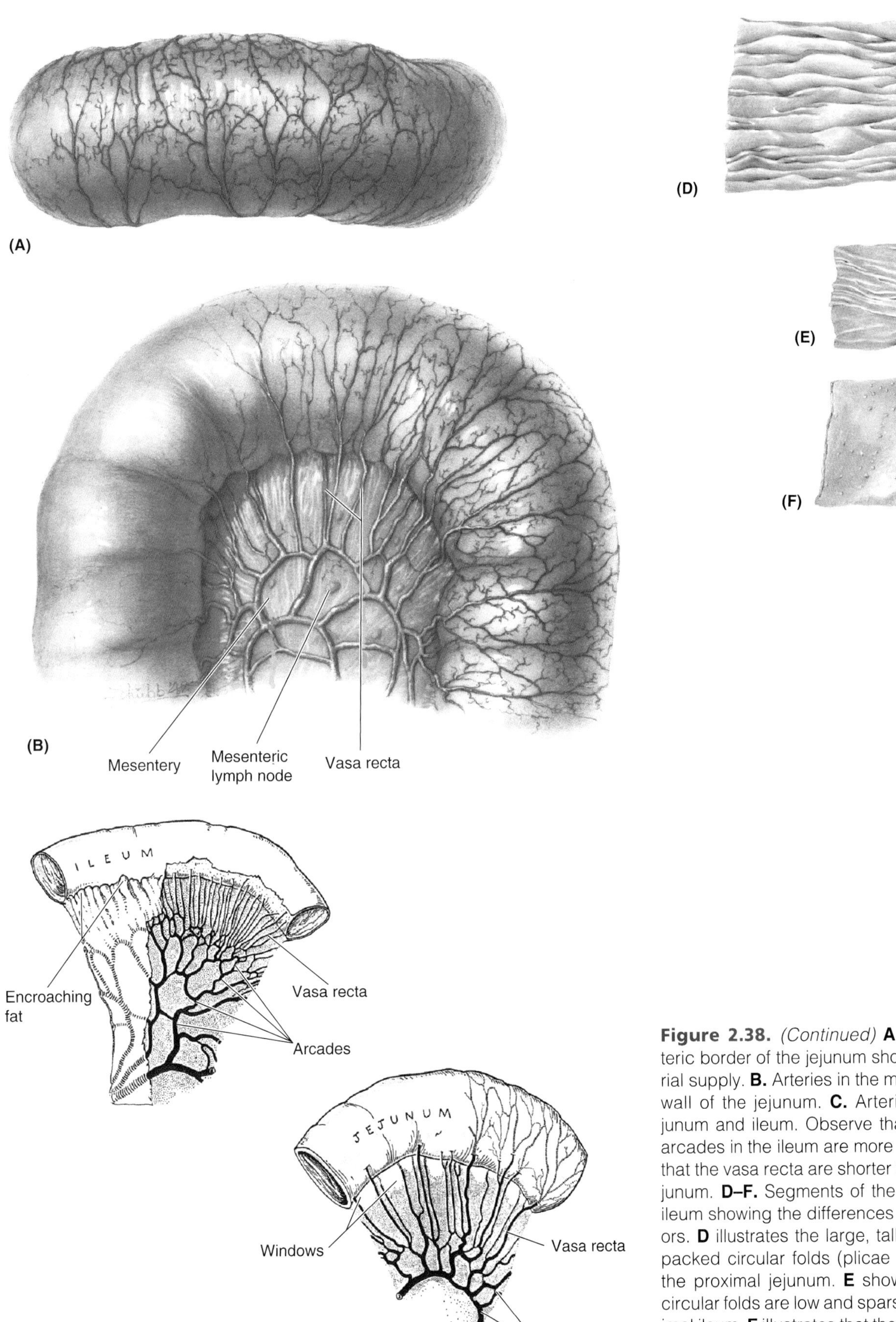

Figure 2.38. *(Continued)* **A.** Antimesenteric border of the jejunum showing its arterial supply. **B.** Arteries in the mesentery and wall of the jejunum. **C.** Arteries of the jejunum and ileum. Observe that the arterial arcades in the ileum are more complex and that the vasa recta are shorter than in the jejunum. **D–F.** Segments of the jejunum and ileum showing the differences in their interiors. **D** illustrates the large, tall, and closely packed circular folds (plicae circulares) in the proximal jejunum. **E** shows that these circular folds are low and sparse in the proximal ileum. **F** illustrates that the circular folds are absent in the terminal ileum and that lymph nodules are visible in its wall.

Table 2.8. **Distinguishing Characteristics of the Jejunum and Ileum in Living Persons**

Characteristic	Jejunum	Ileum
Color	Deeper red	Paler pink
Caliber	2–4 cm	2–3 cm
Wall	Thick and heavy	Thin and light
Vascularity	Greater	Less
Vasa recta	Long	Short
Arcades	A few large loops	Many short loops
Fat in mesentery	Less	More
Circular folds (plicae circulares)	Large, tall, and closely packed	Low and sparse; absent in distal part
Lymphoid nodules (Peyer's patches)	Few	Many

Jejunum and Ileum

The **jejunum** begins at the duodenojejunal flexure and the **ileum** ends at the *ileocecal junction*—the union of the terminal ileum and the cecum (Fig. 2.38). Together, the jejunum and ileum are 6 to 7 meters long, the jejunum constituting approximately two-fifths and the ileum approximately three-fifths. *Most of the jejunum lies in the left upper quadrant, whereas most of the ileum lies in the right lower quadrant.* The terminal ileum usually lies in the pelvis from which it ascends, ending in the medial aspect of the cecum. Although no clear line of demarcation between the jejunum and ileum exists, they have distinctive characteristics that are surgically important (Fig. 2.38, Table 2.8).

The **mesentery**—a fan-shaped fold of peritoneum—attaches the jejunum and ileum to the posterior abdominal wall (Fig. 2.39*A*). The **root (origin) of the mesentery** (approximately 15 cm long) is directed obliquely, inferiorly, and to the right. It extends from the duodenojejunal junction on the left side of the L2 vertebra to the ileocolic junction and the right sacroiliac joint. The average breadth of the mesentery from its root to the intestinal border is 20 cm. *The root of the mesentery crosses (successively) the:*

- Ascending and horizontal parts of the duodenum
- Abdominal aorta
- IVC
- Right ureter
- Right psoas major
- Right testicular or ovarian vessels.

Between the two layers of the mesentery are the superior mesenteric vessels, lymph nodes, a variable amount of fat, and autonomic nerves.

The **SMA** supplies the jejunum and ileum (Fig. 2.39, *A–B*, Table 2.9). The SMA usually arises from the abdominal aorta at the level of the L1 vertebra, approximately 1 cm inferior to the celiac trunk, and runs between the layers of the mesentery, sending 15 to 18 branches to the jejunum and ileum. The arteries unite to form loops or arches—**arterial arcades**—that give rise to straight arteries—the **vasa recta**.

The **SMV** (Fig. 2.39*B*) drains the jejunum and ileum. It lies anterior and to the right of the SMA in the root of the mesentery. The SMV ends posterior to the neck of the pancreas where it unites with the splenic vein to form the portal vein (Fig. 2.36, *B–C*).

The specialized lymphatic vessels that absorb fat—**lacteals**—in the intestinal villi (projections of the mucous membrane 0.15–1.5 mm in length) empty their milklike fluid into the lymphatic plexuses in the walls of the jejunum and ileum. The lymphatic vessels pass between the layers of the mesentery; *the mesenteric lymph nodes* (Fig. 2.40) *are located:*

- Close to the intestinal wall
- Among the arterial arcades
- Along the proximal part of the SMA.

Efferent lymphatic vessels from the mesenteric lymph nodes drain to the **superior mesenteric lymph nodes.** Lymphatic vessels from the terminal ileum follow the ileal branch of the ileocolic artery to the **ileocolic lymph nodes.**

The SMA and its branches are surrounded by a *perivascular nerve plexus* through which the nerves are conducted to the parts of the intestine supplied by this artery. **The sympathetic fibers in the nerves to the jejunum and ileum originate in the T5 through T9 segments of the spinal cord**

Figure 2.39. Arterial supply and mesentery of the intestine. A. Arterial supply to the large intestine. The roots (cut) of the transverse and sigmoid mesocolons and mesentery of the jejunum and ileum are also illustrated. **B.** Arterial supply and venous drainage of the small intestine. Observe the arterial arcades and straight arteries (vasa recta). Observe also that the superior mesenteric artery (SMA) supplies the jejunoileum, and the superior mesenteric vein (SMV) draws blood from the intestine into the portal vein.

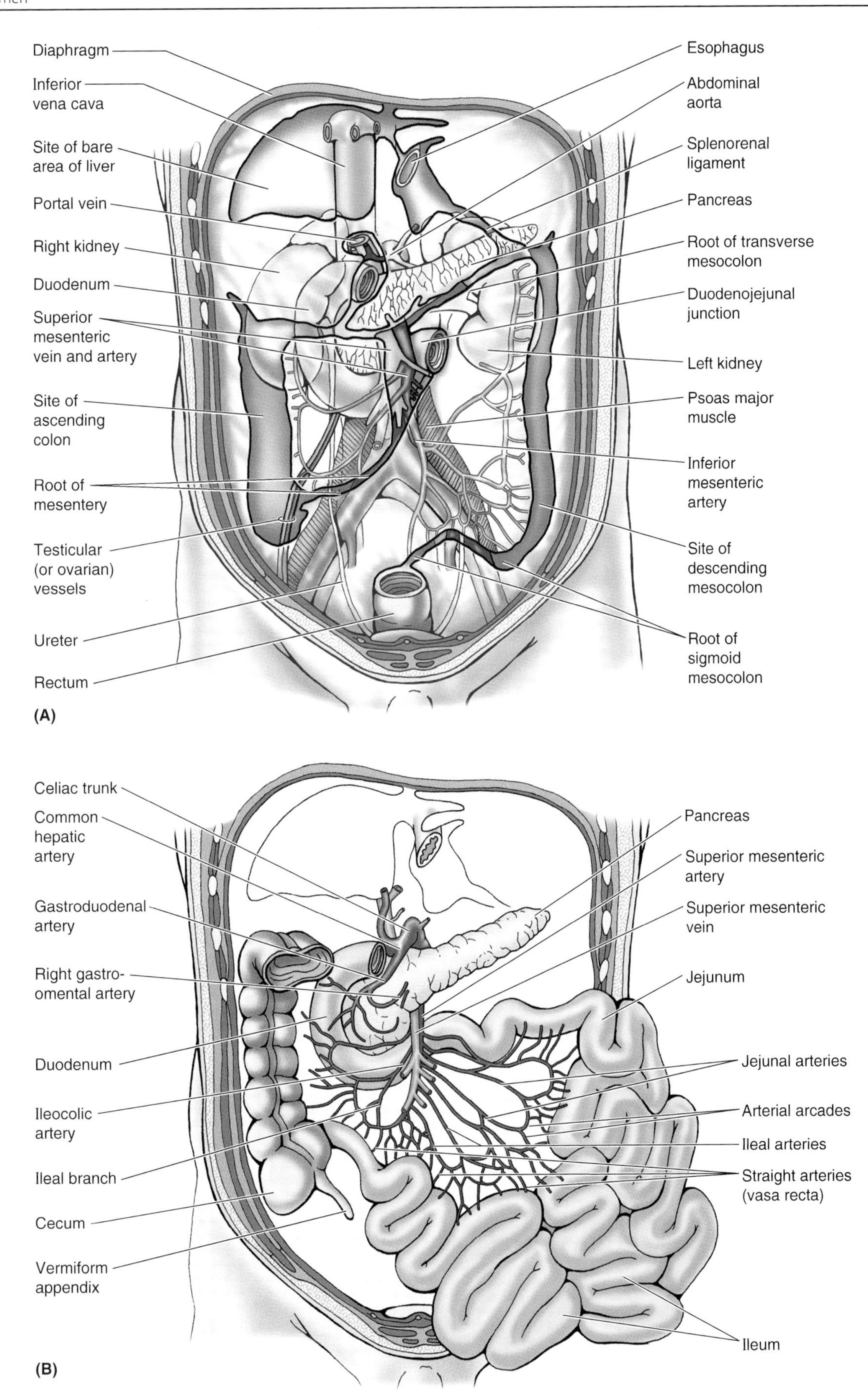
Diaphragm
Inferior vena cava
Site of bare area of liver
Portal vein
Right kidney
Duodenum
Superior mesenteric vein and artery
Site of ascending colon
Root of mesentery
Testicular (or ovarian) vessels
Ureter
Rectum
Esophagus
Abdominal aorta
Splenorenal ligament
Pancreas
Root of transverse mesocolon
Duodenojejunal junction
Left kidney
Psoas major muscle
Inferior mesenteric artery
Site of descending mesocolon
Root of sigmoid mesocolon
(A)
Celiac trunk
Common hepatic artery
Gastroduodenal artery
Right gastro-omental artery
Duodenum
Ileocolic artery
Ileal branch
Cecum
Vermiform appendix
Pancreas
Superior mesenteric artery
Superior mesenteric vein
Jejunum
Jejunal arteries
Arterial arcades
Ileal arteries
Straight arteries (vasa recta)
Ileum
(B)

Figure 2.40. Mesenteric lymph nodes. The superior mesenteric lymph nodes form a system in which the central nodes at the root of the superior mesenteric artery (SMA) receive lymph from the mesenteric, ileocolic, right colic, and middle colic lymph nodes. Observe that the more numerous nodes are located adjacent to the intestine and the less numerous nodes are located along the arteries. The efferent vessels of the celiac and superior mesenteric lymph nodes form the intestinal lymphatic trunk, which usually ends in the left lumbar trunk but ends in the chyle cistern (L. cisterna chyli) in 25% of people.

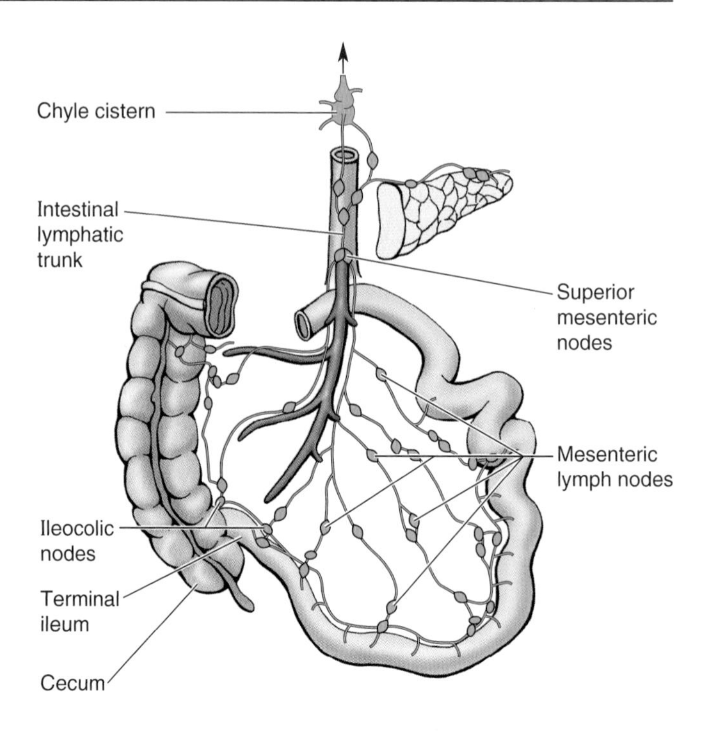

(Fig. 2.41) and reach the *celiac plexus* through the *sympathetic trunks* and *thoracic (greater and lesser) splanchnic nerves*. The presynaptic sympathetic fibers synapse on cell bodies of postsynaptic sympathetic neurons in the *celiac and superior mesenteric (prevertebral) ganglia*. **The parasympathetic fibers in the nerves to the jejunum and ileum derive from the posterior vagal trunks.** The presynaptic parasympathetic fibers synapse with postsynaptic parasympathetic neurons in the *myenteric and submucous plexuses* in the intestinal wall.

In general, sympathetic stimulation reduces motility of the intestine and secretion and acts as a vasoconstrictor, reducing or stopping digestion and making blood (and energy) available for "fleeing or fighting." Parasympathetic stimulation increases motility of the intestine and secretion, restoring digestive activity following a sympathetic reaction. The small intestine also has sensory (visceral afferent) fibers. The intestine is insensitive to most pain stimuli, including cutting and burning; however, it is sensitive to distension that is perceived as **colic** (spasmodic abdominal pains).

Brief Review of the Embryology of the Intestine

An understanding of the development of the intestine clarifies the adult arrangement of the intestines. The primordial gut comprises the foregut, midgut, and hindgut (Moore and Persaud, 1998). Pain arising from foregut derivatives—esophagus, stomach, pancreas, duodenum, liver, and biliary tree—localizes in the epigastric region. Pain arising from midgut derivatives—the small intestine distal to bile duct, cecum, appendix, ascending colon, and most of the transverse colon—localizes in the periumbilical region. Pain arising from hindgut derivatives—the distal part of the transverse colon, descending colon, sigmoid colon, and rectum—localizes in the hypogastric region.

For several weeks the rapidly growing midgut, supplied by the SMA, is physiologically herniated into the proximal part of the umbilical cord (*A*). It is attached to the yolk sac by the yolk stalk. As it returns to the abdominal cavity, the midgut rotates 270° around the axis of the SMA (*B–C*). As the relative size of the liver and kidneys decreases, the midgut returns to the abdominal cavity as increased space becomes available. As the parts of the intestine reach their definitive positions, their mesenteric attachments undergo modification (*D–E*). Some mesenteries shorten and others disappear (e.g., most of duodenal mesentery). Malrotation of the midgut results in several congenital anomalies such as volvulus (twisting) of the intestine. For further details, see Moore and Persaud (1998).

Ischemia of the Intestine

Occlusion of the vasa recta by an embolus (G. embolos, a plug) results in *ischemia*—blood supply deficiency—of the part of the intestine concerned. If the ischemia is severe, *necrosis* of the involved segment results and *ileus* (obstruction of the intestine) of the paralytic type occurs. *Ileus is accompanied by a severe colicky pain*, along with abdominal distension, vomiting, and often fever and dehydration. If the condition is diagnosed early (e.g., using a *superior mesenteric arteriogram)*, the obstructed part of the vessel may be cleared surgically. ▶

Rotation of the Midgut

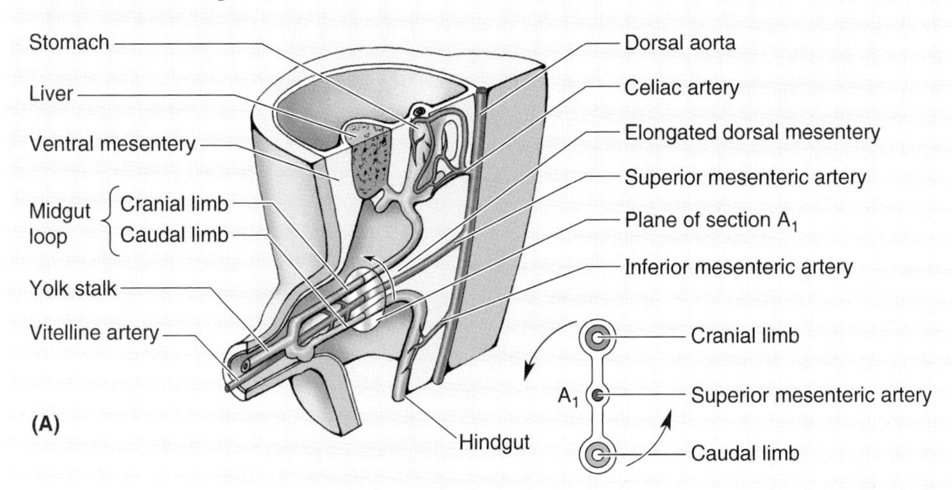

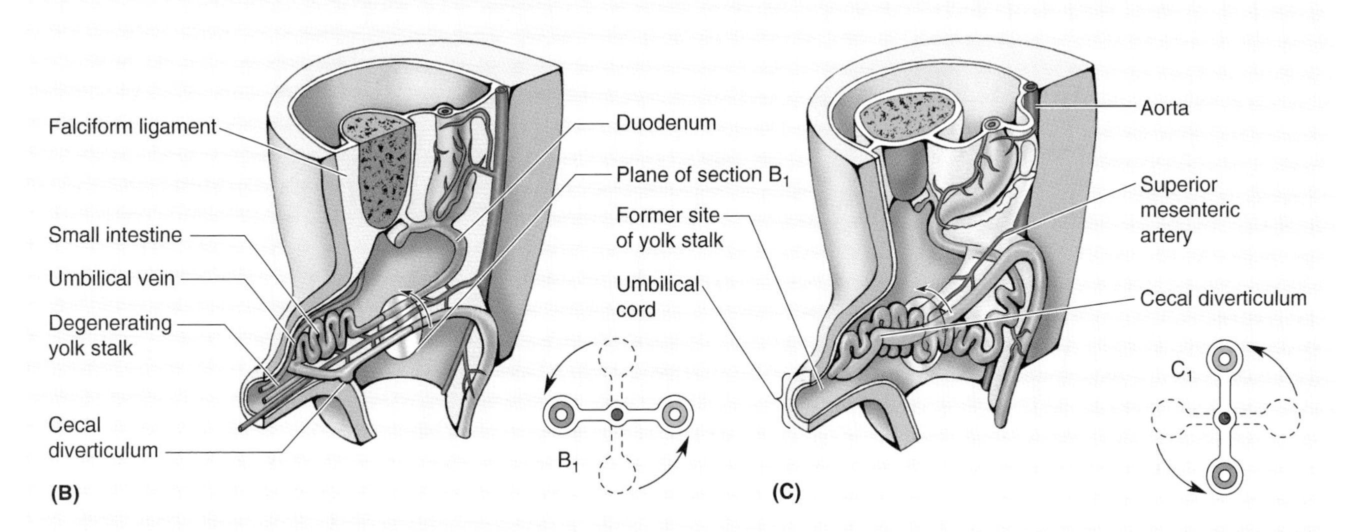

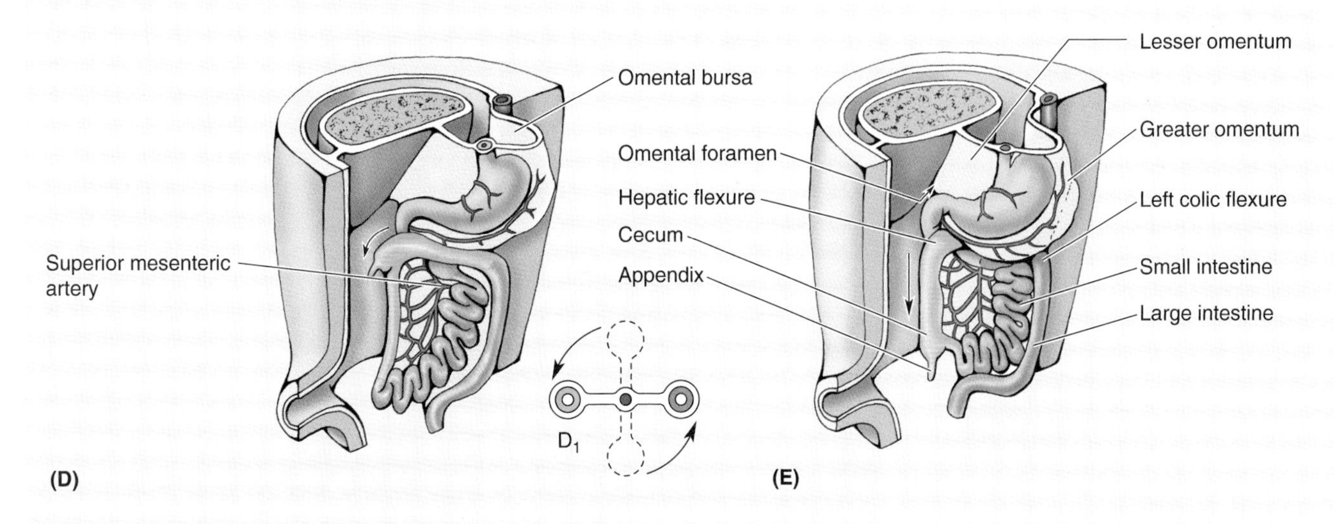

Ileal Diverticulum

An ileal diverticulum (of Meckel) is a congenital anomaly that occurs in 1 to 2% of people (see Fig. 2.43*B*). A remnant of the proximal part of the embryonic yolk stalk, the diverticulum usually appears as a fingerlike pouch (3–6 cm long). It is always on the antimesenteric border of the ileum—the site of attachment of the yolk stalk and the border of the intestine opposite the mesenteric attachment. The diverticulum is usually located approximately 40 cm from the ileocecal junction in infants and 50 cm in adults, and may be free (74%) or attached by a cord to the umbilicus (26%). An ileal diverticulum may become inflamed and produce pain mimicking the pain produced by appendicitis. For discussion and illustrations of the various types of ileal diverticulum, see Fig. 2.43*B*, and Moore et al. (1994) and Moore and Persaud (1998). ⊙

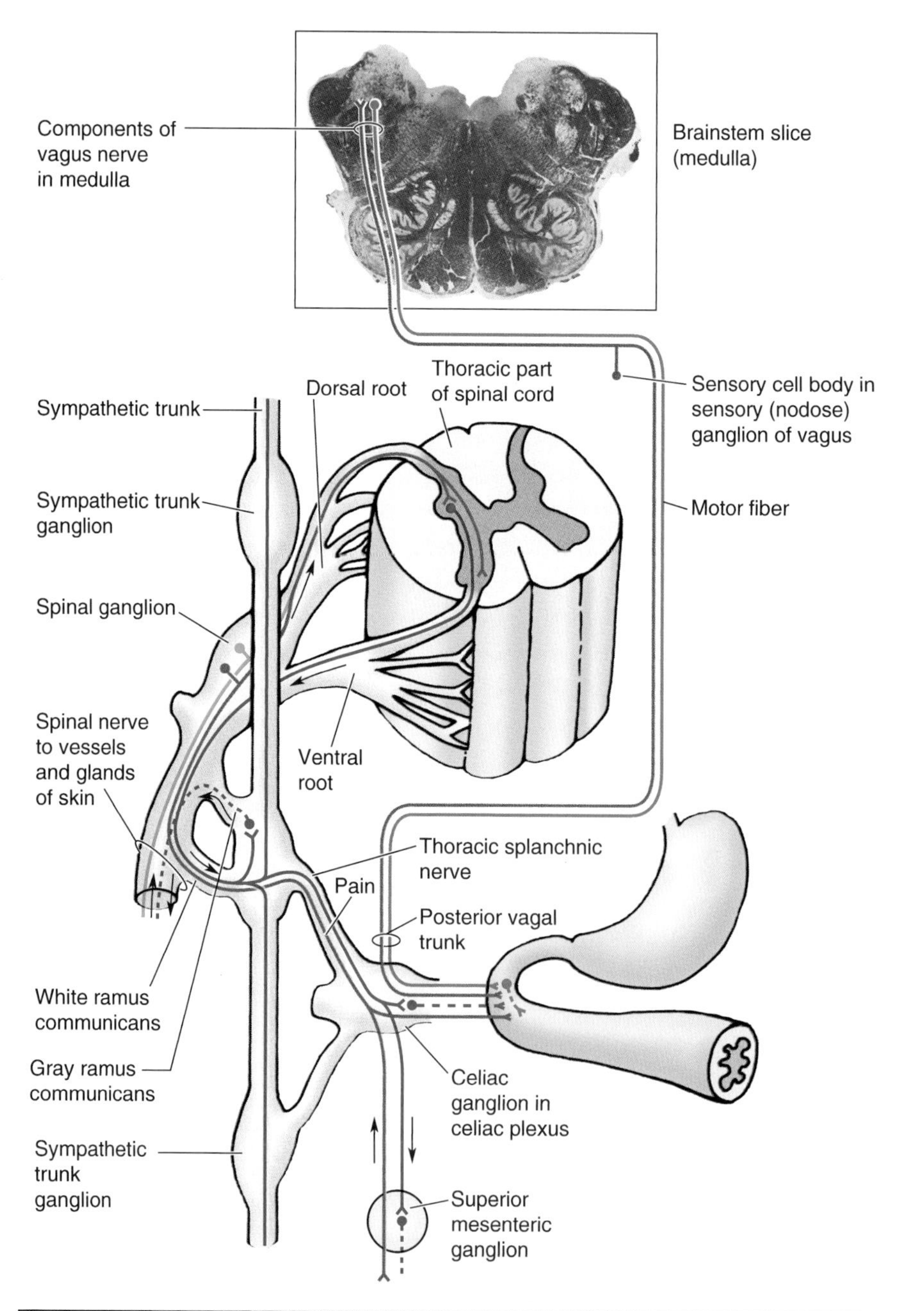

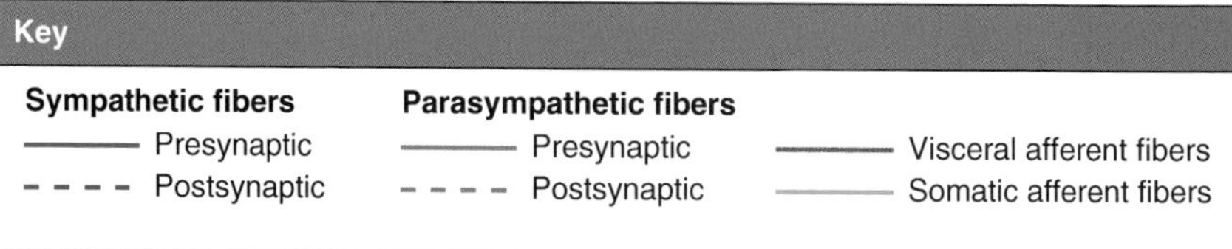

Figure 2.41. Innervation of the small intestine. The *sympathetic nerves* originate in T5 through T9 segments of the spinal cord and reach the celiac plexus through the sympathetic trunks and greater splanchnic nerves. The presynaptic sympathetic fibers synapse in the celiac and superior mesenteric ganglia. The postsynaptic nerve fibers accompany the arteries to the intestine. The afferent fibers contain pain fibers and fibers concerned with reflex regulation of movement and secretion. The intestine is insensitive to most painful stimuli, including cutting and burning; however, it is sensitive to distension, which is perceived as "cramps" or colic. The *parasympathetic (vagus) nerves* originate in the medulla (oblongata). The nerves to the intestine derive from the posterior vagal trunk. The presynaptic fibers synapse with intrinsic postsynaptic neurons in the intestinal wall.

Large Intestine

The large intestine consists of the **cecum;** the **appendix**; the ascending, transverse, descending, and sigmoid **colon;** the **rectum**; and the **anal canal** (Fig. 2.42). *The large intestine can be distinguished from the small intestine by:*

- **Teniae coli**—three thickened bands of muscle
- **Haustra**—sacculations of the colon between the teniae
- **Omental appendices**—small fatty projections of the omentum
- **Caliber**—the internal diameter is much larger.

The three teniae coli (bands of muscle fibers) comprise most of the longitudinal muscle of the large intestine, except in the rectum. Because the teniae are shorter than the intestine, the colon has the typical sacculated shape formed by the haustra. *There are no teniae in the appendix or rectum*; they be-

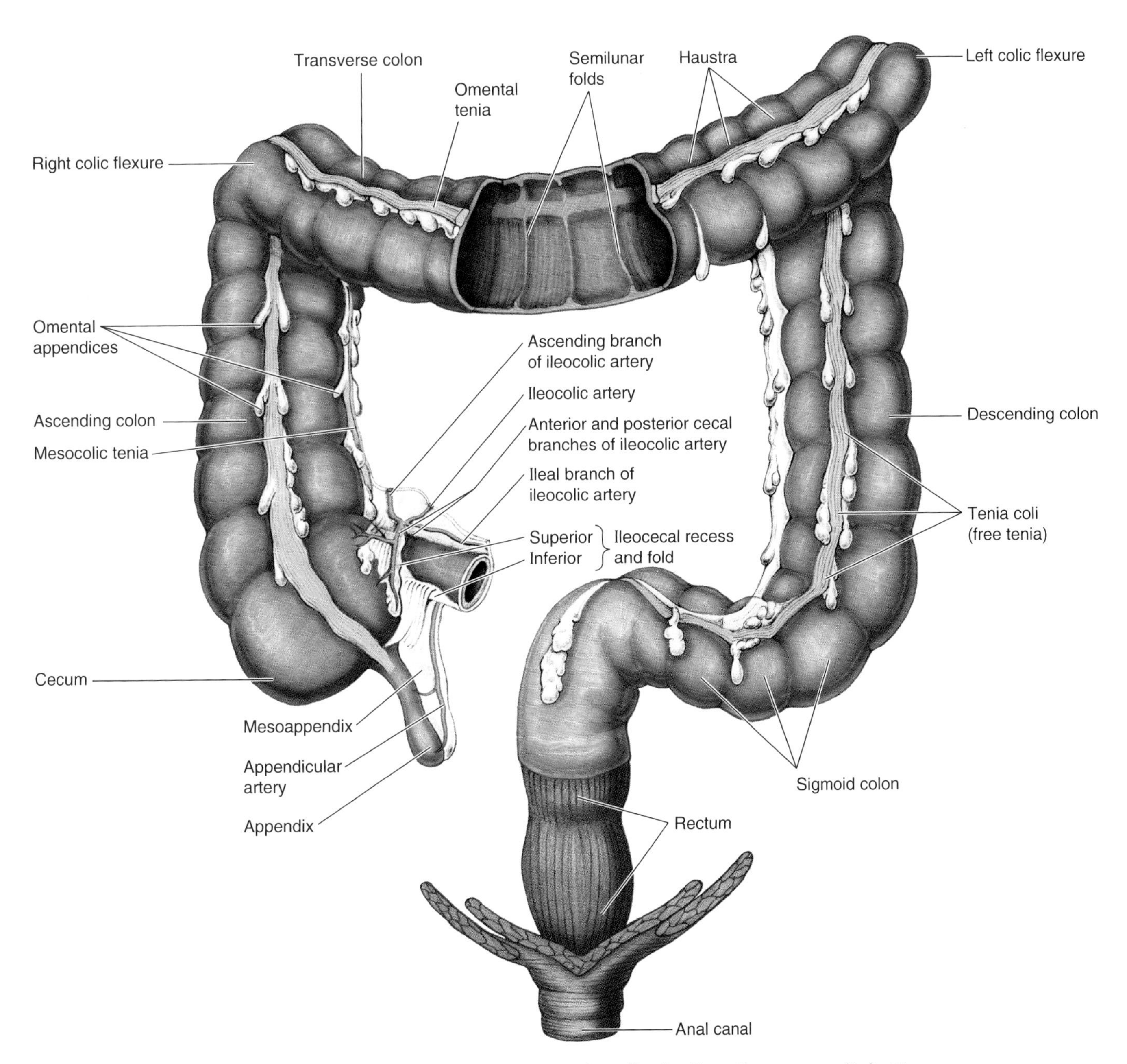

Figure 2.42. Terminal ileum and large intestine (including the appendix). The mucosa, musculature, haustra, and omental (epiploic) appendices are illustrated. Observe that there are no teniae coli (thickened bands of longitudinal muscle) or omental appendices in the rectum. The longitudinal muscle coat forms abruptly, increasingly of broad diffuse bands that become a continuous longitudinal layer of muscle in the rectum. The omental appendices are small masses of fat, enclosed in peritoneum, extending from the external surface of the colon.

gin at the base of the appendix and run through the large intestine to the rectosigmoid junction.

Cecum and Appendix

The **cecum**—the first part of the large intestine that is continuous with the ascending colon—is a *blind intestinal pouch* (approximately 7.5 cm in both length and breadth) in the right lower quadrant, where it lies in the iliac fossa inferior to the junction of the terminal ileum and cecum (Figs. 2.42 and 2.43, *A–D*). If distended with feces or gas, the cecum may be palpable through the anterolateral abdominal wall. The cecum, usually lying within 2.5 cm of the inguinal ligament, is almost entirely enveloped by peritoneum and can be lifted freely; however, *the cecum has no mesentery*. Because of its relative freedom, it may be displaced from the iliac fossa, but it is commonly bound to the lateral abdominal wall by one or more **cecal folds** of peritoneum (Fig. 2.43*C*). The terminal ileum enters the cecum obliquely and partly invaginates into it. Traditionally, based on cadaveric studies, this is said to produce folds superior and inferior to the **ileocecal orifice** that form the **ileocecal valve** (Fig. 2.43*A*). The folds meet laterally to form ridges—the frenula of the valve. When the cecum is distended (and thus, by

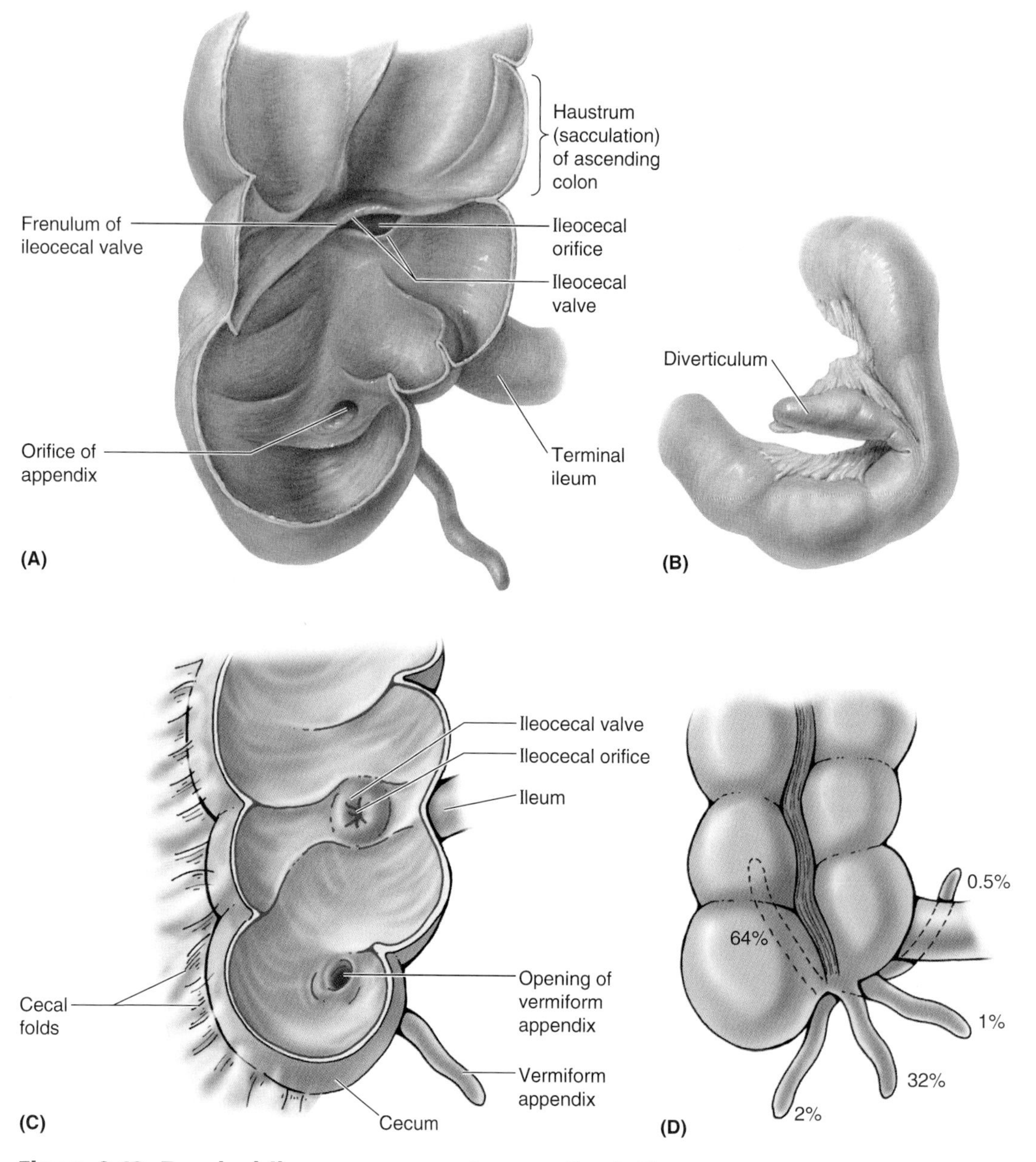

Figure 2.43. Terminal ileum, cecum, and appendix. A. The cecum was filled with air until dry, opened, and varnished. Observe the ileocecal valve and ileocecal orifice. The frenulum is a fold, more evident in cadavers, that runs from the ileocecal valve along the wall at the cecocolic junction. **B.** An ileal diverticulum (Meckel diverticulum) is projecting from the anti-mesenteric side of the intestine (but is shown lying across the mesenteric side in this twisted postion). **C.** Drawing of the interior of the cecum showing the endoscopic (living) appearance of the ileocecal valve. **D.** Diagram showing the approximate incidence of variouslocations of the appendix based on the analysis of 10,000 cases.

assumption, when it contracts), the **frenula** tighten, closing the valve to prevent reflux from the cecum into the ileum. However, direct observation by endoscopy in living persons does not support this description. The circular muscle is poorly developed around the orifice; therefore, the valve is unlikely to have any sphincteric action in terms of controlling passage of the intestinal contents from the ileum into the cecum. The opening is usually closed by tonic contraction, however, giving it a papillary appearance from the cecal side. The valve probably does prevent reflux from the cecum into the ileum as contractions occur to propel contents up the ascending colon and into the transverse colon (Magee and Dalley, 1986).

The vermiform (L. vermis, wormlike) **appendix**, a blind intestinal diverticulum (6–10 cm in length), arises from the posteromedial aspect of the cecum inferior to the ileocecal junction. The wormlike appendix has a short triangular mesentery, the **mesoappendix**, which derives from the posterior side of the mesentery of the terminal ileum (Fig. 2.42). The mesoappendix attaches to the cecum and the proximal part of the appendix. The position of the appendix is variable, but it is usually retrocecal (Fig. 2.43*D*).

The cecum is supplied by the **ileocolic artery**, the terminal branch of the SMA (Fig. 2.44*A*, Table 2.9). The appendix is supplied by the **appendicular artery**, a branch of the ileocolic artery. A tributary of the SMV, the **ileocolic vein**, drains blood from the cecum and appendix (Fig. 2.44*B*).

The *lymphatic vessels* from the cecum and appendix pass to lymph nodes in the mesoappendix and to the **ileocolic lymph nodes** that lie along the **ileocolic artery** (Fig. 2.45*A*). Efferent lymphatic vessels pass to the **superior mesenteric lymph nodes.**

The *nerve supply to the cecum and appendix* derives from the sympathetic and parasympathetic nerves from the **superior mesenteric plexus** (Fig. 2.45*B*). The **sympathetic nerve fibers** originate in the lower thoracic part of the spinal cord, and the **parasympathetic nerve fibers** derive from the vagus nerves. Afferent nerve fibers from the appendix accompany the sympathetic nerves to the T10 segment of the spinal cord.

Position of the Appendix

A *retrocecal appendix* extends superiorly toward the right colic flexure and is usually free; however, it occasionally lies beneath the peritoneal covering of the cecum where it is often fused to the cecum or the posterior abdominal wall. The appendix may project inferiorly toward or across the pelvic brim. The anatomical position of the appendix determines the symptoms and the site of muscular spasm and tenderness when the appendix is inflamed. The base of the appendix lies deep to a point that is one-third of the way along the oblique line, joining the right anterior superior iliac spine to the umbilicus (spinoumbilical or *McBurney's point*). ⊙

Table 2.9. Arterial Supply to Intestines

Artery	Origin	Course	Distribution
Superior mesenteric	Abdominal aorta	Runs in root of mesentery to ileocecal junction	Part of gastrointestinal tract derived from midgut
Intestinal (n = 15–18)	Superior mesenteric artery	Passes between the two layers of mesentery	Jejunum and ileum
Middle colic	Superior mesenteric artery	Ascends retroperitoneally and passes between layers of transverse mesocolon	Transverse colon
Right colic	Superior mesenteric artery	Passes retroperitoneally to reach ascending colon	Ascending colon
Ileocolic	Terminal branch of superior mesenteric artery	Runs along root of mesentery and divides into ileal and colic branches	Ileum, cecum, and ascending colon
Appendicular	Ileocolic artery	Passes between layers of mesoappendix	Vermiform appendix
Inferior mesenteric	Abdominal aorta	Descends retroperitoneally to left of abdominal aorta	Supplies part of gastrointestinal tract derived from hindgut
Left colic	Inferior mesenteric artery	Passes retroperitoneally toward left to descending colon	Descending colon
Sigmoid (n = 3–4)	Inferior mesenteric artery	Passes retroperitoneally toward left to descending colon	Descending and sigmoid colon
Superior rectal	Terminal branch of inferior mesenteric artery	Descends retroperitoneally to rectum	Proximal part of rectum
Middle rectal	Internal iliac artery	Passes retroperitoneally to rectum	Midpart of rectum
Inferior rectal	Internal pudendal artery	Crosses ischioanal fossa to reach rectum	Distal part of rectum and anal canal

Figure 2.44. Blood vessels of the large intestine and appendix. A. *Arteries.* The arteries of the cecum, appendix, ascending colon, and the greater part of the transverse colon are supplied by the superior mesenteric artery (SMA). The inferior mesenteric artery (IMA) supplies the distal part of the transverse colon, the descending colon, and the rectum. **B.** *Veins.* The venous drainage by the superior and inferior mesenteric veins (IMVs) corresponds to the pattern of the arteries they accompany. The IMV is most commonly a tributary of the splenic vein, which then merges with the superior mesenteric vein (SMV) to form the portal vein.

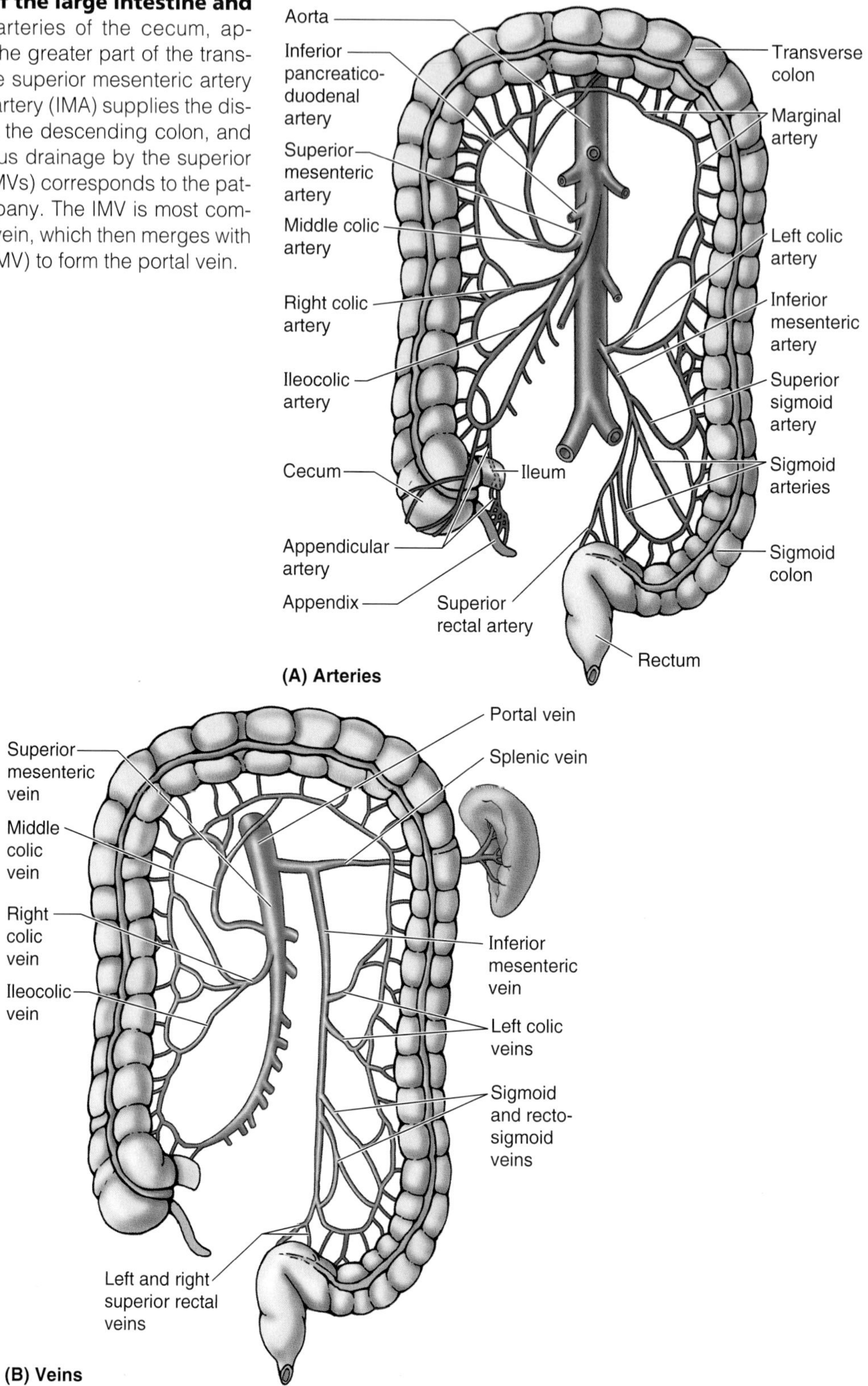

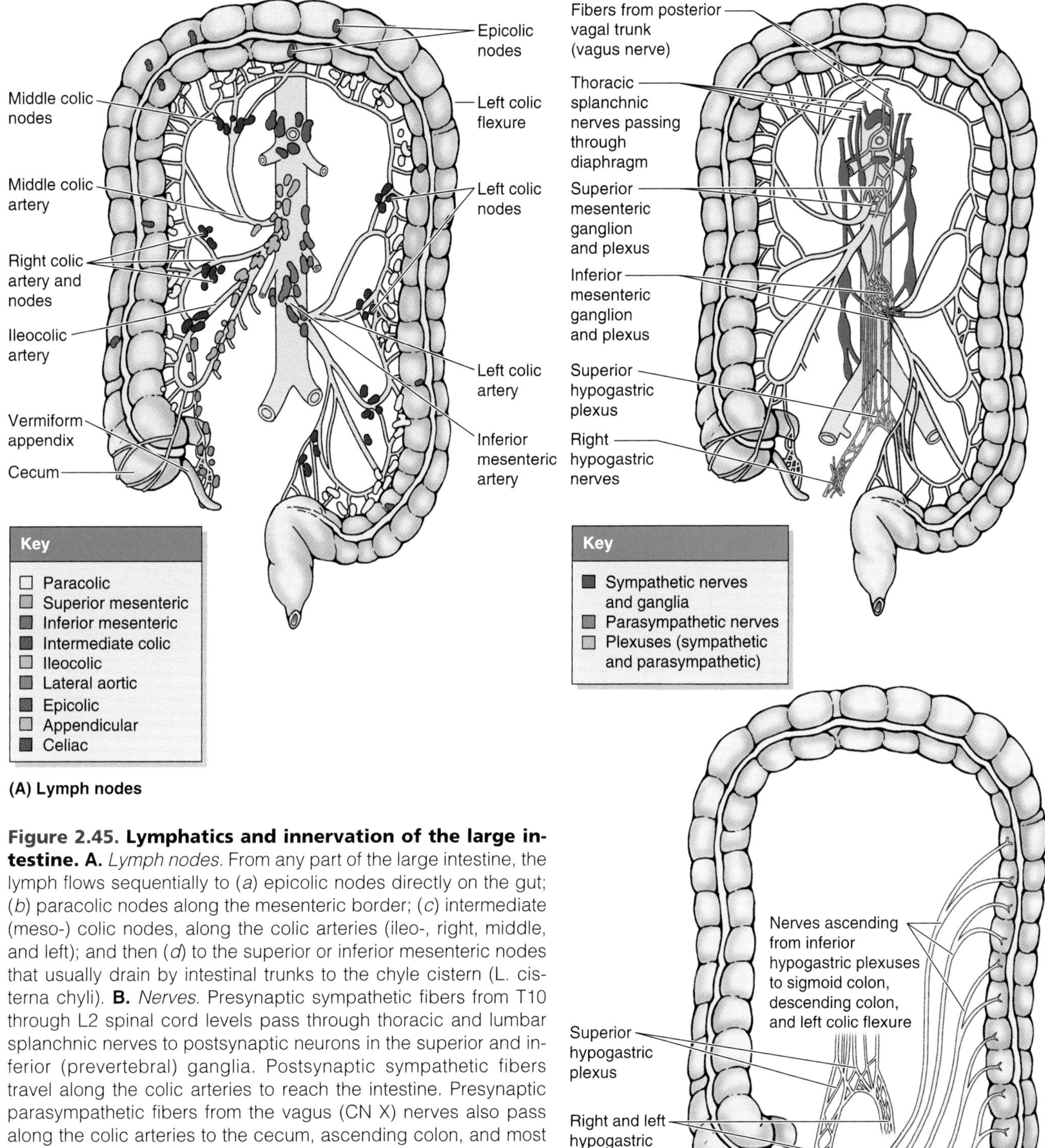

Figure 2.45. Lymphatics and innervation of the large intestine. A. *Lymph nodes.* From any part of the large intestine, the lymph flows sequentially to (*a*) epicolic nodes directly on the gut; (*b*) paracolic nodes along the mesenteric border; (*c*) intermediate (meso-) colic nodes, along the colic arteries (ileo-, right, middle, and left); and then (*d*) to the superior or inferior mesenteric nodes that usually drain by intestinal trunks to the chyle cistern (L. cisterna chyli). **B.** *Nerves.* Presynaptic sympathetic fibers from T10 through L2 spinal cord levels pass through thoracic and lumbar splanchnic nerves to postsynaptic neurons in the superior and inferior (prevertebral) ganglia. Postsynaptic sympathetic fibers travel along the colic arteries to reach the intestine. Presynaptic parasympathetic fibers from the vagus (CN X) nerves also pass along the colic arteries to the cecum, ascending colon, and most of the transverse colon. Presynaptic parasympathetic fibers from S2 through S4 spinal cord levels pass through pelvic splanchnic nerves to the inferior hypogastric (pelvic) plexuses and continue by way of nerves ascending from *both right and left plexuses* (only the right one is shown) to reach the sigmoid colon, descending colon, and most of the distal transverse colon. Pain fibers from most of the large intestine follow the sympathetic fibers retrograde to sensory ganglia of spinal nerves T10 through L2; pain fibers from the distal sigmoid colon and rectum follow the parasympathetic fibers retrograde to sensory ganglia of spinal nerves S2 through S4.

Appendicitis

Acute inflammation of the appendix is a common cause of an *acute abdomen*—severe abdominal pain arising suddenly. Digital pressure over the spinoumbilical (McBurney) point registers maximum abdominal tenderness. Appendicitis in young people is usually caused by hyperplasia of lymphatic follicles in the appendix that occludes the lumen. In older people, the obstruction usually results from a *fecalith* (coprolith), a concretion that forms around a center of fecal matter. When secretions from the appendix cannot escape, the appendix swells, stretching the visceral peritoneum. The pain of appendicitis usually commences as a vague pain in the periumbilical region because afferent pain fibers enter the spinal cord at the T10 level. Later, severe pain in the right lower quadrant results from irritation of the parietal peritoneum lining the posterior abdominal wall. Extending the thigh at the hip joint elicits pain.

Acute infection of the appendix may result in *thrombosis* (clotting of blood) in the appendicular artery, which often results in ischemia, gangrene (death of tissue), and perforation of an acutely inflamed appendix. **Rupture of the appendix** results in infection of the peritoneum (*peritonitis*), increased abdominal pain, nausea and/or vomiting, and *abdominal rigidity* (stiffness of the abdominal muscles). Flexion of the right thigh ameliorates the pain because it causes relaxation of the right psoas muscle, a flexor of the thigh at the hip joint.

Appendectomy

Removal of the appendix is usually performed through a gridiron (muscle-splitting) incision centered at McBurney's point in the right lower quadrant. The incision is usually over this point at a right angle to a line between the anterior superior iliac spine of the ilium and the umbilicus (p. 190). The choice of incision site is at the surgeon's discretion. The following description outlines the clinical anatomy of appendectomy—not the technique. If the appendix is not obvious, one of the teniae coli is traced to its base. The appendix arises from the convergence of the three teniae coli. Following incision of the skin, the external oblique aponeurosis is incised along the lines of its fibers. An opening is then made in the same way in the internal oblique and transverse abdominal muscles and the peritoneum. The cecum is delivered into the surgical wound and the mesoappendix containing the appendicular vessels is firmly ligated and divided. The base of the appendix is tied, the appendix is excised, and its stump is usually cauterized and invaginated into the cecum. The incision is then closed in layers. For details of an appendectomy, refer to a surgical text such as Sabiston and Leyerly (1994) or Skandalakis et al. (1995).

In unusual cases of *malrotation of the intestine,* or failure of descent of the cecum, the appendix is not in the lower right quadrant (Moore and Persaud, 1998). When the cecum is high (*subhepatic cecum),* the appendix is in the right hypochondriac region and the pain localizes there, not in the lower right quadrant.

Laparoscopy

When the diagnosis is unclear, examination of the abdominal contents with a *laparoscope* passed through a small incision in the anterolateral abdominal wall is especially useful in differentiating acute appendicitis from other causes of abdominal pain, including inflammatory pelvic disease (Sabiston and Lyerly, 1994). Laparoscopy has been used for many years by gynecologists in evaluating women with acute lower abdominal pain (Soper, 1993). In addition, laparoscopy is used for removing the gallbladder and appendix and for treating abdominal obstruction. ○

Colon

The colon is described in four parts—ascending, transverse, descending, and sigmoid—that succeed one another in an arch (Fig. 2.42). The colon lies first to the right of the small intestine, then successively superior and anterior to it, to the left of it, and eventually inferior to it.

The **ascending colon**—the second part of the large intestine—passes superiorly on the right side of the abdominal cavity from the cecum to the right lobe of the liver (Fig. 2.42), where it turns to the left at the **right colic flexure** (hepatic flexure). The ascending colon is narrower than the cecum and *lies retroperitoneally* along the right side of the posterior abdominal wall. The ascending colon is covered by peritoneum anteriorly and on its sides; however, in approximately 25% of people it has a short mesentery. The ascending colon is separated from the anterolateral abdominal wall by the greater omentum.

The **arterial supply** to the ascending colon and right colic flexure is from branches of the SMA—the **ileocolic** and **right colic arteries** (Fig. 2.44*A*, Table 2.9). The right branch of the middle colic artery usually anastomoses with the right colic artery. Tributaries of the SMV, the **ileocolic** and **right colic veins,** drain blood from the ascending colon (Fig. 2.44*B*). The lymphatic vessels pass first to the **epicolic** and **paracolic lymph nodes,** next to the **ileocolic** and intermediate **right colic lymph nodes,** and from them to the **superior mesenteric lymph nodes** (Fig. 2.45*A*). The nerves to the ascending colon derive from the **superior mesenteric nerve plexus** (Fig. 2.45*B*).

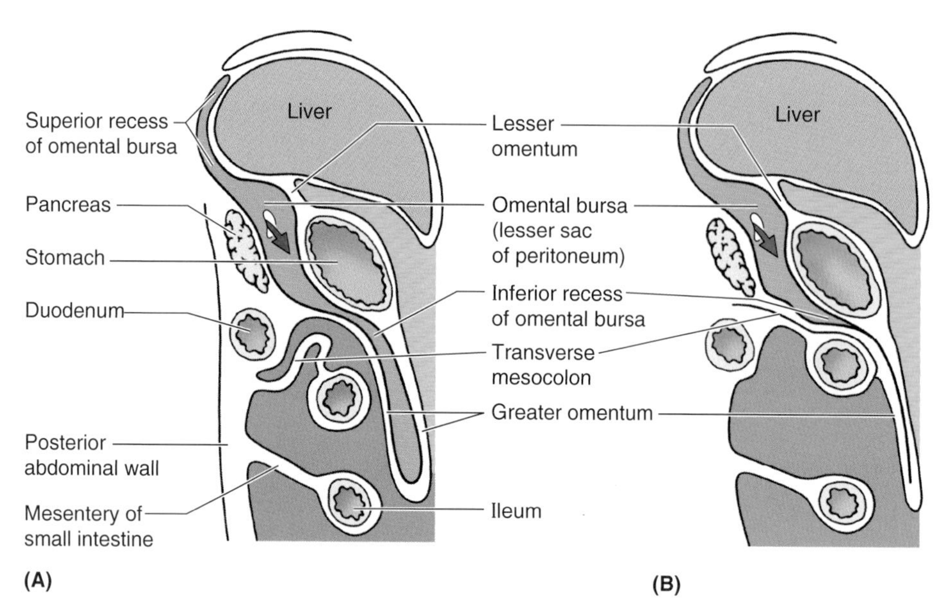

Figure 2.25. Walls and recesses of the omental (epiploic) bursa. Sagittal sections. **A.** Section of the abdomen of an infant showing that the omental bursa (lesser sac) is an isolated part of the peritoneal cavity, lying dorsal to the stomach and extending superiorly to the liver and diaphragm (superior recess of the omental bursa) and inferiorly between the layers of the greater omentum (inferior recess of the omental bursa). **B.** Section of the abdomen of an adult after fusion of the layers of the greater omentum. Observe that the inferior recess of the omental bursa now extends inferiorly only as far as the transverse colon because of the fused layers of the greater omentum. The *red arrows* in (**A**) and (**B**) pass from the greater sac through the omental foramen into the omental bursa (lesser sac).

Abdominal Viscera

The principal viscera of the abdomen are the terminal part of the esophagus, and the stomach, intestines, spleen, pancreas, liver, gallbladder, kidneys, and suprarenal (adrenal) glands (Figs. 2.26 and 2.27). Before studying these organs and their relationship to the omenta, observe that the liver, stomach, and spleen almost fill the domes of the diaphragm, and by thus indenting the thoracic cavity receive protection from the lower thoracic cage. Also observe that the **falciform ligament** normally attaches along a continuous line to the anterior abdominal wall as far inferiorly as the umbilicus and divides the liver into right and left parts. The fat-laden **greater omentum**, when in its "typical" position, conceals almost all the intestine, and the gallbladder projects inferior to the sharp border of the liver (Fig. 2.27*A*).

Food passes from the mouth and pharynx through the **esophagus** to the **stomach**, where it mixes with gastric secretions. *Digestion* mostly occurs in the stomach and duodenum. **Peristalsis**—ringlike contraction waves that begin around the middle of the stomach and move slowly toward the **pylorus**—is responsible for mixing the masticated (chewed) food mass with gastric juices and for emptying the contents of the stomach into the duodenum.

Absorption of chemical compounds occurs principally in the small intestine, a coiled 5- to 6-meter long tube consisting of the **duodenum, jejunum**, and **ileum**. Peristalsis also occurs in the jejunum and ileum; however, it is not forceful unless an obstruction is present. The stomach is continuous with the duodenum, which receives the openings of the ducts from the **pancreas** and **liver** (major glands of the digestive tract).

The large intestine consists of the **cecum**, which receives the terminal part of the ileum; vermiform **appendix**; **colon** (ascending, transverse, descending, and sigmoid); **rectum**; and **anal canal**, which ends at the **anus**. Most reabsorption of water occurs in the ascending colon. Feces (excrement) form in the descending and sigmoid colon and accumulate in the rectum before defecation. The esophagus, stomach, and intestine constitute the digestive tract and are derived from the primordial foregut, midgut, and hindgut.

The *arterial supply to the digestive tract* (Fig. 2.28*A*) is from

Mobile Ascending Colon

When the inferior part of the ascending colon has a mesentery, the cecum and proximal part of the colon are abnormally mobile. This condition, present in approximately 11% of individuals, may cause *volvulus* (L. volvo, to roll) *of the colon*—obstruction of intestine resulting from twisting. *Cecopexy* may avoid volvulus and possible obstruction of the colon. During this anchoring procedure, a tenia coli of the cecum and proximal ascending colon is sutured to the abdominal wall. ⊙

The **transverse colon** (approximately 45 cm long) is the largest and most mobile part of the large intestine (Fig. 2.42). It crosses the abdomen from the *right colic flexure* to the *left colic flexure*, where it bends inferiorly to become the descending colon. The **left colic flexure** (splenic flexure)—usually more superior, more acute, and less mobile than the right colic flexure—lies anterior to the inferior part of the left kidney and attaches to the diaphragm through the *phrenicocolic ligament* (Fig. 2.22). The mesentery of the transverse colon—the **transverse mesocolon**—loops down, often inferior to the level of the iliac crests, and is adherent to or fused with the posterior wall of the omental bursa. The **root of the transverse mesocolon** (Fig. 2.39*A*) lies along the inferior border of the pancreas and is continuous with the parietal peritoneum posteriorly. Being freely movable, the transverse colon is variable in position, usually hanging to the level of the umbilicus. However, in tall thin people, the transverse colon may dip into the pelvis.

The arterial supply of the transverse colon is mainly from the **middle colic artery** (Fig. 2.44*A*, Table 2.9), a branch of the SMA; however, it also is supplied by the **right** and **left colic arteries**. *Venous drainage of the transverse colon* is through the **SMV**. The *lymphatic drainage of the transverse colon* is to the **middle colic lymph nodes**, which in turn drain to the **superior mesenteric lymph nodes** (Fig. 2.45*A*). The *nerves of the transverse colon* arise from the **superior mesenteric nerve plexus** and follow the right and middle colic arteries (Fig. 2.45*B*). These nerves transmit sympathetic and parasympathetic (vagal) nerve fibers. The nerves that derive from the **inferior mesenteric nerve plexus** follow the left colic artery.

The **descending colon** passes retroperitoneally from the left colic flexure into the left iliac fossa, where it is continuous with the sigmoid colon (Fig. 2.42). Peritoneum covers the colon anteriorly and laterally and binds it to the posterior abdominal wall. Although retroperitoneal, the descending colon, especially in the iliac fossa, has a short mesentery in approximately 33% of people; however, it is usually not long enough to cause volvulus of the colon. As it descends, the colon passes anterior to the lateral border of the left kidney. As with the ascending colon, the descending colon has a paracolic gutter on its lateral aspect.

The **sigmoid colon**, characterized by its S-shaped loop of variable length (usually approximately 40 cm), links the descending colon and the rectum (Fig. 2.42). The sigmoid colon extends from the iliac fossa to the third sacral segment where it joins the rectum. The termination of the teniae coli, approximately 15 cm from the anus, indicates the *rectosigmoid junction*. The sigmoid colon usually has a long mesentery and therefore has considerable freedom of movement, especially its middle part. The **root of the sigmoid mesocolon** has an inverted V-shaped attachment, extending first medially and superiorly along the external iliac vessels and then medially and inferiorly from the bifurcation of the common iliac vessels to the anterior aspect of the sacrum. The *left ureter* and the division of the left common iliac artery lie retroperitoneally, posterior to the apex of the root of the sigmoid mesocolon. The *omental appendices* are long in the sigmoid colon; they disappear when the sigmoid mesentery terminates (Fig. 2.42). The teniae coli also disappear as the longitudinal muscle in the wall of the colon broadens to form a complete layer in the rectum.

The *arterial supply of the descending and sigmoid colon* is from the **left colic** and **superior sigmoid arteries**, *branches of the inferior mesenteric artery* (Fig. 2.44*A*, Table 2.9). The sigmoid arteries descend obliquely to the left where they divide into ascending and descending branches. The most superior branch of the superior sigmoid artery anastomoses with the descending branch of the left colic artery, thereby forming a part of the marginal artery of the colon. The **inferior mesenteric vein** (IMV) returns blood from the descending colon and sigmoid colon, flowing into the splenic vein and then the portal vein on its way to the liver.

The *lymphatic vessels from the descending colon and sigmoid* colon pass to the epicolic and paracolic nodes and then through the **intermediate colic lymph nodes** along the left colic artery (Fig. 2.45*A*). Lymph from these nodes passes to the **inferior mesenteric lymph nodes** that lie around the inferior mesenteric artery. However, lymph from the left colic flexure may also drain to the **superior mesenteric lymph nodes.**

The *sympathetic nerve supply of the descending and sigmoid colon* is from the lumbar part of the sympathetic trunk and the **superior hypogastric plexus** through the plexuses on the inferior mesenteric artery and its branches (Fig. 2.45*B*). The *parasympathetic nerve supply* is from the **pelvic splanchnic nerves.**

Rectum and Anal Canal

The rectum—the fixed terminal part of the large intestine—is continuous with the sigmoid colon at the level of S3 vertebra. The junction is at the lower end of the mesentery of the sigmoid colon (Fig. 2.42). The rectum is continuous inferiorly with the anal canal. These parts of the large intestine are described with the pelvis in Chapter 3.

Colitis, Colectomy, Ileostomy, and Colostomy

Chronic inflammation of the colon (*ulcerative colitis*) is characterized by severe inflammation and ulceration of the colon and rectum. In some patients a *colectomy* is performed, during which the terminal ileum and colon, as well as the rectum and anal canal, are removed. An *ileostomy* is then constructed to establish an opening between the ileum and the skin of the anterolateral abdominal wall. Sometimes a *colostomy* is performed to create an artificial cutaneous opening into the colon. A *sigmoidostomy* establishes an artificial anus by creating a cutaneous opening into the sigmoid colon.

Colonoscopy (Coloscopy)

The interior of the colon can be observed with an elongated endoscope, usually a fiberoptic *colonoscope.* The endoscope is a flexible tube that inserts into the colon through the anus and rectum. Most tumors of the large intestine occur in the rectum; approximately 12% of them appear near the rectosigmoid junction. The interior of the sigmoid colon is observed with a *sigmoidoscope* (sigmoscope). ⊙

Spleen

The spleen, the largest of the lymphatic organs, is a mobile organ. Usually a purplish color, the spleen is located intraperitoneally in the left upper abdominal quadrant (Figs. 2.46 and 2.47). The spleen is entirely surrounded by peritoneum except at the **hilum,** where the splenic branches of the splenic artery and vein enter and leave. It is associated posteriorly with the left 9th through 11th ribs and separated from them by the diaphragm and the *costodiaphragmatic recess*—the cleftlike extension of the pleural cavity between the diaphragm and the lower part of the thoracic cage (see Chapter 1). The spleen normally does not descend inferior to the costal (rib) region; it rests on the left colic flexure. **The relations of the spleen are:**

- Anteriorly—the stomach
- Posteriorly—the left part of the diaphragm that separates it from the pleura, lung, and ribs 9 through 11
- Inferiorly—the left colic flexure
- Medially—the left kidney.

The spleen varies considerably in size, weight, and shape; however, it is usually approximately 12 cm long and 7 cm wide—roughly the size and shape of a clenched fist.

The *diaphragmatic surface of the spleen* is convexly curved to fit the concavity of the diaphragm. The anterior and superior borders of the spleen are sharp and often notched, whereas its posterior and inferior borders are rounded.

The spleen normally contains a large quantity of blood that is expelled periodically into the circulation by the action of the smooth muscle in its capsule and trabeculae. The large size of the splenic artery (or vein) indicates the volume of blood that passes through the spleen's capillaries and sinuses.

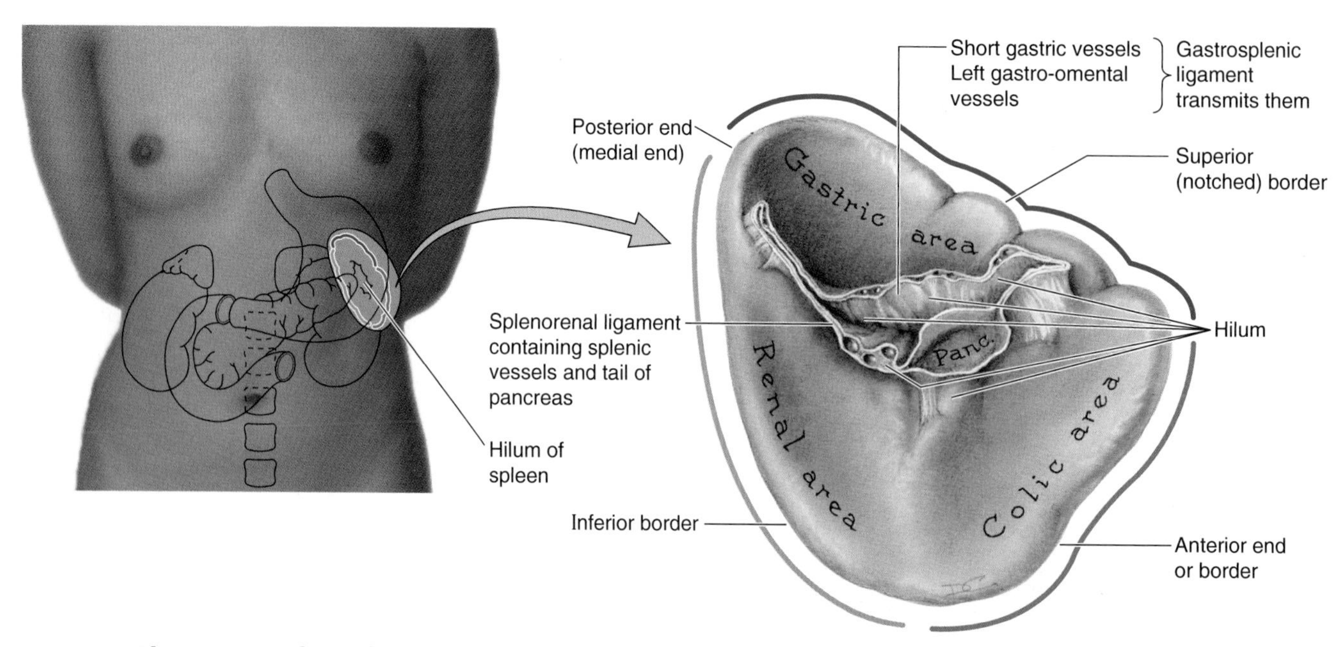

Figure 2.46. Visceral surface of the spleen. The illustration on the left is for orientation and showing the surface anatomy of the spleen and organs associated with it. Observe the notches that are characteristic of the superior border. Note the impressions formed by the structures in contact with the spleen. (*Panc.*, pancreas)

The thin *capsule of the spleen* is composed of dense, irregular fibroelastic connective tissue, containing occasional smooth muscle cells. The capsule is thickened at the hilum of the spleen. Internally the *trabeculae*, arising from the deep aspect of the capsule, carry blood vessels to and from the parenchyma or splenic "pulp"—the substance of the spleen.

The spleen contacts the posterior wall of the stomach and is connected to its greater curvature by the **gastrosplenic ligament** and to the left kidney by the **splenorenal ligament** (p. 216). These ligaments, containing splenic vessels, are attached to the hilum of the spleen on its medial aspect. Except at the hilum where these peritoneal reflections occur, the spleen is intimately covered with peritoneum. The **hilum of the spleen** is often in contact with the tail of the pancreas and constitutes the left boundary of the omental bursa.

The **splenic artery**—the largest branch of the celiac trunk—follows a tortuous course posterior to the omental bursa, anterior to the left kidney, and along the superior border of the pancreas (Figs. 2.47*A* and 2.48*A*). Between the layers of the splenorenal ligament, the splenic artery divides into five or more branches that enter the hilum. The lack of anastomosis of the arterial vessels in the spleen results in the formation of vascular segments—two in 84% of spleens and three in the others—with avascular planes between them.

The **splenic vein** is formed by several tributaries that emerge from the hilum (Figs. 2.47*A* and 2.48*B*). It is joined by the IMV and runs posterior to the body and tail of the pancreas throughout most of its course. The splenic vein unites with the SMV posterior to the neck of the pancreas to form the **portal vein.**

The *splenic lymphatic vessels* leave the lymph nodes in the hilum and pass along the splenic vessels to the **pancreaticosplenic lymph nodes** (Fig. 2.49*A*). These nodes relate to the posterior surface and superior border of the pancreas. The *nerves of the spleen* derive from the **celiac plexus** (Fig. 2.49*B*), are distributed mainly along branches of the splenic artery, and are vasomotor in function.

Rupture of the Spleen

Although well protected by the 9th through 12th ribs, the spleen is the most frequently injured organ in the abdomen when severe blows on the left side fracture one or more of the 9th through 12th ribs. Blunt trauma to other regions of the abdomen that cause a sudden, marked increase in intra-abdominal pressure (e.g., by impalement on a steering wheel of a car or the handlebars of a bicycle) can also rupture the spleen. If ruptured, the spleen bleeds profusely because its capsule is thin and its parenchyma is soft and pulpy. Rupture of the spleen causes severe intraperitoneal hemorrhage and shock.

Splenectomy

Repair of a ruptured spleen is difficult; consequently, splenectomy is often performed to prevent the person from bleeding to death. Partial removal of the spleen is followed by rapid regeneration. Even total splenectomy does not produce serious effects, especially in adults, because its functions are assumed by other reticuloendothelial organs. When the spleen is diseased, resulting from, for example, granulocytic leukemia (high leukocyte and white blood cell count), it may be 10 or more times its normal size (*splenomegaly*—enlargement of the spleen).

Accessory Spleen(s)

One or more small spleens may form near the hilum; they may be embedded partly or wholly in the tail of the pancreas. In most affected individuals, only one accessory spleen is present; externally it often resembles a lymph node. Accessory spleens are common (10%) and usually are approximately 1 cm in diameter (ranging from 0.2 to 10 cm). An accessory spleen is also commonly found between the layers of the gastrosplenic ligament. Awareness of the possible presence of an accessory spleen is important because if not removed during splenectomy, the symptoms that indicated removal of the spleen (e.g., *splenic anemia*) may persist.

Splenic Needle Biopsy and Splenoportography

The relationship of the costodiaphragmatic recess of the pleural cavity to the spleen is clinically important (see Chapter 1). This potential space descends to the level of the 10th rib in the midaxillary line. Its existence must be kept in mind when doing a *splenic needle biopsy*, or when injecting radiopaque material into the spleen for visualization of the portal vein (*splenoportography).* If care is not exercised, this material may enter the pleural cavity, causing *pleuritis* (inflammation of the pleura). ⊙

Pancreas

The pancreas—*an elongated, accessory digestive gland*—lies retroperitoneally and transversely across the posterior abdominal wall, posterior to the stomach between the duodenum on the right and the spleen on the left (Fig. 2.47). The transverse mesocolon attaches to its anterior margin. The pancreas produces

- An exocrine secretion (*pancreatic juice* from the acinar cells) that enters the duodenum through the main and accessory pancreatic ducts
- Endocrine secretions (*glucagon* and *insulin* from the pancreatic islets [of Langerhans]) that enter the blood (Fig. 2.47*D*).

For descriptive purposes the pancreas is divided into four parts: head, neck, body, and tail.

The **head of the pancreas**—the expanded part of the gland—is embraced by the C-shaped curve of the duodenum to the right of the superior mesenteric vessels. The head firmly attaches to the medial aspect of the descending and horizontal parts of the duodenum. The **uncinate process**, a projection

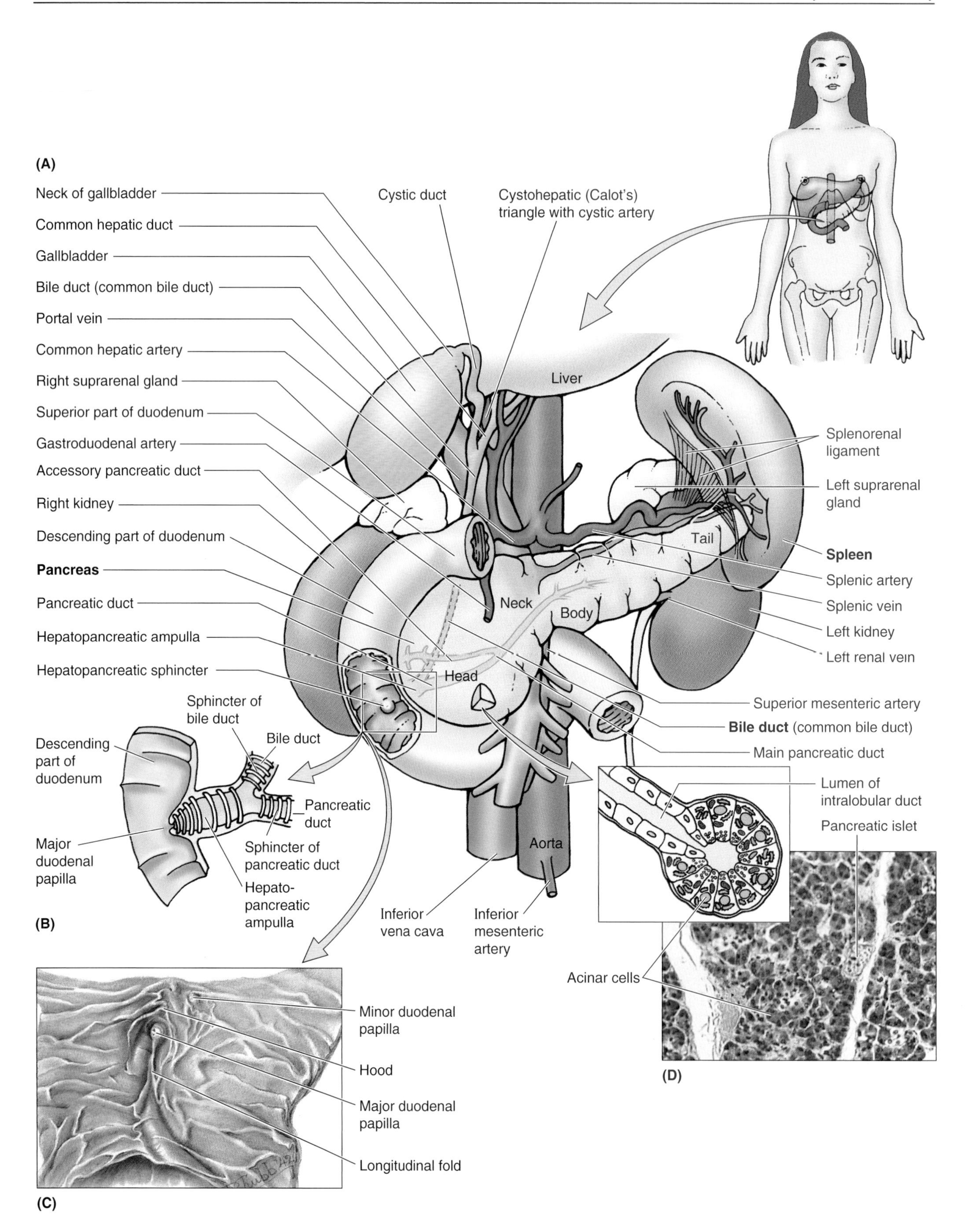
(A)
Neck of gallbladder
Common hepatic duct
Gallbladder
Bile duct (common bile duct)
Portal vein
Common hepatic artery
Right suprarenal gland
Superior part of duodenum
Gastroduodenal artery
Accessory pancreatic duct
Right kidney
Descending part of duodenum
Pancreas
Pancreatic duct
Hepatopancreatic ampulla
Hepatopancreatic sphincter
Cystic duct
Cystohepatic (Calot's) triangle with cystic artery
Liver
Splenorenal ligament
Left suprarenal gland
Tail
Spleen
Splenic artery
Splenic vein
Neck
Body
Left kidney
Left renal vein
Head
Superior mesenteric artery
Bile duct (common bile duct)
Main pancreatic duct
Sphincter of bile duct
Bile duct
Descending part of duodenum
Pancreatic duct
Major duodenal papilla
Sphincter of pancreatic duct
Hepato-pancreatic ampulla
(B)
Aorta
Inferior vena cava
Inferior mesenteric artery
Lumen of intralobular duct
Pancreatic islet
Acinar cells
(D)
Minor duodenal papilla
Hood
Major duodenal papilla
Longitudinal fold
(C)

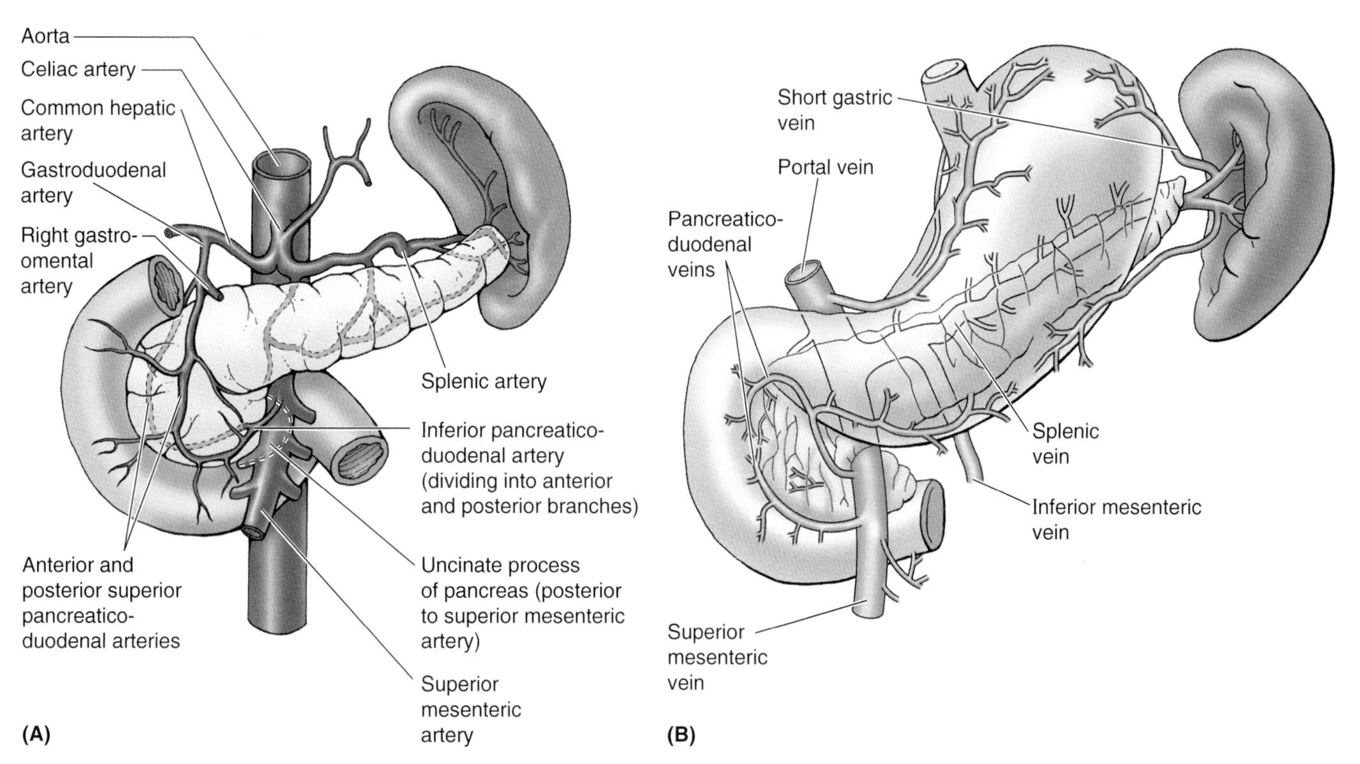

Figure 2.48. Arterial supply and venous drainage of the pancreas. Because of the close relationship of the pancreas and duodenum, their blood vessels are the same in whole or in part. **A.** *Arteries.* **B.** *Veins.* The veins accompany the arteries supplying the pancreas.

from the inferior part of the head, extends medially to the left, posterior to the SMA (Fig. 2.48). *The head of the pancreas rests posteriorly on the:*

- IVC
- Right renal artery and vein
- Left renal vein.

On its way to opening into the descending part of the duodenum, the **bile duct** lies in a groove on the posterosuperior surface of the head or is embedded in its substance (Fig. 2.47*B*).

The **neck of the pancreas** is short (1.5–2 cm) and overlies the superior mesenteric vessels, which form a groove in its posterior aspect (Fig. 2.36, *B–C*). The anterior surface of the neck, covered with peritoneum, is adjacent to the *pylorus* of the stomach. The SMV joins the splenic vein posterior to the neck to form the portal vein.

The **body of the pancreas** continues from the neck and lies to the left of the superior mesenteric vessels, passing over the aorta and L2 vertebra, posterior to the omental bursa. The anterior surface of the body of the pancreas—covered with peritoneum—lies in the floor of the omental bursa and forms part of the stomach bed. The posterior surface of the body of the pancreas (Fig. 2.47*A*) is devoid of peritoneum and is in contact with the:

- Aorta
- SMA
- Left suprarenal gland
- Left kidney and renal vessels.

Figure 2.47. Pancreas, duodenum, biliary (bile) ducts, and spleen. A. Pancreas, extrahepatic bile passages, pancreatic ducts, and duodenum. Observe that the bile duct, after descending posterior to the superior part of the duodenum and the accessory pancreatic duct, joins the main pancreatic duct. **B.** Entry of the bile duct and pancreatic duct into the duodenum through the hepatopancreatic ampulla—the dilation within the major duodenal papilla that normally receives both the bile duct and the main pancreatic duct. Observe the smooth muscle sphincters encircling these bile and pancreatic ducts and the hepatopancreatic ampulla. **C.** Interior of the descending (second) part of the duodenum showing the major and minor duodenal papillae. There is a hood over the larger papilla onto which the hepatopancreatic ampulla is opening. The bile duct and main pancreatic duct open separately on the papilla in approximately 5% of people. The accessory pancreatic duct opens on the minor duodenal papilla. **D.** Drawing illustrating the structure of the acinar (enzyme-producing) tissue. Below the drawing is a photomicrograph of the pancreas displaying secretory acini and a pancreatic islet.

Figure 2.49. Lymphatic drainage and innervation of the pancreas and spleen. A. *Lymph nodes.* The *arrows* indicate lymph flow to the lymph nodes. **B.** *Innervation.* The nerves of the pancreas are autonomic nerves from the celiac and superior mesenteric plexuses. A dense network of nerve fibers passes from the celiac plexus along the splenic artery to the spleen. Most are postsynaptic sympathetic fibers to smooth muscle of the capsule, trabeculae, and intrasplenic vessels.

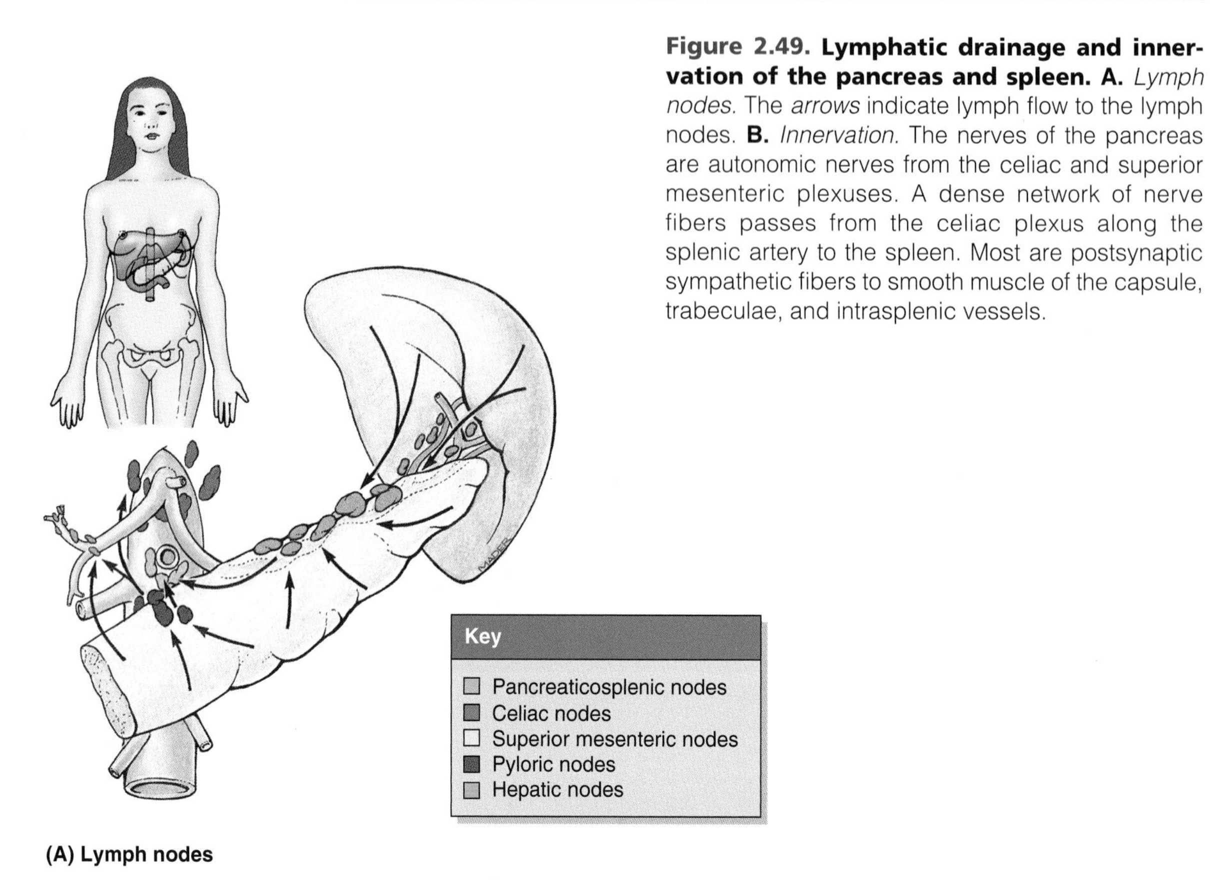

(A) Lymph nodes

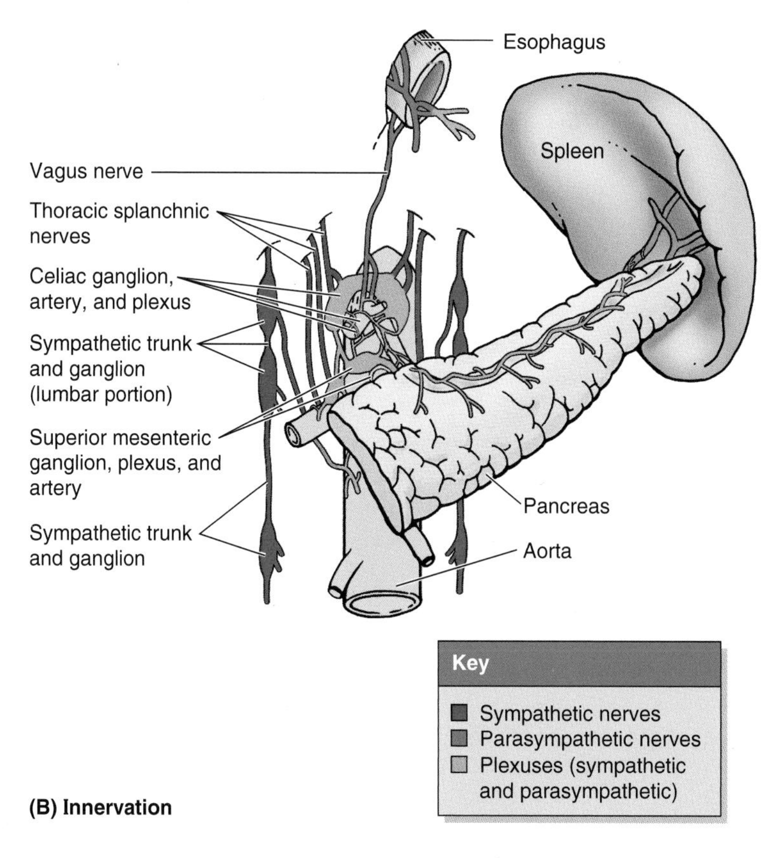

(B) Innervation

The **tail of the pancreas** lies anterior to the left kidney, where it is closely related to the hilum of the spleen and the left colic flexure. The tail is relatively mobile and passes between the layers of the **splenorenal ligament** with the splenic vessels (Figs. 2.46 and 2.47*A*). The tip of the tail is usually blunted and turned superiorly.

The **main pancreatic duct** begins in the tail of the pancreas and runs through the parenchyma of the gland to the head, where it turns inferiorly and is closely related to the bile duct (Fig. 2.47, *A–B*). The main pancreatic duct and the bile duct unite to form a short, dilated **hepatopancreatic ampulla**, which opens into the descending part of the duodenum at the summit of the *major duodenal papilla* (Fig. 2.47, *B–C*). The **sphincter of the pancreatic duct** (around the terminal part of the pancreatic duct), the **sphincter of the bile duct** (around the termination of the bile duct), and the **hepatopancreatic sphincter** (sphincter of Oddi)—around the hepatopancreatic ampulla—are smooth muscle sphincters that control the flow of bile and pancreatic juice into the duodenum.

The **accessory pancreatic duct** (Fig. 2.47*A*) drains the uncinate process and inferior part of the head of the pancreas and opens into the duodenum at the summit of the *minor duodenal papilla* (Fig. 2.47*C*). Usually, the accessory duct (60%) communicates with the main pancreatic duct. In some cases the main pancreatic duct is smaller than the accessory pancreatic duct and the two are not connected. In these people, the accessory duct carries most of the pancreatic juice. These variations of the pancreatic ducts are explainable from their fusion or lack of fusion during pancreatic development (Moore and Persaud, 1998).

The **pancreatic arteries** derive mainly from the branches of the markedly tortuous **splenic artery**, which form several arcades with pancreatic branches of the **gastroduodenal** and **superior mesenteric arteries** (Fig. 2.48*A*). Up to 10 branches of the splenic artery supply the body and tail of the pancreas. The anterior and posterior **superior pancreaticoduodenal arteries**, branches of the gastroduodenal artery, and the anterior and posterior **inferior pancreaticoduodenal arteries**, branches of the SMA, supply the head. The corresponding **pancreatic veins** are tributaries of the splenic and superior mesenteric parts of the portal vein; however, most of them empty into the **splenic vein** (Fig. 2.48*B*).

The **pancreatic lymphatic vessels** follow the blood vessels (Fig. 2.49*A*). Most vessels end in the **pancreaticosplenic nodes** that lie along the splenic artery, but some vessels end in the *pyloric lymph nodes*. Efferent vessels from these nodes drain to the **celiac, hepatic**, and **superior mesenteric lymph nodes.**

The **nerves of the pancreas** are derived from the **vagus** and **thoracic splanchnic nerves** passing through the diaphragm (Fig. 2.49*B*, Table 2.10). The parasympathetic and sympathetic fibers reach the pancreas by passing along the arteries from the **celiac plexus** and **superior mesenteric plexus.** They are vasomotor (sympathetic) and parenchymal (sympathetic and parasympathetic—to pancreatic acinar cells and islets) in their distribution.

Table 2.10. Splanchnic Nerves

Splanchnic Nerves	Autonomic Fiber Type[a]	System	Origin	Destination
A. Cardiopulmonary	Postsynaptic	Sympathetic	Cervical and upper thoracic sympathetic trunk	Thoracic cavity (viscera above level of diaphragm)
B. Abdominopelvic			Lower thoracic and abdominal sympathetic trunk	Abdominopelvic cavity (prevertebral ganglia serving viscera below level of diaphragm)
1. Lower thoracic: a. Greater b. Lesser c. Least	Presynaptic	Sympathetic	Thoracic sympathetic trunk: a. T5–T9 or T10 level b. T10–T11 level c. T12 level	Prevertebral ganglia: a. Celiac ganglia b. Superior mesenteric ganglia c. Aorticorenal ganglia
2. Lumbar			Abdominal sympathetic trunk	Inferior mesenteric ganglia and ganglia of intermesenteric and hypogastric plexuses
C. Pelvic	Presynaptic	Parasympathetic	Ventral rami of S2–S4 spinal nerves	Intrinsic ganglia of descending and sigmoid colon, rectum, and pelvic viscera

[a]Splanchnic nerves also convey visceral afferent fibers, which are not part of the autonomic nervous system.

Blockage of the Hepatopancreatic Ampulla

Because the main pancreatic duct joins the bile duct to form the hepatopancreatic ampulla and pierces the duodenal wall, a *gallstone* passing along the extrahe-patic bile passages may lodge in the constricted distal end of the ampulla, where it opens at the summit of the major duodenal papilla (Fig. 2.47, *A–B*). In this case, both the biliary and pancreatic duct systems are blocked and neither bile nor pancreatic juice can enter the duodenum. However, bile may back up and enter the pancreatic duct. A similar reflux of bile sometimes results from *spasms of the hepatopancreatic sphincter*. Normally, the sphincter of the pancreatic duct prevents reflux of bile into the pancreatic duct; however, if the hepatopancreatic ampulla is obstructed, the weak pancreatic duct sphincter may be unable to withstand the excessive pressure of the bile in the hepatopancreatic ampulla.

Accessory Pancreatic Tissue

It is not unusual for pancreatic tissue to develop in the stomach, duodenum, ileum, and ileal diverticulum; however, the stomach and duodenum are the most common sites. The accessory pancreatic tissue may contain islet cells that produce glucagon and insulin.

Pancreatitis

If the pancreatic duct is blocked, the pancreas may be inflamed (*pancreatitis*). Reflux of bile from the hepatopancreatic ampulla into the pancreatic duct may be another cause of pancreatitis. Swelling of the head of the pancreas (resulting from inflammation) may occlude the main pancreatic duct and result in pancreatitis of the body and tail of the pancreas. If the accessory pancreatic duct connects with the main pancreatic duct and opens into the duodenum, it may compensate for an obstructed main pancreatic duct or spasm of the hepatopancreatic sphincter.

Pancreatectomies

In the treatment of some people with chronic pancreatitis, most of the pancreas is removed. The anatomical relationships and the blood supply of the head of the pancreas, bile duct, and duodenum make it impossible to remove the entire head of the pancreas (Skandalakis et al., 1995). Usually a rim of the pancreas is retained along the medial border of the duodenum to preserve the duodenal blood supply.

Rupture of the Pancreas

Pancreatic injury can result from sudden, severe, forceful compression of the abdomen, such as the force of a seat belt in an automobile accident. Because the pancreas lies transversely, the vertebral column acts like an anvil and the traumatic force may rupture the pancreas. Rupture of the pancreas frequently tears its duct system, allowing pancreatic juice to enter the parenchyma of the gland and to invade adjacent tissues. Digestion of pancreatic and other tissues by pancreatic juice is painful.

Pancreatic Cancer

Cancer involving the pancreatic head accounts for most cases of extrahepatic obstruction of the biliary system. Because of the posterior relationships of the pancreas, cancer of the head often compresses and obstructs the bile duct and/or the hepatopancreatic ampulla. This condition causes *obstructive jaundice*, resulting in the retention of bile pigments, enlargement of the gallbladder, and jaundice (Fr. jaune, yellow)—the yellow staining of most body tissues, skin, mucous membranes, and conjunctiva. Approximately 90% of people with pancreatic cancer have *ductular adenocarcinoma*. Severe pain in the back is frequently present. Cancer of the neck and body of the pancreas may cause portal or inferior vena caval obstruction because the pancreas overlies these large veins (Fig. 2.47*B*). ⊙

Surface Anatomy of the Spleen and Pancreas

The spleen lies superficially in the left hypochondrium between the 9th through 11th ribs; its costal surface is convex to fit the curved bodies of these elongated, flattened bones. In the recumbent position, the long axis of the spleen is roughly parallel to the long axis of the 10th rib; in the left anterior oblique (LAO) photograph, the long axis is parallel to the line of gaze (i.e., directly into the page). Normally the spleen does not extend inferior to the left costal margin; thus, it is seldom palpable through the anterolateral abdominal wall unless it is enlarged. When it is hardened and enlarged to approximately three times its normal size, it moves inferior to the left costal margin and its notched superior border lies inferomedially. The notched border is helpful when palpating an enlarged spleen because when the person takes a deep breath, the notches can often be palpated.

The **neck of the pancreas** overlies the 1st and 2nd lumbar vertebrae in the transpyloric plane. Its head is to the right and inferior to this plane, and its body and tail are to the left and superior to this level. Because the pancreas is deep in the abdominal cavity, posterior to the stomach and omental bursa, it is usually not palpable. ⊙

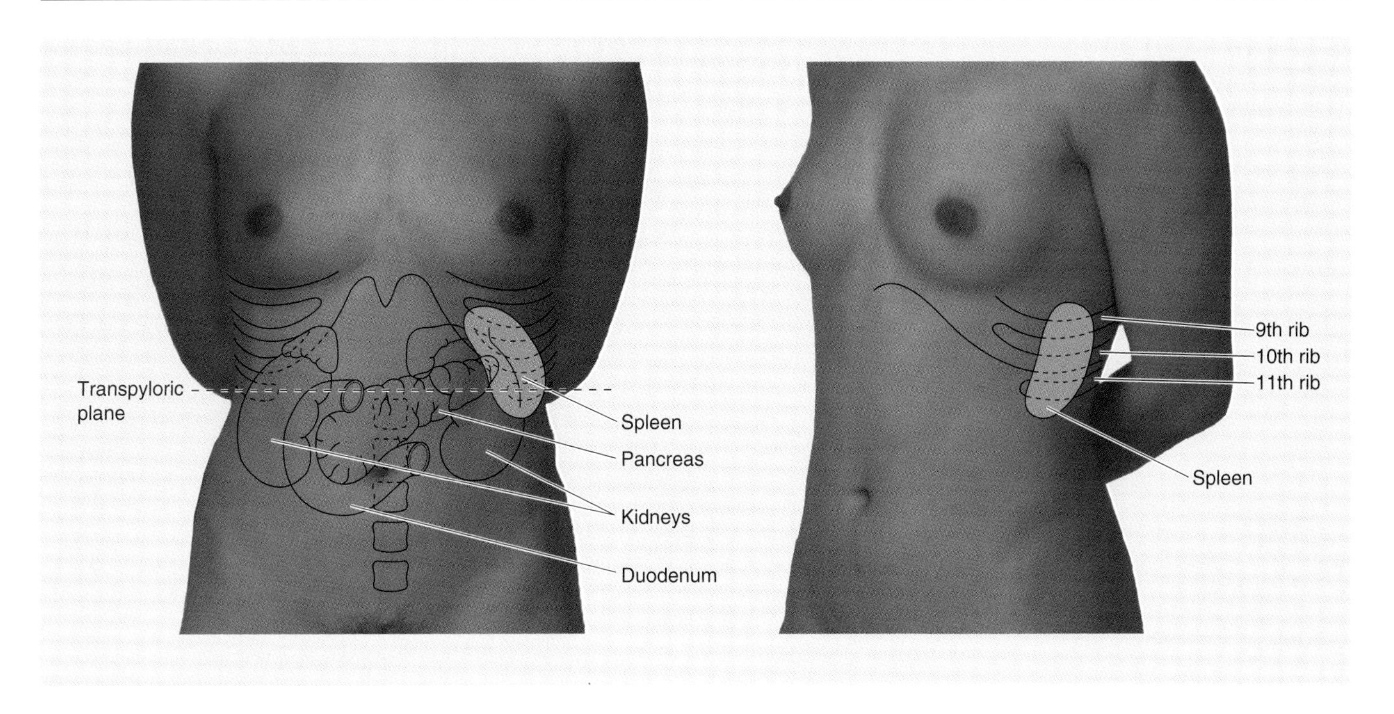

Liver

The liver—*the largest gland in the body*—weighs approximately 1500 gm and accounts for approximately one-fortieth of adult body weight. It lies in the right and left upper quadrants (mainly on the right side), inferior to the diaphragm, which separates it from the pleura, lungs, pericardium, and heart (Fig. 2.50). In addition to its many metabolic activities, the liver stores glycogen and secretes bile. **Bile** passes from the liver in the *right and left hepatic ducts* that join to form the *common hepatic duct*, which unites with the *cystic duct* to form the *bile duct*. In addition to storing bile, the gallbladder concentrates it by absorbing water and salts. When food arrives in the duodenum, the gallbladder sends concentrated bile through the cystic and bile ducts to the duodenum.

Surfaces of the Liver

The liver has a **diaphragmatic surface** (anterior, superior, and some posterior) and a **visceral surface** (posteroinferior), which are separated anteriorly by its sharp inferior border (Fig. 2.50*A*).

The **diaphragmatic surface of the liver** is smooth and dome-shaped where it is related to the concavity of the inferior surface of the diaphragm (Fig. 2.50*C*). However, the diaphragmatic surface is largely separated from the diaphragm by the **subphrenic recesses,** or spaces between the anterior part of the liver and the diaphragm. The subphrenic recesses are separated into right and left recesses by the falciform ligament (Fig. 2.50, *B–D*). The **hepatorenal recess** (hepatorenal pouch; Morison's pouch) is a deep recess of the peritoneal cavity on the right side extending superiorly between the liver anteriorly and the kidney and suprarenal glands posteriorly. The hepatorenal recess is a gravity-dependent part of the peritoneal cavity. When in the supine position, fluid draining from the omental bursa flows into this recess. The hepatorenal recess communicates anteriorly with the right subphrenic recess. Recall that normally all recesses of the peritoneal cavity are potential spaces only, containing only enough peritoneal fluid to lubricate the adjacent peritoneal membranes.

The diaphragmatic surface of the liver is covered with visceral peritoneum except posteriorly in the **bare area of the liver** (Fig. 2.50, *A*, *B*, and *D*), where it lies in contact with the diaphragm. The bare area is demarcated by the reflection of peritoneum from the diaphragm to it as the anterior (upper) and posterior (lower) layers of the **coronary ligament**. These layers meet on the right to form the **right triangular ligament** and diverge toward the left to enclose the triangular bare area. The anterior layer of the coronary ligament is continuous on the left with the right layer of the falciform ligament, and the posterior layer is continuous with the right layer of the lesser omentum. The left layers of the falciform ligament and the lesser omentum meet to form the **left triangular ligament.**

The **visceral surface of the liver** is covered with peritoneum (Fig. 2.50*D*), except at the *bed of the gallbladder* and the **porta hepatis**, where vessels and ducts enter and leave the liver. *The visceral surface is related to the:*

- Right side of the anterior aspect of the stomach—the **gastric and pyloric areas**
- Superior (first) part of the duodenum—the **duodenal area**
- Lesser omentum
- Gallbladder
- Right colic flexure and right transverse colon—the **colic area**
- Right kidney and suprarenal gland—the **renal** and **suprarenal areas.**

Subphrenic Abscesses

Peritonitis may result in the formation of localized abscesses in various parts of the peritoneal cavity. A common site for pus to collect is in a subphrenic recess or space. Subphrenic abscesses occur much more frequently on the right side because of the frequency of ruptured appendices and perforated duodenal ulcers. Because the right and left subphrenic recesses are continuous with the hepatorenal recess (the lowest [most gravity-dependent] parts of the peritoneal cavity when supine), pus from a subphrenic abscess may drain into one of the hepatorenal recesses, especially when patients are bedridden. A subphrenic abscess is often drained by an incision inferior to, or through, the bed of the 12th rib (Ellis, 1992), obviating the formation of an opening in the pleura or peritoneum. An anterior subphrenic abscess is often drained through a subcostal incision located inferior and parallel to the right costal margin. ⊙

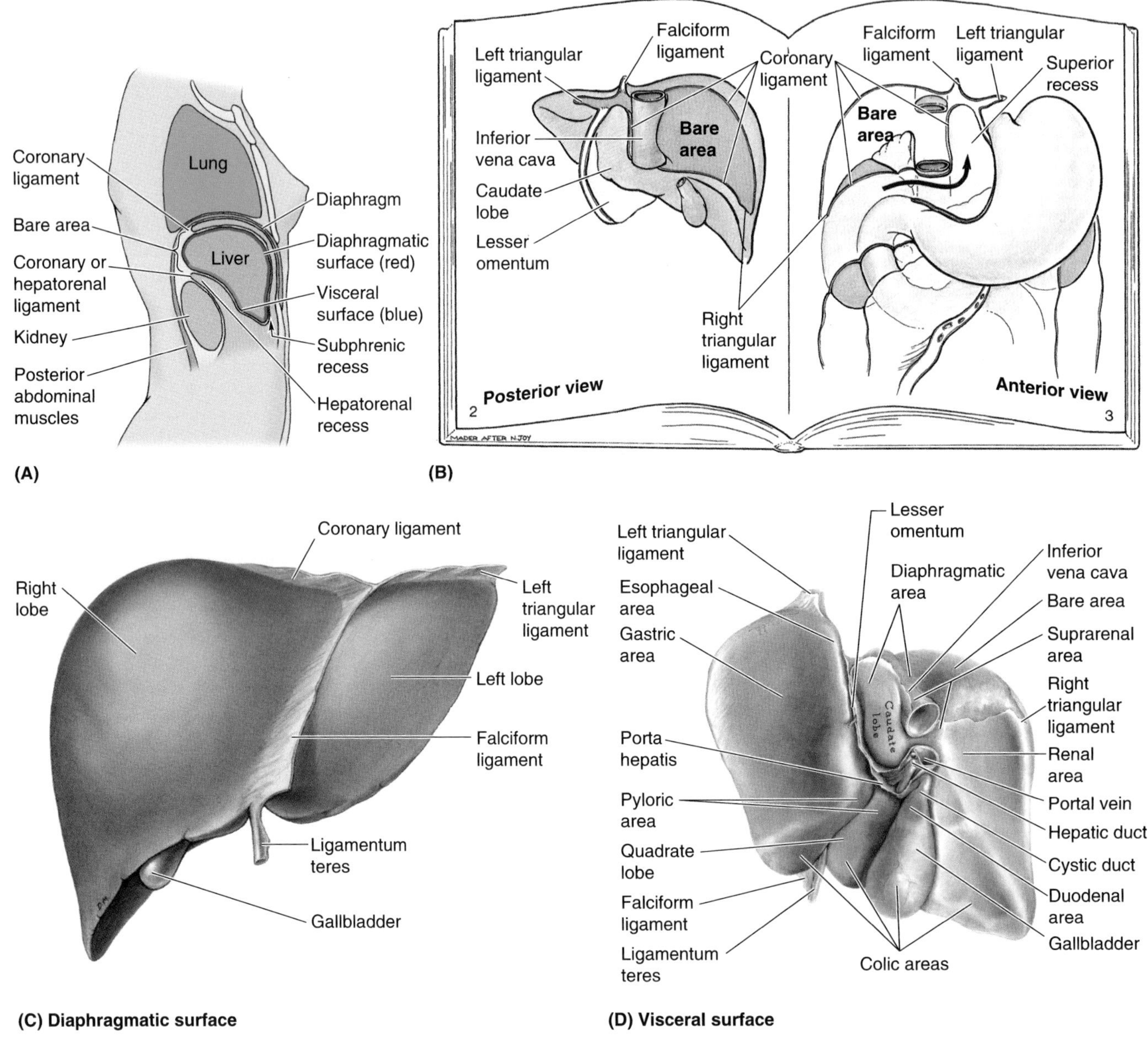

Figure 2.50. Surfaces and peritoneal ligaments of the liver. A. Drawing of a sagittal section through the diaphragm, liver, and right kidney. Observe the surfaces and recesses of the liver. **B.** Diagram of the peritoneal ligaments of the liver. The attachments of the liver are cut through and the liver is turned to the right side, as when turning the page of a book. **C.** Anterior part of the diaphragmatic surface of the liver. The diaphragmatic surface is dome-shaped and conforms to the inferior surface of the diaphragm. This surface is extensive and is divisible into superior, anterior (shown here), right, and posterior parts. **D.** Visceral surface of the liver. In the anatomical position, this surface is directed inferiorly, posteriorly,and to the right. Observe the impressions in this surface of the liver formed by the structures with which it is in contact.

Functional Parts of the Liver

The liver has functionally independent right and left parts (portal lobes) that are approximately equal in size (Fig. 2.51 and see Fig. 2.53*A*). Each part has its own blood supply from the hepatic artery and portal vein and its own venous and biliary drainage. On the visceral surface, the right (part of the) liver is demarcated from the left (part of the) liver by the gallbladder fossa inferiorly and the fossa for the IVC superiorly. An imaginary line over the diaphragmatic surface of the liver that runs from the fundus of the gallbladder to the IVC separates the parts. Both the right and left parts of the liver have medial and lateral divisions; those of the left liver are separated by the falciform ligament (the former terminology considered the lateral segment to be the anatomical left lobe). *In current terminology*, the left liver includes the **caudate lobe** and most of the **quadrate lobe** (Fig. 2.51*A*). The anatomical left lobe is separated from these lobes on the visceral surface by the *fissure for the round ligament* of the liver and the *fissure for the ligamentum venosum* (Fig. 2.51*B*), and on the diaphragmatic surface by the attachment of the falciform ligament.

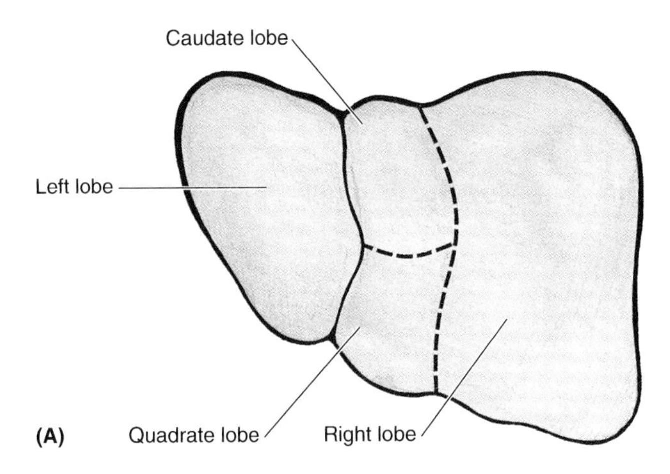

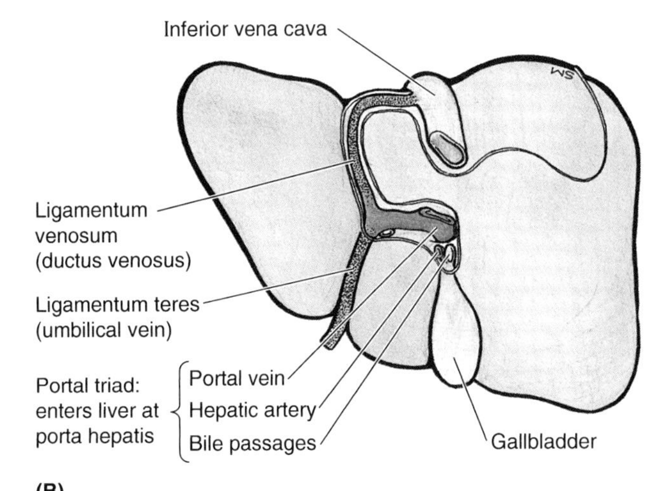

Figure 2.51. Anatomical lobes of the liver. A. Schematic posterior view showing the four (nonfunctional) anatomical lobes of the liver as traditionally described. Current terminology (*Terminologia Anatomica*) refers to the functional right and left (parts of the) liver, demarcated by a line running through the gallbladder fossa to the inferior vena cava (IVC). **B.** Schematic posterior view. The round ligament (L. ligamentum teres) of the liver is the occluded remains of the fetal umbilical vein. The ligamentum venosum is the fibrous remnant of the ductus venosum that shunted blood from the umbilical vein to the IVC.

The **round ligament** of the liver is the fibrous remnant of the *umbilical vein* that carried well-oxygenated and nutrient-rich blood from the placenta to the fetus. The umbilical vein remains patent in infants for a while. In individuals with *portal hypertension* (abnormally increased blood pressure in the portal venous system), there may be enlarged paraumbilical veins that course along the round ligament. The **ligamentum venosum** is the fibrous remnant of the fetal *ductus venosus* that shunted blood from the umbilical vein to the IVC, short-circuiting the liver (Moore and Persaud, 1998).

The **porta hepatis** (hepatic portal; portal fissure) is a transverse fissure on the visceral surface of the liver between the caudate and quadrate lobes (Fig. 2.50*D*), where the portal vein and hepatic artery enter the liver and the hepatic ducts leave. The porta hepatis gives passage to the:

- Portal vein
- Hepatic artery
- Hepatic nerve plexus
- Hepatic ducts
- Lymphatic vessels.

Peritoneal Relations of the Liver

The **lesser omentum**, enclosing the **portal triad** (Fig. 2.52)—bile duct, hepatic artery, and portal vein—passes from the liver to the lesser curvature of the stomach and the first 2 cm of the superior (first) part of the duodenum. The thick, free edge of the lesser omentum—extending between the porta hepatis of the liver and the duodenum (the **hepatoduodenal ligament**)—encloses the portal triad, a few lymph nodes, lymphatic vessels, and the hepatic plexus of nerves. The sheet-like remainder of the lesser omentum, the **hepatogastric ligament**, extends between the groove for the ligamentum venosum of the liver and the lesser curvature of the stomach.

Vessels and Nerves of the Liver

The liver receives blood from two sources (Fig. 2.52):

- The *portal vein* (70%)
- The *hepatic artery* (30%).

The **portal vein**, a short, wide vein, is formed by the superior mesenteric and splenic veins posterior to the neck of the pancreas, ascends anterior to the IVC, and divides at the right end of the porta hepatis into right and left branches that ramify within the liver. The **hepatic artery**, a branch of the celiac trunk (Fig. 2.49*A*), may be divided into the:

- *Common hepatic artery*—from the celiac trunk to the origin of the gastroduodenal artery
- *Hepatic artery proper*—from the origin of the gastroduodenal artery to its bifurcation into right and left branches (arteries).

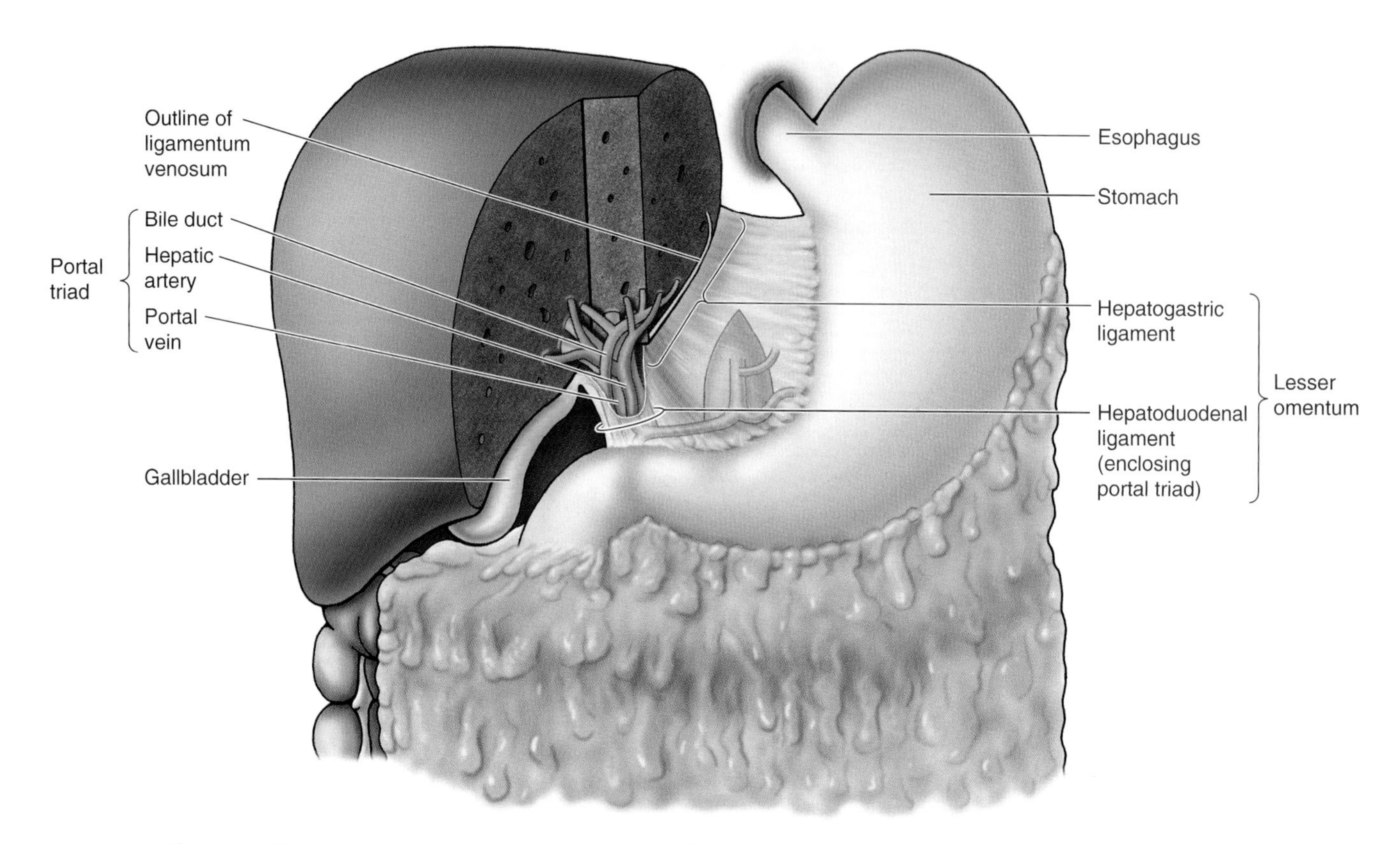

Figure 2.52. Lesser omentum and portal triad. Two sagittal cuts have been made through the liver, and these cuts have been joined by a coronal cut. Observe the portal triad—within the layers of the hepatoduodenal ligament—that enters the liver at the porta hepatis (Fig. 2.50*D*). Observe the hepatic artery passing between the layers of the hepatogastric ligament. When there is bleeding from the cystic artery, a branch of the right hepatic artery, the hemorrhage can be stopped by compressing the hepatic artery (as shown in Fig. 2.24).

The hepatic artery carries well-oxygenated blood from the aorta, and the *portal vein* carries poorly oxygenated but nutrient-rich blood from the GI tract (except for the inferior part of the anal canal) to the sinusoids of the liver. At or close to the **porta hepatis**, the hepatic artery and portal vein terminate by dividing into right and left branches, which supply the right and left parts of the liver, respectively. Within each part the primary branchings of the portal vein and hepatic artery are consistent enough to form **vascular segments** (Fig. 2.53). A horizontal plane through the right lobe and lateral division of the left lobe, plus the caudate lobe, divides the liver into eight vascular segments. The left lobe is composed of segments 1 through 4 and the right lobe segments 5 through 8. Between the segments are the hepatic veins, which are intersegmental in their distribution and function, draining parts of adjacent segments. The **hepatic veins**, formed by the union of the central veins of the liver, open into the IVC just inferior to the diaphragm. The attachment of these veins to the IVC helps to hold the liver in position.

Hepatic Lobectomies and Segmentectomy

When it was discovered that the right and left hepatic arteries and ducts, as well as branches of the right and left portal veins, do not communicate, it became possible to perform **hepatic lobectomies**—removal of the right or left (part of the) liver—without excessive bleeding. Most injuries to the liver involve the right liver. More recently, especially since the advent of cauterizing scalpel and laser surgery, it has become possible to perform **segmentectomies.** If a severe injury or tumor involves one segment or adjacent segments, it may be possible to resect (remove) only the affected segment(s): *segmentectomy.* The intersegmental hepatic veins serve as guides to the interlobular planes; however, they also provide a major source of bleeding with which the surgeon must contend. A more extensive injury that is likely to leave large areas of the liver devascularized may still require lobectomy. ▶

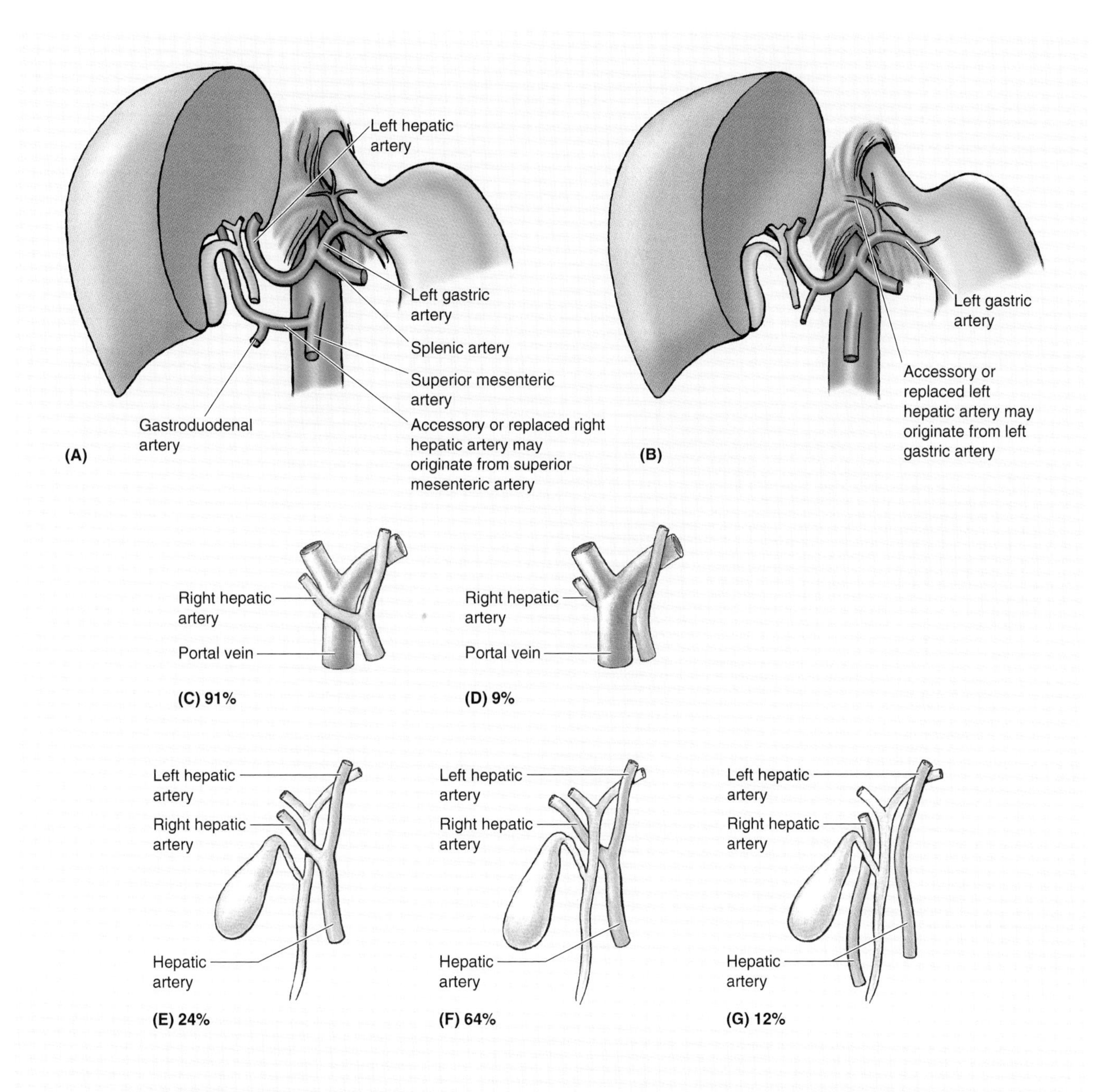

Aberrant Hepatic Arteries

A more common variety of right or left hepatic artery that arises as a terminal branch of the hepatic artery proper may be replaced in part or entirely by an aberrant (accessory or replaced) artery arising from another source (*A–B*). The most common source of an **aberrant left hepatic artery** is the left gastric artery. The most common source of an **aberrant right hepatic artery** is the SMA.

Variations in the Relationships of the Hepatic Arteries

In most people, the right hepatic artery crosses anterior to the portal vein in some people, the artery crosses posterior to the portal vein (*C–D*). In most people, the right hepatic artery runs posterior to the common hepatic duct; however, in some individuals the right hepatic artery crosses anterior to the common hepatic duct (*E–G*).

Unusual Formation of the Portal Vein

Usually, the portal vein forms posterior to the neck of the pancreas by the union of the superior mesenteric and splenic veins and ascends anterior to the IVC. In approximately one-third of individuals, the IMV joins the confluence of the superior mesenteric and splenic veins; hence, all three veins form the portal vein. In most people the IMV enters the splenic vein (60%) or the SMV (40%). ○

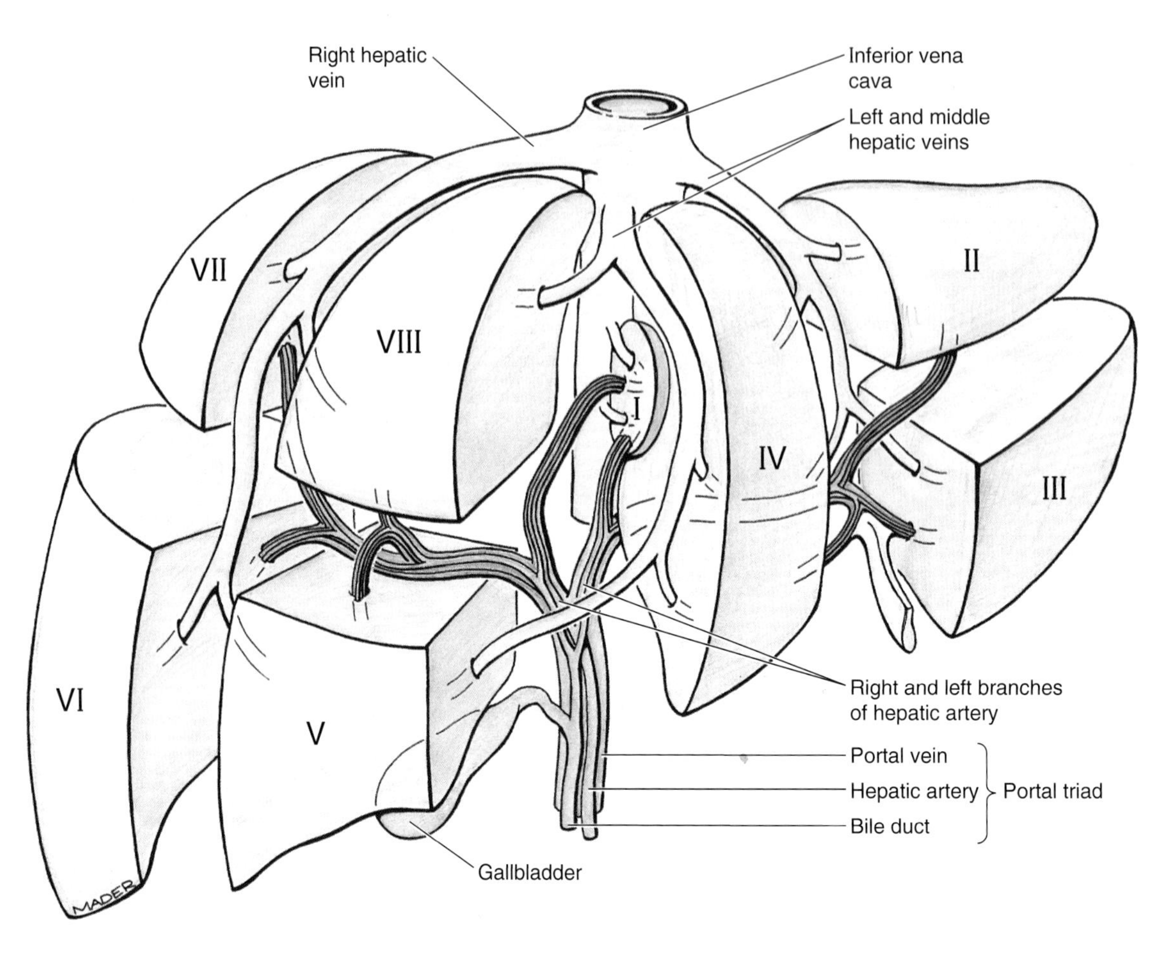

(A)

Schema of Terminology for Subdivisions of the Liver

<table>
<tr><th>Anatomical Term</th><th colspan="3">Right Lobe</th><th>Left Lobe</th><th colspan="2">Caudate Lobe</th></tr>
<tr><td rowspan="4">Functional/
Surgical Term**</td><td colspan="2">Right (part of) Liver
[Right portal lobe*]</td><td colspan="2">Left (part of) Liver
[Left portal lobe^{+}]</td><td colspan="2">Posterior (part of) Liver</td></tr>
<tr><td>Right lateral division</td><td>Right medial division</td><td>Left medial division</td><td>Left lateral division</td><td>[Right caudate lobe*]</td><td>[Left caudate lobe^{+}]</td></tr>
<tr><td>Posterior lateral segment
Segment VII
[Posterior superior area]</td><td>Posterior medial segment
Segment VIII
[Anterior superior area]</td><td rowspan="2">[Medial superior area]
Left medial segment
Segment IV
[Medial inferior area = quadrate lobe]</td><td>Lateral segment
Segment II
[Lateral superior area]</td><td colspan="2" rowspan="2">Posterior segment
Segment I</td></tr>
<tr><td>Right anterior lateral segment
Segment VI
[Posterior inferior area]</td><td>Anterior medial segment
Segment V
[Anterior inferior area]</td><td>Left lateral anterior segment
Segment III
[Lateral inferior area]</td></tr>
</table>

** The labels in the table and figure above reflect the new *Terminologia Anatomica: International Anatomical Terminology.* Previous terminology is in brackets.

**$^{+}$ Under the schema of the previous terminology, the caudate lobe was divided into right and left halves, and

* the right half of the caudate lobe was considered a subdivision of the right portal lobe;

$^{+}$ the left half of the caudate lobe was considered a subdivision of the left portal lobe.

Figure 2.53. Hepatic segmentation. A. The segmentation of the liver is based on the principal divisions of the hepatic artery and portal vein and accompanying hepatic ducts. Each segment of the liver is supplied by a branch of the hepatic artery and portal vein, and drained by a branch of the bile duct. Intersegmental hepatic veins pass between segments on their way to the inferior vena cava (IVC).

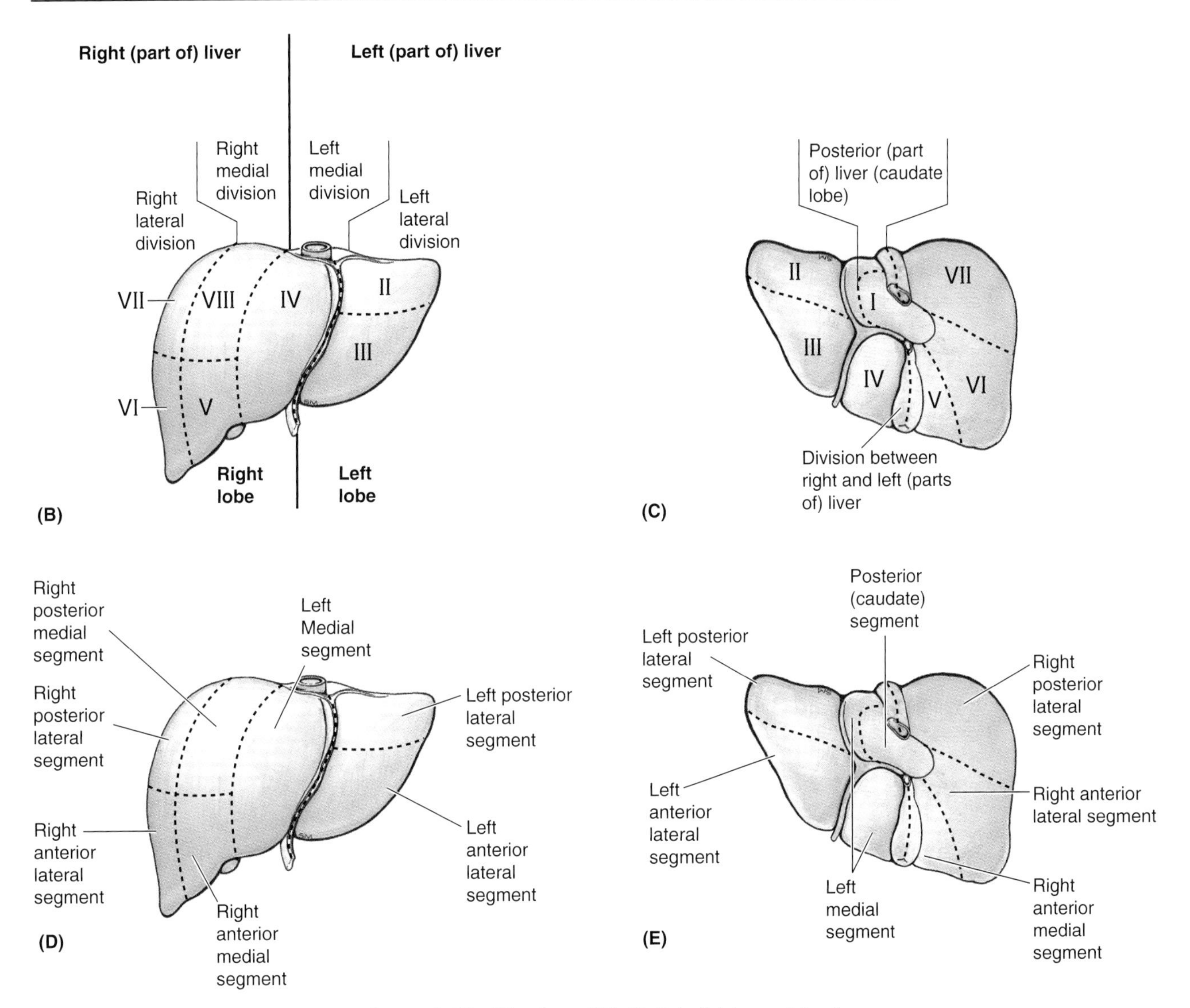

Figure 2.53. *(Continued)* **B–E.** Subdivisions of the liver.

The liver is a major lymph-producing organ: between one-quarter and one-half of the lymph received by the thoracic duct comes from the liver. The **lymphatic vessels of the liver** occur as *superficial lymphatics*, in the subperitoneal fibrous capsule of the liver (Glisson's capsule), which form its outer surface, and as *deep lymphatics* in the connective tissue that accompanies the ramifications of the portal triad and hepatic veins. Superficial lymphatics from the anterior aspects of the diaphragmatic and visceral surfaces and the deep lymphatic vessels accompanying the portal triads converge toward the porta hepatis and drain to the **hepatic lymph nodes** scattered along the hepatic vessels and ducts in the lesser omentum. Efferent lymphatic vessels from the hepatic nodes drain into **celiac lymph nodes**, which in turn drain into the **chyle cistern** (L. cisterna chyli), a dilated sac at the inferior end of the thoracic duct. Superficial lymphatics from the posterior aspects of the diaphragmatic and visceral surfaces of the liver drain toward the bare area of the liver. Here they drain into **phrenic lymph nodes**, or join deep lymphatics that have accompanied the hepatic veins converging on the IVC, and pass with this large vein through the diaphragm to drain into the **posterior mediastinal lymph nodes.** Efferent vessels from these nodes join the right lymphatic and thoracic ducts. *A few lymphatic vessels follow different routes.*

- From the posterior surface of the left lobe toward the esophageal hiatus of the diaphragm to end in **left gastric lymph nodes** (Fig. 2.54*A*)
- From the anterior central diaphragmatic surface along the falciform ligament to **parasternal lymph nodes**
- Along the **round ligament** of the liver to the umbilicus and lymphatics of the anterior abdominal wall.

The **nerves of the liver** are derived from the **hepatic nerve plexus** (Fig. 2.54*B*), the largest derivative of the celiac plexus. The hepatic plexus accompanies the branches of the hepatic artery and portal vein to the liver. It consists of sympathetic fibers from the celiac plexus and parasympathetic fibers from the anterior and posterior vagal trunks. Nerve fibers accompany the vessels and bile ducts of the portal triad in the liver. Other than vasoconstriction, their function is unclear.

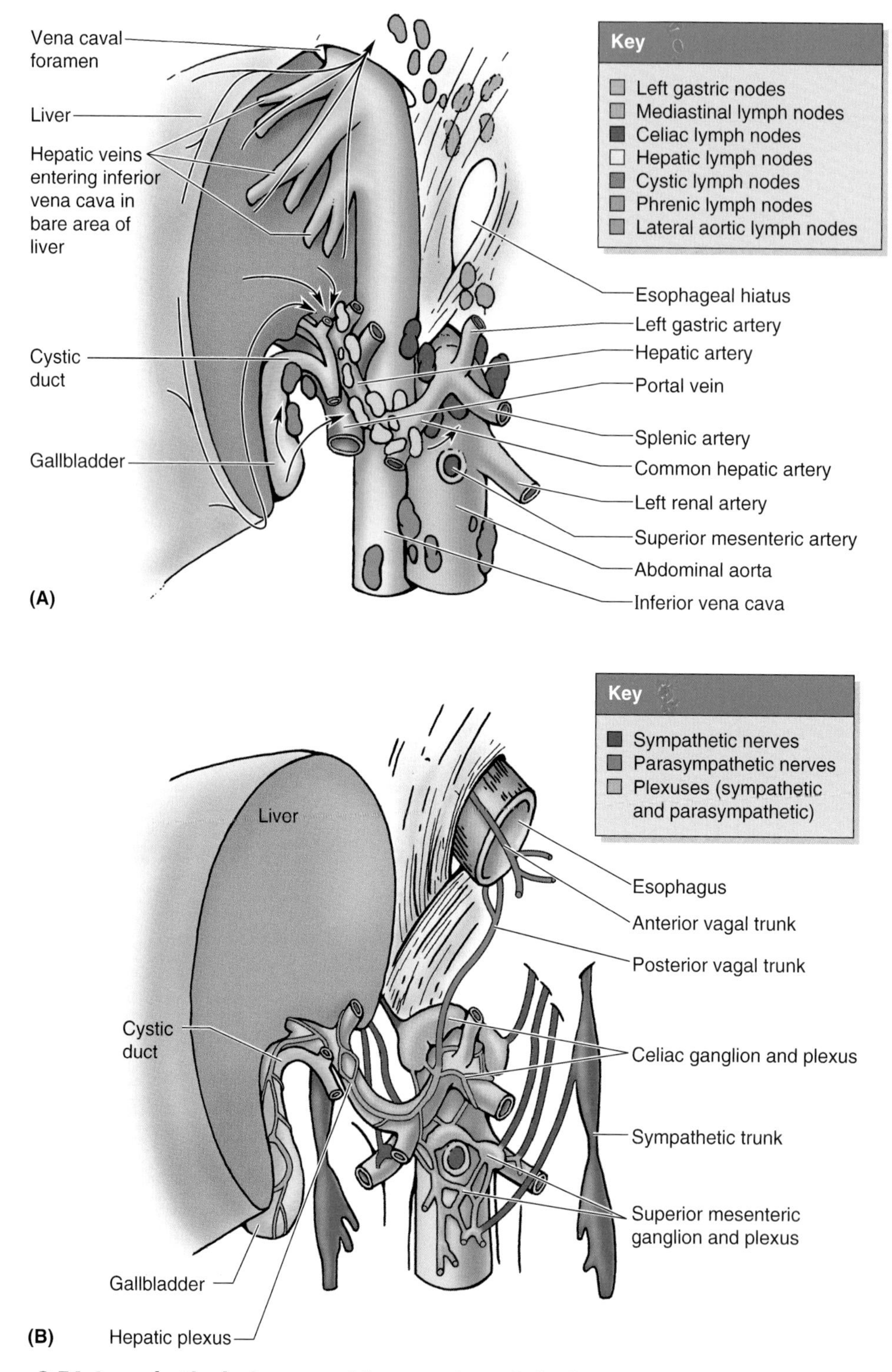

Figure 2.54. Lymphatic drainage and innervation of the liver. A. *Lymph nodes.* Note that lymph from the posterior aspect (superficial and deep) flows toward the bare area to enter phrenic lymph nodes or pass with the inferior vena cava (IVC) through the caval foramen in the diaphragm to enter mediastinal lymph nodes; lymph from the anterior and inferior aspect (superficial and deep) flows toward the porta hepatis to enter hepatic lymph nodes in the lesser omentum. **B.** *Nerves and plexuses.* The nerves of the liver derive from the hepatic plexus, the largest derivative of the celiac plexus. The hepatic plexus accompanies the branches of the hepatic artery and portal vein to the liver. It consists of sympathetic fibers from the celiac plexus and parasympathetic fibers from the anterior and posterior vagal trunks.

Liver Biopsy

Hepatic tissue may be obtained for diagnostic purposes by liver biopsy. The needle puncture commonly goes through the right 10th intercostal space in the midaxillary line. Before the physician takes the biopsy, the person is asked to hold his or her breath in full expiration to reduce the costodiaphragmatic recess (see Fig. 2.60, p. 281) and to lessen the possibility of damaging the lung and contaminating the pleural cavity.

Rupture of the Liver

The liver is easily injured because it is large, fixed in position, and friable (easily crumbled). Often the liver is torn by a fractured rib that perforates the diaphragm. Because of the liver's great vascularity and friability, liver lacerations often cause considerable hemorrhage and right upper quadrant pain. In such cases, the surgeon must decide whether to remove foreign material and the contaminated or devitalized tissue by dissection or to perform a segmental resection.

Hepatomegaly

Many diseases (such as heart failure) cause liver enlargement, or *hepatomegaly*. When the liver is massively enlarged, its inferior edge may reach the pelvic brim in the right lower quadrant of the abdomen. Tumors also enlarge the liver. The liver is a common site of *metastatic carcinoma*—cancer spreading from organs drained by the portal system of veins. Cancer cells may also pass to the liver from the thorax, especially from the right breast, because of the communications between thoracic lymph nodes and the lymphatic vessels draining the bare area of the liver.

Cirrhosis of the Liver

There is progressive destruction of hepatocytes (parenchymal liver cells) in *hepatic cirrhosis* and replacement of them by fibrous tissue. Cirrhosis, the most common of many causes of *portal hypertension*, frequently develops in persons suffering from chronic alcoholism. *Alcoholic cirrhosis* is characterized by enlargement of the liver resulting from fatty changes and fibrosis. The surface of the cirrhotic liver is nodular in appearance, accounting for the colloquial descriptive term "hobnail liver." The fibrous tissue surrounds the intrahepatic blood vessels and biliary ducts, making the liver firm, which impedes the circulation of blood through it. The treatment of cirrhosis may include anastomosing the portal and systemic venous systems. In a *portocaval shunt*, the portal vein is anastomosed to the IVC. For a description of other types of portasystemic shunts and how they are performed, see Skandalakis et al. (1995).

Liver Transplantation

A person with end-stage liver disease—such as *severe alcoholic cirrhosis*—may elect to have his or her liver removed and replaced with a normal liver. This procedure can relieve portal hypertension and restore liver function; however, lifelong drug immunosuppression may be necessary. ⊙

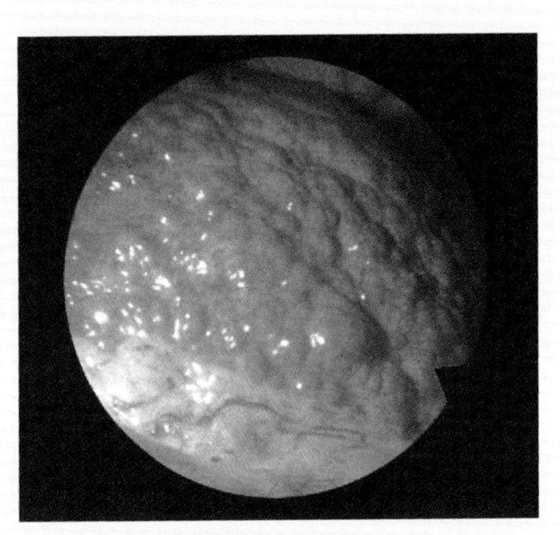

Cirrhosis of the liver
small-node type (endoscopic view)

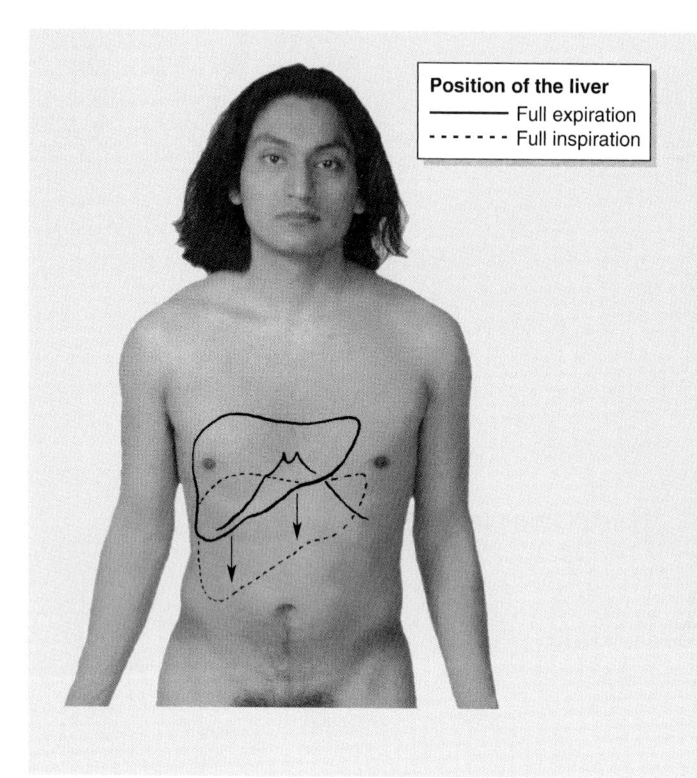

Surface Anatomy of the Liver

The liver lies mainly in the right upper quadrant of the abdomen where it is hidden and protected by the thoracic cage and the diaphragm. The normal liver lies deep to ribs 7 through 11 on the right side and crosses the midline toward the left nipple. Consequently, the liver occupies most of the right hypochondrium, the upper epigastrium, and extends into the left hypochondrium. The liver is located more inferiorly when one is erect because of gravity. Its sharp inferior border follows the right costal margin. When the person is asked to inspire deeply, the liver may be palpated because of the inferior movement of the diaphragm and liver. One method of palpating the liver is to place your left hand posteriorly between the person's right 12th rib and iliac crest. Then, put your right hand on the person's right upper quadrant, parallel and lateral to the rectus abdominis. The person is asked to take a deep breath as the examiner presses posterosuperiorly with the right hand and pulls anteriorly with the left hand. For photographs and other methods used for palpating the liver, see Willms et al. (1994)

Biliary Ducts and Gallbladder

The digestive function of the liver is to produce bile, a yellow–green secretion, for passage to the duodenum. *Bile is produced in the liver and stored in the gallbladder*, which releases it when fat enters the duodenum. Bile emulsifies the fat and distributes it to the distal intestine for further digestion and absorption.

Normal hepatic tissue, when sectioned, is traditionally described as demonstrating a pattern of hexagonal-shaped *liver lobules* (Fig. 2.55*A*). Each lobule has a **central vein** running through its center from which **sinusoids** (large capillaries) and plates of hepatocytes (liver cells) radiate toward an imaginary perimeter extrapolated from surrounding **interlobular portal triads** (portal vein, hepatic artery, and biliary duct). Although commonly said to be the anatomical units of the liver, hepatic "lobules" are not structural entities; instead, the lobular pattern is a physiological consequence of pressure gradients and is altered by disease. Because the bile duct is not central, the hepatic lobule does not represent a functional unit like acini of other glands. However, the hepatic lobule is a firmly established concept and is useful for descriptive purposes. The hepatocytes secrete bile into the **bile canaliculi** formed between them. The canaliculi drain into the small *interlobular biliary ducts* and then into large collecting bile ducts of the intrahepatic portal triad, which merges to form the right and left hepatic ducts (Fig. 2.55*B*). The **right and left hepatic ducts** drain the right and left lobes of the liver, respectively. Shortly after leaving the porta hepatis, the right and left hepatic ducts unite to form the **common hepatic duct**, which is joined on the right side by the **cystic duct** to form the **bile duct** (part of the extrahepatic portal triad of the lesser omentum), which conveys the bile to the duodenum.

Bile Duct

The **bile duct** (formerly, common bile duct) forms in the free edge of the lesser omentum by the union of the **cystic duct** and the **common hepatic duct** (Fig. 2.55*B*). The length of the bile duct varies from 5 to 15 cm, depending on where the cystic duct joins the common hepatic duct. The bile duct descends posterior to the superior (first) part of the duodenum and lies in a groove on the posterior surface of the head of the pancreas. On the left side of the descending (second) part of the duodenum, the bile duct comes into contact with the **main pancreatic duct**. These ducts run obliquely through the wall of this part of the duodenum, where they unite to form the **hepatopancreatic ampulla** (ampulla of Vater)—the dilation within the major duodenal papilla. The distal end of the ampulla opens into the duodenum through the **major duodenal papilla** (Fig. 2.47). The circular muscle around the distal end of the bile duct is thickened to form the **sphincter of the bile duct** (choledochal sphincter). When this sphincter contracts, bile cannot enter the ampulla and the duodenum; hence, bile backs up and passes along the cystic duct to the **gallbladder** for concentration and storage. The *arteries supplying the bile duct* (Fig. 2.56) include the:

- **Cystic artery**, supplying the proximal part of the duct
- **Right hepatic artery**, supplying the middle part of the duct

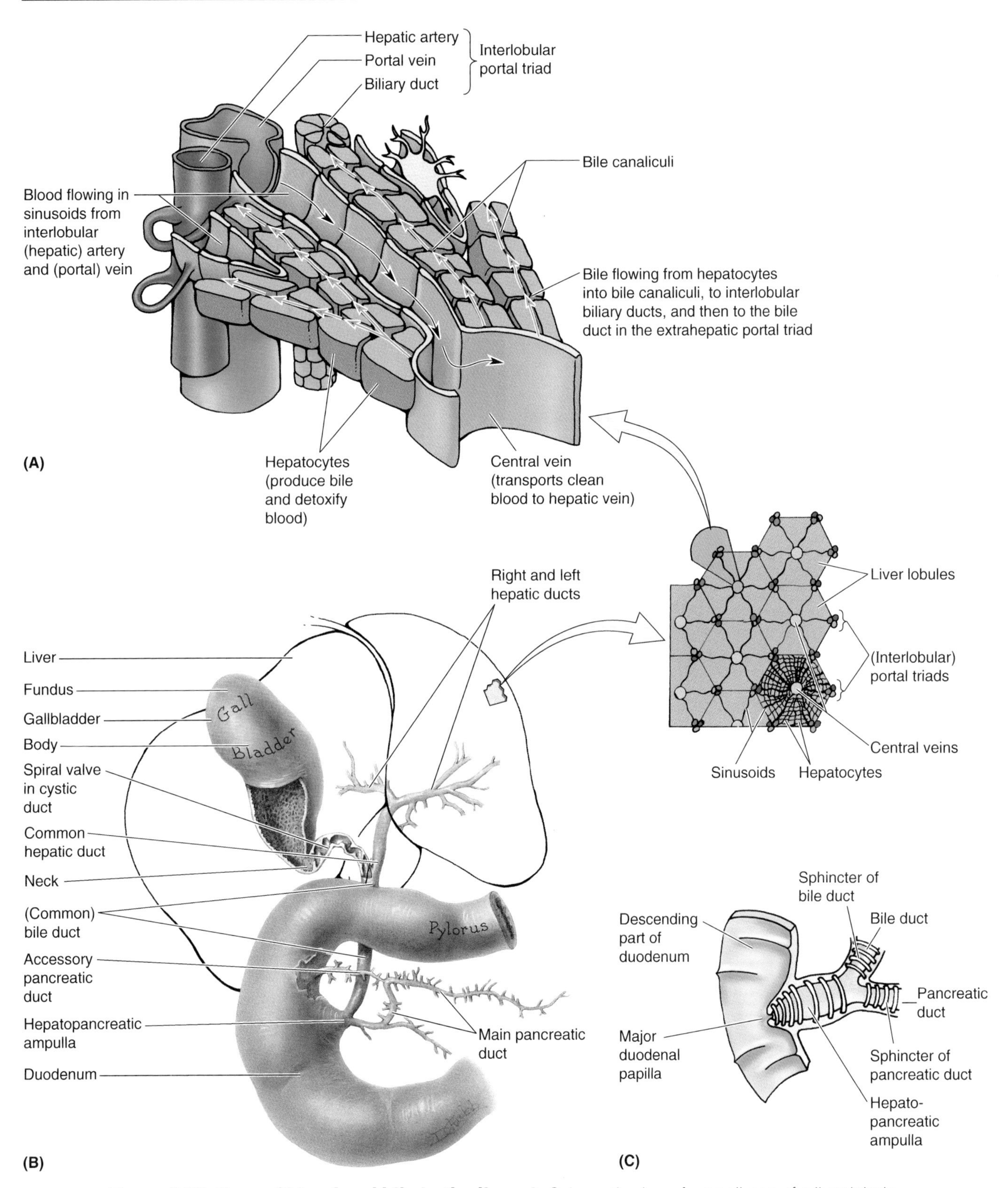

Figure 2.55. Flow of blood and bile in the liver. A. Schematic view of a small part of a liver lobule, illustrating the components of the interlobular portal triad and the positioning of the sinusoids and bile canaliculi. Below this drawing is a schematic view of the cut surface of the liver, showing the hexagonal pattern of lobes. **B.** Extrahepatic bile passages, gallbladder, and pancreatic ducts. **C.** Entry of bile duct and pancreatic duct into the hepatopancreatic ampulla, which opens into the descending part of the duodenum.

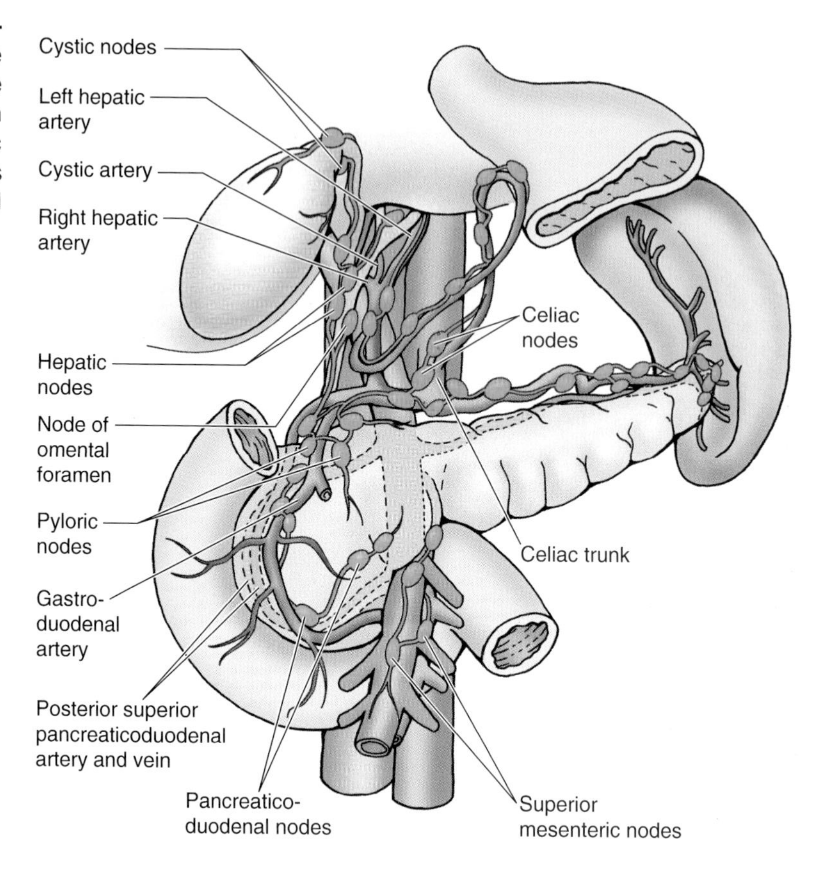

Figure 2.56. Lymphatic drainage of the gallbladder and bile duct. Lymph passes from the cystic and hepatic nodes and the node of the omental (epiploic) foramen to the celiac lymph nodes surrounding the celiac trunk. The lymphatic vessels of the gallbladder and biliary passages anastomose superiorly with those of the liver and inferiorly with those of the pancreas.

- **Posterior superior pancreaticoduodenal artery** and **gastroduodenal artery**, supplying the retroduodenal part of the duct.

The **veins** from the proximal part of the bile duct and the hepatic ducts usually enter the liver directly. The **posterior superior pancreaticoduodenal vein** (Fig. 2.56) drains the distal part of the bile duct and empties into the portal vein or one of its tributaries.

The *lymphatic vessels from the bile duct* pass to the **cystic lymph nodes** near the neck of the gallbladder, the **node of the omental foramen**, and the **hepatic lymph nodes** (Fig. 2.56). Efferent lymphatic vessels from the bile duct pass to the **celiac lymph nodes.**

Gallbladder

The gallbladder (7–10 cm long) lies in the **gallbladder fossa** on the visceral surface of the liver (Figs. 2.50*D*, 2.56, and 2.57). This shallow fossa lies at the junction of the right and left lobes of the liver. The relationship of the gallbladder to the duodenum is so intimate that the superior (first) part of the duodenum is usually stained with bile in the cadaver. The pear-shaped gallbladder has a capacity of up to 50 mL of bile. Peritoneum completely surrounds the fundus of the gallbladder and binds its body and neck to the liver. The hepatic surface of the gallbladder attaches to the liver by connective tissue of the fibrous capsule of the liver. *The gallbladder has three parts* (Figs. 2.55*B* and 2.57).

- The **fundus**, the wide end, projects from the inferior border of the liver and is usually located at the tip of the right 9th costal cartilage in the MCL (Fig.2.26*A*).
- The **body** contacts the visceral surface of the liver, the transverse colon, and the superior part of the duodenum.
- The **neck** is narrow, tapered, and directed toward the porta hepatis. The mucosa of the neck spirals into a fold—the **spiral valve** (Fig. 2.55*B*)—which keeps the cystic duct open so that bile can easily divert into the gallbladder when the distal end of the bile duct is closed by the sphincter of the bile duct and/or the hepatopancreatic sphincter, or when bile can pass to the duodenum as the gallbladder contracts. The neck of the gallbladder makes an S-shaped bend and joins the cystic duct.

The **cystic duct** (approximately 4 cm long) connects the neck of the gallbladder to the common hepatic duct (Figs. 2.55*B* and 2.57). The duct passes between the layers of the lesser omentum, usually parallel to the common hepatic duct, which it joins to form the bile duct.

The **cystic artery** supplying the gallbladder and cystic duct (Figs. 2.56 and 2.58*A*) commonly arises from the right hepatic artery in the angle between the common hepatic duct

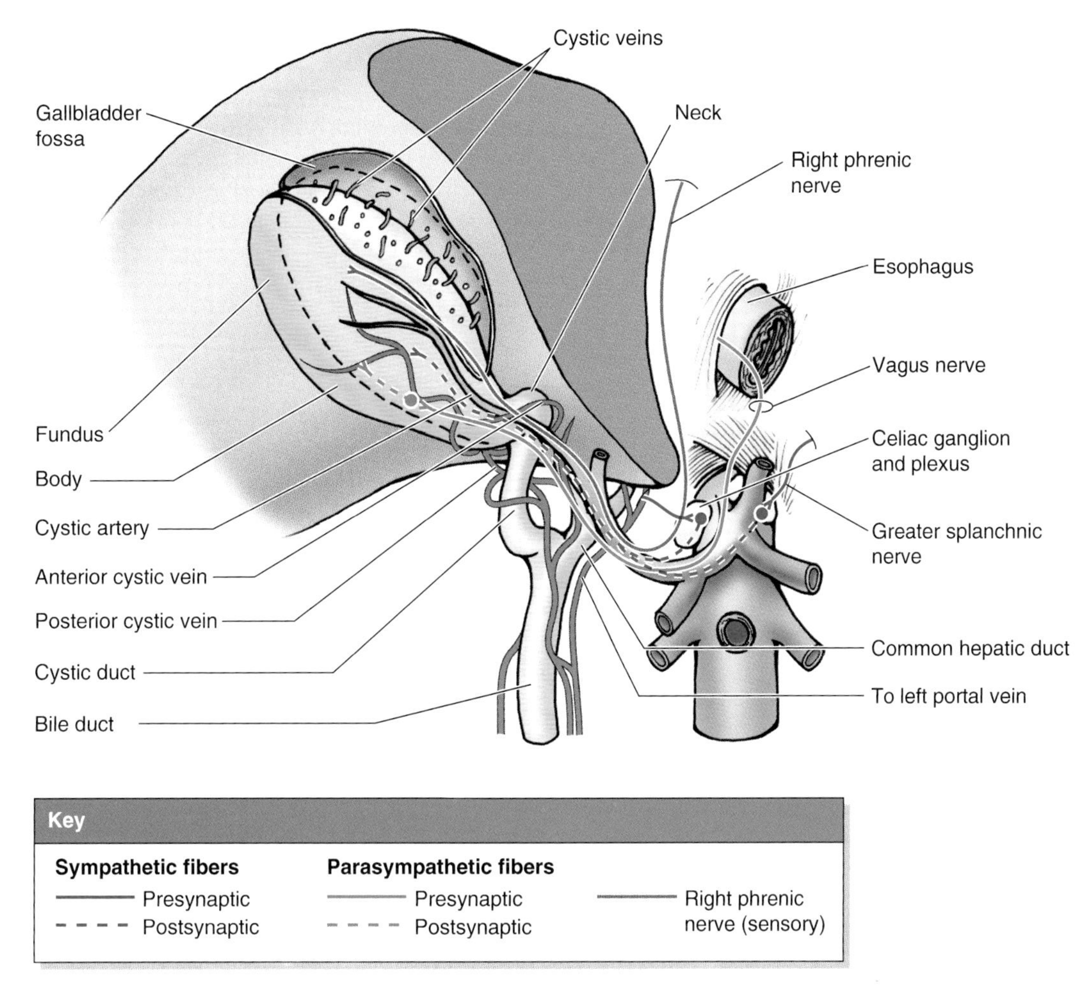

Figure 2.57. Nerves and veins of the liver and biliary system. Nerves are prominent along the hepatic artery, portal vein, and bile duct. The sympathetic nerve supply is vasomotor in the liver and biliary system. The veins of the gallbladder neck communicate with veins along the cystic and bile ducts. Small cystic veins pass from the gallbladder into the liver.

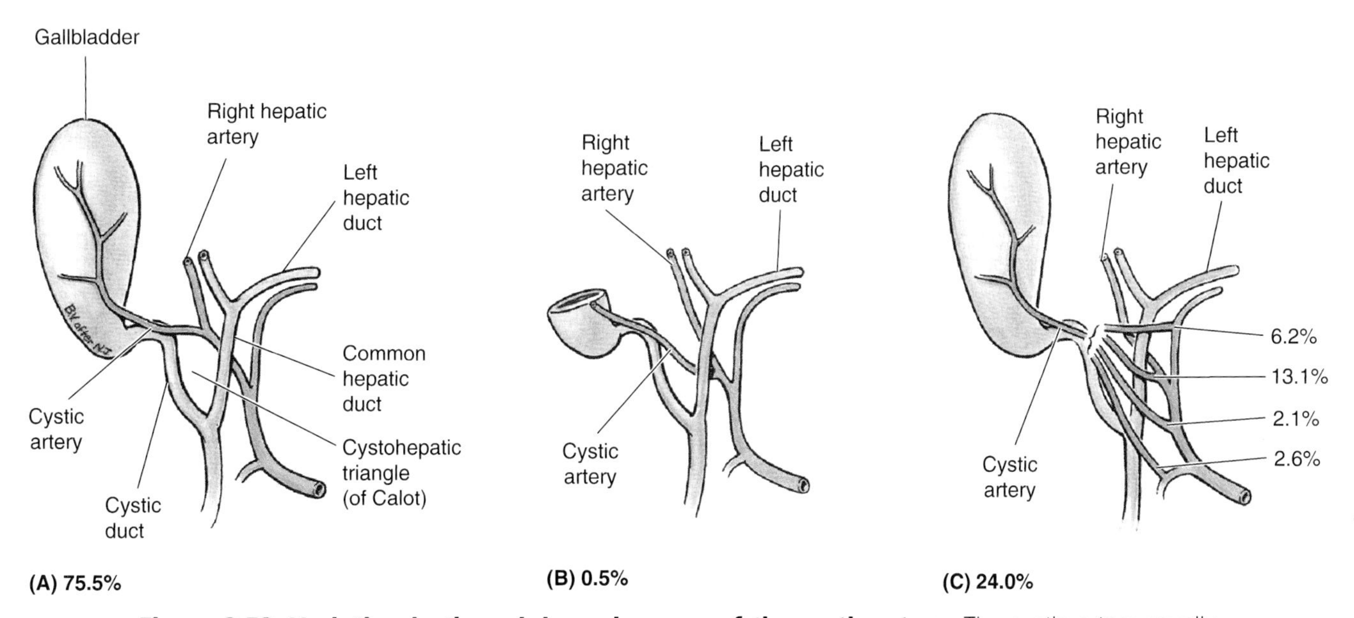

Figure 2.58. Variation in the origin and course of the cystic artery. The cystic artery usually arises from the right hepatic artery in the cystohepatic triangle (of Calot)—bounded by the cystic artery, cystic duct, and common hepatic duct (Fig. 2.47*A*). Variations in the origin and course of the cystic artery occur in 24.5% of people (Daseler et al., 1947), which is of clinical significance during cholecystectomy (surgical removal of the gallbladder).

and the cystic duct. Variations in the origin and course of the cystic artery are common (Fig. 2.58, *B–C*).

The **cystic veins** draining the biliary ducts and neck of the gallbladder enter the liver directly or drain through the portal vein to the liver, or after joining the veins draining the hepatic ducts and upper bile duct (Fig. 2.57). The veins from the fundus and body pass directly into the visceral surface of the liver and drain into the hepatic sinusoids.

The *lymphatic drainage of the gallbladder* is to the **hepatic lymph nodes** (Fig. 2.56), often through **cystic lymph nodes** located near the neck of the gallbladder. Efferent lymphatic vessels from these nodes pass to the **celiac lymph nodes.**

The *nerves to the gallbladder and cystic duct* (Fig. 2.57) pass along the cystic artery from the **celiac plexus** (sympathetic), the **vagus nerve** (parasympathetic), and the **right phrenic nerve** (sensory).

Infundibulum of the Gallbladder

In diseased states of the gallbladder, a dilation or pouch appears at the junction of the neck of the gallbladder and the cystic duct—the *infundibulum of the gallbladder* (Hartmann's pouch). When this pouch is large, the cystic duct arises from its upper left aspect, not from what appears to be the apex of the gallbladder. Gallstones commonly collect in the infundibulum (ampulla). If a peptic duodenal ulcer ruptures, a false passage may form between the infundibulum and the superior part of the duodenum, allowing gallstones to enter the duodenum.

Mobile Gallbladder

The gallbladder has a short mesentery in approximately 4% of individuals. Such gallbladders are subject to vascular torsion and infarction (sudden insufficiency of arterial or venous blood supply).

Variations in the Cystic and Hepatic Ducts

Occasionally, the cystic duct runs alongside the common hepatic duct and adheres closely to it. The cystic duct may be short or even absent. In some people, there is low union of the cystic and common hepatic ducts (*A*); as a result, the bile duct is short and lies posterior to the superior (first) part of the duodenum, or even inferior to it. When there is low union, the two ducts may be joined by fibrous tissue, making clamping the cystic duct difficult without injuring the common hepatic duct. Occasionally there is high union of the cystic and common hepatic ducts near the porta hepatis (*B*). In other cases, the cystic duct spirals anteriorly over the common hepatic duct before joining it on the left side (*C*). Understanding the variations in bile duct formation is important for surgeons when they ligate the cystic duct during *cholecystectomy*—removal of the gallbladder (Gr. cholecyst).

Accessory Hepatic Ducts

Accessory (aberrant) hepatic ducts are common and are in positions of surgical danger during cholecystectomy. An accessory duct is a normal segmental duct that joins the biliary system outside the liver instead of within it (*A–B*). Because it drains a normal segment of the liver, it leaks bile if inadvertently cut during surgery (Skandalakis et al., 1995). Of 95 gallbladders and biliary ducts studied (Agur, 1991), seven had accessory ducts; four joined the common hepatic duct near the cystic ducts; two joined the cystic duct; and one was an anastomosing duct connecting the cystic duct with the common hepatic duct.

Impaction of Gallstones

The distal end of the hepatopancreatic ampulla is the narrowest part of the biliary passages and is the common ▶

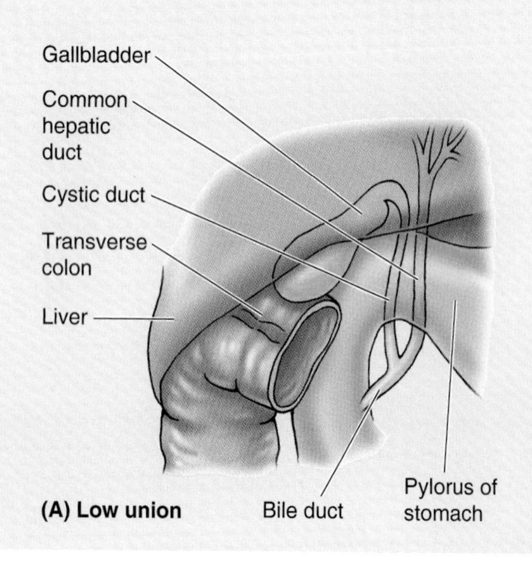

(A) Low union

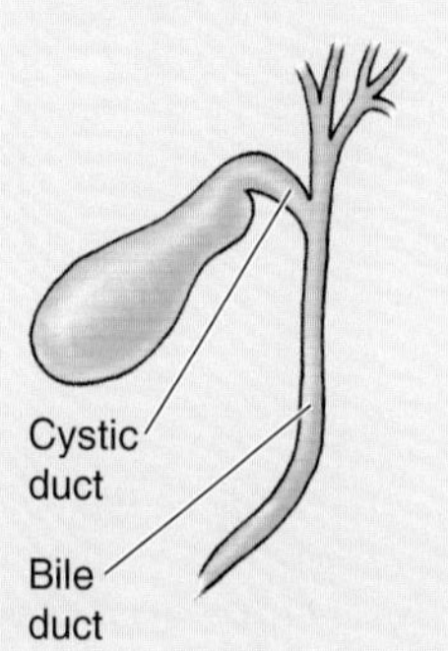

(B) High union

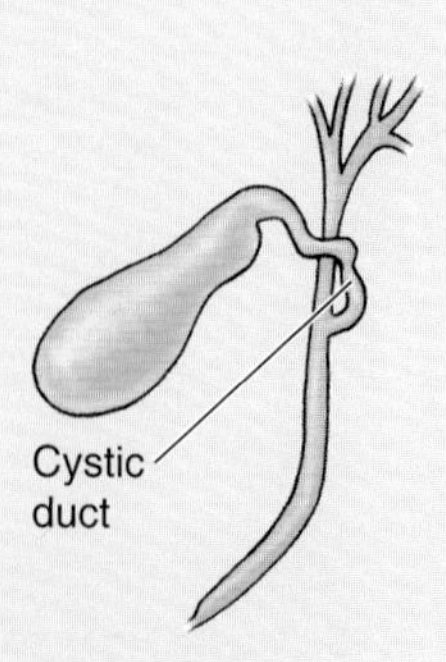

(C) Swerving course

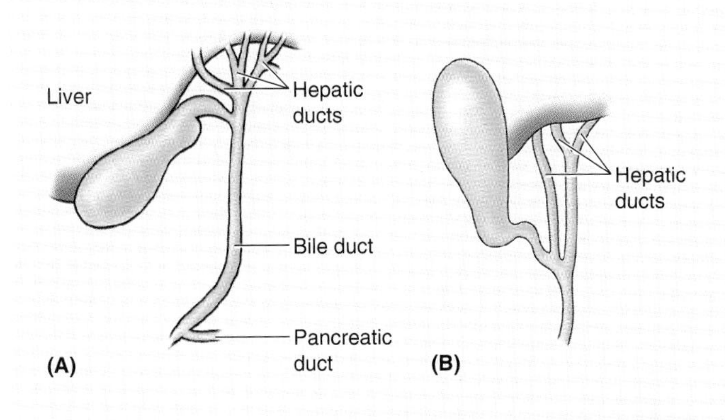

▶ site for impaction of gallstones. The infundibulum of the gallbladder (Hartmann's pouch) is another common site for impaction. Gallstones may also lodge in the hepatic ducts. Ultrasound and CT scans are common noninvasive techniques for locating stones. A stone may also lodge in the cystic duct, causing *biliary colic* (pain in the epigastric region). When the gallbladder relaxes, the stone may pass back into the gallbladder. If the stone blocks the cystic duct, *cholecystitis* (inflammation of the gallbladder) occurs because of bile accumulation, causing enlargement of the gallbladder. Pain develops in the epigastric region and later shifts to the right hypochondriac region at the junction of the 9th costal cartilage and the lateral border of the rectus sheath—indicated by the linea semilunaris. Inflammation of the gallbladder may cause pain in the posterior thoracic wall or right shoulder owing to irritation of the diaphragm. If bile cannot leave the gallbladder, it enters the blood and causes *jaundice* (discussed on p. 262).

Cholecystectomy

People with severe **biliary colic**—intense spasmodic pain in the right upper quadrant of the abdomen resulting from impaction of a gallstone in the cystic duct and acute inflammation of the gallbladder (cholecystitis)—may have their gallbladders removed. *Laparoscopic cholecystectomy* often replaces the open surgical method; for details of this technique, see Skandalakis et al. (1995). Dissection of the **cystohepatic triangle** (Calot's triangle)—bounded by the cystic artery, cystic duct, and common hepatic duct (Fig. 2.58*A*)—early during cholecystectomy safeguards these important structures should there be anatomical variations. *Errors during gallbladder surgery commonly result from failure to appreciate the common variations in the anatomy of the biliary system, especially its blood supply.* Before dividing any structure and removing the gallbladder, surgeons identify all three biliary ducts, as well as the cystic and hepatic arteries. It is usually the right hepatic artery that is in danger during surgery and must be located before ligating the cystic artery. *Bile duct injury is a serious complication of cholecystectomy*, which is estimated to occur in 1 per 600 cases, and the risk appears to be modestly higher for laparoscopic cholecystectomy (Sabiston and Lyerly, 1994). ○

Portal Vein and Portal-Systemic Anastomoses

The **portal vein** is the main channel of the **portal venous system** (Figs. 2.52 and 2.59). It collects poorly oxygenated but nutrient-rich blood from the abdominal part of the GI tract, including the gallbladder, pancreas, and spleen, and carries it to the liver. There it branches to end in expanded capillaries—the **venous sinusoids of the liver** (Fig. 2.55*A*).

The portal venous system communicates with the systemic venous system in the following locations (Fig. 2.59):

- Between the **esophageal veins** draining into either the **azygos vein** (systemic system) or the **left gastric vein** (portal system); when dilated, these are *esophageal varices.*
- Between the **rectal veins**, the inferior and middle draining into the IVC (systemic system), and the superior rectal vein continuing as the IMV (portal system). The submucosal veins involved are normally dilated (varicose in appearance), even in newborns; when the mucosa containing them prolapses, they form *hemorrhoids.* (The varicose appearance of the veins and the occurrence of hemorrhoids is not related to portal hypertension, as is commonly stated.)
- **Paraumbilical veins** of the anterior abdominal wall (portal system) anastomosing with **superficial epigastric veins** (systemic system); when dilated, these veins produce caput medusae—varicose veins radiating from the umbilicus (see the clinical box on portal hypertension, p. 278).
- Twigs of **colic veins** (portal system) anastomosing with retroperitoneal veins (systemic system).

Paraumbilical veins extend along the ligamentum teres in the falciform ligament and the median umbilical ligament in the median umbilical fold. These small veins establish anastomoses between the veins of the anterior abdominal wall and the portal and internal iliac veins.

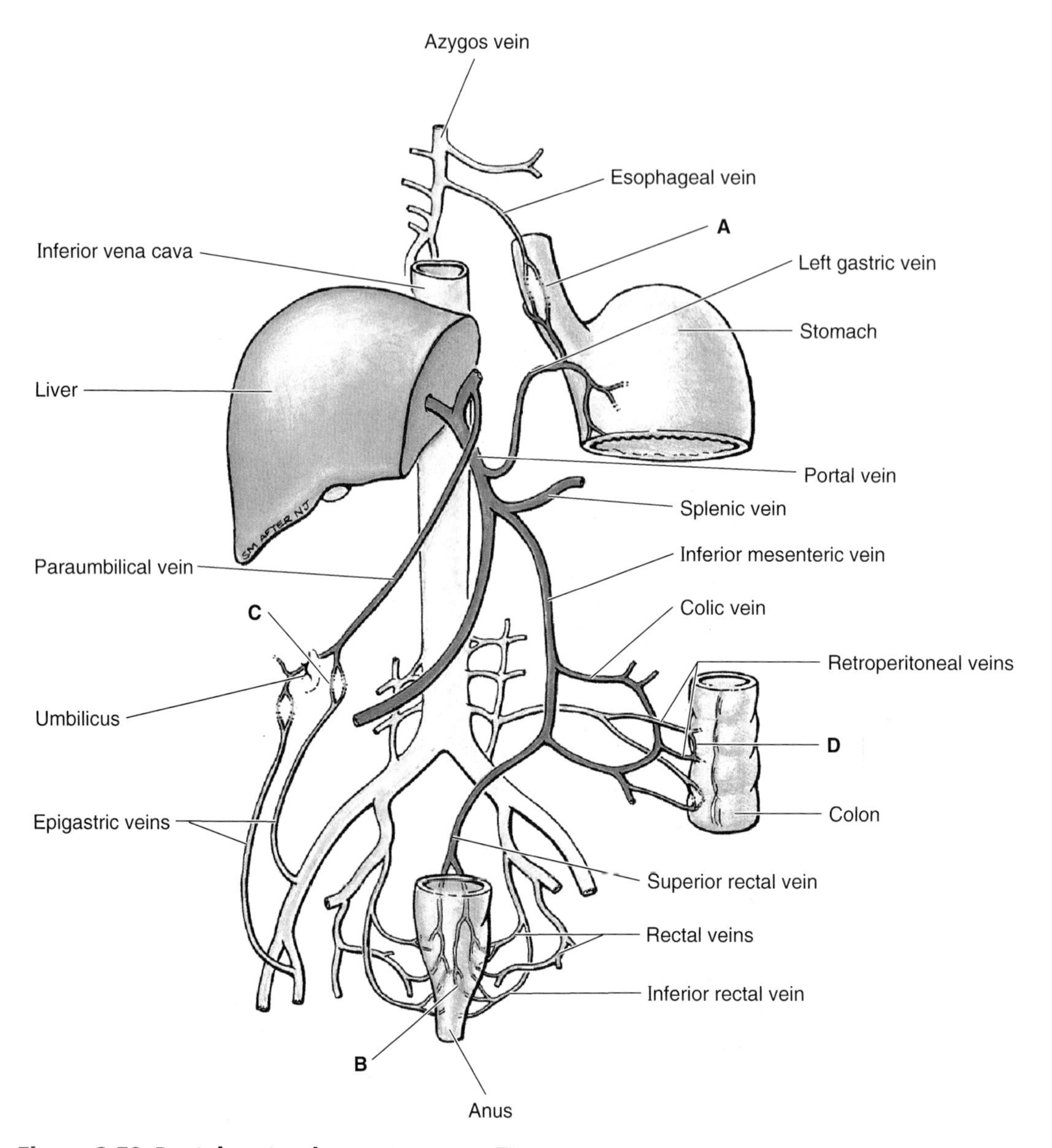

Figure 2.59. Portal-systemic anastomoses. These communications provide a collateral circulation in cases of obstruction in the liver or portal vein. In this diagram, portal tributaries are darker blue and systemic tributaries are lighter blue. **A–D** indicate sites of anastomoses. *A,* between esophageal veins draining into either the azygos vein (systemic) or the left gastric vein (portal); when dilated these are esophageal varices. *B,* between rectal veins, the inferior and middle rectal veins draining into the inferior vena cava (IVC) (systemic) and the superior rectal vein continuing as the inferior mesenteric vein (IMV) (portal); when dilated these are hemorrhoids. *C,* paraumbilical veins (portal) anastomosing with small epigastric veins of the anterior abdominal wall (systemic); may produce the"caput medusae" (p. 279). *D.* Twigs of colic veins (portal) anastomosing with systemic retroperitoneal veins.

Portal-Systemic Anastomoses

The communications between the portal venous system and the systemic venous system are important clinically in the advent of an **intrahepatic** or **extrahepatic portal venous block.** When portal circulation through the liver is diminished or obstructed because of liver disease or physical pressure from a tumor, for example, blood from the GI tract can still reach the right side of the heart through the IVC by way of several collateral routes. These alternate routes are available because the portal vein and its tributaries have no valves; hence, blood can flow in a reverse direction to the IVC.

Portal Hypertension

When scarring and fibrosis from cirrhosis obstruct the portal vein in the liver, pressure rises in the portal vein and ▶

▶ its tributaries—producing *portal hypertension.* At the sites of anastomoses between portal and systemic veins, portal hypertension produces enlarged *varicose veins* and bloodflow from the portal system to the systemic system of veins. The veins may become so dilated that their walls rupture, resulting in hemorrhage.

Bleeding from esophageal varices (dilated veins) at the distal end of the esophagus is often severe and may be fatal. In severe cases of portal obstruction, even the paraumbilical veins may become varicose and look somewhat like small snakes radiating under the skin around the umbilicus. This condition is referred to as *caput medusae* because of its resemblance to the serpents on the head of Medusa, a character in Greek mythology.

Portasystemic Shunts

A common method of reducing portal hypertension is to divert blood from the portal venous system to the systemic venous system by creating a communication between the portal vein and the IVC. This *portacaval anastomosis* or *portasystemic shunt* may be done where these vessels lie close to each other posterior to the liver. Another way of reducing portal pressure is to join the splenic vein to the left renal vein following splenectomy (*splenorenal anastomosis* or *shunt*). For a description of the surgical anatomy of portasystemic shunts, see Skandalakis et al. (1995). ⊙

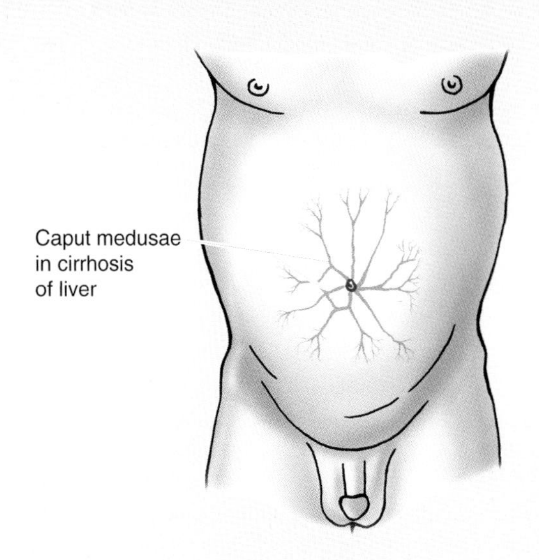

Kidneys, Ureters, and Suprarenal Glands

The **kidneys** lie retroperitoneally on the posterior abdominal wall (Fig. 2.60). These urinary organs remove excess water, salts, and wastes of protein metabolism from the blood while returning nutrients and chemicals to the blood. The kidneys convey the waste products from the blood to the urine through the ureters to the urinary bladder. The **ureters** run inferiorly from the kidneys, passing over the pelvic brim at the bifurcation of the common iliac arteries. They then run along the lateral wall of the pelvis and enter the **urinary bladder**. The superomedial aspect of each kidney normally contacts a **suprarenal (adrenal) gland** enclosed in a fibrous capsule and a cushion of *perirenal fat* (Figs. 2.61 and 2.62). A weak septum of **renal fascia** separates these glands from the kidneys so that they are not actually attached to each other. The suprarenal glands function as part of the endocrine system, completely separate in function from the kidneys. They secrete corticosteroids and androgens and make epinephrine and norepinephrine hormones.

The **perirenal fat**, derived from extraperitoneal fat, is continuous at the **renal hilum** with fat in the **renal sinus** (Fig. 2.61*B*). External to the renal fascia is **pararenal fat**, which is most obvious posterior to the kidney; the renal fascia sends collagen bundles through the fat. The collagen bundles, renal fascia, and perirenal and pararenal fatty tissue, along with the tethering provided by the renal vessels and ureter, hold the kidneys in a relatively fixed position. However, movement of the kidneys occurs during respiration and when changing from the supine to the erect position, and vice versa. Normal renal mobility is approximately 3 cm, approximately the height of one vertebral body. Superiorly, the renal fascia is continuous with the fascia on the inferior surface of the diaphragm (diaphragmatic fascia); thus, the primary attachment of the suprarenal glands is to the diaphragm. Inferiorly the two layers are only loosely united, if attached at all.

Retroperitoneal Pneumography

The loose attachment of the anterior and posterior layers of renal fascia was demonstrated by this clinical radiological procedure, in which air injected into the presacral fatty tissue ascended through the retroperitoneal fat and entered the renal fascia, radiographically outlining the kidney and suprarenal gland. This procedure was largely outdated by modern cross-sectional imaging techniques such as CT, MRI, and ultrasound.

Perinephric Abscess

The attachments of the renal fascia determine the path of extension of a perinephric abscess. For example, fascia at the renal hilum attaches to the renal vessels and ureter, usually preventing the spread of pus to the contralateral side. However, pus from an abscess (or blood from an injured kidney) may force its way into the pelvis between the loosely attached anterior and posterior layers of the pelvic fascia. ▶

Nephroptosis

Because the layers of renal fascia do not fuse firmly inferiorly to offer resistance, abnormally mobile kidneys may descend more than the normal 3 cm when the body is erect. When kidneys descend, the suprarenal glands remain in place because they lie in a separate fascial compartment and are most firmly attached to the diaphragm. *Nephroptosis* ("dropped kidney") is distinguished from an *ectopic kidney* (congenital misplaced kidney) by a ureter of normal length, demonstrating loose coiling or kinks because the distance to the bladder has been reduced. The kinks do not seem to be of significance. Symptoms of intermittent pain in the renal region, relieved by lying down, appear to result from traction on the renal vessels. The lack of inferior support for the kidneys in the lumbar region is one of the reasons transplanted kidneys are placed in the iliac fossa of the greater pelvis. (Other reasons for this placement are the availability of major blood vessels and convenient access to the nearby bladder.) ⊙

Kidneys

The ovoid kidneys lie retroperitoneally on the posterior abdominal wall, one on each side of the vertebral column at the level of T12 through L3 vertebrae (Figs. 2.60 and 2.62). *The right kidney usually lies slightly inferior to the left kidney* because of the large size of the right lobe of the liver. During life the kidneys are reddish-brown and measure approximately 10 cm in length, 5 cm in width, and 2.5 cm in thickness. Superiorly, the kidneys are associated with the diaphragm, which separates them from the pleural cavities and the 12th pair of ribs. More inferiorly, the posterior surfaces of the kidney are related to the *quadratus lumborum muscle* (Figs. 2.60 and 2.61). The subcostal nerve and vessels and the iliohypogastric and ilioinguinal nerves descend diagonally across the posterior surfaces of the kidneys. The liver, duodenum, and ascending colon are anterior to the right kidney (Fig. 2.63). The right kidney is separated from the liver by the **hepatorenal recess**. The *left kidney* is related to the stomach, spleen, pancreas, jejunum, and descending colon.

At the concave medial margin of each kidney is a vertical cleft—**the renal hilum**—where the renal artery enters and the renal vein and renal pelvis leave the renal sinus (Fig. 2.64*A*). At the hilum, the **renal vein** is anterior to the **renal artery**, which is anterior to the renal pelvis. The renal hilum is the entrance to a space within the kidney—**the renal sinus** (Fig. 2.64*B*)—that is occupied by the renal pelvis, calices, vessels and nerves, and a variable amount of fat (Fig. 2.64, *C–D*). Each kidney has anterior and posterior surfaces, medial and lateral margins, and superior and inferior poles. The lateral margin is convex and the medial margin is concave where the renal sinus and renal pelvis are located. The indented medial margin gives the kidney a somewhat kidney bean-shaped appearance.

The **ureters** are muscular ducts (25–30 cm long) with narrow lumina that carry urine from the kidneys to the urinary bladder (Figs. 2.60 and 2.64). The abdominal parts of the ureters adhere closely to the parietal peritoneum and are retroperitoneal throughout their course. As demonstrated radiographically using contrast medium, *the ureters are normally constricted to a variable degree in three places*:

- At the junction of the ureters and renal pelves
- Where the ureters cross the brim of the pelvic inlet
- During their passage through the wall of the urinary bladder.

These constricted areas are potential sites of obstruction by ureteric (kidney) stones.

The **renal pelvis** is the flattened, funnel-shaped expansion of the superior end of the ureter (Fig. 2.64, *B–D*). The apex of the renal pelvis is continuous with the ureter. The renal pelvis receives two or three **major calices** (calyces), each of which divides into two or three **minor calices**. Each minor calix (calyx) is indented by the **renal papilla**, the apex of the *renal pyramid*.

Renal Transplantation

The kidney can be removed from the donor without damaging the suprarenal gland because of the weak septum of renal fascia that separates the kidney from this gland. Renal transplantation is now an established operation for the treatment of selected cases of chronic renal failure. The site for transplanting a kidney is in the iliac fossa of the greater pelvis. The renal artery and vein are joined to the external iliac artery and vein, respectively, and the ureter is sutured into the urinary bladder.

Renal Cysts

Cysts in the kidney, multiple or solitary, are common findings during dissection of cadavers. Adult *polycystic disease of the kidneys* is an important cause of renal failure that is inherited as an autosomal-dominant trait. The kidneys are markedly enlarged and distorted by cysts as large as 5 cm.

Pain in the Pararenal Region

The close relationship of the kidneys to the psoas major muscles explains why extension of the hip joints may increase pain resulting from inflammation in the pararenal areas. These muscles flex the thighs at the hip joints. ⊙

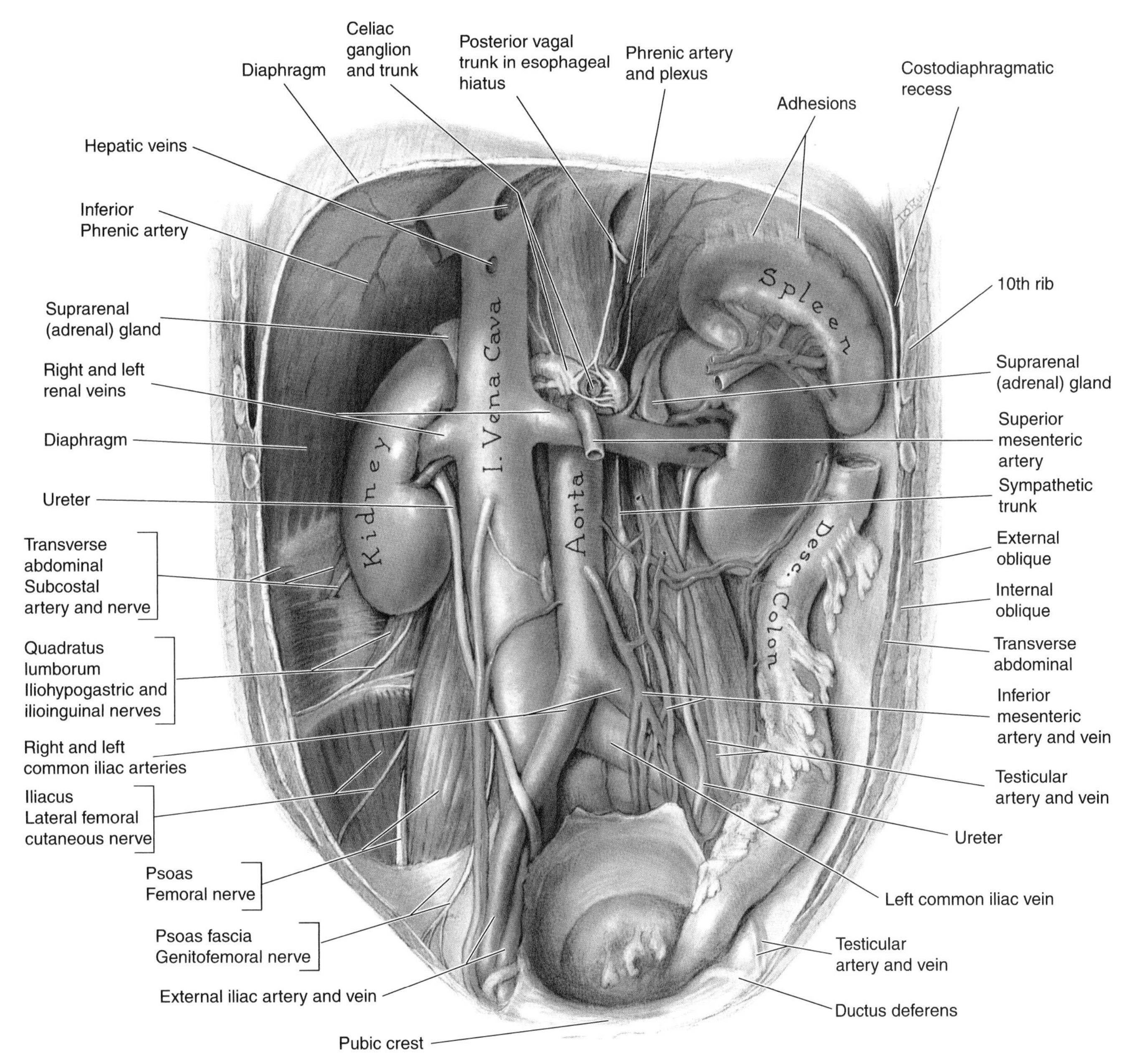

Figure 2.60. Posterior abdominal wall showing great vessels, kidneys, and suprarenal glands. Most of the fascia has been removed. Observe that the ureter crosses the external iliac artery just beyond the common iliac bifurcation and that the testicular vessels cross anterior to the ureter and join the ductus deferens (vas deferens) to enter the inguinal canal. Note that the renal arteries are not seen because they lie posterior to the renal veins. Note also that the left renal vein is compressed between the aorta posteriorly and the superior mesenteric artery (SMA), which bears the weight of the intestine.

Surface Anatomy of the Kidneys and Ureters

The hilum of the left kidney lies near the transpyloric plane, approximately 5 cm from the median plane. The transpyloric plane passes through the superior pole of the right kidney, which is approximately 2.5 cm lower than the left pole. Posteriorly, the superior parts of the kidneys lie deep to the 11th and 12th ribs. The levels of the kidneys change during respiration and with changes in posture. Each kidney moves approximately 3 cm in a vertical direction during the movement of the diaphragm that ▶

In living persons, the renal pelvis and its calices are usually collapsed (empty). The lobes of the kidney are formed by the pyramids and their associated cortex. The lobes are visible on the external surfaces of the kidneys in fetuses, and evidence of the lobes may persist for some time after birth (Moore and Persaud, 1998).

The **arteries to the ureter** (Fig. 2.65) arise mainly from the:

- Renal arteries
- Testicular or ovarian arteries
- Abdominal aorta.

However, arteries may arise from the common iliac, internal iliac, inferior vesical (male), and uterine arteries (female). When the arteries reach the ureter, they divide into ascending and descending branches. The longitudinal anastomosis between these branches on the ureteric wall is good.

The **veins of the ureters** drain into the renal and testicular or ovarian veins (Fig. 2.65). The *lymphatic vessels of the ureter* join the renal collecting vessels or pass directly to **lumbar (aortic) lymph nodes** and the **common iliac lymph nodes** (see Fig. 2.67). Lymph drainage from the pelvic parts of the ureters is into the common, external, and **internal iliac lymph nodes.**

The **nerves of the ureters** derive from the renal, aortic, and superior and inferior hypogastric plexuses (Fig. 2.66*B*). Visceral afferent fibers conveying pain sensation (e.g., resulting from obstruction and consequent distension) follow the sympathetic fibers retrogradely to spinal ganglia and cord segments T11 through L2. Ureteric pain is usually referred to the ipsilateral lower quadrant of the anterior abdominal wall and especially to the groin (see the clinical box "Renal and Ureteric Calculi," p. 288).

▶ occurs with deep breathing. Because the usual surgical approach to the kidneys is through the posterior abdominal wall, it is helpful to know that the inferior pole of the right kidney is approximately a fingersbreadth superior to the iliac crest.

In extremely muscular and/or obese people, the kidneys may be impalpable. In most adults, the inferior pole of the right kidney is palpable by bimanual examination as a firm, smooth, somewhat rounded mass that descends during inspiration. Palpation of the right kidney is possible because it is 1 to 2 cm inferior to the left one. To palpate the kidneys, press the flank—the side part of the body between ribs and pelvis—anteriorly with one hand while palpating deeply at the costal margin with the other. The left kidney is usually not palpable unless it is enlarged or a retroperitoneal mass has displaced it inferiorly. ○

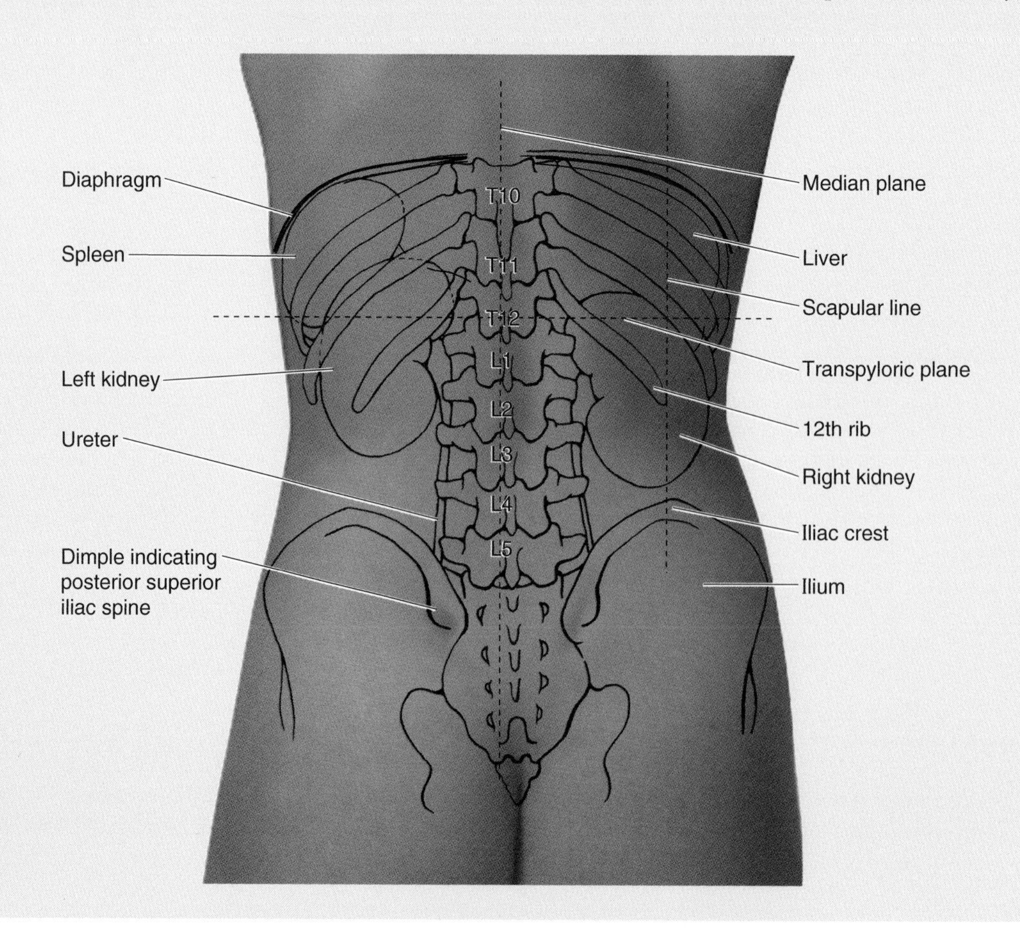

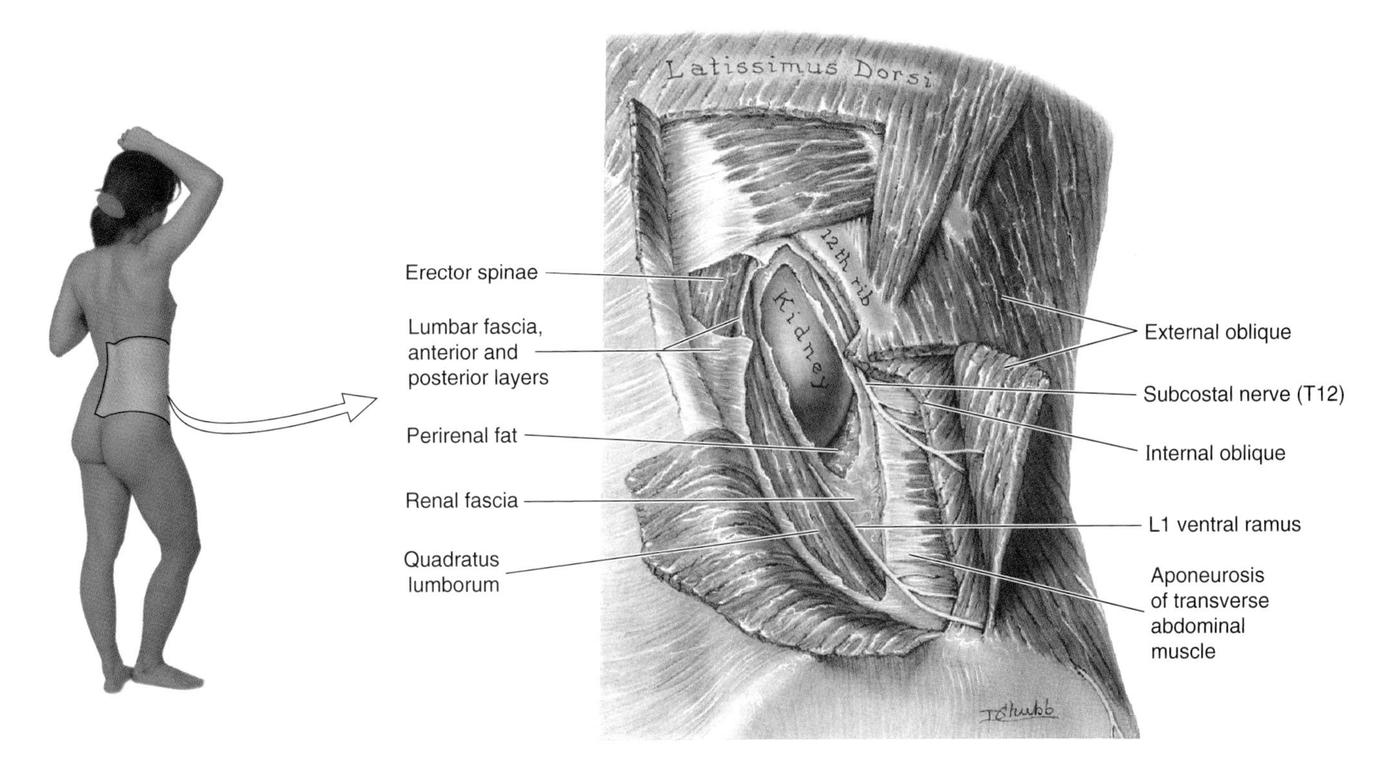

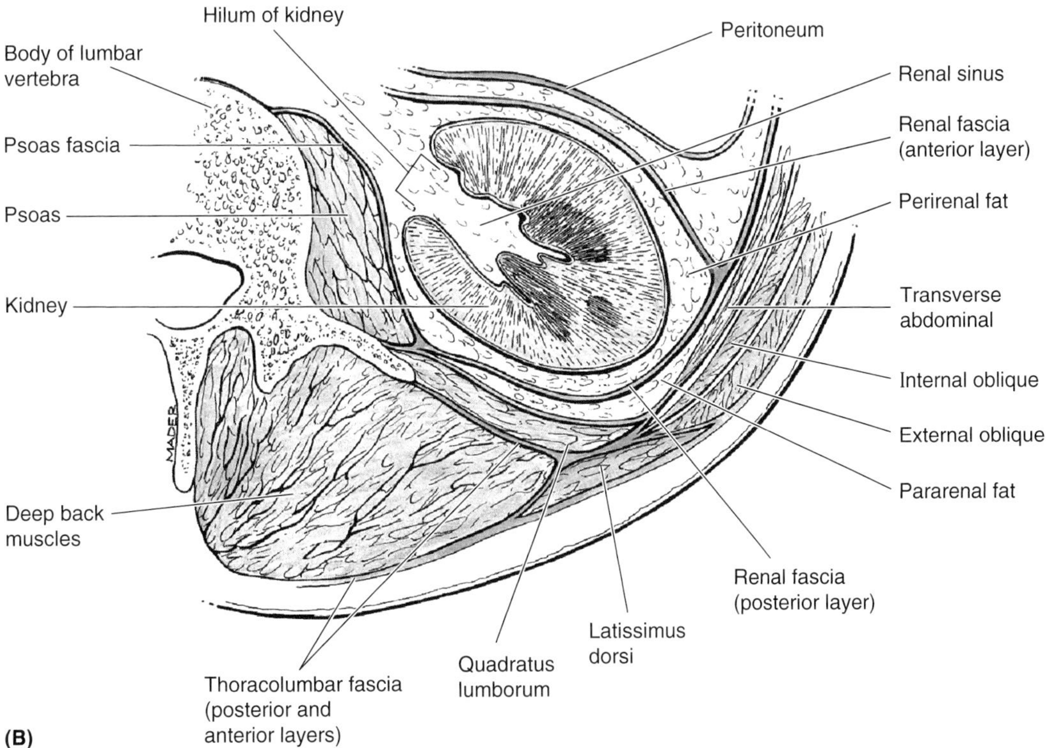

Figure 2.61. External aspect of the right posterior abdominal wall showing the lumbar approach to the kidney and the relationships of the muscles and fascia. A. On dividing the posterior aponeurosis of the transverse abdominal muscle (L. transversus abdominis) between the subcostal and iliohypogastric nerves, and lateral to the oblique lateral border of the quadratus lumborum, the retroperitoneal fat surrounding the kidney is exposed. The renal fascia is within this fat; the fat inside the renal fascia is termed the "fatty renal capsule"(perirenal fat); the fat outside the capsule is pararenal fat (see Fig. 2.73 for an earlier stage of dissection). **B.** Transverse section of the kidney showing the relationships of muscle and fascia. Because the renal fascia surrounds the kidney as a separate sheath, it must be incised in any surgical operation on the kidney, whether from an anterior or a posterior approach.

Figure 2.62. Arterial supply of the kidneys and ureters. Observe the relationship of the abdominal aorta to the vertebral column. Observe also the arterial supply of the kidneys and ureters. Note that there is an accessory left renal artery. Multiple renal arteries are common and usually enter the hilum of the kidney as shown here. Extrahilar renal arteries from the renal artery or aorta may enter the external surface of the kidney, commonly at their poles. Their presence presents a hazard during operations on the kidney.

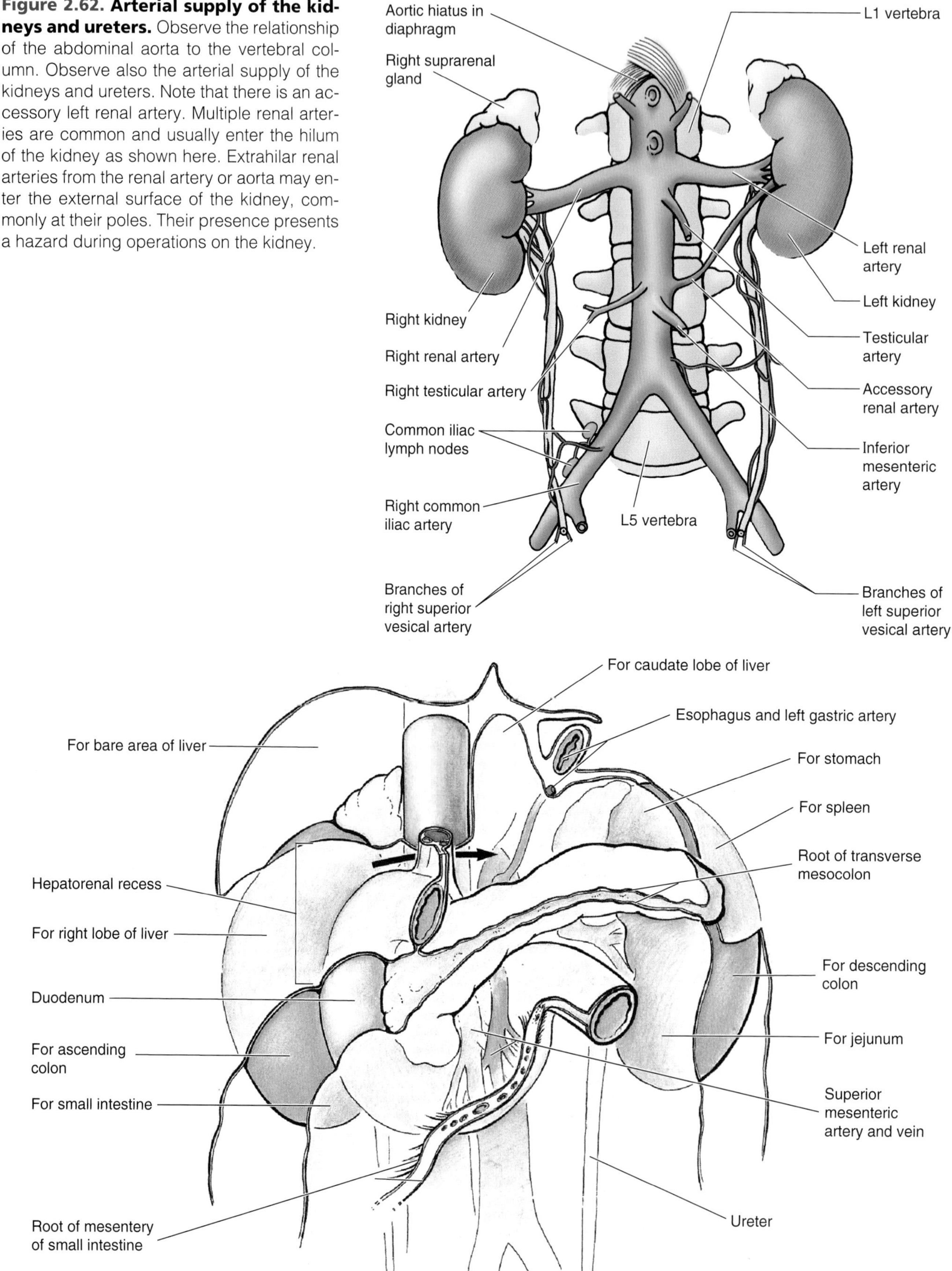

Figure 2.63. Anterior relationships of the kidneys, pancreas, and duodenum. Observe the relationships of the kidneys to the suprarenal glands, liver, colon, duodenum, stomach, pancreas, and spleen. Observe also that the right suprarenal gland is at the level of the omental (epiploic) foramen, indicated by the *black arrow*.

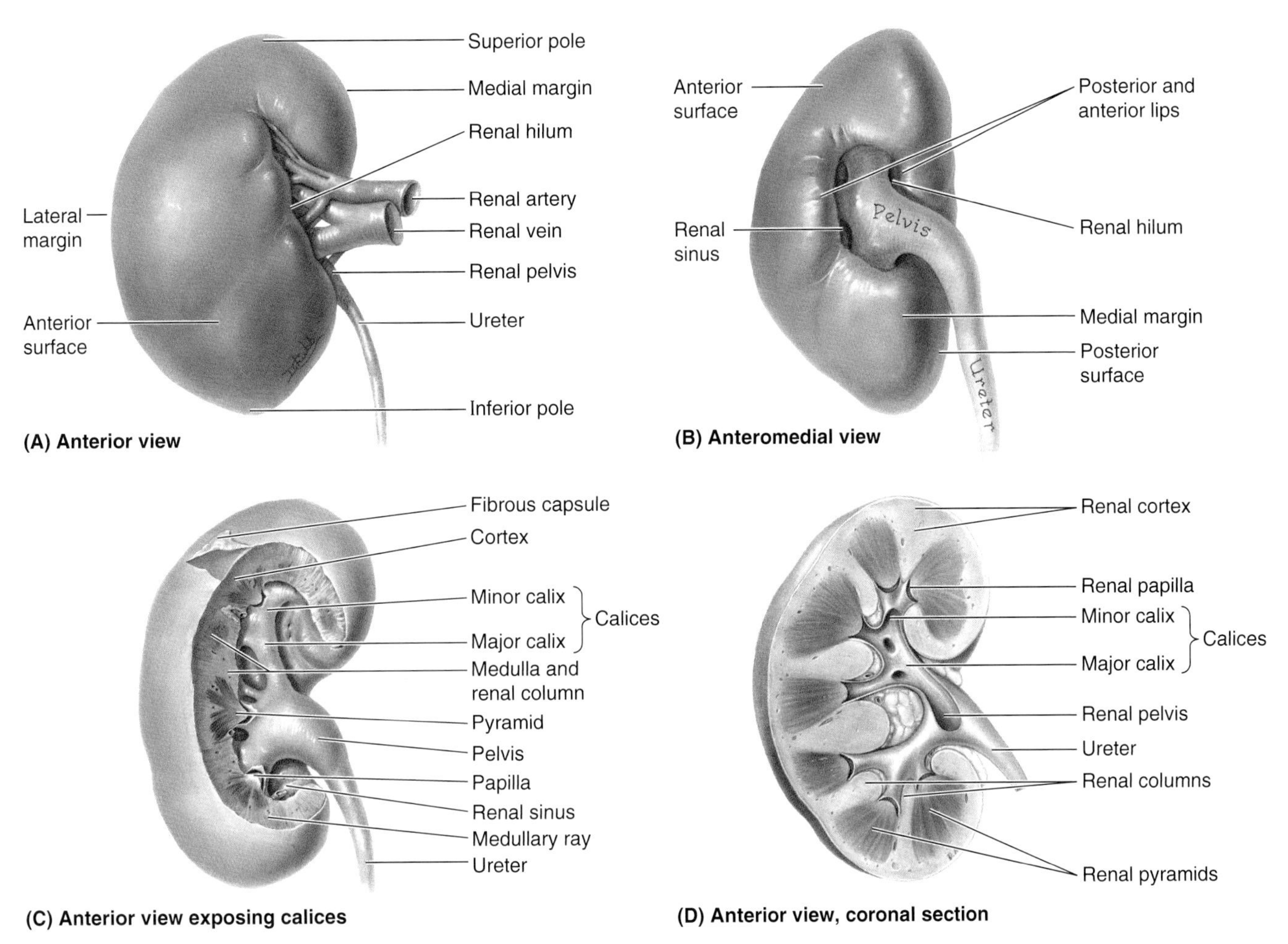

Figure 2.64. External and internal appearance of the kidneys. A. Right kidney, anterior view. Observe the order of the structures at the renal hilum. **B.** Renal sinus as seen through the renal hilum. As shown here and in (**A**), the renal sinus contains the renal pelvis and renal vessels. **C.** The anterior lip of the renal hilum has been cut away to expose the renal pelvis and calices within the renal sinus. **D.** Coronal section of the kidney showing its internal structure. The renal pyramids contain the collecting tubules and form the medulla of the kidney. The renal cortex contains the renal corpuscles. Observe that the blunted apices of the pyramids (the renal papillae) project into the minor calices, into which they discharge urine, which passes into the major calices and renal pelvis.

Suprarenal Glands

The suprarenal (adrenal) glands are located between the superomedial aspects of the kidneys and the diaphragm (Fig. 2.65), where they are surrounded by connective tissue containing considerable perinephric fat. The glands are enclosed by renal fascia by which they are attached to the diaphragm; however, they are separated from the kidneys by fibrous tissue. *The shape and relations of the suprarenal glands differ on the two sides.*

- The triangular right gland lies anterior to the diaphragm and makes contact with the IVC anteromedially (Fig. 2.60) and the liver anterolaterally
- The semilunar left gland is related to the spleen, stomach, pancreas, and the left crus of the diaphragm.

The medial borders of the suprarenal glands are 4 to 5 cm apart. In this area, from right to left, are the IVC, right crus of the diaphragm, celiac ganglion, celiac trunk, SMA, and the left crus of the diaphragm.

Each suprarenal gland, enclosed by a fibrous capsule and a cushion of fat, has two parts: the suprarenal cortex and suprarenal medulla. These parts have different embryological origins and different functions (Moore and Persaud, 1998).

The **suprarenal cortex** derives from mesoderm and secretes corticosteroids and androgens. These hormones cause the kidneys to retain sodium and water in response to stress, increasing the blood volume and blood pressure. They also affect muscles and organs such as the heart and lungs.

The **suprarenal medulla** is a mass of nervous tissue—permeated with capillaries and sinusoids—that derives from neural crest cells associated with the sympathetic nervous system. The *chromaffin cells* of the medulla are related to sympathetic ganglion (postsynaptic) neurons in both derivation (neural

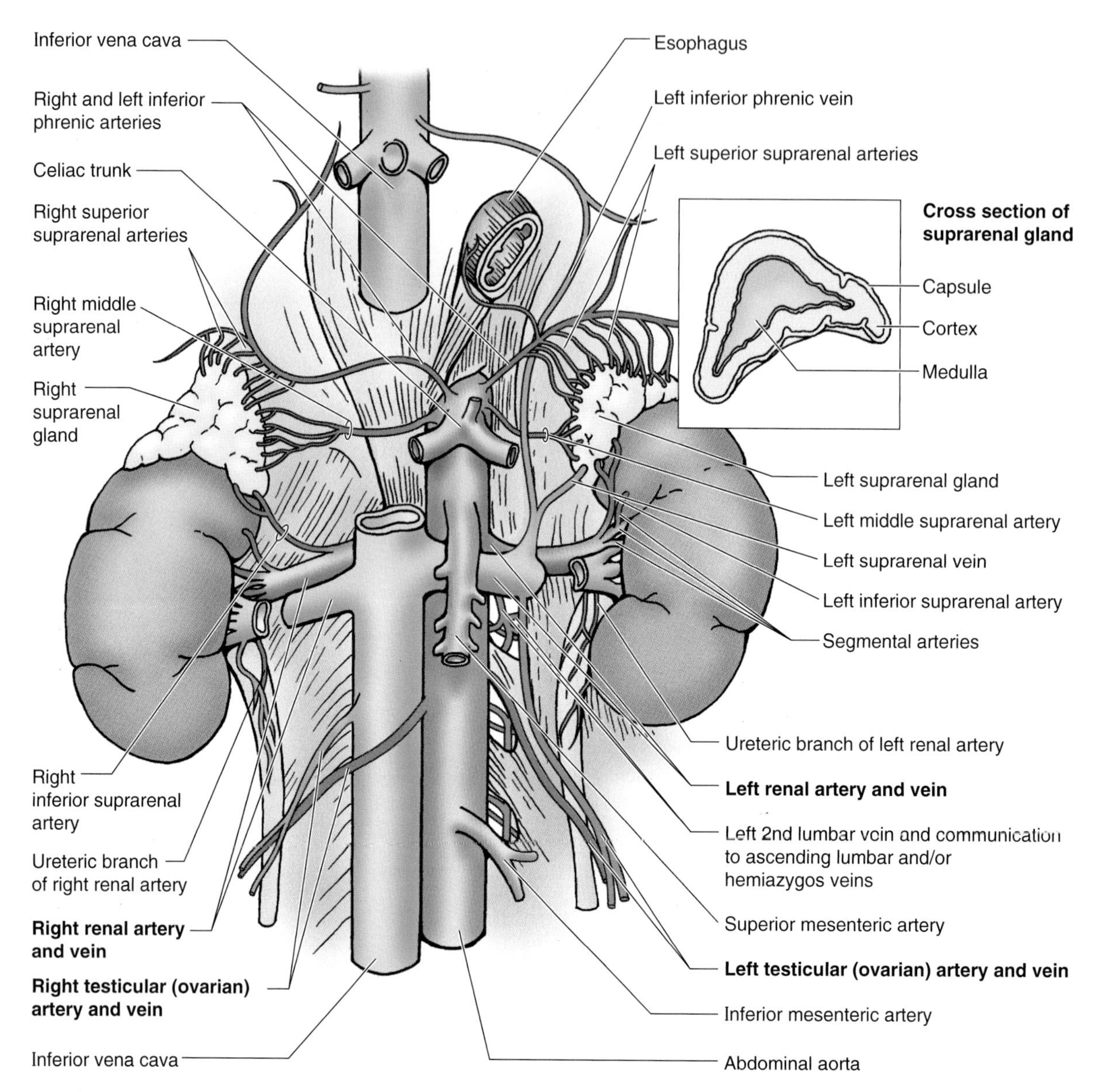

Figure 2.65. Great vessels, suprarenal glands, and kidneys. Observe the celiac trunk; the celiac plexus of nerves and ganglia that surrounds it is removed. The inferior vena cava (IVC) has been transected and the upper part has been elevated from its normal position. The renal veins have been cut and the kidneys have been moved laterally. For the normal relationships of the kidneys and suprarenal glands with the great vessels, see Figure 2.60. Observe the gross structure of the suprarenal glands and their rich arterial supply. The cross section of the suprarenal gland shows that it is composed of two distinct parts, the cortex and medulla, which are two separate endocrine glands that become closely related during embryonic development.

crest cells) and function (Fig. 2.65). These cells secrete catecholamines (mostly epinephrine) into the bloodstream in response to signals from presynaptic neurons (Naftel and Hardy, 1997). The powerful medullary hormones epinephrine (adrenalin) and norepinephrine (noradrenaline) activate the body to a flight-or-fight status in response to traumatic stress. They also increase heart rate and blood pressure, dilate the bronchioles, and change bloodflow patterns for increased alertness. Medullary hormones also play a role in hypertension.

Vessels of the Kidneys and Suprarenal Glands. The renal arteries arise at the level of the IV disc between L1 and L2 vertebrae (Figs. 2.62 and 2.65). The longer **right renal artery** passes posterior to the IVC. Typically, each artery divides close to the hilum into five segmental arteries that are end arteries (i.e., they do not anastomose). *Segmental arteries are distributed to the segments of the kidney* (Fig. 2.66) *as follows*:

- The superior (apical) segment is supplied by the **superior (apical) segmental artery**; the anterosuperior and anteroinfe-

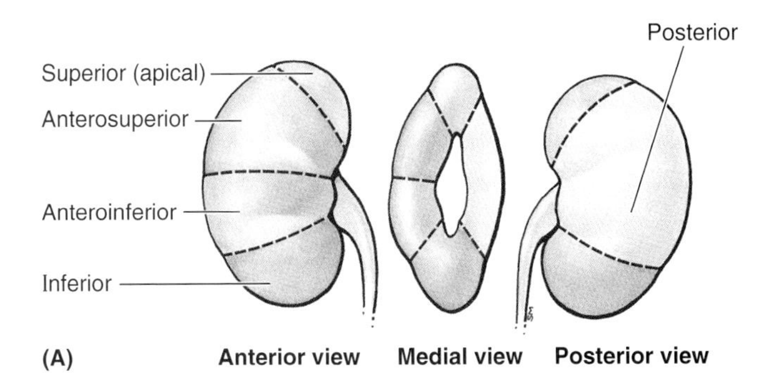

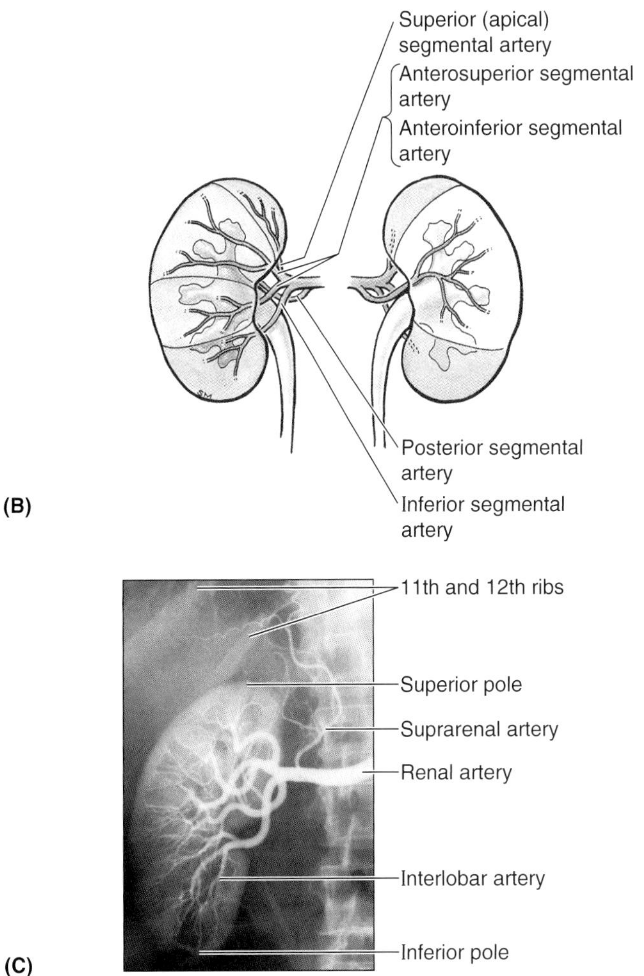

Figure 2.66. Renal segments and segmental arteries. **A.** Segments of the kidney. According to its arterial supply, the kidney has five segments: superior (apical), anterosuperior, anteroinferior, inferior, and posterior. **B.** Segmental arteries. Only the superior and inferior arteries supply the whole thickness of the kidney. The posterior artery crosses superior to the renal pelvis to reach its segment. **C.** Renal arteriogram. Typically, the renal artery divides into five branches, each supplying a segment of the kidney. A segmental artery provides a lobar artery to each pyramid; these divide to provide two or three interlobar arteries that travel between pyramids. Near the junction of the medulla and cortex, arcuate arteries are given off at right angles to the parent stem—these do not anastomose. From the arcuate arteries (and some from the interlobar arteries), interlobular arteries pass into the cortex; the arterioles supplying the glomeruli are mainly from these interlobular arteries. Although the veins of the kidney anastomose freely, segmental arteries are end arteries. (Courtesy of Dr. E.L. Lansdown, Professor of Medical Imaging, University of Toronto, Toronto, Ontario, Canada)

rior segments are supplied by the **anterosuperior segmental** and **anteroinferior segmental arteries**, and the inferior segment is supplied by the **inferior segmental artery**. These arteries originate from the anterior branch of the renal artery.

- The **posterior segmental artery**, which originates from a continuation of the posterior branch of the renal artery, supplies the posterior segment of the kidney.

The endocrine function of the suprarenal glands makes their abundant blood supply necessary. The **suprarenal arteries** (Fig. 2.65) branch freely before entering each gland so that 50 to 60 arteries penetrate the capsule covering the entire surface of the glands. *The suprarenal glands are supplied by:*

- **Superior suprarenal arteries** (six to eight) from the inferior phrenic artery
- **Middle suprarenal arteries** (one or more) from the abdominal aorta near the level of origin of the SMA
- **Inferior suprarenal arteries** (one or more) from the renal artery.

Several *renal veins* drain the kidney and unite in a variable fashion to form the **renal vein.** The renal veins lie anterior to the renal arteries, and the longer left renal vein passes anterior to the aorta. Each renal vein drains into the IVC. *The venous drainage of the suprarenal gland* is into a large **suprarenal vein** (Fig. 2.65). The short right suprarenal vein drains into the IVC, whereas the longer left suprarenal vein, often joined by the inferior phrenic vein, empties into the left renal vein.

The *renal lymphatic vessels* follow the renal veins and drain into the **lumbar (aortic) lymph nodes** (Fig. 2.67*A*). Lymphatic vessels from the superior part of the ureter may join those from the kidney or pass directly to the lumbar nodes. Lymphatic vessels from the middle part of the ureter usually drain into the **common iliac lymph nodes**, whereas vessels from its inferior part of the kidney drain into the common, external, or internal **iliac lymph nodes.** The *suprarenal lymphatic vessels* arise from a plexus deep to the capsule of the gland and from one in its medulla. The lymph passes to the **lumbar lymph nodes.** Many lymphatic vessels leave the suprarenal glands.

Nerves of the Kidneys and Suprarenal Glands. Nerves to the kidneys arise from the renal plexus and consist of sympathetic and parasympathetic fibers (Fig. 2.67*B*). The **renal plexus** is supplied by fibers from the thoracic (especially the least) splanchnic nerves. The suprarenal glands have a rich nerve supply from the **celiac plexus** and **thoracic splanchnic nerves.** The nerves are mainly myelinated presynaptic sympathetic fibers that derive from the lateral horn of the spinal cord and are distributed to the chromaffin cells in the suprarenal medulla (Fig. 2.67*C*).

Accessory Renal Vessels

During their "ascent" to their final site, the embryonic kidneys receive their blood supply and venous drainage from successively more superior vessels (Moore and Persaud, 1998). Usually the inferior vessels degenerate as superior ones take over the blood supply and venous drainage. Failure of these vessels to degenerate results in *accessory renal arteries and veins* (known as "polar arteries and veins" when they enter/exit the poles of the kidneys). Variations in the number and position of these vessels occur in approximately 25% of people.

Renal and Ureteric Calculi

Calculi (L. pebbles) may be located in the calices of the kidneys, ureters, or urinary bladder. A *renal calculus* (kidney stone) may pass from the kidney into the renal pelvis ▶

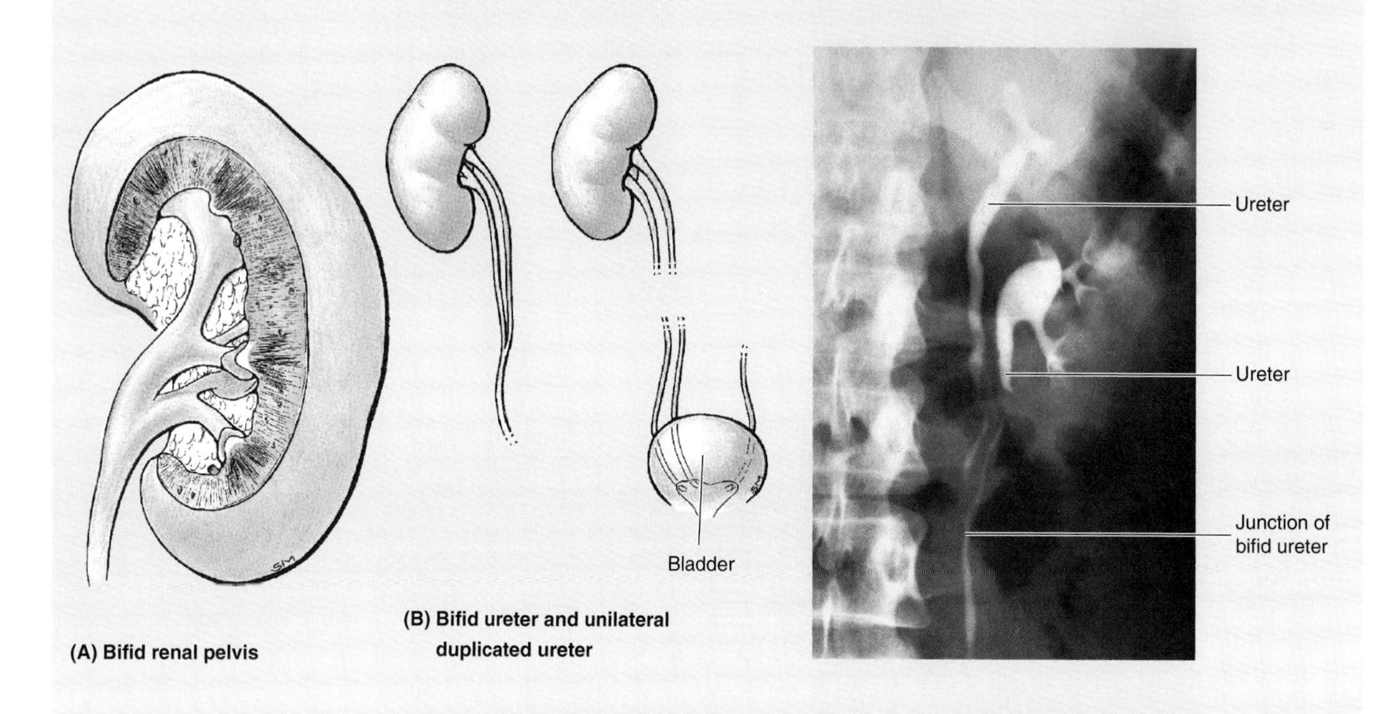

(A) Bifid renal pelvis

(B) Bifid ureter and unilateral duplicated ureter

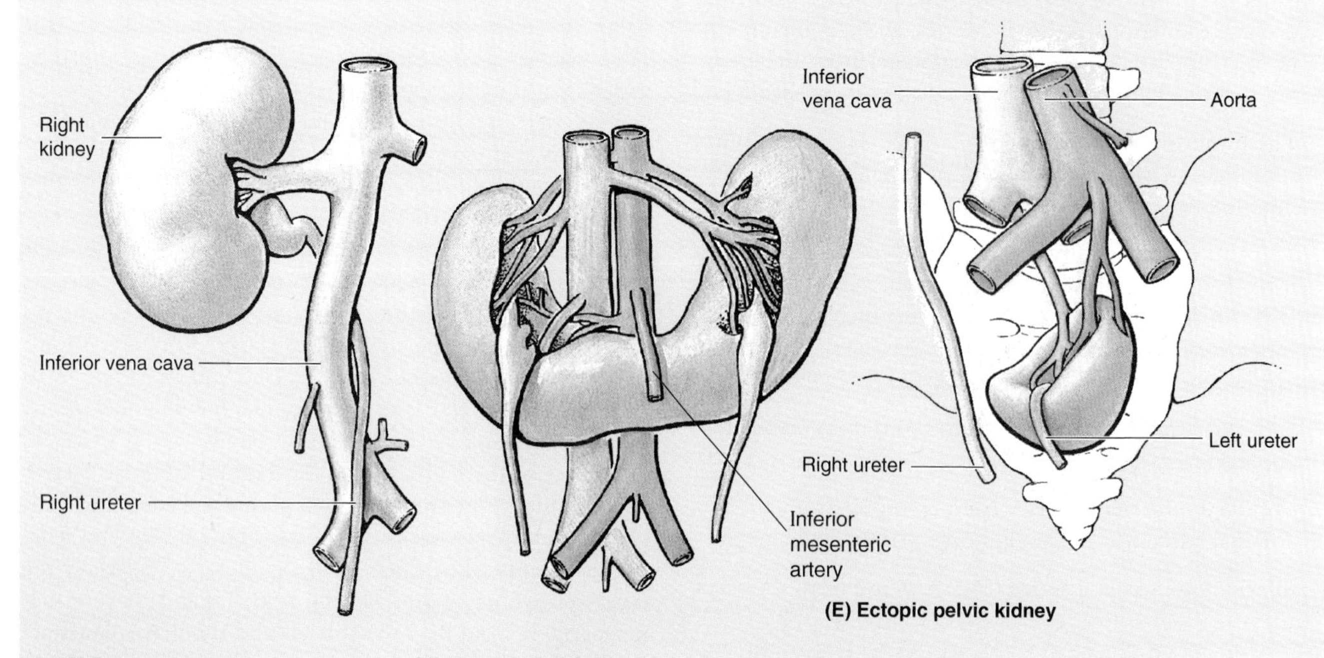

(E) Ectopic pelvic kidney

(C) Retrocaval ureter

(D) Horseshoe kidney

▶ and then into the ureter, causing excessive distension of this muscular tube. A *ureteric calculus* causes severe rhythmic pain—*ureteric colic*—as it is gradually forced down the ureter by waves of contraction. The calculus may cause complete or intermittent obstruction of urinary flow. Depending on the level of obstruction, which changes, the pain may be referred to the:

- Lumbar region
- Hypogastric region
- External genitalia
- Testis.

Ureteric colic is usually a sharp, stabbing pain that follows the course of the ureter.

The pain is referred to the cutaneous areas innervated by spinal cord segments and sensory ganglia, which also supply the ureter—mainly T11 through L2. *The pain passes inferoanteriorly from the loin to the groin.* The loin is the part of the side and back between the ribs and the pelvis, and the groin is the inguinal region. The pain may extend into the proximal anterior aspect of the thigh by projection through the genitofemoral nerve (L1, L2), the scrotum in males, and the labia majora in females. These areas of skin are supplied by the T11 through L2 segments of the spinal cord, which also supply the abdominal skin overlying the ureter.

Ureteric calculi can be observed and removed with a *nephroscope*, an instrument that is inserted through a small incision. Another technique, *lithotripsy*, focuses a shockwave through the body that breaks the stone into small fragments that pass with the urine.

Congenital Anomalies of the Kidneys and Ureters

Bifid renal pelvis and ureter are fairly common (*A–B*). These anomalies result from division of the metanephric diverticulum or ureteric bud—the primordium of the renal pelvis and ureter (Moore and Persaud, 1998). The extent of ureteral duplication depends on the completeness of embryonic division of the ureteric bud. The bifid renal pelvis and/or ureter may be unilateral or bilateral; however, separate openings into the bladder are uncommon. Incomplete division of the ureteric bud results in a *bifid ureter*; complete division results in *supernumerary kidney*. An uncommon anomaly is a *retrocaval ureter* (*C*), which leaves the kidney and passes posterior to the IVC.

When first formed, the kidneys are close together in the pelvis. In approximately 1 in 600 fetuses, the inferior poles (rarely, the superior poles) of the kidneys fuse to form a *horseshoe kidney* (*D*). This U-shaped kidney usually lies at the level of L3 through L5 vertebrae because normal ascent of the abnormal kidney was prevented by the root of the *inferior mesenteric artery*. Horseshoe kidney usually produces no symptoms; however, associated abnormalities of the kidney and renal pelvis may be present, obstructing the ureter.

Sometimes the embryonic kidney on one or both sides fails to ascend to the abdomen and lies anterior to the sacrum. Although uncommon, awareness of the possibility of an *ectopic pelvic kidney* (*E*) should prevent it from being mistaken for a pelvic tumor and removed. A pelvic kidney in a woman also can be injured by or cause obstruction during childbirth. Pelvic kidneys usually receive their blood supply from the common iliac arteries. ⊙

Thoracic Diaphragm

The diaphragm is a dome-shaped, musculotendinous partition separating the thoracic and abdominal cavities. Its mainly convex superior surface faces the thoracic cavity, and its concave inferior surface faces the abdominal cavity (Fig. 2.68*A*). The diaphragm, *the chief muscle of inspiration*, descends during inspiration; however, only its central part moves because its periphery, as the fixed origin of the muscle, attaches to the inferior margin of the thoracic cage and the superior lumbar vertebrae. The **pericardium,** containing the heart, lies on the central part of the diaphragm, depressing it slightly (Fig. 2.68*B*). The diaphragm curves superiorly into right and left domes; normally the right dome is higher than the left. During expiration, the right dome reaches as high as the 5th rib and the left dome ascends to the 5th intercostal space. *The level of the domes of the diaphragm varies according to the:*

- Phase of respiration (inspiration or expiration)
- Posture (e.g., supine or standing)
- Size and degree of distension of the abdominal viscera.

The muscular part of the diaphragm is situated peripherally with fibers that converge radially on the trifoliate central aponeurotic part—the **central tendon**. The central tendon has no bony attachments and is incompletely divided into three leaves, resembling a wide cloverleaf. Although it lies near the center of the diaphragm, the central tendon is closer to the anterior part of the thorax. The **caval foramen** (vena caval foramen), through which the terminal part of the IVC passes to enter the heart, perforates the central tendon. The surrounding muscular part of the diaphragm forms a continuous sheet; however, for descriptive purposes it is divided into three parts, based on the peripheral attachments:

- A **sternal part**, consisting of two muscular slips that attach to the posterior aspect of the xiphoid process; this part is not always present

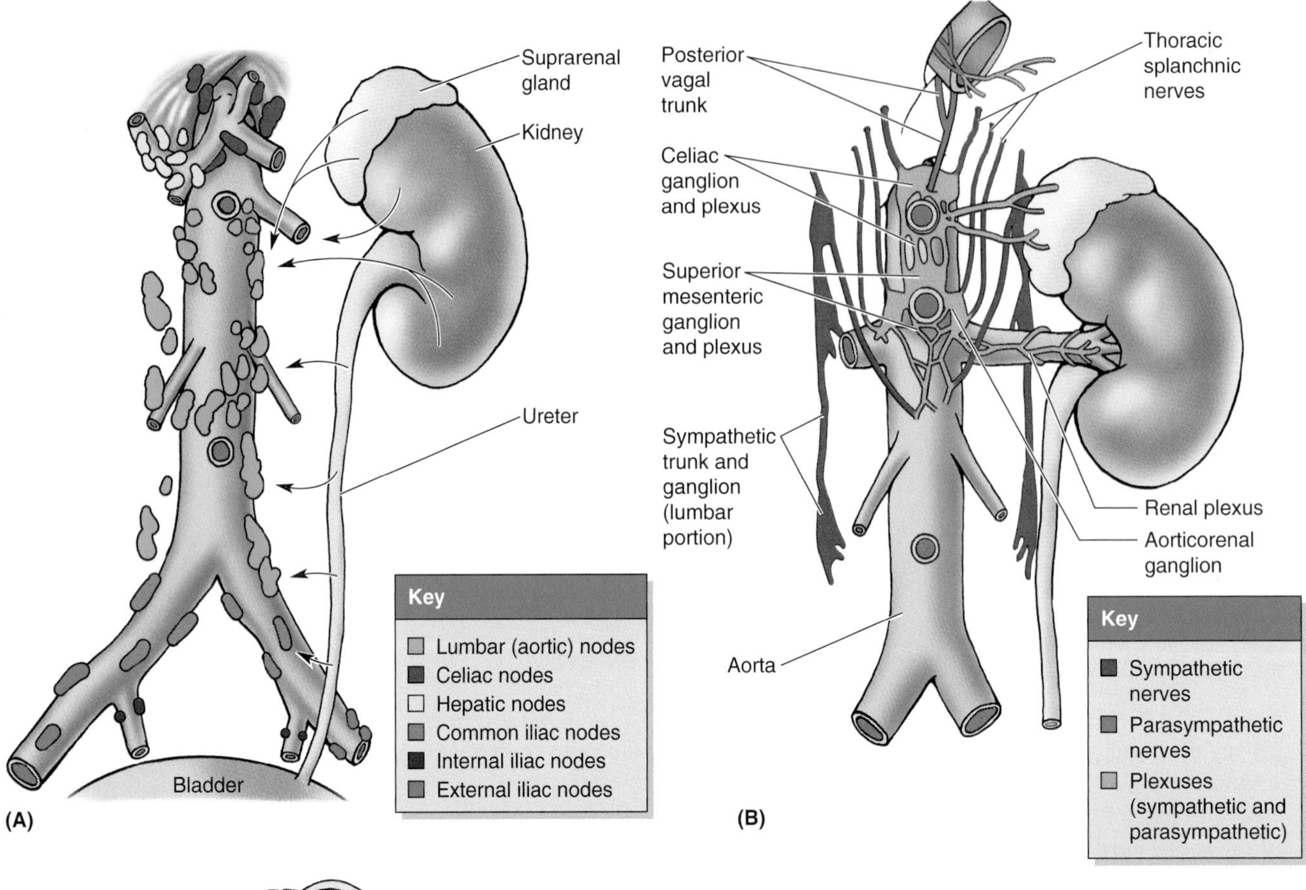

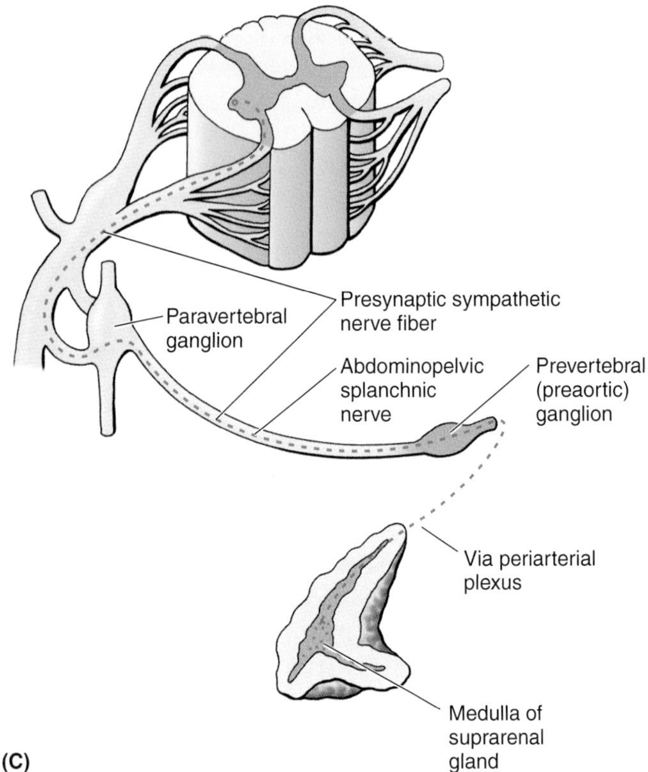

Figure 2.67. Lymphatics and nerves of the kidneys and suprarenal glands. A. The lymphatic vessels of the kidneys form three plexuses: one in the substance of the kidney, one under the fibrous capsule, and one in the perirenal fat. Four or five lymphatic trunks leave the renal hilum and are joined by vessels from the capsule (*arrows*). The lymphatic vessels follow the renal vein to the lumbar (aortic) lymph nodes. Lymph from the suprarenal glands also drains to the lumbar nodes. Lymphatic drainage of the ureters is also illustrated. The lumbar nodes drain through the lumbar lymphatic trunks to the chyle cistern (L. cisterna chyli) . **B.** The nerves of the kidneys and suprarenal glands are derived from the celiac plexus, the lesser and least thoracic splanchnic nerves, and the aorticorenal ganglion. The main efferent innervation of the kidney is vasomotor, autonomic nerves supplying the afferent and efferent arterioles. **C.** The nerves of the suprarenal glands are derived from the thoracic (greater, lesser, and least) splanchnic nerves—some of the abdominopelvic splanchnic nerves conveying presynaptic sympathetic fibers that have passed through the paravertebral ganglia without synapsing to the prevertebral ganglia. Exclusively in the case of the suprarenal medulla, the presynaptic fibers also pass the prevertebral ganglia without synapsing, to end directly on the secretory cells of the suprarenal medulla.

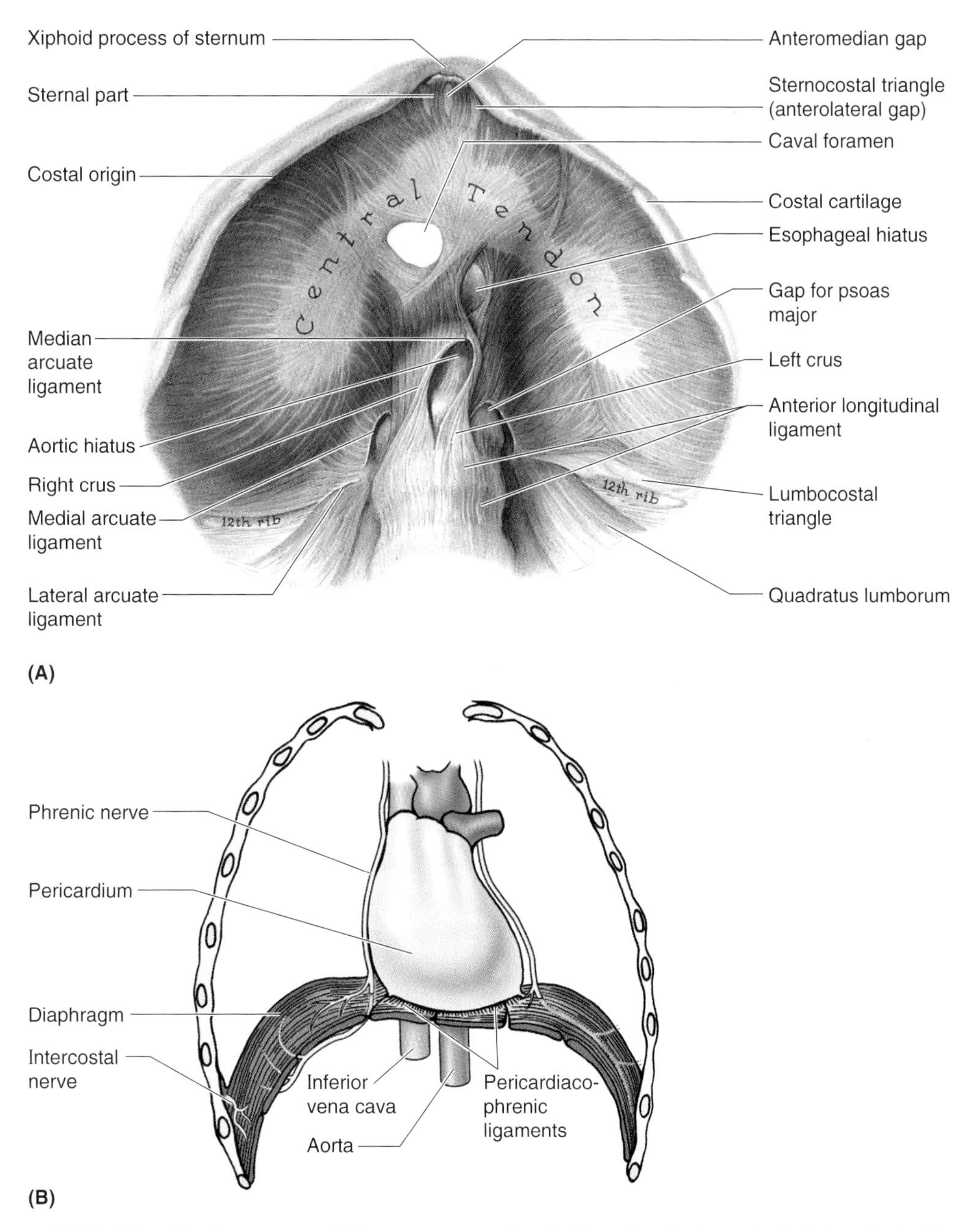

Figure 2.68. The diaphragm and its nerve supply. A. Muscle attachments (inferior view). Observe the fleshy origins of the diaphragm and the trefoil-shaped central tendon that is the aponeurotic insertion of the diaphragmatic muscle fibers. **B.** Nerve supply. Each phrenic nerve (C3 through C5) is the sole motor nerve to its own half of the diaphragm; it is also sensory to its own half, including the pleura superiorly and the peritoneum inferiorly. The lower intercostal nerves are sensory to the peripheral edge of the diaphragm. The pericardiacophrenic ligaments attach the fibrous pericardium to the diaphragm.

- A **costal part**, consisting of wide muscular slips that attach to the internal surfaces of the inferior six costal cartilages and their adjoining ribs on each side; the costal parts form the right and left domes
- A **lumbar part**, arising from two aponeurotic arches—the *medial* and *lateral arcuate ligaments*—and the three superior lumbar vertebrae; the lumbar part forms right and left muscular crura that ascend to the central tendon.

The **crura of the diaphragm** are musculotendinous bundles that arise from the anterior surfaces of the bodies of the superior three lumbar vertebrae, the anterior longitudinal ligament, and the IV discs.

The **right crus**, larger and longer than the left crus, arises from the first three or four lumbar vertebrae; the **left crus** arises from the first two or three. Because it lies to the left of the midline, it is surprising to find that the **esophageal hia-**

tus is a formation in the right crus; however, if the muscular fibers bounding each side of the hiatus are traced inferiorly, it will be seen that they pass to the right of the aortic hiatus. The **aortic hiatus** is formed by the right and left crura and the fibrous **median arcuate ligament**, which unites them as it arches over the anterior aspect of the aorta. The diaphragm is also attached on each side to the **medial** and **lateral arcuate ligaments**, which are thickenings of the fascia covering the psoas major and quadratus lumborum muscles, respectively. **Pericardiacophrenic ligaments** fuse the central tendon of the diaphragm (the thin, strong aponeurotic tendon of all the muscular fibers of the diaphragm) with the inferior surface of the fibrous **pericardium** (Fig. 2.68*B*)—the strong, external part of the fibroserous *pericardial sac* that encloses the heart.

Vessels and Nerves of the Diaphragm

The arteries of the diaphragm form a branchlike pattern on both its superior and inferior surfaces. *The arteries supplying the superior surface of the diaphragm* (Fig. 2.69) *are the*:

- **Pericardiacophrenic** and **musculophrenic arteries**—branches of the internal thoracic artery
- **Superior phrenic arteries**—arising from the thoracic aorta.

The arteries supplying the inferior surface of the diaphragm are the **inferior phrenic arteries**, which typically are the first branches of the *abdominal aorta* (Fig. 2.69*B*, Table 2.11); however, they may arise from the celiac trunk.

The veins draining the superior surface of the diaphragm are the **pericardiacophrenic** and **musculophrenic veins**,

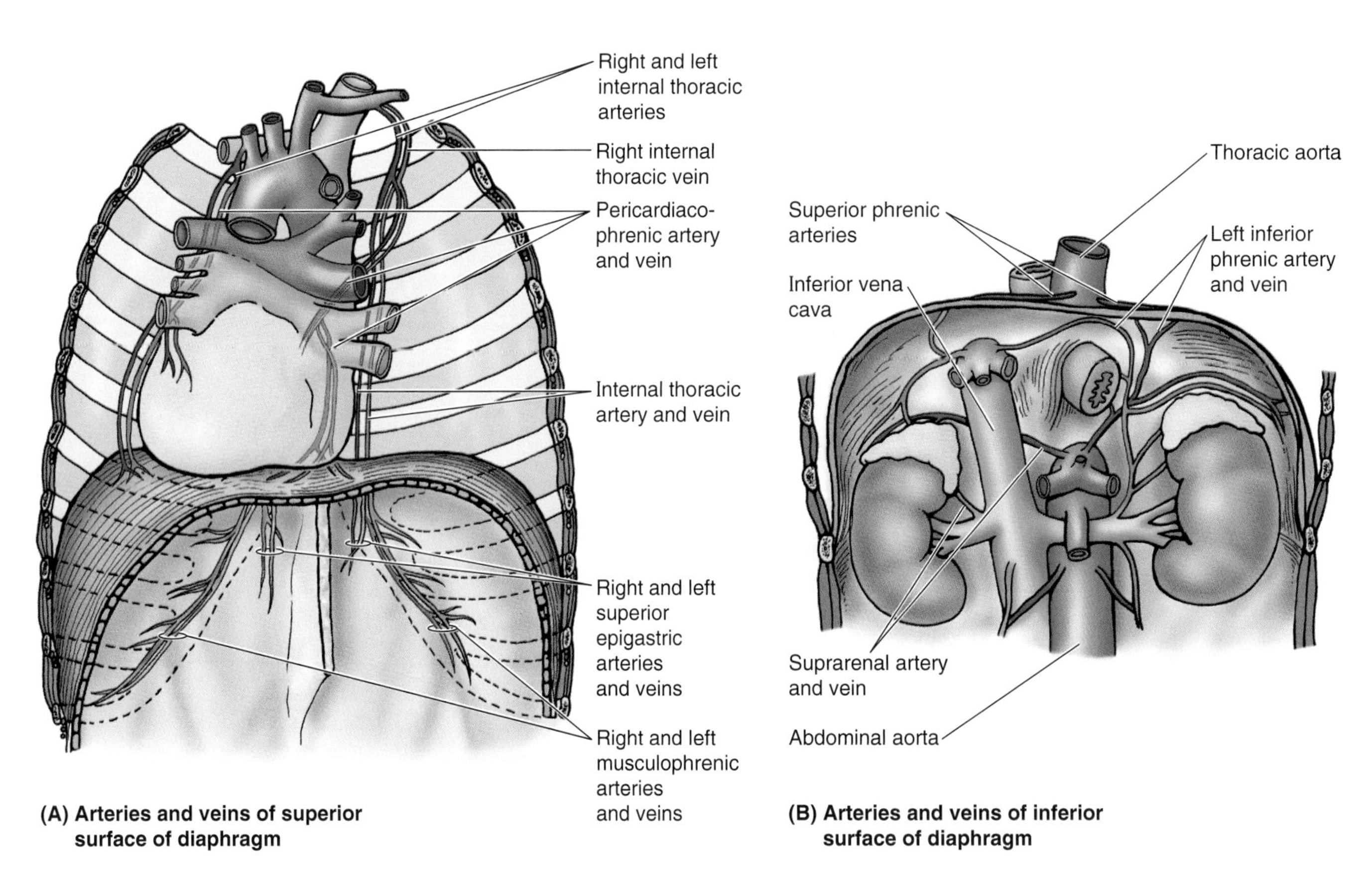

Figure 2.69. Blood vessels of the diaphragm. A. The arteries and veins of the superior surface of the diaphragm are derived from the pericardiacophrenic and musculophrenic arteries and veins (branches of the internal thoracic artery and vein) and from the superior phrenic arteries (small branches, as shown in (**B**), of the aorta). **B.** The inferior phrenic arteries and veins supply and drain blood from the inferior surface of the diaphragm. The inferior phrenic arteries are usually the first branches of the abdominal aorta, but they may arise as branches of the celiac trunk. The right inferior phrenic vein drains to the inferior vena cava (IVC); the left inferior phrenic vein is often double, with one (anterior) branch ending in the IVC and the other (posterior) branch ending in the left renal or suprarenal vein. The two veins may anastomose with each other, as shown here.

Table 2.11. **Vessels and Nerves of the Diaphragm**

Vessels and Nerves	Superior Surface of Diaphragm	Inferior Surface of Diaphragm
Arterial supply	Superior phrenic arteries from thoracic aorta Musculophrenic and pericardiophrenic arteries from internal thoracic arteries	Inferior phrenic arteries from abdominal aorta
Venous drainage	Musculophrenic and pericardiacophrenic veins drain into internal thoracic veins; superior phrenic vein (right side) drains into IVC	Inferior phrenic veins; right vein drains into IVC; left vein is doubled and drains into IVC and suprarenal vein
Lymphatic drainage	Diaphragmatic lymph nodes to phrenic then to parasternal and posterior mediastinal nodes	Superior lumbar lymph nodes; lymphatic plexuses on superior and inferior surfaces communicate freely
Innervation	Motor supply: phrenic nerves (C3–C5) Sensory supply: centrally by phrenic nerves (C3–C5); peripherally by intercostal nerves (T5–T11) and subcostal nerves (T12)	

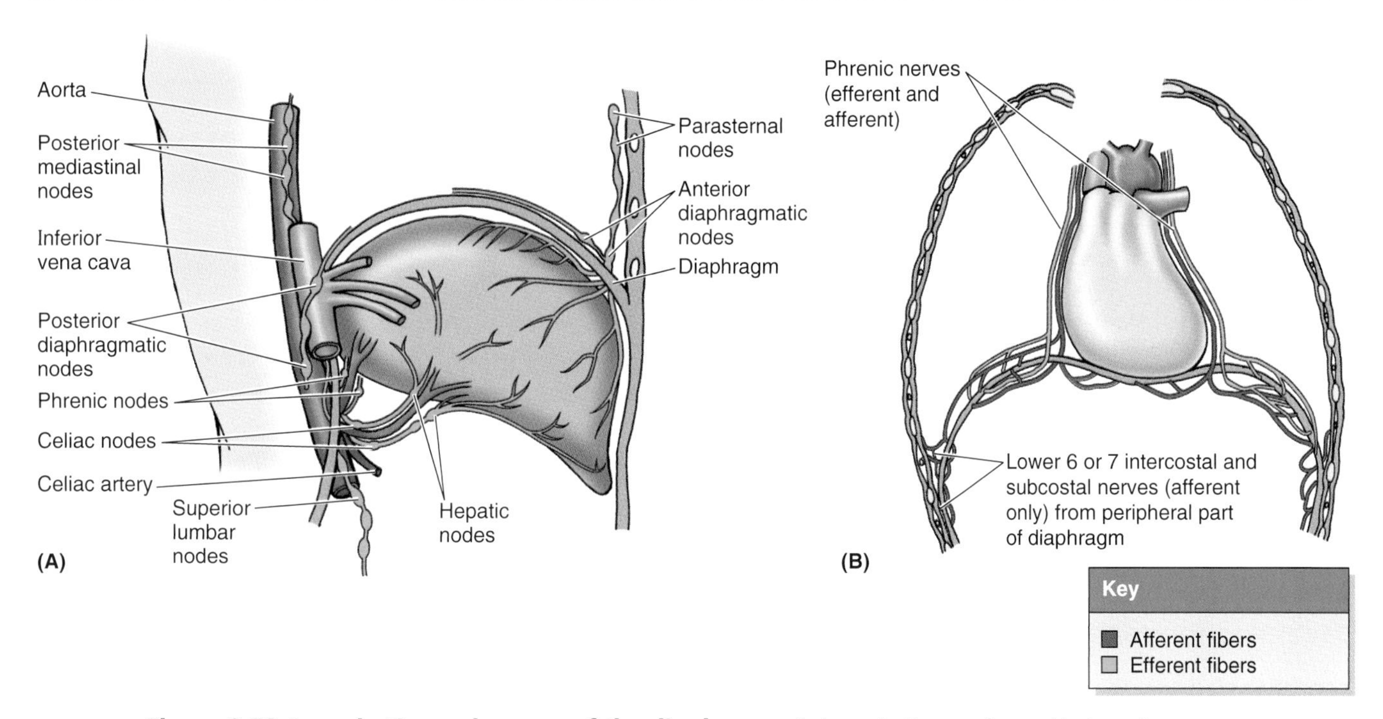

Figure 2.70. Lymphatics and nerves of the diaphragm. A. Lymphatics are formed in two plexuses, one on the thoracic surface of the diaphragm and the other on its abdominal surface; they communicate freely. Lymph from the thoracic surface of the diaphragm passes to the anterior and posterior diaphragmatic nodes; the anterior drainage continues to the parasternal lymph nodes, while posterior drainage passes to phrenic and posterior mediastinal nodes. Lymph drainage from the abdominal surface drains into anterior diaphragmatic, phrenic, and superior lumbar nodes. **B.** The phrenic nerve (from spinal nerves C3 through C5) is the sole motor nerve of the diaphragm. It also carries sensory fibers from the central part of the diaphragm. The lower six or seven intercostal and subcostal nerves carry sensory fibers from the peripheral part of the diaphragm.

which empty into the *internal thoracic veins* and, on the right side, a *superior phrenic vein* that drains into the IVC. Some veins from the posterior curvature of the diaphragm drain into the *azygos and hemiazygos veins* (see Chapter 1). The inferior phrenic veins drain blood from the inferior surface of the diaphragm. The **right inferior phrenic vein** usually opens into the IVC, whereas the **left inferior phrenic vein** is usually double, with one branch passing anterior to the esophageal hiatus to end in the IVC and the other, more posterior branch usually joining the left suprarenal vein.

The **lymphatic plexuses** on the thoracic and abdominal surfaces of the diaphragm communicate freely (Fig. 2.70*A*). The **anterior and posterior diaphragmatic lymph nodes** are on the thoracic surface of the diaphragm. Lymph from these nodes drains into the **parasternal**, **posterior mediastinal**, and **phrenic lymph nodes**. Lymph vessels from the abdominal surface of the diaphragm drain into the anterior diaphragmatic, phrenic, and **superior lumbar lymph nodes**. Lymphatic vessels are dense on the inferior surface of the diaphragm, constituting the primary means for absorption of peritoneal fluid and substances introduced by IP injection.

The entire motor supply to the diaphragm is from the **phrenic nerves**, each of which is distributed to half of the diaphragm and arises from the ventral rami of C3 through C5 segments of the spinal cord (Fig. 2.70*B*). The phrenic nerves also supply sensory fibers (pain and proprioception) to most of the diaphragm. Peripheral parts of the diaphragm receive their sensory nerve supply from the **intercostal nerves** (lower 6 or 7) and the **subcostal nerves.**

Section of a Phrenic Nerve

Section of a phrenic nerve in the neck results in complete paralysis and eventual atrophy of the muscular part of the corresponding half of the diaphragm, except in persons who have an accessory phrenic nerve (see Chapter 8). *Paralysis of a hemidiaphragm* can be recognized radiographically by its permanent elevation and paradoxical movement. Instead of descending on inspiration, it is forced superiorly by the increased intra-abdominal pressure secondary to descent of the opposite unparalyzed hemidiaphragm.

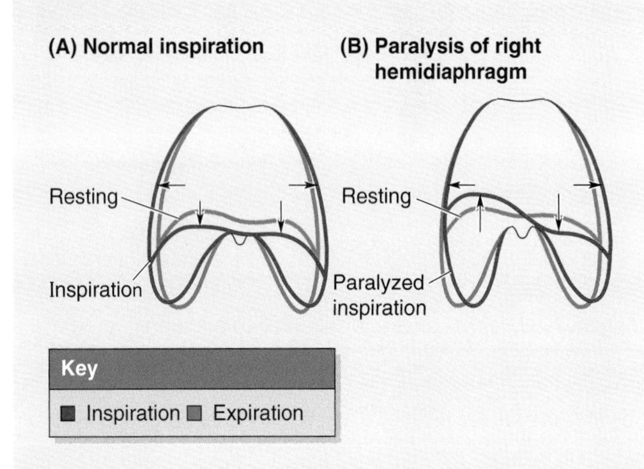

Hiccups

Hiccups (hiccoughs) are involuntary, spasmodic contractions of the diaphragm, causing sudden inhalations that are rapidly interrupted by spasmodic closure of the glottis—the aperture of the larynx—which checks the inflow of air and produces a characteristic sound. Hiccups have many causes, such as indigestion, diaphragm irritation, alcoholism, cerebral lesions, and thoracic and abdominal lesions, all which disturb the phrenic nerves. Hiccups result from irritation of afferent or efferent nerve endings or of medullary centers in the brainstem that control the muscles of respiration, particularly the diaphragm. ⊙

Diaphragmatic Apertures

The diaphragmatic apertures (openings, foramina, hiatus) permit structures (vessels, nerves, and lymphatics) to pass between the thorax and abdomen (Figs. 2.68 and 2.71). *The three large apertures for the IVC, esophagus, and aorta are the:*

- Caval opening
- Esophageal hiatus
- Aortic hiatus.

Caval Opening (Vena Caval Foramen)

The caval opening is an aperture in the central tendon primarily for the IVC. Also passing through the caval opening are terminal branches of the right phrenic nerve and a few lymphatic vessels on their way from the liver to the middle phrenic and mediastinal lymph nodes. The caval opening is located to the right of the median plane at the junction of the tendon's right and middle leaves. The most superior of the three diaphragmatic apertures, the caval opening lies at the level of the disc between the T8 and T9 vertebrae (Fig. 2.71). The IVC is adherent to the margin of the opening; consequently, when the diaphragm contracts during inspiration, it widens the opening and dilates the IVC. These changes facilitate bloodflow to the heart through this large vein.

Esophageal Hiatus (Aperture)

The esophageal hiatus is an oval aperture for the esophagus in the muscle of the right crus of the diaphragm at the level of T10 vertebra. The esophageal hiatus also transmits the anterior and posterior vagal trunks, esophageal branches of the left gastric vessels, and a few lymphatic vessels. The fibers of the right crus of the diaphragm decussate distal to the hiatus,

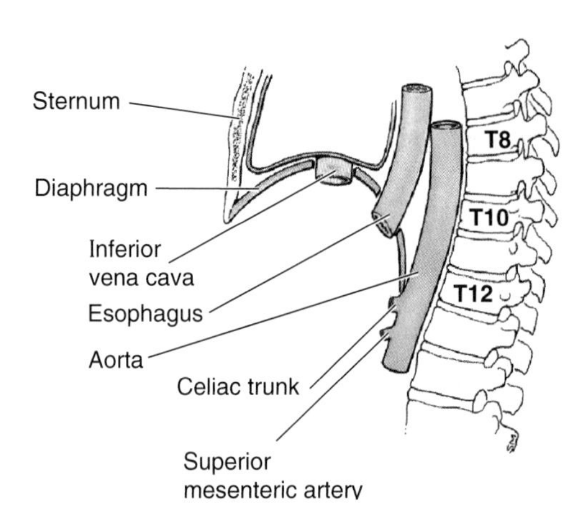

Figure 2.71. Apertures of the diaphragm. There are three large apertures in the diaphragm for major structures to pass to and from the thorax into the abdomen. The caval opening for the inferior vena cava (IVC), most anterior, is at the T8 level and to the right of the midline; the esophageal hiatus, intermediate, is at T10 and to the left of the midline; the aortic hiatus for the aorta passes posterior to the vertebral attachment of the diaphragm in the midline, at T12 (see also Fig. 2.68*A*).

forming a muscular sphincter for the esophagus that constricts it when the diaphragm contracts. The esophageal hiatus is superior to and to the left of the aortic hiatus. In most cases (70%), both margins of the hiatus are formed by muscular bundles of the right crus. In some cases (30%), a superficial muscular bundle from the left crus contributes to the formation of the right margin of the hiatus.

Aortic Hiatus (Aperture)

The aortic hiatus is the opening posterior to the diaphragm for the aorta. Because the aorta does not pierce the diaphragm, bloodflow through it is not affected by its movements during respiration. The aorta passes between the crura of the diaphragm posterior to the median arcuate ligament, which is at the level of the inferior border of the T12 vertebra. The aperture for the aorta also transmits the thoracic duct and sometimes the azygos vein.

Other Apertures in the Diaphragm

In addition to the three main apertures, there is a small opening, the *sternocostal foramen* (triangle), between the sternal and costal attachments of the diaphragm. This foramen transmits the lymphatic vessels from the diaphragmatic surface of the liver and the superior epigastric vessels. The sympathetic trunks pass deep to the medial arcuate ligament. There are two small apertures in each crus of the diaphragm; one transmits the greater and the other the lesser splanchnic nerve.

Actions of the Diaphragm

When the diaphragm contracts, its domes move inferiorly so that the convexity of the diaphragm is somewhat flattened. Although this movement is often described as the "descent of the diaphragm," only the domes of the diaphragm descend; its periphery remains attached to the ribs and cartilages of the inferior six ribs. As the diaphragm descends, it pushes the abdominal viscera inferiorly. This increases the volume of the thoracic cavity and decreases the intrathoracic pressure, resulting in air being taken into the lungs. In addition, the volume of the abdominal cavity decreases slightly and intra-abdominal pressure increases somewhat. Movements of the diaphragm are also important in circulation because the increased intra-abdominal pressure and decreased intrathoracic pressure help to return venous blood to the heart. When the diaphragm contracts, compressing the abdominal viscera, blood in the IVC is forced superiorly into the heart.

The diaphragm is at its most superior level when a person is supine (with the upper body lowered—the *Trendelenburg position*). When a person is supine the abdominal viscera push the diaphragm superiorly in the thoracic cavity. When a person lies on one side, the hemidiaphragm rises to a more superior level because of the greater push of the viscera on that side. Conversely, the diaphragm assumes an inferior level when a person is sitting or standing. For this reason, people with *dyspnea* (difficult breathing) prefer to sit up, not lie down.

Referred Pain from the Diaphragm

Pain from the diaphragm radiates to two different areas because of the difference in the sensory nerve supply of the diaphragm (Table 2.11). Pain resulting from irritation of the diaphragmatic pleura or the diaphragmatic peritoneum is referred to the shoulder region, the area of skin supplied by the C3 through C5 segments of the spinal cord (p. 233). These segments also contribute ventral rami to the phrenic nerves. Irritation of peripheral regions of the diaphragm, innervated by the inferior intercostal nerves, is more localized, being referred to the skin over the costal margins of the anterolateral abdominal wall.

Rupture of the Diaphragm and Herniation of Viscera

Rupture of the diaphragm and herniation of viscera can result from a sudden increase in either the intrathoracic or intra-abdominal pressure. The common cause of this injury is severe trauma to the thorax or abdomen during a motor vehicle accident, for example. Most diaphragmatic ruptures are on the left side (95%) because of that side's congenital weakness (Moore and Persaud, 1998). The gap in the musculature of the diaphragm is usually medial to the most inferior muscular fibers arising from the 12th ribs. In this area, the superior and inferior fasciae form the diaphragm. A defect in this *lumbocostal triangle* (Fig. 2.68*A*) is variable in size. The following structures may herniate into the thorax when there is a *traumatic diaphragmatic hernia*:

- Stomach
- Intestine
- Mesentery
- Spleen.

Hiatal or hiatus hernia, a protrusion of part of the stomach into the thorax through the esophageal hiatus, has been discussed (p. 227). The structures that pass through the esophageal hiatus (vagal trunks, left inferior phrenic vessels, esophageal branches of the left gastric vessels) may be injured in surgical procedures on the esophageal hiatus (e.g., repair of a hiatus hernia).

Congenital Diaphragmatic Hernia

Posterolateral defect of the diaphragm is the only relatively common congenital anomaly of the diaphragm (Moore and Persaud, 1998). This defect occurs approximately once in 2200 newborn infants and is associated with *CDH* (herniation of abdominal contents into the thoracic cavity). Life-threatening breathing difficulties may be associated with this anomaly because of the inhibition of development and inflation of lungs. ⊙

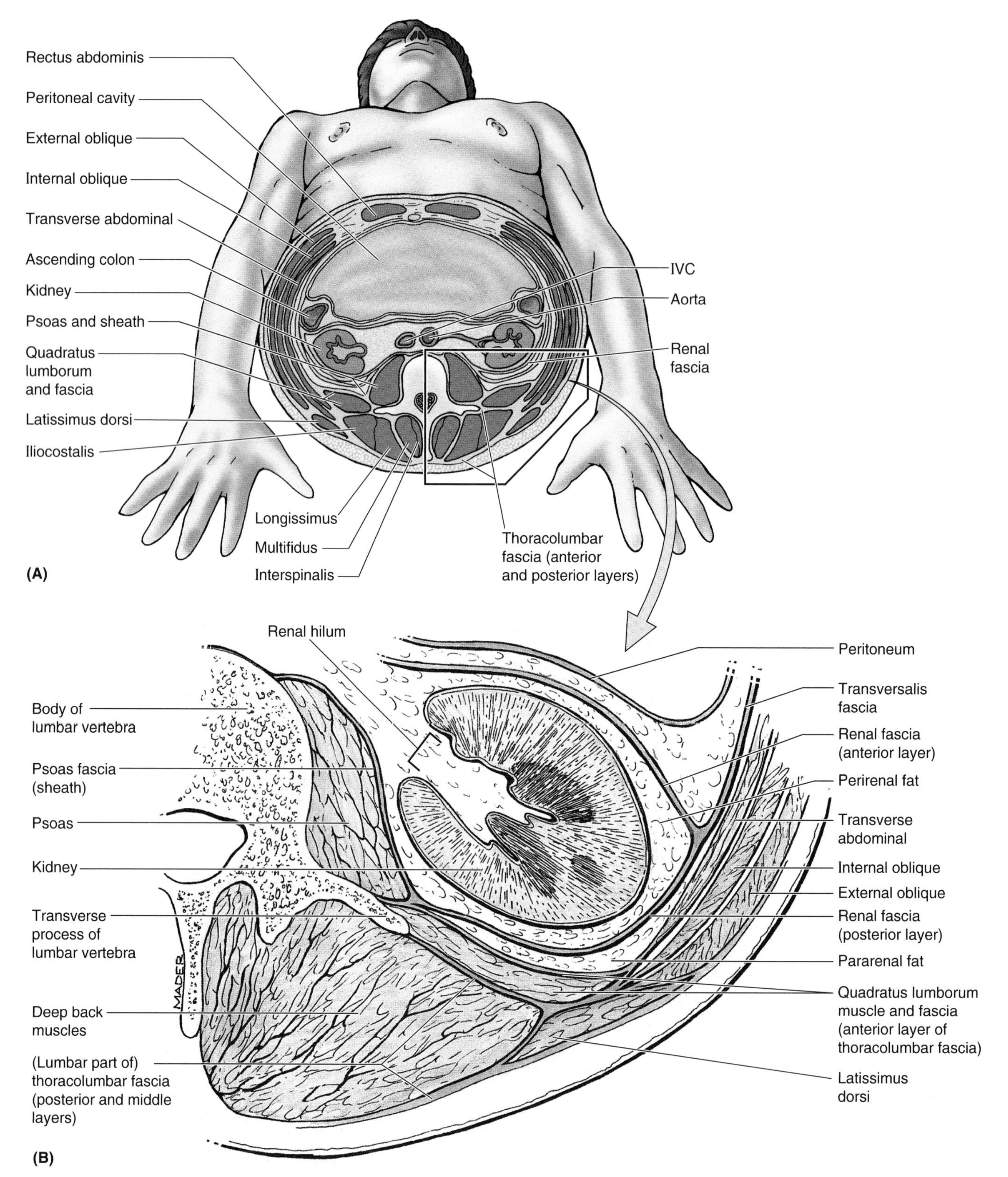

Figure 2.72. Transverse section of the abdomen at the level of the renal hilum. A. The relationships of the muscles, fascia, and abdominal walls are illustrated. The relations of the kidneys are also demonstrated. The anterior wall consists of the rectus abdominis and aponeuroses of the three flat abdominal muscles: external oblique, internal oblique, and transverse abdominal. The lateral walls are formed by the flat abdominal muscles and, in part, by the iliacus muscles and hip bones (see Fig. 2.74*A*). The posterior abdominal wall is formed by the bodies of the five lumbar vertebrae, the intervertebral (IV) discs associated with them, the psoas and quadratus lumborum muscles, and, in part, by the iliacus muscles and iliac bones. **B.** Transverse section of the kidney: relationships of muscles and fascia of the posterior abdominal wall.

Posterior Abdominal Wall

The posterior abdominal wall (Figs. 2.72–2.74) is mainly composed—from deep (posterior) to superficial (anterior)—of the:

- Five lumbar vertebrae and associated IV discs
- Posterior abdominal wall muscles—psoas, quadratus lumborum, iliacus, transverse abdominal, and oblique muscles
- Lumbar plexus, composed of the ventral rami of lumbar spinal nerves
- Fascia, including thoracolumbar fascia
- Diaphragm, contributing to the superior part of the posterior wall
- Fat, nerves, vessels (e.g., aorta and IVC), and lymph nodes.

Fascia of the Posterior Abdominal Wall

The posterior abdominal wall is covered with a continuous layer of endoabdominal fascia that lies between the parietal peritoneum and the muscles (Figs. 2.72 and 2.73). The fascia lining the posterior abdominal wall is continuous with the transversalis fascia that lines the transverse abdominal muscle. It is customary to name the fascia according to the structure it covers.

The **psoas fascia** covering the psoas major (**psoas sheath**) is attached medially to the lumbar vertebrae and pelvic brim (Fig. 2.72). The psoas fascia (sheath) is thickened superiorly to form the *medial arcuate ligament* (Fig. 2.68*A*). The psoas fascia fuses laterally with the **quadratus lumborum** and **thoracolumbar fascias**. Inferior to the iliac crest, the psoas fascia is continuous with the part of the iliac fascia covering the iliacus. The psoas fascia also blends with the fascia covering the quadratus lumborum.

The **quadratus lumborum fascia** covering the quadratus lumborum is a dense membranous layer that is continuous laterally with the anterior layer of the thoracolumbar fascia (Fig. 2.72). The quadratus lumborum fascia attaches to the anterior surfaces of the transverse processes of the lumbar vertebrae, the iliac crest, and the 12th rib and is continuous with the aponeurosis of the transverse abdominal muscle. The quadratus lumborum fascia thickens superiorly to form the *lateral arcuate ligaments* and is adherent inferiorly to the iliolumbar ligaments.

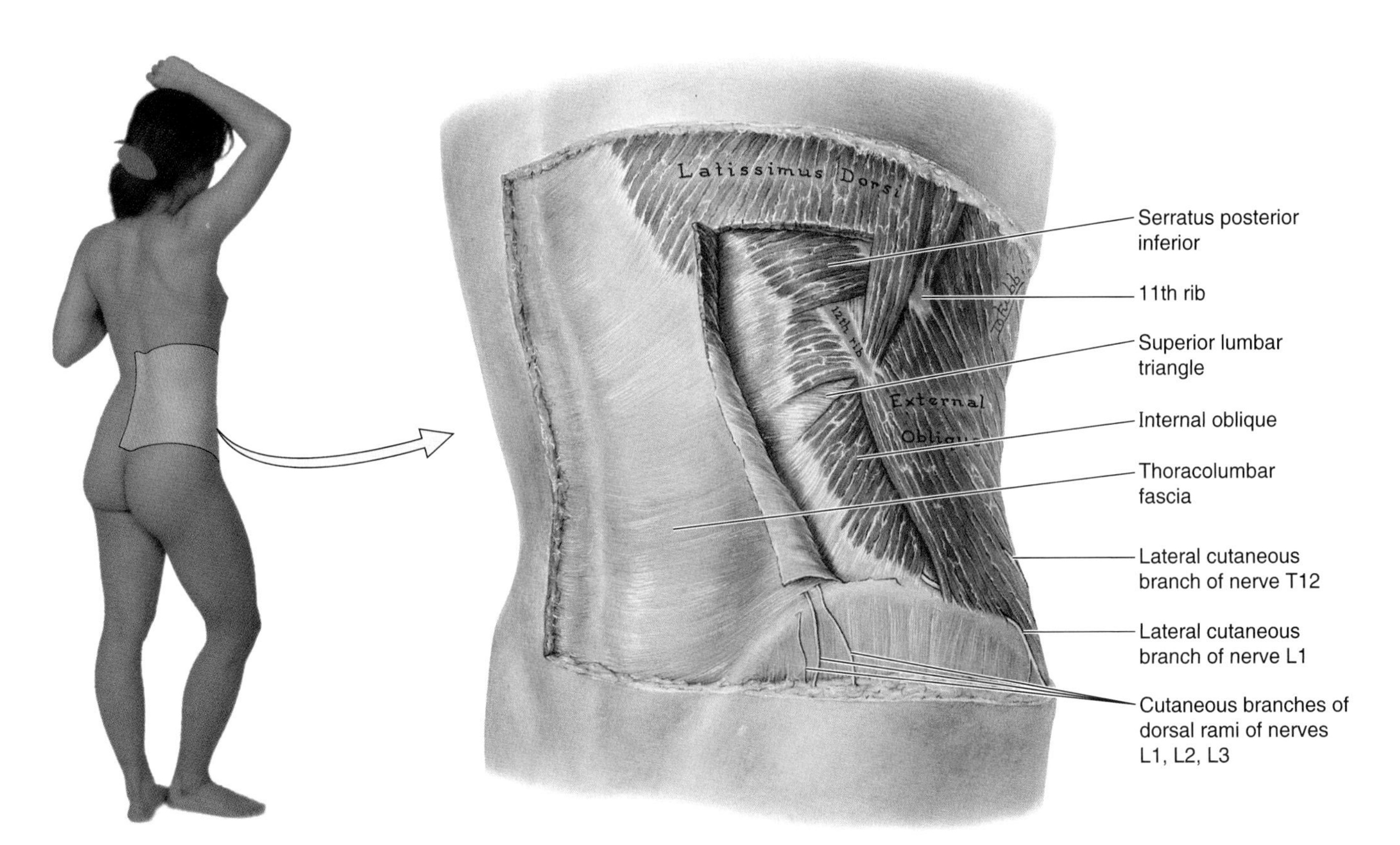

Figure 2.73. Posterolateral view of the posterior abdominal wall. Part of the latissimus dorsi muscle is reflected. Observe the free posterior border of the external oblique muscle that extends from the tip of the 12th rib to the midpoint of the iliac crest. Observe also that the strong, posterior layer of the thoracolumbar fascia covers the deep muscles of the back. The fascia is attached to the angles of the ribs; to the spinous processes of the thoracic, lumbar, and sacral vertebrae; to the transverse processes of the lumbar vertebrae; to the inferior border of the 12th rib; and to the iliac crest, as well as to several ligaments.

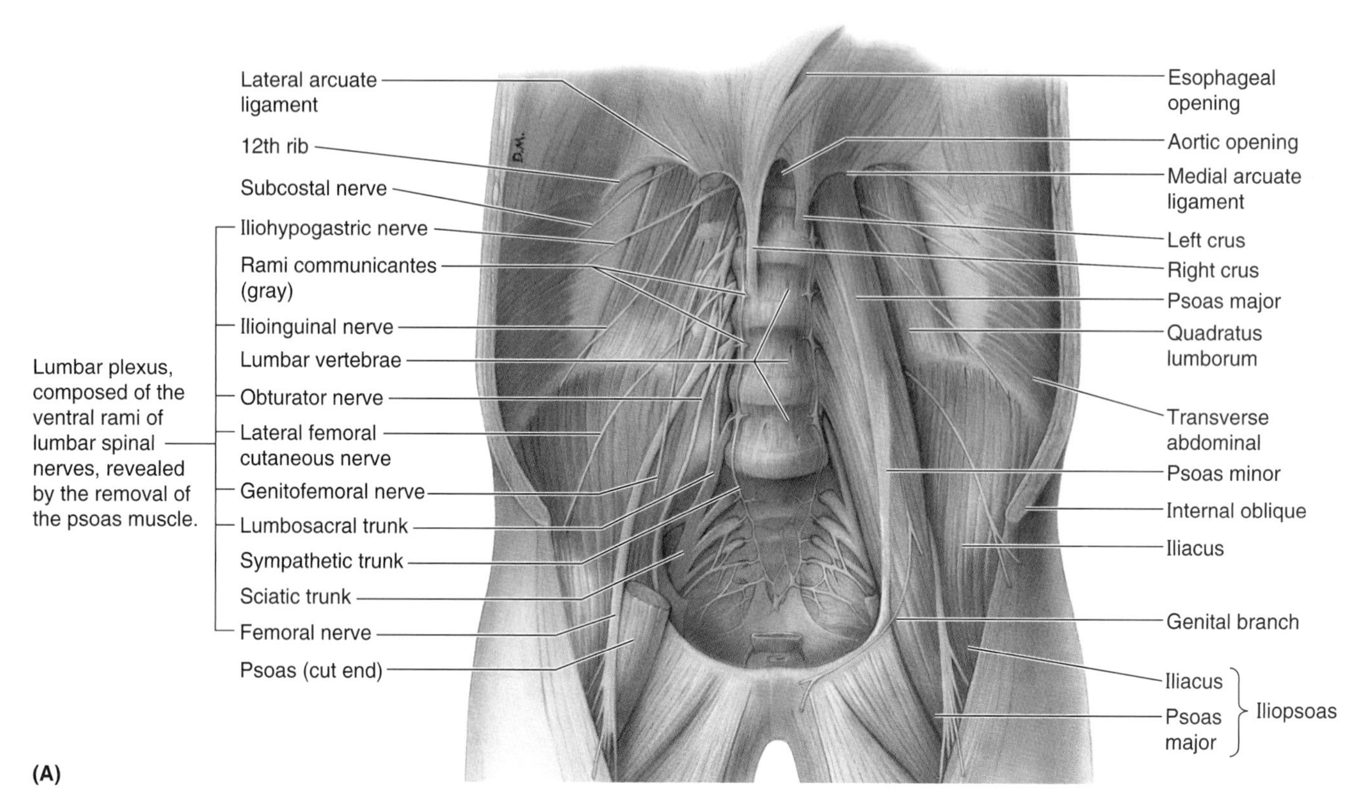
Lateral arcuate ligament
12th rib
Subcostal nerve
Iliohypogastric nerve
Rami communicantes (gray)
Ilioinguinal nerve
Lumbar vertebrae
Obturator nerve
Lateral femoral cutaneous nerve
Genitofemoral nerve
Lumbosacral trunk
Sympathetic trunk
Sciatic trunk
Femoral nerve
Psoas (cut end)
Lumbar plexus, composed of the ventral rami of lumbar spinal nerves, revealed by the removal of the psoas muscle.
Esophageal opening
Aortic opening
Medial arcuate ligament
Left crus
Right crus
Psoas major
Quadratus lumborum
Transverse abdominal
Psoas minor
Internal oblique
Iliacus
Genital branch
Iliacus
Psoas major
Iliopsoas
D.M.
(A)

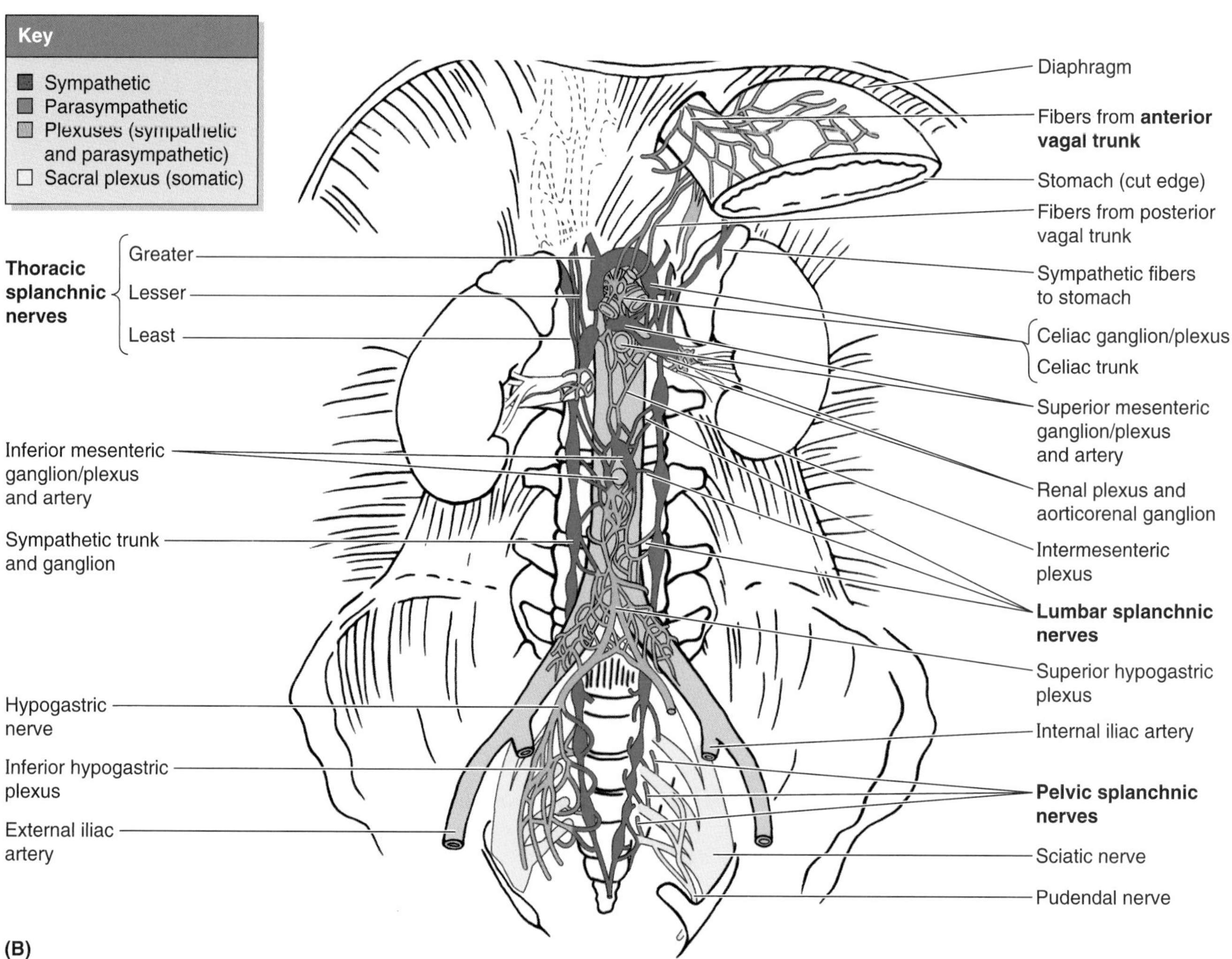
Key
Sympathetic
Parasympathetic
Plexuses (sympathetic and parasympathetic)
Sacral plexus (somatic)
Thoracic splanchnic nerves
Greater
Lesser
Least
Inferior mesenteric ganglion/plexus and artery
Sympathetic trunk and ganglion
Hypogastric nerve
Inferior hypogastric plexus
External iliac artery
Diaphragm
Fibers from **anterior vagal trunk**
Stomach (cut edge)
Fibers from posterior vagal trunk
Sympathetic fibers to stomach
Celiac ganglion/plexus
Celiac trunk
Superior mesenteric ganglion/plexus and artery
Renal plexus and aorticorenal ganglion
Intermesenteric plexus
Lumbar splanchnic nerves
Superior hypogastric plexus
Internal iliac artery
Pelvic splanchnic nerves
Sciatic nerve
Pudendal nerve
(B)

The **thoracolumbar fascia** is an extensive fascial sheet that splits into anterior and posterior layers, enclosing the deep back muscles (Figs. 2.72 and 2.73). It is thin and transparent where it covers the thoracic parts of the deep muscles but is thick and strong in the lumbar region. The lumbar part of the thoracolumbar fascia, extending between the 12th rib and the iliac crest, attaches laterally to the internal oblique and transverse abdominal muscles.

Psoas Abscess

Although the prevalence of *tuberculosis* has been greatly reduced, there is a resurgence of tuberculosis (TB) in Africa and Asia, sometimes in pandemic proportions due to AIDS and drug resistence. TB of the spine is quite common. An infection may spread through the blood to the vertebrae (*hematogenous spread*), particularly during childhood. An abscess resulting from tuberculosis in the lumbar region tends to spread from the vertebrae into the psoas sheath, where it produces a *psoas abscess*. As a consequence, the psoas fascia thickens to form a strong stockinglike tube. Pus from the psoas abscess passes inferiorly along the psoas within this fascial tube over the pelvic brim and deep to the inguinal ligament. The pus usually surfaces in the superior part of the thigh. Pus can also reach the psoas sheath by passing from the posterior mediastinum when the thoracic vertebrae are diseased.

The inferior part of the *iliac fascia* is often tense and raises a fold that passes to the internal aspect of the iliac crest. The superior part of this fascia is loose and may form a pocket—the *iliacosubfascial fossa*—posterior to the above-mentioned fold. Part of the large intestine, such as the cecum and/or appendix on the right side and the sigmoid colon on the left side, may become trapped in this fossa causing considerable pain. ○

Muscles of the Posterior Abdominal Wall

The main paired muscles in the posterior abdominal wall (Fig. 2.74) are the:

- **Psoas major**, passing inferolaterally
- **Iliacus**, lying along the lateral sides of the inferior part of the psoas major
- **Quadratus lumborum**, lying adjacent to the transverse processes of the lumbar vertebrae and lateral to superior parts of the psoas major.

The attachments, nerve supply, and main actions of these muscles are summarized in Table 2.12.

Psoas Major

The long, thick, fusiform psoas major lies lateral to the lumbar vertebrae. *Psoas* is a Greek word meaning "muscle of the loin." (Butchers refer to the psoas of animals as tenderloin.) The psoas passes inferolaterally, deep to the inguinal ligament to reach the lesser trochanter of the femur (see Chapter 4). The lumbar plexus of nerves is embedded in the posterior part of the psoas, anterior to the lumbar transverse processes (Fig. 2.74*A*).

Iliacus

The iliacus is a large triangular muscle that lies along the lateral side of the inferior part of the psoas major. This muscle extends across the sacroiliac joint and attaches to the superior two-thirds of the iliac fossa. Most of its fibers join the tendon of the psoas major; together the psoas and iliacus form the **iliopsoas**—the chief flexor of the thigh. It is also a stabilizer of the hip joint and helps to maintain the erect posture at this joint.

Quadratus Lumborum

The quadrilateral quadratus lumborum forms a thick muscular sheet in the posterior abdominal wall (Figs. 2.72 and 2.74). It lies adjacent to the lumbar transverse processes and is broader inferiorly. Close to the 12th rib, the quadratus lumborum is crossed by the **lateral arcuate ligament**. The *subcostal nerve* passes posterior to this ligament and runs inferolaterally on the quadratus lumborum. *Branches of the lumbar plexus* run inferiorly on the anterior surface of this muscle.

Posterior Abdominal Pain

The iliopsoas has extensive and clinically important relations to the kidneys, ureters, cecum, appendix, sigmoid colon, pancreas, lumbar lymph nodes, and nerves of the posterior abdominal wall. When any of these structures is diseased, movement of the iliopsoas usually causes pain. When intra-abdominal inflammation is suspected, the **iliopsoas test** is performed. The person is asked to lie on the unaffected side and to extend the ▶

Figure 2.74. Muscles and nerves of the posterior abdominal wall. A. Most of the right psoas major has been removed to show that the lumbar plexus of nerves is formed by the ventral rami of the first four lumbar spinal nerves, and that it lies in the substance of the psoas major. **B.** Autonomic nerve supply of the abdomen. Observe that the sympathetic and parasympathetic nerves mingle in the rich tangle of nerve plexuses anterior to the aorta. Both kinds of fiber are distributed by "hitchhiking" on the walls of branches of the abdominal aorta to their destinations.

Table 2.12. **Main Muscles of the Posterior Abdominal Wall**

Muscle	Superior Attachment	Inferior Attachment	Innervation	Main Action
Psoas major	Transverse processes of lumbar vertebrae; sides of bodies of T12–L5 vertebrae and intervening intervertebral discs	By a strong tendon to lesser trochanter of femur	Lumbar plexus via ventral branches of L2–L4 nerves	Acting inferiorly with iliacus, it flexes thigh; acting superiorly it flexes vertebral column laterally; it is used to balance the trunk; when sitting it acts inferiorly with iliacus to flex trunk
Iliacus	Superior two-thirds of iliac fossa, ala of sacrum, and anterior sacroiliac ligaments	Lesser trochanter of femur and shaft inferior to it, and to psoas major tendon	Femoral nerve (L2–L4)	Flexes thigh and stabilizes hip joint; acts with psoas major
Quadratus lumborum	Medial half of inferior border of 12th rib and tips of lumbar transverse processes	Iliolumbar ligament and internal lip of iliac crest	Ventral branches of T12 and L1–L4 nerves	Extends and laterally flexes vertebral column; fixes 12th rib during inspiration

▶ thigh on the affected side against the resistance of the examiner's hand (Swartz, 1994). The elicitation of pain with this maneuver is a *positive psoas sign*. An acutely inflamed appendix, for example, will produce a positive right psoas sign. Because the psoas lies along the vertebral column and the iliacus crosses the sacroiliac joint, disease of the intervertebral and sacroiliac joints may cause *spasm of the iliopsoas*, a protective reflex. *Adenocarcinoma of the pancreas* in advanced stages invades the muscles and nerves of the posterior abdominal wall, producing excruciating pain because of the close relationship of the pancreas to the posterior abdominal wall. ⊙

Nerves of the Posterior Abdominal Wall

There are somatic and autonomic nerves in the posterior abdominal wall.

Somatic Nerves of the Posterior Abdominal Wall

The **subcostal nerves**—the ventral rami of T12—arise in the thorax, pass posterior to the lateral arcuate ligaments into the abdomen, and run inferolaterally on the anterior surface of the quadratus lumborum (Fig. 2.74). They pass through the transverse abdominal and internal oblique muscles to supply the external oblique and skin of the anterolateral abdominal wall.

The **lumbar nerves** pass from the spinal cord through the IV foramina inferior to the corresponding vertebrae, where they divide into dorsal and ventral primary rami. Each ramus contains sensory and motor fibers. The dorsal primary rami pass posteriorly to supply the muscles and skin of the back, whereas the ventral primary rami pass into the psoas major muscles and are connected to the sympathetic trunks by rami communicantes.

The **lumbar plexus of nerves** is in the posterior part of the psoas major, anterior to the lumbar transverse processes (Fig. 2.74*A*). This nerve network is composed of the ventral rami of L1 through L4 nerves. All these rami receive gray rami communicantes (communicating branches) from the sympathetic trunks, and the superior two send white rami communicantes to these trunks. *The following nerves are branches of the lumbar plexus; the three largest are listed first*:

- The **obturator nerve** (L2 through L4) emerges from the medial border of the psoas major and passes through the pelvis to the medial thigh, supplying the adductor muscles.
- The **femoral nerve** (L2 through L4) emerges from the lateral border of the psoas major and innervates the iliacus and passes deep to the inguinal ligament to the anterior thigh, supplying the flexors of the hip and extensors of the knee.
- The **lumbosacral trunk** (L4, L5) passes over the ala (wing) of the sacrum and descends into the pelvis to participate in the formation of the sacral plexus along with the ventral rami of S1 through S4 nerves.
- The **ilioinguinal** and **iliohypogastric nerves** (L1) arise from the ventral ramus of L1 and enter the abdomen posterior to the medial arcuate ligaments and pass inferolaterally, anterior to the quadratus lumborum muscles. They pierce the transverse abdominal muscles near the anterior superior iliac spines and pass through the internal and external oblique muscles to supply the skin of the suprapubic and inguinal regions (both nerves also supply branches to the abdominal musculature).
- The **genitofemoral nerve** (L1, L2) pierces the anterior surface of the psoas major and runs inferiorly on it deep to the psoas fascia; it divides lateral to the common and external iliac arteries into femoral and genital branches.
- The **lateral femoral cutaneous nerve** (L2, L3) runs inferolaterally on the iliacus and enters the thigh posterior to the inguinal ligament, just medial to the anterior superior iliac spine; it supplies skin on the anterolateral surface of the thigh.

Autonomic Nerves of the Posterior Abdominal Wall

The autonomic nerves of the abdomen consist of one cranial nerve (the vagus) and several different splanchnic nerves that deliver presynaptic sympathetic and parasympathetic fibers to the nerve plexuses and sympathetic ganglia along the abdominal aorta and the periarterial extensions of those plexuses that reach the abdominal viscera, where intrinsic parasympathetic ganglia occur (Fig. 2.74*B*, Table 2.10).

The *sympathetic part of the autonomic nervous system in the abdomen* consists of the:

- Abdominopelvic splanchnic nerves from the thoracic and abdominal sympathetic trunks
- Prevertebral sympathetic ganglia
- Abdominal autonomic plexuses
- Periarterial plexuses.

The plexuses are mixed, shared with the parasympathetic nervous system and visceral afferent fibers.

The **abdominopelvic splanchnic nerves** are the source of sympathetic innervation in the abdominopelvic cavity. The presynaptic sympathetic fibers they convey originated from cell bodies in the intermediolateral cell column, or lateral horn, of gray matter of spinal cord segments T7 through L2 or L3. The fibers pass successively through the ventral roots, ventral rami, and white rami communicans of thoracic and upper lumbar spinal nerves to reach the sympathetic trunks. They pass through the paravertebral ganglia of these trunks without synapsing to enter the abdominopelvic splanchnic nerves that convey them to the prevertebral ganglia of the abdominal cavity. *The abdominopelvic splanchnic nerves include the:*

- Lower thoracic splanchnic nerves (greater, lesser, and least splanchnic nerves) from the thoracic part of the sympathetic trunks
- Lumbar splanchnic nerves from the lumbar part of the sympathetic trunks.

The **lower thoracic splanchnic nerves** are the main source of presynaptic sympathetic fibers serving abdominal viscera. The **greater** (from the sympathetic trunk at T5 through T9 or T10 vertebral levels), **lesser** (from T10 and T11 levels), and **least** (from the T12 level) **splanchnic nerves** are the specific thoracic splanchnic nerves that arise from the thoracic part of the sympathetic trunks and pierce the corresponding crus of the diaphragm to convey the presynaptic sympathetic fibers to the celiac, superior mesenteric, and aorticorenal (prevertebral) sympathetic ganglia.

The **lumbar splanchnic nerves** arise from the abdominal part of the sympathetic trunks. The sympathetic trunks extend into the abdomen from the thorax by passing posterior to the medial arcuate ligaments of the diaphragm (Fig. 2.74*A*). They lie on the anterolateral aspects of the bodies of the lumbar vertebrae in a groove formed by the adjacent psoas major muscle. The abdominal part of the trunks is composed of four **lumbar (paravertebral) sympathetic ganglia** and interconnecting fibers. Laterally, the trunks receive white rami communicantes from the ventral rami of the L1, L2, and occasionally L3 spinal nerves, and send gray rami communicantes back to the adjacent ventral rami. Medially, the abdominal sympathetic trunks give off three to four lumbar splanchnic nerves, which pass to the **intermesenteric, inferior mesenteric,** and **superior hypogastric plexuses**, conveying presynaptic sympathetic fibers to the associated prevertebral ganglia.

The cell bodies of postsynaptic sympathetic neurons constitute the major prevertebral ganglia that cluster about the roots of the major branches of the abdominal aorta—the *celiac, aorticorenal, superior mesenteric,* and *inferior mesenteric ganglia*—and minor, unnamed prevertebral ganglia that occur within the intermesenteric and superior hypogastric plexuses. The synapse between presynaptic and postsynaptic neurons occurs in the prevertebral ganglia. Postsynaptic sympathetic nerve fibers pass from the prevertebral ganglia to the abdominal viscera through the periarterial plexuses associated with the branches of the abdominal aorta. Sympathetic innervation in the abdomen, as elsewhere, is primarily involved in producing vasoconstriction. With regard to the digestive tract, it acts to inhibit (slow down or stop) peristalsis.

Visceral afferent fibers conveying pain sensations accompany the sympathetic (visceral motor) fibers, the pain impulses passing retrograde to those of the motor fibers along the splanchnic nerves to the sympathetic trunk, and then through white rami communicantes to the ventral rami of the spinal nerves, and then into the dorsal root to the spinal sensory ganglia.

The *parasympathetic part of the autonomic nervous system in the abdomen* (Fig. 2.74*B*, Table 2.10) consists of the:

- Anterior and posterior vagal trunks
- Pelvic splanchnic nerves
- Abdominal autonomic nerve plexuses
- Periarterial plexuses of nerves
- Intrinsic (enteric) parasympathetic ganglia.

The nerve plexuses are mixed, shared with the sympathetic nervous system and visceral afferent fibers.

The **anterior and posterior vagal trunks** are the continuation of the left and right vagus nerves that emerge from the esophageal plexus and pass through the esophageal hiatus on the anterior and posterior aspects of the esophagus and stomach (Figs. 2.30 and 2.74). The vagus nerves convey presynaptic parasympathetic and visceral afferent fibers (mainly for unconscious sensations associated with reflexes) to the abdominal aortic plexuses and the periarterial plexuses, which extend along the branches of the aorta.

The **pelvic splanchnic nerves** are distinct from other splanchnic nerves in that they:

- Have nothing to do with the sympathetic trunks
- Derive directly from ventral rami of spinal nerves S2 through S4

- Convey presynaptic parasympathetic fibers to the inferior hypogastric (pelvic) plexus.

Presynaptic fibers terminate on the isolated and widely scattered cell bodies of postsynaptic neurons lying on or within the abdominal viscera, constituting intrinsic ganglia.

The presynaptic parasympathetic and visceral afferent reflex fibers conveyed by the vagus nerves extend to intrinsic ganglia of the lower esophagus, stomach, small intestine (including the duodenum), ascending and most of the transverse parts of the colon; those conveyed by the pelvic splanchnic nerves supply the descending and sigmoid parts of the colon, rectum, and pelvic organs. That is, in terms of the digestive tract, the vagus nerves provide parasympathetic innervation of the smooth muscle and glands of the gut as far as the left colic flexure; the pelvic splanchnic nerves provide the remainder.

The **abdominal autonomic plexuses** are nerve networks consisting of both sympathetic and parasympathetic fibers that surround the abdominal aorta and its major branches. The celiac, superior mesenteric, and inferior mesenteric plexuses are interconnected. The **prevertebral sympathetic ganglia** are scattered among the celiac and mesenteric plexuses. The **intrinsic parasympathetic ganglia**, such as the **myenteric plexus** (Auerbach plexus) in the muscular coat of the stomach and intestine, are in the walls of the viscera.

The **celiac plexus** (solar plexus), surrounding the root of the celiac arterial trunk, contains irregular right and left *celiac ganglia* (approximately 2 cm long) that unite superior and inferior to the celiac trunk. The *parasympathetic root* of the celiac plexus is a branch of the *posterior vagal trunk* that contains fibers from the right and left vagus nerves. The *sympathetic roots* of the plexus are the greater and lesser splanchnic nerves.

The **superior mesenteric plexus** and ganglion or ganglia surround the origin of the SMA. The plexus has one median and two lateral roots. The median root is a branch of the celiac plexus and the lateral roots arise from the lesser and least splanchnic nerves, sometimes with a contribution from the first lumbar ganglion of the sympathetic trunk.

The **inferior mesenteric plexus** surrounds the inferior mesenteric artery and gives offshoots to its branches. It receives a medial root from the intermesenteric plexus and lateral roots from the lumbar ganglia of the sympathetic trunks. An *inferior mesenteric ganglion* may also appear just inferior to the root of the inferior mesenteric artery.

The **intermesenteric plexus** is part of the aortic plexus of nerves between the superior and inferior mesenteric arteries. It gives rise to renal, testicular or ovarian, and ureteric plexuses.

The **superior hypogastric plexus** is continuous with the intermesenteric plexus and the inferior mesenteric plexus and lies anterior to the inferior part of the abdominal aorta at its bifurcation. Right and left **hypogastric nerves** join the superior hypogastric plexus to the inferior hypogastric plexus. The superior hypogastric plexus supplies *ureteric and testicular plexuses* and a plexus on each common iliac artery.

The **inferior hypogastric plexus** is formed on each side by a hypogastric nerve from the superior hypogastric plexus. The right and left plexuses are situated on the sides of the rectum, uterine cervix, and urinary bladder. The plexuses receive small branches from the superior sacral sympathetic ganglia and the sacral parasympathetic outflow from S2 through S4 sacral spinal nerves (*pelvic [parasympathetic] splanchnic nerves*). Extensions of the inferior hypogastric plexus send autonomic fibers along the blood vessels, which form visceral plexuses on the walls of the pelvic viscera (e.g., *the rectal and vesical plexuses*).

Partial Lumbar Sympathectomy

The treatment of some patients with arterial disease in the lower limbs may include a *partial lumbar sympathectomy*, the surgical removal of two or more lumbar sympathetic ganglia by division of their rami communicantes. Surgical access to the sympathetic trunks is commonly through a lateral extraperitoneal approach because the sympathetic trunks lie retroperitoneally in the extraperitoneal fatty tissue (Fig. 2.74). The surgeon splits the muscles of the anterior abdominal wall and moves the peritoneum medially and anteriorly to expose the medial edge of the psoas major, along which the sympathetic trunk lies. The left trunk is often overlapped slightly by the aorta. The right sympathetic trunk is covered by the IVC. The intimate relationship of the sympathetic trunks to the aorta and IVC also makes these large vessels vulnerable to injury during lumbar sympathectomy. Consequently, the surgeon carefully retracts them to expose the sympathetic trunks that usually lie in the groove between the psoas major laterally and the lumbar vertebral bodies medially. These trunks are often obscured by fat and lymphatic tissue. Knowing that identification of the sympathetic trunks is not easy, great care is taken not to remove inadvertently part of the:

- Genitofemoral nerve
- Lumbar lymphatics
- Ureter. ⊕

Arteries of the Posterior Abdominal Wall

Most arteries supplying the posterior abdominal wall arise from the **abdominal aorta** (Fig. 2.75*A*); the subcostal arteries arise from the thoracic aorta and distribute inferior to the 12th rib. The abdominal aorta—approximately 13 cm in length—begins at the aortic hiatus in the diaphragm at the level of the T12 vertebra and ends at the level of the L4 vertebra by dividing into the right and left common iliac arter-

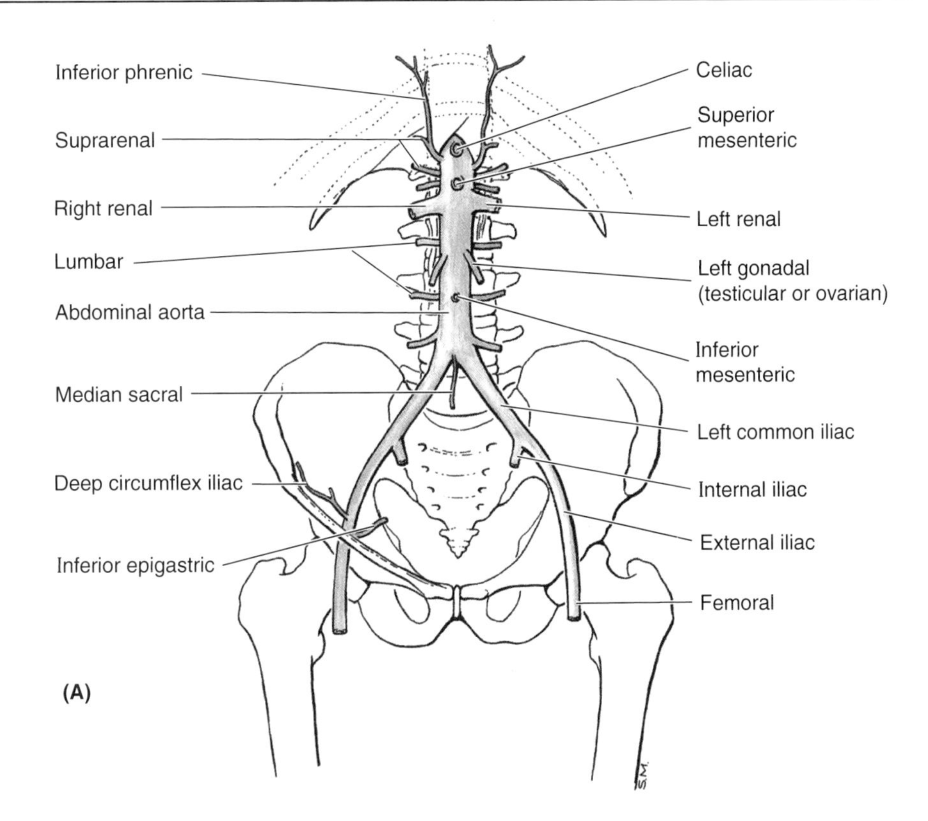

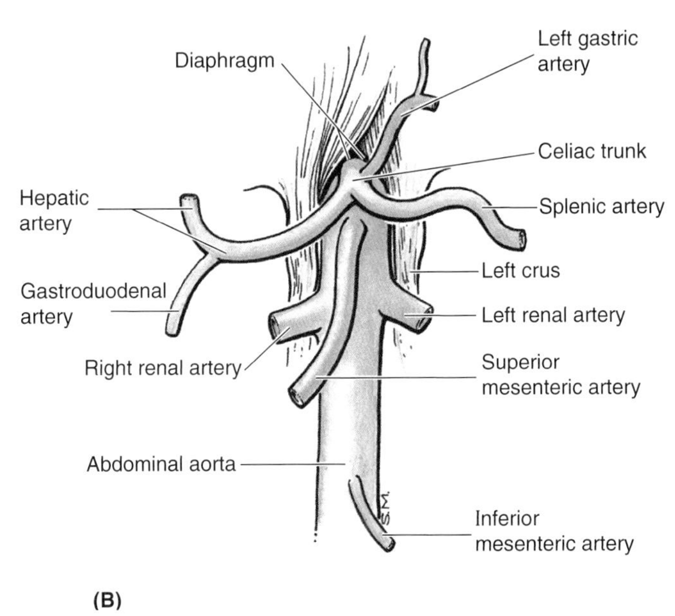

Figure 2.75. Abdominal aorta. A. This diagram shows the relationship of the aorta to the vertebral column. The median sacral artery usually arises from the posterior surface of the aorta. **B.** Major branches of the abdominal aorta. Autonomic nerve plexuses travel with these vessels to the viscera (Fig. 2.74*B*).

ies. The *level of the aortic bifurcation* is 2 to 3 cm inferior and to the left of the umbilicus at the level of the iliac crests. The **common iliac arteries** diverge and run inferolaterally, following the medial border of the psoas muscles to the pelvic brim. Here each common iliac artery divides into the **internal** and **external iliac arteries**. The internal iliac artery enters the pelvis; its course and branches are described in Chapter 3. The external iliac artery follows the iliopsoas muscle. Just before leaving the abdomen, the external iliac artery gives rise to the **inferior epigastric** and **deep circumflex iliac** arteries that supply the anterolateral abdominal wall.

Relations of the Abdominal Aorta

From superior to inferior, the important anterior relations of the abdominal aorta are the:

- Celiac plexus and ganglion (Fig. 2.74)
- Body of the pancreas
- Splenic and left renal veins (Figs. 2.63 and 2.76)
- Horizontal part of the duodenum
- Coils of the small intestine.

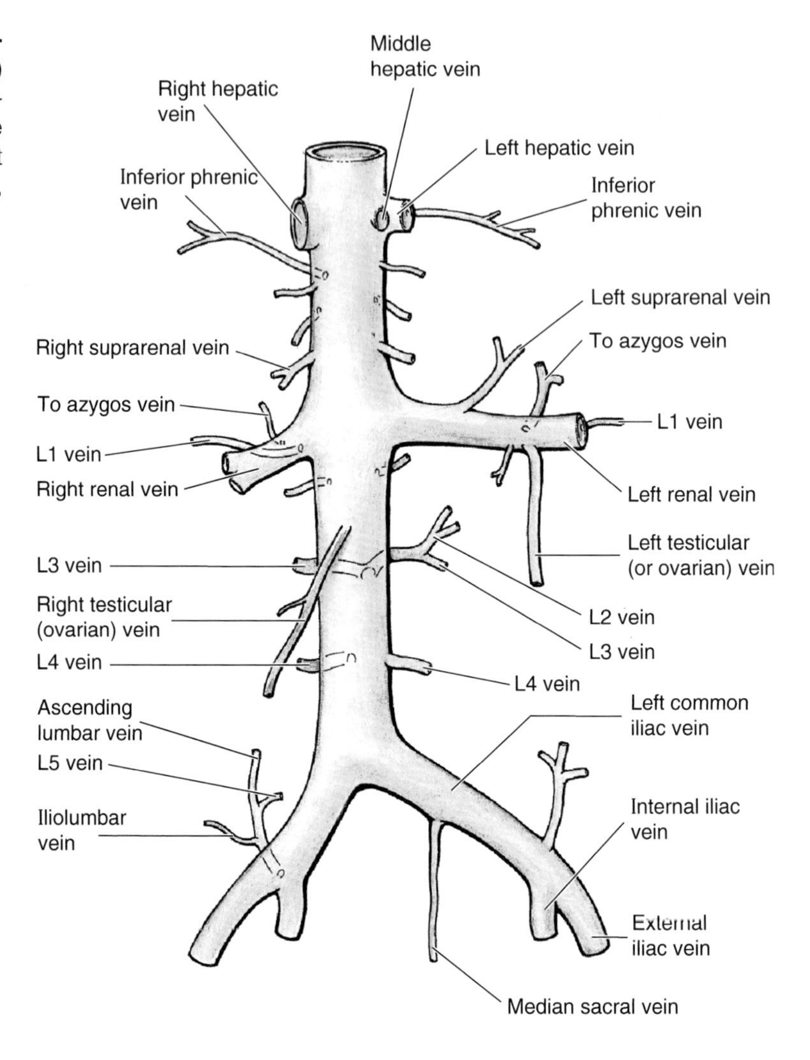

Figure 2.76. Inferior vena cava and its tributaries. Observe that the inferior vena cava (IVC) forms by the junction of the common iliac veins. Observe also that the left renal vein is longer than the right one because it must cross the midline and that the left testicular vein ends in the left renal vein, whereas the right one opens into the IVC.

The abdominal aorta descends anterior to the bodies of T12 through L4 vertebrae. The left lumbar veins pass posterior to the aorta to reach the IVC (Fig. 2.76). *On the right*, the aorta is related to the azygos vein, chyle cistern, thoracic duct, right crus of the diaphragm, and right celiac ganglion. *On the left*, the aorta is related to the left crus of the diaphragm and the left celiac ganglion.

Branches of the Abdominal Aorta

The branches of the abdominal aorta (Fig. 2.75) may be described as visceral or parietal and paired or unpaired.

The **unpaired visceral branches** arise at the following vertebral levels:

- Celiac trunk (T12)
- SMA (L1)
- Inferior mesenteric artery (L3).
- The **paired visceral branches** are the:
- Suprarenal arteries (L1)
- Renal arteries (L1)
- Gonadal arteries, the ovarian or testicular arteries (L2).

The **paired parietal branches** are the:

- Subcostal arteries that enter the abdomen posterior to the lateral arcuate ligaments with the subcostal nerves (T12)
- Inferior phrenic arteries that arise just inferior to the diaphragm
- Lumbar arteries that pass around the sides of the superior four lumbar vertebrae.

The *unpaired parietal branch* is the median sacral artery that arises from the aorta at its bifurcation.

Surface Anatomy of the Abdominal Aorta

The abdominal aorta may be represented on the anterior abdominal wall by a band (approximately 2 cm wide) extending from a median point, approximately 2.5 cm superior to the transpyloric plane to a point slightly inferior to and to the left of the umbilicus. This point indicates the level of bifurcation of the aorta into the common iliac arteries. The site of the aortic ▶

▶ bifurcation is also indicated just to the left of the midpoint of a line joining the highest points of the iliac crests. This line is helpful when examining obese persons in whom the umbilicus is not a reliable surface landmark. ⊕

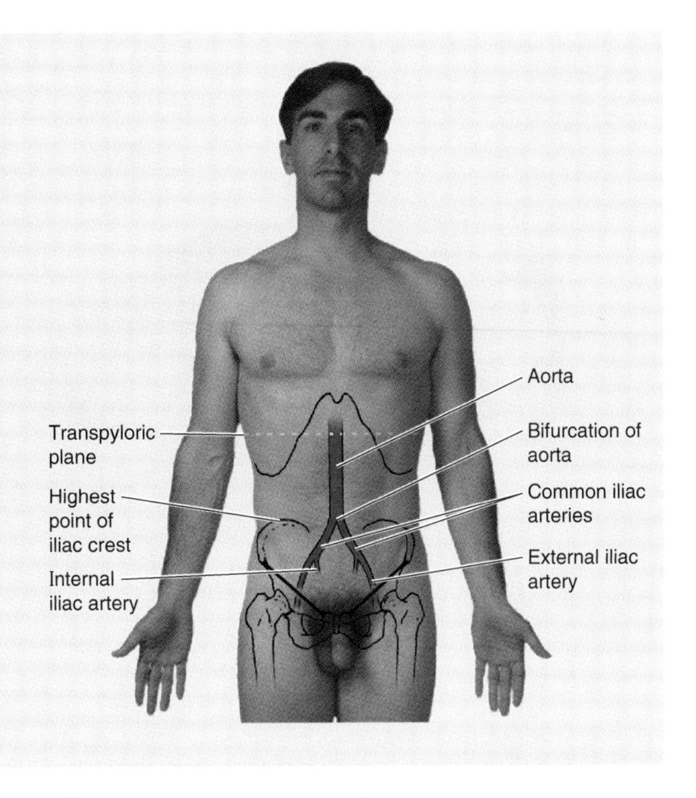

Pulsations of the Aorta and Abdominal Aortic Aneurysm

Because the aorta lies posterior to the pancreas and stomach, a tumor of these organs may transmit pulsations of the aorta that could be mistaken for an **abdominal aortic aneurysm**—a localized enlargement of the aorta (*A*). Deep palpation of the midabdomen can detect an aneurysm (*B–C*), which usually results from a congenital or acquired weakness of the arterial wall. Pulsations of a large aneurysm can be detected to the left of the midline; the pulsatile mass can be moved easily from side to side. Ultrasonography can confirm the diagnosis in doubtful cases.

Acute rupture of an abdominal aortic aneurysm is associated with severe pain in the abdomen or back. If unrecognized, such an aneurysm has a mortality rate of nearly 90% because of heavy blood loss (Swartz, 1994). ▶

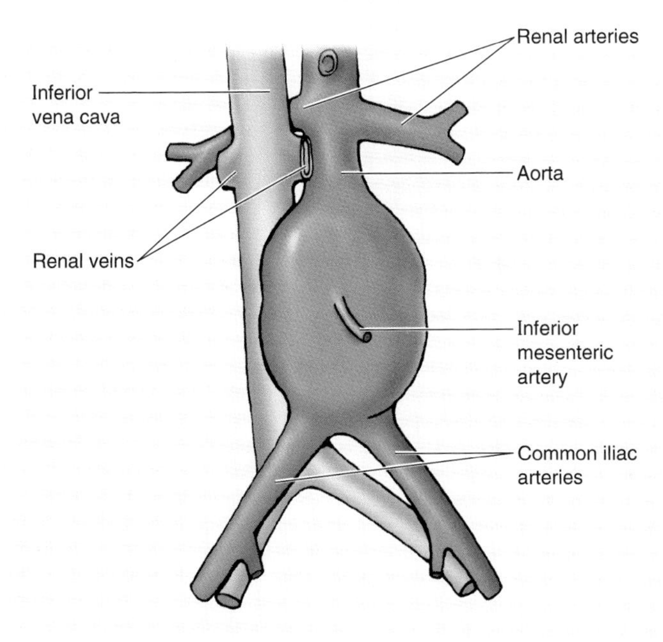

(A)

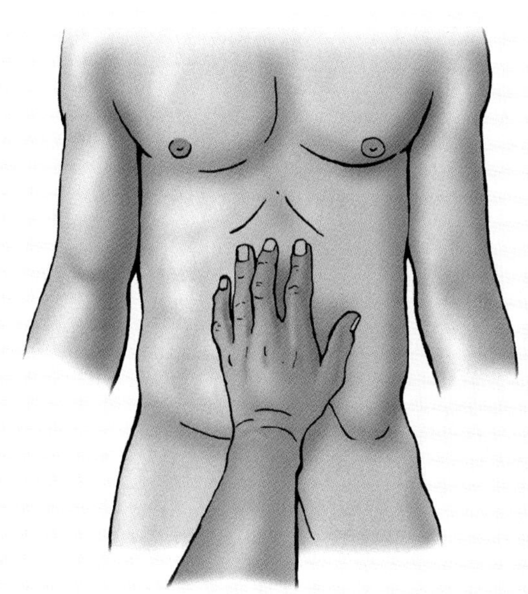

(B)

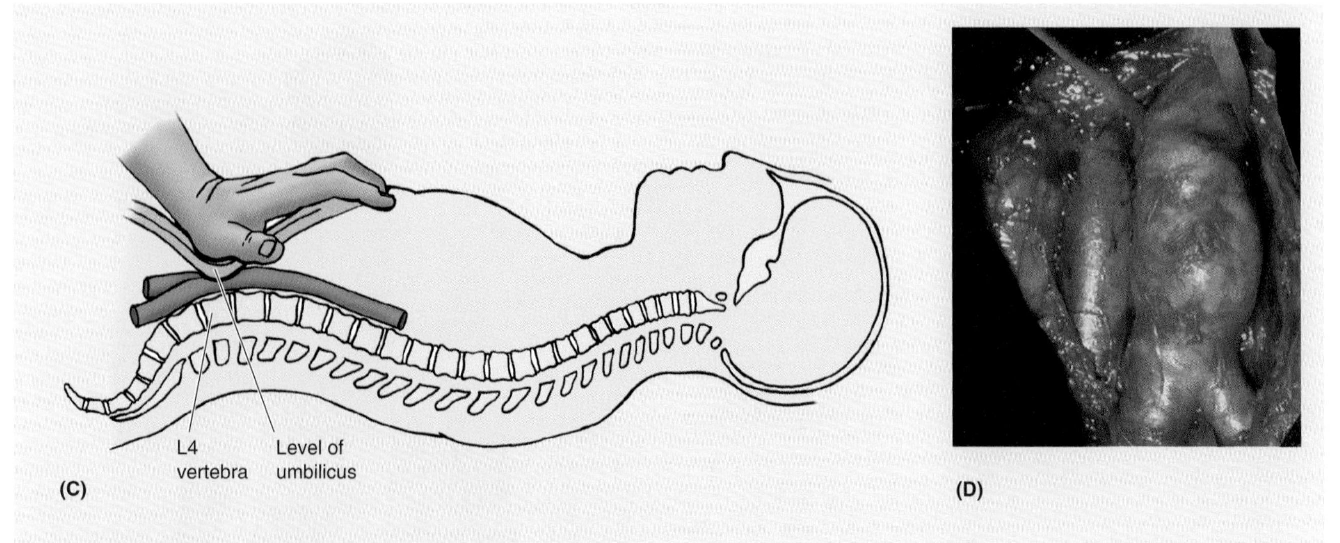

(C)

(D)

▶ Surgeons can repair an aneurysm (*D*) by opening it, inserting a prosthetic graft (such as one made of Dacron), and sewing the wall of the aneurysmal aorta over the graft to protect it.

When the anterior abdominal wall is relaxed, particularly in children and thin adults, the inferior part of the abdominal aorta may be compressed against the body of the L4 vertebra by firm pressure on the anterior abdominal wall, over the umbilicus (*C*). This pressure may be applied to control bleeding in the pelvis or lower limbs. ⊙

Veins of the Posterior Abdominal Wall

The veins of the posterior abdominal wall are tributaries of the IVC, except for the left testicular or ovarian vein that enters the renal vein before entering the IVC (Fig. 2.76). The IVC, the largest vein in the body, has no valves except for a variable, nonfunctional one at its orifice in the right atrium of the heart. The IVC returns poorly oxygenated blood from the lower limbs, most of the back, the abdominal walls, and the abdominopelvic viscera. Blood from the abdominal viscera passes through the *portal venous system* and the liver before entering the IVC through the **hepatic veins**.

Collateral Routes for Abdominopelvic Venous Blood

Three collateral routes, formed by valveless veins of the trunk, are available for venous blood to return to the heart when the IVC is obstructed or ligated.

- The *inferior epigastric veins*, tributaries of the external iliac veins of the inferior caval system, anastomose in the rectus sheath with *superior epigastric veins*, which drain in sequence through the internal thoracic veins of the superior caval system.
- The second collateral route involves the *superficial epigastric* or *superficial circumflex iliac veins*, normally tributaries of the great saphenous vein of the inferior caval system, which anastomoses in the subcutaneous tissues of the anterolateral body wall with one of the tributaries of the axillary vein, commonly the *lateral thoracic vein*. When the IVC is obstructed, this subcutaneous collateral pathway—called the *thoracoepigastric vein*—becomes particularly conspicuous.

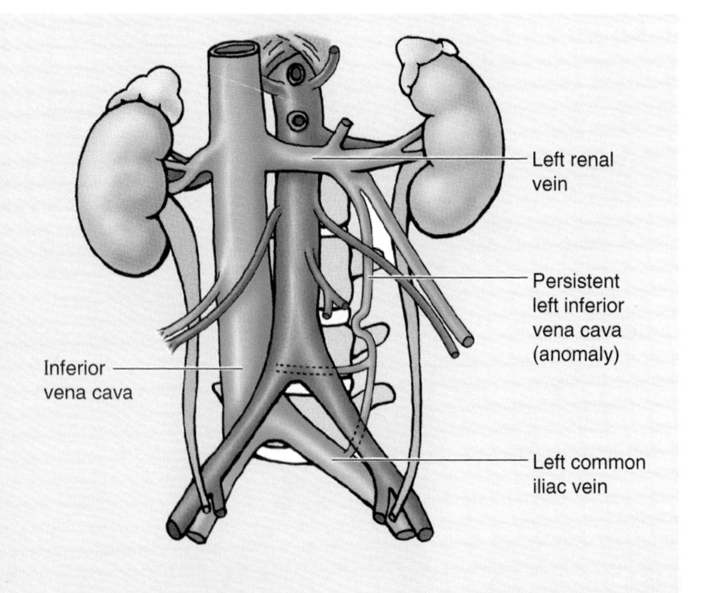

- The third collateral route involves the *epidural venous plexus* inside the vertebral column (illustrated and discussed in Chapter 3), which communicates with the ▶

▶ *lumbar veins* of the inferior caval system, and the tributaries of the *azygos system of veins* that is part of the superior caval system.

The inferior part of the IVC has a complicated developmental history because it forms from parts of three sets of embryonic veins (Moore and Persaud, 1998). Therefore, IVC anomalies are relatively common and most of them, such as a *persisting left IVC*, occur inferior to the renal veins. These anomalies result from the persistence of embryonic veins on the left side that normally disappear. If a left IVC is present, it may cross to the right side at the level of the kidneys. ⊙

The *IVC begins anterior to the L5 vertebra* by the union of the *common iliac veins.* This union occurs approximately 2.5 cm to the right of the median plane, inferior to the bifurcation of the aorta and posterior to the proximal part of the right common iliac artery (Fig. 2.60). The IVC ascends on the right sides of the bodies of the L3 through L5 vertebrae and on the right psoas major to the right of the aorta. The IVC leaves the abdomen by passing through the *caval foramen* in the diaphragm to enter the thorax. The abdominal part of the IVC—approximately 7 cm longer than the abdominal aorta—collects poorly oxygenated blood from the lower limbs and nonportal blood from the abdomen and pelvis. Almost all the blood from the digestive tract is collected by the portal system and passes through the hepatic veins to the IVC.

The **tributaries of the IVC** (Fig. 2.76) correspond to branches of the abdominal aorta:

- Common iliac veins, formed by union of the external and internal iliac veins
- 3rd (L3) and 4th (L4) lumbar veins

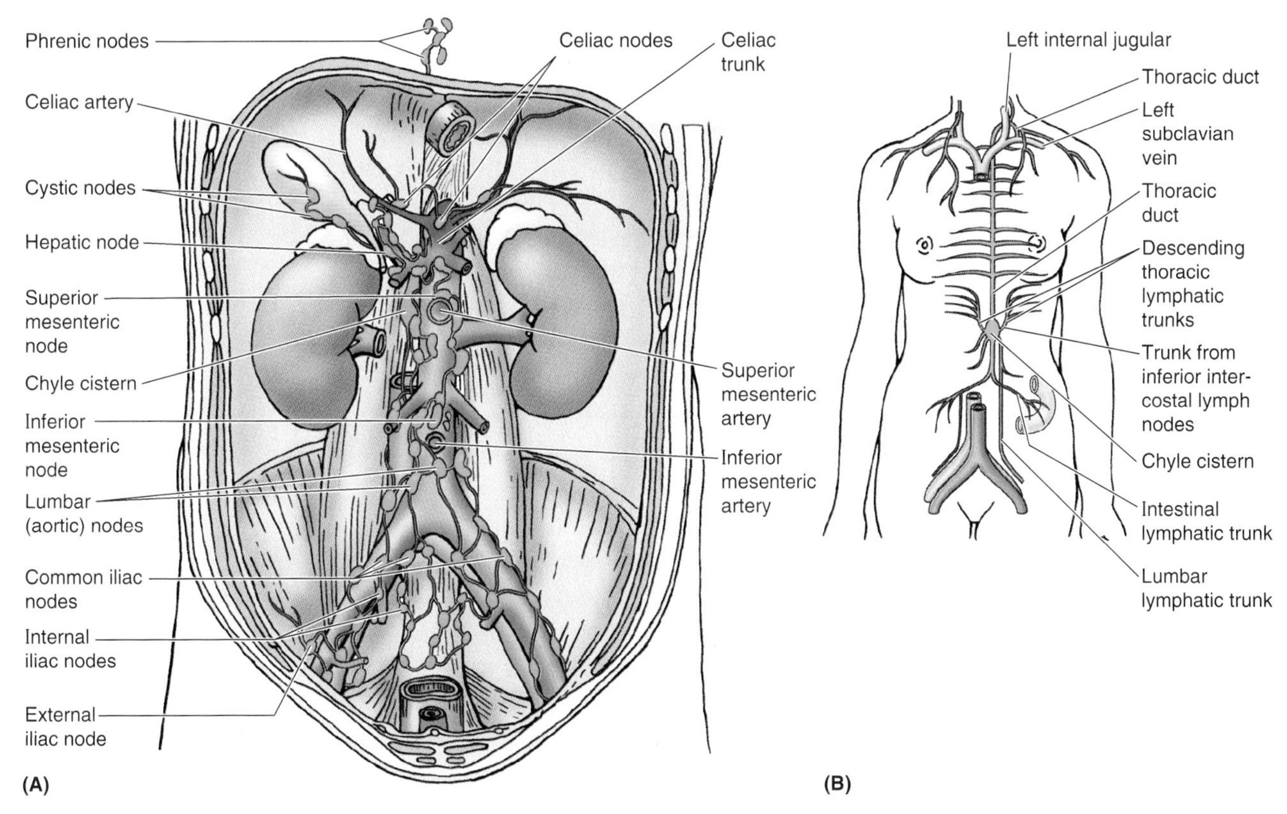

Figure 2.77. Lymphatic drainage of the posterior abdominal wall and lymphatic trunks of the abdomen. A. Parietal lymph nodes. The external iliac, common iliac, and lumbar (aortic) lymph nodes lie in a continuous chain along the abdominal aorta and its terminal branches, receiving lymph from the posterior abdominal wall as well as the efferent drainage of lymph nodes of the abdominal viscera and lower limbs. **B.** Abdominal lymphatic trunks. The thoracic duct begins posterior to the aorta at the aortic hiatus as a convergence of lymphatic trunks, which may or may not take the form of a chyle cistern (L. cisterna chyli). The converging lymphatic trunks include the paired lumbar lymph trunks, the intestinal lymphatic trunk(s), and a pair of descending thoracic lymphatic trunks. Essentially all deep lymphatic drainage from the lower half of the body converges in the abdomen to enter the beginning of the thoracic duct that conducts it into the venous system at the convergence of the left internal jugular and subclavical veins.

- Right testicular or ovarian vein
- Renal veins
- Ascending lumbar (azygos/hemiazygos) veins
- Right suprarenal vein
- Inferior phrenic veins
- Hepatic veins.

The left testicular or ovarian vein and the left suprarenal vein usually drain into the left renal vein. The ascending lumbar and azygos veins connect the IVC and SVC (superior vena cava), either directly or indirectly (see Chapter 1).

Lymphatics of the Posterior Abdominal Wall

Lymphatic vessels and lymph nodes lie along the aorta, IVC, and iliac vessels (Fig. 2.77*A*). The **common iliac lymph nodes** receive lymph from the external and internal iliac lymph nodes. Lymph from the common iliac lymph nodes passes to the **lumbar (aortic) lymph nodes**. Lymph from the digestive tract, liver, spleen, and pancreas passes along the celiac and superior and inferior mesenteric arteries to the **preaortic lymph nodes** (celiac and superior and inferior mesenteric nodes) scattered around the origins of these arteries from the aorta. Efferent vessels from these nodes form the **intestinal lymphatic trunks**, which may be single or multiple, and participate in the confluence of lymphatic trunks that gives rise to the thoracic duct (Fig. 2.77*B*). The **lumbar (aortic) lymph nodes** lie on both sides of the aorta and IVC. These nodes receive lymph directly from the posterior abdominal wall, kidneys, ureters, testes or ovaries, uterus, and uterine tubes. They also receive lymph from the descending colon, pelvis, and lower limbs through the **inferior mesenteric** and **common iliac lymph nodes.** Efferent lymphatic vessels from the large lumbar lymph nodes form the right and left **lumbar lymphatic trunks.**

The inferior end of the **thoracic duct** lies anterior to the bodies of the L1 and L2 vertebrae between the right crus of the diaphragm and the aorta. The thoracic duct begins with the convergence of the main lymphatic ducts of the abdomen, which only in a small proportion of individuals takes the form of the commonly depicted, thin-walled sac or dilation—the **chyle cistern**). Chyle cisterns vary greatly in size and shape. More often there is merely a simple or plexiform convergence at this level of the right and left lumbar lymphatic trunks, the intestinal lymph trunk(s), and a pair of **descending thoracic lymphatic trunks** that carry lymph from the lower six intercostal spaces on each side. Consequently, essentially all the lymphatic drainage from the lower half of the body (deep lymphatic drainage inferior to the level of the diaphragm, and all superficial drainage inferior to the level of the umbilicus) converges in the abdomen to enter the beginning of the thoracic duct. The thoracic duct ascends through the aortic hiatus in the diaphragm into the posterior mediastinum where it collects more parietal and visceral drainage, particularly from the left upper quadrant of the body, and ultimately ends by entering the venous system (Fig. 2.77*B*) at the junction of the **left subclavian and internal jugular veins** (the left venous angle).

Medical Imaging of the Abdomen

Radiographs of the abdomen demonstrate abnormal anatomical relationships of organs such as those resulting from tumors. Radiographs also demonstrate gas and calcifications in the intestine. Abnormal intestinal gas patterns indicate a dilated large intestine and intestinal perforation. The anatomy of the esophagus and GI tract can be demonstrated radiologically following the swallowing of *barium meal*—a contrast medium composed of a mixture of barium sulphate and water. Examinations of the GI tract are ▶

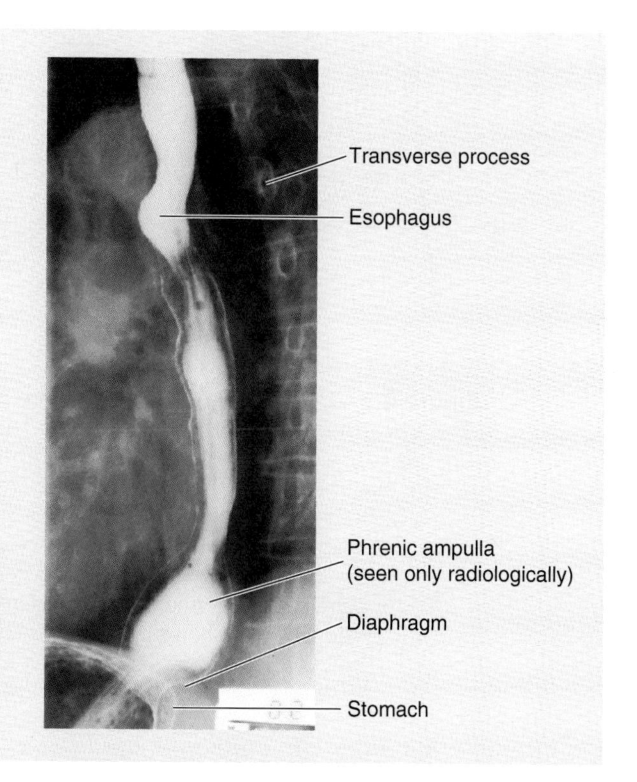

Figure 2.78. Radiograph of the esophagus after swallowing barium. Left posterior oblique (LPO) view. Note normal "constrictions" (impressions) caused by the arch of the aorta and left main bronchus. The phrenic ampulla that is seen only radiologically is the distensible part of the esophagus, lying just superior to the diaphragm. (Courtesy of Dr. E.L. Lansdown, Professor of Medical Imaging, University of Toronto, Toronto, Ontario, Canada)

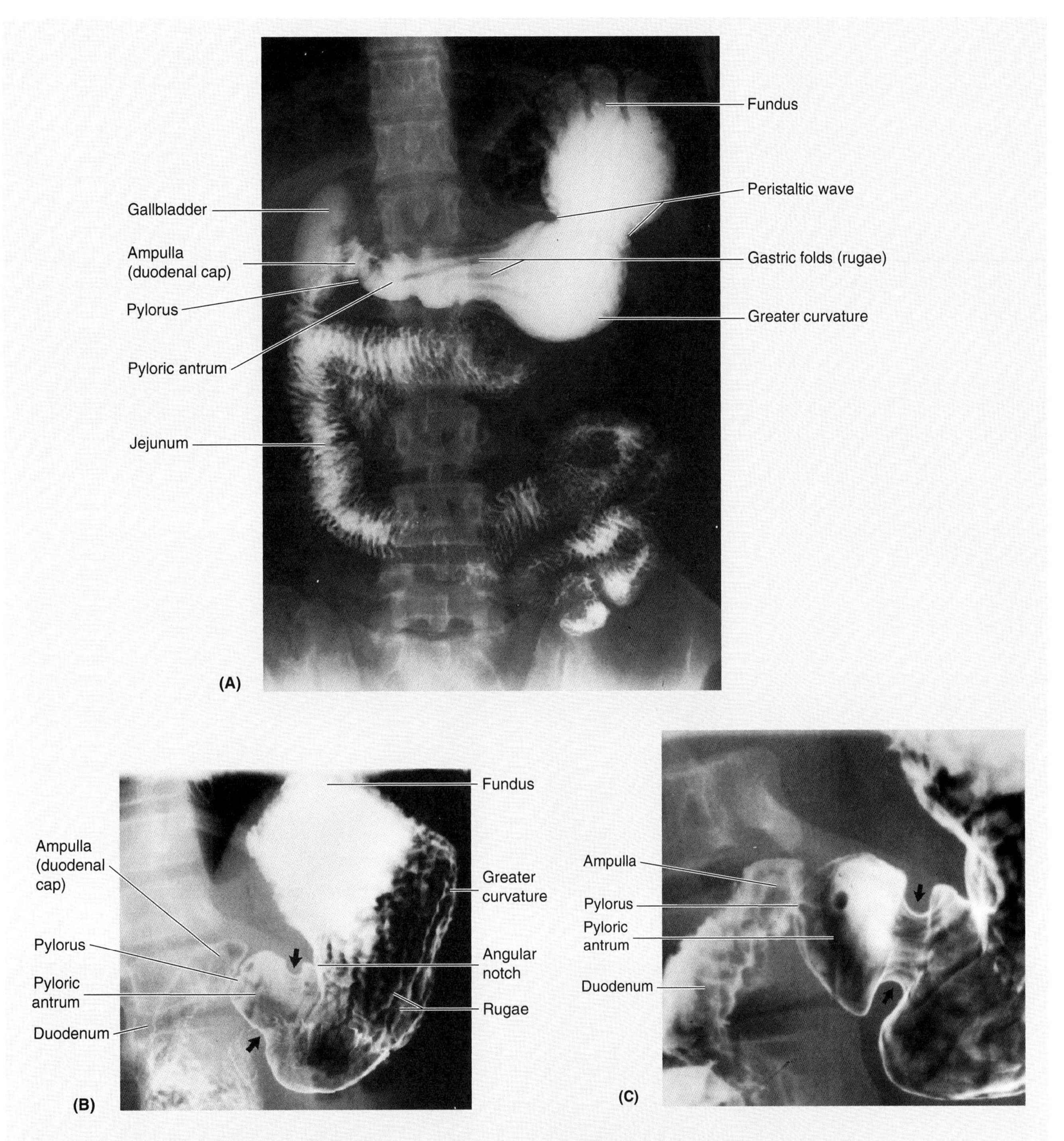

Figure 2.79. Radiographs of the stomach, small intestine, and gallbladder. A. Observe the peristaltic wave in the stomach and the longitudinal gastric folds (rugae) of mucous membrane. **B.** The stomach and small intestine following a barium meal. Observe the peristaltic wave (*arrowheads*), pylorus, duodenal cap, and the feathery appearance of the barium in the small intestine. Also observe the relationship of the gallbladder to the ampulla (duodenal cap) in the superior (first) part of the duodenum. **C.** Radiograph of the pyloric region of the stomach and the superior part of the duodenum. (**A**, Courtesy of Dr. J. Helsin, Toronto, Ontario, Canada; **B–C**, Courtesy of Dr. E.L. Lansdown, Professor of Medical Imaging, University of Toronto, Toronto, Ontario, Canada)

▶ performed with fluoroscopic guidance, supplemented by cineradiography or videotape and radiographs. These procedures enable the radiologist to detect abnormalities of the esophagus, stomach, and intestines, as well as lesions in adjacent structures that displace these organs. Most early knowledge concerning the shape, position, and movements of the esophagus, stomach, and intestines came from barium studies.

Peristaltic waves in the esophagus appear as ringlike contractions that propel the barium toward the stomach (Fig. 2.78). Once the stomach is distended, circular peristaltic waves begin in the body of the stomach and sweep toward the pyloric canal, where they stop. Gas in the fundus of the stomach is clearly visible superior to the barium shadow (Fig. 2.79). When the pyloric sphincter opens and closes, the proximal part of the duodenum expands to form the **ampulla** (duodenal cap). The outline of the barium in the superior part of the duodenum is smooth; in other parts the outline has a feathery appearance because of the circular folds (plicae circulares) in the duodenum.

To examine the colon, a barium enema is given after the bowel is cleared of fecal material by a cleansing enema. Radiographs show the typical **haustra** (Fig. 2.80, *A–B*). The single-contrast examinations of the colon show its various parts; the dilated **rectum** is well illustrated. In chronic stages of *colitis* (inflammation of the colon), the mucosa atrophies and the typical pattern of the haustra disappears. The outline of the haustra is accentuated by the double-contrast method: the barium is evacuated by the patient and air is injected through the anal canal to distend the ▶

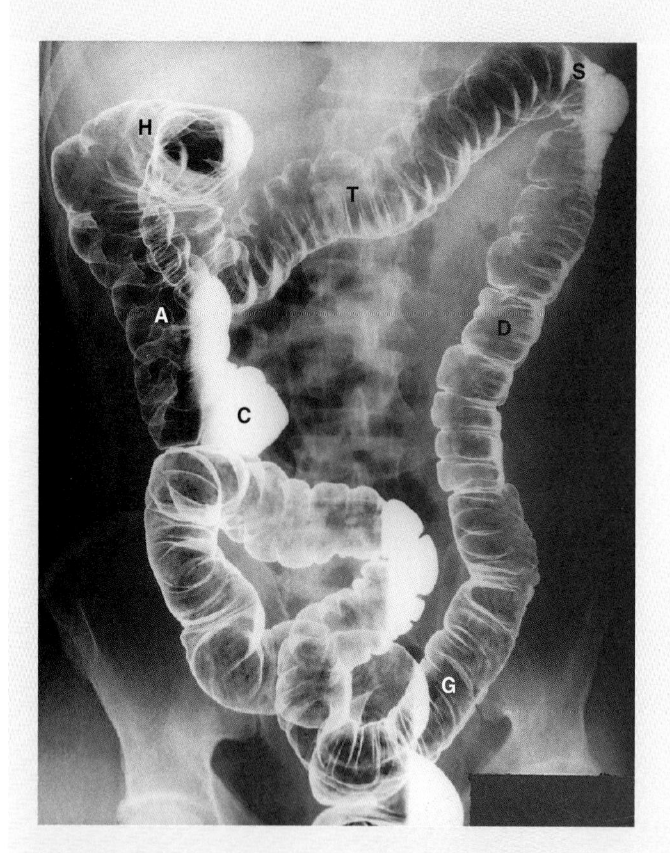

(A) Anteroposterior view

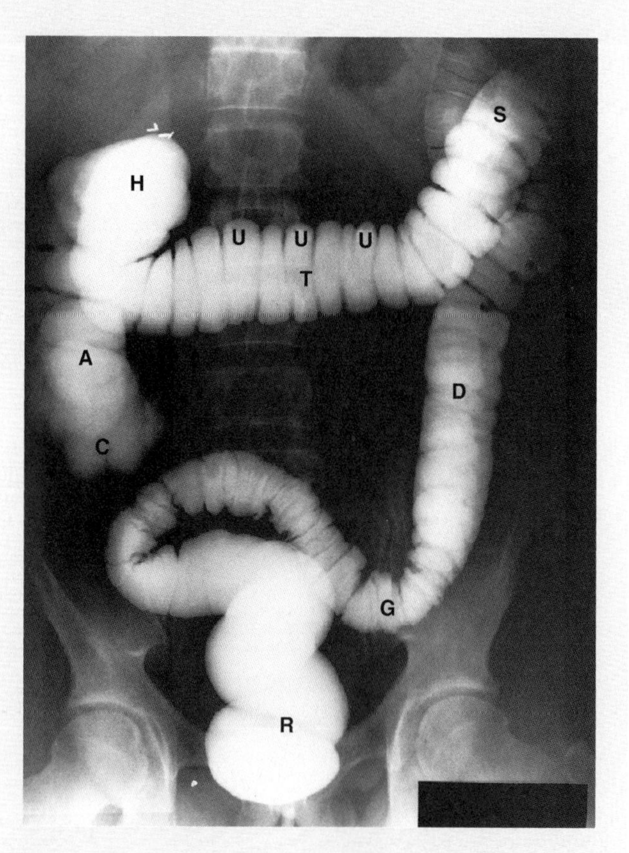

(B)

Figure 2.80. Barium enema examination of the colon and an MRI of the intestine. A. Double contrast. Barium can be seen coating the walls of the colon distended with air, giving a vivid view of the mucosa and haustra. (Courtesy of Dr. C.S. Ho, Professor of Medical Imaging, University of Toronto, Toronto, Ontario, Canada) **B.** Single contrast. A barium enema has filled the colon. Observe the level of the right colic flexure. (Courtesy of Dr. E.L. Lansdown, Professor of Medical Imaging, University of Toronto, Toronto, Ontario, Canada)

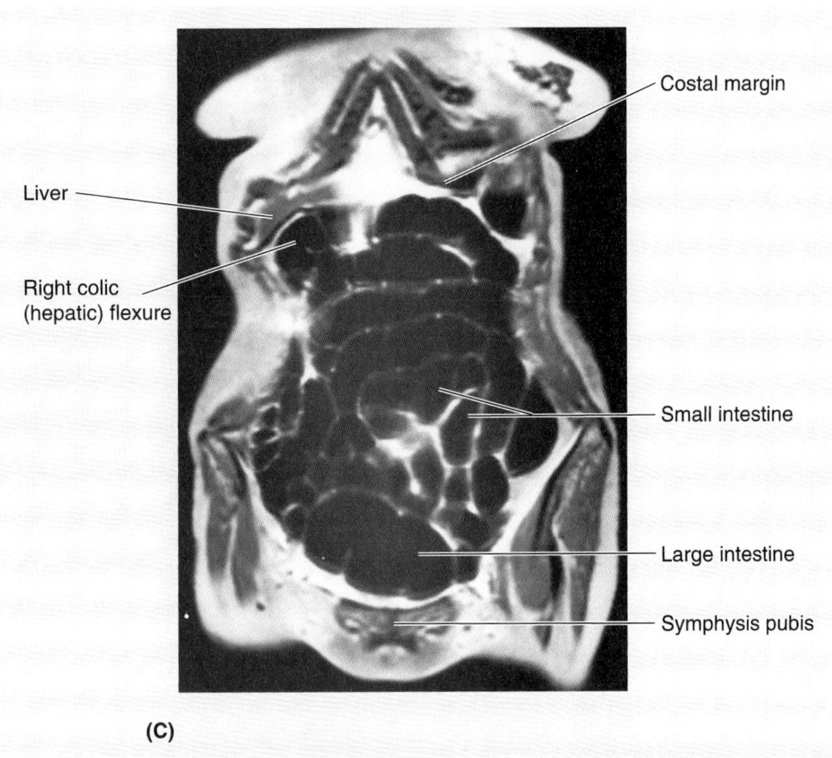

(C)

Figure 2.80. *(Continued)* **C.** Coronal MRI of the lower thoracic cage (rib cage) and abdomen. (Courtesy of Dr. W. Kucharczyk, Professor and Chair of Medical Imaging, University of Toronto, and Clinical Director of Tri-Hospital Resonance Centre, Toronto, Ontario, Canada) *C*, cecum; *A*, ascending colon; *H*, hepatic or right colic flexure; *T*, transverse colon; *S*, splenic or left colic flexure; *D*, descending colon; *G*, sigmoid colon; *R*, rectum; *U*, haustra.

▶ colon that is still lined with a thin layer of barium (Fig. 2.80*A*).

A wide variety of radiographic methods are available for studying the structure and function of the biliary passages and gallbladder clinically, each with its own criteria, advantages, and disadvantages. Radiographic examination of the bile ducts at the time of surgery (peroperative cholangiography—Fig. 2.81*A*) is an important procedure for all patients undergoing **cholecystectomy** to determine the presence and location of stones in the biliary tract, that all calculi have been removed, and that no other obstructions are present. A tube is implanted directly into the biliary duct system to inject radiopaque dye.

Endoscopic retrograde cholangiopancreatography (ERCP—Fig. 2.81*B*) has become a standard procedure for the diagnosis of both pancreatic and biliary disease. Following passage of the fiberoptic endoscope through the mouth, esophagus, and stomach, the duodenum is entered and a cannula is inserted into the major duodenal papilla and advanced under fluoroscopic control into the duct of choice (bile duct or pancreatic duct) for injection of radiographic contrast medium. One radiographic method of studying the anatomy and function of the gallbladder has been used for more than eight decades (Fig. 2.81*C*). This procedure—*oral or intravenous cholecystography*—is a test of gallbladder function, as well as demonstrating its location and shape. This procedure has largely been replaced by *ultrasonography* and cholescintigraphy. The latter is a dynamic function test in many ways similar to cholecystography; however, radioactive tracer and marker compounds are used.

Ultrasound and CT scans of the abdomen, including magnetic resonance images (**MRIs**), are also used to examine the abdominal viscera (Figs. 2.82–2.86). Because MRIs provide better differentiation between soft tissues, its images are more revealing. An image in virtually any plane can be reconstructed after scanning is completed.

Abdominal arteriography—radiography after the injection of radiopaque material directly into the bloodstream—detects abnormalities of the abdominal arteries, such as blood clots (Fig. 2.87). **Nuclear scanning** is useful for detection of lower GI hemorrhage. A positive nuclear scan may be followed by arteriography.

The radiological anatomy of the kidneys and ureters can be examined by *intravenous urography*, or pyelography (Fig. 2.88). An intravenously injected contrast medium is excreted by the kidneys. In *retrograde pyelography*, the contrast medium is injected into the ureters and ascends to fill the pelves and calices of the kidneys. ⭘

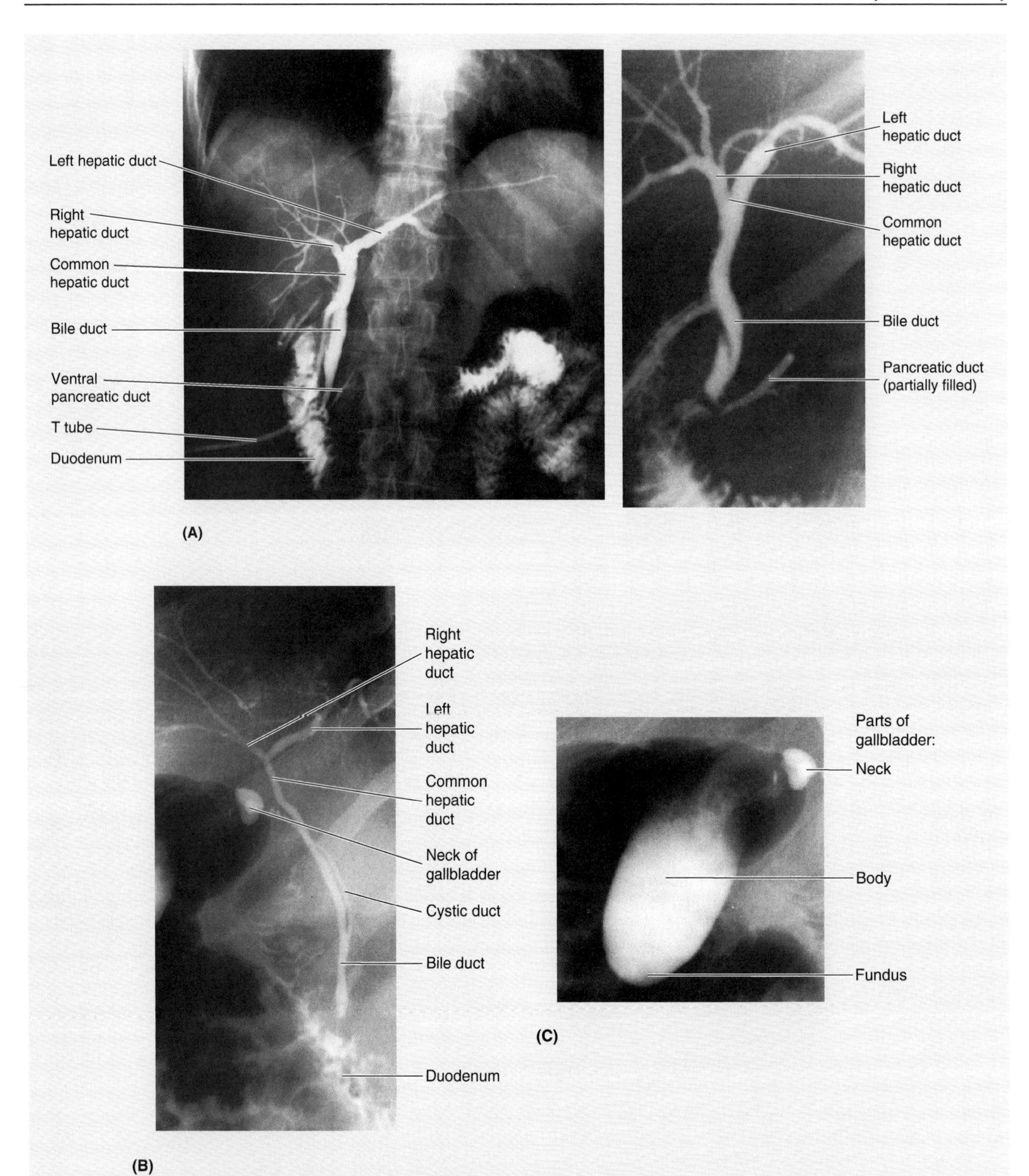

Figure 2.81. Radiographs of biliary passages and the gallbladder. A. Following an open-incision cholecystectomy, contrast medium was injected through a T tube inserted into the bile passages. (Courtesy of Dr. J. Helsin, Toronto, Ontario, Canada) **B–C.** Endoscopic retrograde cholangiography of the gallbladder and biliary passages. The cystic duct usually lies on the right side of the common hepatic duct and joins it just superior to the superior part of the duodenum. The course and length of the cystic duct are variable. (Courtesy of Dr. G.B. Haber, Assistant Professor of Medicine, University of Toronto, Toronto, Ontario, Canada)

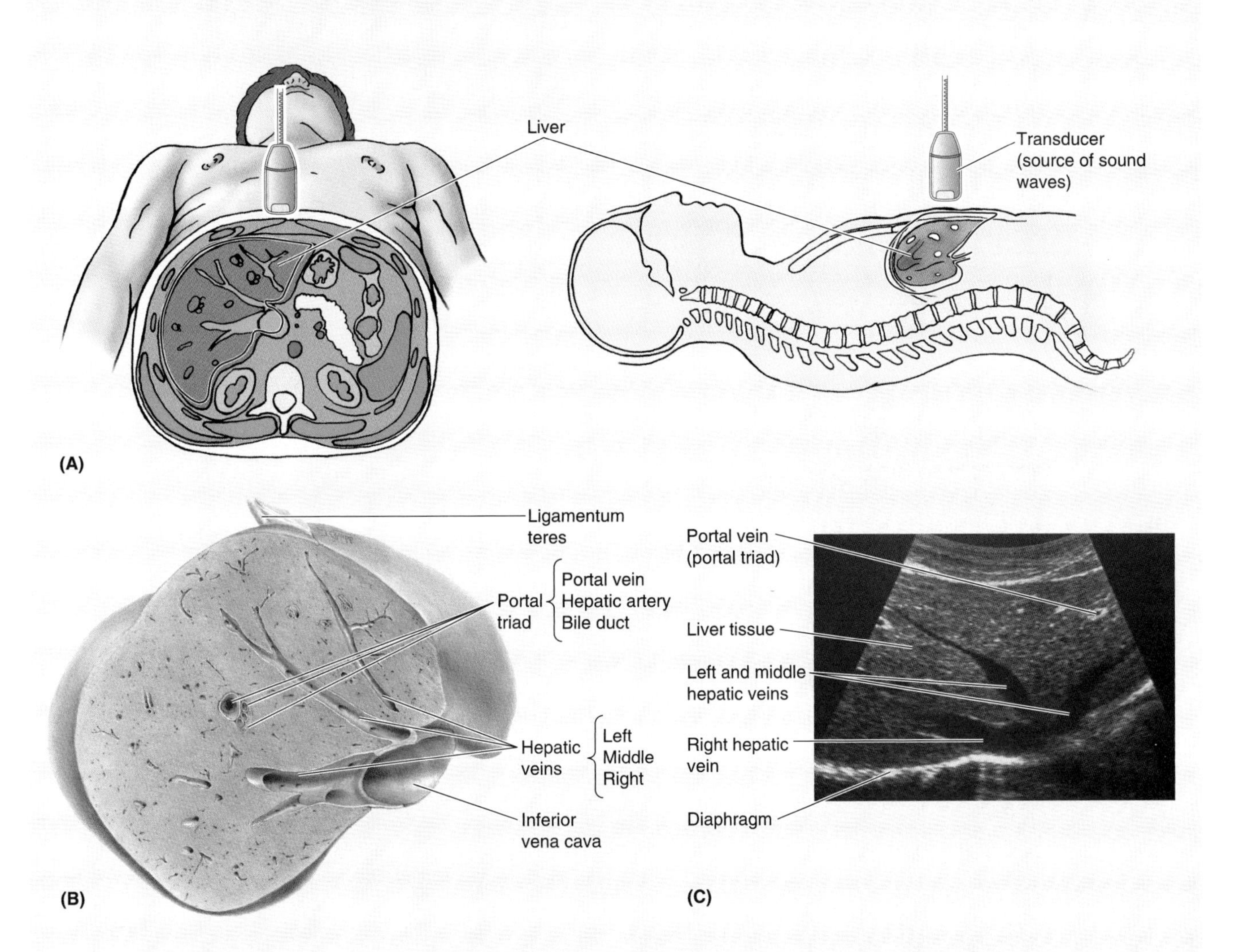

Figure 2.82. Ultrasound scans of the abdomen. Orientation drawings (**A**) demonstrating placement of the transducer on the anterior abdominal wall. Transverse anatomical section (**B**) and ultrasound scan (**C**) of the liver demonstrating the hepatic veins. (Courtesy of Dr. A.M. Arenson, Assistant Professor of Medical Imaging, University of Toronto, Toronto, Ontario, Canada)

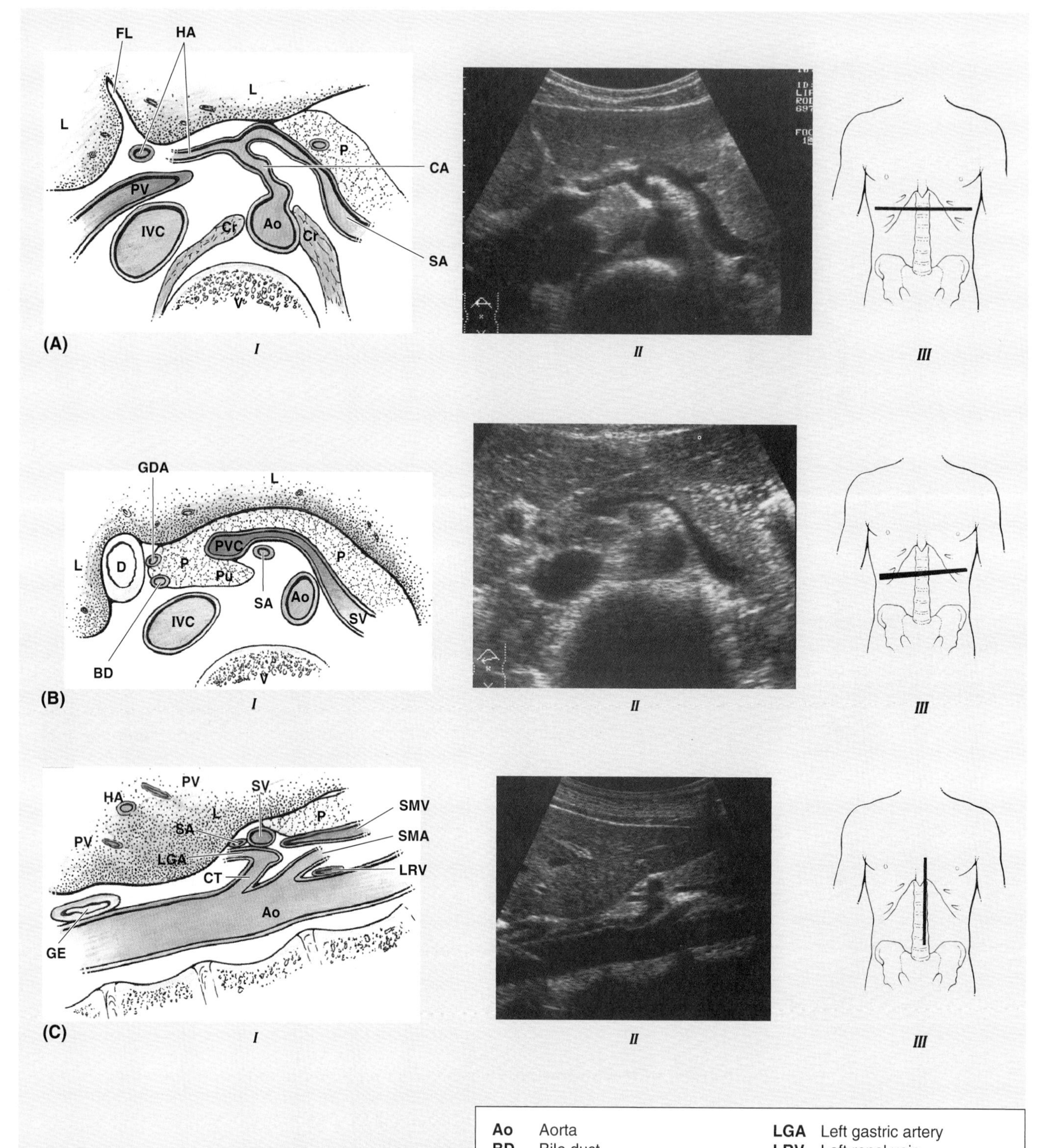

Ao	Aorta	**LGA**	Left gastric artery
BD	Bile duct	**LRV**	Left renal vein
CA	Celiac artery	**P**	Pancreas
Cr	Crus of diaphragm	**Pu**	Uncinate process of pancreas
CT	Celiac trunk		
D	Duodenum	**PV**	Portal vein
FL	Falciform ligament	**PVC**	Portal venous confluence
GDA	Gastroduodenal artery	**SA**	Splenic artery
GE	Gastroesophageal junction	**SMA**	Superior mesenteric artery
HA	Hepatic artery	**SMV**	Superior mesenteric vein
IVC	Inferior vena cava	**SV**	Splenic vein
L	Liver	**V**	Vertebra

Figure 2.83. Ultrasound scans of the abdomen. A. Transverse scan through the celiac trunk. **B.** Transverse scan through the pancreas. **C.** Sagittal scan through the aorta. (Courtesy of Dr. A.M. Arenson, Assistant Professor of Medical Imaging, University of Toronto, Toronto, Ontario, Canada)

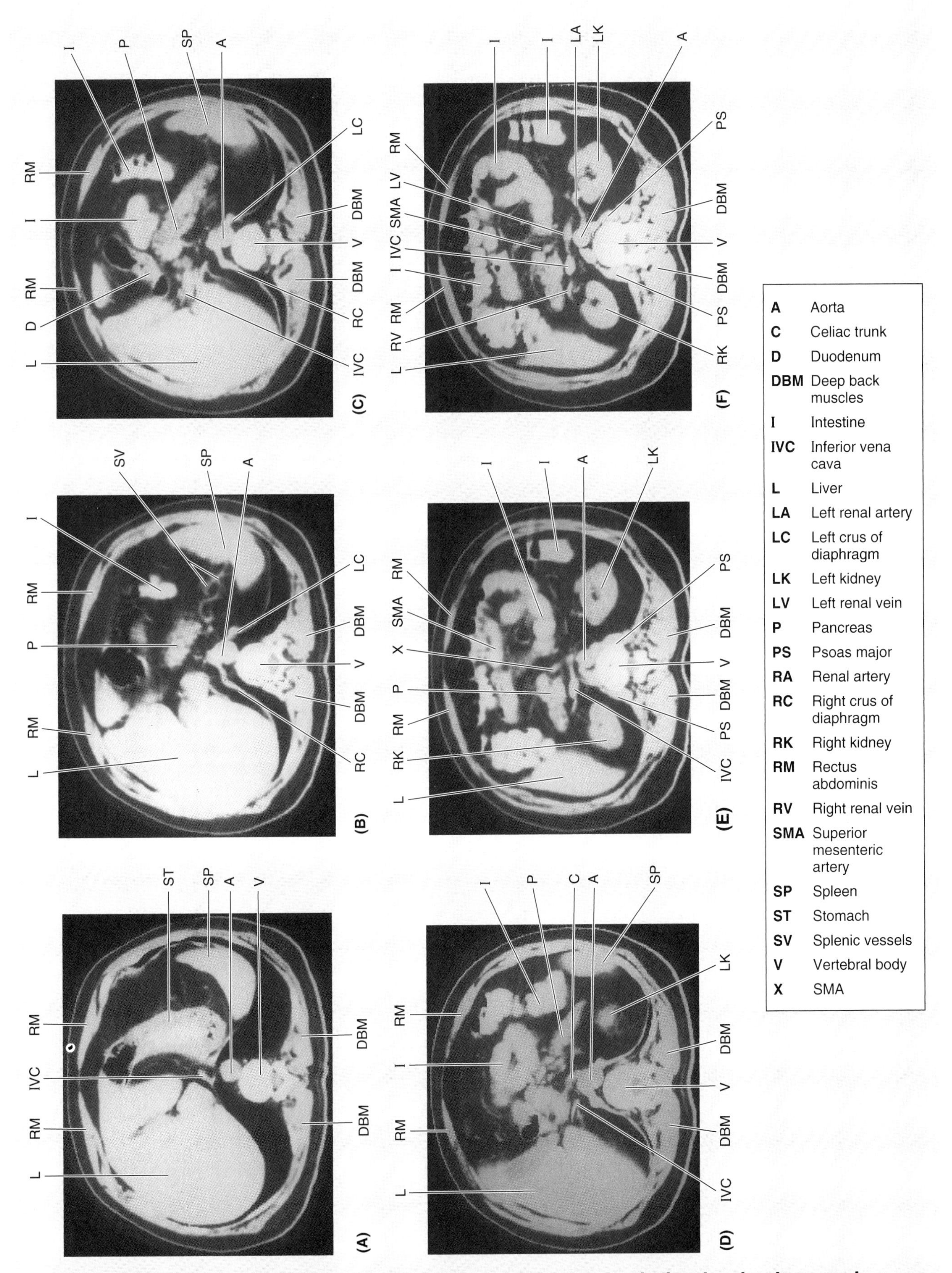

Figure 2.84. CT scans of the abdomen at progressively lower levels showing the viscera and blood vessels. (Courtesy of Dr. Tom White, Department of Radiology, The Health Sciences Center, University of Tennessee, Memphis, TN)

Figure 2.85. Transverse MRIs of the abdomen. (Courtesy of Dr. W. Kucharczyk, Professor and Chair of Medical Imaging, University of Toronto, and Clinical Director of Tri-Hospital Resonance Centre, Toronto, Ontario, Canada)

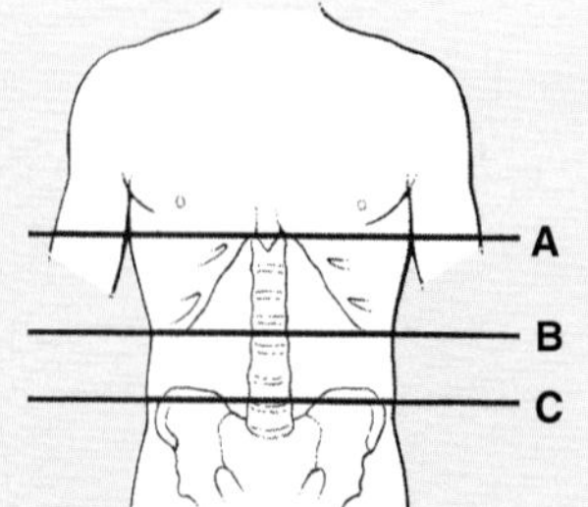

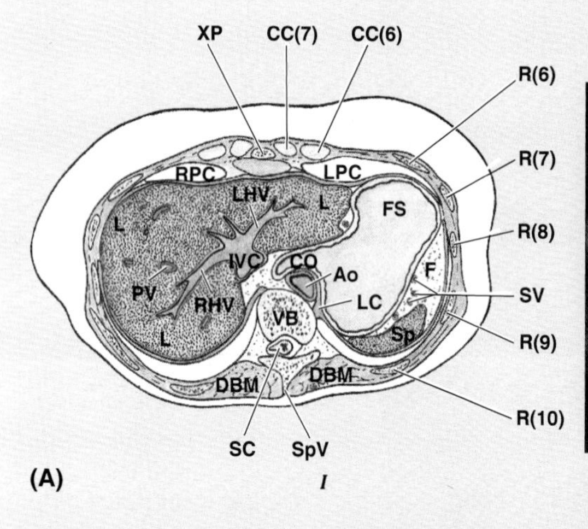

(A) *I*

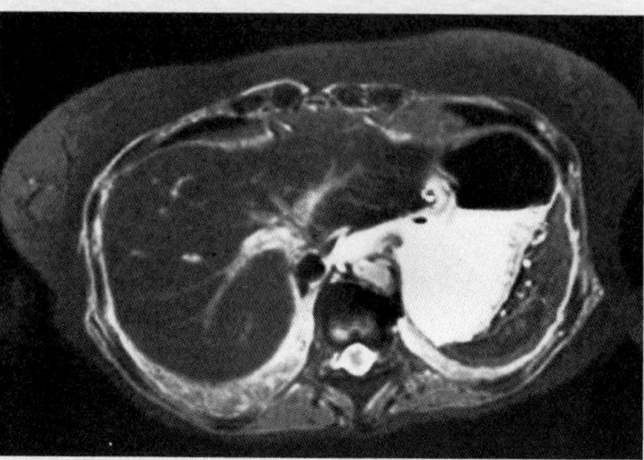

II

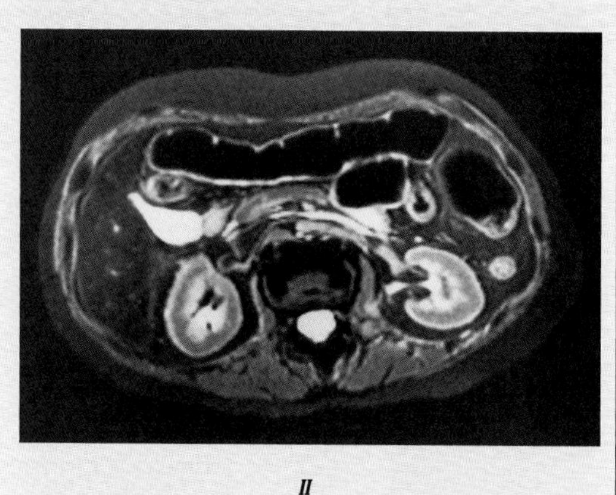

(B) *I* *II*

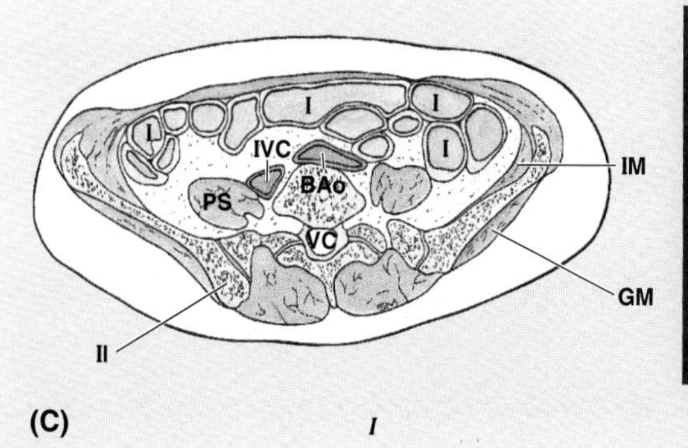

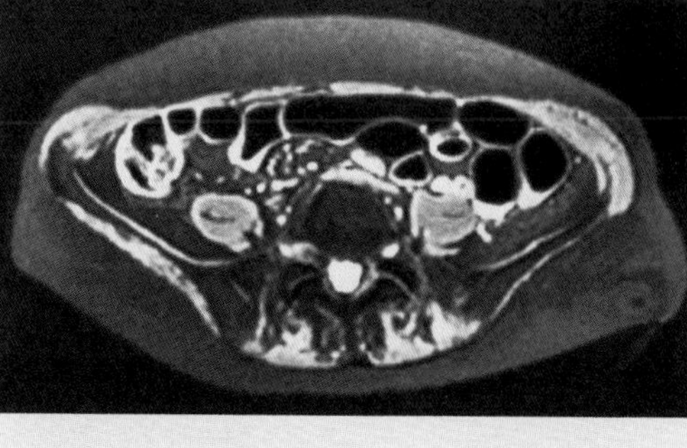

(C) *I* *II*

Ao	Aorta
BAo	Bifurcation of aorta
CC	Costal cartilage
CO	Cardiac orifice of stomach
D	Duodenum
DBM	Deep back muscles
F	Fat
FS	Fundus of stomach
GB	Gallbladder
GM	Gluteus medius muscle
I	Intestine
II	Ilium
IM	Iliacus muscle
IVC	Inferior vena cava
L	Liver
LC	Left crus
LHV	Left hepatic vein
LK	Left kidney
LPC	Left pleural cavity
LRA	Inferior vena cava
LRV	Left renal vein
P	Pancreas
PC	Portal confluence
PF	Perirenal fat
PS	Psoas muscle
PV	Portal vein (triad)
R	Rib
RHV	Right hepatic vein
RK	Right kidney
RLL	Right lobe of liver
RPC	Right pleural cavity
RRV	Right renal vein
SC	Spinal cord
Sp	Spleen
SpV	Spinous process of vertebra
Sv	Splenic vein
SV	Splenic vessels
TC	Transverse colon
VB	Vertebral body
VC	Vertebral canal
XP	Xiphoid process

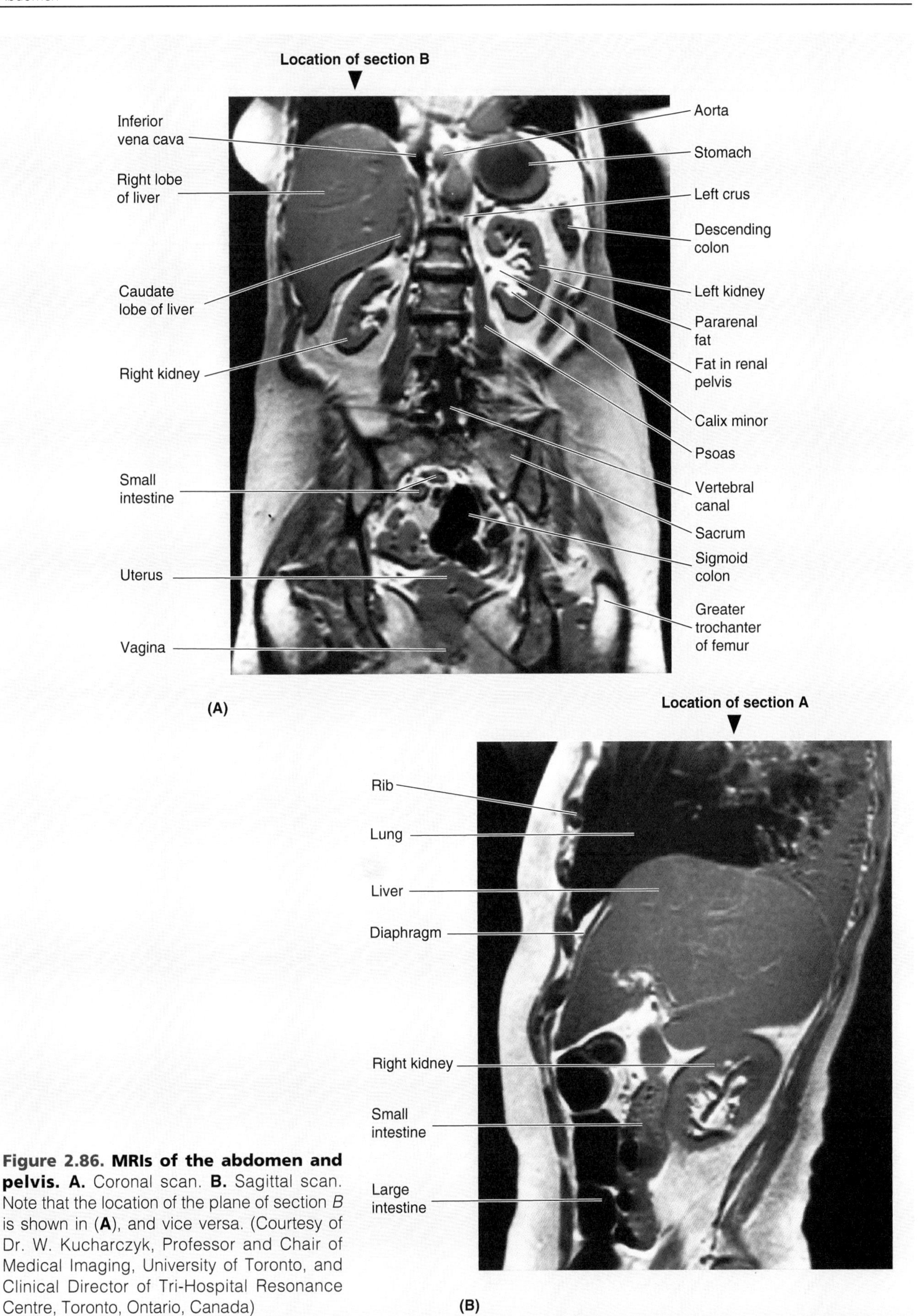

Figure 2.86. MRIs of the abdomen and pelvis. A. Coronal scan. **B.** Sagittal scan. Note that the location of the plane of section *B* is shown in (**A**), and vice versa. (Courtesy of Dr. W. Kucharczyk, Professor and Chair of Medical Imaging, University of Toronto, and Clinical Director of Tri-Hospital Resonance Centre, Toronto, Ontario, Canada)

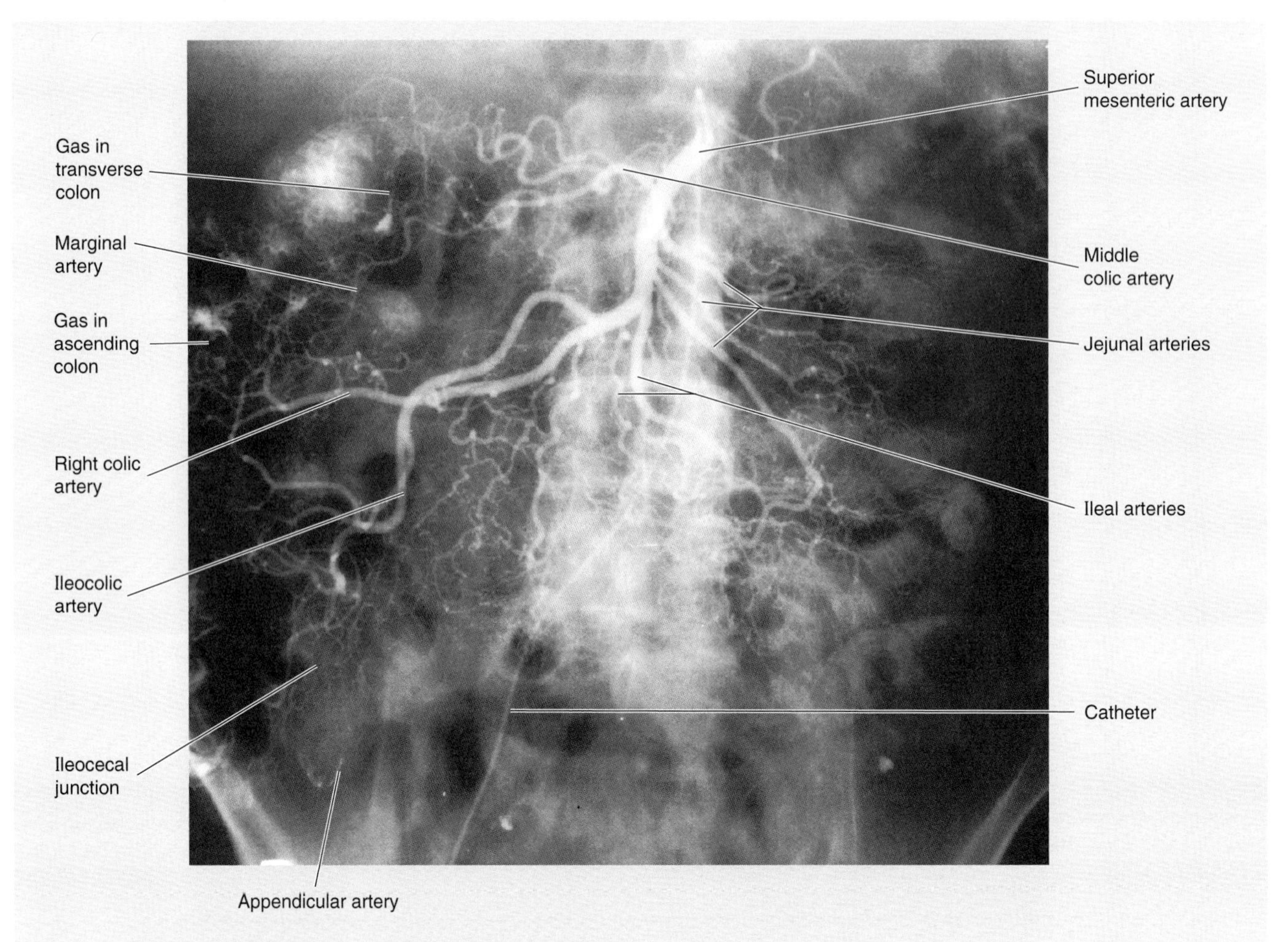

Figure 2.87. Superior mesenteric arteriogram. Radiopaque dye has been injected into the artery by means of the catheter introduced into the femoral artery and advanced through the iliac arteries and aorta to the opening of the superior mesenteric artery (SMA). (Courtesy of Dr. E.L. Lansdown, Professor of Medical Imaging, University of Toronto, Toronto, Ontario, Canada)

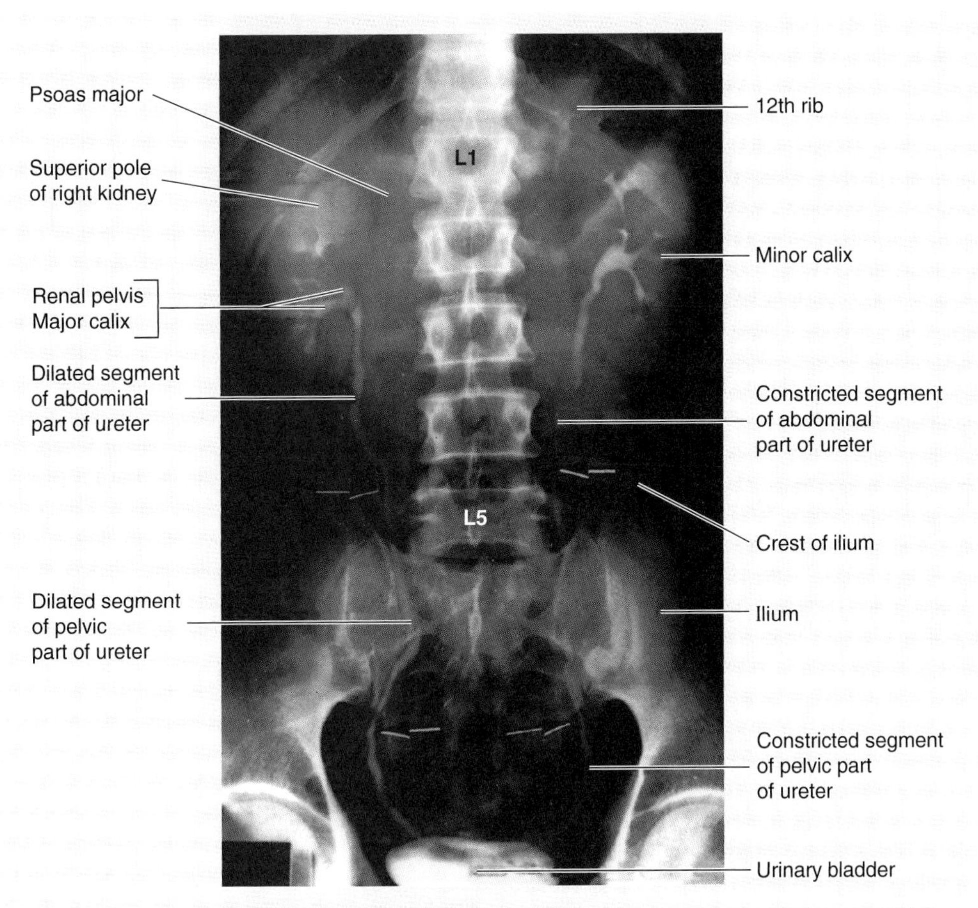

Figure 2.88. Intravenous urogram (pyelogram). The contrast medium was injected intravenously and was concentrated and excreted by the kidneys. This anteroposterior (AP) projection shows the calices, renal pelves, and ureters outlined by the contrast medium filling their lumina. Note the difference in shape and level of the renal pelves and the constrictions and dilations in the ureter resulting from peristaltic contractions of their smooth muscle walls. The *arrows* indicate narrowings of the lumen resulting from peristaltic contractions. (Courtesy of Dr. John Campbell, Department of Medical Imaging, Sunnybrook Medical Centre, University of Toronto, Toronto, Ontario, Canada)

CASE STUDIES

Case 2.1

A third-year medical student was invited by a senior surgeon to observe an abdominal exploratory operation on a patient with intestinal obstruction. While the surgeon and student were scrubbing their hands, the surgeon asked what abdominal skin incision would be appropriate. After a long pause, the student suggested a pararectus incision. The surgeon was not pleased with the suggestion and asked the following questions.

Clinicoanatomical Problems

- Anatomically, why is a pararectus incision undesirable?
- Based on your anatomical knowledge of the structures in the anterolateral abdominal wall, what other type of vertical incision do you think would be better?
- Why do you believe this incision is better?

The problems are discussed on page 325.

Case 2.2

During an appendectomy, the surgical resident asked an attending senior medical student the following questions.

Clinicoanatomical Problems

- When making a transverse incision in the anterolateral abdominal wall for an appendectomy, what nerve must be identified and preserved?
- Where would you expect to find the nerve?
- What may result from sectioning of this nerve?

The problems are discussed on page 325.

Case 2.3

An obese man complained of tingling, burning, and pins-and-needles sensations (*paresthesias*) and some pain on the lateral side of his thigh after a laparoscopic inguinal hernia repair. During surgical rounds the following questions were asked.

Clinicoanatomical Problems

- What nerve supplies the skin on the anterolateral surface of the thigh?
- What anatomical variation of this nerve might explain these paraesthetic symptoms?
- What surgical error could produce the man's pain?

The problems are discussed on page 325.

Case 2.4

Physical examination of an infant with a severe periumbilical infection revealed enlarged lymph nodes in both axillae and inguinal regions.

Clinicoanatomical Problems

- What regions other than the abdomen would you examine for infections?
- How would you explain the lymphadenitis in this infant?

The problems are discussed on page 325.

Case 2.5

During the physical examination of a patient's abdomen, the attending physician asked you what was the key landmark of the abdomen. You said that L1 vertebra is a classical abdominal landmark. The physician asked you the following questions.

Clinicoanatomical Problems

- Why is L1 vertebra so important anatomically and surgically?
- How does one locate the surface level of this vertebra?

The problems are discussed on page 325.

Case 2.6

A man was referred by his primary care physician to a surgeon for a vasectomy. Knowing you were a third-year medical student, the family physician asked you some questions.

Clinicoanatomical Problems

- What constituent of the spermatic cord is easy to palpate?
- Where would you palpate this structure?
- Can this structure always be palpated?

The problems are discussed on page 325.

Case 2.7

During clinical rounds a female patient with an indirect inguinal hernia was visited. The physician stated that indirect inguinal hernias are approximately 20 times more common in males than in females.

Clinicoanatomical Problems

- What is the basis for this sex difference in the frequency of this type of hernia?
- Where does the bulging of the hernia occur in females?

The problems are discussed on page 325.

Case 2.8

While observing a cholecystectomy during open surgery, the surgeon reported severe bleeding. The bloodflow was quickly arrested and the bleeding vessels were treated by electrocautery.

Clinicoanatomical Problems

- What is the quickest way to control hemorrhage during cholecystectomy without using clamps?
- What other surgical procedure is used to remove the gallbladder?

The problems are discussed on page 326.

Case 2.9

A 49-year-old woman complained of recurrent attacks of right upper quadrant pain in her abdomen following fatty meals. In the recent attack, the pain lasted longer than 6 hours and the pain spread over her right shoulder and the tip of her right scapula.

Clinicoanatomical Problems

- How would you explain her right upper quadrant pain?
- What is the basis for the referral of pain to her shoulder and back?

The problems are discussed on page 326.

Case 2.10

A young man who was thrown from his motorcycle complained of sharp pain on his left side and held his hand over his lower ribs. Radiographic studies revealed fractures of the 10th and 11th ribs.

Clinicoanatomical Problems

- What abdominal organ was most likely injured?
- How is severe hemorrhage of this organ controlled?
- Why is this organ so vulnerable to injury?
- Can blunt trauma to other regions of the abdomen injure this organ?

The problems are discussed on page 326.

Case 2.11

A 55-year-old man reported to his primary care physician that he felt a solid swelling in his scrotum. The lump was diagnosed as an advanced carcinoma of the testis.

Clinicoanatomical Problems

- Where would you search for lymphogenous spread of cancer cells from the tumor?
- Would the skin of the scrotum be involved?

The problems are discussed on page 326.

Case 2.12

A 43-year-old woman had symptoms of weight loss, vague abdominal discomfort, obstructive jaundice, and pain penetrating into the back. A diagnosis of pancreatic adenocarcinoma was made.

Clinicoanatomical Problems

- Based on your anatomical knowledge of the relations of the pancreas, in what part of the gland do you think the cancer was located?
- Where would you expect to find metastases of neoplastic cells in this case?

The problems are discussed on page 326.

Case 2.13

A 23-year-old man was admitted to the hospital with severe abdominal pain and a mildly elevated temperature. He complained that he originally experienced generalized abdominal pain. Later he said he felt most pain in the pit of his stomach (epigastrium) and that the pain was more intense around his umbilicus. Further examination revealed right lower quadrant pain and rebound tenderness.

Clinicoanatomical Problems

- Based on anatomical considerations, what do you think is the cause of his pain?
- Explain the change in the location of the pain.
- What is the usual cause of this type of pain?

The problems are discussed on page 326.

Case 2.14

During a brawl outside a bar, a man was struck by a knee in the groin. He doubled up and complained of severe groin pain. He also said that he felt sick to his stomach.

Clinicoanatomical Problems

- Where else would the man feel pain?
- Explain why pain is felt in this location.

The problems are discussed on page 326.

Case 2.15

A 32-year-old accountant complained to her physician about a burning pain in the "pit of her stomach" of approximately 2 weeks' duration. On careful questioning, she revealed that the pain usually began approximately 2 hours after she had eaten and then disappeared when she ate again or drank a glass of milk. Except for mild tenderness in her right upper quadrant, just lateral to the xiphoid process, the physical examination results were normal. Suspecting a peptic ulcer, the physician ordered *Helicobacter pylori* tests, radiographs of the patient's abdomen, and upper GI studies. Endoscopy of the stomach and upper duodenum was requested later.

Radiology Report The radiographs are normal but the upper GI studies revealed a peptic ulcer in a moderately deformed ampulla (duodenal cap).

Bacteriology Report *Helicobacter pylori* bacteria were found in the biopsy of the duodenal mucosa.

Diagnosis Active peptic duodenal ulcer.

Treatment Initially, the patient responded well to medical treatment such as antacids, frequent bland feedings, abstinence from smoking and alcohol, and antibiotic therapy. Although she followed the physician's instructions for approximately 2 months, she began to work long hours again, smoke heavily, and consume excessive amounts of coffee and alcohol. Her symptoms returned and vomiting sometimes occurred when the pain was severe. One evening she developed acute upper abdominal pain, vomited, and fainted. She was rushed to the hospital.

Physical Examination Extreme pain, rigidity of the abdomen, and rebound tenderness were detected. On questioning, the patient revealed that her ulcer had been "acting up" and that she had noticed blood in her vomitus.

Surgical Treatment Emergency surgery was performed and a perforated duodenal ulcer was observed and resected. There was a generalized chemical peritonitis resulting from the escape of bile and the contents of her GI tract into the peritoneal cavity.

Clinicoanatomical Problems

- What structures closely related to the superior part of the duodenum might be eroded by a perforated duodenal ulcer?
- Name the congenital anomaly of the ileum in which a peptic ulcer may develop.
- Explain the anatomical basis for abdominal pain in the right upper and lower quadrants.
- What nerves are vulnerable during surgical procedures designed to reduce acid secretion by the parietal cells of the stomach?

The problems are discussed on page 326.

Case 2.16

On the way home from work, a 42-year-old office worker suddenly experienced a sharp pain in his left side. The pain was so excruciating that he doubled up and moaned in agony. A fellow worker took him to the hospital. When the physician asked him to describe the onset of pain, the patient said that he first felt a slight pain between his ribs and hip bone and that it gradually increased until it was so severe that it brought tears to his eyes. He said that this unbearable pain lasted approximately 30 minutes and then suddenly eased. He explained that the pain comes and goes, but it seemed to be moving toward his groin.

Physical Examination The physician noted some tenderness and guarding (spasm of the muscles) in the left lower quadrant, but no muscular rigidity. While palpating the tender area deeply, he suddenly removed his hand. Instead of wincing, the patient seemed relieved that the probing had stopped (absence of rebound tenderness). By this time, the patient reported that he felt the pain in his left groin and scrotum and along the medial side of his thigh. The physician noted that the left testis was unusually tender and retracted. When asked to produce a urine sample, the patient stated that it was difficult and painful for him to urinate (*dysuria*). The nurse reported that the patient's urine sample contained blood (*hematuria*). Although the physician was certain that the man was passing a ureteric stone, he ordered an abdominal radiograph of the right kidney, ureter, and urinary bladder.

Radiology Report Small ureteric (renal) stones are visible on the superior right part of the left ureter and in the urinary bladder.

Diagnosis Ureteric calculi in the left ureter and renal stones in the urinary bladder.

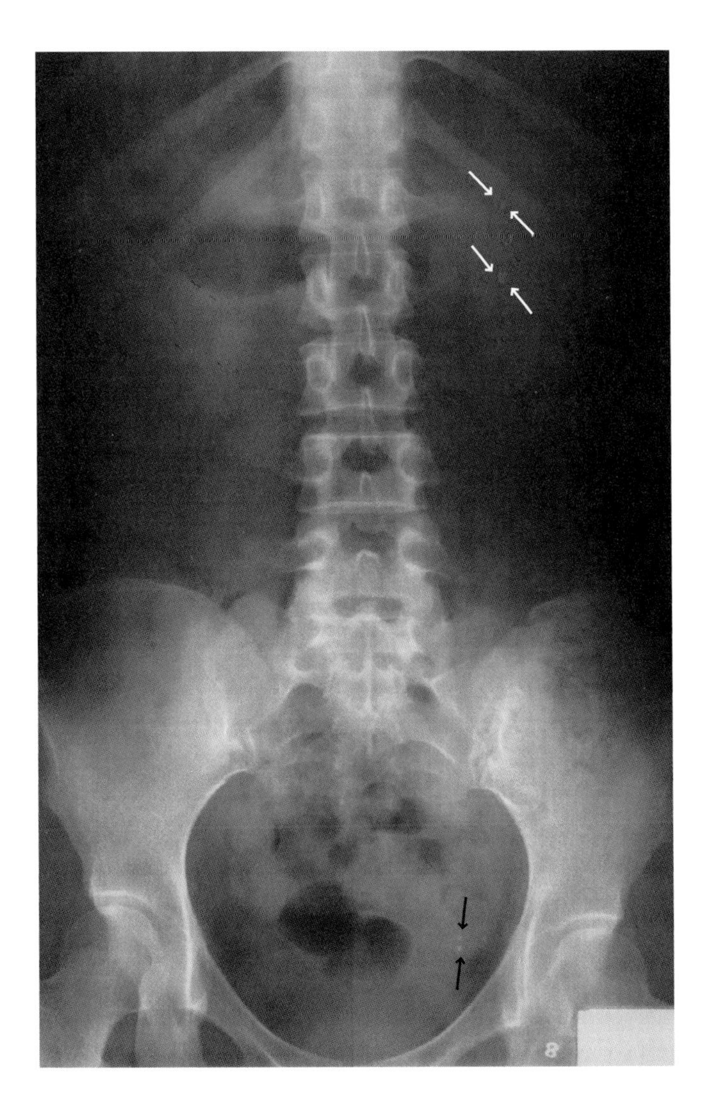

Kidney stones *(arrows)* in ureter and bladder

Clinicoanatomical Problems

- What probably caused the patient's initial attack of excruciating pain?
- Based on the anatomy of the ureter, at what other sites do you think a ureteric calculus is likely to become lodged?
- Explain the intermittent exacerbation of pain and the course taken by the pain.
- Briefly discuss referred pain from the ureter.

The problems are discussed on page 327.

Case 2.17

A 14-year-old boy suffered pain in his right groin while attempting to lift a heavy weight. As soon as he noticed a lump in the region where he felt pain, he decided to lie down. The bulge soon disappeared and he went home. Later, he blew his nose hard and again experienced pain, and the swelling reappeared in his right groin. Fearing that he may have a "rupture," his father made an appointment with their family physician.

Physical Examination The physician inserted his index finger into the boy's scrotum superior to his right testis and along the spermatic cord to the superficial inguinal ring. Nothing was felt until he asked the boy to cough; he then felt an impulse on the tip of his finger. When the patient was in a prone position, the bulge disappeared, but when he was asked to strain, a plum-sized bulge appeared in the right inguinal region superior to the inguinal ligament.

Diagnosis Indirect inguinal hernia.

Clinicoanatomical Problems

- What is an indirect inguinal hernia? Explain the embryological basis of this kind of hernia.
- What layers of the spermatic cord cover the hernial sac?
- What structures are endangered during an operation for repair of an indirect inguinal hernia?
- Does this type of hernia occur in women?

The problems are discussed on page 327.

Case 2.18

A 22-year-old married female medical student woke up one morning not feeling as well as usual. She was anorexic (lacked desire for food) and had crampy abdominal pains. As this coincided with the time of her expected menses, she thought the cramps were the beginning of her usual painful menstruation (*dysmenorrhea*). Because she had missed her last menstrual period, she also thought that she might be having early symptoms of a ruptured *ectopic pregnancy*. She had a slight fever and felt dizzy; therefore, she decided to stay in bed. The pain soon localized around her umbilicus. By evening the site of pain shifted to the right lower quadrant of her abdomen and she suspected acute appendicitis. As she was in considerable pain, her husband decided to take her to the hospital.

Physical Examination The physician observed a slightly elevated temperature and an increased pulse rate. When asked to indicate where the pain began, the patient circled her umbilical area. When asked where she now felt pain, she put her finger on McBurney's point. During gentle palpation of her abdomen, the physician detected localized rigidity (*muscle spasm*) and tenderness in the right lower quadrant. When the physician suddenly removed her palpating hand from the area of McBurney's point, the patient winced in pain (rebound tenderness). The physician ordered a blood count.

Laboratory Report There is an abnormally high white blood cell count (*leukocytosis*).

Diagnosis Acute appendicitis.

Clinicoanatomical Problems

- What types of incision could the surgeon make to expose the appendix?
- Discuss the anatomical basis of these incisions.
- How would you locate the McBurney point used as a guideline for the skin incision? What part of the appendix is usually deep to this point?
- Based on your knowledge of dissection, how do you think the patient's appendix would be exposed?
- Where would her appendix most likely be located?
- What position of an inflamed appendix might give rise to pelvic or rectal pain?
- Discuss referred pain from the appendix.
- Inflammation of what other structure could produce an appendixlike pain in the right lower quadrant?
- Can the appendix be removed by the laparoscopic approach? If so, how do you think it would be performed?

The problems are discussed on page 327.

Case 2.19

A 58-year-old obese man with a history of heartburn, impaired gastric function (*dyspepsia*), and belching after heavy meals complained about recent epigastric pain. He stated that the pain was below his breastbone (*substernal pain*) and in his chest. He said that the pain was most severe after dinner, especially when he stooped down. Fearing the chest pains might be a heart attack, his wife insisted that he consult a physician.

Physical Examination When asked if he had noticed any other abnormalities, the patient stated that he often brought up small amounts of sour or bitter-tasting substances (*gastric reflux*), particularly when he stooped to tie his shoes. He also reported that he recently had been having bouts of hiccups and that he occasionally had difficulty swallowing. The physician ordered an ECG and radiographic and ultrasound studies.

Cardiology Report The ECG shows no evidence of heart disease.

Radiology Report The radiographs of the abdomen were negative; however, a fluoroscopic examination of the thorax showed a round space filled with gas and fluid in the inferior part of the patient's posterior mediastinum. On swallowing barium sulphate, the emulsion was seen to enter this space, which was identified as the gastroesophageal region of the stomach. There was no radiological evidence of peptic gastric or duodenal ulcers. The ultrasound examination showed that part of the stomach passed through the esophageal hiatus in the diaphragm when the patient was asked to touch his toes.

Diagnosis Sliding hiatus hernia.

Clinicoanatomical Problems

- What is a diaphragmatic hernia?
- Does hiatus hernia have an embryological basis? Is it usually present at birth?
- What caused the patient's epigastric chest pain?
- Based on your anatomical knowledge, what structures do you think would be endangered in the surgical repair of a hiatus hernia?

The problems are discussed on page 328.

Case 2.20

A 40-year-old overweight woman was rushed to the hospital with severe colicky pain in the right upper quadrant of her abdomen. When asked where she first felt pain, she pointed to her epigastric region. When asked where the pain was now, she ran her fingers under her right ribs (hypochondriac region) and around her right side to her back. She stated that the pain was felt near the lower end of her shoulder blade (inferior angle of scapula). On questioning, she said the sharp midline pain followed a heavy meal containing fatty foods, after which she felt nauseated and vomited. The pain gradually increased.

Physical Examination During gentle palpation of the patient's abdomen, the physician noted rigidity and tenderness in the right upper quadrant, especially during inspiration. She ordered radiographic studies, including an ultrasound examination.

Radiology Report There is a small calculus (gallstone) in the proximal part of the cystic duct, and the gallbladder is greatly enlarged.

Diagnosis Biliary colic resulting from impaction of a gallstone in the cystic duct.

Clinicoanatomical Problems

- What is a gallstone?
- What materials are in these stones?
- Explain the anatomical basis for the patient's pain in (*a*) the epigastric region, (*b*) the right hypochondrium, and (*c*) the infrascapular region.
- Does peritoneum separate the gallbladder from the liver?
- What structures are endangered during cholecystectomy?
- How do you think a laparoscopic cholecystectomy would be performed?

The problems are discussed on page 328.

Case 2.21

A 54-year-old mechanic was admitted to the hospital because of severe epigastric pain and vomiting of blood (*hematemesis*). It was obvious that he had been drinking heavily.

Physical Examination The blood in his vomitus was bright red. On questioning, the physician learned that the patient had exhibited upper GI bleeding on previous occasions (*ruptured esophageal varices*), but never so profusely. The patient's blood pressure was low, and his pulse rate was high. His skin and conjunctivae were slightly yellow (*jaundice*). His eyes appeared to be slightly sunken. *Spider nevi* (branching arterioles) were present in his cheeks, neck, shoulders, and upper limbs. The patient's abdomen was large, fluid-filled (*ascites*), and pendulous. Palpation of the patient's abdomen revealed some enlargement of the liver (*hepatomegaly*) and spleen (*splenomegaly*). Several bluish, dilated varicose veins radiated from his umbilicus (*caput medusae*). During a proctoscopic examination, internal hemorrhoids were observed. On questioning, the patient said that he sometimes saw blood in his stools (bowel movements), which were black and shiny.

Diagnosis Alcoholic cirrhosis of the liver.

Clinicoanatomical Problems

- Discuss anatomically the basis of the man's hematemesis, esophageal varices, hemorrhoids, bloody stools, and caput medusae.
- What is the likely cause of the ascites and splenomegaly?
- Thinking anatomically, how would you suggest that blood pressure in the portal system be reduced?

The problems are discussed on page 329.

Case 2.22

The initial complaint of a 54-year-old man was oval swelling in his left groin. He stated that this painless swelling enlarged when he coughed and disappeared when he lies down.

Physical Examination During examination with the patient in the standing position, the physician put his finger into the man's left superficial inguinal ring. He noticed a sensation that his finger was going directly into the abdomen rather than along the inguinal canal. When the man coughed, the physician felt a mass strike the side of his finger, which was against the posterior wall of the inguinal canal. When the man was asked to lie down, the mass reduced itself immediately. The physician then placed his fingers over the *inguinal triangle* and instructed the man to hold his nose and blow it. The physician felt a mass protruding from the inferior portion of this triangle.

Diagnosis Direct inguinal hernia.

Clinicoanatomical Problems

- Explain what is meant by the term "direct inguinal hernia."
- How does this hernia differ from an indirect inguinal hernia?
- Does a direct inguinal hernia have an embryological basis?
- Does this type of injury occur at all ages?
- What is the relationship of a direct inguinal hernia to the inferior epigastric artery? Is this relationship different from that of an indirect inguinal hernia?
- Inadvertent injury to which nerves of the abdominal wall during surgery may predispose to the development of an inguinal hernia?

The problems are discussed on page 329.

DISCUSSION OF CASES

Case 2.1

A pararectus incision is undesirable because it cuts across the nerve supply to the rectus abdominis. The segmental nerves enter the deep surface of the muscle near its lateral edge. Rectus paralysis with weakening of the anterior abdominal wall results from section of several of these nerves. The blood supply from the inferior epigastric artery may also be compromised. A paramedian incision is preferable because the rectus muscle is retracted laterally to prevent tension on the nerves and vessels. Release of traction allows the intact muscle to bridge the incision through the rectus sheath (Skandalakis et al., 1995).

Case 2.2

The iliohypogastric nerve must be identified and preserved because it may be cut by a transverse incision for an appendectomy. The nerve perforates the posterior part of the transverse abdominal muscle and divides between this muscle and the internal oblique into lateral and anterior cutaneous branches and muscular branches to both these muscles. The consequent muscular weakness resulting from section of the iliohypogastric nerve may predispose to the development of direct inguinal hernia.

Case 2.3

The lateral femoral cutaneous nerve from the second and third lumbar nerves supplies the skin on the anterolateral and lateral surfaces of the thigh. The nerve passes posterior or through the inguinal ligament, approximately 1 cm medial to the anterior superior iliac spine. The anterior branch pierces the fascia lata and transmits cutaneous sensations from the lateral surface of the thigh. When the nerve passes through the inguinal ligament, a tight retaining strap on the operating table that crosses the anterior superior iliac spine could exert pressure and injure the lateral femoral cutaneous nerve, producing a sensory neuritis known as *meralgia paresthetica* (Rowland, 1995). A suture or surgical clip inserted too close to the anterior superior iliac spine could also injure the lateral femoral cutaneous nerve.

Case 2.4

The upper and lower limbs should be checked for infections because lymph from them drains into the axillary and inguinal lymph nodes, respectively. The lymph vessels from the upper two quadrants of the anterolateral abdominal wall also drain into the pectoral and subscapular axillary lymph nodes. Some lymph from the supraumbilical quadrants drains into the parasternal lymph nodes. The lymph vessels from the lower two quadrants of the anterolateral abdominal wall drain into the superficial inguinal nodes.

Case 2.5

L1 vertebra is a classical abdominal landmark because it indicates the level of the *transpyloric plane*, the key horizontal plane of the abdomen because so many abdominal structures are related to it. The transpyloric plane bisects the line joining the jugular notch and pubic symphysis. This plane passes through L1 vertebra, the tips of the 9th costal cartilages, the fundus of the gallbladder, the neck of the pancreas, the splenic vein, the origin of the SMA, the duodenojejunal junction, the root of the transverse mesocolon, and the hila of the kidneys. Despite its name, the pylorus of the stomach is not usually on this plane in living persons.

Case 2.6

The ductus deferens, or deferent duct, is easily felt in the superior part of the scrotum as a firm cord (2 to 4 mm in diameter) between the thumb and index finger. The duct feels like a firm plastic tube. In some individuals this duct cannot be palpated on one side because of aplasia or dysplasia. This condition has a high association with ipsilateral renal agenesis (absence of the kidney on the same side).

Case 2.7

The processus vaginalis normally disappears. When it persists it forms a potential hernial sac into which a loop of intestine may enter. The sac may enter the scrotum or labium majus. An indirect inguinal hernia is more common in males than in females because the testis passes through the inguinal canals, creating a potentially weak area of the anterior abdominal wall. In females the inguinal canals are not well developed; consequently, herniation of a loop of intestine into them or through them into the labia majora is uncommon.

Case 2.8

The hemorrhage can be controlled by compressing the hepatic artery—the origin of the cystic branch—between the index finger and thumb where it lies in the anterior wall of the omental foramen. Laparoscopic cholecystectomy, in which the gallbladder is removed with instruments and scopes passed through small incisions, is now common. It is currently the operative therapy of choice in most instances (Sabiston and Lyerly, 1994).

Case 2.9

Recurrent attacks of right upper quadrant pain—*biliary colic*—is typical of intermittent obstruction of the cystic duct by a gallstone. The referred pain to the right shoulder and scapula results from *acute cholecystitis* (sudden inflammation of the gallbladder). The fundus of the inflamed gallbladder touches the anterior abdominal wall near the tip of the 9th costal cartilage and irritates the peritoneum on the inferior surface of the diaphragm. As a result, the pain is referred to the shoulder and clavicular area by the supraclavicular nerves (C3 and C4), which supply the skin covering these areas.

Case 2.10

The spleen is the most commonly injured abdominal organ. Severe splenic hemorrhage is easily controlled by splenectomy. The spleen has a thin capsule and is soft, friable, and highly vascular. Hence, it is easily ruptured (e.g., by fractured ribs). Blunt trauma to other regions of the abdomen that cause a sudden, marked increase in intra-abdominal pressure can also rupture the spleen (e.g., resulting from impalement on a steering wheel during a motor vehicle accident).

Case 2.11

Malignant cells from a testicular carcinoma metastasize to the lumbar lymph nodes by lymphogenous dissemination. Because of cross-communications, tumor cells may be present on both sides (Ellis, 1992). The inguinal lymph nodes would not be involved unless the tumor has ulcerated the skin of the scrotum and entered the scrotal lymphatics, which drain to the superficial inguinal nodes. In advanced cases, testicular tumors may metastasize to the liver and to mediastinal and cervical lymph nodes.

Case 2.12

The adenocarcinoma was most likely located in the head of the pancreas because of the presence of obstructive jaundice. Because of the posterior relations of the head, a tumor may produce obstructive jaundice by compressing the bile duct that lies either in a groove in the right part of the gland or embedded in its substance. Metastases from neoplastic cells would be found in the pancreaticosplenic and pyloric lymph nodes. They would most likely also be found in the celiac, superior mesenteric, and lumbar lymph nodes that receive lymph from the head of the pancreas.

Case 2.13

The history and physical findings suggest *acute appendicitis*. Acute inflammation of the appendix is a common cause of acute abdominal pain (*acute abdomen*). Digital pressure over McBurney's point usually registers the maximum abdominal tenderness. *Appendicitis* is usually caused by obstruction of the appendix, most often hardened by fecal material. When its secretions cannot escape, the appendix swells and stretches the visceral peritoneum. The pain of acute appendicitis usually commences as a vague pain in the periumbilical region because afferent pain fibers enter the spinal cord at the T10 level. Later, severe pain develops in the lower right quadrant; this pain is caused by irritation of the parietal peritoneum on the posterior abdominal wall. Pain can be elicited by extending the thigh at the hip joint.

Case 2.14

The nonspecific pain, associated with nausea, would be felt in the abdomen mostly in the periumbilical region because the fetal testes developed as abdominal organs in the upper lumbar region. The testes descended to the inguinal region during the late period and entered the scrotum before birth. The nerve supply of the testis is from approximately the lower three thoracic and first lumbar segments of the spinal cord. The sympathetic nerve supply is from the T10 segment of the spinal cord. Pain from the testes is usually referred to the lower thoracic or upper lumbar regions. The T10 dermatome is in the region of the umbilicus.

Case 2.15

A peptic ulcer is an ulceration of the mucous membrane of the stomach or duodenum. Peptic ulcers are common in the stomach (*gastric ulcers*) and duodenum (*duodenal ulcers*). They are usually found within 3 cm of the pylorus. Peptic gastric and duodenal ulcers tend to bleed. Sometimes organs and vessels adjacent to the duodenum, usually the pancreas, adhere to an ulcer and are eroded by it; for example, a posterior penetrating ulcer may erode the gastroduodenal artery or one of its branches, causing sudden massive hemorrhage, which may be fatal. Peptic ulcers may occur in an ileal (Meckel) diverticulum, a remnant of the yolk stalk attached to the ileum. Gastric tissue that may secrete acid, causing ulcer formation, may be present in the wall of this diverticulum.

The pain resulting from a peptic gastric ulcer is referred to the epigastric and left hypochondriac regions because the stomach is supplied with pain afferents that reach T7 and T8 segments of the spinal cord through the greater splanchnic nerve (Table 2.12). Pain resulting from a peptic duodenal ulcer is referred to the epigastric region of the anterolateral abdominal wall because both the duodenum and this area of skin are supplied by T9 and T10 spinal nerves. When a peptic duodenal ulcer perforates, there may be pain throughout the abdomen. Sometimes the paracolic gutter associated with the ascending colon may act as a watershed and direct the escaping inflammatory material into the right iliac fossa. This explains why pain from an anterior perforation of a duodenal ulcer may cause right upper and lower quadrant pain.

Because the vagus nerves largely control the secretion of acid by the parietal cells of the stomach, section of the vagus nerves (*vagotomy*) as they enter the abdomen is sometimes performed to reduce acid production. Often only the gastric branches of the vagus nerves are cut (*selective vagotomy*), thereby avoiding adverse effects on other organs (e.g., dilation of the gallbladder). Vagotomy may be performed in conjunction with resection of the ulcerated area and the acid-producing part of the stomach. In most cases, the *H. pylori* bacteria causing the ulcer can be eradicated by antibiotics, making surgical treatment unnecessary.

Case 2.16

The patient's initial attack of excruciating pain was almost certainly caused by passage of the kidney stone from the renal pelvis into the superior end of his right ureter. The stones are composed of salts of inorganic or organic acids or of other materials. Calculi that are larger than the lumen of the ureter (approximately 3 mm) cause severe pain when they attempt to pass through it. The pain moves inferomedially as the calculus passes along the ureter. The patient likely experienced the severe pain when the calculus was temporarily impeded because of the angulation of the ureter as it crosses the pelvic brim and, later, when it became wedged in the ureter where it passes through the wall of the urinary bladder. At the inferior end of the ureter is a definite narrowing of the lumen, a common site of obstruction. The pain ceases when it passes into the urinary bladder, although tenderness along the course of the ureter often persists for some time.

Ureteric pain results from passage of the calculus through the ureter. Because the ureter is a muscular tube in which peristaltic contractions normally convey urine from the kidney to the urinary bladder, pain results from distension of the ureter by the calculus and the urine that is unable to pass by it. The smooth muscular coat of the ureter normally undergoes peristaltic contractions from its superior to its inferior end. As the peristaltic wave approaches the obstruction, forceful smooth contraction causes excessive dilation of the ureter between the wave and the stone. It is the ureteric distension that produces the sharp pain. Exacerbation of pain occurs as distension increases.

The afferent pain fibers supplying the ureter are included in the lesser splanchnic nerve. Impulses enter the L1 and L2 segments of the spinal cord, and the pain is felt in the cutaneous areas innervated by the inferior intercostal nerves (T11 and T12), the iliohypogastric and ilioinguinal nerves (L1), and the genitofemoral nerve (L1 and L2). These are the same regions of the spinal cord that supply the ureter (T11 to L2). Consequently, the pain begins in the lateral region and radiates to the groin and scrotum. The retraction of the testis by the cremaster muscle and pain along the medial part of the front of the thigh indicate that the genital and femoral branches of the genitofemoral nerve (L1 and L2) were involved.

Ureteral colic is caused by distension of the ureter, which stimulates pain afferents in its wall. Because there was no peritonitis, no rigidity or rebound tenderness were present. When peritonitis is present, pressing the hand into the abdominal wall and rapidly releasing it causes pain when the abdominal musculature springs back into place, carrying the inflamed peritoneum with it. Hence, the abdominal rebound test is useful in the differentiation of ureteric colic from intestinal colic and appendicitis.

Case 2.17

An indirect inguinal hernia is an outpouching of the peritoneal sac that enters the deep inguinal ring, traverses the inguinal canal, and exits through the superficial inguinal ring. The embryological basis of an indirect inguinal hernia is persistence of all or part of the *processus vaginalis*, an embryonic diverticulum of the peritoneum that pushes through the abdominal wall and forms the inguinal canal. The processus vaginalis evaginates all layers of the abdominal wall before it, and in males the layers become the coverings of the spermatic cord.

The hernial sac (former processus vaginalis) may vary from a short one that does not extend beyond the superficial ring to one that extends into the scrotum or labium majus. In males it is continuous with the tunica vaginalis. A persistent processus vaginalis predisposes a person to indirect inguinal hernia by creating a weakness in the inguinal region of the anterolateral abdominal wall. It also forms a hernial sac into which abdominal contents (usually a loop of intestine) may herniate if the intra-abdominal pressure becomes high, as occurs during straining while lifting a heavy object. Once the deep inguinal ring has been enlarged by a herniation of the intestine, coughing may cause herniation to occur again. This is the basis of the test done during a physical examination, in which the examiner inserts a finger through the superficial ring of the inguinal canal and the patient is asked to cough.

During the surgical repair of an indirect inguinal hernia, the genital branch of the *genitofemoral nerve* is endangered because it traverses the inguinal canal in both sexes and exits through the superficial inguinal ring. The ilioinguinal nerve may also be trapped in a hernia repair. It supplies skin of the superomedial area of the thigh, the skin over the root of the penis, and the part of the scrotum or labia majora with sensory fibers. If this nerve is injured, anesthesia of these areas of skin is a likely result. If the nerve is constricted by a suture, postoperative neuritic pain may also occur in the scrotum. Because the ductus deferens lies immediately posterior to the hernial sac, it may be damaged when the sac is freed, ligated, and excised. Because the hernial sac is within the spermatic cord, the pampiniform plexus of veins and the testicular artery may also be injured, resulting in impaired circulation to the testis. Injury to the vessels of the spermatic cord could also result in atrophy of the testis on that side.

Indirect inguinal hernia, although uncommon, can occur in women if the processus vaginalis persists. The hernia produces a bulge in the labium majus.

Case 2.18

The type of skin incision used for an appendectomy depends on the type of patient, the certainty of the diagnosis, and the preference of the surgeon. It may be a midline incision, a right lower quadrant transverse incision overlying the rectus abdominis, or a gridiron, muscle-splitting incision. The center of the gridiron incision is located at the McBurney point, which is at the junction of the lateral and middle thirds of the line joining the anterior superior iliac spine and the umbilicus. In most people, this point overlies the base of the appendix. Following incision of the skin and the subcutaneous tissue, the aponeurosis of the external oblique is incised in the direction of its fibers. The other two flat muscles of the anterior abdominal wall (internal oblique and transverse abdominal) are then split in the direction of their fibers, lessening the chances of injuring the nerves supplying them. The fibers of the external oblique run inferoanteriorly; the fibers of the internal oblique fan out passing superomedially, medially, and inferomedially. The muscle fibers of the transverse abdominal fan out inferomedially over the ribs, iliac crest, and inguinal ligament and end in an aponeurosis. The gridiron incision is good because each muscle layer can be retracted and split in the direction of its fibers, thus requiring no cutting of fibers. Because each muscle layer runs in a different direction, the incision is well protected when the retracted layers are returned to their normal position.

Next, the transversalis fascia and parietal peritoneum are incised to expose the cecum. The base of the appendix is indicated by the point of convergence of the three teniae coli. Following one of the teniae distally leads to the base of the appendix. The exact location of the appendix is the site where pain and tenderness are usually maximal, which varies from person to person (Sabiston and Lyerly, 1994).

Variations in length and position of the appendix may give rise to varying signs and symptoms in appendicitis. For example, the site of maximum tenderness in cases of retrocecal appendix may be superomedial to the anterior superior iliac spine, even as far superior as the transumbilical plane. If the appendix is long (10–15 cm) and extends into the pelvis minor, the site of pain in a female might suggest peritoneal irritation resulting from a ruptured ectopic pregnancy. As the appendix crosses the psoas major, the person often flexes the right thigh to relieve the pain. Thus, hyperextension of the thigh (*psoas test*) causes pain because it stretches the muscle and its inflamed fascia. Tenderness on the right side during a rectal examination may indicate an inflamed pelvic appendix.

Initially, the pain of typical acute appendicitis is referred to the periumbilical region of the abdomen; later, the pain usually shifts to the right lower quadrant. Afferent nerve fibers from the appendix are carried in the lesser splanchnic nerve, and impulses enter the T10 segment of the spinal cord. As impulses from the skin in the periumbilical region are also sent to this region of the spinal cord, the pain is interpreted as somatic, not visceral, possibly because impulses of cutaneous origin are received more often by the brain.

The shift of pain to the right lower quadrant is caused by irritation of the parietal peritoneum, usually on the posterior abdominal wall. Afferent fibers from this region of peritoneum and skin are carried in the inferior intercostal and subcostal nerves. The pain during palpation results from stimulation of pain receptors in the skin and peritoneum, whereas the increased tenderness detected in the right side of the *rectouterine pouch* (rectovesical pouch in a male) is caused by irritation of the parietal peritoneum in this pouch. When the abdominal wall is depressed and allowed to rebound, the person usually winces because, as the abdominal muscles spring back, the inflamed peritoneum is carried with it.

If the woman had previously had her appendix removed, an inflamed ileal (Meckel) diverticulum could give rise to signs and symptoms similar to appendicitis. An ileal diverticulum represents the remnant of the proximal portion of the yolk stalk and appears as a fingerlike projection from the antimesenteric border of the ileum. The laparoscopic approach may be used for removal of the appendix in selected individuals. (For a description of *laparoscopic appendectomy*, see Skandalakis et al., 1995.) When a diagnosis is uncertain, the laparoscopic approach may also be used to examine other abdominal viscera. If the appendix is inflamed, a trocar (sharp-pointed instrument equipped with a cannula) is inserted through the anterolateral abdominal wall. An endoloop (ligature) is passed through a smaller trocar, and the appendix is snared for retraction and electrocautery of the base of the appendix and the appendiceal vessels.

Case 2.19

A diaphragmatic hernia is a herniation of abdominal viscera into the thoracic cavity through an opening in the diaphragm. *Hiatus hernia* is common, particularly in older people. Usually the gastroesophageal region of the stomach herniates through the esophageal hiatus into the inferior part of the thorax. Hiatus hernia is usually acquired, but a congenitally enlarged esophageal hiatus may be a predisposing factor. Because the right crus passes to the left of the median plane, the esophageal hiatus and hernia are to the left of the midline even though they are within the right crus.

The two main types of hiatal or hiatus hernia are *sliding hiatus hernia* and *paraesophageal hiatus hernia*. Some hernias show features of both types and are referred to as *mixed hiatus hernias*. The thoracic region of the vertebral column becomes shorter with age because of desiccation (dehydration) of the IV discs, and the abdominal fat generally increases during middle age. Both of these occurrences favor development of hiatus hernias. Most of the present man's complaints (heartburn, belching, regurgitation, and epigastric pain) resulted from irritation of the esophageal mucosa by the reflux of gastric juice. The irritant effect of the gastric juice produces esophageal spasm, resulting in dysphagia and retrosternal pain.

Pain endings in the esophagus are stimulated by the forcible contractions of the smooth muscle in the esophageal wall. Pain of gastroesophageal origin is referred to the epigastric and retrosternal regions, the cutaneous areas of reference for these regions of the viscera. Because the esophageal hiatus also transmits the vagus nerves and esophageal branches of the left gastric vessels, these structures must be protected from injury during surgical repair of hiatus hernias.

Case 2.20

Obese middle-aged women who have had several children are most prone to gallbladder disease. *Gallstones* (biliary calculi) are more common in women older than 20 years, but this is not necessarily so after 50 years of age. In approximately 50% of persons, gallstones are "silent" (asymptomatic). A gallstone is a concretion in the gallbladder, cystic duct, or bile duct composed chiefly of cholesterol crystals. The pain is severe when a biliary calculus is lodged in the cystic or bile duct. The sudden severe pain in the epigastric region was caused by a gallstone wedged in the cystic duct. *Biliary colic* is the pain typical of intermittent obstruction of the cystic duct by a calculus.

The pain referred to the right upper quadrant and scapular region results from inflammation of the gallbladder and distension of the cystic duct. The nerve impulses pass centrally in the greater splanchnic nerve on the right side and enter the spinal cord through the dorsal roots of the T7 and T8 nerves. This visceral referred pain is felt in the right upper quadrant of the abdomen and in the right infrascapular region because the source of the stimuli entering this region of the cord is wrongly interpreted as cutaneous. Often the inflamed gallbladder irritates the peritoneum covering the peripheral part of the diaphragm, resulting in a parietal referred pain in the right hypochondriac region (Table 2.7). This part of the peritoneum is supplied by the inferior intercostal nerves. In other cases, the peritoneum covering the diaphragm is irritated and the pain is referred to the right shoulder because the central area of peritoneum is supplied by sensory fibers in the phrenic nerves. The skin of the shoulder region is supplied by the supraclavicular nerves (C3 and C4), the same segments of the cord that receive pain afferents from the central portion of the diaphragm.

When fat enters the duodenum, *cholecystokinin* causes contraction of the gallbladder. In the present case it is likely that the woman's gallbladder contracted vigorously after her fatty meal, squeezing a stone into her cystic duct. *Acute cholecystitis* is associated with a gallstone impacted in the cystic duct in a high percentage of cases. The impacted calculus causes sudden distension of the gallbladder, which compromises its arterial supply and its venous and lymphatic drainage. Usually the peritoneum does not separate the gallbladder from the liver. The gallbladder lies in a fossa on the visceral surface of its right lobe. The peritoneum on this surface of the liver passes over the inferior surface of the gallbladder. The abdominal rigidity detected in the present case resulted from involuntary contractions of the muscles of the anterolateral abdominal wall, particularly the rectus abdominis. This muscle spasm was a reflex response to stimulation of nerve endings in the peritoneum associated with the dilated gallbladder.

Anatomical variations in the gallbladder and cystic duct and in the arteries supplying them are common. Consequently, surgeons must determine the existing anatomical pattern and identify the cystic, bile, and hepatic ducts and the cystic and hepatic arteries before dividing the cystic duct and its artery. Accessory cystic branches from the hepatic arteries may exist; therefore, unexpected hemorrhage may occur during cholecystectomy. Hemorrhage during cholecystectomy may be controlled by compressing the hepatic artery between the finger and thumb where it lies in the anterior wall of the omental foramen.

Laparoscopic cholecystectomy is now used most commonly because patients can be discharged from the hospital in a few days and can usually return to full activity within a week. First, carbon dioxide is infused into the peritoneal cavity to create a peritoneal space in which to work. Four trocars are inserted through the anterolateral abdominal wall in different locations to permit the insertion of surgical instruments (see Skandalakis et al., 1995 for details). Surgeons view the operative field on video monitors, and the cystic duct and artery are located, controlled with clips, and divided. The gallbladder is then dissected from the gallbladder fossa and removed. Extreme care is taken not to injure the bile duct and associated vessels during this procedure.

Case 2.21

Hepatic cirrhosis is a disease characterized by progressive destruction of hepatic parenchymal cells. The cells are replaced by fibrous tissue that contracts and hardens. The fibrous tissue surrounds the intrahepatic blood vessels and biliary radicles (roots). As this process advances, circulation of blood through the branches of the portal vein and of bile through the biliary radicles in the liver is impeded. As pressure in the portal vein rises (*portal hypertension*), the liver becomes more dependent on the hepatic artery for its blood supply and blood pressure in the portal vein rises, reversing bloodflow in the normal portacaval anastomoses and resulting in portal blood entering the systemic circulation. Because these anastomotic veins seldom possess valves, they can conduct blood in either direction. This bloodflow causes enlargement of the veins (*varicose veins*), forming anastomoses at the inferior end of the esophagus (*esophageal varices*), the inferior end of the rectum and anal canal (*hemorrhoids*), and around the umbilicus (*caput medusae*).

Because of pressure during swallowing and defecation, the esophageal varices and hemorrhoids, respectively, may rupture. Rupture results in bloody vomitus and/or bleeding from the anus. Internal hemorrhoids are varicosities of the tributaries of the superior rectal vein. Blood may also pass in a retrograde direction in the paraumbilical veins via the vein in the ligamentum teres. In portal hypertension, the paraumbilical veins may become varicose, forming a *caput medusae*—a radiating venous pattern at the umbilicus.

In *cirrhosis of the liver*, the ramifications of the portal vein are compressed by the contraction of the fibrous tissue in the portal canals. Increased pressure in the splenic and superior and inferior mesenteric veins results. Fluid is forced out of the capillary beds drained by these veins and into the peritoneal cavity. Accumulation of fluid in the peritoneal cavity is called *ascites*. The spleen usually enlarges (*splenomegaly*) in hepatic cirrhosis because of increased pressure in the splenic vein. Because no valves are in the portal system, pressure in the splenic vein is equal to that in the portal vein. A common method of reducing portal pressure is by diverting blood from the portal vein to the IVC through a surgically created anastomosis (*portacaval anastomosis*). Similarly, the splenic vein may be anastomosed to the left renal vein (*splenorenal anastomosis*).

Case 2.22

A *direct inguinal hernia* enters the inguinal canal through its posterior wall, whereas an *indirect inguinal hernia* enters the inguinal canal through the deep inguinal ring. A direct inguinal hernia is much less common than an indirect inguinal hernia; both types occur more often in men than in women. A direct inguinal hernia is usually acquired and commonly occurs in men older than 40 years of age.

The sac of a direct inguinal hernia is formed by the peritoneum lining the anterolateral abdominal wall. The sac protrudes through the inguinal triangle. This triangle is bounded medially by the lateral border of the rectus abdominis, inferiorly by the inguinal ligament, and laterally by the inferior epigastric artery. The hernia may protrude through the abdominal wall and escape from the abdomen on the lateral side of the conjoint tendon to pass through the posterior wall of the inguinal canal. In this case, the hernial sac is covered by transversalis fascia, cremaster muscle, cremasteric fascia, and external spermatic fascia. Occasionally, the hernial sac is forced through the fibers of the conjoint tendon and enters the superficial inguinal ring. When this happens, the hernial sac is covered by transversalis fascia, conjoint tendon, and external spermatic fascia.

Direct inguinal hernias usually protrude anteriorly through the inferior part of the inguinal triangle and extend toward the superficial inguinal ring, but they may pass through this ring and enter the scrotum or labium majus. A direct hernia is acquired and results from weakness of the anterior abdominal wall (e.g, because of weakness of the transversalis fascia and atrophy of the conjoint tendon). No known embryological basis exists for this type of hernia.

The type of inguinal hernia (direct or indirect) can often be determined by the relationship of the hernial sac to the inferior epigastric artery. The pulsations of this artery can be felt by the tip of the examiner's finger in the inguinal canal. In direct inguinal hernias, the neck of the hernial sac is in the inguinal triangle and lies *medial to the inferior epigastric artery*; whereas, in indirect inguinal hernias, the neck of the hernial sac is in the deep inguinal ring and lies lateral to the inferior epigastric artery.

Because the inferior intercostal nerves and the iliohypogastric and ilioinguinal nerves from the first lumbar nerve supply the abdominal musculature, injury to any of them during surgery or an accident could result in weakening of muscles in the inguinal region, predisposing the person to development of direct inguinal hernia. The ilioinguinal nerve also gives motor branches to the fibers of the internal oblique, which are inserted into the lateral border of the conjoint tendon. Division of this nerve paralyzes these fibers and relaxes the conjoint tendon; this may result in a direct inguinal hernia.

References and Suggested Readings

Agur AMR: *Grant's Atlas of Anatomy*, 9th ed. Baltimore, Williams & Wilkins, 1991.

Behrman RE, Kliegman RM, Arvin AM (eds): *Textbook of Pediatrics*, 15th ed. Philadelphia, WB Saunders, 1996.

Daseler EH, Anson BJ, Hambley WC, Reimann AF: The cystic artery and constituents of the hepatic pedical. Surg Gynecol Obstet 85:47, 1947.

Ellis H: A revision and applied anatomy for clinical students. In *Clinical Anatomy*, 8th ed. Oxford, Blackwell Scientific Publications, 1992.

Ger R, Abrahams P, Olson TR: *Essentials of Clinical Anatomy*, 2nd ed. London, Parthenon Publishing Group, 1996.

Haines DE (ed): *Fundamental Neuroscience*. Churchill Livingstone, New York, 1997.

Magee DF, Dalley AF (eds): Digestion and the structure and function of the gut. In *Karger Continuing Education Series*, Volume 8. Basel, S Karger AG, 1986.

Moore KL, Agur AMR: *Essential Clinical Anatomy*. Baltimore, Williams & Wilkins, 1995.

Moore KL, Persaud TVN: *The Developing Human. Clinically Oriented Embryology*, 6th ed. Philadelphia, WB Saunders, 1998.

Naftel JP, Hardy SGP: Visceral motor pathways. *In* Haines DE (ed): *Fundamental Neuroscience*. New York, Churchill Livingstone, 1997.

Rowland LP: *Merritt's Textbook of Neurology*, 9th ed. Baltimore, Williams & Wilkins, 1995.

Sabiston DC Jr, Lyerly H (eds): *Sabiston Essentials of Surgery*, 2nd ed. Philadelphia, WB Saunders, 1994.

Sabiston DC Jr, Lyerly H (eds): *Textbook of Surgery. Pocket Companion*, 2nd ed. Philadelphia, WB Saunders, 1997.

Schochat SJ: Inguinal hernias. *In* Behrman et al. (eds): *Nelson Textbook of Pediatrics*, 15th ed. Philadelphia, WB Saunders, 1996.

Skandalakis JE, Skandalakis PN, Skandalakis JL: *Surgical Anatomy and Technique. A Pocket Manual*. New York, Springer-Verlag, 1995.

Soper DF: Upper genital tract infections. *In* Copeland LJ (ed): *Textbook of Gynecology*. Philadelphia, WB Saunders, 1993.

Swartz MH: History and examination. In *Textbook of Physical Diagnosis*, 2nd ed. Philadelphia, WB Saunders, 1994.

Wakeley CPG: The position of the vermiform appendix as ascertained by the analysis of 10,000 cases. *J Anat* 67:277, 1933.

Williams PL, Bannister LH, Berry MM, Collins P, et al. (eds): The anatomical basis of medicine and surgery. *Gray's Anatomy*, 38th ed. New York, Churchill Livingstone, 1995.

Willms JL, Schneiderman H, Algranati PS: *Physical Diagnosis. Bedside Evaluation of Diagnosis and Function*. Baltimore, Williams & Wilkins, 1994.

c h a p t e r

3 Pelvis and Perineum

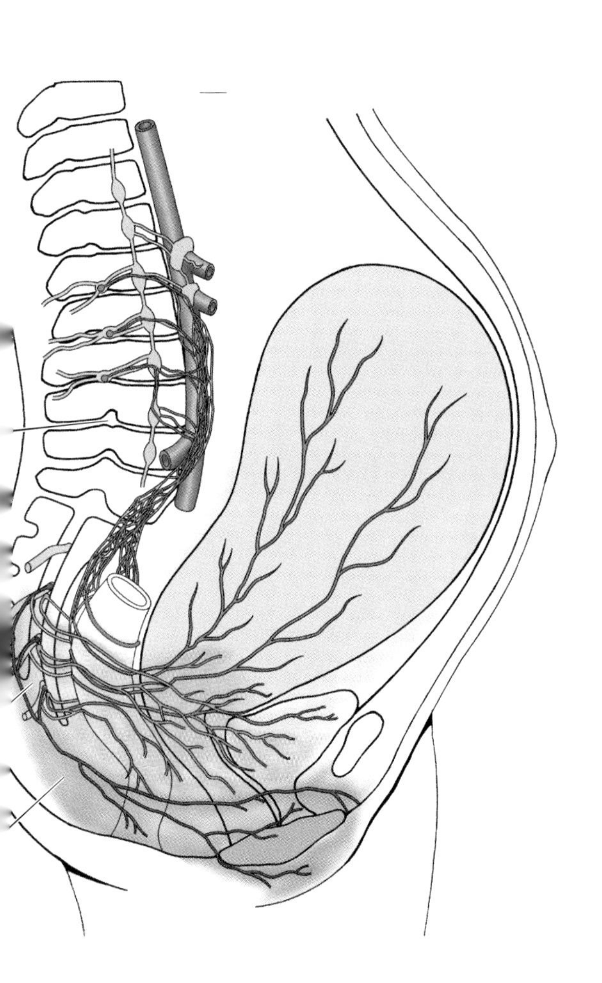

The pelvis (L. basin) is the part of the trunk inferoposterior to the abdomen and is the area of transition between the trunk and the lower limbs (Fig. 3.1*A*). The **bony pelvis** is the basin-shaped ring of bones that protects the distal parts of the intestinal and urinary tracts and the internal genital organs. The **abdominopelvic cavity** extends superiorly into the thoracic cage and inferiorly into the pelvis. Although this cavity is largely protected in its superior and inferior parts, perforating wounds in the thorax and pelvis may involve the abdominopelvic cavity and its contents. The **pelvic axis** is a hypothetical curved line joining the center point of each of the four planes of the pelvis, marking the center of the pelvic cavity at every level. The form of the pelvic axis and its disparity in depth between the anterior and posterior contours of the cavity are important factors in the mechanism of fetal passage through the pelvic canal (Williams et al., 1995).

The **perineum** refers to both the *area* of the trunk between the thighs and buttocks extending from the coccyx to the pubis and the shallow *compartment* lying deep to this area and inferior to the **pelvic floor** formed by the **pelvic diaphragm** (Fig. 3.1, *B* and *C*). In the male this area includes the penis, scrotum, and anus, and in the female, the vulva (external genitalia) and anus.

Pelvis

The pelvis is enclosed by bony, ligamentous, muscular walls. The funnel-shaped **pelvic cavity**—the space bounded at the sides by the bones of the pelvis—is continuous with the abdominal cavity and is angulated posteriorly from it (Fig. 3.1, *A* and *C*). The pelvic cavity contains the urinary bladder, terminal parts of the ureters, pelvic genital organs, rectum, blood vessels, lymphatics, and nerves. The superior bound-

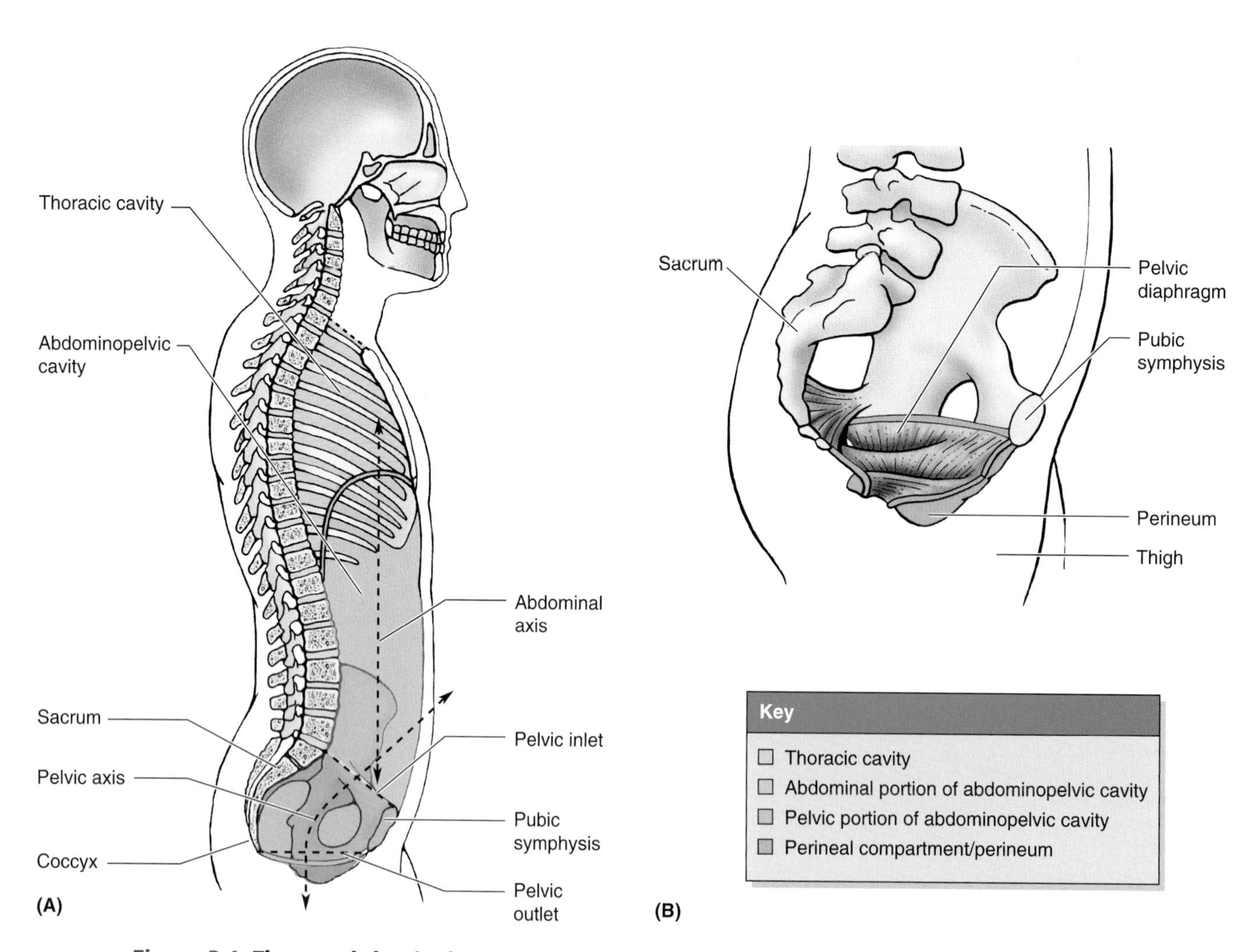

Figure 3.1. Thoracoabdominal cavity. A. Median section of the trunk showing the relationship of the thoracic and abdominopelvic cavities. The pelvic inlet (superior pelvic aperture) is the opening into the lesser pelvis, the true pelvis. The pelvic outlet (inferior pelvic aperture) is the lower opening of the lesser pelvis. **B.** Schematic medial view of the left half of the pelvis, showing the pelvic diaphragm separating the pelvic cavity from the perineum and forming the floor of the pelvis.

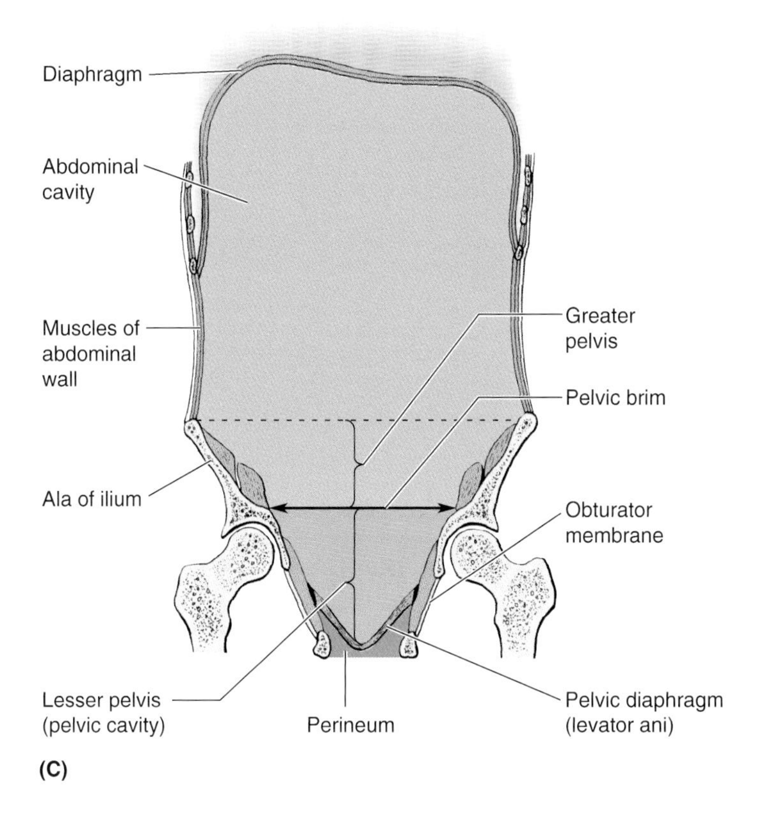

Figure 3.1. *(Continued)* **C.** Schematic coronal section of the abdominopelvic cavity. Observe that the plane of the pelvic brim (*double-headed arrow*) separates the greater pelvis—part of the abdominal cavity—from the lesser pelvis, the pelvic cavity.

ary of the pelvic cavity is the **pelvic inlet** (superior pelvic aperture). The pelvis is limited inferiorly by the **pelvic outlet** (inferior pelvic aperture), which is closed by the musculofascial **pelvic diaphragm** and bounded posteriorly by the **coccyx** and anteriorly by the **pubic symphysis** (Fig. 3.1, *A* and *B*). Although continuous, the abdominal and pelvic cavities are described separately for descriptive and regional purposes.

Bony Pelvis

The bony pelvis (pelvic skeleton) is strong. Its main functions are to transfer the weight of the upper body from the axial to the lower appendicular skeleton and to withstand compression and other forces resulting from its support of body weight and its provision of attachments for powerful muscles (Williams et al., 1995). *In the mature individual the bony pelvis is formed by four bones* (Fig. 3.2):

- **Hip bones**, two large, irregularly shaped bones, each of which develops from the fusion of three bones—*ilium, ischium*, and *pubis*
- **Sacrum**, formed by the fusion of five originally separate sacral vertebrae
- **Coccyx**, formed by the fusion of four rudimentary coccygeal vertebrae; sometimes the first sacral vertebra is separated from the others and thus participates in forming the skeleton of this vestigial tail.

In infants and children, the **hip bones** (coxal bones; innominate bones) consist of three separate bones that are united by cartilage at the **acetabulum** (Fig. 3.2*B*)—the cuplike depression in the lateral surface of the hip bone that articulates with the head of the femur (thigh bone) (Fig. 3.3*A*). At puberty the ilium, ischium, and pubis fuse to form the hip bone.

The hip bones are joined at the **pubic symphysis** (L. symphysis pubis) anteriorly and to the sacrum posteriorly to form the **pelvic girdle** (Fig. 3.2*A*), which

- Articulates with the sacrum at the sacroiliac joints
- Is massively constructed for resistance to stress
- Transmits the thrust between the vertebral column and the lower limbs.

The pelvic girdle—formed by the hip bones and the sacrum—is attached to the lower limbs.

The **ilium** is the superior, flattened, fan-shaped part of the hip bone (Fig. 3.2, *A–C*). The **ala** (L. wing) of the ilium represents the spread of the fan and the body, the handle. The **body** of the ilium helps to form the acetabulum. The **iliac crest**, the rim of the fan, has a curve that follows the contour of the ala between the **anterior** and **posterior superior iliac spines**. The anterior concave part of the ilium forms the **iliac fossa.**

The **ischium** has a body and ramus (L. branch). The **body** of the ischium helps to form the acetabulum and the **ramus** helps to form the **obturator foramen** (Fig. 3.2, *B* and *C*). The large posteroinferior protuberance of the ischium is the **is-**

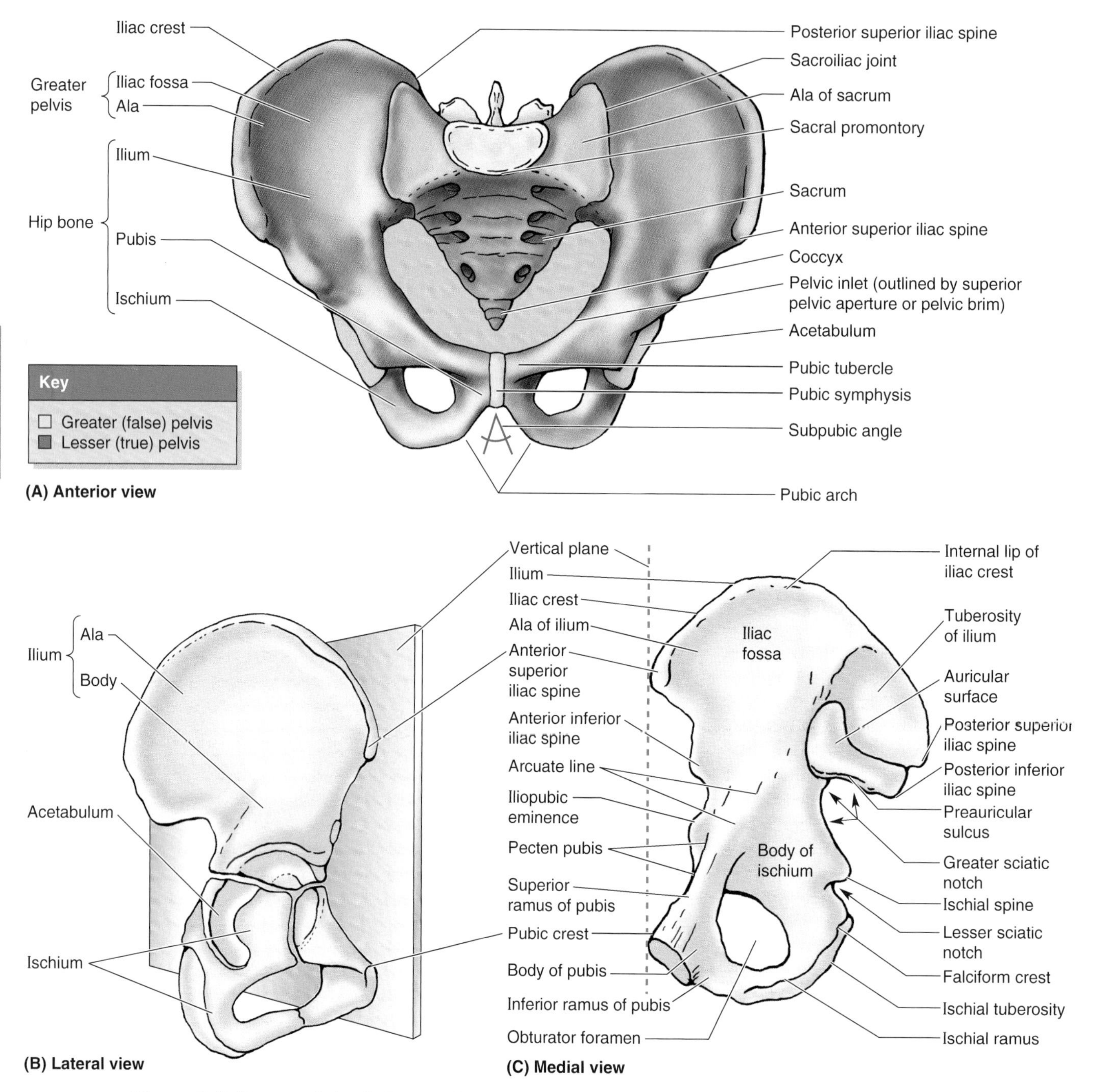

Figure 3.2. Bony pelvis. A. Bones of the pelvis. Observe that the skeleton of the pelvis is formed by the two hip bones anteriorly and laterally, and the sacrum and coccyx posteriorly. **B.** Lateral view of a child's hip bone. Note that in the anatomical position, the anterior superior iliac spine and the anterior aspect of the pubis lie in the same vertical plane. Observe that the hip bone is composed of three bones: ilium, ischium, and pubis, which meet in the cup-shaped acetabulum. Note that the bones are not fused at this age and unite by a triradiate cartilage along a Y-shaped line (*blue*). Fusion is usually complete by age 23. **C.** Medial view of the right hip bone in the anatomical position.

chial tuberosity; the small pointed posterior projection near the junction of the ramus and body is the **ischial spine.** The concavity between the ischial spine and the ischial tuberosity is the **lesser sciatic notch.** The larger concavity, the **greater sciatic notch,** is superior to the ischial spine and is formed in part by the ilium.

The **pubis** is an angulated bone with a **superior ramus** that helps to form the acetabulum and an **inferior ramus** that helps to form the obturator foramen (Fig. 3.2, *B* and *C*). A thickening on the anterior part of the **body** of the pubis is the **pubic crest,** which ends laterally as a prominent bump, the **pubic tubercle** (Fig. 3.2*A*). The lateral part of the superior ra-

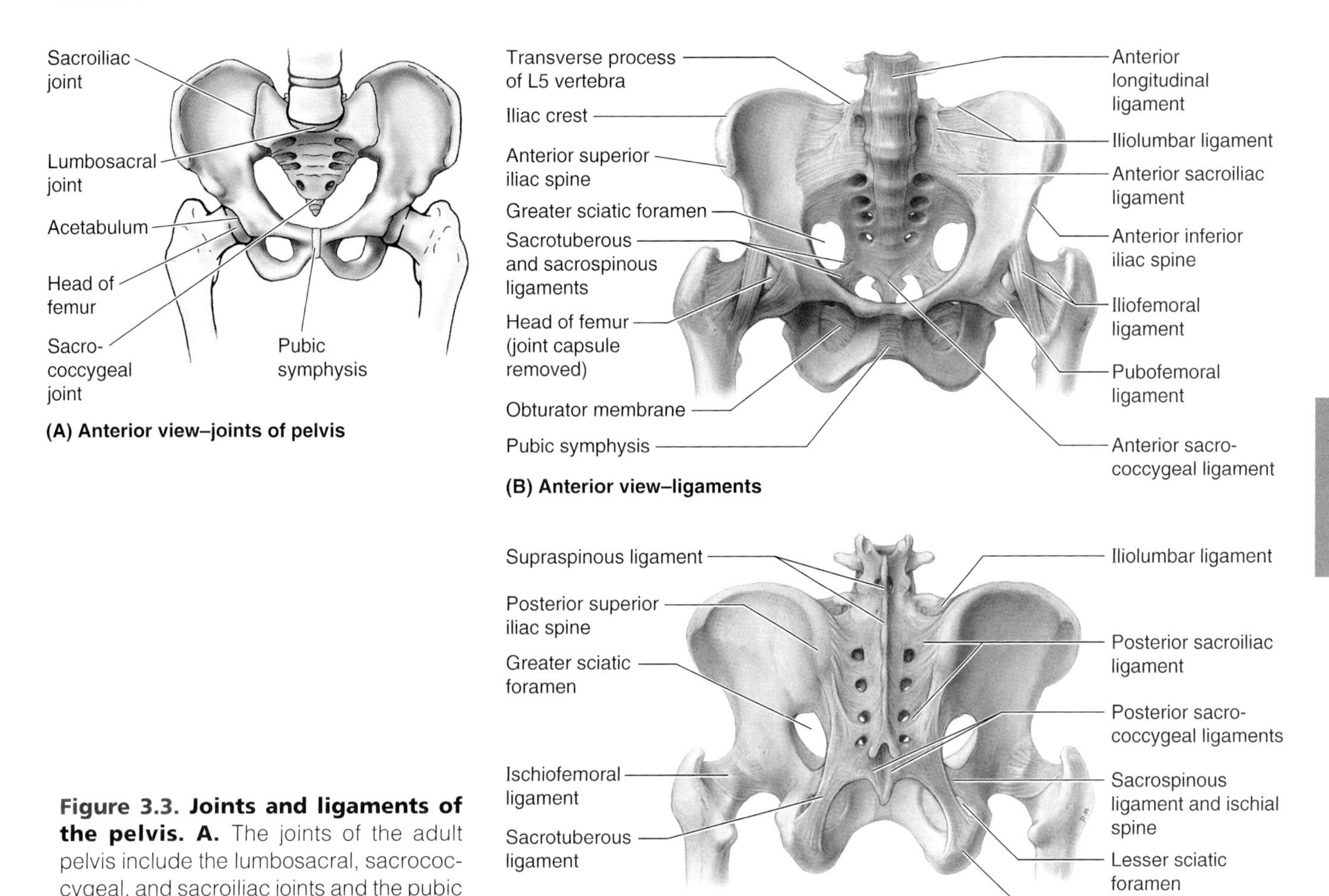

Figure 3.3. Joints and ligaments of the pelvis. A. The joints of the adult pelvis include the lumbosacral, sacrococcygeal, and sacroiliac joints and the pubic symphysis. **B–C.** Ligaments of the pelvis.

mus has an oblique ridge—the **pecten pubis** (pectineal line of pubis).

The bony pelvis is divided into greater and lesser pelves—the false and true pelves—by an oblique plane passing through the **sacral promontory** posteriorly and the **terminal lines** (L. lineae terminales) elsewhere. The **pelvic inlet** is defined by the plane of the terminal lines (Fig. 3.2*A*). The **pubic arch** is formed by the conjoined rami of the pubis and ischium of the two sides. These rami meet at the **pubic symphysis** to form the **subpubic angle** (Table 3.1), which can be measured with the fingers in the vagina during a physical examination.

The **greater pelvis** *(*false pelvis, pelvis major) is:

- Superior to the pelvic inlet
- The location of some abdominal viscera (e.g., the ileum and sigmoid colon)
- Bounded by the abdominal wall anteriorly, the iliac fossae posterolaterally, and L5 and S1 vertebrae posteriorly.

The cavity of the greater pelvis is the inferior part of the abdominal cavity (Fig. 3.1*C*).

The **lesser pelvis** (true pelvis, pelvis minor) is:

- Between the pelvic inlet and the pelvic outlet
- The location of the pelvic viscera—the urinary bladder and reproductive organs such as the uterus and ovaries
- Bounded by the pelvic surfaces of the hip bones, sacrum, and coccyx
- Limited inferiorly by the musculofascial pelvic diaphragm
- Of major obstetrical and gynecological significance.

The cavity of the lesser pelvis is the true pelvic cavity, forming the inferior part of the abdominopelvic cavity (Fig. 3.1*C*).

The **pelvic inlet** separates the greater pelvis from the lesser pelvis (Figs. 3.1, *A* and *C*, and 3.2*A*, Table 3.1). The edge or rim of the pelvic inlet is the **pelvic brim.** The oblique plane of the pelvic brim forms an angle of approximately 55° to the horizontal. This plane coincides with the line joining the **sacral promontory** to the superior margin of the **pubic symphysis.** This line represents the anteroposterior or conjugate diameter of the pelvic inlet. *The pelvic inlet is bounded by the:*

- Superior margin of the pubic symphysis anteriorly
- Posterior border of the pubic crest
- Pecten pubis
- Arcuate line of the ilium

- Anterior border of the ala of the sacrum
- Sacral promontory.

The pelvic outlet is bounded by the:

- Inferior margin of the pubic symphysis anteriorly
- Inferior rami of the pubis and ischial tuberosities anterolaterally
- Sacrotuberous ligaments posterolaterally (Fig. 3.3, *B* and *C*)
- Tip of the coccyx posteriorly.

Table 3.1. Comparison of Male and Female Bony Pelves

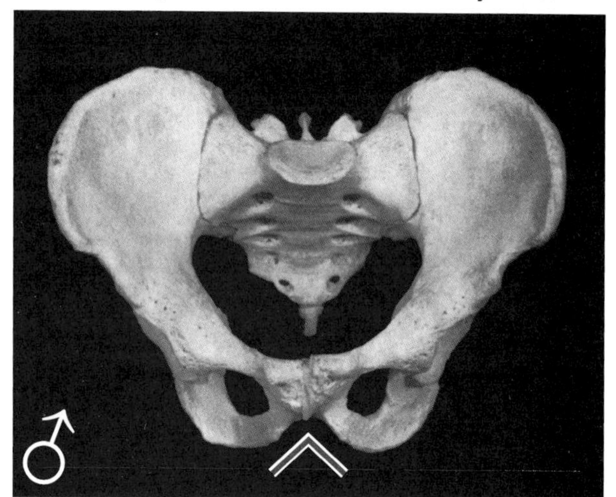

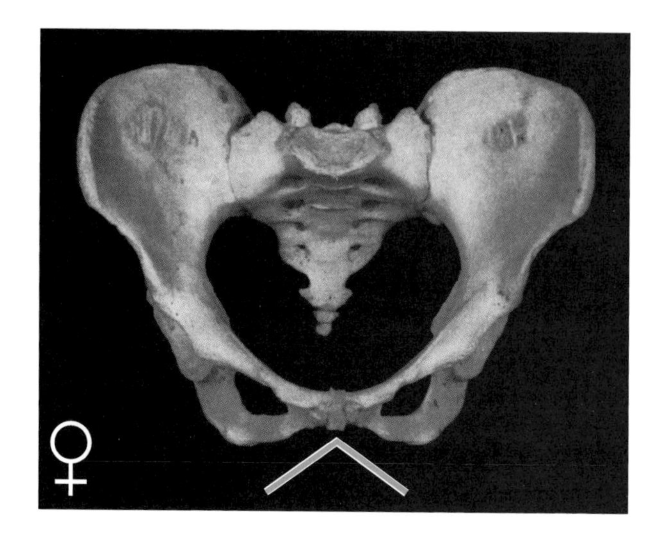

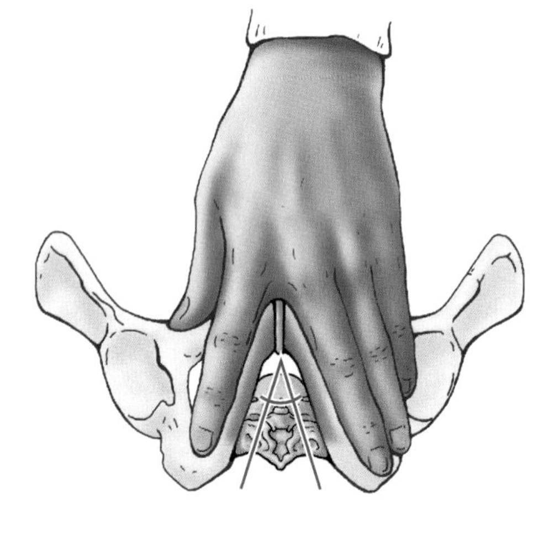

(A)

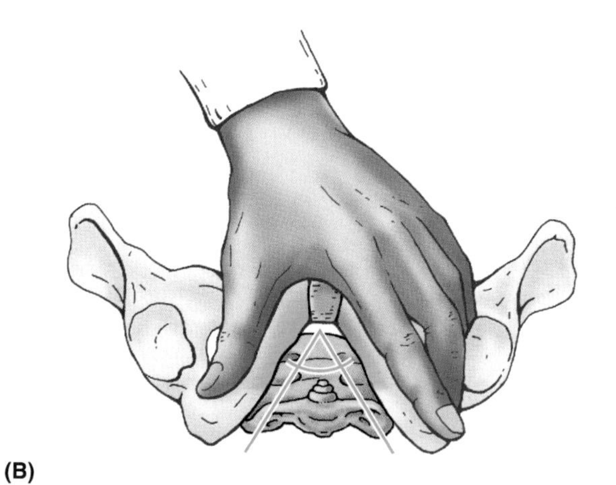

(B)

Bony Pelvis	Male (♂)	Female (♀)
General structure	Thick and heavy	Thin and light
Greater pelvis (pelvis major)	Deep	Shallow
Lesser pelvis (pelvis minor)	Narrow and deep	Wide and shallow
Pelvic inlet (superior pelvic aperture)	Heart-shaped	Oval and rounded
Pelvic outlet (inferior pelvic aperture)	Comparatively small	Comparatively large
Pubic arch and subpubic angle	Narrow	Wide
Obturator foramen	Round	Oval
Acetabulum	Large	Small

Orientation of the Pelvis

When a person is in the anatomical position, the anterior superior iliac spines and the anterior aspect of the pubic symphysis lie in the same vertical plane (Fig. 3.2*B*). Furthermore, the **pelvic canal**—the passage from the pelvic inlet to the pelvic outlet—curves obliquely posteriorly relative to the abdominal cavity and trunk (Fig. 3.1*A*).

Sexual Differences in the Pelves

The pelves of males and females differ in several respects (Table 3.1). These sexual differences are linked to function. *In both sexes* the primary pelvic function is locomotor (pertaining to locomotion)—the ability to move from one place to another. The sexual differences are related mainly to the heavier build and larger muscles of most men and to the adaptation of the pelvis—particularly the lesser pelvis—in women for parturition (childbearing).

The **male pelvis** is heavier and thicker than the female pelvis and usually has more prominent bone markings. The **female pelvis** is wider, shallower, and has a larger pelvic inlet and outlet. The **pubic arch** is formed by the conjoined rami of the pubis and ischium of the two sides. These rami meet at the pubic symphysis to form the **subpubic angle**, which is measurable during a physical examination. The subpubic angle is nearly a right angle in females; it is considerably less in males (approximately 60°). When the vagina admits three fingers side by side, the subpubic angle is sufficient to permit passage of the fetal head after it has passed through the pelvic outlet.

Although anatomical differences between male and female pelves are usually clear-cut, the pelvis of any person may have some features of the opposite sex. The pelvic types shown in *A* and *C* are commonest in males, *B* and *A* in white females, *B* and *C* in black females, while *D* is uncommon in both sexes. The **gynecoid pelvis** (*B*) is the normal female type. An android (masculine or funnel-shaped) pelvis in a woman may present hazards to successful vaginal delivery of a fetus. The size and shape of the pelvic inlet is important because it is through this opening that the fetal head enters the lesser pelvis during labor. The size of the lesser pelvis is particularly important in obstetrics because it is the bony *pelvic canal* ("birth canal") through which the fetus passes during a vaginal birth. To determine the capacity of the female pelvis for childbearing, the diameters of the lesser pelvis are noted radiographically or during a pelvic examination. The typical gynecoid pelvis has a rounded oval shape and the maximum transverse diameter. In all pelves, the ischial spines face each other, and the interspinous distance between them is the narrowest part of the pelvic canal. The anteroposterior or obstetrical diameter (true conjugate) passes from the superior margin of the pubic symphysis to the middle of the sacral promontory.

In **forensic medicine**—dealing with the application of medical and anatomical knowledge for the purposes of law—identification of human skeletal remains usually involves the diagnosis of gender. A prime focus of attention is the pelvis because sexual differences usually are clearly visible. Even parts of the pelvis are useful in making a diagnosis of sex. ▶

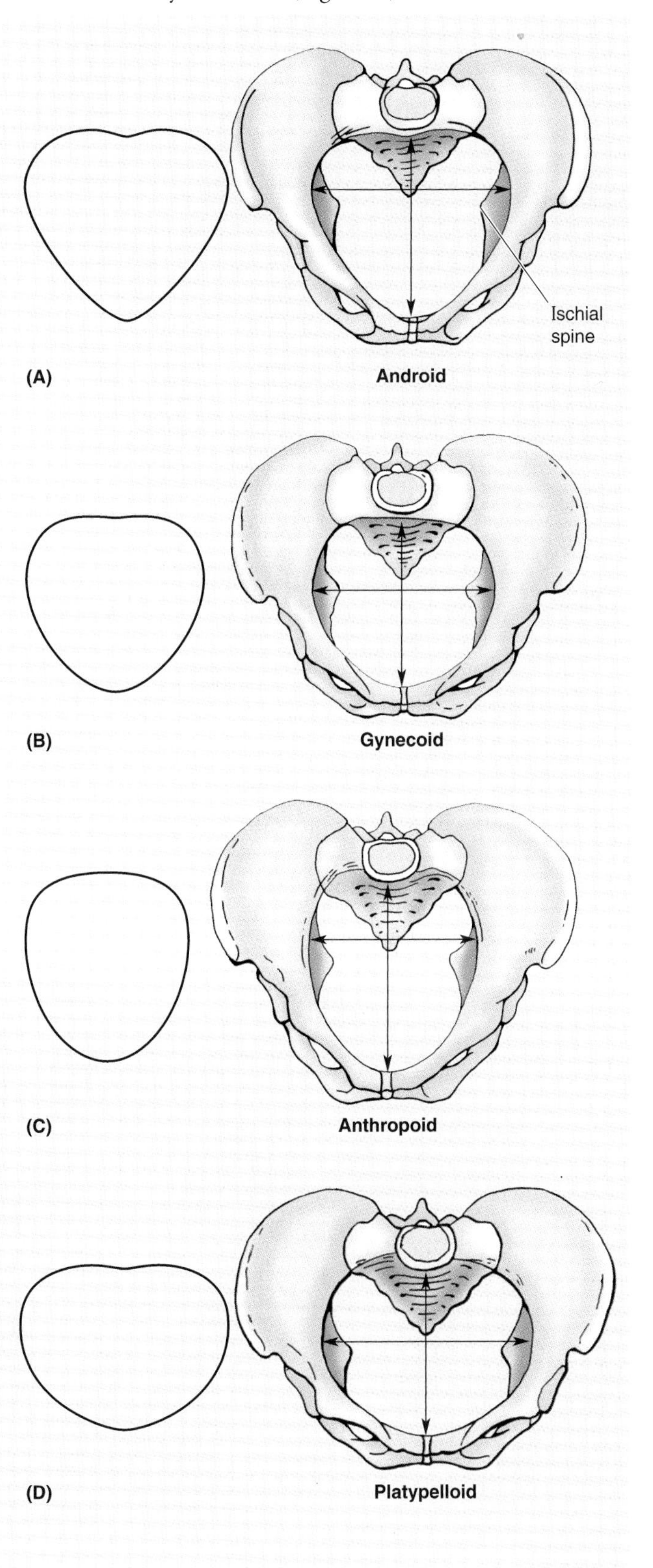

Pelvic Fractures

Anteroposterior compression of the pelvis occurs during "squeezing accidents" (e.g., when a heavy object falls on the pelvis). This type of trauma commonly produces *fractures of the pubic rami.* When the pelvis is compressed laterally, the acetabula and ilia are squeezed toward each other and may be broken. Some pelvic fractures result from the tearing away of bone by the strong posterior pelvic ligaments associated with the sacroiliac joints (Fig. 3.3).

Pelvic fractures can result from direct trauma to the pelvic bones such as occurs during an automobile accident (*A*) or be caused by forces transmitted to these bones from the lower limbs during falls on the feet (*B*). *Weak areas of the pelvis are the*:

- Pubic rami
- Acetabula (or the area immediately surrounding them)
- Region of the sacroiliac joint
- Alae of the ilium.

The fracture usually occurs across a weak part of the pelvis (e.g., across the rami defining the obturator foramen).

Pelvic fractures may cause injury to pelvic soft tissues, blood vessels, nerves, and organs. Fractures in the pubo-obturator area are relatively common and are often complicated because of their relationship to the urinary bladder and urethra, which may be ruptured or torn. *Falls on the feet or buttocks from a high ladder, for example, may produce the following injuries*: ▶

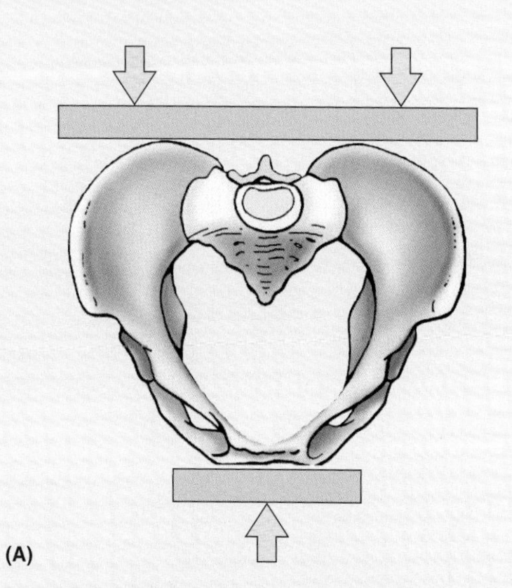

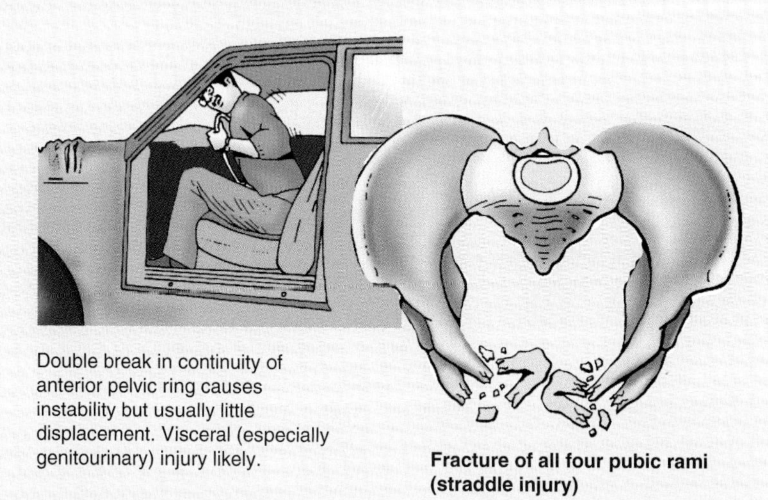

Double break in continuity of anterior pelvic ring causes instability but usually little displacement. Visceral (especially genitourinary) injury likely.

Fracture of all four pubic rami (straddle injury)

(A)

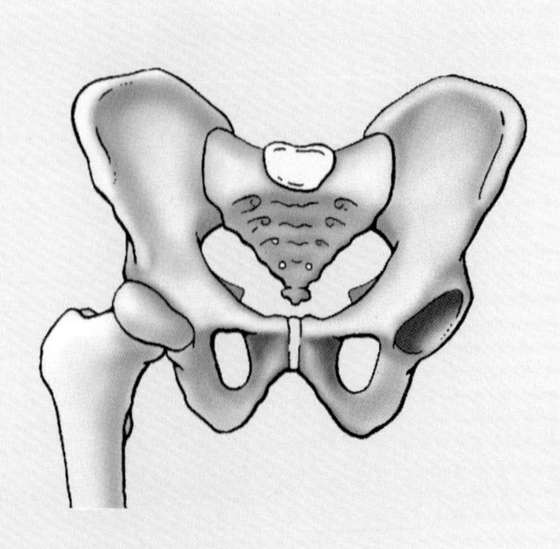

Central fracture of acetabulum with dislocation of femoral head into pelvis.

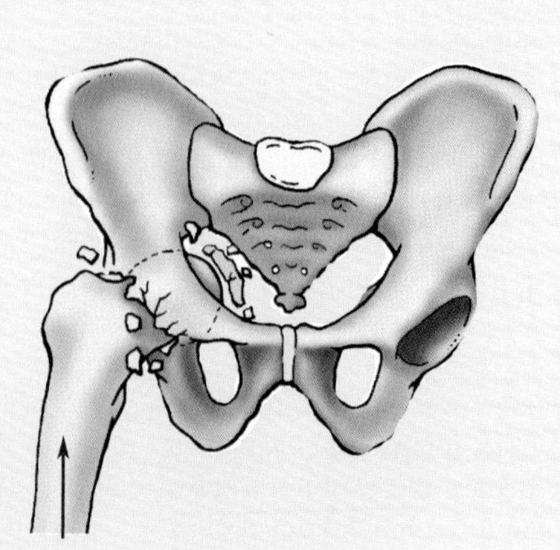

Fracture of acetabulum

(B)

- The pubic rami may be fractured
- The bone or cartilage of the acetabula may be injured
- The head of the femur—the bone that extends from the pelvis to the knee—may be driven through the acetabulum into the pelvic cavity, injuring pelvic viscera, nerves, and vessels.

In persons less than 17 years of age, the acetabulum may fracture through the *triradiate (radiating in three directions) cartilage* into their three developmental parts (Fig. 3.2*B*), or the bony acetabular margins may be torn away. ⊕

Pelvic Joints and Ligaments

The joints of the pelvis are the lumbosacral joints, the sacrococcygeal joint, the sacroiliac joints, and the pubic symphysis (Fig. 3.3). Strong ligaments support and strengthen these joints.

Lumbosacral Joints

L5 and S1 vertebrae articulate at the anterior **intervertebral (IV) joint** formed by the IV disc between their bodies (Fig. 3.3*A*) and at two posterior **zygapophysial joints** (facet joints) between the auricular (articular) processes of these vertebrae. The facets on S1 vertebra face posteromedially, thereby preventing L5 vertebra from sliding anteriorly. **Iliolumbar ligaments** unite the ilia and L5 vertebra (Fig. 3.3, *B* and *C*).

Spondylolysis and Spondylolisthesis

Spondylolysis, a degenerative condition resulting from deficient development of the articulating part of a vertebra, occurs in the inferior lumbar region in approximately 5% of white North American adults. It occurs more frequently in certain races (e.g., the Canadian Inuit people). Persons with spondylolysis have a *defect in the vertebral (neural) arch*—the posterior projection from the vertebral body—between the superior and inferior facets. When bilateral, the defects result in the L5 vertebra being divided into two pieces. If the parts separate, the abnormality is **spondylolisthesis** (anterior displacement of the body of L5 vertebra on the sacrum). The displacement of the anterior piece of L5 vertebra reduces the anteroposterior diameter of the pelvic inlet, which may interfere with parturition. Spondylolisthesis at L5 may also compress the spinal nerves as they enter the sacral canal, causing back pain.

Obstetricians test for spondylolisthesis by running their fingers along the lumbar spinous processes. An abnormally prominent L5 process indicates that the anterior part of L5 vertebra and the vertebral column superior to it may have moved anteriorly relative to the sacrum. Medical images, such as sagittal magnetic resonance imaging (MRI), are taken to confirm the diagnosis and to measure the anteroposterior diameter of the pelvic inlet. ⊕

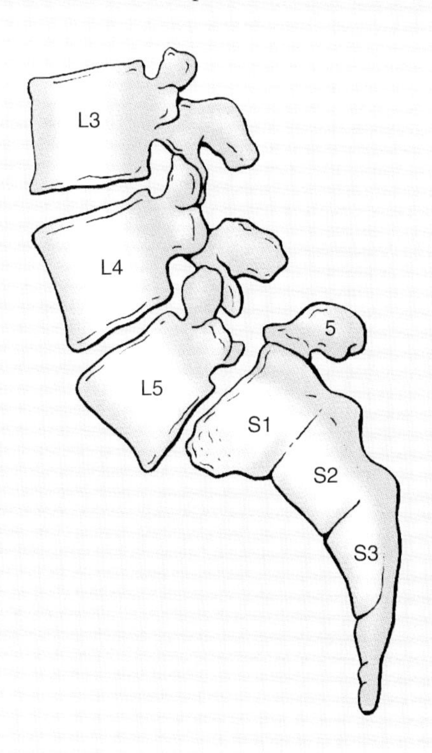

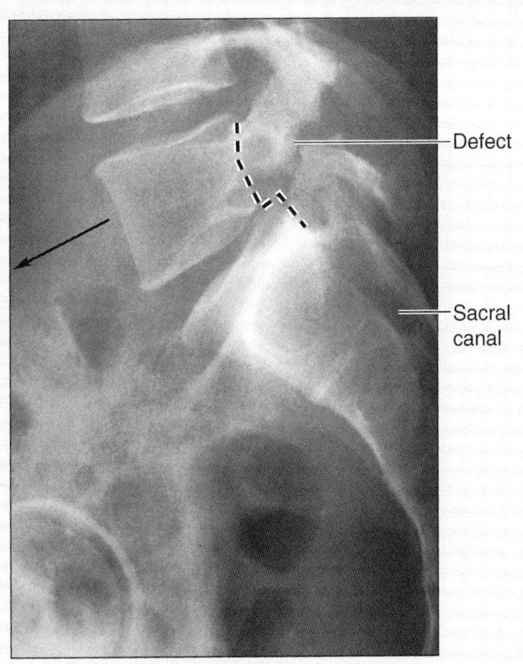

Dotted line follows posterior vertebral margins of L5 and the sacrum

(B)

Sacrococcygeal Joint

This secondary cartilaginous joint (Fig. 3.3*A*) has an IV disc. Fibrocartilage and ligaments join the apex of the sacrum to the base of the coccyx. The anterior and posterior **sacrococcygeal ligaments** are long strands that reinforce the joint, much like the anterior and posterior longitudinal ligaments do for the superior vertebrae (Fig. 3.3, *B* and *C*).

Sacroiliac Joints

These articulations are strong, weightbearing synovial joints between the ear-shaped auricular surfaces of the sacrum and ilium (Figs. 3.3–3.5). These surfaces have irregular elevations and depressions that produce some interlocking of the bones. The sacrum is suspended between the iliac bones and is firmly attached to them by **interosseous** and **sacroiliac ligaments** (Figs. 3.3, *B* and *C*, and 3.4*A*). The sacroiliac joints differ from most synovial joints in that they possess little mobility because of their role in transmitting the weight of most of the body to the hip bones.

Movement of the sacroiliac joints is limited because of the interlocking of the articulating bones and the thick interosseous and posterior **sacroiliac ligaments**. Movement is limited to slight gliding and rotary movements (Fig. 3.4*C*), except when subject to considerable force such as occurs following a high jump. In this case, the force is transmitted through the lumbar vertebrae to the superior end of the sacrum, which tends to rotate anteriorly. This rotation is counterbalanced by the interlocking auricular surfaces and the strong supporting ligaments, especially the sacrotuberous and sacrospinous ligaments, which join the sacrum to the ischium (Figs. 3.3, *B* and *C*, and 3.4, *A–C*). The **sacrotuberous and sacrospinous ligaments** allow only limited upward movement of the inferior end of the sacrum, thereby providing resilience to the sacroiliac region when the vertebral column sustains sudden weight increases (resulting from a jump from a wall, for example).

Pubic Symphysis

This secondary cartilaginous joint is formed by the union of the bodies of the pubic bones in the median plane (Figs. 3.3 and 3.5*A*). The fibrocartilaginous interpubic disc of the pubic symphysis is generally thicker in women than in men. The ligaments joining the bones are thickened superiorly and inferiorly to form the superior pubic ligament and the inferior (arcuate) pubic ligament, respectively. The **superior pubic ligament** connects the superior pubic bodies along their superior surfaces and extends as far laterally as the pubic tubercles. The **inferior pubic ligament** (Fig. 3.5*A*) is a thick arch of fibers that connects the inferior borders of the joint, rounds off the subpubic angle between the inferior rami of the pubic bones, and forms the superior border of the pubic arch (Fig. 3.2*A*). The decussating tendinous fibers of the rectus abdominis and external oblique muscles also strengthen the pubic symphysis anteriorly (see Chapter 2).

Relaxation of Pelvic Joints and Ligaments During Pregnancy

During pregnancy, the pelvic joints and ligaments relax and pelvic movements increase. The relaxation, caused by the increase in sex hormones and the presence of the hormone *relaxin*, permits freer movements between the inferior parts of the vertebral column and the pelvis. The sacroiliac interlocking mechanism is less effective because the relaxation permits greater rotation of the pelvis and a small increase in pelvic diameters during parturition. Loosening of the interpubic disc also occurs, resulting in an increase in the distance between the pubic bones. ▶

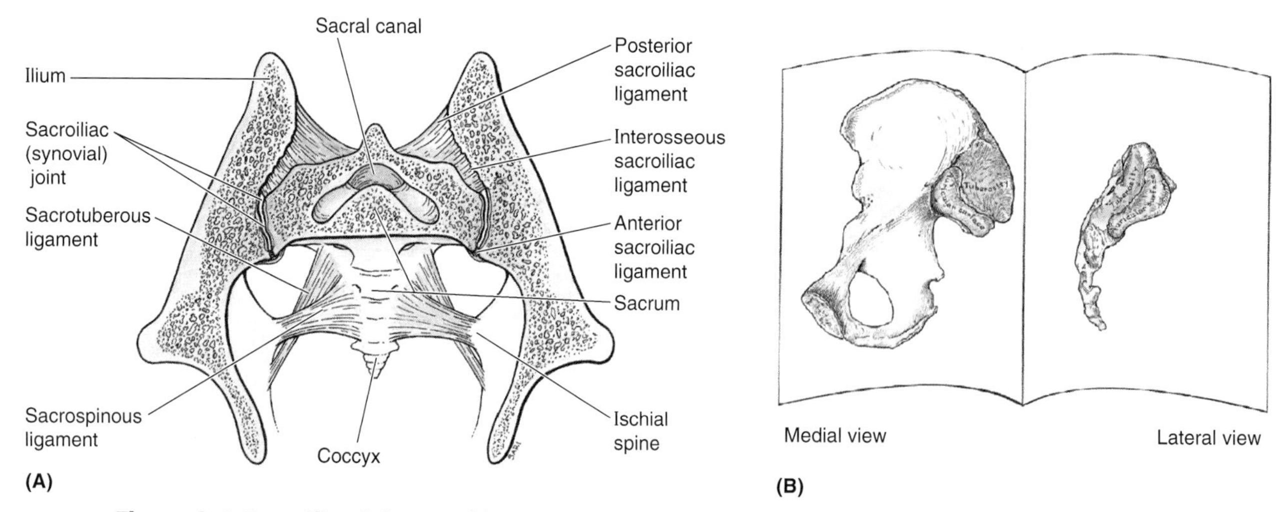

Figure 3.4. Sacroiliac joints and ligaments. A. Anterior view of the posterior half of a coronally sectioned pelvis illustrating these sagittally oriented joints. Note that the strong interosseous sacroiliac ligaments lie deep (anteroinferior) to the posterior sacroiliac ligament and consist of short fibers connecting the tuberosity of the sacrum to the ilium. **B.** Auricular surfaces of the sacroiliac joints.

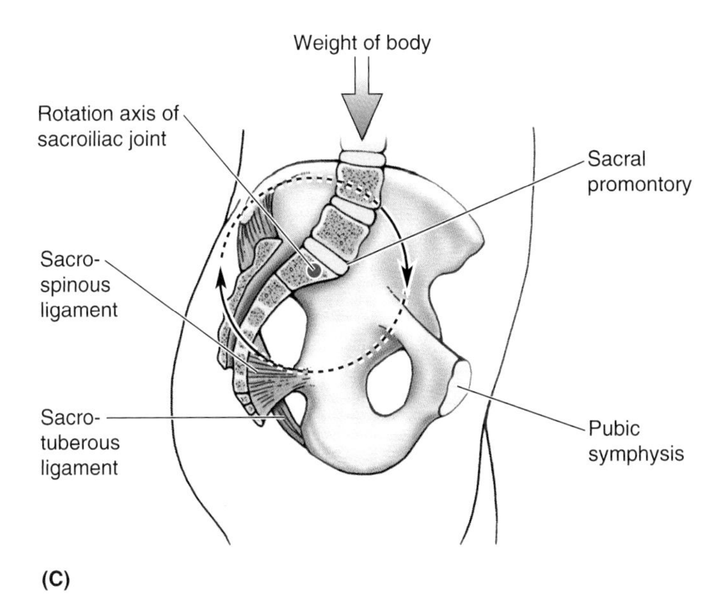

Figure 3.4. *(Continued)* **C.** Drawing illustrating the rotation axis of the sacroiliac joint. The weight of the body is transmitted through the sacrum anterior to the rotation axis, tending to push the upper sacrum inferiorly and thus causing the inferior sacrum to rotate superiorly. This tendency is resisted by the strong sacrotuberous and sacrospinous ligaments. Weight is transferred from the axial skeleton to the ilia, and then to the femurs (thigh bone) during standing and to the ischial tuberosities during sitting.

▶ The coccyx also moves posteriorly during childbirth. All these pelvic changes result in as much as a 10 to 15% increase in diameters (mostly transverse), which facilitate passage of the fetus through the pelvic canal. The one diameter that remains unaffected is the true (conjugate) diameter between the sacral promontory and the posterosuperior aspect of the pubic symphysis. ⊙

Pelvic Walls and Floor

The pelvic wall is divided into an anterior wall, two lateral walls, a posterior wall, and a floor (Figs. 3.1, 3.2, 3.5, and 3.6, Table 3.2).

Anterior Pelvic Wall

The anterior pelvic wall (actually anteroinferiorly placed in the anatomical position) is formed primarily by the bodies and rami of the pubic bones and the pubic symphysis.

Lateral Pelvic Walls

The bony framework of the lateral pelvic walls is formed by the hip bones and the obturator foramen, closed by the obturator membrane. The **obturator internus muscles** cover and thus pad most of the lateral pelvic walls (Figs. 3.5, *C* and *D*, and 3.6*B*). Medial to these muscles are the obturator nerves and vessels and other branches of the internal iliac vessels. Each obturator internus passes posteriorly from the lesser pelvis through the *lesser sciatic foramen* and turns sharply laterally to attach to the greater trochanter of the femur (see Chapter 5).

Posterior Pelvic Wall

The posterior pelvic wall (actually more like a roof or ceiling in the anatomical position) is formed by the sacrum and coccyx, adjacent parts of the ilia, and the sacroiliac joints and their associated ligaments (Fig. 3.5, *A* and *D*). The **piriformis muscles** cover and thus pad this wall posterolaterally. Each muscle leaves the lesser pelvis through the *greater sciatic foramen* to attach to the upper border of the greater trochanter of the femur (see Chapter 5). Medial to the piriformis muscles are the nerves of the *sacral plexus*; these muscles form a "muscular bed" for this nerve network.

Pelvic Floor

The pelvic floor is formed by the funnel-shaped **pelvic diaphragm**, which consists of the levator ani and coccygeus muscles and the fasciae covering the superior and inferior aspects of these muscles (Figs. 3.5*A* and 3.6, *A* and *B*, Table 3.2). *The pelvic diaphragm stretches between the pubis anteriorly and the coccyx posteriorly and from one lateral pelvic wall to the other.* This gives the pelvic diaphragm the appearance of a funnel suspended from these attachments. The levator ani—a broad muscular sheet—is the larger part and most important muscle in the pelvic floor. Its parts are designated according to the direction and attachment of its fibers. The **levator ani** is attached to the internal surface of the lesser pelvis and forms most of the pelvic floor. *The levator ani consists of three parts* (Figs. 3.5, *A* and *D*, and 3.6, Table 3.2):

- The **pubococcygeus,** the main part of the levator ani, arises from the posterior aspect of the body of the pubis and passes back almost horizontally

- The **puborectalis**, consisting of the thickened, medialmost part of the pubococcygeus, unites with its partner to form a U-shaped muscular sling that passes posterior to the anorectal junction (Fig. 3.7)
- The **iliococcygeus**, the posterior part of the levator ani, is thin and often poorly developed.

The levator ani:

- Forms a muscular sling for supporting the abdominopelvic viscera
- Resists increases in intra-abdominal pressure
- Helps to hold the pelvic viscera in position.

Acting together, the parts of the levator ani raise the pelvic floor, thereby assisting the anterolateral abdominal muscles in compressing the abdominal and pelvic contents. This action is an important part of forced expiration, coughing, sneezing, vomiting, urinating, defecating, and fixation of the trunk during strong movements of the upper limbs (e.g., when lifting heavy objects). The levator ani also has important functions in the voluntary control of urination, fecal continence (via the puborectalis), defecation, and uterine support.

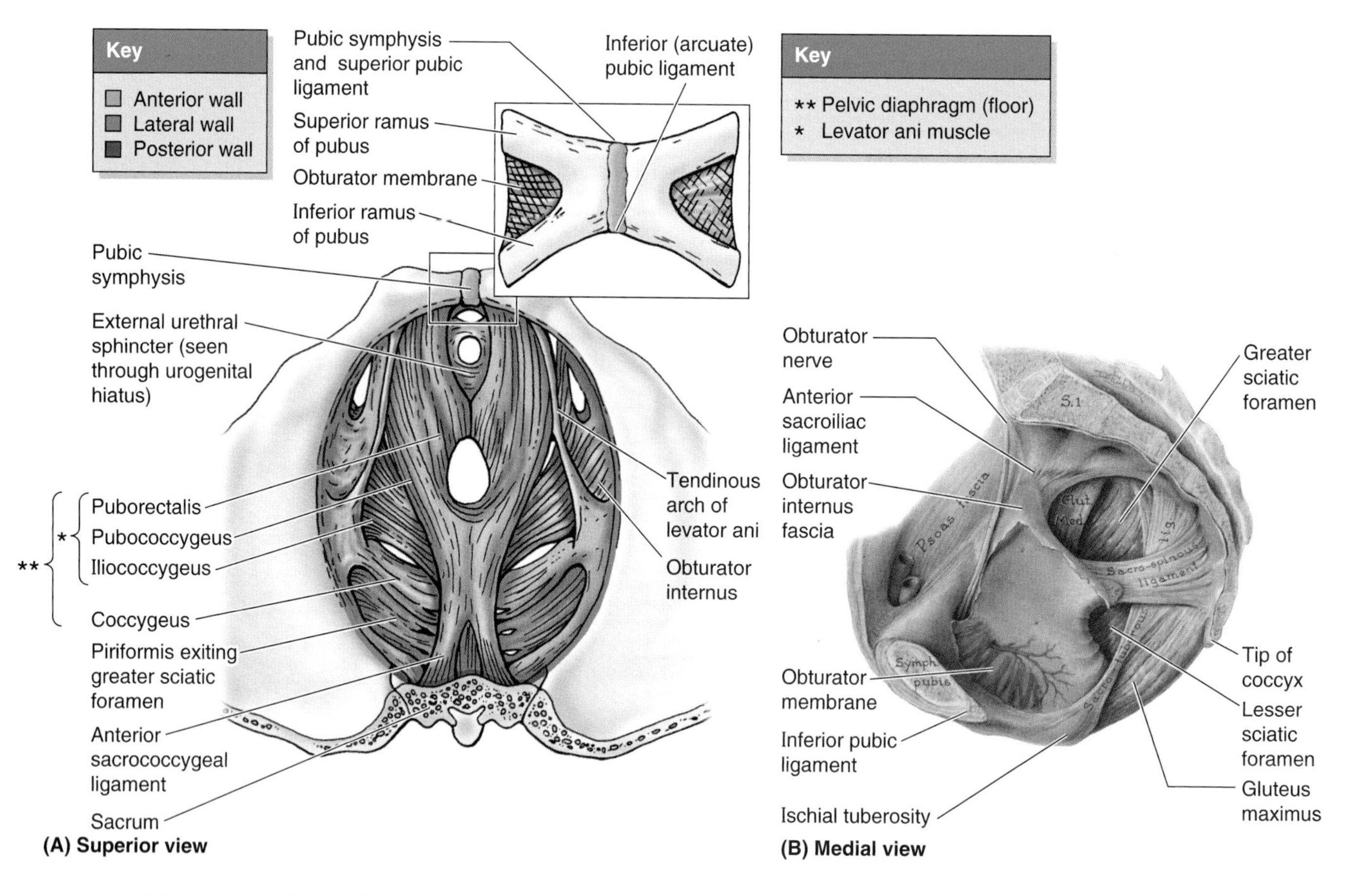

Figure 3.5. Pelvic floor and walls. A. Floor of male pelvis. **B.** Walls of lesser pelvis. Observe: anteriorly, the pubis; posteriorly, the sacrum and coccyx; posterolaterally, the coccyx and inferior part of the sacrum attached to the ischial tuberosity by the sacrotuberous ligament and to the ischial spine by the sacrotuberous ligament; the superior part of the sacrum joined to the ilium by the anterior sacroiliac ligament. The obturator fascia and the obturator internus muscle have been removed to expose the ischium and the obturator membrane composed of strong interlacing fibers filling the obturator foramen.

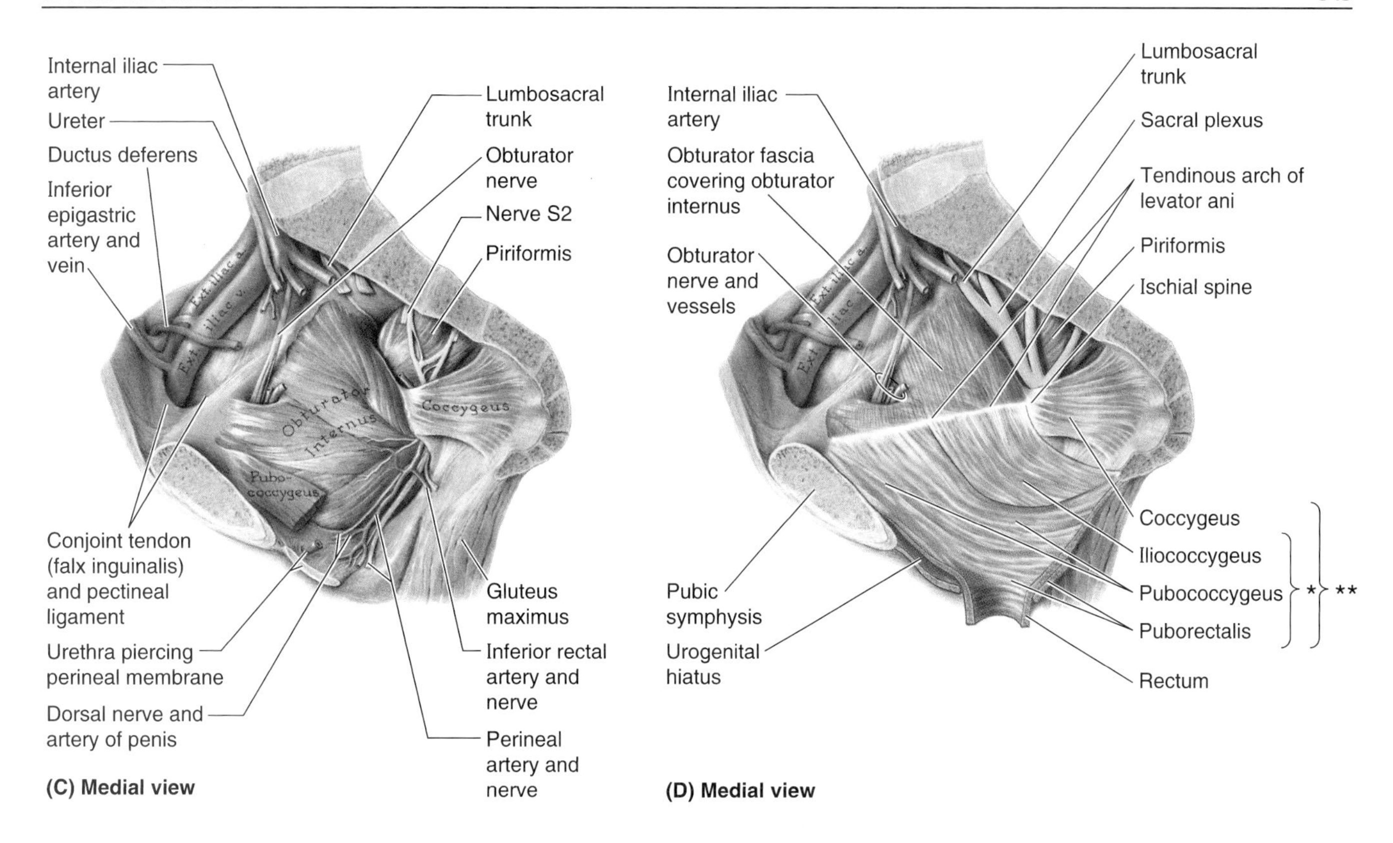

Figure 3.5. *(Continued)* **C.** Muscles of the lesser pelvis. Observe the obturator internus padding the lateral wall of the pelvis and converging to escape posteriorly (see **B**) through the lesser sciatic foramen. **D.** Floor of the male pelvis, showing the pubococcygeus muscle—the main part of the levator ani—arising from the posterior aspect of the body of the pelvis and the tendinous arch of the levator ani, and running posteriorly toward the coccyx.

Table 3.2. **Muscles of the Pelvic Walls**

Muscle	Proximal Attachment	Distal Attachment	Innervation	Main Action
Lateral Wall				
Obturator internus	Pelvic surfaces of ilium and ischium; obturator membrane	Greater trochanter of femur	Nerve to obturator internus (L5, S1, and S2)	Rotates thigh laterally; assists in holding head of femur in acetabulum
Posterior Wall				
Piriformis	Pelvic surface of 2nd and 4th sacral segments: superior margin of greater sciatic notch and sacrotuberous ligament	Greater trochanter of femur	Ventral rami of S1 and S2	Rotates thigh laterally; abducts thigh; assists in holding head of femur in acetabulum
Pelvic Floor				
Levator ani (pubococcygeus, puborectalis, and iliococcygeus)[a]	Body of pubis, tendinous arch of obturator fascia, and ischial spine	Perineal body, coccyx, anococcygeal ligament, walls of prostate or vagina, rectum, and anal canal	Nerve to levator ani (branches of S4) and inferior anal (rectal) nerve and coccygeal plexus	Helps to support the pelvic viscera and resists increases in intra-abdominal pressure
Coccygeus (ischiococcygeus)[a]	Ischial spine	Inferior end of sacrum	Branches of S4 and S5 nerves	Forms small part of pelvic diaphragm that supports pelvic viscera; flexes coccyx

[a] Pelvic diaphragm

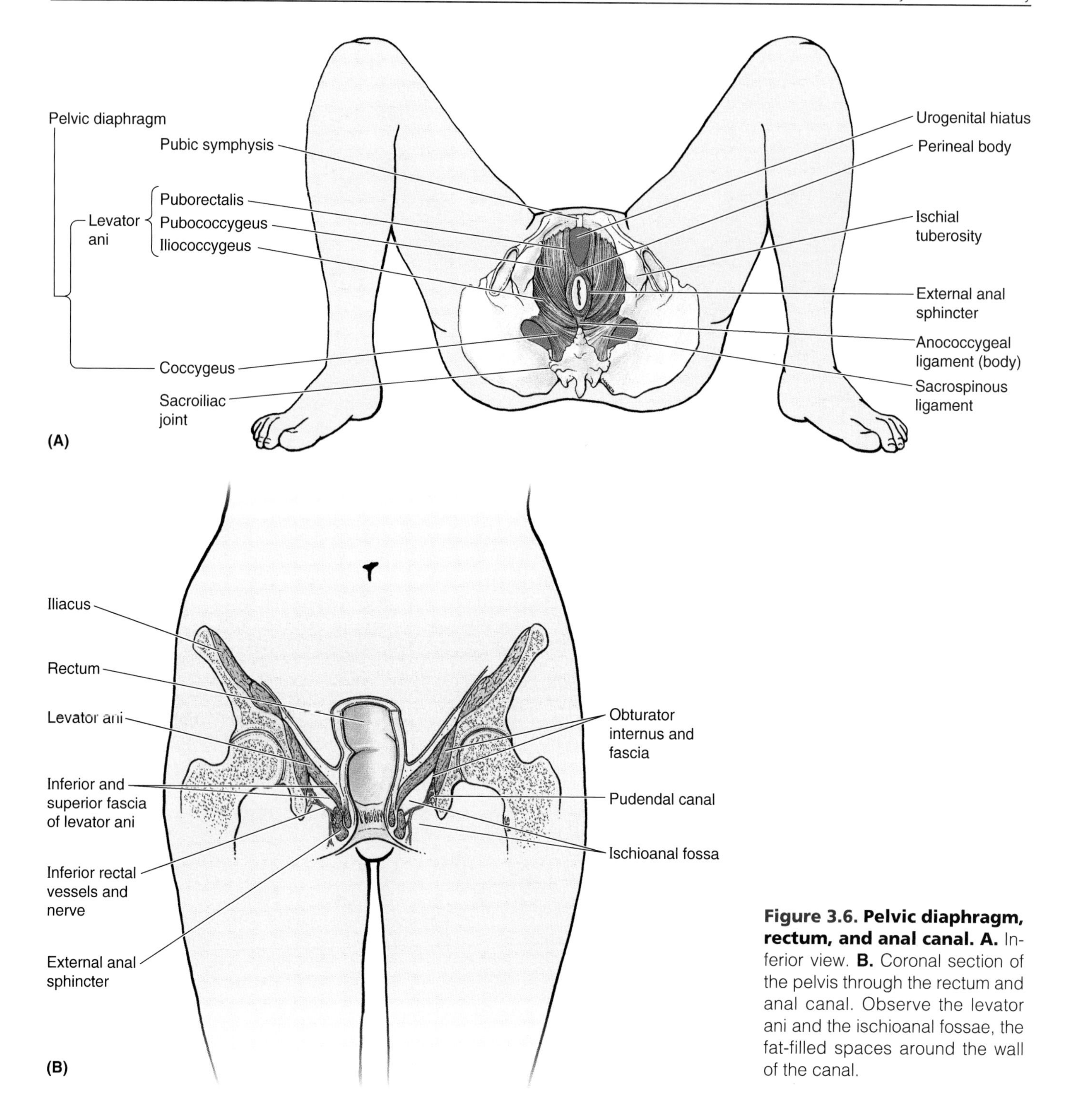

Figure 3.6. Pelvic diaphragm, rectum, and anal canal. A. Inferior view. **B.** Coronal section of the pelvis through the rectum and anal canal. Observe the levator ani and the ischioanal fossae, the fat-filled spaces around the wall of the canal.

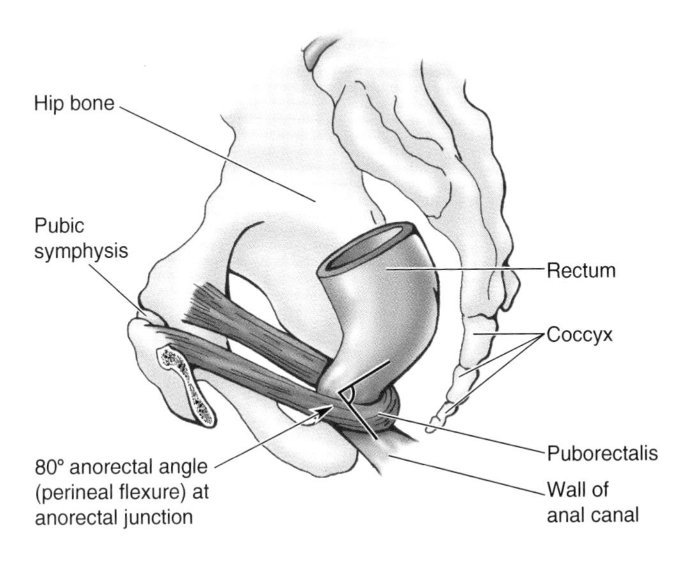

Figure 3.7. Puborectalis muscle. Note that this part of the levator ani unites with its partner to form a U-shaped sling around the anorectal junction. The puborectalis is responsible for the anorectal angle (perineal flexure), which is important in maintaining fecal continence. Relaxation of this muscle during defecation results in straightening of the anorectal junction.

Injury to the Pelvic Floor

During childbirth, the pelvic floor supports the fetal head while the cervix of the uterus is dilating to permit delivery of the fetus. The perineum, levator ani, and pelvic fascia may be injured during childbirth (*A*); it is the pubococcygeus, the main part of the levator ani, that is usually torn (*B*). This part of the muscle is important because it encircles and supports the urethra, vagina, and anal canal. Weakening of the levator ani and pelvic fascia resulting from stretching or tearing during childbirth may alter the position of the neck of the bladder and the urethra. These changes can cause *urinary stress incontinence*, characterized by dribbling of urine when intra-abdominal pressure is raised during coughing and lifting, for instance. ⊕

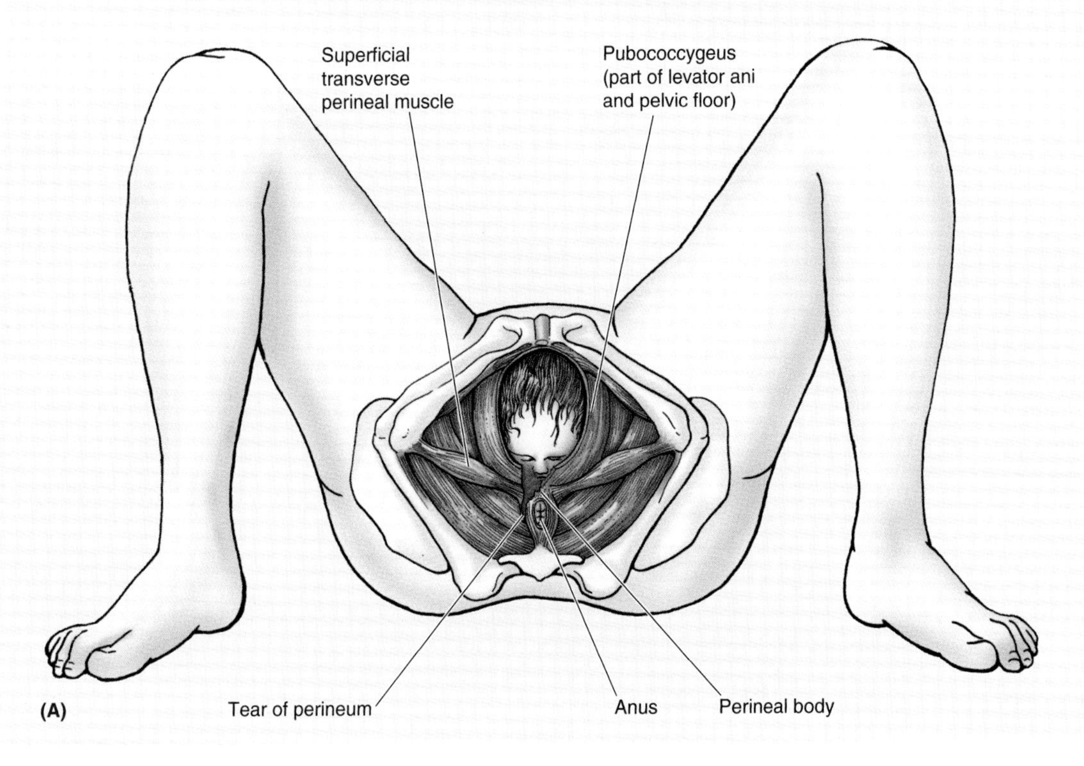

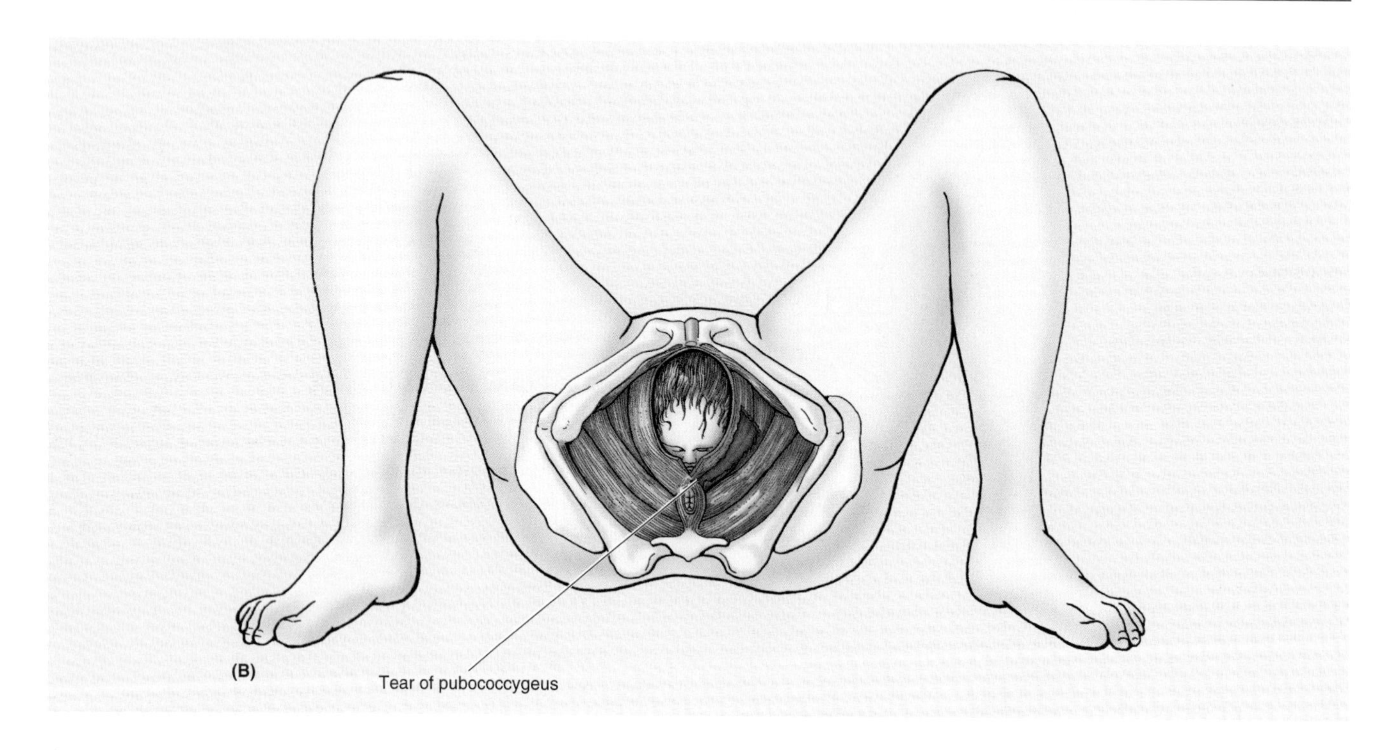

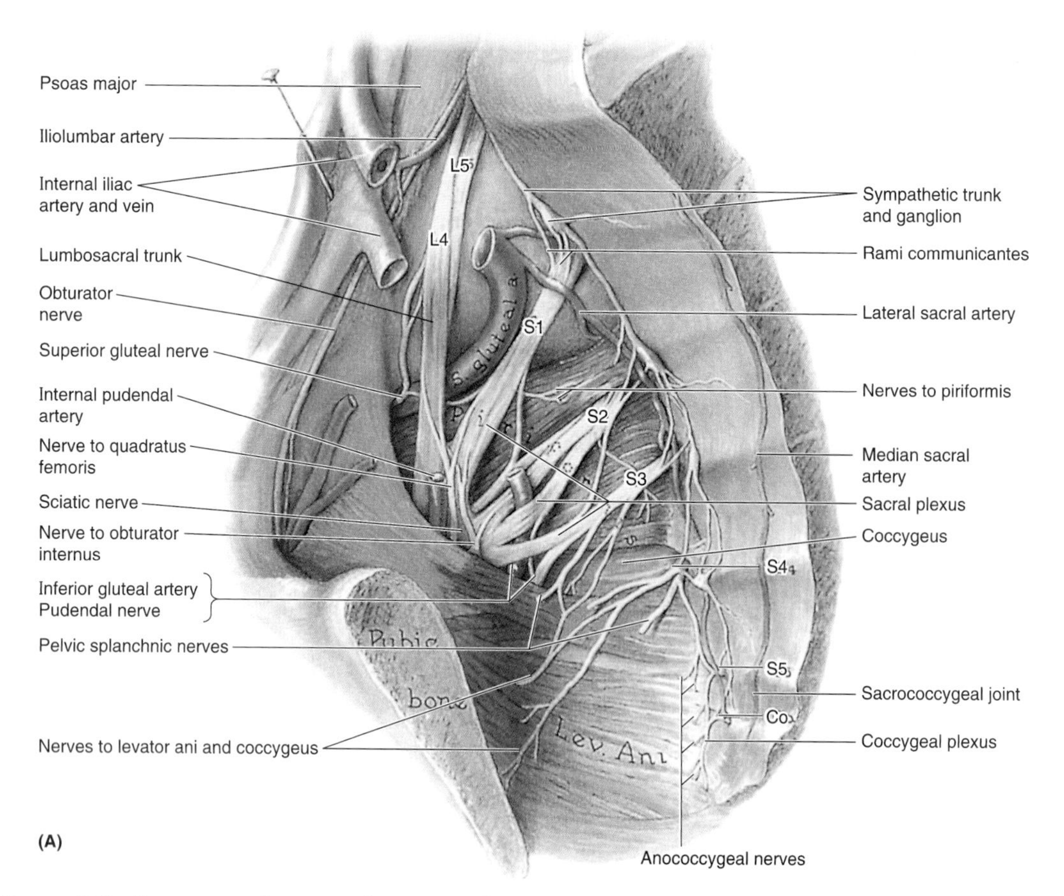

Figure 3.8. Nerves and nerve plexuses of the pelvis. A. Somatic nerves (sacral and coccygeal nerve plexuses) and the pelvic part of the sympathetic trunk. Although located in the pelvis, most of the nerves seen here are involved with the innervation of the lower limb rather than with the pelvic structures.

Pelvic Nerves

The pelvis is innervated mainly by the *sacral and coccygeal nerves* and the pelvic part of the *autonomic nervous system*. The piriformis and coccygeus muscles form a bed for the sacral and coccygeal nerve plexuses (Fig. 3.8*A*). The ventral rami of S2 and S3 nerves emerge between the digitations of these muscles. The descending part of L4 nerve unites with the ventral ramus of L5 nerve to form the thick, cordlike **lumbosacral trunk** (Fig. 3.5*C*). It passes inferiorly, anterior to the ala of the sacrum, where the lumbosacral trunk joins the sacral plexus.

Sacral Plexus

The sacral plexus is located on the posterior wall of the lesser pelvis (Figs. 3.5*D* and 3.8*A*, Table 3.3), where it is closely related to the anterior surface of the piriformis. *The two main nerves of the sacral plexus*—the *sciatic* and *pudendal*—lie external to the parietal pelvic fascia. Most branches of the sacral plexus leave the pelvis through the greater sciatic foramen.

The **sciatic nerve**—the largest and broadest nerve in the body—is formed by the ventral rami of L4 to S3 that converge on the anterior surface of the piriformis. Most commonly, the sciatic nerve passes through the *greater sciatic foramen* inferior to the piriformis to enter the gluteal (buttock) region. It then descends along the posterior aspect of the thigh to supply the posterior aspect of the lower limb (see Chapter 5).

The **pudendal nerve** is derived from the anterior divisions of the ventral rami of S2 through S4. It accompanies the internal pudendal artery and leaves the pelvis through the greater sciatic foramen between the piriformis and coccygeus muscles. *The pudendal nerve—the main nerve of the perineum and the chief sensory nerve of the external genitalia*—hooks around the ischial spine and sacrospinous ligament and enters the perineum through the lesser sciatic foramen. It supplies the skin and muscles of the perineum and, ending as the dorsal nerve of the penis or clitoris, is the chief sensory nerve of the external genitalia.

The **superior gluteal nerve** arises from the posterior divisions of the ventral rami of L4 through S1 and leaves the pelvis through the greater sciatic foramen, superior to the piriformis. It supplies two muscles in the gluteal region—the gluteus medius and minimus—and the tensor of the fascia lata (see Chapter 5).

The **inferior gluteal nerve** arises from the posterior divisions of the ventral rami of L5 through S2 and leaves the pelvis through the greater sciatic foramen, inferior to the piriformis and superficial to the sciatic nerve. It accompanies

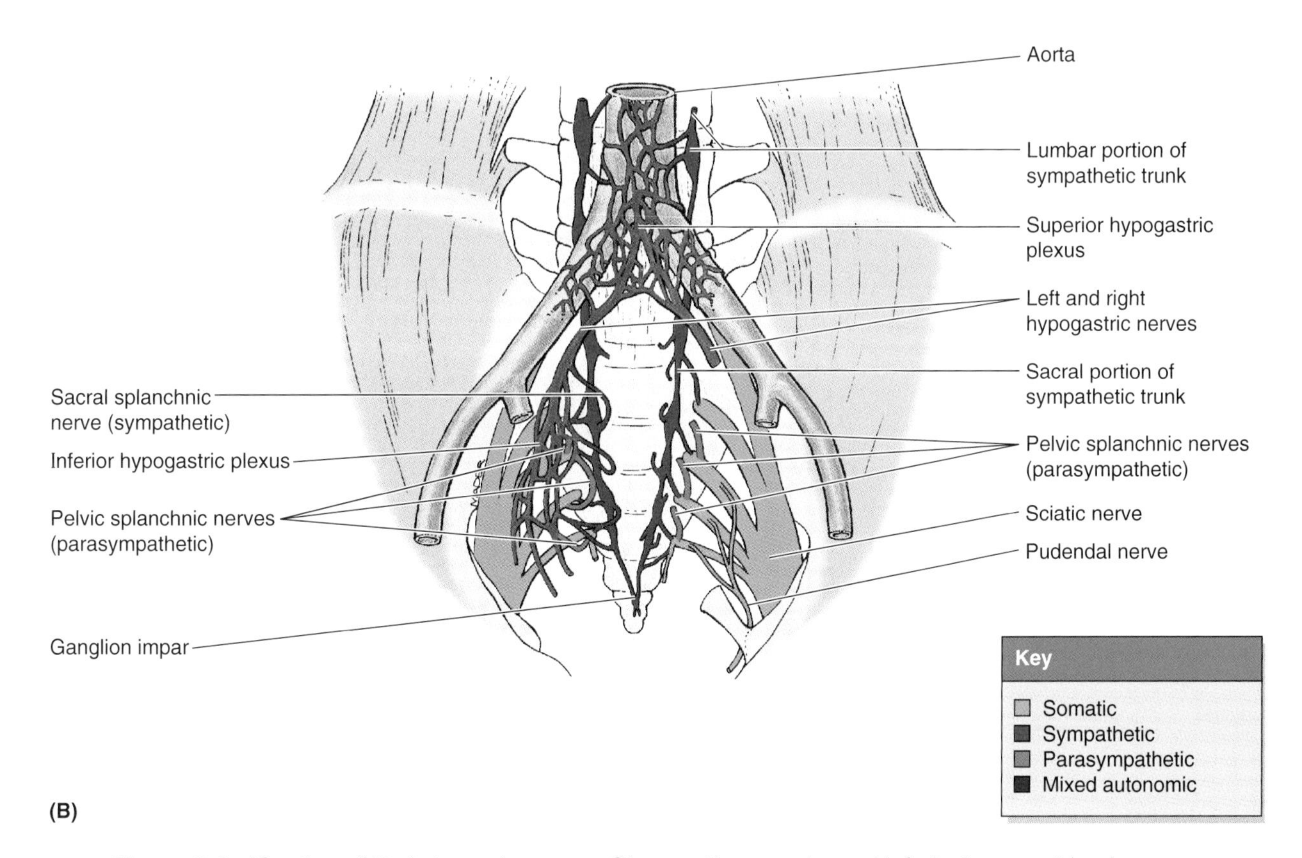

Figure 3.8. *(Continued)* **B.** Autonomic nerves. Observe the superior and inferior hypogastric plexuses and the left and right hypogastric nerves joining the superior hypogastric plexus to the inferior hypogastric plexus. The pelvic plexuses, including the large inferior hypogastric plexus, consist of both sympathetic and parasympathetic fibers.

Table 3.3. **Nerves of Sacral and Coccygeal Plexuses**

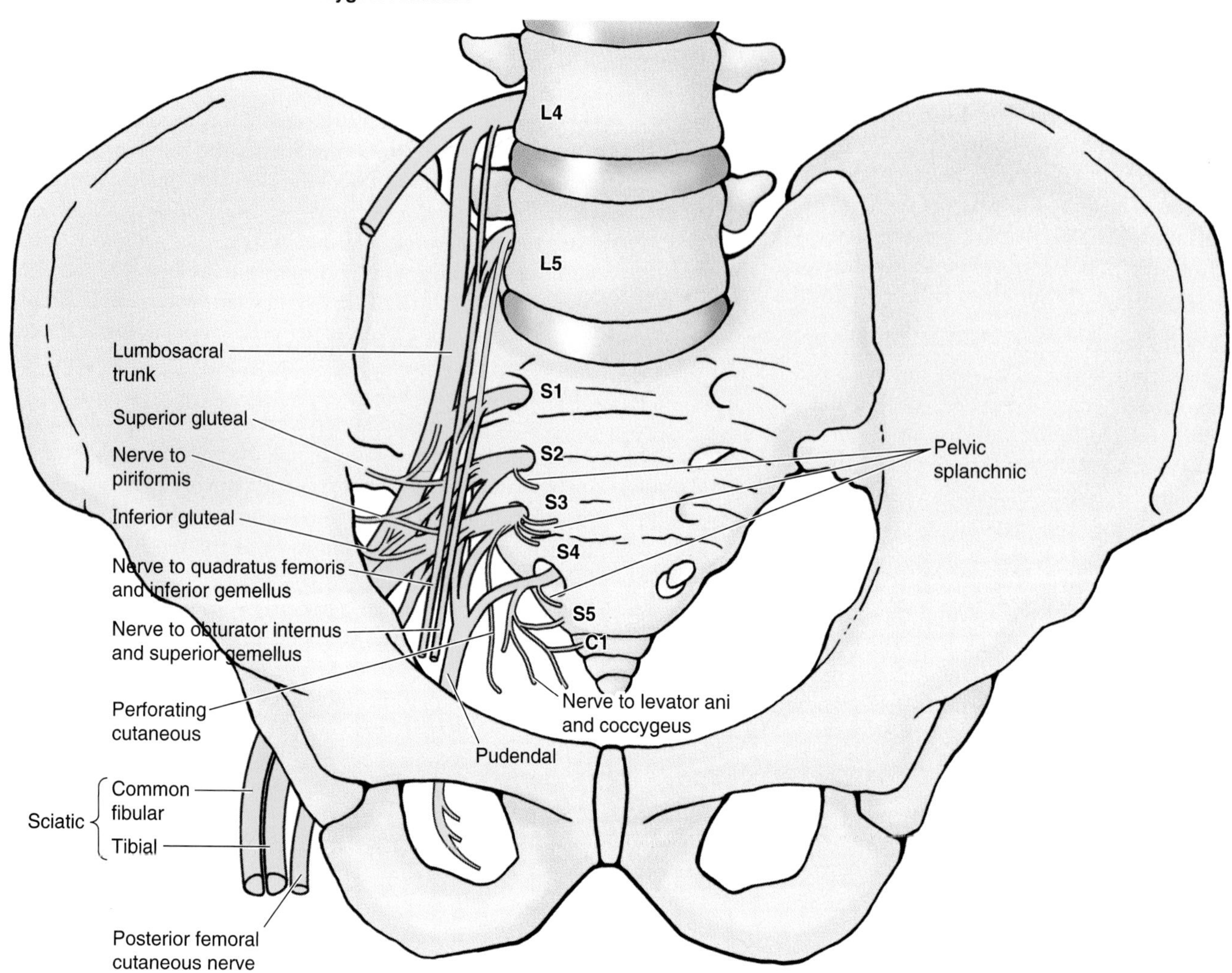

Nerve	Origin	Distribution
Sciatic	L4, L5, S1, S2, S3	Articular branches to hip joint and muscular branches to flexors of knee in thigh and all muscles in leg and foot
Superior gluteal	L4, L5, S1	Gluteus medius and gluteus minimus muscles
Inferior gluteal	L5, S1, S2	Gluteus maximus
Nerve to piriformis	S1, S2	Piriformis muscle
Nerve to quadratus femoris and inferior gemellus	L4, L5, S1	Quadratus femoris and inferior gemellus muscles
Nerve to obturator internus and superior gemellus	L5, S1, S2	Obturator internus and superior gemellus muscles
Pudendal	S2, S3, S4	Structures in perineum: sensory to genitalia; muscular branches to perineal muscles, external urethral sphincter, and external anal sphincter
Nerves to levator ani and coccygeus	S3, S4	Levator ani and coccygeus muscles
Posterior femoral cutaneous	S2, S3	Cutaneous branches to buttock and uppermost medial and posterior surfaces of thigh
Perforating cutaneous	S2, S3	Cutaneous branches to medial part of buttock
Pelvic splanchnic	S2, S3, S4	Pelvic viscera via inferior hypogastric and pelvic plexuses

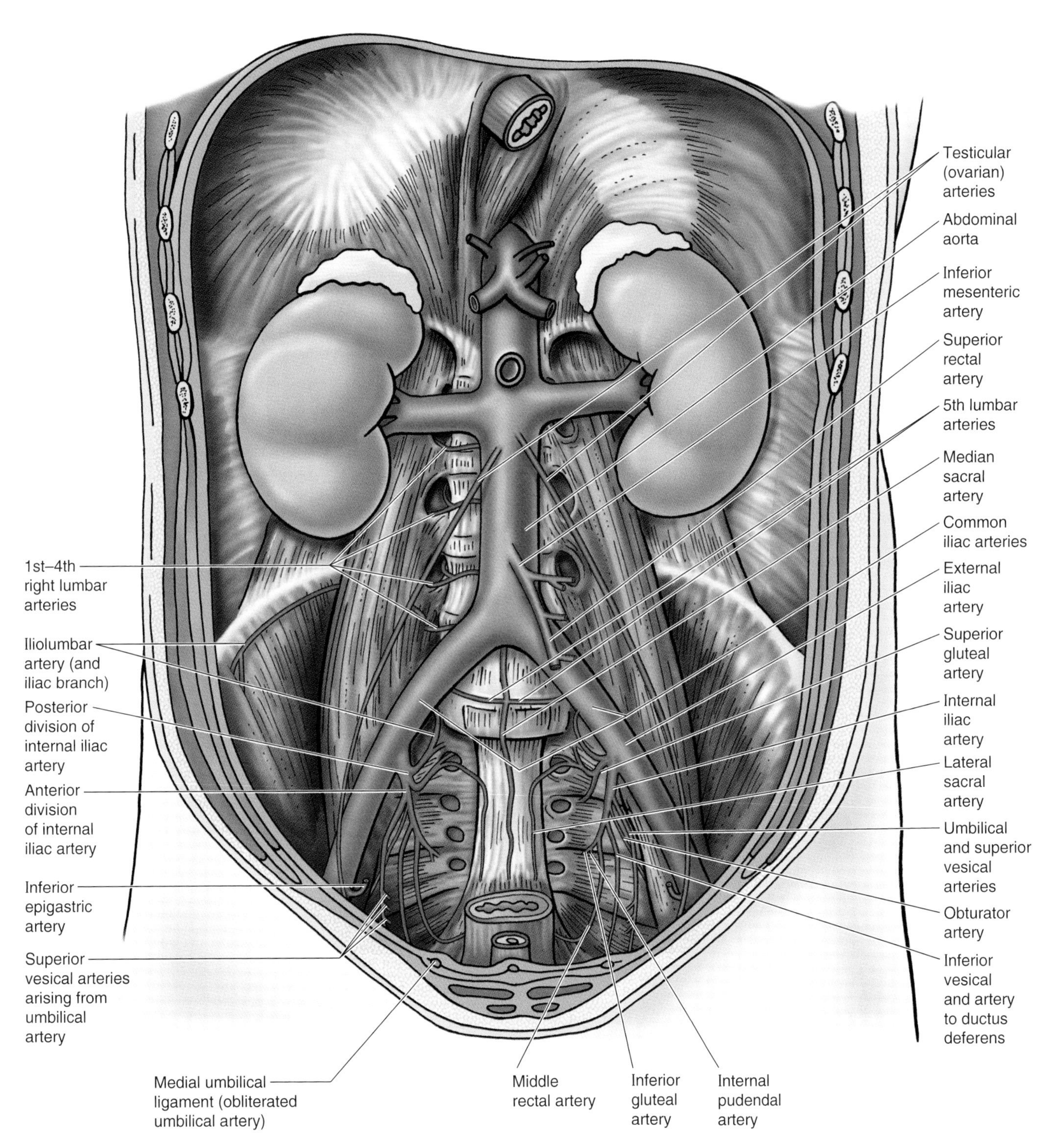

Figure 3.9. Abdominopelvic arteries. Note the high abdominal origin of the gonadal (testicular or ovarian) arteries. The gonads (testes or ovaries) develop near the kidneys and receive their blood supply from the abdominal aorta; they descend to their pelvic/perineal locations before birth, retaining their abdominal blood supply. The internal iliac arteries, providing most of the blood supply to the pelvis, arise from the common iliac arteries anterior to the sacroiliac joints at the level of the L5 and S1 intervertebral (IV) disc. The internal iliac arteries commonly divide into two main divisions, anterior and posterior.

the inferior gluteal artery and breaks up into several branches that supply the overlying gluteus maximus muscle (see Chapter 5).

Obturator Nerve

The obturator nerve arises from the *lumbar plexus* (L2–L4) in the abdomen (greater pelvis) and enters the lesser pelvis (Figs. 3.5*D* and 3.8*A*). It runs in the extraperitoneal fat along the lateral wall of the pelvis to the *obturator canal*, where it divides into anterior and posterior parts that leave the pelvis through this canal and supply the medial thigh muscles (see Chapter 5). The obturator canal is the opening in the obturator membrane (that otherwise fills the obturator foramen) through which the obturator nerves pass from the pelvic cavity into the thigh.

Injury to the Pelvic Nerves

During childbirth the fetal head may compress the nerves of the mother's sacral plexus, producing pain in the lower limbs. The *obturator nerve* is vulnerable to injury during surgery (e.g., during removal of cancerous lymph nodes from the lateral pelvic wall). Injury to this nerve may cause painful spasms of the adductor muscles of the thigh and sensory deficits in the medial thigh region (see Chapter 5). ⊙

Coccygeal Plexus

The coccygeal plexus is a small network of nerve fibers formed by the ventral rami of S4 and S5 and the **coccygeal nerves** (Fig. 3.8*A*). It lies on the pelvic surface of the coccygeus and supplies this muscle, part of the levator ani, and the sacrococcygeal joint. The **anococcygeal nerves** arising from this plexus pierce the sacrotuberous ligament and supply a small area of skin in the coccygeal region.

Pelvic Autonomic Nerves

The **sacral sympathetic trunks** are the inferior continuation of the lumbar sympathetic trunks (Fig. 3.8, *A* and *B*). Each of the sacral trunks is diminished in size from that of the lumbar trunks and usually has four sympathetic ganglia. The sacral trunks descend on the pelvic surface of the sacrum just medial to the pelvic sacral foramina, and converge to form the small median **ganglion impar** (coccygeal ganglion) anterior to the coccyx. The sacral sympathetic trunks descend posterior to the rectum in the extraperitoneal connective tissue and send gray rami communicantes (communicating branches) to each of the ventral rami of the sacral and coccygeal nerves. They also send small branches to the median sacral artery and the inferior hypogastric plexus. *The primary function of the sacral sympathetic trunks is to provide postsynaptic fibers to the sacral plexus for sympathetic innervation of the lower limb* (vasomotor, pilomotor, and sudomotor).

The **hypogastric plexuses** (superior and inferior) are networks of autonomic nerves. The main part of the **superior hypogastric plexus** lies just inferior to the bifurcation of the aorta (Fig. 3.8*B*) and descends into the pelvis. This plexus is the inferior prolongation of the *intermesenteric plexus* (see Chapter 2), which also receives the L3 and L4 splanchnic nerves. Branches from the superior hypogastric plexus enter the pelvis and descend anterior to the sacrum as the left and right **hypogastric nerves.** *In males* these nerves descend lateral to the rectum within the *hypogastric sheaths* (see the subsequent discussion of the pelvic fascia) and then spread in a fanlike fashion as the **inferior hypogastric plexuses.** Extensions of these plexuses, collectively referred to as the *pelvic plexuses*, pass to the prostate, seminal vesicles (seminal glands), and inferolateral surfaces of the urinary bladder. *In females* the cervix of the uterus and the lateral fornices of the vagina take the place of the prostate and seminal vesicles in relation to these plexuses.

The **pelvic splanchnic nerves** (Fig. 3.8*B*, Table 3.3) contain parasympathetic fibers derived from S2, S3, and S4 spinal cord segments and visceral afferent fibers from cell bodies in the spinal ganglia of the corresponding spinal nerves. The contribution from the third sacral nerve is usually the largest. The pelvic splanchnic nerves merge with the hypogastric nerves to form the inferior hypogastric (and pelvic) plexuses. The **inferior hypogastric plexuses** thus contain both sympathetic and parasympathetic fibers, which pass along the branches of the internal iliac arteries to form subplexuses (rectal, vesical, uterovaginal) on the pelvic viscera.

Pelvic Arteries

Four main arteries enter the lesser pelvis. The internal iliac and ovarian arteries are paired, and the median sacral and superior rectal arteries are unpaired.

Internal Iliac Artery

Each internal iliac artery, approximately 4 cm long, begins anterior to the sacroiliac joint at the bifurcation of the **common iliac artery** and descends posteriorly to the greater sciatic foramen. *The internal iliac artery is the artery of the pelvis*; however, it also supplies branches to the buttocks, medial thigh regions, and the perineum. The internal iliac artery supplies most of the blood to the pelvic viscera, as well as supplying the musculoskeletal part of the pelvis and the gluteal region (Figs. 3.9 and 3.10, Table 3.4).

The internal iliac artery begins at the level of the IV disc between L5 and S1 vertebrae, where it is crossed by the ureter (Fig. 3.10*C*). It is separated from the sacroiliac joint by the internal iliac vein and the lumbosacral trunk. The internal iliac artery passes posteromedially into the lesser pelvis, medial to the external iliac vein and obturator nerve, and lateral to the peritoneum. Although variations are frequent, the internal iliac artery commonly ends at the superior edge of the greater sciatic foramen by dividing into anterior and posterior divisions (trunks). The branches of the anterior division are mainly visceral (i.e., they supply the bladder, rectum, and reproductive organs). It also has two parietal branches that pass to the buttock and thigh. The arrangement of the visceral branches is variable.

The following are branches of the anterior division of the internal iliac artery (Figs. 3.9–3.11).

Umbilical Artery. This vessel runs anteroinferiorly between the urinary bladder and the lateral wall of the pelvis. It gives off the **superior vesical artery** (or arteries) that supplies (or arise as) numerous branches to the fundus of the urinary bladder. Before birth the umbilical arteries carry blood from the fetus to the placenta for reoxygenation. Postnatally their distal parts atrophy and form fibrous cords—the *medial umbilical ligaments* (Figs. 3.9 and 3.10, *A* and *B*). They raise folds of peritoneum—the *medial umbilical folds*—that run on the deep surface of the anterior abdominal wall (see Chapter 2).

Obturator Artery. The origin of this vessel is variable; usually it arises close to the umbilical artery, where it is crossed by the ureter. It then runs anteroinferiorly on the obturator fascia on the lateral wall of the pelvis and passes between the obturator nerve and vein. It then leaves the pelvis through the *obturator canal* and supplies muscles of the thigh. Within the pelvis, the obturator artery gives off muscular branches, a nutrient artery to the ilium, and a pubic branch. The *pubic branch* arises just before the obturator artery leaves the pelvis. It ascends on the pelvic surface of the pubis to anastomose with its fellow of the opposite side and the pubic branch of the *inferior epigastric artery*, a branch of the external iliac. In a common variation (20%), *an aberrant obturator artery* arises from the inferior epigastric artery and descends into the pelvis along the usual route of the pubic branch (Williams et al., 1995). The extrapelvic distribution of the obturator artery is described with the lower limb (see Chapter 5).

Inferior Vesical Artery. This vessel occurs only in males (Fig. 3.10*A*). The inferior vesical artery passes to the fundus of the urinary bladder, where it supplies the seminal vesicles, prostate, the fundus of the bladder, and the inferior part of the ureter. The branches to the ductus deferens (vas deferens) and prostate are the *artery of the ductus deferens* and the *prostatic artery*. The artery of the ductus deferens may arise from the superior vesical artery.

Middle Rectal Artery. The middle rectal artery may arise independently from the internal iliac artery or it may arise in common with the inferior vesical artery or the internal pudendal artery. The middle rectal artery supplies the inferior part of the rectum, anastomosing with the superior and inferior rectal arteries, and also supplies the seminal vesicles, prostate, and vagina.

Vaginal Artery. This vessel is the female homologue to the inferior vesical artery in the male. It runs anteriorly and then passes along the side of the vagina (Figs. 3.10*B* and 3.11*B*), where it supplies numerous branches to the anterior and posterior surfaces of the vagina, posteroinferior parts of the urinary bladder, and the pelvic part of the urethra. It anastomoses with the vaginal branch of the uterine artery.

Uterine Artery. This vessel usually arises separately and directly from the internal iliac artery (Fig. 3.10*B*), but it may arise from the umbilical artery. It is the female homologue to the artery of the ductus deferens in the male. It descends on the lateral wall of the pelvis, anterior to the internal iliac artery, and enters the root of the broad ligament where it passes superior to the lateral part of the fornix of the vagina to reach the lateral margin of the uterus (Fig. 3.11, *A* and *B*).

The uterine artery passes anterior and superior to the ureter near the lateral part of the fornix of the vagina. On reaching the side of the cervix, the uterine artery divides into a large superior branch that supplies the body and fundus of the uterus and a smaller vaginal branch that supplies the cervix and vagina. The uterine artery pursues a tortuous course along the lateral margin of the uterus and ends when its ovarian and tubal branches anastomose with the ovarian and tubal branches of the ovarian artery between the layers of the broad ligament.

Internal Pudendal Artery. This vessel, larger in males than in females, passes inferolaterally, anterior to the piriformis muscle and sacral plexus (Fig. 3.10). It leaves the pelvis

Internal Iliac Ligation

Occasionally, the internal iliac artery is ligated to control pelvic hemorrhage. Ligation does not stop bloodflow but reduces blood pressure, allowing hemostasis (arrest of bleeding) to occur. Because of three arterial anastomoses (lumbar to iliolumbar, median sacral to lateral sacral, and superior rectal to middle rectal), bloodflow in the artery is maintained although reversed. This maintains the blood supply to the pelvic viscera, gluteal region, and genital organs.

Iatrogenic Injury to the Ureters

The fact that the uterine artery crosses anterior and superior to the ureter near the lateral part of the fornix of the vagina is clinically important. The ureter is in danger of being inadvertently clamped, ligated, or severed during a **hysterectomy** (excision of uterus) when the uterine artery is tied off. The point of crossing of the uterine artery and ureter lies approximately 2 cm superior to the ischial spine. The left ureter is particularly vulnerable because it passes close to the lateral aspect of the cervix. The ureter is also vulnerable to injury when the ovarian vessels are tied off during an **ovariectomy** (excision of ovary) because these structures lie close to each other where they cross the pelvic brim. ○

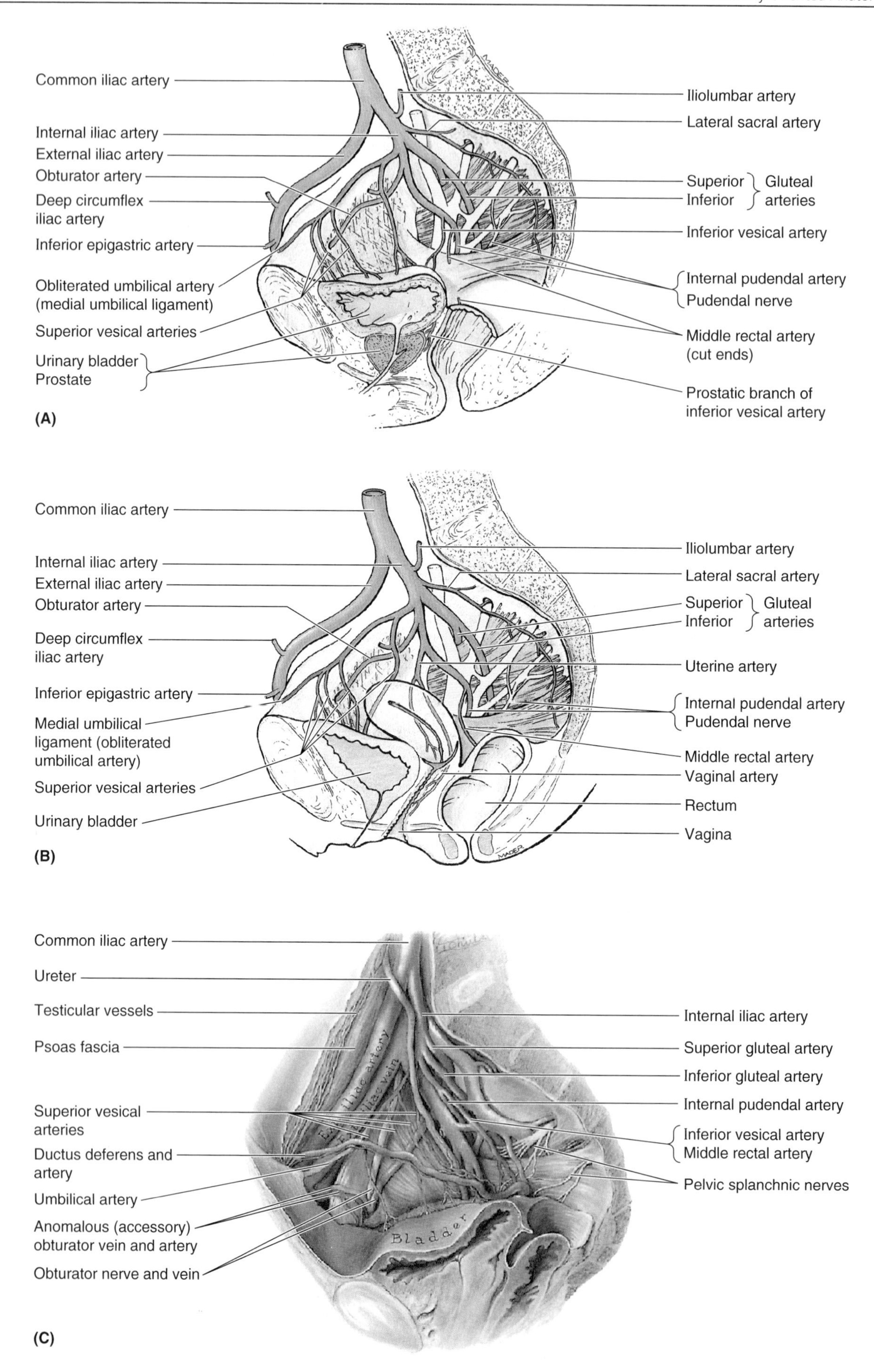
Common iliac artery
Internal iliac artery
External iliac artery
Obturator artery
Deep circumflex iliac artery
Inferior epigastric artery
Obliterated umbilical artery (medial umbilical ligament)
Superior vesical arteries
Urinary bladder
Prostate
(A)
Iliolumbar artery
Lateral sacral artery
Superior
Inferior
Gluteal arteries
Inferior vesical artery
Internal pudendal artery
Pudendal nerve
Middle rectal artery (cut ends)
Prostatic branch of inferior vesical artery
Common iliac artery
Internal iliac artery
External iliac artery
Obturator artery
Deep circumflex iliac artery
Inferior epigastric artery
Medial umbilical ligament (obliterated umbilical artery)
Superior vesical arteries
Urinary bladder
(B)
Iliolumbar artery
Lateral sacral artery
Superior
Inferior
Gluteal arteries
Uterine artery
Internal pudendal artery
Pudendal nerve
Middle rectal artery
Vaginal artery
Rectum
Vagina
Common iliac artery
Ureter
Testicular vessels
Psoas fascia
Superior vesical arteries
Ductus deferens and artery
Umbilical artery
Anomalous (accessory) obturator vein and artery
Obturator nerve and vein
(C)
Internal iliac artery
Superior gluteal artery
Inferior gluteal artery
Internal pudendal artery
Inferior vesical artery
Middle rectal artery
Pelvic splanchnic nerves
Bladder

Table 3.4. **Arteries of the Pelvis**

Artery	Origin	Course	Distribution
Gonadal–testicular (♂) or ovarian (♀)	Abdominal aorta	Descends retroperitoneally; testicular artery passes through inguinal canal into scrotum; ovarian artery crosses pelvic brim, coursing medially in suspensory ligament to ovary	Testis (♂) or ovary (♀)
Superior rectal	Continuation of inferior mesenteric artery	Crosses left common iliac vessels and descends into the pelvis between the layers of the sigmoid mesocolon	Upper part of rectum; anastomoses with middle and inferior rectal arteries
Median sacral	Posterior aspect of abdominal aorta	Descends in median line over L4 and L5 vertebrae and the sacrum and coccyx	Lower lumbar vertebrae, sacrum, and coccyx
Internal iliac	Common iliac	Passes over pelvic brim to reach pelvic cavity	Main blood supply to pelvic organs, gluteal muscles, and perineum
Anterior division of internal iliac	Internal iliac	Passes anteriorly and divides into visceral branches and obturator artery	Pelvic viscera and muscles in medial compartment of thigh
Umbilical	Anterior division of internal iliac	Obliterates becoming medial umbilical ligament after running a short pelvic course during which it gives rise to superior vesical arteries	Superior aspect of urinary bladder; occasionally artery to ductus deferens (males)
Superior vesical	Patent (proximal) part of umbilical	Usually multiple, these arteries pass to the superior aspect of the urinary bladder	Superior aspect of urinary bladder, pelvic portion of ureter
Obturator	Anterior division of internal iliac	Runs anteroinferiorly on lateral pelvic wall to exit pelvis via obturator canal	Pelvic muscles, nutrient artery to ilium, head of femur, muscles of medial compartment of thigh
Inferior vesical (♂)	Anterior division of internal iliac	Passes retroperitoneally to inferior aspect of male urinary bladder	Inferior aspet of urinary bladder, ductus deferens, seminal vesicle, and prostate
Artery to ductus deferens (♂)	Inferior (or superior) vesical	Runs retroperitoneally to ductus deferens	Ductus deferens
Prostatic branches (♂)	Inferior vesical artery	Descends on posterolateral aspect of prostate	Prostate
Uterine (♀)	Anterior division of internal iliac	Runs medially in base of broad ligament superior to cardinal ligament, crossing superior to ureter, to sides of uterus	Uterus, ligaments of uterus, uterine tube, and vagina
Vaginal (♀)	Uterine artery	Arises lateral to ureter and descends inferior to it to lateral aspect of vagina	Vagina; branches to inferior part of urinary bladder and termination of ureter
Internal pudendal	Anterior division of internal iliac	Leaves pelvis through greater sciatic foramen and enters perineum (ischioanal fossa) by passing through lesser sciatic foramen	Main artery to perineum, including muscles and skin of anal and urogenital triangles; erectile bodies
Middle rectal	Anterior division of internal iliac	Descends in pelvis to lower part of rectum	Seminal vesicles and lower part of rectum
Inferior gluteal	Anterior division of internal iliac	Exits pelvis via greater sciatic foramen, passing inferior to piriformis	Pelvic diaphragm (coccygeus and levator ani), piriformis, quadratus femoris, uppermost hamstrings, gluteus maximus, sciatic nerve
Posterior division of internal iliac	Internal iliac	Passes posteriorly and gives rise to parietal branches	Pelvic wall and gluteal region
Iliolumbar	Posterior division of internal iliac	Ascends anterior to sacroiliac joint and posterior to common iliac vessels and psoas major	Psoas major, iliacus and quadratus lumborum muscles, cauda equina in vertebral canal
Lateral sacral (superior and inferior)	Posterior division of internal iliac	Runs on anteromedial aspect of piriformis to send branches into pelvic sacral foramina	Piriformis, structures in sacral canal, erector spinae and overlying skin
Superior gluteal	Posterior division of internal iliac	Exits pelvis via greater sciatic foramen, passing superior to piriformis	Piriformis, all 3 gluteal muscles, tensor fascia lata

Figure 3.10. Iliac arteries and their branches. A. Dissection of the lateral wall of a male pelvis. Observe that the common iliac artery has two terminal branches—external and internal iliac arteries—but no collateral branches. **B.** Internal iliac arteries and their branches in a female. **C.** Dissection of the lateral wall of a male pelvis. Observe the ureter and ductus deferens running a subperitoneal course across the external iliac vessels, umbilical artery, and obturator nerve and vessels. In this specimen the obturator artery is a common variation, taking its origin from the inferior epigastric artery. It is important for surgeons operating in this area to be aware of this occurrence, found in 1 of 5 individuals.

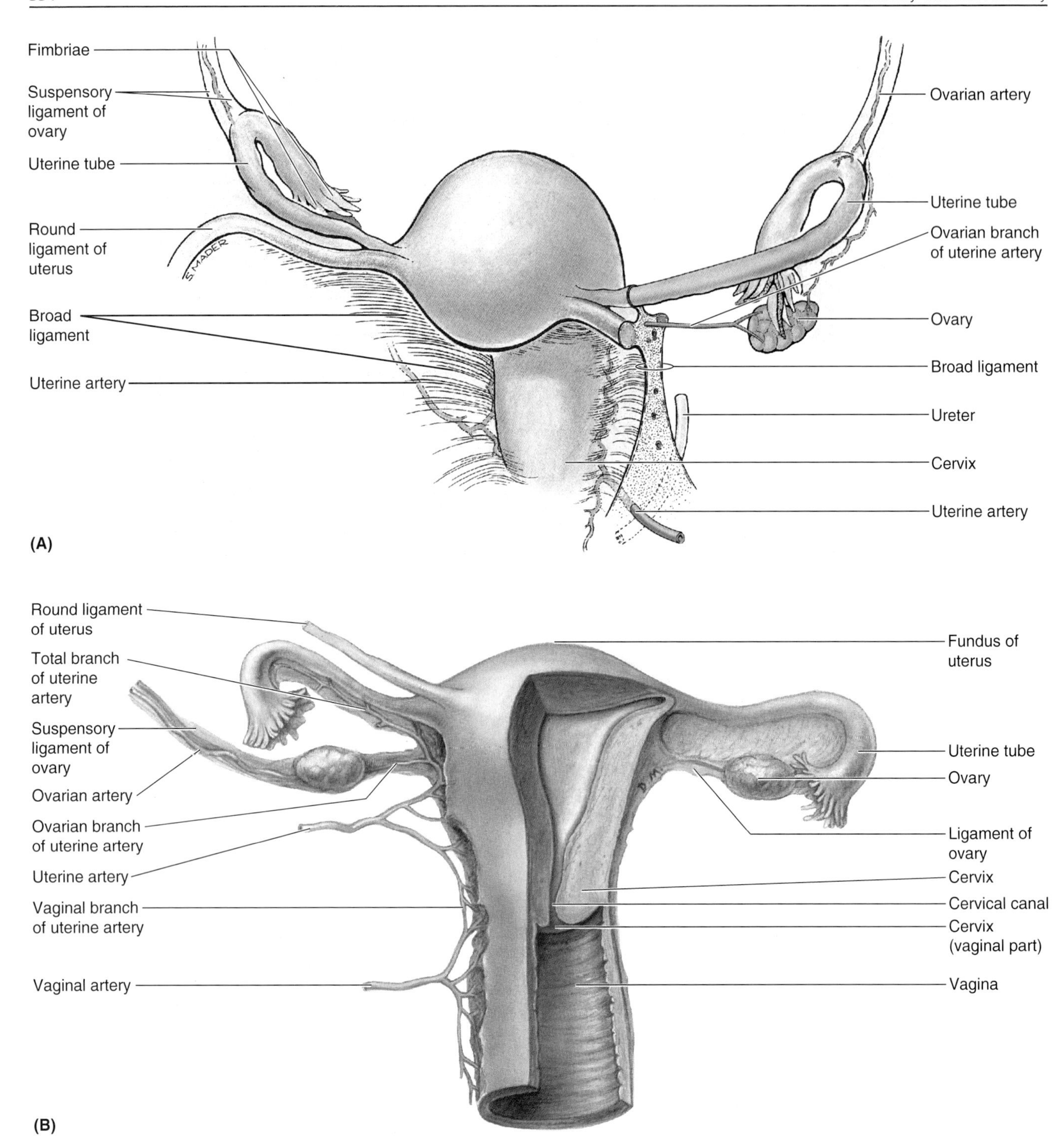

Figure 3.11. Broad ligament and blood supply of the uterus, vagina, and ovaries. **A.** Paramedian section of the left broad ligament to show the structure of this mesentery of the uterus. **B.** Anterior view of the female internal genital organs after removal of most of the broad ligament (i.e., the mesometrium). Part of the uterine and vaginal walls have been cut away to reveal their structure. Note the anastomosis between the ovarian branches of the ovarian and uterine arteries, and between the vaginal branch of the uterine artery and the vaginal artery. These communications occur between the layers of the broad ligament (see (**A**)).

between the piriformis and coccygeus muscles by passing through the inferior part of the *greater sciatic foramen*. The internal pudendal artery then passes around the posterior aspect of the ischial spine or the sacrospinous ligament and enters the *ischioanal fossa* through the lesser sciatic foramen. The internal pudendal artery, along with the internal pudendal veins and branches of the pudendal nerve, passes through the pudendal canal in the lateral wall of the ischioanal fossa (Fig. 3.6*B*). As it exits the pudendal canal, medial to the ischial tuberosity, the internal pudendal artery divides into its terminal branches: *the deep and dorsal arteries of the penis or clitoris*.

Inferior Gluteal Artery. This vessel passes posteriorly between the sacral nerves (usually S2 and S3) and leaves the pelvis through the inferior part of the *greater sciatic foramen*, inferior to the piriformis muscle. It supplies the muscles and skin of the buttock and the posterior surface of the thigh.

The following three arteries are branches of the posterior division of the internal iliac artery.

Superior Gluteal Artery. This large vessel passes posteriorly and runs between the lumbosacral trunk and the ventral ramus of S1 nerve (Fig. 3.10). It leaves the pelvis through the superior part of the *greater sciatic foramen*, superior to the piriformis muscle, to supply the gluteal muscles in the buttocks.

Iliolumbar Artery. This vessel runs superolaterally in a recurrent fashion to the iliac fossa (Fig. 3.10, *A* and *B*). Within the fossa, the iliolumbar artery divides into an *iliac branch* that supplies the iliacus muscle and ilium and a *lumbar branch* that supplies the psoas major and quadratus lumborum muscles.

Lateral Sacral Arteries. These vessels, usually superior and inferior ones on each side, may arise from a common trunk (Fig. 3.10, *A* and *B*). The lateral sacral arteries pass medially and descend anterior to the sacral ventral rami, giving off spinal branches that pass through the anterior sacral foramina and *supply the spinal meninges* enclosing the roots of the sacral nerves. Some branches of these arteries pass from the sacral canal through the posterior sacral foramina and supply the erector spinae muscles and skin overlying the sacrum.

Ovarian Artery

The ovarian artery arises from the abdominal aorta inferior to the renal artery, but considerably superior to the inferior mesenteric artery. As it passes inferiorly, *the ovarian artery adheres to the parietal peritoneum and runs anterior to the ureter on the posterior abdominal wall, usually giving branches to it*. As it enters the lesser pelvis, it crosses the origin of the external iliac vessels. It then runs medially in the suspensory ligament of the ovary and enters the superolateral part of the broad ligament to divide into an ovarian and a tubal branch that supply the ovary and uterine tube, respectively (Fig. 3.11). These branches anastomose with the corresponding branches of the uterine artery.

Median Sacral Artery

This small unpaired artery usually arises from the posterior surface of the abdominal aorta, just superior to its bifurcation, but it may arise from its anterior surface (Fig. 3.9). This vessel runs anterior to the bodies of the last one or two lumbar vertebrae, the sacrum, and the coccyx and ends in a series of anastomotic loops. Before the median sacral artery enters the lesser pelvis, it sometimes gives rise to a pair of *fifth lumbar arteries*. As it descends over the sacrum, the median sacral artery gives off small parietal (lateral sacral) branches that anastomose with the lateral sacral arteries. It also gives rise to small visceral branches to the posterior part of the rectum, which anastomose with the superior and middle rectal arteries. The median sacral artery represents the caudal end of the embryonic dorsal aorta that reduced in size as the tail of the embryo disappeared.

Superior Rectal Artery

This vessel is the direct continuation of the inferior mesenteric artery (Fig. 3.9). It crosses the left common iliac vessels and descends in the sigmoid mesocolon to the lesser pelvis. At the level of S3 vertebra, the superior rectal artery divides into two branches that descend on each side of the rectum and supply it as far inferiorly as the internal anal sphincter. The superior rectal artery anastomoses with branches of the middle rectal artery (a branch of the internal iliac artery) and with the inferior rectal artery (a branch of the internal pudendal artery).

Pelvic Veins

The pelvis is drained mainly by the **internal iliac veins** and their tributaries (Fig. 3.12), but some drainage occurs through the superior rectal, median sacral, and ovarian veins. Some blood from the pelvis also passes to the *internal vertebral venous plexuses* (see Chapter 4).

The **internal iliac vein** joins the external iliac vein to form the **common iliac vein**, which unites with its partner at the level of L5 vertebra to form the **inferior vena cava** (Fig. 3.12*A*). The internal iliac vein lies posteroinferior to the internal iliac artery, and its tributaries are similar to the branches of this artery except that there are no veins accompanying the umbilical arteries between the pelvis and the umbilicus.

The **iliolumbar veins** usually drain into the common iliac veins. The **superior gluteal veins**, the accompanying veins (L. venae comitantes) of the superior gluteal arteries, are the largest tributaries of the internal iliac veins except during pregnancy, when the uterine veins become larger.

Pelvic venous plexuses are formed by the interjoining of veins in the pelvis (Fig. 3.12, *B* and *C*). These *intercommunicating networks of veins* are clinically important. The various plexuses (rectal, vesical, prostatic, uterine, and vaginal) unite and drain mainly into the **internal iliac vein**, but some of them drain through the superior rectal vein into the inferior mesenteric vein or through lateral sacral veins into the internal vertebral venous plexus.

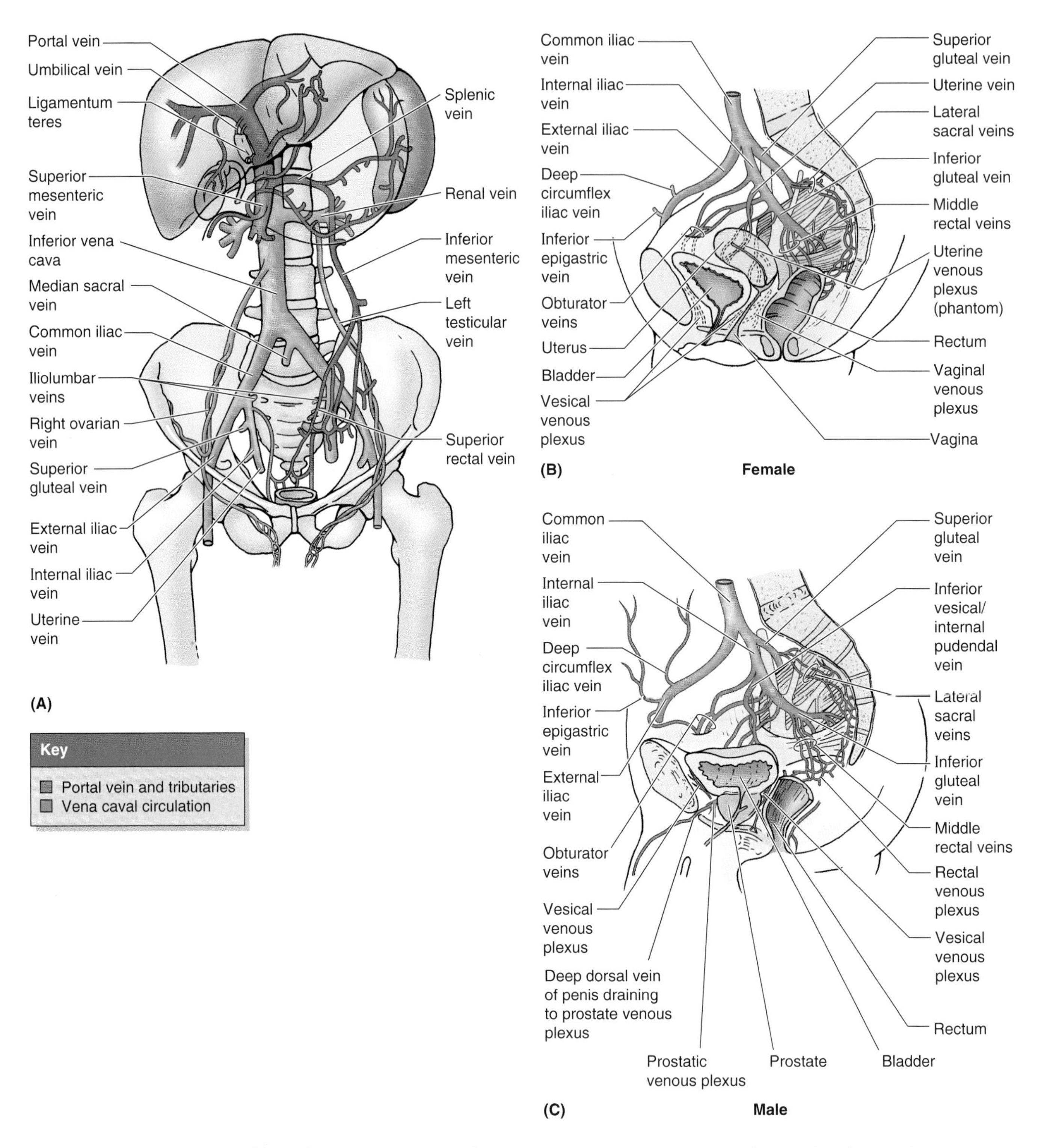

Figure 3.12. Pelvic veins. A. Anterior view of the abdominopelvic cavity showing the portal and systemic (vena caval) venous systems. Because of anastomoses (portal-systemic or portacaval anastomoses) between the two systems, in portal hypertension—occurring in hepatic cirrhosis (see Chapter 2)—the voluminous portal blood, which cannot drain freely through the liver, attempts to pass through these relatively small anastomoses, and the tiny anastomotic veins become engorged, dilated, varicose, and, consequently, may rupture. **B.** Pelvic veins and venous plexuses in a dissection of a female pelvis. **C.** Pelvic veins and venous plexuses in a dissection of a male pelvis.

Viscera of Pelvic Cavity

The pelvic viscera include the urinary bladder and parts of the ureters and reproductive system, as well as the inferior part of the intestinal tract (rectum). Although the sigmoid colon and parts of the small bowel extend into the pelvic cavity, they are not pelvic viscera. The sigmoid colon is continuous with the rectum anterior to S3 vertebra.

Urinary Organs

The pelvic urinary organs are the:

- Ureters, which carry urine from the kidneys
- Urinary bladder, which temporarily stores urine
- Urethra, which conducts urine from the bladder to the exterior.

Ureters

The ureters are muscular tubes, 25 to 30 cm long, that connect the kidneys to the urinary bladder (Fig. 3.13), draining the product of the kidneys (urine) to its temporary storage site. The ureters are retroperitoneal; their superior halves are in the abdomen and their inferior halves lie in the pelvis. As the ureters leave the abdomen and enter the lesser pelvis, they pass over the pelvic brim. The pelvic part of the ureter begins where it crosses the bifurcation of the common iliac artery or the beginning of the external iliac artery (Fig. 3.13*A*). The ureters run posteroinferiorly on the lateral walls of the pelvis, external to the parietal pelvic peritoneum and anterior to the internal iliac arteries. They then curve anteromedially, superior to the levator ani, to enter the urinary bladder. The inferior end of each ureter is surrounded by the **vesical venous plexus** (Fig. 3.12, *B* and *C*). The ureters run obliquely through a gap in the muscular wall of the urinary bladder. Their oblique passage through the bladder wall forms a one-

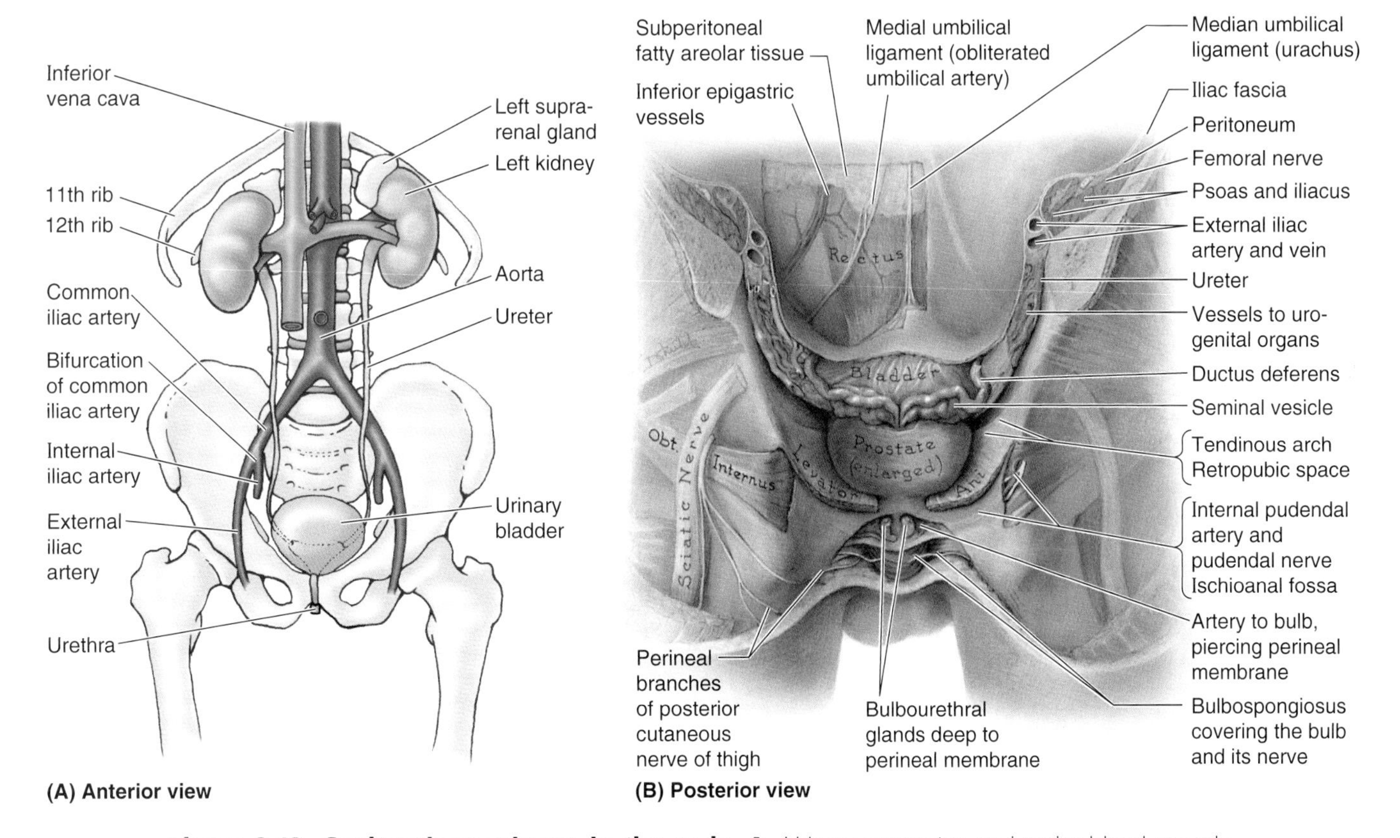

Figure 3.13. Genitourinary viscera in the male. A. Urinary apparatus and major blood vessels. **B.** Coronal section of a dissection of a male pelvis just anterior to the rectum; this a posterior view of the anterior pelvis. Note that the medial umbilical ligament (obliterated umbilical artery) and the median umbilical ligament (remnant of the fetal urachus), like the urinary bladder, are in the subperitoneal fatty-areolar tissue. Examine the levator ani and its fascial coverings separating the pelvic retropubic space from the perineal ischioanal fossae. Observe that the free anterior borders of the levator ani are separated by a gap, the *urogenital hiatus*, through which the urethra (and, in the female, the vagina) passes.

way "flap valve," and contractions of the bladder musculature act as a sphincter, preventing the reflux of urine into the ureters when the bladder empties.

In males the only structure that passes between the ureter and the peritoneum is the **ductus deferens**, the secretory duct of the testis that is the distal continuation of the epididymis (Fig. 3.13*B*). The ureter lies posterolateral to the ductus deferens and enters the posterosuperior angle of the bladder, just superior to the *seminal vesicle*.

In females the ureter passes medial to the origin of the uterine artery and continues to the level of the ischial spine, where it is crossed superiorly by the uterine artery. It then passes close to the lateral part of the fornix of the vagina and enters the posterosuperior angle of the bladder.

Arterial Supply of the Ureters. The common and internal iliac arteries supply the pelvic parts of the ureters (Fig. 3.13, Table 3.4). The most constant arteries supplying pelvic parts of the ureter in females are branches of the **uterine arteries**. The anastomosis between these branches is good. The source of similar branches in males are the **inferior vesical arteries**.

Venous and Lymphatic Drainage of the Ureters. Veins from the ureters accompany the arteries and have corresponding names. Lymph drains into the lumbar (lateral aortic), common iliac, external iliac, and internal iliac lymph nodes (Fig. 3.14*A*).

Innervation of the Ureters. The nerves to the ureters derive from adjacent *autonomic plexuses* (renal, aortic, superior and inferior hypogastric). Afferent (pain) fibers from the ureters follow sympathetic fibers retrogradely to reach the spinal ganglia and spinal cord segments T11 through L1 or L2 (Fig. 3.14*B*). *Ureteric pain* is usually referred to the ipsilateral lower quadrant of the abdomen, especially to the groin.

Ureteric Injuries

The ureters may be injured during gynecologic operations, such as radical hysterectomy (Morris and Burke, 1993), because of their proximity to the internal genital organs. The two common sites of injury are:

- At the pelvic brim where the ureter is close to the ovarian vessels
- Where the uterine artery crosses the ureter at the side of the cervix.

Identification of the ureters during their full course through the pelvis is an important preventative measure. Injury to the ureter may consist of transection, crushing, kinking, ligation, or devascularization of the vascular plexus (Hatch, 1993).

Ureteric Calculi

The ureters are expansile muscular tubes that dilate if obstructed. Acute obstruction usually results from a **ureteric calculus** (L. a pebble or small stone). Although passage of small calculi usually causes little or no pain, larger ones produce severe pain. The symptoms and severity depend on the location, type, and size of the calculus and on whether it is smooth or spiky. The pain caused by a calculus is a *colicky pain* (resembling colonic pain), which results from hyperperistalsis in the ureter, superior to the level of the obstruction. Ureteric calculi may cause complete or intermittent obstruction of urinary flow. The obstruction may occur anywhere along the ureter but it occurs most often where the ureters are normally relatively constricted:

- At the junction of the ureters and renal pelves
- Where they cross the external iliac artery and pelvic brim
- During their passage through the wall of the urinary bladder.

The presence of calculi can often be confirmed by abdominal radiographs or an intravenous urogram (see Chapter 2).

Ureteric calculi may be removed in three ways: open surgery, endourology, and lithotripsy. *Open surgery* is performed uncommonly because the calculi usually can be removed by endourology or lithotripsy. In *endourology*, a cystoscope is passed through the urethra into the bladder (p. 367). The *cystoscope* has a light, an observing lens, and various attachments (e.g., for grasping a stone). Another instrument, a *ureteroscope*, can be inserted into the cystoscope and passed up the ureter to grasp the stone. Calculi may also be broken up with an ultrasonic probe that is inserted through the ureteroscope. *Lithotripsy* uses shock waves to break up a stone into small fragments that can be passed in the urine. ○

Urinary Bladder

The urinary bladder, a hollow viscus with strong muscular walls, is characterized by its distensibility. The urinary bladder is a temporary reservoir for urine that varies in size, shape, position, and relations according to its content and the state of neighboring viscera. When empty, the adult urinary bladder is

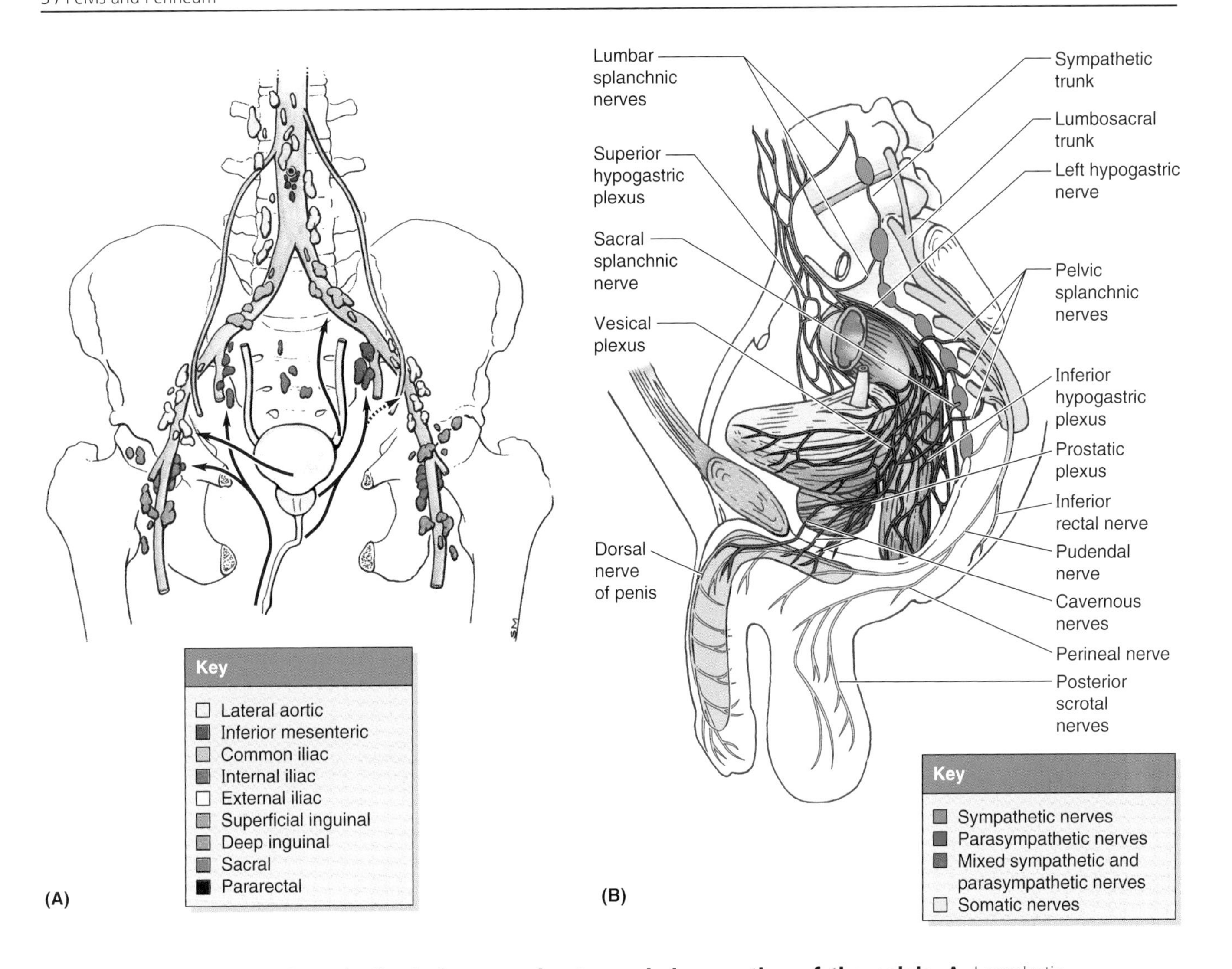

Figure 3.14. Lymphatic drainage and autonomic innervation of the pelvis. A. Lymphatic drainage of the ureters, urinary bladder, and urethra. The *arrows* indicate the direction of lymph flow to the lymph nodes. **B.** Autonomic nerves of the pelvis, with continuations to perineal structures.

in the lesser pelvis, lying posterior and slightly superior to the pubic bones. It is separated from these bones by the potential **retropubic space** and lies inferior to the peritoneum, where it rests on the pelvic floor (Fig. 3.15). The bladder is relatively free within the extraperitoneal subcutaneous fatty tissue except for its neck, which is held firmly by the *puboprostatic ligaments* in males and the *pubovesical ligaments* in females.

In infants and children, the urinary bladder is in the abdomen even when empty (Fig. 3.16*A*). The bladder usually enters the greater pelvis by 6 years of age; however, it is not entirely within the lesser pelvis until after puberty. An empty bladder in an adult lies almost entirely in the lesser pelvis, inferior to the peritoneum, where it rests on the pelvic floor posterior to the pubic symphysis (Fig. 3.16*B*). As the bladder fills, it ascends in the extraperitoneal fatty tissue of the anterior abdominal wall and enters the greater pelvis. A full bladder may ascend to the level of the umbilicus.

The bladder always contains some urine and is usually more or less rounded. When empty, the bladder is somewhat tetrahedral-shaped (Fig. 3.16*B*) and has an:

- Apex
- Body
- Fundus
- Neck
- Uvula.

Study of the empty, contracted bladder in a cadaver helps to describe the bladder's four surfaces: superior, inferolateral, and posterior.

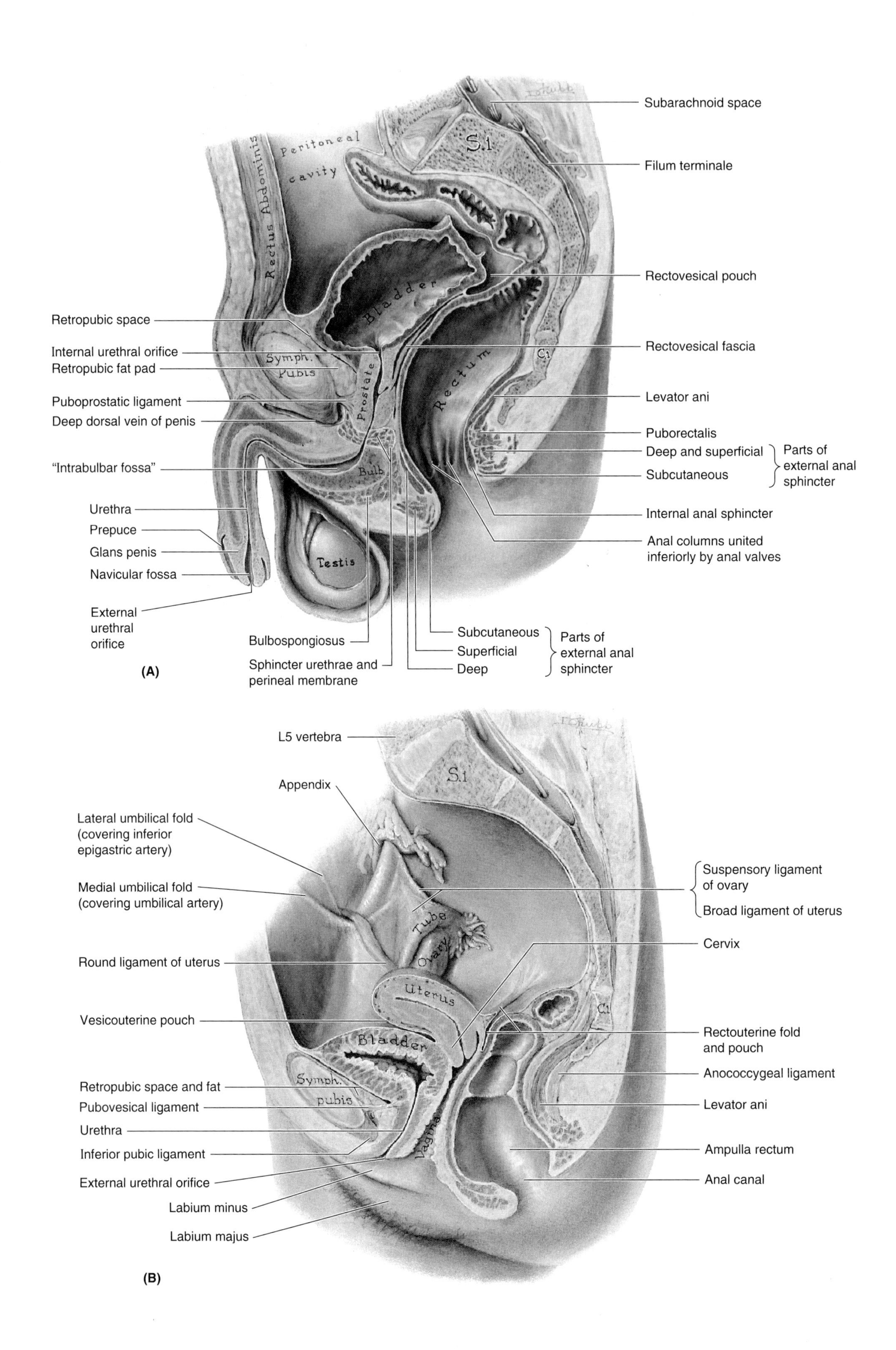
Subarachnoid space
Filum terminale
Peritoneal cavity
Rectus Abdominis
S.1
Bladder
Rectovesical pouch
Retropubic space
Internal urethral orifice
Retropubic fat pad
Symph. Pubis
Prostate
Rectum
Rectovesical fascia
C1
Puboprostatic ligament
Deep dorsal vein of penis
Levator ani
Puborectalis
Deep and superficial
Subcutaneous
Parts of external anal sphincter
"Intrabulbar fossa"
Bulb
Urethra
Prepuce
Glans penis
Navicular fossa
Testis
Internal anal sphincter
Anal columns united inferiorly by anal valves
External urethral orifice
Bulbospongiosus
Sphincter urethrae and perineal membrane
Subcutaneous
Superficial
Deep
Parts of external anal sphincter
(A)
L5 vertebra
S.1
Appendix
Lateral umbilical fold (covering inferior epigastric artery)
Medial umbilical fold (covering umbilical artery)
Suspensory ligament of ovary
Broad ligament of uterus
Tube
Ovary
Cervix
Round ligament of uterus
Uterus
C1
Vesicouterine pouch
Bladder
Rectouterine fold and pouch
Anococcygeal ligament
Retropubic space and fat
Symph. pubis
Pubovesical ligament
Levator ani
Urethra
Vagina
Inferior pubic ligament
Ampulla rectum
External urethral orifice
Anal canal
Labium minus
Labium majus
(B)

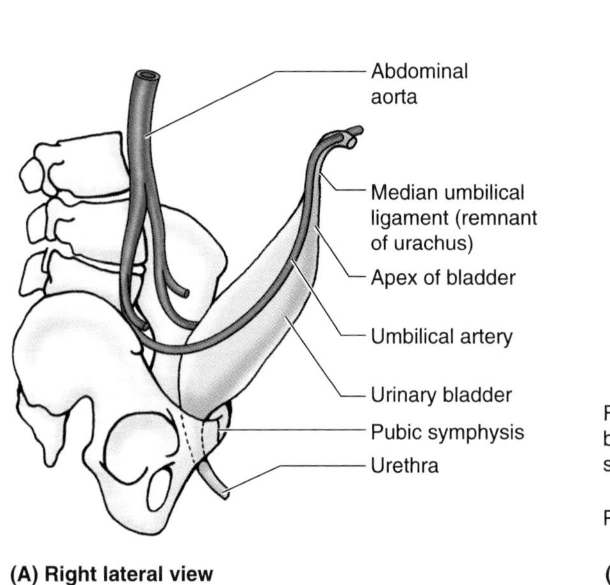

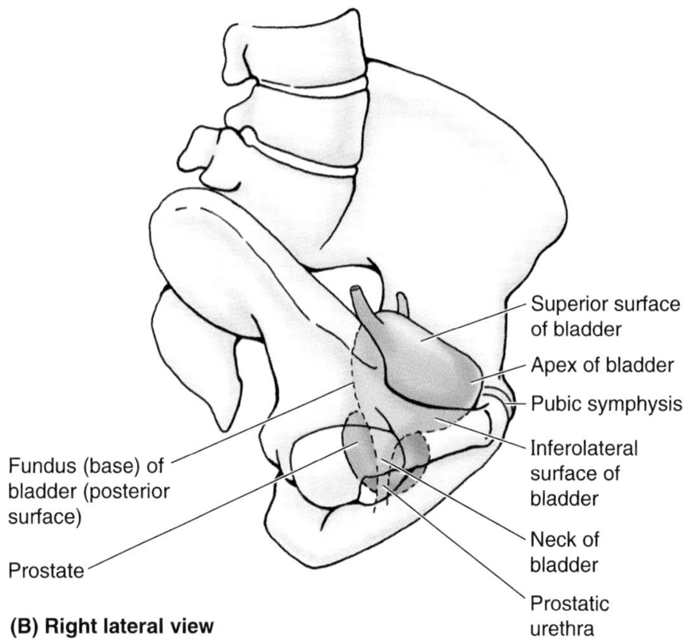

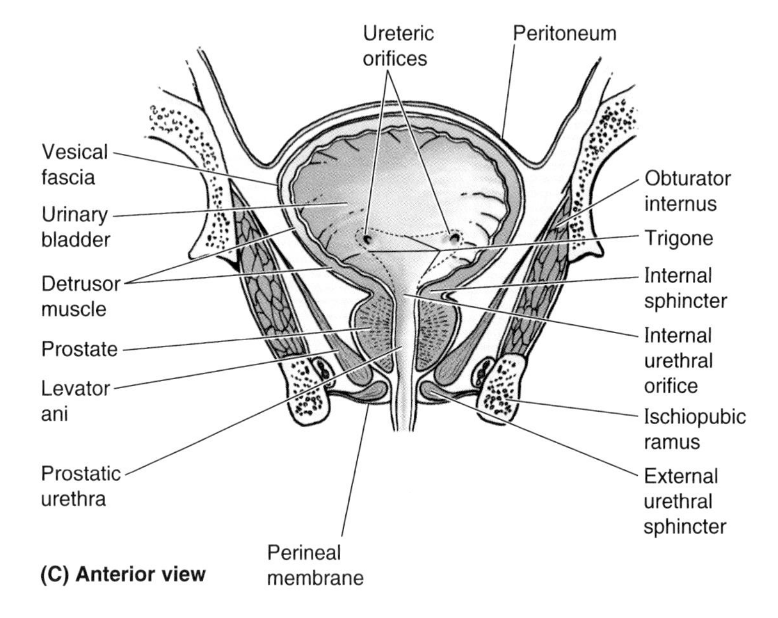

Figure 3.16. Urinary bladder and prostate. A. Empty bladder of an infant. Note its fusiform shape and that it is almost entirely above the pelvic inlet (superior pelvic aperture) in the abdominal cavity. **B.** Lateral view of adult bladder and prostate, demonstrating their pelvic location. The bladder is shown as moderately full. **C.** Anterior view of the posterior part of a male pelvis sectioned coronally in the plane of the prostatic urethra. Observe the obturator internus forming the pelvic walls and the levator ani muscles supporting the prostate and bladder.

Figure 3.15. Median sections of male and female pelves. A. Male pelvis. Observe the moderately distended urinary bladder lying on the rectum and the prostatic urethra descending vertically through a somewhat elongated prostate. The disposition of the bladder, which normally lies anteriorly against the pubic bone (see (**B**)) and extends superiorly between the rectus abdominis and the peritoneum as it fills, is unusual in this individual, creating a large supravesical fossa and a high, posteriorly placed rectovesical pouch (fossa). **B.** Female pelvis. The uterus was sectioned in its own median plane and is depicted as though this coincided with the median plane of the body, which is seldom the case. Observe the uterine tube and ovary and the uterus, bent on itself (anteflexed) at the junction of the body and cervix, and tipped anteriorly (anteverted) so that its mass rests on the bladder. Both (**A**) and (**B**) illustrate cadaveric specimens in which the rectum and anal canal are fixed in a postmortem position because of the absence of muscle tone. This position most closely resembles the position of the living rectum and anal canal during defecation: the puborectalis muscle is relaxed, allowing straightening of the anorectal angle; the rectal ampulla is clearly descended in (**B**) because of the relaxation of the levator ani; and in both specimens the relaxed anal sphincters allow wide gaping of the anal canal. All these occur during the passage of stool (feces) through the anal canal.

The **apex of the bladder** (anterior end) points toward the superior edge of the pubic symphysis. The **body of the bladder** is the part between the apex and fundus. The base, or **fundus of the bladder**, is formed by the posterior wall, which is somewhat convex. The fundus and inferolateral surfaces meet at the **neck of the bladder**. The **uvula of the bladder** is a slight projection of the **trigone of the bladder** (Fig. 3.17); it is usually more prominent in older men.

The two inferolateral surfaces are in contact with the fascia covering the levator ani (Fig. 3.16*C*). *In males* the fundus is related to the rectum (Fig. 3.15*A*); *in females* the fundus is closely related to the anterior wall of the vagina (Fig. 3.15*B*).

The **bladder bed** is formed on each side by the pubic bones and the obturator internus and levator ani (Fig. 3.16*C*) and posteriorly by the rectum or vagina (Fig. 3.15). The bladder is enveloped by loose connective tissue—**vesical fascia**. The wall of the bladder is composed chiefly of the **detrusor muscle**. Toward the neck of the male bladder, the muscle fibers form the involuntary **internal sphincter**. Some fibers run radially and assist in opening the **internal urethral orifice**. In males the muscle fibers in the neck of the bladder are continuous with the fibromuscular tissue of the prostate, whereas in females these fibers are continuous with muscle fibers in the wall of the urethra. The **ureteric orifices** and the internal urethral orifice are at the angles of the **trigone of the bladder** (Figs. 3.16*C* and 3.17). The ureters pass obliquely through the bladder wall in an inferomedial direction. An increase in bladder pressure presses the walls of the ureters together, preventing the pressure in the bladder from forcing urine up the ureters.

In males the peritoneum passes (Fig. 3.15*A*):

- From the anterior abdominal wall
- Superior to the pubic bone
- Along the superior surface of the urinary bladder
- Inferiorly on the posterior surface of the urinary bladder
- On the superior ends of the seminal vesicles
- Posteriorly to line the *rectovesical pouch* and cover the superior part of the rectum
- Posterosuperiorly to become the sigmoid mesocolon.

In females the peritoneum passes (Fig. 3.15*B*):

- From the anterior abdominal wall
- Superior to the pubic bone
- Along the superior surface of the urinary bladder
- From the bladder to the uterus forming the *vesicouterine pouch*
- On the fundus and body of the uterus, posterior fornix, and the wall of the vagina
- Between the rectum and uterus forming the *rectouterine pouch*
- On the anterior and lateral sides of the rectum
- Posterosuperiorly to become the sigmoid mesocolon.

The **vesicouterine pouch** of peritoneum, between the bladder and uterus, is empty except when the uterus is inclined posteriorly (retroverted); in this case it may contain a loop of intestine.

Arterial Supply of the Bladder. The main arteries supplying the bladder are branches of the internal iliac arteries (Fig. 3.10, Table 3.4). The **superior vesical arteries** supply anterosuperior parts of the bladder. In males, **inferior vesical arteries** supply the fundus and neck of the bladder. In females the **vaginal arteries** replace the inferior vesical arteries and send small branches to posteroinferior parts of the bladder (Fig. 3.10*B*). The obturator and inferior gluteal arteries also supply small branches to the bladder.

Venous and Lymphatic Drainage of the Bladder. The names of the veins correspond to the arteries and are tributaries of the internal iliac veins. In males the vesical venous plexus combines with the **prostatic venous plexus** (Fig. 3.12*C*) and envelops the fundus of the bladder and prostate, the seminal vesicles, the ductus deferens, and the inferior ends of the ureters. It receives blood from the deep dorsal vein of the penis, which drains into the prostatic venous plexus. The **vesical venous plexus** mainly drains through the inferior vesical veins into the internal iliac veins; however, it may drain through the sacral veins into the *internal vertebral venous plexuses* (see Chapter 4). In females, the vesical venous plexus envelops the pelvic part of the urethra and the neck of the bladder and receives blood from the dorsal vein of the clitoris and communicates with the *vaginal or uterovaginal venous plexus* (Fig. 3.12*B*).

In both sexes, **lymphatic vessels** leave the superior surface of the bladder and pass to the **external iliac lymph nodes** (Fig. 3.14*A*), whereas those from the fundus pass to the **internal iliac lymph nodes**. Some vessels from the neck of the bladder drain into the sacral or common iliac lymph nodes.

Innervation of the Bladder. Parasympathetic fibers to the bladder are derived from the **pelvic splanchnic nerves** (Fig. 3.14*B*). They are motor to the detrusor muscle and inhibitory to the internal sphincter. Hence, when the visceral afferent fibers are stimulated by stretching, the bladder contracts reflexively, the internal sphincter relaxes, and urine flows into the urethra. With toilet training, we learn to suppress this reflex when we do not wish to void. *Sympathetic fibers* to the bladder are derived from T11 through L2 nerves. The nerves supplying the bladder form the *vesical nerve plexus*, which consists of both sympathetic and parasympathetic fibers. This plexus, one of the several pelvic nerve plexuses, is continuous with the *inferior hypogastric plexus*. Sensory fibers from the bladder are visceral; reflex afferents follow the course of the parasympathetic fibers, as do those transmitting pain sensations (such as results from overdistension) from the inferior part of the bladder. Pain fibers from the superior surface (bladder roof) follow the sympathetic fibers retrogradely to spinal ganglia of T11 through L2.

Male Urethra

The male urethra is a muscular tube (18–20 cm long) that conveys urine from the **internal urethral orifice** of the urinary bladder to the **external urethral orifice** at the tip of the glans penis (Fig. 3.15*A*). The urethra also provides an exit for semen (sperm and glandular secretions). In the flaccid (nonerect) state the urethra has a double curvature. *For descriptive purposes, the urethra is divided into four parts* (Figs. 3.15*A* and 3.17):

- Urethra in the bladder neck (preprostatic urethra)
- Prostatic urethra
- Intermediate part of the urethra (membranous urethra)
- Spongy urethra.

The last two parts are described subsequently with the perineum (p. 403).

The **urethra in the bladder neck** or preprostatic urethra (1–1.5 cm in length) extends almost vertically from the neck of the bladder to the superior aspect of the prostate.

The **prostatic urethra** (4 cm long) is continuous with the urethra in the bladder neck (Figs. 3.16, *B* and *C*, and 3.17) and *descends through the prostate*—closer to the anterior than the posterior surface—forming a gentle curve that is concave anteriorly. The prostatic urethra—*the widest and most dilatable part of the urethra*—ends as the urethra becomes completely encircled by the external urethral sphincter (sphincter urethrae). The internal surface of the posterior wall of the prostatic urethra has notable features (Fig. 3.17). The most prominent part is the **urethral crest**, a median ridge that has a groove—**prostatic si-**

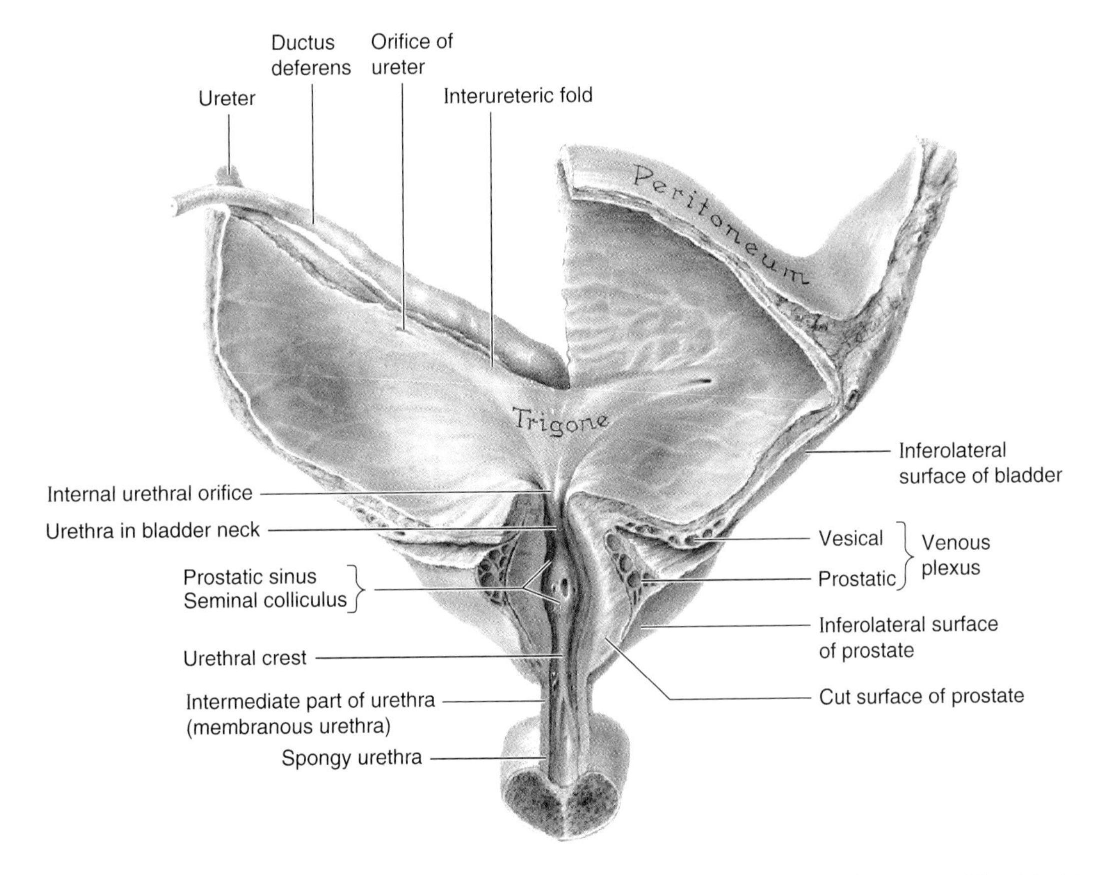

Figure 3.17. Interior of the male bladder and urethra. The anterior parts of the bladder, prostate, and urethra are cut away. The knife was then carried through the posterior wall of the bladder at the superior border of the right ureter and interureteric fold. This fold unites the ureters along the superior limit of the trigone. Observe the slight fullness posterior to the internal urethral orifice (at the tip of the leader line indicating this orifice) that, when exaggerated, becomes the uvula of the bladder (L. uvula vesicae). This small projection is produced by the middle (median) lobe of the prostate. Observe the oval mouth of the prostatic utricle (not labeled) in the summit of the seminal colliculus, and the orifice of an ejaculatory duct on each side of the utricle. Observe also the unusually long urethral crest extending more superior than usual and bifurcating more inferior than usual. Notice the prostatic fascia enclosing the prostatic venous plexus.

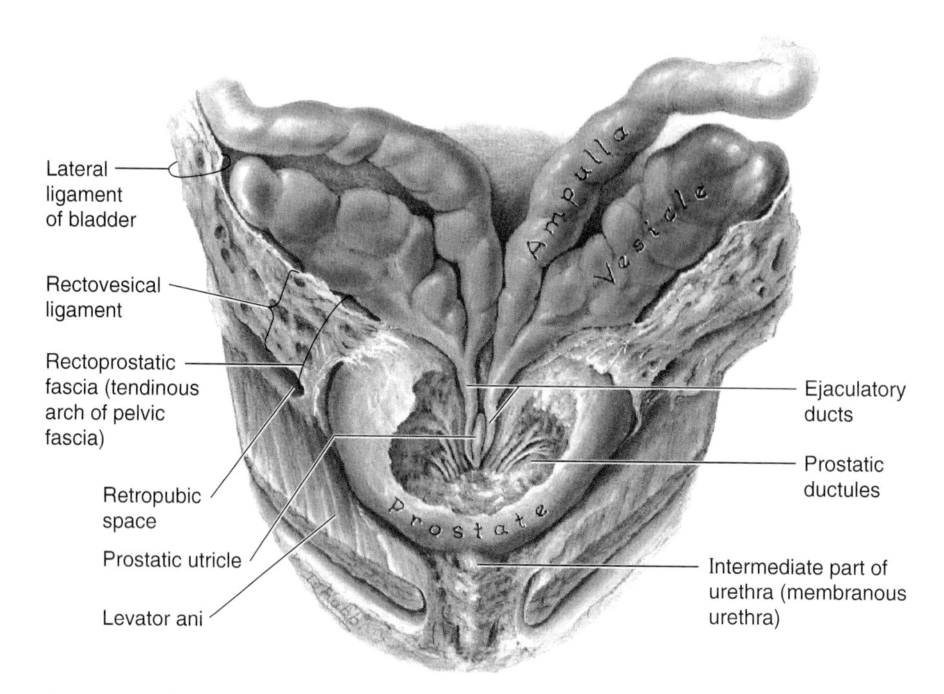

Figure 3.18. Dissection of the posterior part of the prostate, lateral ligament of the bladder, and rectoprostatic fascia. Observe the right and left ejaculatory ducts, each formed by the merger of the duct of the seminal vesicle and the ductus (vas) deferens. Note the vestigial prostatic utricle lying between the ejaculatory ducts.

nus—on each side. Most prostatic ductules open into these sinuses. In the middle part of this crest is the **seminal colliculus**, a rounded eminence with a slitlike orifice, which opens into a small cul-de-sac—the **prostatic utricle** (Fig. 3.18). The prostatic utricle is the vestigial remnant of the embryonic uterovaginal canal, which, in the female, is the primordium of the uterus and a part of the vagina (Moore and Persaud, 1998). There is a minute, slitlike opening of an **ejaculatory duct** on or just within the orifice to the prostatic utricle.

The **intermediate part of the urethra** (membranous part) is the section passing through the external urethral sphincter and the perineal membrane (see Fig. 3.21). The short intermediate part, extending from the prostatic urethra to the spongy urethra, is the narrowest and least distensible part of the urethra.

Arterial Supply of the Male Urethra. The proximal two parts of the urethra are supplied by prostatic branches of the inferior vesical and middle rectal arteries (Fig. 3.10, *A* and *C*).

Venous and Lymphatic Drainage of the Male Urethra. The veins from the proximal two parts of the urethra follow the arteries and have similar names (Fig. 3.12, *A* and *C*). The lymphatic vessels pass mainly to the **internal iliac lymph nodes** (Fig. 3.14*A*); a few vessels drain into the external iliac lymph nodes.

Innervation of the Male Urethra. The nerves are derived from the pudendal nerve and the **prostatic plexus** of the autonomic nervous system (Fig. 3.14*B*). This plexus arises from the inferior part of the inferior hypogastric plexus.

Female Urethra

The female urethra (approximately 4 cm long and 6 mm in diameter) passes anteroinferiorly from the **internal urethral orifice** of the urinary bladder (Fig. 3.15*B*), posterior and then inferior to the pubic symphysis. The **external urethral orifice** is in the vestibule of the vagina directly anterior to the orifice of the vagina. The urethra lies anterior to the vagina; its axis is parallel to that of the vagina. The urethra passes with the vagina through the pelvic diaphragm, external urethral sphincter, and perineal membrane. Urethral glands are present, particularly in the superior part of the urethra. One group of glands on each side—the *paraurethral glands*—are homologues to the prostate. These glands have a common paraurethral duct, which opens (one on each side) near the external urethral orifice. The inferior half of the urethra is in the perineum and is discussed subsequently with this section.

Arterial Supply of the Female Urethra. Blood is supplied by the internal pudendal and vaginal arteries (Fig. 3.10*B*).

Venous and Lymphatic Drainage of the Female Urethra. The veins follow the arteries and have similar names (Fig. 3.12*B*). Most lymphatic vessels from the urethra pass to the **sacral** and **internal iliac lymph nodes** (Fig. 3.14*A*). A few vessels drain into the inguinal lymph nodes.

Innervation of the Female Urethra. The nerves to the urethra arise from the *pudendal nerve*. Most afferents from the urethra run in the pelvic splanchnic nerves as in the male (Fig. 3.14*B*).

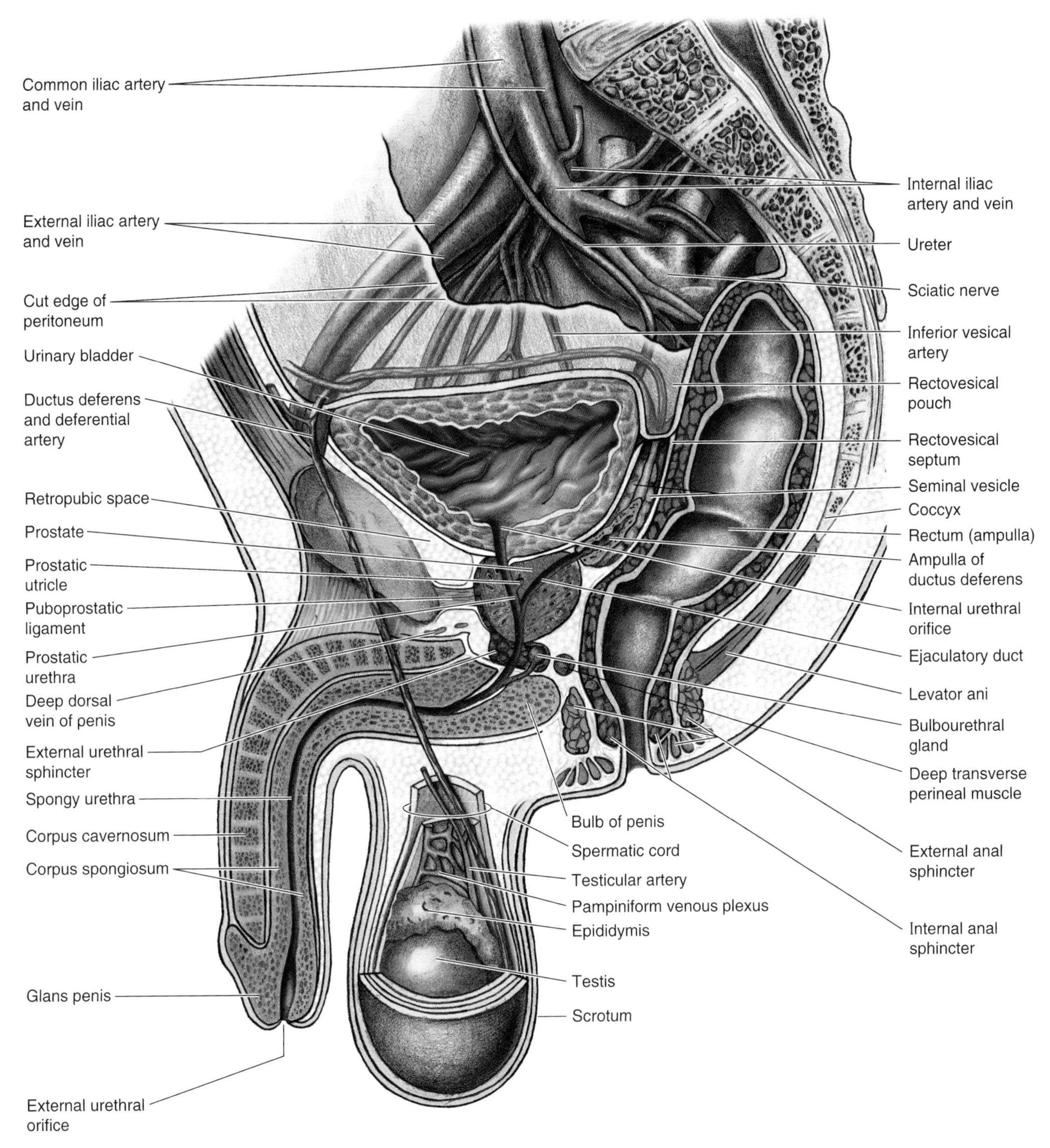

Figure 3.19. Median section of the male pelvis and perineum. Observe the genital organs: testis, epididymis, ductus (vas) deferens, ejaculatory duct, and penis with the accessory glandular structures: seminal vesicle, prostate, and bulbourethral gland. The spermatic cord connects the testis to the abdominal cavity, and the testis lies externally in a musculocutaneous pouch, the scrotum.

Cystocele—Hernia of the Bladder

Loss of bladder support in females by damage to the perineal muscles or their associated fascia can result in herniation of the bladder into the vaginal wall (*A*). A cystocele may also result from prolapse of pelvic viscera secondary to injury to the pelvic floor during childbirth.

Suprapubic Cystotomy

As the bladder fills, it extends superiorly above the pubic symphysis along the anterior abdominal wall. When excessively distended, the bladder may rise to the level of the umbilicus (*B*). In so doing, it inserts itself between the parietal peritoneum and the anterior abdominal wall.

The bladder then lies adjacent to this wall without the intervention of peritoneum. Consequently, the distended bladder may be punctured—*suprapubic cystotomy*—or approached surgically superior to the pubic symphysis for the introduction of in-dwelling catheters or instruments without traversing the peritoneum and entering the peritoneal cavity. Urinary calculi, foreign bodies, and small tumors may also be removed from the bladder through a suprapubic, extraperitoneal incision.

Because of the superior position of the distended bladder, it may be ruptured by injuries to the inferior part of the anterior abdominal wall or by fractures of the pelvis. The rupture may result in the escape of urine extraperitoneally or intraperitoneally. Rupture of the superior part of the bladder frequently tears the peritoneum, resulting in ▶

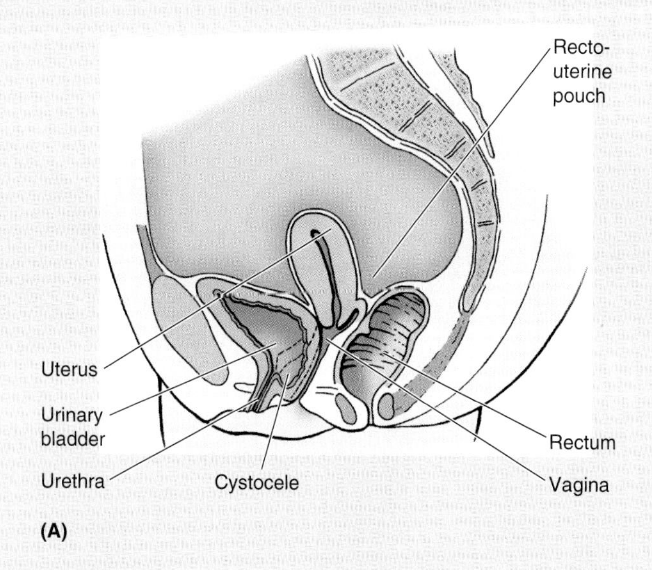

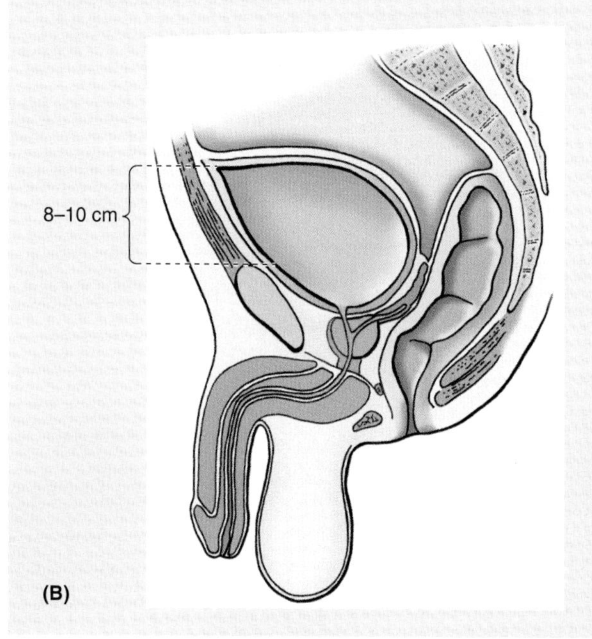

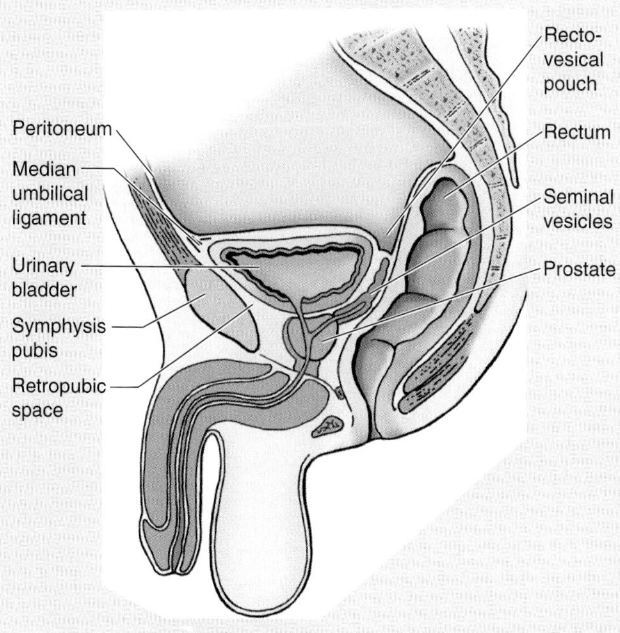

▶ *extravasation (passage) of urine into the peritoneal cavity.* Posterior rupture of the bladder usually results in passage of urine extraperitoneally into the perineum.

Cystoscopy

The interior of the bladder and its three orifices can be examined with a *cystoscope*, a lighted tubular endoscope that is inserted through the urethra. The cystoscope consists of a light, observing lens, and various attachments for grasping, removing, cutting, and cauterizing. During *transurethral resection of a tumor,* an instrument is passed into the bladder through the urethra. Using a high-frequency electrical current, the tumor is removed in small fragments that are washed from the bladder with water. ⊙

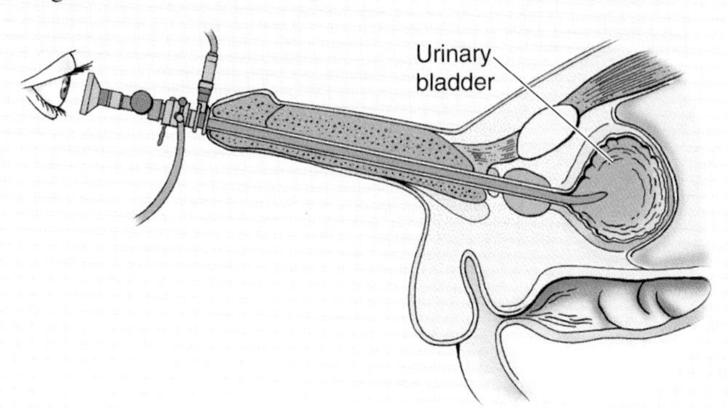

Male Internal Genital Organs

The male internal genital organs include the testes, epididymides (plural of epididymis), ductus deferentes (plural of ductus deferens), seminal vesicles, ejaculatory ducts, prostate, and bulbourethral glands (Fig. 3.19). The testes and epididymides are described in Chapter 2.

Ductus Deferens

The deferent duct (L. ductus deferens, vas deferens) is the continuation of the duct of the epididymis (Fig. 3.19). The ductus deferens:

- Begins in the tail of the epididymis
- Ascends in the spermatic cord
- Passes through the inguinal canal
- Crosses over the external iliac vessels and enters the pelvis
- Passes along the lateral wall of the pelvis, where it lies external to the parietal peritoneum
- Ends by joining the duct of the seminal vesicle to form the ejaculatory duct.

During its course no other structure intervenes between the ductus deferens and the peritoneum.

The ductus deferens crosses superior to the ureter near the posterolateral angle of the bladder, running between the ureter and the peritoneum to reach the fundus of the bladder. The relationship of the ductus deferens to the ureter in the male is similar, although of lesser clinical importance, to that of the uterine artery to the ureter in the female. The developmental basis of this relationship is shown in Figure 3.20. Posterior to the bladder, the ductus deferens at first lies superior to the seminal vesicle, then descends medial to the ureter and the vesicle (Fig. 3.19). Here the ductus deferens enlarges to form the **ampulla of the ductus deferens.** It then narrows and joins the duct of the seminal vesicle to form the *ejaculatory duct.*

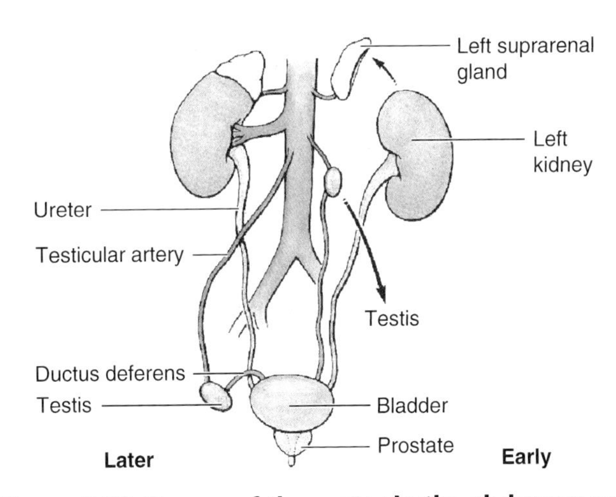

Figure 3.20. Course of the ureter in the abdomen and pelvis. This diagram explains developmentally how the ureter comes to be crossed by testicular vessels in the abdomen and the ductus (vas) deferens in the pelvis.

Arterial Supply of the Ductus Deferens. The tiny *differential artery* usually arises from the inferior vesical artery (Fig. 3.19) and terminates by anastomosing with the testicular artery, posterior to the testis.

Venous and Lymphatic Drainage of the Ductus Deferens. The veins accompany the arteries and have similar names (Fig. 3.12*C*). Lymphatic vessels from the ductus deferens end in the *external iliac lymph nodes* (Fig. 3.14*A*).

Innervation of the Ductus Deferens. The nerves of the ductus deferens are derived from the **inferior hypogastric plexus** (Fig. 3.14*B*). The ductus is richly innervated by autonomic nerve fibers, thereby facilitating its rapid contraction for expulsion of sperms during ejaculation.

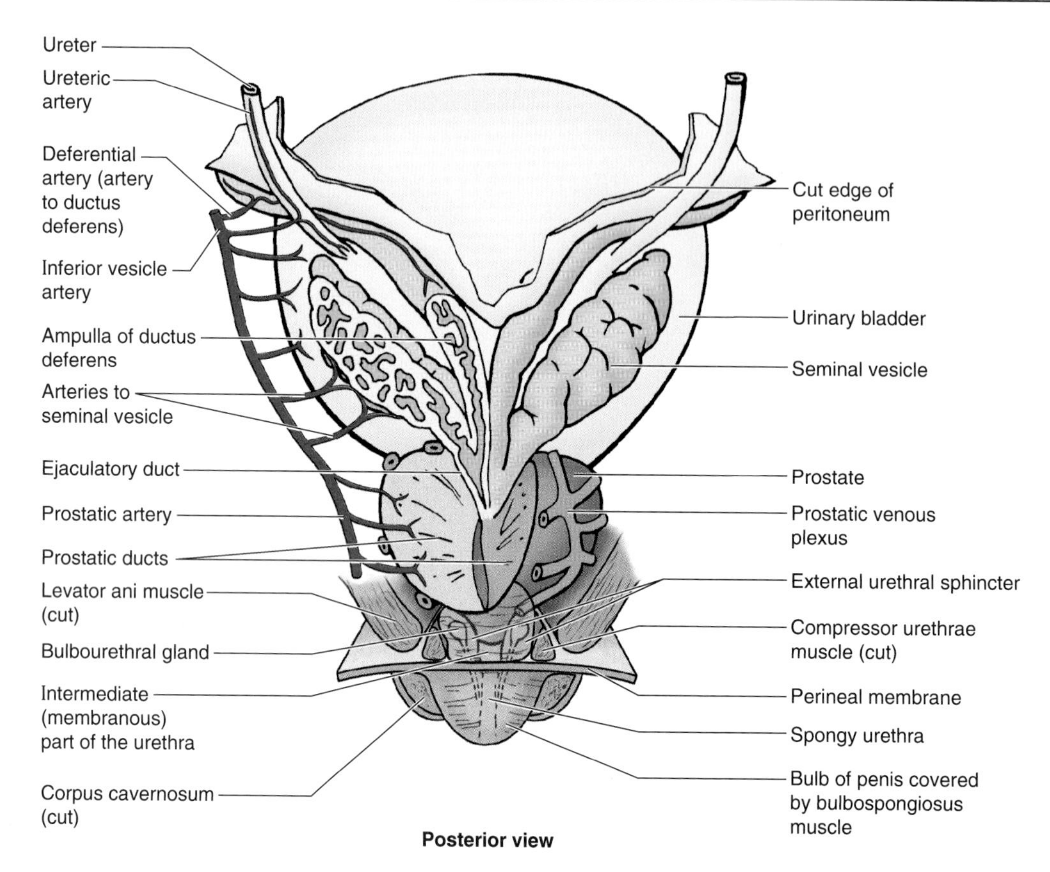

Figure 3.21. Urinary bladder, seminal vesicles, terminal parts of the deferent ducts, and the prostate. The left seminal vesicle and ampulla of the deferent duct (ductus deferens) are dissected free and sliced open. Part of the prostate is also cut away to expose the ejaculatory duct. Observe the perineal membrane lying between the external genitalia and the deep part of the perineum (anterior recess of the ischioanal fossa). It is pierced by the urethra, ducts of the bulbourethral glands, dorsal and deep arteries of the penis, cavernous nerves, and the dorsal nerve of the penis.

Sterilization of Males

The common method of sterilizing males is a **deferentectomy**, popularly called a **vasectomy**. During this procedure part of the ductus deferens is ligated and/or excised through an incision in the superior part of the scrotum. Hence, the subsequent ejaculated fluid from the seminal vesicles, prostate, and bulbourethral glands contains no sperms. The unexpelled sperms degenerate in the epididymis and the proximal part of the ductus deferens. *Reversal of a deferentectomy* is successful in favorable cases (patients under 30 years of age and before 7 years postoperation) in most instances. The ends of the sectioned ductus deferentes are reattached under an operating microscope. ⊙

Seminal Vesicles

Each of the seminal vesicles is an elongated structure (approximately 5 cm long) that lies between the fundus of the bladder and the rectum (Figs. 3.19 and 3.21). The seminal vesicles, obliquely placed glands superior to the prostate, *do not store sperms* as their name implies. They secrete a thick alkaline fluid that mixes with the sperms as they pass into the ejaculatory ducts and urethra. The superior ends of the seminal vesicles are covered with peritoneum and lie posterior to the ureters, where the peritoneum of the *rectovesical pouch* separates them from the rectum. The inferior ends of the seminal vesicles are closely related to the rectum and are separated from it only by the **rectovesical septum**, a membranous partition (Fig. 3.19). The duct of the seminal vesicle joins the ductus deferens to form the *ejaculatory duct.*

Arterial Supply of the Seminal Vesicles. The arteries to the seminal vesicles (Fig. 3.10, *A* and *C*) derive from the inferior vesical and middle rectal arteries.

Venous and Lymphatic Drainage of the Seminal Vesicles. The veins accompany the arteries and have similar names (Fig. 3.12*C*). The iliac lymph nodes, especially the **internal iliac lymph nodes** (Fig. 3.14*A*), receive lymph from the seminal vesicles.

Innervation of the Seminal Vesicles. The walls of these vesicles contain a plexus of nerve fibers and some sympathetic ganglia (Fig. 3.14*B*). The presynaptic sympathetic fibers traverse the *superior lumbar and hypogastric nerves*, and the presynaptic parasympathetic fibers traverse the *pelvic splanchnic nerves* to reach the *inferior hypogastric (pelvic) plexuses.*

Abscesses in the Seminal Vesicles

Localized collections of pus (abscesses) in the seminal vesicles may rupture, allowing pus to enter the peritoneal cavity. Enlarged seminal vesicles can be palpated during a rectal examination. They can also be massaged to release their secretions for microscopic examination to detect *gonococci,* for example (the organisms that cause *gonorrhea*). ⊙

Ejaculatory Ducts

Each ejaculatory duct is a slender tube that arises by the union of the duct of a seminal vesicle with the ductus deferens (Figs. 3.19 and 3.21). The ejaculatory ducts (approximately 2.5 cm long) arise near the neck of the bladder and run close together as they pass anteroinferiorly through the posterior part of the prostate and along the sides of the prostatic utricle. The ejaculatory ducts converge to open on the seminal colliculus by slitlike apertures on, or just within, the opening of the prostatic utricle (Figs. 3.17 and 3.18).

Arterial Supply of the Ejaculatory Ducts. The *deferential arteries,* usually branches of the inferior vesical arteries, supply the ejaculatory ducts (Fig. 3.19).

Venous and Lymphatic Drainage of the Ejaculatory Ducts. The veins join the *prostatic and vesical venous plexuses* (Fig. 3.12*C*). The lymphatic vessels drain into the **external iliac lymph nodes** (Fig. 3.14*A*).

Innervation of the Ejaculatory Ducts. The nerves of the ejaculatory ducts derive from the *inferior hypogastric plexus* (Fig. 3.14*B*).

Prostate

The prostate (approximately 3 cm long) is the largest accessory gland of the male reproductive system (Figs. 3.15*A* and 3.16, *B* and *C*). The glandular part comprises approximately two-thirds of the prostate; the other third is fibromuscular. The firm, walnut-sized prostate, surrounding the prostatic urethra, has a dense fibrous *prostatic capsule* that is surrounded by a fibrous *prostatic sheath,* which is continuous with the puboprostatic ligaments. *The prostate has*:

- A base closely related to the neck of the bladder
- An apex that is in contact with fascia on the superior aspect of the urethral sphincter and deep perineal muscles
- A muscular anterior surface, featuring mostly transversely oriented muscle fibers continuous interiorly with the urethral sphincter, that is separated from the pubic symphysis by retroperitoneal fat in the retropubic space
- A posterior surface that is related to the ampulla of the rectum
- Inferolateral surfaces that are related to the levator ani.

Although not clearly distinct anatomically, the following lobes of the prostate are traditionally described (Figs. 3.19 and 3.21):

- The anterior lobe, or isthmus, lies anterior to the urethra. It is fibromuscular, the muscle fibers representing a superior continuation of the urethral sphincter muscle, and contains little, if any, glandular tissue.
- The posterior lobe lies posterior to the urethra and inferior to the ejaculatory ducts; it is readily palpable by digital rectal examination.
- The lateral lobes on either side of the urethra form the major part of the prostate.
- The middle (median) lobe lies between the urethra and the ejaculatory ducts and is closely related to the neck of the bladder.

Some authors, especially urologists and sonographers, divide the prostate into peripheral and central (internal) zones. The central zone is comparable to the middle lobe. Within each lobe are four lobules, which are defined by the arrangement of the ducts and connective tissue.

The **prostatic ducts** (20 to 30) open chiefly into the grooves—the *prostatic sinuses*—that lie on either side of the seminal colliculus on the posterior wall of the prostatic urethra (Fig. 3.17). Prostatic fluid, a thin, milky fluid, provides approximately 20% of the volume of semen—a mixture of secretions produced by the testes, seminal vesicles, prostate, and bulbourethral glands.

Arterial Supply of the Prostate. The prostatic arteries are mainly branches of the internal iliac artery (Fig. 3.10, *A* and *C*), especially the **inferior vesical arteries** but also the internal pudendal and middle rectal arteries.

Venous and Lymphatic Drainage of the Prostate. The veins join to form a plexus around the sides and base of the prostate (Figs. 3.10*C* and 3.21). This **prostatic venous plexus,** between the fibrous capsule of the prostate and the prostatic sheath, drains into the **internal iliac veins.** The prostatic venous plexus is continuous superiorly with the vesical venous plexus and communicates posteriorly with the *internal vertebral venous plexus.* The **lymphatic vessels** (Fig. 3.14*A*) terminate chiefly in the *internal iliac and sacral lymph nodes.*

Innervation of the Prostate. Parasympathetic fibers arise from the *pelvic splanchnic nerves* (S2 through S4). The sympathetic fibers derive from the *inferior hypogastric plexus* (Fig. 3.14*B*).

Enlargement of the Prostate

The prostate is of considerable medical interest because benign enlargement—*hypertrophy of the prostate*—is common after middle age. An enlarged prostate projects into the urinary bladder and impedes urination by distorting the prostatic urethra. The middle lobe often enlarges the most and obstructs the internal urethral orifice; the more the person strains, the more the prostate occludes the urethra (Williams et al., 1995). ▶

▶ The prostate may be examined by *digital rectal examination*. The palpability of the prostate depends on the fullness of the bladder. A full bladder offers resistance, holding the gland in its place and making it more readily palpable. *Benign prostatic hypertrophy* affects a high proportion of older men and is a common cause of urethral obstruction leading to *nocturia* (need to void during the night), *dysuria* (difficulty and/or pain during urination), and *urgency* (sudden desire to void).

Prostatic cancer is common in men older than age 55. During a rectal examination the malignant prostate feels hard and often irregular. In advanced stages, *cancer cells metastasize to the internal iliac and sacral lymph nodes* and later to distant nodes and bone. Because of the close relationship of the prostate to the prostatic urethra, an obstruction may be relieved with a *resectoscope*, an instrument that is inserted through the external urethral orifice and spongy urethra into the prostatic urethra. During surgical resection of all or part of the prostate (*prostatectomy*), the hypertrophied part of the prostate is removed.

In more serious cases, the entire prostate is removed along with the seminal vesicles, ejaculatory ducts, and the terminal parts of the deferent ducts. Techniques for preserving the nerves and blood vessels to the penis that pass beside the prostate have increased the possibility for patients to retain sexual function after surgery and have improved the likelihood of normal urinary control following this procedure. ⊙

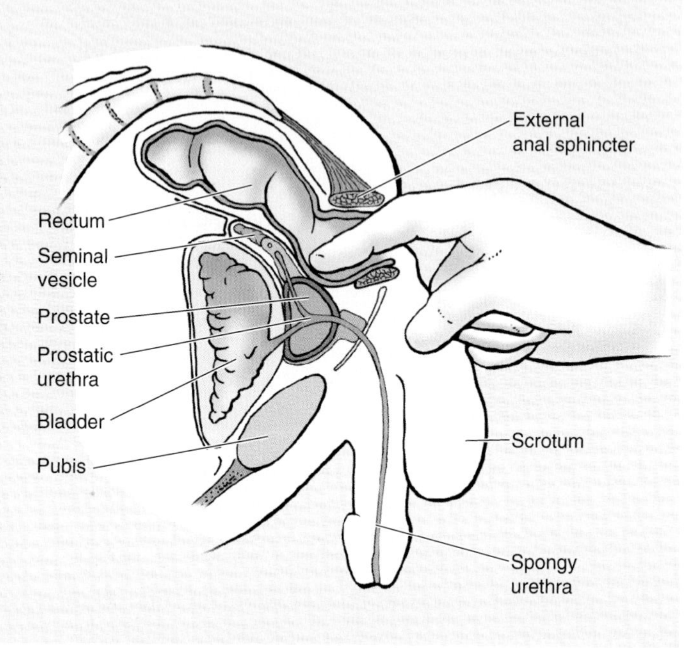

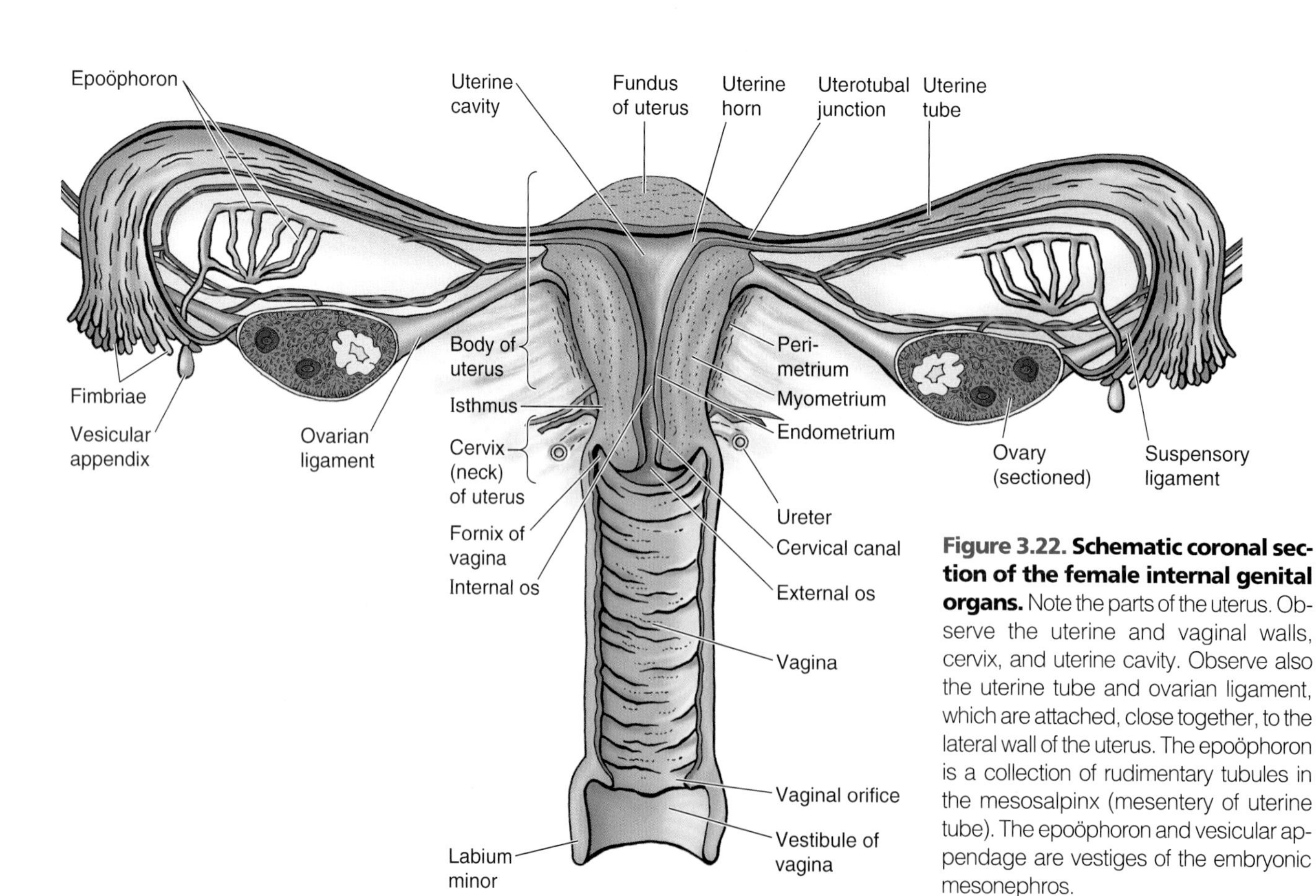

Figure 3.22. Schematic coronal section of the female internal genital organs. Note the parts of the uterus. Observe the uterine and vaginal walls, cervix, and uterine cavity. Observe also the uterine tube and ovarian ligament, which are attached, close together, to the lateral wall of the uterus. The epoöphoron is a collection of rudimentary tubules in the mesosalpinx (mesentery of uterine tube). The epoöphoron and vesicular appendage are vestiges of the embryonic mesonephros.

Bulbourethral Glands

The two pea-sized bulbourethral glands lie posterolateral to the intermediate (membranous) part of the urethra (Figs. 3.19 and 3.21). The ducts of the bulbourethral glands pass through the inferior fascia of the urethral sphincter (perineal membrane) with the urethra and open through minute apertures into the proximal part of the spongy urethra in the bulb of the penis. Their mucuslike secretion enters the urethra during sexual arousal.

Female Internal Genital Organs

The female internal genital organs include the vagina, uterus, uterine tubes, and ovaries.

Vagina

The vagina, a musculomembranous tube (7–9 cm long), extends from the cervix (neck of the uterus) to the **vestibule of the vagina**—the cleft between the labia minora (Figs. 3.22 and 3.23). The superior end of the vagina surrounds the cervix; the lower end passes anteroinferiorly through the pelvic floor to open in the vestibule. The vagina:

- Serves as the excretory duct for menstrual fluid
- Forms the inferior part of the pelvic (birth) canal
- Receives the penis and ejaculate during sexual intercourse
- Communicates superiorly with the *cervical canal*—a fusiform canal extending from the isthmus of the uterus to the external os of the uterus—and inferiorly with the vestibule of the vagina.

The vagina is normally collapsed so that its anterior and posterior walls are in contact except at its superior end, where the cervix holds them apart. The vagina lies posterior to the urethra and urinary bladder and anterior to the rectum (Fig. 3.23), passing between the medial margins of the levator ani muscles. The **vaginal fornix**, the recess around the cervix, is described as having anterior, posterior, and lateral parts. *The posterior*

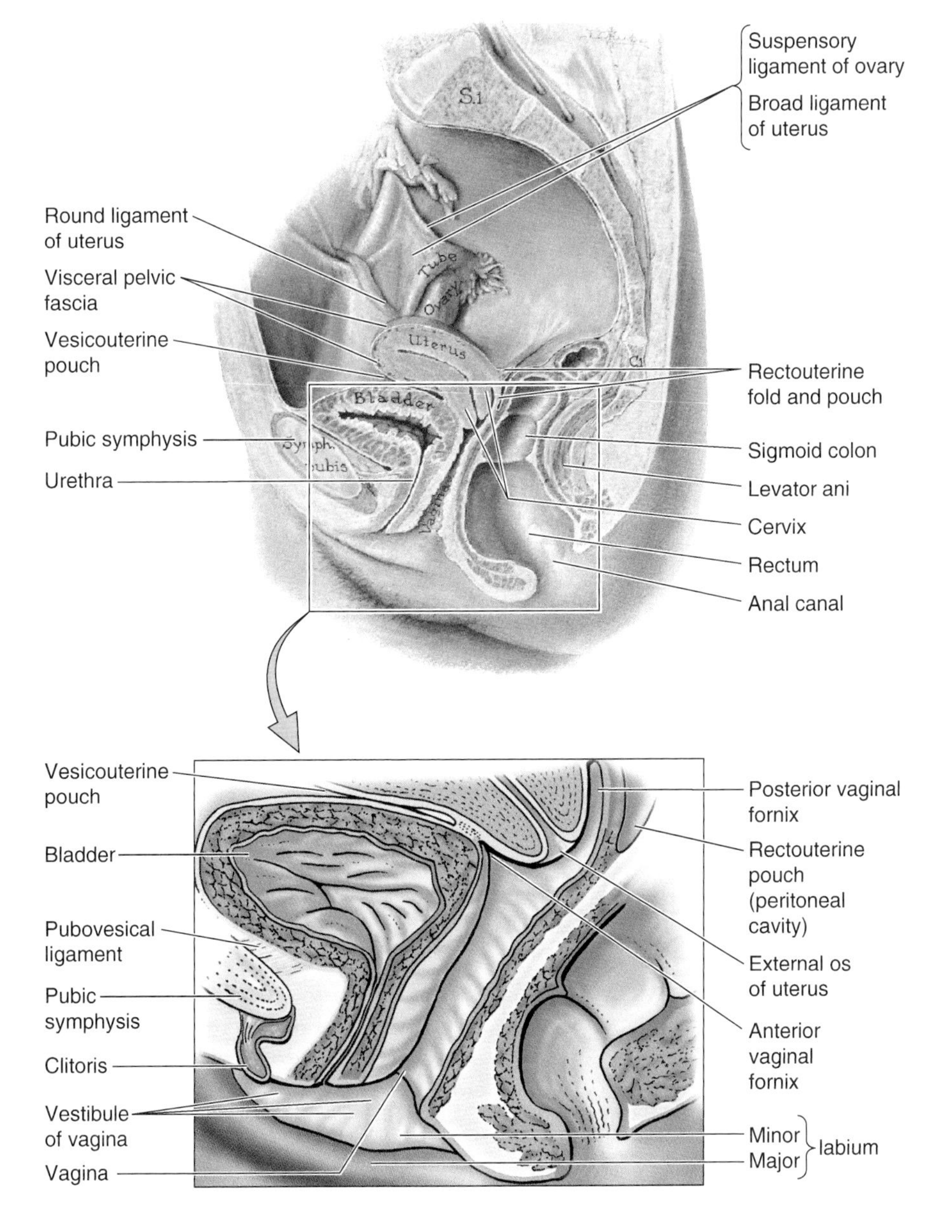

Figure 3.23. Median section of a female pelvis. The *boxed area* shows an enlargement of the urinary bladder, urethra, cervix, anterior and posterior vaginal fornices, vagina, rectouterine pouch, external os of uterus, labia majora and minora, and vestibule of the vagina. Note that the axes of the urethra and vagina are parallel and that the urethra is adherent to the anterior vagina wall. Placing a finger in the vagina can help to direct the insertion of a catheter through the urethra into the bladder.

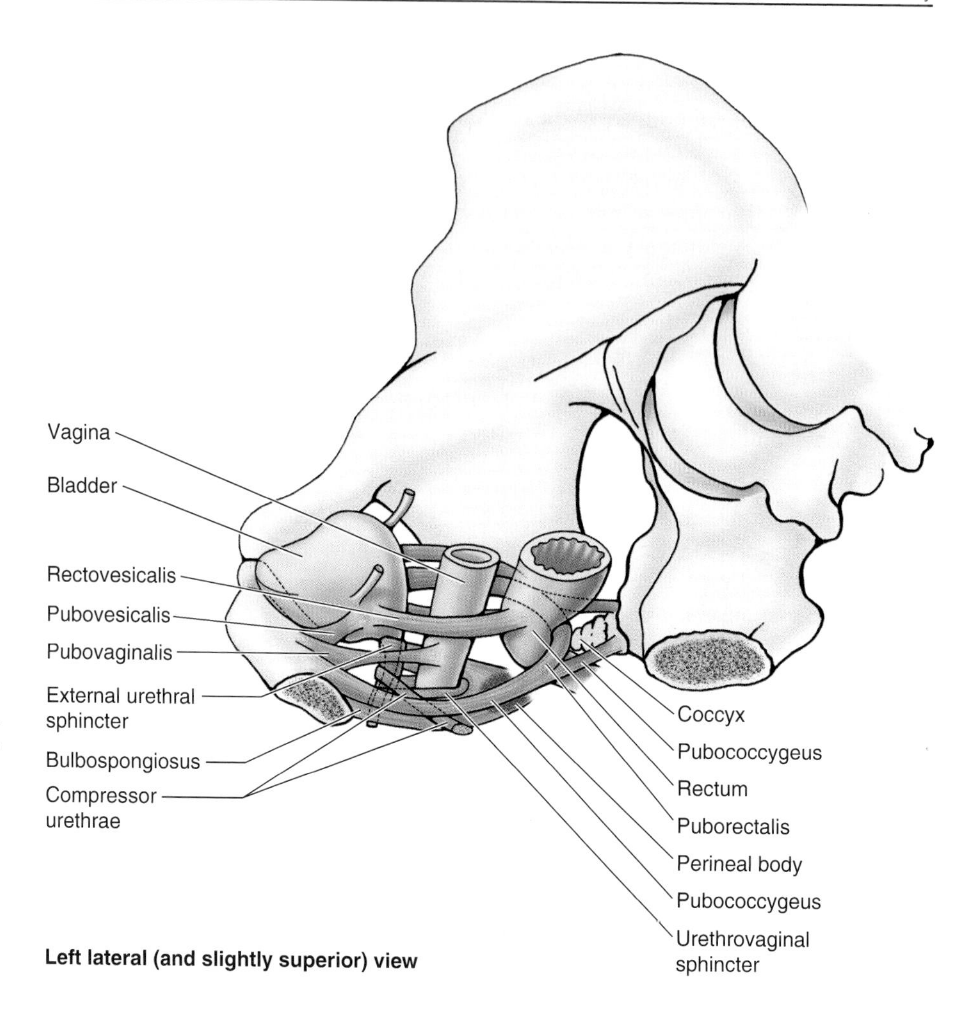

Figure 3.24. Muscles compressing the urethra and vagina. Four muscles compress the vagina and act like sphincters: pubovaginalis, external urethral sphincter (especially its urethrovaginal sphincter part), and bulbospongiosus. The compressor urethrae and external urethral sphincter compress the urethra.

fornix is the deepest part and is closely related to the rectouterine pouch. Four muscles compress the vagina and act like sphincters: pubovaginalis, external urethral sphincter, urethrovaginal sphincter, and bulbospongiosus (Fig. 3.24). *The relations of the vagina are:*

- Anteriorly—the base of the bladder and urethra
- Laterally—the levator ani, visceral pelvic fascia, and ureters
- Posteriorly (inferior to superior)—the anal canal, rectum, and rectouterine pouch.

Arterial Supply of the Vagina. The arteries supplying the superior part of the vagina derive from the **uterine arteries** (Fig. 3.25); the **vaginal arteries** supplying the middle and inferior parts of the vagina derive from the *middle rectal artery* and the *internal pudendal artery.*

Venous and Lymphatic Drainage of the Vagina. The vaginal veins form *vaginal venous plexuses* along the sides of the vagina and within the vaginal mucosa (Fig. 3.25). These veins are continuous with the uterine venous plexus as the *uterovaginal venous plexus* and drain into the internal iliac veins through the uterine vein. This plexus also communicates with the vesical and rectal venous plexuses. *The vaginal lymphatic vessels* (Fig. 3.26*A*) *drain from the parts of the vagina:*

- Superior part into the internal and external iliac lymph nodes
- Middle part into the internal iliac lymph nodes
- Inferior part into the sacral and common iliac nodes, as well as into the superficial inguinal lymph nodes.

Innervation of the Vagina. Most of the vagina (superior three-fourths to four-fifths) is visceral in terms of its innervation (Fig. 3.26*B*). Nerves to this part of the vagina are derived from the **uterovaginal plexus,** which travels with the uterine artery at the junction of the base of the peritoneal broad ligament and the upper part of the fascial transverse cervical ligament. The uterovaginal plexus is one of the pelvic plexuses that extends to the pelvic viscera from the inferior hypogastric plexus. Sympathetic, parasympathetic, and visceral afferent fibers pass through this plexus. *Sympathetic innervation* originates in the lower thoracic spinal cord segments and passes through *lumbar splanchnic nerves* and the intermesenteric/hypogastric series of plexuses. *Parasympathetic innervation* originates in the S2 through S4 spinal cord segments and passes through the *pelvic splanchnic nerves* to the inferior hypogastric/uterovaginal plexus. Afferent fibers from this part of the vagina accompany the parasympathetic fibers through the uterovaginal and inferior hypogastric plexuses and the pelvic splanchnic nerves to cell bodies in the spinal ganglia of S2, S3, and S4.

Only the lower one-fifth to one-fourth of the vagina is somatic in terms of innervation. Innervation of this part of the vagina is from the deep perineal branch of the **pudendal nerve**, which conveys sympathetic and afferent fibers but no parasympathetic fibers. *Only this somatically innervated part is sensitive to touch and temperature*, even though the somatic afferent fibers also have their cell bodies in the spinal ganglia of S2 through S4.

Distension of the Vagina

The vagina can be markedly distended by the fetus during parturition, particularly in an anteroposterior direction. Distension is limited laterally by the ischial spines, which project posteromedially, and the sacrospinous ligaments extending from these spines to the lateral margins of the sacrum and coccyx.

Examination of the Vagina

The interior of the vagina and the cervix can be examined with a *vaginal speculum* (*A* and *C*). The cervix, ischial spines, and sacral promontory can also be palpated with the digits in the vagina or rectum. Pulsations of the uterine arteries may also be felt through the lateral parts of the fornix, as may irregularities of the ovaries (such as cysts).

Culdoscopy, Laparoscopy, and Culdocentesis

A *culdoscope*—an endoscopic instrument—is inserted through the posterior part of the vaginal fornix (*B*) to examine the ovaries or uterine tubes (e.g., for the presence of a tubal pregnancy). Culdoscopy has been largely replaced by *laparoscopy*, which provides greater flexibility for operative procedures and better visualization of pelvic organs. There is also less potential for bacterial contamination of the peritoneal cavity. A pelvic abscess in the rectouterine pouch can be drained through an incision made in the posterior vaginal fornix (*culdocentesis*). Similarly, fluid in the peritoneal cavity (e.g., blood) can be aspirated by this technique. ○

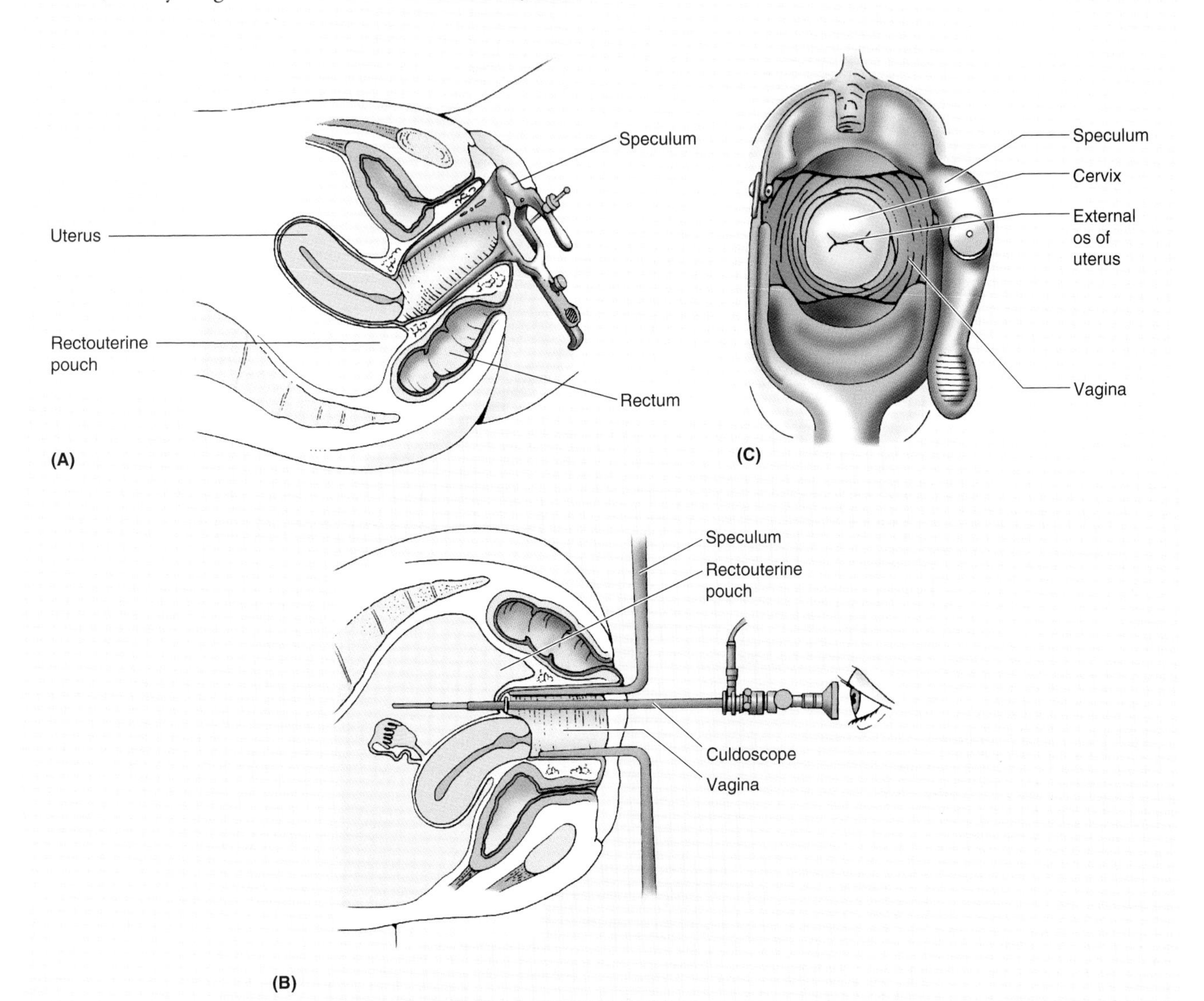

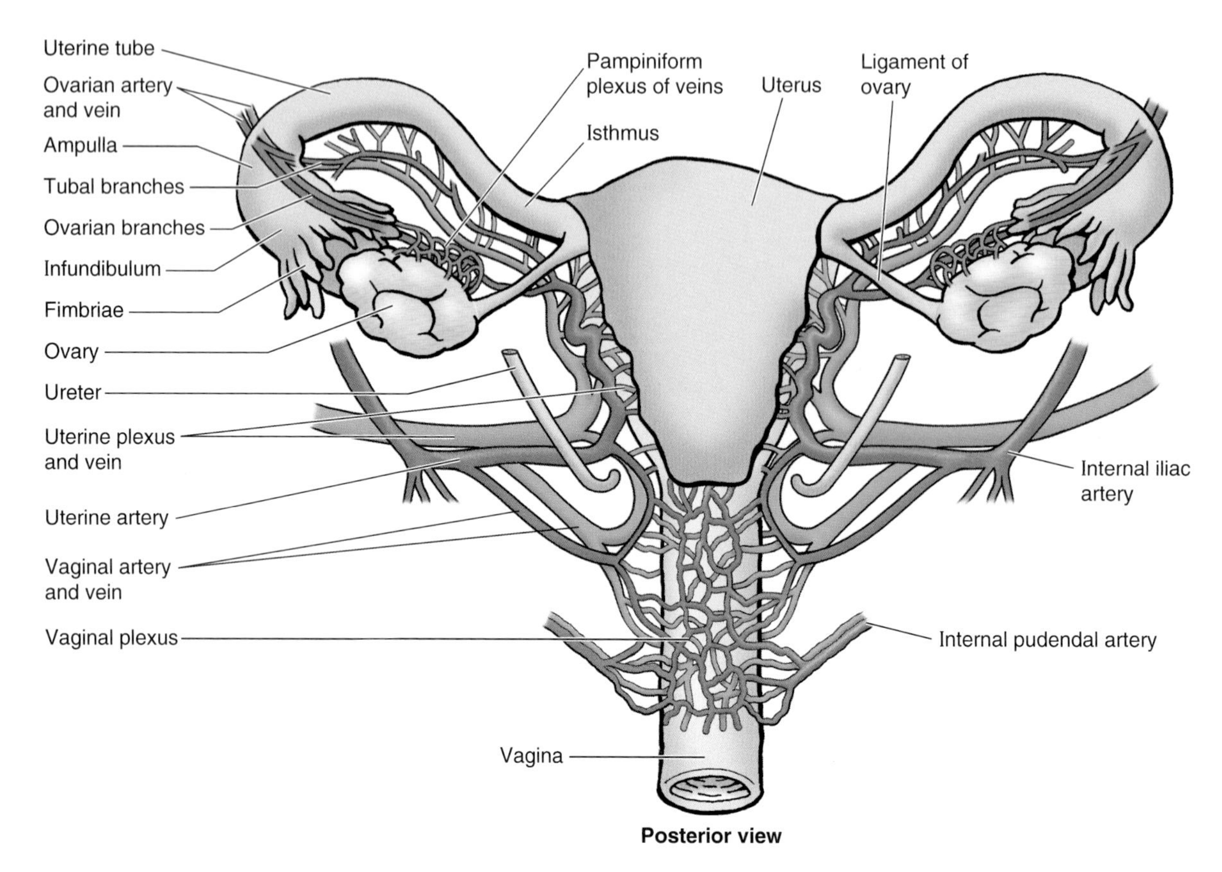

Figure 3.25. Blood supply and venous drainage of the uterus, vagina, and ovaries. The broad ligament of the uterus is removed to show the ovarian artery from the aorta and the uterine artery from the internal iliac artery supplying the ovary, uterine tube, and uterus. Observe also the anastomosing tubal and ovarian branches within the broad ligament (removed). Examine the pampiniform plexus and ovarian vein and the uterine plexus and vein.

Uterus

The uterus—a thick-walled, pear-shaped, hollow muscular organ—lies in the lesser pelvis normally with its body lying on top of the urinary bladder and its neck (cervix) between the urinary bladder and rectum (Figs. 3.23, 3.25, and 3.27). In the adult the uterus is usually *anteverted*—tipped anterosuperiorly relative to the axis of the vagina—and *anteflexed* (the uterine body is flexed or bent anteriorly relative to the cervix) so that its mass lies over the bladder.

The position of the uterus changes with the degree of fullness of the bladder and rectum. Although its size varies considerably, the uterus is approximately 7.5 cm long, 5 cm broad, and 2 cm thick and weighs approximately 90 grams. During pregnancy the uterus enlarges greatly to accommodate

Figure 3.26. Lymphatic drainage and innervation of the female pelvis. A. Lymphatic drainage of the uterus and uterine tubes. The *arrows* indicate the direction of lymph flow to the lymph nodes. **B.** Innervation of the uterus, uterine tubes, and vagina. Observe the parasympathetic pelvic splanchnic nerves arising from the ventral primary rami of S2, S3, and S4, supplying motor fibers to the uterus and vagina and vasodilator fibers to the erectile tissue of the clitoris and bulb of the vestibule. Observe also the presynaptic sympathetic fibers traversing the sympathetic trunk and passing through the lumbar splanchnic nerves to synapse with postsynaptic fibers, which travel through the superior and inferior hypogastric plexuses to reach the pelvic viscera. Afferent fibers conducting pain from intraperitoneal structures (such as the uterine body and fundus) travel with the sympathetic fibers to the T12 through L2 spinal ganglia. Afferent fibers conducting pain from subperitoneal structures such as the cervix and vagina (i.e., the birth canal) travel with parasympathetic fibers to S2 through S4 spinal ganglia. Somatic sensation from the opening of the vagina also passes to S2 through S4 spinal ganglia. Muscular contractions of the uterus are also hormonally induced.

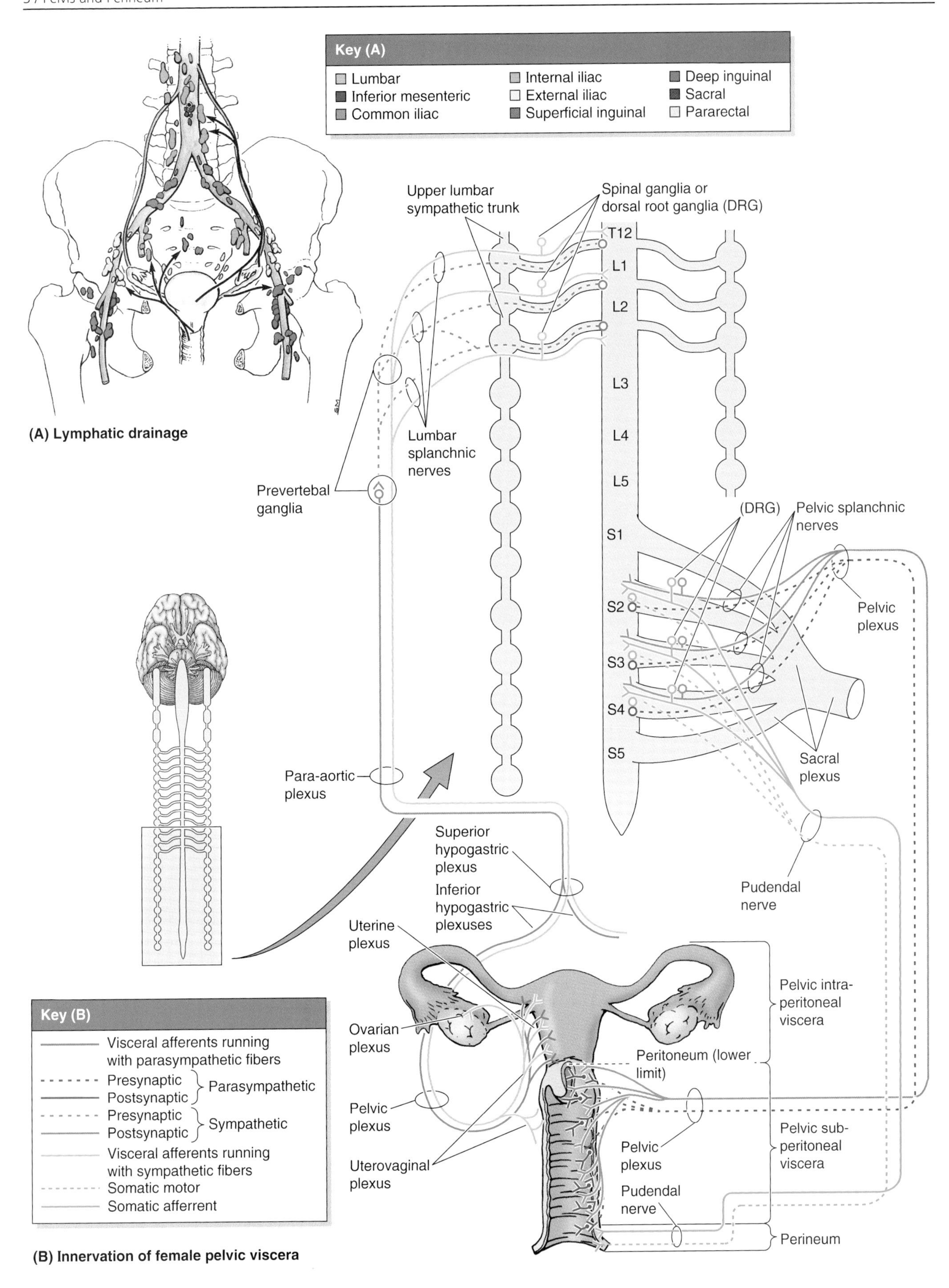

(A) Lymphatic drainage

(B) Innervation of female pelvic viscera

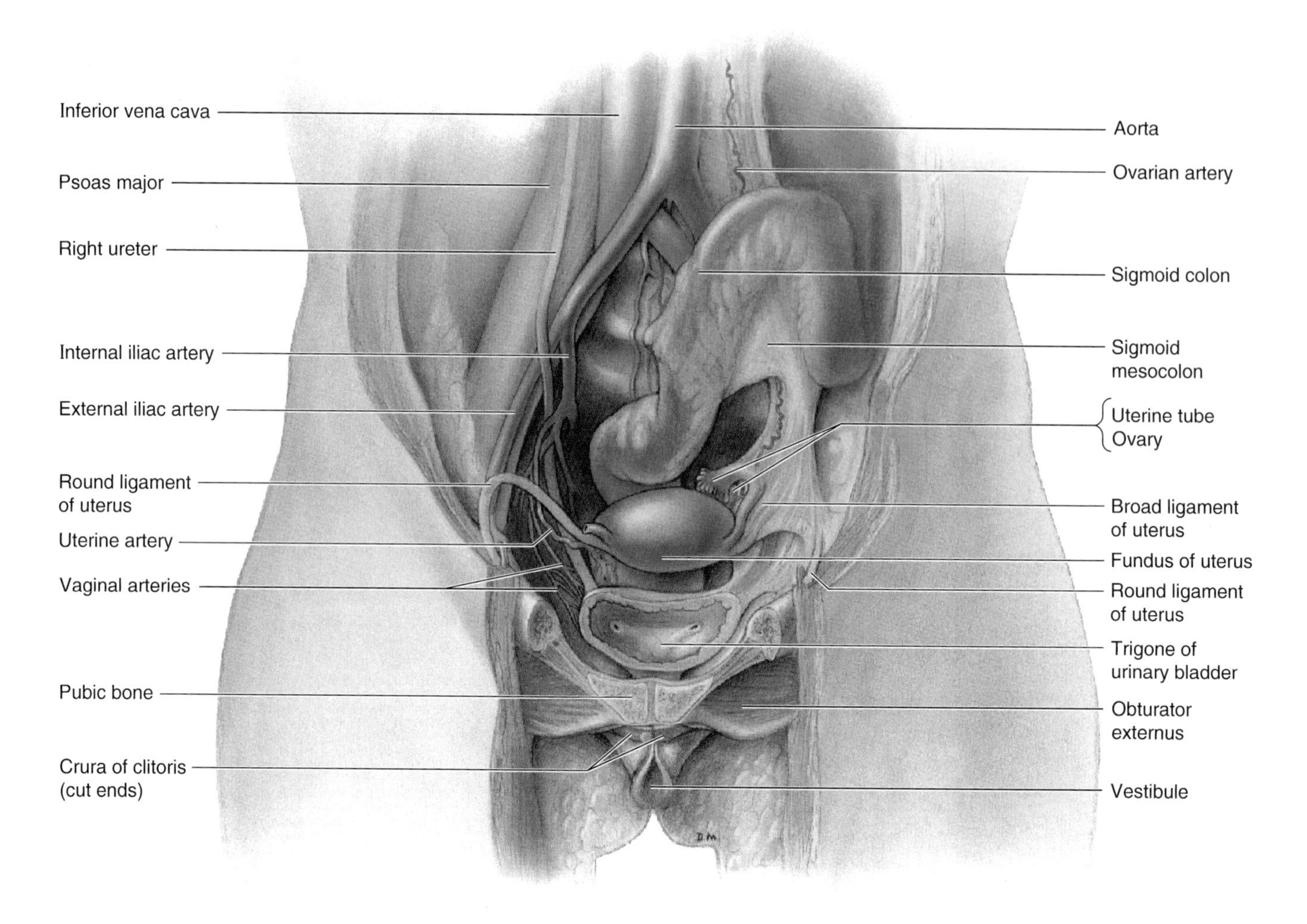

Figure 3.27. Lower abdomen and pelvis of a female. A. Dissection of the female genital organs and coronal section of the bladder and anterior pelvis. Part of the superior ramus and bodies of the pubic bones and the anterior aspect of the bladder have been removed. On the right side, the uterine tube, ovary, broad ligament, and peritoneum covering the lateral wall of the pelvis have been removed. Examine these structures on the left side.

the fetus. *The uterus is divisible into two main parts* (Fig. 3.22)—the body and cervix:

- **Body**, forming the upper two-thirds, has two parts:
- **Fundus**—the rounded part of the body that lies superior to the orifices of the uterine tubes
- **Isthmus**—the relatively constricted region of the body (approximately 1 cm long) just above the cervix
- **Cervix**—the cylindrical, narrow inferior part that protrudes into the uppermost vagina.

The uterus is where the embryo and fetus develop. During pregnancy, the uterus enlarges greatly.

The **body of the uterus** lies between the layers of the broad ligament and is freely movable. It has two surfaces: vesical (related to the bladder) and intestinal. The **uterine horns** (L. cornua) are the superolateral regions where the uterine tubes enter (Fig. 3.27*A*). In the postpubertal premenopausal woman, the body is the pear-shaped superior two-thirds of the uterus. During childhood and postmenopause, the body and cervix are approximately of equal length (height), with the cervix being of greater diameter (thickness). The slitlike **uterine cavity** is approximately 6 cm in length from the *external os* of the uterus to the wall of the fundus.

The **cervix of the uterus**—approximately 2.5 cm long in an adult nonpregnant woman—is divided into supravaginal and vaginal parts. The *supravaginal part of the cervix* is separated from the bladder anteriorly by loose connective tissue and from the rectum posteriorly by the **rectouterine pouch** (Fig. 3.23). The rounded *vaginal part of the cervix* extends into the vagina and communicates with it through the exter-

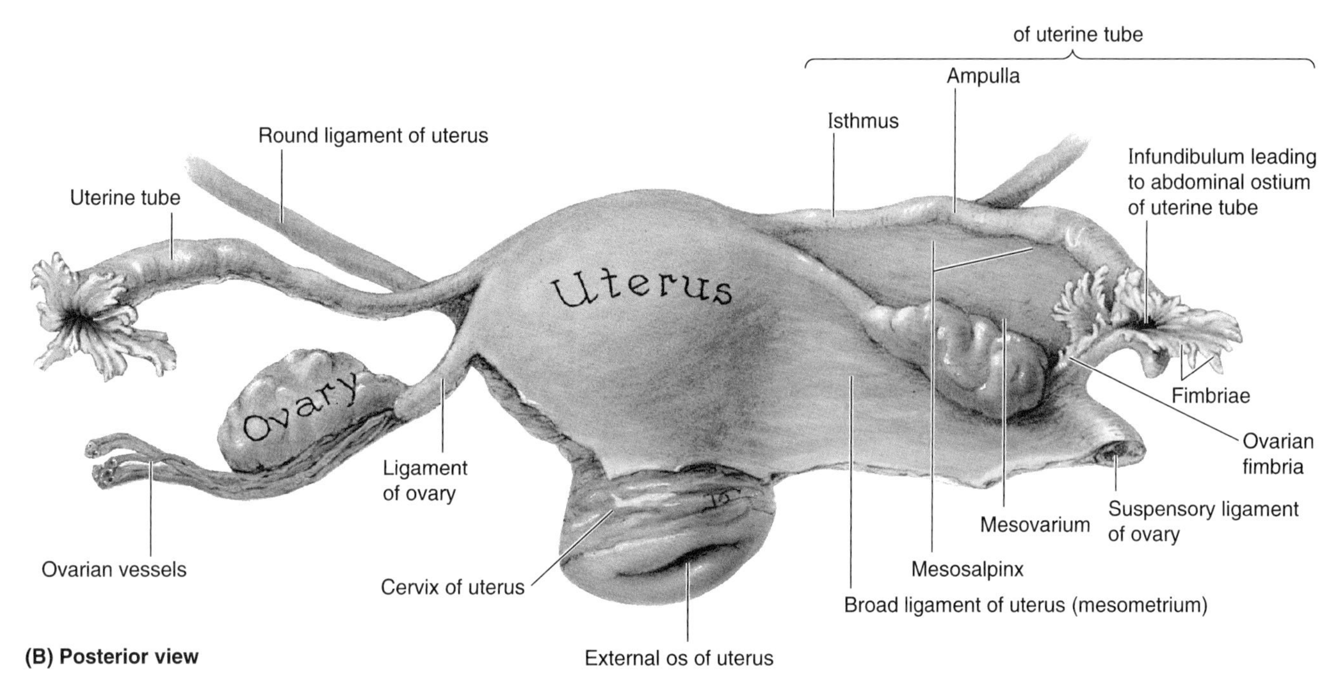

Figure 3.27. *(Continued)* **B.** Posterior view of the female internal genitalia: uterus, ovaries, uterine tubes, and related structures. The broad ligament is removed on the left side.

nal os. The fusiform *cervical canal* extends from the isthmus of the uterine body to the external os of the uterus. The cervical canal is broadest at its middle part and communicates with the uterine cavity through the internal os and with the vagina through the external os (Fig. 3.22).

The **ligament of the ovary** attaches to the uterus posteroinferior to the uterotubal junction. The **round ligament of the uterus** (L. ligamentum teres) attaches anteroinferiorly to this junction. *The wall of the body of the uterus consists of three layers*:

- The **perimetrium**—the outer serous coat—consists of peritoneum supported by a thin layer of connective tissue
- The **myometrium**—the middle muscular coat—becomes greatly distended (more extensive but thinner) during pregnancy; the main branches of the blood vessels and nerves of the uterus are located in the myometrium
- The **endometrium**—the inner mucous coat—is firmly adherent to the underlying myometrium.

The amount of muscular tissue in the cervix is markedly less than in the body of the uterus. The cervix is mostly fibrous and is composed mainly of collagen with a small amount of smooth muscle and elastin.

The **broad ligament of the uterus** is a double layer of peritoneum (mesentery) that extends from the sides of the uterus to the lateral walls and floor of the pelvis (Fig. 3.27). The broad ligament assists in keeping the uterus in position. The two layers of the broad ligament are continuous with each other at a free edge that surrounds the uterine tube. Laterally, the peritoneum of the broad ligament is prolonged superiorly over the vessels as the **suspensory ligament of the ovary** (Fig. 3.27*B*). The *ligament of the ovary* lies posterosuperiorly and the *round ligament of the uterus* lies anteroinferiorly between the layers of the broad ligament. The part of the broad ligament by which the ovary is suspended is the **mesovarium.** The part of the broad ligament forming the mesentery of the uterine tube is the **mesosalpinx**. The major part of the broad ligament, the mesentery of the uterus, or **mesometrium**, is below the mesosalpinx and mesovarium.

The uterus is a dense structure located in the center of the pelvic cavity. The *principal supports of the uterus* holding it in this position are both passive and active (dynamic). Dynamic support is provided by the **pelvic diaphragm.** Its tonus during sitting and standing, and active contraction during periods of increased intra-abdominal pressure (sneezing, coughing, etc.), is transmitted to the uterus through the *surrounding pelvic organs and the endopelvic fascia* in which they are embedded. Passive support of the body of the uterus is provided by the *position of the uterus*—the way in which the normally anteverted and anteflexed uterus "rests" on top of the bladder. When intra-abdominal pressure is increased, the uterus is pressed against the bladder instead of being pushed through the vagina, as the tendency would be if the uterus were upright over the vagina. *The cervix is the least mobile part of the uterus because of the passive support provided by attached condensations of endopelvic fascia ("ligaments"),* which may also contain smooth muscle (Fig. 3.28).

- **Transverse cervical (cardinal) ligaments** extend from the cervix and lateral parts of the fornix of the vagina to the lateral walls of the pelvis.
- **Uterosacral ligaments** pass superiorly and slightly posteriorly from the sides of the cervix to the middle of the sacrum; they are palpable during a rectal examination.

Together these passive and active supports keep the uterus centered in the pelvic cavity and resist the tendency for the uterus to fall or be pushed through the hollow tube formed by the vagina (*uterine prolapse*).

Relations of the Uterus. Peritoneum covers the uterus anteriorly and superiorly (Fig. 3.23), except for the vaginal part of the cervix. The peritoneum is reflected anteriorly from the uterus onto the bladder and posteriorly over the posterior part of the fornix of the vagina to the rectum. Anteriorly, the uterine body is separated from the urinary bladder by the **vesicouterine pouch**, where the peritoneum is reflected from the uterus onto the posterior margin of the superior surface of the bladder. Posteriorly, the uterine body and supravaginal part of the cervix are separated from the sigmoid colon by a layer of peritoneum and the peritoneal cavity and from the rectum by the **rectouterine pouch**. Laterally, the uterine artery crosses the ureter superiorly, near the cervix (Fig. 3.25).

Summary of the relations of the uterus:

- *Anteriorly* (anteroinferiorly in its normal anteverted position), the vesicouterine pouch and superior surface of the bladder; the supravaginal part of the cervix is related to the bladder and is separated from it only by fibrous connective tissue
- *Posteriorly*, the rectouterine pouch containing loops of intestine and the anterior surface of rectum; only the visceral pelvic fascia provides support against increased intra-abdominal pressure
- *Laterally*, the peritoneal broad ligament and fascial transverse cervical (cardinal) ligaments and ureters; the ureters run anteriorly slightly superior to the lateral part of the vaginal fornix and inferior to the uterine arteries, usually approximately 2 cm lateral to the supravaginal part of the cervix.

Arterial Supply of the Uterus. The blood supply of the uterus derives mainly from the **uterine arteries** with an additional supply from the ovarian arteries (Fig. 3.25).

Venous and Lymphatic Drainage of the Uterus. The uterine veins enter the broad ligaments with the arteries and form a **uterine venous plexus** on each side of the cervix (Fig. 3.25). Veins from the uterine plexus drain into the **internal iliac veins**. *The uterine lymphatic vessels follow three main routes* (Fig. 3.26*A*):

- Most vessels from the fundus pass to the lumbar lymph nodes, but some vessels pass to the external iliac lymph nodes or run along the round ligament of the uterus to the superficial inguinal lymph nodes.
- Vessels from the uterine body pass within the broad ligament to the external iliac lymph nodes.
- Vessels from the uterine cervix pass to the internal iliac and sacral lymph nodes.

Innervation of the Uterus. The nerves to the uterus are derived from the uterovaginal plexus, which travels with the uterine artery at the junction of the base of the peritoneal broad ligament and the superior part of the fascial transverse cervical ligament. The **uterovaginal plexus** is one of the pelvic plexuses that extend to the pelvic viscera from the **inferior hypogastric plexus** (Fig. 3.26*B*). Sympathetic, parasympathetic, and visceral afferent fibers pass through this plexus. *Sympathetic innervation* originates in the lower thoracic spinal cord segments and passes through lumbar splanchnic nerves and the intermesenteric/hypogastric series of plexuses. *Parasympathetic innervation* originates in the S2 through S4 spinal cord segments and passes through the **pelvic splanchnic nerves** to the inferior hypogastric/uterovaginal plexus. The afferent innervation of the upper (intraperitoneal—fundus and body) and lower (subperitoneal—mainly cervical) parts of the uterus differ in terms of course and source. Afferent fibers conducting pain impulses from the intraperitoneal uterine fundus and body follow the sympathetic innervation retrogradely, passing through the uterovaginal, inferior and superior hypogastric and intermesenteric plexuses, and then through the lumbar splanchnic nerves, lumbar sympathetic trunks, and gray rami communicantes to reach cell bodies in the lower thoracic/upper lumbar spinal ganglia. Afferent fibers conducting pain impulses from the subperitoneal uterine cervix (as well as all afferent fibers from the uterus not concerned with pain) follow the parasympathetic fibers retrogradely through the uterovaginal and inferior hypogastric plexuses and pelvic splanchnic nerves to reach cell bodies in the spinal ganglia of S2 through S4.

Pelvic Fascia

The pelvic fascia is connective tissue that occupies all the space between the membranous peritoneum and the muscular pelvic walls and floor not occupied by the pelvic organs. This "layer" is a continuation of the comparatively thin endoabdominal fascia that lies between the muscular abdominal walls and the peritoneum superiorly. Traditionally, the pelvic fascia has been described as having parietal and visceral components (Fig. 3.28). This concept is useful only until it gets in the way of understanding or communication.

The **parietal pelvic fascia** is a membranous layer of variable thickness that lines the inner (deep or pelvic) aspect of the muscles forming the walls and floor of the pelvis (Fig. 3.28*A*).

Figure 3.28. Pelvic fascia: endopelvic fascia and fascial ligaments. A. Coronal section of a female pelvis illustrating the pelvic fascia. **B.** Coronal section of a male pelvis demonstrating the pelvic fascia. **C.** Transverse section of the pelvis at the level shown in (**A**), illustrating the female (endopelvic) fascia. **D.** Transverse section of the pelvis at the level shown in (**B**), demonstrating the male (endopelvic) fascia. **E.** Pelvic fascial ligaments adjacent to the male pelvic floor.

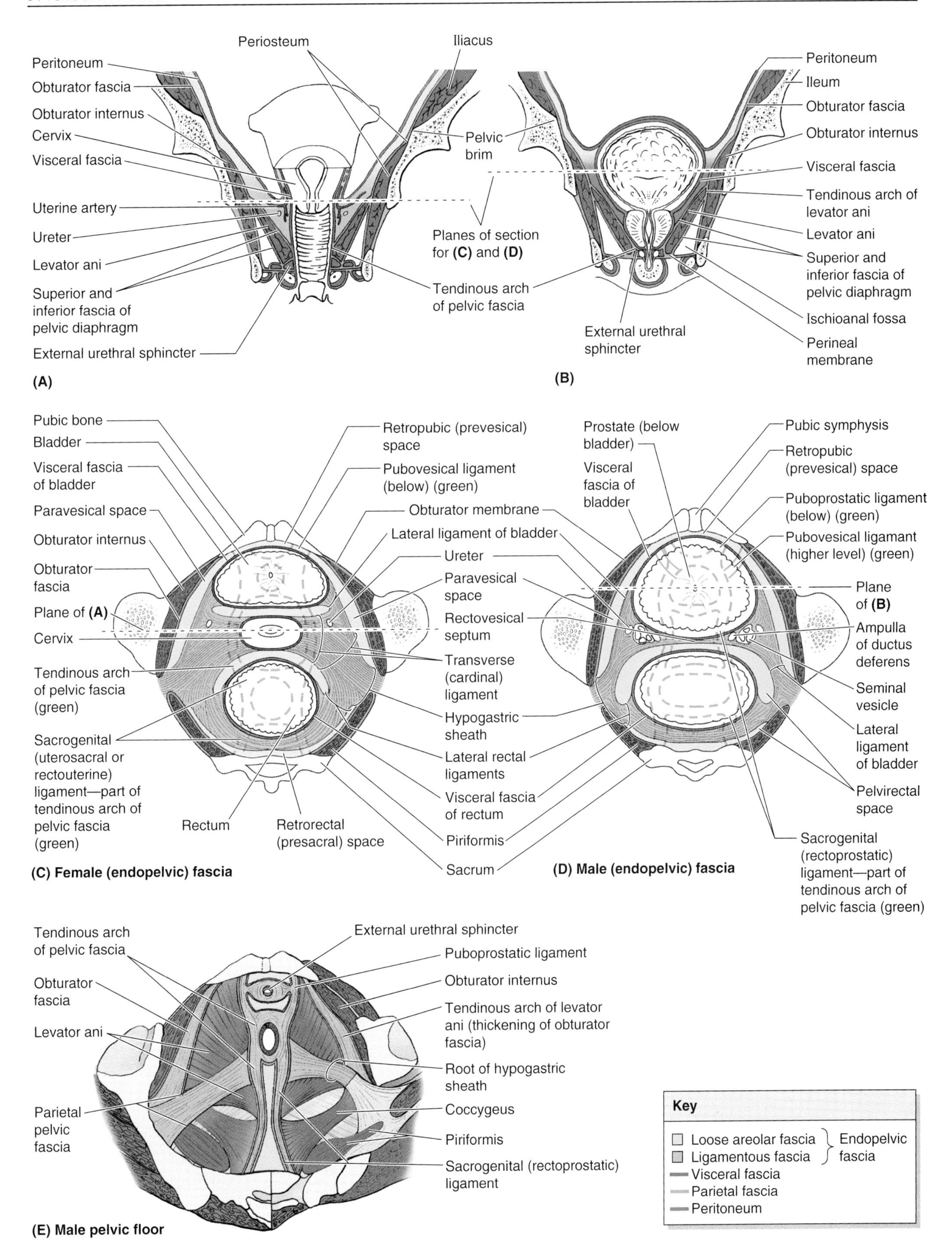
Periosteum
Iliacus
Peritoneum
Obturator fascia
Obturator internus
Cervix
Visceral fascia
Pelvic brim
Uterine artery
Ureter
Levator ani
Superior and inferior fascia of pelvic diaphragm
External urethral sphincter
Planes of section for (C) and (D)
Tendinous arch of pelvic fascia
(A)
Peritoneum
Ileum
Obturator fascia
Obturator internus
Visceral fascia
Tendinous arch of levator ani
Levator ani
Superior and inferior fascia of pelvic diaphragm
Ischioanal fossa
Perineal membrane
External urethral sphincter
(B)
Pubic bone
Bladder
Visceral fascia of bladder
Paravesical space
Obturator internus
Obturator fascia
Plane of (A)
Cervix
Tendinous arch of pelvic fascia (green)
Sacrogenital (uterosacral or rectouterine) ligament—part of tendinous arch of pelvic fascia (green)
Rectum
Retrorectal (presacral) space
Retropubic (prevesical) space
Pubovesical ligament (below) (green)
Obturator membrane
Lateral ligament of bladder
Ureter
Paravesical space
Rectovesical septum
Transverse (cardinal) ligament
Hypogastric sheath
Lateral rectal ligaments
Visceral fascia of rectum
Piriformis
Sacrum
(C) Female (endopelvic) fascia
Prostate (below bladder)
Visceral fascia of bladder
Pubic symphysis
Retropubic (prevesical) space
Puboprostatic ligament (below) (green)
Pubovesical ligamant (higher level) (green)
Plane of (B)
Ampulla of ductus deferens
Seminal vesicle
Lateral ligament of bladder
Pelvirectal space
Sacrogenital (rectoprostatic) ligament—part of tendinous arch of pelvic fascia (green)
(D) Male (endopelvic) fascia
Tendinous arch of pelvic fascia
Obturator fascia
Levator ani
Parietal pelvic fascia
External urethral sphincter
Puboprostatic ligament
Obturator internus
Tendinous arch of levator ani (thickening of obturator fascia)
Root of hypogastric sheath
Coccygeus
Piriformis
Sacrogenital (rectoprostatic) ligament
(E) Male pelvic floor
Key
Loose areolar fascia
Ligamentous fascia
Endopelvic fascia
Visceral fascia
Parietal fascia
Peritoneum

The parietal pelvic fascia therefore covers the pelvic surfaces of the obturator internus, piriformis, coccygeus, levator ani, and part of the urethral sphincter muscles (Fig. 3.28, *C–E*). Specific parts of the parietal fascia are named for the muscle they cover (e.g., obturator fascia).

The **visceral pelvic fascia** includes the membranous fascia that directly ensheathes the pelvic organs, forming the adventitial layer of each. The membranous parietal and visceral layers become continuous where the organs penetrate the pelvic floor. Here the parietal fascia thickens, forming the **tendinous arch of pelvic fascia**, a continuous bilateral band running from the pubis to the sacrum along the pelvic floor adjacent to the viscera. The anteriormost part of this tendinous arch or band (the *puboprostatic ligament* in males or the *pubovesical ligament* in females) connects the prostate to the pubis in the male or the base of the bladder to the pubis in the female. The posteriormost part of the band runs as the **sacrogenital ligaments** from the sacrum around the side of the rectum to attach to the prostate in the male or the vagina in the female.

Most often, the abundant connective tissue remaining between these two membranous layers has been considered part of the visceral fascia, but various authors label parts of it as parietal. It is probably most realistic to consider this remaining fascia simply as extraperitoneal or *subperitoneal endopelvic fascia* (Fig. 3.28, *C* and *D*), which is continuous with both the parietal and visceral membranous fascias. This fascia forms a connective tissue matrix or packing material for the pelvic viscera (Fig. 3.28, *B* and *C*). It varies markedly in density and content. Some of it is extremely loose areolar (fatty) tissue, relatively devoid of all but minor lymphatics and nutrient vessels. In dissection or surgery, the fingers can be pushed into this loose tissue with ease (e.g., between the pubis and the bladder anteriorly, and between the sacrum and rectum posteriorly), creating actual spaces by blunt dissection. These "potential spaces," normally consisting only of a layer of loose, fatty tissue, are the *retropubic* (or *prevesical, extended posterolaterally as paravesical*) and *retrorectal* (or *presacral*) spaces, consecutively. The presence of loose connective tissue here accommodates the expansion of the urinary bladder and rectal ampulla as they fill.

Although these types of endopelvic fascia do not differ much in their gross appearance, other parts of the endopelvic fascia have a much more fibrous consistency, containing an abundance of collagen and elastic fibers and, according to some authors, a scattering of smooth muscle fibers. These parts are often described as "fascial condensations" or pelvic "ligaments." For example, during dissection, if you insert the fingers of one hand into the retropubic space and the fingers of the other hand into the presacral space and attempt to bring them together along the lateral pelvic wall, you will find that they do not meet or pass from one space to the other. They encounter the so-called *hypogastric sheath*, a thick band of condensed pelvic fascia. This fascial condensation is not merely a barrier separating the two potential spaces; it gives passage to essentially all the vessels and nerves passing from the lateral wall of the pelvis to the pelvic viscera, along with the ureters and, in the male, the ductus deferens. As it extends medially from the lateral wall, the hypogastric sheath divides into three laminae ("leaflets" or "wings") that pass to or between the pelvic organs, conveying neurovascular structures and providing support. Because of the latter function, they are also referred to as ligaments. The anteriormost lamina, the *lateral ligament of the bladder*, passes to the bladder, conveying the superior vesical arteries and veins. The posteriormost lamina passes to the rectum, conveying the middle rectal artery and vein.

In the male, the middle lamina forms a relatively thin fascial partition—the *rectovesical septum* (Fig. 3.28*D*)—between the posterior surface of the bladder and the prostate anteriorly and the rectum posteriorly. *In the female*, the middle lamina is the most substantial of the three, passing medially to the uterine cervix and vagina as the *transverse cervical (cardinal) ligament*—also known clinically as the *lateral cervical* or *Mackenrodt's ligament* (Fig. 3.28*C*). In its uppermost portion, at the base of the peritoneal broad ligament, the uterine artery runs transversely toward the cervix while the ureters pass immediately beneath them as they pass on each side of the cervix toward the bladder. This relationship ("water passing under the bridge") is an especially important one for surgeons ligating the uterine artery, as in a hysterectomy (surgical excision of the uterus). The **transverse cervical ligament**, and the way in which the uterus normally "rests" on top of the bladder, provides the main passive support for the uterus. The perineal muscles provide dynamic support for the uterus by contracting during moments of increased intra-abdominal pressure (sneezing, coughing, etc.). Passive and dynamic supports together resist the tendency for the uterus to fall or be pushed through the hollow tube formed by the vagina (uterine prolapse). The transverse cervical ligament has enough fibrous content to anchor wide loops of suture during surgical repairs.

In addition to the ischioanal fossae (Fig. 3.28*B*), there is a surgically important potential *pelvirectal space* in the loose extraperitoneal connective tissue superior to the levator ani (Fig. 3.28*D*). It is divided into anterior and posterior regions by the *lateral rectal ligaments,* which connect the rectum to the parietal pelvic fascia covering the sacrum at the levels of the S2 through S4 pelvic foramina. The rectal arteries and plexuses are embedded in the lateral ligaments.

Cervical Cancer

Until 1940, cervical cancer was the leading cause of death in North American women (Krebs, 1993). The decline in the incidence and number of women dying from cervical cancer is related to the accessibility of the cervix to direct visualization and to cell and tissue study (Papanicolaou smears), leading to the detection of premalignant cervical conditions (Copeland, 1993; Morris and Burke, 1993). ▶

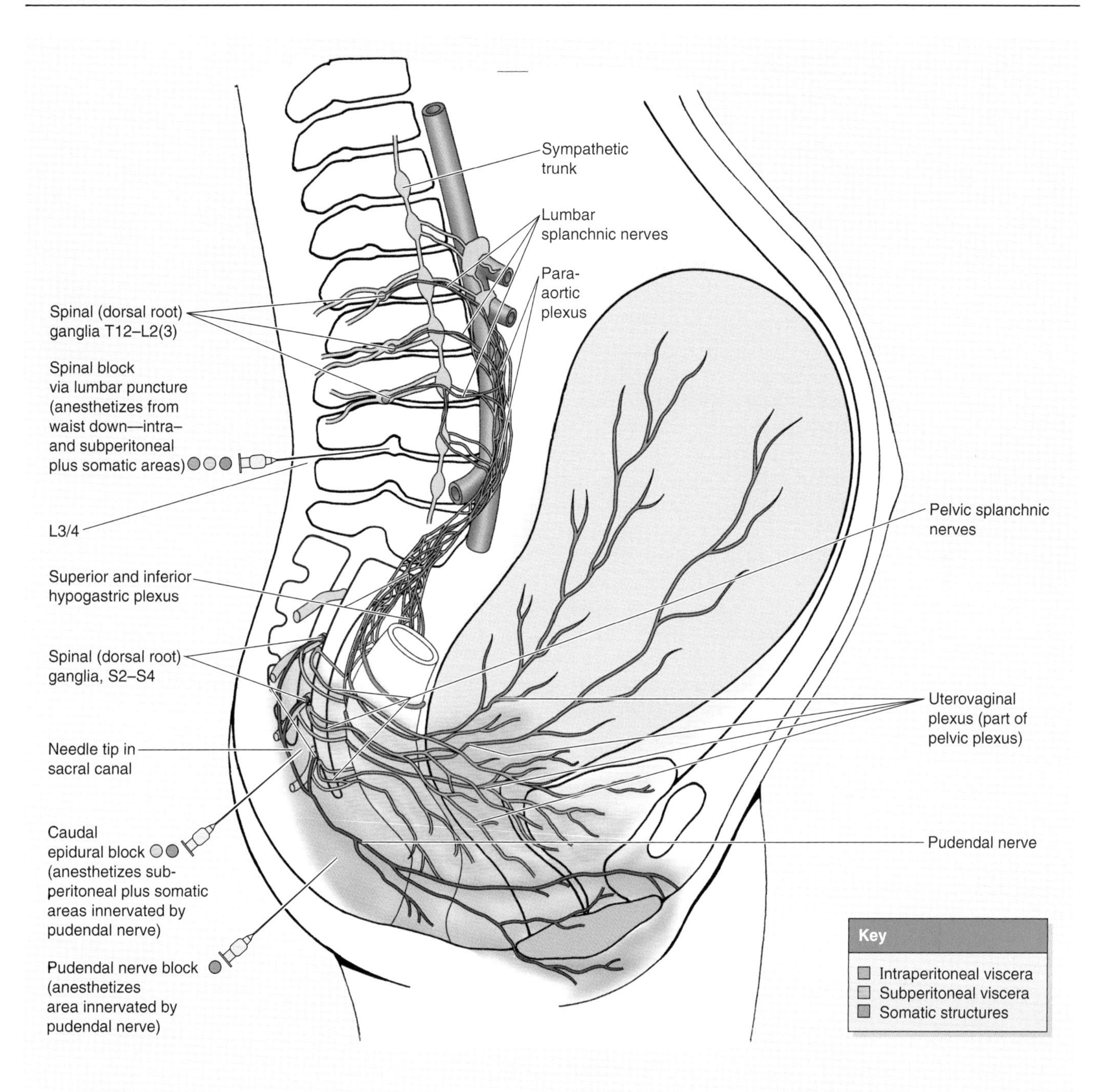

Anesthesia for Childbirth

Several options are available to women to reduce the pain and discomfort experienced during childbirth. **General anesthesia** has advantages for emergency procedures and for women who choose it over regional anesthesia. Because the woman is unconscious, clinicians monitor and regulate maternal respiration and both maternal and fetal cardiac function. Women who choose **regional anesthesia**—such as a spinal, pudendal nerve, or caudal epidural block—often wish to participate actively (e.g., using the Lamaze method) and be conscious of their uterine contractions to "bear down" or push to assist the contractions and expel the fetus, yet do not wish to experience the pain of labor.

General anesthesia renders the mother unconscious; she is unaware of the labor and delivery. Childbirth occurs passively under the control of maternal hormones with the assistance of an obstetrician. The mother is spared pain and discomfort but is unaware of the earliest moments of her baby's life.

A **spinal block**—in which the anaesthetic agent is introduced with a needle into the spinal subarachnoid (leptomeningeal) space at the L3/L4 vertebral level—anesthetizes essentially everything inferior to the waist. The ▶

▶ perineum, pelvic floor, and birth canal are anesthetized, and motor and sensory functions of the entire lower limbs, as well as sensation of uterine contraction, are temporarily eliminated. The mother is conscious but she must depend on electronic monitoring of uterine contractions. If labor is extended or the level of anesthesia was inadequate, it is difficult or may be impossible to readminister the anesthesia. Because the anesthetic agent is heavier than cerebrospinal fluid, it remains in the lower spinal subarachnoid space while the patient is inclined. The anesthetic agent circulates into the cerebral subarachnoid space in the cranial cavity when the patient lies flat following the delivery. A severe headache is a common sequel to spinal anesthesia.

In a **pudendal nerve block**, the anesthetic agent is injected near the pudendal nerve as it exits the greater sciatic foramen and enters the lesser sciatic foramen (i.e., where it passes over the ischial spine). It provides anesthesia specifically over the S2 through S4 dermatomes (the majority of the perineum) and the lower one-fourth of the vagina. It does not block pain from the upper birth canal (uterine cervix and upper vagina), so the mother is able to feel uterine contractions. It can be readministered, but to do so may be disruptive and involve the use of a sharp instrument in close proximity to the infant's head.

The **caudal epidural block** has become a popular choice for participatory childbirth. It must be administered in advance of the actual delivery, which is not possible with a precipitous birth. The anesthetic agent is administered using an in-dwelling catheter in the *sacral canal* (see Chapter 4), enabling administration of more anesthetic agent for a deeper or more prolonged anesthesia if necessary. Within the sacral canal, the anesthesia bathes the S2 through S4 spinal nerve roots, including the pain fibers from the uterine cervix and upper vagina, and the afferent fibers from the pudendal nerve. Thus, the entire birth canal, pelvic floor, and majority of the perineum are anesthetized but the lower limbs usually are not affected. The pain fibers from the uterine fundus and body ascend to the lower thoracic/upper lumbar levels; these and all the fibers superior to them are not affected by the anesthetic, so the mother is aware of her uterine contractions. With epidural anesthesia, no "spinal headache" occurs because the vertebral epidural space is not continuous with the cranial epidural space (see Chapter 4).

Age Changes in the Uterus

When a female baby is born, her uterus is relatively large and has adult proportions (two-thirds body, one-third cervix) because of the prepartum (before childbirth) influence of the adult maternal hormones. Several weeks postpartum (after childbirth), childhood dimensions and proportions are assumed. Because of the small size of the pelvic cavity during infancy, the uterus is mainly an abdominal organ and the cervix is relatively large (approximately 50% of total uterus) throughout childhood. During puberty the uterus (especially its body) grows rapidly. At menopause (46 to 52 years of age), the uterus decreases in size. Incomplete fusion of the embryonic paramesonephric ducts results in a variety of congenital anomalies (e.g., double uterus; see Fig. 3.52*B*) (Moore and Persaud, 1998).

Examination of the Uterus

The uterus may be examined by bimanual palpation. Two fingers of the right hand are passed superiorly in the vagina while the other hand is pressed inferoposteriorly on the pubic region of the anterior abdominal wall. The size and other characteristics of the uterus can be determined in this way (e.g., whether the uterus is in its normal anteverted position). When softening of the uterine isthmus occurs (*Hegar's sign*), the cervix feels as though it were separate from the body. Softening of the uterine isthmus is an early sign of pregnancy.

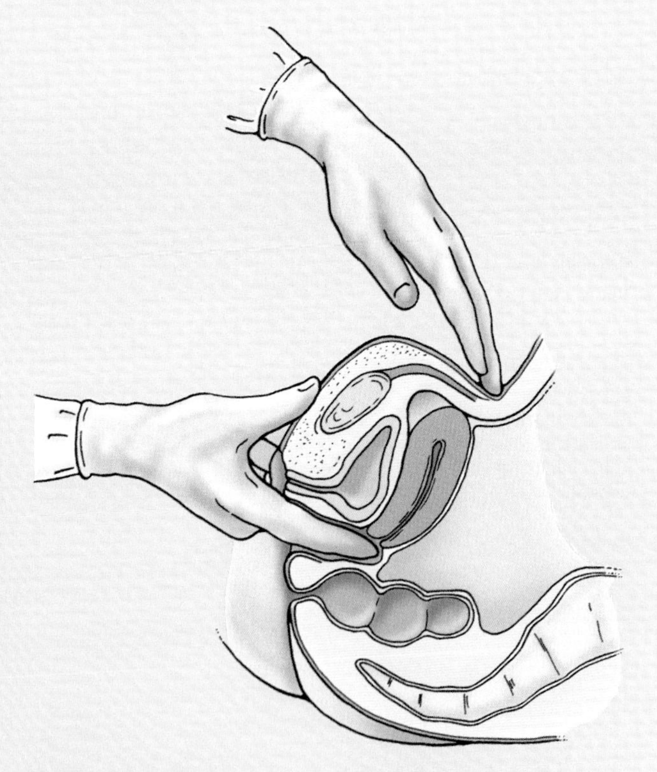

Hysterectomy

Hysterectomy (excision of the uterus) is performed through the anterior abdominal wall or through the vagina. Because the uterine artery crosses anterior and superior to the ureter near the lateral fornix of the vagina, the ureter is in danger of being inadvertently clamped or severed when the artery is tied off. The point of crossing of the artery and the ureter is approximately 2 cm superior to the ischial spine. The left ureter is particularly vulnerable because it runs close to the lateral aspect of the cervix. ⊙

Uterine Tubes

The uterine tubes (formerly fallopian tubes) extend laterally from the uterine horns and open into the peritoneal cavity near the ovaries (Fig. 3.27). The uterine tubes (approximately 10 cm long) lie in the *mesosalpinx* formed by the free edges of the broad ligaments. In the "ideal" disposition as typically illustrated, the tubes extend symmetrically posterolaterally to the lateral pelvic walls, where they arch anterior and superior to the ovaries in the horizontally disposed broad ligament. In reality, as seen on ultrasound examination, the tubes are commonly asymmetrically arranged with one or the other often lying superior and even posterior to the uterus.

The uterine tubes are divisible into four parts, from lateral to medial (Fig. 3.27*B*):

- The **infundibulum** is the funnel-shaped distal end that opens into the peritoneal cavity through the **abdominal ostium**; the fingerlike processes of the fimbriated end of the infundibulum—the **fimbriae**—spread over the medial surface of the ovary; one large *ovarian fimbria* is attached to the superior pole of the ovary.
- The **ampulla**, the widest and longest part, begins at the medial end of the infundibulum; oocytes expelled from the ovaries usually are fertilized in the ampulla.
- The **isthmus**, the thick-walled part, enters the uterus horn.
- The **uterine part** is the short intramural segment that passes through the wall of the uterus (Fig. 3.22) and opens through the **uterine ostium** into the uterine cavity.

Arterial Supply of the Uterine Tubes. The tubal branches arise as anastomosing terminal branches of the **uterine** and **ovarian arteries** (Fig. 3.25).

Venous and Lymphatic Drainage of the Uterine Tubes. The tubal veins drain into the ovarian veins and **uterine venous plexus** (Fig. 3.25). The lymphatic vessels drain to the *lumbar lymph nodes* (Fig. 3.26*A*).

Innervation of the Uterine Tubes. The nerve supply derives partly from the *ovarian plexus* and partly from the *uterine plexus* (Fig. 3.26*B*). Afferent fibers ascend through the ovarian plexus and lumbar splanchnic nerves to cell bodies in the T11 through L1 spinal ganglia.

Infections of the Female Genital Tract

Because the female genital tract communicates with the peritoneal cavity through the abdominal ostia, infections of the vagina, uterus, and tubes may result in peritonitis. Conversely, inflammation of the tube (*salpingitis*) may result from infections that spread from the peritoneal cavity. *A major cause of infertility in women is blockage of the uterine tubes*, often the result of pelvic infection that causes *salpingitis*.

Salpingography

Patency of the uterine tubes may be determined by *salpingography*, a radiographic procedure involving injection of a water-soluble radiopaque material into the uterus—*hysterosalpingography* (see Fig. 3.52*A*). The material enters the uterine tubes, and if the tubes are patent, passes from the abdominal ostium into the peritoneal cavity.

Endoscopy

Patency of the uterine tubes can also be determined by *endoscopy*—examination of the interior of the tubes using a special instrument, an *endoscope*, introduced through the vagina and uterus (*hysteroscopy*).

Ligation of the Uterine Tubes

Ligation of the uterine tubes, a surgical method of birth control, is remarkably safe. Oocytes discharged from the ovaries that enter the tubes of these patients die and soon disappear. Most surgical sterilizations are done by either abdominal tubal ligation or laparoscopic tubal ligation. *Abdominal tubal ligation* is usually performed through a short suprapubic incision made just at the pubic hairline. *Laparoscopic tubal ligation* is done with a laparoscope. It is similar to a small telescope with a powerful light and is inserted through a small incision, usually near the umbilicus.

Ectopic Tubal Pregnancy

In some women, collections of pus may develop in the uterine tube (*pyosalpinx*) and the tube may be partly occluded by adhesions. In these cases—although obviously sperms have done so—the dividing zygote may not be able to pass along the tube to the uterus. The blastocyst may implant in the mucosa of the uterine tube, producing an *ectopic tubal pregnancy*. Although implantation may occur in any part of the tube, the common site is in the ampulla. *Tubal pregnancy is the most common type of ectopic gestation;* it occurs approximately once in every 250 pregnancies in North America (Moore and Persaud, 1998). If not diagnosed early, ectopic tubal pregnancies may result in rupture of the uterine tube and hemorrhage into the abdominopelvic cavity during the first 8 weeks of gestation. Tubal rupture and the associated severe hemorrhage constitute a threat to the mother's life and result in death of the embryo. On the right side, the appendix often lies close to the ovary and uterine tube. This close relationship explains why a *ruptured tubal pregnancy* and the resulting peritonitis may be misdiagnosed as acute appendicitis. In both cases the parietal ▶

▶ peritoneum is inflamed in the same general area, and the pain is referred to the right lower quadrant of the abdomen.

Remnants of the Embryonic Ducts

Occasionally the mesosalpinx between the uterine tube and the ovary contains embryonic remnants (Fig. 3.22). The *epoöphoron* forms from remnants of the mesonephric tubules of the *mesonephros*—the transitory embryonic kidney (Moore and Persaud, 1998). There may also be a persistent *duct of the epoöphoron* (duct of Gartner), a remnant of the mesonephric duct that forms the ductus deferens and ejaculatory duct in the male. It lies between layers of the broad ligament along either side of the uterus and/or vagina. A *vesicular appendage* is sometimes attached to the infundibulum of the uterine tube. It is the remains of the cranial end of the mesonephric duct that forms the *ductus epididymis* (Moore and Persaud, 1998). Although these vestigial structures are mostly of embryological and morphological interest, they occasionally accumulate fluid and form cysts (e.g., Gartner's duct cysts). ○

Ovaries

The ovaries are almond-shaped glands located close to the lateral pelvic walls suspended by the mesovarium of the broad ligament (Fig. 3.27, *A* and *B*). In prepubertal (before puberty) females, the surface of the ovary is covered by a smooth layer of ovarian surface epithelium—a single layer of cuboidal cells—that gives the surface a dull, grayish appearance, contrasting with the shiny surface of the adjacent peritoneal mesovarium with which it is continuous. Following puberty, the surface becomes progressively scarred and distorted because of the repeated rupture of ovarian follicles and discharge of oocytes (ova) that are part of ovulation. The scarring is less in women who have been taking oral contraceptives that inhibit ovulation.

The distal end of the ovary connects to the lateral wall of the pelvis by the **suspensory ligament of the ovary**. This ligament conveys the ovarian vessels, lymphatics, and nerves to and from the ovary, and constitutes the lateral part of the mesovarium of the broad ligament. The ovary also attaches to the uterus by the **ligament of the ovary**, or ovarian ligament, which runs within the mesovarium. This ligament is a remnant of the uppermost part of the ovarian gubernaculum of the fetus (see Chapter 2, Fig. 2.15). The ligament of the ovary connects the proximal (uterine) end of the ovary to the lateral angle of the uterus, just inferior to the entrance of the uterine tube. Because the ovary is suspended in the peritoneal cavity and its surface is not covered by peritoneum, the oocyte expelled at ovulation passes into the peritoneal cavity. However, its intraperitoneal life is short because it is usually trapped by the fimbriae of the infundibulum of the uterine tube and carried into the ampulla, where it may be fertilized.

Arterial Supply of the Ovaries. The ovarian arteries arise from the abdominal aorta (Fig. 3.27*A*) and descend along the posterior abdominal wall. At the pelvic brim, they cross over the external iliac vessels and enter the suspensory ligaments (Fig. 3.27*B*). The **ovarian artery** terminates by bifurcating into ovarian and tubal branches, which pass through the mesovarium to the ovary (Fig. 3.25). Both branches anastomose with corresponding branches of the uterine artery.

Venous and Lymphatic Drainage of the Ovaries. Veins draining the ovary form a vinelike **pampiniform plexus** of veins in the broad ligament near the ovary and uterine tube (Fig. 3.25). The veins of the plexus merge to form usually a singular **ovarian vein**, which leaves the lesser pelvis with the ovarian artery. The right ovarian vein ascends to enter the *inferior vena cava*; the left ovarian vein drains into the *left renal vein*. The **lymphatic vessels** follow the ovarian blood vessels and join those from the uterine tubes and fundus as they ascend to the *lumbar lymph nodes* (Fig. 3.26*A*).

Innervation of the Ovaries. Sympathetic and afferent fibers reach the ovarian plexus by descending along the overian vessels and via communications from the pelvic (uterovaginal) plexus (Fig. 3.26*B*). Parasympathetic fibers from the pelvic splanchnic nerves also utilize the latter route. Afferent fibers from the ovary enter the spinal cord through T10 and T11 nerves.

Injury to the Ureter

The ureter is vulnerable to injury when ovarian vessels are being ligated during an ovariectomy, for example, because these structures lie in close proximity as they cross the pelvic brim. The ureter is medial to the ovarian vessels (Fig. 3.25). ○

Rectum

The rectum is the part of the alimentary tract continuous proximally with the sigmoid colon (Fig. 3.29) and distally with the anal canal. The *rectosigmoid junction* lies anterior to the S3 vertebra. At this point, the teniae of the sigmoid colon spread out to form a continuous outer longitudinal layer of smooth muscle, and the fatty omental appendices are discontinued (see Chapter 2, Fig. 2.42). The rectum follows the curve of the sacrum and coccyx, forming the *sacral flexure* of the rectum. The rectum ends anteroinferior to the tip of the coccyx by turning sharply posteroinferiorly—the *anorectal flexure*—as it perforates the pelvic diaphragm (levator ani) to

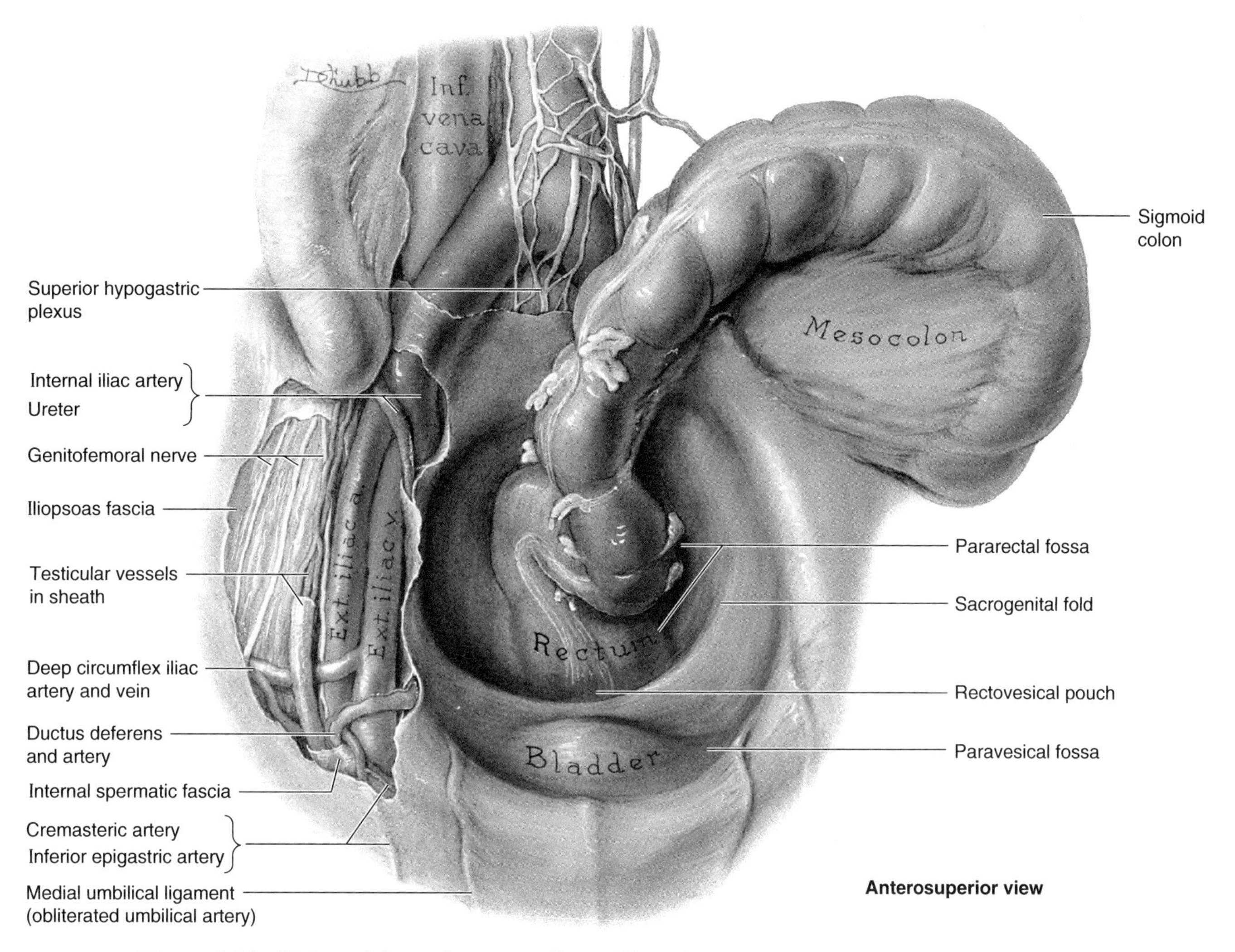

Figure 3.29. Male pelvis and surroundings. Dissection showing the superior hypogastric plexus lying in the bifurcation of the abdominal aorta. Note that the right ureter runs deep (external) to the peritoneum, crosses the external iliac vessels, and descends anterior to the internal iliac artery.

become the anal canal. The roughly 80° **anorectal flexure** is an important mechanism for fecal continence, being maintained during the resting state by the tonus of the puborectalis muscle and by its active contraction during peristaltic contractions if defecation is not to occur. With the flexures of the rectosigmoid junction superiorly and the anorectal junction inferiorly, the rectum has an S-shape when viewed laterally. When viewed anteriorly, the rectum demonstrates three sharp *lateral flexures (superior, intermediate, and inferior)* because of the presence of three internal infoldings (*transverse rectal folds*) of the mucous and submucous coats overlying thickened parts of the circular muscle layer of the rectal wall (Fig. 3.30*B*). The dilated terminal part of the rectum, lying directly above and supported by the pelvic diaphragm (levator ani) and anococcygeal ligament, is the **ampulla of the rectum** (Figs. 3.15*B* and 3.29). The ampulla receives and holds an accumulating fecal mass until it is expelled during defecation. The ability of the ampulla to relax to accommodate the initial and subsequent arrivals of fecal material is another essential element of maintaining fecal continence.

Peritoneum covers the anterior and lateral surfaces of the superior third of the rectum, only the anterior surface of the middle third, and no surface of the inferior third because it is subperitoneal (Table 3.5). In males the peritoneum reflects from the rectum to the posterior wall of the bladder, where it forms the floor of the **rectovesical pouch**. In females the peritoneum reflects from the rectum to the posterior fornix of the vagina, where it forms the floor of the **rectouterine pouch** (cul-de-sac). In both sexes, lateral reflections of peritoneum from the superior one-third of the rectum form **pararectal fossae** (Fig. 3.29), which permit the rectum to distend as it fills with feces.

The rectum rests posteriorly on the inferior three sacral vertebrae and the coccyx, anococcygeal ligament, median sacral vessels, and inferior ends of the sympathetic trunks and sacral plexuses. *In males* the rectum is related anteriorly to the fundus of the urinary bladder, terminal parts of the ureters, ductus deferentes, seminal vesicles, and prostate (Fig. 3.19). The *rectovesical septum* lies between the fundus of the bladder and the ampulla of the rectum and is closely associated with

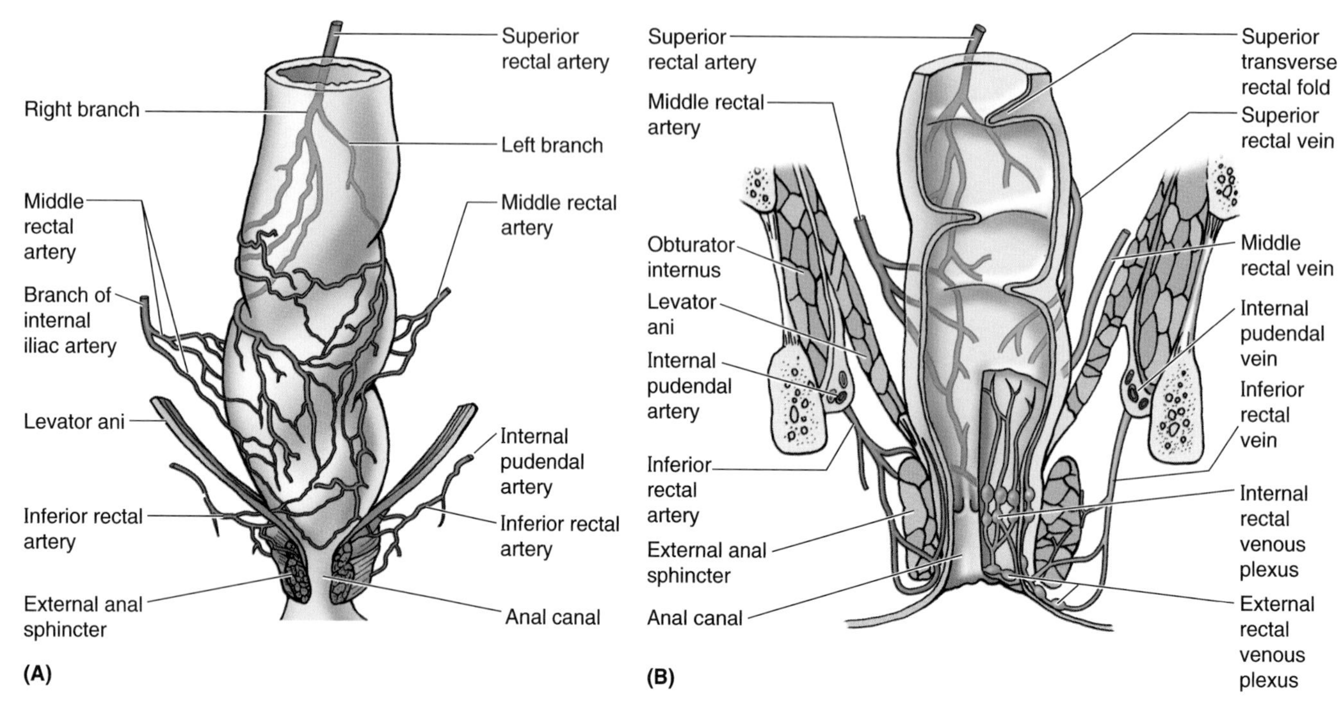

Figure 3.30. Arteries and veins of the rectum and anal canal. A. Anterior view of the arteries. In this specimen there are two right middle rectal arteries. Note that in spite of their name, the inferior rectal arteries, which are branches of the internal pudendal arteries, mainly supply the anal canal. Observe the three sharp lateral flexures of the rectum, which reflect the way in which the lumen navigates the transverse rectal folds (shown in (**B**)) on the internal surface. **B.** Coronal section of the rectum and anal canal showing the arterial supply and venous drainage. Observe the internal and external "rectal" venous plexuses (most directly related to the anal canal). Observe also the flexures and transverse rectal folds that help to support the weight of the feces.

the seminal vesicles and prostate. *In females* the rectum is related anteriorly to the vagina and is separated from its posterior fornix and the cervix by the *rectouterine pouch* (Fig. 3.23). Inferior to this pouch, the weak rectovaginal septum separates the superior half of the posterior wall of the vagina from the rectum.

Arterial Supply of the Rectum. The *superior rectal artery*—the continuation of the inferior mesenteric artery—supplies the proximal part of the rectum (Fig. 3.30). The two *middle rectal arteries*—usually arising from the inferior vesical arteries—supply the middle and inferior parts of the rectum, and the *inferior rectal arteries*—arising from the internal pudendal arteries—supply the anorectal junction and anal canal.

Venous and Lymphatic Drainage of the Rectum. Blood from the rectum drains through the superior, middle, and inferior rectal veins (Fig. 3.30*B*). Anastomoses occur between the portal and systemic veins in the wall of the anal canal. Because the superior rectal vein drains into the portal venous system and the middle and inferior rectal veins drain into the systemic system, these anastomoses are an important area of portacaval anastomosis (see Chapter 2, Fig. 2.59). The submucosal rectal venous plexus surrounds the rectum and communicates with the vesical venous plexus in males and the uterovaginal venous plexus in females. The **rectal venous plexus** consists of two parts (Fig. 3.30*B*), the internal rectal venous plexus just deep to the mucosa of the anorectal junction and the subcutaneous external rectal venous plexus external to the muscular wall of the rectum.

Lymphatic vessels from the superior half of the rectum ascend along the superior rectal vessels to the **pararectal lymph nodes** (Fig. 3.31*A*); they then pass to lymph nodes in the inferior part of the mesentery of the sigmoid colon and from them to the inferior mesenteric and **lumbar lymph nodes.** *Lymphatic vessels from the inferior half of the rectum* ascend with the middle rectal arteries and drain into the **internal iliac lymph nodes.**

Innervation of the Rectum. The nerve supply to the rectum is from the sympathetic and parasympathetic systems (Fig. 3.31*B*). The rectum derives its sympathetic supply from the lumbar part of the sympathetic trunk and the superior hypogastric plexus through plexuses on the branches of the inferior mesenteric artery. The parasympathetic supply derives from the pelvic splanchnic nerves. Fibers pass from these nerves to the left and right inferior hypogastric plexuses to supply the rectum. Visceral afferent or sensory fibers also join these plexuses and reach the spinal cord through the pelvic splanchnic nerves.

Figure 3.31. Lymphatic drainage and innervation of the rectum and anal canal. A. Lymphatic drainage of the rectum and anal canal. The *arrows* indicate the direction of lymph flow to the lymph nodes. **B.** Schematic drawing illustrating the innervation of the rectum and anal canal, with splanchnic nerves and hypogastric plexuses retracted laterally for clarity.

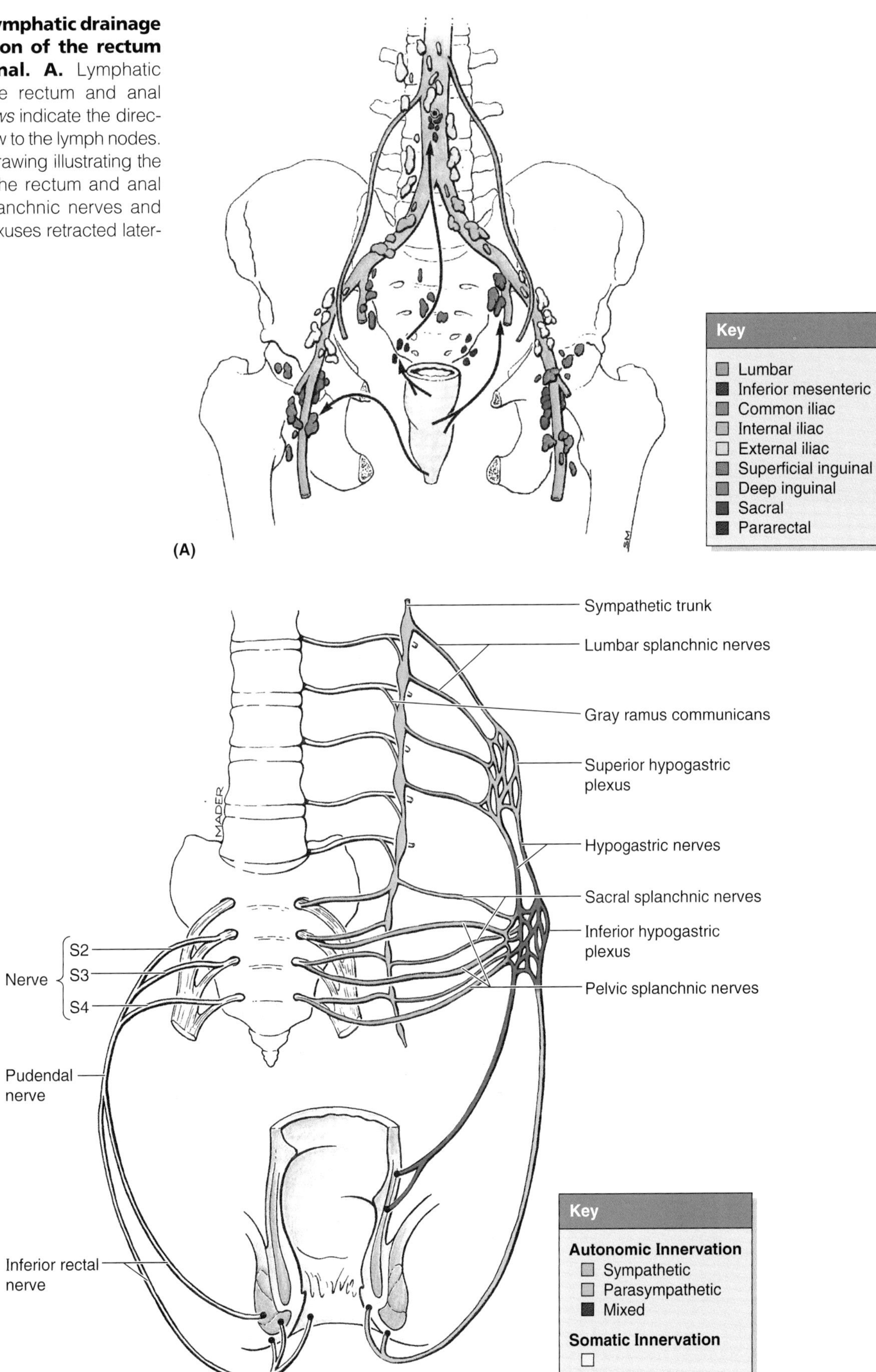

Rectal Examination

Many structures related to the anteroinferior part of the rectum may be palpated through its walls (e.g., the prostate and seminal vesicles in males and the cervix in females). In both sexes the pelvic surfaces of the sacrum and coccyx may be palpated. The ischial spines and tuberosities may also be palpated. Enlarged internal iliac lymph nodes, pathological thickening of the ureters, swellings in the ischioanal fossae (e.g., ischioanal abscesses and abnormal contents in the rectovesical pouch in the male or the rectouterine pouch in the female) may also be palpated. Tenderness of an inflamed appendix may also be detected rectally if it descends into the lesser pelvis (pararectal fossa).

The internal aspect of the rectum can be examined with a *proctoscope*, and biopsies of lesions may be taken through this instrument. During insertion of a *sigmoidoscope*, the curvatures of the rectum and its acute flexion at the rectosigmoid junction have to be kept in mind so that the patient does not undergo unnecessary discomfort. The operator must also know that the *transverse rectal folds, which provide useful landmarks for the procedure,* may temporarily impede passage of these instruments.

Resection of the Rectum

When resecting the rectum in males (e.g., during cancer treatment), the plane of the rectovesical septum (a fascial septum extending superiorly from the perineal body) is located so that the prostate and urethra can be separated from the rectum. In this way these organs are not damaged during the surgery. ⊙

Table 3.5. Peritoneum Covering the Pelvic Organs

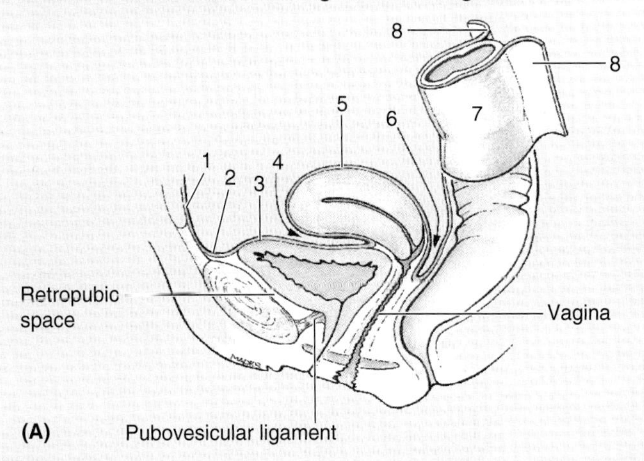

Peritoneal reflections in female pelvis (median section)

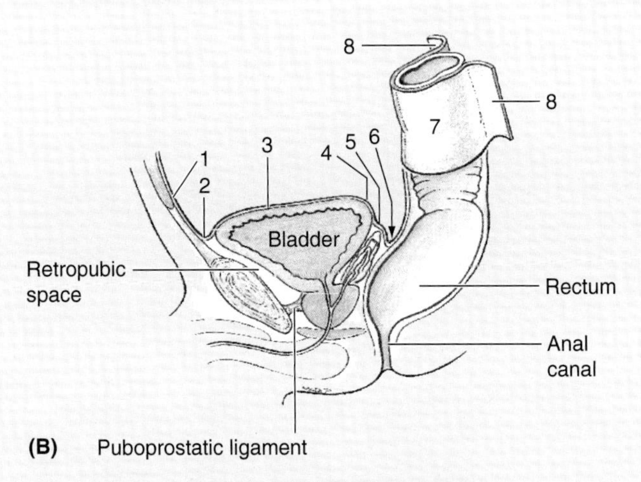

Peritoneal reflections in male pelvis (median section)

Female	Male
Peritoneum passes: • From the anterior abdominal wall *(1)* • Superior to the pubic bone *(2)* • On the superior surface of the urinary bladder *(3)* • From the bladder to the uterus, forming the *vesicouterine pouch (4)* • On the fundus and body of the uterus, posterior firmix, and all of the vagina *(5)* • Between the rectum and uterus, forming the *rectouterine pouch (6)* • On the anterior and lateral sides of the rectum *(7)* • Posteriorly to become the sigmoid mesocolon *(8)*	Peritoneum passes: • From the anterior abdominal wall *(1)* • Superior to the pubic bone *(2)* • On the superior surface of the urinary bladder *(3)* • 2 cm inferiorly on the posterior surface of the urinary bladder *(4)* • On the superior ends of the seminal vesicles *(5)* • Posteriorly to line the *rectovesical pouch (6)* • To cover the rectum *(7)* • Posteriorly to become the sigmoid mesocolon *(8)*

Perineum

The perineum refers to both an external surface area and a shallow "compartment" of the body. The perineal compartment lies inferior to the pelvic outlet and is separated from the pelvic cavity by the pelvic diaphragm, which is formed by the levator ani and coccygeus muscles and has osseofibrous boundaries. In the anatomical position the perineum (perineal area) is the narrow region between the proximal parts of the thighs; however, when the lower limbs are abducted, the perineum is a diamond-shaped area extending from the mons pubis anteriorly, the medial surfaces (insides) of the thighs laterally, and the gluteal folds and upper end of the intergluteal (natal) cleft posteriorly (Fig. 3.32). Some obstetricians apply the term perineum to a more restricted region—the area between the vagina and anus. *The osseofibrous structures marking the boundaries of the perineum (perineal compartment)* (Fig. 3.33) are the:

- Pubic symphysis—anteriorly
- Inferior pubic rami and ischial rami—anterolaterally
- Ischial tuberosities—laterally
- Sacrotuberous ligaments—posterolaterally
- Inferiormost sacrum and coccyx.

A transverse line joining the anterior ends of the ischial tuberosities divides the perineum into two triangles (Fig. 3.33*A*):

- The **anal triangle**, containing the anus, is posterior to this line.

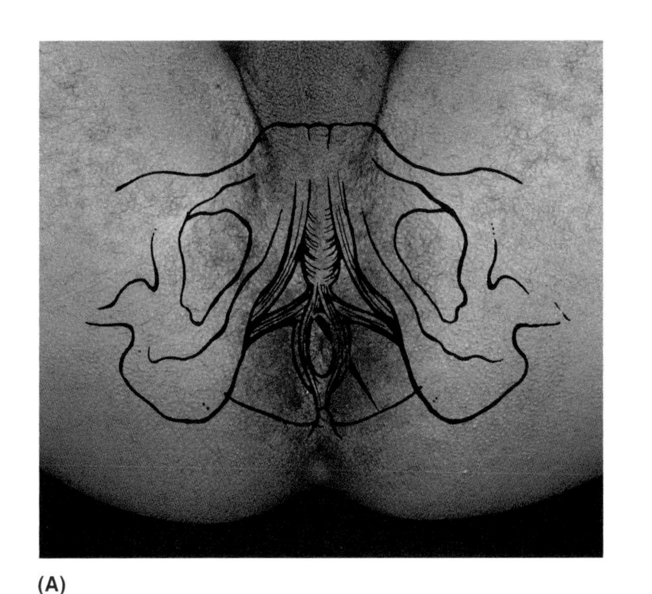

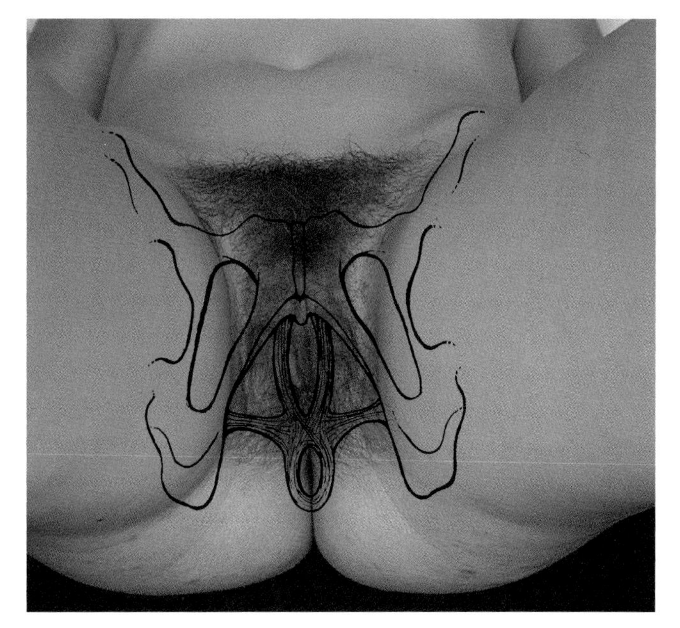

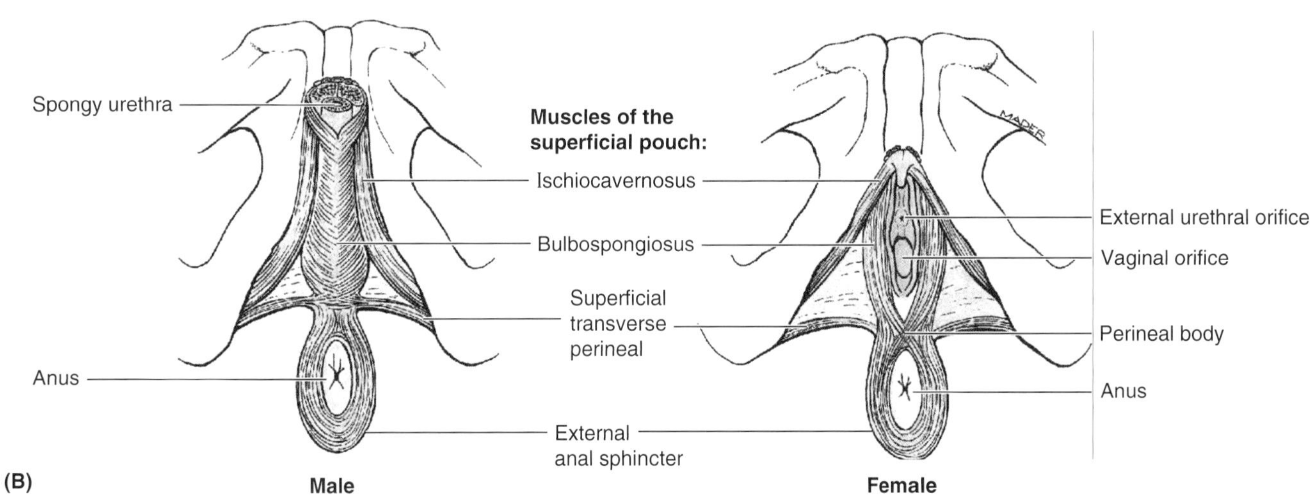

Figure 3.32. Male and female perineum. A. Surface projections of the osseous boundaries and structures of the perineal compartment. The penis (part of the perineum) is not shown. **B.** Muscles of the superficial perineal pouch (space).

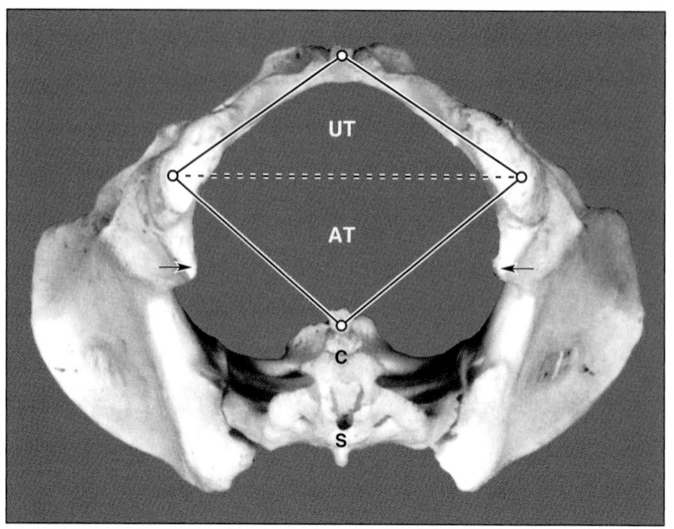

(A) Female

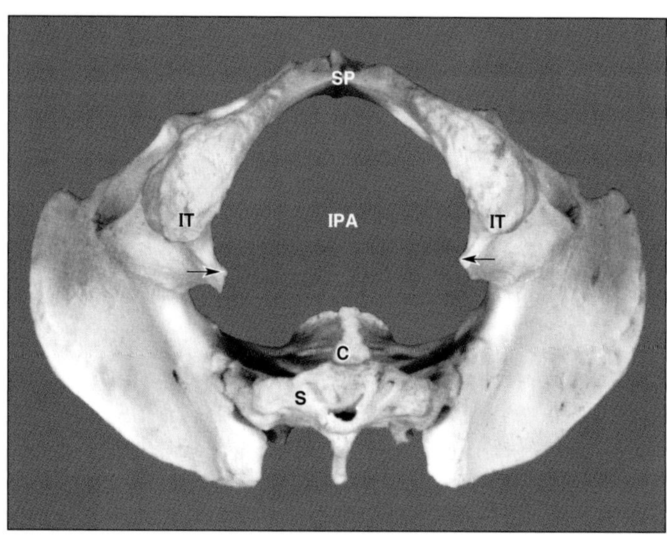

(B) Male

Figure 3.33. Pelvic outlets of female (A) and male pelves (B). Observe the sex difference in the size of the pelvic outlet, or inferior pelvic aperture (*IPA*). The view of the female pelvis is the one that an obstetrician visualizes in his or her "mind's eye" when the patient is on the examining table. At the angles of the IPA are the symphysis pubis (*SP*), coccyx (*C*), and ischial tuberosities (*IT*). In (**A**), the *broken transverse white line* between the right and left ischial tuberosities (*IT*) divides the diamond-shaped perineum into two triangles: the urogenital triangle (*UT*) and the anal triangle (*AT*). The *arrows* indicate the ischial spines. Note that the sacrum (*S*) is wedged between the iliac bones.

- The **urogenital triangle**, containing the root of the scrotum and penis in males or the external genitalia in females (Fig. 3.32), is anterior to this line.

The midpoint of the line joining the ischial tuberosities is the central point of the perineum overlying the **perineal body**, the attachment for the perineal muscles.

A thin sheet of tough, deep fascia, the **perineal membrane** stretches between the two sides of the pubic arch and covers the anterior part of the pelvic outlet (Fig. 3.34*C*). Immediately superior to the perineal membrane (on its superior surface), the *deep transverse perineal muscles* (L. deep transversus perinei) run transversely along the posterior aspect of the perineal membrane. Above the center of the perineal membrane, the external urethral sphincter surrounds the urethra.

The **perineal body** is an irregular fibromuscular mass located in the median plane between the anal canal and the perineal membrane (Figs. 3.32 and 3.34*E*). The perineal body contains collagenous and elastic fibers and both skeletal and smooth muscle. The perineal body is the site of convergence of several muscles (Wendell-Smith, 1995):

- Bulbospongiosus
- External anal sphincter
- Superficial and deep transverse perineal muscles.

The perineal body lies deep to the skin and subcutaneous tissue, posterior to the vestibule of the vagina or bulb of the penis, and anterior to the anus and anal canal. The perineal body attaches to the posterior border of the **perineal membrane** that lies between the external genitalia and the deep perineal pouch or space (Fig. 3.34*D*). The perineal body is variable in size and consistency, with relatively little fat deep to the overlying skin.

Disruption of the Perineal Body

The perineal body is an especially important structure in women because it is the final support of the pelvic viscera. Stretching or tearing of this attachment for the perineal muscles can occur during childbirth, removing support from the inferior part of the posterior wall of the vagina. As a result, *prolapse of the vagina* through the vaginal orifice may occur. The perineal body can also be disrupted by trauma, inflammatory disease, and infection, which can result in the formation of a *fistula* (abnormal canal) that is connected to the vestibule of the vagina. Attenuation of the perineal body, associated with diastasis (separation) of the puborectalis and pubococcygeus parts of the levator ani, may result in the formation of a *rectocele*—hernial protrusion of part of the rectum into the vaginal wall.

Episiotomy

During vaginal surgery and labor, an episiotomy—a surgical incision of the perineum and lower, posterior vaginal wall—is often made to enlarge the vaginal orifice and to prevent a jagged tear of the perineal muscles. The perineal body is the major structure incised during *median episiotomy* for childbirth. Mediolateral episiotomies are also performed. ("Mediolateral" is used inappropriately here. It actually refers to an incision that is initially a median incision that then turns laterally as it proceeds posteriorly.) Although routine prophylactic performance is widely debated, may obstetricians believe that episiotomy decreases the prevalence of excessive perineal body attenuation and decreases trauma to the pelvic diaphragm and perineal musculature. ○

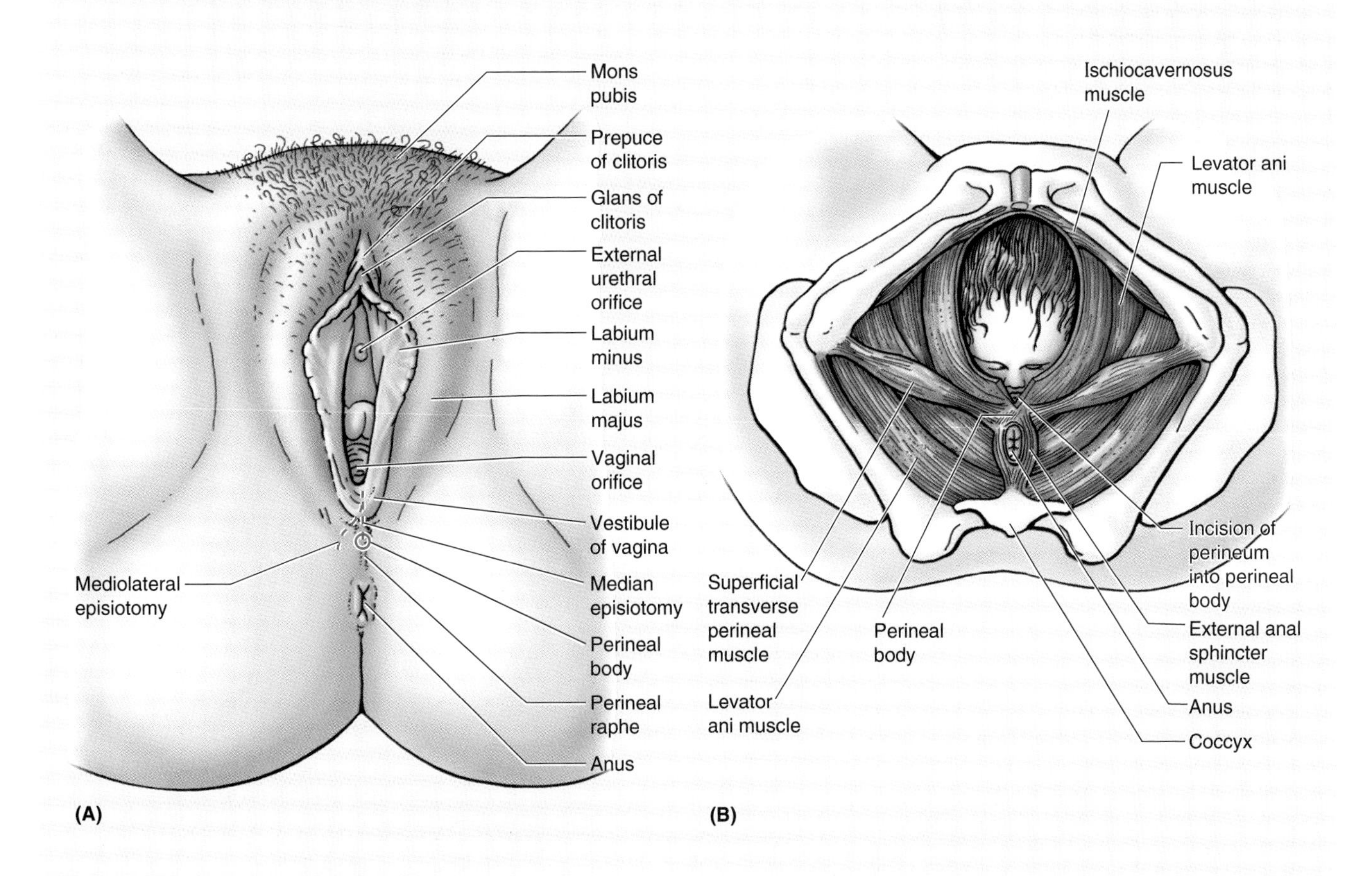

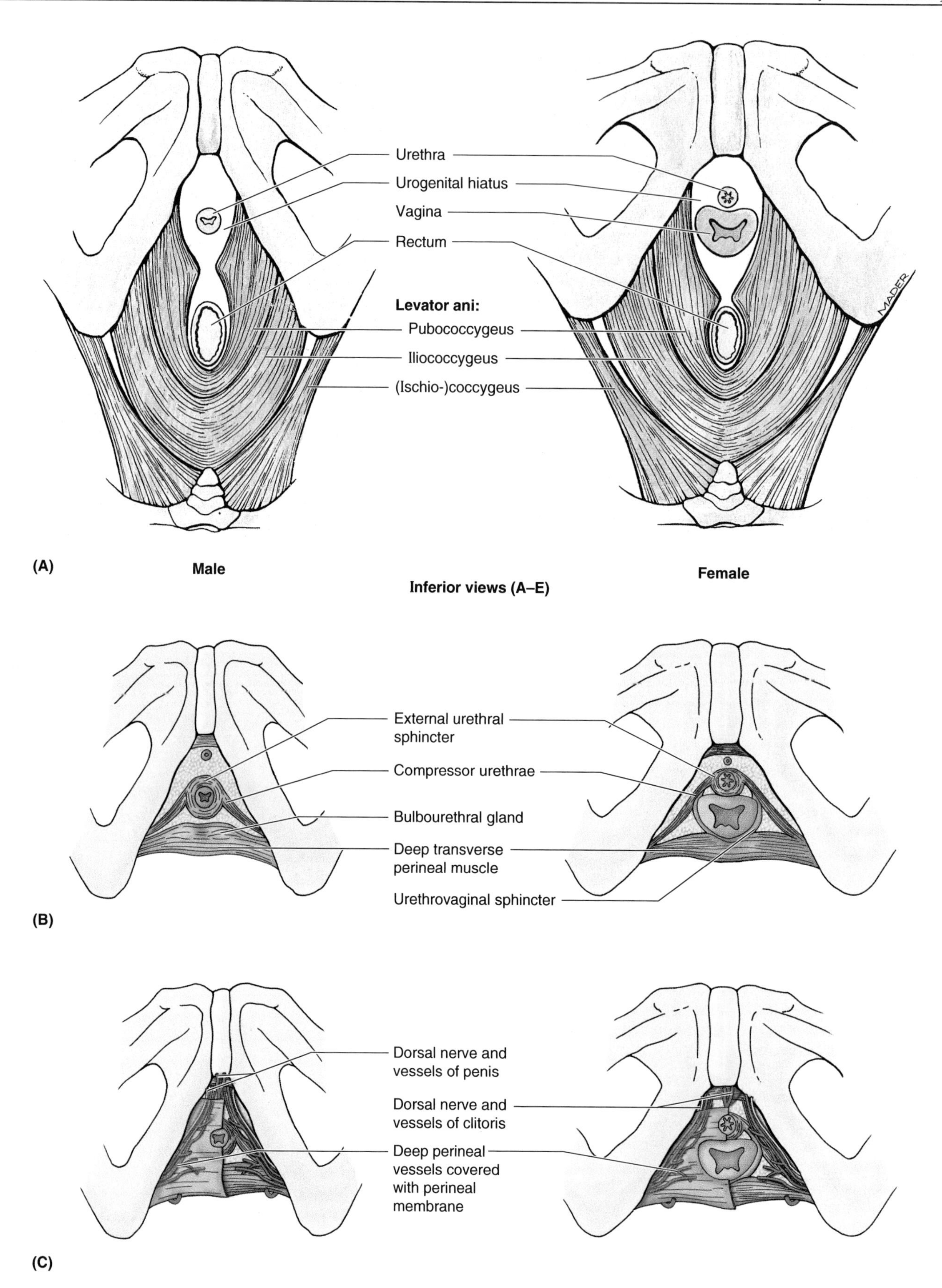

Figure 3.34. Layers of the perineum in the male and female.

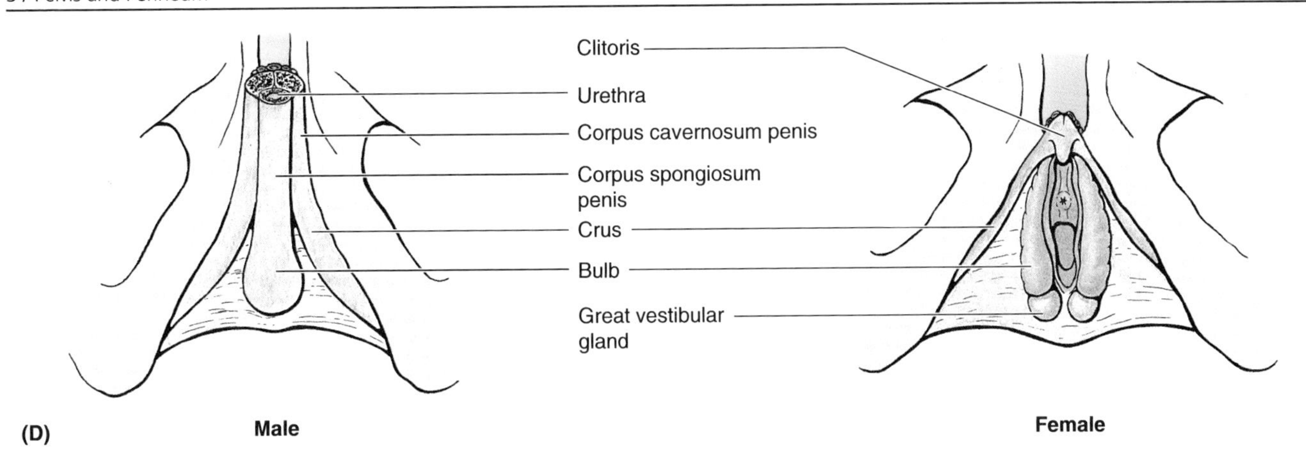

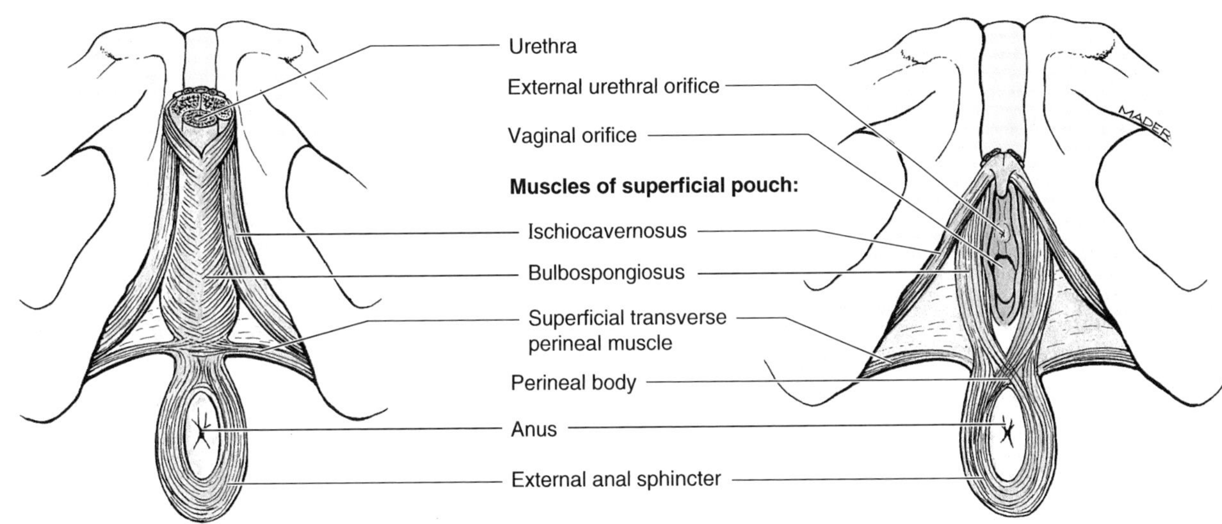

Figure 3.34. (*Continued*) These schematic diagrams show the layers of the perineum being built up from deep (**A**) to superficial (**E**). In (**A**), the pelvic outlet is almost filled by the pelvic diaphragm (levator ani and coccygeus muscles), which forms the roof of the perineal compartment. The urethra (and vagina in females) passes through the urogenital hiatus anteriorly and the rectum posteriorly. The external urethral sphincter and deep transverse perineal muscles (**B**) also lie within the urogenital hiatus, which is sealed inferiorly by the perineal membrane that extends between ischiopubic rami (**C**). Below the perineal membrane, the superficial perineal pouch or space contains the erectile bodies (**D**) and the muscles associated with them (**E**).

Perineal Fascia[1]

The perineal fascia consists of superficial and deep layers (Fig. 3.35). The subcutaneous tissue of the perineum, or *superficial perineal fascia*, like that of the lower anterior abdominal wall (see Chapter 2) consists of a fatty superficial layer and a membranous (deep) layer (Colles' fascia). *In females*, the fatty superficial layer continues anteriorly into the labia majora (Fig. 3.35*A*) and from them into the mons pubis and the fatty superficial layer of the abdomen (Camper's fascia). *In males*, the fatty superficial layer is greatly diminished in the urogenital triangle, being replaced altogether in the penis and scrotum with smooth (dartos) muscle. It is continuous between the scrotum and thighs (Fig. 3.35*B*) with the subcutaneous tissue of the abdomen and posteriorly with a similar layer in the anal region (ischioanal fat pad). The *membranous (deep) layer of superficial perineal fascia* is attached posteriorly to the posterior margin of the perineal membrane and the perineal body. Laterally it is attached to the fascia lata (deep fascia) of the uppermost medial aspect of the thigh. Anteriorly, the membranous layer of superficial perineal fascia is continuous with the dartos fascia in the scrotum; however, on each side of and anterior to the scrotum, the membranous layer becomes continuous with the membranous layer of the abdomen (Scarpa's fascia). In females, the membranous layer passes superior to the fatty layer forming the labia majora and becomes continuous with the membranous layer of subcutaneous fascia of the abdomen.

---

[1] The terminology used in this section is recommended by the *Federative Committee on Anatomical Terminology* (FCAT) in 1998; however, because many clinicians concerned with the perineum use eponyms, the authors have placed commonly used eponyms in parentheses so that the new terms introduced will be understood by all readers.

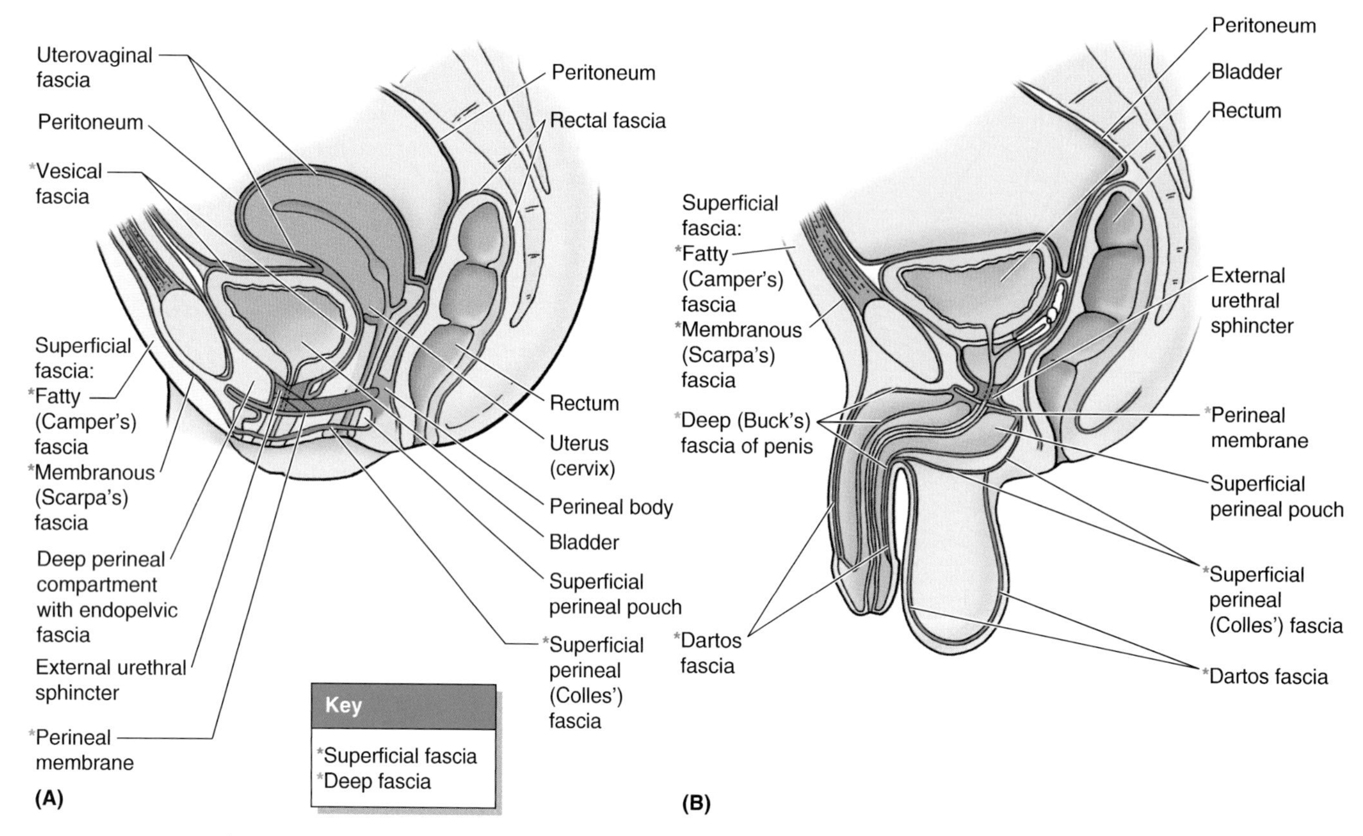

Figure 3.35. Fascia of the pelvis and perineum. Median sections. **A.** Female. **B.** Male.

The *deep perineal fascia* (investing or Gallaudet's fascia) intimately invests the ischiocavernosus, bulbospongiosus, and superficial transverse perineal muscles. It is also attached laterally to the ischiopubic ramus superior to the attachment of the membranous layer of superficial perineal fascia. Anteriorly it is fused to the suspensory ligament of the penis (see Fig. 3.43) and is continuous with the deep fascia covering the external oblique muscle of the abdomen and the rectus sheath. The deep perineal fascia is fused with the suspensory ligament of the clitoris in females and the deep fascia of the abdomen, as in males.

Superficial Perineal Pouch

The superficial perineal pouch (compartment) is a potential space between the membranous layer of the subcutaneous tissue and the perineal membrane (Fig. 3.32). *In males the superficial perineal pouch contains the:*

- Root (bulb and crura) of the penis and the muscles associated with it (ischiocavernosus and bulbospongiosus)
- Proximal part of the spongy urethra
- Superficial transverse perineal muscles
- Branches of the internal pudendal vessels
- Branches of the pudendal nerves (perineal nerves).

In females the superficial perineal pouch contains the:

- Root (crura) of the clitoris and the muscle associated with it (ischiocavernosus)
- Bulbs of the vestibule and surrounding muscle (bulbospongiosus)
- Superficial transverse perineal muscles
- Related vessels and nerves (branches of internal pudendal vessels, perineal nerves)
- Greater vestibular glands.

Deep Perineal Pouch

The deep perineal pouch (space) is not an enclosed compartment; it is open superiorly. This pouch and the deep urogenital muscles are bounded below by the perineal membrane; however, the pouch extends superiorly as the anterior recesses of the ischioanal fossa. *In males the deep perineal pouch contains the:*

- Intermediate (membranous) part of the urethra
- External urethral sphincter muscle
- Bulbourethral glands
- Deep transverse perineal muscles
- Related vessels and nerves.

In females the deep perineal pouch contains the:

- Proximal part of the urethra
- External urethral sphincter muscle
- Deep transverse perineal muscles
- Related vessels and nerves.

Pelvic Diaphragm

The pelvic diaphragm—consisting of the levator ani and coccygeus muscles, together with the fascia above and below them—*separates the pelvic cavity from the perineum* (Fig. 3.36*A*). The pelvic diaphragm forms the funnel-shaped or hammocklike floor of the pelvic cavity and the medial slope of the inverted V-shaped roof of each ischioanal fossa.

Ischioanal Fossae

The ischioanal fossae (formerly ischiorectal fossae) around the wall of the anal canal are large fascia-lined, wedge-shaped spaces between the skin of the anal region and the pelvic diaphragm (Figs. 3.36*B*, 3.37, and Fig. 3.28*A* & *B*). The apex of each fossa lies superiorly where the levator ani muscle arises from the obturator fascia. The ischioanal fossae, wide inferiorly and narrow superiorly, are filled with fat and loose connective tissue. The two ischioanal fossae communicate by means of the *deep postanal space* over the *anococcygeal ligament* (body), a fibrous mass located between the anal canal and the tip of the coccyx (Fig. 3.36*A*). *Each ischioanal fossa is bounded:*

- Laterally by the ischium and the inferior part of the obturator internus, covered with obturator fascia
- Medially by the anal canal to which the levator ani descends and which the external anal sphincter surrounds
- Posteriorly by the sacrotuberous ligament and gluteus maximus
- Anteriorly by the external urethra sphincter and deep transverse perineal muscles and their fasciae. These parts of the fossae, superior to the perineal membrane, are known as the *anterior recesses of the ischioanal fossae.*

The ischioanal fossae are traversed by tough, fibrous bands and filled with fat, forming the *fat bodies of the ischioanal fossae*. These fat bodies support the anal canal but they are readily displaced to permit expansion of the anal canal during the passage of feces.

The lateral walls of the ischioanal fossae contain the *internal pudendal vessels* and the *pudendal nerves* (Fig. 3.37). Posteriorly these vessels and the nerve give rise to the inferior rectal vessels and nerves, respectively, which cross the ischioanal fossae and become superficial as they supply the external anal sphincter and the perianal skin. Two other cutaneous nerves, the perforating branch of S2 and S3 and the perineal branch of S4 nerve, also pass through the ischioanal fossae.

Pudendal Canal

The pudendal canal is a space within the obturator fascia, which covers the medial aspect of the obturator internus and lines the lateral wall of the ischioanal fossa (Fig. 3.37). The internal pudendal artery and vein, the pudendal nerve, and the nerve to the obturator internus enter this canal at the lesser sciatic notch, inferior to the ischial spine.

The **pudendal nerve** supplies most of the innervation to the perineum. Toward the distal end of the pudendal canal, the pudendal nerve splits, giving rise to the perineal nerves (Fig. 3.38) and continuing as the dorsal nerve of the penis or clitoris. These nerves run anteriorly on each side of the continuation of the internal pudendal artery.

The superficial **perineal nerves** give scrotal or labial branches, and the **deep perineal nerve** supplies the muscles of the deep and superficial perineal pouches, the skin of the vestibule of the vagina, and the mucosa of the inferiormost part of the vagina. The **inferior rectal nerve** arises from the pudendal nerve at the entrance to the pudendal canal and crosses the ischioanal fossa to reach the anus. The inferior rectal nerve supplies the external anal sphincter and perianal skin and communicates with the posterior scrotal or labial and perineal nerves. The **dorsal nerve of the penis or clitoris**, a sensory branch of the pudendal nerve, runs through the deep perineal pouch to reach its area of supply.

Anal Canal

The anal canal is the terminal part of the large intestine that extends from the upper aspect of the pelvic diaphragm to the anus (Fig. 3.37). The anal canal (2.5–3.5 cm long) begins where the rectal ampulla narrows at the level of the U-shaped sling formed by the puborectalis muscle (Fig. 3.7). The anal canal ends at the *anus*, the external outlet of the gastrointestinal tract. The anal canal, surrounded by internal and external anal sphincters, descends posteroinferiorly between the anococcygeal ligament and the perineal body. The anal canal is collapsed except during passage of feces. Both sphincters must relax before defecation can occur.

The **external anal sphincter** is a large voluntary sphincter that forms a broad band on each side of the inferior two-thirds of the anal canal (Fig. 3.37). This sphincter blends superiorly with the puborectalis muscle and is supplied mainly by S4 through the inferior rectal nerve (Fig. 3.38).

The **internal anal sphincter** (Fig. 3.37) is an involuntary sphincter surrounding the superior two-thirds of the anal canal. It is a thickening of the circular muscle layer and is innervated by parasympathetic fibers that passed through the *pelvic splanchnic nerves*. This sphincter is tonically contracted most of the time to prevent leakage of fluid or flatus; however, it relaxes in response to the pressure of feces or gas distending the rectal ampulla, requiring voluntary contraction of the puborectalis and external anal sphincters if defecation is not to occur.

In the **interior of the anal canal**, the superior half of the mucous membrane is characterized by a series of longitudinal ridges—**anal columns**. These columns contain the terminal branches of the superior rectal artery and vein. The **anorectal junction**, indicated by the superior ends of the anal columns, is where the rectum joins the anal canal. At this point the wide rectal ampulla abruptly narrows as it traverses the pelvic diaphragm. The inferior ends of the anal columns are joined by **anal valves**. Superior to the valves are small recesses—**anal sinuses**. When compressed by feces, the anal sinuses exude mucus that aids in evacuation of feces from the anal canal. The inferior comb-shaped limit of the anal valves forms an irregular line—the **pectinate line**—that indicates the junction of the superior part of the anal canal (derived from the embryonic

Rupture of the Urethra in Males and Extravasation of Urine

Fractures of the bony pelvis, especially those resulting from separation of the pubic symphysis and puboprostatic ligaments, usually rupture the **intermediate part of the urethra.** Rupture of this part of the urethra results in the extravasation of urine and blood into the deep perineal pouch (*A*), which passes superiorly and extraperitoneally around the prostate and bladder.

The common site of rupture of the spongy urethra and extravasation of urine is in the bulb of the penis (*B*). This injury usually results from a forceful blow to the perineum ("straddle injury"), such as falling on a metal beam or, less commonly, from the incorrect passage ("false passage") of a transurethral catheter or device that fails to negotiate the angle of the urethra in the bulb of the penis. Rupture of the corpus spongiosum and spongy urethra results in urine passing from it (extravasation) into the superficial perineal space. The attachments of the perineal fascia determine the direction of flow of the extravasated urine; hence, urine may pass into the loose connective tissue in the scrotum, around the penis, and superiorly deep to the membranous layer of subcutaneous connective tissue of the lower anterior abdominal wall. The urine cannot pass far into the thighs because the membranous layer of superficial perineal fascia blends with the fascia lata enveloping the thigh muscles, just distal to the inguinal ligament. In addition, urine cannot pass posteriorly into the anal triangle because the superficial and deep layers of perineal fascia are continuous with each other around the superficial perineal muscles and with the posterior edge of the perineal membrane between them. Rupture of a blood vessel into the superficial perineal pouch resulting from trauma would result in a similar containment of blood in the superficial perineal pouch.

Structures in the Deep Perineal Pouch

Traditionally, a trilaminar, triangular urogenital (UG) diaphragm has been described as the main constituent of the deep perineal pouch (p. 397). Although the classical descriptions appear justified when viewing only the superficial aspect of the structures occupying the deep pouch (*A*), the long-held concept of a flat, essentially two-dimensional diaphragm is erroneous. According to this concept, the UG diaphragm consisted of the perineal membrane (inferior fascia of the UG diaphragm) on the bottom and a "superior fascia of the UG diaphragm" on top, between which was a flat muscular sheet composed of a disclike sphincter urethra in front of an equally two-dimensional, transversely oriented deep transverse perineal muscle. The descriptions of the perineal membrane and deep male transverse perineal muscles only appear to be supported by evidence, the female deep transverse perineal muscles being mainly smooth muscle (Wendell-Smith, 1995). The strong *perineal membrane,* extending between the ischiopubic rami and separating superficial and deep perineal pouches, is indeed the final passive support of the pelvic viscera. Immediately superior to the posterior half of the perineal membrane, the flat, sheetlike *deep transverse perineal muscle,* when developed, offers dynamic support for ▶

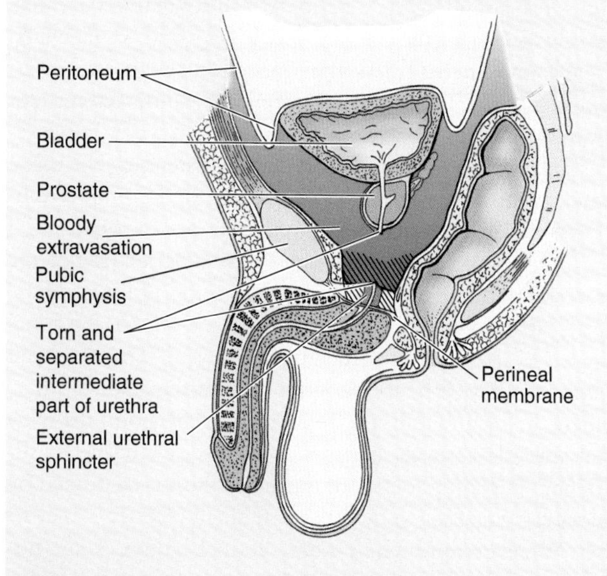

(A)

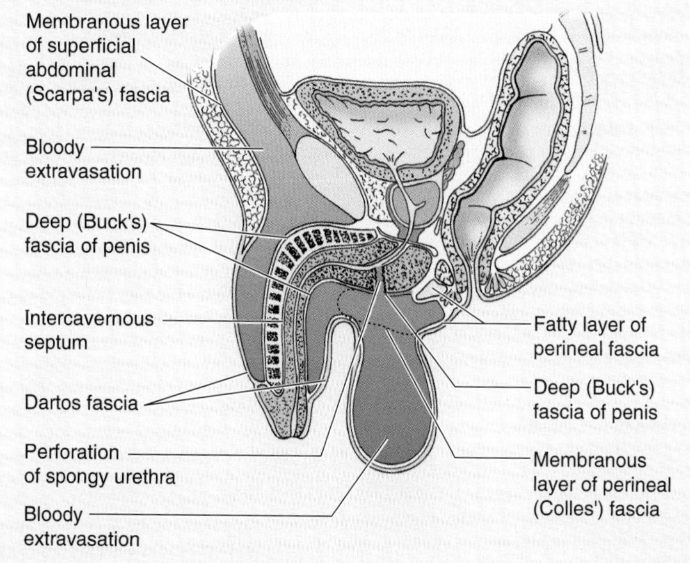

(B)

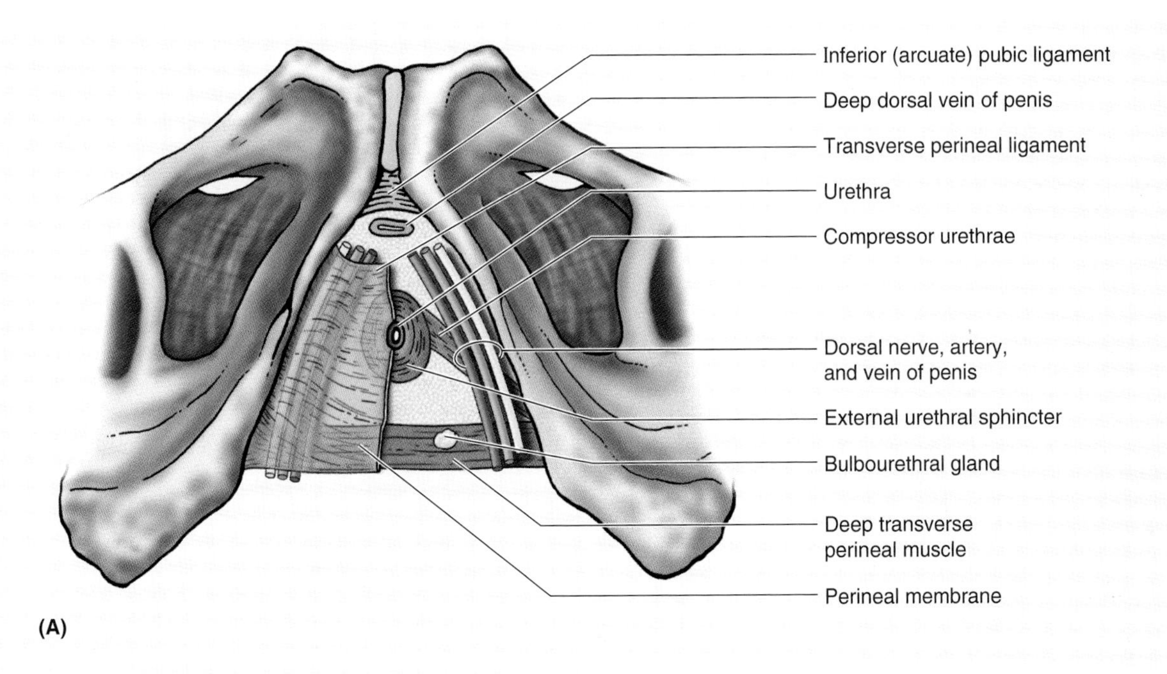
Inferior (arcuate) pubic ligament
Deep dorsal vein of penis
Transverse perineal ligament
Urethra
Compressor urethrae
Dorsal nerve, artery, and vein of penis
External urethral sphincter
Bulbourethral gland
Deep transverse perineal muscle
Perineal membrane
(A)

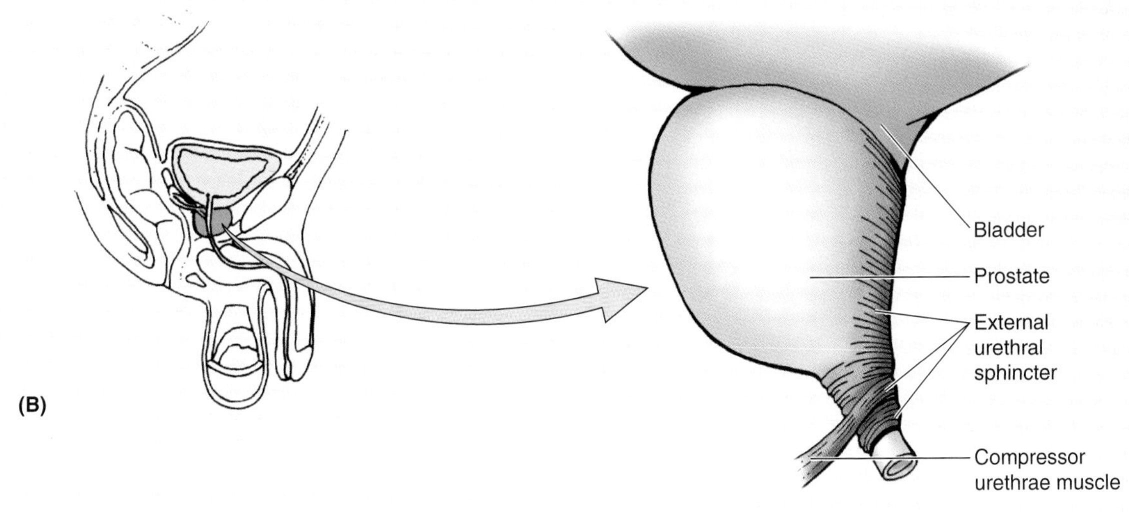
Bladder
Prostate
External urethral sphincter
Compressor urethrae muscle
(B)

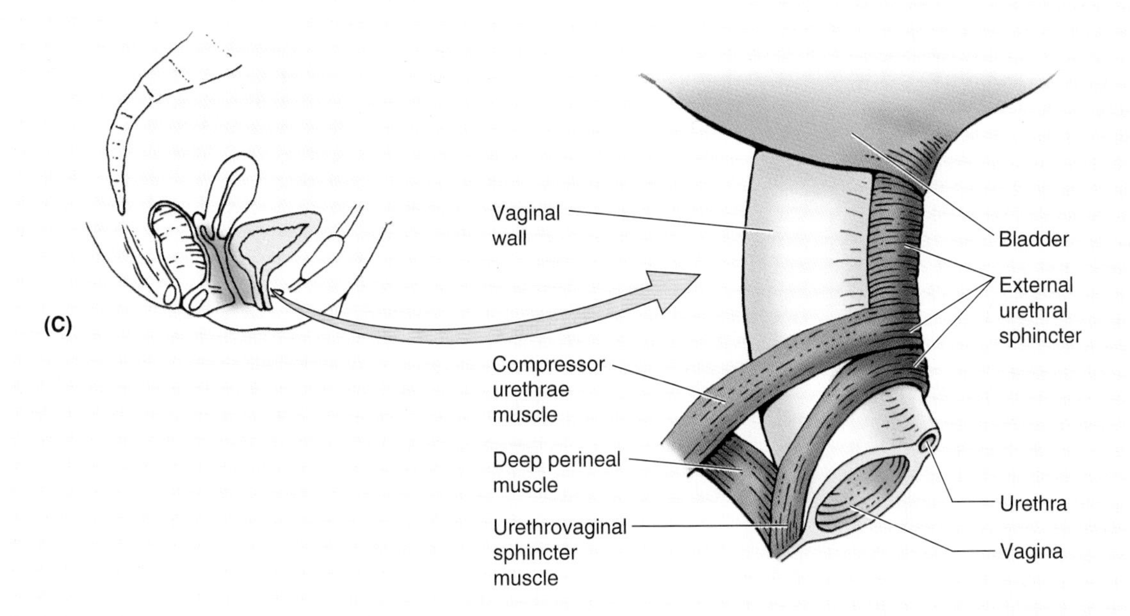
Vaginal wall
Bladder
External urethral sphincter
Compressor urethrae muscle
Deep perineal muscle
Urethrovaginal sphincter muscle
Urethra
Vagina
(C)

▶ the pelvic viscera. As described by Oelrich (1980), however, the urethral sphincter muscle is not a flat, planar structure, and the only "superior fascia" is the intrinsic fascia of the external urethral sphincter muscle.

The *external urethral sphincter* is more tube- and troughlike than disclike, and in the male only a part of the muscle forms a circular investment (a true sphincter) for the intermediate (membranous) part of the urethra inferior to the prostate (*B*). Its larger, troughlike part extends vertically to the neck of the bladder, displacing the prostate and investing the prostatic urethra anteriorly and anterolaterally only. Apparently the muscular primordium is established around the whole length of the urethra before development of the prostate. As the prostate develops from urethral glands, the muscle atrophies or is displaced by the prostate posteriorly and posterolaterally. Whether this part of the muscle compresses or dilates the prostatic urethra is a matter of some controversy.

In the female the external urethral sphincter is more properly a "urogenital sphincter," according to Oelrich (1983). Here, too, he describes a part forming a true anular sphincter around the urethra (*C*), but this having several additional parts extending from it; a superior part, extending to the neck of the bladder; a subdivision described as extending inferolaterally to the ischial ramus on each side (the compressor urethrae muscle); and yet another bandlike part, which encircles both the vagina and urethra (urethrovaginal sphincter). In both the male and female, the musculature described, rather than lying in the plane of the deep perineal muscle, is actually oriented perpendicular to it. Other authors substantiate most of Oelrich's description but dispute the encircling of the urethra in the female, claiming that the muscle is not capable of sphincteric action. Further, they assert that sectioning the perineal nerve supplying the "sphincter" does not result in incontinence, but this may be a consequence of innervation from multiple sources. ○

hindgut) and the inferior part (derived from the embryonic proctodeum). *The anal canal superior to the pectinate line differs from the part inferior to the pectinate line in its arterial supply, innervation, and venous and lymphatic drainage.* These differences result from their different embryological origins of the upper and lower parts of the anal canal (Moore and Persaud, 1998).

Arterial Supply of the Anal Canal. The *superior rectal artery* supplies the anal canal superior to the pectinate line (Fig. 3.30*A*). The two *inferior rectal arteries* supply the inferior part of the anal canal, as well as the surrounding muscles and perianal skin. The *middle rectal arteries* assist with the blood supply to the anal canal by forming anastomoses with the superior and inferior rectal arteries.

Venous and Lymphatic Drainage of the Anal Canal. The internal rectal venous plexus drains in both directions from the level of the pectinate line. *Superior to the pectinate line*, the internal rectal plexus drains chiefly into the **superior**

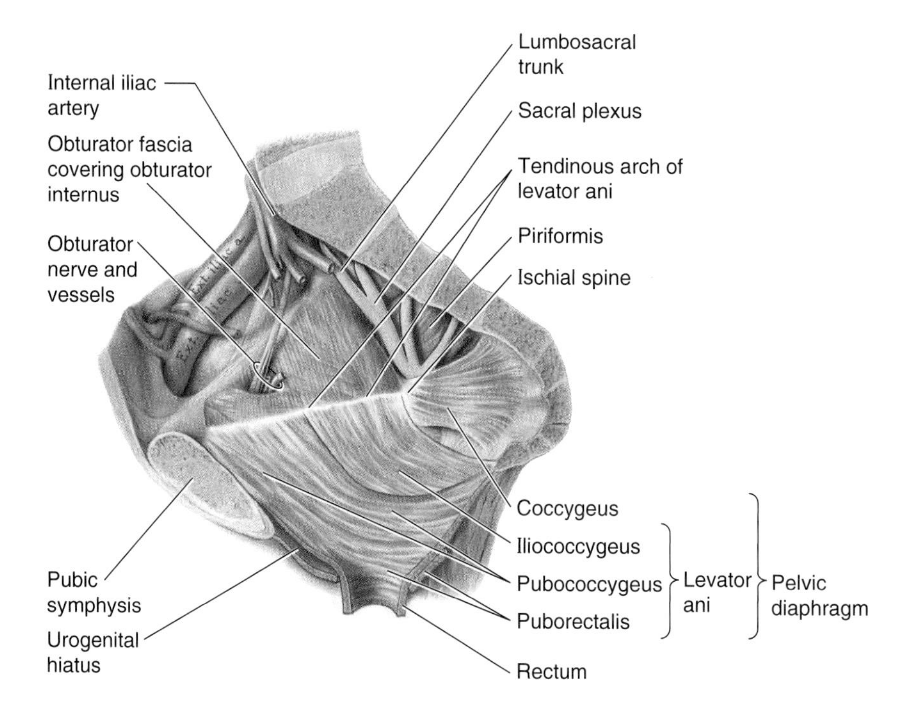

Figure 3.36. Pelvic diaphragm and ischioanal fossae. A. Floor of male pelvis.

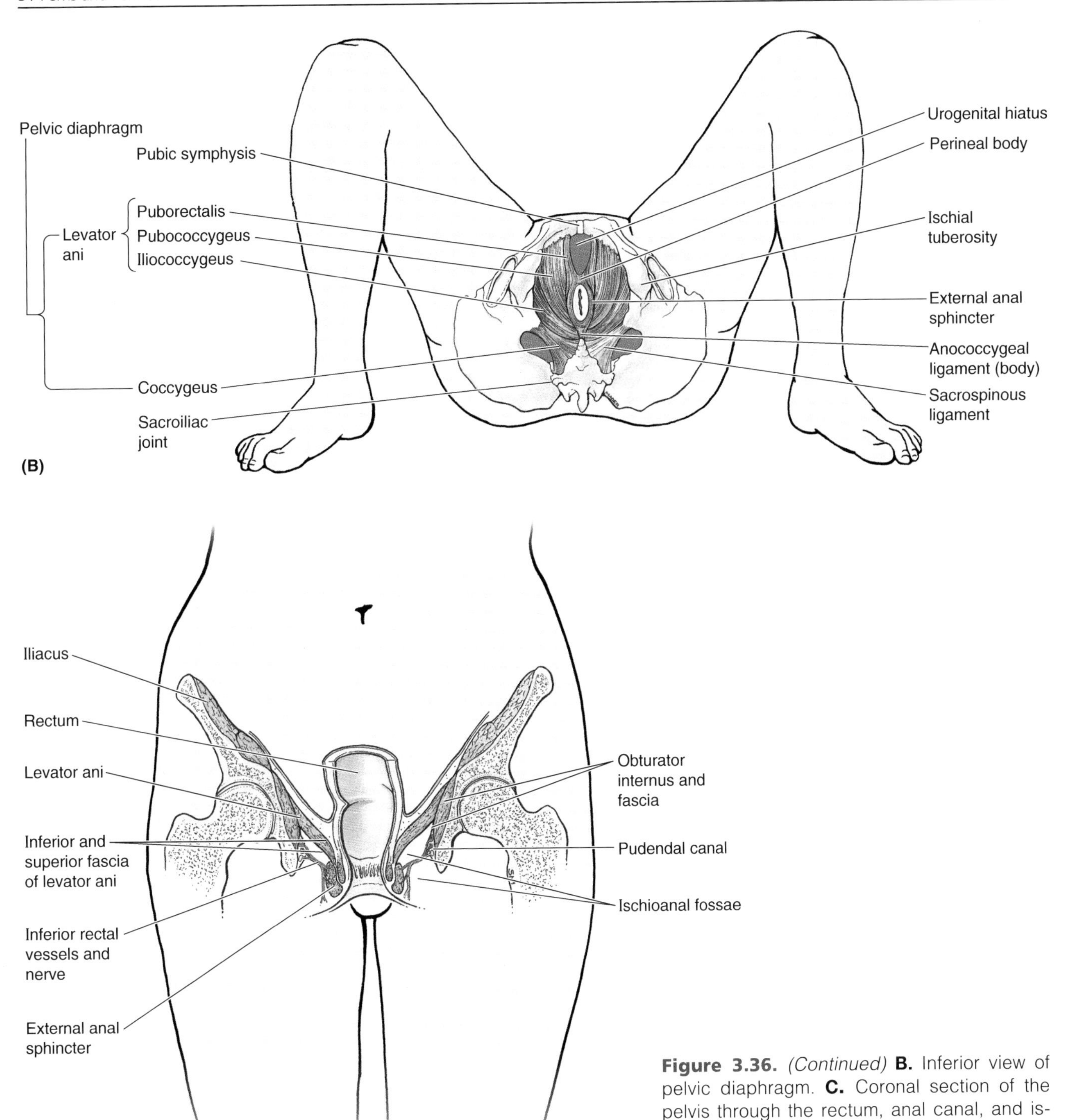

Figure 3.36. *(Continued)* **B.** Inferior view of pelvic diaphragm. **C.** Coronal section of the pelvis through the rectum, anal canal, and ischioanal fossae.

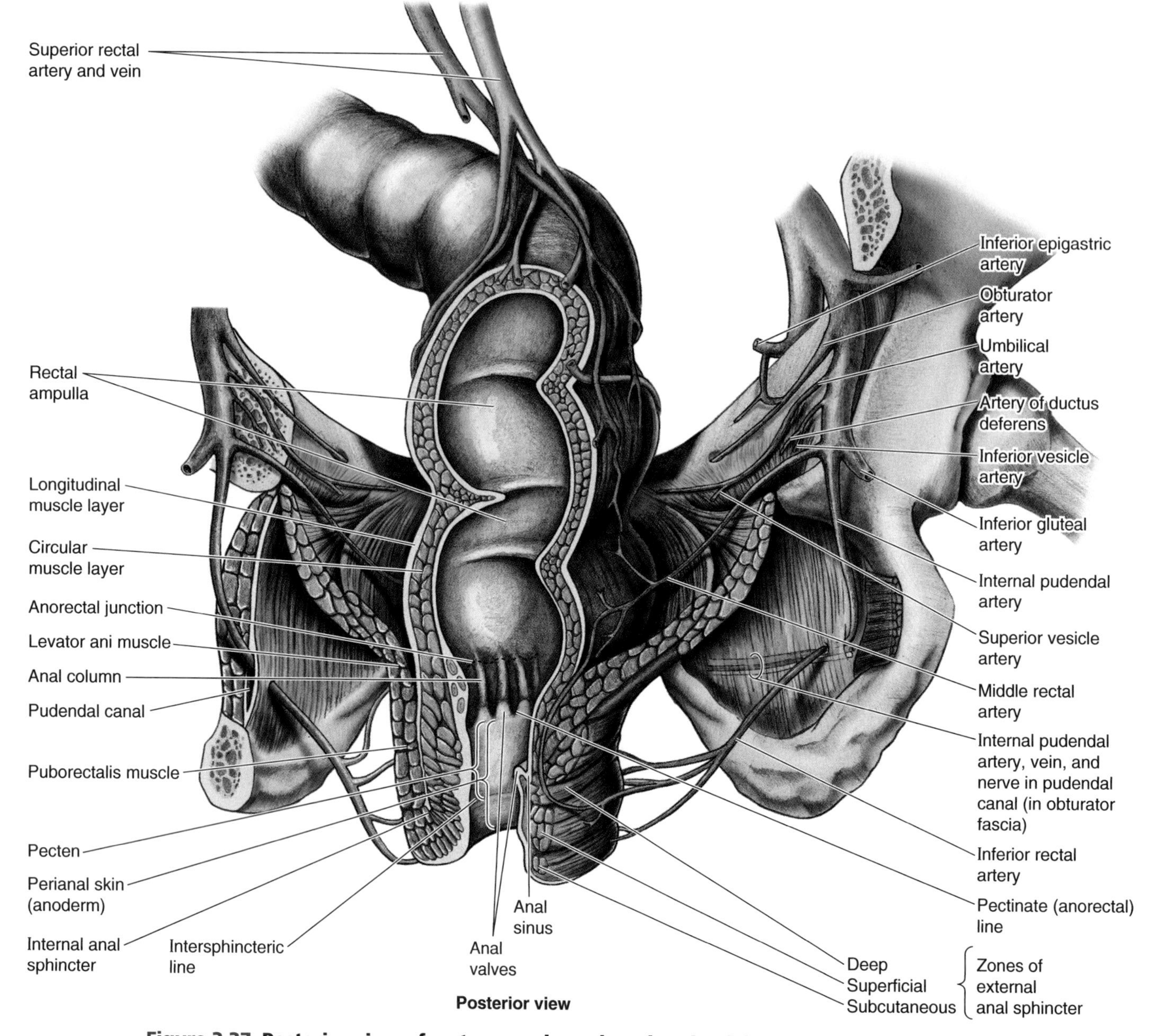

Figure 3.37. Posterior view of rectum, anal canal, and pudendal canal. Observe the pudendal canal—the space within the obturator fascia covering the medial surface of the obturator internus and lining the lateral wall of the ischioanal fossa—that transmits the pudendal vessels and nerves.

rectal vein—a tributary of the inferior mesenteric vein—and the portal system (Fig. 3.30*B*). *Inferior to the pectinate line*, the internal rectal plexus drains into the **inferior rectal veins**—tributaries of the caval venous system—around the margin of the external anal sphincter. The **middle rectal veins**—tributaries of the internal iliac veins—mainly drain the muscularis externa of the ampulla and form anastomoses with the superior and inferior rectal veins.

Superior to the pectinate line, the lymphatic vessels drain into the **internal iliac lymph nodes** and through them into the common iliac and lumbar lymph nodes (Fig. 3.31*A*). *Inferior to the pectinate line*, the lymphatic vessels drain into the *superficial inguinal lymph nodes*.

Innervation of the Anal Canal. *The nerve supply to the anal canal superior to the pectinate line* is visceral innervation from the ***inferior hypogastric plexus*** (sympathetic and parasympathetic fibers [Fig. 3.31*B*]). The superior part of the anal canal is sensitive only to stretching. *The nerve supply of the anal canal inferior to the pectinate line* is somatic innervation derived from the **inferior anal (rectal) nerves**, branches of the pudendal nerve. Therefore, this part of the anal canal is sensitive to pain, touch, and temperature.

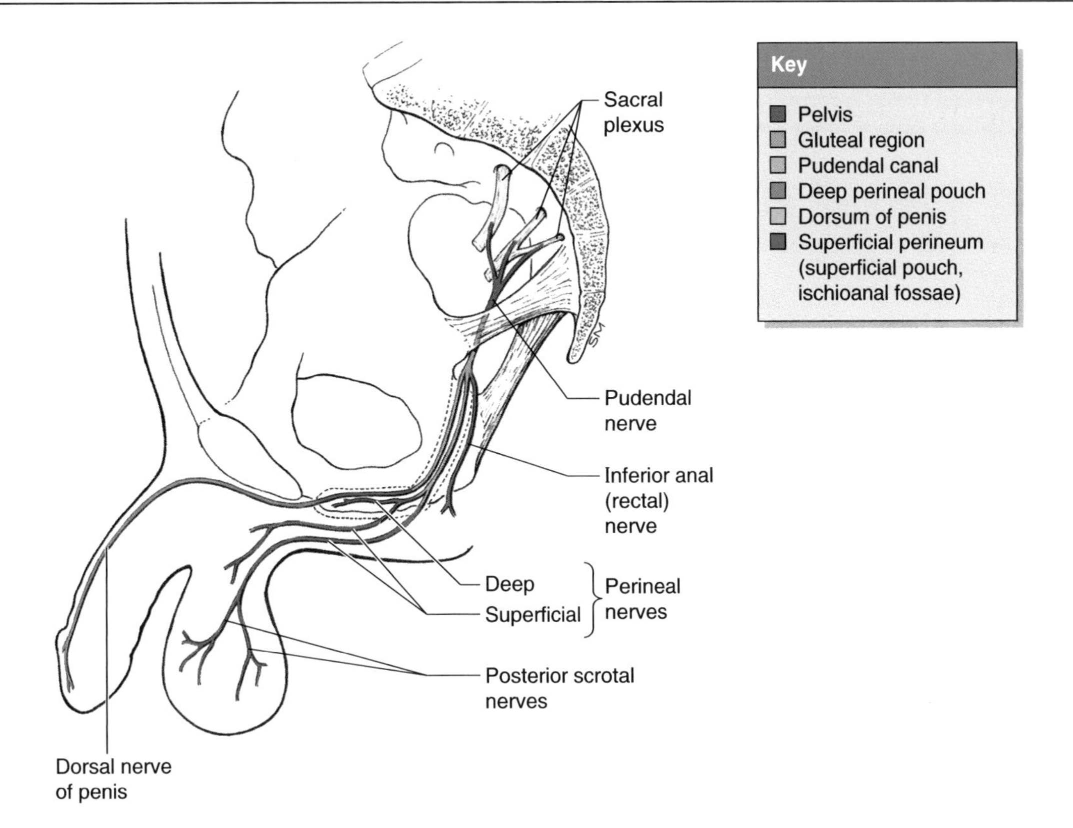

Figure 3.38. Diagram of the pudendal nerve. The five regions in which it runs are shown in color. The pudendal nerve supplies the skin, organs, and muscles of the perineum; it is therefore concerned with micturition, defecation, erection, ejaculation, and, in the female, parturition. Although the pudendal nerve is shown here in the male, its distribution is similar in the female because the parts of the female perineum are homologues of the male.

Clinically Important Landmark

The *pectinate line* (also called the dentate or mucocutaneous line by some clinicians) is a particularly important landmark because it is visible and approximates the level of important anatomical changes mentioned previously (e.g., in the nerve supply to the anal canal).

Anal Fissures and Perianal Abscesses

The ischioanal fossae are occasionally the sites of infection that may result in the formation of *ischioanal abscesses (A)* (p. 402). These collections of pus are annoying and painful. Infections may reach the ischioanal fossae in several ways:

- Following *cryptitis* (inflammation of the anal sinuses)
- Extension from a pelvirectal abscess
- Following a tear in the anal mucous membrane
- From a penetrating wound in the anal region.

Diagnostic signs of an ischioanal abscess are fullness and tenderness between the anus and the ischial tuberosity. A perianal abscess may open spontaneously into the anal canal, rectum, or perianal skin.

Because the ischioanal fossae communicate posteriorly through the *deep postanal space*, an abscess in one fossa may spread to the other one and form a semicircular "horseshoe-shaped" abscess around the posterior aspect of the anal canal. In chronically constipated persons, the anal valves and mucosa may be torn by hard feces. An *anal fissure* (slitlike lesion) is usually located in the posterior midline, inferior to the anal valves. It is painful because this region is supplied by sensory fibers of the inferior rectal nerves. *Perianal abscesses* may follow infection of anal ▶

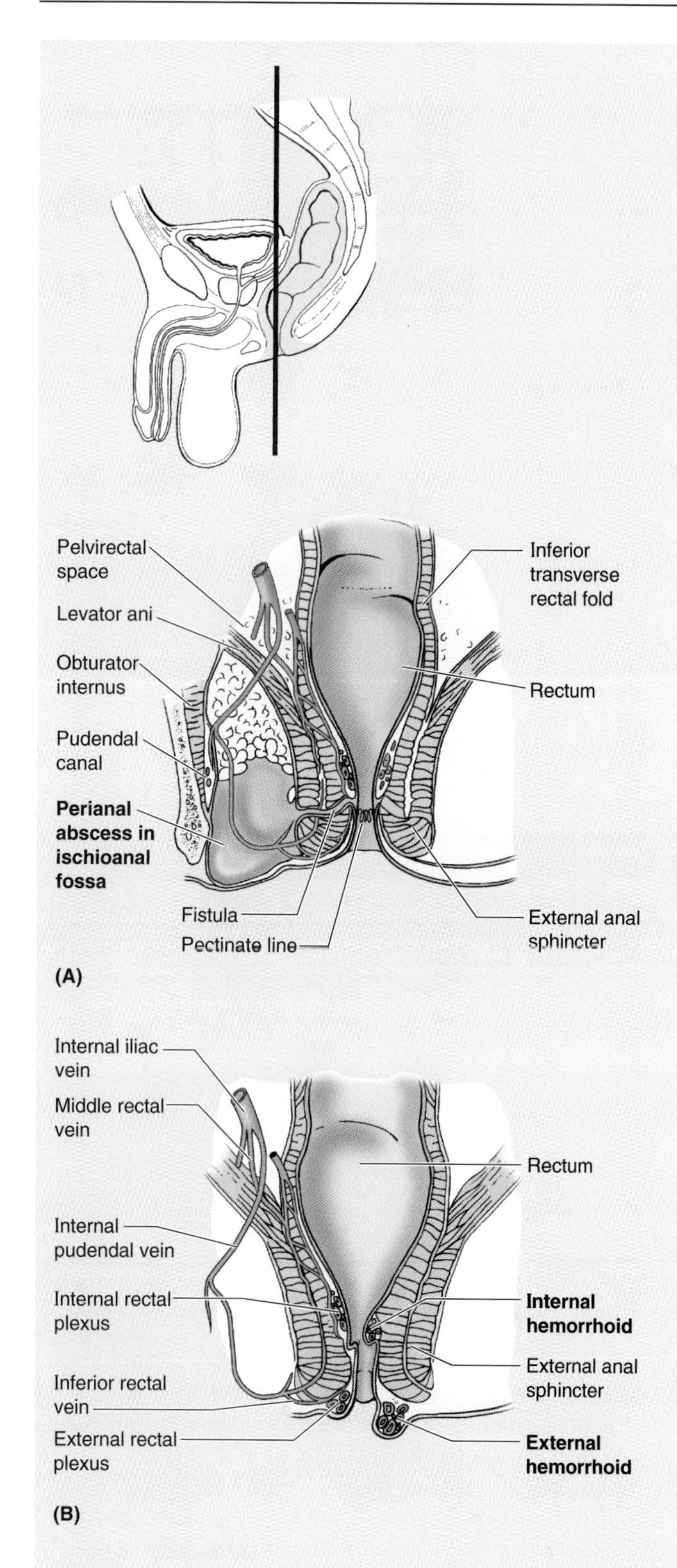

▶ fissures, and the infection may spread to the ischioanal fossae and form *ischioanal abscesses* or spread into the pelvis and form *pelvirectal abscesses.* An *anal fistula* may result from the spread of an anal infection and crypts (*cryptitis*). One end of this abnormal canal (fistula) opens into the *anal canal* and the other end opens into an abscess in the ischioanal fossa or into the perianal skin.

Prolapse of Hemorrhoids

Internal hemorrhoids ("piles") are prolapses of rectal mucosa containing the normally dilated veins of the *internal rectal venous plexus* (*B*). Internal hemorrhoids are thought to result from a breakdown of the muscularis mucosae, a smooth muscle layer deep to the mucosa. Internal hemorrhoids that prolapse through the anal canal are often compressed by the contracted sphincters, impeding bloodflow. As a result they tend to strangulate and ulcerate. *External hemorrhoids* are thromboses (blood clots) in the veins of the *external rectal venous plexus* and are covered by skin. Predisposing factors for hemorrhoids include pregnancy, chronic constipation, and any disorder that results in increased intra-abdominal pressure.

The anastomoses between the superior, middle, and inferior rectal veins form clinically important communications between the portal and systemic venous systems (see Chapter 2, Fig. 2.59). The superior rectal vein drains into the inferior mesenteric vein, whereas the middle and inferior rectal veins drain through the systemic system into the inferior vena cava. Any abnormal increase in pressure in the valveless portal system may cause enlargement of the superior rectal veins, resulting in an increase in bloodflow in the internal rectal venous plexus.

In *portal hypertension* occurring with *hepatic cirrhosis*, the anastomotic veins in the anal canal and elsewhere become varicose. Those in the esophagus are especially prone to rupture. It is important to note that the veins of the rectal plexuses *normally* appear varicose (dilated and tortuous)—even in newborns—and that internal hemorrhoids occur most commonly in the *absence* of portal hypertension.

Because autonomic nerves supply the anal canal superior to the pectinate line, an incision or needle insertion in this region is painless. However, *the anal canal inferior to the pectinate line is sensitive* (e.g., to the prick of a hypodermic needle) because it is supplied by the *inferior anal (rectal) nerves* containing sensory fibers. ⊙

Male Perineum

The male perineum includes the:

- Anal canal
- Intermediate and spongy parts of the urethra
- Root of the penis and the scrotum.

Urethra

The urethra in the bladder neck (preprostatic urethra) and the prostatic urethra, the first two parts of the male urethra, are described with the pelvis (p. 365).

The **intermediate (membranous) part of the urethra** is the shortest (1–2 cm) and narrowest part of the urethra, except for the external urethral orifice. It begins at the apex of the prostate and ends at the bulb of the penis, where it is continuous with the **spongy urethra** (Fig. 3.39). The intermediate part of the urethra traverses the deep perineal pouch, where it is surrounded by the external urethral sphincter and the perineal membrane. Posterolateral to this part of the urethra are the small *bulbourethral glands* (Cowper's glands) and their slender ducts, which open into the proximal part of the spongy urethra.

The **spongy urethra**, the longest part (15–16 cm), passes through the bulb and corpus spongiosum of the penis. It begins at the distal end of the intermediate part of the urethra and ends at the **external urethral orifice**, the narrowest part of the urethra (Fig. 3.39). The lumen of the spongy urethra is approximately 5 mm in diameter; however, it is expanded in the bulb of the penis to form the **intrabulbar fossa** and in the glans penis to form the **navicular fossa.** On each side the slender **ducts of the bulbourethral glands** open into the proximal part of the spongy urethra; the orifices of these ducts are extremely small. There are also many minute openings of the ducts of mucus-secreting **urethral glands** into the spongy urethra.

Arterial Supply of the Distal Two Parts of the Urethra. The arterial supply of the intermediate and spongy parts of the urethra is from branches of the **internal pudendal artery** (Figs. 3.34*C* and 3.37, Table 3.6).

Venous and Lymphatic Drainage of the Intermediate and Spongy Parts of the Urethra. Veins accompany the arteries and have similar names. Lymphatic vessels from the intermediate part of the urethra drain mainly into the **internal iliac lymph nodes** (Fig. 3.40); whereas most vessels from the spongy urethra pass to the **deep inguinal lymph nodes**, but some lymph passes to the external iliac nodes.

Innervation of the Intermediate and Spongy Parts of the Urethra. The nerves to the bulbar and spongy urethra are branches of the **pudendal nerve** (Figs. 3.26 and 3.38). Most afferent fibers from the intermediate part of the urethra run in the **pelvic splanchnic nerves.** Nerves from the *prostatic nerve plexus*, arising from the **inferior hypogastric plexus**, are distributed to this part of the urethra.

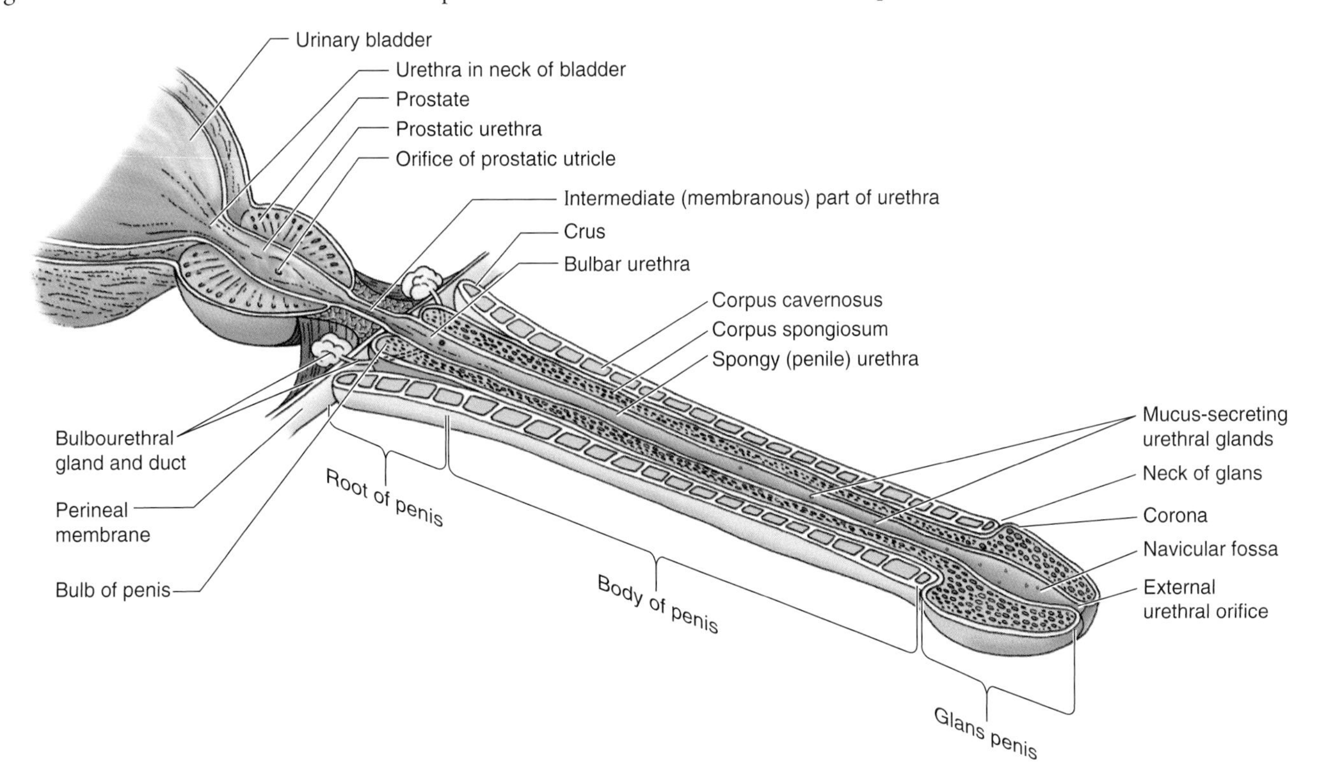

Figure 3.39. Male urethra and associated structures. Observe the four parts of the urethra: urethra in the bladder neck, prostatic urethra, intermediate part (membranous urethra), and spongy urethra. Observe also the ducts of the bulbourethral glands opening to the proximal part of the spongy urethra. Note that the urethra is not uniform in its caliber: the external urethral orifice and intermediate (membranous) part are narrowest. Attempting to approach this "straight-line" position as much as possible facilitates passage of a catheter or other transurethral device.

Table 3.6. Arterial Supply of the Perineum

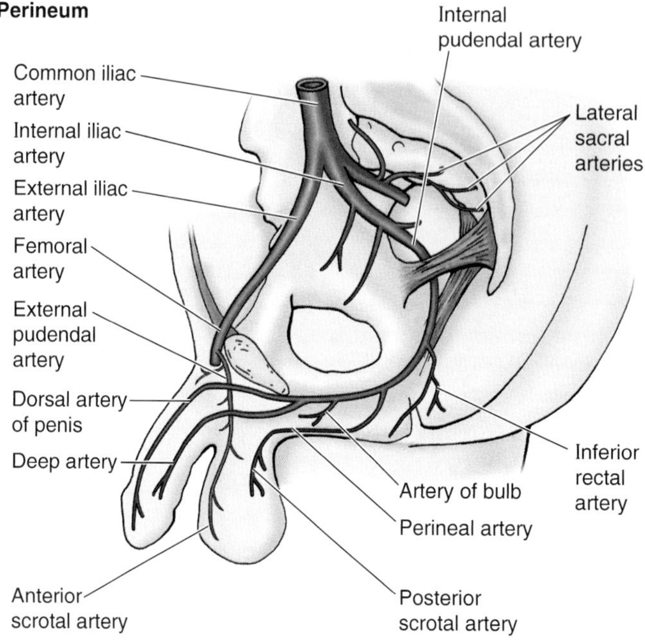

Artery	Origin	Course	Distribution
Internal pudendal	Internal iliac artery	Leaves pelvis through greater sciatic foramen; hooks around ischial spine and enters perineum by way of lesser sciatic foramen and passes to pudendal canal	Perineum and external genital organs
Inferior rectal	Internal pudendal artery	Leaves pudendal canal and crosses ischioanal fossa to anal canal	Distal portion of anal canal
Perineal	Internal pudendal artery	Leaves pudendal canal and enters superficial perineal space	Supplies superficial perineal muscles and scrotum
Posterior scrotal or labial	Terminal branches of perineal artery	Runs in superficial fascia of posterior scrotum or labium majus	Skin of scrotum or labium majus
Artery of bulb of penis or vestibule of vagina	Internal pudendal artery	Pierces perineal membrane to reach bulb of penis or vestibule of vagina	Supplies bulb of penis or vestibule and bulbourethral gland (male) and greater vestibular gland (female)
Deep artery of penis or clitoris	Terminal branch of internal pudendal artery	Pierces perineal membrane to reach corpora cavernosa of penis or clitoris	Supplies erectile tissue of penis or clitoris
Dorsal artery of penis or clitoris	Terminal branch of internal pudendal artery	Pierces perineal membrane and passes through suspensory ligament of penis or clitoris to run on dorsum of penis or clitoris	Skin of penis and erectile tissue of penis or clitoris
External pudendal, superficial, and deep branches	Femoral artery	Pass medially across the thigh to reach the scrotum or labia majora	External genitalia and superomedial part of the thigh

Urethral Catheterization

Urethral catheterization is done to remove urine from a person who is unable to micturate. It is also performed to irrigate the bladder and to obtain an uncontaminated sample of urine. When inserting catheters and urethral sounds—slightly conical instruments for exploring and dilating a constricted urethra—*the curves of the urethra must be considered.* Just distal to the perineal membrane, the spongy urethra is well covered inferiorly and posteriorly by erectile tissue of the bulb of the penis; however, a short segment of the intermediate part of the urethra is unprotected. Because the urethral wall is thin, and because of the angle that must be negotiated to enter the intermediate part of the spongy urethra, it is vulnerable to rupture during the insertion of urethral catheters and sounds. The intermediate part—the least distensible part—runs inferoanteriorly as it passes through the external urethral sphincter. Proximally, the prostatic part takes a slight curve that is concave anteriorly as it traverses the prostate.

Urethral stricture may result from external trauma of the penis or infection of the urethra. *Urethral sounds* are used to dilate the constricted urethra in such cases. The spongy urethra will expand enough to permit passage of an instrument approximately 8 mm in diameter. The external urethral orifice is the narrowest and least distensible part of the urethra; hence, an instrument that passes through this opening normally passes through all other parts of the urethra. ⊙

Scrotum

The scrotum is a cutaneous fibromuscular sac for the testes and associated structures. It is situated posteroinferior to the penis and inferior to the pubic symphysis. The bilateral embryonic formation of the scrotum is indicated by the midline **scrotal raphe** (Fig. 3.41), which is continuous on the ventral surface of the penis with the **penile raphe** and posteriorly along the median line of the perineum with the **perineal raphe**. The contents of the scrotum (testes and epididymides) and their coverings are described with the abdomen (see Chapter 2).

Arterial Supply of the Scrotum. The *external pudendal arteries* supply the anterior aspect of the scrotum and the *internal pudendal arteries* supply the posterior aspect (Table 3.6). The scrotum also receives branches from the testicular and cremasteric arteries.

Venous and Lymphatic Drainage of the Scrotum. The scrotal veins accompany the arteries and join the **external pudendal veins.** *Lymphatic vessels* from the scrotum carry lymph to the *superficial inguinal lymph nodes* (Fig. 3.40).

Innervation of the Scrotum. The anterior aspect of the scrotum is supplied by anterior scrotal nerves derived from the **ilioinguinal nerve**, and by the genital branch of the **genitofemoral nerve.** The posterior aspect of the scrotum is supplied by posterior scrotal nerves, branches of the superficial *perineal nerves* (Fig. 3.38), and by the perineal branch of the **posterior femoral cutaneous nerve.**

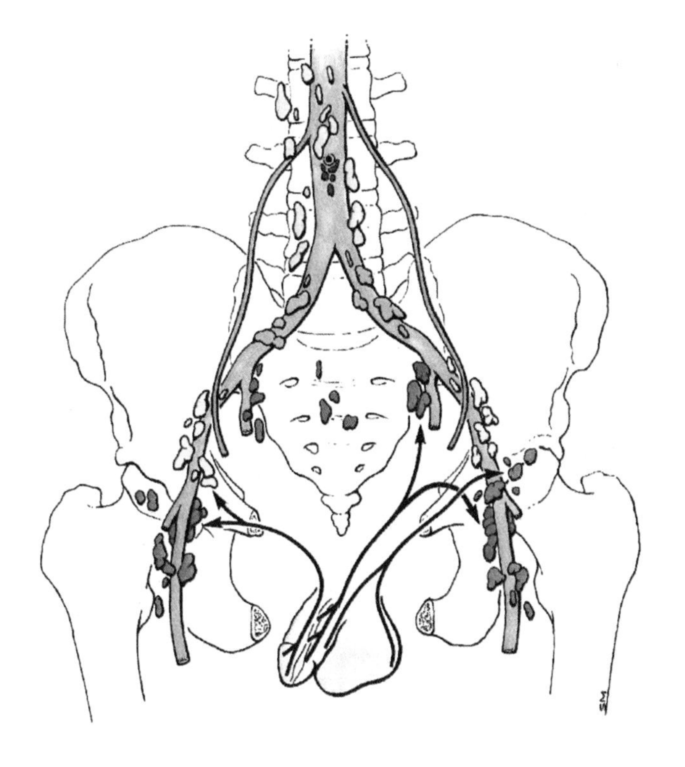

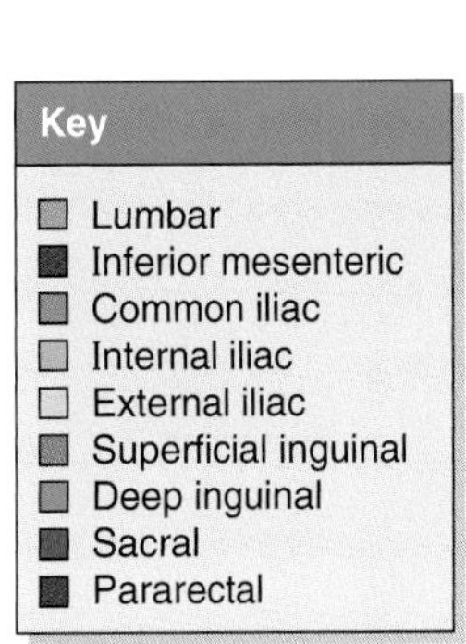

Figure 3.40. Lymphatic drainage of the penis, spongy urethra, and scrotum. The *arrows* indicate the direction of lymph flow to the lymph nodes.

Distension of the Scrotum

The scrotum is easily distended. In persons with large indirect inguinal hernias, for example (see Chapter 2), the intestine may enter the scrotum, making it as large as a soccer ball. Similarly, inflammation of the testes, such as *orchitis*—inflammation of the testis—associated with mumps, or bleeding in the subcutaneous tissue may produce an enlarged scrotum.

Palpation of the Testes

The soft, pliable skin of the scrotum makes it easy to palpate the testes and the structures related to them (e.g., the epididymis and ductus deferens). The left testis commonly lies at a lower level than does the right one. ⊙

Penis

The penis is the male copulatory organ and the common outlet for urine and semen (Figs. 3.39, 3.41, and 3.42). The penis consists of a **root, body,** and **glans penis.** It is composed of three cylindrical bodies of erectile cavernous tissue—the **corpora cavernosa** and **corpus spongiosum**—each of which have an outer fibrous covering or capsule, the **tunica albuginea** (Fig. 3.41*C*). Superficial to the outer covering is the deep fascia of the penis (Buck's fascia), the continuation of the deep perineal fascia that forms a strong membranous covering for the corpora cavernosa and **corpus spongiosum**, binding them together. The corpus spongiosum contains the spongy urethra. The two **corpora cavernosa** are fused with each other in the median plane, except posteriorly where they separate to form the **crura of the penis** (Fig. 3.42*B*).

The **root of the penis**, the attached part, consists of the crura, bulb, and ischiocavernosus and bulbospongiosus mus-

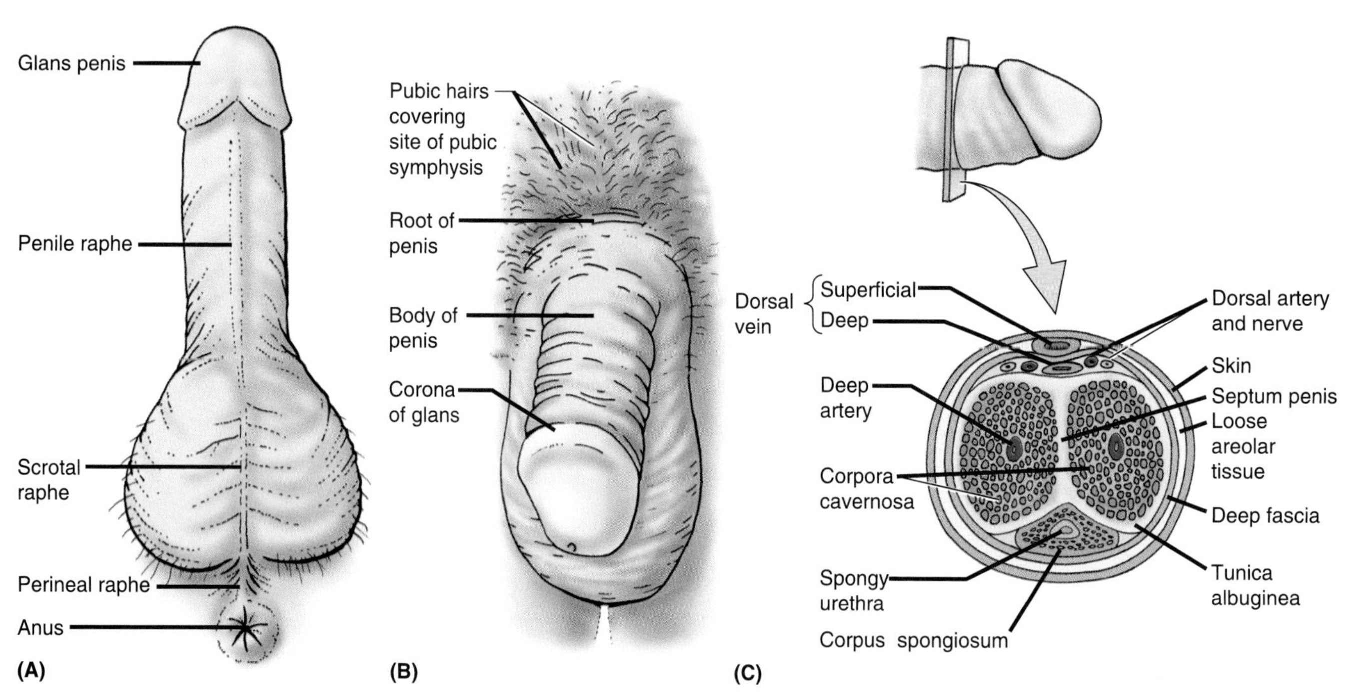

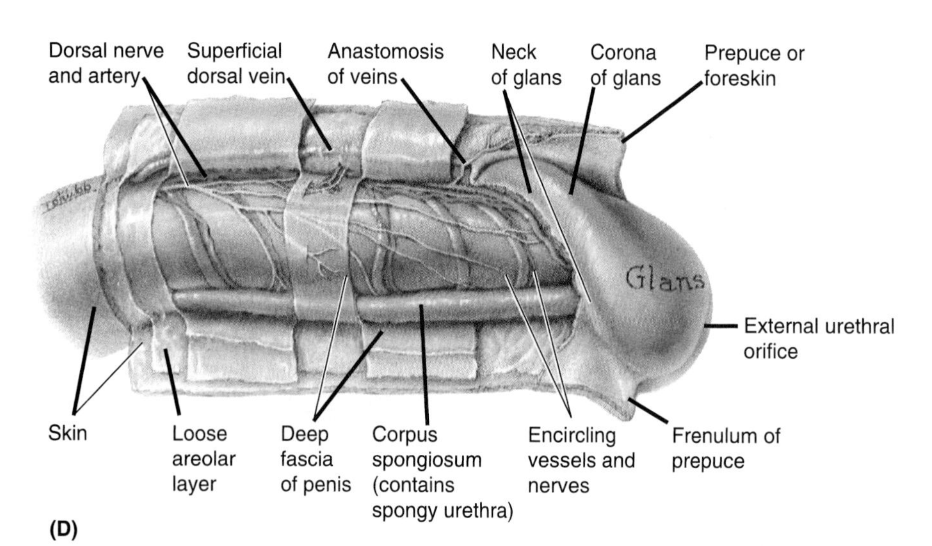

Figure 3.41. Penis and scrotum. **A.** Urethral surface of the penis. The spongy urethra is deep to the cutaneous penile raphe. Observe that the scrotum is also divided into right and left halves by the cutaneous scrotal raphe, which is continuous with the penile and perineal raphae. **B.** Dorsum of the penis and the anterior surface of the scrotum. Observe that the penis comprises a root, body, and glans. **C.** Section of the body of the penis. Observe that it contains three erectile masses: two corpora cavernosa and a corpus spongiosum (containing the spongy urethra). **D.** Lateral view of a dissected penis. Observe that the skin is carried forward as the prepuce. Observe also the vessels and nerves.

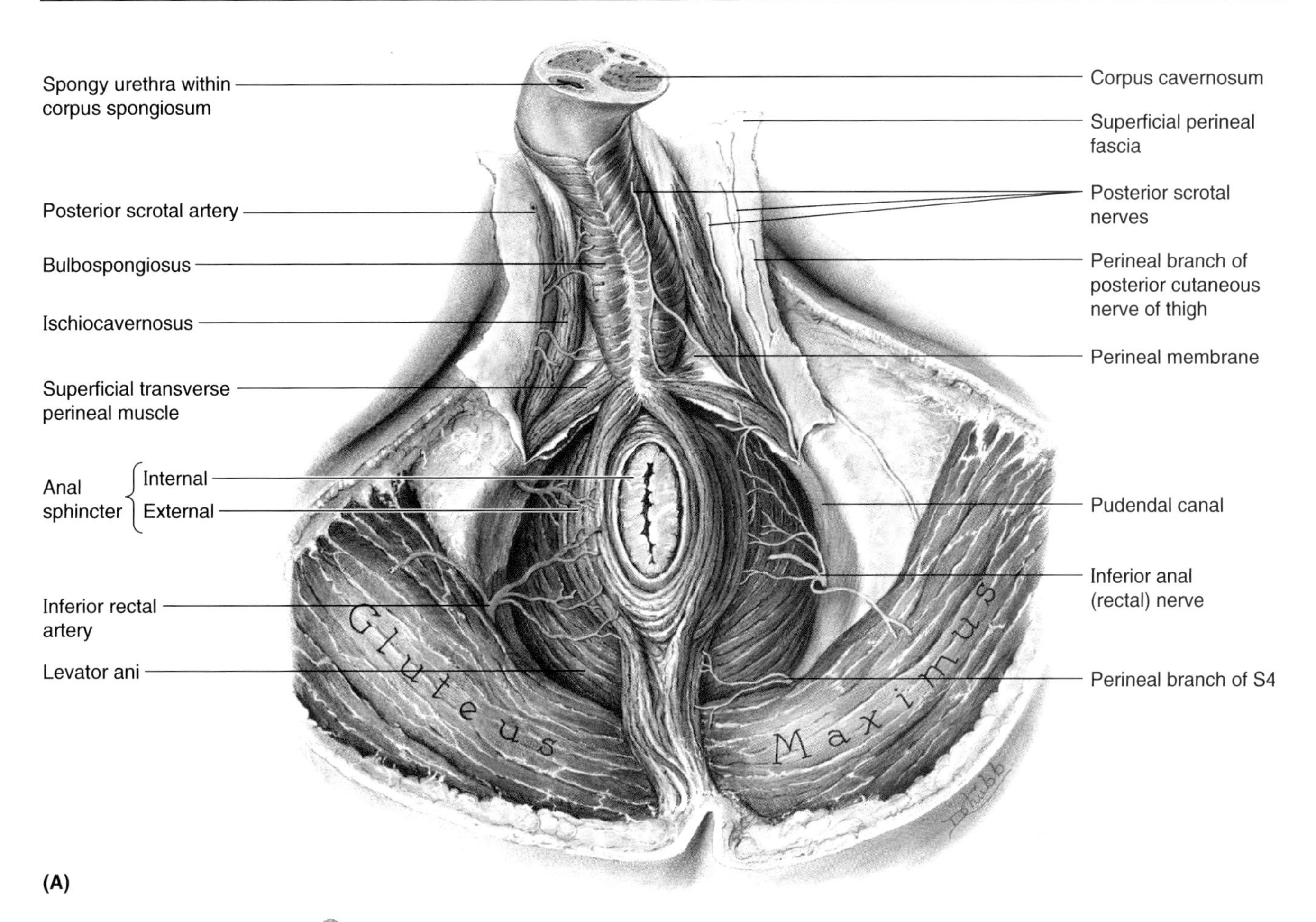

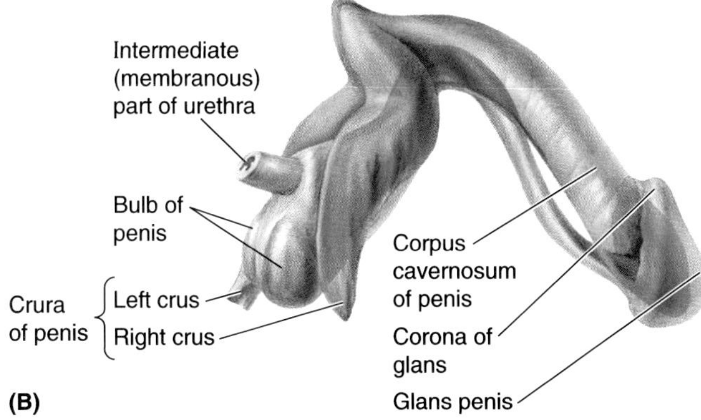

Figure 3.42. Male perineum and structure of the penis. A. Dissection of the perineum. Observe the anus surrounded by the external anal sphincter with an ischioanal fossa on each side. Observe the inferior anal (rectal) nerve that has branched from the pudendal nerve at the entrance to the pudendal canal and, with the perineal branch of S4, is supplying the external anal sphincter. **B.** Structure of the penis. The corpus spongiosum is separated from the corpora cavernosa. The natural flexures of the penis are preserved. Observe that the glans penis fits like a cap on the blunt ends of the corpora cavernosa.

cles. The root is located in the superficial perineal pouch, between the perineal membrane superiorly and the deep perineal fascia inferiorly. The **crura** and **bulb** of the penis contain masses of erectile tissue. Each crus is attached to the inferior part of the internal surface of the corresponding ischial ramus (Fig. 3.34*D*), anterior to the ischial tuberosity. The enlarged posterior part of the **bulb of the penis** is penetrated superiorly by the intermediate part of the urethra (Fig. 3.42*B*).

The **body of the penis** (Figs 3.39 and 3.41*B*) is the free part that is pendulous in the flaccid condition. Except for a few fibers of the bulbospongiosus near the root of the penis and the ischiocavernosus that embrace the crura, the body (shaft) of the penis has no muscles. The penis consists of thin skin, connective tissue, blood and lymphatic vessels, fascia, the corpora cavernosa, and the corpus spongiosum containing the spongy urethra (Fig. 3.41*C*). Distally the corpus spongiosum expands to form the conical **glans penis** (Fig. 3.42*B*), which forms the head of the penis. The margin of the glans projects beyond the ends of the corpora cavernosa to form the **corona of the glans** (Fig. 3.41, *B* and *D*). The corona overhangs an obliquely grooved constriction—the **neck of the glans**—that separates the glans from the body of the penis. The slitlike opening of the spongy urethra, the **external urethral orifice** (meatus), is near the tip of the glans.

The skin of the penis is thin, dark, and connected to the tunica albuginea by loose connective tissue. At the neck of the glans, the skin and fascia of the penis are prolonged as a double layer of skin, the **prepuce** (foreskin), which covers the glans to a variable extent. The **frenulum of the prepuce** is a

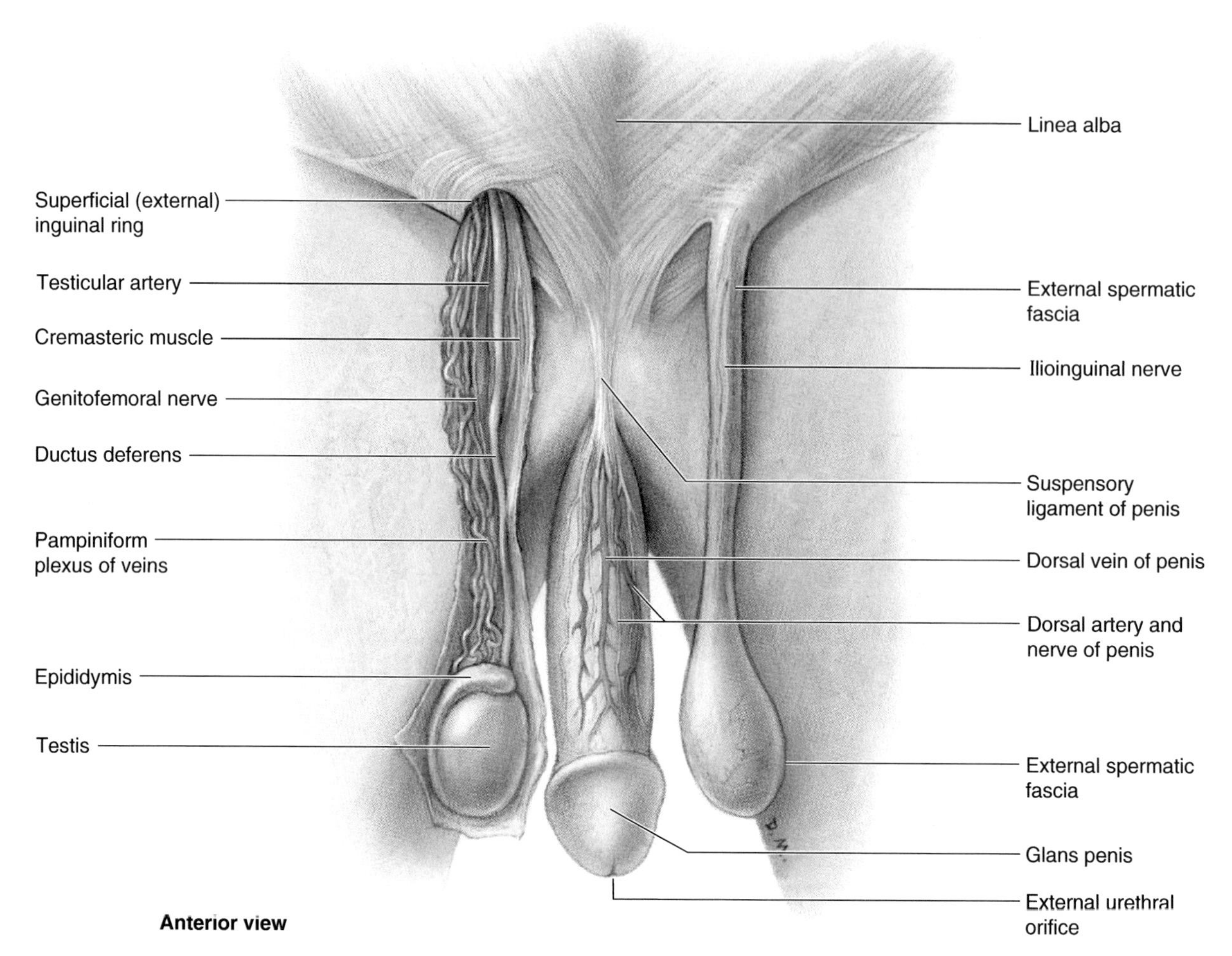

Figure 3.43. Vessels and nerves on the dorsum of the penis and contents of the spermatic cord. Dissection of the penis and spermatic cord showing their vessels and nerves. The skin—including the scrotum—has been removed. The superficial (dartos) fascia covering the penis has also been removed to expose the midline deep dorsal vein and the bilateral dorsal arteries and nerves of the penis. Observe the triangular suspensory ligament of the penis attached to the pubic symphysis and blending with the deep fascia of the penis.

median fold that passes from the deep layer of the prepuce to the urethral surface of the glans (Fig. 3.41*D*).

The **suspensory ligament of the penis** is a condensation of deep fascia that arises from the anterior surface of the pubic symphysis (Fig. 3.43). The ligament passes inferiorly and splits to form a sling that is attached to the deep fascia of the penis at the junction of its root and body. The **fundiform ligament of the penis** is a band of elastic fibers of the subcutaneous tissue that extends from the linea alba superior to the pubic symphysis and splits to surround the penis before attaching to the fascia of the penis.

The **superficial perineal muscles** (Fig. 3.42*A*, Table 3.7) are the:

- Superficial transverse perineal
- Bulbospongiosus
- Ischiocavernosus.

These muscles are in the superficial perineal pouch and are supplied by the perineal nerves.

The **superficial transverse perineal muscles** are slender strips of muscles that pass transversely, anterior to the anus. Each muscle extends from the ischial tuberosity to the perineal body, the region of interdigitation of the fibers of the bulbospongiosus anteriorly, the external anal sphincter posteriorly, and the superficial perineal muscles laterally.

The **bulbospongiosus muscles** lie in the median plane of the perineum, anterior to the anus. The two symmetrical parts are united by a median tendinous raphe inferior to the bulb of the penis. The muscles arise from this raphe and the perineal body. *The bulbospongiosus forms a sphincter that compresses the bulb of the penis and the corpus spongiosum,* thereby aiding in emptying the spongy urethra of residual urine and/or semen. The anterior fibers of the bulbospongiosus, encircling the most proximal part of the body of the penis, also assist erection by increasing the pressure on the erectile tissue in the root of the penis. At the same time, they also compress the deep dorsal vein of the penis, impeding venous drainage of the cavernous spaces and helping to promote enlargement and turgidity of the penis.

Table 3.7. **Muscles of the Perineum**

Muscle	Origin	Insertion	Innervation	Main Action
External anal sphincter	Skin and fascia surrounding anus and coccyx via anococcygeal ligament	Perineal body	Inferior anal nerve	Closes anal canal; works with bulbospongiosus to support and fix perineal body
Bulbospongiosus	Male: median raphe, ventral surface of bulb of penis, and perineal body Female: perineal body	Male: corpora spongiosum and cavernosa and fascia of bulb of penis Female: fascia of corpus cavernosa	Deep branch of perineal nerve, a branch of pudendal nerve	Works with external anal sphincter to support/fix perineal body Male: compresses bulb of penis to expel last drops of urine/semen; assists erection by pushing blood into body of penis and compressing outflow veins Female: "sphincter" of vagina and assists in erection of clitoris
Ischiocavernosus	Internal surface of ischiopubic ramus and ischial tuberosity	Crus of penis or clitoris		Maintains erection of penis or clitoris by compressing outflow veins and pushing blood into body of penis or clitoris
Superficial transverse perineal muscle		Perineal body		Support and fix perineal body (pelvic floor) to support abdominopelvic viscera and resist increased intra-abdominal pressure
Deep transverse perineal muscle		Median raphe, perineal body, and external anal sphincter		
External urethral sphincter	(Compressor urethrae portion only)	Surrounds urethra; in males, also ascends anterior aspect of prostate; in females, some fibers also enclose vagina (urethrovaginal sphincter)		Compresses urethra to maintain urinary continence; in females, urethrovaginal sphincter portion also compresses vagina

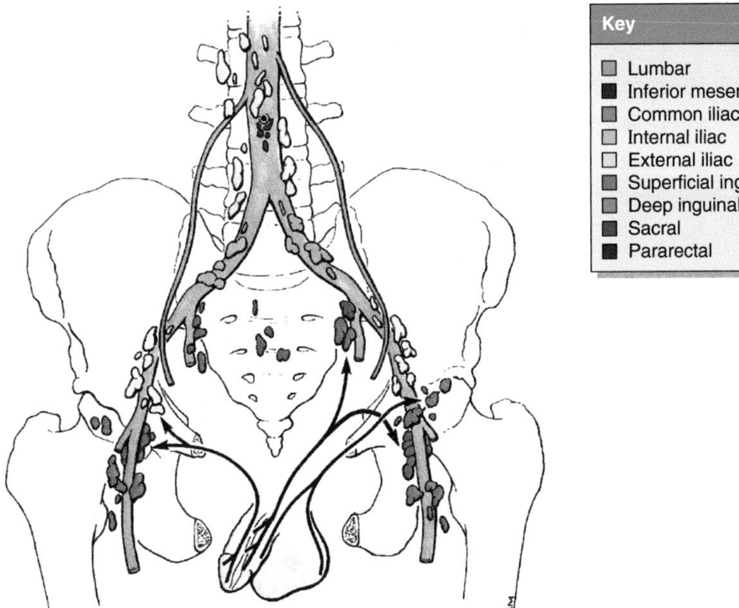

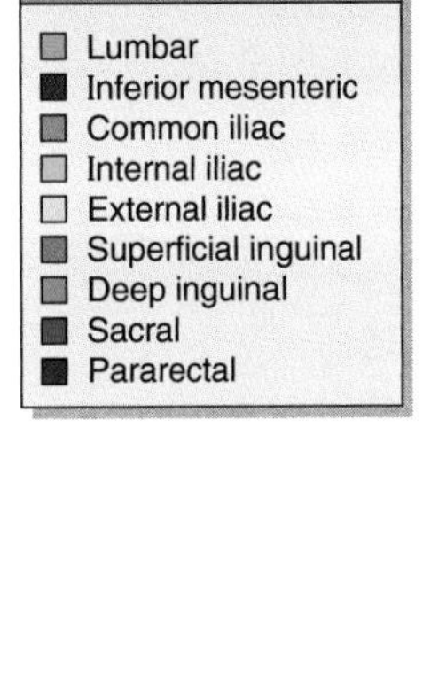

Figure 3.44. Lymphatic drainage of the penis and scrotum. The *arrows* indicate the direction of lymph flow to the lymph nodes.

The **ischiocavernosus muscles** surround the crura in the root of the penis. Each muscle arises from the *internal surface of the ischial tuberosity and ischial ramus* and passes anteriorly on the crus of the penis, where it is inserted into the sides and ventral surface of the crus and the perineal membrane. *The ischiocavernosus muscles force blood from the cavernous spaces in the crura into the distal parts of the corpora cavernosa*; this increases the turgidity of the penis. Contraction of the ischiocavernosus muscles also compresses the deep dorsal vein of the penis as it leaves the crus of the penis, thereby cutting off the venous return from the penis and helping to maintain the erection.

Arterial Supply of the Penis. The penis is supplied mainly by branches of the **internal pudendal arteries** (Figs. 3.41, *C* and *D*, and 3.43, Table 3.6).

- *Dorsal arteries* run in the interval between the corpora cavernosa on each side of the deep dorsal vein, supplying the fibrous tissue around the corpora and the penile skin
- *Deep arteries* pierce the crura and run within the corpora cavernosa, supplying the erectile tissue in these structures
- *Artery of the bulb of the penis* supplies the posterior part of the corpus spongiosum and the bulbourethral gland.

Superficial and deep branches of the **external pudendal arteries** supply the penile skin, anastomosing with branches of the internal pudendal arteries.

The **deep arteries of the penis** are the main vessels supplying the cavernous spaces in the erectile tissue of the corpora cavernosa and are therefore involved in the erection of the penis. They give off numerous branches that open directly into the cavernous spaces. When the penis is flaccid, these arteries are coiled; hence, they are *helicine arteries* (Gr. helix, a coil).

Venous and Lymphatic Drainage of the Penis. Blood from the cavernous spaces is drained by a venous plexus that joins the **deep dorsal vein of the penis** in the deep fascia (Fig. 3.41*C*). This vein passes deep to the arcuate pubic ligament and joins the **prostatic** venous plexus. Blood from the superficial coverings of the penis drain into the **superficial dorsal vein**, which ends in the *superficial external pudendal vein.* Some blood also passes to the lateral pudendal vein. The **superficial inguinal lymph nodes** receive most of the lymph from the penis (Fig. 3.44).

Innervation of the Penis. The nerves derive from the S2 through S4 segments of the spinal cord, passing through the **pudendal nerve** and the pelvic plexuses. The **dorsal nerve of the penis**—a terminal branch of the pudendal nerve (Fig. 3.45)—arises in the pudendal canal and passes anteriorly into the deep perineal pouch. It then runs to the dorsum of the penis, where it passes lateral to the dorsal artery (Fig. 3.41*C*). It supplies both the skin and the glans penis. The penis is richly provided with a variety of sensory nerve endings—especially the glans penis— and thus is sensitive. Branches of the *ilioinguinal nerve* supply the skin at the root of the penis.

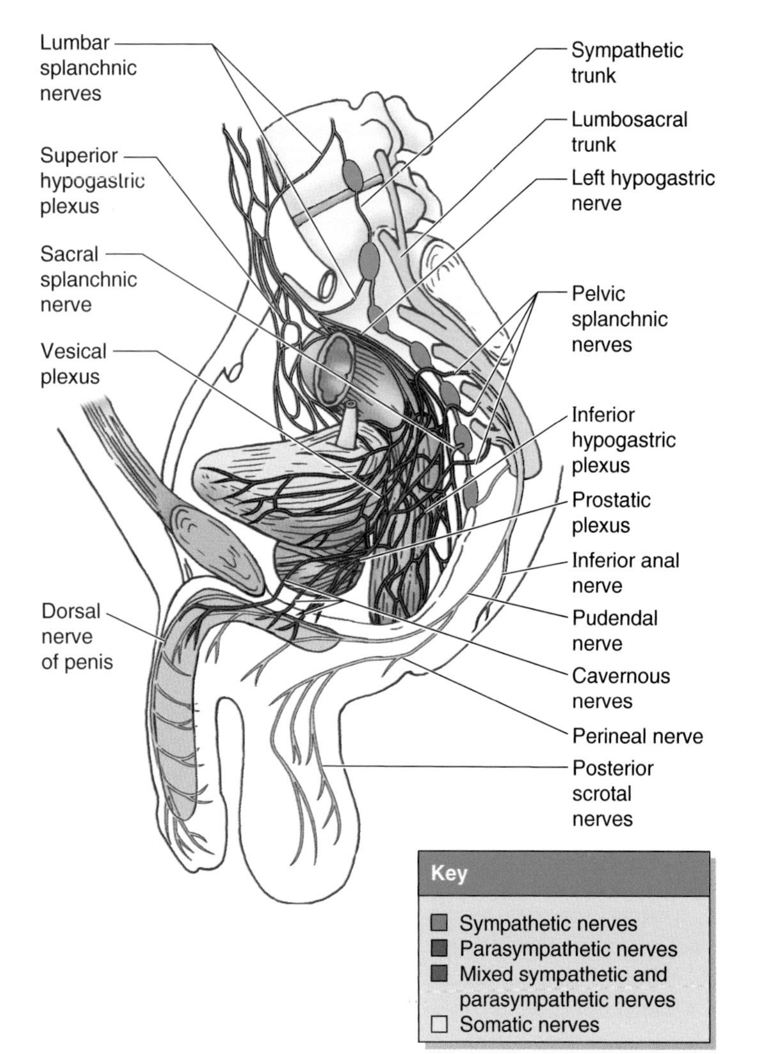

Figure 3.45. Autonomic nerves of the pelvis. Observe the sympathetic trunk, lumbosacral trunk, hypogastric plexuses, sacral and pelvic splanchnic nerves, and the dorsal nerve of the penis. The cavernous nerves course independently of the pudendal nerve and are the only nerves conveying parasympathetic fibers out of the body cavities. They terminate on the arteriovenous anastomoses and helicine arteries, which, when stimulated, produce erection of the penis.

Erection, Emission, and Ejaculation

When a male is stimulated erotically, arteriovenous anastomoses—by which blood is normally able to bypass the "empty" potential spaces or sinuses of the corpora cavernosa—are closed. The smooth muscle in the fibrous trabeculae and coiled arteries relaxes as a result of *parasympathetic stimulation* (S2 through S4 through the cavernous nerves from the *prostatic nerve plexus*). As a result, the helicine arteries straighten, enlarging their lumina and allowing blood to flow into and dilate the cavernous spaces in the corpora of the penis. The bulbospongiosus and ischiocavernosus muscles compress the venous plexuses at the periphery of the corpora cavernosa, impeding the return of venous blood. As a result, the corpora cavernosa and corpus spongiosum become enlarged, rigid, and the penis erects.

During *emission,* semen (sperms and glandular secretions) is delivered to the prostatic urethra through the ejaculatory ducts after peristalsis of the ductus deferentes and seminal vesicles. Emission is a sympathetic response (L1 and L2 nerves). Prostatic fluid is added to the seminal fluid as the smooth muscle in the prostate contracts.

During *ejaculation,* semen is expelled from the urethra through the external urethral orifice. Ejaculation results from:

- Closure of the vesical sphincter at the neck of the bladder—sympathetic (L1 and L2 nerves)
- Contraction of the urethral muscle—a parasympathetic response (S2 through S4 nerves)
- Contraction of the bulbospongiosus muscles—pudendal nerves (S2 through S4).

After ejaculation, the penis gradually returns to a flaccid state, resulting from sympathetic stimulation that causes constriction of the smooth muscle in the coiled arteries. The bulbospongiosus and ischiocavernosus muscles relax, allowing more blood to flow into the veins. Blood is slowly drained from the cavernous spaces in the penile corpora into the deep dorsal vein.

Hypospadias

Hypospadias is a common congenital anomaly of the penis, occurring in 1 in 500 newborns. In the simplest form, *glandular hypospadias,* the external urethral orifice is on the ventral aspect of the glans penis. In other infants the defect is in the skin and ventral wall of the spongy urethra. Hence, the external urethral orifice is on the urethral surface of the penis. Hypospadias results from failure of fusion of the urogenital folds during early fetal development (Moore and Persaud, 1998).

Phimosis, Paraphimosis, and Circumcision

The prepuce of the penis is usually sufficiently elastic for it to be retracted over the glans penis. In some males it fits tightly over the glans and cannot be retracted easily—*phimosis*—if at all. As there are modified sebaceous glands in the prepuce, the oily secretions of cheesy consistency—*smegma*—from them accumulate in the *preputial sac,* located between the glans and prepuce, causing irritation. In some persons, retraction of the prepuce over the glans penis constricts the neck of the glans so much that there is interference with the drainage of blood and tissue fluid. In patients with this condition—*paraphimosis*—the glans penis may enlarge so much that the prepuce cannot be drawn over it. Circumcision is commonly performed in such cases. *Circumcision*—surgical excision of the prepuce—is the most commonly performed minor surgical operation on male infants. Although it is a religious practice in Islam and Judaism, it is often done routinely for nonreligious reasons (mostly related to hygiene) in North America. In adults, circumcision is usually performed when phimosis or paraphimosis is present. ○

Female Perineum

The perineum—the shallow perineal compartment inferior to the pelvic diaphragm—has the same skeletal boundaries as the pelvic outlet (Figs. 3.33*A* and 3.34*A*). *The perineum (perineal area or region) is bounded by the:*

- Mons pubis
- Medial aspects (insides) of the thighs
- Gluteal folds
- Intergluteal (natal) cleft.

The perineum includes the pudendum or vulva (external genitalia) and the anus. Clinically, the term perineum is restricted to the region between the anal and vaginal orifices.

Female External Genitalia

The external genital organs (Figs. 3.46 and 3.47) include the:

- Mons pubis
- Labia majora
- Labia minora

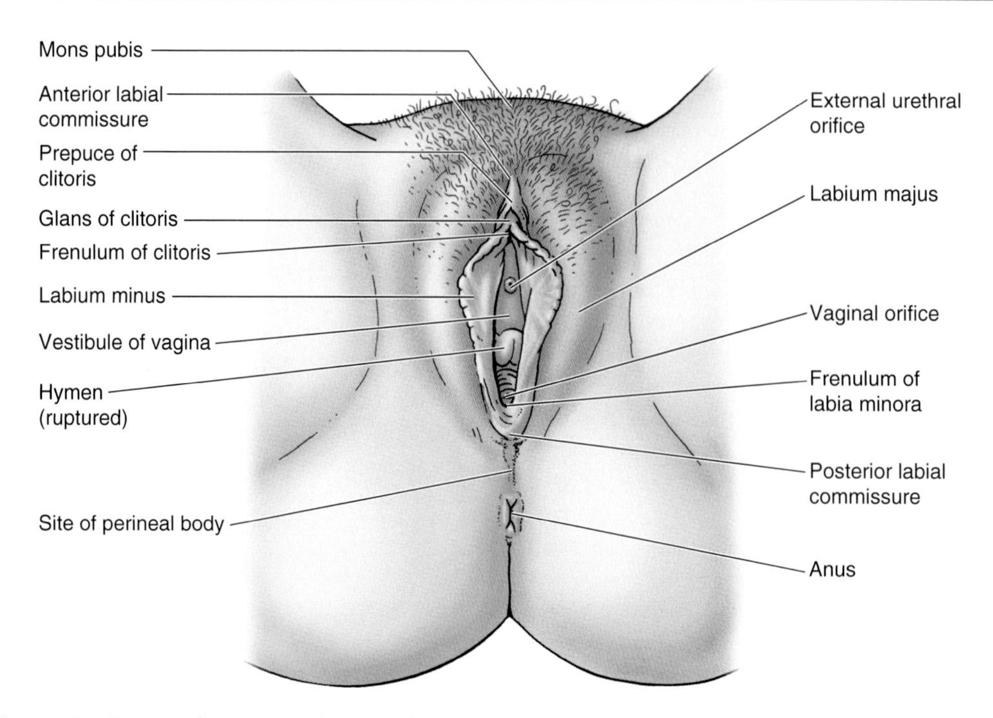

Figure 3.46. Female external genitalia. The labia majora and minora are separated to show the vestibule of the vagina, into which the external urethral orifice and the vaginal orifice open.

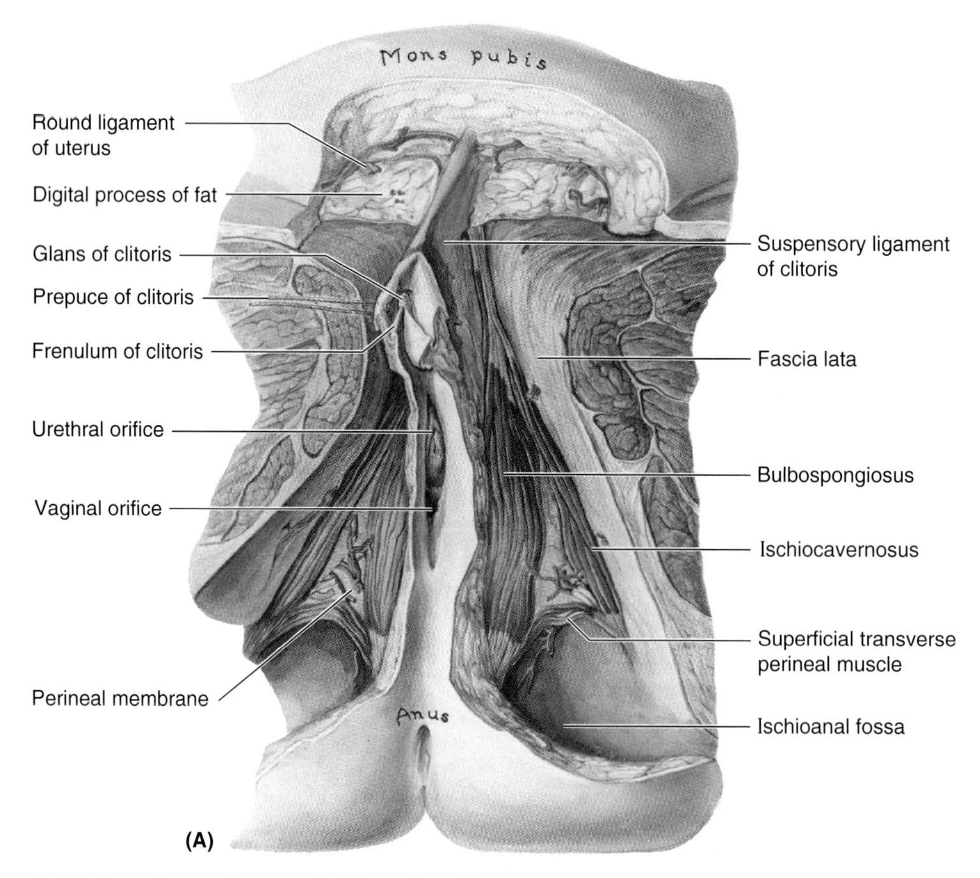

Figure 3.47. Female perineum. A. Dissection. Observe the thickness of the superficial fatty tissue in the mons pubis and the encapsulated digital processes of fat deep to this that largely filled the labia majora. Observe the prepuce of the clitoris forming a hood over the clitoris. Examine the ischioanal fossae lateral to the anal canal.

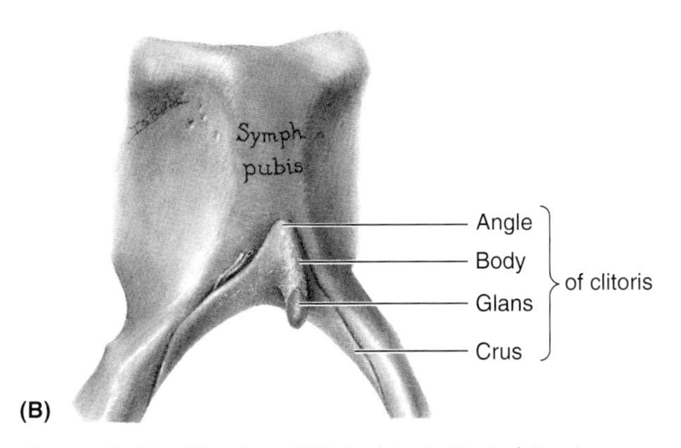

Figure 3.47. *(Continued)* **B.** Isolated clitoris following removal of surrounding soft tissue by dissection.

- Clitoris
- Vestibule of vagina
- Bulbs of vestibule
- Greater vestibular glands.

The synonymous terms **vulva** and **pudendum** include all these parts; the term pudendum is commonly used clinically. The vulva serves:

- As sensory and erectile tissue for sexual arousal and intercourse
- To direct the flow of urine
- To prevent entry of foreign material into the urogenital tract.

Mons Pubis. The mons pubis is the rounded, fatty prominence anterior to the pubic symphysis, pubic tubercles, and superior pubic rami. The eminence is formed by a mass of fatty subcutaneous tissue. The surface of the mons is continuous with the anterior abdominal wall. After puberty the mons pubis is covered with coarse pubic hairs. The amount of fat increases at puberty and decreases after menopause.

Labia Majora. The labia majora are prominent folds of skin that bound the pudendal cleft and indirectly provide protection for the urethral and vaginal orifices. Each labium majus—largely filled with a fingerlike "digital process" of loose subcutaneous tissue containing smooth muscle, the termination of the round ligament of the uterus (see Fig. 3.49), and fat—passes inferoposteriorly from the mons pubis toward the anus. The labia lie at the sides of the **pudendal cleft**, the slit between the labia majora. The external aspects of the labia in the adult are covered with pigmented skin containing many sebaceous glands and are covered with crisp pubic hair. The internal aspects of the labia are smooth, pink, and hairless. The labia are thicker anteriorly where they join to form the **anterior commissure**. Posteriorly, in nulliparous women (never having borne children) they merge to form a ridge—the **posterior commissure**—which overlies the perineal body and is the posterior limit of the vulva. This commissure usually disappears after the first vaginal birth.

Labia Minora. The labia minora are folds of fat-free, hairless skin. They are enclosed in the pudendal cleft within the labia majora, immediately surrounding the **vestibule of the vagina.** They have a core of spongy connective tissue containing erectile tissue and many small blood vessels. The labia minora extend from the clitoris posterolaterally around the external urethral orifice and the orifice of the vagina. In young women, especially virgins, the labia minora are connected by a small fold—the **frenulum of the labia minora** (fourchette). Although the internal surface of each labium minus consists of thin moist skin, it has the typical pink color of mucous membrane and contains many sebaceous glands and sensory nerve endings.

Clitoris. The clitoris is an erectile organ located where the labia minora meet anteriorly. The clitoris consists of a **root** and a **body,** which are composed of two crura, two corpora cavernosa, and a **glans clitoris** that is covered by a prepuce (Fig. 3.48). Together the body and glans of the clitoris are approximately 2 cm in length and less than 1 cm in diameter. The anteriormost part of the labia minora passes anterior to the clitoris and forms the **prepuce** of the clitoris. A more posterior or deeper part of the labia minora passes posterior to the clitoris and forms the **frenulum** of the clitoris. Unlike the penis, the clitoris is not functionally related to the urethra or to urination, and functions solely as an organ of sexual arousal. The clitoris enlarges on tactile stimulation and is highly sensitive. The glans of the clitoris is the most highly innervated part of the clitoris.

Vestibule. The vestibule is the space between the labia minora that contains the openings of the urethra, vagina, and ducts of the greater and lesser vestibular glands (Fig. 3.46). The **external urethral orifice** is located 2 to 3 cm posteroinferior to the glans of the clitoris and anterior to the vaginal orifice. On each side of the external urethral orifice are the openings of the ducts of the *paraurethral glands*. The size and appearance of the **vaginal orifice** vary with the condition of the **hymen**, a thin fold of mucous membrane surrounding the vaginal orifice. After childbirth, only a few remnants of the hymen—*hymenal caruncles* (tags)—are visible. These remnants demarcate the vulva from the vagina.

Bulbs of the Vestibule. The bulbs of the vestibule are paired masses of *elongated erectile tissue*, approximately 3 cm in length (Fig. 3.48). The bulbs lie along the sides of the vaginal orifice under cover of the bulbospongiosus muscles. The bulbs are homologous with the bulb of the penis and the corpus spongiosum.

Vestibular Glands. The greater vestibular glands, approximately 0.5 cm in diameter, are on each side of the vestibule, posterolateral to the vaginal orifice (Fig. 3.48). The **greater vestibular glands** are round or oval and are partly overlapped posteriorly by the **bulbs of the vestibule**, and like the bulbs, are partially surrounded by the bulbospongiosus muscles. The slender ducts of these glands pass deep to the bulbs of the vestibule and open into the vestibule on each side

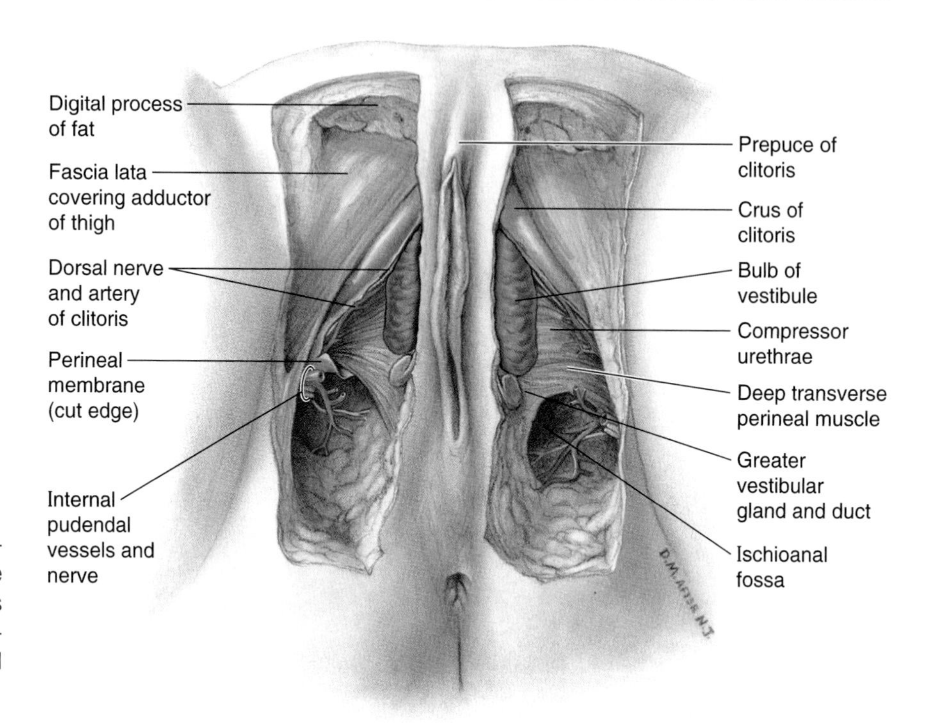

Figure 3.48. Female perineum. Dissection showing the bulbs of the vestibule, the greater vestibular glands and ducts, and the ischioanal fossae. Observe the internal pudendal vessels and nerve.

of the vaginal orifice. These glands secrete mucus into the vestibule during sexual arousal. The **lesser vestibular glands** are small glands on each side of the vestibule that open into it between the urethral and vaginal orifices. These glands secrete mucus into the vestibule, which moistens the labia and vestibule.

Perineal Fascia and Muscles

The *superficial perineal fascia* consists of a fatty layer and a membranous layer of subcutaneous connective tissue. These layers are continuous in the labia majora. The deep layer of fascia attaches medially to the pubic symphysis and laterally to the body of the pubis.

The **superficial perineal muscles** (Fig. 3.47 and Table 3.7) include the:

- Superficial transverse perineal
- Ischiocavernosus
- Bulbospongiosus.

The slender, **superficial transverse perineal** muscle (L. transversus perinei superficialis) passes in the base of the su-

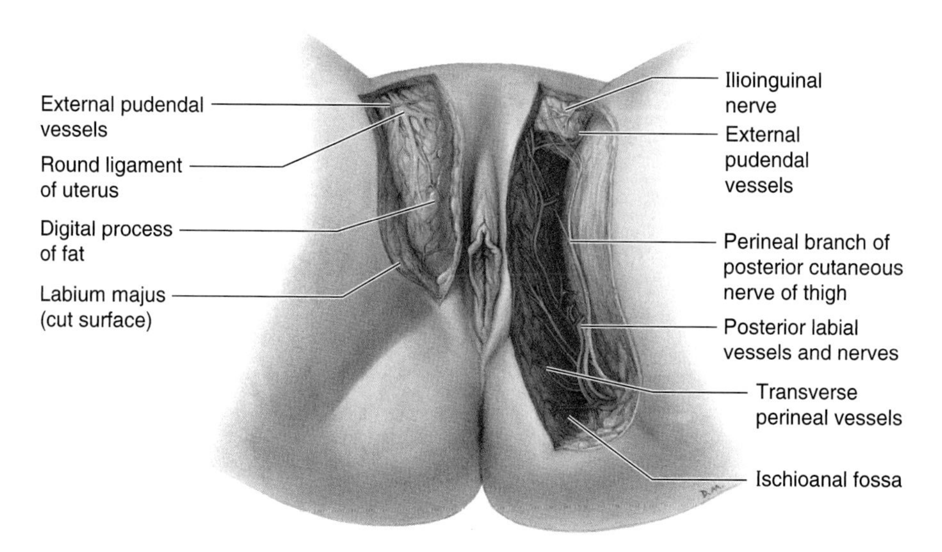

Figure 3.49. Female perineum. Observe the posterior labial vessels and nerves (S2 and S3), joined by the perineal branch of the posterior cutaneous nerve of the thigh (S1 through S3) running anteriorly almost to the mons pubis. Observe the vessels anastomosing here with the external pudendal vessels and the nerves meeting the ilioinguinal nerve (L1).

perficial perineal pouch from the ischial ramus to the perineal body. The **ischiocavernosus**, another slender muscle, attaches to the ischial ramus and partially surrounds the crus of the clitoris. The **bulbospongiosus**, a thin wide muscle, is separated from its contralateral partner by the vagina. It arises from the perineal body, passes around the vagina, and inserts into the clitoris. In its course it covers the bulb of the vestibule and the greater vestibular gland. Acting together, the bulbospongiosus muscles constrict the vagina weakly.

The **perineal body** is a fibromuscular structure that supports the posterior wall of the vagina and is the center of a musculofibrous "cross-member" that *forms the final support of the pelvic viscera*. The perineal body lies between the inferior part of the vagina and the anal canal and is held in position by the attachment of the perineal and levator ani muscles, the other parts of the "cross-member."

Arterial Supply of the Vulva. The abundant arterial supply to the vulva is from the **external pudendal arteries** and one internal pudendal artery on each side (Figs. 3.48 and 3.49). The **internal pudendal artery** supplies the skin, sex organs, and perineal muscles. The labial arteries are branches of the internal pudendal artery, as are those of the clitoris.

Venous and Lymphatic Drainage of the Vulva. The labial veins are tributaries of the **internal pudendal veins** and venae comitantes of the internal pudendal artery. Venous engorgement during the excitement phase of sexual response causes an increase in the size and consistency of the clitoris and the bulbs of the vestibule. The clitoris becomes hard and, in approximately 10% of women, elongates significantly. The vulva contains a rich network of lymphatic vessels that pass laterally to the *superficial inguinal lymph nodes* (Fig. 3.50).

Innervation of the Vulva. The nerves to the vulva are the anterior labial nerves (branches of the **ilioinguinal nerve**); the genital branch of the **genitofemoral nerve**; the perineal branch of the **cutaneous nerve of the thigh**; and the posterior labial nerves (branches of the **perineal nerve**, the larger terminal branch of the **pudendal nerve**) (Fig. 3.49, Table 3.3).

Parasympathetic stimulation produces:

- Increased vaginal secretion
- Erection of the clitoris
- Engorgement of erectile tissue in the bulbs of the vestibule.

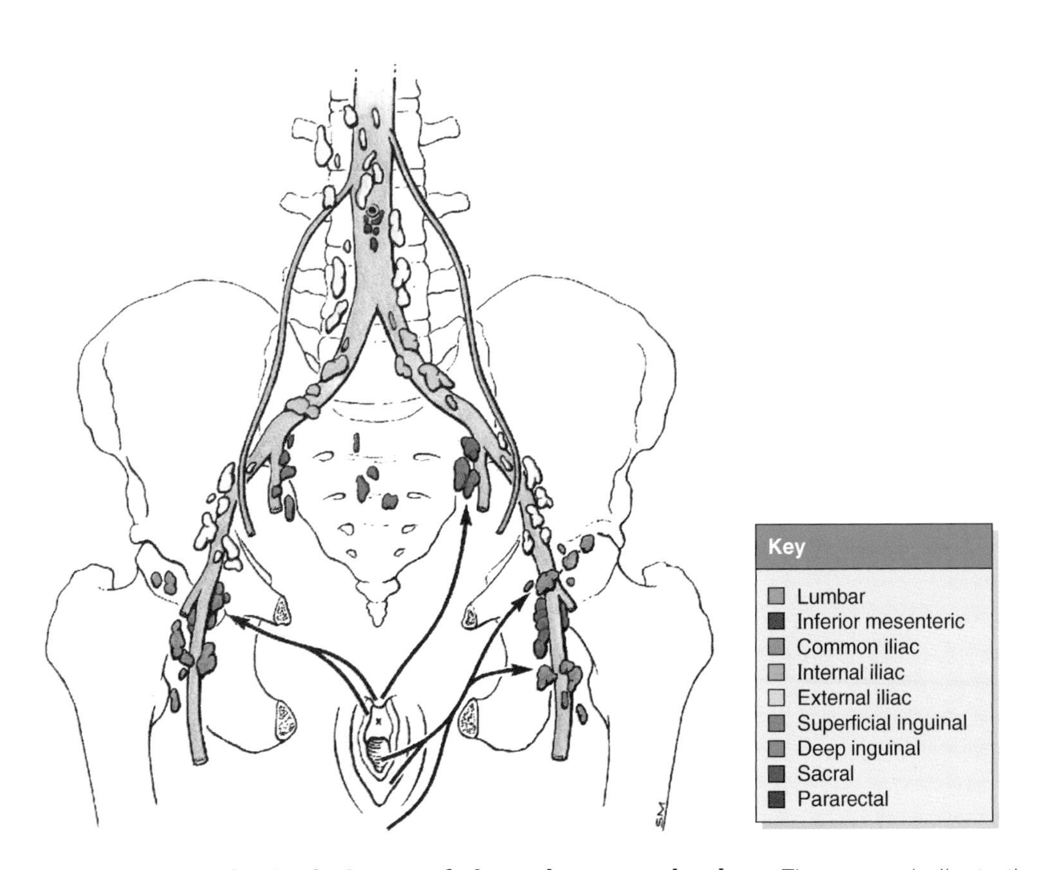

Figure 3.50. Lymphatic drainage of the vulva or pudendum. The *arrows* indicate the direction of lymph flow to the lymph nodes.

Perineal Injuries During Childbirth

Tearing of the perineal body during childbirth may result in a permanent weakness of the pelvic diaphragm. Spontaneous delivery of an infant without the presence of a physician can produce a severe tear in the lower third of the posterior wall of the vagina, perineal body, and overlying skin. When it is obvious that the perineum will tear during childbirth, a surgical incision—*episiotomy*—is performed (p. 391).

Vaginismus

The bulbospongiosus and transverse perineal muscles are thought to be responsible for vaginismus (involuntary spasms of the perivaginal and levator ani muscles). In mild forms it causes *dyspareunia* (painful intercourse); in severe forms it prevents vaginal entry and is reportedly a cause of unconsummated marriages (Fromm, 1993).

Female Circumcision

Although illegal and now being actively discouraged in most countries, female circumcision is widely practiced in many cultures, particularly in Africa. The operation performed during childhood removes the prepuce of the clitoris, but commonly also removes part or all of the clitoris and the labia minora. This disfiguring procedure is thought to inhibit sexual arousal and gratification.

Dilation of the Urethra

The female urethra is distensible because it contains considerable elastic tissue, as well as smooth muscle. It can be easily dilated without injury; consequently, the passage of catheters or cystoscopes is easier in females than in males. The female urethra is easily infected because it is open to the exterior through the vestibule of the vagina.

Infection of the Greater Vestibular Glands

The greater vestibular glands are usually not palpable, but are so when infected. Occlusion of the vestibular gland duct can predispose to infection of the gland. The greater vestibular gland is the site or origin of most vulvar adenocarcinomas. *Bartholinitis*—inflammation of the greater vestibular glands (Bartholin's glands)—may result from a number of pathogenic organisms. Infected glands may enlarge to a diameter of 4 to 5 cm and impinge on the wall of the rectum. Occlusion of the vestibular gland duct without infection can result in the accumulation of mucin (Bartholin's cyst).

Pudendal and Ilioinguinal Nerve Blocks

To relieve the pain experienced during childbirth, *pudendal nerve block anesthesia* may be performed by injecting a local anesthetic agent into the tissues surrounding the pudendal nerve (*A*). The injection is made where the pudendal nerve crosses the lateral aspect of the sacrospinous ligament, near its attachment to the ischial spine. To abolish sensation from the anterior part of the perineum, an *ilioinguinal nerve* block is performed (*B*). When patients continue to complain of pain sensation following proper administration of a pudendal or pudendal and ilioinguinal nerve block, it is usually the result of overlapping innervation by the perineal branch at the posterior cutaneous nerve of the thigh. For a discussion of other types of anesthesia for childbirth, see page 381.

Anorectal Incontinence

Stretching of the pudendal nerve(s) during a traumatic childbirth can result in *pudendal nerve damage* and anorectal incontinence.

Vulvar Trauma

The highly vascular bulbs of the vestibule are susceptible to disruption of vessels as the result of trauma (e.g., athletic injuries, such as running hurdles, sexual assault, and obstetrical injury). These injuries often result in vulvar hematomas in the labia majora, for example. ○

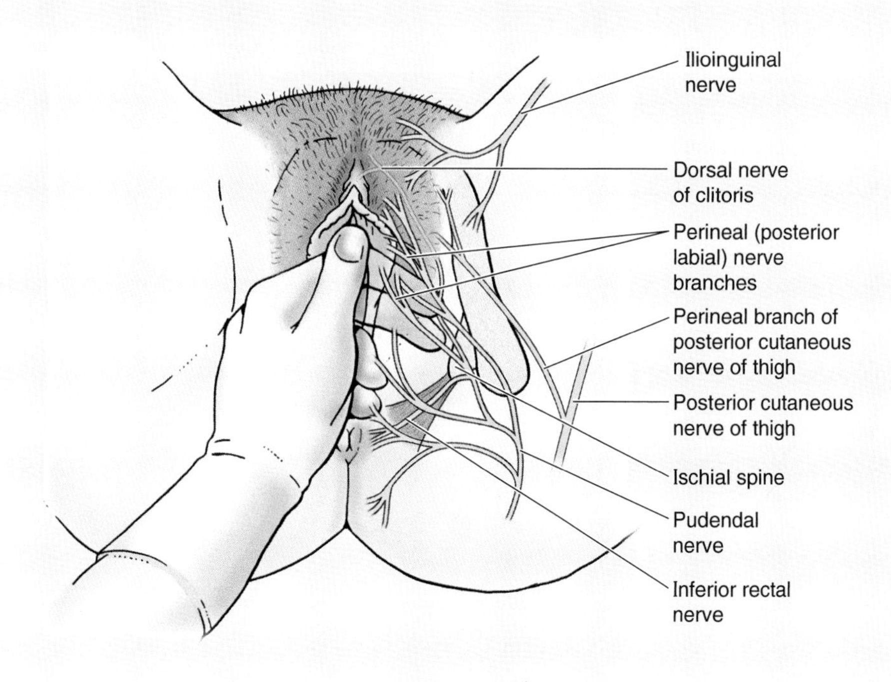
Ilioinguinal nerve
Dorsal nerve of clitoris
Perineal (posterior labial) nerve branches
Perineal branch of posterior cutaneous nerve of thigh
Posterior cutaneous nerve of thigh
Ischial spine
Pudendal nerve
Inferior rectal nerve

(A)

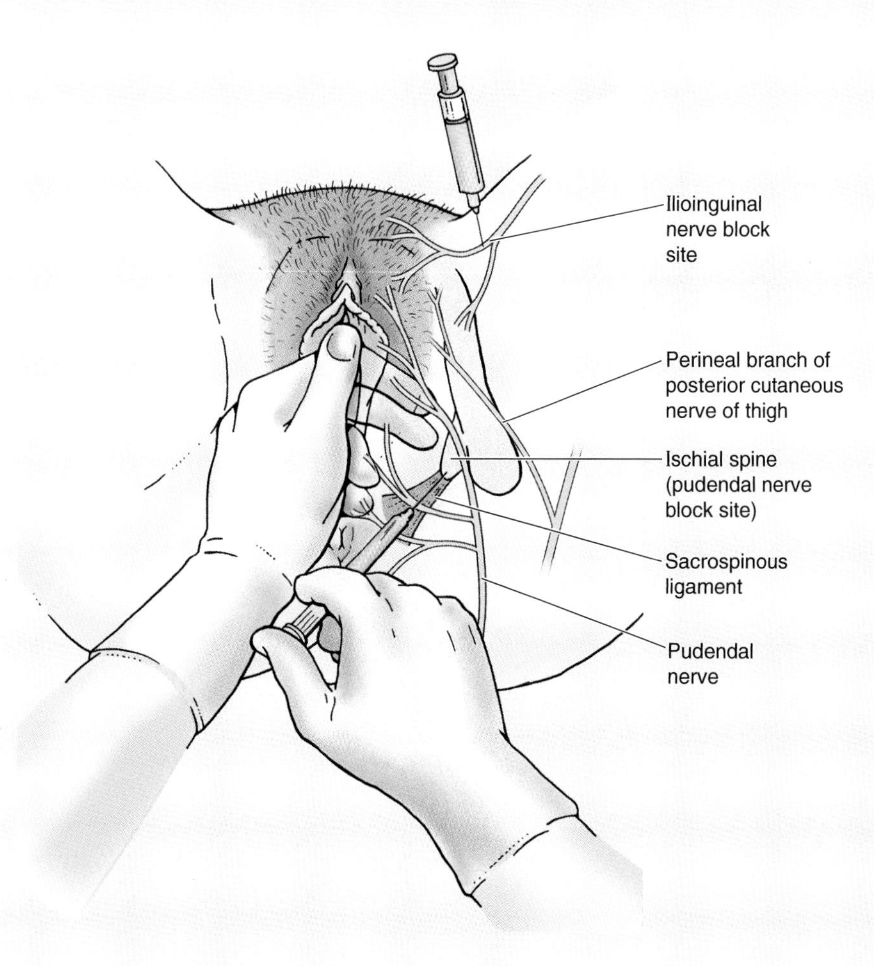
Ilioinguinal nerve block site
Perineal branch of posterior cutaneous nerve of thigh
Ischial spine (pudendal nerve block site)
Sacrospinous ligament
Pudendal nerve

(B)

Medical Imaging of the Pelvis and Perineum

Various diagnostic imaging techniques are used to diagnose pelvic disease and fractures, and to assess congenital and acquired anomalies of the pelvis and pelvic organs.

Radiography

Plain radiographs of the pelvis (Fig. 3.51*A*) are often used as initial screening studies in patients with symptoms of pelvic disease (ureteric calculi and bowel obstruction). They are also used to examine the fetus and to assess the diameter of the pelvic inlet. Contrast agents enhance the visualization of pelvic organs and vessels.

Hysterosalpingography, in which radiopaque dye is injected into the uterine cavity and tubes, is used for demonstrating tubal anatomy and patency (Fig. 3.52*A*) and for detecting uterine and tubal abnormalities (e.g., bicornuate uteri) (Fig. 3.52*B*).

Arteriography, visualization of arteries by X-ray imaging after injection of a radiopaque contrast medium, is used to demonstrate the pelvic arteries (Fig. 3.52*C*). ▶

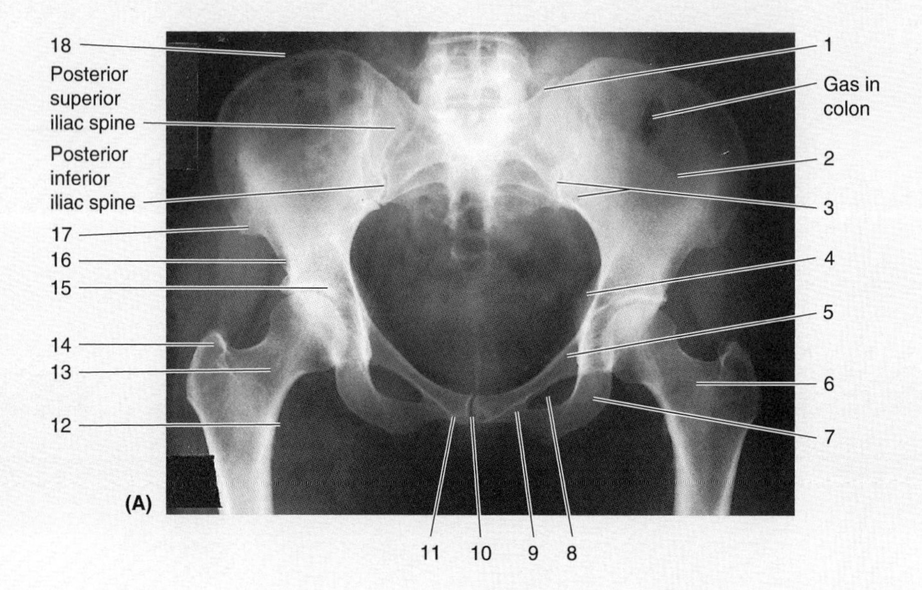

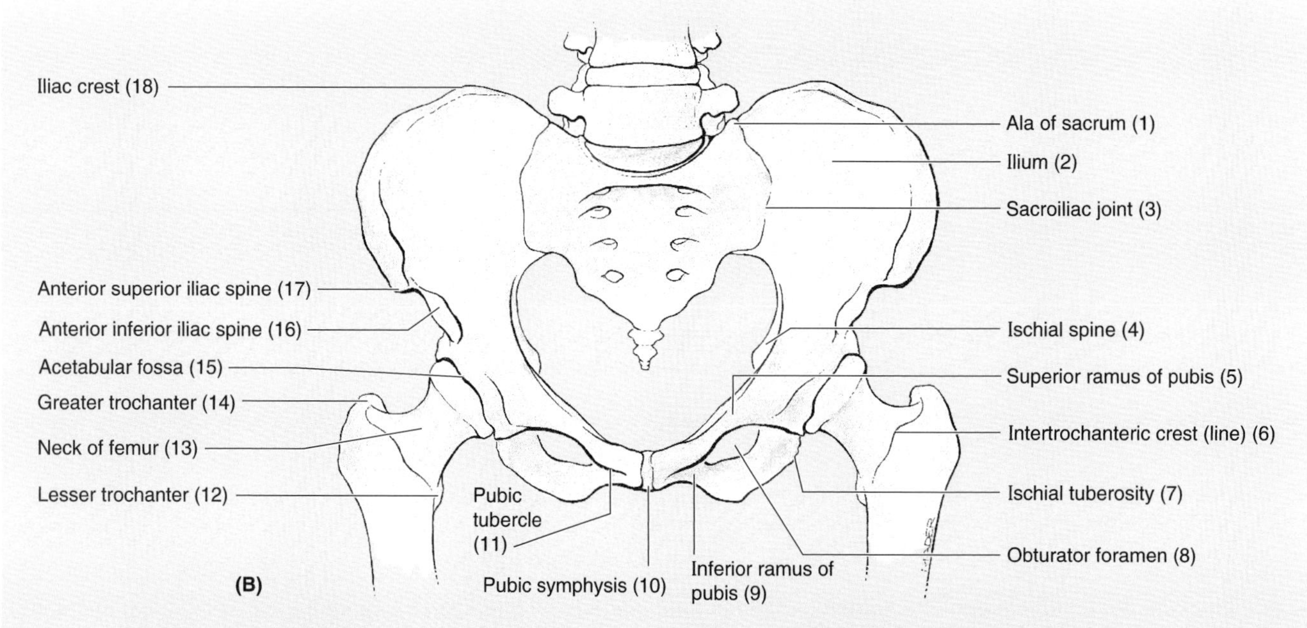

Figure 3.51. Radiograph of the pelvis. A. Anteroposterior (AP) view of a female pelvis. Note the wide subpubic angle and the separation of the ischial spines. (Courtesy of Dr. E.L. Lansdown, Professor of Medical Imaging, University of Toronto, Toronto, Ontario, Canada) **B.** Diagram of the bony pelvis.

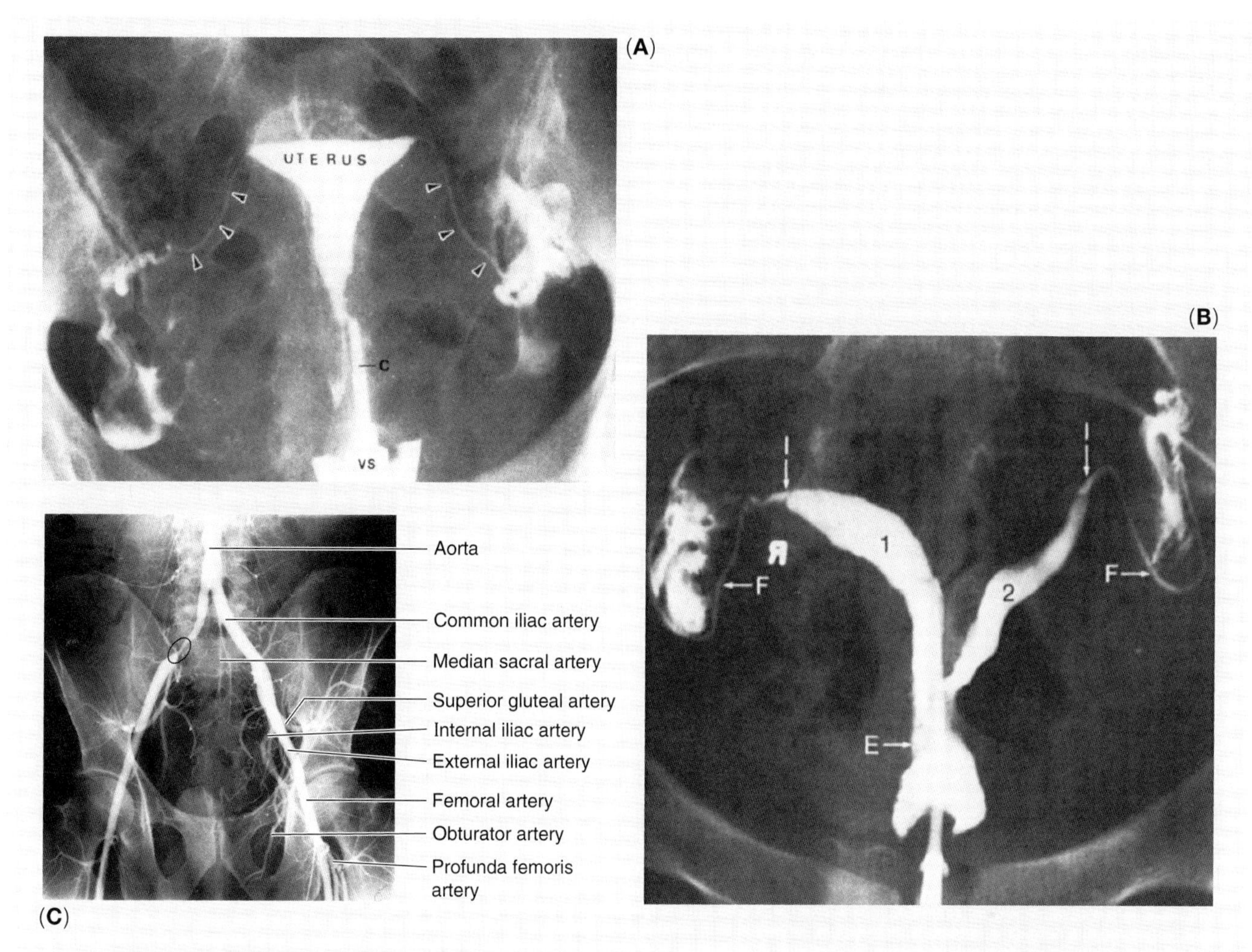

Figure 3.52. Radiograph of the uterus and uterine tubes (*hysterosalpingogram*). Radiopaque material was injected into the uterus through the external os of the uterus (**A**). The contrast medium has traveled through the triangular uterine cavity and uterine tubes (*arrowheads*) and passed into the pararectal fossae of the peritoneal cavity (lateral to the *arrowheads*). (**C**) indicates the catheter in the cervical canal. This radiograph illustrates that the female genital tract is in direct communication with the peritoneal cavity and is therefore a potential pathway for the spread of an infection from the vagina and uterus. **B.** Hysterosalpingogram showing a bicornate uterus. *1* and *2*, uterine cavity; *I*, isthmus of tube; *E*, cervical canal; *F*, uterine tube. (Courtesy of C.E. Stuart and David F. Reid. *In* Copeland LJ: *Textbook of Gynecology*. WB Saunders, Philadelphia, 1993). **C.** Iliac arteriogram. An injection of radiopaque dye has been made into the aorta in the lumbar region. Observe: (*a*) bifurcation of the aorta into right and left common iliac arteries (anterior to L4); (*b*) bifurcation of the common iliacs into internal and external iliac arteries (opposite the sacroiliac joint at the level of the lumbosacral disc); (*c*) the *circled area* on the arteriogram that indicates a site of narrowing (stenosis) of the right common iliac artery. (Courtesy of Dr. D. Sniderman, Associate Professor of Medical Imaging, University of Toronto, Toronto, Ontario, Canada)

Ultrasonography

Ultrasonography (sonography) is used for obstetrical examination and for early evaluation of pelvic problems, such as screening examinations of high-risk patients for carcinoma and evaluation of congenital anomalies. Transabdominal scanning requires a fully distended urinary bladder to displace the bowel loops from the pelvis and to provide an acoustical window through which to observe pelvic anatomy. The introduction of transrectal and transvaginal ultrasonography has resulted in enhanced resolution of pelvic structures (Fig. 3.53). Transvaginal and transrectal ultrasonography enables the placing of the probe closer to the structures of interest, allowing increased resolution of the structures. For example, intrauterine gestations can be distinguished 7 to 10 days earlier than with transabdominal scanning. ▶

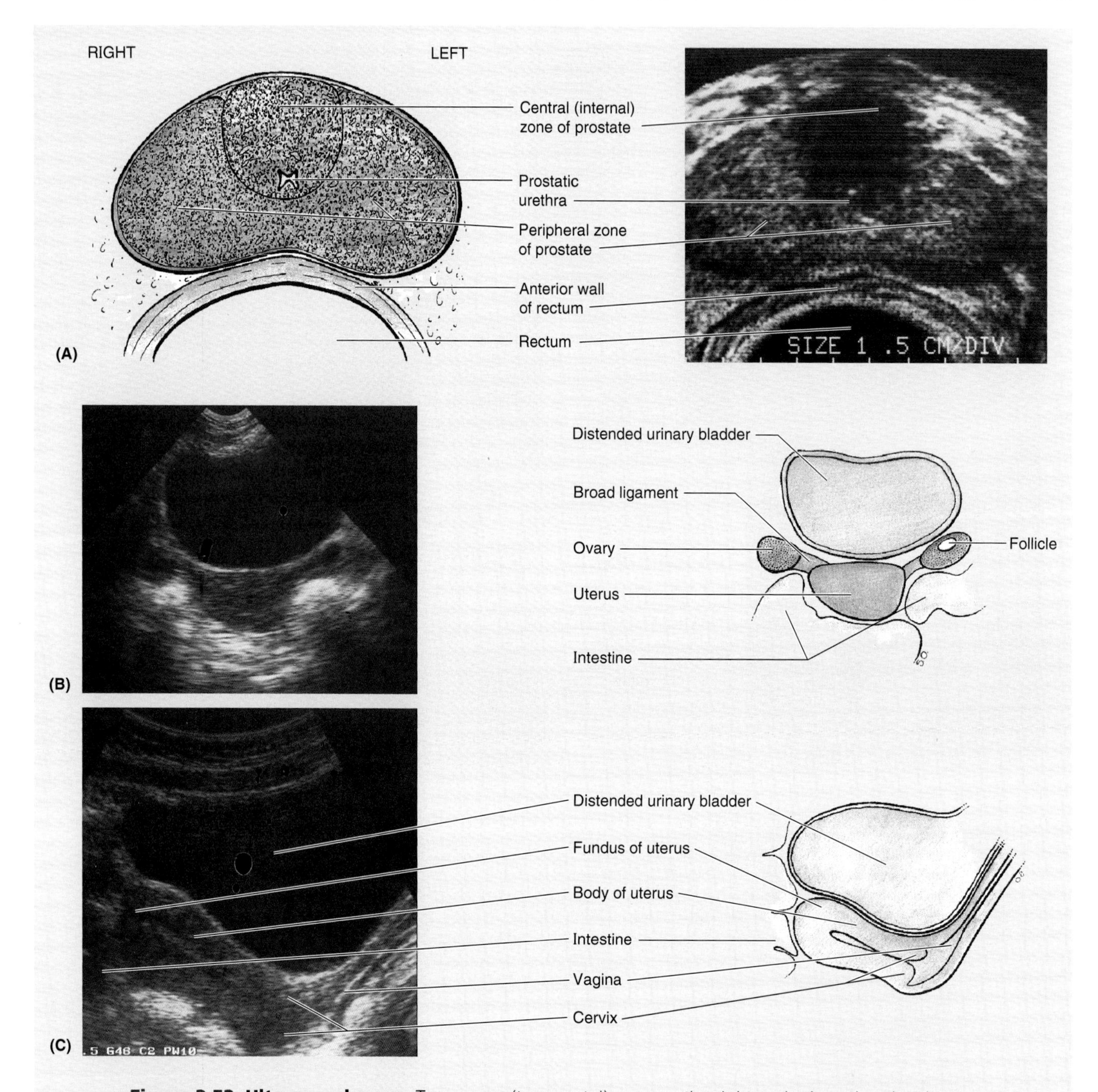

Figure 3.53. Ultrasound scans. Transverse (transrectal) scan on the right and orientation drawing on the left (**A**). The probe was inserted into the rectum to scan the anteriorly located prostate. The ducts of the glands in the peripheral zone open into the prostatic sinuses, whereas the ducts of the glands in the central (internal) zone open into the prostatic sinuses and the seminal colliculus. **B.** Transverse scan of a female's pelvis on the left and diagram of the ultrasound scan on the right. **C.** Sagittal ultrasound scan on the left and diagram of the scan on the right. (Courtesy of Dr. A.M. Arenson, Assistant Professor of Medical Imaging, University of Toronto, Toronto, Ontario, Canada)

Computed Tomography

The anatomy of the pelvis is well demonstrated by computed tomography (CT). The presence of extraperitoneal fat and the relative absence of motion artifacts makes it possible to obtain excellent resolution (Fig. 3.54). *Axial scanning*—a CT scan that is transverse to the axis of the body—is usually satisfactory; however, reformation of images into sagittal and coronal planes can provide additional ▶

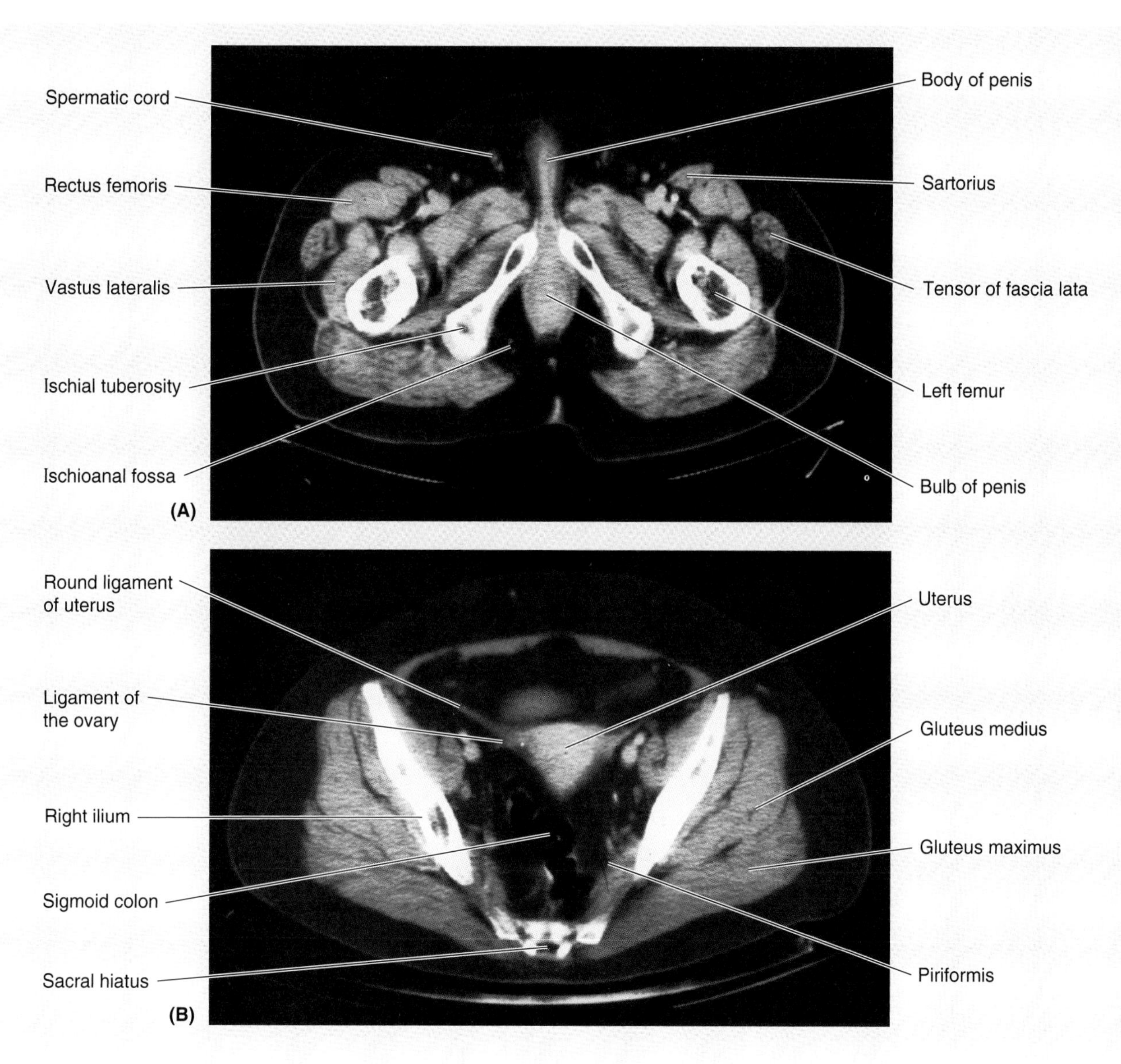

Figure 3.54. Transverse CT images of the pelvis. A. Male. **B.** Female. (Courtesy of Dr. Donald R. Cahill, Department of Anatomy, Mayo Medical School, Rochester, MN)

▶ information. CT is an important modality for assessing local tumor extent and in detecting metastases. It is also accurate in detecting postoperative abscesses.

Magnetic Resonance Imaging

MRI provides excellent evaluation of pelvic structures (Figs. 3.55 and 3.56). Some advantages of MRI are:

- No ionizing radiation
- Multiplanar imaging permits superior spatial resolution in oblique, sagittal, and other anatomical planes
- Soft tissue contrast resolutions are superior to that produced by other techniques
- Vascular structures can be imaged without using contrast materials.

MRI permits excellent delineation of the uterus and ovaries (Fig. 3.57). It also permits the identification of tumors (e.g., a myoma—a benign neoplasm) and congenital anomalies such as bicornuate uterus. ⊙

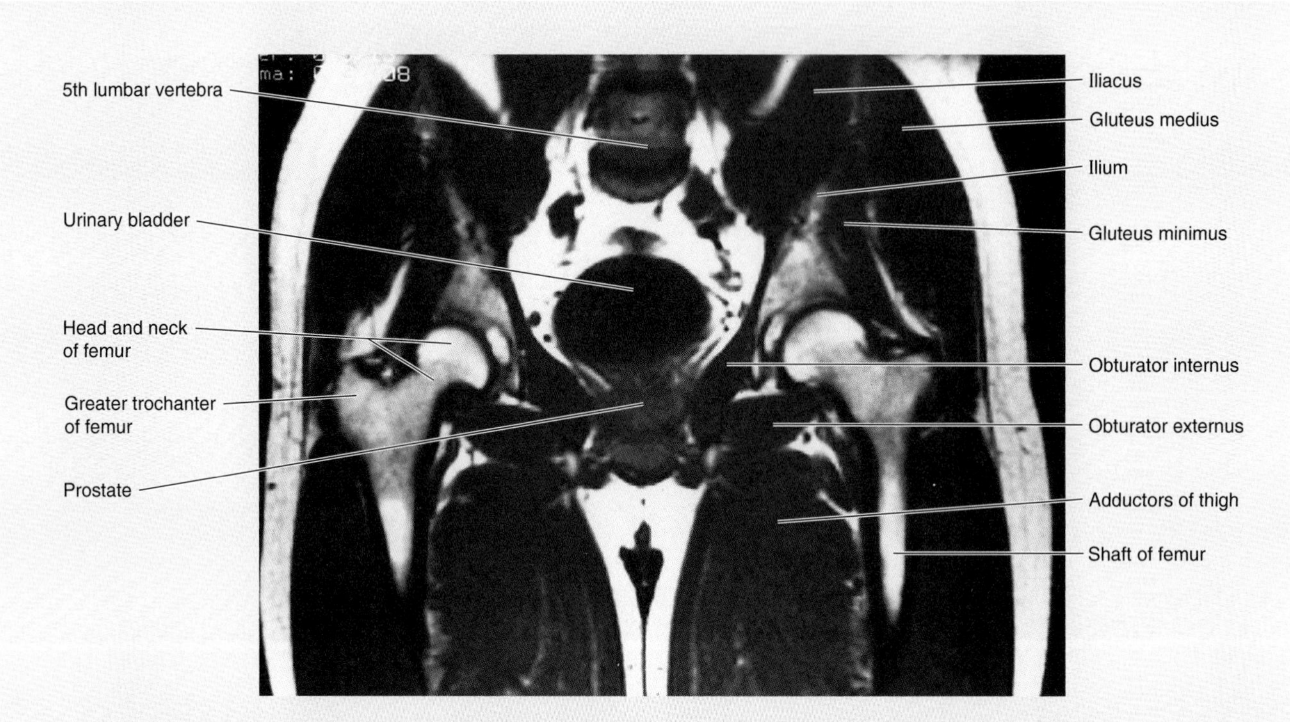

Figure 3.55. Coronal MRI of the male pelvis. (Courtesy of Dr. W. Kucharczyk, Professor and Chair, Department of Medical Imaging, University of Toronto, and Clinical Director of Tri-Hospital Resonence Centre, Toronto, Ontario, Canada)

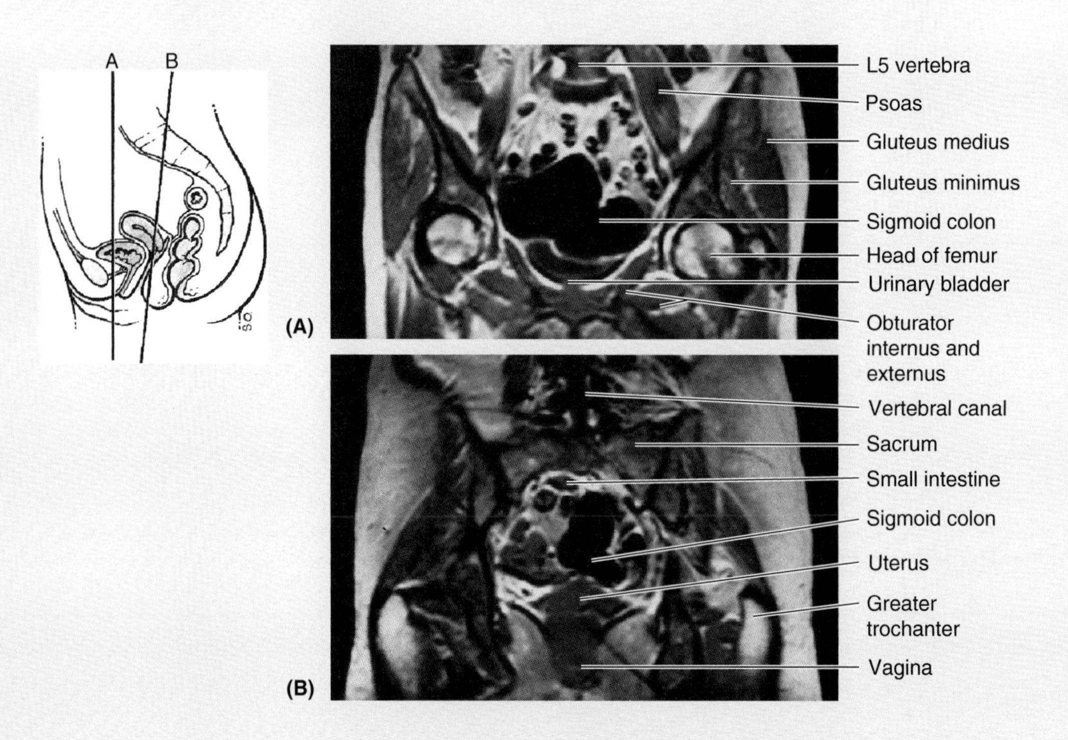

Figure 3.56. MRIs of the female pelvis. Top left, orientation drawing. **A–B.** Coronal MRIs.

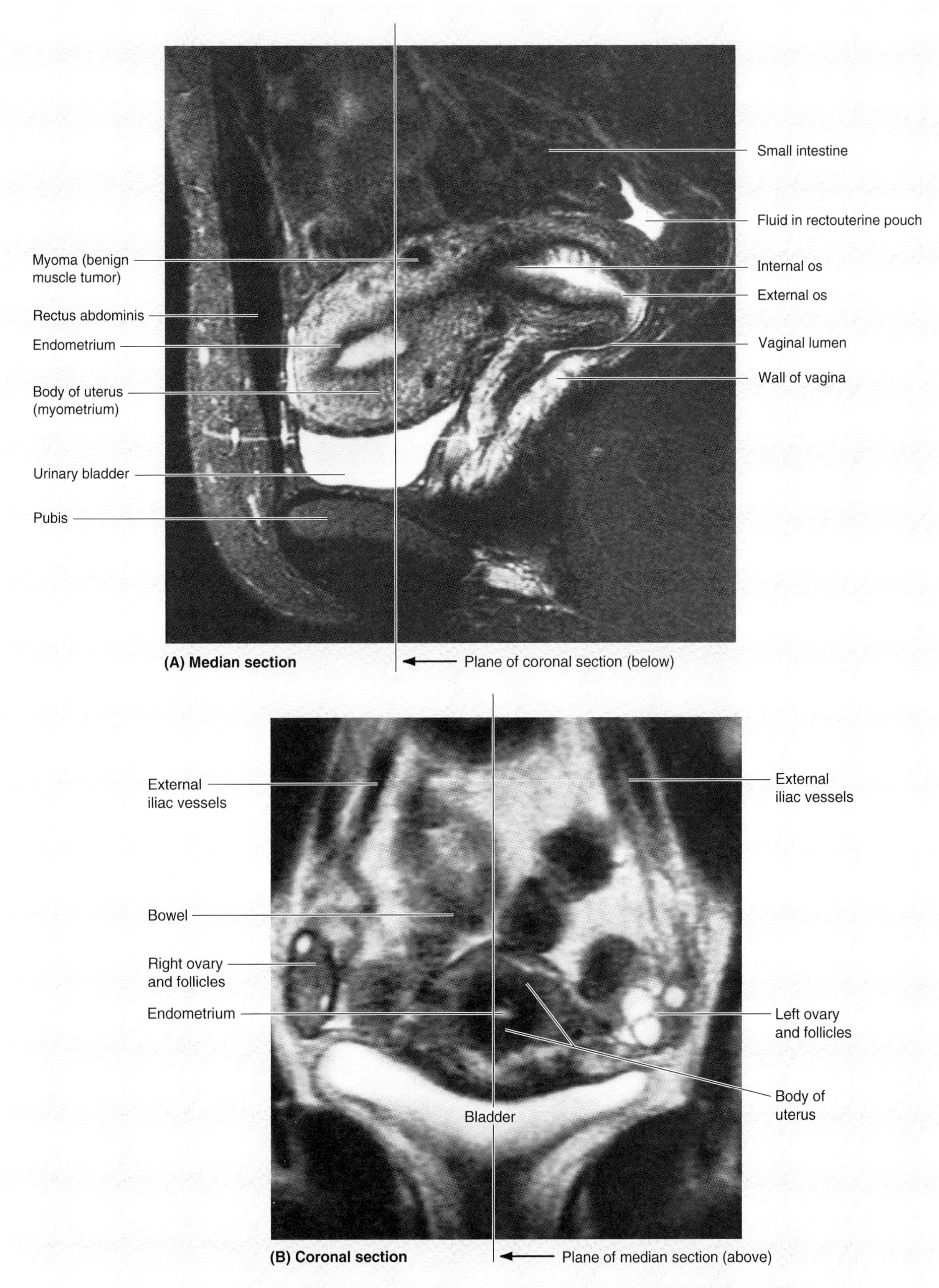

Figure 3.57. MRIs of the female pelvis. Median (above) and coronal (below) MRIs showing the urinary bladder, body of the uterus, vagina, and intestine. Compare coronal section with the dissected specimen of Fig. 3.27*A*. (Courtesy of Dr. Shirley McCarthy, Department of Diagnostic Radiology, Yale University and Yale-New Haven Hospital, New Haven, CT)

CASE STUDIES

Case 3.1

A woman was informed that she has a vulvar malignancy and that removal of all affected lymph nodes would be necessary to avoid the spread of cancer cells to other areas.

Clinicoanatomical Problems

- To which lymph nodes would the malignant cells metastasize?
- If the clitoris was involved in the malignancy, where else might the cancer cells metastasize?

The problems are discussed on page 427.

Case 3.2

A 48-year-old man told his physician that he had "piles" and often passed blood-stained stools. He also said that he felt protrusions from his anus when he strained during defecation. After a physical examination, the physician informed the man that he had *internal hemorrhoids*.

Clinicoanatomical Problems

- What are internal hemorrhoids?
- Explain the anatomical basis of this man's medical problems.

The problems are discussed on page 427.

Case 3.3

A woman in labor was given injections of anesthetic around the pudendal nerve to relieve vulvar pain. She still complained about pain.

Clinicoanatomical Problems

- Assuming the injections were given properly, why do you think the pudendal nerve block might have failed to prevent all the pain experienced by the patient in her labia majora?
- How would complete labial analgesia be obtained?

The problems are discussed on page 427.

Case 3.4

A female gymnast fell during her routine on the balance beam. It was determined later that she had a large labial hematoma.

Clinicoanatomical Problems

- What makes this part of the vulva susceptible to this type of injury?
- What arteries supply the labia majora?

The problems are discussed on page 427.

Case 3.5

A 62-year-old man reported to his physician that his stools were blood-stained and that he was usually unable to completely empty his rectum. He also had a pain along the back of this thigh and weakness of the posterior thigh muscles. Digital examination of the anal canal and lower rectum revealed a tumor in the posterior wall of the rectum.

Clinicoanatomical Problems

- Pressure on what nerve plexus by a rectal tumor could cause pain down the posterior thigh?
- What nerve was likely affected most?

The problems are discussed on page 427.

Case 3.6

A female patient was concerned about her difficulty defecating. She explained to her physician that she could defecate if she pressed posteriorly with her fingers in her vagina. Her history revealed that she had had a difficult childbirth.

Clinicoanatomical Problems

- What type of genital prolapse could produce difficulty in defecation?
- What causes this condition?

The problems are discussed on page 427.

Case 3.7

Shortly after childbirth, a woman consulted her physician about a tender swelling in her perianal region.

Clinicoanatomical Problems

- What fossa related to the anal canal could produce the perianal swelling?
- Disruption of what vessels may cause the collection of blood in this fossa after childbirth?

The problems are discussed on page 427.

Case 3.8

Radiographs of the pelvis of a man involved in a car accident revealed fractures and extravasation of urine. His internal iliac artery was ligated on one side to control severe pelvic hemorrhage.

Clinicoanatomical Problems

- What pelvic organ was likely ruptured?
- Would ligation of the internal iliac artery seriously affect bloodflow to his pelvic viscera?

The problems are discussed on page 427.

Case 3.9

A 25-year-old man who had consumed a large amount of beer was kicked in the lower part of his anterior abdominal wall. He experienced extreme pain and bulging of his abdomen. He was taken to the hospital, where a rectal examination revealed posterior bulging of a pouch related to his rectum.

Clinicoanatomical Problems

- What pouch was bulging and where is it located?
- What caused the bulging of his abdomen?
- What structure was likely injured to cause this bulging?

The problems are discussed on page 428.

Case 3.10

A woman who had had two difficult vaginal births consulted her physician about urinary stress incontinence. Exercises were prescribed for treatment of this condition.

Clinicoanatomical Problems

- What muscles would likely be involved in the physical rehabilitation?
- What usually causes urinary stress incontinence in women who have borne children?

The problems are discussed on page 428.

Case 3.11

A 70-year-old man who had prostate cancer and a prostatectomy 2 years ago complained of pain in his back. Radiographs revealed metastases in his lumbar vertebrae.

Clinicoanatomical Problems

- Using your anatomical knowledge of the prostate, how do you think the cancer cells reached his back?
- Where else would you expect cancer cells to spread?

The problems are discussed on page 428.

Case 3.12

A 68-year-old man with a history of prostatic disease informed the emergency room physician that he had been unable to urinate for 7 hours and was in extreme pain. Because several attempts to catheterize him were unsuccessful, the urologist decided to relieve the bladder pressure by passing a suprapubic tube into his bladder.

Clinicoanatomical Problems

- What caused the buildup of urine in the patient's bladder?
- Through what structures would the suprapubic tube pass?
- Would the tube enter the peritoneal cavity?
- If the bladder ruptured before inserting the tube, where would the urine go?

The problems are discussed on page 428.

Case 3.13

An inebriated 25-year-old man who was involved in an automobile accident complained of severe lower abdominal pain and blood in his urine. A radiographic examination revealed a severe fracture in the pelvic region.

Clinicoanatomical Problems

- What type of fracture do you think the man sustained?
- What injuries do you think he sustained that would explain the hematuria?

The problems are discussed on page 428.

Case 3.14

A 23-year old woman has been in labor for nearly 24 hours. The crown of the fetal head was visible through the vaginal orifice. The obstetrician, fearing the perineal structures might be torn, decided to perform a mediolateral episiotomy to enlarge the inferior opening of the birth canal.

Clinicoanatomical Problems

- What perineal structures would probably be cut during this surgical procedure?
- What structures might have been injured if the perineum had been allowed to tear in an uncontrolled fashion?
- In severe perineal lacerations, what muscle(s) may be torn?

The problems are discussed on page 428.

Case 3.15

A 31-year-old construction worker was walking along a steel beam when he fell, straddling it. He was in severe pain because of trauma to his testes and perineum. Later he observed swelling and discoloration of his scrotum, and when he attempted to urinate, only a few drops of bloody urine appeared. He went to a hospital emergency department. After examining the patient, the physician consulted a urologist who ordered radiographic studies of the patient's urethra and bladder.

Radiology Report The radiographic studies revealed a rupture of the spongy urethra just inferior to the inferior fascia of the external urethral sphincter and deep transverse perineal muscles. Urethrograms showed passage of contrasting material out of the urethra into the surrounding tissues of the perineum.

Diagnosis Rupture of the proximal part of the spongy urethra with extravasation of urine into the surrounding tissues.

Clinicoanatomical Problems

- When the patient tried to urinate, practically no urine came from his external urethral orifice. Where did it go?
- Explain why extravasated urine cannot pass posteriorly, laterally, or into the lesser pelvis.

The problems are discussed on page 428.

Case 3.16

A 49-year-old man complained of tenderness and pain on the right side of his anus. The pain was aggravated by defecation and sitting. Because he had a history of hemorrhoids, he suspected that he might be having a recurrence of this problem. After he explained his symptoms and history, his physician examined his anal canal and rectum. When he asked the patient to strain as if to defecate, prolapsing internal hemorrhoids came into view. During careful digital examination of the anal canal and rectum, the doctor detected some swelling in the patient's right ischioanal fossa. The swelling produced severe pain when it was compressed.

Diagnosis Prolapsing internal hemorrhoids and ischioanal abscess.

Treatment The ischioanal abscess, the more important problem, was drained through an incision in the skin between the anus and the ischial tuberosity.

Clinicoanatomical Problems

- Differentiate between internal and external hemorrhoids.
- What is an ischioanal abscess?
- What nerve is vulnerable to injury during surgical treatment of an ischioanal abscess?
- If this nerve were severed, what structure(s) would be partly denervated?

The problems are discussed on page 428.

Case 3.17

A 40-year-old unconscious woman was rushed to the hospital because she had sustained multiple injuries during an automobile accident. Priority was given to securing a patent airway by inserting an endotracheal tube. Next, emergency care was directed toward controlling bleeding and treating the shock. When the patient's general condition had stabilized, radiographs were taken of the injured regions of her body. Because she had not urinated since being admitted, she was catheterized. The presence of blood in her urine (*hematuria*) suggested rupture of her urinary bladder. Therefore, a sterile dilute contrast solution was injected into her bladder through a catheter and radiographs of the pelvis and abdomen were taken.

Radiology Report There are fractures of the pubic rami on both sides. The cystogram showed extravasation of contrast material from the superior surface of the bladder.

Diagnosis Fractured pelvis and ruptured urinary bladder.

Clinicoanatomical Problems

- Where would the extravasated urine from the bladder go?
- What covers the superior surface of the urinary bladder?
- Thinking anatomically, what route do you think the surgeon would take when repairing the ruptured bladder?

The problems are discussed on page 429.

Case 3.18

A 28-year-old woman was experiencing pregnancy for the first time (*primigravida*). Toward the end of the gestational period, she suffered painful uterine contractions at night that subsided toward morning (*false pains*). When she called her physician, she told her that her labor was imminent. In a few days she observed a discharge of mucus and some blood. When the patient reported that her "pains" (*uterine contractions*) were occurring every 10 minutes, her obstetrician asked her to go to the hospital.

Following admission, the physician palpated the patient's cervix and informed the intern that the external uterine os was open approximately one fingertip and that the patient was still in the *first stage of labor* (period of dilation of the uterine os). Later, a large volume of fluid was expelled (rupture of fetal membranes). When the patient entered the *second stage of labor* (period of expulsive effort beginning with complete dilation of the cervix and ending with delivery of the baby), she began experiencing considerable pain. Although she had wanted to have a natural birth without the use of anesthetics, she was unable to bear the pain. Medication for pain relief was administered as ordered by her physician.

When it was determined that her contractions were 2 minutes apart and lasting 40 to 60 seconds, she was moved to the case room and placed on a delivery table. As the fetal head dilated the cervix, it was obvious that the woman was suffering intense pain. The obstetrician decided to perform a median episiotomy when it appeared that a tear might occur in the patient's perineum. She administered an intradermal injection of an anesthetic agent into the patient's perineum. Although the local anesthetic enabled the incision to be made without pain, it did not alleviate the severe labor pain. The obstetrician decided to perform bilateral pudendal nerve blocks. Thereafter, the patient completed the second stage and proceeded through the *third stage of labor* (beginning after delivery of the child and ending with expulsion of the placenta and fetal membranes).

Clinicoanatomical Problems

- What structures are usually incised during a median episiotomy?
- What is the main structure incised during this procedure and why do you think incision of it might be beneficial?
- Name the structures supplied by the pudendal nerve.
- Based on your knowledge of the anatomy of this nerve, where do you think the obstetrician would inject the anesthetic agent to perform a pudendal nerve block?
- When complete perineal anesthesia is required, branches of what other nerves would have to be blocked?

The problems are discussed on page 429.

Case 3.19

During the examination of a male child, a congenital anomaly of the penis known as hypospadias was detected. The urethra opened just proximal to the site where the frenulum usually attaches the prepuce to the ventral surface of the penis. There was a slight indentation at the site where the external urethral orifice is normally located. In addition, there was a slight ventral curvature of the penis (*chordee*). Micturition was essentially normal except that the child dribbled if he urinated while standing up, wetting his clothing and shoes.

Clinicoanatomical Problems

- What is the embryological basis of hypospadias?
- What type of hypospadias is present?
- Discuss its etiology and other types of hypospadias.
- Do you think this condition would subsequently interfere with reproductive function?

The problems are discussed on page 429.

DISCUSSION OF CASES

Case 3.1

Primary lymphatic drainage of the vulva is to the superficial inguinal lymph nodes. If the carcinoma involved the clitoris, the malignant cells would metastasize to both superficial and deep inguinal lymph nodes.

Case 3.2

Hemorrhoids (piles), occurring in approximately 35% of the population, are mucosal prolapses containing the normally dilated veins of the internal rectal venous plexus located at or near the anorectal junction. During early stages, these hemorrhoids are contained within the anal canal; however, as they enlarge they pass out of the anal canal during defecation and usually return when evacuation is complete. In severe cases the hemorrhoid may remain outside the anus. Internal hemorrhoids are painless because the mucous membrane covering the varicosities is supplied by visceral afferent fibers that accompany autonomic nerves. They are mostly insensitive to touch but may produce an aching sensation during straining. Varicose veins may result from a breakdown or congenital weakness of the muscularis mucosae.

Case 3.3

The rich innervation of the vulva from the ilioinguinal, genitofemoral, and posterior cutaneous nerves of the thigh accounts for the common failure of labial analgesia after a pudendal nerve local anesthetic block. To eliminate most of the pain, these other nerves would have to be blocked.

Case 3.4

The high degree of vascularity of the bulbs of the vestibule makes the labia majora susceptible to trauma. Disruption of the blood vessels results in bleeding and the formation of a hematoma. The labia majora and bulbs of the vestibule are supplied by branches of the external and internal pudendal arteries.

Case 3.5

A rectal tumor (e.g., advanced carcinoma) could exert pressure on the sacral plexus located posterior to the rectum. The sciatic nerve, the large nerve to the limb, arises from the sacral plexus and descends to the posterior aspect of the thigh. The term *sciatica* is used to described the pain produced by irritation or pressure on the sciatic nerve.

Case 3.6

A rectocele could cause difficulty in defecation in a woman whose rectum had herniated into her vagina. This condition results when prolapse of the posterior vaginal wall occurs. The tearing, stretching, and pressure on the pelvic support tissues that often occurs with vaginal deliveries may damage the pelvic floor, especially the perineal body and levator ani, which later results in posterior vaginal wall prolapse.

Case 3.7

The ischioanal fossae are closely related to the anal canal. Pus or blood can collect and form a swelling in the ischioanal fossa. Disruption of the inferior rectal vessels that pass through the ischioanal fossa can produce an ischioanal hematoma. The trauma could have occurred during a difficult childbirth or as the result of an infection resulting from a mediolateral episiotomy.

Case 3.8

Rupture of the urinary bladder commonly occurs when the pelvis is fractured, resulting in urine entering the peritoneal cavity. Internal iliac artery ligation for the control of pelvic hemorrhage does not stop bloodflow; it reduces pulse pressure. Bloodflow is maintained in the artery, although reversed, because of three critical arterial anastomoses (lumbar to iliolumbar, median sacral to lateral sacral, and superior rectal to middle rectal).

Case 3.9

As the bladder fills with urine, it rises from the pelvis into the abdomen. A severe low blow to the anterior abdomen can rupture the urinary bladder, permitting the urine to escape into the peritoneal cavity. When the patient is upright, the urine would accumulate in the rectovesical pouch between the rectum and the bladder. Bulging of the rectovesical pouch can be palpated during a rectal examination.

Case 3.10

Urinary stress incontinence usually results from injury to the fascia and muscles forming the pelvic diaphragm (e.g., the levator ani). This can result from a difficult childbirth that causes stretching of the muscles supporting the neck of the urinary bladder. Alternate contraction and relaxation of the perineal muscles (*Kegel's exercises*) would be involved to strengthen the pelvic diaphragm and the external urethral sphincter. This would increase urethral resting pressure and help to restore the normal urethrovesical angle.

Case 3.11

Cancer cells from the carcinoma of the prostate probably passed through the venous system to the back. The prostatic venous plexus drains into the internal iliac veins and the vertebral venous plexuses, especially the internal vertebral venous plexus. The cancer cells invade the vertebral column and may also pass superiorly in the vertebral venous plexuses and enter the dural venous sinuses and bones of the skull.

Case 3.12

It is obvious that an enlarged prostate was obstructing the man's urethra, preventing urine flow from the bladder. The suprapubic tube would pass superior to the pubis through the layers of the anterior abdominal wall but would not pass through the parietal peritoneum into the peritoneal cavity because as the bladder fills with urine, it rises into the abdomen and strips the peritoneum from the anterior abdominal wall. In most cases the bladder ruptures intraperitoneally. Urine and blood escape into the peritoneal cavity.

Case 3.13

The young man likely fractured his pelvis, and a bony fragment ruptured his urinary bladder. His inebriated condition suggests that his bladder was full and that there was an *extraperitoneal rupture of the bladder*. Hence, the urine would lie between the anterior abdominal wall and the bladder. Blood from the ruptured bladder would enter the urethra and appear as blood in the urine (*hematuria*).

Case 3.14

During a mediolateral episiotomy, the following structures are usually cut: perineal skin, posterior wall of vagina, perineal body, and the attachment of the bulbospongiosus muscle. An episiotomy is performed when a perineal laceration seems inevitable to protect the fascia supporting the urinary bladder, urethra, and rectum. If a tear is allowed to occur spontaneously in whatever direction it may, perineal muscles, the external anal sphincter, the levator ani, and the wall of the rectum may be torn. Episiotomy makes a clean cut away from important structures. If the perineum had been allowed to tear in an uncontrolled fashion, the levator ani forming the pelvic diaphragm might have been torn. This results in poor perineal support for the pelvic organs, which can cause sagging of the pelvic floor in later life. This could lead to difficulty in bladder control (*urinary incontinence*) and could be the basis of subsequent prolapse of the urinary bladder (cystocele).

Case 3.15

Traumatic rupture of the man's spongy urethra in the bulb of his penis resulted in superficial or subcutaneous extravasation of urine when he attempted to urinate. Urine from the torn urethra would pass into the perineum, superficial to the perineal membrane, but deep to the membranous layer of superficial perineal fascia. The urine in the superficial perineal pouch passes inferiorly into the loose connective tissue of the scrotum, anteriorly into the penis, and superiorly into the anterior wall of the abdomen. The perineal membrane and subcutaneous tissue of the perineum are firmly attached to the ischiopubic rami. Therefore, the urine cannot pass posteriorly because the two layers are continuous with each other around the superficial transverse perineal muscles. The urine does not extend laterally because these two layers are connected to the rami of the pubis and ischium. It cannot extend into the lesser pelvis because the opening into this cavity is closed by the perineal membrane. Urine cannot pass into the thighs because the membranous layer of the subcutaneous tissue of the anterior abdominal wall blends with the fascia lata, just distal to the inguinal ligament. The fascia lata is the strong fascia enveloping the muscles of the thigh.

Case 3.16

Internal hemorrhoids are mucosal prolapses containing the normally varicose-appearing veins of the internal rectal venous plexus draining blood from the anal canal. The hemorrhoids occur because of a breakdown of the muscularis mucosae. These veins are usually tributaries of the superior rectal vein. This vein is a tributary of the inferior mesenteric vein and belongs to the portal system of veins. The tributaries of the superior rectal vein arise in the internal rectal plexus that lies in the anal columns. They normally appear varicose (dilated and tortuous), even in newborns. Internal hemorrhoids are covered by mucous membrane. At first they are contained in the anal canal, but as they enlarge they may protrude through the anal canal on straining during defecation. Bleeding from internal hemorrhoids is common.

External hemorrhoids are thromboses (blood clots) in the tributaries of the inferior rectal vein arising from the external rectal plexus, which drains the inferior part of the anal canal. External hemorrhoids are covered by anal skin and are painful but they usually resolve within hours, often by rupturing. Local anesthetics or sitting in a warm bath often brings relief.

Perianal abscesses often result from injury to the anal mucosa by hardened fecal material. Inflammation of the anal sinuses may result, producing a condition called *cryptitis*. The infection may spread through a small crack or lesion in the anal mucosa and pass through the anal wall into the ischioanal fossa, producing an *ischioanal abscess*. The ischioanal fossa is a wedge-shaped space lateral to the anus and levator ani. The main component of the ischioanal fossae is fat. The branches of the nerves and vessels (pudendal nerve, internal pudendal vessels, and the nerve to the obturator internus) enter the ischioanal fossa through the lesser sciatic foramen.

The pudendal nerve and internal pudendal vessels pass in the pudendal canal lying in the lateral wall of the ischioanal fossa. The *inferior rectal nerve* leaves the pudendal canal and runs anteromedially and superficially across the ischioanal fossa. It passes to the external anal sphincter and supplies it. It is vulnerable during surgery in the ischioanal fossa. Damage to the inferior rectal nerve results in impaired action of this voluntary anal sphincter.

Case 3.17

The urine that escaped from the superior surface of the patient's ruptured urinary bladder would pass into the peritoneal cavity. Although pelvic fractures are sometimes complicated by bladder rupture, the radiographs indicated that the rupture was not likely caused by a sharp bone fragment. Probably the bladder was ruptured by the same compressive blow to the region of the pubic symphysis that fractured the pelvis. A full bladder is especially liable to rupture at its superior surface following a nonpenetrating blow.

The superior surface of the bladder is almost completely covered with peritoneum. In the female the peritoneum is reflected onto the uterus at the junction of its body and cervix, forming the *vesicouterine pouch*. In a patient with an intraperitoneal bladder rupture, signs and symptoms of peritoneal irritation are likely to develop. *Septic peritonitis* may develop if pathogenic organisms are present in the urine. As the urine accumulates in the peritoneal cavity, dullness will be detected over the pericolic gutters during percussion of the abdomen. This dullness will disappear from the left side when the patient is rolled onto the right side, and vice versa. This finding indicates free fluid in the peritoneal cavity from a ruptured viscus.

Access to the urinary bladder for surgical repair of its ruptured superior wall would most likely be via the suprapubic route. The bladder is separated from the pubic bones by a thin layer of loose connective tissue, which may contain fat. When the bladder is full, its anteroinferior surface is in contact with the anterior abdominal wall, without the interposition of peritoneum.

Case 3.18

During a median episiotomy, the incision starts at the frenulum of the labia minora and extends through the skin, vaginal mucosa, perineal body, and superficial perineal muscle. The perineal body is the main structure incised during this type of episiotomy. This procedure is believed by many to decrease the prevalence of excessive perineal body attenuation and to decrease trauma to the pelvic and urogenital diaphragms.

The *pudendal nerve* arising from the sacral plexus (S2, S3, and S4) is the main nerve of the perineum. It is both motor and sensory to this region and also carries some postsynaptic sympathetic fibers to the perineum. In the female, the pudendal nerve divides into the perineal nerve and the dorsal nerve of the clitoris. The superficial perineal nerves give off two posterior labial nerves, while the deep perineal nerve supplies both sensory and small terminal muscular branches. The muscular branches enter the superficial and deep perineal pouches to supply the muscles in them and the bulb of the vestibule. The dorsal nerve of the clitoris is sensory and supplies the prepuce and glans of the clitoris and the associated skin.

When the pudendal nerve is blocked via the perineal route, the chief bony landmark is the *ischial spine*. With the patient in the lithotomy position, the ischial spine is palpated. Here, the pudendal nerve enters the pudendal canal before distributing itself over the perineum and so is "collected" at one site. The needle is inserted through the vagina but can also be made through the skin and directed toward the palpating fingertip in the vagina. Because these procedures are most often performed just prior to a vaginal delivery (birth), the palpating finger also serves as a barrier between the needle tip and the baby's scalp, the baby's head now being stationed in the lesser pelvis. When complete perineal anesthesia is required, genital branches of the genitofemoral and ilioinguinal nerves and the perineal branch of the posterior cutaneous nerve of the thigh must also be anesthetized by making an injection along the lateral margin of the labia majora.

Case 3.19

The external urethral orifice in 1 in 300 male infants is on the ventral surface of the penis. Most often the defect is of the glandular type, as in the present case. In other patients, the opening is in the body of the penis (penile hypospadias) or in the perineum (penoscrotal hypospadias).

The embryological basis of glandular and penile hypospadias is failure of the *urogenital folds* to fuse on the ventral surface of the developing penis and form the spongy urethra. Urine is not discharged from the tip of the penis but from an opening on the ventral surface of the penis.

The embryological basis of scrotal hypospadias is failure of the labioscrotal folds to fuse and form the scrotum. The cause of hypospadias is not clearly understood, but it appears to have a *multifactorial etiology* (i.e., genetic and environmental factors are involved). Close relatives of patients with hypospadias are more likely than the general population to have the anomaly. It is generally believed that hypospadias is associated with an inadequate production of androgens by the fetal testes. Differences in the timing and degree of hormonal insufficiency probably account for the different types of hypospadias.

Because the urethral orifice is not located at the tip of the glans and there is ventral bowing of the penis (*chordee*), which is more marked when the penis is erect, reproduction by persons with this malformation is difficult. In some cases the degree of curvature is so severe during erection that *intromission* (insertion of the penis into the vagina) and natural *insemination* are impossible. Surgical correction of chordee to produce a straight penis and repair of the urethra (*urethroplasty*) were recommended in the present case before the boy started school so that it would be possible for him to urinate in the normal standing position and able to reproduce later.

References and Suggested Readings

Ayoub SF: Anatomy of the external anal sphincter in man. *Acta Anat* 105:25, 1979.

Behrman RE, Kliegman RM, Arvin AM (eds): *Nelson Textbook of Pediatrics*, 15th ed. Philadelphia, WB Saunders, 1996.

Cahill DR, Orland MJ, Miller G: *Atlas of Human Cross-Sectional Anatomy*, 3rd ed. New York, Wiley-Liss, 1994.

Copeland LJ (ed): *Textbook of Gynecology*. Philadelphia, WB Saunders, 1993.

Dellenbach P, et al.: The transvaginal method for oocyte retrieval. An update on our experience (1984–1987). *In* Jones HW Jr, Schrader C (eds): *In Vitro Fertilization and Other Assisted Reproduction*. New York, Annals of the New York Academy of Sciences, vol. 541, p. 111, 1988a.

Dellenbach P, Forrier A, Moreau L, Rouard M, Badoc E: Direct intraperitoneal insemination. New treatment for cervical and unexplained infertility. *In* Jones HW Jr, Schrader C (eds): *In Vitro Fertilization and Other Assisted Reproduction*. New York, Annals of the New York Academy of Sciences, vol. 541, p. 761, 1988b.

Ellis H: *Clinical Anatomy. A Revision and Applied Anatomy for Clinical Students*, 8th ed. Oxford, Blackwell Scientific Publications, 1992.

Farrow GA: Urology. *In* Gross A, Gross P, Langer B (eds): *Surgery. A Complete Guide for Patients and Their Families*. Toronto, Harper & Collins, 1989.

Fromm LM: Psychological aspects of gynecology. *In* Copeland LJ (ed): *Textbook of Gynecology*. Philadelphia, WB Saunders, 1993.

Hannah WJ: Obstetrics and gynecology. *In* Gross A, Gross P, Langer B (eds): *Surgery. A Complete Guide for Patients and Their Families*. Toronto, Harper & Collins, 1989.

Hatch K: Urinary tract injury and fistula. *In* Copeland LJ (ed): *Textbook of Gynecology*. Philadelphia, WB Saunders, 1993.

Healey JE Jr, Hodge J: *Surgical Anatomy*, 2nd ed. Toronto, BC Decker, 1990.

Krebs H-B: Premalignant lesions of the cervix. *In* Copeland LJ (ed): *Textbook of Gynecology*. Philadelphia, WB Saunders, 1993.

Moore KL, Persaud TVN: *The Developing Human. Clinically Oriented Embryology*, 6th ed. Philadelphia, WB Saunders, 1998.

Morris M, Burke TW: Cervical cancer. *In* Copeland LJ (ed): *Textbook of Gynecology*. Philadelphia, WB Saunders, 1993.

Oelrich TM: The urethral sphincter muscle in the male. *Am J Anat* 158: 229, 1980.

Oelrich TM: The striated urogenital sphincter muscle in the female. *Anat Rec* 205:223, 1983.

O'Rahilly R: *Gardner-Gray-O'Rahilly Anatomy. A Regional Study of Human Structure*, 5th ed. Philadelphia, WB Saunders, 1986.

Stormont TJ, Cahill DR, King BF, Myers RP: Fascias of the male external genitalia and perineum. *Clin Anat* 7:115, 1994.

Wendell-Smith CP, Wilson PM: The vulva, vagina, and urethra and the musculature of the pelvic floor. *In* Phillip E, Setchell M, Ginsburg J (eds): *Scientific Foundations of Obstetrics and Gynecology*. Oxford, Butterworth-Heinemann, 1991.

Wendell-Smith CP: Muscles and fasciae of the pelvis. *In* Williams PL, Bannister LH, Berry MM, Collins P, Dussek JE, Fergusson MWJ (eds): *Gray's Anatomy*, 38th ed. Edinburgh, Churchill-Livingstone, 1995.

Williams PL, Bannister LH, Berry MM, Collins P, Dussek JE, Fergusson MWJ (eds): *Gray's Anatomy*, 38th ed. Edinburgh, Churchill Livingstone, 1995.

Woodburne RD, Burkel WE: *Essentials of Human Anatomy*, 9th ed. New York, Oxford University Press, 1994.

c h a p t e r

4 Back

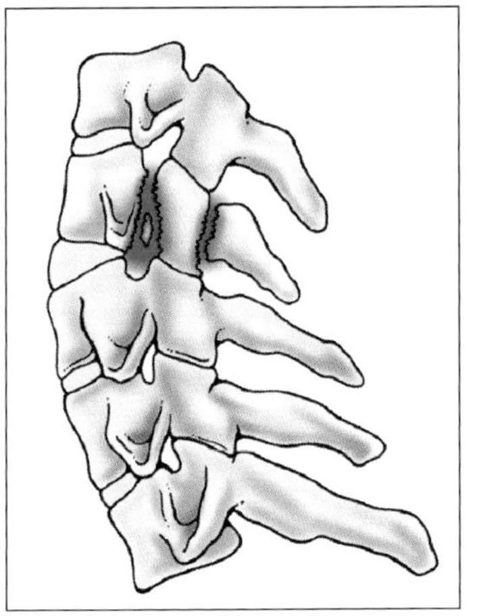

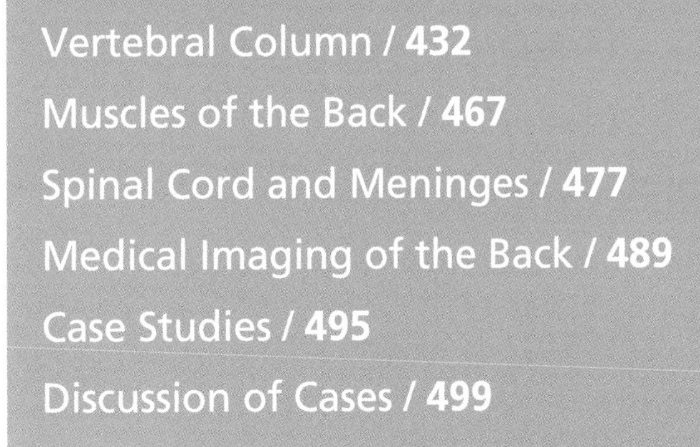

The back—the posterior aspect of the trunk, inferior to the neck and superior to the buttocks—is the region of the body to which the head, neck, and limbs are attached. Because of their close association with the trunk, the back of the neck and the posterior and deep cervical muscles and vertebrae are also described in this chapter. *The term back includes the:*

- Skin
- Subcutaneous tissue—a layer of loose irregular connective tissue consisting of fatty tissue containing cutaneous nerves and vessels, deep fascia, muscles and their vessels and nerves
- Ligaments and vertebral column
- Spinal cord and meninges—three membranes covering the spinal cord
- Ribs (in the thoracic region)
- Various nerves and vessels.

Study of the soft tissues of the back is best preceded by examination of the vertebral column and the fibrocartilaginous intervertebral (IV) discs interposed between the bodies of adjacent vertebrae.

Backache and Back Pain

The parts of the neck and back where the vertebral column has the greatest freedom of movement—the cervical and lumbar regions—are the most frequent sites of disabling pain. Approximately 10% of the population consults a physician or chiropractor each year about back pain. *Backache* is a nonspecific term used to describe back pain. More than 80% of people have back complaints during their lifetime. *Low back pain*—the most common complaint—occurs typically in the 3rd through 6th decades of life. *Back injuries* occur frequently in competitive sports and in industrial and automobile accidents.

The anatomy of the back is complex, and back pain has many causes. A thorough knowledge of the structure and function of the back is required to diagnose and treat back pain. In severe injuries, the examining person must be careful not to cause further damage. For example, if an injured person complains of back pain and is unable to move the limbs, the vertebral column may be fractured. If the neck is flexed or the injured person sits up, the spinal cord may be injured. *Improper handling of an injured person can convert an unstable lesion without a neurological deficit to one with a deficit that produces permanent disability.* ○

Vertebral Column

The vertebral column (backbone, spine)—extending from the skull to the apex (tip) of the coccyx—*forms the skeleton of the neck and back and the main part of the axial skeleton* (the articulated bones of the skull, vertebral column, ribs, and sternum). Most vertebral columns in adults are 72 to 75 cm long, of which approximately one-fourth is formed by the fibrocartilaginous **IV discs**, which separate and bind the **vertebrae** together (Fig. 4.1). *The vertebral column:*

- Protects the spinal cord and spinal nerves
- Supports the weight of the body
- Provides a partly rigid and flexible axis for the body and a pivot for the head
- Plays an important role in posture and locomotion—movement from one place to another.

The vertebral column in an adult typically consists of 33 vertebrae arranged in five regions: 7 cervical, 12 thoracic, 5 lumbar, 5 sacral, and 4 coccygeal (Fig. 4.2).

The **lumbosacral angle** occurs at the junction of, and is formed by, the long axes of the lumbar region of the vertebral column and the sacrum. *Motion occurs between only 24 vertebrae:* 7 cervical, 12 thoracic, and 5 lumbar. The five sacral vertebrae are fused in adults to form the **sacrum**, and the four coccygeal vertebrae are fused to form the **coccyx**. The vertebrae gradually become larger as the vertebral column descends to the sacrum, and then they become progressively smaller toward the apex of the coccyx. These structural differences are related to the fact that the successive vertebrae bear increasing amounts of the body's weight as the column descends, until it is transferred to the pelvic girdle at the sacroiliac joints.

The vertebral column is flexible because it consists of many relatively small bones—the **vertebrae**—that are separated by

Figure 4.1. Vertebral column, spinal cord, and meninges. A. Schematic medial view of the vertebral column showing its normal curvatures; lumbosacral angle; and relationship to the skull, thoracic cage, and hip bone. Observe the intervertebral (IV) foramina where the spinal nerves exit the vertebral (spinal) canal. **B.** Sagittal MRI showing the spinal cord and cauda equina (L. cauda, tail) within the vertebral canal. The medullary cone (L. conus medullaris) is the cone-shaped inferior end of the spinal cord, which typically ends at the L1/L2 level. Note that the dura mater—the external covering of the spinal cord (*gray*)—is separated from the spinal cord by the CSF-filled subarachnoid (leptomeningeal) space (*black*), and from the wall of the vertebral canal by the epidural or extradural space, containing semifluid fat (*white*) and thin-walled veins (not visible here). (Courtesy of Dr. W. Kucharczyk, Chair of Medical Imaging, Faculty of Medicine, University of Toronto, and Clinical Director of Tri-Hospital Resonance Centre, Toronto, Ontario, Canada)

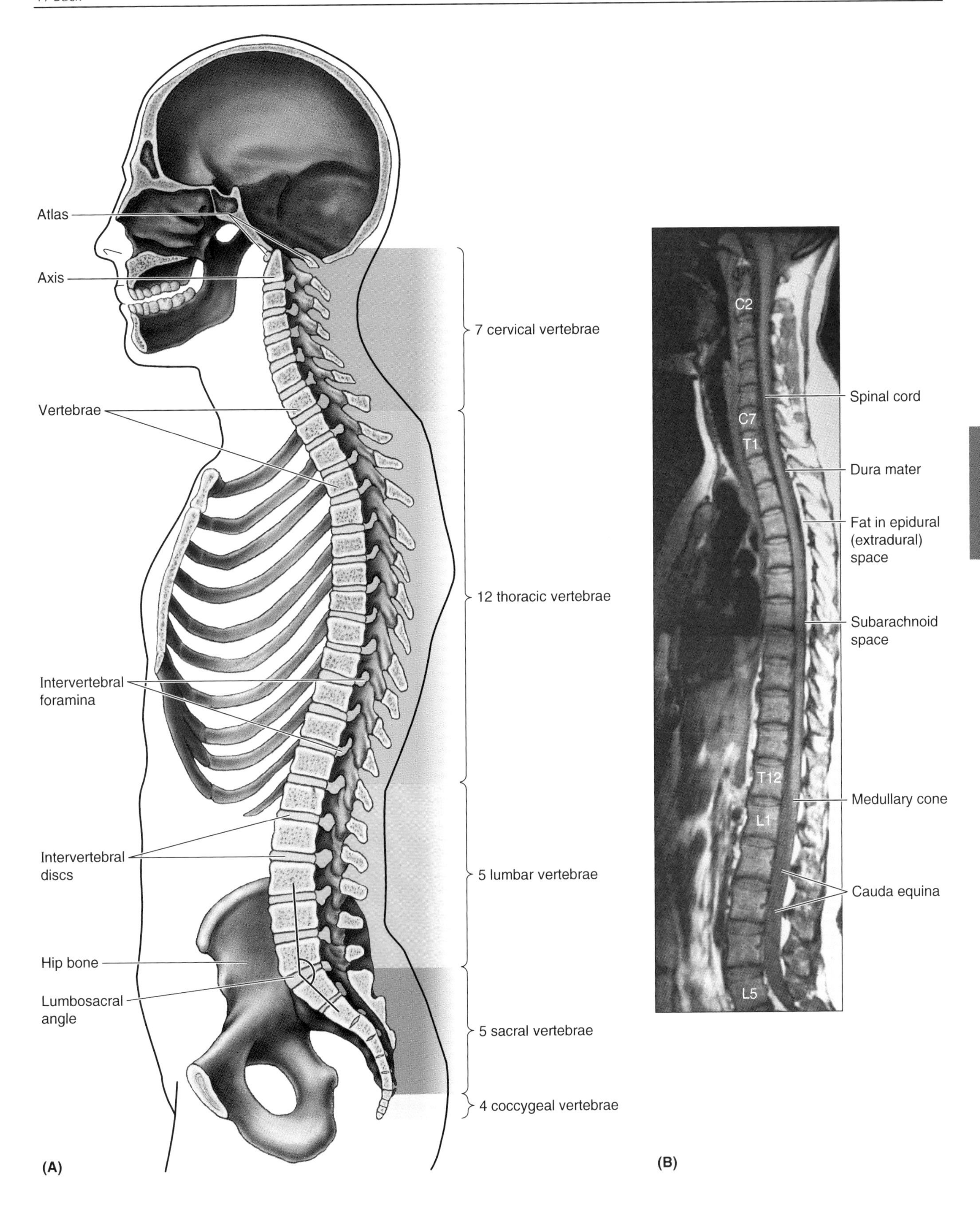
Atlas
Axis
Vertebrae
Intervertebral foramina
Intervertebral discs
Hip bone
Lumbosacral angle
7 cervical vertebrae
12 thoracic vertebrae
5 lumbar vertebrae
5 sacral vertebrae
4 coccygeal vertebrae
(A)
C2
C7
T1
T12
L1
L5
Spinal cord
Dura mater
Fat in epidural (extradural) space
Subarachnoid space
Medullary cone
Cauda equina
(B)

resilient IV discs. The 24 cervical, thoracic, and lumbar vertebrae also articulate at synovial joints that facilitate and control the vertebral column's flexibility. Although the movement between two adjacent vertebrae is small, in aggregate the vertebrae and IV discs uniting them form a column that is remarkably flexible yet necessarily rigid and protective of the spinal cord it surrounds. The shape and strength of the vertebrae and IV discs, ligaments, and muscles provide stability to the vertebral column.

Variations in the Vertebrae

Studies of many skeletons indicate that 88 to 90% of vertebral columns have the normal number of vertebrae superior to the sacrum. An increased length of the presacral region of the vertebral column increases the strain on the inferior part of the lumbar region of the column because of the increased leverage.

Variations in vertebrae are affected by race, sex, and developmental factors (genetic and environmental). Most people have 33 vertebrae, but developmental errors may result in 32 or 34 vertebrae. An increased number of vertebrae occurs more often in males and a reduced number occurs more frequently in females. Some races show more variation in the number of vertebrae. *Variations in the number of vertebrae may be clinically important*; however, most of them are detected in medical images such as radiographs and computed tomographies (CTs) and during dissections and autopsies of persons with no history of back problems.

Caution is necessary, however, when describing an injury (e.g., when reporting the site of a vertebral fracture). When counting the vertebrae, begin at the base of the neck. *The number of cervical vertebrae (seven) is constant*; however, numerical variations of thoracic, lumbar, sacral, and coccygeal vertebrae occur in approximately 5% of otherwise normal people. When considering a numerical variation, the thoracic and lumbar regions must be considered together because some people have more than five lumbar vertebrae and a compensatory decrease in the number of thoracic vertebrae. ⊙

Curvatures of the Vertebral Column

The vertebral column in adults has four curvatures: cervical, thoracic, lumbar, and sacral (Fig. 4.3). The curvatures provide a flexible support (shock-absorbing resilience) for the body. The thoracic and sacral (pelvic) curvatures are concave anteriorly, whereas the cervical and lumbar curvatures are concave posteriorly. The thoracic and sacral curvatures are **primary curvatures** that develop during the fetal period. Primary curvatures are caused by differences in height between the anterior and posterior parts of the vertebrae. Compare the curvatures in Figure 4.3, noting that the primary curvatures are in the same direction as the main curvatures of the fetal vertebral column. The cervical and lumbar curvatures are **secondary curvatures** that begin to appear during the fetal period but do not become obvious until infancy. Secondary curvatures are caused mainly by differences in thickness between the anterior and posterior parts of the IV discs.

The **cervical curvature** becomes prominent when an infant begins to hold its head erect. The **thoracic curvature** results from the slightly wedge-shaped thoracic vertebral bodies. The **lumbar curvature** becomes obvious when an infant begins to walk and assumes the upright posture. This curvature, generally more pronounced in females, ends at the **lumbosacral angle** formed at the junction of L5 vertebra with the sacrum (Fig. 4.1*A*). The **sacral curvature** also differs in males and females.

Abnormal Curvatures of the Vertebral Column

To detect an abnormal curvature of the vertebral column, have the person stand in the anatomical position. Inspect the profile of the vertebral column from the person's side (*A*) and then from the posterior aspect (*D*). With the person bending over, observe the ability to bend directly forward and whether the back is level in the bent position (*E*). Abnormal curvatures in some people result from developmental anomalies; in others, the curvatures result from pathological processes such as *osteoporosis*—reduction in the quality of bone or atrophy of skeletal tissue.

Kyphosis (humpback, hunchback) is characterized by *an abnormal increase in the thoracic curvature*; the vertebral column curves posteriorly (*B*). This abnormality can result from erosion of the anterior part of one or more vertebrae (e.g., because of demineralization resulting from osteoporosis). *Dowager's hump* is a colloquial name for kyphosis in older women caused by wedge fractures of the thoracic vertebrae resulting from osteoporosis; however, kyphosis occurs in both male and female geriatric patients (Swartz, 1994). Progressive erosion and collapse of vertebrae also result in an overall loss of height. Kyphosis results in an increase in the anteroposterior (AP) diameter of the thorax.

Lordosis (hollow back, sway back) is characterized by *an anterior rotation of the pelvis* (the upper sacrum tilts anteroinferiorly) at the hip joints producing an abnormal increase in the lumbar curvature; the vertebral column curves more anteriorly (*C*). This *abnormal extension deformity* is often associated with weakened trunk musculature, especially the anterolateral abdominal muscles. To compensate for alterations to their normal line of gravity, *women develop a temporary lordosis during late pregnancy.* This lordotic curvature may cause low back pain, but the discomfort normally disappears soon after childbirth. *Obesity in both sexes can also cause lordosis and low back pain* because of the increased weight of the abdominal contents (e.g., "potbelly") ante- ▶

Structure and Function of Vertebrae

Vertebrae vary in size and other characteristics from one region of the vertebral column to another and to a lesser degree within each region. *A typical vertebra* (Fig. 4.4) *consists of:*

- A vertebral body
- A vertebral (neural) arch
- Seven processes.

Typical vertebrae vary in size and in other characteristics from one region to another; however, their basic structure is the same.

The **vertebral body** is the anterior, more massive part of the bone that gives strength to the vertebral column and supports body weight (Fig. 4.4*A*). The vertebral bodies, especially from T4 inferiorly, become progressively larger to bear the progressively greater body weight. In dried laboratory and museum skeletal specimens, the hyaline cartilage that covers most of the superior and inferior ends of the vertebral body is absent and the bone appears spongy, except at the periphery where an epiphyseal ring of smooth bone—the **epiphyseal ring** (derived from the anular epiphysis)—fused to the body (Fig. 4.4*B*). As the vertebrae grow, the hyaline epiphyseal plates form the zone from which the vertebral body grows in height (see "Bones" in Introductory Chapter). The *epiphyseal growth plates*, in addition to serving as growth zones, probably provide some protection to the vertebral bodies and permit the diffusion of fluid between the IV disc and the capillaries in the vertebral body. A secondary center of ossification appears around puberty in the margin of each growth plate—forming an epiphyseal ring from the *anular epiphysis* (see Fig. 4.11, *B* and *C*). The superior and inferior epiphyses usually unite with the vertebral body early in adult life (approximately 25 years). The **centrum** is the primary ossification center for the central mass of the vertebral body (Fig. 4.4*B*); therefore, *the terms centrum and body are not synonyms*. The centrum is somewhat less than a vertebral body.

The **vertebral arch** is posterior to the vertebral body and is the part of a vertebra that is formed by the right and left pedi-

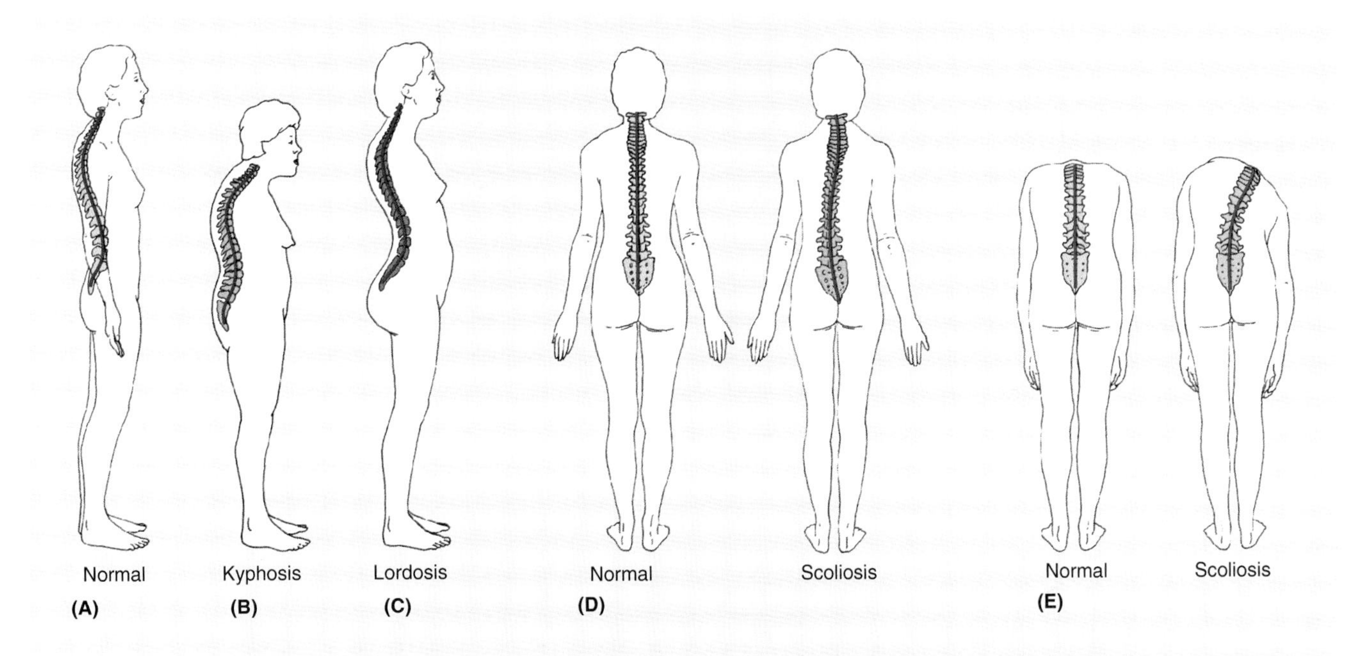

▶ rior to the normal line of gravity. Loss of weight corrects this type of lordosis.

Scoliosis (crooked or curved back) is characterized by *an abnormal lateral curvature that is accompanied by rotation of the vertebrae* (*D* and *E*). The spinous processes turn toward the cavity of the abnormal curvature, and, when bending over, the ribs rotate posteriorly (protrude) on the side of the increased convexity. *Scoliosis is the most common deformity of the vertebral column in pubertal girls.* Asymmetrical weakness of the intrinsic back muscles (*myopathic scoliosis*), failure of half of a vertebra to develop (*hemivertebra*), and a difference in the length of the lower limbs are causes of scoliosis. If the lengths of the lower limbs are not equal, a compensatory pelvic tilt may lead to a functional *static scoliosis.* When a person is standing, an obvious inclination or listing to one side may be a sign of scoliosis that is secondary to a herniated IV disc. *Habit scoliosis* is supposedly caused by habitual standing or sitting in an improper position. When the scoliosis is entirely postural, it disappears with maximum flexion of the vertebral column.

Sometimes there is **kyphoscoliosis**—kyphosis combined with scoliosis—in which an abnormal AP diameter produces a severe restriction of the thorax and lung expansion (Swartz, 1994). ⊙

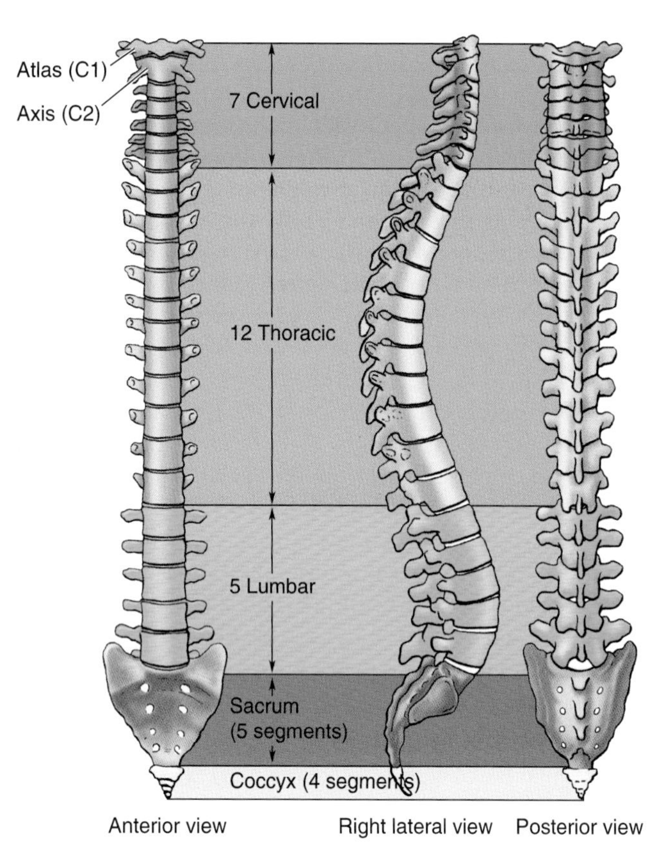

Figure 4.2. Three views of the vertebral column. Observe the five regions of the vertebral column. Observe also that there are 24 separate presacral vertebrae and that the five sacral segments (vertebrae) are fused to form the sacrum. The four coccygeal segments fuse late in life to form the coccyx. The left lateral view shows the normal curvatures of the vertebral column.

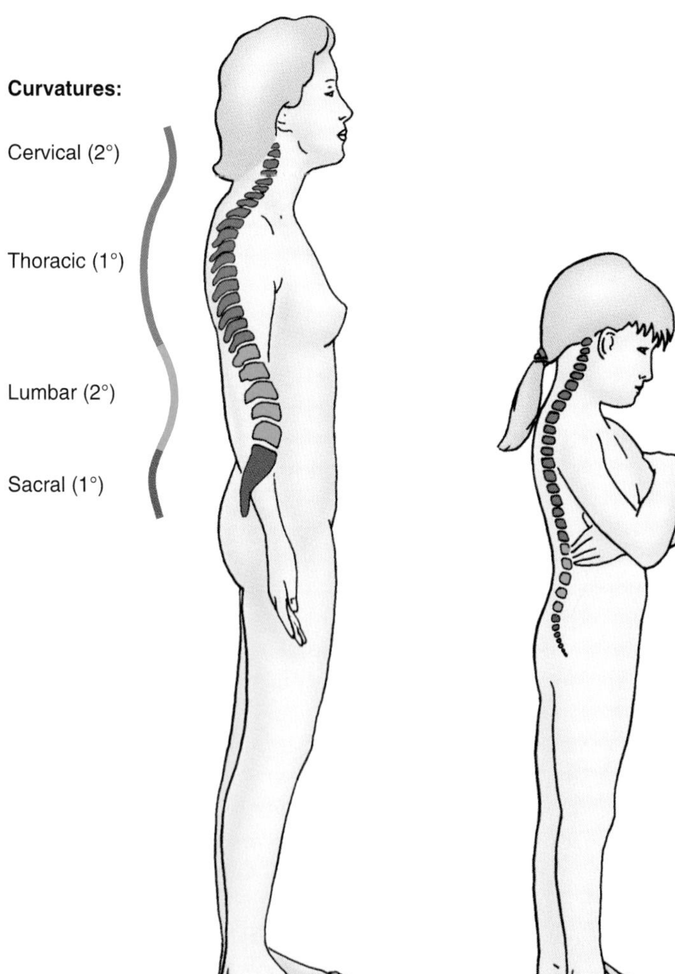

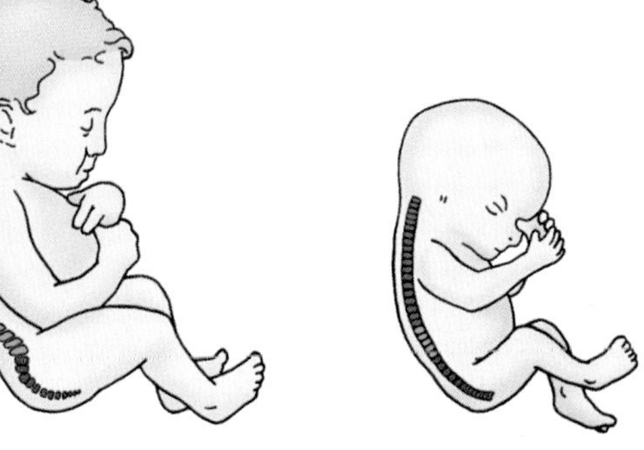

Figure 4.3. Curvatures of the vertebral column. Observe the four curvatures of the adult vertebral column—cervical, thoracic, lumbar, and sacral. Observe also the C-shaped curvature of the vertebral column during fetal life, when only the primary (1°) curvatures exist. Observe also the development of the secondary (2°) curvatures during infancy and childhood.

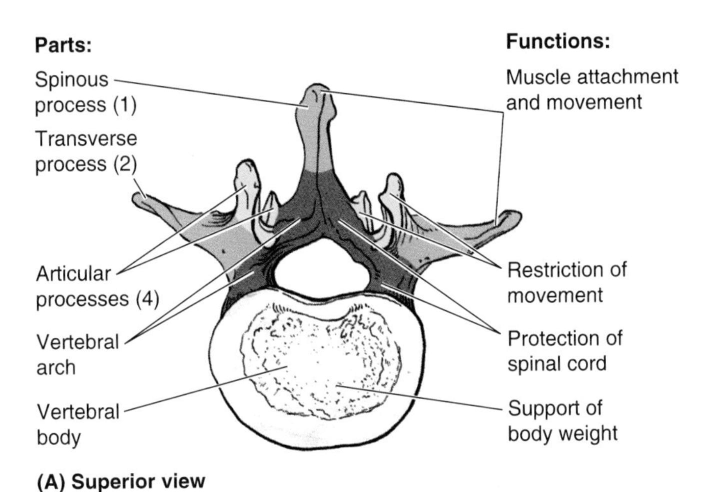

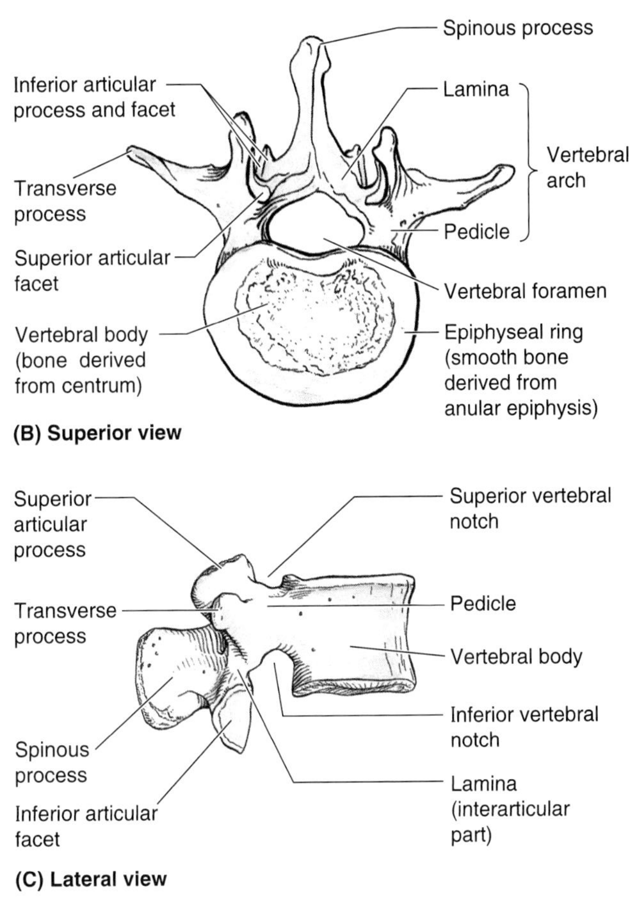

Figure 4.4. Parts and functions of typical vertebrae. **A.** Functional components of a vertebra; the functions of the color-coded components are listed on the figure's right side. **B.** Superior view of a typical vertebra (L2). Primary parts consist of a body, a vertebral (neural) arch, and seven processes. **C.** Lateral view of the L2 vertebra; observe the small vertebral notch superior to the pedicle and the larger inferior vertebral notch. When two vertebrae are in articulation, the adjacent superior and inferior notches of the vertebra, plus the intervertebral (IV) disc that unites them, form an IV foramen (Fig. 4.1*A*) for the passage of the spinal nerve and its accompanying vessels. Observe also that each articular process has an articular facet (articulating surface covered in life with articular cartilage where contact occurs with the articular facets of adjacent vertebrae).

cles and laminae (Fig. 4.4*B*). The **pedicles** are short, stout processes that join the vertebral arch to the vertebral body. The pedicles project posteriorly to meet two broad, flat plates of bone—the **laminae.** The vertebral arch and the posterior surface of the vertebral body form the walls of the **vertebral foramen.** The succession of vertebral foramina in the articulated column forms the **vertebral canal** (spinal canal), which contains the spinal cord, meninges, fat, spinal nerve roots, and vessels (Fig. 4.1*B*). The **vertebral notches** are indentations formed by the projection of the body and articular processes above and below the pedicle (Fig. 4.4*C*). The superior and inferior vertebral notches of adjacent vertebrae contribute to the formation of **IV foramina** (Fig. 4.1*A*), which give passage to spinal nerve roots and accompanying vessels, and contain the spinal ganglia (dorsal root ganglia).

Seven processes arise from the vertebral arch of a typical vertebra (Fig. 4.4*A*):

- A **spinous process** projects posteriorly from the vertebral arch at the junction of the laminae and overlaps the vertebra below.
- Two **transverse processes** project posterolaterally from the junctions of the pedicles and laminae.
- Four **articular processes** (G. zygapophyses)—two superior and two inferior—also arise from the junctions of the pedicles and laminae.

Three processes—two transverse and one spinous—project from the vertebral arch and afford attachments for deep back muscles and form levers that help the muscles to move the vertebrae.

The four articular processes project superiorly and inferiorly respectively from the vertebral arch and are in apposition with corresponding processes of vertebrae superior and inferior to them. Their function is to restrict movements in certain directions (Fig. 4.4*A*) or at least to decree which movements may be permitted. The articular processes also prevent the vertebrae from slipping anteriorly. When one rises from the flexed position, the articular processes bear weight temporarily. The inferior articular processes of L5 vertebra bear weight even in the erect posture.

Regional Characteristics of Vertebrae

Vertebrae in different regions of the vertebral column show some modification from typical vertebrae. The vertebrae in each region can usually be identified because of special characteristics (e.g., cervical vertebrae are characterized by the presence of foramina in their transverse processes). In addition, individual vertebrae have distinguishing characteristics; *C7 vertebra, for example, has a long spinous process that forms a prominence under the skin*, especially when the neck is flexed (see p. 464). Run your finger along the midline of the posterior aspect of your neck until you feel the prominent C7 spinous process.

The **direction of the articular facets** on the articular processes of vertebrae determines the direction of movement of the trunk allowed in any particular region (Fig. 4.4*B*). For example, the articular facets of thoracic vertebrae favor lateral bending and rotation of the vertebral column. Variations in the size and shape of the vertebral canal occur because the spinal cord is enlarged in the cervical and lumbar regions to provide innervation of the limbs.

Cervical Vertebrae

Cervical vertebrae form the bony skeleton of the neck (Figs. 4.1 and 4.2). The smallest of the 24 movable vertebrae, the cervical vertebrae are located between the skull and thorax. The cervical vertebrae are relatively small bones and bear less weight than do the vertebrae inferior to them. The distinctive feature of each cervical vertebra is the oval **foramen of the transverse process** (L. foramen transversarium). These foramina are smaller in C7 than those in other cervical vertebrae; occasionally these foramina are absent (Fig. 4.5, Table 4.1). The *vertebral arteries pass through the transverse foramina*, except those in C7, which transmit only small accessory vertebral veins. The transverse processes of cervical vertebrae end laterally in two projections—the **anterior** and **posterior tubercles**. The large anterior tubercles of C6 are called **carotid tubercles** because the common carotid arteries may be compressed against them to control bleeding from these vessels.

C3 through C7 vertebrae are characterized by large vertebral foramina because of the cervical enlargement of the

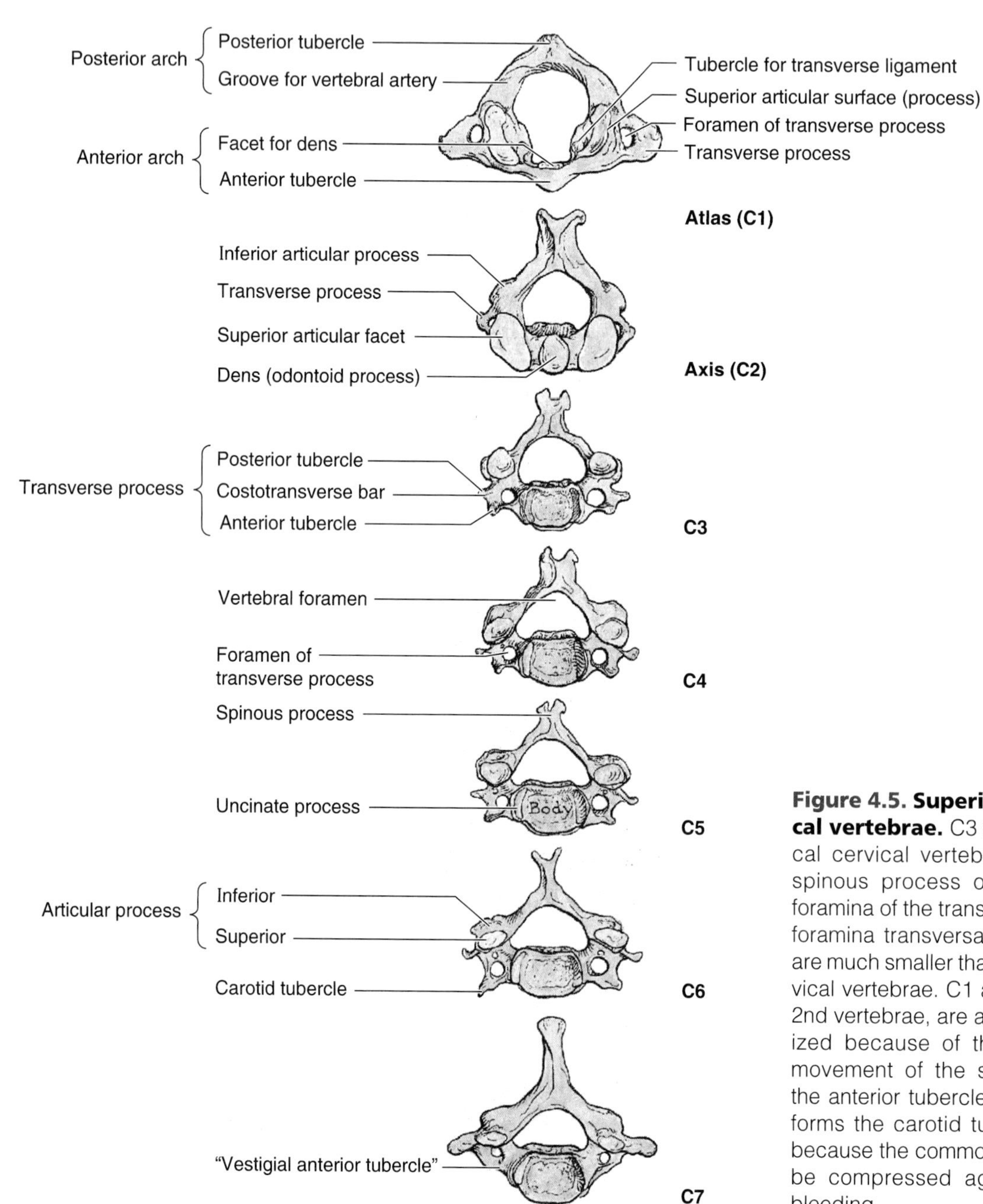

Figure 4.5. Superior views of cervical vertebrae. C3 through C6 are typical cervical vertebrae. Note the long spinous process of C7 and that the foramina of the transverse processes (L. foramina transversaria) in this vertebra are much smaller than those in other cervical vertebrae. C1 and C2, the 1st and 2nd vertebrae, are atypical and specialized because of their relationship to movement of the skull. Observe that the anterior tubercle of C6 is large and forms the carotid tubercle—so named because the common carotid artery can be compressed against it to control bleeding.

Table 4.1. Typical Cervical Vertebrae (C3–C7)[a]

Part	Distinctive Characteristics
Body	Small and wider from side to side than anteroposteriorly; superior surface is concave and inferior surface is convex
Vertebral foramen	Large and triangular
Transverse processes	Transverse foramina (foramina transversaria); small or absent in C7; vertebral arteries and accompanying venous and sympathetic plexuses pass through foramina, except C7, which transmits only small accessory vertebral veins; anterior and posterior tubercles
Articular processes	Superior facets directed superoposteriorly; inferior facets directed inferoanteriorly; obliquely placed facets are most nearly horizontal in this region
Spinous process	Short (C3–C5) and bifid (C3–C5); process of C6 is long but that of C7 is longer (for this reason, C7 is called vertebra prominens)

[a] C1 and C2 vertebrae are atypical.

spinal cord that provides for the innervation of the upper limbs (see Fig. 4.28). The superior borders of the bodies of these vertebrae are raised posteriorly, especially at the sides, but are depressed anteriorly. Their raised margins are *uncinate processes* (see Fig. 4.14). The spinous processes of C3 through C6 vertebrae are short and usually bifid in white persons but usually not in black persons. *C7 is a prominent vertebra that is characterized by a long spinous process*; because of this prominent process, **C7 is called the vertebra prominens.**

C1 and C2 are atypical cervical vertebrae. **C1**—the **atlas**—is a ring-shaped bone. In Figure 4.5 note that the atlas is the widest of the cervical vertebrae. Because it supports the skull, it was named after Atlas who, according to Greek mythology, supported the earth on his shoulders. The kidney-shaped, concave **superior articular surfaces** of C1 (Fig. 4.6*B*) receive the two large protuberances at the sides of the foramen magnum—the **occipital condyles** (Fig. 4.6*A*). Through these condyles, the weight of the head is transmitted to the vertebral column. *The atlas has no spinous process or body*; it consists of **anterior** and **posterior arches**, each of which has a tubercle and a lateral mass. The posterior arch, which corresponds to the lamina of a typical vertebra, has a wide **groove for the vertebral artery** on its superior surface (Fig. 4.6*B*). The 1st cervical nerve also occupies this groove.

C2—the **axis**—is the strongest of the cervical vertebrae because C1, carrying the skull, rotates on it when a person is shaking the head, for example. The axis has two large, flat bearing surfaces, the **superior articular facets**, on which the atlas rotates (Fig. 4.6*C*). The distinguishing feature of the axis is the blunt toothlike **dens** (odontoid process), which projects superiorly from its body. The dens (G. tooth) is held in position by the **transverse ligament of the atlas**, which prevents horizontal displacement of the atlas. C2 has a large bifid spinous process (Fig. 4.6*D*) that can be felt deep in the *nuchal groove*—the posterior groove of the neck.

The reason C1 and C2 vertebrae are atypical is because part of the body of C1 is transferred to the body of C2. The part of the body that remains with C1 is represented by the anterior arch of C1. The part of the body of C1 that was transferred to C2 becomes the dens. It is the pivot around which C1 (carrying the head) rotates.

Dislocation of the Cervical Vertebrae

The bodies of the cervical vertebrae are stacked like books and can be dislocated in neck injuries with less force than is required to fracture them. Because of the large vertebral canal in the cervical region, slight dislocation can occur without damaging the spinal cord. When a cervical vertebra is severely dislocated, it injures the spinal cord (p. 462). However, the vertebra may self-reduce ("slip back into place") so that a radiograph or magnetic resonance image (MRI) may not indicate that the cord has been injured.

Fracture and Dislocation of the Axis

The dens may be fractured as a result of a fall on the head. Displacement of the fractured dens may injure the spinal cord, causing **quadriplegia** (paralysis of all four limbs), or the medulla of the brainstem, causing death. **Fractures through the lamina (pars interarticularis)** of the axis—such as the *hangman's fracture*—with or without subluxation (incomplete dislocation) of the axis on the 3rd cervical vertebra (C3) may also injure the spinal cord and/or medulla causing quadriplegia or death. ⊙

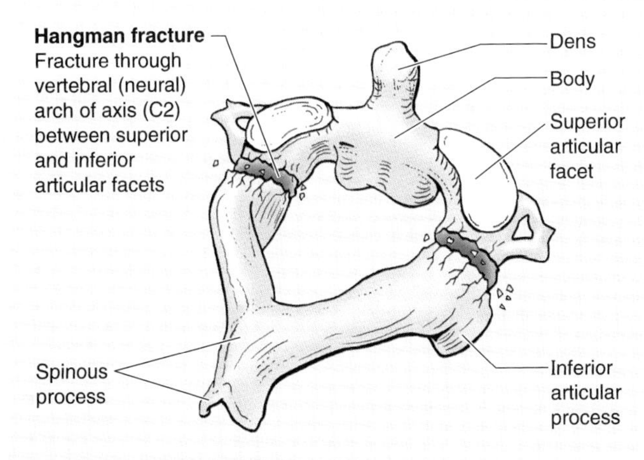

Posterior superior view of axis (C2)

Thoracic Vertebrae

The characteristic features of thoracic vertebrae are the **costal facets** for articulation with ribs (Fig. 4.7, Table 4.2). One or more facets are on each side of the body for articulation with the head of a rib, and one facet is on each transverse process of the superior 10 thoracic vertebrae for the tubercle of a rib.

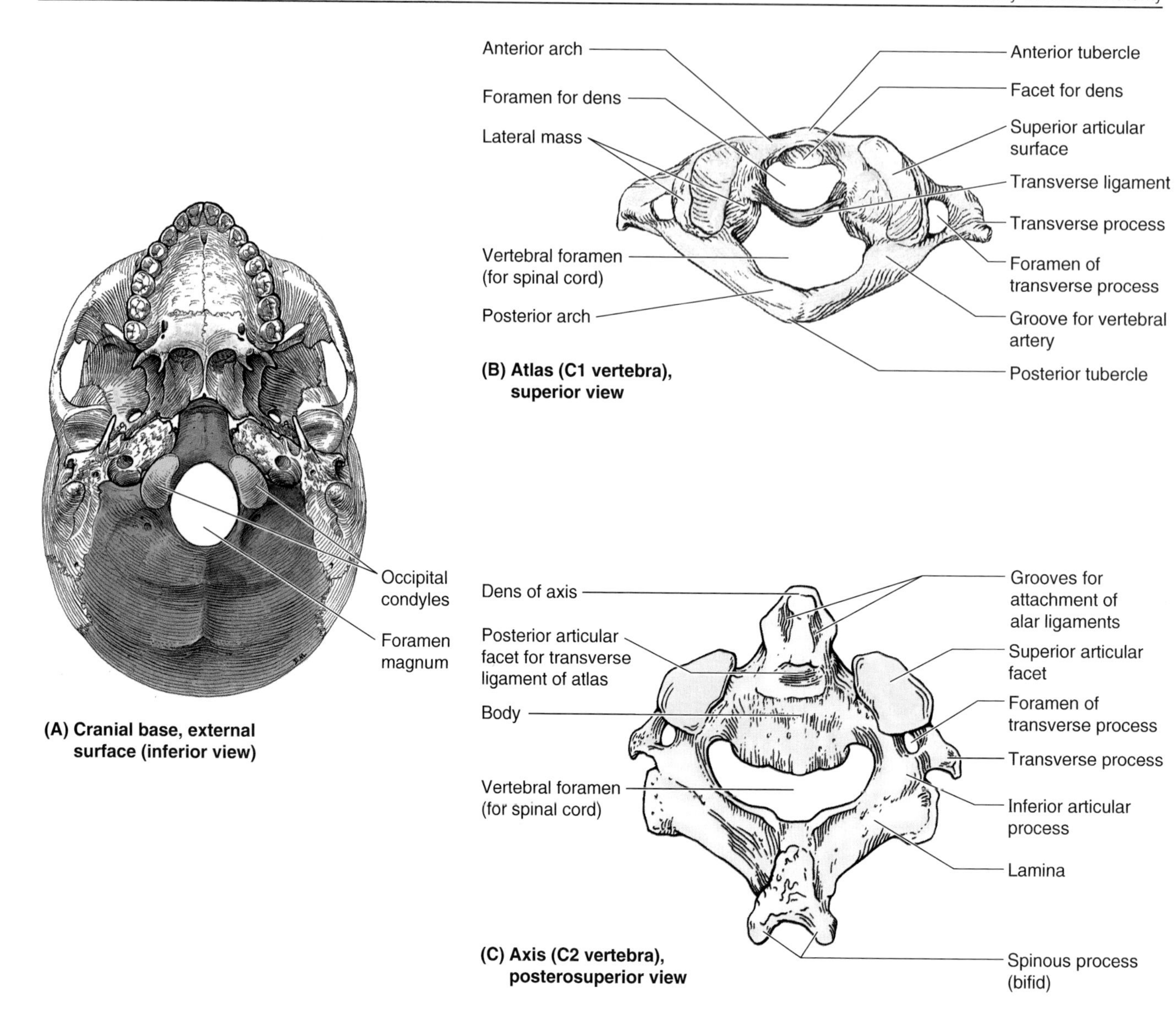

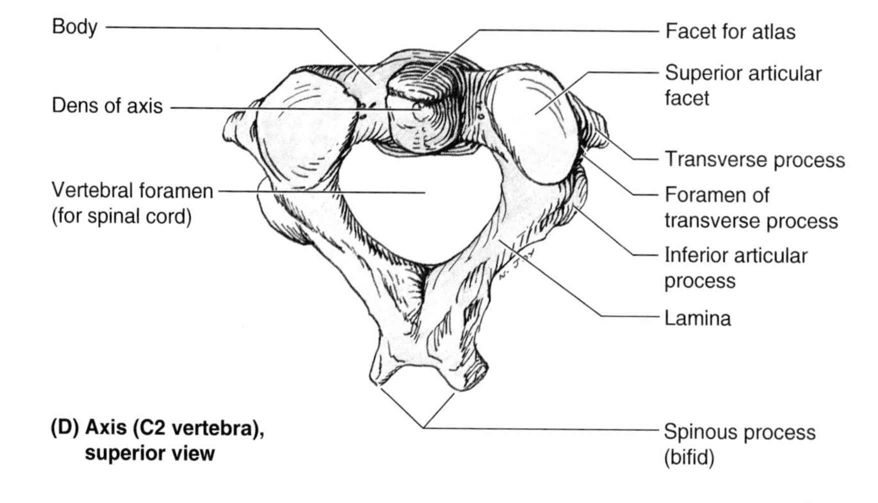

Figure 4.6. Cranial base and the first two cervical vertebrae. A. Observe the occipital condyles that articulate with the superior articular surfaces (facets) of the atlas. **B.** C1 vertebra—the atlas—on which the skull rests. Observe that it has neither a spinous process nor a body. C1 vertebra consists of two lateral masses connected by anterior and posterior arches. **C–D.** The C2 vertebra—the axis—provides a pivot around which the atlas turns and carries the skull. C2 vertebra is characterized by its toothlike dens, projecting superiorly from the vertebral body and articulating anteriorly with the anterior arch of the atlas (see facet for dens in (**B**)) and posteriorly with the transverse ligament of the atlas. C2 vertebra has a thick bifid spinous process.

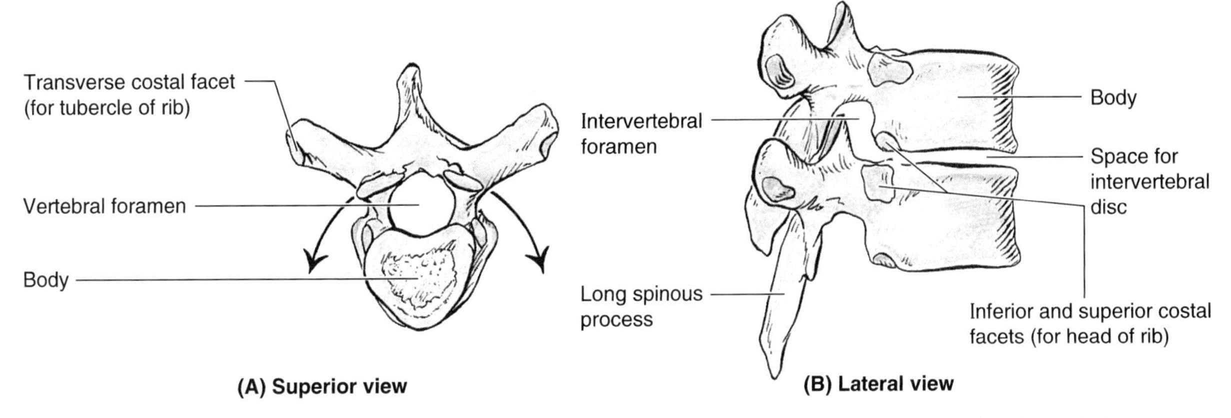

Figure 4.7. Typical thoracic vertebrae. A. Superior view. Observe the heart-shaped body and circular vertebral foramen. Observe also that the transverse processes are long, rounded, and strong. **B.** Lateral view of two articulated vertebrae. During life, an intervertebral (IV) disc occupies the space between their bodies. Note that the spinous processes are long and slender, slope inferoposteriorly, and overlap the spinous process of the thoracic vertebra inferior to it. *Costal facets are unique characteristics of thoracic vertebrae.* The larger superior costal facet, together with the IV disc superior to it, and the smaller inferior costal facet of the suprajacent vertebra form the "socket" for the head of the corresponding rib.

Table 4.2. **Thoracic Vertebrae**

Part	Distinctive Characteristics
Body	Heart-shaped; has one or two costal facets for articulation with head of a rib
Vertebral foramen	Circular and smaller than those of cervical and lumbar vertebrae
Transverse processes	Long and strong and extend posterolaterally; length diminishes from T1–T12 (T1–T10 have transverse costal facets for articulation with tubercle of a rib)
Articular processes	Superior facets directed posteriorly and slightly laterally; inferior facets directed anteriorly and slightly medially; plane of facets lies on an arc centered about vertebral body
Spinous process	Long and slopes posteroinferiorly; tip extends to level of vertebral body below

The **spinous processes** of the thoracic vertebrae are long and slender, and the middle ones are directed inferiorly over the vertebral arches of the vertebrae inferior to them (Figs. 4.1*A* and 4.7*B*).

T1 through T4 vertebrae have some features of cervical vertebrae. T1 is atypical in that it has a long, almost horizontal spinous process that may be nearly as prominent as that of the vertebra prominens. T1 also has a complete costal facet on the superior edge of its body for the 1st rib and a demifacet on its inferior edge that contributes to the articular surface for the 2nd rib. T9 through T12 vertebrae are also atypical in that they have tubercles similar to the accessory and mamillary processes of lumbar vertebrae.

The middle four thoracic vertebrae are typical ones. The outline of their bodies, viewed from the superior aspect, is heart-shaped, and their vertebral foramina are circular (Fig. 4.7*A*). Sometimes an impression is visible on the left sides of the bodies of the thoracic vertebrae, which is produced by the descending thoracic aorta.

Lumbar Vertebrae

These vertebrae are in the lower back ("small of the back") between the thorax and sacrum (Figs. 4.1 and 4.2). Lumbar vertebrae are distinguished by their *massive bodies, sturdy laminae,* and *absence of costal facets* (Fig. 4.8, Table 4.3). These large vertebrae account for much of the thickness of the lower trunk in the median plane. Lumbar vertebrae have massive bodies because the weight they support increases toward the inferior end of the vertebral column. The bodies of lumbar vertebrae, viewed superiorly, are somewhat kidney shaped, and their vertebral foramina vary from oval (L1) to triangular (L5).

L5 is the largest of all movable vertebrae; it carries the weight of the whole upper body. *L5 is characterized by its massive body and transverse processes.* Its body is markedly deeper anteriorly; *therefore, it is largely responsible for the lumbosacral angle* between the long axis of the lumbar region of the vertebral column and that of the sacrum (Fig. 4.1*A*). Body weight is transmitted from L5 vertebra to the **base of the sacrum,** formed by the superior surface of S1 vertebra (see Fig. 4.10*A*). *The hatchet-shaped spinous processes of lumbar vertebrae are thick and broad and point posteriorly* (Figs. 4.1*A* and 4.8). The **articular processes** of lumbar vertebrae facilitate flexion, extension, and lateral bending of the vertebral column; however, they prohibit rotation. The **transverse processes** project somewhat posterosuperiorly as well as laterally. On the posterior surface of the base of each transverse process is a small *accessory process* (Fig. 4.8), which provides an attachment for the medial intertransverse lumborum muscle (see Fig. 4.25*C*). On the posterior surface of the superior articular processes are

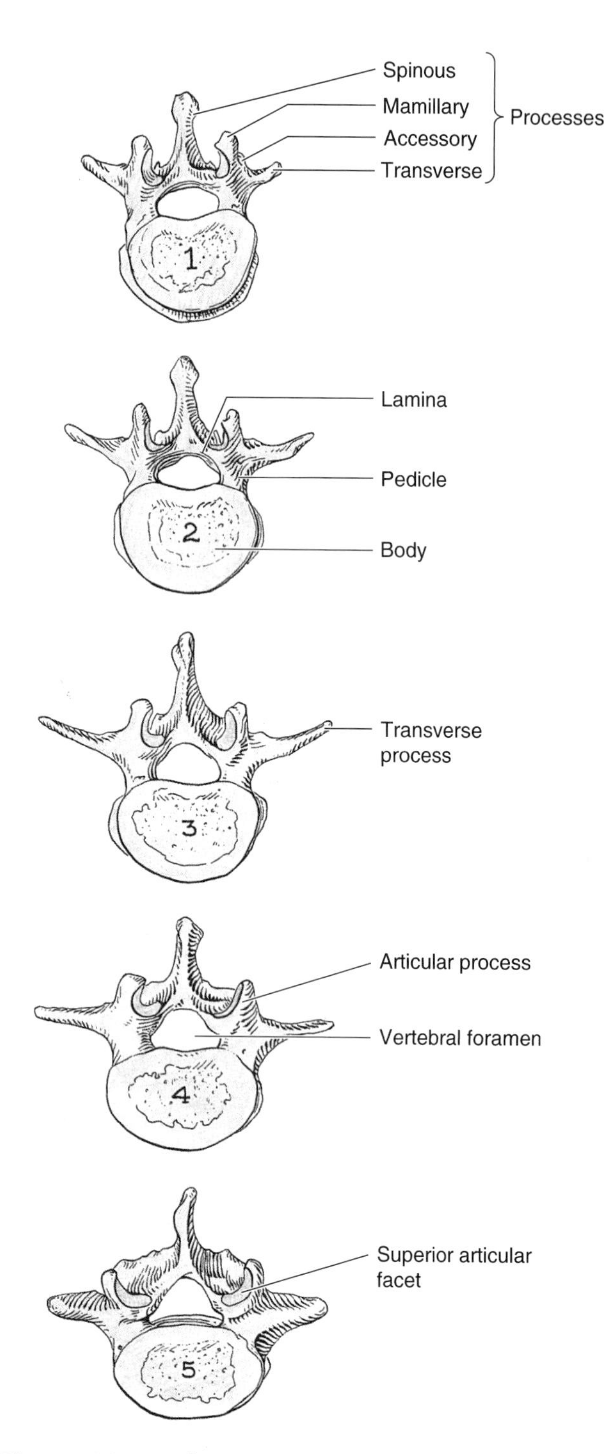

Figure 4.8. Lumbar vertebrae. Superior surfaces of the five lumbar vertebrae showing their massive, kidney-shaped bodies and sturdy vertebral (neural) arches, especially L4 and L5. They are distinguished by their large size, the absence of costal facets, and their long, thin transverse processes, which are homologous to ribs. The quadrilateral, hatchet-shaped spinous processes extend horizontally and posteriorly. In addition to spinous processes, they have mamillary and accessory processes for the attachment of deep back muscles.

Table 4.3. **Lumbar Vertebrae**

Part	Distinctive Characteristics
Body	Massive; kidney-shaped when viewed superiorly
Vertebral foramen	Triangular; larger than in thoracic vertebrae and smaller than in cervical vertebrae
Transverse processes	Long and slender; accessory process on posterior surface of base of each process
Articular processes	Superior facets directed posteromedially (or medially); inferior facets directed anterolaterally (or laterally); mamillary process on posterior surface of each superior articular process
Spinous process	Short and sturdy; thick, broad, and hatchet-shaped

mamillary processes, which give attachment to the multifidus and medial intertransverse muscles.

Sacrum

This large, triangular wedged-shaped bone is usually composed of five fused sacral vertebrae in adults (Fig. 4.9). It is wedged between the hip bones and forms the roof and posterosuperior wall of the posterior pelvic cavity. The triangular shape of the sacrum resulted from the rapid decrease in the size of the lateral masses of the sacral vertebrae during development. The inferior half of the sacrum is not weightbearing, and therefore its bulk is diminished rapidly. The sacrum (L. sacred bone) provides strength and stability to the pelvis and transmits the weight of the body to the pelvic girdle, the bony ring formed by the hip bones and sacrum, to which the lower limbs are attached.

The **sacral canal** (Fig. 4.9)—the continuation of the vertebral canal in the sacrum—*contains the nerve roots of the cauda equina* (L. cauda, tail), the bundle of spinal nerve roots arising from the lumbosacral enlargement and medullary cone (L. conus medullaris) of the spinal cord. The **cauda equina** comprises the roots of all the spinal nerves inferior to L1 vertebra. On the pelvic and dorsal surfaces of the sacrum are typically four pairs of **sacral foramina** between its vertebral components for the exit of the dorsal and ventral rami of the spinal nerves. The anterior (pelvic) sacral foramina are larger than the posterior (dorsal) ones.

The **base of the sacrum** is formed by the superior surface of S1 vertebra. Its superior articular processes articulate with the inferior articular processes of L5 vertebra. The anterior projecting edge of the body of S1 vertebra is the **sacral promontory** (L. mountain ridge)—*an important obstetrical landmark* (see Chapter 3). The **apex of the sacrum**—the tapering inferior end of the sacrum—has an oval facet for articulation with the coccyx.

The sacrum supports the vertebral column and forms the posterior part of the bony pelvis. The sacrum is tilted so that it articulates with L5 vertebra at the **lumbosacral angle** (Fig. 4.1*A*), which varies from 130 to 160°. The sacrum is often wider in proportion to length in the female than in the male, but the body of the 1st sacral vertebra is usually larger in males.

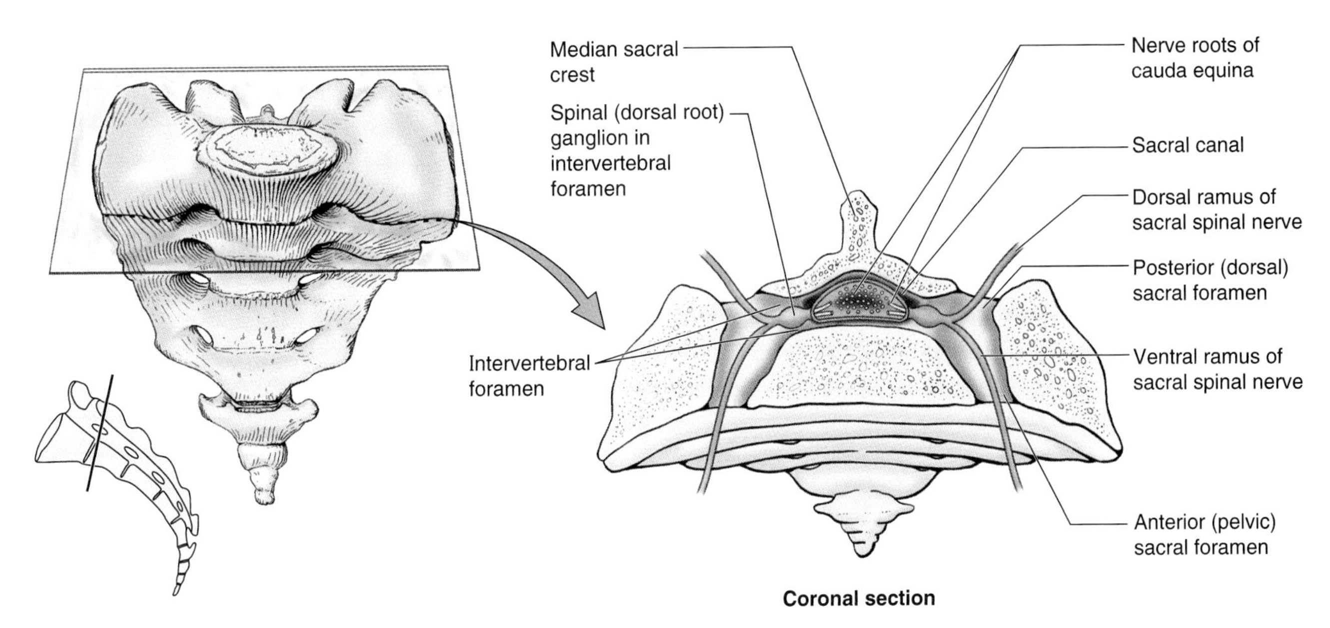

Figure 4.9. Sacrum and coccyx. The orientation drawing of the pelvic surface of the sacrum (viewed anteriorly and inferiorly) shows the level of the coronal section. In the anatomical position, the S1 through S3 vertebrae lie in an essentially horizontal plane, forming a roof for the pelvic cavity. Observe the bundle of spinal nerve roots forming the cauda equina (L. cauda, tail) in the sacral canal—the inferior part of the vertebral (spinal) canal. Observe also the spinal ganglia in the intervertebral (IV) foramina and the dorsal and ventral rami of the spinal nerves exiting from the posterior (dorsal) and anterior (ventral) sacral foramina, respectively.

The **pelvic surface of the sacrum** is smooth and concave (Fig. 4.10*A*). Four transverse lines on this surface of sacra from adults indicate where fusion of the sacral vertebrae occurred. During childhood, the individual sacral vertebrae are connected by hyaline cartilage and separated by IV discs. Fusion of the sacral vertebrae starts after the 20th year; however, most of the IV discs remain unossified up to or beyond middle life (Williams et al., 1995).

The **dorsal surface of the sacrum** (Fig. 4.10*B*) is rough, convex, and marked by five prominent longitudinal ridges. The central one, the **median crest**, represents the fused rudimentary spinous processes of the superior three or four sacral vertebrae—S5 has no spinous process. The **medial (intermediate) crests** represent the fused articular processes, and the **lateral crests** are the tips of the transverse processes of the fused sacral vertebrae. The clinically important features of the dorsal surface of the sacrum are the inverted U-shaped sacral hiatus and the sacral cornua (L. horns).

The **sacral hiatus** results from the absence of the laminae and spinous process of S5 vertebra and sometimes S4. The sacral hiatus leads into the **sacral canal**, the inferior end of the vertebral canal. Its depth varies, depending on how much of the spinous process and laminae of S4 are present. The sacral hiatus contains fatty connective tissue, the **terminal filum** (L. filum terminale)—a connective tissue strand that extends from the inferior end of the spinal cord, the S5 nerve, and the coccygeal nerve (Co.). The **sacral cornua**, representing the inferior articular processes of S5 vertebra, project inferiorly on each side of the sacral hiatus and are a helpful guide to its location.

The superior part of the **lateral surface of the sacrum** looks somewhat like an auricle (L. external ear, dim. of *auris,* ear); because of its shape, this area is named the **auricular surface** (Fig. 4.10*B*). This surface—formed by the fusion of the costal processes (homologs of ribs)—is the *site of the synovial part of the sacroiliac joint,* located between the sacrum and ilium. During life, the auricular surface is covered with hyaline cartilage.

Coccyx

The coccyx (tailbone) is a small triangular bone—usually formed by four rudimentary vertebrae (Fig. 4.10)—but there may be one less or one more. The 1st coccygeal (Co.1) vertebra may be separate. *The coccyx is the remnant of the skeleton of the tail,* which human embryos have until the beginning of the 8th week (Moore and Persaud, 1998). The pelvic surface of the coccyx is concave and relatively smooth, and the dorsal surface has rudimentary articular processes. Co.1 vertebra is the largest and broadest of all the coccygeal bones. Its short transverse processes are connected to the sacrum, and its rudimentary articular processes form **coccygeal cornua**, which articulate with the sacral cornua. The last three coccygeal vertebrae often fuse during middle life, forming a beaklike bone; this accounts for its name (G. coccyx, cuckoo). During old age, Co.1 vertebra often fuses with the sacrum, and the remaining coccygeal vertebrae usually fuse to form a single bone. The coccyx does not participate with the other vertebrae in the support of the body weight; however, it provides attachments for parts of the gluteus maximus and coccygeus muscles and the *anococcygeal ligament,* the median fibrous intersection of the pubococcygeus muscles (see Chapter 3).

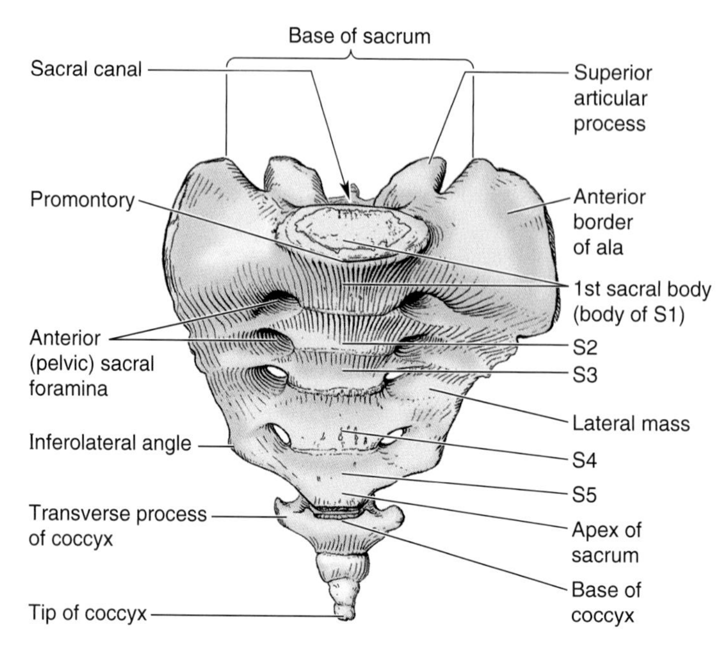

(A) Base and pelvic surface of sacrum

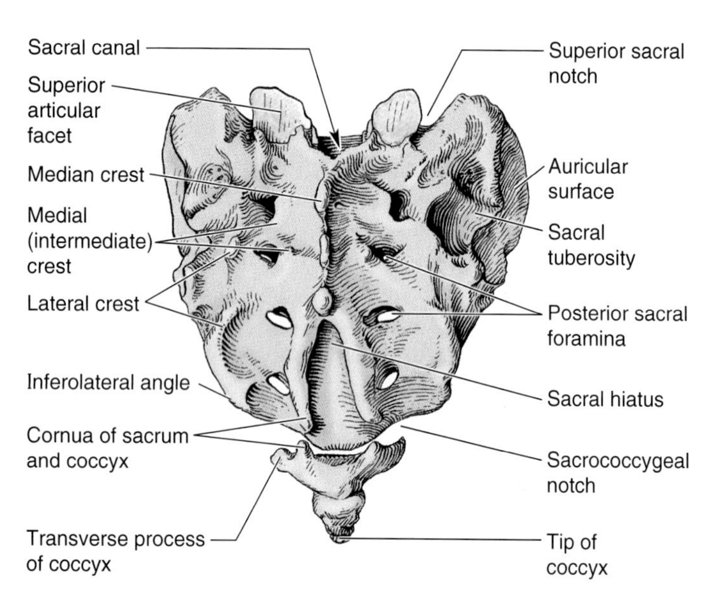

(B) Dorsal and lateral surfaces of sacrum

Figure 4.10. Sacrum and coccyx. A. Observe the base of this adult sacrum and the five sacral vertebral bodies, demarcated by four transverse lines indicating the fusion sites of the sacral vertebrae. The lines end laterally in four pairs of anterior (pelvic) sacral foramina. The base—the anterosuperior surface of the sacrum (compare with the base of an inverted isosceles triangle, opposite the apex)—is characterized by the promontory, the anterior margin of the superior surface of S1 vertebra, the sacral canal, the inferior part of the vertebral (spinal) canal, and the right and left alae (L. wings). The coccyx has four segments (vertebrae). The first segment has a pair of transverse processes. **B.** Observe the dorsal surface of the adult sacrum and coccyx. Note that the absence of the S4 and S5 spinous processes has resulted in the formation of a large sacral hiatus. Observe also the cornua, or horns, of the sacrum and coccyx, which are clinical landmarks when performing a caudal (epidural) anesthesia (p. 445). The superolateral part of the sacrum—lateral to the foramina—is characterized by an irregular, ear-shaped auricular surface, which articulates with the ilium to form the synovial part of the sacroiliac joint. The inferior three coccygeal vertebrae are fused in this specimen.

Caudal Epidural Anesthesia

During life, the sacral hiatus is filled with fatty connective tissue in the epidural or extradural space (*A*). In *epidural anesthesia* or *caudal analgesia,* a local anesthetic agent is injected into the sacral canal (*B*). The anesthetic solution spreads superiorly and extradurally or epidurally, where it acts on the S2 through Co. spinal nerves in the cauda equina. The height to which the anesthetic ascends is controlled by the amount injected and the position of the patient. *Sensation is lost inferior to the epidural block.* As the sacral hiatus is located between the sacral cornua and inferior to the 4th sacral spinous process or median sacral crest, these horns are important bony landmarks for locating the hiatus. Anesthetic agents can also be injected through the posterior sacral foramina—*transsacral (epidural) anesthesia*—into the sacral canal around the spinal nerve roots (*B*). ▶

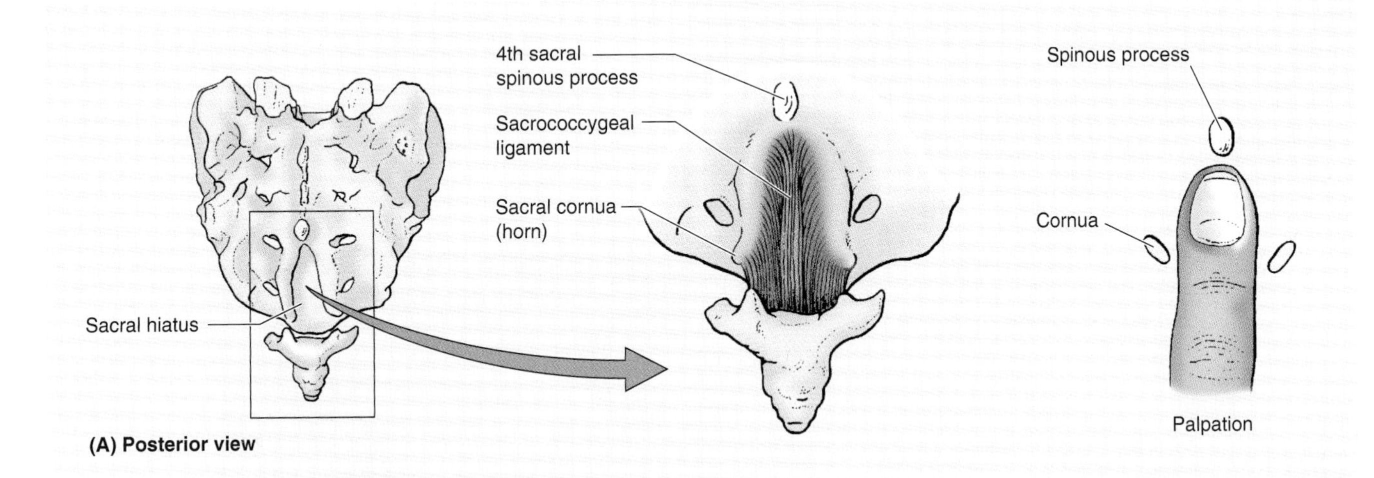
4th sacral
spinous process
Sacrococcygeal
ligament
Sacral cornua
(horn)
Sacral hiatus
Spinous process
Cornua
Palpation
(A) Posterior view

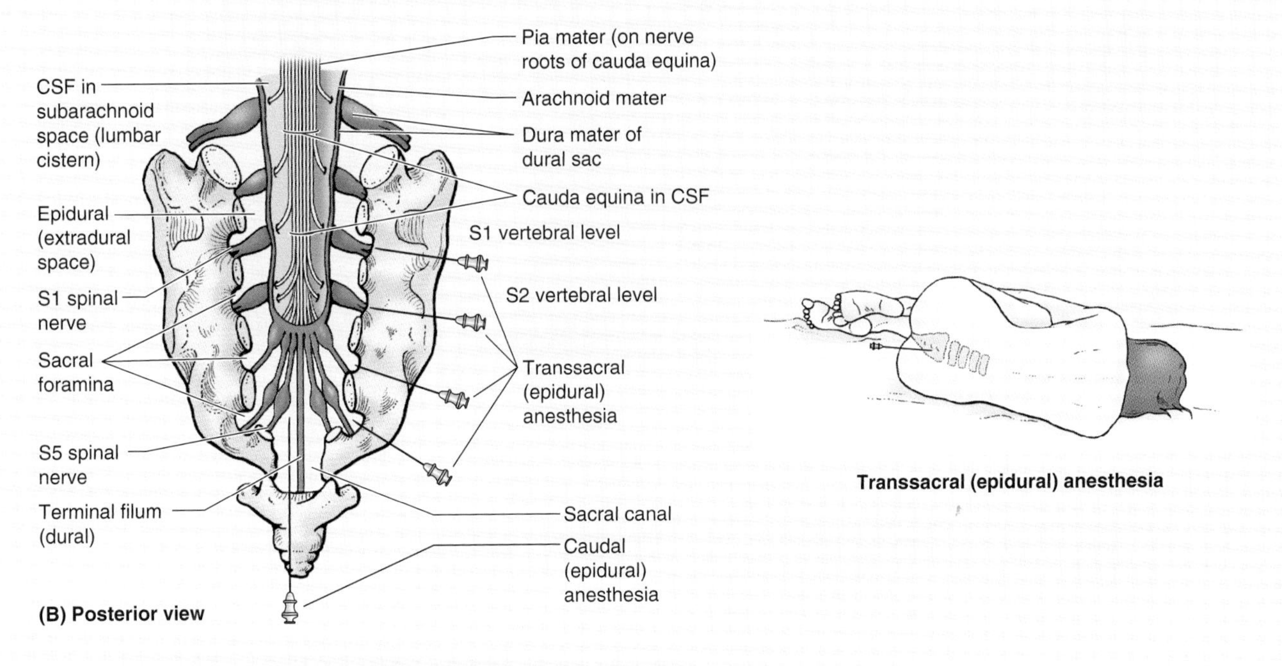
Pia mater (on nerve
roots of cauda equina)
CSF in
subarachnoid
space (lumbar
cistern)
Arachnoid mater
Dura mater of
dural sac
Cauda equina in CSF
Epidural
(extradural
space)
S1 vertebral level
S1 spinal
nerve
S2 vertebral level
Sacral
foramina
Transsacral
(epidural)
anesthesia
S5 spinal
nerve
Terminal filum
(dural)
Sacral canal
Caudal
(epidural)
anesthesia
(B) Posterior view
Transsacral (epidural) anesthesia

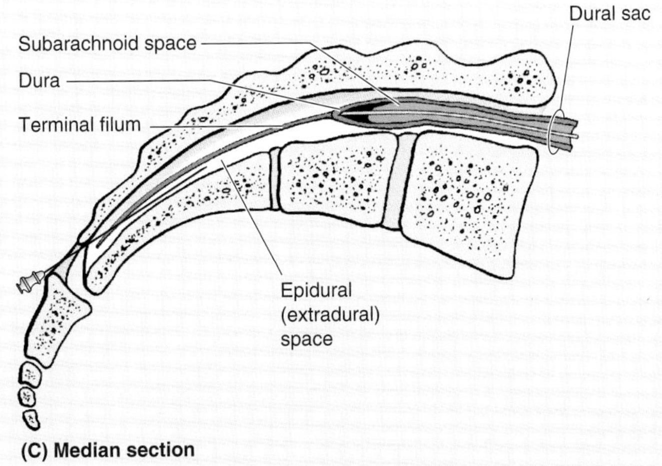
Dural sac
Subarachnoid space
Dura
Terminal filum
Epidural
(extradural)
space
(C) Median section

Abnormal Fusion of Vertebrae

In approximately 5% of people, L5 vertebra is partly or completely incorporated into the sacrum—conditions known as *hemisacralization and sacralization of L5 vertebra*, respectively. In other people, S1 vertebra is more or less separated from the sacrum and is partly or completely fused with L5 vertebra—*lumbarization of S1 vertebra*. When L5 is sacralized, the L5/S1 level is strong and the L4/L5 level degenerates, often producing painful symptoms. ▶

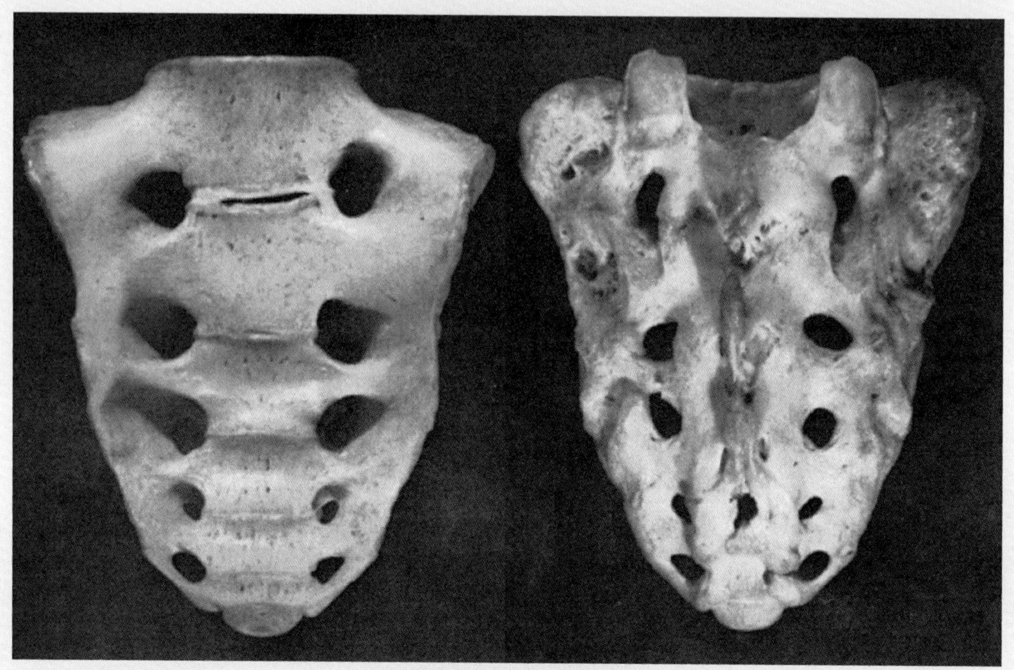

Pelvic surface Dorsal surface

Sacralization of L5 vertebra

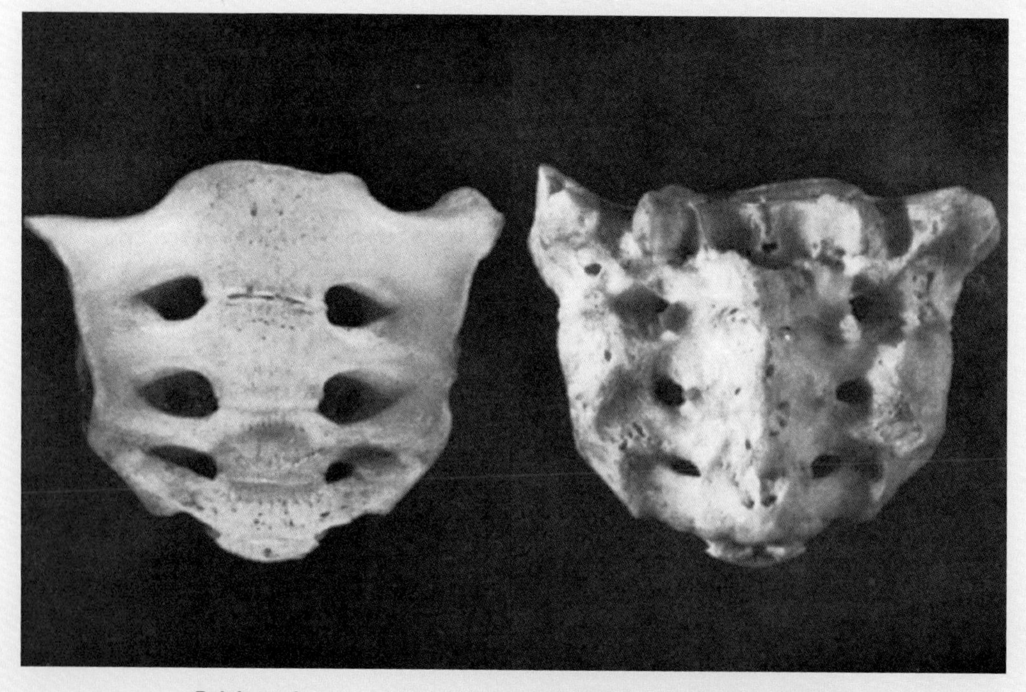

Pelvic surface Dorsal surface

Lumbarization of S1 vertebra

Lumbar Spinal Stenosis

Lumbar spinal stenosis—*stenotic (narrow) vertebral canal* (*A*)—may be a hereditary anomaly, resulting in age-related degenerative changes such as IV disc bulging (Rowland and McCormick, 1995). The narrowing is usually maximal at the level of the IV discs. Congenital stenosis makes a person more vulnerable to these degenerative changes. *Stenosis of L5 vertebra causes compression of the spinal nerve roots* of the cauda equina. *Electromyography* can reveal that the denervation is restricted to muscles innervated by the lumbosacral nerve roots. *Surgical treatment of lumbar stenosis may consist of decompressive laminectomy*—excision of the vertebral laminae or the entire vertebral arch. When IV disc protrusion occurs in a patient with spinal stenosis (*B*) it further compromises a vertebral canal that is already limited, as does arthritic proliferation and ligamentous degeneration (McCormick, 1995). ○

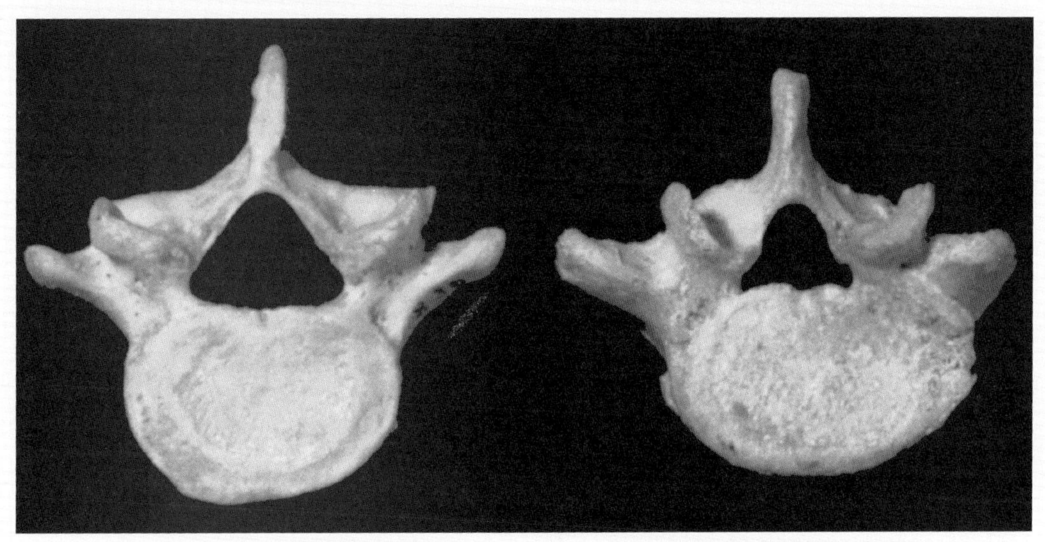

(A) **Normal vertebral foramen** **Stenotic vertebral foramen**

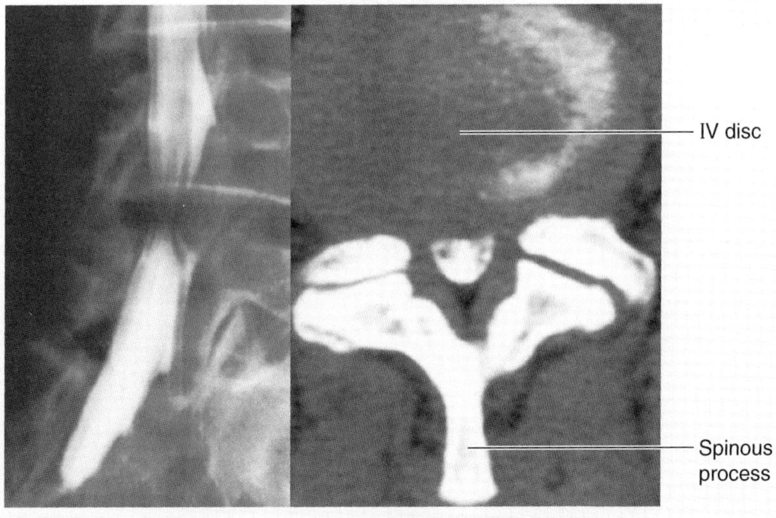

(B) **Lumbar myelogram** (showing high-grade stenosis at the L4–L5 disc space) **CT scan** (demonstrating stenosis caused by IV disc bulging at L4–L5 disc space)

Ossification of Vertebrae

Vertebrae begin to develop during the embryonic period as mesenchymal condensations around the notochord (Moore and Persaud, 1998). Later, these mesenchymal bone models chondrify and cartilaginous vertebrae form. Typical vertebrae begin to ossify toward the end of the embryonic period (8th week) and ossification continues throughout the fetal period.

Three primary ossification centers develop in each cartilaginous vertebra, one in the centrum and one in each half of the vertebral arch (Fig. 4.11*A*). At birth, the inferior sacral vertebrae and all the coccygeal vertebrae are cartilaginous. They begin to ossify during infancy. At birth, the halves of the vertebral arch articulate with the centrum at **neurocentral joints,** which are primary cartilaginous joints. *Each typical vertebra at birth consists of three bony parts* (see (*A*), Clinical Box, *Anomalies of the Vertebrae*) *united by hyaline cartilage.* The halves of the vertebral arch begin to fuse in the cervical region during the lst year, and fusion is usually complete in the lumbar region by the 6th year. The vertebral arch fuses with the centrum during childhood (5 to 8 years).

Five secondary ossification centers develop during puberty in each typical vertebra: one at the tip of the spinous process; one at the tip of each transverse process; and two **anular epiphyses** (ring epiphyses), one on the superior and one on the inferior edge of the centrum (Fig. 4.11, *B* and *C*). The ring epiphyses usually unite with the vertebral body early in the adult period. This union results in the characteristic smooth raised margin around the edges of the superior and inferior surfaces of the body of the vertebra. The body of the vertebra forms mainly from growth of the **centrum,** the primary ossification center of the central mass of the vertebral body. All secondary ossification centers usually unite with the vertebra by the 25th year; however, the times of their union are variable.

Caution must be exercised so that a persistent epiphysis is not mistaken for a vertebral fracture in a radiograph. Exceptions to the typical ossification of vertebrae occur in C1, C2, and C7, the lumbar vertebrae, sacrum, and coccyx. For example, 56 to 58 primary and secondary centers of ossification have been described in the sacrum. (For details about the ossification of vertebrae, see Williams et al., 1995.)

Anomalies of the Vertebrae

At birth, a vertebra consists of three bony parts (*A*, centrum and two parts of the vertebral arch) united by hyaline cartilage (*C*). Beginning at age 2 years, the halves of the vertebral arch begin to fuse with each other, from the lumbar to the cervical region. From approximately age 7 years, the vertebral arches fuse to the centra of the vertebrae in sequence from the cervical to lumbar regions. The centrum of a vertebra is the ossification center of the central mass of the body of a vertebra. The centrum and body of a vertebra are not synonymous; a centrum is somewhat less than a vertebral body (Williams et al., 1995). Sometimes the epiphysis of the transverse process fails to fuse. Its shadow on a radiograph may suggest a fracture of this process (*B*).

The common congenital anomaly of the vertebral column is **spina bifida occulta**, in which the laminae of L5 and/or S1 fail to develop normally and fuse. This bony defect, present in up to 24% of the population (Greer, 1995), usually occurs in the vertebral arch of L5 and/or S1. The defect is concealed by skin, but its loca- ▶

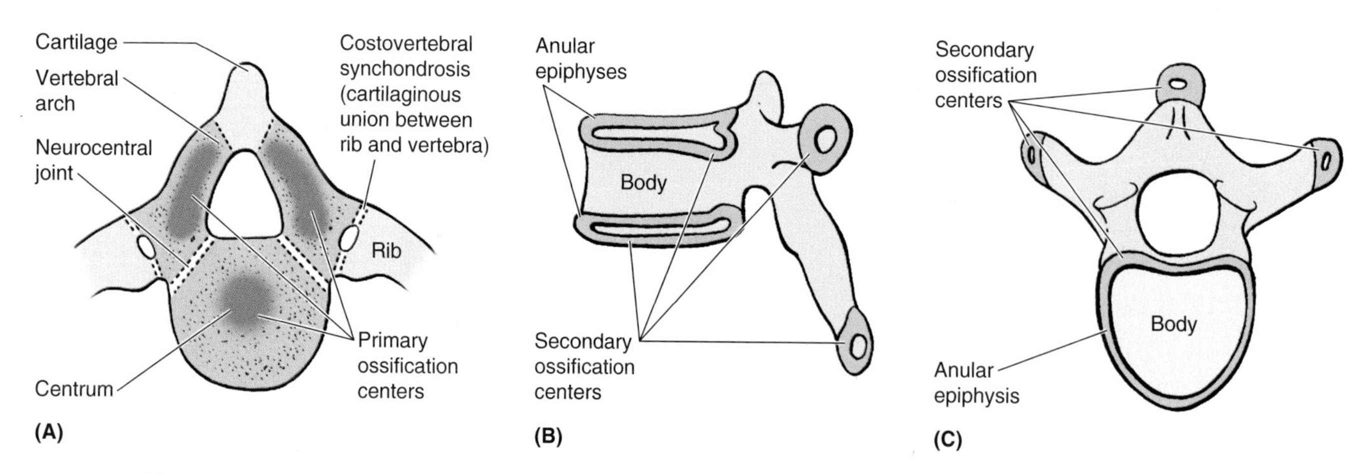

Figure 4.11. Ossification of a typical thoracic vertebra. A. Superior view of the developing vertebra illustrating the three primary ossification centers (*darker blue*) in a cartilaginous vertebra of a 7-week-old embryo. Observe the joints present at this stage. **B–C.** Lateral and superior views, respectively, of a vertebra from an individual at puberty, showing the location of the secondary centers of ossification and the anular epiphyses. These epiphyses form the smooth epiphyseal rings (see Fig. 4.4*B*) on the circumferences of the superior and inferior surfaces of the vertebral bodies.

▶ tion is often indicated by a tuft of hair (Moore and Persaud, 1998). Most people with spina bifida occulta have no back problems. In severe types of spina bifida, **spina bifida cystica**, one or more vertebral arches may almost completely fail to develop. Spina bifida cystica is associated with herniation of the meninges (*meningocele*, a spina bifida associated with a meningeal cyst) and/or the spinal cord (*meningomyelocele*). Usually, neurological symptoms are present in severe cases of meningomyelocele (e.g., paralysis of the limbs and disturbances in bladder and bowel control). Severe forms of spina bifida result from **neural tube defects**—defective closure of the neural tube during the 4th week of embryonic development (Moore and Persaud, 1998). ○

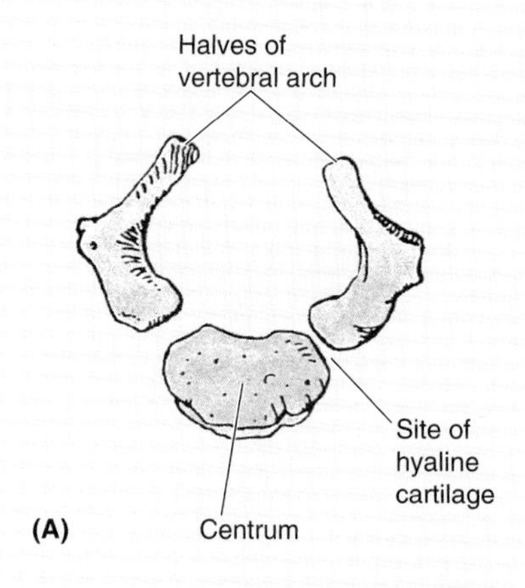

(A)

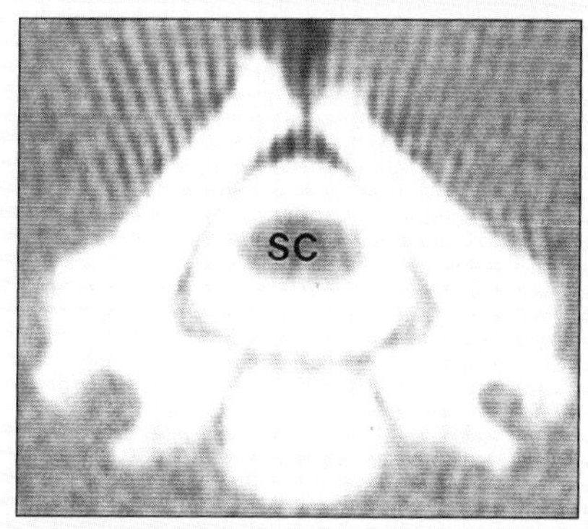

(B) Transverse CT scan (SC, spinal cord)

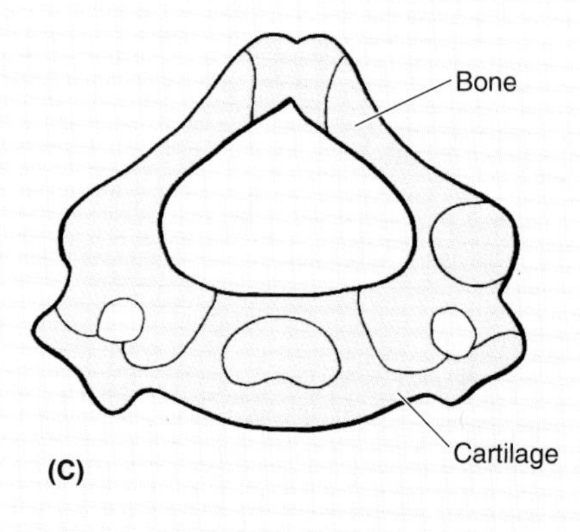

(C)

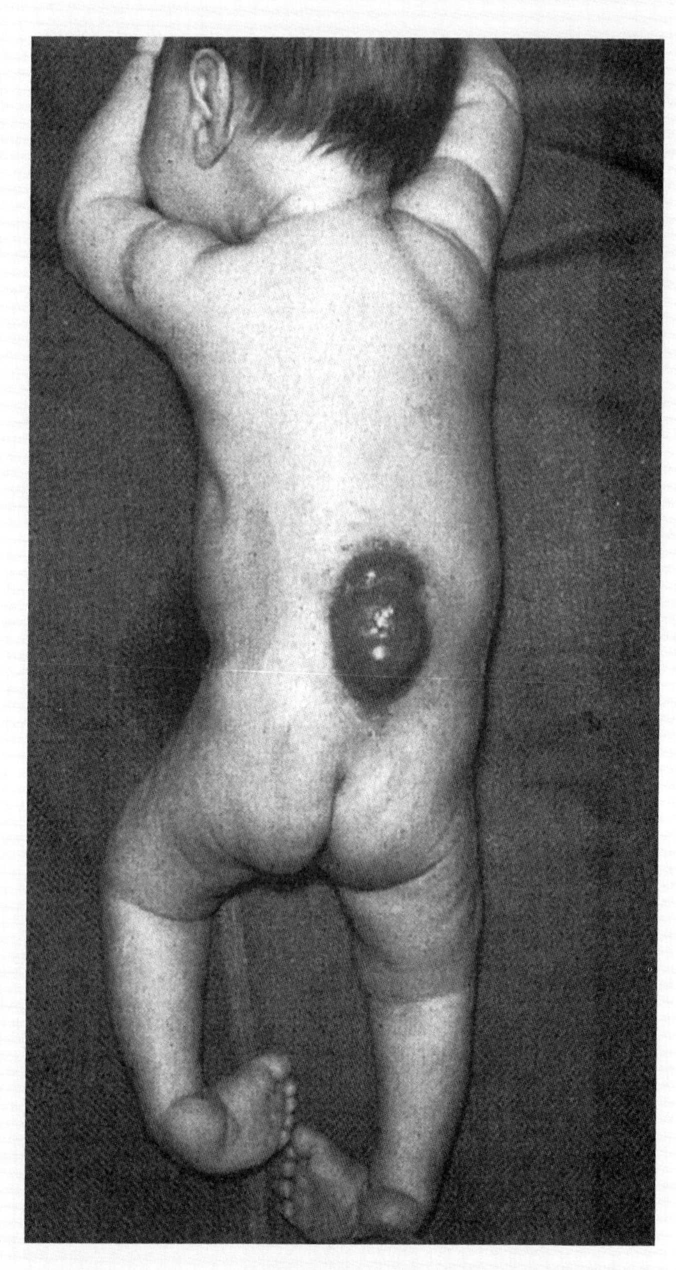

Spina bifida cystica

(Courtesy of Dr. Dwight Parkinson, Department of Surgery and Department of Human Anatomy and Cell Science, University of Manitoba, Winnipeg, Canada)

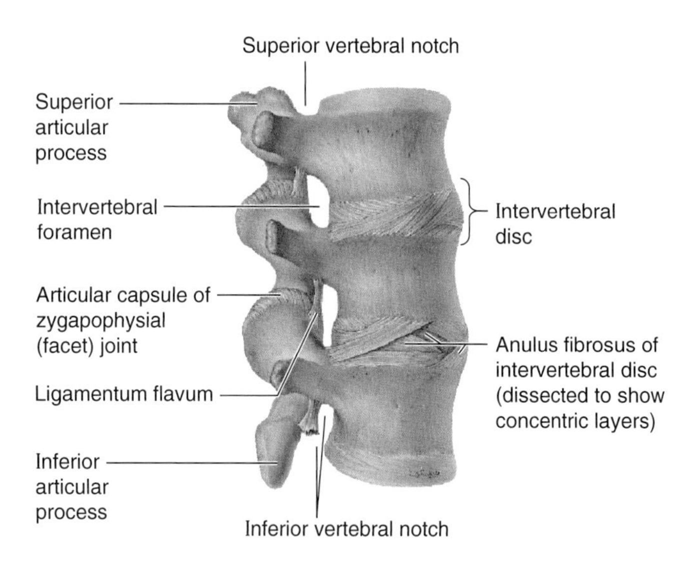

Figure 4.12. Lumbar vertebrae and intervertebral (IV) discs. Lateral view of the superior lumbar region, primarily to show the structure of the anuli fibrosi of the IV discs. Note that the IV disc forms the inferior half of the anterior boundary of the IV foramen. Thus, herniation of the disc will not effect the spinal nerve exiting from the superior bony part of that foramen. Also observe the articular capsules of the zygapophysial (facet) joints.

Joints of the Vertebral Column

The joints of the vertebral column include the:

- Joints of the vertebral bodies
- Joints of the vertebral arches (zygapophysial joints)
- Atlantoaxial joints
- Atlanto-occipital joints
- Costovertebral joints (see Chapter 1)
- Sacroiliac joints (see Chapter 3).

Joints of the Vertebral Bodies

The joints of the vertebral bodies are *secondary cartilaginous joints (symphyses)* designed for weightbearing and strength. The articulating surfaces of adjacent vertebrae are connected by **IV discs** and ligaments (Figs. 4.12 and 4.13). The IV discs provide strong attachments between the vertebral bodies. They also form the inferior half of the anterior border of the IV foramen. The discs act as shock absorbers, and their varying shapes produce the secondary curvatures of the vertebral column. *Each IV disc consists of:*

- An anulus fibrosus—an outer fibrous part—composed of concentric lamellae of fibrocartilage
- A gelatinous central mass—the nucleus pulposus.

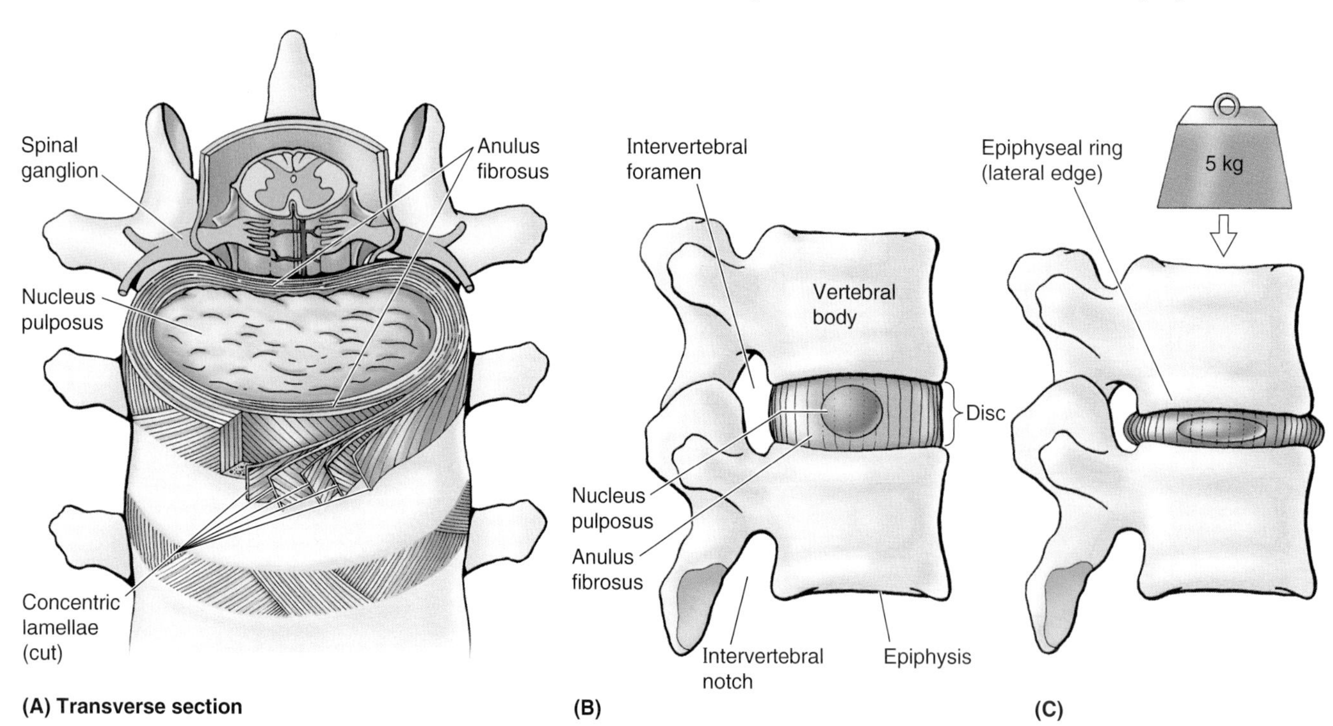

Figure 4.13. Structure of an intervertebral (IV) disc. A. The disc consists of a nucleus pulposus and an anulus fibrosus. The anulus is arranged in concentric layers of parallel fibers that crisscross those of the next layer. The superficial layers of the anulus have been cut and spread apart to show the direction of the fibers. Note that the thickness of the rings of the anulus is diminished posteriorly; that is, the anulus is thinner posteriorly. **B.** The nucleus pulposus—containing fibrogelatinous pulp—occupies the center of the disc and acts as a cushion and shock-absorbing mechanism. **C.** Note that the nucleus flattens and the anulus bulges when weight is applied, as occurs during standing and more so during lifting.

The IV disc is interposed between the bodies of adjacent vertebrae.

The **anulus fibrosus** (Fig. 4.13) is a ring consisting of concentric lamellae of fibrocartilage forming the circumference of the IV disc. The anuli insert into the smooth, rounded epiphyseal rings on the articular surfaces of the vertebral bodies. The fibers forming each lamella run obliquely from one vertebra to another, the fibers of one lamella typically running at right angles to those of the adjacent ones. This arrangement, although allowing some movement between adjacent vertebrae, provides a strong bond between them.

The **nucleus pulposus** (L. pulpa, fleshy) is the central core of the IV disc. It is more cartilaginous than fibrous and is normally highly elastic. The nucleus pulposus is located more posteriorly than centrally and has a high water content that is maximal at birth and decreases with advancing age. It acts like a shock absorber for axial forces and like a semifluid ball bearing during flexion, extension, rotation, and lateral flexion of the vertebral column. It becomes broader when compressed (Fig. 4.13*C*). Because the lamellae of the anulus fibrosus are thinner and less numerous posteriorly than they are anteriorly or laterally, the nucleus pulposus is not centered in the disc but is more posteriorly placed. *The nucleus pulposus is avascular.* It receives its nourishment by diffusion from blood vessels at the periphery of the anulus fibrosus and vertebral body.

There is no IV disc between C1 and C2 vertebrae. The most inferior functional disc is between L5 and S1 vertebrae. The discs vary in thickness in different regions; they are thickest in the lumbar region and thinnest in the superior thoracic region. The discs are thicker anteriorly in the cervical and lumbar regions and more uniform in thickness in the thoracic region.

Uncovertebral "joints" (of Luschka) are between the uncinate processes of C3 through C6 vertebrae and the beveled surfaces of the vertebral bodies superior to them (Fig. 4.14). The "joints" (fissures) are at the lateral and posterolateral margins of the IV discs. These jointlike structures are covered with cartilage and contain a capsule filled with fluid. They are considered to be synovial joints by some; others consider them to be degenerative spaces in the discs that are filled with extracellular fluid. The uncovertebral "joints" are frequent sites of spur formation (projecting processes of bone) that may cause neck pain.

The **anterior longitudinal ligament** is a strong, broad fibrous band that covers and connects the anterolateral aspects of the vertebral bodies and IV discs (Figs. 4.15 and 4.16). The ligament extends from the pelvic surface of the sacrum to the anterior tubercle of C1 and the occipital bone anterior to the foramen magnum. This ligament maintains stability of the joints between the vertebral bodies and helps prevent hyperextension of the vertebral column.

The **posterior longitudinal ligament** is a much narrower, somewhat weaker band than the anterior longitudinal ligament (Figs. 4.15 and 4.16). *The posterior longitudinal ligament runs within the vertebral canal along the posterior aspect of the vertebral bodies.* It is attached to the IV discs and the posterior edges of the vertebral bodies from C2 to the sacrum. This ligament helps prevent hyperflexion of the vertebral column and herniation or posterior protrusion of the discs. It is well provided with nociceptive (pain) nerve endings.

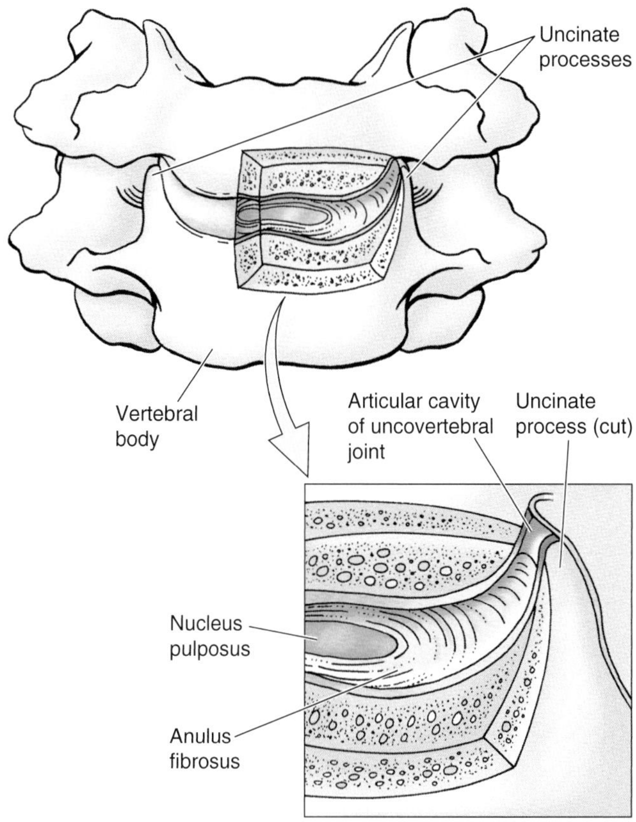

Figure 4.14. Uncovertebral joints. These small, synovial jointlike structures are between the uncinate processes of the lower vertebrae and the beveled surfaces of the vertebral bodies superior to them. These joints are at the posterolateral margins of the intervertebral (IV) discs.

Herniation of the Nucleus Pulposus

Herniation or protrusion of the nucleus pulposus into or through the anulus fibrosus is a well-recognized cause of low back pain (*A*). *Approximately 95% of lumbar disc protrusions occur at the L4/5 or L5/S1 levels.* The IV discs in young persons are strong, and the water content of their nuclei pulposi is high (up to 90%), giving them great turgor (fullness). In young adults, the IV discs are usually so strong that the vertebrae often fracture during a fall before the discs rupture. However, violent hyperflexion of the vertebral column may rupture an IV disc and fracture the adjacent vertebral bodies. ▶

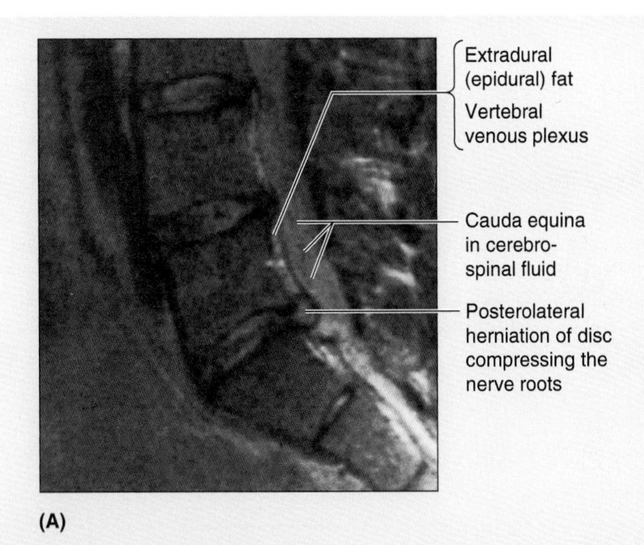

(A)

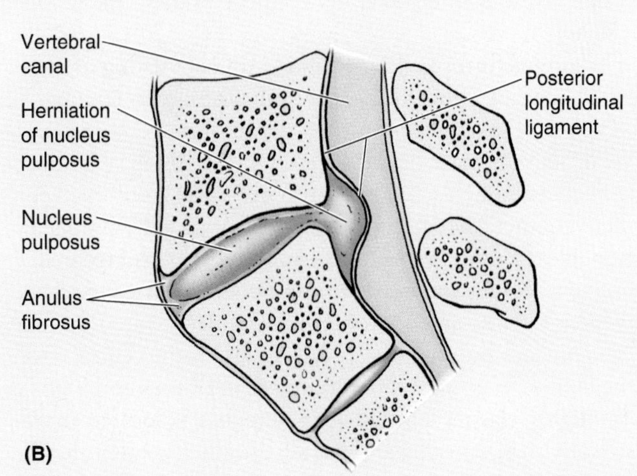

(B)

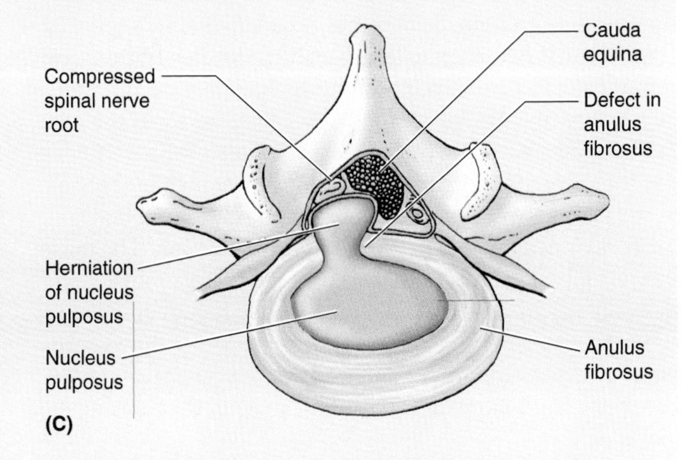

(C)

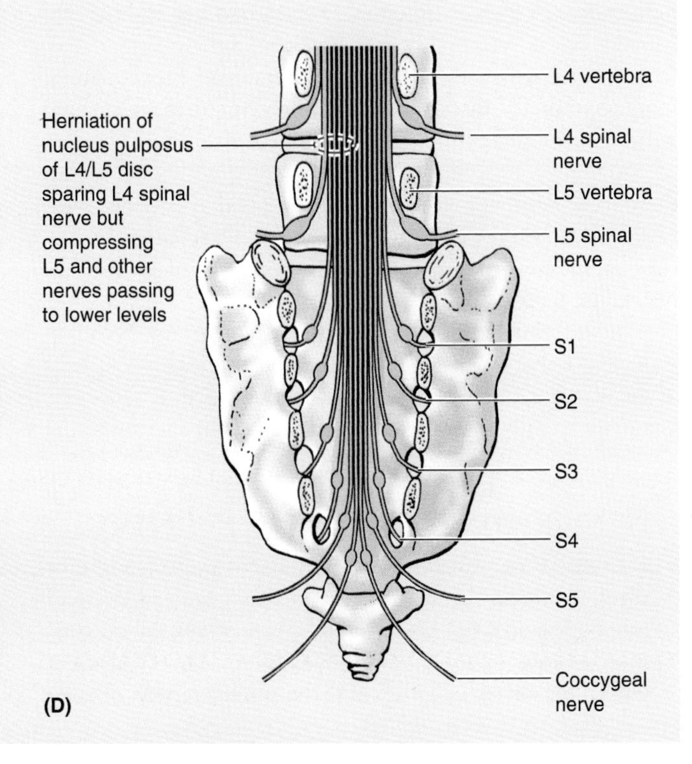

(D)

▶ As people get older, their nuclei pulposi lose their turgor and become thinner because of dehydration and degeneration. These age changes in the IV discs account in part for the slight loss in height that occurs during old age. Decrease in the height of an IV disc also results in narrowing of the IV foramina, which may cause compression of the spinal nerve roots.

Flexion of the vertebral column produces compression anteriorly and stretching or tension posteriorly, pushing the nucleus pulposus further posteriorly toward the thinnest part of the anulus fibrosus. If degeneration of the posterior longitudinal ligament and wearing of the anulus fibrosus has occurred, the nucleus pulposus may herniate into the vertebral canal and compress the spinal cord or the nerve roots of the cauda equina (*B* and *C*). A herniated IV disc is often inappropriately called a "slipped disc" by some people. Sports announcers often call the injury a "ruptured disc."

Protrusions of the nucleus pulposus usually occur posterolaterally, where the anulus fibrosus is relatively thin and poorly supported by either the posterior or anterior longitudinal ligaments. A posterolateral herniated IV disc is more likely to be symptomatic because of the proximity of the spinal nerve roots. The *localized back pain* of a herniated disc results from pressure on the longitudinal ligaments and periphery of the anulus fibrosus and from local inflammation resulting from chemical irritation by substances from the ruptured nucleus pulposus. *Chronic pain* resulting from the spinal nerve roots being compressed by the herniated disc is referred to the area (dermatome) supplied by that nerve. Because the IV discs are largest in the lumbar and lumbosacral regions, where movements are consequently greater, posterolateral herniations of the nucleus pulposus are most common here.

Lumbago—an acute mid and low back pain radiating down the posterolateral aspect of the thigh and leg—is often caused by a posterolateral protrusion of a lumbar ▶

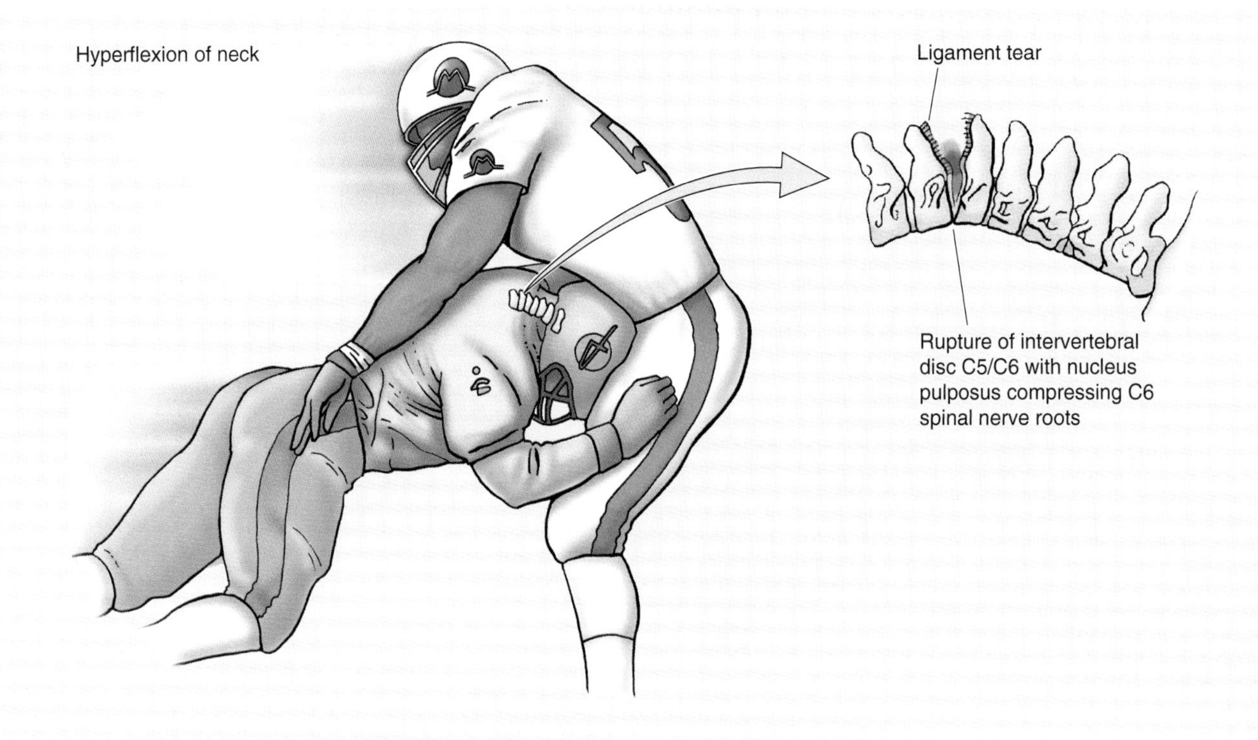

▶ IV disc at the L5/S1 level that affects the S1 component of the sciatic nerve (*C*). The clinical picture varies considerably, but pain of acute onset in the lower back is a common presenting symptom. Because muscle spasm is associated with low back pain, the lumbar region of the vertebral column becomes rigid, and movement is painful. With treatment, the lumbago type of back pain usually begins to fade after a few days; however, it may gradually be replaced by sciatica.

Sciatica is pain in the lower back and hip radiating down the back of the thigh into the leg. Sciatica is caused by a herniated lumbar IV disc that compresses and compromises the L5 or S1 nerve root. The IV foramina in the lumbar region decrease in size and the lumbar nerves increase in size. This phenomenon explains the commonness of sciatica. New bone deposited during *osteoarthritis* narrows the foramina even more, causing shooting pains down the lower limbs. Any maneuver that stretches the sciatic nerve, such as flexing the thigh with the leg extended, may produce or exacerbate (but in some individuals relieves) the pain caused by disc herniation. IV discs may also be damaged by violent rotation or flexing of the vertebral column. The general rule is that when an IV disc protrudes, it may compress the nerve roots numbered one inferior to the disc; for example, L5 nerve is compressed by an L4/L5 IV disc herniation and S1 nerve by a L5/S1 IV disc herniation. Recall that the IV disc forms the inferior half of the anterior border of the IV foramen and that the superior half is formed by the bone of the body of the superior vertebra (Fig. 4.12). The spinal nerve roots descend to the IV foramen from which the spinal nerve formed by their merging will exit. The nerve that exits a given IV foramen passes through the superior bony half of the foramen and thus lies above and is not affected by a herniating disc at that level. However, the nerve roots passing to the IV foramen immediately and further below pass directly across that area of herniation.

Symptom-producing IV disc protrusions occur in the cervical region almost as often as in the lumbar region. As degenerative changes occur, the cervical IV discs thin out and the uncinate processes approach the beveled inferior surfaces of the cervical vertebrae superiorly. This results in encroachment of the IV foramina, pressure on the nerve roots, and neck pain.

A forcible **hyperflexion of the cervical region** during a head-on collision or during illegal head blocking in football, for example, may rupture the IV disc posteriorly without fracturing the vertebral body. The cervical IV discs most commonly ruptured are those between C5/C6 and C6/C7, compressing spinal nerve roots C6 and C7, respectively. IV disc protrusions result in pain in the neck, shoulder, arm, and hand. Any sport in which movement causes downward or twisting pressure on the neck or lower back may produce herniation of the nucleus pulposus. The most common sports involved are bowling, tennis, jogging, football, hockey, weight lifting, and gymnastics.

Hyperextension of the neck, which may occur during head butting in football (*A*), may cause a crush or compression fracture of posterior vertebral arch elements. The *anterior longitudinal ligament* is severely stretched and may be ▶

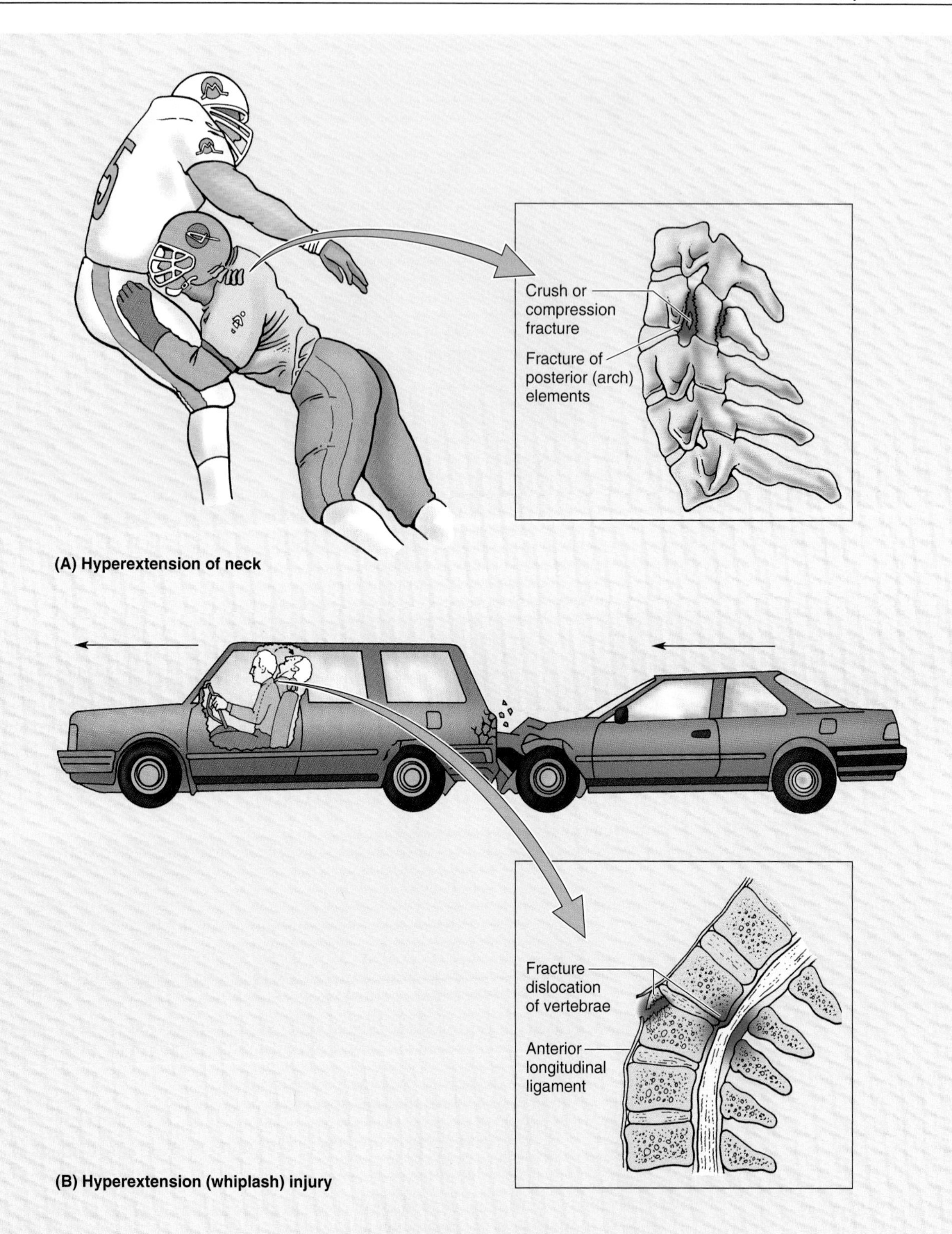

(A) Hyperextension of neck

(B) Hyperextension (whiplash) injury

▶ torn during severe hyperextension of the neck, as occurs during a hyperextension ("whiplash") injury of the neck (*B*). The association of rear-end automobile collision and these injuries is well known, especially to litigation lawyers. *Hyperflexion injury* of the vertebral column may also occur as the head snaps forward onto the thorax. "Facet jumping" or locking of the cervical vertebrae may occur because of dislocation of the vertebral arches. ○

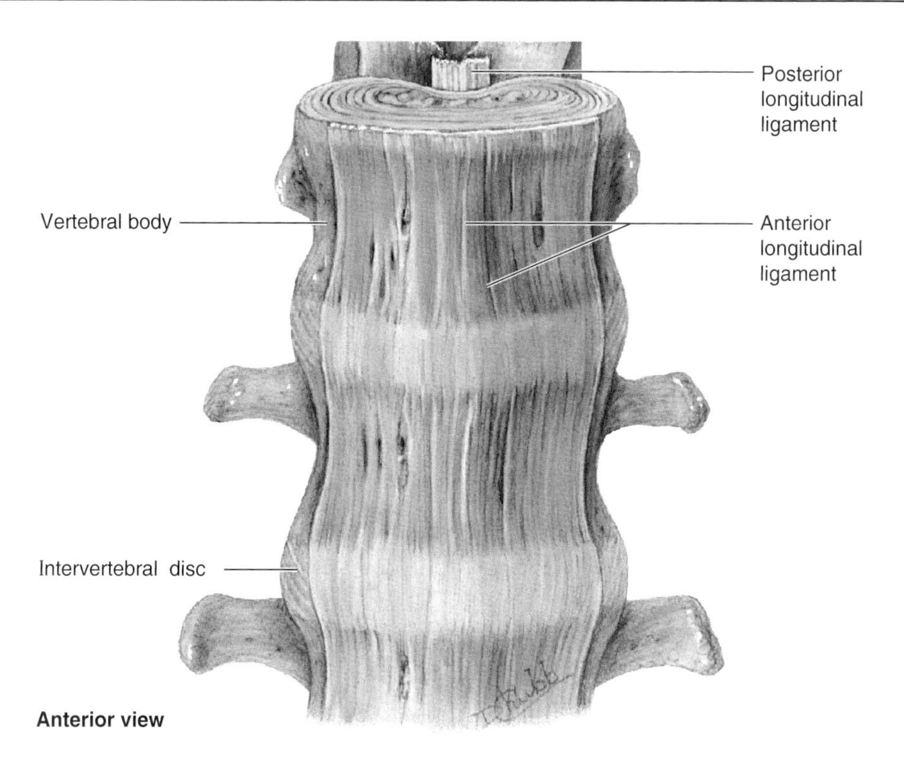

Figure 4.15. Intervertebral (IV) discs and longitudinal ligaments. Parts of the thoracic (T12) and lumbar (L1, L2) regions are shown. Note the broad anterior longitudinal ligament and the narrow posterior longitudinal ligament. The anterior longitudinal ligament is a strong, fibrous band that is attached to the IV discs and adjacent anterolateral parts of the vertebral bodies. Although it is thickest on the anterior aspect of the vertebral bodies (and often only this thickest part is illustrated), it covers both the anterior and lateral aspects of the bodies to the IV foramen, blending between vertebrae with the outermost part of the anulus fibrosus of the IV discs. Observe the foramina in the ligament for arteries and veins passing to and from the vertebral bodies.

Joints of the Vertebral Arches

The joints of the vertebral arches are the **zygapophysial joints** (often called *facet joints* for brevity). These articulations are plane synovial joints between the superior and inferior articular processes (G. zygapophyses) of adjacent vertebrae (Figs. 4.16 and 4.17). Each joint is surrounded by a thin, loose **articular capsule.** Those in the cervical region are especially thin and loose. The capsule is attached to the margins of the articular processes of adjacent vertebrae. Accessory ligaments unite the laminae, transverse processes, and spinous processes and help to stabilize the joints.

The zygapophysial joints permit gliding movements between the vertebrae; the shape and disposition of the articular surfaces determines the type of movement possible. The range (amount) of movement is determined by the size of the IV disc relative to that of the vertebral body. In the cervical and lumbar regions, the zygapophysial joints bear some weight, sharing this function with the IV discs. The zygapophysial joints are innervated by articular branches that arise from the medial branches of the dorsal primary rami of spinal nerves (Fig. 4.16*B*). As these nerves pass posteroinferiorly, they lie in grooves on the posterior surfaces of the medial parts of the transverse processes. Each articular branch supplies two adjacent joints; therefore, each joint is supplied by two nerves.

Injury and Disease of Zygapophysial Joints

The zygapophysial joints are of clinical interest because they are close to the IV foramina through which the spinal nerves emerge from the vertebral canal. When these joints are injured or diseased from *osteoarthritis*, the related spinal nerves are often affected. This causes pain along the distribution patterns of the *dermatomes* and spasm in the muscles derived from the associated *myotomes* (a myotome = all the muscles or parts of muscles receiving innervation from one spinal nerve). The illustration of the lumbar region of the vertebral column shows the innervation of the zygapophysial joints (Fig. 4.16*B*).

Denervation of lumbar zygapophyseal joints is a procedure that may be used for treatment of back pain caused by disease of these joints. The nerves are sectioned near the joints or are destroyed by radiofrequency *percutaneous rhizolysis* (G. rhiza, root + lysis, dissolution). In each procedure, the denervation process is directed at the articular branches of two adjacent dorsal primary rami of the spinal nerves because each joint receives innervation from both the nerve exiting that level and the suprajacent nerve. ⭘

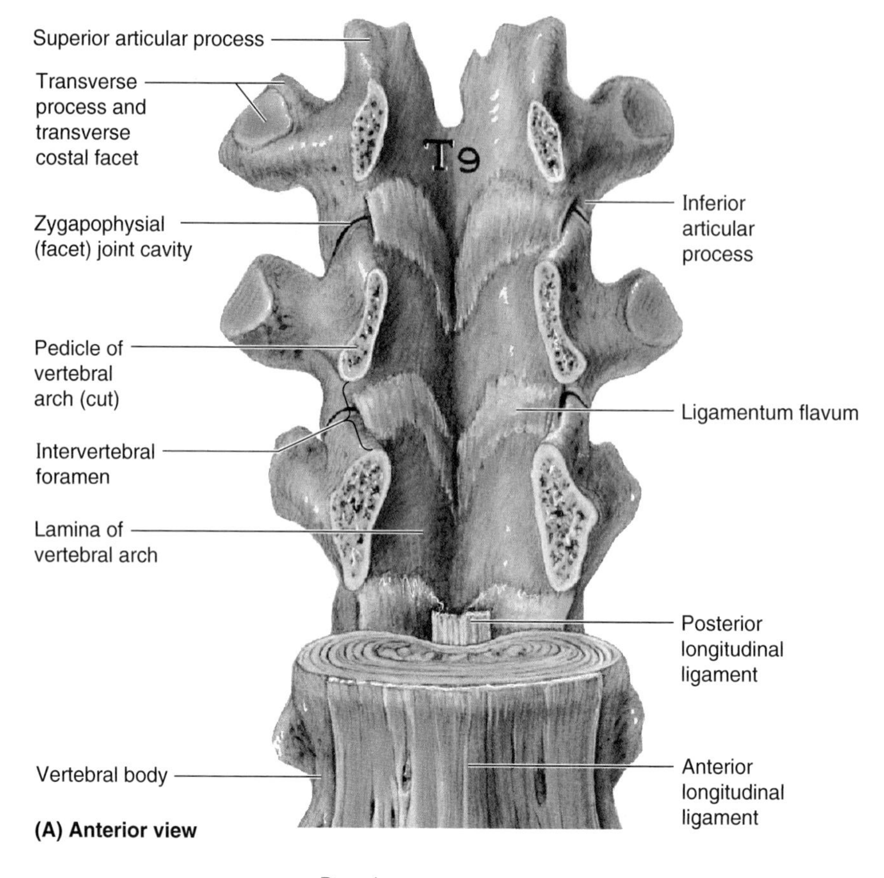

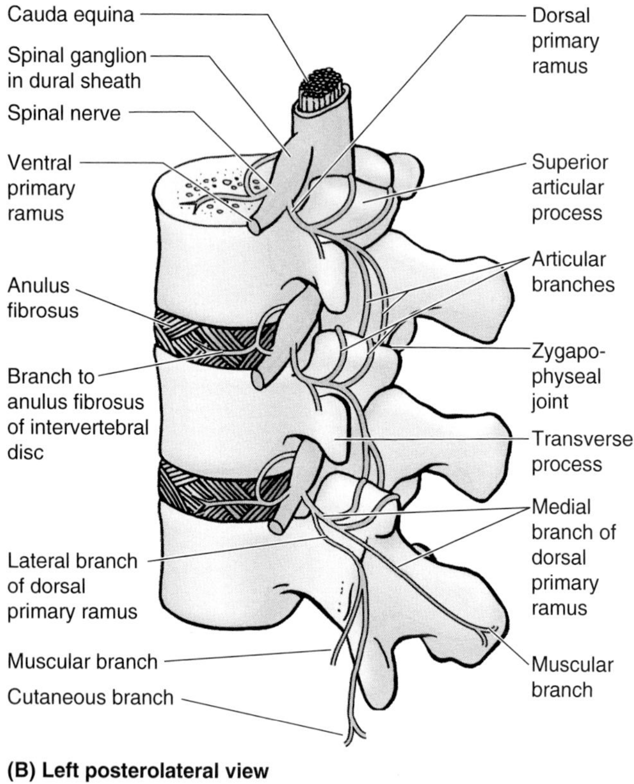

Figure 4.16. Zygapophysial (facet) joints, longitudinal ligaments, ligamenta flava, and innervation of the joints. The zygapophysial articulations are between the articular processes (G. zygapophyses) of the vertebrae. **A.** The pedicles of T9 through T11 vertebrae have been sawn through and their bodies removed to provide this anterior view of the posterior wall of the vertebral (spinal) canal formed by the ligamenta flava composed of yellow elastic fibers, extending between adjacent vertebral laminae. The ligaments of opposite sides meet and blend in the median plane. Between the adjacent left or right pedicles, the inferior and superior articular processes and the zygapophysial joints between them (from which joint capsules have been removed) and the lateralmost extent of the ligamenta flava form the posterior boundaries of intervertebral (IV) foramina. **B.** Observe that the dorsal primary ramus arises from the spinal nerve outside the IV foramen and divides into medial and lateral branches, which supply the zygapophysial joints.

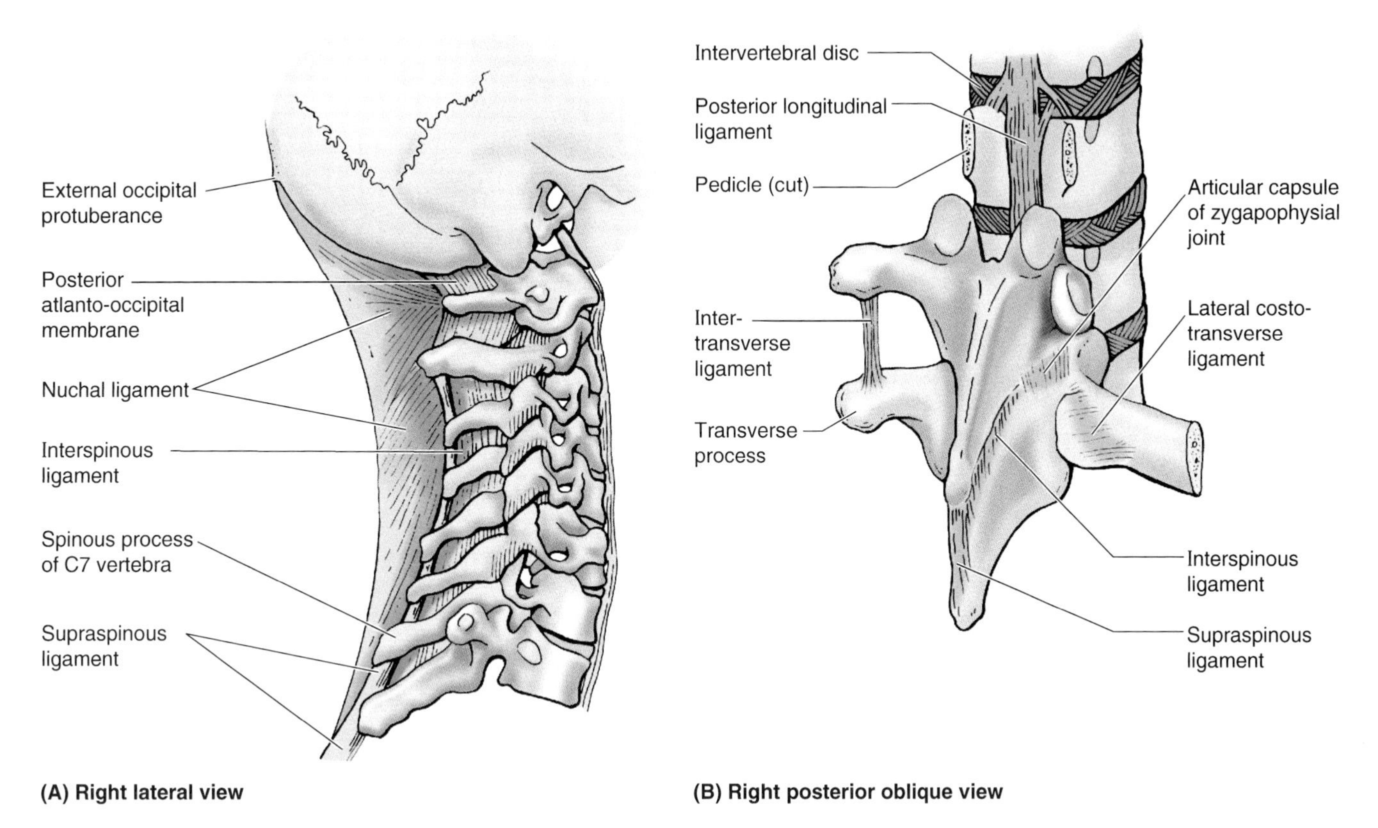

Figure 4.17. Joints and ligaments of the vertebral column. A. Right lateral view of the ligaments in the cervical region. Note the prominent spinous process of C7—the vertebra prominens—and the nuchal ligament (L. ligamentum nuchae). **B.** Posterolateral view of the ligaments in the thoracic region. The pedicles of the upper two vertebrae have been sawn through and the vertebral (neural) arches removed to reveal the posterior longitudinal ligament. Note the intertransverse, supraspinous, and interspinous ligaments associated with the vertebrae with intact vertebral arches.

Accessory Ligaments of the Intervertebral Joints

The laminae of adjacent vertebral arches are joined by broad, yellow elastic fibrous tissue—the **ligamenta flava** (L. flavus, yellow). These *yellow ligaments* extend almost vertically from the lamina above to the lamina below (Fig. 4.16). The ligaments bind the lamina of the adjoining vertebrae together, forming part of the posterior wall of the vertebral canal. The ligamenta flava are long, thin, and broad in the cervical region; thicker in the thoracic region; and thickest in the lumbar region. These ligaments prevent separation of the vertebral lamina, thereby arresting abrupt flexion of the vertebral column and usually preventing injury to the IV discs. The strong elastic ligamenta flava help to preserve the normal curvatures of the vertebral column and assist with straightening of the column after flexing.

Adjoining spinous processes are united by weak, almost membranous interspinous ligaments and strong fibrous supraspinous ligaments (Fig. 4.17). The thin **interspinous ligaments** connect adjoining spinous processes, attaching from the root to the apex of each process. The cordlike **supraspinous ligament**—connecting the apices of the spinous processes from C7 to the sacrum—merges superiorly with the nuchal ligament (L. ligamentum nuchae), the broad, strong median ligament of the neck (Fr. nuque). The **nuchal ligament**—distinct from the interspinous and supraspinous ligaments—is composed of thickened fibroelastic tissue, extending from the external occipital protuberance and posterior border of the foramen magnum to the spinous processes of the cervical vertebrae. Because of the shortness of the C3 through C5 spinous processes, the nuchal ligament substitutes for bone in providing muscular attachments. The **intertransverse ligaments**, connecting adjacent transverse processes, consist of scattered fibers in the cervical region and fibrous cords in the thoracic region (Fig. 4.17, *right*). The intertransverse ligaments in the lumbar region are thin and membranous.

Craniovertebral Joints

There are two craniovertebral joints:

- Atlanto-occipital joint—between the atlas (C1 vertebra) and the occipital bone of the skull
- Atlantoaxial joint—between the atlas and axis (C2 vertebra).

The Greek word *atlanto* refers to the *atlas*. The craniovertebral joints are synovial joints that have no IV discs. They are designed to give a wider range of movement than in the rest

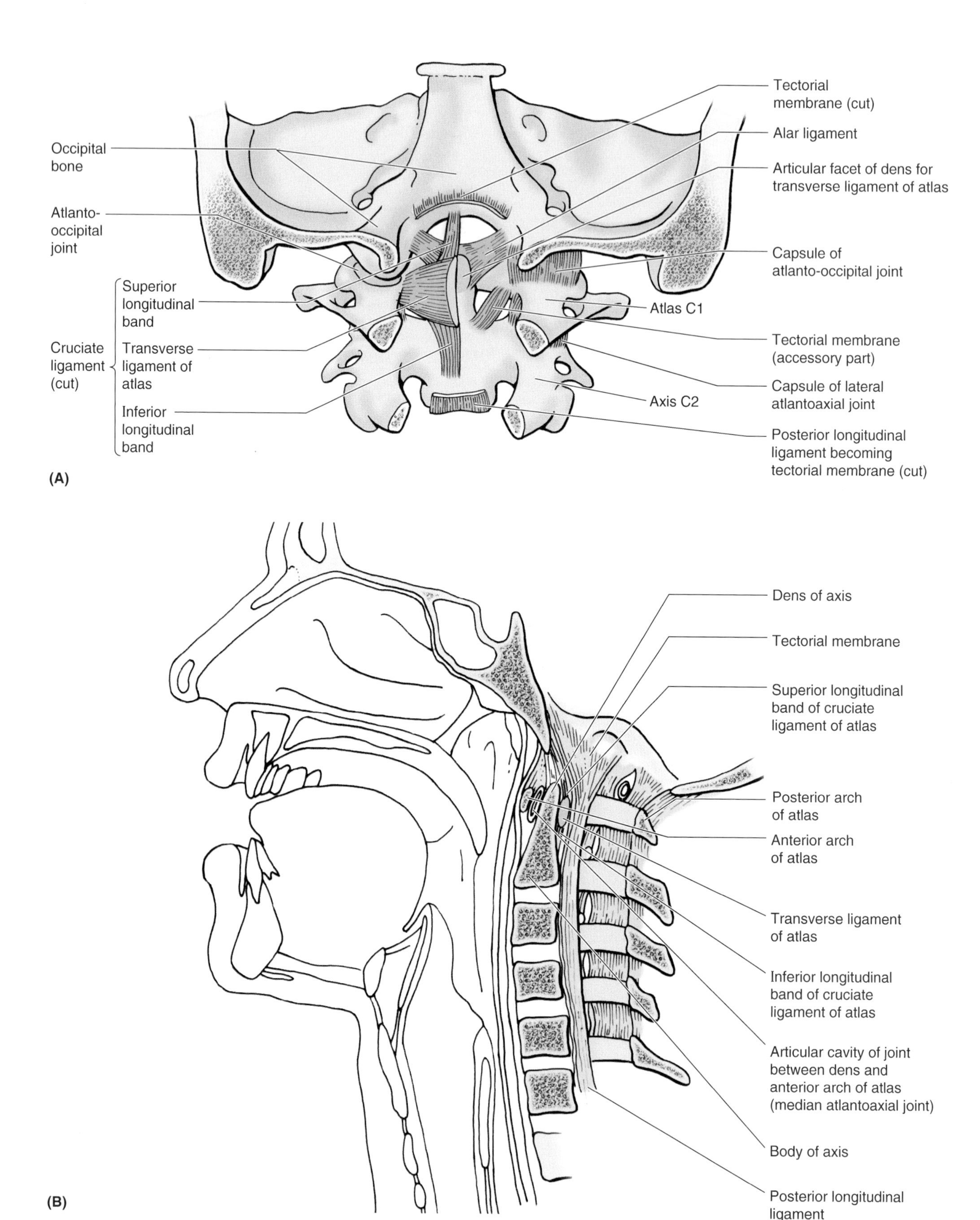

Figure 4.18. Craniovertebral joints and ligaments. A. Posterior view. The tectorial membrane and the transverse ligament of the atlas (C1) have been cut to show the attachment of the alar ligament to the dens of C2 (axis) vertebra. **B.** Median section of the craniovertebral region and neck, showing the joints and ligaments. Observe that the tectorial membrane is the superior continuation of the posterior longitudinal ligament and is attached to (spans between) the bodies of C2 and C3 vertebrae and the superior surface of the basilar part of the occipital bone.

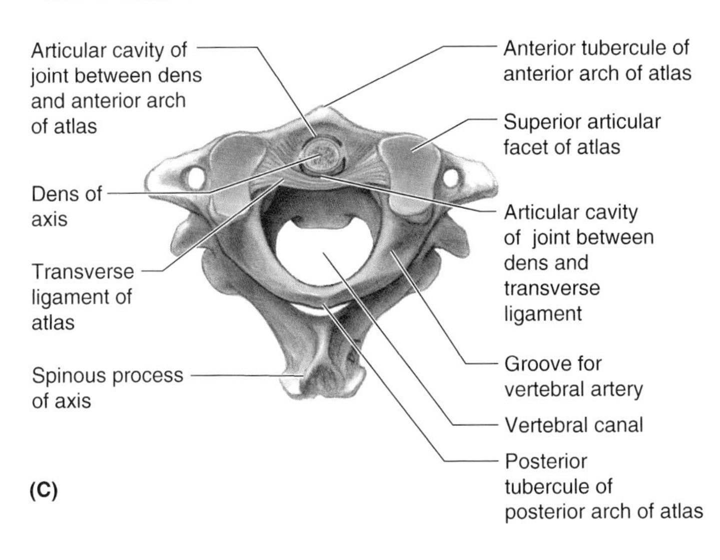

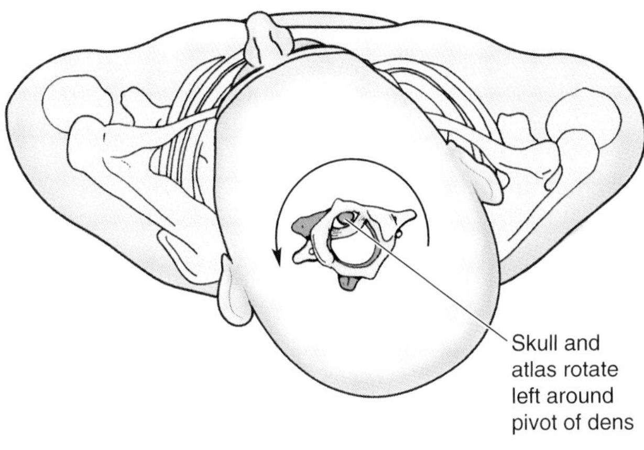

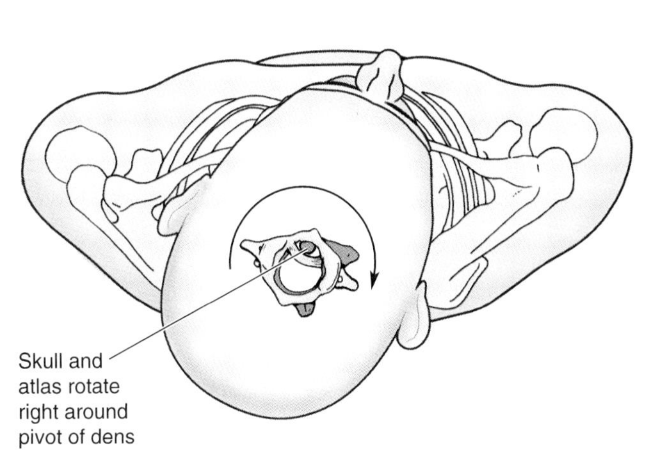

Figure 4.18. *(Continued)* **C.** Superior view of the median atlantoaxial joint. It is formed by the anterior arch and the transverse ligament of the atlas and the dens of the axis. **D.** These drawings illustrate how the skull and atlas rotate around the pivot of the dens when the head is turned.

of the vertebral column. The articulations involve the occipital condyles, atlas, and axis.

Atlanto-Occipital Joints. These articulations between the lateral masses of atlas and the occipital condyles (Fig. 4.18) permit nodding of the head, such as the neck flexion and extension occurring when indicating approval (the "yes" movement). These joints also permit sideways tilting of the head. The main movement is flexion, with a little lateral bending and rotation. They are *synovial joints of the condyloid type* and have thin, loose articular capsules composed of fibrous capsules lined by synovial membranes. The skull and C1 are also connected by anterior and posterior **atlanto-occipital membranes**, which extend from the anterior and posterior arches of C1 to the anterior and posterior margins of the foramen magnum (Fig. 4.19). The anterior membranes are composed of broad, densely woven fibers; the posterior

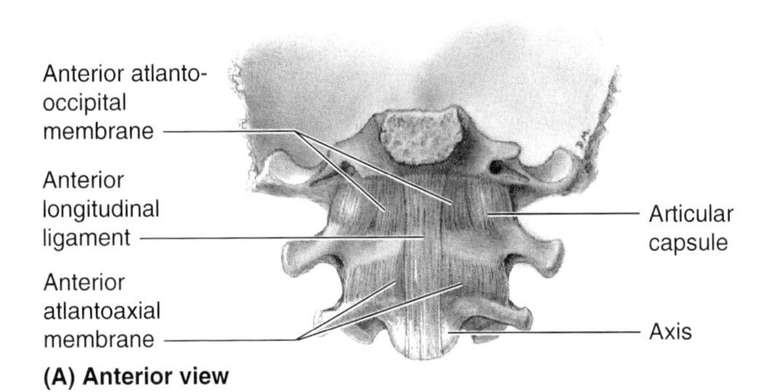

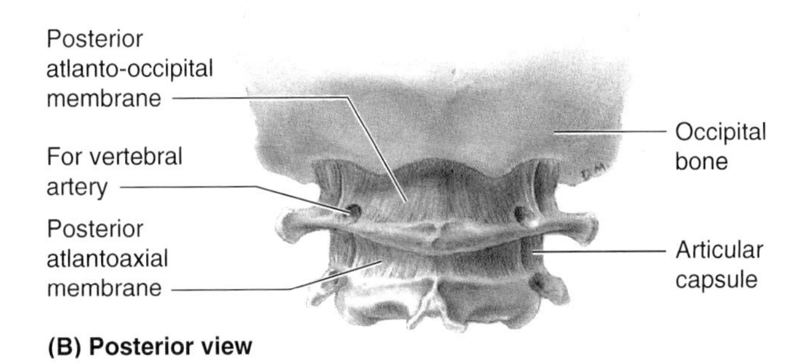

Figure 4.19. Ligaments of the craniovertebral joints. A. Observe the thicker, anteriormost part of the anterior longitudinal ligament blending in the midline with the anterior atlantoaxial membrane and the anterior atlanto-occipital membrane and laterally with the articular capsules of the zygapophysial (facet) joints. **B.** Observe the posterior atlanto-occipital membrane lying between the foramen magnum and the superior surface of the posterior arch of the atlas (C1). Note the opening for the vertebral arteries which, having passed through the transverse foramina of the atlas, pierce the atlanto-occipital membrane before traversing the foramen magnum. Observe also the posterior atlantoaxial membrane lying between the inferior surface of the posterior arch of the atlas and the laminae of the axis (C2).

membranes are broad but relatively weak. The atlanto-occipital membranes help to prevent excessive movement of these joints.

The **transverse ligament of the atlas** (Fig. 4.18, *A–C*) is a strong band extending between the tubercles on the medial aspects of the lateral masses of C1 vertebrae. It holds the dens of C2 against the anterior arch of C1, forming the posterior wall of a socket for the dens. Vertically oriented superior and inferior **longitudinal bands** pass from the transverse ligament to the occipital bone superiorly and to the body of C2 inferiorly. Together, the transverse ligament and the longitudinal bands form the **cruciate ligament** (formerly the cruciform ligament), so named because of its resemblance to a cross.

The **alar ligaments** extend from the sides of the dens to the lateral margins of the foramen magnum (Fig. 4.18*A*). These short, rounded cords, approximately one-half centimeter in diameter—just smaller than a pencil—attach the skull to C1 vertebra and check rotation (side-to-side movements) of the head when it is turned.

The **tectorial membrane** (Fig. 4.18, *A* and *B*) is the strong superior continuation of the posterior longitudinal ligament across the central atlantoaxial joint through the foramen magnum to the central floor of the cranial cavity. It runs from the body of C2 to the internal surface of the occipital bone and covers the alar and transverse ligaments.

Atlantoaxial Joints. There are three atlantoaxial articulations (Fig. 4.18, *A–C*):

- Two lateral atlantoaxial joints
- One median atlantoaxial joint.

These synovial joints are between the inferior facets of the lateral masses of C1 and the superior facets of C2 and between the dens of C2 and the anterior arch of the atlas.

Movement (mainly rotation) at all three atlantoaxial joints permits the head to be turned from side to side (Fig. 4.18, *C* and *D*), as occurs when rotating the head to indicate disapproval (the "no" movement). During this movement, the skull and C1 rotate on C2 as a unit. Excessive rotation at these joints is prevented by the alar ligaments. During rotation of the head, the dens of C2 is the axis or pivot that is held in a socket or collar formed by the anterior arch of the atlas and the transverse ligament of the atlas (Fig. 4.18*C*). The articulation of the dens of C2 with C1 (central atlantoaxial joint) is thus described as a pivot joint, whereas the C1/C2 zygapophyseal joints (lateral atlantoaxial joints) are gliding-type synovial joints.

Fracture of the Dens

The transverse ligament of the atlas is stronger than the dens of C2 vertebra, which usually fractures at its base. Often the fractures do not reunite because of the interposition of the transverse ligament of the atlas (Crockard et al., 1993). Other dens fractures result from abnormal ossification patterns.

Rupture of the Transverse Ligament of the Atlas

When the transverse ligament of the atlas ruptures or is weakened by disease, the dens is set free, resulting in *atlantoaxial subluxation*—incomplete dislocation of the atlantoaxial joint. Pathological softening of the transverse and adjacent ligaments—usually resulting from disorders of connective tissue—may also cause atlantoaxial subluxation (Bogduk and Macintosh, 1984). When complete dislocation occurs, the dens may be driven into the upper cervical region of the spinal cord, causing paralysis of all four limbs (*quadriplegia*), or into the medulla of the brainstem, causing death. Sometimes inflammation in the craniovertebral area may produce softening of the ligaments of the craniovertebral joints and cause dislocation of the atlantoaxial joints. Sudden movement of a patient from a bed to a chair, for example, may produce posterior displacement of the dens and injury to the spinal cord.

Rupture of the Alar Ligaments

The alar ligaments are weaker than the transverse ligament of the atlas. Consequently, combined flexion and rotation of the head may tear one or both alar ligaments. Rupture of an alar ligament results in an increase of approximately 30% in the range of movement to the opposite side (Dvorak et al., 1988).

Ossification of the Posterior Atlanto-Occipital Membrane

The inferior edge of the posterior atlanto-occipital membrane, which bridges the groove for the vertebral artery, may become ossified and convert the groove into a foramen for the vertebral artery and the 1st cervical nerve. This is a familial and genetic trait.

Compression of C2 Spinal Ganglion

Although uncommon, atlantoaxial rotation may compress the 2nd cervical spinal nerve. When the neck is severely hyperextended while the head is turned to the side, the spinal ganglion of C2 nerve on the opposite side may be compressed between C1 and C2 vertebrae. This may be followed by prolonged headaches and cervico-occipital pains that are so severe that suicidal tendencies develop. ○

Movements of the Vertebral Column

The range of movement of the vertebral column varies according to the region and the individual. Movements are extraordinary in some people, such as acrobats who begin to train during early childhood. The mobility of the vertebral column results primarily from the compressibility and elasticity of the IV discs. The following movements of the vertebral column are possible: flexion, extension, lateral bending, and rotation (torsion). The range of movement of the vertebral column is limited by the:

- Thickness, elasticity, and compressibility of the IV discs
- Shape and orientation of the zygapophysial joints
- Tension of the articular capsules of the zygapophysial joints
- Resistance of the back muscles and ligaments (such as the ligamenta flava and the posterior longitudinal ligament).

The back muscles producing movements of the vertebral column are discussed subsequently, but the movements are not produced exclusively by the back muscles. They are assisted by gravity and the action of the anterolateral abdominal muscles (see Chapter 2).

Movements between adjacent vertebrae take place on the resilient nuclei pulposi of the IV discs and at the zygapophysial joints. The orientation of the latter joints permits some movements and restricts others. In the thoracic region, for example, the oblique orientation of the zygapophysial joints, in which extrapolation from the articular surfaces (i.e., from the joint planes) describes an arc centered on the vertebral body, allows some rotation and lateral bending but prevents flexion and posterior sliding of the vertebrae. Although movements between adjacent vertebrae are relatively small, especially in the thoracic region, the summation of all the small movements produces a considerable range of movement of the vertebral column as a whole (e.g., when bending to touch the toes).

Movements of the vertebral column are freer in the cervical and lumbar regions than elsewhere (Fig. 4.20, *A–C*). Flexion, extension, lateral bending, and rotation of the neck are especially free because the:

- IV discs, although thin relative to most other discs, are thick relative to the size of the vertebral bodies at this level
- Articular surfaces of the zygapophysial joints are relatively large and the joint planes are almost horizontal
- Articular capsules of the zygapophysial joints are loose
- Neck is slender (with less surrounding soft tissue bulk).

Flexion of the vertebral column is greatest in the cervical region (Fig. 4.20*A*).

The sagittally oriented joint planes of the lumbar regions are conducive to flexion and extension. *Extension of the vertebral column is most marked in the lumbar region* and is usually more extensive than flexion; however, the interlocking articular processes here prevent rotation. The lumbar region, like the cervical region, has IV discs that are large (the largest discs occur here) relative to the size of the vertebral bodies. *Lateral*

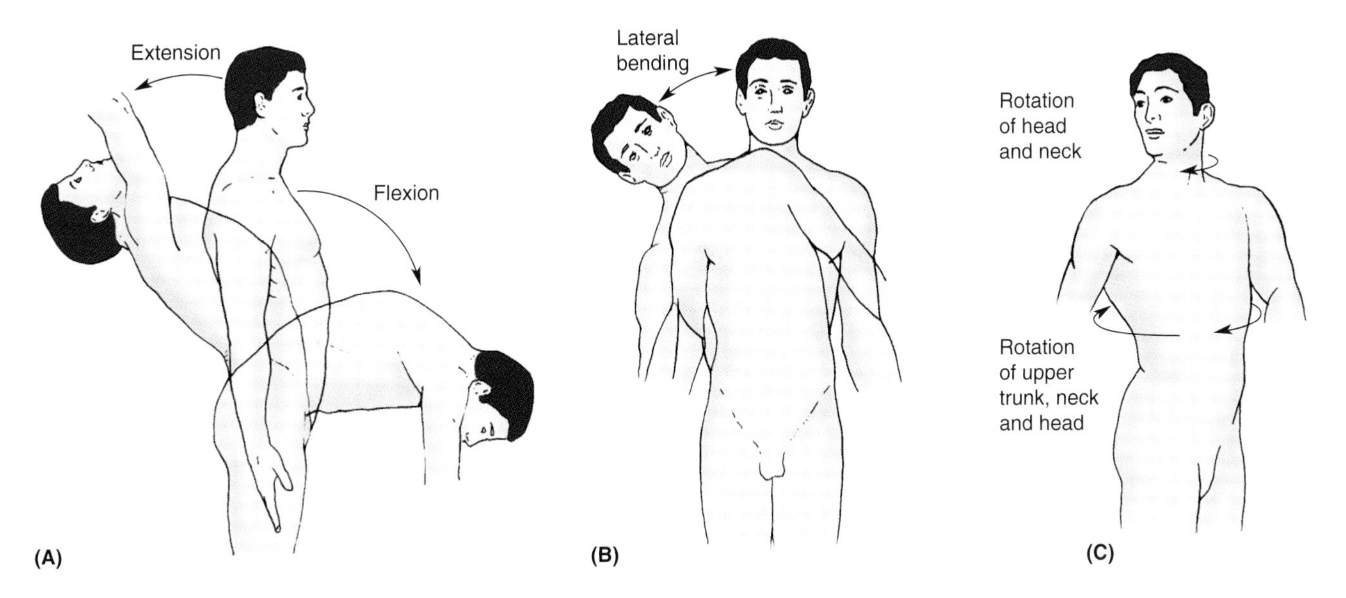

Figure 4.20. Movements of the vertebral column. The movements are flexion (forward bending) and extension (backward bending), both in the median plane (**A**) (flexion and extension are occurring primarily in the cervical and lumbar regions); lateral bending (to the right or left in a coronal plane), also occurring mostly in the cervical and lumbar regions (**B**); rotation around a longitudinal axis, which occurs primarily at the craniovertebral joints (augmented by the cervical region) and the thoracic region (**C**).

Fractures and Dislocations of the Vertebrae

Fractures, dislocations, and fracture-dislocations of the vertebral column usually result from sudden forceful flexion, as occurs in automobile accidents or from a violent blow to the back of the head. *The common injury is a crush or compression fracture of the body of one or more vertebrae.* If violent anterior movement of the vertebra occurs in addition to compression, a vertebra may be displaced anteriorly on the vertebra inferior to it, such as dislocation of C6 or C7 vertebrae. Usually this dislocates and fractures the articular facets between the two vertebrae and ruptures the interspinous ligaments. Irreparable injuries to the spinal cord accompany most severe flexion injuries of the vertebral column.

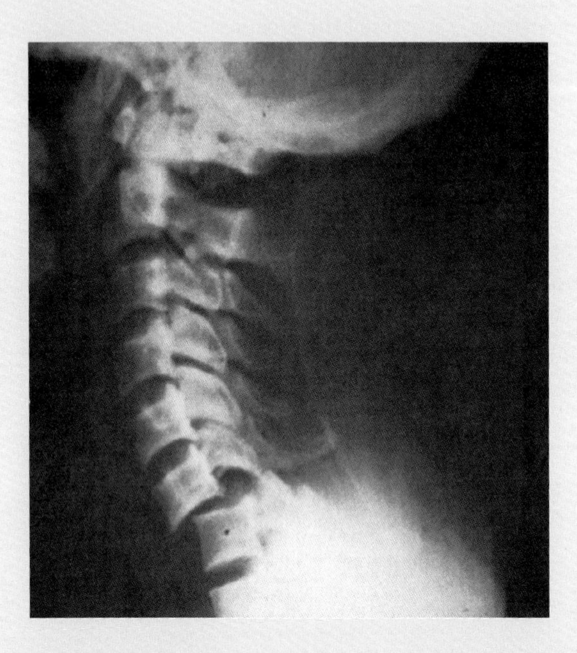

Sudden, forceful extension can also injure the vertebral column and spinal cord. Illegal face blocking in football may lead to an hyperextension injury of the neck (p. 454). Hyperextension of the neck occurs during rear motor vehicle collisions. Severe hyperextension is most likely to injure posterior parts of the vertebrae—the vertebral arches and their processes. *Fractures of cervical vertebrae may radiate pain to the back of the neck and scapular region* because the same spinal sensory ganglia and spinal cord segments receiving pain impulses from the vertebrae are also involved in supplying the levator scapulae, rhomboid, and deep neck muscles(see Fig. 4.23).

Severe hyperextension of the neck may pinch the posterior arch of C1 between the occipital bone and C2. In these cases the atlas usually breaks at one or both grooves for the vertebral arteries. The anterior longitudinal ligament and the adjacent anulus fibrosus of the C2/C3 IV disc may also rupture. If this occurs, the skull, C1, and C2 are separated from the rest of the axial skeleton, and the spinal cord is usually severed. Persons with this severe injury seldom survive.

Dislocation of vertebrae in the thoracic and lumbar regions is uncommon because of the interlocking of their articular processes. Slight dislocation of cervical vertebrae may not damage the spinal cord because the vertebral canal in this region is usually larger than the spinal cord; however, considerable vertebral displacement may cause serious damage to the spinal cord. Football, diving, falls from horses, and motor vehicle accidents cause most fractures of ▶

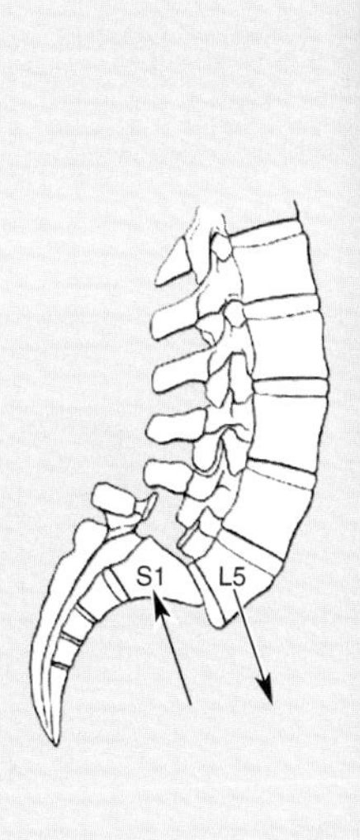

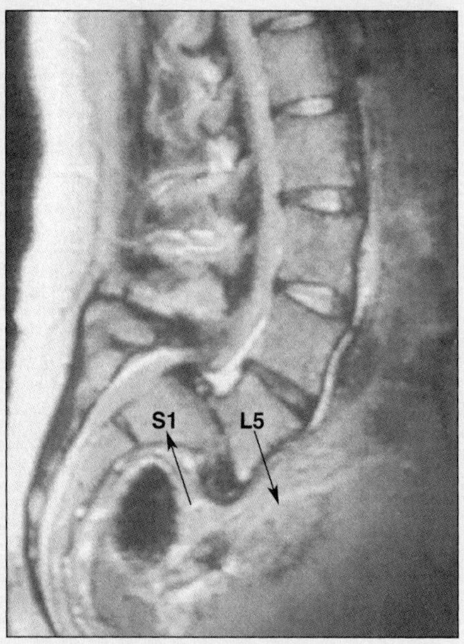

bending of the vertebral column is greatest in the cervical and lumbar regions (Fig. 4.20*B*).

The thoracic region, in contrast, has IV discs that are thin relative to the size of the vertebral bodies. Relative stability is also conferred on this part of the vertebral column through its connection to the sternum by the ribs and costal cartilages. The joint planes here lie on an arc that is centered on the vertebral body, permitting rotation in the thoracic region (Fig. 4.20*C*). This rotation of the upper trunk, in combination with the rotation permitted in the cervical region and that at the atlantoaxial joints, enables the torsion of the axial skeleton that occurs as one looks back over the shoulder. However, flexion is almost nonexistent in the thoracic region, and lateral bending is severely restricted.

▶ the cervical region of the vertebral column. Symptoms range from vague aches to progressive loss of motor and sensory functions.

The transition from the relatively inflexible thoracic region to the much more mobile lumbar region occurs abruptly, unfortunately. Consequently, T11 or T12 are the most commonly fractured noncervical vertebrae (i.e., a "broken back" vs. a "broken neck").

Fracture of the interarticular parts of the vertebral laminae of L5 (spondylolysis) may result in forward displacement of the L5 vertebral body relative to the sacrum (S1 vertebra—spondylolisthesis). The posterior fragment, consisting of most of the vertebral arch, remains in normal relation to the sacrum, but the anterior fragment and the L5 vertebral body may move anteriorly. It is the anterior displacement of most of the vertebral column that constitutes spondylolisthesis. *Spondylolisthesis at the L5/S1 articulation* may result in pressure on the spinal nerves of the cauda equina as they pass into the superior part of the sacrum, causing back and lower limb pain. ○

Surface Anatomy of the Vertebral Column

Because of the cervical curvature, **C7 spinous process** is the only one that is usually evident superficially in the cervical region, hence the name **vertebra prominens** (*large arrow* in photograph on p. 464). The spinous process of C7 may be made more prominent by flexion of the neck. The *spinous process of C2* is the first bony point that can be felt in the midline inferior to the *external occipital protuberance,* a median projection located a little inferior to the bulging part of the posterior aspect of the head at the junction of the head and neck. C1 has no spinous process, and its small posterior tubercle is neither visible nor palpable.

The short bifid spinous processes of C3 through C5 vertebrae may be felt in the *nuchal groove* between the neck muscles, but they are not easy to palpate because they lie deep to the surface from which they are separated by the nuchal ligament. The *spinous process of C6 vertebra* is easily felt when the neck is flexed because of the length of its process. The tips of the spinous processes are normally in line with each other. A shift in their alignment (*small arrow* in photograph on the next page) may be the result of a unilateral dislocation of a zygapophyseal joint; however, this slight malalignment may have resulted from a fracture of the spinous process of T1 vertebra.

The transverse processes of C1 may be felt by deep palpation between the *mastoid processes*—prominences of the temporal bones posterior to the ears—and the angles of the jaws. The *carotid tubercle,* the anterior tubercle of the transverse process of C6 vertebra, may be large enough to be palpable. The *carotid artery* lies anterior to it, and in an emergency this vessel can be compressed by posterior pressure against the tubercle to reduce bleeding from this artery or its branches (bleeding may continue at a slower rate because of the multiple anastomoses of distal branches with adjacent and contralateral branches).

The prominence between C7 and T2 (*large arrow*) is the most prominent *spinous process* in the thoracic region; *T1 spinous process* may be almost as prominent as the process of the vertebra prominens. The spinous processes of the other thoracic vertebrae may be obvious in thin people and in others can be identified by superior to inferior palpation beginning at the T1 spinous process. The *spinous processes of lumbar vertebrae* are large and easy to observe when the trunk is flexed. They can also be palpated in the posteromedian furrow.

The *S2 spinous process* lies at the middle of a line drawn between the posterosuperior iliac spines, indicated by the skin dimples formed by the attachment of the skin and fascia to these spines. *A horizontal line joining the highest points of the iliac crests passes through the tip of the L4 spinous process and the L4/L5 IV disc.* This is a useful landmark when performing a lumbar puncture to obtain a sample of cerebrospinal fluid (CSF) (p. 482).

The *median crest of the sacrum,* formed by the fusion of the spinous processes of the superior three or four sacral vertebrae, can be felt inferior to the L5 spinous process. The *sacral triangle* is formed by the lines joining the two posterosuperior iliac spines and the superior part of the *intergluteal (natal) cleft* between the buttocks (*F*). The sacral triangle outlining the sacrum is a common area of pain resulting from low back sprains. The *sacral hiatus* can be palpated at the inferior end of the sacrum located in the superior part of the intergluteal cleft (p. 445).

The transverse processes of thoracic and lumbar vertebrae are covered with thick muscles and may or may not be palpable. The *coccyx* can be palpated in the intergluteal cleft, inferior to the sacral triangle. The tip or *apex of the coccyx* can be palpated approximately 2.5 cm posterosuperior to the anus. Clinically, the coccyx is examined with a gloved finger in the anal canal. A painful coccyx (*coccydynia*) is usually the result of a direct blow or a fall on a hard surface. ▶

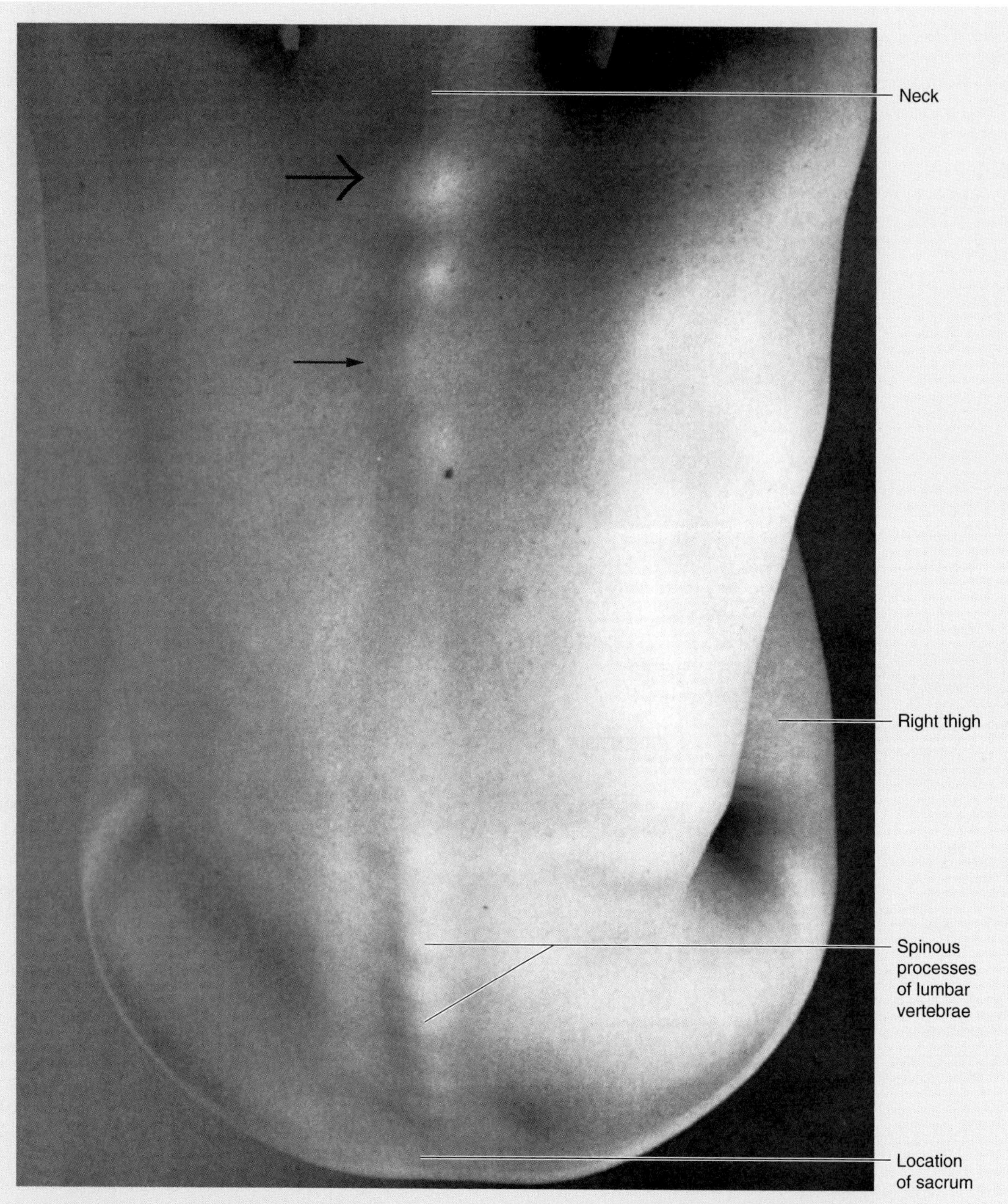

Surface anatomy of the back of a 27-year-old woman. Note the prominence of the spinous process of C7 vertebra—the vertebra prominens (*large arrow*). The spinous process of T2 vertebra (*small arrow*) deviates slightly from the posterior median (midvertebral) line.

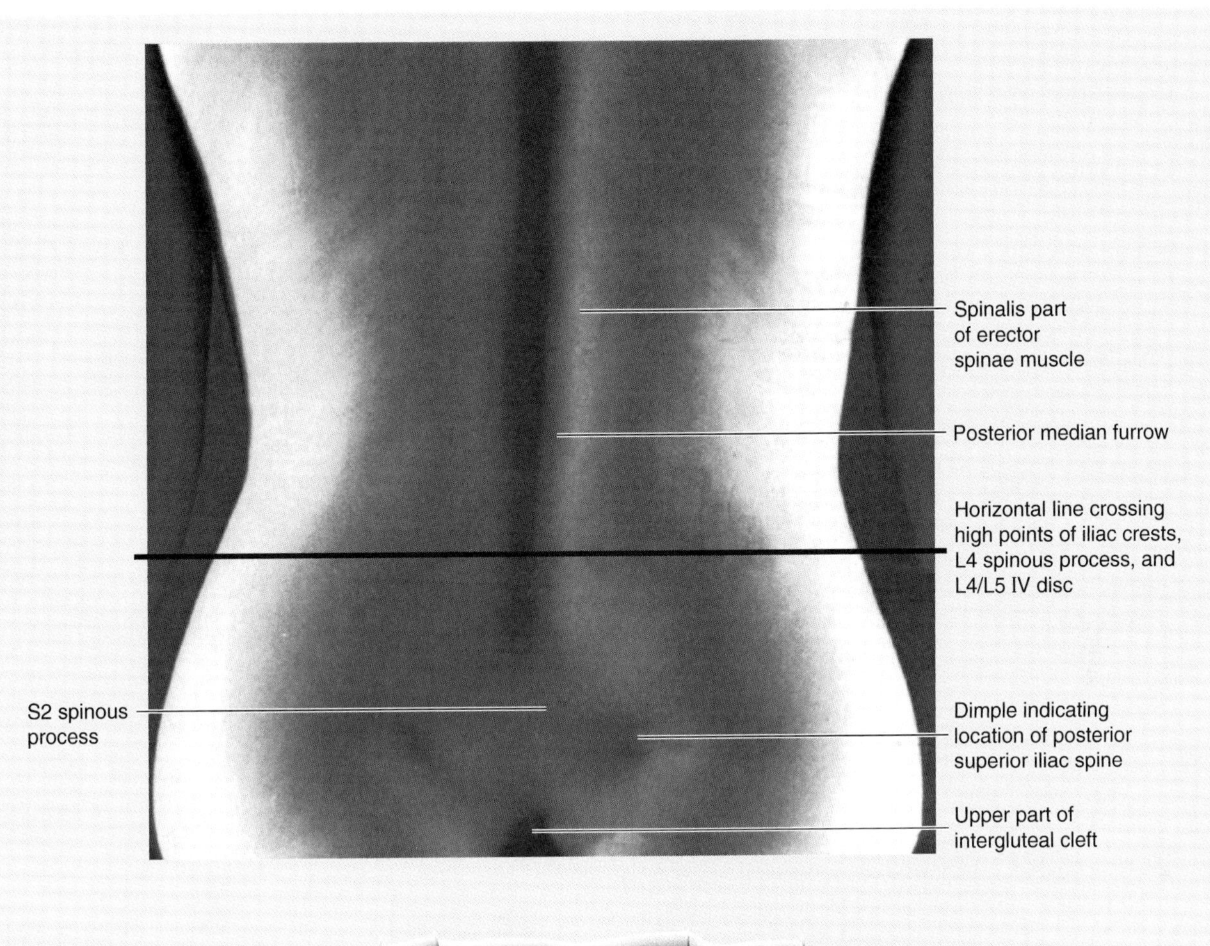
Spinalis part of erector spinae muscle
Posterior median furrow
Horizontal line crossing high points of iliac crests, L4 spinous process, and L4/L5 IV disc
S2 spinous process
Dimple indicating location of posterior superior iliac spine
Upper part of intergluteal cleft

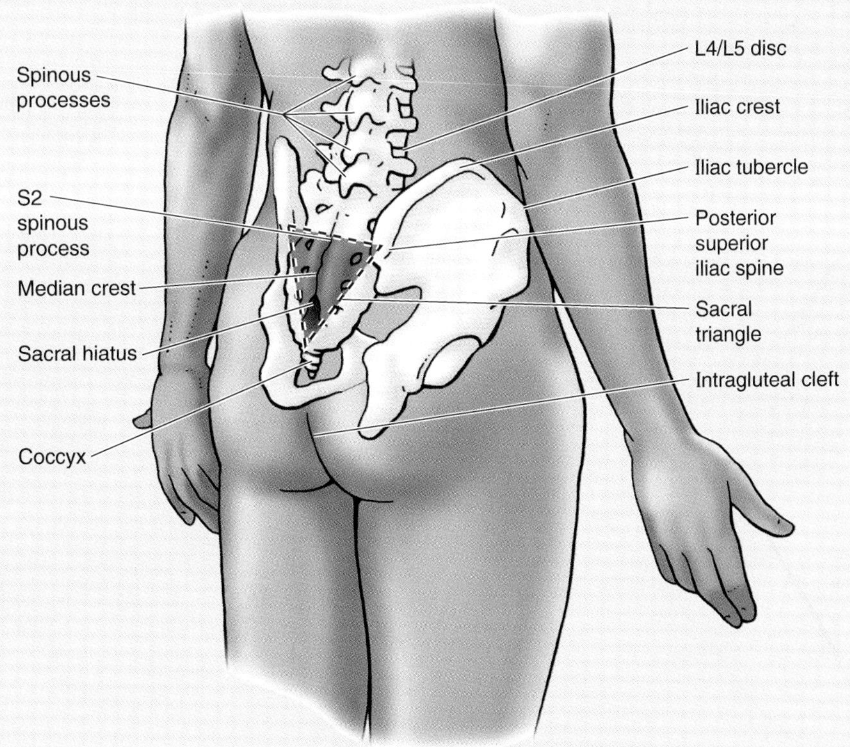
Spinous processes
L4/L5 disc
Iliac crest
Iliac tubercle
S2 spinous process
Posterior superior iliac spine
Median crest
Sacral triangle
Sacral hiatus
Intragluteal cleft
Coccyx

Figure 4.21. Blood supply of a vertebra—viewed superiorly. A typical vertebra is supplied by a segmental artery—here a lumbar artery—that arises from the aorta. Each thoracic and lumbar vertebra is related around its middle to a pair of segmental arteries (intercostal or lumbar). Similar segmental arteries in the cervical region are branches of the vertebral arteries, in the thoracic region are branches of intercostal arteries, and in the sacral region they are branches of the lateral sacral arteries. Observe that the segmental artery supplies twigs to the vertebral body and that dorsal branches supply the spinous process and the back muscles. Note that spinal branches enter the vertebral (spinal) canal through the intervertebral (IV) foramina and supply the bones, periosteum, and ligaments that form the internal aspects of the walls of the vertebral canal. Some spinal branches in the vertebral canal supply the extradural or epidural space.

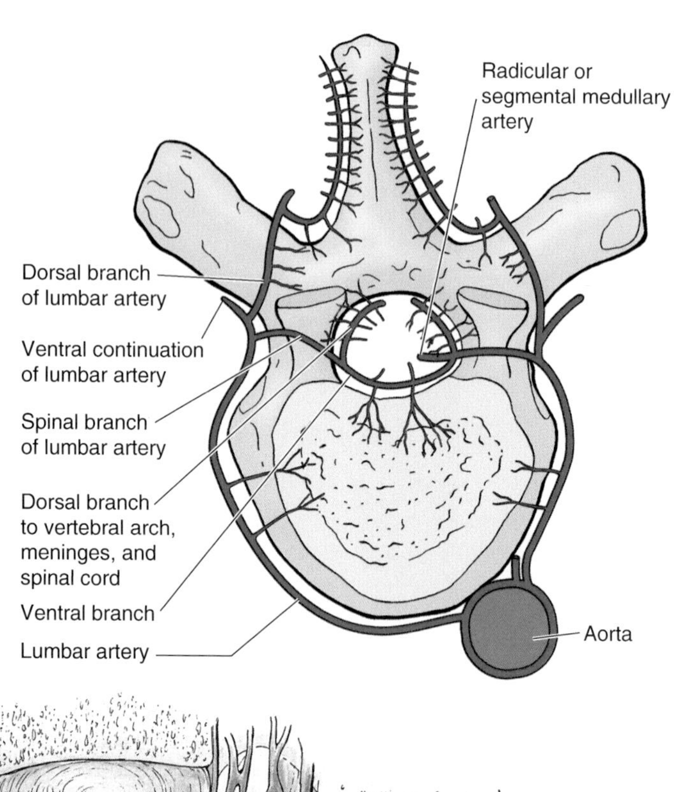

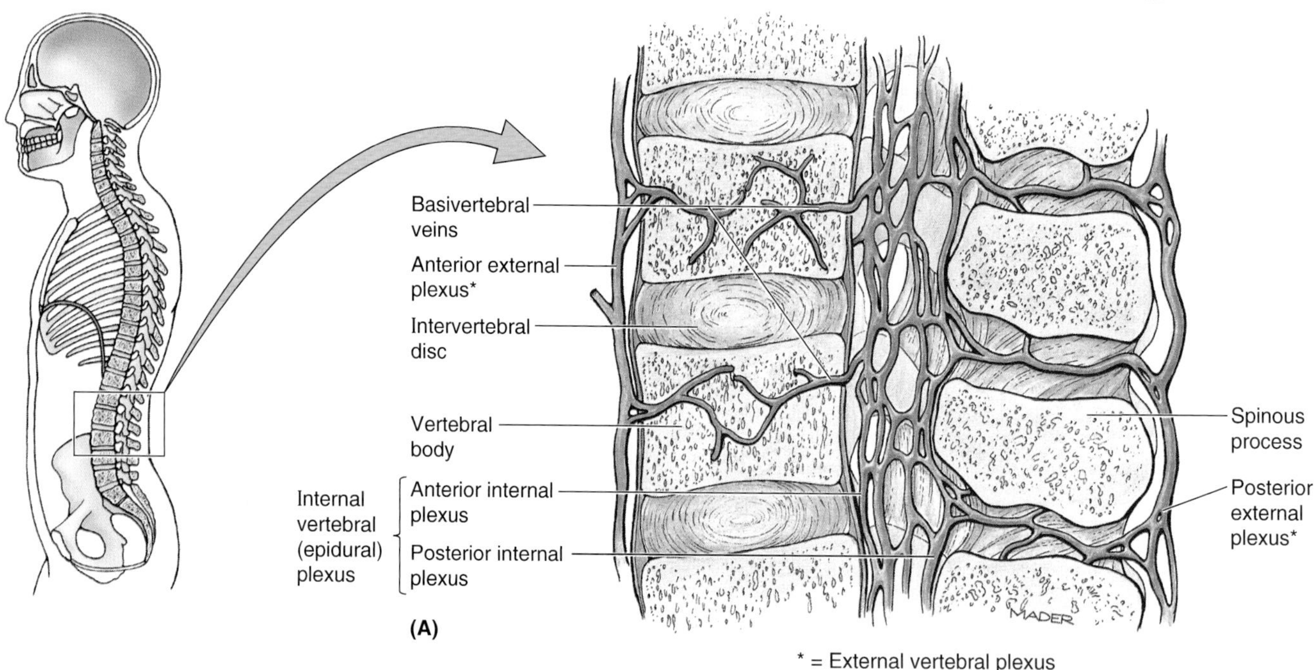

* = External vertebral plexus

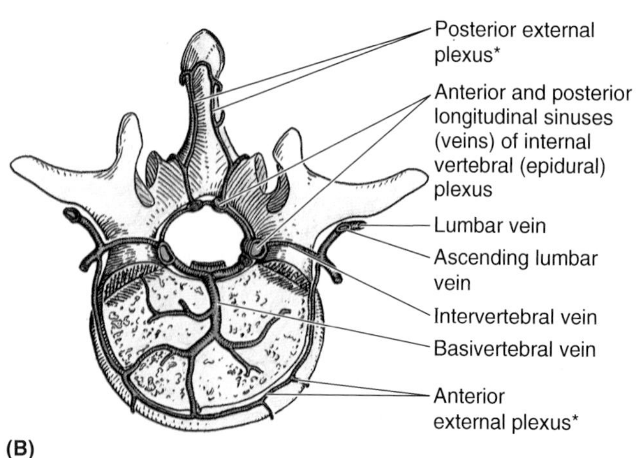

Figure 4.22. Venous drainage of the vertebral column. A. Median section of the vertebral column. The venous drainage parallels the arterial supply and enters the internal vertebral (epidural) venous plexuses that surround the spinal cord. There is also anterolateral drainage from the vertebrae into segmental veins. **B.** Superior view of a lumbar vertebra. The vertebral (spinal) canal contains a dense plexus of thin-walled valveless veins, the internal vertebral venous plexuses, which surround the dura mater. Anterior and posterior longitudinal venous sinuses can be identified in the internal vertebral venous plexus. Superiorly, this plexus communicates through the foramen magnum with the occipital and basilar sinuses of the skull. Veins from the vertebral body—the basivertebral veins—drain primarily into the internal vertebral venous plexus, but they may also drain to the anterior and posterior external plexuses. In the thoracic, lumbar, and sacral regions, the azygos system of veins link the various segments (see Chapter 2).

Vasculature of the Vertebral Column

A typical vertebra is supplied by branches of segmental vessels such as the lumbar arteries, which are closely associated with it (Fig. 4.21). **Spinal arteries** supplying the vertebrae are branches of the:

- Vertebral and ascending cervical arteries in the neck
- Posterior intercostal arteries in the thoracic region
- Subcostal and lumbar arteries in the abdomen
- Iliolumbar and lateral and medial sacral arteries in the pelvis.

Spinal arteries enter the IV foramina and divide mostly into terminal radicular arteries distributed to the dorsal and ventral roots of the spinal nerves and their coverings. Some radicular arteries continue as irregularly spaced medullary segmental arteries that anastomose with arteries of the spinal cord (see Fig. 4.34).

Spinal veins form venous plexuses along the vertebral column both inside and outside the vertebral canal—the **internal vertebral venous plexus** and **external vertebral venous plexuses**, respectively (Fig. 4.22). The large, tortuous **basivertebral veins** are in the substance of the vertebral bodies. They emerge from foramina on the surfaces of the vertebral bodies (mostly the posterior aspect) and drain into the external and especially the internal vertebral venous plexuses. The **IV veins** accompany the spinal nerves through the IV foramina and receive veins from the spinal cord and vertebral plexuses.

Muscles of the Back

Most body weight is anterior to the vertebral column, especially in obese people; consequently, the many strong muscles attached to the spinous and transverse processes are necessary to support and move the vertebral column. *There are three groups of muscles in the back.*

- The superficial and intermediate groups include *extrinsic back muscles* that produce and control limb and respiratory movements, respectively.
- The deep group includes the true or *intrinsic back muscles* that specifically act on the vertebral column, producing its movements and maintaining posture.

Superficial or Extrinsic Back Muscles

The superficial extrinsic back muscles (trapezius, latissimus dorsi, levator scapulae, and rhomboids) connect the upper limbs to the trunk and control limb movements (Fig. 4.23*A*; see also Chapter 6). These muscles, although located in the back region, for the most part receive their nerve supply from the ventral rami of cervical nerves and act on the upper limb. The trapezius receives its motor fibers from a cranial nerve, the accessory nerve (CN XI).

The **intermediate extrinsic back muscles** (serratus posterior) are superficial respiratory muscles and are described with muscles of the thoracic wall (see Chapter 1). The serratus posterior superior lies deep to the rhomboids, and the serratus posterior inferior lies deep to the latissimus dorsi (Fig. 4.24*B*). Both serratus muscles are innervated by intercostal nerves: the superior by the first four intercostals and the inferior by the last four.

Deep or Intrinsic Back Muscles

The deep (true) or intrinsic back muscles are innervated by the dorsal rami of spinal nerves and act to maintain posture and control movements of the vertebral column (Figs. 4.23*B* and 4.24). These muscles—extending from the pelvis to the skull—are enclosed by fascia that attaches medially to the nuchal ligament, the tips of the spinous processes, the supraspinous ligament, and the median crest of the sacrum. The fascia attaches laterally to the cervical and lumbar transverse processes and to the angles of the ribs. The thoracic and lumbar parts of the fascia constitute the **thoracolumbar fascia**, which encloses the deep muscles of the back. It extends laterally from the spinous processes and forms a thin covering for the deep muscles in the thoracic region and a strong thick covering for muscles in the lumbar region (Fig. 4.23*B*). The deep back muscles are grouped according to their relationship to the surface (Table 4.4).

Superficial Layer of Intrinsic Back Muscles

The **splenius muscles** (splenii), thick and flat, lie on the lateral and posterior aspects of the neck, covering the vertical muscles somewhat like a bandage, which explains their name (L. splenion, bandage). The splenii arise from the midline and extend superolaterally to the cervical vertebrae (splenius cervicis) and skull (splenius capitis). The splenii cover and hold the deep neck muscles in position (Fig. 4.24). For more information on their attachments, nerve supply, and actions, see Table 4.4.

Intermediate Layer of Intrinsic Back Muscles

The **erector spinae** (sacrospinalis) lies in a groove on each side of the vertebral column. *The massive erector spinae—the chief extensor of the vertebral column—divides into three columns* (Fig. 4.24, Table 4.4):

- Iliocostalis—lateral column
- Longissimus—intermediate column
- Spinalis—medial column.

Each column is divided regionally into three parts according to its superior attachments (e.g., iliocostalis lumborum, iliocostalis thoracis, and iliocostalis cervicis).

The common origin of the three erector spinae columns is through a broad tendon that attaches inferiorly to the posterior part of the iliac crest, the posterior aspect of the sacrum, the sacroiliac ligaments, and the sacral and inferior lumbar spinous processes. The attachments, nerve supply, and actions of the erector spinae are described in Table 4.4.

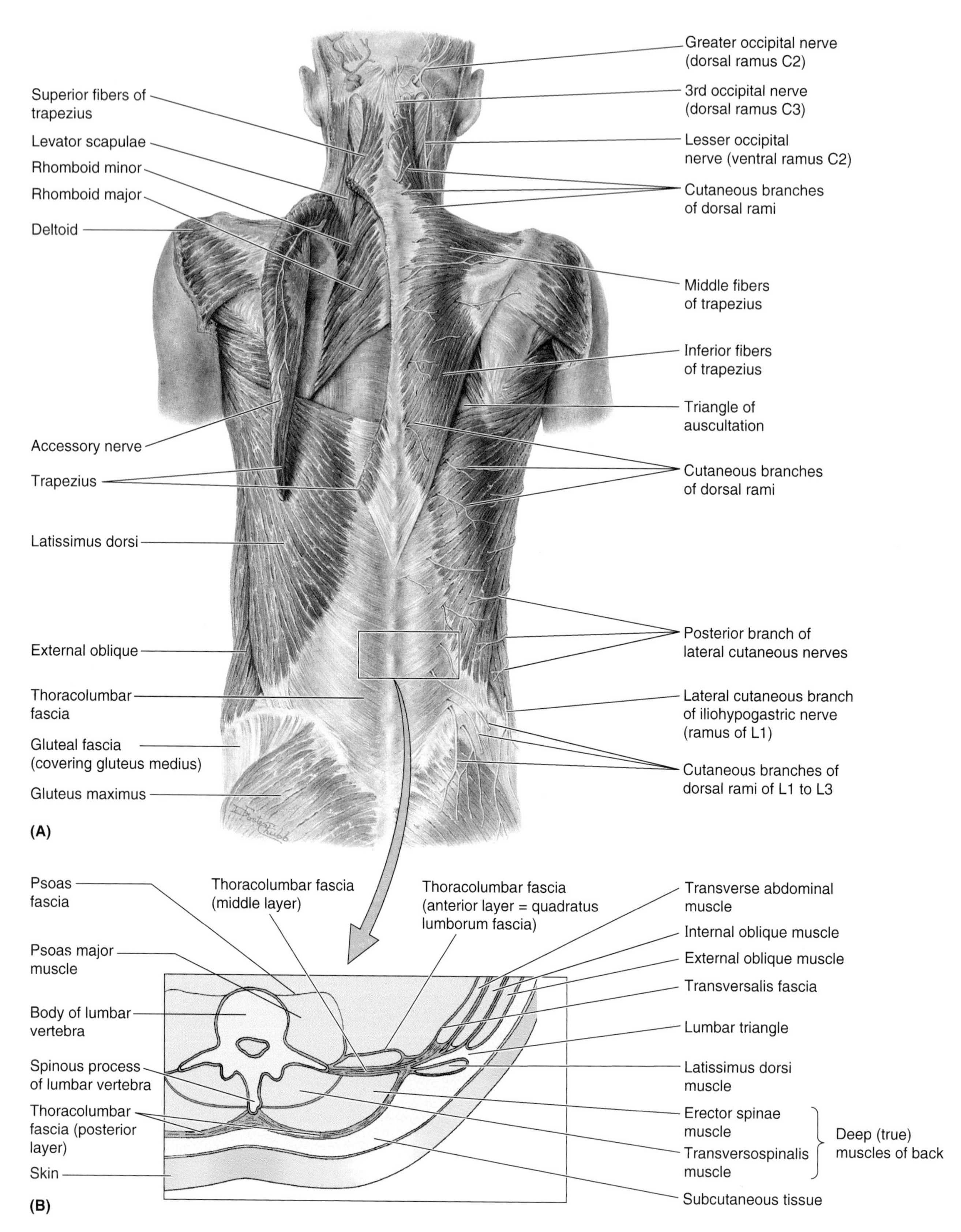

Figure 4.23. Muscles of the back. A. Superficial extrinsic muscles. The trapezius is reflected on the left side to show the accessory nerve (CN XI) coursing on its deep surface, and the levator scapulae and rhomboid muscles. These muscles help to attach the upper limb to the trunk. **B.** Transverse section of part of the back showing the location of the deep (true) back muscles and the layers of fascia associated with them. Observe that the posterior aponeuroses of the transverse abdominal and internal oblique muscles split into two strong sheets—the middle and posterior layers of the thoracolumbar fascia—which enclose the deep muscles.

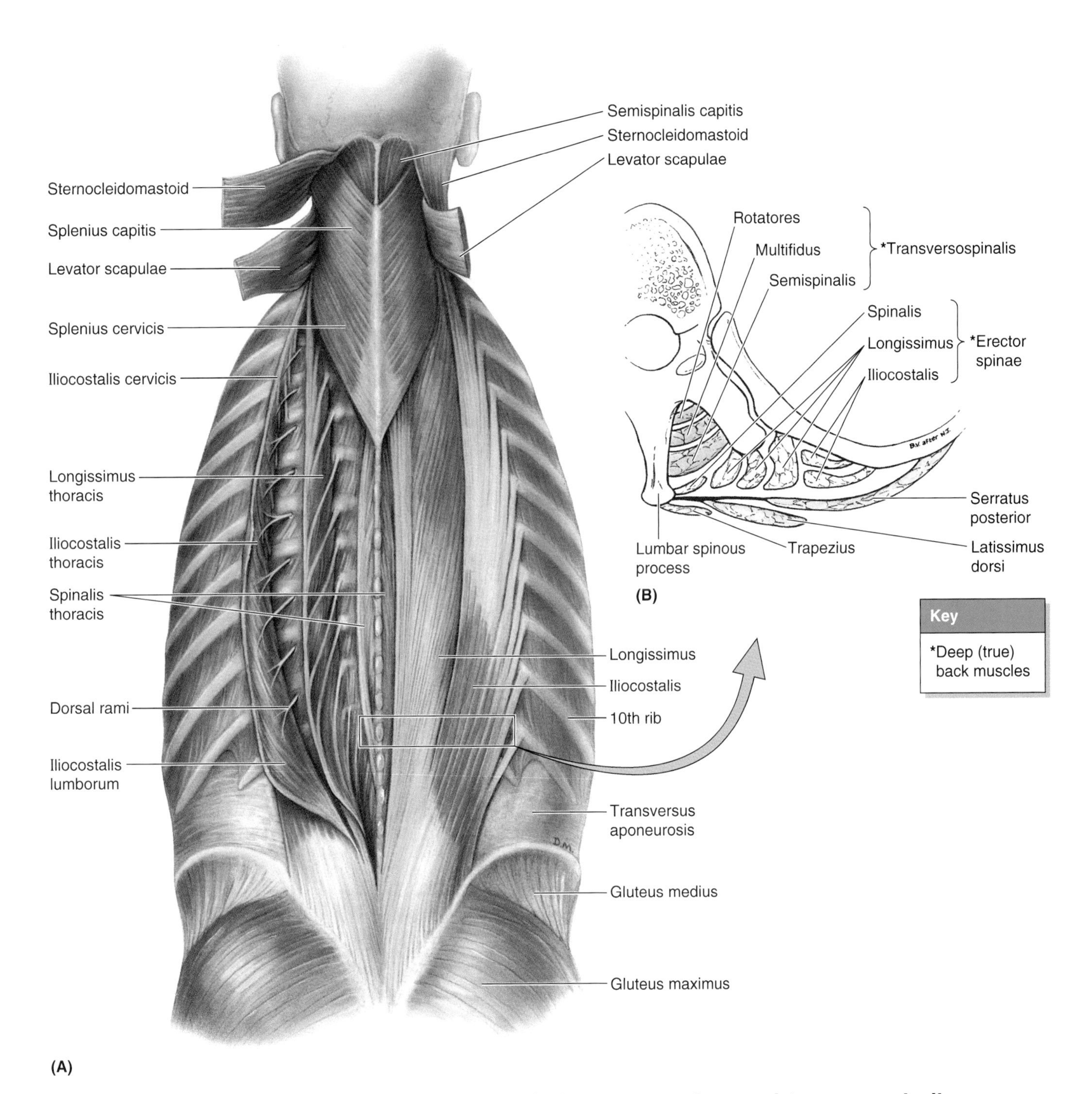

Figure 4.24. Deep muscles of the back: splenius, erector spinae, and transversospinalis. **A.** The sternocleidomastoid and levator scapulae muscles are reflected to reveal the splenius capitis and splenius cervicis muscles. *On the right side* the erector spinae is undisturbed (in situ) and shows the three columns of this massive muscle. *On the left side* the spinalis muscle, the thinnest and most medial of the erector spinae columns, is displayed as a separate muscle by reflecting the longissimus and iliocostalis columns of the erector spinae. **B.** Transverse section of the back showing arrangement of the three columns of the erector spinae and the three layers of the transversospinalis muscle.

Table 4.4. **Deep or Intrinsic Back Muscles**

Muscle	Origin	Insertion	Nerve Supply[a]	Main Action
Superficial layer				
Splenius	Arises from ligamentum nuchae and spinous processes of C7–T3 of T4 vertebrae	Splenius capitis: fibers run superolaterally to mastoid process of temporal bone and lateral third of superior nuchal line of occipital bone Splenius cervicis: tubercles of transverse processes of C1–C3 or C4 vertebrae	Dorsal rami of spinal nerves	*Acting alone,* they laterally bend and rotate head to side of active muscles; *acting together,* they extend head and neck
Intermediate layer				
Erector spine	Arises by a broad tendon from posterior part of iliac crest, posterior surface of sacrum, sacral and inferior lumbar spinous processes, and supraspinous ligament	Iliocostalis—lumborum, thoracis, and cervicis: fibers run superiorly to angles of lower ribs and cervical transverse processes Longissimus—thoracis, cervicis, and capitis: fibers run superiorly to ribs between tubercles and angles, to transverse processes in thoracic and cervical regions, and to mastoid process of temporal bone Spinalis—thoracis, cervicis, and capitis: fibers run superiorly to spinous processes in the upper thoracic region and to skull	Dorsal rami of spinal nerves	*Acting bilaterally,* they extend vertebral column and head; as back is flexed they control movement by gradually lengthening their fibers; *acting unilaterally,* they laterally bend vertebral column
Deep layer				
Transversospinal	Transverse processes: Semispinalis arises from transverse processes of C4–T12 vertebrae Multifidus arises from sacrum and ilium, transverse processes of T1–T3, and articular processes of C4–C7 Rotatores arise from transverse processes of vertebrae; are best developed in thoracic region	Spinous processes: Semispinalis—thoracis, cervicis, and capitis: fibers run superomedially to occipital bone and spinous processes in thoracic and cervical regions, spanning 4–6 segments Multifidus: fibers pass superomedially to spinous processes of vertebrae above, spanning 2–4 segments Rotatores: pass superomedially to attach to junction of lamina and transverse process, or spinous process, of vertebra above their origin, spanning 1–2 segments	Dorsal rami of spinal nerves	Extend head and thoracic and cervical regions of vertebral column and rotate them contralaterally Stabilizes vertebrae during local movements of vertebral column Stabilize vertebrae and assist with local extension and rotary movements of vertebral column; may function as organs of proprioception
Minor deep layer				
Interspinales	Superior surfaces of spinous processes of cervical and lumbar vertebrae	Inferior surfaces of spinous processes of vertebrae superior to vertebrae of origin	Dorsal rami of spinal nerves	Aid in extension and rotation of vertebral column
Intertransversarii	Transverse processes of cervical and lumbar vertebrae	Transverse processes of adjacent vertebrae	Dorsal and ventral rami of spinal nerves[a]	Aid in lateral bending of vertebral column; acting bilaterally, they stabilize vertebral column
Levatores costarum	Tips of transverse processes of C7 and T1–T11 vertebrae	Pass inferolaterally and insert on rib between its tubercle and angle	Dorsal rami of C8–T11 spinal nerves[b]	Elevate ribs, assisting inspiration; assist with lateral bending of vertebral column

[a] Most back muscles are innervated by dorsal rami of spinal nerves, but a few are innervated by ventral rami. Anterior intertransversarii of cervical region are supplied by ventral rami.
[b] Levatores costarum were once said to be innervated by ventral rami, but investigators now agree that they are innervated by dorsal rami.

Deep Layer of Intrinsic Back Muscles

Deep to the erector spinae is an obliquely disposed group of muscles—the **transversospinal muscle group** (L. transversospinalis)—semispinalis, multifidus, and rotatores. These muscles originate from transverse processes of vertebrae and pass to spinous processes of more superior vertebrae. They occupy the "gutter" between the transverse and spinous processes (Figs. 4.24–4.26).

- The semispinalis is superficial
- The multifidus is deeper
- The rotatores are deepest.

The **semispinalis** is the most superficial of the three layers of the transversospinal muscle. As its name indicates, the semispinalis arises from approximately half of the vertebral

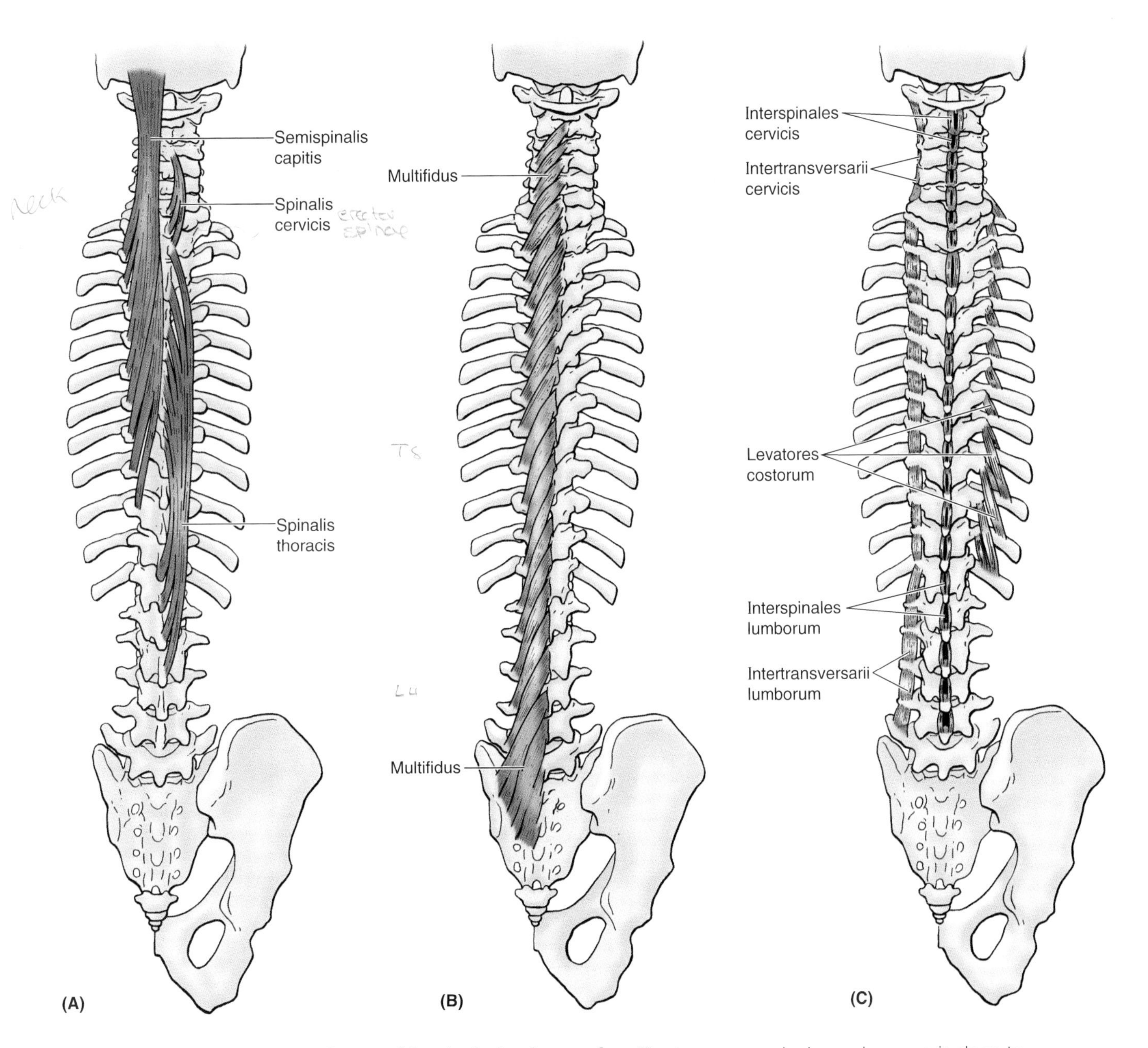

Figure 4.25. Deep layer of intrinsic back muscles. The transversospinal muscle group is deep to the erector spinae. It consists chiefly of a large number of small muscles that run obliquely upward and medially from transverse to spinous processes. **A.** Semispinalis capitis, spinalis cervicis, and spinalis thoracis. **B.** Multifidus—the intermediate layer of the transversospinalis that rotates the vertebral column. **C.** Interspinales, intertransversarii, and levatores costarum. The levator muscles, which elevate the ribs, represent the posterior intertransversarii in the thoracic region.

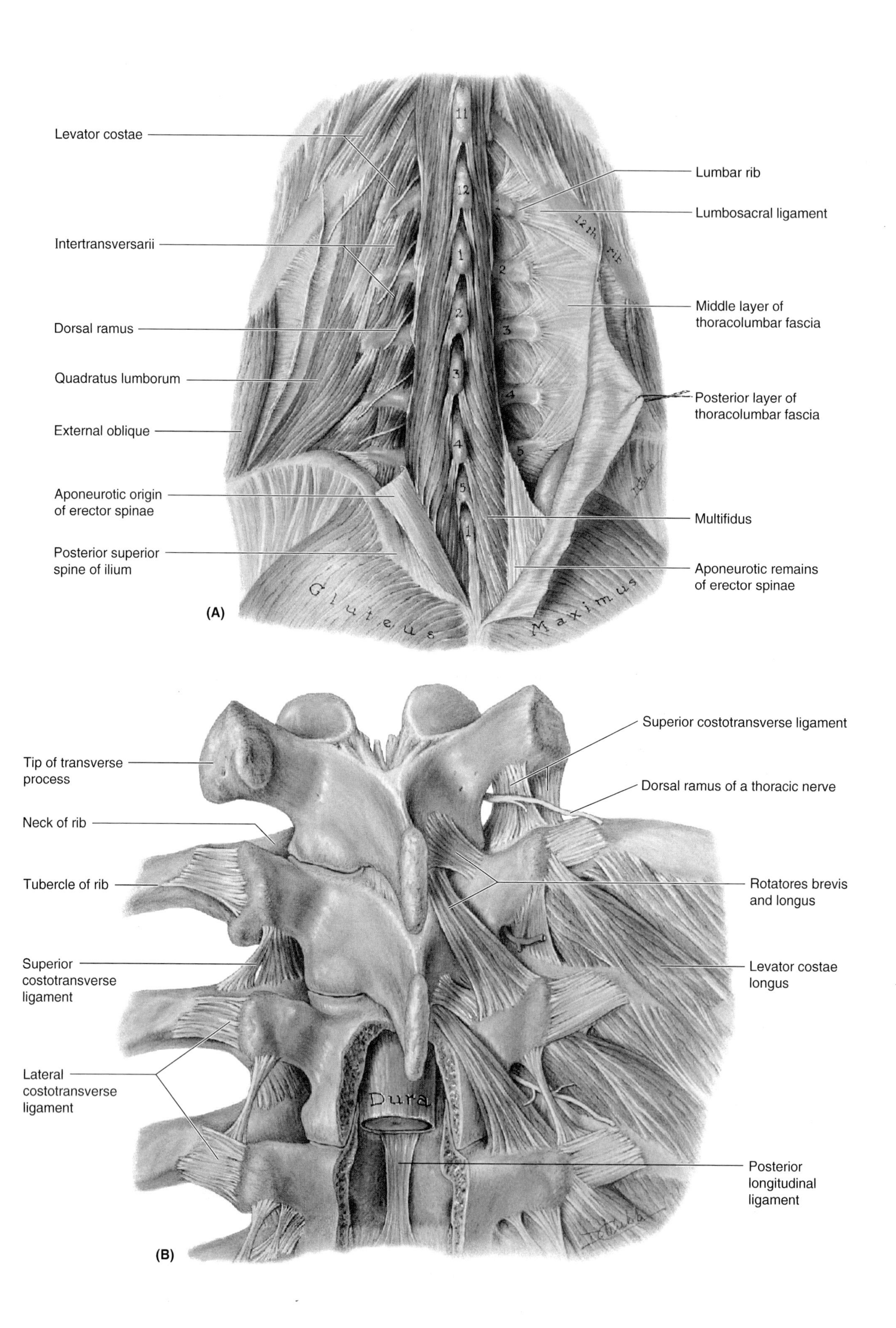
Levator costae
Intertransversarii
Dorsal ramus
Quadratus lumborum
External oblique
Aponeurotic origin of erector spinae
Posterior superior spine of ilium
Lumbar rib
Lumbosacral ligament
Middle layer of thoracolumbar fascia
Posterior layer of thoracolumbar fascia
Multifidus
Aponeurotic remains of erector spinae
12th rib
Gluteus
Maximus
11
12
1
2
3
4
5
1
(A)
Tip of transverse process
Neck of rib
Tubercle of rib
Superior costotransverse ligament
Lateral costotransverse ligament
Superior costotransverse ligament
Dorsal ramus of a thoracic nerve
Rotatores brevis and longus
Levator costae longus
Posterior longitudinal ligament
Dura
(B)

column ("spine"); it is divided into three parts according to their superior attachments:

- Semispinalis capitis
- Semispinalis thoracis
- Semispinalis cervicis.

Semispinalis capitis is responsible for the longitudinal bulge in the back of the neck near the median plane. It ascends from the cervical and thoracic transverse processes to the occipital bone. *Semispinalis thoracis and cervicis* pass superomedially from the transverse processes to the thoracic and cervical spinous processes of more superior vertebrae.

The **multifidus** consists of short, triangular muscular bundles that are thickest in the lumbar region. They arise—caudally to cranially—from the:

- Posterior aspect of the sacrum
- Aponeurosis of the erector spinae
- Posterosuperior iliac spines
- Sacroiliac ligaments
- Mamillary processes of lumbar vertebrae
- Transverse processes of thoracic vertebrae
- Articular processes of the inferior four cervical vertebrae.

Each muscular bundle passes obliquely superiorly and medially and attaches along the whole length of the spinous process of the adjacent superior vertebra.

The **rotatores**, or rotator muscles—best developed in the thoracic region—are the deepest of the three layers of transversospinal muscles. They arise from the transverse process of one vertebra and are inserted into the root of the spinous processes of the next one or two vertebrae superiorly. The attachments, nerve supply, and actions of the transversospinal muscle group (semispinalis, multifidus, and rotatores) are described in Table. 4.4.

The *interspinal* (L. interspinales), *intertransverse* (L. intertransversarii), and elevators of ribs (L. levatores costarum) are minor deep back muscles that are poorly developed in the thoracic region. The interspinal and intertransverse muscles connect spinous and transverse processes, respectively. The elevators of the ribs represent the posterior intertransverse muscles of the neck. Their attachments, nerve supply, and actions are described in Table 4.4.

Muscles Producing Movements of the Intervertebral Joints

The principal muscles producing movements of the cervical, thoracic, and lumbar IV joints are summarized in Tables 4.5 and 4.6. The back muscles are relatively inactive in the stand-easy position, but they act as lateral steadiers of the vertebral column.

Smaller muscles generally have higher densities of *muscle spindles* (sensors of proprioception—the sense of one's position—that are interdigitated among the muscle's fibers) than do large muscles. It has been presumed that this is because small muscles are used for the most precise movements, such as fine postural movements or manipulation, and therefore require more proprioceptive feedback. The movements described for small muscles are assumed from the location of

Table 4.5. Principal Muscles Producing Movements of Cervical Intervertebral Joints

Flexion	Extension	Lateral Bending	Rotation
Bilateral action of: Longus coli Scalene Sternocleidomastoid	Bilateral action of: Splenius capitis Semispinalis capitis and cervicis	Unilateral action of: Iliocostalis cervicis Longissimus capitis and cervicis Splenius capitis and cervicis	Unilateral action of: Rotatores Semispinalis capitis and cervicis Multifidus Splenius cervicis

Table 4.6. Principal Muscles Producing Movements of Thoracic and Lumbar Intervertebral Joints

Flexion	Extension	Lateral Bending	Rotation
Bilateral action of: Rectus abdominis Psoas major Gravity	Bilateral action of: Erector spinae Multifidus Semispinalis thoracis	Unilateral action of: Iliocostalis thoracis and lumborum Longissimus thoracis Multifidus External and internal oblique Quadratus lumborum	Unilateral action of: Rotatores Multifidus External oblique acting synchronously with opposite internal oblique Semispinalis thoracis

Figure 4.26. Deep dissections of the back. A. Multifidus and quadratus lumborum muscles, and the thoracolumbar fascia. Observe the short lumbar rib articulating with the transverse process of L1 vertebra. This common variation does not usually cause a problem; however, persons unfamiliar with its possible presence may think it is a fractured transverse process. **B.** Deeper dissection showing the rotatores and costotransverse ligaments. Observe two of three sets of costotransverse ligaments: superior and lateral. The (medial) costotransverse ligaments extend between the neck of the rib and the transverse process of the vertebra of the same number (not shown).

their attachments and the direction of the muscle fibers and from activity measured by electromyography as movements are performed. Muscles such as the rotatores, however, are so small and are placed in positions of such relatively poor mechanical advantage that their ability to produce the movements described is somewhat questionable. Furthermore, such small muscles are often redundant to other larger muscles having superior mechanical advantage. Hence, it has been proposed (Buxton and Peck, 1989) that the smaller muscles of small–large muscle pairs function more as "kinesiological monitors"—organs of proprioception—and that the larger muscles are the producers of motion.

Back Strains and Sprains

Adequate warmup and stretching prevent most back strains and sprains. *Back strain* is a common back problem in persons who participate in sports. It results from extreme movements of the vertebral column, such as excessive extension or rotation. *Back strain* refers to some degree of stretching or microscopic tearing of muscle fibers and/or ligaments of the back. The muscles usually involved are those producing movements of the lumbar IV joints, especially the columns of the erector spinae. If the weight is not properly balanced on the vertebral column, strain is exerted on the muscles. This is undoubtedly a common cause of low back pain. Using the back as a lever when lifting puts an enormous strain on the vertebral column, its ligaments, and muscles. These strains can be minimized if the lifter crouches, holds the back as straight as possible, and uses the muscles of the buttocks and lower limbs to assist with the lifting.

As a protective mechanism, the back muscles go into spasm following an injury or in response to inflammation of structures in the back, such as ligaments. *A spasm is a sudden involuntary contraction of one or more muscle groups.* Spasms are attended by cramps, pain, and interference with function, producing involuntary movement and distortion. ○

Surface Anatomy of the Back

In the median line of the back is a **posteromedian furrow** that overlies the tips of the spinous processes of the vertebrae. The furrow is continuous superiorly with the **nuchal furrow** in the neck. The posteromedian furrow is deepest in the lower thoracic and upper lumbar regions. The **erector spinae** produces a prominent vertical bulge on each ▶

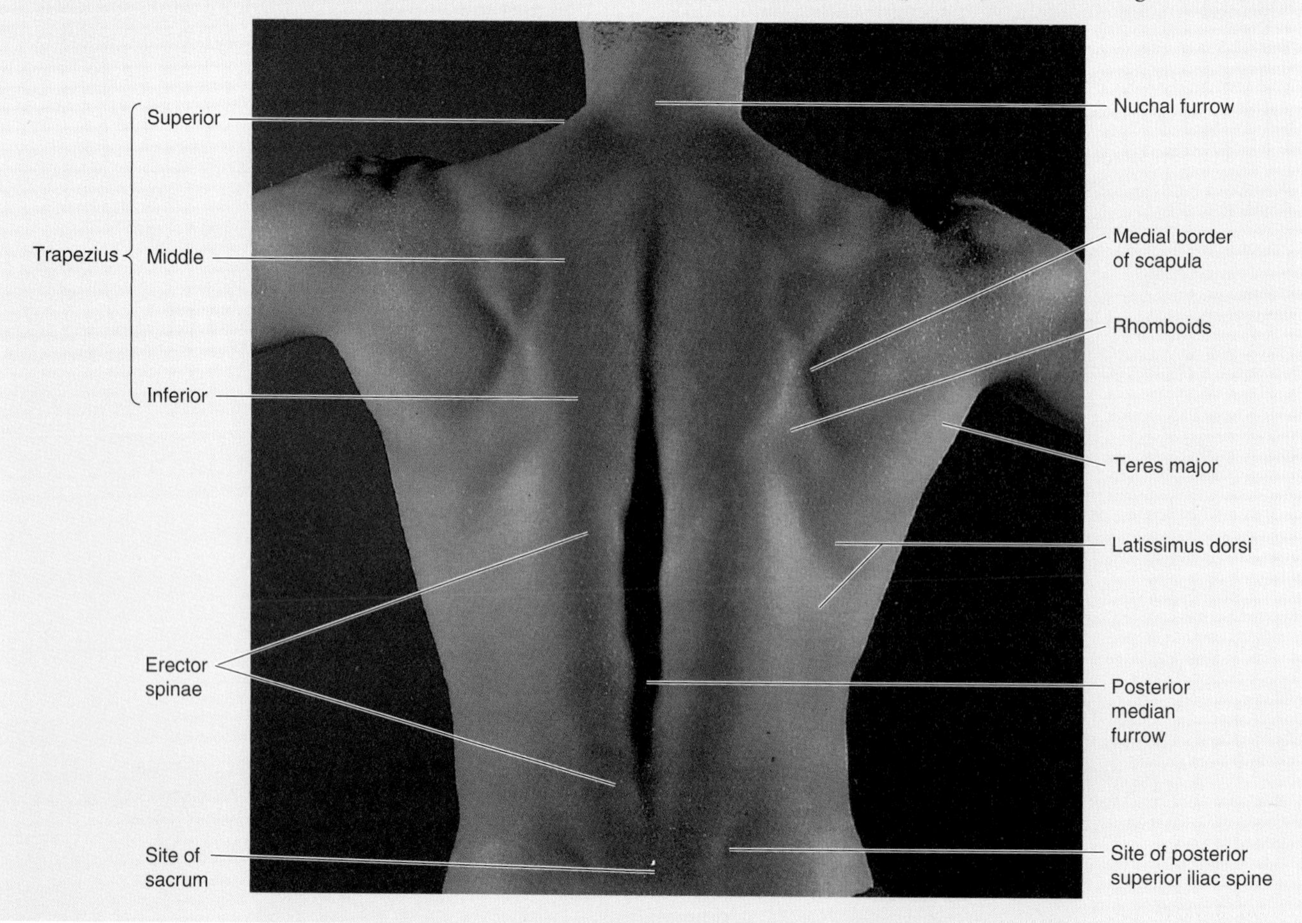

▶ side of the posteromedian furrow. In the lumbar region the erector spinae is readily palpable, and its lateral border is indicated by a groove in the skin. When standing, the lumbar spinous processes may be indicated by depressions in the skin. These processes are visible when the vertebral column is flexed (p. 464). The posteromedian furrow ends in the flattened triangular area covering the sacrum.

When the upper limbs are elevated the scapulae move laterally on the thoracic wall, making the **rhomboids** and **teres major** visible. These muscles are described in Chapter 6. The superficially located **trapezius** and **latissimus dorsi** muscles connecting the upper limbs to the vertebral column are also clearly visible.

Suboccipital and Deep Neck Muscles

The **suboccipital region**—the upper back of the neck—is the triangular area inferior to the occipital region of the head, including the posterior aspects of C1 and C2 vertebrae. The **suboccipital triangle** lies deep to the trapezius and semispinalis capitis muscles (Fig. 4.27). The four small muscles in the suboccipital region, two rectus capitis posterior and two obliquus muscles, are innervated by the dorsal ramus of C1, the **suboccipital nerve.** These muscles are mainly postural muscles, but actions are typically described for each muscle in terms of producing movement of the head. The muscles are considered to act on the head directly or indirectly—as indicated by "capitis" in their name—by extending it on C1 and rotating it on C1 and C2. However, recall the discussion of the small member of the small–large muscle pair functioning as a "kinesiological monitor" for the sense of proprioception (p. 474).

- *Rectus capitis posterior major* arises from the spinous process of C2 and inserts into the lateral part of the inferior nuchal line and the occipital bone.
- *Rectus capitis posterior minor* arises from the posterior tubercle of the posterior arch of C1 and inserts into the medial part of the inferior nuchal line.
- *Inferior oblique of the head* (obliquus capitis inferior) arises from the spinous process of C2 and inserts into the transverse process of C1. The name of this muscle is somewhat misleading; it is the only "capitis" muscle that has no attachment to the cranium.

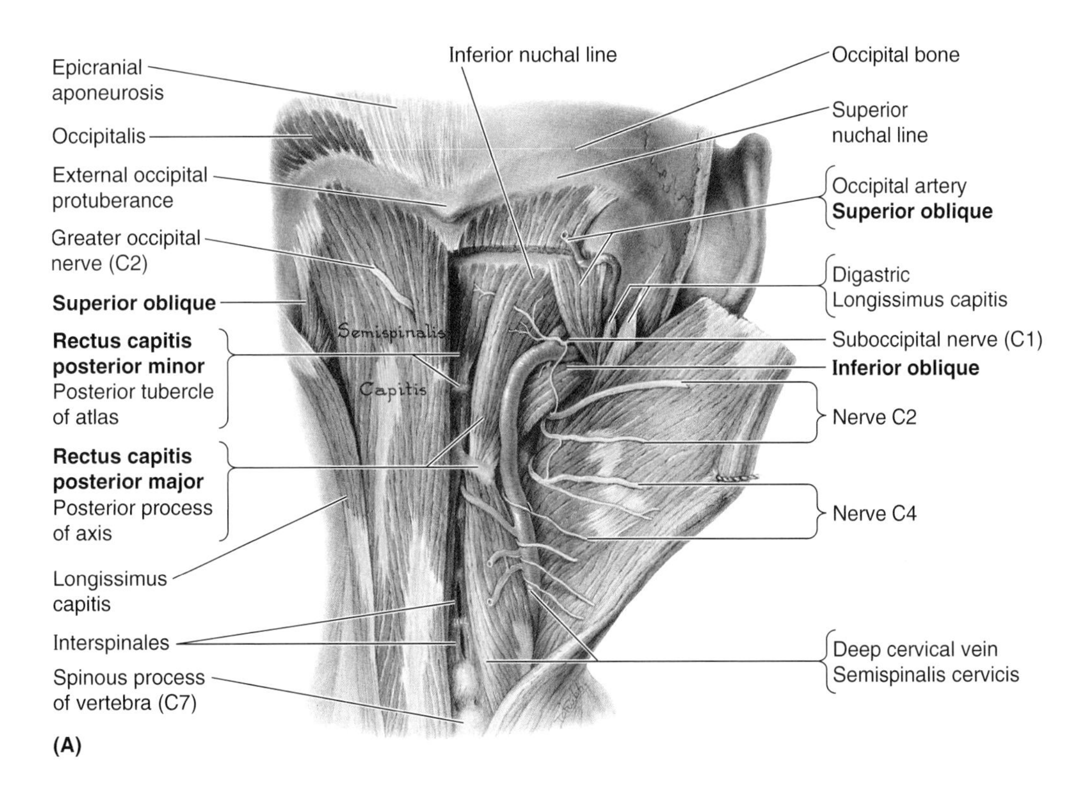

Figure 4.27. Dissections of the suboccipital region. A. The trapezius, sternocleidomastoid, and splenic muscles are removed. The suboccipital triangle is bounded by three muscles: inferior oblique, superior oblique, and rectus capitis posterior major. All nerves are derivatives of dorsal rami.

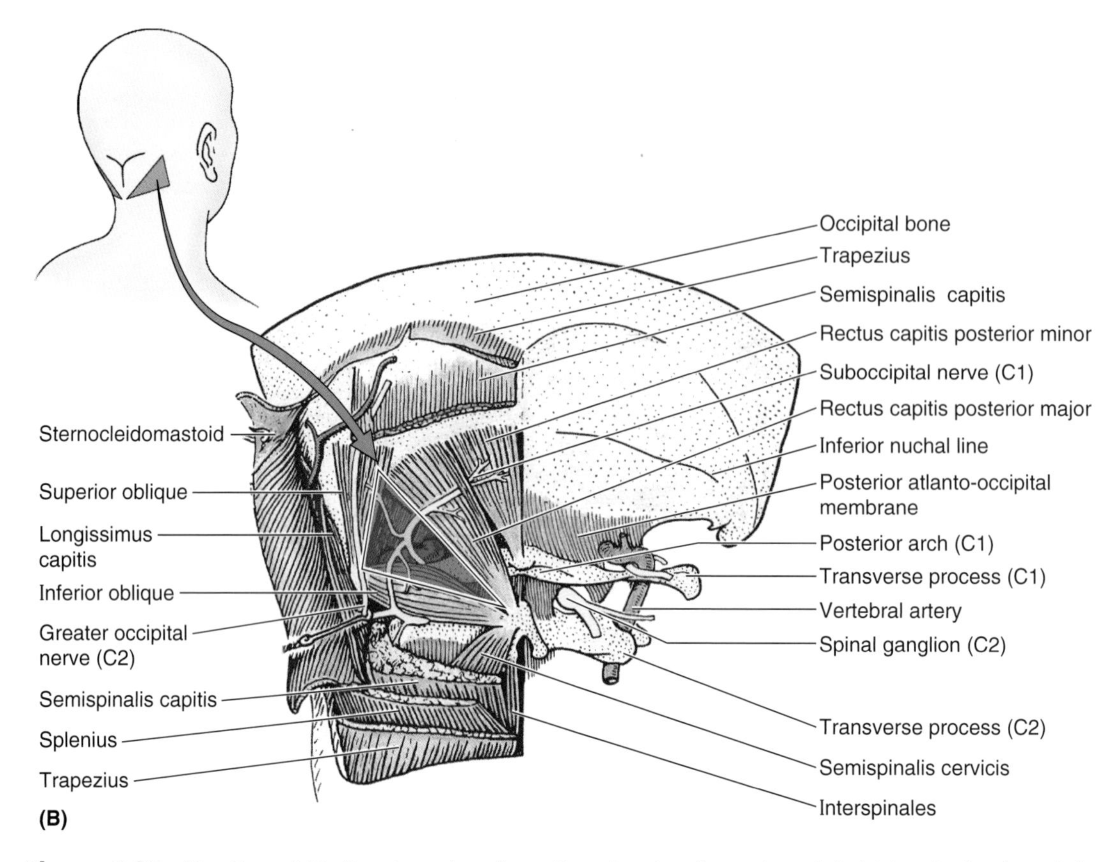

Figure 4.27. *(Continued)* **B.** Drawing of a dissection showing the suboccipital triangle (*red*) and the muscles, nerves (suboccipital-C1, and greater occipital-C2), and vertebrae in the suboccipital region. Observe the vertebral artery winding posterior to the superior articular process of the atlas (C1) to enter the foramen magnum of the skull.

Table 4.7. **Principal Muscles Producing Movements of Atlanto-Occipital Joints**

Flexion	Extension	Lateral Bending
Longus capitis Rectus capitis anterior Anterior fibers of sternocleidomastoid	Rectus capitis posterior major and minor Obliquus capitis superior Semispinalis capitis Splenius capitis Longissimus capitis Trapezius	Sternocleidomastoid Obliquus capitis superior Rectus capitis lateralis Longissimus capitis Splenius capitis

Table 4.8. **Principal Muscles Producing Rotation at Atlantoaxial Joints**[a]

Ipsilateral[b]	Contralateral
Obliquus capitis inferior Rectus capitis posterior, major and minor Longissimus capitis Splenius capitis	Sternocleidomastoid Semispinalis capitis

[a] Rotation is the specialized movement at these joints. Movement of one joint involves the other.
[b] Same side to which head is rotated.

- *Superior oblique of the head* (obliquus capitis superior) arises from the transverse process of C1 and inserts into the occipital bone between the superior and inferior nuchal lines.

The **suboccipital triangle** is the deep triangular area between the rectus capitis posterior major and the superior and inferior oblique capitis muscles (Fig. 4.27). *The boundaries and contents of the suboccipital triangle are:*

- Superomedially—rectus capitis posterior major
- Superolaterally—superior oblique
- Inferolaterally—inferior oblique

Table 4.9. **Nerve Supply of Suboccipital Triangle and Back**

Nerve	Origin	Course	Distribution
Suboccipital	Dorsal ramus of C1 nerve	Runs between skull and 1st cervical vertebra to reach suboccipital triangle	Muscles of suboccipital triangle
Greater occipital	Dorsal ramus of C2 nerve	Emerges inferior to inferior oblique and ascends to back of scalp	Skin over neck and occipital bone
Lesser occipital	Ventral ramus of C2 and C3 nerves	Passes directly to skin	Skin of neck and scalp
Dorsal rami C3–Co.	Spinal nerves C3–Co.	Pass segmentally to muscles and skin	Intrinsic muscles of back and overlying skin (adjacent to vertebral column)

- Floor—posterior atlanto-occipital membrane and posterior arch of C1
- Roof—semispinalis capitis
- Contents—vertebral artery and suboccipital nerve.

The principal muscles producing movements of the craniovertebral joints are summarized in Tables 4.7 and 4.8, and the nerve supply of the muscles in the suboccipital triangle, back, and back of the neck is summarized in Table. 4.9.

Reduced Blood Supply to the Brainstem

The winding course of the vertebral arteries through the suboccipital triangles becomes clinically significant when bloodflow through them is reduced, as occurs with *arteriosclerosis.* Under these conditions, prolonged turning of the head—as occurs when backing up a motor vehicle—may cause dizziness and other symptoms from the interference with the blood supply to the brainstem. ⊙

Spinal Cord and Meninges

The spinal cord, spinal meninges, and related structures are in the **vertebral canal** (Fig. 4.28), formed by successive vertebral foramina. The spinal cord, the major reflex center and conduction pathway between the body and the brain, is a cylindrical structure that is slightly flattened anteriorly and posteriorly. It is protected by the vertebrae and their associated ligaments and muscles, the spinal meninges, and the CSF. The **spinal cord** begins as a continuation of the medulla oblongata, the caudal part of the brainstem. In adults, the spinal cord (42 to 45 cm long) extends from the foramen magnum in the occipital bone to the L2 vertebral level; however, its tapering inferior end—the **medullary cone**—may terminate as high as T12 or as low as L3. Thus, the spinal cord occupies only the superior two-thirds of the vertebral canal. *The spinal cord is enlarged in two regions for innervation of the limbs.*

- The **cervical enlargement** extends from C4 through T1 segments of the spinal cord, and most of the ventral rami of the spinal nerves arising from it form the *brachial plexus of nerves* that innervates the upper limbs (see Chapter 6).
- The **lumbosacral enlargement** extends from T11 through L1 segments of the spinal cord, and the ventral rami of the spinal nerves arising from it make up the *lumbar and sacral plexuses of nerves* that innervate the lower limbs (see Chapter 5). The spinal nerve roots arising from the lumbosacral enlargement and the medullary cone form the **cauda equina**—the bundle of spinal nerve roots running through the *lumbar cistern* (subarachnoid [leptomeningeal] space).

Structure of Spinal Nerves

Thirty-one pairs of spinal nerves are attached to the spinal cord—8 cervical, 12 thoracic, 5 lumbar, 5 sacral, and 1 coccygeal (Fig. 4.28). Several rootlets emerge from the dorsal and ventral surfaces of the spinal cord and converge to form the dorsal and ventral **roots of the spinal nerves** (Fig. 4.29). Each spinal nerve contains afferent fibers that convey sensory input from the periphery and efferent fibers arising from the spinal motor neurons (Haines et al., 1997). Each level or segment of the spinal cord is specified by the IV foramina through which the dorsal and ventral roots attached to that segment exit the vertebral canal (Fig. 4.28).

The **dorsal roots of spinal nerves** contain afferent (or sensory) fibers from skin, from subcutaneous and deep tissues, and often from viscera (Fig. 4.29). The **ventral roots of spinal nerves** contain efferent (or motor) fibers to skeletal muscle and many contain presynaptic autonomic fibers. The cell bodies of somatic axons making up the ventral roots are in the **ventral gray horns** of the spinal cord, whereas the cell bodies of axons making up the dorsal roots are outside the spinal cord in the **spinal ganglia** (dorsal root ganglia) at the distal ends of the dorsal roots. The dorsal and ventral nerve roots unite at their points of exit from the vertebral canal to form a **spinal nerve.** The first cervical nerves lack dorsal roots in 50% of people, and the coccygeal (Co.) nerve may be absent.

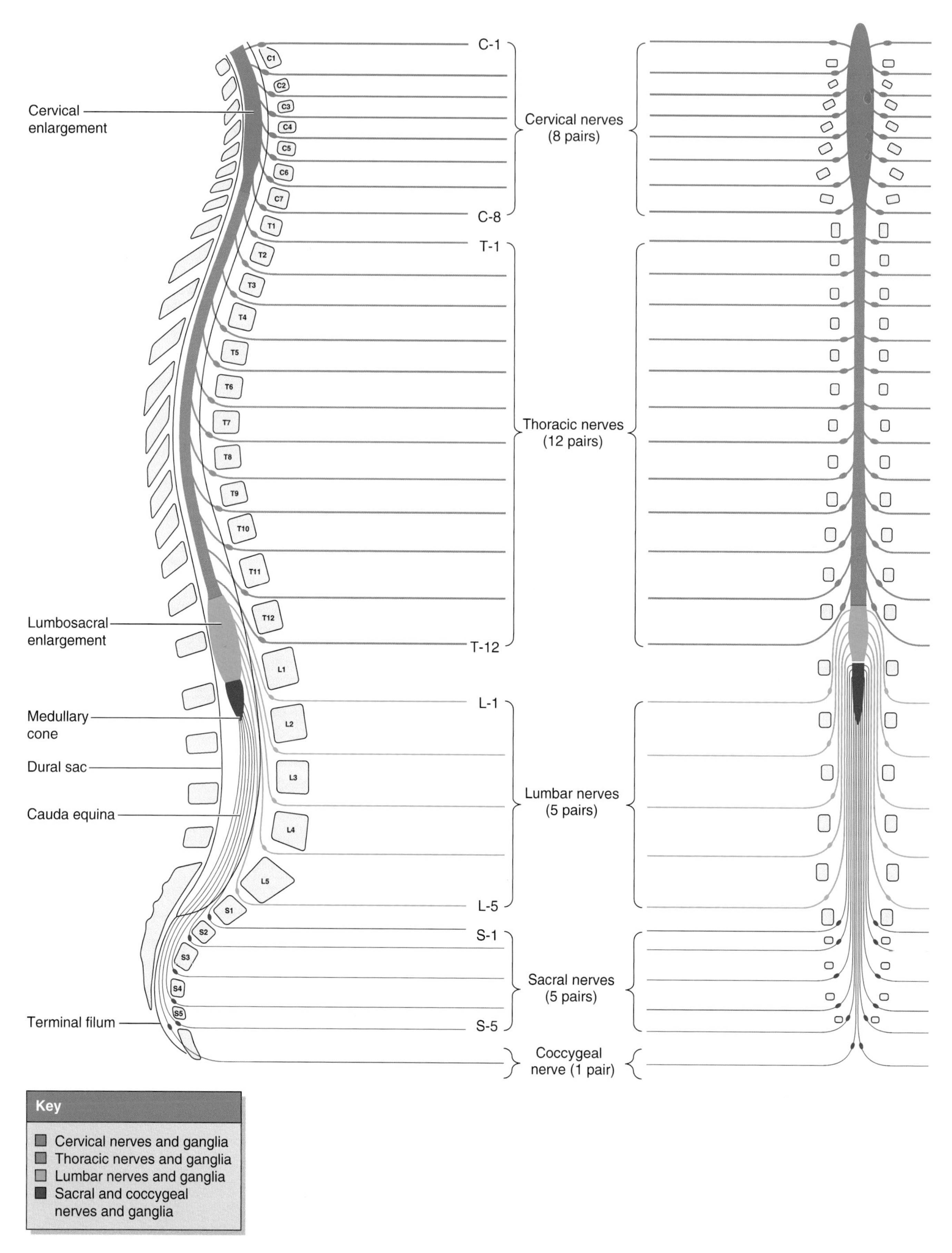

Figure 4.28. Vertebral column, spinal cord, spinal ganglia, and spinal nerves. Lateral and posterior views, illustrating the relation of the spinal cord segments and spinal nerves to the adult vertebral column.

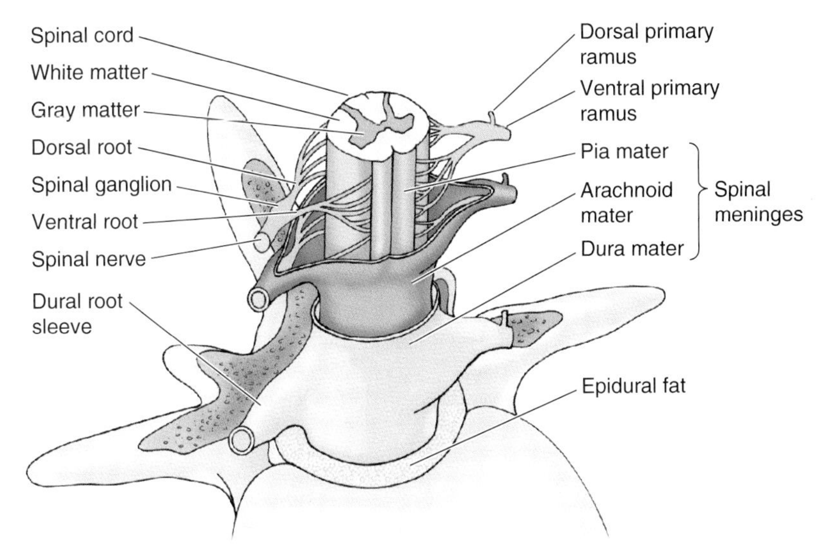

Figure 4.29. Spinal cord, spinal nerves, and spinal meninges. Observe that each nerve root emerges from the spinal cord as a series of rootlets and that each spinal nerve is formed by the union of dorsal and ventral spinal nerve roots. Observe also the three membranes covering the spinal cord (the spinal meninges): dura mater, arachnoid mater, and pia mater. Note that as each spinal nerve approaches an intervertebral (IV) foramen, it enters a dural root sleeve (sheath) that is continuous with the epineurium of the spinal nerve.

Each spinal nerve divides almost immediately into a **dorsal primary ramus** and a **ventral primary ramus**. The dorsal rami supply the skin and true muscles of the back; the ventral rami supply the limbs and the rest of the trunk.

In embryos, the spinal cord occupies the full length of the vertebral canal (Moore and Persaud, 1998) and thus spinal cord segments lie approximately at the vertebral level of the same number; spinal nerves pass laterally to exit the corresponding IV foramen. By the end of the embryonic period (8th week), the tail has disappeared and the number of coccygeal vertebrae is reduced from six to four segments. The spinal cord in the vertebral canal of the coccyx atrophies. During the fetal period, the vertebral column grows faster than the spinal cord; as a result, the cord "ascends" relative to the vertebral canal. At birth, the tip of the spinal cord—the **medullary cone**—is at the L2/L3 level.

In adults, the spinal cord is shorter than the vertebral column; hence, there is a progressive obliquity of the spinal nerve roots (Figs. 4.28 and 4.30). Because of the increasing distance between the spinal cord segments and the corresponding vertebrae, the length of the nerve roots increases progressively as the inferior end of the vertebral column is approached. The lumbar and sacral nerve rootlets are the longest. They descend until they reach their IV foramina of exit in the lumbar and sacral regions of the vertebral column, respectively. The bundle of spinal nerve roots in the **lumbar cistern** (subarachnoid space) within the vertebral canal caudal to the termination of the spinal cord resembles a horse's tail, hence its name—**cauda equina.**

The inferior end of the spinal cord has a conical shape and tapers into the **medullary cone**. From its inferior end, the **terminal filum** descends among the spinal nerve roots in the cauda equina. The terminal filum is the vestigial remnant of the caudal part of the spinal cord that was in the tail of the embryo. Its proximal end consists of vestiges of neural tissue, connective tissue, pia mater, and neuroglial tissue. The terminal filum leaves the inferior end of the dural sac, passes through the sacral hiatus, and attaches to the dorsum of the coccyx. The terminal filum serves as an anchor for the end of the **dural sac**—the continuation of the dura mater inferior to the medullary cone (Fig. 4.28).

Compression of the Lumbar Spinal Nerve Roots

The lumbar spinal nerves increase in size from above downward, whereas the IV foramina decrease in diameter. Consequently, the L5 spinal nerve roots are the thickest and their foramina, the narrowest. This increases the chance that these nerve roots will be compressed if herniation of the nucleus pulposus occurs. ○

Spinal Meninges and Cerebrospinal Fluid

Collectively, the dura mater (dura), arachnoid mater (arachnoid), and pia mater (pia) surrounding the spinal cord form the **spinal meninges** (Fig. 4.29, Table 4.10). These membranes and the CSF in the subarachnoid space surround, support, and protect the spinal cord and spinal nerve roots—including those in the cauda equina.

Dura Mater

The spinal dura mater, composed of tough fibrous and elastic tissue, is the outermost covering membrane of the spinal cord. The spinal dura is separated from the vertebrae by the extradural or epidural space that contains adipose tissue and a venous plexus. The dura forms the **dural sac**, a long tubular sheath within the vertebral canal (Fig. 4.28). The dural sac adheres to the margin of the foramen magnum of the skull, where it is continuous with the cranial dura mater. The dural sac is pierced by the spinal nerves (Fig. 4.30) and is anchored inferiorly to the coccyx by the **terminal filum**. The spinal

Table 4.10. **Spaces Associated with Spinal Meninges**

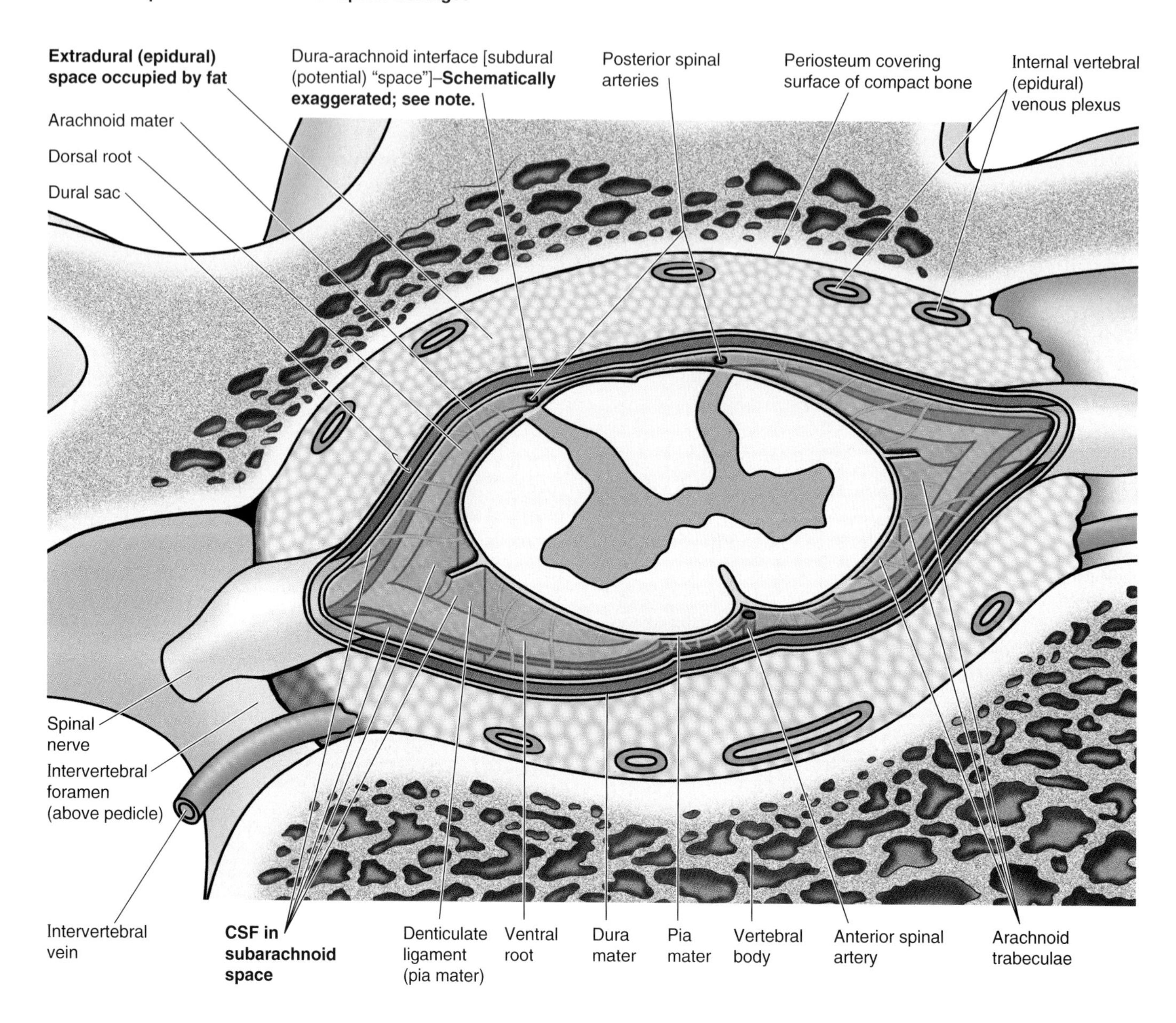

Space	Location	Contents
Extradural (epidural)	Space between periosteum lining bony wall of vertebral canal and dura mater–position of extradural (epidural) herniation	Fat (loose connective tissue), internal vertebral venous plexuses, and inferior to L2 vertebra, the ensheathed roots of spinal nerves
Subarachnoid (lepto-meningeal)	A naturally occuring space between the arachnoid mater and pia mater	CSF, radicular, segmental medullary, spinal arteries, veins, and arachnoid trabeculae

Note: Although it is common to refer to a "subdural space," there is no naturally occuring space at the arachnoid-dura junction (Haines, 1997). Hematomas at this junction are usually caused by extravasated blood, which creates an abnormal space at the dura-arachnoid interface.

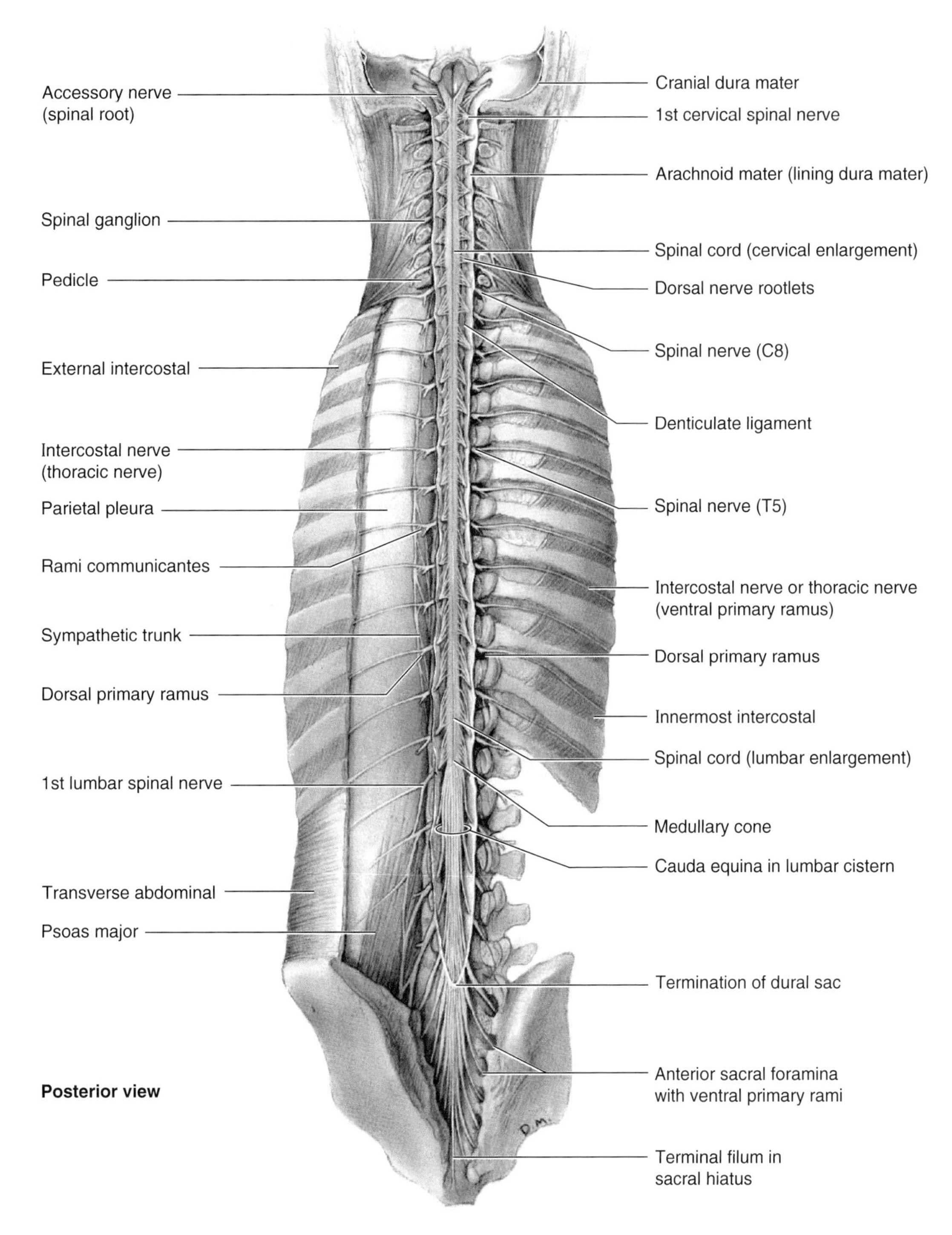

Figure 4.30. Spinal cord in situ. The vertebral (neural) arches and the posterior aspect of the sacrum have been removed to expose the vertebral (spinal) canal. The dural sac has also been cut open posteriorly to reveal the spinal cord and nerve roots, the termination of the spinal cord at the L2 vertebral level, and the termination of the dural sac at the 2nd sacral segment.

dura extends into the IV foramina and along the dorsal and ventral nerve roots distal to the spinal ganglia to form **dural root sleeves** (Fig. 4.29). These sleeves adhere to the periosteum lining the IV foramina and end by blending with the epineurium of the spinal nerves.

Arachnoid Mater

The arachnoid mater is a delicate, avascular membrane composed of fibrous and elastic tissue that lines the dural sac and its dural root sleeves and encloses the CSF-filled subarachnoid space containing the spinal cord, spinal nerve roots, and spinal ganglia. *The arachnoid is not attached to the dura* but is held against the inner surface of the dura by the pressure of the CSF. In a lumbar spinal puncture, the needle traverses the dura and arachnoid simultaneously. Their apposition is the dura–arachnoid interface, often erroneously referred to as the "subdural space." *No such space exists normally.* Bleeding into this space creates a subdural hematoma in the pathological space. In the cadaver—because of the absence of CSF—the arachnoid falls away from the inner surface at the dura and lies loosely on the spinal cord. The arachnoid is separated from the pia mater on the surface of the spinal cord by the **subarachnoid space** containing CSF (Table 4.10). Delicate strands of connective tissue, the **arachnoid trabeculae**, span the subarachnoid space connecting the arachnoid and pia.

Pia Mater

The pia mater, the innermost covering membrane of the spinal cord, consists of flattened cells with long, equally flattened processes that closely follow all the surface features of the spinal cord (Haines, 1997). The pia also directly covers the roots of the spinal nerves and the spinal blood vessels (Fig. 4.31). Inferior to the medullary cone, the pia continues as the terminal filum.

The spinal cord is suspended in the dural sac by the saw-toothed **denticulate ligament** on each side (L. denticulus, a small tooth). These ligaments are lateral extensions from the lateral surfaces of the pia midway between the dorsal and ventral nerve roots (Figs. 4.30–4.32). The 20 to 22 processes, shaped much like shark's teeth, attach to the inner surface of the dural sac (Haines, 1997). The uppermost part of the denticulate ligament attaches to the occipital dura immediately inside the foramen magnum, and the lowermost part spans between the T12 and L1 nerve roots.

Subarachnoid Space

The subarachnoid space is between the arachnoid and pia mater and is filled with CSF. The enlargement of the subarachnoid space in the dural sac, caudal to the medullary cone containing the cauda equina, is the **lumbar cistern** (Figs. 4.30 and 4.33). It extends from L2 vertebra to the second segment of the sacrum. Dural root sleeves (Fig.4.29), enclosing spinal nerve roots in extensions of the subarachnoid space, protrude from the sides of the lumbar cistern.

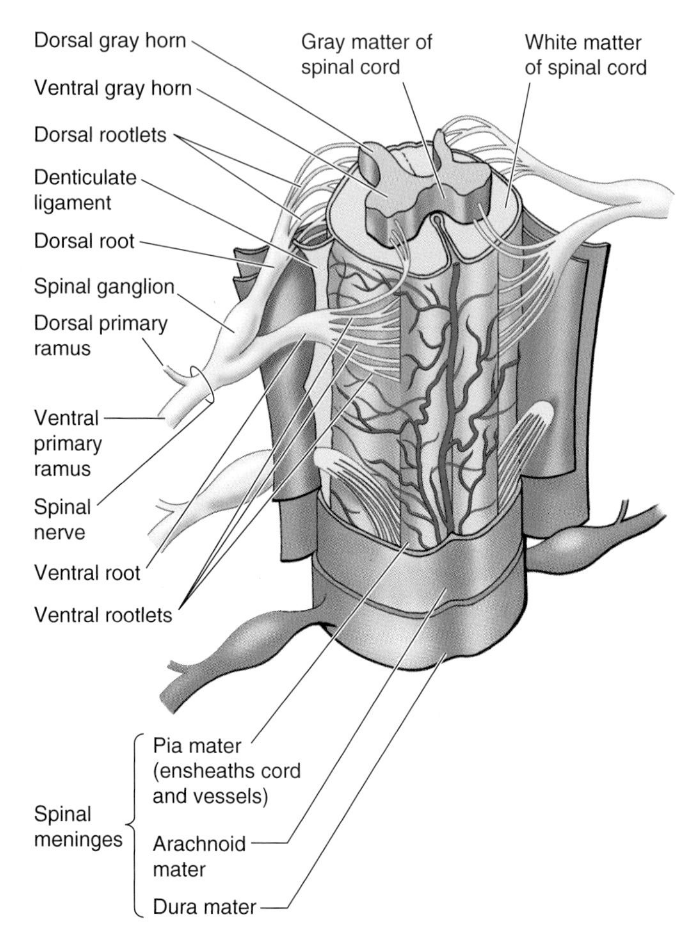

Figure 4.31. Spinal cord, dorsal and ventral roots, spinal nerves, and spinal meninges. This three-dimensional drawing shows the structure of the spinal cord (gray and white matter) and the entrance and exit of nerve fibers. Observe the denticulate ligament, a serrated shelflike extension of the spinal pia mater projecting between the dorsal and ventral roots in a frontal plane from either side of the cervical and thoracic regions of the spinal cord.

Development of the Meninges and Subarachnoid Space

Together, the pia mater and arachnoid mater form the *leptomeninges* (G. slender membranes). They develop as a single layer from the mesenchyme surrounding the embryonic spinal cord. Fluid-filled spaces form within the layer and coalesce to become the *subarachnoid space* (Moore and Persaud, 1998). The origin of the membranes from a single membrane is reflected by the numerous arachnoid trabeculae passing between them. The arachnoid in adults is thick enough to be manipulated with the fingers or forceps. In contrast, the pia mater is barely visible to the unaided eye, although it gives a shiny appearance to the surface of the spinal cord.

Lumbar Spinal Puncture

The retrieval of CSF from the lumbar cistern is an important diagnostic tool for evaluating a variety of central nervous system disorders. Infections of the meninges (*meningitis*) and diseases of the central nervous system may alter ▶

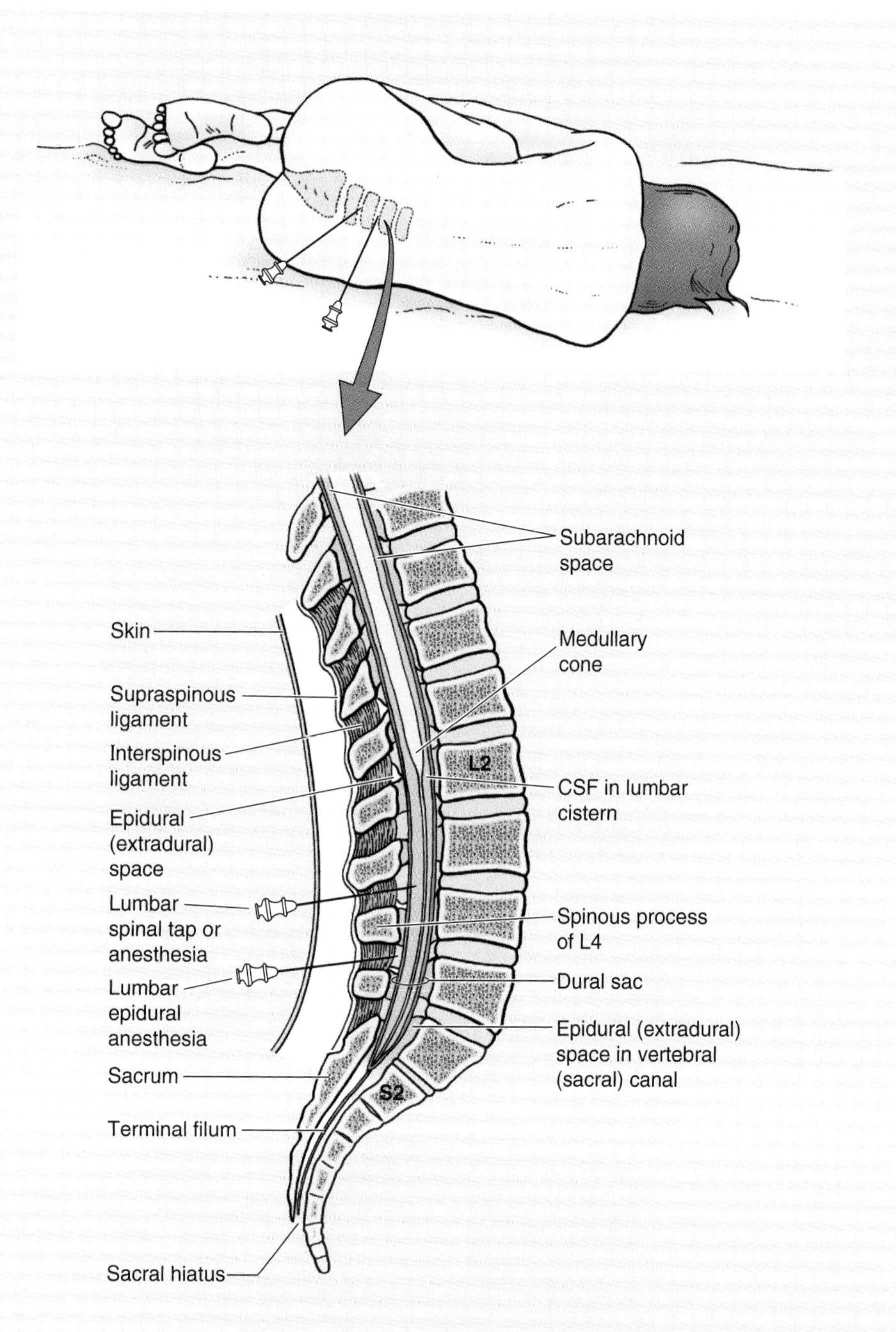

▶ the cells in the CSF or change the concentration of its chemical constituents. Examination of CSF samples can also determine whether blood is in the CSF. Lumbar spinal puncture (spinal tap) is performed with the patient leaning forward or lying on the side with the back flexed. Flexion of the vertebral column facilitates insertion of the needle by stretching the ligamenta flava and spreading the laminae and spinous processes apart.

Under aseptic conditions, the skin covering the lower lumbar vertebrae is anesthetized and a *lumbar puncture needle*, fitted with a stylet, is inserted in the midline between the spinous processes of L3 and L4 (or L4 and L5) vertebrae. Recall that a plane transecting the highest points of the iliac crests—the *supracristal plane*—usually passes through the L4 spinous process. At these levels, there is no danger of damaging the spinal cord. After passing 4 to 6 cm in adults (more in obese persons), the needle punctures the dura and arachnoid and enters the lumbar cistern. As the stylet is removed, CSF normally escapes at the rate of approximately one drop per second. If subarachnoid pressure is high, CSF flows out or escapes as a jet. Lumbar puncture is not performed if examination of the fundus (back) of the interior of the eyeball with an ophthalmoscope reveals high intracranial pressure. The consequent release of pressure in the lumbar region could result in a fatal herniation of the brainstem and cerebellum (L. little brain) into the vertebral canal. ▶

Spinal Block

An anesthetic agent can be injected directly into the CSF. Anesthesia usually takes effect within 1 minute. A headache may follow a *spinal block* and is thought to result from the leakage of CSF through the opening made in the dura and arachnoid. For more information about spinal blocks, see Chapter 3.

Epidural Block

An anesthetic agent can be injected into the extradural or epidural space using the position described for lumbar spinal puncture (lumbar block) or through the foramina and hiatus of the sacrum. The anesthetic has a direct effect on the spinal nerve roots of the cauda equina after they exit the dural sac. The effect of an epidural block usually takes 10 to 20 minutes. An epidural block is used as the sole anesthetic for operations below the diaphragm. With this anesthetic, the patient may be sedated or remain awake. An epidural block is commonly used to prevent pain during childbirth and for cesarean sections because maternal vascular resistance (blood pressure) and thus bloodflow to the fetus through the placenta is more stable (see Chapter 3).

When performing a **caudal epidural block**, the anesthetic agent is administered using an in-dwelling catheter in the sacral canal. The catheter is inserted through the **sacral hiatus,** and the anesthetic expelled bathes the S2 through S4 spinal nerve roots (pp. 381 and 445). The height to which the anesthetic agent ascends is controlled by the amount injected and the position of the patient. If the sacral hiatus is large (as in Figure 4.10*B*), one must be careful not to insert the needle too far because it may enter the lumbar cistern, which extends to the second segment of the sacrum. This could cause excessive anesthesia because the dose of the agent for caudal anesthesia (injection into the epidural space) is much greater than that used for a spinal block (injection into CSF). Anesthetic agents can also be injected through the posterior sacral foramina into the epidural space around specific sacral nerves (p. 445). For a discussion of caudal anesthesia for childbirth, see Chapter 3. ⊙

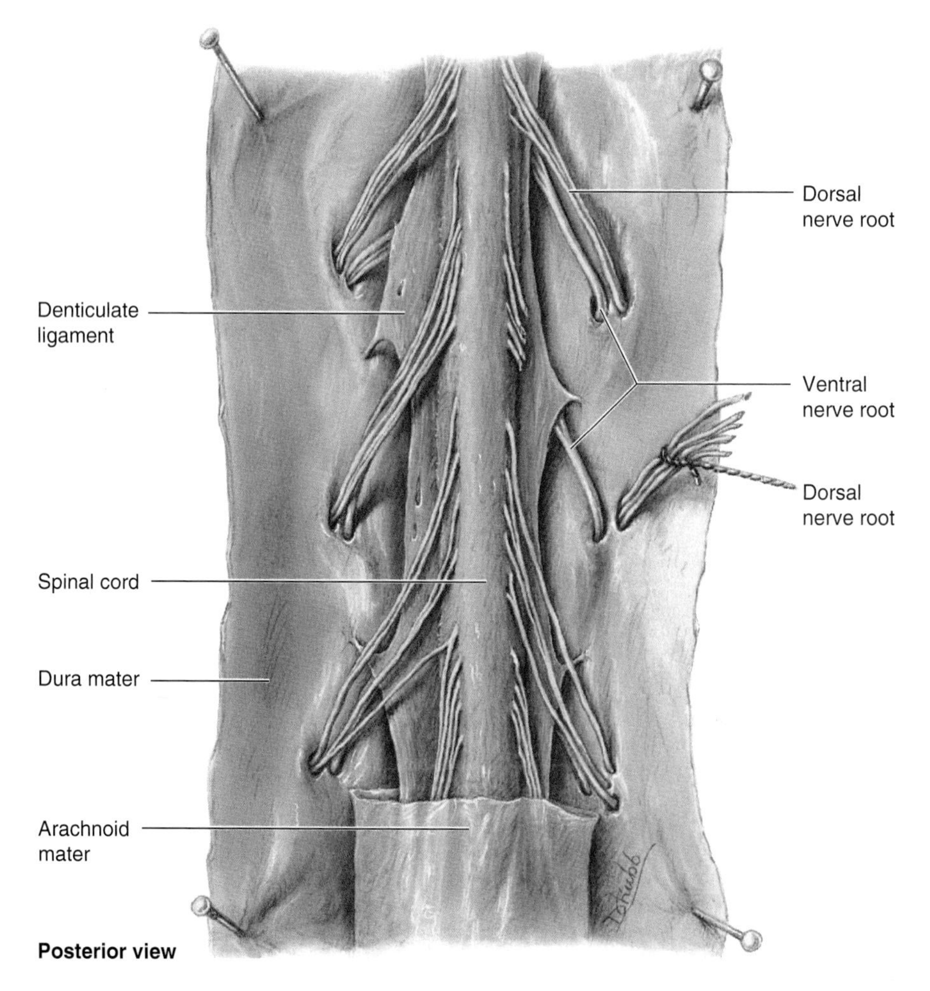

Figure 4.32. Spinal cord within its meninges (membranes). The dura mater and arachnoid mater have been split and pinned to expose the spinal cord and nerve roots. Observe the denticulate ligament, running like a band along each side of the spinal cord and by means of relatively strong, toothlike processes, anchoring the spinal cord to the dura between successive nerve roots.

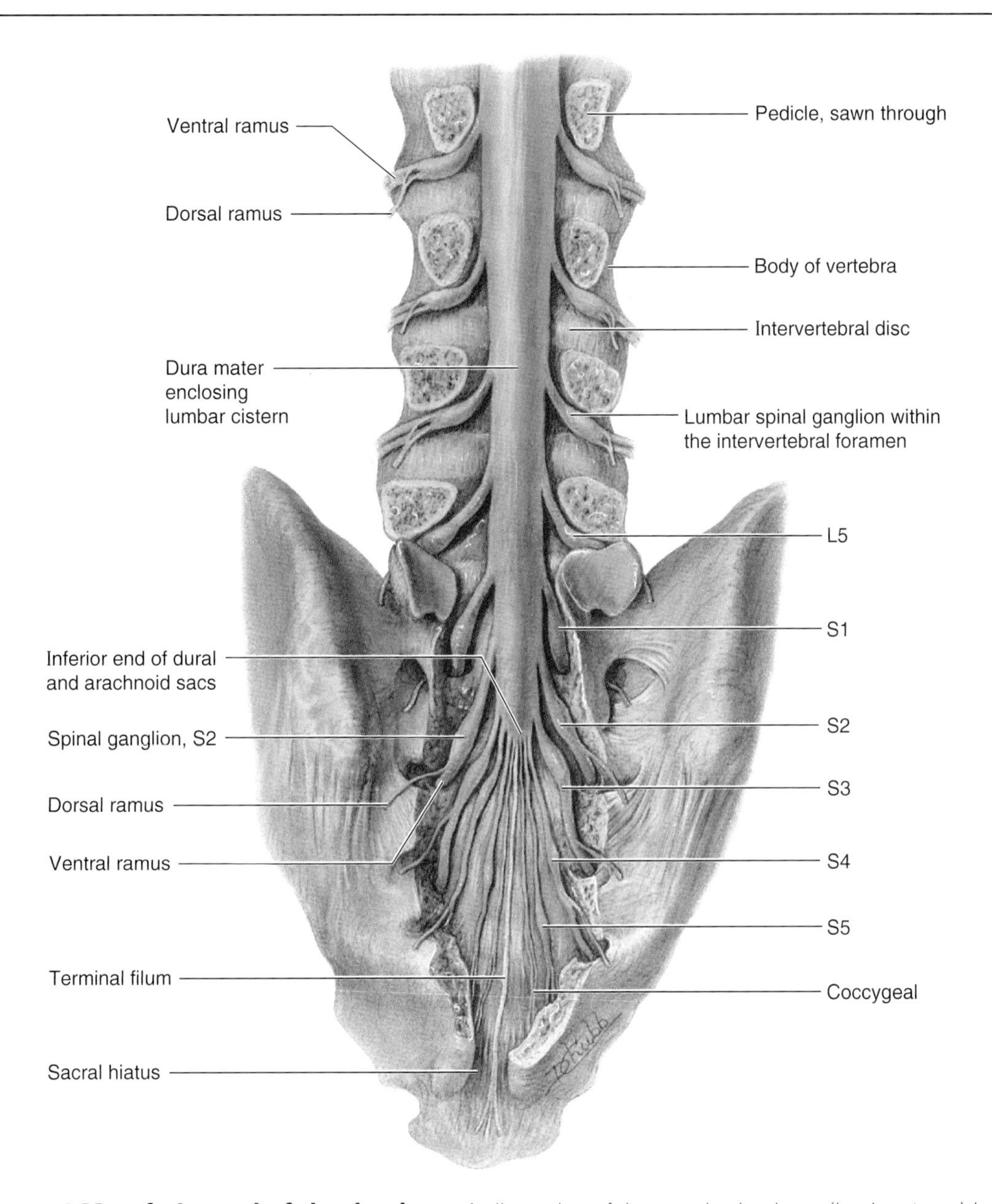

Figure 4.33. Inferior end of the dural sac. A dissection of the vertebral column (laminectomy) has been performed to show the inferior end of the dural sac, which encloses the lumbar cistern containing the cauda equina (L. cauda, tail). Most of the vertebral (neural) arches of the lumbar and sacral vertebrae have been removed. Observe that the lumbar spinal ganglia lie within the intervertebral (IV) foramina and that the sacral spinal ganglia (S1 through S5) are in the sacral canal. Note that in the lumbar region, the nerves exiting an IV foramen pass superior to the IV disc at that level; thus, herniation of the nucleus pulposus tends to impinge on nerves passing to lower levels.

Vasculature of the Spinal Cord

Arteries of the Spinal Cord

The arteries supplying the spinal cord arise from branches of the vertebral, ascending cervical, deep cervical, intercostal, lumbar, and lateral sacral arteries (Fig. 4.34). *Three longitudinal arteries supply the spinal cord*:

- An anterior spinal artery
- Paired posterior spinal arteries.

These arteries run longitudinally from the medulla of the brainstem to the medullary cone of the spinal cord.

The **anterior spinal artery**, formed by the union of branches of the vertebral arteries (Figs. 4.34*A* and 4.35), runs inferiorly in the anteromedian fissure (ventromedian fissure). **Sulcal (central) arteries** arise from the artery and enter the spinal cord through the anteromedian fissure. The sulcal arteries supply approximately two-thirds of the cross-sectional area of the spinal cord (Williams et al., 1995). Each **posterior spinal artery** is a branch of either the vertebral artery or the posteroinferior cerebellar artery. The posterior spinal arteries commonly form anastomosing channels in the pia mater.

By themselves, the anterior and posterior spinal arteries can supply only the short superior part of the spinal cord. The circulation to much of the cord depends on segmental medullary and radicular arteries running along the spinal nerve roots. The anterior and posterior **segmental medullary arteries** are derived from spinal branches of the ascending cervical, deep cervical, vertebral, posterior intercostal, and lumbar arteries. The medullary segmental arteries are located chiefly where the need for a good blood supply to the spinal cord is greatest—the cervical and lumbosacral enlargements. They enter the vertebral canal through the IV foramina.

The **great anterior segmental medullary artery** (medullary artery of Adamkiewicz)—on the left side in 65% of persons—reinforces the circulation to two-thirds of the spinal cord, including the lumbosacral enlargement. The great anterior segmental medullary artery, much larger than the other medullary segmental arteries, usually arises from an inferior intercostal or upper lumbar artery and enters the vertebral canal through the IV foramen at the lower thoracic or upper lumbar level.

The dorsal and ventral roots of the spinal nerves and their coverings are supplied by dorsal and ventral **radicular arteries** that run along the nerve roots. Most spinal nerve roots and proximal spinal nerves and roots are accompanied by radicular arteries that do not reach the posterior, anterior, or spinal arteries. Segmental medullary arteries occur irregularly *in the place of* radicular arteries—they are really just larger vessels that make it all the way to the spinal arteries. Most radicular arteries are small and supply only the nerve roots; however, some of them assist with the supply of superficial parts of the gray matter in the dorsal and ventral horns and the spinal cord.

Veins of the Spinal Cord

In general, the veins of the spinal cord have a distribution similar to that of the spinal arteries. *There are usually three anterior and three posterior spinal veins* (Fig. 4.35*A*). The **spinal veins** are arranged longitudinally, communicate freely with each other, and are drained by up to 12 anterior and posterior medullary and radicular veins. The veins draining the spinal cord join the **internal vertebral (epidural) venous plexus**, lying in the extradural (epidural) space. The internal vertebral venous plexus passes superiorly through the foramen magnum to communicate with dural sinuses and vertebral veins in the skull (see Chapter 7). The internal vertebral plexus also communicates with the external vertebral venous plexus on the external surface of the vertebrae (Fig. 4.22).

Figure 4.34. Arterial supply of the spinal cord. The arteries derive from branches of the vertebral, ascending and deep cervical, intercostal, lumbar and lateral sacral arteries. Three longitudinal arteries supply the spinal cord: an anterior spinal artery and two posterior spinal arteries. These vessels are reinforced by blood from the anterior and posterior segmental medullary arteries. Observe the clinically important great anterior segmental medullary artery (of Adamkiewicz). Segmental medullary arteries supply the nerve roots that they course along, and then contribute to the longitudinal arteries. At levels where segmental medullary arteries do not occur, radicular arteries supply the dorsal and ventral roots of the spinal nerves. (Radicular arteries are only illustrated at cervical and thoracic levels, but they also occur at lumbar and sacral levels.)

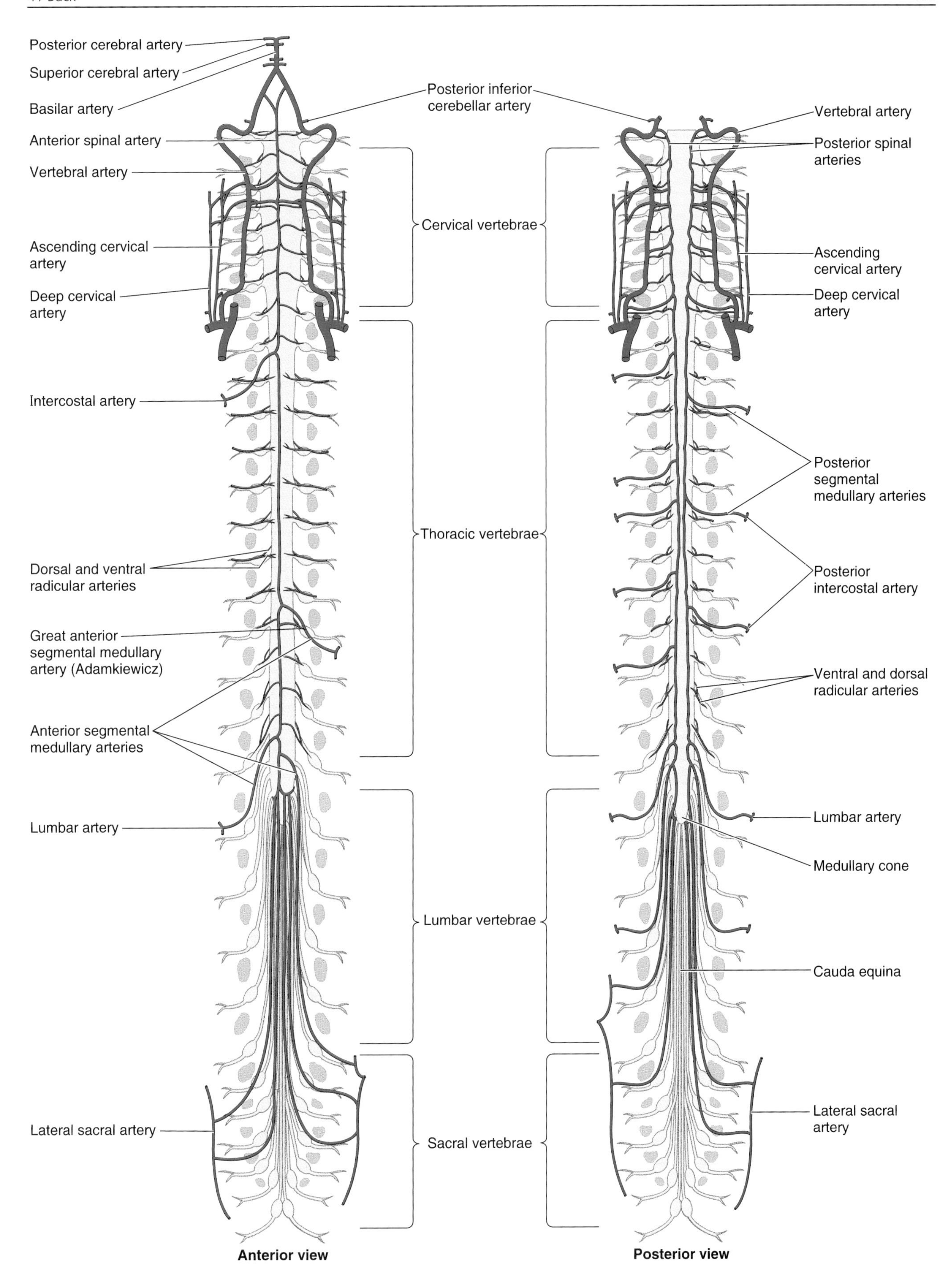
Posterior cerebral artery
Superior cerebral artery
Basilar artery
Anterior spinal artery
Vertebral artery
Ascending cervical artery
Deep cervical artery
Intercostal artery
Dorsal and ventral radicular arteries
Great anterior segmental medullary artery (Adamkiewicz)
Anterior segmental medullary arteries
Lumbar artery
Lateral sacral artery
Posterior inferior cerebellar artery
Cervical vertebrae
Thoracic vertebrae
Lumbar vertebrae
Sacral vertebrae
Vertebral artery
Posterior spinal arteries
Ascending cervical artery
Deep cervical artery
Posterior segmental medullary arteries
Posterior intercostal artery
Ventral and dorsal radicular arteries
Lumbar artery
Medullary cone
Cauda equina
Lateral sacral artery
Anterior view
Posterior view

Ischemia of the Spinal Cord

The segmental reinforcements of the blood supply to the spinal cord from the medullary segmental arteries are important in supplying blood to the anterior and posterior spinal arteries. Fractures, dislocations, and fracture-dislocations may interfere with the blood supply to the spinal cord from the spinal and medullary arteries. *Deficiency of blood supply results in ischemia of the spinal cord* that affects its function and can lead to muscle weakness and paralysis. The spinal cord may also suffer circulatory impairment if the segmental medullary arteries, particularly the great anterior medullary segmental artery (of Adamkiewicz), are narrowed by obstructive arterial disease. Sometimes the aorta is purposely occluded ("cross clamped") during surgery. Patients undergoing such surgeries, and those suffering ruptured aneurysms of the aorta or occlusion of the great anterior medullary segmental artery, may lose all sensation and voluntary movement inferior to the level of impaired blood supply to the spinal cord (*paraplegia*) secondary to death of neurons in the part of the spinal cord supplied by the anterior spinal artery (Fig. 4.34). The occurrence of iatrogenic paraplegia ranges from 0.4 to 40% depending on the age of the patient, the extent of the disease, and the length of time the aorta is cross clamped (Murray et al., 1992).

When systemic blood pressure drops severely for 3 to 6 minutes, bloodflow from the medullary segmental arteries to the anterior spinal artery supplying the midthoracic region of the spinal cord may be reduced or stopped. These patients may also lose sensation and voluntary movement in the areas supplied by the affected level of the spinal cord. ▶

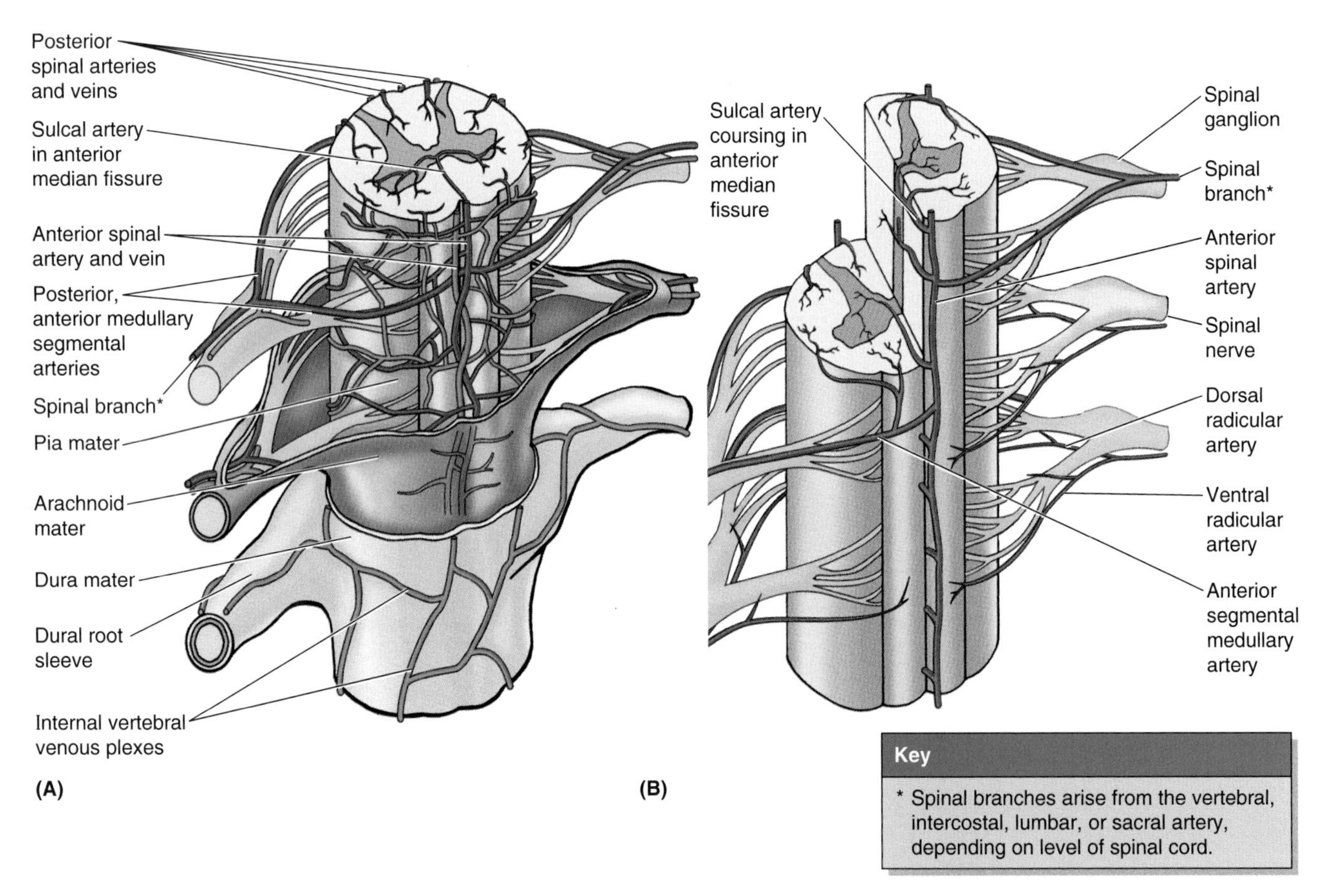

Most proximal spinal nerves and roots are accompanied by **radicular arteries**, which do not reach the posterior, anterior, or spinal arteries. **Medullary segmental arteries** occur irregularly *in the place of* radicular arteries—they are really just larger vessels that make it all the way to the spinal arteries.

Figure 4.35. Venous drainage and arterial supply of the spinal cord and dorsal and ventral nerve roots. A. The veins that drain the spinal cord, as well as the internal vertebral venous plexuses, drain into the intervertebral veins, which in turn drain into segmental veins. **B.** The pattern of the arterial supply of the spinal cord is from three longitudinal arteries: one anterior lying in the anteromedian position and the other two lying posterolaterally. These vessels are reinforced by medullary branches derived from the segmental arteries. The sulcal arteries are small branches of the anterior spinal artery coursing in the anteromedian fissure.

Spinal Cord Injuries

The vertebral canal varies considerably in size and shape from level to level, particularly in the cervical and lumbar regions. A small vertebral canal in the cervical region, into which the spinal cord fits tightly, is potentially dangerous because a minor fracture and/or dislocation of the cervical vertebrae may damage the spinal cord. The protrusion of a cervical IV disc into the vertebral canal after a neck injury may cause "spinal cord shock" associated with paralysis inferior to the site of the lesion. In some patients, no fracture or dislocation of cervical vertebrae can be found. If the patient dies, a softening of the spinal cord may be found at the site of the cervical disc protrusion when an autopsy is performed.

Encroachment of the vertebral canal by a protruding IV disc, by a swollen ligamenta flava, or resulting from *osteoarthritis of the zygapophyseal joints* may exert pressure on one or more of the spinal nerve roots of the cauda equina. Pressure may produce sensory and motor symptoms in the area of distribution of the involved spinal nerve. This group of bone and joint abnormalities—*lumbar spondylosis* (degenerative joint disease)—also causes localized pain and stiffness.

Cervical spondylosis is often accompanied by swollen ligamenta flava and osteoarthritis of the zygapophyseal joints. Under these conditions, encroachment on the IV foramina and/or vertebral canal usually occurs. This may cause pressure on the cervical spinal nerve roots and/or spinal cord, resulting in various neurological signs and symptoms.

In some elderly people, the nuclei pulposi of the IV discs degenerate, the vertebrae come together, and the anuli fibrosi bulge anteriorly, posteriorly, and laterally. This leads to the formation of bony outgrowths—*osteophytes*—that are mostly asymptomatic but may produce pressure on the spinal nerve roots and cause sensory and motor symptoms.

Transection of the spinal cord results in loss of all sensation and voluntary movement inferior to the lesion. *The patient is quadriplegic if the cervical cord superior to C5 is transected*, and the patient may die of respiratory failure if the transected cervical cord is superior to C4. The patient is paraplegic—paralysis of both lower limbs—if the transection is between the cervical and lumbosacral enlargements. ⊙

Medical Imaging of the Back

Radiography

Radiographical examination of the vertebral column usually requires both anteroposterior and lateral views. The anatomical features visible in the regions of the vertebral column are illustrated and described in Figures 4.36 to 4.39. Conventional radiographs are excellent for high-contrast structures such as bone. The advent of *digital radiography* allows improved contrast resolution.

Vertebral Body Osteoporosis

This common metabolic bone disease is often detected during routine radiographical studies. The radiographs ▶

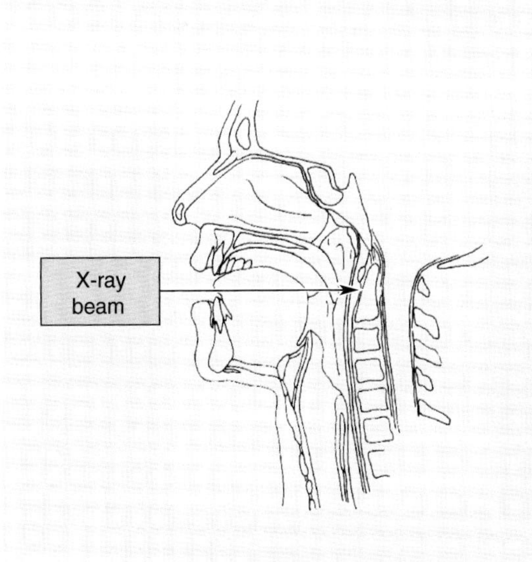

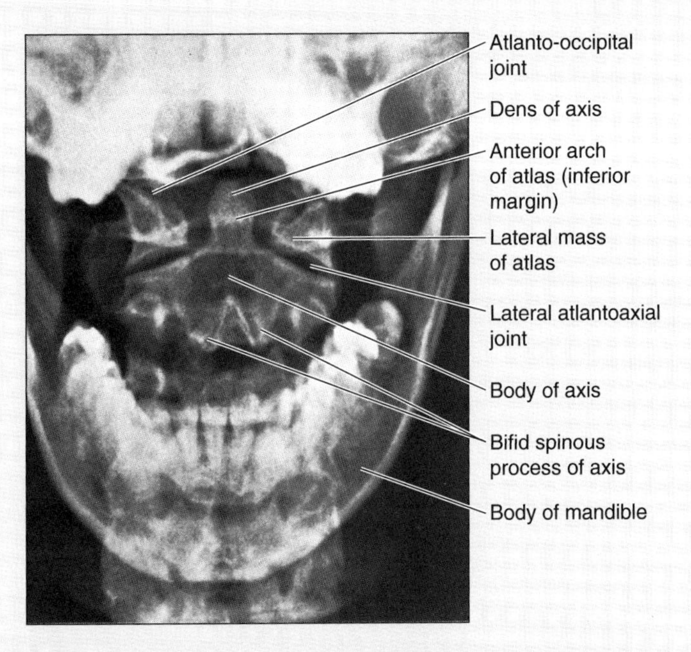

Figure 4.36. Anteroposterior (AP) radiograph of the superior part of the cervical region of the vertebral column. As shown in the orientation drawing, the radiograph was taken through the open mouth. Observe the lateral atlantoaxial joint and the body of the axis (C2) with its dens projecting superiorly between the lateral masses of the atlas (C1).

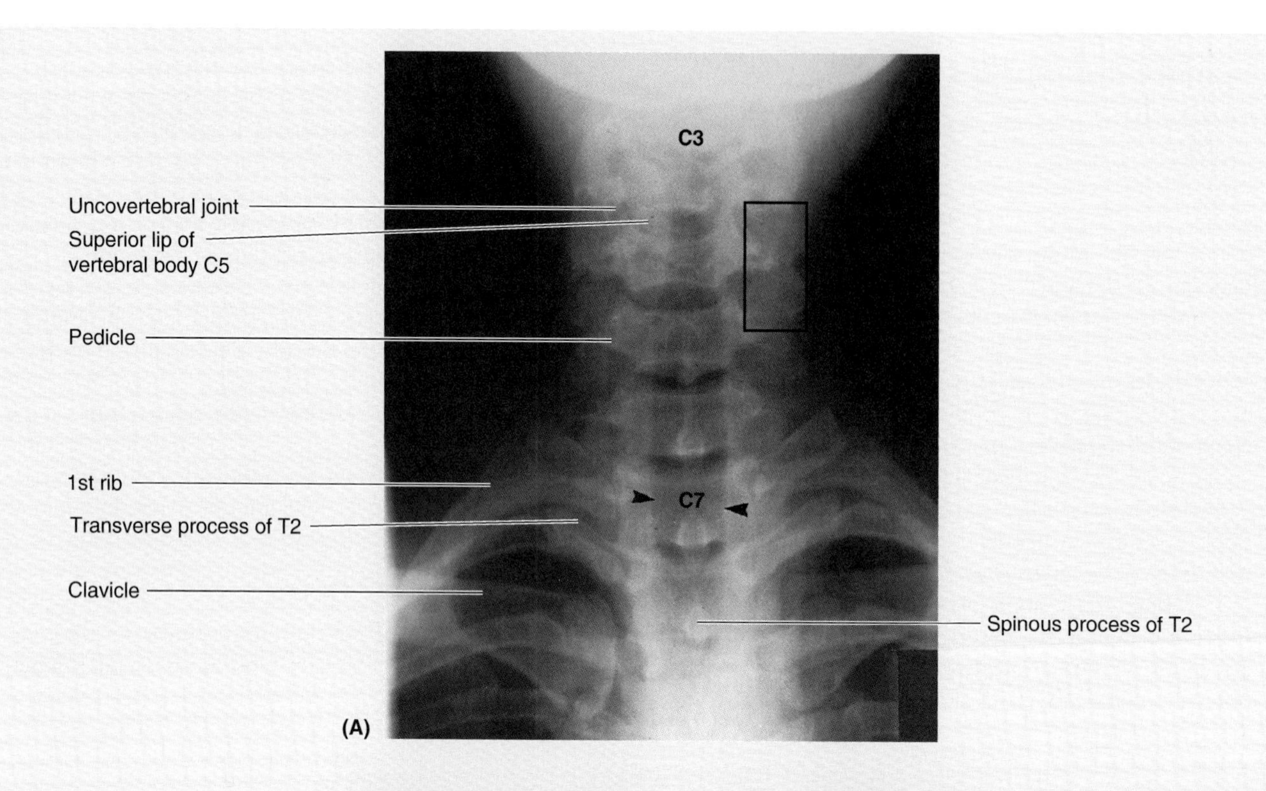

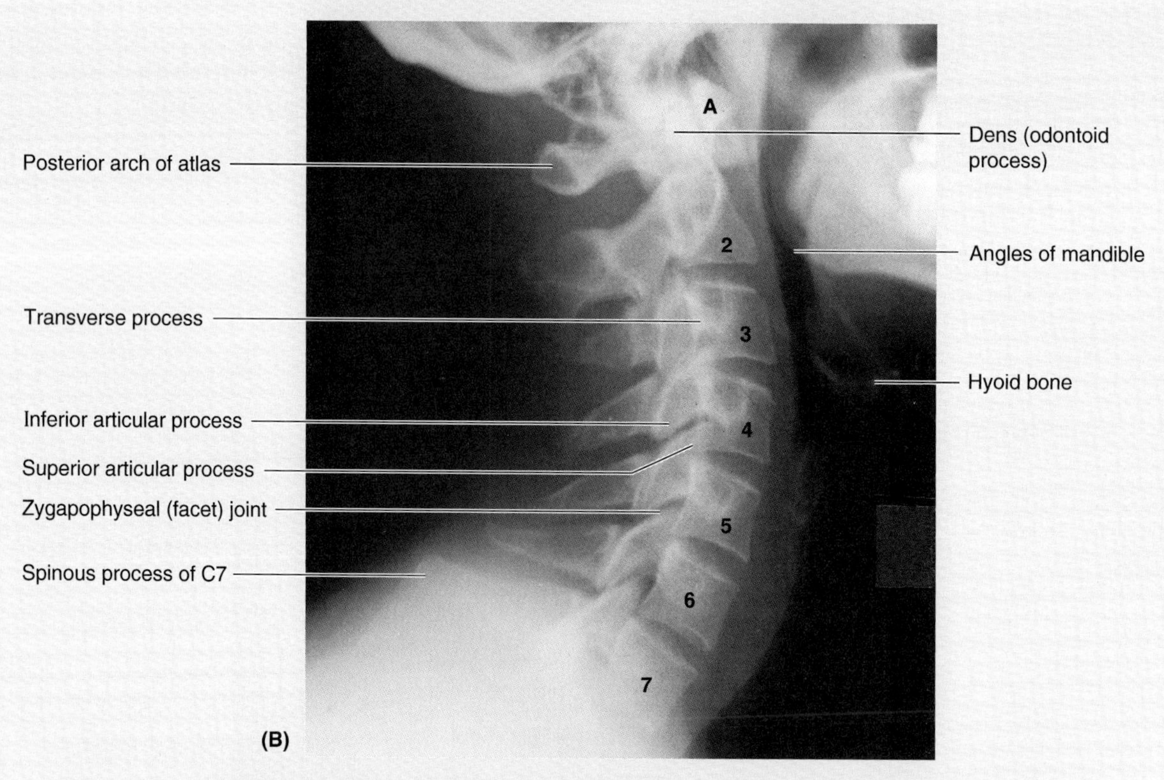

Figure 4.37. Radiographs of the cervical region of the vertebral column. A. Anteroposterior (AP) view. Observe the bifid spinous processes of C2 through C6 vertebrae. The *arrowheads* indicate the margins of the column of air (*black*) in the trachea. The *boxed area* outlines the column of articular processes and the overlapping transverse processes. **B.** Lateral view. Observe that the anterior arch of the atlas (C1) (*A*) is in a plane that is anterior to the curved line joining the anterior borders of the bodies of the vertebrae. The vertebral bodies of C2 through C7 are numbered. Observe also the long spinous process (*C7*) of the vertebra prominens. (Courtesy of Dr. J. Heslin, Toronto, Ontario, Canada)

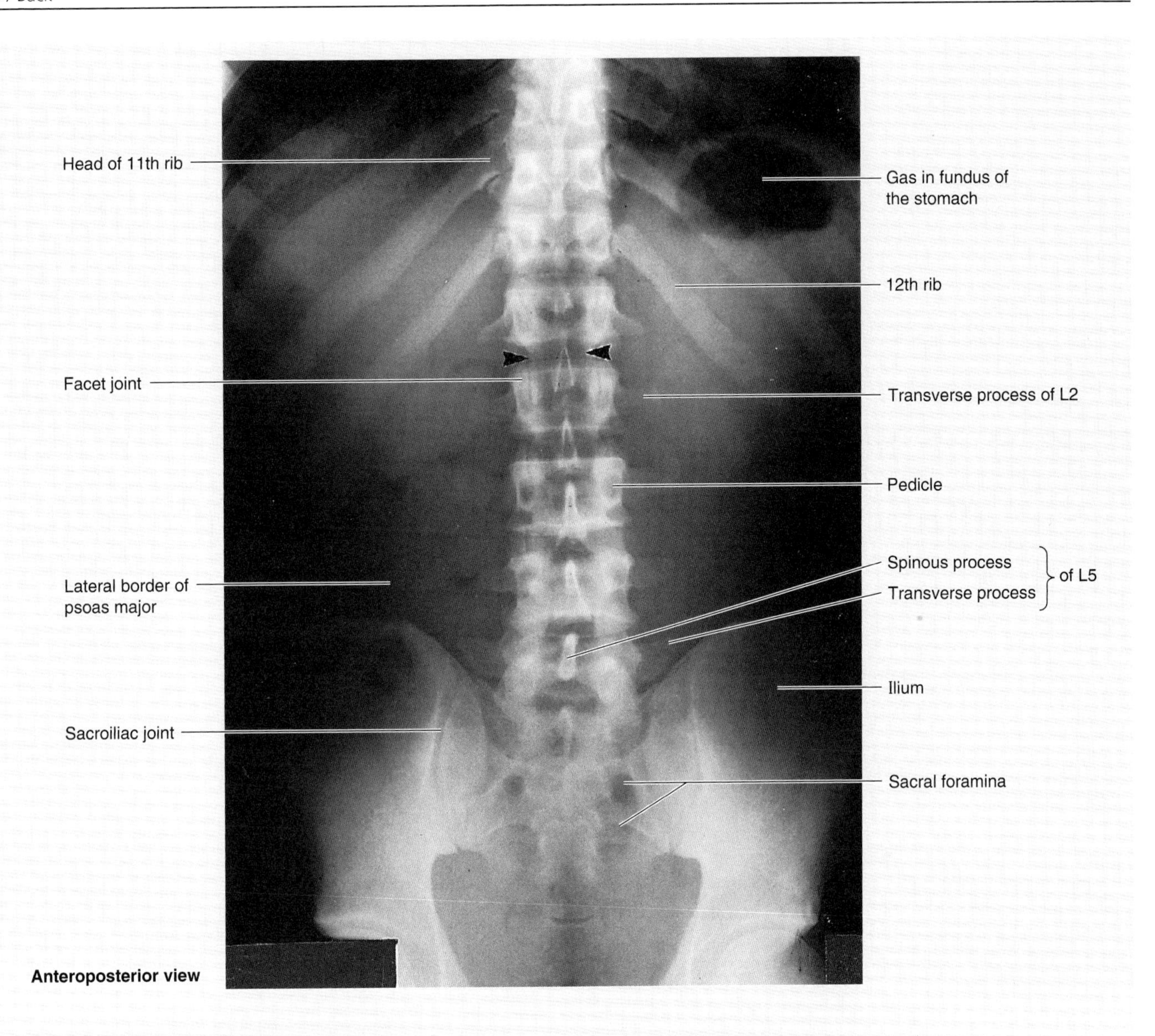

Figure 4.38. Radiograph of the inferior thoracic and lumbosacral regions of the vertebral column. Observe the articulation of the 12th pair of ribs with T12 vertebra. Also observe the vertebral (spinal) canal (*arrowheads*) that contains the spinal cord and meninges (not visible). Observe the midline shadows formed by the spinous processes. The coccyx is not well shown because of its oblique position relative to the x-ray film. In addition, gas and feces in the rectum and sigmoid colon help to obscure the coccyx. (Courtesy of Dr. J. Heslin, Toronto, Ontario, Canada)

▶ show demineralization of the vertebrae and may reveal vertebral collapse and compression fractures of these bones. Vertebral body osteoporosis occurs in all vertebrae but is most common in thoracic vertebrae and is an especially common finding in postmenopausal females.

Myelography

After injection of a radiopaque contrast medium into the subarachnoid space—*myelography*—visualization of the spinal cord and nerve roots is possible (Fig 4.40). CSF is withdrawn by lumbar puncture and replaced with a contrast material. This technique shows the extent of the subarachnoid space and its extensions around the spinal nerve roots within the dural root sleeves (Fig. 4.41). High-resolution MRI with spinal coils has largely supplanted myelography (McCormick and Fetell, 1995).

Computed Tomography

CT differentiates between white and gray matter of the brain and spinal cord. CT has improved the radiological assessment of fractures of the vertebral column, particularly in determining the degree of compression of the ▶

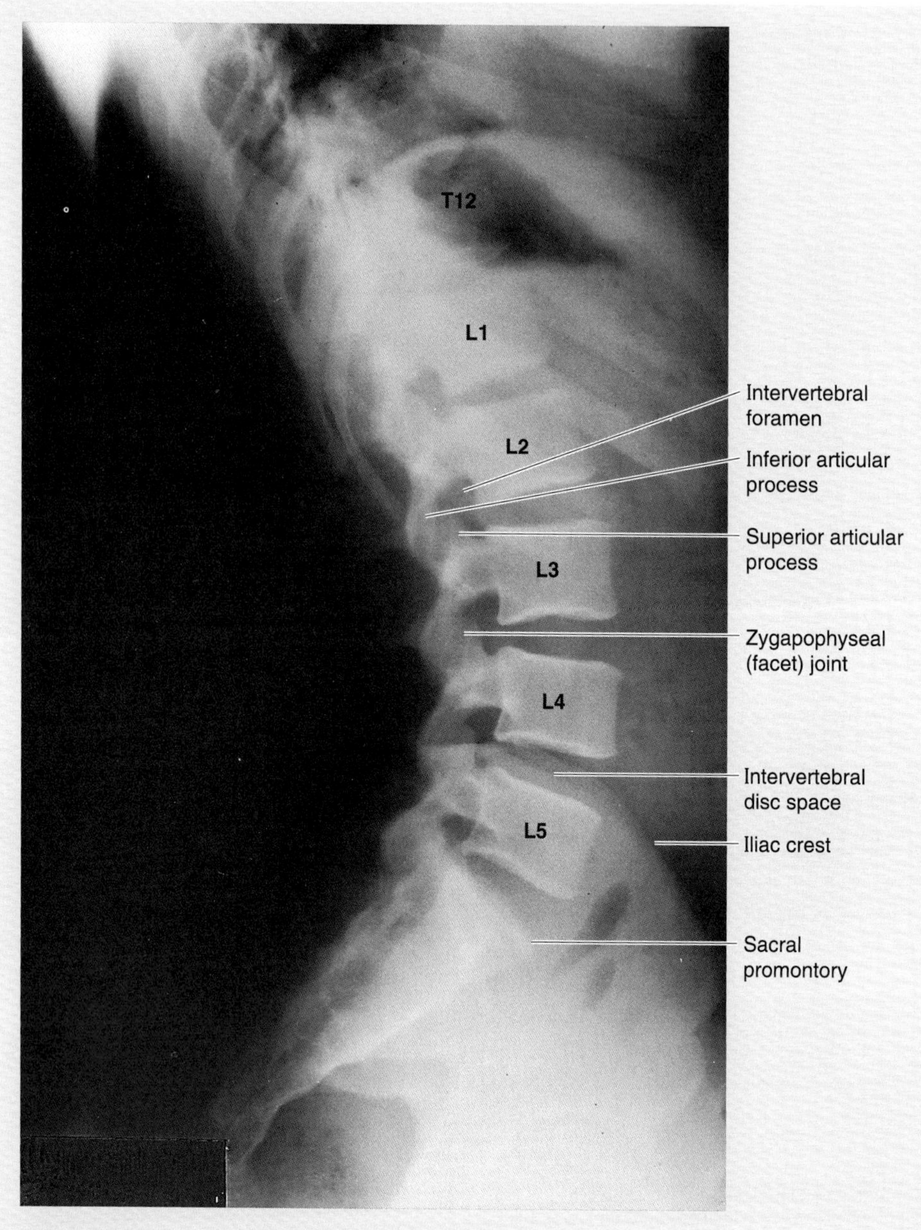

Figure 4.39. Radiograph of the lumbosacral region of the vertebral column. Observe the larger vertebral bodies of the lumbar vertebrae. Also observe the intervertebral (IV) disc spaces, the IV foramina, and the angulation at the lumbosacral junction producing the sacral promontory. Notice that the lumbar IV discs are wedge shaped—especially the L5/S1 disc. Note the zygapophyseal (facet) joint between the superior articular process of L4 and the inferior articular process of L3. (Courtesy of Dr. J. Heslin, Toronto, Ontario, Canada)

▶ spinal cord (McCormick, 1995). CT images of the vertebral column are used to detect lesions and congenital anomalies. The very dense vertebrae attenuate much of the x-ray beam and therefore appear white on the scans (Fig. 4.42). The IV discs have a higher density than the surrounding adipose tissue in the extradural space and the CSF in the subarachnoid space. Herniations of the IV discs are therefore recognizable in CT images, as are displaced fragments of the discs.

Magnetic Resonance Imaging

MRI, like CT, is a computer-assisted imaging procedure, but x-rays are not used as with CT. MRI produces extremely good images of the vertebral column, spinal cord, and CSF (Fig. 4.42). A disadvantage of MRI has been that the person must remain motionless in the scanner for 5 to 10 minutes; however, new technologies are decreasing the time required in the scanner. MRI clearly demonstrates the ▶

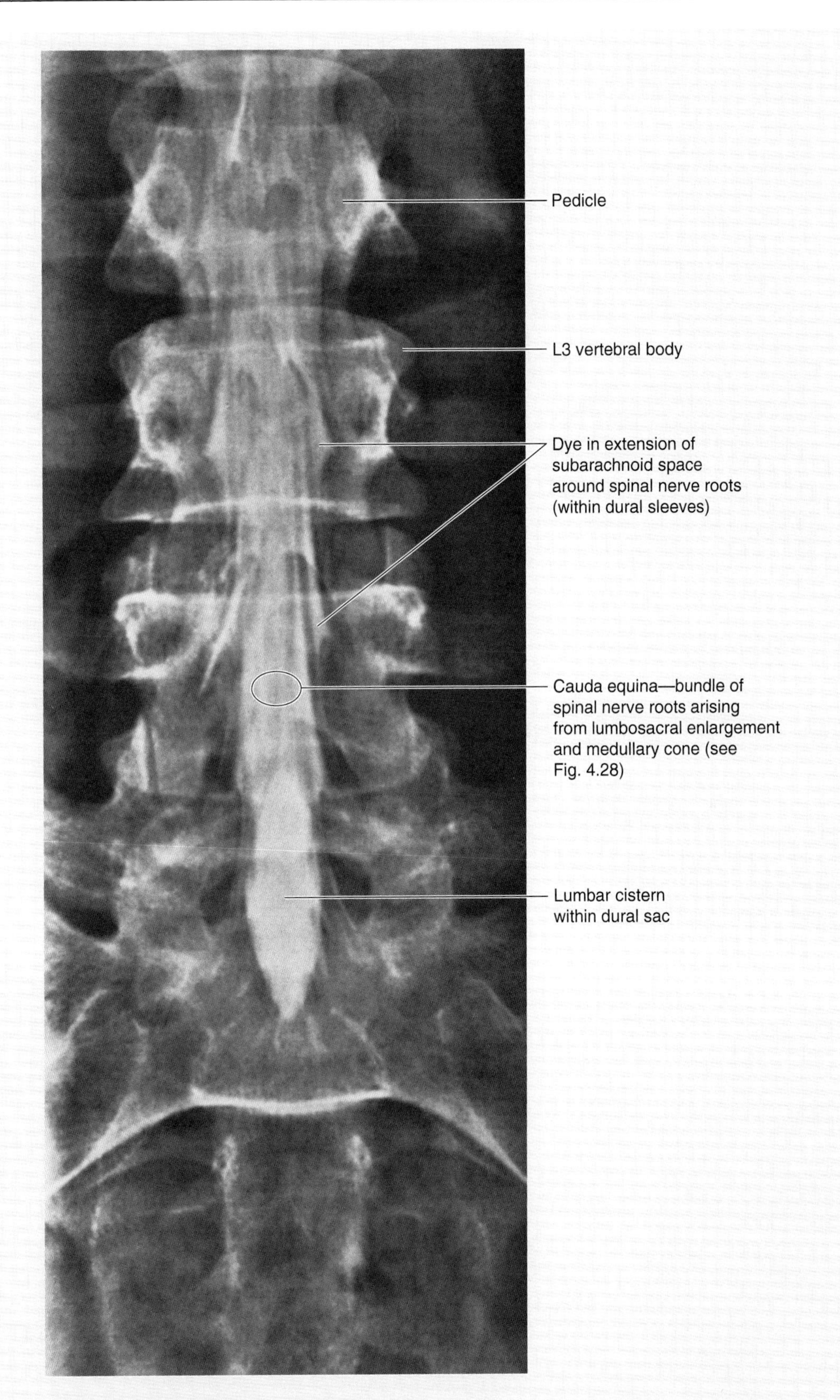

Figure 4.40. Myelogram of the lumbar region of the vertebral column. Contrast medium was injected into the lumbar cistern. The lateral projections indicate the extension of the subarachnoid (leptomeningeal) space in the dural root sleeves around the spinal nerve roots.

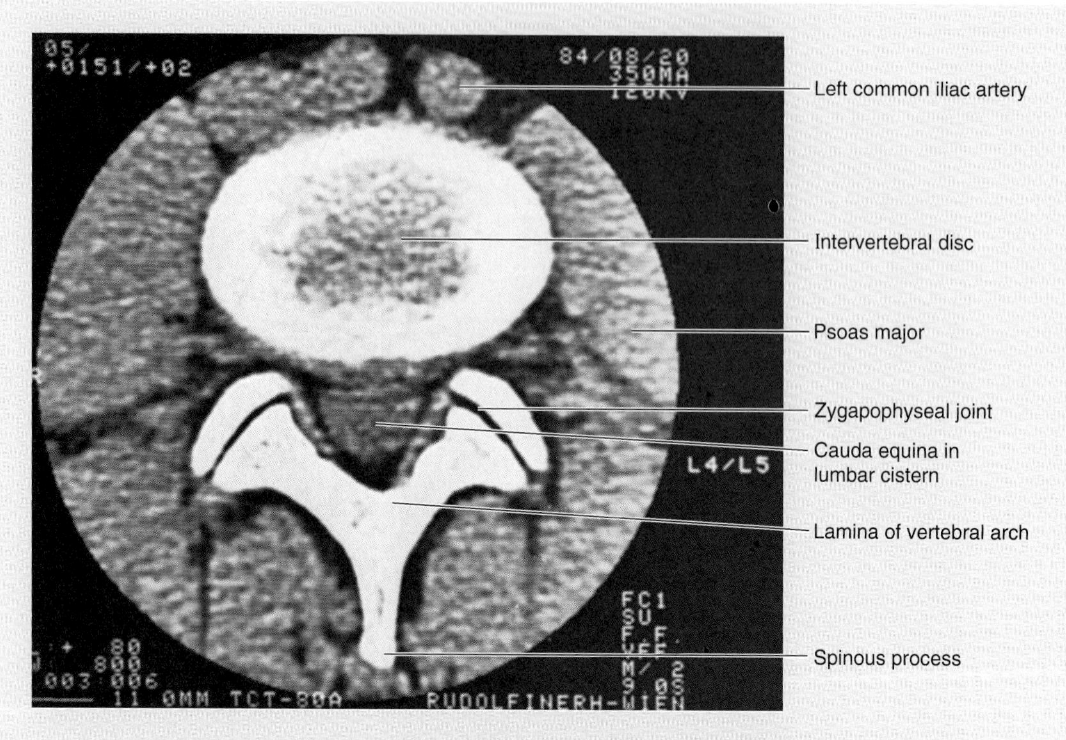

Figure 4.41. Transverse CT image of L4/L5 intervertebral (IV) disc. Observe the cauda equina (L. cauda, tail), the bundle of spinal nerve roots arising from the lumbosacral enlargement and medullary cone (L. conus medullaris) of the spinal cord and running through the lumbar cistern within the vertebral (spinal) canal.

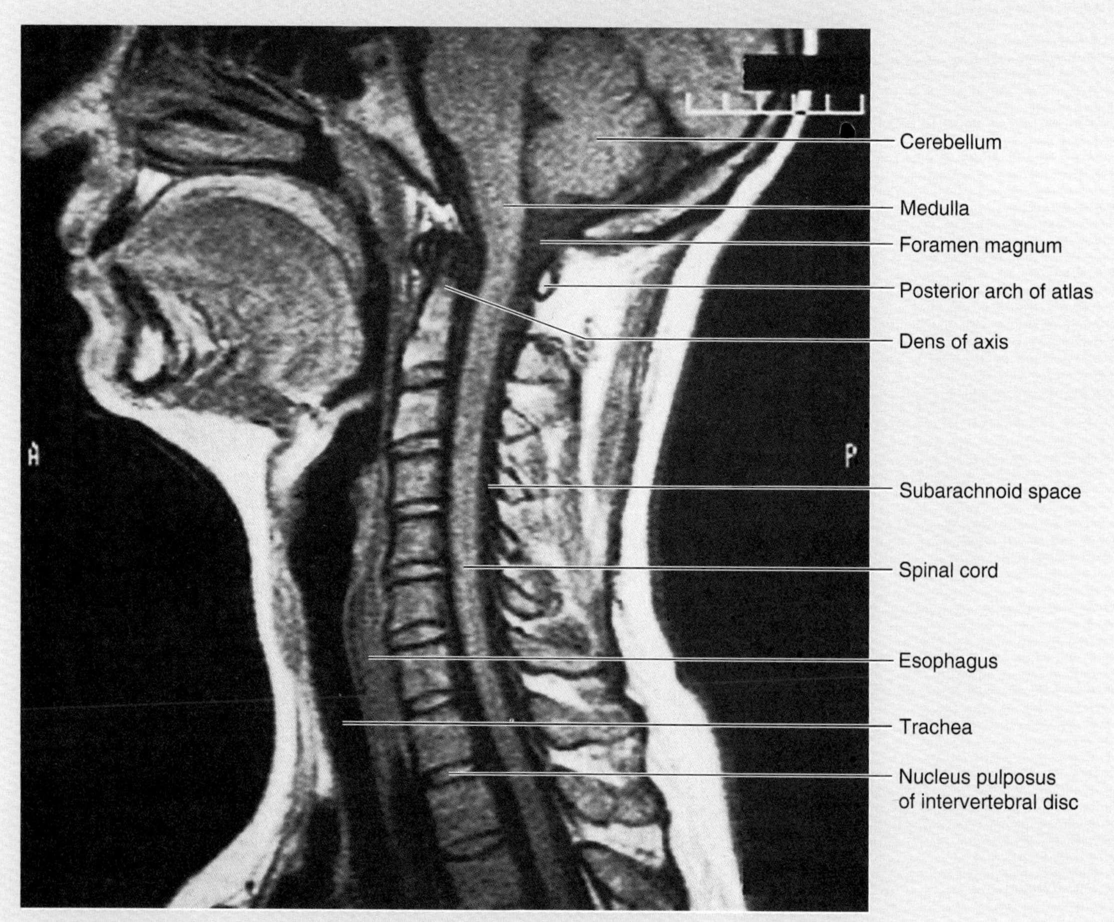

Figure 4.42. Midsagittal MRI of lower head and neck. Observe the cerebellum, medulla, spinal cord, and cervical region of the vertebral column.

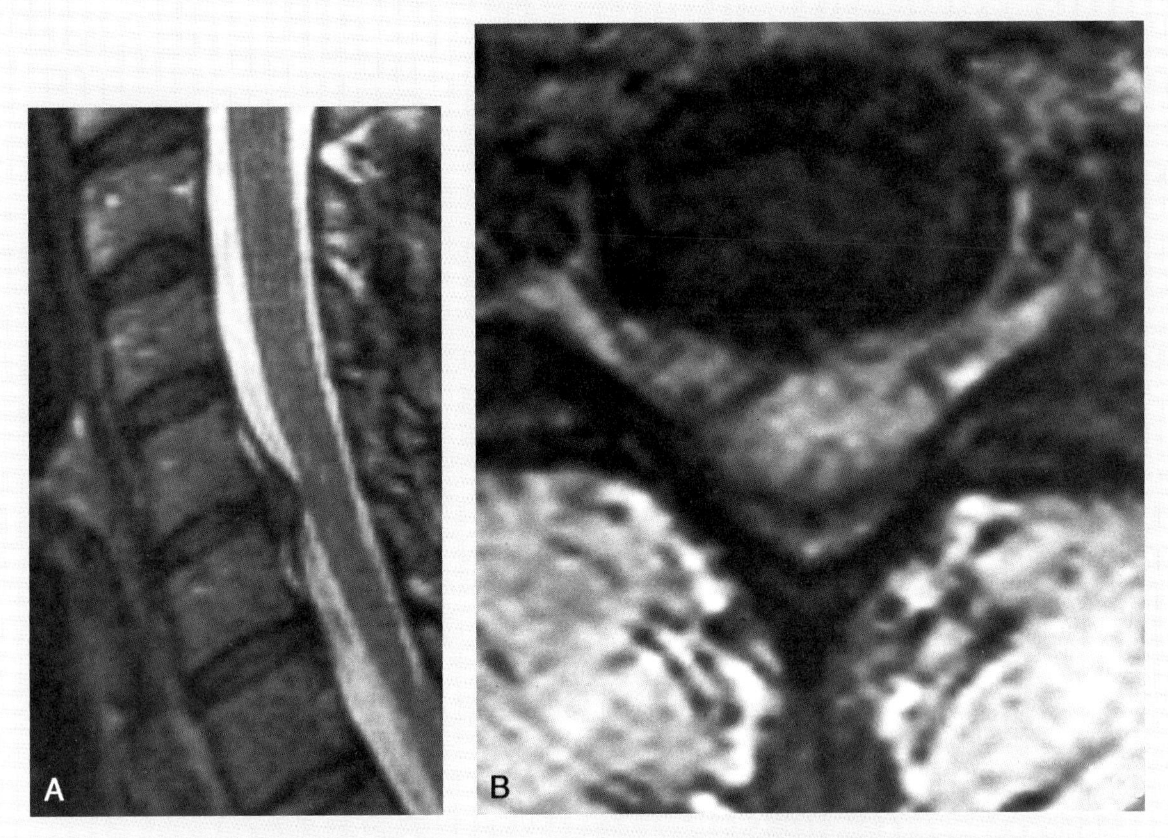

Figure 4.43. Sagittal MRI of the vertebral column. A. T_1-weighted sagittal MRI showing a C5/C6 intervertebral (IV) disc herniation. "T" refers to the time for 63% of longitudinal relaxation to occur; the value is a function of the magnetic field and the chemical environment of the hydrogen nucleus. The image shown here has a bright water signal (see Chan et al. for more information). **B.** Axial MRI demonstrating an IV disc herniation in a patient with radiculopathy—disorder of the spinal nerve roots.

▶ components of the IV discs and shows their relationship to the vertebral bodies and longitudinal ligaments. Herniations of the nucleus pulposus and its relationship to the spinal nerve roots are also well defined. MRI is the imaging procedure of choice for evaluating IV disc disorders and is gradually replacing myelography and CT for the study of these disorders (McCormick, 1995). MRI can identify spinal cord or nerve root compression and demonstrate the degree of degenerative change within the IV disc (Fig. 4.43). MRI is also an ideal screening procedure for the differential diagnosis of structural disorders affecting the spinal cord and spinal nerve roots. ⊙

CASE STUDIES

Case 4.1

A 13-year-old competitive gymnast who practiced 18 to 20 hours a week on her routines complained of low back pain. A physical examination and radiographs revealed that she had a stress fracture of L5 vertebra.

Clinicoanatomical Problems

- What repetitive movement of the vertebral column may result in a stress fracture of the vertebral column?
- What region and vertebra of the column is usually involved?

- What is the clinical name given to this bony defect?
- What activities do you think might produce this type of stress fracture?

The problems are discussed on page 499.

Case 4.2

A weight lifter was preparing for competition by lifting increasingly heavy weights. During hyperextension of his vertebral column, he suddenly he felt a severe pain in his lower back. Following a physical examination, it was decided that he was experiencing acute back pain.

Clinicoanatomical Problems

- What was the likely cause of the man's acute back pain?
- Which back muscles are commonly affected?
- What symptoms and signs would you anticipate in persons with this condition?
- What do you think might prevent this type of back injury?

The problems are discussed on page 499.

Case 4.3

A middle-aged man was lifting a heavy object when he suddenly felt a severe pain in his lower back and hip that radiated down the back of his thigh into his leg, including the dorsum of his foot. MRIs revealed that he had a protruding IV disc at the L5/S1 vertebral level.

Clinicoanatomical Problems

- In which direction does an IV disc usually protrude?
- Why does the nucleus pulposus usually herniate in this direction?
- Why does this cause pain in the lower limb?
- What limb movements do you think would exacerbate the pain?
- What is the anatomical basis of the increased pain?

The problems are discussed on page 499.

Case 4.4

A 52-year-old woman turned her head quickly during a tennis game and suddenly felt a sharp pain in her neck and along her upper limb. Physical examination and medical imaging revealed a herniated degenerated IV disc in the cervical region of her vertebral column.

Clinicoanatomical Problems

- What probably caused the IV disc herniation?
- What causes IV disc degeneration?
- What are the results of disc degeneration?
- What might relieve pain produced by cervical disc degeneration?

The problems are discussed on page 499.

Case 4.5

A 45-year-old man was thrown from a horse. He complained to those who attended him about a severe pain in his neck and a tingling sensation in his upper limbs. A radiograph revealed a dislocated cervical vertebra.

Clinicoanatomical Problems

- What vertebra is usually dislocated in this type of injury?
- What structure may have been compressed if a severe dislocation occurred?
- What is the anatomical basis of the man's injuries?

The problems are discussed on page 499.

Case 4.6

A woman who had just begun labor was experiencing considerable pain. Her obstetrician decided to perform a caudal epidural block.

Clinicoanatomical Problems

- Explain what is meant by a caudal epidural block.
- What nerves are usually anesthetized?
- What are the important bony landmarks used for the administration of caudal anesthesia?

The problems are discussed on page 499.

Case 4.7

A physician was preparing to perform a lumbar puncture to establish the diagnosis in a patient with suspected intracranial bleeding.

Clinicoanatomical Problems

- At what levels would she consider inserting the needle?
- Why are these vertebral levels used?
- Are these levels safe in infants?
- Why is the patient's back flexed as much as possible when performing a lumbar puncture?

The problems are discussed on page 499.

Case 4.8

A patient complaining of neck pain was examined by her physician. After a physical examination and an MRI scan, the physician told her that she had a C5 IV disc herniation.

Clinicoanatomical Problems

- What nerve would a 5th cervical disc protrusion affect?
- Explain the reason for your answer.

The problems are discussed on page 499.

Case 4.9

A 68-year-old man with back and leg pain was examined by a physician. After viewing the MRIs, he told the patient that his discomfort is the result of spinal stenosis.

Clinicoanatomical Problems

- What is spinal stenosis?
- How does stenosis produce back and leg pain?
- What are the causes of spinal stenosis?

The problems are discussed on page 500.

Case 4.10

During a routine radiograph of a young man's back, a common congenital anomaly of the vertebral arch was observed. No signs of back pain were detected.

Clinicoanatomical Problems

- What is the common developmental anomaly of the vertebral column?
- Are the spinal cord and meninges usually normal in these people?
- Does the common anomaly usually cause back pain?

The problems are discussed on page 500.

Case 4.11

During a swarming attack by skinheads, a 16-year-old youth was stabbed with a knife in the posterior aspect of the neck. As he ducked to avoid his attacker, he flexed his neck. Much to the surprise of his assailants, the youth fell to the ground and was completely immobilized from the neck down.

Clinicoanatomical Problems

- How did this serious injury probably occur?
- Using your anatomical knowledge of the vertebral column and its contents, explain the basis of the injury.
- How might this knowledge be used in the diagnosis and treatment of diseases of the nervous system and in the administration of anesthetic agents?

The problems are discussed on page 500.

Case 4.12

A 51-year-old man was waiting for a traffic light to turn green when his car was "rear-ended." His body was pushed forward and his head thrown violently backward. He suffered a slight concussion and felt shaky. When he talked to the man who had hit his car and to the traffic officer, he informed them that he was not badly hurt. The officer noted that the head restraint in his car was not raised to the level that would prevent hyperextension of his neck. The next morning the man's neck was stiff and painful, and he felt pain on the left side of his neck and in his left arm. The neck pain was aggravated by movement of his head. He decided to visit his physician.

Physical Examination The physician observed that the man held his head rigidly and tilted it to the right. She also observed that his chin was pointed to the left and that his neck was slightly flexed. Her palpation of the posterior aspect of his neck revealed some tenderness over the spinous processes of his inferior cervical vertebrae. His biceps reflex was also weak on the left side. She ordered a radiographical study of the cervical region of his vertebral column.

Radiology Report The IV discs between C5 and C6 and between C6 and C7 are thin, and there are small fringes of bone on the opposing edges of the bodies of C5, C6, and C7 vertebrae.

Diagnosis Hyperextension injury of the neck.

Clinicoanatomical Problems

- What is the anatomical basis for the patient's concussion, stiff neck, and pain in his neck and arm?
- What spinal nerve root was probably compressed?
- What muscles were probably affected?
- What probably caused the thinning of the patient's IV discs and the formation of the bony fringes on the edges of his cervical vertebral bodies?

The problems are discussed on page 500.

Case 4.13

While carrying a heavy box of books, a 45-year-old man suddenly experienced severe pain in his lower back. Later he developed a dull ache in the posterolateral aspect of his left thigh that extended along the calf of his leg into his foot. Lateral deviation of the lumbar region of his vertebral column was also observed. He limped when he walked because he did not fully extend his thigh. After consulting his family physician, who recommended back rest, he was referred to a back specialist.

Physical Examination The orthopedist observed that the man's back muscles were in spasm. When asked to indicate the site of the most severe pain, the man pointed to his lower lumbar region. During the examination, no ankle reflex was observed on the left side. He experienced increased pain when the physician raised his extended lower limb on that side. The orthopedist arranged for a radiographical study and MRIs of the man's lower back.

Radiology Report The radiographs show a slight narrowing of the space between the vertebral bodies of L5 and S1. MRI shows that the nucleus pulposus of the L5/S1 IV disc was protruding.

Diagnosis Posterolateral herniation of the nucleus pulposus of the IV disc between L5 and S1.

Clinicoanatomical Problems

- What is the anatomical basis for herniation (protrusion) of an IV disc and the resulting low back pain?
- What produced the lumbar deviation?
- Why did the patient experience pain in the posterolateral aspect of his thigh, leg, and foot?
- Why did the pain increase when the orthopedist raised the patient's extended lower limb?

The problems are discussed on page 500.

Case 4.14

An 18-year-old woman was thrown from a horse and sustained a spinal cord injury as the result of severe hyperextension of her neck. Although she was rushed to the hospital using a proper transport technique, she died after approximately 5 minutes. An autopsy was performed.

Autopsy Report Several cervical vertebrae were fractured. C1 was fractured at both grooves for the vertebral arteries. C6 and C7 vertebrae were broken in several areas. The spinal cord was severely damaged, and extensive bleeding into the soft tissues of the neck occurred.

Diagnosis Transection of the superior end of the spinal cord resulting from multiple fractures of the cervical vertebrae.

Clinicoanatomical Problems

- What are the likely injuries caused by the fractured vertebrae?
- What associated structures of the vertebral column were probably also ruptured?
- Although one would expect the patient to be quadriplegic following a cervical spinal cord transection, what probably caused her death?
- Do cervical fractures usually cause death?

The problems are discussed on page 501.

Case 4.15

A 62-year-old man, who was a heavy drinker and smoker, consulted his physician about feeling a strong pulse in his abdomen. He said it felt like a second heart. He also complained about pain in his abdomen, back, and groin. The physician arranged for radiographical studies, including CT scans.

Radiology Report The plain radiographs show calcium deposits in the wall of the abdominal aorta and an apparent aneurysm. The CT scans revealed an *abdominal aortic aneurysm* that was 11 cm in diameter. Before he could be admitted to the hospital for repair of his abdominal aortic aneurysm, the patient passed out on his way home and was involved in a minor car accident. He was rushed to the hospital and admitted to surgery for repair of the ruptured aneurysm. During surgery, there was extensive mobilization of the aorta, and several segmental arteries were ligated. Although the aorta was successfully repaired using a Dacron graft, the patient was paraplegic and impotent, and his bladder and bowel functions were no longer under voluntary control.

Diagnosis Paraplegia and other neurological deficits resulting in sphincter paralysis in the urinary bladder and anal canal.

Clinicoanatomical Problems

- What arteries supply the spinal cord?
- What is the most likely anatomical basis for the patient's paraplegia, sphincter dysfunctions, and impotence?
- What arteries were probably ligated during surgery?
- Name the important artery supplying the spinal cord that was likely deprived of blood.
- Why is its supply to the spinal cord so important?
- What "bad habits" are known to be associated with the development of aneurysms?

The problems are discussed on page 501.

Case 4.16

A 21-year-old man was involved in a head-on collision. When removed from his sports car, he complained of loss of sensation and voluntary movements in his lower limbs. Upper limb movements also were impaired, particularly in his hands. The patient was kept warm and immobilized until the ambulance arrived. Using a proper transport technique (spine board with the head and neck stabilized), the patient was taken to the emergency department. After examination at the hospital, radiographs of his vertebral columns were taken.

Radiology Report Radiographs show severe dislocation of C6 vertebra on C7 and a chip fracture of the anterosuperior corner of the body of C7.

Diagnosis Dislocation of C6 and C7 vertebrae.

Surgical Treatment Open reduction was carried out, and the spinous processes of C6 and C7 vertebrae were wired together to hold them in normal relation to each other. The reduction was maintained by immobilization of the neck in a plastic collar, thereby allowing the patient to exercise his upper limbs and to sit up within a day or so after the injury.

Clinicoanatomical Problems

- What joints of the cervical region of the vertebral column were dislocated?
- Which ligaments binding the vertebrae together were probably strained and/or torn?
- What was the most likely cause of the patient's paralysis?
- What other physiological functions would no longer be under voluntary control?

The problems are discussed on page 501.

DISCUSSION OF CASES

Case 4.1

Continuous hyperextension of the vertebral column can produce stress fractures of the interarticular parts of the lamina of L5 vertebra, especially in persons whose vertebral columns are immature. The defect—*spondylolysis*—may result in anterior movement of the body of L5 vertebra on the sacrum (*spondylolisthesis*). Persons involved in diving, gymnastics, wrestling, and weight lifting are especially vulnerable to these bony defects. Overhead lifting and repetitive activities performed by painters, carpenters, and electricians can also produce spondylolysis.

Case 4.2

The weight lifter probably experienced a severe back spasm. The back muscles that are usually involved are the deep muscles (e.g., erector spinae and its subdivisions). The usual symptom of muscular spasm is a constant, dull ache in the lower back. The obvious sign is a limited range of motion of the vertebral column and large areas of tenderness. Poor posture and improper lifting mechanics cause most acute back pains. Proper sitting and lifting techniques would be advisable, such as lifting with the lower limbs instead of the back.

Case 4.3

The herniated nucleus pulposus of an IV disc usually protrudes posterolaterally and causes pain that radiates from the sacroiliac and buttock regions to the posterior aspect of the thigh, leg, and dorsum of the foot. The IV disc is supported anterolaterally and directly posteriorly by anterior and posterior longitudinal ligaments, the posterior ligament being weaker and narrower. In addition, the nucleus pulposus is eccentrically placed toward the posterior aspect of the disc. A posterior herniation of the nucleus pulposus through a degenerated anulus fibrosus and weakened posterior longitudinal ligament into the vertebral canal could compress the spinal cord or cauda equina; however, this type of herniation is uncommon. Most herniations occur in the posterolateral direction, passing to either side of—but impinging on—the posterior longitudinal ligament into the IV foramen and thus compressing the spinal nerve roots. Extrusion of disc material can also cause inflammation of the roots owing to chemical irritation from substances released from the nucleus pulposus. Pressure on the lumbar spinal nerve roots causes the pain to radiate through the buttock into the lower limb. *Paresthesia* ("pins and needles" sensation) also occurs in the area of skin involved (dermatome). Exacerbating pain usually occurs (although some patients experience relief from pain) when the extended lower limb is flexed passively at the hip because the spinal nerve root is stretched.

Case 4.4

IV disc degeneration commonly occurs in the cervical region. Disc degeneration is associated with dehydration of the nucleus pulposus. Loss of water and mucopolysaccharide from the nucleus pulposus results in narrowing of the IV space and a reduced capacity of the disc to act as a cushion between the vertebrae. Disc degeneration diminishes stature, limits mobility, and decreases the size of the IV foramina, thereby increasing the possibility of spinal nerve root compression. Superior traction of the head may relieve pain caused by narrowing of the IV foramina and pressure on the zygapophyseal joint capsules.

Case 4.5

C6 vertebra dislocates, sliding anteriorly on C7 vertebra. Because the vertebral canal is relatively large in this region, a slight displacement of a vertebra may not compress the spinal cord but could cause contusion (bruising) of the nerve tissue, resulting in a tingling sensation. However, severe displacement of cervical vertebrae may compress the spinal cord and cause paralysis of the limbs.

Case 4.6

A caudal epidural block means that an anesthetic agent is administered using an in-dwelling catheter in the sacral canal that is inserted through the sacral hiatus. The anesthetic bathes the S2 through S4 spinal nerve roots, including the pain fibers from the uterine cervix and upper vagina. The sacral cornua are the bony landmarks used in locating the sacral hiatus for the administration of caudal anesthesia.

Case 4.7

Lumbar puncture is usually done between L3/L4 or L4/L5 vertebrae. These levels are safe for infants, children, and adults because the spinal cord usually ends at L3 vertebra in infants and the inferior end of L1 vertebra in most adults. Hence, the inferior end of the dural sac contains the cauda equina and no spinal cord. The backs of patients are flexed for lumbar puncture to separate the spinous processes and laminae so the needle may be inserted through the back tissues into the lumbar cistern containing CSF.

Case 4.8

The 6th cervical nerve would be affected because the IV discs are numbered according to the vertebra below which each lies, whereas the cervical nerves are numbered according to the vertebra above which each nerve lies. Consequently, a herniated C5 disc does not affect the 5th cervical nerve but instead affects the 6th nerve.

Case 4.9

Spinal stenosis is a narrowing of the vertebral canal or the IV foramina, usually in the lumbar region of the vertebral column. It may be congenital or it may be caused by degeneration of the vertebrae. The stenosis (narrowing) causes compression of the spinal cord if at cord levels more superior than L1 or compression of the cauda equina at L2 through L5 levels. The symptoms include back pain and paresthesias that result from ischemia of the cauda equina.

Case 4.10

The common developmental anomaly of the vertebral column is *spina bifida occulta.* This results from failure of union of the laminae of the vertebral arch and varies from a slight defect to almost complete failure of the vertebral arch to form. The spinal cord is usually normal, and the defect is not usually accompanied by symptoms. If more than one lumbar vertebra is affected, frequency of back pain may be greater. The spinal cord and meninges are usually normal in people with spina bifida occulta.

Case 4.11

The spinous processes and laminae of the vertebral arches usually protect the spinal cord during injuries to the posterior aspect of the neck, even those resulting from stab wounds. However, when the neck is flexed, the spaces between these processes and laminae of the vertebral arches increase. This could permit a knife to pass between them, enter the vertebral canal, and sever the cervical region of the spinal cord.

Verify this movement of the cervical vertebrae during flexion by placing your hand on the back of your neck and then flexing it as much as possible. Note that the space between the external occipital protuberance and the spinous process of the axis widens, as do the spaces between the spinous processes of C2 through C7 vertebrae. Had the knife severed the spinal cord superior to C4, the lesion would have stopped the patient's breathing. At this site the injury would have interfered with the patient's phrenic outflow (C3, C4, and C5), the nerve supply to the diaphragm. As a result, the patient may have died in a few minutes.

Complete transection of the spinal cord results in loss of all sensation and voluntary movement inferior to the lesion. The patient is quadriplegic when the lesion is superior to C5 segment because the nerves supplying the upper limb are derived from C5 through T1 segments of the spinal cord. Had the youth been stabbed in the same place when his head was erect, he probably would not have been severely injured. The knife likely would have struck the spinous processes and/or laminae of the cervical vertebrae and glanced off without damaging the spinal cord.

Similar gaps exist between the lumbar spinous processes when the back is flexed. The presence of these gaps is important because they enable clinicians to insert a lumbar puncture needle for withdrawal of CSF. In adults, the needle is usually inserted between the spinous processes of L3 and L4 or L4 and L5 vertebrae into the subarachnoid space, inferior to the termination of the spinal cord. This procedure, known as a *lumbar puncture*, is performed to obtain a sample of CSF. These punctures are performed during the investigation of some diseases of the nervous system (e.g., meningitis). Anesthetic solutions can also be injected into the subarachnoid space at this level (lumbar spinal block) or into the epidural space (caudal anesthesia). The extradural or epidural space is a real space that is filled with loose connective tissue, fat, and veins. For epidural anesthesia—called an *epidural block*—the needle may also be inserted through the sacral hiatus at the inferior end of the sacrum, which is inferior to the level of the dural sac.

Case 4.12

The association of rear-end automobile collisions and hyperextension injuries of the soft tissues of the cervical region of the vertebral column is well known. Head restraints ("headrests") and bucket seats have been designed to minimize these injuries. However, a head restraint is useless if it is not raised so that the head will hit it if a rear-end collision occurs.

The mechanism of injury in the present case was primarily one of *rapid hyperextension of the neck.* Because the head restraint was not in the correct position, there was nothing to restrict posterior movement of the head and neck. A *hyperflexion injury of the neck* also may have occurred when the head was secondarily flung forward onto the chest. This would likely have occurred because the muscles of the patient's neck, the chief stabilizers of the cervical region of the vertebral column, would be relatively relaxed when the patient was caught off guard.

During severe hyperextension of the neck, the anterior longitudinal ligament and neck muscles would be severely stretched, and some of its fibers were probably torn, leading to small hemorrhages. The resulting muscle spasms would account for the patient's stiff and painful neck. The concussion probably resulted from the sudden impact of the frontal and sphenoid bones against the frontal and temporal poles of his brain. The pain in the patient's left shoulder and the weakness of his biceps reflex on the left likely resulted from compression of the left C6 spinal nerve root, probably by a posterolateral herniation of the IV disc between the C5 and C6 vertebrae.

The musculocutaneous nerve (C5 and C6) supplies the biceps brachii muscle, and the *biceps reflex* is also mediated through C5 and C6. A hyperextension injury of the neck is popularly called a "*whiplash injury.*" Many doctors consider this term an unacceptable medical designation because no well-defined clinical syndrome or fixed pathological condition is associated with the injury.

The thinning of the IV discs in the cervical region probably resulted from desiccation (loss or removal of fluid) of the nuclei pulposi of the IV discs. *Degenerative disc disease* often occurs with advancing age and may result in bulging of the anuli fibrosi of the IV discs. Formation of fringes of subperiosteal new bone (osteophytes) on the edges of the vertebral bodies occurs in older persons. These fringes of bone can also compress the spinal nerve roots.

Case 4.13

The patient's low back pain and muscle spasm, sometimes called "*lumbago,*" was probably caused by rupture of the posterolateral part of the anulus fibrosus and protrusion of the nucleus pulposus of the IV disc between L5 and S1 vertebrae. This part of the anulus is not supported by either the posterior or anterior longitudinal ligaments.

The lumbar deviation of the patient's vertebral column was produced by spasm of the intrinsic back muscles. Muscle spasm has a protective, splinting effect on the vertebral column. When the man lifted the heavy box of books, the strain on his IV discs and ligaments was so severe that the anulus fibrosus of one of them ruptured, resulting in herniation of the nucleus pulposus of the IV disc. The disc herniation exerted pressure on a nerve root or roots of the sciatic nerve.

Disc protrusions most commonly occur posterolaterally, where the anulus is thin and unsupported by either the anterior or posterior longitudinal ligament. As the dorsal and ventral nerve roots cross the posterolateral region, the protruding nucleus pulposus often affects one or more spinal nerve roots. Some hemorrhage, muscle spasm, and edema (swelling) would be present at the site of the rupture, which probably caused the patient's initial back pain. In the present case, pressure seems to have affected the *S1 component of the sciatic nerve* as it passes inferiorly, posterior to the IV disc between L5 and S1. As a result, the patient experienced pain over the posterolateral region of his thigh and leg. When the physician raised the patient's extended lower limb at the hip joint, the sciatic nerve was stretched. As its S1 component was compressed by the protruding disc, the lower limb pain increased because of stretching of the compressed nerve fibers in that root.

Sciatica is the name given to pain in the area of distribution of the sciatic nerve (L4–S3). Pain is felt in one or more of the following areas: the buttock, especially the region of the greater sciatic notch, the posterior aspect of the thigh; the posterior and lateral aspects of the leg; and usually parts of the lateral aspect of the ankle and foot. The variation in the location of the pain results from the fact that a posterolateral protrusion of a single lumbar disc presses on only one nerve root. However, the sciatic nerve is composed of several inferior lumbar and superior sacral roots.

The *paravertebral muscle spasm and pain* resulted from the muscles being in continuous tonic contraction in an attempt to prevent the vertebrae from moving and causing more severe pain. The narrowing or thinning of the space between the vertebral bodies noted in the radiographs is caused by a greater reduction of the disc material between the adjacent vertebral bodies at the L5/S1 level than normally occurs with advancing age.

Case 4.14

Severe hyperextension of the neck resulting from a fall on the head often causes a fracture of C1 vertebra at one or both grooves for the vertebral arteries. The vertebral arch of C1 may also break at the isthmus between the lateral mass and the inferior articular process. Probably the patient's anterior longitudinal ligament and the anterior part of the IV disc between C2 and C3 were also ruptured. As the patient hit the ground, hyperextending her neck, her skull and C1 and C2 vertebrae were probably separated from the rest of her vertebral column. As a result, her spinal cord was probably torn in the superior cervical region. Patients with this severe injury rarely survive more than a few minutes because the injury to the spinal cord is superior to the *phrenic outflow* (origin of phrenic nerves). Because these nerves are the sole motor supply to the diaphragm, respiration is severely affected; in addition, the actions of the intercostal muscles are lost. Few fatalities are caused by neck fractures, and not all neck fractures result in paralysis. Improper transport techniques, however, can cause paralysis.

Case 4.15

During certain surgical procedures in the abdomen, such as resection of an *aortic aneurysm*, it is necessary to ligate some aortic segmental branches (e.g., lumbar arteries). If the great anterior segmental medullary artery arises from one of the arteries that has been ligated, the blood supply to the *lumbosacral enlargement of the spinal cord* may be severely impaired. As a result, spinal cord infarction, paraplegia, impotence, and loss of sensation inferior to the lesion may follow. Arising more frequently on the left from an inferior intercostal (T6–T12) or lumbar (L1–L3) artery, the great anterior segmental medullary spinal artery enters the vertebral canal through an IV foramen. It supplies blood mainly to the inferior two-thirds of the spinal cord; therefore, it is understandable that function is lost in the lower limbs, bladder, and intestine when this artery and part of the spinal cord are deprived of blood.

The development of an aneurysm is accelerated by smoking; aneurysms occur three times more frequently in smokers than in nonsmokers. For a description of the diagnosis and surgical treatment of abdominal aortic aneurysm, including illustrations of aneurysm repair, see Ameli (1989).

Case 4.16

The patient is paraplegic, and the condition is known as *paraplegia*. Both the IV disc and the *zygapophyseal joints* between the bodies and vertebral arches of C6 and C7, respectively, were dislocated in this patient. Probably the posterior longitudinal and interspinous ligaments, and the anulus fibrosus, ligamenta flava, and articular capsules of the zygapophyseal joints, were severely injured; some of them may have been torn.

The cervical region, being the most mobile part of the vertebral column, is most vulnerable to injuries such as dislocations and fracture-dislocations. Most injuries occur when a person's head moves forward suddenly and violently, as in the present case, or when the back of the head is struck by a hard blow. In *hyperflexion injuries of the neck*, the anterior longitudinal ligament is usually not torn, and when the patient's neck is placed in a position of extension, this ligament tightens and, along with the plastic cervical collar, tends to hold the vertebrae together.

During the surgical treatment of the injury, the spinous processes of C6 and C7 were wired together to help stabilize the vertebral column during the initial part of the rehabilitation program and to promote healing of the strained and/or torn ligaments. Normally, the vertebral bodies are bound together by the longitudinal ligaments and the anuli fibrosi of the IV discs.

The *posterior longitudinal ligament* is a narrower and weaker band than the anterior longitudinal ligament; however, like the anterior longitudinal ligament, it is attached to the IV discs and to the edges of the vertebral bodies. It lies inside the vertebral canal and tends to prevent excessive flexion of the vertebral column. Because dislocation occurred in this case, the posterior longitudinal ligament and the ligamenta flava were severely stretched and were probably torn.

As the anulus fibrosus of the IV disc attaches to the compact bony rims on the articular surfaces of the vertebral discs, its posterior part would also have been stretched and torn at the C6 and C7 level. It is possible that protrusion of the nucleus pulposus of the IV disc between these vertebrae also occurred because these nuclei are semifluid in young adults. Because the vertebral canal in the cervical region is usually larger than the spinal cord, some displacement of the vertebrae can occur without causing damage to the spinal cord. In view of the patient's paraplegia, it is likely that the spinal cord was severely stretched and/or torn. At the moment of impact, the displacement of C6 on C7 was undoubtedly greater than shown in the radiograph. An initial period of *spinal shock* in these cases lasts from a few days to several weeks, during which time all somatic and visceral activity is abolished. On return of reflex activity, there is spasticity of muscles and exaggerated tendon reflexes inferior to the level of the lesion. Bladder and bowel functions are no longer under voluntary control.

References and Suggested Readings

Ameli FW: Vascular surgery. *In* Gross A, Gross P, Langer B (eds): *Surgery. A Complete Guide for Patients and Their Families.* Toronto, Harper & Collins, 1989.

Barr ML, Kiernan JA: *The Human Nervous System: An Anatomical Viewpoint,* 6th ed. Philadelphia, JB Lippincott Company, 1993.

Behrman RE, Kliegman RM: *Nelson Textbook of Pediatrics,* 15th ed. Philadelphia, WB Saunders, 1996.

Bergman RA, Thompson SA, Afifi AK, Saadeh FA: *Compendium of Human Anatomic Variation. Text, Atlas, and World Literature.* Baltimore, Urban & Schwarzenberg, 1988.

Bogduk N, Macintosh JE: Applied anatomy of the thoracolumbar fascia. *Spine* 9:164, 1984.

Buxton DF, Peck D: Neuromuscular spindles relative to joint movement complexities. *Clin Anat* 2(4):211, 1989.

Chan S, Khandji AG, Hilal SK: How to select diagnostic tests. *In* Rowland LP (ed): *Merritt's Textbook of Neurology,* 9th ed. Baltimore, Williams & Wilkins, 1995.

Crockard HA, Heilman AE, Stevens JM: Progressive myelopathy secondary to odontoid fractures: Clinical, radiological, and surgical features. *J Neurosurg* 78:579, 1993.

Dvorak J, Schneider E, Saldinger P, Rahn B: Biomechanics of the craniovertebral region: The alar and transverse ligaments. *J Orthop Res* 6: 452, 1988.

Fishman RA: Lumbar puncture and CSF examination. *In* Rowland LP (ed): *Merritt's Textbook of Neurology,* 9th ed. Baltimore, Williams & Wilkins, 1995.

Greer M: Structural malformations. *In* Rowland LP (ed): *Merritt's Textbook of Neurology,* 9th ed. Baltimore, Williams & Wilkins, 1995.

Griffith HW: *Complete Guide to Sports Injuries.* Los Angeles, Price Stern Sloan, 1986.

Haines DE (ed): *Fundamental Neuroscience.* New York, Churchill Livingstone, 1997.

Haines DE, Mihailoff GA, Yezierski RP: The spinal cord. *In* Haines DE (ed): *Fundamental Neuroscience.* New York, Churchill Livingstone, 1997.

Hew E: Anesthesia. *In* Gross A, Gross P, Langer B (eds): *Surgery. A Complete Guide for Patients and Their Families.* Toronto, Harper & Collins, 1989.

Koenigsberger MR, Kairam R: Birth injuries and development abnormalities. *In* Rowland LP (ed): *Merritt's Textbook of Neurology,* 9th ed. Baltimore, Williams & Wilkins, 1995.

Marotta J: Spinal trauma. *In* Rowland LP (ed): *Merritt's Textbook of Neurology,* 9th ed. Baltimore, Williams & Wilkins, 1995.

McCormick PC: Intervertebral discs and radiculopathy. *In* Rowland LP (ed): *Merritt's Textbook of Neurology,* 9th ed. Baltimore, Williams & Wilkins, 1995.

McCormick PC, Fetell MR: Spinal tumors. *In* Rowland LP (ed): *Merritt's Textbook of Neurology,* 9th ed. Baltimore, Williams & Wilkins, 1995.

Moore KL, Persaud TVN: *The Developing Human: Clinically Oriented Embryology,* 6th ed. Philadelphia, WB Saunders, 1998.

Murray MJ, Bower TC, Carmichael SW: Anatomy of the anterior spinal artery in pigs. *Clin Anat* 5:457, 1992.

O'Rahilly R: *Gardner-Gray-O'Rahilly Anatomy. A Regional Study of Human Structures,* 5th ed. Philadelphia, WB Saunders, 1986.

Rowland LP, McCormick PC: Lumbar spondylosis. *In* Rowland LP (ed): *Merritt's Textbook of Neurology,* 9th ed. Baltimore, Williams & Wilkins, 1995.

Salter R: Orthopedic surgery: pediatric. *In* Gross A, Gross P, Langer B (eds): *Surgery. A Complete Guide for Patients and Their Families.* Toronto, Harper & Collins, 1989.

Sanderson PL, Wood PL: Surgery for lumbar stenosis in old people. *J Bone Joint Surg* 75:393, 1993.

Swartz MH: *Textbook of Physical Diagnosis. History and Examination,* 2nd ed. Philadelphia, WB Saunders, 1994.

Williams PL, Bannister LH, Berry MM, Collins P, Dyson M, Dussek JE, Ferguson MWJ (eds): *Gray's Anatomy,* 38th ed. New York, Churchill Livingstone, 1995.

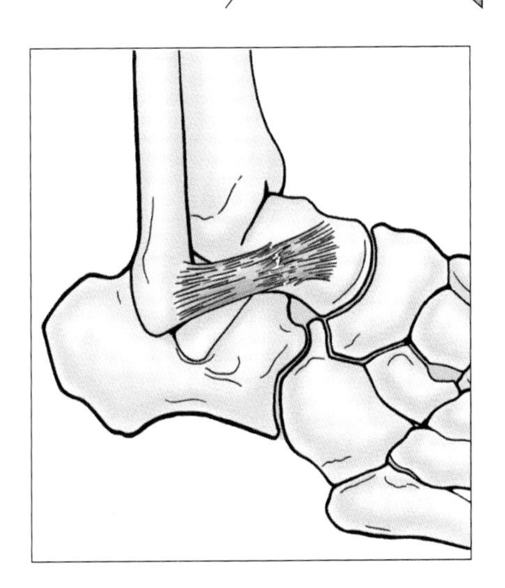

chapter

5 Lower Limb

The lower limb (extremity) is specialized to support body weight and *locomotion*—the ability to move from one place to another—and to maintain *equilibrium*—the condition of being evenly balanced. The lower limbs are connected by the **pelvic girdle** (bony ring formed by the hip bones and sacrum) to the trunk. Some muscles acting on the lower limb arise from the pelvic girdle and vertebral column. It is customary when describing the lower limbs to include regions that are transitional between the trunk and lower limbs, such as the gluteal region (G. gloutos; buttocks). *The lower limb has four parts* (Fig. 5.1):

- **Hip**, the lateral prominence of the pelvis from the *iliac crest* to the thigh containing the *hip bone*, which connects the skeleton of the lower limb to the vertebral column
- **Thigh**, the part between the hip and knee containing the *femur* (thigh bone), which connects the hip and knee; the *patella* (knee cap) covers the anterior surface of the knee
- **Leg**, the part between the knee and ankle containing the *tibia* (shin bone) and *fibula* (calf bone), which connects the knee and ankle
- **Foot**, the distal part containing the *tarsus, metatarsus*, and *phalanges* (toe bones), which connect the ankle and foot (L. pes).

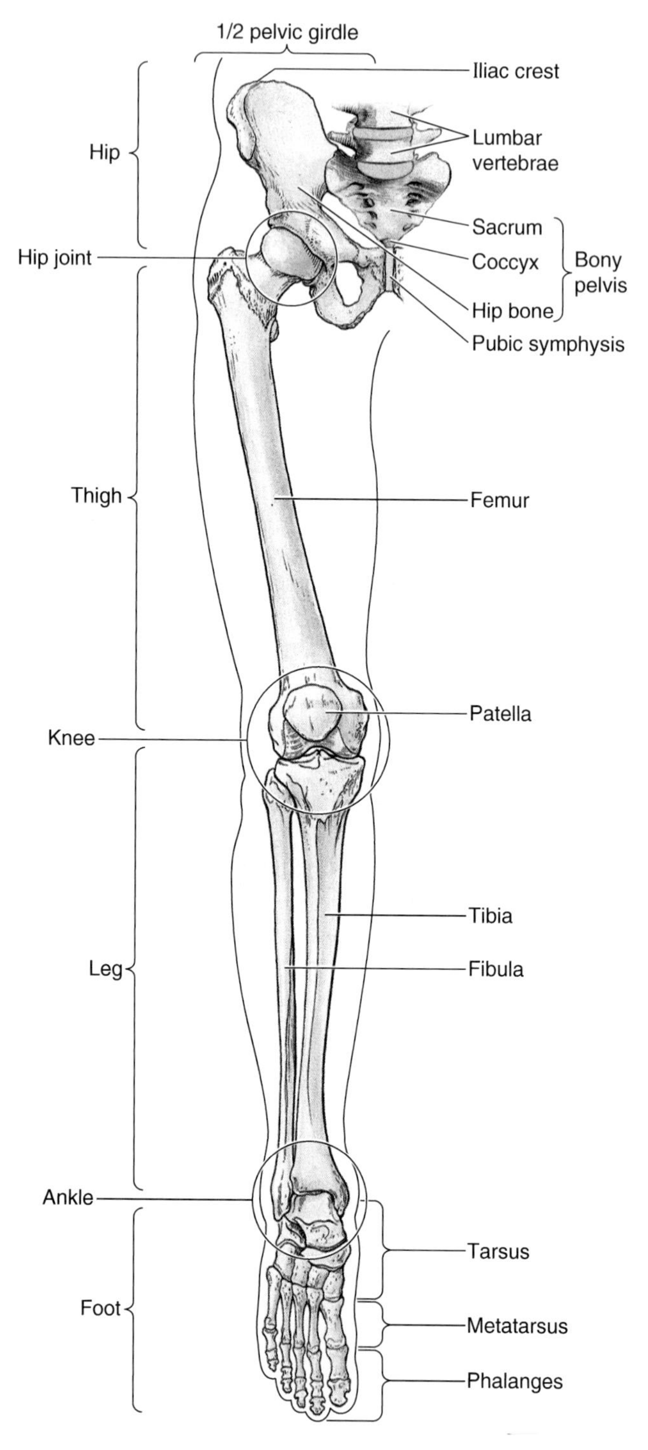

Figure 5.1. Regions and bones of the lower limb. Anterior view. The pelvic girdle is the bony ring formed by the hip bones and sacrum to which the limb bones are attached and through which the weight of the trunk is transferred to the lower limbs.

Lower Limb Injuries

Many recreational and workers' injuries involve the lower limbs. Knee, leg, and foot injuries are most common. Injuries to the hips comprise less than 3% of lower limb injuries. In general, most injuries result from acute trauma during contact sports such as hockey and football and from overuse during endurance sports such as marathon races. Adolescents are most vulnerable to these injuries because of the demands of sports on their slowly maturing musculoskeletal systems. ⊙

Bones of the Lower Limb

The skeleton of the lower limb is composed of the **pelvic girdle**, formed by the two **hip bones**—joined at the **pubic symphysis** (L. symphysis pubis)—and the **sacrum** (Fig. 5.1). The pelvic girdle and sacrum together form the **bony pelvis**. The skeleton of the free limb is attached to the pelvic girdle.

Arrangement of Lower Limb Bones

Body weight is transferred from the vertebral column to the pelvic girdle and from the pelvic girdle through the hip joints to the femurs (L. femora). Observe that each femur is directed inferomedially through the thigh toward the knee joint (Figs. 5.1 and 5.2), where its distal end articulates with the patella and tibia of the leg. *The fibula does not articulate with the femur*. Weight is transferred from the knee joint to the ankle joint by the tibia. The fibula is firmly bound to the tibia inferiorly and forms an important part of the ankle joint. The

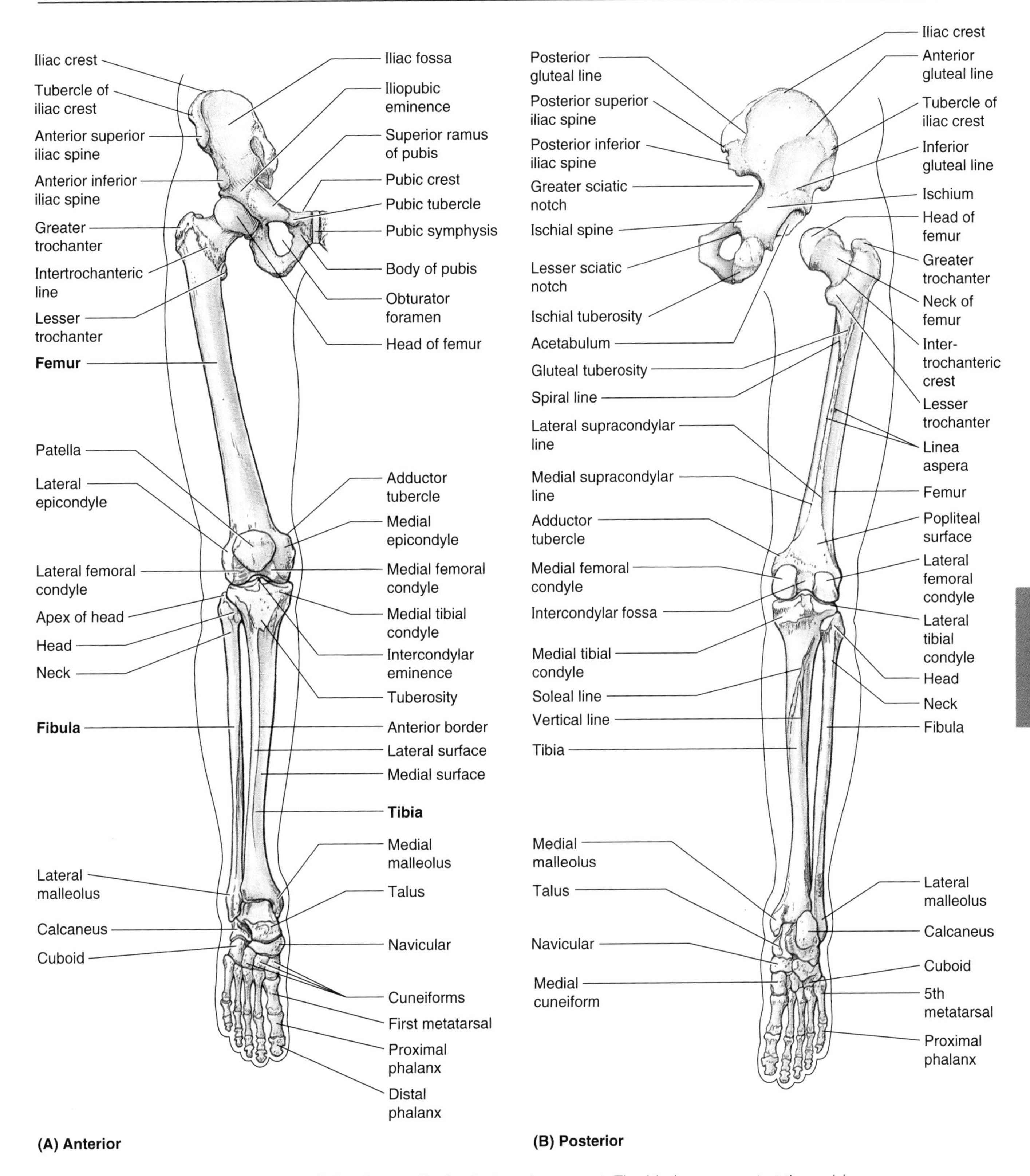

Figure 5.2. Bones of the lower limb. A. Anterior aspect. The hip bones meet at the pubic symphysis (L. symphysis pubis). The femur, or thigh bone, articulates with the hip bone proximally and the tibia (shin bone) distally. The tibia and fibula (calf bone) are the bones of the leg, which join the skeleton of the foot at the ankle. **B.** Posterior aspect. The hip joint is disarticulated to demonstrate the acetabulum of the hip bone with which the head of the femur articulates.

tarsal and metatarsal bones of the foot form a flexible but stable support for the body.

Hip Bone

The mature hip bone (L. os coxae, innominate bone) is the large flat bone formed by the fusion of the **ilium**, **ischium**, and **pubis** at the end of the teenage years. The hip bone forms the bony connection between the trunk (sacrum) and lower limb (femur). The two hip bones, which with the sacrum and coccyx form most of the **bony pelvis**, are united anteriorly by the **pubic symphysis**.

Each of the three bony parts of the immature hip bone forms from its own primary center of ossification. Five secondary centers of ossification appear later. At birth, each hip bone consists of three separate primary bones—*ilium, ischium,* and *pubis*—joined by hyaline cartilage. In infants and children, these large parts of the hip bones are incompletely ossified (Fig. 5.3). At puberty, the three primary bones are still separated by a Y-shaped **triradiate cartilage** centered in the acetabulum (Fig. 5.3*B*). The primary bones begin to fuse at 15 to 17 years; fusion is complete between the 20th and 25th years. Little or no trace of the lines of fusion of the primary bones is visible in older adults (Fig. 5.4). Although the bony components are rigidly fused, their names are still used in adults to describe the three parts of the hip bone.

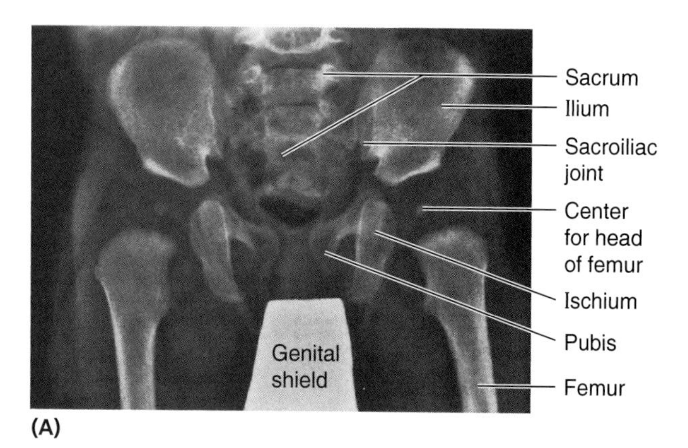

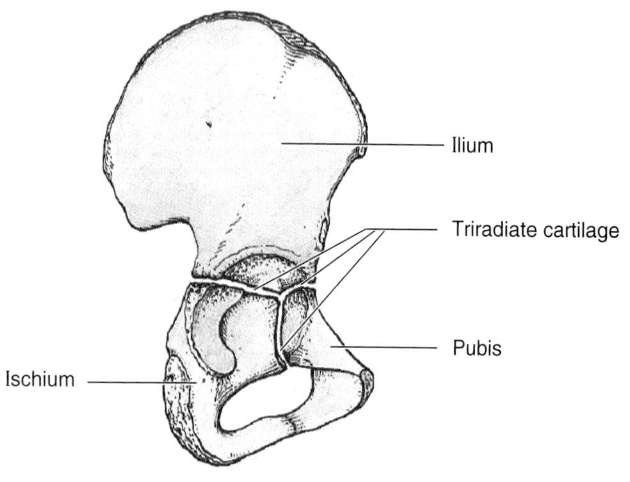

Figure 5.3. Parts of the hip bones. A. Anteroposterior (AP) radiograph of an infant's hip bones. Observe the three parts of the incompletely ossified hip bones (ilium, ischium, and pubis). **B.** Drawing of the right hip bone of a 13-year-old person. Observe the Y-shaped triradiate cartilage extending through the acetabulum and separating the three primary parts of the bone. These bony parts fuse to form the one-part mature hip bone of the adult between the 16th and 18th years.

Ilium

The ilium composes the largest part of the hip bone and contributes *the superior part of the acetabulum* (Fig. 5.4*A*). The ilium has a winglike posterolateral surface—the **ala** (L. wing)—that provides attachment for the gluteal muscles laterally and the iliacus muscle medially. Anteriorly, the ilium has an **anterior superior iliac spine** and inferior to it an **anterior inferior iliac spine.** From the anterior superior iliac spine, the long curved superior border of the ala of the ilium—the **iliac crest**—extends posteriorly, terminating at the **posterior superior iliac spine.** A prominence on the external lip of the crest, the **tubercle of the iliac crest** (iliac tubercle), lies 5 to 6 cm posterior to the anterior superior iliac spine. The posterior superior iliac spine also marks the superior end of the **greater sciatic notch.**

The lateral surface of the ala of the ilium has three rough curved lines—the posterior, anterior, and inferior **gluteal lines**—that separate the proximal attachments of the three large gluteal muscles (muscles of the buttock). Medially, the iliac wing has a large, smooth depression—the **iliac fossa** (Fig. 5.4*B*). The bone forming the superior part of this fossa may be thin and translucent, especially in older women. Posteriorly, the medial aspect of the ilium has a rough ear-shaped area—the **auricular surface** (L. auricula, a little ear)—for articulation with the reciprocally shaped auricular surface of the sacrum at the sacroiliac joint (see Chapter 3).

Ischium

The ischium composes the posteroinferior part of the hip bone. The superior part of the **body of the ischium** fuses with the pubis and ilium, forming the posteroinferior aspect of the acetabulum. The **ramus of the ischium** joins the *inferior ramus of the pubis* to form a bar of bone—the **ischiopubic ramus** (Fig. 5.4*A*)—that constitutes the inferomedial boundary of the **obturator foramen.** The posterior border of the ischium forms the lower margin of a deep indentation—the **greater sciatic notch.** The large, triangular **ischial spine** at the inferior margin of this notch is a sharp demarcation separating the greater sciatic notch from a smaller, rounded inferior indentation—the **lesser sciatic notch.** The rough bony projection at the junction of the inferior end of the body of the ischium and its ramus is the large **ischial tuberosity.** The body's weight rests on this tuberosity in the sitting position.

Pubis

The pubis composes the anteromedial part of the hip bone and contributes the anterior part of the acetabulum. The pubis is divided into a flattened body and two rami, superior and inferior (Fig. 5.4). Medially, the symphyseal surface of the

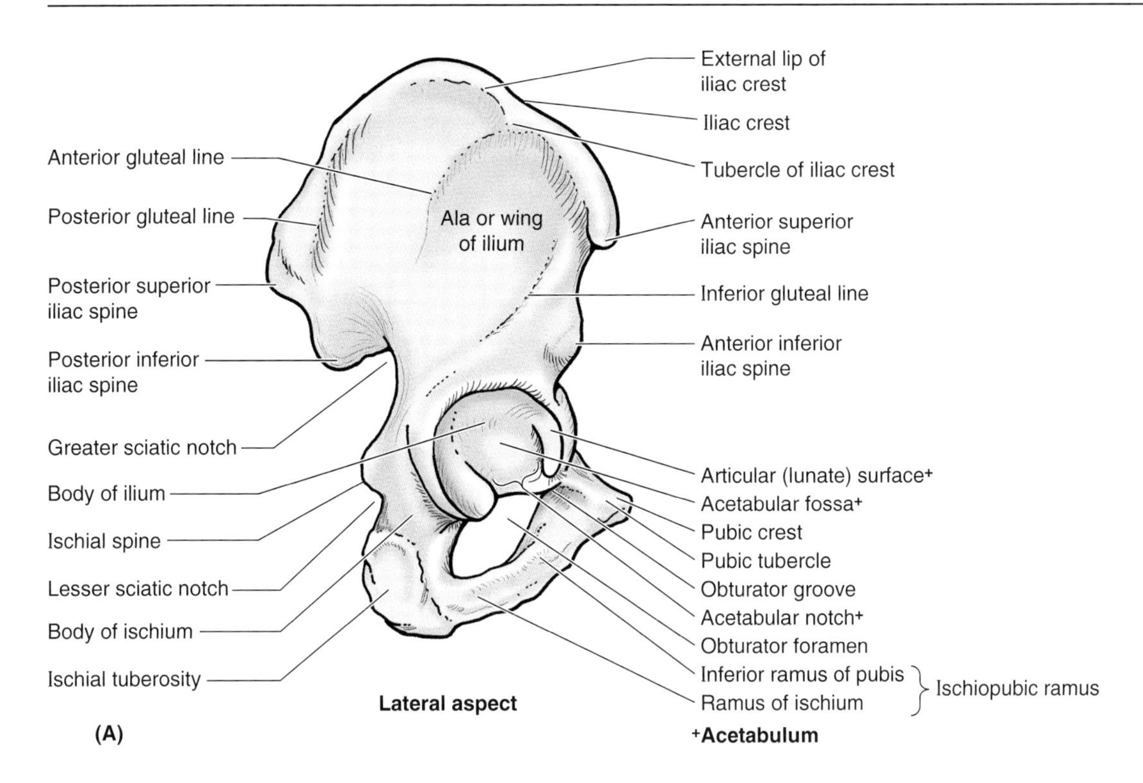

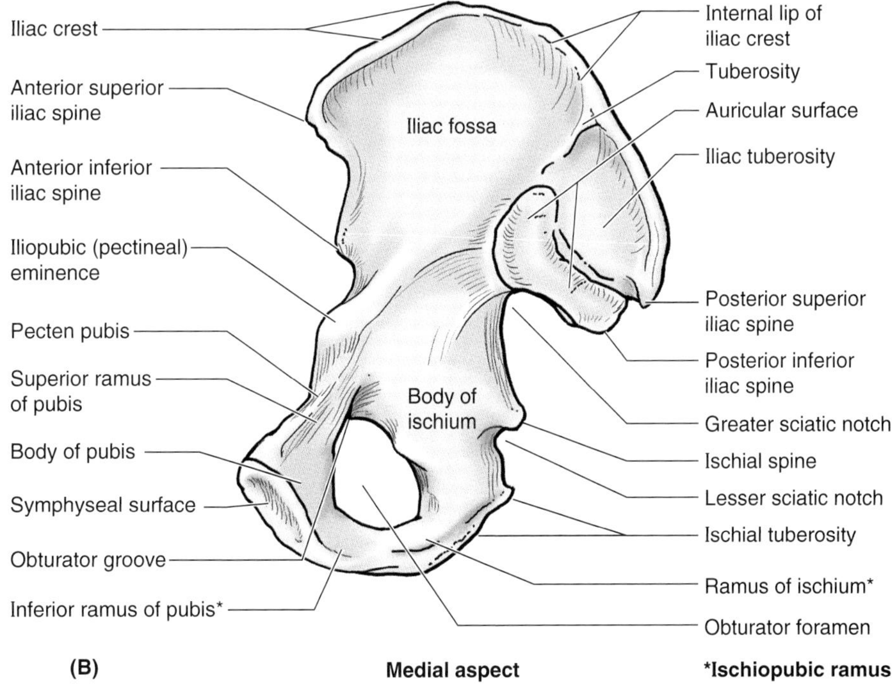

Figure 5.4. The right hip bone of an adult in the anatomical position. A. Lateral aspect. The terminology of the hip bone is based on the anatomical position, in which the articular or symphyseal surface of the pubic symphysis (L. symphysis pubis) is in the median plane and the pubic tubercle and anterior superior iliac spine are in the same coronal plane. Observe that the large hip bone is constricted in the middle and expanded at its superior and inferior ends. The lateral surface has a deep, cup-shaped acetabulum, which forms a socket that receives the head of the femur (thigh bone). Inferior to the acetabulum is the large, oval obturator foramen. **B.** Medial aspect. Observe the superior border of the ilium with its curved iliac crest. Its ends project as anterior and posterior superior iliac spines. Examine the symphyseal surface of the pubis, which articulates with the corresponding surface of the contralateral hip bone. Examine also the auricular surface of the ilium, which articulates with a corresponding surface of the sacrum to form the sacroiliac joint.

body of the pubis articulates with the corresponding surface of the body of the contralateral pubis by means of the *pubic symphysis*. The anterosuperior border of the united bodies and symphysis forms the **pubic crest**. Small projections at the lateral ends of this crest, the **pubic tubercles**, are extremely important landmarks of the inguinal regions. The posterior margin of the **superior ramus** of the pubis has a sharp raised edge—the **pecten pubis** (pectineal line)—which is a part of the pelvic brim (see Chapter 3).

Obturator Foramen

The obturator foramen is a large oval or irregularly triangular aperture in the hip bone. It is bounded by the pubis and ischium and their rami (Fig. 5.4*B*). Except for a small passageway for the obturator nerve and vessels—the *obturator canal*—the obturator foramen is closed by the thin, strong *obturator membrane* covered on both sides by attached muscles.

Acetabulum

The acetabulum (L. shallow vinegar cup) is the large cup-shaped cavity or socket on the lateral aspect of the hip bone, which *articulates with the head of the femur to form the hip joint* (Figs. 5.2*A* and 5.4*A*). All three parts of the hip bone join to form the acetabulum. The margin of the acetabulum is deficient inferiorly at the **acetabular notch**, which makes the fossa resemble a cup with a piece of its lip missing. The rough depression in the floor of the acetabulum extending superiorly from the acetabular notch is the **acetabular fossa**. The acetabular notch and fossa also comprise a deficiency in the smooth articular **lunate surface** of the acetabulum, which actually articulates with the head of the femur.

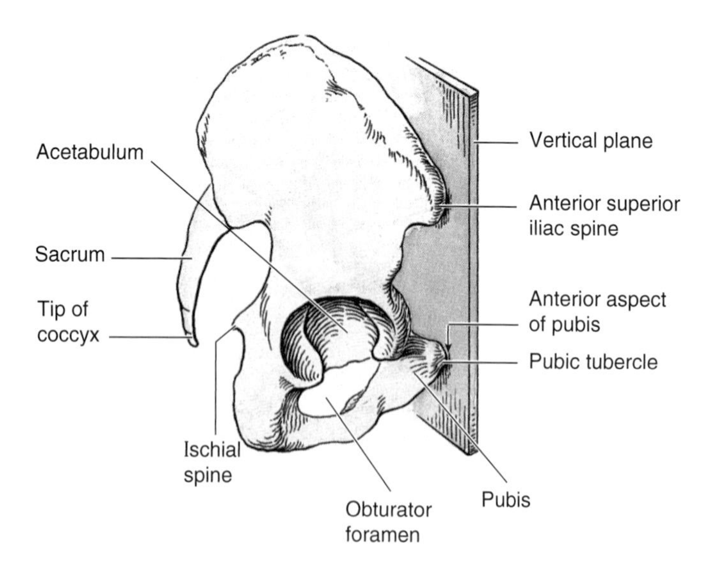

Figure 5.5. Lateral aspect of the right hip bone, sacrum, and coccyx. This drawing demonstrates that, in the anatomical position, the anterior superior iliac spine and the anterior aspect of the pubis lie in the same vertical plane.

Anatomical Position of the Hip Bone

Surfaces and borders of the hip bone are named according to their anatomical position. To place the hip bone in this position, place it so that the acetabulum faces laterally and slightly anteriorly (Fig. 5.5). *When the hip bone is in the anatomical position, the:*

- Anterior superior iliac spine and the anterosuperior aspect of the pubis lie in the same vertical plane
- Ischial spine and superior end of the pubic symphysis are approximately in the same horizontal plane
- Symphyseal surface of the pubis is vertical, parallel to the median plane
- Internal aspect of the body of the pubis faces almost directly superiorly (it essentially forms a floor on which the urinary bladder rests)
- Acetabulum faces inferolaterally, with the acetabular notch directed inferiorly
- Obturator foramen lies inferomedial to the acetabulum
- Tip of the coccyx is typically on a level with the superior half of the body of the pubis.

Hip Injuries

Hip fractures and dislocations are uncommon in contact sports, except with high-energy trauma such as occurs in motor-vehicle, motorcycle, rodeo, and horse-racing accidents. Hip fractures also occur in falls during skiing and ice dancing; however, contusions (bruises) and hematomas (extravasations of blood) are more common in these sports. *Hip fractures* are common in serious vehicular accidents, when severe direct trauma occurs. Anteroposterior (AP) compression of the hip bones commonly fractures the pubic rami. Lateral compression of the pelvis may produce *fractures of the acetabula*, as may falls on the feet (e.g., from a roof). A hip fracture causes severe pain and immediate disability because of the instability of the fracture site.

Avulsion fractures of the hip bone (avulsion or tearing away of the ischial tuberosity) may occur in adolescents and young adults during sports that require sudden acceleration or deceleration forces, such as sprinting or kicking in football, soccer, jumping hurdles, basketball, and martial arts. These fractures occur at *apophyses* (bony projections that lack secondary ossification centers). Avulsion fractures occur where muscles are attached: anterior superior and inferior iliac spines, ischial tuberosities, and ischiopubic rami. A small part of bone with a piece of a tendon or ligament attached is avulsed (torn away). ⊙

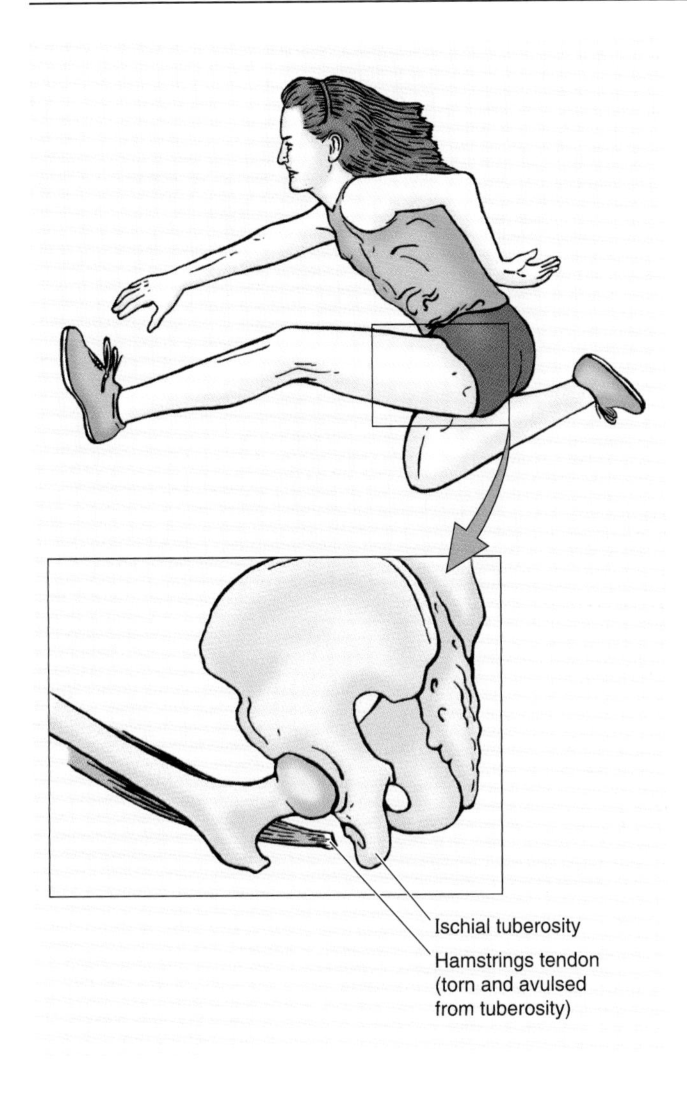

Femur

The femur—*the longest and heaviest bone in the body*—transmits body weight from the hip bone to the tibia when a person is standing (Figs. 5.2 and 5.6). Its length—associated with a striding gait—is approximately a quarter of the person's height (approximately 108 cm or 18 inches). The femur consists of a body (shaft) and two ends, superior and inferior. *The superior end of the femur consists of* a head, neck, and two trochanters (greater and lesser).

The **head of the femur** projects superomedially and slightly anteriorly when articulating with the acetabulum. The head is attached to the femoral body by the **neck of the femur**. The head and neck are at an angle (115 to 140°, averaging 126°) to the long axis of the body of the femur. The angle is widest at birth and diminishes gradually until the adult angle is reached. It is less in females because of the increased breadth of the lesser pelvis (true pelvis, or pelvis minor) and the greater obliquity of the body of the femur (Williams et al., 1995). Although this architecture allows greater mobility of the femur at the hip joint, it imposes considerable strain on the neck of the femur. Consequently, fractures of the femoral neck can occur in older people as a result of a slight stumble.

Where the neck joins the femoral body are two large, blunt elevations—the trochanters. The rounded, conical **lesser trochanter** extends medially from the posteromedial part of the junction of the neck and body. The **greater trochanter** is a large, laterally placed bony mass that projects superiorly and posteriorly where the neck joins the femoral body. The site where the neck joins the body is indicated by the **intertrochanteric line**—a roughened ridge running from the greater to the lesser trochanter (Fig. 5.6, *anterior view*). A similar but smoother ridge, the **intertrochanteric crest**, joins the trochanters posteriorly. The rounded elevation on the crest is the **quadrate tubercle**. In Figure 5.6 (*posterior view*), observe that the greater trochanter is in line with the femoral body and overhangs a deep depression medially—the **trochanteric fossa**.

The **body of the femur** is slightly bowed anteriorly. Most of the body is smoothly rounded, except for a broad, rough line posteriorly—the **linea aspera**. This vertical ridge is especially prominent in the middle third of the femoral body, where it has **medial** and **lateral lips** (margins). Superiorly, the lateral lip blends with the broad, rough **gluteal tuberosity**, and the medial lip continues as a narrow, rough spiral line. The **spiral line** extends toward the lesser trochanter and then passes to the anterior surface of the femur, where it ends in the intertrochanteric line. A prominent intermediate ridge—the **pectineal line**—extends from the central part of the linea aspera to the base of the lesser trochanter. Inferiorly, the linea aspera divides into medial and lateral **supracondylar lines** that lead to the spirally curved medial and lateral condyles (Fig. 5.6). The condyles are separated inferoposteriorly by an **intercondylar fossa** (intercondylar notch).

The **femoral condyles** articulate with the **tibial condyles** to form the knee joint (Fig. 5.2). Anteriorly, the femoral condyles merge at a shallow depression—the **patellar surface** (Fig. 5.6)—where they articulate with the patella. The lateral surface of the lateral condyle has a central projection—the **lateral epicondyle**. The medial surface of the medial condyle has a larger and more prominent **medial epicondyle**, superior to which is another elevation, the **adductor tubercle**. The trochanters, lines, tubercles, and epicondyles are where muscles and ligaments attach.

Evidence of Fetal Viability

The distal end of the femur undergoes ossification just before birth. The visibility of this center of ossification in radiographs is commonly used as medicolegal evidence that a newborn infant found dead was near full term and viable (capable of living at birth). ⊙

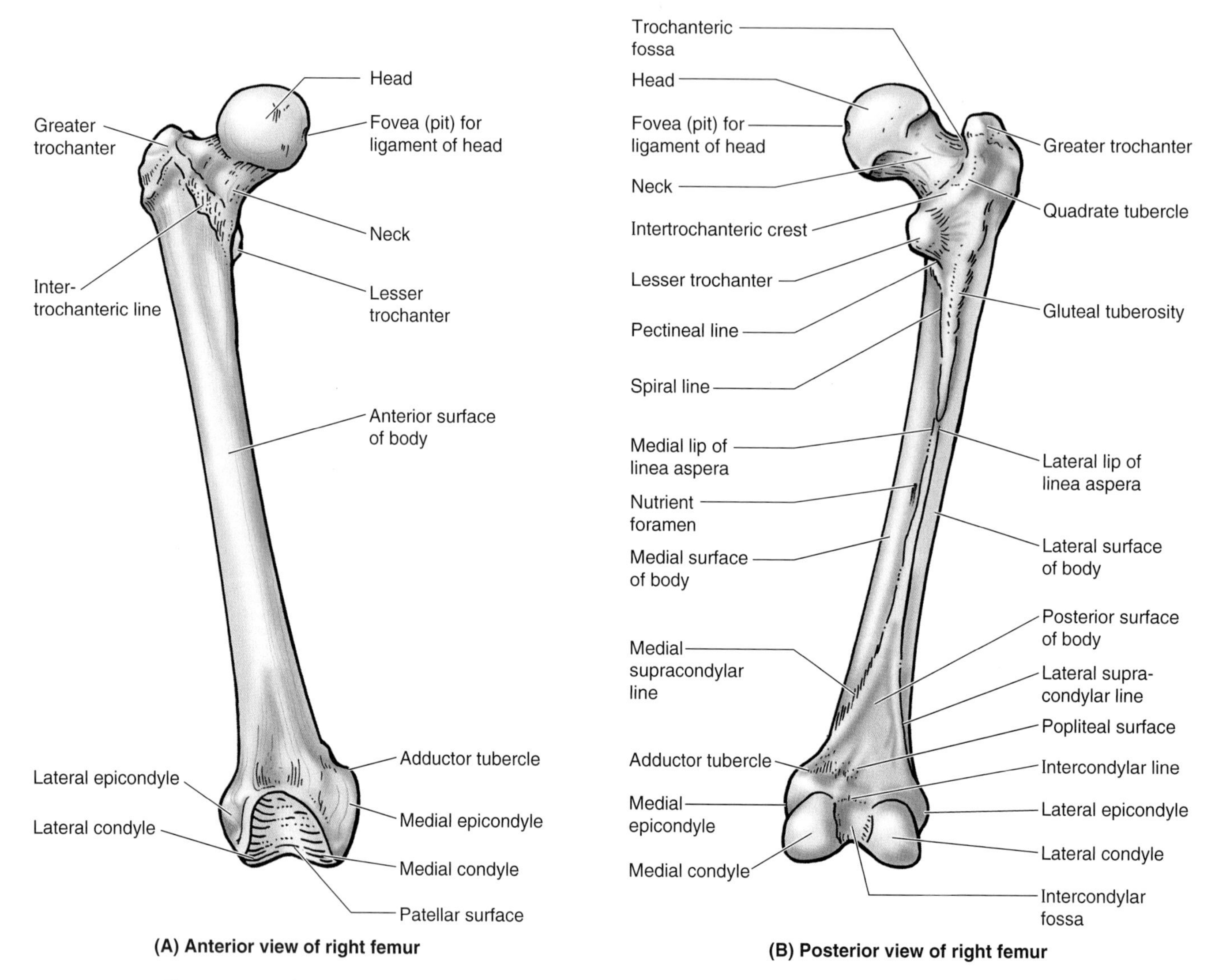

Figure 5.6. The right femur (thigh bone) of an adult. A. Anterior aspect. The body of the femur, the longest and heaviest bone in the body, is almost cylindrical in most of its length. Proximally, the femur has a rounded head, a short neck, and greater and lesser trochanters. The neck is separated from the body of the femur by the intertrochanteric line. The distal end of the femur is massive with medial and lateral condyles. **B.** Posterior aspect. In the middle third of the posterior surface, observe the prominent posterior border—the linea aspera—which has medial and lateral lips. Also, observe the nutrient foramen entering the femoral body near the linea aspera.

Slipped Epiphysis of the Femoral Head

In older children and adolescents (10 to 17 years), the epiphysis of the femoral head may slip away from the femoral neck because of a weakened epiphyseal plate. This injury may be caused by acute trauma or repetitive microtraumas with increased shearing stress on the epiphysis, especially with abduction and lateral rotation of the thigh. The epiphysis often slips slowly and results in a progressive coxa vara. The common initial symptom of the injury is hip discomfort that may be referred to the knee. Radiological examination of the superior end of the femur is usually required to confirm a diagnosis of a slipped (dislocated) epiphysis of the head of the femur.

Coxa Vara and Coxa Valga

The *angle of inclination* that the long axis of the femoral neck makes with the femoral body (*A*) varies with age, sex, and development of the femur (e.g., a congenital defect in the ossification of the femoral neck). It may also change ▶

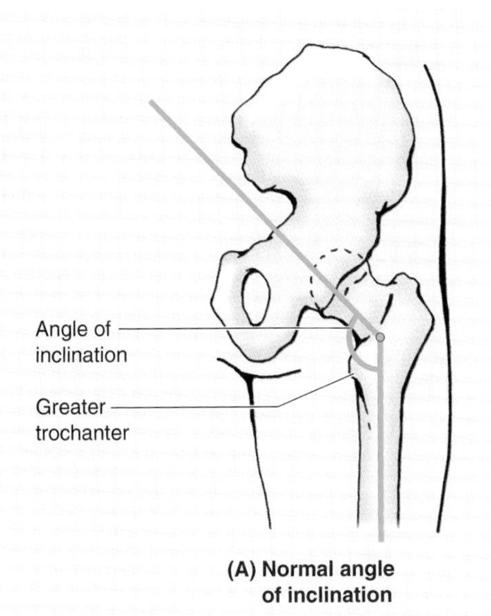

(A) Normal angle of inclination

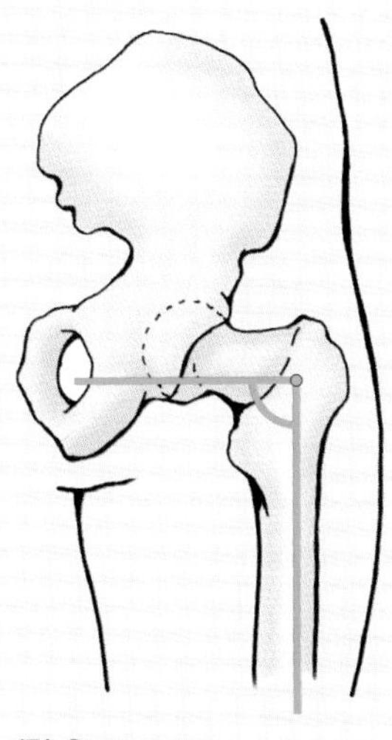
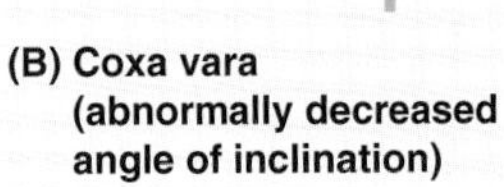
(B) Coxa vara (abnormally decreased angle of inclination)

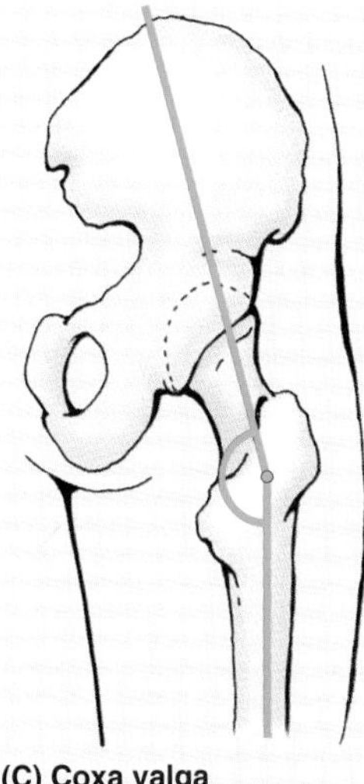
(C) Coxa valga (abnormally increased angle of inclination)

▶ with any pathological process that weakens the neck of the femur (e.g., rickets). When the angle of inclination is decreased, the condition is *coxa vara* (*B*); when it is increased it is *coxa valga* (*C*). Coxa vara causes a mild shortening of the lower limb and limits passive abduction of the hip.

Femoral Fractures

When it is said that an elderly person has a "broken hip," the usual injury is a fracture of the neck of the femur. The neck is frequently fractured when adults older than 60 years stumble. The fracture occurs more frequently in women than in men because it occurs through bone that is weakened by *osteoporosis* (brittle bones resulting from a reduction in bone mass). A *fracture of the femoral neck* is among the most troublesome and problematic of all fractures because of the instability of the fracture site. Furthermore, the periosteum covering the femoral neck is exceedingly thin and has extremely limited powers of *osteogenesis* (bone formation). Because the retinacular arteries arise from the medial circumflex femoral arteries and run parallel to the femoral neck on their way to supply the femoral head, they are vulnerable to injury when the neck of the femur fractures (p. 614). Rupture of these vessels results in degeneration (necrosis) of the femoral head and bleeding into the hip joint. The fractures usually result from indirect violence and often from a trivial mishap, such as slipping on an icy surface or tripping over something (a rug, for instance).

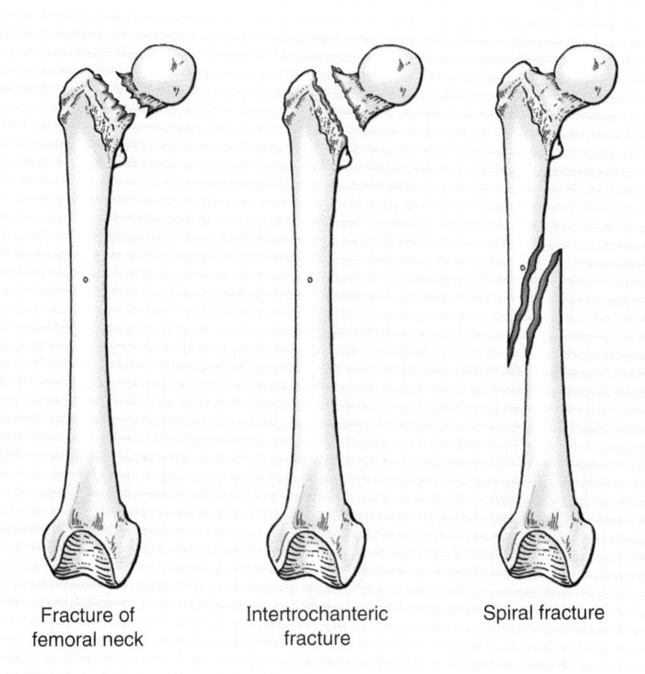

Fractures of the femur between the greater and lesser trochanters *(intertrochanteric fractures)*, or through the trochanters (*pertrochanteric fractures*), are also common in persons older than 60 years. Patients are usually older women who either fall down or are knocked down.

The body of the femur is large and strong; however, a violent direct injury, such as may be sustained in an automobile accident, may fracture it. In some cases a *spiral fracture of the femoral body* occurs, which may be comminuted (broken into several pieces) and the fragments displaced. It may take up to 20 weeks for firm union of the fragments, and union of this serious type of fracture may take up to a year. ⊙

Tibia and Fibula

The tibia and fibula are the bones of the leg (Figs. 5.2 and 5.7). The **tibia** supports the body's weight. It articulates with the condyles of the femur superiorly and the talus (L. ankle; ankle bone) inferiorly. The **fibula** does not transmit body weight; it is mainly for the attachment of muscles, but it also provides stability to the ankle joint. The bodies of the tibia and fibula are connected by an **interosseous membrane** composed of strong oblique fibers.

Tibia

Except for the femur, the large weightbearing tibia is the largest bone in the body. The **tibia** is located on the anteromedial side of the leg, nearly parallel to the fibula. The proximal end of the tibia is large because its **medial** and **lateral condyles** articulate with the large condyles of the femur. The superior surface of the tibia is flat, forming a **tibial plateau** consisting of the medial and lateral tibial condyles and an **intercondylar eminence.** This eminence of the tibia fits into the **intercondylar fossa** (Fig. 5.6*B*) between the femoral condyles. The lateral tibial condyle has a facet inferiorly for the head of the fibula (Fig. 5.7).

The **body of the tibia** is somewhat triangular and has medial, interosseous (lateral), and posterior surfaces. Its anterior border, or crest, has a broad, oblong **tibial tuberosity**, providing a distal attachment for the *patellar ligament*, which stretches from the apex and joins the margins of the patella to the tibial tuberosity. The **anterior border of the tibia** is subcutaneous and is the most prominent border. The body of the tibia is thinnest at the junction of its middle and distal thirds.

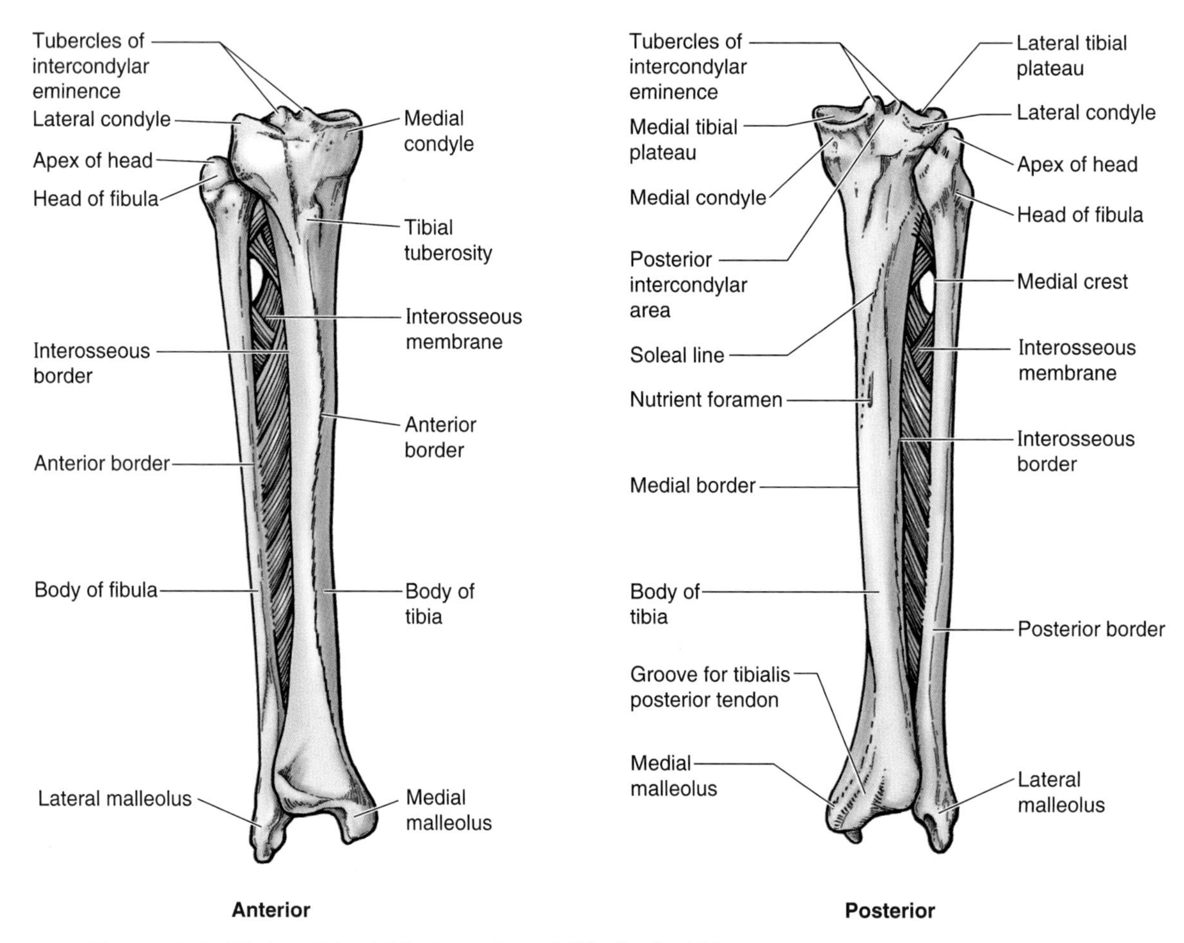

Figure 5.7. Right tibia (shin bone) and fibula (calf bone). Anterior and posterior views. The tibia is located on the anterior and medial (tibial) sides of the leg. Its large proximal end articulates with and receives the body's weight from the large distal end of the femur (thigh bone). The distal end of the tibia has a prominent projection, the medial malleolus. The fibula is located on the lateral (fibular) side of the leg. Its proximal end or head articulates with the posterolateral part of the lateral condyle of the tibia. The fibula has a lateral malleolus that is more prominent than the medial malleolus and extends approximately 1 cm more distally. Observe the interosseous membrane, a dense fibrous layer that connects the interosseous borders of the tibia and fibula. The anterior tibial vessels enter the anterior compartment of the leg through the large oval opening—medial to the medial crest of the fibula—near the proximal end of the membrane.

The distal end of the tibia is smaller than the proximal end and has facets for articulation with the fibula and talus. An inferiorly directed projection from the medial side of the inferior end is the **medial malleolus**, which has a facet on its lateral surface for articulation with the talus. The **interosseous border** of the tibia is sharp where it gives attachment to a dense fibrous membrane—the **interosseous membrane**—uniting the two leg bones (Fig. 5.7).

On the posterior surface of the proximal part of the tibial body is a rough diagonal ridge—the **soleal line**—that runs inferomedially to the medial border, approximately one-third of the way down the body. Immediately distal to the soleal line is an obliquely directed vascular groove, which leads to a large **nutrient foramen**. From it the nutrient canal runs inferiorly in the tibia before it opens into the medullary (marrow) cavity.

Fibula

The slender fibula lies posterolateral to the tibia and serves mainly for muscle attachment (Fig. 5.7). The **fibula** has no function in weightbearing, but its lateral malleolus helps hold the talus in its socket (Fig. 5.2*A*). At its proximal end is the **head of the fibula**, which has a pointed **apex**. The head articulates with the proximal, posterolateral part of the tibia on the inferior aspect of the lateral condyle. The **body** of the fibula is twisted and marked by the sites of muscular attachments. It has three borders (anterior, interosseous, and posterior) and three surfaces (medial, posterior, and lateral). At its distal end, the fibula enlarges to form the **lateral malleolus**, which is more prominent and posterior than the medial malleolus and extends approximately 1 cm more distally. The lateral malleolus articulates with the lateral surface of the talus (Fig. 5.2*A*).

Tibial Fractures

The body of the tibia is narrowest at the junction of its middle and inferior thirds, which is the most frequent site of fracture. The body of the tibia is the most common site for a *compound fracture* (*A*)—one in which the skin is perforated and blood vessels are torn. *Fracture of the tibia through the nutrient canal predisposes to nonunion of the bone fragments* resulting from damage to the nutrient artery.

Transverse stress (march) fractures of the inferior third of the tibia (*B*) are common in people who take long walks when they are not conditioned for this activity. The strain may fracture the anterior cortex of the tibia. Indirect violence applied to the tibial body when the bone turns with the foot fixed during a fall may produce fracture (e.g., when a person is tackled in a football game). In addition, severe torsion during skiing may produce a *diagonal fracture* (*C*) of the tibial body at the junction of the middle and inferior thirds, as well as a *fracture of the fibula*. Frequently during skiing, a fracture results from a high-speed forward fall, which angles the leg over the rigid ski boot, producing a "boot-top fracture." The injury is usually a comminuted fracture in which the tibia is broken into several pieces at the junction of its middle and distal thirds (*D*).

Tibial fractures may also result from direct trauma (e.g., "bumper fractures" caused when the bumper of a car ▶

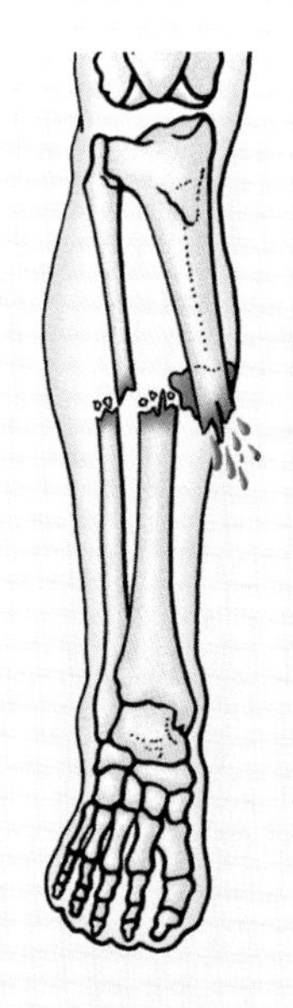

(A) Compound (open) fracture with external bleeding

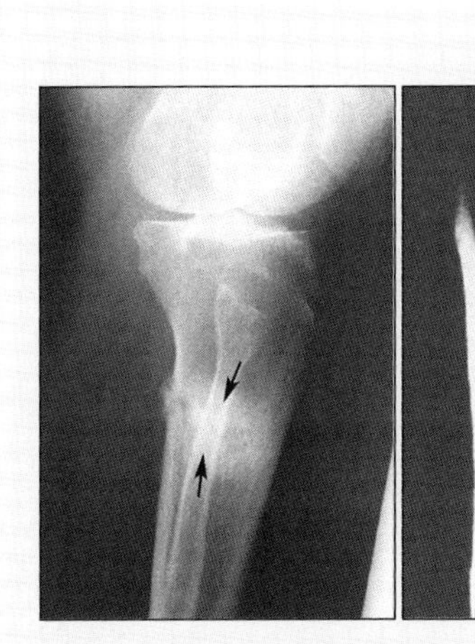

(B) March (stress) fracture of tibia *(arrows)*, most apparent in magnetic resonance image at the right

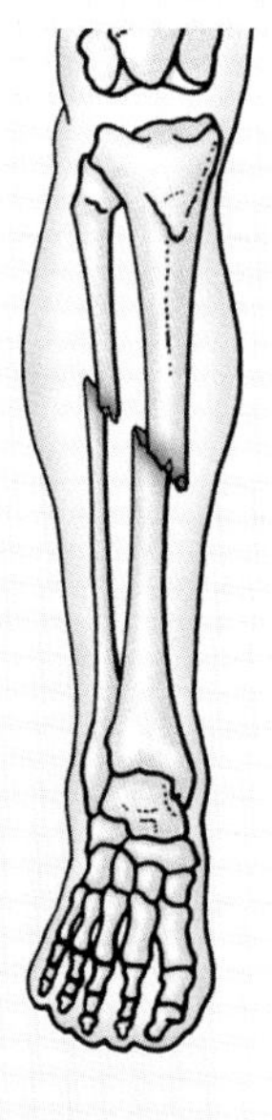

(C) Diagonal fracture with shortening

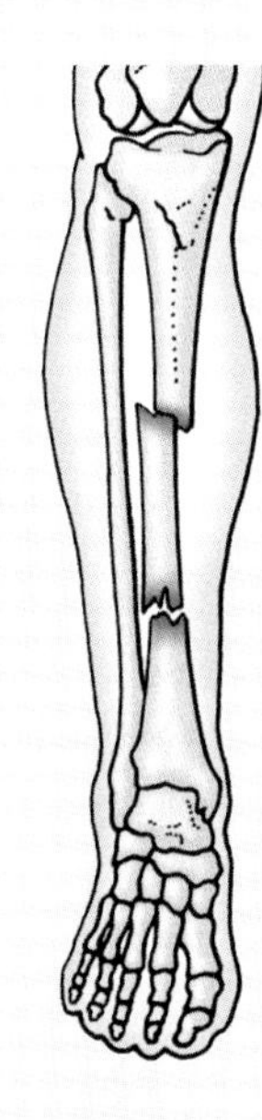

(D) Transverse fractures with fibula intact

▶ strikes the leg). The blow often tears the skin, and because the tibia lies subcutaneously, the bone fragments protrude, producing a *compound fracture*. Because the tibial body is unprotected anteromedially throughout its course and is relatively slender at the junction of its inferior and middle thirds, it is not surprising that the tibia is the most common long bone to fracture and to suffer compound injury. *The tibia has a relatively poor blood supply*; hence, even undisplaced stable fractures may take up to 6 months to heal. Because of its extensive subcutaneous surface, the anterior tibia is accessible for obtaining pieces of bone for grafting in children; it is also used as a site for intramedullary infusion in dehydrated/shocked children.

Fractures Involving the Epiphyseal Plates

The primary ossification center for the superior end of the tibia appears shortly after birth and joins the body of the tibia during adolescence (usually 16 to 18 years). Fractures of the tibia in children are more serious if they involve the epiphyseal plates because continued normal growth of the bone may be jeopardized. The tibial tuberosity usually forms by inferior bone growth from the superior epiphyseal center at approximately 10 years, but a separate center for the tibial tuberosity may appear at approximately 12 years. Disruption of the epiphyseal plate at the tibial tuberosity may cause inflammation of the tibial tuberosity and chronic recurring pain during adolescence (*Osgood-Schlatter disease*), especially in young athletes.

Fibular Fractures

Fractures of the fibula commonly occur 2 to 6 cm proximal to the distal end of the lateral malleolus and are often associated with *fracture-dislocations of the ankle joint*, which are combined with tibial fractures (*F*). When a person slips and the foot is forced into an excessively inverted position, the ankle ligaments tear, forcibly tilting the talus against the lateral malleolus and shearing it off. *Fractures of the lateral and medial malleoli* are relatively common (*G*) in soccer and basketball players.

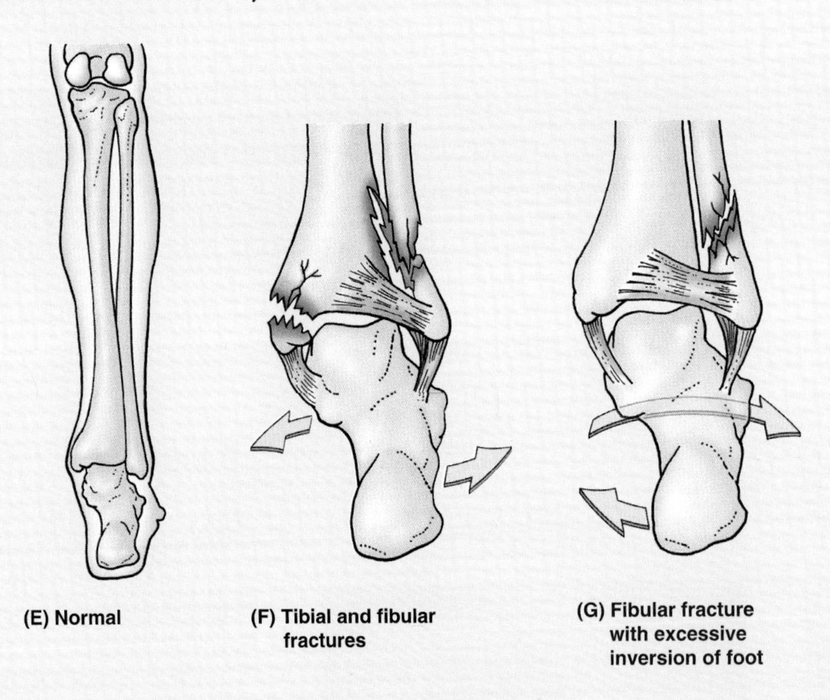

(E) Normal

(F) Tibial and fibular fractures

(G) Fibular fracture with excessive inversion of foot

Bone Grafts

If a part of a major bone is destroyed by injury or disease, the limb becomes useless. Without a bone transplant, the affected part of the limb may have to be amputated. *The fibula is a common source of bone for grafting*. Even after a long piece of the fibula has been removed, walking, running, and jumping can be normal. Free vascularized fibulas have been used to restore skeletal integrity to upper and lower limbs in which congenital bone defects exist or to replace segments of bone following trauma or excision of a malignant tumor. The remaining parts of the fibula usually do not regenerate because the periosteum and nutrient artery are generally removed with the piece of bone so that the graft will remain alive and grow when transplanted to another site. The transplanted piece of fibula, secured in its new site, restores the blood supply of the bone to which it is now attached. Healing proceeds as if merely a fracture were at each of its ends.

Awareness of the location of the nutrient foramen in the fibula is important when performing free vascularized fibular transfers. Because the nutrient foramen is located in the middle third of the fibula in most cases, this segment of the bone is used for transplanting when the graft must include an endosteal, as well as a periosteal, blood supply. ⊙

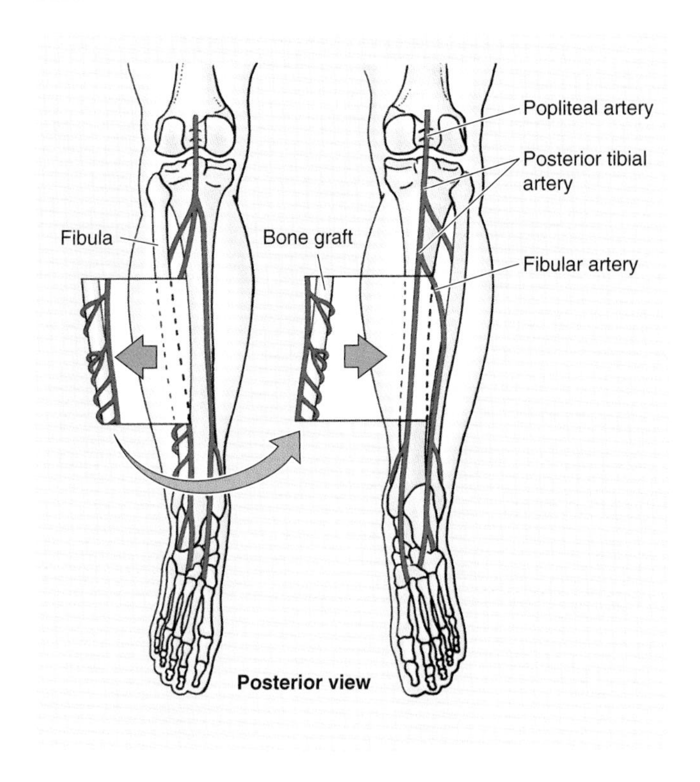

Bones of the Foot

The bones of the foot comprise the tarsus, metatarsus, and phalanges. There are 7 tarsal bones, 5 metatarsal bones, and 14 phalanges. Observe the articulated skeleton of the foot (Figs. 5.1, 5.2, and 5.8), noting that its medial border is almost straight. Note also that the line joining the midpoints of the medial and lateral borders of the foot is oblique and that the metatarsal bones and phalanges are located anterior to this line and the tarsal bones are posterior to it. Although knowledge of the individual characteristics of bones is necessary for an understanding of the structure of the foot, it is important to study the skeleton of the foot as a whole and to identify its principal bony landmarks in the living foot.

Tarsus

The tarsus consists of seven bones (Fig. 5.8, *A* and *B*): calcaneus, talus, cuboid, navicular, and three cuneiforms. Only one bone, the talus, articulates with the leg bones.

The **calcaneus** (calcaneum, heel bone) is the largest and strongest bone in the foot (Fig. 5.8, *A* and *C*). The calcaneus transmits most of the body weight from the talus to the ground. The calcaneus articulates with the talus superiorly and the cuboid anteriorly. The shelflike **sustentaculum tali** (L. support of the talus) (Fig. 5.8, *B* and *D*) projects from the superior border of the medial surface of the calcaneus and supports the talar head. The lateral surface of the calcaneus has an oblique ridge (Fig. 5.8*C*)—the **fibular trochlea** (peroneal trochlea). The posterior part of the calcaneus has a prominence (Fig. 5.8*A*)—the **calcaneal tuberosity** (L. tuber calcanei)—that has medial, lateral, and anterior tubercles (processes). Only the medial tubercle rests on the ground during standing.

The **talus** has a body, neck, and head (Fig. 5.8*C*). The superior surface of the talus bears the weight of the body that is transmitted from the tibia. The talus also articulates with the fibula, calcaneus, and navicular bone. The talus is the only tarsal bone that has no muscular or tendinous attachments. The rounded **talar head** is directed anteromedially and rests on a shelf or bracketlike lateral projection of the calcaneus—the **sustentaculum tali** (Fig. 5.8*D*). The body of the talus is narrow posteriorly and has a groove for a tendon (Fig. 5.8*C*). The groove has a prominent **lateral tubercle** and a less prominent **medial tubercle** (Fig. 5.8*A*).

The **navicular** (L. little ship) is a flattened, boat-shaped bone that is located between the talar head posteriorly and the three cuneiforms anteriorly (Fig. 5.8, *A–D*). The medial surface of the navicular projects inferiorly to form the **navicular tuberosity**. If the tuberosity is too prominent, it may press against the medial part of the shoe and cause foot pain.

The **cuboid**, approximately cubical in shape, is the most lateral bone in the distal row of the tarsus (Fig. 5.8, *A* and *C*). Anterior to the **tuberosity of the cuboid** on the lateral and inferior surfaces of the bone is a groove for the tendon of the fibularis (peroneus) longus muscle.

The three cuneiform bones (Fig. 5.8, *A*, *C*, and *D*) are the medial (1st), intermediate (2nd), and lateral (3rd). The **medial cuneiform** is the largest and the **intermediate cuneiform** is the smallest. Each cuneiform (L. cuneus, wedge shaped) articulates with the navicular bone posteriorly and the base of its appropriate metatarsal anteriorly. The **lateral cuneiform** articulates with the cuboid.

Metatarsus

The metatarsus consists of five metatarsal bones that are numbered from the medial side of the foot (Fig. 5.8*A*). The 1st metatarsal is shorter and stouter than the others. The 2nd metatarsal is the longest. Each metatarsal has a base proximally, a body, and a head distally (Fig. 5.8*C*). *The base of each metatarsal is the larger, proximal end.* The bases of the metatarsals articulate with the cuneiform and cuboid bones and the heads articulate with the proximal phalanges. *The base of the 5th metatarsal has a large tuberosity* that projects over the lateral margin of the cuboid. On the plantar surface of the head of the 1st metatarsal are prominent medial and lateral *sesamoid bones* (not shown); they are embedded in, or covered by, the plantar ligaments (see Fig. 5.44, *B* and *D*).

Phalanges

The 14 phalanges are as follows: the 1st digit (great toe) has two phalanges (proximal and distal); the other four digits have three each—proximal, middle, and distal—(Fig. 5.8, *A* and *C*). Each phalanx consists of a base (proximally), a body, and a head distally. The phalanges of the 1st digit are short, broad, and strong. The middle and distal phalanges of the 5th digit are often fused in elderly people.

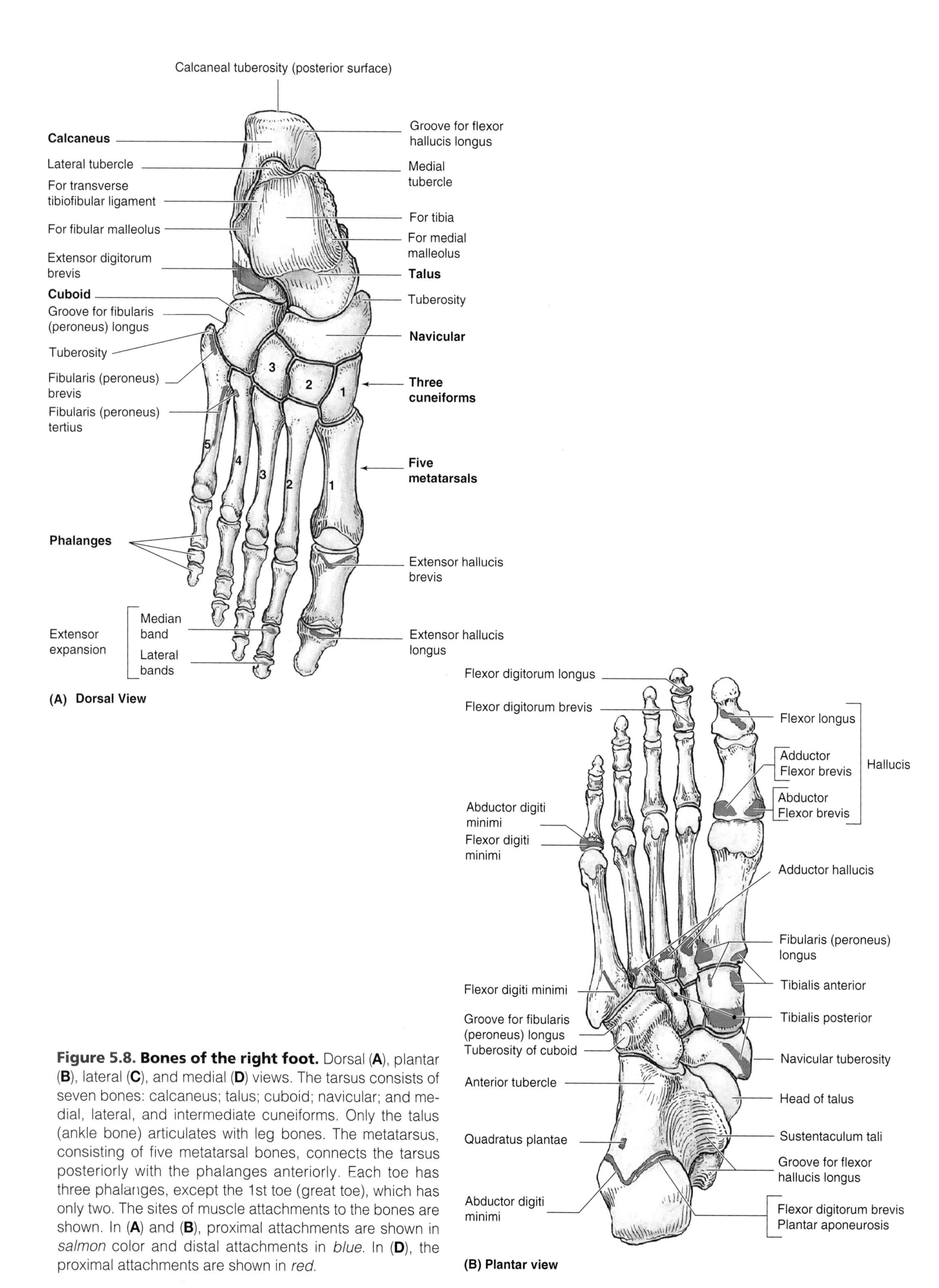

Figure 5.8. Bones of the right foot. Dorsal (**A**), plantar (**B**), lateral (**C**), and medial (**D**) views. The tarsus consists of seven bones: calcaneus; talus; cuboid; navicular; and medial, lateral, and intermediate cuneiforms. Only the talus (ankle bone) articulates with leg bones. The metatarsus, consisting of five metatarsal bones, connects the tarsus posteriorly with the phalanges anteriorly. Each toe has three phalanges, except the 1st toe (great toe), which has only two. The sites of muscle attachments to the bones are shown. In (**A**) and (**B**), proximal attachments are shown in *salmon* color and distal attachments in *blue*. In (**D**), the proximal attachments are shown in *red*.

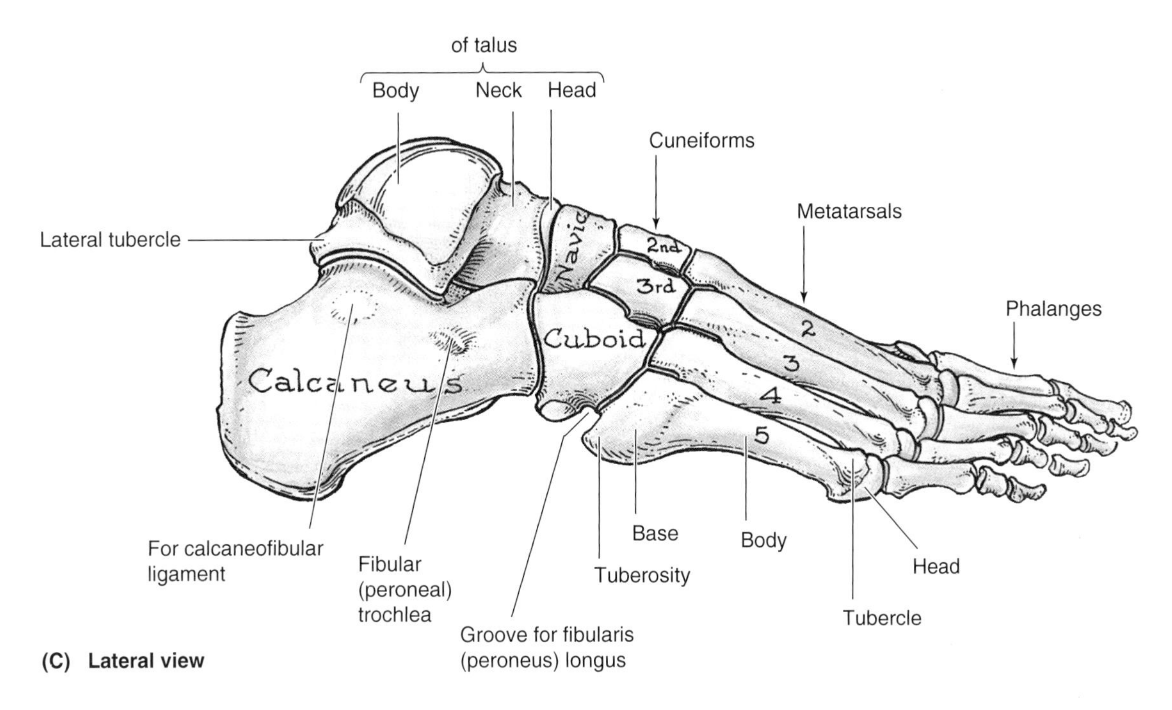

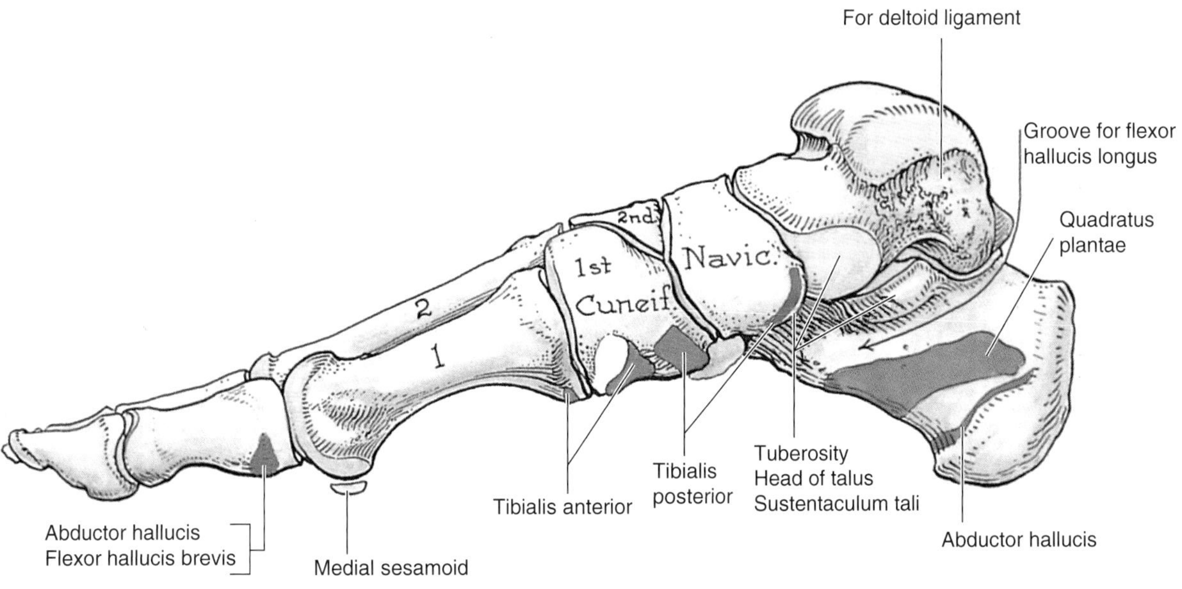

Figure 5.8. Bones of the right foot. *(Continued)*

Surface Anatomy of the Lower Limb Bones

Bony landmarks are helpful during physical examinations and surgery because they can be used to locate structures such as nerves and blood vessels. When your hands are on your hips, they rest on the **iliac crests**, the curved superior borders of the *wings of the ilia* (pleural of ilium). The anterior one-third of these crests are easily palpated because they are subcutaneous. The posterior two-thirds of the iliac crests are often difficult to palpate because they are usually covered with fat. The **supracristal plane** through the highest level of the iliac crests passes through the L4/L5 intervertebral disc (p. 518). Clinically, this level is used as a landmark for inserting a needle into the subarachnoid (leptomeningeal) space when performing a *lumbar puncture* to obtain cerebrospinal fluid (see Chapter 4).

The iliac crest ends anteriorly at the rounded **anterior** ▶

▶ **superior iliac spine,** which is easy to palpate by tracing the iliac crest anteroinferiorly. It is often visible in thin persons. In obese people these spines are covered with fat and may be difficult to locate; however, they are easier to palpate when the person is sitting and the muscles attached to the spines are relaxed. The **iliac tubercle,** 5 to 6 cm posterior to the anterior superior iliac spine, marks the widest point of the iliac crest. To palpate the iliac tubercle, place your thumb on the anterior superior iliac spine and move your fingers posteriorly along the external lip of the iliac crest. The iliac tubercle lies at the level of the spinous process of L5 vertebra.

Approximately a hand's width inferior to the umbilicus, the *bodies and superior rami of the pubic bones* may be palpated (*A*). The **pubic crest,** the rough anterior border of the body of the pubis, may be palpated through the rectus sheath and rectus abdominis (see Chapter 2). The **pubic symphysis** lies in the midline between the bodies of the pubic bones; it may be difficult to palpate because it is covered with fat. The **pubic tubercle** lies anteriorly on each side, a thumb's width (2.5 cm) from the pubic symphysis. The small pubic tubercle at the anterior end of the pubic crest is a guide to the superficial inguinal ring and is an important landmark in the diagnosis and repair of inguinal and femoral hernias (see Chapter 2). The pubic tubercle may also be palpated in males by invaginating the skin of the lateral part of the scrotum with the examining finger (see Chapter 2).

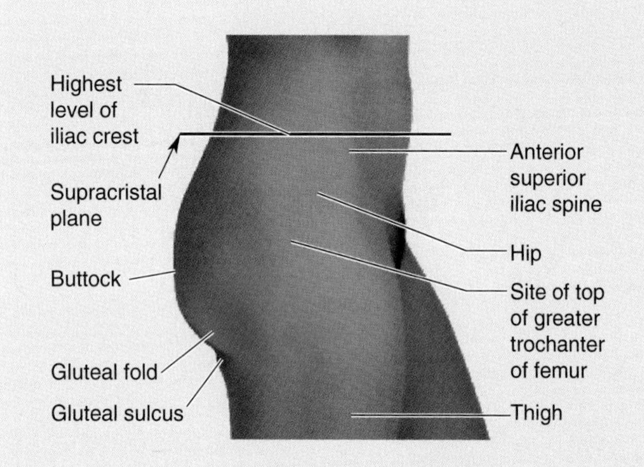

The iliac crest ends posteriorly at the sharp **posterior superior iliac spine** (*B*), which may be difficult to palpate; however, its position is easy to locate because it lies at the bottom of a *skin dimple*, approximately 4 cm lateral to the midline (p. 566). The dimple exists because the skin and underlying fascia attach to the posterior superior iliac spine. A line connecting the dimples lies at the level of the rudimentary S2 spinous process and passes though the center of the *sacroiliac joints*. The skin dimples are useful landmarks when palpating the area of the sacroiliac joints in search of edema (swelling) or local tenderness. These dimples also indicate the termination of the iliac crests from which bone marrow and pieces of bone for grafts can be obtained (e.g., to repair a fractured tibia). The sacroiliac joint is not easy to palpate because of the overhang of the ilium and the presence of strong ligaments.

The **ischial tuberosity** is easily palpated in the inferior part of the buttock when the thigh is flexed. The thick gluteus maximus and fat in the buttock cover and obscure the tuberosity when the thigh is extended (*B*). The **gluteal fold,** a prominent skin fold containing fat, coincides with the inferior border of the gluteus maximus. The **gluteal sulcus,** the skin crease inferior to the gluteal fold, indicates the separation of the buttock from the thigh.

The center of the **head of the femur** (*A*) can be felt deep to a point approximately a thumb's breadth inferior to the midpoint of the inguinal ligament. The **body of the femur** is so covered with muscles that it is not usually palpable. Only the superior and inferior ends of the femur are palpable. The laterally placed **greater trochanter of the femur** ▶

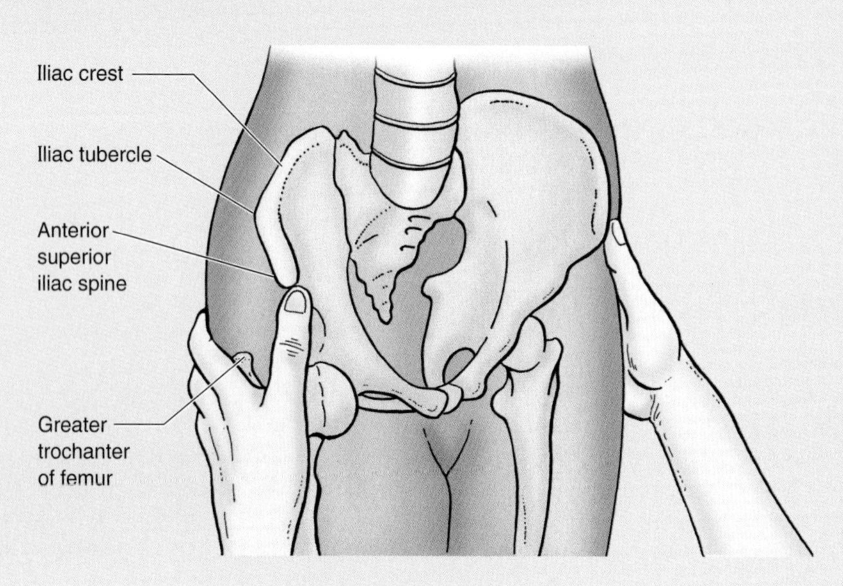

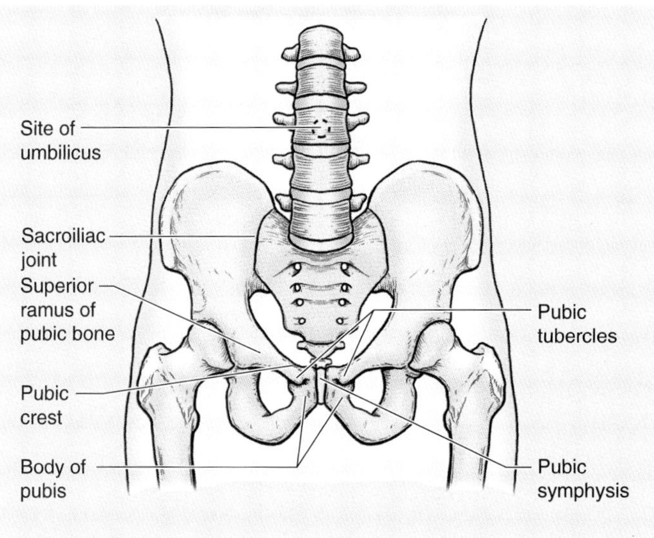

(A) Anterior pelvis

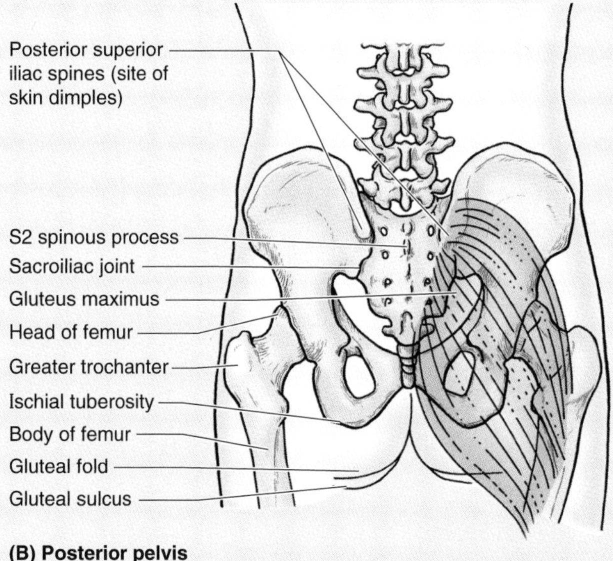

(B) Posterior pelvis

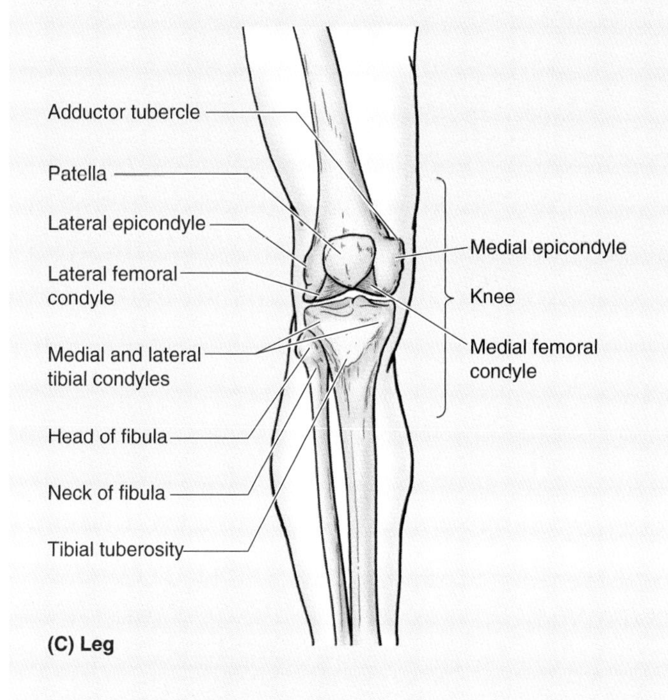

(C) Leg

▶ projects superior to the junction of the body with the femoral neck and can be palpated on the lateral side of the thigh approximately 10 cm inferior to the iliac crest (*B*). The greater trochanter forms a prominence anterior to the hollow on the lateral side of the hip. The prominence of the greater trochanters is responsible for the shape of the adult female pelvis. The posterior edge of the greater trochanter is relatively uncovered and easily palpable. The anterior and lateral parts of the trochanter are covered by fascia and muscle and thus are not easy to palpate. Because it lies close to the skin, the greater trochanter causes discomfort when you lie on your side on a hard surface. In the anatomical position, a line joining the tips of the greater trochanters normally passes through the pubic tubercles and the center of the femoral heads. The lesser trochanter is indistinctly palpable superior to the lateral end of the gluteal fold.

The condyles of the femur are subcutaneous and easily palpated when the knee is flexed or extended (*C*). The **lateral femoral condyle** is superficial and can be palpated on the lateral aspect of the knee. The **medial femoral condyle** can be palpated through the aponeurotic expansion of the vastus (L. great or vast) medialis muscle. At the center of each condyle is a prominent **epicondyle** that is easily palpable. The patellar surface of the femur is where the **patella** slides during flexion and extension of the leg at the knee joint. The lateral and medial margins of the patellar surface can be palpated when the leg is flexed. The **adductor tubercle**, a small prominence of bone, may be felt at the superior part of the medial femoral condyle. The adductor tubercle may be palpated by pushing your thumb inferiorly along the medial side of the thigh until it encounters the tubercle.

The **tibial tuberosity**, an oval elevation on the anterior surface of the tibia, is easily palpated approximately 5 cm distal to the apex of the patella. The subcutaneous, flat anteromedial surface of the tibia is also easy to palpate. The skin covering this surface is freely movable. The **tibial condyles** can be palpated anteriorly at the sides of the patellar ligament, especially when the knee is flexed (p. 532). The **head of the fibula** can be easily palpated at the level of the superior part of the tibial tuberosity because the knoblike head is subcutaneous at the posterolateral aspect of the knee. The **neck of the fibula** can be palpated just distal to the lateral side of the head.

The **medial malleolus**—the prominence on the medial side of the ankle—is also subcutaneous and easy to palpate. Note that its inferior end is blunt and that it does not ▶

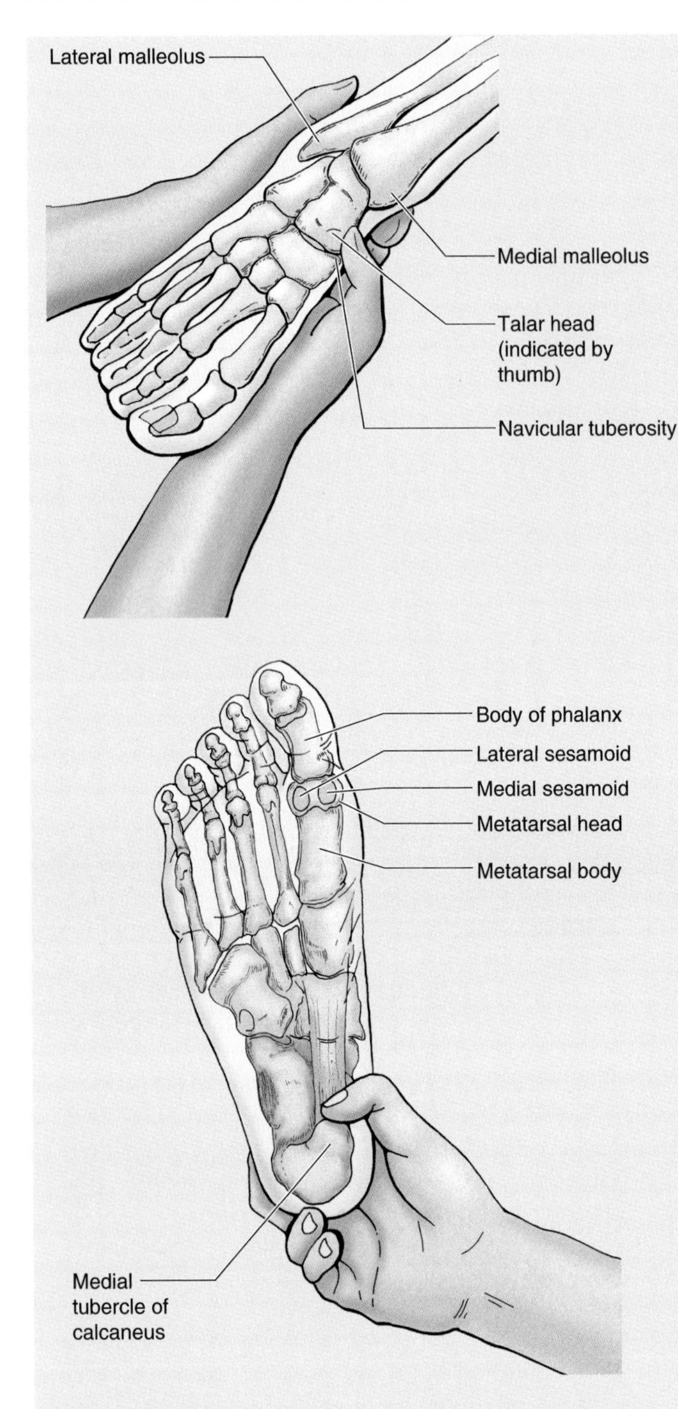

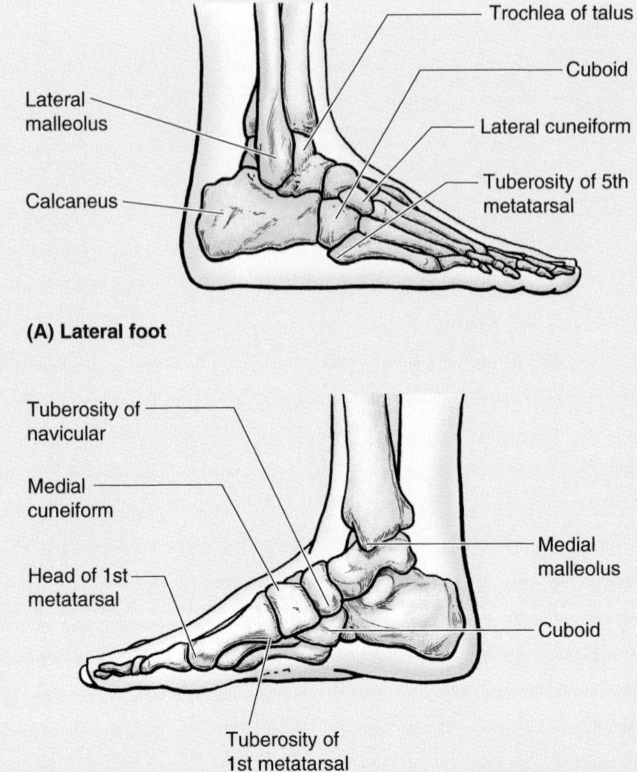

▶ extend as far distally as the lateral malleolus. The medial malleolus lies approximately 1.25 cm proximal to the level of the tip of the lateral malleolus. Only the distal quarter of the body of the fibula is palpable. Feel your **lateral malleolus**, noting that it is subcutaneous and that its inferior end is sharp. Note that the tip of the lateral malleolus extends further distally and more posteriorly than does the tip of the medial malleolus.

The **talar head** is palpable anteromedial to the proximal part of the lateral malleolus when the foot is inverted and anterior to the medial malleolus when the foot is everted. Eversion of the foot makes the head more prominent as it moves away from the navicular. The talar head occupies the space between the sustentaculum tali and the navicular tuberosity. Should the head of the talus be difficult to palpate, draw a line from the tip of the medial malleolus to the navicular tuberosity; the talar head lies deep to the center of this line. When the foot is plantarflexed, the superior surface of the body of the talus can be palpated on the anterior aspect of the ankle, anterior to the inferior end of the tibia.

The weightbearing **medial tubercle of the calcaneus** on the plantar surface of the foot is broad and large, but it is not usually easy to palpate because of the overlying skin and subcutaneous tissue. The **sustentaculum tali**—a shelflike medial extension of the calcaneus that helps support the talus—may be felt as a small prominence approximately a finger's breadth distal to the tip of the medial malleolus. The **fibular trochlea**, a small lateral extension of the calcaneus, may be detectable as a small tubercle on the lateral aspect of the calcaneus, anteroinferior to the tip of the lateral malleolus.

Usually, palpation of bony prominences on the plantar surface of the foot is difficult because of the thick skin, fascia, and pads of fat. The medial and lateral **sesamoid bones** inferior to the head of this metatarsal can be felt to slide when the great (big) toe is moved passively. The heads of the metatarsals can be palpated by placing the thumb on their plantar surfaces and the index finger on their dorsal surfaces. If *callosities* (calluses) or thickenings of the keratin layer of the epidermis are present, the metatarsal heads are difficult to palpate. ▶

▶ The **tuberosity of the 5th metatarsal** forms a prominent landmark on the lateral aspect of the foot (*A*) that can easily be palpated at the midpoint of the lateral border of the foot. The **bodies of the metatarsals and phalanges** can be felt on the dorsum of the foot between the extensor tendons. The **cuboid** can be felt on the lateral aspect of the foot, posterior to the base of the 5th metatarsal. The **medial cuneiform** can be palpated between the tuberosity of the navicular and the base of the 1st metatarsal (*B*). The **head of the 1st metatarsal** forms a prominence on the medial aspect of the foot. The **tuberosity of the navicular** is easily seen and palpated on the medial aspect of the foot (*B*), inferoanterior to the tip of the medial malleolus. The cuboid and cuneiforms are difficult to identify individually by palpation. ⊕

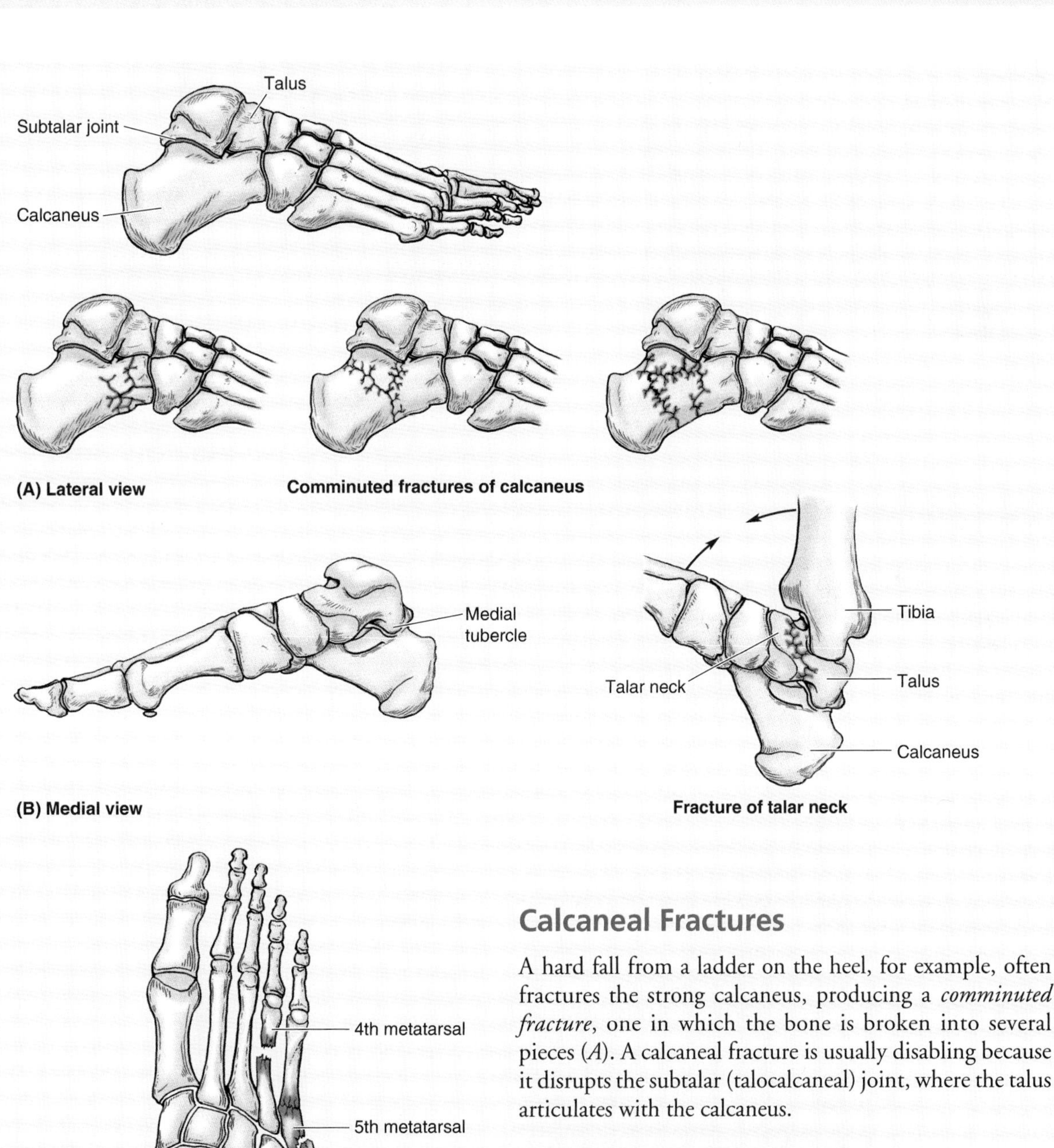

Calcaneal Fractures

A hard fall from a ladder on the heel, for example, often fractures the strong calcaneus, producing a *comminuted fracture*, one in which the bone is broken into several pieces (*A*). A calcaneal fracture is usually disabling because it disrupts the subtalar (talocalcaneal) joint, where the talus articulates with the calcaneus.

Fractures of the Talar Neck

Fractures of the talar neck (*B*) may occur during severe dorsiflexion of the ankle (e.g, when a person is pressing extremely hard on the brake pedal of a vehicle during a head-on collision). In some cases, the body of the talus dislocates posteriorly. ▶

Fractures of the Metatarsals

Metatarsal fractures occur when a heavy object falls on the foot, for example, or it is run over by a heavy object such as a metal wheel (*C*). Metatarsal fractures are also common in dancers, especially female ballet dancers using the demi-pointe technique. The "dancer's fracture" usually occurs when the dancer loses balance, putting the full body weight on the metatarsal and fracturing the bone. *Fatigue fractures of the metatarsals* may result from prolonged walking. These fractures, usually transverse, result from repeated stress on the metatarsals.

When the foot is suddenly and violently inverted, the tuberosity of the 5th metatarsal may be avulsed by the tendon of the fibularis brevis muscle. *Avulsion fractures of the 5th metatarsal tuberosity* (*C*) are common in basketball and tennis players. Part of the tuberosity is pulled off, producing pain and edema at the base of the 5th metatarsal.

Occasionally, the external part of the tuberosity of the 5th metatarsal develops as an accessory bone—the *os vesalianum* (Vesalius' bone); it appears near the base of the metatarsal. When examining radiographs, it is important to be able to recognize a vesalian bone so as not to diagnose it as a fracture of the tuberosity. When the vesalian bone is large, the tuberosity of the 5th metatarsal is small.

Os Trigonum

During ossification of the talus, the lateral tubercle of the talus occasionally fails to unite with the body of the talus. This event results in a bone known as the *os trigonum*, which could be misinterpreted as a fracture by an inexperienced viewer of radiographs.

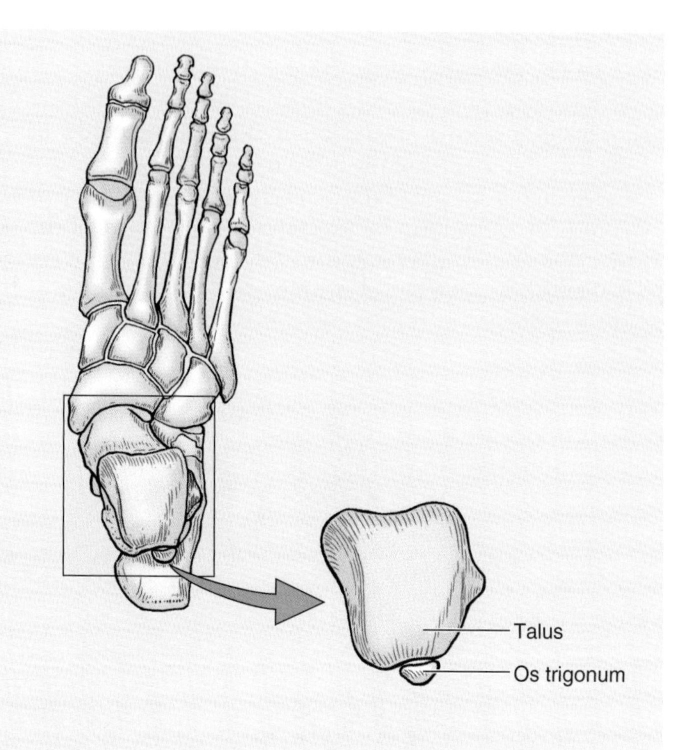

Fracture of the Sesamoid Bones

The *sesamoid bones of the great toe* in the tendon of the flexor hallucis longus *bear the weight of the body*, especially during the latter part of the stance phase of walking. The sesamoids develop before birth and begin to ossify during late childhood. Fracture of the sesamoids may result from a crushing injury (e.g., when a heavy object falls on the great toe). ○

Fascia, Vessels, and Nerves of the Lower Limb

When the connective tissue of the body forms an enveloping sheath it is *fascia* (L. band or bandage), indicating that it binds structures together (Wendell-Smith, 1997). Individual muscles are surrounded by thin fascia—*perimysium*. The fascia (L. fasciae) of the lower limb consists of superficial and deep layers (Fig. 5.9, *A–C*).

The **subcutaneous tissue**, or **superficial fascia**, lies deep to the skin and consists of loose connective tissue that contains a variable amount of fat, cutaneous nerves, superficial veins (great and small saphenous veins and their tributaries), lymphatic vessels, and lymph nodes. The connective tissue fibers blend with those in the dermis so that no distinct plane of cleavage is detectable. The subcutaneous tissue of the hip and thigh is continuous with that of the inferior part of the anterolateral abdominal wall and buttock. At the knee, the subcutaneous tissue loses its fat and blends with the deep fascia, but fat is present in the subcutaneous tissue of the leg.

The **deep fascia** is a dense layer of connective tissue between the subcutaneous tissue and the muscles (Fig. 5.9, *A–C*). It forms fibrous septa that separate muscles from one another and invest them. The deep fascia is especially strong in the lower limb and invests the limb like an elastic stocking. The deep fascia prevents bulging of the muscles during contraction, thereby making muscular contraction more efficient in pumping blood toward the heart. The *deep fascia of the thigh* is called **fascia lata** (L. lata, broad), and the *deep fascia of the leg*, **crural fascia** (L. crus, leg).

The fascia lata attaches:

- Superiorly to the inguinal ligament, pubic arch, body of pubis, and pubic tubercle; the membranous layer of subcutaneous tissue (Scarpa's) of the lower abdominal wall also attaches to the fascia lata approximately a finger's breadth inferior to the inguinal ligament
- Laterally and posteriorly to the iliac crest
- Posteriorly to the sacrum, coccyx, sacrotuberous ligament, and ischial tuberosity.

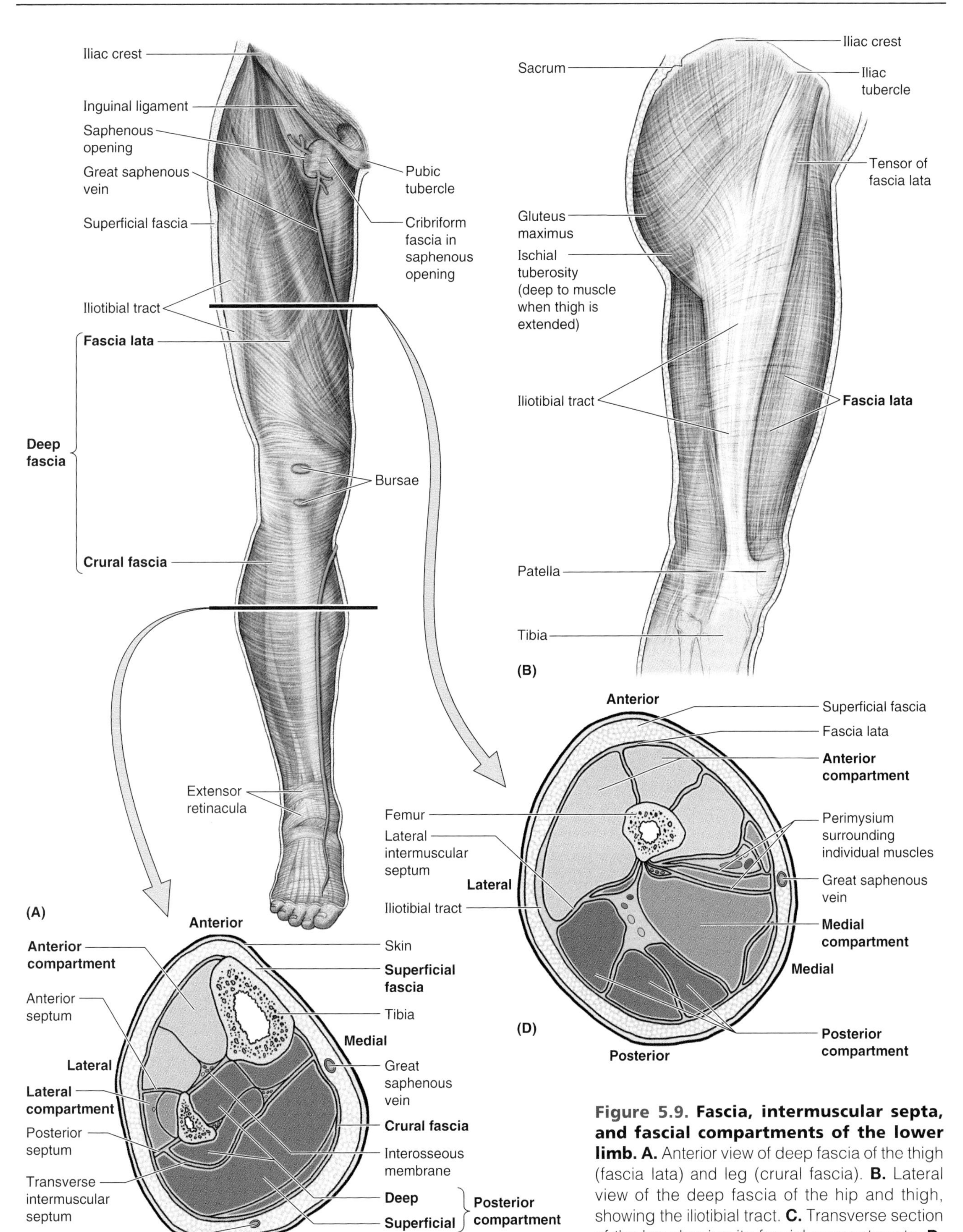

Figure 5.9. Fascia, intermuscular septa, and fascial compartments of the lower limb. A. Anterior view of deep fascia of the thigh (fascia lata) and leg (crural fascia). **B.** Lateral view of the deep fascia of the hip and thigh, showing the iliotibial tract. **C.** Transverse section of the leg showing its fascial compartments. **D.** Transverse section of the thigh showing its fascial compartments.

The **fascia lata** attaches distally to exposed parts of bones around the knee and is continuous with the crural fascia. The fascia lata is substantial because it encloses large thigh muscles, especially laterally where it is thickened and strengthened by additional longitudinal fibers to form the **iliotibial tract** (Fig. 5.9*B*). This broad band of fibers is the conjoint aponeurosis of the **tensor of fascia lata** (L. tensor fasciae latae) and **gluteus maximus** muscles. The iliotibial tract extends from the iliac tubercle to a tubercle on the lateral condyle of the tibia (Gerdy's tubercle).

The thigh muscles are within **three compartments**—anterior, medial, and posterior—the walls of which are formed by *three fascial intermuscular septa* that arise from the deep aspect of the fascia lata and attach to the linea aspera of the femur (Fig. 5.9*D*). The **lateral intermuscular septum** is strong; the other two septa are relatively weak. The lateral intermuscular septum extends from the iliotibial tract to the lateral lip of the linea aspera and lateral supracondylar line of the femur.

The **saphenous opening in the fascia lata** (Fig. 5.9*A*) is a deficiency in the deep fascia lata inferior to the medial part of the inguinal ligament, approximately 4 cm inferolateral to the pubic tubercle. The saphenous opening is usually approximately 3.75 cm in *length* and 2.5 cm in *breadth*, and its long axis is vertical. Its medial margin is smooth but its superior, lateral, and inferior margins form a sharp crescentic edge, the **falciform margin**. This sickle-shaped margin of the saphenous opening is joined at its medial margin by fibrofatty tissue—the **cribriform fascia**. This sievelike fascia (L. cribrum, a sieve), derived from the thin, membranous layer of subcutaneous tissue, spreads over the saphenous opening and closes it. This layer of spongy connective tissue is sievelike because it is pierced by numerous openings for the passage of lymphatic vessels and the great saphenous vein and its tributaries. The **great saphenous vein** passes through the saphenous opening and cribriform fascia to enter the femoral vein (Fig. 5.10*A*). Some efferent lymphatic vessels from the superficial inguinal lymph nodes also pass through the saphenous opening and cribriform fascia to enter the deep inguinal lymph nodes.

The **crural fascia** attaches to the anterior and medial borders of the tibia, where it is continuous with its periosteum. The crural fascia is thick in the proximal part of the anterior aspect of the leg, where it forms part of the proximal attachments of the underlying muscles. Although thin in the distal part of the leg, the crural fascia is thicker where it forms the **extensor retinacula** (Fig. 5.9*A*). Anterior and posterior intermuscular septa pass from the deep surface of the crural fascia and attach to the corresponding margins of the fibula. The **interosseous membrane** and the **crural intermuscular septa** divide the leg into three compartments (Fig. 5.9*C*):

- Anterior (extensor) compartment of leg
- Lateral (fibular) compartment of leg
- Posterior (flexor) compartment of leg.

The muscles in the posterior compartment are subdivided into superficial and deep parts by the **transverse intermuscular septum**.

Venous Drainage of the Lower Limb

The lower limb has superficial and deep veins; the superficial veins are in the subcutaneous tissue, and the deep veins are deep to (beneath) the deep fascia and accompany all major arteries. Superficial and deep veins have valves, but they are more numerous in deep veins. The subcutaneous tissue of the thigh also provides a pathway for superficial veins, lymphatic vessels, and cutaneous nerves to follow.

Superficial Veins of the Lower Limb

The two major superficial veins in the lower limb are the great and small *saphenous veins* (Fig. 5.10, *A* and *B*). Most of their tributaries are unnamed.

The **great saphenous vein** is formed by the union of the dorsal vein of the great toe and the **dorsal venous arch** of the foot. The great saphenous vein:

- Ascends anterior to the medial malleolus
- Passes posterior to the medial condyle of the femur
- Anastomoses freely with the small saphenous vein

Figure 5.10. Veins of the lower limb. A. Medial view. **B.** Posterior view. The veins are subdivided into superficial and deep groups. The superficial veins are in the subcutaneous tissue; the deep veins lie deep to the deep fascia and are usually accompanying veins of the arteries (L. venae comitantes). The main superficial veins are the great and small saphenous veins; most of their tributaries are unnamed. The perforating veins drain blood from the superficial veins to the deep veins, which are subject to compression as the muscles contract, driving blood toward the heart against the pull of gravity. The *inset* drawing demonstrates the valves at the proximal ends of the femoral and great saphenous veins. The veins are opened and spread apart to show the valves. **C.** Anterior view of the deep veins (shown as single veins here); usually they occur as duplicate or multiple accompanying veins. The deep vein of the thigh (L. vena profunda femoris) accompanies the deep artery of the thigh (L. arteria profunda femoris) and joins the femoral vein in the femoral triangle, usually in common with the medial and lateral circumflex femoral veins. **D.** Perforating veins pierce the deep fascia and drain blood from the great saphenous vein to the posterior tibial and fibular veins. **E.** Posterior view of the deep veins of the gluteal region, thigh, leg, and foot. The popliteal vein continues into the thigh as the femoral vein.

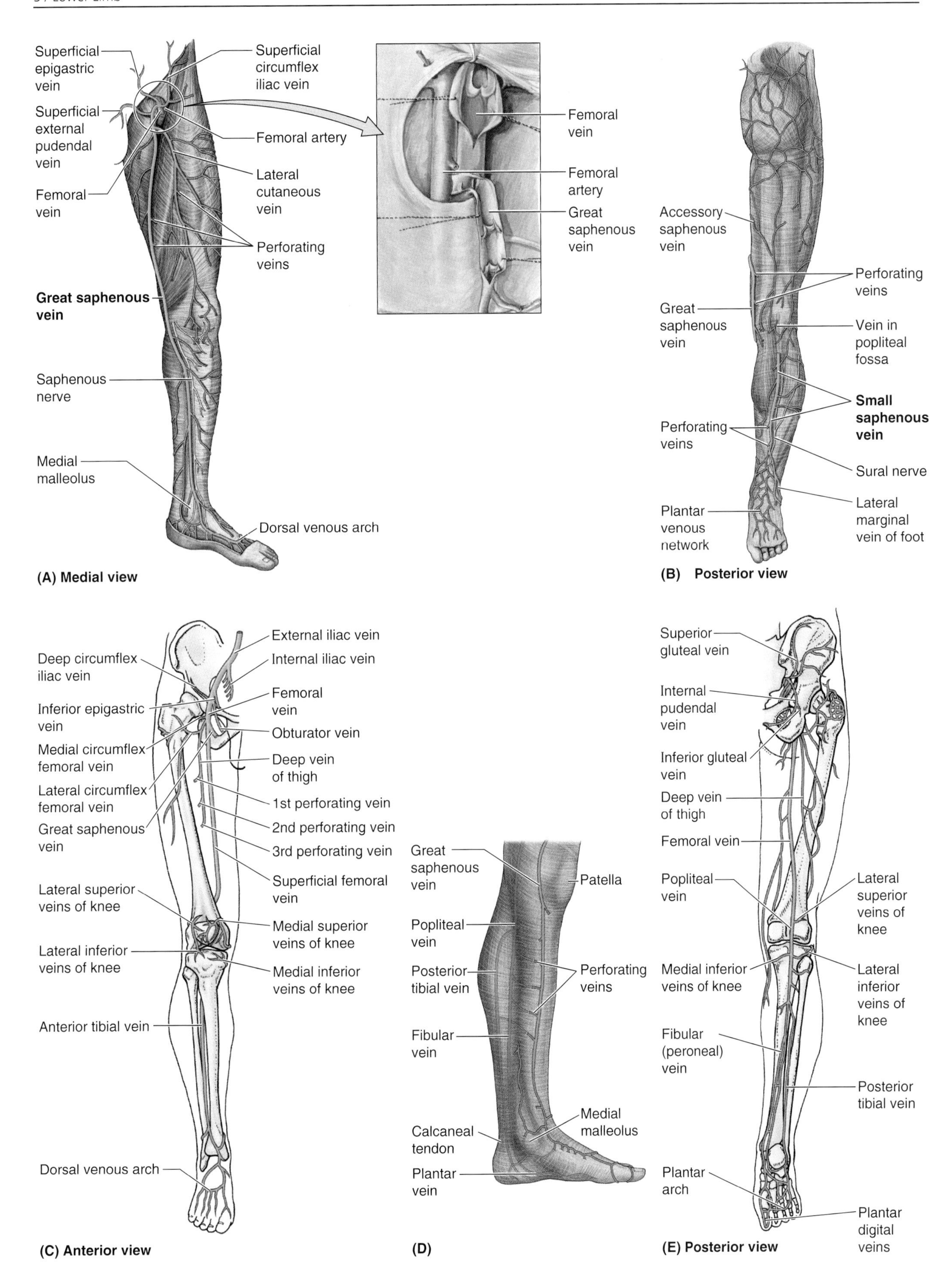

(A) Medial view

(B) Posterior view

(C) Anterior view

(D)

(E) Posterior view

- Traverses the saphenous opening in the fascia lata
- Empties into the femoral vein.

The **great saphenous vein** has 10 to 12 valves, which are more numerous in the leg than in the thigh. These valves are usually located just inferior to the perforating veins (Fig. 5.10*A*). The perforating veins also have valves. Venous valves are cuplike flaps of endothelium that fill from above. When they are full, they occlude the lumen of the vein, thereby preventing reflux of blood distally. This valvular mechanism enables the blood in the saphenous vein to overcome the force of gravity as it passes to the heart.

As it ascends in the leg and thigh, the great saphenous vein receives numerous tributaries and communicates in several locations with the small saphenous vein. Tributaries from the medial and posterior aspects of the thigh frequently unite to form an *accessory saphenous vein* (Fig. 5.10*B*). When present, this vein becomes the main communication between the great and small saphenous veins. Also, fairly large vessels—the *lateral and anterior cutaneous veins*—arise from networks of veins in the inferior part of the thigh and enter the great saphenous vein superiorly, just before it enters the femoral vein. Near its termination, the great saphenous vein also receives the superficial circumflex iliac, superficial epigastric, and external pudendal veins (Fig. 5.10*A*).

The **small saphenous vein** arises on the lateral side of the foot from the union of the dorsal vein of the small (little) toe with the dorsal venous arch. *The small saphenous vein*:

- Ascends posterior to the lateral malleolus as a continuation of the lateral marginal vein
- Passes along the lateral border of the calcaneal tendon
- Inclines to the midline of the fibula and penetrates the deep fascia
- Ascends between the heads of the gastrocnemius muscle
- Empties into the popliteal vein in the popliteal fossa.

Although many tributaries are received by the saphenous veins, their diameter remains remarkably uniform as they ascend the limb. This is possible because the blood they receive is continuously shunted from these superficial veins in the subcutaneous tissue to the deep veins by means of the many perforating veins.

The **perforating veins** penetrate the deep fascia close to their origin from the superficial veins and contain valves that, when functioning normally, only allow blood to flow from the superficial veins to the deep veins. The perforating veins pass through the deep fascia at an oblique angle so that when muscles contract and the pressure increases inside the deep fascia, the perforating veins are compressed. This also prevents blood from flowing from the deep to the superficial veins. This pattern of venous bloodflow—from superficial to deep—is important for proper venous return from the lower limb because it enables muscular contractions to propel blood toward the heart against the pull of gravity (*musculovenous pump*).

Deep Veins of the Lower Limb

The deep veins accompany all the major arteries (L. venae comitantes) and their branches. Instead of occurring as a single vein in the limbs (although they are frequently illustrated as one, and are often referred to as a single vein), the deep veins usually occur as paired, frequently interconnecting veins that flank the artery they accompany (Fig. 5.10*E*). They are contained within the vascular sheath with the artery, whose pulsations also help to compress and move blood in the veins.

The **dorsal digital veins** of the foot receive tributaries from the **plantar venous arch** and join to form common dorsal digital veins that terminate in the **dorsal venous arch**. Medial and lateral plantar veins pass close to the arteries and, after communicating with the great and small saphenous veins, form the **posterior tibial veins** posterior to the medial malleolus. The deep veins communicate with the superficial veins through **perforating veins** (communicating veins), which accompany the perforating arteries from the deep artery of the thigh (L. arteria profunda femoris) (Fig. 5.10, *C* and *D*). The perforating veins drain blood from the thigh muscles and terminate in the **deep vein of the thigh** (L. vena profunda femoris).

Because of the effect of gravity, bloodflow is markedly reduced when a person stands quietly. During exercise, blood received from the superficial veins by the deep veins is propelled by muscular contraction to the femoral and then the external iliac veins. Flow in the reverse direction—away from the heart or from the deep to the superficial veins—is prevented if the venous valves are competent (capable of performing their function). The deep veins are more variable and anastomose much more frequently than the arteries they accompany. Both superficial and deep veins can be ligated with impunity if necessary.

Varicose Veins, Thrombosis, and Thrombophlebitis

Frequently, the great saphenous vein and its tributaries become dilated and varicosed—dilated so that the cusps of their valves do not close. *Varicose veins* are common in the posteromedial parts of the lower limb; they often cause considerable discomfort. Varicose veins form when valves that usually prevent bloodflow from the deep veins through the perforating veins to the superficial veins are incompetent. Further, when the valves within the great saphenous vein itself are incompetent, the pull of gravity on the uninterrupted column of blood results in a higher intraluminal pressure, which also exacerbates varicosities. As a result, the superficial veins become tortuous and dilated. ▶

▶ The veins of the lower limb are subject to *venous thrombosis* (blood clotting) after a bone fractures. *Venous stasis* (stagnation) is an important cause of thrombus formation. Venous stasis can be caused by pressure on the veins from the bedding during a prolonged hospital stay or from a tight cast or bandage and is aggravated by muscular inactivity. As a result of venous thrombosis, inflammation may develop (*thrombophlebitis*) around the vein. *Pulmonary thromboembolism* (obstruction of a pulmonary artery) occurs in a few cases when a thrombus breaks free from a lower limb vein and passes to the lungs. A large embolus may obstruct a main pulmonary artery and cause death (see Chapter 1).

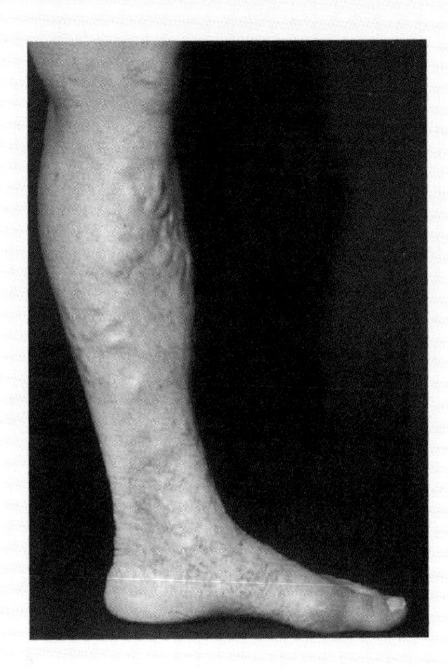

Varicose Veins

Saphenous Vein Grafts

The great saphenous vein is commonly used for coronary arterial bypasses because (*a*) it is readily accessible, (*b*) sufficient distance occurs between the tributaries and perforating veins so that usable lengths can be harvested, and (*c*) its wall contains a higher percentage of muscular and elastic fibers than do other superficial veins. Saphenous vein grafts are used to bypass obstructions in blood vessels (e.g., in an intracoronary thrombus—see Chapter 1). When part of the great saphenous vein is removed for a bypass, the vein is reversed so that the valves do not obstruct bloodflow in the graft. Because there are so many other leg veins, removal of the great saphenous vein rarely produces a significant problem in the lower limb or seriously affects circulation, provided the deep veins are intact.

Saphenous Cutdown and Saphenous Nerve Injury

Even when it is not visible in infants and obese persons, or in patients in shock whose veins are collapsed, the great saphenous vein can always be located by making a skin incision anterior to the medial malleolus. This procedure—a *saphenous cutdown*—is used to insert a cannula for prolonged administration of blood, plasma expanders, electrolytes, or drugs. *The saphenous nerve accompanies the great saphenous vein anterior to the medial malleolus.* Should this nerve be cut during a saphenous cutdown or caught by a ligature during closure of a surgical wound, the patient may complain of pain along the medial border of the foot. ⊙

Lymphatic Drainage of the Lower Limb

The lower limb has superficial and deep lymphatic vessels. The **superficial lymphatic vessels** accompany the saphenous veins and their tributaries. The lymphatic vessels accompanying the great saphenous vein end in the **superficial inguinal lymph nodes** (Fig. 5.11*A*). Most lymph from these nodes passes directly to the **external iliac lymph nodes**—located along the external iliac vein—but lymph may also pass to the **deep inguinal lymph nodes**. The lymphatic vessels accompanying the small saphenous vein enter the **popliteal lymph nodes**, which surround the popliteal vein in the fat of the popliteal fossa (Fig. 5.11*B*). The **deep lymphatic vessels** from the leg accompany deep veins and enter the popliteal lymph nodes. Most lymph from these nodes ascends through deep lymphatic vessels to the **deep inguinal lymph nodes** (Fig. 5.11*A*). These nodes lie under the deep fascia on the medial aspect of the femoral vein. Lymph from the deep nodes passes to the external iliac lymph nodes.

Enlarged Inguinal Lymph Nodes

Lymph nodes enlarge when diseased. *Abrasions and minor sepsis*—pathogenic microorganisms or their toxins in the blood or other tissues—may produce slight enlargement of the superficial inguinal lymph nodes (*lymphadenopathy*) in otherwise healthy people. Because these nodes are in the subcutaneous tissue, they are easy to palpate in normal people. Those who are unaware of this may be concerned when they feel these nodes because they assume they have a serious genital disease, for example. When inguinal lymph nodes are enlarged, their entire field of drainage—the trunk inferior to the umbilicus, including the perineum, as well as the entire lower limb—has to be examined to determine the cause of their enlargement. In female patients, the possibility of metastasis of cancer from the uterus should also be considered because some lymphatic drainage from ▶

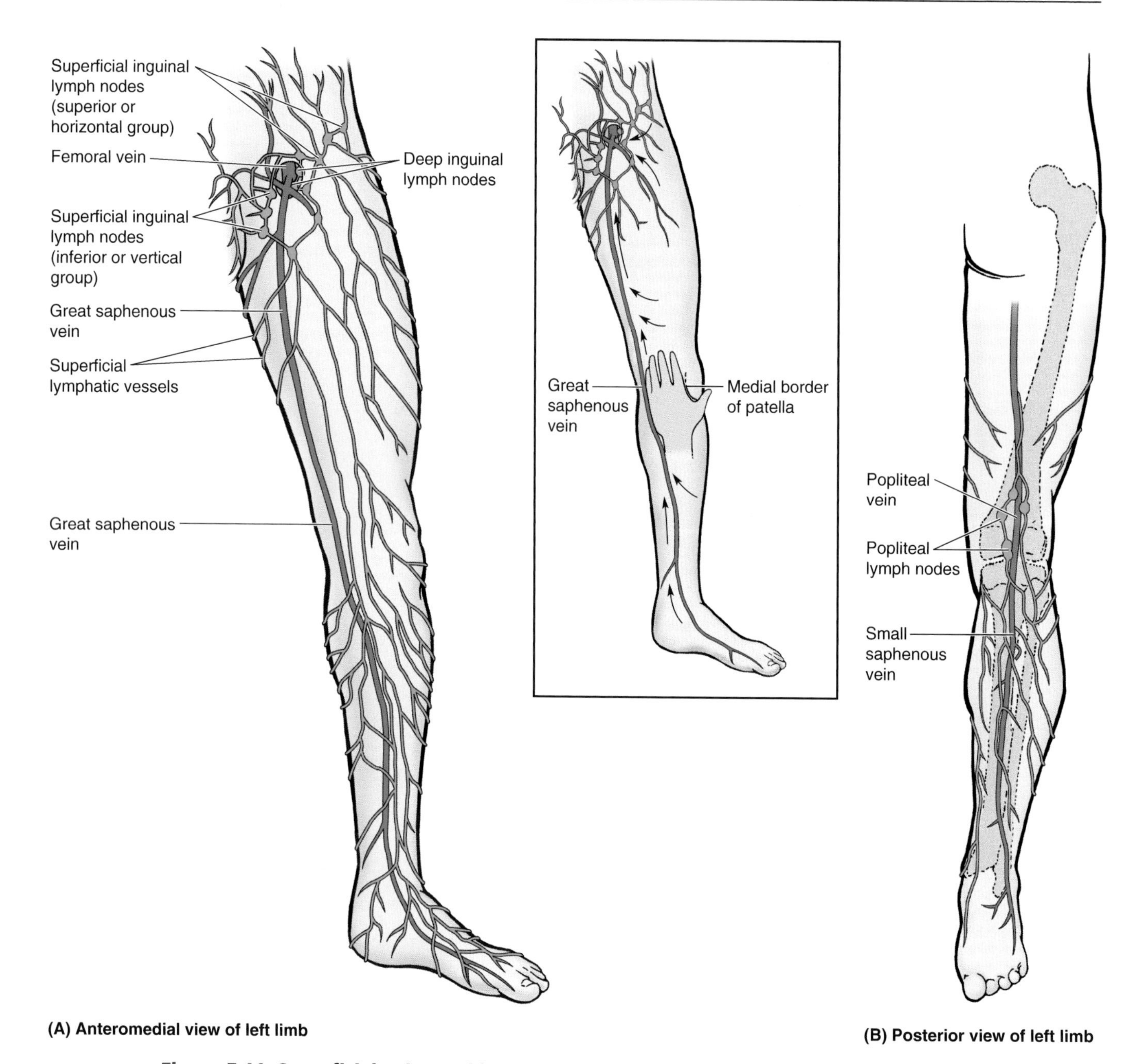

Figure 5.11. Superficial veins and lymphatics of the lower limb. A. Anteromedial view. The great saphenous vein ascends the medial aspect of the limb, passing anterior to the medial malleolus and approximately a hand's breadth posterior to the patella (knee cap). The superficial lymphatic vessels from the medial foot, anteromedial leg, and thigh converge toward and accompany the great saphenous vein, draining into the inferior (vertical) group of superficial inguinal lymph nodes. **B.** Posterior view. The superficial lymphatic vessels of the lateral foot and posterolateral leg accompany the lesser saphenous vein and drain initially into the popliteal lymph nodes, which lie deep to the popliteal fascia. The efferent vessels from these nodes join other deep lymphatics, which accompany the femoral vessels to drain into the deep inguinal lymph nodes.

▶ the uterine fundus may flow along lymphatics accompanying the round ligament of the uterus through the inguinal canal to reach the superficial inguinal lymph nodes (see Chapter 3). ⊙

Cutaneous Innervation of the Lower Limb

Cutaneous nerves in the subcutaneous tissue supply the skin of the lower limb (Fig. 5.12, Table 5.1). These nerves, except for some proximal ones, are branches of the lumbar and sacral plexuses (see Chapters 3 and 4). The area of skin supplied by cutaneous branches from a single spinal nerve is called a **dermatome.** Adjacent dermatomes may overlap except at the **axial line**—the line of junction of dermatomes supplied from discontinuous spinal levels.

Branches of the **subcostal nerve** (T12) descend over the iliac crest toward the anterosuperior iliac spine and enter the superolateral part of the thigh. They supply the skin of the thigh anterior to the greater trochanter of the femur.

The **iliohypogastric nerve** (L1, occasionally T12) divides into lateral and anterior cutaneous branches. The lateral branch supplies the skin over the superolateral part of the buttock (Table 5.1), and the anterior branch supplies skin superior to the pubis.

The **ilioinguinal nerve** (L1, occasionally T12) accompanies the spermatic cord or the round ligament of the uterus through the superficial inguinal ring to the scrotum or labium majus (see Chapter 3). Branches of the ilioinguinal nerve are distributed to the skin over the proximal and medial parts of the thigh and to the scrotum and labium majus through their anterior scrotal and labial branches, respectively.

The **genitofemoral nerve** (L2 and L3) has genital and femoral branches that supply skin just inferior to the middle part of the inguinal ligament.

The **lateral femoral cutaneous nerve** (L2 and L3), a direct branch of the lumbar plexus, runs obliquely toward the anterior superior iliac spine and then passes deep to the inguinal ligament into the thigh, dividing into anterior and posterior branches. The anterior branches become superficial approximately 10 cm distal to the inguinal ligament and supply skin on the lateral and anterior parts of the thigh. The posterior branch passes posteriorly across the lateral and posterior surfaces of the thigh and supplies skin from the level of the greater trochanter to the middle of the area just proximal to the knee.

The **femoral nerve** arises from the 2nd, 3rd, and 4th lumbar nerves in the substance of the psoas major muscle (see Chapter 2) and enters the thigh deep to the inguinal ligament, lateral to the femoral vessels. It sends branches to thigh muscles and anterior femoral cutaneous nerves to the skin on the anterior and medial regions of the thigh.

The **anterior femoral cutaneous nerves** arise from the

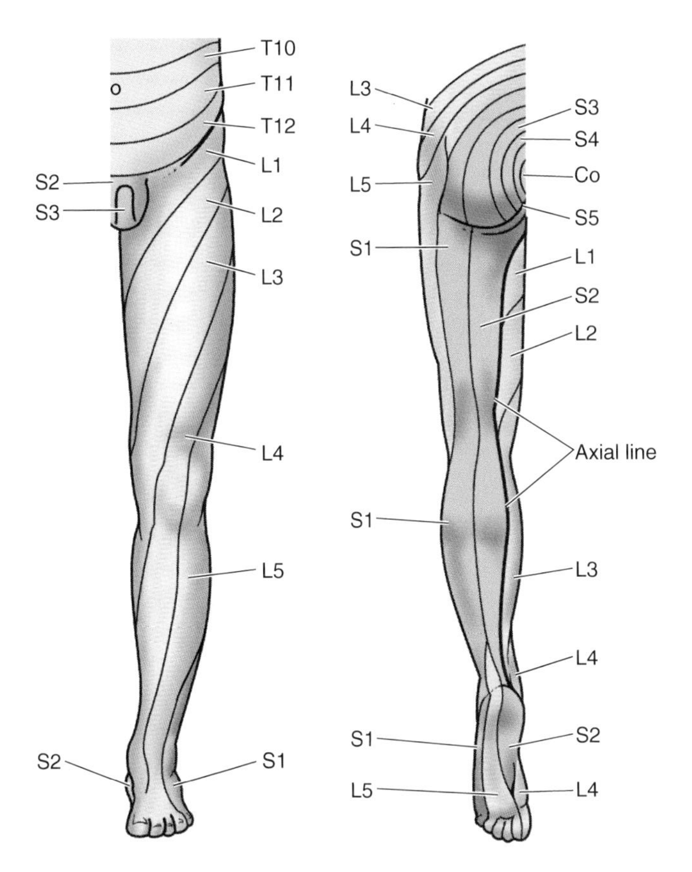

Figure 5.12. Dermatomes of the lower limb. A dermatome is an area of skin supplied by the dorsal (sensory) root of a spinal nerve. The dermatomes L1 through L5 extend as a series of bands from the posterior midline of the trunk into the limbs, passing laterally and inferiorly around the limb to its anterior and medial aspects. Dermatomes S1 and S2 pass inferiorly down the posterior aspect of the limb, separating near the ankle to pass to the lateral and medial margins of the foot. Adjacent dermatomes overlap considerably; that is, each segmental nerve overlaps the territories of its neighbors.

femoral nerve, a branch of the *lumbar plexus*. They arise in the femoral triangle (p. 538), pierce the fascia lata along the path of the sartorius muscle, and supply skin on the medial and anterior aspects of the thigh.

A branch of the **obturator nerve** (L2, L3, and L4) is occasionally present (Table 5.1). It passes to the medial side of the knee, where it communicates with the saphenous nerve and supplies the skin on the anterior, medial, and posterior surfaces of the proximal part of the thigh.

The **posterior femoral cutaneous nerve**, a branch of the *sacral plexus* (S2 and S3), supplies branches to the skin on the posterior aspect of the thigh and over the popliteal fossa.

The **sciatic nerve** arises from the sacral plexus, passes through the greater sciatic foramen into the inferior gluteal region, and then into the posterior thigh. At the apex of the popliteal fossa, the sciatic nerve divides into the common fibular (peroneal) and tibial nerves; their cutaneous branches are discussed with the leg.

Table 5.1. Cutaneous Nerves of the Thigh

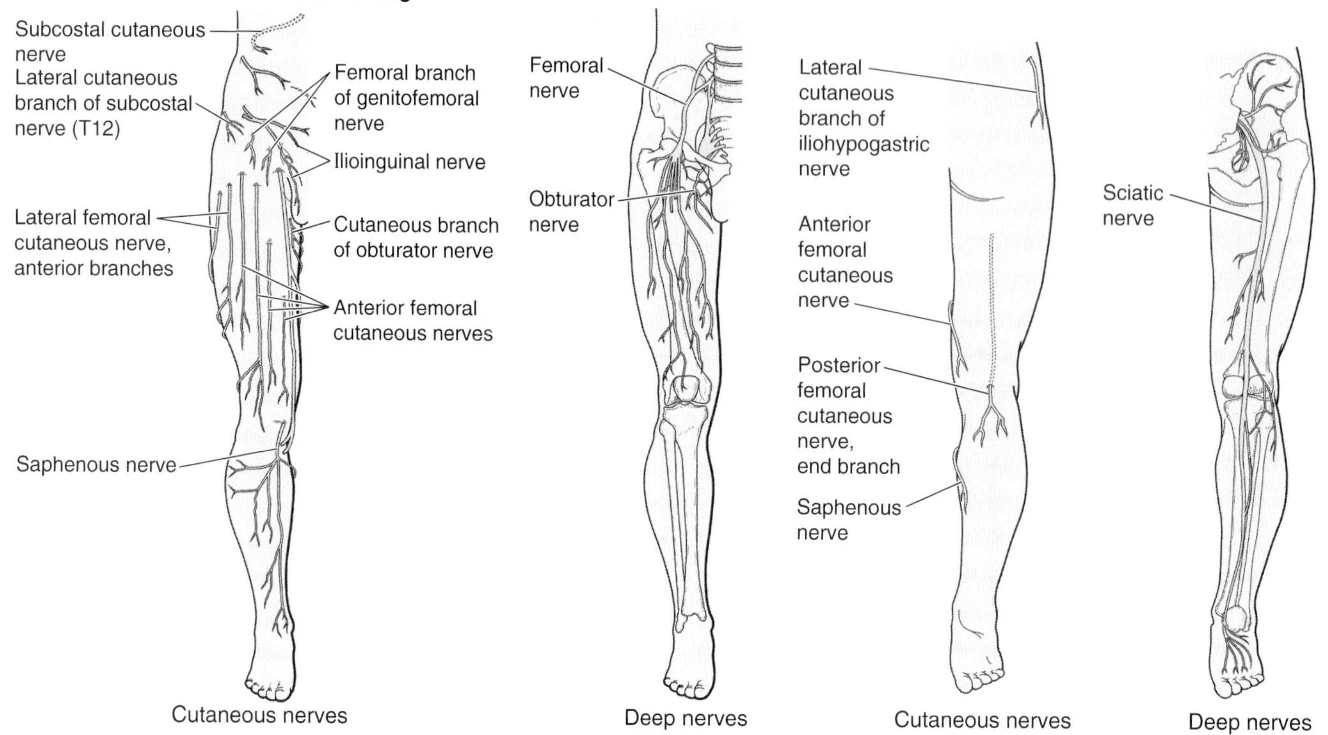

(A) Anterior view **(B) Posterior view**

Nerve	Origin	Course	Distribution
Subcostal	Ventral ramus of T12	Courses along inferior border of 12th rib in same manner as intercostal nerves	Lateral cutaneous branch supplies skin inferior to anterior iliac crest
Iliohypogastric	Lumbar plexus (L1)	Parallels iliac crest to reach inguinal and pubic regions	Lateral cutaneous branch supplies superolateral quadrant of buttock
Ilioinguinal	Lumbar plexus (L1)	Passes through inguinal canal and divides into femoral and scrotal or labial branches	Femoral branch supplies skin over femoral triangle
Genitofemoral	Lumbar plexus (L1 and L2)	Descends on anterior surface of psoas major and divides into genital and femoral branches	Femoral branch supplies skin over femoral triangle; genital branch supplies scrotum or labia majora
Lateral femoral cutaneous	Lumbar plexus (L2 and L3)	Passes deep to inguinal ligament, 2–3 cm medial to anterior superior iliac spine	Supplies skin on anterior and lateral aspects of thigh
Femoral	Lumbar plexus (L2–L4)	Passes deep to midpoint of inguinal ligament, lateral to femoral vessels, and divides into muscular and cutaneous branches	Supplies anterior thigh muscles, hip and knee joints, and skin on antero-medial side of thigh
Anterior femoral cutaneous	Femoral nerve (L2–L4)	Arise in femoral triangle and pierce fascia lata of thigh along path of sartorius muscle	Supply skin on medial and anterior aspects of thigh
Obturator	Lumbar plexus (L2–L4)	Enters thigh through obturator foramen and divides; its anterior branch descends between adductor longus and adductor brevis; its posterior branch descends between adductor brevis and adductor magnus	Anterior branch supplies adductor longus, adductor brevis, gracilis, and pectineus; posterior branch supplies obturator externus and adductor magnus
Posterior femoral cutaneous	Sacral plexus (S1–S3)	Passes through greater sciatic foramen inferior to piriformis, runs deep to gluteus maximus, and emerges from its inferior border	In addition to supplying buttock, supplies skin over posterior aspect of thigh and popliteal fossa
Sciatic	Sacral plexus (L4–S3)	Enters gluteal region through greater sciatic foramen inferior to piriformis, descends along posterior aspect of thigh, and divides proximal to knee into tibial and common fibular peroneal nerves	Innervates hamstrings by its tibial division, except for short head of biceps femoris, which is innervated by its common fibular division; provides articular branches to hip and knee joints

Regional Anesthetic Blocks of the Lower Limbs

The *iliohypogastric and ilioinguinal nerves* can be blocked by injecting an anesthetic agent 4 to 6 cm posterior to the anterior superior iliac spine, along the lateral aspect of the external lip of the iliac crest (see Chapter 2). This is where these nerves perforate the transverse abdominal muscle (L. transversus abdominis). The *femoral nerve* (L2 through L4) can be blocked approximately 2 cm inferior to the inguinal ligament, approximately a finger's breadth lateral to the femoral artery. *Paresthesia* (tingling, burning, tickling) radiates to the knee and over the medial side of the leg if the *saphenous nerve* (terminal branch of femoral) is affected.

Variations of the Cutaneous Nerves

Variations of the cutaneous nerves are common. For example, the iliohypogastric and ilioinguinal nerves may arise from a common trunk, or the ilioinguinal nerve may join the iliohypogastric nerve at the iliac crest. In this case, the iliohypogastric nerve supplies the cutaneous branches for both nerves. When the obturator nerve has a cutaneous branch, the medial cutaneous branch of the femoral nerve is correspondingly small.

Abnormalities of Sensory Function

In most instances, a peripheral nerve sensitizing an area of skin represents more than one segment of the spinal cord. Therefore, to interpret abnormalities of peripheral sensory function, peripheral nerve distribution of the major cutaneous nerves must be interpreted as anatomically different from dermatome distribution of the spinal cord segments—dermatomes are areas of skin supplied by cutaneous branches from a single spinal nerve (Fig. 5.12). Neighboring dermatomes may overlap. Pain sensation is tested by using a safety pin and asking the patient if the pinprick is felt. If sensory loss to pain exists, the spinal cord segment(s) involved can be determined. ⊙

Organization of Thigh Muscles

The thigh muscles are organized into three compartments by intermuscular septa that pass between the muscles from the fascia lata to the femur (Fig. 5.9*D*). The compartments are *anterior, medial,* and *posterior*, so named on the basis of their location, actions, and nerve supply.

Anterior Thigh Muscles

The anterior thigh muscles—**the flexors of the hip and extensors of the knee** (Fig. 5.13)—are in the *anterior compartment of the thigh*. For attachments, nerve supply, and main actions of these muscles, see Table 5.2. *The anterior thigh muscles are*:

- Pectineus
- Iliopsoas
- Tensor of fascia lata
- Sartorius
- Quadriceps femoris.

Pectineus

The pectineus is a flat quadrangular muscle located in the anterior part of the superomedial aspect of the thigh. The pectineus adducts and flexes the thigh and assists in medial rotation of the thigh.

Iliopsoas

The iliopsoas is the *chief flexor of the thigh,* and when the thigh is fixed, it flexes the trunk on the hip. Its broad lateral part, the *iliacus*, and its long medial part, the *psoas major*, arise from the iliac fossa and lumbar vertebrae, respectively. The iliopsoas is also a postural muscle that is active during standing by preventing hyperextension of the hip joint.

Tensor of Fascia Lata

The tensor of the fascia lata is a fusiform muscle approximately 15 cm long that is enclosed between two layers of fascia lata. It inserts into the *iliotibial tract*, joining fibers from the superior part of the gluteus maximus. Although anteriorly placed, the tensor of the fascia lata is actually a gluteal muscle that is usually studied with the anterior thigh muscles. The tensor of the fascia lata receives its nerve supply from the superior gluteal nerve (L4, L5) and is supplied by an inferior branch of the superior gluteal artery. *The tensor of the fascia lata is primarily a flexor of the thigh*; however, it generally does not act independently. To produce flexion, the tensor of the fascia lata acts in concert with the iliopsoas. When the iliopsoas is paralyzed, the tensor of the fascia lata hypertrophies in an attempt to compensate. It also works in conjunction with other muscles (gluteus medius and minimus) to produce medial rotation of the thigh, and it contracts during abduction. It lies too far anteriorly to be a strong abductor and thus probably serves as a synergist or fixator. *The tensor of the fascia lata also tenses the fascia lata and iliotibial tract,* thereby helping to support the femur on the tibia when standing. It has no direct action on the leg.

Sartorius

The sartorius—the "tailor's muscle" (L. sartus, patched or repaired)—is the long, ribbonlike muscle that passes obliquely (lateral to medial) across the superoanterior part of the thigh. This muscle descends inferiorly as far as the side of the knee. The sartorius, the longest muscle in the body, *acts across two joints*: it flexes the hip joint and participates in flexion of the knee. It also weakly abducts the thigh and laterally rotates it. The actions of both sartorius muscles bring the lower limbs

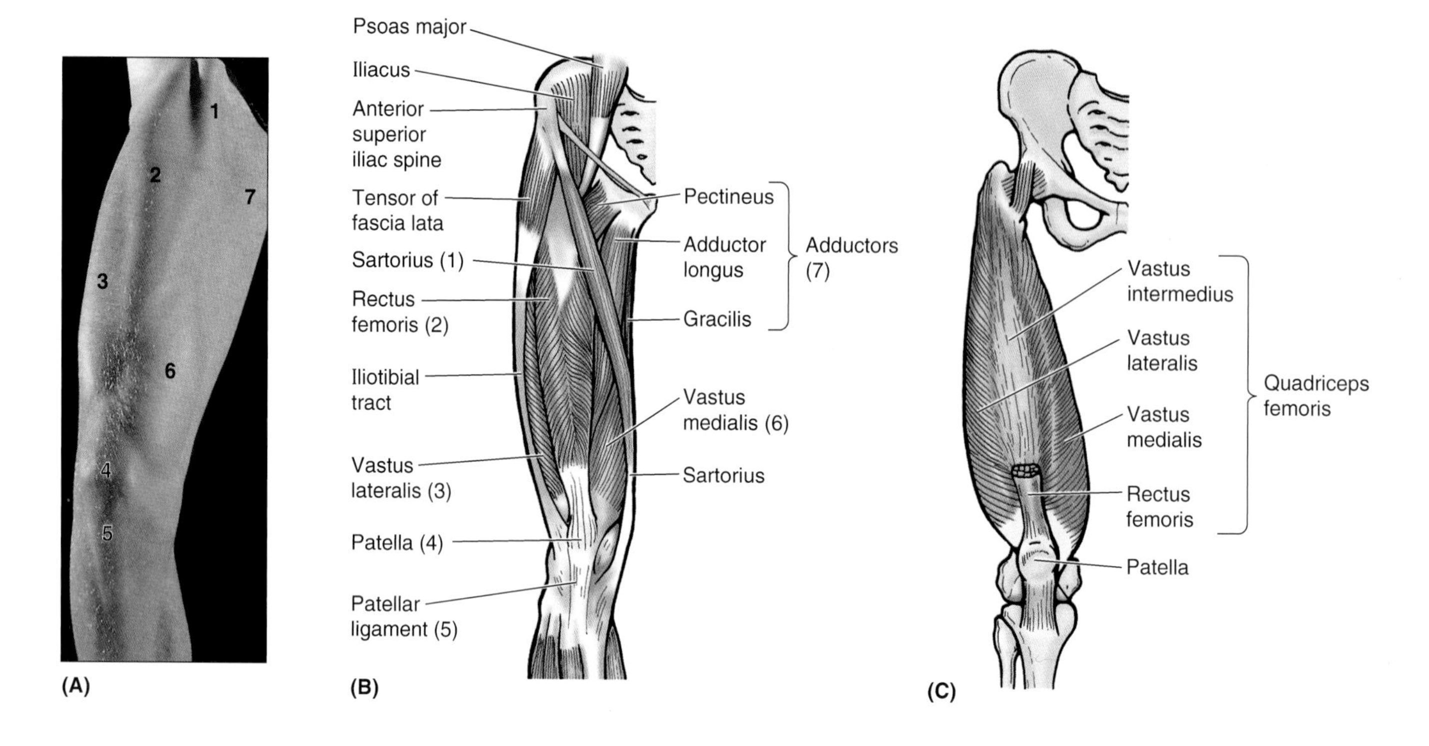

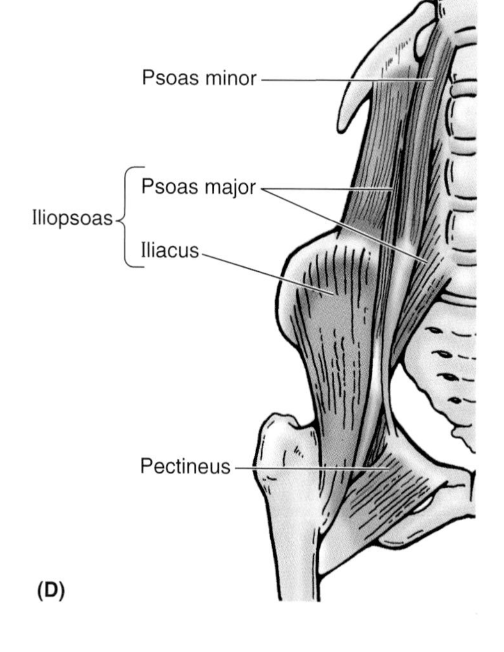

Figure 5.13. Anterior thigh muscles. A. Surface anatomy of the thigh and proximal part of the leg. The numbers following some of the labels in (**A**) refer to structures labeled in parts (**B**) and (**C**). **B.** Muscles in the anterior part of the thigh. Observe the large, four-headed quadriceps femoris, the chief extensor of the leg. In (**C**), most of the rectus femoris has been removed to show the vastus intermedius. **D.** Deep dissection showing the iliopsoas (psoas major and iliacus) and pectineus muscles.

into the cross-legged sitting position, which is still in use by some Asian tailors and jewelers. None of the actions of the sartorius is strong; therefore, other thigh muscles producing these movements are involved.

Quadriceps Femoris

The quadriceps femoris (L. four-headed femoral muscle) forms the main bulk of the anterior thigh muscles and collectively constitutes the largest and one of the most powerful muscles in the body. It covers almost all the anterior aspect and sides of the femur. *The quadriceps consists of four parts*:

- Rectus femoris
- Vastus lateralis
- Vastus intermedius
- Vastus medialis.

The quadriceps is the great extensor of the leg; all four of its parts combine to form a tendinous attachment to the tibia.

The three vastus muscles are separable only with difficulty. The quadriceps is an important muscle during climbing, running, jumping, rising from the sitting position, and walking up and down stairs. The tendons of the four parts of the quadriceps unite in the distal portion of the thigh to form a single, strong, broad **quadriceps tendon** (Fig. 5.13*C*). This tendon is traditionally described as attaching to the *base of the patella*, a large sesamoid bone in the tendon, which in turn is attached through the **patellar ligament** (L. ligamentum patellae) to the tibial tuberosity (Fig. 5.13*B*). However, it is probably more accurate to consider the patellar ligament as the continuation of the quadriceps tendon in which the patella, as a sesamoid bone, is embedded.

Testing the quadriceps is performed with the person in the supine position with the knee partly flexed. The person extends the knee against resistance. During the test, contraction

Table 5.2. Anterior Thigh Muscles

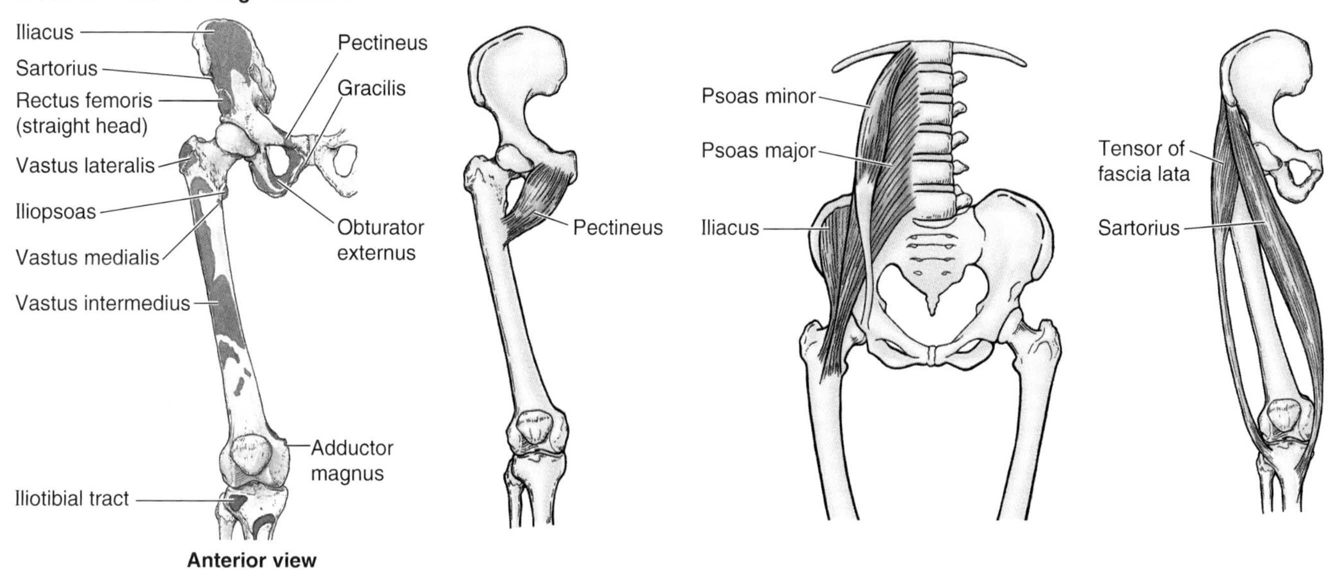

Muscle	Proximal Attachment	Distal Attachment	Innervation[a]	Main Action
Pectineus	Superior ramus of pubis	Pectineal line of femur, just inferior to lesser trochanter	Femoral nerve (**L2** and L3); may receive a branch from obturator nerve	Adducts and flexes thigh; assists with medial rotation of thigh
Iliopsoas				
Psoas major	Sides of T12–L5 vertebrae and discs between them; transverse processes of all lumbar vertebrae	Lesser trochanter of femur	Ventral rami of lumbar nerves (**L1, L2,** and L3)	
Psoas minor	Sides of T12–L1 vertebrae and intervertebral disc	Pectineal line, iliopectineal eminence via iliopectineal arch	Ventral rami of lumbar nerves (L1 and L2)	Act conjointly in flexing thigh at hip joint and in stabilizing this joint[b]
Iliacus	Iliac crest, iliac fossa, ala of sacrum, and anterior sacroiliac ligaments	Tendon of psoas major, lesser trochanter, and femur distal to it	Femoral nerve (**L2** and L3)	
Tensor of fascia lata	Anterior superior iliac spine and anterior part of iliac crest	Iliotibial tract that attaches to lateral condyle of tibia	Superior gluteal (L4 and L5)	Abducts, medially rotates, and flexes thigh; helps to keep knee extended; steadies trunk on thigh
Sartorius	Anterior superior iliac spine and superior part of notch inferior to it	Superior part of medial surface of tibia	Femoral nerve (L2 and L3)	Flexes, abducts, and laterally rotates thigh at hip joint; flexes leg at knee joint[c]

[a] Numbers indicate spinal cord segmental innervation of nerves [e.g., L1, L2, and L3 indicate that nerves supplying psoas major are derived from first three lumbar segments of the spinal cord; boldface type **(L1, L2)** indicates main segmental innervation]. Damage to one or more of these spinal cord segments or to motor nerve roots arising from them results in paralysis of the muscles concerned.

[b] Psoas major is also a postural muscle that helps control deviation of trunk and is active during standing.

[c] Four actions of sartorious (L. sartor, tailor) produce the once common crosslegged sitting position used by tailors–hence the name.

of the rectus femoris should be observable and palpable if the muscle is acting normally, indicating that its nerve supply is intact.

The **patella** provides a bony surface that is able to withstand the compression placed on the quadriceps tendon during kneeling and the friction occurring when the knee is flexed and extended during running. The patella also provides additional leverage for the quadriceps in placing the tendon more anteriorly, further from the joint's axis, causing it to approach the tibia from a position of greater mechanical advantage. The inferiorly directed apex of the patella indicates the level of the joint plane of the knee when the leg is extended and the patellar ligament is taut.

Rectus Femoris. This "kicking muscle" received its name because it runs straight down the thigh (L. rectus, straight). It assists the iliopsoas in flexing the thigh at the hip joint. Because of its attachments to the hip bone and tibia, *the rectus femoris crosses two joints*; hence, it flexes the thigh at the hip joint and extends the leg at the knee joint.

Vastus Muscles. The names of these large muscles indicate their position around the femoral body:

- *Vastus lateralis*—the largest component of the quadriceps—lies on the lateral side of the thigh
- *Vastus medialis* covers the medial side of the thigh
- *Vastus intermedius* lies deep to the rectus femoris, between the vastus medialis and vastus lateralis.

The small, flat **articular muscle of the knee** (L. articularis genus), a derivative of the vastus intermedius (Figs. 5.13 and 5.14), usually consists of a variable number of muscular slips that attach superiorly to the inferior part of the anterior aspect of the femur and inferiorly to the synovial membrane of the knee joint and the wall of the suprapatellar bursa. This articular muscle pulls the synovial capsule superiorly during extension of the leg, thereby preventing folds of the capsule from being compressed between the femur and the patella within the knee joint.

Table 5.2. *(Continued)* **Anterior Thigh Muscles**

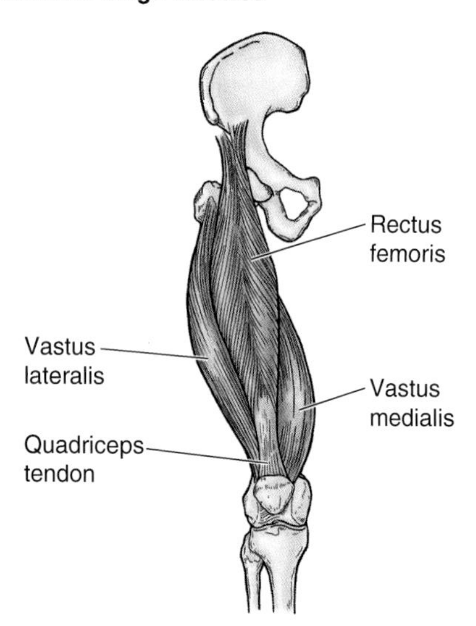

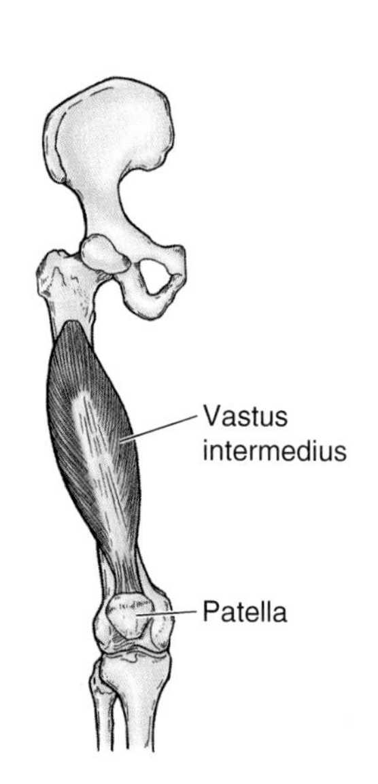

Muscle	Proximal Attachment	Distal Attachment	Innervation	Main Action
Quadriceps femoris		Base of patella and by patellar ligament to tibial tuberosity	Femoral nerve (L2, **L3**, and **L4**)	Extend leg at knee joint; rectus femoris also steadies hip joint and helps iliopsoas to flex thigh
Rectus femoris	Anterior inferior iliac spine and ilium superior to acetabulum			
Vastus lateralis	Greater trochanter and lateral lip of linea aspera of femur			
Vastus medialis	Intertrochanteric line and medial lip of linea aspera of femur			
Vastus intermedius	Anterior and lateral surfaces of body of femur			

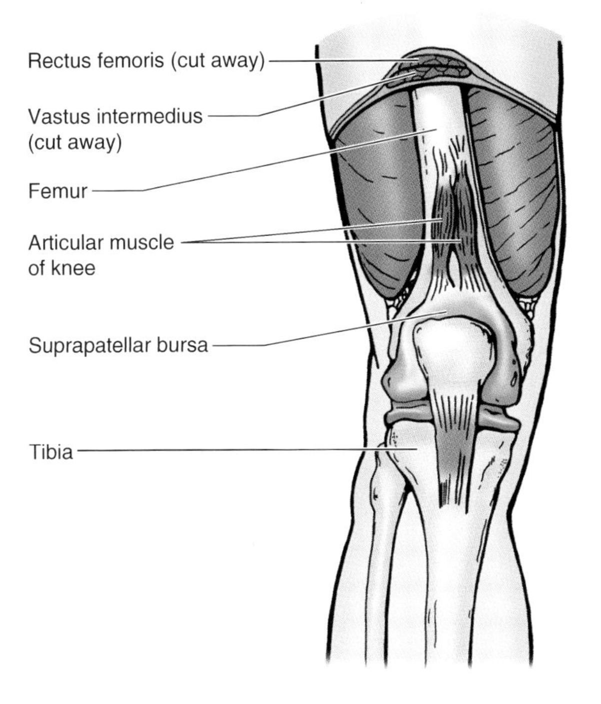

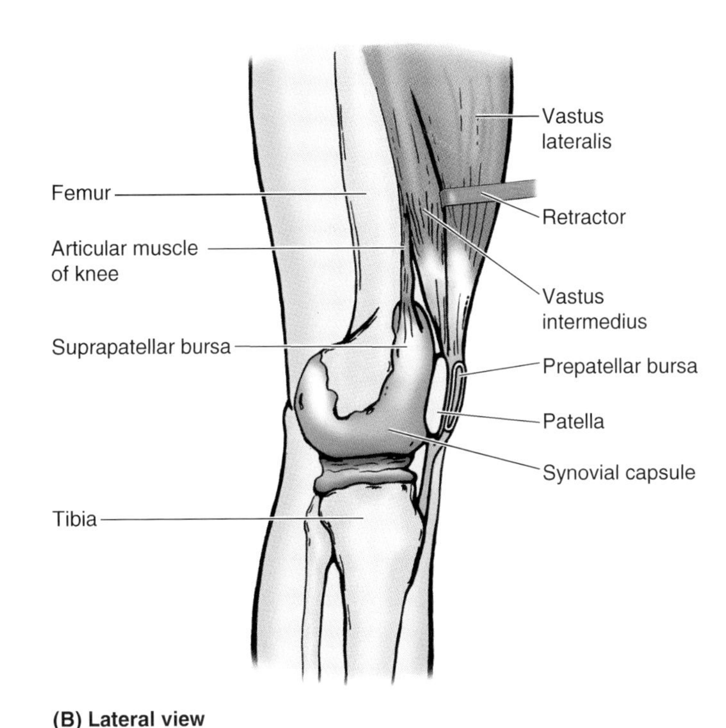

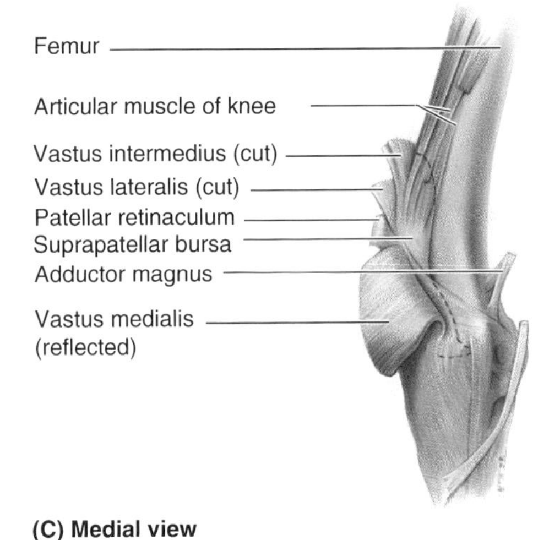

Figure 5.14. Dissection of the distal thigh and knee regions. Anterior (**A**) and lateral (**B**) views. Observe the bursae around the knee and the articular muscle of the knee (L. articularis genus). It is attached proximally to the distal part of the femur and distally to the suprapatellar bursa. This muscle retracts the suprapatellar bursa during extension of the knee. **C.** Medial view of a deep dissection of the knee region, showing the articular muscle of the knee, which often blends with the vastus intermedius muscle. It elevates the synovial membrane during extension of the leg.

Hip and Thigh Contusions

Sports broadcasters and trainers sometimes refer to a "hip pointer." This term refers to a *contusion of the iliac crest*, usually its anterior part (e.g., where the sartorius attaches to the anterior superior iliac spine). Contusions cause bleeding from ruptured capillaries and infiltration of blood into the muscles, tendons, and other soft tissues. A "hip pointer" may also refer to avulsion of bony muscle attachments (e.g., of the iliacus and rectus femoris to the iliac crest); however, these injuries are more accurately called *avulsion fractures*.

Another term sports broadcasters and others use is "charley horse." This term may refer either to the cramping of an individual thigh muscle because of ischemia (inadequate circulation of blood) or to the contusion and tearing of muscle fibers and rupture of blood vessels sufficient enough to form a *hematoma*. The latter injury usually is the consequence of the rupture of some fibers of the rectus femoris; sometimes the quadriceps tendon is also partially torn. The most common site of a thigh hematoma is in the quadriceps. A "charley horse" is associated with localized pain and/or muscle stiffness and commonly follows direct trauma (e.g., a stick slash in hockey or a tackle in football). A common term for this injury in countries where cricket is played is "cricket thigh." ▶

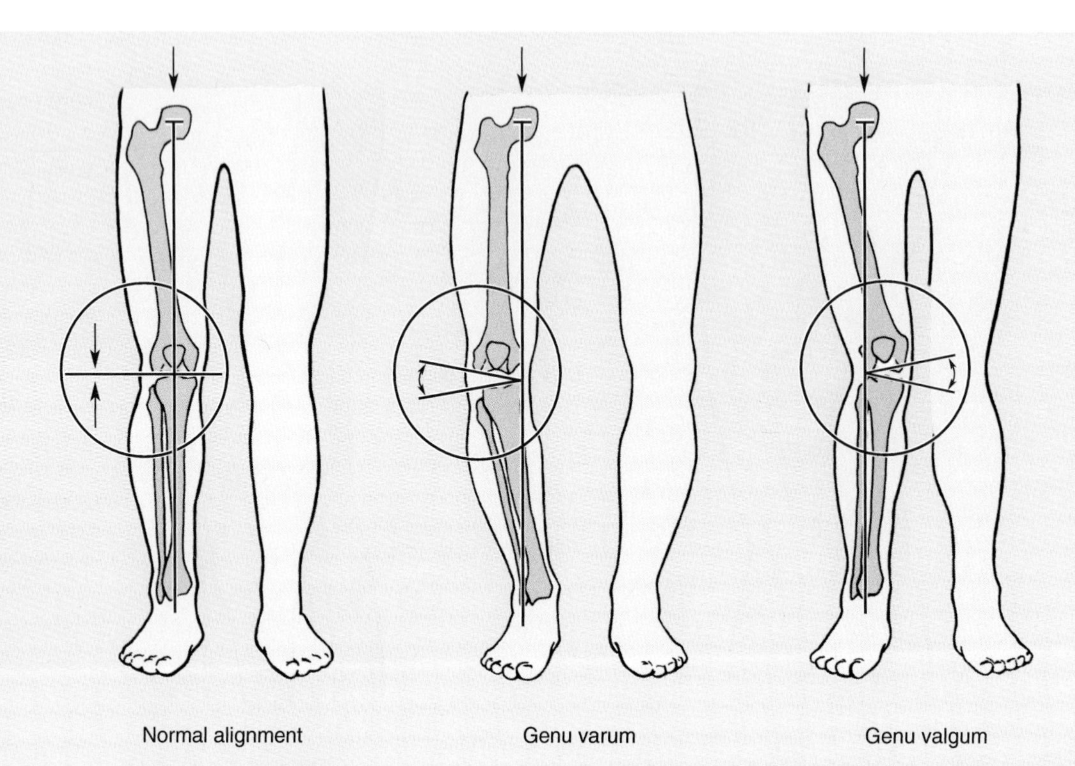

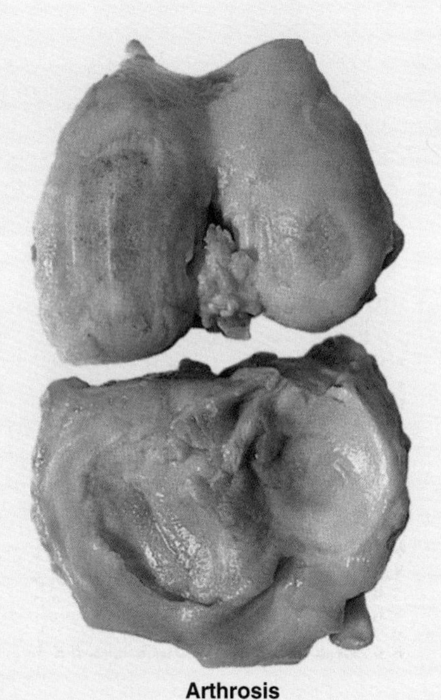
Arthrosis

Psoas Abscess

The psoas major arises in the abdomen from the intervertebral discs, the sides of T12 through L5 vertebrae, and their transverse processes. The *medial arcuate ligament* of the diaphragm arches obliquely over the proximal part of the psoas major (see Chapter 2). The transversalis fascia on the internal abdominal wall is continuous with the *psoas fascia,* where it forms a fascial covering for the psoas major that accompanies the muscle into the anterior region of the thigh. A retroperitoneal pyogenic (pus-forming) infection in the abdomen, characteristically occurring in association with tuberculosis of the vertebral column, may result in the formation of a collection of pus (*psoas abscess*) that, when it passes between the psoas and its fascia to the inguinal and proximal thigh regions, may refer severe pain to the hip, thigh, or knee joints. Consequently, a psoas abscess should always be considered when edema occurs in the proximal part of the thigh.

Genu Valgum and Genu Varum

The femur is set obliquely, creating an angle with the tibia at the knee (*A*). A medial angulation of the leg in relation to the thigh (*B*) is a deformity called *genu varum* (bowleg) that causes unequal weight distribution. All the pressure is taken by the inside of the knee joint, which results in *arthrosis*—destruction of knee cartilages. Because of the exaggerated knee angle in genu varum, the patella tends to move laterally when the leg is extended. This movement is increased by the pull of the vastus lateralis. A lateral angulation of the leg (*C*) in relation to the thigh (exaggeration of the knee angle) is *genu valgum* (knock-knee).

Children commonly appear bowlegged for 1 to 2 years after starting to walk, and knock-knees are frequently observed in children 2 to 4 years of age. Persistence of these abnormal knee angles in late childhood usually means deformities exist that may require correction. Any irregularity of a joint eventually leads to wear and tear of the articular cartilages. ▶

Paralysis of the Quadriceps Femoris

A person with a paralyzed quadriceps cannot extend the leg against resistance and usually presses on the distal end of the thigh during walking to prevent inadvertent flexion of the knee joint. Weakness of the vastus medialis or vastus lateralis, resulting from arthritis or trauma of the knee joint, for example, can result in abnormal patellar movement and loss of joint stability.

Chondromalacia Patellae

A common knee problem for runners is chondromalacia patellae ("runner's knee"). The soreness and aching around or deep to the patella results from *quadriceps imbalance*. Such overstressing of the knee can also occur in running sports such as jogging, basketball, football, or soccer. Chondromalacia patellae may result from excessive running, a blow to the patella, or extreme flexion of the knee (e.g., during squatting when power lifting).

Patellar Dislocation

When dislocation of the patella occurs, the movement is nearly always lateral and it happens more often in women. The lateral dislocation is counterbalanced by the medial, more horizontal pull of the powerful vastus medialis. In addition, the lateral femoral condyle has a more anterior projection and a deeper slope for the larger lateral patellar facet. Hence, a mechanical deterrent to lateral dislocation exists.

Patellar Fractures

A direct blow to the patella may fracture it in two or more fragments. *Transverse patellar fractures* may result from a blow to the knee or from sudden contraction of the quadriceps (e.g., when one slips and attempts to prevent a backward fall). The proximal fragment is pulled superiorly with the quadriceps tendon, and the distal fragment remains with the patellar ligament.

Patellectomy

Removal of the patella (*patellectomy*) because of a comminuted fracture, for example, may require more force by the quadriceps to extend the leg completely. Recently developed surgical techniques can repair the patellar tendon after patellectomy, which minimizes the weakening of knee extension.

Abnormal Ossification of the Patella

The patella is cartilaginous at birth and becomes ossified during the 3rd to 6th years, frequently from more than one ossification center. Although these centers usually coalesce and form a single bone, they may remain separate on one or both sides, giving rise to a bipartite or tripartite patella. An unwary observer might interpret this condition on a radiograph as a comminuted fracture of the patella. Ossification abnormalities are nearly always bilateral; therefore, radiographs should be examined from both sides. If the defects are bilateral, the abnormalities are likely ossification abnormalities.

Patellar Tendon Reflex

Tapping the patellar ligament with a reflex hammer normally elicits the patellar reflex (knee jerk). This reflex is routinely tested during a physical examination by having the patient sit with legs dangling off the side of the examining table. A firm strike on the patellar ligament with a reflex hammer usually causes the leg to extend. If the reflex is normal, a hand on the patient's quadriceps should feel the muscle contract. This tendon reflex tests the L2 through L4 nerves. Tapping the patellar ligament activates muscle spindles in the quadriceps; afferent impulses from these spindles travel in the femoral nerve to the spinal cord (L2 through L4 segments). From here, efferent impulses are transmitted via motor fibers in the femoral nerve to the quadriceps, resulting in a jerklike contraction of the muscle and extension of the leg at the knee joint. *Diminution or absence of the patellar tendon (quadriceps) reflex* may result from any lesion that interrupts the innervation of the quadriceps muscle (e.g., peripheral nerve disease). ⊙

Medial Thigh Muscles

The medial thigh muscles—**the adductor group**—are in the medial compartment of the thigh (Figs. 5.13, *A* and *B*, and 5.15). *The adductor group of thigh muscles consists of:*

- *Adductor longus*
- *Adductor brevis*
- *Adductor magnus*
- *Gracilis*
- *Obturator externus.*

Collectively, these muscles are the adductors of the thigh; however, the actions of some of these muscles are more complex. The attachments, nerve supply, and actions of the muscles are given in Table 5.3.

Adductor Longus

The adductor longus—a large, fan-shaped muscle—is the most anteriorly placed of the adductor group. This triangular *long adductor* arises by a strong tendon from the anterior aspect of the body of the pubis just inferior to the pubic tubercle and expands to attach to the linea aspera of the femur.

Adductor Brevis

The adductor brevis lies deep to the pectineus and adductor longus muscles. This *short adductor*—largely covered by the adductor longus—arises from the body and inferior ramus of the pubis and expands as it passes to its distal attachment to the linea aspera of the femur.

Adductor Magnus

The adductor magnus is the largest muscle in the adductor group. This *large adductor* is a composite, triangular muscle that has *adductor and hamstring parts*. The two parts differ in their attachments, nerve supply, and main actions (Table 5.3).

Gracilis

This long, *straplike muscle* lies along the medial side of the thigh and knee. The gracilis (L. slender) is the most superficial of the adductor group and is the weakest member. It is the only one of the group to cross the knee joint. It adducts the thigh, flexes the knee, and rotates the leg medially.

Obturator Externus

This flat, relatively small, fan-shaped muscle is deeply placed in the superomedial part of the thigh. It extends from the external surface of the obturator membrane and surrounding bone of the pelvis to the posterior aspect of the greater trochanter, passing under the head and neck of the femur.

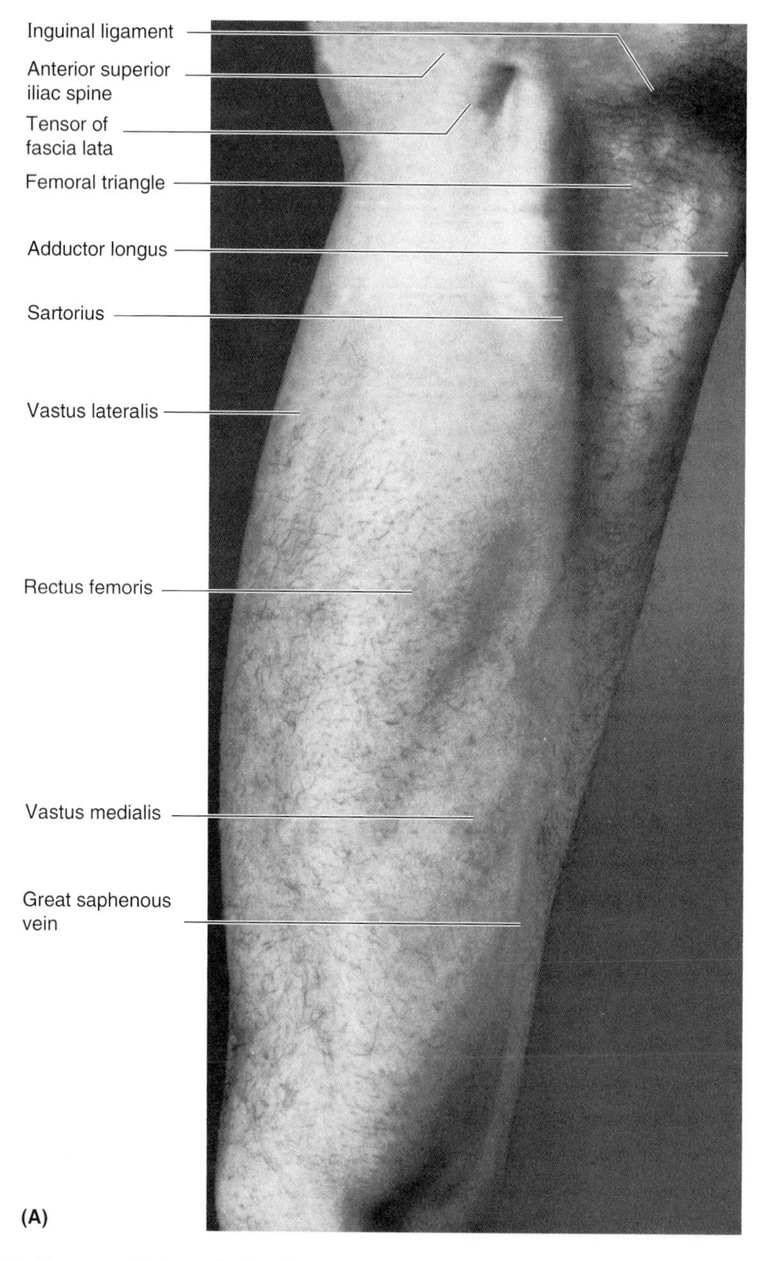

Figure 5.15. Femoral triangle in the superior third of the anterior aspect of the thigh. **A.** Photograph displaying the femoral triangle, adductor longus, and sartorius. To show these structures, the man was asked to flex, abduct, and laterally rotate his thigh. His pelvis is also slightly rotated to the left.

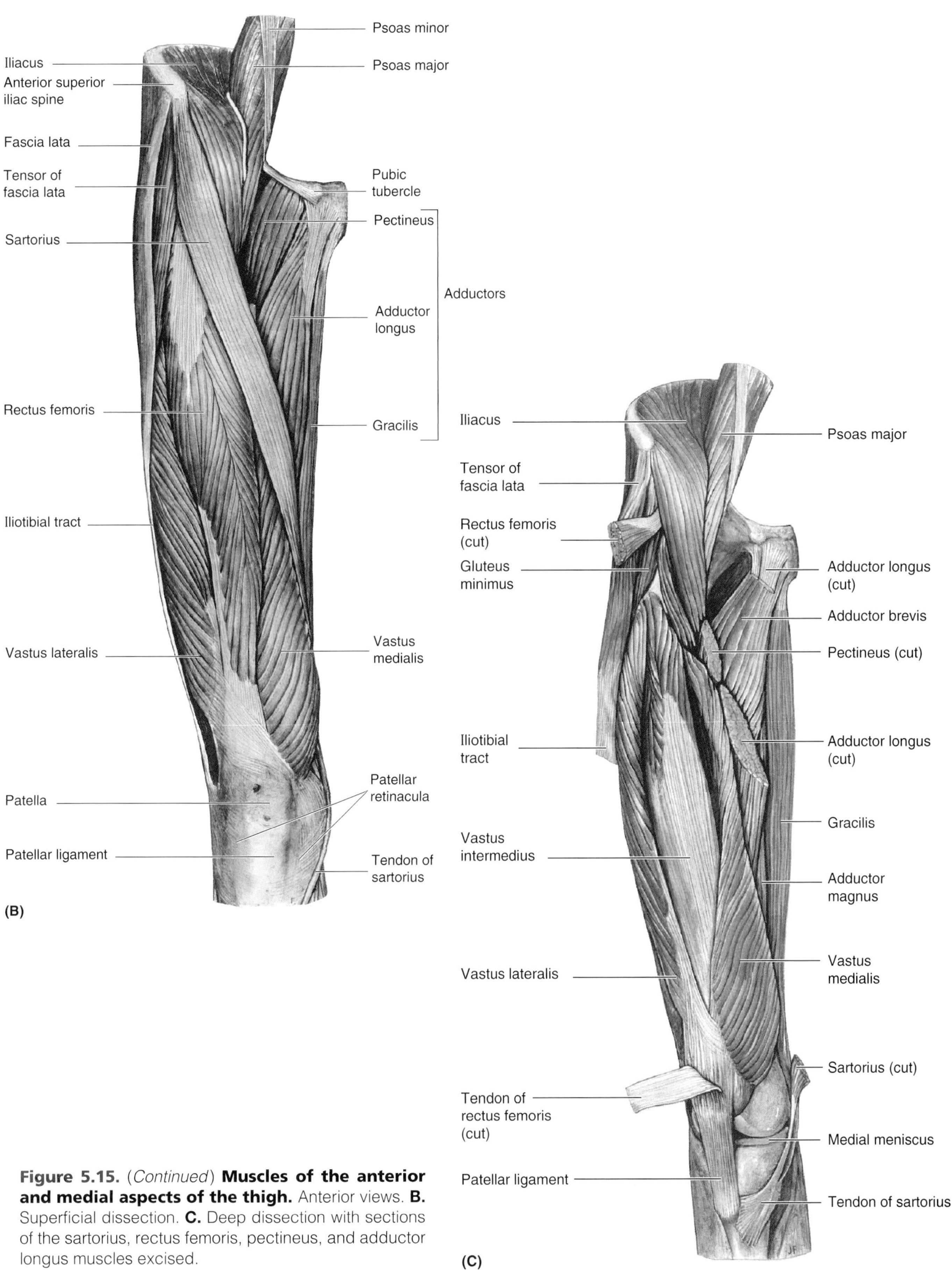

Figure 5.15. (*Continued*) **Muscles of the anterior and medial aspects of the thigh.** Anterior views. **B.** Superficial dissection. **C.** Deep dissection with sections of the sartorius, rectus femoris, pectineus, and adductor longus muscles excised.

Table 5.3. Medial Thigh Muscles

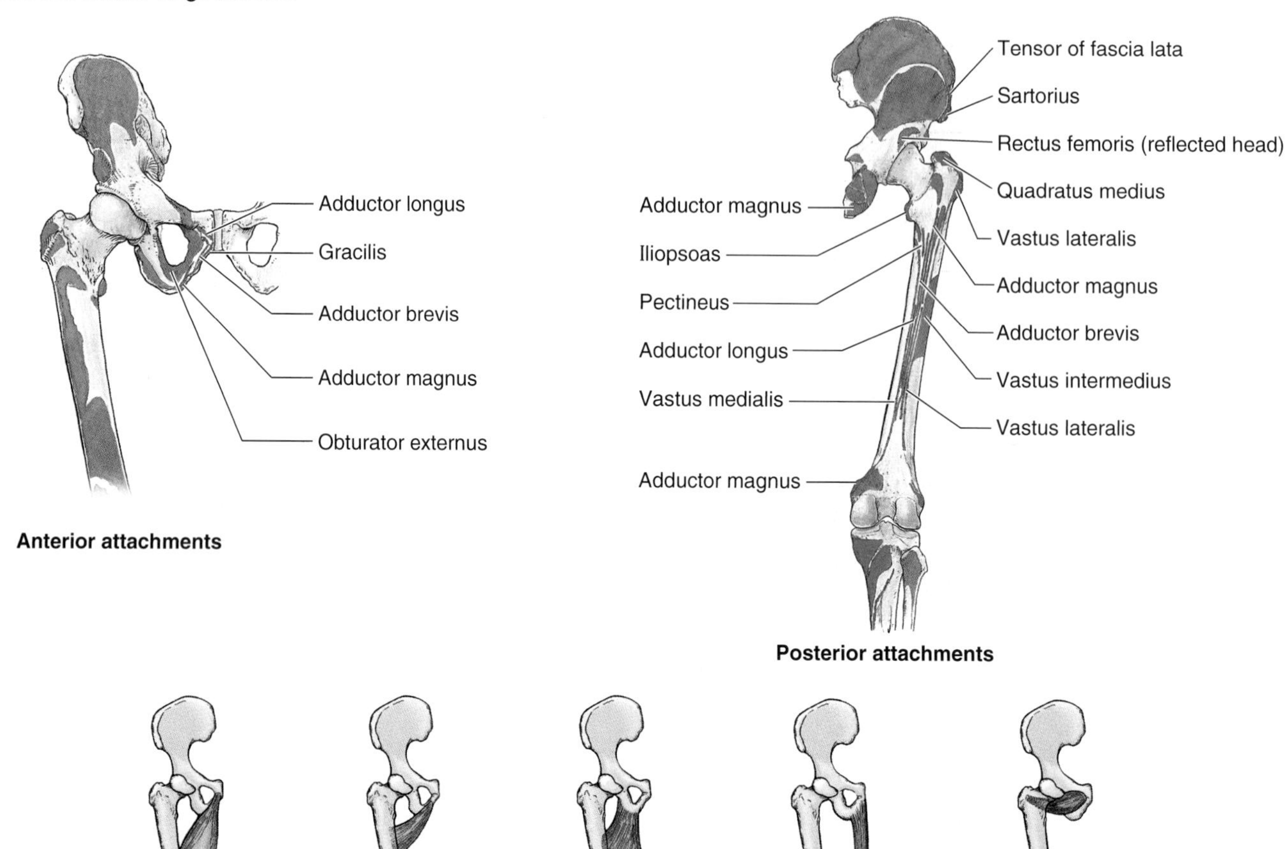

Adductor longus

Adductor brevis

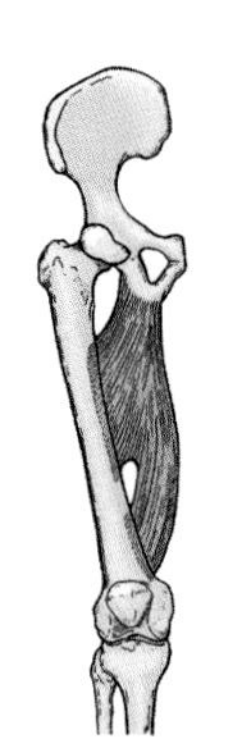

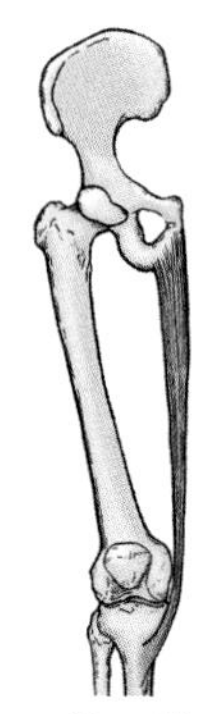

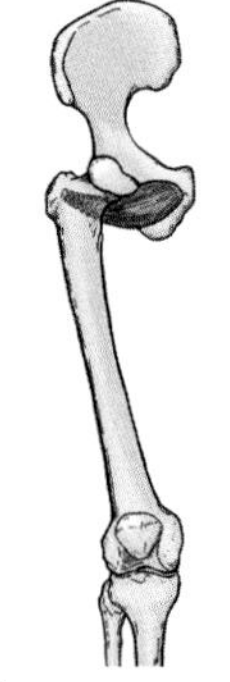

Muscle	Proximal Attachment	Distal Attachment	Innervation[a]	Main Action
Adductor longus	Body of pubis inferior to pubic crest	Middle third of linea aspera of femur	Obturator nerve, branch of anterior division (L2, **L3**, and L4)	Adducts thigh
Adductor brevis	Body and inferior ramus of pubis	Pectineal line and proximal part of linea aspera of femur	Obturator nerve (L2, **L3**, and L4), branch of anterior division	Adducts thigh and to some extent flexes it
Adductor magnus	Adductor part: inferior ramus of pubis, ramus of ischium Hamstrings part: ischial tuberosity	Adductor part: gluteal tuberosity, linea aspera, medial supracondylar line Hamstrings part: adductor tubercle of femur	Adductor part: obturator nerve (L2, **L3**, and **L4**), branches of posterior division Hamstrings part: tibial part of sciatic nerve (**L4**)	Adducts thigh Adductor part: flexes thigh Hamstrings part: extends thigh
Gracilis	Body and inferior ramus of pubis	Superior part of medial surface of tibia	Obturator nerve (**L2** and L3)	Adducts thigh, flexes leg, and helps rotate it medially
Obturator externus	Margins of obturator foramen and obturator membrane	Trochanteric fossa of femur	Obturator nerve (L3 and **L4**)	Laterally rotates thigh; steadies head of femur in acetabulum

Collectively, the five muscles listed are the adductors of the thigh, but their actions are more complex (e.g., they act as flexors of the hip joint during flexion of the knee joint and are active during walking).

[a] See Table 5.1 for explanation of segmental innervation.

Actions of the Adductor Muscle Group. *The main action of the adductor group of muscles is to adduct the thigh.* Three adductors (longus, brevis, and magnus) are used in all movements in which the thighs are adducted (e.g., pressed together when riding a horse). They are also important stabilizing muscles during flexion and extension of the thigh. *Testing of the medial thigh muscles* is performed while the person is lying supine with the knee straight. The person adducts the thigh against resistance, and if the adductors are normal, the proximal ends of the gracilis and adductor longus can easily be palpated.

Adductor Hiatus. An opening in the aponeurotic distal attachment of the adductor magnus—the **adductor hiatus**—transmits the femoral artery and vein from the adductor canal in the thigh to the popliteal fossa posterior to the knee (Table 5.4). The opening is located just superior to the adductor tubercle of the femur. All adductor muscles, except the pectineus and part of the adductor magnus, are supplied by the *obturator nerve* (L2 through L4). The pectineus is supplied by the femoral nerve (L2 through L4) and the "hamstring part" of the adductor magnus is supplied by the tibial part of the sciatic nerve (L4).

Transplantation of the Gracilis

Because the gracilis is a relatively weak member of the adductor group of muscles, it can be removed without noticeable loss of its actions on the leg. Hence, surgeons often transplant the gracilis, or part of it, with its nerve and blood vessels to replace a damaged muscle in the hand, for example. Once the muscle is transplanted, it soon produces good digital flexion and extension.

Groin Pull

Occasionally, sports broadcasters refer to a "pulled groin" or "groin injury." These terms mean that a strain, stretching, and probably some tearing of the proximal attachments of the anteromedial thigh muscles has occurred. The injury usually involves the flexor and adductor thigh muscles. The proximal attachments of these muscles are in the inguinal region (groin), the junction of the thigh and trunk. Groin pulls usually occur in sports that require quick starts such as short-distance racing (e.g., a 60-meter sprint); base stealing in baseball; and quick starts in basketball, football, and soccer.

Injury to the Adductor Longus

Muscle strains of the adductor longus may occur in horseback riders and produce pain ("rider's strain"). Ossification sometimes occurs in the tendons of these muscles because they (horseback riders) actively adduct their thighs to keep from falling from these animals. The ossified tendons are sometimes called "riders' bones." ⊙

Femoral Triangle

The femoral triangle—a junctional region between the trunk and lower limb—is *a triangular fascial space in the superoanterior third of the thigh* (Figs. 5.16 and 5.17). It appears as a triangular depression inferior to the inguinal ligament when the thigh is flexed, abducted, and laterally rotated. *The femoral triangle is bounded*:

- Superiorly by the inguinal ligament
- Medially by the adductor longus
- Laterally by the sartorius.

The *base of the femoral triangle* is formed by the inguinal ligament, and its *apex* is where the lateral border of the sartorius crosses the medial border of the adductor longus. The muscular *floor of the femoral triangle* is formed from lateral to medial by the iliopsoas and pectineus. The *roof of the femoral triangle* is formed by fascia lata and cribriform fascia, subcutaneous tissue, and skin. *The contents of the femoral triangle, from lateral to medial, are the*:

- Femoral nerve and its branches
- Femoral sheath and its contents
- Femoral artery and several of its branches
- Femoral vein and its proximal tributaries, such as the great saphenous and deep femoral veins.

The femoral triangle is bisected by the femoral artery and vein, which leave and enter the **adductor canal** at its apex (Fig. 5.16*B*). The *adductor canal* is the space in the middle third of the thigh between the vastus medialis and adductor muscles, which is converted into a canal by the overlying sartorius muscle.

Femoral Nerve. The femoral nerve (L2 through L4)—*the largest branch of the lumbar plexus*—forms in the abdomen within the psoas major and descends posterolaterally through the pelvis to the midpoint of the inguinal ligament (Figs. 5.16, *A* and *B*). It then passes deep to this ligament and enters the femoral triangle, *lateral to the femoral vessels*. After entering the triangle, the femoral nerve divides into several branches to the anterior thigh muscles. It also sends articular branches to the hip and knee joints and provides several cutaneous branches to the anteromedial side of the thigh (Table 5.1). The terminal cutaneous branch of the femoral nerve—the **saphenous nerve**—descends through the femoral triangle, lateral to the femoral sheath containing the femoral vessels (Fig. 5.17). *The saphenous nerve accompanies the femoral artery and vein through the adductor canal* but then becomes superfi-

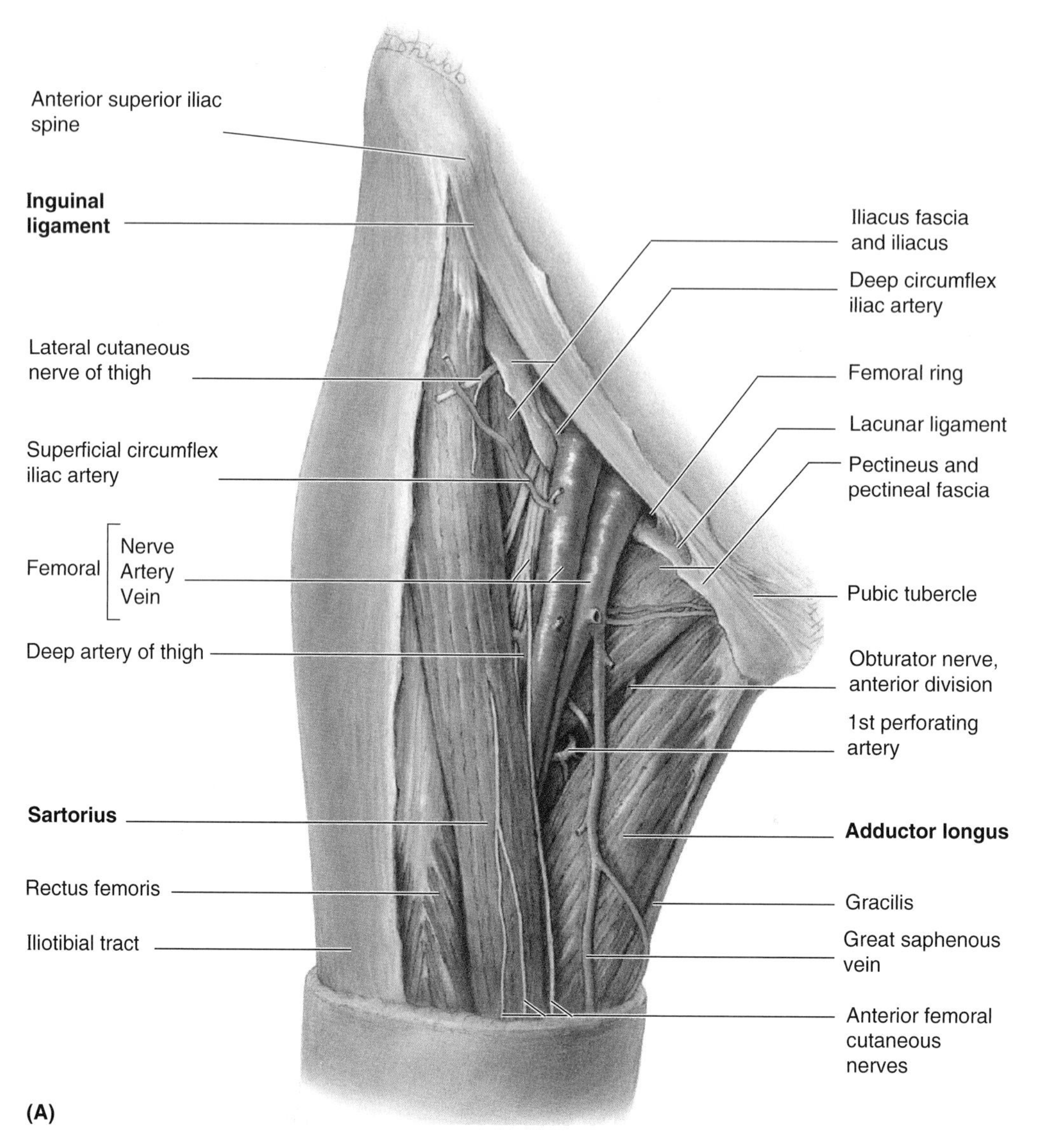

Figure 5.16. A. Dissection of the femoral triangle containing the femoral nerve and vessels. Observe the structures that bound the triangle: the inguinal ligament superiorly, the adductor longus medially, and the sartorius laterally.

cial by passing between the sartorius and gracilis when the femoral vessels traverse the adductor hiatus at the distal end of the canal. It runs anteroinferiorly to supply the skin and fascia on the anteromedial aspects of the knee, leg, and foot.

Femoral Sheath. A *funnel-shaped fascial tube*, the femoral sheath extends 3 to 4 cm inferior to the inguinal ligament and encloses proximal parts of the femoral vessels and the femoral canal (Figs. 5.17 and 5.18). The sheath is formed by an inferior prolongation of transversalis and iliopsoas fascia from the abdomen (see Chapter 2). *The femoral sheath does not enclose the femoral nerve*; the sheath ends by becoming continuous with the adventitia covering the femoral vessels. The medial wall of the femoral sheath is pierced by the great saphenous vein and lymphatic vessels. The femoral sheath allows the femoral artery and vein to glide deep to the inguinal ligament during movements of the hip joint. The femoral sheath is subdivided into three compartments by vertical septa derived from extraperitoneal connective tissue of the abdomen that extends along the vessels. *The compartments of the femoral sheath are the*:

- *Lateral compartment* for the femoral artery
- *Intermediate compartment* for the femoral vein
- *Medial compartment,* which is the femoral canal.

The **femoral canal**—the smallest of the three femoral sheath compartments—is the short (approximately 1.25 cm),

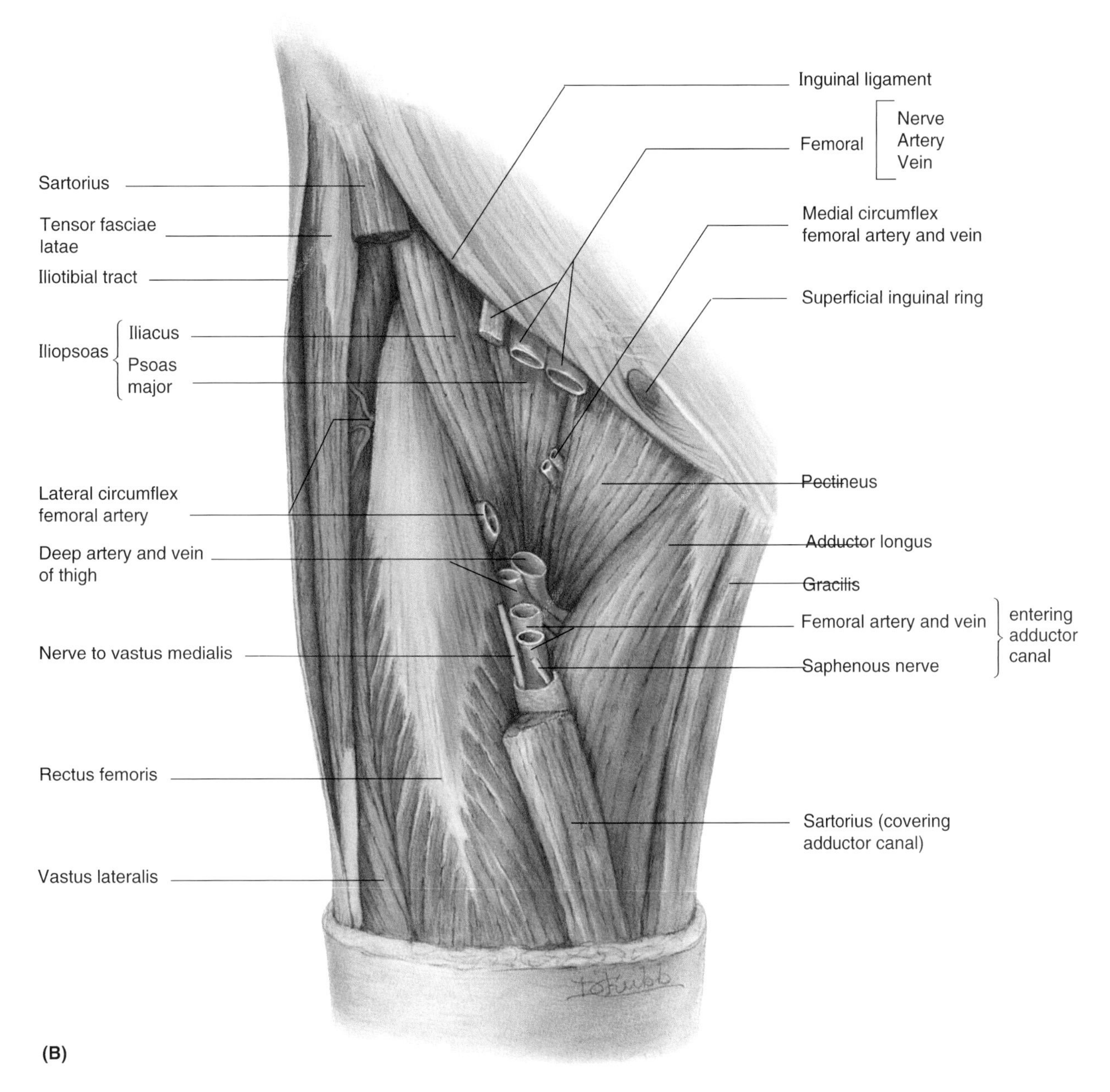

Figure 5.16. *(Continued)* **B.** Deeper dissection showing the floor of the femoral triangle. Sections are removed from the sartorius and the femoral vessels and nerve. Observe the muscles forming the floor of the femoral triangle: the iliopsoas laterally and pectineus medially. At the apex of the femoral triangle, observe four vessels and two nerves in the adductor canal.

conical medial compartment of the femoral sheath that lies between the medial edge of the femoral sheath and the femoral vein. The base of the femoral canal (its abdominal end) is directed superiorly and, although oval shaped, is called the *femoral ring* (Fig. 5.18). **The femoral canal:**

- Extends distally to the level of the proximal edge of the saphenous opening
- Allows the femoral vein to expand when venous return from the lower limb is increased
- Contains loose connective tissue, fat, a few lymphatic vessels, and sometimes a deep inguinal lymph node (Cloquet's node).

The **femoral ring**—the small proximal opening of the femoral canal (approximately 1 cm wide)—is closed by extraperitoneal fatty tissue that forms the **femoral septum**. The abdominal surface of this septum is covered by parietal peritoneum (see Chapter 2). The femoral septum is pierced by lymphatic vessels connecting the inguinal and external iliac lymph nodes. *The boundaries of the femoral ring are:*

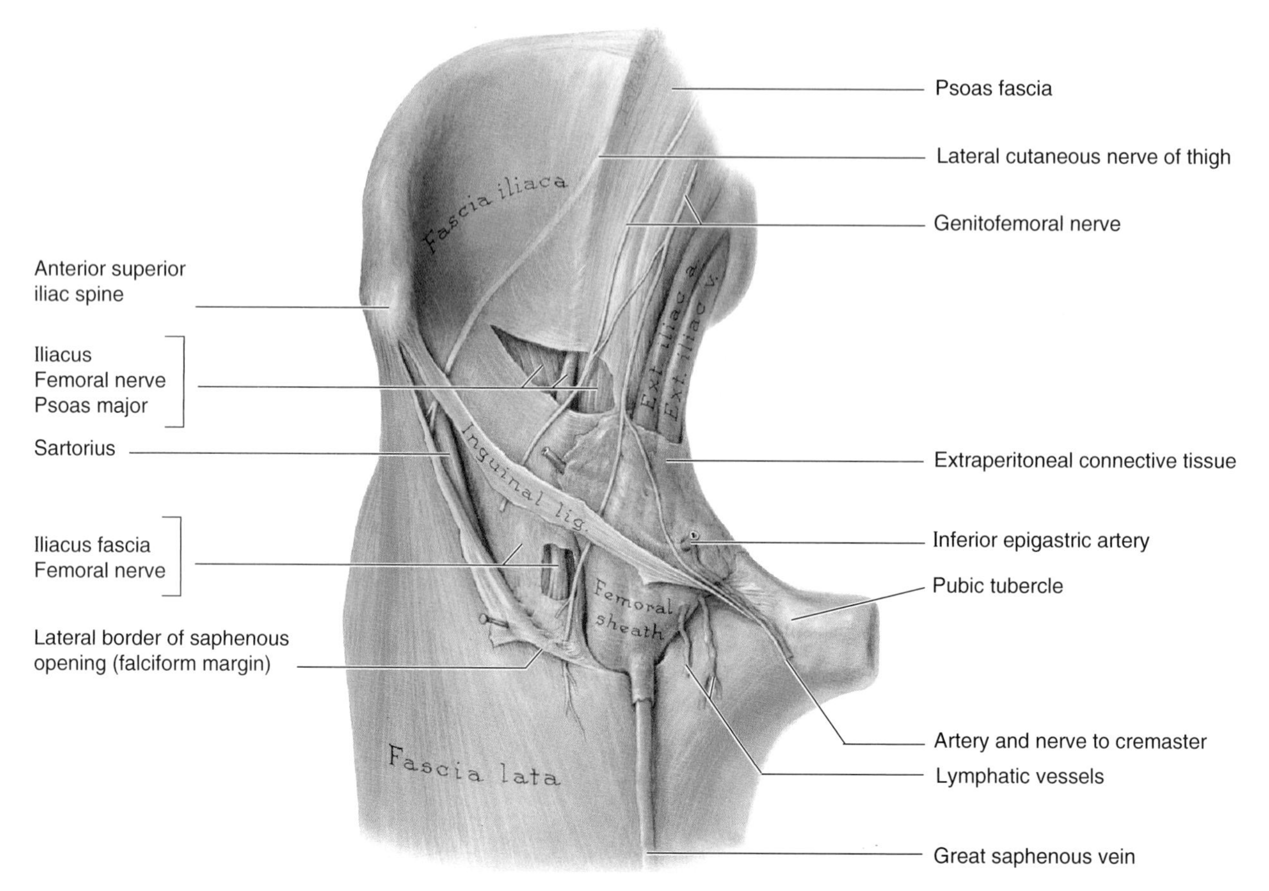

Figure 5.17. Dissection of the femoral sheath in the femoral triangle. The falciform margin of the saphenous opening the fascia lata is cut and reflected. Note that the femoral nerve is external and lateral to the femoral sheath.

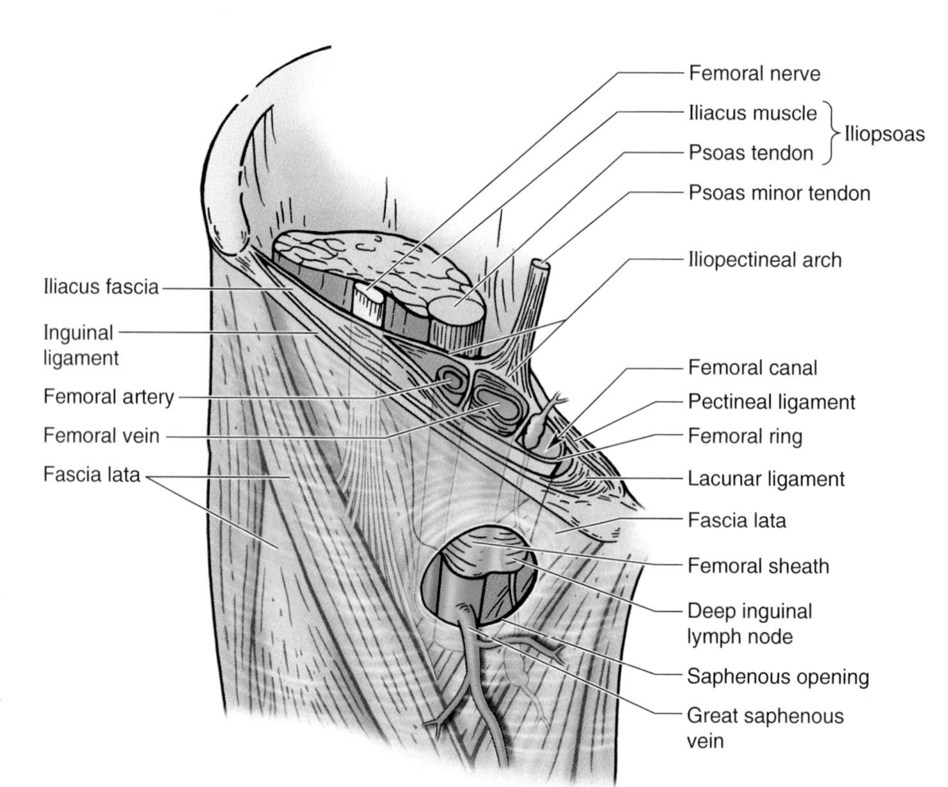

Figure 5.18. Structure of the femoral sheath and its contents. Drawing of a dissection of the superior end of the anterior aspect of the right thigh. The proximal end (opening) of the femoral canal is the femoral ring.

- *Laterally*, the partition between the femoral canal and femoral vein
- *Posteriorly*, the superior ramus of the pubis covered by the pectineus and its fascia
- *Medially*, the lacunar ligament
- *Anteriorly*, the medial part of the inguinal ligament.

Femoral Artery. The femoral artery—*the chief artery to the lower limb*—is the continuation of the external iliac artery (Fig. 5.16*A*, Table 5.4). *The femoral artery*:

- Begins at the inguinal ligament, passing midway between the anterior superior iliac spine and the pubic symphysis
- Enters the femoral triangle deep to the midpoint of the inguinal ligament, *lateral to the femoral vein*
- Lies posterior to the fascia lata and descends on the adjacent borders of the iliopsoas and pectineus forming the floor of the femoral triangle
- Bisects the femoral triangle at its apex and enters the *adductor canal*, deep to the sartorius
- Exits the adductor canal by passing through the *adductor hiatus* and becoming the popliteal artery.

In the proximal part of its course, the femoral artery gives off the superficial epigastric artery, the superficial circumflex iliac artery, and the superficial external pudendal and deep external pudendal arteries.

The **deep artery of the thigh**—the largest branch of the femoral artery and **the chief artery to the thigh**—arises in the femoral triangle from the lateral side of the femoral artery, 1 to 5 cm inferior to the inguinal ligament. It passes deeply into the thigh as it descends so that it lies posterior to the femoral artery and vein on the medial side of the femur (Fig. 5.16, *A* and *B*). The deep artery of the thigh leaves the femoral triangle between the pectineus and adductor longus and descends posterior to the latter muscle, giving off perforating arteries that supply the adductor magnus and hamstrings.

The **circumflex femoral arteries** are usually branches of the deep artery of the thigh, but they may arise directly from the femoral artery. They encircle the thigh, anastomose with each other and other arteries, and supply the thigh muscles and the proximal end of the femur.

The **medial circumflex femoral artery** is especially important because it supplies most of the blood to the head and neck of the femur (Table 5.4). It is often torn when the femoral neck is fractured or the hip joint is dislocated. This artery passes deeply between the iliopsoas and pectineus to reach the posterior part of the thigh.

The **lateral circumflex femoral artery** passes laterally, deep to the sartorius and rectus femoris, and between the branches of the femoral nerve. Here it divides into branches that supply the head of the femur and muscles on the lateral side of the thigh.

The **obturator artery** (Table 5.4) helps the deep artery supply the adductor muscles of the thigh. Arising either from the internal iliac artery or as *an accessory obturator artery* from the inferior epigastric artery, the obturator artery passes through the *obturator foramen*, enters the thigh, and divides into anterior and posterior branches that anastomose. The posterior branch gives off an acetabular branch that supplies the head of the femur.

Palpation of the Femoral Artery

Some vascular surgeons refer to the part of the femoral artery proximal to the branching of the deep artery of the thigh as the *common femoral artery* and to its continuation as the *superficial femoral artery. This terminology is not recommended by the international Federative Committee on Anatomical Terminology* and is not used in this book because these terms may cause misunderstanding. The *femoral pulse* is palpated when the person is lying in the supine position. The *femoral artery begins at the inguinal ligament and runs midway between the anterior superior iliac spine and the pubic symphysis* (*A–B*). The *femoral pulse* can be palpated just inferior to the midpoint of this ligament by pressing firmly. Normally the pulse is strong; however, if the lumina of the common or external iliac arteries are partially occluded, the pulse may be diminished. *Compression of the femoral artery* may also be accomplished at this site by pressing directly posteriorly against the superior pubic ramus, psoas major, and femoral head (*C*). Compression at this point will stop bloodflow through the femoral artery and its branches, such as the deep artery of the thigh.

Cannulation of the Femoral Artery

The femoral artery is easily exposed and cannulated at the base of the femoral triangle—just inferior to the midpoint of the inguinal ligament. For *left cardial (cardiac) angiography*, a long, slender catheter is inserted percutaneously into the artery and passed up the external iliac artery, common iliac artery, and aorta to the left ventricle of the heart (see Chapter 1). The coronary arteries can also be visualized by *coronary arteriography*.

Laceration of the Femoral Artery

The superficial position of the femoral artery in the femoral triangle makes it vulnerable to punctures by lacerations and *gunshot wounds*. Anterior to the femoral artery are the skin, subcutaneous tissue, superficial inguinal lymph nodes, superficial circumflex iliac artery, superficial layer of fascia lata, and the anterior part of the femoral sheath (Fig. 5.17). Commonly, both the femoral artery and vein are lacerated in anterior thigh wounds because they lie so close together. In some cases, an arteriovenous shunt occurs as a result ▶

Table 5.4. Arterial Supply to the Thigh

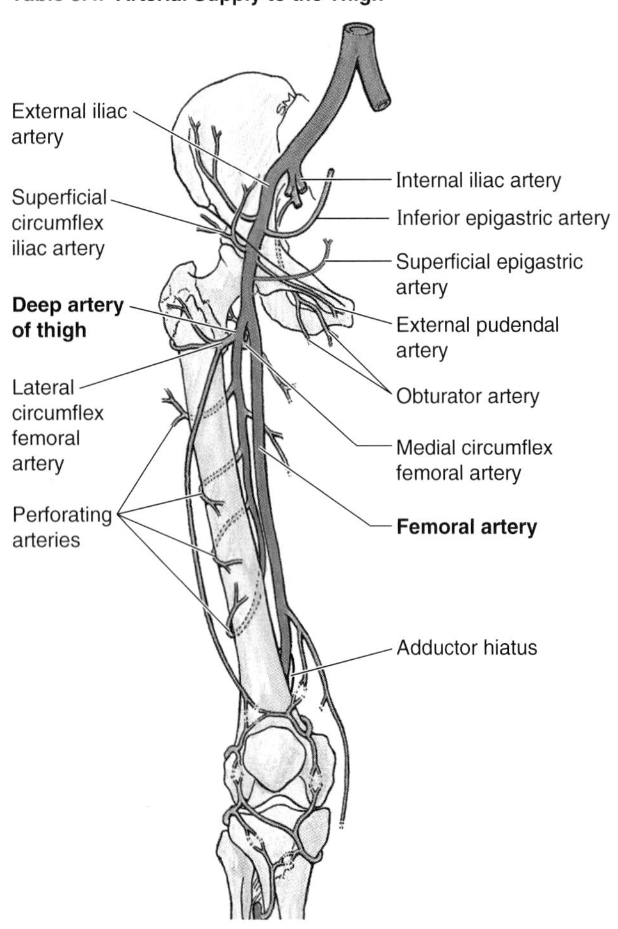

Anterior view

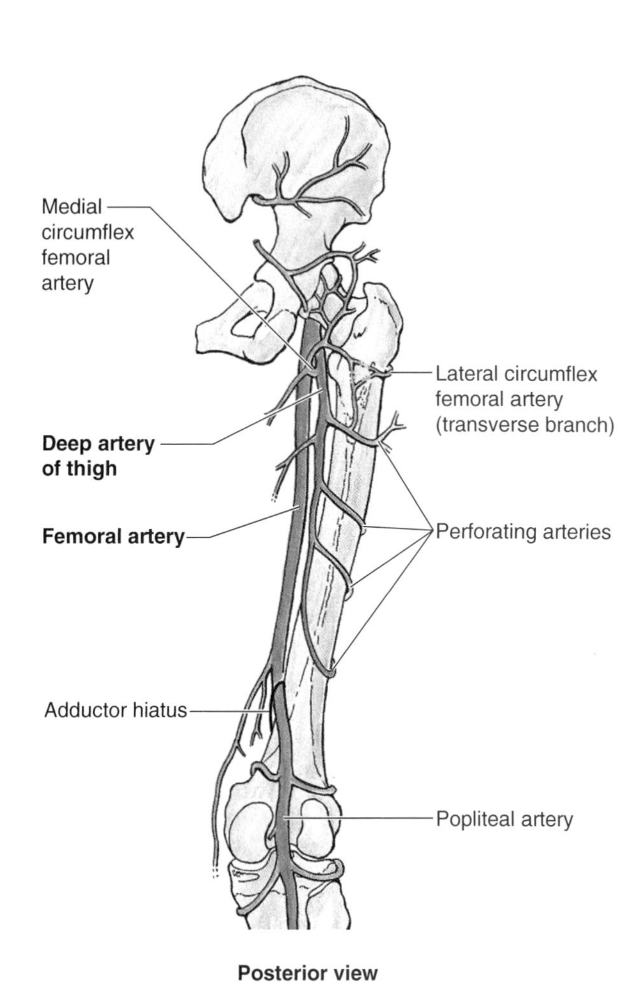

Posterior view

Artery	Origin	Course	Distribution
Femoral	Continuation of external iliac artery distal to inguinal ligament	Descends through femoral triangle, traverses adductor canal, and changes name to "popliteal" at adductor hiatus	Supplies anterior and anteromedial surfaces of thigh
Deep artery of thigh	Femoral artery about 4 cm distal to inguinal ligament	Passes inferiorly, deep to adductor longus	Perforating branches pass through adductor magnus muscle to posterior and lateral part of anterior compartments of thigh
Medial circumflex femoral	Deep artery of thigh; may arise from femoral artery	Passes medially and posteriorly between pectineus and iliopsoas, enters gluteal region, and divides into two branches	Supplies most blood to head and neck of femur; transverse branch takes part in cruciate anastomosis of thigh; ascending branch joins inferior gluteal artery
Lateral circumflex femoral	Deep artery of thigh; may arise from femoral artery	Passes laterally deep to sartorius and rectus femoris and divides into three branches	Ascending branch supplies anterior part of gluteal region; transverse branch winds around femur; descending branch descends to knee and joins genicular anastomoses
Obturator	Internal iliac artery	Passes through obturator foramen, enters medial compartment of thigh, and divides into anterior and posterior branches	Anterior branch supplies obturator externus, pectineus, adductors of thigh, and gracilis; posterior branch supplies muscles attached to ischial tuberosity

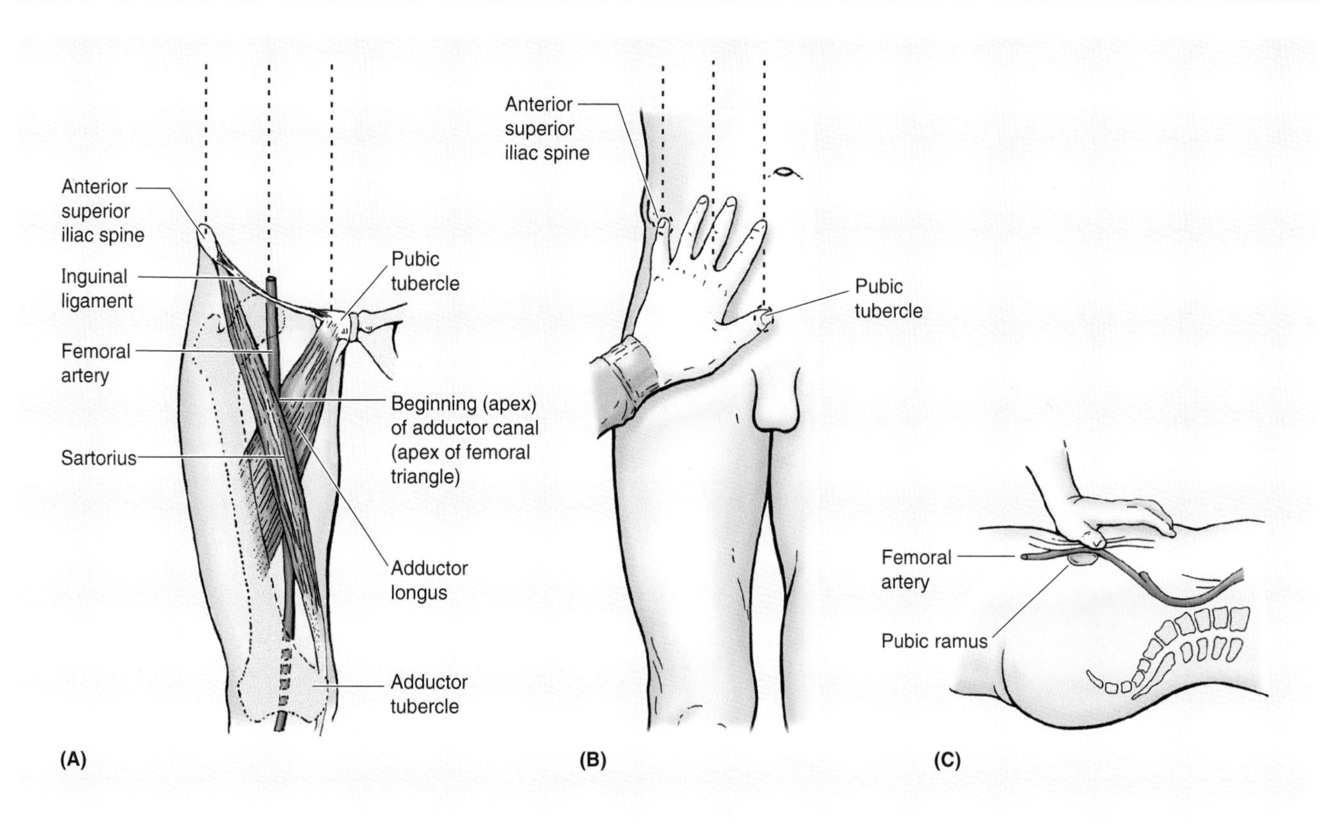

▶ of communication between the injured vessels. When it is necessary to ligate the femoral artery, the *cruciate anastomosis* supplies blood to the lower limb (Table 5.4). This anastomosis, which does not occur as frequently as it is described, is formed by the union of the medial and lateral circumflex femoral arteries with the inferior gluteal artery superiorly and the 1st perforating artery inferiorly.

Accessory Obturator Artery

An enlarged pubic branch of the inferior epigastric artery takes the place of the obturator artery in approximately 20% of people or forms an accessory obturator artery. This artery runs close to or across the femoral ring to reach the obturator foramen. Here it is closely related to the free margin of the lacunar ligament and the neck of a femoral hernia. Consequently, this artery could be involved in a strangulated *femoral hernia* (p. 548). Surgeons placing staples during endoscopic repair of inguinal or femoral hernias must also be vigilant concerning the possible presence of this common arterial variant. ○

Femoral Vein. The femoral vein is *the continuation of the popliteal vein proximal to the adductor hiatus.* As it ascends through the adductor canal, the femoral vein lies posterolateral and then posterior to the femoral artery (Fig. 5.16*B*). The femoral vein enters the femoral sheath lateral to the femoral canal and ends posterior to the inguinal ligament, where it *becomes the external iliac vein.* In the inferior part of the femoral triangle, the femoral vein receives the deep vein of the thigh, the great saphenous vein, and other tributaries. The **deep vein of the thigh**, formed by the union of three or four perforating veins, enters the femoral vein approximately 8 cm inferior to the inguinal ligament and approximately 5 cm inferior to the termination of the great saphenous vein.

Potentially Lethal Misnomer

Some vascular laboratories use the term "superficial femoral vein" in their reports when it is the femoral vein to which they are referring. In addition, most primary care physicians have not been taught and are not aware that the 'superficial' femoral vein is a deep vein and that acute thrombosis of this vessel is potentially life threatening. The adjective "superficial" should not be used because it implies that this vein is superficial, which it is not. *Most pulmonary emboli originate in deep veins*—not in superficial veins—and the risk of embolism can be greatly reduced by anticoagulant treatment. Therefore, anatomical terminology used in clinical reports must be accurate to avoid possible life-threatening situations. ▶

Saphenous Varix

A localized dilation of the terminal part of the great saphenous vein—a *saphenous varix* (L. dilated vein)—may cause edema in the femoral triangle. A saphenous varix may be confused with other groin swellings such as a psoas abscess; however, a varix should be considered when varicose veins are present in other parts of the lower limb.

Location of the Femoral Vein

The femoral vein is not usually palpable, but its position can be located inferior to the inguinal ligament by feeling the pulsations of the femoral artery, which is immediately lateral to the vein. In thin people, the femoral vein may be close to the surface and may be mistaken for the great saphenous vein. *It is therefore important to know that the femoral vein has no tributaries at this level,* except for the great saphenous vein that joins it approximately 3 cm inferior to the inguinal ligament. In *varicose vein operations,* it is obviously important to identify the great saphenous vein correctly and not tie off the femoral vein by mistake.

Cannulation of the Femoral Vein

To secure blood samples and take pressure recordings from the chambers of the right side of the heart and/or from the pulmonary artery, or for *right cardiac angiography,* a long, slender catheter is inserted into the femoral vein as it passes through the femoral triangle. A small cutaneous incision made inferior to the inguinal ligament and medial to the femoral artery allows ready access to the femoral vein for venous puncture, administration of fluids, or catheters. Under fluoroscopic control, the catheter is passed superiorly through the external and common iliac veins into the inferior vena cava and right atrium of the heart.

Femoral Hernia

The femoral ring is a weak area in the anterior abdominal wall that normally admits the tip of the 5th digit. The femoral ring is the usual originating site of a *femoral hernia,* a protrusion of abdominal viscera (often a loop of small intestine) through the femoral ring into the femoral canal. A femoral hernia appears as a mass, often tender, in the femoral triangle, *inferolateral to the pubic tubercle.* The hernia is bounded by the femoral vein laterally and the reflected part of the inguinal ligament—the lacunar ligament—medially (see drawing). The hernial sac compresses the contents of the femoral canal (loose connective tissue, fat, and lymphatics) and distends the wall of the canal. Initially the hernia is small because it is contained within the femoral canal, but it can enlarge by passing inferiorly through the saphenous opening into the subcutaneous tissue of the thigh. A femoral hernia appears inferior to the inguinal ligament and inferolateral to the pubic tubercle, whereas an indirect inguinal hernia is superior to the inguinal ligament and may enter the scrotum (see Chapter 2). Femoral hernias are more common in females, whereas inguinal hernias occur more commonly in males.

Strangulation of a femoral hernia may occur because of the sharp, rigid boundaries of the femoral ring, particularly the concave margin of the *lacunar ligament.* Strangulation of a femoral hernia interferes with the blood supply to the herniated intestine, and this vascular impairment may result in death of the tissues. Sometimes the lacunar ligament has to be incised to release a strangulated hernia. In these cases, an *aberrant obturator artery* passing medial to the hernial sac is vulnerable to injury. ○

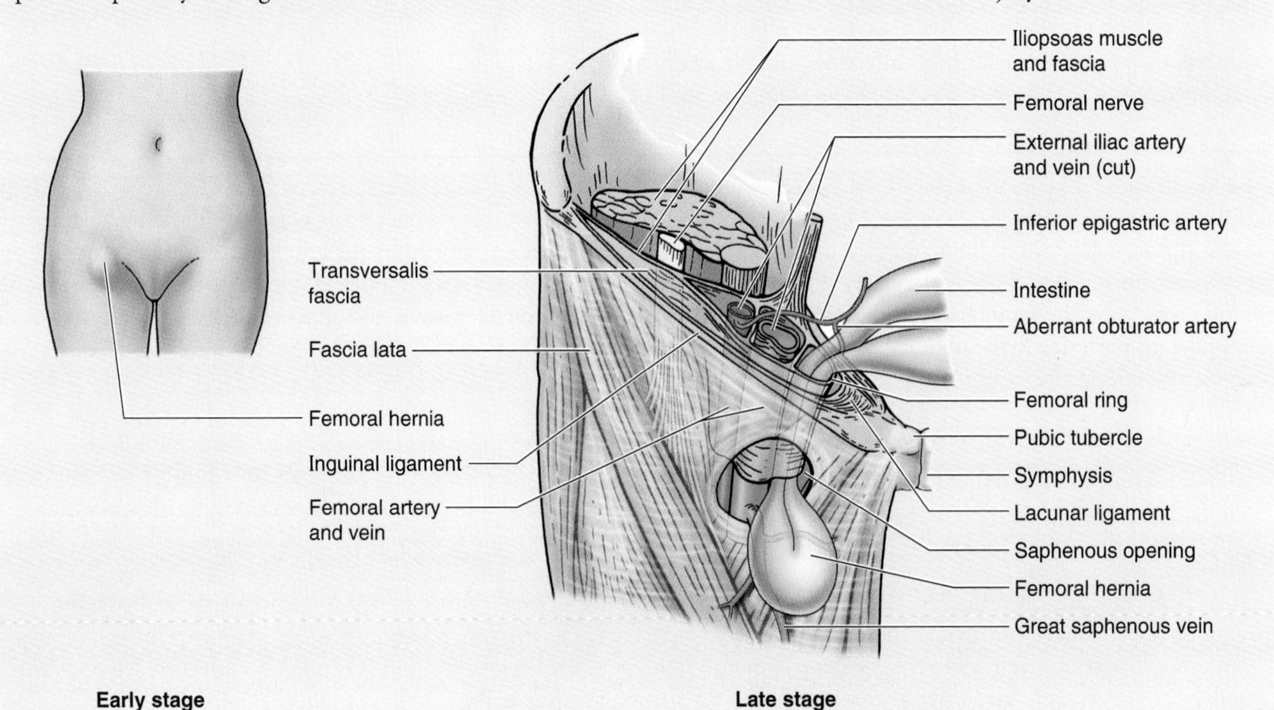

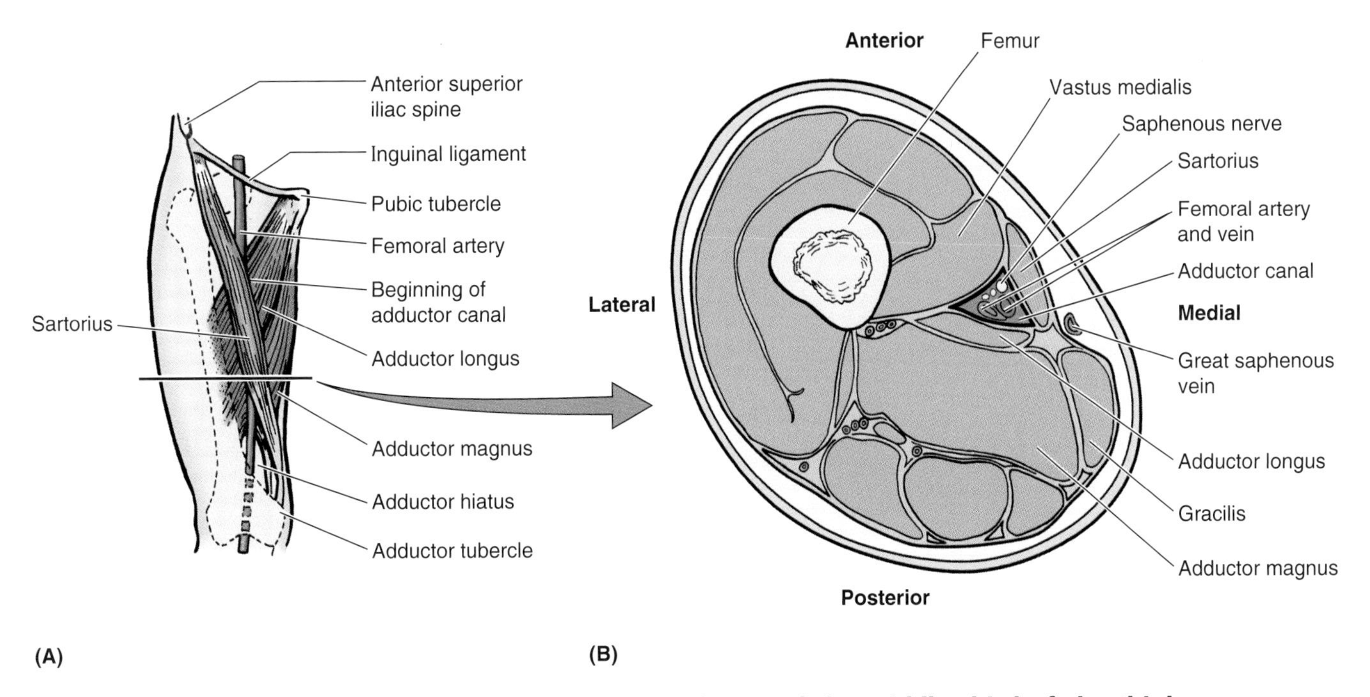

Figure 5.19. Adductor canal in the medial part of the middle third of the thigh. **A.** Orientation drawing showing the level of the section in (**B**). **B.** Transverse section of the thigh showing the adductor canal and its neurovascular contents.

Adductor Canal

The *adductor canal* (subsartorial canal; Hunter's canal)—approximately 15 cm long—is a *narrow fascial tunnel in the thigh running from the apex of the femoral triangle to the adductor hiatus in the tendon of the adductor magnus* (Fig. 5.19, Table 5.4). Located deep to the middle third of the sartorius, the adductor canal provides an intermuscular passage through which the femoral vessels pass to reach the popliteal fossa and become the popliteal vessels. The adductor canal begins where the sartorius crosses over the adductor longus and ends at the adductor hiatus. *The contents of the adductor canal are the*:

- Femoral artery and vein
- Saphenous nerve
- Nerve to vastus medialis.

The adductor canal is bounded:

- Anteriorly and laterally by the vastus medialis
- Posteriorly by the adductors longus and magnus
- Medially by the sartorius.

Gluteal Region

The gluteal region lies posterior to the pelvis between the level of the iliac crests and the inferior borders of the gluteus maximus muscles (Fig. 5.20). The **intergluteal (natal) cleft** separates the buttocks from each other. The *gluteal muscles* (max-

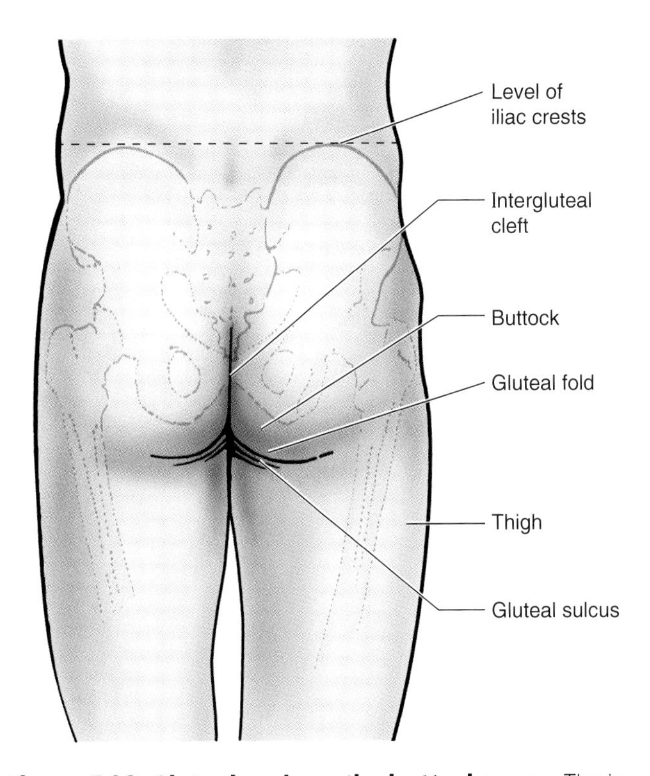

Figure 5.20. Gluteal region—the buttocks area. The intergluteal (natal) cleft separates the buttocks (gluteal prominences). The gluteal sulcus is the groove (crease) beneath the gluteal fold formed by the inferior border of the gluteus maximus. The gluteal sulcus marks the lower limit of the buttock and the upper limit of the thigh.

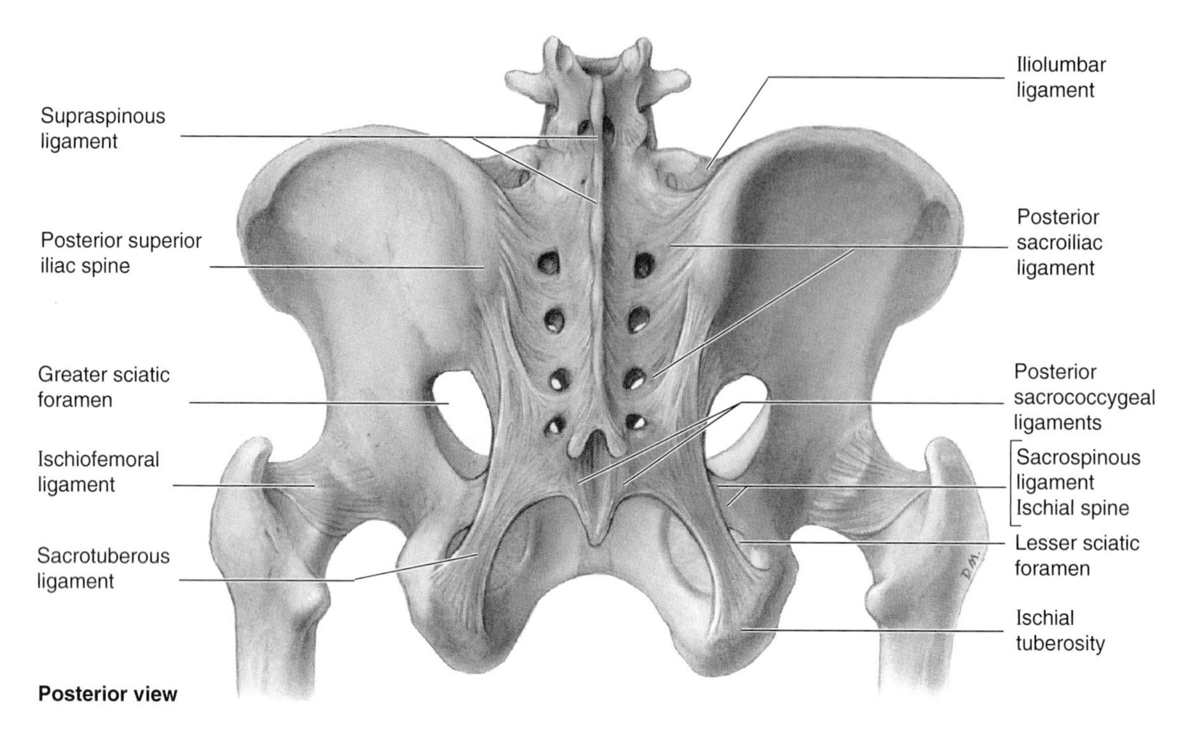

Figure 5.21. Lumbar and pelvic ligaments. The sacrotuberous and sacrospinous ligaments pass from the ischial tuberosity and ischial spine, respectively, to the side of the sacrum and coccyx. These ligaments convert the greater and lesser sciatic notches into foramina.

imus, medius, and minimus) form the bulk of the buttock. The gluteal sulcus lies inferior to the **gluteal fold**, which covers the fat-covered inferior border of the gluteus maximus when the thigh is extended. The **gluteal sulcus** demarcates the inferior boundary of the buttock and the superior boundary of the thigh.

Gluteal Ligaments

The parts of the bony pelvis—hip bones, sacrum, and coccyx—are bound together by dense ligaments (Fig. 5.21). The **sacrotuberous** and **sacrospinous ligaments** convert the sciatic notches in the hip bones into the greater and lesser sciatic foramina. The **greater sciatic foramen** is the passageway for structures entering or leaving the pelvis (e.g., sciatic nerve), whereas the **lesser sciatic foramen** is the passageway for structures entering or leaving the perineum (e.g., pudendal nerve). It is helpful to think of the greater sciatic foramen as the "door" through which all lower limb arteries and nerves leave the pelvis and enter the gluteal region. The *piriformis, occupying a key position in the gluteal region* (Table 5.5), also enters the gluteal region through the greater sciatic foramen and almost fills it.

Gluteal Muscles

The gluteal muscles (Fig. 5.22) consist of:

- Three large glutei (maximus, medius, and minimus), which are mainly extensors and abductors of the thigh
- A deeper group of smaller muscles (piriformis, obturator internus, externus gemelli, and quadratus femoris), which are covered by the inferior half of the gluteus maximus and are the lateral rotators of the thigh. They also stabilize the hip joint by steadying the femoral head in the acetabulum.

For the attachments, innervation, and main actions of these muscles, see Table 5.5.

Gluteus Maximus

The gluteus maximus—the most superficial gluteal muscle—is the largest, heaviest, and most coarsely fibered muscle in the gluteal region. The gluteus maximus covers the other gluteal muscles (Fig. 5.22*A*) except the posterior third of the gluteus medius and forms a pad over the ischial tuberosity. The **ischial tuberosity** can be felt on deep palpation through the inferior part of the muscle, just superior to the medial part of the gluteal fold. When the thigh is flexed, the inferior border of the gluteus maximus moves superiorly, leaving the ischial tuberosity subcutaneous. *You do not sit on your gluteus maximus*; you sit on the fatty fibrous tissue and the ischial bursa that lie between the ischial tuberosity and the skin.

The gluteus maximus slopes inferolaterally at a 45 degree angle from the pelvis to the buttock. The fibers of the superior and larger part of the gluteus maximus and superficial fibers of the inferior part insert into the *iliotibial tract*. Some deep fibers of the inferior part of the muscle attach to the *gluteal tuberosity of the femur* (Fig. 5.6*B*). The inferior gluteal nerve and vessels enter the deep surface of the gluteus maximus at its center. It is supplied by both the inferior and su-

Table 5.5. Muscles of the Gluteal Region

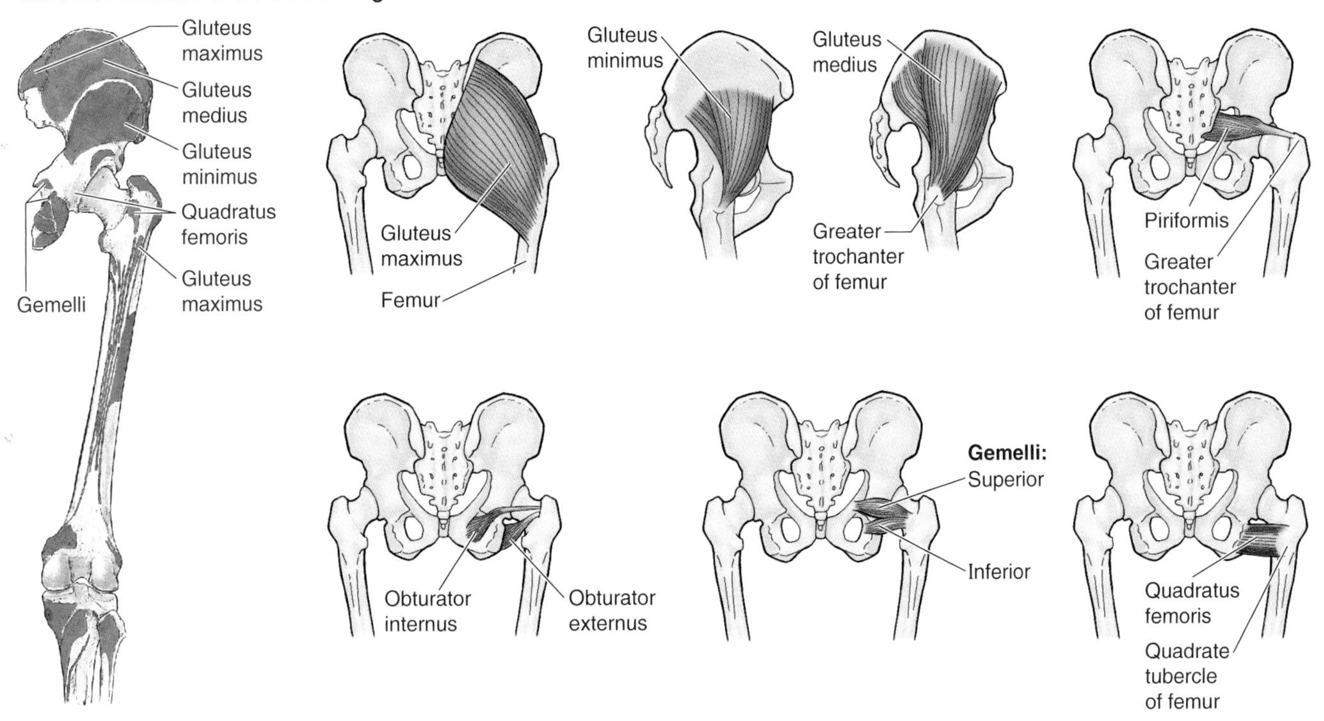

Posterior attachments

Muscle	Proximal Attachment	Distal Attachment	Innervation[a]	Main Action
Gluteus maximus	Ilium posterior to posterior gluteal line, dorsal surface of sacrum and coccyx, and sacrotuberous ligament	Most fibers end in iliotibial tract that inserts into lateral condyle of tibia; some fibers insert on gluteal tuberosity of femur	Inferior gluteal nerve (L5, **S1**, and **S2**)	Extends thigh (especially from flexed position) and assists in its lateral rotation; steadies thigh and assists in rising from sitting position
Gluteus medius	External surface of ilium between anterior and posterior gluteal lines	Lateral surface of greater trochanter of femur	Superior gluteal nerve (**L5** and S1)	Abducts and medially rotates thigh; keeps pelvis level when opposite leg is raised
Gluteus minimus	External surface of ilium between anterior and inferior gluteal lines	Anterior surface of greater trochanter of femur		
Piriformis	Anterior surface of sacrum and sacrotuberous ligament	Superior border of greater trochanter of femur	Branches of ventral rami of **S1** and S2	Laterally rotate extended thigh and abduct flexed thigh; steady femoral head in acetabulum
Obturator internus	Pelvic surface of obturator membrane and surrounding bones	Medial surface of greater trochanter (trochanteric fossa) of femur[b]	Nerve to obturator internus (L5 and **S1**) Superior gemellus: same nerve supply as obturator internus Inferior gemellus: same nerve supply as quadratus femoris	
Gemelli superior and inferior	Superior: ischial spine Inferior: ischial tuberosity			
Quadratus femoris	Lateral border of ischial tuberosity	Quadrate tubercle on intertrochanteric crest of femur and area inferior to it	Nerve to quadratus femoris (L5 and S1)	Laterally rotates thigh[c]; steadies femoral head in acetabulum

[a] See Table 5.1 for explanation of segmental innervation.
[b] Gemelli muscles blend with tendon of obturator internus muscle as it attaches to greater trochanter of femur.
[c] There are six lateral rotators of the thigh: piriformis, obturator internus, gemelli (superior and inferior), quadratus femoris, and obturator externus. These muscles also stabilize the hip joint.

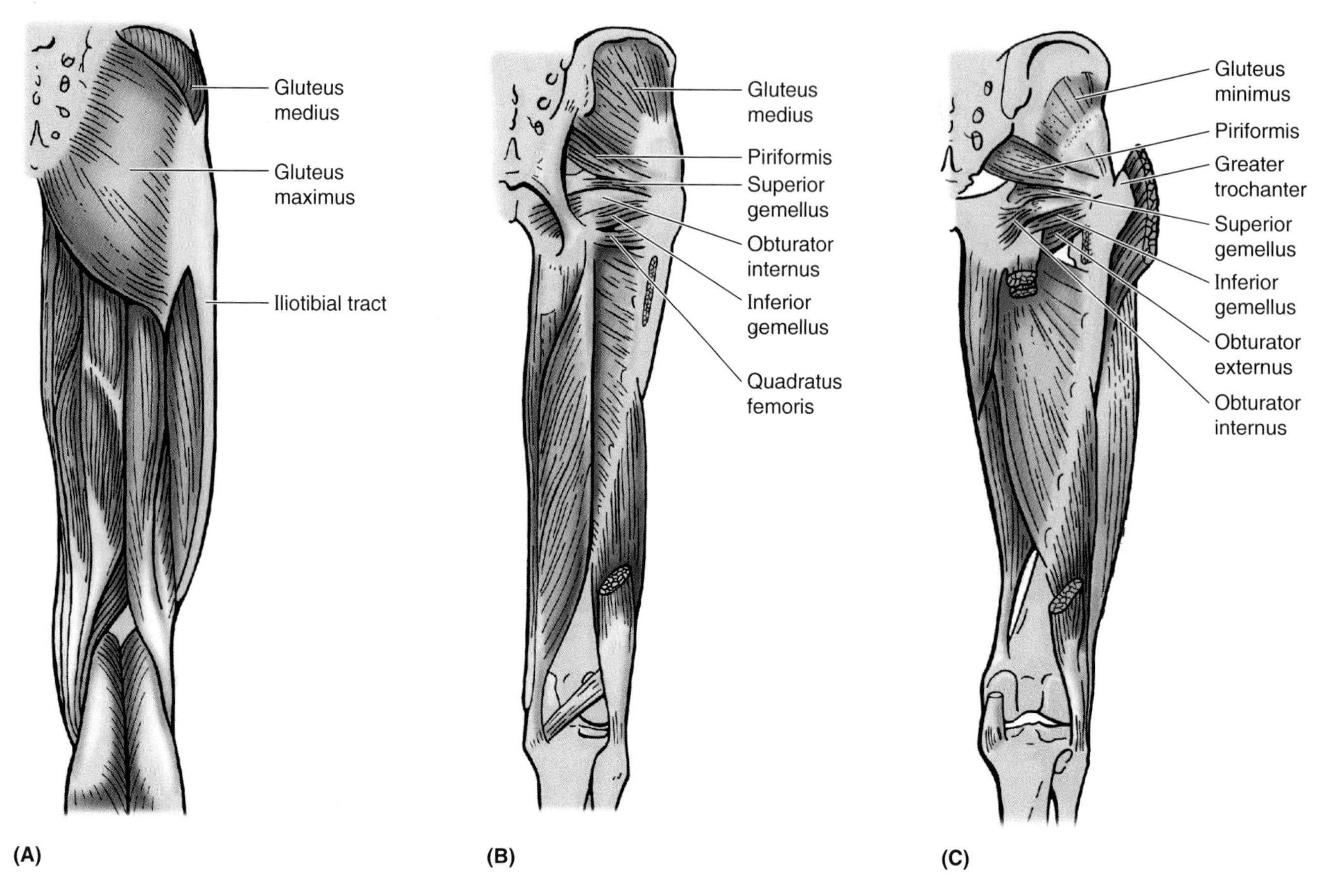

Figure 5.22. Muscles of the gluteal region and the posterior aspect of the thigh. Superficial (**A**), deep (**B**), and deeper (**C**) dissections.

perior gluteal arteries. In the superior part of its course, the **sciatic nerve** passes deep to the gluteus maximus (Fig. 5.23).

The main actions of the gluteus maximus are extension and lateral rotation of the thigh. When the fixed site of the gluteus maximus is at its proximal attachment, the muscle extends the trunk on the lower limb: *the gluteus maximus acts when force is necessary and functions primarily between the flexed and standing (straight) positions of the thigh*, as when rising from the sitting position, straightening from the bending position, walking upstairs, and running. It is used little during casual walking and when standing motionless. Verify this by placing your hand on your buttock when walking slowly. Note that the gluteus maximus contracts little with each step. If you climb stairs and put your hand on your buttock, you will feel the gluteus maximus contract strongly. Because the iliotibial tract crosses the knee and attaches to the tibia in the extended position of the knee, the gluteus maximus is also able to assist in making the knee stable.

Testing the gluteus maximus is performed when the person is prone with the lower limb straight. The person tightens the buttock and extends the hip joint as the examiner observes and palpates the gluteus maximus.

Gluteal bursae (L. purses) separate the gluteus maximus from adjacent structures (Fig. 5.24). Bursae are membranous sacs that are lined by a synovial membrane and contain a capillary layer of slippery fluid resembling egg white. *Bursae are located in areas subject to friction* (e.g., where the iliotibial tract crosses the greater trochanter); their purpose is to reduce friction and permit free movement. Usually three types of bursa are associated with the gluteus maximus:

- The **trochanteric bursae** separate superior fibers of the gluteus maximus from the greater trochanter; they are generally large
- The **ischial bursa** separates the inferior part of the gluteus maximus from the ischial tuberosity; it is often absent
- The **gluteofemoral bursa** separates the iliotibial tract (the fibrous reinforcement of the fascia lata into which most fibers of the gluteus maximus insert) from the superior part of the proximal attachment of the vastus lateralis, a thigh muscle.

Gluteus Medius and Gluteus Minimus

These smaller gluteal muscles are fan-shaped, and their fibers pass in the same direction (Figs. 5.22, *B* and *C*, and 5.23, Table 5.5). They have the same actions and nerve supply and are supplied by the same blood vessel—the superior gluteal artery. The gluteus minimus and most of the gluteus medius

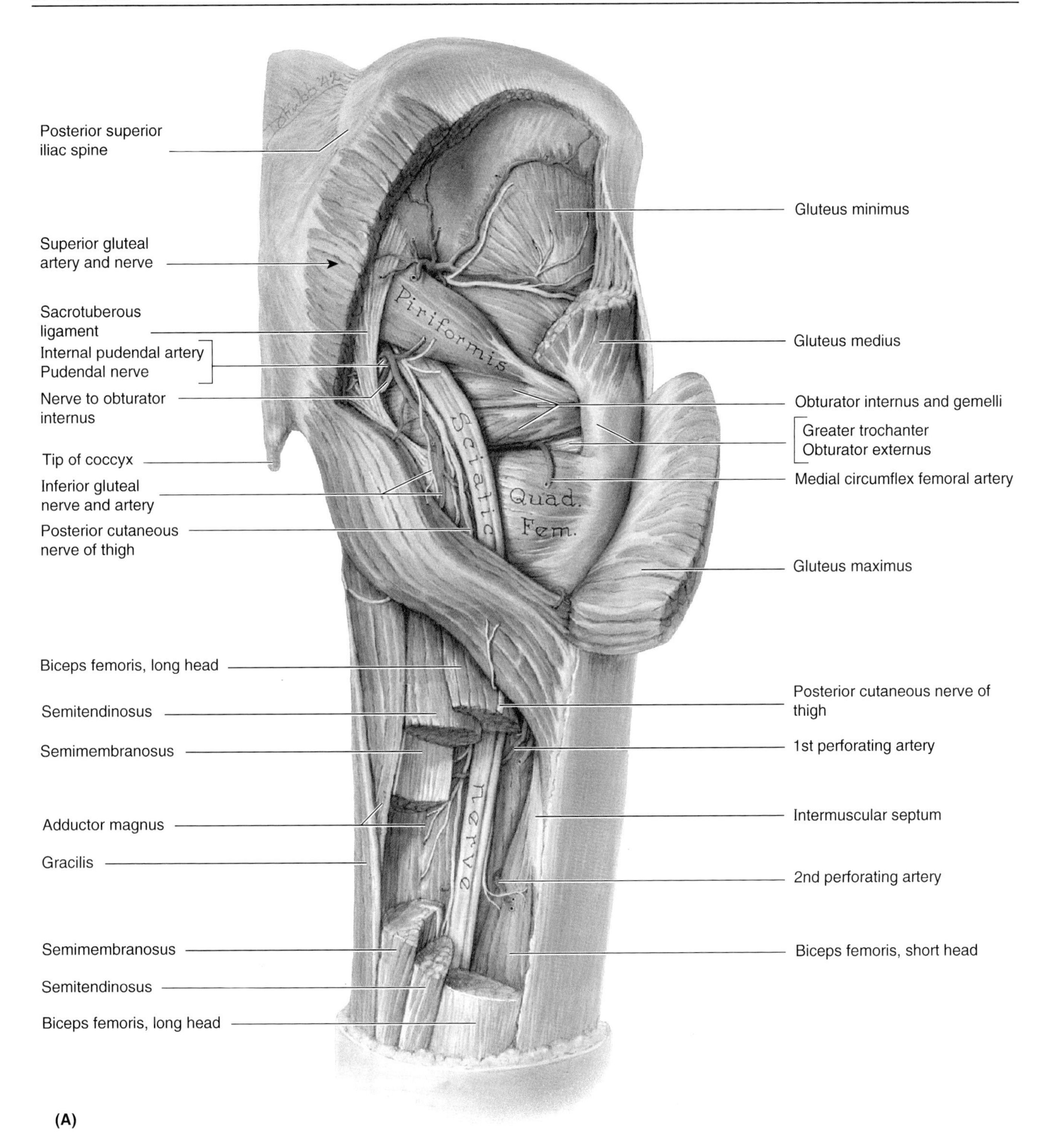

Figure 5.23. Deep dissection of the gluteal region and the posterior aspect of thigh. **A.** Most of the gluteus maximus and medius are removed and segments of the hamstrings are excised. Note that the superior gluteal artery and nerve emerge from the pelvis superior to the piriformis to lie between the gluteus medius and minimus. The inferior gluteal artery and nerve, the sciatic nerve, and the posterior cutaneous nerve of the thigh typically emerge inferior to the piriformis; however, exceptions occur (see Fig. 5.26).

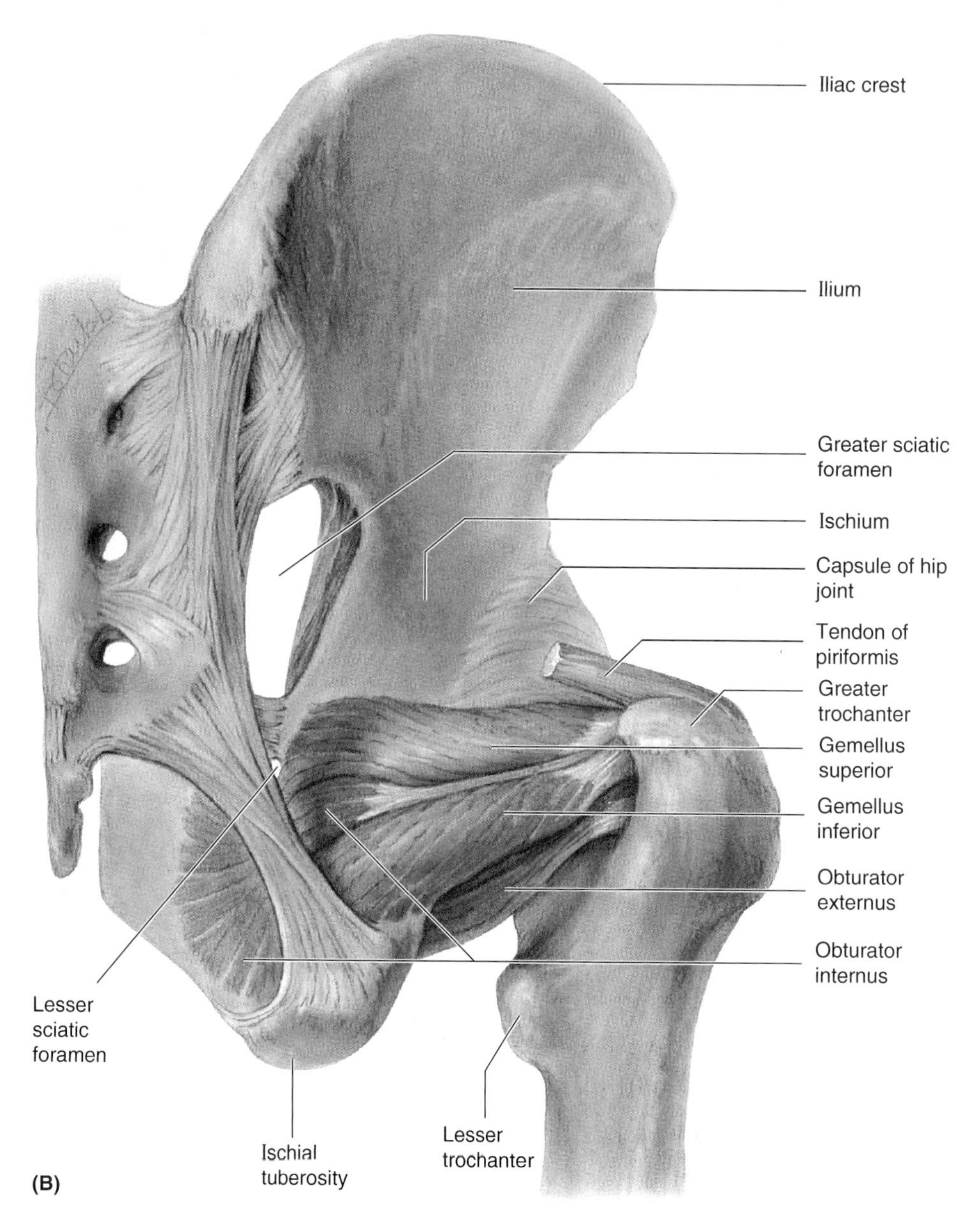

Figure 5.23. *(Continued)* **B.** Dissection of the lateral rotators of the thigh: piriformis (tendon only present), obturator, and gemelli muscles.

lie deep to the gluteus maximus on the external surface of the ilium. *The gluteus medius and minimus abduct the thigh and rotate it medially.* They play an essential role during locomotion and are largely responsible for preventing sagging of the unsupported side of the pelvis during walking (Fig. 5.25). When the left muscles are contracted, the right side is prevented from sagging as the right limb is raised during walking. Keeping the pelvis level enables the nonweightbearing foot to clear the ground as it is brought forward during walking. At this time, with the weightbearing thigh fixed in position, instead of producing medial rotation, the gluteus medius and minimus advance the unsupported side of the pelvis, also helping to advance the free limb. When the opposite foot is raised, the supportive action of the gluteus medius and minimus depends on normal:

- Muscular action and innervation from the superior gluteal nerve
- Articulation of the hip joint components
- Femoral neck (i.e., intact and normally angulated).

Testing the gluteus medius and minimus is performed while the person is prone with the leg flexed to a right angle. The person abducts the thigh against resistance. The gluteus medius can be palpated inferior to the iliac crest, posterior to the tensor of the fascia lata, which is also contracting during abduction of the thigh.

Piriformis

This narrow, pear-shaped (L. pirum, a pear) muscle is located partly on the posterior wall of the lesser pelvis and partly pos-

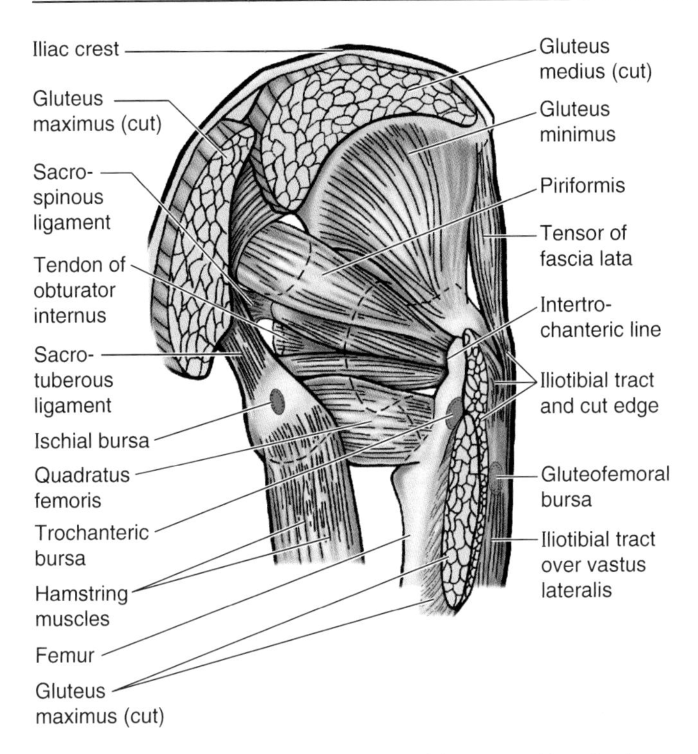

Figure 5.24. Gluteal muscles and bursae. Usually there are three bursae separating the gluteus maximus from underlying structures. Observe the trochanteric bursa, the gluteofemoral bursa, and the ischial bursa.

terior to the hip joint (Figs. 5.22, *B* and *C*, 5.23*B*, and 5.24, Table 5.5). The piriformis leaves the pelvis through the *greater sciatic foramen*, almost filling it, to reach its attachment to the superior border of the greater trochanter. *Because of its key position in the buttock, the piriformis is the landmark of the gluteal region.* The piriformis provides the key to understanding relationships in the gluteal region because it determines the names of the blood vessels and nerves.

- The superior gluteal vessels and nerve emerge superior to it
- The inferior gluteal vessels and nerve emerge inferior to it
- The surface marking of the superior border of the piriformis is indicated by a line joining the skin dimple formed by the posterior superior iliac spine to the superior border of the greater trochanter of the femur (see *A*, p. 567).

Obturator Internus and Gemelli

The obturator internus and the superior and inferior gemelli form a tricipital (three-headed) muscle—sometimes called the *triceps coxae*—which occupies the gap between the piriformis and quadratus femoris (Figs. 5.22, *B* and *C*, and 5.23, Table 5.5). The tricipital tendon of these muscles lies horizontally in the buttock as it passes to the greater trochanter of the femur.

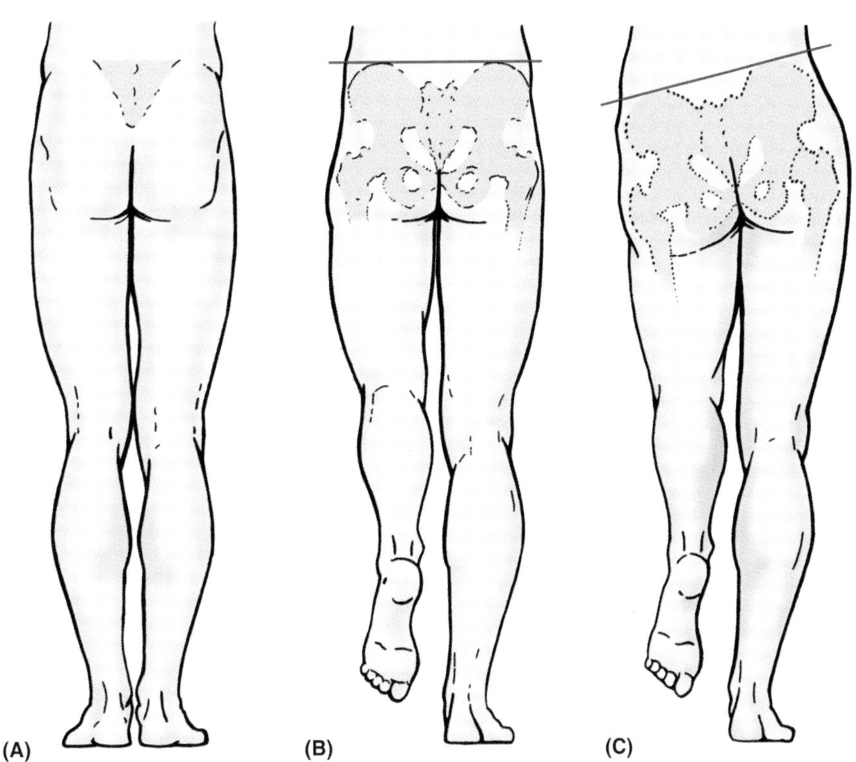

Figure 5.25. Action of abductors of the thigh (gluteus medius and minimus) when walking. Posterior views. **A.** When the weight is on both feet, the pelvis is evenly supported and does not sag. **B.** When the weight is borne by one foot, the muscles on the same side hold the pelvis so that the pelvis will not sag on the side of the raised limb. **C.** When the gluteus medius and minimus are inactive owing to injury of the superior gluteal nerve, the supporting and steadying action of these muscles is lost and the pelvis falls on the side of the raised limb (positive Trendelenburg sign).

The **obturator internus** is located partly in the pelvis, where it covers most of the lateral wall of the lesser pelvis (Fig. 5.23). It leaves the pelvis through the *lesser sciatic foramen*, becoming tendinous as it attaches to the medial surface of the greater trochanter. The *bursa of the obturator internus* allows free movement of the muscle over the posterior border of the ischium, where it forms the lesser sciatic notch.

The **gemelli muscles** assist the obturator internus. The tendon of the obturator internus receives the distal attachments of the small gemelli muscles, which are narrow, triangular extrapelvic reinforcements of the obturator internus. The *superior and inferior gemelli* arise from the ischial spine and ischial tuberosity, respectively.

Quadratus Femoris

This short, flat quadrangular muscle is located inferior to the obturator internus and gemelli (Fig. 5.23*A*). True to its name, the quadratus femoris is a rectangular muscle; it is a strong lateral rotator of the thigh.

Obturator Externus

This muscle was described previously with the medial thigh muscles (Table 5.3). It lies deep in the thigh, posterior to the pectineus and the superior ends of the adductor muscles. The obturator externus is visible only during dissection of the gluteal region (Figs. 5.22*C* and 5.23). The tendon of the obturator externus passes deep to the quadratus femoris on the way to its attachment to the trochanteric fossa of the femur. The obturator externus, with other short muscles around the hip joint, stabilizes the head of the femur in the acetabulum. It is also a lateral rotator of the thigh.

Gluteal Nerves

Several important nerves arise from the sacral plexus and either supply the gluteal region (e.g., superior and inferior gluteal nerves) or pass through it to supply the perineum and thigh (e.g., the pudendal and sciatic nerves, respectively). The *pudendal nerve* enters the gluteal region and then re-enters the pelvis to supply structures in the perineum (see Chapter 3). Table 5.6 describes the origin and distribution of nerves derived from the sacral plexus.

Superficial Gluteal Nerves

The skin of the gluteal region is richly innervated by superior, middle, and inferior **clunial nerves** (L. clunes, buttocks). These nerves supply the skin over the iliac crest, between the posterior superior iliac spines, and over the iliac tubercles. Consequently, these nerves are vulnerable to injury when bone is taken from the ilium for grafting.

- The **superior clunial nerves** are lateral cutaneous branches of the dorsal rami of L1, L2, and L3 nerves. They supply the skin of the gluteal region as far as the greater trochanter.
- The **middle clunial nerves**, lateral branches of the dorsal rami of S1 through S3 nerves, supply the skin and subcutaneous tissue over the sacrum and adjacent area of the buttock.
- The **inferior clunial nerves** are gluteal branches of the *posterior cutaneous nerve* of the thigh, a derivative of the sacral plexus (ventral rami S1 through S3). These nerves curl around the inferior border of the gluteus maximus and supply the inferior half of the buttock.

The **perforating cutaneous nerve** (S2 and S3) passes through the sacrotuberous ligament and the inferior part of the gluteus maximus to supply the skin over the inferior half of the buttock and the medial part of the gluteal fold.

Ischial Bursitis

Recurrent microtrauma resulting from repeated stress (e.g., when cycling or using a weaving machine) overwhelms the ischial bursa's ability to dissipate applied stress. The recurrent trauma results in inflammation of the bursa (*ischial bursitis*); calcification in the bursa may occur with chronic bursitis. Localized pain occurs over the bursa, and the pain increases with movement of the gluteus maximus. Ischial bursitis ("weaver's bottom") is a *friction bursitis* resulting from excessive friction between the ischial bursae and the ischial tuberosities. Weavers extend one lower limb and then the other during weaving. A similar friction bursitis—"paddle soreness"—occurs in cyclists. As the ischial tuberosities bear the body weight during sitting and when supine, these pressure points may lead to *pressure sores* in debilitated people, particularly paraplegic persons with poor nursing care.

Trochanteric Bursitis

Trochanteric bursitis—inflammation of the trochanteric bursae—often results from repetitive actions such as climbing stairs when carrying heavy objects or running on a steeply elevated treadmill. These movements involve the gluteus maximus and move the superior tendinous fibers repeatedly back and forth over the bursae of the greater trochanter. Trochanteric bursitis causes deep diffuse pain in the lateral thigh region. This type of *friction bursitis* is characterized by point tenderness over the great trochanter. *The pain radiates along the iliotibial tract* that extends from the iliac tubercle to the tibia (Fig. 5.22*A*). This thickening of the fascia lata receives tendinous reinforcements from the tensor of the fascia lata and the gluteus maximus. The pain from an inflamed trochanteric bursa, usually localized just posterior to the greater trochanter, is usually elicited by manually resisting abduction and lateral rotation of the thigh while the person is lying on the unaffected side. ⊙

Table 5.6. Nerves of the Gluteal Region

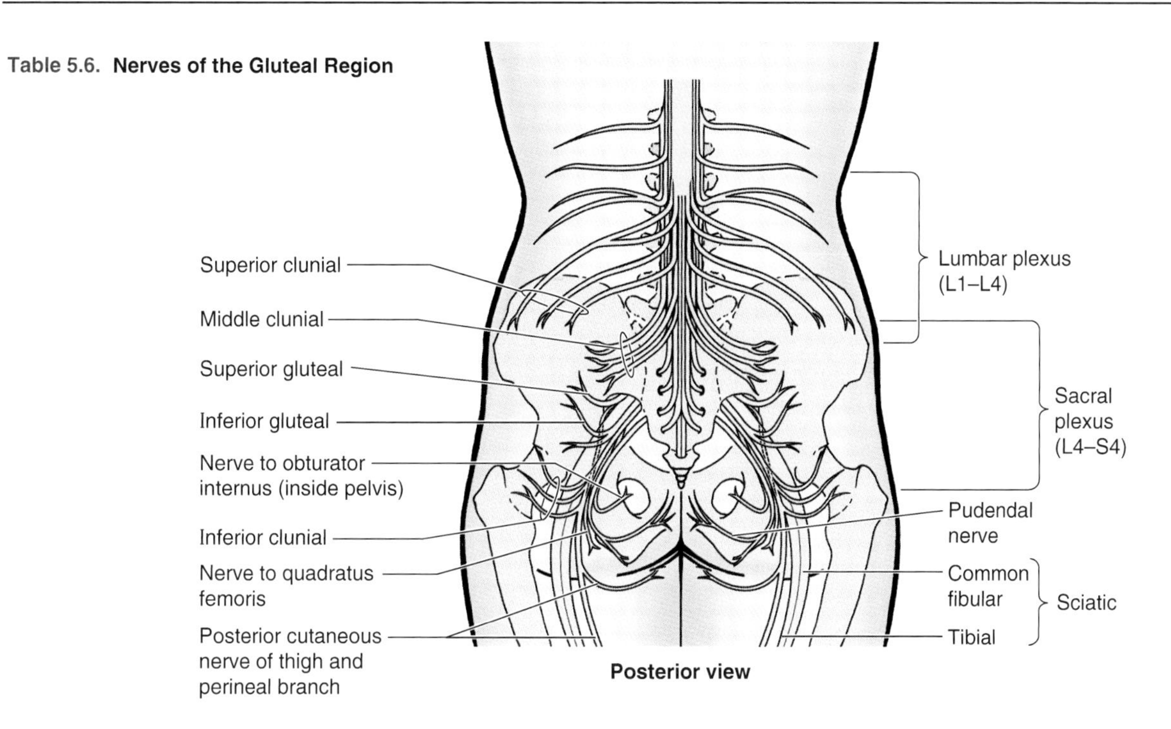

Posterior view

Nerve	Origin	Course	Distribution[a] in Gluteal Region
Clunial (superior, middle, and inferior)	Superior: dorsal rami of L1–L3 nerves Middle: dorsal rami of S1–S3 nerves Inferior: posterior cutaneous nerve of thigh (ventral rami of S2–S3)	Superior nerves cross iliac crest; middle nerves exit through posterior sacral foramina and enter gluteal region; inferior nerves curve around inferior border of gluteal maximus	Supplies skin of buttock or gluteal region as far as greater trochanter
Sciatic	Sacral plexus (L4–S3)	Leaves pelvis through greater sciatic foramen inferior to piriformis and enters gluteal region	Supplies no muscles in gluteal region
Posterior cutaneous nerve of thigh	Sacral plexus (S1–S3)	Leaves pelvis through greater sciatic foramen inferior to piriformis, runs deep to gluteus maximus, and emerges from its inferior border	Supplies skin of buttock through inferior cluneal branches and skin over posterior aspect of thigh and calf; lateral perineum, upper medial thigh via perineal branch
Superior gluteal	Ventral rami of L4–S1 nerves	Leaves pelvis through greater sciatic foramen superior to piriformis and runs between gluteus medius and minimus	Innervates gluteus medius, gluteus minimus, and tensor fasciae latae
Inferior gluteal	Ventral rami of L5–S2 nerves	Leaves pelvis through greater sciatic foramen inferior to piriformis and divides into several branches	Supplies gluteus maximus
Nerve to quadratus femoris	Ventral rami of L4, L5, and S1 nerves	Leaves pelvis through greater sciatic foramen deep to sciatic nerve	Innervates hip joint, inferior gemellus, and quadratus femoris
Pudendal	Ventral rami of S2–S4 nerves	Enters gluteal region through greater sciatic foramen inferior to piriformis; descends posterior to sacrospinous ligament; enters perineum through lesser sciatic foramen	Supplies most innervation to the perineum; supplies no structures in gluteal region
Nerve to obturator internus	Ventral rami of L5, S1, and S2 nerves	Enters gluteal region through greater sciatic foramen inferior to piriformis; descends posterior to ischial spine; enters lesser sciatic foramen and passes to obturator internus	Supplies superior gemellus and obturator internus

[a] See figures 5.10, 5.11, and Table 5.1 for cutaneous innervation of lower limb.

Deep Gluteal Nerves

The deep gluteal nerves are the sciatic, posterior cutaneous nerve of the thigh, superior gluteal, and inferior gluteal nerves; nerve to quadratus femoris; pudendal nerve; and nerve to obturator internus (Fig. 5.23, Table 5.6). All these nerves are branches of the sacral plexus and leave the pelvis through the greater sciatic foramen. Except for the superior gluteal nerve, they all emerge inferior to the piriformis.

Sciatic Nerve. *The sciatic nerve is the largest nerve in the body and forms the greatest part of the sacral plexus.* The sciatic nerve is the continuation of the main part of the sacral plexus, arising from the ventral rami of L4 through S3 (Fig. 5.23, Table 5.6). The rami converge at the inferior border of the piriformis to form the sciatic nerve, a thick, flattened band approximately 2 cm wide. The sciatic nerve passes through the inferior part of the *greater sciatic foramen* and is the most lateral structure emerging inferior to the piriformis. Medial to it are the inferior gluteal nerve and vessels, the internal pudendal vessels, and the pudendal nerve. The sciatic nerve runs inferolaterally under cover of the gluteus maximus, midway between the greater trochanter and ischial tuberosity. The nerve rests on the ischium and then passes posterior to the obturator internus, quadratus femoris, and adductor magnus muscles. The sciatic nerve is so large that it receives its own blood supply from the inferior gluteal artery.

The sciatic nerve supplies no structures in the gluteal region. It supplies the skin of the foot, most of the leg, posterior thigh muscles, and all leg and foot muscles. It also supplies articular branches to all joints of the lower limb. The sciatic nerve is really two nerves, *the tibial and common fibular nerves,* that are loosely bound together in the same connective tissue sheath. *Perone* is Greek for the *fibula*; because of the close relationship of the nerve to the fibular neck, its name has been changed internationally from common peroneal to common fibular. The tibial and common fibular nerves usually separate approximately halfway or more down the thigh (see Fig. 5.32); however, in approximately 12% of cases they separate as they leave the pelvis (Fig. 5.26*B*). In these cases, the tibial nerve passes inferior to the piriformis, and the common fibular nerve pierces this muscle or passes superior to it (Fig. 5.26*C*). The tibial nerve supplies flexor muscles, and the common fibular nerve supplies extensor and abductor muscles.

Posterior Cutaneous Nerve of the Thigh. This nerve arises from the sacral plexus (S1 through S3) and *supplies more skin than any other cutaneous nerve.* The posterior cutaneous nerve of the thigh leaves the pelvis with the inferior gluteal nerve and vessels and the sciatic nerve. It passes through the greater sciatic foramen inferior to the piriformis and descends deep to the gluteus maximus. Its fibers from posterior divisions of S1 and S2 supply the skin of the inferior part of the buttock; those from the anterior divisions supply the skin of the perineum; other branches continue inferiorly, where they supply skin of the posterior thigh and proximal part of the leg. Unlike most nerves bearing the name "cutaneous," the main part of this nerve lies deep to the deep fascia (L. fascia lata), with only its terminal branches penetrating to the subcutaneous tissue for distribution to the skin.

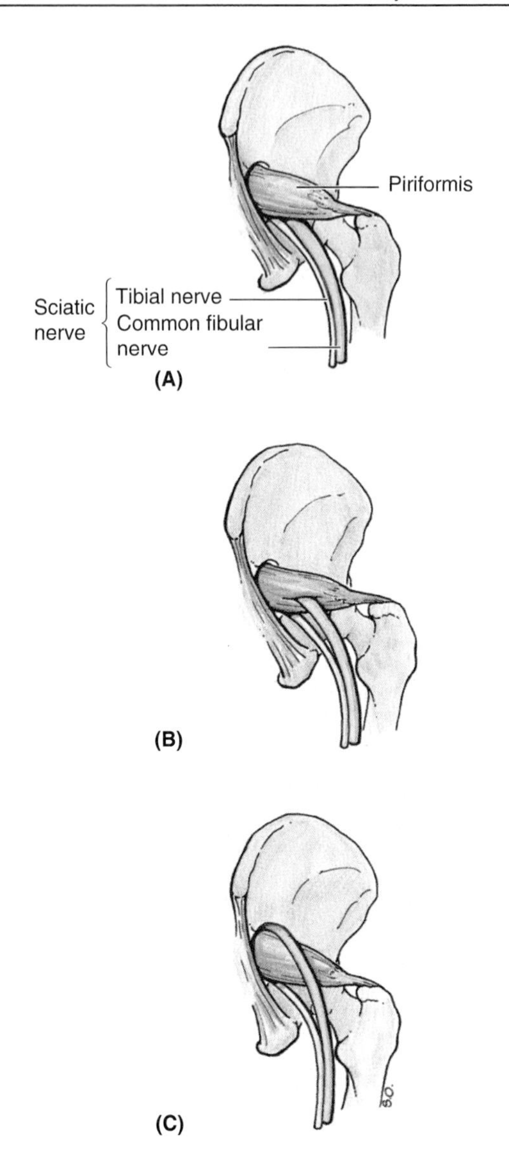

Figure 5.26. Relationship of sciatic nerve to the piriformis. A. Usually, as the sciatic nerve emerges from the greater sciatic foramen, it passes inferior to the piriformis. **B.** In 12.2% of 640 limbs studied, the sciatic nerve divided before it entered the gluteal region, and the common fibular division (*yellow*) passed through the piriformis. **C.** In 0.5% of cases, the common fibular division passed superior to the muscle, where it is especially vulnerable to injury during intragluteal injections.

Superior Gluteal Nerve. This nerve arises from posterior divisions of the ventral rami of L4 through S1. It leaves the pelvis through the greater sciatic foramen, *superior to the piriformis,* and runs laterally between the gluteus medius and minimus with the deep branch of the superior gluteal artery. The superior gluteal nerve divides into a superior branch that supplies the gluteus medius and an inferior branch that passes between the gluteus medius and minimus to supply both of these muscles and the tensor of the fascia lata.

Inferior Gluteal Nerve. This nerve arises from posterior divisions of the ventral rami of L5 through S2 and leaves the pelvis through the greater sciatic foramen, *inferior to the piriformis* and superficial to the sciatic nerve. It divides into several branches that supply the overlying gluteus maximus.

Nerve to Quadratus Femoris. This nerve arises from anterior divisions of the ventral rami of L4, L5, and S1 and leaves the pelvis anterior to the sciatic nerve and obturator internus and passes over the posterior surface of the hip joint. It supplies an articular branch to this joint and innervates the inferior gemellus and quadratus femoris.

Pudendal Nerve. This nerve arises from the anterior divisions of the ventral rami of S2 through S4 nerves and is the most medial structure to pass through the greater sciatic foramen inferior to the piriformis muscle. It passes lateral to the sacrospinous ligament, re-entering the pelvis through the lesser sciatic foramen, to supply structures in the perineum (see Chapter 3); *it supplies no structures in the gluteal region.*

Nerve to Obturator Internus. This nerve arises from the anterior divisions of the ventral rami of L5 through S2 nerves. It leaves the pelvis through the greater sciatic foramen, inferior to the piriformis and medial to the sciatic nerve. It winds around the base of the ischial spine to supply the superior gemellus and then passes posterior to the ischial spine, re-entering the pelvis through the lesser sciatic foramen, and supplies the obturator internus muscle.

Injury to the Superior Gluteal Nerve

Section of the superior gluteal nerve results in a characteristic motor loss, resulting in weakened abduction of the thigh by the gluteus medius, a disabling *gluteus medius limp*, and a *gluteal gait*, a compensatory list of the body to the weakened gluteal side. The compensation occurs to put the center of gravity over the supporting lower limb. Medial rotation of the thigh is also severely impaired. When a person is asked to stand on one leg, the gluteus medius normally contracts as soon as the contralateral foot leaves the floor, preventing tipping of the pelvis on the unsupported side (*A*). When a person with paralysis of the superior gluteal nerve is asked to stand on one leg, the pelvis on the unsupported side descends (*B*), indicating that the gluteus medius on the supported side is weak or nonfunctional. This observation is referred to clinically as a positive *Trendelenburg test.* Other causes of this sign include fracture of the greater trochanter—the distal attachment of gluteus medius. When the pelvis descends on the unsupported side, the lower limb becomes, in effect, "too long" and does not clear the ground when the foot is brought forward in the "swing through" phase of walking. To compensate, the individual leans away from the unsupported side, raising the pelvis to allow adequate room for the foot to come forward. This results in a characteristic "waddling gait." Another way to compensate is to lift the foot higher as it is brought forward—resulting in the so-called "steppage gait"—the same gait adopted in the presence of "foot-drop" from common fibular nerve paralysis (p. 585).

Anesthetic Block of the Sciatic Nerve

The sciatic nerve can be blocked by the injection of an anesthetic agent a few centimeters inferior to the midpoint of the line joining the posterior superior iliac spine and the superior border of the greater trochanter. Paresthesia radiates to the foot because of anesthesia of the plantar nerves—terminal branches of the tibial nerve derived from the sciatic nerve.

Injury to the Sciatic Nerve

A pain in the buttock may result from compression of the sciatic nerve by the piriformis muscle (*piriformis syndrome*). Persons involved in sports that require excessive use of the gluteal muscles and women are more likely to develop this syndrome (e.g., ice and roller skaters, cyclists, and mountain climbers). In approximately 50% of cases, the case histories indicate trauma to the buttock associated with hypertrophy and *spasm of the piriformis*. In approximately 12% of people, the common fibular division of the sciatic nerve passes through the piriformis (Fig. 5.26*B*); in these cases, this muscle may compress the nerve.

Complete section of the sciatic nerve is uncommon. When this occurs, the leg is useless because extension of the hip ▶

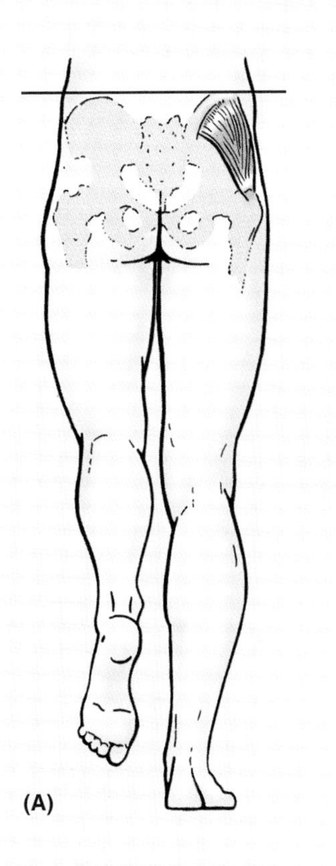

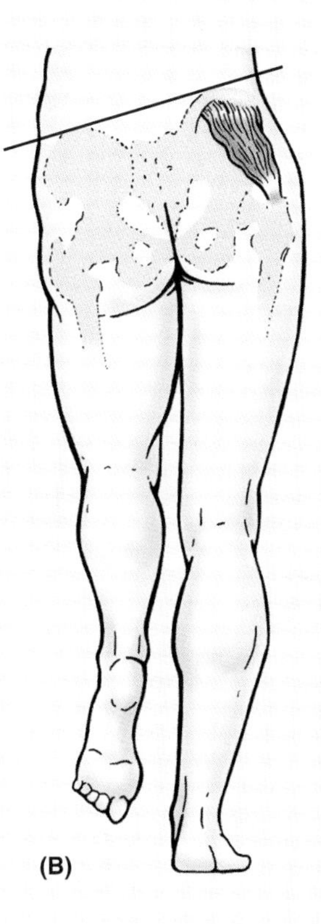

▶ is impaired, as is flexion of the leg. All ankle and foot movements are also lost. Incomplete section of the sciatic nerve from gunshot or stab wounds may also involve the inferior gluteal and/or the posterior femoral cutaneous nerves. *Recovery from a sciatic lesion is slow and usually incomplete.*

With respect to the sciatic nerve, the buttock has a side of safety (its lateral side) and a side of danger (its medial side). Wounds or surgery on the medial side of the buttock are liable to injure the sciatic nerve and its branches to the hamstrings (semitendinosus, semimembranosus, and biceps femoris) on the posterior aspect of the thigh. Paralysis of these muscles results in impairment of thigh extension and leg flexion.

Intragluteal Injections

The gluteal region is a common site for intramuscular injection of drugs. *Gluteal intramuscular injections* penetrate the skin, fascia, and muscles. The gluteal region is a favorable injection site because the muscles are thick and large; consequently, they provide a large surface area for absorption of drugs. Be aware of the extent of the gluteal region and the safe region for giving injections. Some people restrict the area of the buttock to the "cheek," the most prominent part; this is a dangerous concept because the sciatic nerve lies deep to this area.

Injections into the buttock should always be made superior to a line extending from the posterior superior iliac spine to the superior border of the greater trochanter (*A*). Another way to locate the safe injection area is to place the index finger on the anterior superior iliac spine and spread the 3rd digit posteriorly along the iliac crest (*B*). An intragluteal injection can be made safely in the triangular area between the ends of the fingers because it is superior to the sciatic nerve. Injections are safe only in the superolateral part of the buttock. Other areas are dangerous for making injections because many nerves and vessels are present. Intramuscular injections can also be given safely into the anterolateral part of the thigh, where the needle enters the tensor of the fascia lata (Fig. 5.13*B*) as it extends distally from the iliac crest and anterior superior iliac spine. Complications of improper technique include nerve injury, hematoma, and abscess formation. ⊙

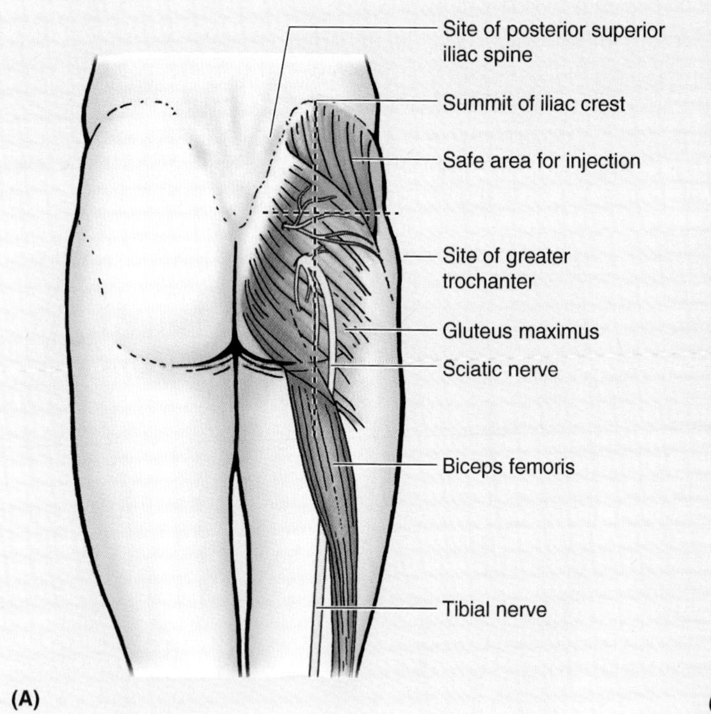

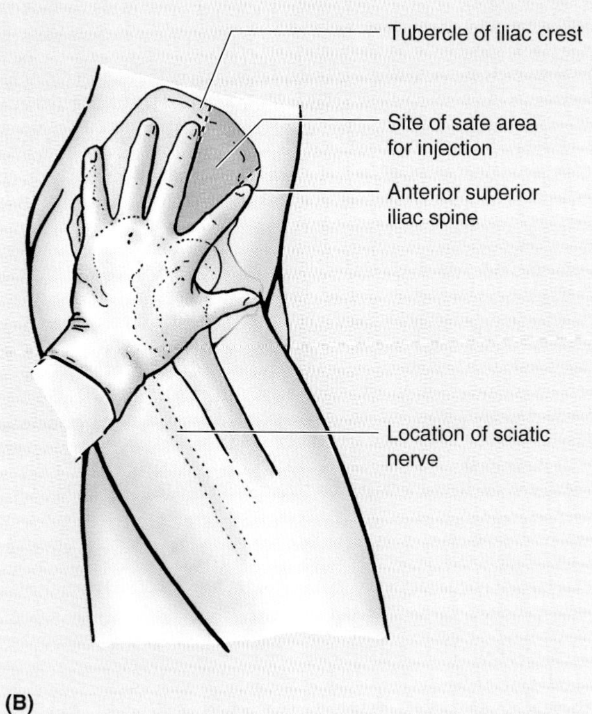

Gluteal Arteries

The gluteal arteries arise, directly or indirectly, from the *internal iliac arteries*, but the patterns of origin of the arteries are variable (Fig. 5.23*A*, Table 5.7). The major gluteal branches of the internal iliac artery are the:

- Superior gluteal artery
- Inferior gluteal artery
- Internal pudendal artery.

Superior Gluteal Artery

The superior gluteal artery—*the largest branch of the internal iliac artery*—passes posteriorly between the lumbosacral trunk and the 1st sacral nerve. The superior gluteal artery leaves the

Table 5.7. **Arteries of the Gluteal Region**

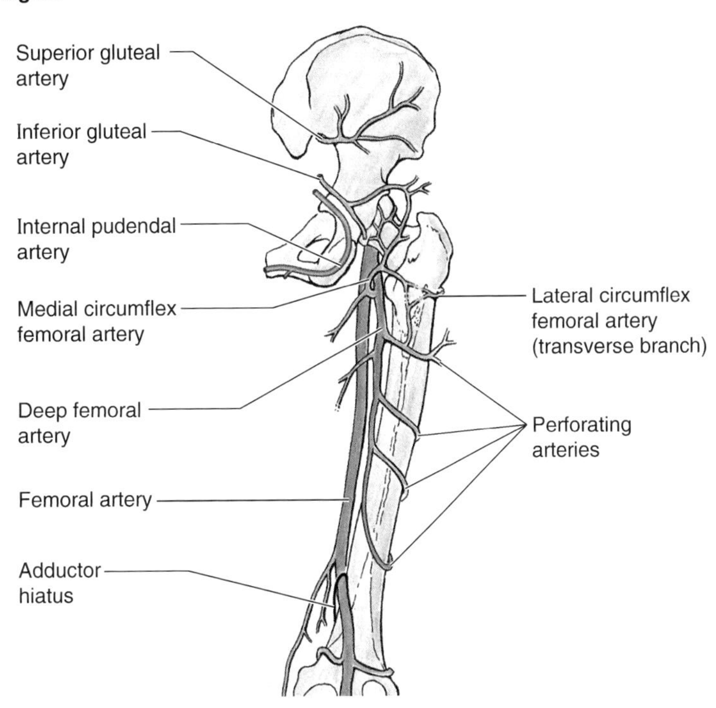

Artery[a]	Course	Distribution
Superior gluteal	Enters gluteal region through greater sciatic foramen superior to piriformis and divides into superficial and deep branches; anastomoses with inferior gluteal and medial circumflex femoral arteries (not shown above)	Superficial branch: supplies gluteus maximus Deep branch: runs between gluteus medius and minimus and supplies them and tensor of fascia lata
Inferior gluteal	Enters gluteal region through greater sciatic foramen inferior to piriformis and descends on medial side of sciatic nerve; anastomoses with superior gluteal artery and participates in cruciate anastomosis of thigh, involving first perforating artery of deep femoral and medial and lateral circumflex femoral arteries (not shown in figure above)	Supplies gluteus maximus, obturator internus, quadratus femoris, and superior parts of hamstrings
Internal pudendal	Enters gluteal region through greater sciatic foramen and descends posterior to ischial spine; enters perineum through lesser sciatic foramen	Supplies external genitalia and muscles in the perineal region; does not supply gluteal region

[a] All these arteries arise from internal iliac artery (see Table 5.4, anterior view).

pelvis through the greater sciatic foramen, superior to the piriformis, and divides immediately into superficial and deep branches. The *superficial branch* supplies the gluteus maximus and skin over the proximal attachment of this muscle; the *deep branch* supplies the gluteus medius, gluteus minimus, and tensor of the fascia lata. The superior gluteal artery anastomoses with the inferior gluteal and medial circumflex femoral arteries.

Inferior Gluteal Artery

This vessel arises from the internal iliac artery and passes posteriorly through the parietal pelvic fascia, between the 1st and 2nd (or 2nd and 3rd) sacral nerves. The inferior gluteal artery leaves the pelvis through the greater sciatic foramen, inferior to the piriformis (Fig. 5.23, Table 5.7). It enters the gluteal region deep to the gluteus maximus, and descends medial to the sciatic nerve. The inferior gluteal artery supplies the gluteus maximus, obturator internus, quadratus femoris, and superior parts of the hamstrings. It anastomoses with the superior gluteal artery and participates in the *cruciate anastomosis of the thigh*, involving the 1st perforating arteries of the deep artery of the thigh and the medial and lateral circumflex femoral arteries.

Internal Pudendal Artery

This vessel arises from the internal iliac artery and lies anterior to the inferior gluteal artery. It accompanies the pudendal nerve and enters the gluteal region through the greater sci-

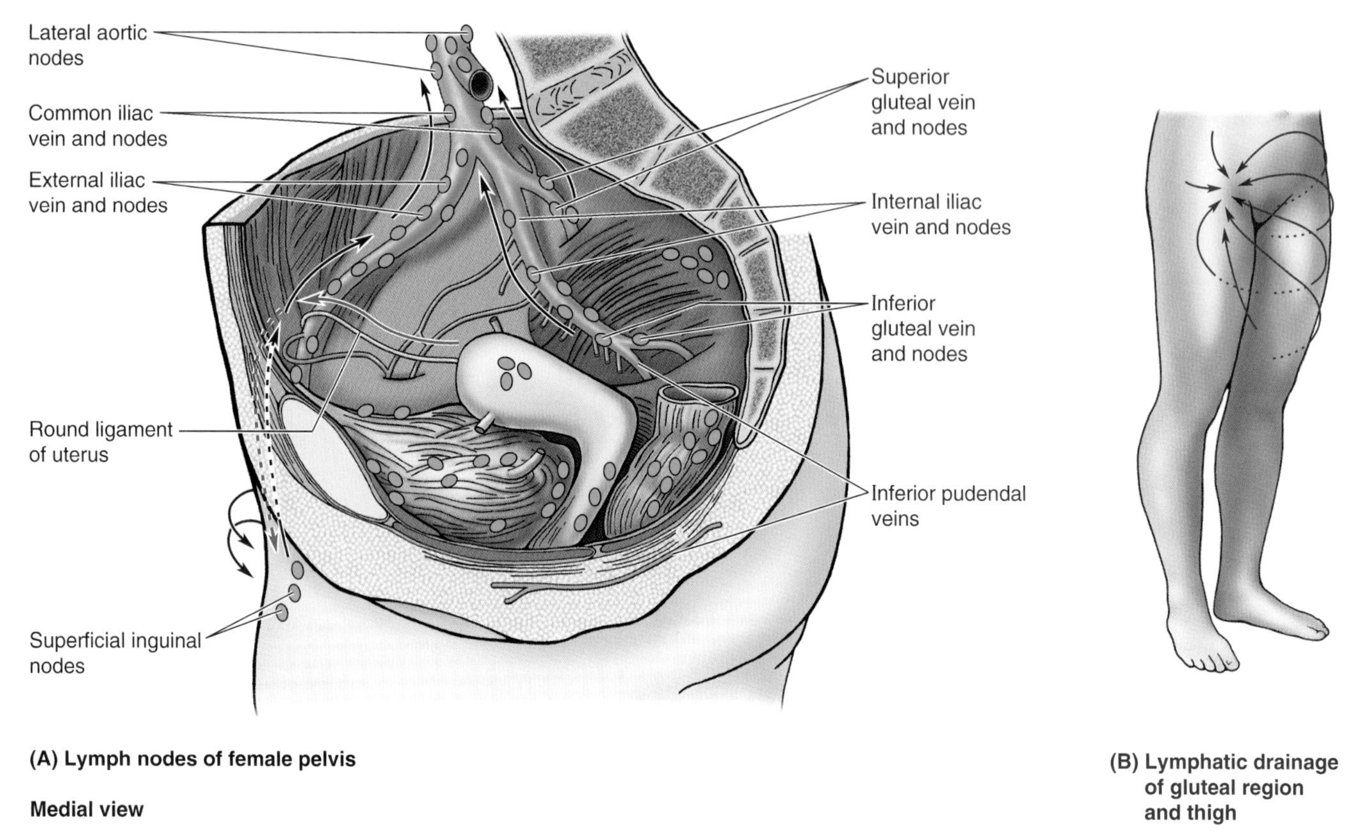

Figure 5.27. Lymphatic drainage of the gluteal region and thigh. A. Lymph from deep tissues of the gluteal region enters the superior and inferior gluteal lymph nodes, and from them it passes to the iliac and lateral aortic lymph nodes. **B.** Lymph from superficial tissues of the gluteal region passes initially to the superficial inguinal nodes, which also receive lymph from the thigh. Lymph from all the superficial inguinal nodes passes through efferent lymph vessels to the external iliac and lateral aortic lymph nodes.

atic foramen, inferior to the piriformis (Fig. 5.23, Table 5.7). *The internal pudendal artery leaves the gluteal region immediately by crossing the ischial spine and re-entering the pelvis through the lesser sciatic foramen.* The artery passes to the perineum with the pudendal nerve and supplies the external genitalia and muscles in the pelvic region. *It does not supply any structures in the gluteal region.*

Gluteal Veins

The gluteal veins—*tributaries of the internal iliac veins*—drain blood from the gluteal region. The **superior and inferior gluteal veins** accompany the corresponding arteries through the greater sciatic foramen, superior and inferior to the piriformis, respectively (Fig. 5.27*A*). They communicate with tributaries of the femoral vein, thereby providing alternate routes for the return of blood from the lower limb if the femoral vein is occluded or has to be ligated. The **internal pudendal veins** accompany the internal pudendal arteries and join to form a single vein that enters the internal iliac vein. These veins drain blood from the external genitalia or pudendum (L. from pudere, to be ashamed).

Lymphatic Drainage of the Gluteal Region and Thigh

Lymph from deep tissues of the buttocks follows the gluteal vessels to the superior and inferior **gluteal lymph nodes** and from them to the internal, external, and common **iliac lymph nodes** (Fig. 5.27) and from them to the **lateral aortic lymph nodes.** Lymph from the superficial tissues of gluteal region enters the **superficial inguinal lymph nodes,** which also receive lymph from the thigh. All the superficial inguinal nodes send efferent lymphatic vessels to the external iliac lymph nodes.

Hematoma of the Buttock

Severe trauma to the buttock usually results from a hard fall (e.g., during figure skating). Because of the large gluteal veins between the gluteus maximus and medius, severe trauma often results in the formation of a large hematoma that results in *ecchymosis,* a purplish patch caused by extravasation of blood into the subcutaneous tissue and skin. Blood from the hematoma may be evacuated by aspiration or by incision and drainage. ○

Posterior Thigh Muscles

The three muscles in the posterior aspect of the thigh are the **hamstrings** (Fig. 5.28, *A* and *B*, Table 5.8):

- Semitendinosus
- Semimembranosus
- Biceps femoris (long head).

These hamstring muscles span the hip and knee joints, arise from the ischial tuberosity deep to the gluteus maximus, and are innervated by the tibial division of the sciatic nerve. The short head of the biceps does not meet these criteria.

The hamstrings are extensors of the thigh and flexors of the leg, especially during walking. Both actions cannot be performed fully at the same time. A fully flexed knee shortens the hamstrings so they cannot contract and extend the thigh. Similarly, a fully extended hip shortens the hamstrings so they cannot act on the knee. When the thighs and legs are fixed, the hamstrings can help to extend the trunk. They are the active thigh extensors when maintaining the relaxed standing posture ("standing at ease"). A person with paralyzed hamstrings tends to fall forward because the gluteus maximus muscles cannot

Table 5.8. Posterior Thigh Muscles

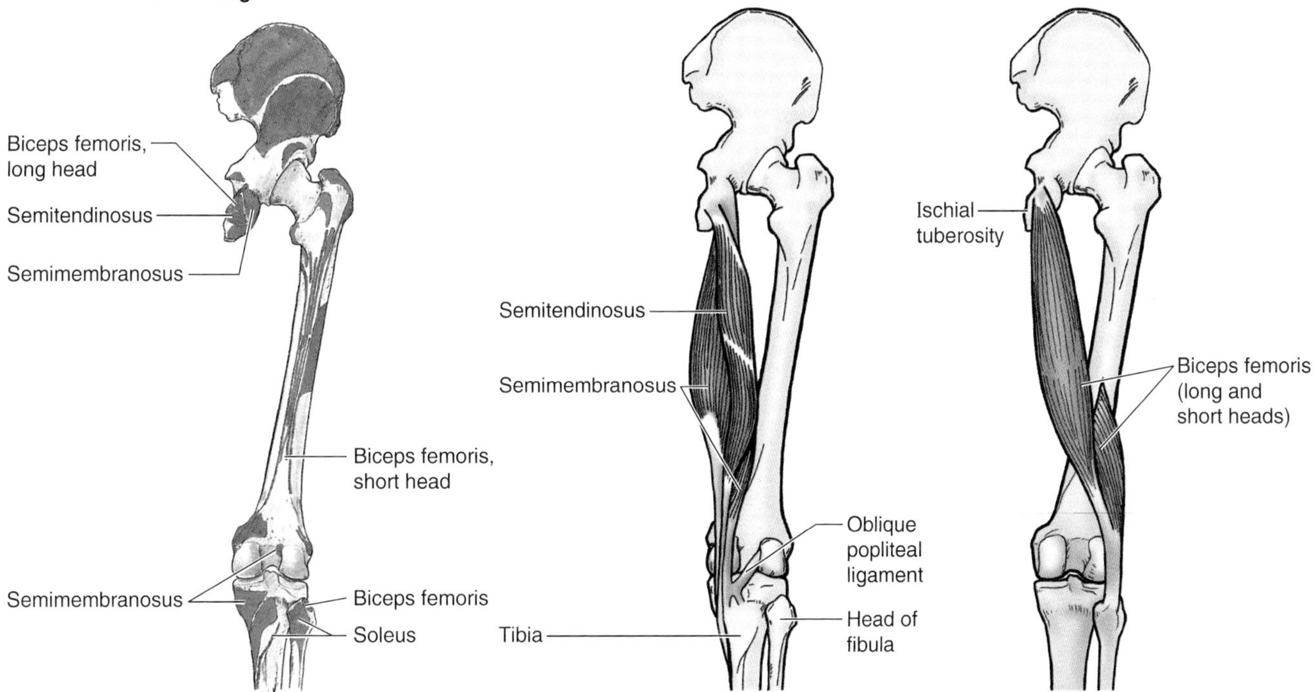

Muscle[a]	Proximal Attachment	Distal Attachment	Innervation[b]	Main Action
Semitendinosus	Ischial tuberosity	Medial surface of superior part of tibia	Tibial division of sciatic nerve (**L5, S1**, and S2)	Extend thigh; flex leg and rotate it medially when knee is flexed; when thigh and leg are flexed, these muscles can extend trunk
Semimembranosus		Posterior part of medial condyle of tibia; reflected attachment forms oblique popliteal ligament (to lateral femoral condyle)		
Biceps femoris	Long head: ischial tuberosity Short head: linea aspera and lateral supracondylar line of femur	Lateral side of head of fibula; tendon is split at this site by fibular collateral ligament of knee	Long head: tibial division of sciatic nerve (L5, **S1**, and S2) Short head: common fibular (peroneal) division of sciatic nerve (L5, **S1**, and S2)	Flexes leg and rotates it laterally when knee is flexed; extends thigh (e.g., when starting to walk)

[a] Collectively these three muscles are known as hamstrings.
[b] See Table 5.1 for explanation of segmental innervation.

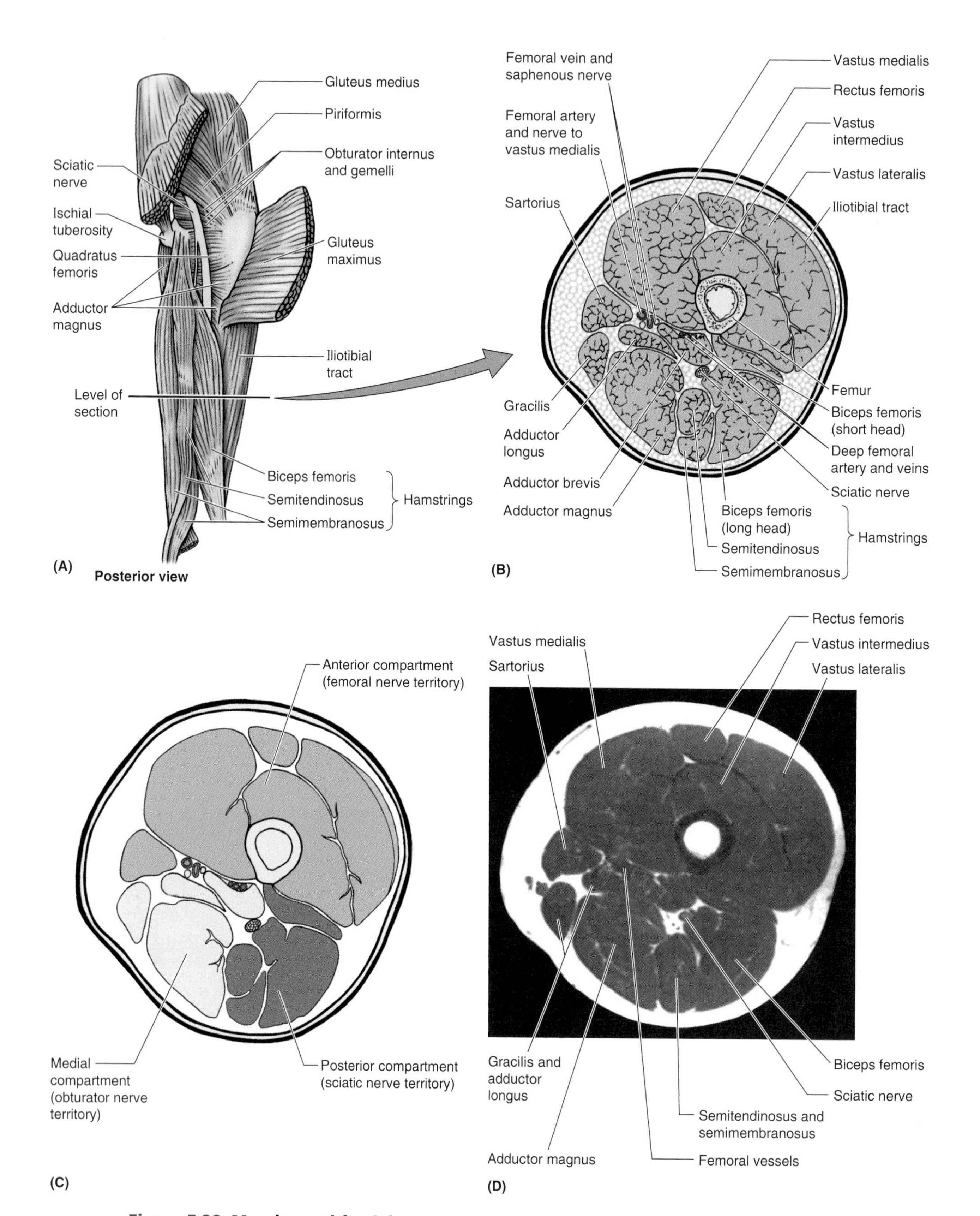

Figure 5.28. Muscles and fascial compartments of the thigh. A. The gluteus maximus has been reflected to show the sciatic nerve entering the proximal thigh and the attachments of the hamstrings. **B.** Transverse section through the thigh, 10 to 15 cm inferior to the inguinal ligament. **C.** Diagram showing the three groups of thigh muscles, each with its own nerve supply and primary function: *anterior*, femoral nerve—flexors of the hip and extensors of the knee joint; *medial,* obturator nerve—adductors of the hip joint; and *posterior*, sciatic nerve—extensors of the hip and flexors of the knee joint. **D.** Transverse MRI of the thigh. (Courtesy of Dr. W. Kucharczyk, Chair of Medical Imaging, Faculty of Medicine, University of Toronto and Clinical Director of the Tri-Hospital Resonance Centre, Toronto, Ontario, Canada)

maintain the necessary muscle tone to stand straight. The hamstrings received their name because it is common to string hams up with a hook around these muscle tendons. This also explains the expression "hamstringing the enemy" by slashing these tendons lateral and medial to the knees.

To test the hamstrings the person flexes the leg against resistance. If acting normally, these muscles—especially their tendons on each side of the popliteal fossa—should be prominent as they bend the knee. The short head of the biceps femoris arises from the lateral lip of the lower third of the linea aspera and supracondylar ridge of the femur. Whereas the hamstrings have a common nerve supply from the tibial division of the sciatic nerve, the short head of the biceps is innervated by the fibular division (Table 5.8).

Semitendinosus

As its name indicates, *this muscle is semitendinous.* It has a fusiform belly and a long, cordlike tendon that begins approximately two-thirds of the way down the thigh and attaches to the medial surface of the superior part of the tibia. Observe that it has a common proximal attachment to the ischial tuberosity with the semimembranosus and the long head of the biceps femoris.

Semimembranosus

This broad muscle is also aptly named because of the *flattened membranosus form of its proximal attachment to the ischial tuberosity.* The tendon of the semimembranosus forms around the middle of the thigh and descends to the posterior part of the medial tibial condyle.

Biceps Femoris

As its name indicates, *this fusiform muscle has two heads,* long and short. In the inferior part of the thigh, the long head becomes tendinous and is joined by the short head. The rounded tendon attaches to the head of the fibula and can easily be seen and felt as it passes the knee, especially when the knee is flexed against resistance (p. 569). Posteriorly, the long head of the biceps femoris crosses and provides protection for the *sciatic nerve* after it descends from the gluteal region into the posterior aspect of the thigh.

Injury to the Nerve Supply of the Biceps Femoris

Because each of the two heads of the biceps femoris has a different nerve supply (from different divisions of the sciatic nerve), a wound in the thigh may sever a nerve, paralyzing one head and not the other.

Variations in the Length of the Hamstrings

The length of the hamstrings varies. In some people they are not long enough to allow them to touch their toes when they flex their backs and keep their knees straight. Furthermore, such people have difficulty making a high kick. However, with practice they can stretch these muscles and tendons. In other people, the hamstrings are long and they can easily touch the floor with their palms and perform high kicks.

Hamstring Injuries

Hamstring strains (pulled and/or torn hamstrings) are common in persons who run and/or kick hard (e.g., in running, jumping, and quick-start sports such as baseball, basketball, football, and soccer). The violent muscular exertion required to excel in these sports may tear part of the proximal tendinous attachments of the hamstrings to the ischial tuberosity. Hamstring strains are twice as common as quadriceps strains (Levandowski and Difiori, 1994).

Usually thigh strains are accompanied by contusion and tearing of muscle fibers, resulting in rupture of the blood vessels supplying the muscles. The resultant *hematoma* is contained by the dense stockinglike fascia lata. Tearing of hamstring fibers is often so painful when the athlete moves or stretches the leg that the person falls and writhes in pain. These injuries often result from inadequate warming up before practice or competition. *Avulsion of the ischial tuberosity* (pp. 508–509) at the proximal attachment of the biceps femoris and semitendinosus ("Hurdler's injury") may result from forcible flexion of the hip with the knee extended (e.g., kicking a football). ⊙

Surface Anatomy of the Gluteal Region and Thigh

The skin of the gluteal region is usually thick and course, especially in men, whereas the skin of the thigh is relatively thin and loosely attached to the underlying subcutaneous tissue. A line joining the highest points of the *iliac crests* crosses the L4/L5 intervertebral disc and is a useful landmark when a lumbar puncture is performed (see Chapter 4). The *intergluteal cleft,* beginning inferior to the apex of the sacrum, is the deep groove between the buttocks. It extends as far superiorly as the 4th or 3rd sacral segment. ▶

▶ The coccyx is palpable in the superior part of the intergluteal cleft. The *posterior superior iliac spines* are located at the posterior extremities of the iliac crests and may be difficult to palpate; however, their position can always be located at the bottom of the permanent skin dimples approximately 3.75 cm from midline. A line joining these dimples, often more visible in women than in men, passes through the S2 spinous process and the middle of the sacroiliac joints.

The location of only two of the gluteal muscles can be observed. The *gluteus maximus* covering most structures in the gluteal region can be felt to contract when straightening up from bending over. The inferior edge of this large muscle is located just superior to the *gluteal fold*, which contains a variable amount of subcutaneous fat. The gluteal fold disappears when the hip joint is flexed. The degree of prominence of the gluteal fold alters in certain abnormal conditions, such as atrophy (wasting) of the gluteus maximus.

An imaginary line drawn from the coccyx to the ischial tuberosity indicates the inferior edge of the gluteus maximus. Another line drawn from the posterior superior iliac spine to a point slightly superior to the greater trochanter indicates the superior edge of this muscle. The *gluteal sulcus*, the skin crease inferior to the gluteal fold, delineates the buttock from the posterior aspect of the thigh. When the thigh is extended as in the photograph, the *ischial tuberosity* is covered by the inferior part of the gluteus maximus; however, it can be felt in the middle of the buttock by deep palpation through the muscle. The tuberosity is ▶

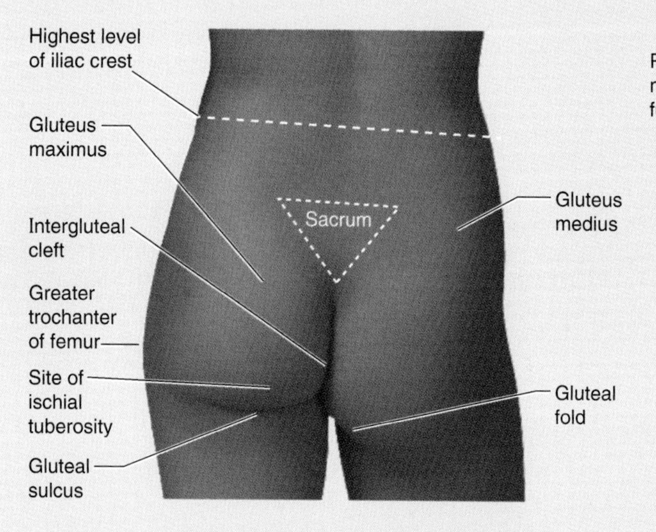

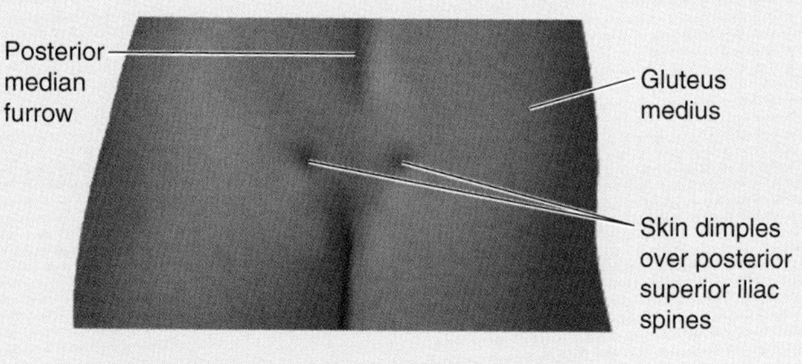

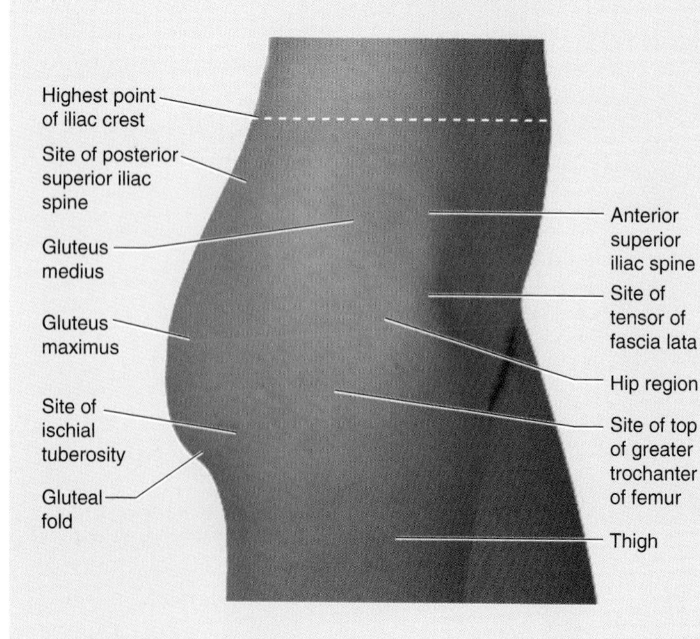

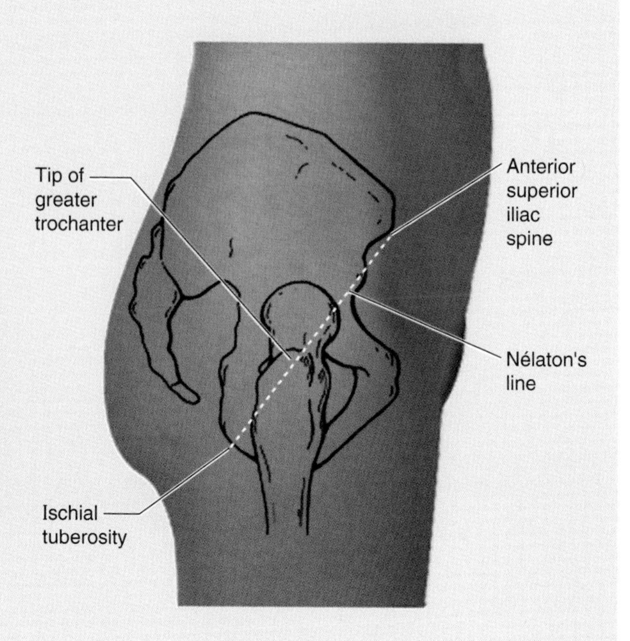

▶ easy to palpate when the thigh is flexed because the gluteus maximus slips superiorly off the tuberosity, which is then subcutaneous. Feel the ischial tuberosity as you bend to sit.

The superior part of the *gluteus medius* can be palpated between the superior part of the gluteus maximus and the iliac crest. The gluteus medius of one buttock can be felt when all the body weight shifts onto the ipsilateral limb—the one on the same side.

The *greater trochanter*, the most lateral bony point in the gluteal region, may be felt on the lateral aspect of the hip, especially its inferior part. It is easier to palpate when you passively abduct your lower limb to relax the gluteus medius and minimus. The top of the trochanter lies approximately a hand's breadth inferior to the tubercle of the iliac crest. The prominence of the trochanter increases when a dislocated hip causes atrophy of the gluteal muscles and displacement of the trochanter. A line drawn from the anterior superior iliac spine to the ischial tuberosity (*Nélaton's line*), passing over the lateral aspect of the hip region, normally passes over or near the top of the greater trochanter. The trochanter can be felt superior to this line in a person with a dislocated hip or a fractured femoral neck. The *lesser trochanter* is palpable with difficulty from the posterior aspect when the thigh is extended and rotated medially.

The *sciatic nerve*, probably the most important structure inferior to the piriformis, is represented by a line from a point midway between the greater trochanter and the ischial tuberosity down the middle of the posterior aspect of the thigh (*A*). The level of the bifurcation of the sciatic nerve into the tibial and common fibular nerves varies. ▶

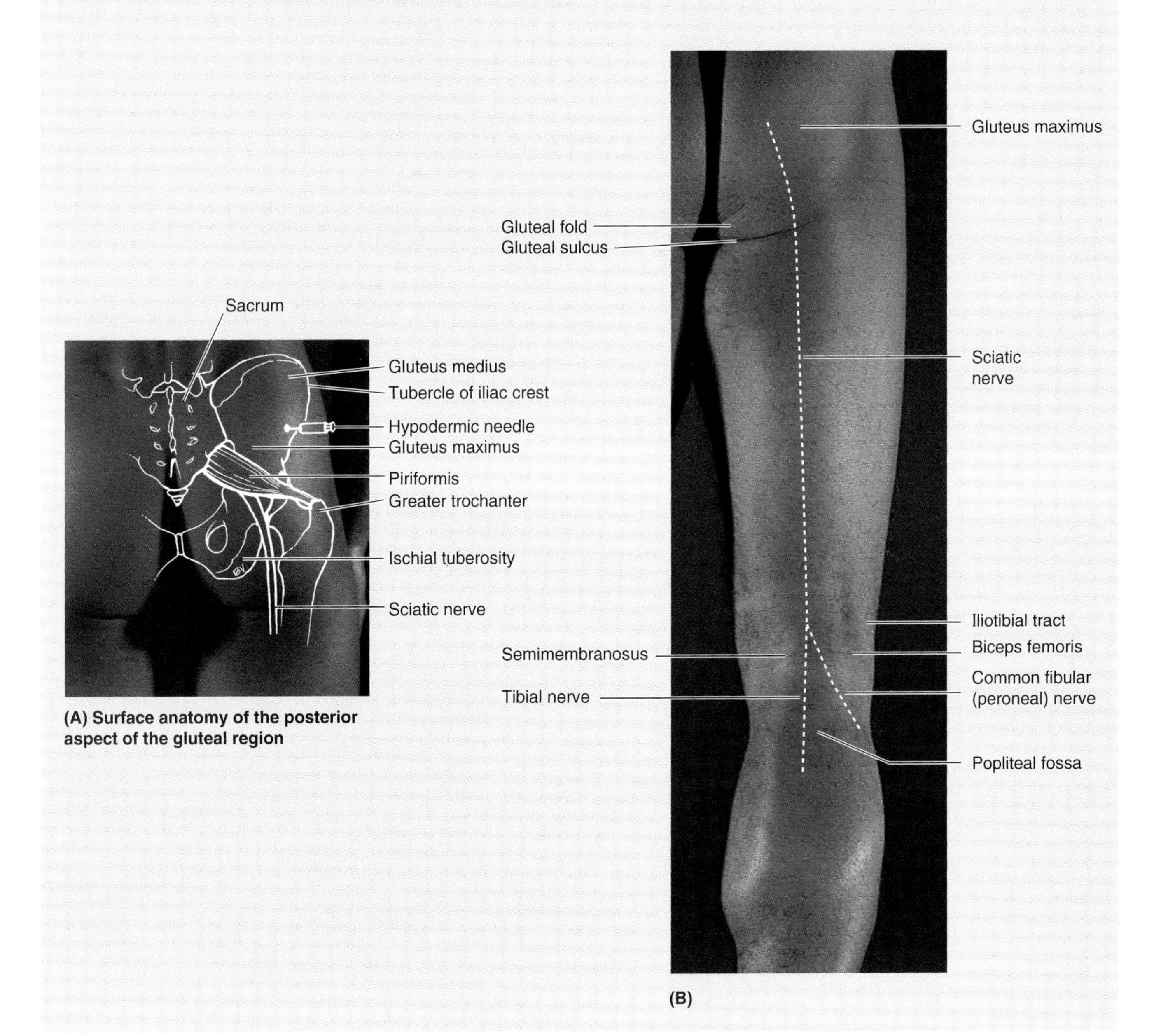

(A) Surface anatomy of the posterior aspect of the gluteal region

(B)

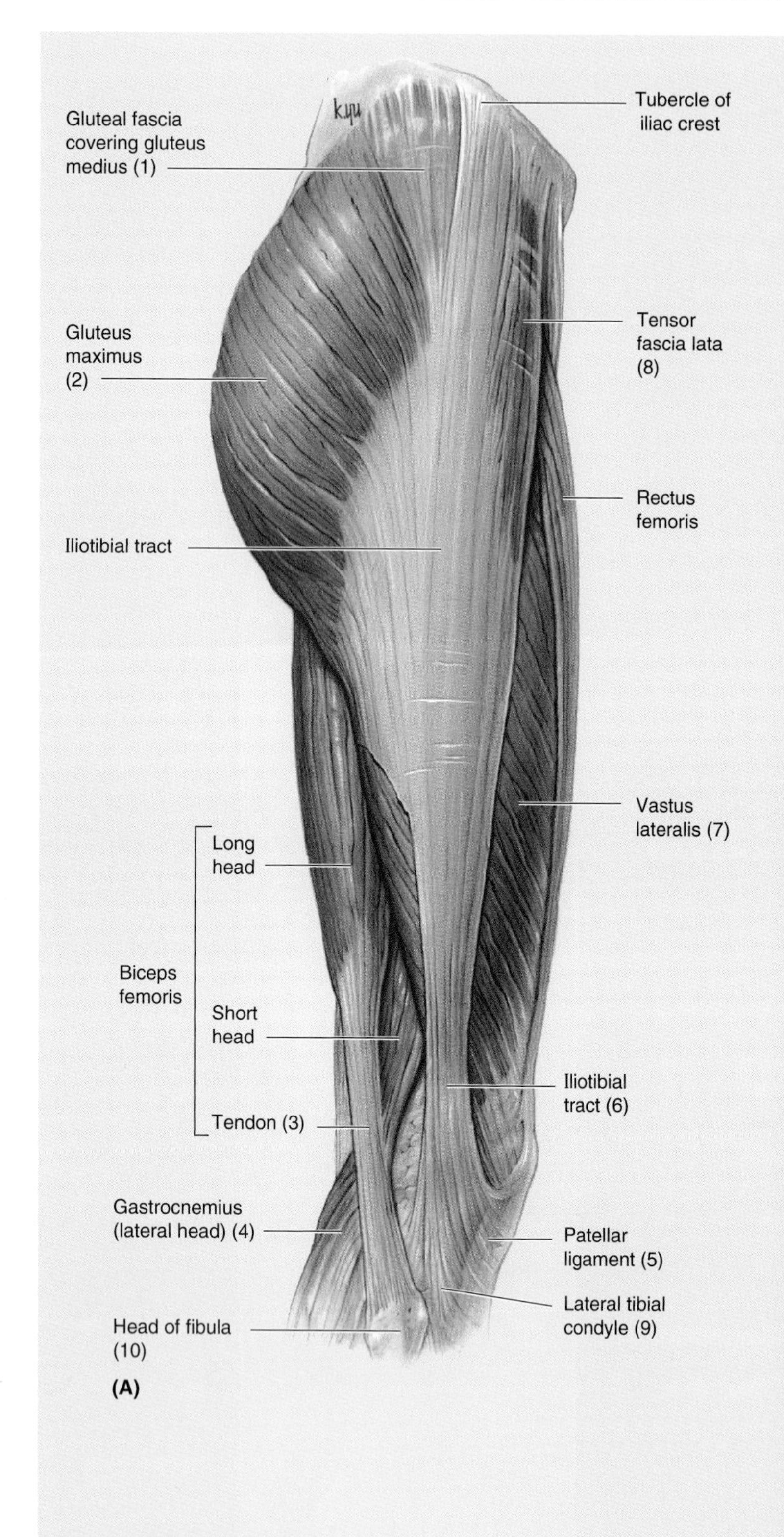

Gluteal fascia covering gluteus medius (1)
Tubercle of iliac crest
Gluteus maximus (2)
Tensor fascia lata (8)
Rectus femoris
Iliotibial tract
Vastus lateralis (7)
Long head
Biceps femoris
Short head
Tendon (3)
Iliotibial tract (6)
Gastrocnemius (lateral head) (4)
Patellar ligament (5)
Lateral tibial condyle (9)
Head of fibula (10)

(A)

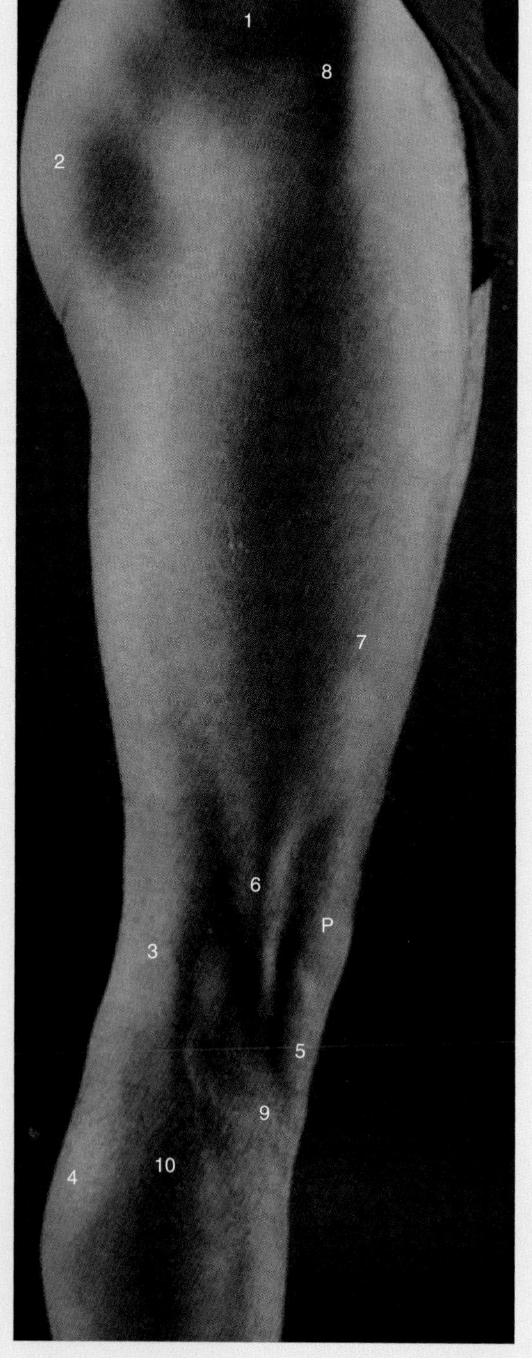

1
8
2
7
6
P
3
5
9
4
10

(B)

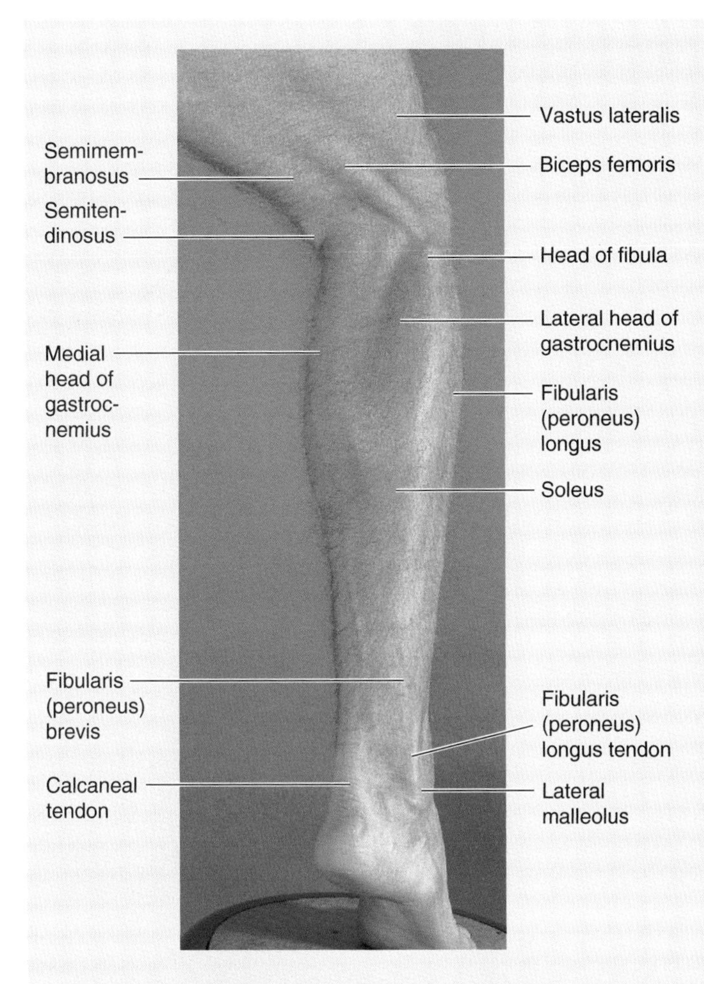

▶ The separation usually occurs in the inferior third of the thigh, but the division of the sciatic nerve may occur as it passes through the sciatic foramen. The *tibial nerve bisects the popliteal fossa* (*B,* p. 567), and the *common fibular nerve* follows the biceps femoris, which covers it. The sciatic nerve stretches when the thigh is flexed and the knee is extended, and it relaxes when the thigh is extended and the knee is flexed.

The hamstrings can be felt as a group as they arise from the ischial tuberosity and extend along the lateral and posterior aspects of the thigh (*A*). The *iliotibial tract,* the fibrous band that reinforces the fascia lata laterally, can be observed on the lateral aspect of the thigh as it passes to the *lateral tibial condyle.* While sitting down with your lower limb extended, raise your heel off the floor and feel the anterior border of the iliotibial tract passing a finger's breadth posterior to the lateral border of the patella (*B*). Note that the iliotibial tract is prominent and taut when the heel is raised and indistinct when the heel is lowered.

The *tendons of the hamstrings* can be observed and palpated at the borders of the popliteal fossa. The *biceps femoris* tendon is on the lateral side of the fossa. The most lateral tendon on the medial side and the most prominent tendon when the knee is flexed against resistance is the *semimembranosus tendon.* While sitting on a chair with your knee flexed, press your heel against the leg of the chair and feel your biceps tendon laterally and trace it to the head of the fibula. Also feel the *semitendinosus tendon* medially, which pulls away from the semimembranosus tendon that attaches to the superomedial part of the tibia.

In fairly muscular individuals, some of the bulky anterior thigh muscles can be observed. The prominent muscles are the quadriceps and sartorius, whereas laterally the tensor of the fascia lata is palpable as is the iliotibial tract to which this muscle attaches. Three of the four parts of the *quadriceps* are visible or can be approximated; the 4th part (vastus intermedius) is deep and almost hidden by the other muscles and cannot be palpated. The *rectus femoris* may be easily observed as a ridge passing down the thigh when the lower limb is raised from the floor while sitting. Observe the large bulges formed by the vastus lateralis and medialis at the knee. The *patellar ligament* is easily observed, especially in thin people, as a thick band running from the patella to the tibial tuberosity. You can also palpate the mass of loose fatty tissue (*infrapatellar fatpads*) on both sides of the patellar ligament.

On the medial aspect of the inferior part of the thigh, the gracilis and sartorius muscles form a well-marked prominence, which is separated by a depression from the large bulge formed by the vastus medialis. Deep in this depressed area, the large tendon of the adductor magnus can be palpated as it passes to its attachment to the adductor tubercle of the femur.

Measurements of the lower limb are taken to detect shortening (e.g., resulting from a femoral fracture). To make such measurements, compare the affected limb with the corresponding limb. Real limb shortening is detected by comparing the measurements from the anterior superior ▶

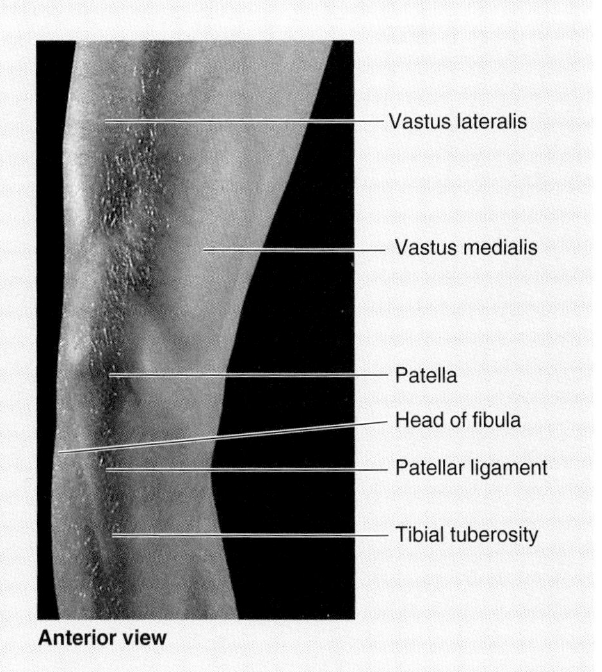

Anterior view

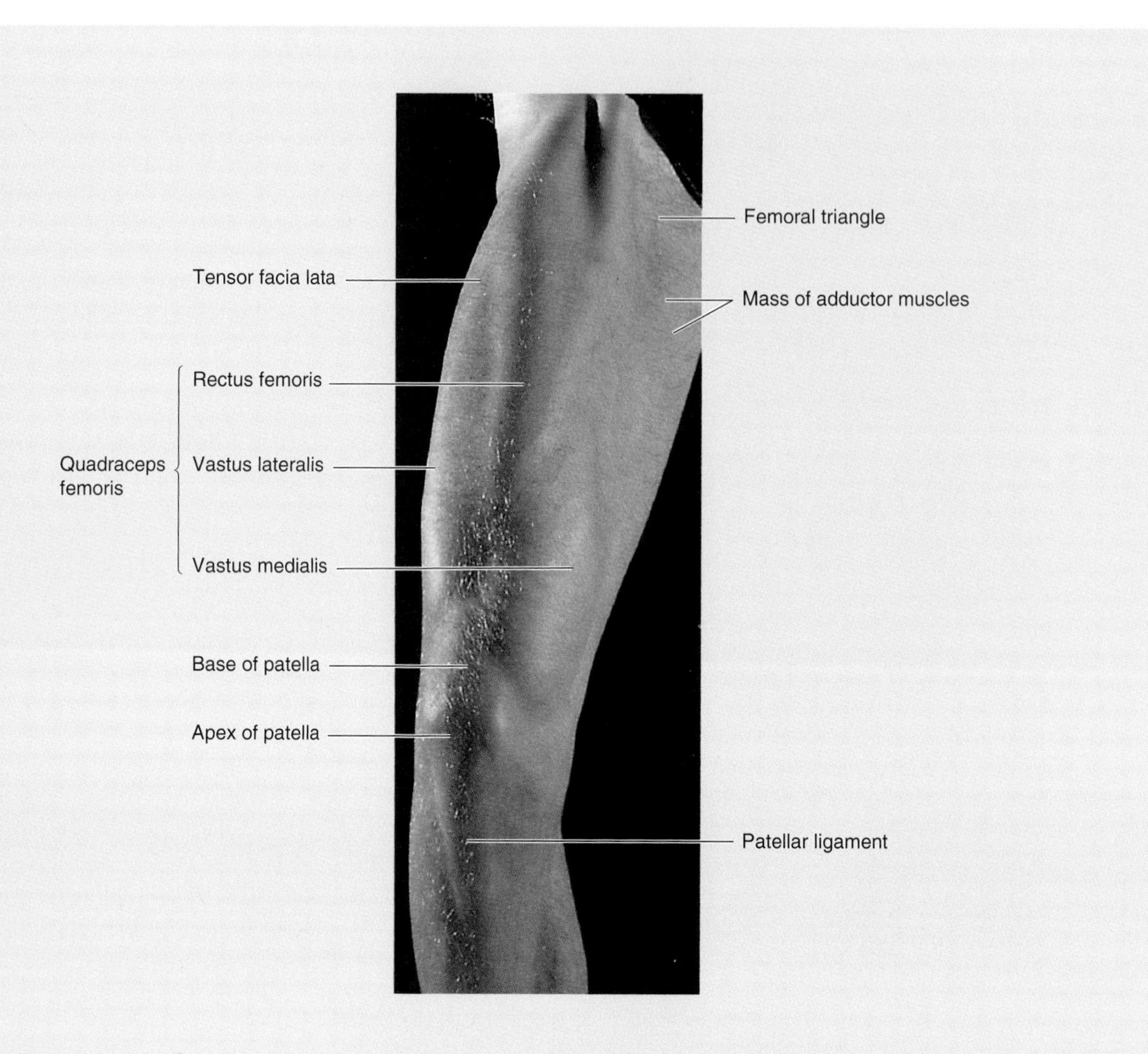

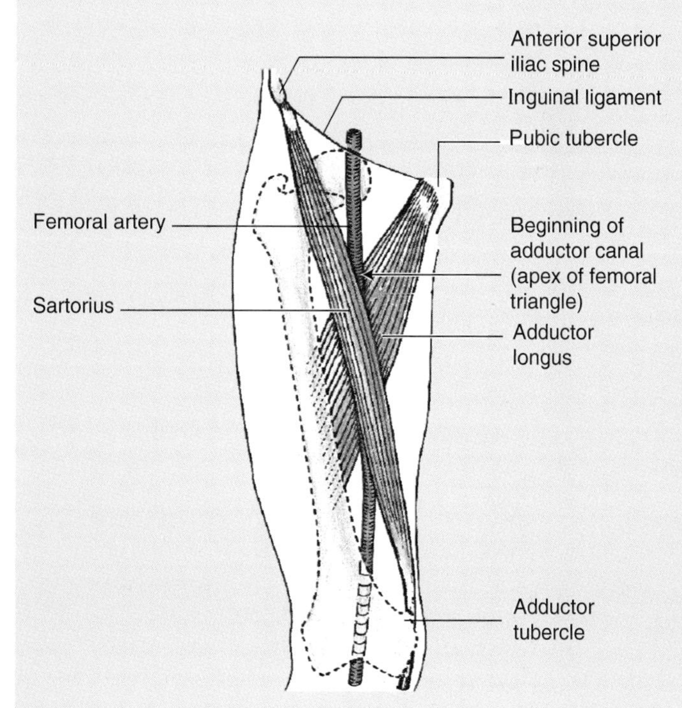

▶ iliac spine to the distal tip of the medial malleolus on both sides. To determine if the shortening is in the thigh, the measurement is taken from the top of the anterior superior iliac spine to the distal edge of the lateral femoral condyle on both sides. Keep in mind that small differences between the two sides—such as a difference of 1.25 cm in total length of the limb—may be normal.

The proximal two-thirds of a line drawn from the midpoint of the inguinal ligament to the adductor tubercle of the femur when the thigh is flexed, abducted, and rotated laterally represents the *course of the femoral artery*. The proximal third of the line represents this artery as it passes through the *femoral triangle*, whereas the middle third represents the artery while it is in the *adductor canal*. Approximately 3.75 cm along this line distal to the inguinal ligament, the *deep artery of the thigh* arises from the femoral artery. The *surface anatomy of the femoral vein*:

- At the base of the femoral triangle (indicated by inguinal ligament), the femoral vein is medial to the femoral artery
- At the apex of the femoral triangle, the vein is posterior to the artery ▶

- In the adductor canal, the vein is posterolateral to the artery.

The *femoral triangle* in the superoanterior aspect of the thigh is not a prominent surface feature in most people. When some people sit cross-legged, the sartorius and adductor longus stand out, delineating the femoral triangle. The surface anatomy of the femoral triangle is clinically important because of its contents. The *femoral artery* can be felt pulsating just inferior at the midinguinal point. When you palpate the femoral pulse, the

- *Femoral vein* is just medial
- *Femoral nerve* is a finger's breadth lateral
- *Femoral head* is just posterior.

The femoral artery runs a 5-cm superficial course through the femoral triangle before it is covered by the sartorius in the adductor canal.

The superficial position of the femoral artery and vein makes them vulnerable to injury (e.g, by a knife or bullet wound). Hemorrhage from the femoral artery and the deep artery of the thigh can be controlled by compressing the femoral artery against the superior pubic ramus and the femoral head. The *great saphenous vein* enters the thigh posterior to the medial femoral condyle and passes superiorly along a line from the adductor tubercle to the *saphenous opening*. The central point of this opening, where the great saphenous vein enters the femoral vein, is located 3.75 cm inferior and 3.75 cm lateral to the pubic tubercle.

Popliteal Fossa

The popliteal fossa is the diamond-shaped depression of the posterior aspect of the knee (Fig. 5.29). The fossa is bounded superiorly by the hamstrings and inferiorly by the two heads of the gastrocnemius and the plantaris. All important vessels and nerves from the thigh to the leg pass through this fossa. *The popliteal fossa is formed*:

- *Superolaterally* by the biceps femoris *(superolateral border)*
- *Superomedially* by the semimembranosus, lateral to which is the semitendinosus (*superomedial border)*
- *Inferolaterally* and *inferomedially* by the lateral and medial heads of the gastrocnemius, respectively (*inferolateral and inferomedial borders*)
- *Posteriorly* by skin and fascia (roof)
- *Anteriorly* by the popliteal surface of the femur, the oblique popliteal ligament, and the popliteal fascia over the popliteus, which together form the floor of the fossa.

The contents of the popliteal fossa (Figs. 5.29*B* and 5.30) *are:*

- Small saphenous vein
- Popliteal arteries and veins
- Tibial and common fibular nerves
- Posterior cutaneous nerve of thigh
- Popliteal lymph nodes and lymphatic vessels.

Fascia of the Popliteal Fossa

The **superficial popliteal fascia** contains fat, the small saphenous vein (although it may penetrate the deep fascia at a more inferior level), and three cutaneous nerves: the terminal branch(es) of the *posterior femoral cutaneous nerve* and the *medial and lateral sural cutaneous nerves*. The **deep popliteal fascia** is a strong sheet of deep fascia that forms a protective covering for neurovascular structures passing from the thigh through the popliteal fossa to the leg. When the leg extends, the popliteal fascia stretches and the semimembranosus moves laterally, providing further protection to the contents of the popliteal fossa.

Blood Vessels in the Popliteal Fossa

The popliteal artery, the continuation of the femoral artery (Figs. 5.29*B* and 5.30), begins when this artery passes through the adductor hiatus. The **popliteal artery** passes inferolaterally through the popliteal fossa and ends at the inferior border of the popliteus by dividing into the anterior and posterior tibial arteries. The deepest structure in the fossa, the popliteal artery, runs close to the articular capsule of the knee joint. Five genicular branches of the popliteal artery supply the articular capsule and ligaments of the knee joint. The **genicular arteries** are the lateral superior, medial superior, middle, lateral inferior, and medial inferior genicular arteries. They participate in the formation of the **genicular anastomosis**, a network of vessels around the knee (Fig. 5.31). Other contributors to this important anastomosis are the:

- Descending genicular branch of the femoral artery, superomedially
- Descending branch of the lateral femoral circumflex artery, superolaterally
- Anterior recurrent branch of the anterior tibial artery, inferolaterally.

The muscular branches of the popliteal artery supply the hamstring, gastrocnemius, soleus, and plantaris muscles. The superior muscular branches of the popliteal artery have clinically important anastomoses with the terminal part of the deep femoral and gluteal arteries.

The **popliteal vein** is formed at the distal border of the popliteus. Throughout its course, the popliteal vein is close to the popliteal artery and lies superficial to and in the same fibrous sheath as the artery. The popliteal vein is at first pos-

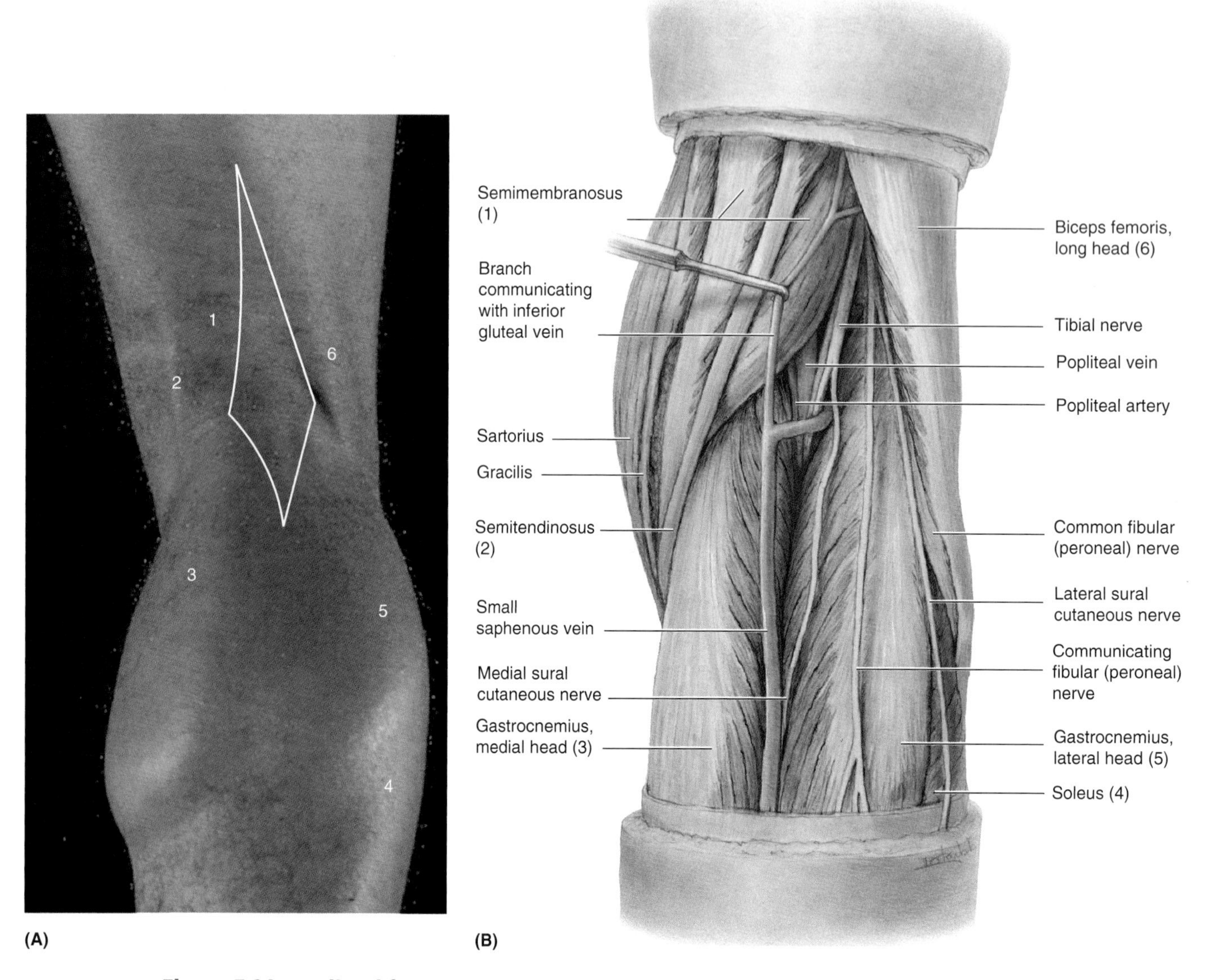

Figure 5.29. Popliteal fossa. A. Surface anatomy. The numbers refer to structures labeled in (**B**) and in Figure 5.30. The diamond-shaped gap in the muscles overlying the fossa is outlined. **B.** Superficial dissection of the fossa. Observe the common fibular nerve following the posterior border of the biceps femoris, and here giving off two cutaneous branches.

teromedial to the artery and lateral to the tibial nerve. More superiorly, the popliteal vein lies posterior to the artery, between this vessel and the overlying tibial nerve. The popliteal vein, which has several valves, ends at the *adductor hiatus*, where it becomes the *femoral vein*. The **small saphenous vein** passes from the posterior aspect of the lateral malleolus to the popliteal fossa, where it pierces the deep popliteal fascia and enters the popliteal vein.

Nerves in the Popliteal Fossa

The **sciatic nerve** usually ends at the superior angle of the popliteal fossa by dividing into the tibial and common fibular nerves (Fig. 5.32). The **tibial nerve**—the medial, larger terminal branch of the sciatic nerve—is the most superficial of the three main central components of the popliteal fossa (i.e., nerve, vein, and artery); however, it is deep and in a protected position. The tibial nerve bisects the fossa as it passes from its superior to its inferior angle. While in the fossa, the tibial nerve gives branches to soleus, gastrocnemius, plantaris, and popliteus muscles. A *medial sural cutaneous* nerve also derives from the tibial nerve, which joins the *lateral sural cutaneous nerve* at a highly variable level to form the **sural nerve.** This nerve supplies the lateral side of the leg and ankle.

The **common fibular nerve**—the lateral, smaller terminal branch of the sciatic nerve—begins at the superior angle of the popliteal fossa and follows closely the medial border of the biceps femoris and its tendon along the superolateral boundary of the popliteal fossa (Fig. 5.32). The common fibular nerve leaves the fossa by passing superficial to the lateral head

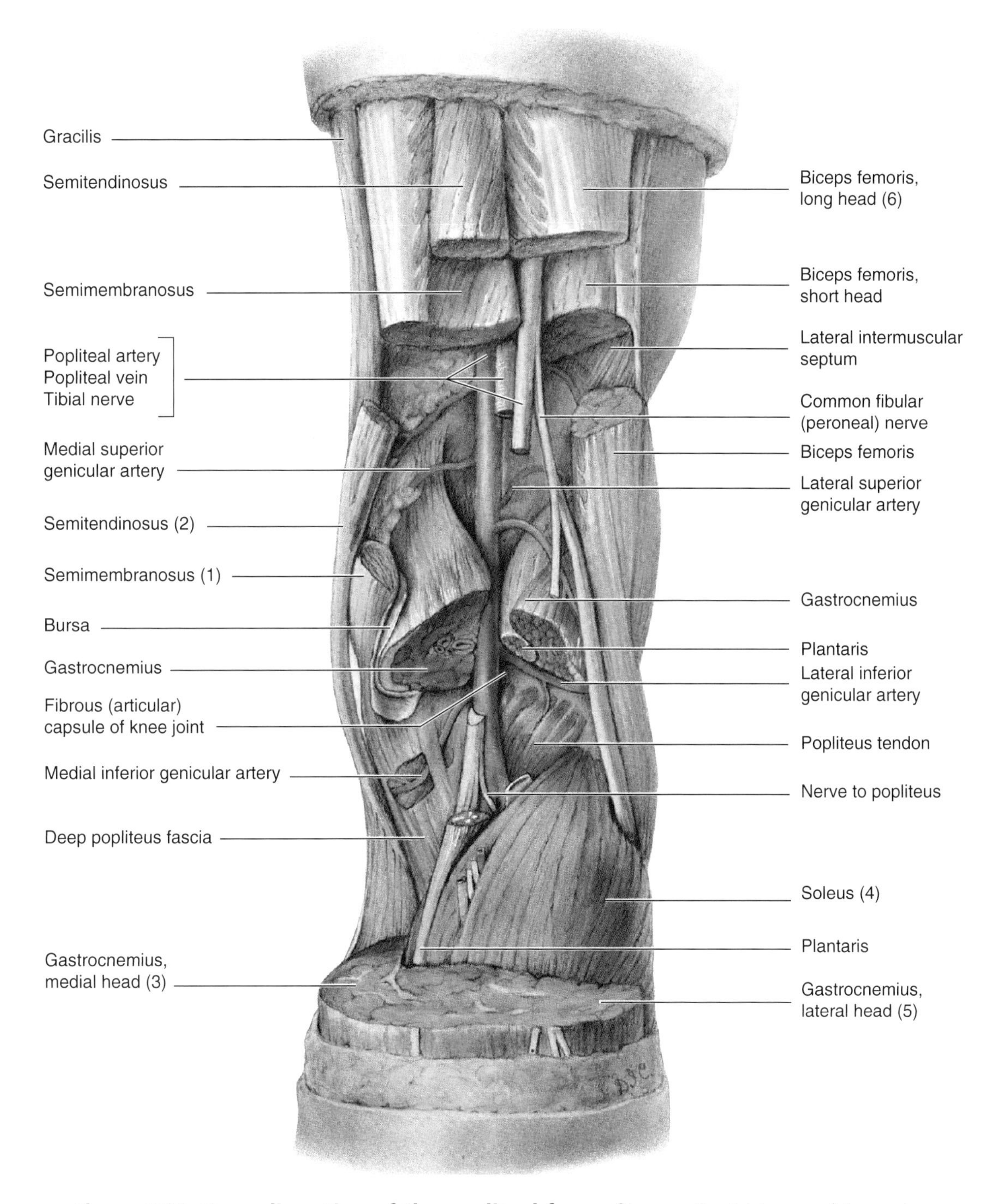

Figure 5.30. Deep dissection of the popliteal fossa. Observe the thickness of the various muscles. Observe also the popliteal artery lying on the floor of the fossa. Note that the floor of the fossa, which extends superiorly to the divergence of the linea aspera of the femur to surround the popliteal face of the femur and inferiorly to the soleal line of the tibia, is much larger than the superficial outline of the popliteal fossa indicated in Figure 5.29*A*.

of the gastrocnemius and then passes over the posterior aspect of the head of the fibula. *The common fibular nerve winds around the fibular neck, where it is susceptible to injury.* Here it divides into its terminal branches.

The posterior cutaneous nerve of the thigh arises from the posterior divisions of the ventral rami of S1 and S2 and the anterior divisions of S2 and S3 nerves. *This nerve supplies more skin than any other cutaneous nerve.* The posterior femoral cutaneous nerve leaves the pelvis with the inferior gluteal nerve and vessels and the sciatic nerve. Its fibers from the posterior divisions of S1 and S2 nerves supply the skin of the inferior part of the buttock; those from the anterior divisions of S2

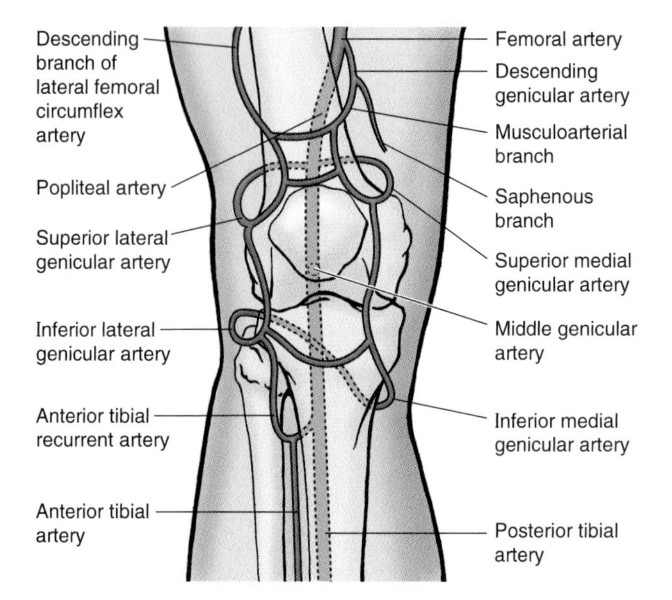

Figure 5.31. Genicular anastomosis around the knee. Observe the many arteries contributing to this anastomosis, which forms an important collateral circulation for bypassing the main (popliteal) vessels when the knee has been maintained too long in a fully flexed position or when the vessels are narrowed or occluded.

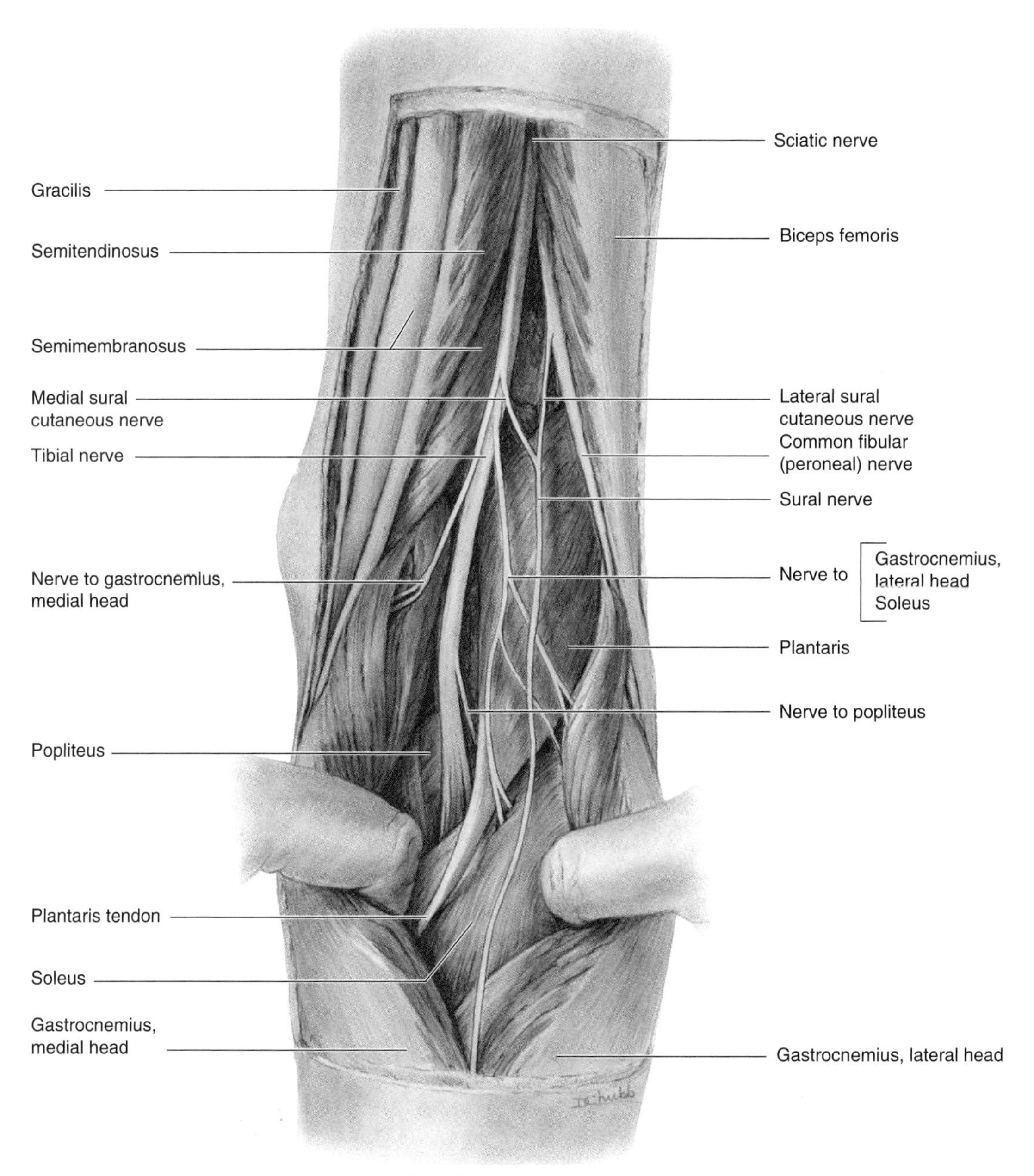

and S3 supply the skin of the perineum; other branches continue inferiorly where they supply the skin of the posterior thigh and proximal part of the leg.

Popliteal Abscesses and Tumors

Because the deep popliteal fascia is strong and limits expansion, pain from an abscess or tumor in the popliteal fossa is usually severe. Popliteal abscesses tend to spread superiorly and inferiorly because of the toughness of the popliteal fascia.

Popliteal Pulse

Because the popliteal artery is deep, it may be difficult to feel the *popliteal pulse*. Palpation of this pulse is commonly performed with the person in the prone position and with the knee flexed to relax the popliteal fascia and hamstrings. The pulsations are best felt in the inferior part of the fossa where the popliteal artery is related to the tibia. Weakening or loss of the popliteal pulse is a sign of a femoral artery obstruction.

Popliteal Aneurysm

A *popliteal aneurysm* (dilation of the popliteal artery) usually causes edema and pain in the popliteal fossa. If the femoral artery has to be ligated, blood can bypass the occlusion through the genicular anastomoses and reach the popliteal artery distal to the ligation.

Injury to the Tibial Nerve

Injury to the tibial nerve is uncommon because of its deep and protected position in the popliteal fossa; however, the nerve may be injured by deep lacerations or wounds in the popliteal fossa. Posterior dislocation of the knee joint may damage the tibial nerve. Severance or damage of the tibial nerve produces paralysis of the flexor muscles in the leg and the intrinsic muscles in the sole of the foot. Persons with a tibial nerve injury are unable to plantarflex their ankle or flex their toes. Loss of sensation also occurs on the sole of the foot. ⊙

Lymph Nodes in the Popliteal Fossa

The **superficial popliteal lymph nodes** are usually small and lie in the popliteal fat. A lymph node lies at the termination of the small saphenous vein and receives lymph from the lymphatic vessels that accompany this vein. The **deep popliteal lymph nodes** receive lymph from the knee joint and the lymphatic vessels that accompany the arteries of the leg. The lymphatic vessels from the popliteal lymph nodes follow the femoral vessels to the *deep inguinal lymph nodes*.

Leg

The leg contains the tibia and fibula, bones that connect the knee and ankle. The **tibia**, the weightbearing bone, is the larger and stronger of the two bones. The tibia articulates with the femoral condyles superiorly and the talus inferiorly. The long, slender **fibula** does not bear weight, and its superior end plays no part in the knee joint; however, it aids in the formation of the ankle joint. The fibula lies posterolateral to the tibia to which it is attached by the **interosseous membrane** (Fig. 5.33). The fibula affords attachment for several muscles and provides stability to the ankle joint.

The leg is divided into three fascial compartments—anterior, lateral, and posterior—by the anterior and posterior *intermuscular septa* and the *interosseous membrane*. The anterior septum separates the anterior and lateral leg muscles, and the posterior septum separates the lateral and posterior muscles; thus, each group has its own compartment.

Compartment Syndromes in the Leg

Because the septa forming the boundaries of the leg compartments are strong, trauma to muscles in the compartments may produce hemorrhage, edema, and inflammation of the muscles. With arterial bleeding, the pressure may reach levels high enough to compress structures in the compartment(s) concerned. Structures distal to the compressed area may become ischemic and permanently injured (e.g., loss of motor function in muscles whose blood supply and/or innervation is affected). Loss of distal pulses is an obvious sign of arterial compression, as is a lowering of the temperature of tissues distal to the compression. A *fasciotomy* (incision of a fascial septum) may be performed to relieve the pressure in the compartment(s) concerned. ⊙

Figure 5.32. Nerves of the popliteal fossa. The two heads of the gastrocnemius muscle are pulled forcibly apart. The sciatic nerve separates into its components at the apex of the popliteal fossa. The common fibular nerve courses along the medial border of the biceps femoris. Observe that all the motor branches arising from the tibial nerve, except one, arise from the lateral side; consequently, it is safer to dissect on the medial side. The level at which the medial and lateral sural nerves merge to form the sural nerve—occurring high here—is highly variable; it may even occur at the level of the ankle.

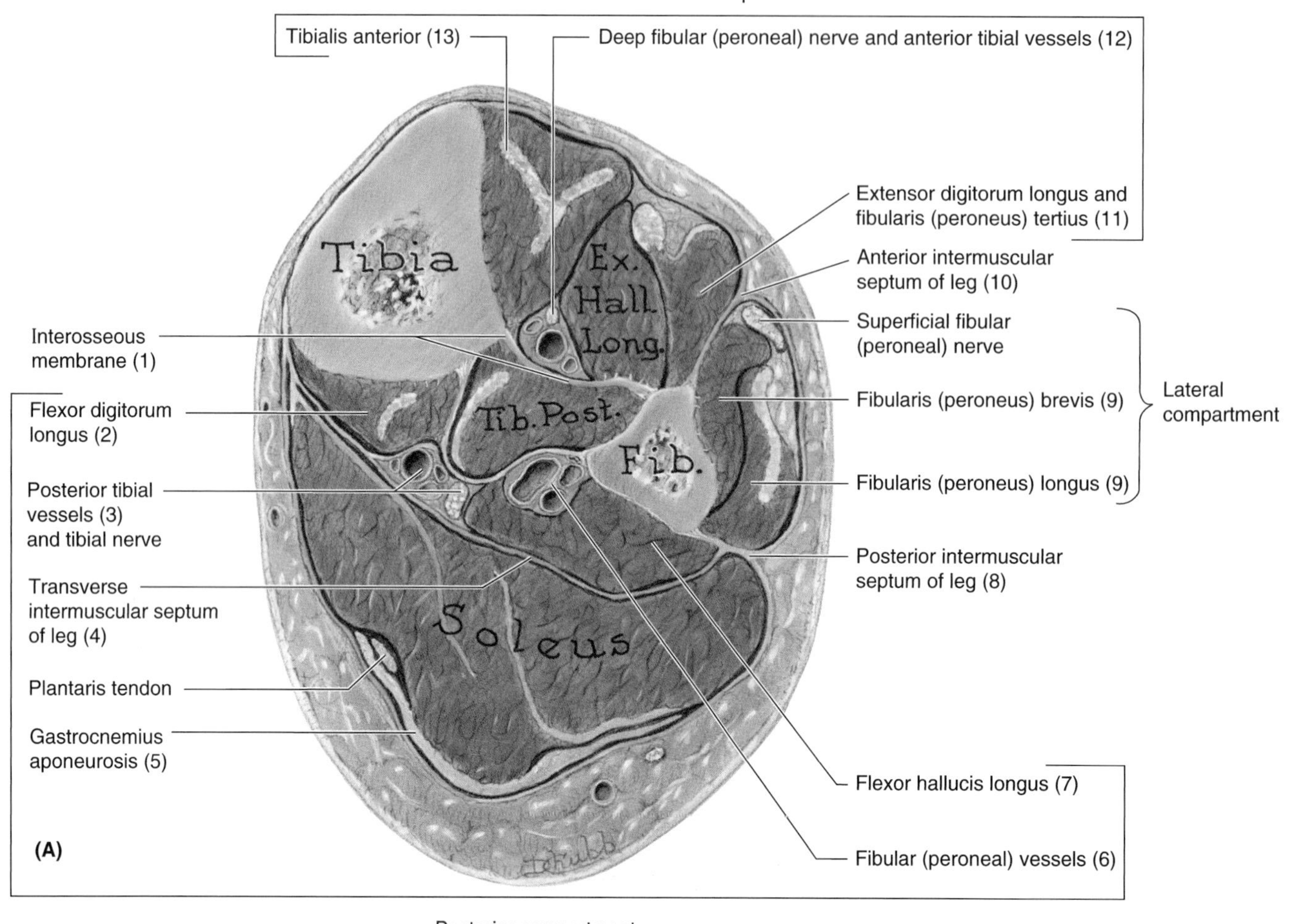

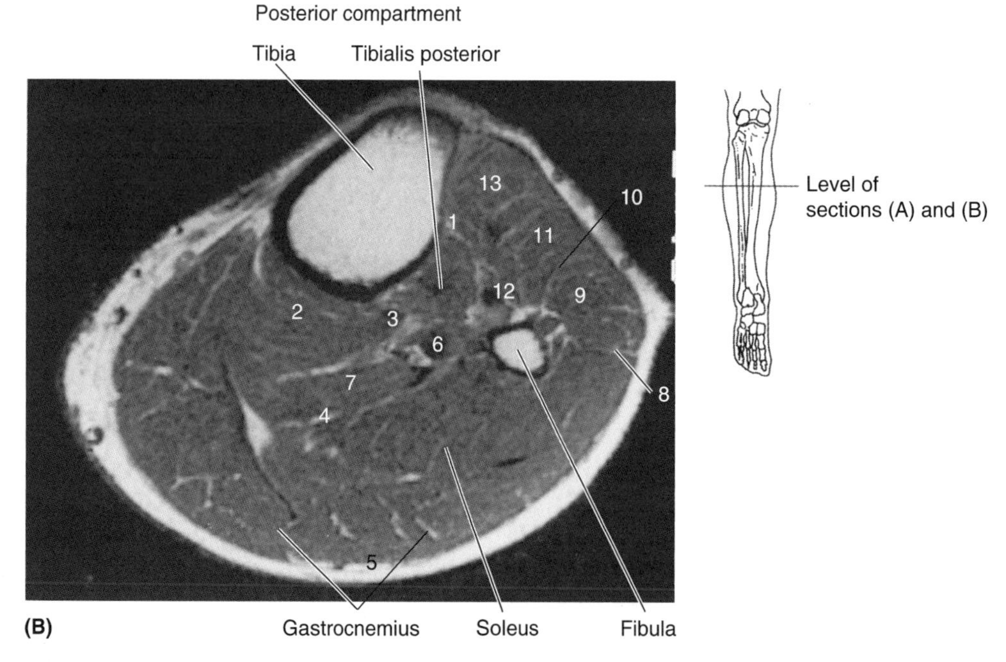

Figure 5.33. Transverse section and an MRI of the leg at the midcalf level. A. Observe the *anterior* (*dorsiflexor extensor*) *compartment* containing four muscles: tibialis anterior, extensor digitorum longus, fibularis (peroneus) tertius, and extensor hallucis longus; *the lateral* (*fibular*) *compartment* containing two muscles: fibularis longus and brevis; and the *posterior* (*flexor*) *compartment* containing seven muscles: gastrocnemius, soleus, and plantaris in the superficial group and popliteus, flexor digitorum longus, flexor hallucis longus, and tibialis posterior in the deep group. **B.** MRI of the leg. The numbers refer to structures labeled in (**A**). (Courtesy of Dr. W. Kucharczyk, Chair of Radiology, Faculty of Medicine, University of Toronto and Clinical Director of the Tri-Hospital Resonance Centre, Toronto, Ontario, Canada)

Table 5.9. **Muscles of the Anterior and Lateral Leg**

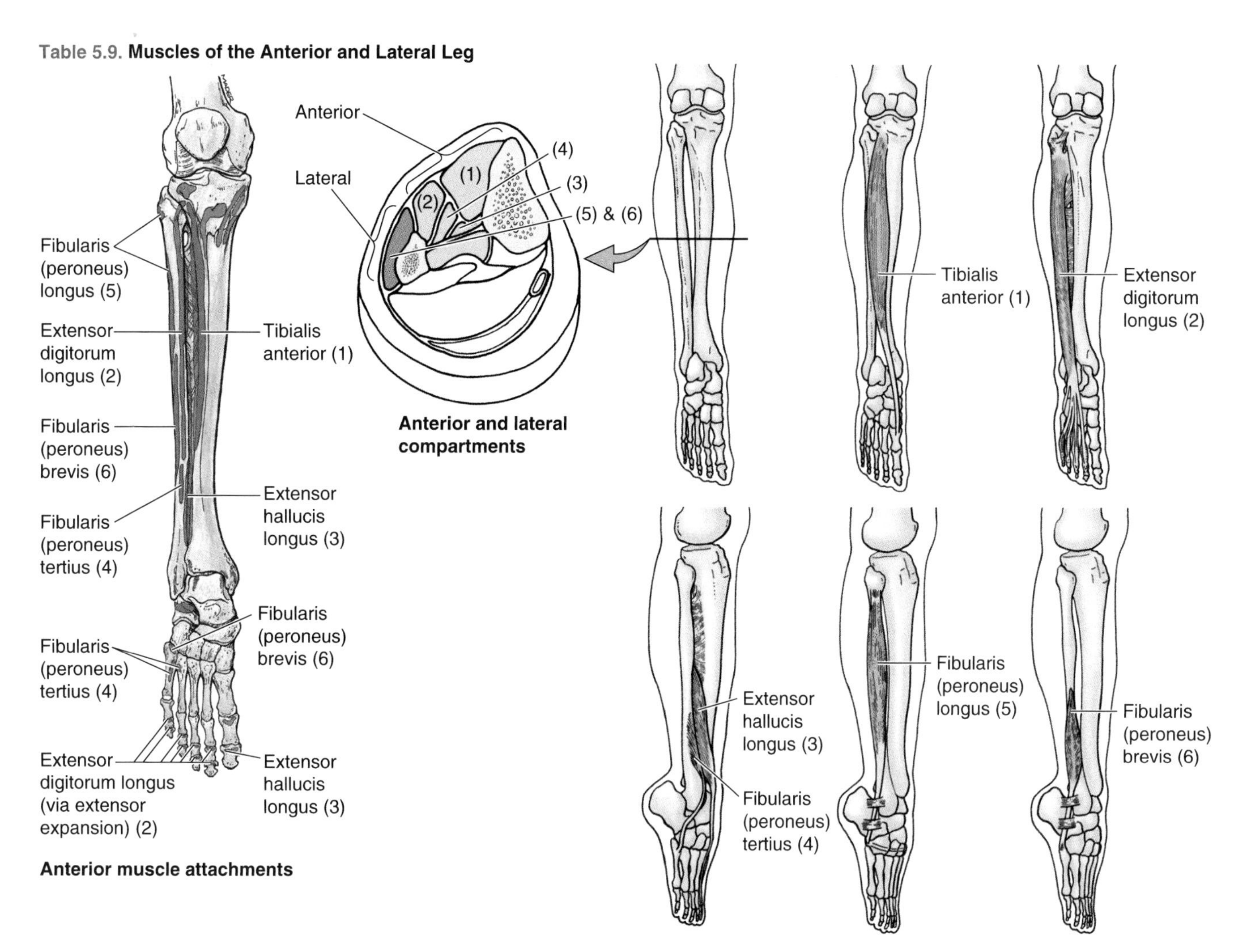

Muscle	Proximal Attachment	Distal Attachment	Innervation	Main Action
Anterior compartment				
Tibialis anterior (1)	Lateral condyle and superior half of lateral surface of tibia and interosseous membrane	Medial and inferior surfaces of medial cuneiform and base of 1st metatarsal	Deep fibular (peroneal) nerve (**L4** and L5)	Dorsiflexes ankle and inverts foot
Extensor digitorum longus (2)	Lateral condyle of tibia and superior three-fourths of medial surface of the fibula and interosseous membrane	Middle and distal phalanges of lateral four digits	Deep fibular (peroneal) nerve (L5 and S1)	Extends lateral four digits and dorsiflexes ankle
Extensor hallucis longus (3)	Middle part of anterior surface of fibula and interosseous membrane	Dorsal aspect of base of distal phalanx of great toe (hallux)		Extends great toe and dorsiflexes ankle
Fibularis (peroneus) tertius (4)	Inferior third of anterior surface of fibula and interosseous membrane	Dorsum of base of 5th metatarsal		Dorsiflexes ankle and aids in eversion of foot
Lateral compartment				
Fibularis (peroneus) longus (5)	Head and superior two-thirds of lateral surface of fibula	Base of 1st metatarsal and medial cuneiform	Superficial fibular (peroneal) nerve (**L5, S1**, and S2)	Everts foot and weakly plantarflexes ankle
Fibularis (peroneus) brevis (6)	Inferior two-thirds of lateral surface of fibula	Dorsal surface of tuberosity on lateral side of base of 5th metatarsal		

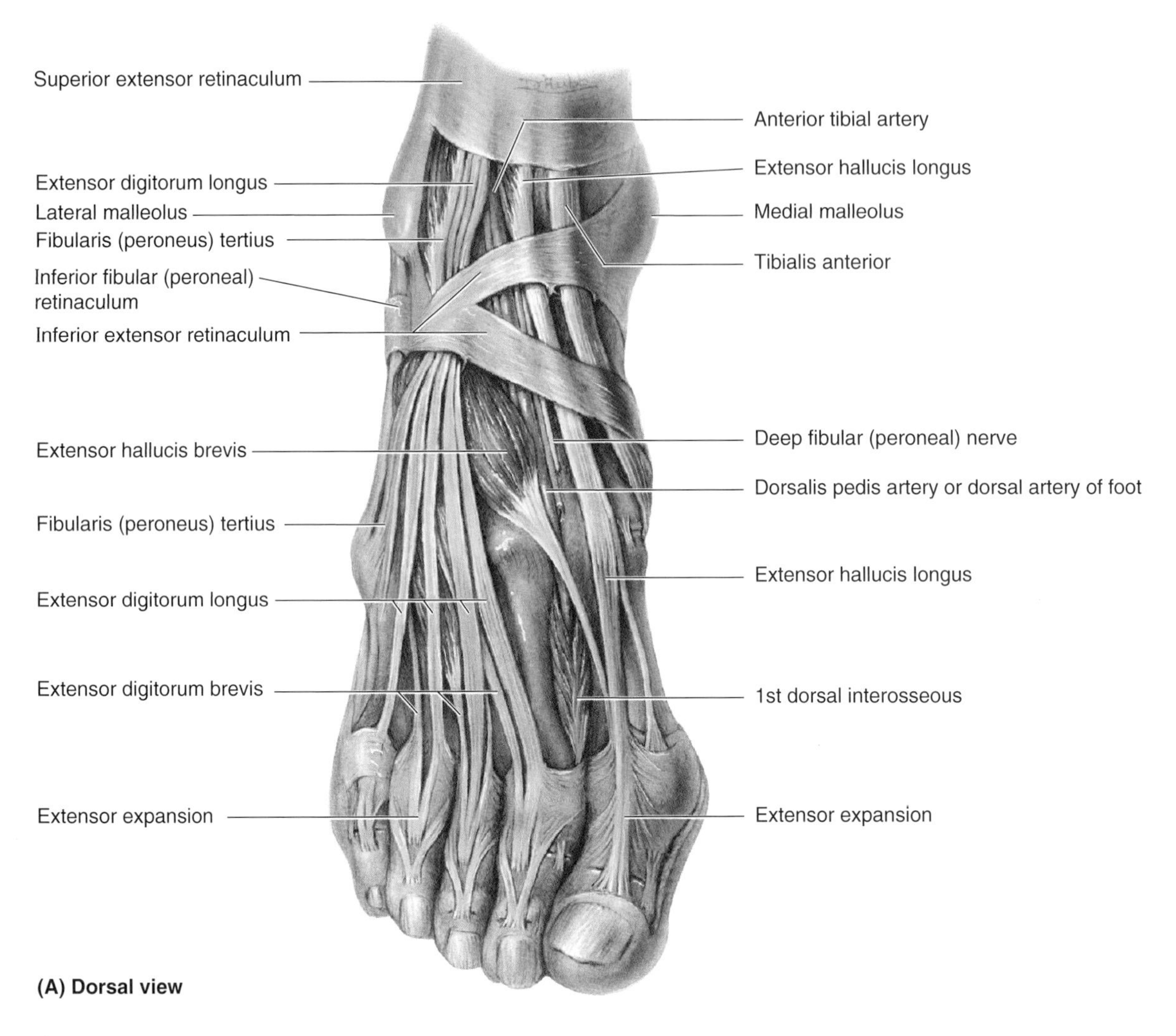

(A) Dorsal view

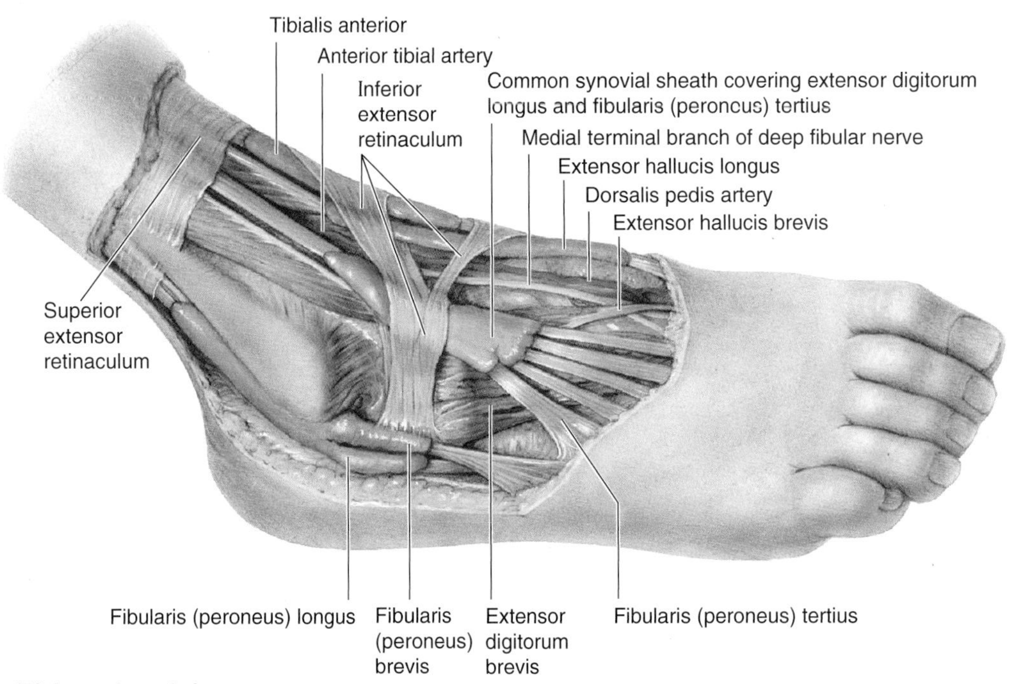

(B) Anterolateral view

Anterior Compartment of the Leg

The anterior compartment—the *dorsiflexor extensor compartment*—is located anterior to the interosseous membrane, between the lateral surface of the tibial body and the anterior intermuscular septum. The anterior compartment is bounded anteriorly by crural fascia and skin.

Muscles in the Anterior Compartment

The four muscles in the anterior compartment (Fig. 5.33, *A* and *B*, Table 5.9) are the:

- Tibialis anterior
- Extensor digitorum longus
- Extensor hallucis longus
- Fibularis tertius.

These muscles are mainly dorsiflexors of the ankle joint and extensors of the toes.

The **superior extensor retinaculum** is a strong, broad band of deep fascia (Fig. 5.34*A*), passing from the fibula to the tibia, proximal to the malleoli. It binds down the tendons of muscles in the anterior compartment, preventing them from bowstringing anteriorly during dorsiflexion of the ankle joint.

The **inferior extensor retinaculum** (Fig. 5.35*A*), a Y-shaped band of deep fascia, attaches laterally to the anterosuperior surface of the calcaneus. It forms a strong loop around the tendons of the fibularis tertius and the extensor digitorum longus muscles.

Tibialis Anterior. The tibialis anterior is a slender muscle that lies against the lateral surface of the tibia (Fig. 5.35, Table 5.9). Its tendon appears in the lower third of the leg on the anterior surface of the tibia. The tendon passes within its own synovial sheath deep to the superior extensor retinaculum (Fig. 5.34*B*) and the inferior extensor retinaculum to its attachment on the medial side of the foot. The tibialis anterior is the strongest dorsiflexor and invertor of the foot. *To test the tibialis anterior*, the foot is dorsiflexed against resistance; if acting normally, its tendon can be seen and palpated.

Extensor Digitorum Longus. The extensor digitorum longus is the most lateral of the anterior leg muscles (Fig. 5.35). A small part of the proximal attachment of the muscle is to the lateral tibial condyle; however, most of it attaches to the medial surface of the fibula and the superior part of the anterior surface of the interosseous membrane (Table 5.9). The muscle becomes tendinous superior to the ankle, and its four tendons attach to the phalanges of the lateral four toes. *To test the extensor digitorum longus*, the lateral four toes are dorsiflexed against resistance; if acting normally, the tendons can be seen and palpated.

A *common synovial sheath* surrounds the four tendons of the extensor digitorum longus as they diverge on the dorsum of the foot and pass to their distal attachments (Fig. 5.34*B*). Each tendon forms a membranous *extensor expansion* (dorsal aponeurosis) over the dorsum of the proximal phalanx of the toe, which divides into two lateral slips and one central slip (Fig. 5.34*A*). The central slip inserts into the base of the middle phalanx and the lateral slips converge to insert into the base of the distal phalanx.

Extensor Hallucis Longus. The extensor hallucis longus is a thin muscle that lies deep at its superior attachment to the fibula and interosseous membrane. It rises to the surface between the tibialis anterior and extensor digitorum longus (Figs. 5.34 and 5.35). *To test the extensor hallucis longus*, the great toe is dorsiflexed against resistance; if acting normally, its tendon can be seen and palpated.

Fibularis Tertius. The fibularis tertius is a separated part of the extensor digitorum longus, which shares its synovial sheath (Figs. 5.34 and 5.35). The extensor digitorum longus and fibularis tertius are fused at their proximal attachments; however, the tendon of the fibularis tertius attaches to the 5th metatarsal—not to a phalanx (Table 5.9). The fibularis tertius is not always present.

Nerve in the Anterior Compartment

The **deep fibular nerve**—the nerve of the anterior compartment (Figs. 5.33*A* and 5.35, Table 5.10)—is one of the two terminal branches of the common fibular nerve. It arises between the fibularis longus muscle and the neck of the fibula. It accompanies the anterior tibial artery, first between the tibialis anterior and extensor digitorum longus and then between the tibialis anterior and extensor hallucis longus.

Artery in the Anterior Compartment

The **anterior tibial artery** supplies structures in the anterior compartment (Figs. 5.33*A* and 5.35*A*, Table 5.11). The smaller terminal branch of the popliteal artery, the anterior tibial artery begins at the inferior border of the popliteus and passes anteriorly through a gap in the superior part of the interosseous membrane and descends on the anterior surface of this membrane between the tibialis anterior and extensor digitorum longus. It ends at the ankle joint, midway between the malleoli, where it becomes the dorsalis pedis artery or dorsal artery of the foot (L. dorsalis pedis artery).

Figure 5.34. Dissections of the right foot. Extensor and fibular retinacula of the ankle and the muscles and tendons of the dorsum of the foot are demonstrated. **A.** The vessels and nerves are cut short. At the ankle, observe that the vessels and the deep fibular nerve lie midway between the malleolus and have tendons of two muscles on each side of them. **B.** Observe the synovial sheaths of the tendons of the ankle.

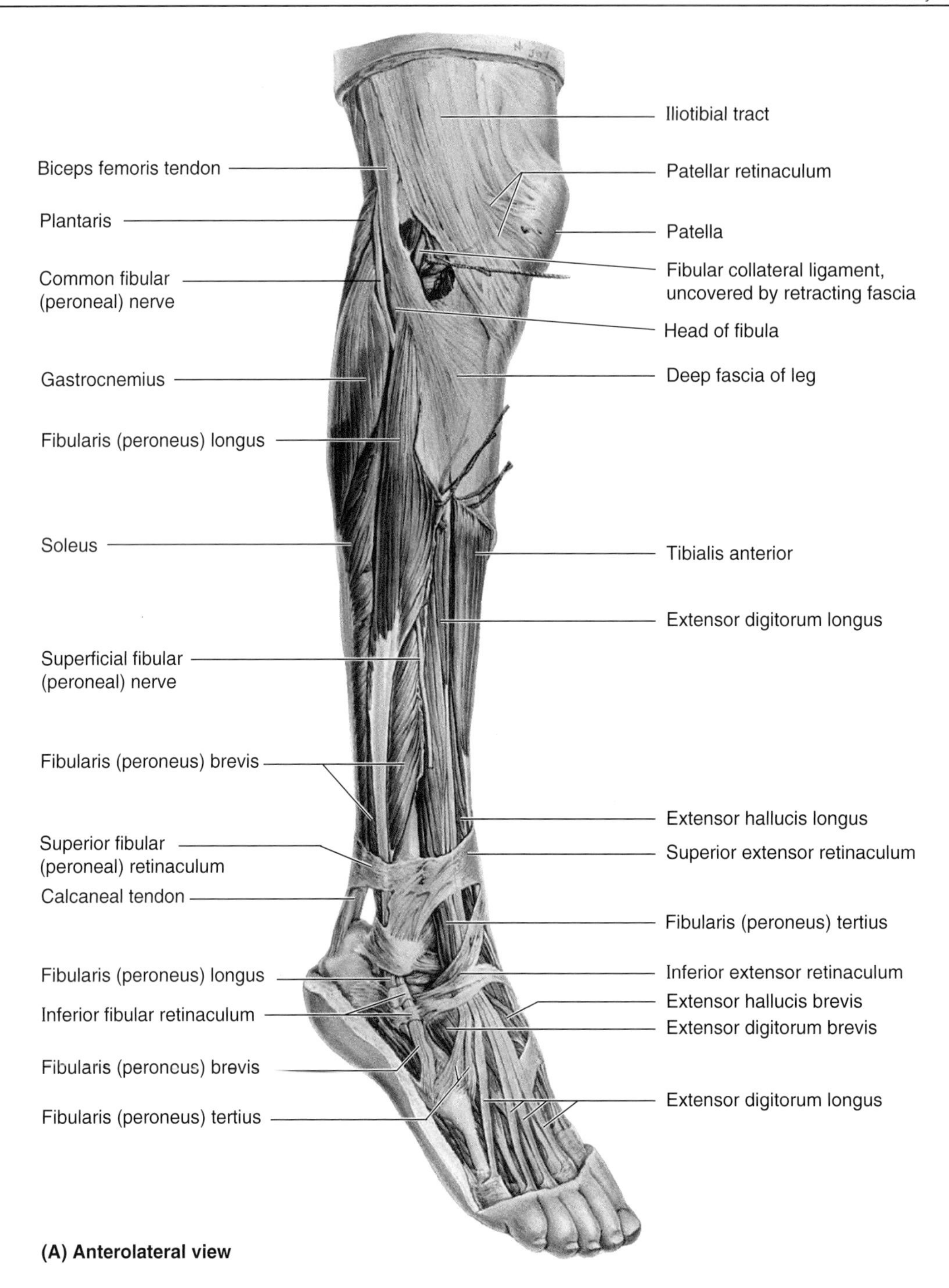

Figure 5.35. Dissections of the right leg. A. Muscles of the leg and foot. Observe the common fibular nerve in contact with the neck of the fibula (calf bone); here it is vulnerable to injury. Observe also the superior and inferior extensor and fibular retinacula.

Anterior Tibialis Strain (Shin Splints)

Shin splints—edema and pain in the area of the distal two-thirds of the tibia—results from repetitive microtrauma of the tibialis anterior and small tears in the periosteum covering the body of the tibia. Shin splints commonly result from traumatic injury or athletic overexertion of muscles in the anterior compartment—especially the tibialis anterior—by untrained persons. Often persons who lead sedentary lives develop shin splints when they participate in walkathons (long-distance walks). Shin splints also occur in trained runners who do not warm up and warm down sufficiently. Muscles in the anterior compartment swell from sudden overuse, and the edema and muscle-tendon ▶

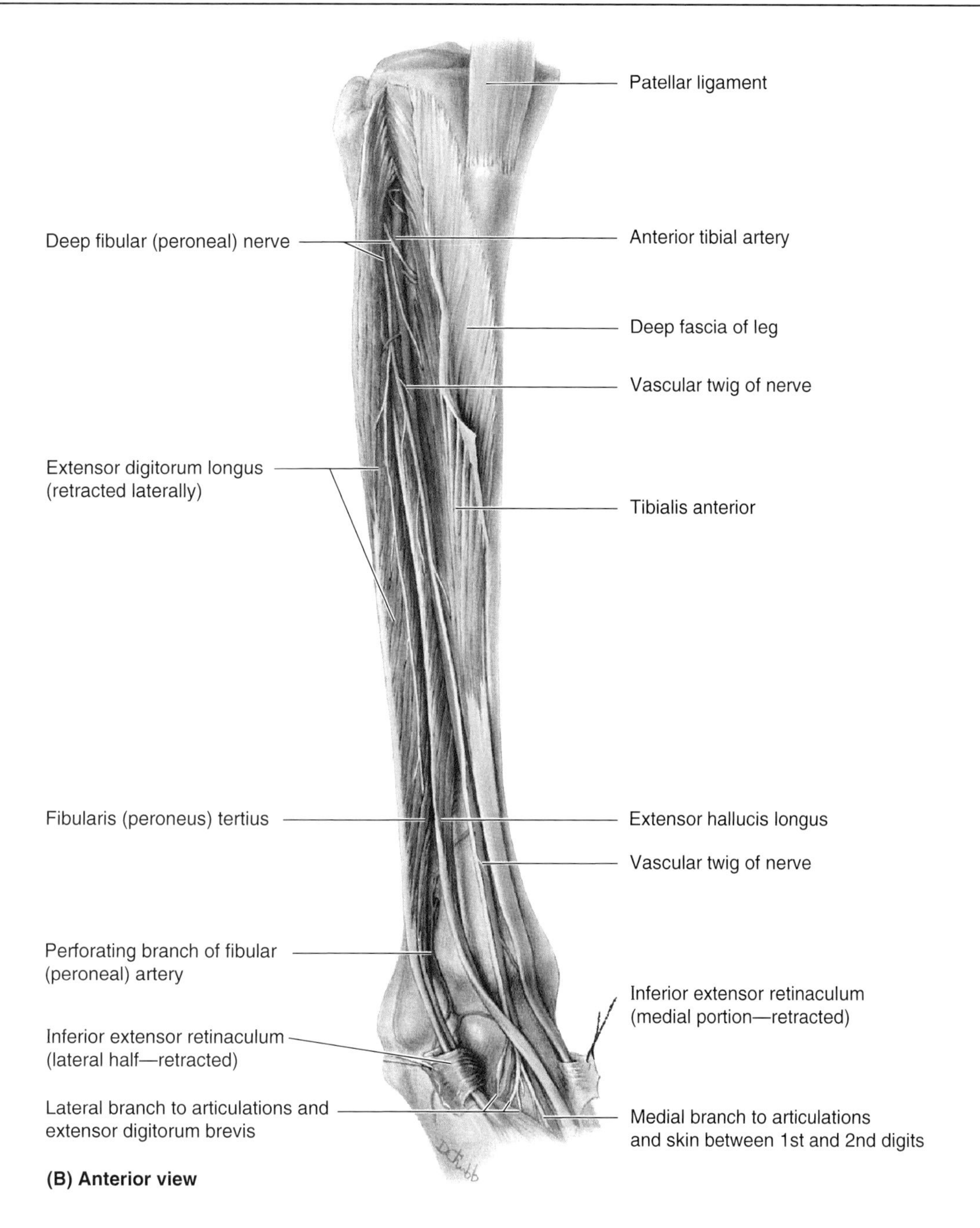

Figure 5.35. *(Continued)* **B.** Dissection of the anterior compartment of the leg. The muscles and inferior extensor retinaculum are separated to display the arteries and nerves of the anterior compartment.

▶ inflammation reduce the bloodflow to the muscles. *Shin splints are a mild form of the anterior compartment syndrome.* The swollen muscles are painful and tender to pressure.

Deep Fibular Nerve Entrapment

Excessive use of muscles supplied by the deep fibular nerve (e.g., during skiing, running, and dancing) may result in muscle injury and edema in the anterior compartment. This entrapment may cause compression of the deep fibular nerve and pain in the anterior compartment. Compression of the nerve by tight-fitting ski boots, for example, may occur where the nerve passes deep to the inferior extensor retinaculum and the extensor hallucis brevis. Pain occurs in the dorsum of the foot and radiates to the web space between the 1st and 2nd digits. Because ski boots are a common cause of this type of nerve entrapment, this condition has been called the "ski boot syndrome;" however, the syndrome in soccer players and runners can result from tight shoes. ⭘

Table 5.10. **Nerves of the Leg**

Nerve	Origin	Course	Distribution in Leg
Saphenous	Femoral nerve	Descends with femoral vessels through femoral triangle and adductor canal and then descends with great saphenous vein	Supplies skin on medial side of leg and foot
Sural	Usually arises from both tibial and common fibular nerves	Descends between heads of gastrocnemius and becomes superficial at the middle of the leg; descends with small saphenous vein and passes inferior to the lateral malleolus to the lateral side of foot	Supplies skin on posterior and lateral aspects of leg and lateral side of foot
Tibial	Sciatic nerve	Forms as sciatic bifurcates at apex of popliteal fossa; descends through popliteal fossa and lies on popliteus; runs inferiorly on the tibialis posterior with the posterior tibial vessels; terminates beneath the flexor retinaculum by dividing into the medial and lateral plantar nerves	Supplies posterior muscles of leg and knee joint
Common fibular	Sciatic nerve	Forms as sciatic bifurcates at apex of popliteal fossa and follows medial border of biceps femoris and its tendon; passes over posterior aspect of head of fibula and then winds around neck of fibula deep to fibularis longus, where it divides into deep and superficial fibular nerves	Supplies skin on lateral part of posterior aspect of leg via its branch, the lateral sural cutaneous nerve; also supplies knee joint via its articular branch
Superficial fibular	Common fibular nerve	Arises between fibularis longus and neck of fibula and descends in lateral compartment of the leg; pierces deep fascia at distal third of leg to become subcutaneous	Supplies fibularis longus and brevis and skin on distal third of anterior surface of leg and dorsum of foot
Deep fibular	Common fibular nerve	Arises between fibularis longus and neck of fibula; passes through extensor digitorum longus and descends on interosseous membrane; crosses distal end of tibia and enters dorsum of foot	Supplies anterior muscles of leg, dorsum of foot, and skin of first interdigital cleft; sends articular branches to joints it crosses

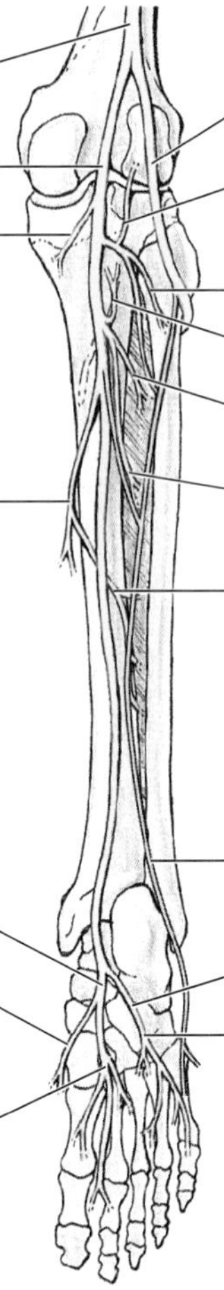

Table 5.11. **Arterial Supply to the Leg**

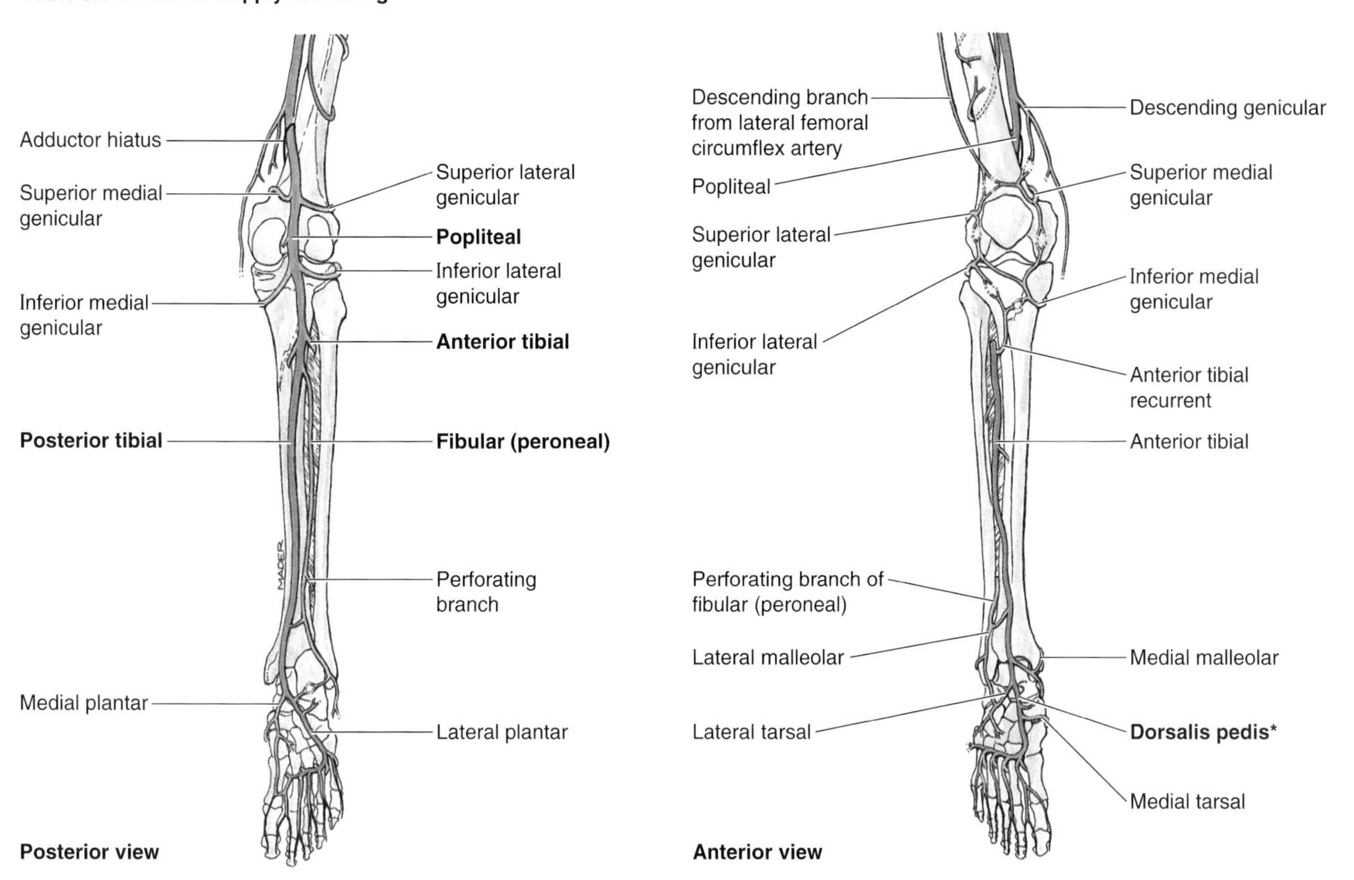

Artery	Origin	Course	Distribution in Leg
Popliteal	Continuation of femoral artery at adductor hiatus in adductor magnus	Passes through popliteal fossa to leg; ends at lower border of popliteus muscle by dividing into anterior and posterior tibial arteries	Superior, middle, and inferior genicular arteries to both lateral and medial aspects of knee
Anterior tibial	Popliteal	Passes between tibia and fibula into anterior compartment through gap in superior part of interosseous membrane and descends this membrane between tibialis anterior and extensor digitorum longus	Anterior compartment of leg
Dorsalis pedis*	Continuation of anterior tibial artery distal to inferior extensor retinaculum	Descends anteromedially to first interosseous space and divides into plantar and arcuate arteries	Muscles on dorsum of foot; pierces first dorsal interosseous muscle as deep plantar artery to contribute to formation of plantar arch
Posterior tibial	Popliteal	Passes through posterior compartment of leg and terminates distal to flexor retinaculum by dividing into medial and lateral plantar arteries	Posterior and lateral compartments of leg; circumflex fibular branch joins anastomoses around knee; nutrient artery passes to tibia
Fibular (peroneal)	Posterior tibial	Descends in posterior compartment adjacent to posterior intermuscular septum	Posterior compartment of leg: perforating branches supply lateral compartment of leg

*Dorsal artery of foot (L. dorsalis pedis)

Lateral Compartment of the Leg

The lateral compartment is bounded by the lateral surface of the fibula, the anterior and posterior intermuscular septa, and the crural fascia (Fig. 5.33, *A* and *B*, Table 5.9).

Muscles in the Lateral Compartment

The lateral compartment contains the fibularis longus and brevis muscles.

Fibularis Longus. The fibularis longus is the longer and more superficial of the two fibularis muscles, and it arises

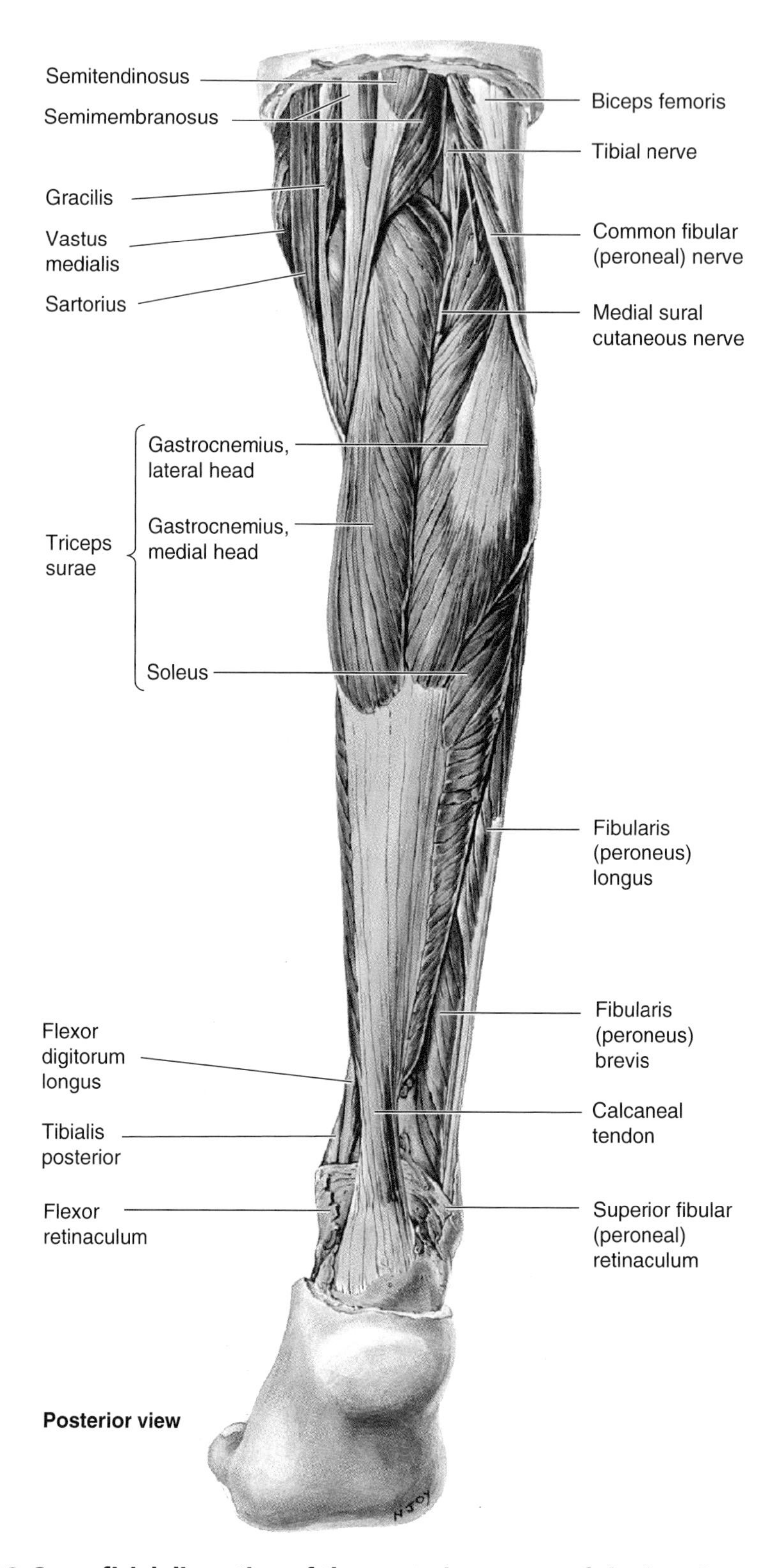

Figure 5.36. Superficial dissection of the posterior aspect of the leg. Observe the nerves and muscles, especially the common fibular nerve, the three-headed triceps surae muscle, and the spiral fibers of the calcaneal tendon.

much more superiorly on the body of the fibula (Figs. 5.33 and 5.35). The narrow fibularis longus extends from the head of the fibula to the sole of the foot. Its tendon can be palpated and observed proximal and posterior to the lateral malleolus. When a person stands on one foot, the fibularis longus helps steady the leg on the foot. The fibularis longus is enclosed in a common synovial sheath with the fibularis brevis (Fig. 5.33*A*). The fibularis longus passes inferior to the fibular trochlea on the calcaneus and enters a groove on the anteroinferior aspect of the cuboid bone. It then crosses the sole of the foot, running obliquely and distally to reach its attachment to the 1st metatarsal and medial cuneiform bones.

Fibularis Brevis. The fibularis brevis is a fusiform muscle that lies deep to the fibularis longus and, true to its name, is shorter than its partner in the lateral compartment (Figs. 5.33 and 5.35). Its broad tendon grooves the posterior aspect of the lateral malleolus and can be palpated inferior to the lateral malleolus, where it lies in a common tendon sheath with the fibularis longus. The narrower tendon of the fibularis longus lies on that of the fibularis brevis and does not contact the lateral malleolus. The tendon of the fibularis brevis can be easily traced to its distal attachment to the base of the 5th metatarsal. The tendons of fibularis longus and brevis are bound down at the malleolus by the **superior fibular retinaculum**, a band of deep fascia that extends from the tip of the lateral malleolus to the calcaneus (Figs. 5.35*A* and 5.36). The tendon of the *fibularis tertius*—a slip of muscle from the extensor digitorum longus—often attaches to the tendon of the fibularis brevis or continues and attaches to the proximal phalanx of this digit. In practice, the primary function of the fibularis muscles is to resist inversion of the foot; the ankle is thus afforded protection because it is most vulnerable in the inverted position. *To test the fibularis longus and brevis*, the foot is everted strongly against resistance; if acting normally, the muscle tendons can be seen and palpated inferior to the lateral malleolus.

Nerves in the Lateral Compartment

The **superficial fibular nerve**, a branch of the common fibular nerve, is *the nerve of the lateral compartment* (Fig. 5.35*A*, Table 5.10). It supplies the skin on the distal part of the anterior surface of the leg and nearly all the dorsum of the foot.

The lateral compartment does not have an artery; the muscles are supplied superiorly by perforating branches of the anterior tibial artery and inferiorly by perforating branches of the fibular artery (Table 5.11).

Superficial Fibular Nerve Entrapment

Chronic ankle sprains may produce recurrent stretching of the superficial fibular nerve that may cause pain along the lateral side of the leg and the dorsum of the ankle and foot. Numbness and paresthesia may be present and increase with activity.

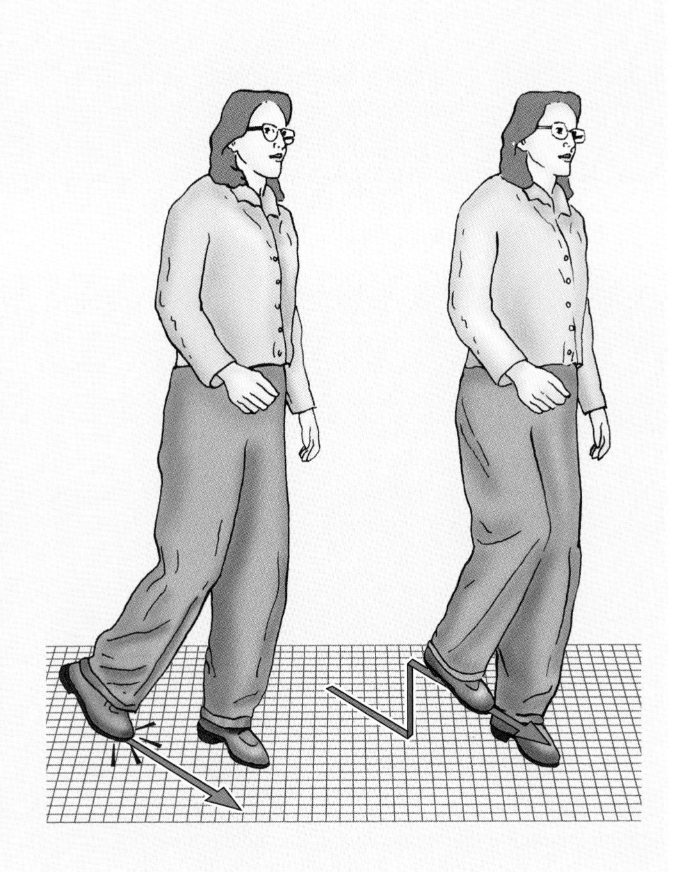

Injury to the Common Fibular Nerve

Because of its superficial position, the common fibular is the nerve most injured in the lower limb, mainly because it winds superficially around the fibular neck. This nerve may be severed during fracture of the fibular neck or severely stretched when the knee joint is injured or dislocated. *Severance of the common fibular nerve* results in paralysis of all muscles in the anterior and lateral compartments of the leg (dorsiflexors of the ankle and evertors of the foot). The loss of eversion of the foot and dorsiflexion of the ankle causes *foot-drop*. The foot drops and the toes drag on the floor when walking. Because it is impossible to make the heel strike the ground first, the patient has a high stepping ("steppage") gait, raising the foot as high as necessary to keep the toes from hitting the ground. In addition, the foot comes down suddenly, producing a distinctive "clop." The person also experiences a variable loss of sensation on the anterolateral aspect of the leg and the dorsum of the foot. ▶

Avulsion of the Tuberosity of the Fifth Metatarsal

Violent inversion of the foot may cause avulsion of the tuberosity of the 5th metatarsal—the distal attachment of fibularis brevis. This *avulsion fracture* is associated with a severely sprained ankle. Injury to the associated superficial fibular nerve causes inversion of the foot because of paralysis of the fibular muscles in the lateral compartment.

Secondary Ossification Centers in the Foot

In some children, a secondary ossification center develops for the lateral surface of the tuberosity of the 5th metatarsal. This condition results in the formation of a chiplike piece of bone that could be mistaken for a "flake" fracture of the tuberosity. The presence of similar secondary centers in both feet usually indicates that a fracture is not present. These centers are not usually observed in adults because by then they have fused with the tuberosity. ⊙

Posterior Compartment of the Leg

The posterior compartment is the largest of the three leg compartments (Fig. 5.33*A*). The *calf muscles* in the posterior compartment are divided into superficial and deep groups by the *transverse intermuscular septum.* The tibial nerve and posterior tibial vessels supply both divisions of the posterior compartment and run between the superficial and deep groups of muscle, just deep to the transverse intermuscular septum.

Superficial Muscle Group in the Posterior Compartment

The superficial group of muscles—*gastrocnemius, soleus, and plantaris*—forms a powerful muscular mass in the calf of the leg that plantarflexes the foot (Fig. 5.33, Table 5.12). The large size of these muscles is a human characteristic that is directly related to our upright stance. These muscles are strong and heavy because they support and move the weight of the body. The three muscles are supplied by the tibial nerve. Together, the two-headed **gastrocnemius** and **soleus** form the three-headed **triceps surae** (L. sura, calf). This large muscle has a common tendon—the **calcaneal tendon** (L. tendo calcaneus, Achilles tendon)—which attaches to the calcaneus (Fig. 5.35*B* and 5.36). A *superficial calcaneal bursa* lies between the skin and the calcaneal tendon, and a *deep calcaneal bursa (retrocalcaneal bursa)* is located between the tendon and the calcaneus.

The triceps surae plantarflexes the ankle joint, raising the heel against the body weight (e.g., when a person is walking, dancing, and/or standing on the toes). "You stroll with the soleus but win the long jump with the gastrocnemius." *To test the triceps surae,* the foot is plantarflexed against resistance (e.g., by "standing on the toes," in which case body weight [gravity] provides resistance). If acting normally, the calcaneal tendon and triceps surae can be seen and palpated.

Gastrocnemius. The gastrocnemius—the most superficial muscle in the posterior compartment—forms part of the prominence of the calf. It is a fusiform, *two-headed, two-joint muscle* with a medial head slightly larger and extending more distally than its lateral head. The heads come together at the inferior margin of the popliteal fossa, where they form the inferolateral and inferomedial boundaries of this fossa. Because its fibers are mainly vertical, contractions of the gastrocnemius produce rapid movements during running and jumping (Table 5.12). Although the gastrocnemius acts on both the knee and ankle joints, it cannot exert its full power on both joints at the same time.

Soleus. The soleus—deep to the gastrocnemius—is powerful. It is a large flat muscle that was given its name because of its resemblance to the sole, a flat fish. Its fibers slope inferomedially. The soleus can be palpated on each side of the gastrocnemius when the subject is standing on the tiptoes. The soleus acts with the gastrocnemius in plantarflexing the ankle joint; it does not act on the knee joint. The soleus is an antigravity muscle that contracts alternately with the extensor muscles of the leg to maintain balance. It is a strong but relatively slow plantarflexor of the ankle joint.

Plantaris. The plantaris is a small muscle with a short belly and long thin tendon (Figs. 5.32 and 5.33*A*). This vestigial muscle, often absent, acts with the gastrocnemius. It has been proposed to be an organ of proprioception for the larger plantar flexors; it has been found to have a high density of muscle spindles (receptors for proprioception). Because of its minor role, the plantaris tendon can be removed for grafting (e.g., during reconstructive surgery of the tendons of the hand) without causing any disability.

Fabella in the Gastrocnemius

Close to its proximal attachment, the lateral head of the gastrocnemius contains a sesamoid bone—the *fabella* (L. bean)—that is visible in lateral radiographs of the knee in 3 to 5% of people (p. 587).

Calcaneal Tendinitis

Inflammation of the calcaneal tendon constitutes 9 to 18% of running injuries. Microscopic tears of collagen fibers in the tendon result in *tendinitis,* which causes pain during walking, especially when wearing rigid-soled shoes. ▶

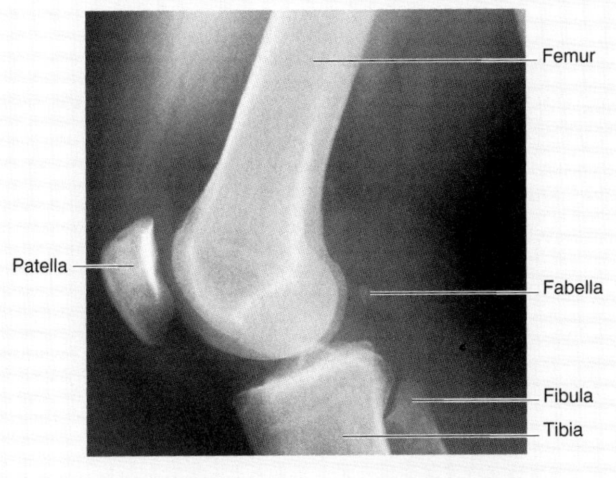

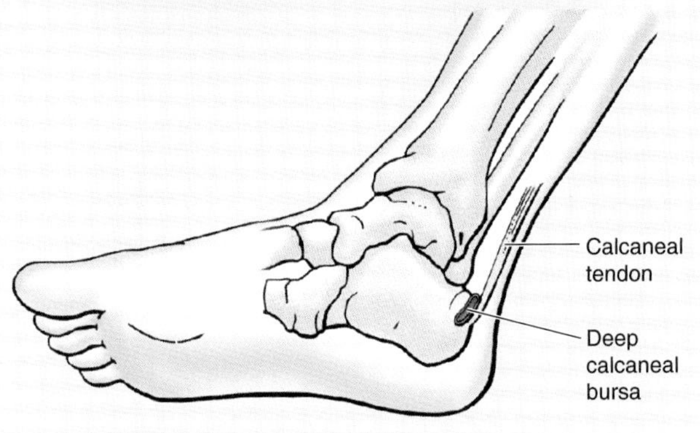

▶ Calcaneal tendinitis often occurs during repetitive activities, especially in persons running after prolonged inactivity.

Ruptured Calcaneal Tendon

This injury often occurs in poorly conditioned 30- to 45-year-old people with a history of calcaneal tendinitis. The usual symptom is sudden calf pain with a audible snap. The injury usually results with sudden dorsiflexion of a plantarflexed foot. In a completely ruptured tendon, a gap is palpable. Persons with this injury cannot use the limb, and a lump appears in the calf owing to shortening of the triceps surae. After complete rupture of the calcaneal tendon, the foot can be dorsiflexed to a greater extent than is normal, but the patient cannot easily plantarflex the foot.

Calcaneal Tendon Reflex

This ankle reflex (jerk) is elicited with the person's legs dangling over the side of the examining table or bed. The calcaneal tendon is struck briskly with a reflex hammer just proximal to the calcaneus. The normal result is plantarflexion of the ankle joint. The calcaneal tendon reflex tests the S1 and S2 nerve roots. If the S1 nerve root is cut or compressed, the ankle reflex is virtually absent.

Gastrocnemius Strain

Gastrocnemius strain ("tennis leg") is a painful fibular injury resulting from partial tearing of the medial belly of the gastrocnemius at or near its musculotendinous junction. It is caused by overstretching the muscle by concomitant full extension of the knee and dorsiflexion of the ankle joint. Usually, an abrupt onset of stabbing pain is followed by edema and spasm of the gastrocnemius.

Calcaneal Bursitis

Calcaneal bursitis (retroachilles bursitis) results from *inflammation of the deep calcaneal bursa* located between the calcaneal tendon and the superior part of the posterior surface of the calcaneus. Calcaneal bursitis causes pain posterior to the heel and is fairly common in long-distance running, basketball, and tennis. It is caused by excessive friction on the bursa as the tendon continuously slides over it. *Inflammation of the superficial calcaneal bursa* also results from repetitive microtrauma from the back of a shoe.

Venous Return from Leg

A venous plexus deep to the triceps surae is involved in the return of blood from the leg. When a person is standing, the venous return from the leg depends largely on the muscular activity of the triceps surae (see pp. 524–526 for a discussion of the *musculovenous pump*). Contraction of the calf muscles pumps blood superiorly in the deep veins. The efficiency of the calf pump is improved by the deep fascia that invests the muscles like an elastic stocking. Normally, blood is prevented from flowing into the superficial veins by the valves in the perforating veins. If these valves are incompetent, blood is forced into the superficial veins during contraction of the triceps surae muscles and by hydrostatic pressure when straining or standing. As a consequence, the vessels become *varicose veins*—dilated and tortuous veins.

Accessory Soleus

An accessory soleus is present in approximately 3% of people. The accessory muscle usually appears as a distal belly medial to the calcaneal tendon (Anne Agur, personal communication 1998). Clinically, accessory soleus muscles are usually associated with pain and edema during prolonged exercise. ○

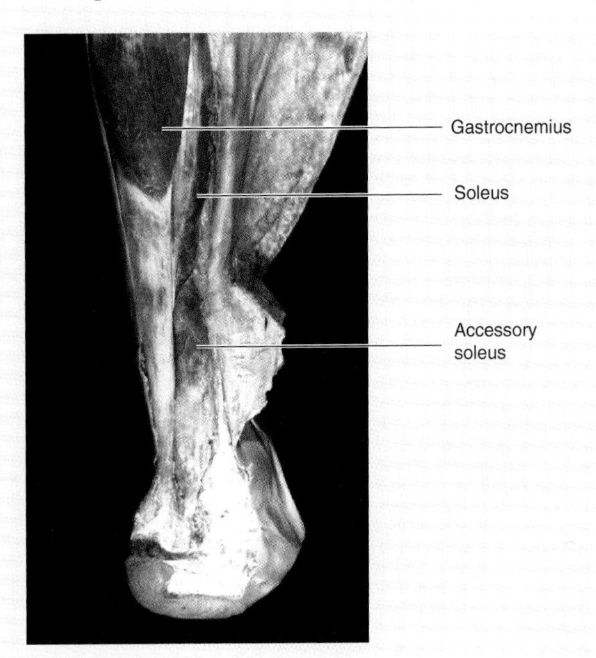

Deep Muscle Group in the Posterior Compartment

Four muscles comprise the deep group in the posterior compartment of the leg (Figs. 5.33 and 5.37*A*, Table 5.12):

- Popliteus
- Flexor digitorum longus
- Flexor hallucis longus
- Tibialis posterior.

The popliteus acts on the knee joint, whereas the other muscles act on the ankle and foot joints.

Popliteus. The popliteus is a thin, triangular muscle that forms the inferior part of the floor of the popliteal fossa (Fig. 5.32). Its tendon, adherent to the articular capsule of the knee joint, lies between the fibrous capsule and the synovial membrane. *The popliteus is a flexor of the knee joint;* when a person is standing with the knee partly flexed, the popliteus contracts to assist the posterior cruciate ligament (PCL) in preventing anterior displacement of the femur on the tibia. The **popliteus bursa** lies deep to the popliteus tendon (Fig. 5.37*B*). When standing with the knees "locked" in the fully extended position, the popliteus acts to rotate the femur laterally 5° on

Table 5.12. Muscles of the Posterior Leg

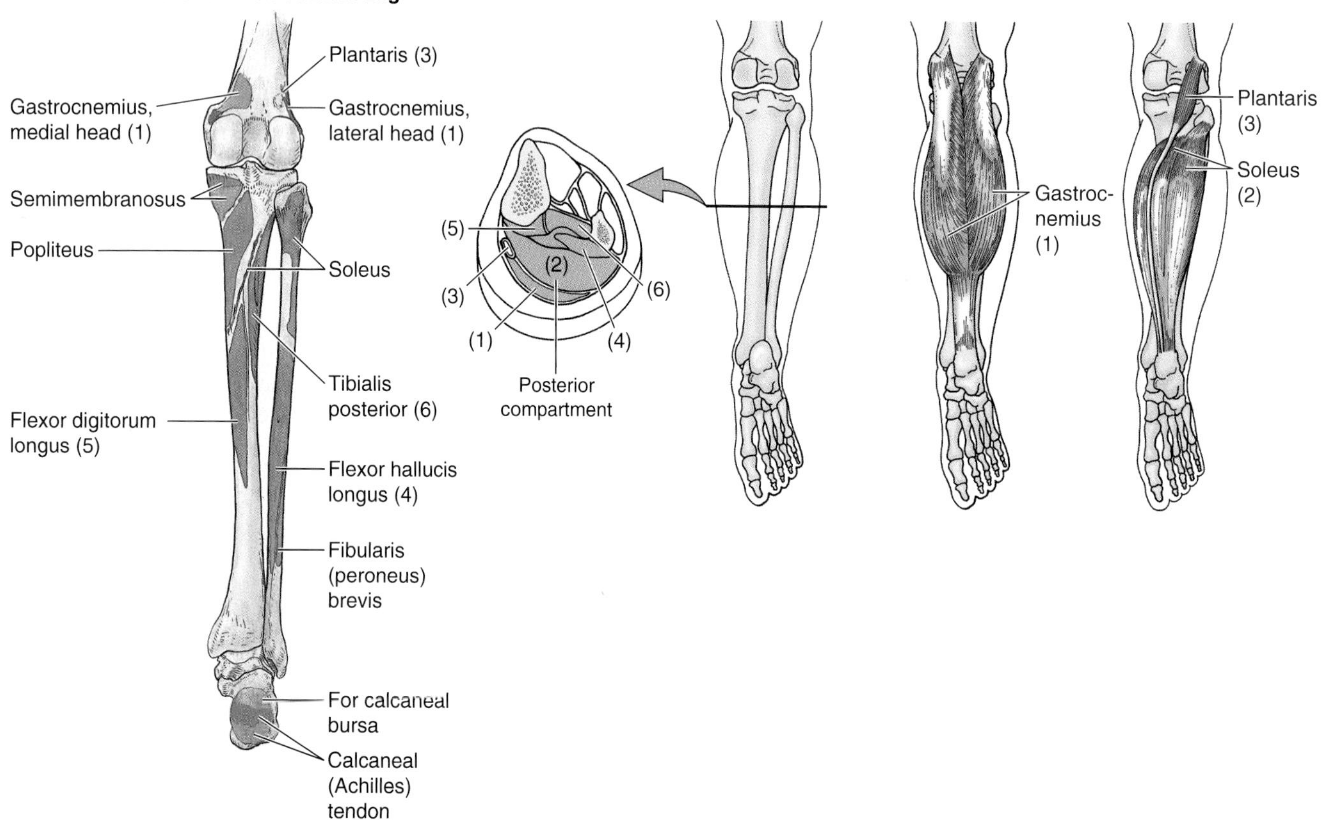

Posterior muscle attachments

Muscle	Proximal Attachment	Distal Attachment	Innervation	Main Action
Superficial muscles				
Gastrocnemius (1)	Lateral head: lateral aspect of lateral condyle of femur Medial head: popliteal surface of femur, superior to medial condyle	Posterior surface of calcaneus via calcaneal tendon	Tibial nerve (S1 and S2)	Plantarflexes ankle when knee is extended, raises heel during walking, and flexes leg at knee joint
Soleus (2)	Posterior aspect of head of fibula, superior fourth of posterior surface of fibula soleal line and medial border of tibia			Plantarflexes ankle independent of position of knee and steadies leg on foot
Plantaris (3)	Inferior end of lateral supracondylar line of femur and oblique popliteal ligament			Weakly assists gastrocnemius in plantarflexing ankle and flexing knee

the tibial plateaus (p. 512), unlocking the knee so that flexion can occur. When the foot is off the ground and the knee is flexed, the popliteus can rotate the tibia medially beneath the femoral condyles.

Flexor Digitorum Longus. The flexor digitorum longus is smaller than the flexor hallucis longus, even though it moves four digits (Fig. 5.37*A*). It passes diagonally into the sole of the foot, superficial to the tendon of the flexor hallucis longus, and divides into four tendons (Fig. 5.39), which pass to the distal phalanges of the lateral four digits. *To test the flexor digitorum longus*, distal phalanges of the lateral four toes are flexed against resistance; if they are acting normally, the tendons of the toes can be seen and palpated.

Flexor Hallucis Longus. The flexor hallucis longus is the powerful "push-off" muscle during walking, running, and jumping (Fig. 5.38). It provides much of the spring to the step. The tendon of the muscle passes posterior to the distal end of the tibia and occupies a shallow groove on the posterior surface of the talus, which is continuous with the groove on the plantar surface of the sustentaculum tali. The tendon then crosses deep to the tendon of the flexor digitorum longus in the sole of the foot. As it passes to the distal phalanx of the great toe, the tendon runs between two *sesamoid bones* in the tendons of the flexor hallucis brevis. These bones protect the tendon from the pressure of the head of the 1st metatarsal bone. *To test the flexor hallucis longus*, the terminal phalanx of

Table 5.12. *(Continued)* **Muscles of the Posterior Leg**

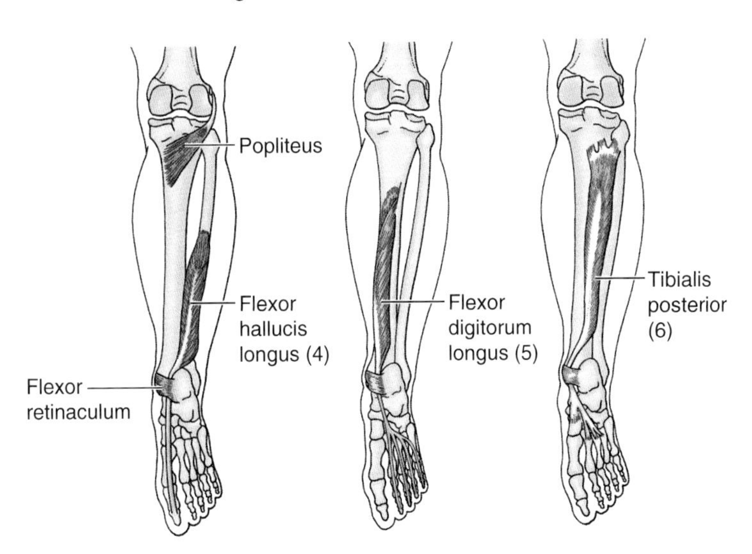

Muscle	Proximal Attachment	Distal Attachment	Innervation	Main Action
Deep muscles				
Popliteus	Lateral surface of lateral condyle of femur and lateral meniscus	Posterior surface of tibia, superior to soleal line	Tibial nerve (L4, L5, and S1)	Weakly flexes knee and unlocks it
Flexor hallucis longus (4)	Inferior two-thirds of posterior surface of fibula and inferior part of interosseous membrane	Base of distal phalanx of great toe (hallux)	Tibial nerve (S2 and S3)	Flexes great toe at all joints and weakly plantarflexes ankle; supports medial longitudinal arches of foot
Flexor digitorum longus (5)	Medial part of posterior surface of tibia inferior to soleal line and by a broad tendon to fibula	Bases of distal phalanges of lateral four digits		Flexes lateral four digits and plantarflexes ankle; supports longitudinal arches of foot
Tibialis posterior (6)	Interosseous membrane, posterior surface of tibia inferior to soleal line, and posterior surface of fibula	Tuberosity of navicular, cuneiform, and cuboid and bases of 2nd, 3rd, and 4th metatarsals	Tibial nerve (L4 and L5)	Plantarflexes ankle and inverts foot

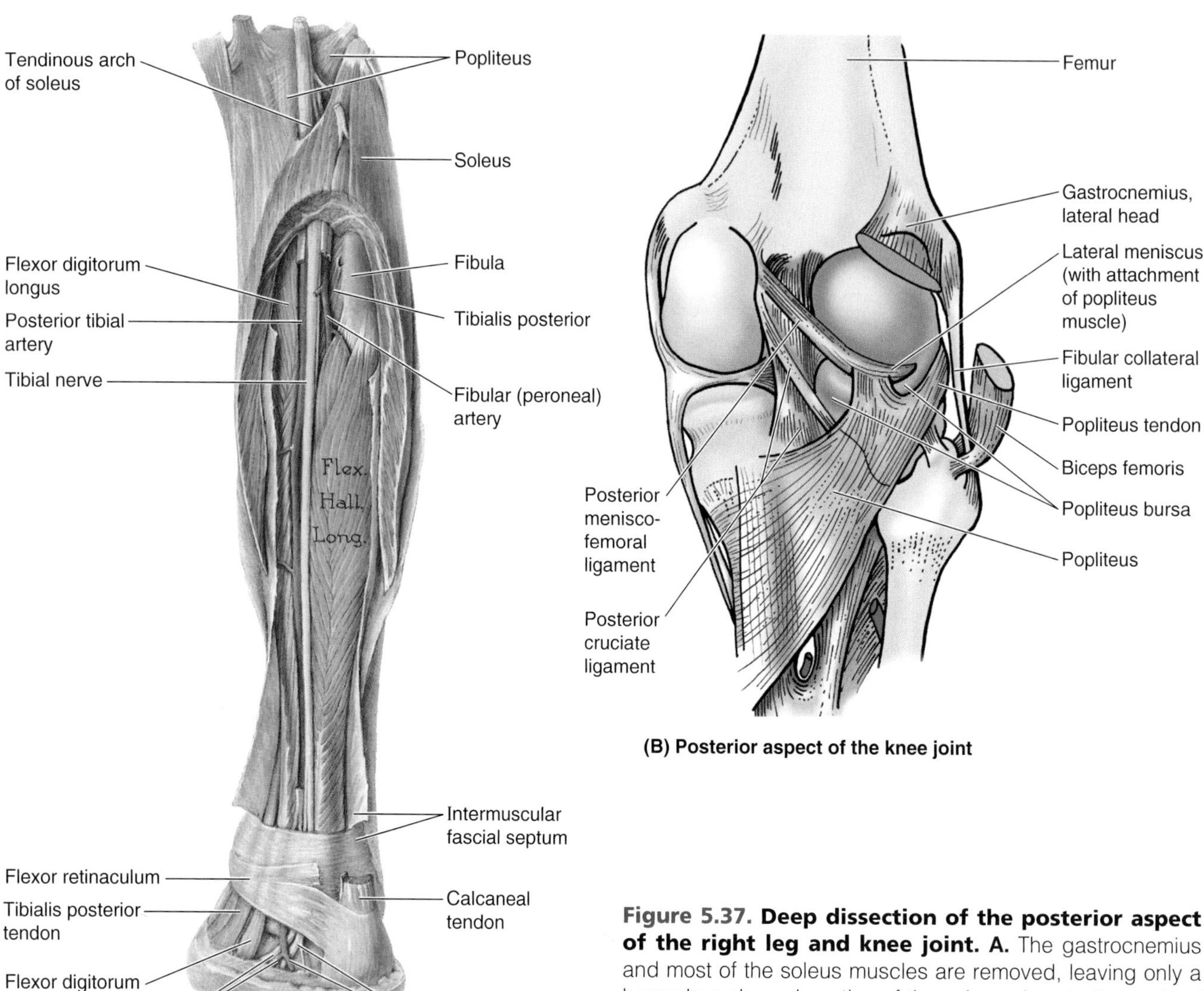

Figure 5.37. Deep dissection of the posterior aspect of the right leg and knee joint. A. The gastrocnemius and most of the soleus muscles are removed, leaving only a horseshoe-shaped section of the soleus close to its proximal attachments and the distal part of the calcaneal tendon. Observe the deep muscles, vessels, and nerves. **B.** Posterior aspect of the knee joint. Observe the popliteus tendon attaching in part to the lateral meniscus and separated from the proximal end of the tibia by the popliteus bursa.

the great toe is flexed against resistance; if it is acting normally, the tendon can be seen and palpated.

Tibialis Posterior. The tibialis posterior, the deepest muscle in the posterior compartment, lies between the flexor digitorum longus and flexor hallucis longus in the same plane as the tibia and fibula (Figs. 5.37*A*, 5.38 and 5.39). It attaches primarily to the navicular bone but has attachments to other tarsals and metatarsals (Fig. 5.37*A*, Table 5.12). *To test the tibialis posterior*, the foot is inverted against resistance with the foot in slight plantarflexion; if it is acting normally, the tendon can be seen and palpated posterior to the medial malleolus.

Nerves in the Posterior Compartment

The **tibial nerve** (L4, L5, and S1 through S3) is the larger of the two terminal branches of the sciatic nerve. It leaves the popliteal fossa between the heads of the gastrocnemius and *supplies all muscles in the posterior compartment of the legs* (Fig. 5.37*A*, Table 5.10). The tibial nerve descends in the median plane of the fibula, deep to the soleus. At the ankle the nerve lies between the tendons of the flexor hallucis longus and the flexor digitorum longus. *Posteroinferior to the medial malleolus, the tibial nerve divides into the medial and lateral plantar nerves.* A branch of the tibial nerve, the *medial sural cutaneous nerve*, usually unites with the communicating branch of the common fibular nerve to form the **sural nerve.** This nerve supplies the skin of the lateral and posterior part of the inferior third of the leg and the lateral side of the foot. Articular branches of the tibial nerve supply the knee joint, and medial calcaneal branches supply the skin of the heel.

Arteries in the Posterior Compartment

The **posterior tibial artery**, the larger terminal branch of the popliteal artery, provides the main blood supply to the foot (Figs. 5.37*A* and 5.40, Table 5.11). It begins at the distal bor-

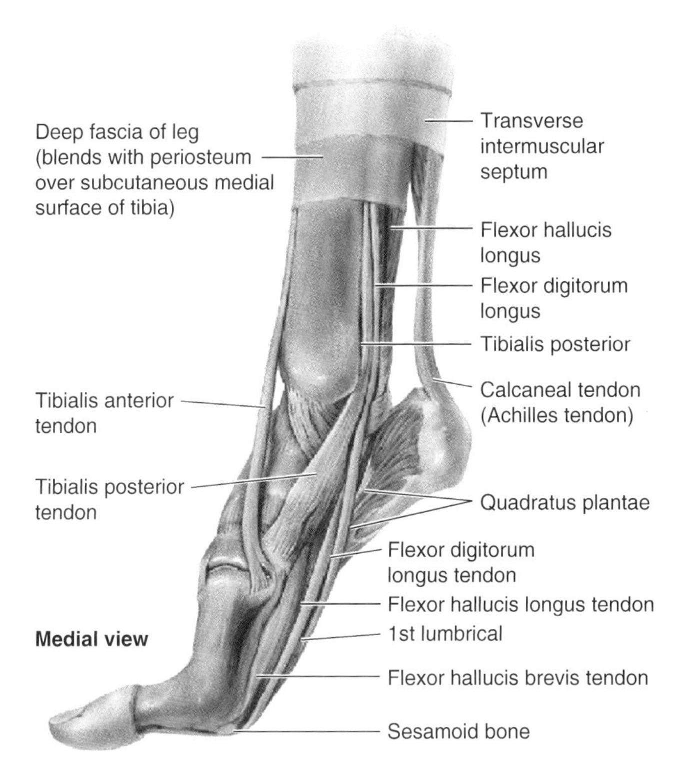

Figure 5.38. Dissection of the lower leg and the foot. The foot is raised as in the "push-off" phase of walking. Observe the sesamoid bone acting as a "foot stool" for the 1st metatarsal, giving it extra height and protecting the flexor hallucis longus tendon.

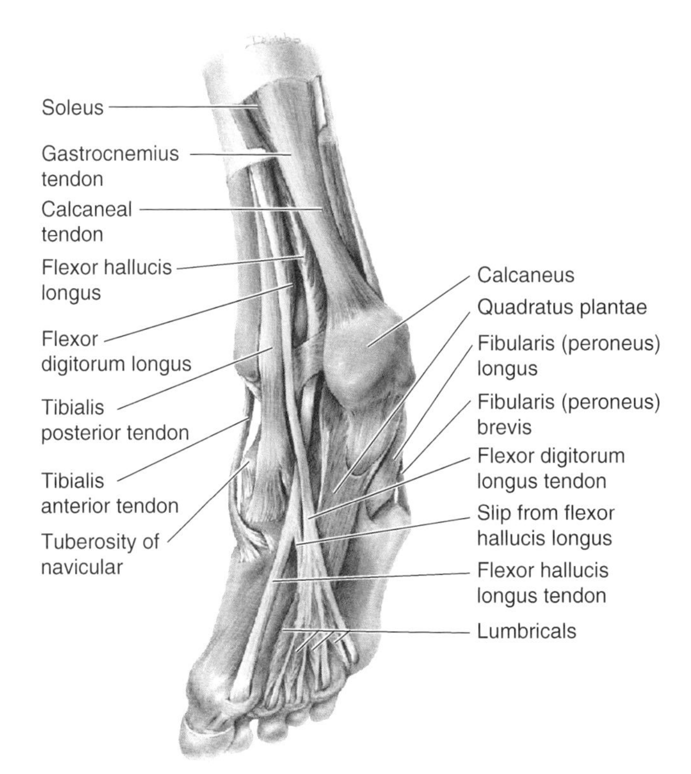

Figure 5.39. Second layer of plantar muscles. This layer of foot muscles includes the tendons of the flexor hallucis longus and flexor digitorum longus, four lumbrical (L. lumbricus, earth worm) muscles, and the quadratus plantae.

der of the popliteus muscle and passes deep to the origin of the soleus. After giving off the **fibular artery**, its largest branch, the posterior tibial artery passes inferomedially on the posterior surface of the tibialis posterior. During its descent, it is accompanied by the tibial nerve and veins. The posterior tibial artery runs posterior to the medial malleolus, from which it is separated by the tendons of the tibialis posterior and flexor digitorum longus. Inferior to the medial malleolus, it runs between the tendons of the flexor hallucis longus and flexor digitorum longus. Deep to the flexor retinaculum and the origin of the abductor hallucis, the posterior tibial artery divides into *medial and lateral plantar arteries*.

The **fibular artery**, the largest and most important branch of the tibial artery, begins inferior to the distal border of the popliteus and the tendinous arch of the soleus (Fig. 5.37*A*, Table 5.11). It descends obliquely toward the fibula and passes along its medial side, usually within the flexor hallucis longus. The fibular artery gives muscular branches to the popliteus and other muscles in the posterior and lateral compartments of the leg. It also supplies a nutrient artery to the fibula. The fibular artery usually pierces the interosseous

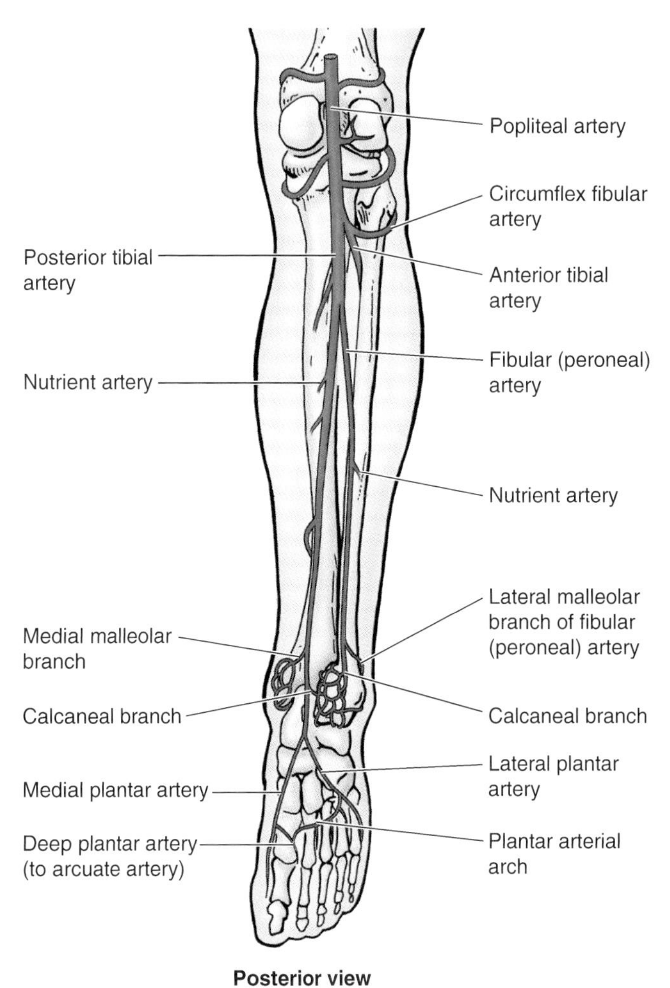

Figure 5.40. Arteries of the knee, leg, and foot. The popliteal artery bifurcates into anterior and posterior tibial arteries; the latter gives rise to the fibular artery and terminates as it enters the foot, bifurcating into medial and lateral plantar arteries.

membrane and passes to the dorsum of the foot, where it anastomoses with the arcuate artery. The *circumflex fibular artery* arises from the origin of the anterior or posterior tibial artery at the knee and passes laterally over the neck of the fibula to the anastomoses around the knee.

The **nutrient artery of the tibia**, the largest nutrient artery in the body, *arises from the posterior tibial artery near its origin.* It pierces the tibialis posterior, to which it supplies branches, and enters the nutrient foramen in the proximal third of the posterior surface of the tibia. The *calcaneal arteries* supply the heel, and a *malleolar branch* joins the network of vessels on the medial malleolus.

Posterior Tibial Pulse

The posterior tibial pulse can usually be palpated between the posterior surface of the medial malleolus and the medial border of the calcaneal tendon. Because the posterior tibial artery passes deep to the flexor retinaculum, it is important when palpating this pulse to have the person relax the retinaculum by inverting the foot. Failing to do this may lead to the erroneous conclusion that the pulse is absent. Both arteries are examined simultaneously for equality of force. *Palpation of the posterior tibial pulse is essential for examining patients with occlusive peripheral arterial disease. Intermittent claudication,* characterized by leg cramps, develops during walking and disappears after rest. This painful condition results from ischemia of the leg muscles caused by narrowing or occlusion of the leg arteries. Although posterior tibial pulses are absent in approximately 15% of normal young people, absence of posterior tibial pulses is considered to be a sign of occlusive *peripheral arterial* disease in people older than 60 years. ○

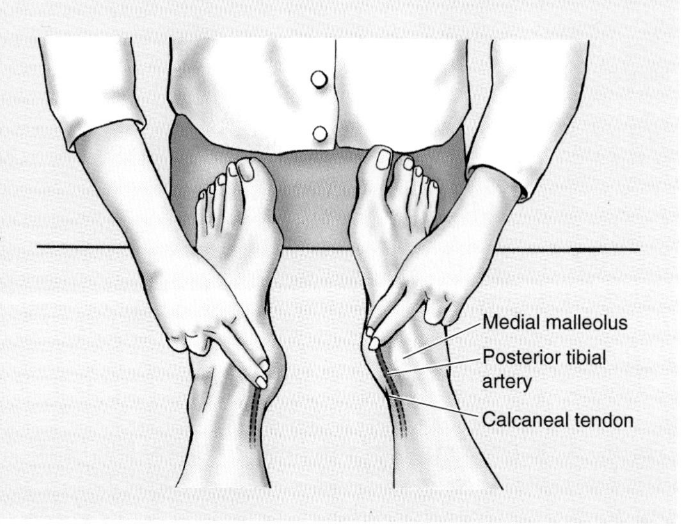

Surface Anatomy of the Leg

The **tibial tuberosity** is an easily palpable elevation on the anterior aspect of the proximal part of the tibia, approximately 5 cm distal to the apex of the patella (p. 519). The tibial tuberosity indicates the level of the head of the fibula and the bifurcation of the popliteal artery into the anterior and posterior tibial arteries. The **patellar ligament** may be felt as it extends from the inferior border of the apex of the patella. It is most easily felt when the knee is extended. When the knee flexes to a right angle, a depression may be felt on each side of the patellar ligament. The joint cavity is superficial in these depressions. The **head of the fibula** is subcutaneous and may be palpated at the posterolateral aspect of the knee, at the level of the tibial tuberosity. The **neck of the fibula** can be palpated just distal to the head. The **tendon of the biceps femoris** may be traced by palpating its distal attachment to the lateral side of the head of the fibula. This tendon and the neck of the fibula guide the examining finger to the *common fibular nerve* (Fig. 5.36). The nerve is indicated by a line along the biceps femoris tendon, posterior to the head of the fibula, and around the lateral aspect of the fibular neck to its anterior aspect, just distal to the fibular head. The common fibular nerve can usually be palpated just posterior to the fibular head and rolled against the fibular neck with the fingertips.

The **anterior border of the tibia** ("shin") is sharp, subcutaneous, and easily followed distally by palpation from the tibial tuberosity to the medial malleolus. It is not usually perfectly straight. The medial surface of the *body of the tibia* is also subcutaneous, except at its proximal end. Its inferior third is crossed obliquely by the great saphenous vein as it passes proximally to the medial aspect of the knee.

The **tibialis anterior** lies superficially and is easily palpable just lateral to the anterior border of the tibia. As the foot is inverted and dorsiflexed, the large **tendon of the tibialis anterior** can be seen and palpated as it runs distally and slightly medially over the anterior surface of the ankle joint to the medial side of the foot. If the 1st digit is dorsiflexed, the tendon of the **extensor hallucis longus** can be palpated just lateral to the tendon of the tibialis anterior. The *tendon of the extensor hallucis brevis* may also be visible.

As the toes are dorsiflexed, the **tendons of the extensor digitorum longus** can be palpated lateral to the extensor hallucis longus and followed to the four lateral digits. The *tendon of the fibularis tertius* may be palpable lateral to the tendons of the extensor digitorum longus, especially when the foot is dorsiflexed and everted.

The **body of the fibula** is subcutaneous only in its distal part, proximal to the lateral malleolus. This part is the ▶

▶ common site of fractures. The **medial** and **lateral malleoli** are subcutaneous and prominent. Palpate them, noting that the tip of the lateral malleolus extends farther distally and posteriorly than the medial malleolus.

The **fibularis longus** is subcutaneous throughout its course. The tendons of this muscle and the fibularis brevis are palpable when the foot is everted as they pass around the posterior aspect of the lateral malleolus. These tendons may be followed anteriorly along the lateral side of the foot. The tendon of the fibularis longus runs as far anteriorly as the cuboid and then disappears by turning into the sole of the foot. The **tendon of the fibularis brevis** may be traced to its attachment to the base of the 5th metatarsal.

The **calcaneal tendon** can be easily followed to its attachment to the posterior part of the calcaneus. The ankle joint is fairly superficial in the depression on each side of the calcaneal tendon. The **heads of the gastrocnemius** are easily recognizable in the superior part of the fibula. The **soleus** can be palpated deep to and at the sides of the superior part of the calcaneal tendon. Both the soleus and gastrocnemius are easier to palpate when the foot is plantarflexed and when standing on the toes. The deep muscles of the fibula are not easily palpated, but their tendons can be observed just posterior to the medial malleolus, especially when the foot is inverted and the toes are flexed. ⊙

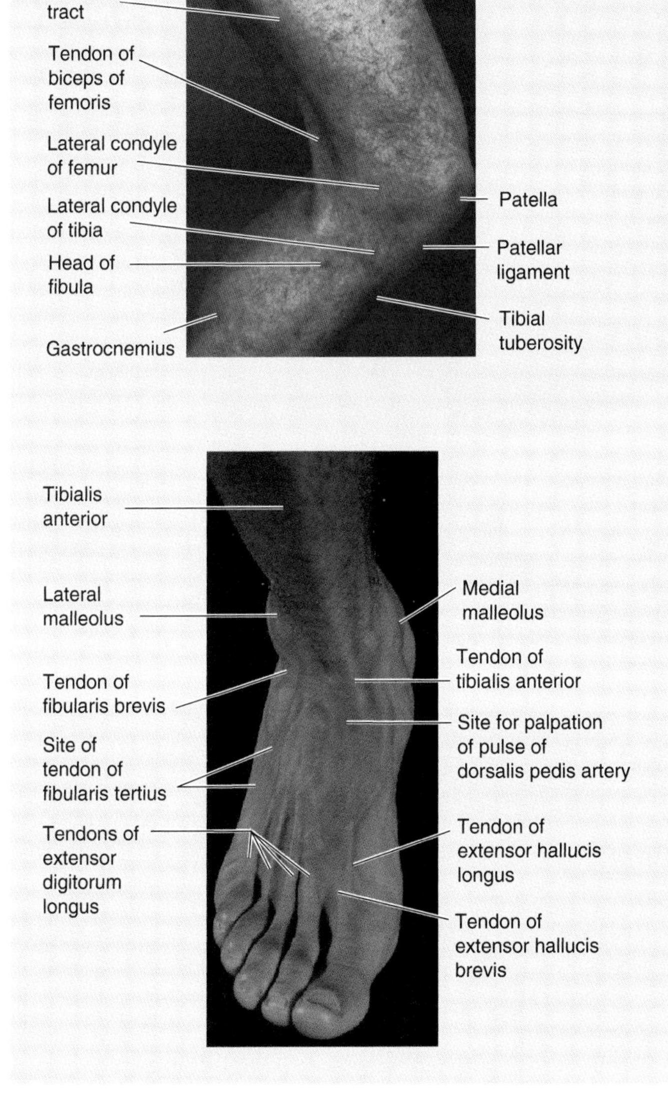

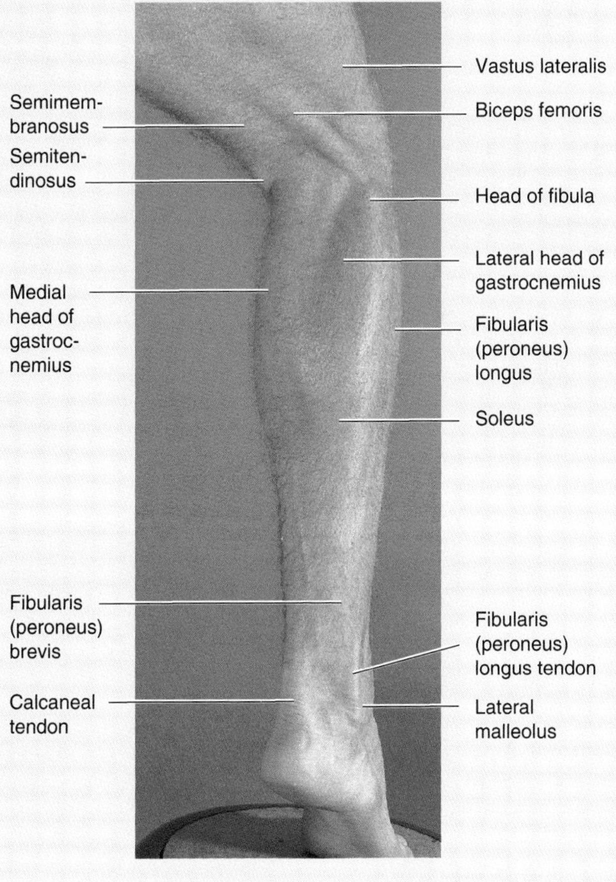

Foot

The **ankle** refers to the angle between the leg and foot and the ankle joint. The **foot**, distal to the leg, supports the weight of the body and has an important role in locomotion. The skeleton of the foot consists of 7 tarsal bones, 5 metatarsals, and 14 phalanges (Fig. 5.41). *The foot and its bones are divided into three parts*:

- The *hindfoot*—talus and calcaneus
- The *midfoot*—navicular, cuboid, and cuneiforms
- The *forefoot*—metatarsals and phalanges.

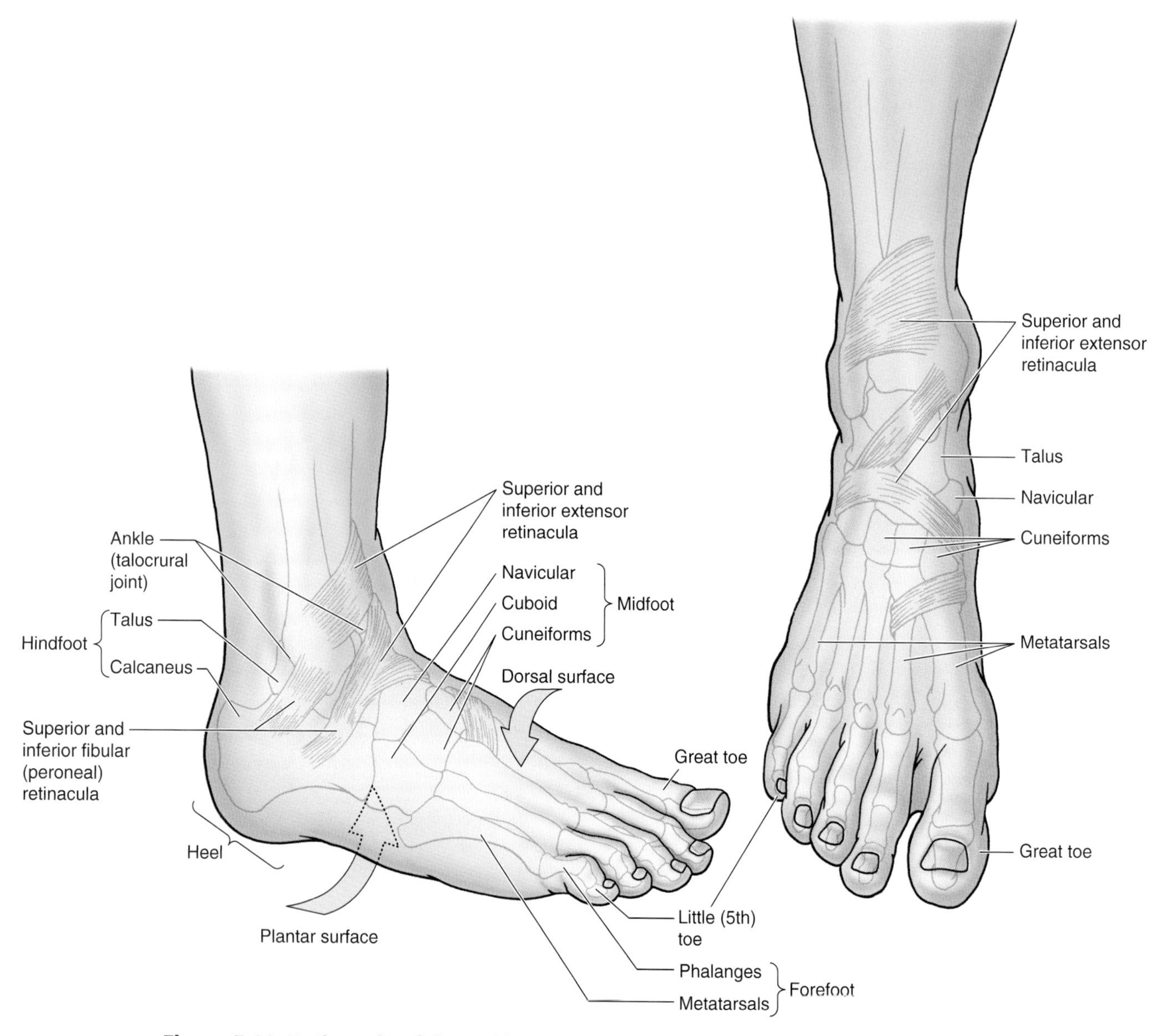

Figure 5.41. Retinacula of the ankle and parts of the foot. Observe the bones of the foot and the superior and inferior extensor and fibular retinacula.

The clinical importance of the foot is indicated by an estimate that physicians devote 20% of their practice to foot problems, and the practice of *podiatry* is concerned with the diagnosis and treatment of diseases, injuries, and abnormalities of the foot.

The part of the foot facing the floor or ground is the *plantar surface* (sole of the foot) and the part facing superiorly is the *dorsal surface* (dorsum of the foot). The part of the sole of the foot underlying the calcaneus is the *heel* and the part of the sole underlying the heads of the metatarsals is the *ball of the foot*. The great toe is the *hallux* or 1st digit; the small toe is the 5th digit (L. digitus minimus).

Skin of the Foot

The skin of the dorsal surface of the foot is much thinner and less sensitive than skin on the plantar surface. The subcutaneous tissue is loose deep to the dorsal skin; therefore, edema (G. oidēma, a swelling) is most marked over this surface, especially anterior to and around the medial malleolus. The skin over the major weightbearing areas of the sole of the foot—heel, lateral margin of foot, ball of great toe—is thick. The subcutaneous tissue in the sole is more fibrous than in other areas of the foot. Fibrous septa divide this tissue into fat-filled areas, making it a shock-absorbing pad, especially over the heel. The fibrous septa also anchor the skin to the underlying plantar aponeurosis, improving the "grip" of the sole. The plantar skin is hairless and sweat glands are numerous; the entire sole of the foot is sensitive ("ticklish").

Deep Fascia of the Foot

The deep fascia is thin on the dorsum of the foot, where it is continuous with the **inferior extensor retinaculum** (Fig.

5.42*A*). Over the lateral and posterior aspects of the foot, the deep fascia is continuous with the **plantar fascia**—deep fascia of the sole (Fig. 5.42, *B* and *C*). The thick, central part of this fascia forms the strong **plantar aponeurosis**—longitudinally arranged bands of dense fibrous connective tissue—which has a thick central part and weaker medial and lateral parts. The plantar fascia:

- Holds the parts of the foot together
- Helps protect the plantar surface of the foot from injury
- Helps support the longitudinal arches of the foot.

The plantar aponeurosis arises posteriorly from the calcaneus and divides into five bands that split to enclose the digital tendons that attach to the margins of the fibrous digital sheaths and the sesamoid bones of the great toe.

From the margins of the central part of the plantar

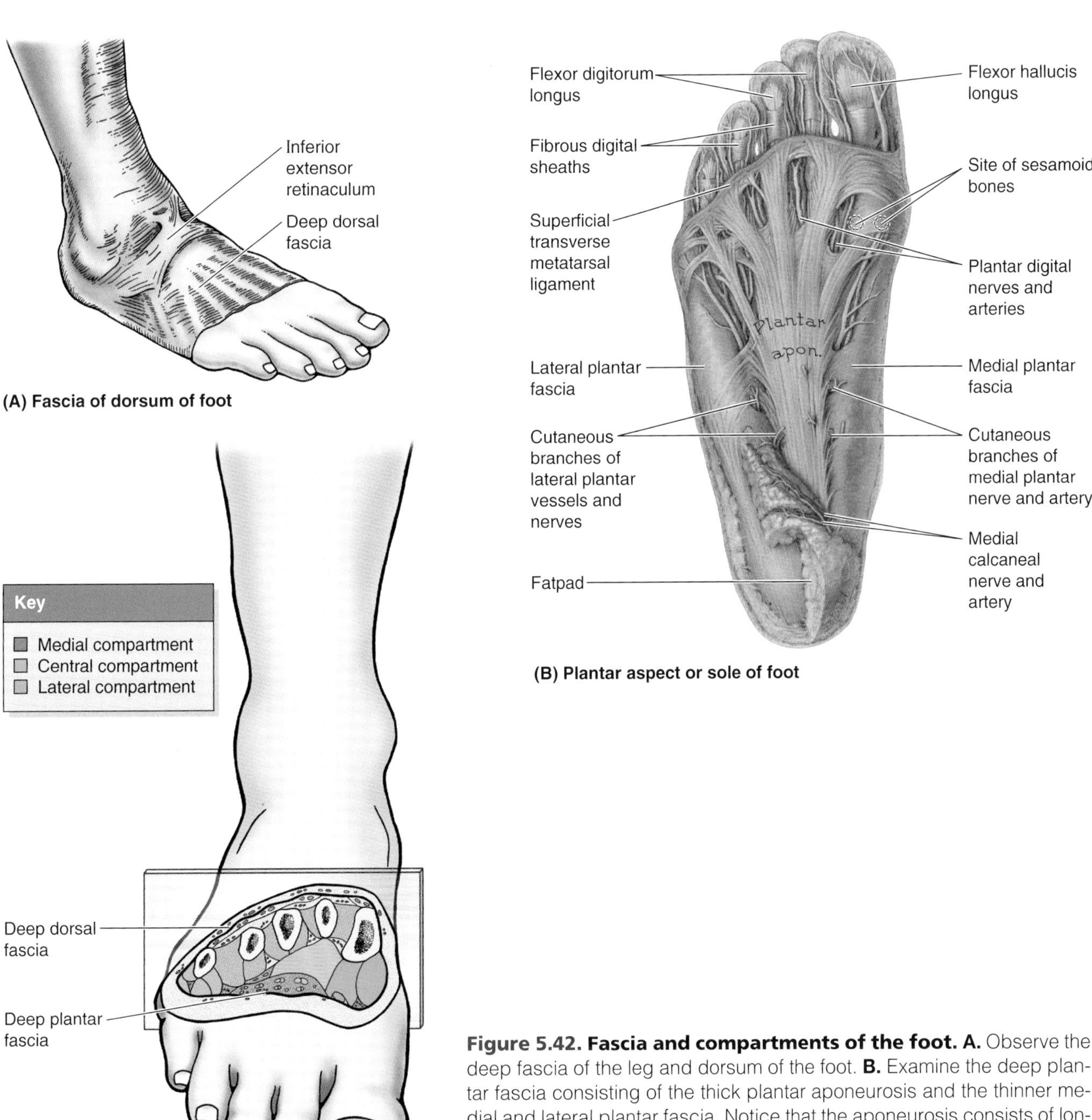

Figure 5.42. Fascia and compartments of the foot. A. Observe the deep fascia of the leg and dorsum of the foot. **B.** Examine the deep plantar fascia consisting of the thick plantar aponeurosis and the thinner medial and lateral plantar fascia. Notice that the aponeurosis consists of longitudinal bands of dense fibrous connective tissue. Observe also the plantar digital vessels and nerves. **C.** Note the deep dorsal fascia, deep plantar fascia, and large central and smaller medial and lateral compartments of the foot.

aponeurosis, vertical septa extend deeply to form three compartments of the sole of the foot (Fig. 5.42C):

- *Medial compartment*—containing the abductor hallucis, flexor hallucis brevis, and medial plantar nerve and vessels
- *Central compartment*—containing the flexor digitorum brevis, flexor digitorum longus, quadratus plantae, lumbricals, proximal part of the tendon flexor hallucis longus, and the lateral plantar nerve and vessels
- *Lateral compartment*—containing the abductor and flexor digiti minimi brevis.

The muscles, nerves, and vessels in the sole are described according to these compartments; however, the muscles are more easily dissected in layers than by compartments.

Plantar Fasciitis

Straining and inflammation of the plantar aponeurosis may result from running and high-impact aerobics, especially when inappropriate footwear such as worn-out shoes are worn. Plantar fasciitis causes pain on the plantar surface of the heel and on the medial aspect of the foot. The pain is often most severe after sitting and when beginning to walk in the morning. Point tenderness is located at the proximal attachment of the aponeurosis to the medial tubercle of the calcaneus and on the medial surface of the bone. The pain increases with passive dorsiflexion of the great toe. If a *calcaneal spur* (abnormal bony process) protrudes from the medial calcaneal tubercle, the plantar fasciitis may produce the "heel spur syndrome." Usually a bursa develops at the end of the spur that may also become inflamed and tender. ⊙

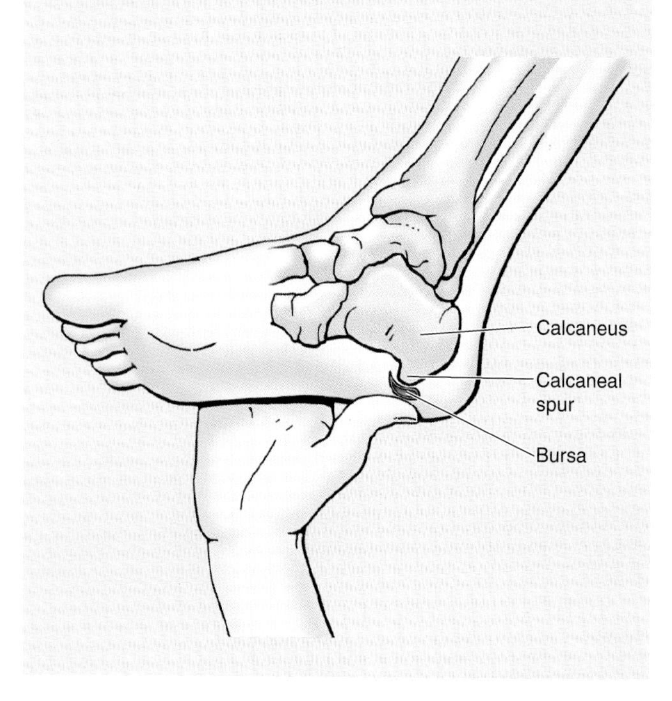

Muscles of the Foot

The four muscular layers in the sole of the foot (Figs. 5.43 and 5.44, Table 5.13) help maintain the arches of the foot and enable one to stand on uneven ground. The muscles are of little importance individually because fine control of the individual toes is not important to most people.

The 1st layer of plantar muscles consists of:

- Abductor hallucis—the abductor of the great toe
- Flexor digitorum brevis—the flexor of the lateral four toes
- Abductor digiti minimi—the abductor of the small toe.

The 2nd layer of plantar muscles consists of:

- Quadratus plantae (flexor accessorius)—the flexor of the lateral four toes
- Tendons of flexor hallucis longus and flexor digitorum longus—the long flexors of the toes
- Lumbricals (L. lumbricus, earthworm)—the wormlike flexors of the proximal phalanges and extensors of the middle and distal phalanges of lateral four toes.

The 3rd layer of plantar muscles consists of:

- Flexor hallucis brevis—the flexor of the proximal phalanx of the great toe
- Adductor hallucis (transverse and oblique heads)—the adductor of the great toe
- Flexor digiti minimi brevis—the flexor of the proximal phalanx of the small toe.

The 4th layer of plantar muscles consists of:

- Three plantar interossei—the adductors of digits 2 to 4 and the flexors of the metatarsophalangeal joints
- Four dorsal interossei—the abductors of digits 2 to 4 and the flexors of the metatarsophalangeal joints.

In Table 5.13, note that the:

- **P**lantar interossei **AD**duct (**PAD**) and arise from a single metatarsal
- **D**orsal interossei **AB**duct (**DAB**) and arise from two metatarsals.

Two neurovascular planes are in the foot (Fig. 5.43):

- A superficial one between the 1st and 2nd muscular layers
- A deep one between the 3rd and 4th muscular layers.

Figure 5.43. Layers of plantar muscles. A. First layer, consisting of the abductors of the large and small toes, and the short flexor of the toes. **B.** Second layer, consisting of the long flexor tendons and associated muscles: four lumbricals and the quadratus plantae. **C.** Third layer, consisting of the flexor of the small (little) toe and the flexor and adductor of the large (big) toe. **D.** Fourth layer, consisting of the dorsal and plantar interosseous muscles.

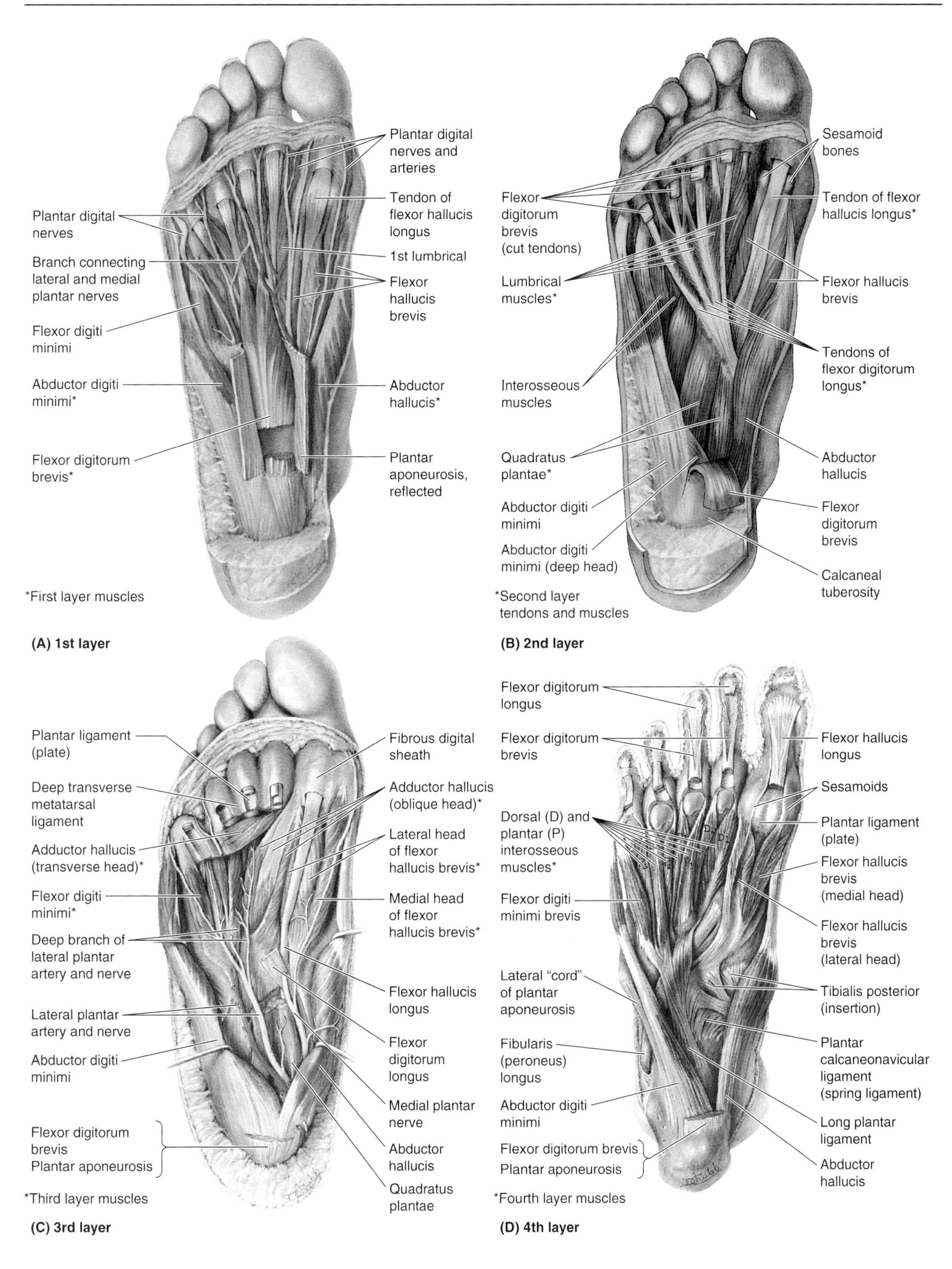
Plantar digital nerves and arteries
Tendon of flexor hallucis longus
Plantar digital nerves
Branch connecting lateral and medial plantar nerves
1st lumbrical
Flexor hallucis brevis
Flexor digiti minimi
Abductor digiti minimi*
Abductor hallucis*
Flexor digitorum brevis*
Plantar aponeurosis, reflected
*First layer muscles
(A) 1st layer
Sesamoid bones
Flexor digitorum brevis (cut tendons)
Tendon of flexor hallucis longus*
Lumbrical muscles*
Flexor hallucis brevis
Tendons of flexor digitorum longus*
Interosseous muscles
Quadratus plantae*
Abductor hallucis
Abductor digiti minimi
Flexor digitorum brevis
Abductor digiti minimi (deep head)
Calcaneal tuberosity
*Second layer tendons and muscles
(B) 2nd layer
Plantar ligament (plate)
Fibrous digital sheath
Deep transverse metatarsal ligament
Adductor hallucis (oblique head)*
Lateral head of flexor hallucis brevis*
Adductor hallucis (transverse head)*
Flexor digiti minimi*
Medial head of flexor hallucis brevis*
Deep branch of lateral plantar artery and nerve
Flexor hallucis longus
Lateral plantar artery and nerve
Flexor digitorum longus
Abductor digiti minimi
Medial plantar nerve
Flexor digitorum brevis
Plantar aponeurosis
Abductor hallucis
Quadratus plantae
*Third layer muscles
(C) 3rd layer
Flexor digitorum longus
Flexor digitorum brevis
Flexor hallucis longus
Sesamoids
Dorsal (D) and plantar (P) interosseous muscles*
Plantar ligament (plate)
Flexor hallucis brevis (medial head)
Flexor digiti minimi brevis
Flexor hallucis brevis (lateral head)
Lateral "cord" of plantar aponeurosis
Tibialis posterior (insertion)
Plantar calcaneonavicular ligament (spring ligament)
Fibularis (peroneus) longus
Abductor digiti minimi
Long plantar ligament
Flexor digitorum brevis
Plantar aponeurosis
Abductor hallucis
*Fourth layer muscles
(D) 4th layer

Table 5.13. **Muscles in the Sole of the Foot**

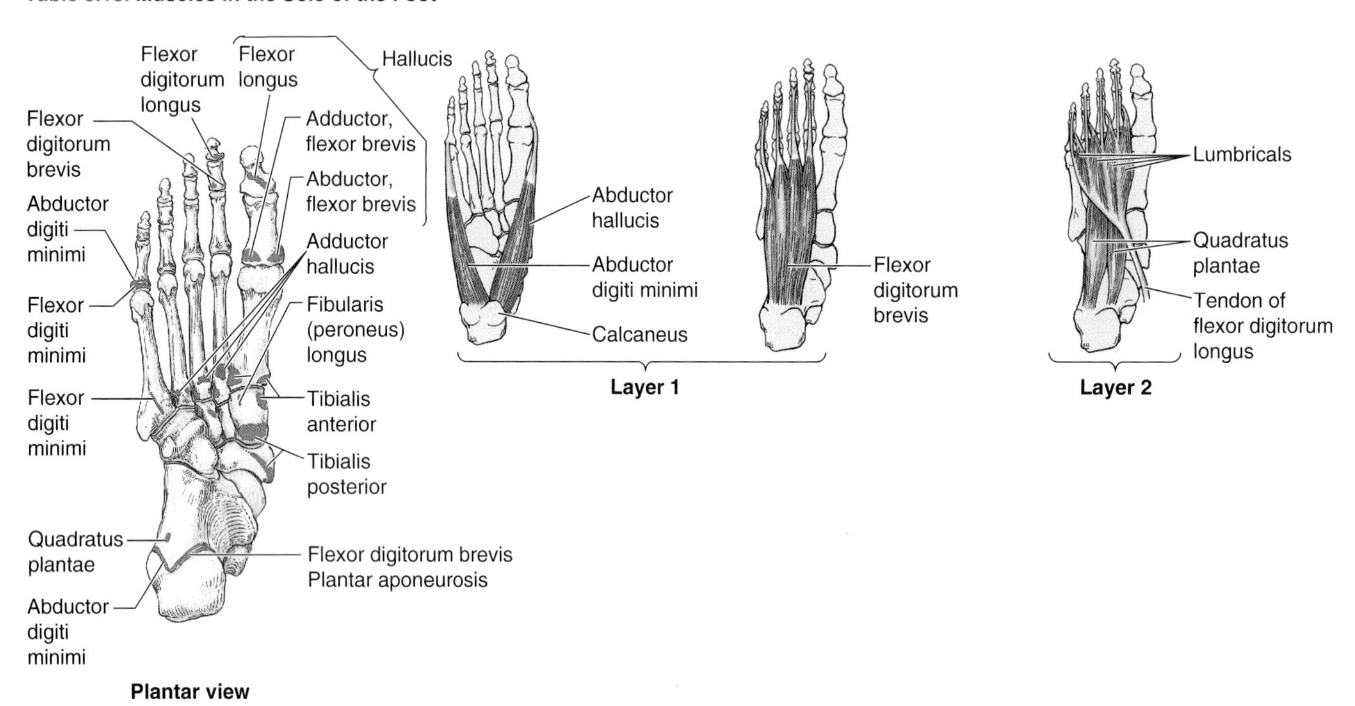

Muscle	Proximal Attachment	Distal Attachment	Innervation	Main Action[a]
Layer 1				
Abductor hallucis	Medial tubercle of tuberosity of calcaneus, flexor retinaculum, and plantar aponeurosis	Medial side of base of proximal phalanx of 1st digit	Medial plantar nerve (S2 and S3)	Abducts and flexes 1st digit (great toe, hallux)
Flexor digitorum brevis	Medial tubercle of tuberosity of calcaneus, plantar aponeurosis, and intermuscular septa	Both sides of middle phalanges of lateral four digits		Flexes lateral four digits
Abductor digiti minimi	Medial and lateral tubercles of tuberosity of calcaneus, plantar aponeurosis, and intermuscular septa	Lateral side of base of proximal phalanx of 5th digit	Lateral plantar nerve (S2 and S3)	Abducts and flexes 5th digit
Layer 2				
Quadratus plantae	Medial surface and lateral margin of plantar surface of calcaneus	Posterolateral margin of tendon of flexor digitorum longus	Lateral plantar nerve (S2 and S3)	Assists flexor digitorum longus in flexing lateral four digits
Lumbricals	Tendons of flexor digitorum longus	Medial aspect of expansion over lateral four digits	Medial one: medial plantar nerve (S2 and S3) Lateral three: lateral plantar nerve (S2 and S3)	Flex proximal phalanges and extend middle and distal phalanges of lateral four digits

[a] In spite of individual actions, the primary function of the intrinsic muscles of the sole of the foot is to resist flattening or maintain the longitudinal arch of the foot.

Table 5.13. *(Continued)* **Muscles in the Sole of the Foot**

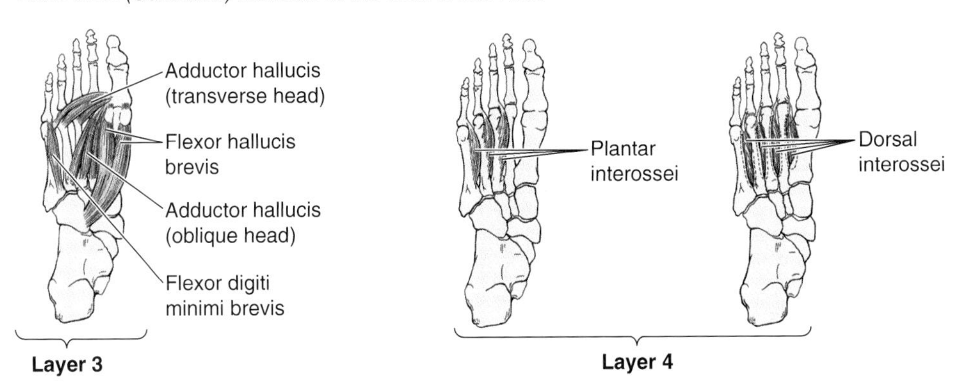

Muscle	Proximal Attachment	Distal Attachment	Innervation	Main Action[a]
Layer 3				
Flexor hallucis brevis	Plantar surfaces of cuboid and lateral cuneiforms	Both sides of base of proximal phalanx of 1st digit	Medial plantar nerve (S2 and S3)	Flexes proximal phalanx of 1st digit
Adductor hallucis	Oblique head: bases of metatarsals 2–4 Transverse head: plantar ligaments of metatarso-phalangeal joints	Tendons of both heads attach to lateral side of base of proximal phalanx of 1st digit	Deep branch of lateral plantar nerve (S2 and S3)	Adducts 1st digit; assists in maintaining transverse arch of foot
Flexor digiti minimi brevis	Base of the 5th metatarsal	Base of proximal phalanx of 5th digit	Superficial branch of lateral plantar nerve (S2 and S3)	Flexes proximal phalanx of 5th digit, thereby assisting with its flexion
Layer 4				
Plantar interossei (three muscles)	Bases and medial sides of metatarsals 3–5	Medial sides of bases of proximal phalanges of 3rd to 5th digits	Lateral plantar nerve (S2 and S3)	Adduct digits (2–4) and flex metatarsophalangeal joints
Dorsal interossei (four muscles)	Adjacent sides of metatarsals 1–5	First: medial side of proximal phalanx of 2nd digit Second to fourth: lateral sides of 2nd to 4th digits		Abduct digits (2–4) and flex metatarsophalangeal joints

The lateral plantar artery and nerve course laterally between the muscles of the 1st and 2nd layers of plantar muscles. Their deep branches course medially between the muscles of the 3rd and 4th layers.

Two closely connected muscles on the dorsum of the foot are the *extensor digitorum brevis* and *extensor hallucis brevis*. These thin broad muscles form a fleshy mass on the lateral part of the dorsum of the foot, anterior to the lateral malleolus (Fig. 5.34). The extensor hallucis brevis is part of the extensor digitorum brevis. Its small fleshy belly may be felt when the toes are extended. The extensor digitorum brevis extends digits 2 to 4 at the metatarsophalangeal joints, and the extensor hallucis brevis extends the great toe. Both these muscles help the long extensors to extend the toes.

Contusion of the Extensor Digitorum Brevis

Functionally, the extensor digitorum brevis and extensor hallucis brevis muscles are relatively unimportant. Clinically, knowing the location of the belly of the extensor digitorum brevis is important for distinguishing it from an abnormal edema. Contusion and tearing of the muscle's fibers and associated blood vessels result in a *hematoma*, producing edema anteromedial to the lateral malleolus. Most people who have not seen this inflamed muscle assume they have a badly sprained ankle. ⊙

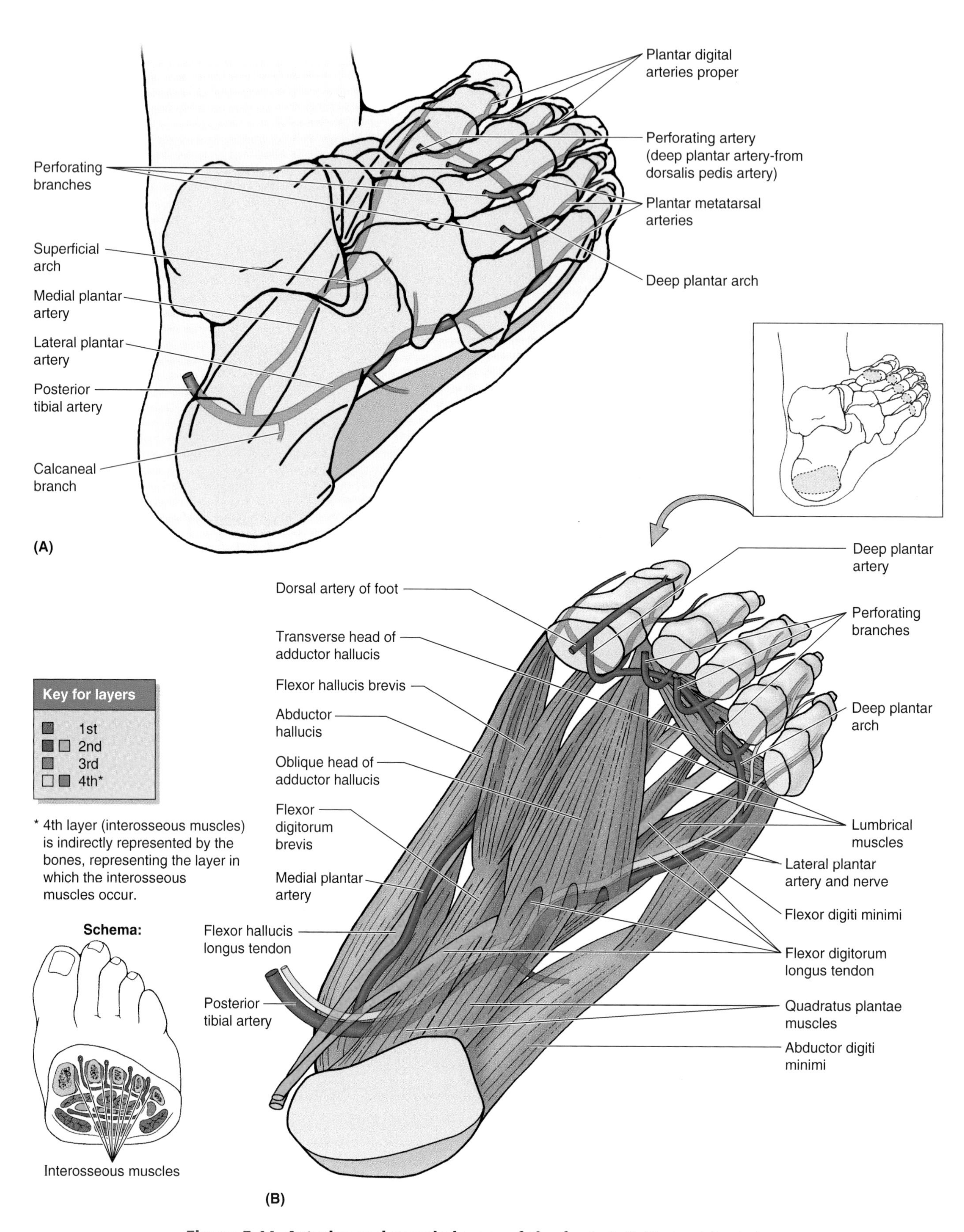

Figure 5.44. Arteries and muscle layers of the foot. A–B. Posterolateral views. Observe the posterior tibial artery terminating as it enters the foot by dividing into the medial and lateral plantar arteries. Observe also the distal anastomoses of these vessels with the deep plantar artery from the dorsal artery of the foot (L. dorsalis pedis artery) and the perforating branches to the arcuate artery on the dorsum of the foot (Fig. 5.45). Examine the relationship of the plantar arteries to four muscle layers in the foot.

Nerves of the Foot

The **tibial nerve** divides posterior to the medial malleolus into the *medial and lateral plantar nerves* (Table 5.14). These nerves supply the intrinsic muscles of the foot, except for the extensor digitorum brevis, which is supplied by the **deep fibular nerve**. *The cutaneous innervation of the foot is supplied by the:*

- Saphenous nerve—medial side of the foot as far as the head of 1st metatarsal
- Superficial and deep fibular nerves—dorsum of the foot
- Medial and lateral plantar nerves—sole of the foot
- Sural nerve—lateral aspect of the foot, including part of the heel
- Calcaneal branches of the tibial and sural nerves—heel.

Saphenous Nerve

The saphenous nerve is the largest cutaneous branch of the femoral nerve. In addition to supplying the skin and fascia on the anterior and medial sides of the leg, the saphenous nerve passes anterior to the medial malleolus to the dorsum of the foot, where it supplies skin along the medial side of the foot as far anteriorly as the head of the 1st metatarsal.

Superficial and Deep Fibular Nerves

The superficial fibular nerve lies between the fibular muscles and leaves them around the junction of the middle two-thirds and the inferior third of the lateral side of the leg. The nerve ends by supplying the skin on the dorsum of the foot. The **deep fibular nerve** passes deep to the extensor retinaculum and supplies the skin on the contiguous sides of the 1st and 2nd toes.

Medial Plantar Nerve

The medial plantar nerve, the larger of the two terminal branches of the tibial nerve, passes deep to the abductor hallucis and runs anteriorly between this muscle and the flexor digitorum brevis on the lateral side of the medial plantar artery. The medial plantar nerve terminates near the bases of the metatarsals by dividing into three sensory branches, which supply cutaneous branches to the medial three and a half digits and motor branches to the abductor hallucis, flexor digitorum brevis, flexor hallucis brevis, and the most medial lumbrical muscles.

Lateral Plantar Nerve

The lateral plantar nerve, the smaller of the two terminal branches of the tibial nerve, begins deep to the flexor retinaculum and the abductor hallucis and runs anterolaterally, medial to the lateral plantar artery and between the 1st and 2nd layers of plantar muscles. The lateral plantar nerve terminates by dividing into superficial and deep branches. The superficial branch divides into two digital nerves that send cutaneous branches to the lateral one and a half digits. The superficial and deep branches of the lateral plantar nerve supply motor branches to muscles of the sole that are not supplied by the medial plantar nerve.

Sural Nerve

The sural nerve is formed by union of branches from the tibial and common fibular nerves (Table 5.10). The level of junction of these branches is variable; sometimes these nerves do not join and therefore no sural nerve forms. In these people, the skin normally innervated by the sural nerve is supplied by the tibial and fibular branches. *The sural nerve accompanies the small saphenous vein and enters the foot posterior to the lateral malleolus* to supply skin along the lateral margin of the foot and the lateral side of the 5th digit (Table 5.14).

Sural Nerve Grafts

Pieces of the sural nerve are often used for nerve grafts in procedures such as repairing nerve defects resulting from wounds. Because of the variations in the level of formation of the sural nerve, the union of the tibial and common fibular nerves may be high (in the popliteal fossa) or low (proximal to heel). Sometimes the medial sural cutaneous nerve does not unite with the communicating branch of the common fibular nerve; thus, no sural nerve is present.

Anesthetic Block of the Superficial Fibular Nerve

After the superficial fibular nerve pierces the deep fascia, it divides into medial and lateral branches. In thin people, these branches can often be seen or felt as ridges under the skin when the foot is plantarflexed. Injections of an anesthetic agent around these branches anesthetizes the skin on the dorsum of the foot for any superficial surgery.

Plantar Reflex

The plantar reflex (L4, L5, S1, and S2 nerve roots) is a deep tendon reflex that is routinely tested during neurological examinations. The lateral aspect of the sole of the foot is stroked with a blunt object, such as a tongue depressor, beginning at the heel and crossing to the base of the great toe. The motion is firm and continuous but neither painful nor ticklish. Flexion of the toes is a normal response. Slight fanning of the lateral four toes and *dorsiflexion of the great toe is an abnormal response* (*Babinski sign*), indicating brain injury or cerebral disease, except in infants. Because the corticospinal tracts are not fully developed in newborns, a Babinski sign is usually elicited and may be present until children are 4 years of age. ▶

Medial Plantar Nerve Entrapment

Compressive irritation of the medial plantar nerve as it passes deep to the flexor retinaculum or curves deep to the abductor hallucis may cause aching, burning, numbness, and tingling (paresthesia) on the medial side of the sole and in the region of the navicular tuberosity. *Medial plantar nerve compression* may occur during repetitive eversion of the foot (e.g., during gymnastics and running). Because of its frequency in runners, these symptoms have been called "jogger's foot."

Table 5.14. **Nerves of the Foot**

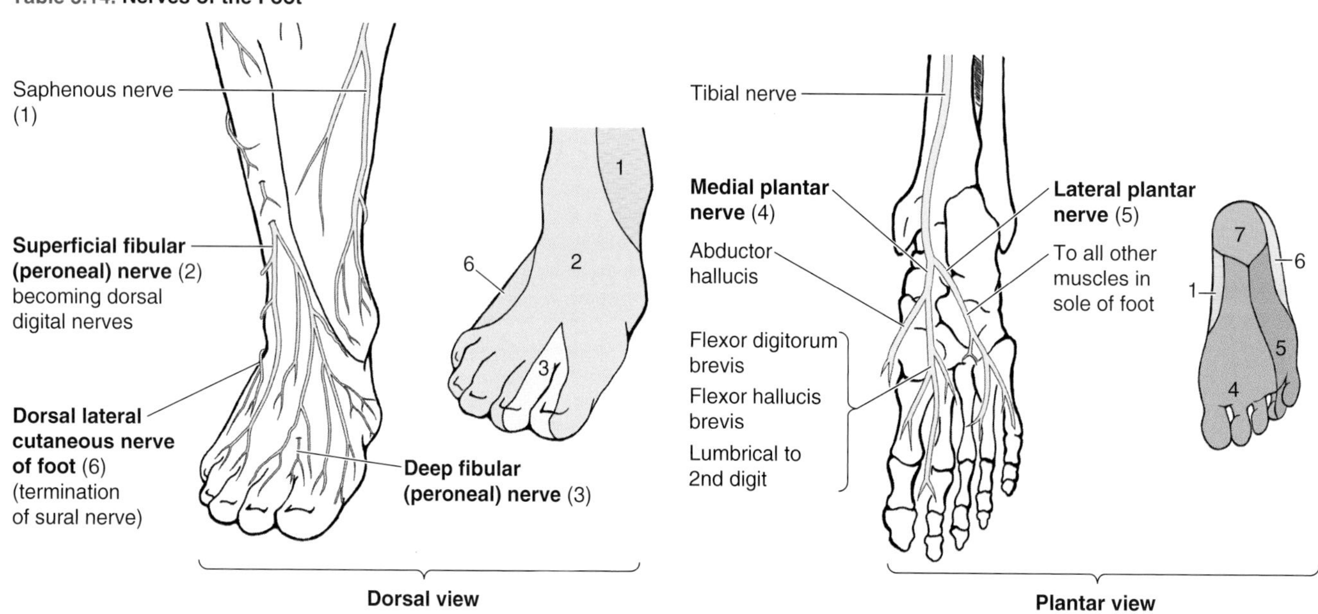

Nerve	Origin	Course	Distribution in the Foot
Saphenous (1)	Femoral nerve	Arises in femoral triangle and descends through thigh and leg; accompanies great saphenous vein anterior to medial malleolus and ends on medial side of foot	Supplies skin on medial side of foot as far anteriorly as head of 1st metatarsal
Superficial fibular (2)	Common fibular (peroneal) nerve	Pierces deep fascia in distal third of leg to become cutaneous and send branches to foot and digits	Supplies skin on dorsum of foot and all digits, except lateral side of 5th and adjoining sides of the 1st and 2nd digits
Deep fibular (3)		Passes deep to extensor retinaculum to enter dorsum of foot	Supplies extensor digitorum brevis and skin on contiguous sides of 1st and 2nd digits
Medial plantar (4)	Larger terminal branch of the tibial nerve	Passes distally in foot between abductor hallucis and flexor digitorum brevis and divides into muscular and cutaneous branches	Supplies skin of medial side of sole of foot and sides of first three digits; also supplies abductor hallucis, flexor digitorum brevis, flexor hallucis brevis, and first lumbrical nerve
Lateral plantar (5)	Smaller terminal branch of the tibial nerve	Passes laterally in foot between quadratus plantae and flexor digitorum brevis muscles and divides into superficial and deep branches	Supplies quadratus plantae, abductor digiti minimi, and flexor digiti minimi brevis; deep branch supplies plantar and dorsal interossei, lateral three lumbricals, and adductor hallucis; supplies skin on sole lateral to a line splitting 4th digit
Sural (6)	Usually arises from both tibial and common fibular (peroneal) nerves	Passes inferior to the lateral malleolus to lateral side of foot	Lateral aspect of foot
Calcaneal branches (7)	Tibial and sural nerves	Pass from distal part of the posterior aspect of leg to skin on heel	Skin of heel

Arteries of the Foot

The arteries of the foot are terminal branches of the anterior and posterior tibial arteries (Fig. 5.45), respectively the dorsal and plantar arteries.

Dorsal Artery of the Foot

The dorsal artery of the foot—the *major source of blood supply to the toes*—is the direct continuation of the anterior tibial artery (Fig. 5.45*A*). The dorsal artery begins midway between the malleoli and runs anteromedially, deep to the inferior extensor retinaculum between the extensor hallucis longus and the extensor digitorum longus tendons on the dorsum of the foot. The dorsal artery passes to the 1st interosseous space, where it divides into a deep plantar artery that passes to the sole of the foot, where it joins the plantar arch and the 1st *dorsal metatarsal artery*.

Lateral Tarsal Artery. The lateral tarsal artery, a branch of the dorsal artery of the foot, runs laterally in an arched course beneath the extensor digitorum brevis to supply this muscle and the underlying tarsals and joints. It anastomoses with other branches such as the *arcuate artery*. The **deep plantar artery** passes deeply through the 1st interosseous space to participate in the formation of the *deep plantar arch* by joining the lateral plantar artery.

First Dorsal Metatarsal Artery. The 1st dorsal metatarsal artery divides into branches that supply both sides of the great toe and the medial side of the 2nd toe.

Arcuate Artery. The arcuate artery runs laterally across the bases of the lateral four metatarsals, deep to the extensor tendons, where it gives off the 2nd, 3rd, and 4th dorsal **metatarsal arteries**. These vessels run to the clefts of the toes, where each of them divides into two **dorsal digital arteries** for the sides of adjoining toes. The metatarsal arteries are connected to the plantar arch and to the plantar metatarsal arteries by *perforating arteries*.

Palpation of the Dorsalis Pedis Pulse

The *dorsalis pedis pulse*—the pulse of the dorsalis pedis artery, or dorsal artery of the foot—is evaluated during a physical examination of the peripheral vascular system. Dorsalis pedis pulses may be palpated with the ▶

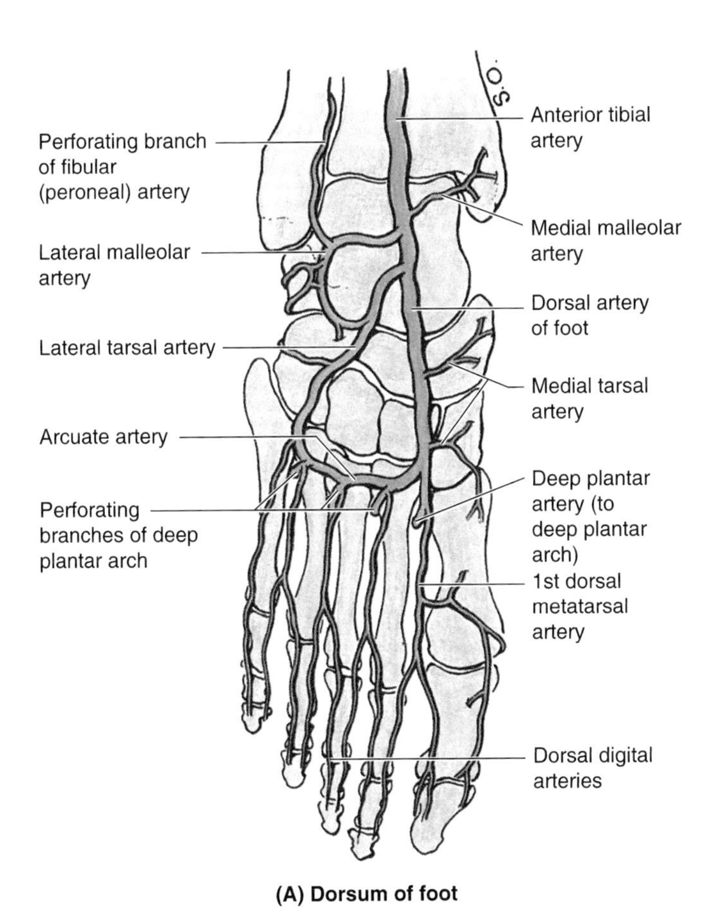

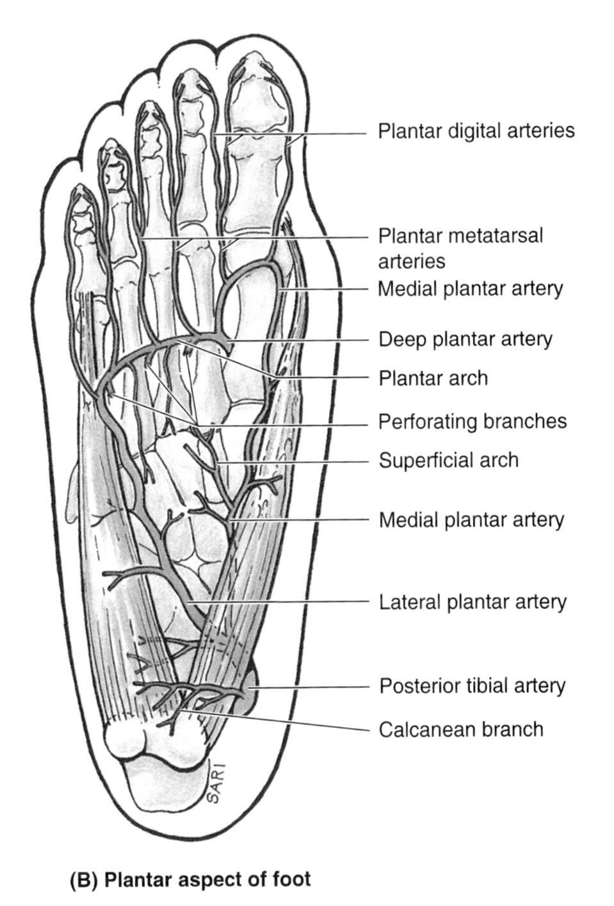

Figure 5.45. Arteries of the foot. A. Observe the anterior tibial artery becoming the dorsal artery of the foot (L. dorsalis pedis artery) and the arcuate artery. **B.** Observe the posterior tibial artery and its terminal branches—the medial and lateral plantar arteries. The deep plantar arch is formed by the lateral plantar artery. In (**A**) and (**B**), note the anastomoses between the dorsal and plantar arteries through the deep plantar artery and perforating branches of the deep plantar arch.

▶ feet slightly dorsiflexed. The pulses are usually easy to palpate because the arteries are subcutaneous and pass along a line from the extensor retinaculum to a point just lateral to the extensor hallucis longus tendons (Swartz, 1994). Some healthy adults—and even children—have *congenitally nonpalpable dorsalis pedis pulses*; the anomaly is usually bilateral. In these cases, the dorsalis pedis artery is replaced by an enlarged perforating fibular artery. *A diminished or absent dorsalis pedis pulse usually suggests vascular insufficiency resulting from arterial disease. Five "P signs" of acute arterial occlusion are* pain, pallor, paresthesia, paralysis, and pulselessness. ⊙

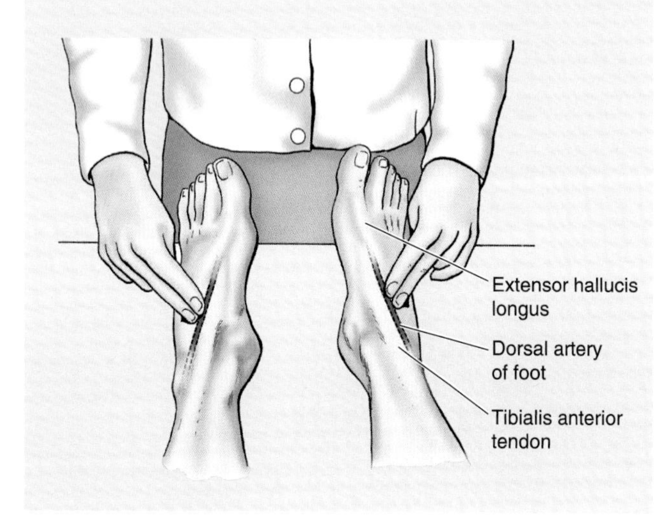

Sole of the Foot

The sole of the foot has a prolific blood supply. The arteries derive from the **posterior tibial artery** (Fig. 5.45*B*), which divides deep to the abductor hallucis to form the medial and lateral *plantar arteries.* They run parallel to the similarly named nerves.

Medial Plantar Artery. The medial plantar artery is small and supplies mainly the muscles of the great toe; however, most *plantar digital arteries* arise from this vessel. The superficial branch of the medial plantar artery helps supply the skin on the medial side of the sole and has digital branches that accompany digital branches of the medial plantar nerve; however, these digital arterial branches contribute little to the circulation of the toes.

Lateral Plantar Artery. The lateral plantar artery, much larger than the medial plantar artery, accompanies the nerve of the same name. It runs laterally and anteriorly, at first deep to the abductor hallucis and then deep to the flexor digitorum brevis (Fig. 5.45*B*). The lateral plantar artery arches medially across the foot with the deep branch of the lateral plantar nerve to form the **deep plantar arch**, which is completed by the *medial plantar artery.* The deep plantar arch begins opposite the base of the 5th metatarsal and is completed medially by union with the **deep plantar artery** (Fig. 5.45*A*), a branch of the dorsal artery of the foot. As it crosses the foot, the deep plantar arch gives off four **plantar metatarsal arteries** (Fig. 5.45*B*) and three **perforating arteries** and many branches to the skin, fascia, and muscles in the sole of the foot. These arteries join with the superficial branches of the medial and lateral plantar arteries to form the **plantar digital arteries**, supplying the adjacent digits.

Puncture Wounds of the Sole of the Foot

Puncture wounds of the sole of the foot involving the deep plantar arch and its branches usually result in severe bleeding. Ligature of the arch is difficult because of its depth and the structures that surround it.

Infections of the Foot

Foot infections are common, and infected areas may be drained according to their location. The palmar fascial spaces are usually incised and drained on the medial side of the foot so that a painful scar will not be in a weightbearing area. ⊙

Venous Drainage of the Foot

Dorsal digital veins running along the dorsum of each toe are continuous with the **dorsal metatarsal veins**, which join to form the **dorsal venous arch** in the subcutaneous tissue (Fig. 5.46*A*). The dorsal venous arch communicates with the **plantar venous arch.** Veins leave the dorsal venous arch and converge medially to form the **great saphenous vein** and laterally to form the **small saphenous vein** (Fig. 5.46*B*). The superficial veins of the sole unite to form a **plantar venous network** from which efferent vessels pass to medial and lateral **marginal veins** that join the great and small saphenous veins. The deep veins of the sole begin as **plantar digital veins** on the plantar aspects of the digits that communicate with the **dorsal digital veins** through perforating veins (Fig. 5.46*A*). Most blood returns from the foot through deep veins that accompany the arteries.

Lymphatic Drainage of the Foot

The lymphatics of the foot begin in subcutaneous plexuses (Fig. 5.47*A*). The collecting vessels consist of superficial and deep lymphatic vessels that follow the veins. Superficial lymphatic vessels are most numerous in the sole. They leave the foot medially along the *great saphenous vein* and laterally along the small saphenous vein. The **medial superficial lymphatic vessels**, larger and more numerous than the lateral ones, drain the dorsum and medial side of the dorsum of the foot and the sole. These vessels converge on the great saphenous vein and accompany it to the distal group of **superficial inguinal lymph nodes** (Fig. 5.47, *A* and *B*), located along the termination of the great saphenous vein.

The superficial inguinal lymph nodes drain mainly into

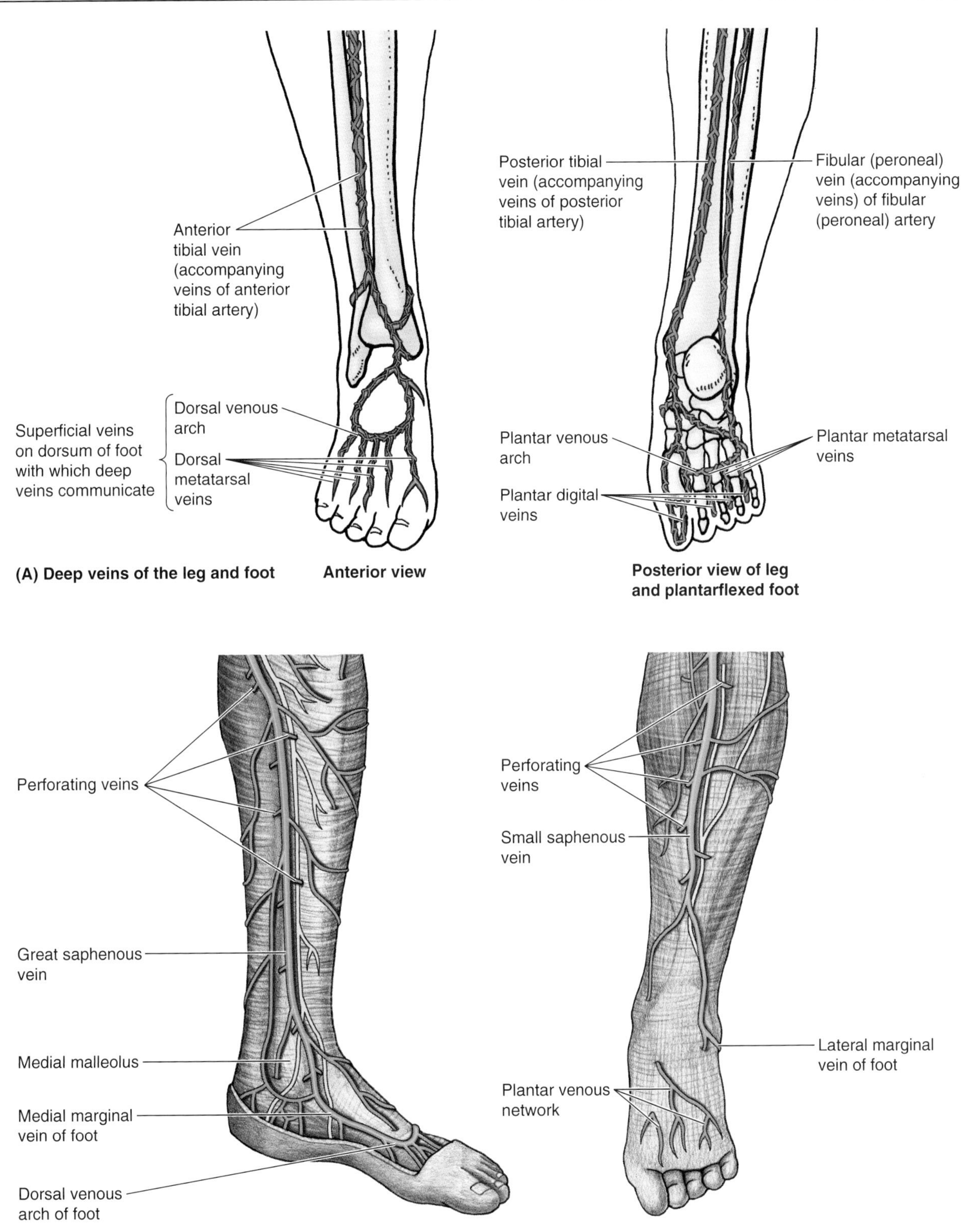

Figure 5.46. Veins of the leg and foot. A. The deep veins accompany the arteries and their branches (L. venae comitantes); they anastomose frequently and have numerous valves. **B.** The main superficial veins are the great and small saphenous veins, which drain into the deep veins by means of perforating veins so that muscular compression can propel blood toward the heart against the pull of gravity.

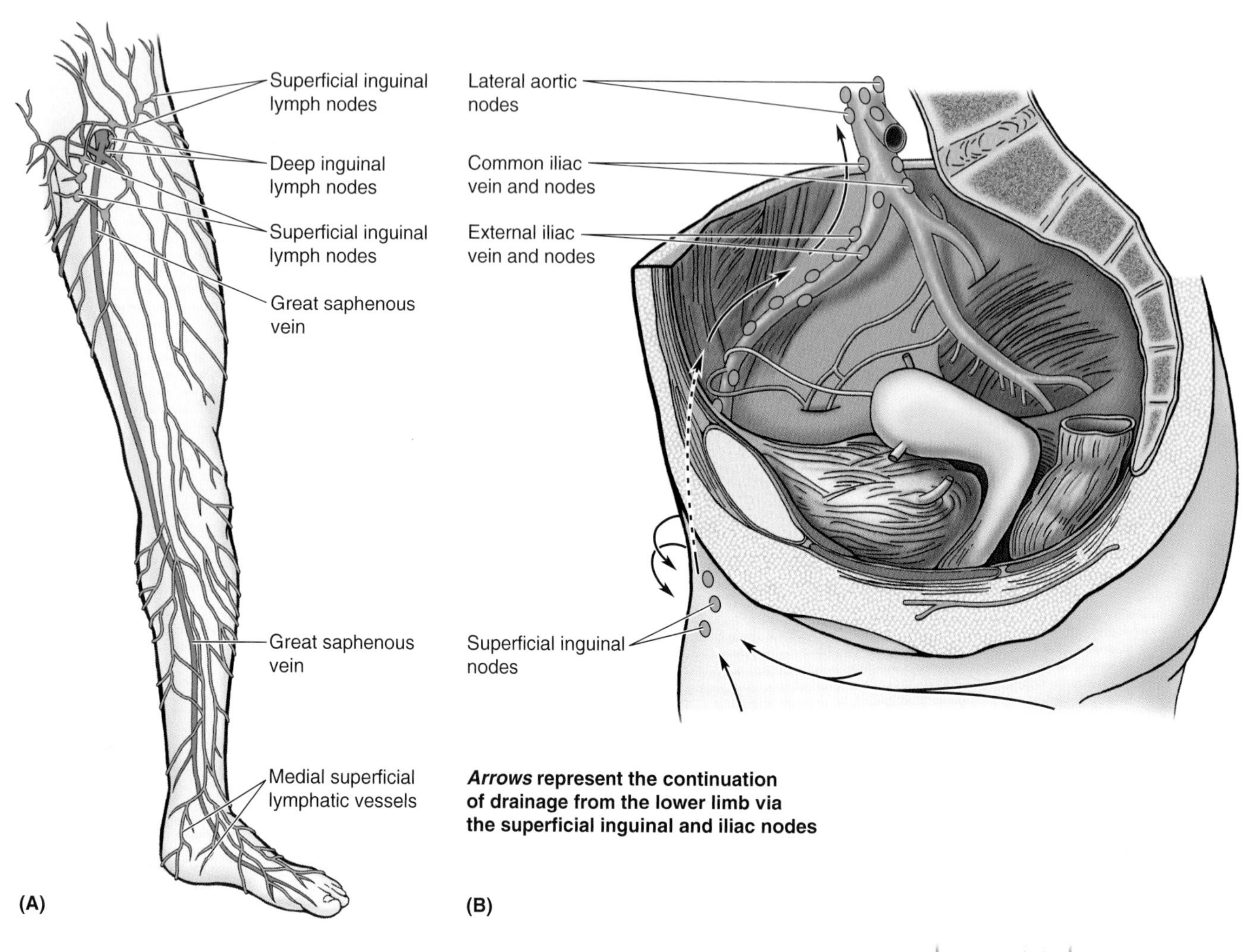

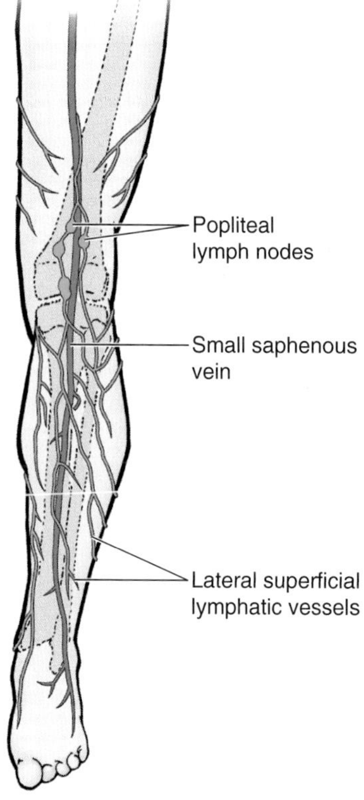

Figure 5.47. Lymphatic drainage of the lower limb. A. Superficial lymphatic vessels drain to the superficial inguinal lymph nodes. **B.** The *arrows* show drainage from the superficial inguinal lymph nodes to the external and common iliac lymph nodes on its way to the lateral aortic nodes. **C.** Lymphatic vessels from the lateral foot and posterolateral leg accompany the small saphenous vein and enter the popliteal lymph nodes. Deep lymphatic vessels then accompany the femoral vein, passing to the deep inguinal lymph nodes. Lymph from these nodes then joins the superficial drainage in traversing the external and common iliac nodes before entering the lateral aortic lymph nodes.

the **external iliac lymph nodes**, but some nodes drain into the deep inguinal nodes. The **lateral superficial lymphatic vessels** drain the lateral side of the foot and sole (Fig. 5.47*C*). Most of these vessels pass posterior to the lateral malleolus and accompany the *small saphenous vein* to the popliteal fossa, where they enter the popliteal lymph nodes. The **deep lymphatic vessels** follow the main blood vessels: anterior and posterior tibial, fibular, popliteal, and femoral. The deep vessels from the foot drain into the **popliteal lymph nodes**. Lymphatic vessels from them follow the femoral vessels, carrying lymph to the **deep inguinal lymph nodes**. From them lymph passes to the external iliac lymph nodes, common iliac nodes, and lateral aortic nodes.

Lymphadenopathy

Infections of the foot may spread proximally, causing enlargement of popliteal and inguinal lymph nodes (*lymphadenopathy*). Inflammation of the popliteal nodes often results from *lateral lesions of the heel* (Williams et al., 1995). Infections on the lateral side of the foot and sole initially produce enlargement of popliteal lymph nodes (*popliteal lymphadenopathy*); later, the inguinal lymph nodes may enlarge. *Inguinal lymphadenopathy* can result from infection of skin on the leg and/or foot; however, enlargement of these nodes can also result from an infection or tumor in the vulva, penis, scrotum, perineum, and gluteal region and from terminal parts of the urethra, anal canal, and vagina.

Lymphangiography

Because of the numerous large lymphatic vessels in the subcutaneous tissue on the dorsum of the foot, a contrast medium injected into the foot is transported by these vessels to the inguinal and iliac lymph nodes. A radiograph shows the lymph nodes outlined by contrast medium so that their size and number can be studied to help determine the cause of the lymphadenopathy. *Lymphangiography is gradually being replaced by other diagnostic imaging techniques such as computed tomography (CT) and magnetic resonance imaging (MRI)*, which do not necessitate the injection of contrast media. ⊙

Joints of the Lower Limb

The joints of the lower limb include the joints of the pelvic girdle, lumbosacral joints, sacroiliac joints, and pubic symphysis (Fig. 5.48), which are discussed in Chapter 3. The remaining joints of the lower limb are the hip joint, knee joint, tibiofibular joints, ankle joint, and foot joints.

Hip Joint

The hip joint forms the connection between the lower limb and the pelvic girdle. It is a strong and stable multiaxial ball-and-socket type of synovial joint—the femoral head is the ball and the acetabulum is the socket (Fig. 5.49). The hip joint is designed for stability as well as for a wide range of movement. Next to the shoulder joint, it is the most movable of all joints. During standing, the entire weight of the upper body is transmitted through the hip bones to the heads and necks of the femurs. The hip joint is mechanically most stable when a person is bearing weight, when lifting a heavy object, for example.

Articular Surfaces of the Hip Joint

The round head of the femur articulates with the cuplike acetabulum of the hip bone (Figs. 5.48–5.50). Because the depth of the acetabulum is increased by the fibrocartilaginous **acetabular labrum** (L. labrum, lip), which attaches to the bony rim of the acetabulum and the **transverse acetabular ligament**, more than half of the head fits within the acetabulum. The acetabular labrum "grasps" the femoral head beyond its equator. The head of the femur is covered with articular cartilage, except for the pit (L. fovea) for the ligament of the femoral head. The central and inferior part of the acetabulum, the **acetabular fossa**, is thin, nonarticular, and often translucent.

Articular Capsule of the Hip Joint

The strong, loose **fibrous capsule** permits free movement of the hip joint; it attaches proximally to the acetabulum and transverse acetabular ligament (Figs. 5.49 and 5.50). Some parts of the fibrous capsule are thicker than others and are called ligaments—the iliofemoral ligament, for example (Fig. 5.51). *The fibrous capsule attaches distally to the neck of the femur only anteriorly at the intertrochanteric line and root of the greater trochanter* (Fig. 5.51*B*). Posteriorly, the fibrous capsule crosses to the neck proximal to the *intertrochanteric crest* but is not attached to it (Fig. 5.51*C*). Most capsular fibers take a spiral course from the hip bone to the intertrochanteric line, but some deep fibers pass circularly around the neck, forming the **orbicular zone** (Fig. 5.51*C*). These fibers form a collar around the neck that constricts the capsule and helps hold the femoral head in the acetabulum.

Some deep longitudinal fibers of the capsule form **retinacula**, which reflect superiorly along the femoral neck as longitudinal bands that blend with the periosteum (Fig. 5.50). The retinacula contain retinacular blood vessels (branches of the medial [and a few from the lateral] femoral circumflex artery) that supply the head and neck of the femur (see Fig. 5.53). Thick parts of the fibrous capsule form the **ligaments of the hip joint**, which pass in a spiral fashion from the pelvis to the femur. They allow considerable flexion of the hip joint but restrict extension of the joint to 10 to 20° beyond the vertical position.

The ligaments of the hip joint are as follows:

- The fibrous capsule is reinforced anteriorly by the strong, Y-shaped **iliofemoral ligament** (of Bigelow) (Fig. 5.51, *A*

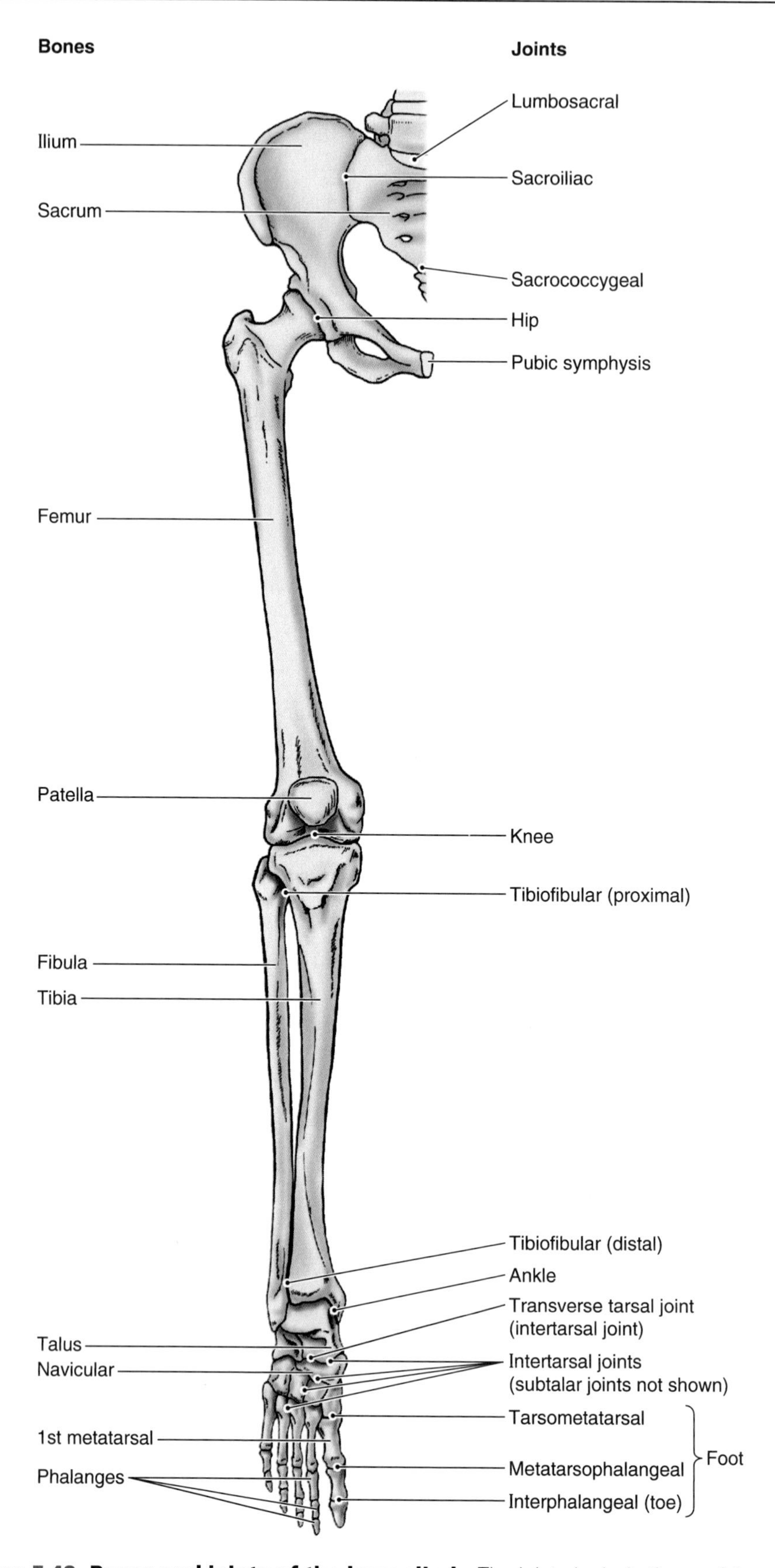

Figure 5.48. Bones and joints of the lower limb. The joints include those of the pelvic girdle connecting the lower limbs to the vertebral column.

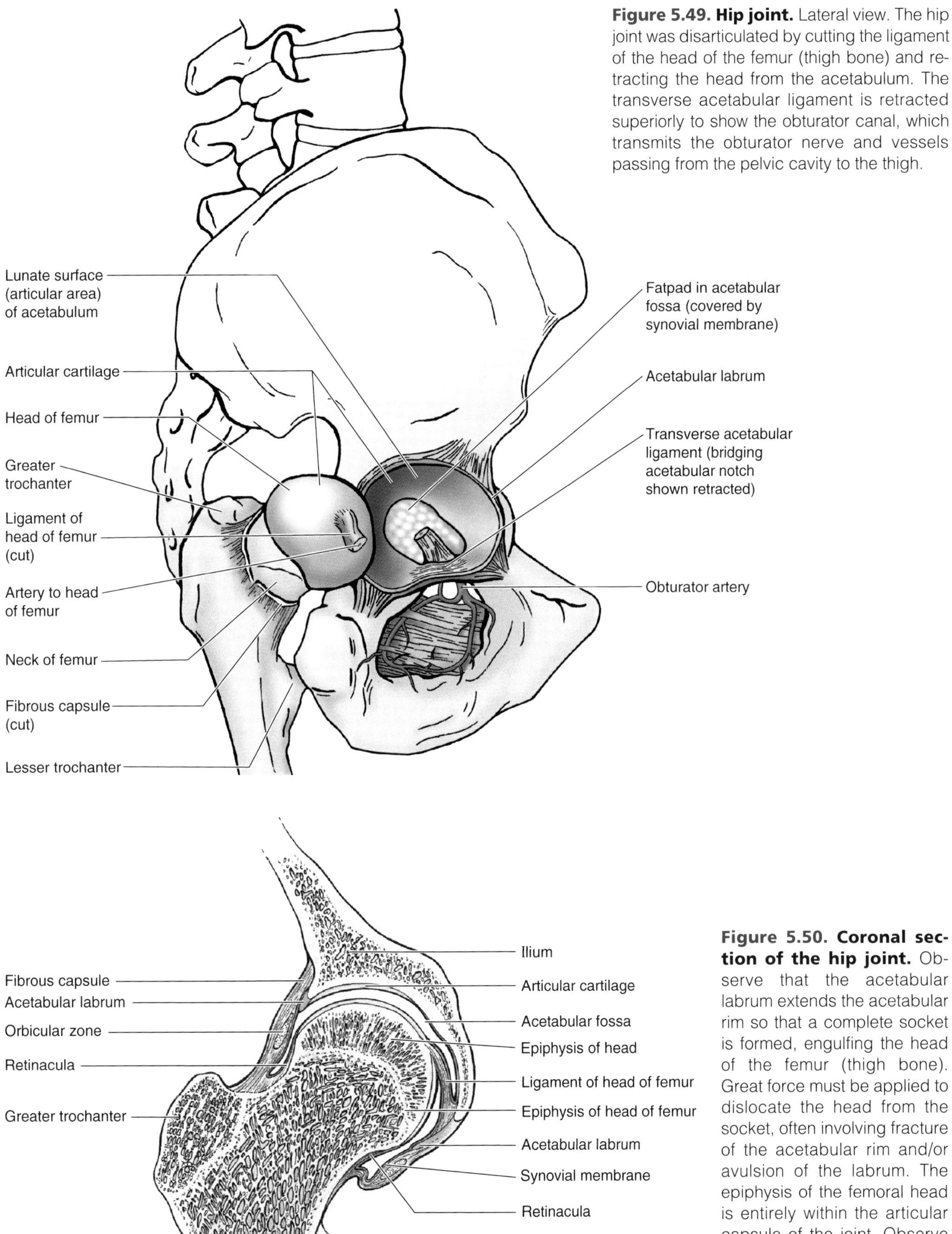

Figure 5.49. Hip joint. Lateral view. The hip joint was disarticulated by cutting the ligament of the head of the femur (thigh bone) and retracting the head from the acetabulum. The transverse acetabular ligament is retracted superiorly to show the obturator canal, which transmits the obturator nerve and vessels passing from the pelvic cavity to the thigh.

Figure 5.50. Coronal section of the hip joint. Observe that the acetabular labrum extends the acetabular rim so that a complete socket is formed, engulfing the head of the femur (thigh bone). Great force must be applied to dislocate the head from the socket, often involving fracture of the acetabular rim and/or avulsion of the labrum. The epiphysis of the femoral head is entirely within the articular capsule of the joint. Observe also that the ligament of the head is surrounded by a tube of synovial membrane. It transmits medial epiphyseal vessels to the femur, which may or may not persist in adults.

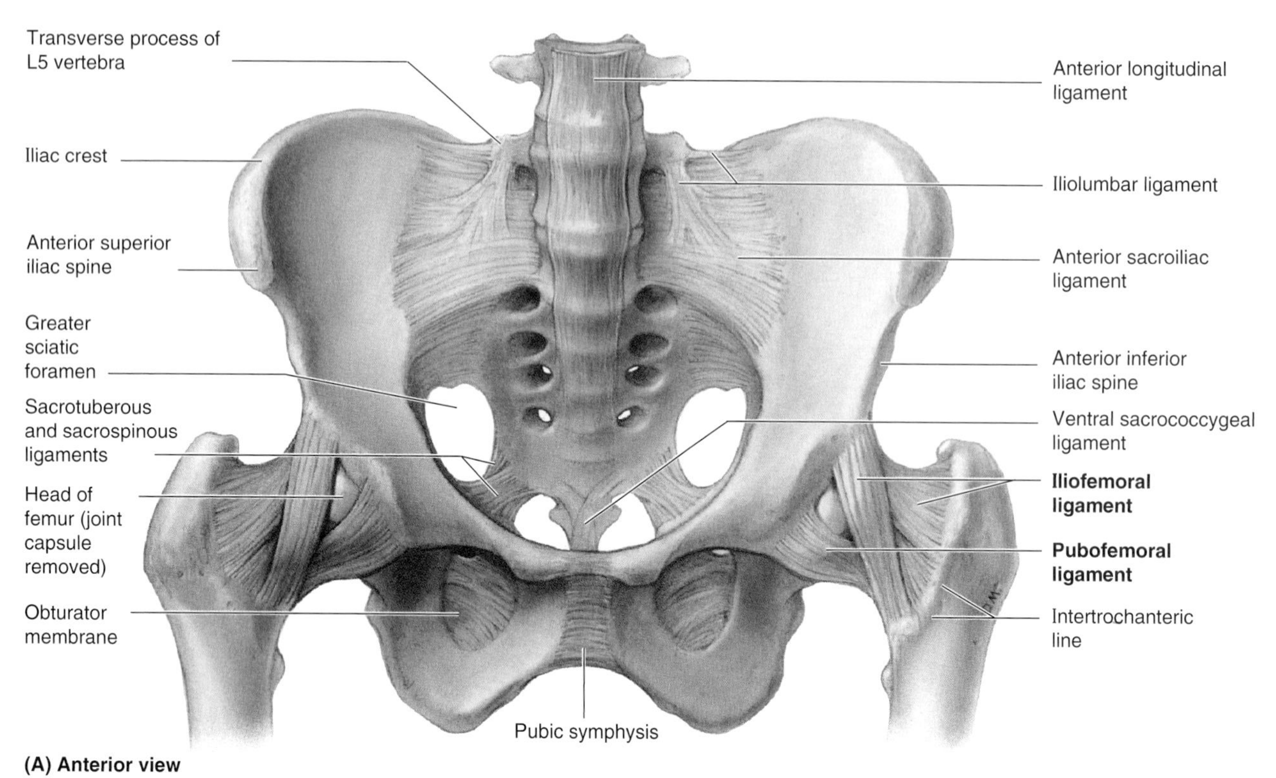

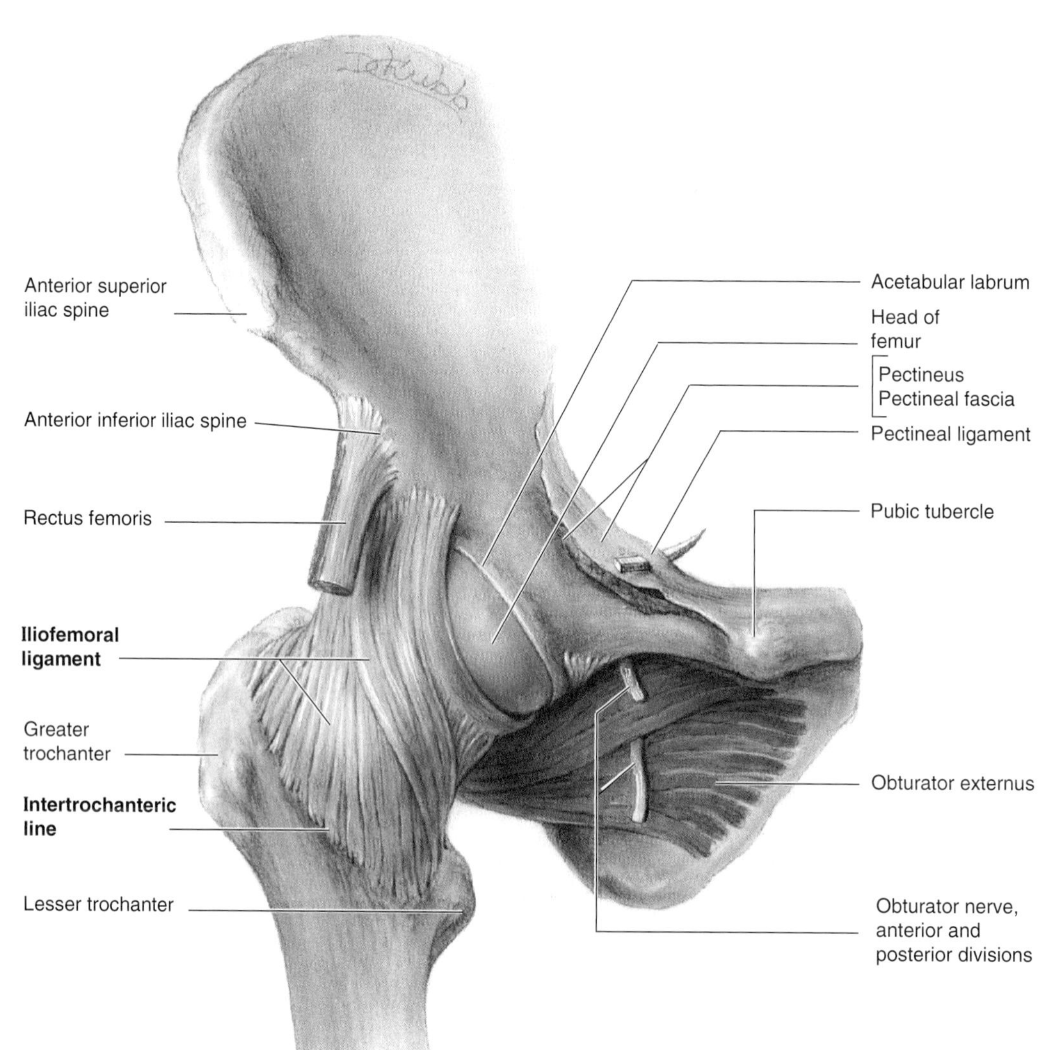

Figure 5.51. Ligaments of the pelvis and hip joint. A–B. Anterior views. Observe that the strong, triangular iliofemoral ligament attaches at its apex to the anterior inferior iliac spine and the rim of the acetabulum and at its base to the anterior intertrochanteric line of the femur (thigh bone). Observe also that the pubofemoral ligament, a thickened part of the fibrous capsule, extends from the superior ramus of the pubis to the intertrochanteric line of the femur, passing deep to the iliofemoral ligament.

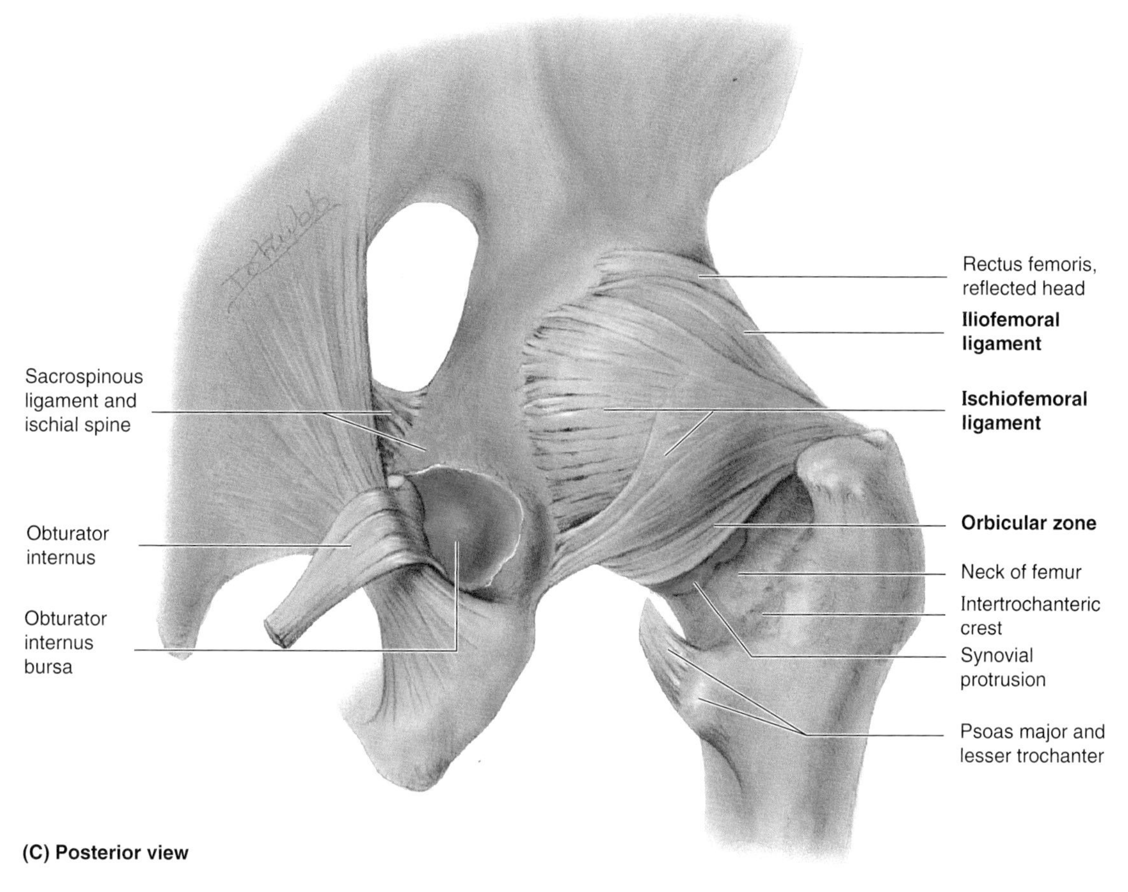

Figure 5.51. *(Continued)* **C.** Observe the ischiofemoral ligament, also a thickened part of the fibrous capsule, passing superolaterally from the ischium over the neck of the femur. The fibrous capsule of the hip joint does not attach to the posterior aspect of the femur. Observe that the synovial membrane protruding inferior to the fibrous capsule forms the obturator externus bursa to facilitate movement of the tendon of the obturator externus (muscle shown in (**B**)) over the bone.

and *B*), which attaches to the anterior inferior iliac spine and the acetabular rim proximally and the intertrochanteric line distally. *The iliofemoral ligament prevents hyperextension of the hip joint during standing* by screwing the femoral head into the acetabulum.

- The fibrous capsule is reinforced inferiorly and anteriorly by the **pubofemoral ligament** that arises from the obturator crest of the pubic bone and passes laterally and inferiorly to merge with the fibrous capsule of the hip joint (Fig. 5.51*A*). This ligament blends with the medial part of the iliofemoral ligament and tightens during extension and abduction of the hip joint. *The pubofemoral ligament prevents overabduction of the hip joint.*
- The fibrous capsule is reinforced posteriorly by the **ischiofemoral ligament** that arises from the ischial part of the acetabular rim (Fig. 5.51*C*) and spirals superolaterally to the neck of the femur, medial to the base of the greater trochanter. *The ischiofemoral ligament tends to screw the femoral head medially into the acetabulum, preventing hyperextension of the hip joint.*

The **synovial membrane of the hip joint** (Fig. 5.50) lines the fibrous capsule and covers the:

- Neck of the femur between the attachment of the fibrous capsule and the edge of the articular cartilage of the head
- Nonarticular area of the acetabulum, providing a covering for the ligament of the femoral head.

A **synovial protrusion** beyond the free margin of the fibrous capsule onto the posterior aspect of the femoral neck (Fig. 5.51*C*) forms a bursa for the obturator externus tendon.

The **ligament of the head of the femur** (Figs. 5.49 and 5.50) is weak and of little importance in strengthening the hip joint. Its wide end attaches to the margins of the acetabular notch and the transverse acetabular ligament; its narrow end attaches to the pit in the head of the femur. Usually the ligament contains a small artery to the head of the femur. The *fat-pad in the acetabular fossa* (covered with synovial membrane)

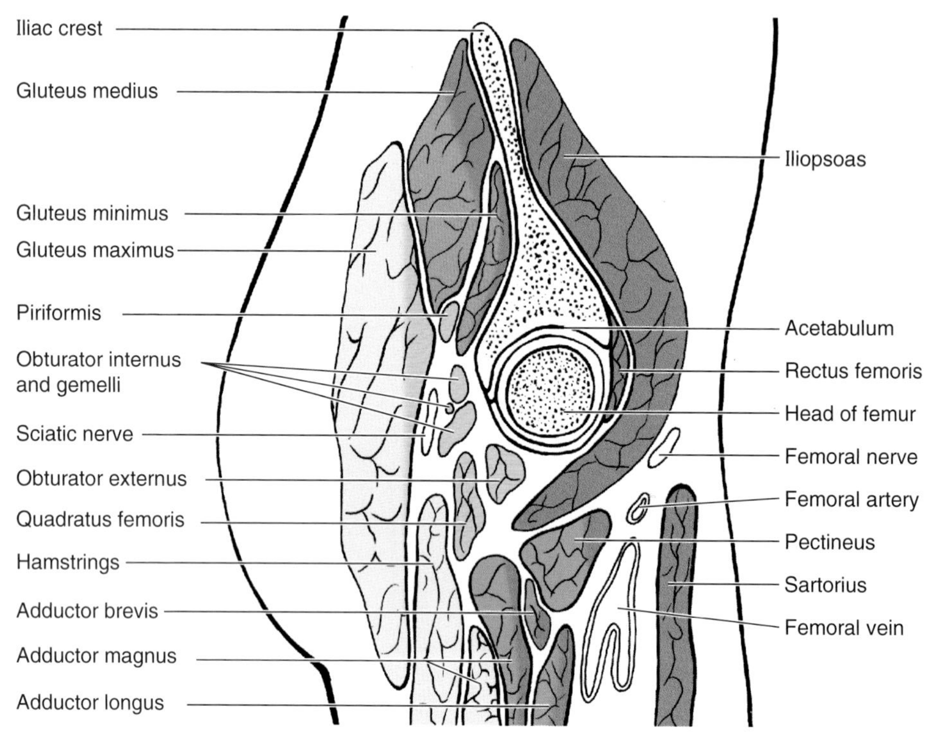

(A) Sagittal section through femoral head

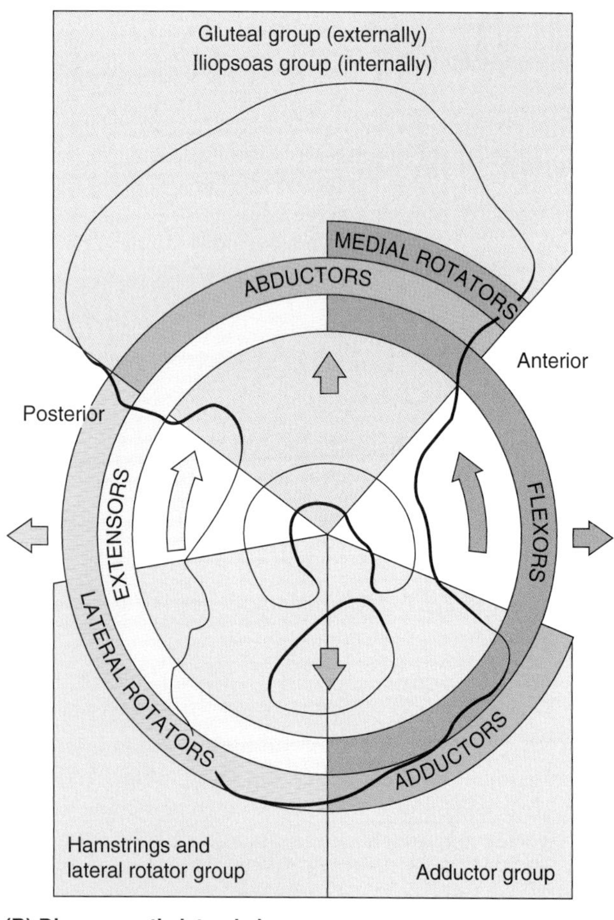

(B) Diagrammatic lateral view

Circular Zones =
The zones represent the position of origin of functional groups relative to center of humeral head in acetabulum (point of rotation). Pull is applied on the femur (femoral trochanters or shaft) from these positions.

Colored Arrows =
The arrows show the direction of rotation of femoral head caused by activity of functional groups.

Functional groups of muscles acting at hip joint

Flexors
Iliopsoas Sartorius Tensor of fascia lata Rectus femoris Pectineus Adductor longus Adductor brevis Adductor magnus-anterior part Gracilis
Adductors
Adductor longus Adductor brevis Adductor magnus Gracilis Pectineus Obturator externus
Lateral rotators
Obturator externus Obturator internus Gemelli Piriformis Quadratus femoris Gluteus maximus
Extensors
Hamstrings: Semitendinosus Semimembranosus Long head, biceps femoris Adductor magnus-posterior part Gluteus maximus
Abductors
Gluteus medius Gluteus minimus Tensor of fascia lata
Medial rotators
Gluteus medius } Anterior parts Gluteus minimus } Tensor of fascia lata

fills the part of the acetabular fossa that is not occupied by the femoral head. The malleable nature of the fat pad permits it to change shape to accommodate the varying shape of the head during joint movements.

Movements of the Hip Joint

Hip movements are flexion-extension, abduction-adduction, medial-lateral rotation, and circumduction (Fig. 5.52). Movements of the trunk at the hip joints are also important, such as those occurring when a person lifts the trunk from the supine position during sit-ups, for example.

The degree of flexion and extension of the hip joint depends on the position of the knee. If the knee is flexed relaxing the hamstrings, the thigh can be moved toward the anterior abdominal wall. Not all this movement occurs at the hip joint; some results from flexion of the vertebral column. During extension of the hip joint, the fibrous capsule, especially the iliofemoral ligament, is taut; therefore, the hip can usually be extended only slightly beyond the vertical.

Abduction of the hip joint is usually somewhat freer than adduction. Rotation of the hip can be carried through approximately one-sixth of a circle when the thigh is extended and more when it is flexed. Lateral rotation is much more powerful than medial rotation. The *main muscles producing movements of the hip joint* (Fig. 5.52*B*) are:

- **Flexion**—Iliopsoas (the strongest flexor), sartorius, tensor of fascia lata, rectus femoris, pectineus, adductor longus, adductor brevis, adductor magnus (anterior part), and gracilis
- **Extension**—Hamstrings (semitendinosus, semimembranosus, and long head of biceps femoris), adductor magnus—posterior part, and gluteus maximus; the gluteus maximus is relatively inactive from the straight (standing) position to the fully extended position unless forceful extension is required. It acts mostly from the fully flexed to the straight position, as in climbing stairs or in rising from a sitting position
- **Abduction**—Gluteus medius and minimus, and tensor of fascia lata
- **Adduction**—Adductor longus, adductor brevis, adductor magnus, gracilis, pectineus, and obturator externus
- **Rotation**—*Medial rotators*: anterior fibers of gluteus medius, gluteus minimus, and tensor of fascia lata; *lateral rotators*: obturator externus, obturator internus, gemelli, piriformis, quadratus femoris, and gluteus maximus.

Blood Supply of the Hip Joint

The arteries supplying the hip joint (Fig. 5.53) are:

- Medial and lateral **circumflex femoral arteries**—usually branches of the *deep artery of the thigh* but occasionally arising as branches of the *femoral artery*
- *Artery to the head of femur*—a branch of the obturator artery—enters through the ligament of the head.

The main blood supply of the hip joint is from branches of the circumflex femoral arteries (especially the medial circumflex femoral artery) that travel in the retinacula (reflections of the capsule along the neck of the femur toward the head). These retinacular vessels may be damaged in femoral neck fractures and result in avascular necrosis of the femoral head.

Nerve Supply of the Hip Joint

The nerve supply of the hip joint (Figs. 5.51*B* and 5.52*A*) is from the:

- Femoral nerve or its muscular branches (anteriorly)
- Accessory obturator nerve, if present (anteriorly)
- Obturator nerve (anterior division) (inferiorly)
- Superior gluteal nerve (superiorly and posteriorly)
- Nerve to quadratus femoris (posteriorly).

Pain in the hip may be misleading because pain can be referred from the vertebral column.

Relations of the Hip Joint

Structures related to the hip joint (Fig. 5.52*A*) are:

- *Anteriorly*—Pectineus, iliopsoas, subtendinous iliac bursa, and femoral artery and nerve
- *Laterally*—Rectus femoris anterior to iliofemoral ligament, iliotibial tract, and gluteus minimus
- *Inferiorly*—Obturator externus crosses inferior to the femoral head and runs posterior to the femoral neck
- *Superiorly*—Gluteus minimus, gluteus medius, and overlying gluteus maximus
- *Posteriorly*—Piriformis, obturator externus, obturator internus and gemelli, superior border of quadratus femoris, and sciatic nerve.

The large *subtendinous iliac bursa*, which may communicate with the cavity of the hip joint, separates the iliopsoas tendon from the articular capsule of this joint.

Figure 5.52. Relations of hip joint and the muscles producing movements of the joint. **A.** Sagittal section of the hip joint showing the muscles, vessels, and nerves related to it. The muscles are color coded to indicate muscular functions. **B.** Diagrammatic illustration showing the relative positions of the muscles producing movements of the hip joint and the direction of the movement.

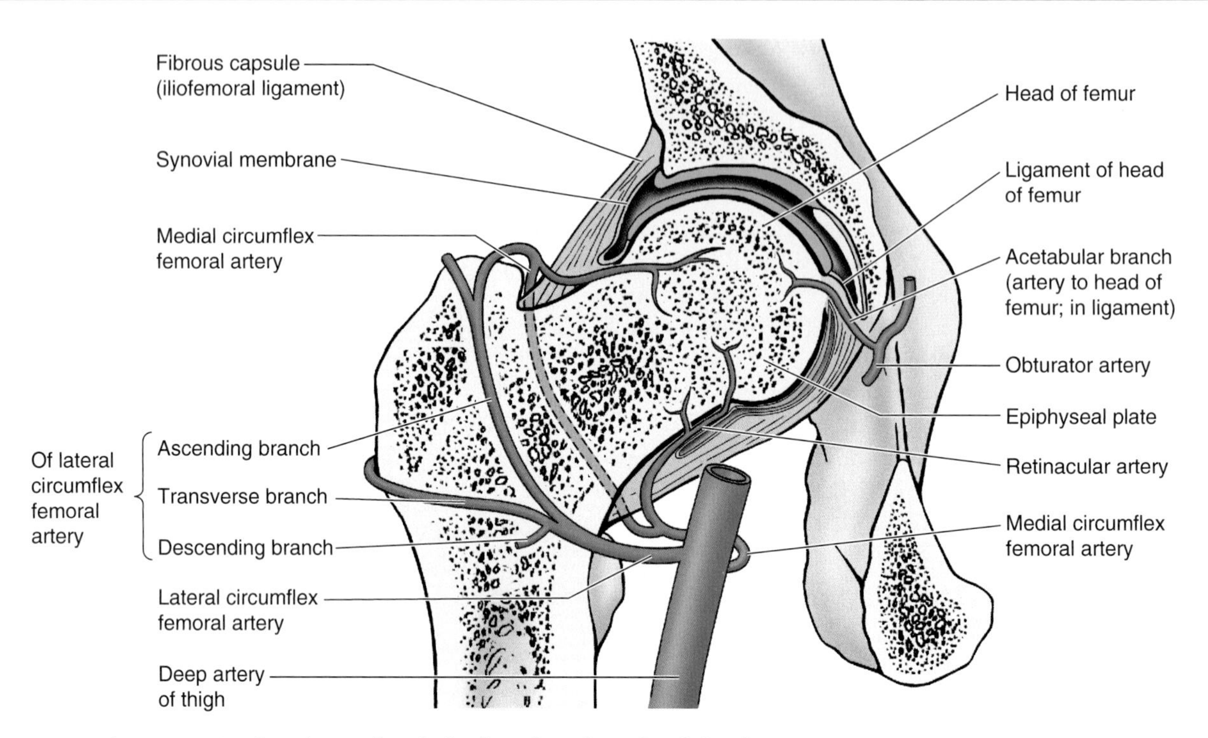

Figure 5.53. Blood supply of the head and neck of the femur. Branches of the medial and lateral circumflex femoral arteries, branches of the deep artery of the thigh (L. arteria profunda femoris), and the artery to the femoral head (a branch of the obturator artery) supply the head and neck of the femur. In the adult, the medial circumflex femoral artery is the most important source of blood to the femoral head and adjacent (proximal) neck.

Fractures of the Femoral Neck

Fractures of the neck of the femur are uncommon in most contact sports because the participants are usually young and the femoral neck is strong in persons younger than 40 years. However, these fractures may occur with high-energy impacts (e.g., during race-car accidents, figure skating, long jumping, and equestrian events) when the lower limb is straight and the force of the impact is transmitted to the hip joint. *Fractures of the femoral neck are intracapsular,* and realignment of the neck fragments requires internal skeletal fixation. Femoral neck fractures are among the most troublesome and problematic of all fractures (Salter, 1998). These fractures are especially common in persons older than 60 years, especially in women because their bones are often weak and brittle as a result of *osteoporosis*. The femoral neck may also fracture when traumatic force is applied to the foot and ankle. For example, ▶

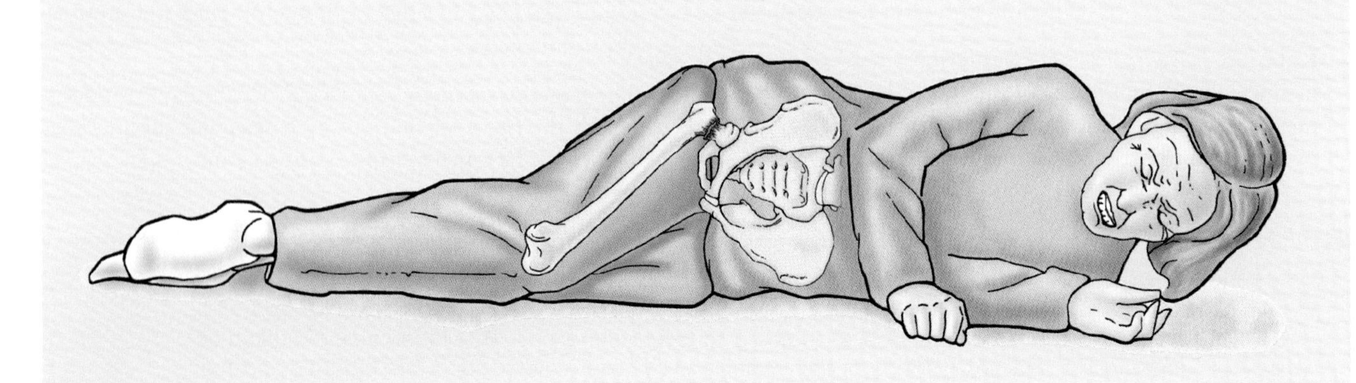

▶ if the foot is firmly braced with the knee locked during a head-on collision, the force of the impact may be transmitted superiorly and produce a femoral neck fracture.

Fractures of the femoral neck often disrupt the blood supply to the head of the femur. The medial circumflex femoral artery is clinically important because it supplies most of the blood to the head and neck of the femur (Fig. 5.53). It is often torn when the femoral neck is fractured or the hip joint is dislocated. In some femoral neck fractures, the blood from the artery of the ligament of the head may be the only blood that the proximal fragment of the femoral head receives; consequently, if the artery to the head of the femur is ruptured, the fragment of bone may receive no blood and will undergo *aseptic vascular necrosis*.

Surgical Hip Replacement

Although the hip joint is strong and stable, it is subject to severe traumatic injury and degenerative disease. Therefore, it is the 1st joint for which a replacement prosthesis was successfully developed. *Osteoarthritis of the hip joint*—characterized by pain, edema, limitation of motion, and erosion of articular cartilage—is a common cause of disability. During hip replacement, a metal prosthesis anchored to the person's femur by bone cement replaces the femoral head and neck. A plastic socket cemented to the hip bone replaces the acetabulum.

Normal hip

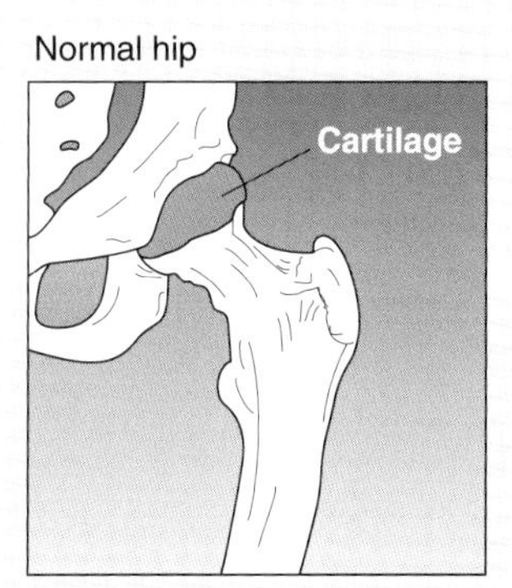

Hip with moderate arthritis

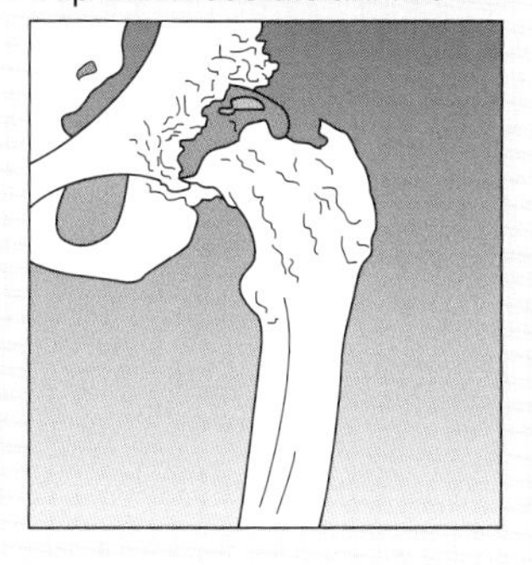

Necrosis of the Femoral Head in Children

When the femoral head loses its blood supply by disruption of the artery of the head of the femur, *post-traumatic avascular necrosis of the head* occurs. The head may be compressed, and slippage of the femoral head along the epiphyseal plate may occur. This injury, common in children 3 to 9 years of age, causes hip pain that may radiate to the knee (Salter, 1998).

Dislocation of the Hip Joint

Congenital dislocation of the hip joint is common, occurring in approximately 1.5 per 1000 livebirths; it is bilateral in approximately half the cases. Girls are affected at least eight times as often as boys (Salter, 1989). Dislocation occurs when the femoral head is not properly located in the acetabulum. Inability to abduct the thigh is characteristic of congenital dislocation. In addition, the affected limb appears shorter because the dislocated femoral head is more superior than on the normal side. Approximately 25% of all cases of arthritis of the hip in adults are the direct result of residual defects from congenital dislocation of the hip.

Acquired dislocation of the hip joint is uncommon because this articulation is so strong and stable. Nevertheless, dislocation may occur during an automobile accident when the hip is flexed, adducted, and medially rotated—the position of the lower limb when a person is riding in a car. Posterior dislocations (as illustrated on p. 616) are most common. A head-on collision that causes the knee to strike the dashboard (*A*) may dislocate the hip when the femoral head is forced out of the acetabulum. The fibrous capsule ruptures inferiorly and posteriorly, allowing the femoral head to pass through the tear in the capsule and over the posterior margin of the acetabulum onto the lateral surface of the ilium, shortening and medially rotating the affected limb (*B*). Because of the close relationship of the *sciatic nerve* to the hip joint (Fig. 5.52*A*), it may be injured (stretched and/or compressed) during posterior dislocations or fracture-dislocations of the hip joint. This kind of injury may result in paralysis of the hamstrings and muscles distal to the knee and supplied by the sciatic nerve. Sensory changes may also occur in the skin over the posterolateral aspects of the leg and over much of the foot because of injury to sensory branches of the sciatic nerve.

Anterior dislocation of the hip joint results from a violent injury that forces the hip into extension, abduction, and lateral rotation. In these cases, the femoral head is inferior to the acetabulum. Often, the acetabular margin fractures, producing a *fracture-dislocation of the hip joint*. When the femoral head dislocates, it usually carries the acetabular bone fragment and acetabular labrum with it. ○

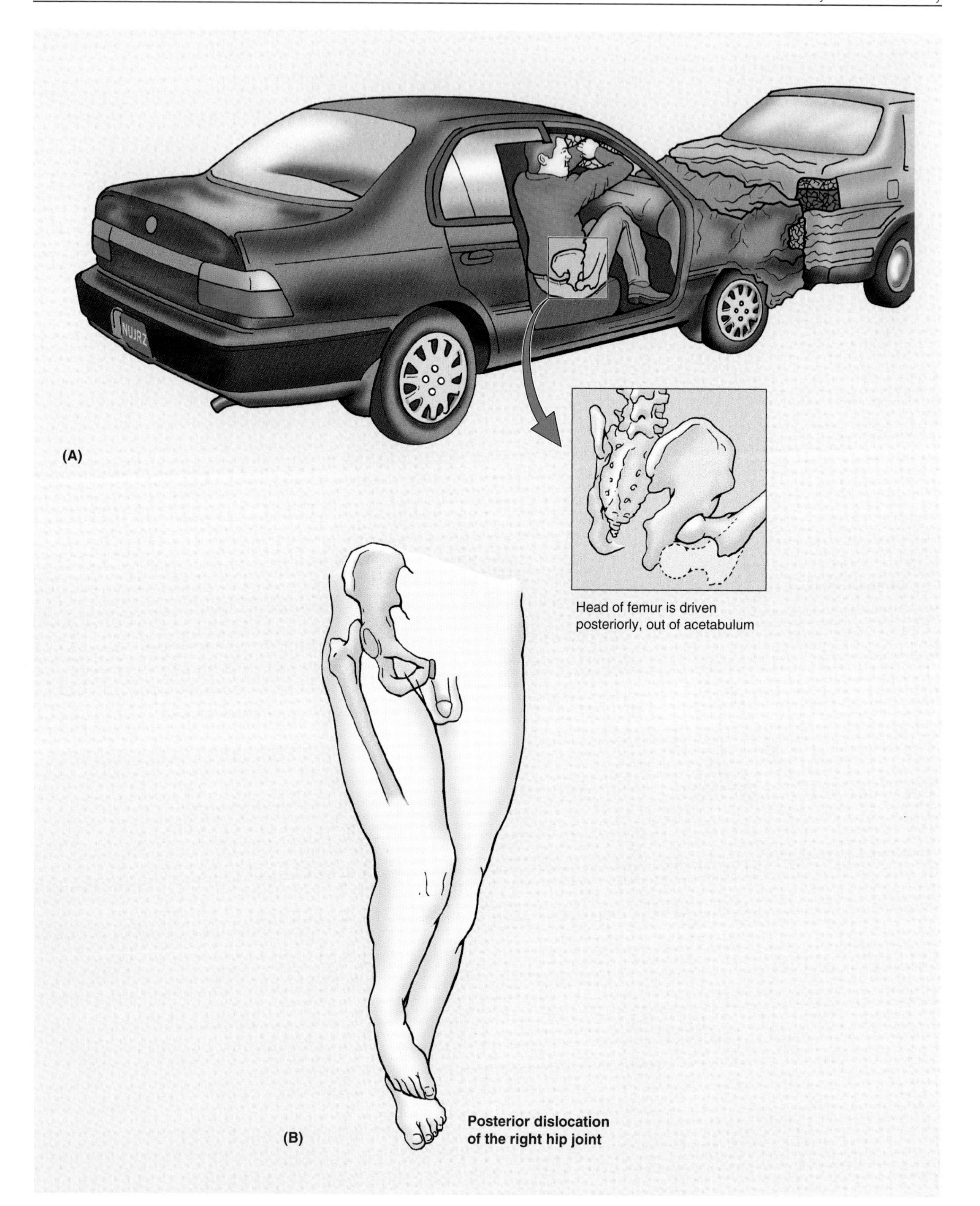
NUJRZ
(A)
Head of femur is driven posteriorly, out of acetabulum
(B)
Posterior dislocation of the right hip joint

Knee Joint

The knee is primarily a hinge type of synovial joint allowing flexion and extension; however, the hinge movements are combined with gliding and rolling and with rotation about a vertical axis. Although the knee joint is well constructed, its function is commonly impaired when it is hyperextended (e.g., in body contact sports).

Articular Surfaces of the Knee Joint

The articular surfaces of the knee joint are characterized by their large size and their complicated and incongruent shapes. The femur slants medially at the knee, whereas the tibia is almost vertical. The knee joint consists of three articulations (Figs. 5.54 and 5.55):

- Lateral and medial articulations between the femoral and tibial condyles
- Intermediate articulation between the patella and femur; this articulation is sometimes referred to as the patellofemoral joint.

The fibula is not involved in the knee joint. The smooth superior surfaces of the tibia with which the medial and lateral femoral condyles articulate are the medial and lateral **tibial plateaus** (Fig. 5.55*C*).

The knee joint is relatively weak mechanically because of the configurations of its articular surfaces. *The stability of the knee joint* (Fig. 5.55, *A* and *B*) *depends on the*:

- Strength and actions of the surrounding muscles and their tendons
- Ligaments that connect the femur and tibia.

Of these supports, the muscles are most important; therefore, many sport injuries are preventable through appropriate conditioning and training. The most important muscle in stabilizing the knee joint is the large **quadriceps femoris**, partic-

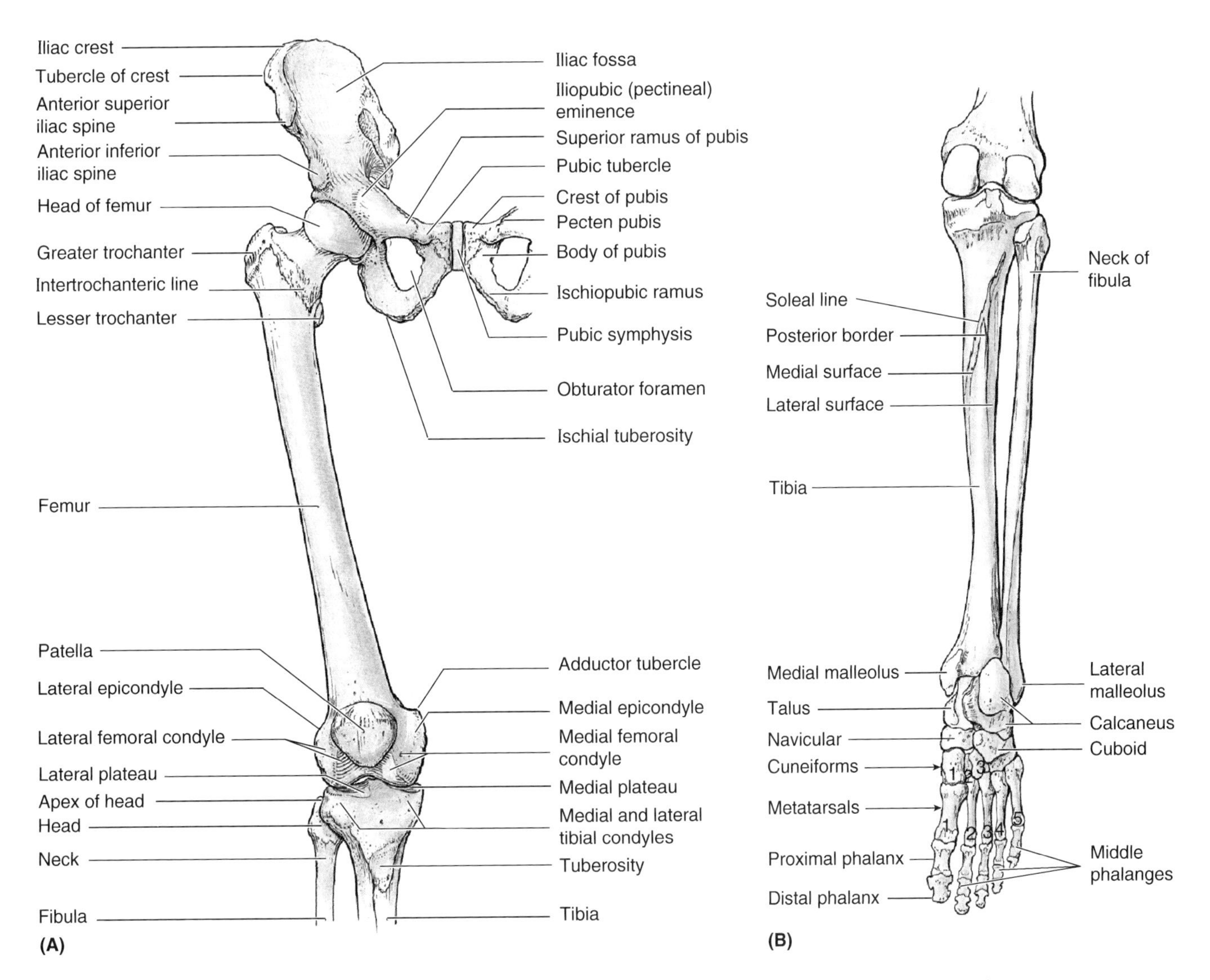

Figure 5.54. Bones of the hip and knee joints. Articular cartilages are colored *blue*. **A.** Anterior view of the bones articulating at the hip and knee joints. **B.** Bones and bony features of the knee, ankle, and joints of the foot.

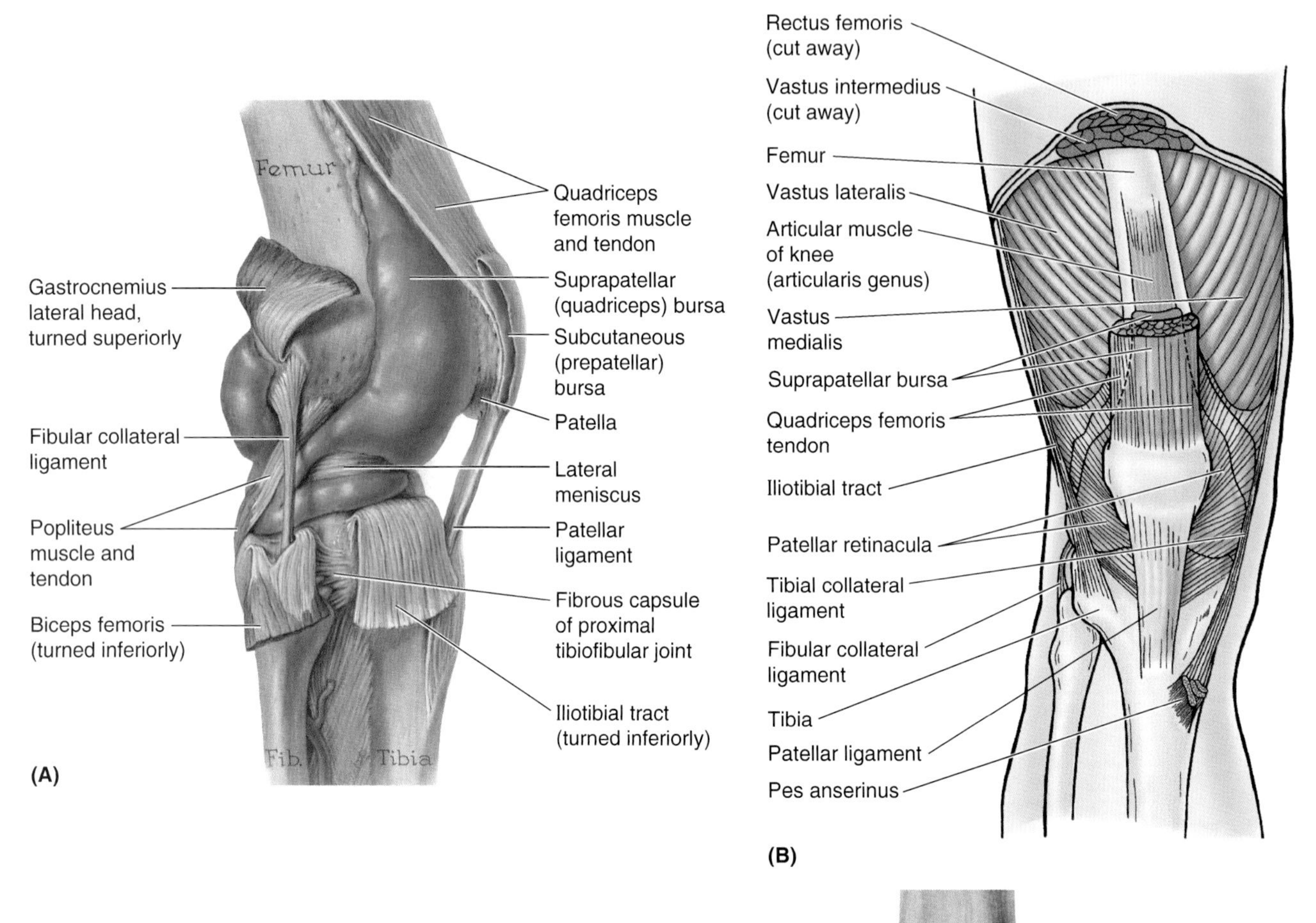

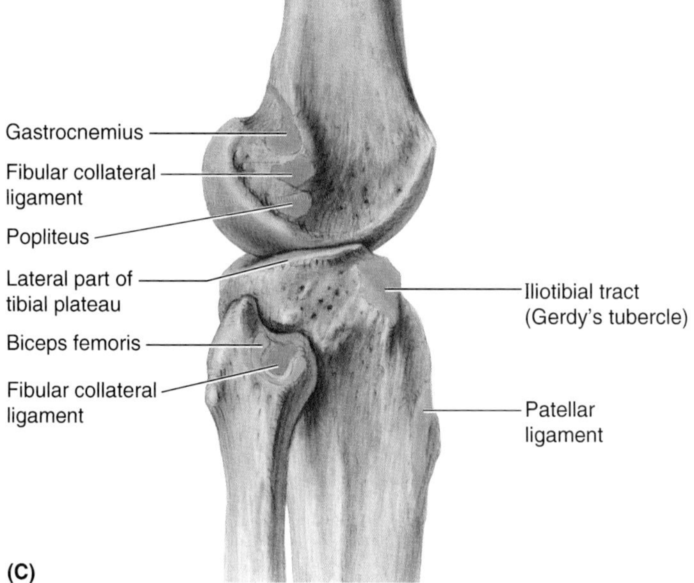

Figure 5.55. Right knee joint in extension. **A.** Lateral view. Blue latex was injected into the joint cavity to demonstrate the extensive synovial capsule. Observe the extent of the synovial capsule deep to the quadriceps, forming the suprapatellar bursa. **B.** Anterior view, demonstrating the bursae around the knee. **C.** Lateral view of the bones of the right knee region showing the attachment sites of muscles and ligaments. Proximal attachments of muscles are shown in *salmon color* and distal attachments are in *blue*. The attachment sites of the fibular collateral ligament are shown in *green*.

ularly inferior fibers of the vastus medialis and lateralis. The knee joint will function surprisingly well following a ligament strain if the quadriceps is well conditioned.

Articular Capsule of the Knee Joint

The articular capsule investing the joint is thin and is deficient in some areas. The strong **fibrous capsule** (Figs. 5.55*A* and 5.56) attaches to the femur superiorly, just proximal to the articular margins of the condyles and also to the *intercondylar fossa* posteriorly. The fibrous capsule is deficient on the lateral condyle to allow the tendon of the popliteus to pass out of the joint to attach to the tibia. Inferiorly, the fibrous capsule attaches to the articular margin of the tibia, except where the tendon of the popliteus crosses the bone. The patella and patellar ligament serve as a capsule anteriorly.

The extensive **synovial membrane** lines the internal aspect

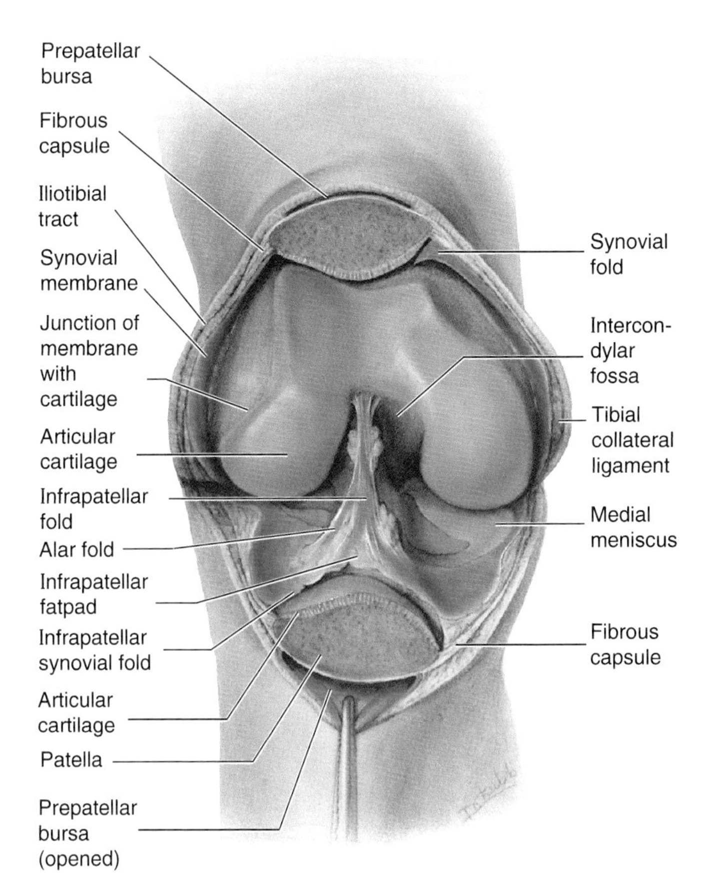

Figure 5.56. Synovial cavity of the knee joint, opened anteriorly. The patella (knee cap) is sawn through, the skin and joint capsule are cut through, and the joint is flexed. The infrapatellar fold of synovial membrane encloses the cruciate ligaments, excluding them from the joint cavity. All internal surfaces not covered with or made of articular cartilage (*blue*, or *gray* in the case of the menisci) are lined with synovial membrane (*purple*, except where it is covering nonarticular surfaces of the femur).

of the fibrous capsule and attaches to the periphery of the patella and the edges of the menisci (Fig. 5.56)—the fibrocartilaginous discs between the tibial and femoral articular surfaces. The synovial membrane reflects from the posterior aspect of the joint onto the cruciate ligaments (see Fig. 5.59*A*). The reflection of the membrane between the tibia and patella covers the *infrapatellar fat pad*. The synovial membrane covering the fat pad and cruciate ligaments separates them from the joint cavity. The median *infrapatellar synovial fold* extends posteriorly from the fat pad to the intercondylar fossa of the femur. *Alar folds* project from the *synovial fold* to the lateral edges of the patella. The *infrapatellar synovial fold* is a fold of synovial membrane extending from below the level of the articular surface of the patella to the anterior part of the intercondylar fossa.

The knee joint cavity extends superior to the patella as the **suprapatellar bursa**, which lies deep to the articular muscle of the knee and vastus intermedius (Fig. 5.55*A*). The synovial membrane of the articular capsule is continuous with the synovial lining of this bursa. This large bursa usually extends approximately 5 cm superior to the patella; however, it may extend halfway up the anterior aspect of the femur.

Extracapsular Ligaments of the Knee Joint

The fibrous capsule is strengthened by five extracapsular ligaments (Figs. 5.55, 5.57, and 5.58):

- Patellar ligament
- Fibular collateral ligament
- Tibial collateral ligament
- Oblique popliteal ligament
- Arcuate popliteal ligament.

They are sometimes called external ligaments to differentiate them from internal ligaments such as the cruciate ligaments (Fig. 5.59*A*).

The **patellar ligament**, the distal part of the quadriceps tendon, is a strong, thick fibrous band passing from the apex and adjoining margins of the patella to the tibial tuberosity (Fig. 5.55, *A–C*). The patellar ligament is the anterior ligament of the knee joint. It blends with the medial and lateral **patellar retinacula** (Fig. 5.55*B*), which are aponeurotic expansions of the vastus medialis and lateralis and the overlying deep fascia. The retinacula support the articular capsule of the knee laterally.

The **fibular collateral ligament** (lateral collateral ligament), rounded and cordlike, is strong. It extends inferiorly from the lateral epicondyle of the femur to the lateral surface of the head of the fibula (Fig. 5.55, *A–C*). The tendon of the popliteus passes deep to the fibular collateral ligament, separating it from the lateral meniscus. The tendon of the biceps femoris is also split into two parts by this ligament (Fig. 5.55*A*).

The **tibial collateral ligament** (medial collateral ligament) is a strong flat band that extends from the medial epicondyle of the femur to the medial condyle and the superior part of the medial surface of the tibia (Figs. 5.55*B*, 5.57*A*, and 5.59*A*). *At its midpoint, the deep fibers of the tibial collateral ligament are firmly attached to the medial meniscus.* The tibial collateral ligament, weaker than the fibular collateral ligament, is more often damaged. As a result, the tibial collateral ligament and medial meniscus are commonly torn during contact sports such as football.

The **oblique popliteal ligament** (Fig. 5.58) is an expansion of the tendon of the semimembranosus that strengthens the fibrous capsule posteriorly. It arises posterior to the medial tibial condyle and passes superolaterally to attach to the central part of the posterior aspect of the fibrous capsule.

The **arcuate popliteal ligament** also strengthens the fibrous capsule posteriorly. It arises from the posterior aspect of the fibular head (Fig. 5.58), passes superomedially over the tendon of the popliteus, and spreads over the posterior surface of the knee joint.

Intra-Articular Ligaments of the Knee Joint

The intra-articular ligaments within the knee joint consist of the cruciate ligaments and menisci (semilunar cartilages). The popliteal tendon is also intra-articular during part of its course.

The **cruciate ligaments** (L. crux, a cross) join the femur and tibia, crisscrossing within the articular capsule of the joint but outside the synovial joint cavity (Fig. 5.59*A*). The cruciate ligaments are located in the center of the joint and cross each other obliquely like the letter X, providing stability to the knee joint.

The **anterior cruciate ligament** (ACL), the weaker of the two cruciate ligaments, arises from the anterior intercondylar area of the tibia, just posterior to the attachment of the medial meniscus. It extends superiorly, posteriorly, and laterally to attach to the posterior part of the medial side of the lateral condyle of the femur. The ACL has a relatively poor blood supply. It is slack when the knee is flexed and taut when it is fully extended, preventing posterior displacement of the femur on the tibia and hyperextension of the knee joint. When the joint is flexed at a right angle, the tibia cannot be pulled anteriorly because it is held by the ACL.

The **PCL**, the stronger of the two cruciate ligaments, arises from the posterior intercondylar area of the tibia (Fig. 5.59, *A* and *B*). The PCL passes superiorly and anteriorly on the medial side of the ACL to attach to the anterior part of the lateral surface of the medial condyle of the femur (Fig. 5.59*A*). The PCL tightens during flexion of the knee joint, preventing anterior displacement of the femur on the tibia or posterior displacement of the tibia on the femur. It also helps prevent hyperflexion of the knee joint. In the weightbearing flexed knee, the PCL is the main stabilizing factor for the femur (e.g., when walking downhill).

The **menisci of the knee joint** are crescentic plates of fibrocartilage on the articular surface of the tibia that deepen the surface and act like shock absorbers (Figs. 5.59 and 5.60). The Greek word *meniskos* means crescent. The menisci are thicker at their external margins and taper to thin, unattached edges in the interior of the joint. Wedge shaped in transverse

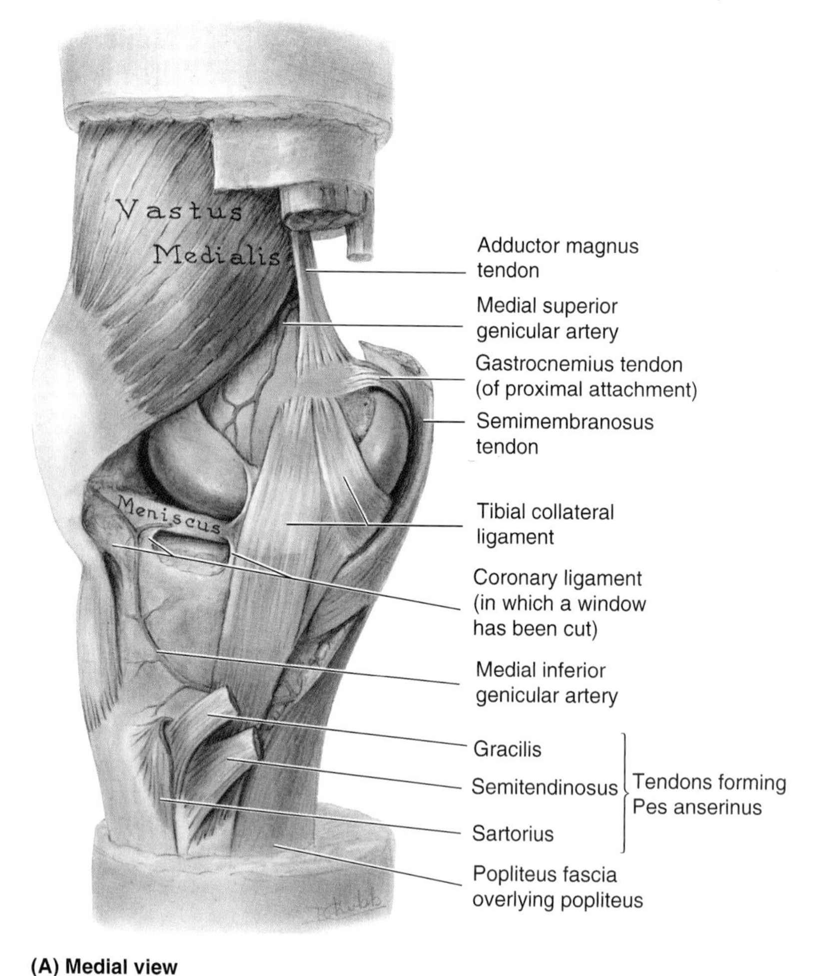

Figure 5.57. Dissection of the right knee joint, medial views. A. Observe the bandlike part of the tibial collateral ligament (isolated here from the fibrous capsule of the knee, of which it is a part), which attaches to the medial epicondyle, almost in line with the adductor magnus tendon, crossing the insertion of the semimembranosus muscle, and passing deep to the pes anserinus (L. "goose's foot")—the combined attachment sites of the tendinous distal attachment of the gracilis, semitendinosus, and sartorius muscles.

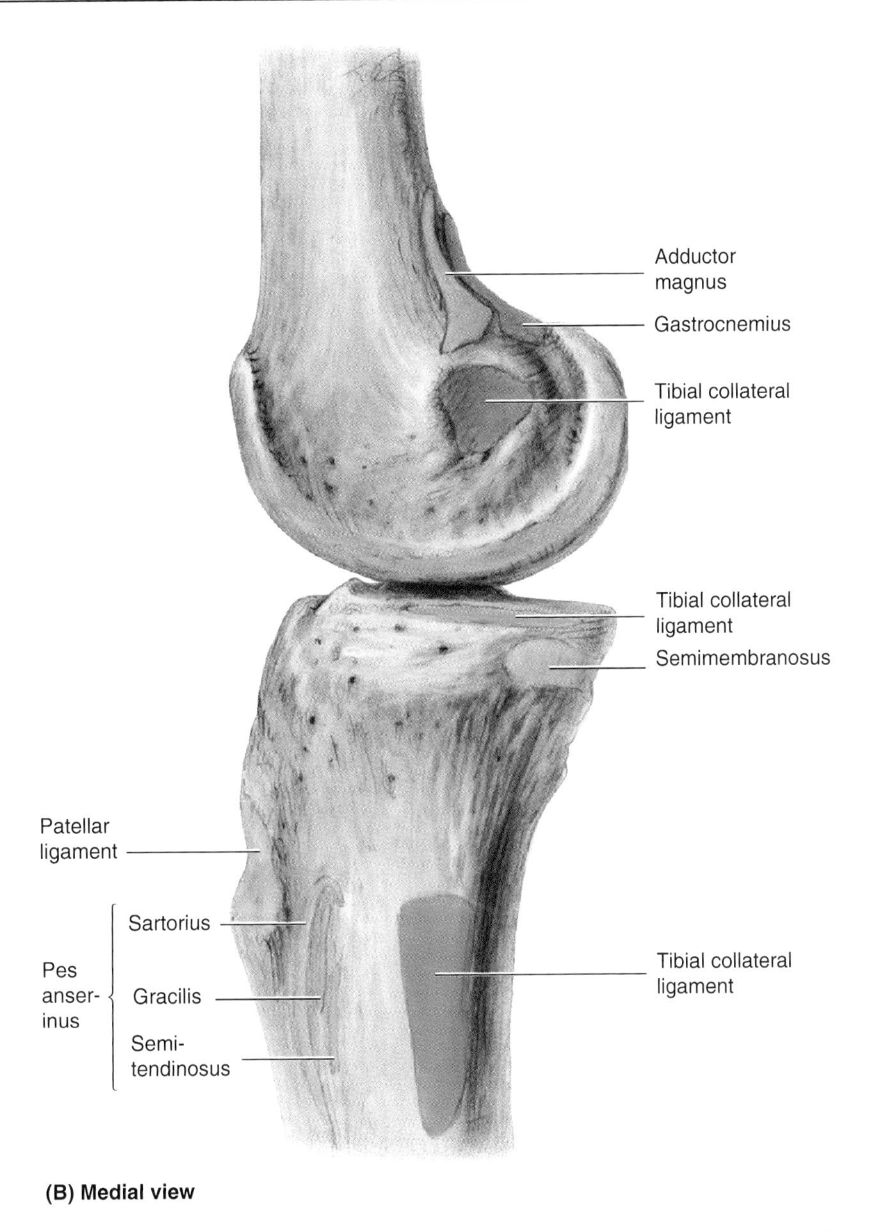

Figure 5.57. *(Continued)* **B.** Bones of the knee showing muscle and ligament attachments.

section, the menisci are firmly attached at their ends to the intercondylar area of the tibia. Their external margins attach to the fibrous capsule of the knee joint. The **coronary ligaments** are capsular fibers that attach the margins of the menisci to the tibial condyles (Fig. 5.59*A*). A slender fibrous band—the **transverse ligament of the knee**—joins the anterior edges of the menisci (Fig. 5.60*A*), allowing them to move together during knee movements.

The **medial meniscus** is C shaped and broader posteriorly than anteriorly. Its anterior end (horn) attaches to the anterior intercondylar area of the tibia, anterior to the attachment of the ACL (Fig. 5.59*B*). Its posterior end (horn) attaches to the posterior intercondylar area, anterior to the attachment of the PCL. The medial meniscus firmly adheres to the deep surface of the tibial collateral ligament (Fig. 5.60, *A* and *D*).

The **lateral meniscus** is nearly circular and is smaller and more freely movable than the medial meniscus. The tendon of the popliteus separates the lateral meniscus from the fibular collateral ligament (Fig. 5.59*A*). A strong tendinous slip, the **posterior meniscofemoral ligament**, joins the lateral meniscus to the PCL and the medial femoral condyle (Fig. 5.59*D*).

Movements of the Knee Joint

Flexion and extension are the main knee movements; some rotation occurs when the knee is flexed. When the knee is fully extended with the leg and foot on the ground, the knee "locks" because of medial rotation of the femur on the tibia. This position makes the lower limb a solid column and more adapted for weightbearing. When the knee is "locked," the thigh and leg muscles can relax briefly without making the knee joint too unstable. To "unlock" the knee the popliteus contracts, rotating the femur laterally so that flexion of the

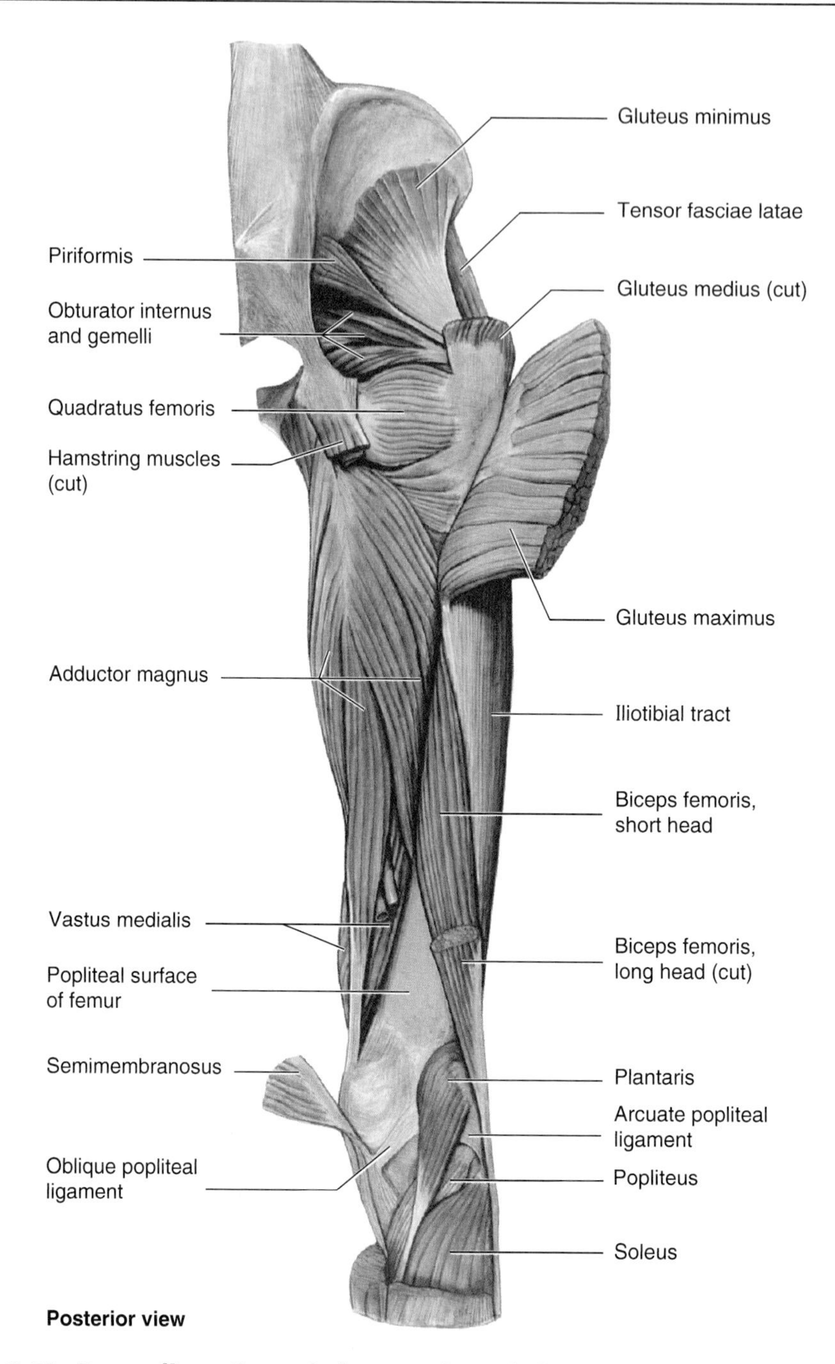

Figure 5.58. Deep dissection of the muscles of the right gluteal and posterior thigh regions. Most of the hamstring muscles and the posterior intermuscular septum have been cut and removed to expose the adductor magnus and the floor of the popliteal fossa, formed by the popliteal surface of the femur, the oblique popliteal ligament, and the popliteus fascia (removed to expose the popliteus).

knee can occur. *The main movements of the knee joint and the muscles producing them are*:

- *Flexion*—principally by the hamstrings—movement is limited by contact between the calf and thigh
- *Rotation*—possible when the knee is partially flexed
- *Medial rotation*—popliteus, semitendinosus, and slightly by the semimembranosus—movement is checked by the cruciate ligaments
- *Lateral rotation*—biceps femoris—movement is checked by the collateral ligaments
- *Extension*—principally by quadriceps—movement is limited as the cruciate and collateral ligaments become taut.

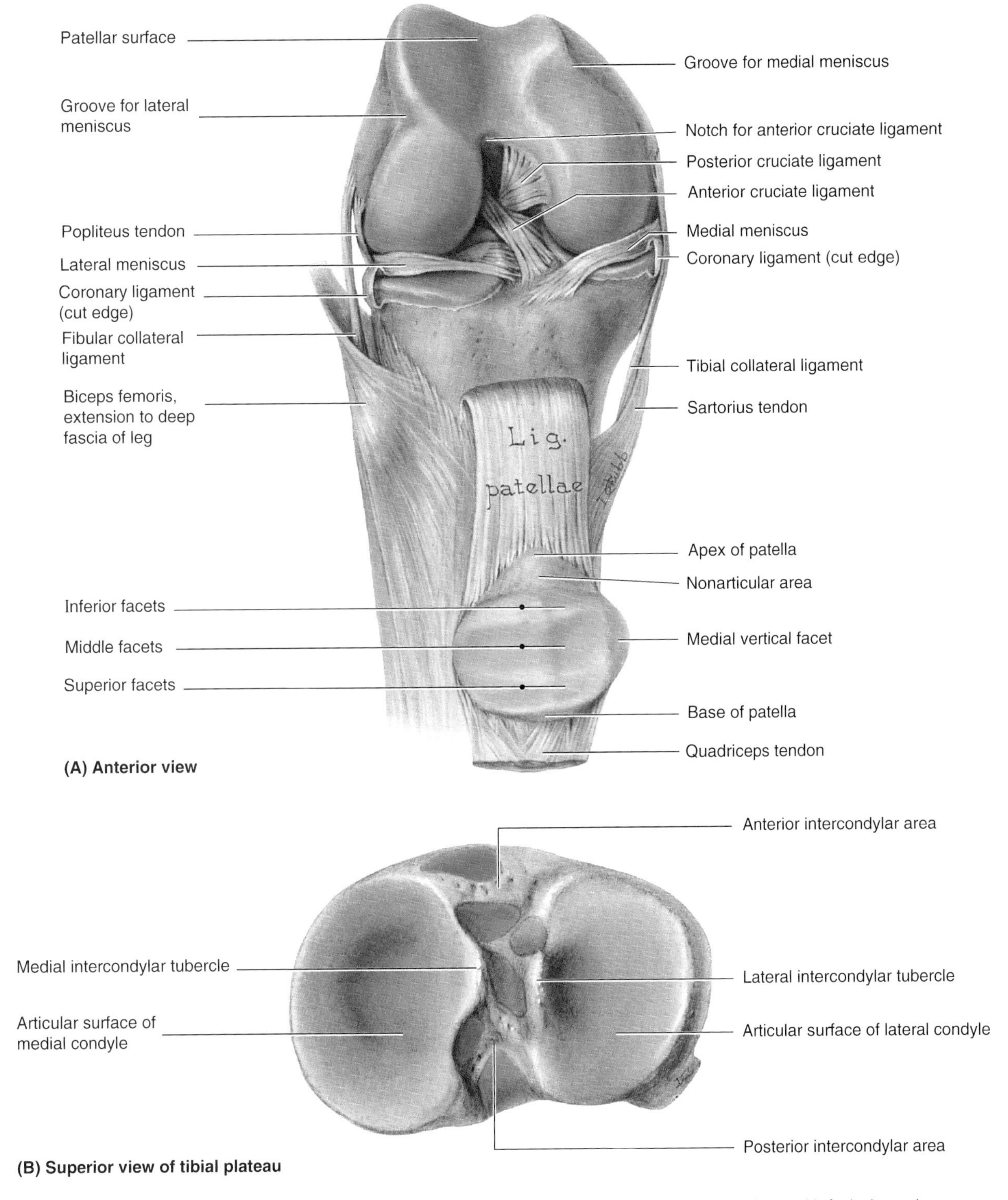

Figure 5.59. Ligaments of the knee joint. A. Anterior view. The patella is reflected inferiorly and the joint is flexed to demonstrate the cruciate ligaments. **B.** Superior aspect of the proximal end of the tibia (tibial plateau), showing the medial and lateral condyles (articular surfaces) and the intercondylar eminence between them. The sites of attachment of the cruciate ligaments are colored *green*; those of the medial meniscus, *purple*; and those of the lateral meniscus, *orange*.

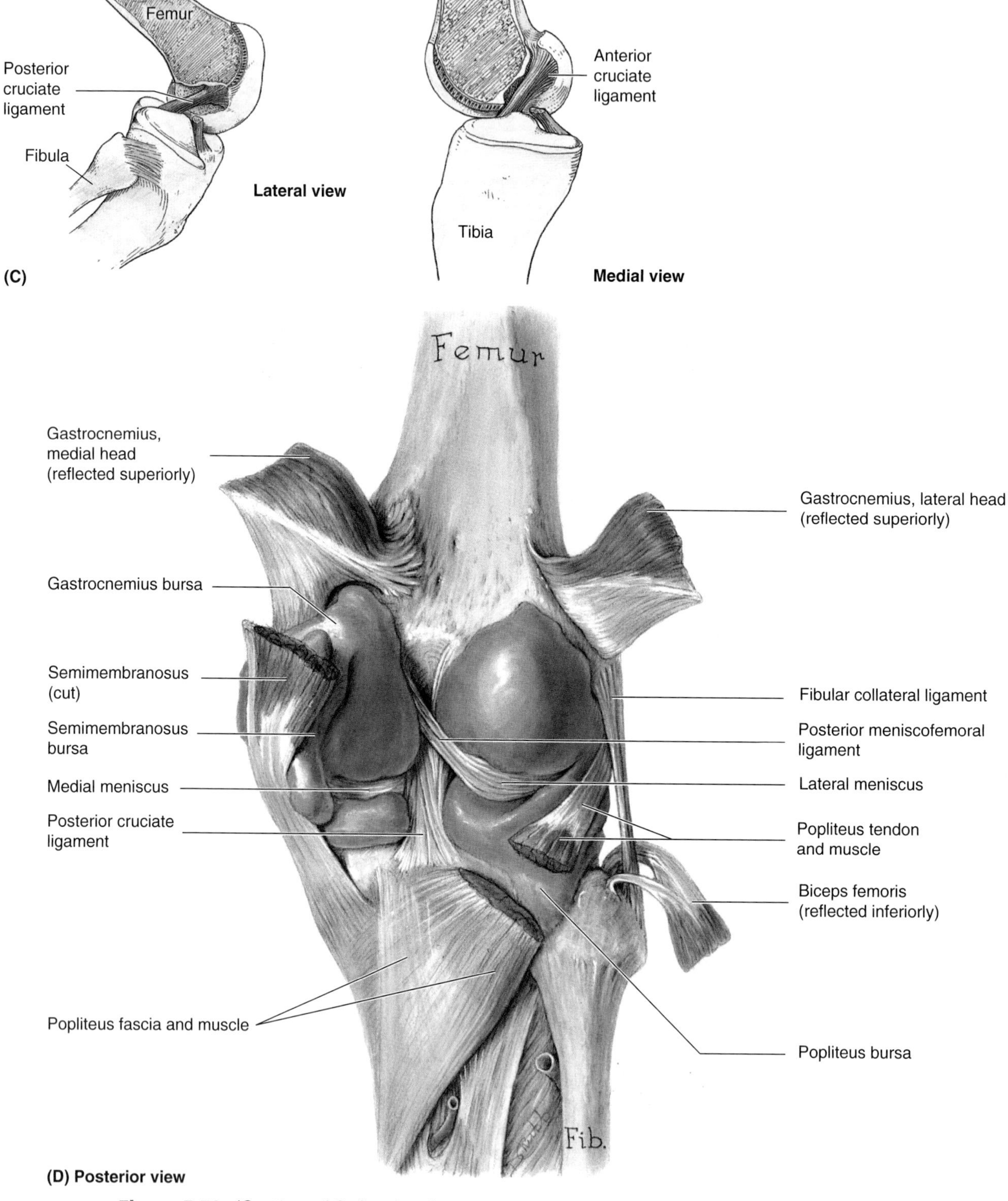

Figure 5.59. *(Continued)* **C.** Cruciate ligaments. In each drawing, the femur has been sectioned longitudinally and the near half has been removed with the proximal part of the corresponding cruciate ligament. The *right figure* demonstrates how the anterior cruciate ligament resists posterior displacement of the femur on the tibial plateau; the *left figure* demonstrates how the posterior cruciate ligament resists anterior displacement of the femur on the tibial plateau. **D.** Posterior view of the extended knee joint. Both heads of the gastrocnemius are reflected superiorly, and the biceps femoris is reflected inferiorly. Observe the bursae, most of which are in continuity with the joint cavity.

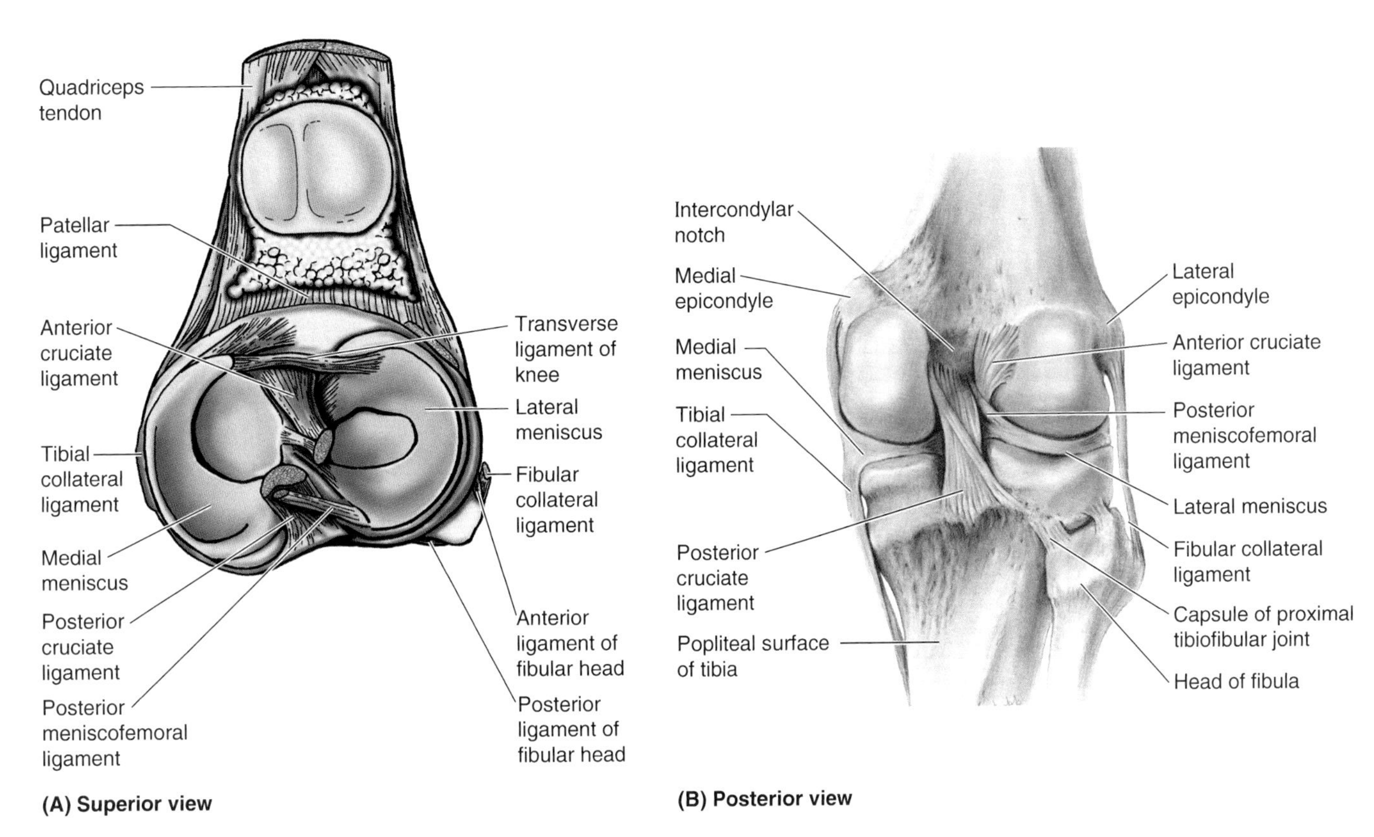

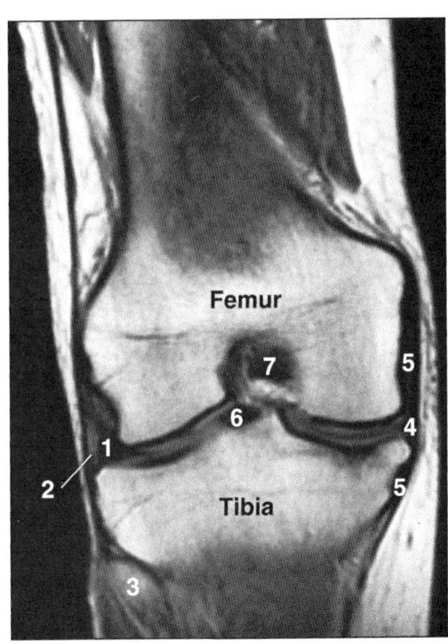

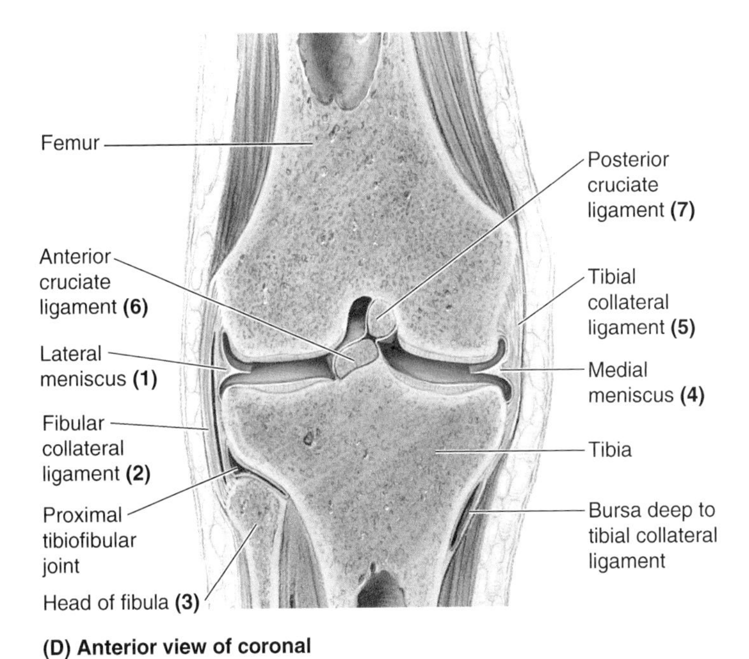

Figure 5.60. Cruciate ligaments and menisci of the knee joint. A. The quadriceps tendon is cut and the patella and patellar ligament are reflected inferiorly and anteriorly. The menisci, their attachments to the intercondylar area of the tibia, and the tibial attachments of the cruciate ligaments are shown. **B.** Observe that the bandlike tibial collateral ligament is attached to the medial meniscus and that the cordlike fibular collateral ligament is separated from the lateral meniscus. **C.** Coronal MRI of the knee. The numbers on the MRI refer to the structures labeled in the corresponding anatomical coronal section (**D**). (Courtesy of Dr. W. Kucharczyk, Chair of Medical Imaging, University of Toronto, and Clinical Director of Tri-Hospital Magnetic Resonance Centre, Toronto, Ontario, Canada)

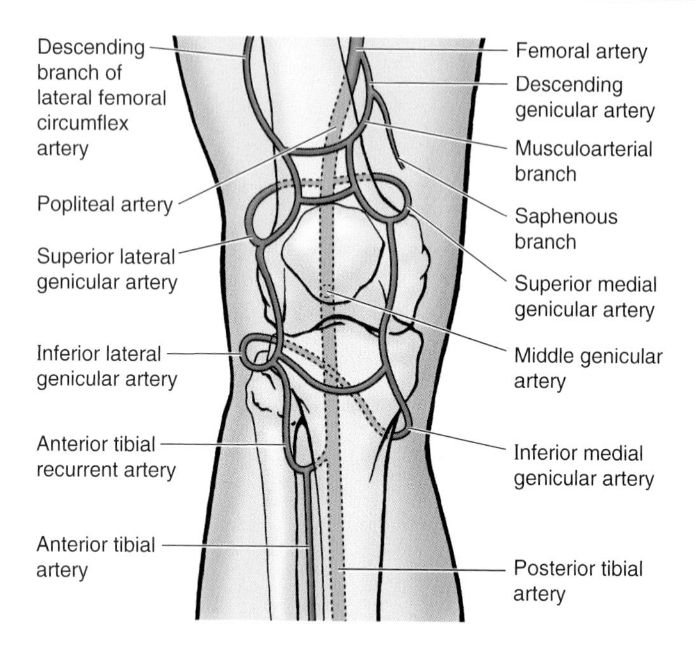

Figure 5.61. Arterial anastomoses around the knee. Anterior view of the communicating genicular arteries.

Blood Supply of the Knee Joint

The arteries supplying the knee joint are genicular branches of the femoral, popliteal, and anterior and posterior recurrent branches of the anterior tibial recurrent and circumflex fibular arteries, which form the **genicular anastomosis around the knee** (Figs. 5.61 and 5.62*B*). The middle genicular branches of the popliteal artery penetrate the fibrous capsule of the knee joint and supply the cruciate ligaments, synovial membrane, and peripheral margins of the menisci.

Innervation of the Knee Joint

The nerves of the knee joint (Fig. 5.62*D*) are branches of the obturator, femoral, tibial, and common fibular nerves.

Bursae Around the Knee Joint

Many bursae are around the knee joint (12 or more) because most tendons run parallel to the bones and pull lengthwise across the joint during knee movements. Subcutaneous bursae—**prepatellar and infrapatellar bursae**—are also at the convex surface of the joint because the skin must be able to move freely during movements of the knee.

Four bursae communicate with the synovial cavity of the knee joint: suprapatellar bursa, popliteus bursa, anserine bursa, and gastrocnemius bursa. The most important bursae around the knee are illustrated in Figure 5.14. The large **suprapatellar bursa** (Fig. 5.55, *A* and *B*) is especially important because an infection in it may spread to the knee cavity. Although it develops separately from the knee joint, it becomes continuous with it. For further information on the bursae, see Williams et al. (1995).

Knee Joint Injuries

Knee joint injuries are common because the knee is a mobile, weightbearing joint, and its stability depends almost entirely on its associated ligaments and muscles. The knee joint is essential for everyday activities such as standing, walking, and climbing stairs. It is also a main joint for sports that involve running, jumping, kicking, and changing directions. To perform these activities, the knee joint must be mobile; however, this mobility makes it susceptible to injuries in contact and noncontact sports. The most common knee injuries in contact sports are *ligament sprains*, which occur when the foot is fixed in the ground. If a force is applied against the knee when the foot cannot move, ligament injuries are likely to occur. The tibial and fibular collateral ligaments normally prevent disruption of the sides of the knee joint. They are tightly stretched when the leg is extended and usually prevent rotation of the tibia laterally or the femur medially. Because the collateral ligaments are slack during flexion of the leg, they permit some rotation of the tibia on the femur in this position.

The firm attachment of the tibial collateral ligament to the medial meniscus is of considerable clinical significance because tearing of the tibial collateral ligament frequently results in concomitant tearing of the medial meniscus. The damage is frequently caused by a blow to the lateral side of the knee. Injury to the medial meniscus results from the twisting strain on the knee when it is flexed. Because the meniscus is firmly adherent to the tibial collateral ligament, twisting strains of this ligament may tear and/or detach the medial meniscus from the fibrous capsule (*A*). This injury is common in athletes who twist their flexed knees while running (e.g., in football and volleyball). The ACL may tear when the tibial collateral ligament ruptures. First, the tibial collateral ligament ruptures, opening the joint on the medial side and possibly tearing the medial meniscus and ACL. This "unhappy triad of injuries" can result from clipping in football.

Severe force directed anteriorly with the knee semiflexed may also tear the ACL. **ACL ruptures**, one of the most common knee injuries in skiing accidents, for example, allow the tibia to slide anteriorly from the femur—*the anterior drawer sign* (*B*). The ACL may tear away from the femur or tibia; however, tears commonly occur in the midportion of the ligament.

Although strong, **PCL ruptures** may occur when a player lands on the tibial tuberosity with the knee flexed (e.g., when knocked to the floor in basketball). PCL ruptures usually occur in conjunction with tibial or fibular ligament tears (*C*). These injuries can also occur in head-on collisions when seatbelts are not worn and the proximal end of the tibia strikes the dashboard. PCL ruptures allow the tibia to slide posteriorly from the femur—*the posterior drawer sign.* ▶

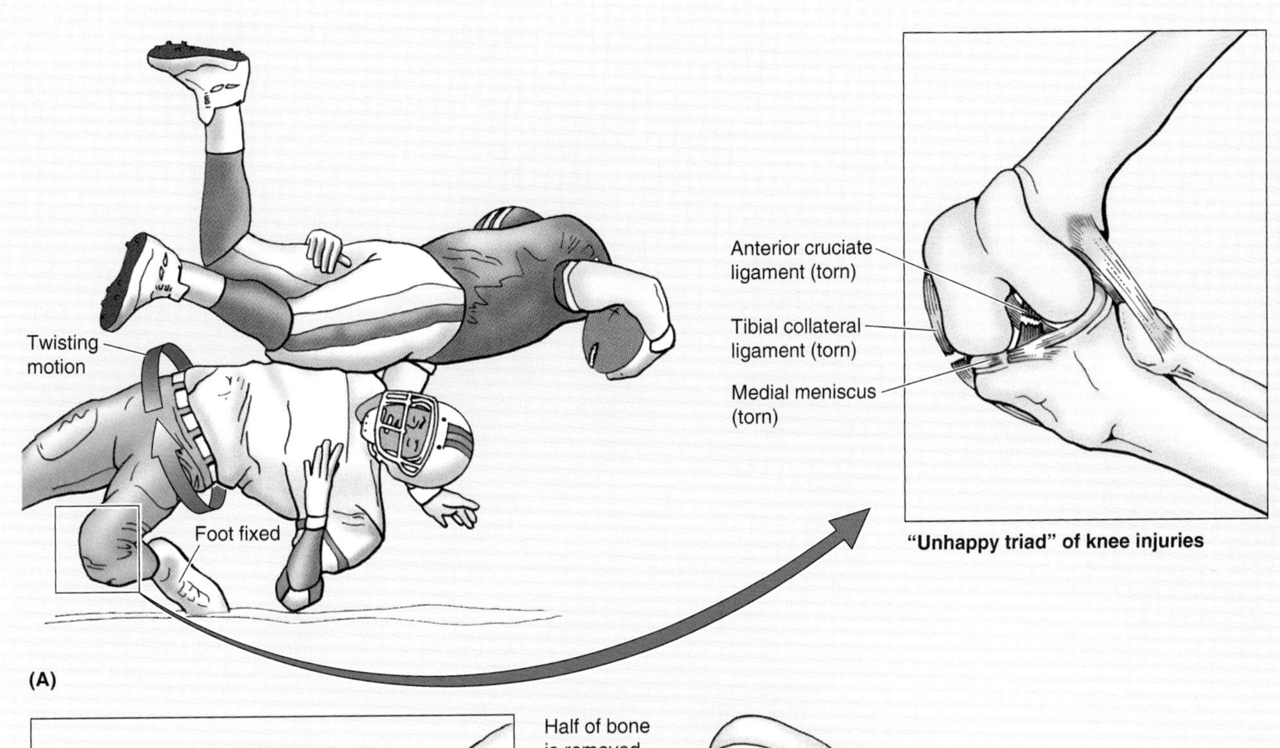

"Unhappy triad" of knee injuries

(A)

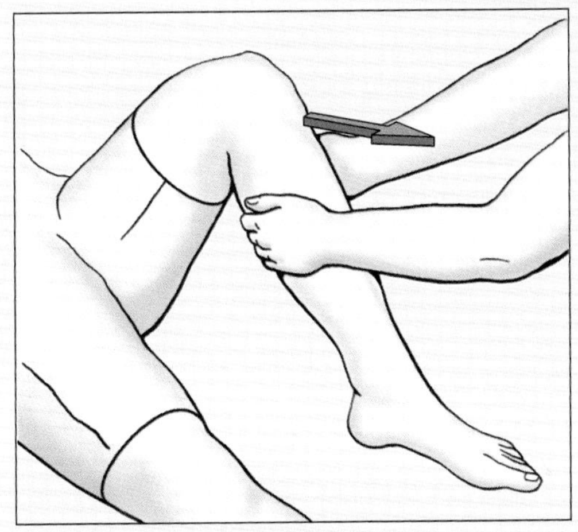

(B) Anterior drawer sign (ACL)

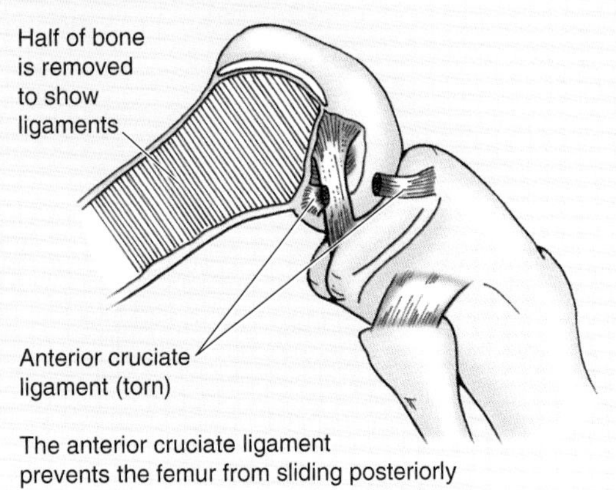

The anterior cruciate ligament prevents the femur from sliding posteriorly on the tibia and hyperextension of the knee and limits medial rotation of the femur when the foot is on the ground, and the leg is flexed.

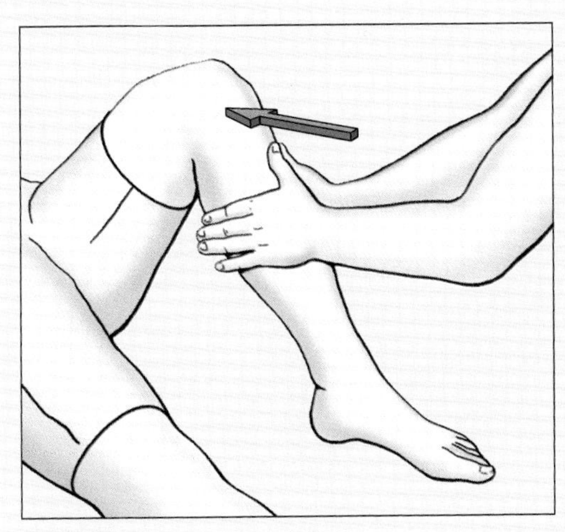

(C) Posterior drawer sign (PCL)

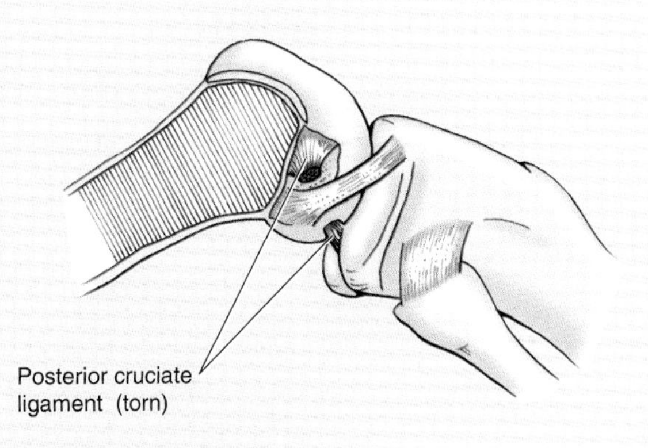

The posterior cruciate ligament prevents the femur from sliding anteriorly on the tibia, particularly when the knee is flexed.

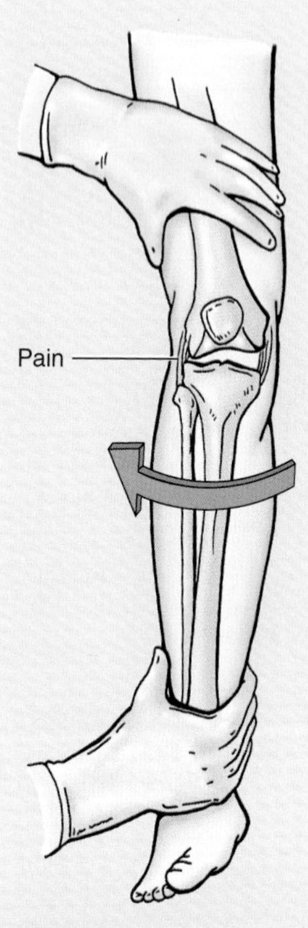

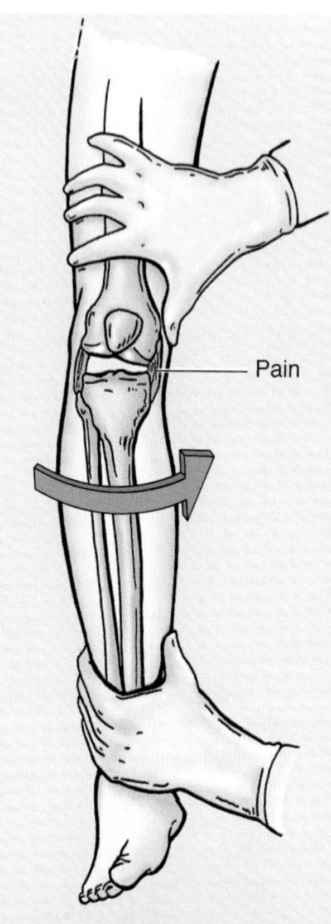

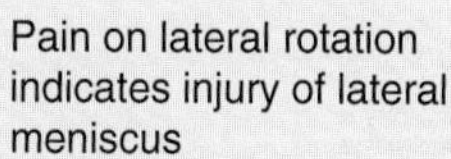

Pain on lateral rotation indicates injury of lateral meniscus

Pain on medial rotation indicates injury of medial meniscus

▶ *Meniscal tears commonly involve the medial meniscus.* The lateral meniscus does not usually tear because of its mobility. *Pain on lateral rotation of the tibia on the femur indicates injury of the lateral meniscus, whereas pain on medial rotation of the tibia on the femur indicates injury of the medial meniscus.* Most meniscal tears occur when the tibial collateral ligament and/or the ACL is torn. Peripheral meniscal tears can often be repaired or will heal on their own because of the generous blood supply to this area. Meniscal tears that do not heal or cannot be repaired are usually removed (e.g., by arthroscopic surgery). Knee joints from which the menisci have been removed suffer no loss of mobility; however, the tibial plateaus often undergo inflammatory reactions.

Arthroscopy of the Knee Joint

The arthroscope is a useful diagnostic and surgical tool that allows visualization of the interior of the joint cavity. This technique allows removal of torn menisci, loose bodies in the joint such as bone chips, and debridement—excision of devitalized articular cartilaginous material in advanced cases of arthritis. Although general anesthesia is preferable, knee arthroscopy can be performed using local or regional anesthesia. For further information see Soames (1995).

Patellofemoral Syndrome

Pain deep to the patella often results from excessive running, especially downhill; hence, this type of pain is often called "runner's knee." The pain results from repetitive microtrauma caused by abnormal tracking of the patella with the patellar surface of the femur. The patellofemoral syndrome may also result from a direct blow to the patella and from *osteoarthritis of the patellofemoral compartment* (degenerative wear and tear of articular cartilages). In some cases, strengthening of the vastus medialis corrects *patellofemoral dysfunction.* This muscle tends to prevent lateral dislocation of the patella because its fibers attach to the medial border of the patella. Hence, weakness of the vastus medialis predisposes to the patellofemoral dysfunction and patellar dislocation.

Aspiration of the Knee Joint

Fractures of the distal end of the femur or lacerations of the anterior thigh may involve the suprapatellar bursa and result in infection of the knee joint. When the knee joint is infected and inflamed, the amount of synovial fluid may increase. *Joint effusions*—escape of fluid from blood or lymphatic vessels—results in increased amounts of fluid in the joint cavity. Because the suprapatellar bursa communicates freely with the synovial cavity of the knee joint, fullness of the thigh in the region of the suprapatellar bursa may indicate increased synovial fluid. This bursa can be aspirated to remove the fluid. Direct aspiration of the knee joint is usually performed with the patient sitting on a table with the knee flexed. The joint is approached laterally, using three bony points as landmarks for needle insertion: the tip of the lateral tibial condyle, the lateral epicondyle, and the apex of the patella. The needle is inserted into the joint through the triangle formed by the above bony points. In addition to being the route for aspiration of serous and sanguineous (bloody) fluid, this triangular area also lends itself to drug injection for treating pathology of the knee joint.

Bursitis in the Knee Region

Prepatellar bursitis is usually a friction bursitis caused by friction between the skin and the patella; however, the bursa may be injured by compressive forces resulting from a direct blow or from falling on the flexed knee (Anderson and Hall, 1995). If the inflammation is chronic, the bursa becomes distended with fluid and forms a swelling anterior to the knee. This condition has been called "housemaid's knee"; however, other people who work on their knees without knee pads, such as hardwood floor and rug installers, also develop prepatellar bursitis. *Subcutaneous infrapatellar bursitis* results from excessive friction between the skin and tibial tuberosity; the edema occurs over the ▶

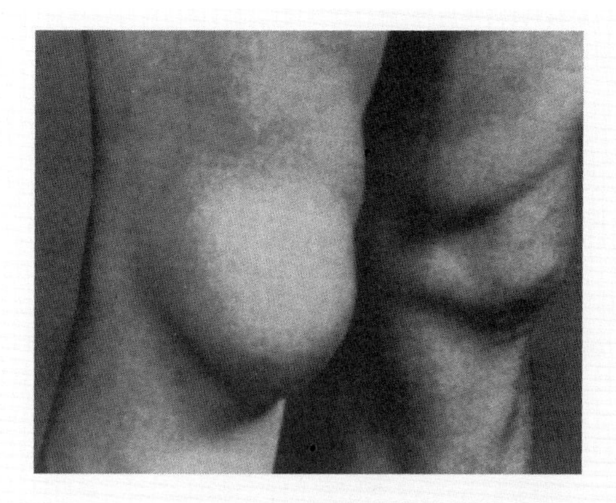

▶ proximal end of the tibia. This condition was formerly called "clergyman's knee" because of frequent genuflecting (L. genu, knee); however, it occurs more commonly in roofers and floor tilers who do not wear knee pads.

Deep infrapatellar bursitis results in edema between the patellar ligament and the tibia, superior to the tibial tuberosity. Inflammation of this bursa is usually caused by overuse and subsequent friction between the patellar tendon and structures posterior to it—infrapatellar fat pad and tibia (Anderson and Hall, 1995). Enlargement of the deep infrapatellar bursa obliterates the dimples on each side of the patellar ligament when the leg is extended.

Abrasions or penetrating wounds may result in *suprapatellar bursitis* caused by bacteria entering the bursa from the torn skin. This infected bursa differs from acute bursitis because of the localized redness and enlarged popliteal and inguinal lymph nodes. The infection may spread to the knee joint.

Popliteal Cysts

Popliteal cysts are fluid-filled herniations of the synovial membrane of the knee joint, or distensions of the gastrocnemius or semimembranosus bursa. A popliteal cyst is almost always a complication of chronic knee joint effusion (Slaby et al., 1994). Synovial fluid may escape from the knee joint (*synovial effusion*) or a bursa around the knee and collect in the popliteal fossa. Here it forms a new synovial-lined sac or *popliteal cyst* (Baker's cyst). The cyst may communicate with the synovial cavity of the knee joint by a narrow stalk, which suggests that some cysts result from herniation of the synovial membrane through the fibrous capsule into the popliteal fossa. Popliteal cysts are common in children but seldom cause symptoms. In adults, ▶

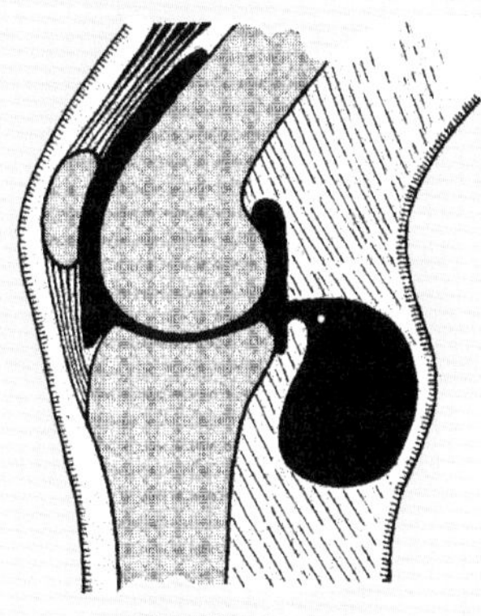

Popliteal cyst

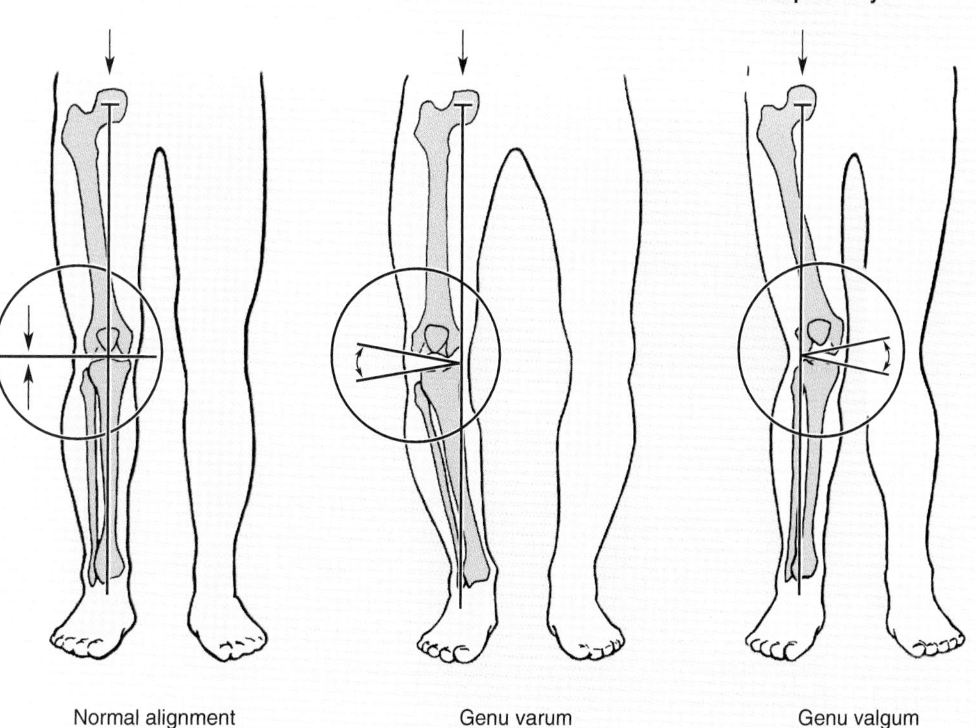

Normal alignment Genu varum Genu valgum

popliteal cysts can be large and may extend as far as the midcalf. In some cases, the cyst may interfere with knee movements.

Knee Deformities

Wear and tear of the knee joints is part of the normal aging process; however, *osteoarthritis accelerates the degenerative wear and tear of the menisci.* Osteoarthritis usually results from some predisposing factor, such as a deformity or an injury. Any irregularity of the knee joint results in wear and tear of the menisci. A torn meniscus can no longer serve as a shock absorber, and localized pressure and damage to the joint result.

Genu varum and genu valgum result in deviation of the tibia from the midline. In *genu varum* the tibia is diverted medially, and in *genu valgum* the tibia is diverted laterally; these deformities cause unequal weight distribution. In the varum deformity, the medial side of the knee takes all the pressure, leading to wear and tear of the medial meniscus. These deformities can be corrected by knee realignment (Gross, 1989).

Knee Replacement

If the patient's entire knee is diseased, resulting from osteoarthritis, for example, an artificial knee joint may be inserted (*total knee replacement arthroplasty*). The artificial knee consists of plastic and metal components that are cemented to the bone ends. The combination of metal and plastic mimics the smoothness of cartilage on cartilage and produces good results in "low-demand" people who have a relatively sedentary life. In "high-demand" people who are active in sports, the bone-cement junctions may break down, and the artificial knee components may loosen (Gross, 1989). For more information see Soames (1995).

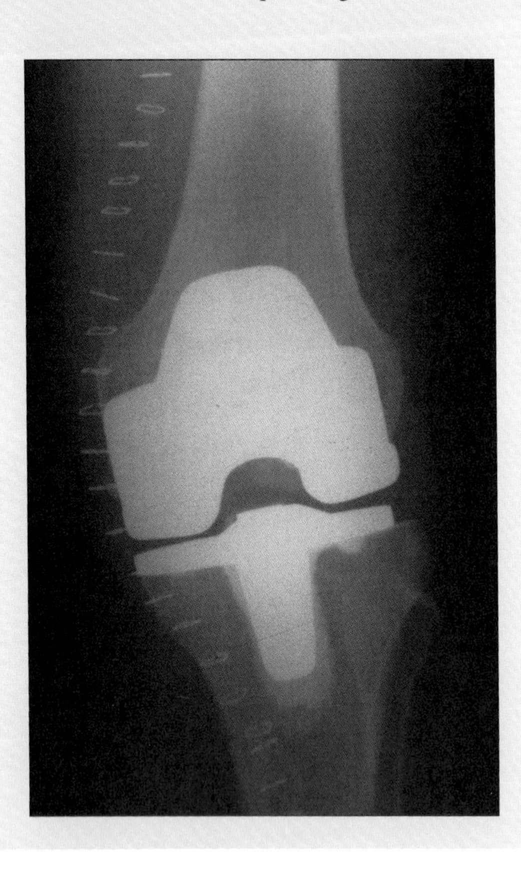

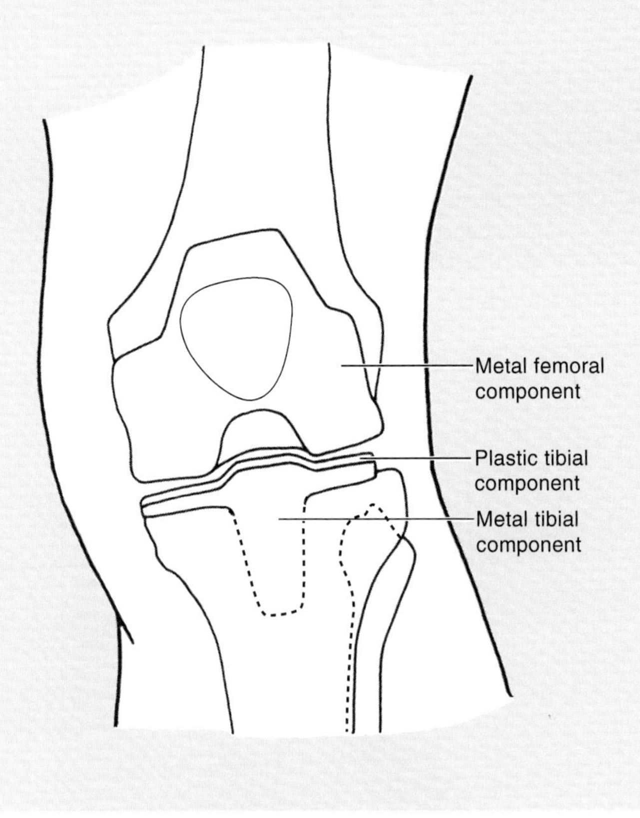

Tibiofibular Joints

The tibia and fibula are connected by two joints: the *proximal tibiofibular joint* and the *distal tibiofibular joint.* In addition, an interosseous membrane joins the bodies of the two bones (Fig. 5.62*A*). The anterior tibial vessels pass through a hiatus at the superior end of the membrane. At the inferior end of the interosseous membrane is a smaller hiatus through which

Figure 5.62. Tibiofibular joints and arteries and nerves of the leg. A. Anterior view. Diagram of the tibiofibular articulations. **B.** Anterior view of the lower thigh, knee, leg, and foot, showing the arterial supply of the joints. **C.** Anterior view of the nerve supply of the leg. **D.** Posterior view of the knee and tibiofibular joints and the joints of the foot.

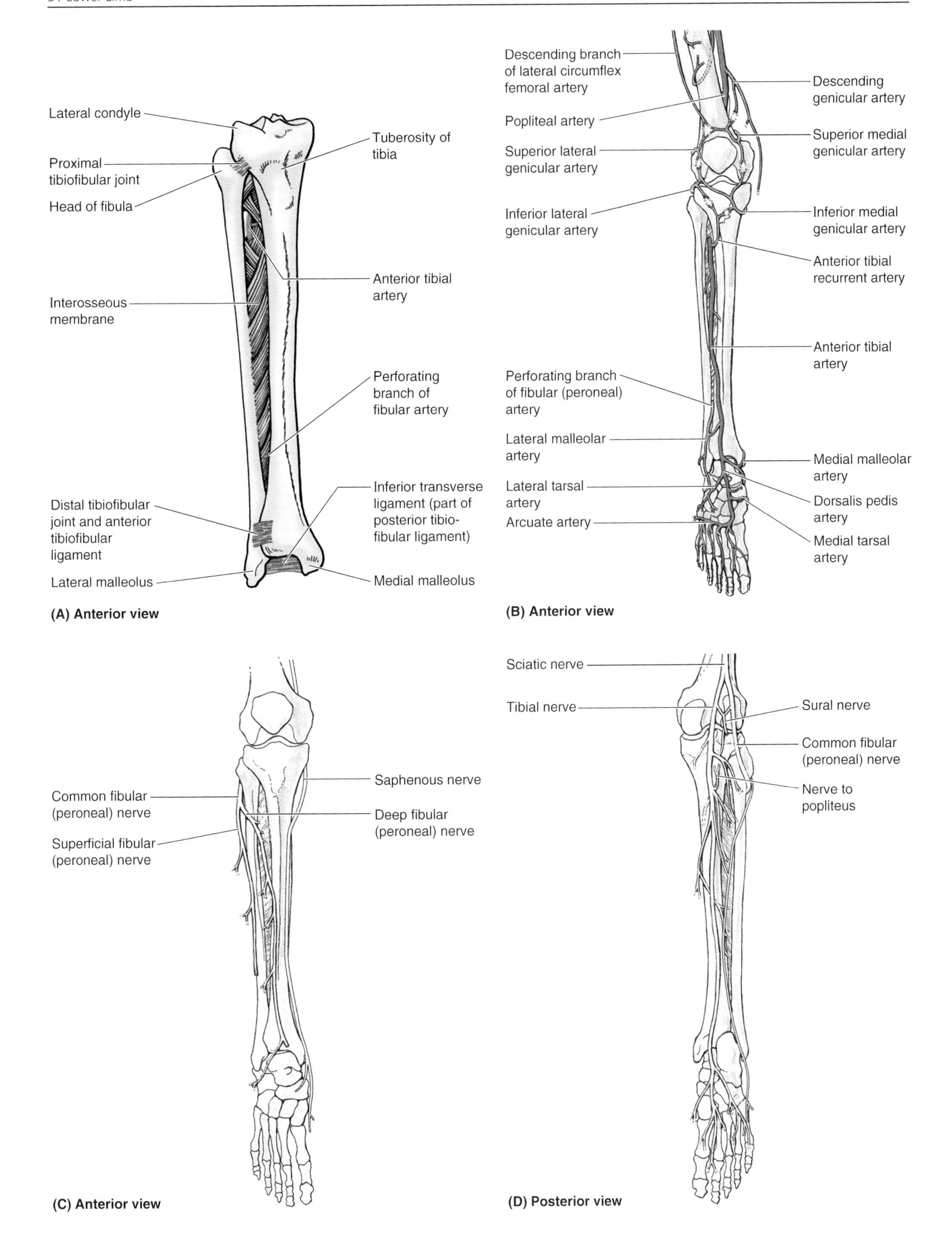

(A) Anterior view

(B) Anterior view

(C) Anterior view

(D) Posterior view

the perforating branch of the fibular artery passes. Movement at the proximal tibiofibular joint is impossible without movement at the distal joint.

Proximal Tibiofibular Joint

The proximal tibiofibular joint (Figs. 5.60, *B* and *D*, and 5.62*A*) is a plane type of synovial joint between the head of the fibula and the lateral condyle of the tibia.

Articular Surfaces. The flat facet on the fibular head articulates with a similar articular facet located posterolaterally on the lateral tibial condyle.

Articular Capsule and Ligaments. The *fibrous capsule* surrounds the joint and attaches to the margins of the articular surfaces of the fibula and tibia. The capsule is strengthened by anterior and posterior tibiofibular ligaments. The *anterior ligament of the head of the fibula* consists of two or three broad flat bands that pass superomedially from the anterior aspect of the fibular head to the anterior part of the lateral tibial condyle. The *posterior ligament of the head of the fibula* is a single broad band that passes superomedially from the posterior aspect of the fibular head to the posterior part of the lateral tibial condyle. The thick band is covered by the tendon of the popliteus. The *synovial membrane* lines the fibrous capsule. A pouch of synovial membrane, the **popliteus bursa** (Table 5.15), passes between the tendon of the popliteus and the lateral condyle of the tibia and may communicate with the synovial cavity of the knee joint.

Movements. Slight movement of the joint occurs during dorsiflexion and plantarflexion of the foot.

Blood Supply. The arteries of the joint (Fig. 5.62*B*) are from the inferior lateral genicular and anterior tibial recurrent arteries.

Nerve Supply. The nerves of the joint (Fig. 5.62, *C* and *D*) are from the common fibular nerve and the nerve to the popliteus.

Distal Tibiofibular Joint

The distal tibiofibular joint is a fibrous joint (syndesmosis). The integrity of this articulation is essential for the stability of the ankle joint because it keeps the lateral malleolus firmly against the lateral surface of the talus.

Articular Surfaces and Ligaments. The rough, triangular articular area on the medial surface of the inferior end of the fibula articulates with a facet on the inferior end of the tibia (Fig. 5.62*A*). The strong **interosseous ligament**, continuous superiorly with the interosseous membrane, forms the principal connection between the tibia and fibula. The joint is also strengthened anteriorly and posteriorly by the strong anterior and posterior **inferior tibiofibular ligaments**. The distal deep continuation of the posterior inferior tibiofibular ligament—the *inferior transverse (tibiofibular) ligament*—forms a strong connection between the distal ends of the tibia (medial malleolus) and fibula (lateral malleolus). It contacts the talus and forms the posterior "wall" of a three-sided "socket" for the trochlea of the talus, the lateral and medial walls being formed by the respective malleoli.

Movements. Slight movement of the joint occurs to accommodate the talus during dorsiflexion of the foot.

Blood Supply. The arteries are from the perforating branch of the fibular artery and from medial malleolar branches of the anterior and posterior tibial arteries (Figs. 5.61 and 5.62*B*).

Nerve Supply. The nerves are from the deep fibular, tibial, and saphenous nerves (Fig. 5.62, *C* and *D*).

Ankle Joint

The ankle joint (talocrural articulation) is a hinge type of synovial joint. It is located between the distal ends of the tibia and fibula and the superior part of the talus. The ankle joint can be felt between the tendons on the anterior surface of the ankle as a slight depression, approximately 1 cm proximal to the tip of the medial malleolus.

Articular Surfaces of the Ankle Joint

The distal ends of the tibia and fibula (along with the inferior transverse part of the posterior tibiofibular ligament) form a *mortise* (deep socket) into which the pulley-shaped **trochlea of the talus** fits (Fig. 5.63). The trochlea (L. pulley) is the rounded superior articular surface of the talus. The medial surface of the lateral malleolus articulates with the lateral surface of the talus. The tibia articulates with the talus in two places:

- Its inferior surface forms the roof of the mortise
- Its medial malleolus articulates with the medial surface of the talus.

The malleoli grip the talus tightly as it rocks anteriorly and especially posteriorly in the mortise during movements of the joint.

The grip of the malleoli on the trochlea is strongest during dorsiflexion of the foot because this movement forces the wider, anterior part of the trochlea posteriorly, spreading the tibia and fibula slightly apart. This spreading is limited by the strong interosseous ligament and by the transverse, anterior, and posterior tibiofibular ligaments that unite the tibia and fibula (Fig. 5.62*A*). The ankle joint is relatively unstable during plantarflexion because the trochlea is narrower posteriorly and therefore lies relatively loosely within the mortise. It is during plantarflexion that most injuries of the ankle occur (usually as a result of sudden, unexpected—and therefore adequately resisted—inversion of the foot).

Articular Capsule of the Ankle Joint

The *fibrous capsule* is thin anteriorly and posteriorly but is supported on each side by strong collateral ligaments (Figs. 5.64 and 5.65). It is attached superiorly to the borders of the articular surfaces of the tibia and to the malleoli and inferiorly to the talus.

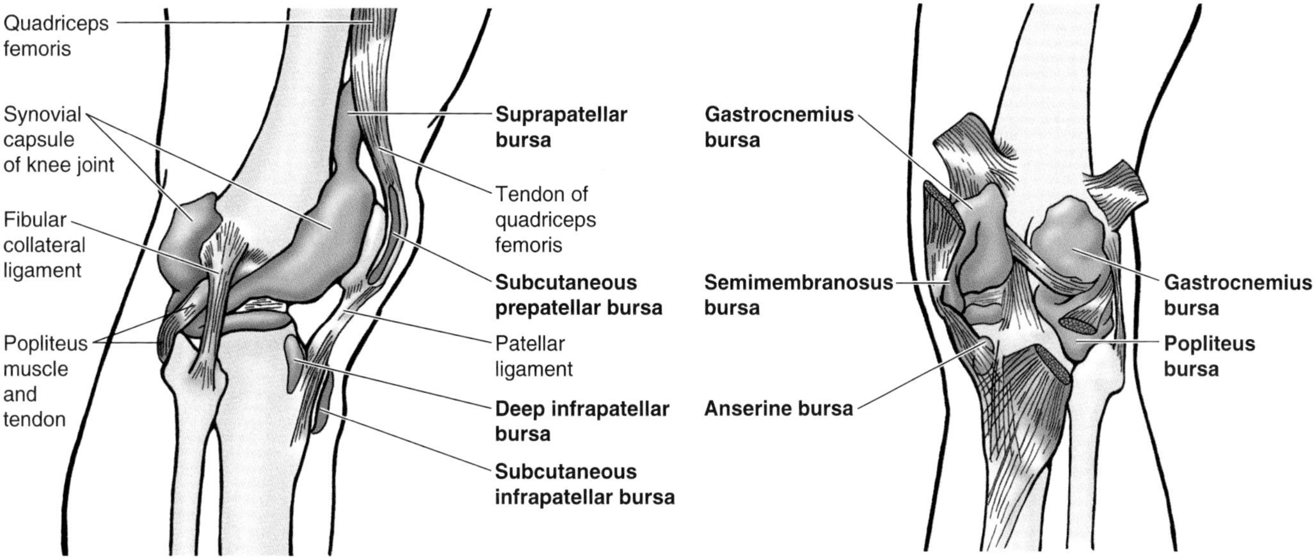

(A) Lateral view

(B) Posterior view

Bursae	Locations	Comments
Suprapatellar (quadriceps)	Located between femur and tendon of quadriceps femoris	Held in position by articularis genu muscle; communicates freely with (superior extension of) synovial cavity of knee joint
Popliteus	Located between tendon of popliteus and lateral condyle of tibia	Opens into synovial cavity of knee joint inferior to lateral meniscus
Anserine	Separates tendons of sartorius, gracilis, and semitendinosus from tibia and tibial collateral ligament	Is area where tendons of these muscles attach to tibia; resembles the foot of a goose (pes anserinus–L. pes, foot; L. anser, goose)
Gastrocnemius	Lies deep to proximal attachment of tendon of medial head of gastrocnemius	Is an extension of synovial cavity of knee joint
Semimembranosus	Located between medial head of gastrocnemius and semimembranosus tendon	Related to the distal attachment of semimembranosus
Subcutaneous prepatellar	Lies between skin and anterior surface of patella	Allows free movement of skin over patella during movements of leg
Subcutaneous infrapatellar	Located between skin and tibial tuberosity	Helps knee to withstand pressure when kneeling
Deep infrapatellar	Lies between patellar ligament and anterior surface of tibia	Separated from knee joint by infrapatellar fatpad

Ligaments of the Ankle Joint

The fibrous capsule is reinforced by the **lateral ligament** (weaker than the medial ligament), which consists of three parts (Fig. 5.64):

- The *anterior talofibular ligament*—a flat, weak band that extends anteromedially from the lateral malleolus to the neck of the talus
- The *posterior talofibular ligament*—a thick, fairly strong band that runs horizontally medially and slightly posteriorly from the malleolar fossa to the lateral tubercle of the talus
- The *calcaneofibular ligament*—a round cord that passes posteroinferiorly from the tip of the lateral malleolus to the lateral surface of the calcaneus.

The three discrete ligaments are collectively referred to as the lateral ligament.

The *fibrous capsule* is reinforced medially by the large, strong **medial ligament** (deltoid ligament) that attaches prox-

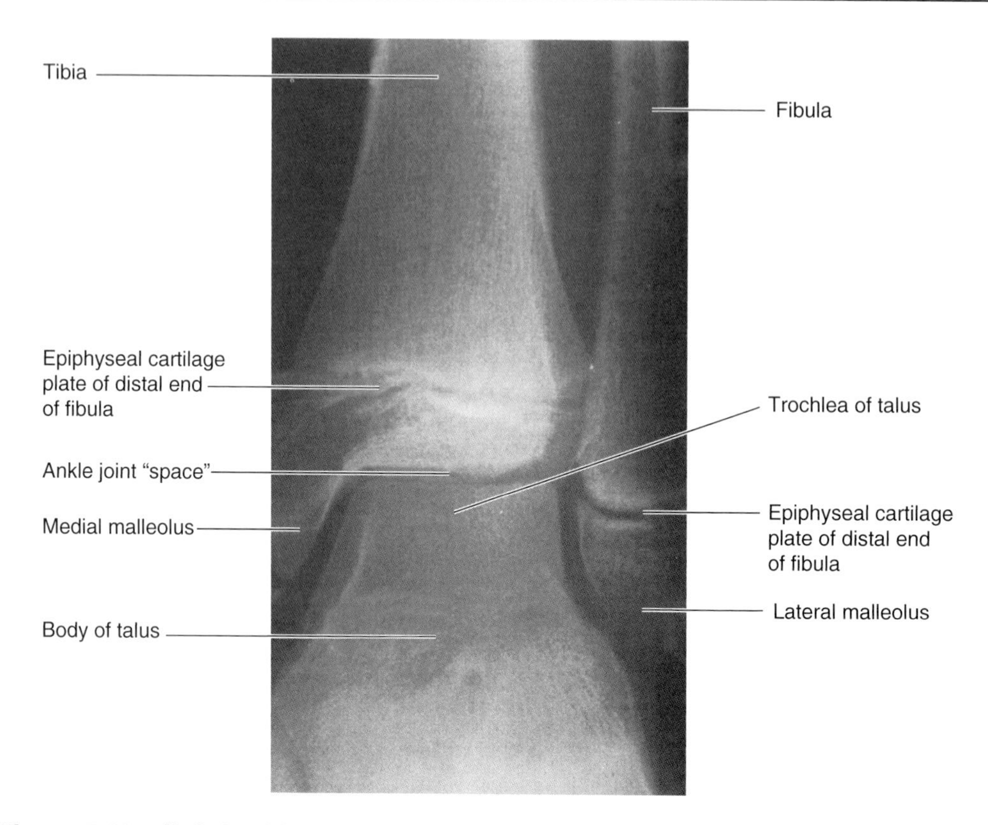

Figure 5.63. Slightly oblique frontal radiograph of the ankle joint of a 14-year-old boy. Observe how the trochlea of the body of the talus (ankle bone) fits into the mortise formed by the medial and lateral malleoli.

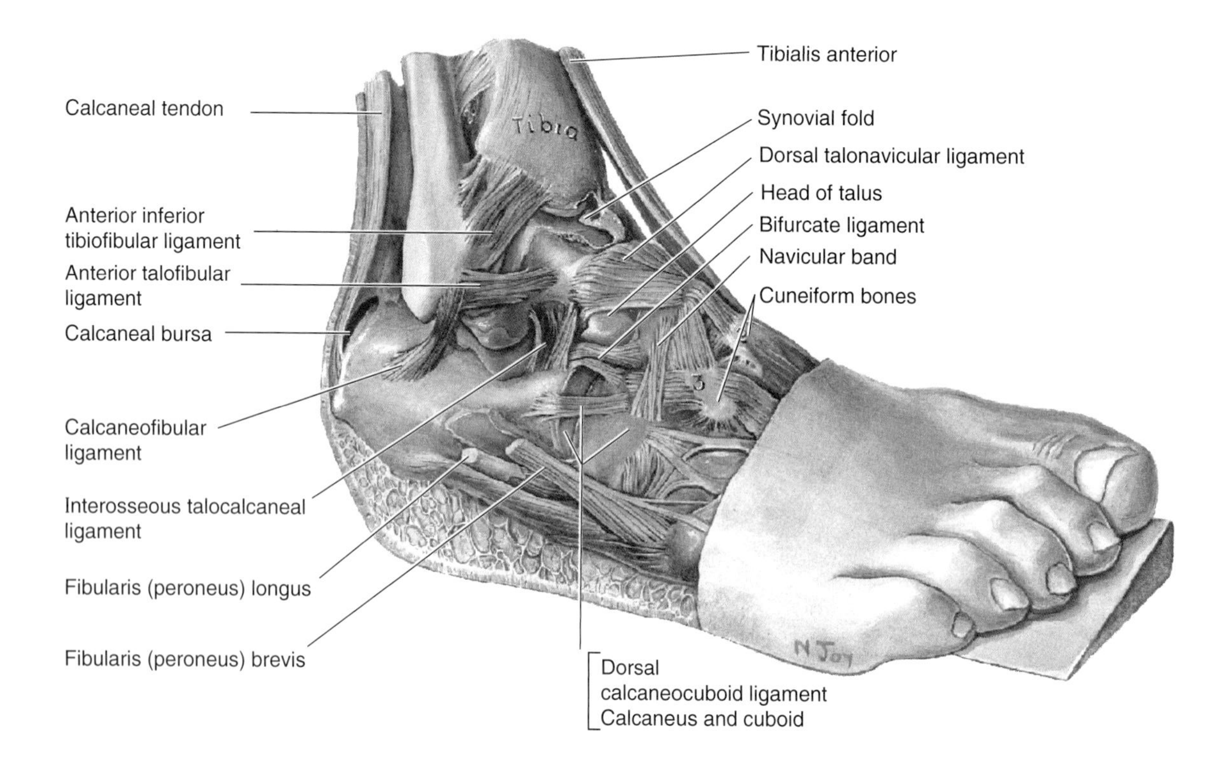

Figure 5.64. Dissection of the right ankle joint and the joints of inversion and eversion. Lateral view. The foot has been inverted to demonstrate the articular areas and the lateral ligaments that become taut during inversion of the foot.

imally to the medial malleolus (Fig. 5.65). The medial ligament has fibers that fan out from the malleolus and attach distally to the talus, calcaneus, and navicular—forming the:

- Tibionavicular ligament
- Anterior and posterior tibiotalar ligaments
- Tibiocalcaneal ligament.

The medial ligament stabilizes the ankle joint during eversion and prevents subluxation (partial dislocation) of the joint.

The **synovial membrane** is loose and lines the fibrous capsule (Fig. 5.64). The synovial cavity often extends superiorly between the tibia and fibula as far as the interosseous ligament of the distal tibiofibular joint.

Movements of the Ankle Joint

The main movements are dorsiflexion and plantarflexion. When the foot is plantarflexed, some rotation, abduction, and adduction of the ankle joint are possible.

- *Dorsiflexion of the ankle* is produced by the muscles in the anterior compartment of the leg (Table 5.9, p. 577). Dorsiflexion is usually limited by the passive resistance of the triceps surae to stretching and by tension in the medial and lateral ligaments.
- *Plantarflexion of the ankle* is produced by the muscles in the posterior compartment of the leg (Table 5.12, p. 588). In toe dancing by ballet dancers, for example, the dorsum of the foot is in line with the anterior surface of the leg.

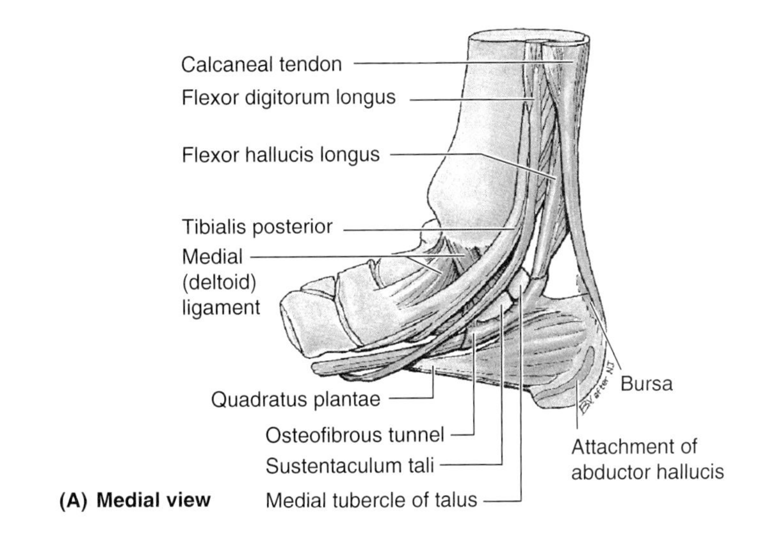

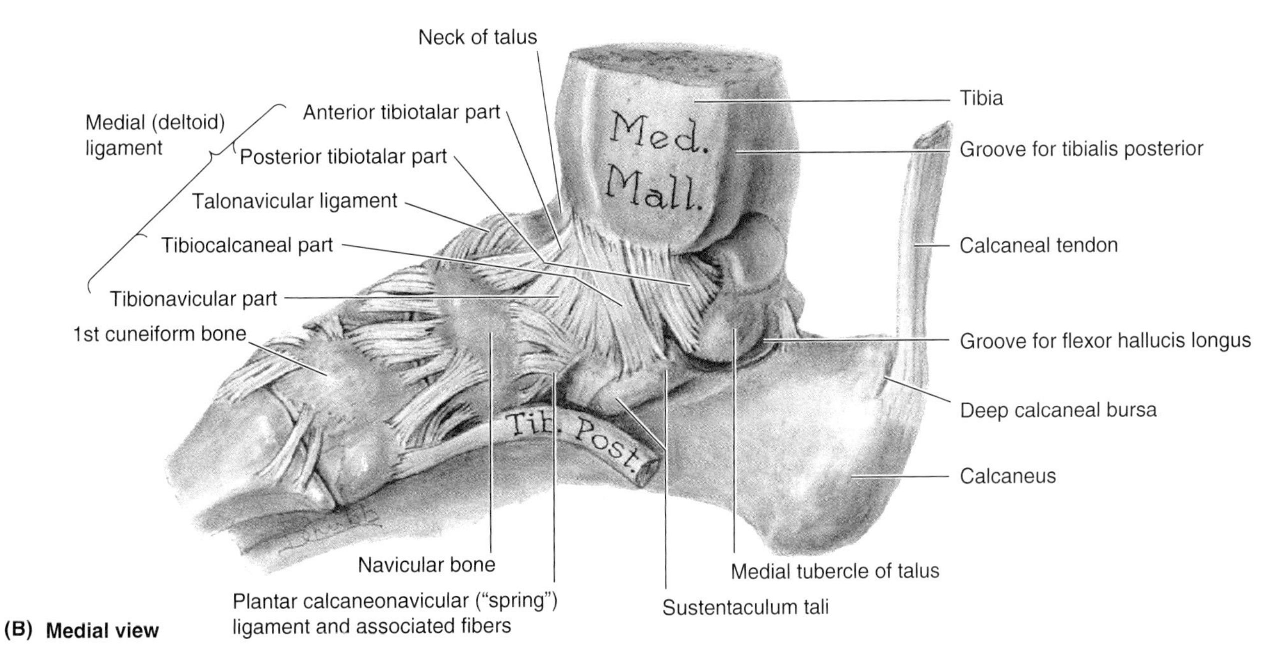

Figure 5.65. Ankle and tarsal joints. Medial views. **A.** Tendons of the medial aspect of the ankle. **B.** Dissection of the ankle and tarsal joints. Observe the four parts of the medial ligament of the ankle.

Blood Supply of the Ankle Joint

The arteries are derived from malleolar branches of the fibular and anterior and posterior tibial arteries (Figs. 5.61 and 5.62*B*).

Nerve Supply of the Ankle Joint

The nerves are derived from the tibial nerve and the deep fibular nerve, a division of the common fibular nerve (Fig. 5.62, *C* and *D*).

Ankle Injuries

The ankle is the most frequently injured major joint in the body. *Ankle sprains* (tearing fibers of ligaments) are most common. A sprained ankle is nearly always an *inversion injury*, involving twisting of the weightbearing foot. The person steps on an uneven surface and the foot is forcibly inverted. *Lateral ligament sprains* occur in sports in which running and jumping are common, particularly basketball (70% to 80% of players have had at least one sprain). The lateral ligament is often injured because it is much weaker than the medial ligament. Many fibers of the *anterior talofibular ligament*—part of the lateral ligament—are torn during ankle sprains, either partially or completely, resulting in instability of the ankle joint. The *calcaneofibular ligament* may also be torn. In severe sprains, the lateral malleolus of the fibula may be fractured. *Shearing injuries fracture the lateral malleolus* at or superior to the ankle joint. *Avulsion fractures* break the malleolus inferior to the ankle joint; a fragment of bone is pulled off by the attached ligament(s).

A *Pott's fracture-dislocation of the ankle* occurs when the foot is forcibly everted. This action pulls on the extremely strong medial ligament, often tearing off the medial malleolus. The talus then moves laterally, shearing off the lateral malleolus or, more commonly, breaking the fibula superior to the inferior tibiofibular joint. If the tibia is carried anteriorly, the posterior margin of the distal end of the tibia is also sheared off by the talus, producing a *"trimalleolar fracture."* In this injury, the distal end of the tibia is considered to be a "malleolus."

Tibial Nerve Entrapment

The tibial nerve leaves the posterior compartment of the leg by passing deep to the flexor retinaculum in the interval between the medial malleolus and calcaneus. Entrapment and compression of the tibial nerve (*tarsal tunnel syndrome*) occurs when there is edema and tightness in the ankle involving the synovial sheaths of the tendons of muscles in the posterior compartment of the leg. The area involved is from the medial malleolus to the calcaneus, and the heel pain results from compression of the tibial nerve by the flexor retinaculum. ⊙

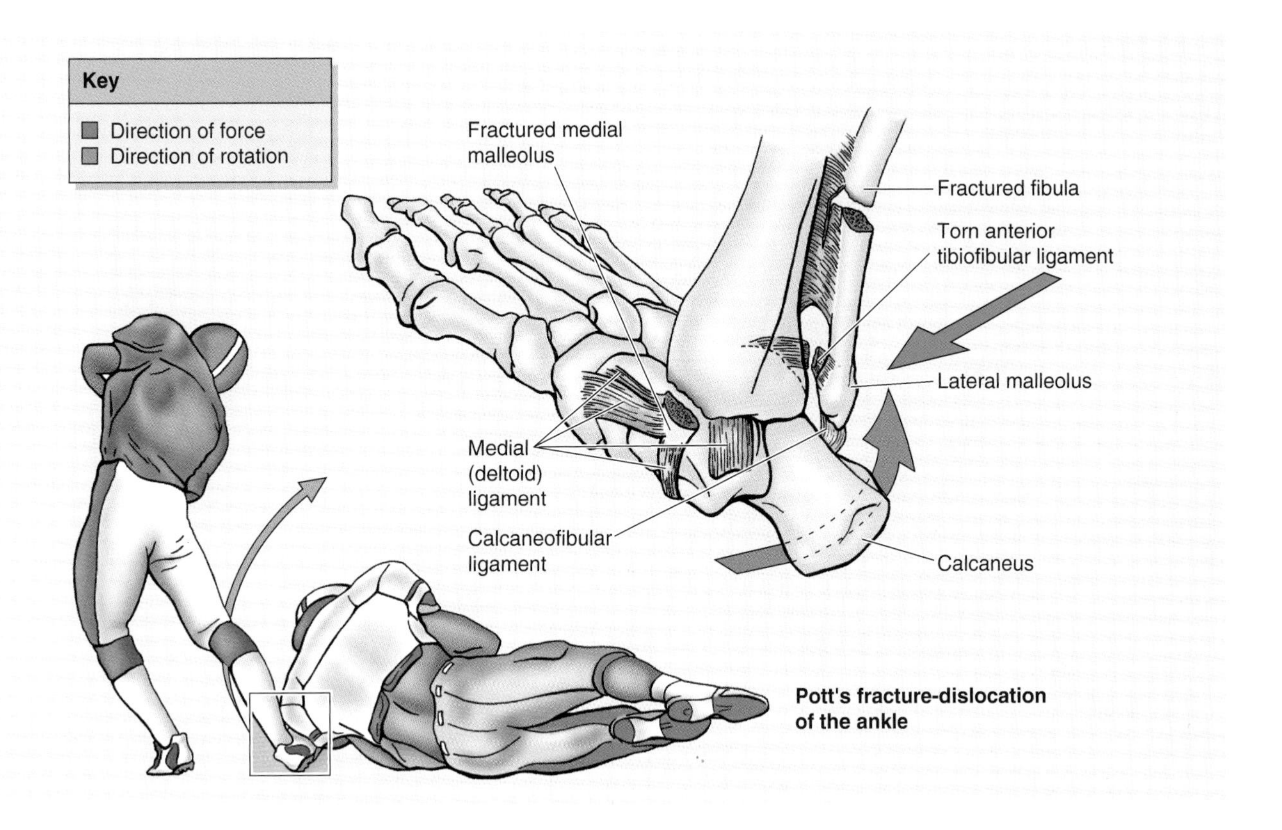

Foot Joints

The many joints of the foot involve the tarsals, metatarsals, and phalanges (Table 5.16). The important intertarsal joints are the **transverse tarsal joint** (calcaneocuboid and talonavicular joints) and the **subtalar joint**. Inversion and eversion of the foot are the main movements involving these joints. The other joints in the foot are relatively small and are so tightly joined by ligaments that only slight movement occurs between them (Table 5.17). All foot bones are united by dorsal and plantar ligaments.

The **transverse tarsal joint** is formed by the combined talonavicular part of the talocalcaneonavicular and calcaneocuboid joints; two separate joints align transversely. Transection across the transverse tarsal joint is a standard method for *surgical amputation of the foot*.

The **subtalar joint** occurs where the talus rests on and articulates with the calcaneus. The subtalar joint is a synovial joint that is surrounded by an articular capsule, which is attached near the margins of the articular facets. The fibrous capsule is weak but is supported by medial, lateral, posterior, and interosseous *talocalcaneal ligaments* (Fig. 5.64). Orthopaedic surgeons often use the term "subtalar joint" as a functional term where most inversion and eversion movements occur).

Ligaments of the Foot

The major ligaments of the foot (Fig. 5.66) are the:

- **Plantar calcaneonavicular ligament** (spring ligament) that extends from the sustentaculum tali to the posteroinferior surface of the navicular. This ligament plays an important role in maintaining the longitudinal arch of the foot.
- **Long plantar ligament** that passes from the plantar surface of the calcaneus to the groove on the cuboid. Some of its fibers extend to the bases of the metatarsals, thereby forming a tunnel for the tendon of the fibularis longus (Fig. 5.66*A*). The long plantar ligament is important in maintaining the arches of the foot.
- **Plantar calcaneocuboid ligament** (short plantar ligament) that is deep to the long plantar ligament (Fig. 5.66*B*). It extends from the anterior aspect of the inferior surface of the calcaneus to the inferior surface of the cuboid.

Arches of the Foot

The tarsal and metatarsal bones are arranged in longitudinal and transverse arches that add to the weightbearing capabilities and resiliency of the foot. They act as shock absorbers for supporting the weight of the body and for propelling it during movement. The resilient arches of the foot make it adaptable to surface and weight changes. The weight of the body is transmitted to the talus from the tibia. Then it is transmitted posteroinferiorly to the calcaneus and anteroinferiorly to the heads of the 2nd to 5th metatarsals and the sesamoids of the 1st digit (Fig. 5.67). Between these weightbearing points are

Table 5.16. **Joints of the Foot**

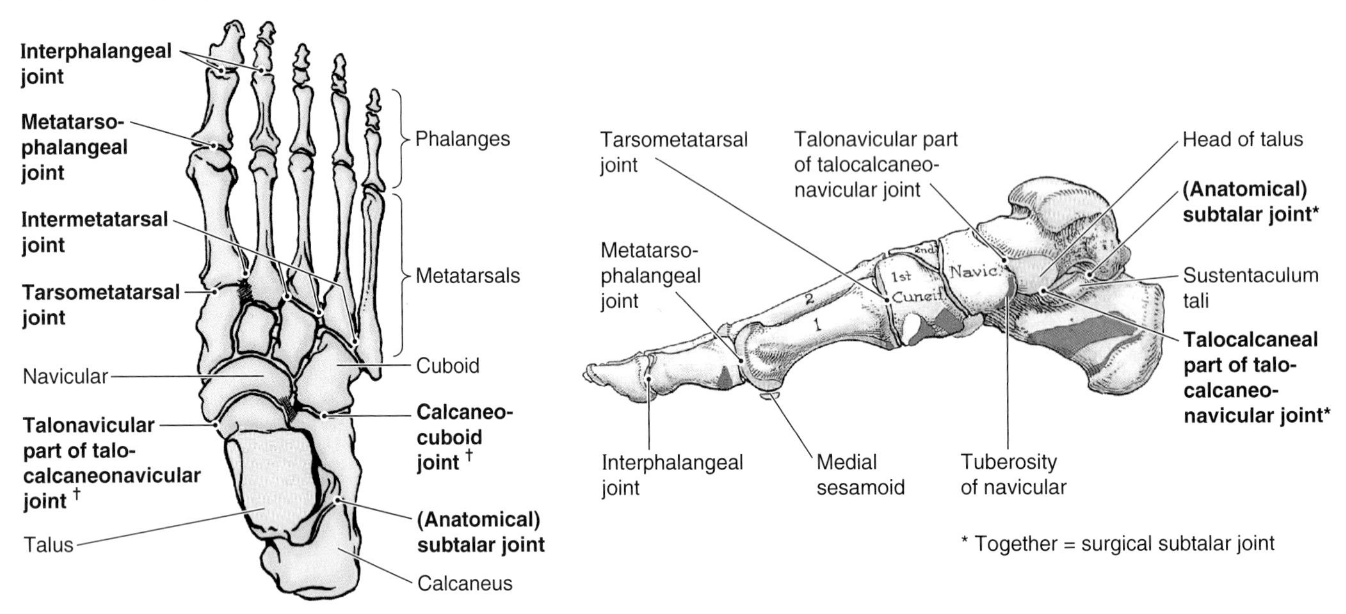

Joint	Type	Articular Surface	Articular Capsule	Ligaments	Movements	Blood Supply	Nerve Supply
Subtalar (talo-calcaneal)	Plane type of synovial joint	Inferior surface of body of talus (posterior calcanean articular facet) articulates with superior surface (posterior talar articular surface) of calcaneus	Fibrous capsule is attached to margins of articular surfaces	Medial, lateral, and posterior talocalcaneal ligaments support capsule; interosseous talocalcaneal ligament binds bones together	Inversion and eversion of foot	Posterior tibial and fibular arteries	
Talocalcaneo-navicular	Synovial joint; talonavicular part is ball and socket type	Head of talus articulates with calcaneus and navicular bones	Fibrous capsule incompletely encloses joint	Plantar calcaneo-navicular (spring) ligament supports head of talus	Gliding and rotatory movements are possible	Anterior tibial artery via lateral tarsal artery	Plantar aspect: medial or lateral plantar nerves; Dorsal aspect: deep fibular nerve
Calcaneo-cuboid	Plane type of synovial joint	Anterior end of calcaneus articulates with posterior surface of cuboid	Fibrous capsule encloses joint	Dorsal calcaneocuboid ligament, plantar calcaneocuboid ligament, and long plantar ligament support fibrous capsule	Inversion and eversion of foot	Anterior tibial artery via lateral tarsal artery	
Tarso-metatarsal	Plane type of synovial joint	Anterior tarsal bones articulate with bases of metatarsal bones	Fibrous capsule encloses each joint	Dorsal, plantar, and interosseous ligaments	Gliding or sliding	Lateral tarsal artery, a branch of dorsalis pedis artery	Deep fibular medial and lateral plantar, and sural nerves
Intermetatarsal	Plane type of synovial joint	Bases of meta-tarsal bones articulate with each other	Fibrous capsule encloses each joint	Dorsal, plantar, and interosseous ligaments bind bones together	Little individual movement of bones possible	Lateral meta-tarsal artery, a branch of dorsalis pedis artery	
Metatarso-phalangeal	Condyloid type of synovial joint	Heads of meta-tarsal bones articulate with bases of proximal phalanges	Fibrous capsule encloses each joint	Collateral ligaments support capsule on each side; plantar ligament supports plantar part of capsule	Flexion, extension, and some abduction, adduction, and circumduction	Lateral tarsal artery, a branch of dorsalis pedis artery	Digital nerves
Inter-phalangeal	Hinge type of synovial joint	Head of one phalanx articulates with base of one distal to it	Fibrous capsule encloses each joint	Collateral and plantar ligaments support joints	Flexion and extension	Digital branches of plantar arch	

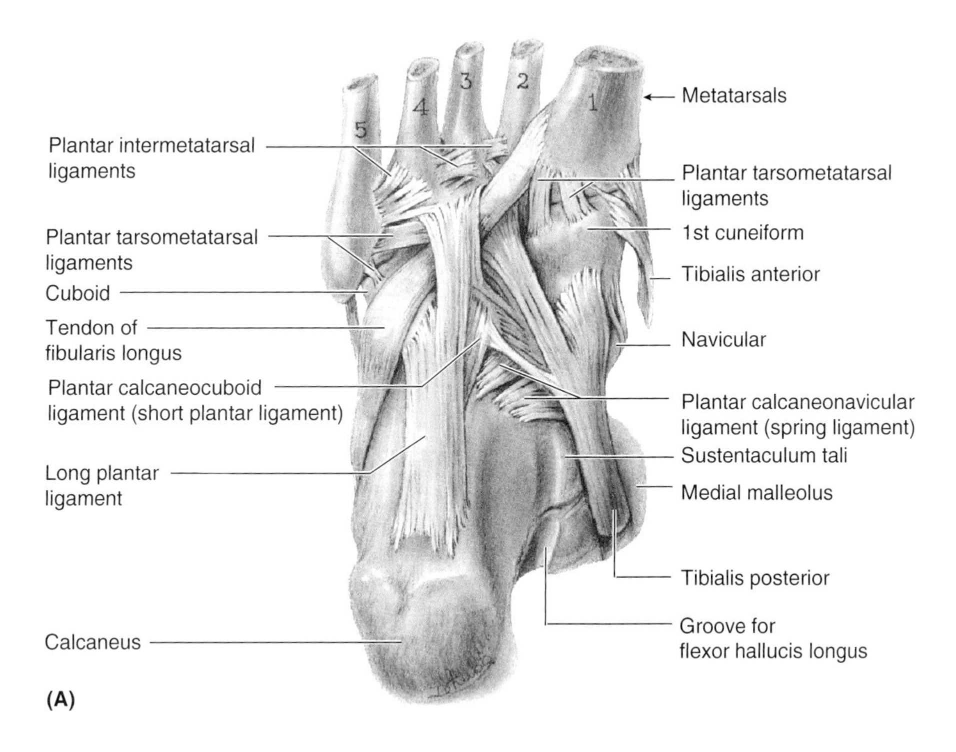

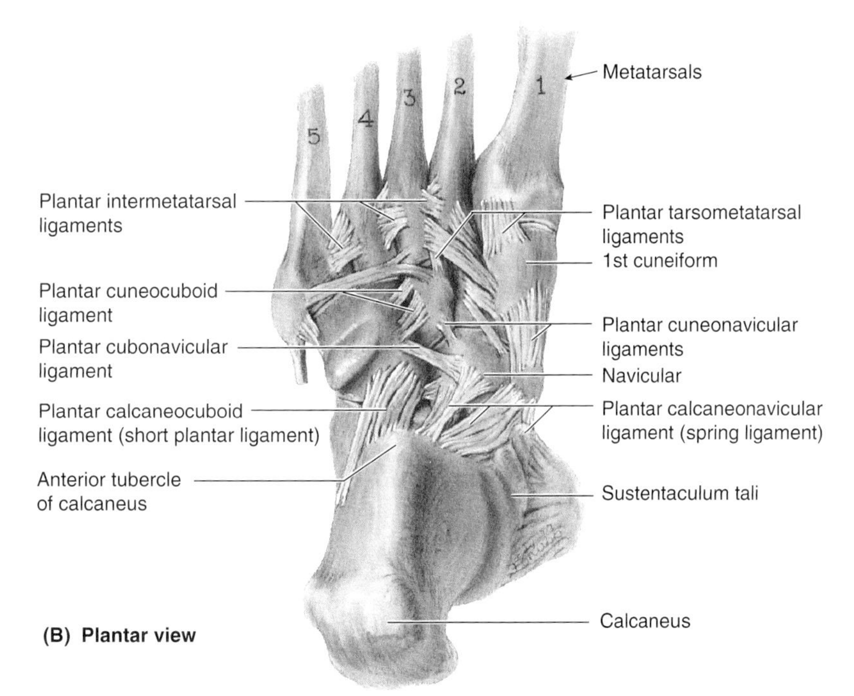

Figure 5.66. Plantar ligaments. A–B. Deep dissection of the sole of the right foot showing the attachments of the ligaments and the long tendons of the leg muscles. The main ligaments from this view are the plantar calcaneonavicular and the long and short plantar ligaments.

Table 5.17. Main Muscles Moving the Metatarsophalangeal and Interphalangeal Joints[a]

Movement	Muscles
Metatarsophalangeal joints	
Flexion	**Flexor digitorum brevis** **Lumbricals** **Interossei** **Flexor hallucis brevis** **Flexor hallucis longus** Flexor digiti minimi brevis Flexor digitorum longus
Extension	**Extensor hallucis longus** **Extensor digitorum longus** **Extensor digitorum brevis**
Abduction	**Abductor hallucis** **Abductor digiti minimi** **Dorsal interossei**
Adduction	**Adductor hallucis** **Plantar interossei**
Interphalangeal joints	
Flexion	**Flexor hallucis longus** **Flexor digitorum longus** **Flexor digitorum brevis** Quadratus plantae
Extension	**Extensor hallucis longus** **Extensor digitorum longus** **Extensor digitorum brevis**

[a] **Boldface** indicates the muscles that are chiefly responsible for the movement; the other muscles assist them.

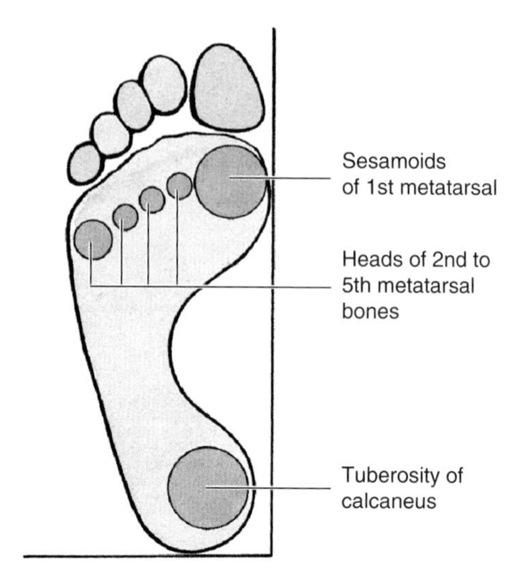

Figure 5.67. Weightbearing areas of the foot. Body weight is divided approximately equally between the calcaneus (heel bone) and the heads of the metatarsals. The anterior part of the foot has five points of contact with the ground: a large medial one that includes the two sesamoid bones associated with the head of the 1st metatarsal and the heads of the lateral four metatarsals. The 1st metatarsal supports a double load.

the relatively elastic arches of the foot that become slightly flattened by body weight during standing, but they normally resume their curvature (recoil) when body weight is removed (e.g., during sitting).

The **longitudinal arch of the foot** is composed of medial and lateral parts (Fig. 5.68). Functionally, both parts act as a unit with the transverse arch, spreading the weight in all directions. The **medial longitudinal arch** is higher and more important than the lateral longitudinal arch (Fig. 5.68*A*). The medial longitudinal arch is composed of the calcaneus, talus, navicular, three cuneiforms, and three metatarsals. *The talar head is the keystone of the medial longitudinal arch.* The tibialis anterior, attaching to the 1st metatarsal and medial cuneiform, helps strengthen the medial longitudinal arch. The fibularis longus tendon, passing from lateral to medial, also helps support this arch (Fig. 5.66*A*). The **lateral longitudinal arch** is much flatter than the medial part of the arch and rests on the ground during standing (Fig. 5.68*B*). It is composed of the calcaneus, cuboid, and lateral two metatarsals.

The **transverse arch of the foot** runs from side to side (Fig. 5.68*C*). It is formed by the cuboid, cuneiforms, and bases of the metatarsals. The medial and lateral parts of the longitudinal arch serve as pillars for the transverse arch. The tendon of the fibularis longus, crossing the sole of the foot obliquely (Fig. 5.66*A*), helps maintain the curvature of the transverse arch. *The integrity of the bony arches of the foot is maintained by the*:

- Shape of the interlocking bones
- Strength of the plantar ligaments, especially the plantar calcaneonavicular (spring) ligament and the long and short plantar ligaments
- Plantar aponeurosis (central part of plantar fascia)
- Action of muscles through the bracing action of their tendons.

Of these factors, the plantar ligaments and the plantar aponeurosis bear the greatest stress and are most important in maintaining the arches.

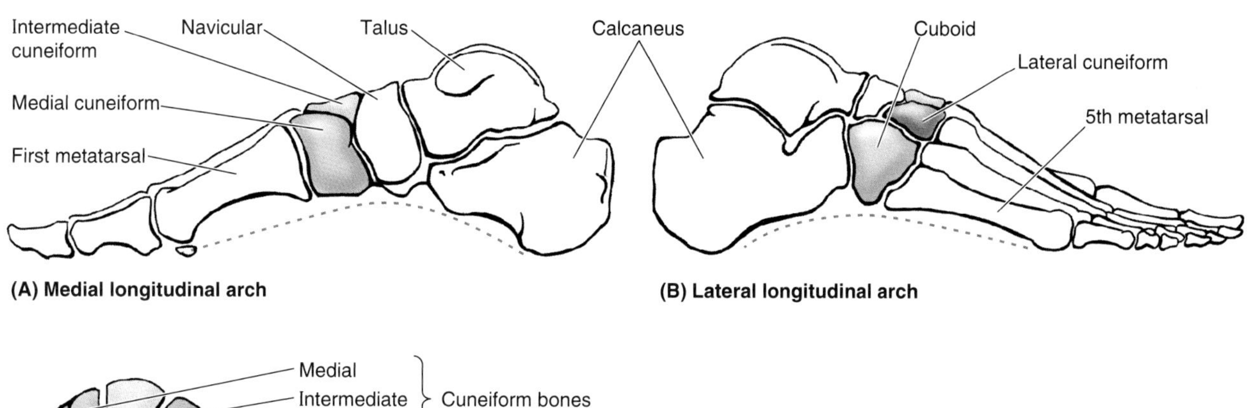

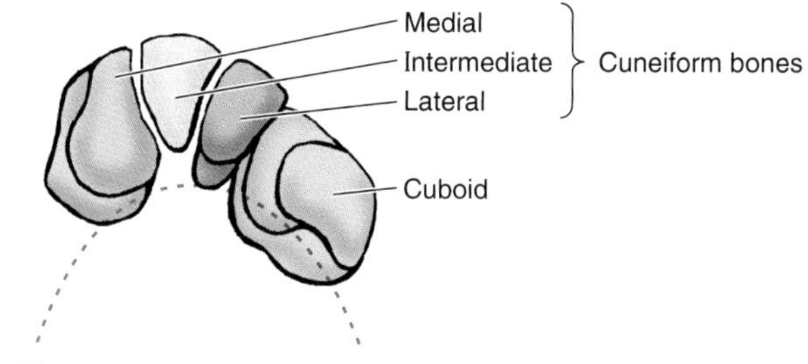

Figure 5.68. Arches of the foot. A–B. Medial and lateral longitudinal arches. **C.** Transverse arch.

Hallux Valgus

Hallux valgus is a foot deformity characterized by *lateral deviation of the great toe.* The **L** in valgus indicates *lateral deviation*. In some people the deviation is so great that the toe overlaps the 2nd toe (*A*). These persons cannot move their 1st digit away from their 2nd digit because *the sesamoids under the head of the 1st metatarsal are usually displaced* and lie in the space between the heads of the 1st and 2nd metatarsals (*B*). The 1st metatarsal bone shifts laterally and the sesamoid bones shift medially. Often the surrounding tissues swell and the resultant pressure and friction against the shoe cause a bursa to form; when tender and inflamed, the bursa is called a *bunion* (*A*). Often *hard corns* (inflamed areas of thick skin) also form over the proximal interphalangeal joints, especially the small toe.

Hammer Toe

Hammer toe is a deformity in which the proximal phalanx is permanently flexed at the metatarsophalangeal joint and the middle phalanx is plantarflexed at the interphalangeal joint. The distal phalanx is also flexed or extended, giving the digit (usually the 2nd) a hammerlike appearance (p. 642). This deformity of one or more toes may result from weakness of the lumbricals and interossei, which flex the metatarsophalangeal joints and extend the interphalangeal joints. A *callosity* or callus—hard thickening of the keratin layer of the skin—often develops where the dorsal surface of the toe repeatedly rubs on the shoe.

Claw Toes

Claw toes are characterized by hyperextension of the metatarsophalangeal joints and flexion of the distal ▶

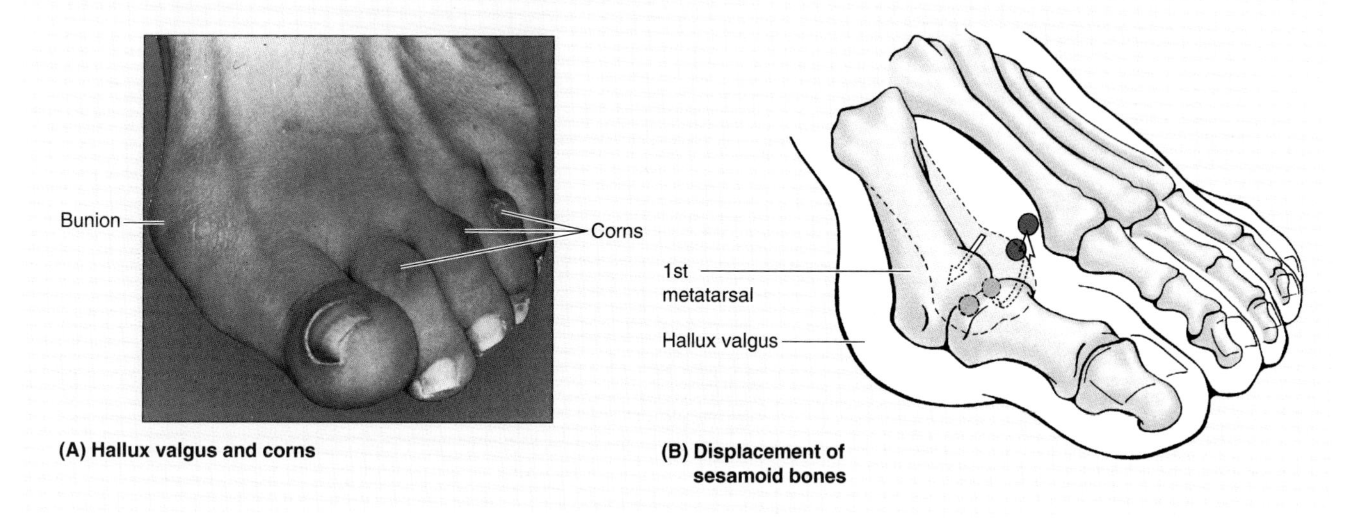

(A) Hallux valgus and corns

(B) Displacement of sesamoid bones

▶ interphalangeal joints. Usually all the lateral four toes are involved. Callosities develop on the dorsal surfaces of the toes because of pressure of the shoe. They may also form on the plantar surfaces of the metatarsal heads and the toe tips because they bear extra weight when claw toes are present.

Pes Planus (Flatfeet)

The flat appearance of infants' feet is normal and results from the thick subcutaneous fat pads in the soles of their feet. The arches of the feet are present at birth but they do not become visible until after the infant has walked for a few months. Pes planus or flatfeet in adolescents and adults results from "fallen arches," usually the medial parts of the longitudinal arches. When a person is standing, the plantar ligaments and plantar aponeurosis stretch under the body weight (*A*). If these ligaments become abnormally stretched during long periods of standing, the plantar calcaneonavicular ligament can no longer support the head of the talus. Consequently, the talar head displaces inferomedially and becomes prominent (*B*). Observe that the head of the talus is displaced inferomedially (*red arrow*). As a result, some flattening of the medial part of the longitudinal arch occurs, along with lateral deviation of the forefoot.

In the common type of flatfoot, the foot resumes its arched form when the weight is removed from it. *Flatfeet are common in older persons*, particularly if they undertake much unaccustomed standing or gain weight rapidly, adding stress on the muscles and increasing strain on the ligaments supporting the arches.

Clubfoot (Talipes)

Clubfoot refers to a foot that is twisted out of position. Of the several types, all are congenital (present at birth). *Talipes equinovarus*, the common type (2 per 1000 livebirths), involves the subtalar joint; boys are affected twice as often as girls. The foot is inverted, the ankle is plantarflexed, and the forefoot is adducted. The foot assumes the position of a horse's hoof, hence the prefix "equino" (L. equinus, horse). In half of those affected, both feet are malformed. A person with an uncorrected clubfoot cannot put the heel and sole of the foot flat and must bear the weight on the lateral surface of the forefoot. Consequently, walking is painful. The main abnormality is shortness and tightness of the muscles, tendons, ligaments, and articular capsules on the medial side and posterior aspect of the foot and ankle. ⊙

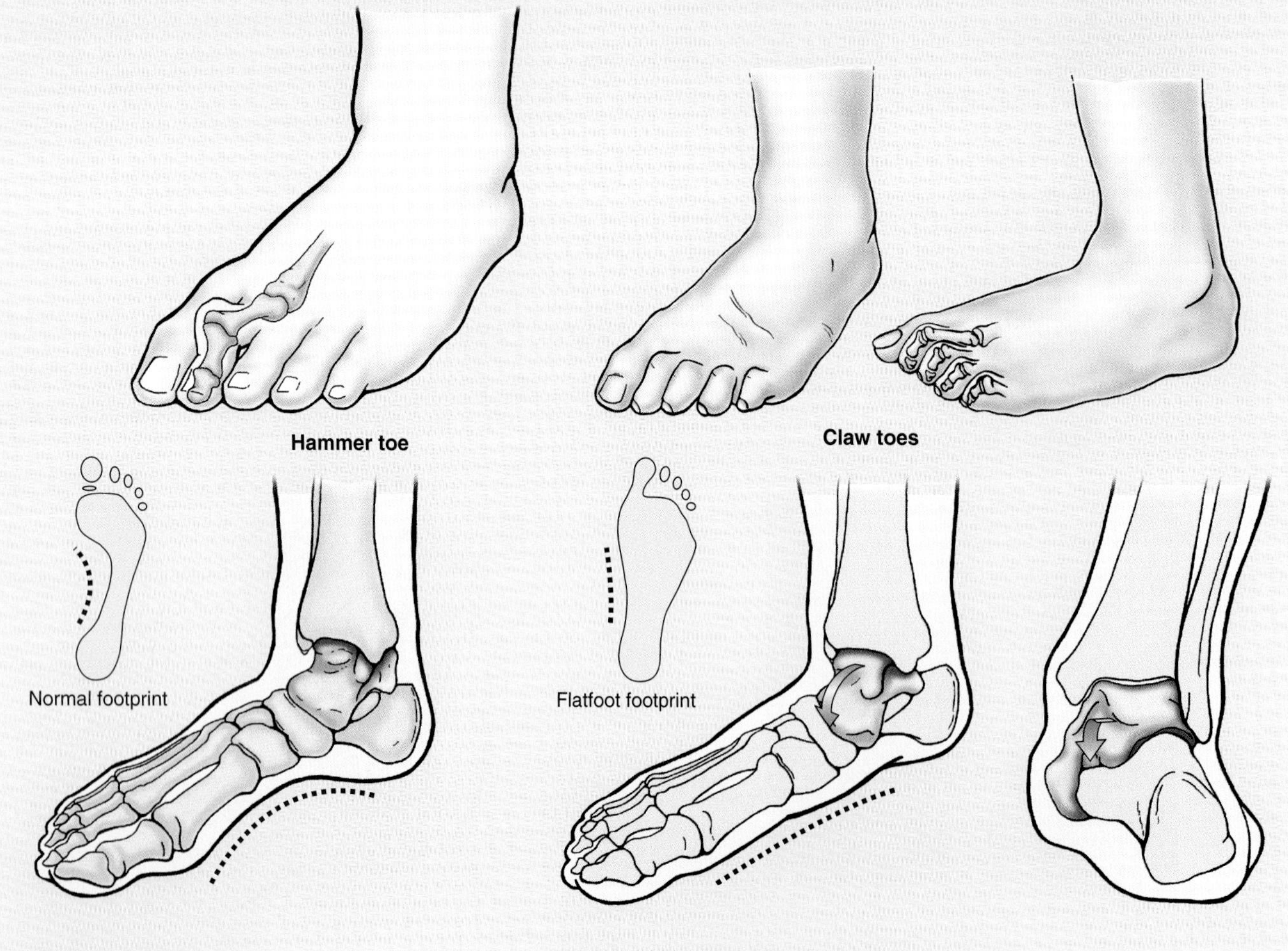

(A) View of normal arch

(B) View of fallen arch

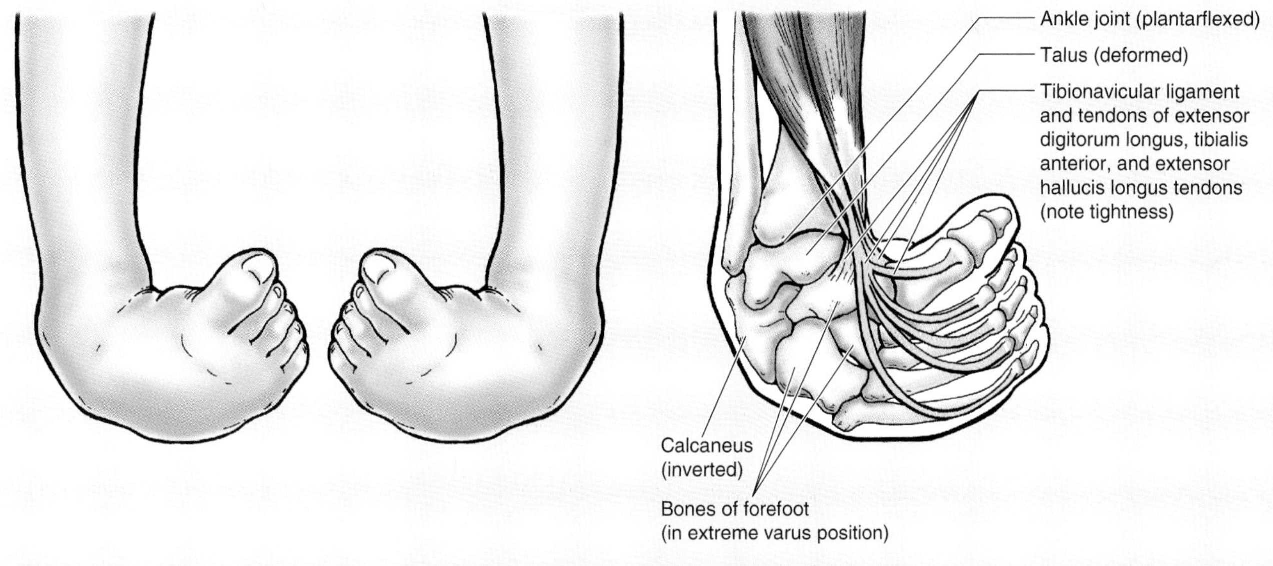

(A) Clubfeet or talipes equinovarus **(B)**

Surface Anatomy of the Ankle and Foot

The *medial* and *lateral malleoli* are subcutaneous and prominent. They can be easily palpated. Note that the tip of the lateral malleolus is farther distally and posteriorly than the medial malleolus. The *tuberosity of the navicular* is the most prominent and important landmark in the foot (*A*). It is easily seen and palpated on the medial aspect of the foot, inferoanterior to the tip of the medial malleolus. If the foot is actively inverted, the tendon of the tibialis posterior may be palpated as it passes posterior and ▶

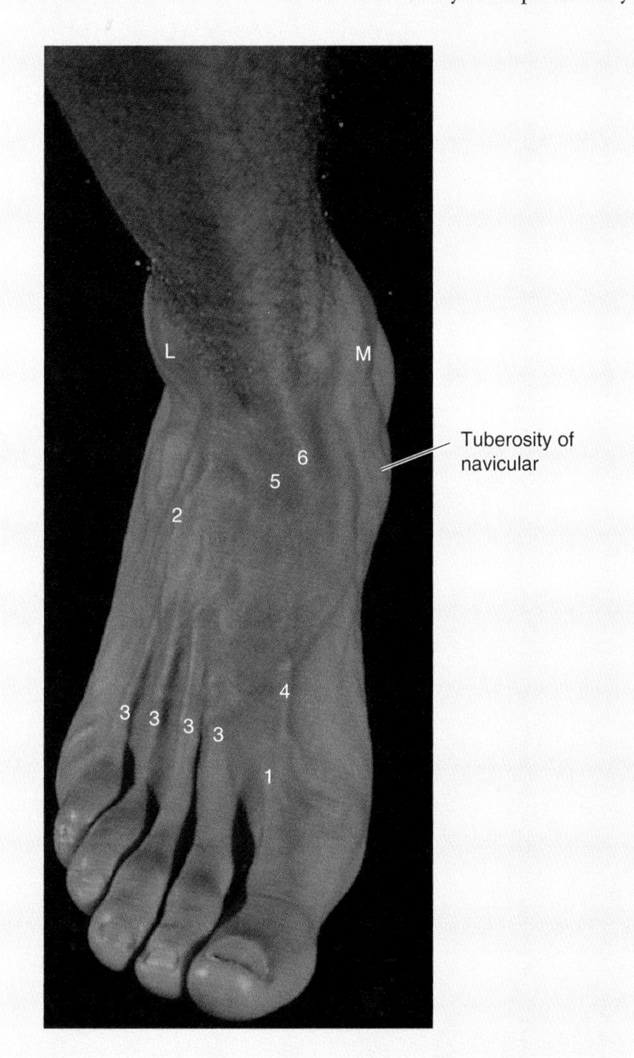

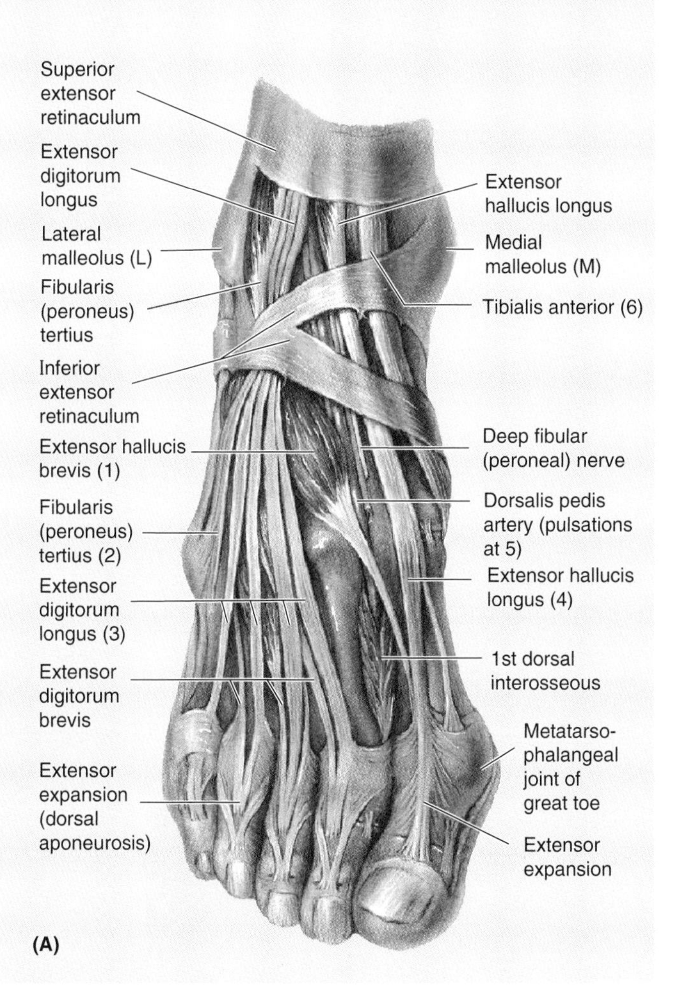

(A)

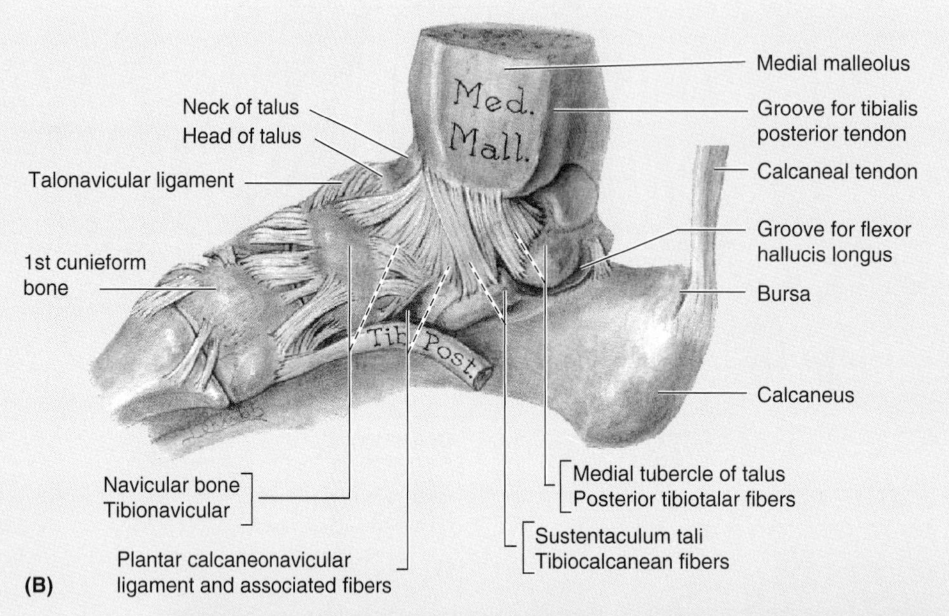

(B)

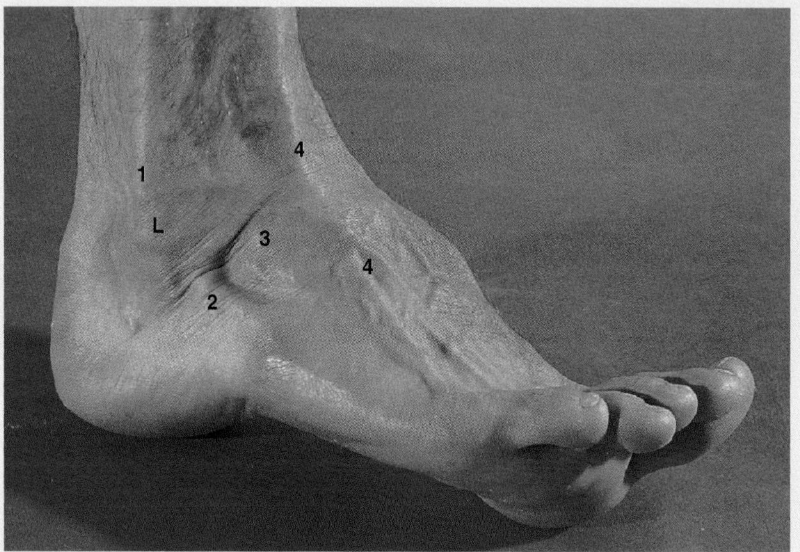

Lateral view

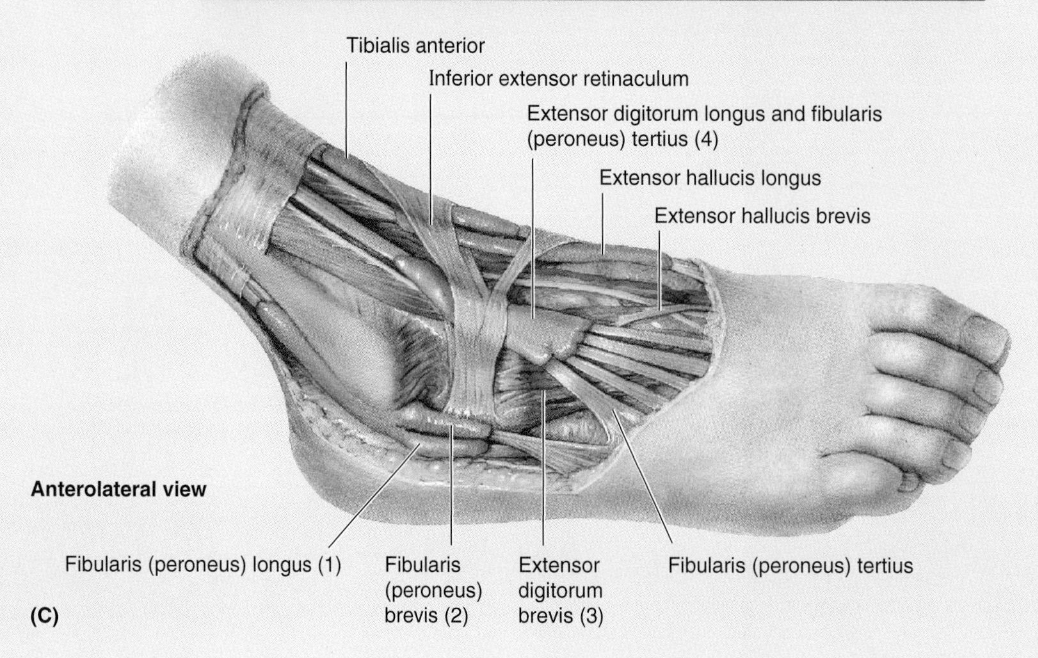

Anterolateral view

(C)

▶ distal to the medial malleolus, then superior to the sustentaculum tali, to reach its attachment to the tuberosity of the navicular. Hence, *the tibialis posterior tendon is the guide to the navicular.* The tendon of the tibialis posterior also indicates the site for palpating the *posterior tibial pulse* (halfway between the medial malleolus and the calcaneal tendon).

The *sustentaculum tali* forms a small bony prominence approximately 2 cm distal to the tip of the medial malleolus (*B*). It is best felt by palpating it from below where it is somewhat obscured by the tendon of the flexor digitorum longus, which crosses it. When the foot is inverted, the anterior surface of the calcaneus is uncovered and palpable. This indicates the site of the *calcaneocuboid joint.* When the foot is plantarflexed, the *head of the talus* is exposed. Palpate it dorsal to where the anterior surface of the calcaneus was felt. The *calcaneal tendon* at the posterior aspect of the ankle is easily palpated and traced to its attachment to the posterior part of the calcaneus. At the depression on each side of the tendon, the ankle joint is superficial. When the joint is overfilled with fluid, these depressions may be obliterated.

The tendons in the ankle region can be identified satisfactorily only when their muscles are acting. The *tendons of the fibularis longus and brevis* (peroneus longus and brevis) may be followed distally, posterior and inferior to the lateral malleolus, and then anteriorly along the lateral aspect of the foot (*C*). The fibularis longus tendon can be palpated as far as the cuboid, and then it disappears as it turns into the sole. The fibularis brevis tendon can easily be traced to its attachment to the dorsal surface of the tuberosity on the base of the 5th metatarsal. This tuberosity is located at the middle of the lateral border of the foot. With toes actively extended, the small fleshy belly of the *extensor digitorum brevis* (*C*) may be seen and palpated anterior to the lateral malleolus. Its position should be observed and palpated so that it may not be mistaken for an abnormal edema.

The *metatarsophalangeal joint of the great toe* lies distal to the knuckle formed by the head of the 1st metatarsal. *Gout,* a metabolic disorder, commonly causes edema and tenderness of this joint, as does *osteoarthritis* (degenerative joint disease). Severe pain in the 1st metatarsophalangeal joint is called *podagra* (G. fr. *pous* + *agra*, a seizure). Often the 1st metatarsophalangeal joint is the first one affected by arthritis.

The tendons on the anterior aspect of the ankle (from medial to lateral side) are easily palpated when the foot is dorsiflexed.

- The large tendon of the tibialis anterior (*A*, p. 643) leaves the cover of the superior extensor tendon, from which level the tendon is invested by a continuous synovial sheath; the tendon may be traced to its attachment to the 1st cuneiform and the base of the 1st metatarsal
- The tendon of the *extensor hallucis longus,* obvious when the great toe is dorsiflexed against resistance (*A*), may be followed to its attachment to the base of the distal phalanx of the great toe
- The tendons of the *extensor digitorum longus* (*C*) may be followed easily to their attachments to the lateral four toes
- The tendon of the *fibularis tertius* may also be traced to its attachment to the base of the 5th metatarsal. This muscle is of minor importance and may be absent.

The *transverse tarsal joint* is indicated by a line from the posterior aspect of the tuberosity of the navicular to a point halfway between the lateral malleolus and the tuberosity of the 5th metatarsal. ○

Posture and Gait

When a person is standing at ease with the feet slightly apart and rotated laterally so the toes point outward, few of the back and lower limb muscles are active. The mechanical arrangement of the joints and muscles are such that a minimum of muscular activity is required to keep from falling. In the stand-easy position, the hip and knee joints are extended and are in their most stable positions. The ankle joint is less stable than the hip and knee joints. Consequently, when forward swaying occurs, the calf muscles contract to prevent falling. Lateral stability is dependent on the fascia lata, iliotibial tract, fibular collateral ligament of the knee joint, and tibialis anterior. The latter muscle checks lateral sway at the ankle.

Locomotion is a complex function; however, it is important to understand because disturbances of gait are indicative of disorders of the nervous system. The movements of the lower limbs during walking on a level surface may be divided into swing and stance phases (Fig. 5.69). The *swing phase* begins with push-off when the foot is off the ground and lasts until the heel strikes the ground. The *stance phase* begins when the foot is on the ground and is bearing body weight. The swing phase occupies approximately one-third of the walking cycle and the stance phase two-thirds. The push-off to start the swing phase usually results from plantarflexion of the foot.

If the triceps surae is paralyzed or the calcaneal tendon is ruptured, the push-off can be accomplished less effectively by the gluteus maximus and hamstrings that extend the thigh at the hip joint. The swing phase also involves almost simultaneous flexion of the hip and knee joints, followed by dorsiflexion of the ankle joint. Usually the hip swings forward concurrently (rotation of hip to opposite side), involving the tensor fascia lata, pectineus, and sartorius. As the forward swing continues, the foot is dorsiflexed by the action of mus-

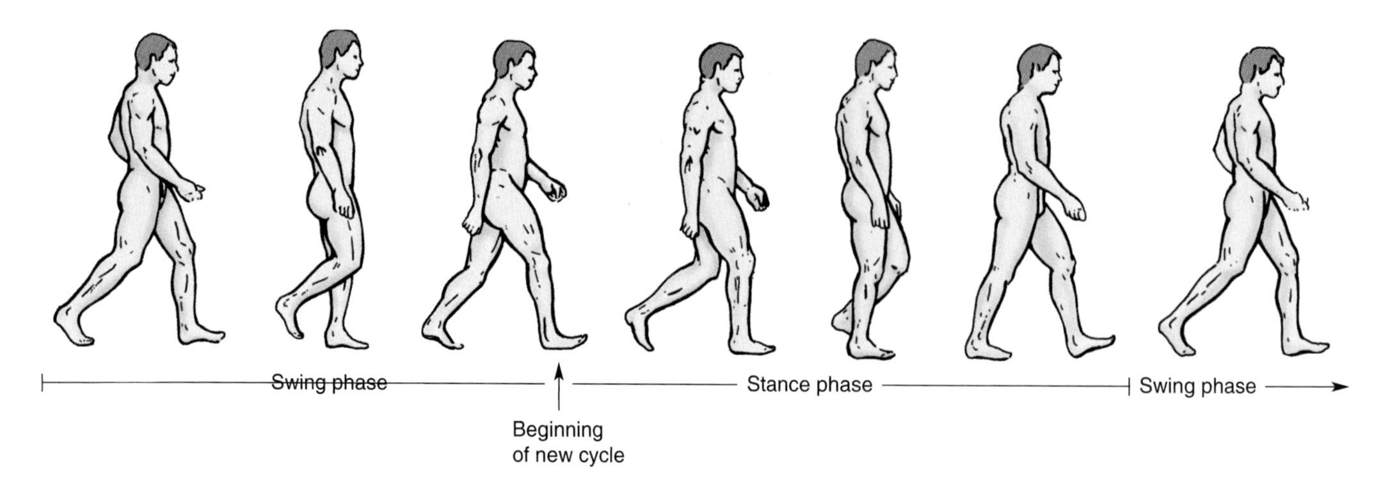

Figure 5.69. Mechanics of gait. Observe the swing and stance phases of the right lower limb.

cles in the anterior compartment of the leg, and the quadriceps contract to begin extension of the leg. The quadriceps and gluteus maximus muscles contribute little to level walking but are active during walking up and down hills. During the late part of the stance phase, the toes tend to flex and grip the ground. The long extensors and intrinsic muscles of the foot stabilize the toes and provide fixed distal attachments so that the long flexors and extensors can act on the leg.

Stabilization is important during walking; control of pelvic tilt is executed by the gluteus medius. When weight is borne on one lower limb during the stance phase, the pelvis tilts toward the ground on the free or swing side because of gravity. The hip abductors (gluteus medius and minimus) on the stance side minimize this tilt. They contract strongly, acting on the pelvis from a fixed femur. During walking, the pelvic tilt alternates from side to side. The invertors and evertors of the foot are the principal stabilizers of the foot during the stance phase. They also help support the arches of the foot during the stance phase, as do the intrinsic muscles of the sole.

Medical Imaging of the Lower Limb

Radiography

Pelvic Girdle and Hip Joint

Radiographs of the pelvis and hip show bone and joint abnormalities or malalignment. In an *AP projection of the hip joint* (Fig. 5.70*A*), the patient is in the supine position on the radiographic table with the foot pointing straight up. The central x-ray beam is centered over the hip joint (see orientation). Examine the anatomical features (labeled and unlabeled). Superimposed on the femoral head is the posterior rim of the acetabulum (PR). At the junction of the femoral neck and body, the greater and lesser trochanters may be seen. Between the trochanters is an oblique line cast by the superimposed intertrochanteric line and crest (IC). Observe the architecture of the trabecular bone of the head, neck, and proximal body of the femur. The strength of the angled bone depends on this structure. Note the tension and pressure lines related to the weightbearing function of this bone. Observe that the dense compact bone appears transparent (white), whereas the less dense spongy bone appears dark. To obtain a complete view of the femoral neck when a fracture is suspected, several views have to be taken (Fig. 5.70*B*).

Knee Joint

Multiple radiographic projections (AP, lateral, and oblique) are necessary to evaluate the knee properly. In an AP projection of the knee joint (Fig. 5.71*B*), the patient is in the supine position with the knee extended. The central x-ray beam is directed through the joint cavity. Identify the femoral and tibial condyles and the shadows of the patella overlying the distal femur. Observe the joint cavity that appears large because the menisci are not visible. They can be visualized if air or opaque fluid are injected into the knee cavity. Observe the prominent adductor tubercle just proximal to the medial epicondyle. The lateral femoral epicondyle appears more prominent than the medial one. The intercondylar fossa is opposite the medial and lateral tubercles of the intercondylar eminence of the proximal tibia. Note that the articular surfaces of the tibial condyles are concave.

In the lateral projection of the arthrogram of the knee joint (Fig. 5.72), the knee is slightly flexed. A contrast medium was injected through the articular capsule to show the joint cavity and the extent of the synovial membrane. The articular cartilage on the femoral condyle is clear, as ▶

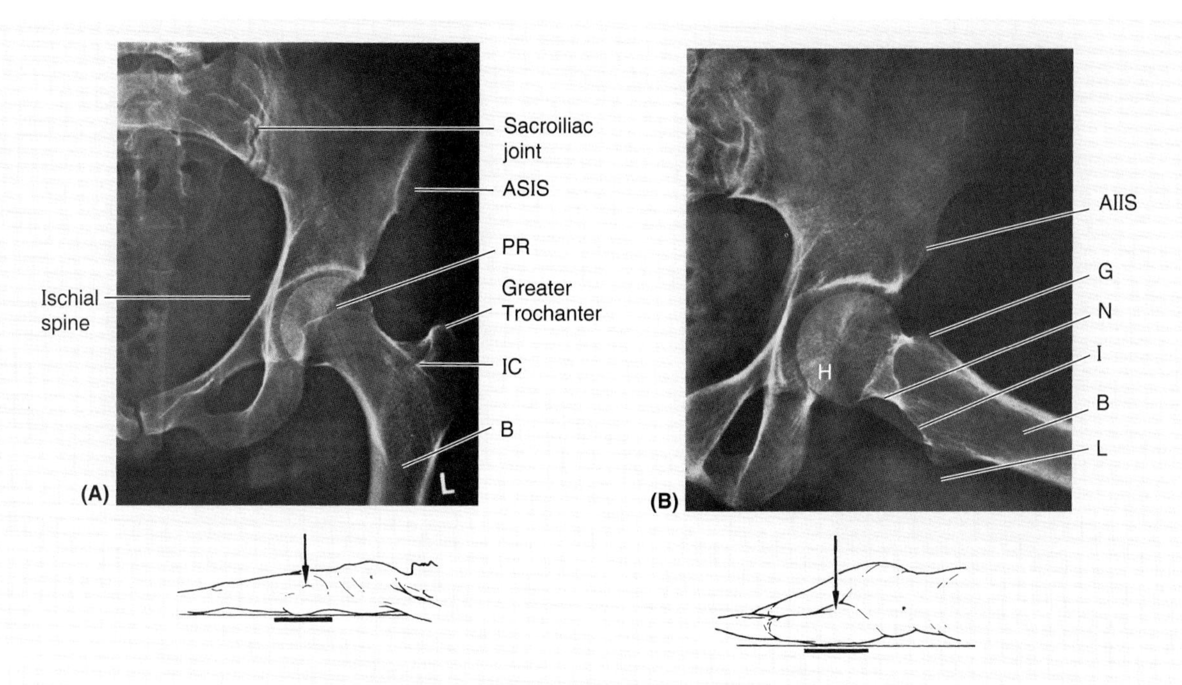

Figure 5.70. Radiograph of a normal hip joint. A. Anteroposterior (AP) projection of the left hip joint. *On the femur* (thigh bone) observe the head, neck, greater trochanter, intertrochanteric crest (*IC*), and body, or shaft (*B*). *On the hip bone* observe the lunate surface of the acetabulum, the posterior rim of the acetabulum (*PR*), the anterior superior iliac spine (*ASIS*), the ischial spine, and the sacroiliac joint. **B.** Left thigh abducted. Observe the acetabular fossa, anterior inferior iliac spine (*AIIS*), head of femur (*H*), neck of femur (*N*), greater trochanter (*G*), lesser trochanter (*L*), intertrochanteric crest (*I*), and body of femur (*B*).

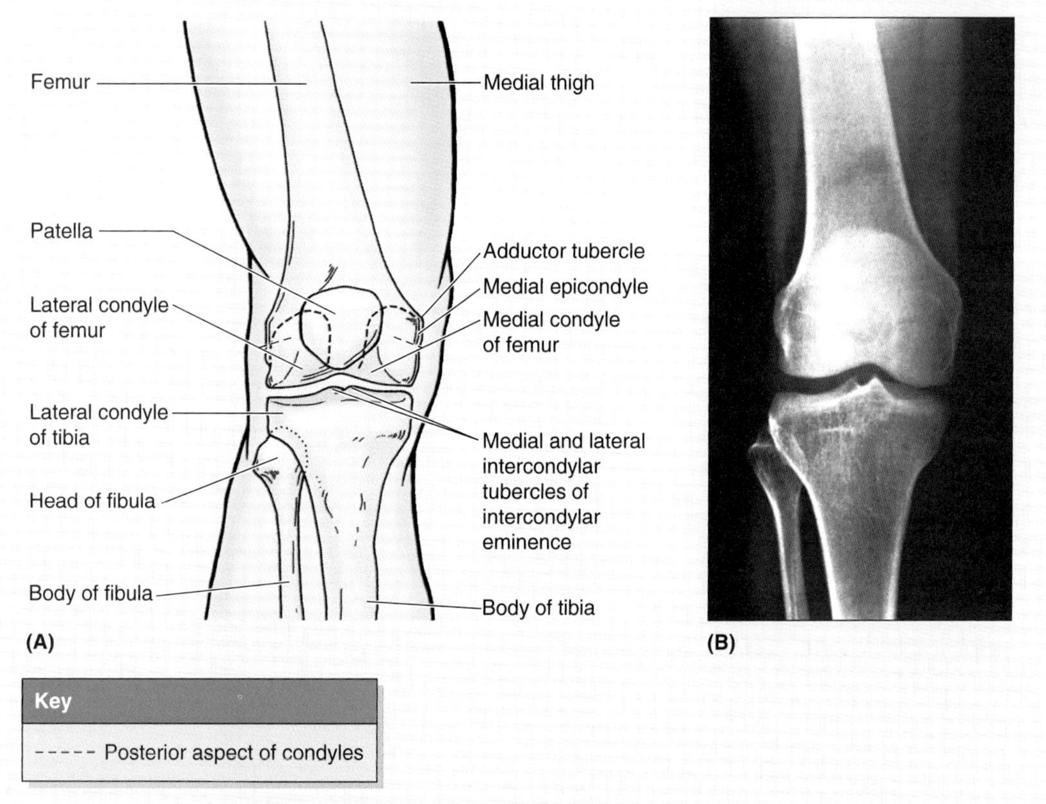

Figure 5.71. Knee joint. A. Orientation drawing. **B.** Radiograph of the right knee joint; anteroposterior (AP) projection.

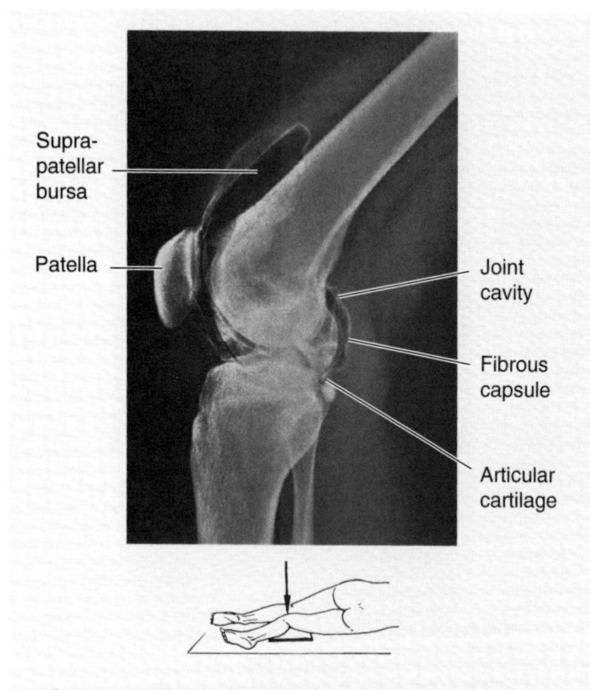

Figure 5.72. Arthrogram of the knee joint. Lateral projection with the knee slightly flexed. Observe the large suprapatellar bursa in communication with the knee joint cavity.

▶ is the fibrous capsule lined with synovial membrane. Observe that the large suprapatellar bursa is continuous with the joint cavity. Although MRI has largely replaced arthrography, it is sometimes useful for detecting loose articular bodies if MRI is inconclusive.

Ankle Joint and Foot

The common radiographs are lateral and AP. A lateral radiograph is taken with the lateral malleolus placed against the x-ray film cassette (Fig. 5.73*A*). Observe the convex surface of the trochlea of the talus (T) articulating with the malleoli of the tibia and fibula (shadows of the malleoli are visible). Also observe the neck (N) and head (H) of the talus, the disc-shaped navicular (Na), and the talonavicular joint. The calcaneus (Ca) and cuboid (C) articulate at the calcaneocuboid joint. The *tarsal sinus* (TS)—the space between the calcaneus and talus—contains the interosseous talocalcaneal ligament. ▶

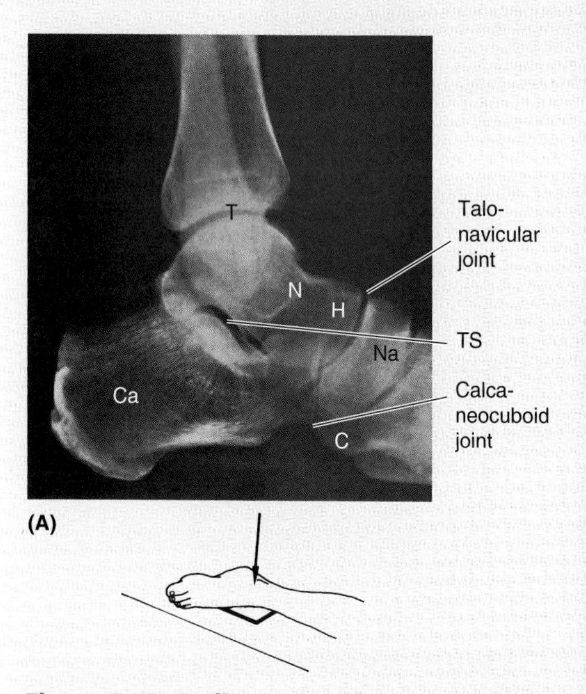

Figure 5.73. Radiographs of the lower leg, ankle, and foot. A. Left ankle, lateral projection. **B.** Left foot, dorsoplantar projection.

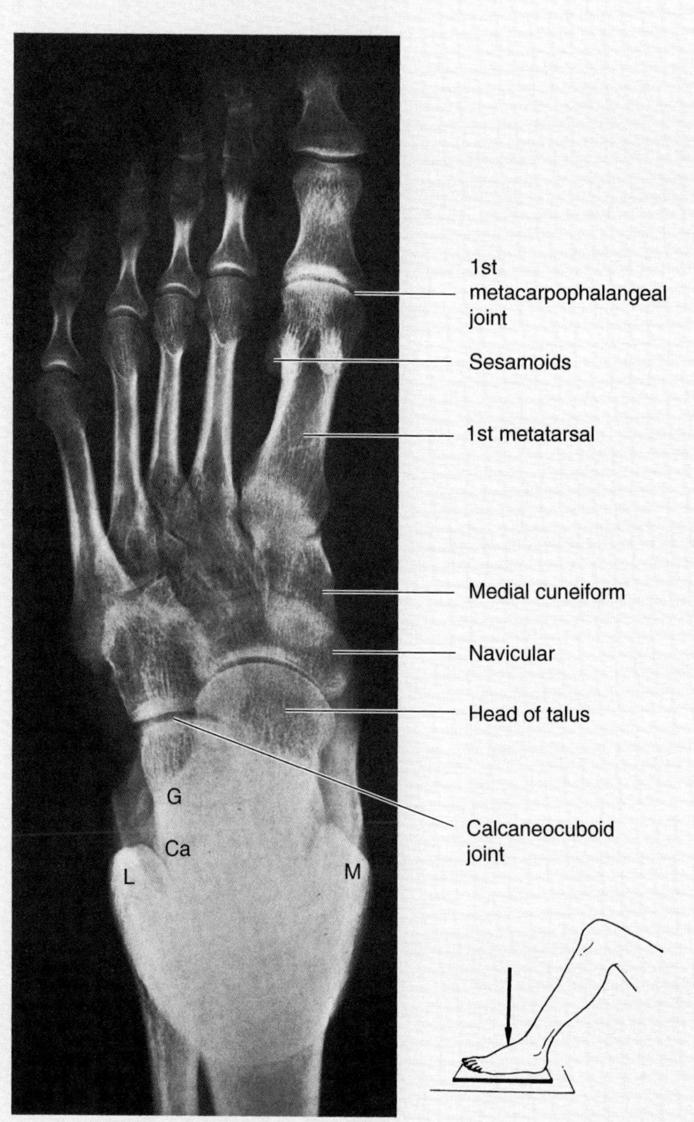

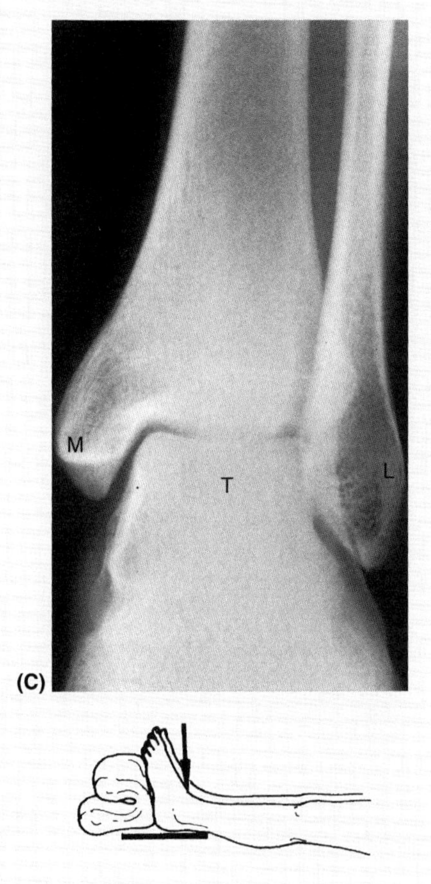

Figure 5.73. *(Continued)* **C.** Left ankle, AP projection. *C*, cuboid; *Ca*, calcaneus; *T*, talus; *TS*, tarsal sinus (tunnel); *N*, neck of talus; *H*, head of talus; *Na*, navicular; *L*, lateral malleolus; *M*, medial malleolus. (**C** Courtesy of Drs. P. Bobechko and E. Becker, Department of Medical Imaging, University of Toronto, Toronto, Ontario, Canada)

▶ In the dorsoplantar projection, the patient is supine with the knee flexed and the plantar surface of the foot placed on the x-ray film holder (Fig. 5.73*B*). The central x-ray beam is centered on the base of the 3rd metatarsal. Observe the phalanges and interphalangeal joints. Because the toes are in slight flexion, the interphalangeal joints of the 2nd to 5th toes are not clear. Observe the sesamoids on the plantar surface of the head of the 1st metatarsal and that the base of the 1st metatarsal articulates with the medial cuneiform and the base of the 2nd metatarsal. The bases of the 2nd to 5th metatarsals overlap such that the intermetatarsal joints are easy to see. The tarsal bones overlap somewhat because of the normal curvature of the foot. Consequently, all the tarsal joints do not show clearly. Only anterior parts of the talus and calcaneus are visible because of the overlap of the malleoli. An AP radiograph (*C*) is taken with the person in the supine position with the foot dorsiflexed to a right angle and the great toe pointed slightly medially. To visualize all the bones and joints of the ankle and foot, other projections would be required. ▶

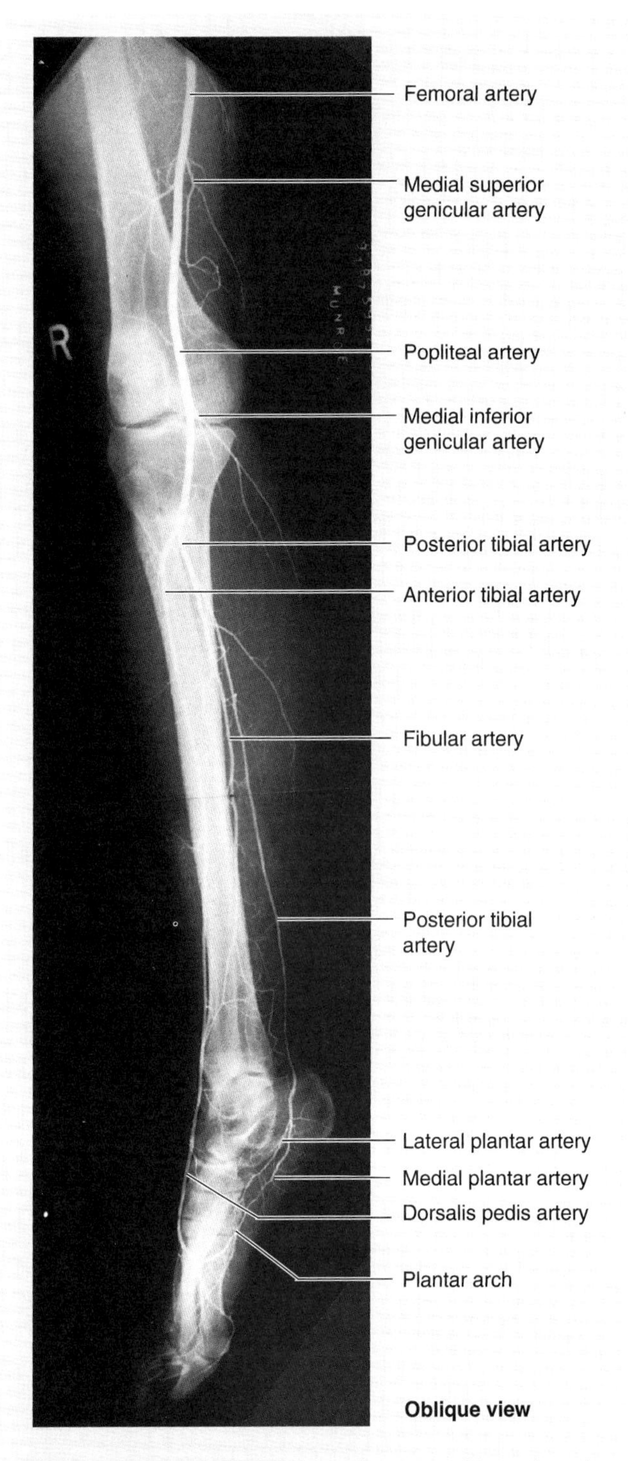

Figure 5.74. Popliteal arteriogram. Note that the popliteal artery begins at the site of the adductor hiatus (where it may be compressed) and then lies successively on the distal end of the femur, articular capsule of the knee joint, and popliteus muscle (not visible) before dividing into the anterior and posterior tibial arteries at the inferior angle of the popliteus fossa. Here it is subject to entrapment as it passes beneath the tendinous arch of the soleus muscle. (Courtesy of Dr. K. Sniderman, Associate Professor of Medical Imaging, University of Toronto, Toronto, Ontario, Canada)

Arteriography

Visualization of arteries by x-ray imaging after injection of a radiographic contrast medium is a helpful way of studying selected arteries to determine the existence of abnormalities such as *popliteal aneurysm* (circumscribed dilation of the popliteal artery). In a **popliteal arteriogram** (Fig. 5.74), the radiopaque material is injected into the femoral artery and spreads through the popliteal artery and its branches.

Computed Tomography

CT also uses x-rays that pass through the limb at various angles; however, the images are created by computer reconstructions of the data. CT scans can be set to display soft tissues or bone. Hypodense areas on CT scans suggest strains (swelling), and hyperdense areas suggest hematomas. CT arthrotomography is reliable in the assessment of the cruciate ligaments, menisci, patellar cartilage, and localization of osteochondral defects and loose bodies (Cahill et al., 1994; Levandowski, 1994).

Magnetic Resonance Imaging

MRI produces images of exquisite resolution of limbs without the use of radiation. MRI scanning requires the patient to keep the limbs motionless for 5 to 10 min. MRIs show much more detail in the soft tissues than radiographs or CTs do (Figs. 5.75–5.78).

Hip

In Figure 5.75, *A* and *B*, observe that the fibrous capsule of the joint is thick near the iliofemoral ligament and thin posterior to the psoas bursa and tendon. The *inset* is an orientation drawing showing the level of section in *A* and the MRI in *B*. In the sections, observe that the femoral sheath, which encloses the femoral artery, vein, lymph nodes, lymph vessels, and fat, is free except posteriorly where, between the psoas and pectineus, it is attached to the capsule of the hip joint. Note that the femoral artery is separated from the joint by the tough psoas tendon. Also observe the femoral vein at the interval between the psoas, pectineus, and femoral nerve lying between the iliacus and its fascia. ▶

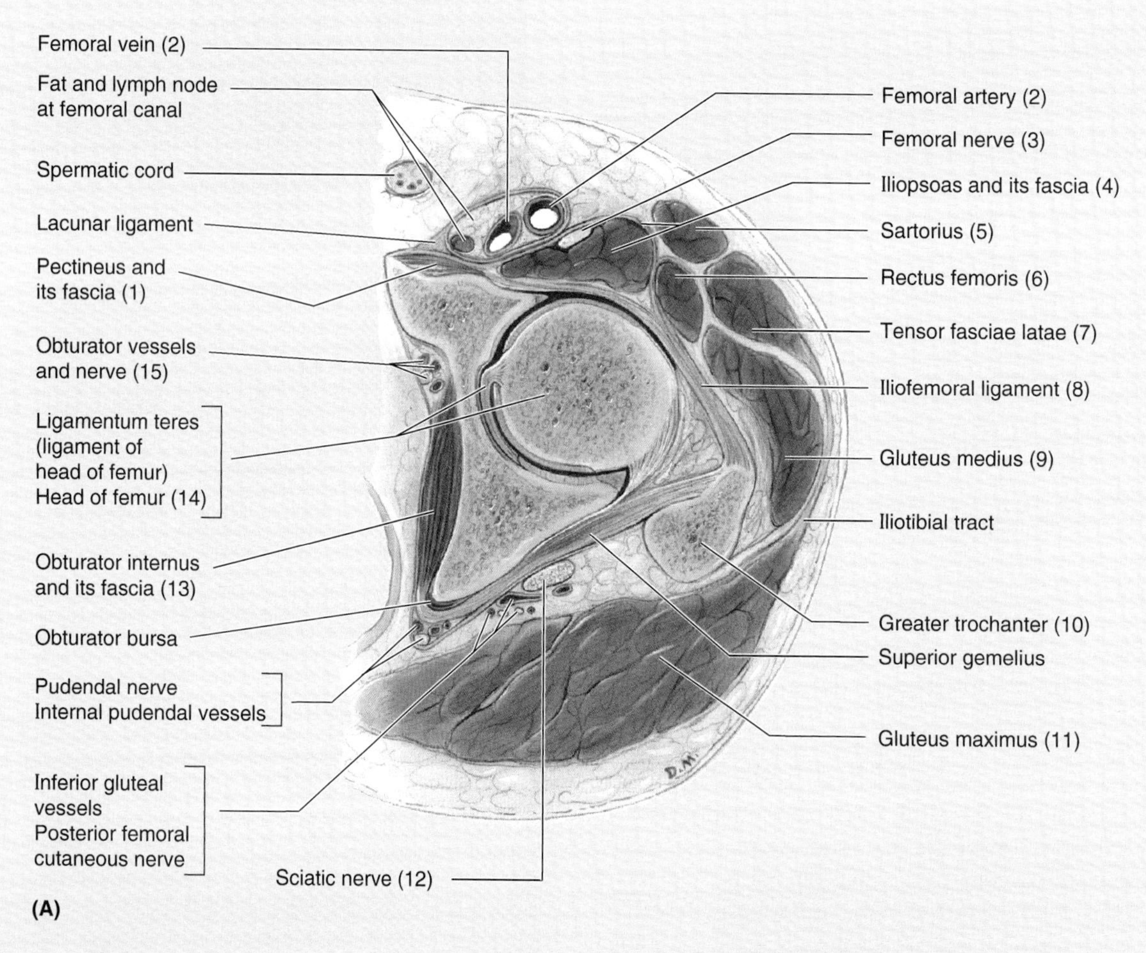

Figure 5.75. Transverse section and MRI through the thigh at the level of the hip joint. A. Drawing of the section in **B.** Inset **C.** Orientation sketch.

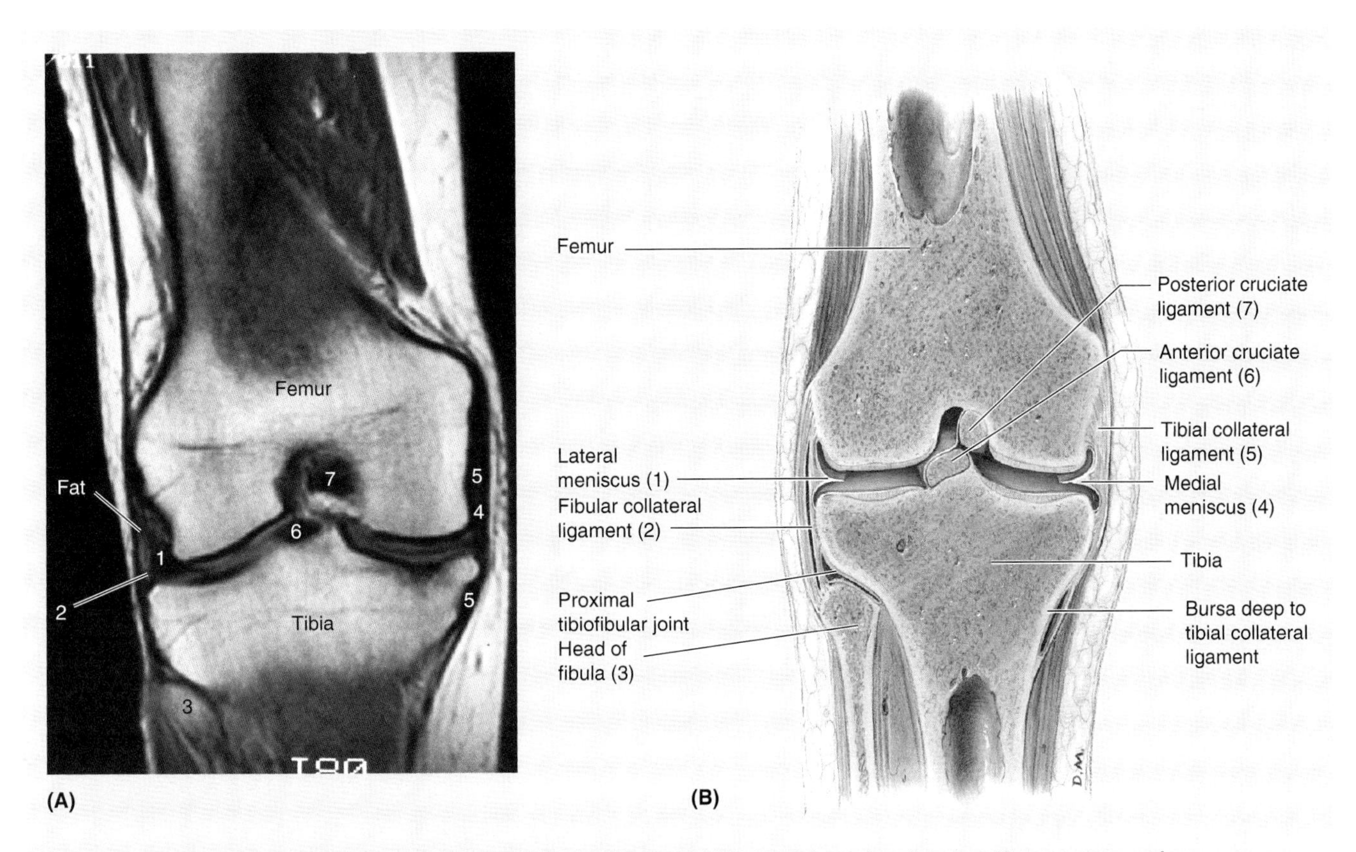

Figure 5.76. MRI of the knee joint. A. Coronal MRI. **B.** Orientation drawing. The numbers refer to structures labeled in (**A**). (Courtesy of Dr. W. Kucharczyk, Professor and Chair of Medical Imaging, University of Toronto, and Clinical Director of Tri-Hospital Magnetic Resonance Centre, Toronto, Ontario, Canada)

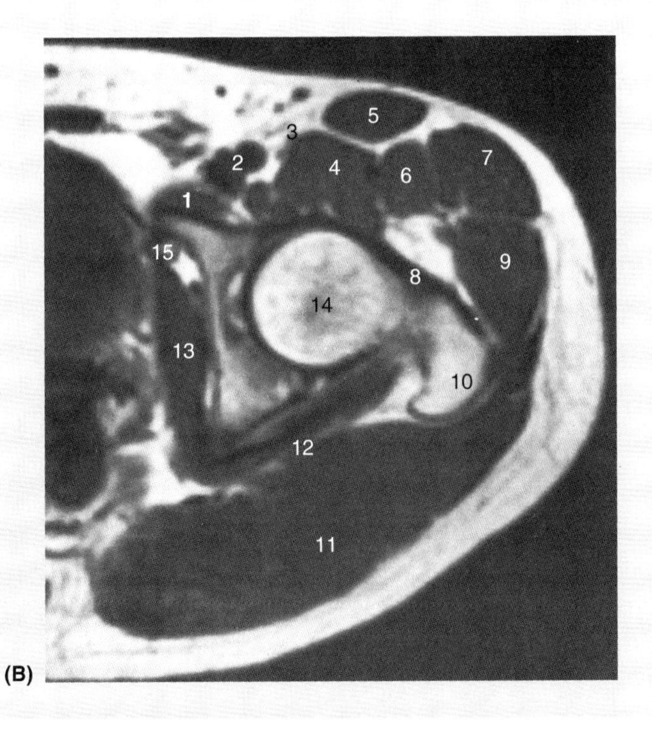

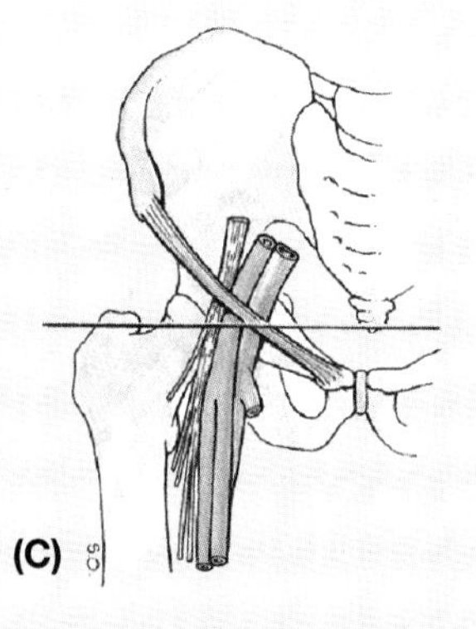

Figure 5.75. (*Continued*) **B.** MRI. The numbers refer to structures labeled in (**A**). (Courtesy of Dr. W. Kucharczk, Professor and Chair of Medical Imaging, University of Toronto, and Clinical Director of Tri-Hospital Magnetic Resonance Centre, Toronto, Ontario, Canada)

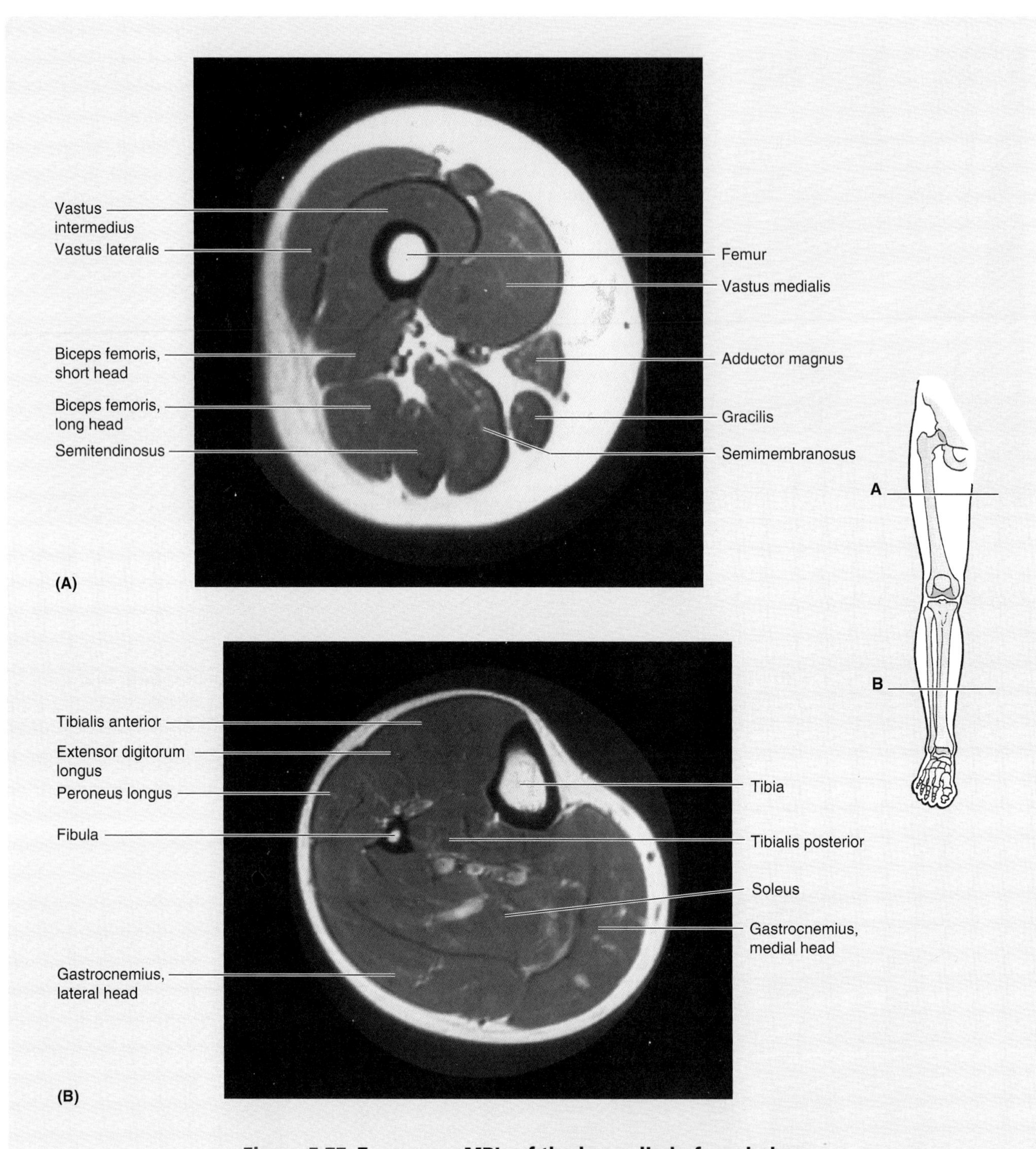

Figure 5.77. Transverse MRIs of the lower limb, from below.

Knee

MRIs are helpful in evaluating the menisci, collateral ligaments, and cruciate ligaments of the knee joint (Fig. 5.76). It is the procedure of choice for assessing internal derangements of the knee.

Ankle

MRIs are helpful for identifying the soft tissue and intraosseous and neurovascular anatomy of the ankle when evaluating ankle injuries. In Figure 5.78, observe the tibia resting on the talus and the talus resting on the calcaneus. Between the calcaneus and skin, observe several encapsulated cushions of fat. Also observe the lateral malleolus descending much farther inferiorly than the medial malleolus because of the weak interosseous tibiofibular ligament. Note the interosseous band between the talus and calcaneus separating the subtalar joint from the talocalcaneonavicular joint. Observe the sustentaculum tali acting as a pulley for the flexor hallucis longus and giving attachment to the calcaneotibial band of the medial ligament.

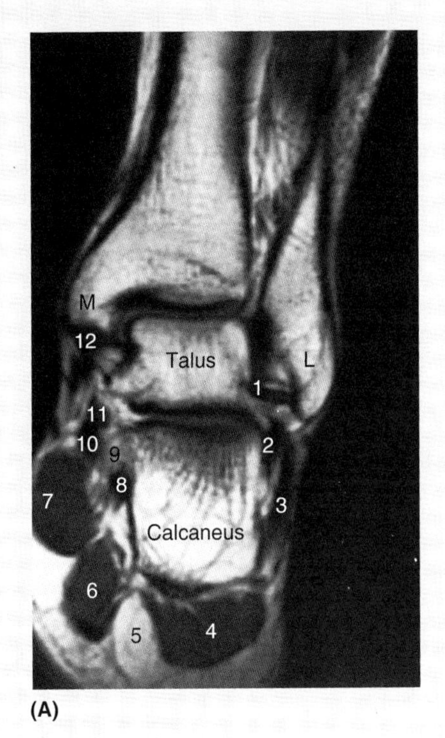

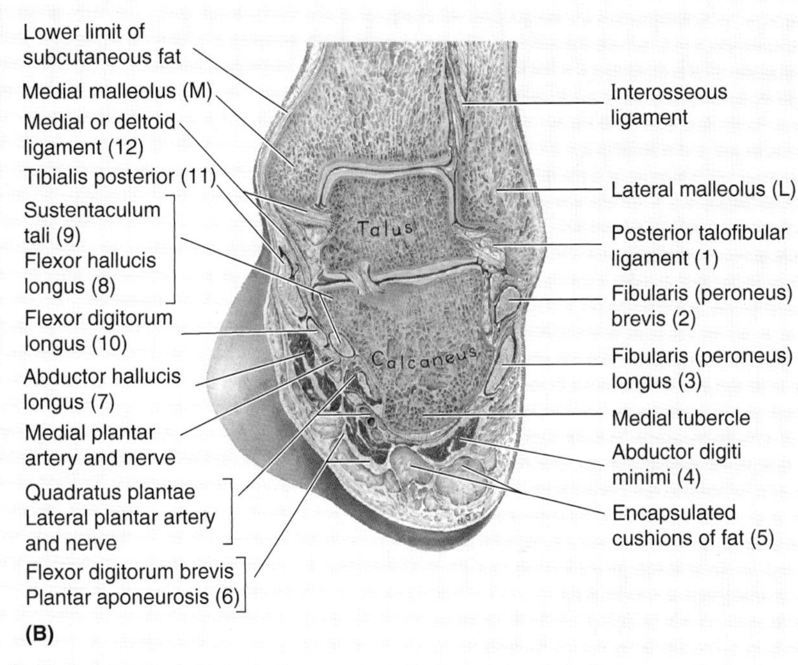

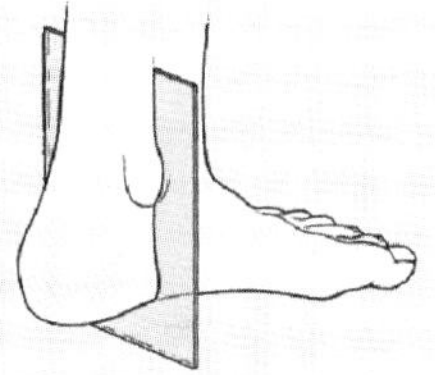

Figure 5.78. Coronal MRI of the ankle region. A. MRI. The numbers refer to structures in (**B**), an orientation drawing for the MRI. (Courtesy of Dr. W. Kucharczyk, Professor and Chair of Medical Imaging, University of Toronto, and Clinical Director of Tri-Hospital Magnetic Resonance Centre, Toronto, Ontario, Canada)

CASE STUDIES

Case 5.1

A primary care physician asked her patient if he would let you palpate his popliteal pulse. He agreed and after several attempts, you were unable to detect a pulse. The physician offered some advice.

Clinicoanatomical Problems

- What do you think would be the most likely reason for your inability to palpate the popliteal pulse?
- How would you position the patient to attempt again to palpate the pulse?

The problems are discussed on page 658.

Case 5.2

During clinical rounds you were asked if you wished to observe a *lumbar puncture*. You enthusiastically said, "Yes." Before beginning the procedure, the physician asked you three questions.

Clinicoanatomical Problems

- What parts of the ilia are important landmarks when preparing to perform a lumbar puncture?
- Where does the spinal cord usually terminate in adults?
- Where would you insert the lumbar puncture needle?

The problems are discussed on page 658.

Case 5.3

A teenager was stabbed with a jackknife approximately 5 cm proximal to the base of the patella. The wound did not bother him too much so he did not seek medical attention. Two days later his lower thigh became swollen and tender. He came to an emergency clinic in considerable pain.

Clinicoanatomical Problems

- What fluid-filled structure lies deep in this area of the thigh?
- How is a wound in this region related to the knee joint?

The problems are discussed on page 658.

Case 5.4

During a soccer game a player was kicked hard on the lateral surface of the knee. Knowing you were a medical student, the trainer asked you to palpate the head of the man's fibula.

Clinicoanatomical Problems

- How would you palate the fibular head?
- What bony landmark would you use?
- Why is it important to be able to palpate the fibular head?

The problems are discussed on page 658.

Case 5.5

A 52-year-old woman was concerned about a plum-sized bulge in her upper thigh, just inferior to her inguinal ligament. A physical examination revealed that the swelling was in the femoral triangle and that it was herniating through the saphenous opening.

Clinicoanatomical Problems

- What is the position of the saphenous opening relative to the pubic tubercle?
- Why is this relationship important in the differential diagnosis of hernias?
- Is a femoral hernia found more frequently in women than in men?

The problems are discussed on page 658.

Case 5.6

A young man received a superficial slash wound on the superomedial side of his knee. A rounded cordlike tendon, closely related to the medial head of the gastrocnemius, was cut.

Clinicoanatomical Problems

- What tendon was probably cut?
- What is the relationship of this tendon to the medial femoral condyle?
- How is the muscle to which this tendon belongs tested?

The problems are discussed on page 659.

Case 5.7

A young woman injured her ankle during a basketball game. She said, "I turned my ankle in but it's only a slight sprain!" A preliminary examination by the trainer indicated that she had a severely sprained ankle.

Clinicoanatomical Problems

- What ankle ligament did she most likely tear?
- What other ligaments may have been torn?
- What bone might she have fractured?

The problems are discussed on page 659.

Case 5.8

A patient spoke to her primary care physician about the pain in her feet. She was also concerned about the uneven wearing of the soles of her shoes. The physician examined her feet while she was standing. He observed that her longitudinal arch was flatter than normal.

Clinicoanatomical Problems

- What do you think the physician's diagnosis would be?
- What is the cause of this condition?
- Where would the person's shoes show excessive wear?

The problems are discussed on page 659.

Case 5.9

During a football game, a 20-year-old wide receiver was illegally blocked by a linebacker, who threw himself against the posterolateral aspect of the runner's legs. The wide receiver grasped his knee and was in obvious pain.

Clinicoanatomical Problems

- What injury can occur from a block against the lateral aspect of the knee if the foot is planted on the ground?
- In what other sport does this knee injury frequently occur?
- What is the mechanism of the knee injury?

The problems are discussed on page 659.

Case 5.10

When his knee was flexed at a right angle, a football player was tackled in such a way that his tibia was driven in an anterior direction. The trainer examined his knee and observed the anterior drawer sign.

Clinicoanatomical Problems

- What is the anterior drawer sign?
- Application of a severe force that drives the tibia in an anterior direction when the knee is flexed at a right angle may cause what kind of knee injury?
- In which sport does this injury commonly occur?
- What other knee injuries are often associated with this knee injury?

The problems are discussed on page 659.

Case 5.11

A young man involved in a street brawl was kicked hard on the anterolateral surface of his leg. The pain in his leg increased severely after several hours, and his leg became swollen. He was unable to extend his toes or dorsiflex his foot on the side of the injury. He also had a foot-drop when he tried to walk. Neither the tibia nor the fibula was fractured.

Clinicoanatomical Problems

- Explain the anatomical basis of these signs and symptoms?
- What is this concurrence of signs and symptoms called?
- What other sign would you expect to find?
- How could the severe pain in his leg be relieved?

The problems are discussed on page 659.

Case 5.12

An elderly woman was knocked down by a man attempting to steal her purse. She was unable to get up, not only because of pain but also because of a fracture of the proximal end of her femur. Her thigh was swollen because of bleeding from ruptured blood vessels.

Clinicoanatomical Problems

- Why is this type of fracture associated with swelling of the thigh when femoral neck fractures do not exhibit such swelling?
- What would be the position of her lower limb?
- Is good union of the bone fragments in this region of the femur possible by closed treatment (i.e., is internal surgical fixation of the fragments required)?
- If good union is possible by closed treatment, to what do you attribute this fact?

The problems are discussed on page 659.

Case 5.13

A jogger who started to run several miles a day also ran up several flights of stairs during his run. He complained of point tenderness just posterior to his right greater trochanter and radiation of pain along his thigh. He said the pain increased when the physician manually resisted abduction and lateral rotation of his thigh.

Clinicoanatomical Problems

- What do you think may be causing the hip pain: inflammation of the hip joint or trochanteric bursa?
- What are the anatomical reasons for your tentative diagnosis?
- Why did resistance to abduction and lateral rotation of the thigh exacerbate the pain?

The problems are discussed on page 659.

Case 5.14

A poorly conditioned 40-year-old man started to play squash, particularly on weekends. At first he felt pain in his calcaneal tendon, and then he felt sudden calf pain. When the pain disappeared, he continued to play vigorously. He again felt calf pain and heard an audible snap. He had difficulty tiptoeing and climbing the stairs, but he found it easy to dorsiflex his ankle.

Clinicoanatomical Problems

- What do you think caused the audible snap and calf pain?
- What physical sign would you expect to observe?
- Why did he have trouble tiptoeing and climbing stairs?

The problems are discussed on page 659.

Case 5.15

Your grandmother slipped on the polished floor in the front hall. As you approached her, she was lying on her back in severe pain. She said she heard a loud snap as she fell. Her right lower limb immediately attracted your attention because it was laterally rotated and noticeably shorter than her left limb. She was unable to get up or lift her limb off the floor, and when she attempted to do so, she experienced excruciating pain. You called an ambulance, and she was taken to an emergency clinic.

Physical Examination The right lower limb is noticeably shorter than the left one and is laterally rotated. On palpation, the hip region is tender, but no swelling is obvious. Passive movement of the thigh causes extreme pain. A radiological examination of the hip region was requested.

Radiology Report An intracapsular fracture of the femoral neck is present, with the distal part of the femur rotated laterally and shifted proximally.

Diagnosis Fracture of the femoral neck.

Clinicoanatomical Problems

- What is the common fracture site of the femur in elderly people?
- Why is this part of the bone so fragile in elderly people?
- Explain anatomically why her injured limb was shorter than the other one.
- What are the anatomical reasons for the complications (nonunion and avascular necrosis) commonly associated with these fractures?

The problems are discussed on page 659.

Case 5.16

While playing in an old-timer's hockey game, a 55-year-old man was accidentally kicked with a skate on the lateral surface of his right leg just inferior to the knee. The trainer treated the superficial laceration but the man was unable to continue playing because of pain in the region of the wound and loss of power in his leg and foot. He also experienced numbness and tingling on the lateral surface of his leg and on the dorsum of his foot. When he removed his skates, he found that he was unable to dorsiflex his right foot or his toes. The trainer advised him to see his primary care physician immediately.

Physical Examination As the man walked into the examining room, the physician observed that he had an *abnormal gait*—he raised his right foot higher than usual and brought it down suddenly, making a flapping noise. During the physical examination, the physician detected tenderness over the head and neck regions of the patient's fibula and a sensory deficit on the lateral side of the distal part of his leg, including the dorsum of his foot. Radiographs of the knee region were requested.

Radiology Report A fracture of the neck of the fibula is visible.

Diagnosis Fracture of the neck of the fibula and peripheral nerve injury.

Clinicoanatomical Problems

- What is the anatomical basis of the loss of sensation and impaired function in the patient's foot?
- What nerve appears to have been injured?
- What is the relationship of this nerve to the neck of the fibula?
- If the skate blade had not severed the nerve, what probably would have injured it if he had continued to play?
- What is the name of the foot condition exhibited by the patient when he walked?

The problems are discussed on page 660.

Case 5.17

While a 26-year-old worker was loading a heavy crate, it fell on his knee. He experienced severe pain and was unable to get up. The first aid team carried him to the physician's office on a stretcher. Following a physical examination, the physician requested radiographs of the man's knee.

Radiology Report A comminuted fracture of the proximal end of the tibia is visible; the fibular neck is also fractured.

Diagnosis Fracture of the proximal end of the tibia and neck of the fibula associated with a peripheral nerve injury.

Clinicoanatomical Problems

- What artery or arteries might have been torn by the tibial bone fragments?
- Using your anatomical knowledge, where would you check the patient's pulse to determine whether these arteries have sustained damage?
- What nerve may have been injured by the fracture of the fibula?

The problems are discussed on page 660.

Case 5.18

A 32-year-old man slipped on a patch of ice and fell. After he was helped up, he was unable to bear weight on his right foot. When he noticed that his ankle was beginning to swell, he hailed a cab and went to the hospital for treatment of what he thought was a badly "sprained ankle."

Physical Examination On examination, the physician noted that the patient could barely move his ankle because of pain and tenderness, especially over the lateral malleolus. Radiographs of the ankle were requested.

Radiology Report A transverse fracture of the lateral malleolus is visible at the level of the superior articular surface of the talus.

Diagnosis Fracture of the lateral malleolus and a severely sprained ankle with tearing of ligaments.

Clinicoanatomical Problems

- What excessive movement usually results in a sprained ankle?
- Discuss the meaning of the term *sprain*.
- Explain anatomically how this fracture probably occurred.
- What structures were probably torn or ruptured?
- Does the patient have what is usually referred to as a *Pott's fracture*?

The problems are discussed on page 660.

Case 5.19

A 22-year-old woman was a front seat passenger in a car that was involved in a head-on collision. Although she sustained head injuries, her chief complaint was a sore right hip that prevented her from standing up. Believing that she might have broken her hip, the paramedics at the scene rushed her to the nearest hospital.

Physical Examination The physical examination revealed that her lower limb was slightly flexed, adducted, and medially rotated and that it appeared shorter than the other limb. Radiographs of her hip were requested.

Radiology Report Posterior dislocation of the right hip joint with fracture of the posterior margin of the acetabulum.

Diagnosis Traumatic posterior dislocation of the hip joint.

Clinicoanatomical Problems

- Explain anatomically how this injury probably occurred.
- What nerve may have been injured?
- When paralysis of this nerve is complete, what muscles are paralyzed?
- Where may cutaneous sensation be lost?

The problems are discussed on page 661.

Case 5.20

A 62-year-old man complained to his family physician about an aching pain in his left buttock that extended along the posterior aspect of his left thigh.

Physical Examination During the examination, the patient pointed to the area where he felt most pain, which was in the region of the greater sciatic notch. Pain was also elicited by pressure along a line beginning midway between the tip of the greater trochanter of his femur and the ischial tuberosity to the midline of the thigh approximately halfway to the knee. When seated, the patient was unable to extend his left leg fully because of severe pain. With the patient in the supine position, the physician grasped the patient's left ankle and placed his other hand on the anterior aspect of the knee to keep the leg straight. He then slowly raised the left lower limb; when it reached an approximate 75° angle, the man grimaced with pain. Even more pain was elicited when the patient's foot was dorsiflexed. MRIs of the lower back were requested.

Radiology Report Herniation of the L5/S1 intervertebral disc is visible.

Diagnosis Herniation of L5/S1 intervertebral disc and compression of the 1st sacral nerve root.

Clinicoanatomical Problems

- What nerve is involved in this case?
- From which segments of the spinal cord does this nerve arise?
- Why does the *straight leg-raising test* elicit pain?
- Why did the pain increase when his foot was dorsiflexed?
- What back lesion probably produced the pain in the buttock and posterior thigh region?
- Thinking anatomically, what other lesions—resulting from disease or injury—do you think might have caused the patient's symptoms?

The problems are discussed on page 661.

Case 5.21

In a football game, a player was clipped (blocked from behind) by a lineman as he was about to tackle the ball carrier. The lineman's hip hit the runner's knee from the side. It was evident on the slow-motion videotape replay that the runner's knee was slightly flexed and his foot was firmly planted in the turf when he was hit. As he lay on the ground clutching his knee, it was obvious from his face that he was in severe pain. While he was being helped to the sidelines, you said to your friend, "I'm afraid he has torn knee ligaments." Not knowing much about the functioning of the knee joint, your friend said, "Which knee ligaments are probably torn?"

Clinicoanatomical Problems

- How would you explain the knee injury to your friend, assuming that he has little knowledge of the anatomy of the knee joint?
- What ligament was probably ruptured?
- What ligament may have been torn?
- Would the menisci be injured?

The problems are discussed on page 661.

Case 5.22

A 55-year-old man became concerned because of a globular *swelling in his right groin.* He stated that the swelling became smaller when he lay down but never completely disappeared. He also said that the mass occasionally became large and bulged under the skin on the anterior aspect of his thigh. When this occurred, he said he felt a pain down the inside of his thigh.

Physical Examination On examination, the physician noted that the swelling was inferior to the medial third of the inguinal ligament and *lateral to the pubic tubercle.* When he inserted his index finger into the superficial ring of the patient's inguinal canal and asked him to cough, the physician felt no mass or protruding gut; however, he observed a slight increase in the size of the swelling. When the physician asked the patient to point to the site where the swelling first appeared, he placed his finger over the site of the femoral ring. When asked in which direction the swelling "came down" when he felt pain, he ran his finger along his thigh to the region of the *saphenous opening.* The physician applied extremely gentle manual pressure to the swelling with the thigh flexed and medially rotated; however, he was unable to reduce the protrusion.

Diagnosis Irreducible complete femoral hernia.

Clinicoanatomical Problems

- Define the terms: femoral ring, femoral canal, and femoral hernia.
- What are the usual contents of the femoral canal?
- Use your anatomical knowledge to explain why a femoral hernia curves superiorly.
- Can you think of any anatomical reason why femoral hernias are more common in females than in males?
- Explain anatomically why strangulation of this type of hernia is common.
- Enlargement of what structure in the femoral canal might be mistaken for a femoral hernia?

The problems are discussed on page 661.

DISCUSSION OF CASES

Case 5.1

You were probably unable to palpate the popliteal pulse because the patient's knee was extended, which tenses the popliteal fascia. You should hold the person's knee in a mild degree of flexion and place your thumbs on his patella and the fingers of both hands in his popliteal fossa. The person should not be asked to flex his leg because this will tighten the muscles and again make the pulse difficult to palpate. Both hands should squeeze the popliteal fossa firmly to feel the pulse. Because this pulse is often difficult to palpate, it is necessary in some cases to ask the person to lie in a prone position with the knee flexed to a right angle. Deep pressure in the fossa should compress the popliteal artery against the popliteal surface of the femur and make the pulse palpable.

Case 5.2

The highest points of the iliac crests, palpated from the posterior aspect, are important in the performance of a lumbar puncture. The line joining these points lies opposite the L4–L5 intervertebral space, which is a landmark for inserting the needle into the subarachnoid space to obtain a cerebrospinal fluid sample. The spinal cord usually ends at the disc between L1 and L2 vertebrae, but it may terminate at the disc between T12 and L1 or at the disc between L2 and L3. The lumbar puncture needle is usually inserted in the interspinous space L3 and L4 vertebrae or between L4 and L5 vertebrae.

Case 5.3

The suprapatellar bursa lies deep to this area. The suprapatellar bursa almost invariably communicates freely with the knee joint and surgically forms part of the joint cavity. Consequently, a wound in the anterior aspect of the distal part of the thigh may infect the suprapatellar bursa and the infection may spread to the knee joint.

Case 5.4

The head of the fibula is subcutaneous at the posterolateral aspect of the knee at the level of the tibial tuberosity. The fibular head is best palpated from the posterior aspect. It is important to be able to palpate the fibular head because the fibular neck is commonly fractured when there is a hard blow to the lateral side of the knee. Furthermore, the common fibular nerve winds around the fibular neck and may be injured by a blow to the knee and severed if the fibular neck is fractured.

Case 5.5

The saphenous opening is approximately 4 cm inferolateral to the pubic tubercle. A *femoral hernia* enters the femoral canal and may enlarge by passing inferiorly through the saphenous opening into the subcutaneous tissue of the thigh. A femoral hernia forms a bulge inferolateral to the pubic tubercle, whereas an *indirect in-*

guinal hernia forms a bulge superior to the pubic tubercle as the hernia exits the superficial inguinal ring and enters the scrotum. Consequently, a swelling in the femoral triangle cannot be formed by an inguinal hernia. Femoral hernias are more frequently found in women because the femoral ring is larger in women than in men, their femoral vessels are smaller, and changes occur in the associated tissues during pregnancy.

Case 5.6

Undoubtedly, the tendon of the semitendinosus was cut. It lies superficial to the tendon of the semimembranosus. The tendon of the semitendinosus passes posterior to the medial femoral condyle. To test the muscle during a physical examination (not in this case), ask the person to lie in the prone position. Then ask him to flex his knee against the resistance of your hand. If the muscle is normal, the tendon can be palpated proximal to the knee on the medial side.

Case 5.7

In most lateral compartment ligament sprains, the anterior talofibular ligament is torn. In severe sprains, the calcaneofibular and posterior talofibular ligaments may also be torn and accompanied by an avulsion fracture of the distal fibula.

Case 5.8

Obviously, the person has flat feet. People with flat feet wear down their soles on the medial side, extending to the tip of the shoe. The most important ligament in the foot is the plantar calcaneonavicular ligament because of the support it gives to the medial longitudinal arch of the foot. Its principal attachments are to the sustentaculum tali and the tuberosity of the navicular. A line joining these two bony points indicates the location of this ligament on the surface of the foot.

Case 5.9

Injury to the tibial collateral ligament can result from this type of tackle or block. Tearing of this ligament may also cause detachment of the medial meniscus and anterior cruciate ligament. These ligamentous injuries can also occur when a skier traps one ski in the snow and momentum carries the person forward.

Case 5.10

The anterior drawer sign indicates an unstable knee that moves anteriorly when the flexed leg is pulled. Forcefully driving the tibia anteriorly with respect to the femur usually causes *tearing of the anterior cruciate ligament*, one of the most common injuries in sports, especially skiing. This ligament is usually torn as part of a more complex knee injury. The fibular collateral ligament and the menisci may also be torn.

Case 5.11

These signs and symptoms indicate an acute *anterior compartment syndrome* resulting from hemorrhage from the anterior tibial artery. Compartmental pressure builds up within the anterior fascial compartment, compressing the deep fibular nerve in this compartment, which explains why the man could not extend his toes or dorsiflex his foot. Severe compression of the tibialis anterior and deep fibular nerve also results in foot-drop and a stepping gait. Compression and hemorrhage of the anterior tibial artery also causes *loss of the dorsalis pedis pulse* because the dorsal artery of the foot is the terminal branch of the anterior tibial artery. The severe pain could be relieved by performing a *fasciotomy*, an incision through the anterior crural intermuscular septum to relieve the compartmental pressure.

Case 5.12

The woman's fracture was most likely between the greater and lesser trochanters or through the trochanters. These fractures commonly occur through bone that is markedly weakened by *osteoporosis*, and the fractures are often comminuted. Bleeding from the abundant blood vessels supplying the trochanteric region is into the groin or thigh because these fractures are outside the articular capsule (*extracapsular fractures*). Fractures of the femoral neck all occur within the articular capsule (*intracapsular fractures*); consequently, the blood cannot pass into the groin or thigh. The blood accumulates in the cavity of the hip joint. The lower limb of a person with an extracapsular trochanteric fracture is laterally rotated, as it is with fracture of the femoral neck. Because of the abundant blood supply to the proximal end of the femur, trochanteric fractures usually unite well and can usually be reduced by closed treatment using continuous traction.

Case 5.13

The signs and symptoms suggest *trochanteric bursitis* (inflammation of trochanteric bursa). Often this injury results from repetitive actions such as climbing stairs and mountain climbing. These movements involve the gluteus maximus and move the tendinous fibers of its superior part repeatedly back and forth over the trochanteric bursa. This type of bursitis is characterized by point tenderness just posterior to the greater trochanter. The pain radiates along the side of the thigh, the course of the iliotibial tract that receives tendinous reinforcements from the gluteus maximus and tensor fascia lata. Manually resisting abduction and lateral rotation of the thigh puts pressure on the trochanteric bursa and causes pain.

Case 5.14

The man ruptured his calcaneal tendon. This was apparent because of the site of the pain and weakness of plantarflexion of his ankle. The calf pain likely resulted from calcaneal tendinitis. After the calcaneal tendon ruptures, the prominence of the calf increases because of shortening (release) of the triceps surae. He had trouble tiptoeing and climbing stairs because the soleus, the strong plantar flexor of the ankle, attaches to the calcaneus through the calcaneal tendon.

Case 5.15

A fracture of the neck of the femur is a common injury in elderly women. This fracture is often wrongly referred to as a "fractured hip," implying that the hip bone is broken. As your grandmother stumbled and tried to catch herself, she probably exerted a torsional force on one hip, producing a fracture of the femoral neck, the most fragile part of the bone. She fell when the bone fractured; hence, the fracture was the cause of her fall, not the result of it.

Lateral rotation and shortening of the injured limb are characteristic clinical features following fractures of the femoral neck. The rotation results from the change in the axis of the limb because of the separation of the body and head of the femur. The shortening of the lower limb results from the superior pull of the muscles connecting the femur to the hip bone. Spasm of the muscles (sudden involuntary muscular contractions) causes the pull.

In a skeletal disorder called *osteoporosis*, the body's total bone mass becomes progressively diminished with advancing age. The problem is not failure of new bone to become adequately calcified. The disorder results primarily because the amount of new bone matrix produced is insufficient to keep pace with bone resorption. As a result, the body's total bone mass becomes progressively diminished. Because of osteoporosis, the femoral neck becomes weak. Fractures of the proximal part of the femur can result from little or no trauma. In this bone disorder of postmenopausal women and elderly men, absorption of bone is greater than bone formation.

The blood vessels to the proximal part of the femur derive mostly from the medial and lateral *circumflex femoral arteries*. Branches of these arteries run in the retinacula of the fibrous capsule of the hip joint. A branch of the obturator artery supplies a variable amount of blood to the femoral head—the *artery of the ligament of the head*. The ligament may be ruptured during fractures of the femoral neck. The artery of this ligament is often not patent in elderly people because they commonly have *arteriosclerosis* (hardening of arteries). Sometimes other blood vessels supplying the femoral head are torn when the femoral neck fractures. Generally, the more proximal the fracture, the greater the chances of interrupting the vascular supply.

A poor blood supply may result in nonunion and *avascular necrosis* of the femoral head (death and collapse of proximal bone fragment owing to poor blood supply). *Intracapsular fractures* (high in the neck) almost always present healing problems because they usually interfere with the blood supply to the proximal bone fragment. The importance of preserving the blood supply to the proximal part of the femur is one reason that patients with this type of injury are handled with extreme care; another is that this injury is painful.

Case 5.16

The close relationship of the *common fibular nerve* to the neck of the fibula makes it vulnerable to injury when this region of the bone is fractured. The nerve lies on the lateral aspect of the neck of the fibula and can be easily injured by superficial lacerations. This patient's signs and symptoms make it obvious that the common fibular nerve was injured. Superficial wounds, prolonged pressure by hard objects (e.g., the sharp edge of a bed during sleep), or compression by a tight plaster cast may present similar clinical features.

Injury to the common fibular nerve affects muscles in the lateral compartment of the leg (fibularis longus and brevis supplied by superficial fibular nerve) and in the anterior compartment of the leg supplied by the deep fibular nerve. Consequently, eversion and dorsiflexion of the foot and extension of the toes are impaired. This patient showed a characteristic *foot-drop* (plantarflexion and slight inversion) and stepping gait. As the patient walked, his toes dragged and his foot slapped the floor. In an attempt to prevent this from happening, he raised his foot higher than usual.

The *dysesthesia* (impairment of sensation) on the patient's leg and foot resulted from injury to cutaneous branches of the common fibular nerve. The injury to the nerve resulted from the skate grazing the nerve or the nerve being compressed or torn by the bone fragments. Although the fibula is not a weightbearing bone, fractures of its proximal end cause pain on walking because the pull of muscles attached to it causes the fragments to move, which is painful.

Case 5.17

Because the *popliteal artery* lies deep in the popliteal fossa against the fibrous capsule of the knee joint, it could have been torn by fragments from the comminuted fractures of the proximal ends of the tibia and fibula. The popliteal artery divides into its terminal branches (*anterior and posterior tibial arteries*) at the inferior end of the popliteal fossa; therefore, these vessels may also have been torn when the bones fractured. Undoubtedly, one or more of the *genicular arteries*, branches of the popliteal artery, were also torn. They supply the articular capsule and ligaments of the knee joint.

Pulsations of the posterior tibial artery may be felt halfway between the medial malleolus and heel. The dorsal artery of the foot—the continuation of the anterior tibial artery—can also be palpated where it passes over the navicular and cuneiform bones in the foot. These places are good for taking the pulse of the arteries because they are superficial and can be compressed against the bones. Loss of a pulse in these arteries in the present case would have suggested a torn popliteal and/or tibial artery.

There is also a chance that the *tibial nerve* was injured in this patient, as it is the most superficial of the three main structures in the popliteal fossa. Severance of the tibial nerve results in paralysis of the popliteus and muscles of the calf (gastrocnemius, soleus, flexor hallucis longus, and tibialis posterior), together with those in the sole of the foot. Probably some of the genicular branches (articular nerves) to the knee joint were also severed. The close relationship of the common fibular nerve to the neck of the fibula makes it vulnerable to injury when this part of the bone is fractured. For the signs and symptoms resulting from severance of this nerve, see the discussion of Case 5.2.

Case 5.18

The usual sprained ankle results from excessive inversion of the weightbearing foot, which ruptures the anterolateral part of the fibrous capsule of the ankle joint and calcaneofibular and talofibular ligaments. The term *sprain* is used to indicate some degree of tearing of the ligaments. Severe sprains involve the tearing of many fibers of the ligaments, often resulting in considerable instability of the ankle joint. In the present case, the severe sprain and fracture occurred when the patient slipped in such a way that his foot was forced into an excessively inverted position. His body weight then caused forceful inversion of the ankle joint. Probably the calcaneofibular and anterior talofibular ligaments were partly or completely torn. Normally, the deep mortise formed by the distal end of the tibia and malleoli holds the talus firmly in position. When the ankle ligaments tear, the talus is forcibly tilted against the lateral malleolus, shearing it off.

Had the man's ankle been forced in the opposite direction (i.e., in an extremely everted position), the strong *medial ligament* may have avulsed the medial malleolus. As the force continued, it might have tilted the talus, moving it and the lateral malleolus laterally. Because the interosseous ligament acts as a pivot, the fibula breaks proximal to the proximal tibiofibular joint.

This patient's ankle injury was probably not a *Pott's fracture*. The term "Pott's fracture-dislocation" is often loosely used to include most fractures and fracture-dislocations involving the malleoli of the ankle. Thus, a *first degree injury* involves two malleoli (or the malleolus and one ligament), and a *third degree injury* includes both malleoli and the posterior margin of the tibia (or two malleoli and one ligament).

Case 5.19

Dislocation of the hip joint is uncommon because of the stability of this articulation. The head of the femur sits deep in the acetabulum and is held there by an exceedingly strong fibrous capsule. Traumatic dislocations of the hip joint may occur during automobile accidents when the hip joint is flexed and the thigh is adducted and medially rotated. This patient's knee likely struck the dashboard when her right lower limb was in the position just described. Consequently, the force was transmitted along the femur, driving its head and the posterior margin of the acetabulum posteriorly. Because the femoral head in this position is covered posteriorly by capsule and not bone, the articular capsule probably ruptured inferiorly and posteriorly. This permitted the head to dislocate posteriorly and carry the fractured posterior margin of the acetabulum and acetabular labrum with it. As a result, the femoral head came to lie on the gluteal surface of the ilium.

The close relationship of the *sciatic nerve* (L4 through S3) to the posterior surface of the hip joint makes it vulnerable to injury in posterior dislocations. If the paralysis is complete, which is rarely the case, the hamstring muscles and those distal to the knee will all be paralyzed. In addition, anesthesia will probably be in the lower leg and foot, except for the skin on the medial side, which is supplied by the *saphenous nerve* (L3 and L4), a terminal branch of the femoral nerve.

Case 5.20

The site of the patient's pain and its course down the posterior aspect of the thigh clearly indicates compression of the roots of the *sciatic nerve*, the largest branch of the sacral plexus. It arises from spinal cord segments L4 through S3. The sciatic nerve leaves the pelvis through the inferior part of the greater sciatic notch and extends from the inferior border of the piriformis to the distal third of the thigh, along the course clearly indicated by the patient's pain. The *straight leg–raising test* elicits pain because the sciatic nerve stretches when the limb is raised. Dorsiflexion of the foot further increases the pull on the sciatic nerve and its roots.

A posterolateral *protrusion of an intervertebral disc* is a common cause of sciatica, most often affecting the 1st sacral nerve roots. A protruding L5/S1 disc exerts pressure on the dorsal and ventral roots, producing *sciatica*, which may be accompanied by low back pain. Sciatic pain can also result from pressure (e.g., a tumor) on the sciatic nerve or its components in the pelvis, gluteal region, or thigh. Pain can also be produced by irritation of the sciatic nerve resulting from inflammation of the nerve (neuritis) and its sheath or compression of the nerve.

Case 5.21

The knee joint is one of the most secure joints in the body, particularly when it is extended. Although these bones are bound together by strong ligaments, the knee is subject to a wide range of injuries because of the severe strain that is placed on its attachments, especially in contact sports like hockey and football. The blow on the lateral side of the player's knee by the lineman's hip occurred while he was running and at a time when his foot was fixed in the ground. As he was bearing weight on his leg, the hard blow bent the runner's knee medially relative to the fixed tibia. This severely stressed the *tibial collateral ligament*. Some fibers of this ligament may have ruptured, and he may only have a sprain; however, because of the severity of the blow, the entire ligament probably ruptured near its attachment to the medial femoral epicondyle. Because the *medial meniscus* is attached to this ligament, it was probably torn.

This description may have been the full extent of the athlete's injury, but he may also have ruptured his *anterior cruciate ligament*. This ligament, which prevents posterior displacement of the femur and hyperextension of the knee joint, sometimes tears when the knee is hit hard from the lateral side.

In summary, the forced abduction and lateral rotation of the runner's leg by the block probably resulted in the simultaneous rupture of three structures: tibial collateral ligament, medial meniscus, and anterior cruciate ligament—the so-called "unhappy triad of injuries."

Case 5.22

A *femoral hernia* is a protrusion of fat, peritoneum, omentum, and usually a loop of intestine through the femoral ring into the femoral canal. Bowel sounds may be heard with a stethoscope when the intestine is in the hernial sac.

The *femoral canal* is a short, blind, potential space in the medial compartment of the *femoral sheath*, which is a prolongation of the fascial lining of the interior of the abdomen (fascia transversalis anteriorly and fascia iliaca posteriorly). Normally, this space contains lymphatic vessels and at least one lymph node embedded in connective tissue.

The femoral canal is a source of weakness in the abdominal wall; thus, when intra-abdominal pressure rises very high (as may occur when a chronically constipated person attempts to defecate), abdominal contents may be forced through the femoral ring into the femoral canal. The intestine carries a pouch of peritoneum before it as it descends along the femoral canal and through the saphenous opening. Being prevented from extending further down by the fascia lata of the thigh, the hernial sac travels anteriorly and then superiorly, forming a swelling inferior to the inguinal ligament. While the hernia is in the femoral canal (*incomplete femoral hernia*) it is usually small, but after it passes anteriorly through the saphenous opening into the loose connective tissue of the thigh, it becomes much larger (*complete femoral hernia*).

The differential diagnosis between indirect inguinal hernia and complete femoral hernia is at times difficult because advanced types of femoral hernia may produce a swelling superior to the inguinal ligament. The swelling produced by femoral hernia is inferolateral to that caused by an indirect inguinal hernia.

The *pubic tubercle* is the important bony landmark in differentiating an inguinal from a femoral hernia. The neck of an *inguinal hernial sac is superomedial to the pubic tubercle* at the superficial inguinal ring, whereas the neck of a *femoral hernial sac is inferolateral to the pubic tubercle* at this site. In addition, if a hernia does not present in the inguinal canal during the *invagination test*, as in the present case, the hernia cannot be an indirect inguinal hernia.

If the hand is placed gently over the hernia and eased inferiorly, the fold of the groin produced by the inguinal ligament will be seen passing superior to a femoral hernia, whereas if the hand is eased superiorly, the fold of the groin will be seen passing inferior to an inguinal hernia.

Femoral hernia is more common in women than in men (approximately 3:1) because the femoral ring is larger in women than in men because of the greater breadth of the female pelvis, the smaller size of their femoral vessels, and the changes that occur in the associated tissues during pregnancy.

Strangulation of a complete femoral hernia is common. The tendency for this type of hernia to strangulate (i.e., compress the

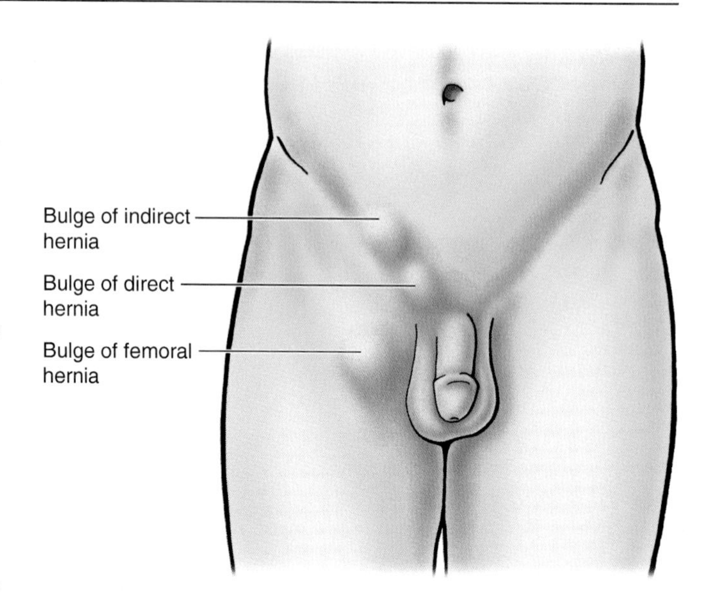

vessels of the hernia) results from the sharp boundaries of the femoral ring (e.g., the inguinal ligament anteriorly and the lacunar ligament medially). Stricture of the hernia may also be produced by the sharp edges of the saphenous opening.

Because of the relatively small size of the femoral ring and saphenous opening and the rigidity of the surrounding structures, blockage of venous return from the protruded loop of intestine often occurs. As arterial blood continues to pass into the loop, it becomes engorged with blood, and circulation through it soon stops. Early surgical intervention is required to prevent necrosis of the strangulated loop of intestine.

A soft enlarged lymph node in the femoral canal could be mistaken for a femoral hernia, although the node would likely be firmer. Cancer or infections in the areas drained by these nodes could cause one or more of them to enlarge.

References and Suggested Readings

Anderson MK, Hall SJ: *Sports Injury Management.* Baltimore, Williams & Wilkins, 1995.

Behrman RE, Kliegman RM, Arvin AM: *Nelson Textbook of Pediatrics*, 15th ed. Philadelphia, WB Saunders, 1996.

Birrer RB (ed): *Sports Medicine for the Primary Care Physician*, 2nd ed. Boca Raton, CRC Press, 1994.

Cahill DR, Orland MJ, Miller G: *Atlas of Human Cross-Sectional Anatomy*, 3rd ed. New York, Wiley-Liss, 1994.

Clemente CD: *Gray's Anatomy of the Human Body*, 30th American Edition. Philadelphia, Lea & Febiger, 1985.

Crelin ES: An experimental study of hip stability in human newborn cadavers. *Yale J Biol Med* 49:109, 1976.

Devinsky O, Feldman E: *Examination of Cranial and Peripheral Nerves.* New York, Churchill Livingstone, 1988.

Ellis A: *Clinical Anatomy. A Revision and Applied Anatomy for Clinical Students*, 8th ed. Oxford, Blackwell Scientific Publications, 1992.

Ger R, Sedlin E: The accessory soleus muscle. *Clin Orthop* 116:200, 1976.

Griffith HW: *Complete Guide to Sports Injuries.* Los Angeles, Price Stern Sloan, 1986.

Gross A: Orthopedic surgery adult. *In* Gross A, Gross P, Langer B (eds): *Surgery: A Complete Guide for Patients and Their Families.* Toronto, Harper & Collins, 1989.

Healey JE Jr, Hodge J: *Surgical Anatomy*, 2nd ed. Toronto, BC Decker, 1990.

Jenkins DB: *Functional Anatomy of the Limbs and Back*, 6th ed. Philadelphia, WB Saunders, 1990.

Levandowski R: Knee injuries. *In* Birrer RB (ed): *Sports Medicine for the Primary Care Physician*, 2nd ed. Boca Raton, CRC Press, 1994.

Levandowski R, Difiori JP: Thigh injuries. *In* Birrer RB (ed): *Sports Medicine for the Primary Care Physician*, 2nd ed. Boca Raton, CRC Press, 1994.

McKee NH, Fish JS, Manktelow RT, McAvoy GV, Young S, Zuker RM: Gracilis muscle anatomy as related to function of a free functioning muscle transplant. *Clin Anat* 3:87–92, 1990.

McMinn RMH: *Last's Anatomy. Regional and Applied*, 8th ed. New York, Churchill Livingstone, 1990.

Salter RB: *Textbook of Disorders and Injuries of the Musculoskeletal System*, 3rd ed. Baltimore, Williams & Wilkins, 1998.

Salter RB: Orthopedic surgery pediatric. *In* Gross A, Gross P, Langer B (eds): *Surgery: A Complete Guide for Patients and Their Families*. Toronto, Harper & Collins, 1989.

Slaby FJ, McCune SK, Summers RW: *Gross Anatomy in the Practice of Medicine*. Baltimore, Lea & Febiger, 1994.

Soames RW: Arthroscopy of the knee. *In* Williams PH, Bannister LH, Berry MM, Collins P, Dyson M, Dussek JE, Ferguson MWJ (eds): *Gray's Anatomy*, 38th ed. New York, Churchill Livingstone, 1995.

Swartz MH: *Textbook of Physical Diagnosis*, 2nd ed. Philadelphia, WB Saunders, 1994.

Wendell-Smith CP: Fascia: An illustrative problem in international terminology. *Surg Radiol Anat* 19:273, 1997.

Williams PH, Bannister LH, Berry MM, Collins P, Dyson M, Dussek JE, Ferguson MWJ: *Gray's Anatomy*, 38th ed. New York, Churchill Livingstone, 1995.

Woodburne RT, Burkel WE: *Essentials of Human Anatomy*, 9th ed. New York, Oxford University Press, 1994.

chapter

6 Upper Limb

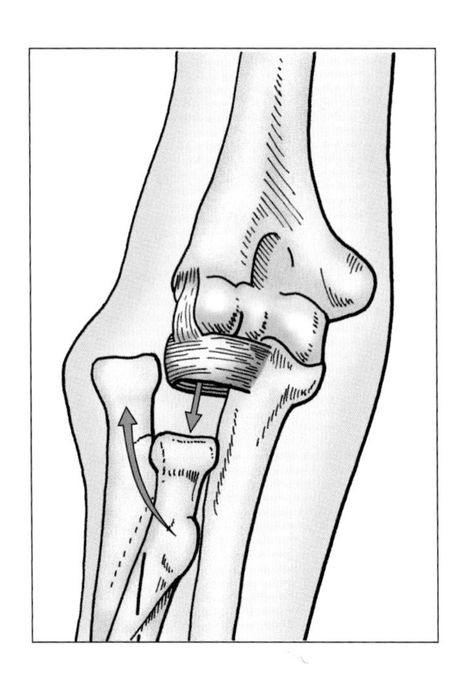

The upper limb (extremity) is characterized by its mobility and ability to grasp and manipulate. These characteristics are especially marked in the hand (L. manus) when performing manual activities such as buttoning a shirt. Because the upper limb is not usually involved in weightbearing, its stability has been sacrificed to gain mobility. The digits (fingers including the thumb) are the most mobile, but other parts are still more mobile than comparable parts of the lower limb. *The upper limb consists of four segments* (Fig. 6.1):

- *Pectoral girdle*—the bony ring, incomplete posteriorly, formed by the scapulae and clavicles, which is completed anteriorly by the manubrium of the sternum (breast bone)

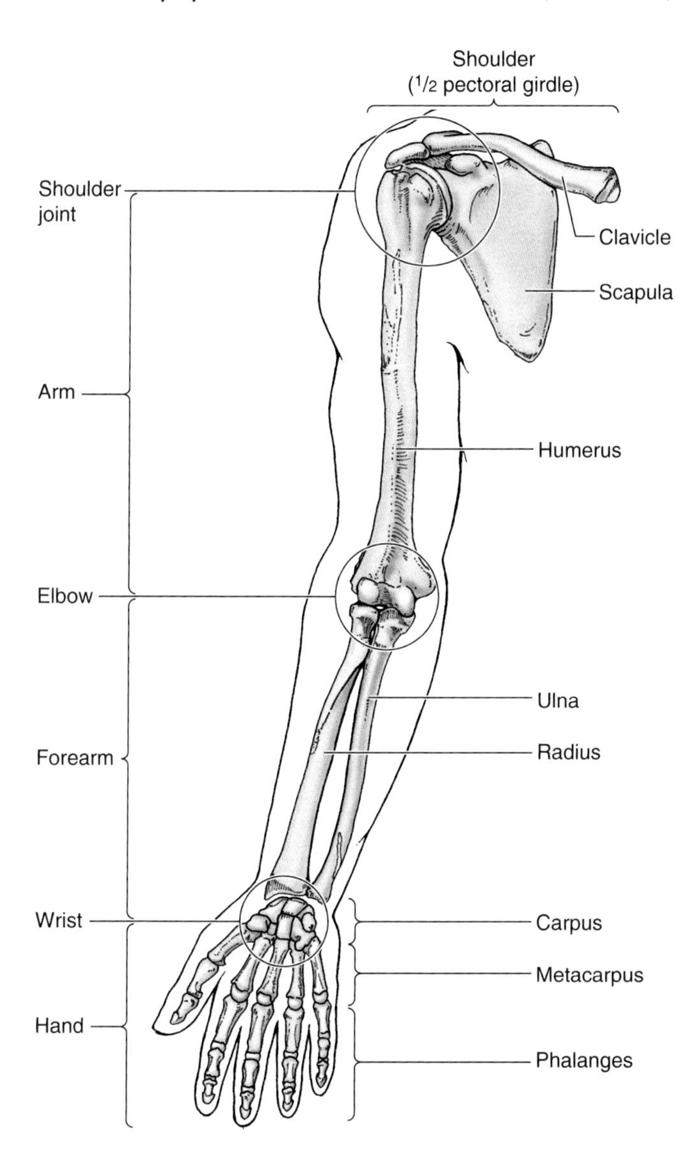

Figure 6.1. Regions and bones of the upper limb. Anterior view. The joints divide the superior appendicular skeleton—and thus the limb itself—into four main regions: shoulder, arm (L. brachium), forearm, and hand (L. manus). The pectoral girdle (shoulder girdle) is a bony ring, incomplete posteriorly, that provides attachment and support of the upper limbs and protection of upper thoracic, lower neck, and axillary structures.

- *Arm*—the part between the shoulder and elbow containing the humerus, which connects the shoulder and the elbow
- *Forearm*—the part between the elbow and wrist containing the ulna and radius, which connect the elbow and wrist
- *Hand*—the part of the upper limb distal to the forearm containing the carpus, metacarpus, and phalanges, which is composed of the wrist, palm, dorsum of hand, and fingers including the thumb.

Upper Limb Injuries

Because the disabling effects of an injury to the upper limb, particularly the hand, are far out of proportion to the extent of the injury, a sound understanding of the structure and function of the upper limb is of the highest importance. Knowledge of its structure without an understanding of its functions is almost useless clinically because the aim of treating an injured limb is to preserve or restore its functions. ○

Bones of the Upper Limb

The pectoral girdle (shoulder girdle) and bones of the free part of the upper limb form the superior free part of the **appendicular skeleton** (Fig. 6.2); the pelvic girdle and bones of the free part of the lower limb form the inferior part. The **pectoral girdle**, formed by the scapulae and clavicles and joined to the manubrium of the sternum, connects the free parts of the upper limbs to the **axial skeleton** (bones in the head, neck, and trunk). Although extremely mobile, the pectoral girdle is supported and stabilized by muscles that are attached to the ribs, sternum, and vertebrae.

Clavicle

The clavicle (collar bone), a doubly curved long bone, connects the upper limb to the trunk (Fig. 6.3). Its *sternal (medial) end* is enlarged and triangular where it articulates with the manubrium of the sternum at the *sternoclavicular (SC) joint*. Laterally, the clavicle articulates with the acromion of the scapula (Fig. 6.2). Its *acromial (lateral) end* is flat where it articulates with the acromion at the *acromioclavicular (AC) joint* (Figs. 6.2*B* and 6.3). The medial two-thirds of the body (shaft) of the clavicle are convex anteriorly, whereas the lateral third is flattened and concave anteriorly. These curvatures increase the resilience of the clavicle and give it the appearance of an elongated capital "S." The clavicle:

- Serves as a strut (rigid support) from which the scapula and free limb are suspended, keeping them away from the thorax so that the arm (L. brachium) has maximum freedom of motion; the strut is movable and allows the scapula to move on the thoracic wall (at the conceptual "scapulothoracic joint"), increasing the range of motion of the limb;

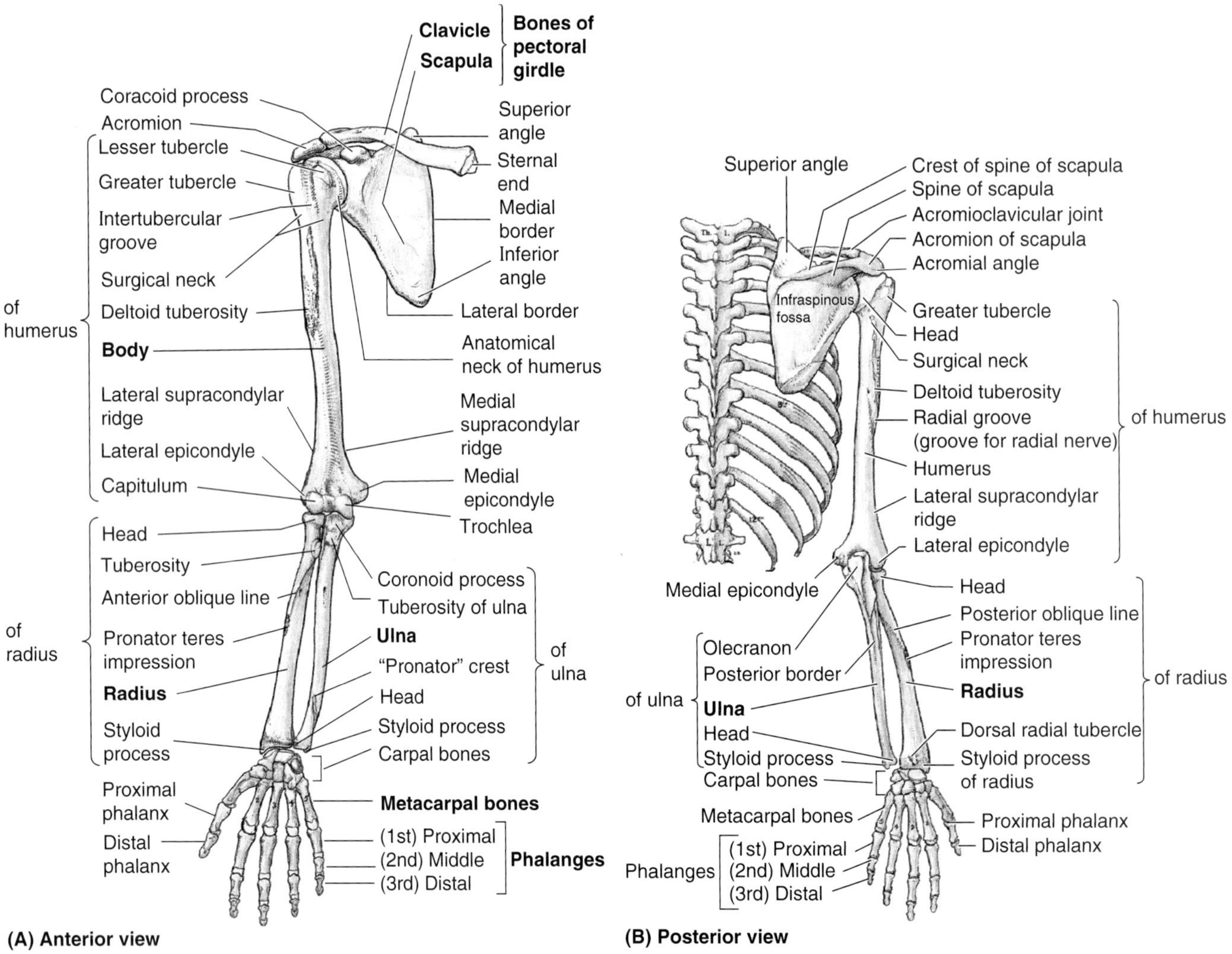

Figure 6.2. Bones of the upper limb. A. Anterior view of the superior appendicular skeleton. The clavicle (collar bone) and scapula (L. shoulder blade) of the right and left sides form the pectoral girdle (shoulder girdle); the remaining bones form the skeleton of the free limb. **B.** Posterior view of the superior appendicular and thoracic part of the axial skeleton. Note the relationship of the scapula to the ribs (L. costae) and thoracic vertebrae. The scapula overlaps parts of the 2nd through 7th ribs. The upper limb is supported and stabilized by thoracoappendicular muscles, which extend from the ribs and vertebrae (axial skeleton) to the limb (appendicular skeleton), forming a conceptual "scapulothoracic joint" between the scapula and the thoracic (chest) wall.

fixing the strut in position—especially following its elevation—enables elevation of the ribs for deep respiration (inspiration)

- Forms one of the bony boundaries of the *cervicoaxillary canal* (passageway between the neck and arm), affording protection to the neurovascular bundle supplying the upper limb
- Transmits shocks (traumatic impacts) from the upper limb to the axial skeleton.

Although designated as a long bone, the clavicle has no medullary (marrow) cavity. It consists of spongy (cancellous) bone with a shell of compact bone.

The **superior surface of the clavicle**, lying just deep to the skin and platysma (G. flat plate) muscle in the subcutaneous tissue, is smooth. The **deltoid tubercle** is the prominence indicating the attachment of the deltoid, the muscle that gives the rounded contour to the shoulder.

The **inferior surface of the clavicle** is rough because strong ligaments bind it to the 1st rib (L. costa) near its sternal end and suspend the scapula from its acromial end. The **conoid tubercle**, near the acromial end of the clavicle (Fig. 6.3), gives attachment to the **conoid ligament**—the medial part of the coracoclavicular ligament. The **subclavian groove** in the medial third of the clavicle is the site of attachment of the subclavius muscle. More medially is the **impression for the costoclavicular ligament**, which binds the 1st rib to the clavicle. Near the acromial end of the clavicle is the **trapezoid line**, to which the trapezoid ligament attaches; it is the lateral part of the coracoclavicular ligament.

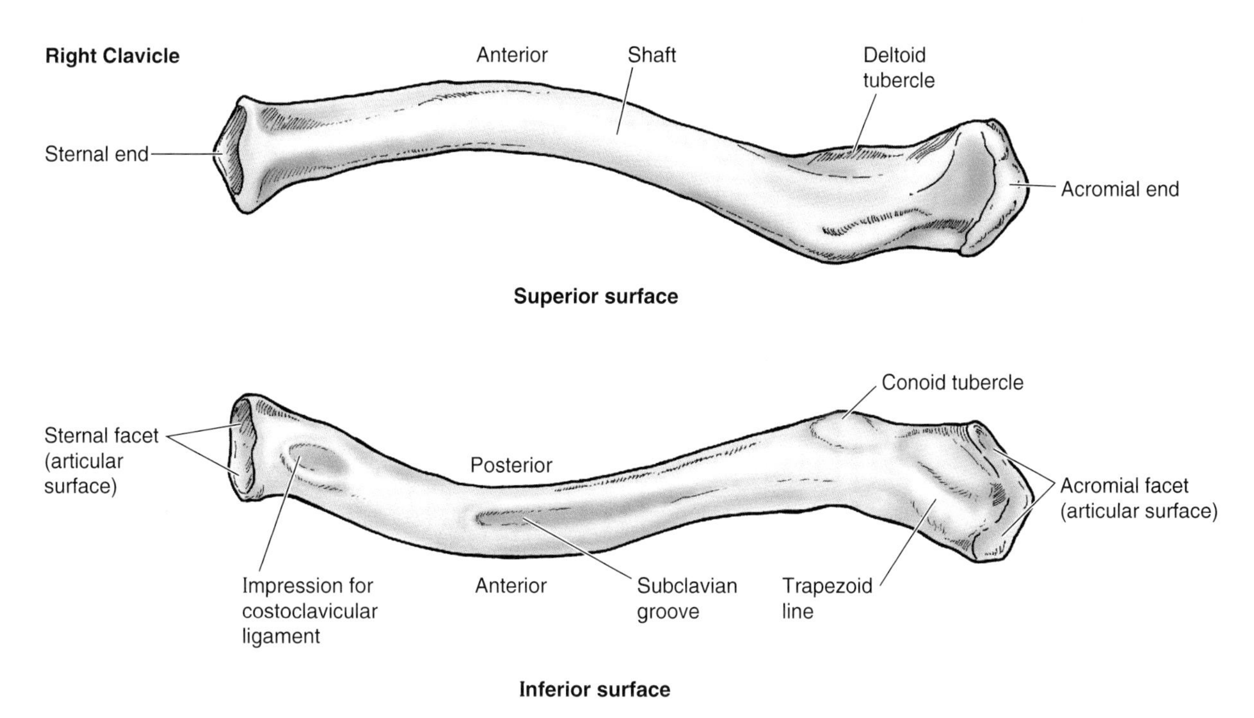

Figure 6.3. Right clavicle (collar bone). Superior and inferior surfaces showing their prominent features. The clavicle acts as a mobile strut, connecting the trunk to the upper limb by extending from the manubrium of the sternum (breast bone) to the acromion of the scapula (L. shoulder blade) (Fig. 6.2).

Variations of the Clavicle

The clavicle varies more in shape than most other long bones. Occasionally, the clavicle is pierced by a branch of the supraclavicular nerve. The clavicle is thicker and more curved in manual workers, and the sites of muscular attachments are more marked. The right clavicle is stronger than the left and is usually shorter.

Fracture of the Clavicle

The clavicle is commonly fractured, often by an indirect force resulting from violent impacts to the outstretched hand during a fall—transmitted through the bones of the forearm and arm to the shoulder—or by falls directly onto the shoulder itself. The relatively strong clavicles of adults are less frequently fractured than the slender ones of children. Because the clavicle is the area of bony contact with the manubrium of the sternum, that is, it is "the link between the appendicular and axial parts of the skeleton," fractures of the clavicle are relatively common. *The weakest part of the clavicle is the junction of its middle and lateral thirds.* Although a main function of the clavicle is to transmit forces from the upper limb to the axial skeleton, if the force during a fall on the shoulder is greater than the strength of the clavicle, a fracture will result.

After fracture of the clavicle, the sternocleidomastoid muscle elevates the medial fragment of bone. Because the trapezius muscle is unable to hold the lateral fragment up because of the weight of the upper limb, the shoulder drops. The strong *coracoclavicular ligament* usually prevents dislocation of the AC joint. People with fractured clavicles support the sagging limb with the other limb. In addition to being depressed, the lateral fragment of the clavicle may be pulled medially by the adductor muscles of the arm, such as the pectoralis major. Overriding of the bone fragments shortens the clavicle.

The slender clavicles of newborn infants may be fractured during delivery of broad-shouldered babies; however, they usually heal quickly. A fracture of the clavicle is often incomplete in children; that is, it is a *greenstick fracture* in which one side of a bone is broken and the other is bent. This fracture was so named because the parts of the bone do not separate; it resembles a tree branch (greenstick) that is sharply bent but the parts do not disconnect.

Ossification of the Clavicle

The clavicle is the first long bone to ossify, beginning during the 5th and 6th embryonic weeks in condensed mesenchyme (*intramembranous ossification*) from two primary centers, medial and lateral, close together in the body of the clavicle. The two ends of the clavicle later pass through a cartilaginous phase (*endochondral ossification*); the cartilages form growth zones similar to those of other long bones. A secondary ossification center appears at the sternal end and forms a scalelike epiphysis that begins to fuse with the body (diaphysis) between the 18th and 25th years and is completely fused to it between the 25th and 31st ▶

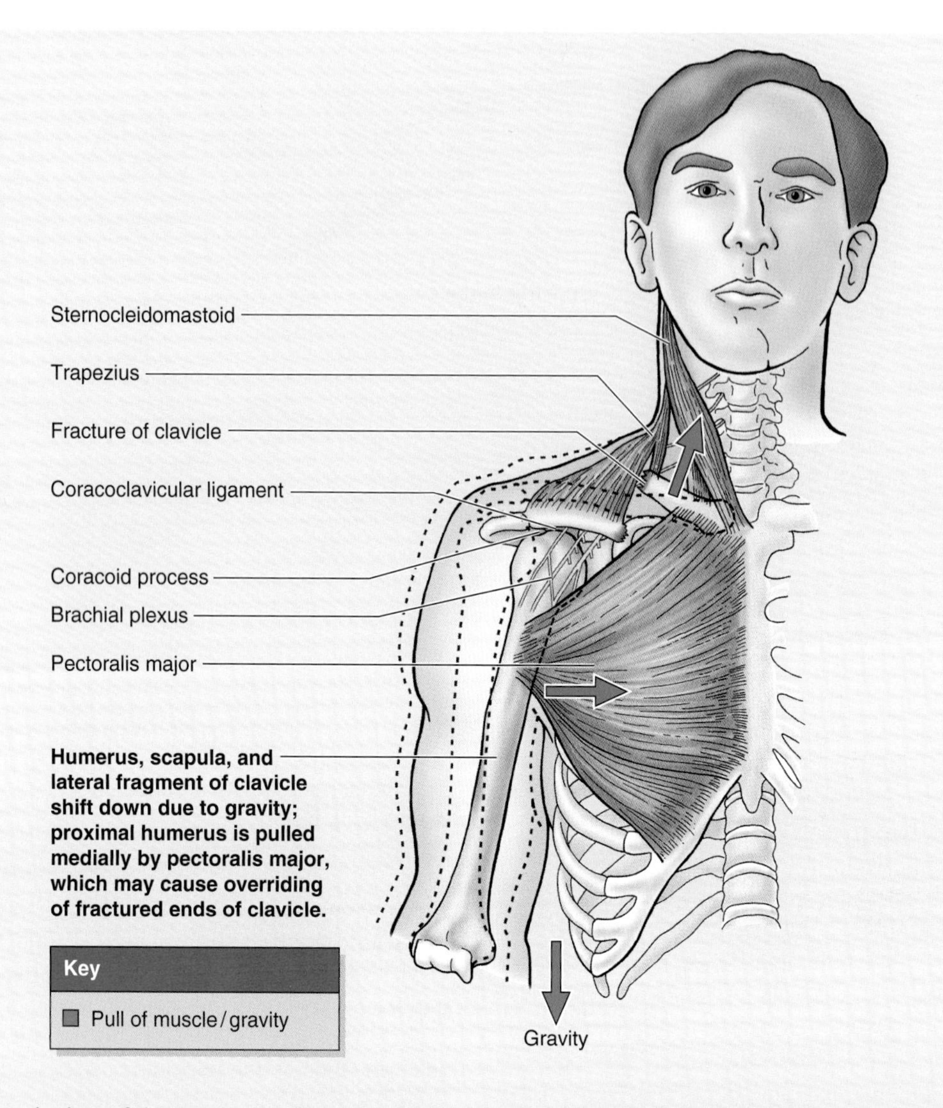

▶ years. This is the last of the epiphyses of long bones to fuse. An even smaller scalelike epiphysis may be present at the acromial end of the clavicle; it must not be mistaken for a fracture.

Abnormal Ossification of the Clavicle

Abnormal ossification of the clavicles may result in the absence of most of both clavicles, which enables the person to bring the shoulders together. Sometimes fusion of the two ossification centers of the clavicle fails to occur; as a result, a bony defect occurs between the lateral and medial two-thirds of the clavicle. Awareness of this possible congenital defect should prevent diagnosis of a fracture in an otherwise normal clavicle. When doubt exists, both clavicles are radiographed because this defect is usually bilateral (Ger et al., 1996). ○

Scapula

The scapula (shoulder blade) is a triangular flat bone that lies on the posterolateral aspect of the thorax, overlying the 2nd through 7th ribs (Fig. 6.2*B*). The concave **costal surface** of most of the scapula forms a large **subscapular fossa** (Fig. 6.4); the convex **posterior surface** of the scapula is unevenly divided by the spine of the scapula (Fig. 6.2*B*) into a small **supraspinous fossa** and a much larger **infraspinous fossa**. The broad bony surfaces of the three fossae provide attachments for fleshy muscles. The triangular **body of the scapula** is thin and translucent superior and inferior to the spine, although its borders, especially the lateral one, are somewhat thicker. The **spine of the scapula**, a thick projecting ridge of bone, continues laterally as the flat expanded **acromion** (G. akros, point), which forms the subcutaneous "point of the shoulder" and articulates with the acromial end of the clavicle. Superolaterally, the lateral surface of the scapula forms the **glenoid cavity** (G. socket), which articulates with the head of the humerus at the glenohumeral (shoulder) joint (Fig. 6.2*B*).

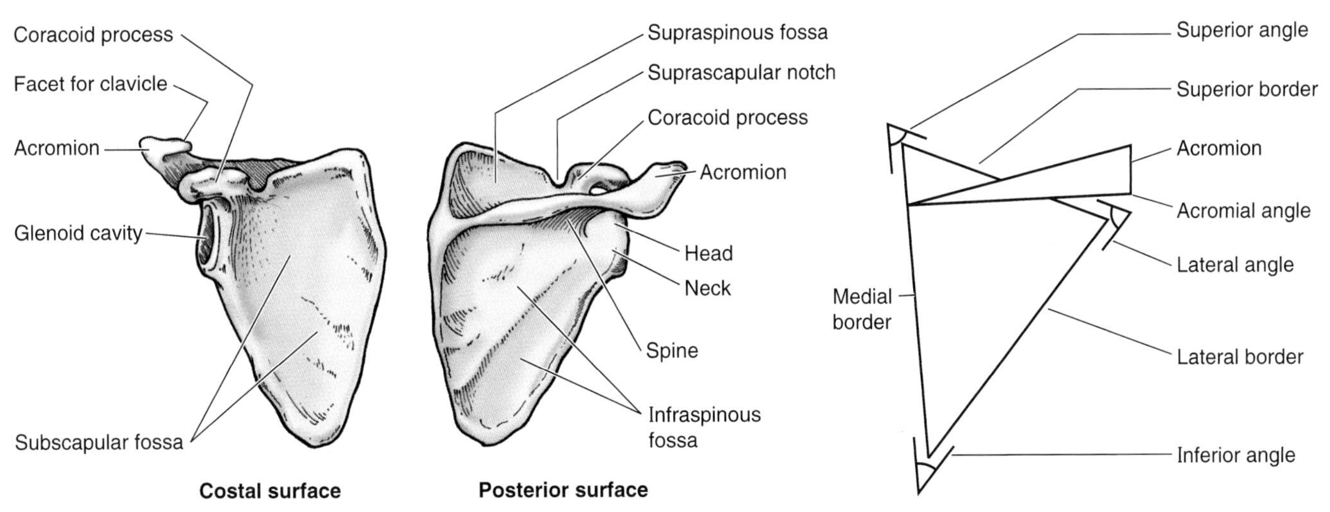

Figure 6.4. Right scapula (L. shoulder blade). Costal (anterior) and posterior surfaces. The diagram on the right illustrates the borders and angles of the scapula. The scapula is suspended from the clavicle (collar bone) and forms the actual skeletal connection between the clavicle and the humerus (arm bone). Although not a "true" joint, the scapula "articulates" with the posterosuperior thoracic wall via the conceptual "scapulothoracic joint" (Fig. 6.2) and truly articulates with the head of the humerus.

The beaklike **coracoid process** (G. korakōdēs, "like a crow's beak") is superior to the glenoid cavity and projects anterolaterally (Fig. 6.4). This process also resembles in size, shape, and direction a bent finger pointing to the shoulder.

The scapula has medial, lateral, and superior borders and superior, lateral, and inferior angles. When the scapular body is in the anatomical position, the thin **medial border of the scapula** runs parallel to and approximately 5 cm lateral to the spinous processes of the thoracic vertebrae (Fig. 6.2*B*); hence, it is often called the vertebral border. From the inferior angle, the **lateral border of the scapula** runs superolaterally toward the apex of the axilla; hence, it is often called the axillary border. The lateral border contains a thick bar of bone that prevents buckling of this stress-bearing region of the scapula. The lateral border terminates in the truncated **lateral angle** of the scapula, the thickest part of the bone where the glenoid cavity is located; the adjacent broadened process is the **head** of the scapula. The constriction between the head and body is the **neck** of the scapula. The **superior border** of the scapula is marked near the junction of its medial two-thirds and lateral third by the **suprascapular notch** (scapular notch). The notch is located where the superior border joins the base of the coracoid process. The superior border is the thinnest and shortest of the three borders. The **glenoid cavity** for reception of the head of the humerus is a shallow, concave, oval fossa (L. fossa ovalis) approximately 4 cm long and 2 to 3 cm wide; it faces anterolaterally and slightly superiorly. *The scapula is capable of considerable movement on the thoracic wall at the conceptual scapulothoracic joint.* These movements, enabling the arm to move freely, are discussed subsequently with those that move the scapula. In addition to giving attachment to muscles, the glenoid cavity of the scapula forms the socket of the shoulder joint.

Fracture of the Scapula

Most of the scapula is well protected by muscles and by its association with the thoracic wall; consequently, most fractures of the scapula involve the protruding subcutaneous acromion. ⊙

Humerus

The humerus (arm bone), the largest bone in the upper limb, articulates with the scapula at the scapulohumeral (shoulder) joint and the radius and ulna at the elbow joint (Fig. 6.2). The proximal end of the humerus has a head, a neck, and greater and lesser tubercles. The ball-shaped **head** of the humerus articulates with the glenoid cavity of the scapula. The **anatomical neck of the humerus** is formed by the groove circumscribing the head and separating it from the greater and lesser tubercles. The junction of the head and neck with the **body of the humerus** is indicated by the greater and lesser tubercles (tuberosities), which provide attachment and leverage to some scapulohumeral muscles. The **greater tubercle** is at the lateral margin of the humerus, whereas the **lesser tubercle** projects anteriorly from the bone. The **intertubercular groove** (bicipital groove) separates the tubercles. The **surgical neck of the humerus** is the narrow part distal to the tubercles and the crests descending from them, flanking the intertubercular groove. *The surgical neck is a common fracture site of the humerus.*

The body of the humerus has two prominent features: the **deltoid tuberosity**, laterally, for attachment of the deltoid muscle, and the oblique **radial groove**, posteriorly, in which the radial nerve and deep artery of the arm (L. arteria pro-

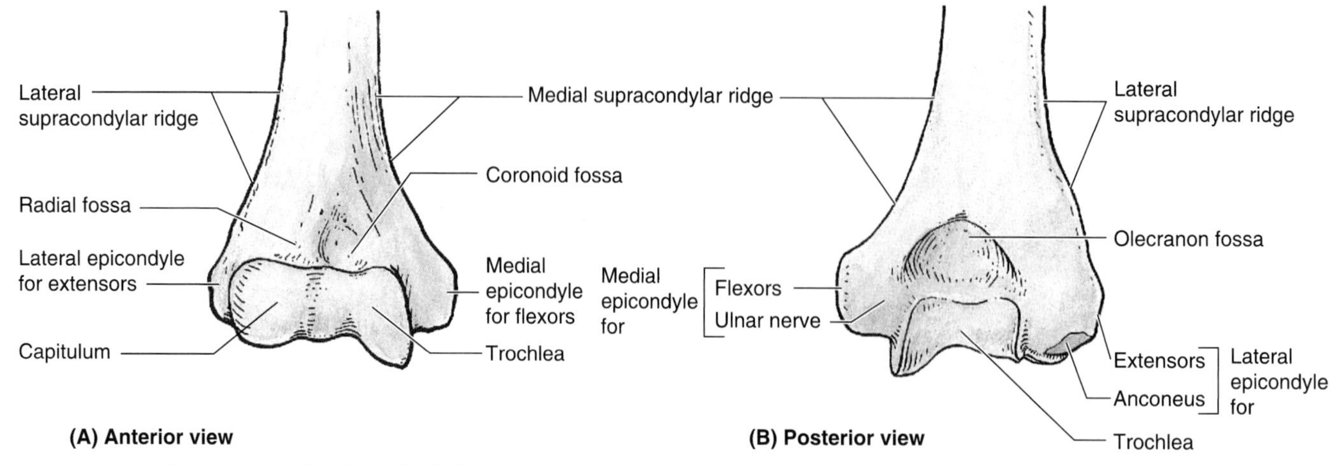

Figure 6.5. Distal end of the right humerus. Anterior (**A**) and posterior (**B**) views illustrating the lateral and medial epicondyles and the condyle consisting of the capitulum and trochlea (L. pulley), which articulate with the radius and ulna, respectively, at the elbow joint (see Fig. 6.6).

funda brachii) lie as they pass between the medial and the long and then the lateral heads of the triceps brachii muscle. The inferior end of the humerus widens as the sharp medial and lateral **supracondylar ridges** form and then end distally in medial and especially prominent lateral extensions, the **epicondyles,** providing for muscle attachment. The distal end of the humerus—including the epicondyles, trochlea (L. pulley), and capitulum and the olecranon, coronoid, and radial fossae—constitutes the **condyle of the humerus** (Fig. 6.5). The condyle has two articular surfaces: a lateral **capitulum** (L. little head) for articulation with the head of the radius and a medial **trochlea** for articulation with the proximal end (trochlear notch) of the ulna (Fig. 6.1). Superior to the pulleylike trochlea anteriorly is the **coronoid fossa,** which receives the coronoid process of the ulna during full flexion of the elbow, and posteriorly is the **olecranon fossa,** which accommodates the olecranon of the ulna during full extension of the elbow. Superior to the capitulum anteriorly, a shallow **radial fossa** accommodates the edge of the head of the radius when the forearm is fully flexed.

Fracture of the Humerus

Most injuries of the proximal end of the humerus are *fractures of the surgical neck.* These injuries are common in elderly people, especially those with *osteoporosis* because their demineralized bones are brittle. Humeral fractures are often ones in which one fragment is driven into the spongy bone of the other fragment (*impacted fractures*). The injuries usually result from a minor fall on the hand, with the force being transmitted up the forearm bones of the extended limb. Because of impaction of the fragments, the fracture site is stable and the patient is able to move the arm passively with little pain. Radiographs or computed tomographic (CT) scans reveal the humeral fracture.

Avulsion fractures of the greater tubercle of the humerus (*A*) are relatively common, especially in middle-aged and elderly people. They usually result from a fall on the point of the shoulder—the acromion. In younger people, an *avulsion fracture of the greater tubercle* (pulling tubercle away from head) usually results from a fall on the hand when the arm is abducted. Muscles (especially the subscapularis) remaining attached to the humerus pull the limb into medial rotation. An inferior dislocation of the shoulder joint often occurs in the absence of the muscle attachments to the greater tubercle.

Transverse fractures of the body of the humerus frequently result from a direct blow to the arm. The pull of the deltoid carries the proximal fragment of the fractured humerus laterally (*B*). Indirect injury resulting from a fall on the outstretched hand may produce a *spiral fracture of the humeral body.* Overriding of the oblique ends of a spirally fractured bone may result in foreshortening. Because the humerus is surrounded by muscle and has a well-developed periosteum, the bone fragments usually unite well. When describing *fractures of the condyle of the humerus,* the terms medial and lateral condyles may be used. *Intercondylar fractures of the humerus* result from a severe fall on the "point" of the flexed elbow. The olecranon of the ulna is driven like a wedge into the condyle of the humerus, separating one or both parts from the humeral body.

The following parts of the humerus are in direct contact with the indicated nerves:

- Surgical neck—axillary nerve
- Radial groove—radial nerve
- Distal end of humerus—median nerve
- Medial epicondyle—ulnar nerve.

These nerves may be injured when the associated part of the humerus is fractured. These injuries are discussed after these nerves are described. ⊙

Ulna

The ulna—*the stabilizing bone of the forearm*—is the medial and longer of the two forearm bones (Figs. 6.6 and 6.7). Its proximal end has two prominent projections, the **olecranon,** which projects proximally from its posterior aspect, and the **coronoid process,** which projects anteriorly. The anterior surface of the olecranon forms the posterior wall of the **trochlear notch,** which articulates with the trochlea of the humerus. On the lateral side of the **coronoid process** is a smooth, rounded concavity, the **radial notch,** which articulates with the head of the radius. Inferior to the coronoid process is the **tuberosity of the ulna** (Fig. 6.6*A*) for attachment of the tendon of the biceps brachii muscle. The proximal end of the ulna looks somewhat like a pipe wrench, with the olecranon representing the upper jaw (maxilla) and the coronoid process representing the lower jaw (mandible); the trochlear notch is the "mouth of the wrench." The olecranon and coronoid process grasp the trochlea of the humerus somewhat like a pipe wrench clasps a pipe (Fig. 6.6, *B* and *C*). Inferior to the radial notch on the lateral surface of the ulna is a prominent ridge—the **supinator crest.** Between it and the distal part of the coronoid process is a concavity—the **supinator fossa.** The deep part of the supinator muscle attaches to the supinator crest and fossa. The **body of the ulna** is thick and cylindrical proximally, but it tapers, diminishing in diameter as it continues

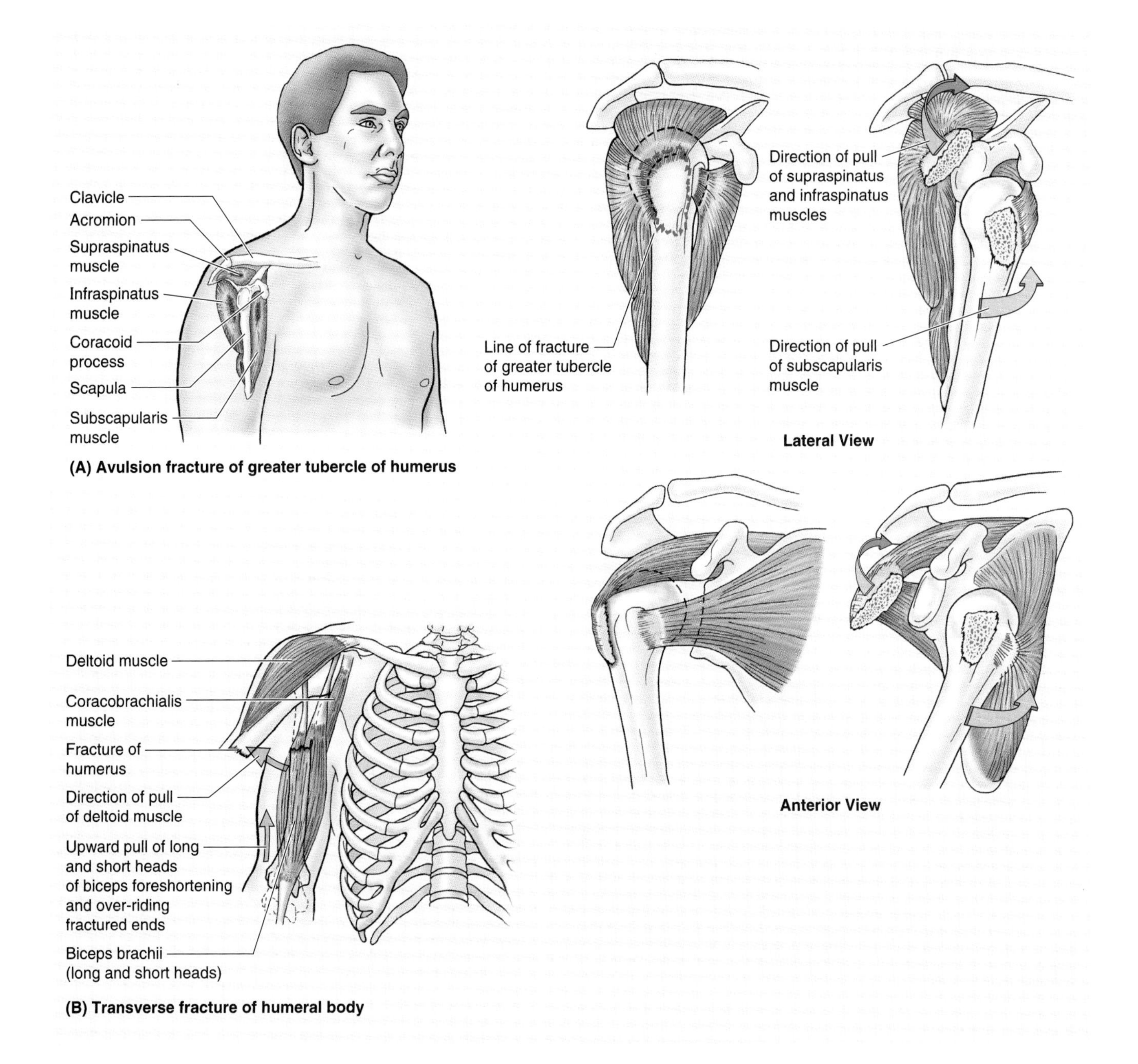

distally (Fig. 6.7*A*). At its narrow distal end is a somewhat abrupt enlargement forming a disclike **head** and a small, conical **styloid process**. Note that the head of the ulna lies distally (i.e., at the wrist). The articulation between the ulna and humerus allows primarily only flexion and extension of the elbow joint, although a small amount of abduction-adduction "wobble" occurs during pronation and supination of the forearm.

Radius

The radius is the lateral and shorter of the two forearm bones. Its proximal end consists of a short cylindrical (or thick disclike) head, a neck, and a medially directed tuberosity (Fig. 6.7, *A* and *B*). Proximally, the smooth superior aspect of the **head of the radius** is concave for articulation with the capitulum of the humerus during flexion and extension of the elbow joint. The head also articulates peripherally with the radial notch of the ulna; thus, the head is covered with articular cartilage. The **neck of the radius** is relatively constricted between the head—which overhangs it—and the tuberosity. The oval **radial tuberosity** separates the proximal end (head and neck) of the radius from the body. The **body of the radius** has a lateral convexity and gradually and progressively enlarges in girth as it passes distally. The distal end of the radius is essentially rectangular when sectioned transversely. Its medial aspect forms a concavity, the **ulnar notch,** which accommodates the head of the ulna. Extending from its lateral aspect is the **radial styloid process.** The **dorsal tubercle,** projecting dorsally (Fig. 6.7, *A–C*), lies between grooves for the passage of the tendons of forearm muscles (Fig. 6.7*D*). *The radial styloid process is much larger than the ulnar styloid process and extends approximately a finger's breadth further distally.* This relationship is of clinical importance when the ulna and/or the radius are fractured.

Bones of the Hand

The skeleton of the wrist—the **carpus**—is composed of eight **carpal bones** (carpals) arranged in two rows of four each (Fig. 6.8). These small bones give flexibility to the wrist. The carpus is markedly convex from side to side posteriorly and concave anteriorly. Augmenting movement at the radiocarpal joint or wrist, the two rows of carpals glide on each other; in addition, each bone glides on those adjacent to it. The carpals are attached to each other by interosseous ligaments. From lateral to medial, *the four bones in the proximal row of carpals are the:*

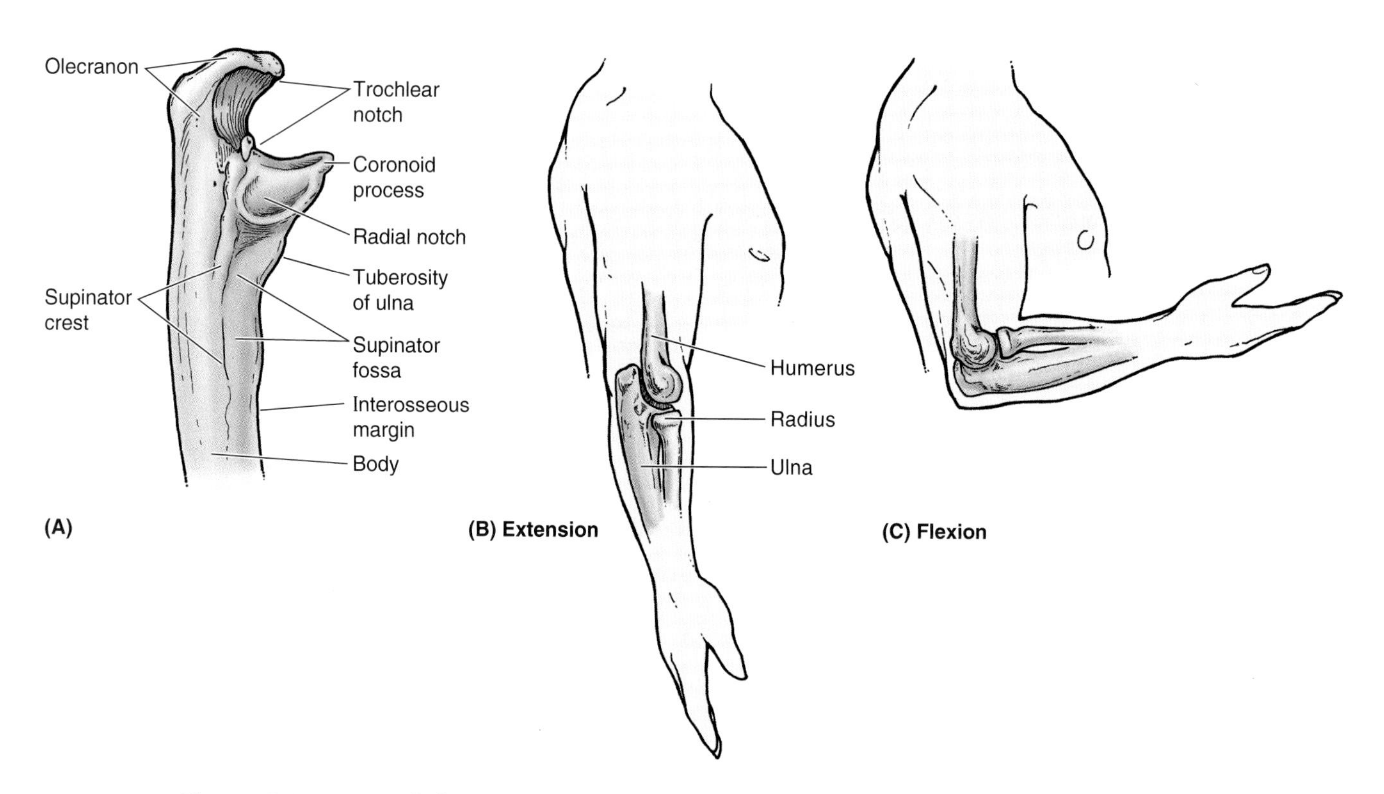

Figure 6.6. Bones of the right elbow region. A. Lateral view of the proximal part of the ulna. **B.** Lateral view of the bones of the elbow region, showing the relationship of the humerus (arm bone), ulna, and radius during extension of the elbow joint. **C.** Relationship of the humerus and forearm bones during flexion of the elbow joint.

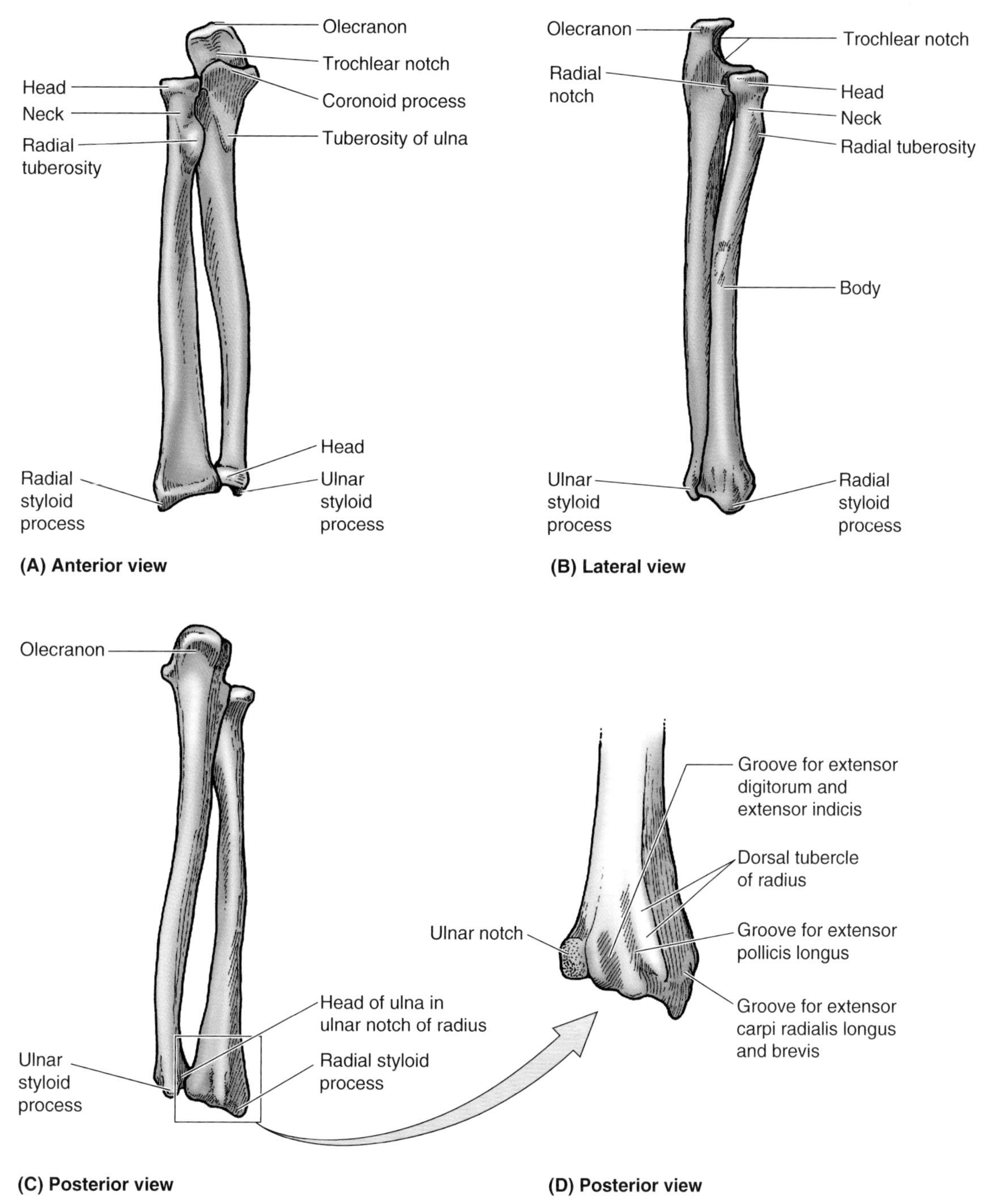

Figure 6.7. Various views of the right radius and ulna. A. Anterior view. **B.** Lateral view. **C.** Posterior view. **D.** Enlarged drawing of the dorsal aspect of the distal end of the radius. Observe the ulnar notch where the articular surface of the head of the ulna articulates with the radius. Notice the grooves and prominent dorsal tubercle of the radius formed in relationship to the tendons of some forearm extensor muscles passing to the hand (L. manus). The dorsal tubercle serves as a trochlea (L. pulley) for the extensor pollicis longus (EPL) tendon.

Fracture of the Radius and Ulna

Fractures of both bones are usually the result of severe injury. A direct injury usually produces transverse fractures at the same level, usually in the middle third of the bones. Isolated fractures of the radius or ulna may occur. Because the bodies of these bones are firmly bound together by the *interosseous membrane*, a fracture of one bone is likely to be associated with dislocation of the nearest joint; for example, a fracture of the distal third of the radius may be associated with dislocation of the distal radioulnar joint. ▶

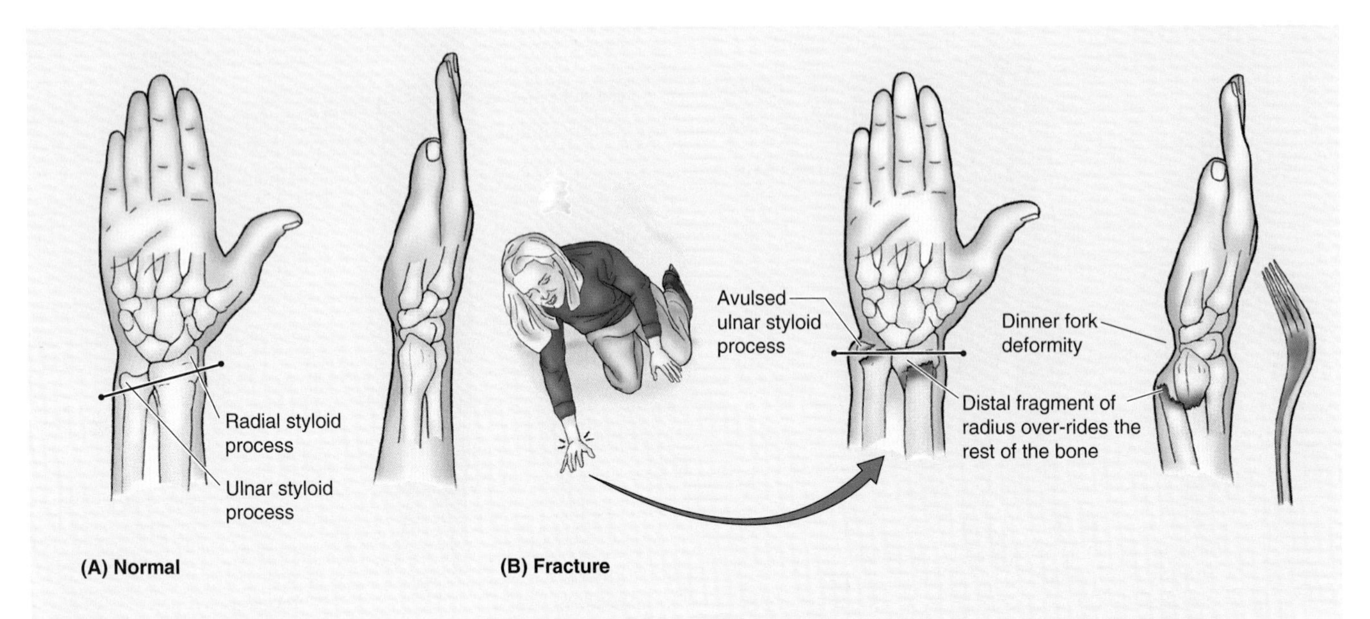

▶ *Fracture of the distal end of the radius is the most common fracture in adults older than 50 years* and occurs more frequently in women because their bones are more commonly weakened by *osteoporosis*—reduction in the quantity of bone or atrophy of skeletal tissue. A complete transverse fracture within the distal 2 cm of the radius—*Colles' fracture*—is the most common fracture of the forearm. The distal fragment is displaced dorsally and is often comminuted (broken into pieces). The fracture results from forced dorsiflexion of the hand, usually as the result of trying to ease a fall by outstretching the upper limb (*A*). In approximately 40% of cases, the ulnar styloid process is avulsed (broken off). Normally, the radial styloid process projects further distally than the ulnar styloid; consequently, when a Colles' fracture occurs, this relationship is reversed because of shortening of the radius (*B*). The clinical deformity is often referred to as a "dinner fork (silver fork) deformity" because a "jog" occurs in the forearm just proximal to the wrist, which is produced by the posterior displacement and tilt of the distal fragment of the radius. *The typical history of a person with a Colles' fracture* includes slipping (as on ice) or tripping (as on a rug) and, in an attempt to break the fall, landing on the outstretched limb with the forearm and hand pronated. Because of the rich blood supply to the distal end of the radius, bony union is usually good.

When the distal end of the radius fractures in children, the fracture line may extend through the distal epiphyseal plate. *Epiphyseal plate injuries* are common in older children because of frequent falls in which the forces are transmitted from the hand to the radius and ulna. The healing process may result in malalignment of the epiphyseal plate and disturbance of radial growth. ⊙

- **Scaphoid**—a boat-shaped bone that articulates proximally with the radius and has a prominent tubercle
- **Lunate**—a moon-shaped bone that articulates proximally with the radius and is broader anteriorly than posteriorly
- **Triquetrum**—a three-cornered (L. triquetrus) pyramidal bone that articulates proximally with the articular disc of the distal radioulnar joint
- **Pisiform**—a small, pea-shaped bone that lies on the palmar surface of the triquetrum.

From lateral to medial, *the four bones in the distal row of carpals are the*:

- **Trapezium**—which is four sided
- **Trapezoid**—which is wedge shaped
- **Capitate**—which has a rounded head
- **Hamate**—which is wedge-shaped and has a hooked process, the *hook of the hamate*.

The proximal surfaces of the distal row of bones articulate with the proximal row of carpals, and their distal surfaces articulate with the metacarpals.

The skeleton of the hand between the carpus and phalanges—the **metacarpus**—is composed of five **metacarpal bones** (metacarpals). Each metacarpal consists of a **body** and **two ends**. The distal ends or **heads of the metacarpals** articulate with the proximal phalanges and form the knuckles of the fist; the proximal ends or **bases of the metacarpals** articulate with the carpal bones. The 1st metacarpal (of the thumb) is the thickest and shortest of the metacarpals. The 3rd metacarpal is distinguished by a **styloid process** on the lateral side of its base.

Each digit has three **phalanges** except the first (the thumb), which has only two (however, they are stouter than

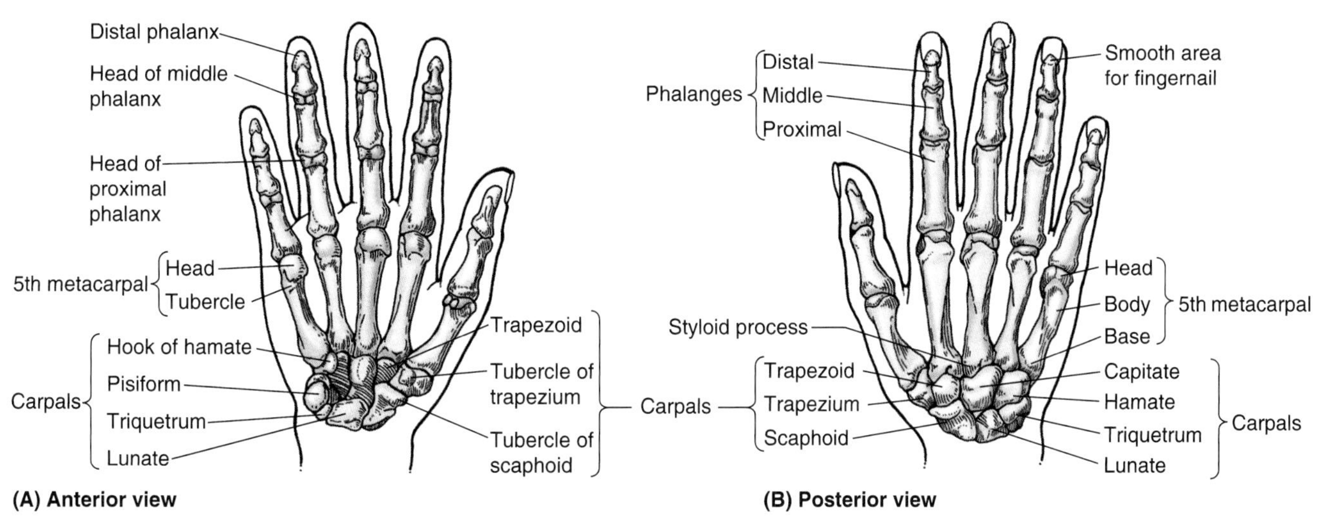

Figure 6.8. Bones of the right hand. Anterior (**A**) and posterior (**B**) views showing that the skeleton of the hand consists of three segments: carpals of the wrist, metacarpals of the palm, and phalanges of the digits. The heads of the metacarpals form the knuckles of the dorsum of the hand, and the heads of the phalanges form the knuckles of the digits.

those in the fingers). Each phalanx has a **base** proximally, a **head** distally, and a **body** between the base and the head (Fig. 6.8). The proximal phalanges are the largest, the middle ones are intermediate in size, and the distal ones are the smallest. Each terminal phalanx is flattened and expanded at its distal end to form the nailbed.

Fracture of the Scaphoid

The scaphoid is the most frequently fractured carpal bone, and *fracture of the scaphoid is the most common injury of the wrist*, especially as a result of a fall onto the palm when the hand is abducted. Pain occurs primarily on the lateral (radial) side of the wrist, especially during dorsiflexion and abduction of the hand. Initial radiographs may not reveal a fracture of the scaphoid; often, an apparent *severely sprained wrist* is subsequently diagnosed as a fractured scaphoid after repeated radiographs 2 to 3 weeks later reveal a fracture site because bone resorption has occurred there. Because of poor blood supply to the proximal part of the scaphoid, union of the fractured parts may take several months. *Avascular necrosis of the proximal fragment of the scaphoid* (pathological death of bone resulting from inadequate blood supply) may occur and produce *degenerative joint disease of the wrist*. In some cases, it is necessary to fuse the carpals surgically (*arthrodesis*).

Fracture of the Hamate

Fracture of the hamate may result in nonunion of the fractured bony parts because of the traction produced by the ▶

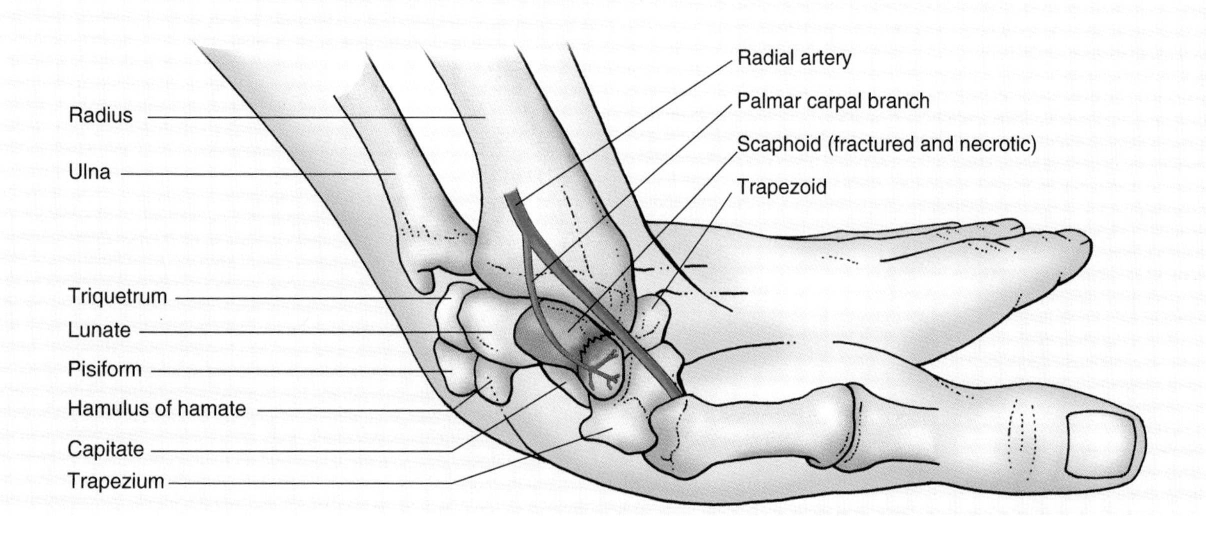

▶ attached muscles. Because the ulnar nerve is in close proximity to the hook of the hamate, the nerve may be injured by a hamate fracture, causing decreased grip strength of the hand.

Fracture of the Metacarpals

The metacarpals (except the first) are closely bound together; hence, isolated fractures tend to be stable. Furthermore, these bones have a good blood supply, and fractures heal rapidly. *Fracture of the necks of the 1st and 2nd metacarpals* are often referred to as "boxer's fractures." In unskilled street fighters, the neck of the more mobile 5th metacarpal is commonly fractured when they strike a blow with the fist clenched. *Severe crushing injuries of the hand* may produce multiple metacarpal fractures, resulting in instability of the hand.

Fracture of the Phalanges

Crushing injuries of the distal phalanges are common (e.g., when a finger is caught in a car door). Because of the highly developed sensation in the fingers, these injuries are extremely painful. A *fracture of a distal phalanx* is usually comminuted and a painful hematoma (local collection of blood) develops. Fractures of the proximal and middle phalanges are usually the result of crushing or hyperextension injuries. Because of the close relationship of phalangeal fractures to the flexor tendons, the bone fragments must be carefully realigned to restore normal function of the fingers. ○

Surface Anatomy of the Upper Limb Bones

The **clavicle** is subcutaneous and can be easily palpated throughout its length. Its sternal end projects superior to the manubrium. Between the elevated sternal ends of the clavicles is the **jugular notch** (suprasternal notch). The acromial end of the clavicle often rises higher than the acromion, forming a palpable elevation at the **AC joint** (*A*). The acromial end can be palpated 2 to 3 cm medial to the lateral border of the acromion, particularly when the upper limb is swung back and forth. Either or both ends of the clavicle may be prominent; when present, this condition is usually bilateral. Note the elasticity of the skin over the clavicle, observing how easily it can be pinched up and moved around. This property of the skin is useful when ligating the third part of the subclavian artery: the skin lying superior to the clavicle is pulled down onto the clavicle and then incised; after the incision is made, the skin is allowed to return to its position superior to the clavicle, where it overlies the artery (which was thus not endangered during the incision).

The **acromion of the scapula** is easily felt (*B*) and often visible. The superior surface of the acromion is subcutaneous and may be traced medially to the AC joint. The lateral and posterior borders of the acromion meet to form the **acromial angle**—the point from which the length of the upper limb is measured. Inferior to the acromion, the **deltoid** forms the rounded curve of the shoulder. The **crest of the scapular spine** is subcutaneous throughout and is easily palpated. *When the upper limb is in the anatomical position*:

- The superior angle of the scapula lies at the level of T2 vertebra
- The medial end of the root of the scapular spine is opposite the spinous process of T3 vertebra
- The inferior angle of the scapula lies at the level of vertebra T7, near the inferior border of the 7th rib and 7th intercostal space.

The **medial border** of the scapula is palpable inferior to the root of the spine of the scapula, as it crosses the 2nd to 7th ribs (*B*); the **lateral border** is not easily palpated because it is covered by the teres major (L. teres, round) muscle.

When the upper limb is abducted and the hand is placed on the back of the head, the scapula is rotated, elevating the glenoid cavity such that the medial border of the scapula parallels and thus can be used to estimate the position of the 6th rib and, deep to the rib, the oblique fissure of the lung. The **inferior angle** of the scapula is easily felt and is often visible. It is grasped when testing movements of the glenohumeral joint to immobilize the scapula.

The **coracoid process of the scapula** (p. 678) can be felt by palpating deeply at the lateral side of the **deltopectoral triangle**. The head of the humerus is surrounded by muscles, except inferiorly; consequently, it can be palpated only by pushing the fingers well up into the axilla. The arm should not be fully abducted, otherwise the fascia in the axilla will be tense and impede palpation of the head. When the arm is moved and the scapula is immobilized, the head can be palpated.

The **greater tubercle of the humerus** may be felt with the person's arm by the side on deep palpation through the deltoid, inferior to the lateral border of the acromion. In this position, the greater tubercle is the most lateral bony point of the shoulder and, along with the deltoid, gives the shoulder its rounded contour. When the arm is abducted, the greater tubercle disappears beneath the acromion and is no longer palpable.

The **lesser tubercle of the humerus** may be felt with difficulty by deep palpation through the deltoid on the an- ▶

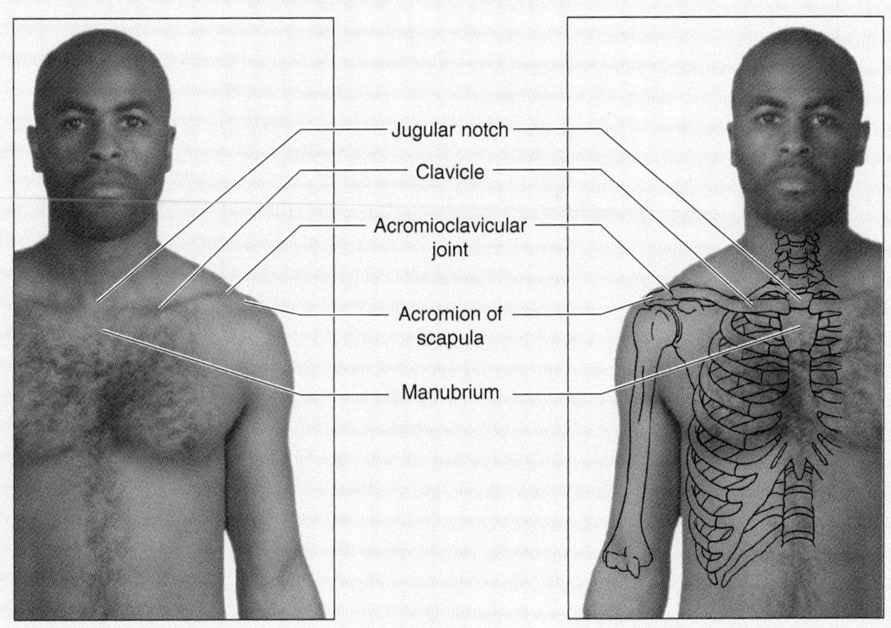

(A) Anterior view

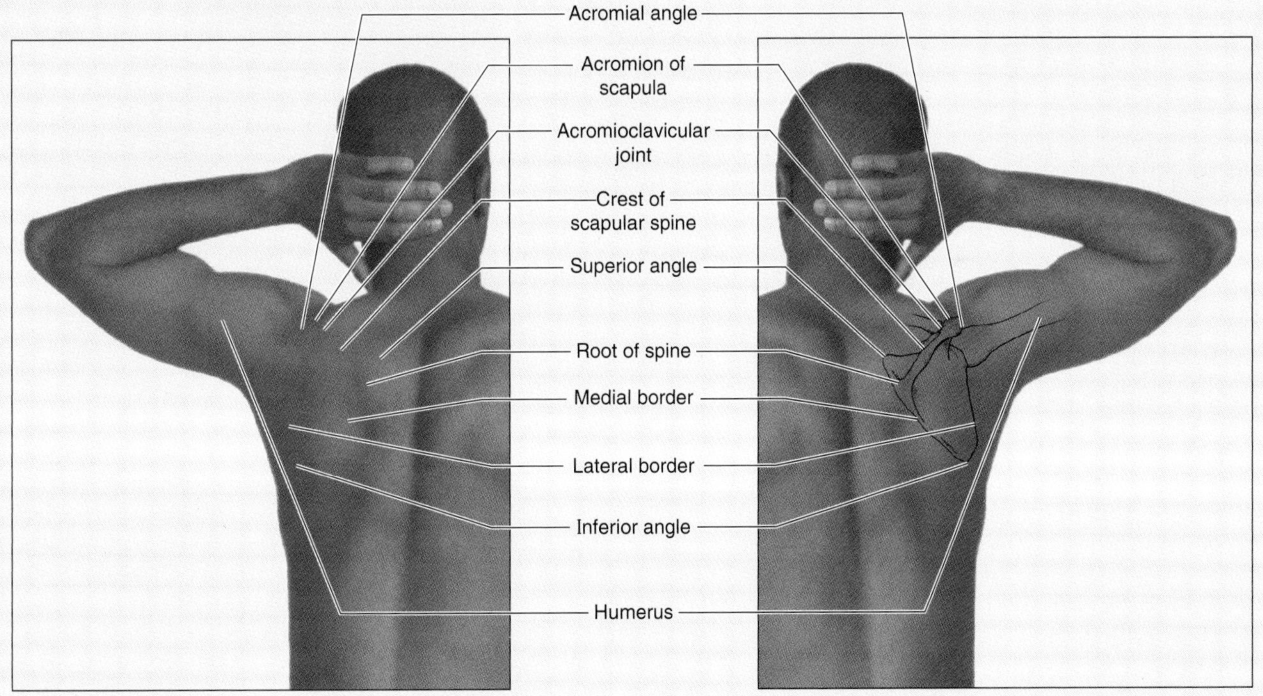

(B) Posterior view

▶ terior aspect of the arm, approximately 1.25 cm lateral and slightly inferior to the tip of the coracoid process. Rotation of the arm facilitates palpation of the lesser tubercle. *The location of the intertubercular groove*, between the greater and lesser tubercles, is identifiable during flexion and extension of the elbow joint by palpation of the tendon of the long head of the biceps brachii as it moves through the intertubercular groove.

The **body of the humerus** may be felt with varying distinctness through the muscles surrounding it. No part of the proximal part or the humeral body is subcutaneous. The medial and lateral epicondyles of the humerus are subcutaneous and easily palpated at the proximal parts of the medial and lateral aspects of the elbow region (p. 679). The knoblike medial epicondyle, projecting posteromedially, is more prominent than the **lateral epicondyle.** When the elbow joint is partially flexed, the lateral epicondyle is visible. When the elbow joint is fully extended, the lateral epicondyle can be palpated but not seen at the bottom of the depression on the posterolateral aspect of the elbow. ▶

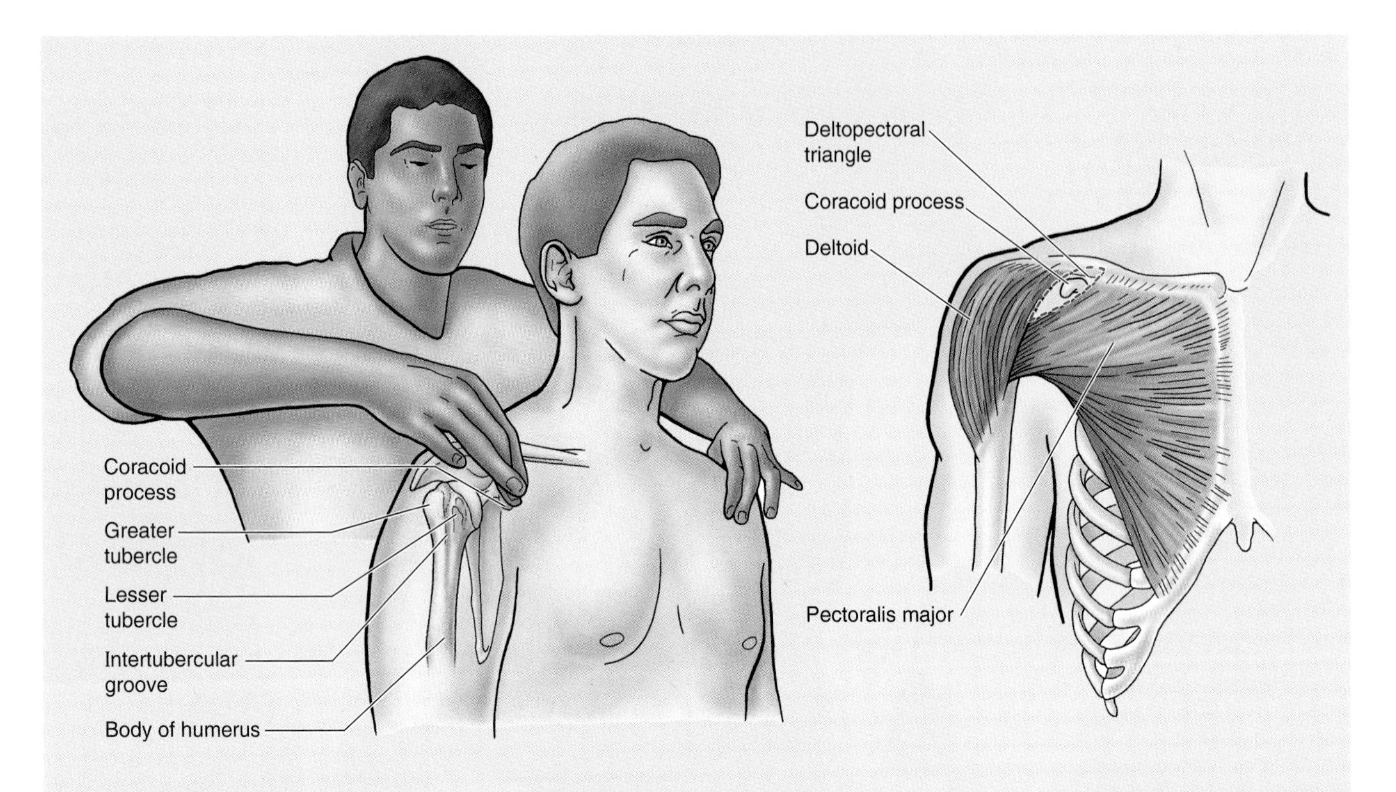

▶ The **olecranon** and posterior border of the ulna can be easily palpated. When the elbow joint is extended, observe that the tip of the olecranon and the humeral epicondyles lie in a straight line (*A–B* on p. 679), whereas when the elbow is flexed, the olecranon descends until its tip forms the apex of an approximately equilateral triangle, of which the epicondyles form the angles at its base (*C*). These normal relationships are important in the diagnosis of certain elbow injuries (e.g., dislocation of the elbow joint).

The **head of the radius** can be palpated and felt to rotate in the depression on the posterolateral aspect of the extended elbow joint, just distal to the lateral epicondyle of the humerus. The head can also be palpated as it rotates during pronation and supination of the forearm. The **ulnar nerve** feels like a thick cord where it passes posterior to the medial epicondyle of the humerus; pressing the nerve here evokes an unpleasant "crazy bone" sensation.

The **radial styloid process** can be easily palpated in the anatomical snuff box on the lateral side of the wrist; it is larger and approximately 1 cm more distal than the ulnar styloid process. The radial styloid process is easiest to palpate when the thumb is abducted. It is overlaid by the tendons of the thumb muscles. Because the radial styloid process extends more distally than the ulnar styloid process, more ulnar deviation than radial deviation of the wrist is possible. The relationship of the radial and ulnar processes is important in the diagnosis of certain wrist injuries (e.g., Colles' fracture). Proximal to the radial styloid process, the anterior, lateral, and posterior surfaces of the radius are palpable for several centimeters. The **dorsal radial tubercle** is easily felt around the middle of the dorsal aspect of the distal end of the radius. The dorsal tubercle acts as a pulley for the long extensor tendon of the thumb, which passes medial to it.

The **head of the ulna** forms a large, rounded subcutaneous prominence that can be easily seen and palpated on the medial side of the dorsal aspect of the wrist, especially when the hand is pronated. The subcutaneous **ulnar styloid process** may be felt slightly distal to the ulnar head when the hand is supinated.

The **pisiform** can be felt on the anterior aspect of the medial border of the wrist and can be moved from side to side when the hand is relaxed. The **hook of the hamate** can be palpated on deep pressure over the medial side of the palm, approximately 2 cm distal and lateral to the pisiform. The **tubercles of the scaphoid and trapezium** can be palpated at the proximal end of the thenar eminence (ball of thumb) when the hand is extended.

The **metacarpals,** although covered by the long extensor tendons of the digits, can be palpated on the dorsum of the hand. The heads of these bones form the knuckles of the fist; the 3rd metacarpal head is most prominent. The **styloid process** on the lateral surface of the base of the 3rd metacarpal can be palpated approximately 3.5 cm from the dorsal radial tubercle. The dorsal aspects of the phalanges can be easily palpated. The knuckles of the fingers are formed by the heads of the proximal and middle phalanges. ⊙

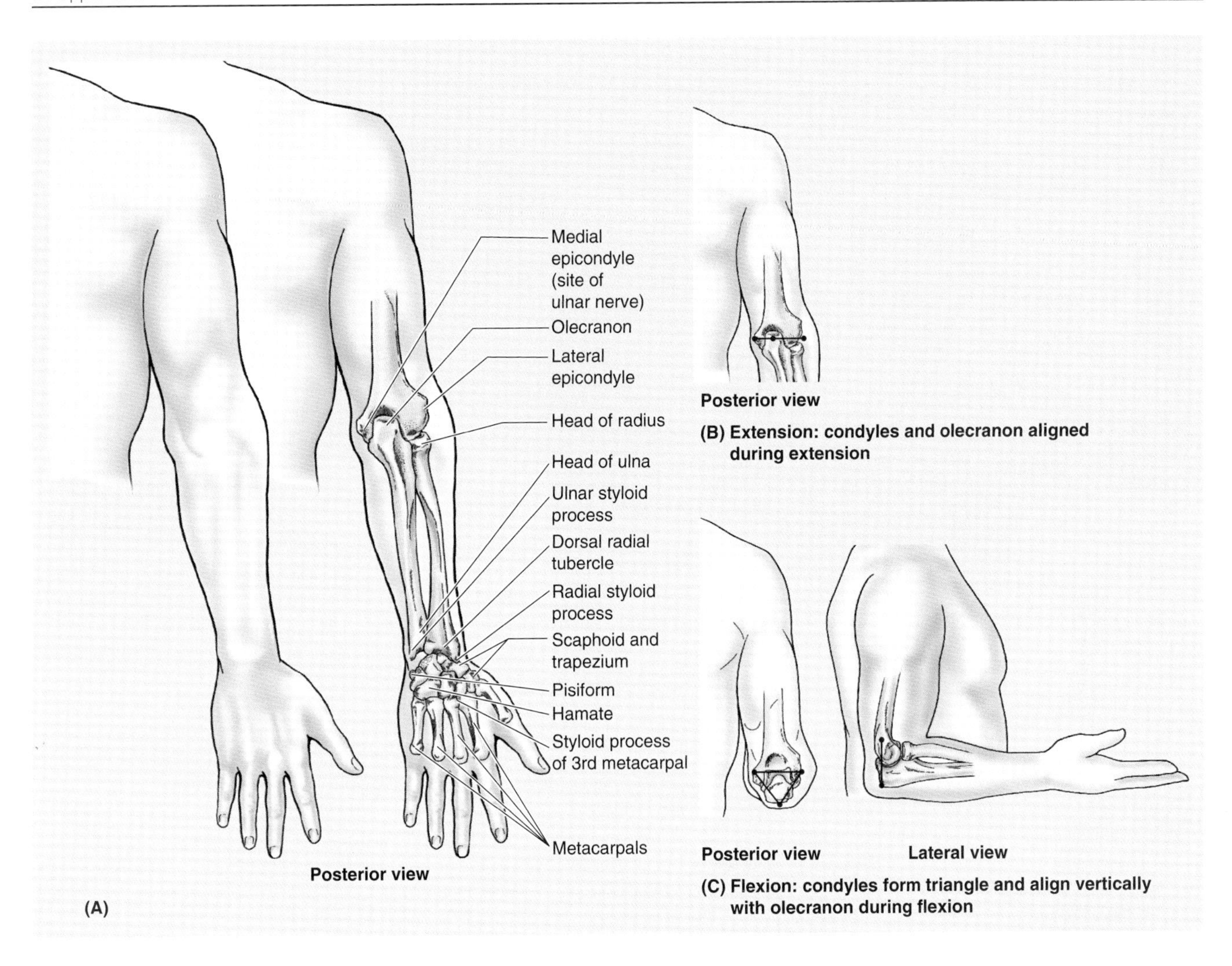

Superficial Structures of the Upper Limb

Deep to the skin is subcutaneous tissue (superficial fascia) containing fat and deep fascia surrounding the muscles. If nothing (no muscle, tendon, or bursa, for example) intervenes between the skin and bone, the deep fascia is usually attached to bone.

Fascia of the Upper Limb

The fascia of the pectoral region is attached to the clavicle and sternum. The **pectoral fascia** invests the pectoralis major and is continuous inferiorly with the fascia of the anterior abdominal wall. The pectoral fascia leaves the lateral border of the pectoralis major and becomes the **axillary fascia** (Fig. 6.9*A*), which forms the floor of the axilla (armpit). A fascial layer—the **clavipectoral fascia**—extends from the axillary fascia, encloses the pectoralis minor and subclavius muscles, and then attaches to the clavicle (Fig. 6.9*B*). The part of the clavipectoral fascia superior to the pectoralis minor—the **costocoracoid membrane**—is pierced by the lateral pectoral nerve that primarily supplies the pectoralis major. The part of the clavipectoral fascia inferior to the pectoralis minor—the **suspensory ligament of the axilla**—supports the axillary fascia and pulls it and the skin inferior to it upward during abduction of the arm, forming the axilla, or "armpit."

A sheath of deep fascia—the **brachial fascia**—encloses the arm (L. brachium) like a sleeve (Fig. 6.10); it is continuous superiorly with the pectoral and axillary layers of fascia. The brachial fascia is attached inferiorly to the epicondyles of the humerus and the olecranon of the ulna and is continuous with the **antebrachial fascia**, the deep fascia of the forearm. Two intermuscular septa—the **medial** and **lateral intermuscular septa**—extend from the deep surface of the brachial fascia to the medial and lateral supracondylar ridges of the humerus, dividing the arm into **anterior** (**flexor**) and **posterior** (**extensor**) **fascial compartments**, each of which contains muscles serving similar functions, nerves, and blood vessels that supply them. These fascial compartments are separated by an **interosseous membrane** connecting the radius and ulna.

The **antebrachial fascia** thickens posteriorly over the distal ends of the radius and ulna to form a transverse band, the **extensor retinaculum**, which retains the extensor tendons in position. The antebrachial fascia also forms an anterior thickening,

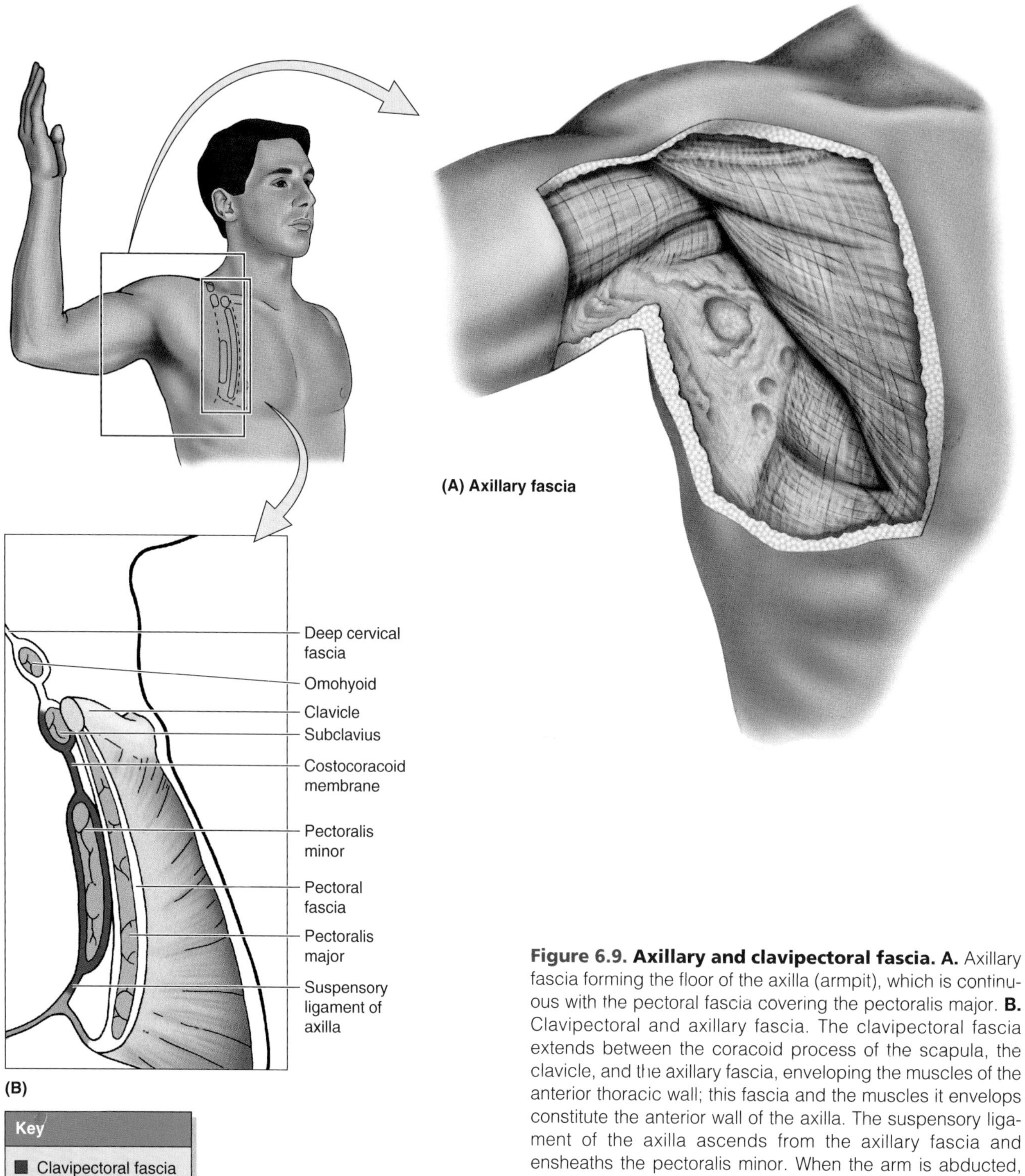

Figure 6.9. Axillary and clavipectoral fascia. A. Axillary fascia forming the floor of the axilla (armpit), which is continuous with the pectoral fascia covering the pectoralis major. **B.** Clavipectoral and axillary fascia. The clavipectoral fascia extends between the coracoid process of the scapula, the clavicle, and the axillary fascia, enveloping the muscles of the anterior thoracic wall; this fascia and the muscles it envelops constitute the anterior wall of the axilla. The suspensory ligament of the axilla ascends from the axillary fascia and ensheaths the pectoralis minor. When the arm is abducted, traction by the suspensory ligament produces the hollow of the axilla.

which is continuous with the extensor retinaculum but is officially unnamed; some authors identify it as the *palmar carpal ligament.* Immediately distal but at a deeper level to the latter, the antebrachial fascia is also continued as the **flexor retinaculum** (transverse carpal ligament). This fibrous band extends between the anterior prominences of the outer carpal bones and converts the anterior concavity of the carpus into a **carpal tunnel** through which the flexor tendons and median nerve pass.

The **deep fascia of the hand** is continuous through the extensor and flexor retinacula with the antebrachial fascia. The

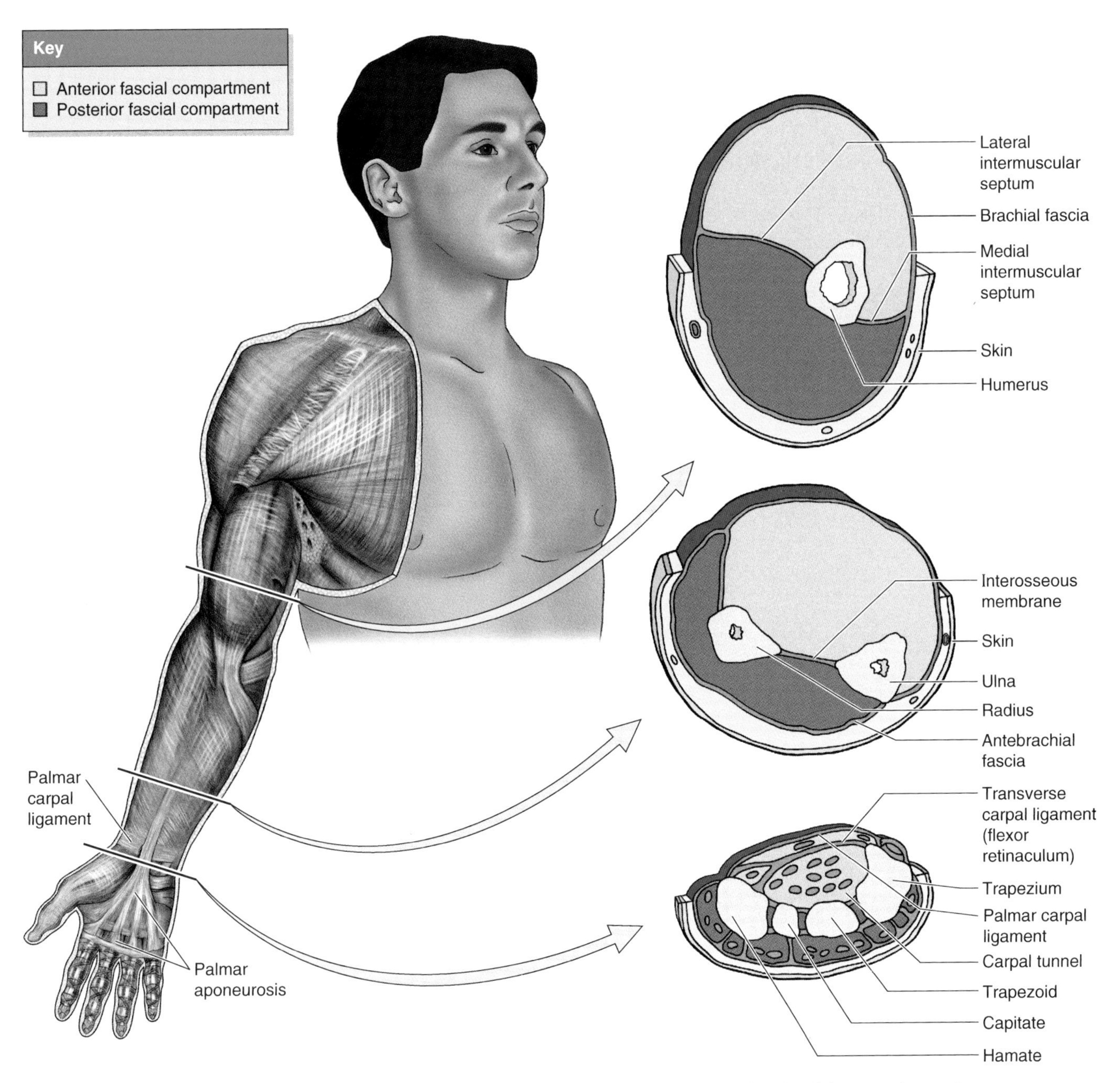

Figure 6.10. Fascia of the upper limb. The *brachial fascia*, the deep fascia of the arm, is continuous superiorly with the pectoral and axillary layers of fascia. Medial and lateral intermuscular septa extend from the deep aspect of the brachial fascia to the humerus, dividing the arm into anterior and posterior compartments, each of which contains muscles serving similar functions, and the nerves and vessels supplying them. The *antebrachial fascia*, surrounding the forearm muscles, is continuous with the brachial fascia and the deep fascia of the hand. The *interosseous membrane* and the bones it connects (radius and ulna) separate the forearm into anterior and posterior compartments. Over the distal ends of the radius and ulna, the deep fascia of the forearm thickens to form the *extensor retinaculum* posteriorly and a corresponding (but officially unnamed) thickening anteriorly—identified by some authors as the "palmar carpal ligament." Immediately distal and at a deeper level to the latter—but also continuous with the antebrachial fascia—a ligamentous formation—the *flexor retinaculum*—extends between anterior prominences of the outer carpal bones, converting the anterior concavity of the carpus into an osseofibrous *carpel tunnel*. This tunnel provides passage for the median nerve and tendons of the flexor muscles passing from the forearm to the hand. The deep fascia of the hand is continuous through the retinacula with the antebrachial fascia. The central part of the *palmar fascia* thickens to form the *palmar aponeurosis*.

central part of the palmar fascia—the **palmar aponeurosis**—is thick, tendinous, and triangular; it overlies the central compartment of the palm with its apex proximally, which is continuous with the tendon of the palmaris longus (when this muscle is present). The aponeurosis forms four distinct thickenings that radiate to the bases of the fingers and become continuous with the fibrous tendon sheaths of the digits. The bands are traversed distally by the superficial *transverse metacarpal ligament* that forms the base of the triangular palmar aponeurosis. Innumerable minute, strong skin ligaments (L. retinacula cutis) extend from the palmar aponeurosis to the skin. These ligaments hold the skin close to the aponeurosis, allowing little sliding movement of the skin.

Cutaneous Nerves of the Upper Limb

The cutaneous nerves of the upper limb follow a general pattern that is easy to understand if it is noted that developmen-

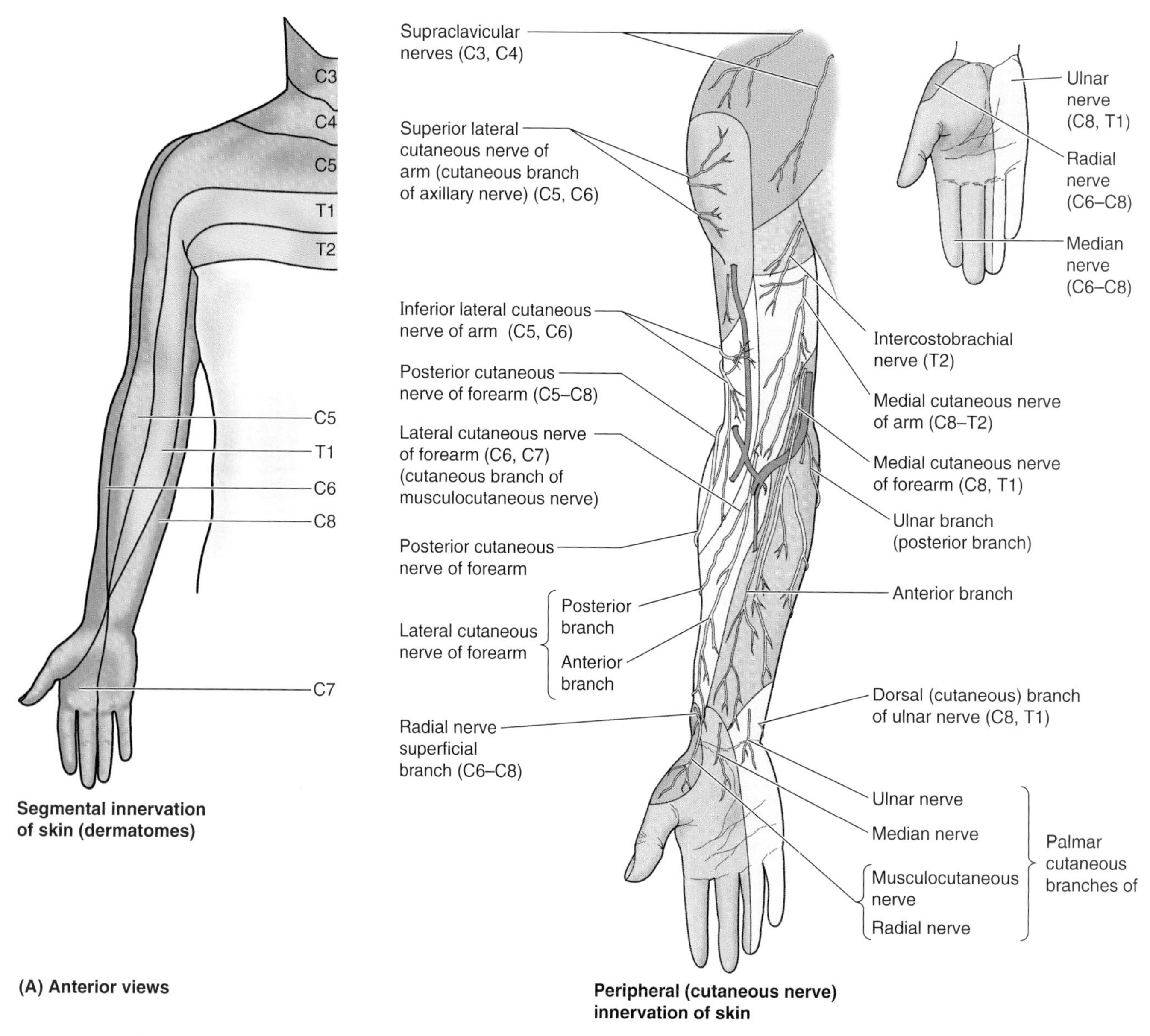

Figure 6.11. Segmental (dermatomal) and peripheral (cutaneous nerve) innervation of the upper limb. Anterior (**A**) and posterior (**B**) views showing the distribution of the peripheral (named cutaneous) nerves, which usually are branches of nerve plexuses and therefore contain fibers from more than one spinal nerve or spinal cord segment. The segmental (dermatomal) pattern of innervation of the limbs is also shown. A *dermatome* is the area of skin receiving sensory innervation from the dorsal root of a single spinal nerve or spinal cord segment through the dorsal and ventral rami (root) of that spinal nerve. Neighboring dermatomes usually overlap extensively so that loss of innervation to the skin consequent to damage to a single dorsal root may involve only a small part of the indicated dermatome.

tally the limbs grow as lateral protrusions of the trunk, with the 1st digit (thumb or great toe) located on the cranial side (thumb is directed superiorly). Thus, the lateral surface of the upper limb is more cranial than the medial surface (Moore and Persaud, 1998).

Observe the progression of the segmental innervation of the various cutaneous areas around the limb (Fig. 6.11*A* and *B*):

- C3 and C4 nerves supply the region at the base of the neck extending laterally over the shoulder
- C5 nerve supplies the arm laterally (i.e., on the superior aspect of the outstretched limb)
- C6 nerve supplies the forearm laterally and the thumb
- C7 nerve supplies the middle and ring fingers and the middle of the posterior surface of the limb
- C8 nerve supplies the little finger, the medial side of the hand, and the forearm (i.e., the inferior aspect of the outstretched limb)
- T1 nerve supplies the middle of the forearm to the axilla
- T2 nerve supplies a small part of the arm and the skin of the axilla.

Most cutaneous nerves of the upper limb are derived from the **brachial plexus**—a major nerve network formed by the ventral rami of the 5th cervical to the 1st thoracic spinal nerves (see Table 6.4). The nerves to the shoulder, however, are derived from the **cervical plexus**—a nerve network consisting of a series of nerve loops formed between adjacent ventral primary rami of the first four cervical nerves, which also receives gray communicating rami from the superior cervical ganglion. The cervical plexus lies deep to the sternocleido-

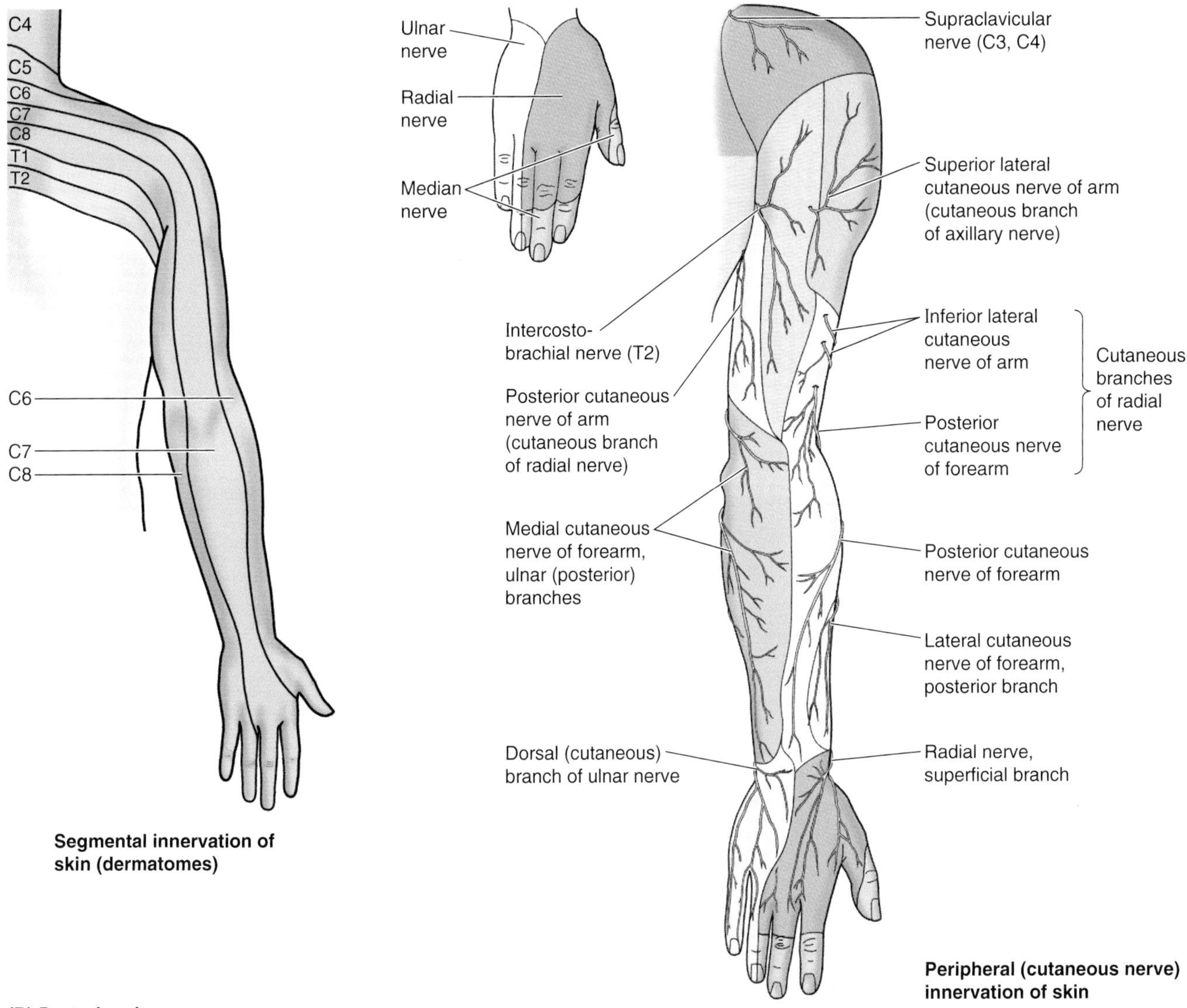

Figure 6.11. *(Continued)*

mastoid on the anterolateral aspect of the trunk. *The cutaneous nerves of the arm and forearm are as follows:*

- The *supraclavicular nerves* (C3, C4) pass anterior to the clavicle, immediately deep to the platysma, and supply the skin over the clavicle and the superolateral aspect of the pectoralis major.
- The *posterior cutaneous nerve of the arm,* a branch of the radial nerve, supplies the skin on the posterior surface of the arm.
- The *posterior cutaneous nerve of the forearm,* also a branch of the radial nerve, supplies the skin on the posterior surface of the forearm.
- The *superior lateral cutaneous nerve of the arm,* the terminal branch of the axillary nerve, emerges from beneath the posterior margin of the deltoid to supply skin over the lower part of this muscle and on the lateral side of the midarm for a short distance inferior to its distal attachment to the lateral side of the arm a little above its middle.
- The *inferior lateral cutaneous nerve of the arm,* a branch of the radial nerve, supplies the skin over the inferolateral aspect of the arm; it is frequently a branch of the posterior cutaneous nerve of the forearm.
- The *lateral cutaneous nerve of the forearm,* the terminal cutaneous branch of the musculocutaneous nerve, supplies the skin on the lateral side of the forearm.
- The *medial cutaneous nerve of the arm* arises from the medial cord of the brachial plexus, unites in the axilla with the lateral cutaneous branch of the 2nd intercostal nerve, and supplies the skin on the medial side of the arm.
- The *intercostobrachial nerve,* a lateral cutaneous branch of the 2nd intercostal nerve from T2, also contributes to the innervation of the skin on the medial surface of the arm.
- The *medial cutaneous nerve of the forearm* arises from the medial cord of the brachial plexus and supplies the skin of the anterior and medial surfaces of the forearm.

Note that like the brachial plexus, which has posterior, lateral, and medial—but no anterior—cords, the upper limb has posterior, lateral, and medial—but no anterior—cutaneous nerves.

Superficial Veins of the Upper Limb

The main superficial veins of the upper limb—the **cephalic** and **basilic veins**—originate in the subcutaneous tissue on the dorsum of the hand from the **dorsal venous network** (Fig. 6.12). The **perforating veins** form communications between the superficial veins and the deep veins.

The **cephalic vein** (G. kephalé, head) ascends from the lateral aspect of the dorsal venous network, proceeding along the lateral border of the wrist and the anterolateral surface of the forearm and arm. Anterior to the elbow it communicates with the **median cubital vein,** which passes obliquely across the anterior aspect of the elbow and joins the basilic vein. Superiorly, the cephalic vein courses along the *deltopectoral groove* (between the deltoid and pectoralis major) and enters the **deltopectoral triangle** (Fig. 6.12*B*), where it pierces the clavipectoral fascia (costocoracoid membrane) and joins the axillary vein.

The **basilic vein** ascends from the medial end of the dorsal venous network along the medial side of the forearm and the inferior part of the arm. It then passes deeply, piercing the brachial (deep) fascia, and runs superiorly parallel to the brachial artery to the axilla, where it merges with the accompanying veins (L. venae comitantes) of the axillary artery to form the **axillary vein.**

The highly variable and commonly absent **median antebrachial vein** begins at the base of the dorsum of the thumb, curves around the lateral side of the wrist, ascends in the middle of the anterior aspect of the forearm between the cephalic and basilic veins, and may join the basilic vein in the cubital fossa. Sometimes the median antebrachial vein divides into intermediate (median) cephalic and basilic veins, which drain into the cephalic and basilic veins, respectively, and may replace the median cubital vein when located on the anterior aspect of the elbow (p. 756).

Venipuncture of the Upper Limb

Because of the prominence and accessibility of the superficial veins of the upper limb, they are commonly used for venipuncture (puncture of a vein to draw blood or inject a solution). These veins may be embedded with the subcutaneous tissue (fat), making them difficult to see; however, by applying a tourniquet to the arm, the venous return is occluded and the veins distend and are usually visible and/or palpable. The *median cubital vein* is commonly used for venipuncture for drawing blood and inserting a catheter for *right cardiac catheterization* (see Chapter 1). Considerable variation occurs in the connection of the basilic and cephalic veins in the *cubital fossa*—the depression on the anterior surface of the elbow. If the median cubital vein is very large, most blood from the cephalic vein of the forearm enters the basilic vein (of the arm). In these cases, the superior cephalic vein (of the arm) may be diminished or absent. The veins forming the *dorsal venous network* and the cephalic and basilic veins arising from it are commonly used for long-term introduction of fluids (*intravenous feeding*). ○

Lymphatic Drainage of the Upper Limb

Superficial lymphatic vessels arise from **lymphatic plexuses** in the digits, palm, and dorsum of the hand and ascend mostly with superficial veins such as the cephalic and basilic veins (Fig. 6.13). Some vessels accompanying the basilic vein enter

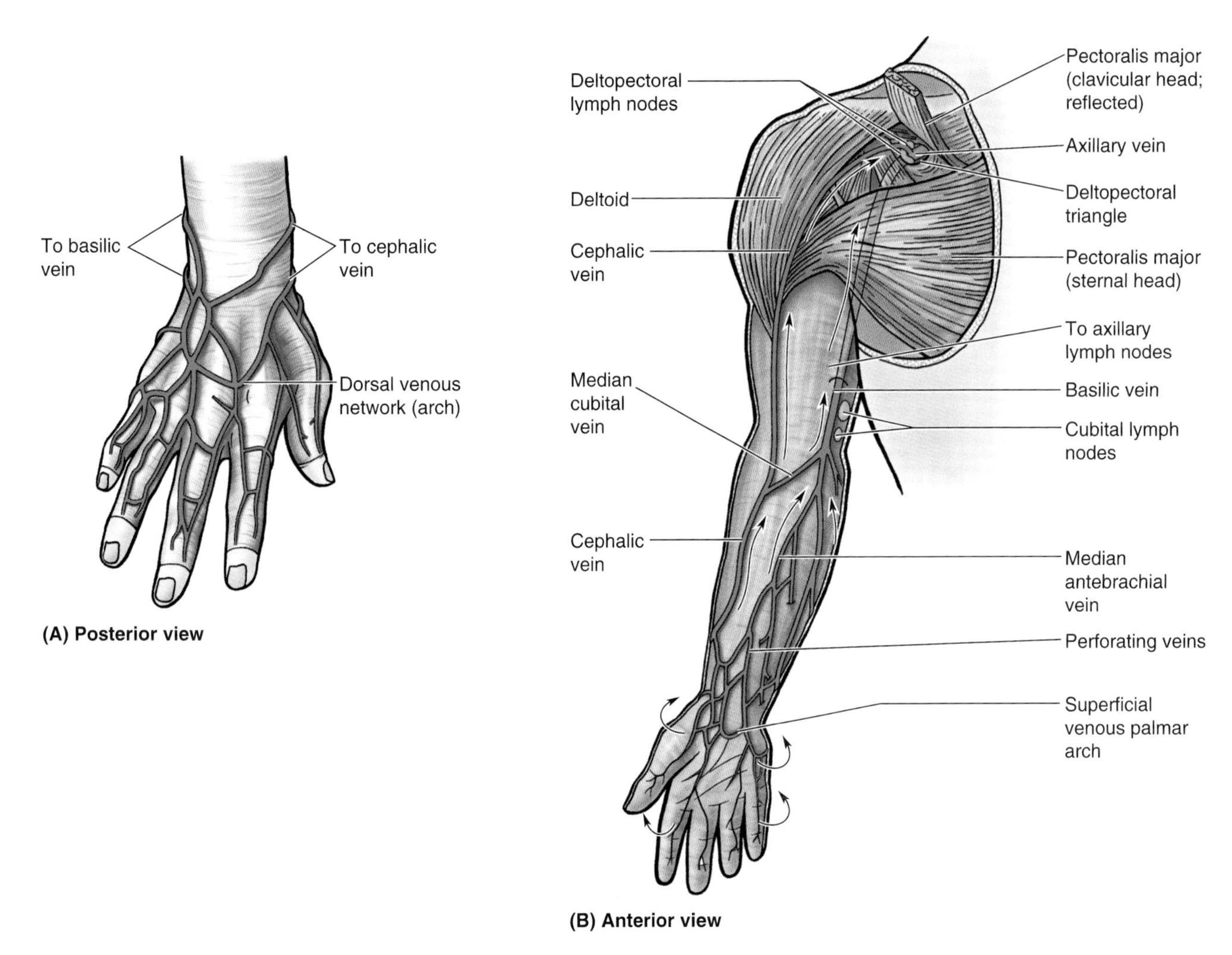

Figure 6.12. Veins and lymph nodes of the upper limb. A. Posterior view (dorsum) of the hand. The digital veins drain into the dorsal venous network (arch), which leads to two prominent superficial vessels, the *cephalic* and *basilic veins.* **B.** Anterior view of the right upper limb illustrating the superficial venous drainage. The main veins, the cephalic and basilic, ultimately drain into the axillary vein. The cephalic vein ascends to the anterior aspect of the shoulder and passes deep in the *deltopectoral triangle* to join the axillary vein. The basilic vein penetrates the deep fascia on the medial side of the middle part of the arm and then joins with the brachial veins to form the *axillary vein.* The median cubital vein is the communication between the basilic and cephalic veins in the *cubital fossa,* the hollow area on the anterior surface of the elbow. Perforating veins connect the superficial veins to the deep veins. *Arrows* indicate the flow of lymph to the cubital and axillary lymph nodes.

the **cubital nodes**, located proximal to the medial epicondyle and medial to the basilic vein. Efferent vessels from these lymph nodes ascend in the arm and terminate in the humeral (lateral) **axillary lymph nodes.** Most lymphatic vessels accompanying the cephalic vein cross the proximal part of the arm and anterior aspect of the shoulder to enter the apical group of axillary nodes; however, some vessels previously enter the **deltopectoral nodes.** *Deep lymphatic vessels,* less numerous than superficial vessels, accompany the major deep veins in the upper limb and also terminate in the humeral group of axillary nodes.

Anterior Thoracoappendicular Muscles of the Upper Limb

Four anterior thoracoappendicular (pectoral) muscles move the pectoral girdle: pectoralis major, pectoralis minor, subclavius, and serratus anterior. The attachments, nerve supply, and main actions of these muscles are given in Figure 6.14 and Table 6.1.

The **pectoralis major**, large and fan shaped, covers the superior part of the thorax. It has *clavicular* and *sternocostal*

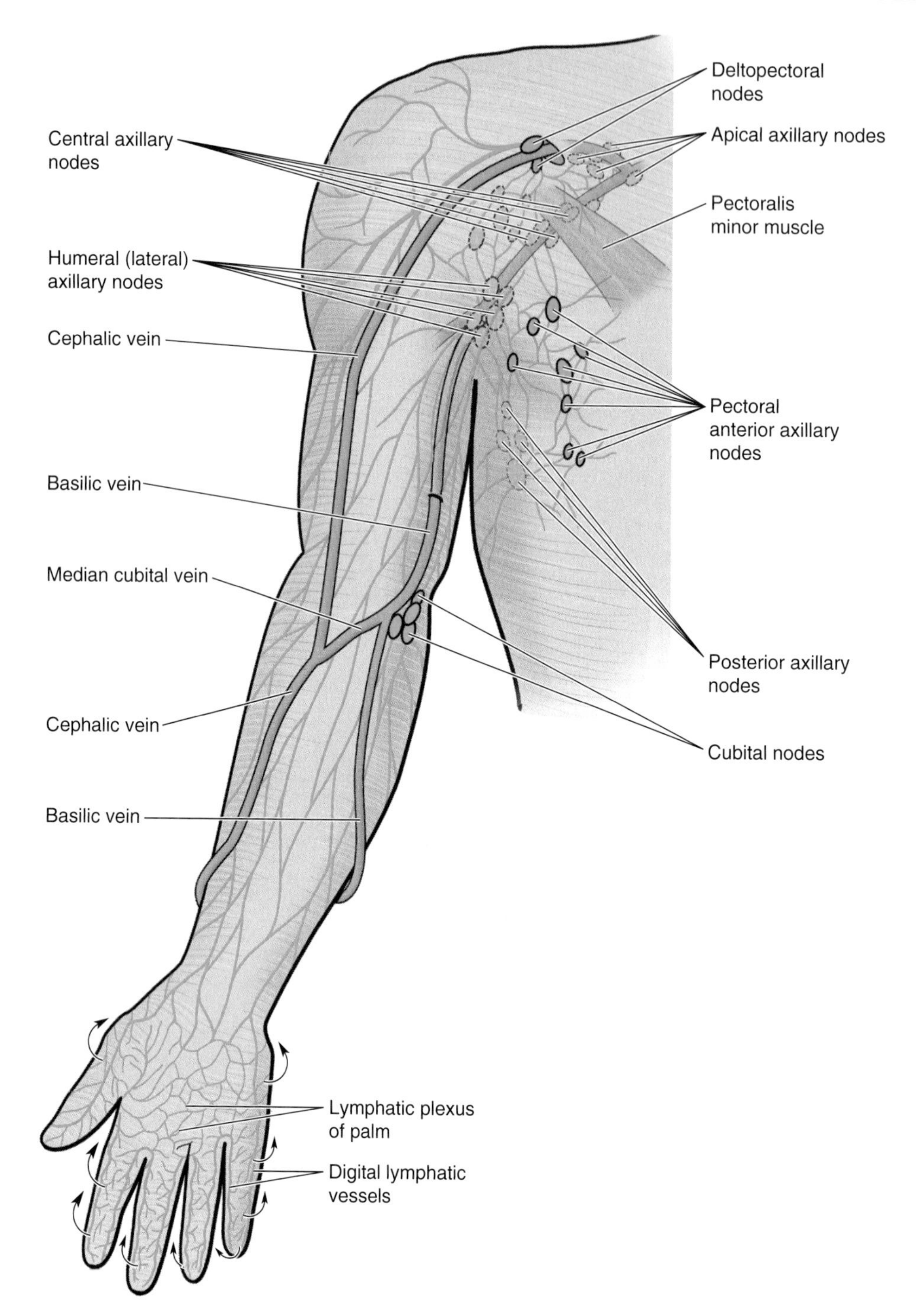

Figure 6.13. Lymphatic drainage of the upper limb. Superficial lymphatic vessels begin in the skin of the hand from digital lymphatic vessels and the lymphatic plexus of the hand. Most drainage from the palm passes to the dorsum of the hand (*arrows*). The vessels ascend through the forearm and arm, converging toward the cephalic vein and especially the basilic vein to reach the *axillary lymph nodes*. Some lymph first passes through the *cubital lymph nodes* in the elbow region or *deltopectoral nodes* in the shoulder region. Deep lymphatic vessels (less numerous than superficial vessels) follow the main neurovascular bundles and drain lymph from the joint capsules, periosteum, tendons, nerves, and muscles. A few lymph nodes are present along their course, which end primarily in the humeral (lateral) and then the central axillary lymph nodes.

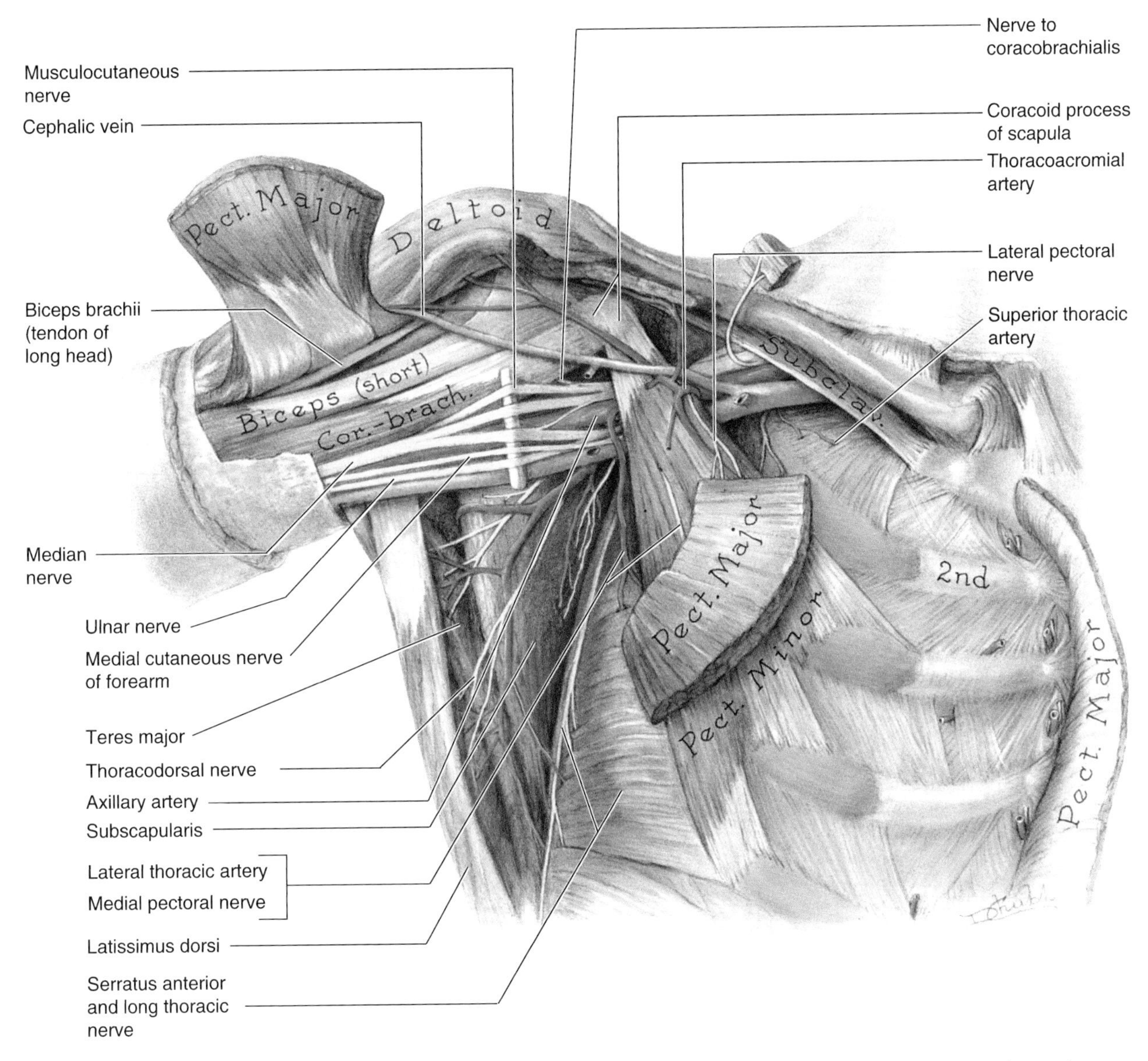

Figure 6.14. Structures of the axilla. Most of the anterior wall of the axilla and the axillary fat pad have been removed, revealing the axilla's medial and posterior walls and neurovascular contents. Of the structures forming the anterior wall, only portions of the pectoralis major (attaching ends, a central part overlying the pectoralis minor, and a cube of muscle reflected superior to the clavicle) and the pectoralis minor remain. All the clavipectoral fascia has been removed, as has the axillary sheath surrounding the neurovascular bundle. Observe the *axillary artery* emerging from the cervicoaxillary canal inferior to the clavicle and subclavius muscle and then passing a finger's breadth inferior to the coracoid process of the scapula. As the axillary artery passes through the axilla, it is surrounded by the *brachial plexus of nerves.* The major nerves arising from the lateral and medial cords (anterior divisions) of the plexus have been elevated by an applicator stick.

heads. The latter head is much larger, and its lateral border is responsible for the muscular mass that forms most of the anterior wall of the axilla, with its inferior border forming the anterior axillary fold. The pectoralis major and adjacent deltoid form the narrow *deltopectoral groove*, in which the cephalic vein runs; however, they diverge slightly from each other superiorly and, along with the clavicle, form the **deltopectoral triangle** (Fig. 6.12). *The pectoralis major is a powerful adductor of the arm and a medial rotator of the humerus.* The two parts of the pectoralis major can act independently: the clavicular head flexing the humerus and the sternocostal head extending it (Table 6.1). *To test the clavicular head of pectoralis major*, the arm is abducted 90° and then the person moves the arm anteriorly against resistance. If acting normally, the clavicular head can be seen and palpated. *To test the sternocostal head of the pectoralis major*, the arm is raised 60°

Table 6.1. **Anterior Thoracoappendicular Muscles**

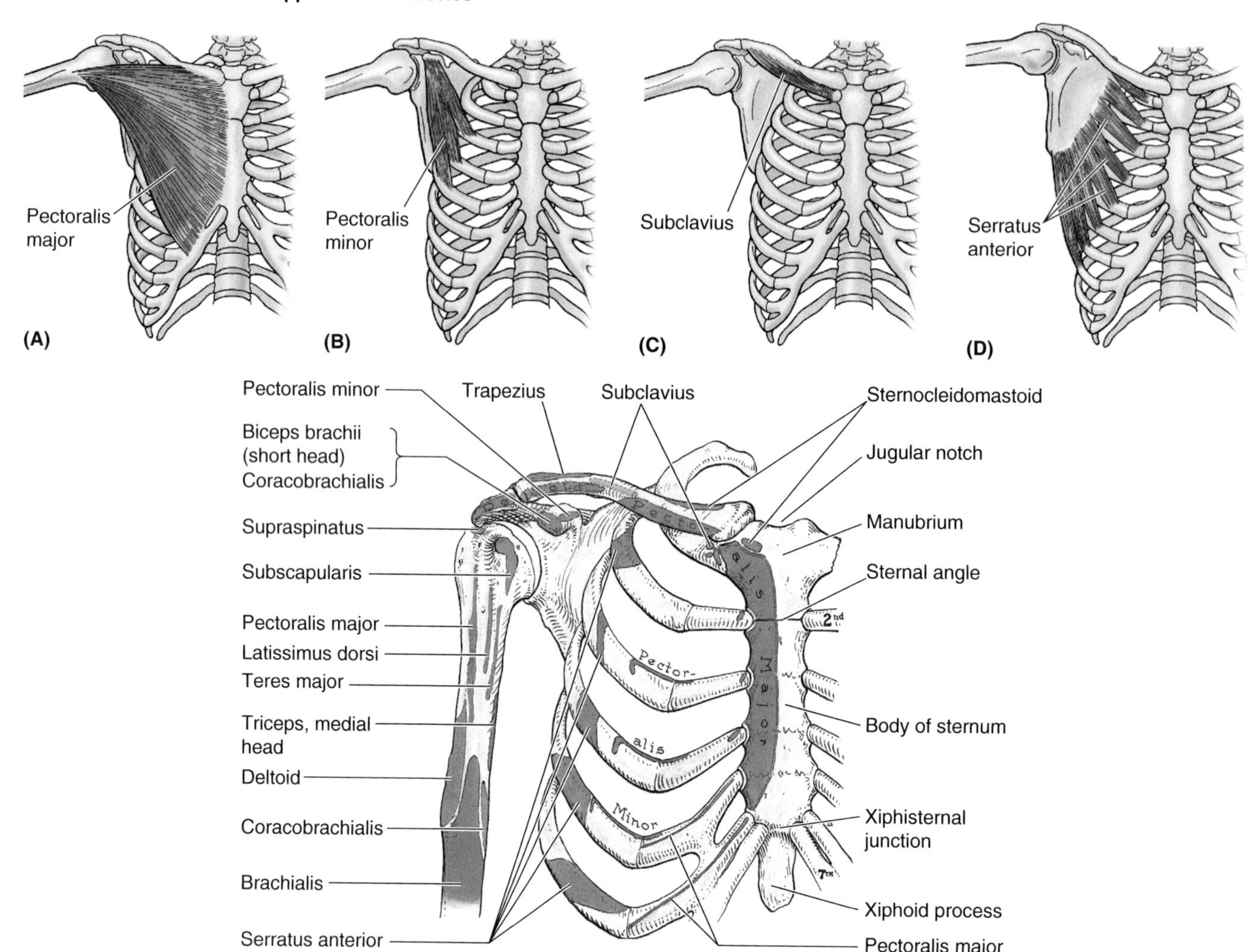

Muscle	Proximal Attachment	Distal Attachment	Innervation[a]	Main Action
Pectoralis major	Clavicular head: anterior surface of medial half of clavicle Sternocostal head: anterior surface of sternum, superior six costal cartilages, and aponeurosis of external oblique muscle	Lateral lip of intertubercular groove of humerus	Lateral and medial pectoral nerves; clavicular head (C5 and **C6**), sternocostal head (**C7**, **C8**, and T1)	Adducts and medially rotates humerus; draws scapula anteriorly and inferiorly Acting alone: clavicular head flexes humerus and sternocostal head extends it from the flexed position.
Pectoralis minor	3rd to 5th ribs near their costal cartilages	Medial border and superior surface of coracoid process of scapula	Medial pectoral nerve (C8 and T1)	Stabilizes scapula by drawing it inferiorly and anteriorly against thoracic wall
Subclavius	Junction of 1st rib and its costal cartilage	Inferior surface of middle third of clavicle	Nerve to subclavius (**C5** and C6)	Anchors and depresses clavicle
Serratus anterior	External surfaces of lateral parts of 1st to 8th ribs	Anterior surface of medial border of scapula	Long thoracic nerve (C5, **C6**, and **C7**)	Protracts scapula and holds it against thoracic wall; rotates scapula

[a] Numbers indicate spinal cord segmental innervation (e.g., C5 and C6 indicate that nerves supplying the clavicular head of pectoralis major are derived from 5th and 6th cervical segments of spinal cord). **Boldface** numbers indicate the main segmental innervation. Damage to these segments, or to motor nerve roots arising from them, results in paralysis of muscles concerned.

and then adducted against resistance. If acting normally, the sternocostal head can be seen and palpated.

The **pectoralis minor** lies in the anterior wall of the axilla, where it is largely covered by the much larger pectoralis major. The pectoralis minor is triangular in shape: its base (proximal attachment) is formed by fleshy slips attached to the anterior ends of the 3rd through 5th ribs near their costal cartilages; its apex (distal attachment) is on the coracoid process of the scapula. Variations in the costal attachments of the muscle are common. *The pectoralis minor stabilizes the scapula and is used when stretching the arm forward to touch an object that is just out of reach.* The pectoralis minor also assists in elevating the ribs for deep inspiration when the pectoral girdle is fixed or elevated. *The pectoralis minor is a useful anatomical and surgical landmark for structures in the axilla* (e.g., the axillary artery). With the coracoid process, the pectoralis minor forms a "bridge" under which vessels and nerves must pass to the arm (Fig. 6.14).

The **subclavius** lies almost horizontally when the arm is in the anatomical position. This small, round muscle is located inferior to the clavicle and affords some protection to the subclavian artery when the clavicle fractures. It may also prevent the jagged ends of a fractured clavicle from injuring the adjacent subclavian vessels and the superior trunk of the brachial plexus. The subclavius anchors and depresses the clavicle, stabilizing it during movements of the upper limb. It also helps resist the tendency for the clavicle to dislocate at the SC joint, for example, when pulling hard during a tug-of-war game.

The **serratus anterior** overlies the lateral part of the thorax and forms the medial wall of the axilla. This broad sheet of thick muscle was named because of the saw-toothed appearance of its fleshy slips or digitations (L. serratus, a saw). The muscular slips pass posteriorly and then medially to converge on the whole length of the anterior surface of the medial border of the scapula, including its inferior angle. *The serratus anterior—one of the most powerful muscles of the pectoral girdle—is a strong protractor of the scapula that is used when punching or reaching anteriorly* (some people call it the "boxer's muscle"). Its strong inferior part rotates the scapula, elevating its glenoid cavity so the arm can be raised above the shoulder. By keeping the scapula closely applied to the thoracic wall, it anchors this flat bone enabling other muscles to use it as a fixed bone for movements of the humerus. The serratus anterior holds the scapula against the thoracic wall when doing push-ups or when pushing against resistance (pushing a car, for example). Thus, *to test the serratus anterior (or the function of the long thoracic nerve that supplies it)*, the hand of the outstretched limb is pushed against a wall. If the muscle is acting normally, several digitations of the muscle can be seen and palpated.

Absence of the Pectoral Muscles

Absence of part of the pectoralis major, usually its sternocostal part, is uncommon, but when it occurs, no disability usually results. However, the *anterior axillary fold*—formed by the skin and fascia overlying the inferior border of the pectoralis major—is absent on the affected side, and the nipple is more inferior than usual. In the *Poland syndrome*, both the pectoralis major and minor are absent; breast hypoplasia and absence of two to four rib segments are also seen.

Paralysis of the Serratus Anterior

When the serratus anterior is paralyzed because of *injury to the long thoracic nerve* (Fig. 6.14), the medial border of the scapula moves laterally and posteriorly away from the thoracic wall, giving the scapula the appearance of a wing, especially when the person leans on the hand or presses the upper limbs against a wall. When the arm is raised, the medial border and inferior angle of the scapula pull markedly away from the posterior thoracic wall—consequently the term *winged scapula*. In addition, the arm cannot be abducted above the horizontal position because the serratus anterior is unable to rotate the glenoid cavity superiorly (face upward) to allow complete abduction of the arm. Although protected when the limbs are at one's sides, the long thoracic nerve is exceptional in that it courses on the superficial aspect of the serratus anterior muscle, which it supplies. Thus, when the limbs are elevated—as in a knife fight—the nerve is especially vulnerable. Weapons, including missiles (bullets) directed toward the thorax, are a common source of injury. ○

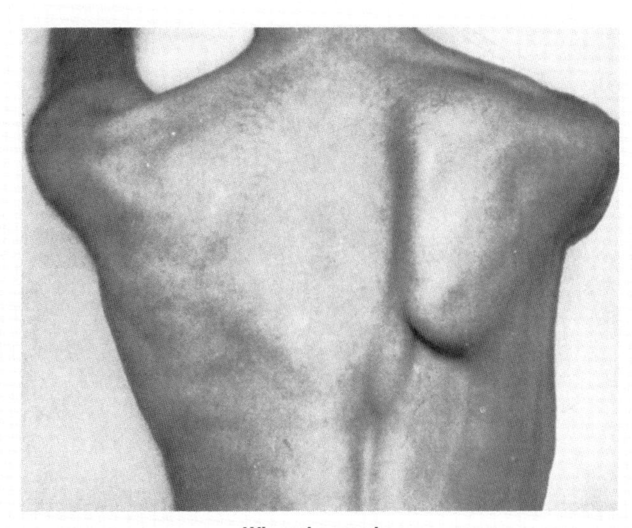

Winged scapula

Table 6.2. Scapulohumeral and Posterior Thoracoappendicular Muscles

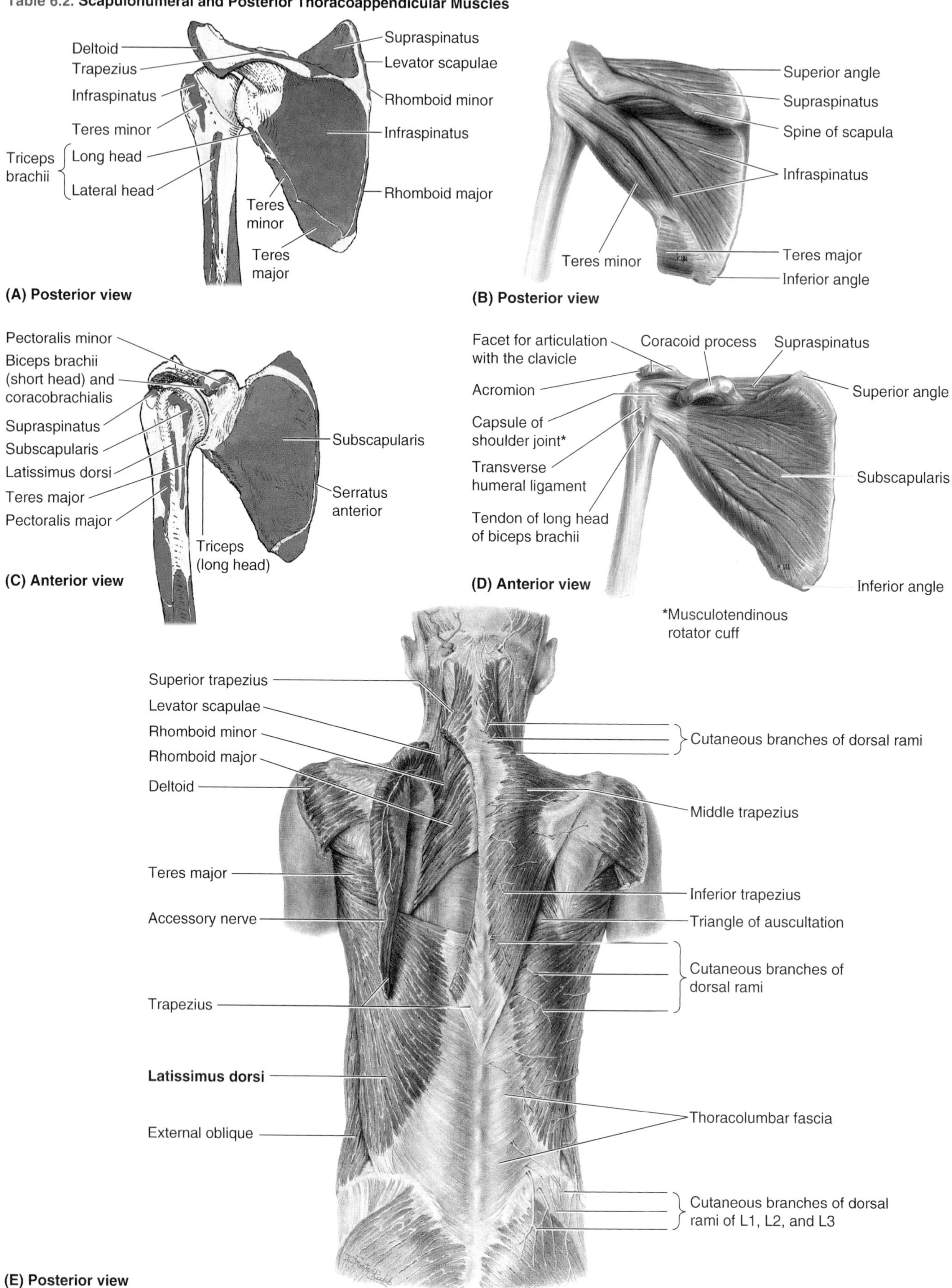

(A) Posterior view

(B) Posterior view

(C) Anterior view

(D) Anterior view

(E) Posterior view

Table 6.2. (*Continued*) **Scapulohumeral and Posterior Thoracoappendicular Muscles**

Muscle	Proximal Attachment	Distal Attachment	Innervation[a]	Main Action
Trapezius	Medial third of superior nuchal line; external occipital protuberance, nuchal ligament, and spinous processes of C7–T12 vertebrae	Lateral third of clavicle, acromion, and spine of scapula	Spinal root of accessory nerve (CN XI) (motor) and cervical nerves (C3 and C4) (pain and proprioception)	Elevates, retracts, and rotates scapula; superior fibers elevate, middle fibers retract, and inferior fibers depress scapula; superior and inferior fibers act together in superior rotation of scapula
Latissimus dorsi	Spinous processes of inferior 6 thoracic vertebrae, thoracolumbar fascia, iliac crest, and inferior 3 or 4 ribs	Floor of intertubercular groove of humerus	Thoracodorsal nerve (**C6**, **C7**, and C8)	Extends, adducts, and medially rotates humerus; raises body toward arms during climbing
Levator scapulae	Posterior tubercles of transverse processes of C1–C4 vertebrae	Superior part of medial border or scapula	Dorsal scapular (C5) and cervical (C3 and C4) nerves	Elevates scapula and tilts its glenoid cavity inferiorly by rotating scapula
Rhomboid minor and major	*Minor:* nuchal ligament and spinous processes of C7 and T1 vertebrae *Major:* spinous processes of T2–T5 vertebrae	Medial border of scapula from level of spine to inferior angle	Dorsal scapular nerve (C4 and **C5**) rotate	Retract scapula and rotate it to depress glenoid cavity; fix scapula to thoracic wall
Deltoid	Lateral third of clavicle, acromion, and spine of scapula	Deltoid tuberosity of humerus	Axillary nerve (**C5** and C6)	Anterior part: flexes and medially rotates arm Middle part: abducts arm Posterior part: extends and laterally rotates arm
Supraspinatus[b]	Supraspinous fossa of scapula	Superior facet on greater tubercle of humerus	Suprascapular nerve (C4, **C5**, and C6)	Initiates and assists deltoid in abduction of arm and acts with rotator cuff muscles[b]
Infraspinatus[b]	Infraspinous fossa of scapula	Middle facet on greater tubercle of humerus	Suprascapular nerve (**C5** and C6)	Laterally rotate arm; help to hold humeral head in glenoid cavity of scapula
Teres minor[b]	Superior part of lateral border of scapula	Inferior facet on greater tubercle of humerus	Axillary nerve (**C5** and C6)	
Teres major	Dorsal surface of inferior angle of scapula	Medial lip of intertubercular groove of humerus	Lower subscapular nerve (**C6** and C7)	Adducts and medially rotates arm
Subscapularis[b]	Subscapular fossa	Lesser tubercle of humerus	Upper and lower subscapular nerves (C5, **C6**, and C7)	Medially rotates arm and adducts it; helps to hold humeral head in glenoid cavity

[a] Numbers indicate spinal cord segmental innervation (e.g., C5 and C6 indicate that nerves supplying the teres minor muscle are derived from 5th and 6th cervical segments of spinal cord). **Boldface** numbers indicate main segmental innervation. Damage to these segments, or to motor nerve roots arising from them, results in paralysis of muscles concerned.
[b] Collectively, the supraspinatus, infraspinatus, teres minor, and subscapularis muscles are referred to as the rotator cuff muscles. Their prime function during all movements of shoulder joint is to hold head of humerus in glenoid cavity of scapula.

Posterior Thoracoappendicular and Scapulohumeral Muscles

The posterior thoracoappendicular muscles (superficial and intermediate groups of *extrinsic back muscles)* attach the superior appendicular skeleton (of the upper limb) to the axial skeleton (in the trunk). *The intrinsic back muscles,* which maintain posture and control movements of the vertebral column, are described in Chapter 4. The shoulder muscles are divided into three groups (Table 6.2):

- *Superficial posterior thoracoappendicular (extrinsic shoulder) muscles*: trapezius and latissimus dorsi

- *Deep posterior thoracoappendicular (extrinsic shoulder) muscles*: levator scapulae and rhomboids
- *Scapulohumeral (intrinsic shoulder) muscles*: deltoid, teres major, and four rotator cuff muscles.

Superficial Posterior Thoracoappendicular (Extrinsic Shoulder) Muscles

The superficial thoracoappendicular muscles are the trapezius and latissimus dorsi. The attachments, nerve supply, and main actions of these muscles are given in Table 6.2.

Trapezius

The trapezius provides a direct attachment of the pectoral girdle to the trunk. This large, triangular muscle covers the posterior aspect of the neck and the superior half of the trunk (Fig. 6.15). It was given its name because the muscles of the two sides form a *trapezium* (G. irregular, four-sided figure). The trapezius attaches the pectoral girdle to the skull and vertebral column and assists in suspending the upper limb. *The fibers of the trapezius are divided into three parts that have different actions* at the conceptual scapulothoracic joint between the scapula and the thoracic wall:

- Superior fibers—elevate the scapula (e.g., when squaring the shoulders)
- Middle fibers—retract the scapula (i.e., pull it posteriorly)
- Inferior fibers—depress the scapula and lower the shoulder.

The trapezius also braces the shoulders by pulling the scapulae posteriorly and superiorly, fixing them in position on the thoracic wall with tonic contraction; consequently, weakness of this muscle causes drooping of the shoulders.

Superior and inferior trapezius fibers act together in rotating the scapula on the thoracic wall in different directions, twisting it like a wingnut. *To test the trapezius (or the function of the accessory nerve [CN XI] that supplies it),* the shoulder is shrugged against resistance (the patient attempts to raise the shoulders as the physician or physical therapist presses down on them). If acting normally, the superior border of the muscle can be easily seen and palpated.

Latissimus Dorsi

The Latin name for this muscle, meaning "widest of the back," was well chosen because it covers a wide area of the back (Fig. 6.16, Table 6.2*E*). This large, *fan-shaped muscle* passes from the trunk to the humerus and acts directly on the glenohumeral joint and indirectly on the pectoral girdle (scapulothoracic joint). The latissimus dorsi extends, retracts, and rotates the humerus medially (e.g., when folding the arms behind the back or scratching the skin over the opposite

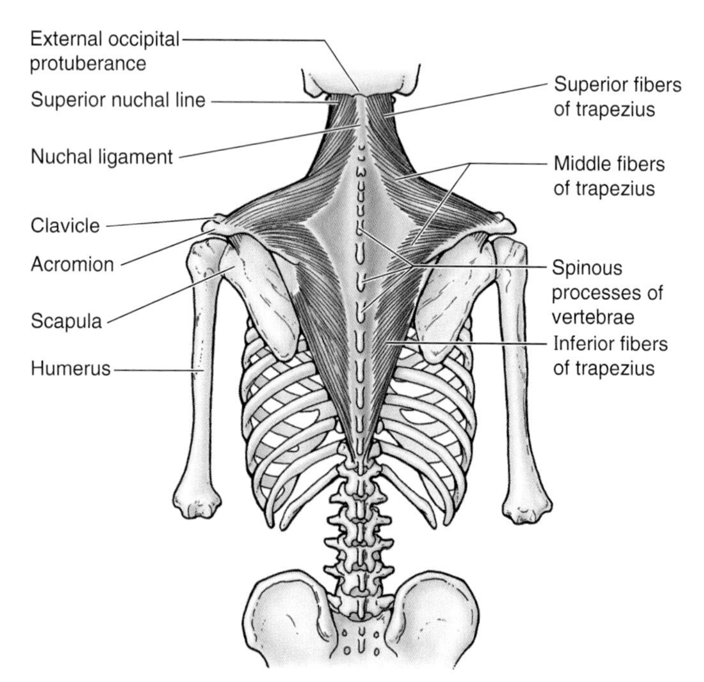

Figure 6.15. Trapezius muscle. This large, superficial, triangular muscle is responsible for the lateral slope between the neck and shoulder. It assists in suspending the pectoral girdle (shoulder girdle) and elevates, retracts, and rotates the scapula.

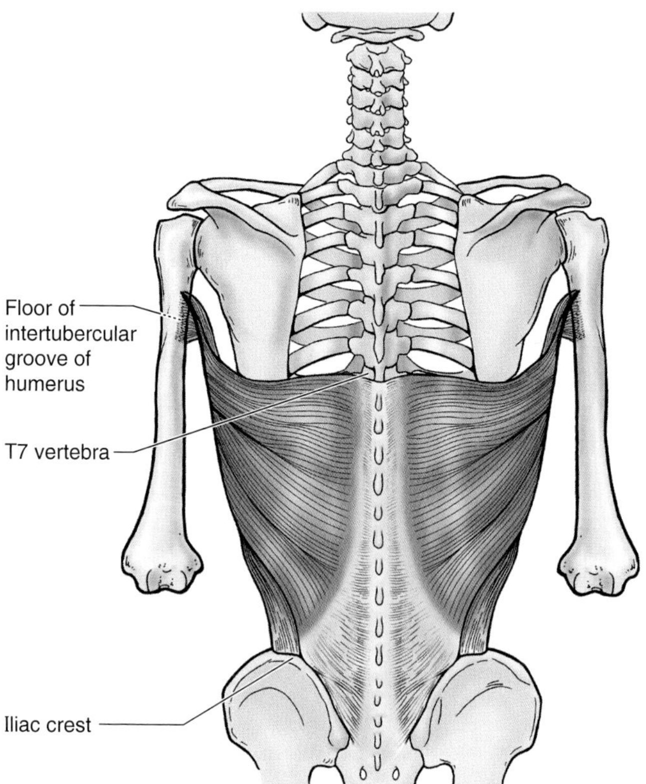

Figure 6.16. Latissimus dorsi muscle. This broad, triangular, mostly superficial muscle extends, adducts, and medially rotates the humerus. It is a powerful adductor and extensor of the arm and raises the body toward the arm during climbing.

scapula). In combination with the pectoralis major, the latissimus dorsi is a powerful adductor of the humerus. It is also useful in restoring the upper limb from abduction superior to the shoulder; hence, *the latissimus dorsi is the important climbing muscle.* In conjunction with the pectoralis major, the latissimus dorsi raises the trunk to the arm, which occurs when performing chin-ups (hoisting oneself on an overhead bar) or climbing a tree, for example. These movements are also used when chopping wood, paddling a canoe, and swimming (particularly during the crawl stroke). *To test the latissimus dorsi (or the function of the thoracodorsal nerve that supplies it),* the arm is abducted 90° and then adducted against resistance provided by the examiner. If acting normally, the anterior border of the muscle can be seen and easily palpated in the posterior axillary fold. The muscle can also be felt to contract when a person is asked to cough.

Triangle of Auscultation

Near the inferior angle of the scapula is a small triangular gap in the musculature. The superior horizontal border of the latissimus dorsi, the medial border of the scapula, and the inferolateral border of the trapezius form a *triangle of auscultation* (Table 6.2*E*). This triangular gap in the thick back musculature is a good place to examine posterior segments of the lungs with a stethoscope. When the scapulae are drawn anteriorly by folding the arms across the chest and the trunk is flexed, the auscultatory triangle enlarges and parts of the 6th and 7th ribs and 6th intercostal space are subcutaneous.

Injury of the Thoracodorsal Nerve

During surgical operations in the inferior part of the axilla, the thoracodorsal nerve (C6, C7, and C8) supplying the latissimus dorsi is vulnerable to injury. This nerve passes inferiorly along the posterior wall of the axilla and enters the medial surface of the latissimus dorsi close to where it becomes tendinous. The nerve is also vulnerable to injury during operations on scapular lymph nodes because its terminal part lies anterior to them and the subscapular artery (p. 694). The latissimus dorsi and the inferior part of the pectoralis major form an anteroposterior muscular sling between the trunk and the arm; however, the latissimus dorsi forms the more powerful part of the sling. *With paralysis of the latissimus dorsi, the person is unable to raise the trunk as occurs during climbing (pulling one's self up by the arms).* Furthermore, the person cannot use an axillary crutch because the shoulder is pushed superiorly by it. ⊙

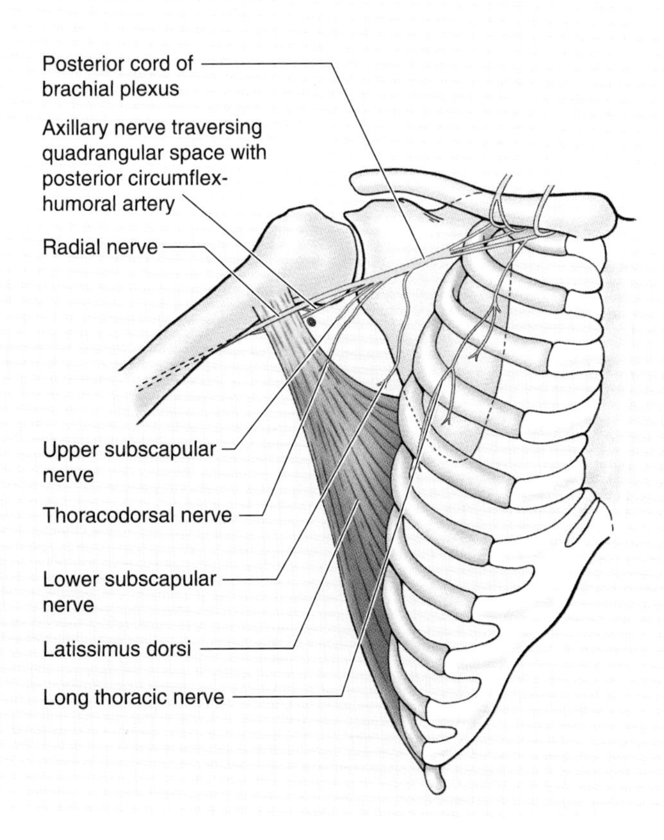

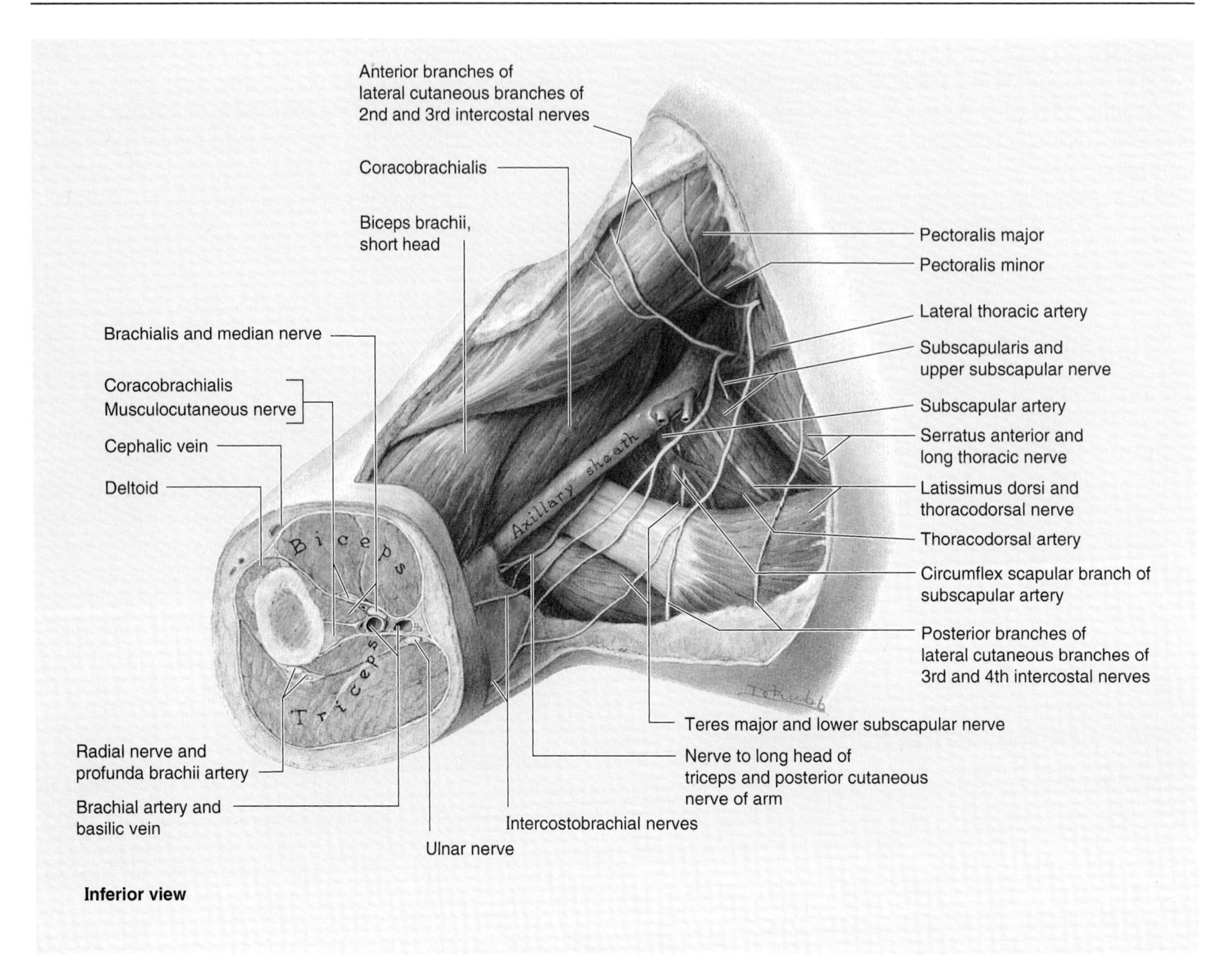

Deep Thoracoappendicular (Extrinsic Shoulder) Muscles

The deep thoracoappendicular muscles are the levator scapulae and rhomboids. These muscles provide direct attachment of the appendicular skeleton to the axial skeleton. The attachments, nerve supply, and main actions are given in Table 6.2.

Levator Scapulae

The superior third of the straplike levator scapulae lies deep to the sternocleidomastoid; the inferior third is deep to the trapezius. From the transverse processes of the upper cervical vertebrae, the fibers of the levator scapulae pass inferiorly to the superomedial border of the scapula (Fig. 6.17). True to its name, *the levator scapulae elevates and rotates the scapula, depressing the glenoid cavity (tilting it inferiorly).* It also assists in retracting the scapula and fixing it against the trunk and in flexing the neck laterally.

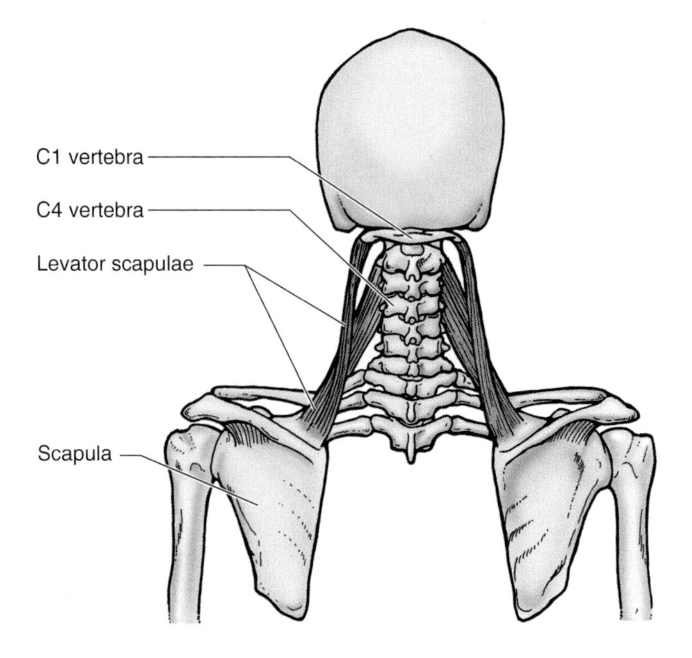

Rhomboids

The two rhomboid muscles—not always clearly separated from each other—have a rhomboid appearance; they form an oblique equilateral parallelogram (Fig. 6.18). The **rhomboid major and minor** lie deep to the trapezius and form broad parallel bands that pass inferolaterally from the vertebrae to the medial border of the scapula. The thin, flat rhomboid major is approximately two times wider than the thicker rhomboid minor lying superior to it. *The rhomboids retract and rotate the scapula*, depressing its glenoid cavity. They also assist the serratus anterior in holding the scapula against the thoracic wall and fixing the scapula during movements of the upper limb. The rhomboids are used when forcibly lowering the raised upper limbs (e.g., when driving a stake with a sledge hammer). *To test the rhomboids (or the function of the dorsal scapular nerve that supplies them)*, the patient is asked to place the hands posteriorly on the hips and to push the elbows posteriorly against resistance provided by the examiner. If the rhomboids are acting normally, they can be palpated along the medial borders of the scapulae. Because they lie deep to the trapezius, the rhomboids are not always visible during testing.

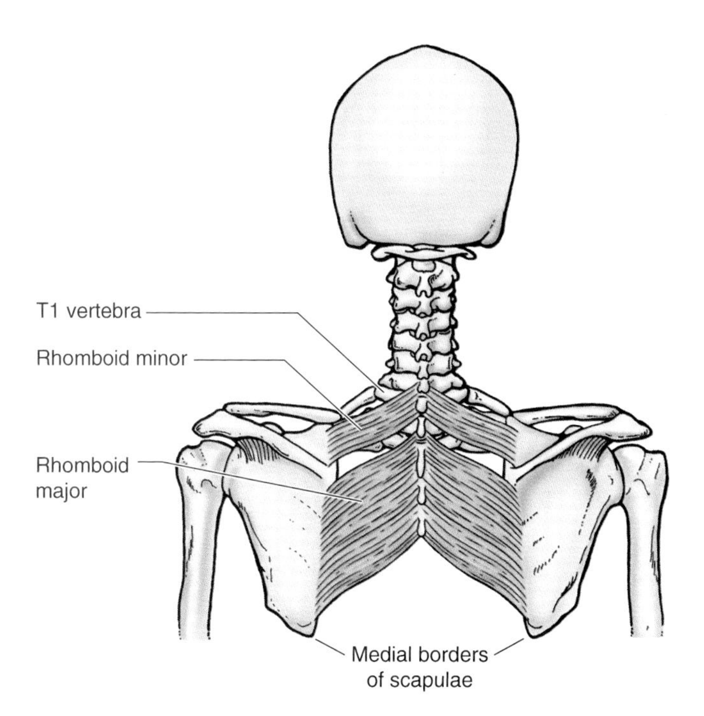

Figure 6.18. Rhomboid muscles, posterior view. The rhomboids (major and minor) retract the scapula and rotate it to depress the glenoid cavity. They also fix the scapula to the thoracic wall.

Figure 6.17. Levator scapulae muscle, posterior view. This thick, straplike muscle descends from the first four cervical vertebrae and attaches to the medial border of the superior angle of the scapula. It elevates and rotates the scapula, tilting (depressing) its glenoid cavity inferiorly.

Injury to the Dorsal Scapular Nerve

Injury to the dorsal scapular nerve, the nerve to the rhomboids, affects the actions of these muscles. If the rhomboids of one side are paralyzed, the scapula on the affected side is located further from the midline than that on the normal side. ⊙

Scapulohumeral (Intrinsic Shoulder) Muscles

The *six scapulohumeral muscles* (deltoid, teres major, supraspinatus, infraspinatus, subscapularis, and teres minor) are relatively short muscles that pass from the scapula to the humerus and act on the glenohumeral (shoulder) joint. The attachments, nerve supply, and main actions of these intrinsic shoulder muscles are summarized in Table 6.2.

Deltoid

The deltoid is a thick, powerful, coarse-textured muscle covering the shoulder and forming its rounded contour (Fig. 6.19, Table 6.2*E*). As its name indicates, the deltoid is shaped like the inverted Greek letter delta (Δ). The muscle is divided into unipennate anterior and posterior parts and a multipennate middle part; the parts of the deltoid can act separately or as a whole. When all three parts contract simultaneously, the arm is abducted. The anterior and posterior parts act like guy ropes to steady the arm as it is abducted. To initiate movement during the 1st 15° of abduction, the deltoid is assisted

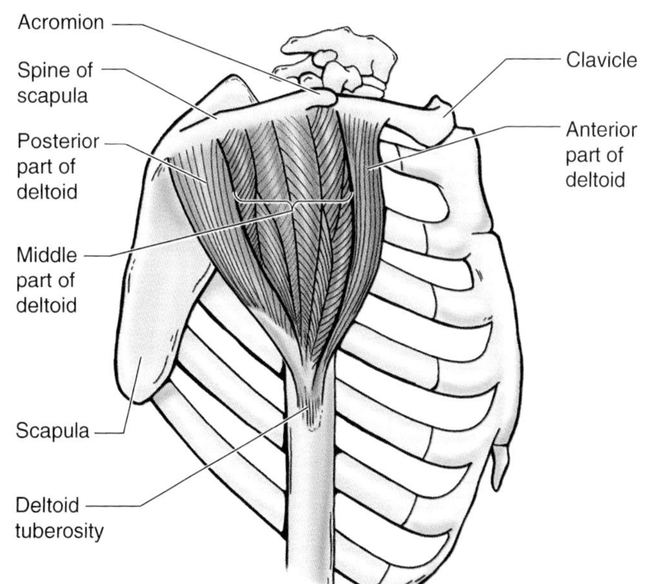

Figure 6.19. Deltoid muscle. This thick, coarse-textured, triangular muscle covers the glenohumeral (shoulder) joint and forms the rounded contour of the shoulder. The middle, multipennate part of the deltoid is the principal abductor of the arm; the anterior part flexes and medially rotates the arm, and the posterior part extends and laterally rotates the arm.

by the *supraspinatus* (Table 6.2*B*). When the arm is fully adducted, the line of pull of the deltoid coincides with the axis of the humerus; thus, it pulls directly upward on the bone and cannot initiate or produce abduction. It is, however, able to act as a shunt muscle, resisting inferior displacement of the head of the humerus from the glenoid cavity, as when lifting and carrying suitcases or buckets of water. From the fully adducted position, abduction must be initiated by the supraspinatus or by leaning to the side, allowing gravity to do so. The deltoid becomes fully effective as an abductor following the initial 15° of abduction.

The anterior and posterior parts of the deltoids are used to swing the limbs during walking. The anterior part assists the pectoralis major in flexing the arm, and the posterior part assists the latissimus dorsi in extending the arm. *The deltoid also helps stabilize the shoulder joint* and hold the head of the humerus in the glenoid cavity during arm movements. *To test the deltoid (or the function of the axillary nerve that supplies it)*, the arm is abducted—starting from approximately 15°—against resistance (Fig. 6.20). If acting normally, the muscle can easily be seen and palpated. The influence of gravity is avoided when the person is supine.

Injury to the Axillary Nerve

The deltoid atrophies when the axillary nerve (C5 and C6) is severely damaged (e.g., following fracture of the surgical neck of the humerus). As the deltoid atrophies, the rounded contour of the shoulder disappears. This gives the shoulder a flattened appearance and produces a slight hollow inferior to the acromion. Because it winds around the surgical neck of the humerus, the axillary nerve is usually injured during fracture of the proximal end of the humerus. It may also be damaged during dislocation of the shoulder joint. In addition to atrophy of the deltoid, a loss of sensation may occur ▶

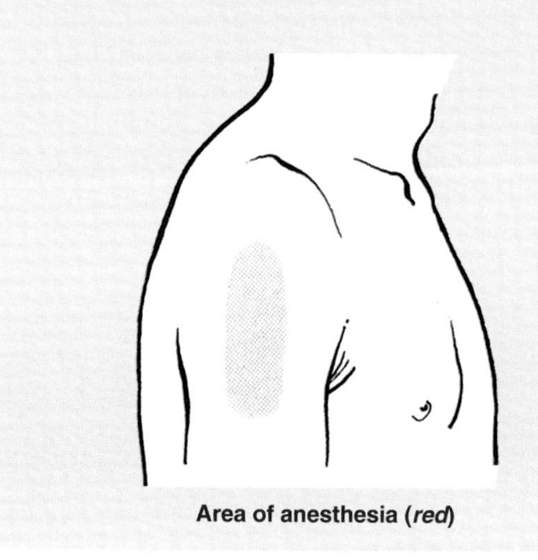

Area of anesthesia (*red*)

Figure 6.20. Testing the deltoid muscle. The examiner resists the patient's abduction of the limb by the deltoid. If acting normally, contraction of the middle part of the muscle can be palpated.

▶ over the lateral side of the proximal part of the arm (shown in *red*). The deltoid is a common site for the intramuscular injection of drugs. The axillary nerve, supplying the deltoid, runs transversely under cover of the deltoid and winds around the surgical neck of the humerus. Awareness of its location avoids injury to it during injections. ⊙

Teres Major

The teres major is a thick, rounded muscle that forms a raised oval area on the inferolateral third of the scapula when the arm is adducted against resistance (Fig. 6.21*A*, Table 6.2). The inferior border of teres major forms the inferior border of the lateral part of the posterior wall of the axilla. *The teres major adducts and medially rotates the arm* (Fig. 6.21*B*). It can also help extend it from the flexed position. The teres major is also an important stabilizer of the humeral head in the glenoid cavity; that is, it steadies the head in its socket. With the teres minor, the teres major holds the humeral head against the pull of the deltoid during abduction of the arm. *To test the teres major (or the lower subscapular nerve that supplies it)*, the abducted arm is adducted against resistance. If acting normally, the muscle can be easily seen and palpated in the posterior axillary fold.

Rotator Cuff Muscles

Four of the scapulohumeral muscles (intrinsic muscles of the shoulder)—*s*upraspinatus, **i**nfraspinatus, **t**eres minor, and **s**ubscapularis (referred to as SITS muscles for brevity)—(Fig. 6.22) are called **rotator cuff muscles** because they form a musculotendinous rotator cuff around the glenohumeral joint. All except the supraspinatus are rotators of the

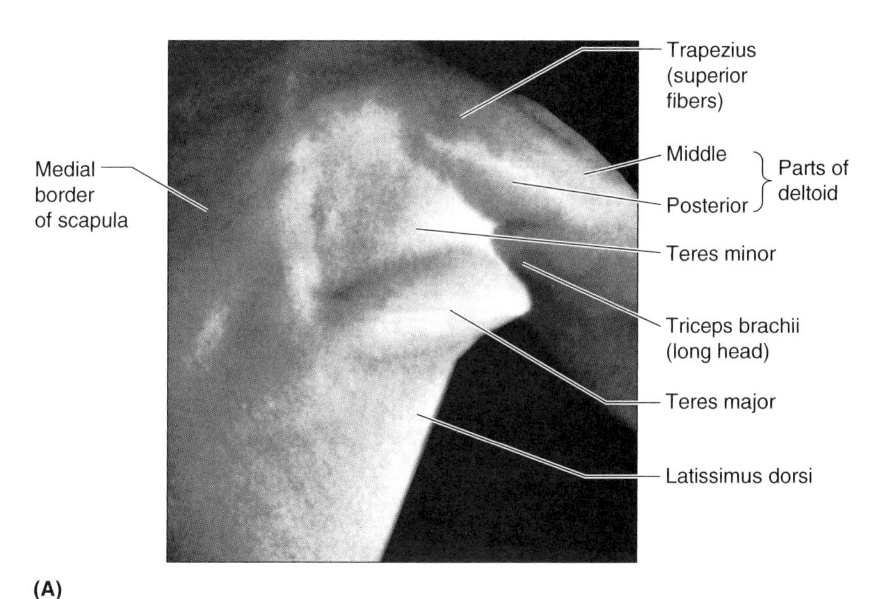

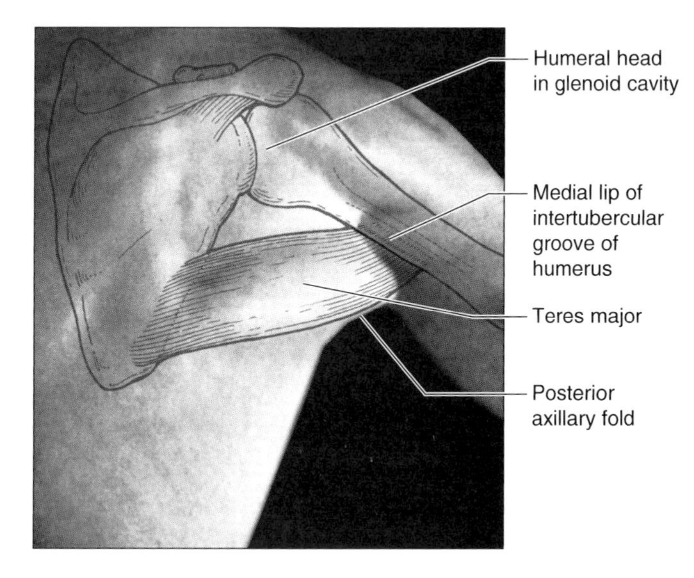

Figure 6.21. Scapulohumeral muscles. Posterior views. These muscles pass from the scapula to the humerus (arm bone) and act on the shoulder joint. **A.** Surface anatomy of the scapular muscles and the latissimus dorsi. **B.** Overlay drawings illustrating the attachments of the teres major (L. teres, round), a thick, rounded muscle that adducts and medially rotates the arm. The latissimus dorsi and teres major form the posterior axillary fold. When the arm is adducted against resistance, this fold is accentuated as in (**A**).

humerus; the supraspinatus, besides being part of the rotator cuff, initiates and assists the deltoid in the first 15° of abduction of the arm. The tendons of the four rotator cuff muscles blend with the articular capsule of the glenohumeral joint, reinforcing it as the **rotator cuff**, which protects the joint and gives it stability, with their tonic contraction holding the relatively large head of the humerus in the small, shallow glenoid cavity of the scapula during arm movements. The attachments, nerve supply, and main actions of the rotator cuff muscles are given in Table 6.2.

The **supraspinatus** occupies the supraspinous fossa of the scapula. A bursa separates it from the lateral fourth of the fossa. See the discussion of this muscle's cooperative action with the description of the deltoid (p. 695). *To test the supraspinatus*, the arm is abducted from the fully adducted position against resistance, and the muscle is palpated superior to the spine of the scapula.

The **infraspinatus** occupies the medial three-fourths of the infraspinous fossa and is partly covered by the deltoid and trapezius. In addition to helping stabilize the shoulder joint, the infraspinatus is a powerful lateral rotator of the humerus. *To test the infraspinatus*, the person is asked to flex the elbow and adduct the arm. The arm is then laterally rotated against resistance. If acting normally, the muscle can be palpated inferior to the scapular spine. *To test the function of the suprascapular nerve*, which supplies the supraspinatus and infraspinatus, both muscles must be tested as described above.

The **teres minor** is a narrow, elongate muscle that is completely hidden by the deltoid and is often not clearly delineated from the infraspinatus. The teres minor rotates the arm and assists in its adduction.

The **subscapularis** is a thick, triangular muscle that lies on the costal surface of the scapula and forms part of the posterior wall of the axilla. It crosses the anterior aspect of the scapulohumeral joint on its way to the humerus. *The subscapularis is the primary medial rotator of the arm and also adducts it.* It also joins the other rotator cuff muscles (supraspinatus, infraspinatus, and teres minor) in holding the

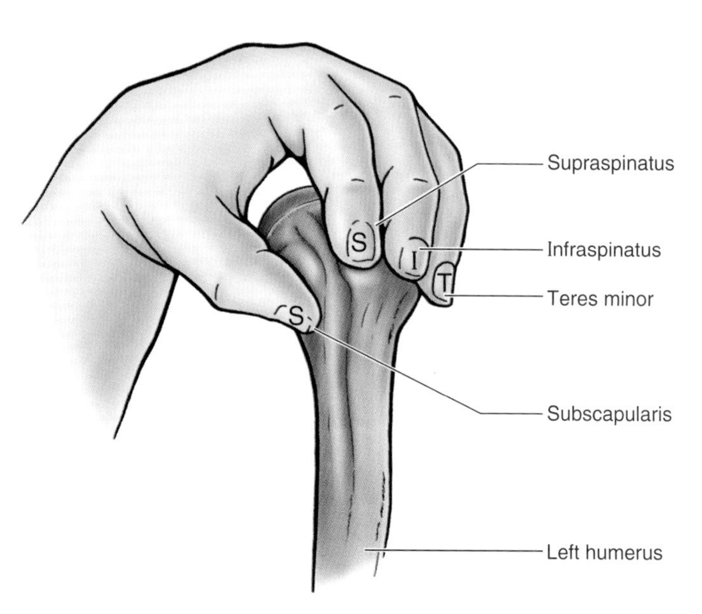

Figure 6.22. Demonstration of the position of the rotator cuff muscles. Anterior view. The primary combined function of these four scapulohumeral muscles is to hold the relatively large head of the humerus (arm bone) in the smaller, shallow, glenoid cavity of the scapula. The tendons of the muscle blend with the fibrous capsule of the glenohumeral joint to form a musculotendinous rotator cuff, which reinforces the capsule on three sides (anteriorly, superiorly, and posteriorly) as it provides active support for the glenohumeral joint.

Rotator Cuff Injuries

Injury or disease may damage the musculotendinous rotator cuff, producing instability of the glenohumeral (shoulder) joint. Trauma may tear or rupture one or more of the tendons of the muscles forming the rotator cuff. Acute tears may occur when the arm is violently pushed into abduction, such as when a hockey player is pushed into the boards while using the upper limbs to cushion the impact. The person may tear the rotator cuff and report a sharp pain in the anterosuperior part of the shoulder. Rotator cuff injuries are also common in baseball pitchers and third basemen who throw the ball hard. Rotator cuff tears also follow dislocation of the shoulder.

Degenerative tendonitis of the rotator cuff is common, especially in old people. To test for this disease, the person is asked to lower the fully abducted limb slowly and smoothly. From an approximately 90° abduction, the limb will suddenly drop to the side in an uncontrolled manner if the rotator cuff (especially the supraspinatus part) is ▶

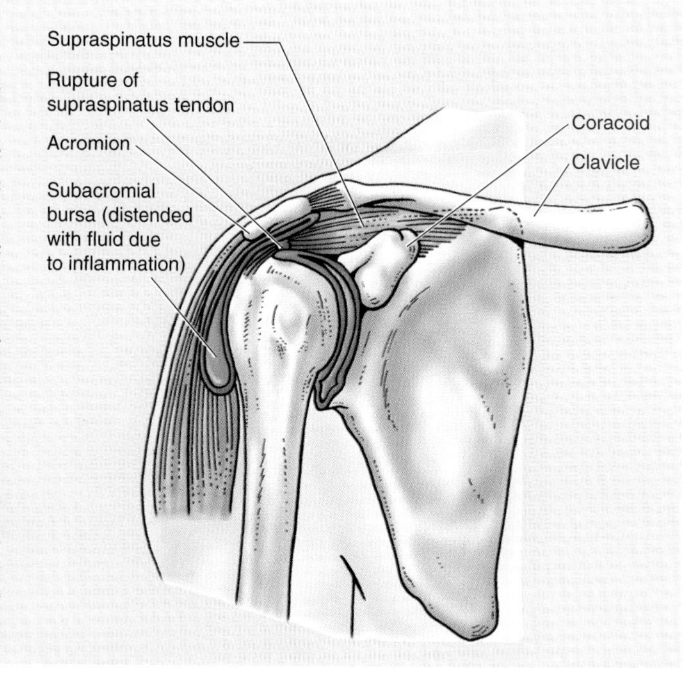

▶ diseased and torn. The supraspinatus tendon is the most commonly torn part of the rotator cuff, probably because it is relatively avascular. When the tendon tears acutely, or when it is eroded by chronic abrasion, especially in older persons, the two associated bursae communicate. The injury often results from an indirect force to the abducted arm, such as a fall during skiing in a person older than 45 years. Acute tears are uncommon in young persons. This injury causes tenderness around the greater tubercle of the humerus and pain during 45° of passive abduction.

Subacromial Bursitis

The tendon of the supraspinatus is separated from the coracoacromial ligament, acromion, and deltoid by the *subacromial bursa.* When this bursa is inflamed (*subacromial bursitis*), abduction of the arm is extremely painful during the arc of 50 to 130° (*painful arc syndrome*). The pain may radiate as far distally as the hand. Acute pain is also felt lateral to the acromion. ⊙

head of the humerus in the glenoid cavity during all movements of the scapulohumeral joint (i.e., it helps stabilize this joint during movements of the elbow, wrist, and hand).

Axilla

The axilla (armpit) is the pyramidal space inferior to the glenohumeral joint and superior to the axillary fascia at the junction of the arm and thorax (Fig. 6.23). The shape and size of the axilla varies depending on the position of the arm; it almost disappears when the arm is fully abducted. *The axilla provides a passageway for vessels and nerves to reach the upper limb.* The axilla has an apex, a base, and four walls, three of which are muscular:

- *Apex of axilla*—the entrance from neck to axilla—lies between the 1st rib, clavicle, and superior edge of subscapularis; the arteries, veins, lymphatics, and nerves pass from the neck to the axilla through the *cervicoaxillary canal*—the superior opening to the axilla—to reach the arm
- *Base of axilla* is formed by the concave skin, subcutaneous tissue, and axillary (deep) fascia extending from the arm to the thoracic wall
- *Anterior wall of axilla* is formed by the pectoralis major and pectoralis minor and the pectoral and clavicopectoral fascia associated with them
- *Posterior wall of axilla* is formed chiefly by the scapula and subscapularis on its anterior surface and inferiorly by the teres major and latissimus dorsi
- *Medial wall of axilla* is formed by the thoracic wall (1st to 4th ribs and intercostal muscles) and the overlying serratus anterior
- *Lateral wall of axilla* is a narrow bony wall formed by the *intertubercular groove* in the humerus.

The axilla contains axillary blood vessels (axillary artery and its branches, axillary vein and its tributaries), lymphatic vessels, and several groups of *axillary lymph nodes* (Fig. 6.23*C*). The axilla also contains large nerves that comprise the cords and branches of the *brachial plexus*, a network of interjoining nerves that pass from the neck to the upper limb. Proximally, these neurovascular structures are ensheathed in a fascial sleeve, the *axillary sheath* (Fig. 6.24*A*).

Axillary Artery

The axillary artery *begins at the lateral border of the 1st rib* as the continuation of the subclavian artery and *ends at the inferior border of the teres major* (Table 6.3). It passes posterior to the pectoralis minor into the arm and *becomes the brachial artery* when it passes distal to the inferior border of the teres major, at which point it usually has reached the humerus. For descriptive purposes, *the axillary artery is divided into three parts by the pectoralis minor* (the part number also indicates its number of branches):

- **First part of the axillary artery**—located between the lateral border of the 1st rib and the medial border of the pectoralis minor—is enclosed in the **axillary sheath** and has one branch, the *superior thoracic artery* (Fig. 6.24*B*, Table 6.3)
- **Second part of the axillary artery**—lies posterior to pectoralis minor and has two branches, the *thoracoacromial* and *lateral thoracic arteries*, which pass medial and lateral to the muscle, respectively
- **Third part of the axillary artery**—extends from the lateral border of pectoralis minor to the inferior border of teres major; it has three branches: the *subscapular*—the largest branch of the axillary artery—opposite which the *anterior circumflex humeral* and *posterior circumflex humeral arteries* arise.

The **superior thoracic artery** (highest thoracic artery) is a small vessel that arises from the first part of the axillary artery, just inferior to the subclavius (Fig. 6.24*B*). It runs inferomedially posterior to the axillary vein and supplies muscles in the 1st and 2nd intercostal spaces and the serratus anterior. It anastomoses with the intercostal arteries.

The **thoracoacromial artery**, a short wide trunk, is usually the first branch of the second part of the axillary artery (Fig. 6.25), deep to the pectoralis minor. It pierces the costocoracoid membrane, part of the *clavipectoral fascia*, and then di-

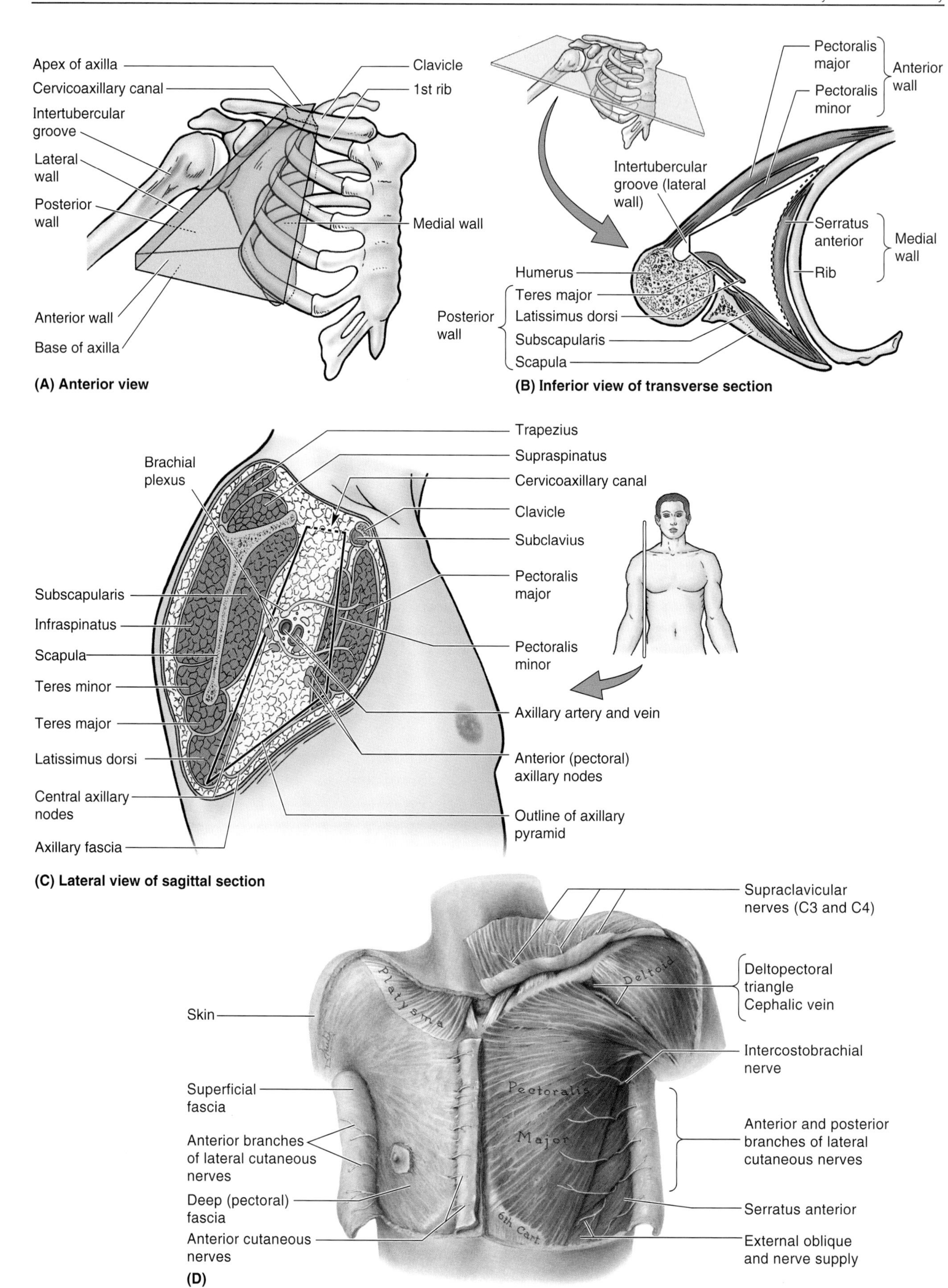
Apex of axilla
Cervicoaxillary canal
Intertubercular groove
Lateral wall
Posterior wall
Anterior wall
Base of axilla
Clavicle
1st rib
Medial wall
(A) Anterior view
Pectoralis major
Pectoralis minor
Anterior wall
Intertubercular groove (lateral wall)
Serratus anterior
Rib
Medial wall
Humerus
Teres major
Latissimus dorsi
Subscapularis
Scapula
Posterior wall
(B) Inferior view of transverse section
Brachial plexus
Trapezius
Supraspinatus
Cervicoaxillary canal
Clavicle
Subclavius
Pectoralis major
Pectoralis minor
Subscapularis
Infraspinatus
Scapula
Teres minor
Teres major
Latissimus dorsi
Central axillary nodes
Axillary fascia
Axillary artery and vein
Anterior (pectoral) axillary nodes
Outline of axillary pyramid
(C) Lateral view of sagittal section
Supraclavicular nerves (C3 and C4)
Deltoid
Platysma
Deltopectoral triangle
Cephalic vein
Skin
Intercostobrachial nerve
Pectoralis
Major
Superficial fascia
Anterior and posterior branches of lateral cutaneous nerves
Anterior branches of lateral cutaneous nerves
Deep (pectoral) fascia
Serratus anterior
6th Cart.
Anterior cutaneous nerves
External oblique and nerve supply
(D)

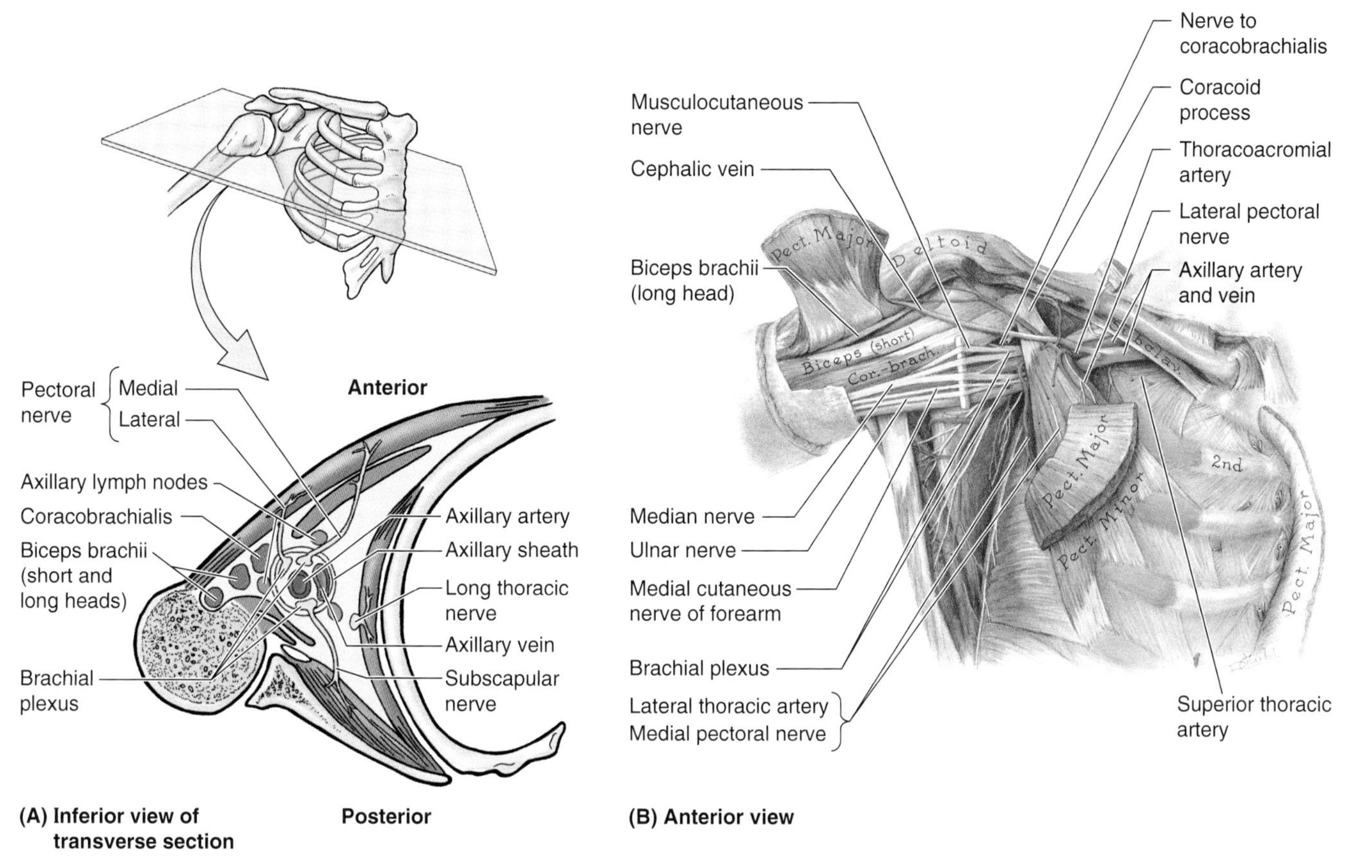

Figure 6.24. Neurovascular structures in the axilla (armpit). A. Inferior view of a transverse section of the axilla. Examine the contents of the axilla, especially the *axillary sheath* enclosing the axillary artery and vein and the three cords of the brachial plexus. **B.** Structures of the axilla. The pectoralis major is mostly—and the clavipectoral fascia and axillary sheath are completely—removed. Observe the *brachial plexus* shown here surrounding the axillary artery on its lateral and medial aspects (appearing here to be its superior and inferior aspects because the limb is abducted) as well as on its posterior aspect (which cannot be seen from this view).

vides into four branches (acromial, deltoid, pectoral, and clavicular), deep to the clavicular head of the pectoralis major.

The **lateral thoracic artery** has a variable origin. It usually arises as the second branch of the second part of the axillary artery and descends along the lateral border of the pectoralis minor (Fig. 6.24*B*); however, it may arise from the thoracoacromial, suprascapular, or subscapular arteries. The lateral thoracic artery supplies the pectoral muscles, the axillary lymph nodes, and the breast; it is an important source of blood to the lateral part of the mammary gland in women.

The **subscapular artery**, the largest branch of the axillary artery (Table 6.3), arises from its third part and descends along the lateral border of the subscapularis on the posterior axillary wall. It soon divides into the circumflex scapular and thoracodorsal arteries and supplies the subscapularis, teres major, serratus anterior, and latissimus dorsi muscles.

Figure 6.23. Location, boundaries, and contents of the axilla. A. Drawing illustrating that the axilla is a pyramidal space inferior to the glenohumeral joint and superior to the skin of the axillary fossa at the junction of the arm and thorax. Observe its apex, base, and walls. **B.** Transverse section of the axilla illustrating its three muscular walls. The small lateral or bony wall of the axilla is the intertubercular (bicipital) groove of the humerus. **C.** Sagittal section of the shoulder showing the contents of the axilla and the scapular and pectoral muscles forming its posterior and anterior walls, respectively. Observe that the *axillary artery* is surrounded by the *brachial plexus,* a major nerve plexus formed by the ventral primary rami of the 5th cervical to the 1st thoracic spinal nerves for innervation of the upper limb. **D.** Superficial dissection of the pectoral region of a man. The subcutaneous platysma muscle, descending from the neck to the 2nd or 3rd rib, is cut short on the right side and reflected on the left side together with the supraclavicular nerves. The inferior border of the pectoralis major forms the anterior axillary fold.

Table 6.3. **Arteries of the Proximal Upper Limb**

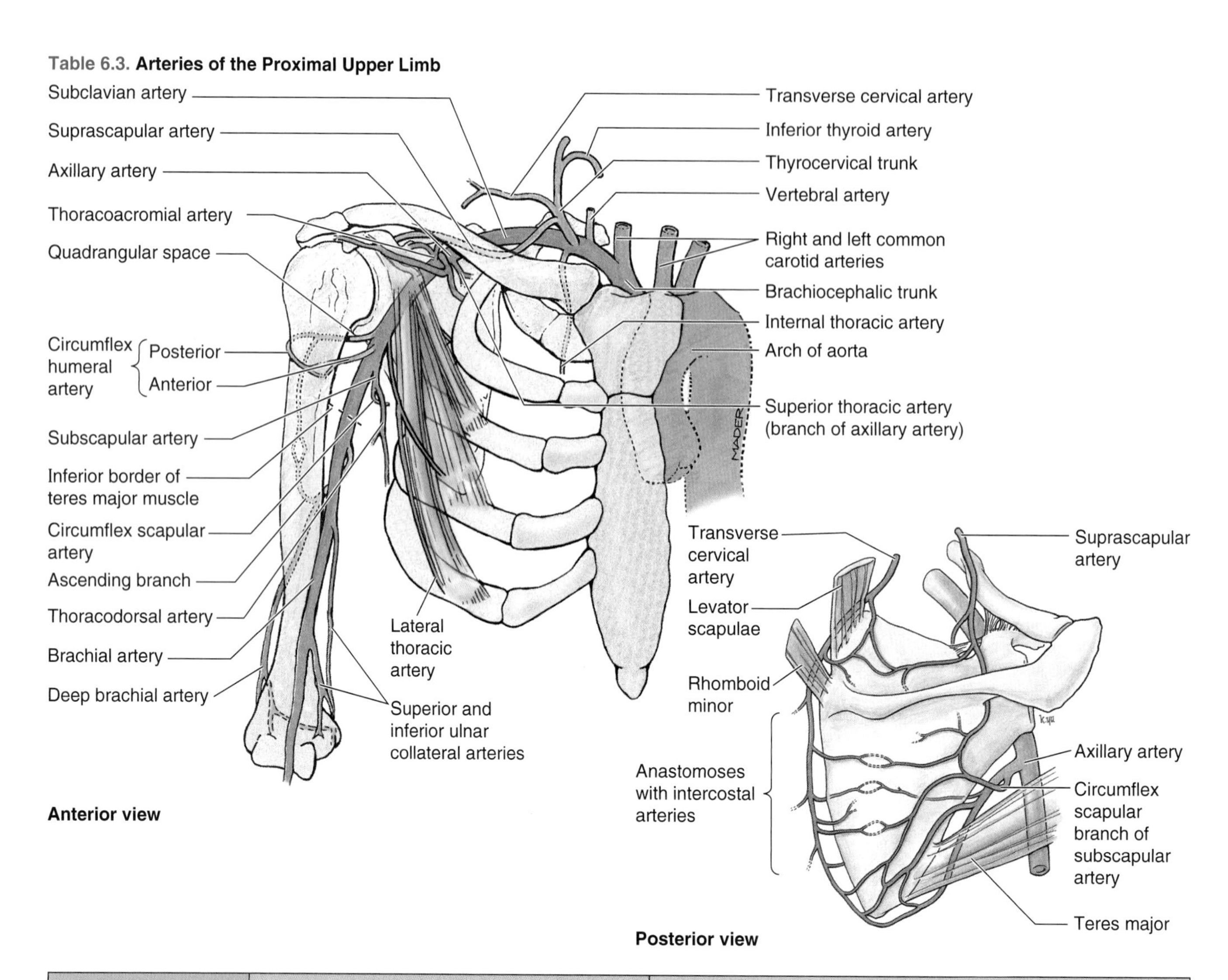

Artery	Origin	Course
Internal thoracic	Inferior surface of subclavian artery	Descends, inclining anteromedially, posterior to sternal end of clavicle and costal cartilage and enters thorax
Thyrocervical trunk	Anterior aspect of first part of subclavian artery	Ascends as a short, wide trunk and gives rise to four branches: suprascapular, transverse cervical, ascending cervical, and inferior thyroid arteries
Suprascapular	Thyrocervical trunk	Passes inferolaterally over anterior scalene muscle and phrenic nerve, crosses subclavian artery and brachial plexus, and runs laterally posterior and parallel to clavicle; it then passes to posterior aspect of scapula and supplies supraspinatus and infraspinatus muscles
Superior thoracic	Only branch of first part of axillary artery	Runs anteromedially along superior border of pectoralis minor and then passes between it and pectoralis major to thoracic wall; helps to supply 1st and 2nd intercostal spaces and superior part of serratus anterior
Thoracoacromial	Second part of axillary artery deep to pectoralis minor	Curls around superomedial border of pectoralis minor, pierces clavipectoral fascia, and divides into four branches
Lateral thoracic	Second part of axillary artery	Descends along axillary border of pectoralis minor and follows it onto thoracic wall
Subscapular	Third part of axillary artery	Descends along lateral border of subscapularis and axillary border of scapula to its inferior angle, where it passes onto thoracic wall
Circumflex scapular artery	Subscapular artery	Curves around axillary border of scapula and enters infraspinous fossa

Table 6.3. (*Continued*) **Arteries of the Proximal Upper Limb**

Artery	Origin	Course
Thoracodorsal	Subscapular artery	Continues course of subscapular artery and accompanies thoracodorsal nerve to latissimus dorsi
Anterior and posterior circumflex humeral	Third part of axillary artery	These arteries anastomose to form a circle around surgical neck of humerus; larger posterior circumflex humeral artery passes through quadrangular space with axillary nerve
Deep brachial	Brachial artery near its origin	Accompanies radial nerve through radial groove in humerus and takes part in anastomosis around elbow joint
Ulnar collateral (superior and inferior)	Superior ulnar collateral artery arises from brachial artery near middle of arm; inferior ulnar collateral artery arises from brachial artery just superior to elbow	Superior ulnar collateral artery accompanies ulnar nerve to posterior aspect of elbow; inferior ulnar collateral artery divides into anterior and posterior branches; both ulnar collateral arteries take part in anastomosis around elbow joint

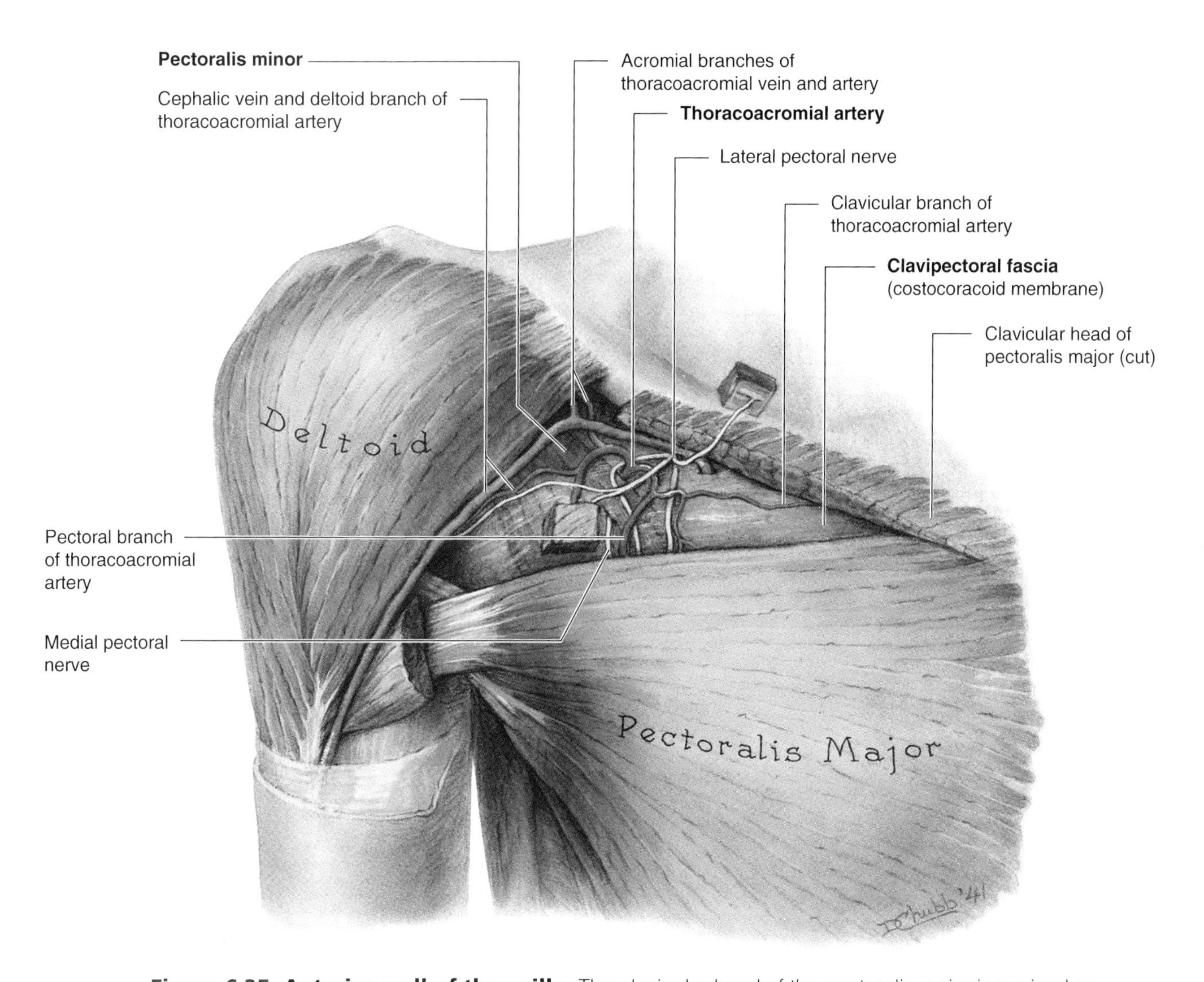

Figure 6.25. Anterior wall of the axilla. The clavicular head of the pectoralis major is excised except for its clavicular and humeral attaching ends and two cubes that remain to identify its nerves. The anterior wall of the axilla is formed by the pectoralis major and minor (and the pectoral and clavipectoral fascia that envelops them). The pectoralis major covers the whole of this wall and forms the anterior axillary fold. Extending between the superior border of the pectoralis minor and the clavicle is the costocoracoid membrane, part of the clavipectoral fascia.

Compression of the Axillary Artery

The axillary artery can be palpated in the inferior part of the lateral wall of the axilla. Compression of the third part of this artery against the humerus may be necessary when profuse bleeding occurs (e.g., resulting from an extensive stab or bullet wound in the axilla). If compression is required at a more proximal site, the axillary artery can be compressed at its origin (as the subclavian artery crosses the 1st rib) by exerting downward pressure in the angle between the clavicle and the attachment of the sternocleidomastoid.

Arterial Anastomoses Around the Scapula

Many arterial anastomoses (communications between arteries) occur around the scapula. Several vessels join to form networks on the anterior and posterior surfaces of the scapula—the dorsal scapular, suprascapular, and subscapular (via the circumflex scapular). The importance of the *collateral circulation* that is possible through these anastomoses becomes apparent when ligation of a lacerated subclavian or axillary artery is necessary. For example, the axillary artery may have to be ligated between the 1st rib and subscapular artery; in other cases, *vascular stenosis* (narrowing) of the axillary artery may result from an atherosclerotic lesion that causes reduced bloodflow. In either case, the direction of bloodflow in the subscapular artery is reversed, enabling blood to reach the third part of the axillary artery. Note that the subscapular artery receives blood through several anastomoses with the suprascapular artery, transverse cervical artery, and intercostal arteries.

Slow occlusion of the axillary artery (e.g., resulting from disease or trauma) often enables sufficient collateral ▶

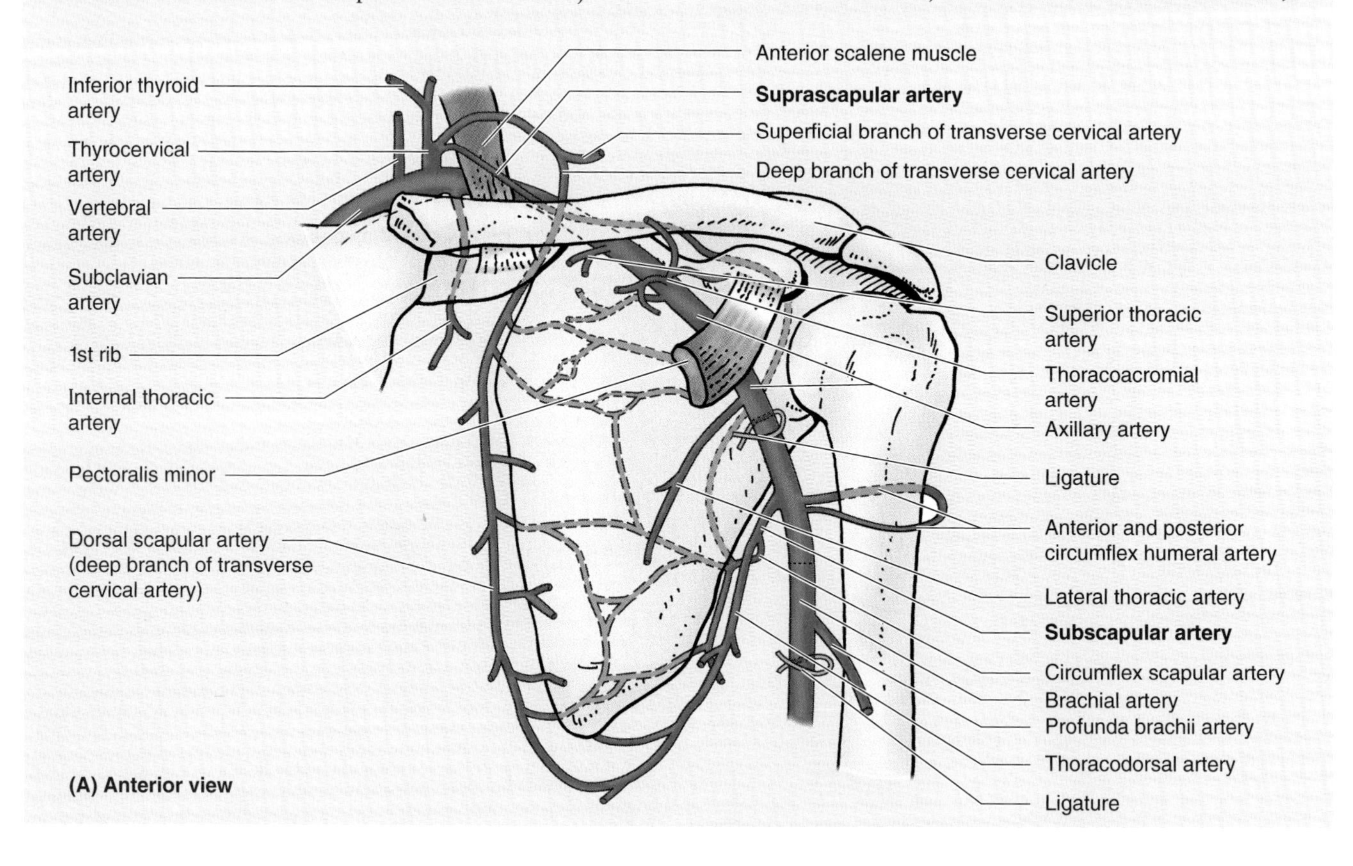

(A) Anterior view

The **circumflex scapular artery**, the larger branch of the subscapular, curves posteriorly around the axillary border of the scapula, passing between the subscapularis and teres major to supply muscles on the dorsum of the scapula. It participates in the anastomoses around the scapula.

The **thoracodorsal artery** continues the general course of the subscapular to the inferior angle of the scapula and supplies adjacent muscles, principally latissimus dorsi. It also participates in the arterial anastomoses around the scapula.

The **circumflex humeral arteries** usually arise from the third part of the axillary artery opposite the subscapular artery and pass around the surgical neck of the humerus to anastomose with each other. The smaller **anterior circumflex humeral artery** passes laterally, deep to the coracobrachialis and biceps brachii. It gives off an ascending branch that supplies the shoulder. The larger **posterior circumflex humeral artery** passes through the posterior wall of the axilla via the **quadrangular space** (Table 6.3) with the axillary nerve to supply surrounding muscles (e.g., the deltoid, teres major and minor, and long head of the triceps).

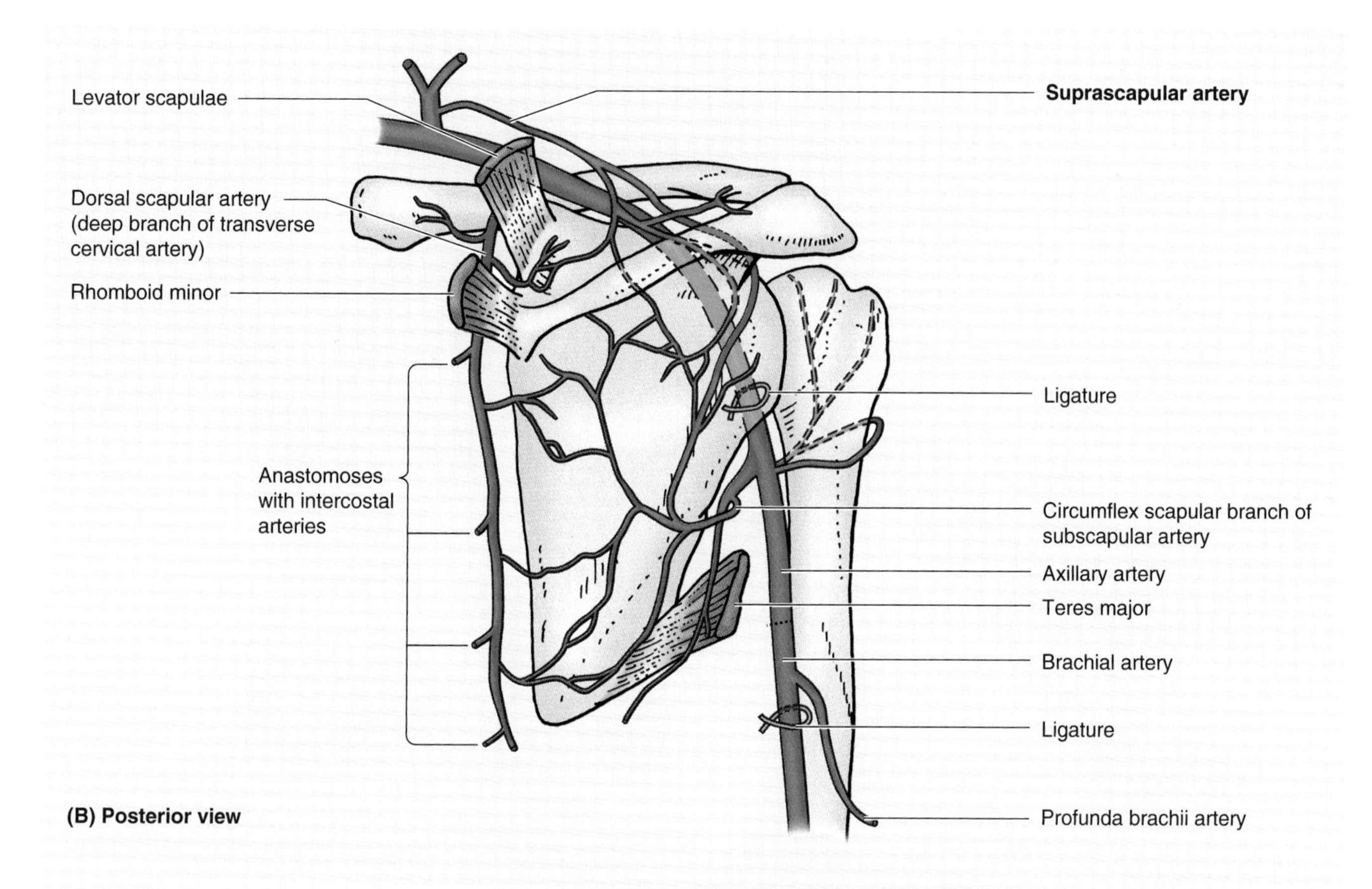

(B) Posterior view

▶ circulation to develop, preventing *ischemia* (deficiency of blood). Sudden occlusion usually does not allow sufficient time for a good collateral circulation to develop; as a result, an inadequate supply of bloodflow to the arm, forearm, and hand. *Ligation of the axillary artery distal to the subscapular artery and proximal to the deep artery of the arm cuts off the blood supply to the arm because the collateral circulation is inadequate.*

Aneurysm of the Axillary Artery

The first part of the axillary artery may enlarge (*aneurysm of the axillary artery*) and compress the trunks of the brachial plexus, causing pain and anesthesia (loss of feeling or sensation) in the areas skin supplied by the affected nerves. Aneurysm of the axillary artery occurs in baseball pitchers and must be repaired so they can continue to play. ⭘

Axillary Vein

The axillary vein lies on the medial side of the axillary artery (Fig. 6.26). This large vein is *formed by the union of the brachial veins*—accompanying veins of the brachial artery—and the basilic vein at the inferior border of the teres major. The axillary vein ends at the lateral border of the 1st rib, where it becomes the **subclavian vein**. Although the veins of the axilla are more abundant than the arteries, are highly variable, and frequently communicate (anastomose), the axillary vein receives tributaries that generally correspond to branches of the axillary artery with a few major exceptions.

- The veins corresponding to the branches of the thoracoacromial artery do not merge to enter by a common tributary; some enter independently into the axillary vein, but others empty into the cephalic vein, which—superior to the pectoralis minor—also enters the axillary vein close to its transition to the subclavian vein.
- The axillary vein receives—directly or indirectly—the *thoracoepigastric vein(s)*, which is formed by the anastomoses of superficial veins from the inguinal (groin) region with tributaries of the axillary vein—usually the lateral thoracic vein—constituting a collateral route that enables venous return in the presence of obstruction of the inferior vena cava.

Injuries to the Axillary Vein

Wounds in the axilla often involve the axillary vein because of its large size and exposed position. When the arm is fully abducted, the axillary vein overlaps the axillary artery anteriorly. A wound in the proximal part of the axillary vein is particularly dangerous not only because of profuse bleeding but also because of the risk of air entering the vessel and producing *air emboli* (air bubbles in the blood). ▶

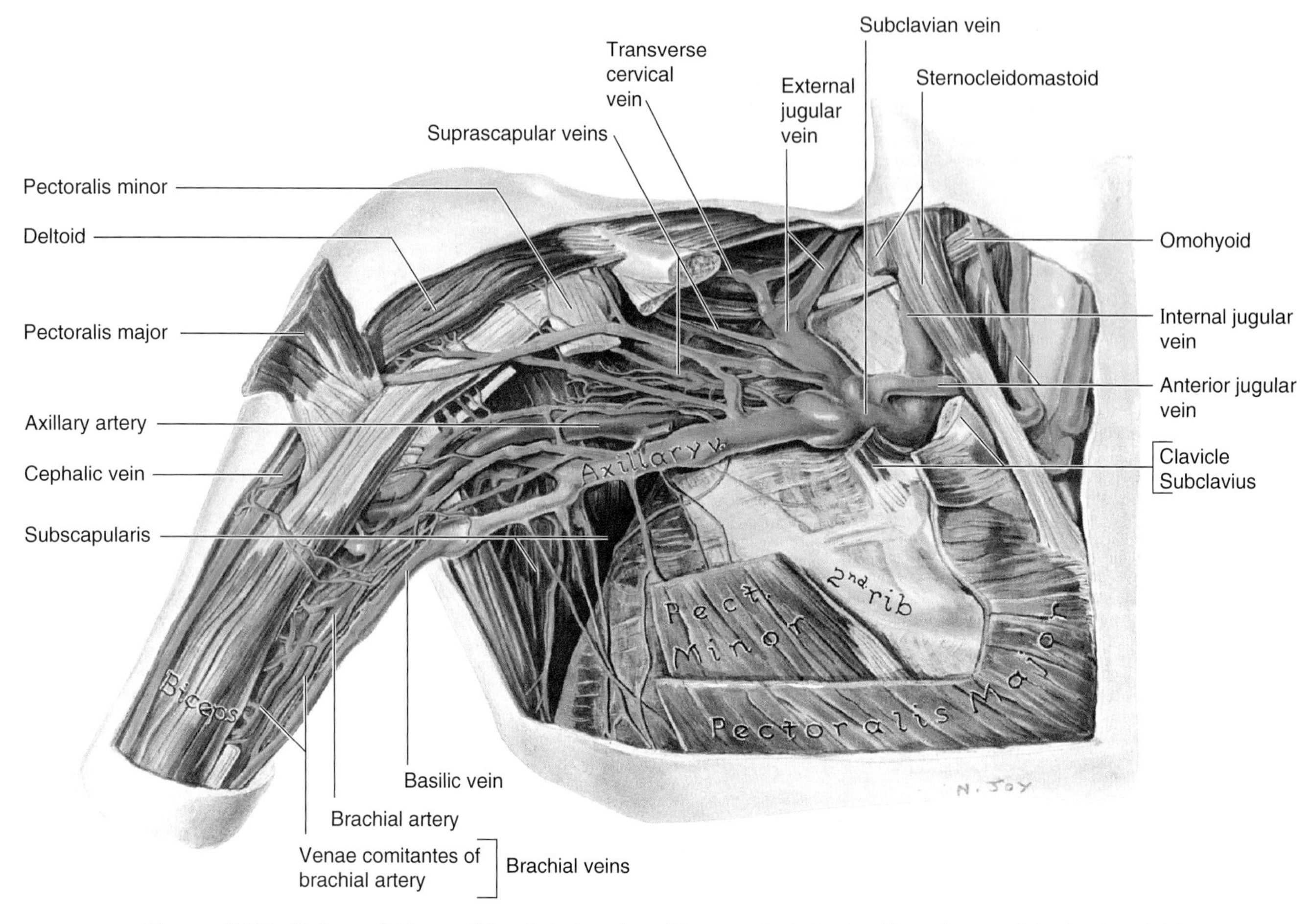

Figure 6.26. Veins of the axilla. Anterior view. Observe that the basilic vein parallels the brachial artery to the axilla, where it merges with the accompanying veins (L. venae comitantes) of the axillary artery to form the axillary vein. Notice also the large number of highly variable veins in the axilla, which are also tributaries of the axillary vein.

Expansion of the Axillary Vein

Because the axillary sheath does not enclose the second and third parts of the axillary vein, they are free to expand when bloodflow increases or the first part compresses. ⊙

Axillary Lymph Nodes

The fibrofatty connective tissue of the axilla has many lymph nodes. The axillary lymph nodes are arranged in five principal groups: apical, pectoral, subscapular, humeral, and central (see Fig. 1.27*A*).

The **apical group of axillary lymph nodes** consists of lymph nodes at the apex of the axilla, located along the medial side of the axillary vein and the first part of the axillary artery. The *apical group receives lymph from all other groups of axillary lymph nodes* as well as from lymphatics accompanying the proximal cephalic vein. Efferent vessels from the apical group of nodes unite to form the **subclavian lymphatic trunk**, which may join the jugular and bronchomediastinal trunks on the right side to form the **right lymphatic duct**, or it may enter the right venous angle independently. On the left side, the subclavian trunk most commonly joins the **thoracic duct** (Fig. 6.27, *A* and *B*).

The **pectoral (anterior) group of axillary lymph nodes** consists of three to five lymph nodes that lie along the medial wall of the axilla, around the lateral thoracic vein and the inferior border of the pectoralis minor. The pectoral group of nodes receives lymph mainly from the anterior thoracic wall including the breast. Efferent lymphatic vessels from these nodes pass to the central and apical groups of axillary lymph nodes.

The **subscapular (posterior) group of axillary lymph nodes** consists of six or seven lymph nodes that lie along the posterior axillary fold and subscapular blood vessels. This group of lymph nodes receives lymph from the posterior aspect of the thoracic wall and scapular region. Efferent lymphatic vessels pass from these nodes to the central and apical groups of axillary lymph nodes.

The **humeral (lateral) group of axillary lymph nodes** consists of four to six lymph nodes that lie along the lateral wall of the axilla, medial and posterior to the axillary vein. This

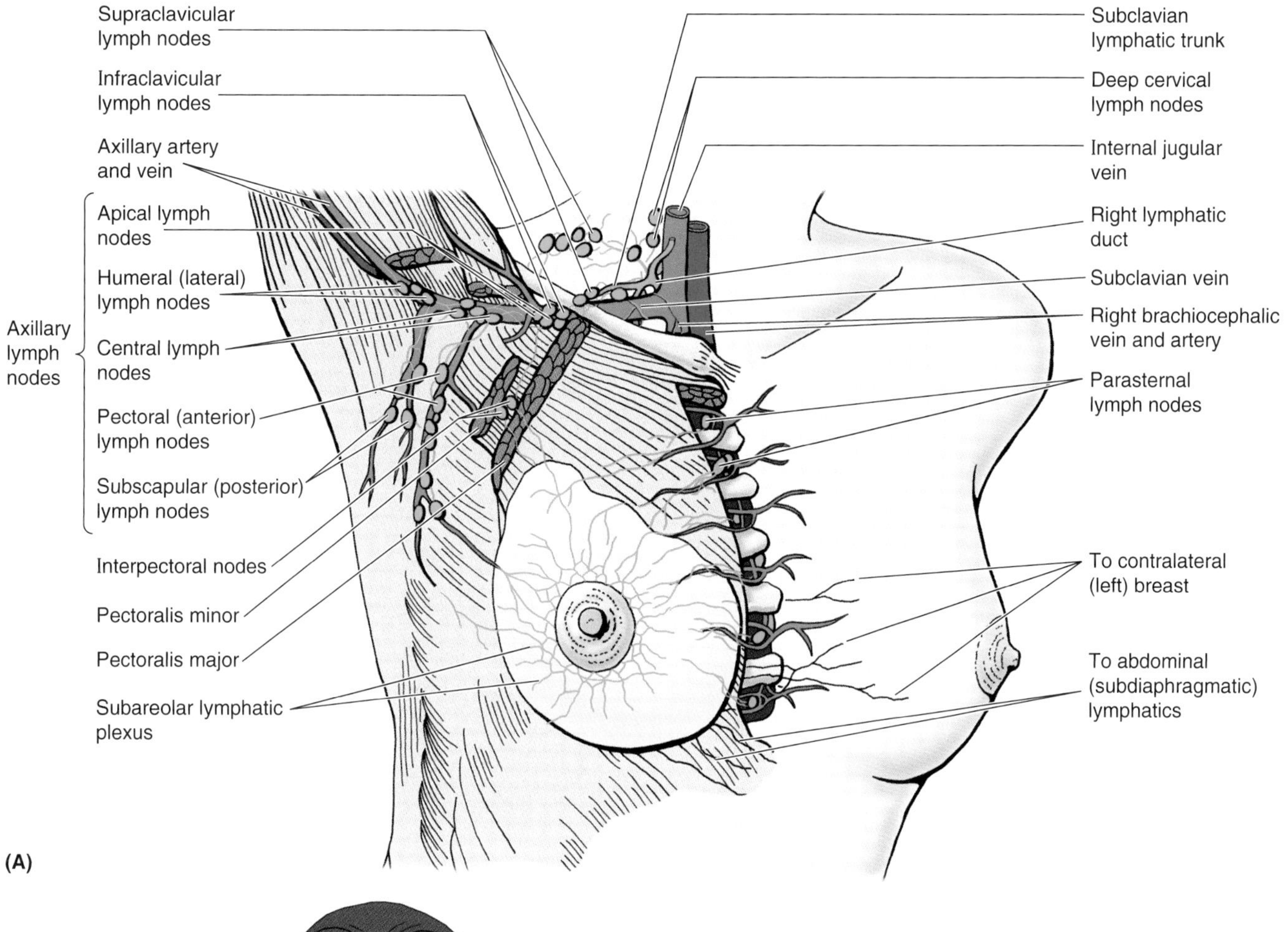

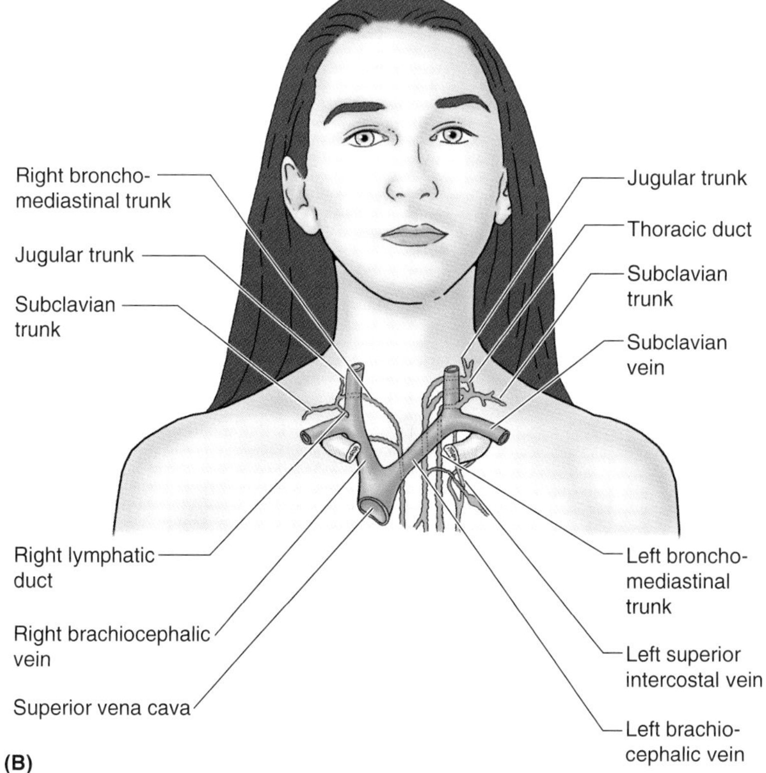

Figure 6.27. Axillary lymph nodes and lymphatic drainage of the right upper limb and breast. A. Observe the five groups of axillary lymph nodes. Most lymphatic vessels of the upper limb terminate in the humeral (brachial or lateral) and central lymph nodes, but those accompanying the upper part of the cephalic vein terminate in the apical lymph nodes. **B.** Lymph passing through the axillary nodes enters efferent lymphatic vessels that form the *subclavian lymphatic trunk*, which usually empties into the junction of the internal jugular and subclavian veins. Occasionally, on the right side, it merges with the jugular lymphatic and/or bronchomediastinal trunks to form a very short *right lymphatic duct*; usually on the left side, it enters the termination of the *thoracic duct*. The lymphatics of the breast are discussed in Chapter 1.

group of lymph nodes receives nearly all the lymph from the upper limb, except that carried by lymphatic vessels accompanying the cephalic vein, which drains to the central and apical axillary nodes.

The **central group of axillary lymph nodes** consists of three or four large lymph nodes situated deep to the pectoralis minor near the base of the axilla, in association with the second part of the axillary artery. As its name indicates, *the central group receives lymph from the pectoral, subscapular, and humeral groups of axillary lymph nodes.* Efferent vessels from the central group pass to the apical group of lymph nodes.

Enlargement of the Axillary Nodes

The axillary lymph nodes enlarge and become tender when infections of the upper limb occur. The humeral group of nodes is the first one to be involved in *lymphangitis* (inflammation of lymphatic vessels, e.g., resulting from a hand infection). Lymphangitis is characterized by the development of red, warm, tender streaks in the skin. Infections in the pectoral region and breast, including the superior part of the abdomen, can also produce enlargement of axillary lymph nodes. In carcinoma (cancer) of the apical group, the lymph nodes often adhere to the axillary vein, which may necessitate excision of part of this vessel. Enlargement of the apical group of lymph nodes may obstruct the cephalic vein superior to the pectoralis minor.

Axillary Lymph Node Dissection

Excision and pathologic analysis of axillary lymph nodes is often necessary for staging and treatment of a malignancy such as breast cancer (see Chapter 1). During axillary lymph node dissection, two nerves are in danger of injury. During the surgery, the *long thoracic nerve* is identified and maintained against the thoracic wall. As discussed previously, cutting the long thoracic nerve to the serratus anterior causes a winged scapula (p. 689). If the *thoracodorsal nerve* to the latissimus dorsi is cut, medial rotation and adduction of the arm are weakened, but deformity does not result. If the lymph nodes around this nerve are obviously malignant, sometimes it has to be sacrificed as the nodes are resected to increase the likelihood of complete removal of all malignant cells. ⊙

Brachial Plexus

Most nerves in the upper limb arise from the **brachial plexus**—a major nerve network supplying the upper limb—which begins in the neck and extends into the axilla. Almost all branches of the brachial plexus arise in the axilla (after it has crossed the 1st rib). *The brachial plexus is formed by the union of the ventral rami of C5 through C8 nerves and the greater part of the ventral ramus of T1* (Fig. 6.28, Table 6.4). The ventral rami of the last four cervical and the first thoracic nerves form the **roots of the brachial plexus;** they usually pass through the gap between the anterior and middle scalene (L. scalenus medius) muscles with the subclavian artery. The sympathetic fibers carried by each root of the plexus are received from the gray rami of the middle and inferior cervical ganglia as they pass between the scalene muscles (Fig. 6.29). In the inferior part of the neck, *the roots of the brachial plexus unite to form three trunks* (Fig. 6.28):

- **Superior trunk**—from the union of the C5 and C6 roots
- **Middle trunk**—a continuation of the C7 root
- **Inferior trunk**—from the union of the C8 and T1 roots.

Each trunk of the brachial plexus divides into anterior and posterior divisions as the plexus passes posterior to the clavicle (through the cervicoaxillary canal). Anterior divisions supply anterior (flexor) compartments of the upper limb, and posterior divisions supply posterior (extensor) compartments.

The divisions of the brachial plexus form three cords:

- Anterior divisions of the superior and middle trunks unite to form the **lateral cord**
- Anterior division of the inferior trunk continues as the **medial cord**
- Posterior divisions of all three trunks unite to form the **posterior cord**.

The cords of the brachial plexus bear the relationship to the second part of the axillary artery that is indicated by their names (e.g., the lateral cord is lateral to the axillary artery, although it may appear to lie superior to the artery because it is most easily seen when the limb is abducted).

The brachial plexus is divided into **supraclavicular** and **infraclavicular parts** by the clavicle (Table 6.4).

- *Supraclavicular branches of the brachial plexus* arise from the roots (ventral rami) and trunks of the brachial plexus (dorsal scapular nerve, long thoracic nerve, nerve to subclavius, and suprascapular nerve) and are approachable through the neck.
- *Infraclavicular branches of the brachial plexus* arise from the cords of the brachial plexus and are approachable through the axilla.

Supraclavicular Branches of the Brachial Plexus

The **dorsal scapular nerve** arises chiefly from the posterior aspect of the ventral ramus of C5 with a frequent contribution from C4. It pierces the middle scalene, runs deep to the levator scapulae (providing a variable supply to it), and enters the deep surface of the rhomboids, supplying them.

The **long thoracic nerve** arises from the posterior aspect of the ventral rami of C5, C6, and C7 and passes through the apex of the axilla (cervicoaxillary canal) posterior to the other brachial plexus components to supply the *serratus anterior.*

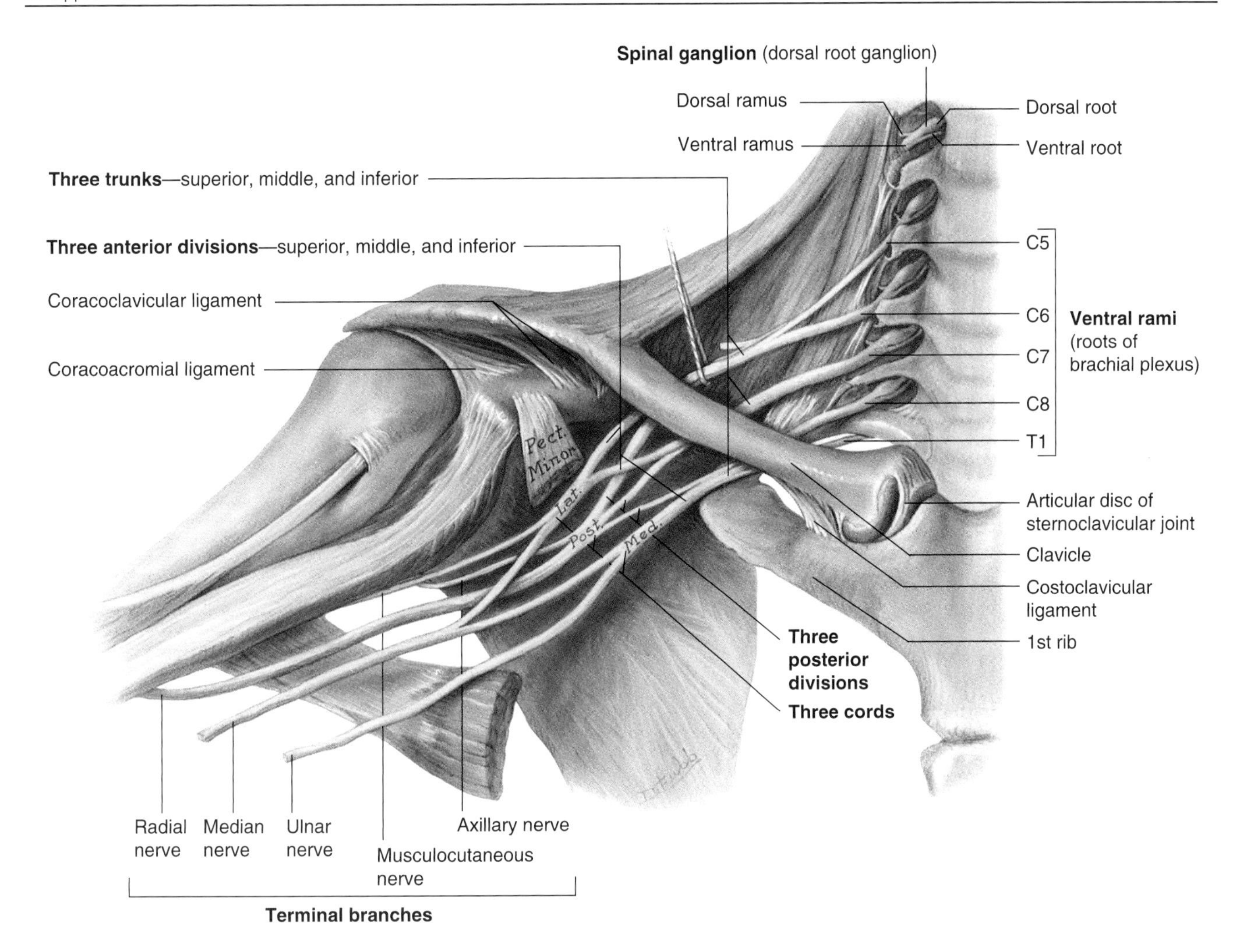

Figure 6.28. Formation of the brachial plexus. This large nerve network provides innervation to the upper limb and shoulder region. The brachial plexus is formed by the ventral rami of the 5th through 8th cervical nerves and the greater part of the ramus of the 1st thoracic nerve (the roots of the brachial plexus). Small contributions may be made by the 4th cervical and 2nd thoracic nerves. Observe the merging and continuation of certain roots of the plexus to three trunks, the separation of each trunk into anterior and posterior divisions, the union of the divisions to form three cords, and the derivation of the main terminal branches from the cords.

The roots from C5 and C6 pierce the middle scalene, and the root from C7 passes anterior to this muscle.

The **nerve to the subclavius**, a slender nerve, arises from the anterior aspect of the superior trunk of the brachial plexus. It receives fibers chiefly from C5, with occasional additions from C4 and C6. It descends posterior to the clavicle and anterior to the brachial plexus to supply the subclavius.

The **suprascapular nerve** arises from the posterior aspect of the superior trunk of the brachial plexus, receiving fibers from C5, C6, and often C4. It supplies the *supraspinatus* and *infraspinatus* and the glenohumeral joint. To reach the muscles, the suprascapular nerve passes laterally across the posterior triangle of the neck, superior to the brachial plexus (Fig. 6.29), and passes through the scapular notch (Table 6.4). The articular branches to the capsule of the glenohumeral joint arise from the intramuscular parts of the muscular branches.

Infraclavicular Branches of the Brachial Plexus

The **lateral cord of the brachial plexus**, carrying fibers primarily from C5 through C7, *has three branches* (Fig. 6.30, *A* and *B*, Table 6.4):

- One side branch—the lateral pectoral nerve
- Two terminal branches—the musculocutaneous nerve and the lateral root of the median nerve.

The **lateral pectoral nerve** (C5, **C6**, and C7) pierces the clavipectoral fascia to supply the pectoralis major (Fig. 6.30*A*). It also sends a branch to the medial pectoral nerve that supplies the pectoralis minor. The lateral pectoral nerve may arise from the lateral cord or from the anterior divisions of the superior and middle trunks of the brachial plexus, accompanying the lateral cord into the axilla.

Table 6.4. **Brachial Plexus and Nerves of the Upper Limb**

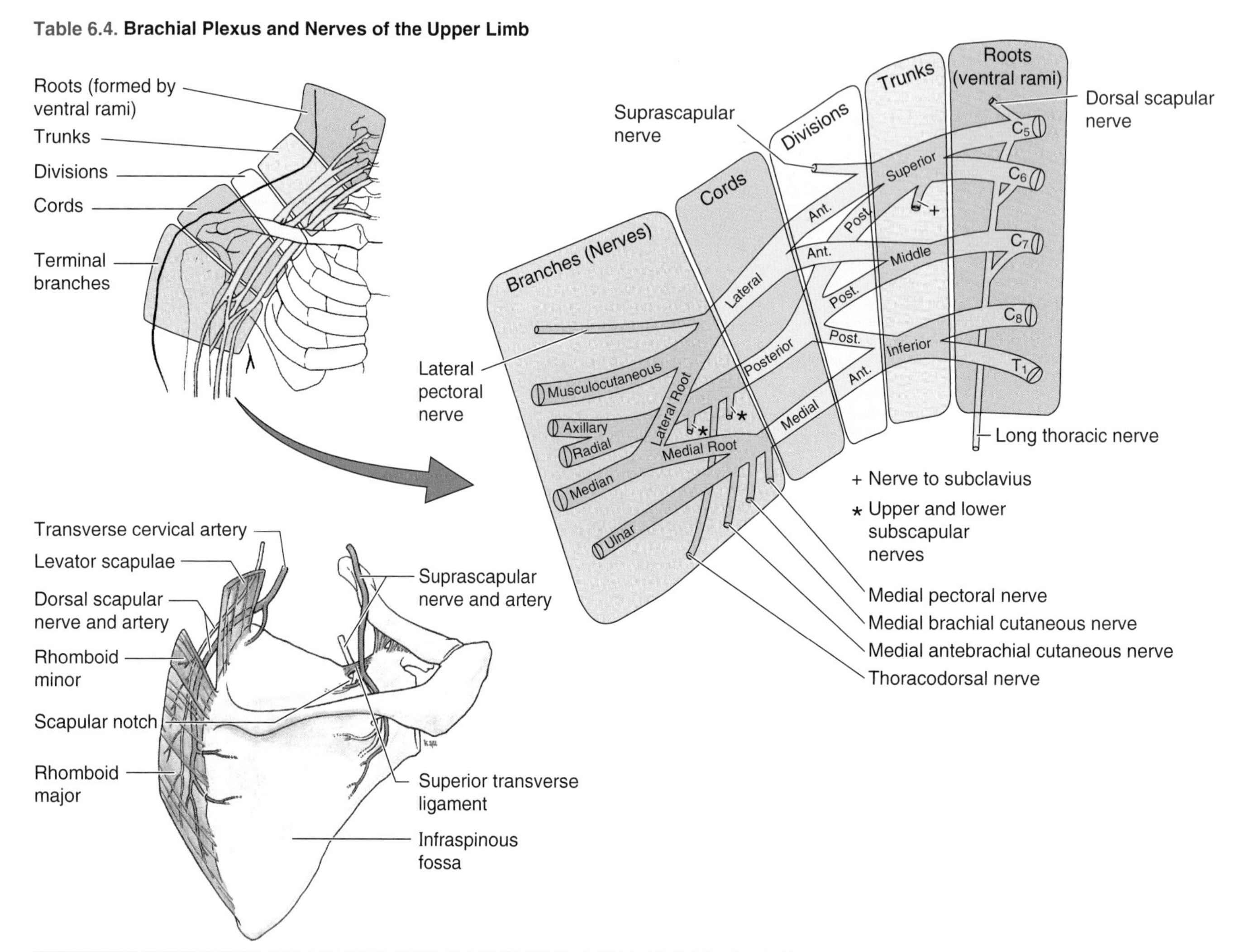

Nerve	Origin	Course	Distribution
Supraclavicular branches			
Dorsal scapular	Ventral ramus of C5 with a frequent contribution from C4	Pierces scalenus medius, descends deep to levator scapulae, and enters deep surface of rhomboids	Innervates rhomboids and occasionally supplies levator scapulae
Long thoracic	Ventral rami of C5–C7	Descends posterior to C8 and T1 rami and passes distally on external surface of serratus anterior	Innervates serratus anterior
Nerve to subclavius	Superior trunk, receiving fibers from C5 and C6 and often C4	Descends posterior to clavicle and anterior to brachial plexus and subclavian artery	Innervates subclavius and sternoclavicular joint
Suprascapular	Superior trunk, receiving fibers from C5 and C6 and often C4	Passes laterally across posterior triangle of neck, through scapular notch under superior transverse scapular ligament	Innervates supraspinatus, infraspinatus, and glenohumeral (shoulder) joint

Table 6.4. (*Continued*) **Brachial Plexus and Nerves of the Upper Limb**

Nerve	Origin	Course	Distribution
Infraclavicular branches			
Lateral pectoral	Lateral cord, receiving fibers from C5–C7	Pierces clavipectoral fascia to reach deep surface of pectoral muscles	Primarily supplies pectoralis major but sends a loop to medial pectoral nerve that innervates pectoralis minor
Musculocutaneous	Lateral cord, receiving fibers from C5–C7	Enters deep surface of coraco-brachialis and descends between biceps brachii and brachialis	Innervates coracobrachialis, biceps brachii, and brachialis; continues as lateral antebrachial cutaneous nerve
Median	Lateral root is a continuation of lateral cord, receiving fibers from C6 and C7; medial root is a continuation of medial cord receiving fibers from C8 and T1	Lateral root joins medial root to form median nerve lateral to axillary artery	Innervates flexor muscles in forearm (except flexor carpi ulnaris, ulnar half of flexor digitorum profundus) and five hand muscles
Medial pectoral	Medial cord, receiving fibers from C8 and T1	Passes between axillary artery and vein and enters deep surface of pectoralis minor	Innervates the pectoralis minor and part of pectoralis major
Medial brachial cutaneous	Medial cord, receiving fibers from C8 and T1	Runs along the medial side of axillary vein and communicates with inter-costobrachial nerve	Supplies skin on medial side of arm
Medial antebrachial cutaneous	Medial cord, receiving fibers from C8 and T1	Runs between axillary artery and vein	Supplies skin over medial side of forearm
Ulnar	A terminal branch of medial cord, receiving fibers from C8 and T1 and often C7	Passes down medial aspect of arm and runs posterior to medial epicondyle to enter forearm	Innervates one and one-half flexor muscles in forearm, most small muscles in hand, and skin of hand medial to a line bisecting 4th digit (ring finger)
Upper subscapular	Branch of posterior cord, receiving fibers from C5 and C6	Passes posteriorly and enters subscapularis	Innervates superior portion of subscapularis
Thoracodorsal	Branch of posterior cord, receiving fibers from C6–C8	Arises between upper and lower subscapular nerves and runs infero-laterally along posterior axillary wall to latissimus dorsi	Innervates latissimus dorsi
Lower subscapular	Branch of posterior cord, receiving fibers from C5 and C6	Passes inferolaterally, deep to subscapular artery and vein, to sub-scapularis and teres major	Innervates inferior portion of subscapularis and teres major
Axillary	Terminal branch of posterior cord, receiving fibers from C5 and C6	Passes to posterior aspect of arm through quadrangular space[a] in company with posterior circumflex humeral artery and then winds around surgical neck of humerus; gives rise to lateral brachial cutaneous nerve	Innervates teres minor and deltoid, shoulder joint, and skin over inferior part of deltoid
Radial	Terminal branch of posterior cord, receiving fibers from C5–C8 and T1	Descends posterior to axillary artery; enters radial groove with deep brachial artery to pass between long and medial heads of triceps	Innervates triceps brachii, anconeus, brachioradialis, and extensor muscles of forearm; supplies skin on posterior aspect of arm and forearm via posterior cutaneous nerves of arm and forearm

[a] Quadrangular space is bounded superiorly by subscapularis and teres minor, inferiorly by teres major, medially by long head of triceps, and laterally by humerus.

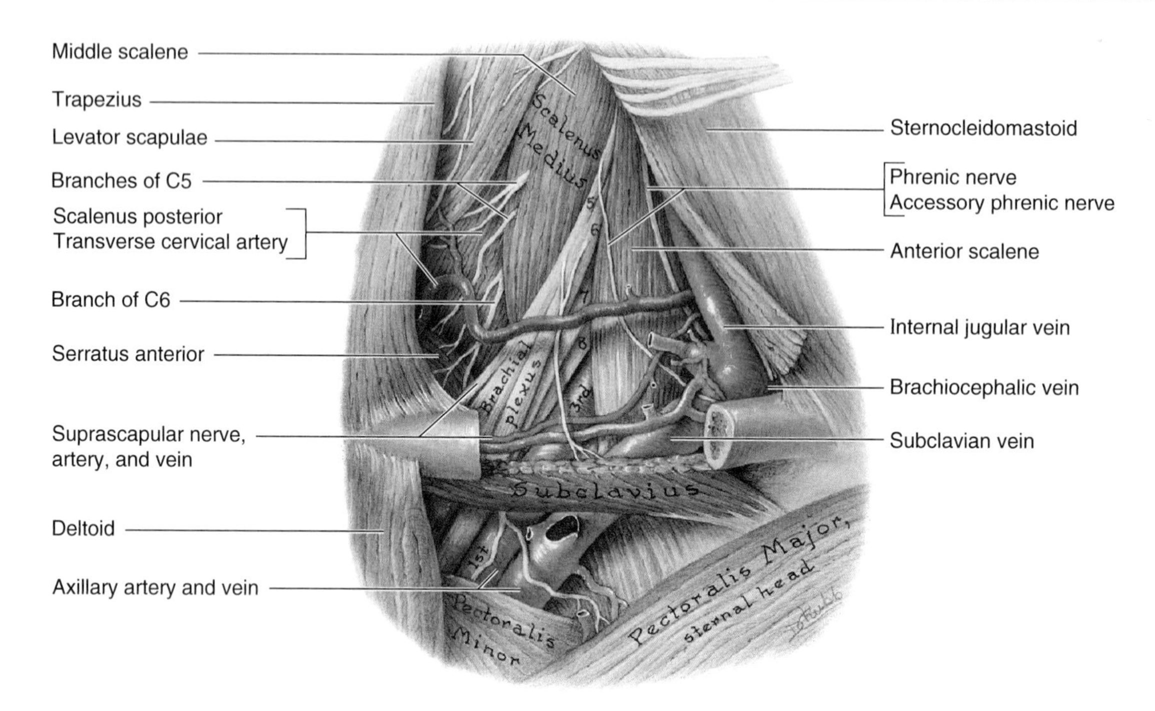

Figure 6.29. Dissection of the right posterior triangle of the neck. Anterolateral or anterior oblique view. Observe the ventral rami constituting the roots of the brachial plexus (numbered), the plexus itself (trunks and divisions at this level), and the subclavian artery emerging with the plexus between the middle scalene (L. scalenus medius) and anterior scalene (L. scalenus anterior) muscles. The inferior root of the plexus (T1) is concealed by the third part of the subclavian artery. As the neurovascular structures pass posterior to the clavicle, they traverse the cervicoaxillary canal connecting the neck and axilla. The subclavius—not of major importance as a muscle—affords some protection to the underlying neurovascular structures when the clavicle is fractured in its middle third.

The **musculocutaneous nerve** (C5 through C7) exits the axilla by piercing the coracobrachialis—supplying the muscle as it traverses it—and then passes between the biceps brachii and the brachialis (Figs. 6.28 and 6.30*B*), supplying both. Thus, the musculocutaneous nerve supplies all the muscles in the anterior compartment of the arm. It continues as the lateral cutaneous nerve of the forearm.

The **median nerve** arises by lateral and medial roots from the lateral and medial cords of the brachial plexus, respectively. The median nerve supplies primarily flexor muscles in the anterior compartment of the forearm, skin of part of the hand, and five muscles of the hand.

The **medial cord of the brachial plexus**, carrying fibers from C8 and T1, *has five branches*:

- Three side branches—medial pectoral nerve, medial cutaneous nerve of the arm, and medial cutaneous nerve of the forearm
- Two terminal branches—ulnar nerve and medial root of the median nerve.

The **medial pectoral nerve (C8,** T1) is a slender nerve (Fig. 6.30*A*) that passes through the pectoralis minor, supplying it and then continuing to supply the pectoralis major. Although it is called the medial pectoral nerve (because it arises from the medial cord of the brachial plexus), *it is located lateral to the lateral pectoral nerve*.

The **medial cutaneous nerve of the arm** (C8, T1) is a slender nerve that supplies skin on the medial side of the arm and the superior part of the forearm.

The **medial cutaneous nerve of the forearm** (C8, T1) is a much larger nerve that runs between the axillary artery and vein and supplies skin on the medial side of the forearm. Because it is close to the ulnar nerve in size and initially in position (Fig. 6.11), the medial cutaneous nerve of the forearm is often mistaken for the ulnar nerve and has been dubbed the "fool's nerve."

The **ulnar nerve** (C8, T1, and sometimes C7) traverses the arm to the forearm without branching (Figs. 6.28 and 6.30*C*). It supplies one and a half muscles in the anterior compartment of the forearm (flexor carpi ulnaris and the ulnar part of flexor digitorum profundus [FDP]) and then continues into the hand, where it supplies most intrinsic muscles and the skin on the medial side of the hand.

The **medial root of the median nerve** unites with the lateral root of the median nerve to form the **median nerve**, the distribution of which has been described.

The **posterior cord of the brachial plexus** (Figs. 6.28, 6.30*D*, and 6.31), carrying fibers from C5 through T1, also has *five branches*:

- Three side branches—upper subscapular, thoracodorsal, and lower subscapular nerves
- Two terminal branches—axillary and radial nerves.

The **upper subscapular nerve** (C5, C6) supplies the subscapularis; the **thoracodorsal nerve (C6, C7,** C8) supplies the latissimus dorsi; and the **lower subscapular nerve** (C6, C7) supplies the teres major; as well as the inferior part of the subscapularis.

The **axillary nerve** (C5, C6), a terminal branch of the posterior cord (Fig. 6.30*D*), supplies the teres minor as it exits the axilla through the *quadrangular space* (Table 6.4). It then supplies the deltoid from its deep posterior aspect and continues as the *superior lateral cutaneous nerve*, supplying the skin over the inferior half of the deltoid.

The **radial nerve** (C5 through C8, T1), the other terminal branch of the posterior cord, is *the largest branch of the brachial plexus* (Fig. 6.30*D*). It supplies all the extensor muscles of the posterior compartments of the upper limb and skin on the posterior aspect of the arm and forearm. In the axilla, the radial nerve lies posterior to the axillary artery and anterior to the subscapularis, teres major, and latissimus dorsi muscles. As it leaves the axilla, the radial nerve runs posteroinferiorly and laterally between the long and medial heads of the triceps. It enters the radial groove on the humerus, where it is vulnerable to injury when the humerus fractures.

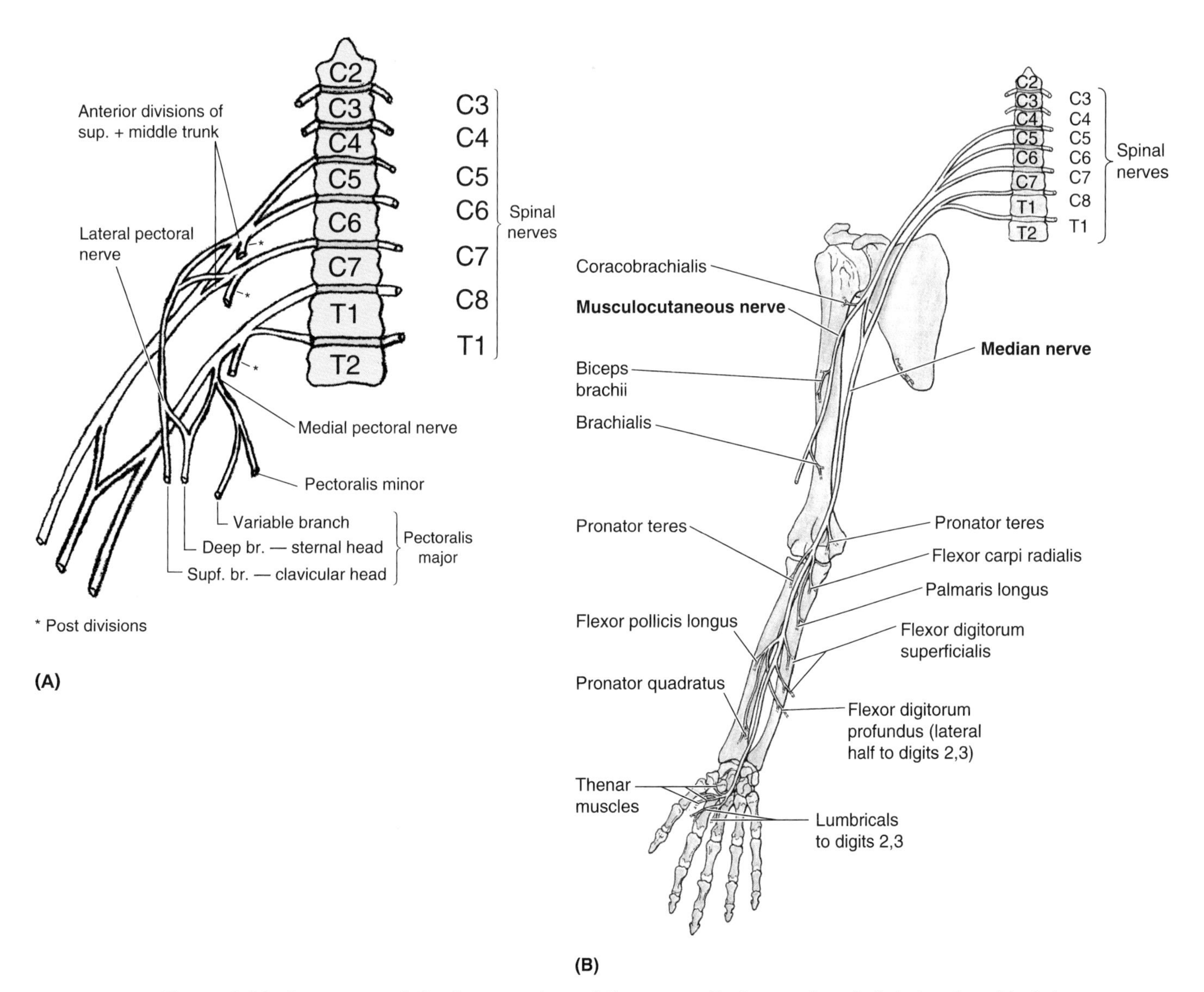

Figure 6.30. Summary of the innervation of the upper limb muscles. A. Anterior view. *Medial and lateral pectoral nerves.* **B.** Anterior view. *Median and musculocutaneous nerves.* The average levels at which the motor branches leave the stems of the main nerves are shown.

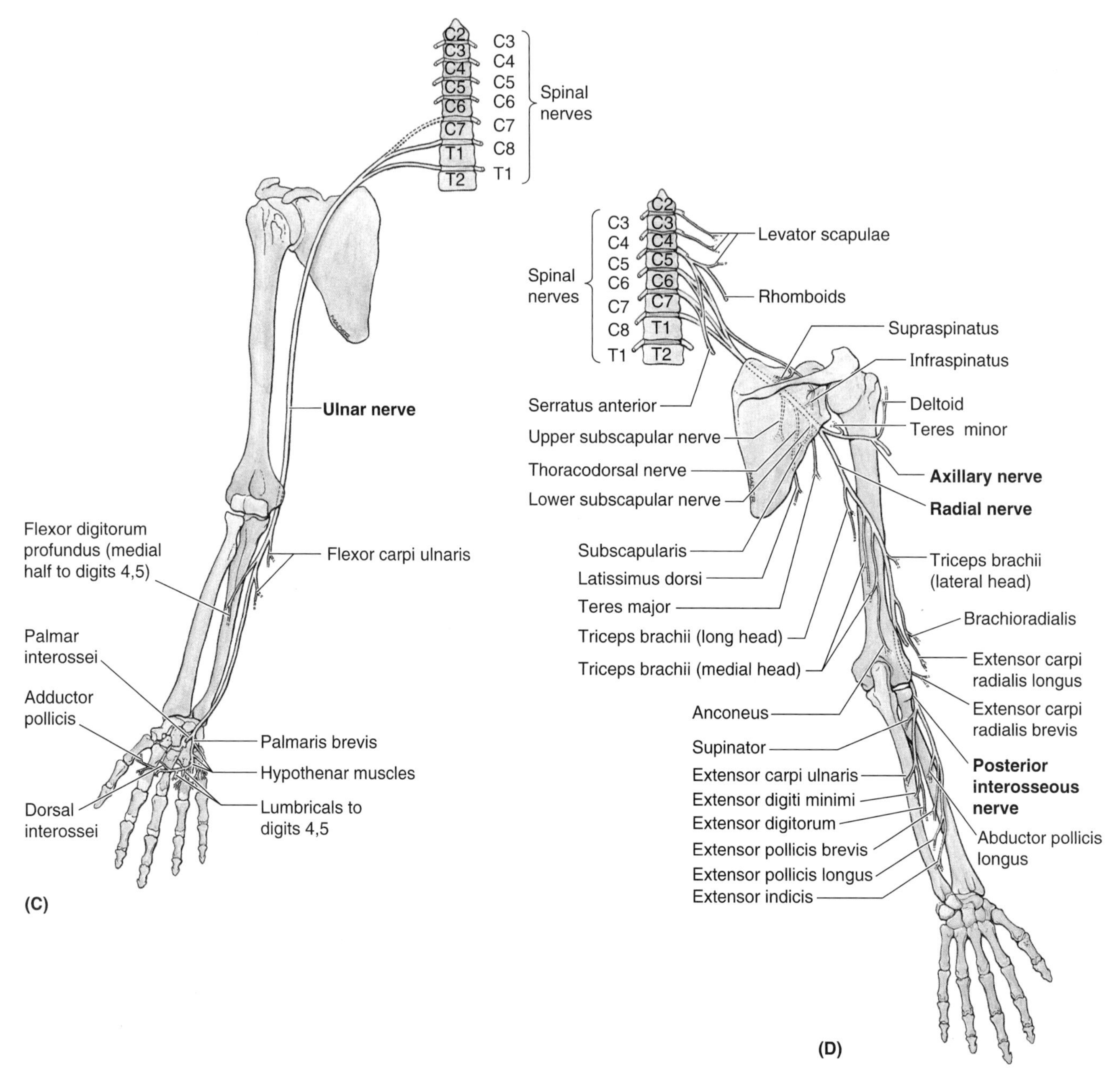

Figure 6.30. *(Continued)* **C.** Anterior view. *Ulnar nerve.* The average levels of origin of the motor branches are shown. **D.** Posterior view. *Axillary and radial nerves.* The posterior interosseous nerve, the continuation of the deep branch of the radial nerve, supplies the abductor pollicis longus (APL), extensor pollicis brevis (EPB), extensor pollicis longus (EPL), and extensor indicis. The dorsum of the hand has no fleshy muscle fibers; therefore, it has no motor nerves.

Variations of the Brachial Plexus

Variations in the formation of the brachial plexus are common (Bergman et al., 1988). In addition to the five ventral rami (C5 through C8 and T1) that form the roots of the brachial plexus, small contributions may be made by the ventral rami of C4 or T2. When the superiormost root (ventral ramus) of the plexus is C4 and the inferiormost root is C8, it is a *prefixed brachial plexus.* Alternately, when the superior root is C6 and the inferior root is T2, it is a *postfixed brachial plexus.* In the latter type, the inferior trunk of the plexus may be compressed by the 1st rib, producing neurovascular symptoms in the upper limb. *Variations may also occur in the:*

- Formation of trunks, divisions, and cords
- Origin and/or combination of branches
- Relations to axillary artery and scalene muscles.

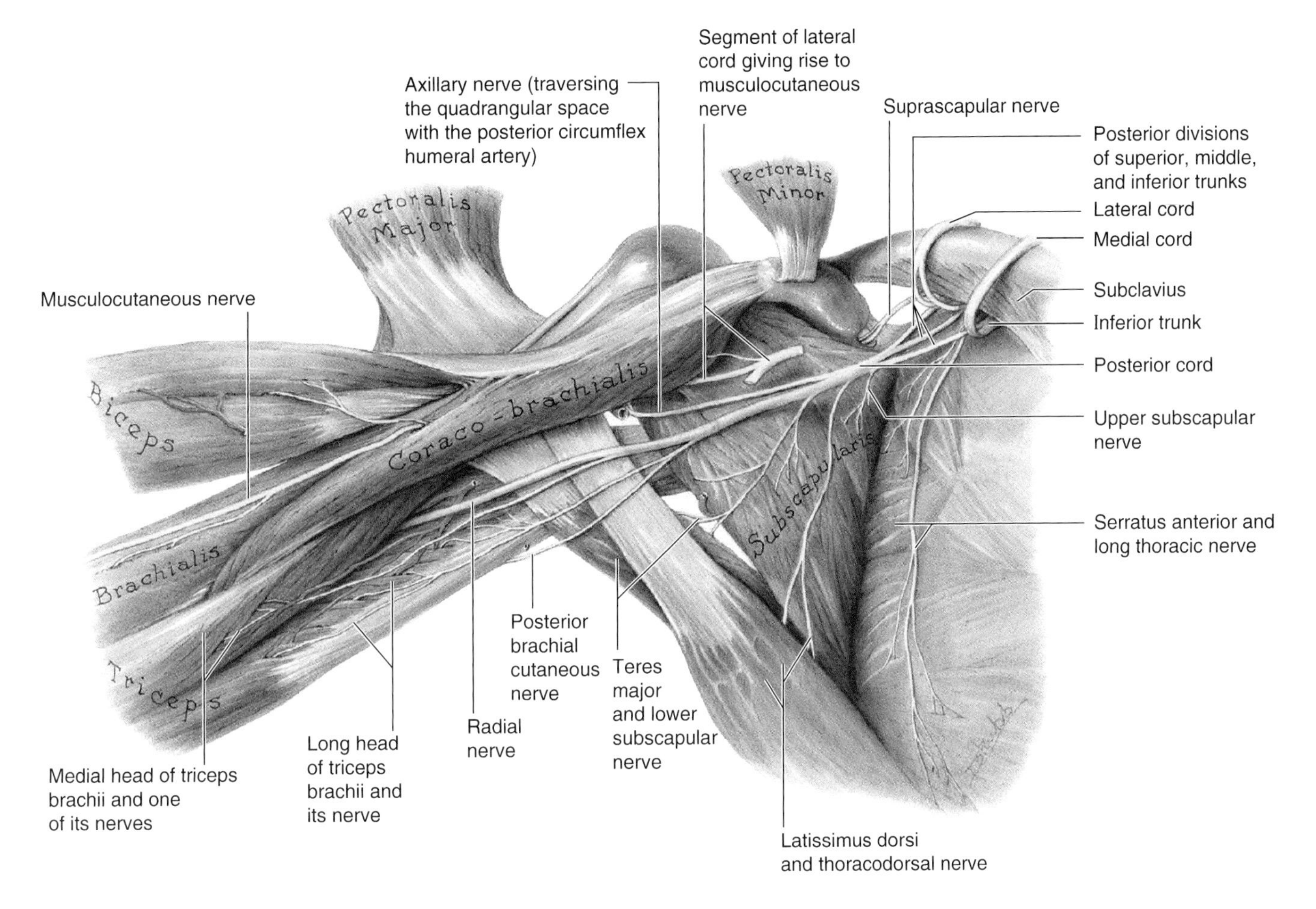

Figure 6.31. Posterior wall of the axilla, musculocutaneous nerve, and posterior cord of the brachial plexus. The pectoralis major and minor muscles are reflected superolaterally, and the lateral and medial cords of the brachial plexus are reflected superomedially. All major vessels and the nerves arising from the medial and lateral cords of the brachial plexus (except for the musculocutaneous nerve from the lateral cord) are removed. The posterior cord, formed by the merging of the posterior divisions of all three trunks of the brachial plexus, gives rise to five nerves: radial, axillary, upper and lower subscapular, and thoracodorsal. Note that the musculocutaneous nerve traverses the coracobrachialis muscle and that the lower subscapular nerve supplies the teres major as well as the subscapularis.

▶ In some individuals, trunk divisions or cord formations may be absent in one or other parts of the plexus; however, the makeup of the terminal branches is unchanged. In addition, the lateral or medial cords may receive fibers from ventral rami inferior or superior to the usual levels, respectively.

Because each peripheral nerve is a collection of nerve fibers bound together by connective tissue, it is understandable that the median nerve, for instance, may have two medial roots instead of one (i.e., the nerve fibers are simply grouped differently). This results from the fibers of the medial cord of the brachial plexus dividing into three branches, two forming the median nerve and the third forming the ulnar nerve. Sometimes it may be more confusing when the two medial roots are completely separate; however, understand that although the median nerve may have two medial roots, the components of the nerve are the same (i.e., the impulses arise from the same place and reach the same destination whether they go through one or two roots).

Brachial Plexus Injuries

Injuries to the brachial plexus are important because they affect movements and cutaneous sensations in the upper limb. Disease, stretching, and wounds in the posterior triangle of the neck or in the axilla may produce brachial plexus injuries. Signs and symptoms depend on which part of the plexus is involved. Injuries to the brachial plexus result in *paralysis* and *anesthesia*. Testing the patient's ability ▶

▶ to perform movements assesses the degree of paralysis. In *complete paralysis,* no movement is detectable. In incomplete paralysis, not all muscles are paralyzed; therefore, the patient can move, but the movements are weak compared with those on the normal side. Determining the ability of the person to feel pain (e.g., from a pinprick of the skin) tests the degree of anesthesia.

Injuries to superior parts of the brachial plexus (C5 and C6) usually result from an excessive increase in the angle between the neck and the shoulder. These injuries can occur in a person who is thrown from a motorcycle or a horse and lands on the shoulder in a way that widely separates the neck and shoulder (*A*). When thrown, the person's shoulder often hits something (e.g., a tree or the ground) and stops, but the head and trunk continue to move. This stretches or tears (avulses) superior parts of the brachial plexus. Injury to the superior trunk of the plexus is apparent by the characteristic position of the limb ("waiter's tip position"), in which the limb hangs by the side in medial rotation (*B*). *Upper brachial plexus injuries* can also occur in a newborn when excessive stretching of the neck occurs during delivery (*C*).

Various terms are used to describe injuries to superior parts of the brachial plexus: *Erb palsy* (paralysis), *Erb-Duchenne palsy* (paralysis), *Duchenne-Erb palsy* (paralysis), and *upper radicular syndrome* (radicular refers to the roots of the plexus). In all cases, paralysis of the muscles of the shoulder and arm supplied by the C5 and C6 spinal nerves occurs: deltoid, biceps, brachialis, and brachioradialis. The usual clinical appearance is an upper limb with an adducted shoulder, medially rotated arm, and extended elbow. The lateral aspect of the upper limb also experiences loss of sensation. Chronic microtrauma to the superior trunk of the brachial plexus from carrying a heavy backpack can produce motor and sensory deficits in the distribution of the musculocutaneous and radial nerves. The upper brachial plexus injury may produce muscle spasms and a severe disability in hikers (*backpacker's palsy*) who carry heavy backpacks for long periods.

Acute brachial plexus neuritis (brachial plexus neuropathy) is a neurological type of disorder of unknown cause that is characterized by the sudden onset of severe pain, usually around the shoulder (Rowland, 1995). Usually the pain begins at night and is soon followed by muscle weakness and sometimes muscular atrophy (*neurologic amyotrophy*). Inflammation of the brachial plexus (*brachial neuritis*) is often preceded by some event (e.g., upper respiratory infection, vaccination, or nonspecific trauma). The nerve fibers involved are usually derived from the superior trunk of the brachial plexus.

Compression of cords of the brachial plexus may result from prolonged hyperabduction of the arm during performance of manual tasks over the head, such as painting or plastering a ceiling. The cords are impinged or compressed between the coracoid process of the scapula and the pectoralis minor tendon. *Common neurological symptoms* are pain running down the arm, numbness, paresthesia (tingling), erythema (redness of skin caused by capillary dilation), and weakness of the hands. These signs and symptoms result from *compression of the axillary vessels and nerves.* Compression of the axillary artery and vein causes ischemia of the upper limb and distension of the superficial veins. These signs and symptoms are part of the *hyperabduction syndrome* of the upper limb.

Injuries to inferior parts of the plexus are much less common. Lower brachial plexus injuries may occur when the upper limb is suddenly pulled superiorly—for example, when a person grasps something to break a fall (*D*) or a baby's upper limb is pulled excessively during delivery (*E*). Inferior parts of the brachial plexus may also be injured ▶

(A)

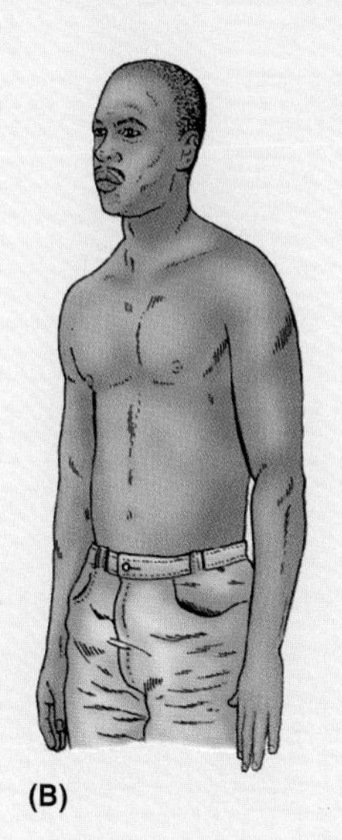

(B)

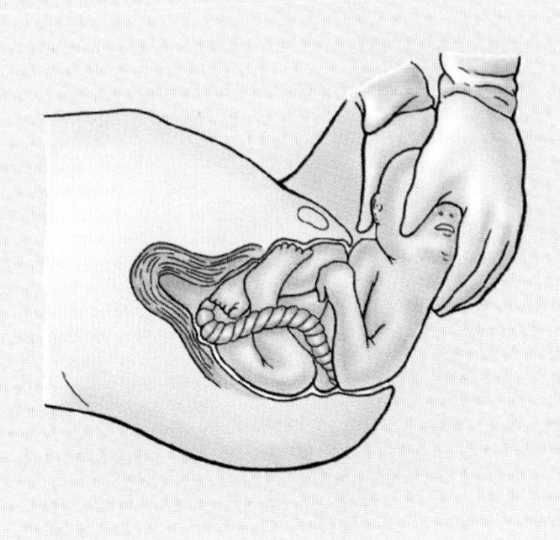

(C)

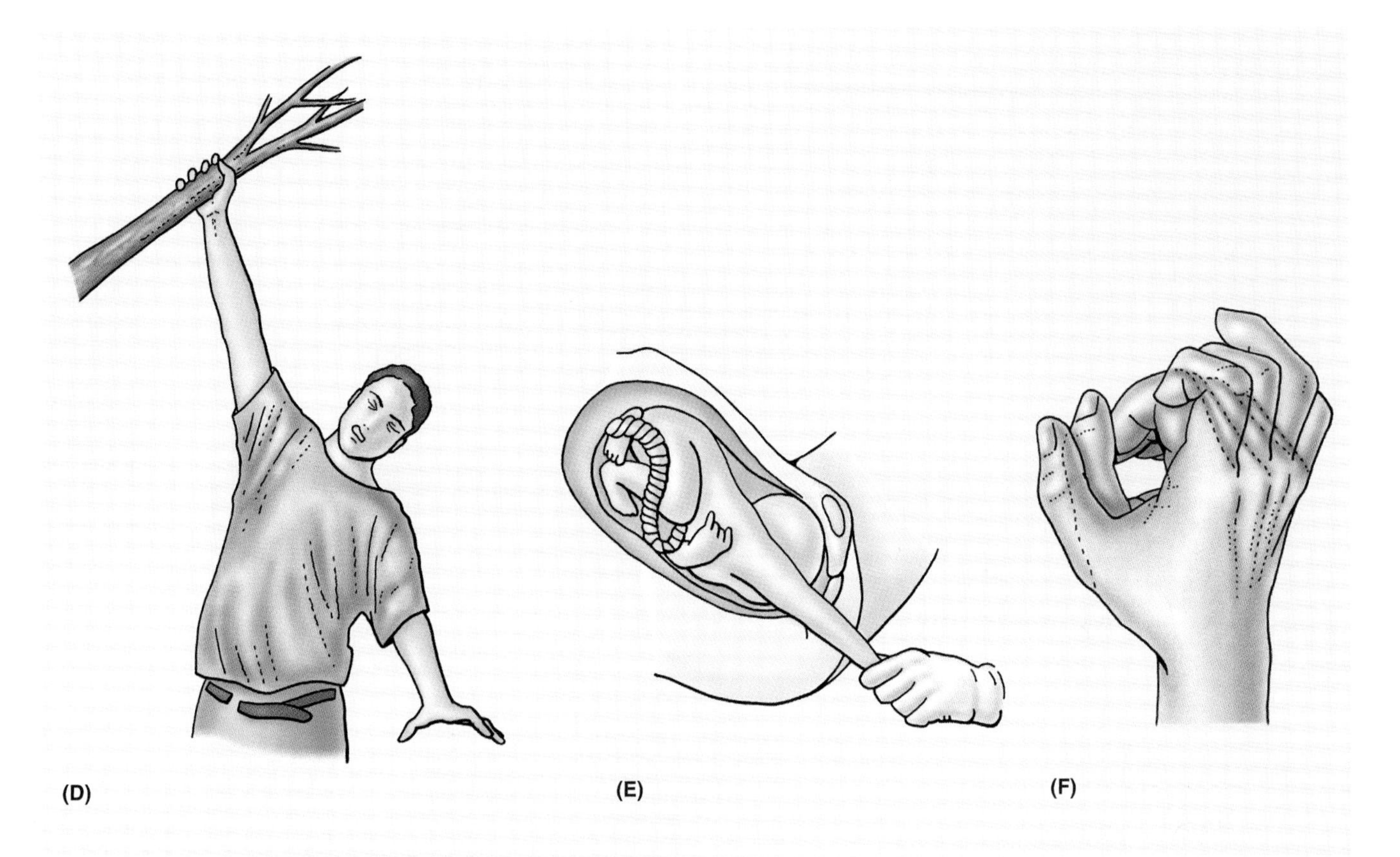

▶ during a breech birth when the infant's limbs are pulled over the head. These events injure the inferior trunk of the brachial plexus (C8 and T1) and may pull (avulse) the dorsal and ventral roots of the spinal nerves from the spinal cord. The short muscles of the hand are affected, and a *clawhand* results (*F*).

Brachial Plexus Block

Injection of an anesthetic solution into the angle between the posterior border of the sternocleidomastoid and the clavicle surrounds the thin axillary sheath (Fig. 6.24*A*) containing the cords of the brachial plexus and axillary vessels. The anesthetic interrupts nerve impulses and produces anesthesia of the structures supplied by the branches of the cords of the plexus, thereby rendering insensitive all the deep structures of the upper limb and the skin distal to the middle of the arm. Combined with an occlusive tourniquet technique, this procedure enables surgeons to operate on the upper limb without using a general anesthetic. ○

Surface Anatomy of the Pectoral and Scapular Regions

The clavicle is the boundary demarcating the root of the neck from the thorax. As the clavicle passes laterally, its medial part can be felt to be convex anteriorly. The large vessels and nerves to the upper limb pass posterior to this convexity. The flattened acromial end of the clavicle does not reach the point of the shoulder, formed by the lateral tip of the acromion of the scapula. The acromion is palpable and may be obvious when the deltoid contracts against resistance.

The **deltopectoral triangle** (infraclavicular fossa) is the slightly depressed area just inferior to the lateral part of the clavicle. This triangle is bounded by the clavicle superiorly, the deltoid laterally, and the pectoralis major medially. The *cephalic vein* ascending from the upper limb enters the deltopectoral triangle and pierces the clavipectoral fascia to enter the *axillary vein*. The **coracoid process** of the scapula is not subcutaneous; it is covered by the anterior border of the deltoid; however, the tip of the process can be felt on deep palpation in the deltopectoral triangle. *The coracoid process is used as a bony landmark when performing a brachial plexus block*, and its position is of importance in diagnosing shoulder dislocations.

While lifting a weight, palpate the anterior sloping ▶

▶ border of the trapezius and where its superior fibers attach to the lateral third of the clavicle. When the arm is abducted and then adducted against resistance, the sternocostal part of the pectoralis major can be seen and palpated. If the **anterior axillary fold** bounding the axilla is grasped between the finger and thumb, the inferior border of the sternocostal head of the pectoralis major can be felt. Several digitations of the **serratus anterior** are visible inferior to the anterior axillary fold. The **posterior axillary fold** is composed of skin and muscular tissue (latissimus dorsi and teres major) bounding the axilla posteriorly.

The lateral border of the acromion may be followed posteriorly with the fingers until it meets the **acromial angle.** Clinically, the length of the arm is measured from the acromial angle to the lateral condyle of the humerus. The **spine of the scapula** is subcutaneous throughout and is easily palpated as it extends medially and slightly inferiorly from the acromion. The **root of the scapular spine** (medial end) is located opposite the T3 spinous process when the arm is adducted. The **medial border of the scapula** may be palpated inferior to the root of the spine as it crosses ribs 2 through 7. It may be visible in some people, ▶

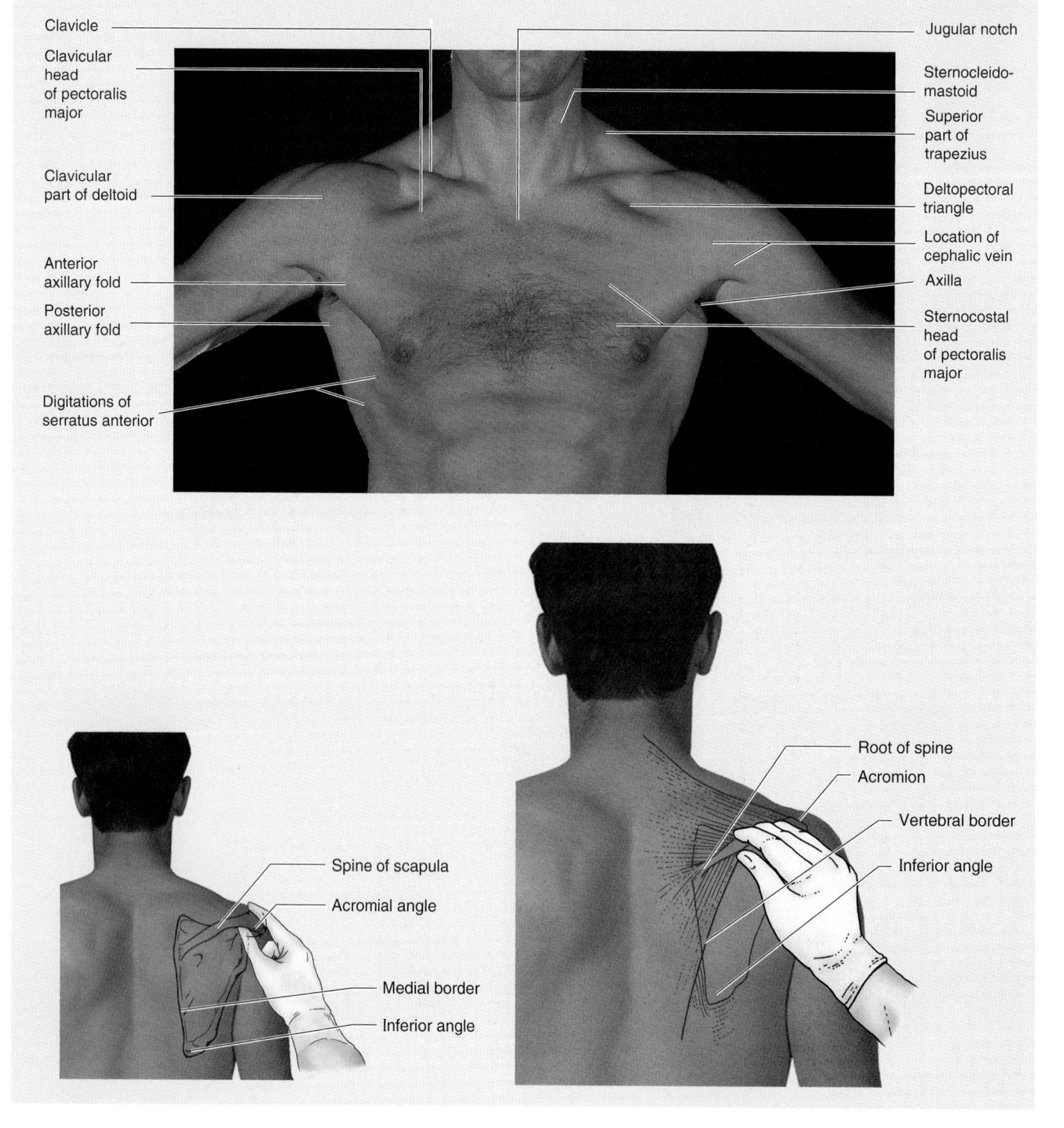

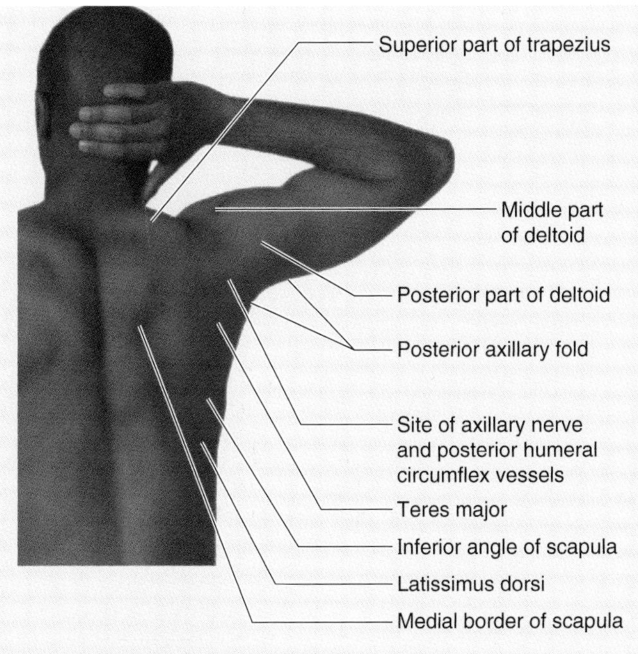

▶ especially thin people. The **inferior angle of the scapula** is easily palpated and is usually visible. Grasp the inferior scapular angle with the thumb and fingers and move the scapula up and down. When the arm is adducted, the inferior scapular angle is opposite the T7 spinous process and lies over the 7th rib or intercostal space.

The **deltoid** covering the proximal part of the humerus *forms the rounded muscular contour of the shoulder*. The greater tubercle of the humerus is the most lateral bony point in the shoulder when the arm is adducted and may be felt on deep palpation through the deltoid inferior to the lateral border of the acromion. When the arm is abducted, observe that the greater tubercle disappears beneath the acromion and is no longer palpable. The borders and parts of the deltoid are usually visible when the arm is abducted against resistance. Loss of the rounded muscular appearance of the shoulder and the appearance of a surface depression distal to the acromion are characteristic of a *dislocated shoulder*—dislocation of the glenohumeral joint. The depression results from displacement of the humeral ▶

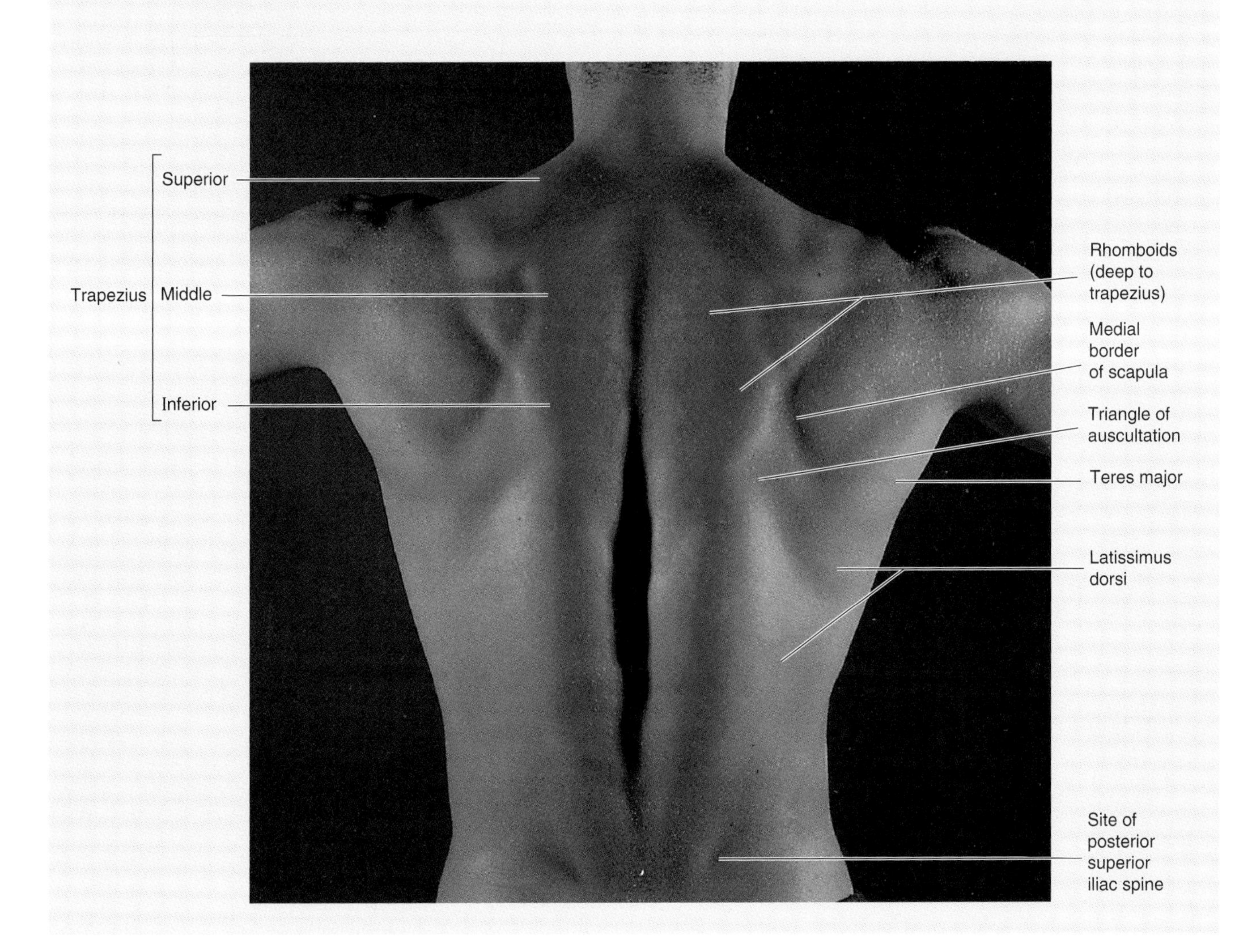

▶ head. The **teres major** is prominent when the abducted arm is adducted against resistance.

When the upper limbs are abducted, the scapulae move laterally on the thoracic wall, enabling the rhomboids to be palpated. Being deep to the trapezius, the rhomboids are not always visible. If the rhomboids of one side are paralyzed, the scapula on the affected side remains farther from the midline than on the normal side because the paralyzed muscles are unable to retract it. ⊙

Arm

The arm extends from the shoulder to the elbow. Two types of movement occur between the arm and forearm at the elbow joint: flexion-extension and pronation-supination. The muscles performing these movements are clearly divided into anterior and posterior groups. The chief action of both groups is at the elbow joint, but some muscles also act at the glenohumeral joint. The superior part of the humerus provides attachments for tendons of the shoulder muscles.

Muscles of the Arm

Of the four arm (brachial) muscles, *three flexors* (biceps brachii, brachialis, and coracobrachialis) are in the *anterior compartment* and supplied by the musculocutaneous nerve (Figs. 6.31 and 6.32) and *one extensor* (triceps brachii) is in the *posterior compartment* and supplied by the radial nerve. The *anconeus muscle*, at the posterior aspect of the elbow, is partly blended with—and is essentially a distally placed continuation of—the triceps.

Biceps Brachii

As its name "biceps" indicates, the proximal attachment of this fusiform muscle has *two heads* (Bi, two + L. caput, head). The two muscle bellies unite just distal to the middle of the arm (Fig. 6.31, Table 6.5). *The biceps is located in the anterior compartment* of the arm (Fig. 6.32). When the elbow is extended, the biceps is *a simple flexor of the forearm*; however, when the elbow is flexed and more power is needed against resistance, the biceps is *the primary (most powerful) supinator of the forearm* (e.g., when right-handed persons drive a screw into hard wood). It is also used when inserting a corkscrew and pulling the cork from a wine bottle. The biceps barely operates during flexion of the prone forearm. The rounded tendon of the long head of the biceps crosses the head of the humerus within the cavity of the glenohumeral joint. The tendon, surrounded by synovial membrane, descends in the intertubercular groove of the humerus. A broad band, the

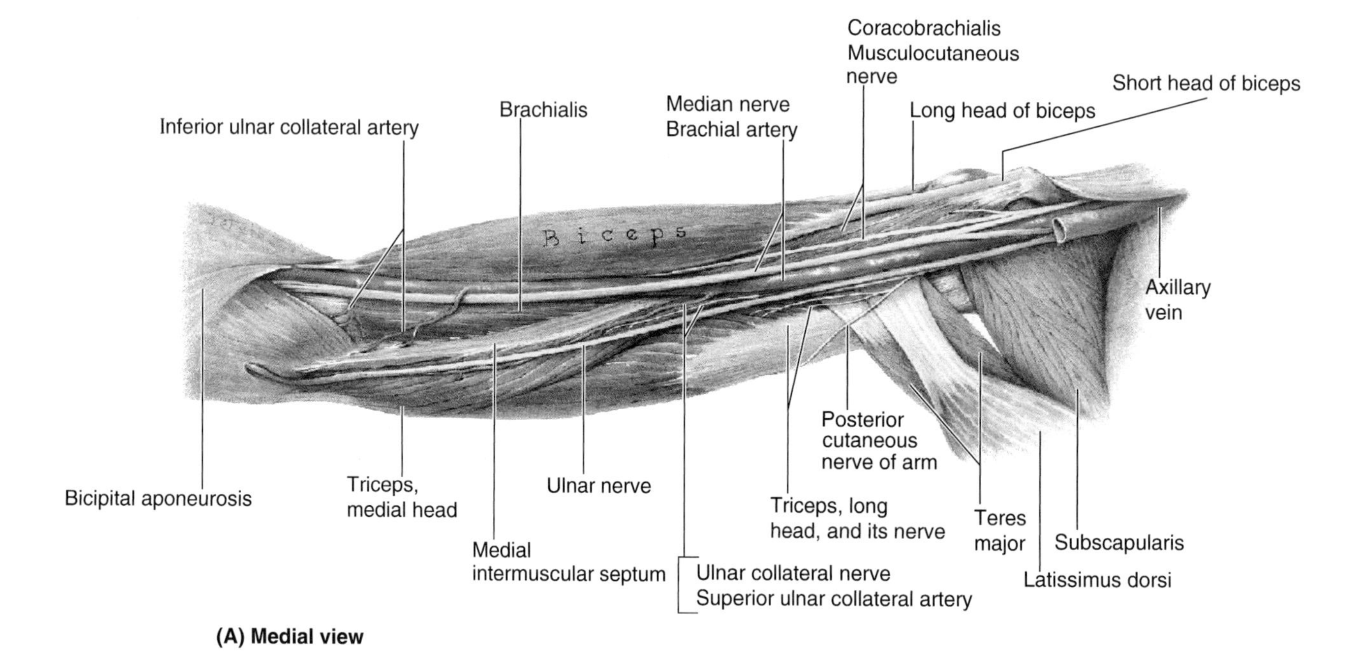

Figure 6.32. Muscles and neurovascular structures of the arm. A. Medial view. Dissection of the right arm. The veins have been removed, except for the proximal part of the axillary vein. Observe the course of the median, musculocutaneous, and ulnar nerves on the medial (protected) aspect of the arm. Observe the biceps, brachialis, and coracobrachialis lying in the *anterior compartment* of the arm and the triceps lying in the *posterior compartment.* Note the medial intermuscular septum separating these two compartments (muscle groups) in the distal two-thirds of the arm.

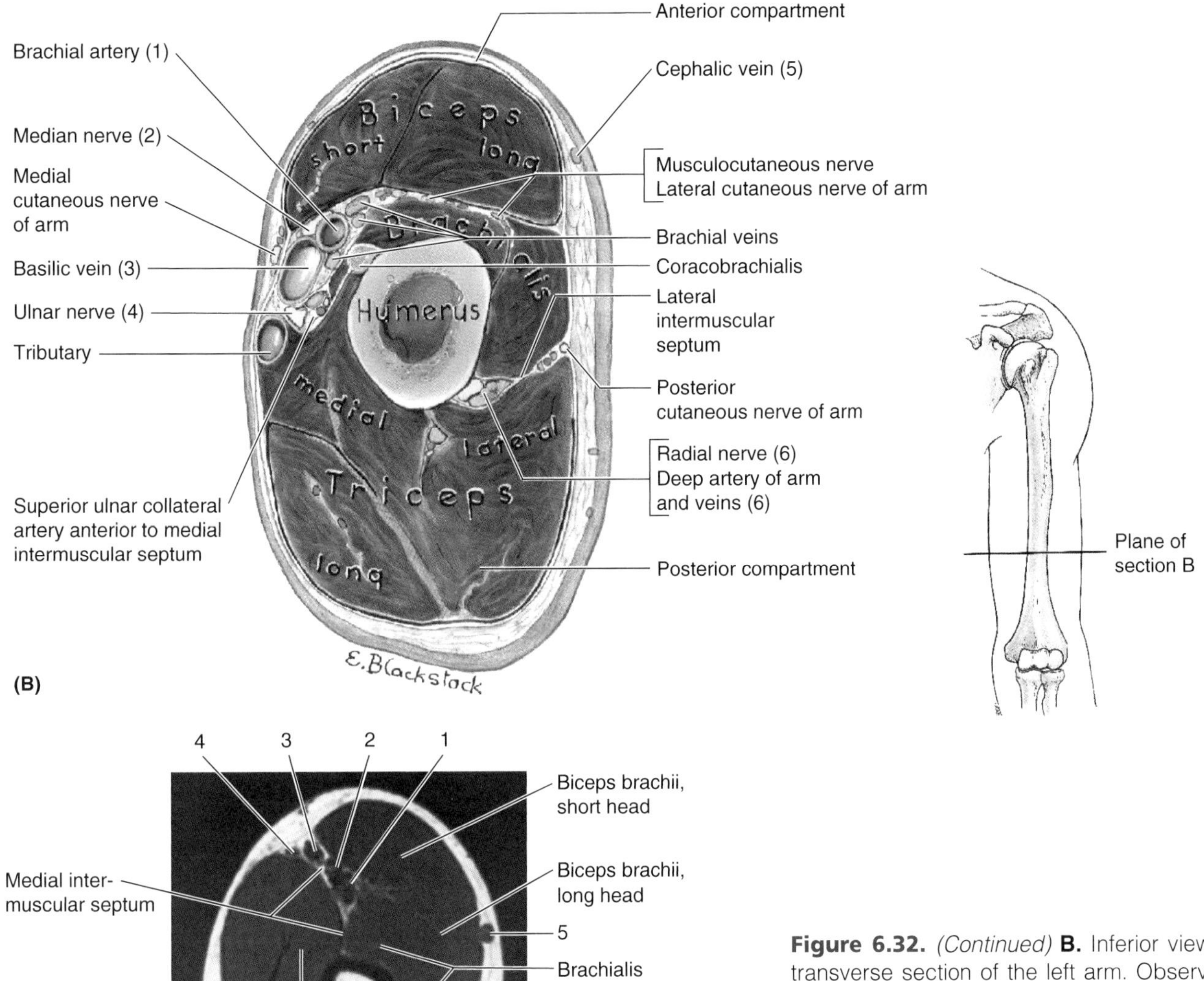

Figure 6.32. *(Continued)* **B.** Inferior view of a transverse section of the left arm. Observe the three heads of the triceps in the posterior compartment. Notice that the radial nerve and its companion vessels are in contact with the humerus. **C.** Transverse MRI demonstrating the features of (**B**) in a living person. (Courtesy of Dr. W. Kucharczyk, Chair of Medical Imaging and Clinical Director of Tri-Hospital Resonance Centre, Toronto, Ontario, Canada)

transverse humeral ligament, passes from the lesser to the greater tubercle of the humerus and converts the intertubercular groove into a canal. The ligament holds the tendon of the long head of the biceps in the groove. Distally, the tendon attaches to the tuberosity of the radius. The *bicipitoradial bursa* separates the biceps tendon from—and reduces abrasion against—the anterior part of the radial tuberosity.

The biceps continues distally as the **bicipital aponeurosis**, a triangular membranous band that runs from the biceps tendon across the cubital fossa and merges with the antebrachial (deep) fascia covering the flexor muscles in the medial side of the forearm (Table 6.5) and attaching by means of the antebrachial fascia with the subcutaneous border of the ulna. The proximal part of the bicipital aponeurosis can be easily felt where it passes obliquely over the brachial artery and median nerve (Fig. 6.33). This aponeurosis affords protection for these and other structures in the cubital fossa. It also helps lessen the pressure of the biceps tendon on the radial tuberosity during pronation and supination of the forearm. Approximately 10% of people have a 3rd head to the biceps, arising at the superomedial part of the brachialis (with which it is blended). In most, the 3rd head lies posterior to the brachial artery. *To test the biceps*, the elbow joint is flexed against resistance when the forearm is supinated. If acting normally, the muscle forms a prominent bulge on the anterior aspect of the arm that is easily palpated.

Table 6.5. **Muscles of the Arm**

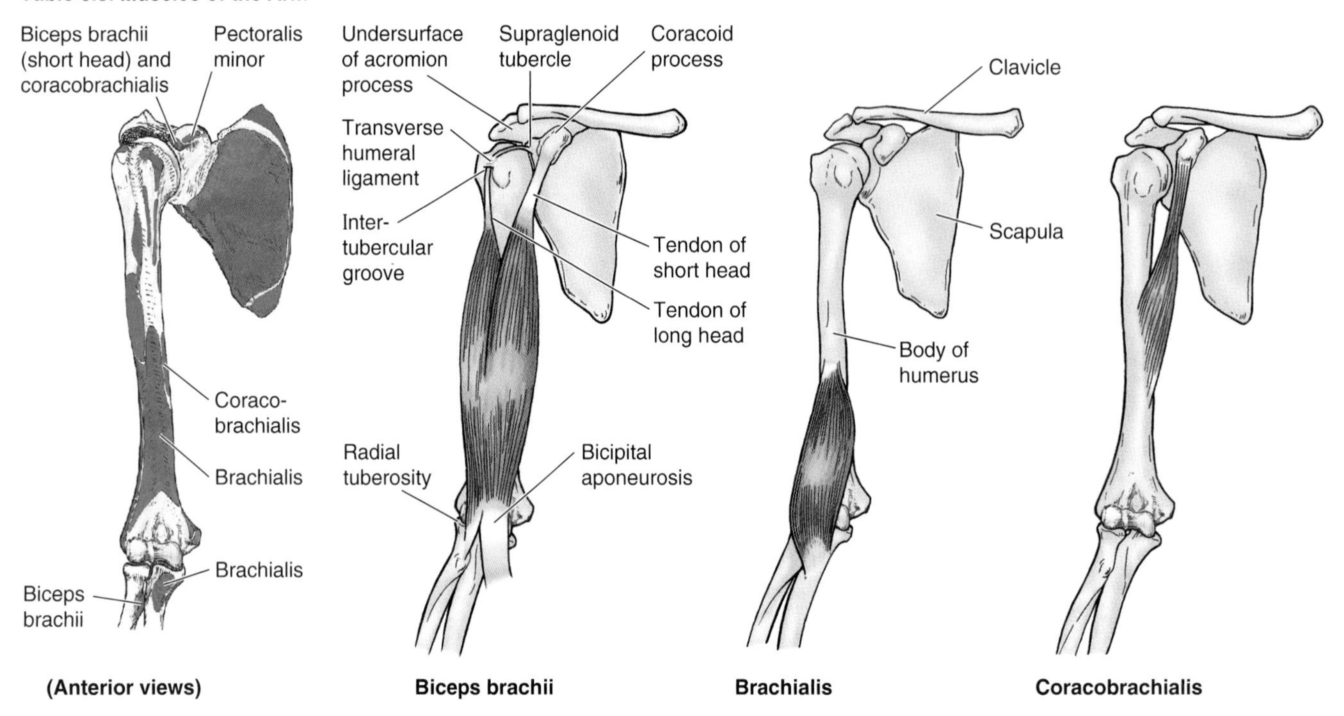

Muscle	Proximal Attachment	Distal Attachment	Innervation[a]	Main Action
Biceps brachii	Short head: tip of coracoid process of scapula Long head: supraglenoid tubercle of scapula	Tuberosity of radius and fascia of forearm via bicipital aponeurosis	Musculocutaneous nerve[b] (C5 and **C6**)	Supinates forearm and, when it is supine, flexes forearm
Brachialis	Distal half of anterior surface of humerus	Coronoid process and tuberosity of ulna		Flexes forearm in all positions
Coracobrachialis	Tip of coracoid process of scapula	Middle third of medial surface of humerus	Musculocutaneous nerve (C5, **C6**, and C7)	Helps to flex and adduct arm

[a] Numbers indicate spinal cord segmental innervation (e.g., C5 and **C6** indicate that nerves supplying the biceps brachii muscle are derived from the 5th and 6th cervical segments of spinal cord). **Boldface** numbers indicate the main segmental innervation. Damage to these segments, or to motor nerve roots arising from them, results in paralysis of muscles concerned.
[b] Some of the lateral part of the brachialis is innervated by a branch of the radial nerve.

Biceps Tendinitis

The tendon of the long head of the biceps, enclosed by a synovial sheath, moves back and forth in the intertubercular groove of the humerus. Wear and tear of this mechanism is a common cause of shoulder pain. Inflammation of the tendon (*biceps tendinitis*), usually the result of repetitive microtrauma, is common in sports involving throwing (e.g., baseball and cricket) and use of a racquet (e.g., tennis). A tight, narrow, and/or rough intertubercular groove may irritate and inflame the tendon, producing tenderness and *crepitus* (crackling sound).

Dislocation of the Tendon of the Long Head of the Biceps

Sometimes the tendon is partially or completely dislocated from the intertubercular groove in the humerus. This painful condition may occur in young persons during traumatic separation of the proximal epiphysis of the humerus. The injury also occurs in older athletes with a history of biceps tendinitis. Usually a sensation of popping or catching is felt during arm rotation. ▶

Table 6.5. (*Continued*) **Muscles of the Arm**

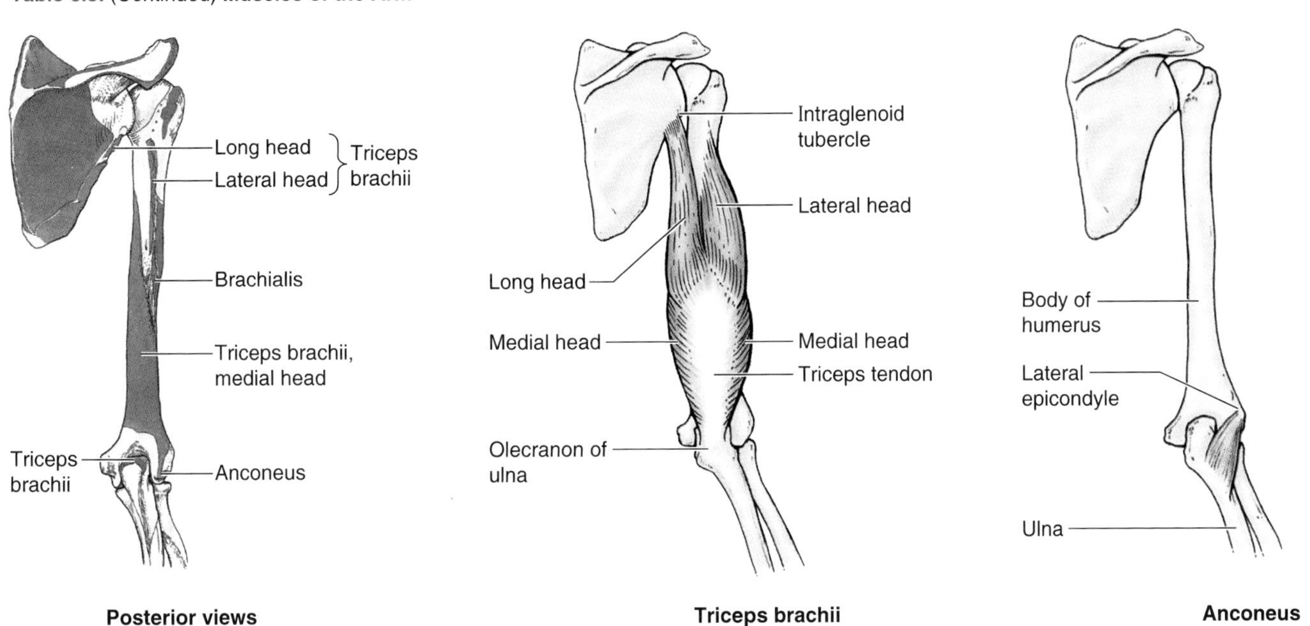

Muscle	Proximal Attachment	Distal Attachment	Innervation[a]	Main Action
Triceps brachii	Long head: infraglenoid tubercle of scapula Lateral head: posterior surface of humerus, superior to radial groove Medial head: posterior surface of humerus, inferior to radial groove	Proximal end of olecranon of ulna and fascia of forearm	Radial nerve (C6, **C7**, and **C8**)	Extends the forearm; it is chief extensor of forearm; long head steadies head of abducted humerus
Anconeus	Lateral epicondyle of humerus	Lateral surface of olecranon and superior part of posterior surface of ulna	Radial nerve (C7, C8, and T1)	Assists triceps in extending forearm; stabilizes elbow joint; abducts ulna during pronation

Brachialis

This flattened fusiform muscle lies posterior (deep) to the biceps. Its distal attachment covers the anterior part of the elbow joint (Figs. 6.31–6.34, Table 6.5). *The brachialis is the main flexor of the forearm; it flexes the forearm in all positions and during slow and quick movements.* When the forearm is extended slowly, the brachialis steadies the movement by slowly relaxing (e.g., you use it to pick up and put down a teacup carefully). The brachialis always contracts during flexion of the elbow joint and is primarily responsible for maintaining flexion. Because of its many functions, it is regarded as the workhorse of the elbow flexors.

Coracobrachialis

This elongated muscle in the superomedial part of the arm is a useful landmark (Figs. 6.31 and 6.32, Table 6.5). For example, the musculocutaneous nerve pierces it, and the distal part of its attachment indicates the location of the nutrient foramen of the humerus. The coracobrachialis helps flex and adduct the arm and stabilize the glenohumeral joint. With the deltoid and long head of the triceps, it serves as a shunt muscle, resisting downward dislocation of the head of the humerus. The median nerve and/or the brachial artery may run deep to the coracobrachialis and be compressed by it.

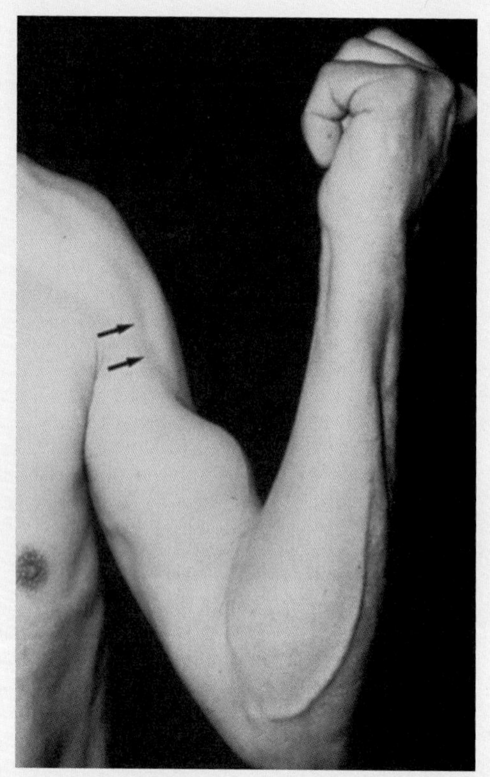

Rupture of biceps tendon

Rupture of the Tendon of the Long Head of the Biceps

Rupture of the tendon usually results from wear and tear of an inflamed tendon as it moves back and forth in the intertubercular groove of the humerus. Rupture of the tendon commonly occurs in athletes older than 33 years (e.g., baseball pitchers). Usually the tendon is torn from its attachment to the supraglenoid tubercle of the scapula. The rupture is commonly dramatic and is associated with a snap or pop. The detached muscle belly forms a ball near the center of the distal part of the anterior aspect of the arm ("popeye deformity"). Rupture of the biceps tendon may result from forceful flexion of the arm against excessive resistance, as occurs in weight lifters (Anderson and Hall, 1995). However, the tendon ruptures more often as the result of prolonged tendinitis that weakens it. The rupture results from repetitive overhead motions, such as occurs in swimmers, which tear the weakened tendon where it passes over the head of the humerus.

Fracture-Dislocation of the Proximal Humeral Epiphysis

A direct blow or indirect injury of the shoulder of a child or adolescent may produce a fracture-dislocation of the proximal humeral epiphysis because the articular capsule of the shoulder joint is stronger than the epiphyseal plate. In severe fractures, the body of the humerus is markedly displaced, but the humeral head retains its normal relationship with the glenoid cavity of the scapula. ⊙

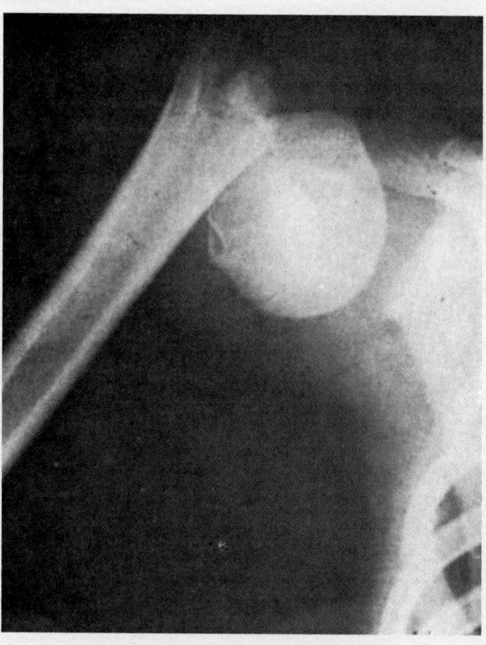

Separation of humeral epiphysis

Triceps Brachii

This large fusiform muscle is located in the posterior compartment of the arm (Figs. 6.32 and 6.35, Table 6.5). As indicated by its name, the triceps has three heads: long, lateral, and medial. *The triceps is the main extensor of the elbow joint.* Because its long head crosses the shoulder joint, *the triceps helps stabilize the adducted glenohumeral joint* by serving as a shunt muscle, resisting inferior displacement of the head of the humerus. The triceps also aids in extension and adduction of the arm. Just proximal to its distal attachment is a friction-reducing *subtendinous olecranon bursa* between the triceps tendon and the olecranon. To *test the triceps (or to determine the level of a radial nerve lesion)*, the arm is abducted 90° and then the flexed forearm is extended against resistance provided by the examiner. If acting normally, the triceps can be seen and palpated. Its strength should be comparable with the contralateral muscle, given consideration for lateral dominance (right- or left-handedness).

Anconeus

The anconeus is a small, relatively unimportant triangular muscle on the posterolateral aspect of the elbow; it is usually partially blended with the triceps (Table 6.5). The anconeus helps the triceps extend the forearm and *resists adduction of the*

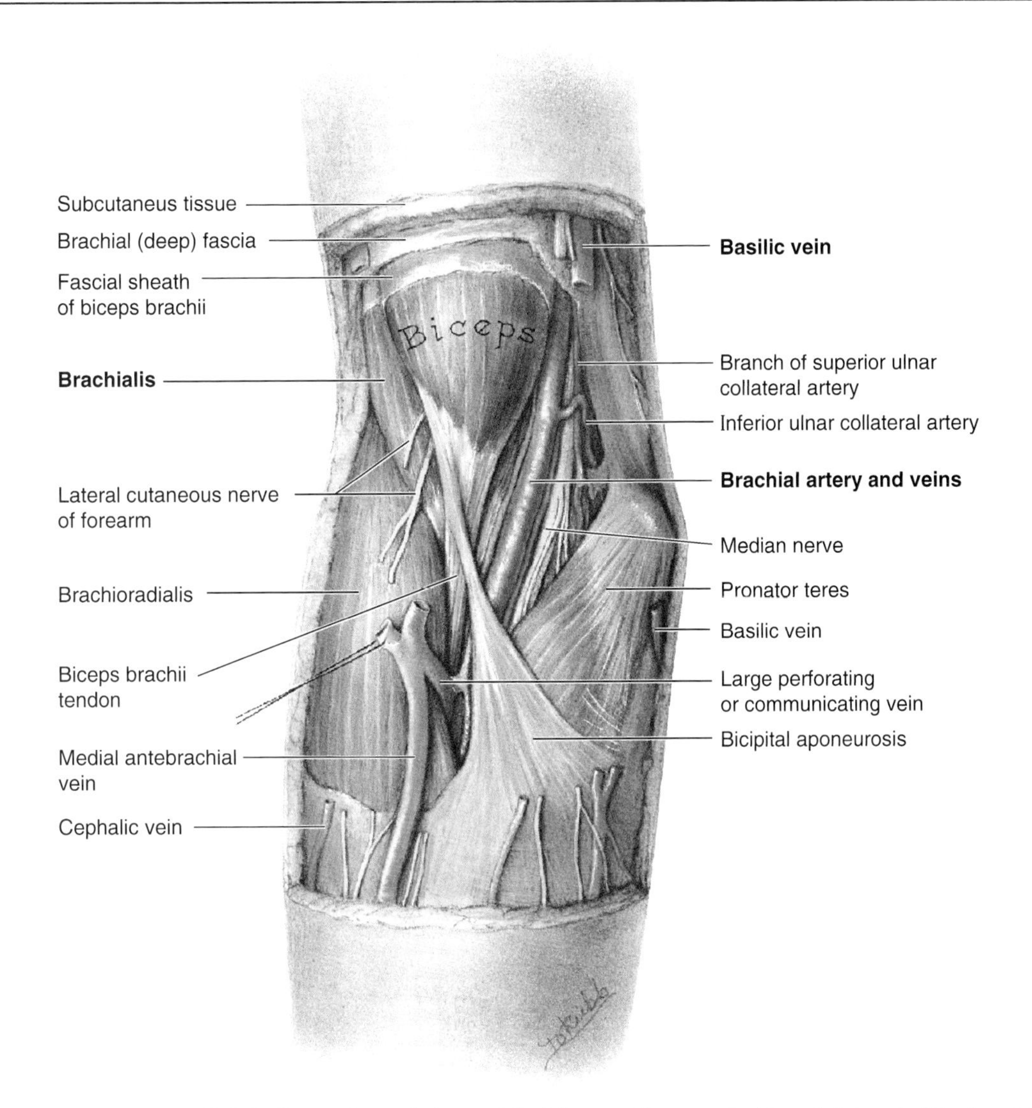

Figure 6.33. Superficial dissection of the cubital fossa. Observe that the distal end of the brachial artery lies medial to the biceps tendon. This is where a stethoscope is placed when listening to pulsations of the brachial artery during the measurement of blood pressure. The median nerve, which lies lateral to the artery in the proximal arm (Fig. 6.32*A*), has crossed anterior to the artery during its course in the arm so that it now lies on its medial aspect. Both the artery and nerve are afforded protection by the overlying *bicipital aponeurosis*. Although needles are commonly used to withdraw blood from the median cubital vein—which lies superficial to the aponeurosis—the artery and nerve are rarely traumatized. (Compare with figures at bottom of p. 732.)

ulna during pronation of the forearm. It is also said to be a tensor of the capsule of the elbow joint, preventing its being pinched during extension of the joint.

Brachial Artery

The brachial artery provides the main arterial supply to the arm (Fig. 6.36). The brachial artery, *the continuation of the axillary artery*, begins at the inferior border of the teres major (Fig. 6.32*A*) and *ends in the cubital fossa* opposite the neck of the radius (Fig. 6.33). Under cover of the bicipital aponeurosis, *the brachial artery divides into the radial and ulnar arteries* (Fig. 6.34). The brachial artery, superficial and palpable throughout its course, lies anterior to the triceps and brachialis. At first it lies medial to the humerus, and then anterior to it. As it passes inferolaterally, *the brachial artery accompanies the median nerve*, which crosses anterior to the artery (Figs. 6.32*A* and 6.37). During its course through the arm, the brachial artery gives rise to many unnamed muscular branches and humeral nutrient arteries, which arise from its lateral aspect.

The main (named) branches of the brachial artery arising from its medial aspect are the **deep artery of the arm** and the superior and inferior **ulnar collateral arteries** (Figs. 6.32*B* and 6.34). The collateral arteries help form the **arterial anastomoses of the elbow region** (Fig. 6.36). Other arteries involved are recurrent branches, sometimes double, from the radial, ulnar, and interosseous arteries, which run superiorly anterior and posterior to the elbow joint. These arteries anastomose with descending articular branches of the deep artery of the arm and the ulnar collateral arteries.

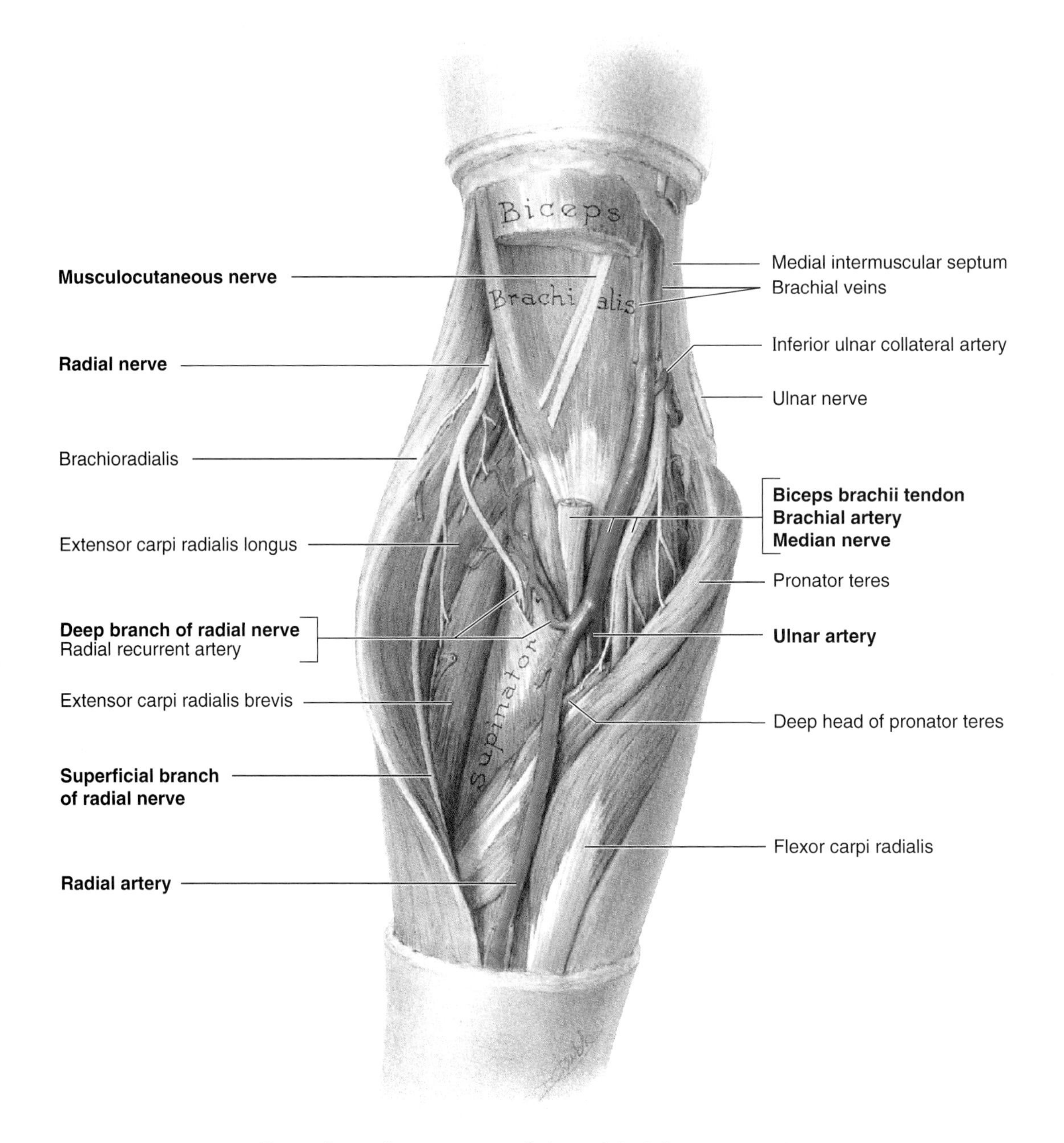

Figure 6.34. Deep dissection of structures of the cubital fossa. Part of the biceps brachii muscle is excised and the cubital fossa is opened widely by retracting the forearm extensor muscles laterally and the flexor muscles medially. Observe the brachialis and supinator muscles in the floor of the cubital fossa and the musculocutaneous nerve passing between the biceps and brachialis. The *radial nerve*, which has just left the posterior compartment of the arm (L. brachium) by piercing the lateral intermuscular septum, emerges between the brachialis and brachioradialis muscles and divides into a superficial (sensory) and a deep (motor) branch. The superficial branch of the radial nerve remains under the protection of the brachioradialis, but the deep branch of the radial nerve pierces the supinator to return to the posterior aspect of the limb. Observe also the *brachial artery* lying between the biceps brachii tendon and the median nerve, and then mostly commonly divides into two nearly equal branches: the *ulnar artery*, which runs deeply, and the radial artery, which remains superficial, making pulsations of it palpable throughout the forearm.

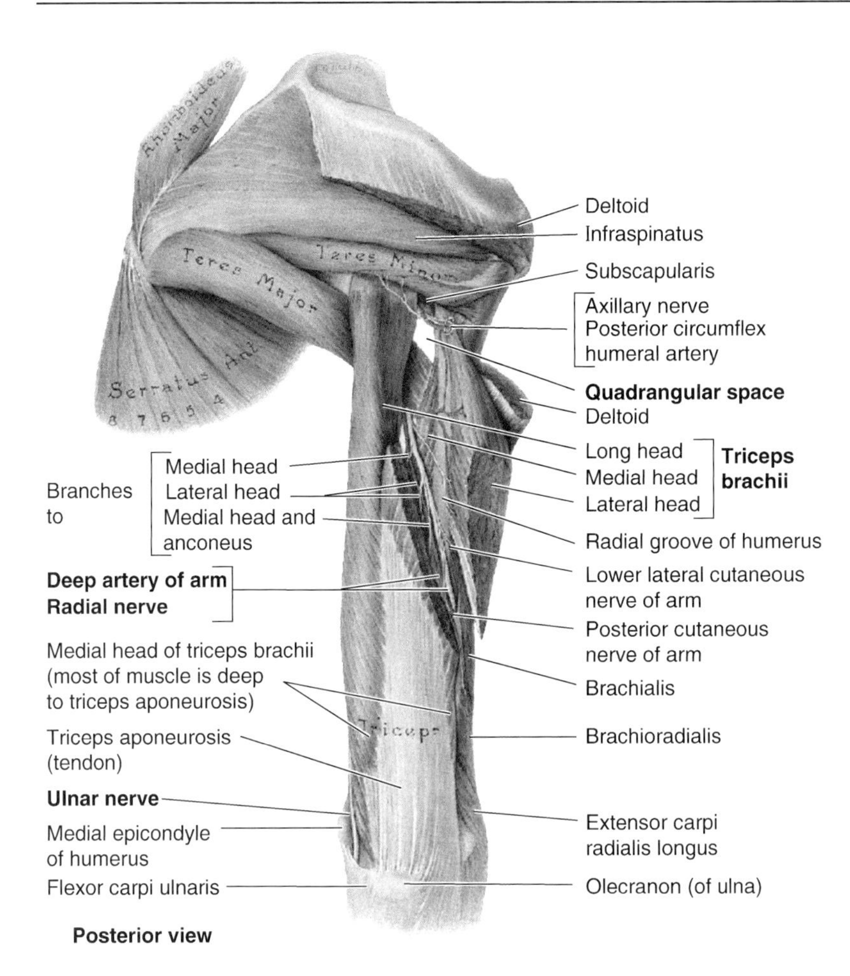

Figure 6.35. Dissection of the triceps brachii. The lateral head of the muscle is divided and displaced to show the *radial nerve* and the deep artery of the arm (L. arteria profunda brachii), which accompanies the nerve as it passes posterior to—and in direct contact with—the radial (spiral) groove of the humerus. This is the most common site of injury for the radial nerve, usually as a result of fractures of the middle third of the humerus. The "bare" bone of the radial groove separates the humeral attachments of the lateral and medial heads of the triceps. Observe also the *quadrangular space* and the structures passing through it. Note the ulnar nerve passing posterior to the medial epicondyle of the humerus where this nerve is most commonly injured.

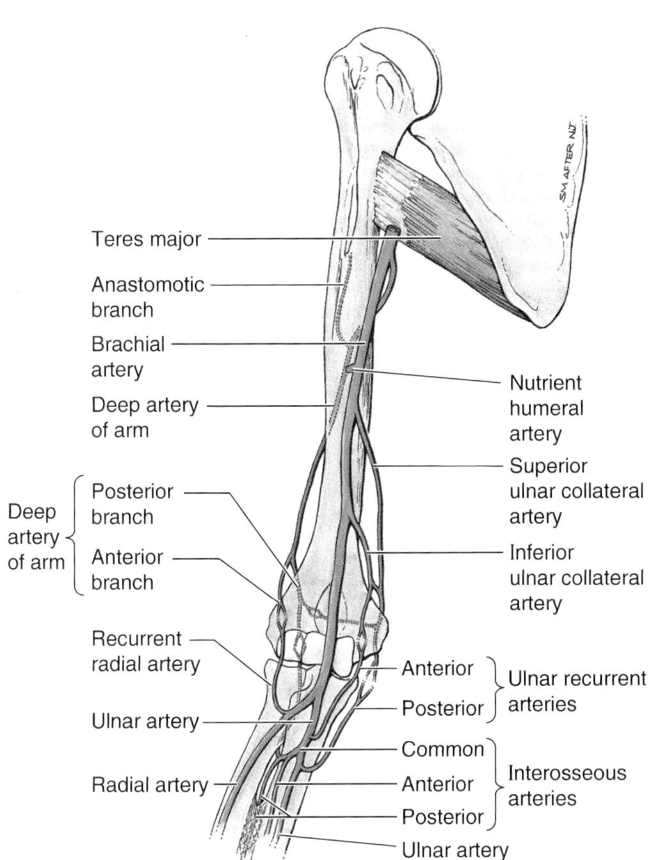

Figure 6.36. Arterial supply of the arm and the proximal part of the forearm. Observe the functionally and clinically important arterial anastomoses around the elbow. These pathways for collateral circulation allow blood to reach the forearm when flexion of the elbow compromises flow through the terminal part of the brachial artery.

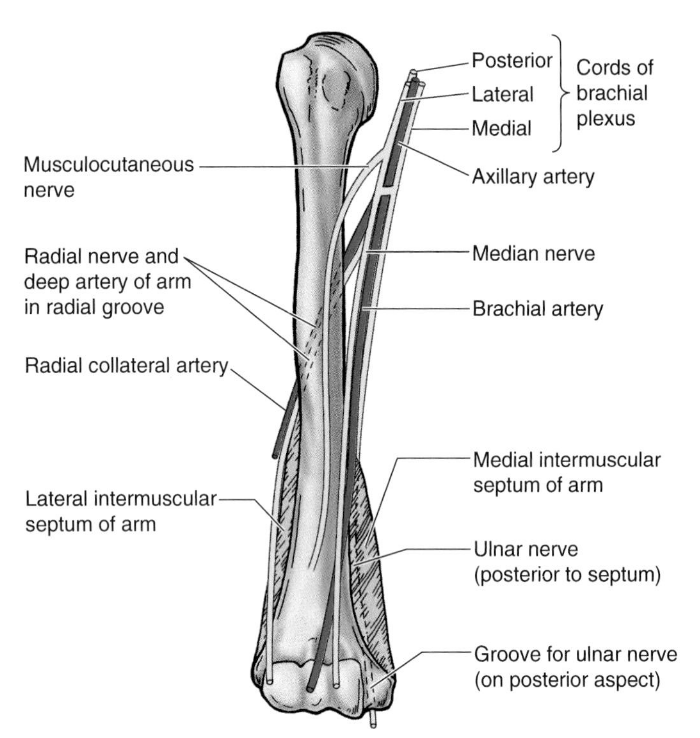

Figure 6.37. Relationship of the arteries and nerves of the arm to the humerus. Observe the cords of the brachial plexus surrounding the axillary artery. The radial nerve and accompanying deep artery of the arm (L. arteria profunda brachii) wind posteriorly around—and directly on the surface of—the humerus in the radial groove. The ulnar nerve pierces the medial intermuscular septum to enter the posterior fascial compartment and then lies in the groove for the ulnar nerve on the posterior aspect of the medial epicondyle of the humerus. Both nerves are most commonly injured at these sites. The median nerve descends in the arm to the medial side of the cubital fossa, where it is well protected and rarely injured (Fig. 6.33).

Deep Artery of the Arm

The deep artery of the arm—the largest branch of the brachial artery—has the most superior origin (Fig. 6.36). *The deep artery accompanies the radial nerve through the radial groove* and passes around the body of the humerus (Figs. 6.35 and 6.37). The deep artery divides into anterior and posterior descending branches that participate in the arterial anastomoses around the elbow.

Nutrient Humeral Artery

This artery arises from the brachial artery around the middle of the arm (Fig. 6.36) and enters the *nutrient canal* on the anteromedial surface of the humerus. The artery runs distally in the canal toward the elbow.

Superior Ulnar Collateral Artery

This artery arises from the medial aspect of the brachial artery near the middle of the arm and accompanies the ulnar nerve posterior to the medial epicondyle of the humerus (Figs. 6.32*A* and 6.36). Here it anastomoses with the posterior (or posterior branch of the) ulnar recurrent artery and inferior ulnar collateral artery, another branch of the brachial artery arising distally.

Inferior Ulnar Collateral Artery

This artery arises from the brachial artery approximately 5 cm proximal to the elbow crease (Figs. 6.32*A*, 6.33, 6.34, and 6.36). It then passes inferomedially anterior to the medial epicondyle of the humerus and joins the anastomoses of the elbow region by anastomosing with the anterior (or anterior branch of the) ulnar recurrent artery.

Measuring Blood Pressure

Health care workers routinely measure arterial blood pressure levels using a *sphygmomanometer*. A cuff is placed around the arm and inflated with air until it compresses the brachial artery against the humerus and occludes it. A stethoscope is placed over the artery in the cubital fossa, the pressure in the cuff is gradually released, and the examiner detects the sound of blood beginning to spurt through the artery. The first audible spurt indicates *systolic blood pressure*. As the pressure is completely released, the point at which the pulse can no longer be heard is the *diastolic blood pressure*.

Palpation of the Brachial Pulse

Detection of the pulse of the brachial artery may be difficult. Usually the belly of the biceps has to be pushed laterally to detect pulsations of the artery. To locate a brachial pulse push laterally, not deeper.

Compressing the Brachial Artery

The best place to compress the brachial artery to control hemorrhage is near the middle of the arm. Because the arterial anastomoses around the elbow provide a functionally and surgically important collateral circulation, the brachial artery may be clamped distal to the deep artery of the arm without producing tissue damage. The anatomical basis for this is that the ulnar and radial arteries still receive sufficient blood through the anastomoses around the elbow. *Ischemia of the elbow and forearm* results from clamping of the brachial artery proximal to the deep artery of the arm for an extended period.

Occlusion or Laceration of the Brachial Artery

Although collateral pathways confer some protection against gradual temporary and partial occlusion, sudden complete occlusion or laceration of the brachial artery creates a surgical emergency because paralysis of muscles results from ischemia within a few hours. ▶

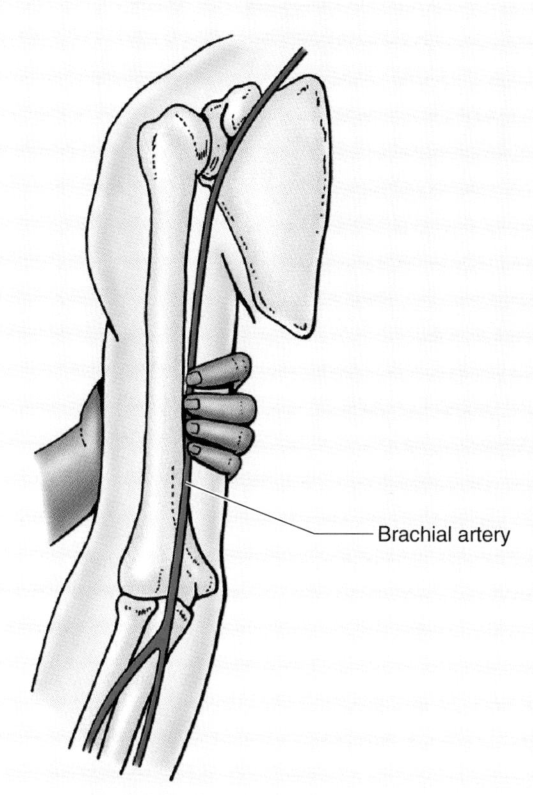

Compression of brachial artery

▶ Muscles and nerves can tolerate up to 6 hours of ischemia (Salter, 1998); after this, fibrous scar tissue replaces necrotic tissue and causes the involved muscles to shorten permanently, producing a flexion deformity—an *ischemic compartment syndrome* (Volkmann's contracture, ischemic contracture). Contraction of the digits and sometimes the wrist results in loss of hand power as a result of irreversible necrosis of the forearm flexor muscles.

Fracture of the Humeral Body

The common mechanism of a humeral fracture is a direct blow. Usually the fracture is transverse and somewhat comminuted. A *midhumeral fracture* may injure the radial nerve in the radial groove as it winds around the humeral body. Consequently, the possibility of a radial nerve lesion should always be considered with a midhumeral fracture. Although the radial nerve is damaged, the fracture is not likely to paralyze the triceps because of the high origin of the nerves to two of three heads of this muscle. A fracture of the distal part of the humerus, near the supracondylar ridges, is a *supracondylar fracture*. The distal bone fragment may be displaced anteriorly or posteriorly. The actions of the brachialis and triceps tend to pull the distal fragment over the proximal fragment, foreshortening the limb. Any of the nerves or branches of the brachial vessels related to the humerus may be injured. An injury to the brachial artery necessitates arterial repair. It may be completely or incompletely divided by a displaced fracture fragment. ○

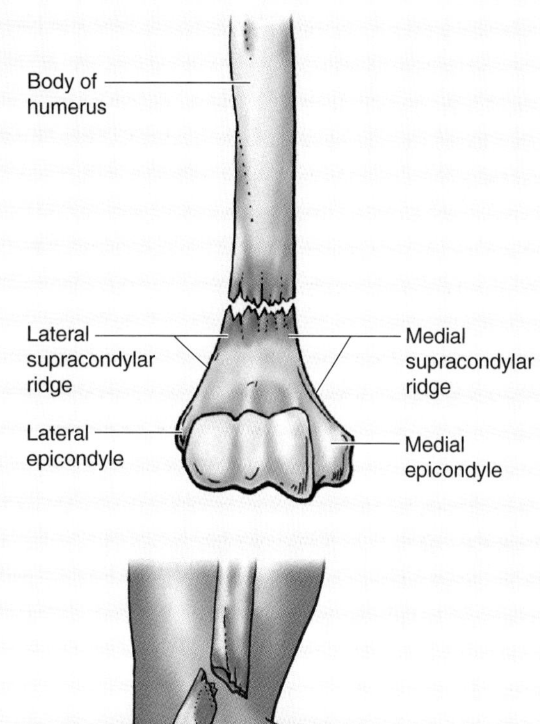

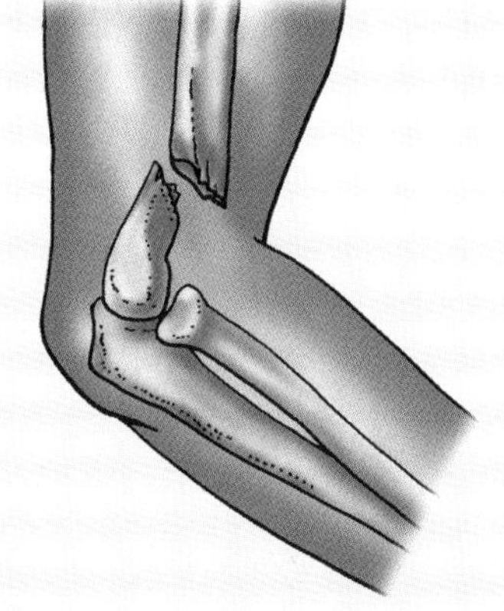

Veins of the Arm

Two sets of veins, *superficial* and *deep*, anastomose freely with each other. The superficial veins are in the subcutaneous tissue, and the deep veins accompany the arteries. Both sets of veins have valves, but they are more numerous in the deep veins than in the superficial veins.

Superficial Veins

The two main superficial veins of the arm are the cephalic and basilic veins (Figs. 6.32*B* and 6.33). The **cephalic vein** is located in the subcutaneous tissue along the anterolateral surface of the proximal forearm and arm and is often visible through the skin (Fig. 6.38). The cephalic vein passes superiorly between the deltoid and pectoralis major in the deltopectoral groove and then in the deltopectoral triangle, where it *empties into the termination of the axillary vein.* The **basilic vein** is also located in the subcutaneous tissue and passes on the medial side of the inferior part of the arm; often it is also visible through the skin. Near the junction of the middle and inferior thirds of the arm, the basilic vein pene-

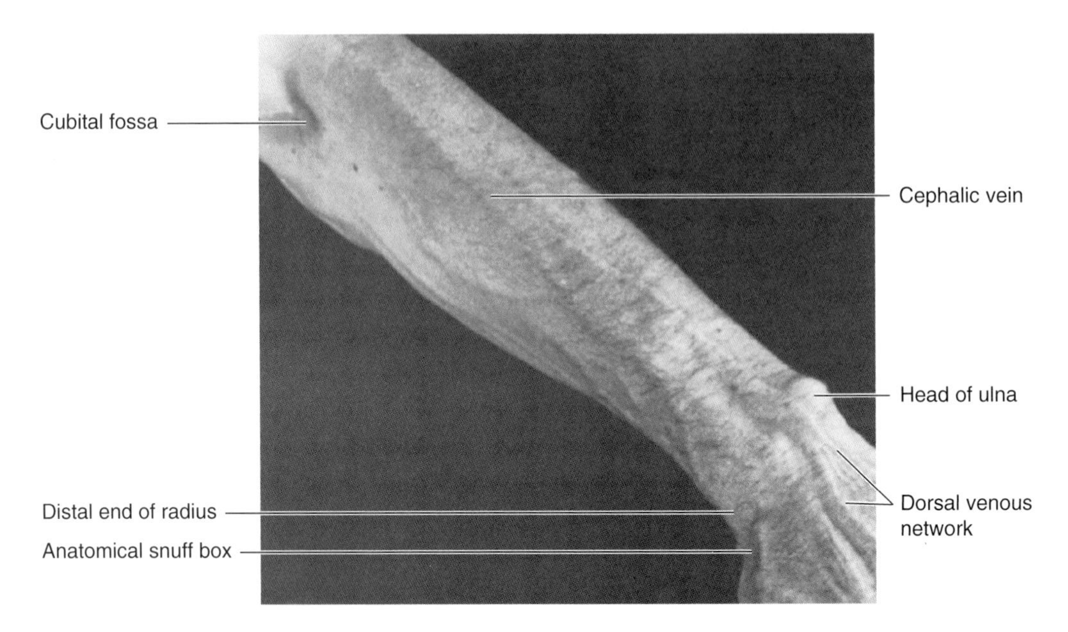

Figure 6.38. Surface anatomy of the cubital fossa, forearm, and hand. The forearm and hand are pronated. The cubital fossa is a triangular intermuscular depression, which contains the biceps tendon, the termination of the brachial artery, and the formation of its accompanying veins, the commencement of the radial and ulnar arteries, and parts of the median and radial nerves.

trates the brachial (deep) fascia and runs superiorly into the axilla, where it merges with the accompanying veins of the brachial artery to form the *axillary vein.*

Deep Veins

The paired veins accompanying the brachial artery (Figs. 6.32*B*) and their frequent connections encompass the artery, forming an anastomotic network within a common vascular sheath. The pulsations of the brachial artery help move the blood through this venous network. The brachial veins begin at the elbow by union of the accompanying veins of the ulnar and radial arteries and end by merging with the basilic vein to form the **axillary vein.** The brachial veins contain valves. Not uncommonly, the deep veins join to form one brachial vein during part of their course.

Nerves of the Arm

Four main nerves pass through the arm: median, ulnar, musculocutaneous, and radial (Fig. 6.37, Table 6.4). *The median and ulnar nerves supply no branches to the arm.*

Musculocutaneous Nerve

The musculocutaneous nerve, one of the terminal branches of the lateral cord of the brachial plexus (Figs. 6.28, 6.30–6.32, Table 6.4), supplies all the muscles in the anterior (flexor) compartment of the arm. The musculocutaneous nerve begins opposite the inferior border of the pectoralis minor, pierces the coracobrachialis, and continues distally between the biceps and brachialis (Fig. 6.34). In the interval between the biceps and brachialis, after supplying all three of these muscles, *the musculocutaneous nerve becomes the lateral cutaneous nerve of the forearm,* which continues across the anterior aspect of the elbow to supply a large area of forearm skin (Fig. 6.33).

Radial Nerve

The radial nerve, *the direct continuation of the posterior cord of the brachial plexus* (Figs. 6.28, 6.30*D*, 6.32, and 6.34, Table 6.4), supplies all the muscles in the posterior compartment of the arm. The radial nerve enters the arm posterior to the brachial artery, medial to the humerus, and anterior to the long head of the triceps (Fig. 6.35). The radial nerve descends inferolaterally with the deep brachial artery and *passes around the humeral body in the radial groove* (Fig. 6.37). Before entering the groove, it gives branches to the long and lateral heads of the triceps. The branch to the medial head arises within the radial groove. When it reaches the lateral border of the humerus, the radial nerve pierces the lateral intermuscular septum and continues inferiorly in the anterior compartment of the arm between the brachialis and brachioradialis to the level of the lateral epicondyle of the humerus. *The radial nerve then divides into deep and superficial branches* (Fig. 6.34).

- The *deep branch of the radial nerve* is entirely muscular and articular in its distribution.
- The *superficial branch of the radial nerve* is entirely cutaneous in its distribution, supplying sensation to the dorsum of the hand and digits.

Median Nerve

This major nerve of the arm is formed in the axilla by the union of a lateral root from the lateral cord and a medial root from the medial cord of the brachial plexus (Figs. 6.28, 6.30*B*, and 6.32, Table 6.4). The median nerve runs distally in the arm, initially on the lateral side of the brachial artery until it reaches the middle of the arm (Fig. 6.37), where it crosses to the medial side and contacts the brachialis. The median nerve then descends to the cubital fossa, where it lies deep to the bicipital aponeurosis and median cubital vein. *The median nerve has no branches in the axilla or the arm*, but it supplies articular branches to the elbow joint.

Ulnar Nerve

This is the larger of the two terminal branches of the medial cord of the brachial plexus (Figs. 6.28, 6.30*C*, and 6.32, Table 6.4). It passes distally, anterior to the triceps, on the medial side of the brachial artery. Around the middle of the arm it pierces the medial intermuscular septum with the superior ulnar collateral artery and descends between the septum and the medial head of the triceps. The ulnar nerve passes posterior to the medial epicondyle and medial to the olecranon to enter the forearm (Fig. 6.37). Posterior to the medial epicondyle—where the ulnar nerve is referred to in lay terms as the "crazy bone"—it is superficial, easily palpable, and vulnerable to injury. The ulnar nerve has no branches in the arm, but it supplies articular branches to the elbow joint.

Cubital Fossa

The cubital fossa is the triangular hollow area on the anterior aspect of the elbow (Figs. 6.33, 6.34, and 6.38). *The boundaries of the cubital fossa are*:

- Superiorly—an imaginary line connecting the medial and lateral epicondyles
- Medially—the pronator teres
- Laterally—the brachioradialis.

The *floor of the cubital fossa* is formed by the brachialis and supinator muscles of the arm and forearm, respectively. The *roof of the cubital fossa* is formed by deep fascia—reinforced by the bicipital aponeurosis—subcutaneous tissue, and skin.

The contents of the cubital fossa (Figs. 6.33, 6.34, and 6.36) *are the*:

- **Terminal part of the brachial artery** and the commencement of its terminal branches, the *radial* and *ulnar arteries*; the brachial artery lies between the biceps tendon and the median nerve
- **(Deep) accompanying veins** of the arteries
- **Biceps brachii tendon**
- **Median nerve.**

Injury to the Musculocutaneous Nerve

Injury to the musculocutaneous nerve in the axilla (uncommon in this protected position) is typically inflicted by a weapon and results in paralysis of the coracobrachialis, biceps, and brachialis. Consequently, flexion of the elbow joint and supination of the forearm are greatly weakened. Loss of sensation may occur on the lateral surface of the forearm supplied by the lateral antebrachial cutaneous nerve.

Injury to the Radial Nerve

Injury to the radial nerve superior to the origin of its branches to the triceps brachii results in *paralysis of the triceps, brachioradialis, supinator, and extensor muscles of the wrist and digits*. Loss of sensation in areas of skin supplied by this nerve also occurs. When the nerve is injured in the radial groove, the triceps is usually not completely paralyzed but only weakened because only the medial head is affected; however, the muscles in the posterior compartment of the forearm that are supplied by more distal branches of the nerve are paralyzed. The characteristic clinical sign of radial nerve injury is *wrist-drop* (inability to extend the wrist and the digits at the metacarpophalangeal joints as shown in (**A**); instead the wrist is flexed due to unopposed tonus of flexor muscles and gravity (**B**). ⊕

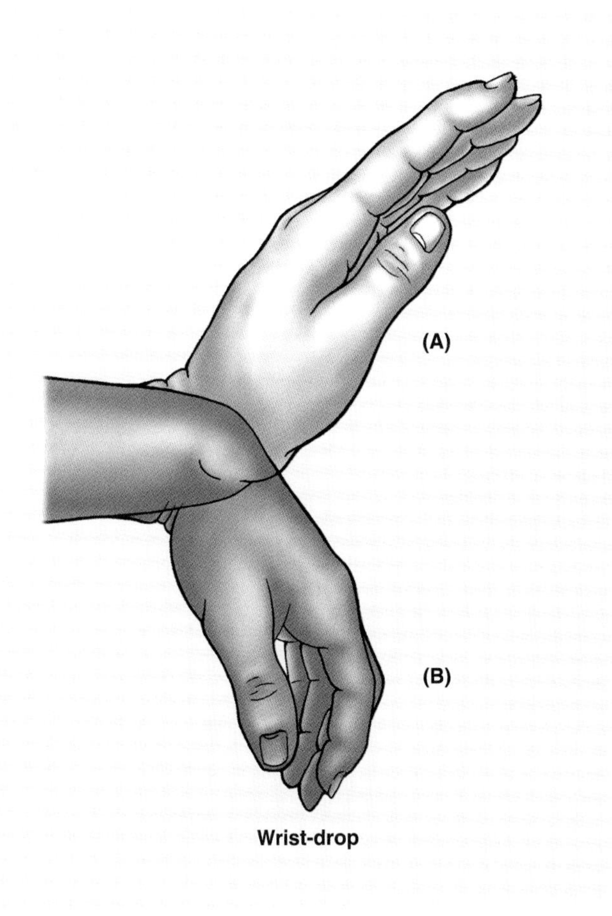

Superficially, in the subcutaneous tissue overlying the fossa are the:

- **Median cubital vein,** lying anterior to the brachial artery
- **Medial** and **lateral antebrachial cutaneous nerves**, related to the basilic and cephalic veins.

The deep and superficial branches of the radial nerve are within the floor of the fossa.

Venipuncture

The cubital fossa is the common site for sampling and transfusion of blood and intravenous injections because of the prominence and accessibility of veins. Considerable variation occurs in the connection of the basilic and cephalic veins. Usually the median cubital vein or basilic vein is selected. The median cubital vein crosses the bicipital aponeurosis, which separates it from the underlying brachial artery and median nerve. The cubital veins are also a site for the introduction of cardiac catheters to secure blood samples from the great vessels and chambers of the heart. These veins are also used for *cardioangiography* (see Chapter 1). A tourniquet is placed around the midarm to distend the veins in the cubital fossa. Once the vein is punctured, the tourniquet is removed so that when the needle is removed the vein will not bleed extensively. ⊙

Surface Anatomy of the Arm and Cubital Fossa

The borders of the deltoid are visible when the arm is abducted against resistance. When the arm is fully adducted, the **greater tubercle of the humerus** can usually be felt on deep palpation through the deltoid, inferior to the lateral border of the acromion. Normally, the greater tubercle is the most lateral bony point of the shoulder. The **distal attachment of the deltoid** can be palpated on the lateral surface of the humerus. The **three heads of the triceps** (long, lateral, and medial) form a bulge on the posterior aspect of the arm and are identifiable when the forearm is extended from the flexed position against resistance. The **olecranon,** to which the triceps tendon attaches distally, is easily palpated. It is separated from the skin only by the *olecranon bursa*. The **triceps tendon** is easily felt as it descends along the posterior aspect of the arm to the olecranon. The fingers can be pressed inwards on each side of the tendon, where the elbow joint is superficial. An abnormal collection of fluid in the elbow joint or in the **triceps bursa** is palpable at these sites; the bursa lies deep to the triceps tendon.

The **biceps brachii** forms a bulge on the anterior aspect of the arm; its belly becomes more prominent when the elbow is flexed and supinated against resistance. The **biceps brachii tendon** can be palpated in the cubital fossa, immediately lateral to the midline, especially when the elbow is flexed ▶

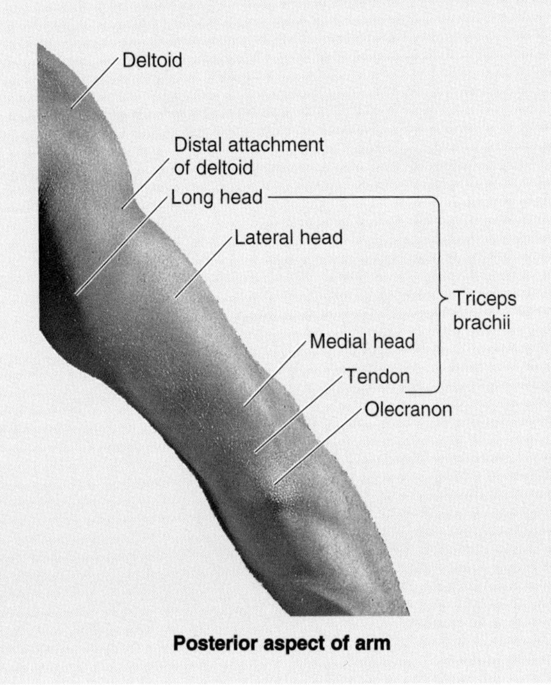

Posterior aspect of arm

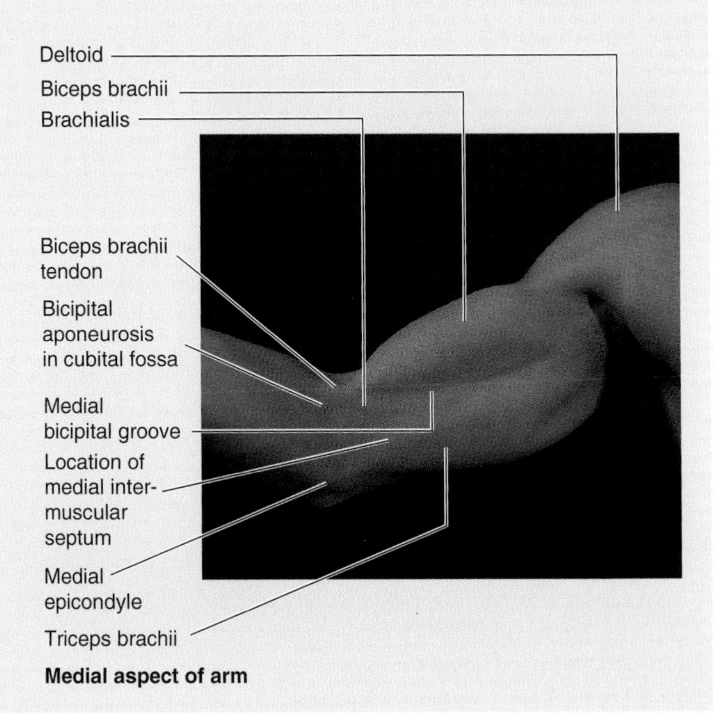

Medial aspect of arm

▶ against resistance. The proximal part of the **bicipital aponeurosis** can be palpated where it passes obliquely over the brachial artery and median nerve. Medial and lateral **bicipital grooves** separate the bulges formed by the biceps and triceps and indicate the location of the medial and lateral intermuscular septa. The cephalic vein runs superiorly in the lateral bicipital groove, and the basilic vein ascends in the medial bicipital groove.

No part of the body of the humerus is subcutaneous; however, it can be palpated with varying distinctness through the muscles surrounding it. The **head of the humerus** is surrounded by muscles on all sides, except inferiorly; thus, it can be palpated by pushing the fingers well up into the axilla. The arm should be close to the side so the axillary fascia is loose. The humeral head can be identified by its movements when the arm is moved and the inferior angle of the scapula is held so the scapula does not move. The **brachial artery** may be felt pulsating deep to the medial border of the biceps (p. 729). The medial and lateral **epicondyles of the humerus** are subcutaneous and can be easily palpated at the medial and lateral aspects of the elbow. The medial epicondyle is more prominent.

In the cubital fossa, the **cephalic and basilic veins** in the subcutaneous tissue are clearly visible when a tourniquet is applied to the arm, as is the **median cubital vein**, which crosses the bicipital aponeurosis as it runs superomedially connecting the cephalic to the basilic. The median cubital vein often receives the **median antebrachial vein** (median vein of forearm); however, the median antebrachial vein may bifurcate to form a *median cephalic vein* and a *median basilic vein*. In these cases, no median cubital vein is present. ⊕

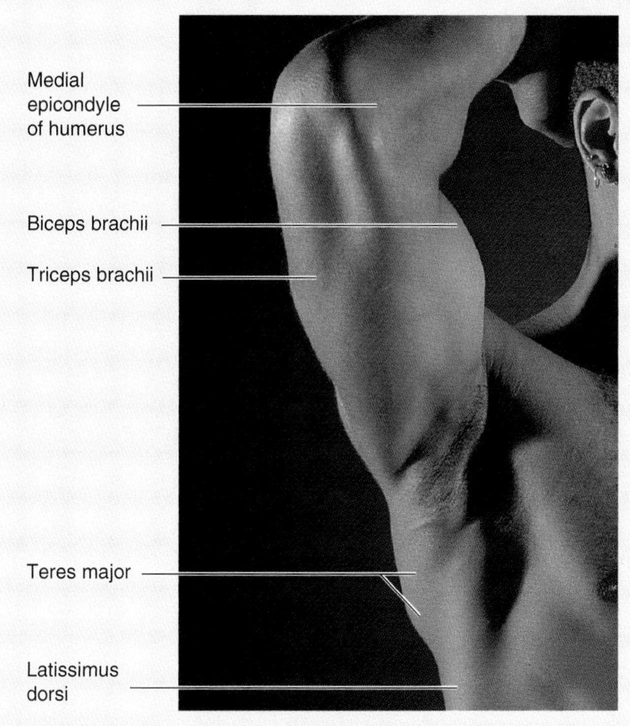

Elbow, arm, and axilla

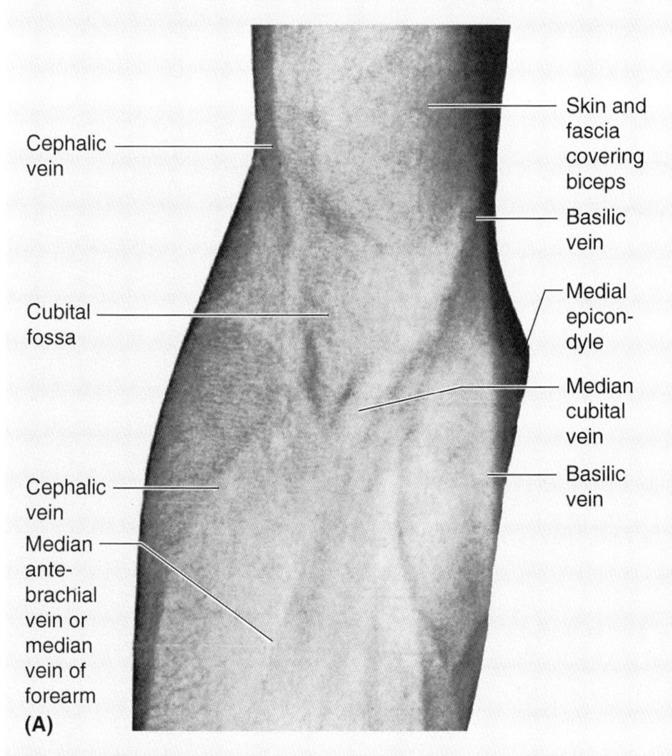

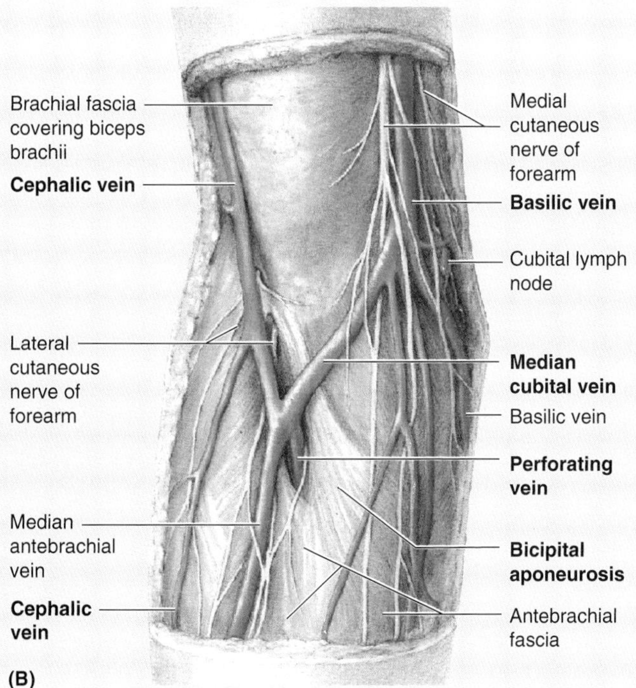

Forearm

The forearm extends from the elbow to the wrist and contains two bones, the **radius** and **ulna** (Fig. 6.39, *A–C*), which are joined by an interosseous membrane. Although thin, this fibrous membrane is strong. In addition to tying the forearm bones together, the **interosseous membrane** provides attachment for some deep forearm muscles. The head of the ulna is at the distal end of the forearm, whereas the head of the radius is at its proximal end.

Compartments of the Forearm

Although the proximal boundary of the forearm is defined by the joint plane of the elbow, functionally the forearm includes the distal humerus. For the distal forearm, wrist, and hand to have minimal bulk to maximize their functionality, they are operated by "remote control" by extrinsic muscles having their bulky, fleshy, contractile parts located proximally in the forearm—distant from the site of action—with long, slender tendons extending distally to the operative site, like long ropes reacting to distant pulleys. Furthermore, because the structures on which the muscles and tendons act (wrist and fingers) have an extensive range of motion, a large amount of contraction is needed, requiring that the muscles have long contractile parts as well as a long tendon(s). The forearm proper is not in fact long enough to provide the required length and sufficient area for attachment proximally, so the proximal attachments (origins) of the muscles must occur proximal to the elbow—in the arm—and be provided by the humerus. Generally, flexors lie anteriorly and extensors posteriorly; however, the anterior and posterior aspects of the distal humerus are occupied by the flexors and extensors of the elbow. Thus, to provide the required attachment sites, medial and lateral extensions (epicondyles and epicondylar ridges) have developed from the distal humerus. The medial epicondyle and epicondylar ridge provide attachment for the forearm flexors, and the lateral formations provide attachment for the forearm extensors. Thus, rather than lying strictly anteriorly and posteriorly, the proximal parts of the "anterior" (flexor-pronator) compartment of the forearm lie anteromedially, and the posterior (extensor-supinator) compartment lies posterolaterally (Fig. 6.39*D* and see 6.42*B*). Spiraling gradually over the length of the forearm, the compartments become truly anterior and posterior in position in the distal forearm and wrist. These functional, fascial compartments are demarcated by the subcutaneous border of the ulna posteriorly (in the proximal forearm) and then medially (distal forearm) and by the radial artery anteriorly and then laterally. These structures are palpable (the artery by its pulsations) throughout the forearm. Because neither boundary is crossed by motor nerves, they also provide sites for surgical incision.

The flexors and pronators of the forearm are in the anterior compartment and are served mainly by the median nerve; the one and a half exceptions are innervated by the ulnar nerve. The extensors and supinators of the forearm are in the posterior compartment and are all served by the radial nerve (directly or by its deep branch). The fascial compartments of the limbs generally end at the joints, and so fluids and infections in compartments are usually contained and cannot readily spread to other compartments. The anterior compartment of the forearm is exceptional in this regard because it communicates with the central compartment of the palm through the carpal tunnel (p. 775).

Muscles of the Forearm

The muscles of the forearm act on the joints of the elbow, wrist, and digits. In the proximal part of the forearm, the muscles form fleshy masses extending inferiorly from the medial and lateral epicondyles of the humerus (Fig. 6.39, *C–E*). The tendons of these muscles pass through the distal part of the forearm and continue into the wrist, hand, and digits.

Flexor-Pronator Muscles of the Forearm

The flexor muscles of the forearm are in the anterior (flexor-pronator) compartment of the forearm and are separated from the extensor muscles of the forearm by the radius and ulna and the interosseous membrane that connects them (Fig. 6.39, *B* and *C*). The tendons of most flexor muscles are located on the anterior surface of the wrist and are held in place by the **palmar carpal ligament** and the **flexor retinaculum (transverse carpal ligament)**, thickenings of the antebrachial fascia (Figs. 6.39*C* and 6.40). *The flexor muscles are arranged in four layers and are divided into two groups, superficial and deep* (Table 6.6):

- **A superficial group of five muscles** (pronator teres, flexor carpi radialis, palmaris longus, flexor carpi ulnaris, and flexor digitorum superficialis [FDS]); these muscles all attach, at least in part, by a *common flexor tendon* from the medial epicondyle of the humerus—the *common flexor attachment*
- **A deep group of three muscles** (flexor digitorum profundus [FDP], flexor pollicis longus, and pronator quadratus).

The five superficial muscles cross the elbow joint; the three deep muscles do not. With the exception of the pronator quadratus, the more distally placed a muscle's distal attachment lies, the more distally and deeply placed is its proximal attachment.

All muscles in the anterior compartment of the forearm are supplied by the median and/or ulnar nerves (most by the median—only one and a half exceptions are supplied by the ulnar). Functionally, the brachioradialis is a flexor of the forearm, but it is located in the posterior (posterolateral) or extensor compartment and is thus supplied by the radial nerve; therefore, this muscle is a major exception to the rule that the radial nerve supplies only extensor muscles and that all flexors lie in the anterior (flexor) compartment. The **long flexors of the digits** (FDS and FDP) also flex the metacarpophalangeal and wrist joints. The FDP flexes the digits in slow action; this action is reinforced by the FDS when speed

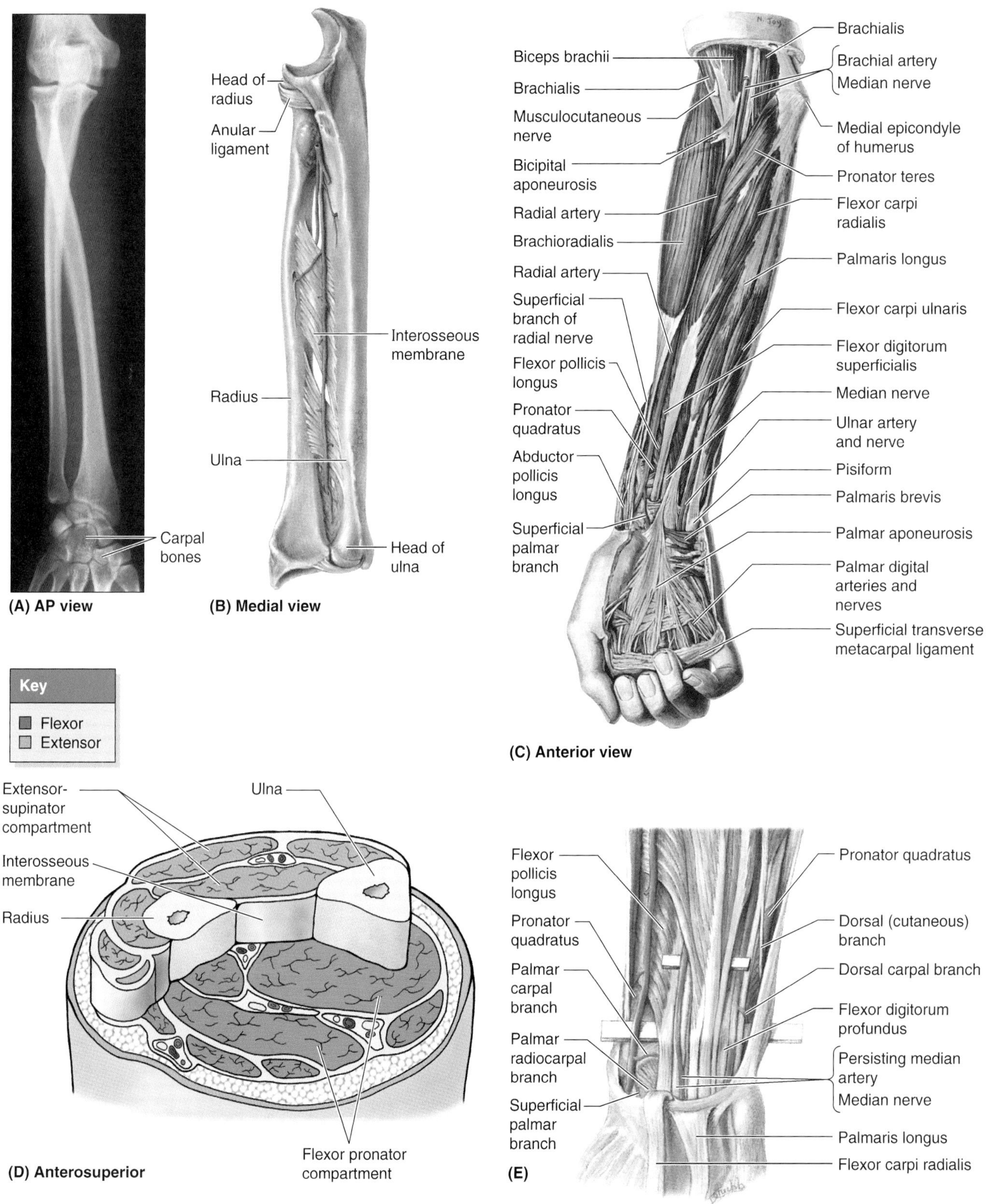

Figure 6.39. Bones, muscles, and compartments of the forearm. A. Anteroposterior (AP) radiograph of the forearm in pronation. (Courtesy of Dr. J. Heslin, Toronto, Ontario, Canada) **B.** Medial view of the bones of the forearm and the radioulnar ligaments. **C.** Anterior view. Superficial muscles of the forearm and the palmar aponeurosis. **D.** Anterosuperior view of a stepped transverse section, demonstrating the compartments of the forearm. **E.** Anterior view. Flexor digitorum superficialis (FDS) and related structures. Observe the ulnar artery descending obliquely posterior to the FDS to meet and accompany the ulnar nerve.

Table 6.6. Muscles of the Anterior Compartment of the Forearm

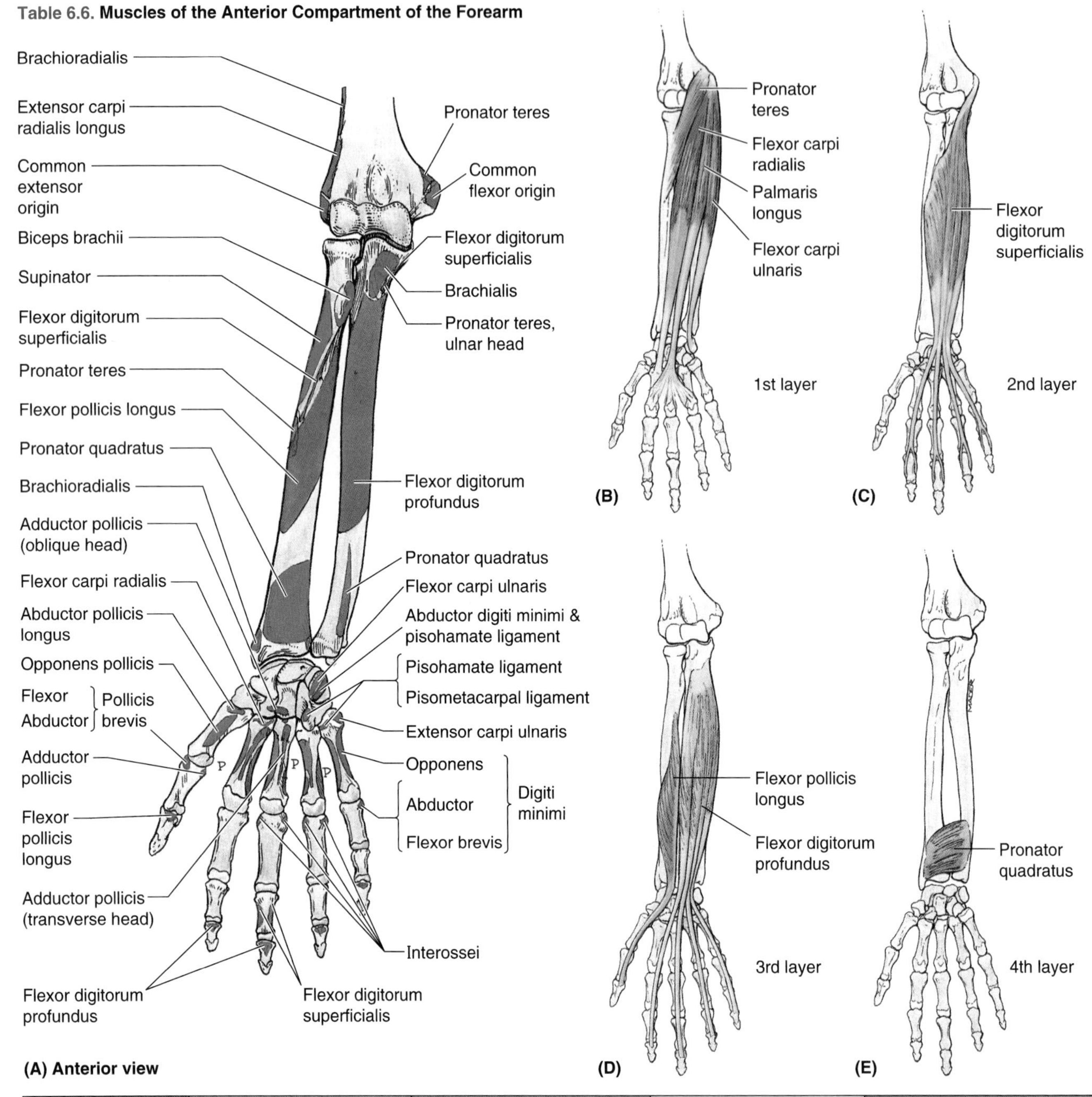

(A) Anterior view (B) (C) (D) (E)

Muscle	Proximal Attachment	Distal Attachment	Innervation [a]	Main Action
Pronator teres	Medial epicondyle of humerus and coronoid process of ulna	Middle of lateral surface of radius	Median nerve (C6 and **C7**)	Pronates and flexes forearm (at elbow)
Flexor carpi radialis	Medial epicondyle of humerus	Base of 2nd metacarpal bone		Flexes and abducts hand (at wrist)
Palmaris longus		Distal half of flexor retinaculum and palmar aponeurosis	Median nerve (C7 and C8)	Flexes hand (at wrist) and tightens palmar aponeurosis
Flexor carpi ulnaris	Humeral head: medial epicondyle of humerus Ulnar head: olecranon and posterior border of ulna	Pisiform bone, hook of hamate bone, and 5th metacarpal bone	Ulnar nerve (C7 and **C8**)	Flexes and adducts hand (at wrist)

Table 6.6. (*Continued*) **Muscles of the Anterior Compartment of the Forearm**

Muscle	Proximal Attachment	Distal Attachment	Innervation [a]	Main Action
Flexor digitorum superficialis	Humeroulnar head: medial epicondyle of humerus, ulnar collateral ligament, and coronoid process of ulna Radial head: superior half of anterior border of radius	Bodies of middle phalanges of medial four digits	Median nerve (C7, C8, and T1)	Flexes middle phalanges at proximal interphalangeal joints of medial four digits; acting more strongly, it also flexes proximal phalanges at metacarpophalangeal joints and hand
Flexor digitorum profundus	Proximal three-fourths of medial and anterior surfaces of ulna and interosseous membrane	Bases of distal phalanges of medial four digits	Medial part: ulnar nerve (**C8** and T1) Lateral part: median nerve (**C8** and T1)	Flexes distal phalanges at distal interphalangeal joints of medial four digits; assists with flexion of hand
Flexor pollicis longus	Anterior surface of radius and adjacent interosseous membrane	Base of distal phalanx of thumb	Anterior interosseous nerve from median (**C8** and T1)	Flexes phalanges of 1st digit (thumb)
Pronator quadratus	Distal fourth of anterior surface of ulna	Distal fourth of anterior surface of radius		Pronates forearm; deep fibers bind radius and ulna together

[a] Numbers indicate spinal cord segmental innervation (e.g., C5 and **C7** indicate that nerves supplying the pronator teres muscle are derived from the 5th and 7th cervical segments of spinal cord). **Boldface** numbers indicate main segmental innervation. Damage to these segments, or to motor nerve roots arising from them, results in paralysis of muscles concerned.

and flexion against resistance are required. When the wrist is flexed at the same time that the metacarpophalangeal and interphalangeal joints are flexed, the long flexor muscles of the fingers are operating over a shortened distance between attachments, and the action resulting from their contraction is consequently weaker. Extending the wrist increases their operating distance, and thus their contraction is more efficient in producing a strong grip (see Fig. 6.52). Tendons of the long flexors of the digits pass through the distal part of the forearm, wrist, and palm and continue to the medial four digits: FDS flexes the middle phalanges, and FDP flexes the distal phalanges.

The following are the superficial (and intermediate) layers of the forearm flexor muscles.

Pronator Teres (Figs. 6.39 and 6.40, Table 6.6). This fusiform muscle is *a pronator of the forearm and a flexor of the elbow joint.* It has two heads of proximal attachment, one of which is the common flexor tendon. Its distal attachment is at the radius' most lateral point (which occurs approximately in the middle of its curved body) to provide maximal leverage. The pronator teres is prominent when the forearm is strongly flexed and pronated; its lateral border forms the medial boundary of the cubital fossa. *To test the pronator teres,* the person's forearm is pronated from the supine position against resistance provided by the examiner. If acting normally, the muscle can be seen and palpated at the medial margin of the cubital fossa.

Flexor Carpi Radialis (Figs. 6.39 and 6.40, Table 6.6). This long, fusiform muscle is located *medial to the pronator teres.* In the middle of the forearm, its fleshy belly is replaced by a long, flattened tendon that becomes cordlike as it approaches the wrist. The flexor carpi radialis produces flexion (when acting with the flexor carpi ulnaris) and abduction of the wrist (when acting with the extensors carpi radialis longus and brevis). When acting alone, it produces a combination of flexion and abduction simultaneously at the wrist so that the hand moves anterolaterally. To reach its distal attachment, its tendon passes through a canal in the lateral part of the flexor retinaculum and through a vertical groove in the trapezium. The tendon of the flexor carpi radialis is a good guide to the radial artery, which lies just lateral to it (Fig. 6.39*C*). *To test the flexor carpi radialis,* the person is asked to flex the wrist against resistance. If acting normally, the tendon can be easily seen and palpated.

Palmaris Longus (Figs. 6.39 and 6.40, Table 6.6). This small fusiform muscle is absent on one or both sides (usually the left) in approximately 14% of people, but its actions are not missed. It has a short belly and a long tendon that passes superficial to the flexor retinaculum and attaches to it and the apex of the palmar aponeurosis. *The palmaris longus tendon is a useful guide to the median nerve at the wrist.* The median nerve lies deep and slightly medial to this tendon before it passes deep to the transverse carpal ligament. *To test the palmaris longus,* the wrist is flexed and the pads of the little finger and thumb are pinched together. If present and acting normally, the tendon can be easily seen and palpated.

Flexor Carpi Ulnaris (Figs. 6.39 and 6.40, Table 6.6). This is the most medial of the superficial flexor muscles. The *flexor carpi ulnaris flexes and adducts the hand* at the wrist simultaneously if acting alone. It flexes the wrist when it acts with the flexor carpi radialis and adducts it when acting with the extensor carpi ulnaris. It has two heads (humeral—the common flexor tendon—and ulnar) of proximal attachment between which the ulnar nerve passes distally in the forearm. This muscle is exceptional among muscles of the anterior compartment, being fully innervated by the ulnar nerve.

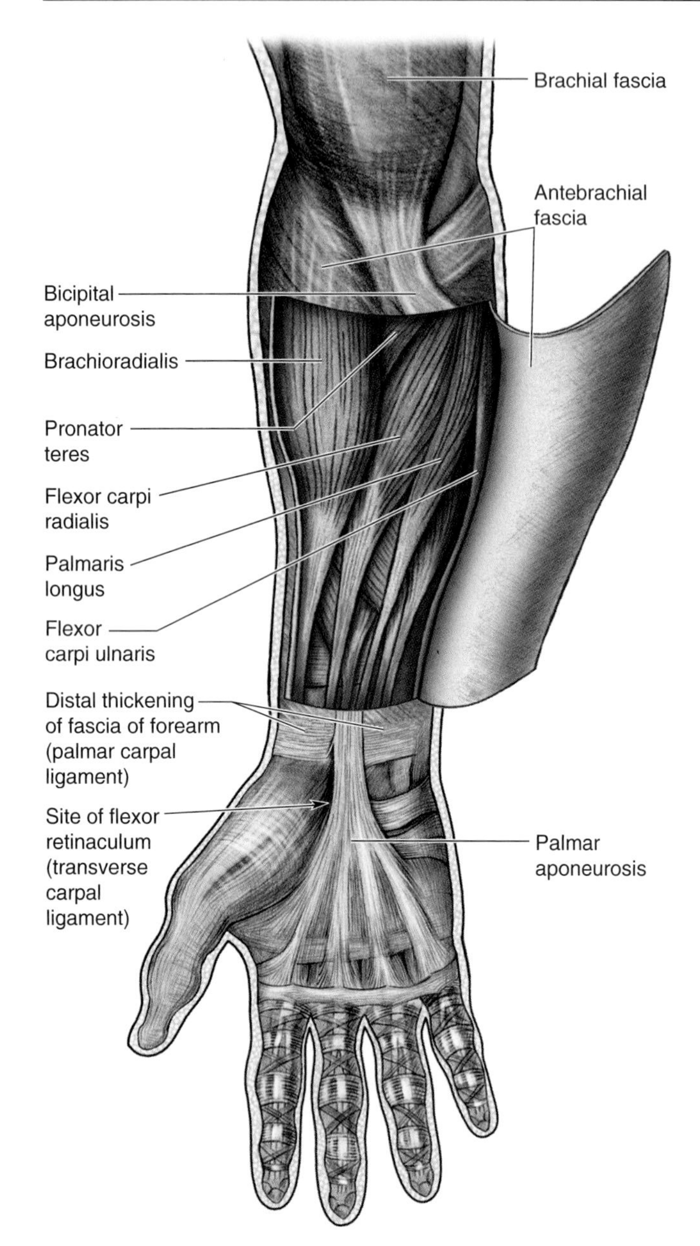

Figure 6.40. Superficial muscles of the forearm and palmar aponeurosis. The brachioradialis, representing the lateral group of muscles, slightly overlaps the radial artery. The four superficial muscles of the anterior (flexor/pronator) compartment of the forearm (pronator teres, flexor carpi radialis, palmaris longus, and flexor carpi ulnaris) radiate from the medial epicondyle of the humerus (arm bone) (Fig. 6.39*B*). Over the distal ends of the radius and ulna, the deep fascia of the forearm thickens to form the extensor retinaculum posteriorly and a corresponding (but officially unnamed) thickening anteriorly, identified by some authors as the "palmar carpal ligament." Immediately distal and at a deeper level to the latter thickening, the *flexor retinaculum* is a heavy fibrous band also continuous with the distal part of the antebrachial fascia, which extends between the anterior prominences of the outer carpal bones and completes the carpal tunnel. The tendons of the flexor muscles of the hand (L. manus) and fingers pass through this tunnel. The *palmar aponeurosis* is a strong, triangular membrane overlying the tendons in the palm. Its apex is continuous with the tendon of the palmaris longus and is anchored to the anterior aspect of the flexor retinaculum.

The tendon of the flexor carpi ulnaris is a guide to the ulnar nerve and artery, which are on its lateral side at the wrist (Fig. 6.39*B*). *To test the flexor carpi ulnaris*, the person is asked to put the posterior aspect of the forearm and hand on a flat table. The person is then asked to flex the wrist against resistance while the examiner palpates the muscle and its tendon.

Flexor Digitorum Superficialis (Fig. 6.39, Table 6.6). The *superficial flexor of the digits* is the largest superficial muscle in the forearm. Although included here with the superficial muscles of the forearm, which have an attachment to the common flexor attachment and therefore cross the elbow, the FDS actually forms an intermediate layer between the superficial and deep groups of forearm muscles. *The FDS has two heads*—humeroulnar and radial—between which the median nerve and ulnar artery pass (Fig. 6.39*D*). Near the wrist, the FDS gives rise to four tendons, which pass deep to the flexor retinaculum through the carpal tunnel to the digits. The four tendons for the digits are enclosed (along with the four tendons of the deep flexor of the digits [L. profundus, deep]) in a **common flexor synovial sheath**. *The FDS flexes the middle phalanges of the medial four digits (fingers) at the proximal interphalangeal joints.* In continued action, the FDS also flexes the proximal phalanges at the metacarpophalangeal joints and the wrist joint. *To test the FDS*, one finger is flexed at the proximal interphalangeal joint against resistance and the other three fingers are held in an extended position to inactivate the FDP.

The following muscles form the deep layer of forearm flexor muscles.

Flexor Digitorum Profundus (Figs. 6.39*E*, Table 6.6). The thick FDP is the only muscle that can flex the distal interphalangeal joints of the digits. This **deep flexor of the digits** has an extensive proximal attachment to the ulna and interosseous membrane and "clothes" the anterior aspect of the ulna. *The FDP flexes the distal phalanges of the medial four digits after the FDS has flexed their middle phalanges* (i.e., it "curls the fingers" and assists with flexion of the hand—making a fist). Each tendon is capable of flexing two interphalangeal joints, the metacarpophalangeal joint, and the wrist joint. *The FDP divides into four parts that end in four tendons*, which pass posterior to the tendons of the FDS and the flexor retinaculum. The part of the muscle going to the 2nd digit usually separates from the rest of the muscle relatively early in the distal part of the forearm. Each tendon enters the fibrous sheath of its digit, posterior to the tendon of the FDS. The lateral part of the muscle serving digits 2 and 3 is innervated by the median nerve, and the medial (or ulnar) part of the muscle serving digits 4 and 5 is innervated by the ulnar nerve. *To test the FDP*, the proximal interphalangeal joint is held in the extended position while the person attempts to flex the distal interphalangeal joint. The integrity of the median nerve in the proximal forearm can be tested by performing this test using the index finger, and that of the ulnar nerve can be assessed by using the little finger.

Flexor Pollicis Longus (Fig. 6.39, *C* and *E*, Table 6.6). This long flexor of the thumb (L. pollex, thumb) lies lateral to the FDP, where it clothes the anterior aspect of the radius distal to the attachment of the supinator. Its flat tendon passes deep to the flexor retinaculum, enveloped in its own synovial sheath on the lateral side of the common flexor synovial sheath. *The flexor pollicis longus flexes the distal phalanx of the 1st digit (thumb)* and, second, the proximal phalanx and 1st metacarpal bone. The flexor pollicis longus is the only muscle that flexes the interphalangeal joint of the thumb. It also flexes the metacarpophalangeal and carpometacarpal joints of the thumb and may assist in flexion of the wrist joint. *To test the flexor pollicis longus,* the proximal phalanx of the thumb is held and the distal phalanx is flexed against resistance.

Pronator Quadratus (Fig. 6.39, *C* and *E*, Table 6.6). As its name states, this small muscle is quadrangular and pronates the forearm. It cannot be palpated or observed, except in dissections, because it is the deepest muscle in the anterior aspect of the forearm. It clothes the distal fourth of the radius and ulna and the interosseous membrane between them. The pronator quadratus is the only muscle that attaches only to the ulna at one end and only to the radius at the other end. *The pronator quadratus pronates the forearm at the radioulnar joints and also at the "intermediate" (radioulnar)*

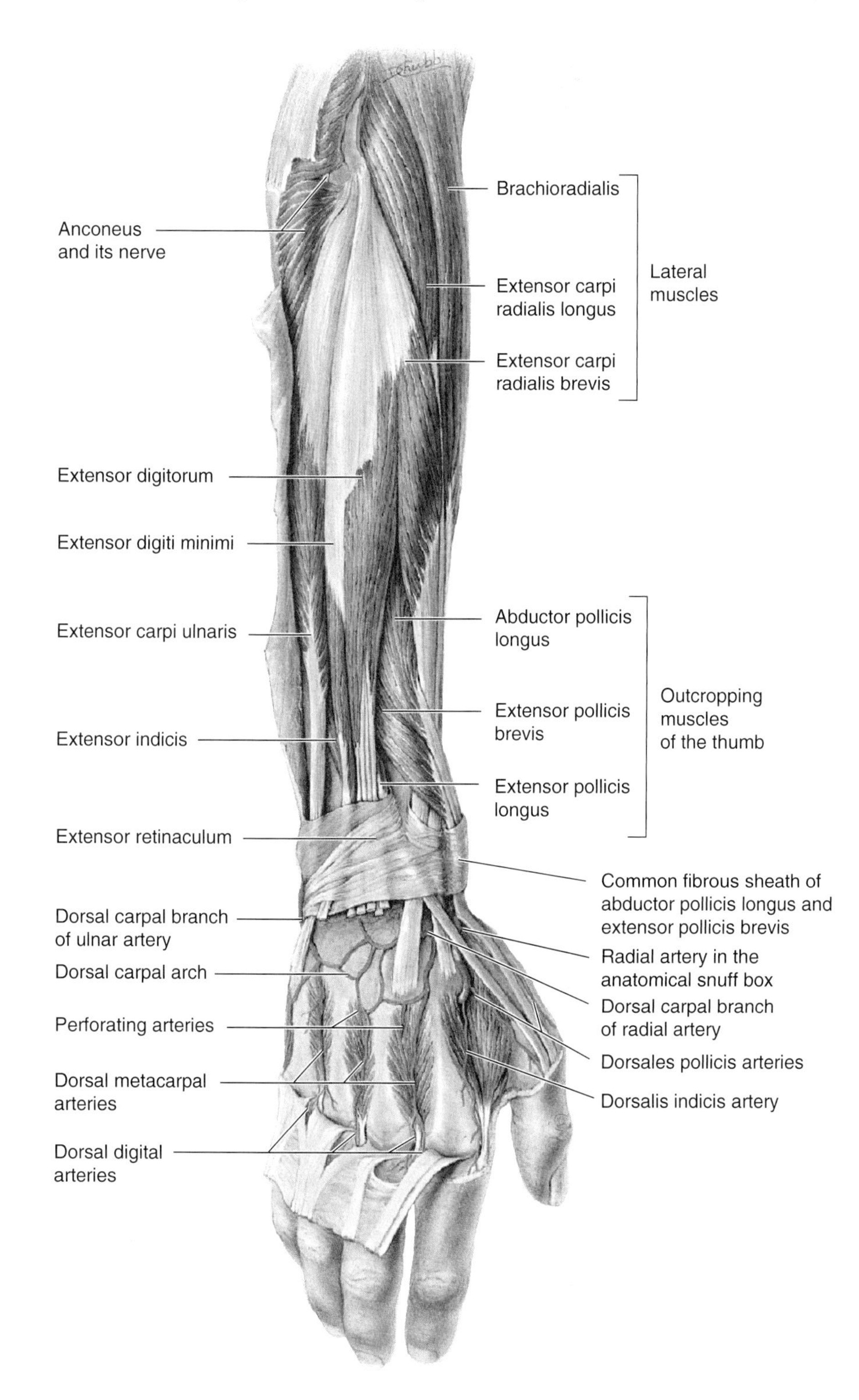

Figure 6.41. Extensor muscles of the right forearm and arteries on the dorsum of the hand. The distal extensor tendons have been removed from the dorsum of the hand without disturbing the arteries because they lie on the skeletal plane. Observe that the fascia on the posterior aspect of the forearm is thickened to form the extensor retinaculum, which is anchored to the radius and ulna.

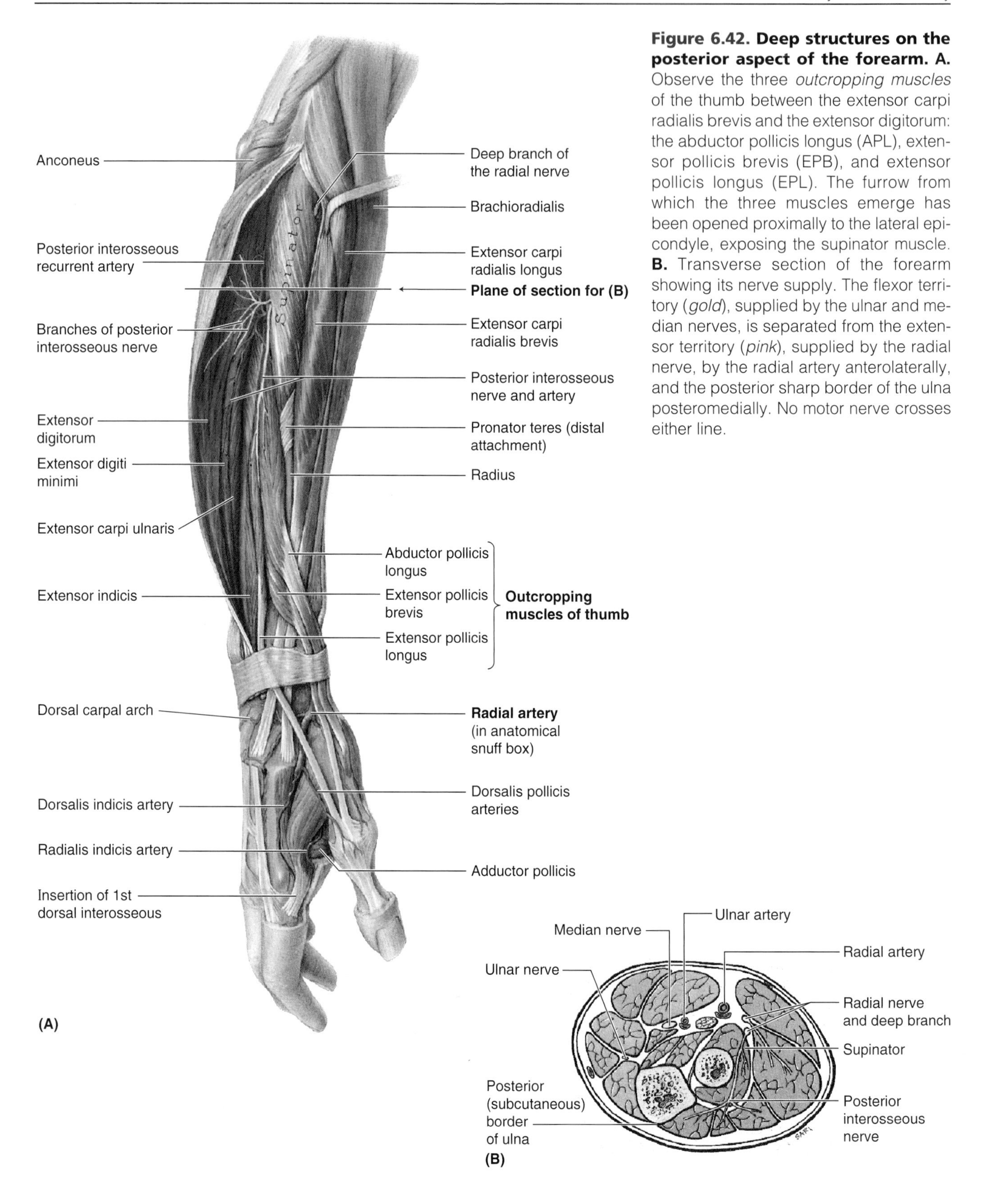

Figure 6.42. Deep structures on the posterior aspect of the forearm. A. Observe the three *outcropping muscles* of the thumb between the extensor carpi radialis brevis and the extensor digitorum: the abductor pollicis longus (APL), extensor pollicis brevis (EPB), and extensor pollicis longus (EPL). The furrow from which the three muscles emerge has been opened proximally to the lateral epicondyle, exposing the supinator muscle. **B.** Transverse section of the forearm showing its nerve supply. The flexor territory (*gold*), supplied by the ulnar and median nerves, is separated from the extensor territory (*pink*), supplied by the radial nerve, by the radial artery anterolaterally, and the posterior sharp border of the ulna posteromedially. No motor nerve crosses either line.

syndesmosis; it is the prime mover in pronation. The pronator quadratus initiates pronation; it is assisted by the pronator teres when more speed and power are needed. The pronator quadratus also helps the interosseous membrane hold the radius and ulna together, particularly when upward thrusts are transmitted through the wrist (e.g., during a fall on the hand).

Extensor Muscles of the Forearm

The extensor muscles are in the posterior (extensor-supinator) compartment of the forearm, and all are innervated by the radial nerve (Figs. 6.41–6.43, Table 6.7). These muscles can be organized into **three functional groups**:

- Muscles that extend and abduct or adduct the hand at the wrist joint (extensor carpi radialis longus, extensor carpi radialis brevis, and extensor carpi ulnaris)
- Muscles that extend the medial four digits (extensor digitorum, extensor indicis, and extensor digiti minimi)
- Muscles that extend or abduct the 1st digit, or thumb (abductor pollicis longus [APL], extensor pollicis brevis [EPB], and extensor pollicis longus [EPL]).

The extensor tendons are held in place in the wrist region by the **extensor retinaculum**, which prevents bowstringing of

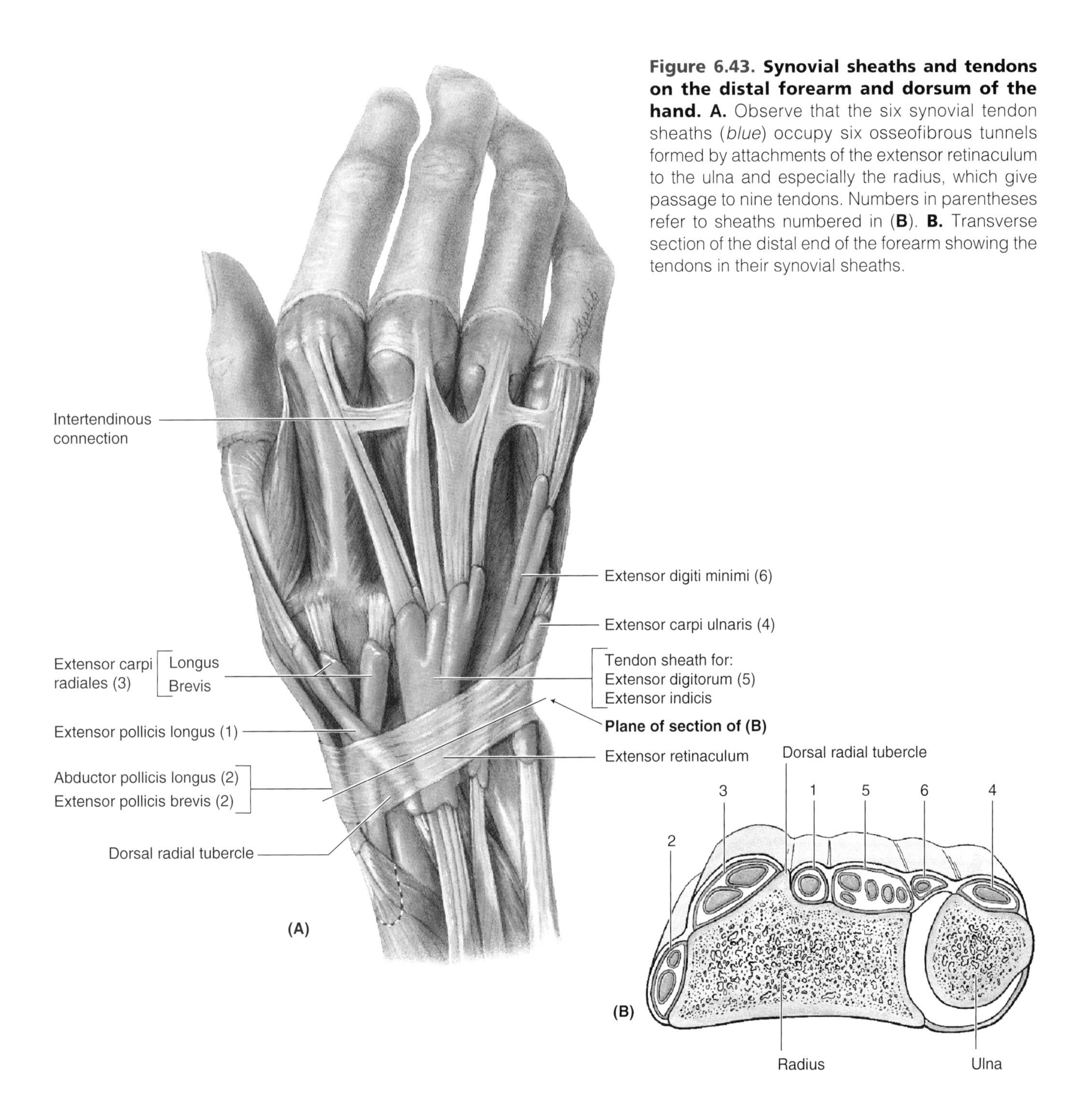

Figure 6.43. Synovial sheaths and tendons on the distal forearm and dorsum of the hand. A. Observe that the six synovial tendon sheaths (*blue*) occupy six osseofibrous tunnels formed by attachments of the extensor retinaculum to the ulna and especially the radius, which give passage to nine tendons. Numbers in parentheses refer to sheaths numbered in (**B**). **B.** Transverse section of the distal end of the forearm showing the tendons in their synovial sheaths.

Table 6.7. **Muscles of the Posterior Compartment of the Forearm**

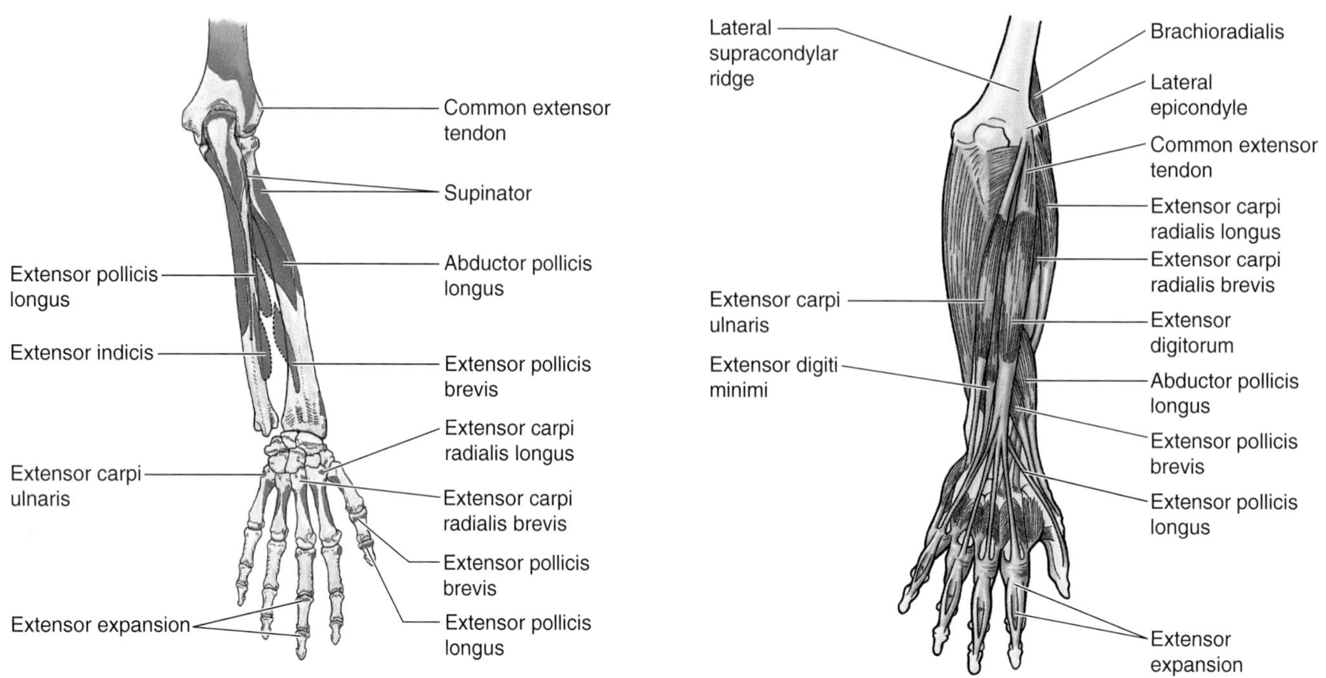

Posterior views

<table>
<tr><th>Muscle</th><th>Proximal Attachment</th><th>Distal Attachment</th><th>Innervation[a]</th><th>Main Action</th></tr>
<tr><td>Brachioradialis</td><td>Proximal two-thirds of lateral supracondylar ridge of humerus</td><td>Lateral surface of distal end of radius</td><td>Radial nerve (C5, C6, and C7)</td><td>Flexes forearm</td></tr>
<tr><td>Extensor carpi radialis longus</td><td>Lateral supracondylar ridge of humerus</td><td>Base of 2nd metacarpal</td><td>Radial nerve (C6 and C7)</td><td rowspan="2">Extend and abduct hand at wrist joint</td></tr>
<tr><td>Extensor carpi radialis brevis</td><td rowspan="3">Lateral epicondyle of humerus</td><td>Base of 3rd metacarpal</td><td>Deep branch of radial nerve (C7 and C8)</td></tr>
<tr><td>Extensor digitorum</td><td>Extensor expansions of medial four digits</td><td rowspan="3">Posterior interosseous nerve (C7 and C8), the continuation of deep branch of radial nerve</td><td>Extends medial four digits at metacarpo-phalangeal joints; extends hand at wrist joint</td></tr>
<tr><td>Extensor digiti minimi</td><td>Extensor expansion of 5th digit</td><td>Extends 5th digit at metacarpophalangeal and interphalangeal joints</td></tr>
<tr><td>Extensor carpi ulnaris</td><td>Lateral epicondyle of humerus and posterior border of ulna</td><td>Base of 5th metacarpal</td><td>Extends and adducts hand at wrist joint</td></tr>
<tr><td>Supinator</td><td>Lateral epicondyle of humerus, radial collateral and anular ligaments, supinator fossa, and crest of ulna</td><td>Lateral, posterior, and anterior surfaces of proximal third of radius</td><td>Deep branch of radial nerve (C5 and C6)</td><td>Supinates forearm (i.e., rotates radius to turn palm anteriorly)</td></tr>
<tr><td>Abductor pollicis longus</td><td>Posterior surfaces of ulna, radius, and interosseous membrane</td><td>Base of 1st metacarpal</td><td>Posterior interosseous nerve (C7 and C8), the continuation of deep branch of radial nerve</td><td>Abducts thumb and extends it at carpometacarpal joint</td></tr>
</table>

Table 6.7. (*Continued*) **Muscles of the Posterior Compartment of the Forearm**

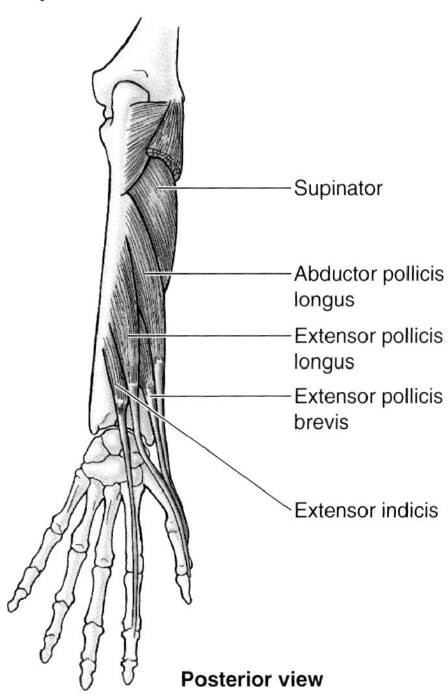

Muscle	Proximal Attachment	Distal Attachment	Innervation[a]	Main Action
Extensor pollicis brevis	Posterior surface of radius and interosseous membrane	Base of proximal phalanx of thumb	Posterior interosseous nerve (C7 and **C8**), the continuation of deep branch of radial nerve	Extends proximal phalanx of thumb at carpometacarpal joint
Extensor pollicis longus	Posterior surface of middle third of ulna and interosseous membrane	Base of distal phalanx of thumb		Extends distal phalanx of thumb at metacarpophalangeal and interphalangeal joints
Extensor indicis	Posterior surface of ulna and interosseous membrane	Extensor expansion of 2nd digit		Extends 2nd digit and helps to extend hand

[a] Numbers indicate spinal cord segmental innervation (e.g., C5, **C6**, and C7 indicate that nerves supplying the brachioradialis muscle are derived from fifth to seventh cervical segments of spinal cord). **Boldface** numbers indicate main segmental innervation. Damage to these segments, or to motor nerve roots arising from them, results in paralysis of muscles concerned.

the tendons when the hand is hyperextended at the wrist joint. As the tendons pass over the dorsum of the wrist, they are provided with **synovial tendon sheaths** that reduce friction between the extensor tendons and bones (Fig. 6.43).

The extensor muscles of the forearm may also be divided into superficial and deep groups. Four of the *superficial extensors* (extensor carpi radialis brevis, extensor digitorum, extensor digiti minimi, and extensor carpi ulnaris) are attached by a *common extensor tendon* to the lateral epicondyle (Figs. 6.41 and 6.42, Table 6.7). The proximal attachment of the other two muscles in the superficial group (brachioradialis and extensor carpi radialis longus) is to the lateral supracondylar ridge of the humerus and adjacent lateral intermuscular septum. The four flat tendons of the extensor digitorum pass deep to the extensor retinaculum to the medial four digits. The common tendons of the index and little fingers are joined on their medial sides near the knuckles by the respective tendons of the extensor indicis and extensor digiti minimi (extensors of index and little fingers, respectively). The extensor indicis tendon enters the hand in the same tunnel as the tendons of the extensor digitorum. The tendon of the extensor digiti minimi has its own tunnel. Usually three oblique bands unite the four tendons of the extensor digitorum proximal to the knuckles, restricting independent actions of the fingers (especially the ring finger). Consequently, normally no digit can remain fully flexed as the other ones are fully extended.

On the distal ends of the metacarpals and along the phalanges, the extensor tendons flatten to form **extensor expansions** (Fig. 6.44, *A–C*). Each extensor digital expansion (dorsal expansion, dorsal hood) is a triangular, tendinous

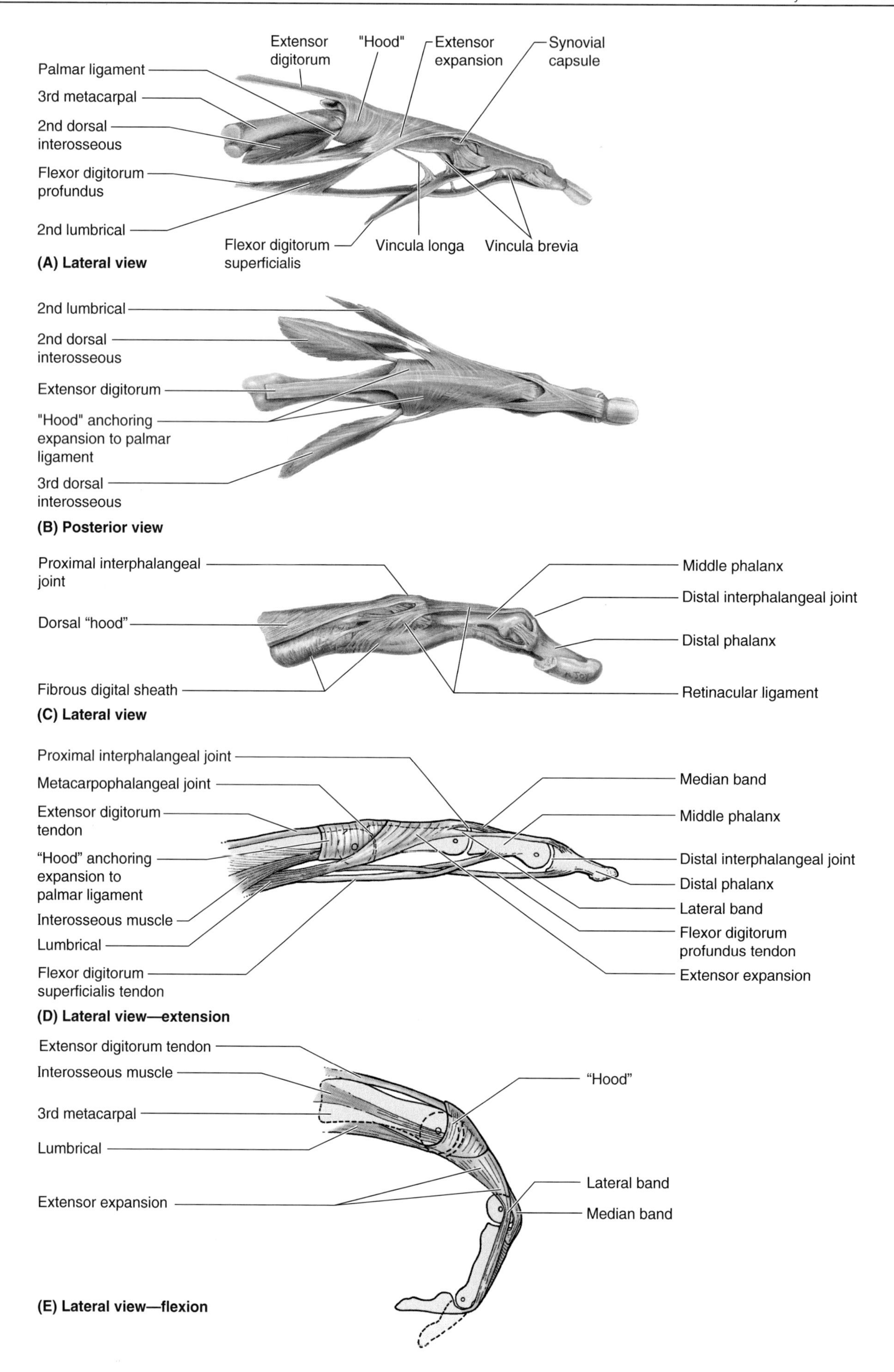
Extensor digitorum
"Hood"
Extensor expansion
Synovial capsule
Palmar ligament
3rd metacarpal
2nd dorsal interosseous
Flexor digitorum profundus
2nd lumbrical
Flexor digitorum superficialis
Vincula longa
Vincula brevia
(A) Lateral view
2nd lumbrical
2nd dorsal interosseous
Extensor digitorum
"Hood" anchoring expansion to palmar ligament
3rd dorsal interosseous
(B) Posterior view
Proximal interphalangeal joint
Dorsal "hood"
Fibrous digital sheath
Middle phalanx
Distal interphalangeal joint
Distal phalanx
Retinacular ligament
(C) Lateral view
Proximal interphalangeal joint
Metacarpophalangeal joint
Extensor digitorum tendon
"Hood" anchoring expansion to palmar ligament
Interosseous muscle
Lumbrical
Flexor digitorum superficialis tendon
Median band
Middle phalanx
Distal interphalangeal joint
Distal phalanx
Lateral band
Flexor digitorum profundus tendon
Extensor expansion
(D) Lateral view—extension
Extensor digitorum tendon
Interosseous muscle
3rd metacarpal
Lumbrical
Extensor expansion
"Hood"
Lateral band
Median band
(E) Lateral view—flexion

aponeurosis that wraps around the dorsum and sides of a head of the metacarpal and proximal phalanx. The visorlike **hood** formed by the extensor expansion over the head of the metacarpal, holding the extensor tendon in the middle of the digit, is anchored on each side to the **palmar ligament** (Fig. 6.44*B*). The extensor expansion divides into a median band that passes to the base of the middle phalanx (Fig. 6.44, *D* and *E*) and two lateral bands that pass to the base of the distal phalanx. The interosseous and lumbrical (L. lumbricus, earthworm) muscles of the hand attach to the lateral bands of the extensor expansion (Figs. 6.43 and 6.44). The **retinacular ligament** is a delicate fibrous band that runs from the proximal phalanx and fibrous digital sheath obliquely across the middle phalanx and two interphalangeal joints (Fig. 6.44*C*). It joins the extensor expansion to the distal phalanx. On flexing the distal interphalangeal joint, the retinacular ligament becomes taut and pulls the proximal joint into flexion. Similarly, on extending the proximal joint, the distal joint is pulled by the retinacular ligament into nearly complete extension.

Brachioradialis (Figs. 6.41–6.43, Table 6.7). This fusiform muscle lies superficially on the anterolateral surface of the forearm and forms the lateral border of the cubital fossa (Fig. 6.33). As mentioned previously, the brachioradialis is exceptional among muscles of the "posterior" (extensor-supinator) compartment in that it flexes the forearm at the elbow, especially when quick movement is required and when a weight is lifted during slow flexion of the forearm. The brachioradialis and the supinator are the only muscles of the compartment that do not cross—and therefore are incapable of acting at—the wrist. As it descends, the brachioradialis overlies the radial nerve and artery where they lie together on the supinator, pronator teres tendon, FDS, and flexor pollicis longus. The distal part of the tendon is covered by the APL and extensor pollicis brevis as they pass to the thumb. *To test the brachioradialis,* the elbow joint is flexed against resistance with the forearm in the midprone position. If acting normally, the muscle can be seen and palpated.

Extensor Carpi Radialis Longus (Figs. 6.41–6.43, Table 6.7). This fusiform muscle is partly overlapped by the brachioradialis, with which it often blends. It passes distally, posterior to the brachioradialis. Its tendon is crossed by the APL and EPB. *The extensor carpi radialis longus extends and abducts the wrist*; it is indispensable when clenching the fist. *To test the extensor carpi radialis longus,* the wrist is extended and abducted with the forearm pronated. If acting normally, the muscle can be palpated inferoposterior to the lateral side of the elbow. Its tendon can be palpated proximal to the wrist.

Extensor Carpi Radialis Brevis (Figs. 6.41–6.43, Table 6.7). As its name indicates, this fusiform muscle is shorter than the extensor carpi radialis longus. As it passes distally, it is covered by the extensor carpi radialis longus. *The extensor carpi radialis brevis extends and abducts the hand at the wrist joint.* This muscle and the extensor carpi radialis longus act together to steady the wrist during flexion of the medial four digits.

Extensor Digitorum (Figs. 6.41–6.44, Table 6.7). The extensor digitorum, *the principal extensor of the medial four digits*, occupies much of the posterior surface of the forearm. Its four tendons proximal pass through a **common synovial sheath**, deep to the extensor retinaculum with the tendon of the extensor indicis (Fig. 6.43, *A* and *B*). On the dorsum of the hand, the tendons spread out as they run toward the fingers. Adjacent tendons are linked by **intertendinous connections**. Commonly, the 4th tendon is fused initially with the tendon to the ring finger and reaches the little finger by a tendinous band. *The extensor digitorum extends the proximal phalanges* and, through its collateral reinforcements, the middle and distal phalanges as well. It also helps extend the hand at the wrist joint after exerting its traction primarily on the digits. *To test the extensor digitorum,* the forearm is pronated and the fingers are extended. The person attempts to keep the fingers extended at the metacarpophalangeal joints as the examiner exerts pressure on the proximal phalanges by attempting to flex them. If acting normally, the extensor digitorum can be palpated in the forearm, and its tendons can be seen and palpated on the dorsum of the hand.

Extensor Digiti Minimi (Figs. 6.41–6.43, Table 6.7). This fusiform slip of muscle is a partially detached part of the extensor digitorum. The tendon of this *extensor of the little finger* runs through a separate compartment deep to the extensor retinaculum and then divides into two slips; the lateral one is joined to the tendon of the extensor digitorum. *The extensor digiti minimi extends the proximal phalanx of the 5th digit* at the metacarpophalangeal joint and assists with exten-

Figure 6.44. Dorsal "hood" and extensor expansion of the 3rd digit. Lateral (**A**) and posterior (**B**) views of the dorsal "hood" and extensor expansion. Observe the interosseous muscles inserted in part into the base of the proximal phalanx and in part into the extensor expansion and the lumbrical muscle (L. lumbricus, earth worm), which inserts into the radial side of the extensor expansion. **C–D.** Lateral views of the *extensor expansion* in extension. The extensor expansion is a delicate fibrous band that runs from the proximal phalanx and fibrous digital sheath obliquely across the middle phalanx and two interphalangeal joints to join the dorsal "hood" to the distal phalanx. **E.** Lateral view of the finger in flexion. On flexing the distal interphalangeal joint (**C**), the extensor expansion (particularly its lateral bands) becomes taut, pulling the proximal joint into flexion (**E**). Similarly, on flexing the metacarpophalangeal joint, the proximal and distal joints are pulled by the extensor expansions (lateral bands) into nearly complete extension (the so-called "Z-movement").

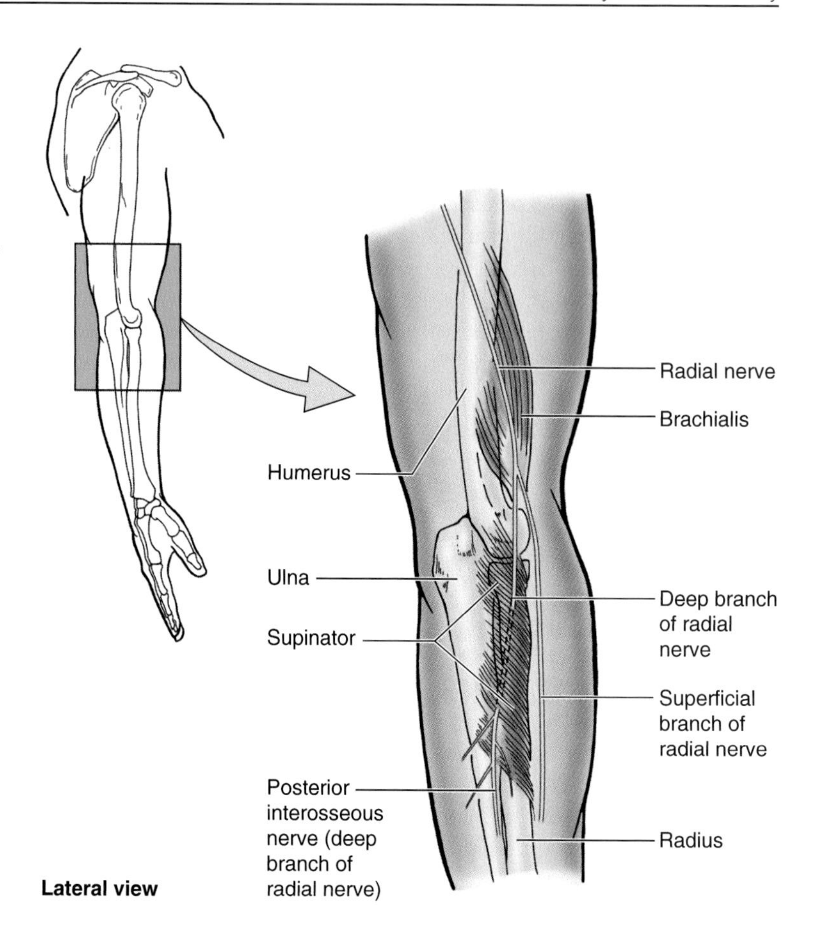

Figure 6.45. Relationship of the radial nerve to the brachialis and supinator muscles. When the radial nerve reaches the distal third of the humerus, it passes from the posterior to the anterior fascial compartment by piercing the lateral intermuscular septum. The nerve then runs between the brachialis and brachioradialis muscles across the anterior aspect of the lateral epicondyle. The *radial nerve* divides in the cubital fossa into motor (deep) and sensory (superficial) branches. The deep branch penetrates the supinator muscle to reach the posterior compartment of the forearm. Beginning at the inferior end of the supinator, the deep branch of the radial nerve is called the *posterior interosseous nerve* as it begins to course with the artery of the same name.

sion of its interphalangeal joints. It also assists with extension of the hand after exerting its traction, primarily on the 5th digit.

Extensor Carpi Ulnaris (Figs. 6.41–6.43, Table 6.7). This long fusiform muscle, located on the medial border of the forearm, has two heads. Distally, its tendon runs in a groove between the ulnar head and the styloid process, within a separate compartment of the extensor retinaculum. *The extensor carpi ulnaris extends and adducts the hand at the wrist joint simultaneously when acting independently.* Acting with the extensor carpi radialis, it extends the hand; acting with the flexor carpi ulnaris, it adducts the hand. Like the extensor carpi radialis longus, it is indispensable when clenching the fist. *To test the extensor carpi ulnaris,* the forearm is pronated and the fingers are extended. The extended wrist is then adducted against resistance. If acting normally, the muscle can be seen and palpated in the proximal part of the forearm and the tendon can be felt proximal to the head of the ulna.

Supinator (Figs. 6.42 and 6.45, Table 6.7). This muscle lies deep in the cubital fossa and, along with the brachialis, forms its floor. The humeral and ulnar heads of attachment of the supinator envelop the neck and proximal part of the body of the radius. The deep branch of the radial nerve passes between the two parts of the muscle as it leaves the cubital fossa to enter the posterior part of the arm; as it exits the muscle and joins the posterior interosseous artery, it may be referred to as the posterior interosseous nerve. *The supinator—the prime mover in supination—supinates the forearm by rotating the radius.* The biceps brachii also supinates the forearm, especially during rapid and forceful supination when resistance is required and the forearm is flexed (e.g., when a right-handed person drives a screw).

Elbow Tendinitis or Lateral Epicondylitis

Elbow tendinitis (golfer's or tennis elbow) is a painful musculoskeletal condition that may follow repetitive use of the superficial extensor muscles of the forearm. Pain is felt over the lateral epicondyle and radiates down the posterior surface of the forearm. People with elbow tendinitis often feel pain when they open a door or lift a glass. Repeated forceful flexion and extension of the wrist strain the attachment of the common tendon, producing inflammation of the periosteum of the lateral epicondyle (*lateral epicondylitis*) and the common extensor attachment of the muscles. ▶

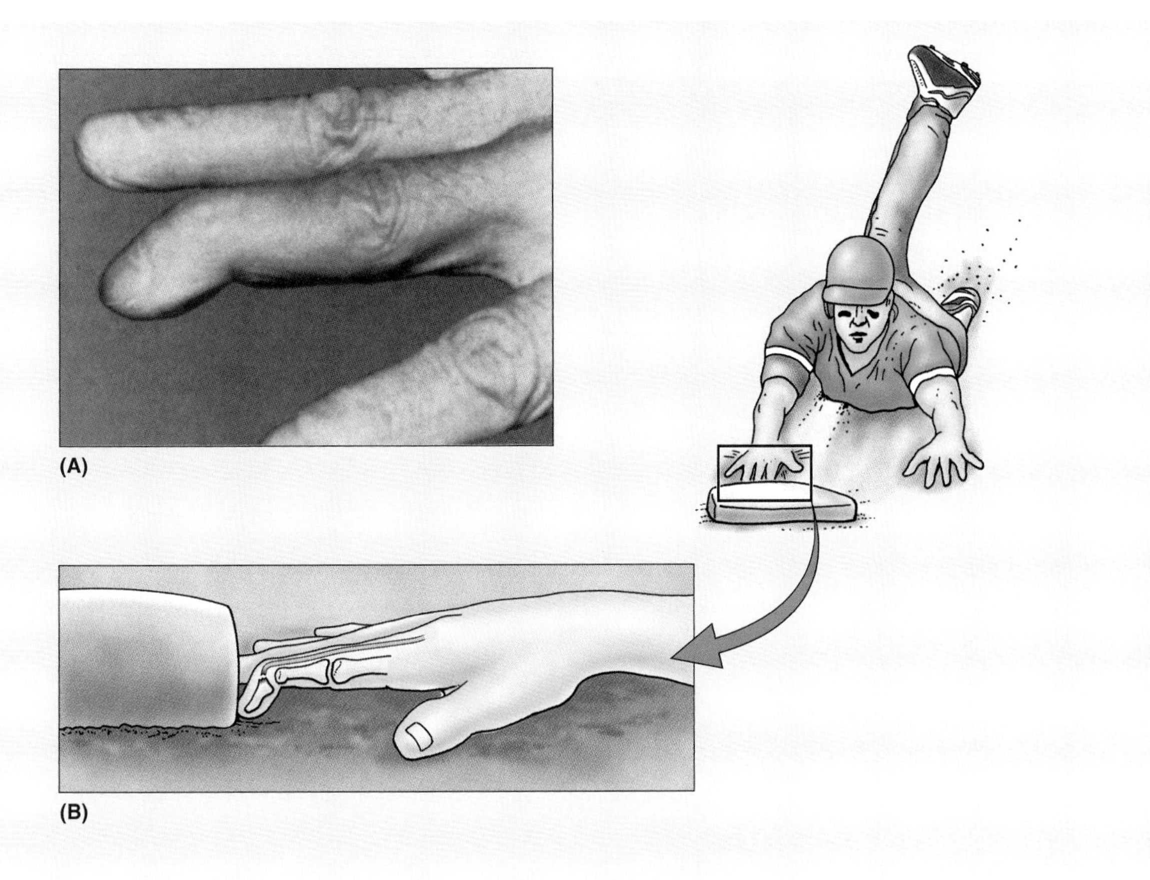

Mallet or Baseball Finger

Sudden severe tension on a long extensor tendon may avulse part of its attachment to the phalanx. The most common result of the injury is a mallet or baseball finger (*A*). This deformity results from the distal interphalangeal joint suddenly being forced into extreme flexion (hyperflexion)—for example, when a baseball is miscaught or a finger is jammed into the base pad (*B*). These actions avulse the attachment of the tendon into the base of the distal phalanx. As a result, the patient cannot extend the distal interphalangeal joint. The resultant deformity bears some resemblance to a mallet.

Fracture of the Olecranon (Fractured Elbow)

This fracture is common because the olecranon is subcutaneous. The common mechanism of injury is a fall on the elbow combined with sudden powerful contraction of the triceps. The fractured olecranon is pulled apart and the injury is often considered to be an avulsion fracture (Salter, 1998). This is a serious fracture requiring the services of an orthopaedic surgeon. Because of the traction produced by the tonus of the triceps on the olecranon fragment, pinning is usually required. Healing occurs slowly, and patients often wear a cast for nearly a year. ○

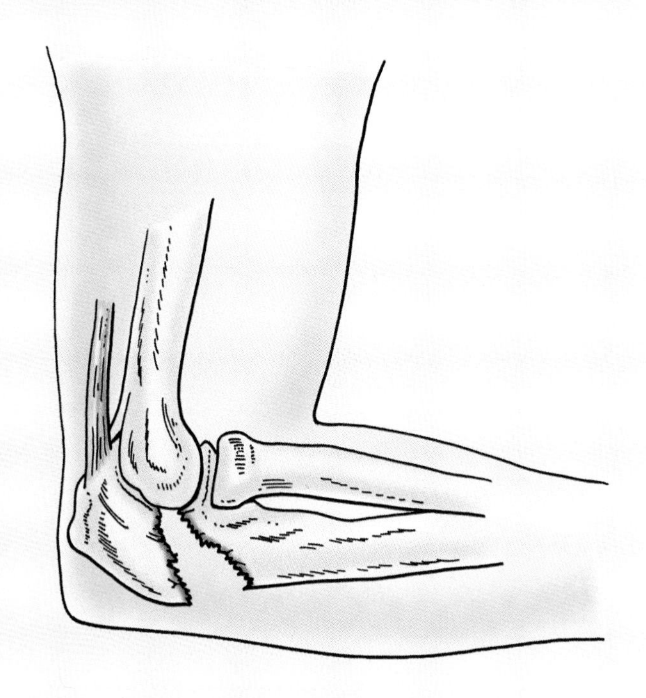

The **deep extensors of the forearm** (APL, EPB, and EPL) act on the thumb, and the extensor indicis helps extend the index finger (Figs. 6.41–6.43, Table 6.7). The three muscles acting on the thumb are deep to the superficial extensors and crop out from the furrow in the lateral part of the forearm that divides the extensors. Because of this characteristic, they are referred to as *outcropping (thumb) muscles.*

Abductor Pollicis Longus (Figs. 6.41–6.43, Table 6.7). The long, fusiform belly of the *abductor of the thumb* lies just distal to the supinator and is closely related to the EPB. Its tendon—and sometimes its belly—is commonly split into two parts, one of which may attach to the trapezium instead of the usual site at the base of the 1st metacarpal. *The APL abducts and extends the thumb at the carpometacarpal joint.* It acts with the abductor pollicis brevis during abduction of the thumb and with the extensor pollicis during extension of this digit. Although deeply situated, the APL emerges at the wrist as *one of the outcropping muscles.* Its tendon passes deep to the extensor retinaculum in a common synovial sheath with the tendon of the EPB. *To test the APL,* the thumb is abducted against resistance at the metacarpophalangeal joint. If acting normally, the tendon of the muscle can be seen and palpated at the lateral side of the snuff box and on the lateral side of the adjacent EPB tendon.

Extensor Pollicis Brevis (Figs. 6.41–6.43, Table 6.7). The belly of this fusiform *short extensor of the thumb* lies distal to the long abductor of the thumb (APL) and is partly covered by it. Its tendon lies parallel and immediately medial to that of the abductor pollicis longus but extends further, reaching the base of the proximal phalanx. *The EPB extends the proximal phalanx of the thumb at the metacarpophalangeal joint* and helps extend the distal phalanx. In continued action, it helps extend the 1st metacarpal. It also helps extend and abduct the hand. *To test the EPB,* the thumb is extended against resistance at the metacarpophalangeal joint. If acting normally, the tendon of the muscle can be seen and palpated at the lateral side of the snuff box and on the medial side of the adjacent APL tendon (Fig. 6.43).

Extensor Pollicis Longus (Figs. 6.41–6.43, Table 6.7). This *long extensor of the thumb* is larger and its tendon is longer than that of the EPB. The tendon passes medial to the dorsal tubercle of the radius, using it as a trochlea (pulley) changing its line of pull as it proceeds to the base of the distal phalanx of the thumb. The gap thus created between the long extensor tendons of the thumb is the anatomical snuff box. *The EPL extends the distal phalanx of the thumb* and, in continued action, extends the metacarpophalangeal

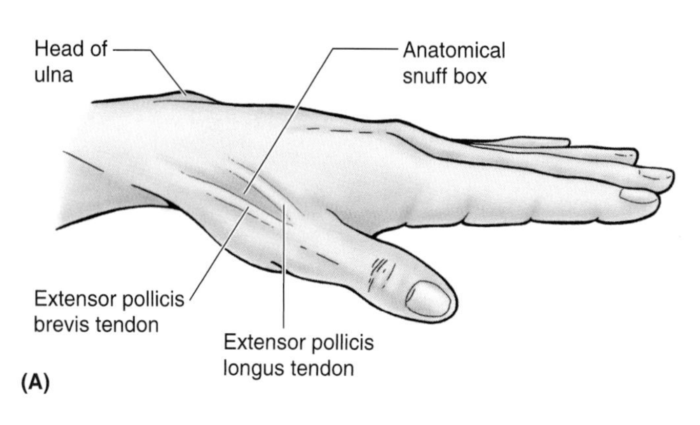

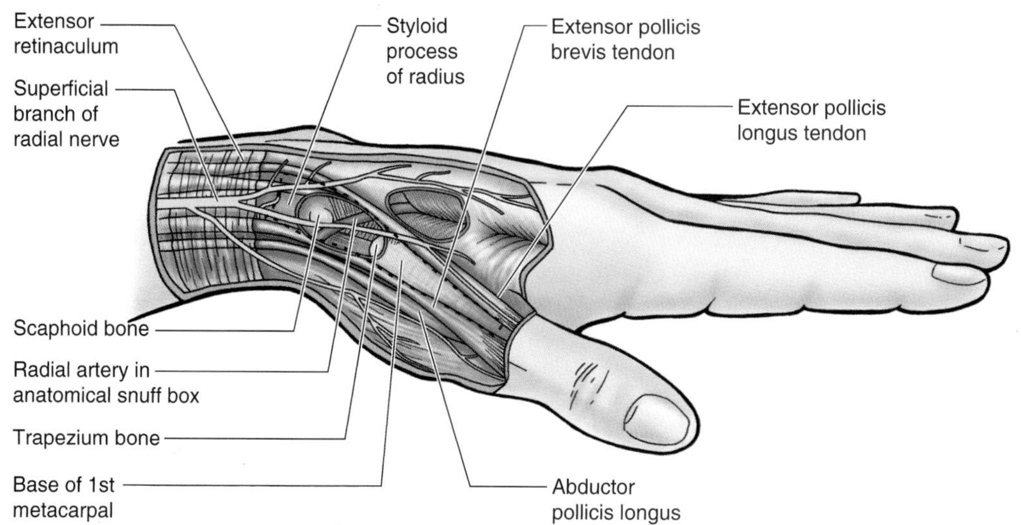

Figure 6.46. Radial artery in the anatomical snuff box. A. When the thumb is abducted and extended, a triangular hollow appears between the tendon of the extensor pollicis longus (EPL) medially and the tendons of the extensor pollicis brevis (EPB) and abductor pollicis longus (APL) laterally. **B.** The floor of the snuff box is formed by the scaphoid and trapezium bones. The floor is crossed by the radial artery as it passes diagonally from the anterior surface of the radius to the dorsal surface of the hand.

and interphalangeal joints of the thumb. It also adducts the extended thumb and rotates it laterally. *To test the EPL*, the thumb is extended against resistance at the interphalangeal joint. If acting normally, the tendon of the muscle can be seen and palpated on the medial side of the anatomical snuff box.

The tendons of the APL and EPB bound the **anatomical snuff box** anteriorly, and the tendon of the EPL bounds it posteriorly (Figs. 6.43 and 6.46). The snuff box is visible when the thumb is fully extended; this draws the tendons up and produces a concavity between them. Observe that the:

- *Radial artery* lies in the floor of the snuff box
- *Radial styloid process* can be palpated proximally and the base of the 1st metacarpal can be palpated distally in the snuff box
- *Scaphoid and trapezium* can be felt in the floor of the snuff box between the radial styloid process and the 1st metacarpal.

Extensor Indicis (Figs. 6.41–6.43, Table 6.7). The narrow, elongated belly of the *extensor of the index finger* lies medial to and alongside that of the EPL. This muscle confers independence to the index finger in that the extensor indicis may act alone or together with the extensor digitorum to *extend the index finger at the proximal interphalangeal joint*, as in pointing. It also helps extend the hand.

Fracture of the Scaphoid

The scaphoid and trapezium lie in the floor of the anatomical snuff box. *The scaphoid is the most frequently fractured carpal bone.* Injury to this bone results in localized tenderness in the snuff box. Initial radiographs may not reveal a fracture of the scaphoid; however, repeated radiographs 2 to 3 weeks later may reveal a fracture because resorption of bone has occurred at the fracture site.

Synovial Cyst of the Wrist

Sometimes a nontender cystic swelling appears on the hand, most commonly on the dorsum of the wrist. Usually the cyst is the size of a small grape, but it varies in size and may be as large as a plum. The thin-walled cyst contains clear mucinous fluid. The cause of the cyst is unknown, but it may result from mucoid degeneration (Salter, 1998). Flexion of the wrist makes the cyst enlarge, and it may be painful. Clinically, this type of swelling is called a "ganglion" (G. swelling or knot); anatomically, a ganglion refers to a collection of nerve cells (e.g., a spinal ganglion). These cystic swellings are close to and often communicate with the synovial sheaths on the dorsum of the wrist. The distal attachment of the extensor carpi radialis brevis tendon into the base of the 3rd metacarpal is a common site for such a cystic swelling. A cystic swelling of the common flexor synovial sheath on the anterior aspect of the wrist can enlarge enough to produce compression of the median nerve by narrowing the carpal tunnel (*carpal tunnel syndrome*). This syndrome produces pain and paresthesia in the sensory distribution of the median nerve and clumsiness of finger movements. ⊙

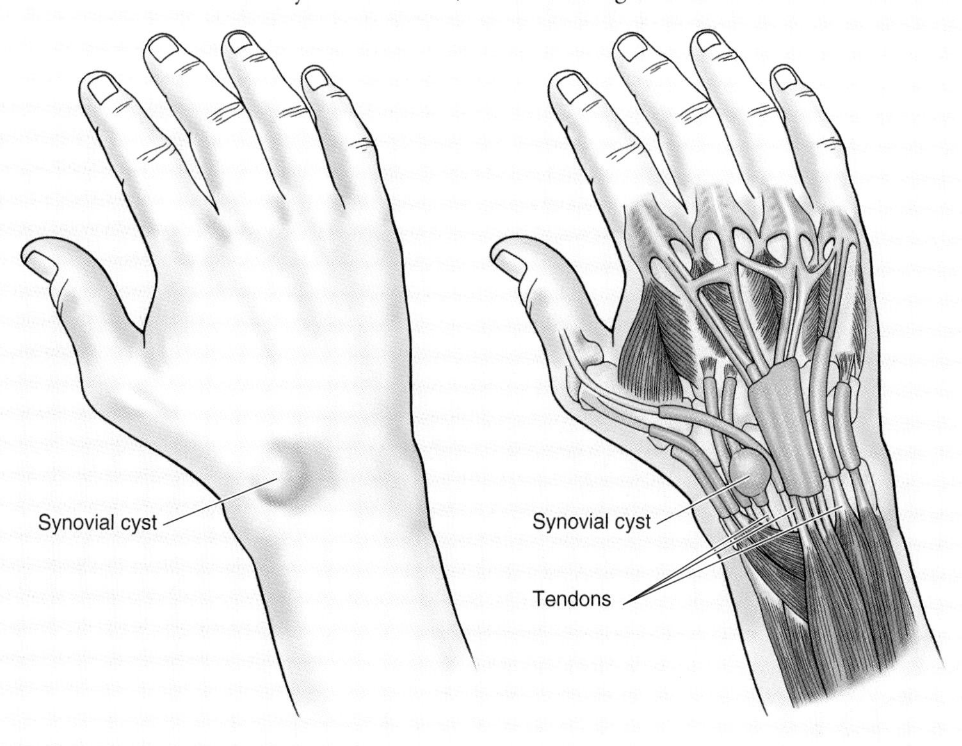

Table 6.8. **Arteries of the Forearm and Hand**

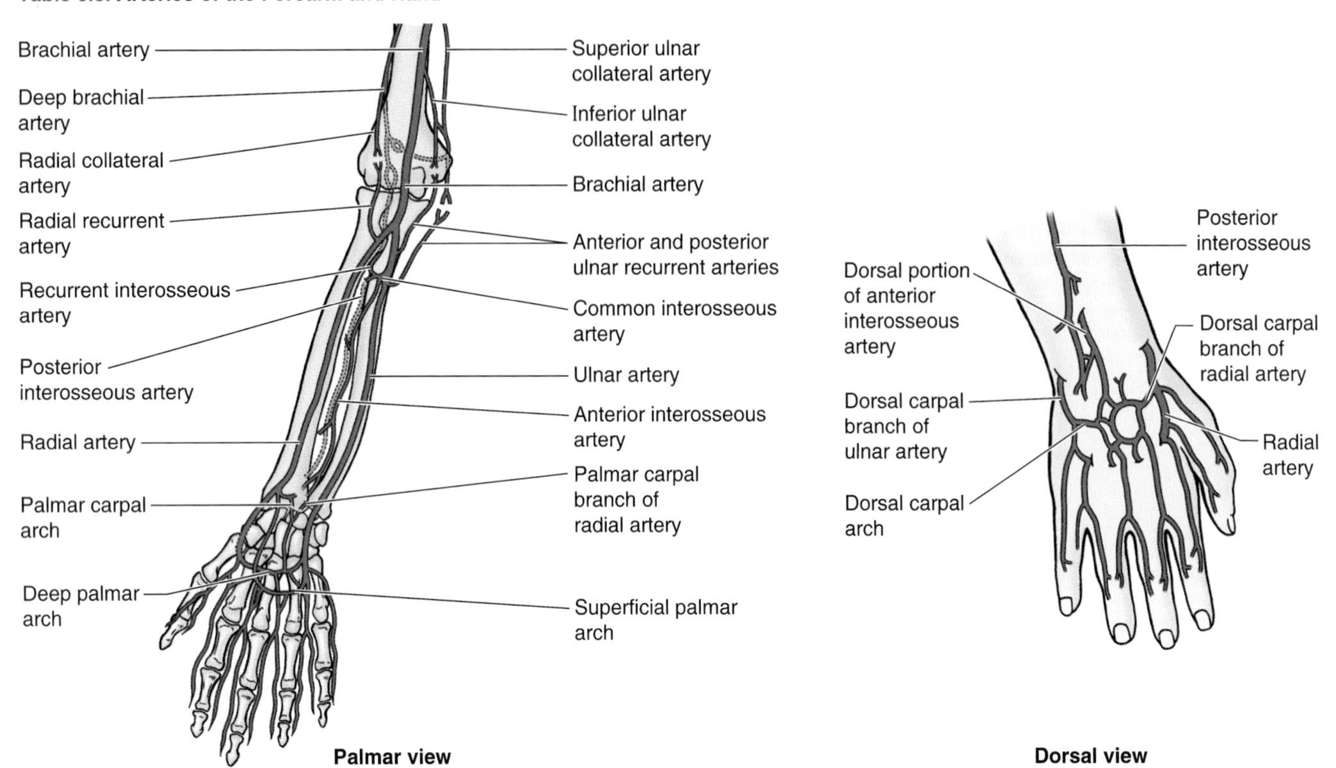

Artery	Course	Distribution
Ulnar	Larger terminal branch of brachial artery in cubital fossa	Passes inferomedially and then directly inferiorly, deep to pronator teres, palmaris longus, and flexor digitorum superficialis to reach medial side of forearm; passes superficial to flexor retinaculum at wrist and gives a deep palmar branch to deep arch and continues as superficial palmar arch
Anterior and posterior ulnar recurrent	Ulnar artery, just distal to elbow joint	Anterior ulnar recurrent artery passes superiorly and posterior ulnar recurrent artery passes posteriorly to anastomose with ulnar collateral and interosseous recurrent arteries
Common interosseous	Ulnar artery, just distal to bifurcation of brachial artery	After a short course, terminates by dividing into anterior and posterior interosseous arteries
Anterior and posterior interosseous	Common interosseous artery	Pass to anterior and posterior sides of interosseous membrane: anterior interosseous artery supplies both anterior and posterior compartments in distal forearm; the posterior interosseous artery gives off the recurrent interosseous artery, which participates in the arterial anastomoses around the elbow
Dorsal and palmar carpal branches	Ulnar artery at level of wrist	Anastomose with corresponding branches of radial artery to form dorsal and palmar carpal arches, providing collateral circulation at wrist
Radial	Smaller terminal division of brachial artery in cubital fossa	Runs inferolaterally under cover of brachioradialis and distally lies lateral to flexor carpi radialis tendon; winds around lateral aspect of radius and crosses floor of anatomical snuff box to pierce fascia; ends by forming deep palmar arch with deep branch of ulnar artery
Radial recurrent	Lateral side of radial artery, just distal to its origin	Ascends on supinator and then passes between brachioradialis and brachialis
Dorsal and palmar carpal branches	Radial artery at level of wrist	Anastomose with corresponding branches of ulnar artery to form dorsal and palmar carpal arches, providing collateral circulation at wrist

Arteries of the Forearm

The main arteries of the forearm are the ulnar and radial arteries. The brachial artery usually ends opposite the neck of the radius in the inferior part of the cubital fossa. Here it divides into its terminal branches, the ulnar and radial arteries.

Ulnar Artery

The ulnar artery, the larger of the two terminal branches of the brachial artery, usually *begins in the cubital fossa near the neck of the radius*, just medial to the biceps tendon (Table 6.8). It descends through the anterior compartment of the forearm, deep to the pronator teres. The ulnar artery then passes distally over the anterior aspect of the wrist to the palm. Pulsations of the artery can be palpated on the lateral side of the flexor carpi ulnaris tendon, where it lies anterior to the ulnar head. The ulnar nerve is on the medial side of the ulnar artery.

Branches of the Ulnar Artery in the Forearm (Table 6.8). These branches supply medial muscles in the forearm and hand, the common flexor synovial sheath, and the ulnar nerve.

- The *anterior ulnar recurrent artery* arises from the ulnar artery just inferior to the elbow joint and runs superiorly between the brachialis and pronator teres. It supplies these

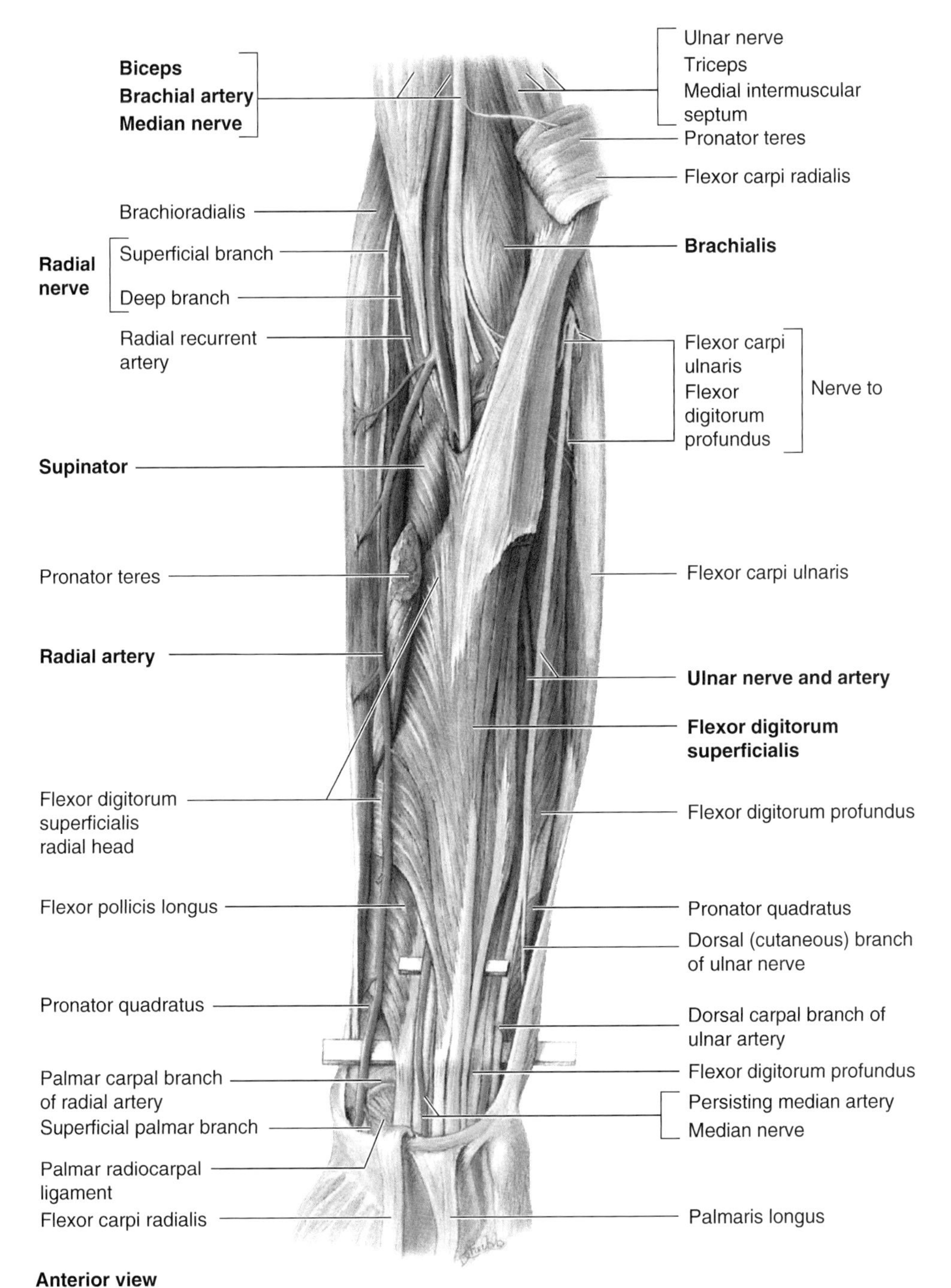

Figure 6.47. Flexor digitorum superficialis (FDS) and related structures. Observe the ulnar artery descending obliquely, posterior to the FDS, to meet and accompany the ulnar nerve. Observe also the ulnar nerve descending vertically near the medial border of the FDS, which is exposed by splitting the septum between the FDS and the flexor carpi ulnaris.

muscles and *anastomoses with the inferior ulnar collateral artery*, a branch of the brachial, thereby participating in the arterial anastomoses around the elbow.

- The *posterior ulnar recurrent artery* arises from the ulnar artery distal to the anterior ulnar recurrent artery. It passes superiorly, posterior to the medial epicondyle, where it lies deep to the tendon of the flexor carpi ulnaris. It supplies adjacent muscles and then participates in the arterial anastomoses around the elbow. The anterior and posterior arteries may be present as anterior and posterior branches of a (common) ulnar recurrent artery.
- The *common interosseous artery*, a short branch of the ulnar artery, arises in the distal part of the cubital fossa and divides almost immediately into anterior and posterior interosseous arteries.
- The *anterior interosseous artery* passes distally on the anterior aspect of the interosseous membrane to the proximal border of the pronator quadratus. Here it pierces the interosseous membrane and continues distally into the wrist on the posterior aspect of the interosseous membrane.

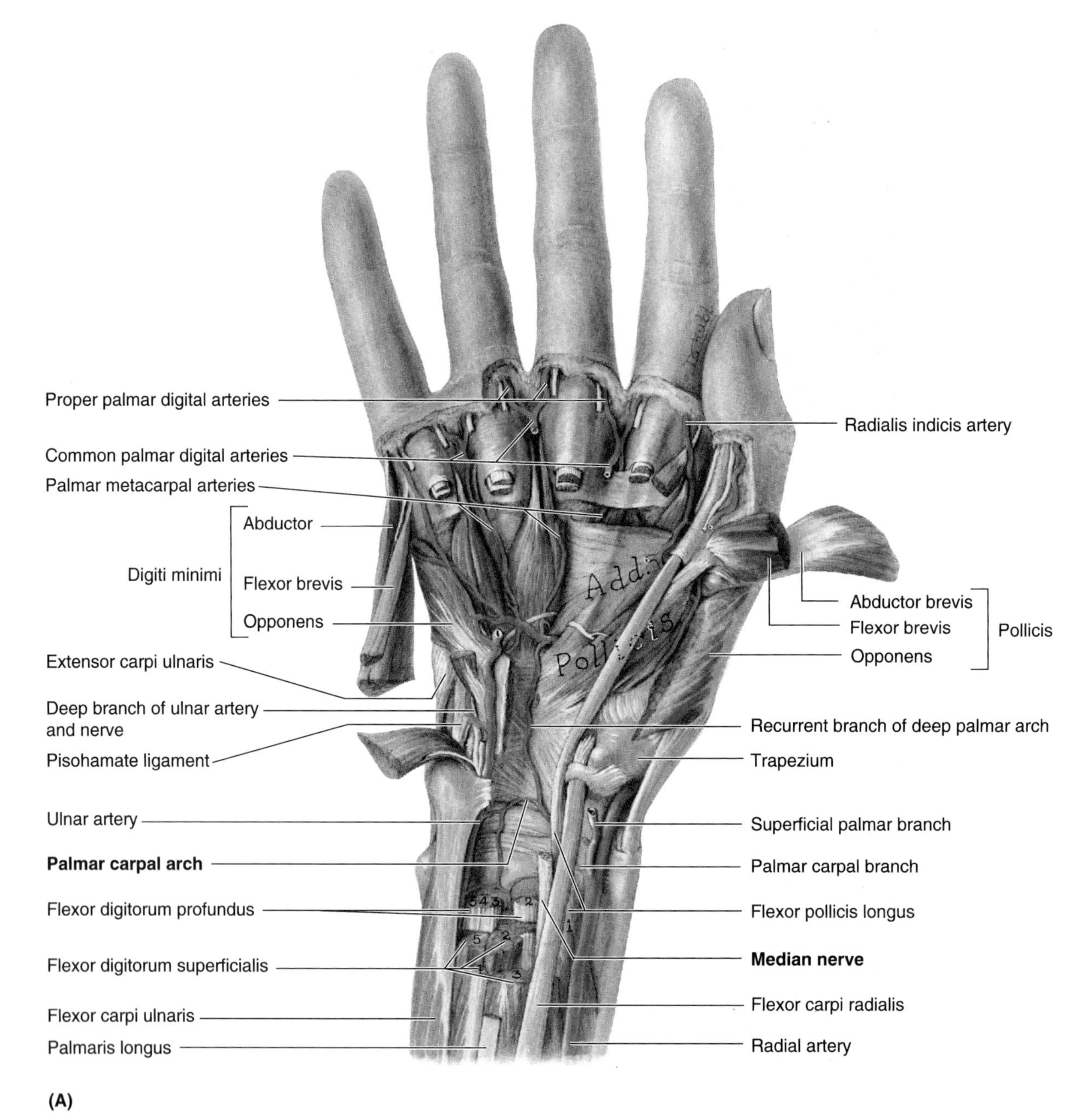

Figure 6.48. Muscles and arteries of the distal forearm and hand. A. Deep dissection of the palm of the right hand. Observe the anastomosis of the palmar carpal branch of theradial artery with the palmar carpal branch of the ulnar artery to form the palmar carpal arch.

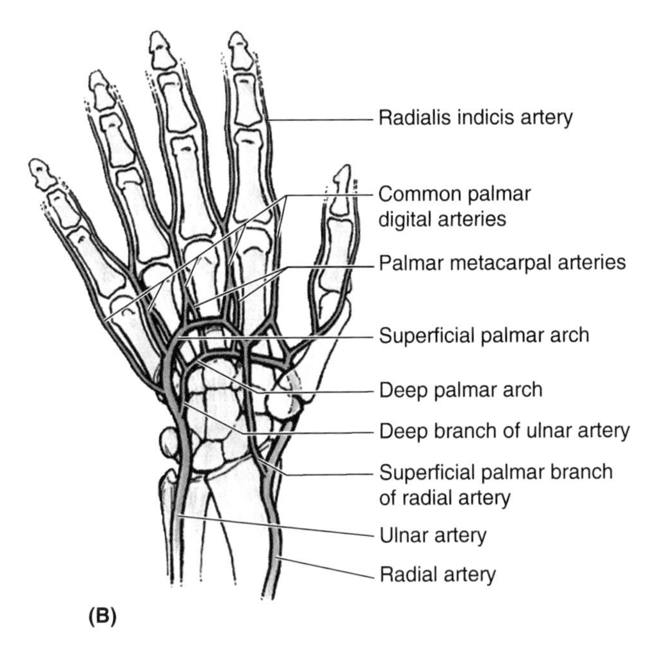

Figure 6.48. (*Continued*) **B.** Diagram of the palmar arterial arches. Observe that the deep palmar arch lies at the level of the bases of the metacarpal bones and that the superficial palmar arch is located more distally.

- The *posterior interosseous artery* passes posteriorly between the radius and ulna, just proximal to the interosseous membrane. It supplies adjacent muscles and then gives off the *posterior interosseous recurrent artery*, which passes superiorly, posterior to the lateral epicondyle and participates in the arterial anastomoses around the elbow. It does not run on the interosseous membrane but courses between the superficial and deep layers of the extensor muscles in the company of the posterior interosseous nerve. It is mostly exhausted in the distal forearm and is replaced by the anterior interosseous artery.
- *Muscular branches of the ulnar artery* supply muscles on the medial side of the forearm, mainly those in the flexor-pronator group.
- The *palmar carpal branch of the ulnar artery* is a small branch that runs across the anterior aspect of the wrist, deep to the tendons of the FDP. This branch anastomoses with the palmar carpal branch of the radial artery, forming the **palmar carpal arch** (Figs. 6.47 and 6.48).
- The *dorsal carpal branch of the ulnar artery* arises just proximal to the pisiform. It passes across the dorsal surface of the wrist, deep to the extensor tendons, where it anastomoses with the dorsal carpal branch of the radial artery, forming the **dorsal carpal arch** (Table 6.8).
- The *superficial branch of the ulnar artery* continues into the palm as the *superficial palmar arch* (Fig. 6.48).
- The *deep palmar branch of the ulnar artery* passes deeply in the hand, where it anastomoses with the radial artery and completes the **deep palmar arch.**

Radial Artery

The radial artery, smaller than the ulnar artery, *begins in the cubital fossa near the neck of the radius.* It passes inferolaterally deep to the brachioradialis, but its pulsations can be felt throughout the forearm, making it useful as an anterolateral demarcation of the flexor and extensor compartments of the forearm. When the brachioradialis is pulled laterally, the entire length of the artery is visible (Fig. 6.47, Table 6.8). The radial artery lies on muscle until it reaches the distal part of the forearm. Here it lies on the anterior surface of the radius and is covered only by skin and fascia. The course of the radial artery in the forearm is represented by a line joining the midpoint of the cubital fossa to a point just medial to the radial styloid process. The radial artery leaves the forearm by winding around the lateral aspect of the wrist and crosses the floor of the anatomical snuff box (Fig. 6.46). In the hand, the ulnar and radial arteries anastomose, forming the superficial and deep *palmar arterial arches.*

The **branches of the radial artery** in the forearm are listed below.

- *Muscular branches* of the radial artery supply both flexor and extensor muscles on the lateral side of the forearm (Fig. 6.47, Table 6.8).
- The *radial recurrent artery* arises from the lateral side of the radial artery just distal to its origin and ascends between the brachioradialis and brachialis. It supplies these muscles and the elbow joint and anastomoses with the *radial collateral artery* (Table 6.8), a branch of the deep artery of the arm, thereby participating in the arterial anastomoses around the elbow.
- The *superficial palmar branch of the radial artery* (Fig. 6.48) arises at the distal end of the radius, just proximal to the wrist. It passes through, sometimes over, the thumb muscles and supplies them. The superficial palmar branch commonly anastomoses with the terminal part of the ulnar artery to form the **superficial palmar arch.**
- The *palmar carpal branch of the radial artery* (Table 6.8) is a small vessel that arises near the distal border of the pronator quadratus (Fig. 6.48*A*). It runs across the wrist deep to the flexor tendons and anastomoses with the carpal branch of the ulnar artery and recurrent branches of the deep palmar arch to form the **palmar carpal arch.**
- The *dorsal carpal branch of the radial artery* runs medially across the dorsal surface of the wrist, deep to the extensor tendons (Table 6.8), where it anastomoses with the dorsal carpal branch of the ulnar artery and terminal branches of the anterior and posterior interosseous arteries to form the **dorsal carpal arch.**

High Division of the Brachial Artery

Sometimes the brachial artery divides at a more proximal level than usual. In this case, the ulnar and radial arteries begin near the middle of the arm and the median nerve passes between them. The musculocutaneous and median nerves commonly communicate as shown here.

Superficial Ulnar Artery

In approximately 3% of people, the ulnar artery descends superficial to the flexor muscles. Pulsations of the superficial ulnar artery can be felt and may be visible. *This variation must be kept in mind when performing venesections for withdrawing blood or making intravenous injections.* If an aberrant ulnar artery is mistaken for a vein, it may be damaged and produce bleeding. If certain drugs are injected into the aberrant artery, the result could be disastrous. ⊙

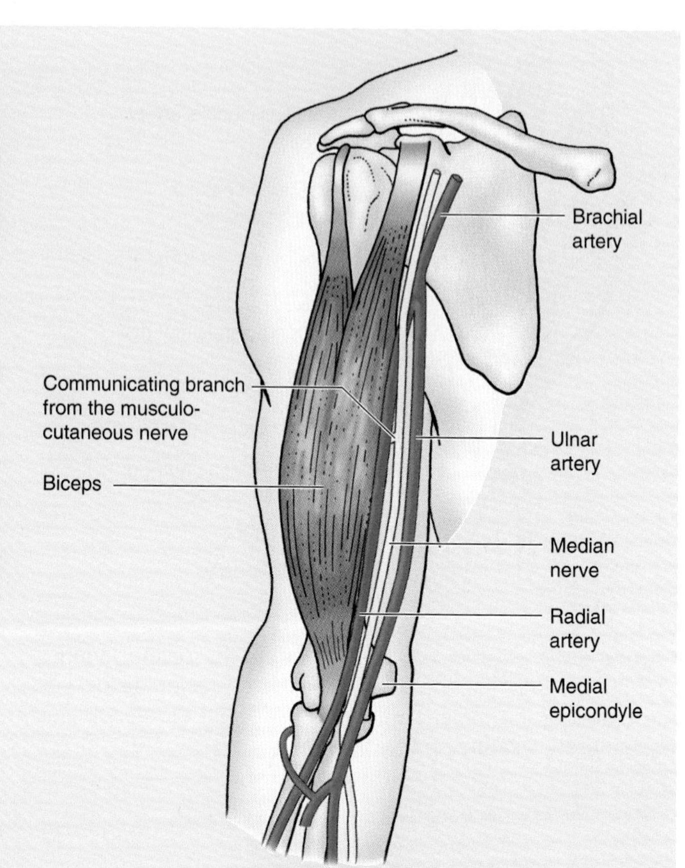

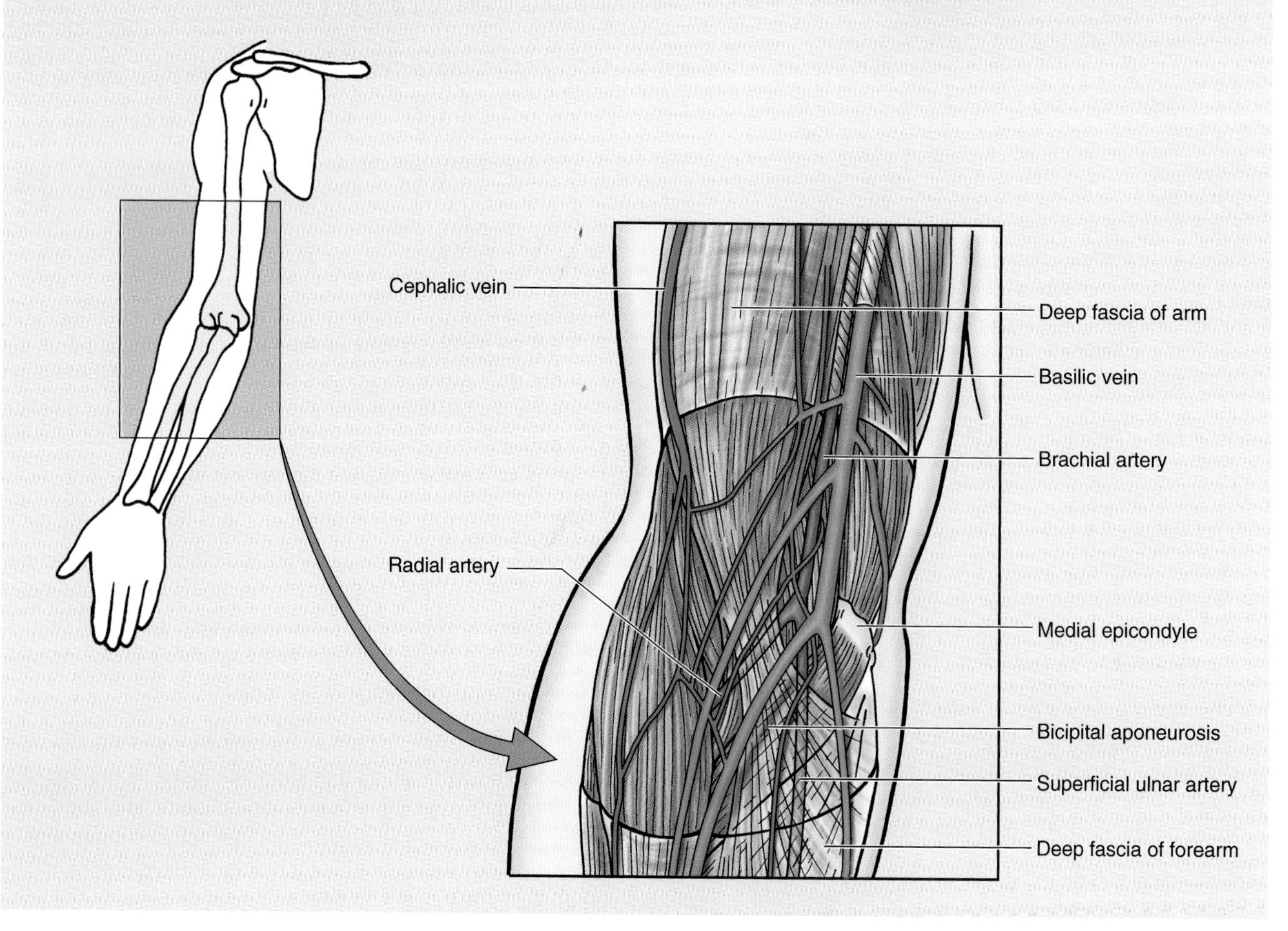

Measuring the Pulse Rate

The common place for measuring the pulse rate is where the radial artery lies on the anterior surface of the distal end of the radius, lateral to the tendon of the flexor carpi radialis. Here it is covered only by fascia and skin. Approximately 4 cm of this artery can be compressed against the distal end of the radius, where it lies between the tendons of the flexor carpi radialis and APL. When measuring the pulse rate, the pulp of the thumb should not be used because it has its own pulse, which could obscure the patient's pulse. If a pulse cannot be felt, try the other wrist because an *aberrant radial artery* on one side may make the pulse difficult to palpate. A radial pulse may also be felt by pressing lightly in the anatomical snuff box.

Variations in the Origin of the Radial Artery

The origin of the radial artery may be more proximal than usual; it may be a branch of the axillary artery or the brachial artery. Sometimes the radial artery is superficial to the deep fascia instead of deep to it. When a superficial vessel is pulsating in the forearm, it may be a superficial radial or ulnar artery and is vulnerable to laceration. ⊙

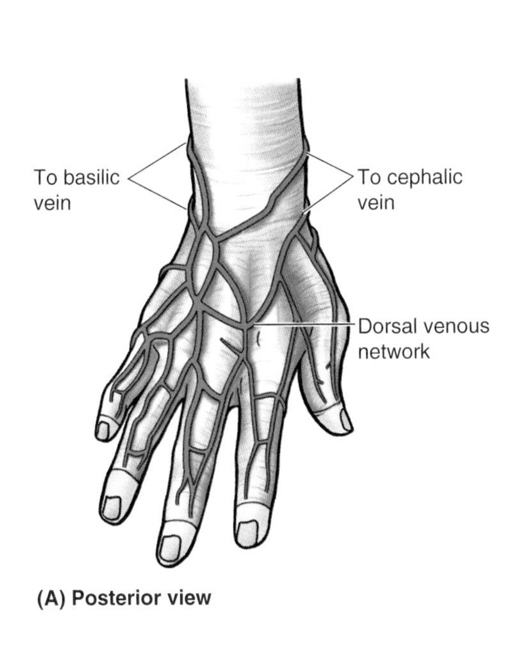

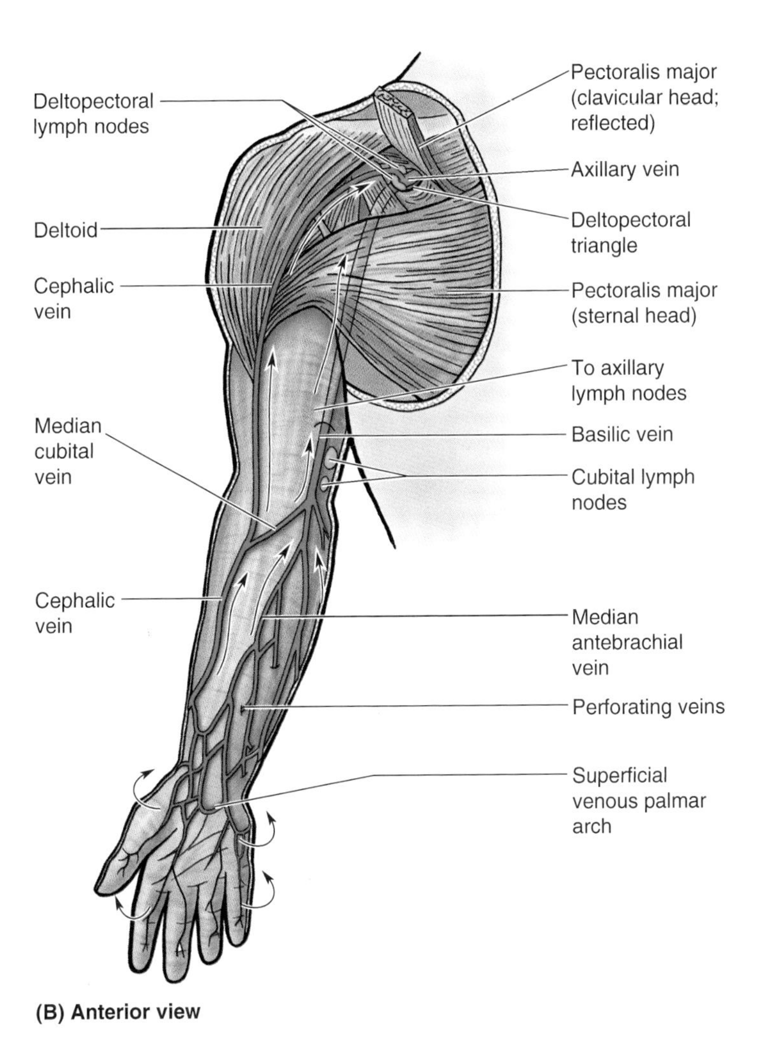

Figure 6.49. Superficial venous and lymphatic drainage of the upper limb. The superficial veins are the cephalic, basilic, and median antebrachial (median vein of the forearm). **A.** Dorsum of the hand. Veins from the fingers (dorsal digital veins) end in the dorsal venous network (arch) opposite the middle of the dorsum of the hand. **B.** Anterior aspect of the upper limb. The median antebrachial vein drains the superficial palmar venous arch and ascends to end in the basilic vein. The perforating veins carry blood from the deep veins to the superficial veins. The superficial lymphatic vessels (*arrows*) mostly accompany the superficial veins; for example, the lymphatic vessels from the wrist follow the median antebrachial vein and enter the cubital lymph nodes. Lymph from the upper limb eventually enters the axillary lymph nodes.

Veins of the Forearm

In the forearm, as in the arm, are superficial and deep veins. The superficial veins ascend in the subcutaneous tissue. The deep veins accompany the deep forearm (antebrachial) arteries.

Superficial Veins

The main superficial veins of the forearm are the cephalic, basilic, median cubital, and antebrachial veins and their tributaries (Fig. 6.49). The **cephalic vein** forms over the anatomical snuff box from the tributaries that arise from the lateral side of the **dorsal venous network** (arch). The cephalic vein ascends along the lateral border of the forearm and communicates with the basilic vein through the **median cubital vein.** The cephalic vein then ascends along the lateral side of the arm and empties into the axillary vein. The **basilic vein** arises from the medial side of the dorsal venous network (Fig. 6.49) and ascends posteromedially in the forearm, reaching the anterior surface just distal to the elbow, where it is joined by the median cubital vein. The **median antebrachial vein** drains subcutaneous tissue in the anterior aspect of the wrist and forearm. It begins in the superficial **venous palmar arch** and usually ends in the basilic vein.

Deep Veins

Deep veins accompanying the arteries are plentiful in the forearm (Fig. 6.50). These accompanying veins arise from a **deep venous arcade** (a series of anastomosing venous arches) in the hand. From the lateral side of the arcade, paired **radial**

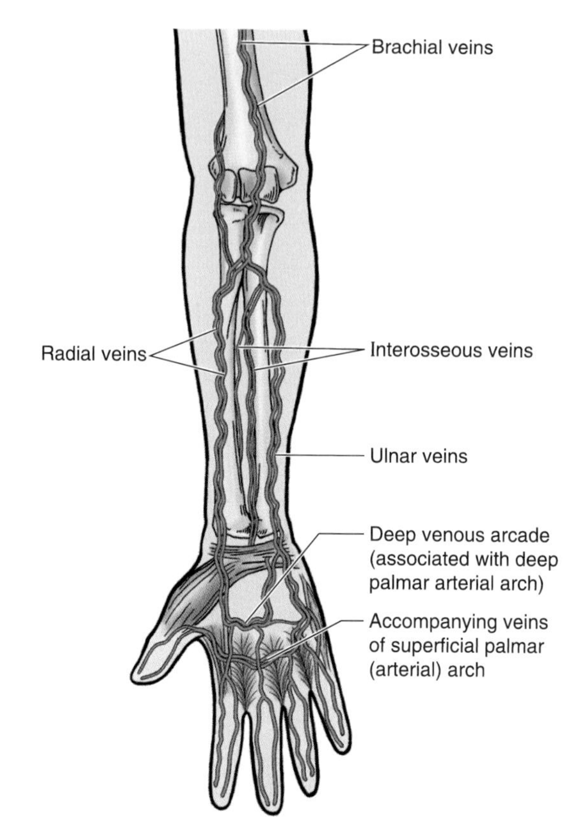

Figure 6.50. Deep venous drainage of the upper limb. The deep veins follow the arteries as their companions. They are relatively small and usually paired and are connected at intervals by transverse branches.

Variation of Veins in the Cubital Fossa

The pattern of veins in the cubital fossa varies greatly. In approximately 20% of people, the median antebrachial vein (median vein of the forearm) divides into a *median basilic vein* that joins the basilic vein and a *median cephalic vein* that joins the cephalic vein. In these cases, a clear "M" *formation* is produced by the cubital veins. It is important to observe and remember that either the median cubital vein or median basilic vein, whichever pattern is present, crosses superficial to the brachial artery, from which it is separated by the bicipital aponeurosis. These veins are good sites for drawing blood but are not ideal for injecting an irritating drug because of the danger of injecting it into the brachial artery. ⊙

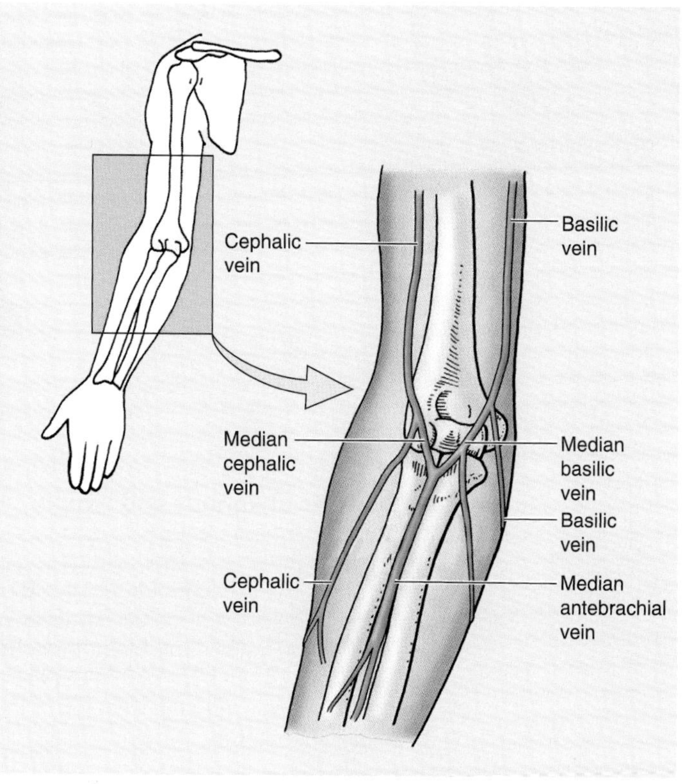

veins arise and accompany the radial artery; from the medial side, paired **ulnar veins** arise and accompany the ulnar artery. The veins accompanying each artery anastomose freely with each other. The radial and ulnar veins drain the forearm but carry relatively little blood from the hand. The deep veins ascend in the forearm along the sides of the corresponding arteries, receiving tributaries from veins leaving the muscles with which they are related. Deep veins communicate with the superficial veins. The deep **interosseous veins**, which accompany the interosseous arteries, unite with the accompanying veins of the radial and ulnar arteries. The deep veins in the cubital fossa are connected to the **median cubital vein**, a superficial vein. These **deep cubital veins** also unite with the accompanying veins of the brachial artery.

Nerves of the Forearm

The nerves of the forearm are the median, ulnar, and radial. The median nerve is the principal nerve of the anterior (flexor-pronator) compartment of the forearm (Fig. 6.51). Although the radial nerve appears in the cubital region, it soon enters the posterior (extensor-supinator) compartment of the forearm. Aside from cutaneous branches, the nerves of the anterior aspect of the forearm are only two in number: the median and ulnar nerves. Their origins, courses, and distributions are described in Tables 6.4 and 6.9.

Median Nerve

The median nerve is the principal nerve of the anterior compartment of the forearm (Fig. 6.51, Table 6.9). It enters the forearm with the brachial artery and lies on its medial side. It leaves the cubital fossa by passing between the heads of the pronator teres, giving branches to them. The median nerve then passes deep to the FDS and continues distally through the middle of the forearm, between the FDS and the FDP. Near the wrist, the median nerve becomes superficial by passing between the tendons of FDS and flexor carpi radialis, deep to the palmaris longus tendon (Fig. 6.51).

Branches of the Median Nerve. The median nerve has no branches in the arm, other than small twigs to the brachial artery (Fig. 6.51). The branches of the median nerve arise in the forearm and hand as follows (Table 6.9):

- *Articular branches* pass to the elbow joint as the median nerve passes it.
- *Muscular branches* supply the pronator teres, pronator quadratus, and all the flexor muscles except the flexor carpi ulnaris and the medial half of the FDP. The *nerve to the pronator teres* usually arises at the elbow and enters the lateral border of the muscle. A broad bundle of nerves pierces the superficial flexor group of muscles and innervates the *flexor carpi radialis*, the *palmaris longus*, and the FDS.

Communications Between the Median and Ulnar Nerves

Occasionally, communications occur between the median and ulnar nerves in the forearm. These branches are usually represented by slender nerves, but the communications are important clinically because even with a complete lesion of the median nerve, some muscles may not be paralyzed. This may lead to an erroneous conclusion that the median nerve has not been damaged.

Median Nerve Injury

When the median nerve is severed in the elbow region, flexion of the proximal interphalangeal (PIP) joints of digits 1 to 3 is lost and flexion of digits 4 and 5 is weakened. Flexion of the distal interphalangeal (DIP) joints of the 2nd and 3rd digits is also lost. Flexion of the distal interphalangeal joints of the 4th and 5th digits is not affected because the medial part of the FDP, which produces these movements, is supplied by the ulnar nerve. The ability to flex the metacarpophalangeal joints of the 2nd and 3rd digits will be affected because the digital branches of the median nerve supply the 1st and 2nd lumbricals. Thus, when the patient attempts to make a fist, digits 2 and 3 remain partially extended ("hand of benediction"). Thenar muscle function is also lost as in carpal tunnel syndrome (p. 761).

Pronator Syndrome

This nerve entrapment syndrome is caused by compression of the median nerve near the elbow. The nerve may be compressed between the heads of the pronator teres owing to trauma, muscular hypertrophy, or fibrous bands. Patients are first seen clinically with pain and tenderness in the proximal aspect of the anterior forearm. Symptoms often follow activities that involve repeated elbow movements. ⊙

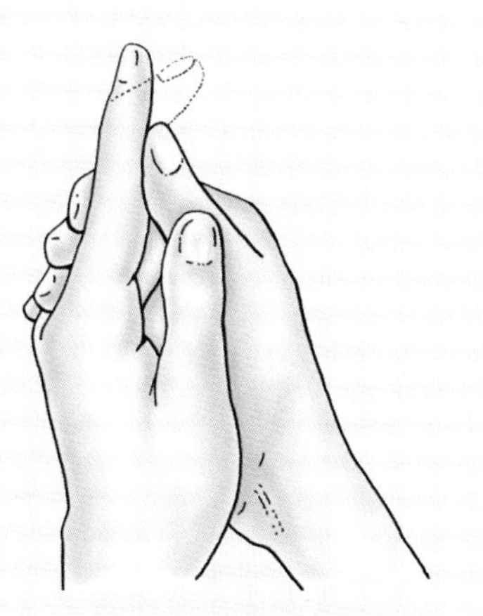

Testing flexion of DIP joint of index finger

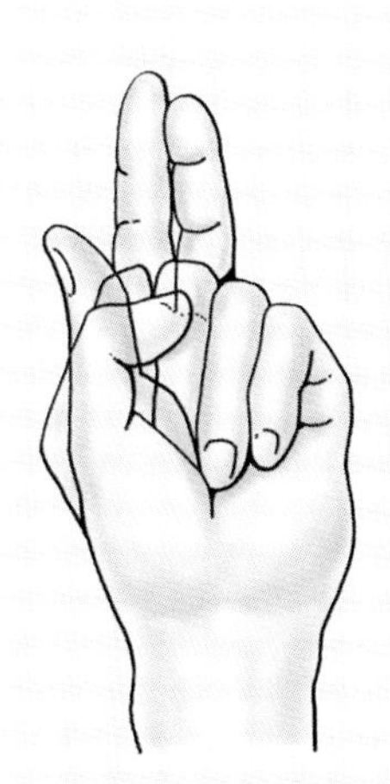

"Hand of benediction"

Table 6.9. **Nerves of the Forearm**

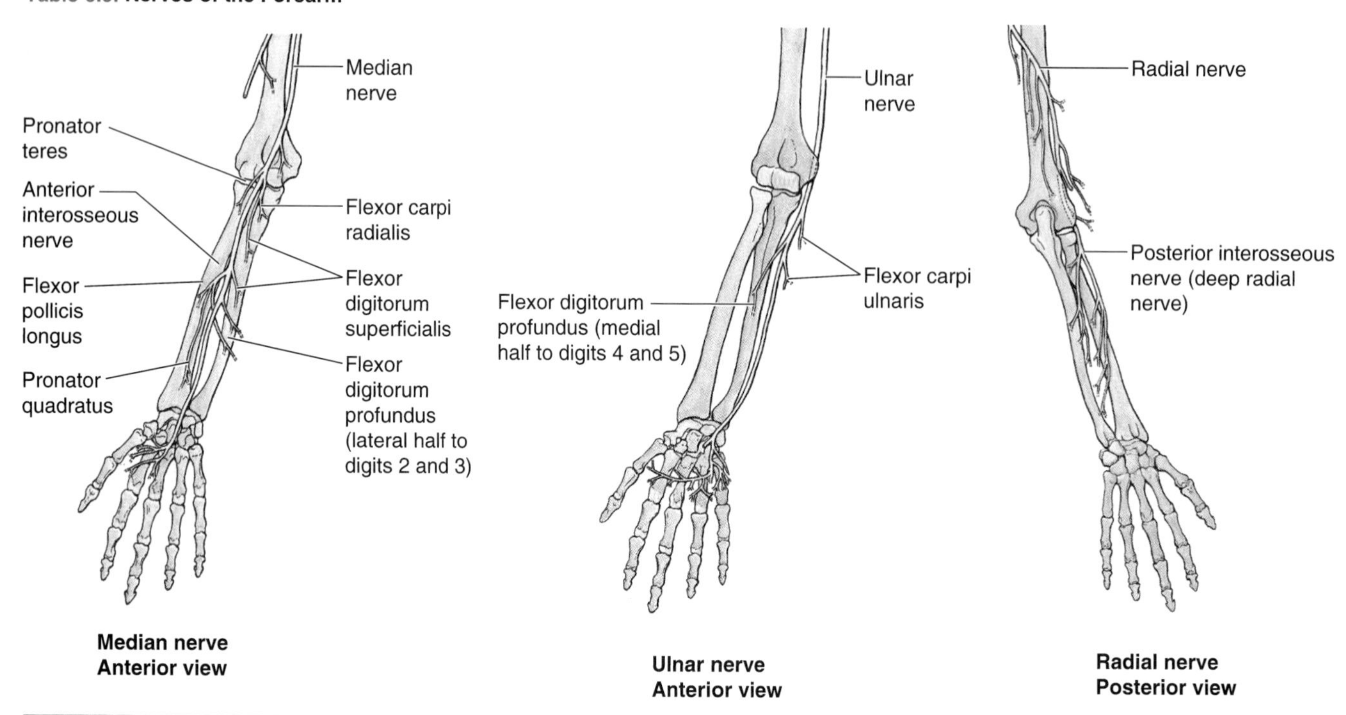

Nerve	Origin	Course
Median	By two roots from lateral (C6 and C7) and medial (C8 and T1) cords of brachial plexus	Enters cubital fossa medial to brachial artery, passes between heads of pronator teres, descends between flexor digitorum superficialis and flexor digitorum profundus, and passes close to flexor retinaculum as it passes through carpal tunnel to reach hand
Anterior interosseous	Median nerve in distal part of cubital fossa	Passes inferiorly on interosseous membrane to supply flexor digitorum profundus, flexor pollicis longus, and pronator quadratus
Palmar cutaneous branch of median	Median nerve just proximal to flexor retinaculum	Passes between tendons of palmaris longus and flexor carpi radialis and runs superficial to flexor retinaculum
Ulnar	Medial cord of brachial plexus (C8 and T1), but it often receives fibers from ventral ramus of C7	Passes posterior to medial epicondyle of humerus and enters forearm between heads of flexor carpi ulnaris; descends through forearm between flexor carpi ulnaris and flexor digitorum profundus; becomes superficial in distal part of forearm and passes superficial to flexor retinaculum
Palmar cutaneous branch of ulnar nerve	Ulnar nerve near middle of forearm	Descends on ulnar artery and perforates deep fascia in the distal third of forearm
Radial	Posterior cord of brachial plexus (C5–C8 and T1)	Passes into cubital fossa and descends between brachialis and brachioradialis; at level of lateral epicondyle of humerus, it divides into superficial and deep branches
Superficial branch of radial nerve	Continuation of radial nerve after deep branch is given off	Passes distally, anterior to pronator teres and deep to brachioradialis; pierces deep fascia at wrist and passes onto dorsum of hand
Deep branch of radial nerve	Arises from radial nerve just distal to elbow	Winds around neck of radius in supinator; enters posterior compartment to supply muscles shown in diagram
Posterior interosseous	Terminal branch of deep branch of radial nerve	Passes deep to extensor pollicis longus and ends on interosseous membrane

- *The anterior interosseous nerve* arises from the median nerve in the distal part of the cubital fossa and passes distally on the interosseous membrane with the anterior interosseous branch of the ulnar artery. It runs between the FDP and flexor pollicis longus to reach the pronator quadratus. It supplies these muscles; however, the ulnar nerve supplies half of the FDP (the ulnar [medial] part of the muscle sending tendons to digits 4 and 5). The anterior interosseous nerve supplies the lateral part of the muscle (which sends tendons to digits 2 and 3) and passes deep to

Table 6.9. (*Continued*) **Nerves of the Forearm**

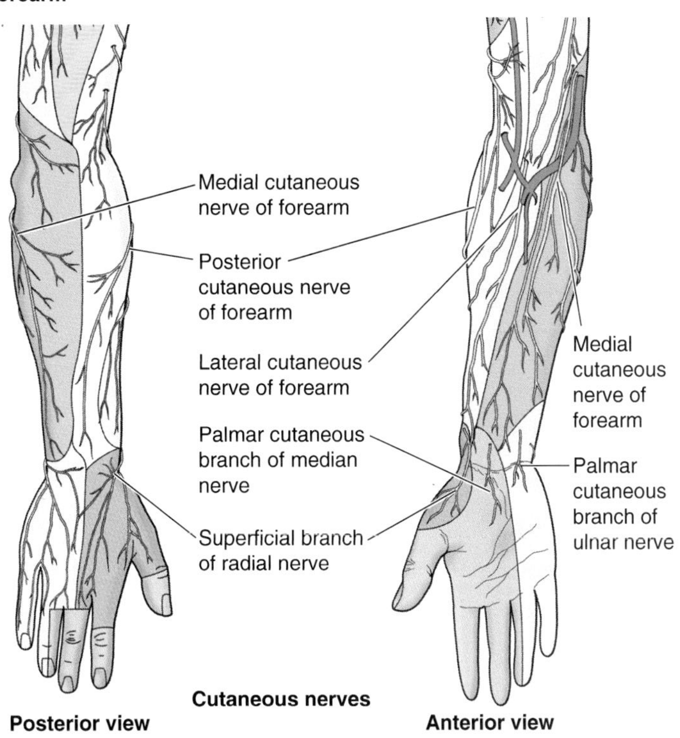

Nerve	Origin	Course
Posterior antebrachial cutaneous nerve	Arises in arm from radial nerve	Perforates lateral head of triceps and descends along lateral side of arm and posterior aspect of forearm to wrist
Lateral antebrachial cutaneous nerve	Continuation of musculocutaneous nerve	Descends along lateral border of forearm to wrist
Medial antebrachial cutaneous nerve	Medial cord of brachial plexus, receiving fibers from C8 and T1	Runs down arm on medial side of brachial artery; pierces deep fascia in cubital fossa and runs along medial aspect of forearm

the pronator quadratus, then ends by sending articular branches to the wrist joint.

- The *recurrent branch of the median nerve* (C8 and T1) arises from the median nerve as soon as it passes distal to the flexor retinaculum. It loops around the distal border of this retinaculum to supply the thenar muscles.
- *The palmar cutaneous branch of the median nerve* arises just proximal to the flexor retinaculum and becomes cutaneous between the tendons of the palmaris longus and flexor carpi radialis (Table 6.9). It passes superficial to the flexor retinaculum to supply the skin of the lateral part of the palm.

Ulnar Nerve

After passing posterior to the medial epicondyle of the humerus, the ulnar nerve enters the forearm by passing between (and in doing so, sending motor branches to) the heads of the flexor carpi ulnaris (Table 6.9). It then passes inferiorly between the flexor carpi ulnaris and FDP, supplying the ulnar (medial) part of the muscle that sends tendons to digits 4 and 5. The ulnar nerve becomes superficial at the wrist and supplies skin on the medial side of the hand (Fig. 6.51). It then runs on the medial side of the ulnar artery and the lateral side of the flexor carpi ulnaris tendon. The ulnar nerve and artery emerge from beneath the flexor carpi ulnaris tendon just proximal to the wrist and pass superficial to the flexor retinaculum to enter the hand. The ulnar nerve and artery lie on the retinaculum and then pass through a groove between the pisiform and the hook of the hamate. A band of fibrous tissue from the flexor retinaculum bridges the groove to form a small canal known clinically as the *canal of Guyon.*

Branches of the Ulnar Nerve. The ulnar nerve has no branches in the arm. They arise in the forearm and hand as follows:

- *Articular branches* pass to the elbow joint while the nerve is between the olecranon and medial epicondyle
- *Muscular branches* supply the flexor carpi ulnaris and the medial half of the FDP
- The *palmar cutaneous branch* arises from the ulnar nerve near the middle of the forearm and descends under the antebrachial fascia anterior to the ulnar artery; the nerve pierces the deep fascia in the distal third of the forearm to supply skin on the medial part of the palm

Ulnar Nerve Injury

Ulnar nerve injury commonly occurs where the nerve passes posterior to the medial epicondyle of the humerus. The injury results when the lateral part of the elbow hits a hard surface, fracturing the medial epicondyle ("funny bone").

Compression of the ulnar nerve at the elbow (*ulnar nerve entrapment*) is common. It usually produces numbness and tingling of the medial part of the palm and the medial one ▶

- The *dorsal cutaneous branch* arises from the ulnar nerve in the distal half of the forearm and passes posteroinferiorly between the ulna and the flexor carpi ulnaris to supply the posterior surface of the medial part of the hand and the digits
- The *deep branch* arises from the ulnar nerve at the wrist and supplies the hypothenar muscles (muscles of little finger); it also supplies the interosseous muscles and the 3rd and 4th lumbrical muscles.

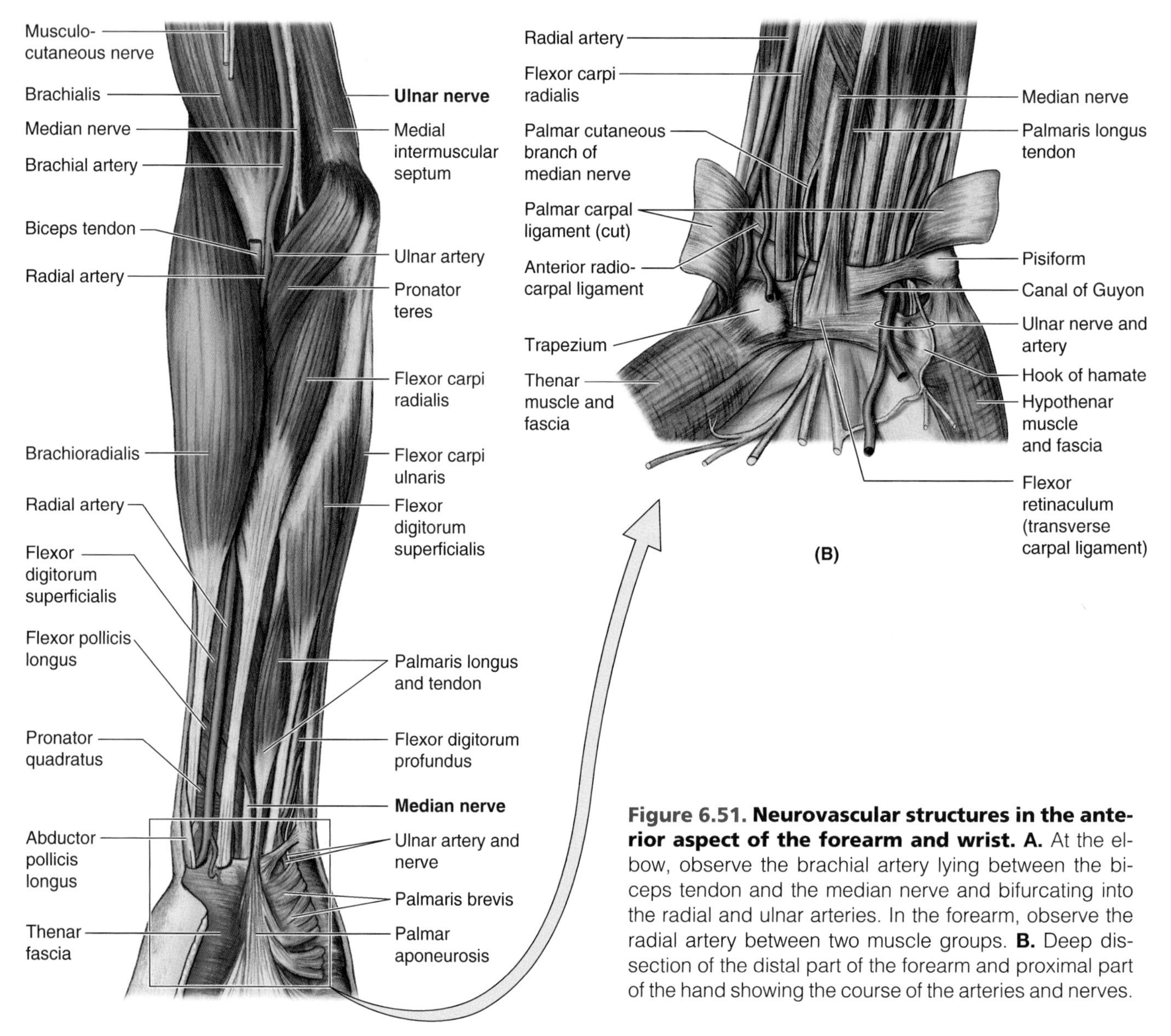

Figure 6.51. Neurovascular structures in the anterior aspect of the forearm and wrist. A. At the elbow, observe the brachial artery lying between the biceps tendon and the median nerve and bifurcating into the radial and ulnar arteries. In the forearm, observe the radial artery between two muscle groups. **B.** Deep dissection of the distal part of the forearm and proximal part of the hand showing the course of the arteries and nerves.

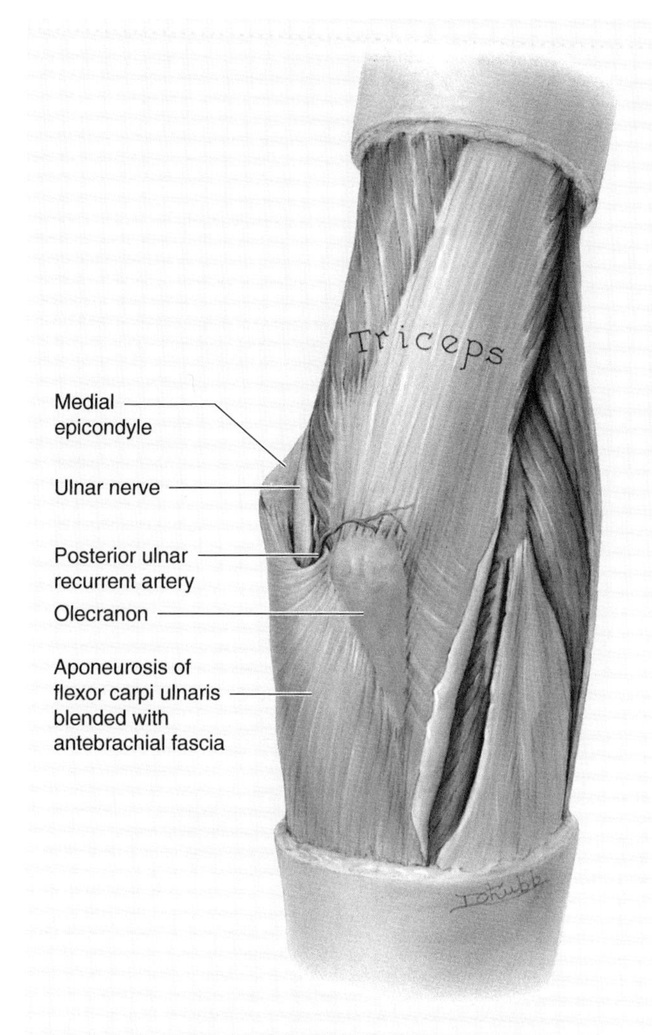

Vulnerable position of ulnar nerve

▶ and a half digits. Pluck your ulnar nerve at the posterior aspect of your elbow with your finger and you may feel tingling in these fingers. Severe compression may also produce elbow pain that radiates distally. Uncommonly, the ulnar nerve is compressed as it passes through the canal of Guyon.

Ulnar nerve injury can result in extensive motor and sensory loss to the hand. An injury to the nerve in the distal part of the forearm denervates most intrinsic hand muscles. Power of adduction is impaired, and when an attempt is made to flex the wrist joint, the hand is drawn to the lateral side by the flexor carpi radialis (supplied by the median nerve). Following ulnar nerve injury, patients have difficulty making a fist because they cannot flex their 4th and 5th digits at the distal interphalangeal joints. This characteristic appearance of the hand resulting from a distal lesion of the ulnar nerve is known as *clawhand* (main en griffe). The deformity results from atrophy of the interosseous muscles of the hand supplied by the ulnar nerve. The "claw" is produced by the opposed action of the extensors and FDP.

Cubital Tunnel Syndrome

The ulnar nerve may be compressed in the cubital tunnel formed by the tendinous arch joining the humeral and ulnar heads of attachment of the flexor carpi ulnaris (Table 6.6). The signs and symptoms are the same as an ulnar nerve lesion in the ulnar groove on the posterior aspect of the medial epicondyle of the humerus (see the discussion at the beginning of this box). ⊙

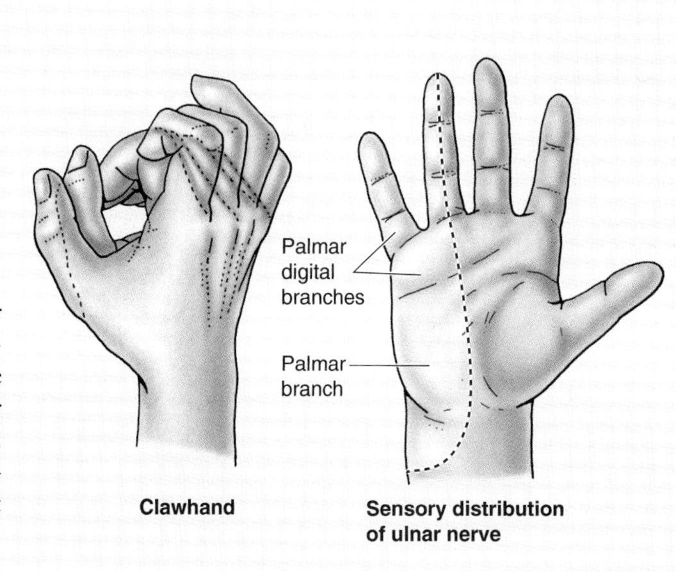

Clawhand **Sensory distribution of ulnar nerve**

Radial Nerve

The radial nerve leaves the posterior compartment of the arm to cross the anterior aspect of the lateral epicondyle of the humerus (Table 6.9). It appears in the cubital fossa between the brachialis and brachioradialis. Soon after it enters the forearm, the radial nerve divides into deep and superficial branches (Fig. 6.45). The deep branch arises anterior to the lateral epicondyle of the humerus and pierces the supinator.

The **superficial branch of the radial nerve** is a cutaneous and articular nerve that descends in the forearm under cover of the brachioradialis. It emerges in the distal part of the forearm and crosses the roof of the anatomical snuff box (Fig. 6.46) and is distributed to skin on the dorsum of the hand and to a number of joints in the hand (Table 6.9).

The **deep branch of the radial nerve,** the larger of the two terminal branches, is the direct continuation of the radial nerve. After it pierces the supinator, it winds around the lateral aspect of the neck of the radius and enters the posterior compartment of the forearm (Fig. 6.45). The *posterior interosseous nerve* is the continuation of the deep branch of the radial nerve.

The **posterior cutaneous nerve of the forearm** (a branch of the radial) descends along the posterior aspect of the forearm to the wrist, supplying the skin during its course.

Radial Nerve Injury

The radial nerve is usually injured in the arm by a fracture of the humeral body. This injury is proximal to the branches to the extensors of the wrist, and so wrist-drop is the primary clinical manifestation of an injury at this level (p. 731). However, injury to the deep branch of the radial nerve may occur when wounds of the forearm are deep (penetrating). Severance of the deep branch of the radial nerve results in an inability to extend the thumb and the metacarpophalangeal (MP) joints of the other digits. Loss of sensation does not occur because the deep branch of the radial is entirely muscular and articular in distribution. See Table 6.7 to determine the muscles that are paralyzed (e.g., extensor digitorum [p. 745]) when this nerve is severed. When the superficial branch of the radial nerve, a cutaneous nerve, is severed, sensory loss is usually minimal. Commonly, a coin-shaped area of anesthesia is distal to the bases of the 1st and 2nd metacarpals. The reason the area of sensory loss is less than expected from the illustration in Table 6.9 is because of the considerable overlap from cutaneous branches of the median and ulnar nerves. ⊙

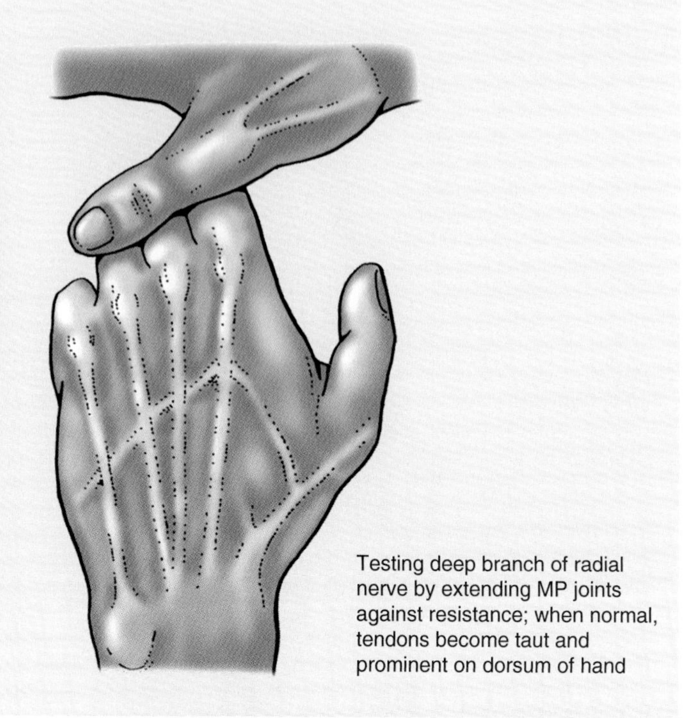

Testing deep branch of radial nerve by extending MP joints against resistance; when normal, tendons become taut and prominent on dorsum of hand

Surface Anatomy of the Forearm

Three bony landmarks are easily palpated at the elbow: the **medial** and **lateral epicondyles** of the humerus and the **olecranon** of the ulna. In the hollow located posterolaterally when the forearm is extended, the head of the radius can be palpated distal to the lateral epicondyle. Supinate and pronate your forearm and feel the movement of the radial head. The posterior border of the ulna can be palpated distally from the olecranon along the entire length of the bone, which is subcutaneous. The **cubital fossa**, the triangular hollow area on the anterior surface of the elbow, is bounded medially by the prominence formed by the flexor-pronator group of muscles that are attached to the medial epicondyle. To estimate the position of these muscles, put your thumb posterior to your medial epicondyle and then place your fingers on your forearm as shown in the photograph on the next page. The black dot on the dorsum of the hand indicates the position of the medial epicondyle.

The **head of the ulna** is at its distal end and is easily seen and palpated. It appears as a rounded prominence at the wrist when the hand is pronated. The **ulnar styloid process** can be palpated just distal to the ulnar head. The larger *radial styloid process* can be easily palpated on the lateral side of the wrist when the hand is supinated, particularly when the tendons covering it are relaxed. The **radial styloid process** is located approximately 1 cm more distal than the ulnar styloid process. This relationship of the styloid processes is important in the diagnosis of certain injuries in the wrist region (e.g., fracture of the distal end of the radius). Proximal to the radial styloid process, the surfaces of the radius are palpable for a few centimeters. The lateral surface of the distal half of the radius is easy to palpate. ⊙

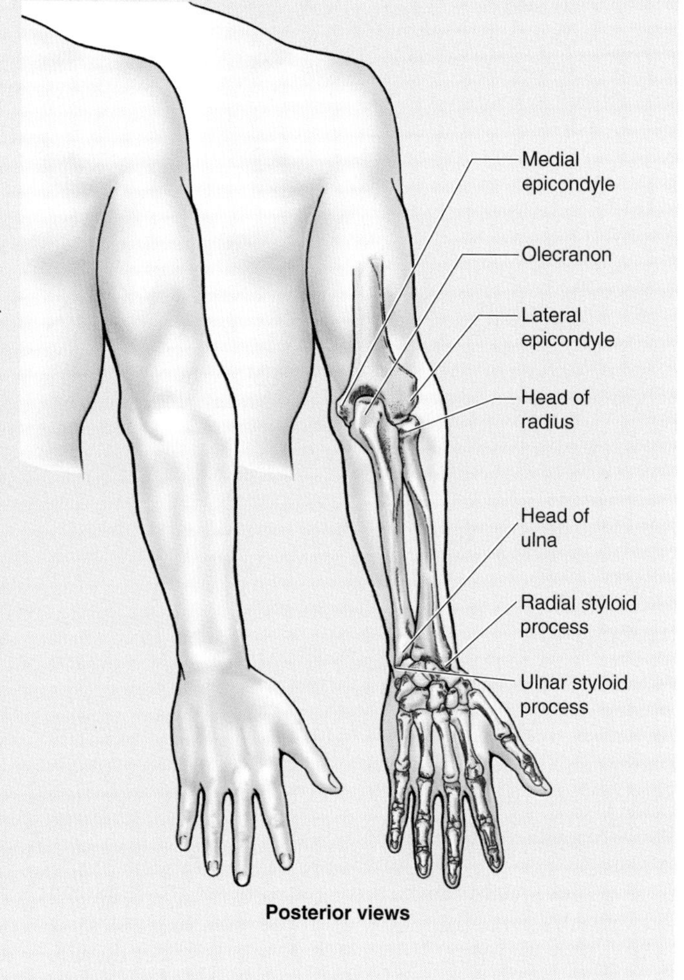

Posterior views

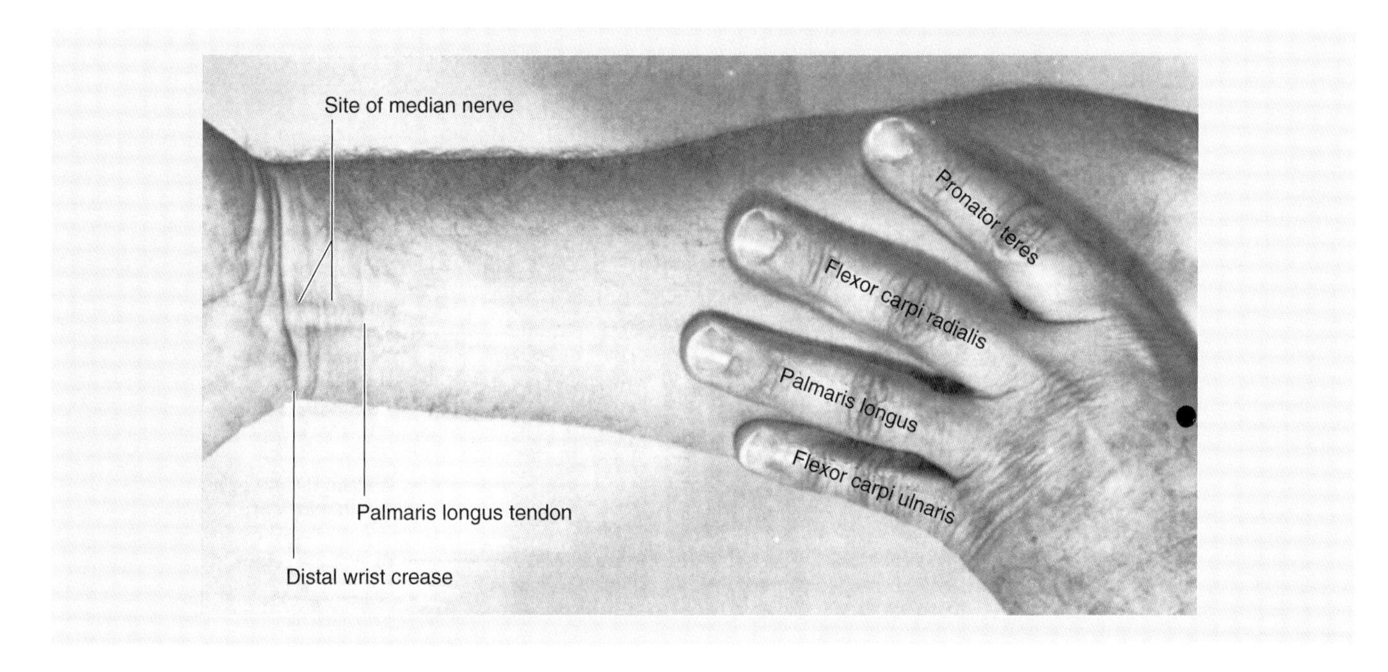

Hand

The wrist is at the junction of the forearm and hand. Movement of the hand occurs at the wrist joint. The hand (L. manus) is the manual part of the upper limb distal to the forearm. The *skeleton of the hand* consists of carpals in the wrist, metacarpals in the hand proper, and phalanges in the digits. The digits are numbered from one to five, beginning with the thumb and ending with the little finger. Because of the importance of manual dexterity in occupational and recreational activities, a good understanding of the structure and function of the hand is essential for all persons involved in maintaining or restoring its activities: free motion, power grasping, precision handling, and pinching.

The **power grip** (palm grasp) refers to forcible motions of the digits acting against the palm; the fingers are wrapped around an object with counter pressure from the thumb—for example, when grasping a cylindrical structure (Fig. 6.52). The power grip involves the long flexor muscles to the fingers, the intrinsic muscles in the palm, and the extensors of the wrist (see the discussion on p. 809). A *hook grip* is the posture of the hand that is used when carrying a brief case. This grip involves primarily the long flexors of the fingers, which are flexed to a varying degree depending on the size of the object that is grasped. The *precision handling grip* involves a change in the position of a handled object that requires fine control of the movements of the fingers and thumb (e.g., holding a pen or winding a watch). In a precision grip, the wrist and fin-

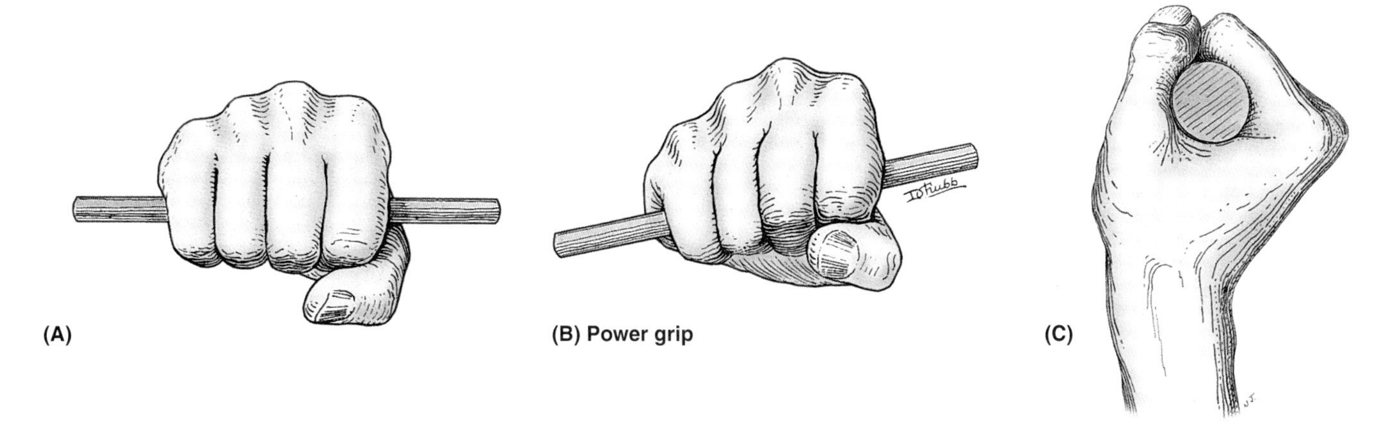

Figure 6.52. Power grip of grasping hand (L. manus). A. Loosely held. **B–C.** Firmly gripped. The 2nd and 3rd carpometacarpal joints are rigid and stable, but the 4th and 5th are hinge joints permitting flexion and extension. **C.** When grasping an object, the metacarpophalangeal and interphalangeal joints are flexed, but the wrist and transverse carpal joints are extended. "Cocking the wrist" in this way increases the distance over which the flexors of the fingers act, producing the same result as a more complete muscular contraction. Without this extension, the grip is feeble and insecure.

gers are held firmly by the long flexor and extensor muscles, and the intrinsic hand muscles perform fine movements of the digits (e.g., when threading a needle or buttoning a shirt or blouse).

Pinching refers to compression of something between the thumb and index finger (e.g., handling a teacup) or between the thumb and adjacent two fingers (e.g., snapping the fingers). The *position of rest* is assumed by an inactive hand (e.g., when the forearm and hand are laid on a table). This position is often used when it is necessary to immobilize the wrist and hand in a cast to stabilize a fracture.

Fascia of the Palm

The fascia of the palm is continuous with the antebrachial fascia and the fascia of the dorsum of the hand (Fig. 6.10). The **palmar fascia** is thin over the thenar and hypothenar eminences, but it is thick centrally where it forms the fibrous palmar aponeurosis and in the digits where it forms the digital sheaths (Fig. 6.53). The **palmar aponeurosis**, a strong, well-defined part of the deep fascia of the palm, covers the soft tissues and overlies the long flexor tendons. The proximal end or apex of the triangular palmar aponeurosis is continuous with the flexor retinac-

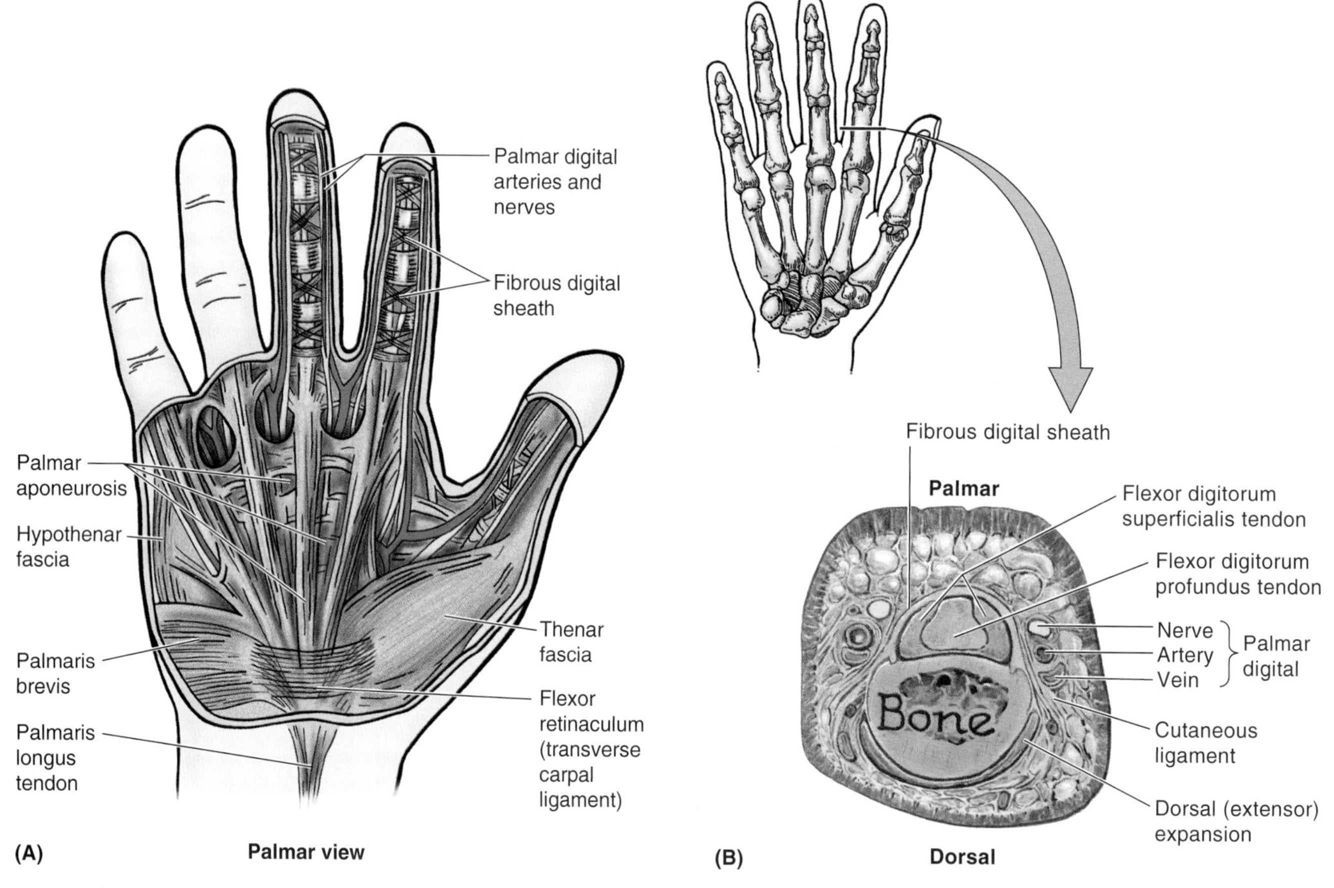

Figure 6.53. Palmar fascia and fibrous digital sheaths. The palmar fascia is continuous with the antebrachial fascia. The fascia investing the muscles and associated vessels and nerves of the little finger is called *hypothenar fascia*. Similarly, the fascia investing the muscles and associated vessels and nerve of the thumb is called *thenar fascia*. Between the medial and lateral muscle masses, the central compartment of the palm is roofed by the palmar aponeurosis, an extension of the palmaris longus tendon (when present [84% of people]). **A.** Observe the fibrous digital sheaths (shown here for the middle and index fingers), which are continuous with the longitudinal fiber bundles of the palmar aponeurosis and form the coverings of the tendons of the flexor digitorum superficialis (FDS) and flexor digitorum profundus (FDP). The fibrous digital sheaths prevent the tendons from pulling away ("bowstringing") from the phalanges. **B.** Transverse section of the index finger. Observe that the palmar digital nerve, artery, and vein are adjacent to the fibrous digital sheath, not to the phalanx (bone of finger). The dorsal digital neurovascular structures are exhausted as they reach the middle third of the middle phalanges, the palmar nerves, arteries, and veins serving all (palmar and dorsal aspects) of the digits distally.

ulum and the palmaris longus tendon. When the muscle is present, the palmar aponeurosis is the expanded tendon of the palmaris longus. Distal to the apex, the palmar aponeurosis forms four longitudinal digital bands that radiate from the apex and attach distally to the bases of the proximal phalanges and become continuous with the fibrous digital sheaths.

The **fibrous digital sheaths of the hand** (fibrous sheaths of digits) are ligamentous tubes that enclose the synovial sheaths, the superficial and deep flexor tendons (Figs. 6.44 and 6.53), and the tendon of the flexor pollicis longus in their passage along the palmar aspect of their respective digits. The flexor digital sheaths are composed of five anular and four cruciform (cross-shaped) parts or "pulleys."

A **medial fibrous septum** extends deeply from the medial border of the palmar aponeurosis to the 5th metacarpal. Medial to this septum is the medial or **hypothenar compartment** containing the hypothenar muscles (Fig. 6.54*A*). Similarly, a **lateral fibrous septum** extends deeply from the lateral border of the palmar aponeurosis to the 3rd metacarpal. Lateral to this septum is the lateral or **thenar compartment** containing the thenar muscles. Between the hypothenar and thenar compartments is the **central compartment** containing the flexor tendons and their sheaths, the lumbricals, the superficial palmar arterial arch, and the digital vessels and nerves. The deepest muscular plane of the palm is the **adductor compartment** containing the adductor pollicis muscle.

Between the flexor tendons and the fascia covering the deep palmar muscles are two potential spaces, the **thenar space** and the **midpalmar space** (Fig. 6.54, *A* and *B*). The spaces are bounded by fibrous septa passing from the edges of the palmar aponeurosis to the metacarpals. Between the two spaces is the especially strong lateral fibrous septum that is attached to the 3rd metacarpal.

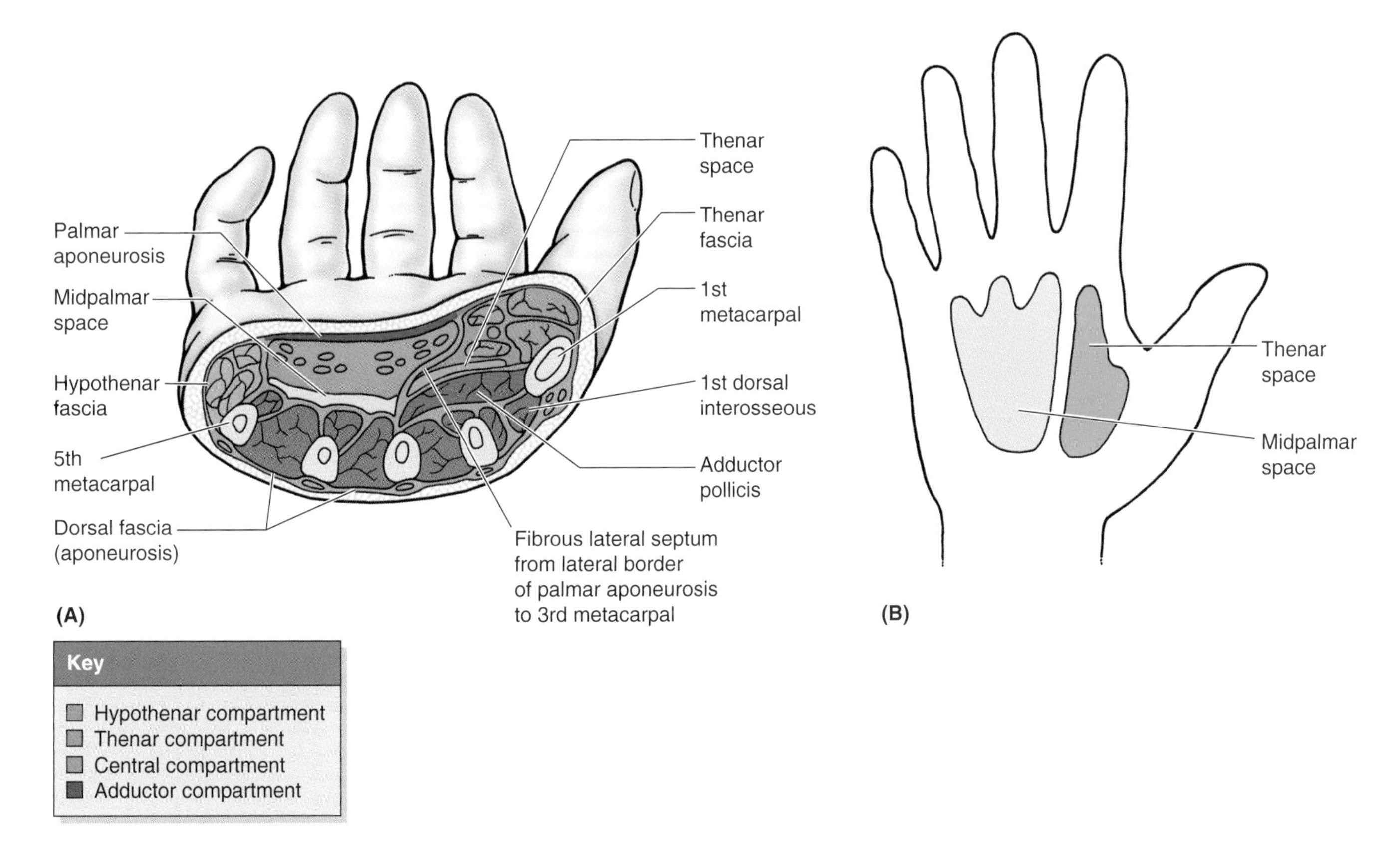

Figure 6.54. Compartments, spaces, and fascia of the palm. A. Transverse section through the middle of the palm illustrating the fascial compartments of the hand. The *hypothenar fascia*, attached to the lateral side of the 5th metacarpal, bounds the hypothenar compartment. Similarly, the *thenar fascia* attaches to the palmar aspect of the 1st metacarpal and bounds, with the metacarpal, the thenar compartment. The *central compartment* of the palm is covered by the palmar aponeurosis. The *adductor compartment*, the deepest muscular plane of the palm, contains the adductor pollicis muscle. **B.** The *midpalmar (central) space* underlies the central compartment of the palm and is bounded medially by the hypothenar compartment, related distally to the synovial tendon sheaths of digits 3, 4, and 5 and proximally to the common flexor sheath as it emerges from the carpal tunnel. The *thenar space* underlies the thenar compartment and is related distally to the synovial tendon sheath of the index finger and proximally to the common flexor sheath distal to the carpal tunnel.

Dupuytren's Contracture of Palmar Fascia

Dupuytren's contracture is a progressive shortening, thickening, and fibrosis of the palmar fascia and aponeurosis. The fibrous degeneration of the longitudinal bands of the palmar aponeurosis on the medial side of the hand pulls the ring and little fingers into partial flexion at the metacarpophalangeal and proximal interphalangeal joints (*A*). The contracture is frequently bilateral and is common in men older than 50 years; its cause is unknown, but evidence points to a hereditary predisposition. The disease first manifests itself as painless nodular thickenings of the palmar aponeurosis that adhere to the skin. Gradually, progressive contracture of the longitudinal bands produces raised ridges in the palmar skin that extend from the proximal part of the hand to the base of the ring and little fingers (*B*). Treatment of Dupuytren's contracture usually involves surgical excision of all fibrotic parts of the palmar fascia to free the fingers (Salter, 1998).

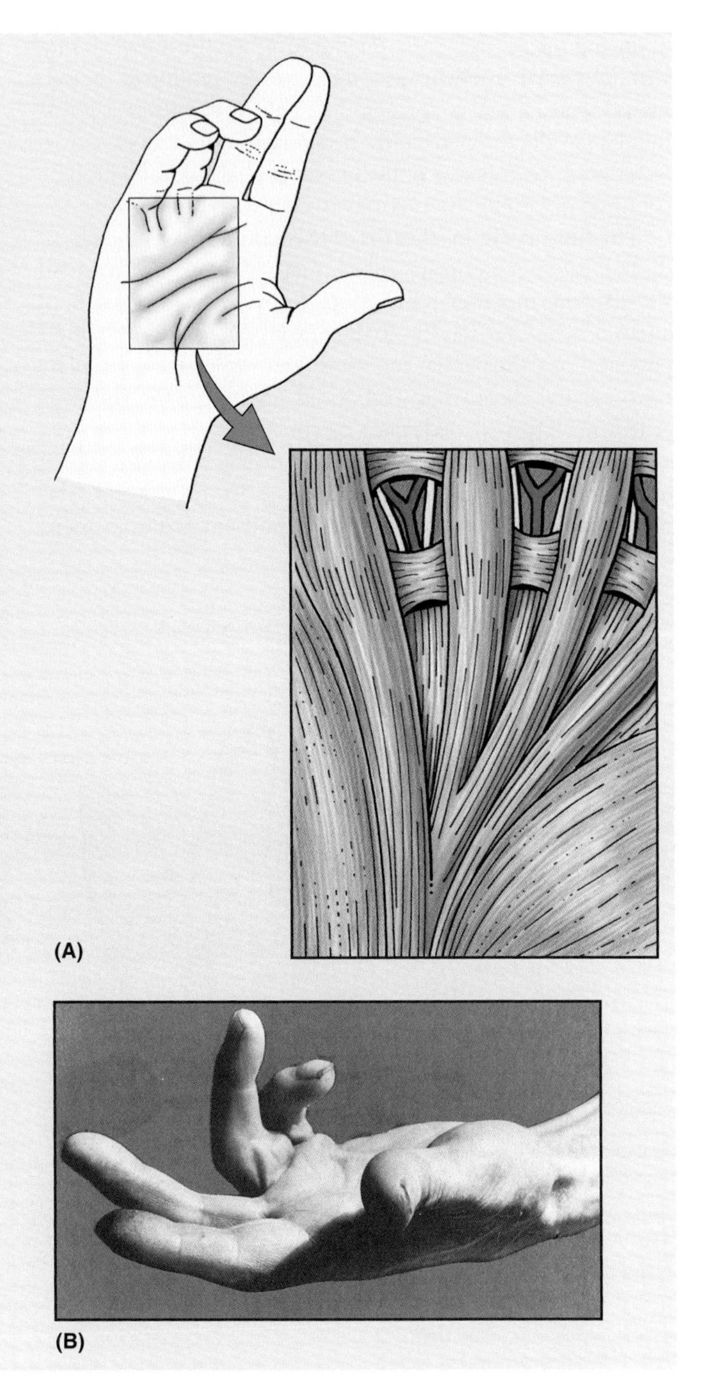

Hand Infections

Because the palmar fascia is thick and strong, swellings resulting from hand infections usually appear on the dorsum of the hand, where the fascia is thinner. The potential fascial spaces of the palm are important because they may become infected. The fascial spaces determine the extent and direction of the spread of pus formed by these infections. Depending on the site of infection, pus will accumulate in the thenar, hypothenar, or adductor compartments. Owing to the widespread use of antibiotics today, infections are rarely encountered that spread from one of these fascial compartments, but an untreated infection can spread proximally through the carpal tunnel into the forearm, anterior to the pronator quadratus and its fascia. ⊙

Muscles of the Hand

The intrinsic muscles of the hand are in four compartments (Figs. 6 54 and 6.55, Table 6.10):

- Thenar muscles in the **thenar compartment**: abductor pollicis brevis, flexor pollicis brevis, and opponens pollicis
- Adductor pollicis in the **adductor compartment**
- Hypothenar muscles in the **hypothenar compartment**: abductor digiti minimi, flexor digiti minimi, and opponens digiti minimi
- Short muscles of the hand: the lumbricals are in the **central compartment** and the interossei are between the metacarpals.

Thenar Muscles

The thenar muscles form the *thenar eminence* on the lateral surface of the palm and are *chiefly responsible for opposition of the thumb.* This complex movement begins with the thumb in the extended position and initially involves a medial rotation of the 1st metacarpal (cupping the palm) produced by the action of the opponens pollicis muscle at the carpometacarpal joint and then abduction, flexion, and usually adduction (Fig. 6.56). The reinforcing action of the *adductor pollicis* and flexor pollicis longus increases the pressure that the opposed thumb can exert on the fingertips. Normal movement of the thumb is important for the precise activities of the hand. Because the 1st metacarpal is more mobile than in other digits, several muscles are required to control its freedom of movement (Williams et al., 1995):

- *Abduction*: APL and abductor pollicis brevis
- *Adduction*: adductor pollicis and 1st dorsal interosseous
- *Extension:* EPL, EPB, and APL
- *Flexion:* flexor pollicis longus and flexor pollicis brevis

The above movements occur at the metacarpophalangeal joint

- *Opposition:* opponens pollicis; the movement referred to here occurs at the carpometacarpal joint and results in a "cupping" of the palm.

Opposition—bringing the tip of the thumb into contact with the little finger or any of the other digits—also involves abduction and flexion and adduction at the metacarpophalangeal joint.

Abductor Pollicis Brevis (Fig. 6.55, Table 6.10). This *short abductor of the thumb* forms the anterolateral part of the thenar eminence. *The abductor pollicis brevis abducts the thumb at the carpometacarpal joint* and assists the opponens pollicis during the early stages of opposition by rotating its proximal phalanx slightly medially. *To test the abductor pollicis brevis,* abduct the thumb against resistance. If acting normally, the muscle can be seen and palpated.

Flexor Pollicis Brevis (Fig. 6.55*A*, Table 6.10). This *short flexor of the thumb* is located medial to the abductor pollicis brevis. Its tendon usually contains a sesamoid bone. *The flexor pollicis brevis flexes the thumb at the carpometacarpal and metacarpophalangeal joints* and aids in opposition of the thumb. *To test the flexor pollicis brevis,* flex the thumb against resistance. If acting normally, the muscle can be seen and palpated; *however,* keep in mind that the flexor pollicis longus also flexes the thumb.

Opponens Pollicis (Fig. 6.55*B*, Table 6.10). This quadrangular muscle lies deep to the abductor pollicis brevis and lateral to the flexor pollicis brevis. *The opponens pollicis opposes the thumb—the most important thumb movement*—that is, it

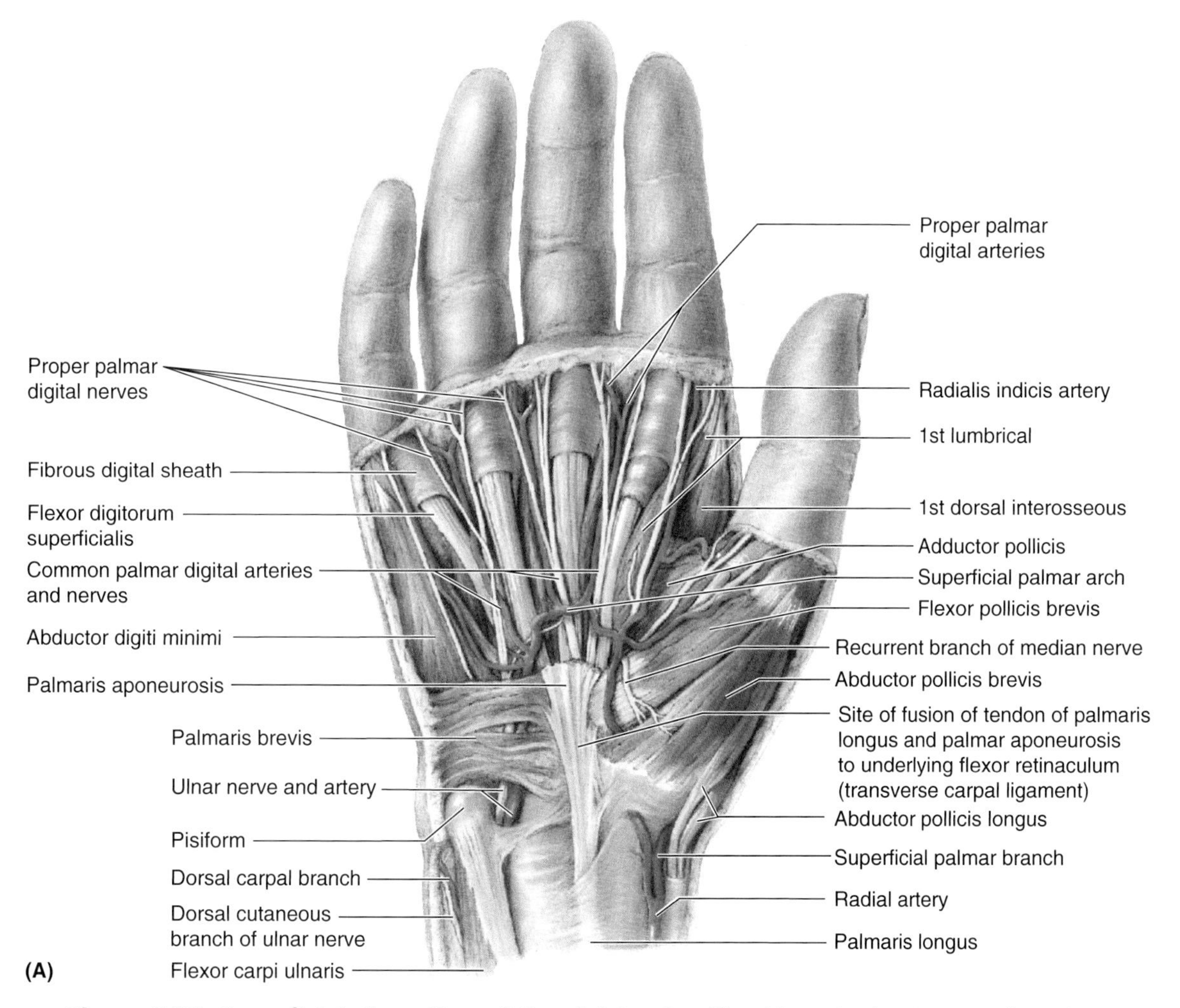

Figure 6.55. Superficial dissection of the right palm. The skin and subcutaneous tissue have been removed, as have the palmar aponeurosis and the thenar and hypothenar fasciae. **A.** Observe the superficial palmar arch located superficial to the long flexor tendons. This arterial arch gives rise to the common palmar digital arteries. In the digits, observe a digital artery (e.g., radialis indicis) and nerve lying on the medial and lateral sides of the fibrous digital sheath. Note that the pisiform bone protects the ulnar nerve and artery as they pass into the palm.

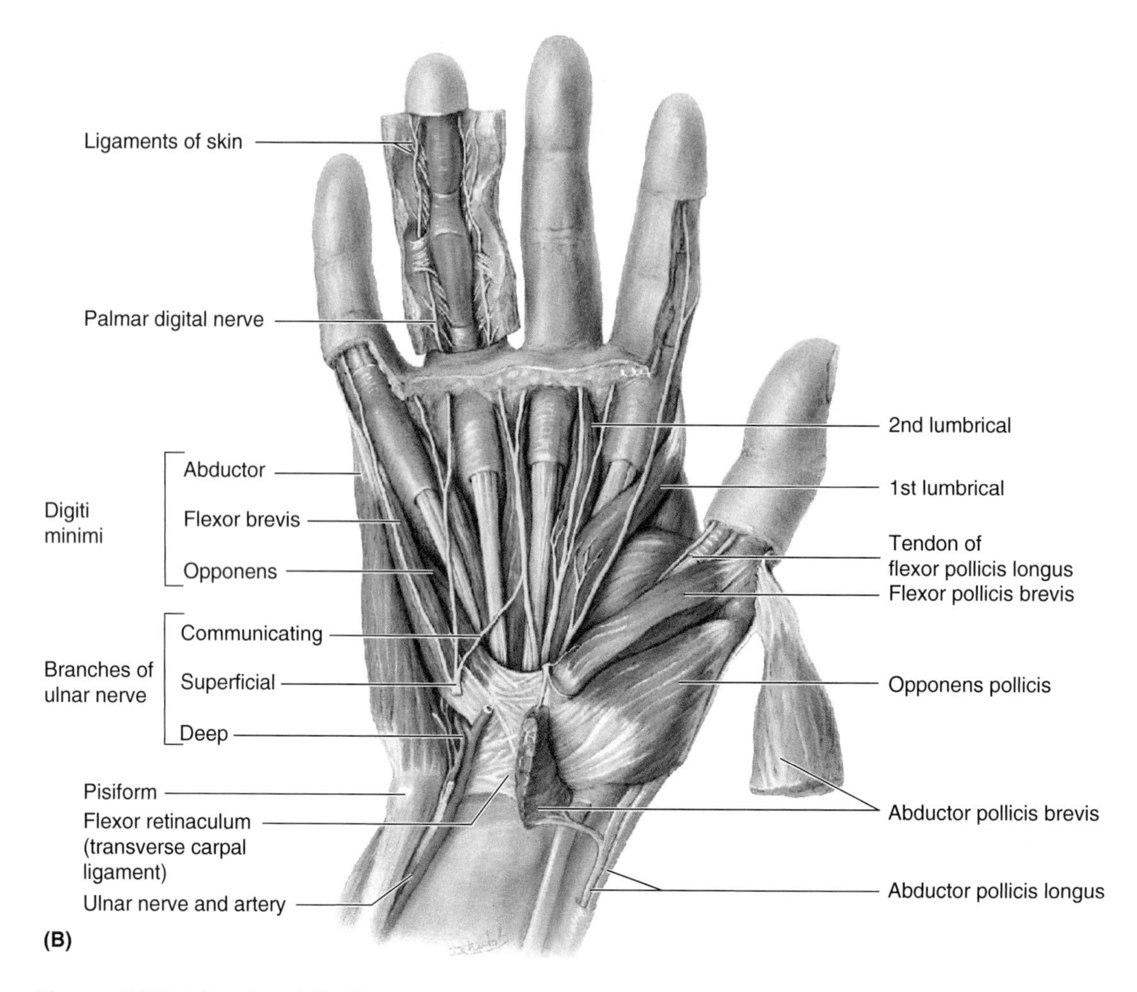

Figure 6.55. (*Continued*) **B.** Observe the three thenar and three hypothenar muscles attaching to the flexor retinaculum and to the four marginal carpal bones united by the retinaculum

flexes and rotates the 1st metacarpal medially at the carpometacarpal joint during opposition; this movement occurs when picking up an object. In Figure 6.56, note that the tip of the thumb is brought into contact with the pad of the little finger.

Adductor Pollicis

The deeply placed, fan-shaped *adductor of the thumb* is located in the *adductor compartment* of the hand (Figs. 6.54*A* and 6.55*A*). The adductor pollicis has two heads of origin that are separated by the radial artery as it enters the palm to form the deep palmar arch (Fig. 6.55*A*). Its tendon usually contains a sesamoid bone. *The adductor pollicis adducts the thumb*—moves the thumb to the palm of the hand, thereby giving power to the grip.

Hypothenar Muscles

The hypothenar muscles (abductor digiti minimi, flexor digiti minimi brevis, and opponens digiti minimi) produce the *hypothenar eminence* on the medial side of the palm and move the little finger. They are in the hypothenar compartment with the 5th metacarpal (Figs. 6.54*A* and 6.55, Table 6.10).

Abductor Digiti Minimi. This *abductor of the little finger* is the most superficial of the three muscles forming the hypothenar eminence. *The abductor digiti minimi abducts the 5th digit* and helps flex its proximal phalanx.

Flexor Digiti Minimi Brevis. This *short flexor of the little finger* is variable in size; it lies lateral to the abductor digiti minimi. *The flexor digiti minimi brevis flexes the proximal phalanx of the 5th digit at the metacarpophalangeal joint.*

Opponens Digiti Minimi. This quadrangular muscle lies deep to the abductor and flexor muscles of the 5th digit. *The opponens digiti minimi draws the 5th metacarpal bone anteriorly and rotates it laterally,* thereby deepening the hollow of the palm and bringing the 5th digit into opposition with the thumb.

Palmaris Brevis

The palmaris brevis is a small, thin muscle in the subcutaneous tissue of the hypothenar eminence (Fig. 6.55*A*); *it is not in the hypothenar compartment.* The palmaris brevis wrinkles the skin of the hypothenar eminence and deepens the hollow of the palm, thereby aiding the palmar grip. *The palmaris brevis covers and protects the ulnar nerve and artery.* It is attached proximally to the medial border of the palmar aponeurosis and to the skin on the medial border of the hand.

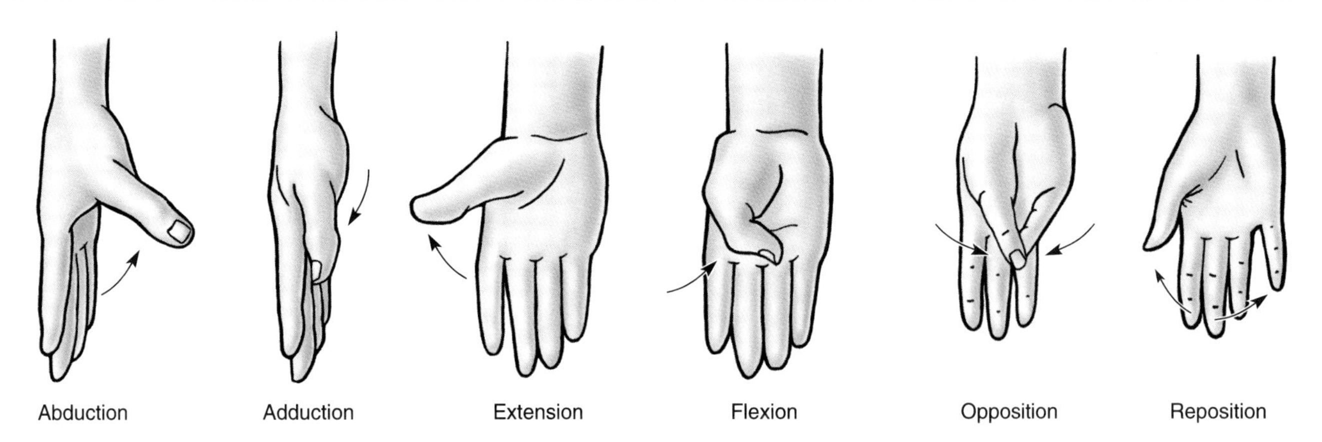

Figure 6.56. Movements of the thumb. Opposition, the action bringing the tip of the thumb in contact with the pads of the other fingers (e.g., with the little finger), is the most complex movement. The components of opposition are medial rotation at the carpometacarpal joint and abduction and flexion of the metacarpophalangeal joint.

Table 6.10. **Intrinsic Muscles of the Hand**

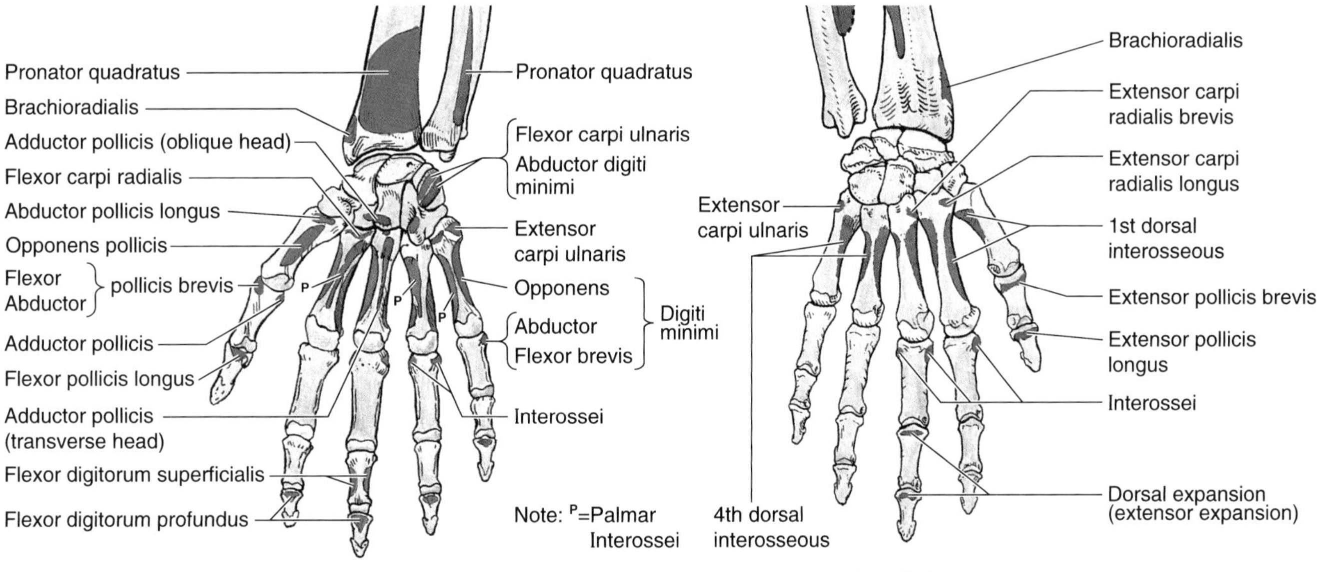

Muscle	Proximal Attachment	Distal Attachment	Innervation[a]	Main Action
Thenar muscles Abductor pollicis brevis	Flexor retinaculum and tubercles of scaphoid and trapezium	Lateral side of base of proximal phalanx of thumb	Recurrent branch of median nerve (**C8** and T1)	Abducts thumb and helps oppose it
Flexor pollicis brevis				Flexes thumb
Opponens pollicis		Lateral side of 1st metacarpal		Draws 1st metacarpal bone medially to oppose thumb toward center of palm and rotates it medially

[a] Numbers indicate spinal cord segmental innervation (e.g., C8 and T1 indicate that nerves supplying the thenar muscles are derived from C8 and T1 segments of spinal cord). **Boldface** numbers indicate main segmental innervation. Damage to these segments, or to motor nerve roots rising from them, results in paralysis of muscles concerned.

Table 6.10. (*Continued*) **Intrinsic Muscles of the Hand**

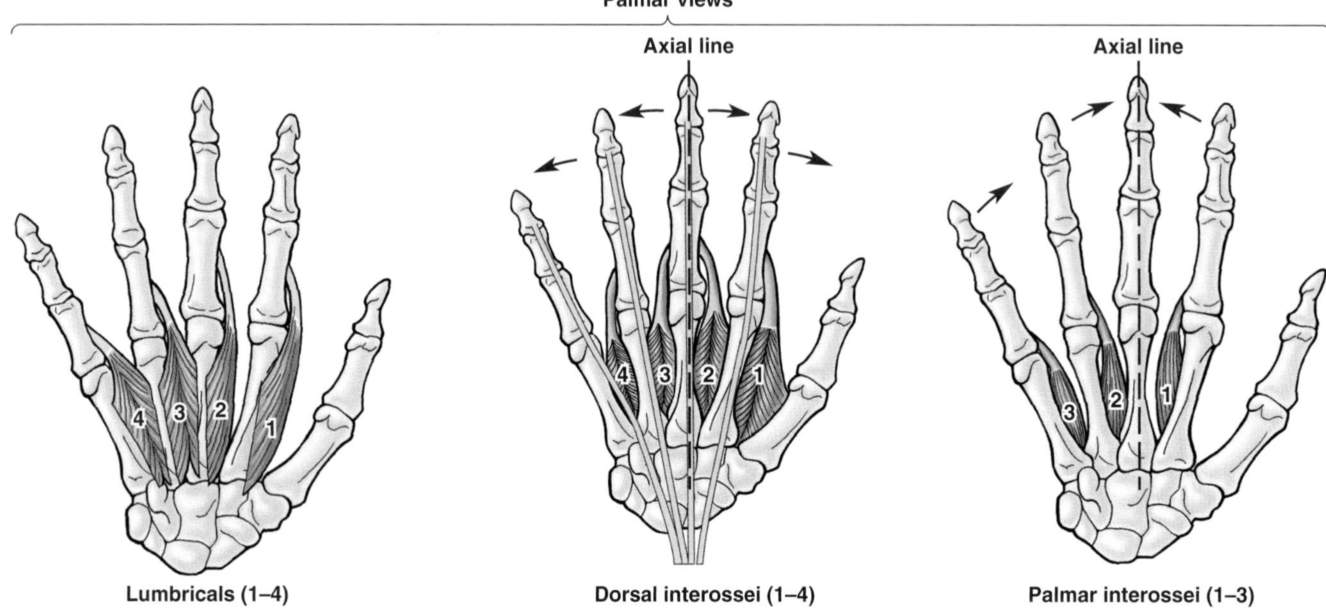

Muscle	Proximal Attachment	Distal Attachment	Innervation[a]	Main Action
Adductor pollicis	Oblique head: bases of 2nd and 3rd metacarpals, capitate, and adjacent carpals Transverse head: anterior surface of body of 3rd metacarpal	Medial side of base of proximal phalanx of thumb	Deep branch of ulnar nerve (C8 and **T1**)	Adducts thumb toward middle digit
Hypothenar muscles Abductor digiti minimi	Pisiform	Medial side of base of proximal phalanx of little finger		Abducts digit 5
Flexor digiti minimi brevis	Hook of hamate and flexor retinaculum			Flexes proximal phalanx of digit 5
Opponens digiti minimi		Medial border of 5th metacarpal		Draws 5th metacarpal anteriorly and rotates it, bringing digit 5 into opposition with thumb
Short muscles Lumbricals 1 and 2	Lateral two tendons of flexor digitorum profundus (unipennate muscles)	Lateral sides of extensor expansions of digits 2–5	Median nerve (C8 and **T1**)	Flex digits at metacarpophalangeal joints and extend interphalangeal joints
Lumbricals 3 and 4	Medial three tendons of flexor digitorum profundus (bipennate muscles)		Deep branch of ulnar nerve (C8 and **T1**)	
Dorsal interossei 1–4	Adjacent sides of two metacarpals (bipennate muscles)	Extensor expansions and bases of proximal phalanges of digits 2–4	Deep branch of ulnar nerve (C8 and **T1**)	Abduct digits from axial line and act with lumbricals to flex metacarpophalangeal joints and extend interphalangeal joints
Palmar interossei 1–3	Palmar surfaces of 2nd, 4th, and 5th metacarpals (unipennate muscles)	Extensor expansions of digits and bases of proximal phalanges of digits 2, 4, and 5		Adduct digits toward axial line and assist lumbricals in flexing metacarpophalangeal joints and extending interphalangeal joints

Short Muscles of the Hand

The short muscles of the hand are the lumbricals and interossei.

Lumbricals. The four slender lumbrical muscles were named because of their wormlike form (L. lumbricus, earthworm). *The lumbricals flex the digits at the metacarpophalangeal joints and extend the interphalangeal joints.*

Interossei. The four dorsal interosseous muscles are located between the metacarpals; the three palmar interosseous muscles are on the palmar surfaces of the metacarpal bones. The 1st dorsal interosseous muscle is easy to palpate; oppose the thumb firmly against the index finger and it can be easily felt. Some authors describe four palmar interossei; in so doing, they are including the deep head of the flexor pollicis brevis because of its similar innervation and placement on the thumb (Table 6.10).

The four dorsal interossei abduct the digits, and the three palmar interossei adduct them. A mnemonic device is to make acronyms of **D**orsal **AB**duct (**DAB**) and **P**almar **AD**uct (**PAD**). Acting together, the dorsal and palmar interossei and the lumbricals produce flexion at the metacarpophalangeal joints and extension of the interphalangeal joints (the so-called "Z-movement"). This occurs because of their attachment to the lateral bands of the extensor expansions. Understanding the *Z-movement* is useful because it is the opposite of clawhand, which occurs in ulnar paralysis when the interossei and the 3rd and 4th lumbricals are incapable of acting together to produce the Z-movement (p. 761).

Flexor Tendons of Extrinsic Hand Muscles

The tendons of the FDS and FDP enter the **common flexor synovial sheath** (ulnar bursa) deep to the flexor retinaculum (Fig. 6.57*A*). The tendons enter the central compartment of the hand and fan out to enter their respective **digital synovial sheaths.** The flexor and digital sheaths enable the tendons to slide freely over each other during movements of the digits. Near the base of the proximal phalanx, the tendon of FDS splits and surrounds the tendon of FDP (Fig. 6.57*B*). The halves of the FDS tendon are attached to the margins of the anterior aspect of the base of the middle phalanx. The tendon of FDP, after passing through the split in the FDS tendon (Camper's chiasm), passes distally to attach to the anterior aspect of the base of the distal phalanx.

The **fibrous digital sheaths** are the strong ligamentous tunnels containing the flexor tendons and their synovial sheaths (Fig. 6.57*C*). The fibrous digital sheaths extend from the heads of the metacarpals to the bases of the distal phalanges. These sheaths prevent the tendons from pulling away from the digits ("bowstringing"). The fibrous digital sheaths combine with the bones to form **osseofibrous tunnels** through which the tendons pass to reach the digits. The **anular** and **cruciform parts** or pulleys are thickened reinforcements of the fibrous digital sheaths (Fig. 6.57*B*).

The long flexor tendons are supplied by small blood vessels that pass within synovial folds (*vincula*) from the periosteum of the phalanges. *The tendon of the flexor pollicis longus* passes deep to the flexor retinaculum to the thumb within its own synovial sheath. At the head of the metacarpal, the tendon runs between two *sesamoid bones*, one in the combined tendon of the flexor pollicis brevis and abductor pollicis brevis and the other in the tendon of the adductor pollicis.

Tenosynovitis

Injuries such as a puncture of the palm by a rusty nail can cause infection of the synovial sheaths. When inflammation of the tendon and synovial sheath occurs (*tenosynovitis*), the digit swells and movement becomes painful. Because the tendons of the 2nd, 3rd, and 4th digits nearly always have separate synovial sheaths, the infection is usually confined to the infected digit. In neglected infections, however, the proximal ends of these sheaths may rupture, allowing the infection to spread to the midpalmar space. Because the synovial sheath of the little finger is usually continuous with the common flexor sheath, tenosynovitis in this digit may spread to the common sheath and thus through the palm and carpal tunnel to the anterior forearm. Likewise, tenosynovitis in the thumb may spread via the continuous synovial sheath of the flexor pollicis longus (radial bursa). Just how far an infection spreads from the digits depends on variations in their connections with the common flexor sheath.

The tendons of the APL and EPB are in the same tendinous sheath on the dorsum of the wrist. Excessive friction of these tendons on their common sheath results in fibrous thickening of the sheath and stenosis of the osseofibrous tunnel. The excessive friction is caused by repetitive forceful use of the hands during gripping and wringing (e.g., squeezing water out of clothes). This condition—*de Quervain's tenovaginitis stenosans*—causes pain in the wrist that radiates proximally to the forearm and distally toward the thumb. Local tenderness is felt over the common fibrous sheath on the lateral side of the wrist.

Thickening of a fibrous digital sheath on the palmar aspect of the digit produces stenosis of the osseofibrous tunnel of the finger or thumb. This narrowing of the tunnel results from repetitive forceful use of the fingers. If the tendons of FDS and FDP enlarge proximal to the tunnel, the person is unable to extend the finger. When the finger is extended passively, a snap is audible. Flexion produces another snap as the thickened tendon moves. This condition is called *digital tenovaginitis stenosans* ("trigger finger" or "snapping finger"). ○

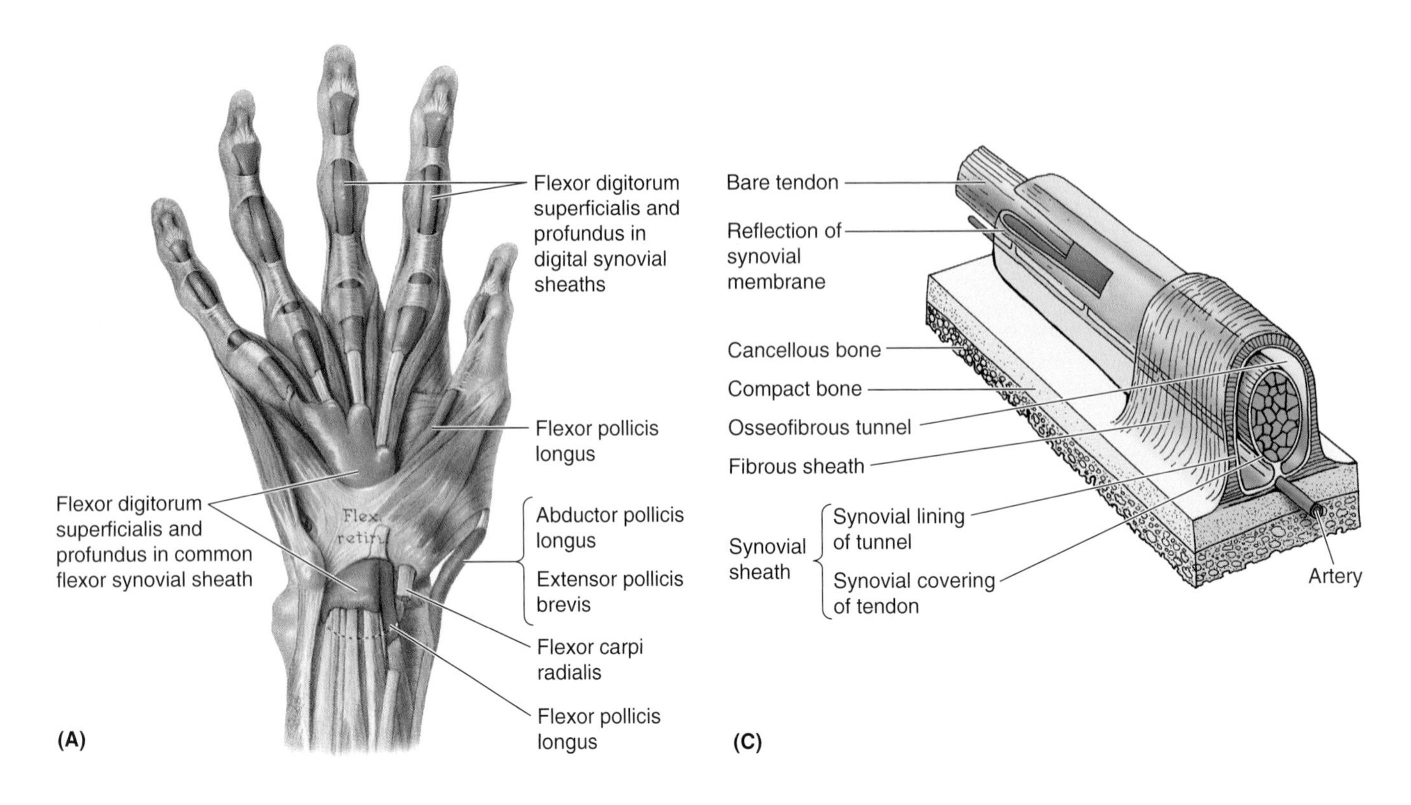

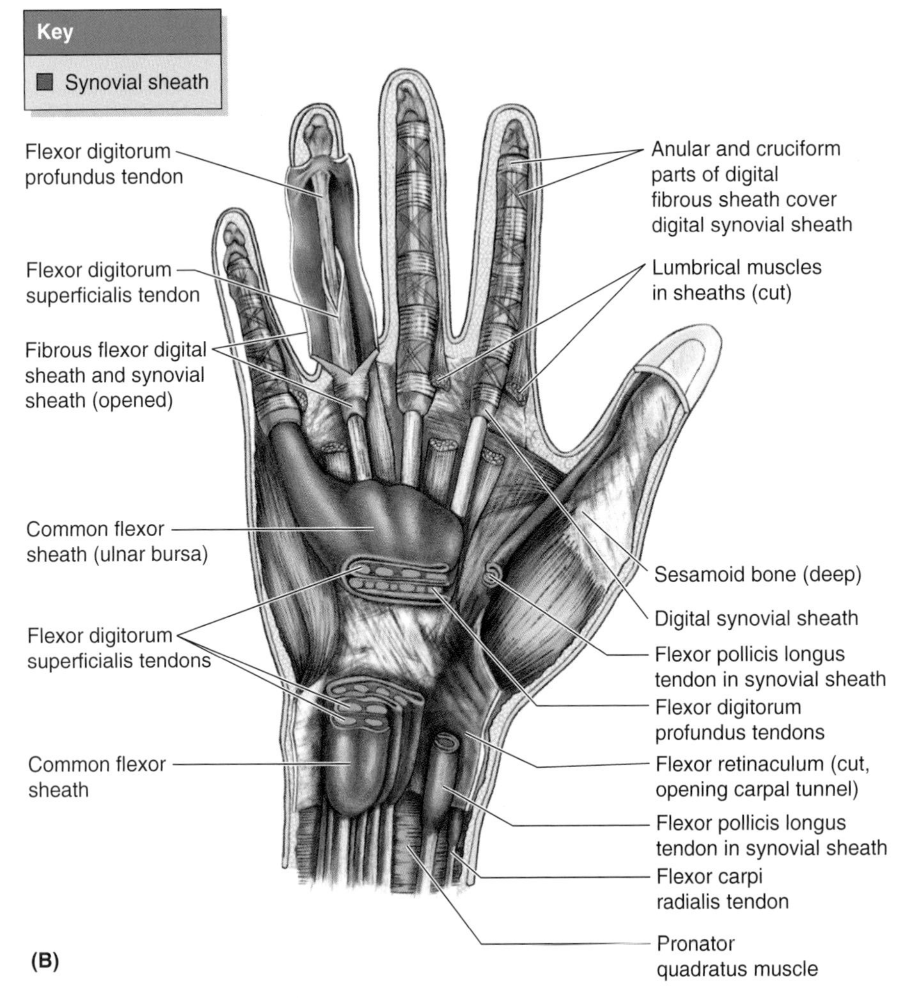

Figure 6.57. Flexor tendons, common flexor sheath, fibrous digital sheaths, and synovial sheaths of the digits. A. Dissection of the anterior aspect of the distal forearm and hand showing the digital synovial sheaths of the long flexor tendons to the digits. Observe the two sets: (*a*) proximal or carpal, posterior to the flexor retinaculum, and (*b*) distal or digital, within to the fibrous sheaths of the digital flexors. **B.** Dissection of the palm of the hand illustrating the tendons and fibrous digital tendon sheaths. The fibrous sheaths of the digits are strong coverings of the flexor tendons, which extend from the heads of the metacarpals to the base of the distal phalanges. They prevent the tendons from pulling away from the bones of the digits. They attach along the borders of the proximal and middle phalanges, to capsules of the interphalangeal joints, and to the surface of the distal phalanx. **C.** Drawing of the osseofibrous tunnel of a finger containing a tendon. Within the fibrous sheath, observe the synovial sheath consisting of the synovial lining of the fibrous sheath and the synovial covering of the tendon. These layers of the synovial sheath are actually separated only by a capillary layer of synovial fluid, which forms a lubricating system for the tendon.

Arteries of the Hand

The ulnar and radial arteries and their branches provide all the blood to the hand (Table 6.11).

Ulnar Artery

The ulnar artery enters the hand anterior to the flexor retinaculum between the pisiform bone and the hook of the hamate (Guyon's canal). The ulnar artery lies lateral to the ulnar nerve (Fig. 6.55). It divides into two terminal branches, the superficial palmar arch and the deep palmar branch. The **superficial palmar arch**, the main termination of the ulnar artery, gives rise to three **common palmar digital arteries** (Fig. 6.55*A*) that anastomose with **palmar metacarpal arteries** from the deep palmar arch. Each common palmar digital artery divides into a pair of **proper palmar digital arteries** that run along the adjacent sides of the 2nd through 4th digits.

Radial Artery

The radial artery curves dorsally around the scaphoid and trapezium in the floor of the anatomical snuff box (Fig. 6.55, Table 6.11) and enters the palm by passing between the heads of the 1st dorsal interosseous muscle. It then turns medially and passes between the heads of the adductor pollicis. The ra-

Table 6.11 Arteries of the hand

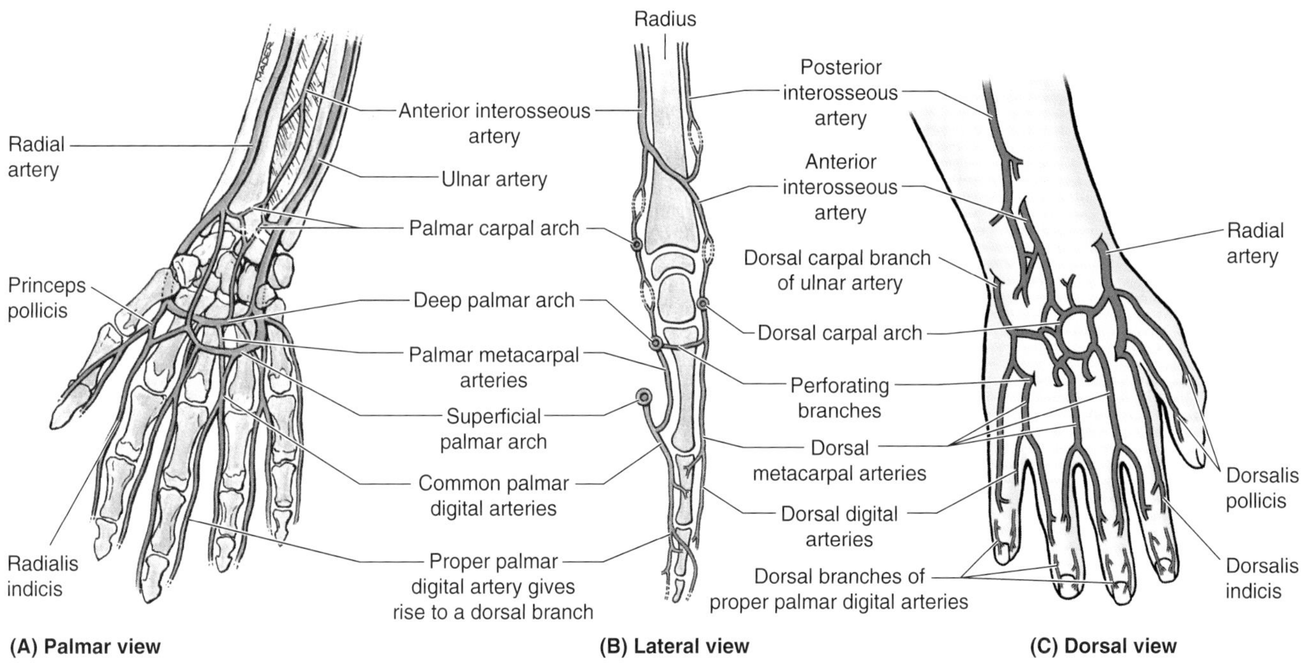

Artery	Origin	Course
Superficial palmar arch	Direct continuation of ulnar artery; arch is completed on lateral side by superficial branch of radial artery or another of its branches	Curves laterally deep to palmar aponeurosis and superficial to long flexor tendons; curve of arch lies across palm at level of distal border of extended thumb
Deep palmar arch	Direct continuation of radial artery arch is completed on medial side by deep branch of ulnar artery	Curves medially, deep to long flexor tendons and is in contact with bases of metacarpals
Common palmar digitals	Superficial palmar arch	Pass distally on lumbricals to webbings of digits
Proper palmar digitals	Common palmar digital arteries	Run along sides of digits 2–5
Princeps pollicis	Radial artery as it turns into palm	Descends on palmar aspect of first metacarpal and divides at the base of proximal phalanx into two branches that run along sides of thumb
Radialis indicis	Radial artery, but may arise from princeps pollicis artery	Passes along lateral side of index finger to its distal end
Dorsal carpal arch	Radial and ulnar arteries	Arches within fascia on dorsum of hand

dial artery anastomoses with the deep branch of the ulnar artery to form the deep palmar arch. The **deep palmar arch**, formed mainly by the radial artery, lies across the metacarpals just distal to their bases (Table 6.11). The deep arch gives rise to three **palmar metacarpal arteries** and the princeps pollicis arteries, which supply the palmar surface and sides of the thumb.

Laceration of the Palmar Arterial Arches

Bleeding is usually profuse when the palmar arterial arches are lacerated. It may not be sufficient to ligate only one forearm artery when the arches are lacerated because these vessels usually have numerous communications in the forearm and hand and thus bleed from both ends. To obtain a bloodless surgical operating field for treating complicated hand injuries, it may be necessary to compress the brachial artery and its branches proximal to the elbow (e.g., using a pneumatic tourniquet). This procedure prevents blood from reaching the ulnar and radial arteries through the anastomoses around the elbow.

Ischemia of the Fingers

Intermittent bilateral attacks of ischemia of the fingers, marked by pallor and often accompanied by paresthesia and pain, is characteristically brought on by cold and emotional stimuli. The condition may result from an anatomical abnormality or an underlying disease. When the cause of the condition is idiopathic (unknown) or primary, it is called *Raynaud's disease*. The arteries of the upper limb are innervated by sympathetic nerves. Postsynaptic fibers from the sympathetic ganglia enter nerves that form the brachial plexus and are distributed to the digital arteries through branches arising from the plexus. When treating ischemia resulting from Raynaud's disease, it may be necessary to perform a cervicodorsal *presynaptic sympathectomy*—excision of a segment of a sympathetic nerve—to dilate the digital arteries. ⊙

Veins of the Hand

The superficial and deep palmar arterial arches are accompanied by superficial and deep **palmar venous arches**, respectively (Figs. 6.49 and 6.50). The dorsal digital veins drain into three dorsal metacarpal veins, which unite to form a **dorsal venous network**. Superficial to the metacarpus, this network is prolonged proximally on the lateral side as the **cephalic vein**. The **basilic vein** arises from the medial (ulnar) side of the dorsal venous network.

Nerves of the Hand

The median, ulnar, and radial nerves supply the hand (Fig. 6.58, Table 6.12). Branches or communications from the lateral and posterior cutaneous nerves may contribute some fibers to supply the dorsum of the hand.

Median Nerve

The median nerve enters the hand through the carpal tunnel, deep to the flexor retinaculum, along with the nine tendons of the flexors digitorum superficialis and profundus and the flexor pollicis longus (Fig. 6.58). The **carpal tunnel** is the passageway deep to the flexor retinaculum between the tubercles of the scaphoid and trapezoid bones on the lateral side and the pisiform and hook of the hamate on the medial side. Distal to the carpal tunnel, the *median nerve supplies the three thenar muscles and the 1st and 2nd lumbricals* (Table 6.12). It also sends sensory fibers to the skin on the entire palmar surface, the sides of the 1st three digits, the lateral half of the 4th digit, and the dorsum of the distal halves of these digits. Note, however, that the palmar branch, which supplies the central palm, arises proximal to the carpal tunnel and does not traverse the tunnel (i.e., it runs superficial to the flexor retinaculum). Thus, although this skin lies distal to the tunnel, it does not lose sensation in the carpal tunnel syndrome.

Lesions of the Median Nerve

Lesions of the median nerve occur in two places: in the forearm and at the wrist. The most common site is where the nerve passes through the carpal tunnel.

Carpal Tunnel Syndrome

Carpal tunnel syndrome results from any lesion (e.g., inflammation of synovial sheaths) that significantly reduces the size of the carpal tunnel (*A*). Fluid retention, infection, and excessive exercise of the fingers may cause swelling of the tendons or their synovial sheaths. *The median nerve is the most sensitive structure in the carpal tunnel* (*B–D*) and therefore is the most affected. This nerve has two terminal sensory branches that supply the skin of the hand; hence, paresthesia, *hypoesthesia* (diminished sensation), or *anesthesia* may occur in the lateral three and a half digits. The nerve also has one terminal motor branch—the thenar or recurrent branch—which serves three thenar muscles.

Progressive loss of coordination and strength in the thumb (owing to weakness of the abductor pollicis brevis and opponens pollicis) may occur if the cause of the *median nerve compression* is not alleviated. Persons with median nerve compression are unable to oppose the thumb (*E*). As the condition progresses, sensory changes radiate into the forearm and axilla. Symptoms ▶

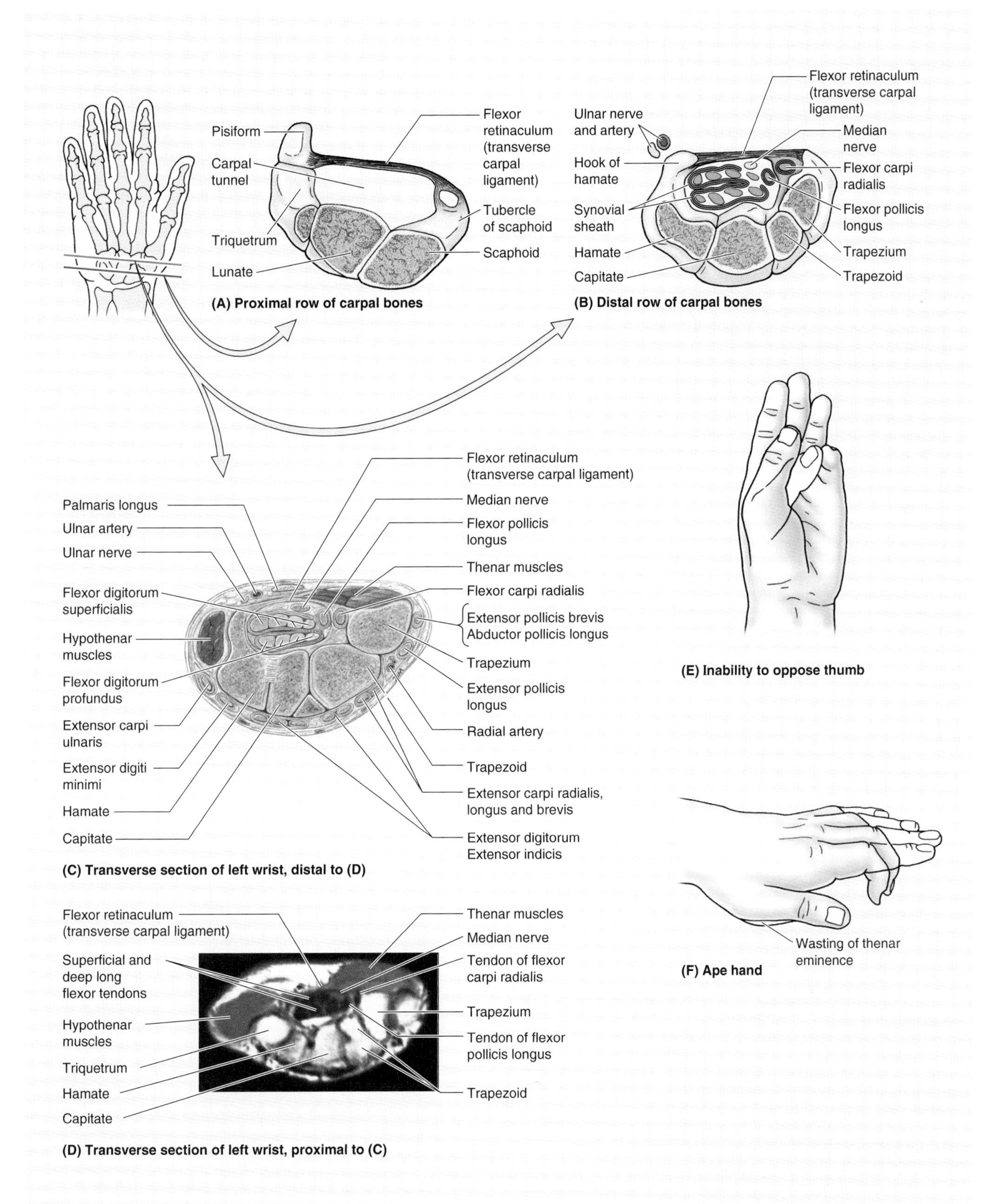
Pisiform
Carpal tunnel
Triquetrum
Lunate
Flexor retinaculum (transverse carpal ligament)
Tubercle of scaphoid
Scaphoid
(A) Proximal row of carpal bones
Ulnar nerve and artery
Hook of hamate
Synovial sheath
Hamate
Capitate
Flexor retinaculum (transverse carpal ligament)
Median nerve
Flexor carpi radialis
Flexor pollicis longus
Trapezium
Trapezoid
(B) Distal row of carpal bones
Palmaris longus
Ulnar artery
Ulnar nerve
Flexor digitorum superficialis
Hypothenar muscles
Flexor digitorum profundus
Extensor carpi ulnaris
Extensor digiti minimi
Hamate
Capitate
Flexor retinaculum (transverse carpal ligament)
Median nerve
Flexor pollicis longus
Thenar muscles
Flexor carpi radialis
Extensor pollicis brevis
Abductor pollicis longus
Trapezium
Extensor pollicis longus
Radial artery
Trapezoid
Extensor carpi radialis, longus and brevis
Extensor digitorum
Extensor indicis
(C) Transverse section of left wrist, distal to (D)
Flexor retinaculum (transverse carpal ligament)
Superficial and deep long flexor tendons
Hypothenar muscles
Triquetrum
Hamate
Capitate
Thenar muscles
Median nerve
Tendon of flexor carpi radialis
Trapezium
Tendon of flexor pollicis longus
Trapezoid
(D) Transverse section of left wrist, proximal to (C)
(E) Inability to oppose thumb
Wasting of thenar eminence
(F) Ape hand

▶ of median nerve compression can be reproduced by compression of the median nerve with your finger at the wrist for approximately 30 seconds. Persons with carpal tunnel syndrome have difficulty performing fine movements of the thumb (e.g., buttoning a shirt or blouse as well as gripping things such as a hairbrush. To relieve symptoms of the syndrome, partial or complete surgical division of the flexor retinaculum—*carpal tunnel release*—may be necessary.

Laceration of the Wrist

Accidental laceration of the wrist often causes median nerve injury because this nerve is relatively close to the surface. In *attempted suicides by wrist slashing,* the median nerve is commonly injured just proximal to the flexor retinaculum. This results in paralysis of the thenar muscles and the first two lumbricals. Hence, opposition of the thumb is not possible, and fine control movements of the 2nd and 3rd digits are impaired. Sensation is also lost over the thumb and adjacent two and a half digits.

Trauma to the Median Nerve

Median nerve injury resulting from a perforating wound in the elbow region results in loss of flexion of the proximal and distal interphalangeal joints of the 2nd and 3rd digits. The ability to flex the metacarpophalangeal joints of these digits is also affected because digital branches of the median nerve supply the 1st and 2nd lumbricals. *Ape hand* (*F,* p. 775) refers to a deformity that is marked by thumb movements being limited to flexion and extension of the thumb in the plane of the palm because of the inability to oppose—and limited abduction of—the thumb. The *recurrent (thenar) branch of the median nerve* (Fig. 6.58*A*) supplying the thenar muscles lies subcutaneously and may be severed by relatively minor lacerations involving the thenar eminence. Severance of this nerve paralyzes the thenar muscles, and the thumb loses much of its usefulness. The incision for *carpal tunnel release* is made toward the medial side of the wrist and flexor retinaculum to avoid possible injury to the recurrent branch of the median nerve.

Most nerve injuries in the upper limb affect opposition of the thumb. Undoubtedly, injuries to the nerves supplying the intrinsic muscles of the hand, especially the median nerve, have the most severe effects on this complex movement. If the median nerve is severed in the forearm or at the wrist, the thumb cannot be opposed; however, the APL and adductor pollicis (supplied by the posterior interosseous and ulnar nerves, respectively) may imitate opposition. ⊕

Ulnar Nerve

The ulnar nerve leaves the forearm by emerging from deep to the tendon of the flexor carpi ulnaris (Figs. 6.55 and 6.58, Table 6.12). It passes distally to the wrist, where it is bound by fascia to the anterior surface of the flexor retinaculum. It then passes alongside the lateral border of the pisiform; the ulnar artery is on its lateral side. Just proximal to the wrist, the ulnar nerve gives off a **palmar cutaneous branch** that passes superficial to the flexor retinaculum and palmar aponeurosis; it supplies skin on the medial side of the palm. The ulnar nerve also gives off a **dorsal cutaneous branch** that supplies the medial half of the dorsum of the hand, the 5th digit, and the medial half of the 4th digit (Fig. 6.58).

The ulnar nerve ends at the distal border of the flexor retinaculum by dividing into superficial and deep branches (Fig. 6.55*B*). The **superficial branch of the ulnar nerve** supplies cutaneous branches to the anterior surfaces of the medial one and a half digits. The **deep branch of the ulnar nerve** supplies the hypothenar muscles, the medial two lumbricals, the adductor pollicis, and all the interossei. The deep branch also supplies several joints (wrist, intercarpal, carpometacarpal, and intermetacarpal). *The ulnar nerve is referred to as the nerve of fine movements because it innervates muscles that are concerned with intricate hand movements* (Table 6.12).

Ulnar Nerve Injury

Ulnar nerve injuries usually occur in four places:

- Posterior to the medial epicondyle of the humerus
- In the cubital tunnel formed by the tendinous arch connecting the humeral and ulnar heads of the flexor carpi ulnaris
- At the wrist
- In the hand.

More than 27% of nerve lesions of the upper limb affect the ulnar nerve (Rowland, 1995). The ulnar nerve is frequently injured by gunshot wounds, stab wounds, and fractures of the distal end of the humerus, olecranon, or head of the radius (Lange et al., 1995).

Ulnar nerve injury commonly occurs where the nerve passes posterior to the medial epicondyle of the humerus. Often the injury occurs when the elbow hits a hard surface, fracturing the epicondyle. The ulnar nerve may be compressed at the elbow during sleep or as an *occupa-* ▶

► *tional neuritis* in workers who rest their elbows on hard surface for long periods (Lange et al., 1995). Ulnar nerve injury may result in extensive motor and sensory loss to the hand with accompanying impaired power of adduction. On flexing the wrist joint, the hand is drawn to the lateral (radial) side by the flexor carpi radialis in the absence of the "balance" provided by the flexor carpi ulnaris. *After ulnar nerve injury, patients are likely to have difficulty making a fist because of paralysis of most intrinsic hand muscles* (Table 6.10). In addition, their metacarpophalangeal joints become hyperextended, and they cannot flex their 4th and 5th digits at the distal interphalangeal joints when they try to make a fist; nor can they extend their interphalangeal joints when they try to straighten their fingers. This results in a characteristic *clawhand appearance* of the hand (p. 761).

Guyon's Canal Syndrome

Compression of the ulnar nerve may occur at the wrist where it passes between the pisiform and the hook of the hamate. The depression between these bones is converted by the *pisohamate ligament* into an osseofibrous tunnel (Guyon's canal). Compression of the ulnar nerve in this tunnel may result in hypoesthesia in the medial one and a half digits and weakness of the intrinsic muscles of the hand.

Cyclist's Palsy

Persons who ride long distances on bicycles with their hands in an extended position against the hand grips put pressure on the hooks of their hamates, which compresses the ulnar nerve. Because of this, this type of nerve compression is called *handlebar neuropathy*. This injury results in sensory loss on the medial side of the hand and weakness of the intrinsic hand muscles. ⊙

Radial Nerve

The radial nerve supplies no hand muscles (Table 6.12). Its terminal branches, superficial and deep, arise in the cubital fossa. The *superficial branch of the radial* is the direct continuation of the radial nerve along the anterolateral side of the forearm and is entirely sensory (Fig. 6.58*A*). It travels under cover of the brachioradialis and then pierces the deep fascia near the dorsum of the wrist to supply the skin and fascia over the lateral two-thirds of the dorsum of the hand, the dorsum of the thumb, and the proximal parts of the lateral one and a half digits.

Radial Nerve Injury

Although the radial nerve supplies no muscles in the hand, radial nerve injury in the arm can produce serious disability of the hand. The characteristic handicap is inability to extend the wrist resulting from *paralysis of extensor muscles of the forearm* (Table 6.7). The hand is flexed at the wrist and lies flaccid, a condition known as *wrist-drop* (p. 731). The digits also remain in the flexed position at the metacarpophalangeal joints. The interphalangeal joints can be extended weakly through the action of the intact lumbricals and interossei, which are supplied by the median and ulnar nerves (Table 6.10). The radial nerve has only a small area of exclusive cutaneous supply on the hand. The extent of anesthesia is minimal, even in serious radial nerve injuries, and is usually confined to a small area on the lateral part of the dorsum of the hand. ⊙

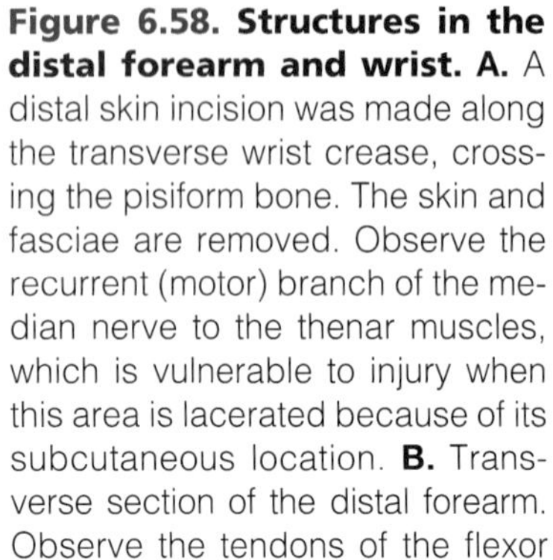

Figure 6.58. Structures in the distal forearm and wrist. A. A distal skin incision was made along the transverse wrist crease, crossing the pisiform bone. The skin and fasciae are removed. Observe the recurrent (motor) branch of the median nerve to the thenar muscles, which is vulnerable to injury when this area is lacerated because of its subcutaneous location. **B.** Transverse section of the distal forearm. Observe the tendons of the flexor carpi radialis, palmaris longus, and flexor carpi ulnaris forming a surface layer of flexors of the wrist. Observe also the ulnar nerve and artery under cover of the flexor carpi ulnaris; this is why the pulse of this artery cannot be felt here. **C.** Orientation drawing indicating the plane of the section shown in (**B**).

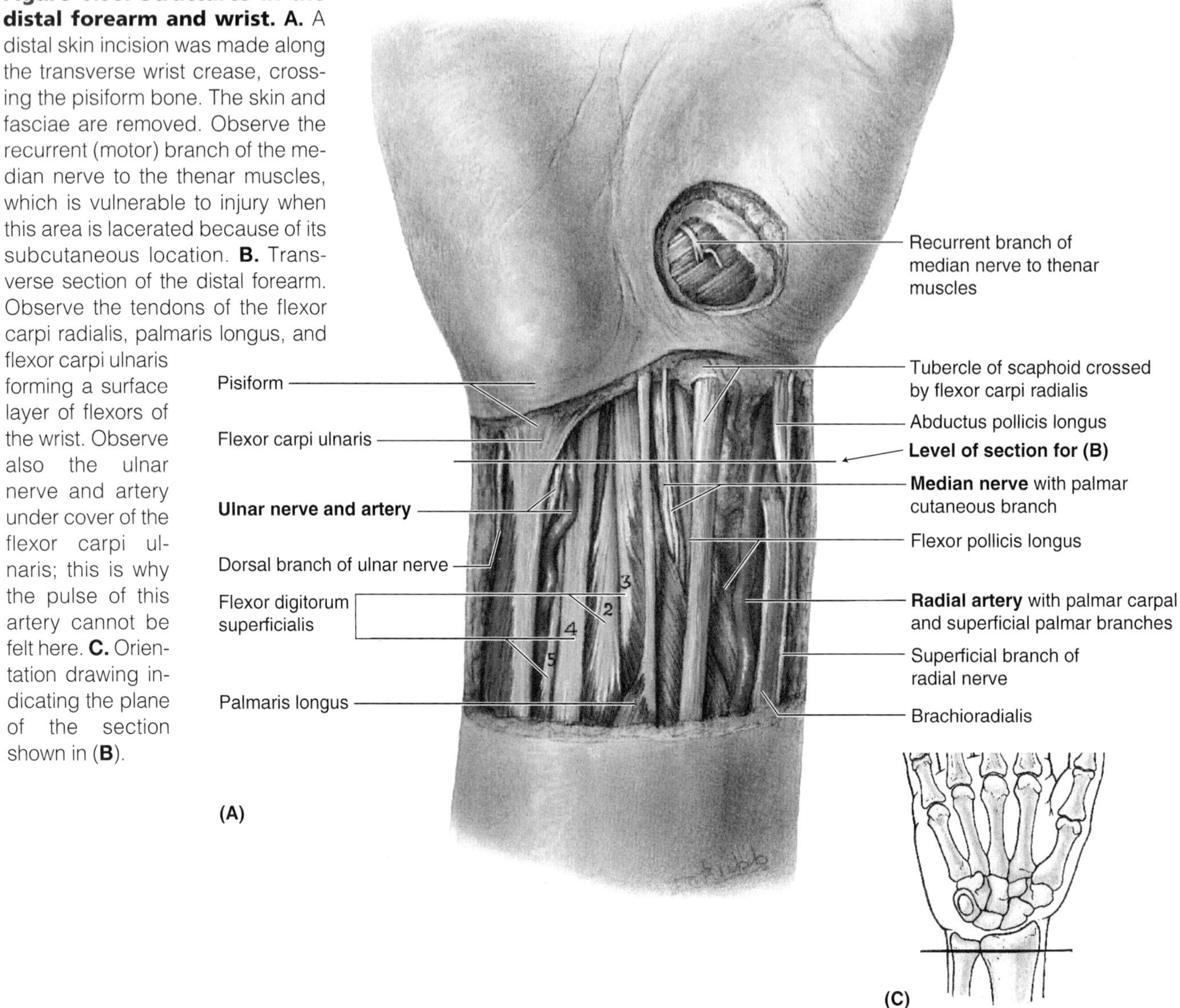

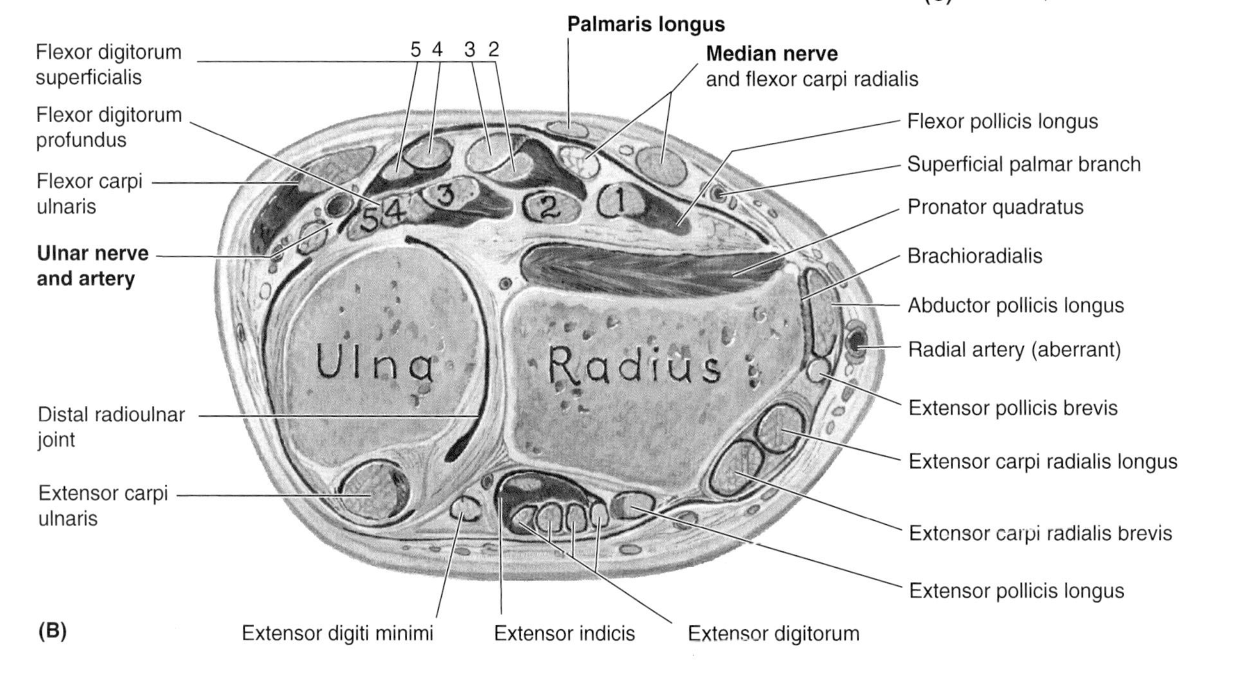

Surface Anatomy of the Hand

A common place for measuring the pulse rate is where the radial artery lies on the anterior surface of the distal end of the radius, lateral to the tendon of the flexor carpi radialis. Here the artery can be compressed against the radius, where it lies between the tendons of the flexor carpi radialis and APL. The **radial pulse**, like other palpable pulses, is a peripheral reflection of cardiac action. The tendons of flexor carpi radialis and palmaris longus can be palpated and usually observed by flexing the closed fist against resistance (*A*). *The tendon of flexor carpi radialis* may be seen ▶

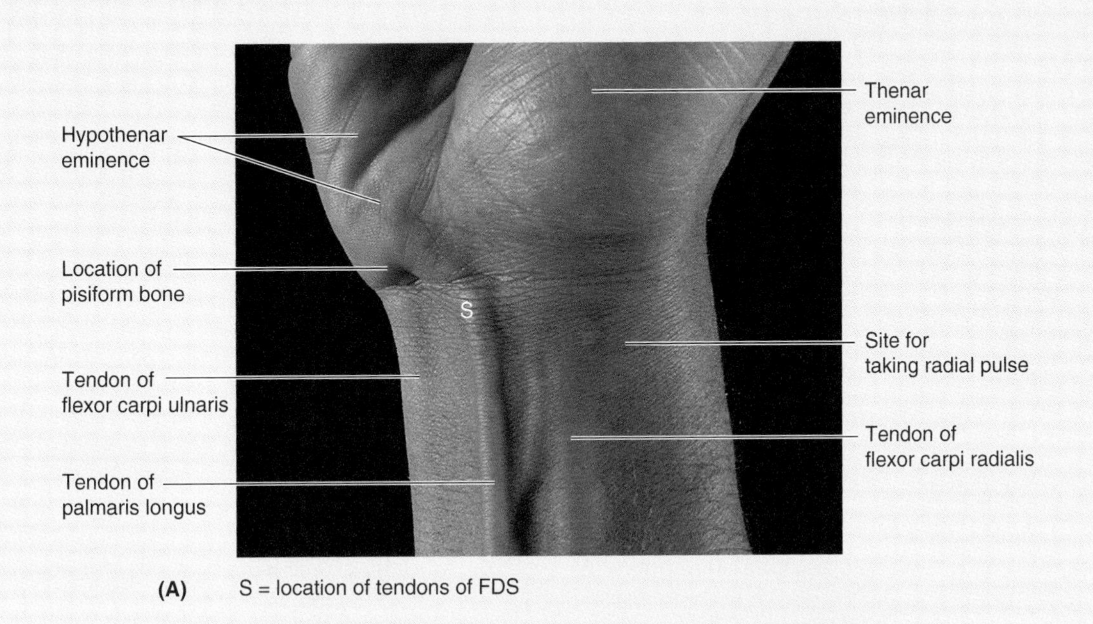

(A) S = location of tendons of FDS

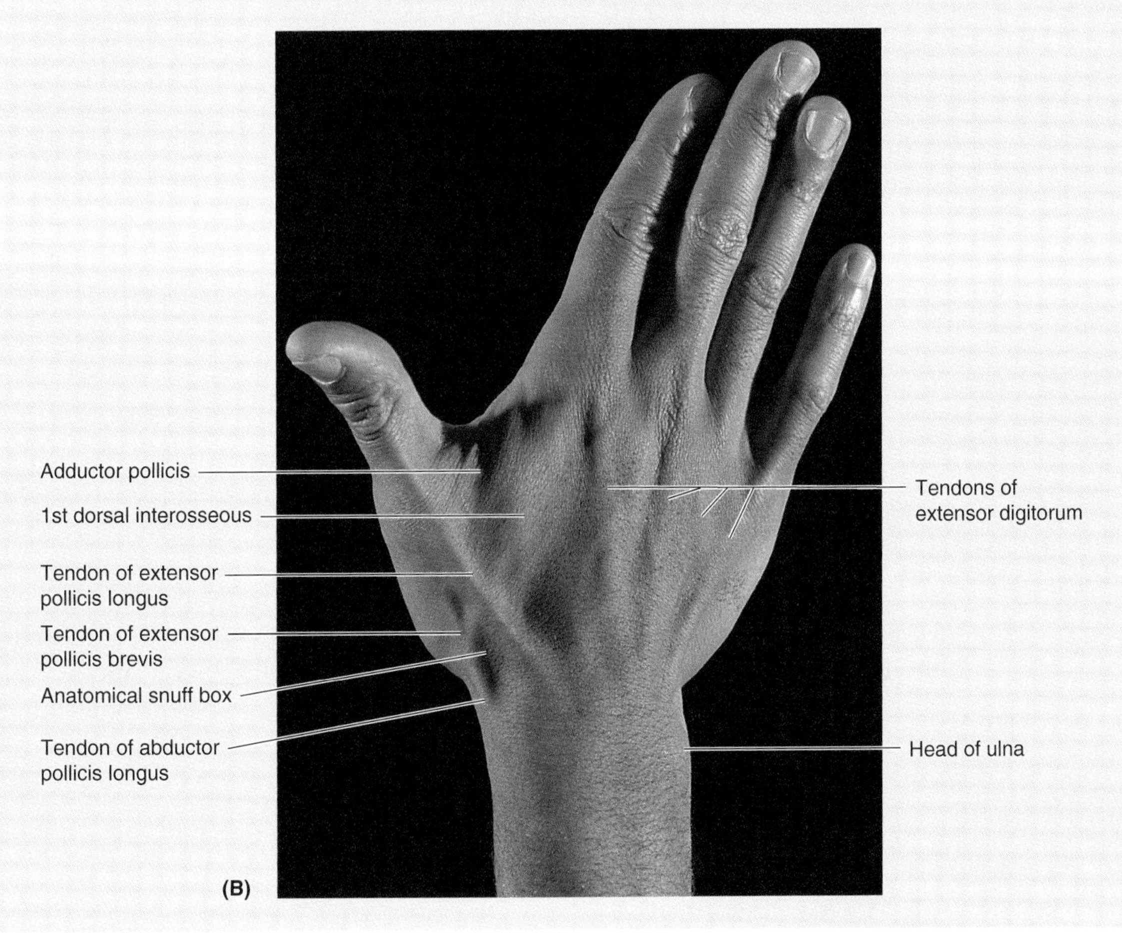

(B)

▶ and palpated anterior to the wrist, a little lateral to its middle. This tendon serves as a *guide to the radial artery*, which can be felt pulsating just lateral to the tendon.

The tendon of palmaris longus can be seen and palpated in the middle of the anterior aspect of the wrist. It is smaller than the tendon of the flexor carpi radialis and is not always present. *The palmaris longus tendon serves as a guide to the median nerve, which lies deep to it* (Fig. 6.58*B*). *The tendon of the flexor carpi ulnaris* can be palpated as it crosses the anterior aspect of the wrist near the medial side and inserts into the pisiform. *The flexor carpi ulnaris tendon serves as a guide to the ulnar nerve and artery.* The tendons of the FDS can be palpated as the digits are alternately flexed and extended. The *ulnar pulse* is often difficult to palpate. The tendons of the APL and EPB muscles indicate the anterior boundary of the *anatomical snuff box* (*B*, p. 779), and the tendon of the EPL indicates the posterior boundary of the box. The radial artery passes through the snuff box, where its pulsations may be felt. The scaphoid and trapezium lie in the floor of the snuff box.

The skin covering the dorsum of the hand is thin and loose when the hand is relaxed. The looseness of the skin results from the mobility of the subcutaneous tissue and from the relatively few skin ligaments that are present. Hair is present on the dorsum of the hand and on the proximal parts of the digits, especially in men. If the dorsum of the hand is examined with the wrist extended against resistance and the digits abducted, the tendons of the extensor digitorum to the fingers usually stand out, particularly in thin persons. These tendons are not visible far beyond the knuckles because they flatten here to form the extensor expansions. The knuckles that become visible when a fist is made are produced by the heads of the metacarpals. Under the loose subcutaneous tissue and extensor tendons on the dorsum of the hand, the metacarpals can be palpated. A prominent feature of the dorsum of the hand is the *dorsal venous network*.

The skin on the palm is thick because it is required to withstand the wear and tear of work and play. It is richly supplied with sweat glands but contains no hair or sebaceous glands. The *superficial palmar arch* lies across the center of the palm, level with the distal border of the fully extended thumb. The main part of the arch ends at the thenar eminence. The *deep palmar arch* runs across the palm approximately 1 cm proximal to the superficial palmar arch. The palmar skin presents several more or less constant *flexion creases* where the skin is firmly bound to the deep fascia. The **distal wrist crease** indicates the proximal border of the flexor retinaculum. The transverse palmar creases indicate where the skin folds during flexion of the hand. The longitudinal creases deepen when the thumb is opposed, and the transverse creases deepen when the metacarpophalangeal joints are flexed. The **radial longitudinal crease** (the "life line" of palmistry) partially ▶

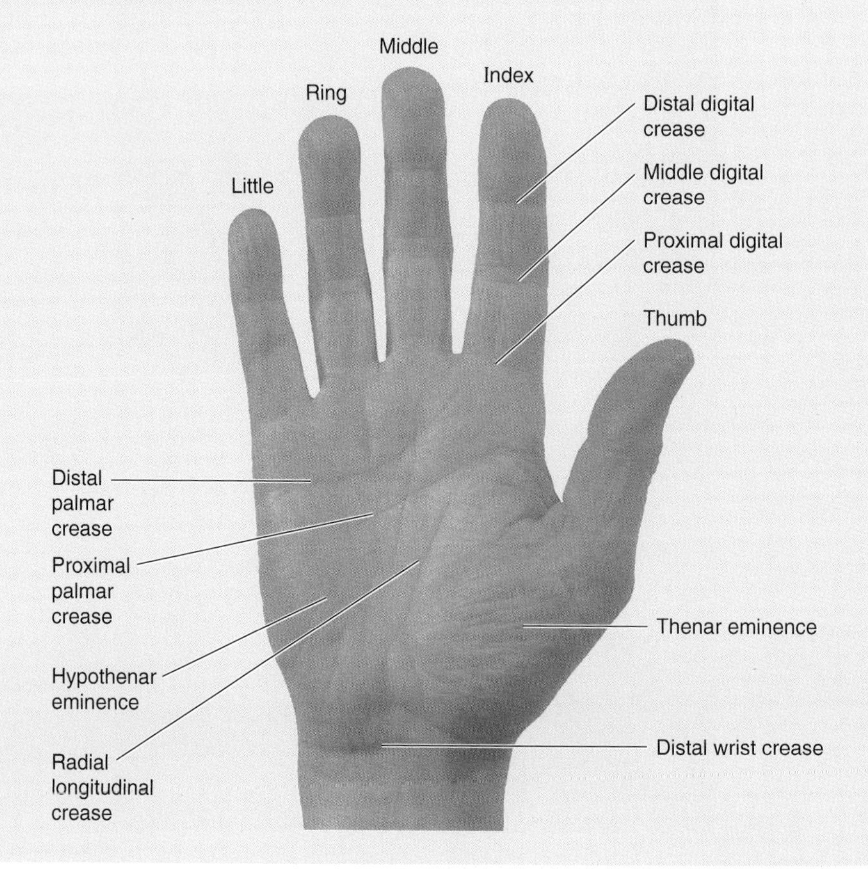

▶ encircles the **thenar eminence**, formed by the short muscles of the 1st digit. The **proximal palmar crease** commences on the lateral border of the palm, superficial to the head of the 2nd metacarpal. It extends medially and slightly proximally across the palm, superficial to the bodies of the 3rd through 5th metacarpals. The **distal palmar crease** begins at or near the cleft between the index and middle fingers and crosses the palm with a slight convexity, superficial to the heads of the 2nd through 4th metacarpals.

Each of the medial four digits usually has three transverse flexion creases. The **proximal digital crease** is located at the root of the digit, approximately 2 cm distal to the metacarpophalangeal joint. The proximal digital crease of the thumb crosses obliquely, proximal to the 1st metacarpophalangeal joint. The **middle digital crease** lies over the proximal interphalangeal joint, and the **distal digital crease** lies proximal to the distal interphalangeal joint. The thumb, having two phalanges, has only two flexion creases. Like other digital creases, they deepen when the thumb is flexed. The skin ridges on the pads of the digits—*fingerprints*—are used for identification because of their unique patterns. The anatomical function of the *epidermal ridge patterns* is to reduce slippage when grasping objects. ⊕

Dermatoglyphics

The science of studying ridge patterns of the palm—*dermatoglyphics*—is a valuable extension of the conventional physical examination of patients with certain congenital anomalies and genetic diseases. For example, persons with trisomy 21 (Down syndrome) often have only one transverse palmar crease (simian crease); however, approximately 1% of the general population has this crease with no other clinical features of the syndrome.

Palmar Wounds and Incisions

The superficial and deep palmar arterial arches are not palpable, but their surface markings are visible:

- The *superficial palmar arch* is at the level of the distal border of the fully extended thumb
- The *deep palmar arch* lies approximately 1 cm proximal to the superficial palmar arch.

The location of these arches should be borne in mind in wounds of the palm and when palmar incisions are made. Furthermore, it should be borne in mind that the superficial palmar arch is at the same level as the distal extremity of the common flexor sheath. Incisions or wounds along the medial surface of the thenar eminence may injure the recurrent branch of the median nerve to the thenar muscles. ⊕

Joints of the Upper Limb

The *pectoral girdle* involves the SC, AC, and glenohumeral joints (Fig. 6.59). Generally, these joints move at the same time. Functional defects in any of the joints impair movements of the pectoral girdle. Mobility of the scapula is essential for free movement of the upper limb. The clavicle forms a strut that holds the shoulder away from the thorax so it can move freely. The clavicle is the radius through which the shoulder moves at the SC joint. The 15 to 20° of movement at the AC joint permits movement of the glenoid cavity that is necessary for arm movements. When testing the range of motion of the pectoral girdle, both scapulothoracic (movement of the scapula on the thoracic wall) and glenohumeral movements must be considered. When elevating the arm, the movement occurs in a 2:1 ratio; for every 3° of elevation, approximately 2° occurs at the glenohumeral joint and 1° at the scapulothoracic movement. *The important movements of the pectoral girdle are scapular movements* (Fig. 6.60):

- Elevation and depression
- Protraction (lateral or forward movement of the scapula) and retraction (medial or backward movement of the scapula)
- Rotation of the scapula.

During scapular movements, the acromion is held away from the thorax by the clavicle—elevating or depressing the glenoid cavity of the scapula.

Sternoclavicular Joint

The SC joint is a saddle type of synovial joint but functions as a ball and socket joint (Fig. 6.59). The SC joint is divided into two compartments by an **articular disc**. The articular disc is firmly attached to the anterior and posterior **SC ligaments**—thickenings of the fibrous capsule of the joint—as well as the **interclavicular ligament**. The great strength of the SC joint is a consequence of these attachments. Thus, although the articular disc serves as a shock absorber of forces transmitted along the clavicle from the upper limb, dislocation of the clavicle is rare, whereas fracture of the clavicle is common. The SC joint—the only articulation between the upper limb and the axial skeleton—can be readily palpated because the sternal end of the clavicle lies superior to the manubrium of the sternum.

Articulation of the Sternoclavicular Joint

The sternal end of the clavicle articulates with the manubrium of the sternum and the 1st costal cartilage. The articular surfaces are covered with fibrocartilage.

Table 6.12. **Nerves of the Hand**

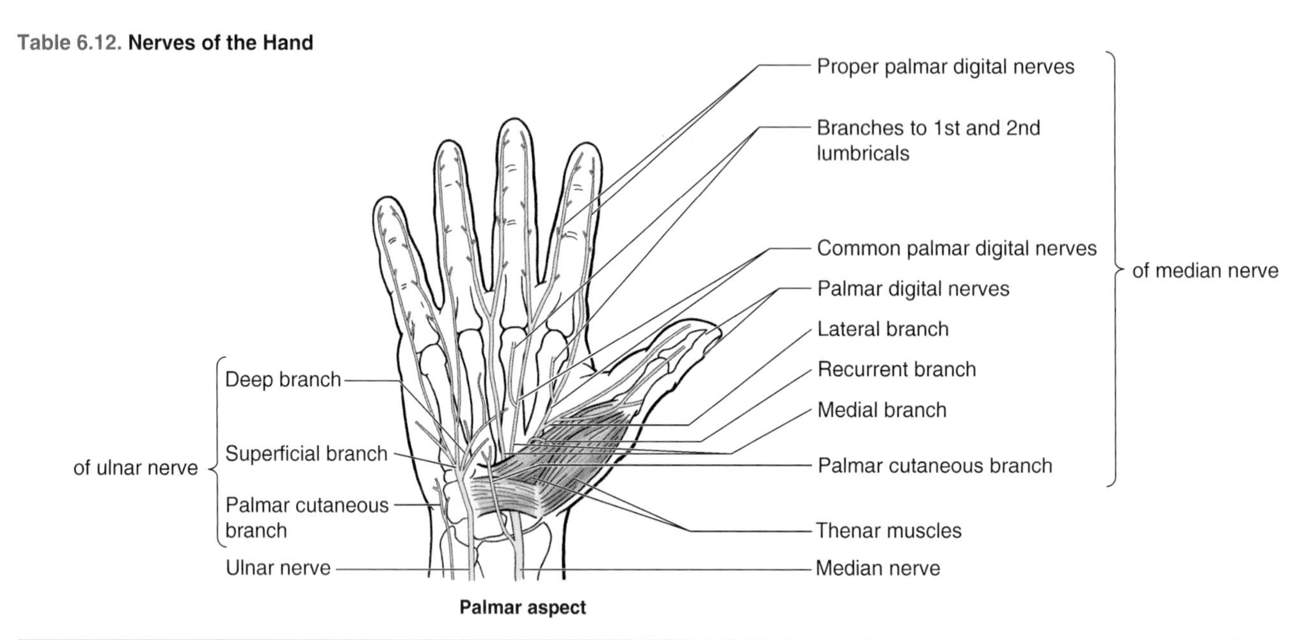

Nerve	Origin	Course	Distribution
Median nerve	Arises by two roots, one from the lateral cord of the brachial plexus (C6 and C7 fibers) and one from the medial cord (C8 and T1 fibers)	Becomes superficial proximal to the wrist and passes deep to the flexor retinaculum (transverse carpal ligament) as it passes through the carpal tunnel to the hand	Thenar muscles (except adductor pollicis and deep head of flexor pollicis brevis), lateral lumbricals (for digits 2 and 3); provides sensation to the skin of the palmar and distal dorsal aspects of the lateral (radial) 3 1/2 digits and adjacent palm
Recurrent (thenar) branch of median nerve	Arises from median nerves as soon as it has passed distal to the flexor retinaculum	Loops around distal border of flexor retinaculum and enters thenar muscles	Abductor pollicis brevis, opponens pollicis, and superficial head of flexor pollicis brevis
Lateral branch of median nerve	Arises as the lateral division of median nerve as it enters palm of hand	Runs laterally to palmar thumb and radial side of index finger	1st lumbrical and skin of palmar and distal dorsal aspects of thumb and radial half of index finger
Medial branch of median nerve	Arises as the medial division of median nerve as it enters palm of hand	Runs medially to adjacent sides of index, middle, and ring fingers	2nd lumbrical and skin of palmar and distal dorsal aspects of adjacent sides of index, middle, and ring fingers
Palmar cutaneous branch of median nerve	Arises from median nerve just proximal to flexor retinaculum	Passes between tendons of palmaris longus and flexor carpi radialis and runs superficial to flexor retinaculum	Skin of central palm

Articular Capsule of the Sternoclavicular Joint

The *fibrous part of the articular capsule* surrounds the SC joint, including the epiphysis at the sternal end of the clavicle, and is attached to the margins of the articular surfaces, including the periphery of the articular disc (Fig. 6.59). A *synovial membrane* lines the fibrous part of the articular capsule and both surfaces of the articular disc.

Ligaments of the Sternoclavicular Joint

Anterior and posterior **SC ligaments** reinforce the capsule anteriorly and posteriorly. The **interclavicular ligament** strengthens the capsule superiorly (Fig. 6.59). It extends from the sternal end of one clavicle and passes to the sternal end of the other clavicle. In between, it is also attached to the superior border of the manubrium of the sternum. The **costoclavicular ligament** anchors the inferior surface of the sternal end of the clavicle to the 1st rib and its costal cartilage, limiting elevation of the pectoral girdle.

Movements of the Sternoclavicular Joint

Although the SC joint is extremely strong, it is significantly mobile to allow movements of the pectoral girdle and upper limb (Figs. 6.61 and 6.62). During full elevation of the limb, the clavicle is raised to approximately a 60° angle. *The SC joint moves in several directions*: anteriorly, posteriorly, and inferiorly; up to 25 to 30° along its long axis.

Blood Supply of the Sternoclavicular Joint

The SC joint is supplied by internal thoracic and suprascapular arteries (Table 6.3).

Nerve Supply of the Sternoclavicular Joint

Branches of the medial supraclavicular nerve and the nerve to the subclavius supply the SC joint (Table 6.4).

Table 6.12. (*Continued*) **Nerves of the Hand**

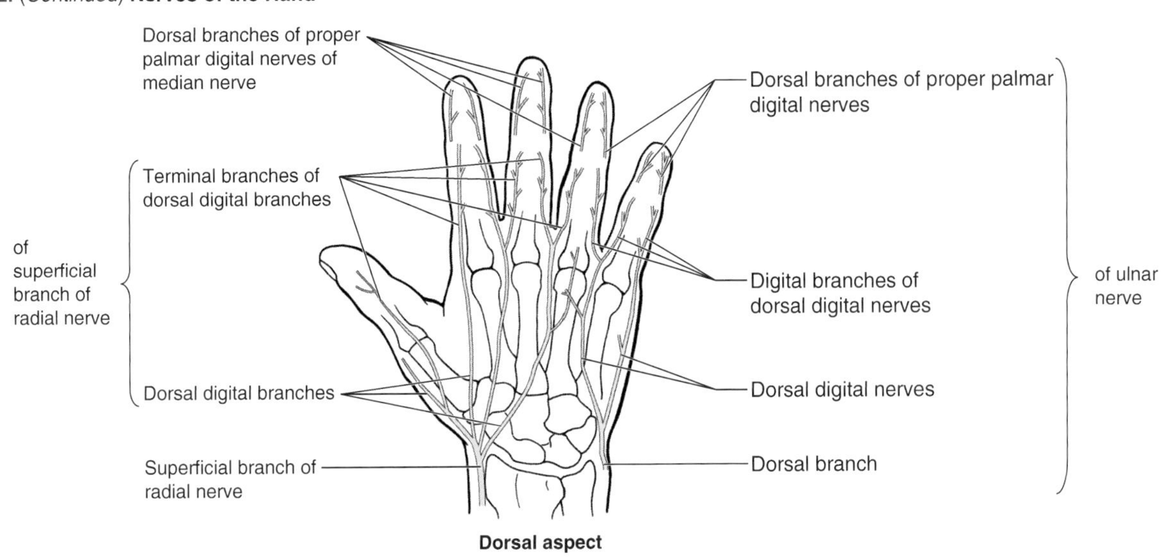

Dorsal aspect

Nerve	Origin	Course	Distribution
Ulnar nerve	Terminal branch of medial cord of brachial plexus (C8 and T1 fibers; often also receives C7 fibers)	Becomes superficial in distal forearm, passing superficial to the flexor retinaculum (transverse carpal ligament) to enter the hand	The majority of intrinsic muscles of hand (hypothenar, interosseous, adductor pollicis, and deep head of flexor pollicis brevis, plus the medial lumbricals [for digits 4 and 5]); provides sensation to the palmar and distal dorsal aspects of medial (ulnar) 1 1/2 digits and adjacent palm
Palmar cutaneous branch of ulnar nerve	Arises from ulnar nerve near middle of forearm	Descends on ulnar artery and perforates deep fascia in the distal third of forearm	Skin at base of medial palm, overlying the medial carpals
Dorsal branch of ulnar nerve	Arises from ulnar nerve about 5 cm proximal to flexor retinaculum	Passes distally deep to flexor carpi ulnaris, then dorsally to perforate deep fascia and course along medial side of dorsum of hand, dividing into 2 to 3 dorsal digital nerves	Skin of medial aspect of dorsum of hand and proximal portions of little and medial half of ring finger (occasionally also adjacent sides of proximal portions of ring and middle fingers)
Superficial branch of ulnar nerve	Arise from ulnar nerve at wrist as they pass between pisiform and hamate bones	Passes palmaris brevis and divides into two common palmar digital nerves	Palmaris brevis and sensation to skin of the palmar and distal dorsal aspects of digit 5 and of the medial (ulnar) side of digit 4 and proximal portion of palm
Deep branch of ulnar nerve		Passes between muscles of hypothenar eminence to pass deeply across palm with deep palmar (arterial) arch	Hypothenar muscles (abductor, flexor, and opponens digiti minimi), lumbricals of digits 4 and 5, all interossei, adductor pollicis, and deep head of flexor pollicis brevis
Radial nerve Superficial branch	Arises from radial nerve in cubital fossa	Courses deep to brachioradialis, emerging from beneath it to pierce the deep fascia lateral to distal radius	Skin of the lateral (radial) half of dorsal aspect of the hand and thumb, the proximal portions of the dorsal aspects of digits 2 and 3, and of the lateral (radial) half of digit 4

Dislocation of the Sternoclavicular Joint

The rarity of dislocation of the SC joint attests to its strength. When a blow is received to the acromion of the scapula, or when a force is transmitted to the pectoral girdle during a fall on the outstretched hand, the force of the blow is usually transmitted along the long axis of the clavicle. The clavicle may break near the junction of its middle and lateral thirds, but it is uncommon for the SC joint to dislocate. Most dislocations of the SC joint in persons younger than 25 years result from fractures through the epiphyseal plate because the epiphysis at the sternal end of the clavicle does not close until 23 to 25 years (Halpern, 1994). ▶

Ankylosis of the Sternoclavicular Joint

Movement at the SC joint is critical to movement of the shoulder. When ankylosis (fixation) of the joint occurs, or is necessary, a section of the center of the clavicle is removed, creating a pseudo- or "flail" joint to permit scapular movement. ⊙

Acromioclavicular Joint

The AC joint is a plane type of synovial joint (Fig. 6.59). It is located 2 to 3 cm from the point of the shoulder formed by the lateral part of the acromion (Fig. 6.62, *A* and *B*).

Articulation of the Acromioclavicular Joint

The acromial end of the clavicle articulates with the acromion of the scapula. The articular surfaces, covered with fibrocartilage, are separated by an incomplete wedge-shaped articular disc.

Articular Capsule of the Acromioclavicular Joint

The sleevelike, relatively loose *fibrous capsule* is attached to the margins of the articular surfaces (Fig. 6.62*A*). A *synovial membrane* lines the fibrous capsule. Although relatively weak, the capsule is strengthened superiorly by fibers of the trapezius.

Ligaments of the Acromioclavicular Joint

The **AC ligament**, a fibrous band extending from the acromion to the clavicle (Figs. 6.59 and 6.63), strengthens the AC joint superiorly; however, the integrity of the joint is maintained by extrinsic ligaments, distant from the joint itself. The **coracoclavicular** ligament is a strong pair of bands that unites the coracoid process of the scapula to the clavicle, anchoring the clavicle to the coracoid process. The coracoclavicular ligament consists of two ligaments, the conoid and trapezoid ligaments, which are often separated by a bursa. The vertical **conoid ligament** is an inverted triangle (cone), which has its apex inferiorly where it is attached to the root of the *coracoid process* in front of the scapular notch. Its wide attachment (base of the triangle or cone) is to the *conoid tubercle* on the inferior surface of the clavicle. The nearly horizontal **trapezoid ligament** is attached to the superior surface of the coracoid process and extends laterally to the trapezoid line on the inferior surface of the clavicle. In ad-

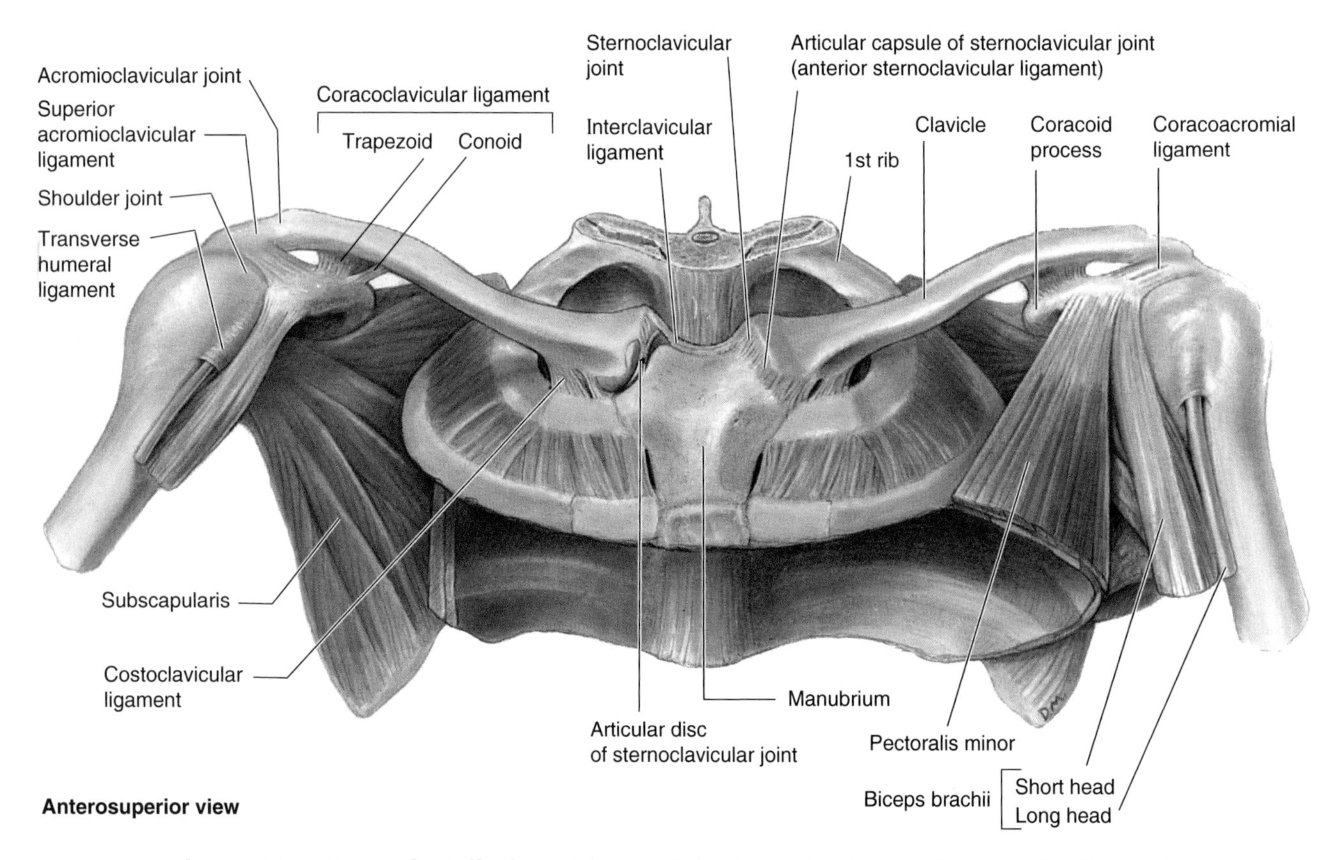

Figure 6.59. Pectoral girdle (shoulder girdle). The pectoral girdle, the bony ring (incomplete posteriorly), is formed by the manubrium of the sternum, the clavicle, and the scapulae. Observe the joints associated with these bones: sternoclavicular (SC), acromioclavicular (AC), and glenohumeral. The girdle serves for the attachment and support of the upper limbs.

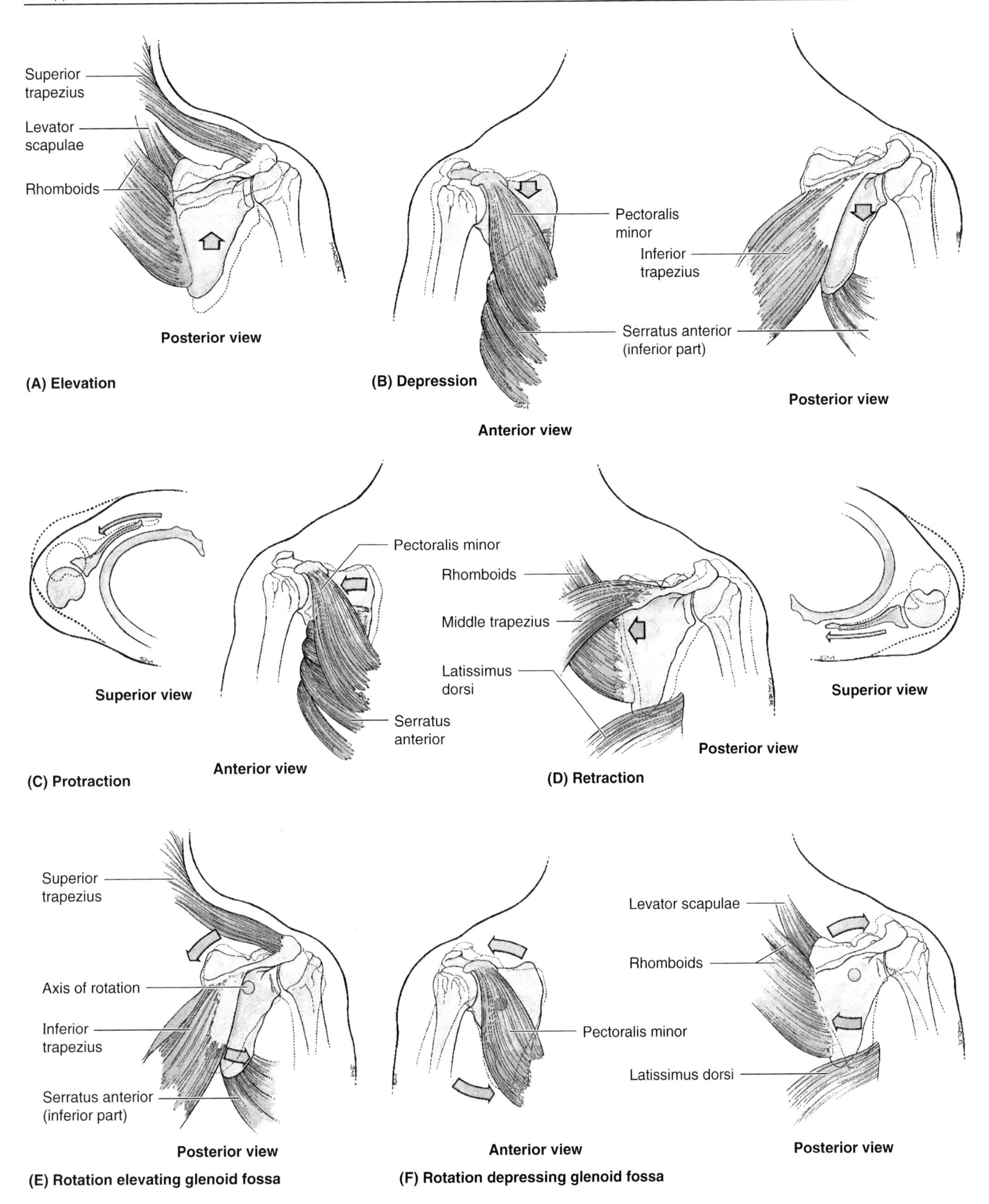

Figure 6.60. Scapular movements. The scapula moves on the thoracic (chest) wall of the conceptual "scapulothoracic joint". **A.** Elevation. **B.** Depression. **C.** Protraction. **D.** Retraction. **E.** Rotation elevating the glenoid fossa. **F.** Rotation depressing the glenoid fossa. The *dotted outlines* represent the starting position of each of the movements.

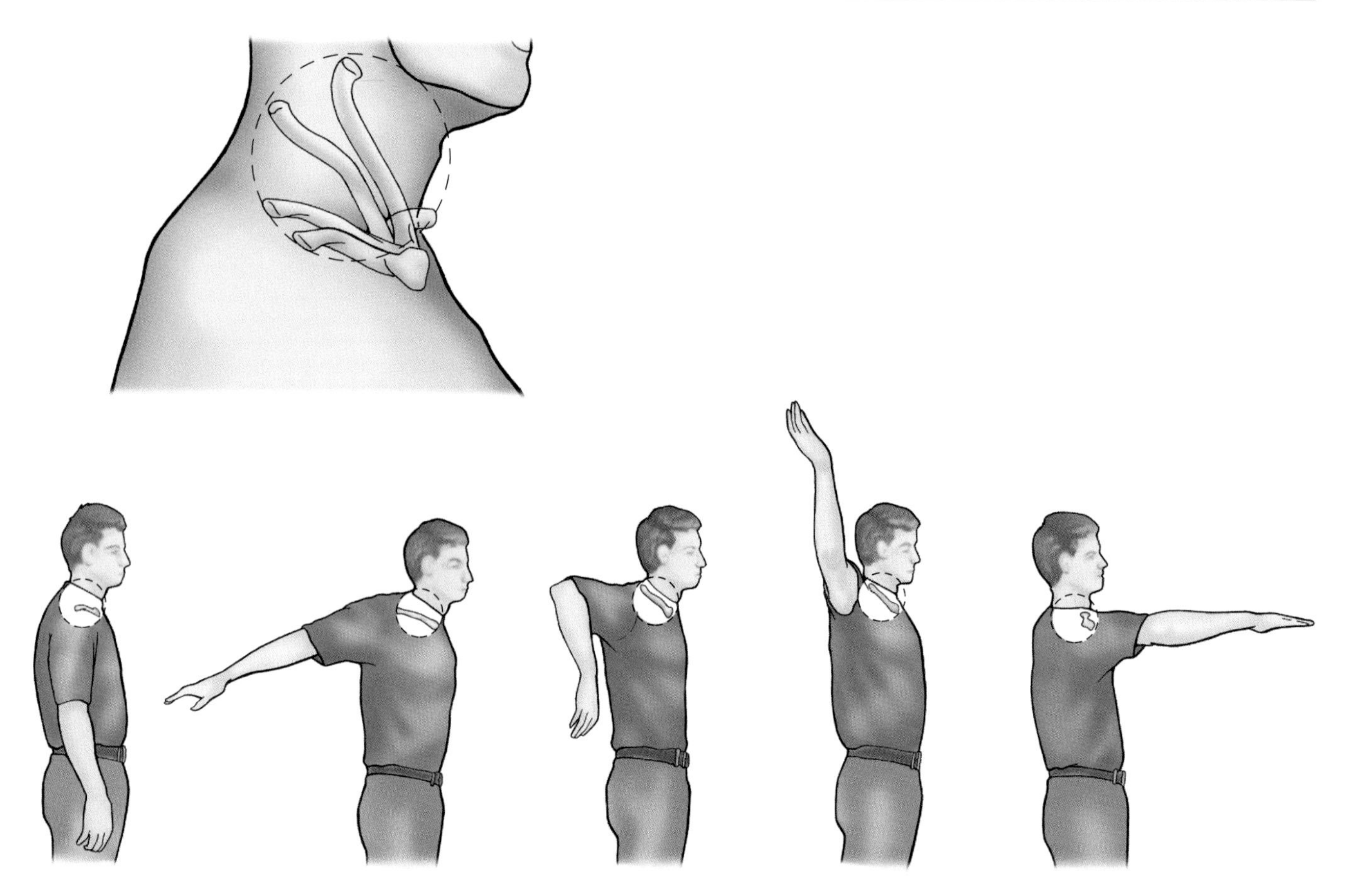

Figure 6.61. Movements of the upper limb at the joints of the pectoral girdle (shoulder girdle). Mobility of the clavicle at the sternoclavicular (SC) joint is essential for the freedom of movement of the upper limb. Observe that the clavicle moves clockwise as the limb is moved over the shoulder and inferiorly to the horizontal position.

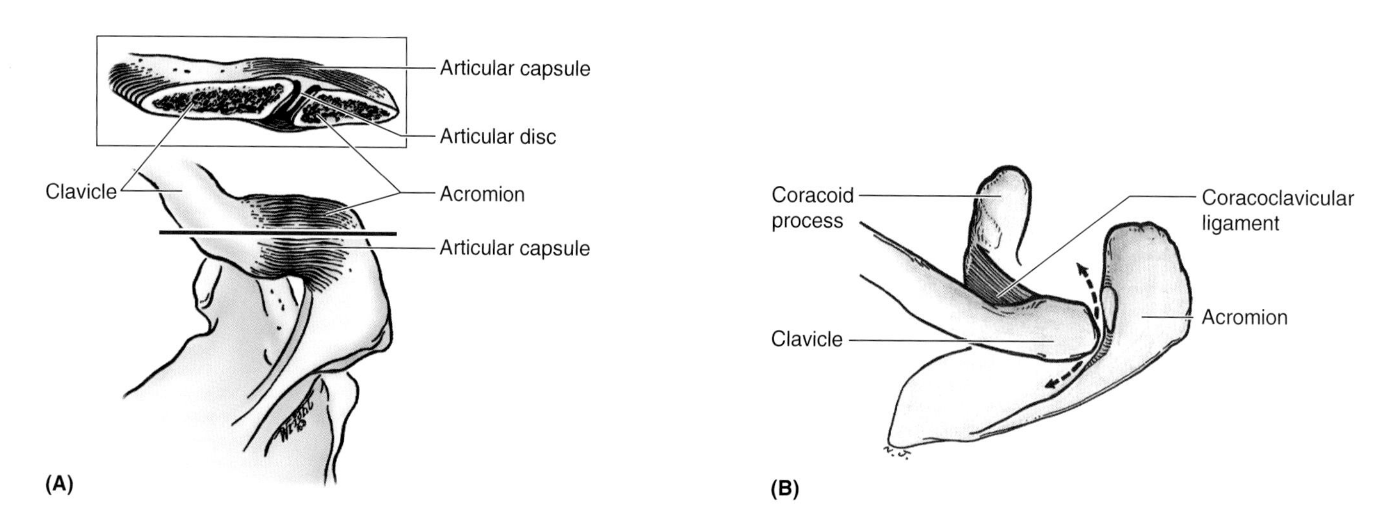

Figure 6.62. Acromioclavicular (AC) and sternoclavicular (SC) joints. A. Superior view of the right AC joint. *Inset*, coronal section of the AC joint showing its articular capsule and disc. **B.** Diagram demonstrating function of the coracoclavicular ligament. As long as the coracoclavicular ligament is intact, the acromion cannot be driven inferior to the clavicle. The ligament, however, does permit protraction and retraction of the acromion.

dition to augmenting the AC joint, the coracoclavicular ligament provides the means by which the scapula and free limb are (passively) suspended from the clavicular strut.

Movements of the Acromioclavicular Joint

The acromion of the scapula rotates on the acromial end of the clavicle. These movements are associated with motion at the conceptual scapulothoracic joint (Figs. 6.61 and 6.62). *No muscles connect the articulating bones to move the AC joint*; the thoracoappendicular muscles that attach to and move the scapula cause the acromion to move on the clavicle.

Blood Supply of the Acromioclavicular Joint

The AC joint is supplied by the suprascapular and thoracoacromial arteries (Table 6.3).

Nerve Supply of the Acromioclavicular Joint

Supraclavicular, lateral pectoral, and axillary nerves supply the AC joint (Table 6.4).

Dislocation of the Acromioclavicular Joint

Although its extrinsic (coracoclavicular) ligament is strong, the AC joint itself is weak and easily injured by a direct blow. In contact sports such as football, soccer, and hockey—or the martial arts—it is not uncommon for *dislocation of the AC joint* to result from a hard fall on the shoulder with the impact taken by the acromion or from a fall on the outstretched upper limb. Dislocation of the AC joint can also occur when a hockey player is driven into the boards or when a person receives a severe blow to the superolateral part of the back. The AC injury, often called a "shoulder separation," is severe when both the AC and coracoclavicular ligaments are torn. When the coracoclavicular ligament tears, the shoulder separates (falls away) from the clavicle because of the weight of the upper limb. *Rupture of the coracoclavicular ligament* allows the fibrous capsule of the joint to also be torn so that the acromion can pass inferior to the acromial end of the clavicle. Dislocation of the AC joint makes the acromion more prominent, and the clavicle may move superior to this process.

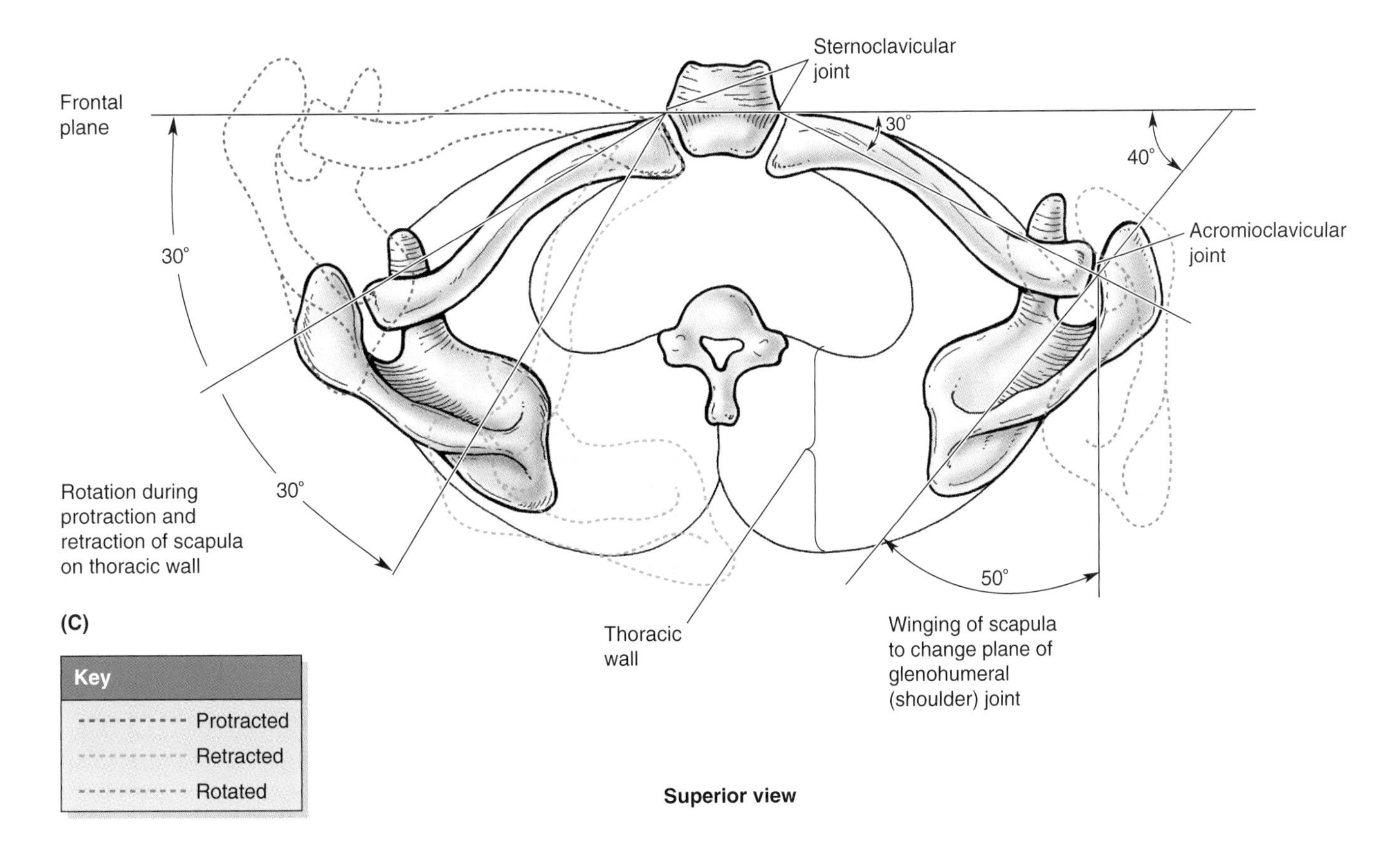

Figure 6.62. (*Continued*) **C.** Clavicular movements at the SC and AC joints during rotation, protraction, and retraction of the scapula on the thoracic wall and winging of the scapula.

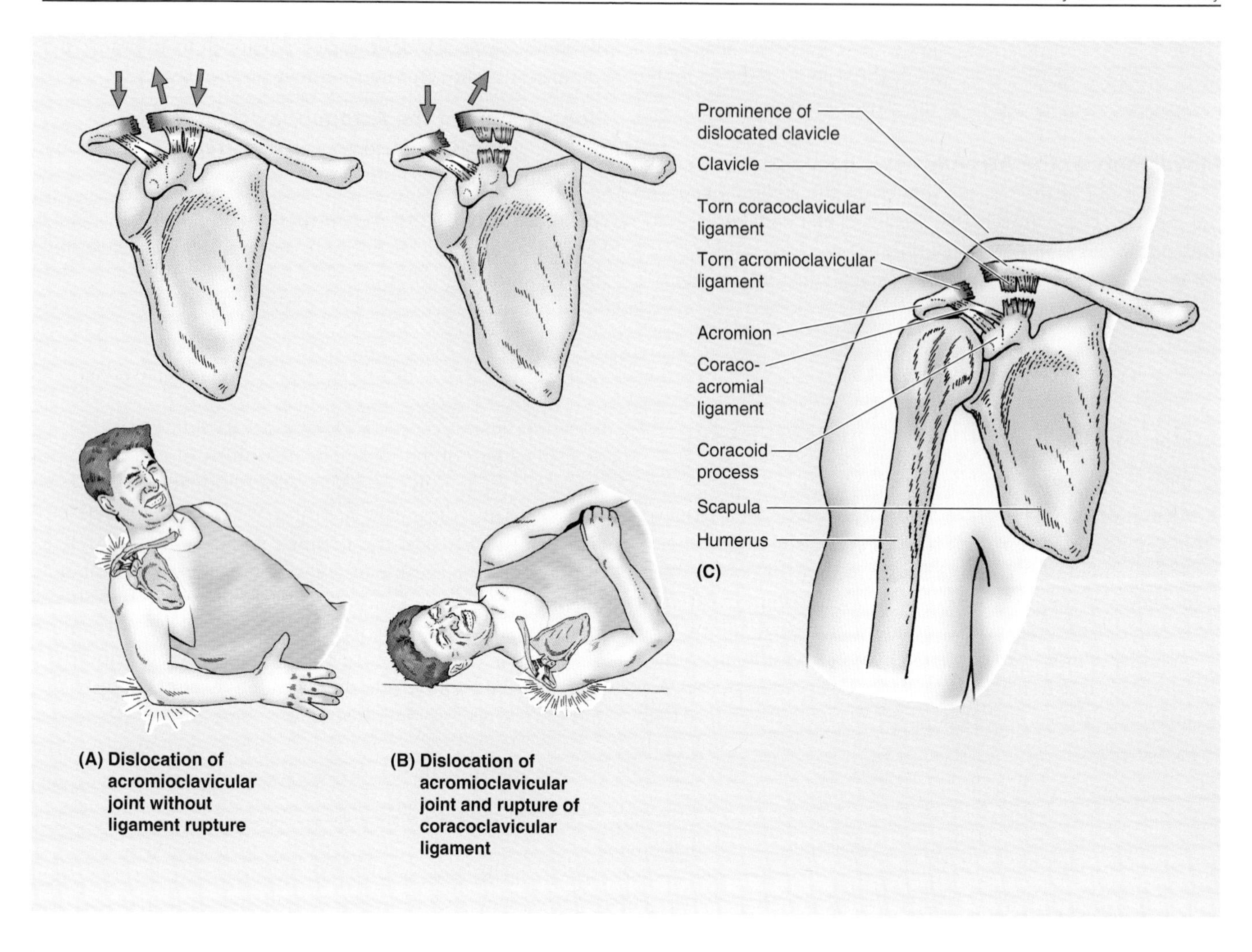

(A) Dislocation of acromioclavicular joint without ligament rupture

(B) Dislocation of acromioclavicular joint and rupture of coracoclavicular ligament

Glenohumeral (Shoulder) Joint

The glenohumeral joint is a ball-and-socket type of synovial joint that permits a wide range of movement; however the mobility makes the joint relatively unstable.

Articulation of the Glenohumeral Joint

The large, round **humeral head** articulates with the relatively shallow **glenoid cavity** of the scapula (Figs. 6.64 and 6.65), which is deepened slightly but effectively by the ringlike, fibrocartilaginous **glenoid labrum** (L. lip). Both articular surfaces are covered with hyaline cartilage. The glenoid cavity accepts little more than a third of the humeral head, which is held in the cavity by the *tonus of the rotator cuff (SITS) muscles*: Supraspinatus, Infraspinatus, Teres minor, and Subscapularis (Table 6.2).

Articular Capsule of the Glenohumeral Joint

The loose **fibrous capsule** surrounds the glenohumeral joint and is attached medially to the margin of the glenoid cavity and laterally to the anatomical neck of the humerus (Fig. 6.65, *A* and *B*). Superiorly, this part of the articular capsule encroaches on the root of the coracoid process so that the fibrous capsule encloses the proximal attachment of the long head of the biceps brachii—supraglenoid tubercle of scapula—within the joint. *The articular capsule of the glenohumeral joint has two apertures*:

- The opening between the tubercles of the humerus is for passage of the tendon of the long head of the biceps brachii (Fig. 6.63)
- The other opening situated anteriorly, inferior to the coracoid process, allows communication between the subscapular bursa and the synovial cavity of the joint.

The inferior part of the articular capsule—the only part not reinforced by the rotator cuff muscles—is its weakest area. Here the capsule is particularly lax and lies in folds when the arm is adducted; however, it becomes taut when the arm is abducted. The **synovial membrane** lines the fibrous capsule and reflects from it onto the glenoid labrum and neck of the humerus, as far as the articular margin of the head (Figs. 6.64*B* and 6.65). The synovial membrane also forms a tubular sheath for the tendon of the long head of the biceps brachii, where it passes into the joint cavity and lies in the intertubercular groove, extending as far as the surgical neck of the humerus (Fig. 6.63).

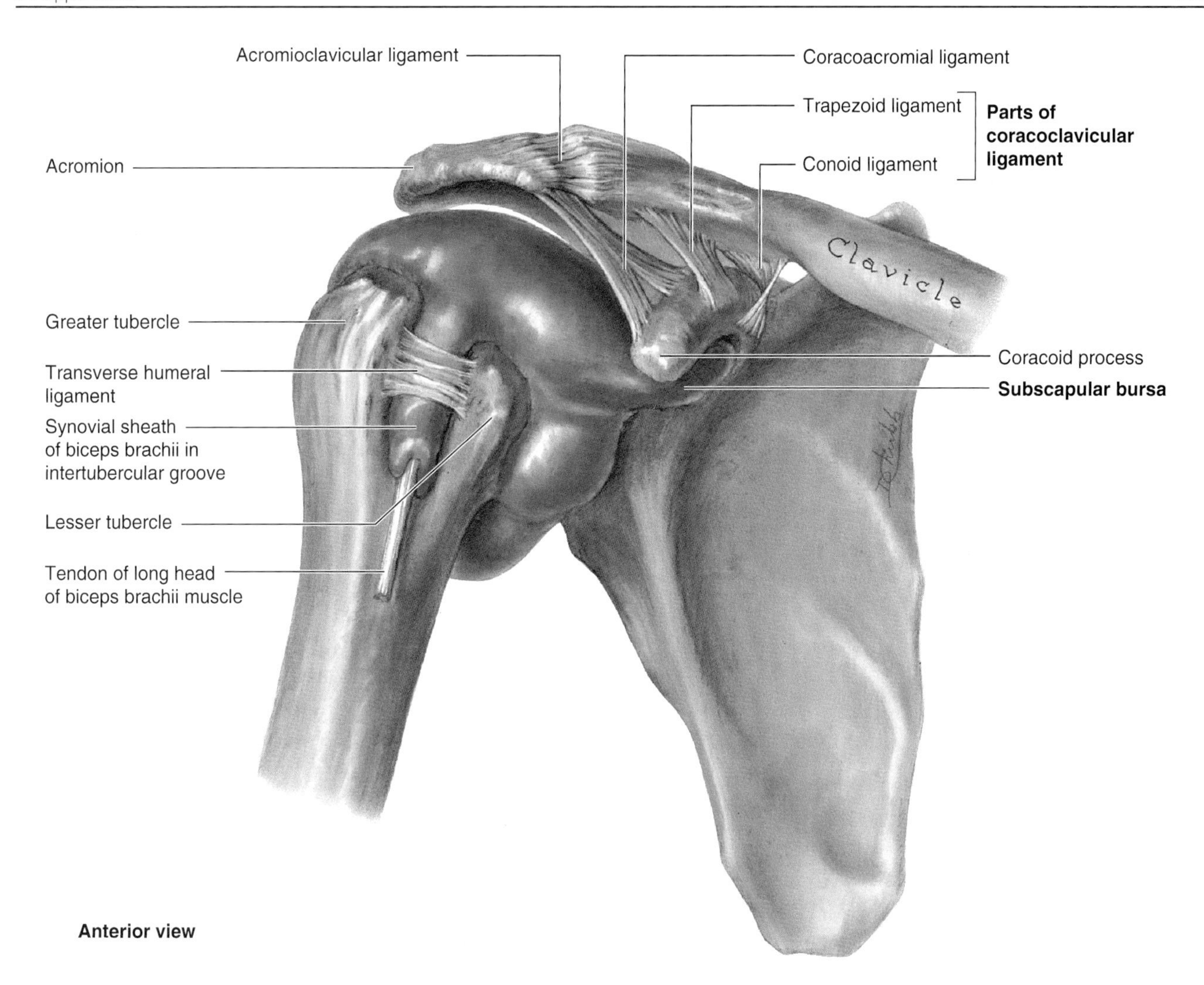

Figure 6.63. Glenohumeral joint. Anterior view of the synovial capsule of the right glenohumeral joint and the ligaments at the lateral end of the clavicle. Note that the synovial capsule has two extensions: (1) where it forms a synovial sheath for the tendon of the long head of the biceps brachii muscle in the intertubercular groove of the humerus and (2) inferior to the coracoid process where it forms the subscapular bursa between the subscapularis tendon and the margin of the glenoid cavity.

Ligaments of the Glenohumeral Joint

The **glenohumeral ligaments**, which strengthen the anterior aspect of the articular capsule of the joint, and the **coracohumeral ligament,** which strengthens the capsule superiorly, are intrinsic ligaments—part of the fibrous capsule (Fig. 6.65, *B* and *C*). The capsule is also thickened by the **transverse humeral ligament**; it strengthens the capsule and bridges the gap between the greater and lesser tubercles of the humerus.

The **glenohumeral ligaments** are three fibrous bands—evident only on the internal aspect of the capsule—that reinforce the anterior part of the articular capsule; they radiate laterally and inferiorly from the glenoid labrum at the supraglenoid tubercle of the scapula and blend distally with the fibrous capsule as it attaches to the anatomical neck of the humerus.

The **coracohumeral ligament** is a strong, broad band that passes from the base of the coracoid process to the anterior aspect of the greater tubercle of the humerus.

The **transverse humeral ligament** is a broad fibrous band that runs more or less obliquely from the greater to the lesser tubercle of the humerus, bridging over the intertubercular groove. The ligament converts the groove into a canal that holds the synovial sheath and tendon of the biceps brachii in place during movements of the glenohumeral joint.

The **coracoacromial arch** is an extrinsic, *protective structure* formed by the smooth inferior aspect of the *acromion* and the *coracoid process* of the scapula, with the **coracoacromial ligament** spanning between them (Fig. 6.65*B*). This osseoligamentous structure forms a protective arch that overlies the head of the humerus, preventing its superior displacement

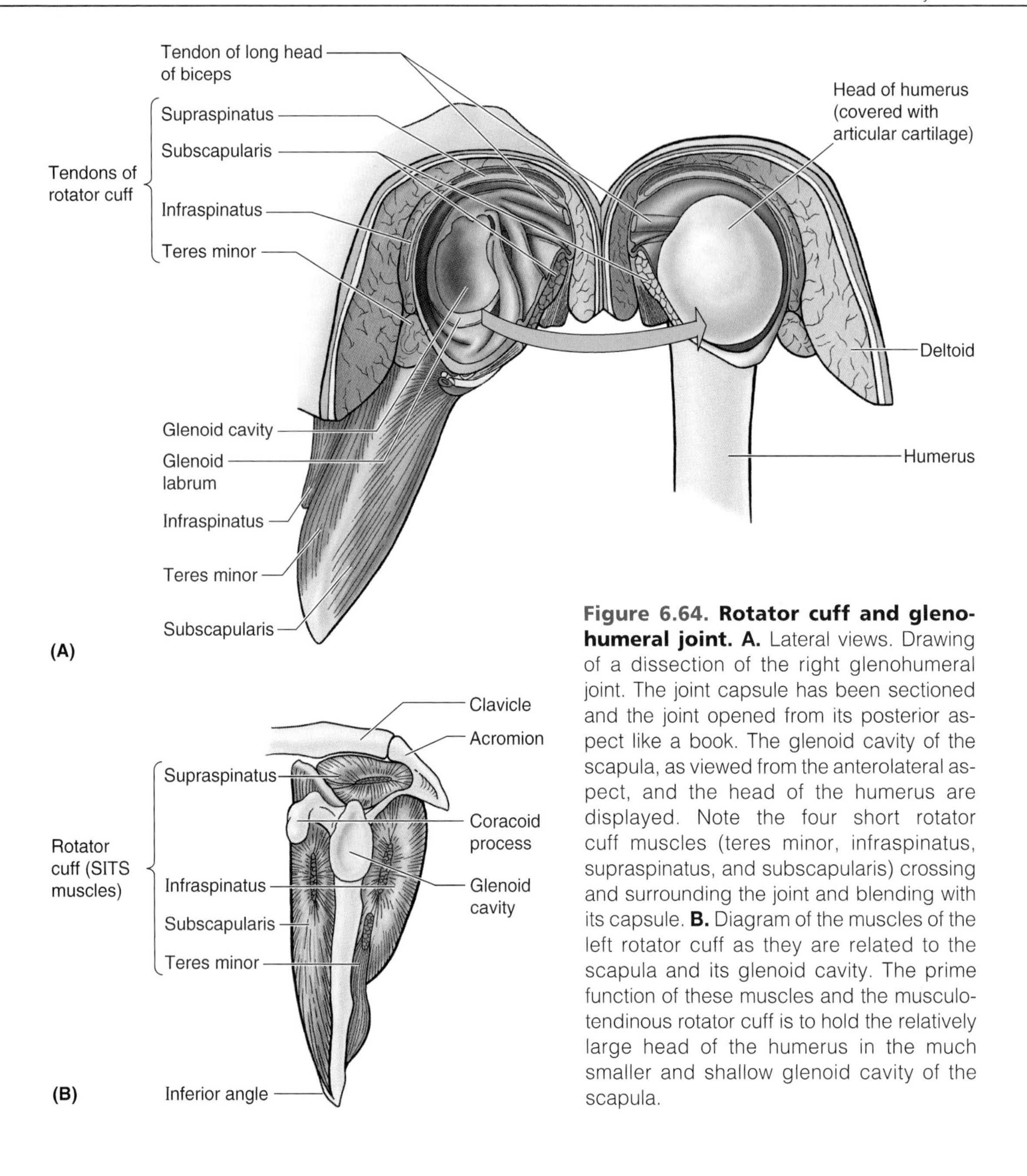

Figure 6.64. Rotator cuff and glenohumeral joint. A. Lateral views. Drawing of a dissection of the right glenohumeral joint. The joint capsule has been sectioned and the joint opened from its posterior aspect like a book. The glenoid cavity of the scapula, as viewed from the anterolateral aspect, and the head of the humerus are displayed. Note the four short rotator cuff muscles (teres minor, infraspinatus, supraspinatus, and subscapularis) crossing and surrounding the joint and blending with its capsule. **B.** Diagram of the muscles of the left rotator cuff as they are related to the scapula and its glenoid cavity. The prime function of these muscles and the musculotendinous rotator cuff is to hold the relatively large head of the humerus in the much smaller and shallow glenoid cavity of the scapula.

from the glenoid cavity of the scapula. *The coracoacromial arch is so strong that a forceful superior thrust of the humerus will not fracture it; the humeral body or clavicle fractures first.* Transmitting force superiorly along the humerus (e.g., when standing at a desk and partly supporting the body with the outstretched limbs), the humeral head presses against the coracoacromial arch. The supraspinatus muscle passes under this arch and lies deep to the deltoid muscle as its tendon blends with the articular capsule of the glenohumeral joint as part of the rotator cuff (Fig. 6.64*A*). Movement of the **supraspinatus tendon**, passing to the greater tubercle of the humerus, is facilitated as it passes under the arch by the **subacromial bursa**, which lies between the arch and the tendon and tubercle.

Movements of the Glenohumeral Joint

The glenohumeral joint has more freedom of movement than any other joint in the body. This freedom results from the laxity of its articular capsule and the large size of the humeral head compared with the small size of the glenoid cavity. *The glenohumeral joint allows movements around three axes and permits:*

- Flexion-extension
- Abduction-adduction
- Rotation (medial and lateral) of the humerus
- Circumduction.

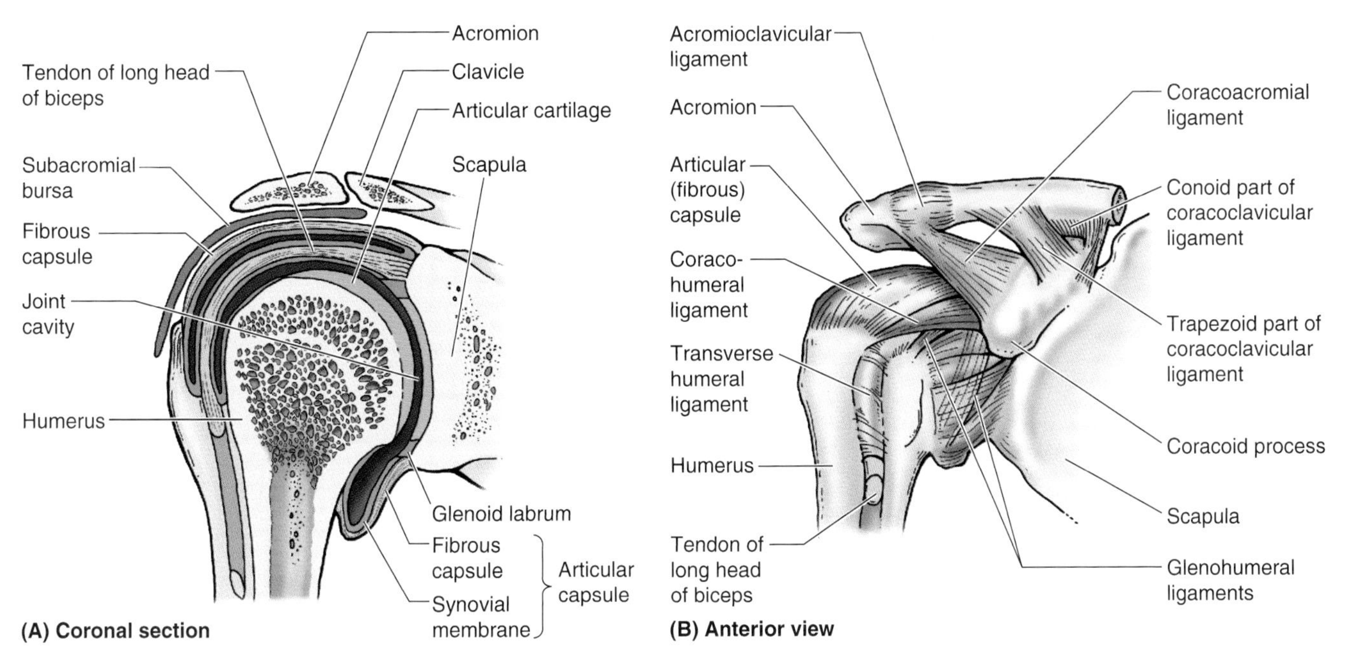

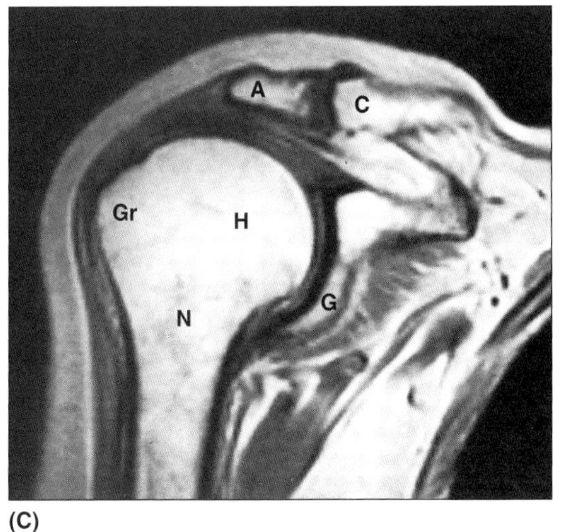

Figure 6.65. Glenohumeral joint. A. Coronal section of the shoulder region illustrating the articulating bones, the articular capsule and cartilage, and the subacromial bursa. **B.** Drawing of an anterior view of a dissection of the acromioclavicular (AC), coracohumeral, and glenohumeral ligaments. The glenohumeral ligaments strengthen the anterior aspect of the capsule of the glenohumeral joint, and the coracohumeral ligament strengthens the capsule superiorly. **C.** Coronal MRI of the right glenohumeral and AC joints. *A*, acromion; *C*, clavicle; *Gr*, greater tubercle of the humerus; *H*, head of humerus; *G*, glenoid cavity; *N*, surgical neck of humerus. (Courtesy of Dr. W. Kucharczyk, Chair of Medical Imaging and Clinical Director of Tri-Hospital Resonance Centre, Toronto, Ontario, Canada)

Circumduction is an orderly sequence of flexion, abduction, extension, and adduction—or the reverse.

Muscles Moving the Glenohumeral Joint

The muscles producing movements of the glenohumeral joint are the *thoracoappendicular muscles*, which may act indirectly on the joint (i.e., act on the pectoral girdle), and the *scapulohumeral muscles*, which act directly on the glenohumeral joint (Fig. 6.60, Table 6.13):

- *Chief flexors of the glenohumeral joint*—pectoralis major (clavicular part) and deltoid (anterior fibers), assisted by the coracobrachialis and biceps brachii
- *Chief extensor of the glenohumeral joint*—latissimus dorsi
- *Chief abductor of the glenohumeral joint*—deltoid, especially the central fibers (following initiation of the movement by the supraspinatus)
- *Chief adductors of the glenohumeral joint*—pectoralis major and latissimus dorsi
- *Chief medial rotator of the glenohumeral joint*—subscapularis
- *Chief lateral rotator of the glenohumeral joint*—infraspinatus.

Other muscles serve the glenohumeral joint as shunt muscles, acting to resist dislocation without producing movement at the joint. The tonus of the rotator cuff muscles holds the large head of the humerus in the relatively shallow glenoid cavity. The coracobrachialis, short head of the biceps, and long head of the triceps all assist the deltoid (acting as a whole) in resisting downward dislocation of the joint (e.g., when carrying heavy suitcases or buckets of water).

Table 6.13. **Movements of the Glenohumeral (Shoulder) Joint**

Movement (Function)	Prime Mover(s) (from pendent position)	Synergists	Note
Flexion	Pectoralis major (clavicular head) Deltoid (anterior part)	Coracobrachialis (assisted by biceps)	From fully extended position to its own (coronal) plane, the sternocostal head of pectoralis major is a major force
Extension	Deltoid (posterior part)	Teres major	The latissimus dorsi, sternocostal head of pectoralis major, and long head of the triceps act from the fully flexed position to their own (coronal) planes
Abduction	Deltoid (as a whole, but especially central part)	Supraspinatus	Supraspinatus is particularly important in initiating the movement; also, upward rotation of the scapula occurs throughout movement, making a significant contribution
Adduction	Pectoralis major and latissimus dorsi	Subscapularis, infraspinatus, and teres minor	In upright position and in the absence of resistance, gravity is the prime mover
Medial rotation	Subscapularis	Pectoralis major, deltoid (anterior fibers), latissimus dorsi	With arm elevated, "synergists" become more important than prime movers
Lateral rotation	Infraspinatus	Teres minor, deltoid (posterior fibers)	
Tensors of articular capsule (to hold head of humerus against glenoid cavity)	Supscapularis and infraspinatus (simultaneously)	Supraspinatus and teres minor	The "rotator cuff" (SITS) muscles acting together; when "resting," their tonus adequately maintains integrity of the joint
Resisting downward dislocation (shunt muscles)	Deltoid (as a whole)	Long head of triceps Coracobrachialis and short head of biceps	Used especially when carrying heavy objects (suitcases, buckets)

Blood Supply of the Glenohumeral Joint

The glenohumeral joint is supplied by the anterior and posterior **circumflex humeral arteries** and branches of the **suprascapular artery**.

Innervation of the Glenohumeral Joint

The suprascapular, axillary, and lateral pectoral nerves supply the glenohumeral joint.

Bursae Around the Glenohumeral Joint

Several bursae containing capillary films of synovial fluid are in the vicinity of the glenohumeral joint. Bursae are located where tendons rub against bone, ligaments, or other tendons and where skin moves over a bony prominence. The bursae around the glenohumeral joint are of special clinical importance. Some of them communicate with the joint cavity (e.g., the subscapular bursa); hence, opening a bursa may mean entering the cavity of the shoulder joint.

Subscapular Bursa (Fig. 6.63). This bursa is located between the tendon of the subscapularis and the neck of the scapula. The bursa protects the tendon where it passes inferior to the root of the coracoid process and over the neck of the scapula. It usually communicates with the cavity of the shoulder joint through an opening in the fibrous capsule; thus, it is really an extension of the glenohumeral joint cavity.

Subacromial Bursa (Fig. 6.65*A*). Sometimes referred to as the *subdeltoid bursa*, this large bursa lies between the deltoid, the supraspinatus tendon, and the fibrous capsule of the glenohumeral joint. Its size varies, but it does not normally communicate with the cavity of the shoulder joint. The subacromial bursa is located inferior to the acromion and coracoacromial ligament, between them and the supraspinatus. This bursa facilitates movement of the supraspinatus tendon under the coracoacromial arch and of the deltoid over the fibrous capsule of the shoulder joint and the greater tubercle of the humerus.

Calcific Supraspinatus Tendinitis

Inflammation and calcification of the subacromial bursa result in pain, tenderness, and limitation of movement of the glenohumeral joint. This condition is also known as *calcific scapulohumeral bursitis*. Deposition of calcium in the supraspinatus tendon is common. This causes increased local pressure that often causes excruciating pain during abduction of the arm; the pain may radiate ▶

▶ as far as the hand. The calcium deposit may irritate the overlying subacromial bursa, producing an inflammatory reaction known as *subacromial bursitis.* As long as the glenohumeral joint is adducted, no pain usually results because in this position the painful lesion is away from the inferior surface of the acromion. In most patients, the pain occurs during 50 to 130° of abduction because during this arc the supraspinatus tendon is in intimate contact with the inferior surface of the acromion. This condition is sometimes referred to as the *painful arc syndrome.* The pain usually develops in males 50 years and older after unusual or excessive use of the shoulder (e.g., during a tennis game).

Rotator Cuff Injuries

The musculotendinous rotator cuff is commonly injured during repetitive use of the upper limb above the horizontal (e.g., during throwing and racquet sports, swimming, and weight lifting). *Recurrent inflammation of the rotator cuff,* especially the relatively avascular area of the supraspinatus tendon, is a common cause of shoulder pain and tears of the musculotendinous rotator cuff. Repetitive use of the rotator cuff muscles (e.g., by baseball pitchers) may allow the humeral head and rotator cuff to impinge on the coracoacromial arch, producing irritation of the arch and inflammation of the rotator cuff. Tendon degeneration and rupture of the cuff may occur.

When an older person strains to lift something—a window sash that is stuck, for example—a previously degenerated musculotendinous rotator cuff may rupture. A fall on the shoulder may also tear a previously degenerated rotator cuff. Often the intracapsular part of the tendon of the long head of the biceps brachii becomes frayed—even worn away—leaving it adherent to the intertubercular groove. As a result, shoulder stiffness occurs. Because they fuse together, the integrity of the fibrous capsule of the glenohumeral joint is usually compromised (tears) when the rotator cuff is injured. As a result, the joint cavity communicates with the subacromial bursa. Because the supraspinatus muscle is no longer functional with a complete tear of the rotator cuff, the person cannot initiate abduction of the upper limb. If the arm is passively abducted 15° or more, the person cannot maintain or even continue the abduction using the deltoid.

Dislocation of the Glenohumeral Joint

Because of its freedom of movement and instability, the glenohumeral joint is commonly dislocated by direct or indirect injury. Because of the presence of the coracoacromial arch and the support of the rotator cuff, most dislocations of the humeral head occur in the downward (inferior) direction; however, they are described clinically as anterior or (more rarely) posterior dislocations, indicating whether the humeral head had descended anterior or ▶

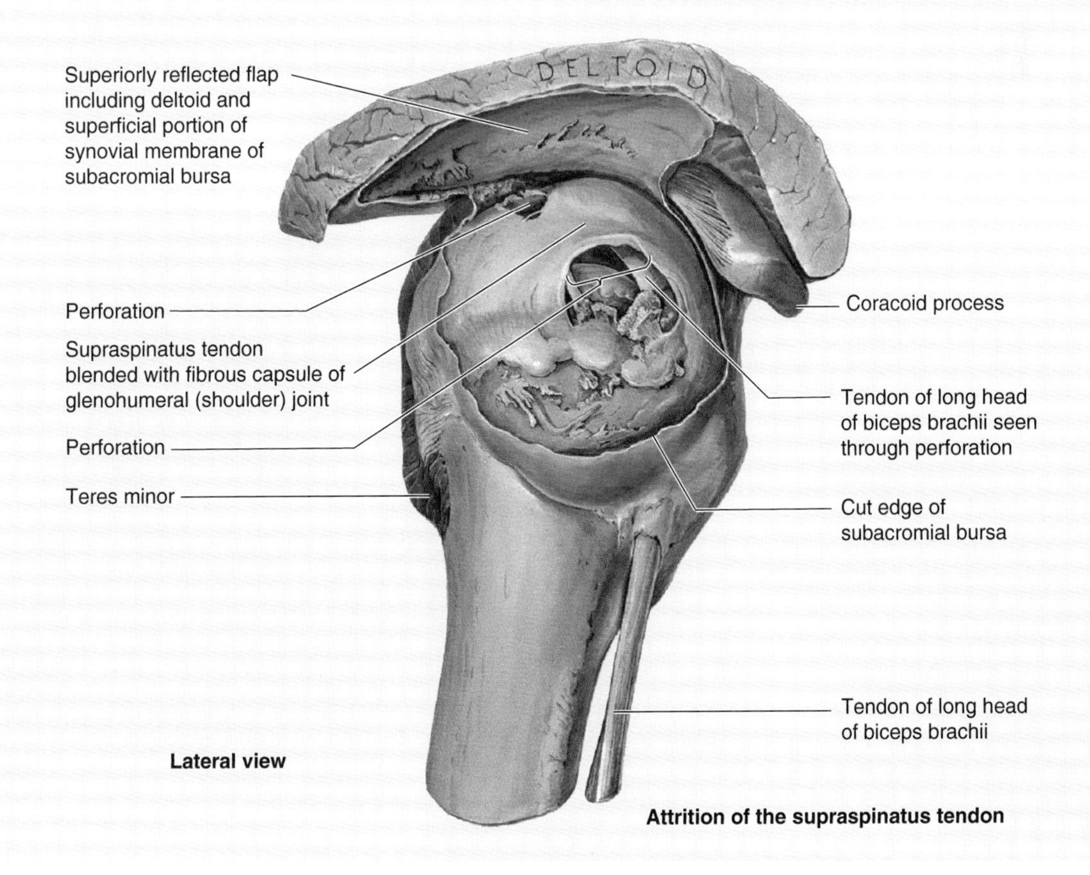

Attrition of the supraspinatus tendon

▶ posterior to the infraglenoid tubercle and the long head of the triceps and ends up lying in front of or behind the glenoid cavity.

Anterior dislocation of the glenohumeral joint occurs most often in young adults, particularly athletes. It is usually caused by excessive extension and lateral rotation of the humerus. The head of the humerus is driven inferoanteriorly, and the fibrous capsule and glenoid labrum may be stripped from the anterior aspect of the glenoid cavity in the process. A hard blow to the humerus when the glenohumeral joint is fully abducted tilts the head of the humerus inferiorly onto the inferior weak part of the capsule. This may tear the capsule and dislocate the shoulder so that the humeral head comes to lie inferior to the glenoid cavity and anterior to the infraglenoid tubercle. The strong flexor and abductor muscles of the glenohumeral joint usually subsequently pull the humeral head anterosuperiorly into a subcoracoid position. Unable to use the arm, the patient commonly supports it with the other hand.

Axillary Nerve Injury

The axillary nerve may be injured when the glenohumeral joint dislocates because of its close relation to the inferior part of the articular capsule of this joint. The subglenoid displacement of the head of the humerus into the quadrangular space damages the axillary nerve. *Axillary nerve injury is indicated by paralysis of the deltoid* and loss of sensation in a small area of skin covering the central part of the deltoid.

Glenoid Labrum Tears

Tearing of the fibrocartilaginous glenoid labrum commonly occurs in athletes who throw a baseball or football or in those who have shoulder instability and partial dislocation (subluxation) of the glenohumeral joint. The tear often results from sudden contraction of the biceps or forceful subluxation of the humeral head over the glenoid labrum. Usually, a tear occurs in the anterosuperior part ▶

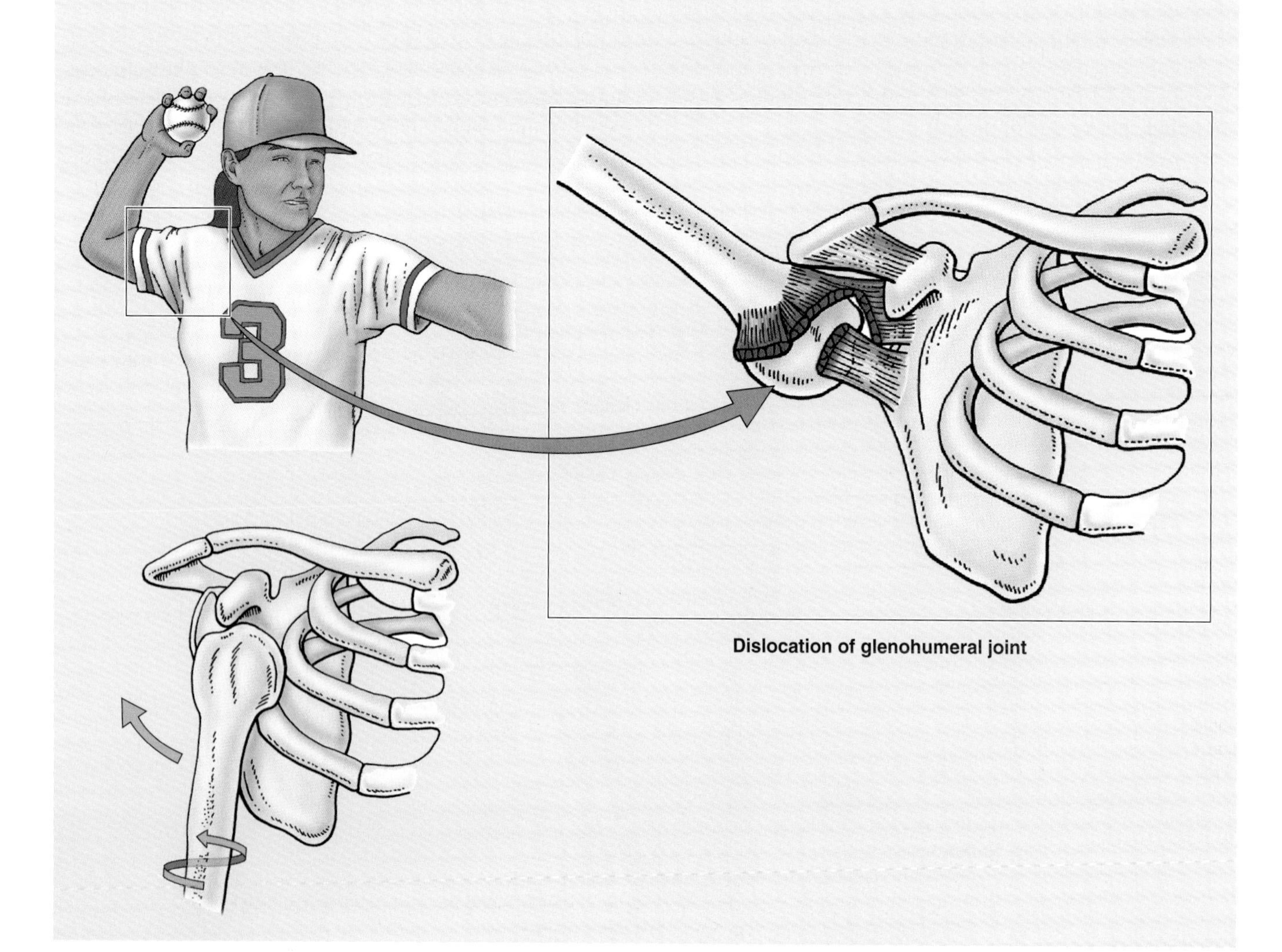

Dislocation of glenohumeral joint

▶ of the glenoid labrum (Halpern, 1994). The usual symptom is pain while throwing, especially during the acceleration phase, but a sense of popping or snapping may be felt in the glenohumeral joint during abduction and lateral rotation of the arm.

Adhesive Capsulitis of the Glenohumeral Joint

Adhesive fibrosis and scarring between the inflamed articular capsule of the glenohumeral joint, rotator cuff, subacromial bursa, and deltoid usually cause *adhesive capsulitis* ("frozen shoulder"). A person with this condition has difficulty abducting the arm. They can obtain an apparent abduction of up to 45° by elevating and rotating the scapula. Because of the lack of movement of the glenohumeral joint, strain is placed on the AC joint, which may be painful during other movements (e.g., elevation [shrugging] of the shoulder). Injuries that may initiate acute capsulitis, usually in 40- to 60-year-old persons, are glenohumeral dislocations, calcific supraspinatus tendinitis, partial tearing of the rotator cuff, and bicipital tendinitis (Salter, 1998). ⊙

Elbow Joint

The elbow joint—a hinge type of synovial joint—is located 2 to 3 cm inferior to the epicondyles of the humerus (Fig. 6.66).

Articulation of the Elbow Joint

The spool-shaped **trochlea** and spheroidal **capitulum** of the humerus articulate with the **trochlear notch** of the ulna and the slightly concave superior aspect of the **head of the radius**, respectively; therefore, there are humeroulnar and humeroradial articulations. The articular surfaces, covered with hyaline cartilage, are most fully congruent (in contact) when the forearm is in a position midway between pronation and supination and is flexed to a right angle.

Articular Capsule of the Elbow Joint

The **fibrous capsule** surrounds the elbow joint (Fig. 6.66, *A* and *C*). It is attached to the humerus at the margins of the lateral and medial ends of the articular surfaces of the capitulum and trochlea. Anteriorly and posteriorly it is carried superiorly, proximal to the coronoid and olecranon fossae. The **synovial membrane** lines the fibrous capsule and the intracapsular parts of the humerus and is continuous inferiorly with the synovial membrane of the proximal radioulnar joint. The articular capsule is weak anteriorly and posteriorly but is strengthened on each side by collateral ligaments.

Ligaments of the Elbow Joint

The collateral ligaments of the elbow joint are strong triangular bands that are medial and lateral thickenings of the fibrous capsule (Figs. 6.66*A* and 6.67). The lateral, fanlike **radial collateral ligament** extends from the lateral epicondyle of the humerus and blends distally with the *anular ligament of the radius*, which encircles and holds the head of the radius in the radial notch of the ulna, forming the proximal radioulnar joint and permitting pronation and supination of the forearm. The medial, triangular **ulnar collateral ligament**—ex-

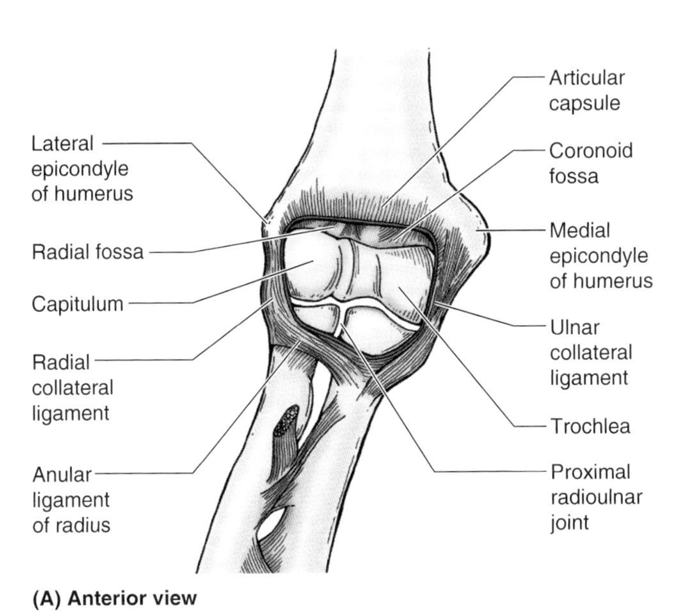

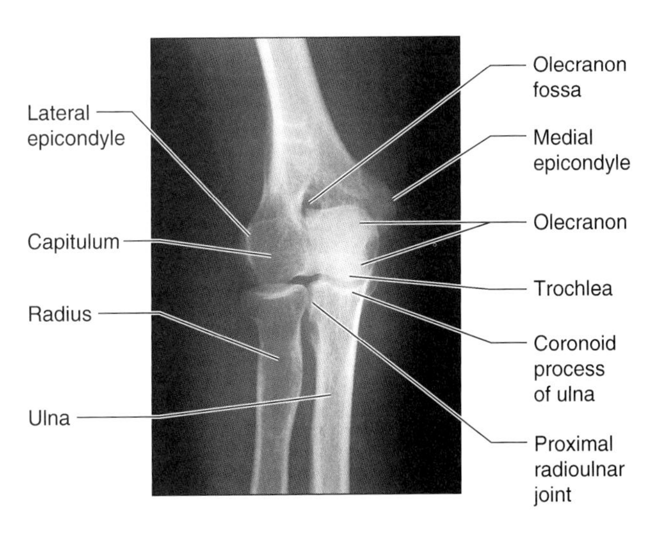

Figure 6.66. Elbow and proximal radioulnar joints. A. Anterior view of a dissection showing the bones, ligaments, and articular capsule of the elbow joint. **B.** Anteroposterior (AP) radiograph of the extended elbow joint.

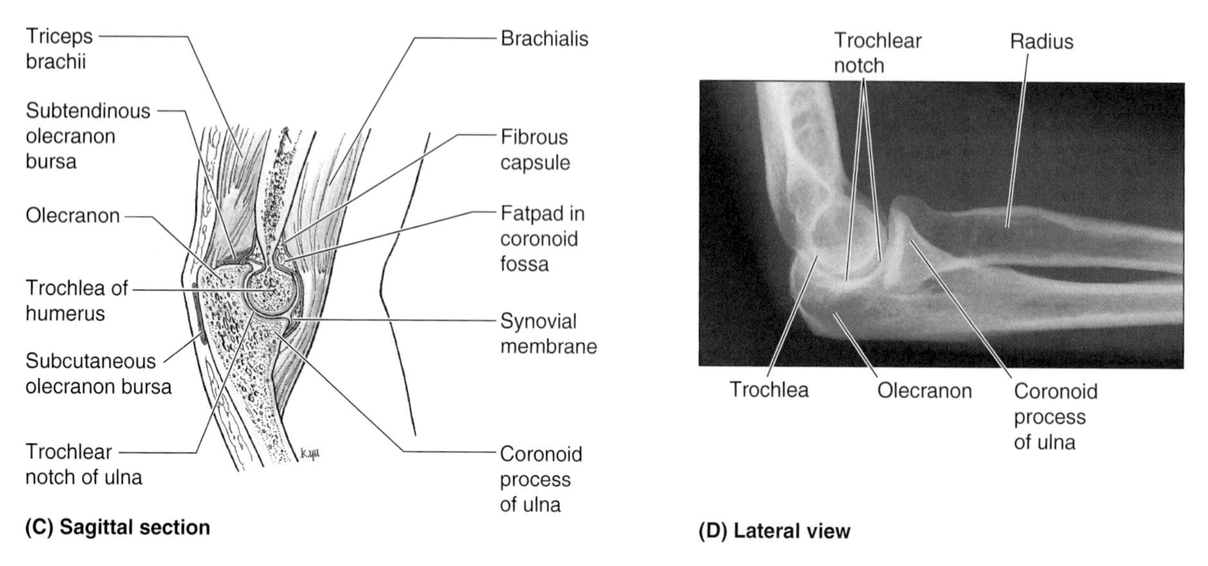

Figure 6.66. (*Continued*) **C.** Sagittal section of the elbow region, showing the fibrous capsule, synovial membrane, subtendinous and subcutaneous olecranon bursae, and articulation of the ulna. **D.** Lateral radiograph of the flexed elbow joint. (**B** and **D** Courtesy of Dr. E. Becker, Associate Professor of Medical Imaging, University of Toronto, Toronto, Ontario, Canada)

tending from the medial epicondyle of the humerus to the coronoid process and olecranon of the ulna—consists of three bands:

- The *anterior cordlike band* is the strongest
- The *posterior fanlike band* is the weakest
- The slender *oblique band* deepens the socket for the trochlea of the humerus.

Movements of the Elbow Joint

The movements of the elbow joint are flexion and extension. The long axis of the fully extended ulna makes an angle of approximately 170° with the long axis of the humerus—the **carrying angle** (Fig. 6.68). The carrying angle received its name because of the way the forearm angles away from the body when something is carried, such as a pail of water. The obliquity of the ulna (of the carrying angle) is more pro-

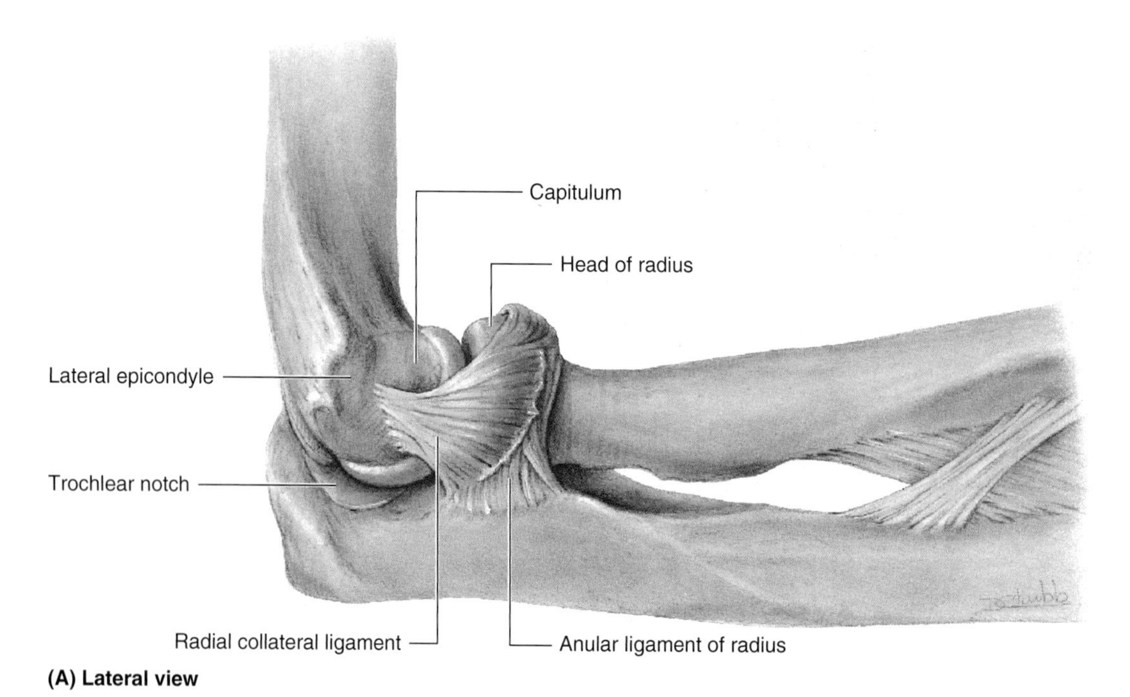

Figure 6.67. Collateral ligaments of the elbow. A. Radial (lateral) collateral ligament. The fan-shaped lateral ligament is attached to the anular ligament of the radius, but its superficial fibers continue on to the ulna.

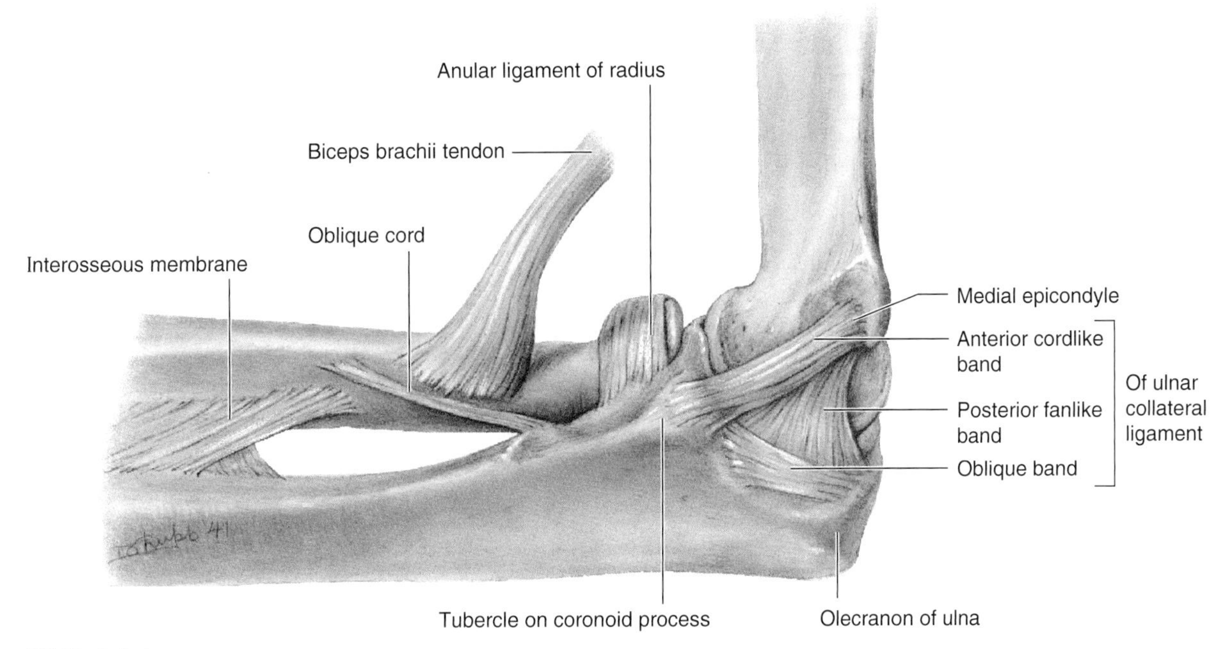

Figure 6.67. (*Continued*) **B.** Ulnar (medial) collateral ligament. The anterior band (part), a strong, round cord, is taut when the elbow joint is extended; the posterior band is a weak, fanlike ligament that is taut in flexion of the joint; the oblique fibers merely deepen the socket for the trochlea of the humerus.

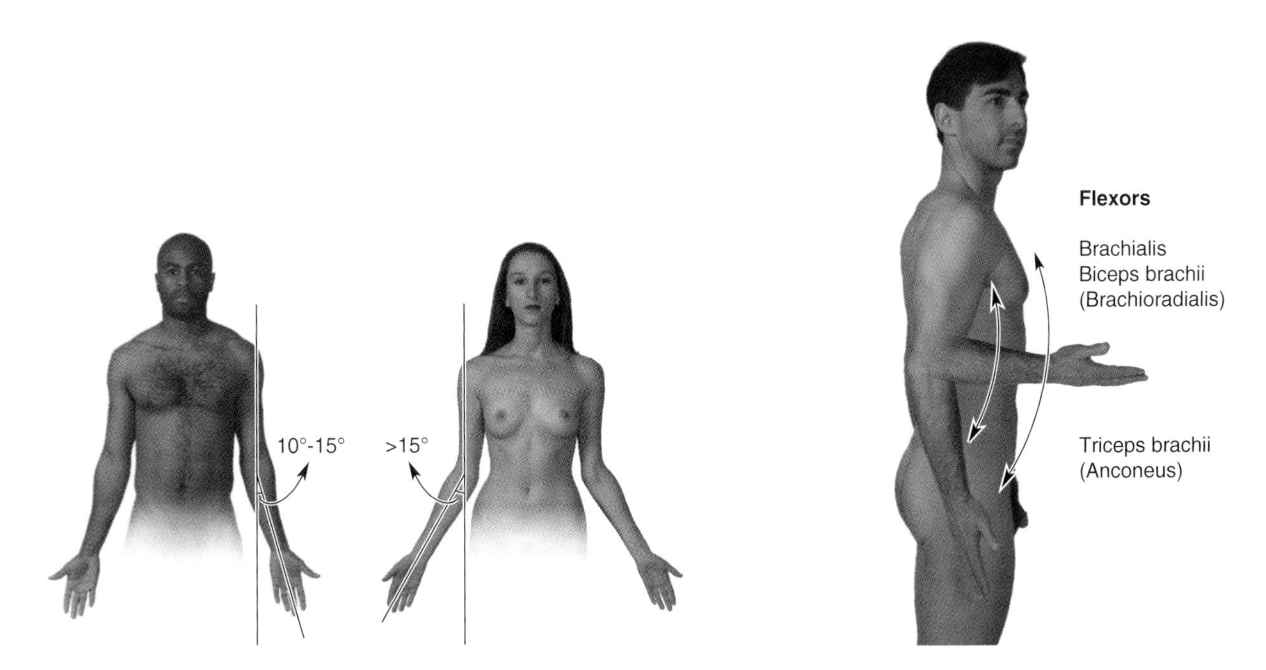

Figure 6.68. Carrying angle of the elbow joint. The carrying angle is the angle made by the axes of the arm and forearm when they are fully extended and supinated. Note that the forearm diverges laterally, forming an angle that is greater in the woman; however, no significant functional difference exists.

Figure 6.69. Flexor and extensor muscles of the elbow joint. During slow flexion or maintenance of flexion against gravity, the brachialis and biceps brachii are mainly involved. With increasing speed, the brachioradialis becomes involved. The triceps brachii is the main extensor of the forearm at the elbow joint. The anconeus muscle and gravity assist the triceps in extending the elbow joint.

nounced (the angle is approximately 10° more acute) in women than men. In the anatomical position, the elbow is against the waist. The carrying angle disappears when the forearm is pronated.

Muscles Moving the Elbow Joint

Several muscles cross the elbow and extend to the forearm and hand:

- *Chief flexors of the elbow joint* (Fig. 6.69)—brachialis, biceps brachii, and brachioradialis, in order of decreasing strength, assisted by pronator teres when flexion is resisted
- *Chief extensors of the elbow joint*—triceps brachii, especially medial head, assisted by anconeus.

Blood Supply of the Elbow Joint

The arteries supplying the elbow joint are derived from the anastomosis around the elbow joint.

Nerve Supply of the Elbow Joint

The elbow joint is supplied by the musculocutaneous, radial, and ulnar nerves.

Bursae Around the Elbow Joint

Only some of the bursae around the elbow joint are clinically important. The three **olecranon bursae** (Figs. 6.66*C* and 6.70) are the:

- *Intratendinous olecranon bursa,* which is sometimes present in the tendon of triceps brachii
- *Subtendinous olecranon bursa,* which is located between the olecranon and the triceps tendon, just proximal to its attachment to the olecranon
- *Subcutaneous olecranon bursa,* which is located in the subcutaneous connective tissue over the olecranon.

The *radioulnar bursa* lies between the extensor digitorum, the radiohumeral articulation, and the supinator muscle. This bursa lies posterior to the supinator, lateral to the tendon of the biceps, and medial to the ulna. The *bicipitoradial bursa* (biceps bursa) lies between the biceps tendon and the anterior part of the radial tuberosity.

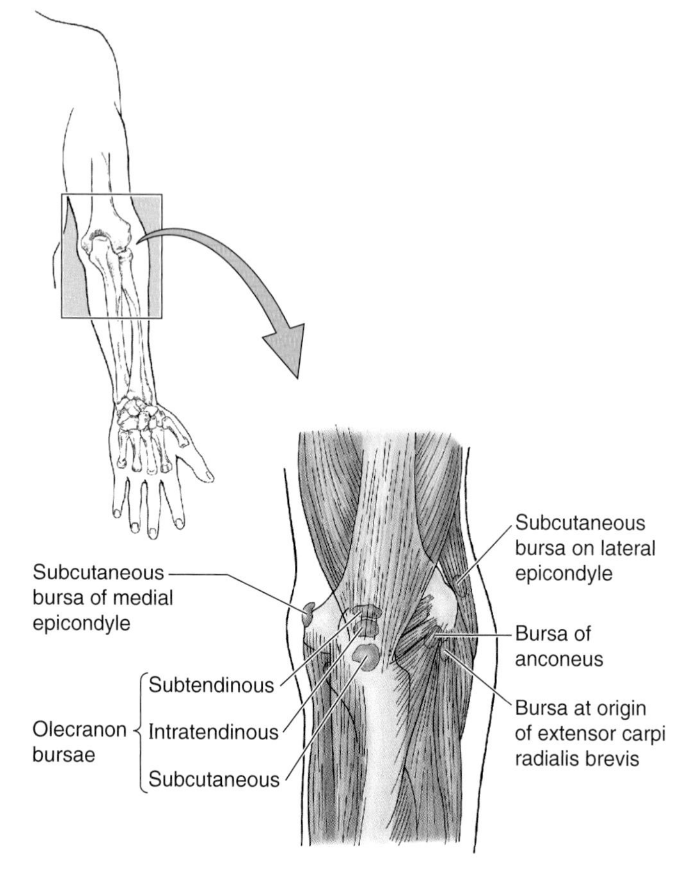

Figure 6.70. Bursae of the elbow joint. Several bursae are around the elbow joint. The olecranon bursae are most important clinically. Trauma of these bursae often produces bursitis (inflammation of the bursa).

Bursitis of the Elbow

The *subcutaneous olecranon bursa* is exposed to injury during falls on the elbow and to infection from abrasions of the skin covering the olecranon. Repeated excessive pressure and friction, as occurs in wrestling, for example, may cause this bursa to become inflamed, producing a friction *subcutaneous olecranon bursitis* (e.g., "student's elbow"). This type of bursitis is also known as "dart thrower's elbow" and "miner's elbow." Occasionally, the bursa becomes infected and the area over the bursa becomes inflamed. *Subtendinous olecranon bursitis* is much less common. It results from excessive friction between the triceps tendon and olecranon, for example, resulting from repeated flexion-extension of the forearm as occurs during certain assembly line jobs. The pain is most severe during flexion of the forearm because of pressure exerted on the inflamed subtendinous olecranon bursa by the triceps tendon. ▶

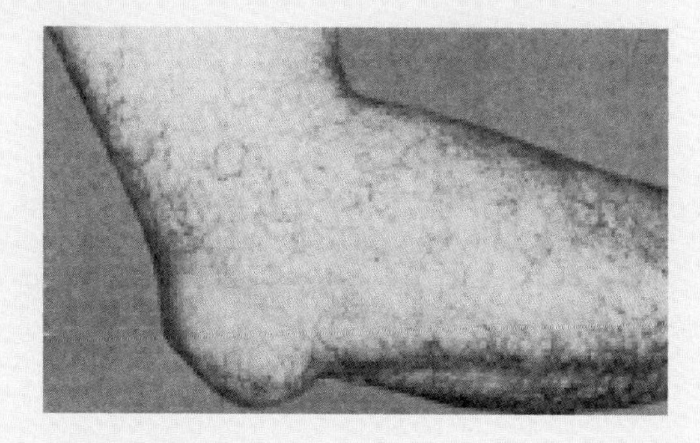

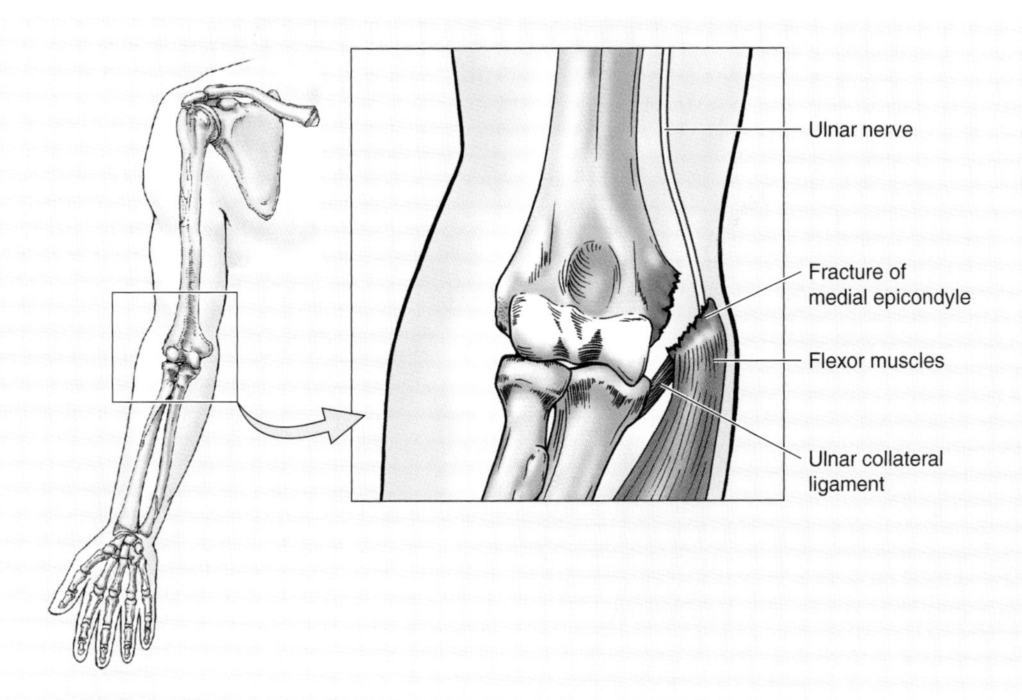

▶ *Bicipitoradial bursitis (biceps bursitis)* results in pain when the forearm is pronated because this action compresses the bicipitoradial bursa against the anterior half of the tuberosity of the radius.

Avulsion of the Medial Epicondyle

Avulsion of the medial epicondyle in children can result from a fall that causes severe abduction of the extended elbow, an abnormal movement of this articulation. The resulting traction on the ulnar collateral ligament pulls the medial epicondyle distally. The anatomical basis of avulsion of the medial epicondyle is that the epiphysis for the medial epicondyle may not fuse with the distal end of the humerus until up to 20 years of age. Usually fusion is complete radiographically at age 14 years in females and age 16 years in males. *Traction injury of the ulnar nerve* is a frequent complication of the abduction type of avulsion of the medial epicondyle. The anatomical basis for this stretching of the ulnar nerve is that it passes posterior to the medial epicondyle before entering the forearm.

Dislocation of the Elbow Joint

Posterior dislocation of the elbow joint may occur when children fall on their hands with their elbows flexed. Dislocations of the elbow may result from hyperextension or a blow that drives the ulna posterior or posterolateral (Anderson and Hall, 1995). The distal end of the humerus is driven through the weak anterior part of the fibrous capsule as the radius and ulna dislocate posteriorly. The ulnar collateral ligament is often torn, and an associated fracture of the head of the radius, coronoid process, or olecranon process of the ulna may occur. Injury to the ulnar nerve may occur, resulting in numbness of the little finger and weakness of flexion and adduction of the wrist. ○

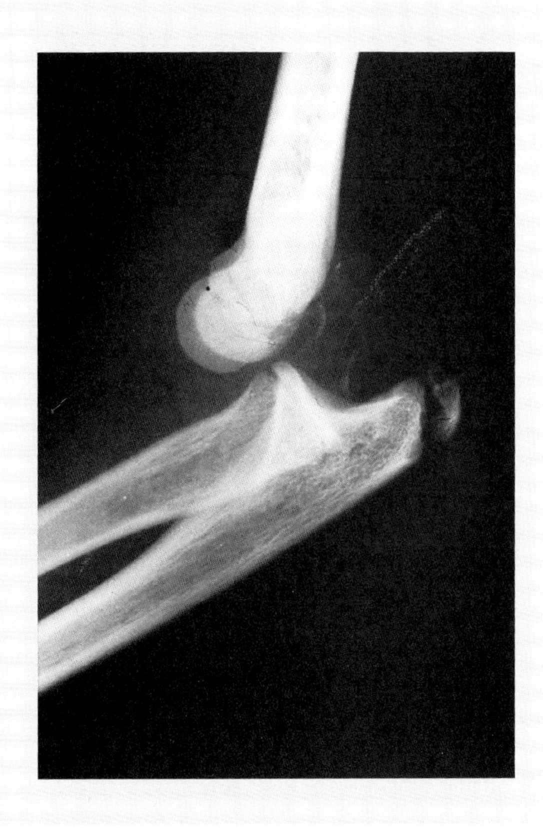

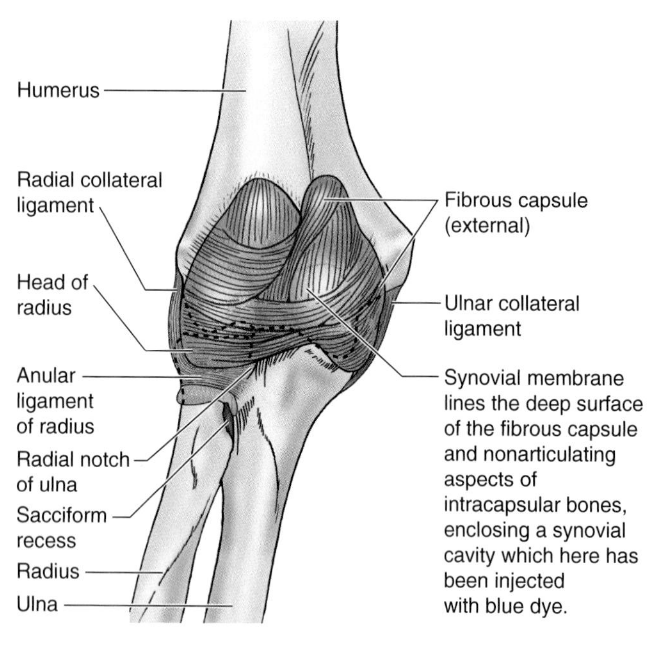

Figure 6.71. Proximal radioulnar joint. This is a pivot type of synovial joint between the head of the radius and the radial notch of the ulna. The anular ligament attaches to the radial notch of the ulna and surrounds the joint, forming a collar around the head of the radius.

Proximal Radioulnar Joint

The proximal (superior) radioulnar joint is a pivot type of synovial joint that allows movement of the head of the radius on the ulna (Fig. 6.71).

Articulation of the Proximal Radioulnar Joint

The head of the radius articulates with the radial notch of the ulna. The radial head is held in position by the anular ligament.

Articular Capsule of the Proximal Radioulnar Joint

The **fibrous capsule** encloses the joint and is continuous with that of the elbow joint. The **synovial membrane** lines the deep surface of the fibrous capsule and nonarticulating aspects of the bones. The synovial membrane is an inferior prolongation of the synovial capsule of the elbow joint.

Ligaments of the Proximal Radioulnar Joint

The **anular ligament**, attached to the ulna anterior and posterior to its radial notch (Figs. 6.71 and 6.72), surrounds the articulating bony surfaces and forms a collar or loop that, with the radial notch, forms a ring that completely *encircles the head of the radius*. The deep surface of the anular ligament is lined with synovial membrane, which continues distally as a *sacciform recess* on the neck of the radius. This arrangement

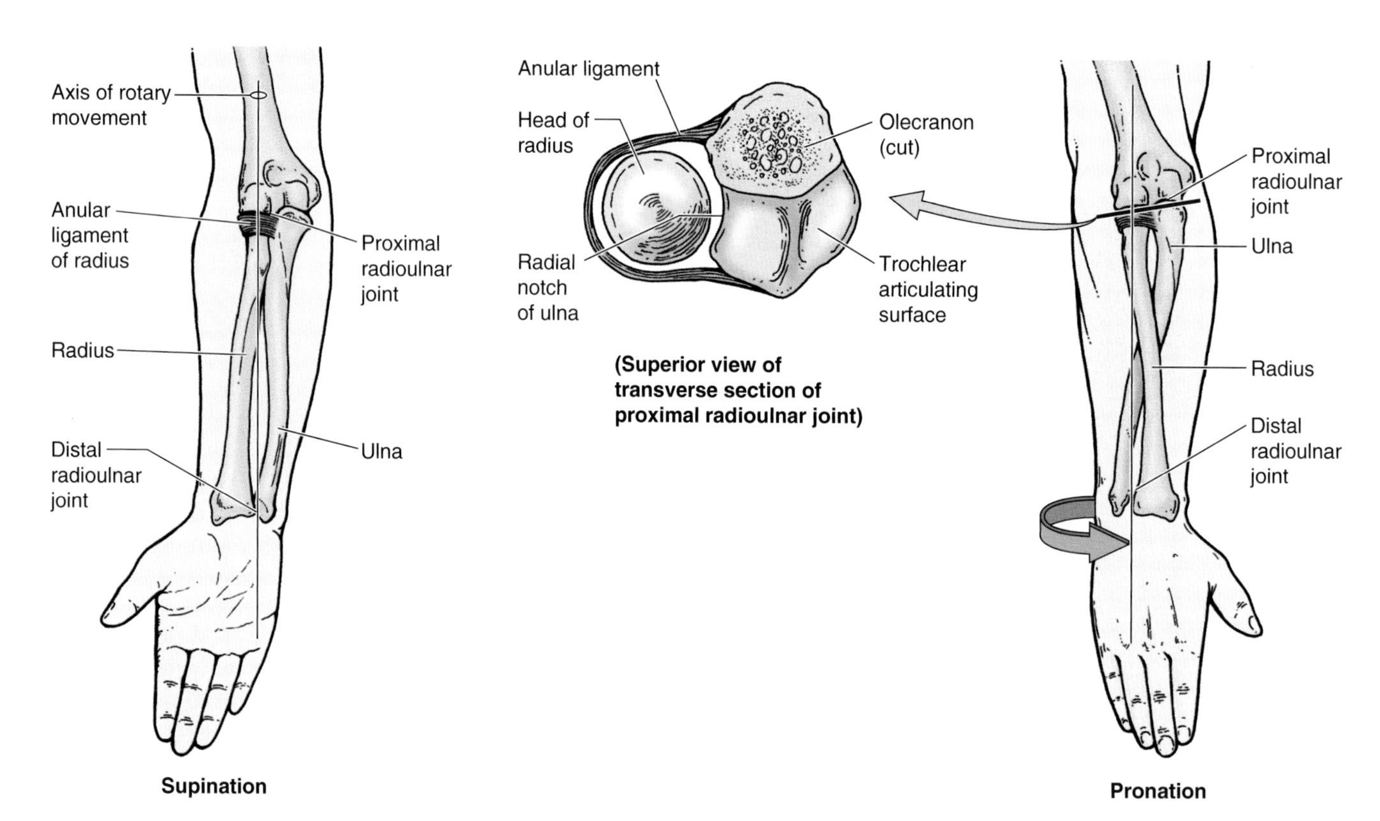

Figure 6.72. Supination and pronation of the forearm. Supination is the movement of the forearm and hand that rotates the radius laterally around its longitudinal axis so that the dorsum of the hand faces posteriorly and the palm faces anteriorly. Pronation is the movement of the forearm and hand that rotates the radius medially around its longitudinal axis so that the palm of the hand faces posteriorly and its dorsum faces anteriorly. The central drawing illustrates the rotation of the head of the radius in the radial notch of the ulna.

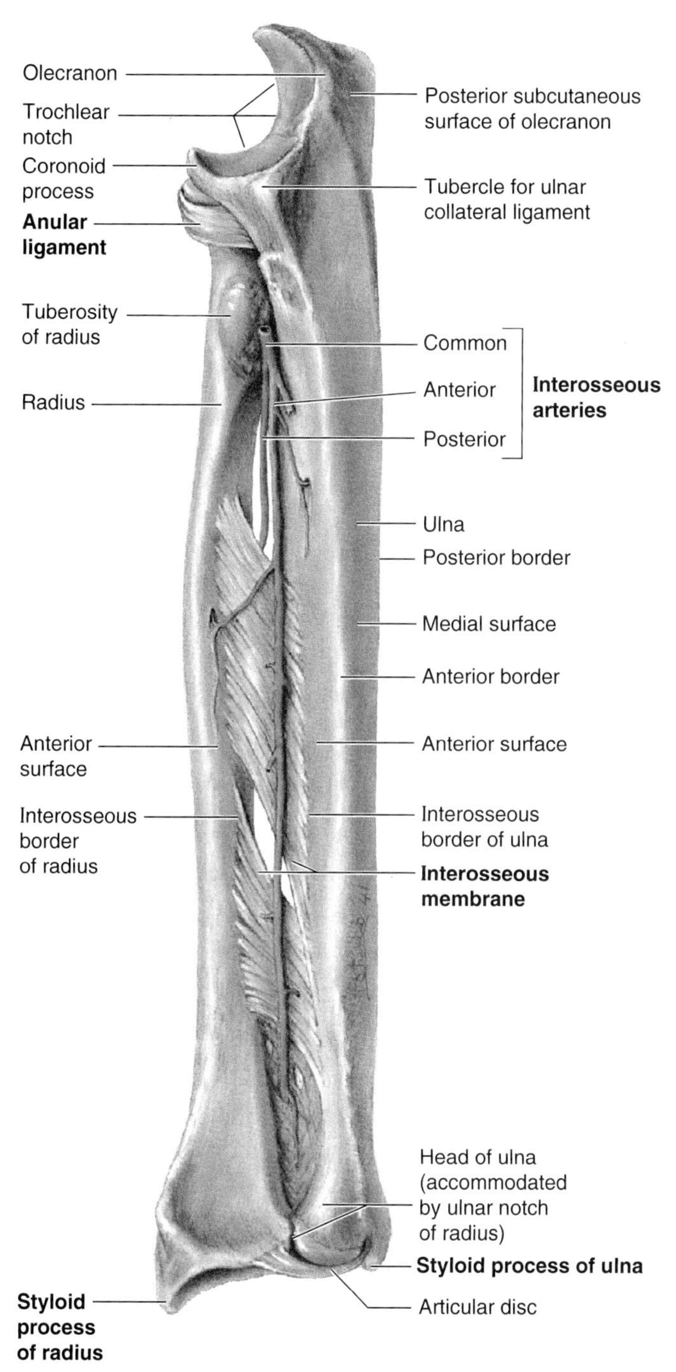

Figure 6.73. Radioulnar ligaments and interosseous arteries. Medial view. The ligament of the proximal radioulnar joint is the anular ligament. The ligament of the distal radioulnar joint is the articular disc. The interosseous membrane connects the interosseous margins of the radius and ulna, forming the radioulnar syndesmosis. The general direction of the fibers of the interosseous membrane is such that a superior thrust to the hand (L. manus) is received by the radius and is transmitted to the ulna.

allows the radius to rotate within the anular ligament without binding, stretching, or tearing the synovial capsule.

Movements of the Proximal Radioulnar Joint

During pronation and supination of the forearm, the head of the radius rotates within the ring formed by the anular ligament and the radial notch of the ulna. **Supination** turns the palm anteriorly (or superiorly when the forearm is flexed). **Pronation** turns the palm posteriorly (or inferiorly when the forearm is flexed). The axis for these movements passes proximally through the center of the head of the radius and distally through the site of attachment of the apex of the articular disc ("triangular ligament") to the head (styloid process) of the ulna. During pronation and supination, it is the radius that rotates; its head rotates within the cup-shaped ring formed by the anular ligament and the radial notch on the ulna. Distally, the end of the radius rotates around the head of the ulna. Almost always, supination and pronation are accompanied by synergic movements of the glenohumeral and elbow joints that produce simultaneous movement of the ulna, except when the elbow is flexed.

Muscles Moving the Proximal Radioulnar Joint

Supination is produced by the supinator (when resistance is absent) and biceps brachii (when power is required because of resistance)—with some assistance from the EPL and extensor carpi radialis longus. *Pronation is produced by the pronator quadratus (primarily) and pronator teres (secondarily)*—with some assistance from the flexor carpi radialis, palmaris longus, and brachioradialis (when the forearm is in the midprone position).

Blood Supply of the Proximal Radioulnar Joint

The proximal radioulnar joint is supplied by the anterior and posterior interosseous arteries (Fig. 6.73).

Innervation of the Proximal Radioulnar Joint

The proximal radioulnar joint is supplied mainly by the musculocutaneous, median, and radial nerves. Pronation is essentially a function of the median nerve, whereas supination is a function of the musculocutaneous and radial nerves.

Subluxation and Dislocation of the Radial Head

Preschool children, particularly girls, are vulnerable to transient subluxation (incomplete dislocation) of the head of the radius ("pulled elbow"). The history of these cases is typical. The child is suddenly lifted (jerked) by the upper limb while the forearm is pronated (e.g., lifting a child into a bus). The child may cry out and refuse to use the limb, which is protected by holding it with the elbow flexed and the forearm pronated. The sudden pulling of the upper limb tears the distal attachment of the anular ligament, where it is ▶

▶ loosely attached to the neck of the radius. The radial head then moves distally, partially out of the torn anular ligament. The proximal part of the torn ligament may become trapped between the head of the radius and the capitulum of the humerus. The source of pain is the pinched anular ligament. Treatment of the subluxation consists of supination of the child's forearm while the elbow is flexed (Salter, 1998). The tear in the anular ligament soon heals when the limb is placed in a sling for 2 weeks. ⊙

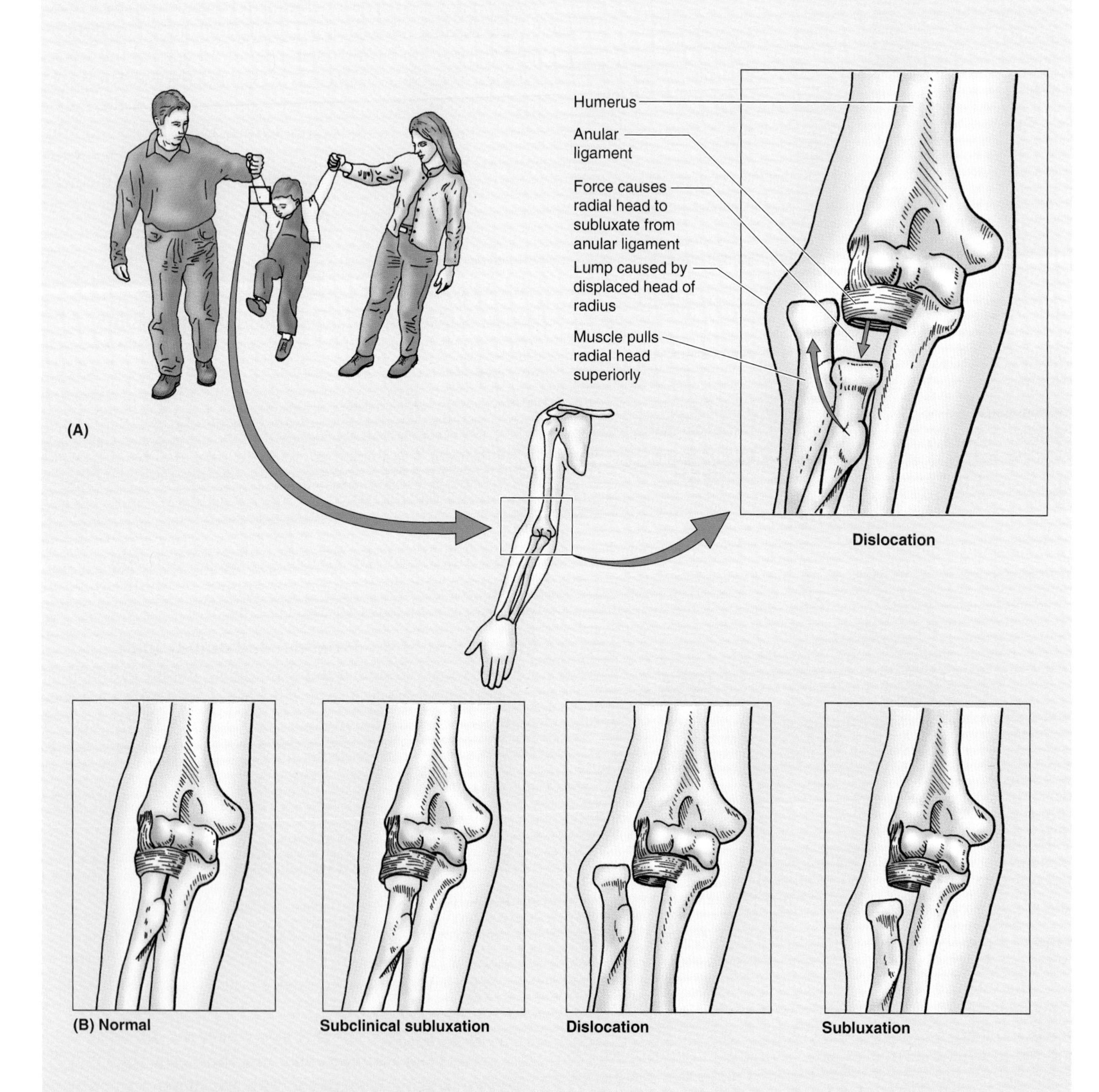

Distal Radioulnar Joint

The distal (inferior) radioulnar joint is a pivot type of synovial joint. The radius moves around the relatively fixed distal end of the ulna.

Articulation of the Distal Radioulnar Joint

The rounded head of the ulna articulates with the ulnar notch on the medial side of the distal end of the radius. A fibrocartilaginous **articular disc** ("triangular ligament") binds the ends of the ulna and radius together and is the main uniting structure of the joint (Figs. 6.74 and 6.75*A*). The base of the articular disc is attached to the medial edge of the ulnar notch of the radius, and its apex is attached to the lateral side of the base of the styloid process of the ulna. The proximal surface of this triangular disc articulates with the distal aspect of the head of the ulna. Hence, the joint cavity is L-shaped in a coronal section, with the vertical bar of the "L" between the radius and the ulna and the horizontal bar between the ulna and the articular disc. The articular disc separates the cavity of the distal radioulnar joint from the cavity of the wrist joint.

Articular Capsule of the Distal Radioulnar Joint

The **fibrous capsule** encloses the distal radioulnar joint but is deficient superiorly. The **synovial membrane** extends superiorly between the radius and ulna to form the *sacciform recess*. This redundancy of the synovial capsule accommodates the twisting of the capsule that occurs when the distal end of the radius travels around the relatively fixed distal end of the ulna during pronation of the forearm.

Ligaments of the Distal Radioulnar Joint

Anterior and posterior ligaments strengthen the fibrous capsule of the distal radioulnar joint. These relatively weak transverse bands extend from the radius to the ulna across the anterior and posterior surfaces of the joint.

Movements of the Distal Radioulnar Joint

During pronation and supination of the forearm and hand, the distal end of the radius moves anteriorly and medially, crossing the ulna anteriorly.

Muscles Moving the Distal Radioulnar Joint

The muscles producing movements of the distal radioulnar joint are discussed with the proximal radioulnar joint.

Blood Supply of the Distal Radioulnar Joint

The anterior and posterior *interosseous arteries* supply the distal radioulnar joint (Fig. 6.73).

Innervation of the Distal Radioulnar Joint

The anterior and posterior *interosseous nerves* supply the distal radioulnar joint.

Wrist Joint

The wrist (radiocarpal) joint is a condyloid type of synovial joint. The position of the joint is indicated approximately by a line joining the styloid processes of the radius and ulna (Fig. 6.74, *A* and *B*).

Articulation of the Wrist Joint

The distal end of the radius and the articular disc of the distal radioulnar joint articulate with the proximal row of carpal bones, except for the pisiform.

Articular Capsule of the Wrist Joint

The **fibrous capsule** surrounds the wrist joint and is attached to the distal ends of the radius and ulna and the proximal row

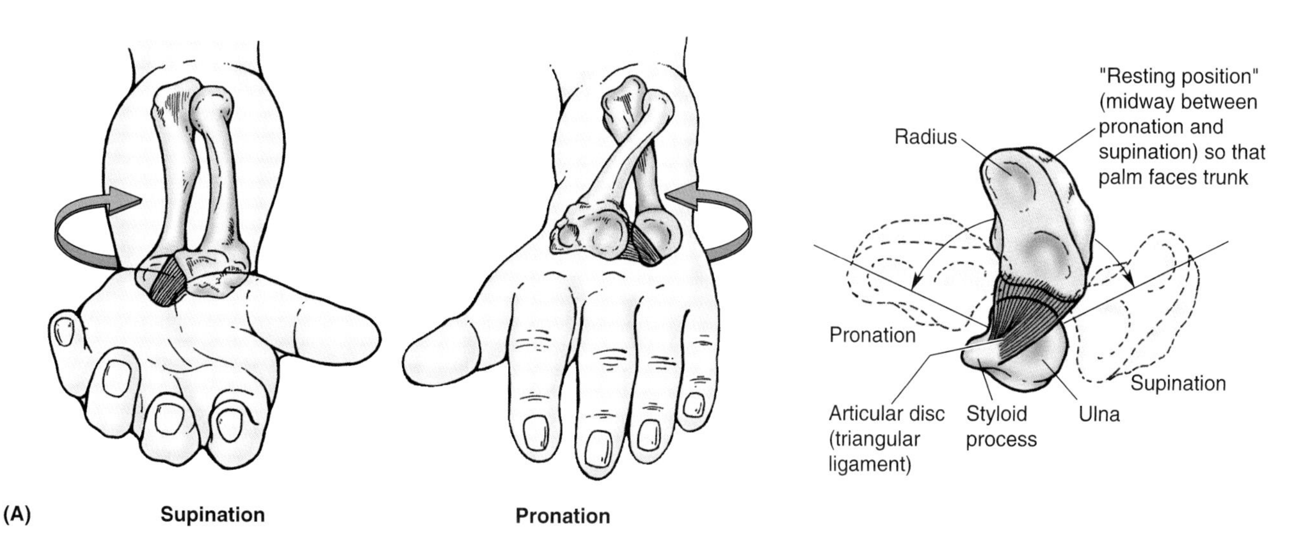

Figure 6.74. Movements of the distal radioulnar joint and joints of the wrist and hand. A. The distal radioulnar joint is the pivot type of synovial joint between the head of the ulna and the ulnar notch of the radius. The inferior end of the radius moves around the relatively fixed end of the ulna during supination and pronation of the hand. The two bones are firmly united distally by the articular disc—referred to clinically as the "triangular ligament" of the distal radioulnar joint. It has a broad attachment to the radius but a narrow attachment to the styloid process of the ulna, which serves as the pivot point for the rotary movement.

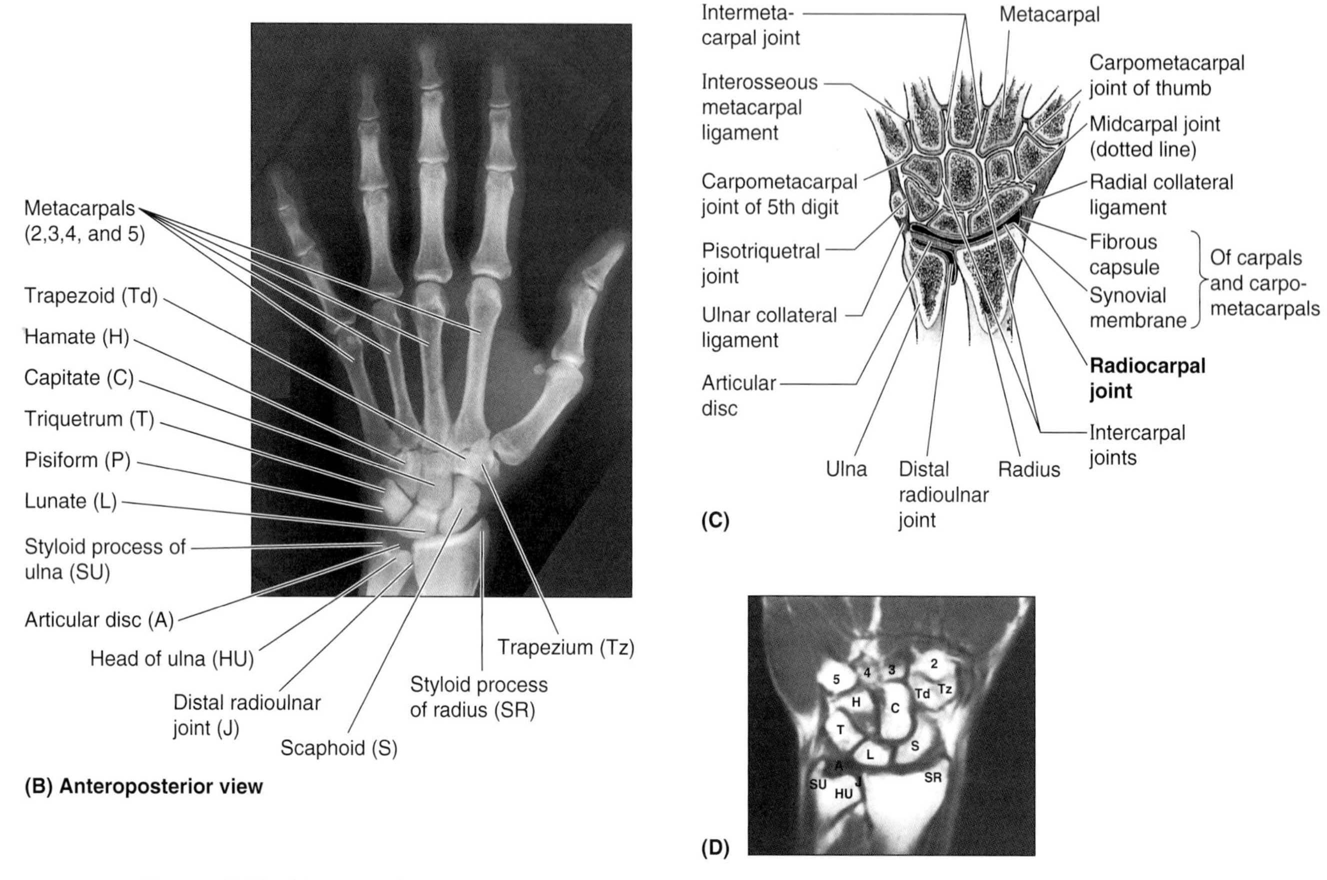

Figure 6.74. (*Continued*) **B.** Radiograph of the wrist and hand. Notice the wide "joint space" at the distal end of the ulna because of the radiolucent articular disc. (Courtesy of Dr. E.L. Lansdown, Professor of Medical Imaging, University of Toronto, Toronto, Ontario, Canada) **C.** Coronal section of the right wrist and hand. Observe the distal radioulnar, wrist, intercarpal, carpometacarpal, and intermetacarpal joints. Note that—although they appear to be continuous when viewed radiographically in (**B**) and (**D**)—the cavities of the distal radioulnar and wrist joints are separated by the articular disc of the distal radioulnar joint. **D.** Coronal MRI of the wrist and hand. (Courtesy of Dr. W. Kucharczyk, Chair of Medical Imaging and Clinical Director of Tri-Hospital Magnetic Resonance Centre, Toronto, Ontario, Canada)

of carpals (scaphoid, lunate, and triquetrum). The **synovial membrane** lines the fibrous capsule and is attached to the margins of the articular surfaces (Fig. 6.75*A*). Numerous synovial folds are present.

Ligaments of the Wrist Joint

The fibrous capsule is strengthened by strong dorsal and palmar radiocarpal ligaments. The **palmar radiocarpal ligaments** pass from the radius to the two rows of carpals (Fig. 6.75*B*). They are strong and directed so that hand follows the radius during supination of the forearm. The **dorsal radiocarpal ligaments** take the same direction so that the hand follows the radius during pronation of the forearm. The fibrous capsule is also strengthened medially by the **ulnar collateral ligament** that is attached to the ulnar styloid process and triquetrum (Figs. 6.74*C* and 6.75*B*). The capsule is also strengthened lat-

Figure 6.75. Articular surfaces of the wrist (radiocarpal) joint. A. Ligaments of the distal radioulnar, radiocarpal, and intercarpal joints. The hand is forcibly extended but the joint is intact. Observe the anterior or palmar radiocarpal ligaments, passing from the radius to the two rows of carpal bones. These strong ligaments are directed so that the hand follows the radius during supination. **B.** The joint is opened from its anterior aspect, with the dorsal radiocarpal ligaments serving as a hinge. Observe the nearly equal proximal articular surfaces of the scaphoid and lunate bones and that the lunate articulates with both the radius and articular disc. Only during adduction of the wrist does the triquetrum (L. triquetrus) articulate with the disc.

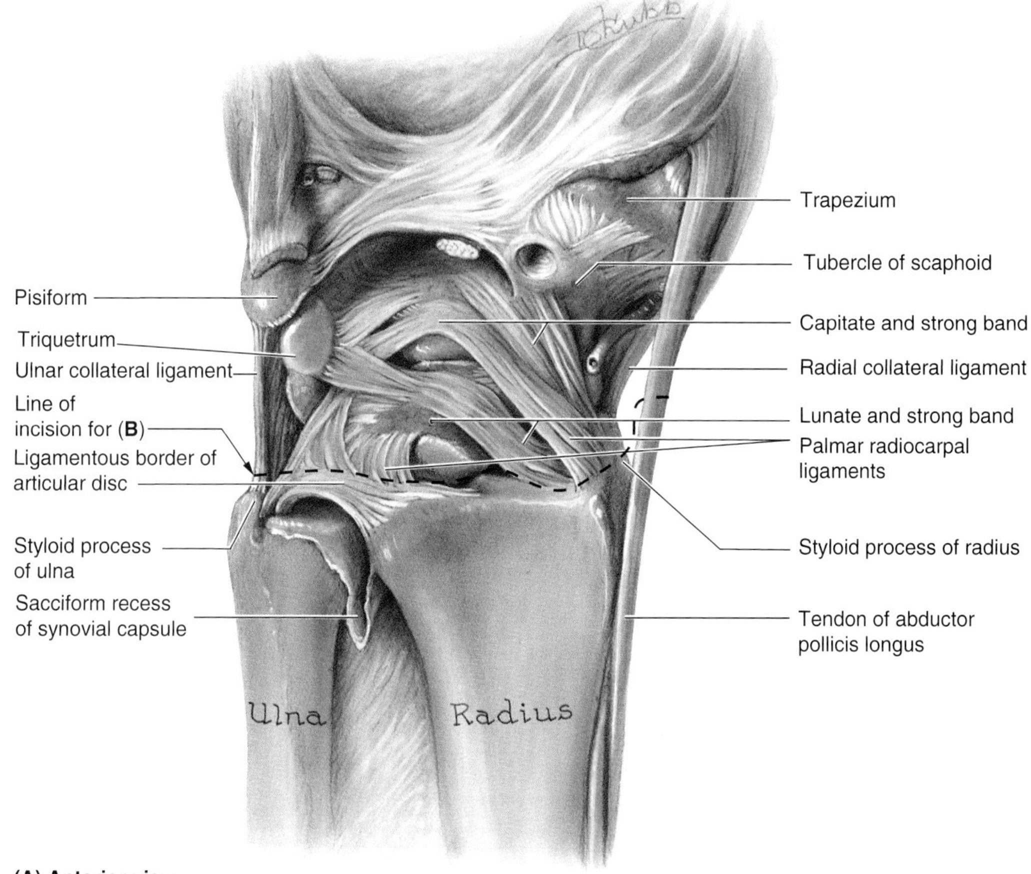

(A) Anterior view

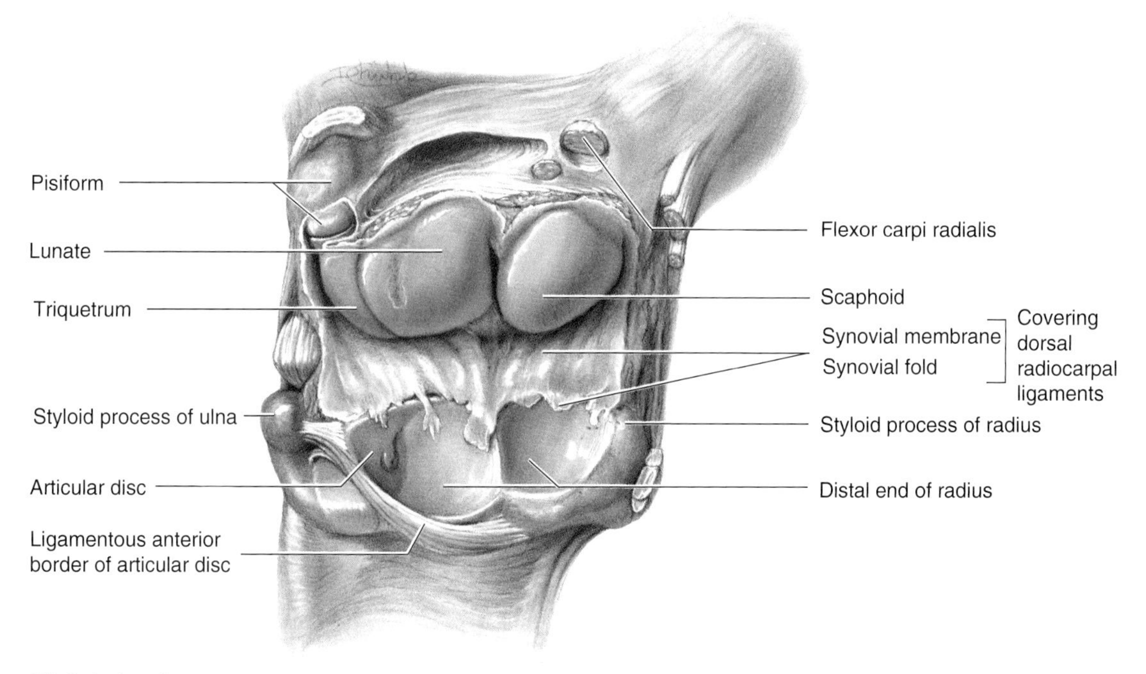

(B) Anterior view

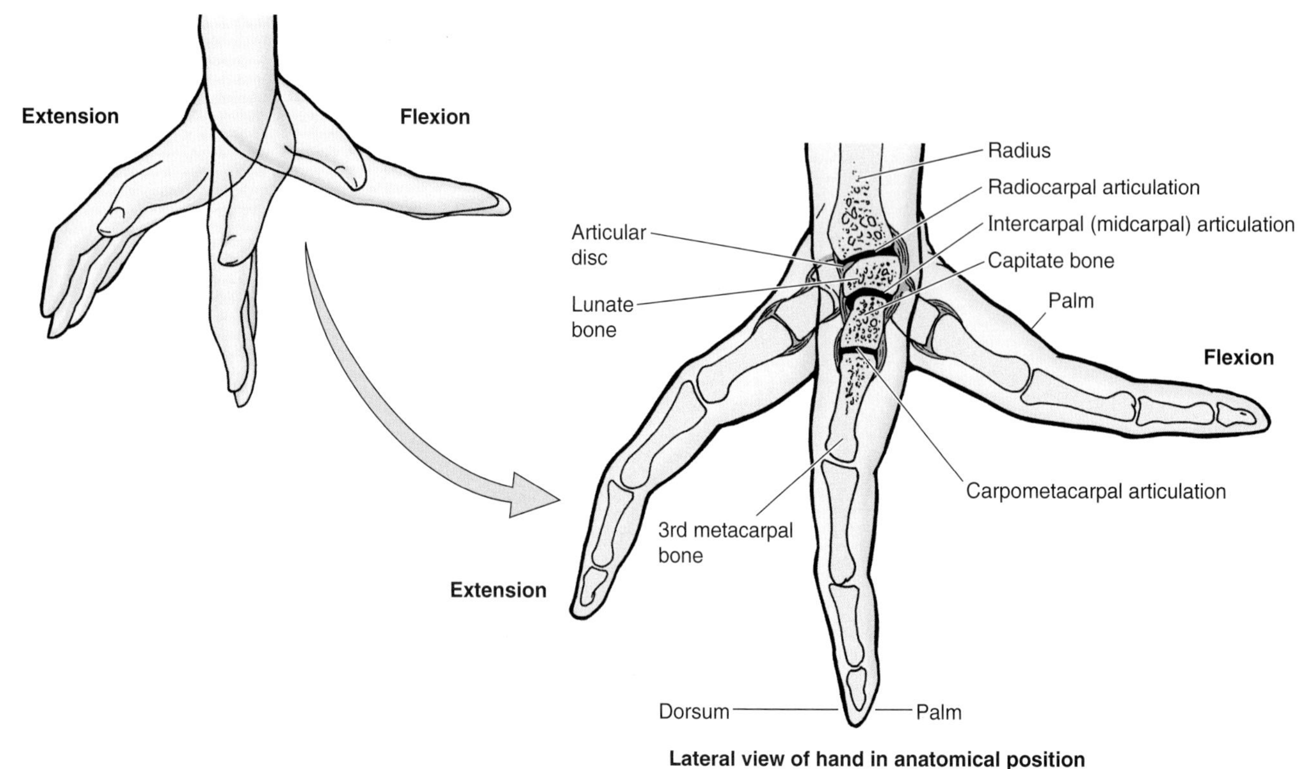

Figure 6.76. Sagittal section of the wrist and hand during extension and flexion. Observe the radiocarpal, intercarpal (midcarpal), and carpometacarpal articulations. Most movement occurs at the radiocarpal joint, with additional movement taking place at the midcarpal joint during full flexion and extension. Virtually no flexion or extension occurs at the carpometacarpal joints of the 2nd and 3rd digits; the 4th is slightly mobile and the 5th extremely mobile, flexing considerably when the fist is clenched (p. 763).

erally by the **radial collateral ligament** that is attached to the radial styloid process and scaphoid.

Movements of the Wrist Joint

The movements at the wrist joint (Fig. 6.76) may be augmented by additional smaller movements at the intercarpal and midcarpal joints. The movements of the wrist joint are flexion-extension, abduction-adduction (radial deviation–ulnar deviation), and circumduction. The hand can be flexed on the forearm more than it can be extended; these movements are accompanied by similar movements at the midcarpal joint between proximal and distal rows of carpal bones. Adduction of the hand is greater than abduction. Most adduction occurs at the wrist joint. Abduction from the neutral position occurs at the midcarpal joint. Circumduction of the hand consists of successive flexion, adduction, extension, and abduction.

Muscles Moving the Wrist Joint

- *Flexion of the wrist* is produced by the flexor carpi radialis and flexor carpi ulnaris with assistance from the flexors of the fingers and thumb, and palmaris longus, and APL
- *Extension of the wrist* is produced by the extensor carpi radialis longus and brevis and extensor carpi ulnaris with assistance from the extensors of the fingers and thumb
- *Abduction of the wrist* is produced by the APL, the flexor carpi radialis, and the extensor carpi radialis longus and brevis; abduction is limited to approximately 15° because of the projecting radial styloid process
- *Adduction of the wrist* is produced by simultaneous contraction of extensor and flexor carpi ulnaris.

Blood Supply of the Wrist Joint

The arteries supplying the wrist joint are branches of the dorsal and palmar carpal arches.

Innervation of the Wrist Joint

The nerves to the wrist joint are derived from the anterior interosseous branch of the *median nerve*, the posterior interosseous branch of the *radial nerve*, and the dorsal and deep branches of the *ulnar nerve*.

Wrist Fractures

Fracture of the distal end of the radius (*Colles' fracture*), the most common fracture in persons older than 50 years, is discussed on page 674. Much less common than Colles' fracture is a *Smith fracture of the radius,* usually occurring in young men. The transverse fracture results from a fall or a blow on the dorsal aspect of the flexed wrist and produces a ventral angulation of the wrist. *Fracture of the scaphoid* is relatively common in young adults, especially males. The injury usually results from a fall on the open hand with the wrist extended and abducted. Because the blood supply of the scaphoid is precarious, union of the fracture is usually slow, requiring at least 3 months (Salter, 1998).

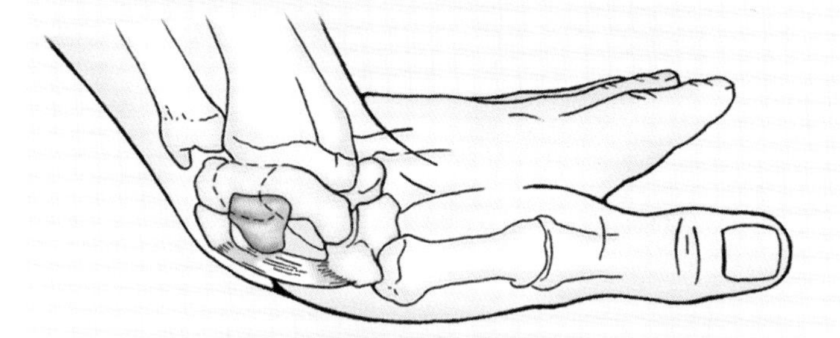

Anterior dislocation of the lunate is an uncommon injury that usually results from a fall on the dorsiflexed wrist. The lunate is pushed out of its place in the floor of the carpal tunnel toward the palmar surface of the wrist. The displaced lunate may compress the median nerve and lead to *carpal tunnel syndrome* (discussed previously). Because of its poor blood supply, *avascular necrosis of the lunate* may occur. In some cases, excision of the lunate may be required. In *degenerative joint disease of the wrist,* surgical fusion of carpals (*arthrodesis*) may be necessary to relieve the severe pain.

Fracture-separation of the distal radial epiphysis is common in children because of frequent falls in which forces are transmitted from the hand to the radius. In a lateral radiograph of a child's wrist, dorsal displacement of the distal radial epiphysis is obvious (*B*). A small fracture of the body of the radius is also visible (not obvious). When the epiphysis is placed in its normal position during reduction, the prognosis for normal bone growth is good. ○

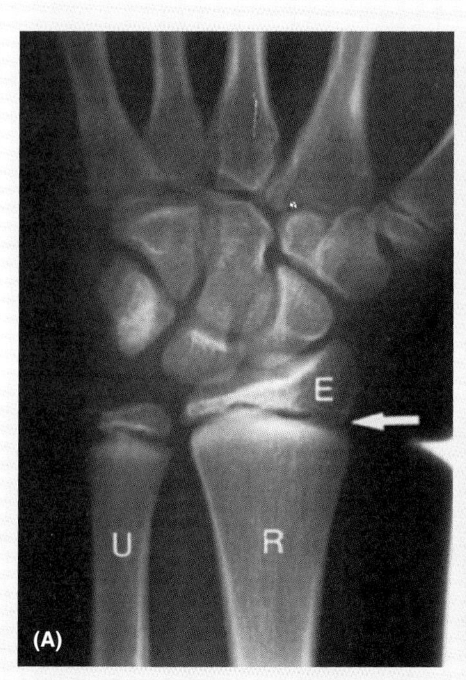

E, epiphysis at distal end of radius (*R*). The *arrow* indicates the epiphyseal plate

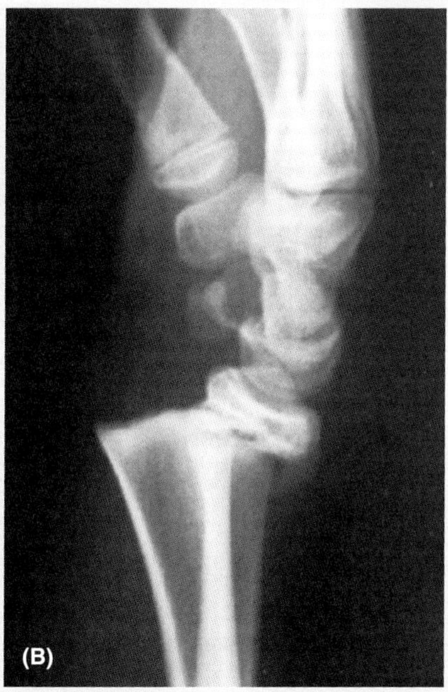

Displacement of epiphysis

Intercarpal Joints

The intercarpal joints, interconnecting the carpal bones, are plane synovial joints (Fig. 6.74, *B–D*), which may be summarized as:

- Joints between the carpal bones of the proximal row
- Joints between the carpal bones of the distal row
- A complex joint—*the midcarpal joint*—between the proximal and distal rows of carpal bones
- The pisiform articulates with the palmar surface of the triquetrum to form a small joint, the *pisotriquetral joint.*

Articular Capsule of the Intercarpal Joints

A continuous, common articular cavity is formed by the intercarpal and carpometacarpal joints, with the exception of the carpometacarpal joint of the thumb, which is independent. The wrist joint is also independent. The continuity of the articular cavities—or the lack of it—is significant in relation to the spread of infection and also to arthroscopy, in which a flexible fiberoptic scope is inserted into the joint cavity to view its internal surfaces and features. The **fibrous capsule** surrounds these joints, which helps unite the carpals. The **synovial membrane** lines the fibrous capsule and is attached to the margins of the articular surfaces of the carpals.

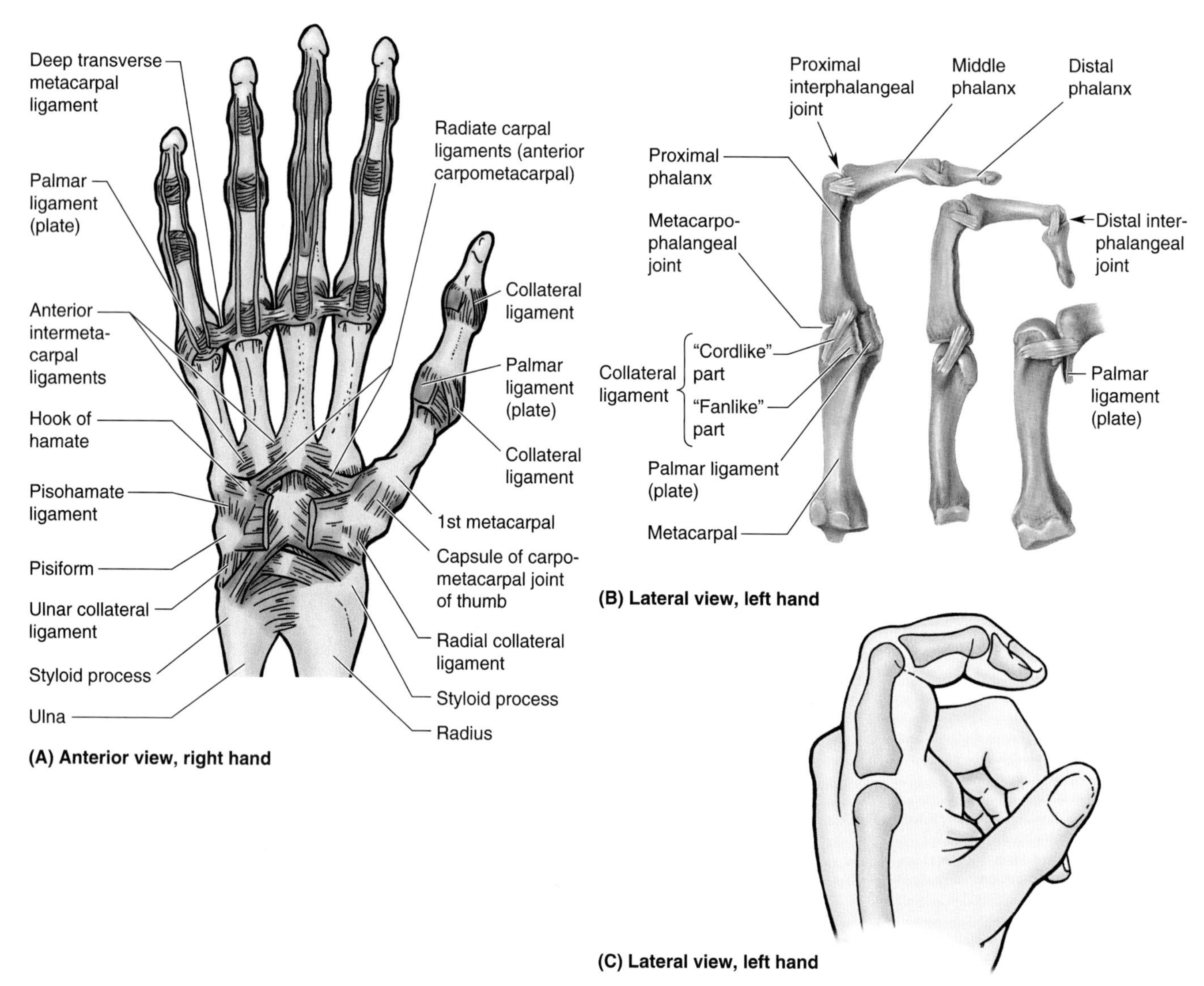

Figure 6.77. Joints of the carpus and digits. A. Anterior (palmar) view of the distal right forearm and hand showing the palmar ligaments of the radioulnar, radiocarpal, intercarpal, carpometacarpal, and interphalangeal joints. **B.** Lateral view of the left metacarpophalangeal and interphalangeal joints. Observe the palmar ligament (fibrocartilaginous plate), a modification of the anterior aspect of the metacarpophalangeal joint capsule, attached distally to the base of the proximal phalanx. It is fixed in place anterior to the head of the metacarpal by the weaker, fanlike part of the collateral ligament and moves like a visor across the metacarpal head. It is concave anteriorly and incorporates cartilage to facilitate passage of the overlying long flexor tendons. Observe also the extremely strong cordlike parts of the collateral ligaments of this joint, being eccentrically attached to the metacarpal heads; they are slack during extension and taut during flexion; hence, the fingers cannot usually be spread (abducted) when the metacarpophalangeal joints are fully flexed (i.e., unless the hand is open). Observe that the interphalangeal joints have corresponding ligaments but that the distal ends of the proximal and middle phalanges, being flattened anteroposteriorly and having two small condyles, permit neither adduction or abduction. **C.** Lateral view of a drawing of the flexed left index finger showing its phalanges (bones) and the position of the metacarpophalangeal and interphalangeal joints. Note that the knuckles are formed by the heads of the bones, with the joint plane lying distally.

Ligaments of the Intercarpal Joints

The carpals are united by anterior, posterior, and interosseous ligaments (Figs. 6.75 and 6.77*A*).

Movements at the Intercarpal Joints

A small amount of gliding movement is possible between the carpals. Movement at the intercarpal joints augments movements at the wrist joint, increasing the range of movement. The *midcarpal joint*—between the proximal and distal rows of carpals—is mainly involved in flexion and abduction of the hand (Figs. 6.74*A* and 6.76).

Blood Supply of the Intercarpal Joints

The arteries supplying the intercarpal joints are derived from the dorsal and palmar carpal arches.

Innervation of the Intercarpal Joints

The intercarpal joints are supplied by the anterior interosseous branch of the *median nerve* and the dorsal and deep branches of the *ulnar nerve*.

Carpometacarpal and Intermetacarpal Joints

These articulations are the plane type of synovial joint, except for the carpometacarpal joint of the thumb, which is a saddle joint (Fig. 6.74, *B–D*).

Articulation of the Carpometacarpal and Intermetacarpal Joints

Like the carpals, the metacarpals articulate with each other at their bases; the carpometacarpal joint of the thumb is between the trapezium and the base of the 1st metacarpal and has a separate joint cavity.

Articular Capsule of the Carpometacarpal and Intermetacarpal Joints

The medial four carpometacarpal joints and three intermetacarpal joints are enclosed by a fibrous capsule on the palmar and dorsal surfaces. The synovial membrane lines the fibrous capsule. The **fibrous capsule** of the carpometacarpal joint of the thumb surrounds the joint and is attached to the margins of the articular surfaces. The **synovial membrane** lines the capsule. The looseness of the capsule facilitates free movement of the joint.

Ligaments of the Carpometacarpal and Intermetacarpal Joints

The bones are united by anterior carpometacarpal and metacarpal ligaments (Fig. 6.77*A*), as well as by the posterior and interosseous ligaments.

Movements at the Carpometacarpal and Intermetacarpal Joints

Flexion-extension, abduction-adduction, and circumduction occur at the carpometacarpal joint of the thumb (Fig. 6.74*C*). This joint permits angular movements in any plane and a restricted amount of axial rotation. Almost no movement occurs at the carpometacarpal joints of the 2nd and 3rd digits, the 4th digit is slightly mobile, and the 5th digit is extremely mobile. This 5th carpometacarpal joint flexes considerably during a tight grasp and rotates slightly (Fig. 6.52).

Blood Supply of the Carpometacarpal and Intermetacarpal Joints

These joints are supplied by dorsal and palmar metacarpal arteries and deep carpal and deep palmar arches (see Fig. 6.82).

Innervation of the Carpometacarpal and Intermetacarpal Joints

These joints are supplied by the anterior interosseous branch of the *median nerve*, posterior interosseous branch of the *radial nerve*, and dorsal and deep branches of the *ulnar nerve*.

Metacarpophalangeal and Interphalangeal Joints

The metacarpophalangeal articulations are the condyloid type of synovial joint that permit movement in two planes: flexion-extension and adduction-abduction. The interphalangeal articulations are the hinge type of synovial joint—flexion-extension only (Fig. 6.77*B*).

Articulations of the Metacarpophalangeal and Interphalangeal Joints

The heads of the metacarpals articulate with the bases of the proximal phalanges in the metacarpophalangeal joints, and the heads of the phalanges articulate with the bases of more distally located phalanges in the interphalangeal joints.

Articular Capsule of the Metacarpophalangeal and Interphalangeal Joints

A **fibrous capsule** encloses each joint and a **synovial membrane** lines each fibrous capsule that is attached to the margins of each joint.

Ligaments of the Metacarpophalangeal and Interphalangeal Joints

Each fibrous capsule is strengthened by two **collateral ligaments** that pass distally from the heads of the metacarpals and phalanges to the bases of the phalanges (Fig. 6.77, *A* and *B*). The collateral ligaments fuse to form the anterior part of each capsule, which is a thick, densely fibrous or fibrocartilaginous plate—the **palmar ligament** (plate). The 2nd through 5th metacarpophalangeal joints are united by deep transverse metacarpal ligaments that hold the heads of the metacarpals together.

Movements of the Metacarpophalangeal and Interphalangeal Joints

Metacarpophalangeal joints—flexion-extension, abduction-adduction, and circumduction of the 2nd through 5th digits; flexion-extension of the thumb occurs but abduction-adduction is limited. *Interphalangeal joints*—flexion-extension.

Blood Supply of the Metacarpal and Interphalangeal Joints

Deep digital arteries that arise from the superficial palmar arches supply these joints (see Fig. 6.82).

Innervation of the Metacarpal and Interphalangeal Joints

Digital nerves that arise from the ulnar and median nerves supply these joints.

Bullrider's Thumb

This injury refers to a sprain of the radial collateral ligament and an avulsion fracture of the lateral part of the proximal phalanx of the thumb. This is a common injury in persons who ride mechanical bulls.

Skier's or Gamekeeper's Thumb

This injury refers to rupture or chronic laxity of the collateral ligament of the 1st metacarpophalangeal joint. The injury results from hyperabduction of the metacarpophalangeal joint of the thumb, which occurs when the thumb is held by the ski pole while the rest of the hand hits the ground or enters the snow. In severe injuries, the head of the metacarpal has an avulsion fracture. ○

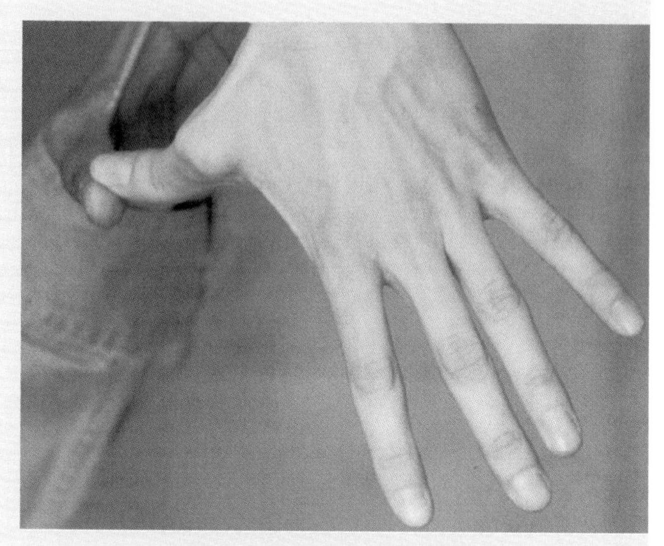

Skier's (gamekeeper's) thumb

Medical Imaging of the Upper Limb

Radiography

Radiographs are often taken when an injury produces bone tenderness, angulation, rotation, or instability. Anteroposterior (AP), lateral, axial, and oblique views are usually sufficient to detect bony and other injuries. Radiological examinations of the upper limb focus mainly on bony structures because muscles, tendons, and nerves are not well visualized. When examining radiographs of the upper limb, it is essential to know the median times of appearance of postnatal ossification centers and when fusion of epiphyses is complete radiographically in males and females. Without this knowledge, an epiphyseal line could be mistaken for a fracture.

Pectoral Girdle and Joints

The usual radiograph is an AP view with the humerus laterally rotated so that its epicondyles are parallel with the x-ray film cassette (Fig. 6.78*A*). The humerus is viewed in its anatomical position. Because some obliquity of the glenoid cavity occurs normally, it appears oval in shape. In this view, the lateral two-thirds of the clavicle is visible. Observe the articulation of the acromial end of the clavicle with the acromion of the scapula at the *AC joint.* This is a frequent site of subluxation of the joint (often referred to as a separated shoulder). Observe the surgical neck of the humerus, approximately 2.5 cm distal to the greater and lesser tubercles, which is a common site of fracture. Keep in mind that the axillary nerve is in contact with the surgical neck and vulnerable to injury. Examine the body of the humerus, noting its dense cortical bone. Observe that this layer thins out proximally and becomes extremely thin over the head of the humerus. Be mindful that the radial nerve runs inferolaterally on the posterior surface of the humerus and is vulnerable to injury in a midhumeral fracture.

An axial projection of the shoulder is examined to obtain another view of the AC and glenohumeral joints. To obtain an axial projection, the patient is asked to abduct the arm and extend the shoulder over the x-ray film cassette. In Figure 6.78*B*, observe the acromion, glenoid cavity, humeral head, coracoid process, and suprascapular notch.

Elbow Joint

In an AP projection, the extended elbow is placed on the x-ray film cassette with the forearm supinated. The x-ray beam is directed perpendicular to the cassette and at the elbow joint. Figure 6.79*A* shows the distal end of the humerus as it flares out and ends as the prominent medial epicondyle and the more flattened lateral epicondyle. Keep in mind that the ulnar nerve is in contact with the medial epicondyle and may be injured when it is fractured or the epiphysis for it is dislocated. Distal to the epicondyles are the convex capitulum and pulley-shaped trochlea. Between the epicondyles is a radiolucent area representing the superimposed olecranon and coronoid fossae. Observe the disc-shaped radial head distal to the capitulum, noting its concave articular surface. Fractures of the radial head are relatively common in young adults after a fall on the hand. Observe the radial notch in the ulna where the radial head articulates at the proximal radioulnar joint. The olecranon of the ulna is superimposed on the distal end of the humerus where it lies in the olecranon fossa. Just dis- ▶

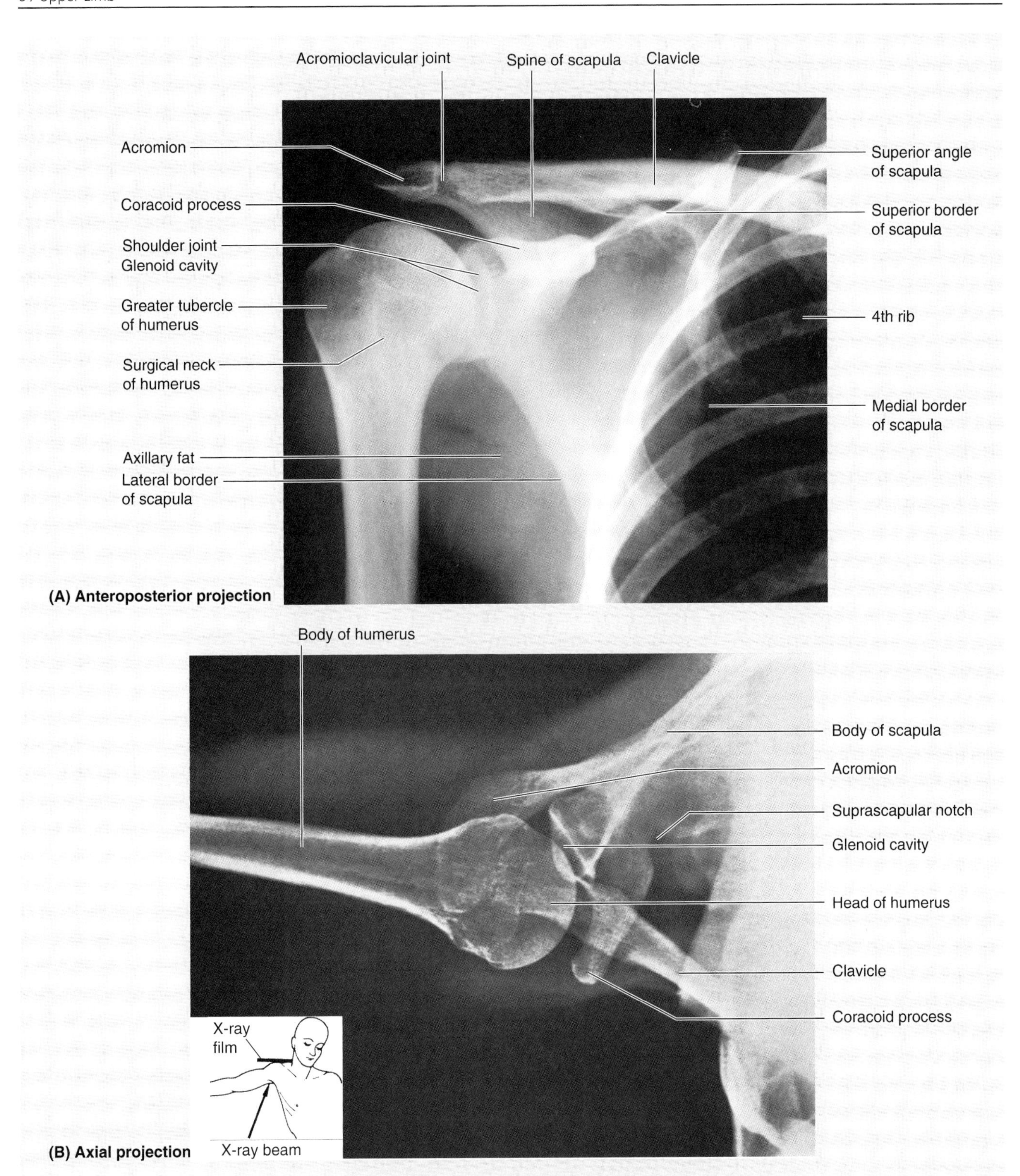

Figure 6.78. Radiographs of the glenohumeral joint. A. Anteroposterior (AP) projection. (Courtesy of Dr. E. L. Lansdown, professor of Medical Imaging, University of Toronto, Toronto, Ontario, Canada) **B.** Axial projection. The orientation drawing shows how the radiograph was taken.

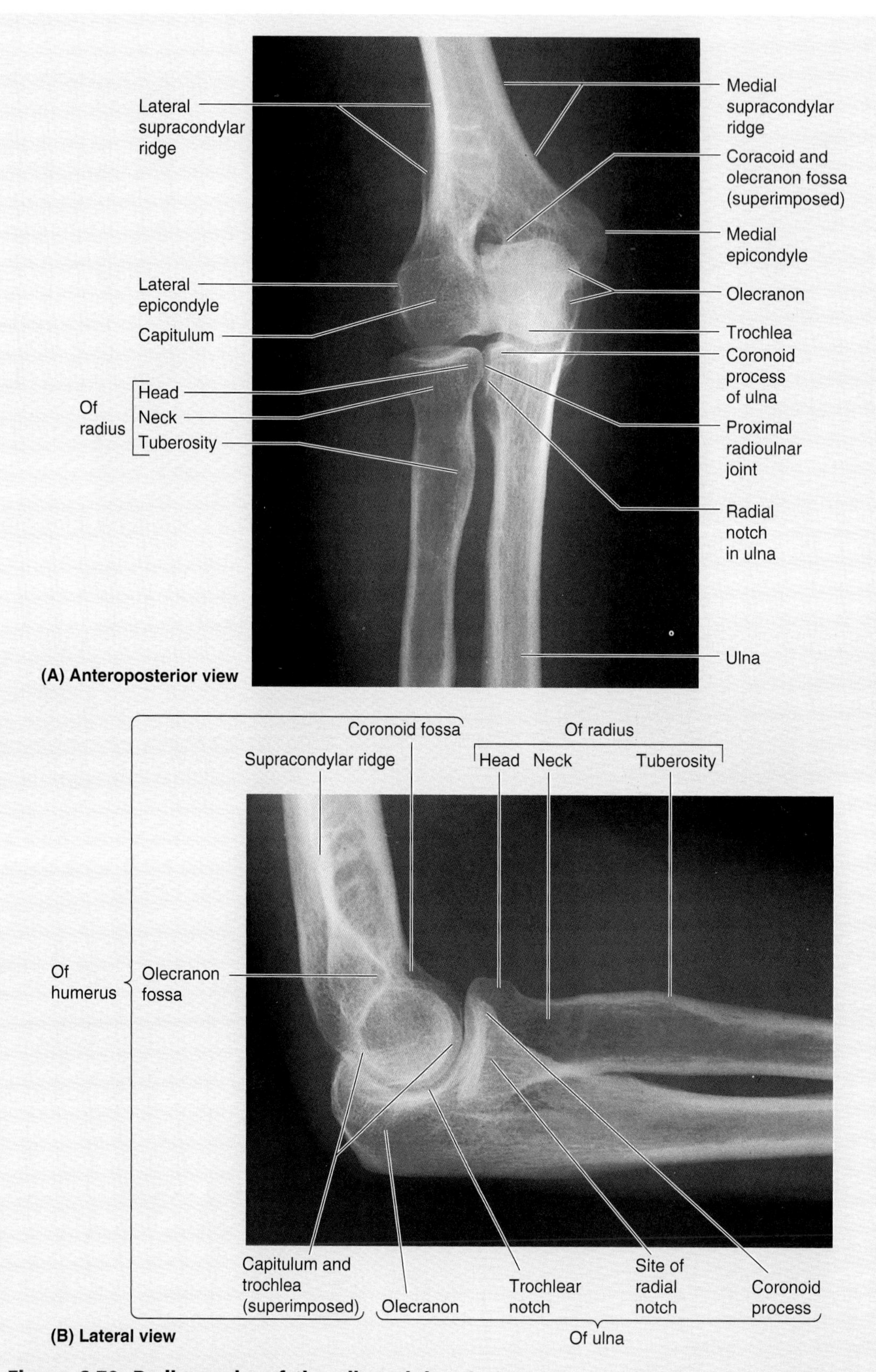

Figure 6.79. Radiographs of the elbow joint. A. Anteroposterior (AP) view of extended joint. **B.** Lateral view of joint flexed to a 90° angle. Observe the articular surfaces of the bones of the elbow. Also observe the proximal radioulnar joint between the head of the radius and the radial notch of the ulna. This joint is involved in pronation and supination. (Courtesy of Dr. E. Becker, Association Professor of Medical Imaging, University of Toronto, Toronto, Ontario, Canada)

▶ tal to the trochlea of the humerus is the coronoid process of the ulna.

In a lateral projection of the elbow (Fig. 6.79*B*), the elbow is flexed 90° and the arm and forearm are placed on the x-ray film cassette with the thumb extended. In this projection, the epicondyles, capitulum, and trochlea are superimposed. The superimposed supracondylar ridges form a dense radiopaque shadow. The coronoid process of the ulna is partly obscured by the radial head.

Forearm

To obtain the AP view shown in Figure 6.80*A*, the forearm is positioned so that it is supinated and the elbow joint is fully extended. Note that both the elbow and wrist joints are shown. Observe the translucent area between the capitulum and radial head and between the trochlea and coronoid process. This area is where the articular cartilages of the articulating bones are located. Observe how the radius and ulna bow as they extend distally from the elbow. Also observe that the bones separate from each other in their distal two-thirds. The interosseous membrane is invisible because it is radiolucent. To detect fractures and verify their correct reduction (restoration to the normal position), keep in mind that the radial styloid process normally ends more distally than the ulnar styloid process. In Figure 6.80*B*, observe that the radius crosses the ulna when ▶

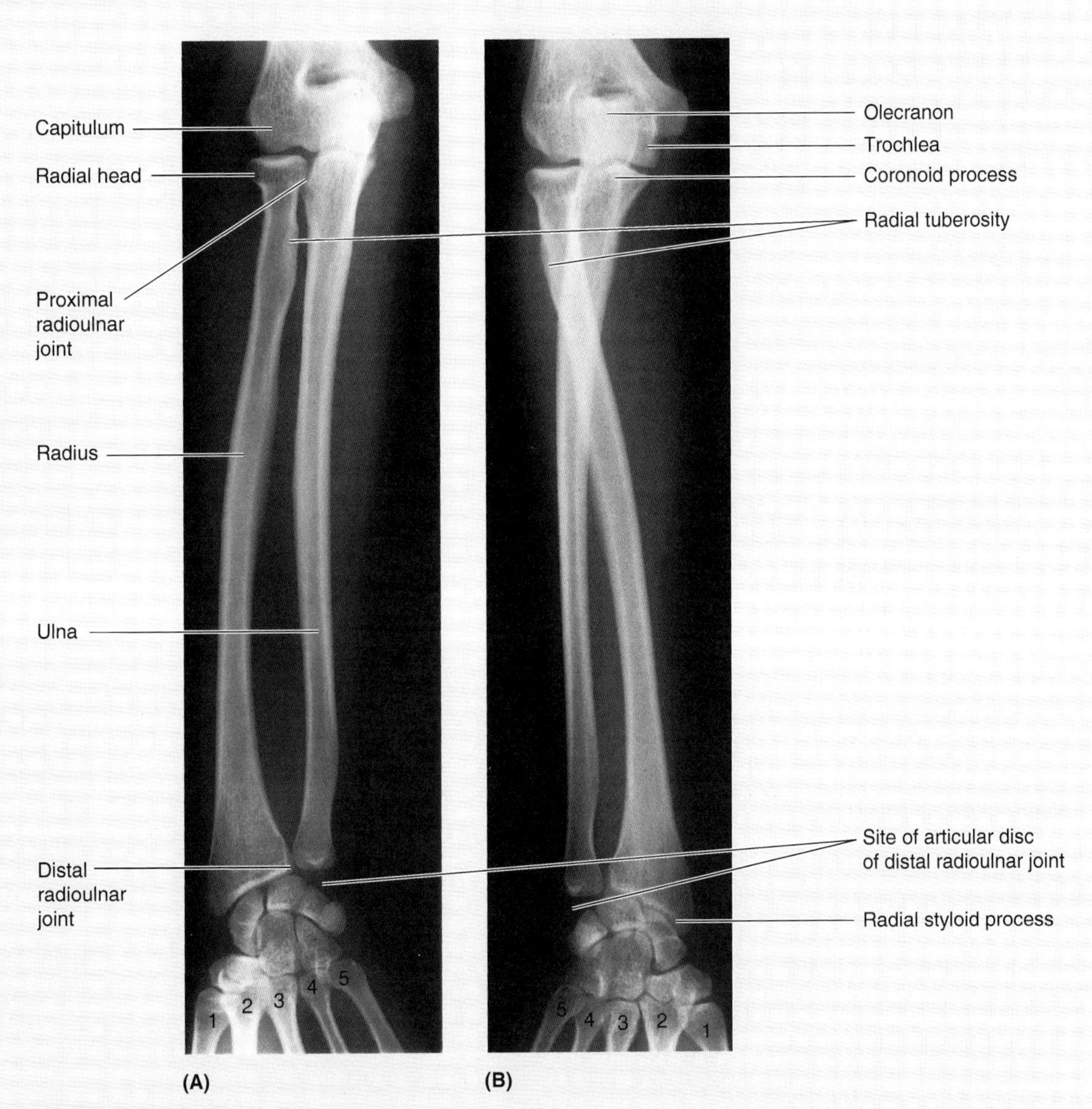

Figure 6.80. Radiographs of the radioulnar joints. The movements between the radius and ulna are supination and pronation. **A.** Anteroposterior (AP) views of the right forearm in supination. **B.** AP view of the right forearm in pronation. Observe that the radius crosses the ulna. During pronation the inferior end of the radius moves anteriorly and medially around the inferior end of the ulna, carrying the hand (L. manus) with it. (Courtesy of Dr. J. Heslin, Toronto, Ontario, Canada)

▶ the forearm is pronated. Examine the proximal and distal radioulnar joints.

Wrist and Hand

Radiographs of the wrist and hand are commonly used to assess skeletal age. For clinical studies, the radiographs are compared with a series of standards in a radiographic atlas of skeletal development to determine the age of the child. *Epiphyseal fractures* are more common in young children than in adolescents because tight fusion of the epiphyses with the bodies of the bones has not occurred. Fusion of the distal radial and ulnar epiphyses is complete radiographically at 16 years in females and at 18 years in males. The carpal bones are superimposed on each other; however, when compared with a skeleton of the wrist and hand, the bones can be identified. A lateral projection of the wrist and hand (Fig. 6.81*A*) is of great importance because in some instances it is the only projection that reveals certain fractures. A fall on the outstretched hand may result in *fracture of the scaphoid*, generally across its narrow part ("waist"). A radiologist looks for a scaphoid fracture if the physician reported tenderness over the scaphoid in the anatomical snuff box; however, it may not be detectable even in several oblique views until approximately 2 weeks after the injury (p. 675). ▶

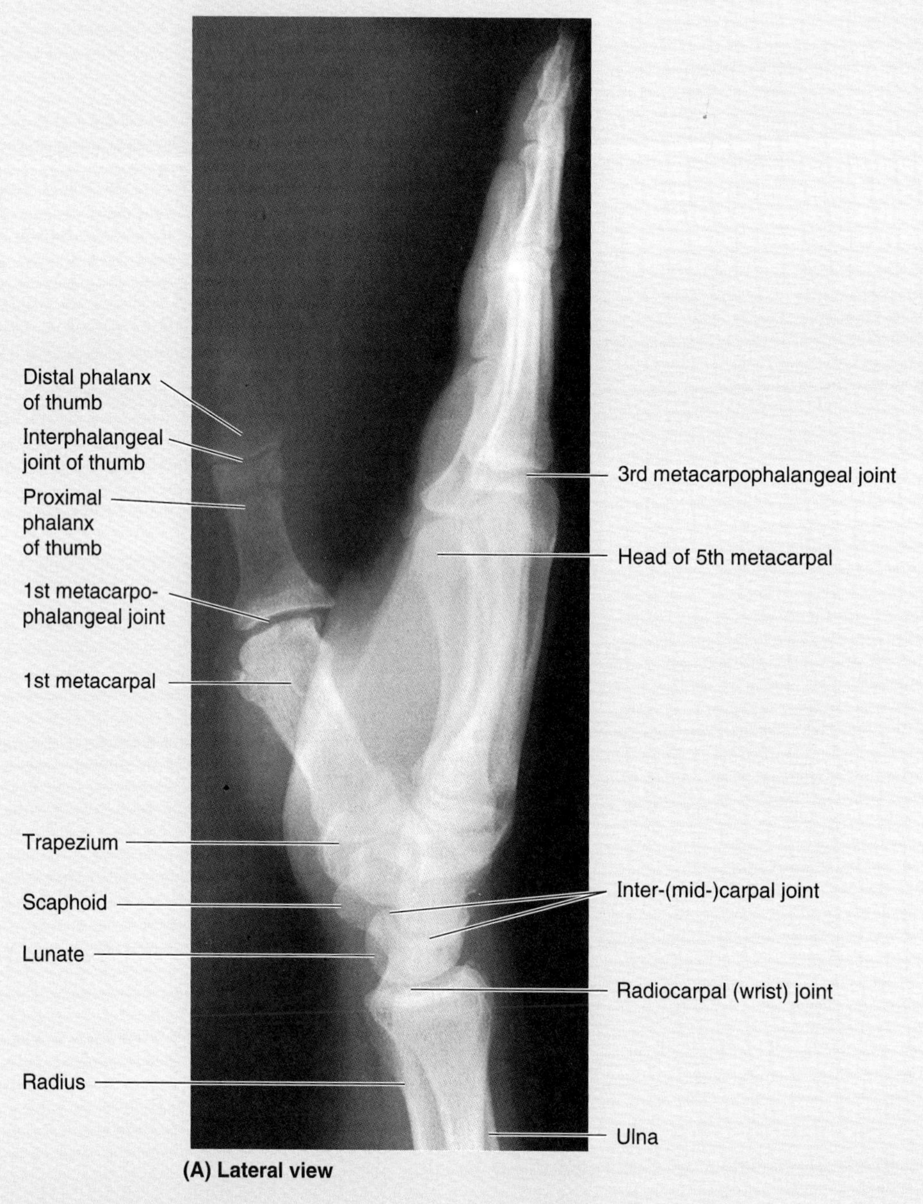

Figure 6.81. Radiographs of the wrist and hand. A. Lateral view of an adult hand. (Courtesy of Dr. E. L. Lansdown, Professor of Medical Imaging, University of Toronto, Toronto, Ontario, Canada)

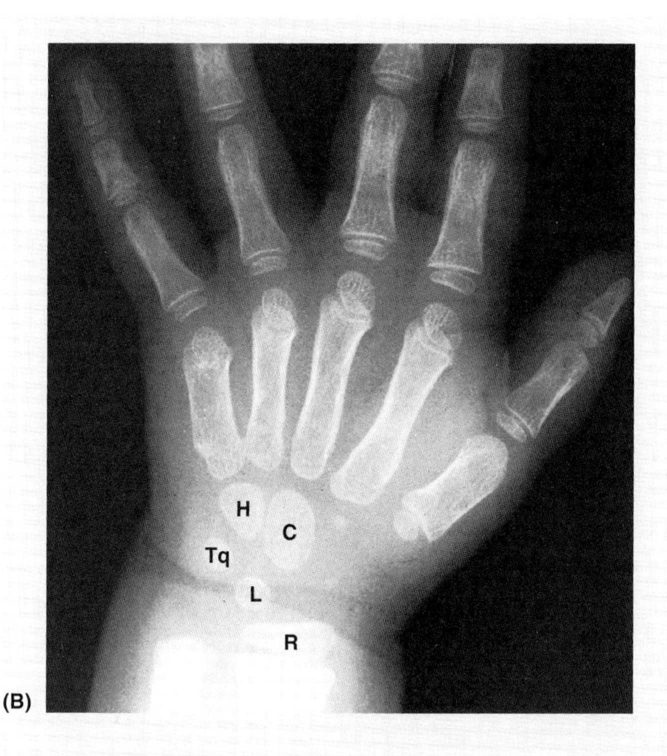

(B)

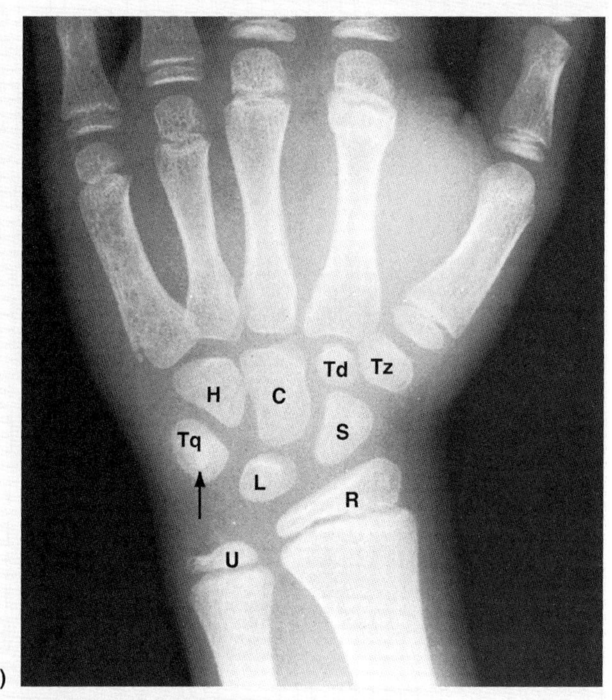

(C)

Posteroanterior view

Figure 6.81. (*Continued*) **B.** Posteroanterior view of the distal end of the forearm and the hand of a 2½-year-old child. Ossification centers of only four carpal bones appear. Radiographs of the hand and wrist are often used to assess skeletal age. Observe the distal radial epiphysis. *C*, capitate; *H*, hamate; *Tq*, triquetrum; *L*, lunate. **C.** Posteroanterior view of the inferior end of the forearm and hand of an 11-year-old child. Ossification centers of all carpal bones are apparent. The *arrowhead* indicates the pisiform. Observe that the distal epiphysis of the ulna has ossified but that all epiphyseal plates (lines) "remain open" (i.e., they are still unossified). *S*, scaphoid; *Td*, trapezoid; *Tz*, trapezium. (Courtesy of Dr. D. Armstrong, Associate Professor of Medical Imaging, University of Toronto, Toronto, Ontario, Canada)

▶ *Anterior dislocation of the lunate* is a fairly common wrist injury. Each carpal usually ossifies from one center postnatally. The centers for the capitate and hamate appear first and usually are obvious during the 1st year; however, they may appear before birth. You should be aware that *accessory ossicles* (small bones) are sometimes observed between the usual carpals, and their possible occurrence should be kept in mind when examining radiographs of the carpus. *Carpal fusions* (e.g., between the lunate and triquetrum) may occur and should be regarded as variations.

Except for the thumb, the metacarpals and phalanges of the fingers are superimposed, and little information can be obtained by examining them (Fig. 6.81*A*). If a particular finger is injured and suspected of having a fracture, it is viewed laterally with the other fingers flexed. The body of each metacarpal begins to ossify during fetal life, and centers appear postnatally in the heads of the four medial metacarpals (Fig. 6.81*B*) and in the base of the 1st metacarpal. *Pseudoepiphyses* (accessory centers) are occasionally seen in the head of the 1st metacarpal and in the base of the 2nd metacarpal.

Ultrasonography

Ultrasonography is highly accurate for rotator cuff tears (Halpern, 1994). *Doppler ultrasound* is used to visualize blood flowing through blood vessels of the limbs and to measure its velocity. *Color doppler ultrasound* technology allows red and blue colors to be superimposed on the standard grayscale images.

Arteriography

Arteriography (visualization of arteries) is used to detect vascular injuries, ischemia (deficiency of blood supply), and variations of arteries. *Arteriograms* are produced by injecting a dye into an artery as a radiograph is taken (Fig. 6.82). When interest is in a specific artery, the dye is injected directly into it, usually through a catheter. Arteriography may be used to determine the patency of arteries before and after surgical procedures in the upper limb to repair injuries resulting from trauma. ▶

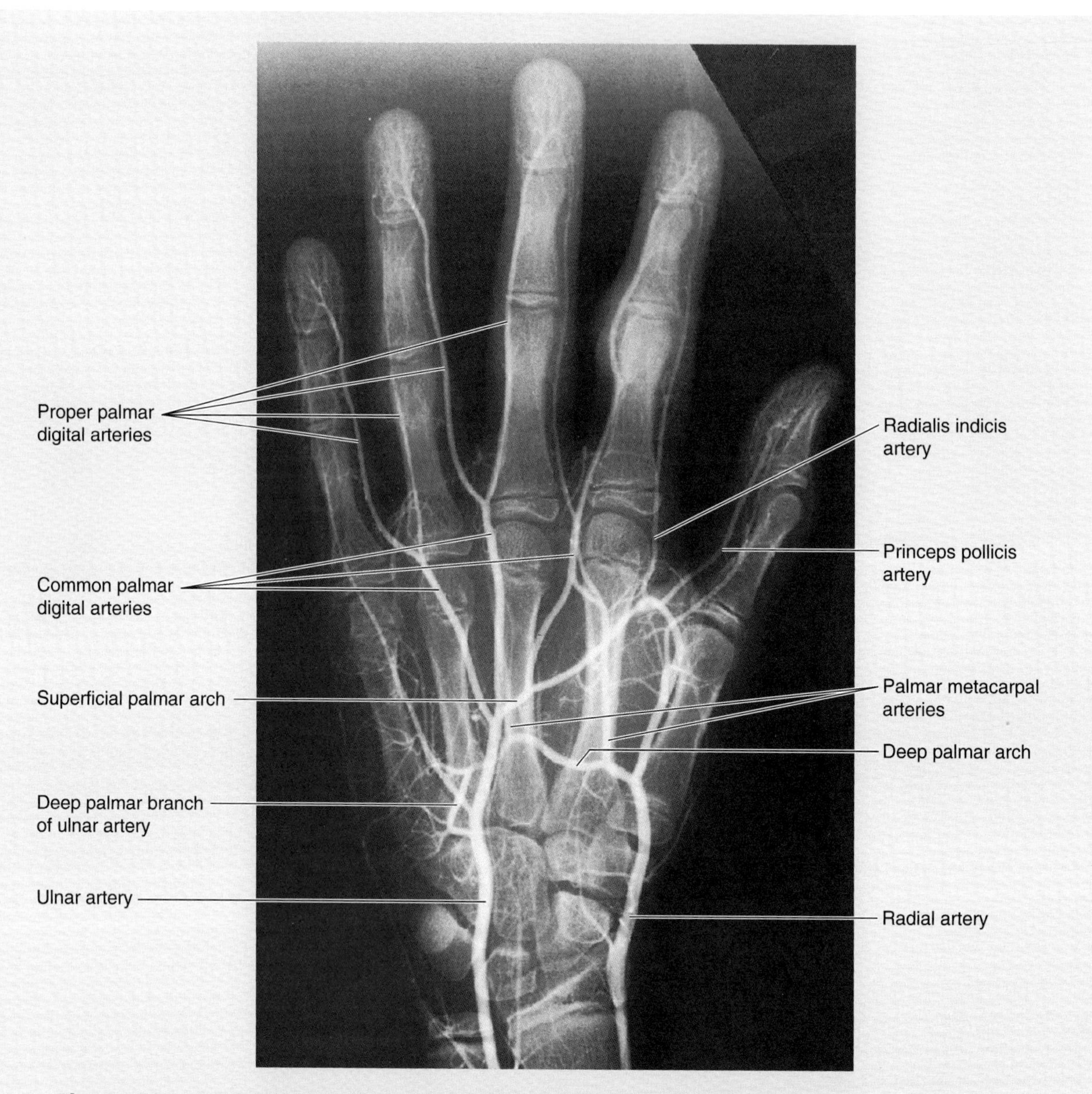

Figure 6.82. Angiogram of the wrist and hand. Observe the superficial and deep palmar arterial arches. The superficial palmar arch is formed primarily by the ulnar artery and is usually completed by the superficial palmar branch of the radial artery. In this teenage hand, the carpal bones are fully ossified, but the epiphyseal lines of the long bones are still open. Closure will occur when growth is complete, usually at the end of the teenage years. (Courtesy of Dr. D. Armstrong, Associate Professor of Medical Imaging, University of Toronto, Toronto, Ontario, Canada)

Computed Tomography

CT scans of the upper limb are produced by x-rays, but the anatomical images are created by computer reconstructions of the electronic data. The reconstructed images look like cross sections of the limb, and they can be reconstructed in the coronal and sagittal planes. Because the patient only has to be in the CT scanner for brief periods, fast (2-second) CT scans can be produced; consequently, involuntary movement produces minimal distortion of the images. *Arthrotomography* enhanced by CT demonstrates subtle fractures (e.g., of the rim of the glenoid cavity) and reactive bone changes around a recurrently dislocating shoulder joint (Halpern, 1994).

Magnetic Resonance Imaging

Magnetic resonance imaging (MRI) produces images with good resolution of the various soft tissues of the limbs because adults can keep their limbs motionless for the 5 to 10 ▶

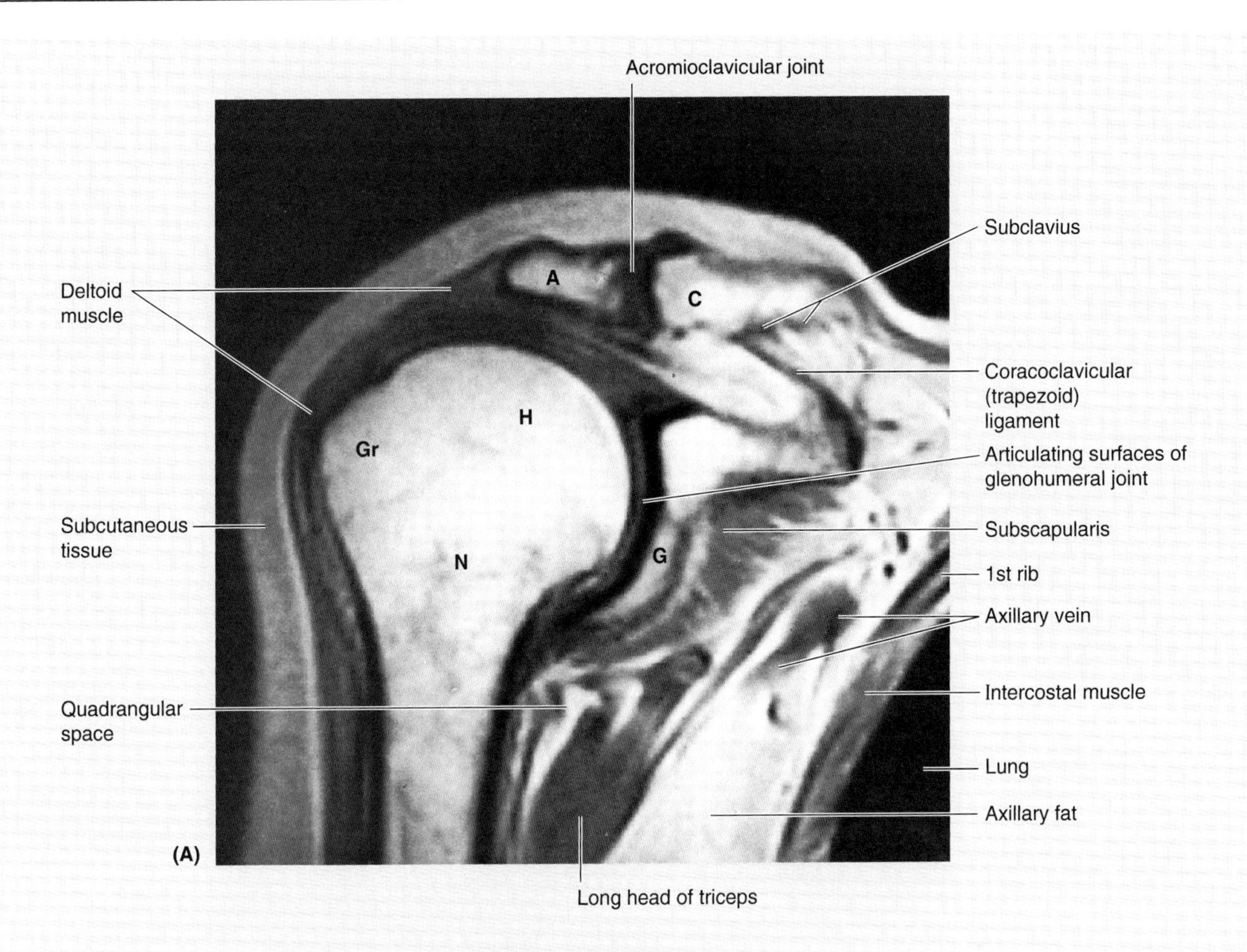

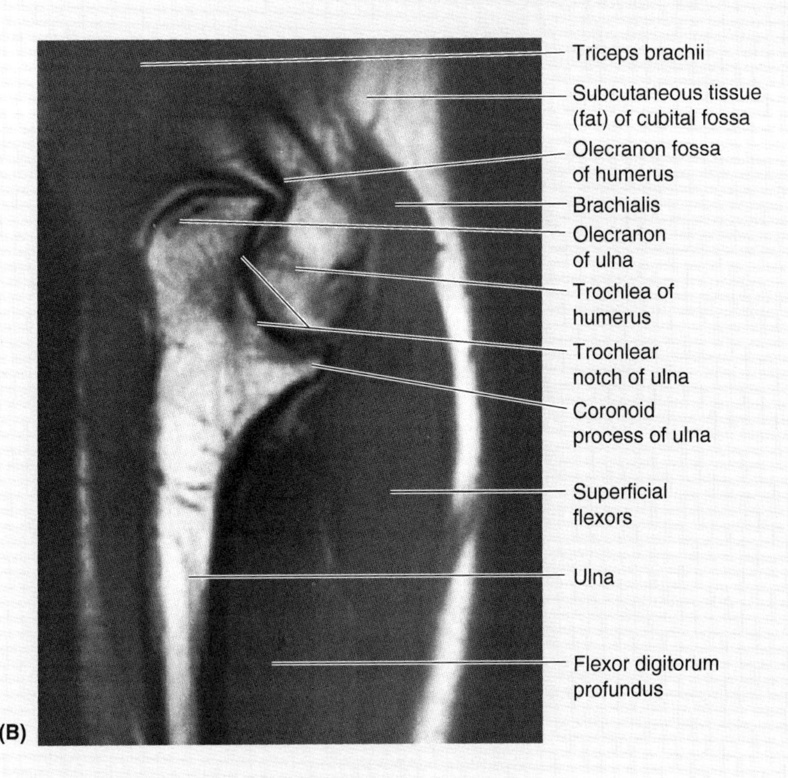

Figure 6.83. MRIs of the upper limb. **A.** Coronal MRI of the glenohumeral and acromioclavicular (AC) joints. The "white" (signal intense) parts of the identified bones are the fatty matrix of cancellous bone; the thin black outlines (absence of signal) of the bones are the compact bone that forms their outer surface. *A*, acromion; *C*, clavicle; *Gr*, greater tubercle of humerus; *N*, surgical neck of humerus; *G*, glenoid cavity; *H*, head of humerus. **B.** Sagittal MRI of the elbow.

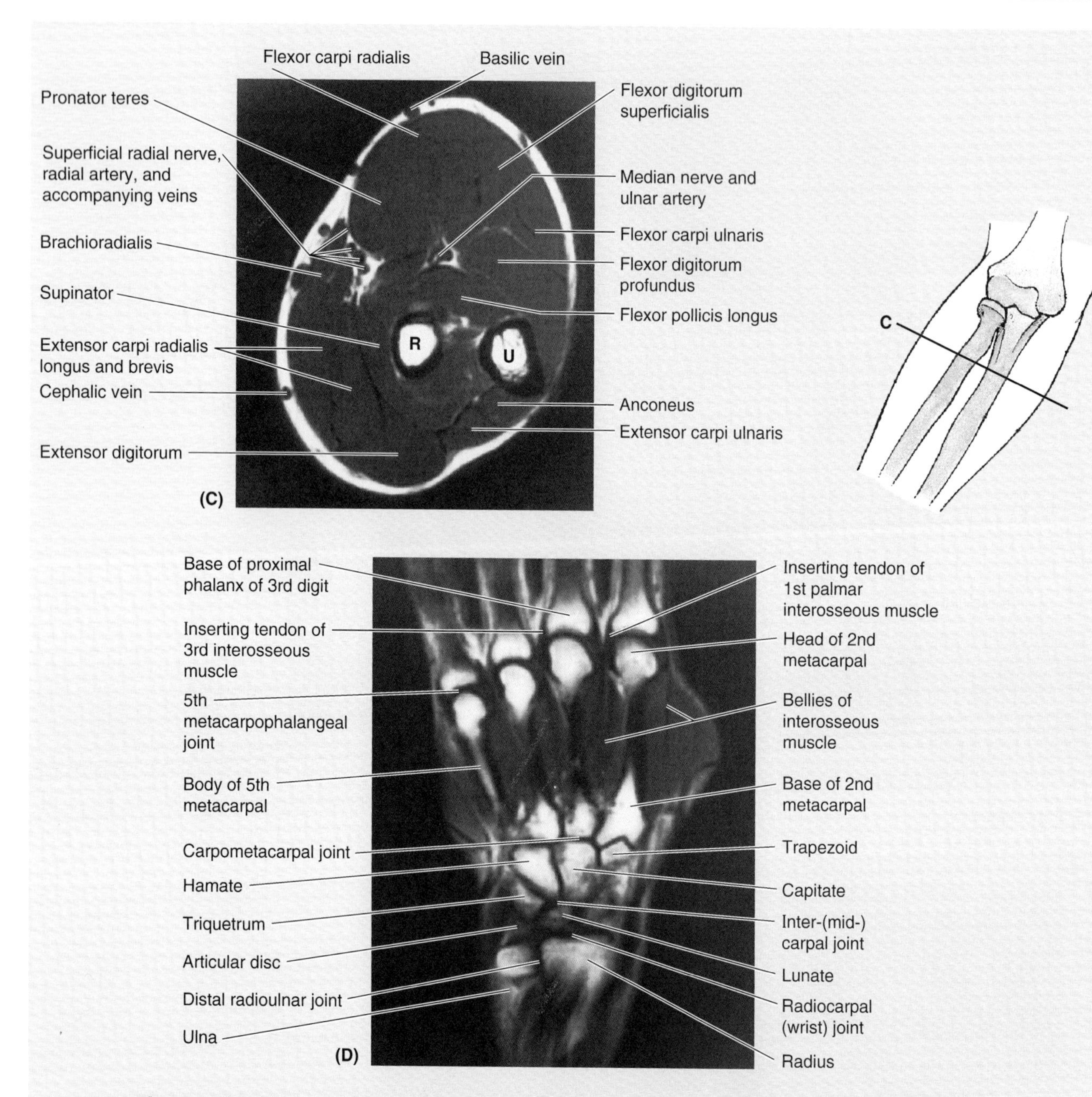

Figure 6.83. (*Continued*) **C.** Transverse MRI of the proximal part of the forearm. The orientation drawing shows the plane of the MRI. The subcutaneous border of the ulna and the radial artery demarcate the anteromedial flexor and posterolateral extensor compartments. *U*, ulna; *R*, radius. **D.** Coronal MRI of the wrist and hand (L. manus). Observe the bones, joints, and interosseous muscles. (Courtesy of Dr. W. Kucharczyk, Chair of Medical Imaging and Director of Tri-Hospital Magnetic Resonance Centre, Toronto, Ontario, Canada)

▶ minutes required for scanning. This technique produces cross-sectional, coronal, and sagittal images and has the advantage that it does not involve the use of x-rays. Because MRI requires the patient to remain still for several minutes, this technique is difficult to use with children. In the MRIs shown in Figure 6.83, observe the good resolution of the muscles, bones, and joints. MRI is helpful in the diagnosis of subluxation, dislocation, impingement of nerves, cartilage tears, and rotator cuff tears of the shoulder joint (Halpern, 1994). Coronal MRIs are helpful in demonstrating ruptured tendons in the shoulder, elbow, hand, and wrist regions. ⊙

CASE STUDIES

Case 6.1

A 52-year-old woman was riding her bicycle along a gravel path. Suddenly she lost her balance and fell on her outstretched upper limb. She said she heard a distinct cracking noise and felt sudden pain in her shoulder region. Her husband, a physician, observed a deformity of her clavicle at the junction of its lateral and intermediate thirds and realized that she had broken her clavicle. He also noted that the lateral part of her shoulder was slumped inferomedially and that the medial fragment of the fractured clavicle was elevated. He made a sling for her limb with his T-shirt.

Clinicoanatomical Problems

- Where does the clavicle commonly fracture?
- Are clavicular fractures more common in adults than in children?
- Why did her shoulder slump inferomedially?
- Why did her clavicle fracture without dislocating her AC joint?
- Why did her clavicle fracture rather than her wrist?

The problems are discussed on page 824.

Case 6.2

A 35-year-old baseball pitcher told his catcher and coach that he felt a gradual onset of shoulder pain. He continued to pitch but had to stop because of pain and weakness, especially during abduction and lateral rotation of his arm. When examined by the team physician, supraspinatus tenderness was detected near the greater tubercle of his humerus. An MRI revealed a tear in the pitcher's rotator cuff.

Clinicoanatomical Problems

- What is the rotator cuff of the shoulder?
- What usually causes rotator cuff strain?
- What part of the rotator cuff usually tears?
- Do these injuries occur only in baseball pitchers?
- What shoulder movement is weak and causes pain?

The problems are discussed on page 824.

Case 6.3

Several weeks after returning home following surgical dissection of her right axilla for the removal of lymph nodes for staging and treatment of her breast cancer, a 44-year-old woman was told by her husband that her right scapula protruded abnormally when she pushed against the wall during her stretching exercises. She also experienced difficulty in raising her right arm above her head when she was combing her hair. During her return visit with her surgeon, she was told that a nerve was accidentally injured during the diagnostic surgical procedure and that this produced her scapular abnormality and inability to raise her arm normally.

Clinicoanatomical Problems

- What nerve was probably injured?
- Why did this injury cause "winging of her scapula" and difficulty in raising her arm?
- If these scapular abnormalities were observed in an automobile accident victim, what fractures might have caused the nerve injury?
- During removal of axillary lymph nodes, what other nerves are vulnerable to injury?
- What abnormalities of arm movement would likely be present?
- Would any areas have cutaneous anesthesia?

The problems are discussed on page 825.

Case 6.4

During a pickup game of football on artificial turf, a 38-year-old ball carrier was slammed to the ground by a linebacker. He landed hard on his right shoulder and indicated that he felt moderate pain that became worse when he attempted to raise his arm. During the physical examination of his shoulder by an orthopaedist, slight superior displacement of the acromial end of his clavicle was noted. Inferior pressure on the clavicle revealed tenderness and some mobility of the clavicle where it articulates with the acromion. Abduction of the arm beyond 90° caused severe pain and abnormal movement of the acromion and clavicle at the AC joint.

Clinicoanatomical Problems

- What part of the AC joint hit the hard artificial turf?
- Based on your anatomical knowledge of the AC joint, what injury do you think resulted from the fall on his shoulder?
- What do sportswriters call this injury?
- What ligament(s) do you think would be disrupted and torn?
- Would the articular capsule be injured?
- If the player had fumbled the ball and landed on his open hand, what bone do you think might have been fractured?

The problems are discussed on page 825.

Case 6.5

A 32-year-old woman who was learning to play tennis practiced daily for approximately 2 weeks. She reported to her coach that she experienced pain over the lateral region of her elbow that radiated down her forearm. Familiar with this complaint by beginners, he asked her to hold the tennis racket and extend her hand at the wrist. She felt no pain until he resisted extension of her hand. When he asked her to pinpoint the area of most pain, she placed her finger over her lateral epicondyle. When he put pressure on the epicondyle, she pulled her elbow away because of the severe pain. The coach compressed the common extensor tendon, and she again experienced intense pain.

Clinicoanatomical Problems

- What elbow injury do you think she has sustained?
- What are the mechanisms of this injury?
- Does this injury occur only in tennis players?
- Where is the discrete point of local tenderness in these injuries?
- Why did the woman experience radiation of pain along the posterolateral aspect of her forearm?

The problems are discussed on page 825.

Case 6.6

A 65-year-old man consulted his physician about a dull ache in his right shoulder that often awakens him during the night, especially when he raises his arm above his shoulder when asleep. He said the severe pain sometimes runs down his right arm and that a friend told him the pain might be a symptom of heart disease. During the physical examination, the physician noted that he kept his shoulder in the adducted position. As she pressed his shoulder firmly just lateral to his acromion, she noted extreme tenderness. She asked him to raise his upper limb slowly. When his limb reached approximately 50° of abduction, he said he felt extreme pain (Fig. 6.84). The pain persisted until his limb was well above his shoulder (approximately 130° of abduction). As his limb approached his head, he said the pain disappeared.

Clinicoanatomical Problems

- Using your knowledge of the rotator cuff of the shoulder joint, what inflammatory condition do you think would cause this arc of painful abduction?
- Inflammation of what synovial sac would exacerbate the pain?
- Why was pressure on the shoulder lateral to the acromion so painful?
- How would you explain anatomically the painful arc syndrome?

The problems are discussed on page 825.

Case 6.7

During the difficult birth of a baby, the physician applied force to the baby's upper limb while the shoulder was still in the birth

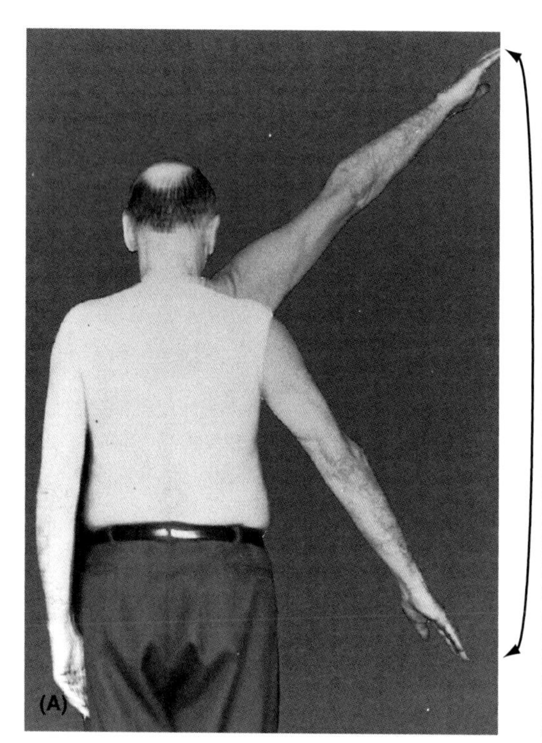

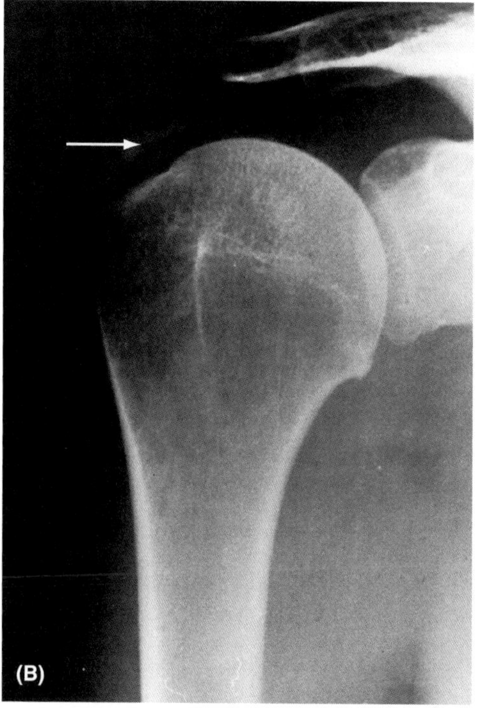

Figure 6.84. Arc of painful abduction. A. Double exposure photograph of a middle-aged man demonstrating the *painful arc syndrome* associated with calcific supraspinatus tendinitis in the right shoulder. Abduction of the shoulder joint from approximately 50 to 130° (*two-headed arrow*) causes severe pain because of tendinitis and subacromial bursitis. **B.** Radiograph of the right shoulder joint showing calcium (*arrow*) deposits in the supraspinatus tendon of the musculotendinous rotator cuff, close to its insertion into the humerus (arm bone).

throwing (e.g., a baseball) injures the rotator cuff and may tear it. The relatively avascular supratendinous part of the rotator cuff near the greater tubercle of the humerus is the part that usually tears initially.

Rotator cuff injuries can occur in anyone who throws something or falls with the arm abducted. The injury is common in people older than 45 years who strain themselves on weekends or holidays performing activities such as skiing, body surfing, and weight lifting. The mechanism of rotator cuff injury is indirect force to the abducted arm and repetitive microtrauma to the shoulder joint. When the rotator cuff is torn, abduction of the arm is weak and pain is severe during 70 to 120° of arm abduction.

Case 6.3

The *long thoracic nerve* to the serratus anterior was obviously injured. During axillary dissection, it is normally identified and maintained against the thoracic wall while the lymph nodes are excised. However, the nerve may be accidentally damaged during removal of nodes. Injury to the long thoracic nerve causes *paralysis of the serratus anterior*, the muscle that keeps the medial border of the scapula in firm apposition with the thoracic wall. The serratus anterior, also powerful, assists the trapezius in rotating the scapula laterally and superiorly when raising the arm over the shoulder. This explains why the patient had difficulty combing her hair.

Injuries to the long thoracic nerve and paralysis of the serratus anterior frequently result from weapons (knives, gunshots); however, they may occur during severe automobile accidents, or when a person is run over by a motor vehicle. Scapular fractures and injury to the long thoracic nerve are usually associated with rib fractures. The *thoracodorsal nerve* (nerve to latissimus dorsi) is in danger during operations on the inferior part of the axilla. The nerve runs inferolaterally along the posterior wall of the axilla and enters the latissimus dorsi at the level of the 2nd and 3rd ribs. A person with paralysis of latissimus dorsi would have difficulty adducting the arm and rotating it medially.

During axillary dissections, care must also be taken to avoid injury to the pectoral nerves supplying the pectoralis major. Paralysis of this muscle would seriously affect adduction and weaken medial rotation of the arm. The *intercostobrachial nerves*, the lateral cutaneous branches of the 2nd intercostal nerve, sometimes have to be sacrificed in radical axillary dissections because these nerves pass close to the axillary lymph nodes to reach the arm. Injury to these nerves causes anesthesia of the skin of the axilla and the posteromedial aspect of the arm.

Case 6.4

The acromion of the scapula likely struck the hard artificial turf first. The acromion forms the bony tip of the shoulder. The medial end of the acromion articulates with the acromial end of the clavicle. Likely the injury was a subluxation of the AC joint. Sportswriters refer to this injury as a shoulder separation, but it should be referred to as an *AC subluxation.*

If marked widening of the AC joint and considerable instability of the acromial end of the clavicle with a "high-riding clavicle" was seen, an *AC dislocation* would most likely have occurred. In this case, the *AC ligament* would be completely torn (i.e., disrupted), and parts of the coracoclavicular ligament would also be torn. The articular capsule would be severely stretched and the superior part of it would probably be torn where the clavicle was displaced superiorly.

If the player had landed on his open hand, he might have fractured his clavicle. The mechanism of injury is that the traumatic force is transmitted through the forearm and arm to the pectoral girdle. In this case, the force is transmitted to the clavicle and does not usually injure the AC joint.

Case 6.5

The symptoms clearly indicate *lateral epicondylitis* (tennis elbow), the most common painful condition in the elbow region. This elbow injury usually results from repetitive microtrauma of the common extensor origin of the forearm extensor muscles. Lateral epicondylitis represents approximately 70% of all sports injuries in persons 40 to 50 years of age. Lateral epicondylitis also occurs in baseball, swimming, gymnastics, fencing, and golf—that is, in sports that involve heavy use of the forearm extensors. Lateral epicondylitis can also result from direct trauma to the lateral epicondyle (e.g., a slash with a hockey stick on an unpadded elbow).

The discrete point of local tenderness in lateral elbow injuries is just distal to the inflamed lateral epicondyle; however, as the *inflammation of the common extensor tendon* develops, the pain is referred distally along the extensor tendon. In chronic cases, radiographs may reveal dystrophic calcification in the area of degeneration of the extensor muscle origin.

Case 6.6

Calcification in the musculotendinous rotator cuff, particularly in its supraspinatus part, is usually the cause of the painful arc syndrome. In *acute calcific supraspinatus tendinitis*, calcium deposited in the substance of the supraspinatus tendon causes excruciating throbbing pain that is not relieved by rest. *Inflammation of the subacromial bursa*, resulting from irritation by calcification in the rotator cuff, causes an aggravation of the shoulder pain. Pressure on the shoulder lateral to the acromion causes pain because this is close to the greater tubercle of the humerus, where calcium deposits are usually located in the supraspinatus tendon.

The *painful arc syndrome* can be explained by the fact that at 50 to 60° of abduction, the inflamed area of the supraspinatus tendon comes into contact with the undersurface of the acromion. Because of this, the syndrome is also referred to as the *subacromial impingement syndrome.*

Case 6.7

The ventral rami of C5 and C6 nerves, which join to form the superior trunk of the brachial plexus, are torn or severely stretched by pulling on the baby's head when the shoulder is not delivered. The clinical condition is called Erb palsy or Erb-Duchenne palsy. The paralyzed upper limb hangs limply by the side and is adducted and medially rotated; the elbow joint is extended and the forearm is pronated. Consequently, the palmar surface of the hand, instead of facing medially, faces posteriorly when the limb rests at the side in the adducted position.

An upper brachial plexus injury in young adults usually occurs when they fall on the shoulder in such a fashion that the shoulder is forcefully depressed and the head and neck are forcefully flexed to the other side of the body (p. 716). These injuries commonly result from a fall from a motorcycle. An upper brachial plexus injury may occur during a football game when one tackler is pulling on a ballcarrier's arm and another is pulling on the person's face mask.

Years ago, it was considered poor taste to put your hand out for a tip; consequently, porters and waiters held their open palms facing posteriorly so that the tip could be tucked into their hands from behind their backs so the receipt of tips was not so obvious. In Great Britain you will still see this porter's tip position, which at first may make you think all porters were injured during birth.

The muscles partially or completely paralyzed by tearing of C5 and C6 roots of the brachial plexus are those whose sole or major innervation is from C5 and/or C6 (e.g., supraspinatus, deltoid, biceps, and supinator). Atrophy of the deltoid results in the shoulder losing its rounded contour and appearing to droop.

Case 6.8

The long head of the man's biceps brachii was likely injured.

The popping sensation and tenderness over the bicipital groove suggest that the tendon of the long head of the biceps was ruptured. Weakness of flexion and supination of the forearm support the view that the biceps is injured because it participates in both of these upper limb movements. Undoubtedly the bulge was formed by the tonically contracted "fallen" belly of the ruptured long head of the biceps. An avulsion fracture of the supraglenoid tubercle of the scapula, the proximal attachment of the long head of the biceps, may occur, allowing the biceps tendon to dislocate from the intertubercular groove.

Biceps tendinitis, the second most common cause of a painful shoulder, occurs in a wide range of throwing, swimming, and racquet sports (Halpern, 1994). This injury results from repetitive microtrauma of the biceps tendon and its synovial sheath as it slides to and fro in the bicipital groove of the humerus during flexion and supination of the forearm. The constant use of the muscle may cause fraying and eventual rupture of the tendon. Rupture of the biceps tendon may also result from sudden forceful contraction against resistance (e.g., during a checked swing in baseball, a fast-pitched softball, or an arm tackle on a quarterback's passing arm).

Case 6.9

The symptoms and signs suggest that the golfer has an elbow injury called *medial epicondylitis* (golfer's elbow). The mechanism of injury resulting in medial epicondylitis is usually medial tension overload of the elbow from repeated microtrauma of the flexor-pronator group of forearm muscles at the *common flexor origin*, the medial epicondyle.

The probable cause of medial epicondylitis is inflammation of the medial epicondyle and common flexor tendon of the flexor-pronator group of forearm muscles. A sprain of the ulnar collateral ligament of the elbow could also cause pain in the medial part of the elbow. This ligament passes from the medial epicondyle to the coronoid process and olecranon of the ulna. Despite its common name (golfer's elbow), medial epicondylitis occurs primarily in tennis players using the twist serve. It also occurs during the acceleration phase in throwing sports.

Case 6.10

The median nerve was compressed at the wrist as it passed beneath the flexor retinaculum. This fairly common condition—*carpal tunnel syndrome*—can be caused by a variety of conditions (Salter, 1998): edema of acute and chronic trauma, synovial cyst in the carpal tunnel, osteophytes (bony outgrowths of carpals), lipoma, tenosynovitis (inflammation of a tendon and its synovial sheath), and excessive exercise. The symptoms of carpal tunnel syndrome (e.g., sensory changes over the lateral side of the hand, including the lateral two and a half fingers) are aggravated by movements of the wrist. If untreated, ulnar nerve compression might produce objective findings of sensory loss (e.g., elicited by pinpricks) and weakness and atrophy of hand muscles (Fig. 6.85).

Case 6.11

Undoubtedly, the pain and paresthesia in her forearm and hand resulted from compression of her radial nerve in the axilla because all the symptoms relate to the distribution of this nerve. Prolonged and faulty use of the axillary type of crutches results in most of the weight being taken through the axilla rather than through the hands. This produces intermittent compression of the radial nerve as it leaves the axilla. The patient should be instructed on how to use the axillary type of crutches, that is, bearing the weight through the hands rather than the axilla. However, it would be better to provide the persons with elbow-length crutches. If radial nerve compression ("crutch palsy") is not eliminated, paralysis of the finger and wrist extensors might occur (Table 6.7). This nerve injury is completely reversible, provided the cause is eliminated.

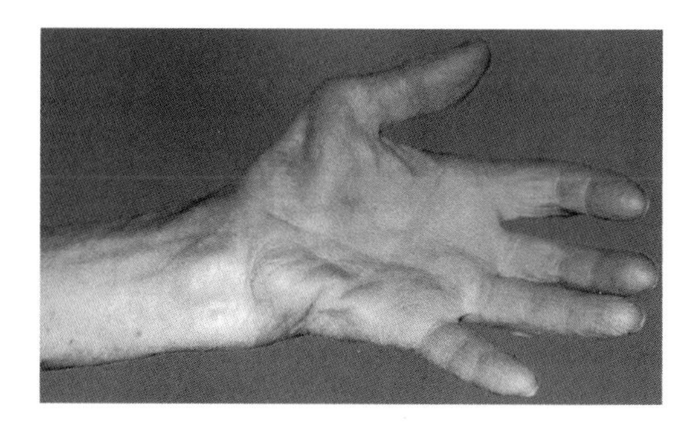

Figure 6.85. Atrophy of the thenar muscles in a 50-year-old woman. This condition resulted from long-standing compression of the median nerve at the wrist (carpal tunnel syndrome).

Case 6.12

Compression of the axillary artery and vein and the cords of the brachial plexus could produce the symptoms described by the patient:

- *Compression of the axillary nerve*—paresthesia over the shoulder and proximal arm and weakness of arm abduction
- *Compression of the axillary artery*—weakening of the radial pulse
- Compression of *the axillary vein*—edema of hand and distension of superficial limb veins.

The axillary structures could become impinged (Fig. 6.86) or compressed between the coracoid process of the scapula and the pectoralis minor tendon during prolonged hyperabduction of the arm.

This type of compression syndrome is called the *hyperabduction syndrome of the arm.* The signs and symptoms depend on which cord(s) of the brachial plexus is compressed.

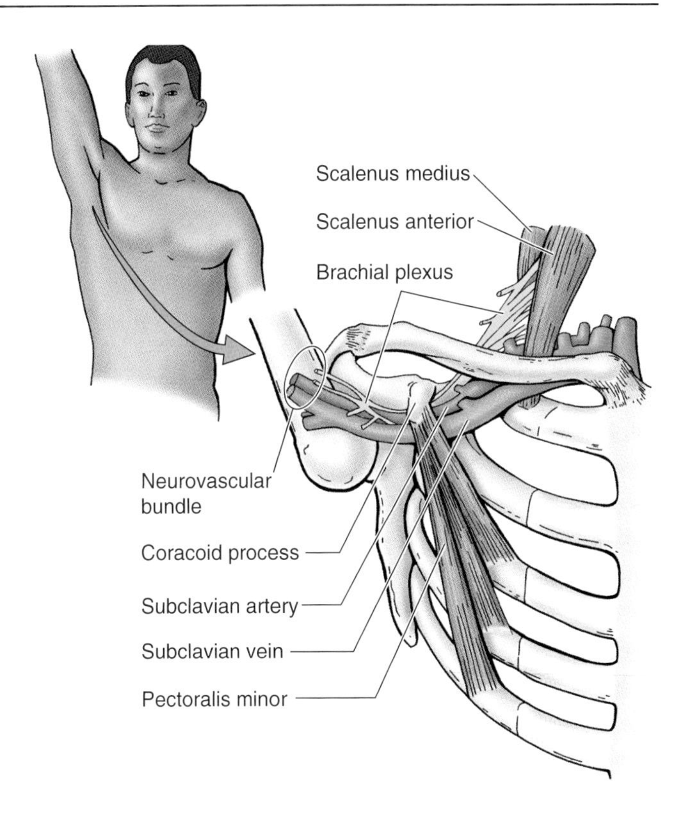

Figure 6.86. Compression of the neurovascular bundle in the axilla (armpit). This condition causes ischemia and paresthesia in the upper limb, exacerbated by elevation of the limb above the head.

Case 6.13

The scaphoid lies in the floor of the anatomical snuff box and is the bone that is commonly fractured in hyperextension of the wrist joint. *Fractures of the scaphoid are difficult to detect*, especially right after an injury, because they are often hairline fractures that are usually not displaced. Subsequent bone resorption at the fracture site makes the break more radiolucent and more apparent after 10 to 14 days.

Because of the position of the scaphoid in the wrist and its relatively small size, it is difficult to immobilize. Continued movement of the wrist often results in nonunion of the bone fragments. Usually, displacement and tearing of ligaments occurs that may interfere with the blood supply to one of the scaphoid fragments. *Ischemic necrosis* (death) of part of the scaphoid may result. Usually the bone is supplied by two nutrient arteries, one to the proximal half and one to the distal half. Occasionally both vessels supply the distal half; consequently, the separated proximal half receives no blood. The resulting necrosis may cause a delay or lack of union of the bone fragments.

Case 6.14

The typical clinical abnormality of the wrist in a Colles' fracture is a dinner (silver) fork deformity resulting from the posterior displacement and tilt of the distal radial fragment. The hand is typically laterally deviated because of displacement of the distal part of the radius. Colles' fracture is more common in people older than 50 years, especially women because their bones are often weakened by postmenopausal osteoporosis. The distal radioulnar joint may be subluxated when a Colles' fracture is not well united, and movements of the wrist joint may be limited.

Case 6.15

The lateral flexion of the baby's trunk and neck severely injured the entire brachial plexus and probably tore one or more of its trunks. The expression *iatrogenic injury* means a harmful condition produced "unwittingly and inadvertently by a physician or surgeon" (Salter, 1998). It is important to understand that had the obstetrician not acted quickly during delivery, he might not have been able to save the baby's life because it was cyanosed (dark bluish coloration of skin resulting from deficient oxygenation of blood).

This severe type of injury involving the entire brachial plexus is not amenable to surgery (Salter, 1998) because it is likely that some of the ventral rami were completely avulsed from the spinal cord. Drooping of the upper eyelid (*ptosis*) was caused by injury of sympathetic fibers in the 1st thoracic nerve root. This is part of the *Horner syndrome*; other characteristics are a constricted pupil (*meiosis*) and absence of sweating of the injured limb (*anhydrosis*).

Case 6.16

When the young man was thrown from his motorcycle and hit a tree, his right shoulder was pulled violently away from his head. This pulled on the superior trunk of his brachial plexus, stretching or tearing the ventral primary rami of C5 and C6 spinal nerves. As a result, the nerves arising from these rami and the superior trunk are affected and the muscles supplied by them are paralyzed. The muscles involved would be the deltoid, biceps brachii, brachialis, brachioradialis, supraspinatus, infraspinatus, teres minor, and supinator.

The patient's arm was medially rotated because the infraspinatus and teres minor (lateral rotators of the shoulder) were paralyzed. His forearm was pronated because the supinator and biceps were paralyzed. Flexion of his elbow was weak because of paralysis of the brachialis and biceps brachii. The inability of the patient to flex his humerus resulted from paralysis of the deltoid and coracobrachialis and probably the clavicular head of the pectoralis major. Loss of abduction of the humerus resulted from paralysis of the supraspinatus and deltoid.

The paralysis of his limb muscles would be permanent if the nerve rootlets forming the C5 and C6 rami were pulled from the spinal cord. As these rootlets cannot be sutured into the spinal cord, the axons of the nerves would not regenerate and the muscles supplied by them would soon undergo atrophy (wasting). Movements of the shoulder and elbow would be greatly affected; for example, the person will always have difficulty lifting a glass to his mouth with his right arm. The loss of sensation in his arm resulted from damage to sensory fibers of C5 and C6 that are conveyed in the upper lateral brachial cutaneous nerve (from the axillary nerve), the lower lateral brachial cutaneous nerve (from the radial nerve), and the lateral antebrachial cutaneous nerve (from the musculocutaneous nerve).

Case 6.17

A shoulder pointer is a sportswriter's term for contusion (bruise) of the acromion (point of shoulder) and the AC joint. To explain a shoulder separation, first you should make a simple diagram of the scapula and clavicle, showing the ligaments attached to them. You should emphasize that it is the coracoclavicular ligament that provides most stability to the AC joint. You should also explain that the scapula and clavicle are parts of the upper limb and make up what is called the pectoral girdle. Explain that the clavicle articulates laterally with the acromion to form the AC joint. Also explain that the scapula and clavicle are held together by the AC and coracoclavicular ligaments.

When your friend hit his shoulder on the boards, the AC and coracoclavicular ligaments were torn (p. 788). As a result, the shoulder fell under the weight of the limb, and the acromion was pulled inferiorly, relative to the clavicle. Also, the lateral end of the clavicle was displaced superiorly, relative to the acromion, producing an obvious prominence. Stress that the expression "separation of the shoulder" is misleading. Explain that it is the AC joint that is dislocated (separated), not the shoulder joint. Rupture of the AC ligament alone is not a serious injury; however, when combined with rupture of the coracoclavicular ligament, dislocation of the AC joint is complicated because the scapula and clavicle are separated and the scapula and upper limb are displaced inferiorly.

Case 6.18

Undoubtedly, the ulnar nerve was injured by the displaced epiphysis of the medial epicondyle. The medial epicondyle does not completely fuse with the side of the diaphysis until 16 years of age in males (14 years in females). Although an epiphyseal separation is sometimes called an "epiphyseal fracture," or a fracture-dislocation, it is best to refer to this injury as a *separation of the epiphysis of the medial epicondyle*. Had this accident occurred in a person older than 16 years, a fracture of the medial epicondyle might have occurred.

Because the epiphyseal plate is weaker than the surrounding bone in children, a direct blow that causes a fracture in adolescents and adults is likely to cause an *epiphyseal plate injury* in children. Because the ulnar nerve passes posterior to the medial epicondyle, between it and the olecranon, it is vulnerable to injuries at the elbow. This kind of injury causes paralysis of muscles and some loss of sensation in the area of skin supplied by the ulnar nerve.

Appreciation of light touch is usually lost over the medial one and a half digits, and response to pinpricks is lost over the 5th digit and the medial border of the palm. Knowing that the in-

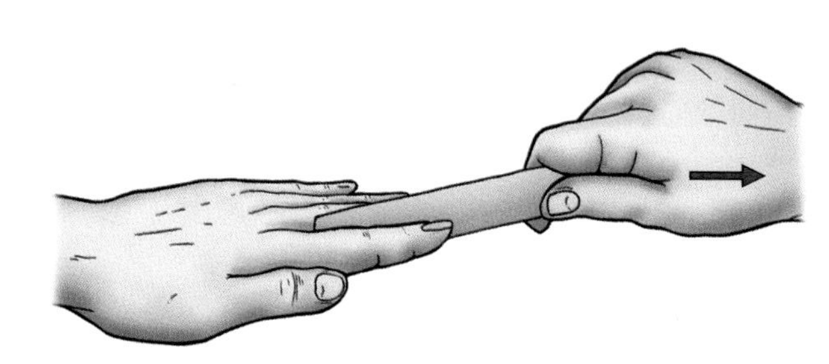

(A) Adduction

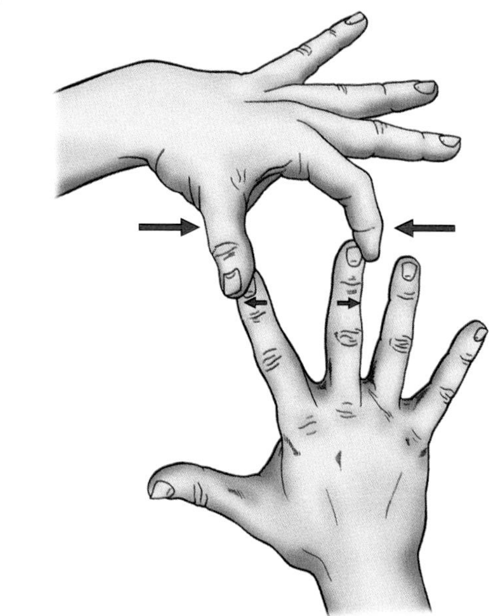

(B) Abduction

Figure 6.87. Testing the interosseous muscles. A. Testing adduction of the fingers. The examiner is attempting to pull the tongue depressor from between the patient's fingers as the patient attempts resistance (*arrow*). Inability to adduct the fingers is a sign of paralysis of the palmar interossei resulting from an ulnar nerve injury. **B.** Testing abduction of the fingers. The examiner is attempting to adduct the abducted 2nd and 3rd digits as the patient attempts to resist (*arrows*). Weakness of abduction indicates incomplete paralysis of the dorsal interossei, which also results from an ulnar nerve injury.

terosseous muscles are supplied by the ulnar nerve, the pediatrician tested them for weakness by placing a piece of paper between the boy's fully extended digits and asking him to grip it as tightly as possible when he pulled on it (Fig. 6.87). Inability to adduct the digits is a classic sign of paralysis of the palmar interosseous muscles and ulnar nerve injury. Loss of other muscle movements would likely have occurred:

- Inability to abduct the digits (*paralysis of dorsal interossei*)
- Loss of adduction of the thumb (*paralysis of adductor pollicis*)
- Weakness of flexion of the 4th and 5th digits at the metacarpophalangeal joints (*paralysis of medial two lumbricals*)
- Impaired flexion and adduction of the wrist (*paralysis of flexor carpi ulnaris*)
- Poor grasp in the 4th and 5th digits (*paralysis of palmar interossei*)
- Inability to flex the distal interphalangeal joints of the 4th and 5th digits (*paralysis of lumbricals, interossei, and part of FDP*).

Because all but five of the intrinsic muscles of the hand are supplied by the ulnar nerve, injury of this nerve at the elbow has its primary effect in the hand.

Because the ulnar nerve was crushed and not severed, it does not require suturing because new axons can grow into the part of the nerve distal to the injury within the original endoneural sheaths and neurolemmal sheaths and reinnervate the paralyzed muscles. Hence, after a crushed nerve injury, as in this case, restoration of function should occur in a few months' time with appropriate physiotherapy.

Case 6.19

The inability of the young man to extend his hand at the wrist indicates injury to the radial nerve. Because the fracture is in the middle of the humerus, it is likely that the radial nerve was damaged where it passes diagonally across the humerus in the radial groove. The nerve is particularly susceptible to injury in this location because of its close relationship to the humerus. *Section of the radial nerve* paralyzes the extensor muscles of the forearm and hand (Table 6.7). As a result, extension of the wrist is impossible, and the hand assumes the flexed position referred to clinically as *wrist-drop*. The radial nerve supplies no muscles in the hand, but it supplies muscles whose tendons pass into the hand; hence, the patient is unable to extend his metacarpophalangeal joints. Because the lumbricals (supplied by the median and ulnar nerves) and interossei (supplied by the ulnar nerve) are intact, the patient is able to flex his metacarpophalangeal joints and extend his interphalangeal joints. However, he would not have normal power of extension of his digits.

Elbow flexion would be painful and would be weakened when the forearm is in the position midway between pronation and supination. Recall that the radial nerve innervates the brachioradialis, a strong flexor of the elbow in this position. The area of sensory loss is often minimal following radial nerve injury because its area of exclusive supply is small. The degree of sensory loss varies from patient to patient, depending on the extent to which the territory is overlapped by adjacent nerves. Sometimes no loss of sensation is detectable.

The shortening of the patient's arm occurred because the proximal and distal fragments of bone overlapped. Contraction of the deltoid abducts the proximal part of the humerus. The proximal contraction of the triceps, biceps, and coracobrachialis pulls the distal fragment superiorly. Although the deep brachial artery accompanies the radial nerve through the radial groove and may be severed by bone fragments, the muscles and structures supplied by this artery (e.g., the humerus) are not likely to cause ischemia because the radial recurrent artery anastomoses with the profunda brachii artery (Fig. 6.26). This communication should provide sufficient blood for the structures supplied by the damaged artery.

Case 6.20

The lateral bones of the carpus, the scaphoid and trapezium, lie in the floor of the anatomical snuff box. This depression at the base of the thumb is limited proximally by the radial styloid process and distally by the base of the 1st metacarpal.

Fracture of the scaphoid is the most common carpal injury and usually results from a fall on the hand. No other fracture in adults is more frequently overlooked at the time of injury (Salter, 1983). Because of the position of the scaphoid and its relatively small size, it is a difficult bone to immobilize. Continued movement of the wrist often results in nonunion of the fragments of bone. Usually displacement and tearing of ligaments occurs that may interfere with the blood supply to one of the fragments.

Ischemic necrosis of the proximal half of the scaphoid may result. Usually the bone is supplied by two nutrient arteries, one to the proximal half and one to the distal half. Occasionally both vessels supply the distal half; the separated proximal half receives no blood. The resulting ischemia may cause a delay or lack of union of the fragments.

Case 6.21

Obviously the patient had not cut her wrist deeply on the left side; the slight bleeding was probably from severed superficial veins. On the right side, she would have certainly cut the tendon of her palmaris longus. She probably also cut the tendon of her flexor carpi radialis. In view of the clinical findings, it is obvious that her *median nerve was severed* or severely injured. At the wrist, this nerve lies deep to and lateral to the tendon of the palmaris longus. The slight spurting of blood in her right wrist suggests that she probably cut the superficial palmar branch of her radial artery. This branch arises from the radial artery just proximal to the wrist. Had she severed her radial artery, the bleeding would have been severe.

Cutting the median nerve at her wrist resulted in paralysis of her thenar muscles and first two lumbricals. Paralysis of the thenar muscles explains her inability to oppose her thumb. Because the posterior interosseous nerve (branch of radial) was unaffected, she could abduct her thumb with her APL, but some impairment of this movement would result because of paralysis of the abductor pollicis brevis, supplied by the recurrent branch of the median nerve. The patient could extend her thumb normally using her EPL and EPB muscles. Because the nerve supply to her adductor pollicis by the deep branch of the ulnar is intact, she could also adduct her thumb. Because of paralysis of her first two lumbricals and the loss of sensation over the thumb and adjacent two and a half digits and the radial two-thirds of her palm, fine control of movements of her 2nd and 3rd digits is lacking. Thus, *cutting the median nerve produces a serious disability of the hand.* In a few weeks, atrophy of the thenar muscles will occur.

Cutting the tendons of the palmaris longus and flexor carpi radialis would weaken flexion of her wrist. In addition, if she attempted to flex her wrist, her hand would be pulled to the ulnar side by the flexor carpi ulnaris, which is unaffected because it is supplied by the ulnar nerve.

Case 6.22

The common injury of the wrist in persons older than 50 years, particularly women, is fracture of the distal end of the radius (*Colles' fracture*). The distal fragment of the radius tilts posteriorly, producing the typical dinner fork deformity of the wrist. The styloid processes of the ulna and radius are at the same level, instead of the radial styloid being more distal than the ulnar styloid process, as is normal. The distal radioulnar joint is also subluxated.

References and Suggested Readings

Anderson MK, Hall SJ: *Sports Injury Management.* Baltimore, Williams & Wilkins, 1995.

Barr ML, Kiernan JA: *The Human Nervous System. An Anatomical Viewpoint*, 6th ed. Philadelphia, JB Lippincott, 1993.

Behrman RE, Kliegman RM, Arvin AM (eds): *Nelson Textbook of Pediatrics*, 15th ed. Philadelphia, WB Saunders, 1996.

Bergman RA, Thompson SA, Afifi AK, Saadeh FA: *Compendium of Human Anatomic Variation: Text, Atlas and World Literature.* Baltimore, Urban & Schwarzenberg, 1988.

Birrer RB (ed): *Sports Medicine for the Primary Care Physician*, 2nd ed. Boca Raton, CRC Press, 1994.

Dravinsky O, Feldman E: *Examination of the Cranial and Peripheral Nerves.* New York, Churchill Livingstone, 1988.

Ellis H: *Clinical Anatomy: A Revision and Applied Anatomy for Clinical Students*, 8th ed. Oxford, Blackwell Scientific Publications, 1992.

Ger R, Abrahams P, Olson T: *Essentials of Clinical Anatomy*, 3rd ed. New York, Parthenon Publishing Group, 1996.

Griffith HW: *Complete Guide to Sports Injuries.* Los Angeles, Price Stern Sloan, 1986.

Haines DE (ed): *Fundamental Neuroscience.* New York, Churchill Livingstone, 1997.

Halpern BC: Shoulder injuries. *In* Birrer RB (ed): *Sports Medicine for the Primary Care Physician*, 2nd ed. Boca Raton, CRC Press, 1994.

Healey JE Jr, Hodge J: *Surgical Anatomy*, 2nd ed. Toronto, BC Decker, 1980.

Lange DJ, Trojaborg W, Rowland LP: Peripheral and cranial nerve lesions. *In* Rowland LP (ed): *Merritt's Textbook of Neurology*, 9th ed. Baltimore, Williams & Wilkins, 1995.

McMinn RM: *Last's Anatomy: Regional and Applied*, 8th ed. Edinburgh, Churchill Livingstone, 1990.

McVay CB: *Anson and McVay Surgical Anatomy*, 6th ed. Philadelphia, WB Saunders, 1984.

Moore KL, Persaud TVN: *The Developing Human: Clinically Oriented Embryology*, 6th ed. Philadelphia, WB Saunders, 1998.

Rowland LP (ed): *Merritt's Textbook of Neurology*, 9th ed. Baltimore, Williams & Wilkins, 1995.

Sabiston DC Jr, Lyerly HK: *Sabiston Essentials of Surgery*, 2nd ed. Philadelphia, WB Saunders, 1994.

Salter RB: *Textbook of Disorders and Injuries of the Musculoskeletal System*, 3rd ed. Baltimore, Williams & Wilkins, 1998.

Swartz MH: *Textbook of Physical Diagnosis. History and Examination*, 2nd ed. Philadelphia, WB Saunders, 1994.

Williams PL, Bannister LH, Berry MM, Collins P, Dyson M, Dussek JE, Fergusson MWJ (eds): *Gray's Anatomy*, 38th ed. Edinburgh, Churchill Livingstone, 1995.

Willms JL, Schneiderman H, Algranati PS: *Physical Diagnosis: Bedside Evaluation of Diagnosis and Function.* Baltimore, Williams & Wilkins, 1994.

Woodburne RT, Burkel WE: *Essentials of Human Anatomy*, 9th ed. New York, Oxford University Press, 1994.

c h a p t e r

7 Head

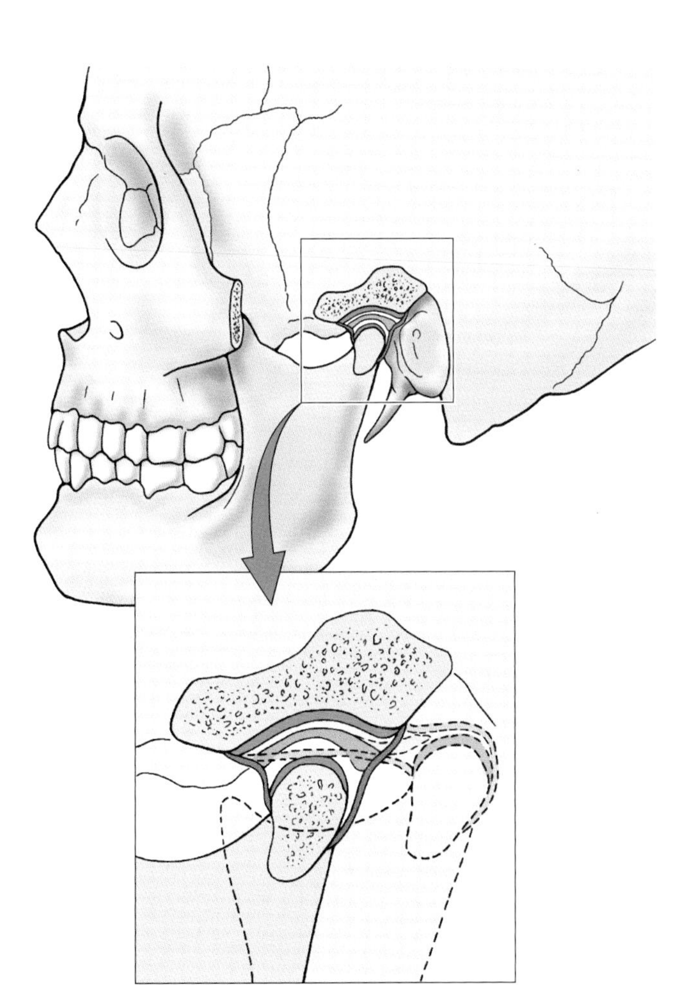

The head consists of the skull, face, scalp, teeth, brain, cranial nerves, meninges, special sense organs, and other structures such as blood vessels, lymphatics, and fat. It is also the site where food is ingested and air is inspired and expired. Diseases of important structures in the head form the bases of many medical, dental, and surgical specialties—dentistry, maxillofacial surgery, neurology, neuroradiology, neurosurgery, ophthalmology, oral surgery, otology, psychiatry, and rhinology.

Head Injuries

Head injuries are a major cause of death and disability. The complications of head injuries include vascular lesions such as hemorrhage, infection (e.g., osteomyelitis, or inflammation of bone marrow and adjacent bone), and injury to the brain and cranial nerves (Rowland, 1995). Disturbance of consciousness is the most common symptom of head injury.

Few complaints are more common than headache and facial pain. Headache is usually a benign symptom and only occasionally the manifestation of a serious illness such as a brain tumor (Raskin, 1995). *Neuralgias*—pain of a severe throbbing or stabbing character in the course of a nerve caused by a demyelinative lesion of nerves—are a common cause of facial pain. Terms such as *facial neuralgia* describe diffuse painful sensations. Localized aches have specific names such as *earache* (otalgia) and *toothache* (odontalgia). Headache often accompanies fever, tension, and fatigue but it may indicate a serious intracranial problem such as a brain tumor, subarachnoid hemorrhage, or meningitis. Consequently, a sound knowledge of the anatomy of the head helps in understanding the causes of headaches and facial pains.

Head injuries are a scourge of industrialized society and current lifestyles (Rowland, 1995). They are a major cause of death, especially in young adults, and of disability. Few other conditions exceed the cost in human misery and dollars that head injuries cause. Almost 10% of all deaths in the United States are caused by head injury, and approximately half of traumatic deaths involve the brain (Rowland, 1995). Head injuries occur at all ages but mostly in young persons between the ages of 15 and 24. Men are affected three or four times as often as women. The major cause of brain injury varies but motor vehicle and motorcycle accidents are prominent. ⊙

Skull

The skull is the skeleton of the head; a series of bones form its two parts, the *neurocranium* and *facial skeleton* (Fig. 7.1). The **neurocranium** ("brain box" or cranial vault) provides a case for the brain and cranial meninges (membranes covering the brain), proximal parts of the cranial nerves, and blood vessels. The term *cranium* (L. skull) is sometimes restricted to a skull without the mandible. The cranium has a domelike roof—the **calvaria** (skullcap)—and a floor or **cranial base** (*basicranium)* consisting of the ethmoid bone and parts of the occipital and temporal bones. The facial skeleton consists of the bones surrounding the mouth and nose and contributing to the orbits (eye sockets, orbital cavities).

The **neurocranium** in adults is formed by a series of eight bones (Fig. 7.1, *A* and *C*):

- A frontal bone
- Paired parietal bones
- Paired temporal bones
- An occipital bone
- A sphenoid bone
- An ethmoid bone.

Most of these bones are largely flat, curved, and united by fibrous interlocking sutures. During childhood, some bones are united by hyaline cartilage (*synchondroses*) between the occipital and sphenoid bones. A number of irregular bones form the framework of the face and cranial base.

The **facial skeleton** (viscerocranium or splanchnocranium) forms the anterior part of the skull containing the orbits and nasal cavities and includes the maxilla and mandible (upper and lower jaws). *The facial skeleton consists of 14 irregular bones* (Fig. 7.1, *A–C*):

- Lacrimal bones (2)
- Nasal bones (2)
- Maxillae (2)
- Zygomatic bones (2)
- Palatine bones (2)
- Inferior nasal conchae (2)
- Mandible (1)
- Vomer (1).

The maxillae and mandible house the teeth; that is, they provide the sockets and supporting bone for the maxillary and mandibular teeth. The maxillae form the skeleton of the upper jaw, which is fixed to the cranial base. In general, the maxillae contribute to the greater part of the upper facial skeleton. The mandible forms the skeleton of the lower jaw, which is movable because it articulates with the cranial base at the temporomandibular joints (TMJs).

In the *anatomical position*, the skull is oriented so that the inferior margin of the orbit and the superior margin of the external acoustic meatus (auditory canal) of both sides lie in the same horizontal plane (Fig. 7.1*A*). This standard craniometric reference is the *orbitomeatal plane* (Frankfort horizontal plane).

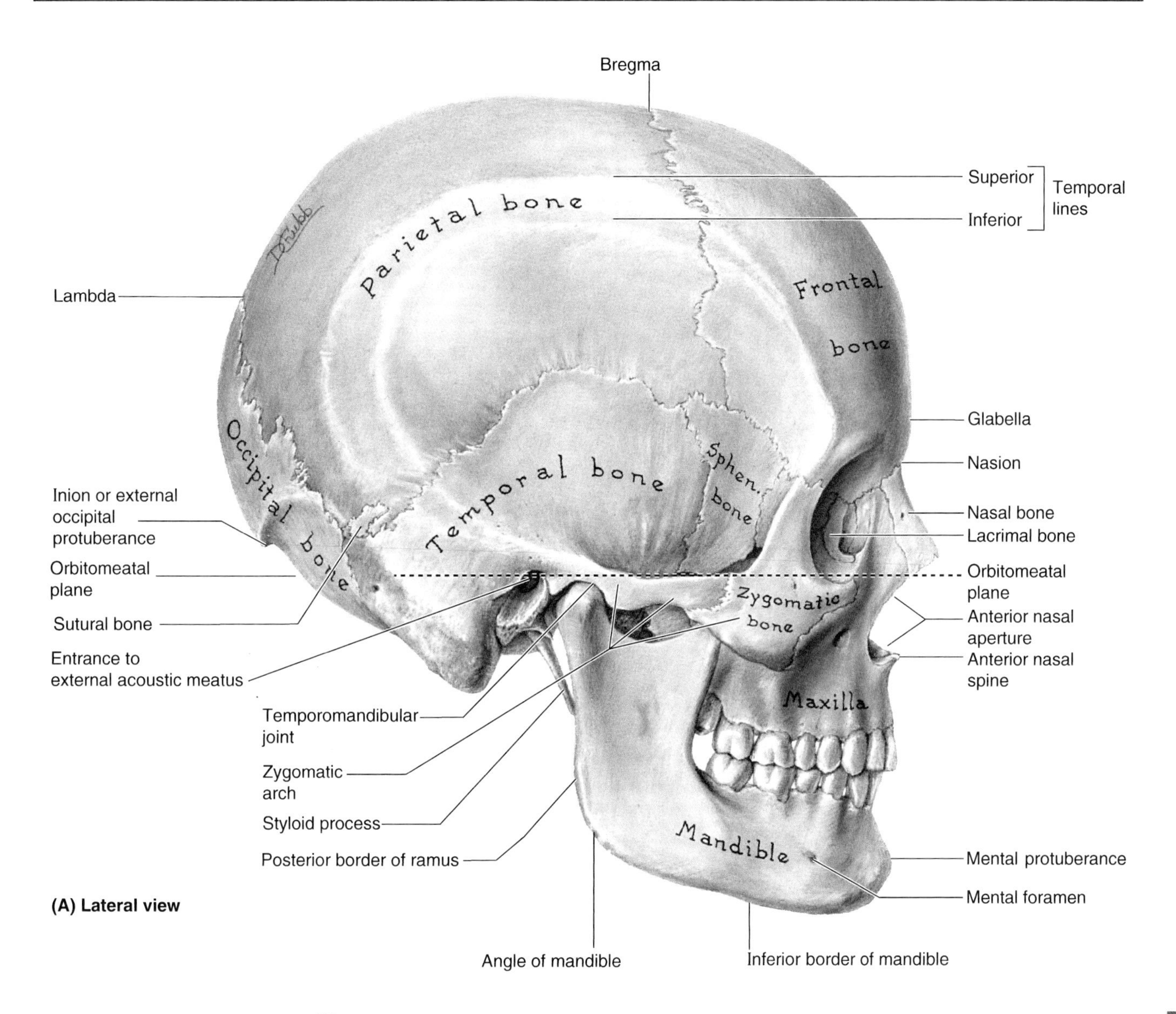

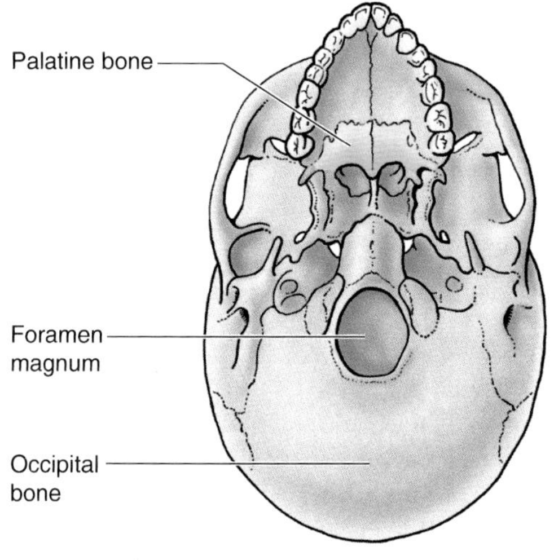

Figure 7.1. Three views of an adult skull. A. Lateral view in the anatomical position. Notice that the *orbitomeatal plane* (Frankfort horizontal plane), a standard craniometric reference, is horizontal in this position. Observe the *temporal fossa* (temple) on the side of the skull superior to the zygomatic arch and inferior to the temporal lines. **B.** Inferior view. Observe the palatine bone, an irregularly shaped bone that contributes to the formation of the nasal cavity, the hard palate (bony plate), and a small part of the orbit (eye socket). Note the large opening in the basal part of the occipital bone—the *foramen magnum*—through which the spinal cord is continuous with the medulla oblongata (medulla) of the brain.

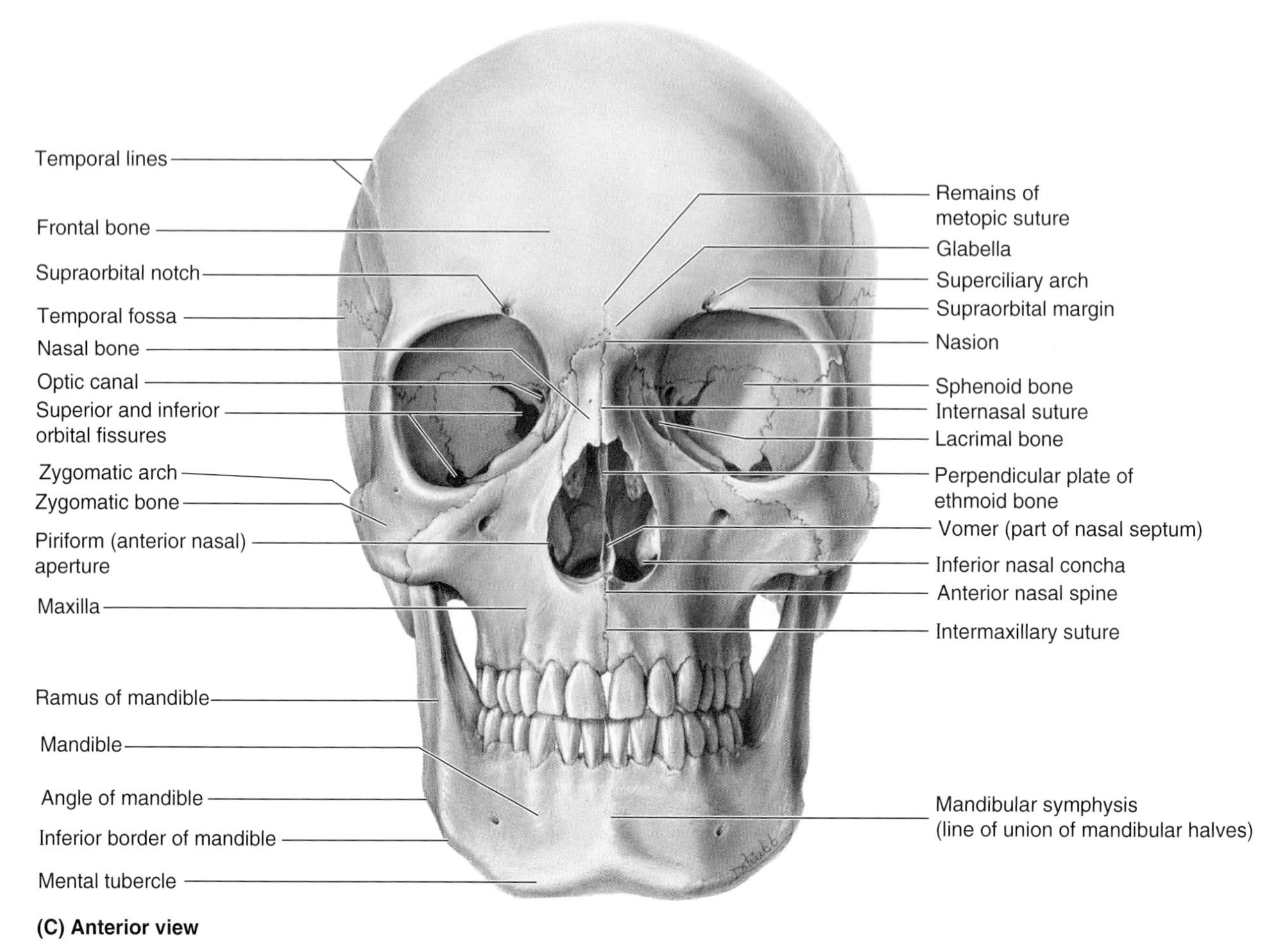

Figure 7.1. *(Continued)* **C.** Anterior view. Observe that the supraorbital notch, the infraorbital foramen, and the mental foramen are approximately in a vertical line.

Anterior Aspect of the Skull

Features of the anterior aspect of the skull are the frontal and zygomatic bones, orbits, nasal region, maxillae, and mandible (Fig. 7.1*C*).

The **frontal bone**—specifically its squamous (flat) part—forms the skeleton of the forehead, articulating inferiorly with the nasal and zygomatic bones. In the fetal skull, the two halves of the frontal bone are separated by the *frontal suture* (p. 847), and they remain separate until approximately 6 years of age. In some adults the separation line persists as the **metopic suture** in the midline of the **glabella** (Fig. 7.1*C*), the smooth, slightly depressed area between the superciliary arches. A *persistent metopic suture* in radiographic images must not be mistaken for a fracture line. The intersection of the frontal and nasal bones is the **nasion** (L. nasus, nose), which in most people is related to a distinctly depressed area ("bridge of the nose"). The frontal bone also articulates with the lacrimal, ethmoid, and sphenoid bones, and a horizontal portion of bone (the orbital part [plate] of the frontal bone) forms both the roof of the orbit and part of the floor of the cranial cavity. The **supraorbital margin** (arch), the angular boundary between the squamous and orbital parts, has a **supraorbital notch** or a **foramen** in some skulls (Fig. 7.2*A*) for passage of the supraorbital nerve and vessels. Just superior to the supraorbital margin is a ridge—the **superciliary arch**—that extends laterally on each side from the glabella. The prominence of this ridge, deep to the eyebrows, is generally greater in males. Within the orbits are the superior and inferior **orbital fissures** and **optic canals** (Fig. 7.1*C*).

The **zygomatic bones** (zygoma bones, cheek bones, malar bones)—forming the prominences of the cheeks (L. buccae)—lie on the inferolateral sides of the orbits and rest on the maxillae. The anterolateral rims, walls, floor, and much of the infraorbital margins of the orbits are formed by these bones. A small **zygomaticofacial foramen** pierces the lateral aspect of each zygomatic bone (Fig. 7.2, *A* and *B*). The zygomatic bones articulate with the frontal, sphenoid, and temporal bones and the maxillae. Inferior to the nasal bones are the pear-shaped **piriform apertures**, or the anterior nasal apertures (Fig. 7.1, *A* and *C*). The bony **nasal septum** can be observed through this opening, dividing the nasal cavity into right and left parts. On the lateral wall of each nasal cavity are curved bony plates, the **nasal conchae** (Figs. 7.1*C* and 7.2*A*).

The **maxillae** form the upper jaw; their **alveolar processes** include the sockets (alveoli) and constitute the supporting bone for the **maxillary teeth.** The maxillae surround most of the piriform apertures and form the infraorbital margins medially. They have a broad connection with the zygomatic bones laterally and have an **infraorbital foramen** inferior to each orbit for the infraorbital nerve and vessels (Fig. 7.2*A*). The two maxillae are united at the **intermaxillary suture** in the median plane.

The **mandible** is a U-shaped bone with alveolar processes that house the *mandibular teeth.* Inferior to the second premolar teeth are the **mental foramina** for the mental nerve and vessels (Fig. 7.2*A*). The **mental protuberance**—forming the prominence of the chin—is a triangular elevation of bone inferior to the **mandibular symphysis** (L. symphysis menti), the region where the halves of the fetal mandible fuse (Fig. 7.1*C*). The mandible is described in more detail later in this chapter.

Injury to the Superciliary Arches

The superciliary arches are relatively sharp ridges of bone; consequently, a blow to them (e.g., during a boxing match) may lacerate the skin and cause profuse bleeding. Bruising of the skin surrounding the orbit causes tissue fluid and blood to accumulate in the surrounding connective tissue, which gravitates into the superior (upper) eyelid and around the eye ("black eye").

Malar Flush

The zygomatic bone was once called the malar bone; consequently, you will hear the clinical term *malar flush.* This redness of the skin covering the zygomatic prominence (malar eminence) is associated with a rise in temperature in various fevers occurring with certain diseases, such as tuberculosis. ▶

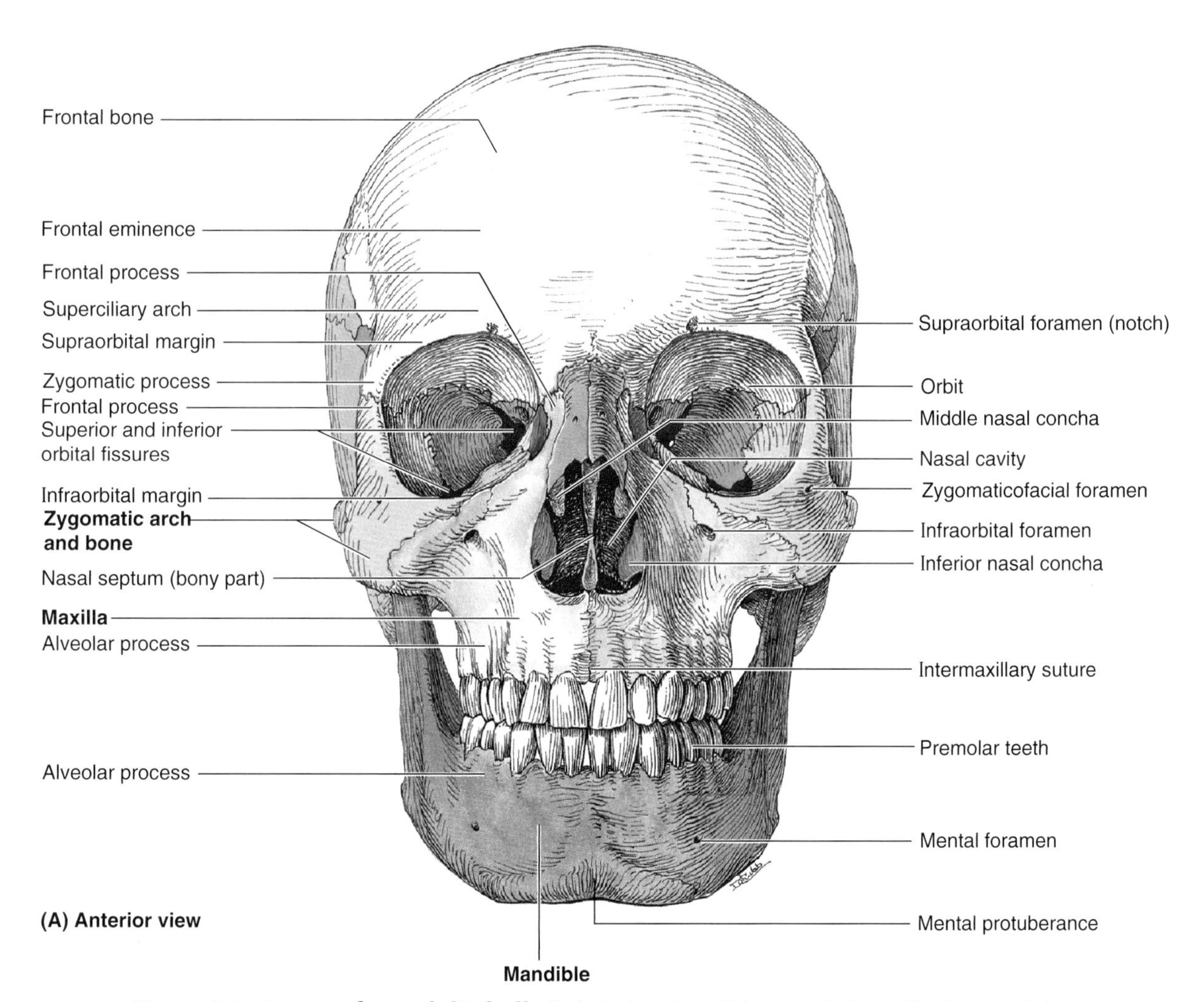

Figure 7.2. Bones of an adult skull. A. Anterior view. This aspect shows the forehead, the orbits (eye sockets), the bony part of the external nose, and the upper (maxilla) and lower (mandible) jaws.

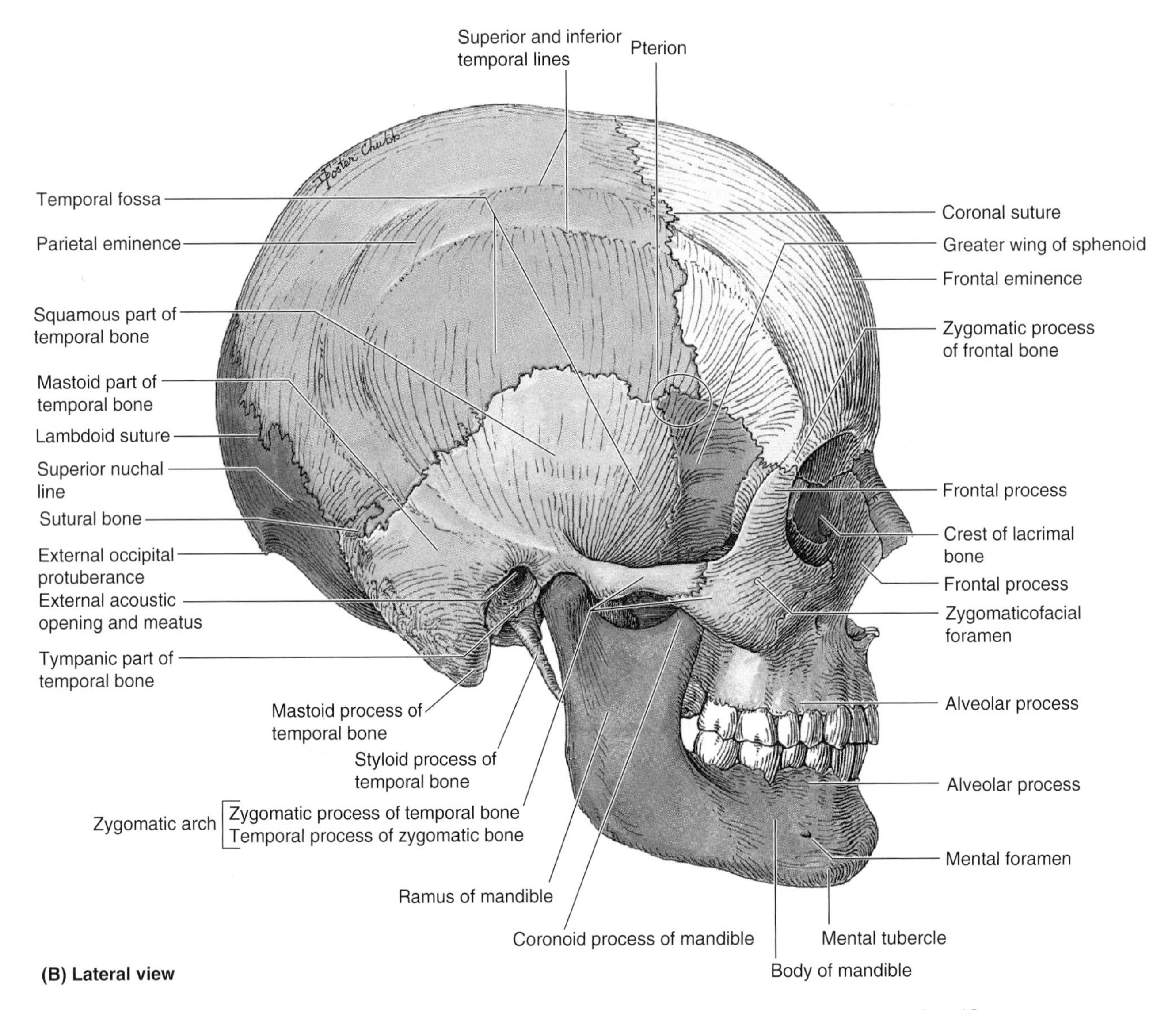

Figure 7.2. *(Continued)* **B.** Lateral view. Within the temporal fossa, observe the **pterion** (G. pteron, wing), a craniometric point at the junction of the greater wing of the sphenoid, the squamous temporal bone, the frontal, and the parietal bones. It intersects the course of the anterior division of the middle meningeal artery (p. 839). Notice that the skull is in the anatomical position when the inferior margin of the orbit and the superior margin of the external acoustic meatus (auditory canal) lie in the same horizontal (orbitomeatal) plane.

Fractures of the Maxillae and Associated Bones

Dr. Le Fort, a Paris surgeon and gynecologist, classified the common variants of fractures of the maxillae. The three types of fracture are:

- A *Le Fort I fracture* (remarkably constant) is a horizontal one of the maxillae, located just superior to the alveolar process, crossing the bony nasal septum and the pterygoid plates of the sphenoid.
- A *Le Fort II fracture* passes from the posterolateral parts of the maxillary sinuses (cavities in the maxillae) superomedially through the infraorbital foramina, lacrimals, or ethmoids to the bridge of the nose. As a result, the entire central part of the face, including the hard palate and alveolar processes, is separated from the rest of the skull.
- A *Le Fort III fracture* is a horizontal one that passes through the superior orbital fissures, the ethmoid and nasal bones, and extends laterally through the ▶

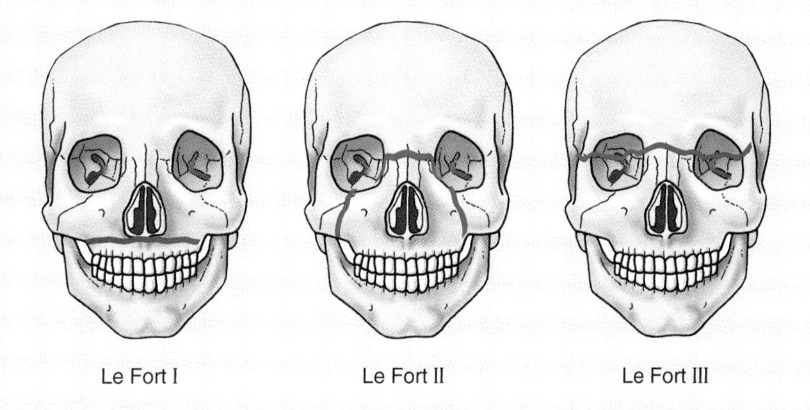

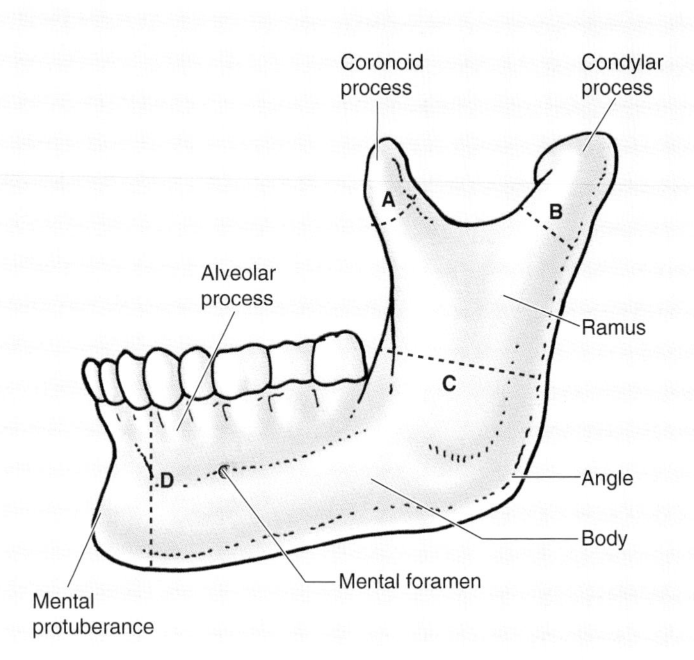

▶ greater wings of the sphenoid bone and the frontozygomatic sutures. Concurrent fracturing of the zygomatic arches causes the maxillae and zygomatic bones to separate from the rest of the skull.

Fractures of the Mandible

A fracture of the mandible usually involves two fractures, and they frequently occur on opposite sides of the mandible; thus, if one fracture is observed, a search should be made for another. For example, a hard blow to the jaw often fractures the neck of the mandible and its body in the region of the opposite canine tooth. Fractures of the coronoid process (*A*) are uncommon and usually single; *fractures of the neck of the mandible* (*B*) are often transverse and may be associated with dislocation of the TMJ on the same side. Fractures of the angle of the mandible (*C*) are usually oblique and may involve the alveolus of the 3rd molar tooth. *Fractures of the body of the mandible* (*D*) frequently pass through the alveolus of a canine tooth.

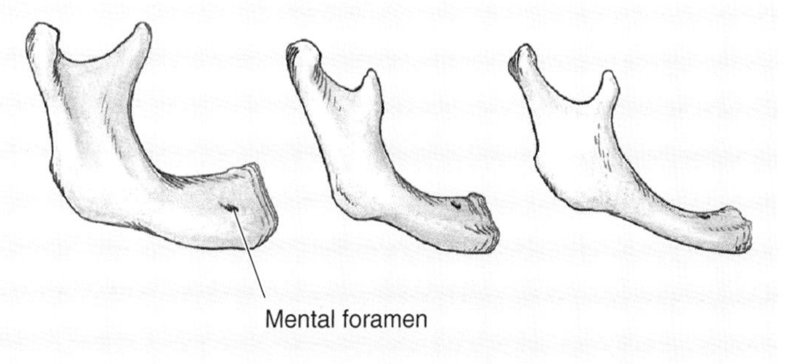

Resorption of Alveolar Bone

Removal of the teeth causes the bone of the alveolar process to resorb in the affected region(s). Following complete loss of maxillary teeth, the alveoli (tooth sockets) begin ▶

► to fill in with bone and the alveolar process begins to resorb. Similarly, removal of mandibular teeth causes the alveolar process of the mandible to resorb. Gradually the mental foramen lies near the superior border of the body of the mandible (p. 837). In extreme cases, the mental foramina disappear, exposing the mental nerves to injury. *Pressure from a dental prosthesis* (e.g., a denture resting on an exposed nerve) may produce pain during eating. Loss of all the teeth results in a decrease in the vertical facial dimension and *mandibular prognathism* (overclosure). Deep creases in the facial skin also appear that pass posteriorly from the corners of the mouth. ○

Lateral Aspect of the Skull

The lateral aspect of the skull is formed by cranial and facial bones (Fig. 7.2*B*). The main features of the cranial part include the **temporal fossa**, the opening of the **external acoustic meatus** (canal), and the mastoid region of the temporal bone. The main features of the facial part include the infratemporal fossa, zygomatic arch, and lateral aspects of the maxilla and mandible. The **temporal fossa** (Fig. 7.1*C*) is bounded superiorly and posteriorly by the **temporal lines**, anteriorly by the frontal and zygomatic bones, and inferiorly by the **zygomatic arch** (Fig. 7.2, *A* and *B*). The superior border of this arch corresponds to the inferior limit of the cerebral hemisphere of the brain. The zygomatic arch is formed by the union of the temporal process of the zygomatic bone and the zygomatic process of the temporal bone. In the anterior part of the temporal fossa, 3 to 4 cm superior to the midpoint of the zygomatic arch, is a clinically important area of bone junctions—the **pterion** (G. pteron, wing) (Fig. 7.2*B*, Table 7.1). It is usually indicated by an H-shaped formation of sutures that unite the frontal, parietal, sphenoid (greater wing), and temporal bones. Less commonly, the frontal and temporal bones articulate; sometimes all four bones meet at a point. The **external acoustic opening** is the entrance to the **external acoustic meatus**, which leads to the tympanic membrane (eardrum). The **mastoid process** of the temporal bone is posteroinferior to the opening of the external acoustic meatus. Anteromedial to the mastoid process is the slender **styloid process**. The *mandible* consists of a horizontal part, the **body,** and a vertical part, the **ramus** (Fig. 7.2*B*). The *infratemporal fossa* is an irregular space inferior and deep to the **zygomatic arch** and the mandible, and posterior to the maxilla.

Fractures of the Calvaria

The convexity of the calvaria—the domelike part of the skull—distributes and thereby minimizes the effects of a blow to it. However, hard blows to the head in thin areas of the cranium are likely to produce *depressed fractures,* in which a fragment of bone is depressed inward to compress or injure the brain. *Linear skull fractures,* the most frequent type, usually occur at the point of impact, but fracture lines often radiate away from it in two or more directions. In *comminuted fractures,* the bone is broken into several pieces. If the area of the calvaria is thick at the site of impact, the bone usually bends inward without fracturing; however, a fracture may occur some distance from the site of direct trauma where the calvaria is thinner. In a *contrecoup (counterblow) fracture,* no fracture occurs at the point of impact but one occurs on the opposite side of the skull.

The pterion is an important clinical landmark because it overlies the anterior branches of the middle meningeal vessels, which lie in grooves on the internal aspect of the lateral wall of the calvaria (Fig. 7.4). The pterion is two fingers' breadth superior to the zygomatic arch and a thumb's breadth posterior to the frontal process of the zygomatic ►

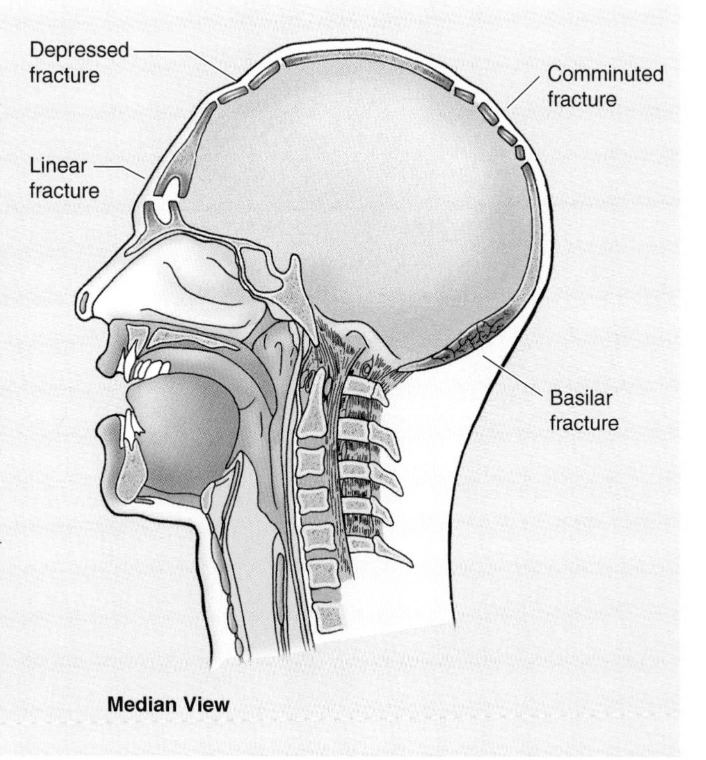

▶ bone (*B*). A blow to the side of the head may fracture the thin bones forming the pterion (Fig. 7.1*A*), rupturing the anterior branch of the middle meningeal artery crossing the pterion (*A*). The resulting *hematoma* (collection of blood) exerts pressure on the underlying cerebral cortex. Untreated *middle meningeal artery hemorrhage* may cause death in a few hours. ⊙

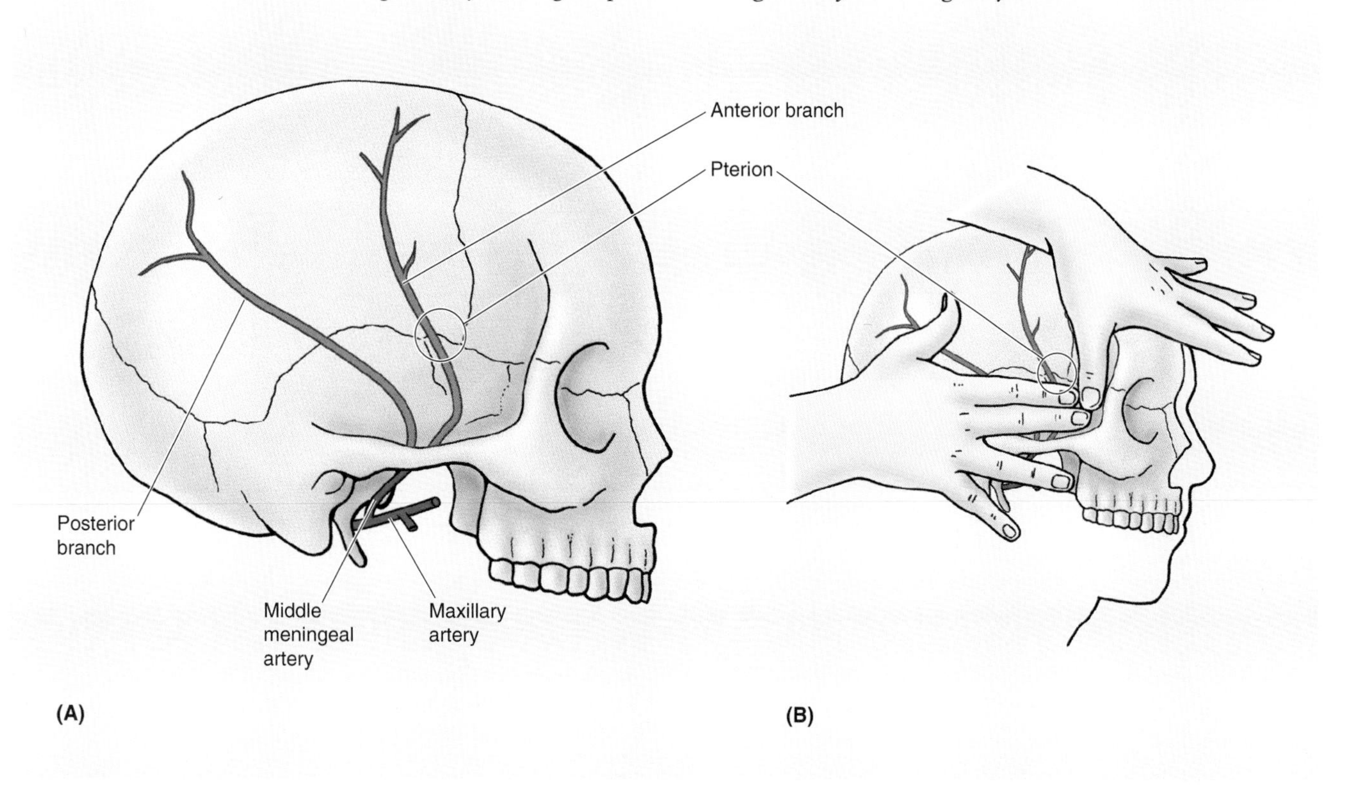

Posterior Aspect of the Skull

The posterior aspect of the skull, or **occiput** (L. back of head), is typically ovoid or round in outline (Fig. 7.3*A*). It is formed by the occipital bone, parts of the parietal bones, and mastoid parts of the temporal bones. The **external occipital protuberance** (also known as the **inion**) is usually an easily palpable elevation in the median plane; however, occasionally (especially in females) it may be inconspicuous. The **external occipital crest** descends from the external occipital protuberance toward the **foramen magnum**—the large opening in the basal part of the occipital bone (Figs. 7.1*B* and 7.3*B*). The **superior nuchal line**, marking the superior limit of the neck, extends laterally from each side of the external occipital protuberance; the inferior nuchal line is less distinct. In the center of the occiput, the **lambda** indicates the junction of the sagittal and lambdoid sutures (Fig. 7.3*A*, Table 7.1). The lambda can sometimes be felt as a depression. One or more *sutural bones* (accessory bones) may be located at the lambda or near the mastoid process of the temporal bone (Fig. 7.2*B*).

Superior Aspect of the Skull

The superior aspect of the skull, usually somewhat oval in form, broadens posterolaterally at the **parietal eminences** (Figs. 7.2*B* and 7.3*B*). In some people the **frontal eminences** are also prominent, giving the skull an almost square appearance. The bones forming the calvaria are visible from this aspect: the frontal bone anteriorly, the right and left parietal bones laterally, and the occipital bone posteriorly. The **coronal suture** separates the frontal and parietal bones (Fig. 7.4*A*); the **sagittal suture** separates the parietal bones, and the **lambdoid suture** separates the parietal and temporal bones from the occipital bone (Fig. 7.3*A*). The **bregma** is the landmark formed by the intersection of the sagittal and coronal sutures (Fig. 7.4*A*, Table 7.1). The **vertex**—the most superior point of the skull—is near the midpoint of the sagittal suture. The **parietal foramen** is a small, inconstant aperture (which may be paired) located posteriorly in the parietal bone near the sagittal suture (Fig. 7.4*B*). The parietal foramen transmits an emissary vein, connecting the scalp to a venous sinus of the cranial cavity (see Fig. 7.14).

External Aspect of the Cranial Base

The external surface of the cranial base shows the alveolar arch of the maxillae (the free border of the **alveolar process** surrounding and supporting the maxillary teeth), the **palatine processes** of the maxillae, and the palatine, sphenoid, vomer, temporal, and occipital bones (Fig. 7.5, *A* and *B*). The **hard**

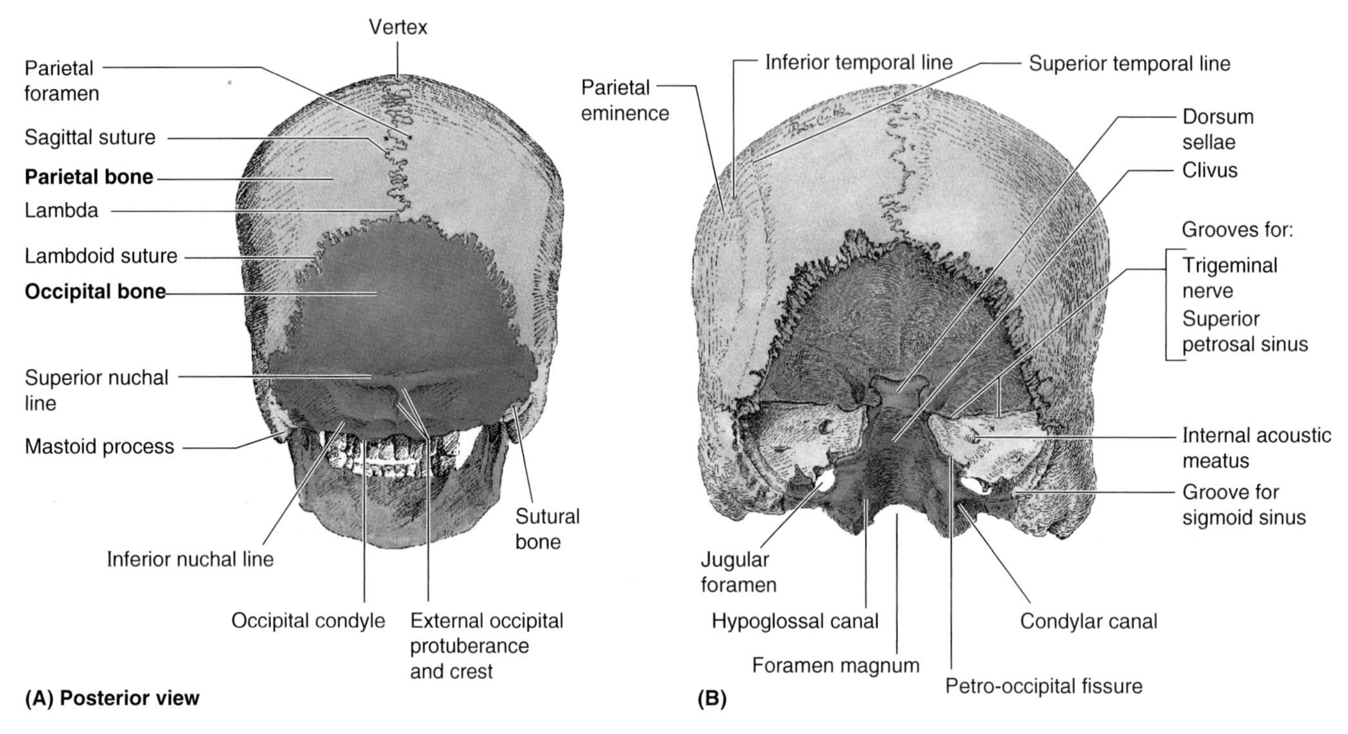

Figure 7.3. Posterior views of an adult skull. A. The posterior aspect of the skull, or occiput (L. back of head), is composed of parts of the parietal bones, the occipital bone, and the mastoid parts of the temporal bones. The sagittal and lambdoid sutures meet at the *lambda*, which can often be felt as a depression in living persons. **B.** Most of the occipital bone (the squamous part) has been removed. Exposed within the cranial cavity, observe the *dorsum sellae ("back of the saddle")*, a squarish plate of bone rising from the body of the sphenoid bone, often broken off in dry skull specimens.

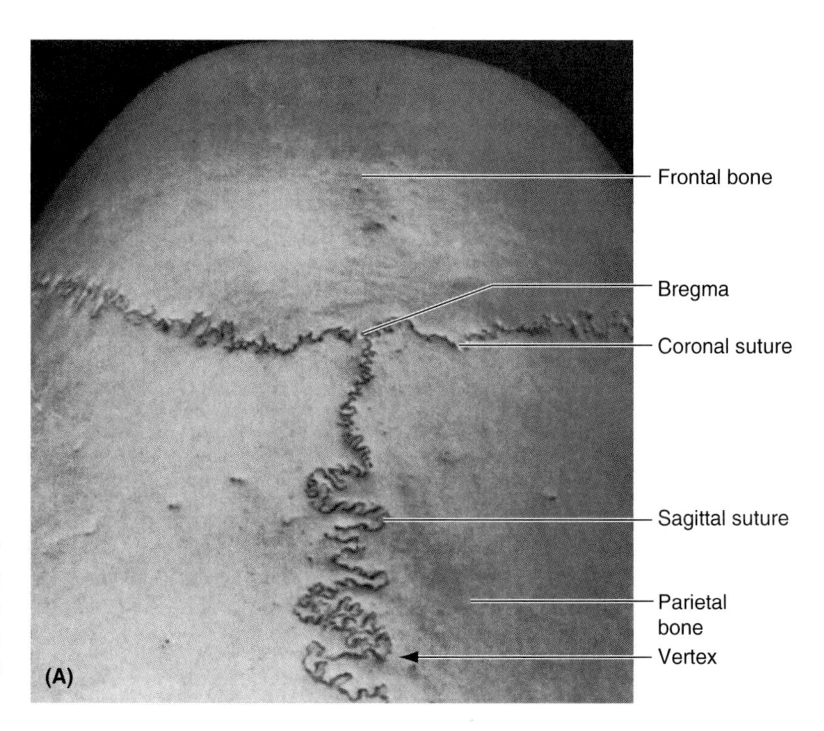

Figure 7.4. Superior aspect of the anterior part of an adult skull. A. External superior view of the calvaria (skullcap) demonstrating the *bregma* where the coronal and sagittal sutures meet, and the vertex, the superior-most point of the skull.

palate (bony palate) is formed by the *palatine processes of the maxillae* anteriorly and the *horizontal plates of the palatine bones* posteriorly. The free posterior border of the hard palate projects posteriorly in the median plane as the **posterior nasal spine.** Posterior to the central incisor teeth is a depression, the **incisive fossa**, through which the nasopalatine nerves pass from the nose through a variable number of *incisive canals* and *foramina.* Posterolaterally are the **greater** and **lesser palatine foramina.** Superior to the posterior edge of the palate are two large openings, the **choanae** (posterior nasal apertures), which are separated from each other by the **vomer**, a thin, flat unpaired bone that makes a major contribution to the bony nasal septum.

Wedged between the frontal, temporal, and occipital bones is the **sphenoid bone** (Fig. 7.5*A*), an irregular unpaired bone that consists of a body and three pairs of processes: greater wings, lesser wings, and pterygoid processes. The **greater** and **lesser wings** spread laterally from the body of the bone (Fig. 7.5*C*). The **pterygoid processes,** consisting of lateral and medial **pterygoid plates** (Fig. 7.5*A*), extend inferiorly on each side of the sphenoid from the junction of the body and greater wings. The groove for the cartilaginous part of the pharyngotympanic (auditory) tube lies medial to the **spine of the sphenoid.** Depressions in the temporal bone—the **mandibular fossae** (Fig. 7.5*B*)—accommodate the condyles of the mandible when the mouth is closed.

The cranial base is formed posteriorly by the occipital bone, which articulates with the sphenoid bone anteriorly. The four parts of the occipital bone are arranged around the **foramen magnum**, the most conspicuous feature of the cranial base. The major structures passing through this large foramen are the spinal cord and its coverings (meninges), the vertebral arteries, the anterior and posterior spinal arteries, and the accessory nerve (CN XI). On the lateral parts of the occipital bone are two large protuberances, the **occipital condyles** (Fig. 7.5*B*), by which the skull articulates with the vertebral column. The large opening between the occipital bone and the petrous part of the temporal bone is the **jugular foramen,** from which the internal jugular vein (IJV) and several cranial nerves (CN IX through XI) emerge from the skull (Fig. 7.5*C*). Superolateral to the jugular foramen is the **internal acoustic meatus** (for CN VII and CN VIII). The entrance to the **carotid canal** for the internal carotid artery is just anterior to the jugular foramen (Fig. 7.5*B*). The **mastoid process** is ridged because it provides for muscle attachment. The **stylomastoid foramen**, transmitting the facial nerve (CN VII) and stylomastoid artery, lies posterior to the base of the styloid process (Fig. 7.5, *A* and *B*).

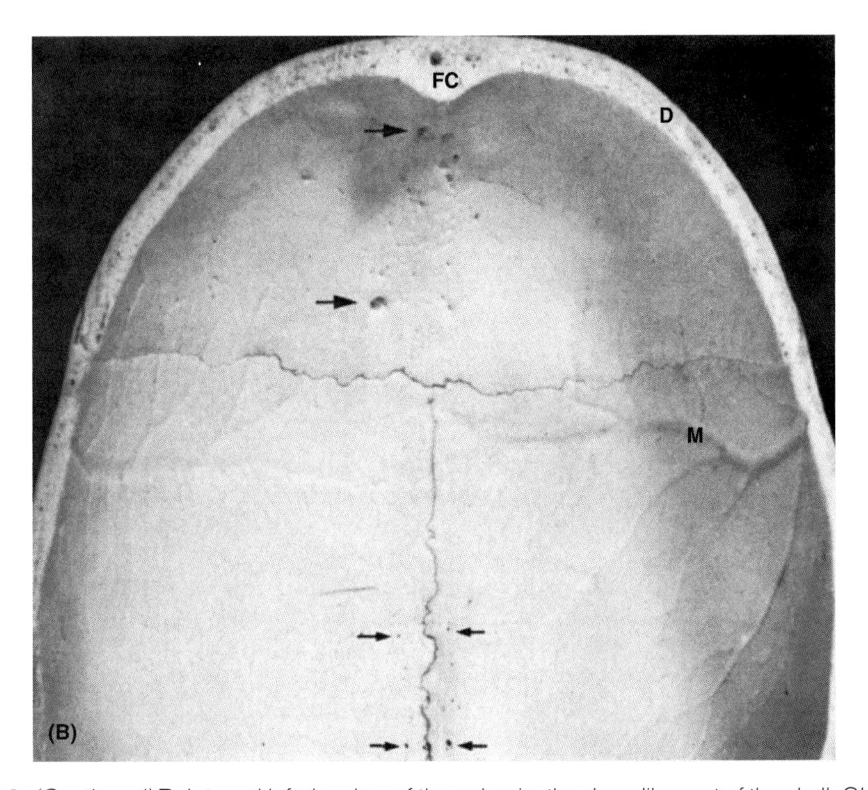

Figure 7.4. *(Continued)* **B.** Internal inferior view of the calvaria, the domelike part of the skull. Observe the pits on the frontal bone (*large arrows*) produced by arachnoid granulations (tufted prolongations [Fig. 7.14] of pia-arachnoid [parts of the coverings of the brain]). On each side of the sagittal suture note the parietal foramina (*smaller arrows*) through which emissary veins pass between the superior sagittal sinus and veins in the diploë (*D*) and scalp. The spongy diploë of cancellous bone contains red marrow in life. Note also the sinuous vascular groove (*M*) formed by the frontal branch of the middle meningeal artery. Note the frontal crest (*FC*) to which the cerebral falx (L. falx cerebri) was attached. The falx is a short process of dura mater, the outer covering of the brain.

Table 7.1. Bony Landmarks of the Skull

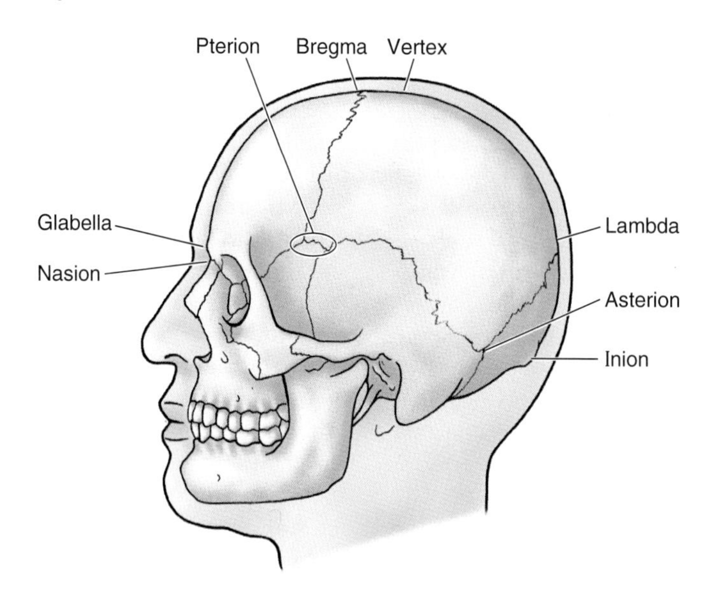

Landmark	Shape and Location
Pterion (G. wing)	Junction of the greater wing of the sphenoid, squamous temporal, frontal, and parietal bones; overlies course of anterior division of middle meningeal artery
Lambda (G. the letter L)	Point on calvaria at junction of lambdoid and sagittal sutures
Bregma (G. forepart of head)	Point on calvaria at junction of coronal and sagittal sutures
Vertex (L. whirl, whorl)	Superior point of neurocranium in the midline with skull oriented in anatomical (orbitomeatal or Frankfort) plane
Asterion (G. asterios, starry)	Star-shaped; located at junction of three sutures: parietomastoid, occipitomastoid, and lambdoid
Glabella (L. smooth, hairless)	Smooth prominence, most marked in males, on the frontal bone superior to root of nose; most anterior projecting part of forehead
Inion (G. back of head)	Most prominent point of external occipital protuberance
Nasion (L. nose)	Point on skull where frontonasal and internasal sutures meet

Internal Aspect of the Cranial Base

The internal surface of the cranial base has three large, distinct depressions that lie at different levels—the **anterior**, **middle**, and **posterior cranial fossae** (Fig. 7.5*D*)—which form the bowl-shaped floor of the cranial cavity. The anterior cranial fossa is at the highest level and the posterior cranial fossa is at the lowest level.

Anterior Cranial Fossa

The inferior and anterior parts of the frontal lobes (frontal poles) of the brain occupy the anterior cranial fossae, the shallowest of the three fossae. The anterior cranial fossa is formed by the frontal bone anteriorly, the ethmoid bone in the middle, and the body and lesser wings of the sphenoid posteriorly (Fig. 7.5, *C* and *D*). The greater part of the anterior cranial fossa is formed by the ridged **orbital parts of the frontal bone** (Fig. 7.5*C*), which support the frontal lobes of the brain and form the roofs of the orbits. This surface shows sinuous impressions (brain markings) of the orbital gyri of the frontal lobes (Fig. 7.5, *C* and *D*). The **frontal crest** is a median bony extension of the frontal bone. At its base is the **foramen cecum** of the frontal bone, which is insignificant postnatally but gives passage to vessels during development. The **crista galli** (L. cock's comb) is a median ridge of bone posterior to the foramen cecum that projects superiorly from the ethmoid. On each side of the crista galli is the sievelike **cribriform plate of the ethmoid**. The olfactory nerves (CN I) from the olfactory areas of the nasal cavities pass through the foramina in the cribriform plate to reach the olfactory bulbs of the brain that lie on this plate (see Fig. 7.70).

Middle Cranial Fossa

The middle cranial fossa is butterfly-shaped, composed of large, deep depressions on each side of the much smaller **sella turcica** (L. Turkish saddle) centrally on the body of the sphenoid bone. The bones forming the middle cranial fossa are the greater wings of the sphenoid and squamous parts of the temporal bones laterally and the petrous (rocklike) parts of the temporal bones posteriorly (Fig. 7.5, *C* and *D*). The middle cranial fossa is posteroinferior to the anterior cranial fossa, separated from it by the sharp sphenoidal crests laterally and the sphenoidal limbus medially. The middle cranial fossa supports the temporal lobes of the brain. The boundary between the middle and posterior cranial fossae is the petrous crests of the temporal bones laterally and a flat plate of bone, the **dorsum sellae** of the sphenoid, medially.

The **sella turcica**—the saddlelike bony formation on the upper surface of the body of the sphenoid bone—is surrounded by the **anterior** and **posterior clinoid processes.** *The sella turcica is composed of three parts* (Fig. 7.5*C*):

- *Tuberculum sellae* (saddle horn)—a slight, olive-shaped swelling anterior to the hypophysial fossa
- *Hypophysial fossa* (pituitary fossa)—a depression (seat of the saddle) for the pituitary gland (L. hypophysis cerebri) in the middle
- *Dorsum sellae* ("back of the saddle")—a square part of bone on the body of the sphenoid posterior to the sella turcica.

The sharp posterior margins (sphenoidal ridges) of the **lesser wings of the sphenoid** overhang the middle cranial

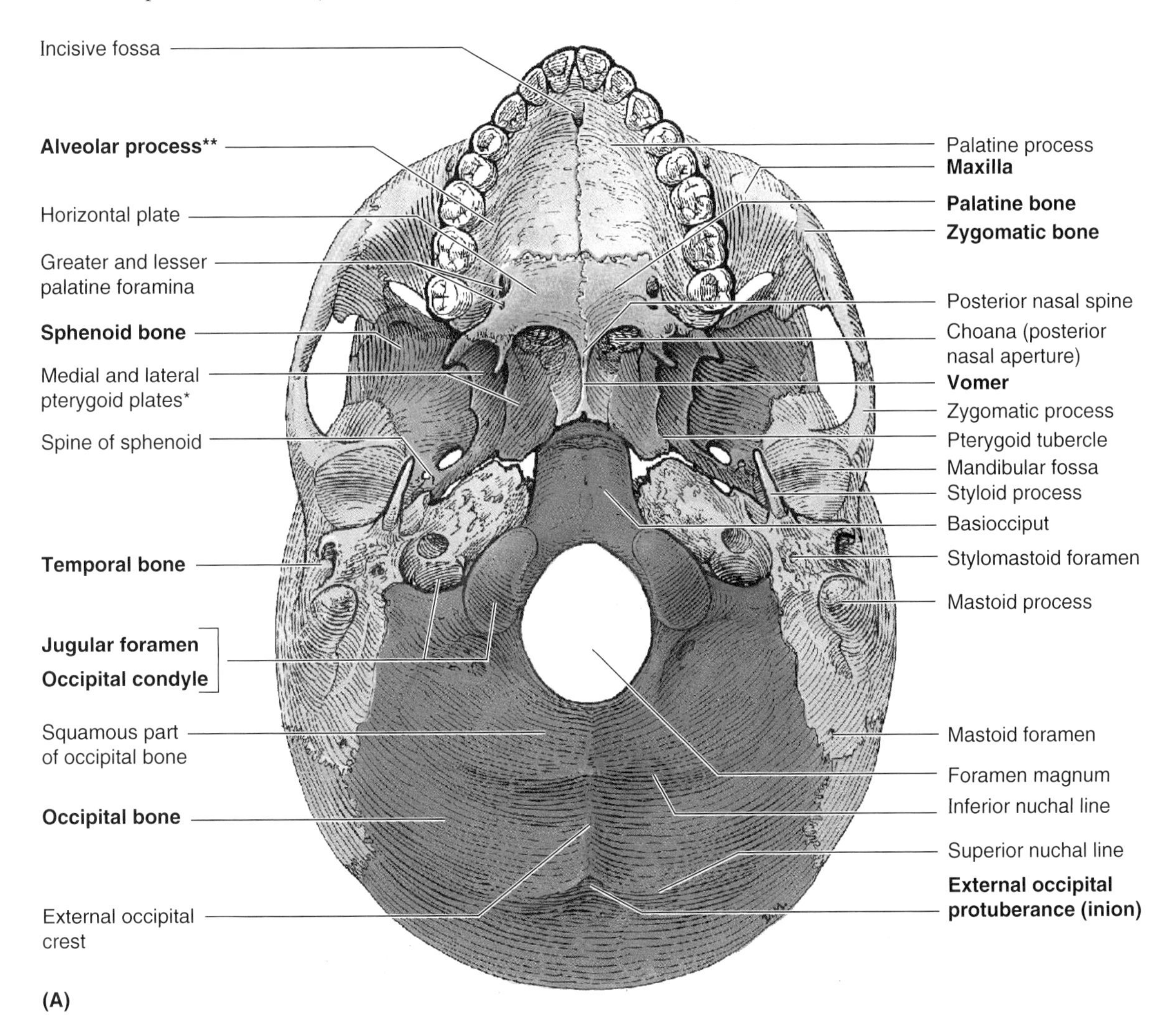

Figure 7.5. Inferior aspect (base) of an adult skull. A. Inferior view of the external aspect of the skull. This surface is formed posteriorly by the occipital bone, which becomes continuous anteriorly with the sphenoid bone following synostosis (bony fusion replacing the synchondrosis uniting the bones during childhood). Observe that the *foramen magnum* is located midway between—and on a level with—the mastoid processes. Observe also the bony palate or skeleton of the hard (bony) palate, which forms both a part of the roof of the mouth and the floor of the nasal cavity.

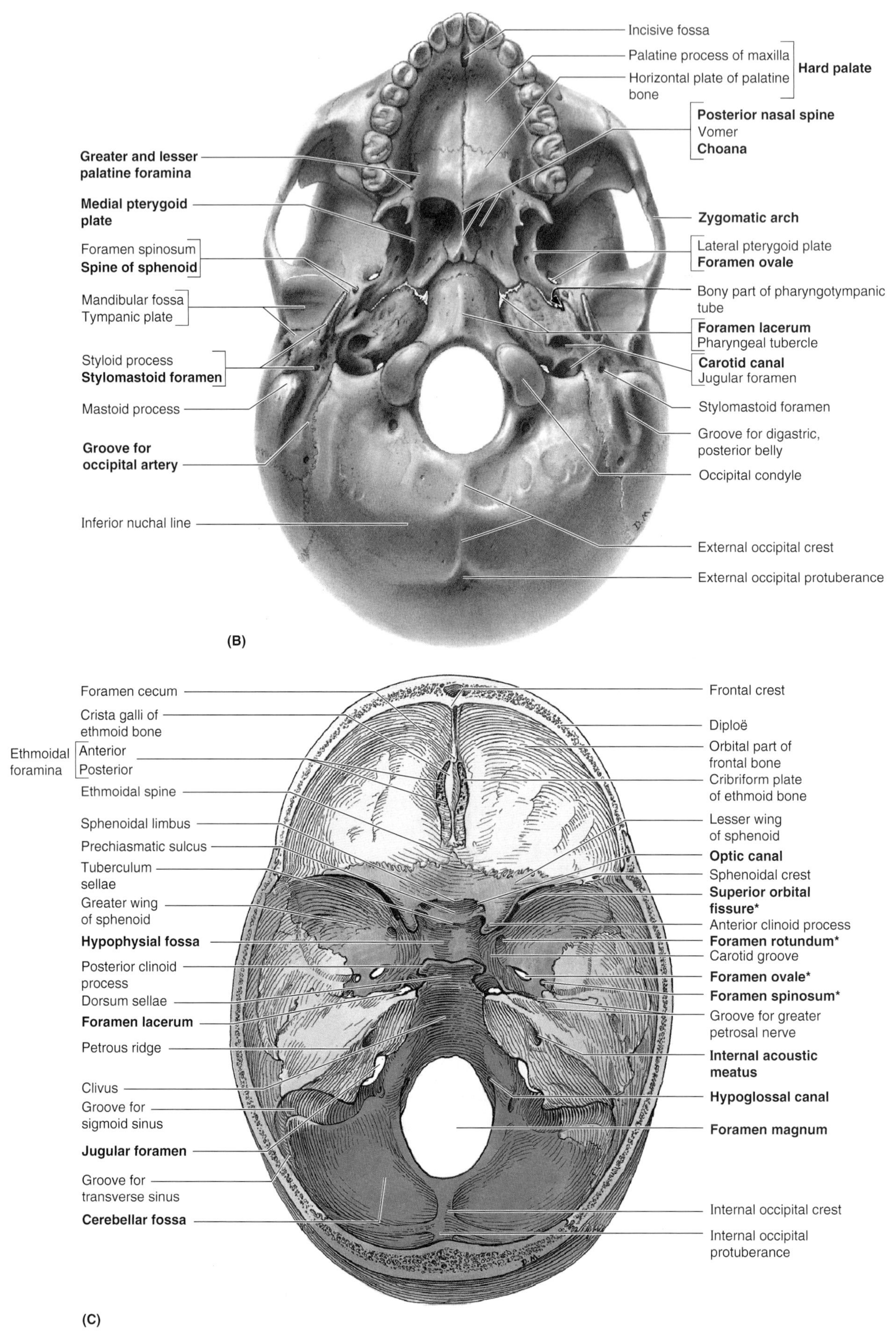

*Form crescent of foramina

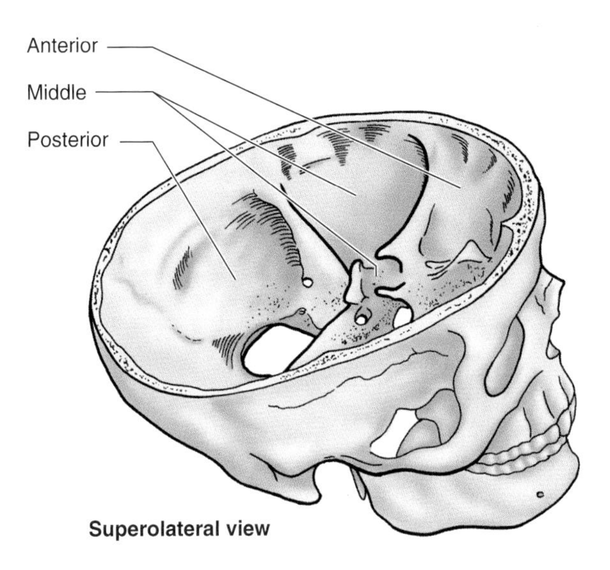

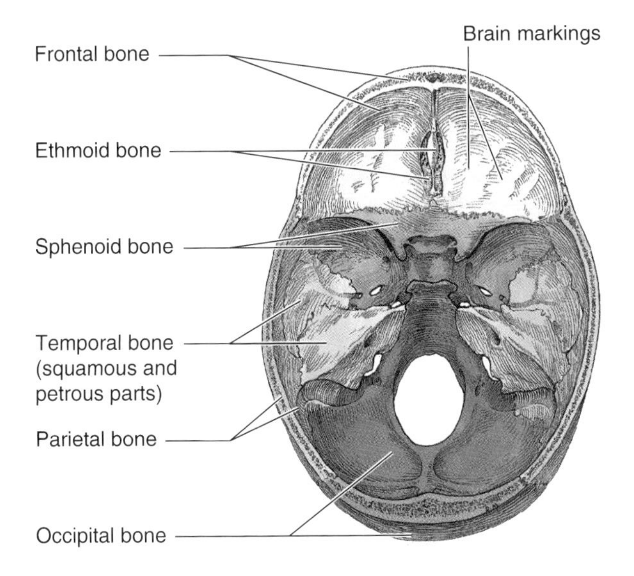

Figure 7.5. *(Continued)* **B.** Inferior view of the external aspect of the skull. Observe the *incisive fossa*, the depression in the midline of the bony palate. Notice the large *choanae* (posterior nasal apertures) on each side of the vomer. **C.** Superior view of the interior of the base of the skull. Observe the three bones contributing to the *anterior cranial fossa*: the orbital part (plate) of the frontal bone, the cribriform plate of ethmoid bone, and the lesser wing of the sphenoid bone. Observe the prechiasmatic sulcus (optic groove) leading from one optic canal to the other. Notice also the tuberculum sellae (saddle horn), hypophysial fossa, and dorsum sellae ("back of the saddle") of the sphenoid bone, which collectively form the sella turcica (L. Turkish saddle) that houses the pituitary gland in living persons. **D.** Right superolateral and superior views. Observe that the floor of the cranial cavity is divisible into three levels (steps): the anterior, middle, and posterior cranial fossae. The lower illustration shows the bones forming these fossae.

fossa. The lesser wings end medially in two projections—the **anterior clinoid processes.** "Clinoid" means "bedpost," and the four clinoid processes surround the "bed" of the pituitary gland (hypophysial fossa) like the posts in a four-poster bed.

In the middle cranial fossa on each side of the base of the body of the sphenoid bone is a **crescent of four foramina**; the foramina perforate the root of the greater wing of the sphenoid bone (Fig. 7.5*C*, Table 7.2):

- The **superior orbital fissure** is between the greater and lesser wings. This fissure provides communication with the orbit and transmits the ophthalmic veins and nerves entering the orbit (CN III, CN IV, CN V_1, CN VI, and sympathetic fibers).
- The **foramen rotundum** (L. round) is posterior to the medial end of the superior orbital fissure; it transmits the maxillary nerve (CN V_2) that supplies the skin, teeth, and mucosa related to the maxillary bone (i.e., the upper jaw and maxillary sinus) of the cheek.
- The **foramen ovale** (L. oval) is a large foramen posterolateral to the foramen rotundum; it opens inferiorly into the infratemporal fossa and transmits the mandibular nerve (CN V_3) and a small accessory meningeal artery.
- The **foramen spinosum** (L. spinous), posterolateral to the foramen ovale, transmits the middle meningeal vessels and the meningeal branch of the mandibular nerve.

The **foramen lacerum**—not part of the crescent of foramina—is a ragged foramen that lies posterolateral to the hypophysial fossa; it is an artifact of a dried skull. In life it is closed by a plate of cartilage. Nothing is transmitted vertically through the foramen lacerum. The internal carotid artery and its accompanying sympathetic and venous plexuses pass across the superior aspect of the cartilage, and some nerves traverse the cartilage horizontally. Extending posteriorly and laterally from the foramen lacerum is a narrow **groove for the greater petrosal nerve** on the anterior surface of the petrous part of the temporal bone. There is also a small groove for the lesser petrosal nerve.

Posterior Cranial Fossa

The posterior cranial fossa, the largest and deepest of the three cranial fossae (Fig. 7.5, *C* and *D*), lodges the cerebellum (L. little brain), pons, and medulla oblongata (medulla). The posterior cranial fossa is formed largely by the occipital bone, but the dorsum sellae of the sphenoid marks its anterior boundary centrally and the petrous and mastoid parts of the temporal bones contribute its anterolateral "walls." From the dorsum sellae, the **clivus** is a marked incline in the center of the anterior part of the posterior cranial fossa leading to the **foramen magnum.** Posterior to this large foramen, the posterior cranial fossa is partly divided by the **internal occipital crest** into two large concave impressions—the **cerebellar fossae.** The internal occipital crest ends in the **internal occipital protuberance** formed in relationship to a merging of dural venous sinuses, the confluence of the sinuses (p. 881). Broad grooves show the horizontal course of the **transverse** and

Table 7.2. **Foramina and Other Apertures in the Cranial Fossae and Their Contents**

Foramina/Apertures	Contents
Anterior cranial fossa	
Foramen cecum	Nasal emissary vein (1% of population)
Foramina in cribriform plate	Axons of olfactory cells in olfactory epithelium that form olfactory nerves
Anterior and posterior ethmoidal foramina	Vessels and nerves with same names
Middle cranial fossa	
Optic canals	Optic nerves (CN II) and ophthalmic arteries
Superior orbital fissures	Ophthalmic veins, ophthalmic nerve (CN V_1), CN III, IV, and VI, and sympathetic fibers
Foramen rotundum	Maxillary nerve (CN V_2)
Foramen ovale	Mandibular nerve (CN V_3) and accessory meningeal artery
Foramen spinosum	Middle meningeal artery and vein and meningeal branch of CN V_3
Foramen lacerum*	*Internal carotid artery and its accompanying sympathetic and venous plexuses
Groove or hiatus of greater petrosal nerve	Greater petrosal nerve and petrosal branch of middle meningeal artery
Posterior cranial fossa	
Foramen magnum	Medulla and meninges, vertebral arteries, spinal roots of CN XI, dural veins, anterior and posterior spinal arteries
Jugular foramen	CNs IX, X, and XI, superior bulb of internal jugular vein, inferior petrosal and sigmoid sinuses, and meningeal branches of ascending pharyngeal and occipital arteries
Hypoglossal canal	Hypoglossal nerve (CN XII)
Condylar canal	Emissary vein that passes from sigmoid sinus to vertebral veins in neck
Mastoid foramen	Mastoid emissary vein from sigmoid sinus and meningeal branch of occipital artery

*Structures actually pass *across* (rather than through) the area of foramen lacerum, an artifact of dry skulls, which is closed by cartilage in life.

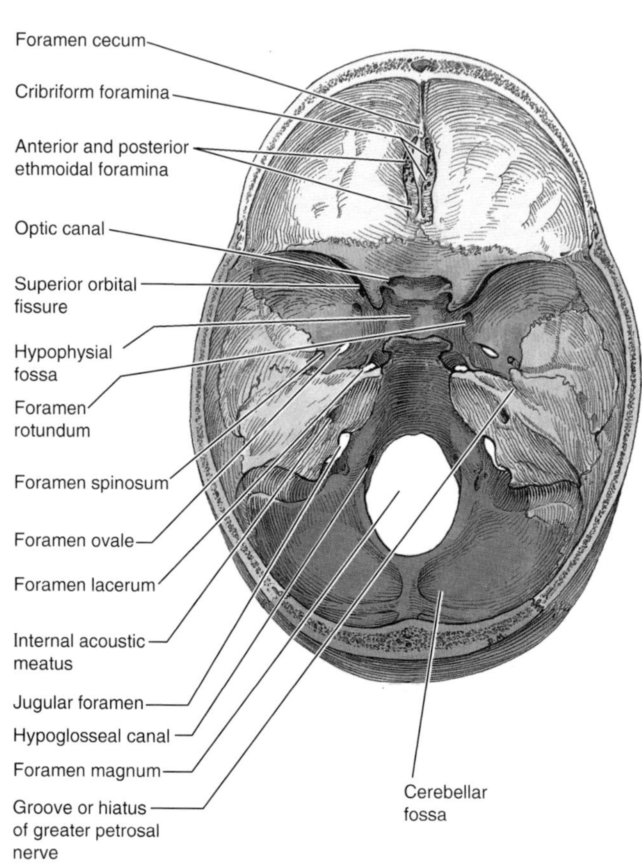

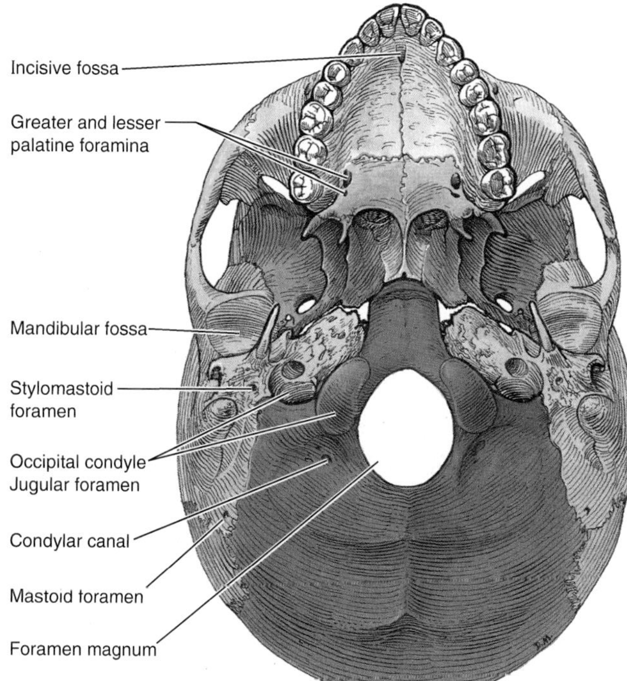

S-shaped **sigmoid sinuses** (dural venous sinuses). At the base of the petrous crest (ridge) of the temporal bone is the **jugular foramen**, which transmits several cranial nerves in addition to the sigmoid sinus, which exits the skull as the IJV (Fig. 7.5C, Table 7.2). Anterosuperior to the jugular foramen is the **internal acoustic meatus** for the facial and vestibulocochlear nerves (CN VIII) and the labyrinthine artery. The **hypoglossal canal** for the hypoglossal nerve (CN XII) is superior to the anterolateral margin of the foramen magnum.

Walls of the Cranial Cavity

The walls of the cranial cavity vary in thickness in different regions. The walls are usually thinner in females than in males and thinner in children and elderly people. The bone tends to be thinnest in areas that are well covered with muscles, such as the squamous part of the temporal bone and the posteroinferior part of the skull posterior to the foramen magnum. You can observe these thin areas of bone by holding a skull up to a bright light. Most bones of the calvaria consist of internal and external tables of compact bone, separated by spongy **diploë** (Fig. 7.5C). The diploë is cancellous bone containing red bone marrow during life, through which run canals formed by diploic veins. Examine the diploë in the calvaria of a laboratory specimen. It is not red in a dried cranium because the protein was removed during preparation of the specimen. Also observe that the inner table of bone is thinner than the outer table and that in some areas there is a thin plate of compact bone with no diploë.

Development of the Skull

The bones forming the calvaria and some parts of the cranial base develop by *intramembranous ossification*, whereas most parts of the cranial base develop by *endochondral ossification*. For more information on the early development of the skull, see Moore and Persaud (1998). At birth, the bones of the calvaria are smooth and unilaminar; no diploë is present. The frontal and parietal eminences are especially prominent. *The skull of a newborn infant is disproportionately large compared with other parts of the skeleton; however, the facial skeleton is small compared with the calvaria*, forming approximately one-eighth of the skull; in the adult the facial skeleton forms one-third of the skull. The large size of the newborn's calvaria results from precocious growth and development of the brain. The smallness of the face results from the rudimentary development of the maxillae, mandible, and paranasal sinuses (cavities in facial bones), the absence of erupted teeth, and the small size of the nasal cavities. The rudimentary development of the face makes the orbits appear relatively large.

Observe that the halves of the frontal bone are separated by a **frontal suture** and that the maxillae and mandibles are separated by an intermaxillary suture and a mandibular symphysis (intermandibular suture), respectively. Note the absence of mastoid and styloid processes. Because of the absence of mastoid processes at birth, *the facial nerves are close to the surface when they emerge from the stylomastoid foramina*. As a result, the facial nerves may be injured by forceps during a difficult delivery or by an ▶

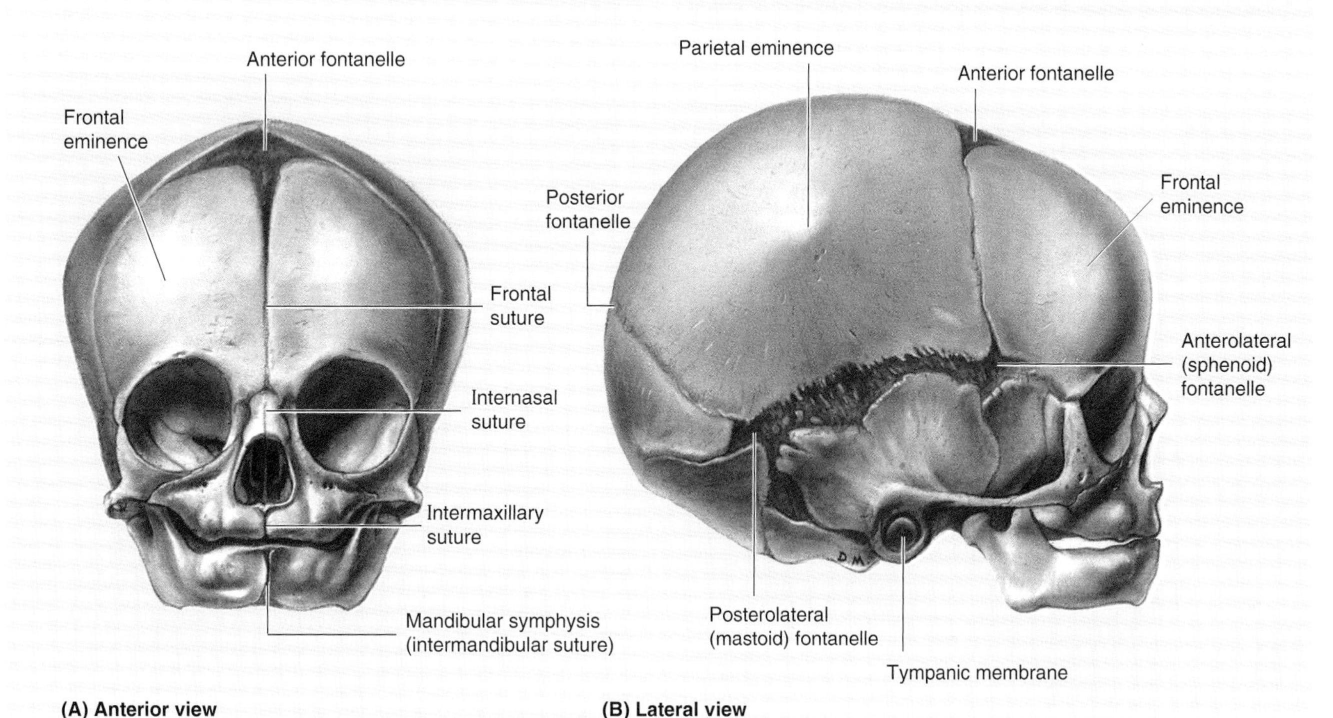

▶ incision posterior to the auricle (L. audire, to hear; external ear or pinna) in infants. The mastoid processes form during the 1st year as the sternocleidomastoid muscles complete their development and pull on the petromastoid parts of the temporal bones.

The bones of the calvaria of a newborn infant are separated by areas of fibrous tissue membrane—the **fontanelles**—which represent parts of unossified bones. *There are six fontanelles*: two are in the median plane—anterior and posterior—and two pairs are on each side—the anterolateral or *sphenoidal fontanelles* and the posterolateral or *mastoid fontanelles*. Palpation of the fontanelles during infancy (commonly referred to as a baby's "soft spot"), especially the anterior and posterior ones, enables physicians to determine the:

- Progress of growth of the frontal and parietal bones
- Degree of hydration of the infant (a depressed fontanelle indicates dehydration)
- Level of intracranial pressure (a bulging fontanelle indicates increased pressure on the brain). The *anterior fontanelle*, the largest one, is diamond or star-shaped; it is bounded by the halves of the frontal bone anteriorly and the parietal bones posteriorly.

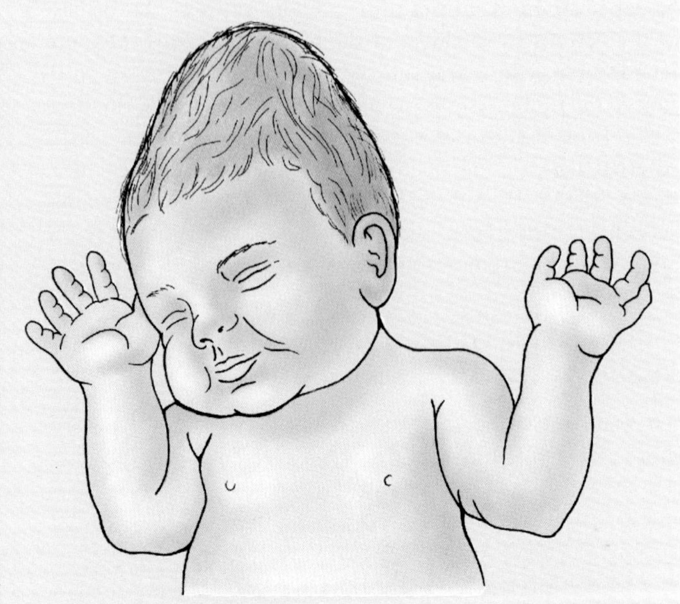

Molding of the calvaria

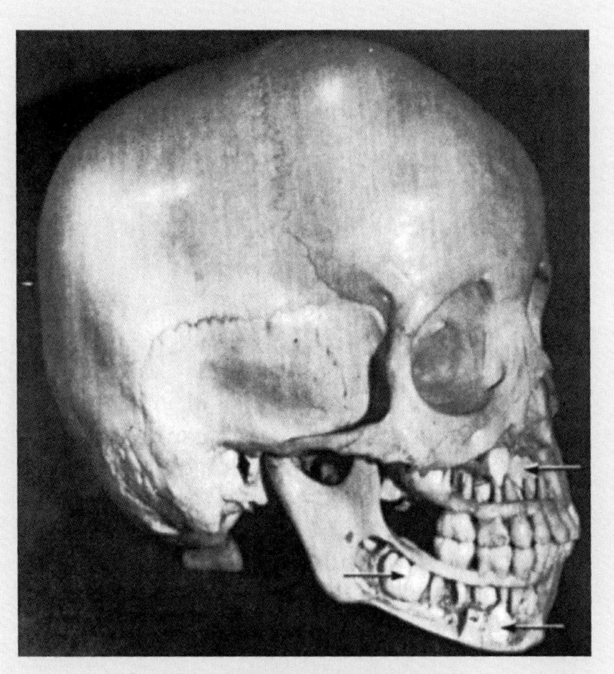

***Arrows* indicate secondary teeth**

The **anterior fontanelle** is located at the junction of the sagittal, coronal, and frontal sutures, the future site of the *bregma*. By 18 months of age, the surrounding bones have fused and the anterior fontanelle is no longer clinically palpable. Union of the halves of the frontal bone begins in the 2nd year. In most cases, the **frontal suture** (interfrontal suture) is obliterated in the 8th year; however, in approximately 8% of people, a remnant of it—the **metopic suture**—persists. Be aware that the halves of the frontal bone may remain partly separated so that a persistent suture will not be interpreted as a fracture in a radiograph or other medical image.

The **posterior fontanelle** is triangular and bounded by the parietal bones anteriorly and the occipital bone posteriorly. It is located at the junction of the lambdoid and sagittal sutures, the future site of the **lambda** (Fig. 7.3*A*). The posterior fontanelle begins to close during the first few months after birth, and by the end of the 1st year it is small and no longer clinically palpable.

The *anterolateral* (sphenoidal) and *posterolateral* (mastoid) *fontanelles*, overlain by the temporal (L. temporalis) muscle, fuse in infancy and are less important clinically than the anterior and posterior fontanelles. The halves of the mandible fuse early in the 2nd year. The two maxillae and nasal bones usually do not fuse.

The softness of the bones and their loose connections at the sutures enable the calvaria to undergo changes of shape, or *molding of the calvaria*, during birth. During passage of the baby through the birth canal, the frontal bone becomes flat, the occipital bone is drawn out, and one parietal bone slightly overrides the other. Within a few days after birth, the shape of the calvaria returns to normal. The resilience of the bones of the fetal skull allows it to resist forces that would produce a fracture in adults. The fibrous sutures of the calvaria also permit the skull to enlarge during infancy and childhood. The increase in the size of the calvaria is greatest during the first 2 years, the period ▶

▶ of most rapid brain development (Moore and Persaud, 1998). A person's calvaria normally increases in capacity until 15 or 16 years of age. After this, the calvaria usually increases slightly in size for 3 to 4 years because of bone thickening.

Age Changes in the Face

The mandible is the most dynamic of our bones; its size and shape and the number of teeth it normally bears undergo considerable change with age. In the newborn, the mandible consists of two halves united in the median plane by a fibrous tissue joint, the *mandibular symphysis*. The former site of this articulation is evident in the adult skull as a bony ridge. The *mental protuberance* (chin) begins to develop in the 2nd year but is not fully developed until after puberty. The two halves of the mandible begin to fuse during the 1st year and are fused by the end of the 2nd year.

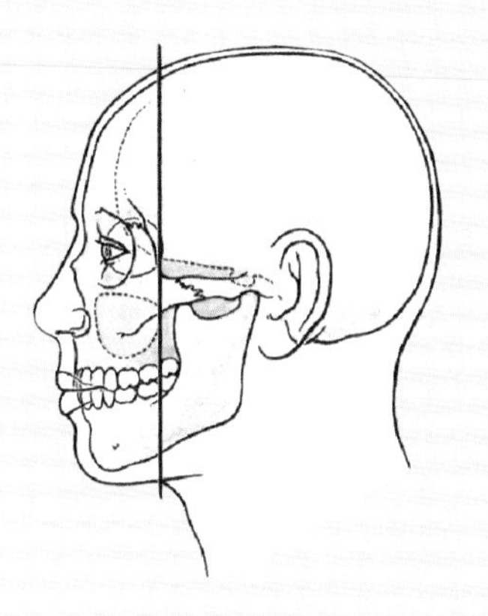

The body of the mandible in newborn infants is a mere shell lacking an alveolar process, each half enclosing five primary (deciduous) teeth. These teeth usually begin to erupt in infants of approximately 6 months of age. The body of the mandible elongates, particularly posterior to the mental foramen, to accommodate the development and then the bearing of eight secondary (permanent) teeth, which begin to erupt during the 6th year of life. Eruption of these permanent teeth is not complete until early adulthood. Rapid growth of the face during infancy and early childhood coincides with the eruption of primary teeth. Vertical growth of the upper face results mainly from dentoalveolar development. These changes are more marked after the secondary teeth erupt. Following complete loss of teeth in old age (or younger if care is neglected), the alveoli begin to fill in with bone and the alveolar processes begin to resorb.

Concurrent enlargement of the frontal and facial regions is associated with the increase in the size of the *paranasal sinuses*—air-filled extensions of the nasal cavities in certain cranial bones. Most paranasal sinuses are rudimentary or absent at birth. Growth of the paranasal sinuses is important in altering the shape of the face and in adding resonance to the voice. ▶

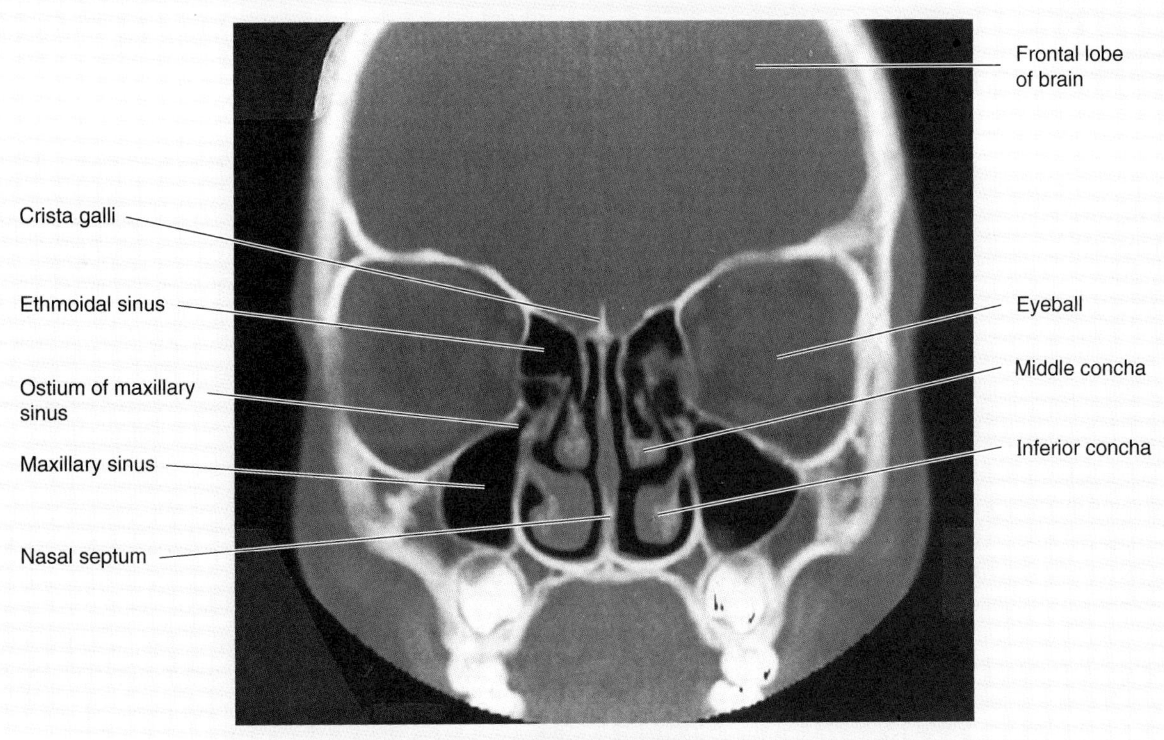

CT of child's head

Obliteration of the Cranial Sutures

The obliteration of sutures between the bones of the calvaria usually begins between the ages of 30 and 40 on the internal surface and approximately 10 years later on the external surface. Obliteration of sutures usually begins at the bregma and continues sequentially in the sagittal, coronal, and lambdoid sutures.

Age Changes in the Skull

As people age the skull bones normally become progressively thinner and lighter, and the diploë gradually become filled with a gray gelatinous material. In these individuals the bone marrow has lost its blood cells and fat, giving it a gelatinous appearance.

Craniosynostosis and Skull Deformities

Premature closure of the sutures of the skull (*primary craniosynostosis*) results in several skull deformities. The incidence of primary craniosynostosis is approximately 1 per 2000 births (Behrman et al., 1996). The cause of craniosynostosis is unknown, but genetic factors appear to be important. The prevailing hypothesis is that abnormal development of the cranial base creates exaggerated forces on the dura mater (outer covering membrane of the brain) that disrupt normal cranial suture development. These deformities are much more common in males than in females and are often associated with other skeletal anomalies. The type of deformed skull that forms depends on which sutures close prematurely.

Premature closure of the sagittal suture, in which the anterior fontanelle is small or absent, results in a long, narrow, and wedge-shaped skull—*scaphocephaly.* Scaphocephaly, constituting approximately one-half the cases of craniosynostosis, does not produce abnormal neurological development.

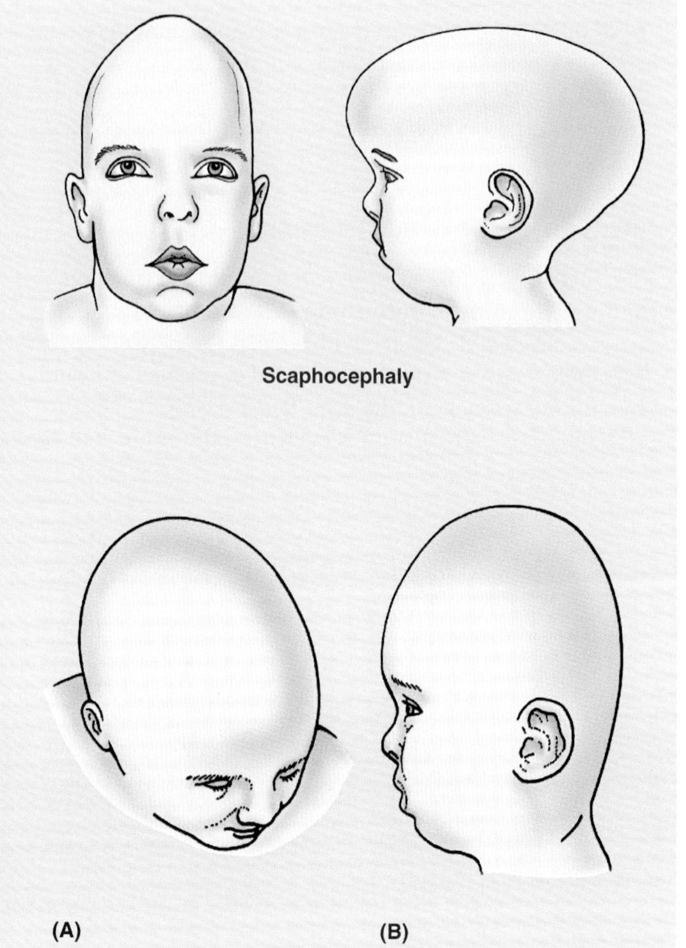

When premature closure of the coronal or the lambdoid suture occurs on one side only, the skull is twisted and asymmetrical, a condition known as *plagiocephaly* (*A*). Another 30% of cases of skull deformity involve *premature closure of the coronal suture,* resulting in a high, towerlike skull—*oxycephaly* or *turricephaly* (*B*). This type of skull deformity is more common in females. ○

Face

The face is the anterior aspect of the head from the forehead to the chin and from one ear to the other ear. The basic shape of the face is determined by the underlying bones. The buccal fatpads in the cheeks and the facial muscles contribute to the final shape of the face. The relatively large size of the **buccal fatpads** in infants, which prevent collapse of the cheeks during sucking, produce their pudgy appearance. The muscles of the face are in the subcutaneous tissue; most of them attach to the skull bones and the skin or mucous membrane. The overlying skin is connected to the bones by skin ligaments (L. retinacula cutis)—bands of connective tissue (p. 14). Growth of the facial bones takes longer than those of the calvaria. The ethmoid bone, orbital cavities, and superior parts of the nasal cavities have nearly completed their growth by the 7th year. Expansion of the orbits and growth of the nasal septum carry the maxillae inferoanteriorly. Considerable facial growth occurs during childhood as the paranasal sinuses develop (p. 849) and the permanent teeth erupt.

Muscles of the Face

The facial muscles or muscles of facial expression are subcutaneous. They move the skin and change facial expressions to convey mood. These muscles are in the anterior and posterior scalp, face, and neck. Most muscles attach to bone or fascia and produce their effects by pulling the skin. For a summary of the facial muscles, their attachments and actions, see Table 7.3.

Table 7.3. **Muscles of the Scalp and Face**

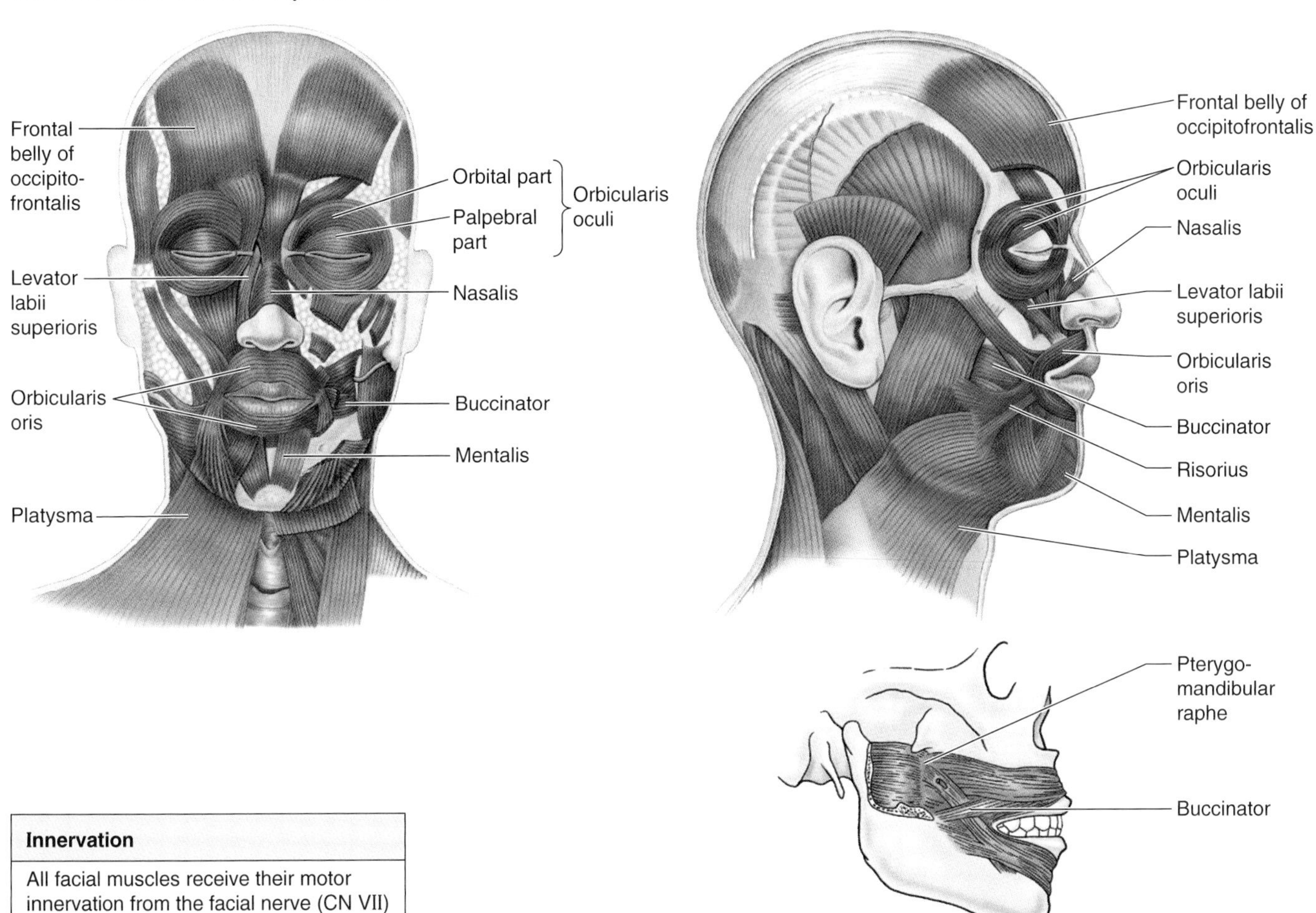

Innervation
All facial muscles receive their motor innervation from the facial nerve (CN VII)

Muscle	Origin	Insertion	Main Action
Frontal belly of occipitofrontalis	Epicranial aponeurosis	Skin of forehead and eyebrows	Elevates eyebrows and skin of forehead
Orbicularis oris	Some fibers arise near median plane of maxilla superiorly and mandible inferiorly; other fibers arise from deep surface of skin	Mucous membrane of lips	As sphincter of oral opening, compresses and protrudes lips (e.g., purses them during whistling and sucking)
Levator labii superioris	Frontal process of maxilla and infraorbital region	Skin of upper lip and alar cartilage of nose	Elevates upper lip and dilates nostril
Mentalis	Incisive fossa of mandible	Skin of chin	Elevates and protrudes lower lip
Buccinator	Mandible, pterygomandibular raphe, and alveolar processes of maxilla and mandible	Angle of mouth	Presses cheek against molar teeth, thereby aiding chewing; expels air from oral cavity as occurs when playing a wind instrument; draws mouth to one side when acting unilaterally
Orbicularis oculi	Medial orbital margin, medial palpebral ligament, and lacrimal bone	Skin around margin of orbit; tarsal plate	Closes eyelids: palpebral part gently closes lids; orbital part tightly closes them
Nasalis	Superior part of canine ridge of maxilla	Nasal cartilages	Draws ala (side) of nose toward nasal septum
Platysma	Superficial fascia of deltoid and pectoral regions	Mandible, skin of cheek, angle of mouth, and orbicularis oris	Depresses mandible and tenses skin of lower face and neck

All muscles of facial expression develop from the 2nd pharyngeal arch (Moore and Persaud, 1998) *and are supplied by its nerve, the 7th cranial nerve (CN VII).* The muscles of facial expression surround the orifices of the mouth, eyes, and nose and act as sphincters and dilators, which close and open the orifices. The facial muscles develop from mesenchyme (embryonic connective tissue) of the 2nd pharyngeal arch and are part of a subcutaneous muscular sheet that spreads over the neck and face during embryonic development, carrying branches of the facial nerve with it. Because of their common embryological origin, the platysma (G. flat plate) and facial muscles are often fused and their fibers frequently intermingled. Functionally the muscular sheet differentiates into muscles that surround the facial orifices, such as the mouth. A sphincter and dilator mechanism for each orifice also produces facial expressions (Fig. 7.7).

Facial Lacerations and Incisions

Because the face has no distinct deep fascia and the subcutaneous tissue between the cutaneous attachments of the facial muscles is loose, *facial lacerations* tend to gape (part widely). Consequently, the skin must be sutured with great care to prevent scarring. The looseness of the subcutaneous tissue also enables fluid and blood to accumulate in the loose connective tissue following bruising of the face. Similarly, facial inflammation causes considerable swelling (e.g., a bee sting on the bridge of the nose may close both eyes). *As a person ages, the skin loses its resiliency* (elasticity). As a result, ridges and wrinkles occur in the skin perpendicular to the direction of the facial muscle fibers. Incisions along these cleavage or wrinkle lines (Langer's lines) heal with minimal scarring (p. 14). ⊙

Muscle of the Forehead

The *frontalis* is the anterior part (frontal belly) of the scalp muscle—the **occipitofrontalis**. The frontal belly of this muscle arises from the anterior part of the **epicranial aponeurosis** (L. aponeurosis epicranialis) and is attached to the skin of the eyebrows (Figs. 7.6, *A* and *B*, and 7.7, Table 7.3). *The frontalis has no bony attachments*; it elevates the eyebrows, giving the face a surprised look, and produces transverse wrinkles in the forehead when one frowns.

Muscles of the Mouth, Lips, and Cheeks

Several muscles alter the shape of the mouth and lips (e.g., during speaking, singing, whistling, and mimicry). The shape of the mouth and lips is controlled by a complex three-dimensional group of muscular slips (Williams et al., 1995), which include:

- Elevators, retractors, and evertors of the upper lip
- Depressors, retractors, and evertors of the lower lip
- A compound sphincter around the mouth
- The buccinator (L. trumpeter) in the cheek.

At rest, the lips are in gentle contact and the teeth are close together.

The **orbicularis oris** is the sphincter of the mouth (Fig. 7.6, *A* and *B*, Table 7.3) and is the first of the series of sphincters associated with the digestive tract. Its fibers encircle the mouth and are within the lips. Normally when the orbicularis oris is tonically contracted, the mouth is closed; active (phasic) contraction causes the mouth to become narrow as the lips pucker, as occurs in whistling (Fig. 7.7). The orbicularis oris is important during articulation (speech) and compresses the lips against the teeth, working with the tongue (L. lingua; G. glossa) to hold food between the teeth during mastication (chewing).

The **dilator muscles** radiate from the lips like the spokes of a wheel. Except for the buccinator, they are of little practical importance to most health care professionals (except dentists who prefer to have the mouth opened widely); however, the dilator muscles are commonly used by actors and mimics. Only the main features and functions of the muscles of the mouth, lips, and cheeks are mentioned here.

The **levator labii superioris alaeque nasi** ("elevator" of the upper lip and wings [L. alae] of the nose) attaches superiorly to the maxilla. It divides into two slips that attach to the alar cartilage of the nose and the upper lip and elevate both these structures.

The **mentalis** is a small muscle arising from the mandible and ascending to the skin of the chin. It raises the skin of the chin during the expression of doubt (Fig. 7.7).

The **buccinator** is a thin, flat, rectangular muscle that attaches laterally to the alveolar processes of the maxillae and mandible, opposite the molar teeth (Fig. 7.6, Table 7.3), and to the *pterygomandibular raphe*—a tendinous thickening of the buccopharyngeal fascia separating and giving origin to the buccinator muscle anteriorly. The buccinator, active in smiling, also keeps the cheek taut, thereby preventing it from folding and being injured during chewing. The fibers of the buccinator mingle medially with those of the orbicularis oris. *The buccinator aids mastication by pressing the cheeks against the molar teeth during chewing.* Thus, working together with the tongue on the lingual aspect and the orbicularis oris anteriorly, food is held between the occlusal surfaces of the teeth for

Figure 7.6. Muscles of the scalp, auricle (external ear, pinna), face, and neck. The muscles of facial expression may be grouped as the muscles of the scalp and auricle; the muscles around the opening of the orbit (eye socket); the muscles of the nose; and the muscles of the mouth and the platysma (G. flat plate). **A.** Anterior view of a dissection of the head and neck. **B.** Lateral view of the facial muscles.

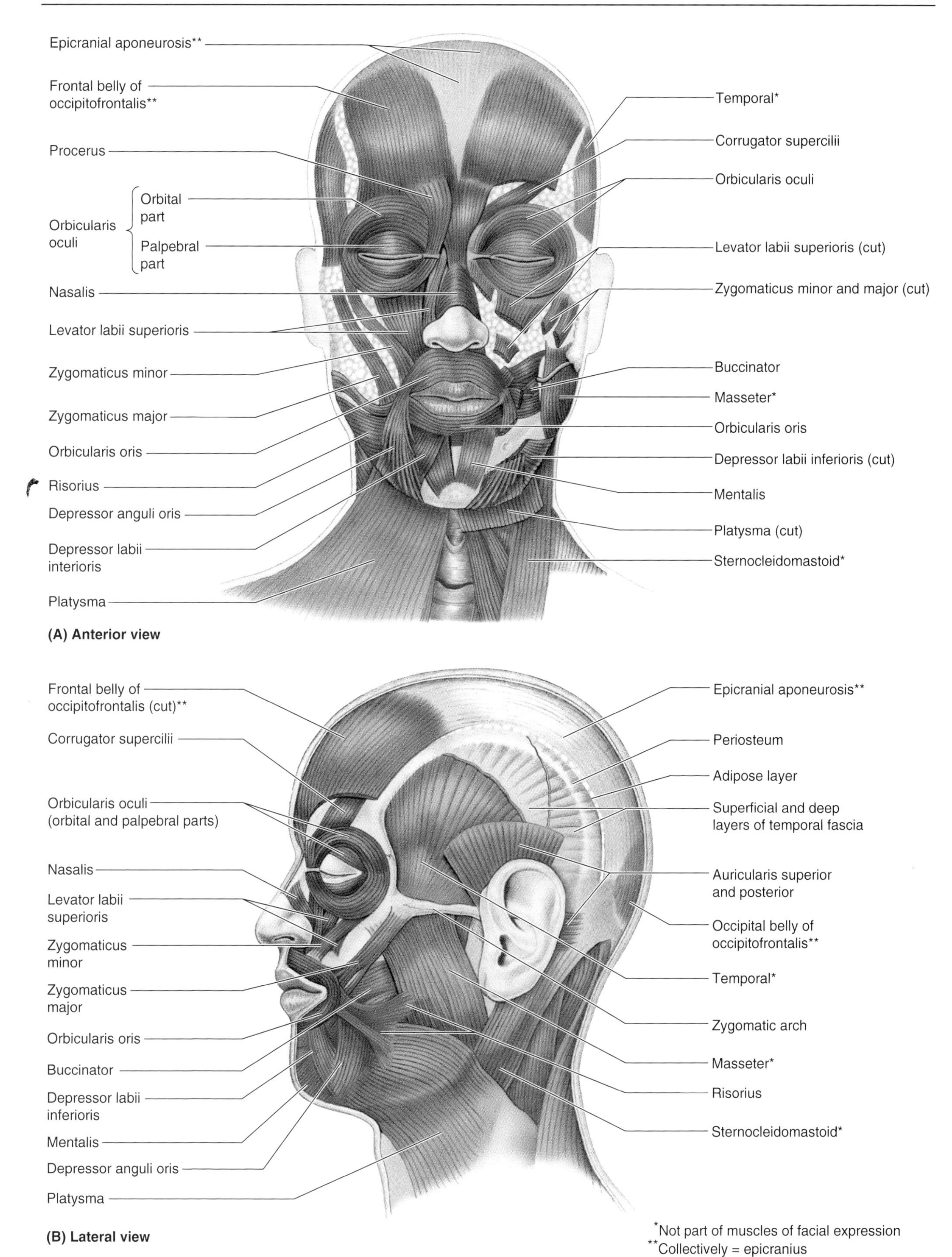
Epicranial aponeurosis**
Frontal belly of occipitofrontalis**
Procerus
Orbicularis oculi
Orbital part
Palpebral part
Nasalis
Levator labii superioris
Zygomaticus minor
Zygomaticus major
Orbicularis oris
Risorius
Depressor anguli oris
Depressor labii interioris
Platysma
Temporal*
Corrugator supercilii
Orbicularis oculi
Levator labii superioris (cut)
Zygomaticus minor and major (cut)
Buccinator
Masseter*
Orbicularis oris
Depressor labii inferioris (cut)
Mentalis
Platysma (cut)
Sternocleidomastoid*
(A) Anterior view
Frontal belly of occipitofrontalis (cut)**
Corrugator supercilii
Orbicularis oculi (orbital and palpebral parts)
Nasalis
Levator labii superioris
Zygomaticus minor
Zygomaticus major
Orbicularis oris
Buccinator
Depressor labii inferioris
Mentalis
Depressor anguli oris
Platysma
Epicranial aponeurosis**
Periosteum
Adipose layer
Superficial and deep layers of temporal fascia
Auricularis superior and posterior
Occipital belly of occipitofrontalis**
Temporal*
Zygomatic arch
Masseter*
Risorius
Sternocleidomastoid*
(B) Lateral view
*Not part of muscles of facial expression
**Collectively = epicranius

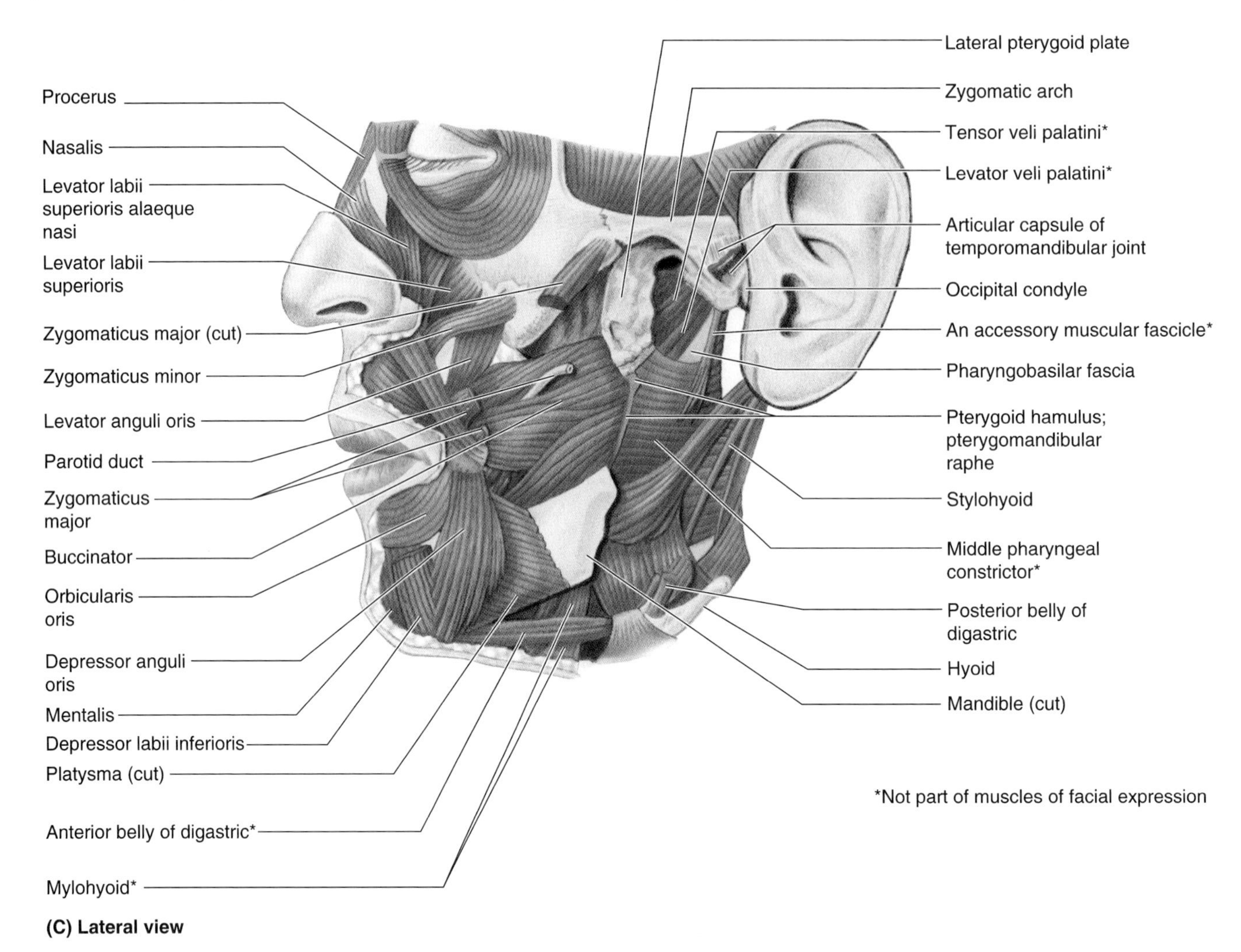

Figure 7.6. *(Continued)* **C.** Lateral view of a deeper dissection of the head and upper neck. The superficial muscles have been cut away to show the full extent of the buccinator (L. trumpeter) and its posterior attachments, which include the pterygoid hamulus, the pterygomandibular raphe, and the mandible (lower jaw). The pterygomandibular raphe is an attachment shared with the middle superior pharyngeal constrictor muscle.

chewing. The buccinator is also used during whistling and sucking by forcing the cheeks against the teeth. The buccinator was given its name because it compresses the cheeks during blowing (e.g., when a musician plays a wind instrument). Some trumpeters (notably Dizzy Gillespie) stretch their buccinator and other cheek muscles so much that their cheeks balloon out when they blow forcibly on their instruments.

The **depressor anguli oris**, as its name indicates, depresses the angles of the mouth (Fig. 7.7) as in frowning. Posterior fibers of the platysma assist with this movement.

The **levator anguli oris** attaches superiorly to the infraorbital margin and inferiorly to the angle of the mouth. It elevates the corner of the mouth.

The **zygomaticus major**, extending from the zygomatic bone to the angle of the mouth, draws the angle of the mouth superolaterally as occurs during smiling or laughing (Fig. 7.7).

The **zygomaticus minor**, a narrow slip of muscle, passes obliquely from the zygomatic bone to the orbicularis oris. It helps to raise the upper lip when showing contempt or to deepen the nasolabial sulcus when showing sadness.

The **levator labii superioris** (L. labii, lips), descending from the infraorbital margin to the upper lip, raises and everts the upper lip. It also helps the minor zygomatic to deepen the nasolabial sulcus when showing sadness.

The **depressor labii inferioris**, lateral to the mentalis, attaches inferiorly to the mandible and merges superiorly with its contralateral partner and the orbicularis oris. It draws the lip inferiorly and slightly laterally, as when showing impatience.

The **risorius,** a variable muscle, arises from the platysma and fascia of the masseter. It attaches to the fascia covering the parotid gland inferoanterior to the ear and to the angle of the mouth. It draws the corner of the mouth laterally when grinning (Fig. 7.7).

The **platysma** is a broad, thin sheet of muscle in the sub-

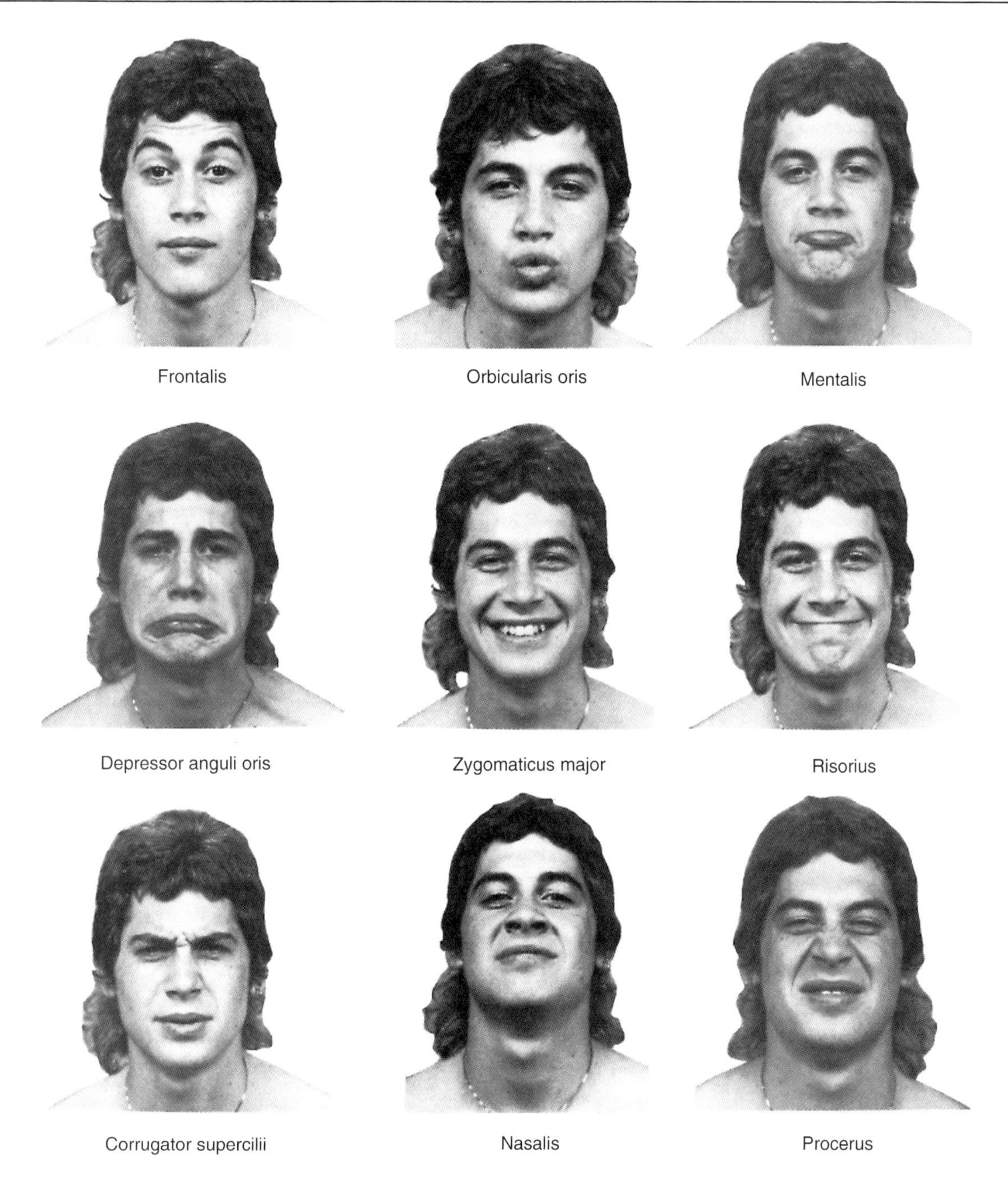

Figure 7.7. Muscles of facial expression in action. These muscles are superficial sphincters and dilators of the orifices of the head. The facial muscles—supplied by the facial nerve (CN VII)—are attached to and move the skin of the face, producing various facial expressions.

cutaneous tissue of the neck (Fig. 7.6, *A* and *B*). It arises in the fascia covering the superior parts of the deltoid and pectoralis major muscles and sweeps superomedially over the clavicle to the inferior border of the mandible (Fig. 7.6*B*). The anterior borders of the two muscles decussate over the chin and blend with the facial muscles. Acting from its superior attachment, the platysma tenses the skin, producing vertical skin ridges covering great stress, and releases pressure on the superficial veins. Acting from its inferior attachment, the platysma helps to depress the mandible and draw the corners of the mouth inferiorly, as in a grimace. The platysma is supplied by the cervical branch of the facial nerve.

Muscles Around the Orbital Opening

The function of the eyelids (L. palpebrae) is to protect the eye from injury and excessive light. The eyelids also keep the cornea moist by spreading the tears.

The **orbicularis oculi** closes the eye and wrinkles the forehead vertically (Figs. 7.6, *A* and *B*, and 7.8, Table 7.3). Its fibers attach primarily to the medial orbital margins and me-

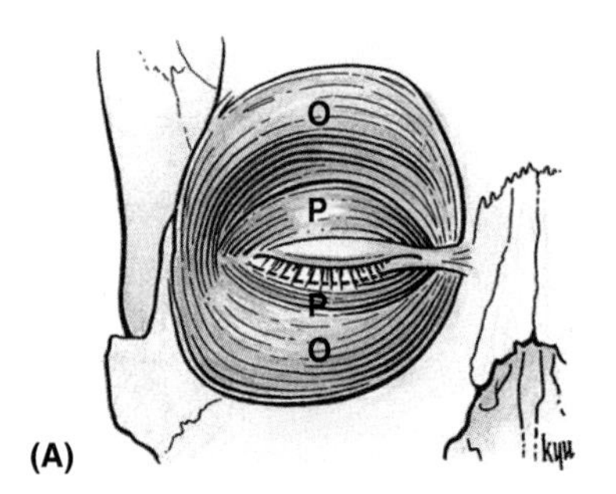

(A)

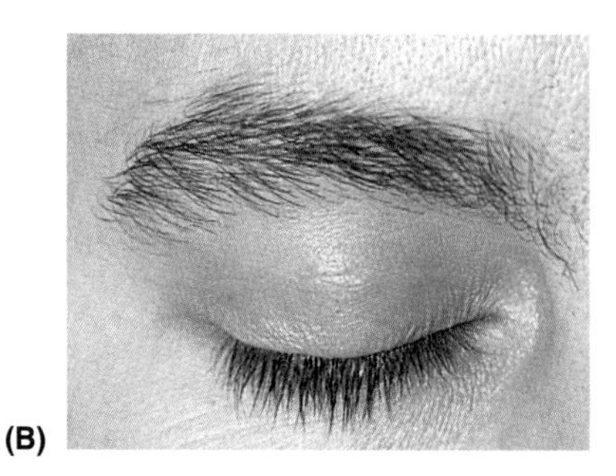
(B)

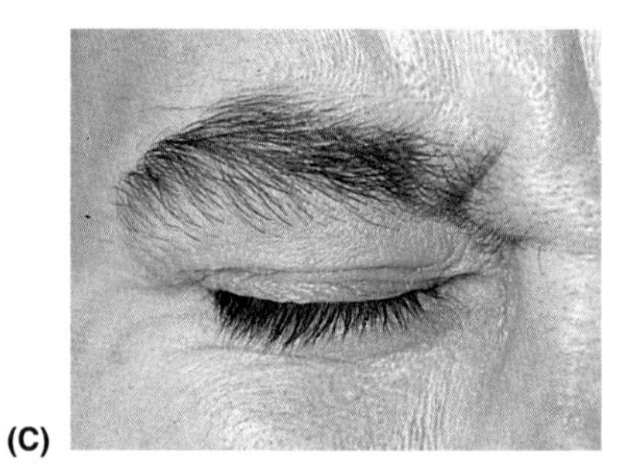
(C)

Figure 7.8. Disposition and actions of the orbicularis oculi muscle. A. Diagram showing the *palpebral (P) and orbital (O) parts* of the orbicularis oculi. **B.** The palpebral part gently closes the eyelids (L. palpebrae). **C.** The orbital part tightly closes the eyelids. The lacrimal part (not shown) passes posterior to the lacrimal sac to attach to the tarsal plates and the lateral margin of the lacrimal sac and aids in the spread of lacrimal secretions by holding the eyelids close to the eyeballs.

dial palpebral ligament; they sweep in concentric circles around the orbital margin and eyelids. Contraction of these fibers narrows the palpebral fissure (aperture between the eyelids) and assists the flow of lacrimal fluid (tears) by bringing the lids together laterally first, closing the palpebral fissure in a lateral to medial direction. *The orbicularis oculi consists of three parts:*

- The *lacrimal part* draws the eyelids and *lacrimal puncta* medially, pressing the latter into the lacrimal lake (L. lacus lacrimalis) so that capillary action may drain lacrimal fluid from it (pp. 901–902).
- The *palpebral part* gently closes the eyelids—as in blinking or in sleep—to keep the cornea from drying.
- The *orbital part* strongly closes the lids, as in squinting (narrowing the fissure), to protect against glare and dust. This squinting causes a radiating folding of the skin lateral to the lateral angle of the lids, which may develop into the permanent "crow's feet" wrinkles of older age.

The lacrimal part of the orbicularis oculi lies deep to the palpebral part and is often considered a part of it. It may aid the flow of lacrimal fluid by holding the eyelids close to the eyeballs. The lacrimal part is also said to exert traction on the lacrimal fascia, thereby dilating the lacrimal sac; however, a proposed pumping action on the lacrimal gland is unlikely. When all three parts of the orbicularis oculi contract, the eyes are firmly closed (Fig. 7.8*C*) and the adjacent skin is wrinkled. Similar wrinkling occurs when a person scrutinizes something (squints). The orbicularis oculi is supplied by a *zygomatic branch of the facial nerve.*

The ***corrugator supercilii*** (Fig. 7.6*A*) arises from the orbital part of the orbicularis oculi and nasal prominence and inserts into the skin of the eyebrow. It draws the medial end of the eyebrow downward and wrinkles the forehead vertically (Fig. 7.7), demonstrating concern.

Muscles Around the Nose

All muscles around the nose are supplied by the facial nerve. The procerus and depressor septi muscles are relatively unimportant to most health care professionals. The **nasalis,** *the main muscle of the nose* (Fig. 7.6, *A* and *B*, Table 7.3), consists of transverse (compressor naris) and alar (dilator naris) parts. The *transverse part* arises from the superior part of the canine ridge on the anterior surface of the maxilla, superior to the incisor teeth, and passes superomedially to the dorsum of the nose. It compresses the nostril. The *alar part* arises from the maxilla superior to the transverse part and attaches to the alar cartilages of the nose. It widens the anterior nasal aperture, flaring the nostrils (Fig. 7.7). The alar part also draws the nostril down, as occurs during fright and anger. Both parts of the nasalis are supplied by a buccal branch of the facial nerve.

The **procerus,** a small slip of muscle that is continuous with the frontalis muscle (Fig. 7.6*A*), passes from the forehead over the bridge of the nose. The procerus draws the medial part of the eyebrow inferiorly, producing transverse wrinkles over the bridge of the nose (Fig. 7.7). This action probably attempts to reduce the glare of bright sunlight; it is also used when frowning. The procerus is supplied by a buccal branch of the facial nerve.

The **depressor septi** arises from the maxilla superior to the central incisor tooth and inserts into the mobile part of the nasal septum. It assists the alar (dilator naris) part of the nasalis muscle to widen the nostril during deep inspiration. It is supplied by a buccal branch of the facial nerve.

Flaring of the Nostrils

The actions of the nasalis have generally been held as insignificant; however, observant clinicians study its action because of its diagnostic value. For example, true *nasal breathers* can flare their nostrils distinctly. Habitual mouth breathing, caused by chronic nasal obstruction, for example, diminishes and sometimes eliminates the ability to flare the nostrils. Children who are chronic *mouth breathers* often develop dental malocclusion (improper ▶

▶ bite). Recently, antisnoring devices have been developed, which attach to the nose to flare the nostrils and maintain a more patent air passageway.

Injury to the Facial Nerve

Injury to the facial nerve or its branches produces paralysis of some or all facial muscles on the affected side. *The most common nontraumatic cause of facial paralysis is inflammation of the facial nerve near the stylomastoid foramen.* This produces edema (swelling) and compression of the nerve in the facial canal. The loss of tonus of the orbicularis oculi causes the lower lid to evert (fall away from the surface of the eye) so that the cornea on the affected side is not adequately hydrated—lubricated or flushed with lacrimal fluid—making it vulnerable to ulceration. Patients cannot whistle, blow a wind instrument, or chew effectively. The palsy weakens or paralyzes the buccinator and orbicularis oris, the cheek and lip muscles that aid chewing by holding food between oclusive surfaces and out of the gutter between the teeth and cheek. Food accumulates during chewing and often must be continually removed with a finger. Displacement of the mouth (drooping of its corner) is produced by contraction of unopposed contralateral facial muscles, resulting in food and saliva dribbling out of the side of the mouth. Patients frequently dab their eyes and mouth with a handkerchief to wipe the fluid (tears and saliva), which runs from the drooping lid and mouth; the fluid and constant wiping results in localized skin irritation.

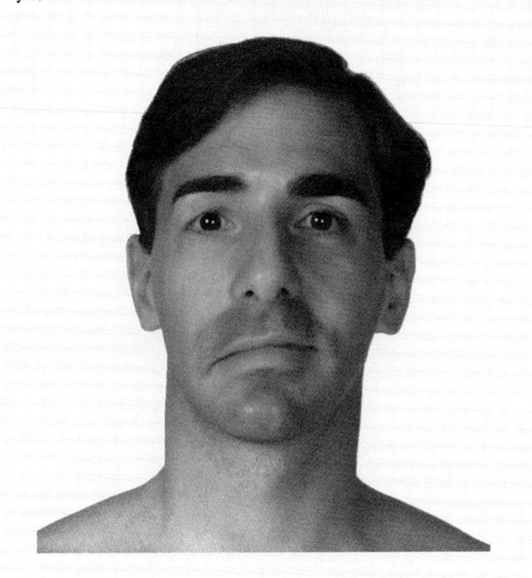

Facial nerve palsy has many causes. It may be idiopathic—occurring without a known cause (Bell's palsy)—but it often follows exposure to cold, as occurs when riding in a car or sleeping with a window open. Patients with *idiopathic facial paralysis* demonstrate the clinical manifestations described previously. Facial paralysis may be a complication of surgery; consequently, identification of the facial nerve is essential during surgery of the parotid gland. The facial nerve is visible as it emerges from the stylomastoid foramen; if necessary, electrical stimulation may be used for confirmation. Facial nerve palsy may also be associated with dental manipulation, vaccination, pregnancy, HIV infection, Lyme disease (inflammatory disorder causing headache and stiff neck), and infections of the middle ear (otitis media). ○

Nerves of the Face

The cutaneous nerves of the neck overlap those of the face. Cutaneous branches of cervical nerves from the *cervical plexus* extend over the ear, the posterior aspect of the neck, and much of the parotid region of the face (area overlying the angle of the jaw).

The trigeminal nerve (CN V) is the sensory nerve for the face and is the motor nerve for the muscles of mastication and several small muscles (Fig. 7.9, Table 7.4). The peripheral processes of the trigeminal ganglion constitute the:

- Ophthalmic nerve (CN V_1)
- Maxillary nerve
- Sensory component of the mandibular nerve.

These nerves are named according to their main areas of termination—the eye, maxilla, and mandible, respectively. The first two divisions (CN V_1 and CN V_2) are wholly sensory; the mandibular division is also largely sensory but contains fibers of the motor root of CN V.

Ophthalmic Nerve

CN V_1—the superior division of the trigeminal nerve—is the smallest of the three divisions. It arises from the trigeminal ganglion as a wholly sensory nerve and supplies the area of skin derived from the embryonic *frontonasal prominence* (Moore and Persaud, 1998). The ophthalmic nerve enters the orbit through the *superior orbital fissure* and supplies branches to the eyeball and the superior part of the nasal cavity; it then leaves the orbit to supply the face. On entering the orbit, the ophthalmic nerve divides into three branches (Fig. 7.9):

- Nasociliary nerve
- Frontal nerve
- Lacrimal nerve.

The cutaneous nerves of these branches of the ophthalmic nerve are the:

- External nasal nerve
- Infratrochlear nerve
- Supratrochlear nerve

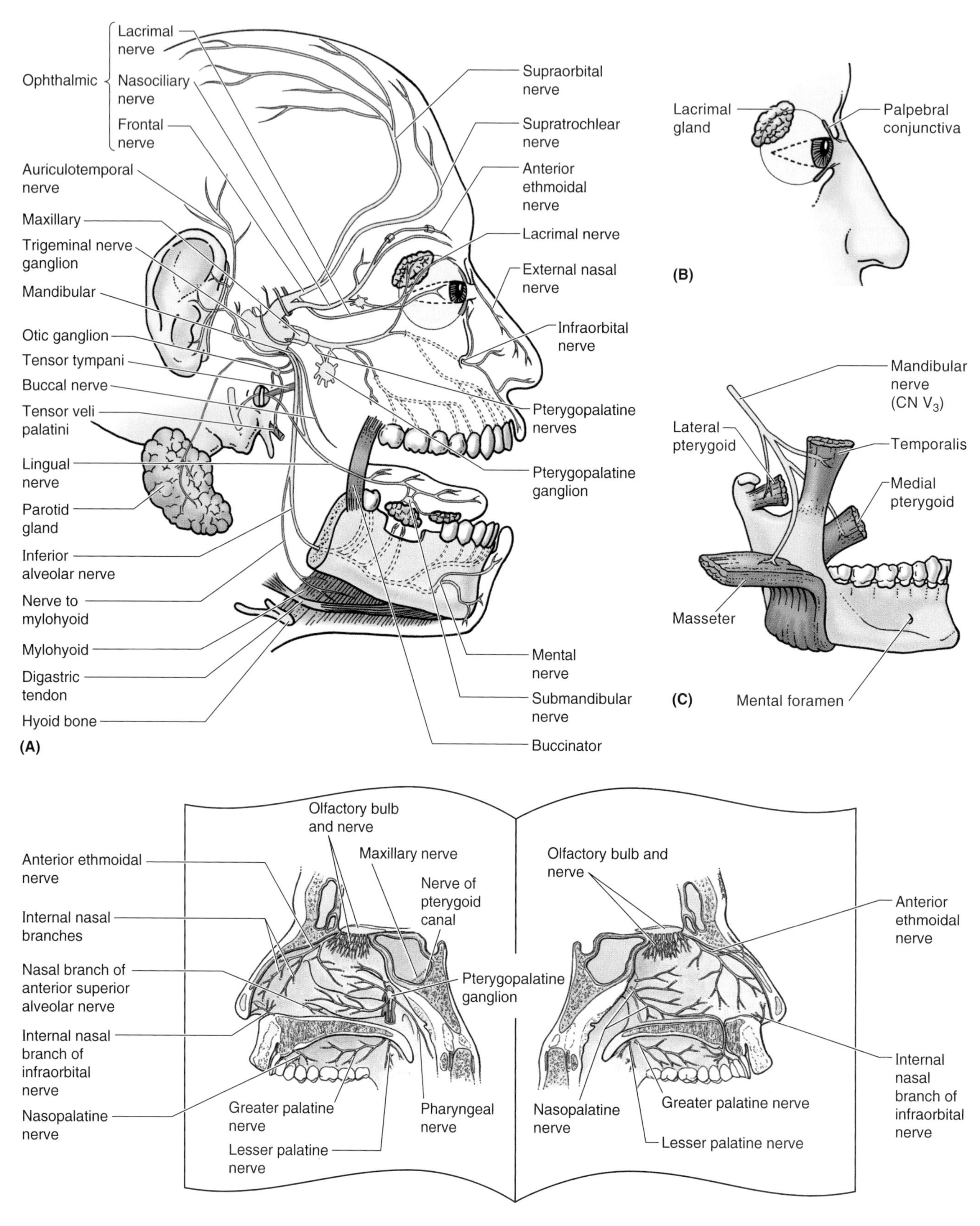
Ophthalmic
Lacrimal nerve
Nasociliary nerve
Frontal nerve
Auriculotemporal nerve
Maxillary
Trigeminal nerve ganglion
Mandibular
Otic ganglion
Tensor tympani
Buccal nerve
Tensor veli palatini
Lingual nerve
Parotid gland
Inferior alveolar nerve
Nerve to mylohyoid
Mylohyoid
Digastric tendon
Hyoid bone
(A)
Supraorbital nerve
Supratrochlear nerve
Anterior ethmoidal nerve
Lacrimal nerve
External nasal nerve
Infraorbital nerve
Pterygopalatine nerves
Pterygopalatine ganglion
Mental nerve
Submandibular nerve
Buccinator
Lacrimal gland
Palpebral conjunctiva
(B)
Mandibular nerve (CN V_3)
Lateral pterygoid
Temporalis
Medial pterygoid
Masseter
(C)
Mental foramen
Olfactory bulb and nerve
Maxillary nerve
Nerve of pterygoid canal
Anterior ethmoidal nerve
Internal nasal branches
Nasal branch of anterior superior alveolar nerve
Pterygopalatine ganglion
Internal nasal branch of infraorbital nerve
Nasopalatine nerve
Greater palatine nerve
Lesser palatine nerve
Pharyngeal nerve
Olfactory bulb and nerve
Anterior ethmoidal nerve
Internal nasal branch of infraorbital nerve
Nasopalatine nerve
Greater palatine nerve
Lesser palatine nerve
(D)
Lateral wall
Nasal septum

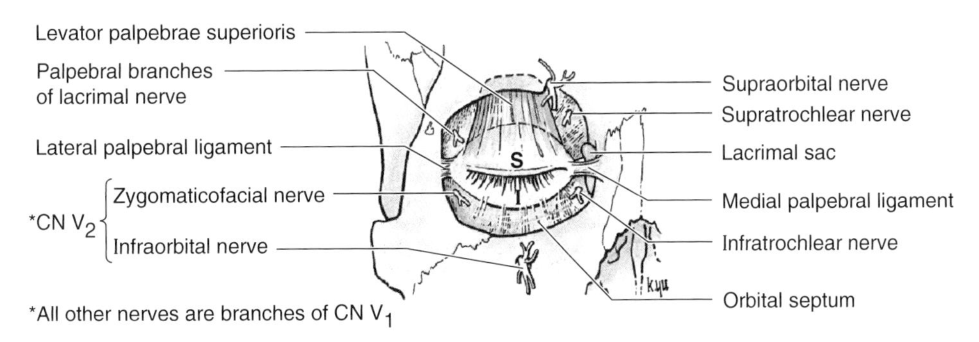

Figure 7.10. Innervation of the eyelids (L. palpebrae). This diagram shows the cutaneous nerves serving the orbital region in relation to the skeleton of the eyelids. The skeleton of the lids is formed by the superior (*S*) and inferior (*I*) tarsal plates and their attachments, the medial and lateral palpebral ligaments, and the orbital septum (palpebral fascia in the eyelid). The fan-shaped aponeurosis of the levator palpebrae superioris is attached to the superior tarsal plate. The skin of the superior (upper) eyelid is supplied by branches of the ophthalmic nerve (CN V_1) while the inferior (lower) eyelid is supplied mainly by branches of the maxillary nerve (CN V_2).

- Supraorbital nerve
- Lacrimal nerve.

The **nasociliary nerve** divides into the posterior ethmoidal, anterior ethmoidal, and infratrochlear nerves. It supplies the tip of the nose through the **external nasal nerve,** a branch of the anterior ethmoidal nerve and the root of the nose through the infratrochlear nerve (Table 7.4). The **infratrochlear nerve**, a terminal branch of the nasociliary nerve, supplies skin on the medial part of the upper eyelid and passes superior to the medial palpebral ligament to the side of the nose (Fig. 7.10). It supplies the lacrimal sac and skin over the bridge of the nose.

The **frontal nerve**, the direct continuation of CN V_1, divides within the orbit into two branches: the supratrochlear and supraorbital nerves. The **supratrochlear nerve** passes superiorly on the medial side of the supraorbital nerve and divides to supply skin in the middle of the forehead to the hairline. The **supraorbital nerve**, the continuation of the frontal nerve, emerges through the supraorbital notch or foramen in the supraorbital margin formed by the frontal bone. As it passes superiorly onto the forehead, the nerve breaks up into several small branches that supply the mucous membrane of the frontal sinus and the upper eyelid (palpebral conjunctiva). Branches of the supraorbital nerve also supply the subcutaneous tissue and skin of the forehead and scalp as far as the vertex of the skull. The **lacrimal nerve**, the smallest of the ophthalmic branches (Table 7.4), supplies a small area of skin over—and conjunctiva deep to—the lateral part of the upper eyelid and, with fibers "borrowed" from the maxillary nerve through a communicating branch, the lacrimal gland.

Maxillary Nerve

CN V_2—*the intermediate division of the trigeminal nerve—also arises as a wholly sensory nerve* (Fig. 7.9, Table 7.4). CN V_2 passes from the **trigeminal ganglion** and leaves the skull through the foramen rotundum in the base of the greater wing of the sphenoid bone (Fig. 7.5*B*). It enters the pterygopalatine fossa, where it gives off branches to the **pterygopalatine ganglion** (Fig. 7.9) and continues forward to give off the **zygomatic nerve** (CN V_2) in the infraorbital foramen, through which it passes and gives rise to the *zygomaticotemporal* and *zygomaticofacial nerves.* It also sends a communicating branch to the lacrimal nerve. After giving off palatine and nasal branches and branches to the posterior teeth, CN V_2 terminates as the infraorbital nerve. *The major cutaneous branches of the maxillary nerve are the* (Table 7.4):

- Infraorbital nerve
- Zygomaticotemporal nerve
- Zygomaticofacial nerve.

These nerves supply the area of skin derived from the embryonic maxillary prominence (Moore and Persaud, 1998).

The **infraorbital nerve** is the continuation of CN V_2 after it has entered the orbit through the inferior orbital fissure.

Figure 7.9. Distribution of the trigeminal nerve (CN V). A. The three divisions of CN V (V_1, V_2, and V_3) arise from the large trigeminal ganglion. Note that the maxillary nerve (CN V_2) gives off two pterygopalatine nerves by which the small pterygopalatine ganglion is suspended. **B.** Illustrates the lacrimal gland and the palpebral conjunctiva lining the eyelids (L. palpebrae). **C.** Shows the branches of the mandibular nerve (CN V_3) passing to the muscles of mastication (the chewing muscles). **D.** "Opened book" view of the lateral wall and the septum of the right part of the nasal cavity, demonstrating the distribution of the olfactory nerves (CN I) and branches of the ophthalmic (CN V_1) and maxillary nerves.

Table 7.4. **Nerves of the Face and Scalp**

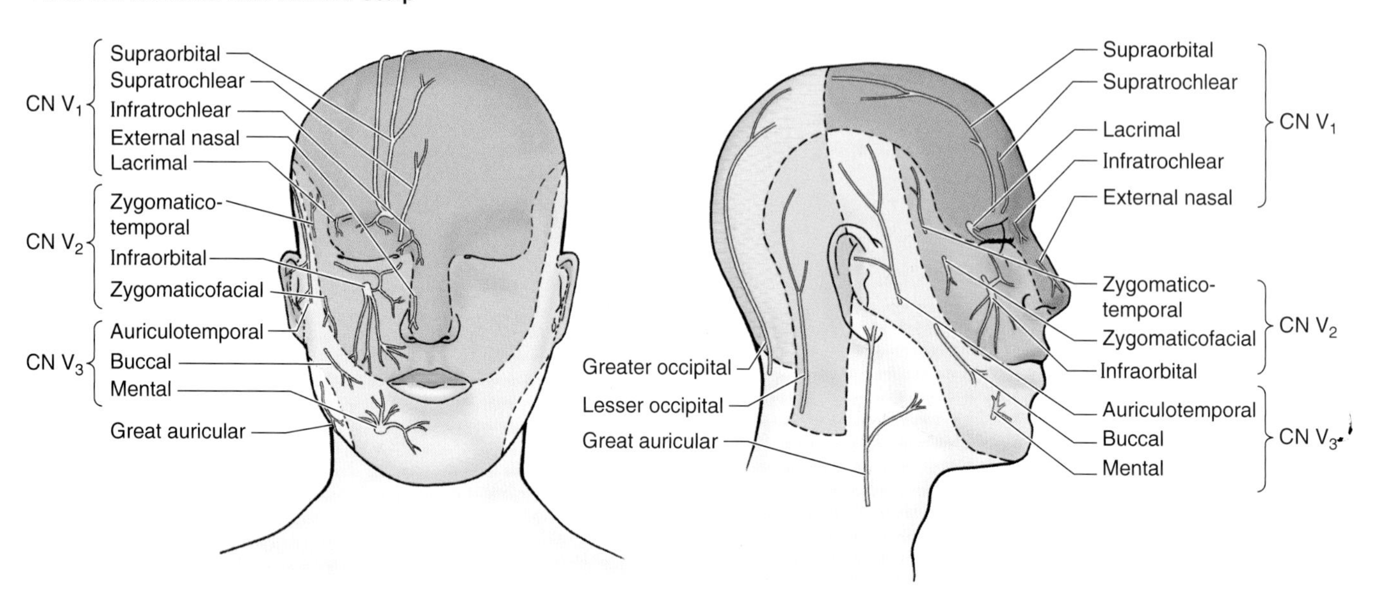

Nerve	Origin	Course	Distribution
Frontal	Ophthalmic nerve (CN V_1)	Crosses orbit on superior aspect of levator palpebrae superioris; divides into supraorbital and supratrochlear branches	Skin of forehead, scalp, upper eyelid, and nose; conjunctiva of upper lid and mucosa of frontal sinus
Supraorbital	Continuation of frontal nerve (CN V_1)	Emerges through supraorbital notch, or foramen, and breaks up into small branches	Mucous membrane of frontal sinus and conjunctiva (lining) of upper eyelid; skin of forehead as far as vertex
Supratrochlear	Frontal nerve (CN V_1)	Passes superiorly on medial of supraorbital nerve and divides into two or more branches	Skin in middle of forehead to hairline
Infratrochlear	Nasociliary nerve (CN V_1)	Follows medial wall of orbit to upper eyelid	Skin and conjunctiva (lining) of upper eyelid
Lacrimal	Ophthalmic nerve (CN V_1)	Passes through palpebral fascia of upper eyelid near lateral angle (canthus) of eye	Lacrimal gland and small area of skin and conjunctiva of lateral part of upper eyelid
External nasal	Anterior ethmoidal nerve (CN V_1)	Runs in nasal cavity and emerges on face between nasal bone and lateral nasal cartilage	Skin on dorsum of nose, including tip of nose
Zygomatic	Maxillary nerve (CN V_2)	Arises in floor of orbit, divides into zygomaticofacial and zygomaticotemporal nerves, which traverse foramina of same name	Skin over zygomatic arch and anterior temporal region; carries postsynaptic parasympathetic fibers from pterygo-palatine ganglion to lacrimal nerve
Infraorbital	Terminal branch of maxillary nerve (CN V_2)	Runs in floor of orbit and emerges at infraorbital foramen	Skin of cheek, lower lid, lateral side of nose and inferior septum and upper lip, upper premolar incisors and canine teeth; mucosa of maxillary sinus and upper lip
Auriculotemporal	Mandibular nerve (CN V_3)	From posterior division of CN V_3, it passes between neck of mandible and external acoustic meatus to accompany superficial temporal artery	Skin anterior to ear and posterior temporal region, tragus and part of helix of auricle, and roof of exterior acoustic meatus and upper tympanic membrane
Buccal	Mandibular nerve (CN V_3)	From the anterior division of CN V_3 in infratemporal fossa, it passes anteriorly to reach cheek	Skin and mucosa of cheek, buccal gingiva adjacent to 2nd and 3rd molar teeth
Mental	Terminal branch of inferior alveolar nerve (CN V_3)	Emerges from mandibular canal at mental foramen	Skin of chin and lower lip and mucosa of lower lip

The infraorbital nerve exits through infraorbital foramen and breaks up into branches that supply the skin of the upper cheek, the mucosa of the maxillary sinus, the maxillary incisor, canine, and premolar teeth and adjacent upper gingiva, the skin and conjunctiva of the inferior eyelid, part of the nose, and the skin and mucosa of the upper lip.

The **zygomaticotemporal nerve**, a branch of the zygomatic nerve, emerges from the zygomatic bone through a small foramen of the same name. It enters the temporal fossa and supplies a small area of skin over the anterior part of the temple (the hairless part).

The **zygomaticofacial nerve**, the smaller branch of the zygomatic nerve, emerges from the zygomatic bone through a small foramen with the same name. It supplies the skin of the face over the zygomatic prominence of the zygomatic bone.

Mandibular Nerve

CN V_3—*the inferior and largest division of the trigeminal nerve*—is formed by the union of sensory fibers from the sensory ganglion and the motor root of the trigeminal nerve in the *foramen ovale*, through which the mandibular nerve emerges from the skull (Fig. 7.9, Table 7.4). CN V_3 has three sensory branches that supply the area of skin derived from the embryonic mandibular prominence (Moore and Persaud, 1998). It also supplies motor fibers to the muscles of mastication (Fig. 7.9*C*). *CN V_3 is the only division of CN V that carries motor fibers.* The major cutaneous branches of CN V_3 are the:

- Auricotemporal nerve
- Buccal nerve
- Mental nerve.

These sensory nerves are distributed to the auricle, external acoustic meatus, tympanic membrane, temporal region, cheek, and skin overlying the mandible, except at its angle.

The **auriculotemporal nerve**, usually arising by two roots embracing the middle meningeal artery, passes through the parotid gland conveying secretomotor fibers and then passes superiorly anterior to the ear to the temporal region. As its name suggests, it supplies parts of the auricle, external acoustic meatus, external surface of the tympanic membrane, and skin superior to the auricle.

The **buccal nerve**, a relatively small sensory branch of CN V_3, emerges from deep to the ramus of the mandible and runs anteriorly on the buccinator muscle, piercing but not supplying this muscle. It sends branches to a thumb-sized area of skin over the cheek and supplies the mucous membrane lining the cheek and the posterior part of the buccal surface of the gingiva (gum).

The ***mental nerve***, a large cutaneous branch of the *inferior alveolar nerve*—one of the major branches of CN V_3—arises in the mandibular canal and emerges from the **mental foramen** in the mandible (Fig. 7.9, *A* and *C*). It divides into three branches that radiate away from the mental foramen: one descends to the skin of the chin, the other two supply the skin and mucous membrane of the lower lip and the inferior labial gingiva.

Infraorbital Nerve Block

For local anesthesia of the inferior part of the face, the infraorbital nerve is infiltrated with an anesthetic agent. For treating wounds such as those of the upper lip and cheek or, more commonly, for repairing the maxillary incisor teeth, *the site of injection is the infraorbital foramen.* To determine where the infraorbital nerve emerges, exert pressure on the maxilla in the region of the infraorbital foramen. Pressure on the nerve causes considerable pain. Care must be exercised when performing an infraorbital nerve block because companion infraorbital vessels leave the infraorbital foramen with the nerve. Careful aspiration of the syringe during injection prevents inadvertent injection of anesthetic fluid into a blood vessel. Because the orbit is located just superior to the injection site, a careless injection could result in passage of anesthetic fluid into the orbit, causing *temporary paralysis of the extraocular muscles.*

Inferior Alveolar Nerve Block

Dentists often anesthetize the inferior alveolar nerve before repairing or removing mandibular teeth. Because the mental and incisive nerves are its terminal branches, one's chin and lower lip on the affected side also lose sensation. *The site of anesthetic injection is the mandibular foramen*, the mouth of the mandibular canal, located on the medial aspect of the ramus of the mandible. "Of all the routine injections attempted in the dental office, the inferior alveolar block is perhaps the most difficult to perform effectively" (Liebgott, 1986). If the needle goes too far posteriorly, it may enter the parotid gland and anesthetize branches of the facial nerve, producing transient unilateral facial paralysis.

Mental and Incisive Nerve Blocks

Occasionally it is desirable to anesthetize one side of the skin and mucous membrane of the lower lip and the skin of the chin (e.g., to suture a severe laceration of the lip). An injection of anesthetic agent into the mental foramen will block the mental nerve that supplies the skin and mucous membrane of the lower lip from the mental foramen to the midline, including the skin of the chin.

Buccal Nerve Block

To anesthetize the skin and mucous membrane of the cheek (e.g., to suture a knife wound), an anesthetic injection can be made into the mucosa covering the retromolar fossa, located posterior to the 3rd mandibular molar. ▶

Trigeminal Neuralgia

Trigeminal neuralgia (tic douloureux) is a sensory disorder of the sensory root of CN V that is characterized by sudden attacks of excruciating, lighteninglike jabs of facial pain. A *paroxysm* (sudden sharp pain) can last for 15 minutes or more (Rowland, 1995). "The maxillary nerve is most frequently involved, then the mandibular nerve, and least frequently the ophthalmic nerve" (Barr and Kiernan, 1993). The pain is often initiated by touching a sensitive *trigger zone of the skin*. The cause of trigeminal neuralgia is unknown; however, some investigators believe that most affected persons have an anomalous blood vessel that compresses the nerve (Lange et al., 1995). When the aberrant artery is moved away from the sensory root of CN V, the symptoms usually disappear. In some cases it is necessary to section the sensory root of CN V for relief of trigeminal neuralgia.

Lesions of the Trigeminal Nerve

Lesions of the entire trigeminal nerve cause widespread anesthesia involving the:

- Corresponding anterior half of the scalp
- Face, except for an area around the angle of the mandible, the cornea, and conjunctiva
- Mucous membranes of the nose, mouth, and anterior part of the tongue.

Paralysis of the muscles of mastication also occurs.

Herpes Zoster

Herpes zoster (shingles) is an infection caused by a herpes virus that produces lesions in the spinal or cranial ganglia. The infection is characterized by an eruption of groups of vesicles (p. 87) following the course of the affected nerve (e.g., *herpes zoster ophthalmicus* involves CN V_1). Involvement of the trigeminal ganglion occurs in approximately 20% of cases (Berg and Klebanoff, 1995). Any division of CN V may be involved, but the ophthalmic division is the one most commonly affected. Usually the cornea is involved, often resulting in painful *corneal ulceration*.

Testing the Sensory Function of CN V

The sensory function of the trigeminal nerve is tested by asking the patient to close his or her eyes and respond when feeling a touch. A piece of gauze or test tubes filled with warm and cold fluid are applied to one cheek and then to the corresponding position on the other side. This test is then performed on the lower jaw. The patient is also asked if one side feels the same as or different from the other side. The testing is then repeated with the gentle touch of a sharp pin, alternating sides. ⊙

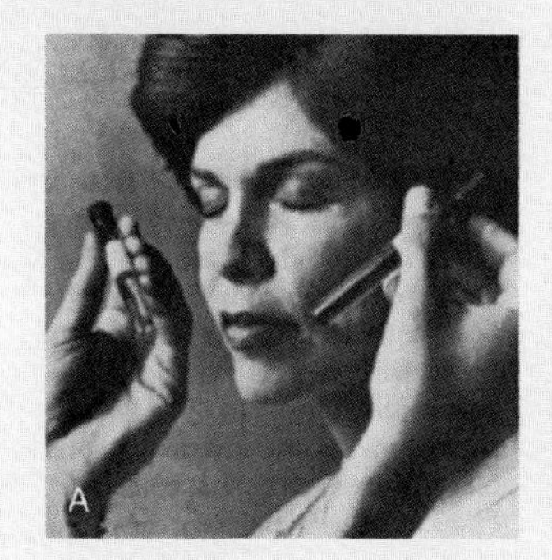
A

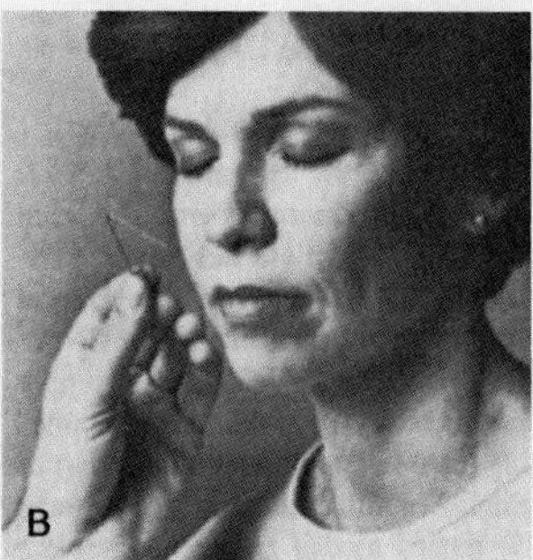
B

Motor Nerves of the Face

The motor nerves of the face are the *facial nerve* to the muscles of facial expression and the *motor root of the mandibular nerve* to the muscles of mastication (masseter, temporal, medial and lateral pterygoids). These nerves also supply some more deeply placed muscles described later in relation to the mouth, middle ear, and neck (Fig. 7.9*A*).

Facial Nerve

The facial nerve has motor and sensory roots (Fig. 7.11). The **motor root** supplies the muscles of facial expression, including the superficial muscle of the neck (platysma), auricular (ear) muscles, scalp muscles, and certain other muscles derived from mesenchyme in the embryonic 2nd pharyngeal arch (Moore and Persaud, 1998). *CN VII is the sole motor supply to the muscles of facial expression* and is sensory to the taste buds in the anterior two-thirds of the tongue. The facial nerve has no sensory fibers in the face. It conveys general sensation from a small area around the external acoustic meatus and is the parent nerve for secretomotor fibers to the submandibular and sublingual salivary glands and small lingual glands.

Following a circuitous route through the temporal bone, the facial nerve emerges from the skull through the **stylomastoid foramen** located between the mastoid and styloid processes (Fig. 7.5*A*). It immediately gives off the **posterior auricular nerve** that passes posterosuperior to the ear to supply the auricularis posterior and occipital belly of the occipitofrontalis muscle (Fig. 7.11). *The main trunk of the facial nerve runs anteriorly and is engulfed by the parotid gland,* in which it forms the ***parotid plexus***, which gives rise to the five terminal branches of the facial nerve:

- Temporal
- Zygomatic
- Buccal

- Marginal mandibular
- Cervical.

The names of the branches refer to the regions they supply.

The **temporal branch of CN VII** emerges from the superior border of the parotid gland and crosses the zygomatic arch to supply the auricularis superior and auricularis anterior, the frontal belly of the occipitofrontalis, and—most importantly—the superior part of the orbicularis oculi (Figs. 7.6*A* and 7.11*A*).

The **zygomatic branch of CN VII** passes via two or three branches superior and mainly inferior to the eye to supply the inferior part of the orbicularis oculi and other facial muscles inferior to the orbit.

The **buccal branch of CN VII** passes external to the buccinator to supply this muscle and muscles of the upper lip (upper parts of the orbicularis oris and inferior fibers of the levator labii superioris).

The **marginal mandibular branch of CN VII** supplies the risorius and muscles of the lower lip and chin. It emerges from the inferior border of the parotid gland and crosses the

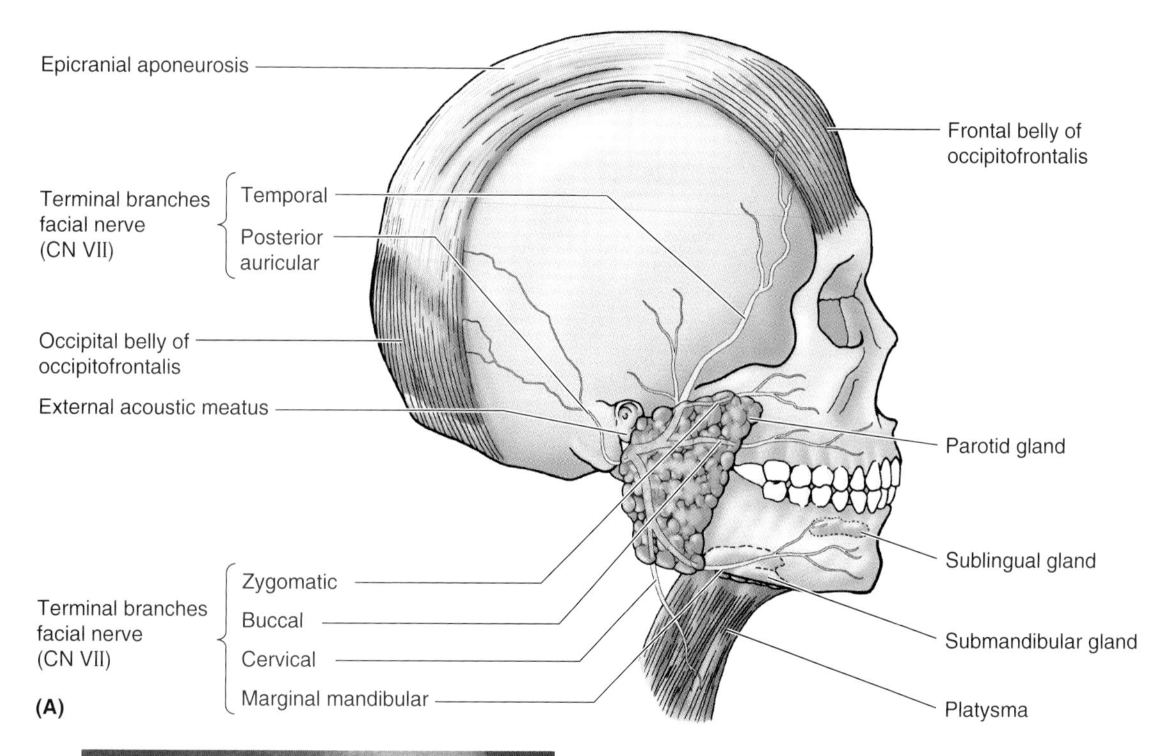

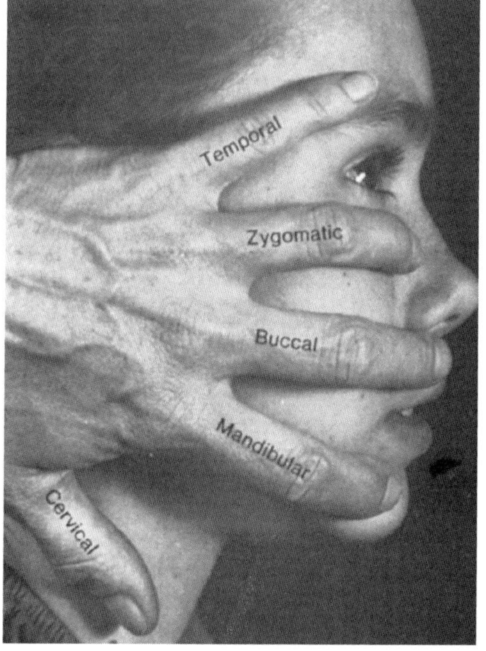

Figure 7.11. Branches of the facial nerve (CN VII). A. Lateral view of the skull. The terminal branches of the facial nerve arise from the parotid plexus within the parotid gland, emerging from the gland under cover of its lateral surface, and radiating in a generally anterior direction across the face. Although intimately related to the parotid gland (and often contacting the submandibular gland via one or more of its lower branches), the facial nerve does not deliver nerve fibers to the salivary glands. Two muscles representing the extremes of the distribution of the facial nerve—the occipitofrontalis and platysma (G. flat plate)—are also shown. **B.** Lateral view of the face of a 12-year-old girl, illustrating a simple method for demonstrating and remembering the general course of the five terminal branches of the facial nerve to the face and neck.

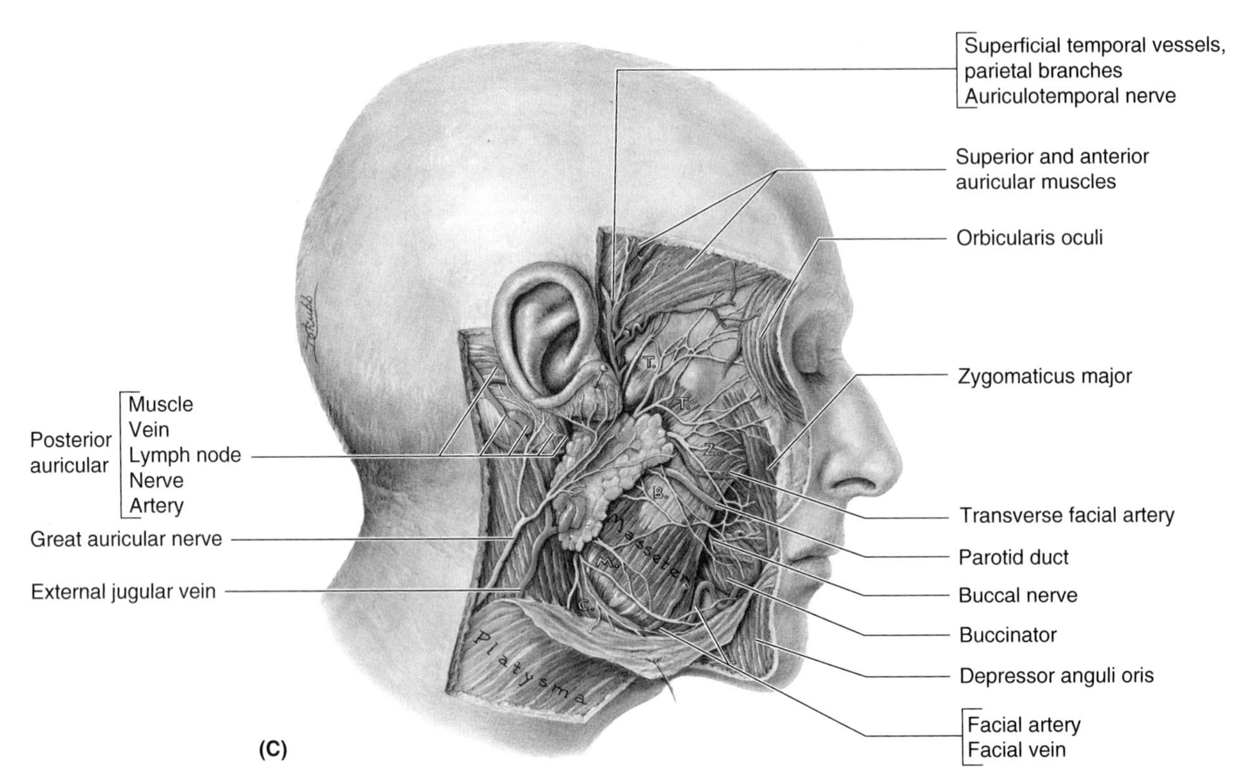

Figure 7.11. *(Continued)* **C.** Lateral view of a dissection of the right side of the head showing the great auricular nerve (C2 and C3)—which supplies the parotid sheath and skin over the angle of the mandible (lower jaw)—and terminal branches of the facial nerve that supply the muscles of facial expression. *T*, temporal; *Z*, zygomatic; *B*, buccal; *M*, mandibular; and *C*, cervical.

Injuries to Branches of CN VII

Injury to branches of the facial nerve causes paralysis of the facial muscles, with or without loss of taste on the anterior two-thirds of the tongue or altered secretion of the lacrimal and salivary glands. Lesions near the origin of CN VII or proximal to the origin of the greater petrosal nerve (in the region of the geniculate ganglion) result in loss of motor, gustatory (taste), and autonomic functions. Lesions distal to the geniculate ganglion, but before the origin of the chorda tympani nerve, produce the same dysfunction except that lacrimal secretion is not affected. Lesions near the stylomastoid foramen result in loss of motor function only (e.g., facial paralysis).

Because the branches of the facial nerve are superficial, they are subject to injury by stab and gunshot wounds, cuts, and injury at birth. Injury of the facial nerve often results from *fracture of the temporal bone*, and facial muscle paralysis is evident soon after the injury. If the nerve is sectioned, the chances of complete or even partial recovery are remote. Improvement of muscular movements is the rule when the nerve damage is associated with head trauma; however, recovery may not be complete (Rowland, 1995).

A lesion of the zygomatic branch of CN VII causes paralysis, including loss of tonus of the orbicularis oculi, in the lower eyelid; thus, the lower lid droops (falls away from the surface of the eyeball). As a result, tears do not spread over the cornea and the dry cornea ulcerates. The resultant corneal scar impairs vision.

Paralysis of the buccal branch of CN VII prevents the emptying of food from the vestibule of the cheeks. The food lodges in the vestibule and cannot be maintained in position between the teeth for chewing.

Paralysis of the marginal mandibular branch of *CN VII* may occur when an incision is made along the inferior border of the mandible. Injury to this branch (e.g., during a surgical approach to the submandibular gland) results in an unsightly drooping of the corner of the mouth.

Idiopathic paralysis of the facial nerve occurring without any known cause is discussed on page 857. This type of transient paralysis can occur at any age but is slightly most common in the 30s through 50s (Rowland, 1995). ⊙

inferior border of the mandible deep to the platysma to reach the face. In approximately 20% of people, the marginal mandibular branch passes inferior to the angle of the mandible.

The **cervical branch of CN VII** passes inferiorly from the inferior border of the parotid gland and runs posterior to the mandible to supply the platysma, the superficial muscle of the neck.

Vasculature of the Face

The face is richly supplied by arteries, the terminal branches of which anastomose freely. Most *facial arteries* are branches of the *external carotid artery*; for the origin, course, and distribution of these arteries, see Table 7.5. The *facial veins* anastomose freely and are drained by veins that accompany the arteries of the face. As with most superficial veins, they are subject to many variations; a common pattern is shown in Table 7.6. The venous return from the face is essentially superficial.

Facial Artery

The facial artery provides the major arterial supply to the face. It arises from the external carotid artery and winds its way to the inferior border of the mandible, just anterior to the masseter (Fig. 7.11*C*, Table 7.5). The artery lies superficially here, immediately deep to the platysma. The facial artery crosses the mandible, buccinator, and maxilla as it courses over the face to the *medial angle (canthus) of the eye*, where the upper and lower eyelids meet. The facial artery lies deep to the zygomaticus major and levator labii superioris muscles. Near the termination of its sinuous course through the face, the facial artery passes approximately a finger's breadth lateral to the angle of the mouth. It sends branches to the upper and lower lips—the **superior** and **inferior labial arteries**—ascends along the side of the nose, and joins the dorsal nasal branch of the ophthalmic artery. Distal to its superior branch, the terminal of the facial artery is called the **angular artery**.

Superficial Temporal Artery

The superficial temporal artery is the smaller terminal branch of the external carotid artery; the other branch is the maxillary artery. The superficial temporal artery emerges on the face between the TMJ and the ear, enters the temporal fossa, and ends in the scalp by dividing into frontal and parietal branches.

Transverse Facial Artery

The transverse facial artery arises from the superficial temporal artery within the parotid gland and crosses the face superficial to the masseter (Fig. 7.11*C*), approximately a finger's breadth inferior to the zygomatic arch. It divides into numerous branches that supply the parotid gland and duct, the masseter, and the skin of the face. It anastomoses with branches of the facial artery.

Supratrochlear Vein

The supratrochlear vein begins in the forehead from a network of veins connected to the frontal tributaries of the superficial temporal vein (Table 7.6). The **supratrochlear vein** descends near the median plane, communicating with its fellow on the other side, and passes to the nose. The supratrochlear vein joins the supraorbital vein to form the **angular vein**, which becomes the facial vein near the medial angle of the eye.

Supraorbital Vein

The supraorbital vein begins in the forehead where it connects with the tributaries of the supratrochlear and the superficial and middle temporal veins. The **supraorbital vein** passes medially superior to the orbit and joins the supratrochlear vein to form the angular vein near the medial angle of the eye. A branch of the supraorbital vein passes through the supraorbital notch, or foramen, and communicates with the superior ophthalmic vein.

Facial Vein

The two facial veins provide the venous drainage of the face. Each vein begins near the medial angle of the eye and the inferior border of the orbit as the continuation of the angular vein. At the side of the nose, the facial vein receives external nasal veins and inferior palpebral veins from the nose and lower eyelid, respectively. The facial vein runs inferoposteriorly through the face, posterior to the facial artery. Tributaries of the facial vein include the deep facial vein, which drains the *pterygoid venous plexus* of the infratemporal fossa anteriorly, and the superior and inferior labial veins, which drain the upper and lower lips. Inferior to the margin of the mandible, the facial vein is joined by the anterior (communicating) branch of the **retromandibular vein**. The facial vein drains directly or indirectly into the **IJV**. At the medial angle of the eye, the facial vein communicates with the **superior ophthalmic vein**, which drains into the **cavernous sinus**. The facial vein also communicates with the *pterygoid venous plexus* (Table 7.6).

Superficial Temporal Vein

The superficial temporal vein begins from a widespread network of veins in the scalp and along the zygomatic arch. The facial vein drains the scalp and forehead and receives tributaries from the veins of the temple and face. Near the auricle, the superficial temporal vein enters the parotid gland.

Compression of the Facial Artery

The facial artery can be occluded by pressure against the mandible where the vessel crosses it (Table 7.5). Because of the numerous anastomoses between the branches of the facial artery and other arteries of the face, *compression of the facial artery on one side does not stop all bleeding from a lacerated facial artery* or one of its branches. In lacerations of the lip, pressure must be applied on both sides of the cut to stop the bleeding. In general, facial wounds bleed freely and heal quickly.

Pulses of the Arteries of the Face

The pulses of the superficial temporal and facial arteries are often measured when it is inconvenient to measure the pulse of other arteries. For example, anesthesiologists at the head of the operating table often take the *temporal pulse* just anterior to the auricle where the superficial temporal artery crosses the root of the zygomatic process of the temporal bone. They may also palpate the *facial pulse* where the facial artery winds around the inferior border of the mandible (Table 7.5). ⊙

Table 7.5 **Arteries of the face and scalp**

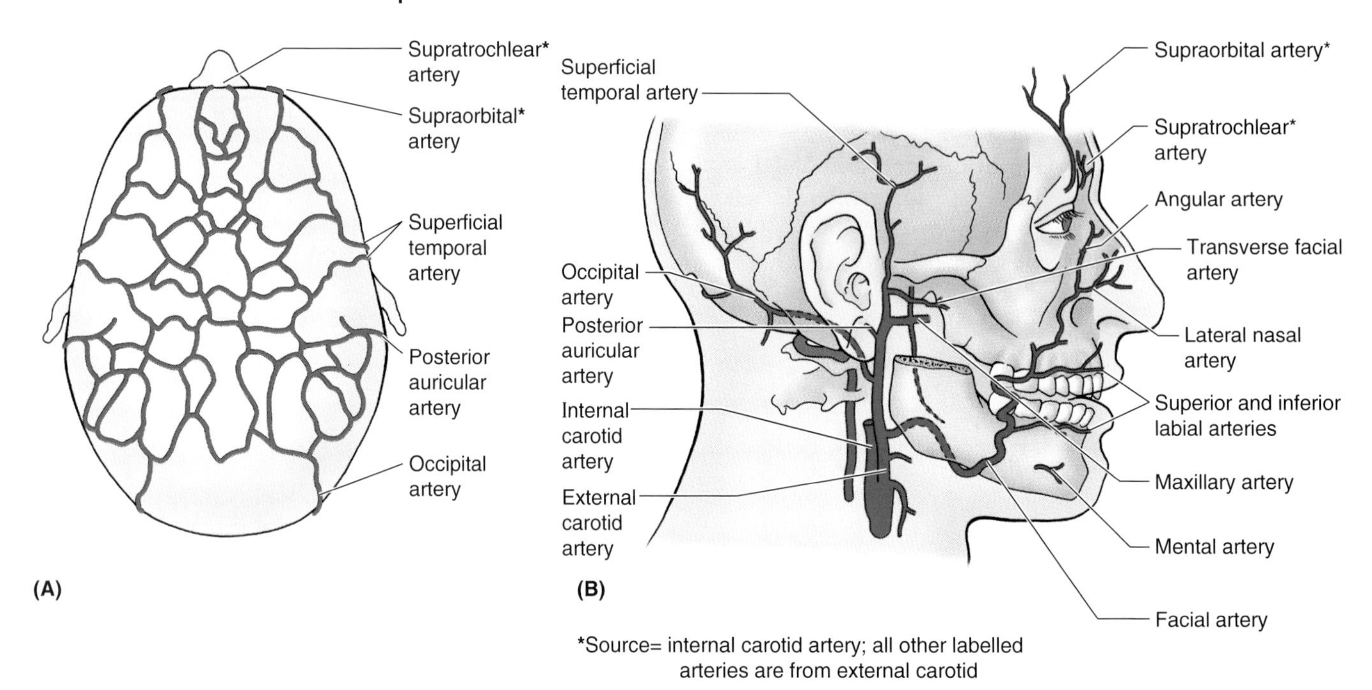

*Source= internal carotid artery; all other labelled arteries are from external carotid

Artery	Origin	Course	Distribution
Facial	External carotid artery	Ascends deep to submandibular gland, winds around inferior border of mandible and enters face	Muscles of facial expression and face
Inferior labial	Facial artery near angle of mouth	Runs medially in lower lip	Lower lip and chin
Superior labial		Runs medially in upper lip	Upper lip and ala (side) and septum of nose
Lateral nasal	Facial artery as it ascends alongside nose	Passes to ala of nose	Skin on ala and dorsum of nose
Angular	Terminal branch of facial artery	Passes to medial angle (canthus) of eye	Superior part of cheek and lower eyelid
Occipital	External carotid artery	Passes medial to posterior belly of digastric and mastoid process; accompanies occipital nerve in occipital region	Scalp of back of head, as far as vertex
Posterior auricular		Passes posteriorly, deep to parotid, along styloid process between mastoid process and ear	Scalp posterior to auricle and auricle
Superficial temporal	Smaller terminal branch of external carotid artery	Ascends anterior to ear to temporal region and ends in scalp	Facial muscles and skin of frontal and temporal regions
Transverse facial	Superficial temporal artery within parotid gland	Crosses face superficial to masseter and inferior to zygomatic arch	Parotid gland and duct, muscles and skin of face
Mental	Terminal branch of inferior alveolar artery	Emerges from mental foramen and passes to chin	Facial muscles and skin of chin
*Supraorbital	Terminal branch of ophthalmic artery, a branch of internal carotid artery	Passes superiorly from supraorbital foramen	Muscles and skin of forehead and scalp
*Supratrochlear		Passes superiorly from supratrochlear notch	Muscles and skin of scalp

Table 7.6. Veins of the Face

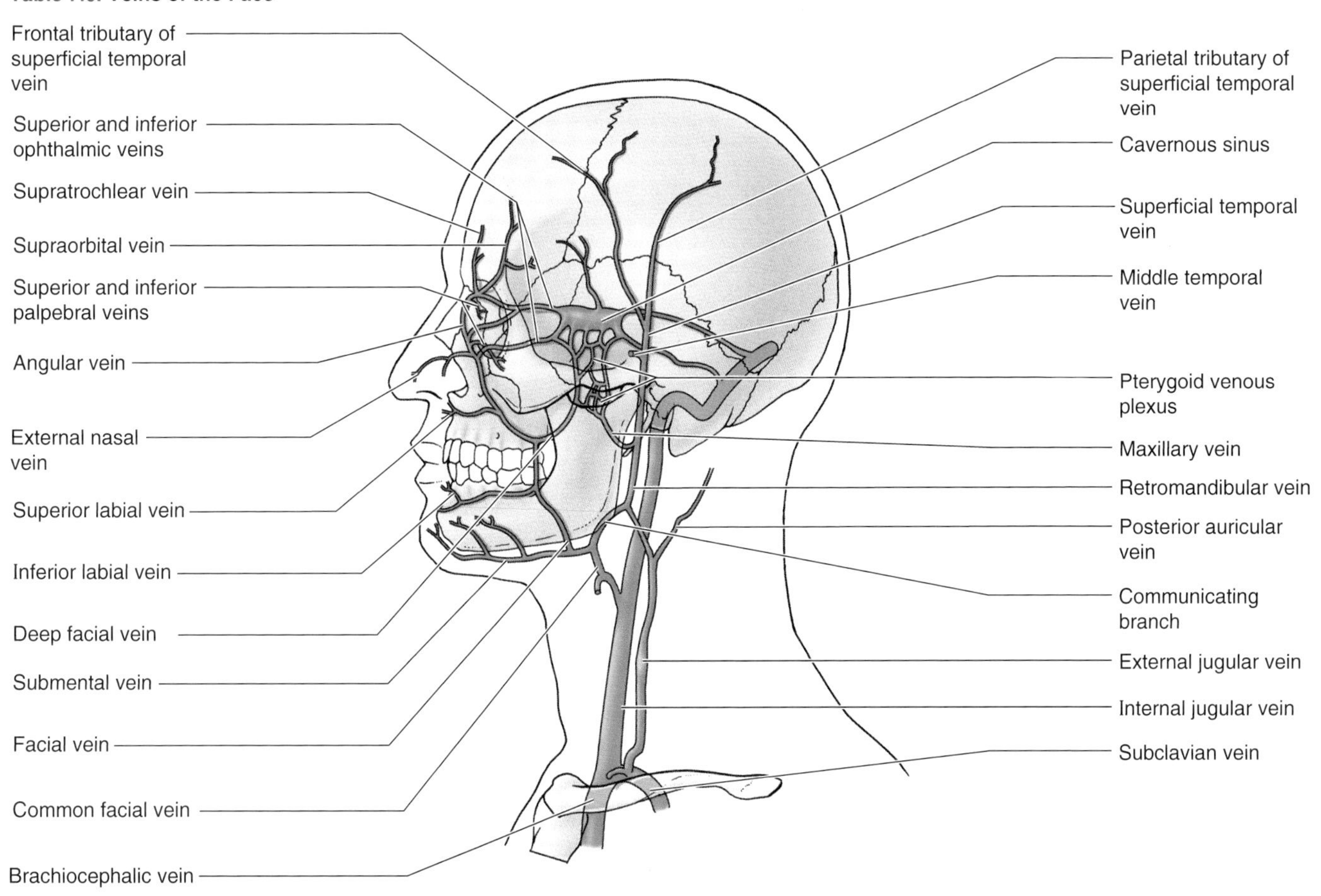

<table>
<tr><th>Vein</th><th>Origin</th><th>Course</th><th>Termination</th><th>Area Drained</th></tr>
<tr><td>Supra-trochlear</td><td>Begins from a venous plexus on the forehead and scalp, through which it communicates with the frontal branch of the superficial temporal vein, its contralateral partner, and the supraorbital vein</td><td>Descends near the midline of the forehead to the root of the nose where it joins the supraorbital vein</td><td rowspan="2">Angular vein at the root of the nose</td><td rowspan="2">Anterior part of scalp and forehead</td></tr>
<tr><td>Supraorbital</td><td>Begins in the forehead by anastomosing with a frontal tributary of the superficial temporal vein</td><td>Passes medially superior to the orbit and joins the supratrochlear vein; a branch passes through the supraorbital notch and joins with the superior ophthalmic vein</td></tr>
<tr><td>Angular</td><td>Begins at root of nose by union of supratrochlear and supraorbital veins</td><td>Descends obliquely along the root and side of the nose to the inferior margin of the orbit</td><td>Becomes the facial vein at the inferior margin of the orbit</td><td>In addition to above, drains upper and lower lids and conjunctiva; may receive drainage from cavernous sinus</td></tr>
</table>

Table 7.6. *(Continued)* **Veins of the Face**

Vein	Origin	Course	Termination	Area Drainage
Facial	Continuation of angular vein past inferior margin of orbit	Descends along lateral border of the nose, receiving external nasal and inferior palpebral veins, then obliquely across face to cross inferior border of mandible; receives communication from retro-mandibular vein, after which it is some-times called the common facial vein	Internal jugular vein opposite or inferior to the level of the hyoid bone	Anterior scalp and forehead, eyelids, external nose, anterior cheek, lips, chin, and submandibular gland
Deep facial	Pterygoid venous plexus	Runs anteriorly on maxilla above buccinator and deep to masseter, emerging medial to anterior border of masseter onto face	Enters posterior aspect of facial vein	Infratemporal fossa (most areas supplied by maxillary artery)
Superficial temporal	Begins from a widespread plexus of veins on the side of the scalp and along the zygomatic arch	Its frontal and parietal tributaries unite anterior to the auricle; it crosses the temporal root of the zygomatic arch to pass from the temporal region and enter the substance of the parotid gland	Joins the maxillary vein posterior to the neck of the mandible to form the retromandibular vein	Side of the scalp, superficial aspect of the temporal muscle, and external ear
Retro-mandibular	Formed anterior to the ear by the union of the super-ficial temporal and maxillary veins	Runs posterior and deep to the ramus of the mandible through the substance of the parotid gland; communicates at its inferior end with the facial vein	Unites with the posterior auricular vein to form the external jugular vein	Parotid gland and masseter muscle

Thrombophlebitis of the Facial Vein

The facial vein makes clinically important connections (*A*) with the:

- *Cavernous sinus*—a venous sinus of the dura mater covering the brain—through the superior ophthalmic vein
- *Pterygoid venous plexus*—a network of small veins ▶

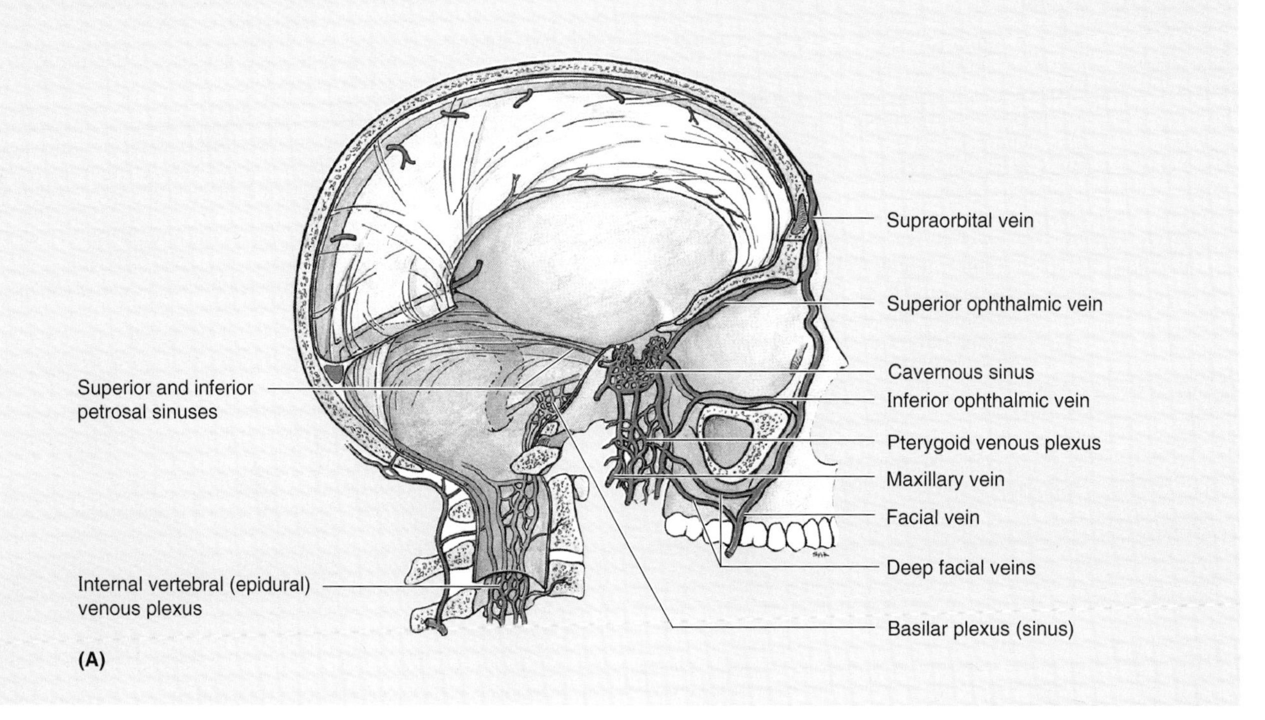

(A)

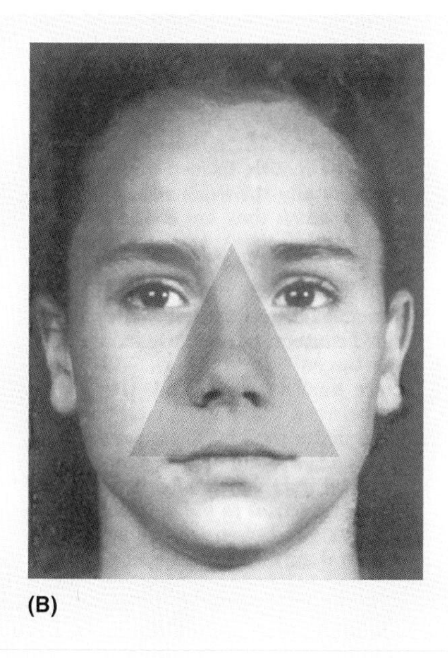

(B)

▶ within the *infratemporal fossa*—through the inferior ophthalmic and deep facial veins (see Fig. 7.21).

Because of these connections, an infection of the face may spread to the cavernous sinus and pterygoid venous plexus.

Blood from the medial angle of the eye, nose, and lips usually drains inferiorly through the facial vein, especially when a person is erect. *Because the facial vein has no valves*, blood may pass through it in the opposite direction; consequently, venous blood from the face may enter the cavernous sinus. In patients with *thrombophlebitis of the facial vein*—inflammation of the facial vein with secondary thrombus (clot) formation—pieces of an infected clot may extend into the intracranial venous system, such as the cavernous sinus, and produce *thrombophlebitis of the cavernous sinuses*. Infection of the facial veins spreading to the dural venous sinuses may result from lacerations of the nose or be initiated by squeezing pustules (pimples) on the side of the nose and upper lip. Consequently, the triangular area from the upper lip to the bridge of the nose is considered as the *danger triangle of the face* (*B*). ⭘

Retromandibular Vein

The retromandibular vein, *formed by the union of the superficial temporal and maxillary veins*, runs posterior to the ramus of the mandible and descends through the parotid gland, superficial to the external carotid artery and deep to the facial nerve. The retromandibular vein divides into an anterior branch that unites with the facial vein and a posterior branch that joins the posterior auricular vein to form the **external jugular vein** (Table 7.6).

Lymphatic Drainage of the Face

The lymphatic vessels of the face accompany other facial vessels. Superficial lymphatic vessels accompany veins and deep lymphatics accompany arteries. All lymph from the head and neck eventually drains into the **deep cervical group of lymph nodes** (Fig. 7.12), a chain of nodes surrounding the IJV in the neck. Lymph from the deep cervical lymph nodes passes to the *jugular lymphatic trunk*, which joins the **thoracic duct** on the left side and the **IJV** or **brachiocephalic vein** on the right side. *Summary of the lymphatic drainage of the face follows below.*

- Lymph from the lateral part of the face, including the eyelids, drains inferiorly to the **parotid lymph nodes**.
- Lymph from the deep parotid nodes drains into the **deep cervical lymph nodes**.
- Lymph from the upper lip and lateral parts of the lower lip drains into the **submandibular lymph nodes**.
- Lymph from the chin and central part of the lower lip drains into the **submental lymph nodes**.

Squamous Cell Carcinoma of the Lip

Squamous cell carcinoma (cancer) of the lip usually involves the lower lip. Overexposure to sunshine over many years, as occurs in outdoor workers, is a common factor in these cases. Chronic irritation from pipe smoking also appears to be a contributing factor. Cancer cells from the central part of the lower lip (as in the photograph), the floor of the mouth, and the apex (tip) of the tongue spread to the *submental lymph nodes*, whereas cancer cells from lateral parts of the lower lip drain to the *submandibular lymph nodes*. ⭘

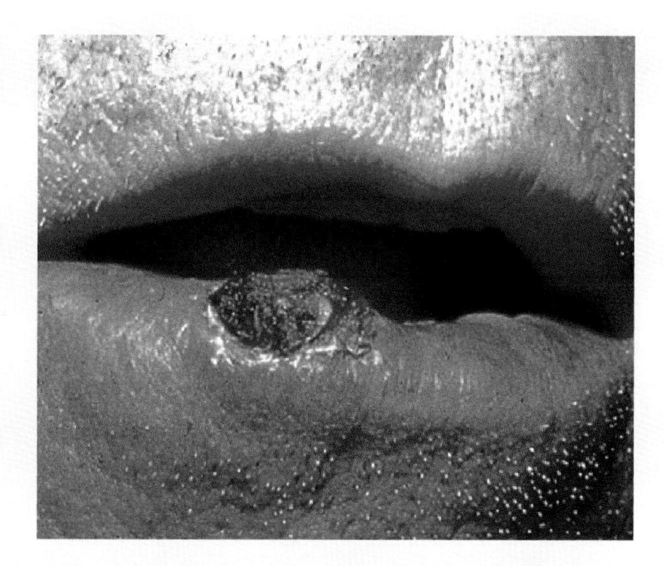

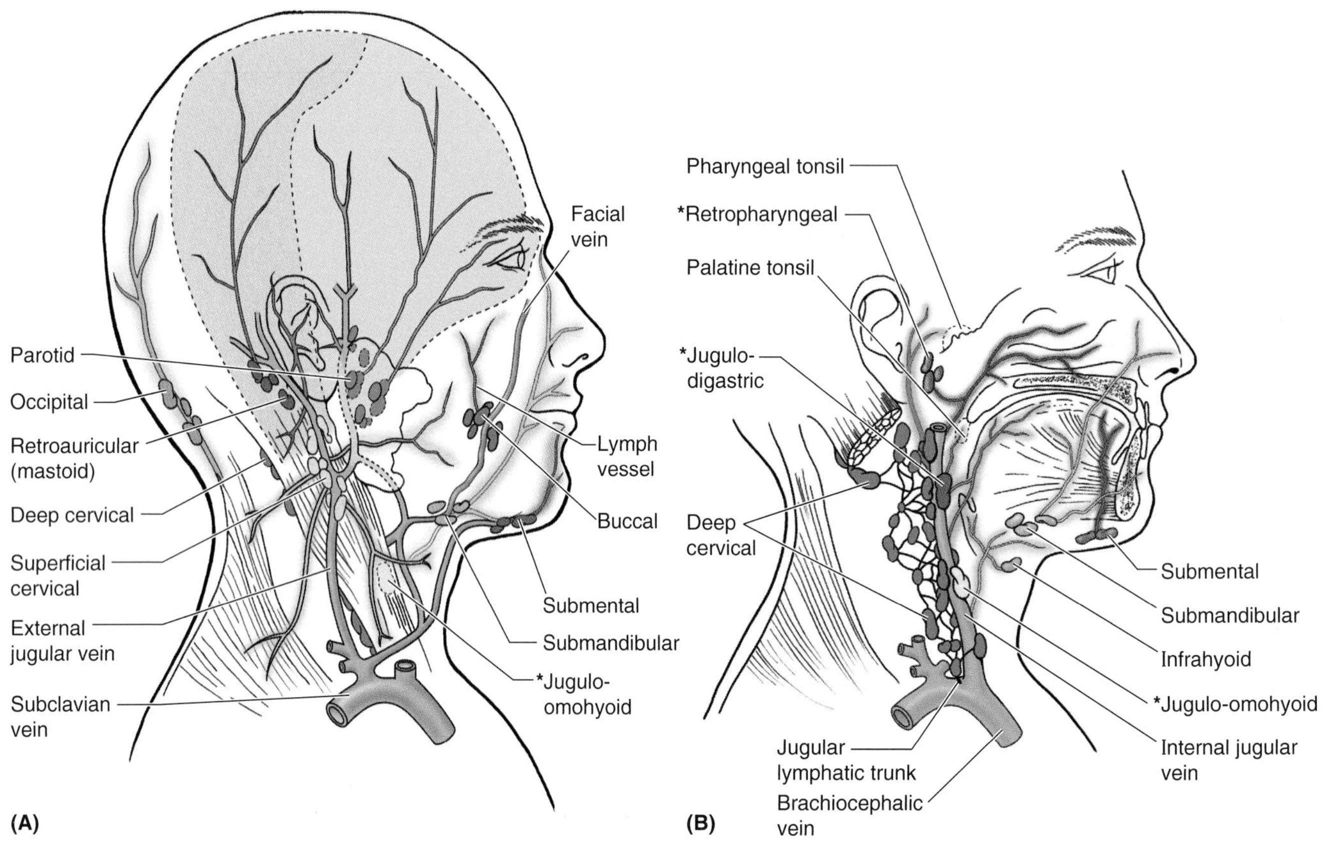

Figure 7.12. Lymphatic drainage of the face and scalp. A. Superficial drainage. A pericervical collar of superficial nodes is formed at the junction of the head and neck by the submental, submandibular, parotid, retroauricular (mastoid), and occipital nodes. These nodes initially receive most of the lymph drainage from the face and scalp. **B.** Deep drainage. All lymphatic vessels from the head and neck ultimately drain into the deep cervical lymph nodes, either directly from the tissues or indirectly after passing through an outlying group of nodes.

Parotid Gland

The parotid gland—the largest of the three paired salivary glands—is enclosed within a tough fascial capsule, the **parotid sheath**, derived from the investing layer of deep cervical fascia (Fig. 7.13). The parotid gland has an irregular shape because the area occupied by the gland, the **parotid bed**, is anteroinferior to the external acoustic meatus, where it is wedged between the ramus of the mandible and the mastoid process. The apex of the parotid gland is posterior to the angle of the mandible, and its base is related to the zygomatic arch. The lateral surface of the parotid gland is almost flat and is subcutaneous. The **parotid duct** passes horizontally from the anterior edge of the gland. At the anterior border of the masseter, the duct turns medially, pierces the buccinator, and enters the oral cavity through a small orifice opposite the 2nd maxillary molar tooth.

Structures Within the Parotid Gland

The structures within the parotid gland, from superficial to deep, are the **facial nerve** and its branches (Figs. 7.11, *A* and *C*, and 7.13), the **retromandibular vein**, and the **external carotid artery**. On the parotid sheath and within the gland are **parotid lymph nodes** (Figs. 7.12*A* and 7.13), which receive lymph from the forehead, lateral parts of the eyelids, temporal region, lateral surface of the auricle, anterior wall of the external acoustic meatus, and the middle ear. Lymph from the parotid nodes drains into the superficial and deep **cervical lymph nodes**.

Innervation of Parotid Gland and Related Structures

The **auriculotemporal nerve**, a branch of CN V_3, is closely related to the parotid gland and passes superior to it with the superficial temporal vessels. The **great auricular nerve** (C2 and C3), a branch of the cervical plexus, innervates the parotid sheath (Fig. 7.13). The parasympathetic component of the **glossopharyngeal nerve** (CN IX) supplies secretory fibers to the parotid gland, which are conveyed by the auriculotemporal nerve from the **otic ganglion** (Fig. 7.9*A*). Stimulation of these fibers produces a thin, watery saliva. Sympathetic fibers are derived from the cervical ganglia through the **external carotid nerve plexus** on the external carotid artery (Fig. 7.13). Sensory nerve fibers pass to the gland through the great auricular and auriculotemporal nerves.

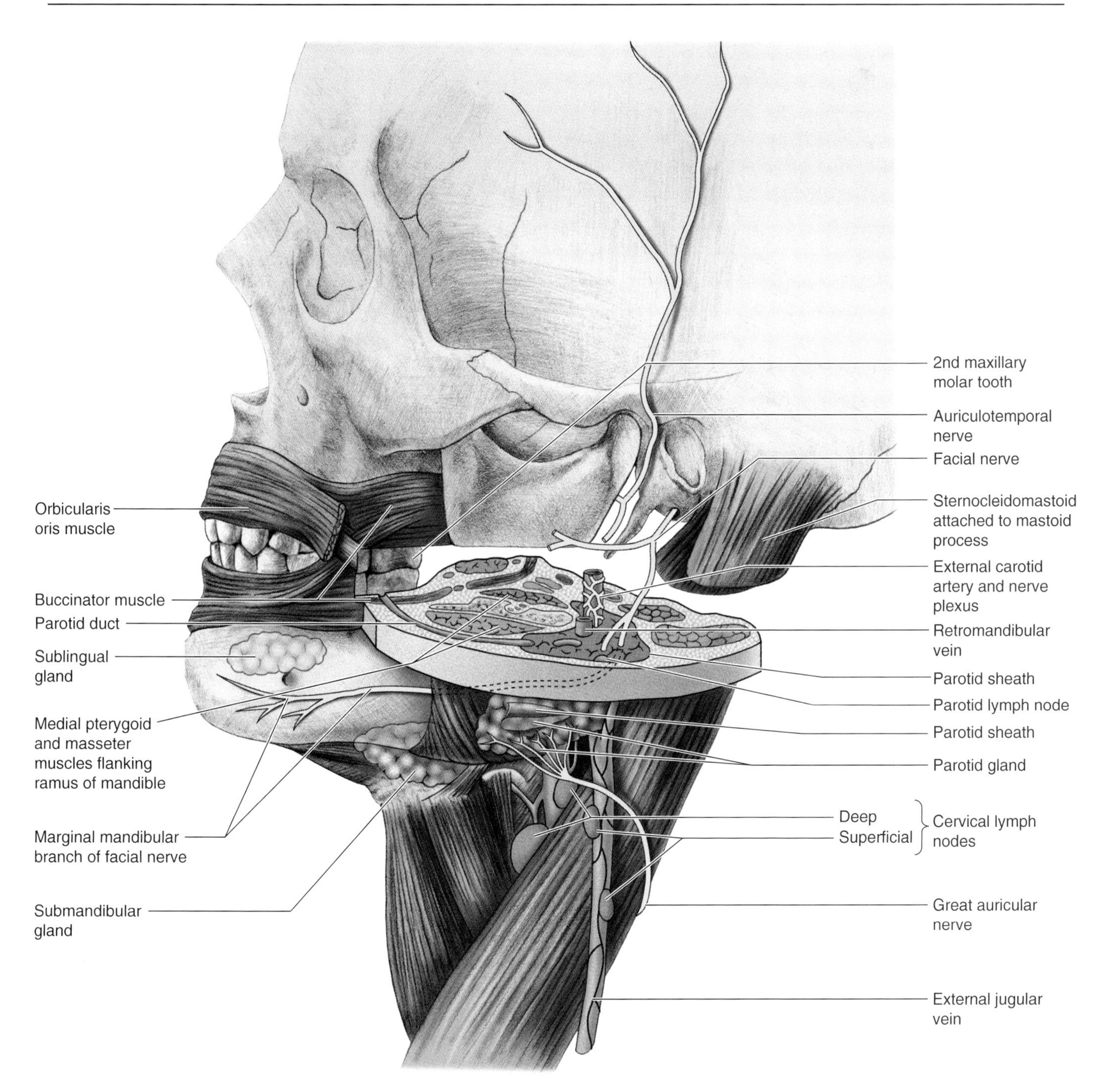

Figure 7.13. Relationships of the parotid gland—left lateral view. A transverse slice through the bed of the parotid gland demonstrates the relationship of the gland to the surrounding structures. Note that the gland passes deep between the ramus of the mandible (lower jaw)—flanked by the muscles of mastication (the chewing muscles)—anteriorly, and the mastoid process and sternocleidomastoid muscle posteriorly. The dimensions of the parotid bed change with movements of the mandible. Observe the structures embedded within the gland itself. Note that the parotid duct turns medially at the anterior border of the masseter muscle and pierces the buccinator (L. trumpeter) muscle.

Injury to the Facial Nerve During Surgery

Because branches of CN VII pass through the parotid gland, they are in jeopardy during surgery of the parotid. An important step in *parotidectomy* (surgical excision of the parotid gland) is the identification, dissection, isolation, and preservation of the facial nerve.

Infection of the Parotid Gland

The parotid gland may become infected through the bloodstream as occurs in *mumps*, an acute communicable viral disease. Infection of the parotid gland causes inflammation (*parotiditis*) and swelling of the gland. Severe pain occurs because the parotid sheath limits swelling. Often the pain is worse during chewing because the enlarged gland is wrapped around the posterior border of the ramus of the mandible and is compressed against the mastoid process of the temporal bone when the mouth is open. *The mumps virus may also cause inflammation of the parotid duct,* producing redness of the parotid papilla, the small, nipplelike projection marking the orifice of the duct in the mucous membrane of the cheek into the oral cavity. Because the pain produced by mumps may be confused with a toothache, redness of the *parotid papilla* is often an early sign that the disease involves the gland and not a tooth. *Parotid gland disease* often causes pain in the auricle, external acoustic meatus, temporal region, and TMJ because the auriculotemporal nerve, from which the parotid gland receives sensory fibers, also supplies sensory fibers to the skin over the temporal fossa and the auricle.

Sialography

A radiopaque fluid can be injected into the duct system of the parotid gland through a cannula inserted through the orifice of the parotid duct in the mucous membrane of the cheek. This technique, followed by radiography of the parotid gland, is called sialography. *Parotid sialograms* (G. sialon, saliva + grapho, to write) demonstrate parts of the parotid duct system that may be displaced or dilated by disease.

Blockage of the Parotid Duct

The parotid duct may be blocked by a calcified deposit—a *sialolith* or calculus (L. pebble). The resulting pain in the parotid gland is made worse by eating. Sucking a lemon slice is painful because of the buildup of saliva in the proximal part of the blocked duct.

Accessory Parotid Gland

Sometimes an accessory parotid gland lies on the masseter muscle between the parotid duct and zygomatic arch. Several ducts open from this accessory gland into the parotid duct. ○

Scalp

The scalp consists of skin (normally hair-bearing) and subcutaneous tissue, which cover the calvaria, from the superior nuchal lines on the occipital bone to the supraorbital margins of the frontal bone. Laterally, the scalp extends over the temporal fascia to the zygomatic arches.

Layers of the Scalp

The scalp consists of five layers of tissue, the first three layers of which are connected intimately and move as a unit (e.g., when wrinkling the forehead and thus moving the scalp). Each bold-faced letter in the term **scalp** serves as a memory key for one of its five layers (Fig. 7.14):

- Skin, thin except in the occipital region, contains many sweat and sebaceous glands and hair follicles; it has an abundant arterial supply and good venous and lymphatic drainage
- Connective tissue, forming the thick, dense, richly vascularized subcutaneous layer that is well supplied with cutaneous nerves
- Aponeurosis—the *epicranial aponeurosis*—a strong tendinous sheet that covers the calvaria between the occipitalis, superior auricular, and frontalis muscles (collectively, these structures constitute the **epicranius muscle**). The frontalis pulls the scalp anteriorly, wrinkles the forehead, and elevates the eyebrows; the occipitalis pulls the scalp posteriorly and wrinkles the skin at the back of the neck.
- Loose connective tissue, somewhat like a sponge because of its many potential spaces that may distend with fluid resulting from injury or infection; this layer allows free movement of the **scalp proper** (the first three layers—skin, connective tissue, and epicranial aponeurosis) over the underlying calvaria.
- Pericranium, a dense layer of connective tissue that forms the external periosteum of the calvaria; it is firmly attached but can be stripped fairly easily from the calvaria of living persons, except where the pericranium is continuous with the fibrous tissue in the cranial sutures.

Nerves of the Scalp

Innervation of the scalp anterior to the auricles is through branches of all three divisions of CN V, the **trigeminal nerve** (Fig. 7.15). Posterior to the auricles, the nerve supply of the scalp is from spinal cutaneous nerves (C2 and C3).

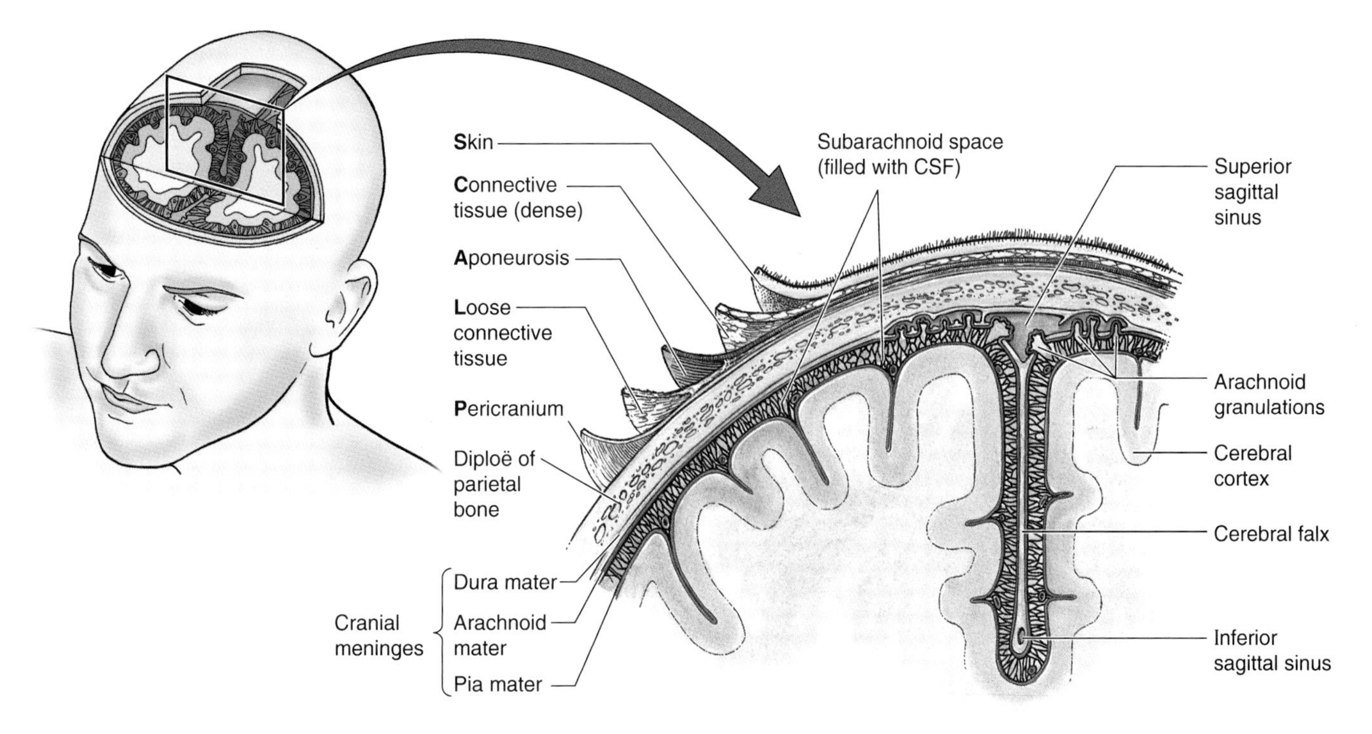

Figure 7.14. The scalp, calvaria (skullcap), and meninges. Anterior view of the coronal section. Observe the layers of the scalp. The skin is bound tightly to the epicranial aponeurosis (L. aponeurosis epicranialis), which moves freely over the pericranium and skull because of the intervening loose connective tissue. The aponeurosis is the flat intermediate tendon of the occipitofrontalis muscle. Observe the cranial meninges and the subarachnoid (leptomeningeal) space (filled with cerebrospinal fluid [CSF]).

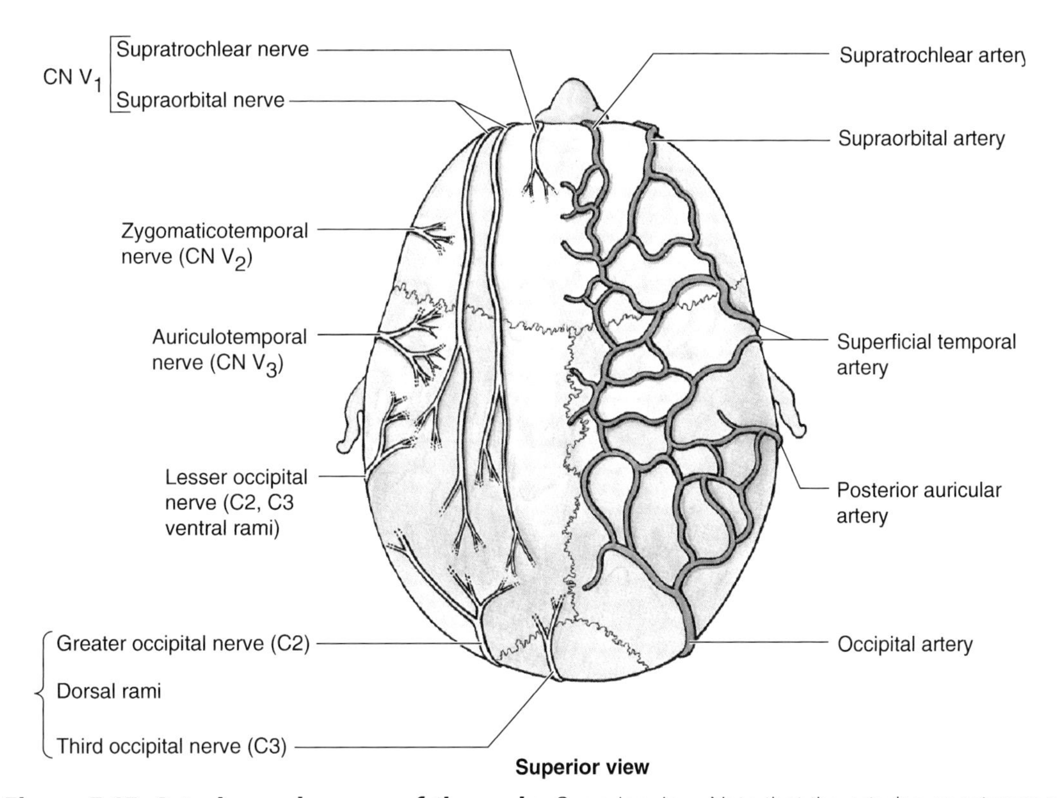

Figure 7.15. Arteries and nerves of the scalp. Superior view. Note that the arteries anastomose freely and thus bleed from both ends when cut. The supraorbital and supratrochlear arteries are branches of the internal carotid artery through the ophthalmic artery; the other arteries are branches of the external carotid artery. The nerves appear in sequence: CN V_1, CN V_2, CN V_3, ventral rami of C2 and C3, and dorsal rami of C2 and C3.

Vasculature of the Scalp

The blood vessels run in layer two of the scalp—the dense subcutaneous layer between the skin and the epicranial aponeurosis. They are held by the dense connective tissue in such a way that they tend to remain open when cut. Consequently, bleeding from scalp wounds is profuse.

Arteries of the Scalp

The arterial supply of the scalp is from the **external carotid arteries**—through the occipital, posterior auricular, and superficial temporal arteries—and from the **internal carotid arteries** through the supratrochlear and supraorbital arteries (Fig. 7.15, Table 7.5). *The scalp has a rich blood supply, and the arteries anastomose freely with one another in layer two of the scalp,* the dense subcutaneous connective tissue layer.

Veins of the Scalp

The venous drainage of superficial parts of the scalp is through the accompanying veins of the scalp arteries—the **supraorbital** and **supratrochlear veins** (Table 7.6), which begin in the forehead and descend to unite at the medial angle of the eye to form the angular vein that becomes the *facial vein* at the inferior margin of the orbit. The **superficial temporal veins** and **posterior auricular veins** drain the scalp anterior and posterior to the auricles, respectively. The **posterior auricular vein** often receives a *mastoid emissary vein* from the sigmoid sinus, a dural venous sinus. The **occipital veins** drain the occipital region of the scalp. Venous drainage of deep parts of the scalp in the temporal region is through *deep temporal veins* that are tributaries of the *pterygoid venous plexus.*

Lymphatic Drainage of the Scalp

There are no lymph nodes in the scalp. Lymph from this region drains into the **superficial ring (pericervical collar) of lymph nodes**—submental, submandibular, parotid, mastoid or retroauricular, and occipital—that is located at the junction of the head and neck (Fig. 7.12*A*). Lymph from these nodes drains into the **deep cervical lymph nodes** along the IJV (Fig. 7.12*B*).

Scalping Injury

Because the scalp arteries arising at the sides of the head are well protected by dense connective tissue and anastomose freely, a *partially detached scalp* may be replaced with a reasonable chance of healing as long as one of the vessels supplying the scalp remains intact (Williams et al., 1995). During an *attached craniotomy*—removal of a segment of the calvaria with a soft tissue scalp flap to expose the cranial cavity—the incisions are usually made convex upward, and the superficial temporal artery is included in the tissue flap.

The *scalp proper*—the first three layers of the scalp—are often regarded clinically as a single layer because they remain together when a scalp flap is made during a craniotomy and when part of the scalp is torn off (e.g., during vehicular or industrial accidents). Nerves and vessels of the scalp enter inferiorly and ascend through layer two to the skin. Consequently, surgical pedicle scalp flaps are made so that they remain attached inferiorly to preserve the nerves and vessels, thereby promoting good healing. *The arteries of the scalp supply little blood to the calvaria*; the bones forming it are supplied by the middle meningeal artery. Hence, loss of the scalp does not produce necrosis of the calvarial bones.

Scalp Infections

The loose connective tissue layer (layer four) of the scalp is the *danger area of the scalp* because pus or blood spreads easily in it. Infection in this layer can also pass into the cranial cavity through *emissary veins* that pass through parietal foramina in the calvaria (Fig. 7.4*B*) and infect the intracranial structures such as the brain and meninges. An infection cannot pass into the neck because the occipitalis muscle attaches to the occipital bone and mastoid parts of the temporal bones. Neither can a scalp infection spread laterally beyond the zygomatic arches because the epicranial aponeurosis is continuous with the temporal fascia that attaches to these arches. An infection or fluid (e.g., pus or blood) can enter the eyelids and the root of the nose because the frontalis inserts into the skin and subcutaneous tissue and does not attach to the bone. Conse- ▶

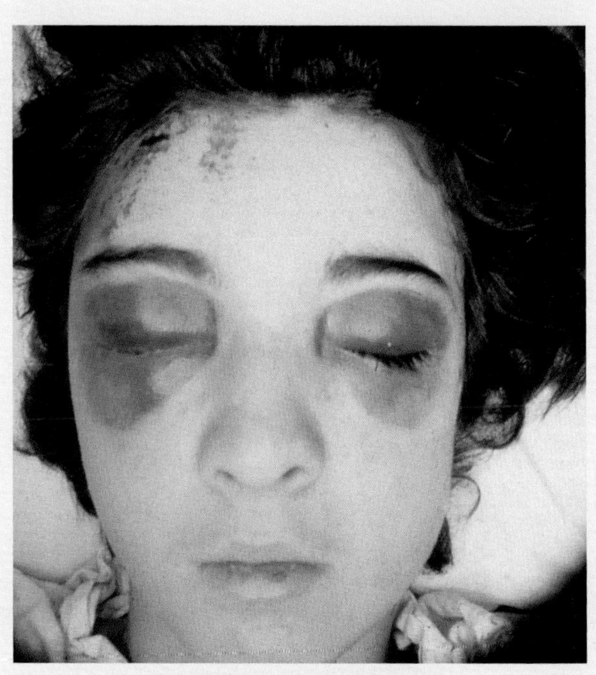

(Courtesy of Dr. Ralph Ger, NYC)

Cranial Meninges

The cranial meninges are internal to the skull (Figs. 7.14 and 7.16). *The cranial meninges*:

- Protect the brain
- Form the supporting framework for arteries, veins, and venous sinuses
- Enclose a fluid-filled cavity, the *subarachnoid (leptomeningeal) space*, which is vital to the normal function of the brain (Haines, 1997).

The cranial meninges consist of three layers:

- **Dura mater** (dura)—an external thick, *dense fibrous membrane*
- **Arachnoid mater** (arachnoid)—an intermediate, *delicate membrane*
- **Pia mater** (pia)—an internal delicate, *vascular membrane.*

The meninges also enclose the **cerebrospinal fluid** (CSF) and help to maintain the balance of extracellular fluid in the brain. CSF is a clear liquid similar to blood in constitution—it provides nutrients but has less protein and a different ion concentration. CSF is formed by the **choroid plexuses** of the four ventricles of the brain (Fig. 7.16). CSF leaves the ventricular system and enters the **subarachnoid space** between the arachnoid and pia mater, where it cushions and nourishes the brain.

Dura Mater

The dura mater (pachymeninx—G. pachy, thick + menix, membrane) is adherent to the internal surface of the skull and is described as a two-layered membrane (Figs. 7.16 and 7.17):

- An *external periosteal layer*, formed by the periosteum covering the internal surface of the calvaria
- An *internal meningeal layer*, a strong fibrous membrane

▶ quently, black eyes can result from an injury to the scalp or forehead. Most blood enters the upper eyelid but some may also enter the lower one.

Scalp Lacerations

Scalp lacerations are the most common type of head injury requiring surgical care. These wounds bleed profusely because the arteries entering the periphery of the scalp bleed from both ends because of abundant anastomoses. They do not retract when lacerated because they are held open by the dense connective tissue in layer two of the scalp. Hence, *unconscious patients may bleed to death from scalp lacerations if bleeding is not controlled* (e.g., by sutures).

The epicranial aponeurosis—layer three of the scalp—is clinically important. Because of its strength, a superficial laceration in the skin does not gape because the margins of the wound are held together by this aponeurosis. Furthermore, when suturing *superficial scalp wounds*, deep sutures are not necessary because the epicranial aponeurosis does not allow wide separation of the skin. *Deep scalp wounds gape widely when the epicranial aponeurosis is split or lacerated in the coronal plane* because of the pull of the frontal and occipital bellies of the epicranius muscle in opposite directions (anteriorly and posteriorly).

Stenosis of the Internal Carotid Artery

At the medial angle of the eye, an anastomosis occurs between the facial branch of the external carotid artery and cutaneous branches of the internal carotid artery. During old age the internal carotid artery may become narrow (stenotic) because of atherosclerotic thickening of the intima of the arteries. Because of the arterial anastomosis, intracranial structures—such as the brain—can receive blood from the connection of the facial artery to the dorsal nasal branch of the ophthalmic artery.

Sebaceous Cysts

The ducts of sebaceous glands associated with hair follicles in the scalp may become obstructed, resulting in the retention of secretions and the formation of sebaceous cysts (wens). Because they are in the skin of the scalp, *sebaceous cysts move with the scalp.* Hair follicles in the scalp go through alternate growing and resting phases; hairs grow and eventually drop out of their follicles during combing, for example. After a while, new hairs normally begin to grow in the same follicles.

Cephalohematoma

Sometimes during a difficult birth, bleeding occurs between the baby's pericranium and calvaria, usually over one parietal bone. The bleeding results from rupture of multiple, minute periosteal arteries that nourish the bones of the calvaria. The resulting collection of blood developing several hours after birth is a *cephalohematoma.*

Bone Flaps

Because the adult pericranium has poor osteogenic properties, little regeneration occurs after bone loss (e.g., when pieces of bone are removed during repair of a comminuted skull fracture; p. 838). *Surgically produced bone flaps* are usually put back into place and wired to other parts of the calvaria. Large defects in the adult calvaria resulting from severe trauma usually do not "fill in"; consequently, the insertion of a metal or plastic plate is necessary to protect the area of the brain that is related to the defect. ⊙

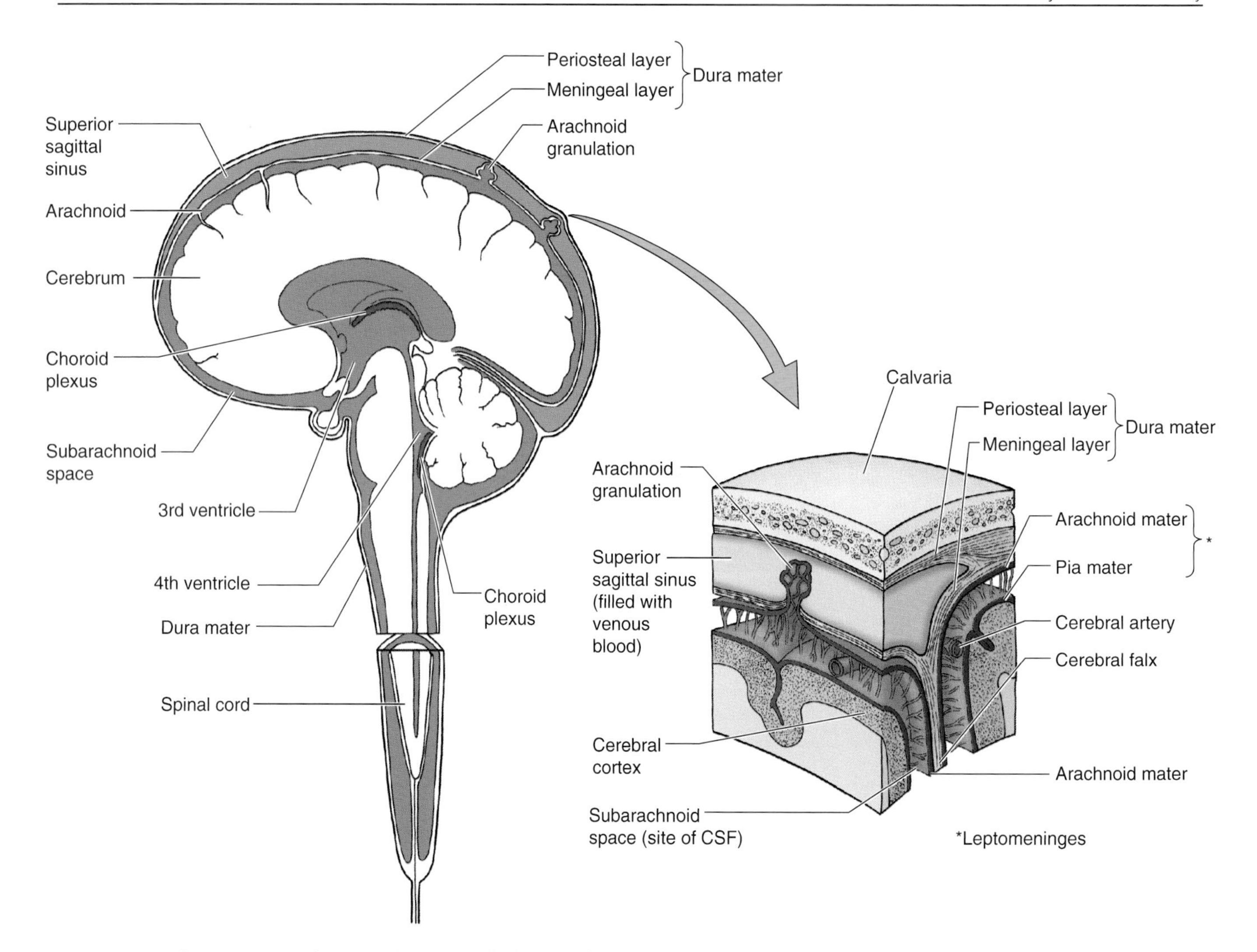

Figure 7.16. The meninges and their relationship to the calvaria (skullcap), brain, and spinal cord. Observe that the dura mater and subarachnoid (leptomeningeal) space (*purple*) surround the brain and are continuous with that around the spinal cord. The two layers of dura separate to form the dural venous sinuses, such as the superior sagittal sinus. The cerebrum (L. brain) includes mainly the cerebral hemispheres. Arachnoid granulations are formed by tufted prolongations of the arachnoid that protrude through the meningeal layer of the dura into the dural venous sinuses and effect transfer of cerebrospinal fluid (CSF) to the venous system.

that is continuous at the foramen magnum with the spinal dura mater covering the spinal cord.

Understand that the periosteal layer of the dura is the periosteum lining the calvaria. It adheres to the internal surface of the skull, and its attachment is tenacious along the suture lines and in the cranial base (Haines, 1997).

The external periosteal layer is continuous at the cranial foramina with the periosteum on the external surface of the calvaria; *it is not continuous with the dura mater of the spinal cord.* The meningeal layer is intimately fused with the periosteal layer and cannot be separated from it. The fused external and internal layers of dura over the calvaria can be easily stripped from the cranial bones (e.g., when the superior part of the calvaria is removed at autopsy). A blow to the head can detach the periosteal layer from the calvaria without fracturing the cranial bones. In the cranial base the two dural layers are firmly attached and difficult to separate from the bones. Consequently, a fracture of the cranial base usually tears the dura and results in leakage of CSF. The internal part of the meningeal layer of dura—*the dural border cell layer*—is composed of flattened fibroblasts that are separated by large extracellular spaces. This layer constitutes a plane of structural weakness at the dura-arachnoid junction (Haines, 1997).

Dural Infoldings or Reflections

The internal meningeal layer of dura draws away from the external periosteal layer of dura to form dural infoldings (reflections), which separate the regions of the brain from each other (Fig. 7.18). The largest of these septa is the **cerebral falx** (L.

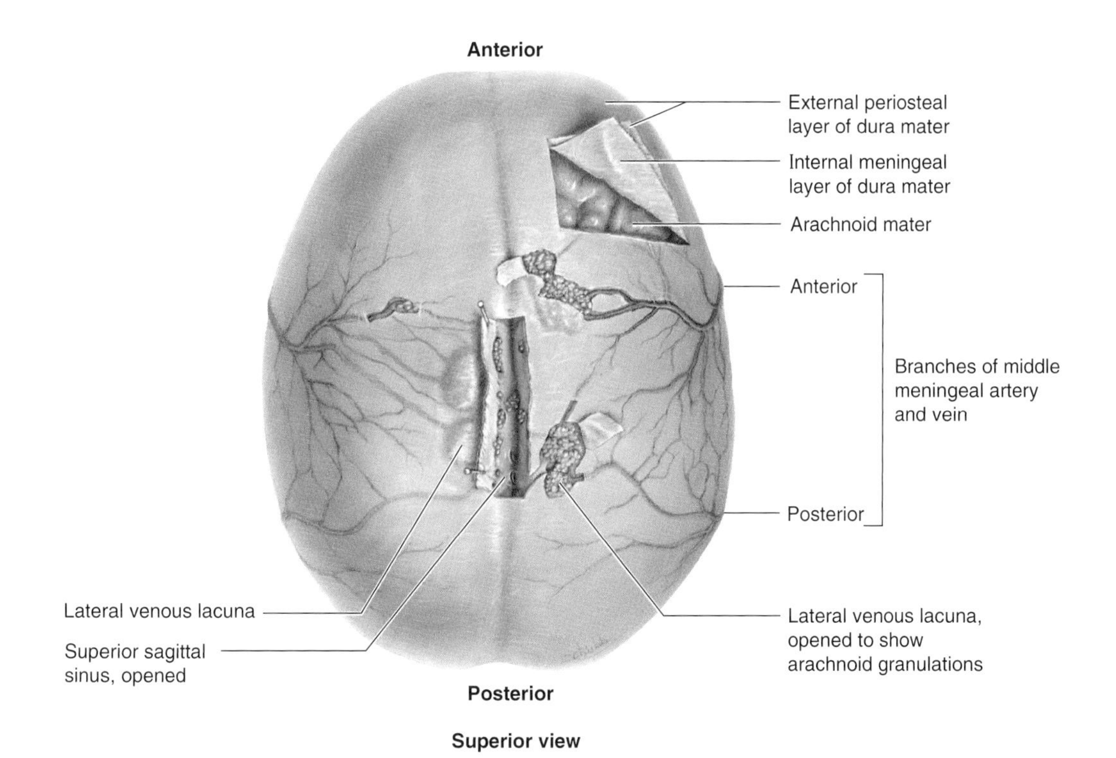

Figure 7.17. External surface of the dura mater: arachnoid granulations. Superior view. The calvaria (skullcap) is removed. In the median plane, a part of the thick roof of the superior sagittal sinus has been incised and retracted; laterally, the thin roofs of two lateral lacunae are reflected to demonstrate the abundant arachnoid granulations, responsible for absorption of cerebrospinal fluid (CSF). On the right, an angular flap of dura is turned anteriorly; the convolutions of the cerebral cortex are visible through the arachnoid mater.

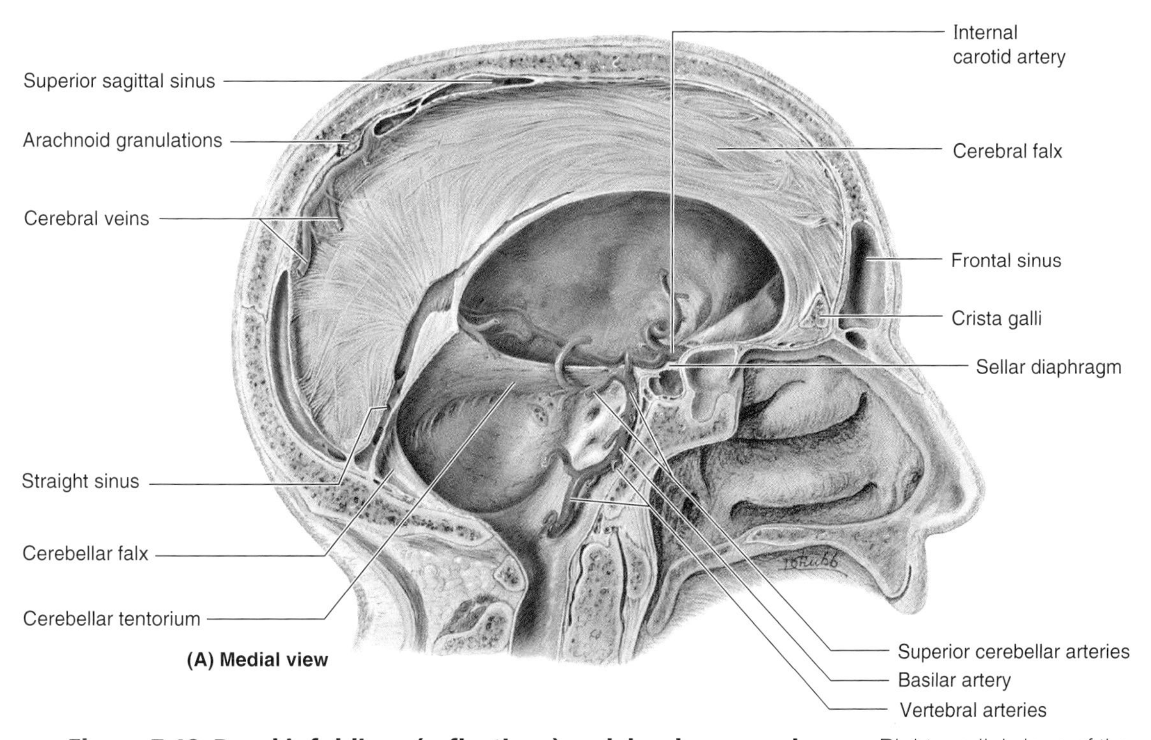

Figure 7.18. Dural infoldings (reflections) and dural venous sinuses. Right medial views of the left half of the head. **A.** Observe the two sickle-shaped dural folds (septae), the cerebral falx (L. falx cerebri) and cerebellar falx (L. falx cerebelli), which are vertically oriented in the median plane. Observe also the two rooflike folds, the cerebellar tentorium (L. tentorium cerebelli) and the small sellar diaphragm (L. diaphragma sellae), which lie horizontally. Notice the left members of the paired arteries that supply the brain—the internal carotid and vertebral arteries.

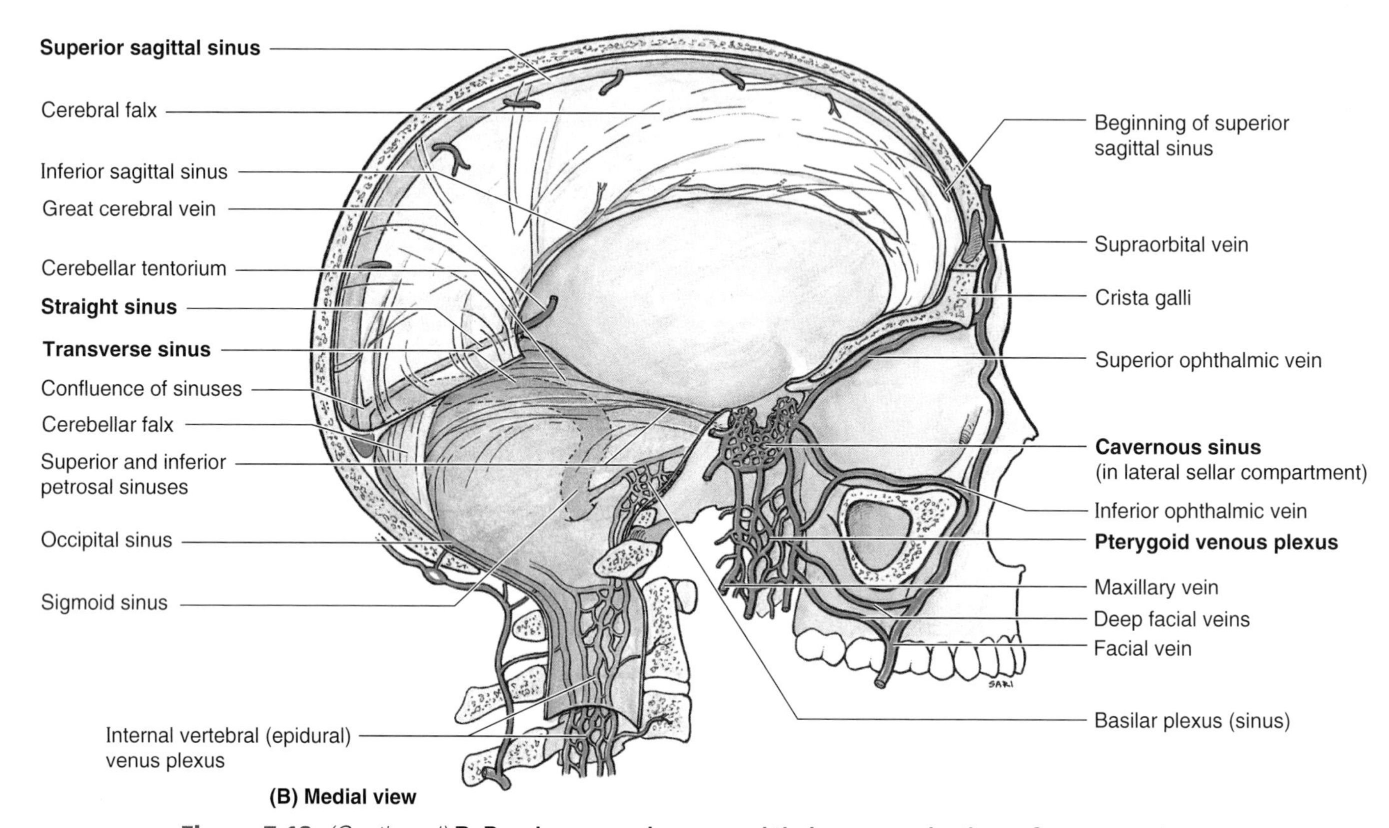

Figure 7.18. *(Continued)* **B. Dural venous sinuses and their communications.** Cavernous sinuses lie on each side of the body of the sphenoid bone. Note that the cavernous sinus communicates anteriorly with the facial vein through the superior and inferior ophthalmic veins. The cavernous sinus also communicates inferiorly with the pterygoid venous plexus and posteriorly with the basilar plexus, which in turn communicates with the internal vertebral (epidural) venous plexus.

falx cerebri). The dural infoldings divide the cranial cavity into compartments that support parts of the brain. The dural infoldings include the:

- Cerebral falx (L. falx cerebri)
- Cerebellar tentorium (L. tentorium cerebelli)
- Sellar diaphragm (L. diaphragma sellae).

Falx is the Latin term for a sickle; thus the falces are sickle-shaped structures.

The **cerebral falx**, the largest dural reflection, lies in the longitudinal fissure and separates the right and left cerebral hemispheres. The cerebral falx attaches in the median plane to the internal surface of the calvaria, from the *frontal crest* of the frontal bone and crista galli of the ethmoid bone anteriorly to the *internal occipital protuberance* posteriorly (Figs. 7.18 and 7.20). The cerebral falx ends by becoming continuous with the cerebellar tentorium.

The **cerebellar tentorium**, the second largest dural infolding, is a wide crescentic septum that separates the occipital lobes of the cerebral hemispheres from the cerebellum. The cerebellar tentorium attaches rostrally to the clinoid processes of the sphenoid bone, rostrolaterally to the petrous part of the temporal bone, and posterolaterally to the internal surface of the occipital bone and part of the parietal bone. The cerebral falx attaches to the cerebellar tentorium and holds it up, giving it a tentlike appearance (L. tentorium, tent). The tentlike shape of the cerebellar tentorium divides the cranial cavity into supratentorial and infratentorial compartments. The supratentorial compartment is divided into right and left halves by the cerebral falx. Its concave anteromedial border is free, producing a gap—the **tentorial notch,** or incisura of tentorium (see Fig. 7.21, *A* and *B*)—through which the brainstem extends from the posterior into the middle cranial fossa.

The **cerebellar falx** is a vertical dural infolding that lies inferior to the cerebellar tentorium in the posterior part of the posterior cranial fossa; it partially separates the cerebellar hemispheres.

The **sellar diaphragm**, the smallest dural infolding, is a circular sheet of dura that is suspended between the clinoid processes forming a roof over the hypophysial fossa in the sphenoid bone (Fig. 7.18*A*). The sellar diaphragm covers the pituitary gland in this fossa and has an aperture for passage of the infundibulum (pituitary stalk) and hypophysial veins.

Tentorial Herniation

The tentorial notch is the opening in the cerebellar tentorium for the brainstem—slightly larger than is necessary ▶

▶ to accommodate the midbrain. Hence, *space-occupying lesions*—such as tumors in the supratentorial compartment—produce increased intracranial pressure and may cause part of the adjacent temporal lobe of the brain to herniate through the tentorial notch. During *tentorial herniation*, the temporal lobe may be lacerated by the tough cerebellar tentorium and the oculomotor nerve (CN III) may be stretched, compressed, or both. *Oculomotor lesions* may produce paralysis of the extrinsic eye muscles supplied by CN III.

Bulging of the Sellar Diaphragm

Pituitary tumors may extend superiorly through the aperture in the sellar diaphragm and/or cause bulging of it. These tumors often expand the sellar diaphragm, producing endocrine symptoms early or late (i.e., before or after enlargement of the sellar diaphragm). Superior extension of a tumor may cause visual symptoms owing to pressure on the optic chiasm (L. chiasma opticum), the place where the optic nerve fibers cross. ○

Dural Venous Sinuses

The dural venous sinuses are endothelium-lined spaces between the periosteal and meningeal layers of the dura; they form where the dural septa attach (Figs. 7.18 and 7.19). Large veins from the surface of the brain empty into these sinuses and all blood from the brain ultimately drains through them into the IJVs (Fig. 7.19).

The **superior sagittal sinus** lies in the convex attached border of the cerebral falx (Fig. 7.18). It begins at the crista galli and ends near the internal occipital protuberance (Fig. 7.20) at the **confluence of sinuses,** a meeting place of the superior sagittal, straight, occipital, and transverse sinuses (Fig. 7.19). The superior sagittal sinus receives the superior cerebral veins and communicates on each side through slitlike openings with the **lateral venous lacunae**—lateral expansions of the superior sagittal sinus (Fig. 7.17).

Arachnoid granulations (collections of arachnoid villi) are tufted prolongations of the arachnoid that protrude through the meningeal layer of the dura mater into the dural venous sinuses, especially the lateral lacunae, and effect transfer of CSF to the venous system (Figs. 7.17 and 7.18A). These small, fleshy looking elevations may be large enough to erode bone, forming pits in the calvaria (Fig. 7.4*B*). They are usually observed in the vicinity of the superior sagittal, transverse, and some other dural venous sinuses. Arachnoid villi or granulations are structurally adapted for the transport of CSF from the subarachnoid space to the venous system (p. 891). For more information, see Haines (1997).

The **inferior sagittal sinus** (Fig. 7.18*B*), much smaller than the superior sagittal sinus, runs in the inferior concave free border of the cerebral falx and ends in the straight sinus.

The **straight sinus** (L. sinus rectus) is formed by the union

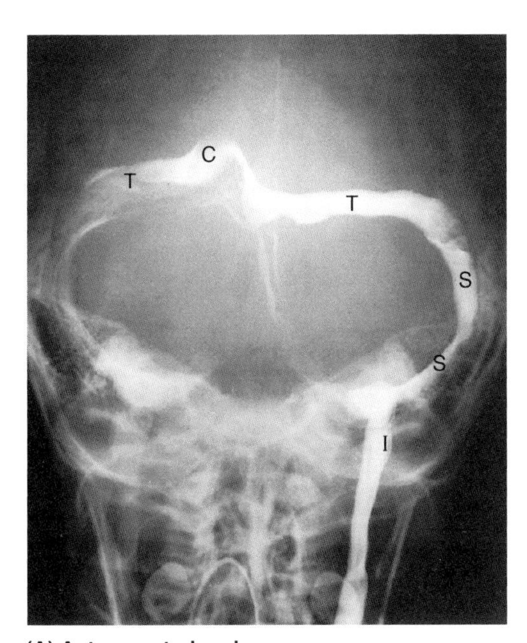

(A) Anteroposterior view

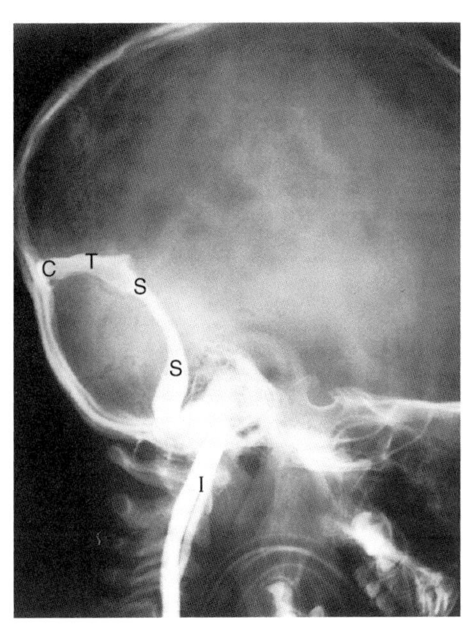

(B) Lateral view

Figure 7.19. Venograms of the dural sinuses. Following an injection of radiopaque dye into the arterial system, sufficient time has elapsed for the dye to circulate through the capillaries of the brain and be collected in the dural venous sinuses when anteroposterior (AP) and lateral radiographic studies were performed. *C*, confluence of sinuses; *T*, transverse sinus; *S*, sigmoid sinus; *I*, internal jugular vein. In the AP view (**A**), notice the left-sided dominance in the drainage of the confluence of sinuses. (Courtesy of Dr. D. Armstrong, Associate Professor of Medical Imaging, University of Toronto, Toronto, Ontario, Canada)

of the inferior sagittal sinus with the **great cerebral vein**. It runs inferoposteriorly along the line of attachment of the cerebral falx to the cerebellar tentorium, where it joins the confluence of sinuses.

The **transverse sinuses** pass laterally from the confluence of sinuses, grooving the occipital bones and the posteroinferior angles of the parietal bones (Figs. 7.19–7.21). The transverse sinuses course along the posterolateral attached margins of the cerebellar tentorium and then become the sigmoid sinuses as they approach the posterior aspect of the petrous temporal bones. Blood received by the **confluence of sinuses** is drained by the transverse sinuses, but rarely equally. Usually the left sinus is dominant (larger).

The **sigmoid sinuses** follow S-shaped courses in the posterior cranial fossa, forming deep grooves in the temporal and occipital bones (Figs. 7.20 and 7.21*A*). Each sigmoid sinus turns anteriorly and then continues inferiorly as the IJV after traversing the jugular foramen (Fig. 7.21*B*).

The **occipital sinus** lies in the attached border of the cerebellar falx (Fig. 7.18*B*) and ends superiorly in the confluence of sinuses (Fig. 7.19). The occipital sinus communicates inferiorly with the internal vertebral venous plexus.

The **cavernous sinus,** or lateral sellar compartment, is situated bilaterally on each side of the sella turcica on the upper surface of the body of the hollow sphenoid bone, which contains the sphenoidal (air) sinus (Figs. 7.18*B* and 7.21). Each sellar compartment contains a cavernous sinus consisting of a venous plexus of extremely thin-walled veins and

- Extends from the superior orbital fissure anteriorly to the apex of the petrous part of the temporal bone posteriorly
- Receives blood from the superior and inferior ophthalmic veins, superficial middle cerebral vein, and sphenoparietal sinus.

The venous channels in these sinuses communicate with each other through venous channels anterior and posterior to the stalk of the pituitary gland—the **intercavernous sinuses**

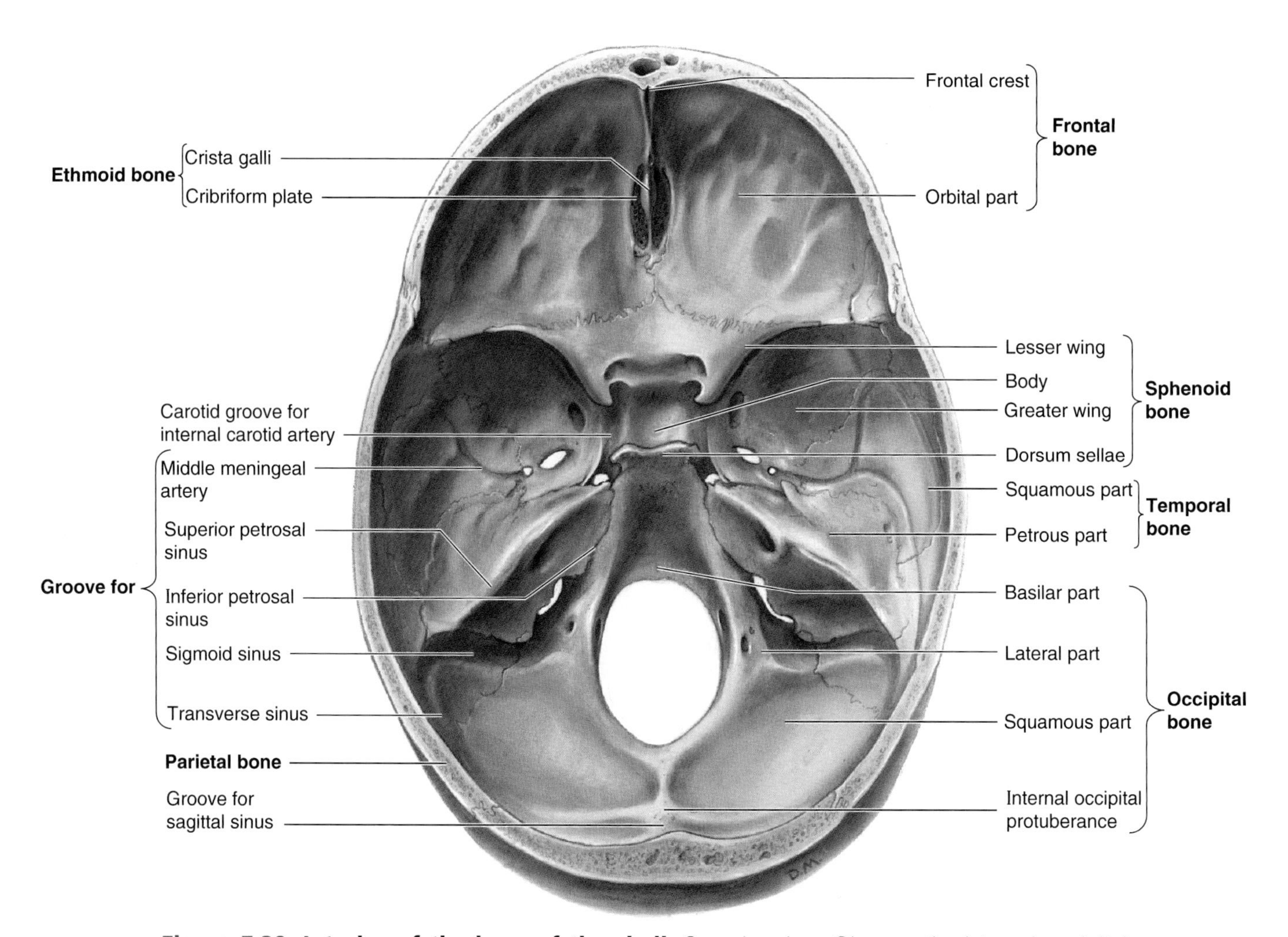

Figure 7.20. Interior of the base of the skull. Superior view. Observe the internal occipital protuberance, formed in relationship to the confluence of sinuses, and the grooves formed in the base of the skull by the dural venous sinuses (e.g., the sigmoid sinus). The grooves for the transverse sinuses and superior petrosal sinuses are also the sites of attachment for the cerebellar tentorium (L. tentorium cerebelli).

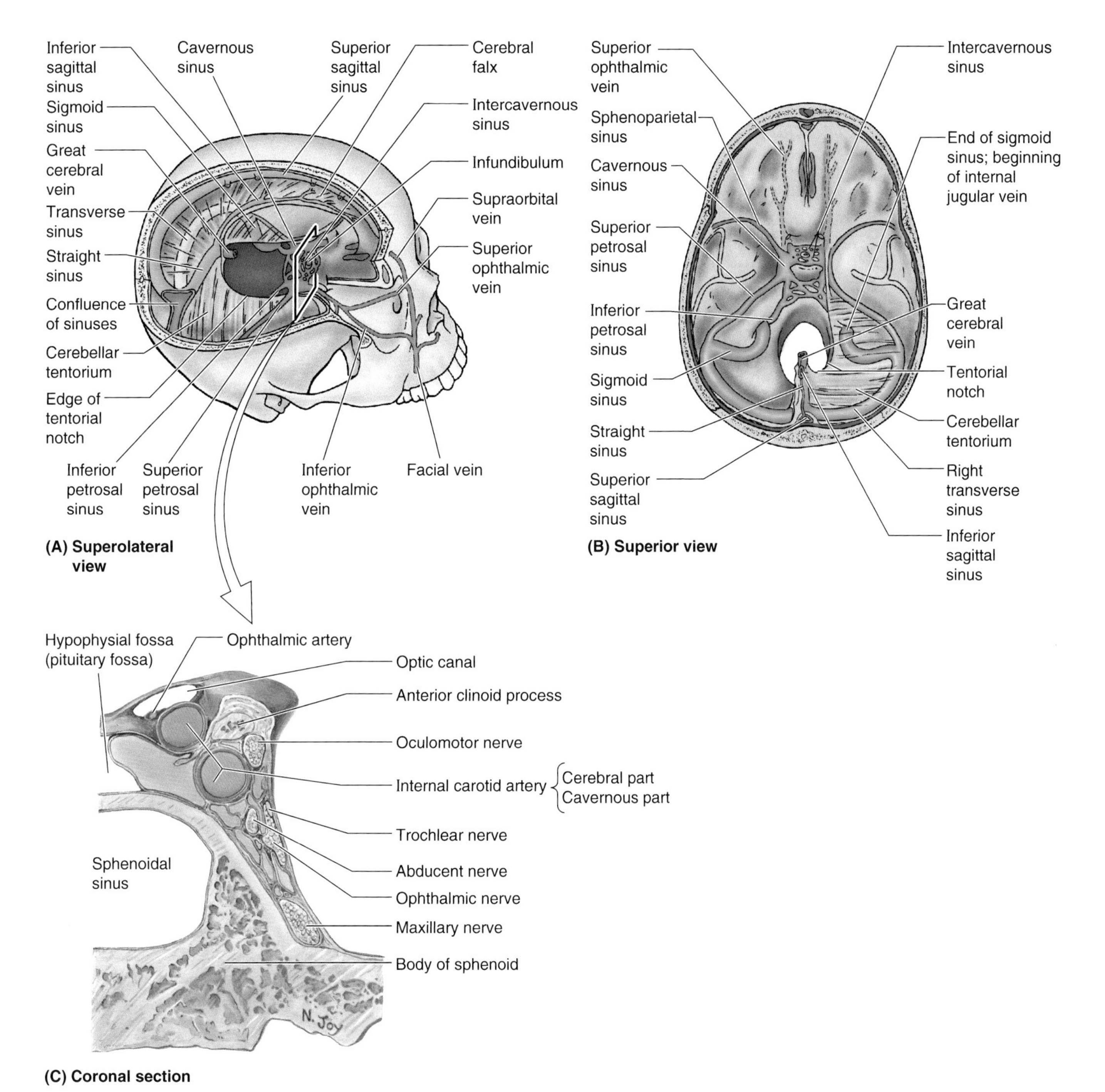

Figure 7.21. Venous sinuses of the dura mater. Blood from the brain drains into the sinuses that are formed within the dura. **A.** Right superolateral view. The brain and part of the calvaria (skullcap) are removed to demonstrate the sinuses related to the cerebral falx (L. falx cerebri) and the cerebellar tentorium (L. tentorium cerebelli). **B.** Superior view of the interior of the base of the skull, demonstrating most communications of the cavernous sinuses (inferior communication with pterygoid plexus is a notable exception) and drainage of the confluence of sinuses. Observe that the ophthalmic veins drain into the cavernous sinus. **C.** Anterior view of a coronal section of the right cavernous sinus and the body of the sphenoid. This venous sinus is situated bilaterally at the lateral aspect of the body of the sphenoid—containing the sphenoidal (air) sinus—and the hypophyseal fossa (pituitary fossa) above. Observe that cranial nerves III, IV, V_1, and V_2 are in a sheath in the lateral wall of the sinus and that the internal carotid artery is surrounded by the internal carotid plexus of sympathetic nerves (not drawn) as the artery and the abducent nerve (CN VI) pass through the cavernous sinus. Notice that the internal carotid artery, having made an acute bend, is cut twice; the inferior section of the artery is from the artery as it passes anteriorly along the carotid groove toward the acute bend (cavernous part); the superior section is from the artery as it passes posteriorly from the bend to join the cerebral arterial circle (of Willis—cerebral part). The ophthalmic artery has just branched from the latter.

(Fig. 7.21, *A* and *B*)—and sometimes through veins inferior to the pituitary gland. The cavernous sinuses drain posteroinferiorly through the superior and inferior **petrosal sinuses** and emissary veins to the **pterygoid plexuses** (Figs. 7.18*B* and 7.21*B*). Inside each sinus or compartment is the **internal carotid artery** with its small branches (Fig. 7.21*C*), surrounded by the carotid plexus of sympathetic nerve(s), and the **abducent nerve** (CN VI). *From superior to inferior, the lateral wall of the cavernous sinus incorporates the*:

- Oculomotor nerve
- Trochlear nerve (CN IV)
- Trigeminal nerve (CN V_1 and, rarely, CN V_2).

These cranial nerves are ensheathed in the lateral wall of the sinus.

The **superior petrosal sinuses** run from the posterior ends of the veins comprising the cavernous sinus to the transverse sinuses at the site where these sinuses curve inferiorly to form the sigmoid sinuses (Fig. 7.21*B*). Each superior petrosal sinus lies in the anterolateral attached margin of the cerebellar tentorium, which attaches to the superior border (crest or ridge) of the petrous part of the temporal bone (Fig. 7.20).

The **inferior petrosal sinuses** also commence at the posterior end of the cavernous sinus inferiorly. Each inferior petrosal sinus runs in a groove between the petrous part of the temporal bone and the basilar part of the occipital bone (Fig. 7.20). The inferior petrosal sinuses drain the veins of the lateral cavernous sinus directly into the origin of the IJVs.

The **basilar plexus** (sinus) connects the inferior petrosal sinuses and communicates inferiorly with the internal vertebral venous plexus (Fig. 7.18*B*).

Emissary veins connect the dural venous sinuses with veins outside the cranium. Although they are valveless and blood may flow in both directions, flow in the emissary veins is usually away from the brain. The size and number of emissary veins vary. A *frontal emissary vein* is present in children and some adults. It passes through the foramen cecum of the skull, connecting the superior sagittal sinus with veins of the frontal sinus and nasal cavities. A *parietal emissary vein*, which

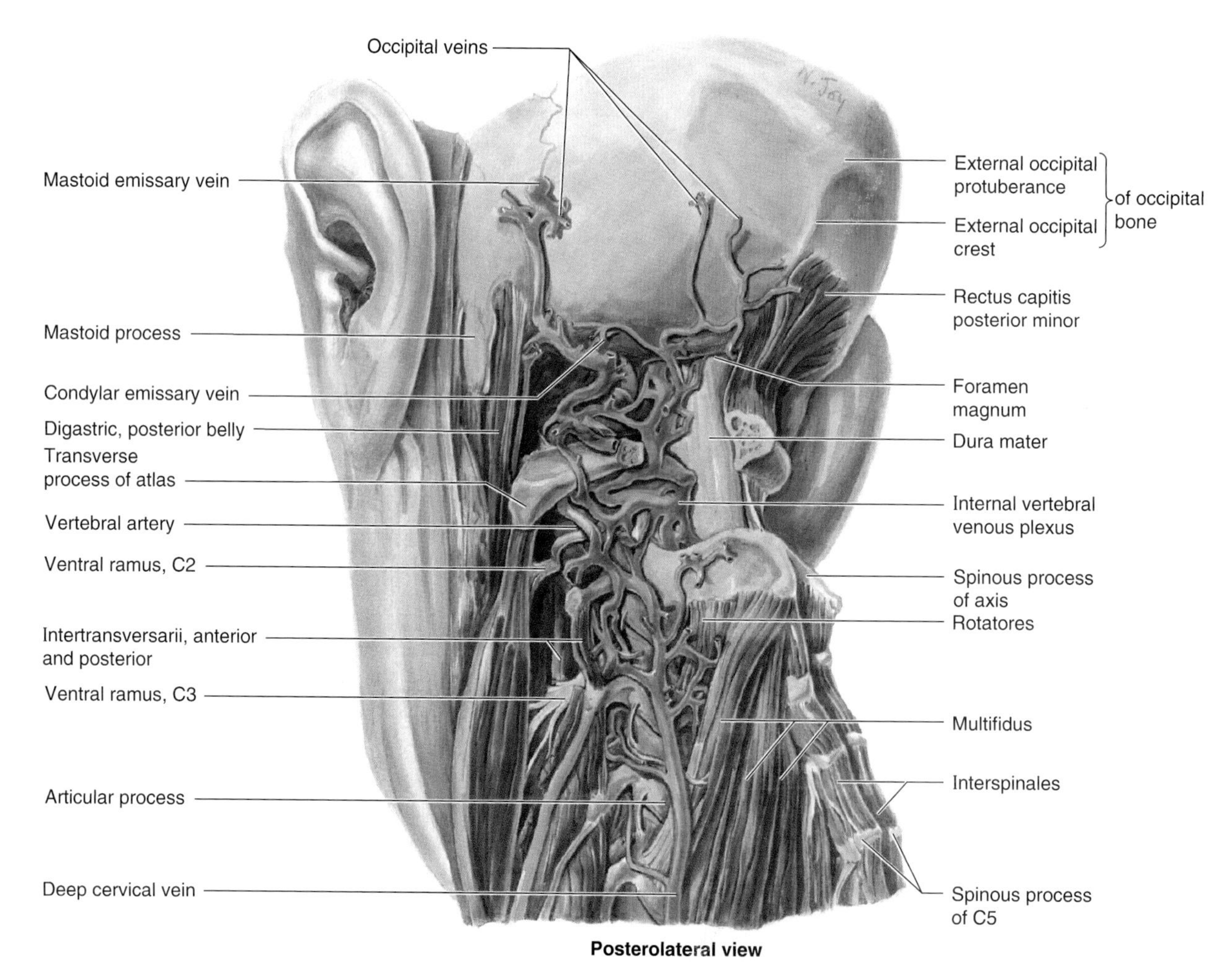

Figure 7.22. Deep dissection of the suboccipital region. Observe the vertebral venous system of veins and its numerous intercommunications and connections (e.g., through the foramen magnum and the mastoid foramen and condylar canal with the intracranial venous sinuses) between the laminae and through the intervertebral (IV) foramina, the internal vertebral (epidural) venous plexus; communicating with the veins of the scalp, with the veins around the vertebral artery and through the deep cervical vein, and with the brachiocephalic vein inferiorly.

may be paired bilaterally, passes through the parietal foramen in the calvaria, connecting the superior sagittal sinus with the veins external to it, particularly those in the scalp. A **mastoid emissary vein** (Fig. 7.22) passes through the mastoid foramen and connects each sigmoid sinus with the occipital or posterior auricular vein. A *posterior condylar emissary vein* may also be present and pass through the condylar canal, connecting the sigmoid sinus with the suboccipital plexus of veins.

Occlusion of Cerebral Veins and Dural Venous Sinuses

Occlusion of cerebral veins and dural venous sinuses may result from a thrombus, thrombophlebitis, or tumors (e.g., meningiomas). The dural sinuses most frequently thrombosed are the transverse, cavernous, and superior sagittal sinuses (Fishman, 1995a).

The facial veins make clinically important connections with the cavernous sinus through the superior ophthalmic veins (Fig. 7.18*B*). *Cavernous sinus thrombosis* usually results from infections in the orbit, nasal sinuses, and superior part of the face (the danger triangle [p. 869]). In patients with thrombophlebitis of the facial vein, pieces of an infected clot may extend into the cavernous sinus, producing *thrombophlebitis of the cavernous sinus.* The infection usually involves only one sinus initially but may spread to the opposite side through the intercavernous sinuses. Septic thrombosis of the cavernous sinus often results in the development of *acute meningitis.*

Metastasis of Tumor Cells to the Dural Sinuses

The basilar and occipital sinuses communicate through the foramen magnum with the *internal vertebral (epidural) venous plexuses* (Fig. 7.18*B*). Because these venous channels are valveless, compression of the thorax, abdomen, or pelvis—as occurs during heavy coughing and straining—may force venous blood from these regions into the internal vertebral venous system and from it into the dural venous sinuses. As a result, pus in abscesses and tumors in these regions may spread to the vertebrae and brain.

Fractures of the Cranial Base

In fractures of the cranial base, the internal carotid artery may tear within the cavernous sinus, producing an *arteriovenous fistula.* Arterial blood rushes into the cavernous sinus, enlarging it and forcing blood into the connecting veins, especially the ophthalmic veins. As a result, the eye protrudes (*exophthalmos*) and the conjunctiva becomes engorged (*chemosis*). The protruding eye pulsates in synchrony with the radial pulse, a phenomenon known as *pulsating exophthalmos.* Because CN III, CN IV, CN V_1, CN V_2, and CN VI lie in or close to the lateral wall of the cavernous sinus, these nerves may also be affected when the sinus is injured. ⊙

Vasculature of the Dura Mater

The **arteries of the dura** supply more blood to the calvaria than to the dura. The largest of the meningeal arteries, the **middle meningeal artery** (Fig. 7.17), is a branch of the maxillary artery. It enters the floor of the middle cranial fossa through the **foramen spinosum**, runs laterally in the fossa, and turns superoanteriorly on the greater wing of the sphenoid, where it divides into anterior and posterior branches. The **anterior branch of the middle meningeal artery** runs superiorly to the pterion and then curves posteriorly to ascend toward the vertex of the skull. A blow to the side of the head may fracture the bones forming the pterion and rupture branches of the middle meningeal artery. The **posterior branch of the middle meningeal artery** runs posterosuperiorly and ramifies over the posterior aspect of the skull. Small areas of dura are supplied by other arteries: meningeal branches of the ophthalmic arteries, branches of the occipital arteries, and small branches of the vertebral arteries.

The **veins of the dura** accompany the meningeal arteries, often in pairs (Fig. 7.17), and may also be torn in fractures of the calvaria. The **middle meningeal veins** accompany the middle meningeal artery, leave the cranial cavity through the foramen spinosum or foramen ovale, and drain into the **pterygoid venous plexus** (Fig. 7.18*B*).

Nerve Supply of the Dura Mater

The nerve supply of the dura of the anterior and middle cranial fossa dura is largely from branches of the trigeminal nerve. The *anterior meningeal branches* of the **ethmoidal nerves** (CN V_1) and the meningeal branches of the **maxillary** and **mandibular nerves** supply the dura of the anterior cranial fossa (Fig. 7.23). The dura of the middle cranial fossa is supplied mainly by branches of CN V_2 and CN V_3. The dura of the posterior cranial fossa is supplied by the **tentorial nerve** (a branch of the ophthalmic nerve) and by sensory branches from dorsal roots of C1 through C3 nerves and may receive some innervation from the vagus nerve (CN X). Sensory endings are more numerous in the dura along each side of the superior sagittal sinus and in the cerebellar tentorium than they are in the floor of the cranium. Pain fibers are numerous where arteries and veins course in the dura.

Dural Origin of Headaches

The dura is sensitive to pain, especially where it is related to the dural venous sinuses and meningeal arteries. Consequently, pulling on arteries at the base of the skull or veins near the vertex where they pierce the dura causes pain. Although the causes of headache are numerous (Raskin, 1995), distension of the scalp or meningeal vessels (or both) is believed to be one cause. Many headaches appear to be dural in origin, such as the headache occurring after a *lumbar spinal puncture for removal of CSF* (see Chapter 4) that is thought to result from stimulation of sensory nerve endings in the dura. When CSF is removed, the brain sags slightly, pulling on the dura; this may cause pain and headache. For this reason, patients are asked to keep their heads down after a lumbar puncture to minimize or prevent headaches. ⊙

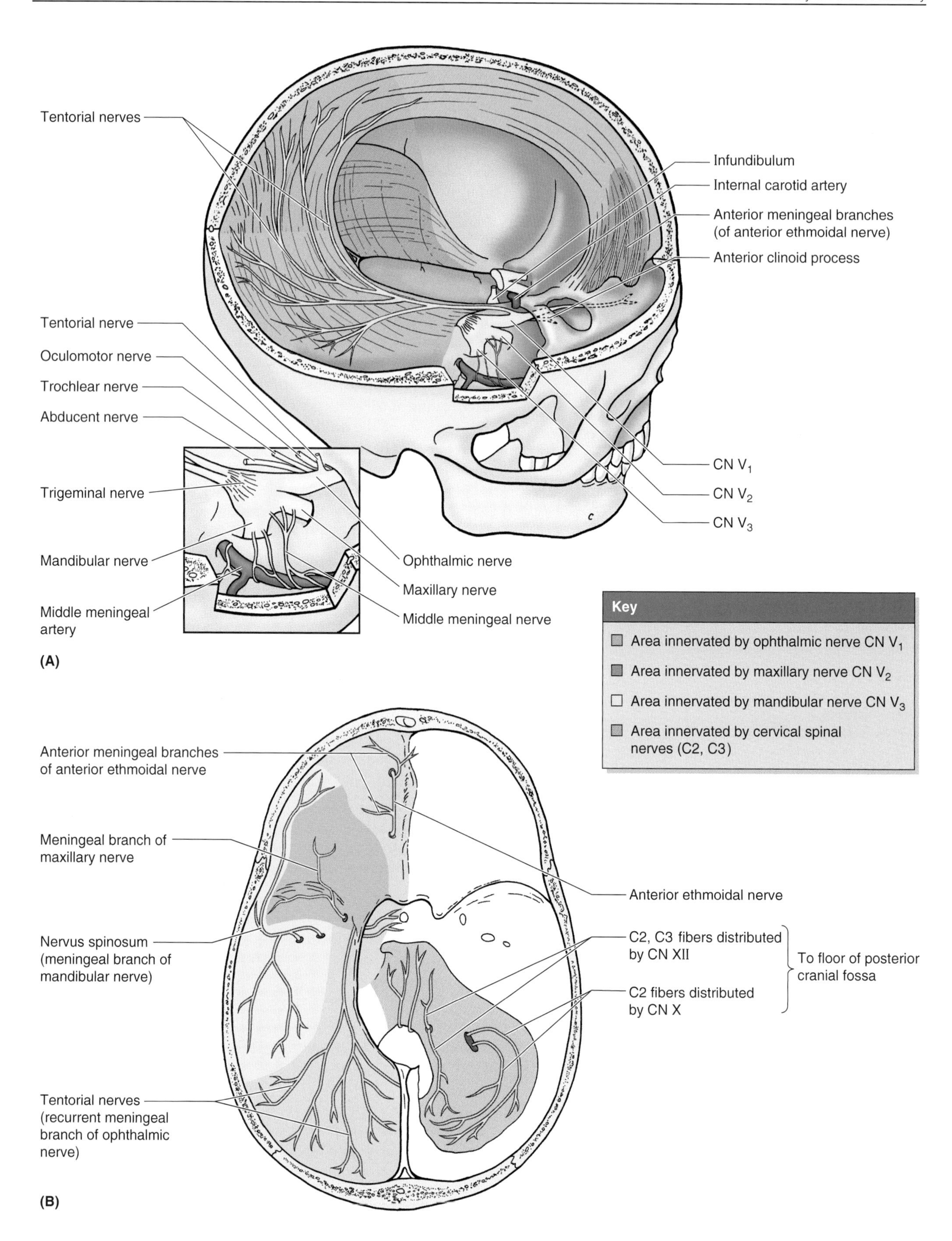
Tentorial nerves
Infundibulum
Internal carotid artery
Anterior meningeal branches (of anterior ethmoidal nerve)
Anterior clinoid process
Tentorial nerve
Oculomotor nerve
Trochlear nerve
Abducent nerve
Trigeminal nerve
Mandibular nerve
Middle meningeal artery
CN V_1
CN V_2
CN V_3
Ophthalmic nerve
Maxillary nerve
Middle meningeal nerve
(A)
Key
Area innervated by ophthalmic nerve CN V_1
Area innervated by maxillary nerve CN V_2
Area innervated by mandibular nerve CN V_3
Area innervated by cervical spinal nerves (C2, C3)
Anterior meningeal branches of anterior ethmoidal nerve
Meningeal branch of maxillary nerve
Anterior ethmoidal nerve
Nervus spinosum (meningeal branch of mandibular nerve)
C2, C3 fibers distributed by CN XII
C2 fibers distributed by CN X
To floor of posterior cranial fossa
Tentorial nerves (recurrent meningeal branch of ophthalmic nerve)
(B)

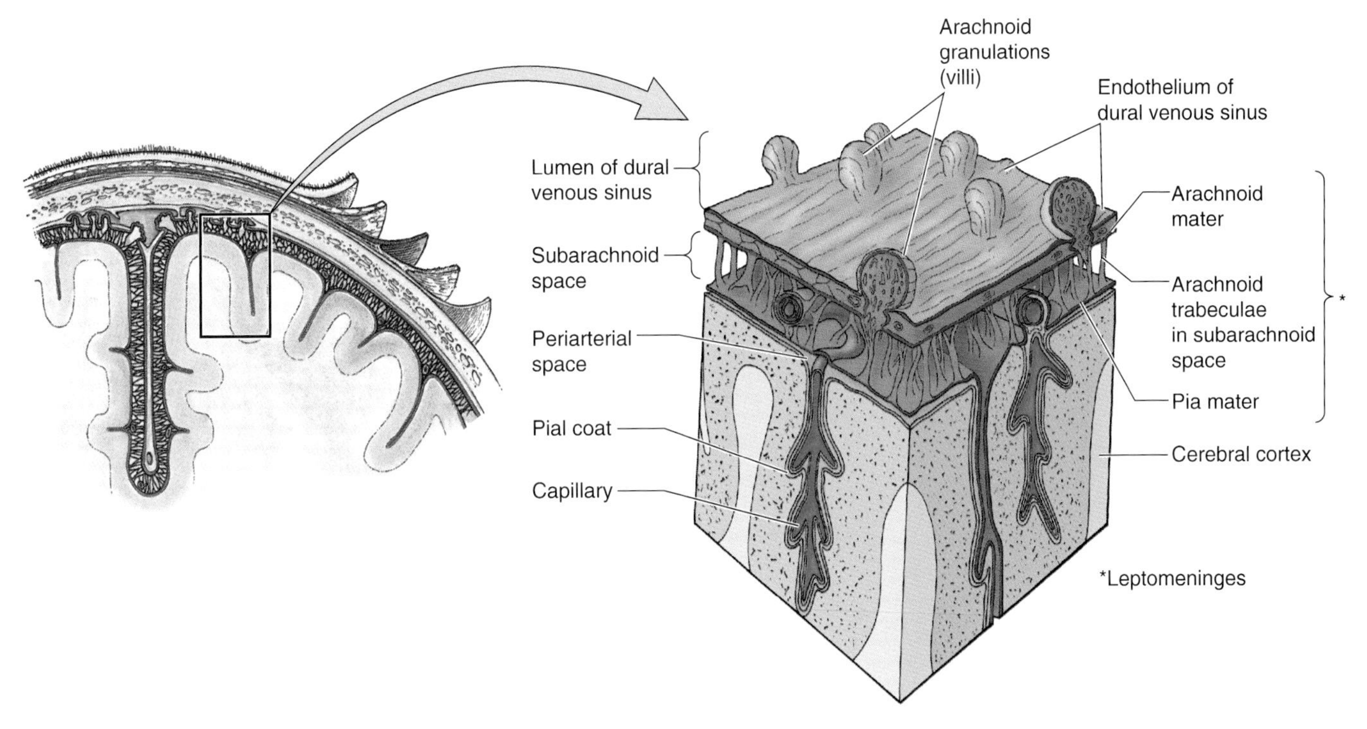

Figure 7.24. Coronal section of the vertex of the skull illustrating the meninges and subarachnoid (leptomeningeal) space. Observe that the subarachnoid space separates the arachnoid mater and pia mater. The main site of absorption of cerebrospinal fluid (CSF) into the venous system is through the arachnoid granulations (villi) projecting into the dural venous sinuses, especially the superior sagittal sinus and adjacent venous lacunae (see Fig. 7.17).

Pia-Arachnoid

The pia-arachnoid (combined pia mater and arachnoid mater) develops from a single layer of mesenchyme surrounding the embryonic brain (Moore and Persaud, 1998). Even in the adult, the arachnoid and pia are actually the "visceral" and parietal parts of the same layer, the *leptomeninges* (G. slender membranes). Fluid-filled spaces form within this layer and coalesce to form the **subarachnoid space** (Fig. 7.24). The derivation of the pia-arachnoid from a single embryonic layer is evident in the adult by the numerous weblike **arachnoid trabeculae** passing between the arachnoid and pia, which give the arachnoid its name (G. resembling a spider's web). The trabeculae are composed of flattened, irregularly shaped fibroblasts that bridge the subarachnoid space (Haines, 1997). The pia and arachnoid are in continuity immediately proximal to the nerve's exit from the dura mater.

The **arachnoid mater** contains fibroblasts, collagen fibers, and some elastic fibers. Although thin, the arachnoid is thick enough to be manipulated with forceps. The avascular arachnoid, closely applied to the meningeal layer of the dura, is not attached to the dura; it is held against the inner surface of the dura by the pressure of the CSF.

The **pia mater** is an even thinner membrane that is highly vascularized by a network of fine blood vessels. The pia is difficult to see but it gives the surface of the brain a shiny appearance. The pia adheres to the surface of the brain and follows all its contours. When the cerebral arteries penetrate the cerebral cortex, the pia follows them for a short distance, forming a **pial coat** and a periarterial space.

Figure 7.23. Innervation of the dura mater. A. The right side of the calvaria (skullcap) and brain is removed, and the trigeminal nerve (CN V) is dissected. Observe that the dura of the anterior cranial fossa is supplied by the ophthalmic nerve (CN V_1) through meningeal branches of the ethmoidal nerves. The dura of the middle cranial fossa is innervated by meningeal branches of the maxillary (CN V_2) and mandibular (CN V_3) nerves. The dura of the roof of the posterior cranial fossa (tentorium cerebelli) is supplied by tentorial branches of the ophthalmic nerve. Observe that autonomic nerve fibers pass to blood vessels (e.g., the middle meningeal artery). **B.** Superior view of the internal aspect of the base of the skull, showing that the dura is supplied by branches of the trigeminal and cervical nerves. Observe that the dura on the floor of the posterior cranial fossa receives meningeal branches of the vagus (CN X) and hypoglossal (CN XII) nerves, both of which contain cervical spinal fibers (C2 and C3).

Meningeal Spaces

Three meningeal spaces are related to the cranial meninges:

- *The dura-skull interface (extradural* or *epidural "space")* is normally not an actual space but only a potential one between the cranial bones and the external periosteal layer of the dura because the dura is attached to the bones. It becomes a real space only pathologically, for example, when blood from torn meningeal vessels pushes the periosteum from the skull and accumulates.
- *The dura-arachnoid junction or interface ("subdural space")* is likewise normally only a potential space that may develop in the dural border cell layer of the dura after a blow to the head (Haines, 1997).
- *The subarachnoid space,* between the arachnoid and pia, is an actual space that contains CSF, trabecular cells, arteries, and veins.

Although it is commonly stated that the brain "floats" in CSF, the brain is suspended in the CSF-filled subarachnoid space by the arachnoid trabeculae.

Leptomeningitis

Leptomeningitis is an inflammation of the leptomeninges resulting from pathogenic micro-organisms. The infection and inflammation are usually confined to the subarachnoid space and the pia-arachnoid (Miller and Jubelt, 1995). The bacteria may enter the subarachnoid space through the blood (*septicemia*—"blood poisoning") or spread from an infection of the heart, lungs, or other viscera. Micro-organisms may also enter the subarachnoid space from a compound skull fracture or a fracture through the nasal sinuses. *Acute purulent meningitis* can result from infection with almost any pathogenic bacteria (e.g., *meningococcal meningitis*). Inflammation of the meninges consisting of or associated with pus is called purulent meningitis.

Head Injuries and Intracranial Hemorrhage

Extradural or *epidural hemorrhage* is arterial in origin. Blood from torn branches of a middle meningeal artery collects between the external periosteal layer of the dura and the calvaria—usually following a blow to the head—and forms an *extradural* or *epidural hematoma.* Typically, a brief *concussion* (loss of consciousness) occurs, followed by a lucid interval of some hours. Later, drowsiness and coma (profound unconsciousness) occur. Compression of the brain occurs as the blood mass increases, necessitating evacuation of the blood and occlusion of the bleeding vessels.

A dural border hematoma is classically called a *subdural hematoma*; however, this is a misnomer because there is no naturally occurring space at the dura-arachnoid junction. "Hematomas at this junction are usually caused by extravasated blood that splits open the dural border cell layer. This blood does not collect within preexisting space, but rather creates a space at the dura-arachnoid junction" (Haines, 1997). *Dural border hemorrhage* usually follows a blow to the head that jerks the brain inside the skull and injures it. The precipitating trauma may be trivial or forgotten. Displacement of the brain is greatest in elderly people in whom some shrinkage of the brain has occurred. *Dural border hemorrhage is typically venous in origin and commonly results from tearing of a cerebral vein* (Fig. 7.18*B*) as it enters the superior sagittal sinus (Haines, 1991). Although the dura and arachnoid are normally adjacent and are usually encountered as two surfaces of a single membrane, blood may collect in the abnormal space that forms when trauma separates them.

Subarachnoid hemorrhage is an extravasation (escape) ▶

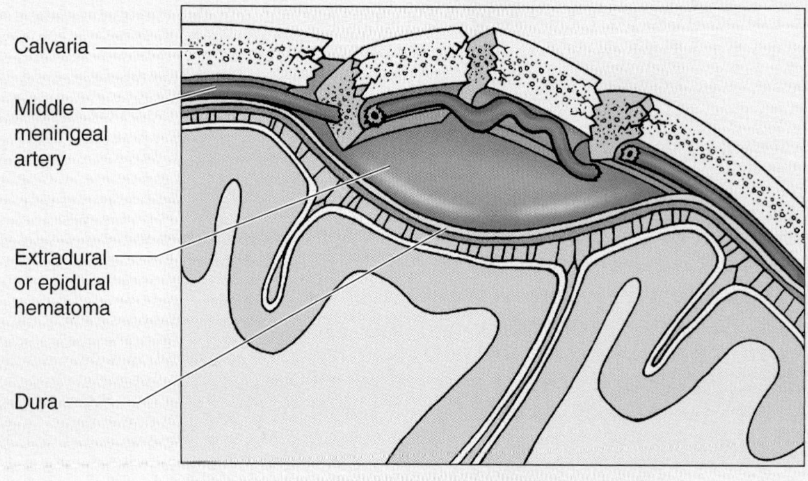

Extradural or epidural hemorrhage

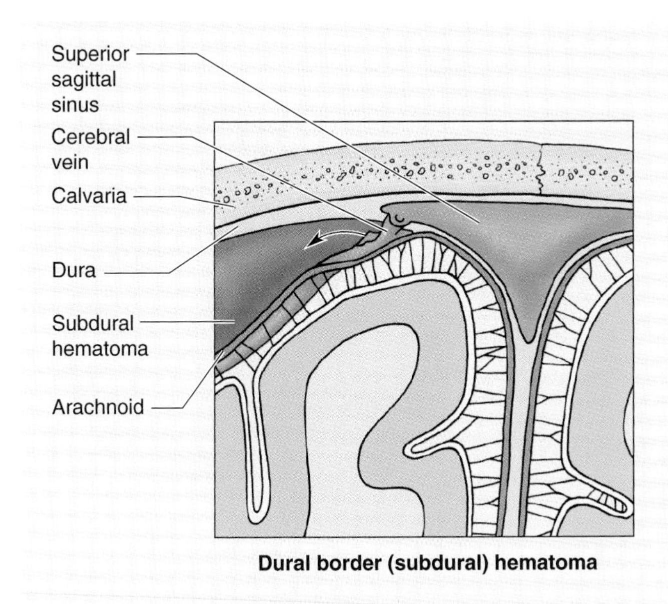

Dural border (subdural) hematoma

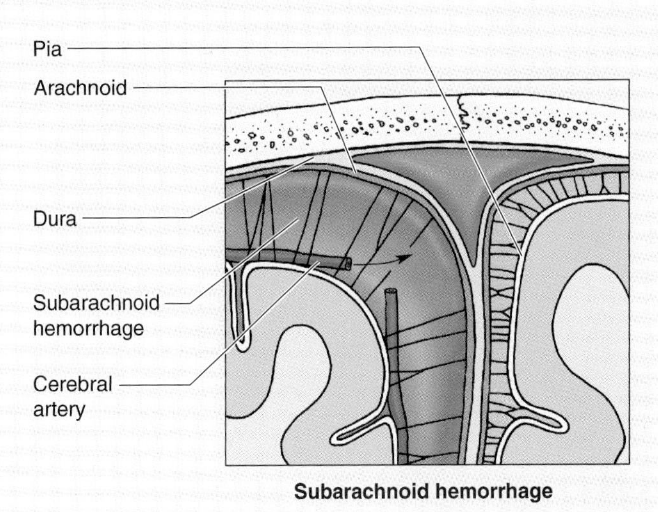

Subarachnoid hemorrhage

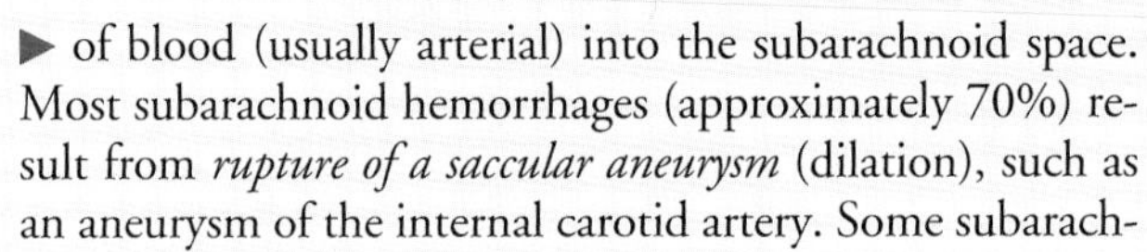

▶ of blood (usually arterial) into the subarachnoid space. Most subarachnoid hemorrhages (approximately 70%) result from *rupture of a saccular aneurysm* (dilation), such as an aneurysm of the internal carotid artery. Some subarachnoid hemorrhages are associated with head trauma involving skull fractures and cerebral lacerations. Bleeding into the subarachnoid space results in meningeal irritation, a severe headache, stiff neck, and often loss of consciousness. ⭘

Brain

The brain, composed of the *cerebrum*, *cerebellum*, and *brainstem* (midbrain, pons, and medulla oblongata), is lodged in the cranial cavity—the space within the skull occupied by the brain, meninges, and CSF. The roof of the cranial cavity is formed by the calvaria, and its floor is formed by the cranial base. The following brief discussion of the gross structure of the brain shows how the brain relates to the cranium, cranial nerves, meninges, and CSF.

Parts of the Brain

When the calvaria and dura are removed, *gyri* (folds), *sulci* (grooves), and *fissures* of the cerebral cortex are visible through the delicate pia-arachnoid layer (Fig. 7.25). The fissures and sulci of the brain are distinctive landmarks that subdivide the cerebral hemispheres into smaller areas (lobes and gyri). The **cerebrum** (L. brain)—the principal part of the brain—includes the cerebral hemispheres and diencephalon but not the brainstem (medulla, pons, and midbrain).

- The **cerebral hemispheres** form the largest part of the brain, occupying the anterior and middle cranial fossae and extending posteriorly over the cerebellar tentorium and cerebellum (i.e., the entire supratentorial cranial cavity). The cavity in each cerebral hemisphere, a *lateral ventricle*, is part of the ventricular system of the brain (Fig. 7.26).
- The **diencephalon**—composed of the epithalamus, dorsal thalamus, and hypothalamus—forms the central core of the brain and *surrounds the 3rd ventricle* (Fig. 7.25*B*); the cavity between the right and left halves of the diencephalon forms this narrow ventricle.
- The **midbrain** (mesencephalon)—the rostral part of the brainstem—lies at the junction of the middle and posterior cranial fossae; the cavity of the midbrain forms a narrow canal—the **cerebral aqueduct**—that conducts CSF from the lateral and 3rd ventricles to the 4th ventricle.
- The **pons**—the part of the brainstem between the midbrain rostrally and the medulla oblongata caudally—lies in the anterior part of the posterior cranial fossa; the cavity in the pons forms the superior part of the 4th ventricle.
- The **medulla oblongata**—the most caudal subdivision of the brainstem that is continuous with the spinal cord—lies in the posterior cranial fossa; the cavity of the medulla forms the inferior part of the 4th ventricle.
- The **cerebellum**—the large brain mass lying dorsal to the pons and medulla and ventral to the posterior part of the cerebrum—lies beneath the cerebellar tentorium in the posterior cranial fossa; it consists of two lateral hemispheres united by a narrow middle part, the vermis.

Eleven of twelve cranial nerves arise from the brain; they all exit the cranial cavity. They have motor, parasympathetic, and/or sensory functions. Generally, these nerves are surrounded by a dural sheath as they leave the cranium; the dural sheath becomes continuous with the connective tissue of the epineurium. For a summary of the cranial nerves, see Chapter 9.

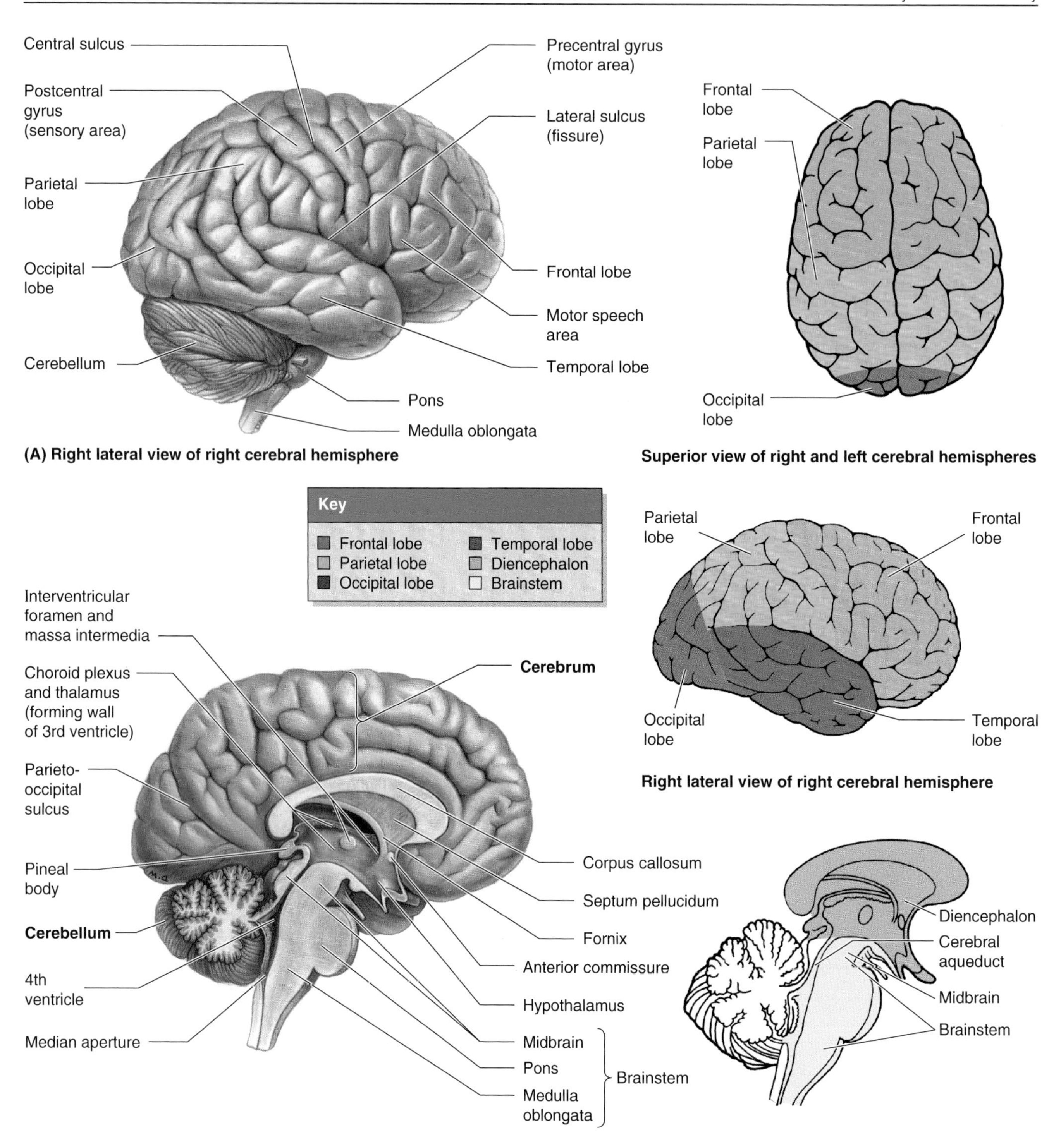

Figure 7.25. Structure of the brain. A. Right lateral and superior views of the surface of the brain. Observe the gyri (folds) and sulci (grooves) of the frontal, parietal, and occipital lobes. **B.** Left median section of the brain. Again notice that the cerebral cortex consists of gyri and sulci. Observe the parts of the brainstem and the ventricles of the brain. On the right, observe the lobes of the brain and the diencephalon, midbrain, and brainstem. Examine the cerebral aqueduct (aqueduct of midbrain) connecting the 3rd and 4th ventricles.

Cerebral Injuries

Cerebral concussion is an abrupt, brief loss of consciousness immediately after head injury. Consciousness may be lost for only 8 to 10 seconds, as occurs in a knockout during boxing. With a more severe injury, such as that in an automobile accident, consciousness may be lost for hours and even days. If a patient recovers consciousness within 6 hours, the long-term outcome is excellent (Rowland, 1995). If the coma lasts longer than 6 hours, brain tissue injury usually results. Professional boxers are especially at risk for *chronic traumatic encephalopathy—*the *punchdrunk syndrome—*a brain injury characterized by weakness in the lower limbs, unsteady gait, slowness of muscular movements, tremors of the hands, hesitancy of speech, and slow cerebration (use of one's brain). The injuries result from acceleration and deceleration of the head that shears or stretches axons (*diffuse axonal injury*). The sudden stopping of the moving head results in the brain hitting the suddenly stationary skull.

Cerebral contusion (bruising) results from brain trauma in which the pia is stripped from the injured surface of the brain and may be torn, allowing blood to enter the subarachnoid space. The bruising results either from the sudden impact of the still-moving brain against the suddenly stationary skull, or from the suddenly moving skull against the still-stationary brain. Cerebral contusion may result in an extended loss of consciousness.

Cerebral lacerations (tearing of neural tissue) are often associated with depressed skull fractures or gunshot wounds. Lacerations result in rupture of blood vessels and bleeding into the brain and subarachnoid space, increasing intracranial pressure, and cerebral compression. *Cerebral compression may be produced by*:

- Intracranial collections of blood
- Obstruction of CSF circulation or absorption of CSF
- Intracranial tumors or abscesses
- Edema of the brain, such as swelling associated with a head injury.

Brain swelling resulting from a head injury may be caused in part by cerebral edema. *Brain edema* is defined as an increase in brain volume resulting from an increase in water and sodium content (Fishman, 1995b). Brain edema accompanies a wide variety of pathological processes. ⊙

Ventricular System of the Brain

The ventricular system of the brain consists of *two lateral ventricles* and the midline *3rd and 4th ventricles* connected by the *cerebral aqueduct* (Fig. 7.26). CSF—largely secreted by the **choroid plexuses** of the ventricles—fills these cavities and the subarachnoid spaces of the brain and spinal cord. The ventricles, the choroid plexuses, and the CSF produced by the plexuses are essential elements in the normal function of the brain (Corbett et al., 1997).

Ventricles of the Brain

The **lateral ventricles**—the 1st and 2nd ventricles—-are the largest cavities of the ventricular system and occupy large areas of the cerebral hemispheres. Each lateral ventricle opens through an **interventricular foramen** into the 3rd ventricle. The **3rd ventricle**—a slitlike cavity between the right and left halves of the diencephalon—is continuous posteroinferiorly with the **cerebral aqueduct** (aqueduct of midbrain), a narrow channel in the midbrain connecting the 3rd and 4th ventricles (Fig. 7.26*B*). The **4th ventricle** in the posterior part of the pons and medulla extends inferoposteriorly. It is continuous through the central canal in the inferior part of the medulla with the **central canal in the spinal cord** (Fig. 7.26*A*). CSF drains from the 4th ventricle through a single **median aperture** and paired **lateral apertures** into the subarachnoid space. These apertures are the only means by which CSF enters the subarachnoid space. If they are blocked, CSF accumulates and the ventricles distend, producing compression of the cerebral hemispheres. For further information about the ventricles of the brain, see Corbett et al. (1997).

Subarachnoid Cisterns

In certain places, primarily at the base of the brain, the arachnoid and pia are widely separated by **subarachnoid cisterns** (Fig. 7.26*B*). These *large pools of CSF* contain arteries, veins, and the roots of cranial nerves in some cases (e.g., the **chiasmatic cistern** contains the **optic chiasm** and the roots of the optic nerves [CN II]). The cisterns are usually named according to the structures related to them. *Major subarachnoid cisterns include the:*

- **Posterior cerebellomedullary cistern** (L. cisterna magna)—the largest of the subarachnoid cisterns—is between the cerebellum and the medulla and receives CSF from the apertures of the 4th ventricle.
- **Pontine cistern** (prepontine cistern)—an extensive space ventral to the pons—is continuous inferiorly with the spinal subarachnoid space.
- **Interpeduncular cistern** (basal cistern)—between the cerebral peduncles of the midbrain and structures of the interpeduncular fossa—contains the *cerebral arterial circle* (of Willis [Fig. 7.2]).
- **Chiasmatic cistern** (cistern of optic chiasm)—inferior and anterior to *optic chiasm—*the point of crossing or decussation of optic nerve fibers.
- **Quadrigeminal cistern** (cistern of the great cerebral vein, superior cistern)—between the posterior part of the corpus callosum and the superior surface of the cerebellum.

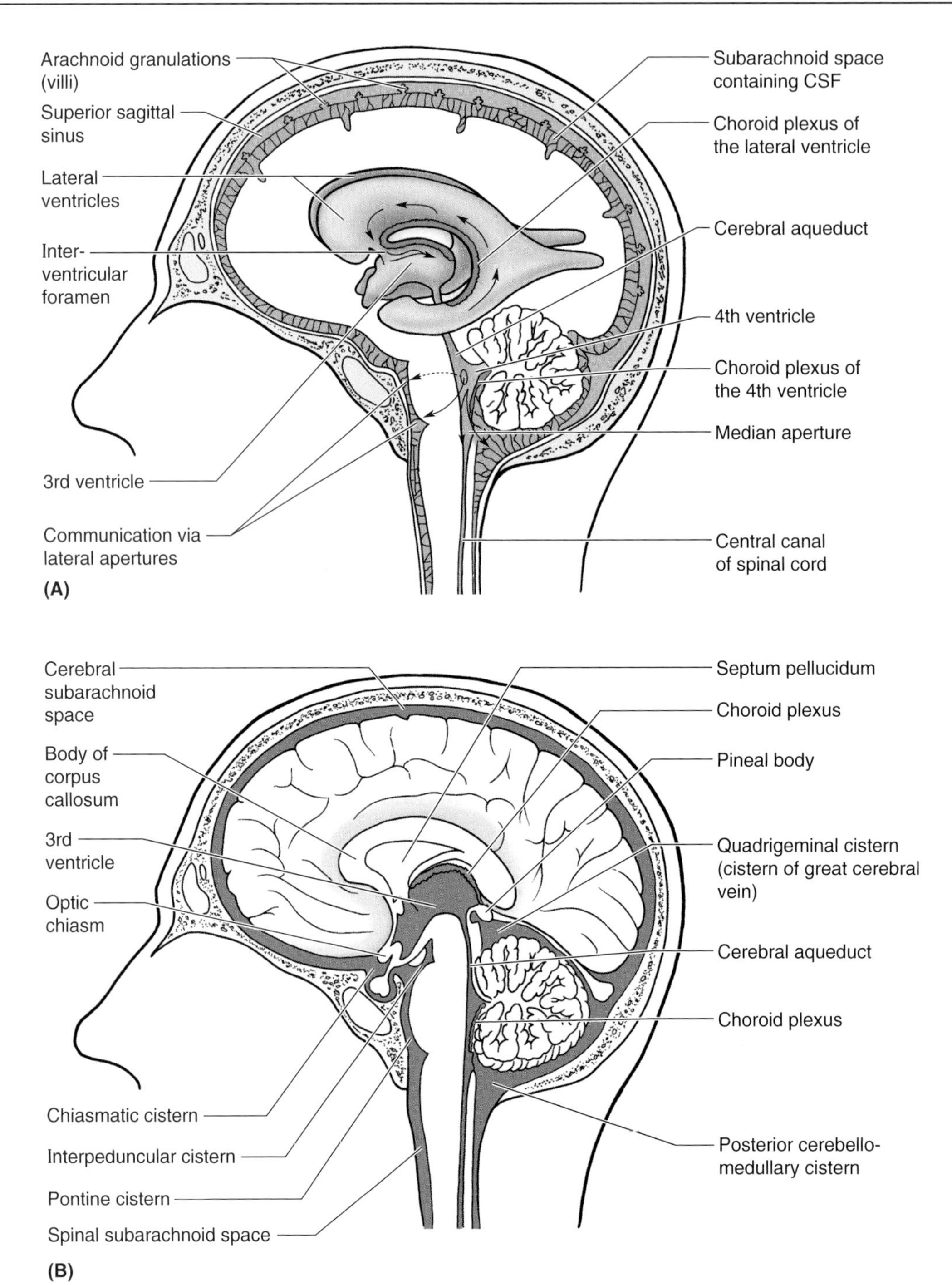

Figure 7.26. Subarachnoid (leptomeningeal) spaces, ventricles, and subarachnoid cisterns. Diagrammatic views of median sections of the head. **A.** Ventricular system and circulation of cerebrospinal fluid (CSF). The production of CSF is mainly by the choroid plexuses of the lateral, 3rd, and 4th ventricles. The plexuses in the lateral ventricles are the largest and most important. The main site of absorption of CSF into the venous system is through the arachnoid granulations projecting into the dural venous sinuses. **B.** Observe the subarachnoid cisterns—expanded regions of the subarachnoid space—that contain more substantial amounts of CSF. The posterior cerebellomedullary cistern (L. cisterna magna) is clinically important as a site for cisternal puncture (p. 891).

For a description of other subarachnoid cisterns, see Haines (1997).

Secretion of Cerebrospinal Fluid

CSF is secreted by choroidal epithelial cells (modified ependymal cells) of the **choroid plexuses** in the lateral, 3rd, and 4th ventricles (Figs. 7.25*B* and 7.26); 400 to 500 mL of CSF are secreted daily by the choroid plexuses (Corbett et al., 1997). The choroid plexuses consist of vascular fringes of pia mater (tela choroidea) covered by cuboidal epithelial cells and are invaginated into the roofs of the 3rd and 4th ventricles and on the floors of the bodies and inferior horns of the lateral ventricles.

Circulation of Cerebrospinal Fluid

CSF leaves the lateral ventricles through the **interventricular foramina** (Fig. 7.26*A*) and enters the 3rd ventricle. From it, CSF passes through the **cerebral aqueduct** into the 4th ventricle. It leaves this ventricle through its **median** and **lateral apertures** and enters the subarachnoid space, which is continuous around the spinal cord and posterosuperiorly over the cerebellum. However, most CSF flows into the interpeduncular and quadrigeminal cisterns. CSF from the various subarachnoid cisterns flows superiorly through the sulci and fissures on the medial and superolateral surfaces of the cerebral hemispheres. CSF also passes into the extensions of the subarachnoid space around the cranial nerves, the most important of which are those surrounding the optic nerves. For further information on the movement of CSF through the ventricular system and the subarachnoid space, see Corbett et al. (1997).

Absorption of Cerebrospinal Fluid

The main site of CSF absorption into the venous system is through the **arachnoid granulations** (Figs. 7.24 and 7.26*A*)—tiny *tuftlike protrusions of arachnoid villi* into the walls of the dural venous sinuses, especially the superior sagittal sinus and its lateral lacunae (Fig. 7.17). The subarachnoid space containing CSF extends into the cores of the arachnoid granulations. CSF enters the venous system through two routes (Corbett et al., 1997):

- Most CSF enters the venous system by transport through the cells of the arachnoid granulations into the dural venous sinuses
- Some CSF moves between the cells making up the arachnoid granulations.

Approximately 330 to 380 mL of CSF enters the venous circulation each day.

Functions of Cerebrospinal Fluid

Along with the meninges and calvaria, *CSF protects the brain* by providing a cushion against blows to the head. The CSF in the subarachnoid space provides the buoyancy necessary for preventing the weight of the brain from compressing the cranial nerve roots and blood vessels against the internal surface of the skull (Corbett et al., 1997). Because the brain is slightly heavier than the CSF, the gyri on the basal surface of the brain are in contact with the cranial fossae in the floor of the cranial cavity when a person is standing erect. In many places at the base of the brain, only the cranial meninges intervene between the brain and the cranial bones. In the erect position the CSF is in the subarachnoid cisterns and sulci on the superior and lateral parts of the brain; therefore, CSF and dura normally separate the superior part of the brain from the calvaria.

Small, rapidly recurring changes take place in *intracranial pressure* owing to the beating heart; slow recurring changes result from unknown causes. Momentarily large changes in pressure occur during coughing and straining and during changes in position (erect vs. supine). Any change in the volume of the intracranial contents (e.g., a brain tumor, an accumulation of ventricular fluid because of blockage of the cerebral aqueduct, or blood from a ruptured aneurysm) will be reflected by a change in intracranial pressure. This rule is called the *Monro-Kellie doctrine*, which states that the cranial cavity is like a closed rigid box and that a change in the quantity of intracranial blood can occur only through the displacement or replacement of CSF.

Cisternal Puncture

CSF may be obtained from the posterior cerebellomedullary cistern through a *cisternal puncture*. The needle is carefully inserted through the posterior atlanto-occipital membrane into the cistern (see Chapter 4). The subarachnoid space or the ventricular system may also be entered for measuring or monitoring CSF pressure, injecting antibiotics, or administering contrast media for medical imaging. The cerebellomedullary cistern is the site of choice in infants and young children; the lumbar cistern is used most frequently in adults (see Chapter 4).

Hydrocephalus

Overproduction of CSF, obstruction of CSF flow, or interference with CSF absorption results in excess fluid in the ventricles and enlargement of the head—*obstructive hydrocephalus* (*A* [p. 892]). The excess CSF dilates the ventricles, thins the cerebral cortex, and separates the bones of the calvaria in infants. Although an obstruction can occur any place, *the blockage usually occurs in the cerebral aqueduct* (*B*) or at the interventricular foramina (Moore and Persaud, 1998). *Aqueductal stenosis* (narrow aqueduct) may be caused by a nearby tumor in the midbrain or by cellular debris following intraventricular hemorrhage or bacterial and fungal infections of the CNS (Corbett et al., 1997).

Blockage of CSF circulation results in dilation of the ventricles superior to the point of obstruction and in pressure on the cerebral hemispheres. This condition squeezes the brain between the ventricular fluid and the bones of the calvaria. In infants the internal pressure results in expansion of the brain and calvaria because the ▶

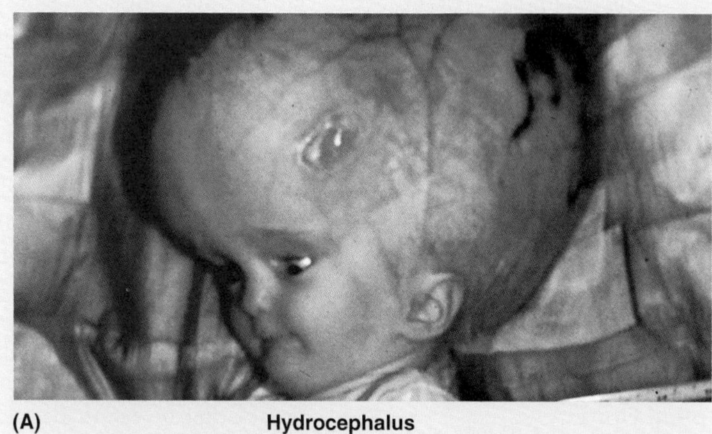

(A) Hydrocephalus

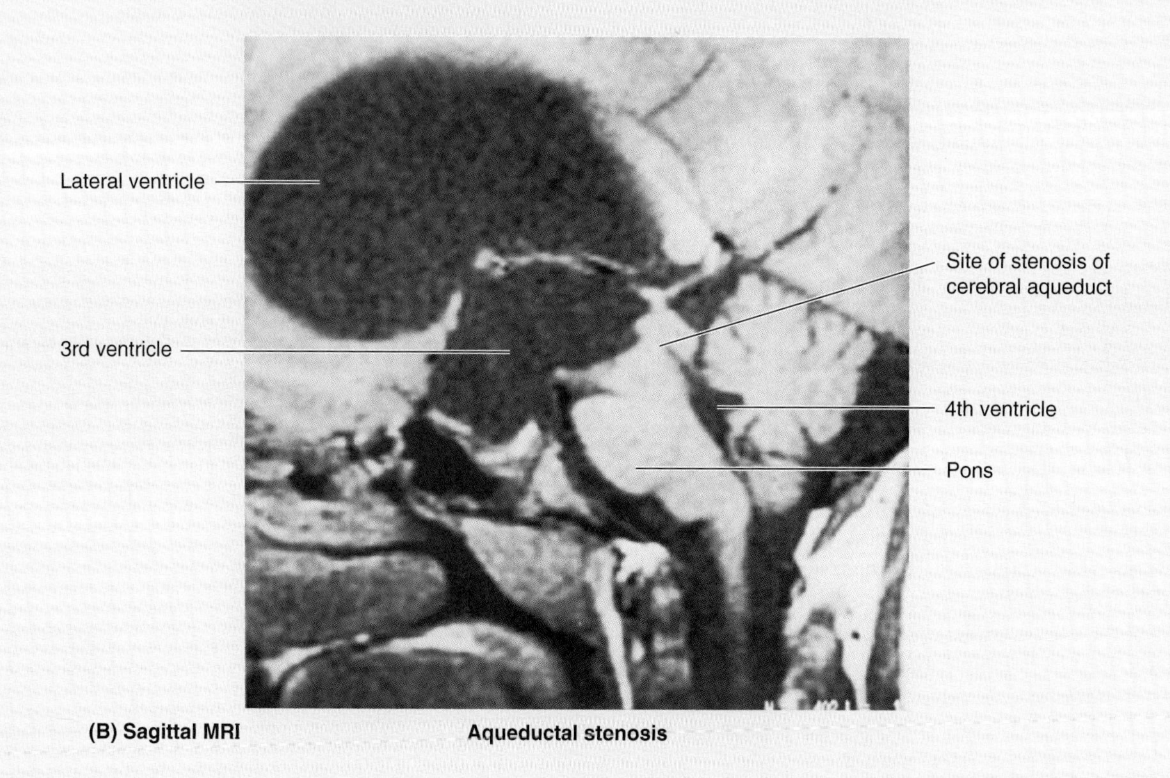

(B) Sagittal MRI Aqueductal stenosis

▶ sutures and fontanelles are still open. It is possible to produce an artificial drainage system to bypass the blockage and allow CSF to escape, thereby lessening damage to the brain.

In *communicating hydrocephalus,* the flow of CSF through the ventricles and into the subarachnoid space is not impaired; however, movement of CSF from this space into the venous system is partly or completely blocked. The blockage may be caused by the congenital absence of arachnoid granulations or the villi in granulations may be blocked by red blood cells as the result of a subarachnoid hemorrhage (Corbett et al., 1997).

Leakage of Cerebrospinal Fluid

Fractures in the floor of the middle cranial fossa may result in CSF leakage from the ear (*CSF otorrhea*) if the meninges superior to the middle ear are torn and the tympanic membrane is ruptured. Fractures in the floor of the anterior cranial fossa may involve the cribriform plate of the ethmoid, resulting in CSF leakage through the nose (*CSF rhinorrhea*). CSF otorrhea and rhinorrhea may be the primary indication of a skull base fracture and present the risk of *meningitis* because an infection could spread to the meninges of the brain from the ear or nose. ⊙

Blood Supply of the Brain

The blood supply to the brain derives from the internal carotid and vertebral arteries (Figs. 7.27 and 7.28, Table 7.7), which lie in the subarachnoid space.

Internal Carotid Arteries

The internal carotid arteries arise in the neck from the common carotid arteries, ascend vertically to the base of the skull, and enter the cranial cavity through the **carotid canals** in the temporal bones (Figs. 7.27 and 7.29). In addition to the carotid arteries, the carotid canals contain venous plexuses and *carotid plexuses of sympathetic nerves.* Within the carotid canals, the carotid arteries change direction, passing anteriorly and medially; they then run forward through the *cavernous sinus,* lying in the carotid groove on the side of the body of the sphenoid (Fig. 7.20). The terminal branches of the internal carotids are the **anterior** and **middle cerebral arteries.** Clinically, the internal carotids and their branches are often referred to as the *anterior circulation of the brain.* The **anterior cerebral arteries** are connected by the **anterior communicating artery.** Near their termination the internal carotid arteries are joined to the posterior cerebral arteries by

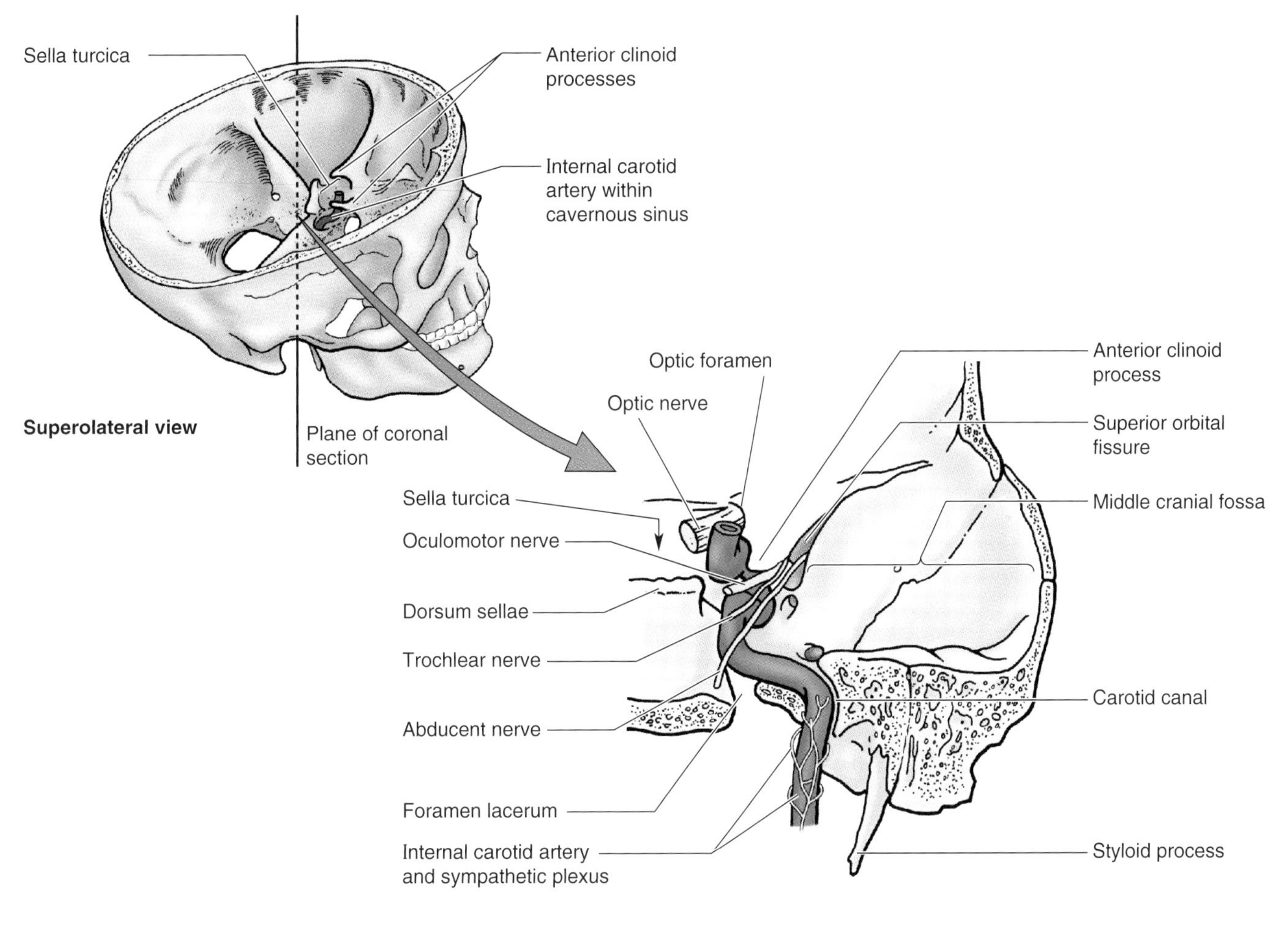

Figure 7.27. Course of the internal carotid artery. The orientation drawing shows the plane of the coronal section through the carotid canal in the lower figure, which shows the internal carotid artery in the carotid canal and the region of the cavernous sinus. The artery ascends vertically in the neck (cervical part) to the entrance of the carotid canal in the petrous (rocklike) temporal bone. In the carotid canal, the artery turns horizontally and medially toward the apex of the petrous bone (petrous part). It emerges from the canal superior to the foramen lacerum—which is closed in life by a cartilage plate—and enters the cranial cavity where it takes an S-shaped course. The artery runs anteriorly across the cartilage of the foramen lacerum and then along the carotid grooves on the lateral side of the body of the sphenoid, coursing through the cavernous sinus (cavernous part). Inferior to the anterior clinoid process, the artery makes a 180° turn, heading posteriorly to join the cerebral arterial circle (of Willis—cerebral part).

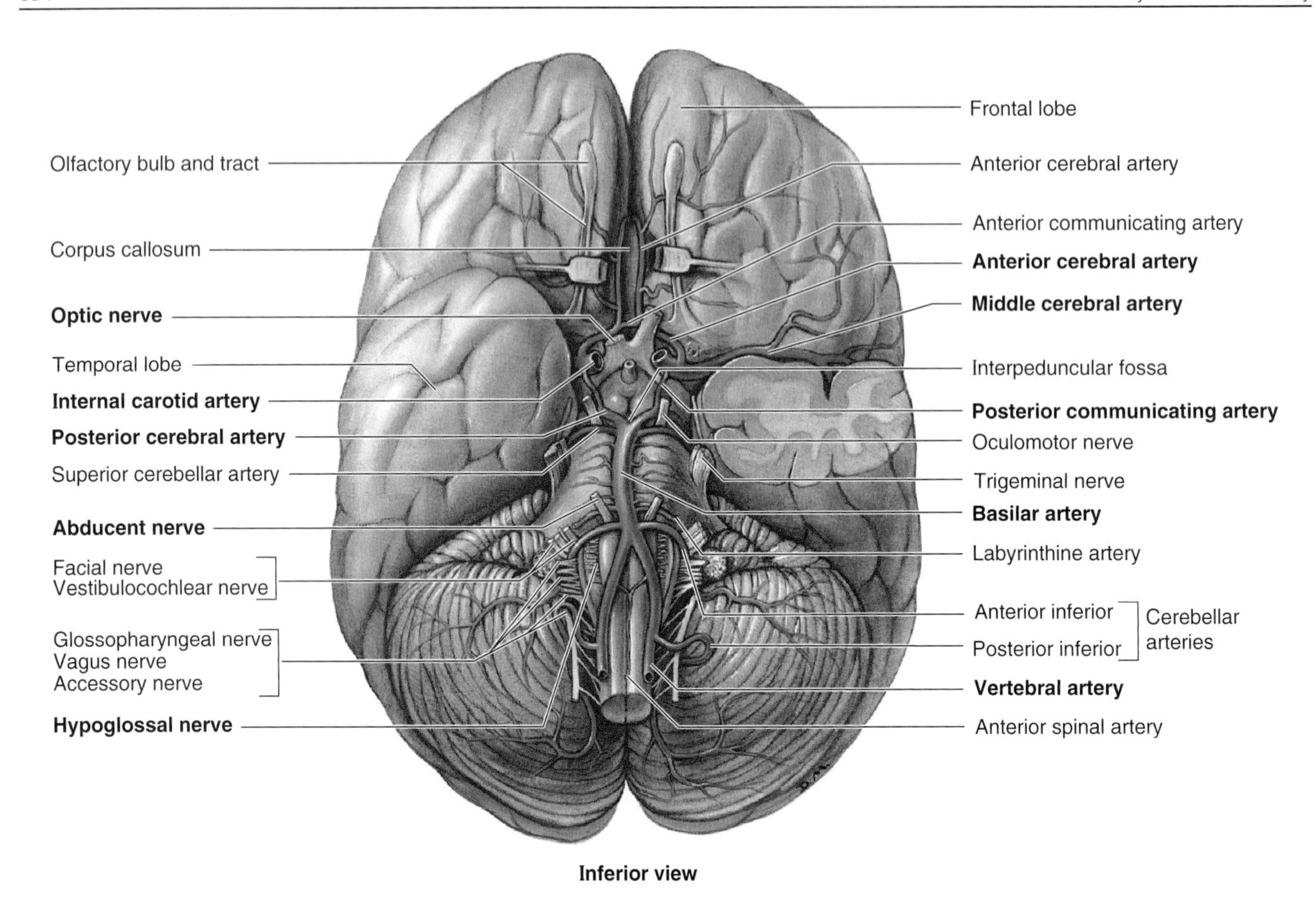

Figure 7.28. Base of the brain illustrating the cerebral arterial circle (of Willis). Inferior view. Observe that three arterial stems ascend to supply the brain: right and left internal carotid arteries and the basilar artery, which results from the union of the two vertebral arteries. Observe that the three arterial stems form an arterial circle through the linkages provided by the anterior communicating artery and the two posterior communicating arteries. The cerebellum (L. little brain) is mainly supplied by branches from the vertebral and basilar arteries. The left temporal pole is removed to enable visualization of the middle cerebral artery in the lateral fissure of the brain. The frontal lobes are separated to expose the anterior cerebral arteries.

the **posterior communicating arteries**, completing the **cerebral arterial circle** around the *interpeduncular fossa*—the deep depression on the inferior surface of the midbrain between the cerebral peduncles.

Vertebral Arteries

The vertebral arteries begin in the root of the neck as the first branches of the first part of the subclavian arteries. The two vertebral arteries are usually unequal in size, the left being larger than the right. The transversarial parts of the arteries pass through the transverse foramina of the first six cervical vertebrae (p. 476). The suboccipital parts of the vertebral arteries perforate the dura and arachnoid and pass through the foramen magnum. The vertebral arteries unite at the caudal border of the pons to form the basilar artery. The vertebrobasilar arterial system and its branches are often referred to clinically as the *posterior circulation of the brain*.

The **basilar artery**, so-named because of its close relationship to the base of the skull, ascends the **clivus** (Fig. 7.5C)—the sloping surface from the dorsum sellae to the foramen magnum—through the pontine cistern to the superior border of the pons, and ends by dividing into the two **posterior cerebral arteries.**

Cerebral Arteries

Each cerebral artery supplies a surface and a pole of the brain (Table 7.7):

- The **anterior cerebral artery** supplies most of the medial and superior surfaces of the brain and the frontal pole
- The **middle cerebral artery** supplies the lateral surface of the brain and the temporal pole
- The **posterior cerebral artery** supplies the inferior surface of the brain and the occipital pole.

Table 7.7. **Arterial Supply of the Cerebral Hemisphere**

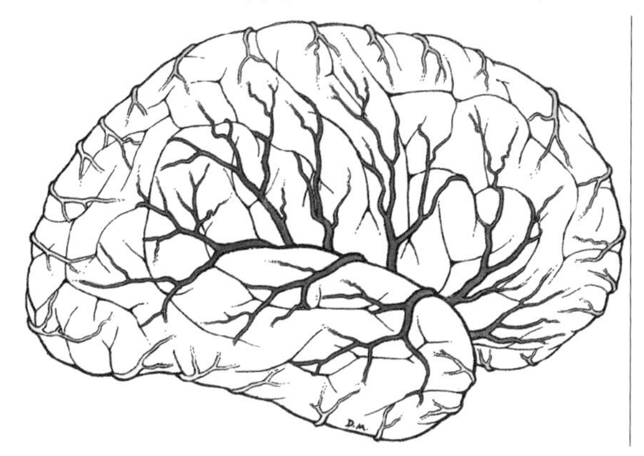

(A) Right lateral view of right hemisphere

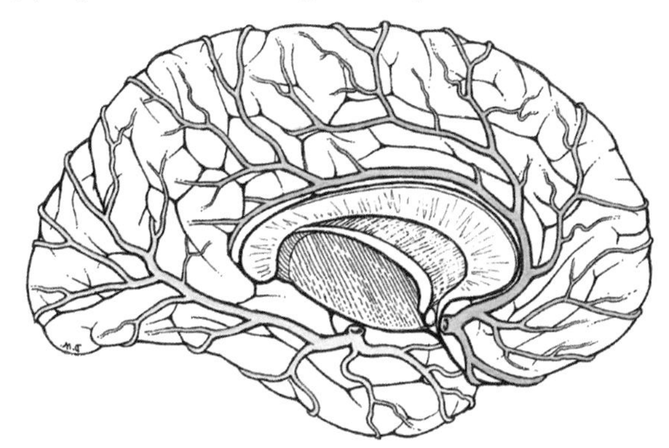

(B) Medial view of left hemisphere

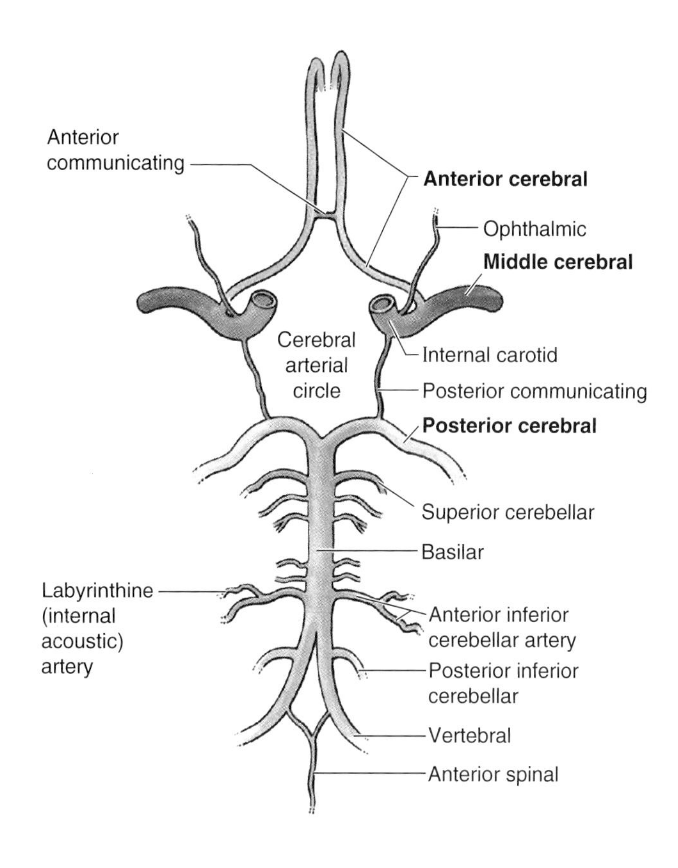

(C) Inferior view

Artery	Origin	Distribution
Internal carotid	Common carotid artery at superior border of thyroid cartilage	Gives branches to walls of cavernous sinus, pituitary gland, and trigeminal ganglion and provides primary supply to brain
Anterior cerebral	Internal carotid artery	Cerebral hemispheres, except for occipital lobes
Anterior communicating	Anterior cerebral artery	Cerebral arterial circle
Middle cerebral	Continuation of the internal carotid artery distal to anterior cerebral artery	Most of lateral surface of cerebral hemispheres
Vertebral	Subclavian artery	Cranial meninges and cerebellum
Basilar	Formed by union of vertebral arteries	Brain stem, cerebellum, and cerebrum
Posterior cerebral	Terminal branch of basilar artery	Inferior aspect of cerebral hemisphere and occipital lobe
Posterior communicating	Posterior cerebral artery	Optic tract, cerebral peduncle, internal capsule, and thalamus

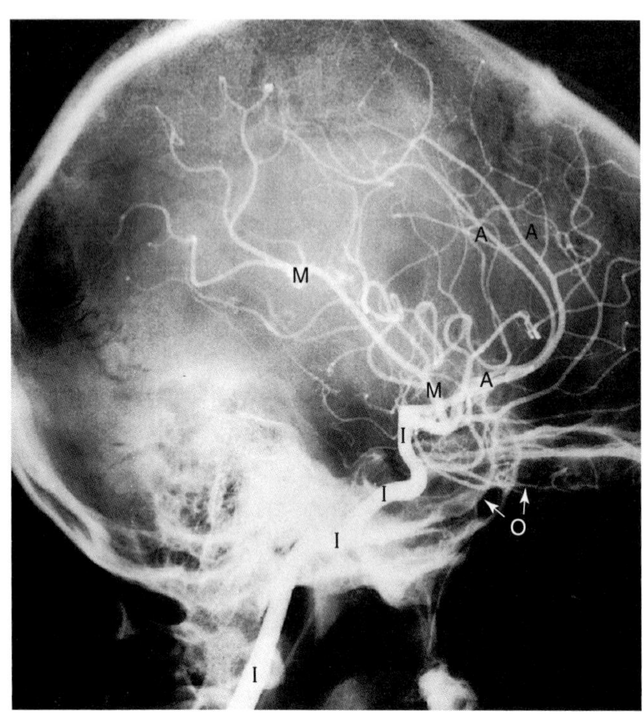

Lateral view

Figure 7.29. Carotid arteriogram. Lateral view. Radiopaque dye was injected into the carotid arterial system before the radiograph was taken. The four letter *I*'s indicate the parts of the internal carotid artery (from inferior to superior): transversarial or cervical, within the neck (before entering the skull); petrous (rocklike), within the carotid canal in the petrous temporal bone; cavernous, within the cavernous sinus; and cerebral, within the cranial subarachnoid (leptomeningeal) space. Observe the anterior cerebral artery and its branches (*A*), the middle cerebral artery and its branches (*M*), and the ophthalmic artery (*O*). (Courtesy of Dr. D. Armstrong, Associate Professor of Medical Imaging, University of Toronto, Ontario, Canada)

Cerebral Arterial Circle

The cerebral arterial circle is an important anastomosis at the base of the brain between the four arteries (vertebrals and internal carotids) that supply the brain (Fig. 7.28, Table 7.7). *The cerebral arterial circle is formed by the*:

- Posterior cerebral arteries
- Posterior communicating arteries
- Internal carotid arteries
- Anterior cerebral arteries
- Anterior communicating arteries.

The various components of the cerebral arterial circle give numerous small branches to the brain.

Anastomoses of Cerebral Arteries and Cerebral Embolism

Branches of the three cerebral arteries anastomose with each other on the surface of the brain; however, if a cerebral artery is blocked by a *cerebral embolism* (e.g., a blood clot), these microscopic anastomoses are not capable of providing enough blood for the area of cerebral cortex concerned. Consequently, *cerebral ischemia* and *infarction* may occur, and an area of necrosis (tissue death) may result. Small cerebral emboli may temporarily occlude small cerebral vessels, producing a *transient ischemic attack* (TIA) and sudden loss of neurological function that usually disappears within minutes. Large cerebral emboli occluding major cerebral vessels may cause severe neurological problems and death (Rowland, 1995; Haines, 1997). ⊙

Variations of the Cerebral Arterial Circle

Variations in the size of the vessels forming the circle are common. The posterior communicating arteries are absent in some individuals; in others the anterior communicating artery may be double. In approximately 1 in 3 persons, one posterior cerebral artery is a major branch of the internal carotid artery; this variation is usually unilateral. One of the anterior cerebral arteries often is small in the proximal part of its course; the anterior communicating artery is larger than usual in these individuals.

Vascular Strokes

A stroke denotes the sudden development of focal neurological deficits that are usually related to *impaired cerebral bloodflow*. Strokes are one of the four main causes of death and are responsible for many neurological disorders. Stroke is the most common neurologic disorder in adults in the United States (Sacco, 1995); it is more often disabling than fatal. *The cardinal feature of stroke is the sudden onset of neurological symptoms.*

The cerebral arterial circle is an important means of collateral circulation in the event that one of the major arteries forming the circle is gradually obstructed. Although capable of providing true collateral circulation in children, if one or more of the four contributing arteries to the cerebral arterial circle is blocked in adults, little exchange of blood usually takes place between the main arteries through the slender communicating arteries. Sudden occlusion—even if only partial—results in embarrassment of the cerebrum. In elderly persons, the anastomoses are often inadequate when a large artery (e.g., the internal carotid) becomes occluded, even if the occlusion is gradual (in which case function is ▶

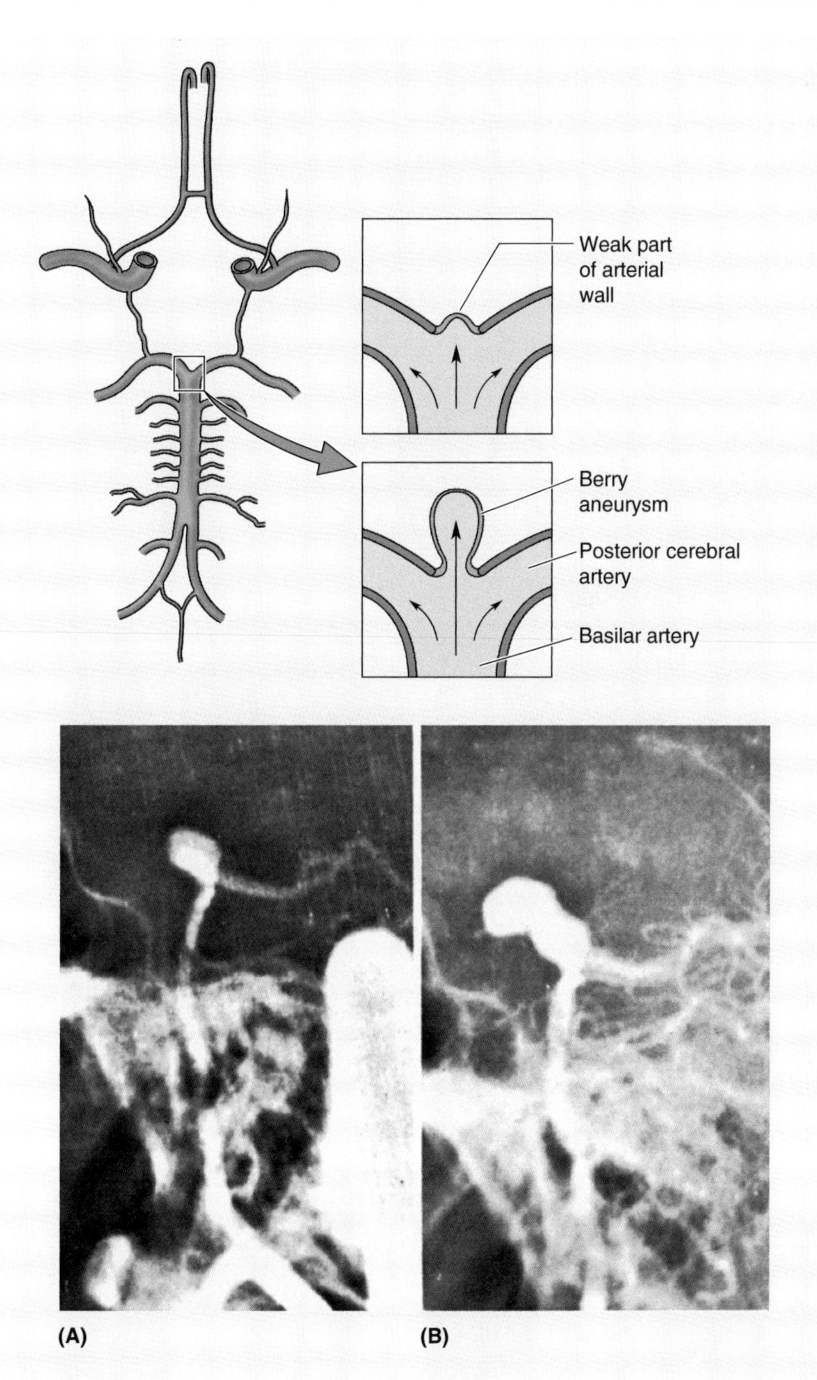

▶ impaired at least to some degree); if the occlusion is sudden, a *vascular stroke* results. The most common causes of strokes are *spontaneous cerebrovascular accidents* such as cerebral thrombosis, cerebral hemorrhage, cerebral embolism, and subarachnoid hemorrhage (Rowland, 1995).

Hemorrhagic stroke follows the rupture of an artery or an aneurysm (balloonlike sac that forms on the weak part of the wall of an artery). The most common type of aneurysm is a *berry aneurysm*, occurring in the vessels of or near the cerebral arterial circle and the medium-sized arteries at the base of the brain (*A*). An aneurysm also occurs at the bifurcation of the basilar artery into the two posterior cerebral arteries. In time, especially in persons with high blood pressure (hypertension), the weak part of the vessel wall expands and may rupture (*B*), allowing blood to enter the subarachnoid space. Sudden rupture of an aneurysm usually produces a severe, almost unbearable headache with a stiff neck. These symptoms result from gross bleeding into the subarachnoid space.

Brain Infarction

An *atherosclerotic plaque* at a bend of an artery (e.g., the bifurcation of a common carotid artery into external and internal carotids) results in progressive narrowing (stenosis) of the artery, producing increasingly severe functional deficits. ▶

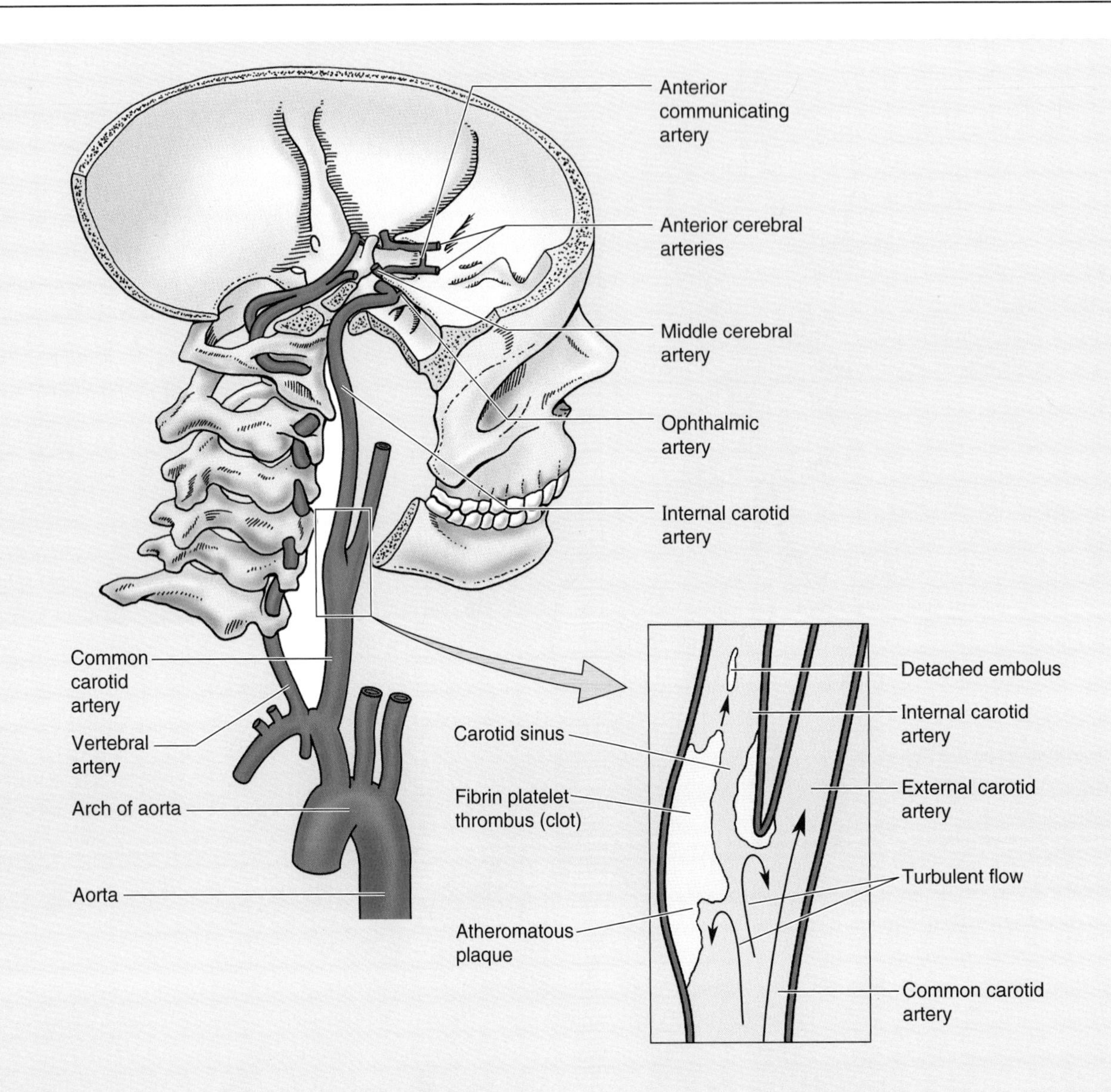

▶ A *detached embolus* (clot) is carried through the blood until it lodges in an artery, usually an intracranial branch that is too small to allow its passage. This event usually results in *acute cortical infarction*—sudden insufficiency of arterial blood to the brain (e.g., of the left frontal and parietal lobes). An interruption of blood supply for 30 seconds alters a person's brain metabolism. After 1–2 minutes the victim may lose neural function; after 5 minutes, lack of oxygen (anoxia) can result in cerebral infarction. Quickly restoring oxygen to the blood supply may reverse the brain damage (Sacco, 1995).

Transient Ischemic Attacks

As mentioned previously, a TIA refers to neurological symptoms resulting from ischemia—deficient blood supply to the brain. Most attacks last a few minutes only, but some persist for up to an hour. With major carotid or vertebrobasilar stenosis, the TIA tends to last longer and cause distal closure of intracranial vessels. The symptoms of TIA may be ambiguous: staggering, dizziness, lightheadedness, fainting, and paresthesias (e.g. tingling in a limb). Patients with TIAs "are at increased risk for myocardial infarction and major stroke" (Brust, 1995). ⊙

Venous Drainage of the Brain

Venous blood from superficial and deep veins of the brain enters the **dural venous sinuses** (Figs. 7.17–7.19), which drain into the **IJVs**. The cerebral veins on the superolateral surface of the brain drain into the superior sagittal sinus; **cerebral veins** on the posteroinferior aspect drain into the straight, transverse, and superior petrosal sinuses, as do the superior cerebellar veins and transverse sinuses. For more information on the cerebral veins, see Haines (1997).

Orbit

The orbit is a pyramidal, bony cavity in the facial skeleton with its base anterior and its apex posterior (Fig. 7.30). The orbits contain and protect the eyeballs and their muscles, nerves, and vessels, together with most of the lacrimal apparatus. The bones forming the orbit are lined with **periorbita** (periosteum of the orbit), which forms the **fascial sheath of the eyeball** (Fig. 7.31*A*). The periorbita is continuous at the **optic canal** and **superior orbital fissure** with the periosteal layer of dura. The periorbita is also continuous over the orbital margins and through the **inferior orbital fissure** with the periosteum covering the external surface of the skull (pericranium).

The orbit has four walls and an apex (Fig. 7.30):

- The **superior wall** (roof) is approximately horizontal and is formed mainly by the **orbital part of the frontal bone,** which separates the orbital cavity from the anterior cranial fossa. Near the apex of the orbit, the superior wall is formed by the lesser wing of the sphenoid. The lacrimal gland occupies the **fossa for the lacrimal gland** (lacrimal fossa) in the orbital part of the frontal bone.
- The **medial wall** is formed primarily by the **ethmoid bone,** along with contributions from the frontal, lacrimal, and sphenoid bones; anteriorly, the *paper-thin medial wall* is indented by a **lacrimal fossa for the lacrimal sac** and the proximal part of the nasolacrimal duct. The medial walls of the two orbits are essentially parallel, separated by the ethmoidal sinuses and the upper nasal cavity.
- The **inferior wall** (floor) is formed mainly by the maxilla and partly by the zygomatic and palatine bones; the thin floor is partly separated from the lateral wall of the orbit by the **inferior orbital fissure.** The inferior wall slants inferiorly from the apex to the inferior orbital margin.
- The **lateral wall** is formed by the frontal process of the zygomatic bone and the greater wing of the sphenoid; the lateral wall is thick, especially its posterior part, which separates the orbit from the middle cranial fossa. The lateral walls of the two orbits are nearly perpendicular to each other.
- The **apex of the orbit** is at the optic canal, in the lesser wing of the sphenoid just medial to the **superior orbital fissure.**

Fracture of the Orbit

Because of the thinness of the medial and inferior walls of the orbit, a blow to the eye may fracture the orbit. Indirect traumatic injury that displaces the orbital walls ▶

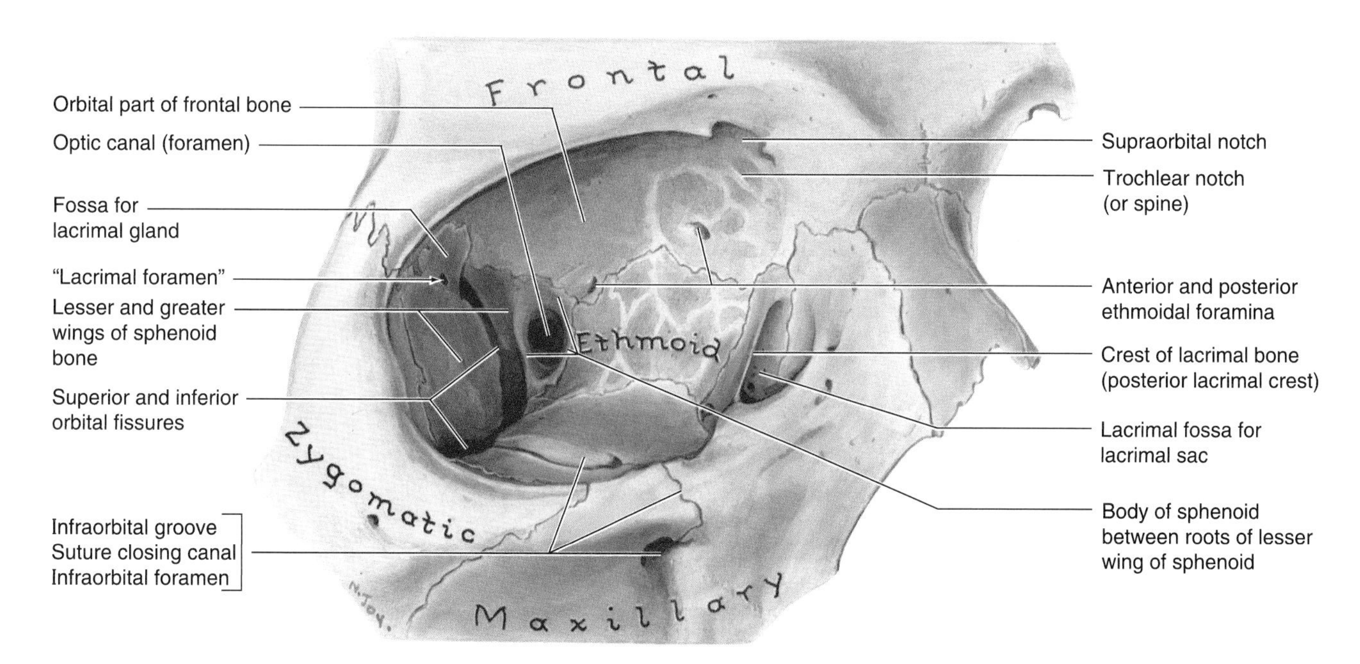

Figure 7.30. Orbit or orbital cavity (eye socket). Anterior view. Observe the bony walls of the orbit. The word "ethmoid" has been printed directly on the thin orbital lamina (plate) that separates the orbit from the ethmoidal sinuses. Observe the optic canal situated at the apex (deepest part) of the pyramidal orbital cavity and placed between the body of the sphenoid bone and the two roots of the lesser wing of this bone. A straight probe must pass along the lateral wall of the cavity if it is to traverse the canal.

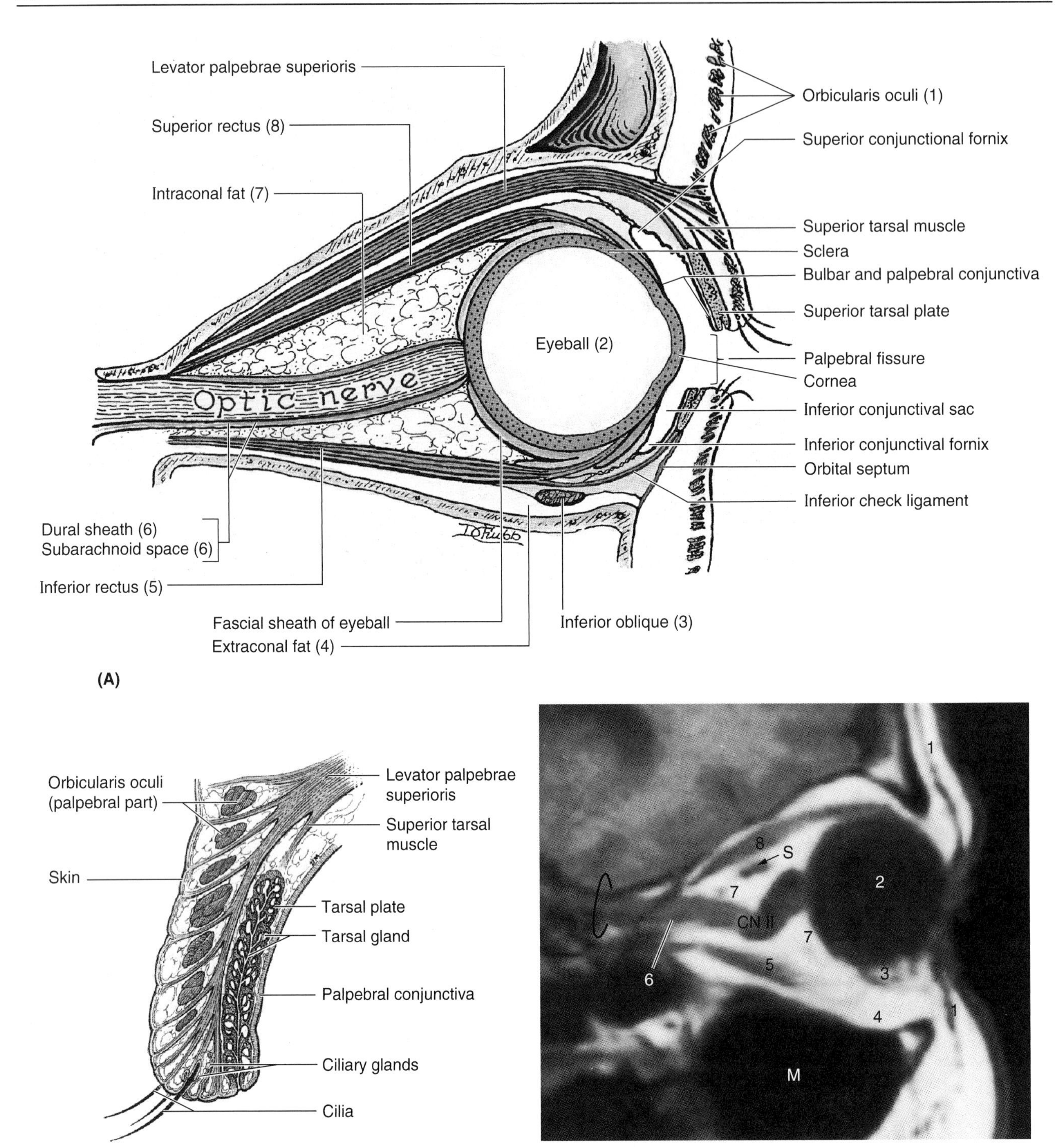

Figure 7.31. The orbit (eye socket, orbital cavity) and eyelids (L. palpebrae). A. Sagittal section of the orbit showing its contents. The numbers in parentheses refer to structures in (**C**). Observe the subarachnoid (leptomeningeal) space around the optic nerve (CN II), which is continuous with the space between the arachnoid and pia covering the brain. **B.** Sagittal section of the superior (upper) eyelid. Notice the tarsal plate that forms the skeleton of the eyelid and contains tarsal glands. **C.** Sagittal MRI through the optic nerve and eyeball. *S*, superior ophthalmic vein; *M*, maxillary sinus; *circle*, optic foramen. (Courtesy of Dr. W. Kucharczyk, Chair of Medical Imaging, University of Toronto, and Clinical Director of Tri-Hospital Resonance Centre, Toronto, Ontario, Canada)

▶ is called a "blowout" fracture. Fractures of the medial wall may involve the ethmoidal and sphenoidal sinuses, whereas fractures of the inferior wall may involve the maxillary sinus. Although the superior wall is stronger than the medial and inferior walls, it is thin enough to be translucent and may be readily penetrated. Thus, a sharp object may pass through it and enter the frontal lobe of the brain.

Orbital fractures often result in intraorbital bleeding, which exerts pressure on the eyeball, causing exophthalmos. Any trauma to the eye may affect adjacent structures:

- Bleeding into the maxillary sinus
- Displacement of maxillary teeth
- Fracture of nasal bones resulting in hemorrhage, airway obstruction, and infection that could spread to the cavernous sinus through the ophthalmic vein.

Periorbital Ecchymosis

Blows to the periorbital region often cause significant swelling and hemorrhage into the eyelids and extravasation of blood into the periorbital skin (ecchymosis). This type of injury is common in boxers and basketball players.

Orbital Tumors

Because of the closeness of the optic nerve to the sphenoidal and posterior ethmoidal sinuses, a malignant tumor in these sinuses may erode the thin bony walls of the orbit and compress the optic nerve and orbital contents. Tumors in the orbit produce *exophthalmos*—protusion of the eyeball. The easiest entrance to the orbital cavity for a tumor in the middle cranial fossa is through the superior orbital fissure; tumors in the temporal or infratemporal fossa gain access to this cavity through the inferior orbital fissure. Although the lateral wall of the orbit is nearly as long as the medial wall because it extends laterally and anteriorly, it does not reach as far anteriorly as the medial wall does, which occupies essentially a sagittal plane. Thus, nearly 2.5 cm of the eyeball is exposed when the pupil is turned medially as far as possible. This is why the lateral side affords a good approach for operations on the eyeball. ⊙

Eyelids and Lacrimal Apparatus

The eyelids and lacrimal fluid (tears)—secreted by the lacrimal glands—protect the cornea and eyeball from injury (e.g., dust and small particles).

Eyelids

When closed, the eyelids cover the eyeball anteriorly, thereby protecting it from injury and excessive light. As mentioned, they also keep the cornea moist by spreading the lacrimal fluid. The eyelids are movable folds that are covered externally by thin skin and internally by **palpebral conjunctiva** (Fig. 7.31*A*). The palpebral conjunctiva is reflected onto the eyeball, where it is continuous with the bulbar conjunctiva. This part of the conjunctiva is thin and transparent and attaches loosely to the anterior surface of the eye. The **bulbar conjunctiva**, loose and wrinkled over the sclera (where it contains small, apparent blood vessels), is adherent to the periphery of the cornea (Fig. 7.33). The lines of reflection of the palpebral conjunctiva onto the eyeball form deep recesses, the superior and inferior **conjunctival fornices** (Fig. 7.31*A*).

The superior and inferior eyelids are strengthened by dense bands of connective tissue—the superior and inferior **tarsal plates**, which form the "skeleton" of the eyelids (Figs. 7.31*B* and 7.32). Fibers of the **orbicularis oculi** are in the connective tissue superficial to these plates and deep to the skin of the eyelids. Embedded in the tarsal plates are **tarsal glands**, the lipid secretion of which lubricates the edges of the eyelids and prevents them from sticking together when they close. The lipid secretion also forms a barrier that lacrimal fluid does not cross when produced in normal amounts. When production is excessive, it spills over the barrier as tears. The **eyelashes** (L. cilia) are in the margins of the lids. The large sebaceous glands associated with the eyelashes are **ciliary glands**. The place where the eyelids meet is the *angle* of the eye (G. kanthos, corner of eye). Thus, each eye has medial and lateral angles, or canthi (Fig. 7.32).

Between the nose and the medial angle of the eye is the **medial palpebral ligament**, which connects the tarsal plates to the medial margin of the orbit (Fig. 7.32). The orbicularis oculi muscle originates and inserts onto this ligament. A similar **lateral palpebral ligament** attaches the tarsal plates to the lateral margin of the orbit but does not provide for direct muscle attachment. The **orbital septum** is a weak membrane that spans from the tarsal plates to the margins of the orbit, where it becomes continuous with the periosteum. It contains orbital fat and can limit the spread of infection to and from the orbit.

Lacrimal Apparatus

The lacrimal apparatus (Figs. 7.31 and 7.33) consists of:

- *Lacrimal glands*, which secrete lacrimal fluid
- *Lacrimal ducts,* which convey lacrimal fluid from the lacrimal glands to the conjunctival sac
- *Lacrimal canaliculi* (L. small canals), each commencing at a *lacrimal punctum* (opening) on the *lacrimal papilla* near the medial angle of the eye, which conveys the lacrimal fluid from the *lacrimal lake*—a triangular space at the medial angle of the eye where the tears collect—in the *lacrimal sac,* the dilated superior part of the nasolacrimal duct

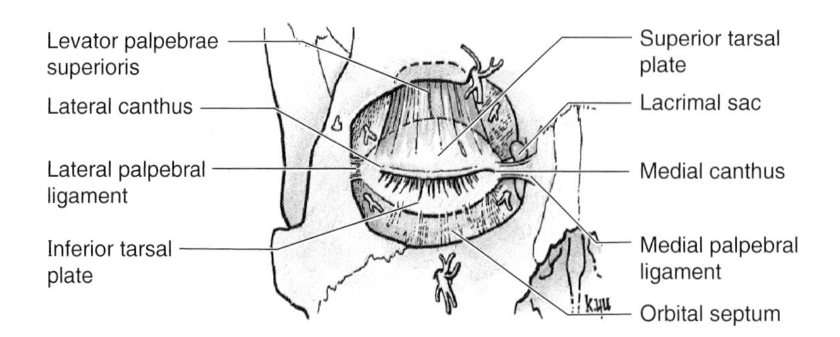

Figure 7.32. Skeleton of the eyelids (L. palpebrae). This diagram shows the superior and inferior tarsal plates and their attachments. Their ciliary margins are free but peripherally the margins are attached to the orbital septum (palpebral fascia in the eyelid). The angles are anchored by medial and lateral palpebral ligaments. The fan-shaped aponeurosis of the levator palpebrae superioris is attached to the anterior surface and superior edge of the superior tarsal plate.

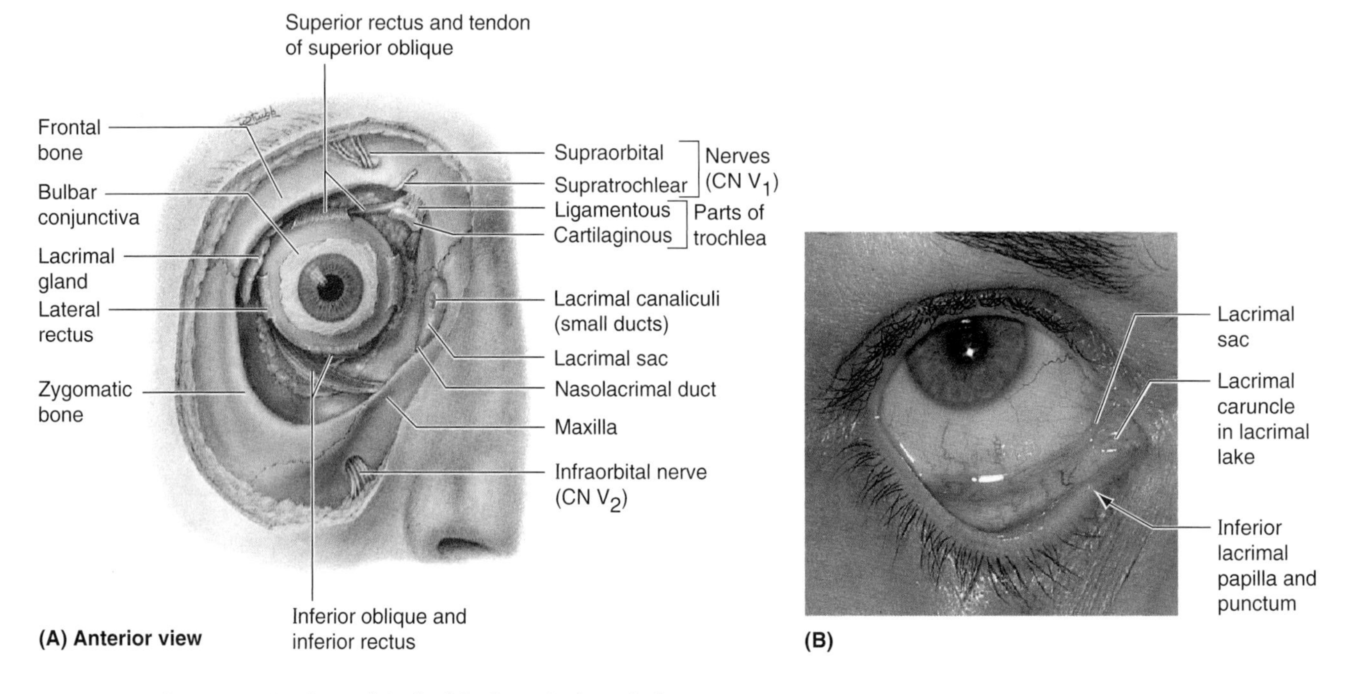

Figure 7.33. The orbit (orbital cavity) and the eye. A. Anterior view. Anterior dissection of the orbit. The eyelids (L. palpebrae), orbital septum, levator palpebrae superioris, and some fat are removed. Observe the bulbar conjunctiva, loose and wrinkled over the sclera but adherent to the cornea. Examine the aponeurotic attachments of the four recti muscles inserted 6 to 8 mm posterior to the sclerocorneal junction. Notice that the tendon of the superior oblique muscle moves on a cartilaginous pulley or trochlea, which is fixed by ligamentous fibers just posterior to the superomedial angle of the orbital margin. Observe the inferior oblique muscle, the only muscle attaching to the anterior aspect of the orbit. The inferior branch of the oculomotor nerve (CN III) to the muscle is shown entering its posterior border. The lacrimal gland can also be seen lying between the bony orbital wall laterally and the eyeball and lateral rectus muscle medially. **B.** Surface features of the eye. Observe the tough white fibrous outer coat of the eyeball, the sclera; the central transparent cornea, through which can be seen the pigmented iris with its aperture, the pupil. Notice that the superior (upper) and inferior (lower) lids meet at the medial and lateral angles (L. canthi). The inferior lid has been everted to show the reflection of conjunctiva from the anterior surface of the eyeball to the inner surface of the eyelid. An *arrow* points to the inferior lacrimal punctum (opening). Observe near the medial angle a vertical fold of conjunctiva, the semilunar fold (L. plica semilunaris) at the lacrimal caruncle (see also p. 915).

- *Nasolacrimal duct*, which conveys the lacrimal fluid to the nasal cavity (p. 916).

The lacrimal fluid usually flows from the lacrimal ducts across the eye to the *lacrimal lake*. The tears enter the *lacrimal sac* (via the canaliculi) and then the nasal cavity, where they flow to the back of the *nasal cavity* and are swallowed. However, when the tears increase as a result of emotion or other causes, they flow anteriorly over the lipid barrier on the edge of the lids and onto the cheeks.

The **lacrimal gland** (Fig. 7.33*A*), almond-shaped and approximately 2 cm long, lies in the **fossa for the lacrimal gland** in the superolateral part of each orbit (Fig. 7.30). The gland is divided into superior (orbital) and inferior (palpebral) parts by the lateral expansion of the tendon of the **levator palpebrae superioris** (Fig. 7.31*B*). *Accessory lacrimal glands* are also present; they are more numerous in the superior lid than in the inferior lid.

Lacrimal fluid—the production of which is stimulated by parasympathetic impulses from CN VII—enters the conjunctival sac through up to 12 **lacrimal ducts** that open into the *superior conjunctival fornix*—the superior line of reflection of the palpebral conjunctiva to the eyeball (Fig. 7.31*A*). The **conjunctival sac** is the space bound by the conjunctival membrane between the *palpebral* and *bulbar conjunctiva*. After passing over the eyeball—due in large part to the way the eye closes from lateral to medial—the tears enter the **lacrimal lake** at the medial angle of the eye from which they drain by capillary action through the **lacrimal puncta** and **lacrimal canaliculi** to the lacrimal sac. From this sac, the tears pass to the nasal cavity through the **nasolacrimal duct** (Fig. 7.33*A*). When the cornea becomes dry, the eye blinks and the eyelids carry a film of fluid over the cornea, somewhat like the car wipers wash the windshield. In this way, foreign material such as dust is carried to the medial angle of the eye where it can be removed.

The **nerve supply of the lacrimal gland** is both sympathetic and parasympathetic. The presynaptic, parasympathetic secretomotor fibers are conveyed from the facial nerve by the *greater petrosal nerve* and then by the *nerve of the pterygoid canal* to the *pterygopalatine ganglion*, where they synapse with the cell body of the postsynaptic fiber. Vasoconstrictive, postsynaptic sympathetic fibers brought from the *superior cervical ganglion* by the *internal carotid plexus* and deep petrosal nerve join the parasympathetic fibers to form the nerve of the pterygoid canal and traverse the pterygopalatine ganglion. The zygomatic nerve (from the maxillary nerve) brings both types of fiber to the lacrimal branch of the ophthalmic nerve, by which they enter the gland.

Injury to the Nerves Supplying the Eyelids

Because it supplies the levator palpebrae superioris, a *lesion of the oculomotor nerve* causes paralysis of the muscle, and the upper eyelid droops (*ptosis*). *Damage to the facial nerve* involves paralysis of the orbicularis oculi, preventing the eyelids from fully closing. Normal rapid protective blinking of the eye is also lost. The loss of tonus of the muscle in the lower eyelid causes the lid to fall away (become everted) from the surface of the eye, leading to drying of the cornea, which leaves it unprotected from dust and small particles. Thus, irritation of the unprotected eyeball results in excessive but inefficient *lacrimation* (tear formation). Excessive tears also form when the lacrimal drainage apparatus is obstructed, thereby preventing the tears from reaching the inferior part of the eye. People often dab their eyes constantly to wipe the tears, resulting in further irritation.

Inflammation of the Palpebral Glands

Any of the glands in the eyelid may become inflamed and swollen from infection or obstruction of their ducts. If the ducts of the ciliary glands become obstructed, a painful red, suppurative (pus-producing) swelling—a *sty* (hordeolum)—develops on the eyelid. *Cysts of the sebaceous glands* of the eyelid—*chalazia*—may also form. Obstruction of a tarsal gland produces an inflammation—a *tarsal chalazion*—that protrudes toward the eyeball and rubs against it as the eyelids blink. Chalazia usually are more painful than sties.

Hyperemia of the Conjunctiva

The bulbar conjunctiva is colorless, except when its vessels are dilated and congested ("bloodshot eyes"). *Hyperemia* of the conjunctiva is caused by local irritation (e.g., from dust, chlorine, or smoke). *An inflamed conjunctiva—conjunctivitis* ("pinkeye")—is a common, contagious infection of the eye.

Subconjunctival Hemorrhages

Subconjunctival hemorrhages are common and are manifested by bright or dark red patches deep to and in the bulbar conjunctiva. The hemorrhages may result from injury or inflammation. A blow to the eye, excessively hard blowing of the nose, and paroxysms of coughing or violent sneezing can cause hemorrhages resulting from rupture of small subconjunctival capillares.

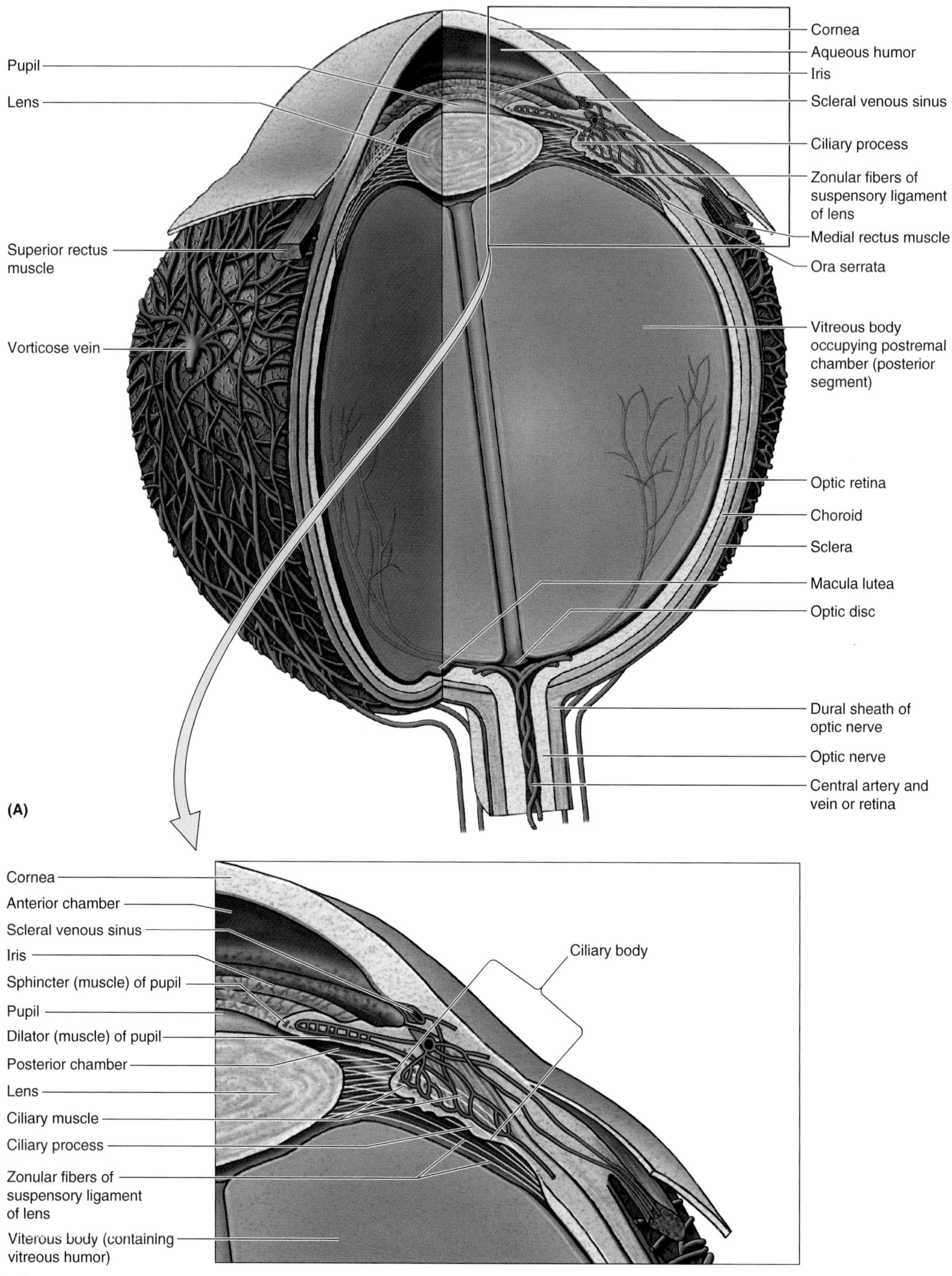
Cornea
Aqueous humor
Iris
Scleral venous sinus
Ciliary process
Zonular fibers of suspensory ligament of lens
Medial rectus muscle
Ora serrata
Pupil
Lens
Superior rectus muscle
Vorticose vein
Vitreous body occupying postremal chamber (posterior segment)
Optic retina
Choroid
Sclera
Macula lutea
Optic disc
Dural sheath of optic nerve
Optic nerve
Central artery and vein or retina
(A)
Cornea
Anterior chamber
Scleral venous sinus
Iris
Sphincter (muscle) of pupil
Pupil
Dilator (muscle) of pupil
Posterior chamber
Lens
Ciliary muscle
Ciliary process
Zonular fibers of suspensory ligament of lens
Viterous body (containing vitreous humor)
Ciliary body
(B)

Orbital Contents

The contents of the orbit are the eyeball, optic nerve, ocular muscles, fascia, nerves, vessels, fat, lacrimal gland, and conjunctival sac. *The eyeball has three layers* (Fig. 7.34):

- The outer **fibrous layer**—the sclera and cornea
- The middle **vascular (pigmented) layer**—the choroid, ciliary body, and iris
- The **inner layer**—the retina, consisting of optic and nonvisual parts.

Outer Fibrous Layer of the Eyeball

The **sclera** is the opaque part of the fibrous coat of the eyeball covering the posterior five-sixths of the eyeball. The anterior part of the sclera is visible through the transparent bulbar conjunctiva as "the white of the eye." The **cornea** is the transparent part of the fibrous coat covering the anterior one-sixth of the eyeball.

Middle Vascular Layer of the Eyeball

The **choroid**—a dark brown membrane between the sclera and retina—forms the largest part of the vascular layer of the eyeball and lines most of the sclera. It terminates anteriorly in the **ciliary body**. The choroid attaches firmly to the pigment layer of the retina, but it can easily be stripped from the sclera.

The **ciliary body**—which is muscular as well as vascular—connects the choroid with the circumference of the iris. Folds on its internal surface—the **ciliary processes** (Fig. 7.34*B*)—*secrete aqueous humor*, which fills the anterior and posterior **chambers of the eye**—fluid-filled spaces. The *anterior chamber of the eye* is the space between the cornea anteriorly and the iris/pupil posteriorly (Fig. 7.34*B*). The *posterior chamber of the eye* is between the iris/pupil anteriorly and the lens and ciliary body posteriorly.

The **iris**, which literally lies on the anterior surface of the lens, is a thin, contractile diaphragm with a central aperture—the **pupil**—for transmitting light. When a person is awake, the size of the pupil varies continually to regulate the amount of light entering the eye. Two muscles control the size of the pupil: the *sphincter pupillae* closes the pupil and the *dilator pupillae* opens it.

Inner Layer of the Eyeball (Retina)

Grossly, the retina consists of three parts:

- Optic part
- Ciliary part
- Iridial part.

The **optic part of the retina**, which receives the visual light rays, has two layers: a neural layer and pigment cell layer. The *neural layer* is the light-receptive part. The *pigmented layer* consists of a single layer of cells that reinforces the light-absorbing property of the choroid in reducing the scattering of light in the eye. The **ciliary** and **iridial parts of the retina** are anterior continuations of the pigmented layer and a layer of supporting cells over the ciliary body and the posterior surface of the iris, respectively.

In the **fundus** (posterior part) of the eye is a circular depressed area—the **optic disc** (optic papilla)—where the optic nerve enters the eyeball (Fig. 7.34*A*). Because it contains nerve fibers and no photoreceptors, *the optic disc is insensitive to light*. Just lateral to this "blind spot" of the retina is the **macula lutea** (L. yellow spot). The yellow color of the macula is apparent only when the retina is examined with red-free light. *The macula lutea—a small oval area of the retina with special photoreceptor cones—is specialized for acuity of vision*; it is not normally observed with an *ophthalmoscope*, a device for viewing the interior of the eyeball through the pupil. At the center of the macula lutea is a depression—the *fovea centralis* (L. central pit)—*the area of most acute vision*. The fovea is approximately 1.5 mm in diameter; its center—the *foveola*—does not have the capillary network visible elsewhere deep to the retina.

The functional optic part of the retina terminates anteriorly along the *ora serrata* (L. serrated edge), an irregular border slightly posterior to the ciliary body (Fig. 7.34). The ora serrata marks the anterior termination of the light-receptive

Pupillary Light Reflex

The pupillary light reflex is tested using a penlight during neurological examinations. This reflex, involving CN II and CN III, is the constriction of the pupil in response to light. When light enters one eye, both pupils constrict because each retina sends fibers into the optic tracts of both sides. The *sphincter pupillae muscle* is innervated by parasympathetic fibers; consequently, interruption of these fibers causes dilation of the pupil because of the unopposed action of the *dilator pupillae muscle*. The first sign of compression of the oculomotor nerve is ipsilateral slowness of the pupillary response to light (Barr and Kiernan, 1993). ○

Figure 7.34. Drawing of a dissection of the eyeball. A. Observe the parts of the eyeball. The inner part of the retina is supplied by the central artery of the retina, whereas the outer part of the retina is nourished by the choriocapillary layer (see Fig. 7.38). Observe that the central retinal artery courses through the optic nerve (CN II) and at the optic disc divides into superior and inferior branches. The branches of the central artery do not anastomose with each other or with any other vessel. **B.** Inset drawing showing the structural details of the ciliary region. Observe the ciliary muscle and process, the sphincter pupillae and dilator pupillae muscles, and the scleral venous sinus (L. sinus venosus sclerae, canal of Schlemm).

part of the retina. Except for the cones and rods of the neural layer, the retina is supplied by the **central artery of the retina**, a branch of the ophthalmic artery. The cones and rods of the outer neural layer receive nutrients from the *choriocapillaris*, or choriocapillary layer (see Fig. 7.38)—the finest vessels on the inner surface of the choroid against which the retina is pressed. A corresponding system of retinal veins unites to form the **central vein of the retina**.

Development of the Retina

The retina and optic nerve develop from the *optic cup*, a derivative of an outgrowth of the embryonic forebrain—the *optic vesicle*. As it evaginates from the brain, the optic vesicle carries the developing meninges with it (Moore and Persaud, 1998). Hence, the optic nerve is invested with cranial meninges and an extension of the subarachnoid space. The central artery and vein of the retina cross the subarachnoid space to run within the distal part of the optic nerve. The pigment cell layer of the retina develops from the outer layer of the optic cup, and the neural layer develops from the inner layer of the cup.

Ophthalmoscopy

Physicians view the fundus of the eye with an *ophthalmoscope*, sometimes called a funduscope (*A*). The retinal arteries and veins radiate over the fundus from the optic disc. Observe the pale, oval optic disc with retinal vessels radiating from its center in this view of the retina through an ophthalmoscope. *Pulsation of the retinal arteries* is usually visible. The fovea appears darker than the reddish hue of surrounding areas of the retina because the black melanin pigment in the choroid and pigment cell layer is not screened by capillary blood.

Papilledema

An increase in CSF pressure slows venous return from the retina, causing *edema of the retina* (fluid accumulation). Edema of the retina appears during ophthalmoscopy as swelling of the optic disc—*papilledema*. Normally the optic disc is flat and does not form a papilla. Papilledema results from increased intracranial pressure and increased CSF pressure in the extension of the subarachnoid space around the optic nerve.

Detachment of the Retina

The layers of the retina are separated in the embryo by an intraretinal space. During the early fetal period, the embryonic layers fuse, obliterating the intraretinal space. Although the pigment cell layer becomes firmly fixed to the choroid, its attachment to the neural layer is not firm. Consequently, detachment of the retina may follow a blow to the eye (*B*), which restores the condition that existed in the embryo. A detached retina usually results from seepage of fluid between the neural and pigment layers of the retina, perhaps days or even weeks after trauma to the eye. Persons with a retinal detachment may complain of flashes of light going on and off (Anderson and Hall, 1995). ⊙

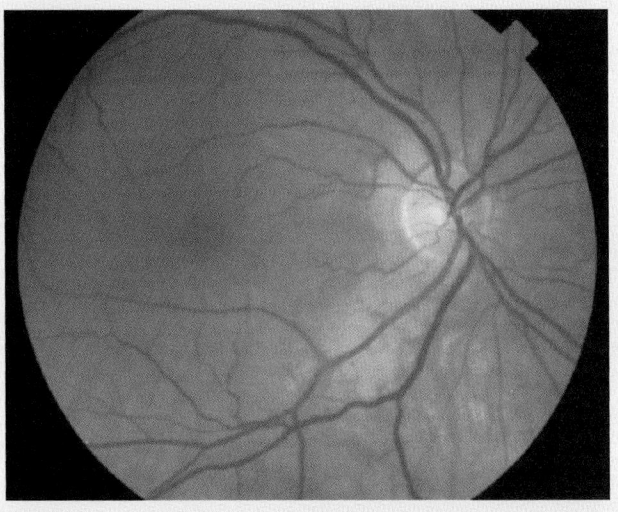

(A)

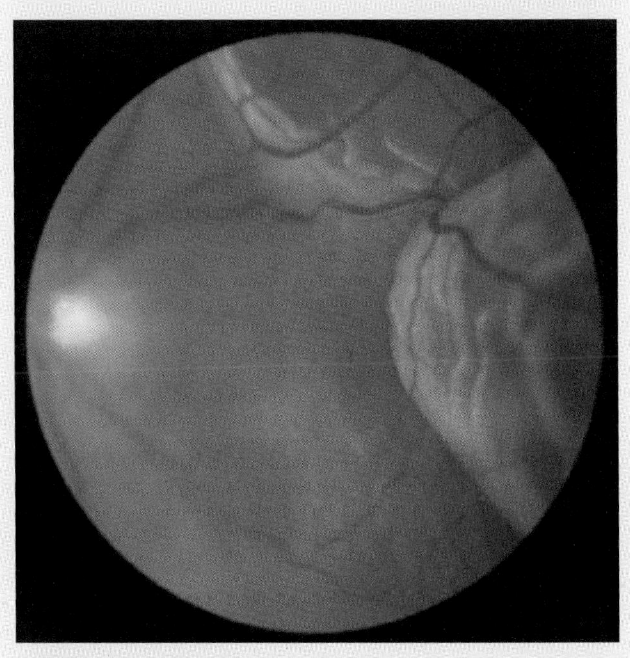

(B) Retinal detachment

Refractive Media of the Eye

On their way to the retina, light waves pass through the *refractive media of the eye*: cornea, aqueous humor, lens, and vitreous humor (Fig. 7.34*B*).

The **cornea** is the circular area of the anterior part of the outer fibrous layer of the eyeball; it is largely responsible for refraction of the light that enters the eye. It is transparent, avascular, and sensitive to touch. The cornea is supplied by the ophthalmic nerve and is nourished by aqueous humor, tears, and oxygen absorbed from the air.

The **aqueous humor** in the anterior and posterior chambers of the eye is produced by the ciliary processes. This clear watery solution provides nutrients for the avascular cornea and lens. After passing through the pupil from the posterior chamber into the anterior chamber, the aqueous humor drains into the scleral venous sinus (L. sinus venosus sclerae, canal of Schlemm) at the iridocorneal angle.

The **lens**, posterior to the iris and anterior to the vitreous humor of the vitreous body, is a transparent biconvex structure enclosed in a capsule. The *lens capsule* is anchored by the suspensory ligament of the lens to the ciliary body and encircled by the ciliary processes. The convexity of the lens, particularly its anterior surface, constantly varies to focus near or distant objects on the retina. The ciliary muscle in the ciliary body changes the shape of the lens (Fig. 7.34*B*); in this way the isolated unattached lens assumes a nearly spherical shape. Stretched within the circle of the relaxed ciliary body, the attachments around its periphery pull the lens relatively flat so that its refraction enables far vision. When parasympathetic stimulation causes the smooth muscle of the circular ciliary body to contract, the circle—like a sphincter—becomes smaller in size and the tension on the lens is reduced, allowing the lens to round up. The increased convexity makes its refraction suitable for near vision. In the absence of parasympathetic stimulation, the ciliary muscles relax again and the lens is pulled into its flatter, far vision shape.

The **vitreous humor** is a watery fluid enclosed in the meshes of the **vitreous body**, a transparent jellylike substance in the posterior four-fifths of the eyeball posterior to the lens (postremal or vitreous chamber, or posterior segment). In addition to transmitting light, the vitreous humor holds the retina in place and supports the lens.

Corneal Abrasions and Lacerations

Foreign objects such as dirt and sand produce *corneal abrasions* (scratches) that cause sudden, stabbing eye pain and excess tears. Opening and closing the eyelids is also painful. *Corneal lacerations* are caused by sharp objects such as fingernails and skate blades. ▶

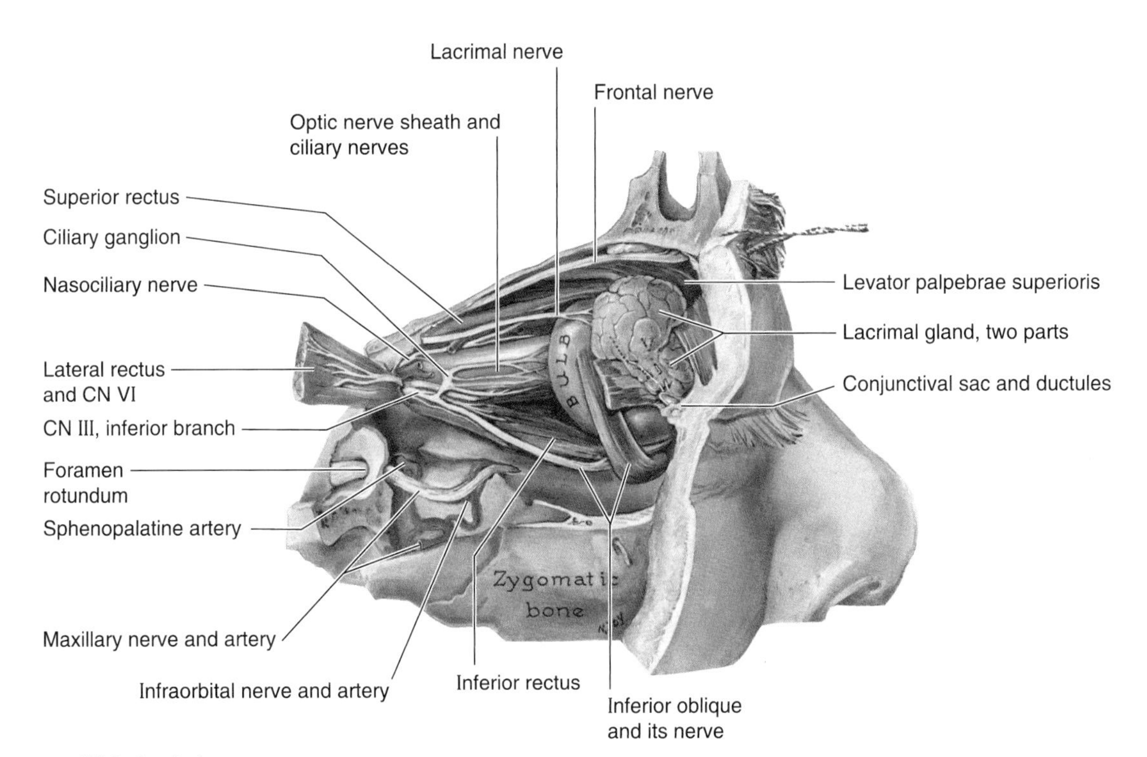

Figure 7.35. Dissection of the orbit (orbital cavity). A. Note in particular the ciliary ganglion that receives sensory fibers from the nasociliary branches of CN V_1, sympathetic fibers from the internal carotid plexus traveling around the ophthalmic artery, and parasympathetic fibers (which synapse in the ganglion) from the inferior branch of the oculomotor nerve (CN III). Observe the 8 to 10 short ciliary nerves passing to the eyeball. Observe also the orbital muscles and the lacrimal gland.

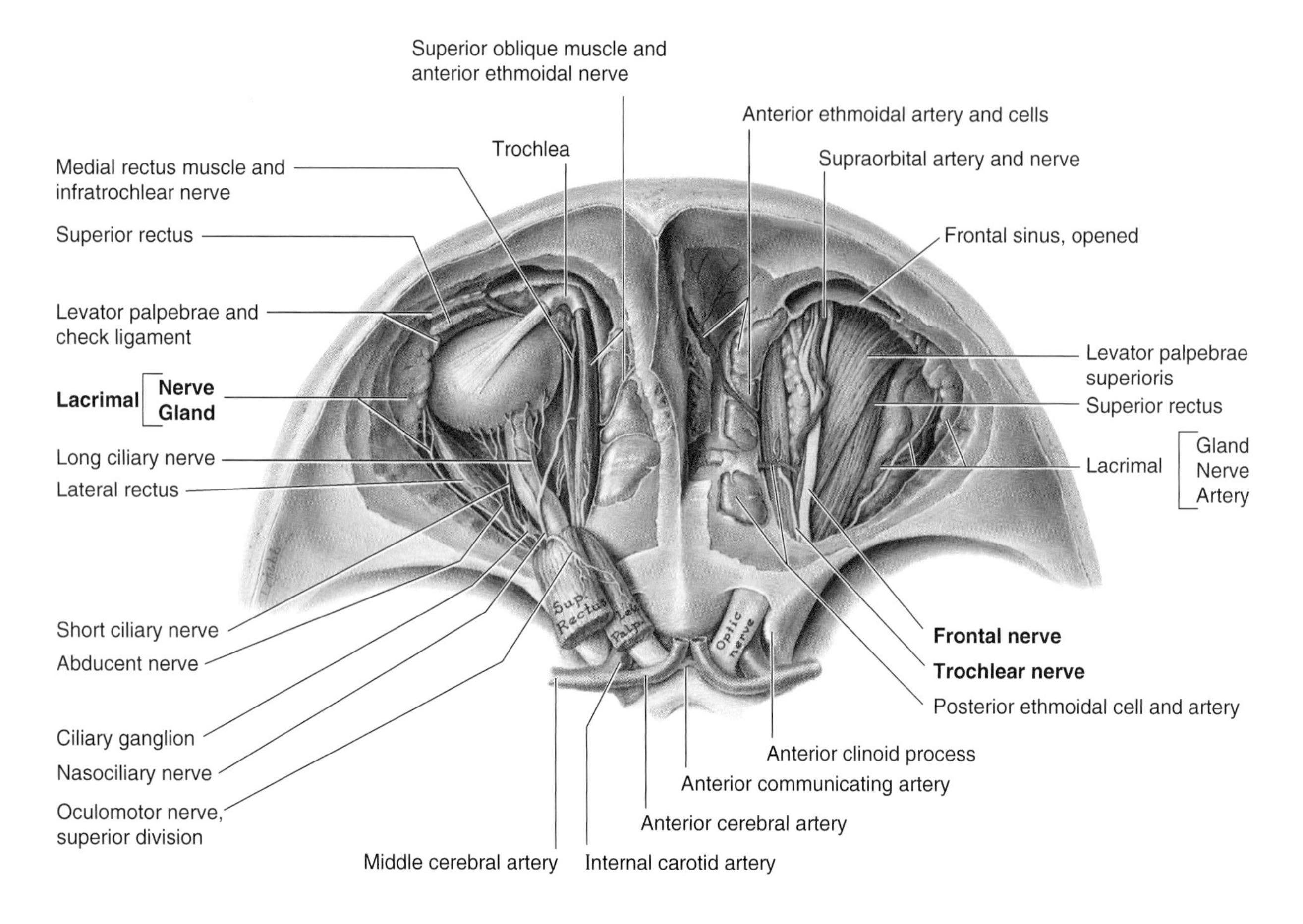

(B) Superior view

Figure 7.35. *(Continued)* **B.** The orbital part of the frontal bone is removed. Observe the three nerves applied to the roof of the orbital cavity—trochlear, frontal, and lacrimal.

Corneal Ulcers and Transplants

Damage to the sensory innervation of the cornea from CN V_1 leaves the cornea vulnerable to injury by foreign particles. Persons with scarred or opaque corneas may receive *corneal transplants* from donors. Corneal implants of nonreactive plastic material are also used.

Presbyopia and Cataracts

As people get older, their lenses become harder and more flattened. These changes gradually reduce the focusing power of the lenses—*presbyopia* (G. presbyos, old). Some elderly people also experience a loss of transparency (*cataracts*) of the lens from areas of opaqueness. *Cataract extraction* is a common eye operation.

Hemorrhage Into the Anterior Chamber

Hemorrhage into the anterior chamber of the eye (*hyphema*) usually results from blunt trauma to the eyeball, such as from a squash or racquet ball or a hockey puck. Initially the anterior chamber is tinged red, but blood soon accumulates in this cavity. The initial hemorrhage usually stops in a few days, and recovery is usually good. ⊙

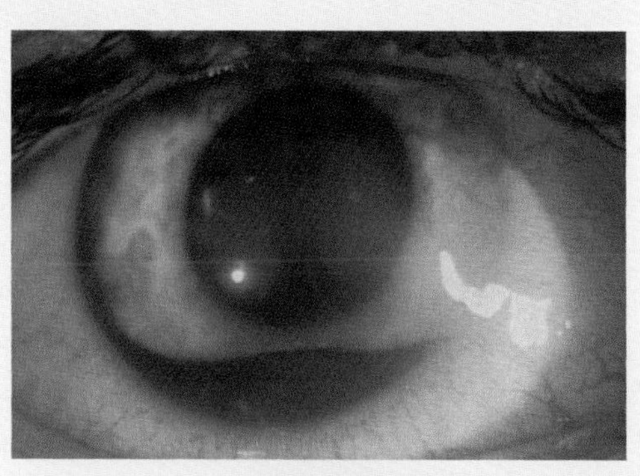

Hyphema

Muscles of the Orbit

The muscles of the orbit are the:

- Levator palpebrae superioris
- Four recti (superior, inferior, medial, and lateral)
- Two oblique (superior and inferior).

These muscles work together to move the upper lids and eyes. The attachments, nerve supply, and actions of the orbital muscles are illustrated in Figures 7.35 and 7.36 and in Table 7.8.

Levator Palpebrae Superioris

This thin, flat *elevator muscle of the superior eyelid* broadens into a wide aponeurosis as it approaches its distal attachment to the tarsal plate. This muscle is the opponent of the orbicularis oculi, the sphincter of the palpebral fissure.

Recti and Oblique Muscles

The four recti arise from a fibrous cuff (Fig. 7.36*B*), the **common tendinous ring**, that surrounds the optic canal and part of the superior orbital fissure (Fig. 7.37). Structures that enter the orbit through this canal and the adjacent part of the fissure lie at first in the cone of recti. The lateral and medial recti lie in the same horizontal plane, and the superior and inferior recti lie in the same vertical plane. All four **recti muscles** attach to the sclera on the anterior half of the eyeball. *The actions of the recti muscles are* (Table 7.8):

- Medial and lateral recti rotate the pupil medially and laterally, respectively
- Superior rectus rotates the pupil superiorly (elevation)
- Inferior rectus rotates the pupil inferiorly (depression).

Neither the superior rectus nor the inferior rectus pulls directly parallel to the long axis of the eyeball. As a result, both recti tend to rotate the pupil medially (adduction). This medial pull of the superior and inferior recti is normally balanced by a similar tendency of the oblique muscles to rotate the pupil laterally (abduction).

The **inferior oblique** directs the pupil laterally and superiorly; therefore, when it works synergistically with the superior rectus, superior movement of the eyeball occurs. Similarly, the **superior oblique** directs the pupil inferiorly and laterally; therefore, when it works synergistically with the inferior rectus, an inferior movement results.

Fascial Sheath of the Eyeball

The fascial sheath (bulbar sheath, Tenon's capsule) envelops the eyeball from the optic nerve to the corneoscleral junction, forming the actual socket for the eyeball (Fig. 7.31*A*). The fascial sheath is pierced by the tendons of the extraocular muscles and is reflected onto each of them as a tubular sheath. There are triangular expansions from the sheaths of the medial and lateral rectus muscles called the medial and lateral **check ligaments** (Fig. 7.35*B*), which are attached to the lacrimal and zygomatic bones, respectively, limiting abduc-

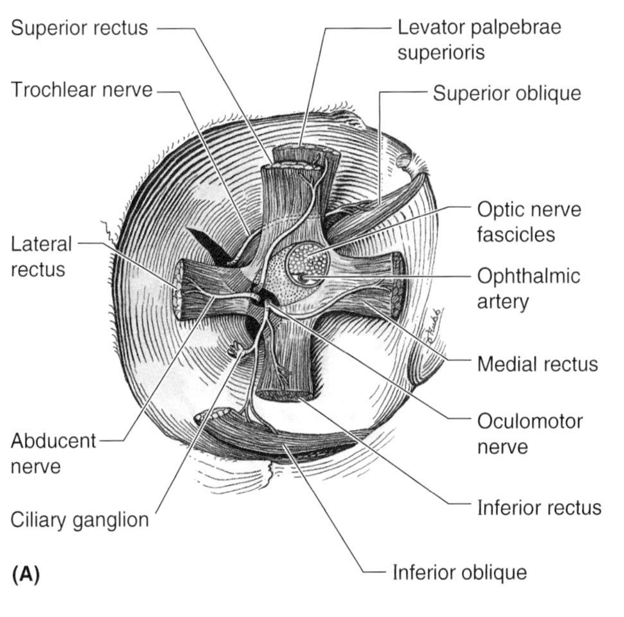

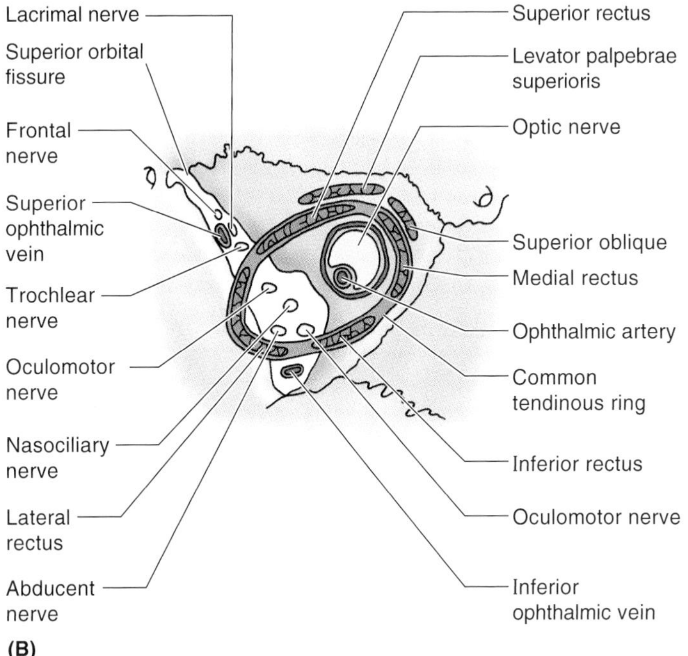

Figure 7.36. Nerves of the orbit (orbital cavity). A. Dissection of the orbit showing the common tendinous ring and the motor nerves of the orbit. Note the four rectus muscles arising from the fibrous cuff and the common tendinous ring. This ring encircles the dural sheath of the optic nerve (CN II), the abducent nerve (CN VI), and the superior and inferior divisions of the oculomotor nerve (CN III). **B.** Structures in the apex of the orbit. Observe the nerves that enter the orbit through the superior orbital fissure and supply the muscles of the eyeball (oculomotor [CN III], trochlear [CN IV], and abducent [CN VI]).

Table 7.8. **Muscles of Orbit**

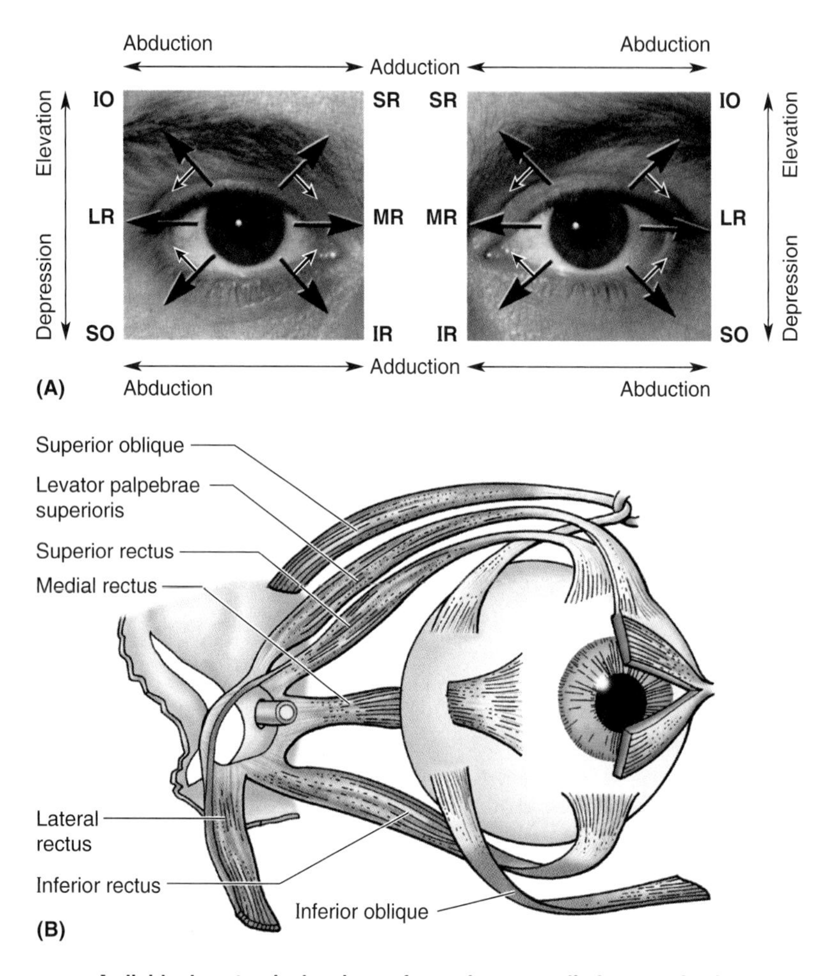

Individual anatomical actions of muscles as studied anatomically

It is essential to appreciate that all muscles are continuously involved in eye movements; thus, the individual actions are not usually tested clinically. **SR**, superior rectus (CN III); **LR**, lateral rectus (CN VI); **IR**, inferior rectus (CN III); **IO**, inferior oblique (CN III); **MR**, medial rectus (CN III); **SO**, superior oblique (CN IV).

Muscle	Origin	Insertion	Innervation	Main Action
Levator palpebrae superioris	Lesser wing of sphenoid bone, superior and anterior to optic canal	Tarsal plate and skin of superior (upper) eyelid	Oculomotor nerve; deep layer (superior tarsal muscle) is supplied by sympathetic fibers	Elevates superior (upper) eyelid
Superior rectus	Common tendinous ring	Sclera just posterior to cornea	Oculomotor nerve	Elevates, adducts, and rotates eyeball medially
Inferior rectus				Depresses, adducts, and rotates eyeball laterally
Lateral rectus			Abducent nerve	Abducts eyeball
Medial rectus			Oculomotor nerve	Adducts eyeball
Superior oblique	Body of sphenoid bone	Its tendon passes through a fibrous ring or trochlea, changes its direction, and inserts into sclera deep to superior rectus muscle	Trochlear nerve	Abducts, depresses, and medially rotates eyeball
Inferior oblique	Anterior part of floor of orbit	Sclera deep to lateral rectus muscle	Oculomotor nerve	Abducts, elevates, and laterally rotates eyeball

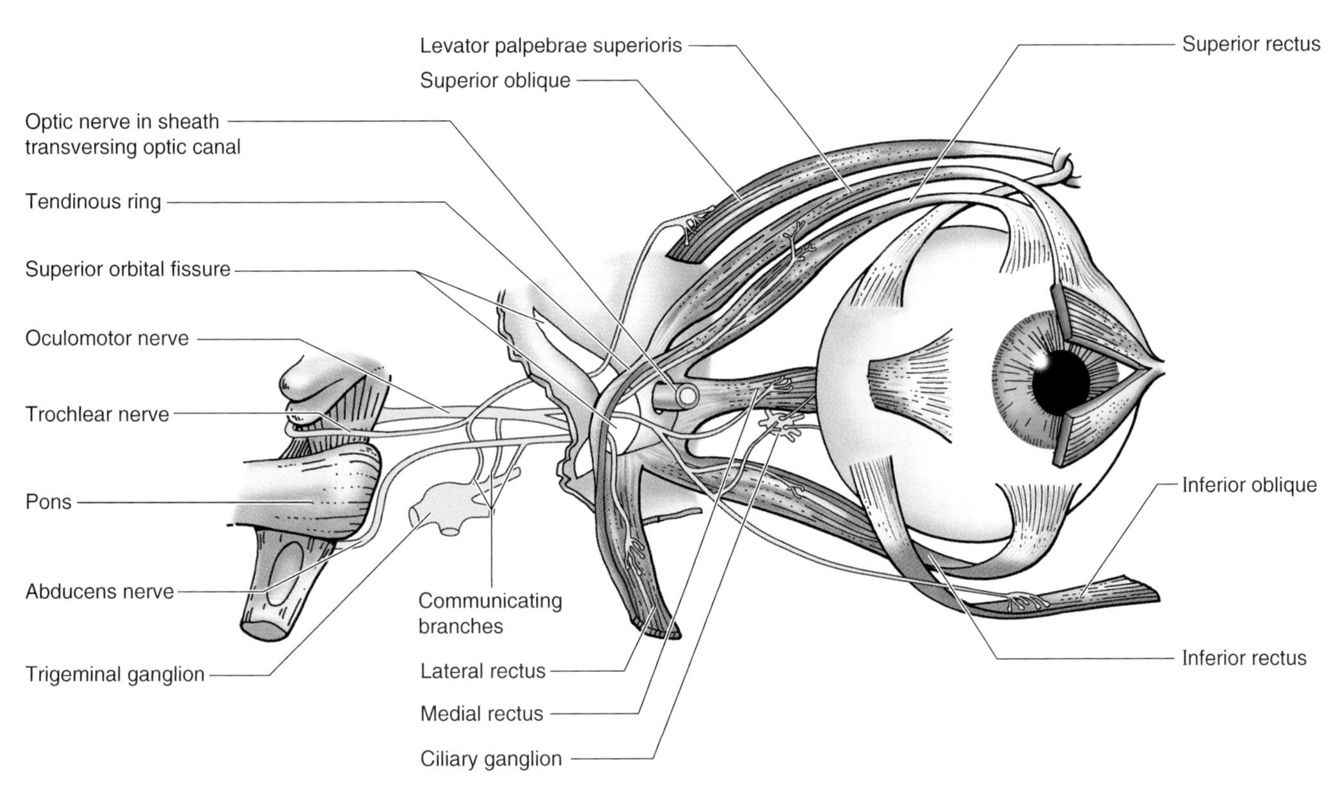

Figure 7.37. Innervation of the muscles of the eyeball. Observe the distribution of the oculomotor (CN III), trochlear (CN IV), and abducent (CN VI) nerves to the muscles of the eyeball. The nerves enter the orbit (eye socket, orbital cavity) through the superior orbital fissure. Note that CN IV supplies the superior oblique, CN VI supplies the lateral rectus, and CN III supplies the remaining five muscles.

tion and adduction. A blending of the check ligaments with the fascia of the inferior rectus and inferior oblique muscles forms a hammocklike sling, the **suspensory ligament** (Fig. 7.34), which supports the eyeball. A potential space between the eyeball and the fascial sheath allows the eyeball to move inside the cuplike sheath. A similar check ligament from the fascial sheath of the inferior rectus retracts the lower eyelid when the gaze is directed downward (Fig. 7.31*A*). Because the sheaths of the superior rectus and levator palpebrae superioris are fused, the upper eyelid is elevated when the gaze is directed upward.

Artificial Eye

The cuplike fascial sheath of the eyeball forms a socket for an artificial eye when the eyeball is removed (*enucleated*). After this operation, the eye muscles cannot retract too far because their fascial sheaths remain attached to the fascial sheath of the eyeball. Thus, some coordinated movement of a properly fitted artificial eye is possible. Because the suspensory ligament supports the eyeball, it is preserved when surgical removal of the bony floor of the orbit is carried out (e.g., during the removal of a tumor). ✪

Innervation of the Orbit

In addition to the *optic nerve*, the nerves of the orbit include those that enter through the *superior orbital fissure* and supply the ocular muscles: oculomotor, *III*; trochlear, *IV*; and abducent, *VI* (Figs. 7.35 and 7.36, Table 7.8).

- CN III supplies the levator palpebrae superioris, superior rectus, medial rectus, inferior rectus, and inferior oblique.
- CN IV supplies the superior oblique.
- CN VI supplies the lateral rectus.

In summary, all muscles of the orbit are supplied by CN III, except for the superior oblique and lateral rectus, which are supplied by CN IV and VI, respectively. A memory device is: **LR_6, SO_4, AO_3** (**L**ateral **R**ectus, CN **VI**, **S**uperior **O**blique, CN **IV**, **A**ll **O**thers, CN **III**).

Several branches of the **ophthalmic nerve** (CN V_1) pass through the superior orbital fissure and supply structures in the orbit. The **lacrimal nerve** arises in the lateral wall of the cavernous sinus and passes to the lacrimal gland, giving branches to the conjunctiva and skin of the superior eyelid and providing secretomotor fibers conveyed to it from the zygomatic nerve (CN V_2). The **frontal nerve** divides into the supraorbital nerve and supratrochlear nerve, which supply the upper eyelid, forehead, and scalp. The **nasociliary nerve**, the sensory nerve to the eye, supplies several branches to the orbit. The **infratrochlear nerve**, a terminal branch of the nasociliary nerve, supplies the eyelids, conjunctiva, skin of the nose, and lacrimal sac. The **ethmoidal nerves**, also branches of the nasociliary nerve, supply the mucous membrane of the sphenoidal and ethmoidal sinuses and the nasal cavities, and the dura of the anterior cranial fossa.

The **short ciliary nerves**, branches of the ciliary ganglion (Fig. 7.35*B*), carry parasympathetic and sympathetic fibers to the ciliary body and iris. The **ciliary ganglion** is a small group of nerve cell bodies between the optic nerve and lateral rectus toward the posterior limit of the orbit. The short ciliary nerves consist of postsynaptic parasympathetic fibers originating in the ciliary ganglion, afferent fibers from the nasociliary nerve that pass through the ganglion, and postsynaptic sympathetic fibers that also pass through it. The **long ciliary nerves**, branches of the nasociliary nerve (CN V_1), transmit postsynaptic sympathetic fibers to the dilator pupillae and afferent fibers from the iris and cornea.

Oculomotor Nerve Palsy

Complete oculomotor nerve palsy affects most of the ocular muscles, the levator palpebrae superioris, and the sphincter pupillae. The superior eyelid droops and cannot be raised voluntarily because of the unopposed orbicularis oculi, supplied by the facial nerve. Paralysis of CN VII does not cause ptosis but prevents wrinkling of the eyelids. The pupil is also fully dilated and nonreactive because of the unopposed dilator pupillae. The pupil is fully abducted and depressed ("down and out") because of the unopposed lateral rectus and superior oblique, respectively.

Horner Syndrome

Interruption of a cervical sympathetic trunk results in paralysis of the superior tarsal muscle supplied by sympathetic fibers, causing ptosis. Other signs of the Horner syndrome are a constricted pupil; sinking, redness, and dryness of the eye; and increased temperature of the face on the affected side.

Paralysis of the Extraocular Muscles

One or more extraocular muscles may be paralyzed by disease in the brainstem or by head injury, resulting in *diplopia* (double vision). Paralysis of a muscle is apparent by the limitation of eye movement in the field of action of the muscle, and by the production of two images when one attempts to use the muscle. When the abducent nerve supplying only the lateral rectus is paralyzed, the patient cannot abduct the pupil on the affected side. The pupil is fully adducted by the unopposed pull of the medial rectus. ⊙

Vasculature of the Orbit

Arteries of the Orbit

The blood supply of the orbit is mainly from the **ophthalmic artery**; the **infraorbital artery** also contributes blood to this region (Figs. 7.35 and 7.36, Table 7.9). The **central artery of the retina**, a branch of the ophthalmic artery inferior to the optic nerve, runs within the dural sheath of the optic nerve until it approaches the eyeball (Fig. 7.34). The central artery pierces the optic nerve and runs within it to emerge at the optic disc. Branches of the central artery spread over the internal surface of the retina. The terminal branches of the central artery are end arteries that provide the only blood supply to the retina. The retina is also supplied by the capillary layer of the choroid (Fig. 7.38), the **choriocapillary layer** (lamina). Of the eight or so posterior ciliary arteries—also branches of the ophthalmic artery—six short posterior ciliary arteries directly supply the choroid, which nourishes the outer nonvascular layer of the retina. Two long **posterior ciliary arteries**, one on each side of the eyeball, pass between the sclera and choroid to anastomose with the anterior ciliary arteries—continuations of the muscular branches of the ophthalmic artery—supplying the ciliary plexus.

Veins of the Orbit

Venous drainage of the orbit is through the **superior** and **inferior ophthalmic veins** that pass through the superior orbital fissure and enter the cavernous sinus (Fig. 7.39). The **central vein of the retina** (Fig. 7.38) usually enters the cavernous sinus directly, but it may join one of the ophthalmic veins. The vortex or **vorticose veins** from the vascular layer drain into the inferior ophthalmic vein. The **scleral venous sinus** is a vascular structure encircling the anterior chamber of the eye through which the aqueous humor is returned to the blood circulation.

Glaucoma

When drainage of aqueous humor through the scleral venous sinus decreases significantly, pressure builds up in the anterior and posterior chambers of the eye—a condition called *glaucoma*. Blindness can result from compression of the neural layer of the retina and the retinal blood supply if aqueous humor production is not reduced to maintain normal intraocular pressure.

Blockage of the Central Retinal Artery

Because terminal branches of the central retinal artery are end arteries, obstruction of them by an embolus results in instant and total blindness. Blockage of the artery is usually unilateral and occurs in older people. ▶

Table 7.9. **Arteries of Orbit**

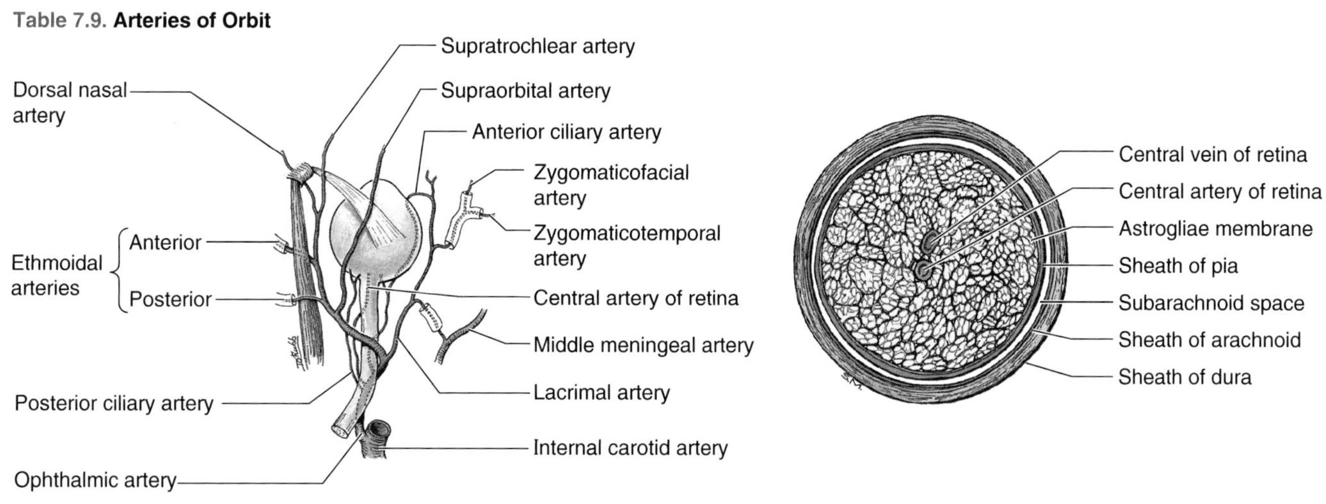

(A) Superior view

(B) Transverse section of optic nerve

Artery	Origin	Course and Distribution
Ophthalmic	Internal carotid artery	Traverses optic foramen to reach orbital cavity
Central artery of retina	Ophthalmic artery	Runs in dural sheath of optic nerve and pierces nerve near eyeball; appears at center of optic disc; supplies optic retina (except cones and rods)
Supraorbital		Passes superiorly and posteriorly from supraorbital foramen to supply forehead and scalp
Supratrochlear		Passes from supraorbital margin to forehead and scalp
Lacrimal		Passes along superior border of lateral rectus muscle to supply lacrimal gland, conjunctiva, and eyelids
Dorsal nasal		Courses along dorsal aspect of nose and supplies its surface
Short posterior ciliaries		Pierce sclera at periphery of optic nerve to supply choroid, which in turn supplies cones and rods of optic retina
Long posterior ciliaries		Pierce sclera to supply ciliary body and iris
Posterior ethmoidal		Passes through posterior ethmoidal foramen to posterior ethmoidal cells
Anterior ethmoidal		Passes through anterior ethmoidal foramen to anterior cranial fossa; supplies anterior and middle ethmoidal cells, frontal sinus, nasal cavity, and skin on dorsum of nose
Anterior ciliary	Muscular (rectus) branches of ophthalmic artery	Pierces sclera at attachments of rectus muscles and forms network in iris and ciliary body
Infraorbital	Third part of maxillary artery	Passes along infraorbital groove and foramen to face

Blockage of the Central Retinal Vein

Because the central vein of the retina enters the cavernous sinus, *thrombophlebitis* of this sinus may result in the passage of thrombi to the central retinal vein and produce clotting in the small retinal veins. Blockage of the central retinal vein usually results in slow, painless loss of vision. ⊙

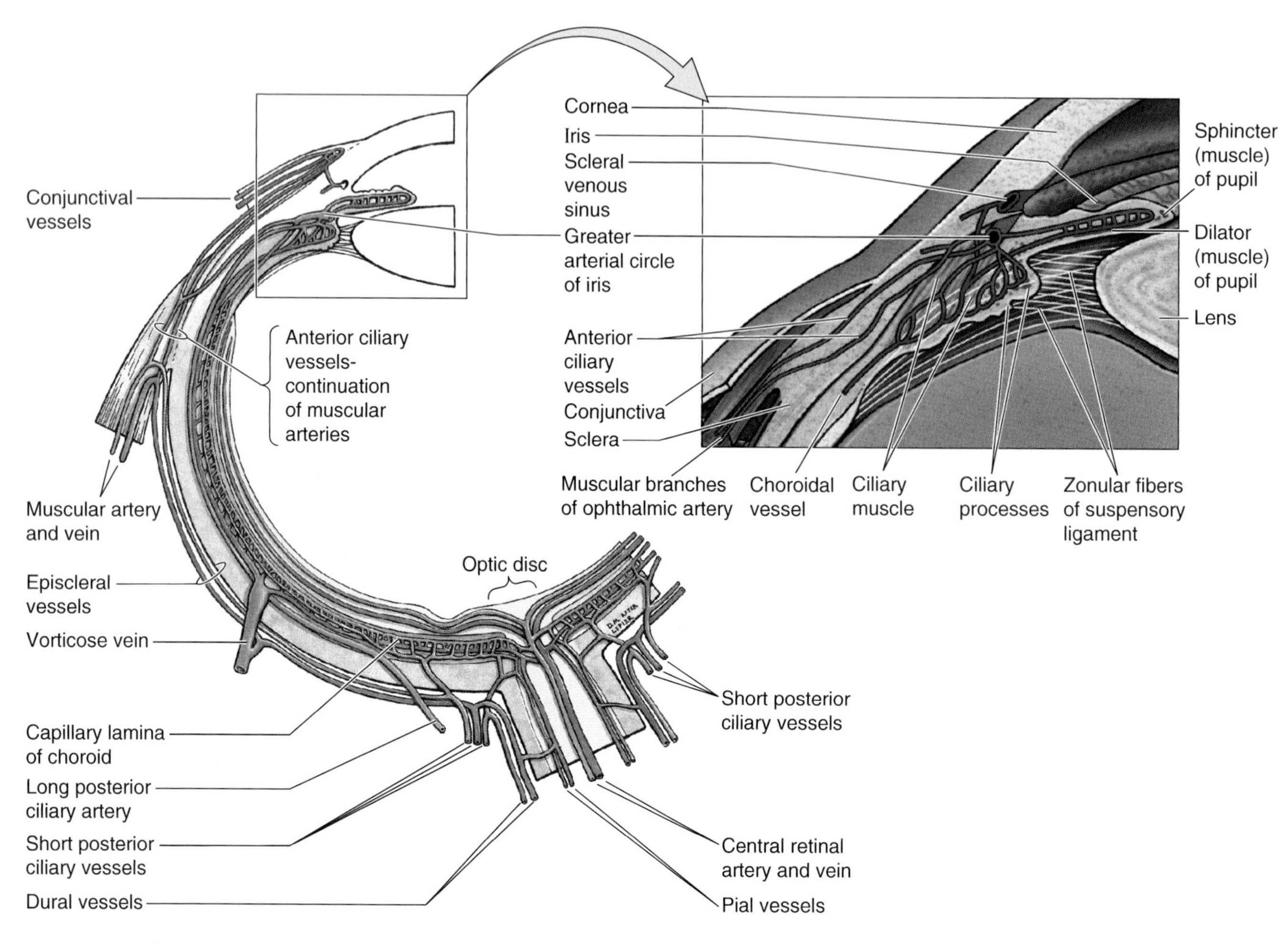

Figure 7.38. Partial horizontal section of the right eyeball. Observe the central artery of the retina, a branch of the ophthalmic artery. It is an end artery. Of the eight or so posterior ciliary arteries, six supply the choroid, which in turn nourishes the outer nonvascular layer of the retina. Two long posterior ciliary arteries, one on each side of the eyeball, run between the sclera and the choroid to anastomose with the anterior ciliary arteries, which are derived from muscular branches. Examine the vorticose vein (one of four to five), which drains venous blood from the choroid into the posterior ciliary and ophthalmic veins (see Fig. 7.39). The scleral venous sinus (L. sinus venosus sclerae, canal of Schlemm) returns the aqueous humor, secreted into the anterior chamber by the ciliary processes, to the venous circulation.

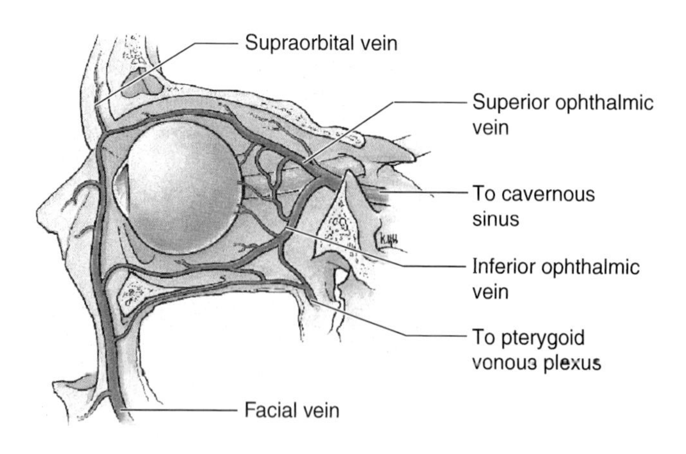

Figure 7.39. Ophthalmic veins. Observe the superior and inferior ophthalmic veins that empty into the cavernous sinus and the pterygoid venous plexus posteriorly and communicate with the facial and supraorbital veins anteriorly. The superior ophthalmic vein accompanies the ophthalmic artery and its branches. These venous communications allow infection or thrombophlebitis to readily extend to the cranial cavity (cavernous sinus) and deep face (pterygoid venous plexus) (p. 878).

Surface Anatomy of the Eyeball, Eyelids, and Lacrimal Apparatus

The opening between the superior and inferior eyelids is the **palpebral fissure**. When the eyelids are closed, the palpebral fissure is nearly in the horizontal plane, except in certain races (e.g., Asian). These individuals have a slight superior slant of the palpebral fissure toward the nose because the medial ends of the upper eyelids project superomedially. Furthermore, the medial angles of their eyeballs are covered by an *epicanthal fold* that varies in size. In a lateral view (*A*), most of the visible part of the eyeball ▶

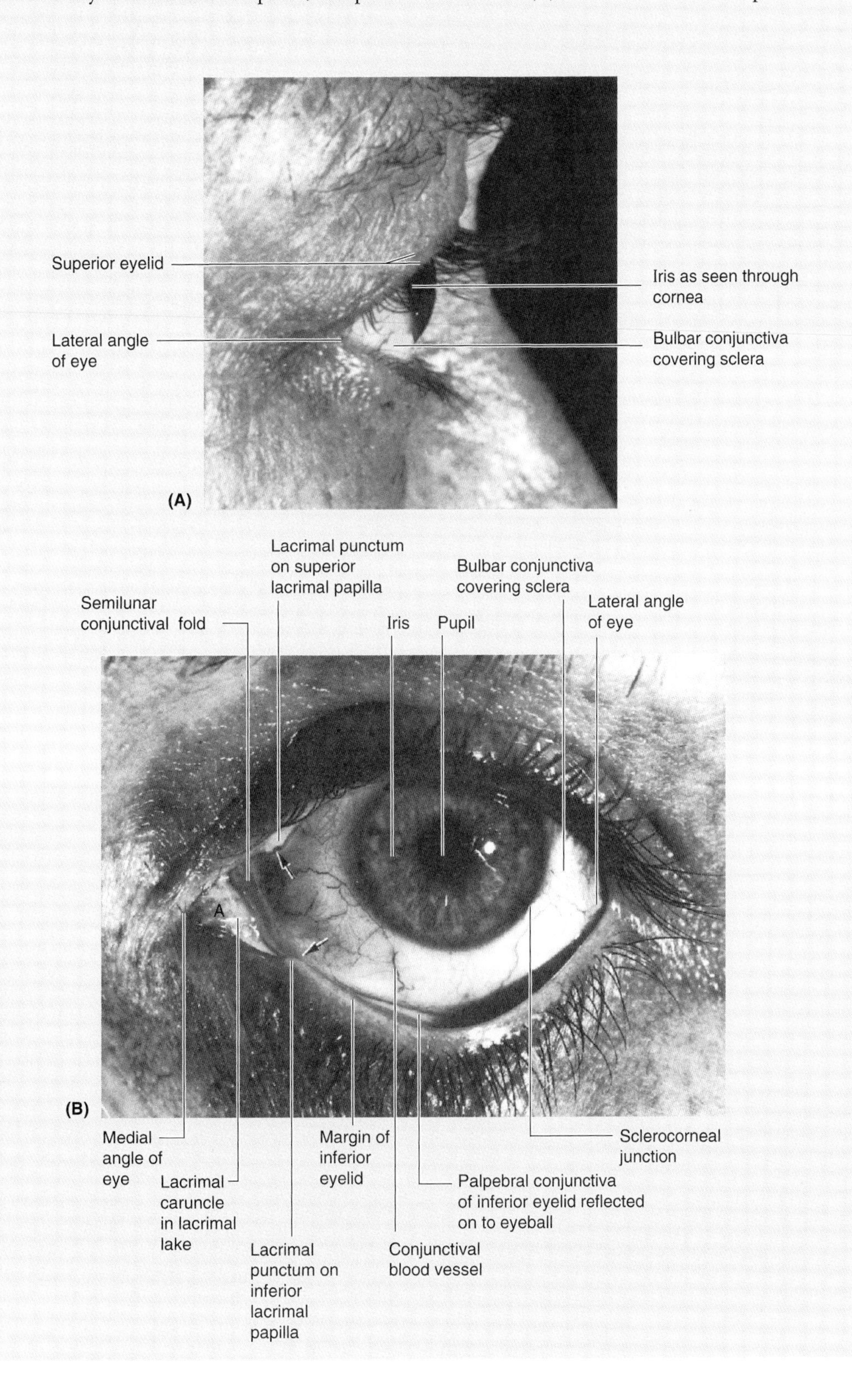

▶ protrudes slightly between the eyelids (i.e., through the palpebral fissure). In an anterior view, most of the eyeball appears to lie within the orbit. The anterior part of the **sclera** is covered by the transparent bulbar conjunctiva. The tough, opaque sclera appears slightly blue in infants and children and has a yellow hue in many older people. The anterior transparent part of the eye is the **cornea**, which is continuous with the sclera at its margins. However, as is apparent when viewed laterally, the cornea has a greater curvature than that of the rest of the eyeball (the part covered by sclera); thus, an angle occurs at the *sclerocorneal junction*, or limbus of cornea. The dark circular opening through which light enters the eye—the **pupil**—is surrounded by the **iris**, a circular, pigmented diaphragm.

The **bulbar conjunctiva** is reflected from the sclera onto the deep surface of the eyelid. The **palpebral conjunctiva** is normally red and vascular and is commonly examined in cases of suspected *anemia*, a blood condition commonly manifested by pallor of the mucous membranes. As the bulbar conjunctiva is continuous with the anterior epithelium of the cornea and with the palpebral conjunctiva, it forms the **conjunctival sac.** The *palpebral fissure* is the "mouth" of the conjunctival sac. When the eyelids are closed, the bulbar and palpebral conjunctivae form a closed conjunctival sac.

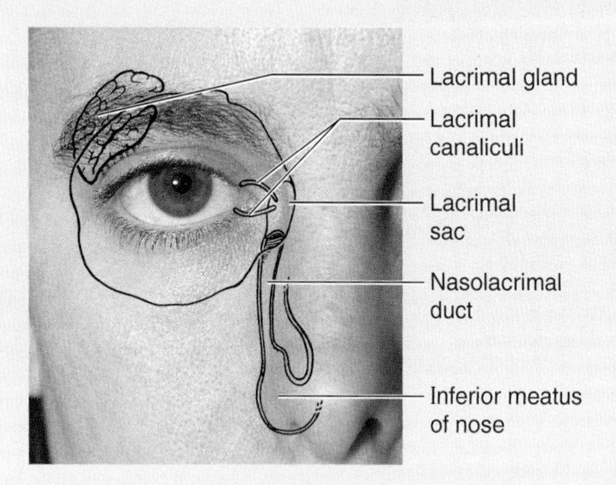

In the **medial angle** of the eye, observe a reddish, shallow reservoir of tears—the **lacrimal lake** (see *B* on p. 915). Within the lake is the *lacrimal caruncle*, a small mound of moist modified skin. Lateral to the caruncle is a **semilunar conjunctival fold**, which slightly overlaps the eyeball. The semilunar fold is a rudiment of the nictitating membrane of birds and reptiles. When the edges of the eyelids are everted, a small pit—the **lacrimal punctum**—is visible at its medial end on the summit of a small elevation—the **lacrimal papilla** (see *B* on p. 915). A similar punctum and papilla are on the superior eyelid.

The **lacrimal gland** in the superolateral part of the orbit secretes lacrimal fluid; its ducts convey the fluid to the surface of the eyeball. When the cornea begins to dry, the eyelids blink, carrying a film of fluid over the cornea. Each lacrimal punctum is the opening of a slender canal, the **lacrimal canaliculus**, which carries the fluid to the **lacrimal sac.** From here the fluid passes through the **nasolacrimal duct** to the **inferior meatus of the nose**—a passage along the lateral wall of the nasal cavity formed by the projection of the inferior nasal concha. The fluid drains to the posterior aspect of the nasal cavity to the nasopharynx, where it is swallowed. ○

Temporal Region

The temporal region includes the temporal and infratemporal fossae—superior and inferior to the zygomatic arch, respectively (Fig. 7.40).

Temporal Fossa

The temporal fossa (Fig. 7.40*A*), in which the temporal muscle is located, is bounded:

- Posteriorly and superiorly by the temporal lines
- Anteriorly by the frontal and zygomatic bones
- Laterally by the zygomatic arch
- Inferiorly by the infratemporal crest (Fig. 7.40*B*).

The *floor of the temporal fossa* is formed by parts of the four bones that form the *pterion*: frontal, parietal, temporal, and greater wing of the sphenoid. The fan-shaped temporal muscle arises from the floor (i.e., the temporal fossa extending to the inferior temporal line) and the overlying temporal fascia (Fig. 7.41), which comprises the *roof of the temporal fossa.* This tough fascia covers the temporal muscle, extending to

Figure 7.40. Bony boundaries of the temporal and infratemporal fossae. A. Lateral view of the intact skull. The lateral wall of the infratemporal fossa is formed by the ramus of the mandible (lower jaw). The space is deep to the zygomatic arch and is traversed by the temporal (L. temporalis) muscle and the deep temporal nerves and vessels. Through this interval, the temporal fossa communicates inferiorly with the infratemporal fossa. **B.** Lateral view of the skull with the zygomatic arch and the ramus of the mandible removed, showing the roof and three walls of the infratemporal fossa. The fossa is an irregularly shaped space posterior to the maxilla (anterior wall). The roof of the fossa is formed by the infratemporal surface of the greater wing of the sphenoid. The medial wall is formed by the lateral pterygoid plate, and the posterior wall is formed by the tympanic plate, styloid process, and mastoid process of the temporal bone. The infratemporal fossa communicates with the pterygopalatine fossa through the pterygomaxillary fissure.

the superior temporal line. Inferiorly, the fascia splits into two layers that attach to the superior margin of the zygomatic arch. The temporal fascia also tethers the zygomatic arch superiorly. When the powerful masseter—attached to the inferior border of the arch—contracts, exerting a strong downward pull on the arch, the temporal fascia provides resistance.

Infratemporal Fossa

The infratemporal fossa is an irregularly shaped space deep and inferior to the zygomatic arch, deep to the ramus of the mandible, and posterior to the maxilla (Fig. 7.40*B*). It communicates with the temporal fossa through the interval be-

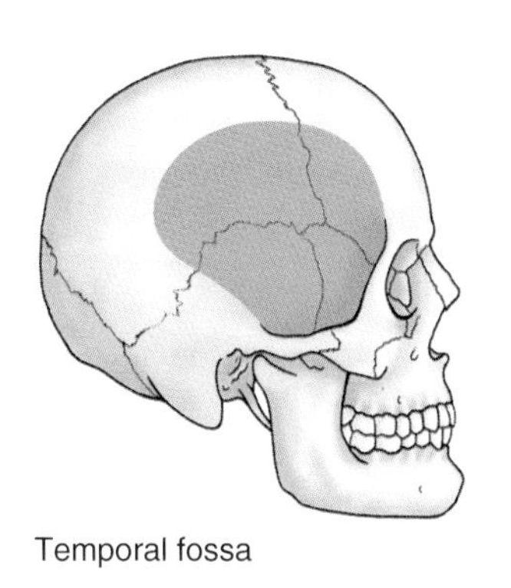

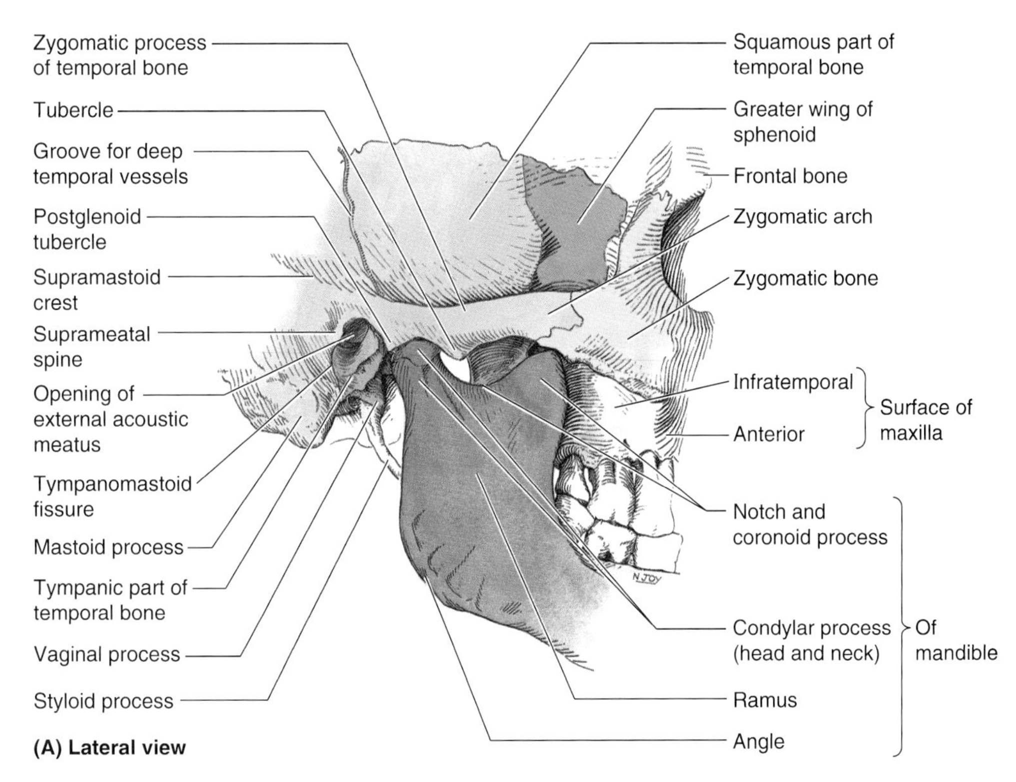

(A) Lateral view

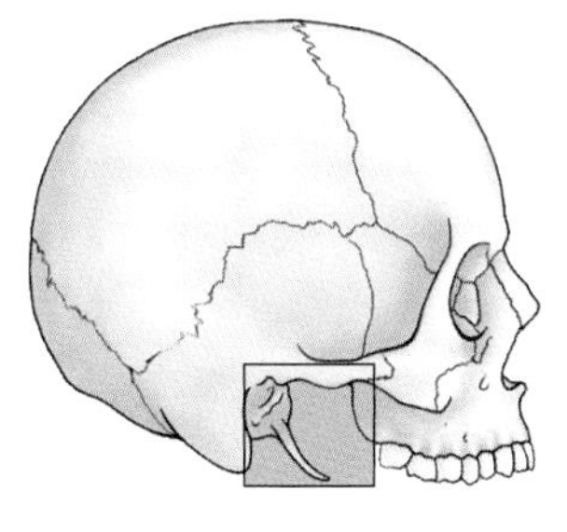

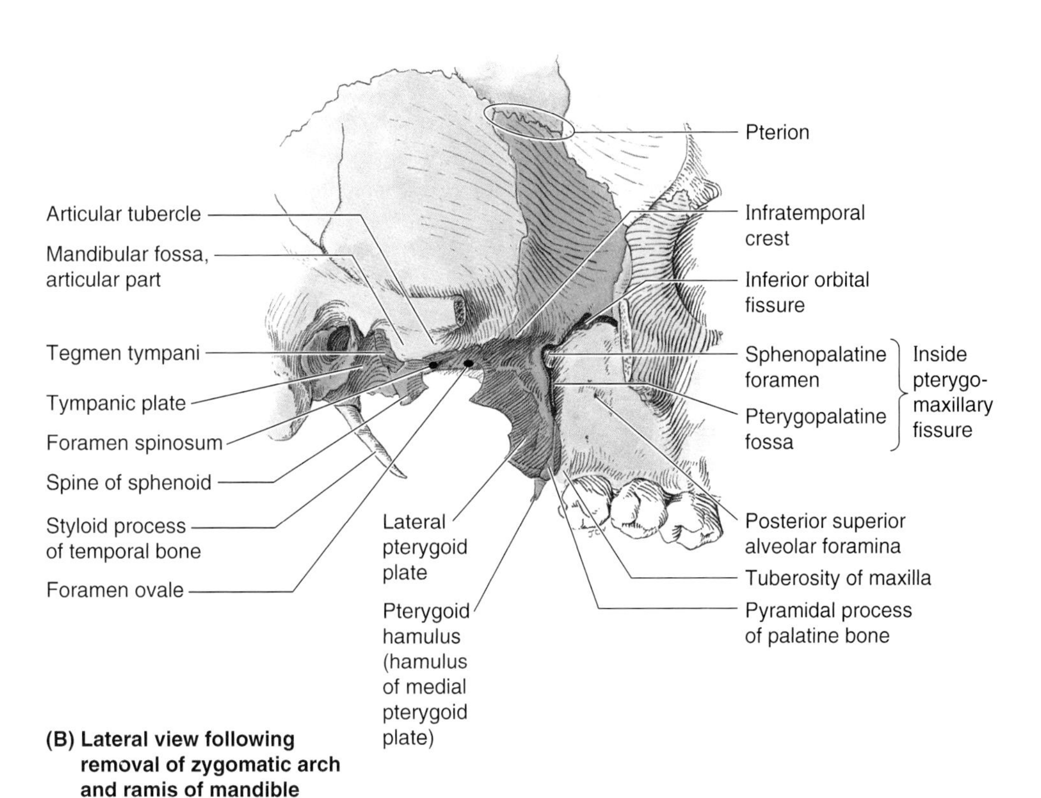

(B) Lateral view following removal of zygomatic arch and ramis of mandible

tween (deep to) the zygomatic arch and (superficial to) the cranial bones.

Boundaries of the Infratemporal Fossa

The boundaries of the infratemporal fossa (Fig. 7.40, *A* and *B*) are:

- Laterally—ramus of the mandible
- Medially—lateral pterygoid plate
- Anteriorly—posterior aspect of the maxilla
- Posteriorly—tympanic plate and the mastoid and styloid processes of the temporal bone
- Superiorly—inferior (infratemporal) surface of the greater wing of the sphenoid bone
- Inferiorly—where the medial pterygoid muscle attaches to the mandible near its angle (Fig. 7.41, *A* and *B*).

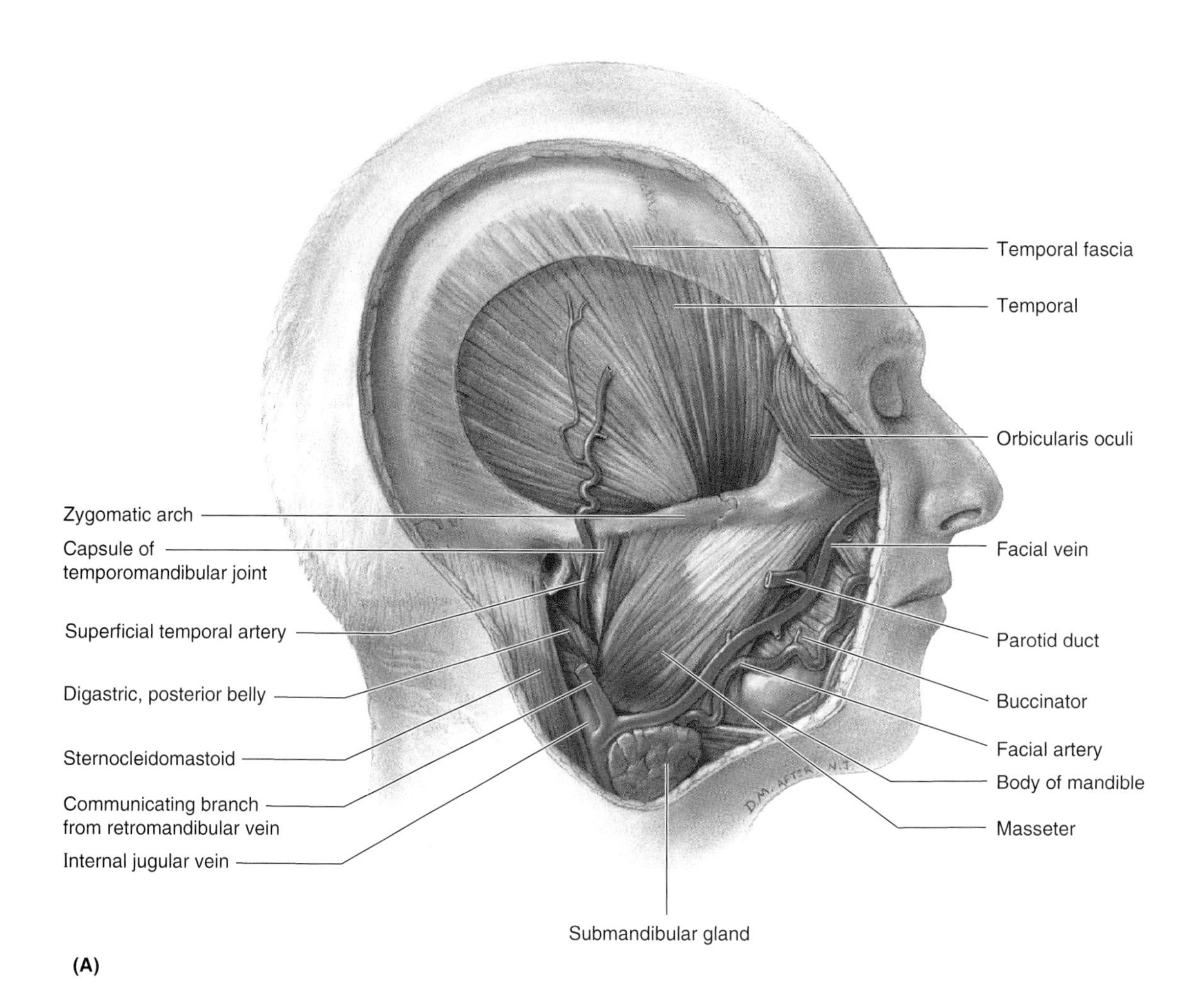

Figure 7.41. Dissections of the temporal and infratemporal regions. Lateral views. **A.** Lateral view of a superficial dissection of the great muscles on the side of the skull. The parotid gland and most of the temporal fascia have been removed. Observe the temporal (L. temporalis) and masseter muscles; both are supplied by the trigeminal nerve (CN V) and both close the jaw. Observe also the submandibular gland, with the facial artery passing deep to it and the facial vein passing superficial to it. **B.** Superficial dissection of the infratemporal region. Most of the zygomatic arch and attached masseter muscle, the coronoid process and adjacent parts of the ramus of the mandible (lower jaw), and the inferior half of the temporal muscle have been removed. Observe the first part of the maxillary artery—the larger of the two end branches of the external carotid—running anteriorly, deep to the neck of the mandible and then passing deeply between the lateral and medial pterygoid muscles. **C.** Deep dissection of the infratemporal region. More of the ramus of the mandible, the lateral pterygoid muscle, and most branches of the maxillary artery have been removed. Observe the second part of the maxillary artery, branches of the mandibular nerve (CN V_3) including the auriculotemporal nerve passing between the sphenomandibular ligament and the neck of the mandible.

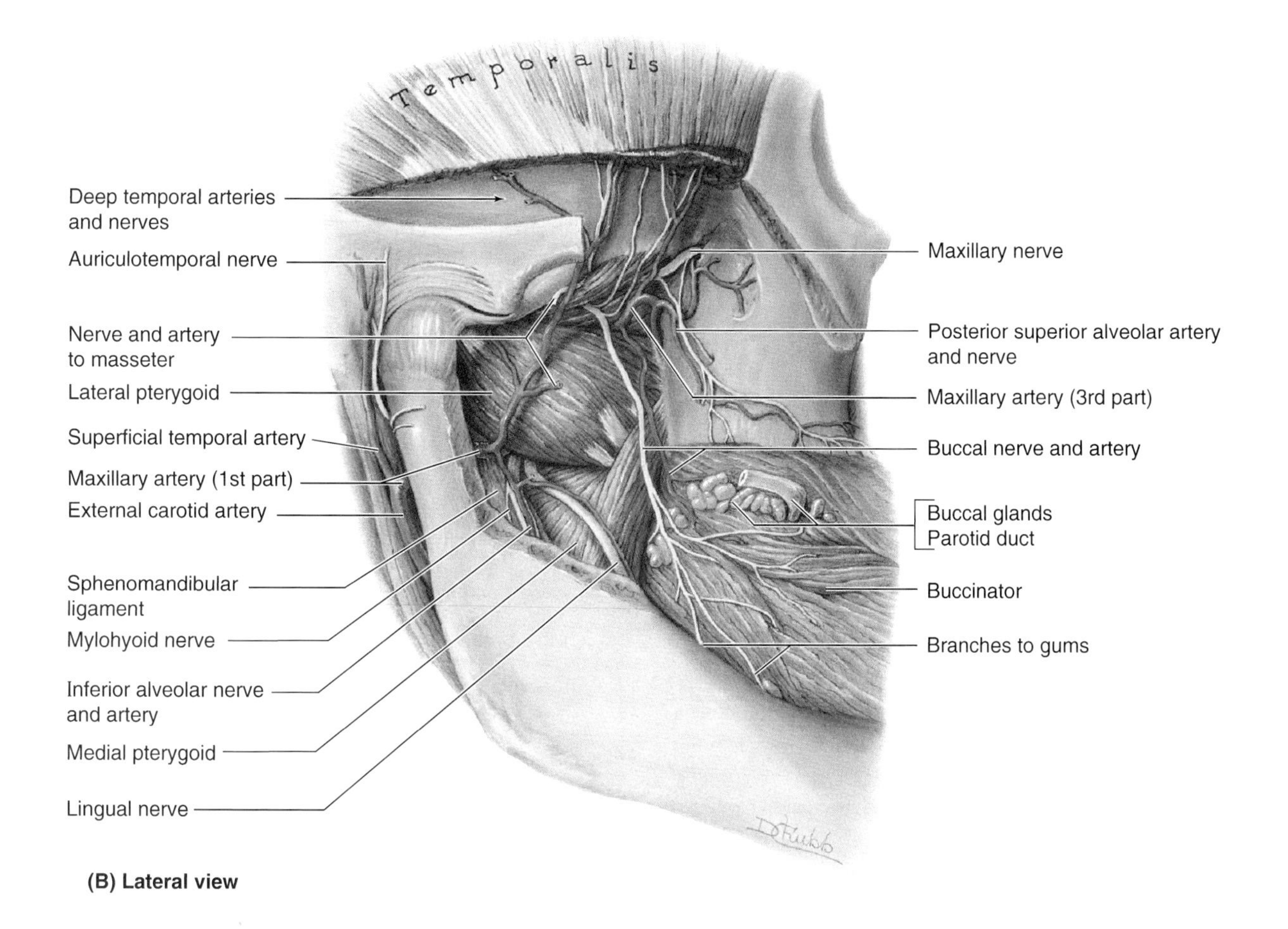

(B) Lateral view

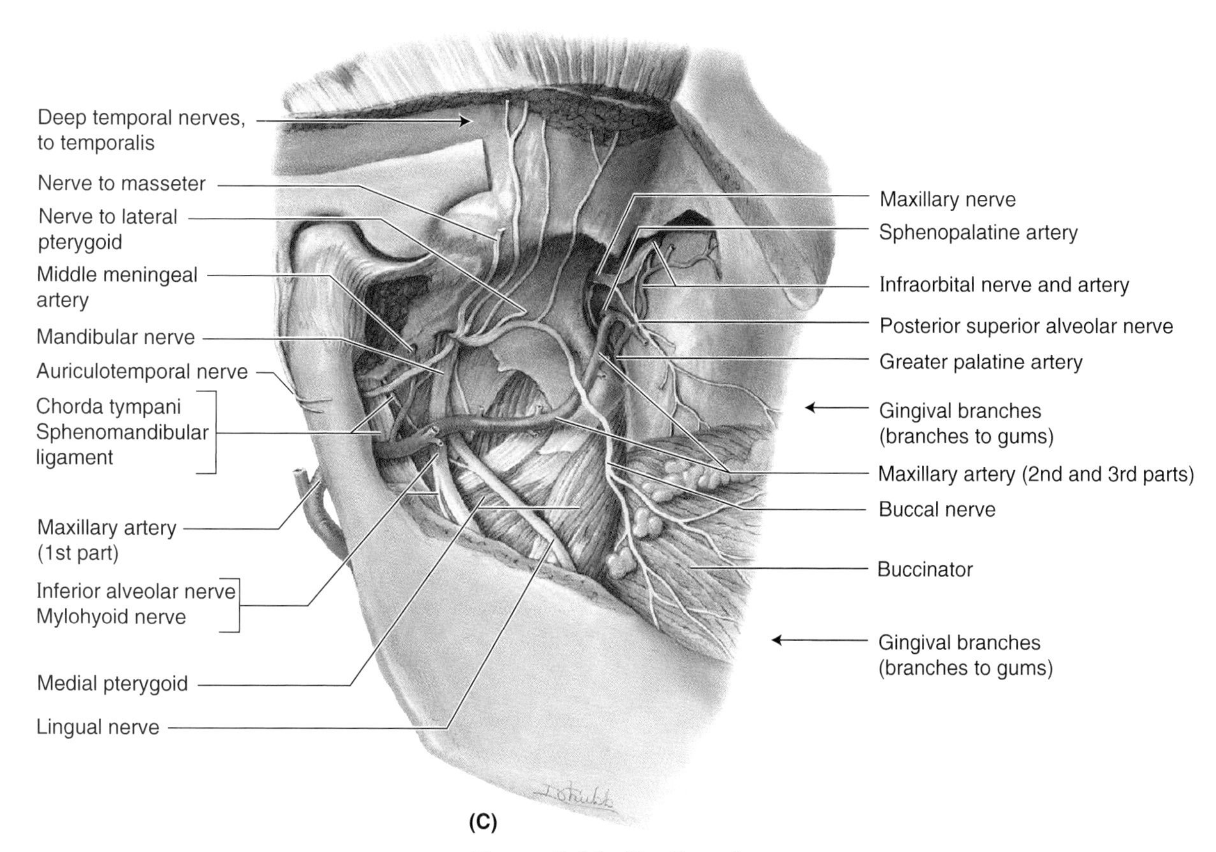

(C)

Figure 7.41. *(Continued)*

Contents of the Infratemporal Fossa

The infratemporal fossa contains the:

- Inferior part of the temporal muscle
- Lateral and medial pterygoid muscles
- Maxillary artery
- Pterygoid venous plexus
- Mandibular, inferior alveolar, lingual, buccal, and chorda tympani nerves, and otic ganglion.

The **temporal muscle** (Fig. 7.41, Table 7.10) is attached proximally to the temporal fossa and distally to the coronoid process and anterior border of the ramus of the mandible. The temporal muscle elevates the mandible (closes lower jaw); its posterior fibers retrude (retract) the protruded mandible.

The **lateral pterygoid muscle** (Fig. 7.41, Table 7.10), which has two heads of origin, passes posteriorly and its upper head attaches to the capsule and disc of the TMJ; the lower head attaches primarily to the pterygoid fovea at the condylar process (neck) of the mandible. The lateral pterygoid muscles protrude the mandible (move it forward) to enable depression of the chin (produced in large part by gravity.

The **medial pterygoid muscle** lies on the medial aspect of the ramus of the mandible (Fig. 7.41*C*). Its two heads embrace the inferior head of the lateral pterygoid and then unite. The medial pterygoid passes inferoposteriorly and attaches to the medial surface of the mandible near its angle. The medial pterygoid elevates the mandible, closing the jaw.

The **maxillary artery**—*the larger of the two terminal branches of the external carotid artery* (Fig. 7.42):

- Arises posterior to the neck of the mandible
- Passes anteriorly, deep to the neck of the mandibular condyle (first or mandibular part)
- Passes superficial or deep to the lateral pterygoid muscle (second or pterygoid part)
- Disappears through the pterygomaxillary fissure to enter the infratemporal fossa (third or pterygopalatine part).

The maxillary artery is thus divided into three parts by the lateral pterygoid muscle (Figs. 7.41, *A* and *B*, and 7.42).

The branches of the first or mandibular part of the maxillary artery are:

- Deep auricular artery to the external acoustic meatus
- Anterior tympanic artery to the tympanic membrane
- Middle meningeal artery to the dura mater and calvaria
- Accessory meningeal arteries to the cranial cavity
- Inferior alveolar artery to the mandible, gingivae, and teeth.

The branches of the second or pterygoid part of the maxillary artery are:

- Deep temporal arteries, anterior and posterior, which supply the temporal muscle
- Pterygoid arteries, which supply the pterygoid muscles

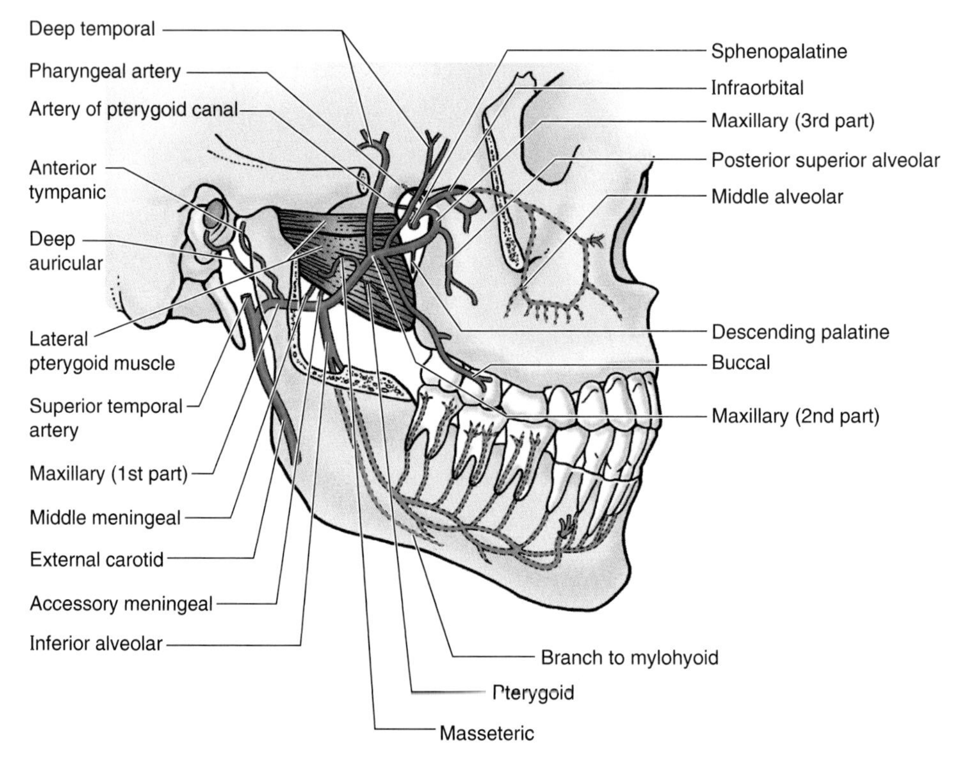

Figure 7.42. Maxillary artery and its branches. Observe the maxillary artery arising at the neck of the mandible (lower jaw). Notice that it is divided into three parts in relation to the lateral pterygoid muscle; it may pass medial or lateral to this muscle. Examine the branches of the first (mandibular) part that pass through the foramina or canals: deep auricular to external acoustic meatus (auditory canal), anterior tympanic to the tympanic cavity, middle and accessory meningeal to the cranial cavity, and inferior alveolar to the mandible and teeth. Observe that the branches of the second (pterygoid) part of the artery supply muscles by masseteric, deep temporal, pterygoid, and buccal branches. The branches of the third (pterygopalatine) part arise just before and within the pterygopalatine fossa: posterior superior alveolar, infraorbital, descending palatine, and sphenopalatine arteries.

Table 7.10. **Muscles Acting on the Temporomandibular Joint (TMJ)**

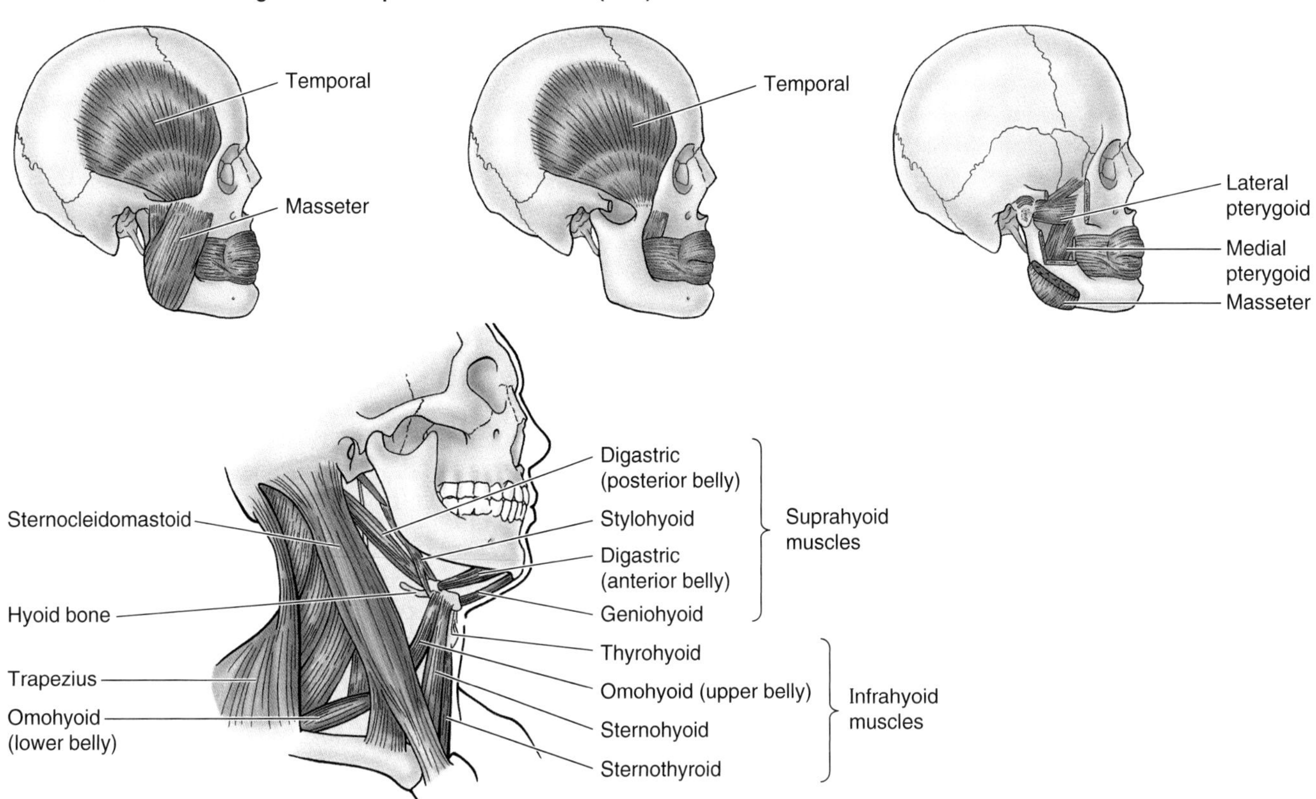

Muscle	Proximal Attachment	Distal Attachment	Innervation	Main Action
Temporal	Floor of temporal fossa and deep surface of temporal fascia	Tip and medial surface of coronoid process and anterior border of ramus of mandible	Deep temporal branches of mandibular nerve	Elevates mandible closing jaws; its posterior fibers retrude mandible after protrusion
Masseter	Inferior border and medial surface of zygomatic arch	Lateral surface of ramus of mandible and its coronoid process	Mandibular nerve (CN V_3) via masseteric nerve, which enters its deep surface	Elevates and protrudes mandible, thus closing jaws; deep fibers retrude it
Lateral pterygoid	Superior head: infratemporal surface and infratemporal crest of greater wing of sphenoid bone Inferior head: lateral surface of lateral pterygoid plate	Neck of mandible (pterygoid fovea); articular disc and capsule of temporomandibular joint	Mandibular nerve via lateral pterygoid nerve from anterior trunk, which enters its deep surface	Acting together, they protrude mandible and depress chin; acting alone and alternately, they produce side-to-side movements of mandible
Medial pterygoid	Deep head: medial surface of lateral pterygoid plate and pyramidal process of palatine bone Superficial head: tuberosity of maxilla	Medial surface of ramus of mandible, inferior to mandibular foramen	Mandibular nerve via medial pterygoid nerve	Acting together, they help to elevate mandible, closing jaws; they help to protrude mandible; acting alone, they protrude side of jaw; acting alternately, they produce a grinding motion

- Masseteric artery, which supplies the deep surface of the masseter muscle
- Buccal artery, which supplies the buccinator muscle.

The branches of the third or pterygopalatine part of the maxillary artery are:

- Posterior superior alveolar (dental) artery, which supplies the maxillary molar and premolar teeth, the lining of the maxillary sinus, and the gingiva
- Infraorbital artery, which supplies the inferior eyelid, lacrimal sac, the side of the nose, and the superior lip
- Descending palatine artery, which supplies the maxillary gingiva, palatine glands, and the mucous membrane of the roof of the mouth
- Artery of pterygoid canal, which supplies the superior part of the pharynx, the pharyngotympanic tube, and the tympanic cavity
- Pharyngeal artery, which supplies the roof of the pharynx, the sphenoidal sinus, and the inferior part of the pharyngotympanic tube
- Sphenopalatine artery, the termination of the maxillary artery, which supplies the lateral nasal wall, the nasal septum, and the adjacent paranasal sinuses.

The pterygoid venous plexus (Fig. 7.21) is partly between the temporal and pterygoid muscles. The plexus has connections with the facial vein through the cavernous sinus.

The **mandibular nerve** descends through the foramen ovale into the infratemporal fossa and divides into sensory and motor branches (Fig. 7.43). The branches of CN V_3 are the auriculotemporal, inferior alveolar, lingual, and buccal nerves. Branches of the mandibular nerve also supply the four muscles of mastication but not the buccinator, which is supplied by the facial nerve.

The **otic ganglion** (parasympathetic) is in the infratemporal fossa, just inferior to the foramen ovale, medial to the mandibular nerve and posterior to the medial pterygoid muscle. Presynaptic parasympathetic fibers, derived mainly from the glossopharyngeal nerve, synapse in the otic ganglion.

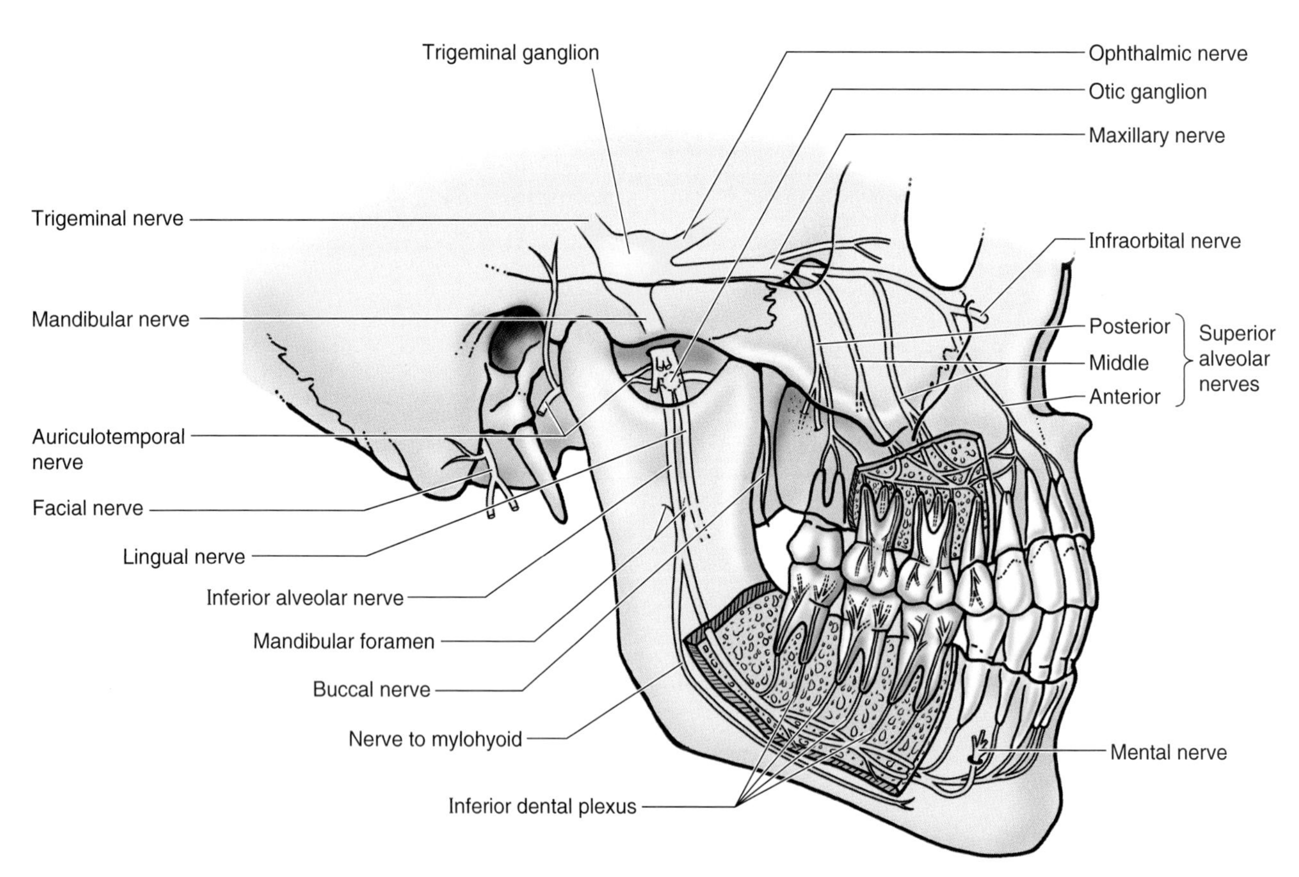

Figure 7.43. Innervation of the teeth. The upper teeth are innervated by the superior alveolar nerves from the maxillary nerve (CN V_2). The superior alveolar nerves form a superior dental plexus within the upper jaw (maxilla) that sends dental branches to the roots of each maxillary tooth. The posterior and middle alveolar nerves arising from the maxillary nerve, and the anterior from the infraorbital nerve, supply the maxillary molar teeth. Notice that the lower teeth are innervated by the inferior alveolar branch of the mandibular nerve (CN V_3), the nerve entering the mandibular foramen on the medial surface of the ramus of the mandible (lower jaw). As in the upper jaw, an inferior dental plexus is formed that sends dental branches to the roots of each mandibular tooth.

Postsynaptic parasympathetic fibers, which are secretory to the parotid gland, pass from the otic ganglion to this gland through the auriculotemporal nerve.

The **auriculotemporal nerve** (Fig. 7.43) encircles the middle meningeal artery and divides into numerous branches, the largest of which passes posteriorly, medial to the neck of the mandible, and supplies sensory fibers to the auricle and temporal region. The auricotemporal nerve also sends articular fibers to the *temporomandibular joint* and parasympathetic secretomotor fibers to the parotid gland (p. 870).

The **inferior alveolar nerve** enters the mandibular foramen and passes through the mandibular canal forming the inferior *dental plexus*, which sends dental branches to all mandibular teeth on its side (Fig. 7.43). Another branch of the plexus—the **mental nerve**—passes through the mental foramen and supplies the skin and mucous membrane of the lower lip, the skin of the chin, and the vestibular gingiva of the mandibular incisor teeth.

The **lingual nerve** lies anterior to the inferior alveolar nerve (Fig. 7.41, *B* and *C*). It is sensory to the anterior two-thirds of the tongue, the floor of the mouth, and the lingual gingivae. It enters the mouth between the medial pterygoid and the ramus of the mandible and passes anteriorly under cover of the oral mucosa, just inferior to the 3rd molar tooth.

The **chorda tympani nerve**, a branch of CN VII carrying taste fibers from the anterior two-thirds of the tongue, joins the lingual nerve in the infratemporal fossa (Fig. 7.41*B*). The chorda tympani also carries secretomotor fibers for the submandibular and sublingual salivary glands.

Mandibular Nerve Block

To produce mandibular nerve block, an anesthesiologist applies an anesthetic agent to the mandibular nerve where it enters the infratemporal fossa. In the extraoral approach, the needle passes through the mandibular notch of the ramus of the mandible into the infratemporal fossa, where the anesthetic is injected and usually anesthetizes the auriculotemporal, inferior alveolar, lingual, and buccal branches of CN V_3.

Inferior Alveolar Nerve Block

An inferior alveolar nerve block—commonly used by dentists when repairing the mandibular teeth—anesthetizes the inferior alveolar nerve, a branch of CN V_3. The injection of anesthetic is around the mandibular foramen, the opening into the mandibular canal on the medial surface of the ramus of the mandible, which gives passage to the inferior alveolar nerve, artery, and vein. When this nerve block is successful, all mandibular teeth are anesthetized to the median plane. The skin and mucous membrane of the lower lip, the labial alveolar mucosa and gingivae, and the skin of the chin are also anesthetized because they are supplied by the mental branch of this nerve. If the needle travels too far posteriorly, it could penetrate the parotid gland and produce transient paralysis of branches of the facial nerve. ⊙

Temporomandibular Joint (TMJ)

The TMJ is a modified hinge type of synovial joint (Fig. 7.44). The articular surfaces involved are the condyle of the mandible, the articular tubercle of the temporal bone, and the mandibular fossa. The articular capsule of the TMJ is loose. The **fibrous capsule** attaches to the margins of the articular area on the temporal bone and around the neck of the mandible (Fig. 7.44*C*). The joint has two synovial membranes:

- The *superior synovial membrane* lines the fibrous capsule superior to the articular disc
- The *inferior synovial membrane* lines the capsule inferior to the disc.

The **articular disc** divides the joint into two separate compartments. The gliding movements of protrusion and retrusion (translation) occur in the superior compartment; the hinge movements of depression and elevation occur in the inferior compartment. The thick part of the articular capsule forms the intrinsic **lateral ligament** (temporomandibular ligament), which strengthens the TMJ laterally and, with the postglenoid tubercle, acts to prevent posterior dislocation of the joint.

Two extrinsic ligaments and the lateral ligament connect the mandible to the cranium. The **stylomandibular ligament**—actually a thickening in the fibrous capsule of the parotid gland—runs from the styloid process to the angle of the mandible (Fig. 7.44, *C* and *D*). It does not contribute significantly to the strength of the joint. The **sphenomandibular ligament** runs from the spine of the sphenoid to the lingula of the mandible. It is the primary passive support of the mandible, although the tonus of the muscles of mastication usually bears the mandible's weight. However, the ligament does serve as a "swinging hinge" for the mandible, serving both as a fulcrum and as a check ligament for the movements of the mandible at the TMJs.

Muscles (or forces) producing movements of the mandible at the TMJs are (Fig. 7.45, Table 7.11):

- *Depression* (open mouth)
  - Gravity (prime mover)
  - Suprahyoid and infrahyoid muscles

Note: Protrusion must occur for all but minimal depression (see below).

- *Elevation* (close mouth)
  - Temporal
  - Masseter
  - Medial pterygoid
- *Protrusion* (protraction of chin)
  - Lateral pterygoid (prime mover)

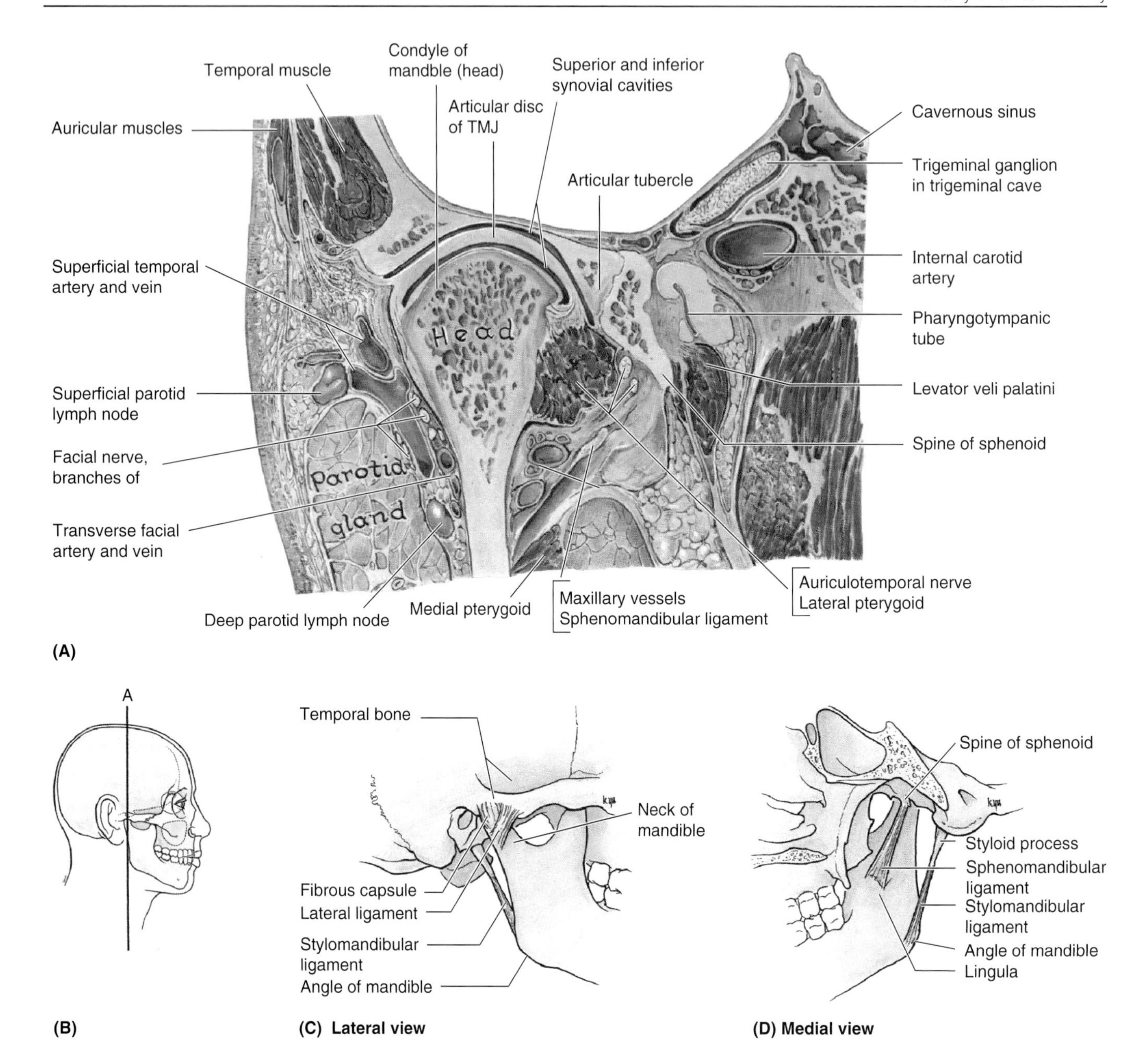

Figure 7.44. Temporomandibular joint (TMJ). A. Anterior view of a coronal section of the right TMJ. Observe the articular disc dividing the joint cavity into superior and inferior compartments. **B.** Orientation drawing showing the plane of the coronal section in (**A**). **C–D.** The TMJ and the extrinsic stylomandibular and sphenomandibular ligaments. The sphenomandibular ligament passively bears the weight of the lower jaw (mandible) and is the "swinging hinge" of the mandible, permitting protrusion and retrusion as well as elevation and depression.

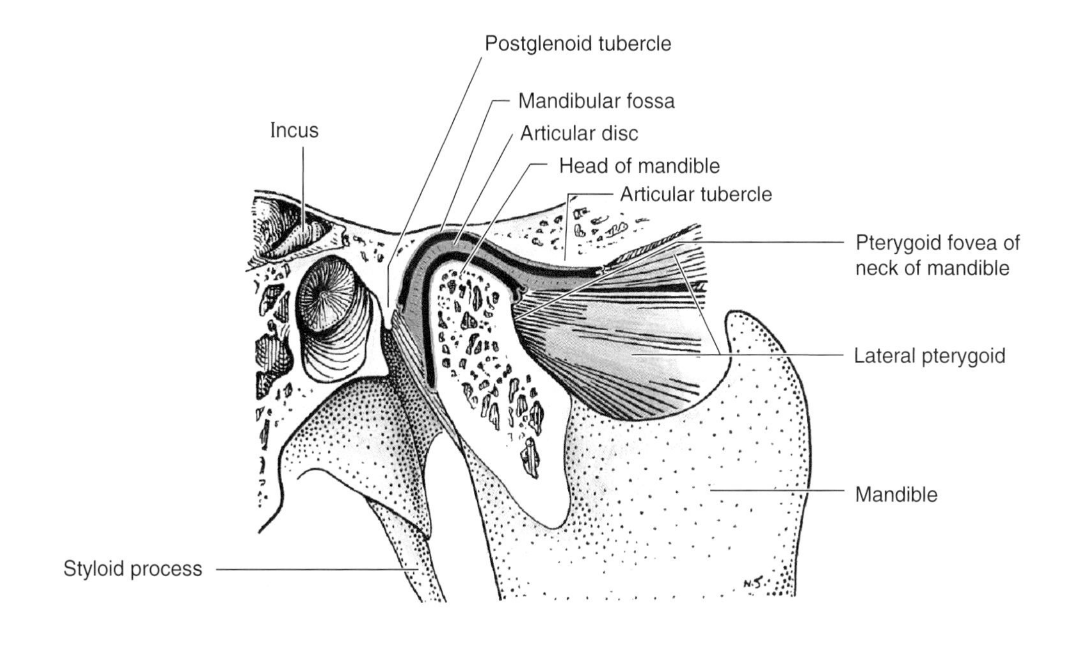

(A) Sagittal section

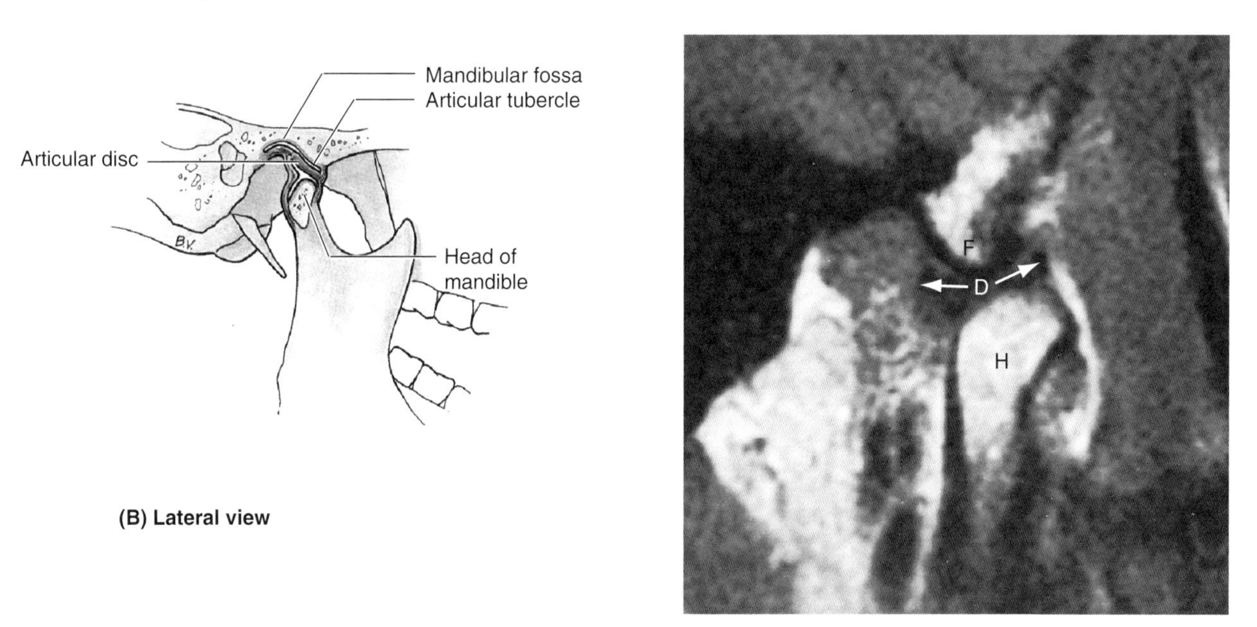

(B) Lateral view

(C) Sagittal view

Figure 7.45. Temporomandibular joint (TMJ). A. Lateral view of a sagittal section of the right TMJ. Note the articular disc dividing the articular cavity into superior and inferior compartments, and the lateral pterygoid muscle with the tendon of its upper belly inserted into the anterior aspect of the disc and fibrous capsule of the joint (not shown); the lower belly of the muscle is seen inserting onto a depression, the pterygoid fovea, on the anterior aspect of the neck of the mandible (lower jaw). **B.** Lateral view of a sagittal section of the TMJ (mouth open). **C.** Sagittal MRI of the TMJ (mouth open). In both (**B**) and (**C**), note the position of the articular disc (*D*) and the head of the mandible (*H*) in relation to the mandibular fossa and the articular tubercle of the temporal bone (*F*). (MRI scan courtesy of Dr. W. Kucharczyk, Chair of Medical Imaging, University of Toronto, and Clinical Director of Tri-Hospital Resonance Centre, Toronto, Ontario, Canada)

Table 7.11. **Movements of the Temporomandibular Joint**

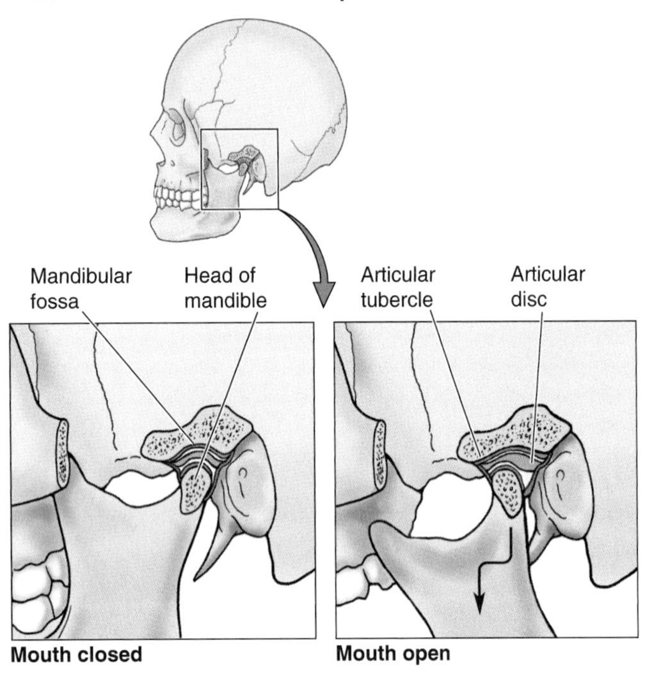

Movements	Muscle(s)
Elevation (close mouth)	Temporal, masseter, and medial pterygoid
Depression (open mouth)	Lateral pterygoid and suprahyoid and infrahyoid muscles*
Protrusion (protrude chin)	Lateral pterygoid, masseter, and medial pterygoid**
Retrusion (retrude chin)	Temporal (posterior oblique and near horizontal fibers) and masseter
Lateral movements (grinding and chewing)	Temporal of same side, pterygoids of opposite side, and masseter

*Prime mover normally is gravity—these muscles are mainly active against resistance

**The lateral pterygoid is the prime mover here, with very secondary roles played by masseter and medial pterygoid

- Masseter (oblique [superficial] fibers only—secondary synergist)
- Medial pterygoid (secondary synergist)

- *Retrusion* (retraction of chin)
  - Temporal (middle [oblique] and posterior [nearly horizontal] fibers only—prime mover)
  - Masseter (vertical [deep] fibers only—secondary synergist)
- *Lateral movement* (side-to-side grinding and chewing)
  - Retractors of same side (see above)
  - Protruders of opposite side (see above).

To enable more than a small amount of depression of the mandible (Fig. 7.45*B*), that is, to open the mouth wider than just to separate the upper and lower teeth, the head of the mandible and articular disc must move anteriorly on the articular surface until the head lies inferior to the articular tubercle (a movement referred to as "translation" by dentists). If this anterior gliding occurs unilaterally, the head of the contralateral mandible rotates (pivots) on the inferior surface of the articular disc, permitting simple side-to-side chewing or grinding movements over a small range. During protrusion and retrusion of the mandible, the head and articular disc slide anteriorly and posteriorly on the articular surface of the temporal bone, with both sides moving together.

TMJ movements are produced chiefly by the muscles of mastication. The attachments, nerve supply, and actions of these muscles are described in Table 7.11.

- The **temporal** (L. temporalis) is a fan-shaped muscle, with its wide part covering the temporal region and its narrow part attaching to the coronoid process of the mandible. The main action of this muscle is to elevate the mandible—closing the mouth and approximating the teeth; however, its middle, oblique, and posterior (nearly horizontal) fibers constitute the major retractor of the mandible.
- The **masseter**, a quadrangular muscle, covers the lateral aspect of the ramus and coronoid process of the mandible. The main action of the masseter is to elevate the mandible and occlude the teeth for biting and chewing.
- The **medial pterygoid**, a quadrilateral muscle, is deep to the ramus of the mandible; its two heads of origin embrace the inferior head of the lateral pterygoid. The main action of the medial pterygoid is to also elevate the mandible.
- The **lateral pterygoid** is almost triangular in shape: the base of the triangle is formed by the anterior attachment of its two heads, and its apex is formed by its posterior attachments to the capsule and disc of the TMJ and to the neck of the mandible. The main action of the lateral pterygoid is protrusion—anterior translation, or moving the jaw forward.

Generally, depression of the mandible is produced by gravity. The *suprahyoid* and *infrahyoid muscles*—straplike muscles on either side of the neck (see Chapter 8)—are primarily used to raise and depress the hyoid bone and larynx, respectively, during swallowing, for example. Indirectly they can also help to depress the mandible, especially when opening the mouth suddenly or against resistance.

Dislocation of the TMJs

Sometimes during yawning or taking a large bite, excessive contraction of the lateral pterygoids may cause the heads of the mandible to dislocate anteriorly (pass anterior to the articular tubercles). In this position, the mandible remains depressed and the person is unable ▶

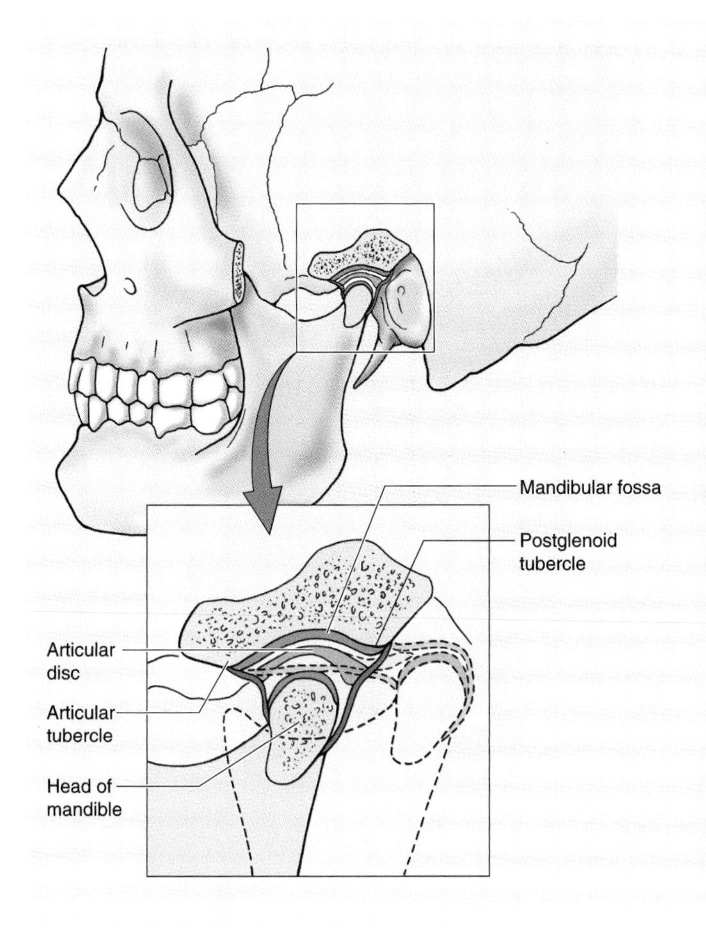

▶ to close their mouth. Most commonly, a sideways blow to the chin when the mouth is open dislocates the TMJ on the side that received the blow. Dislocation of the TMJ may also accompany *fractures of the mandible* (p. 837). Posterior dislocation is uncommon, being resisted by the presence of the postglenoid tubercle and the strong intrinsic lateral or temporomandibular ligament. Usually in falls on or direct blows to the chin, the neck of the mandible fractures before dislocation occurs. Because of the close relationship of the facial and auriculotemporal nerves to the TMJ, care must be taken during surgical procedures to preserve both the branches of the facial nerve overlying it and the articular branches of the auriculotemporal nerve that enter the posterior part of the joint. Injury to articular branches of the auriculotemporal nerve supplying the TMJ—associated with traumatic dislocation and rupture of the articular capsule and lateral ligament—leads to laxity and instability of the TMJ.

Arthritis of the TMJ

The TMJ may become inflamed from *degenerative arthritis*, for example (Liebgott, 1986). Abnormal function of the TMJ may result in structural problems such as dental malocclusion and joint clicking (crepitus). The clicking is thought to result from delayed anterior disc movements during mandibular depression and elevation. ⊙

Oral Region

The oral region includes the oral cavity (mouth), teeth, gingivae (gums), tongue, palate, and the region of the palatine tonsils. The oral cavity is where food is ingested and prepared for digestion in the stomach and small intestine. Food is chewed by the teeth, and saliva from the salivary glands facilitates the formation of a manageable *food bolus* (L. lump). Deglutition (swallowing) is voluntarily initiated in the oral cavity. The voluntary phase of the process pushes the bolus from the oral cavity into the pharynx—the expanded part of the digestive tract—where the automatic phase of swallowing occurs.

Oral Cavity

The oral cavity consists of two parts: the *oral vestibule* and the *oral cavity proper* (Fig. 7.46). The oral cavity is associated with several functions, but the most pleasurable for most people are those associated with eating and drinking. It is in the oral cavity that food and drinks are tasted and savored.

The **oral vestibule** is the slitlike space between the teeth and buccal gingiva and the lips and cheeks. The vestibule communicates with the exterior through the mouth; the size of this orifice is controlled by the circumoral muscles such as the orbicularis oris (the sphincter of the oral aperture [p. 852]), the buccinator, risorius, and depressors and elevators of the lips (dilators of the aperture).

The **oral cavity proper** is the space between the upper and lower dental arches (Fig. 7.46). It is limited laterally and anteriorly by the maxillary and mandibular alveolar arches housing the teeth. The roof of the oral cavity is formed by the *palate*. Posteriorly, the oral cavity communicates with the oropharynx (oral part of the pharynx). When the mouth is closed and at rest, the oral cavity is fully occupied by the tongue.

Lips, Cheeks, and Gingivae

Lips and Cheeks

The **lips** (Figs. 7.47 and 7.48)—mobile, muscular folds surrounding the mouth—contain the orbicularis oris and superior and inferior labial muscles, vessels, and nerves. The lips are covered externally by skin and internally by mucous membrane. The lips are used for grasping food, sucking liquids, clearing food from the labial vestibule, forming speech, and osculation (kissing).

Examine the skin of the external aspect of the lip and the **vermilion border** that indicates the abrupt beginning of the **transitional zone** of the lip (Fig. 7.47*A*). The skin of the transitional zone is hairless and so thin that it appears red because of the underlying capillary bed.

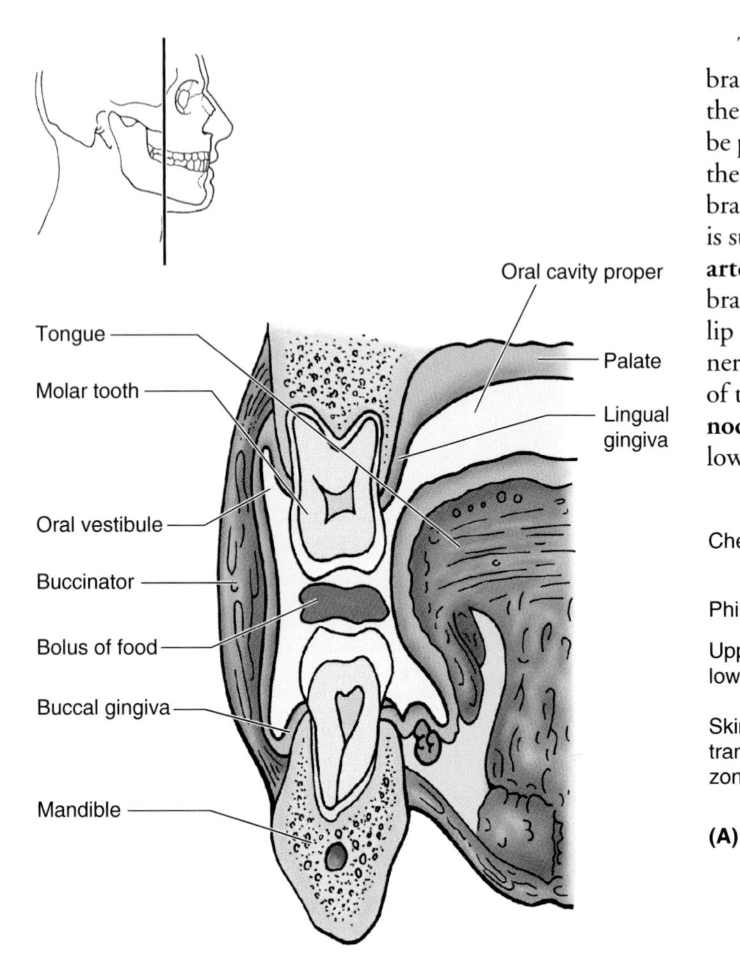

Figure 7.46. Coronal section of the mouth region. The orientation drawing shows the plane of the section. During chewing (mastication), the tongue (L. lingua; G. glossa) and the buccinator (L. trumpeter)—and the orbicularis oris, anteriorly—work together to retain the bolus of food between the occlusive surface of the molar teeth.

The **upper lip** lies between the nose and the orifice of the oral cavity. Laterally the lips are separated from the cheeks by the **nasolabial grooves** that extend from the nose and pass approximately 1 cm lateral to the angles of the mouth. These grooves are easiest to observe when smiling. The upper lip has an infranasal depression, the **philtrum** (G. love-charm), that extends from the external nasal septum, separating the nostrils, to the vermilion border. The **lower lip** lies between the mouth and the **labiomental groove**, which separates the lower lip from the chin. The upper and lower lips are continuous at the angles of the mouth and are separated from the cheeks by the nasolabial grooves. The transitional zone of the lips, ranging from brown to red, continues into the oral cavity where it is continuous with the mucous membrane. This membrane covers the intraoral, vestibular part of the lips (Fig. 7.48). The **labial frenula** are free-edged folds of mucous membrane in the midline, extending from the vestibular gingiva to the mucosa of the upper and lower lips; the one extending to the lower lip is smaller. Other smaller frenula sometimes appear laterally in the premolar vestibular regions.

The superior and inferior **labial arteries** (Table 7.5), branches of the facial arteries, anastomose with each other in the lips to form an arterial ring. The pulse of these arteries can be palpated by grasping the upper or lower lip lightly between the first two digits. The upper lip is supplied by superior labial branches of the **facial** and **infraorbital arteries.** The lower lip is supplied by inferior labial branches of the **facial** and **mental arteries.** The upper lip is supplied by the superior labial branches of the **infraorbital nerves** (of CN V_2), and the lower lip is supplied by the inferior labial branches of the mental nerves (of CN V_3). Lymph from the upper lip and lateral parts of the lower lip passes primarily to the **submandibular lymph nodes** (Fig. 7.47*B*), whereas lymph from the medial part of the lower lip passes initially to the **submental lymph nodes.**

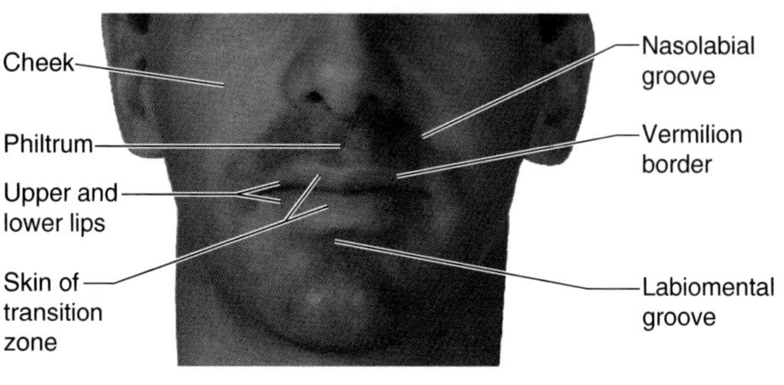

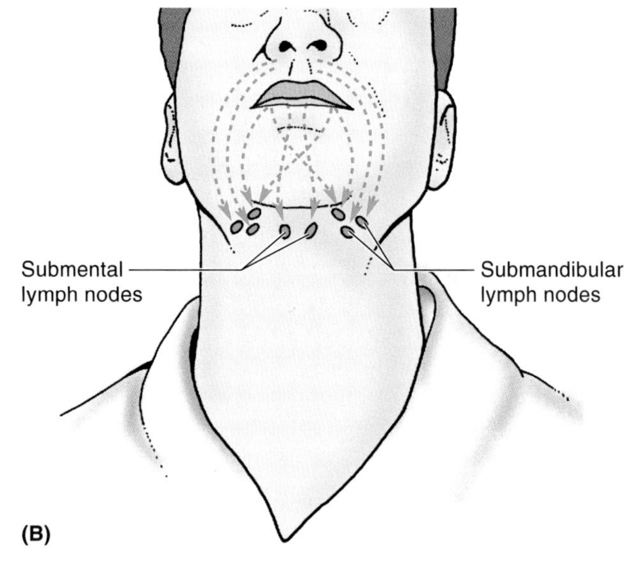

Figure 7.47. Cheeks (L. buccae), lips, and chin. A. Surface anatomy. The median part of the upper lip is marked externally by a shallow groove, the philtrum. The junction between the cheeks and lips is marked on each side by a nasolabial groove, which extends inferolaterally from the nose to approximately 1 cm lateral to the angle of the mouth. The lower lip lies between the mouth and the labiomental groove, which separates the lower lip from the chin. The vermillion border marks the beginning of the red transitional zone between the skin and the mucous membrane of the lip. **B.** Lymphatic drainage of the lips. Observe that lymph from the upper lip and lateral parts of the lower lip drains to the submandibular lymph nodes; lymph from the middle part of the lower lip drains to the submental lymph nodes.

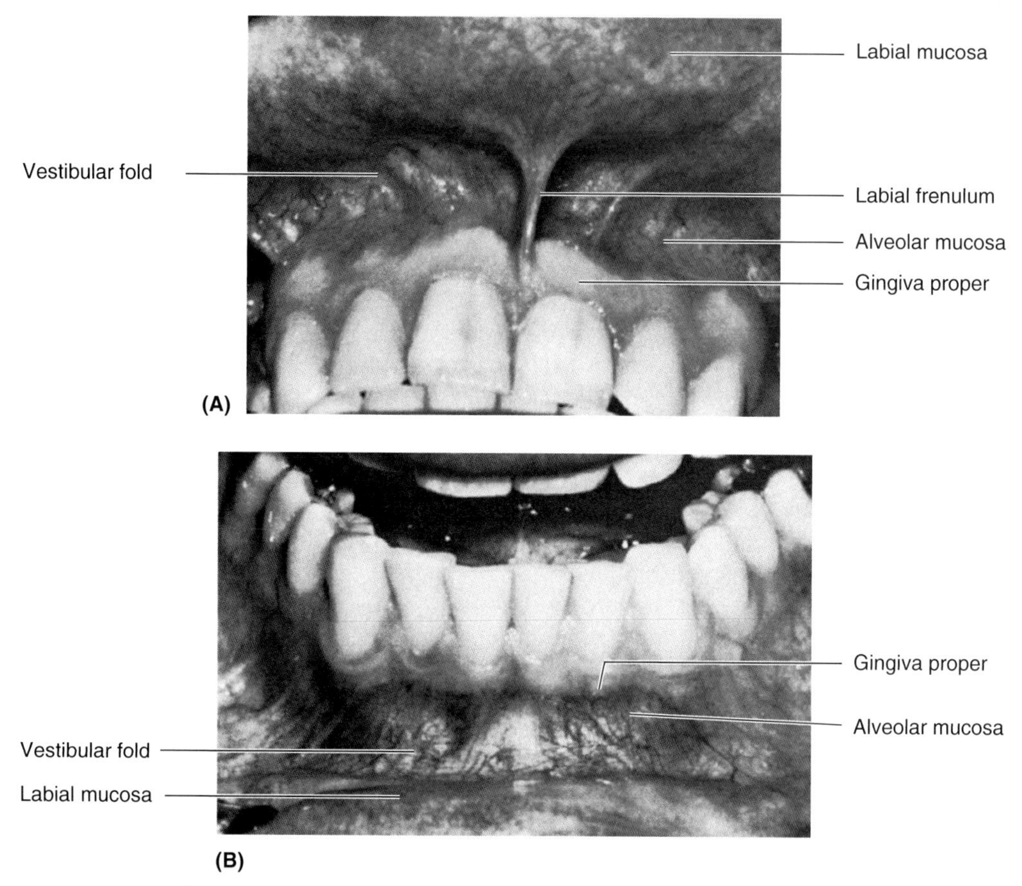

Figure 7.48. Vestibule and gingivae (gums) of the oral cavity. A. Vestibule and gingivae of the maxilla. **B. Vestibule and gingivae of the mandible.** Observe the gingiva and its relationship to the teeth. As the alveolar mucosa approaches the necks of the teeth, it changes in texture and color to become the gingiva proper.

Cleft Lip

Cleft lip ("harelip") is a congenital anomaly of the upper lip that occurs once in 1000 births (Moore and Persaud, 1998); 60 to 80% of affected infants are males. The clefts vary from a small notch in the transitional zone and vermilion border to ones that extend through the lip into the nose. In severe cases, the cleft extends deeper and is continuous with a cleft in the palate. Cleft lip may be unilateral or bilateral.

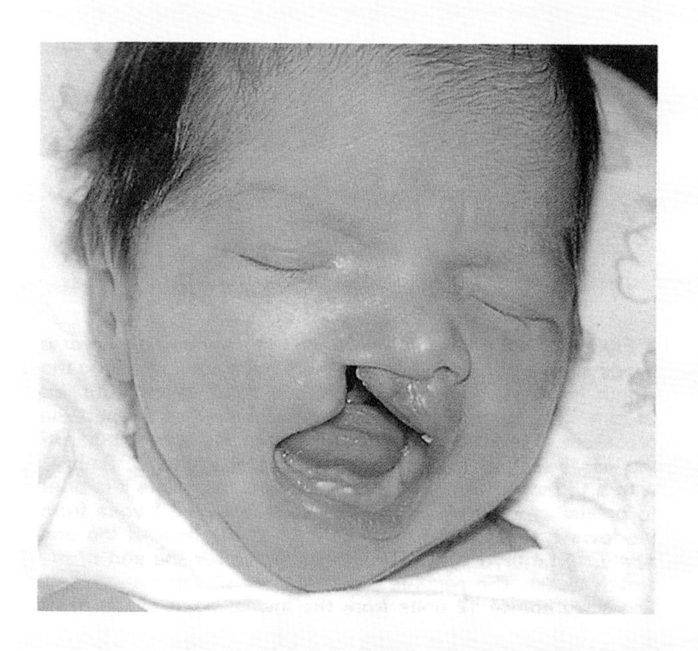

Carcinoma of the Lip

Carcinoma of the lip usually involves the lower lip (p. 869). These tumors metastasize through lymph vessels to the submandibular and submental lymph nodes. The submental lymph nodes enlarge when malignant cells metastasize from the medial part of the lower lip.

Cyanosis of the Lips

Cyanosis is a dark-bluish or purplish coloration of the lips and mucous membranes resulting from deficient oxygenation of capillary blood. The common blue discoloration of the lips in a cold environment results from the decreased blood supply and increased extraction of oxygen. Simple warming up reverses nonpathological, vasoconstrictive cyanosis in cold lips. ▶

Large Labial Frenula

Excessively large labial frenula in children may cause a space between the central incisor teeth (Liebgott, 1986). Recision of the frenulum and a bundle of dense connective tissue between the incisors (*frenulectomy*) allows approximation of the teeth, which may require an orthodontic appliance ("braces"). Large lower labial frenula in adults may pull on the labial gingiva and contribute to *gingival recession,* which results in abnormal exposure of the roots of the teeth. ⊕

Gingivitis

Improper oral hygiene results in food and bacterial deposits in tooth and gingival crevices that may cause *inflammation of the gingivae* (gingivitis). The gingivae swell and redden as a result. If untreated, the disease spreads to other supporting structures including alveolar bone, producing *periodontitis*—inflammation and destruction of alveolar bone and the *periodontal membrane,* the connective tissue that surrounds the root of the tooth and attaches it to its alveolus (bony socket). *Dentoalveolar abscesses* (collections of pus resulting from death of inflamed tissues) may drain to the oral cavity and lips. ⊕

The **cheeks** form the lateral movable walls of the oral cavity and the zygomatic prominences of the cheeks over the zygomatic bones. The cheeks have essentially the same structure as the lips with which they are continuous. *The principal muscle of the cheek is the buccinator* (Fig. 7.46). Numerous small buccal glands lie between the mucous membrane and the buccinator. Superficial to the buccinator is an encapsulated collection of fat; this **buccal fatpad** is proportionately much larger in infants, presumably to reinforce the cheeks and keep them from collapsing during sucking. The lips and cheeks function as an oral sphincter that pushes food from the vestibule into the mouth proper. The tongue and buccinator work together to keep the food between the occlusal surfaces of the molar teeth during chewing. The tonic contraction of the buccinator and especially of the orbicularis oris provides a gentle but continual resistance to the tendency of the teeth to tilt in an outward direction. In the presence of a short upper lip, or retractors that remove this force, crooked or protrusive ("buck") teeth develop. The cheeks are supplied by buccal branches of the maxillary artery and innervated by buccal branches of the mandibular nerve.

Gingivae

The gingivae are composed of fibrous tissue covered with mucous membrane. The *gingiva proper* ("attached gingiva") is firmly attached to the alveolar processes of the jaws and the necks of the teeth (Fig. 7.48); it is normally pink, stippled, and keratinizing. The *alveolar mucosa* ("loose gingiva") is normally shiny red and nonkeratinizing. The *lingual gingivae* (related to the tongue) of the maxillary incisor and canine teeth are supplied by branches of the nasopalatine nerves and vessels. The lingual gingivae of the maxillary premolar and molar teeth are supplied by the greater palatine nerves and vessels. The mandibular labial and buccal gingivae (related to the lips and cheek) of the mandibular incisor, canine, and premolar teeth are supplied by the branches of the inferior alveolar nerve and arteries (Figs. 7.42 and 7.43). The buccal gingivae of the mandibular molar teeth are supplied by the buccal nerve. The lingual gingivae of all mandibular teeth are supplied by the lingual nerve and vessels.

Teeth

Teeth are hard conical structures set in the alveoli of the upper and lower jaws that are used in mastication and in assisting in articulation.

Types of Teeth and Their Orientation in the Jaw

A tooth is identified and described on the basis of whether it is primary or secondary, the type of tooth, and its proximity to the midline or front of the mouth (e.g., medial and lateral incisors; the lst molar is anterior to the 2nd). The surfaces of a tooth are described on the basis of the tooth's orientation in the jaw.

Children have 20 deciduous (primary, or milk) teeth (Table 7.12). In each jaw on either side, the deciduous teeth (and the usual ages of eruption) are:

- A medial and a lateral incisor (erupting at approximately 6 and 8 months, respectively)
- One canine (erupting at 10 months)
- Two premolars (erupting at 20 to 24 months).

Adults normally have 32 secondary teeth (Fig. 7.49, Table 7.9). In each jaw on either side, the permanent teeth (and the usual ages of eruption) are:

- A medial and a lateral incisor (erupting at 7 to 8 years)
- One canine (erupting at 10 years)
- Two premolars (bicuspids) (erupting at 9 to 11 years)
- Three molars (lst and 2nd molars erupting at 6 years and in early teens; 3rd molars ["wisdom teeth"] erupting during the late teens or early twenties).

The types of teeth are identified by their characteristics:

- **Incisors**: thin cutting edge
- **Canines**: single prominent cone

Table 7.12. Primary and Secondary Dentition

Deciduous Teeth	Medial Incisor	Lateral Incisor	Canine	First Molar	Second Molar
Eruption (months)[a]	6 to 8	8 to 10	16 to 20	12 to 16	20 to 24
Shedding (years)	6 to 7	7 to 8	10 to 12	9 to 11	10 to 12

[a]In some normal infants, the first teeth (medial incisors) may not erupt until 12 to 13 months of age.

Permanent Teeth	Medial Incisor	Lateral Incisor	Canine	First Premolar	Second Premolar	First Molar	Second Molar	Third Molar
Eruption (years)	7 to 8	8 to 9	10 to 12	10 to 11	11 to 12	6 to 7	12	13 to 25

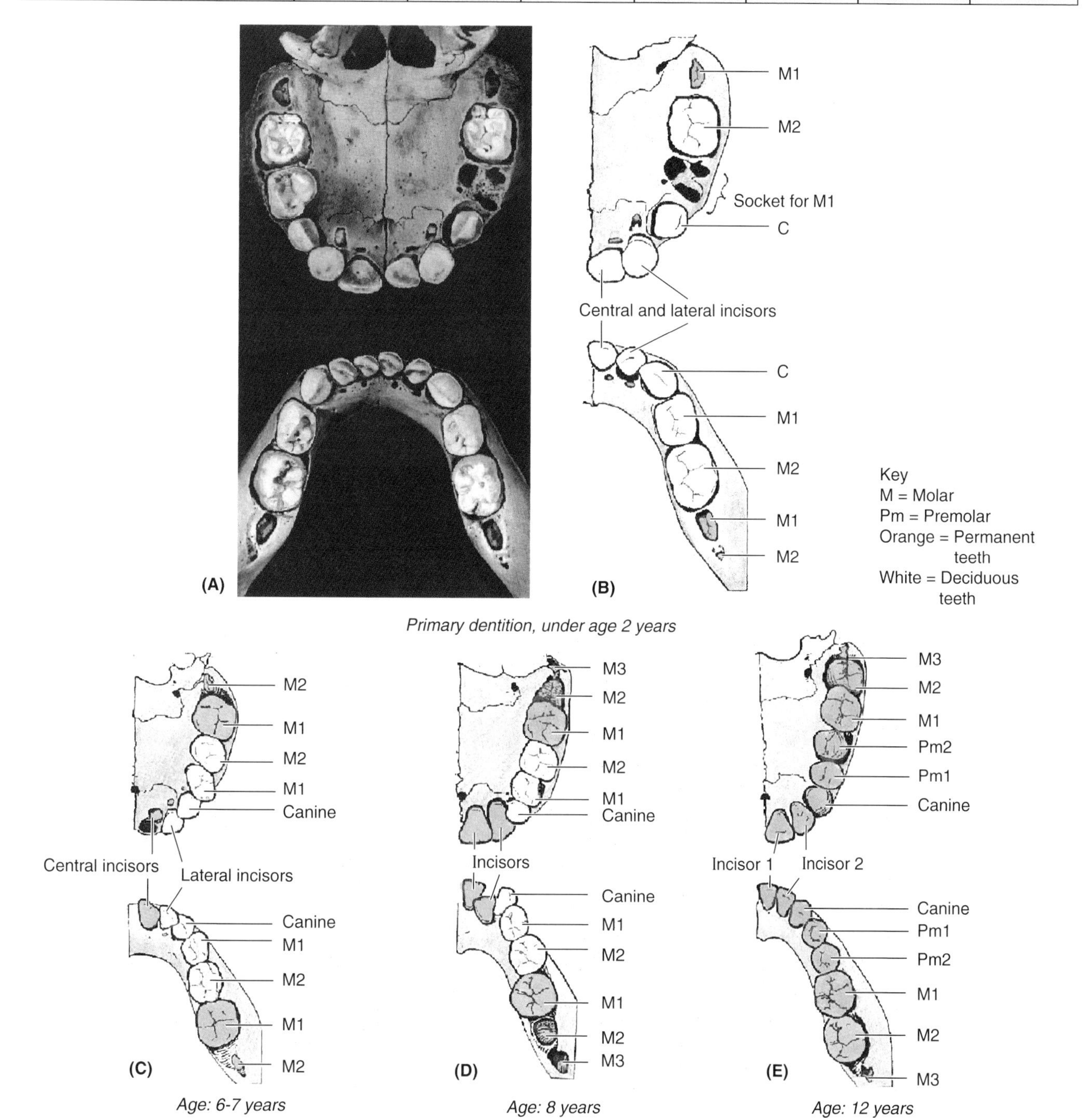

Primary dentition, under age 2 years

Age: 6-7 years *Age: 8 years* *Age: 12 years*

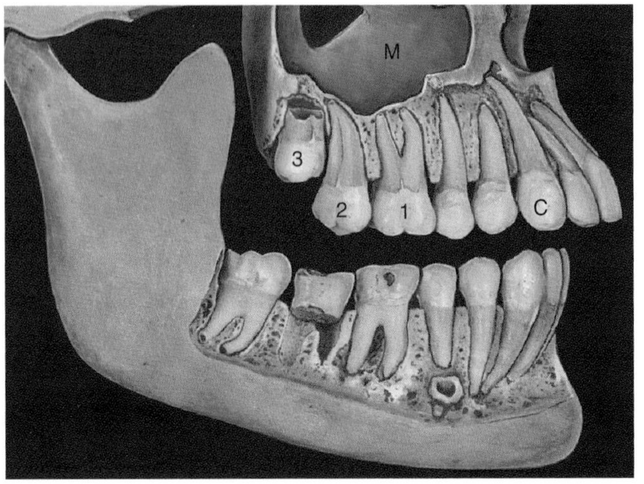

(A)

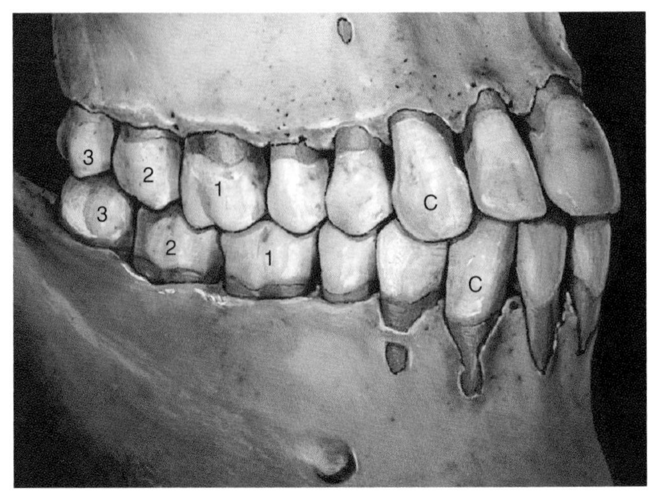

(B)

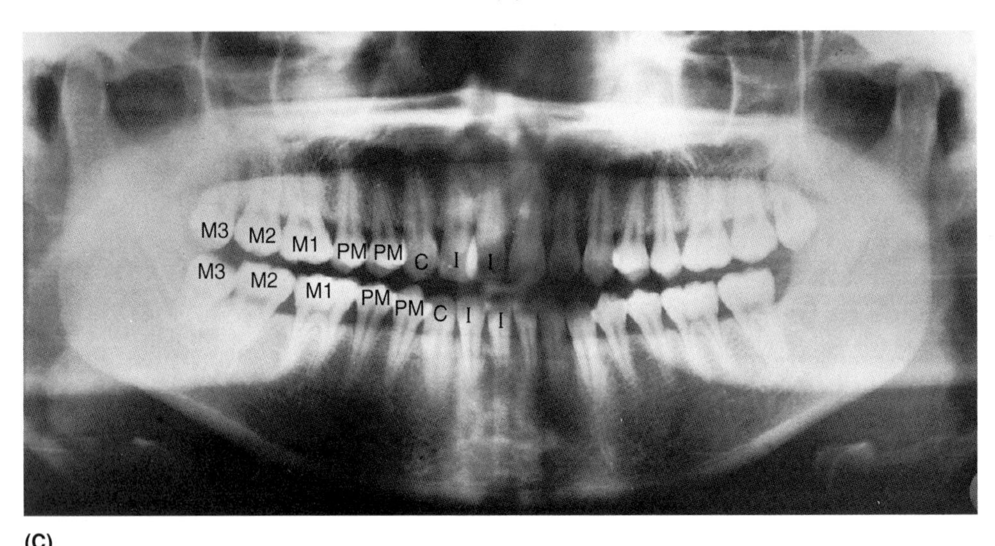

(C)

Figure 7.49. Secondary (permanent) teeth of an adult. A. The alveolar bone is ground away to expose the roots of the teeth. Note that the maxillary canine tooth (*C*) has the longest root and that the roots of the three maxillary molars are close to the floor of the maxillary sinus (*M*). **B.** The teeth in occlusion. **C.** Pantomographic radiograph of the mandible (lower jaw) and maxilla. *I*, incisors; *C*, canine; *PM*, premolars; *M1*, *M2*, and *M3*, molars. Left lower M3 is not present. (Courtesy of M.J. Pharoah, Associate Professor of Dental Radiology, Faculty of Dentistry, University of Toronto, Toronto, Ontario, Canada)

- **Premolars** (bicuspids): two cusps divided by a sagittal groove
- **Molars:** 3 or more cusps.

The *vestibular (labial or buccal) surface* of each tooth is directed outwardly or superficially, and the *oral* (*lingual or palatal*) *surface* is directed inwardly or deeply. The *mesial (proximal) surface* is directed toward the median plane of the facial skull; its *distal surface* is directed away from this plane. Both are *contact surfaces*, that is, surfaces that contact other teeth. The masticatory surface is the *occlusal surface*.

Parts and Structure of the Teeth

A tooth has a crown, neck, and root (Fig. 7.50). Each type of tooth has a characteristic appearance. The **crown** projects from the gingiva. The **neck** is the part of the tooth between the crown and root. The **root** is fixed in the alveolus by the fibrous *periodontal membrane*; the number of roots varies (Fig. 7.49). Most of the tooth is composed of **dentin**, which is covered by **enamel** over the crown and **cement** (L. cementum) over the root. The **pulp cavity** (tooth cavity) contains connective tissue, blood vessels, and nerves. The **root canal** transmits the nerves and vessels to and from the pulp cavity through the apical (root) foramen.

Tooth sockets (L. alveoli dentalis) occur within the alveolar processes of the maxillae and mandible and are the skeletal features that display the greatest change during a lifetime. Adjacent sockets, or alveoli, are separated by **interalveolar septa;** within the socket, the roots of teeth with more than

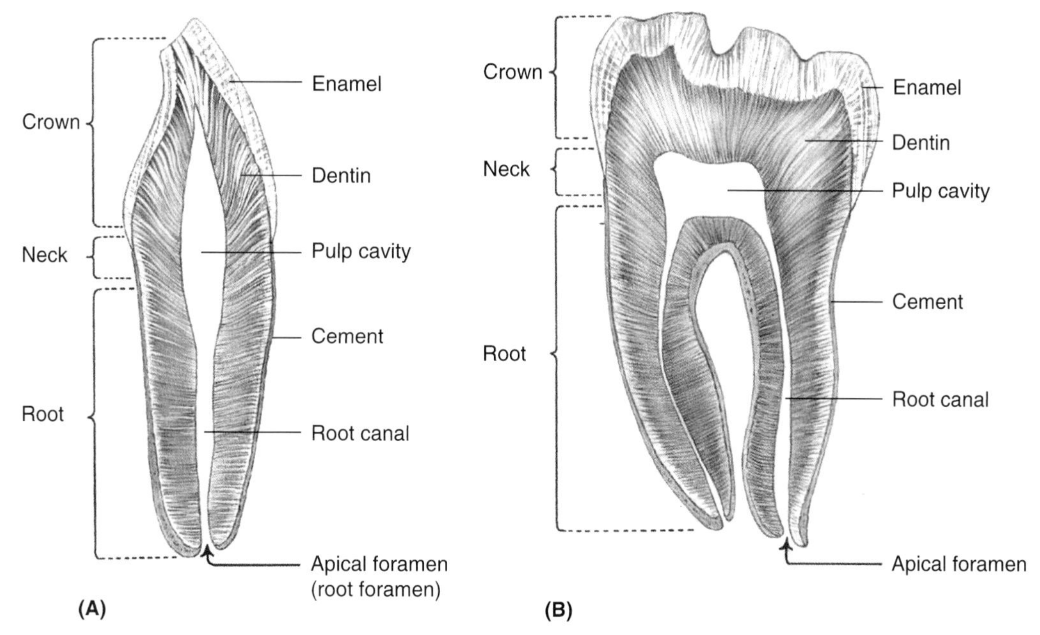

Figure 7.50. Longitudinal sections of the teeth. A. Incisor tooth. **B.** Molar tooth. In living persons the pulp cavity (tooth cavity) is a hollow space within the crown and neck of the tooth. It contains pulpal soft tissues, connective tissue, vessels, and nerves. The pulp cavity narrows down to the root canal in a single-rooted tooth or to one canal per root of a multirooted tooth. The vessels and nerves enter or leave through the apical (root) foramen.

one root are separated by **interradicular septa**. The bone of the socket has a thin cortex separated from the adjacent labial and lingual cortices by a variable amount of trabeculated bone: the labial wall of the socket is particularly thin over the incisor teeth; the reverse is true for the molars, where the lingual wall is thinner. Thus, the labial surface commonly is broken to extract incisors, and the lingual surface is broken to extract molars. The lingual nerve is closely related to the medial aspect of the 3rd mandibular molars; therefore, take caution to avoid injuring this nerve during their extraction.

The **periodontium** includes the cement and the periodontal membrane. The roots of the teeth are connected to the bone of the socket by a springy suspension forming a special type of fibrous joint termed a **gomphosis**. The collagenous fibers that extend between the cement of the root and the periosteum of the alveolar wall is the **periodontal ligament** or membrane. It is abundantly supplied with tactile, pressoreceptive nerve endings, lymph capillaries, and glomerular blood vessels that act as hydraulic cushioning to curb axial masticatory pressure. *Pressoreceptive nerve endings* are capable of receiving changes in pressure as stimuli.

Vasculature of the Teeth

The superior and inferior **alveolar arteries**, branches of the maxillary artery, supply both the maxillary and mandibular teeth, respectively (Figs. 7.41*B* and 7.42). **Veins** with the same names and distribution accompany the arteries. **Lymphatic vessels** from the teeth and gingivae pass mainly to the submandibular lymph nodes (Fig. 7.47).

Innervation of the Teeth

The superior and inferior **alveolar nerves**, branches of CN V_2 and CN V_3, respectively, arise from the *dental plexuses* to supply the maxillary and mandibular teeth (Fig. 7.43).

Dental Caries

Decay of the hard tissues of a tooth results in the formation of dental caries (cavities). Treatment involves removal of the decayed tissue and restoration of the anatomy of the tooth (*restorative dentistry*) with a dental material.

Pulpitis and Tooth Abscess

Neglected dental caries eventually invade and inflame tissues in the pulp cavity. Invasion of the pulp by a deep carious lesion results in infection and irritation of the tissues. This condition causes an inflammatory process known as *pulpitis*. Because the pulp cavity is a rigid space, the swollen pulpal tissues cause considerable pain (*toothache*). If untreated, the small vessels in the root canal may die from the pressure of the swollen tissue, and the infected material may pass through the apical canal into the periodontal tissues. *Infection of the pulp cavity* of a tooth may lead to infection of the periodontal membrane, destroying it and the compact layer of bone lining the alveolus. An infective process develops and spreads through the root ▶

▶ canal to the alveolar bone, producing an abscess. A *dentoalveolar abscess* causes swelling of the adjacent soft tissues ("gum boil"). Pus from abscesses of the maxillary molar teeth may extend into the nasal cavity or the maxillary sinus. The roots of the maxillary molar teeth are closely related to the floor of this sinus. As a consequence, *infection of the pulp cavity may also cause sinusitis,* or sinusitis may stimulate nerves entering the teeth and simulate a toothache. Pus from abscesses of the maxillary canine teeth often penetrates the facial region just inferior to the medial angle of the eye, and the swelling may obstruct drainage from the angular vein and allow infected material to pass through the superior ophthalmic vein to the cavernous sinus.

Extraction of the Teeth

Sometimes it is not practical to restore a tooth because of extreme tooth destruction. The only alternative is oral surgery (tooth removal). A tooth may lose its blood supply as a result of trauma. The blow to the tooth disrupts the blood vessels entering and leaving the apical foramen. It is not always possible to save the tooth. *Unerupted 3rd molars are a common dental problem*; these teeth are the last to erupt, usually when people are in their late teens or early twenties. Often there is not enough room for them to erupt and they become lodged (impacted) under the 2nd molar (see photograph below). If impacted 3rd molars become painful, they are usually removed. When doing so, be careful not to injure the alveolar nerves.

Gingival Recession

As people age, their teeth appear to get longer because of the slow recession of their gingivae; hence the expression "long in the tooth." Gingival recession exposes the sensitive cement of the teeth. This process occurs faster in persons who do not have *tartar* (yellowish-brown deposit) removed from their teeth by a procedure called *scaling.* The periodontal membrane is exposed as a consequence of gingival recession, allowing micro-organisms to invade and destroy it.

Periodontal Disease

As stated earlier, poor oral hygiene and neglect of carious teeth result in *gingivitis.* If untreated, the disease spreads to the other supporting structures and the alveolar bone, causing *periodontitis.* Untreated periodontitis over a variable period of time results in increasing mobility and eventual loss of teeth. ○

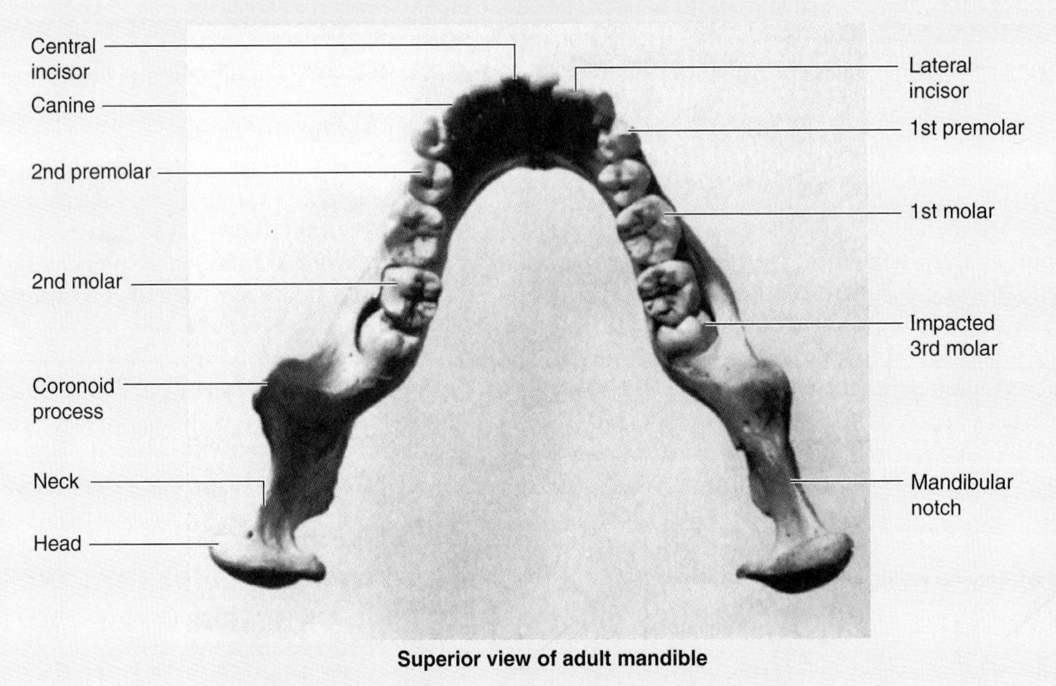

Superior view of adult mandible

Palate

The palate forms the arched roof of the mouth and the floor of the nasal cavities. It separates the oral cavity from the nasal cavities and the nasopharynx—the part of the pharynx lying superior to the soft palate (Fig. 7.51). The superior (nasal) surface of the palate is covered with *respiratory mucosa,* and the inferior (oral) surface is covered with *oral mucosa.* The palate consists of two regions—the *hard palate* anteriorly and the *soft palate* posteriorly.

Hard Palate

The hard palate is vaulted (concave); this space is filled by the tongue when it is at rest (Fig. 7.46). The anterior two-thirds

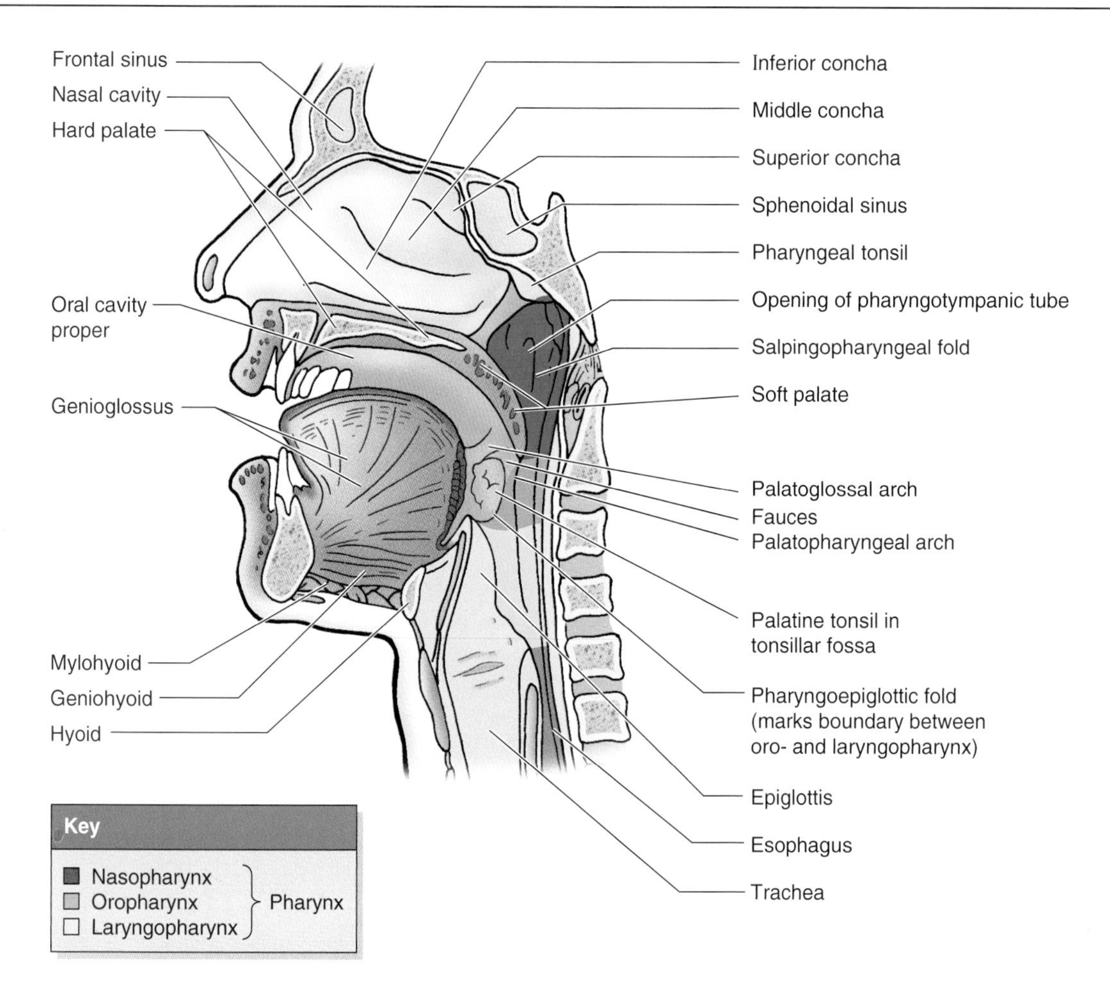

Figure 7.51. Median section of the head and neck. Observe the nasal cavity, palate, tongue (L. lingua; G. glossa), fauces (L. throat), palatine tonsil, the three parts of the pharynx, esophagus, and trachea. Note that the airway and food passageway cross in the pharynx and that the soft palate acts as a valve, elevating to seal the pharyngeal isthmus connecting the nasal cavity/nasopharynx with oral cavity/oropharynx.

of the palate has a bony skeleton formed by the palatine processes of the maxillae and the horizontal plates of the palatine bones (Fig. 7.52*A*). Three foramina are open on the oral aspect of the hard palate: the incisive fossa and the greater and lesser palatine foramina. The **incisive fossa** of the maxilla is a slight depression posterior to the central incisor teeth. The nasopalatine nerves (Fig. 7.52*B*) pass from the nose through a variable number of incisive canals and foramina that open into the incisive fossa. Medial to the 3rd molar tooth, the **greater palatine foramen** pierces the lateral border of the bony palate. The *greater palatine vessels and nerve* emerge from this foramen and run anteriorly on the palate. The **lesser palatine foramina** posterior to the greater palatine foramen pierces the pyramidal process of the palatine bone. The lesser palatine foramina transmit the *lesser palatine nerves and vessels* to the soft palate and adjacent structures (Fig. 7.52, *A* and *B*).

Soft Palate

The soft palate is the movable posterior third of the palate, which is suppended from the posterior border of the hard palate. The soft palate has no bony skeleton; however, it includes an anterior membranous *"aponeurotic palate"* that attaches to the posterior edge of the hard palate and a posterior fibromuscular *"muscular palate"* (Fig. 7.52*B*). The soft palate extends posteroinferiorly as a curved free margin from which hangs a conical process, the **uvula** (Figs. 7.52 and 7.53*B*). The soft palate is strengthened by the **palatine aponeurosis** (aponeurotic palate), formed by the expanded tendon of the tensor veli palatini (Fig. 7.52*B*). The aponeurosis, attached to the posterior margin of the hard palate, is thick anteriorly and thin posteriorly. When a person swallows, the soft palate initially is tensed to allow the tongue to press against it, squeezing the bolus of food to the back of the mouth. The soft palate

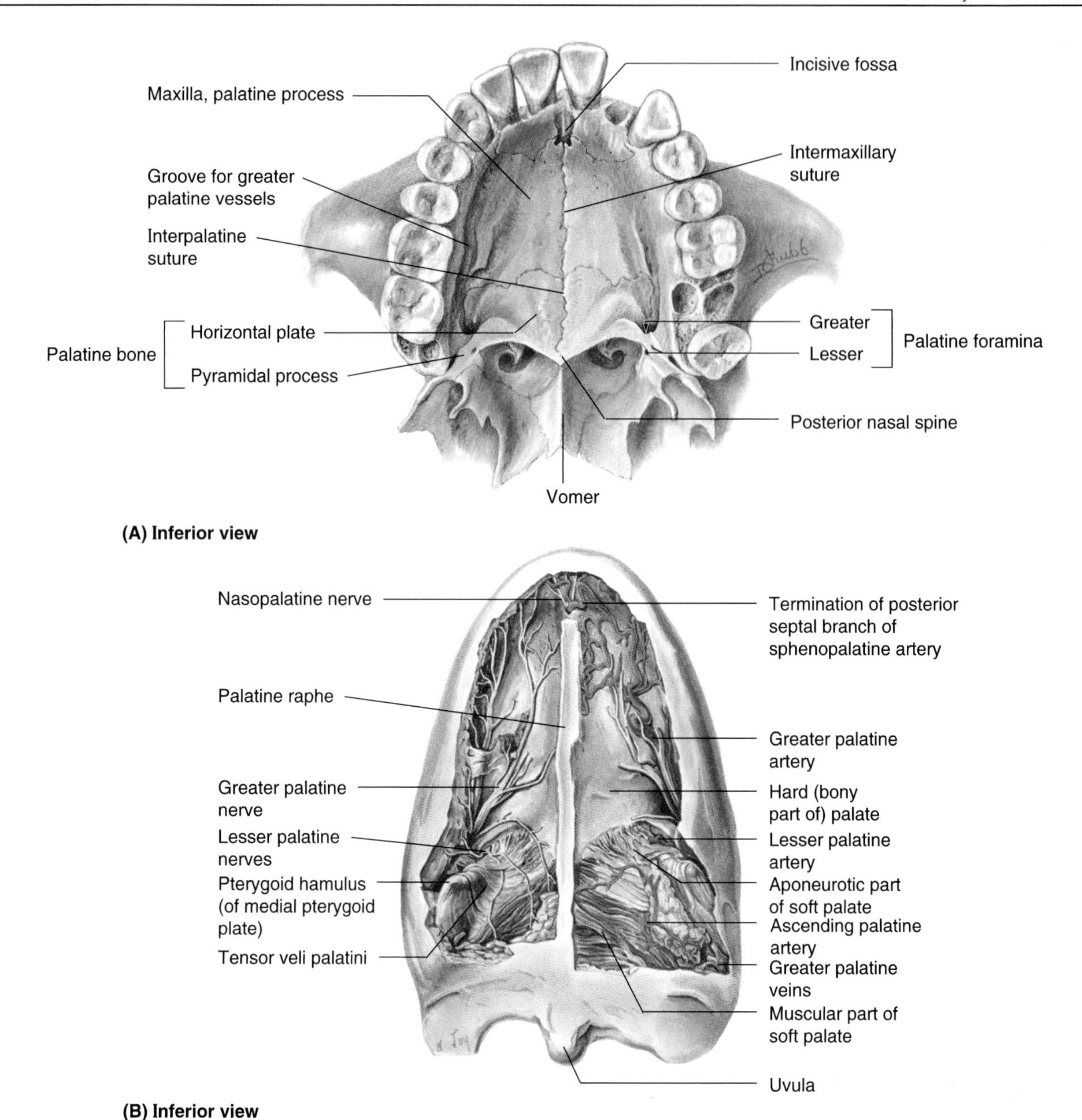

Figure 7.52. Palate. A. Inferior view of the bones of the hard palate (bony plate). **B.** Inferior view of the nerves and vessels of the palate. Observe that the palate has bony, aponeurotic, and muscular parts. The mucosa has been removed on each side of the central palatine raphe, demonstrating a branch of the greater palatine nerve on each side and the artery on the lateral side. Observe the lateral branches of the nerve expended mainly on the gums (gingivae), the medial branch on the hard palate, the nasopalatine nerve in the incisive region, and the lesser palatine nerves in the soft palate. Examine the four palatine arteries, two on the hard palate and two on the soft palate: greater palatine, terminal branch of sphenopalatine artery (posterior nasal septal), lesser palatine, and ascending palatine.

then is elevated posteriorly and superiorly against the wall of the pharynx, thereby preventing passage of food into the nasal cavity. Laterally, the soft palate is continuous with the wall of the pharynx and is joined to the tongue and pharynx by the **palatoglossal** and **palatopharyngeal arches**, respectively (Figs. 7.51 and 7.53*B;* see Blue Box, p. 946).

The **fauces** (L. throat) is the passage from the mouth to the pharynx, including the lumen and its boundaries. The fauces is bounded superiorly by the soft palate, inferiorly by the root (base) of the tongue, and laterally by the *pillars of the fauces*: the palatoglossal and palatopharyngeal arches (Fig. 7.51). The *isthmus of the fauces* is the short constricted space that establishes the connection between the oral cavity proper and the oropharynx. The isthmus is bounded anteriorly by the palatoglossal folds and posteriorly by the palatopharyngeal folds.

The **palatine tonsils**, often referred to as "the tonsils," are masses of lymphoid tissue, one on each side of the oropharynx (Fig. 7.51). Each tonsil is in a **tonsillar fossa** (tonsillar bed), bounded by the palatoglossal and palatopharyngeal arches and the tongue.

Superficial Features of the Palate

The mucosa of the hard palate is tightly bound to the underlying bone (Fig. 7.53); consequently, submucous injections here are extremely painful. The **lingual gingiva**—the part of the gingiva covering the lingual surface of the teeth and the alveolar process—is continuous with the mucosa of the palate; therefore, injection of an anesthetic agent into the gingiva of a tooth anesthetizes the adjacent palatal mucosa. Deep

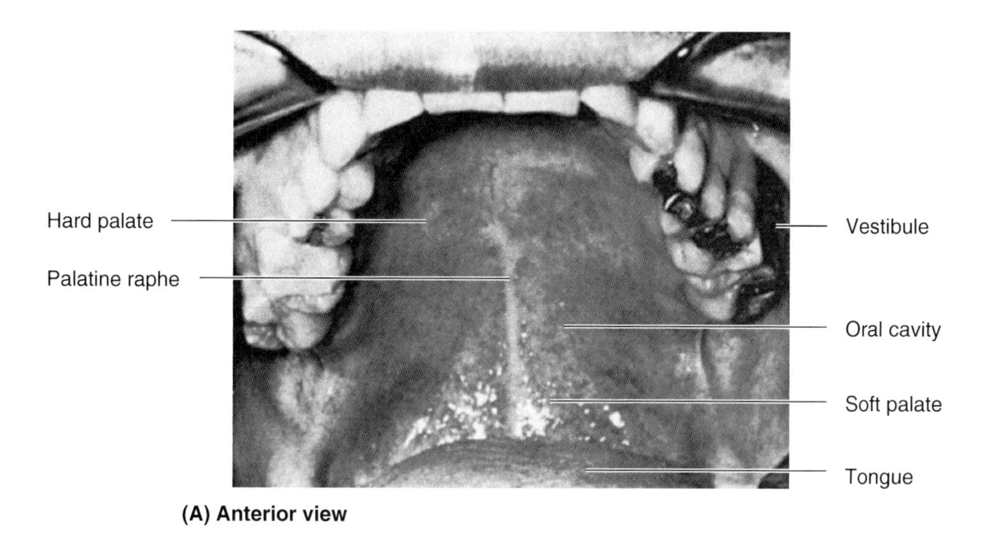

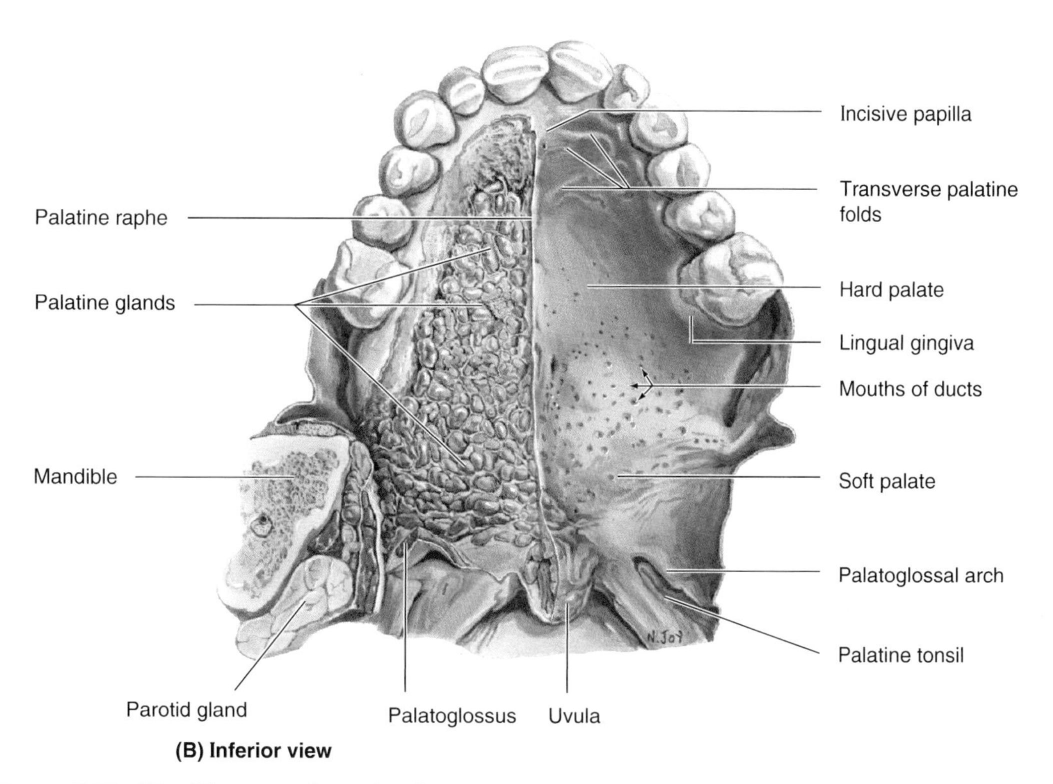

Figure 7.53. Maxillary teeth and palate. A. Anterior view of the maxillary teeth and the mucosa covering the hard palate (bony plate) in a living person. **B.** Inferior view of the mucous membrane and glands of the palate. Observe the orifices of the ducts of the palatine glands, which give the mucous membrane an orange-skin appearance. Note that the palatine glands form a thick layer in the soft palate, a thin one in the hard palate, and are absent in the region of the incisive fossa and the anterior part of the palatine raphe.

to the mucosa are mucus-secreting **palatine glands** (Fig. 7.53*B*). The orifices of the ducts of these glands give the palatine mucosa a pitted or "orange-peel" appearance. In the midline, posterior to the maxillary incisor teeth, is the **incisive papilla.** This elevation of the mucosa lies directly anterior to the underlying **incisive fossa** of the bony palate into which the incisive canals open. The incisive foramina are the openings of the incisive canals that transmit the nasopalatine nerves and vessels (Fig. 7.52*B*). *The incisive papilla is the injection site for anesthetizing the nasopalatine nerve.*

Radiating laterally from the incisive papilla are several parallel **transverse palatine folds**, or palatine rugae (Fig. 7.53*B*). These folds assist with manipulation of food during mastication. Passing posteriorly in the midline of the palate from the incisive papilla is a narrow whitish streak, the **palatine raphe.** It may present as a ridge anteriorly and a groove posteriorly (Fig. 7.53, *A* and *B*). The palatine raphe marks the site of fusion of the embryonic palatal processes, or shelves (Moore and Persaud, 1998).

Muscles of the Soft Palate

The five muscles of the soft palate arise from the base of the cranium and descend to the palate. The soft palate may be elevated so that it is in contact with the posterior wall of the pharynx. The soft palate may also be drawn inferiorly so that it is in contact with the posterior part of the tongue. Tensing the soft palate pulls it tight at an intermediate level. For attachments, nerve supply, and actions of the muscles of the soft palate, refer to Table 7.13. *The muscles of the soft palate are listed below.*

- The **levator veli palatini** or levator muscle of the soft palate is a cylindrical muscle that runs inferoanteriorly, spreading out in the soft palate where it attaches to the superior surface of the palatine aponeurosis.
- The **tensor veli palatini** or tensor muscle of the soft palate is a muscle with a triangular belly that passes inferiorly; the tendon formed at its apex hooks around the pterygoid hamulus—the hamulus of the medial pterygoid plate (Fig. 7.52*B*)—before spreading out as the palatine aponeurosis.
- The **palatoglossus** is a slender slip of muscle that is covered with mucous membrane; it forms the *palatoglossal arch* (Fig. 7.53*B*).
- The **palatopharyngeus** is a thin flat muscle also covered with mucous membrane (Fig. 7.51); it forms the *palatopharyngeal arch.*
- The **musculus uvulae** or uvular muscle inserts into the mucosa of the uvula.

Vasculature and Innervation of the Palate

The palate has a rich blood supply chiefly from the **greater palatine artery** on each side, a branch of the descending palatine artery (Figs. 7.52*B* and 7.54). The greater palatine artery passes through the greater palatine foramen and runs anteromedially. The **lesser palatine artery**—a smaller branch of the descending palatine artery—enters the palate through the lesser palatine foramen and anastomoses with the **ascending palatine artery**, a branch of the facial artery (Fig. 7.52*B*). The **veins of the palate** are tributaries of the pterygoid venous plexus.

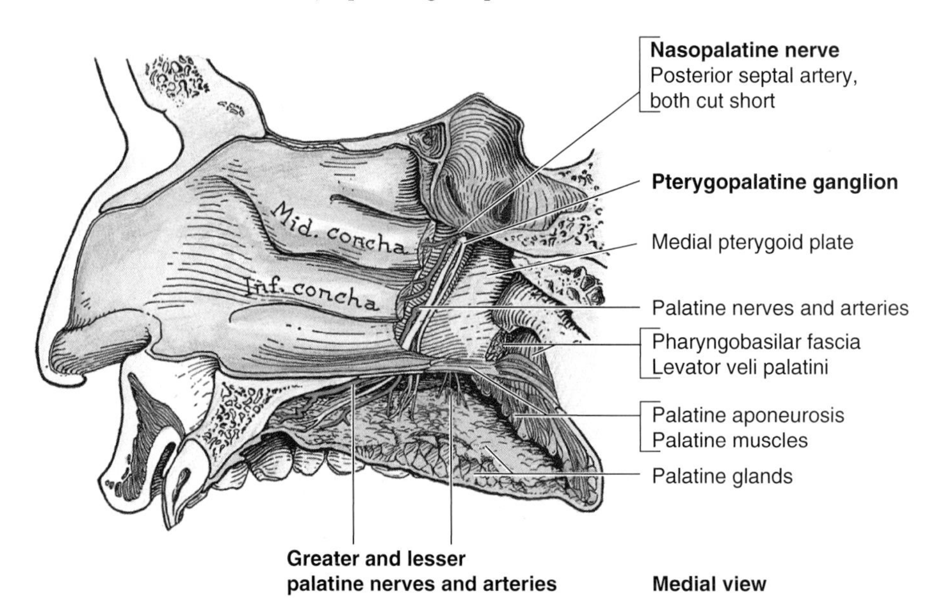

Figure 7.54. Palatine nerves and vessels. Dissection of the posterior part of the lateral wall of the nasal cavity and the palate. The mucous membrane of the palate, containing a layer of mucous glands, has been separated from the hard and soft regions of the palate by blunt dissection. The posterior ends of the middle and inferior conchae are cut through; these and the mucoperiosteum are pulled off the side wall of the nose as far as the posterior border of the medial pterygoid plate. The perpendicular plate of the palatine bone was broken through to expose the palatine nerves and arteries descending from the pterygopalatine fossa in the palatine canal.

Table 7.13. **Muscles of the Soft Palate**

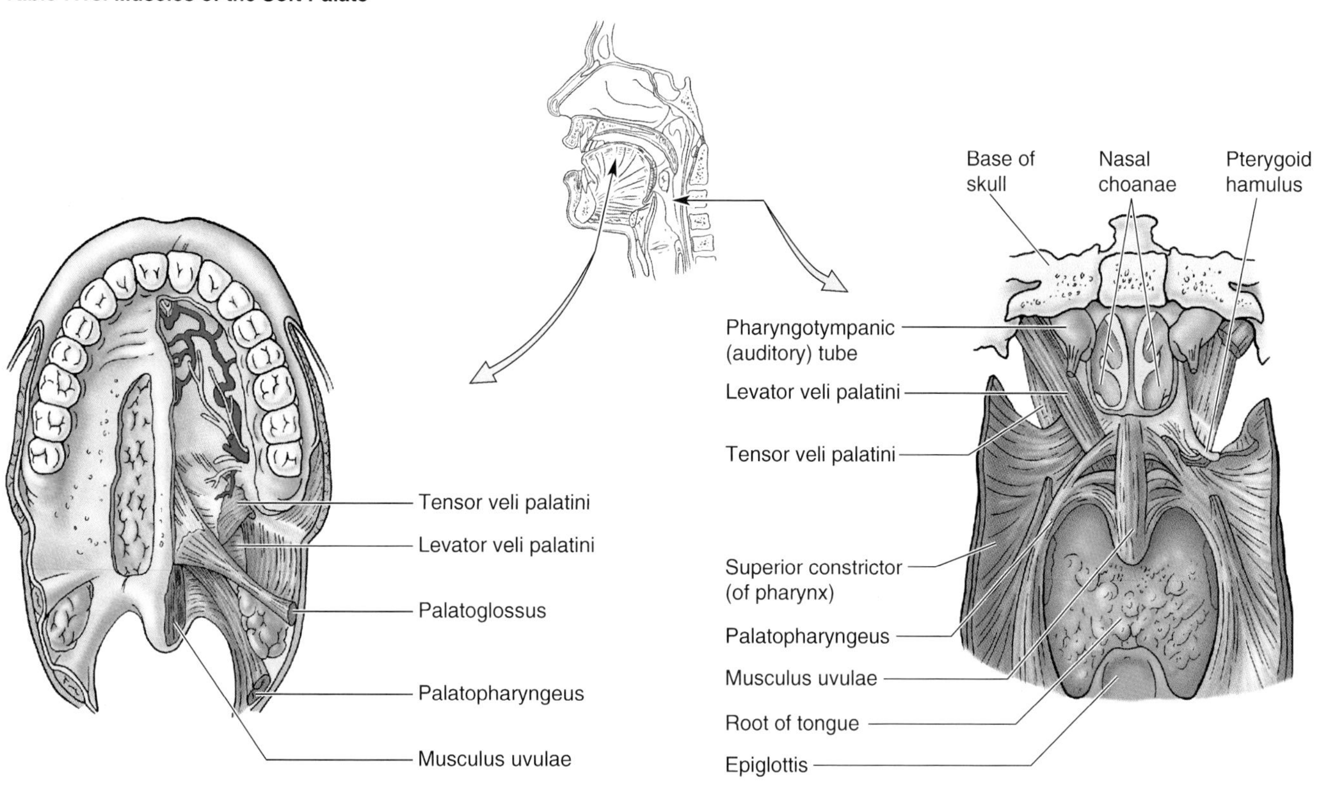

<table>
<tr><th>Muscle</th><th>Superior Attachment</th><th>Inferior Attachment</th><th>Innervation</th><th>Main Action</th></tr>
<tr><td>Tensor veli palatini</td><td>Scaphoid fossa of medial pterygoid plate, spine of sphenoid bone, and cartilage of pharyngotympanic (auditory) tube</td><td rowspan="2">Palatine aponeurosis</td><td>Medial pterygoid nerve (a branch of mandibular nerve—CN V_3) via otic ganglion</td><td>Tenses soft palate and opens mouth of auditory tube during swallowing and yawning</td></tr>
<tr><td>Levator veli palatini</td><td>Cartilage of pharyngotympanic (auditory) tube and petrous part of temporal bone</td><td rowspan="4">Cranial part of CN XI through pharyngeal branch of vagus nerve (CN X) via pharyngeal plexus</td><td>Elevates soft palate during swallowing and yawning</td></tr>
<tr><td>Palatoglossus</td><td>Palatine aponeurosis</td><td>Side of tongue</td><td>Elevates posterior part of tongue and draws soft palate onto tongue</td></tr>
<tr><td>Palatopharyngeus</td><td>Hard palate and palatine aponeurosis</td><td>Lateral wall of pharynx</td><td>Tenses soft palate and pulls walls of pharynx superiorly, anteriorly, and medially during swallowing</td></tr>
<tr><td>Musculus uvulae</td><td>Posterior nasal spine and palatine aponeurosis</td><td>Mucosa of uvula</td><td>Shortens uvula and pulls it superiorly</td></tr>
</table>

The sensory nerves of the palate are branches of the pterygopalatine ganglion (Fig. 7.54). The **greater palatine nerve** supplies the gingivae, mucous membrane, and glands of most of the hard palate. The **nasopalatine nerve** supplies the mucous membrane of the anterior part of the hard palate. The **lesser palatine nerves** supply the soft palate. The palatine nerves accompany the arteries through the greater and lesser palatine foramina, respectively. Except for the tensor veli palatini supplied by CN V_2, all muscles of the soft palate are supplied through the *pharyngeal plexus of nerves* (described in Chapter 8).

Nasopalatine Nerve Block

The nasopalatine nerve can be anesthetized by injecting anesthetic into the incisive fossa in the hard palate. The needle is inserted posterior to the incisive papilla. Both nasopalatine nerves can be anesthetized by the same injection where they emerge through the incisive fossa. The affected tissues are the palatal mucosa, the lingual gingivae and alveolar bone of the six maxillary teeth, and the hard palate.

Greater Palatine Nerve Block

The greater palatine nerve can be anesthetized by injecting anesthetic into the greater palatine foramen. The nerve emerges between the 2nd and 3rd molar teeth. This nerve block anesthetizes, on the side concerned, all the palatal mucosa and lingual gingivae posterior to the maxillary canine teeth and the underlying bone of the palate.

Cleft Palate

Cleft palate, with or without cleft lip, occurs approximately once in 2500 births and is more common in females than in males (Moore and Persaud, 1998). The cleft may involve only the uvula, giving it a fish-tail appearance, or it may extend through the soft and hard regions of the palate. In severe cases associated with cleft lip, the cleft palate extends through the alveolar process of the maxilla and the lips on both sides. The *embryological basis of cleft palate* is failure of mesenchymal masses in the lateral palatine processes (palatal shelves) to meet and fuse with each other, with the nasal septum, and/or with the posterior margin of the median palatine process. ○

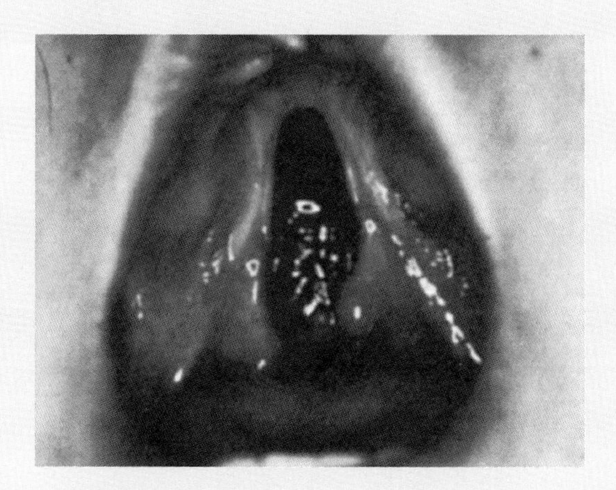

Tongue

The tongue is a mobile muscular organ that can assume a variety of shapes and positions. The tongue is partly in the oral cavity and partly in the pharynx. At rest, it occupies essentially all the oral cavity proper. The tongue is involved with mastication, taste, deglutition, articulation, and oral cleansing; two of these, however, are its main functions:

- Forming words during speaking
- Squeezing food into the pharynx when swallowing.

Parts and Surfaces of the Tongue

The tongue has a root, a body, an apex, a curved dorsal surface or dorsum, and an inferior surface. Some authors describe the **root of the tongue** as the inferior, relatively fixed part attached to the hyoid bone and mandible and in proximity to the geniohyoid and mylohyoid muscles in between (Williams et al., 1995). Most commonly the root is defined as the posterior (postsulcal) third of the tongue. The **body of the tongue** is the remaining part of the tongue; hence, its definition varies with the definition of the root. Most commonly the body is defined as the anterior (presulcal) two-thirds of the tongue (Fig. 7.55). The **apex of the tongue** is usually the pointed anterior part of the body. The body and the apex of the tongue are extremely mobile. The **dorsum of the tongue** is the posterosuperior surface of the tongue, which includes a V-shaped groove—the **terminal sulcus** (L. sulcus terminalis)—the apex of which points posteriorly to the **foramen cecum**, a small pit that is the nonfunctional remnant of the proximal part of the embryonic thyroglossal duct from which the thyroid gland developed (Moore and Persaud, 1998). The terminal sulcus divides the dorsum of the tongue into the anterior part (presulcal or oral part), which lies in the oral cavity proper, and the posterior part (postsulcal or pharyngeal part), which lies in the oropharynx. The anterior and posterior parts both include rough, irregular surfaces.

The mucous membrane on the anterior part of the tongue is rough because of the presence of numerous small **lingual papillae** (Fig. 7.55):

- *Vallate papillae* are large and flat-topped; they lie directly anterior to the terminal sulcus and are surrounded by deep moatlike trenches, the walls of which are studded by taste buds; the ducts of serous glands open into these trenches.
- *Foliate papillae* are small lateral folds of the lingual mucosa; they are poorly developed in humans.

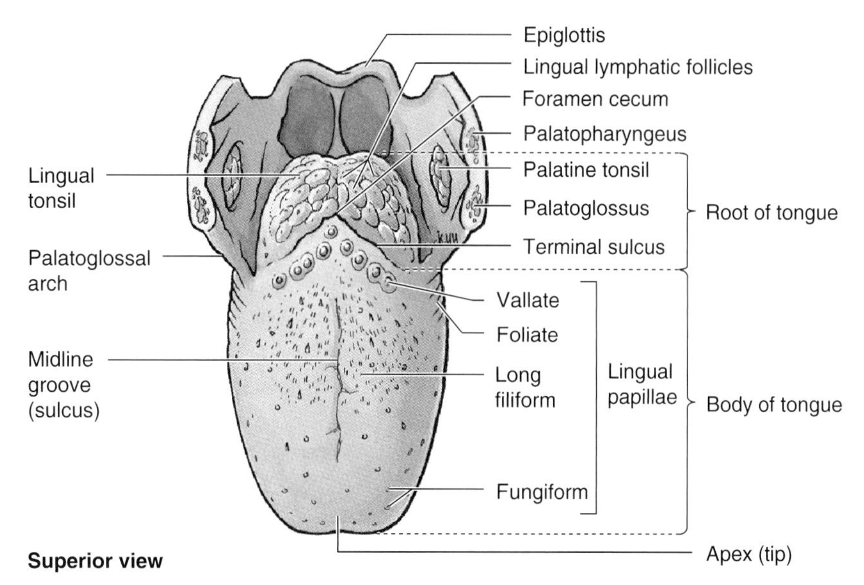

Figure 7.55. Dorsum of the tongue (L. lingua; G. glossa). Superior view. Observe the parts of the tongue: body and root, separated by the terminal sulcus (L. sulcus terminalis) and foramen cecum. The *brackets* indicate the parts of the tongue and do not embrace specific labels. Note also the placement of the various lingual papillae.

- *Filiform papillae* are long and numerous and contain afferent nerve endings that are sensitive to touch; these scaly, threadlike papillae are pinkish-grey and are arranged in V-shaped rows that are parallel to the terminal sulcus, except at the apex of the tongue where they tend to be arranged transversely.
- *Fungiform papillae* are mushroom-shaped and appear as pink or red spots; they are scattered among the filiform papillae but are most numerous at the apex and sides of the tongue.

The vallate, foliate, and most of the fungiform papillae contain taste receptors in the *taste buds*. A few taste buds are also in the epithelium covering the oral surface of the soft palate, the posterior wall of the oropharynx, and the epiglottis. Incidentally, it is the scaly filiform papillae that make the tongues of some animals (e.g., cats) so raspy.

The mucous membrane of the dorsum is thin over the anterior part of the tongue and is closely attached to the underlying muscle. A depression on the dorsal surface, the **midline groove** (median sulcus), divides the tongue into right and left halves; it also indicates the site of fusion of the embryonic distal tongue buds (Moore and Persaud, 1998). Deep to the midline groove is a fibrous **lingual septum** (Table 7.14) that divides the tongue into right and left halves.

The **posterior part of the tongue** is that part located posterior to the terminal sulcus and the palatoglossal arches (structures that demarcate the posterior boundary of the oral cavity). Its mucous membrane is thick and freely movable. It has no lingual papillae but the underlying nodules of *lingual lymphatic follicles* give this part of the tongue its irregular, cobblestone appearance. The nodular masses of **lingual follicles** (Fig. 7.55) are collectively known as the *lingual tonsil.*

The **inferior surface of the tongue** (sublingual surface) is covered with a thin, transparent mucous membrane through which one can see the underlying veins. With the tongue raised, observe the **lingual frenulum** (Fig. 7.56), a large midline fold of mucosa that passes from the gingiva covering the lingual aspect of the anterior alveolar ridge to the posteroinferior surface of the tongue. The frenulum connects the tongue to the floor of the mouth while allowing the anterior part of the tongue to move freely. On each side of this mucosal fold, a deep lingual vein is visible through the thin mucous membrane. Also observe the **sublingual caruncle**, a papilla on each side of the base of the lingual frenulum that is the location of the *opening of the submandibular duct* from the submandibular salivary gland.

Muscles of the Tongue

The tongue is essentially a mass of muscles that is mostly covered by mucous membrane (Fig. 7.55). Although it is traditional to do so, providing descriptions of the actions of tongue muscles ascribing a single action to a specific muscle—or implying that a particular movement is the consequence of a single muscle acting—greatly oversimplify the actions of the tongue and are misleading. The muscles of the tongue do not act in isolation, and some muscles perform multiple actions with parts of one muscle capable of acting independently, producing different—even antagonistic—actions. In general, however, *extrinsic muscles alter the position of the tongue and intrinsic muscles alter its shape.* The four intrinsic and four extrinsic muscles in each half of the tongue are separated by the fibrous **lingual septum** that merges posteriorly with the aponeurosis of the tongue. For attachments, nerve supply, and actions of the intrinsic and extrinsic muscles of the tongue, see Table 7.14.

Table 7.14. **Muscles of the Tongue**

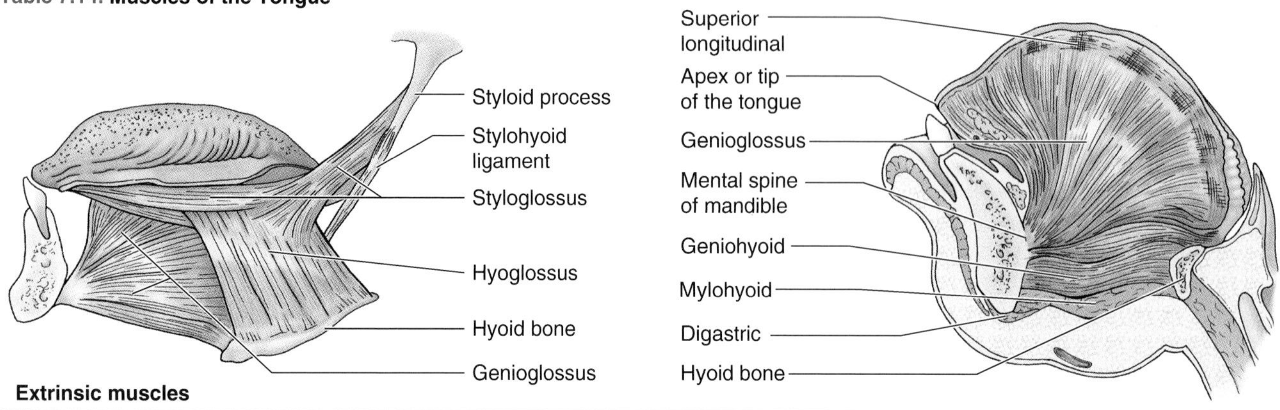

Extrinsic muscles

Muscle	Origin	Insertion	Innervation	Main Action
Genioglossus	Superior part of mental spine of mandible	Dorsum of tongue and body of hyoid bone	Hypoglossal nerve (CN XII)	Depresses tongue; its posterior part pulls tongue anteriorly for protrusion*
Hyoglossus	Body and greater horn of hyoid bone	Side and inferior aspect of tongue		Depresses and retracts tongue
Styloglossus	Styloid process and stylohyoid ligament	Side and inferior aspect of tongue		Retracts tongue and draws it up to create a trough for swallowing
Palatoglossus	Palatine aponeurosis of soft palate	Side of tongue	Cranial root of CN XI via pharyngeal branch of CN X and pharyngeal plexus	Elevates posterior part of tongue

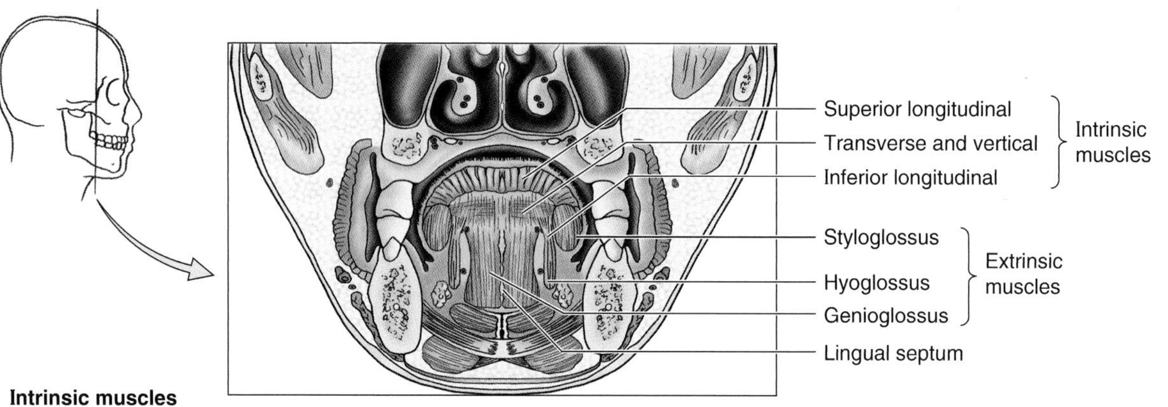

Intrinsic muscles

Muscle	Origin	Insertion	Innervation	Main Action
Superior longitudinal	Submucous fibrous layer and median fibrous septum	Margins of tongue and mucous membrane	Hypoglossal nerve (CN XII)	Curls tip and sides of tongue superiorly and shortens tongue
Inferior longitudinal	Root of tongue and body of hyoid bone	Apex of tongue		Curls tip of tongue inferiorly and shortens tongue
Transverse	Median fibrous septum	Fibrous tissue at margins of tongue		Narrows and elongates the tongue*
Vertical	Superior surface of borders of tongue	Inferior surface of borders of the tongue		Flattens and broadens the tongue*

* Act simultaneously to protrude tongue

Extrinsic Muscles of the Tongue. The extrinsic muscles (genioglossus, hyoglossus, styloglossus, and palatoglossus) originate outside the tongue and attach to it. They mainly move the tongue but they can alter its shape as well.

The **genioglossus**—a fan-shaped muscle—contributes most of the bulk of the tongue. It arises by a short tendon from the superior part of the mental spine of the mandible. It fans out as it enters the tongue inferiorly and its fibers attach to the entire dorsum of the tongue. Its most inferior fibers insert into the body of the hyoid bone and pull the root of the tongue anteriorly, which may be done as an element of the complex action of protruding the tongue. Acting bilaterally, the genioglossus muscles depress, especially the central part of the tongue, creating a central furrow or groove. Acting unilaterally, the genioglossus will deviate ("wag") the tongue toward the contralateral side. When the tongue has been protruded, contraction of the anterior part of the muscle retracts the apex.

The **hyoglossus**—a thin quadrilateral muscle—arises from the body and greater horn of the hyoid bone and passes superoanteriorly to insert into the side and inferior aspect of the tongue. The hyoglossus depresses the tongue, especially pulling its sides inferiorly; it also aids in retrusion (retraction) of the tongue.

The **styloglossus**—a small short muscle—arises from the anterior border of the styloid process near its tip and the stylohyoid ligament. It passes inferoanteriorly to insert into the side and inferior aspect of the tongue. Its fibers interdigitate with those of the hyoglossus muscle. The styloglossus retrudes the tongue and curls its sides, acting with the genioglossus to create a trough during swallowing.

The **palatoglossus** originates in the palatine aponeurosis of the soft palate and inserts into the side of the tongue. It enters the lateral part of the tongue with the styloglossus but passes almost transversely through the tongue with the intrinsic transverse muscle fibers. The palatoglossus is in its derivation, innervation, and function actually more "palatal" than "glossal"; that is, it is more a part of the soft palate than of the tongue. Although it is capable of elevating the posterior part of the tongue, it may pull down the soft palate; most commonly, the palatoglossus muscles act bilaterally, approximating the palatoglossal folds to constrict the isthmus of the fauces. The palatoglossus enters the lateral part of the tongue with the styloglossus but passes almost transversely through the tongue with the intrinsic transverse muscle fibers.

Intrinsic Muscles of the Tongue. The superior and inferior longitudinal, transverse, and vertical muscles are confined to the tongue. They have their attachments entirely within the tongue and are not attached to bone (Table 7.14).

The **superior longitudinal muscle of the tongue** forms a thin layer deep to the mucous membrane on the dorsum of the tongue, running from its apex to its root. It arises from the submucous fibrous layer and the lingual septum and inserts mainly into the mucous membrane. The superior longitudinal muscle curls the apex of the tongue, pushing it against the palate, or turning the protruded apex upward toward the nose, making the dorsum of the tongue concave longitudinally.

The **inferior longitudinal muscle of the tongue** consists of a narrow band close to the inferior surface of the tongue. It extends from the apex to the root of the tongue; some of its fibers attach to the hyoid bone. The inferior longitudinal muscle curls the apex of the tongue inferiorly, making the dorsum of the tongue convex. The superior and inferior longitudinal muscles act together to make the tongue short and thick and in retracting the protruded tongue.

The **transverse muscles of the tongue** lie deep to the superior longitudinal muscle. They arise from the lingual septum and run lateral to its right and left margins. Its fibers insert into the submucous fibrous tissue. The transverse muscles narrow and increase the height of the tongue.

The **vertical muscle of the tongue** runs inferolaterally

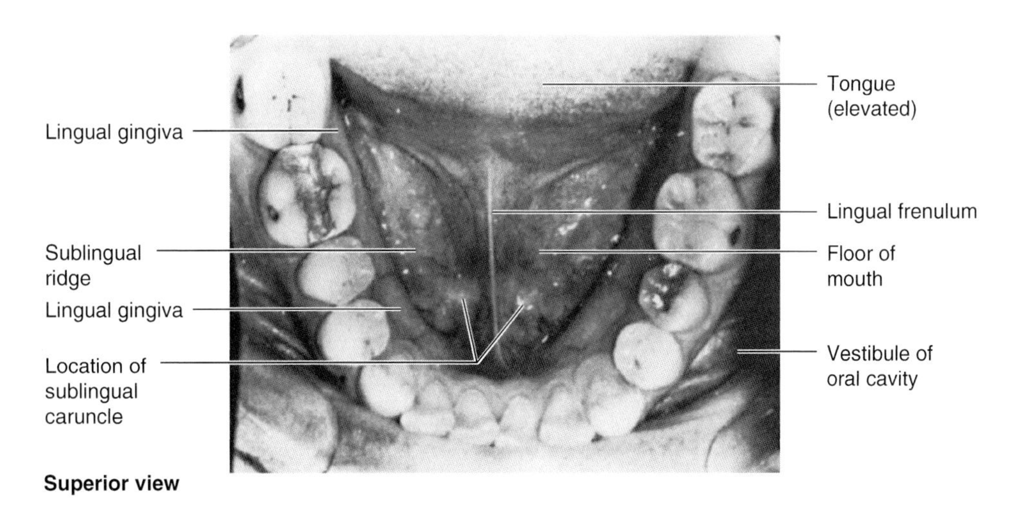

Figure 7.56. Superior view of the floor of the mouth and the vestibule of the oral cavity. The tongue (L. lingua; G. glossa) is elevated and retracted superiorly. Observe the mucous membrane, the lingual gingiva, and the lingual frenulum.

from the dorsum of the tongue. It flattens and broadens the tongue. Acting with the transverse muscles, they make the tongue long and narrow, which may push the tongue against the front teeth or protrude the tongue from the open mouth (especially when acting with the posterior, inferior part of the genioglossus).

Innervation of the Tongue

All muscles of the tongue *except for the palatoglossus* (supplied by the *pharyngeal plexus*—formed by fibers from the cranial root of CN XI carried by CN X) are supplied by CN XII, the **hypoglossal nerve** (Fig. 7.57). For general sensation (touch and temperature), the mucosa of the anterior two-thirds of the tongue is supplied by the **lingual nerve** (Fig. 7.43), a branch of CN V_3. For special sensation (taste), this part of the tongue, except for the vallate papillae, is supplied through the **chorda tympani nerve**, a branch of CN VII (Fig. 7.41*C*). The chorda tympani joins the lingual nerve and runs anteriorly in its sheath. The mucous membrane of the posterior one-third of the tongue and the vallate papillae are supplied by the lingual branch of the **glossopharyngeal nerve** for both general and special sensation. Twigs of the **internal laryngeal nerve**, a branch of the vagus nerve, supply mostly general but some special sensation to a small area of the tongue just anterior to the epiglottis. These mostly sensory nerves also carry **parasympathetic secretomotor fibers** to serous glands in the tongue. Parasympathetic fibers from the chorda tympani nerve travel with the lingual nerve to the submandibular and sublingual salivary glands. These nerve fibers synapse in the submandibular ganglion (see Fig. 7.60) that hangs from the lingual nerve.

The four basic taste sensations—sweet, salty, sour, and bitter—are detected on the tongue as follows:

- Sweetness—apex (tip)
- Saltiness—lateral margins
- Sourness and bitterness—posterior part.

Vasculature of the Tongue

The *arteries of the tongue* derive from the **lingual artery**, which arises from the **external carotid artery** (Fig. 7.58). On entering the tongue, the lingual artery passes deep to the hyoglossus muscle. **The main branches of the lingual artery are the:**

- *Dorsal lingual arteries,* which supply the posterior part and send a tonsillar branch to the palatine tonsil
- *Deep lingual artery,* which supplies the anterior part
- *Sublingual artery,* which supplies the sublingual gland and the floor of the mouth.

The dorsal lingual arteries communicate with each other near the apex of the tongue. The other branches of the lingual artery are prevented from communicating by the fibrous *lingual septum,* which separates the tongue into right and left halves. The **veins of the tongue** (Fig. 7.59) are the:

- *Dorsal lingual veins,* which accompany the lingual artery
- *Deep lingual veins* (ranine veins), which begin at the apex of the tongue and run posteriorly beside the lingual frenulum to join the *sublingual vein.*

All these veins terminate, directly or indirectly, in the IJV. **Lymph draining from the tongue takes four routes** (Fig. 7.60):

- Lymph from the posterior third drains into the *superior deep cervical lymph nodes* on both sides
- Lymph from the medial part of the anterior two-thirds drains directly to the *inferior deep cervical lymph nodes*
- Lymph from the lateral parts of the anterior two-thirds drains to the *submandibular lymph nodes*
- The apex drains to the *submental lymph nodes*
- The posterior third and the area near the midline drain bilaterally.

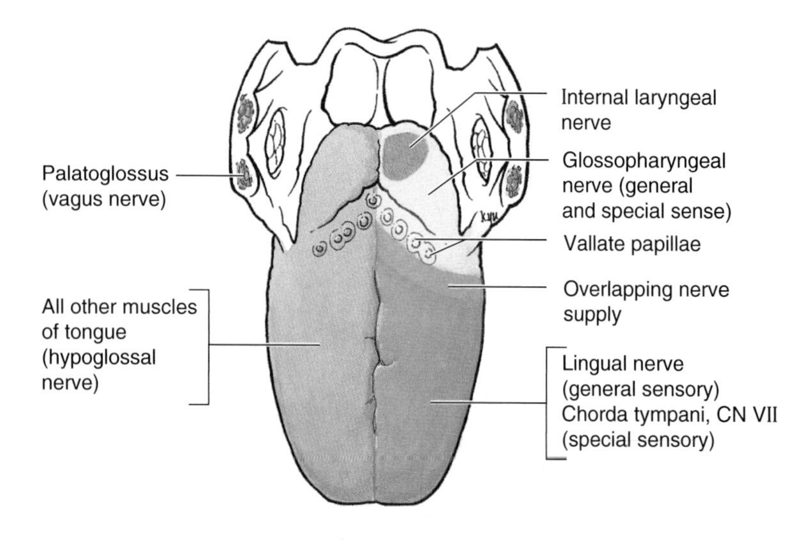

Figure 7.57. Nerve supply of the tongue (L. lingua; G. glossa). Superior view. The anterior two-thirds are supplied by the lingual nerve (CN V_3) for general sensation and the chorda tympani—a branch of the facial nerve (CN VII) transferring nerve fibers to the lingual nerve—for taste. The posterior third of the tongue and the vallate papillae are supplied by the lingual branch of the glossopharyngeal nerve (CN IX) for both general sensation and taste. Other contributions are from a small lingual branch of the facial nerve (taste) and from the internal laryngeal branch of the vagus (CN X) for general sensation and taste. Hence, CN VII, CN IX, and CN X convey nerve fibers for taste.

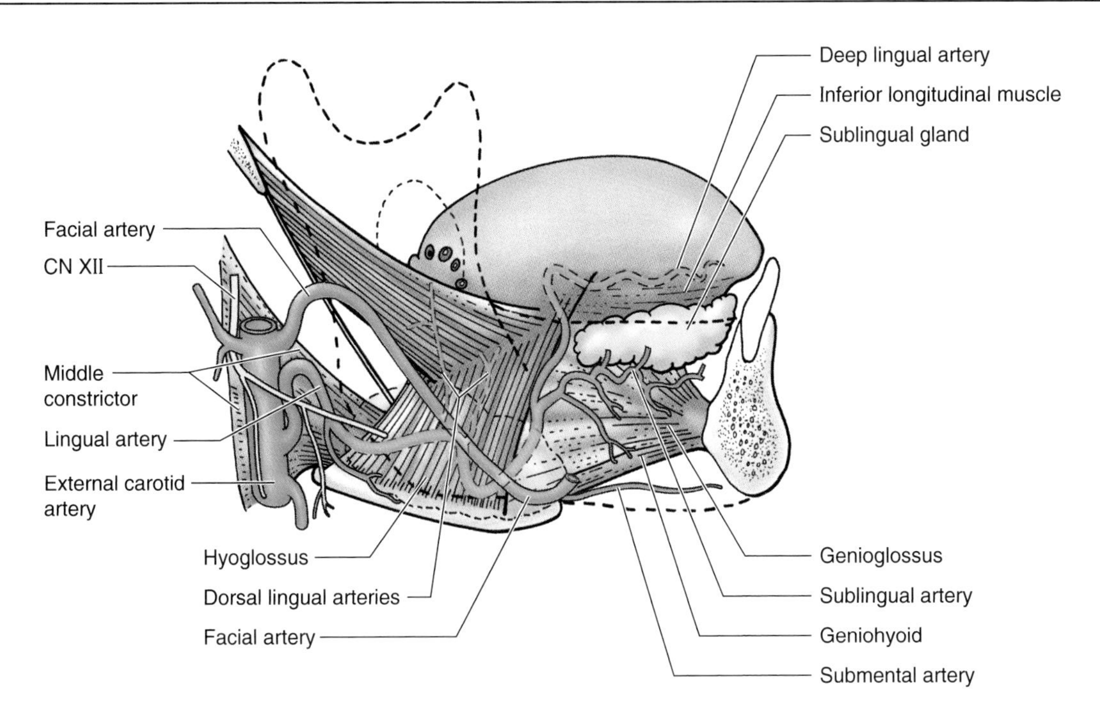

Figure 7.58. Blood supply of the tongue. Lateral view. The main artery to the tongue is the lingual, a branch of the external carotid artery. The *dorsal lingual artery* provides the blood supply to the root of the tongue, while the *deep lingual artery* supplies the body. The dorsal lingual artery also sends a branch to the palatine tonsil. The sublingual branch of the lingual artery provides the blood supply to the floor of the mouth, including the sublingual gland.

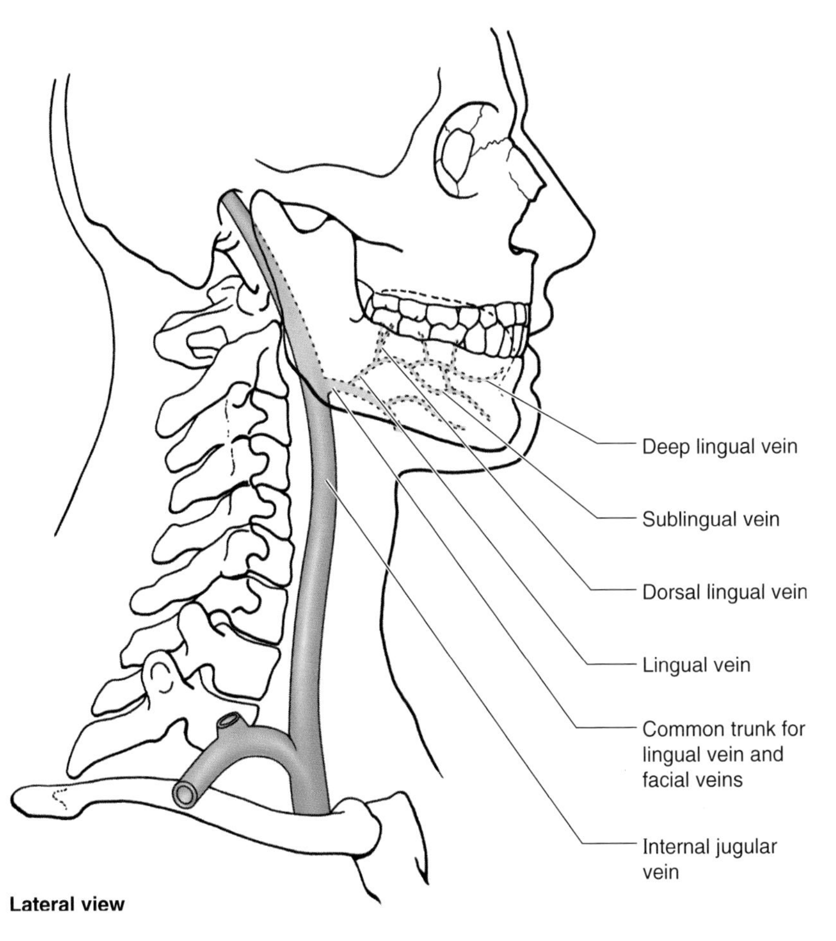

Figure 7.59. Venous drainage of the tongue. The tongue is drained by two lingual veins that accompany the lingual arteries and receive the dorsal lingual veins. The deep lingual vein runs posteriorly under the mucous membrane of the underside of the tongue at the side of the lingual frenulum (where it can be observed through the mucosa). It unites with the sublingual vein from the floor of the mouth and the sublingual salivary gland, and then receives the dorsal lingual vein. All these veins drain directly or indirectly into the internal jugular vein (IJV).

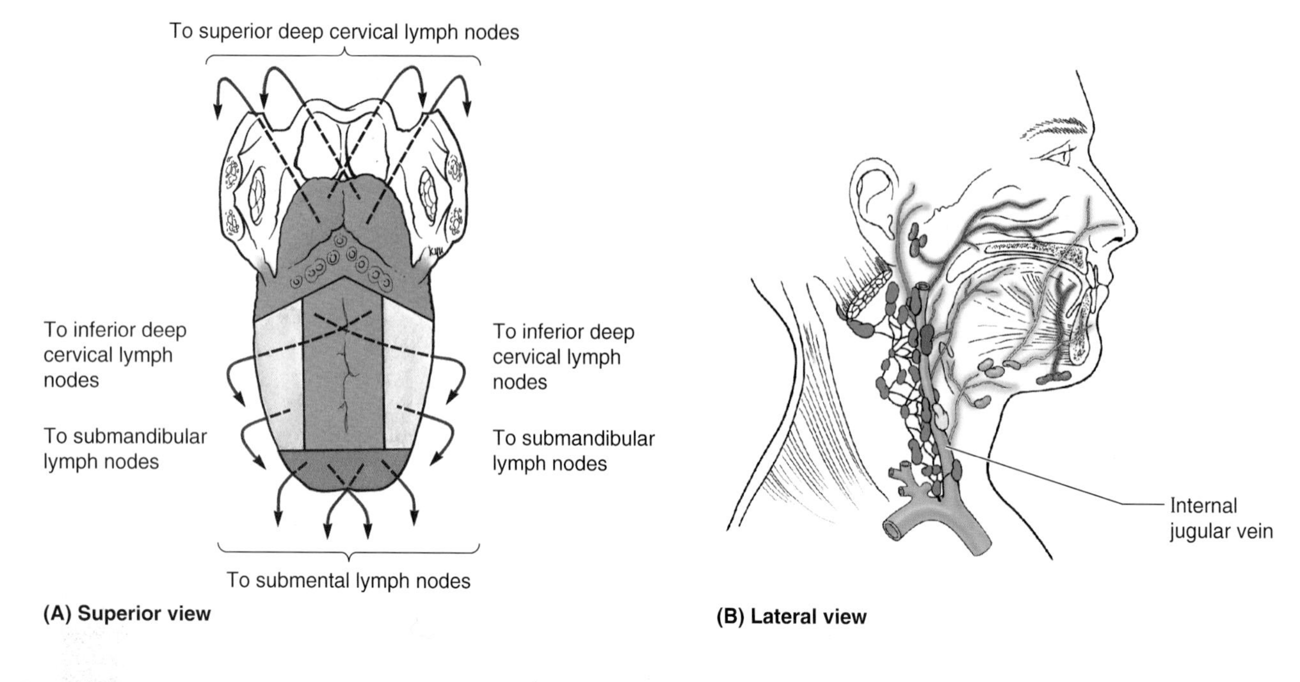

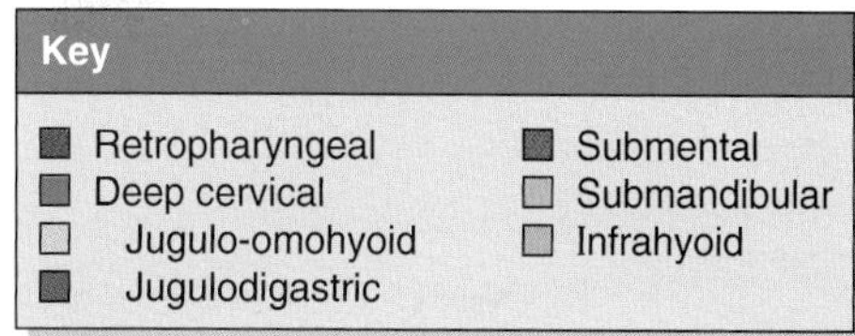

Figure 7.60. Lymphatic drainage of the tongue. A. Superior view of the dorsum of the tongue. **B.** Lateral view of the head and neck. Lymph drains to the submental, submandibular, and superior and inferior deep cervical lymph nodes, including the jugulodigastric and jugulo-omohyoid nodes. Extensive communications occur across the midline of the tongue.

Gag Reflex

One may touch the anterior part of the tongue without feeling discomfort; however, when the posterior part is touched, one gags. CN IX and CN X are responsible for the muscular contraction of each side of the pharynx. Glossopharyngeal branches provide the afferent limb of the *gag reflex*.

Paralysis of the Genioglossus

When the genioglossus muscle is paralyzed, the tongue has a tendency to fall posteriorly, obstructing the airway and presenting the risk of suffocation. Total relaxation of the genioglossus muscles occurs during general anesthesia; therefore, the tongue of an anesthetized patient must be prevented from relapsing by inserting an airway.

Injury to the Hypoglossal Nerve

Trauma, such as a fractured mandible, may injure the hypoglossal nerve, resulting in paralysis and eventual atrophy of one side of the tongue. The tongue deviates to the paralyzed side during protrusion because of the action of the unaffected genioglossus muscle on the other side.

Sublingual Absorption of Drugs

For quick absorption of a drug—for instance, when nitroglycerin is used as a vasodilator in angina pectoris (chest pain)—the pill or spray is put under the tongue where it dissolves and enters the deep lingual veins in less than ▶

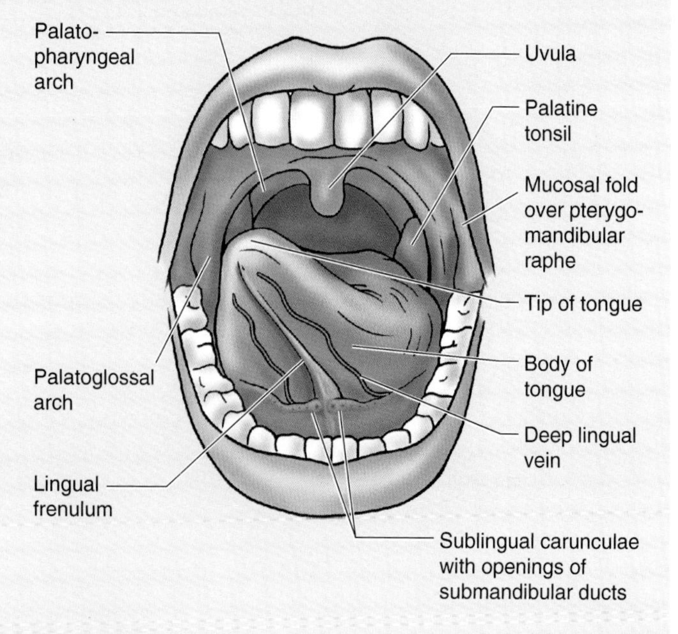

▶ a minute. The sublingual veins in older people are often varicose (enlarged and tortuous), but they do not bleed and have no clinical significance (Swartz, 1994).

Lingual Carcinoma

Lymphatic drainage of the tongue is of particular importance because of the common occurrence of lingual carcinoma. Malignant tumors in the posterior part of the tongue metastasize to the superior deep cervical lymph nodes on both sides, whereas tumors in the anterior part usually do not metastasize to the inferior deep cervical lymph nodes until late in the disease. Because these nodes are closely related to the IJV, metastatic carcinoma from the tongue may be widely distributed through the submental and submandibular regions and along the IJVs in the neck.

Frenectomy

An overly large lingual frenulum may interfere with tongue movements and affect speech. A short lingual frenulum (tongue-tie) rarely interferes with eating or speech. In unusual cases, a *frenectomy* (cutting the frenulum) may be necessary in infants to free the tongue for normal speech.

Thyroglossal Duct Cyst

A cystic remnant of the thyroglossal duct—associated with development of the thyroid gland (Moore and Persaud, 1998)—may be found in the root of the tongue and be connected to a sinus that opens at the foramen cecum. However, most *thyroglossal duct cysts* lie close or just inferior to the body of the hyoid bone and produce a painless midline swelling in the neck. Occasionally these cysts open spontaneously onto the skin of the neck, producing a nonhealing sore—a *thyroglossal fistula (abnormal passage)*. Surgical excision is required for healing to occur.

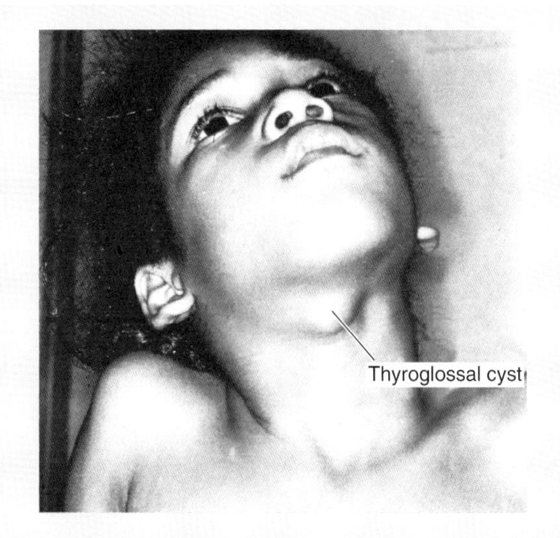

Aberrant Thyroid Gland

Aberrant thyroid glandular tissue may be found anywhere along the path of descent of the embryonic thyroglossal duct. Although uncommon, the thyroglossal duct carrying thyroid-forming tissue at its distal end may fail to descend to its definitive position in the neck. *Aberrant thyroid tissue* may be in the root of the tongue, just posterior to the foramen cecum, or in the neck (*A*). Cystic remnants of the thyroglossal duct may be differentiated from an undescended thyroid by radioisotope scanning (*B*). An aberrant thyroid gland may be the only thyroid tissue the person has; if so, removal will require the patient be continually medicated with thyroid hormone. ⊙

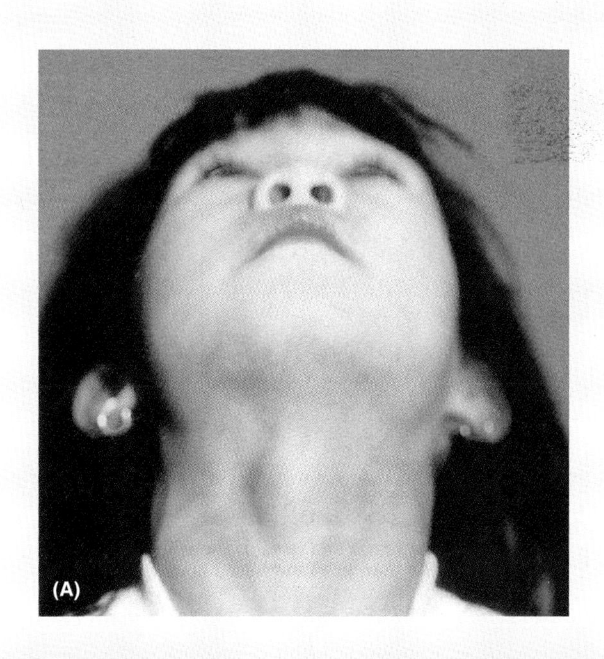

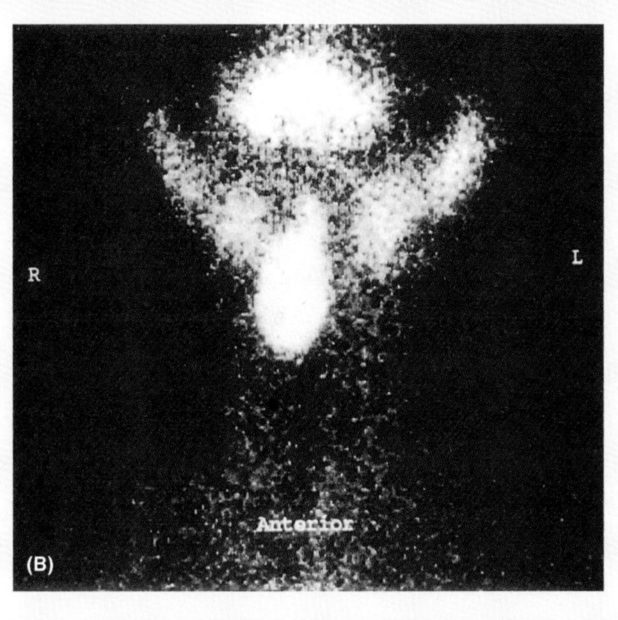

Salivary Glands

The salivary glands are the parotid, submandibular, and sublingual glands. The clear, tasteless, odorless viscid fluid (saliva) secreted by these glands and the mucous glands of the oral cavity:

- Keep the mucous membrane of the mouth moist
- Lubricate the food during mastication
- Begin the digestion of starches
- Serve as an intrinsic "mouthwash"
- Play significant roles in the prevention of tooth decay and in the ability to taste.

In addition to the main salivary glands, small accessory salivary glands are scattered over the palate, lips, cheeks, tonsils, and tongue.

Parotid Glands

The parotid glands are the largest of the three pairs of salivary glands. The parotid gland (Fig. 7.61) has an irregular shape because it occupies the gap between the ramus of the mandible and the styloid process of the temporal bone. The purely serous secretion of these glands empties into the vestibule of the oral cavity. In addition to its digestive function, it washes food particles into the mouth proper. The **arterial supply** of the parotid gland and duct is from branches of the *external carotid* and *superficial temporal arteries*. The **veins** from the parotid gland drain into the *retromandibular veins*. The **lymphatic vessels** from the parotid gland end in the *superficial* and *deep cervical lymph nodes*. For a discussion of the **nerve supply of the parotid gland**, see page 870.

Submandibular Glands

Each **submandibular gland** lies along the body of the mandible, partly superior and partly inferior to the posterior half of the mandible, and partly superficial and partly deep to the mylohyoid muscle (Fig. 7.61). The **submandibular duct**, approximately 5 cm long, arises from the portion of the gland that lies between the mylohyoid and hyoglossus. Passing from lateral to medial, the **lingual nerve** loops under the duct that runs anteriorly to open by one to three orifices on a small sublingual papilla beside the base of the lingual frenulum. The orifice of the submandibular duct is visible, and saliva can often be seen trickling from it (or spraying from it during yawning). The **arterial supply** of the submandibular gland is from the *submental artery* (Fig. 7.58). The **veins** accompany the arteries. The submandibular gland is supplied by presynaptic parasympathetic, secretomotor fibers conveyed from the facial nerve to the lingual nerve by the chorda tympani nerve, which synapse with postsynaptic neurons in the *submandibular ganglion* (Fig. 7.61*A*). The latter fibers accompany arteries to reach the gland, along with vasoconstrictive postsynaptic sympathetic fibers from the superior cervical ganglion. The **lymphatic vessels** of the submandibular gland drain into the *deep cervical lymph nodes*, particularly the *jugulo-omohyoid node* (Fig. 7.60*A*).

Sublingual Glands

The sublingual glands are the smallest and most deeply situated of the salivary glands (Fig. 7.61). Each almond-shaped gland lies in the floor of the mouth between the mandible and the genioglossus muscle. The glands from each side unite to form a horseshoe-shaped mass around the lingual frenulum. Numerous small *sublingual ducts* open into the floor of the mouth along the sublingual folds. The **arterial supply** of the sublingual glands is from the *sublingual* and *submental* arteries, branches of the lingual and facial arteries (Fig. 7.58). The **nerves** of the sublingual glands accompany those of the submandibular gland. Presynaptic parasympathetic secretomotor fibers are conveyed by the facial, chorda tympani, and lingual nerves to synapse in the submandibular ganglion as described for the submandibular glands (Fig. 7.61*A*).

Parotiditis

Parotiditis (parotitis) is an inflammation of the parotid gland. *Mumps* is an acute, generalized viral infection that causes enlargement of the salivary glands, chiefly the parotid glands. Mumps is painful because of enlargement of the parotid gland within its tight fibrous capsule.

Abscess in the Parotid Gland

A *bacterial infection* localized in the parotid gland produces an abscess. The infection could result from extremely poor dental hygiene and spread to the gland through the parotid ducts. Physicians and dentists must determine whether a swelling of the cheek results from infection of the parotid gland or from an abscess of dental origin.

Excision of the Submandibular Gland

Excision of a submandibular gland because of a calculus (stone) in its duct or a tumor in the gland is not uncommon. The skin incision is made at least 2.5 cm inferior to the angle of the mandible to avoid the mandibular branch of the facial nerve.

Sialography

The parotid and submandibular salivary glands may be examined radiographically following injection of a contrast medium into their ducts. This special type of radiograph (*sialogram*) demonstrates the salivary ducts and some secretory units. Because of the small size of the ducts of the sublingual glands and their multiplicity, one cannot usually inject contrast medium into the sublingual ducts. ⊙

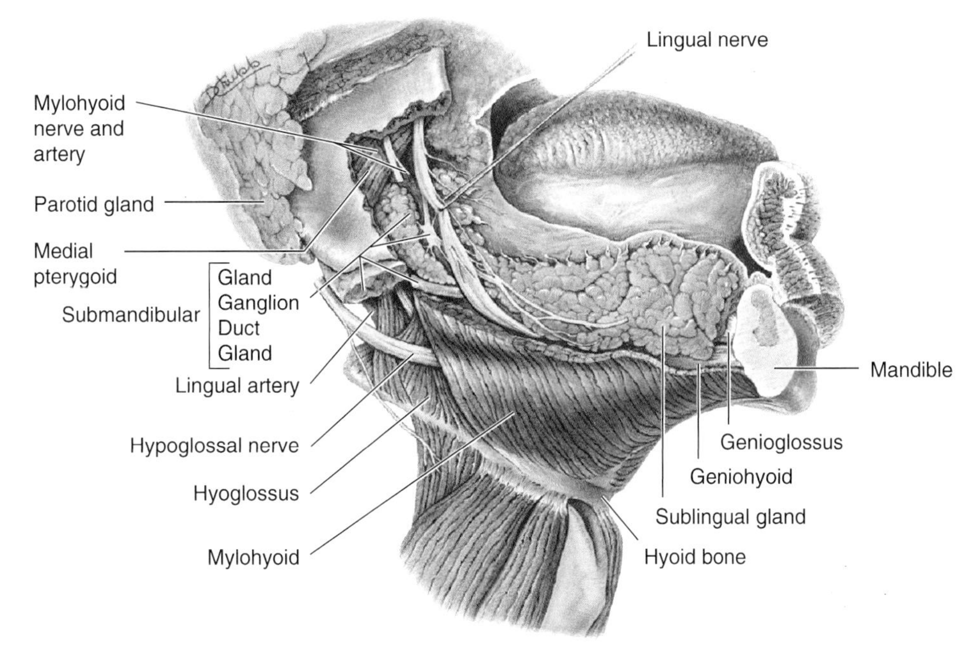

(A) Right lateral view

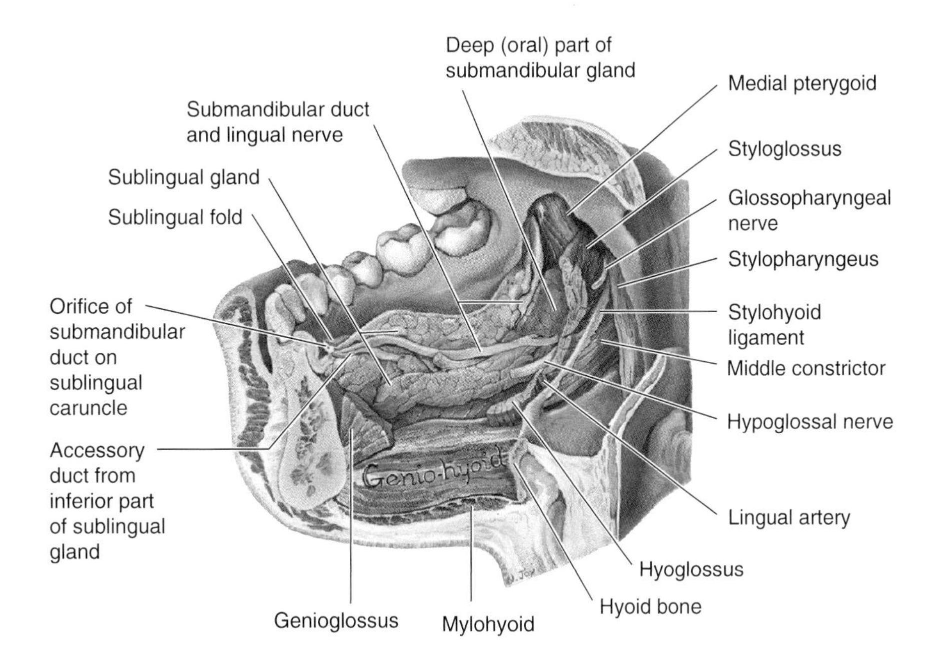

(B) Medial view

Figure 7.61. Parotid, submandibular, and sublingual salivary glands. A. Right lateral view. The body and parts of the ramus of the mandible (lower jaw) have been removed. Observe that the parotid gland is in contact with the deep part of the submandibular gland posteriorly. Notice the fine ducts passing from the superior border of the sublingual gland to open on the sublingual fold. **B.** Medial view of the right lower jaw and the floor of the mouth; the tongue has been excised. Observe the deep or oral part of the submandibular gland in the angle between the lingual nerve and the submandibular duct, which separates it from the sublingual gland. The orifice of the duct is seen at the anterior end of the sublingual fold. Examine the submandibular duct adhering to the medial side of the sublingual gland and here receiving, as it sometimes does, a large accessory duct from the inferior part of the sublingual gland.

Pterygopalatine Fossa

The pterygopalatine fossa—*a small pyramidal space inferior to the apex of the orbit* (Fig. 7.62)—lies between the pterygoid process of the sphenoid bone posteriorly and the posterior aspect of the maxilla anteriorly. The fragile vertical plate of the palatine bone forms its medial wall. The incomplete *roof of the pterygopalatine fossa* is formed by the **greater wing of the sphenoid bone.** The *floor of the pterygopalatine fossa* is formed by the **pyramidal process of the palatine bone.** Its superior, larger end opens into the *inferior orbital fissure*; its inferior end is closed except for the palatine foramina. *The pterygopalatine fossa communicates*:

- Laterally with the *infratemporal fossa* through the *pterygomaxillary fissure*
- Medially with the nasal cavity through the *sphenopalatine foramen*
- Anterosuperiorly with the orbit through the *inferior orbital fissure*
- Posterosuperiorly with the middle cranial fossa through the *foramen rotundum* and *pterygoid canal* (Fig. 7.63).

Contents of the Pterygopalatine Fossa

The contents of the pterygopalatine fossa (Fig. 7.64, *A–C*) are the:

- Terminal (third or pterygopalatine) part of the maxillary artery and the initial parts of its branches
- Maxillary nerve
- Nerve of the pterygoid canal
- Pterygopalatine ganglion.

Maxillary Artery

The maxillary artery, a terminal branch of the external carotid artery, passes anteriorly and traverses the infratemporal fossa. It passes over the lateral pterygoid muscle and enters the pterygopalatine fossa. The *pterygopalatine part of the maxillary artery* (Fig. 7.64*A*), its third part, passes through the *pterygomaxillary fissure* (Fig. 7.62) and enters the pterygopalatine fossa, where it lies anterior to the *pterygopalatine ganglion* (Fig. 7.64*B*). The artery gives rise to branches that accompany all nerves in the fossa with the same names.

The branches of the pterygopalatine part of the maxillary artery (Figs. 7.42, 7.53B, *and* 7.64A) *are the:*

- Posterior superior alveolar artery
- Descending palatine artery, which divides into the greater and lesser palatine arteries
- Artery of the pterygoid canal
- Sphenopalatine artery, which divides into posterior lateral nasal branches to the lateral wall of the nasal cavity and its associated paranasal sinuses, and the posterior septal branches

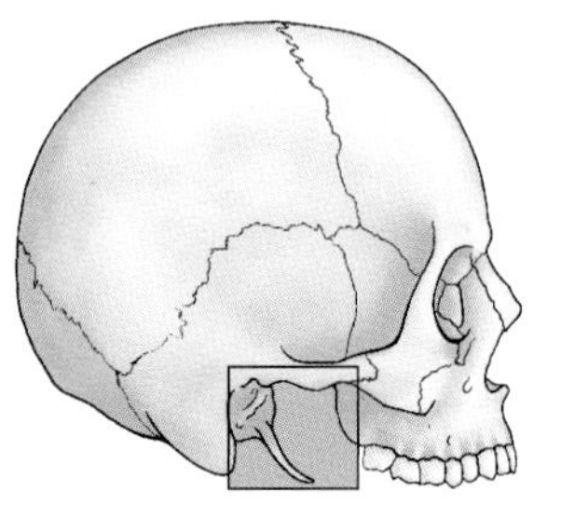

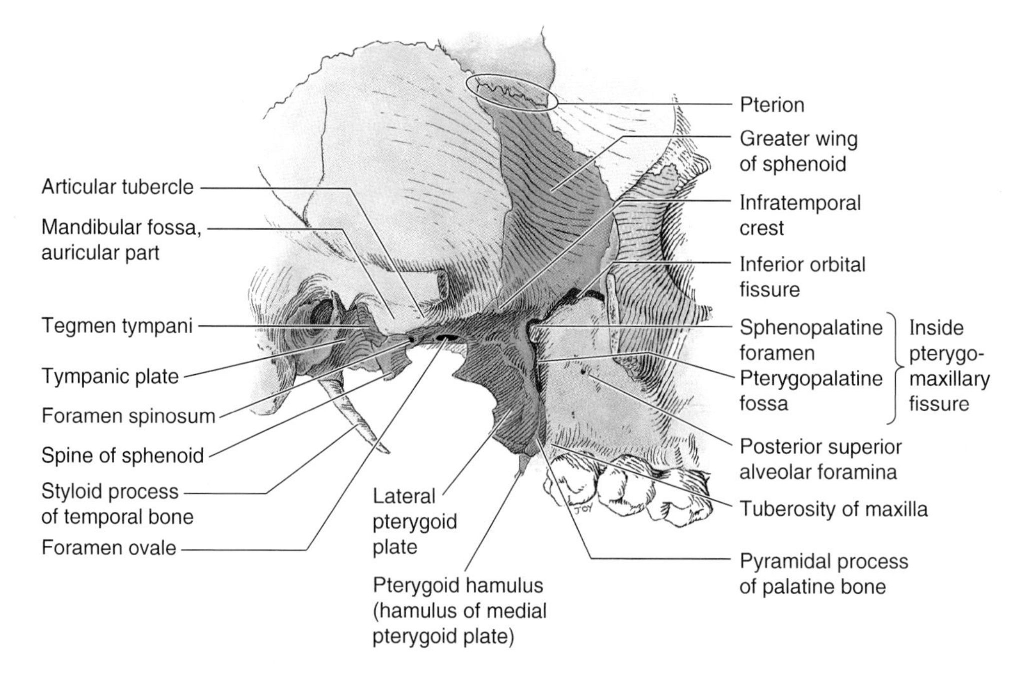

Figure 7.62. Infratemporal and pterygopalatine fossae. Lateral view. The pterygopalatine fossa is a small pyramidal space housing the pterygopalatine ganglion (see Fig. 7.54), which is seen through the pterygomaxillary fissure between the pterygoid process and the maxilla. The sphenopalatine foramen is an opening into the nasal cavity at the top of the palatine bone.

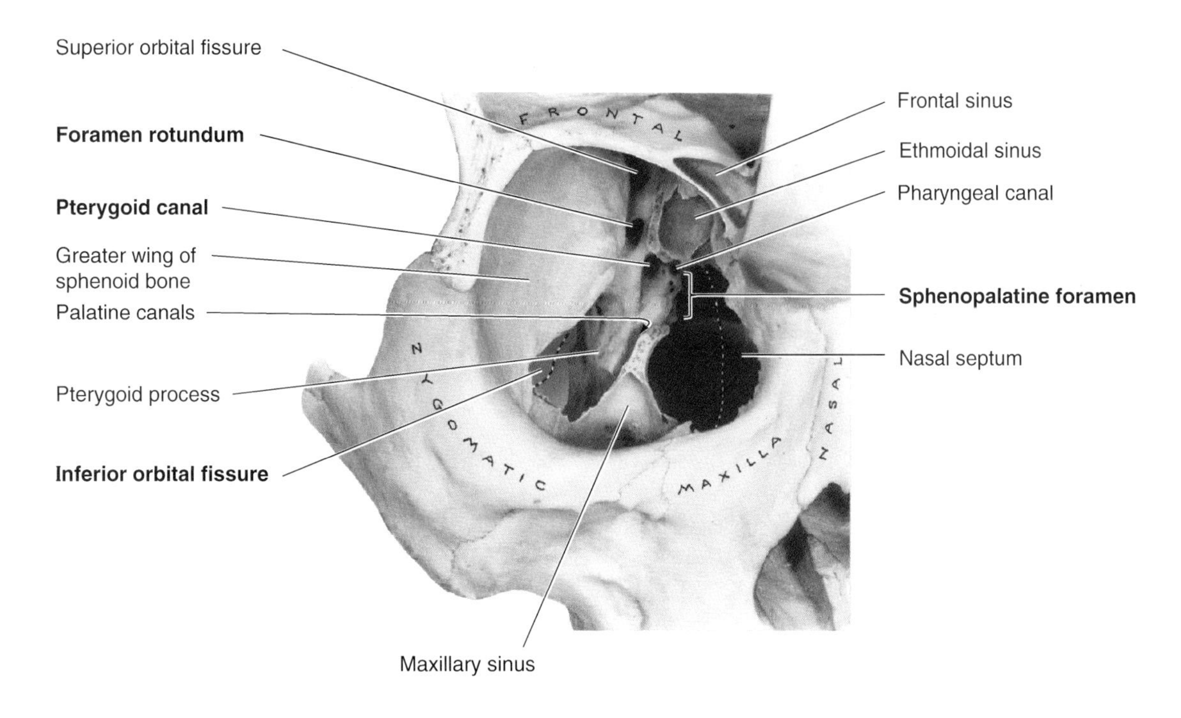

Figure 7.63. Pterygopalatine fossa. Anterior view of the fossa that has been exposed through the floor of the orbit and maxillary sinus. The foramen rotundum (L. round), pterygoid canal, and pharyngeal canal are openings in the posterior wall of the pterygopalatine fossa.

- Infraorbital artery, which gives rise to the anterior superior alveolar artery and terminates as branches to the lower eyelid, nose, and upper lip.

Maxillary Nerve

The maxillary nerve enters the pterygopalatine fossa through the *foramen rotundum* (Fig. 7.63) and runs anterolaterally in the posterior part of the fossa. Within the pterygopalatine fossa, the maxillary nerve gives off the *zygomatic nerve* (Fig. 7.64*C*), which divides into zygomaticofacial and zygomaticotemporal nerves. These nerves emerge from the zygomatic bone through cranial foramina of the same name and supply general sensation to the lateral region of the cheek and temple. The *zygomaticotemporal nerve* also gives rise to a communicating branch, which conveys parasympathetic secretomotor fibers to the lacrimal gland by way of the heretofore purely sensory lacrimal nerve from CN V_1 (Fig. 7.64*C*). While in the pterygopalatine fossa, the maxillary nerve also gives off the two *pterygopalatine nerves* that suspend the parasympathetic *pterygopalatine ganglion* in the superior part of the pterygopalatine fossa (Fig. 7.64, *B* and *C*). The pterygopalatine nerves convey general sensory fibers of the maxillary nerve, which pass through the pterygopalatine ganglion without synapsing and supply the nose, palate, tonsil, and gingivae. The maxillary nerve leaves the pterygopalatine fossa through the inferior orbital fissure, after which it is known as the **infraorbital nerve**.

The parasympathetic fibers to the pterygopalatine ganglion come from the facial nerve by way of its first branch, the *greater petrosal nerve* (Fig. 7.64*C*). This nerve joins the *deep petrosal nerve* as it passes through the foramen lacerum to form the *nerve of the pterygoid canal*, which passes anteriorly through this canal to the pterygopalatine fossa. The parasympathetic fibers of the greater petrosal nerve synapse in the pterygopalatine ganglion.

The **deep petrosal nerve** is a sympathetic nerve from the *internal carotid plexus*. Its postsynaptic fibers are from nerve cell bodies in the superior cervical sympathetic ganglion. Thus, the fibers do not synapse in the pterygopalatine ganglion but pass directly to join the branches of the maxillary nerve (Fig. 7.64*C*). The postsynaptic parasympathetic and the sympathetic fibers pass to the lacrimal gland, the palatine glands, and the mucosal glands of the nasal cavity and upper pharynx.

Nose

The nose is the part of the respiratory tract superior to the hard palate and contains the peripheral organ of smell. It is divided into right and left cavities by the **nasal septum.** Each nasal cavity is divisible into an *olfactory area* and a *respiratory area. The functions of the nose and nasal cavities are*:

- Olfaction (smelling)
- Respiration (breathing)

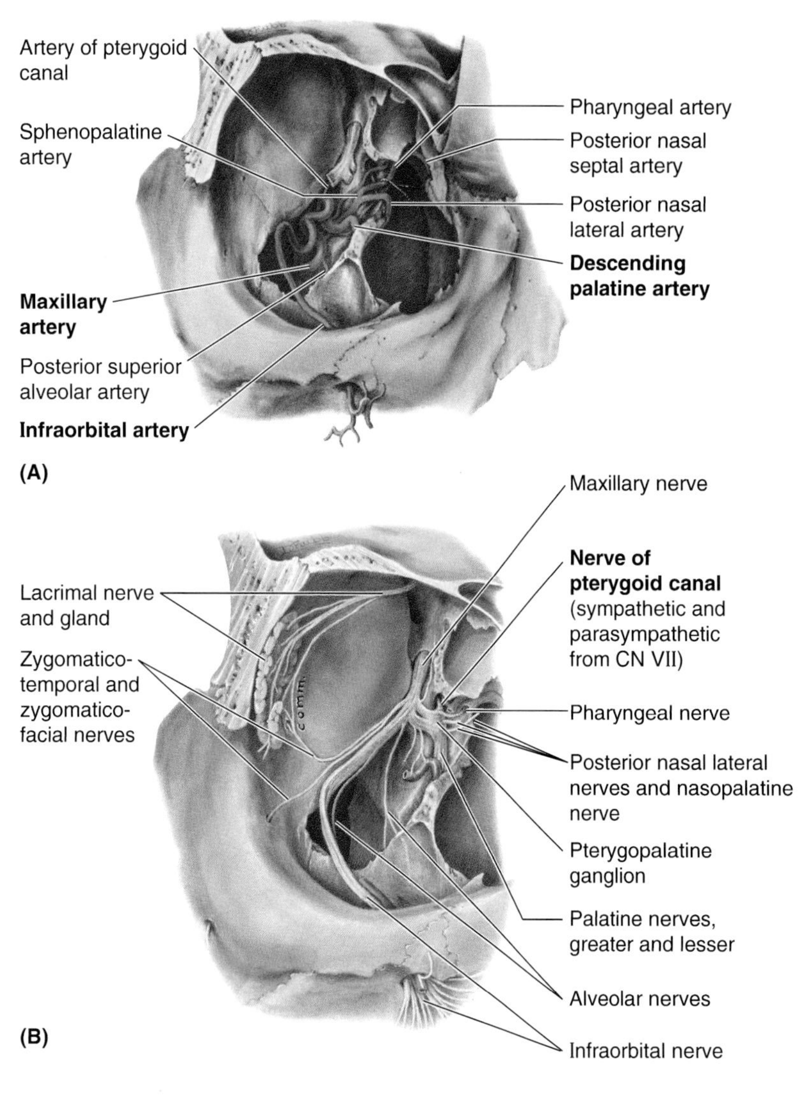

Figure 7.64. Maxillary artery, maxillary nerve (CN V_2), and facial nerve (CN VII). A. Maxillary artery, third (pterygopalatine) part. The course of the maxillary artery is divided into three parts by the lateral pterygoid muscle (see Fig. 7.42). The branches of the third part arise just before and within the pterygopalatine fossa. **B.** Maxillary nerve. The lateral and medial branches are separated by the maxillary sinus. **C.** Diagram of the source of autonomic nerve fibers distributed by the maxillary nerve. The greater petrosal nerve (carrying presynaptic parasympathetic fibers from the facial nerve) merges with the deep petrosal nerve (carrying postsynaptic sympathetic fibers conveyed from the superior cervical ganglion via the internal carotid plexus) to form the nerve of the pterygoid canal in the area of the foramen lacerum. The latter nerve thus brings both sympathetic and parasympathetic fibers to the pterygopalatine ganglion, where only the parasympathetic fibers synapse. The postsynaptic fibers of both divisions of the autonomic nervous system enter all the branches of the maxillary nerve that pass from the ganglion to the palatine, nasal, and uppermost pharyngeal mucosa. Some parasympathetic fibers pass through the pterygopalatine nerves and travel "retrogradely" through the maxillary nerve to reach the zygomatic nerve (CN V_2) for distribution to the lacrimal gland.

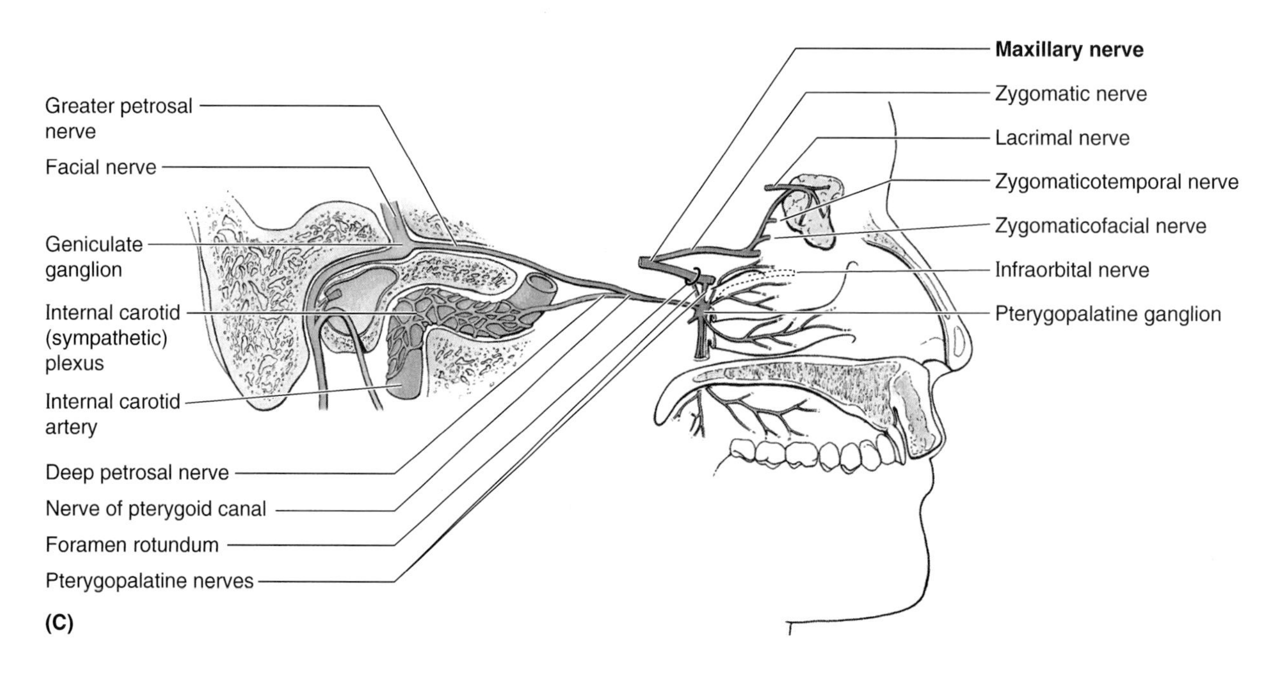

- Filtration of dust
- Humidification of inspired air
- Reception of secretions from the paranasal sinuses and nasolacrimal ducts.

External Nose

The external nose projects from the face; its skeleton is mainly cartilaginous (Fig. 7.65). Noses vary considerably in size and shape, mainly because of differences in the nasal cartilages. The dorsum of the nose extends from its superior angle or **root** to the **apex of the nose.** The inferior surface of the nose is pierced by two piriform (L. pear-shaped) openings, the **nares** (nostrils, anterior nasal apertures), which are bound laterally by the **alae of the nose** and are separated from each other by the fleshy part of skin overlying the **nasal septum** (Figs. 7.65 and 7.66). The superior bony part of the nose, including its root, is covered by thin skin. The skin over the cartilaginous part of the nose is covered with thicker skin that contains many sebaceous glands. The skin extends into the **vestibule of the nose** (Fig. 7.66), where it has a variable number of stiff hairs (*vibrissae*). The junction of the skin and mucous membrane is beyond the hair-bearing area.

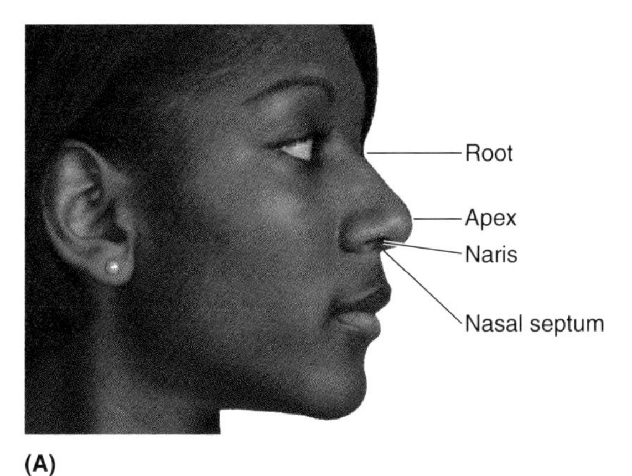

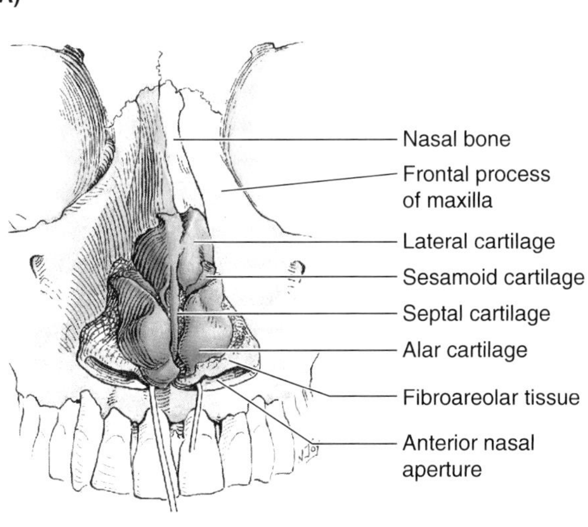

Figure 7.65. External nose. A. Surface anatomy. Observe the apex (free tip), ala (wing), naris (nostril), and nasal septum. The nose is attached to the forehead by its root (bridge). The rounded border between the apex and the bridge is the dorsum of the nose. **B.** Cartilages of the nose. The cartilages are being retracted inferiorly to expose the sesamoid cartilages. The lateral nasal cartilages are fixed by suture to the nasal bones and are continuous with the septal cartilage.

Skeleton of the External Nose

The supporting skeleton of the nose is composed of bone and hyaline cartilage. *The bony part of the nose* (Figs. 7.65*B* and 7.66) *consists of the*:

- Nasal bones
- Frontal processes of maxillae
- Nasal part of the frontal bone and its nasal spine.

The nasal septum has a hard bony part and a soft mobile part. The *cartilaginous part of the nose* (Fig. 7.66) consists of five main cartilages: two lateral cartilages, two alar cartilages, and a septal cartilage. The U-shaped alar nasal cartilages are free and movable. They dilate or constrict the nares when the muscles acting on the nose contract.

Nasal Septum

The partly bony and cartilaginous nasal septum divides the chamber of the nose into two nasal cavities. *The main components of the nasal septum* (Figs. 7.65 and 7.66) *are the*:

- Perpendicular plate of ethmoid
- Vomer
- Septal cartilage.

The thin **perpendicular plate of the ethmoid**, forming the superior part of the nasal septum, descends from the cribriform plate and is continued superior to this plate as the crista galli. The **vomer**, a thin flat bone, forms the posteroinferior part of the nasal septum, with some contribution from the nasal crests of the maxillary and palatine bones. The **septal cartilage** has a tongue-and-groove articulation with the edges of the bony septum.

Nasal Fractures

Fractures of the nasal bones are common facial fractures in sports and automobile accidents because of the prominence of the nose. With a fracture, deformity of the nose is usually present, particularly when a lateral force is applied by someone's elbow, for example. *Epistaxis* (nosebleed) usually occurs. In severe fractures, disruption of the bones and cartilages results in displacement of the nose. When the injury results from a direct blow (e.g., from a hockey stick), the cribriform plate of the ethmoid bone may also fracture. Severe fractures of the nose are potentially dangerous because the cranial meninges may be torn and bacteria in the nasal mucosa (mucous membrane) can enter the cranial cavity, producing *meningitis.* ▶

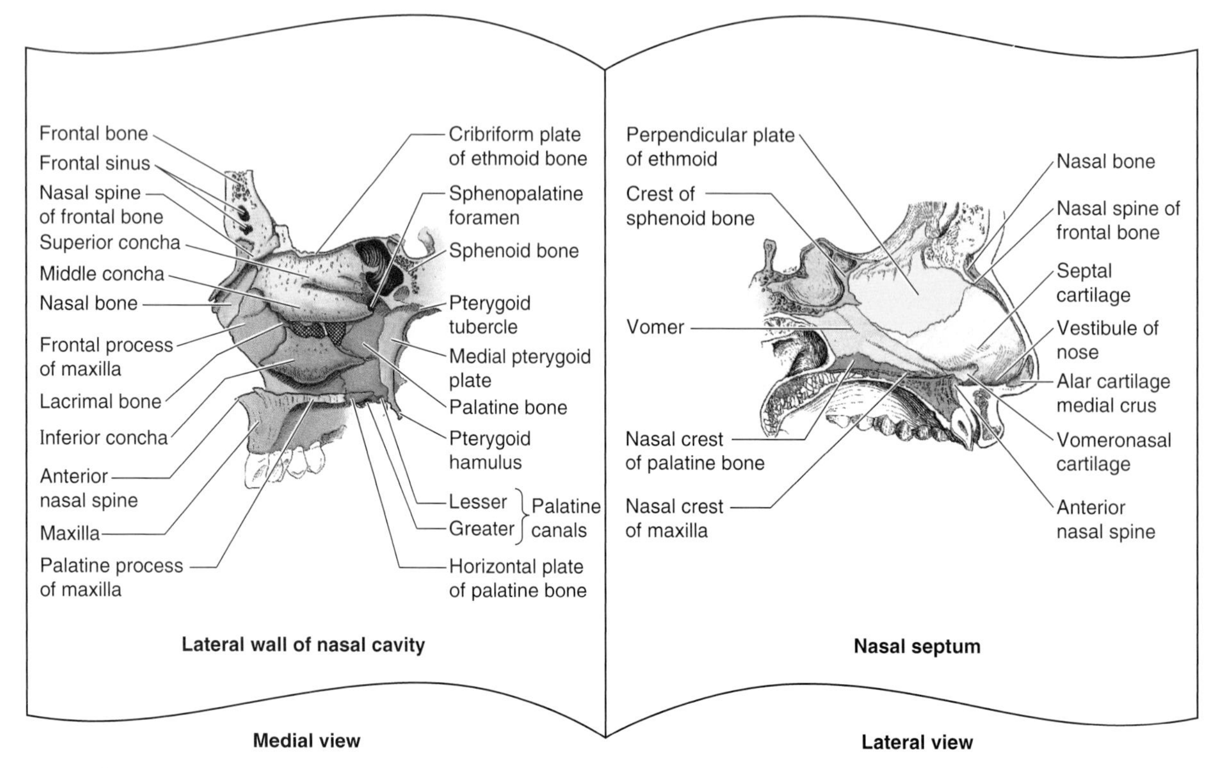

Figure 7.66. Lateral and medial (septal) walls of the right side of the nasal cavity. The walls are separated and demonstrated like adjacent pages of a book. **Left.** Medial view of the right lateral wall of the nasal cavity. **Right.** Right lateral view of the nasal septum. Observe that the nasal septum has a hard (bony) part located deeply (posteriorly) where it is protected, and a soft or mobile part located superficially (anteriorly) mostly in the more vulnerable external nose.

Deviation of the Nasal Septum

The nasal septum may be displaced or deviate from the median plane as the result of a birth injury. More often, however, the deviation results during adolescence and adulthood from postnatal trauma (e.g., during a fist fight). Sometimes the deviation is so severe that the nasal septum is in contact with the lateral wall of the nasal cavity. If this obstructs breathing, surgical repair of the septum may be necessary.

CSF Rhinorrhea

Although nasal discharges are common with upper respiratory tract infections, a clear nasal discharge after a head injury may be CSF. CSF rhinorrhea results from fracture of the cribriform plate, tearing of the cranial meninges, and leakage of CSF from the nose. The CSF—in the subarachnoid space—is in close proximity to the external environment at this site. The fluid drains from the nose within 48 hours of the injury. Persistent rhinorrhea increases the risk of *meningitis* (Rowland, 1995). ○

Nasal Cavities

The nasal cavities, entered anteriorly through the nares, open posteriorly into the nasopharynx through the **choanae** (p. 843). Mucosa lines the nasal cavities except for the **vestibule**, which is lined with skin; vibrissae grow from this skin (Fig. 7.67*A*). The *nasal mucosa* is firmly bound to the periosteum and perichondrium of the supporting bones and cartilages of the nose.

Figure 7.67. Lateral wall of the nasal cavity. Medial views of the right half of the head. **A.** Observe the inferior and middle conchae, curving medially and inferiorly from the lateral wall, dividing it into three nearly equal parts and covering the inferior and middle meatus, respectively. Observe also the superior concha, small and anterior to the sphenoidal sinus; the middle concha, with an angled inferior border, ending inferior to the sphenoidal sinus; the inferior concha, with a slightly curved inferior border, ending inferior to the middle concha approximately 1 cm anterior to the orifice of the pharyngotympanic (auditory) tube, that is, approximately the width of the medial pterygoid plate. **B.** Dissection of the lateral wall of the nasal cavity. Parts of the superior, middle, and inferior conchae are cut away. Observe the sphenoidal sinus in the body of the sphenoid bone; its orifice, superior to the middle of its anterior wall, opens into the sphenoethmoidal recess. The orifices of posterior ethmoidal cells open into the superior meatus; those of middle ethmoidal cells open into the middle meatus.

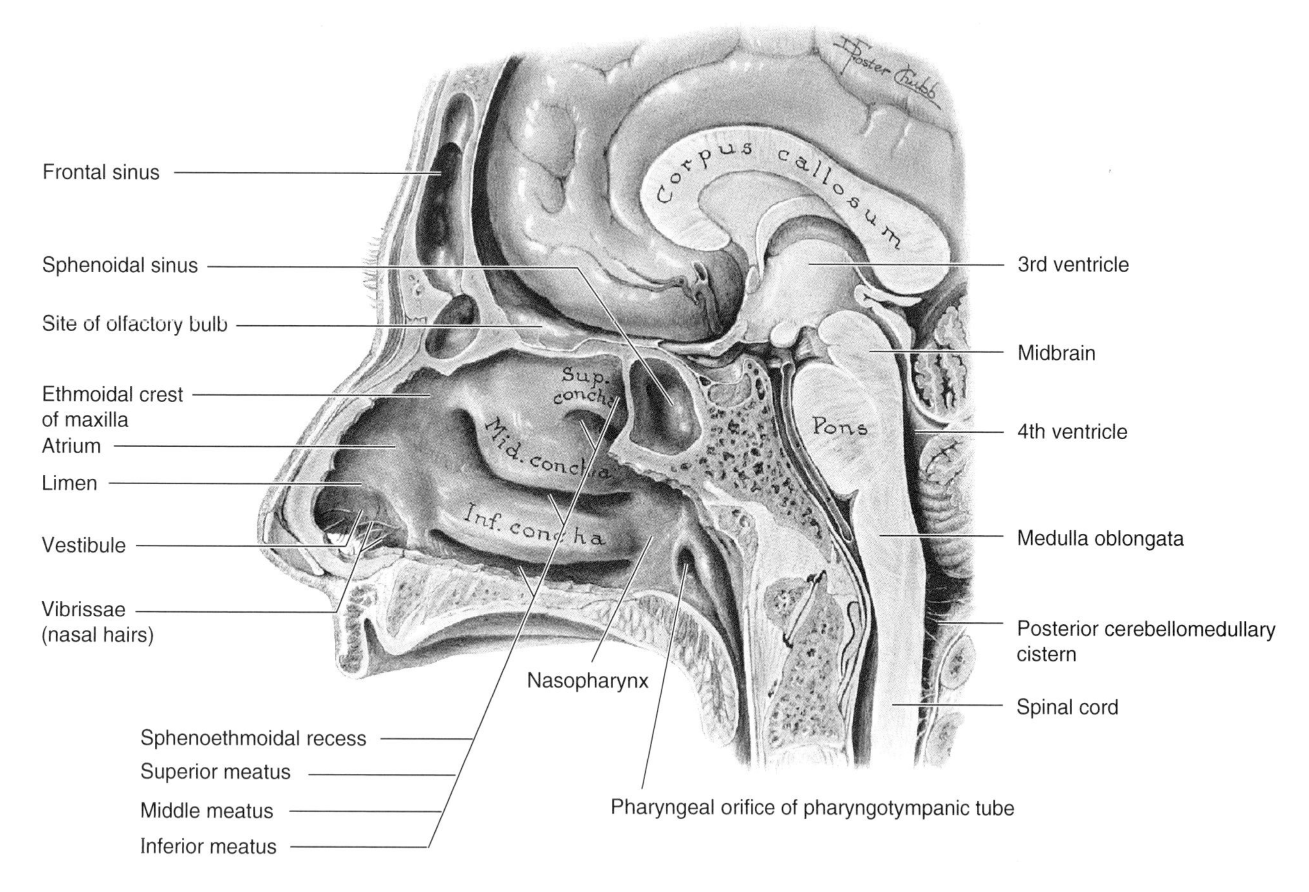

(A) Medial view

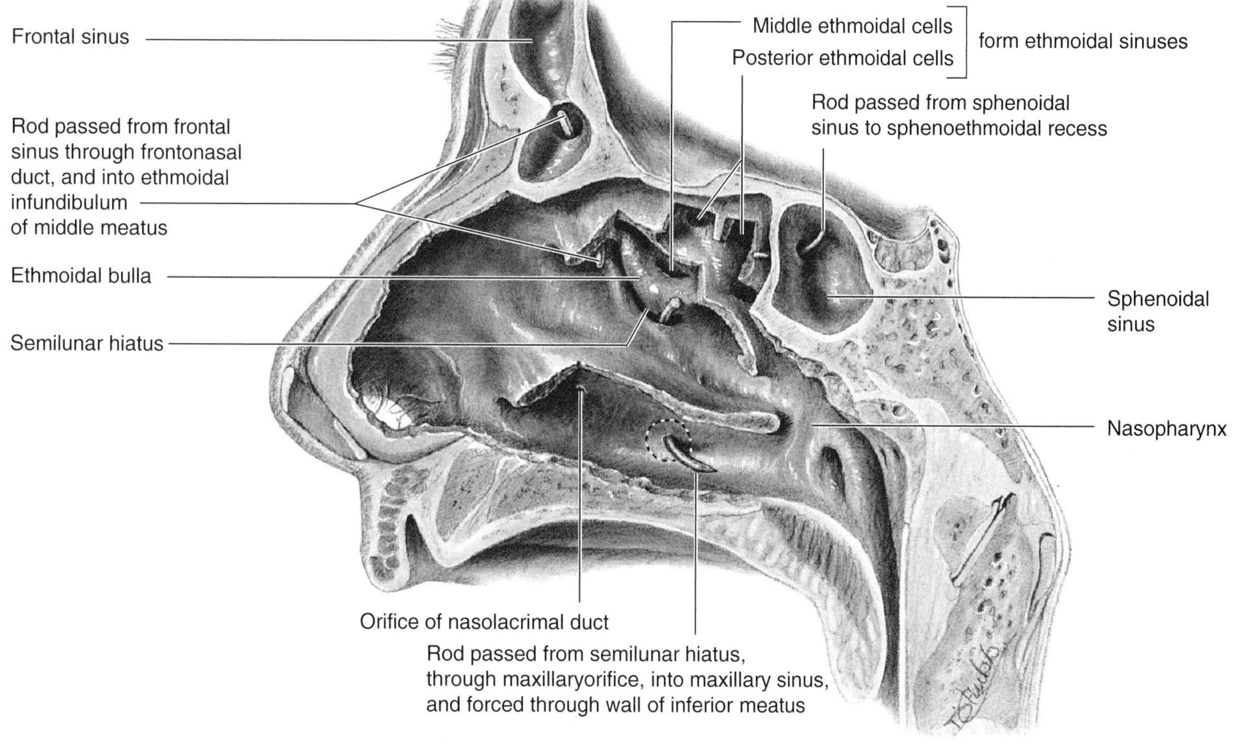

(B) Medial view

The mucosa is continuous with the lining of all the chambers with which the nasal cavities communicate: the nasopharynx posteriorly, the paranasal sinuses (frontal, ethmoidal, sphenoidal, and maxillary) superiorly and laterally, and the lacrimal sac and conjunctiva superiorly.

The inferior two-thirds of the nasal mucosa is the *respiratory area* and the superior one-third is the *olfactory area.* Air passing over the respiratory area is warmed and moistened before it passes through the rest of the upper respiratory tract to the lungs. The *olfactory area* contains the peripheral organ of smell; sniffing draws air to the area. The central processes of *olfactory cells* in the olfactory epithelium unite to form nerve bundles that pass through the cribriform plate and enter the *olfactory bulb* of the brain (Fig. 7.70).

Boundaries of the Nasal Cavity

- The *roof of the nasal cavity* is curved and narrow, except at its posterior end; the roof is divided into three parts (frontonasal, ethmoidal, and sphenoidal) that are named from the bones that form them (Fig. 7.66).
- The *floor of the nasal cavity*, wider than the roof, is formed by the palatine process of the maxilla and the horizontal plate of the palatine bone.
- The *medial wall of the nasal cavity* is formed by the nasal septum.
- The *lateral wall of the nasal cavity* is irregular because of three scroll-shaped elevations—the *nasal conchae*—that project inferiorly like scrolls (Figs. 7.66 and 7.67).
- The conchae curve inferomedially, each forming a roof for a groove, or *meatus*—a passage in the nasal cavity.

The **nasal conchae** (superior, middle, and inferior) divide the nasal cavity into four passages:

- Sphenoethmoidal recess
- Superior meatus
- Middle meatus
- Inferior meatus.

The inferior concha is the longest and broadest and is covered by mucous membrane that contains large vascular spaces that can enlarge to control the caliber of the nasal cavity. When infected, the mucosa may swell rapidly, "blocking the nose."

The **sphenoethmoidal recess,** lying superoposterior to the superior concha, receives the opening of the sphenoidal sinus. The **superior meatus** (Fig. 7.67*A*) is a narrow passage between the superior and middle nasal conchae into which the posterior ethmoidal sinuses open by one or more orifices. The long **middle meatus** is wider than the superior one. The anterosuperior part of this passage leads into a funnel-shaped opening, the **ethmoidal infundibulum**, through which it communicates with the frontal sinus (Fig. 7.68). The passage that leads inferiorly from each frontal sinus to the infundibulum is the *frontonasal duct*. The **semilunar hiatus** (L. hiatus semilunaris) is a semicircular groove into which the frontal sinus opens (Fig. 7.67*B*). The **ethmoidal bulla** (L. bubble)—a rounded elevation located superior to the hiatus—is visible when the middle concha is removed. The bulla is formed by middle ethmoidal cells that form the *ethmoidal sinuses*. The maxillary sinus also opens into the posterior end of the semilunar hiatus.

The **inferior meatus** is a horizontal passage inferolateral to the inferior nasal concha. The *nasolacrimal duct* draining tears from the lacrimal sac opens into the anterior part of this meatus.

Vasculature and Innervation of the Nasal Cavity

The **arterial supply** of the medial and lateral walls of the nasal cavity (Fig. 7.69) is from branches of the:

- Sphenopalatine artery
- Anterior and posterior ethmoidal arteries
- Greater palatine artery
- Superior labial artery and lateral nasal branches of the facial artery.

On the anterior part of the nasal septum is an area rich in capillaries (Kiesselbach's area) where all five arteries supplying the septum anastomose. Thus, this area is often where profuse bleeding from the nose occurs.

A *rich plexus of veins* deep to the nasal mucosa drains into the sphenopalatine, facial, and ophthalmic veins. This venous plexus is an important part of the body's thermoregulatory system, exchanging heat and warming air before it enters the lungs.

The **nerve supply** of the posteroinferior half to two-thirds of the nasal mucosa is chiefly from the **maxillary nerve** by way of the **nasopalatine nerve** to the nasal septum, and posterior lateral nasal branches of the **greater palatine nerve** to the lateral wall (Fig. 7.70). Its anterosuperior part is supplied by the **anterior** and **posterior ethmoidal nerves**, branches of the nasociliary nerve from CN V_1.

The **olfactory nerves**, concerned with smell only, arise from cells in the *olfactory epithelium* in the superior part of the lateral and septal walls of the nasal cavity. The central processes of these cells (forming the olfactory nerve) pass through the *cribriform plate* and end in the *olfactory bulb*, a forebrain structure.

Rhinitis

The nasal mucosa becomes swollen and inflamed (*rhinitis*) during upper respiratory infections and allergic reactions (e.g., hayfever). Swelling of the mucosa occurs readily because of its vascularity. *Infections of the nasal cavities may spread to the*:

- Anterior cranial fossa through the cribriform plate
- Nasopharynx and retropharyngeal soft tissues ▶

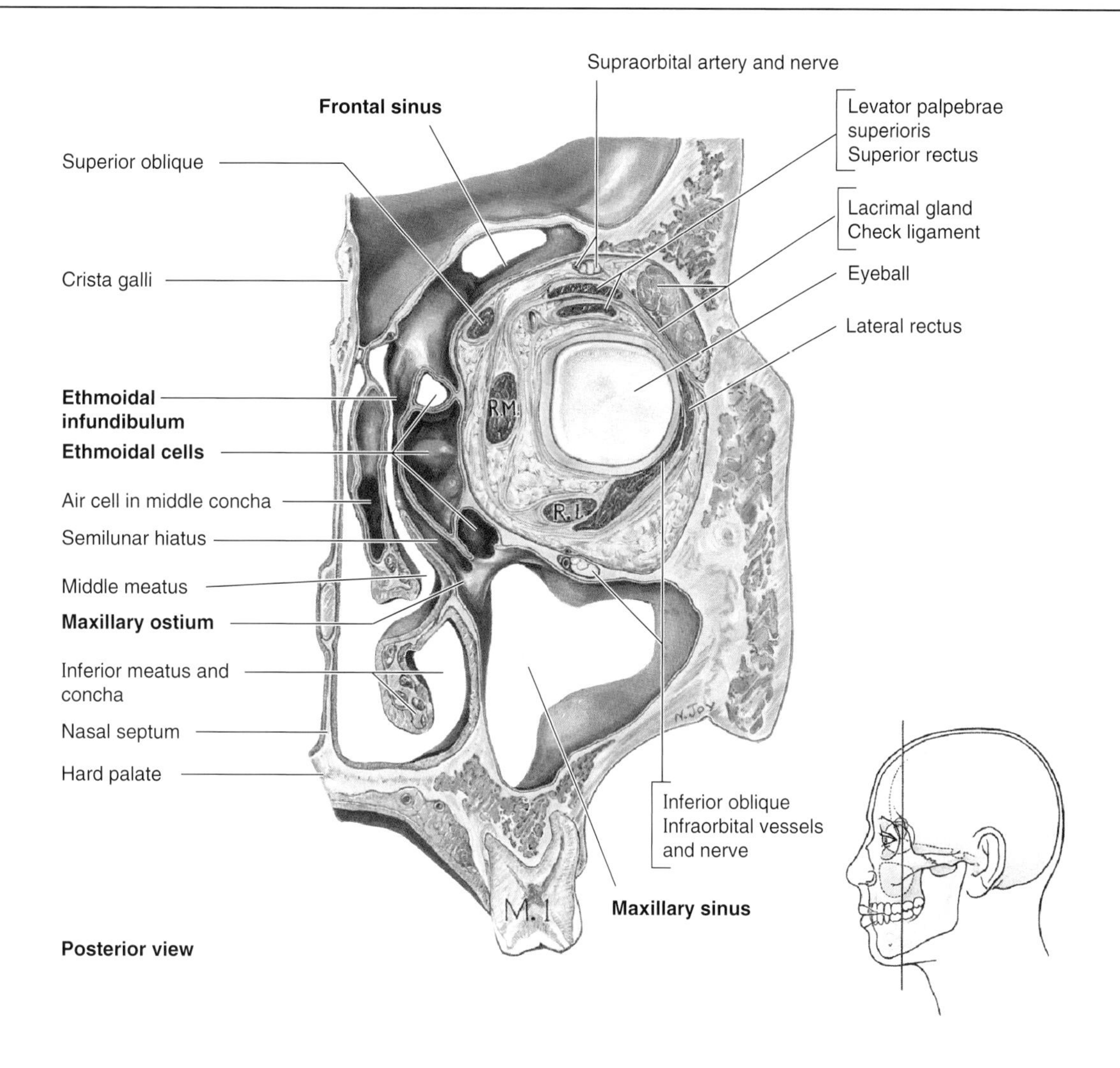

Figure 7.68. Coronal section of the right half of the head. Posterior view. Observe the relationship of the orbit, nasal cavity, and paranasal sinuses. Observe also the orbital contents, including the four recti and the fascia uniting them forming a circle (a cone when viewed in three dimensions) around the posterior part (fundus) of the eyeball. *R.M.*, rectus medius or medial rectus; *R.I.*, rectus inferior or inferior rectus; *M.I.*, first molar. The orientation drawing illustrates the plane of the section.

- Middle ear through the pharyngotympanic tube
- Paranasal sinuses
- Lacrimal apparatus and conjunctiva.

Epistaxis

Epistaxis is relatively common because of the rich blood supply to the nasal mucosa. In most cases the cause is trauma and the bleeding is located in the anterior third of the nose (Kiesselbach's area). Epistaxis is also associated with infections and hypertension. Spurting of blood from the nose results from rupture of arteries. Mild epistaxis often results from *nose picking*, which tears veins in the vestibule of the nose. ⭘

Paranasal Sinuses

The paranasal sinuses are air-filled extensions of the respiratory part of the nasal cavity into the following cranial bones: frontal, ethmoid, sphenoid, and maxilla. They are named according to the bones in which they are located.

Frontal Sinuses

The frontal sinuses (Figs. 7.67, 7,68, and 7.71*A*) are between the outer and inner tables of the frontal bone, posterior to the superciliary arches and the root of the nose. Frontal sinuses are usually detectable in children by 7 years of age. Each sinus drains through the *frontonasal duct* into the *infundibulum*, which opens into the *semilunar hiatus* of the middle meatus. The frontal sinuses are innervated by branches of the *supraorbital nerves* (CN V_1).

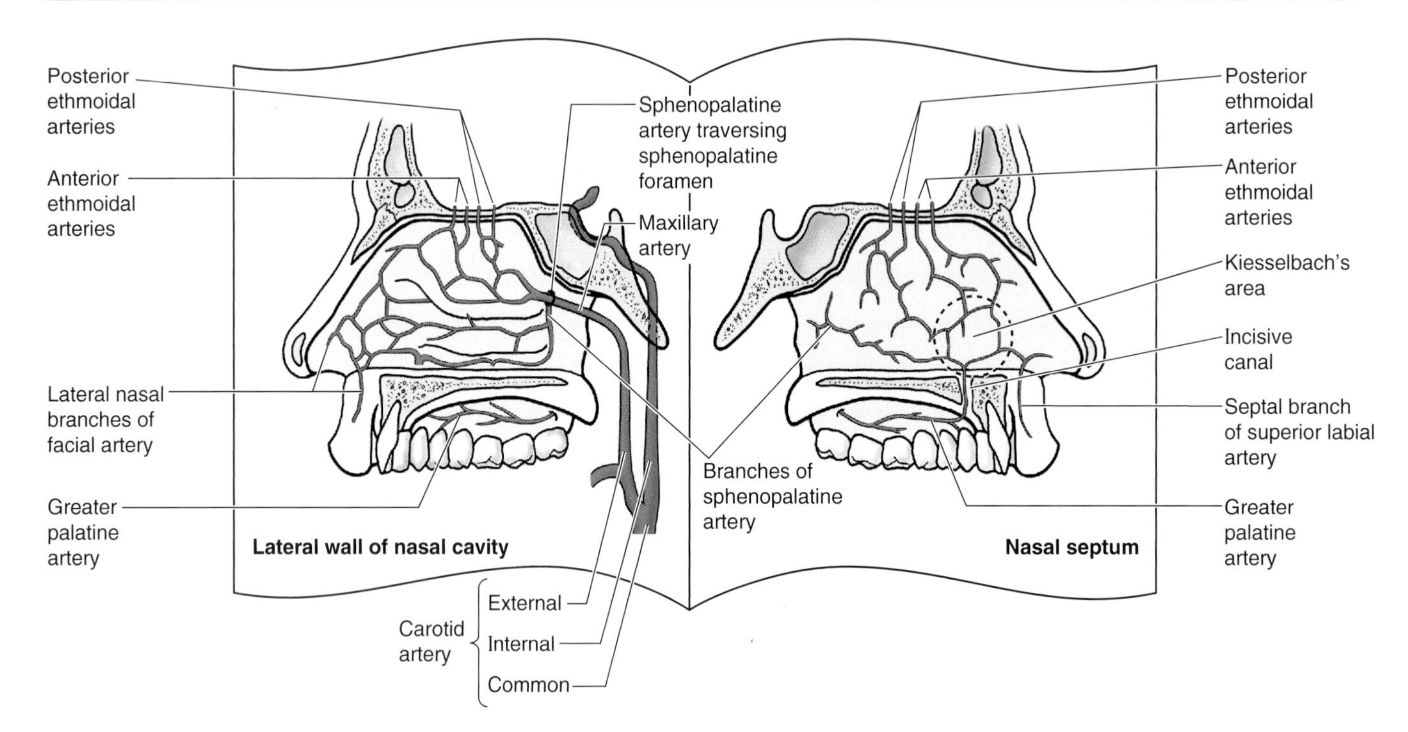

Figure 7.69. Arterial supply of the nasal cavity. "Opened book" view of the lateral and medial (septal) walls of the right side of the nasal cavity. The left "page" shows the lateral wall of the nasal cavity. The sphenopalatine artery (a branch of the maxillary) and the anterior ethmoidal artery (a branch of the ophthalmic) are the most important arteries to the nasal cavity. The right "page" shows the nasal septum. An anastomosis of four to five named arteries supplying the septum occurs in the anteroinferior portion of the nasal septum (Kiesselbach's area), an area commonly involved in chronic epistaxis (nosebleeds).

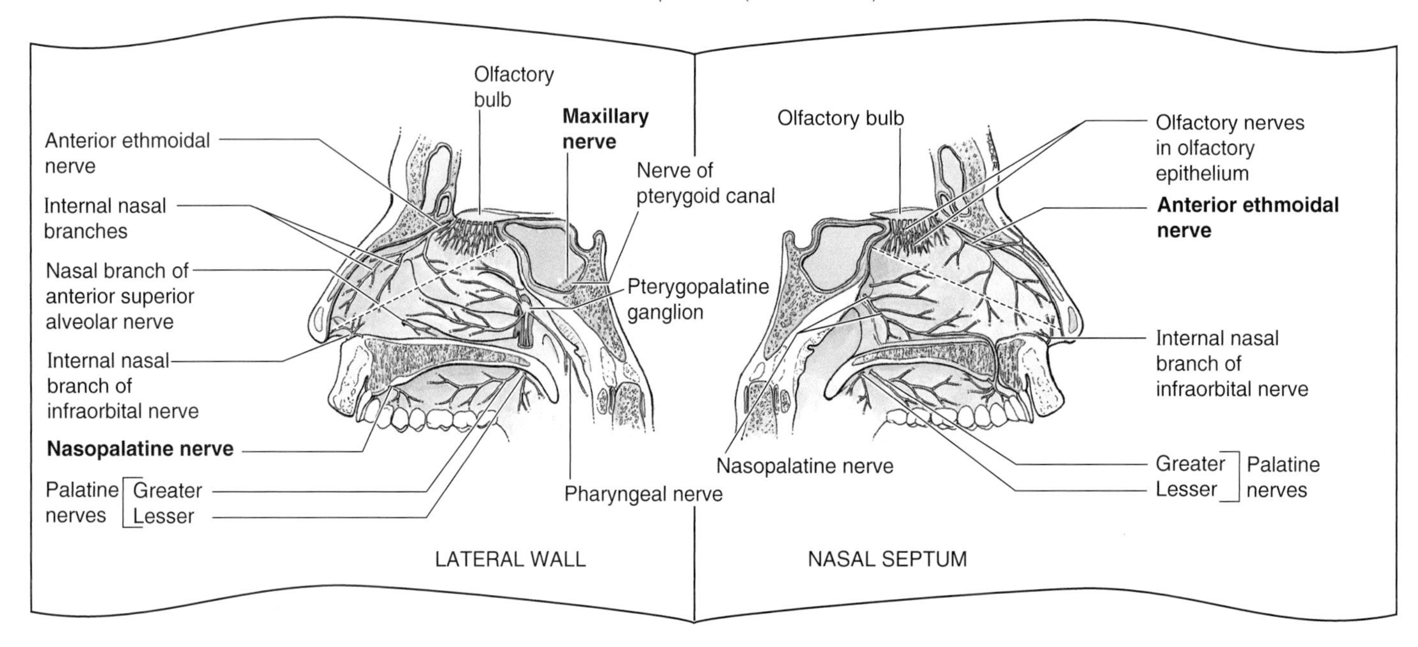

Figure 7.70. Innervation of the nasal cavity. "Opened book" view of the lateral and medial (septal) walls of the right side of the nasal cavity. A line extrapolated approximately from the sphenoethmoidal recess to the tip of the nose demarcates the territories of the ophthalmic (CN V_1) and maxillary (CN V_2) nerves for supplying general sensation to both the lateral wall and the nasal septum. The olfactory nerve (CN I) is distributed to the olfactory mucosa superior to the level of the superior concha on both the lateral wall and the nasal septum. The pterygopalatine ganglion sends the nasopalatine nerve through the sphenopalatine foramen, the greater and lesser palatine nerves through canals of the same name, and the pharyngeal nerve through the pharyngeal canal.

Sinusitis

Because the paranasal sinuses are continuous with the nasal cavities through apertures that open into them, infection may spread from the nasal cavities, producing inflammation and swelling of the mucosa of the sinuses (*sinusitis*) and local pain. Sometimes several sinuses are inflamed (*pansinusitis*), and the swelling of the mucosa may block one or more openings of the sinuses into the nasal cavities.

Variation of the Frontal Sinuses

The right and left frontal sinuses are rarely of equal size, and the septum between the right and left sinuses usually is not situated entirely in the median plane. The frontal sinuses vary in size from approximately 5 mm to large spaces extending laterally into the greater wings of the sphenoid bone. Often a frontal sinus has two parts: a vertical part in the squamous part of the frontal bone and a horizontal part in the orbital part of the frontal bone. One or both parts may be large or small. When the supraorbital part is large, its roof forms the floor of the anterior cranial fossa and its floor forms the roof of the orbit. The frontal sinuses may be multiple on each side, and each of them may have a separate frontonasal duct. ⊙

Ethmoidal Sinuses

The ethmoidal sinuses comprise several cavities—**ethmoidal cells**—that are located in the lateral mass of the ethmoid bone between the nasal cavity and orbit (Figs. 7.68 and 7.71). The ethmoidal sinuses usually are not visible in plain radiographs before 2 years of age but are recognizable in computerized tomography (CT) scans. The *anterior ethmoidal cells* drain directly or indirectly into the middle meatus through the infundibulum. The *middle ethmoidal cells* open directly into the middle meatus and are sometimes called "bullar cells" because they form the *ethmoidal bulla*, a swelling on the superior border of the semilunar hiatus (Fig. 7.67*B*). The *posterior ethmoidal cells* open directly into the superior meatus. The ethmoidal sinuses are supplied by the anterior and posterior ethmoidal branches (Fig. 7.70) of the *nasociliary nerves* (CN V_1).

Infection of the Ethmoidal Cells

If nasal drainage is blocked, infections of the ethmoidal cells may break through the fragile medial wall of the orbit. Severe infections from this source may cause blindness because some posterior ethmoidal cells lie close to the optic canal, which gives passage to the optic nerve and ophthalmic artery. Spread of infection from these cells could also affect the dural nerve sheath of the optic nerve, causing *optic neuritis*. ⊙

Sphenoidal Sinuses

The sphenoidal sinuses (Figs. 7.67 and 7.71*A*), unevenly divided and separated by a bony septum, are in the body of the sphenoid; they may extend into the wings of this bone. Because of these sinuses, the body of the sphenoid is fragile. Only thin plates of bone separate the sinuses from several important structures: the optic nerves and optic chiasm, the pituitary gland, the internal carotid arteries, and the cavernous sinuses. The sphenoidal sinuses derive from a posterior ethmoidal cell that begins to invade the sphenoid bone at approximately 2 years of age. In some people, several posterior ethmoidal cells invade the sphenoid bone, giving rise to multiple sphenoidal sinuses that open separately into the *sphenoethmoidal recess*. The posterior ethmoidal arteries (Fig. 7.69) and posterior ethmoidal nerve supply the sphenoidal sinuses.

Maxillary Sinuses

The maxillary sinuses are the largest of the paranasal sinuses. These large pyramidal cavities occupy the bodies of the maxillae (Figs. 7.68 and 7.71).

- The *apex of the maxillary sinus* extends toward and often into the zygomatic bone
- The *base of the maxillary sinus* forms the inferior part of the lateral wall of the nasal cavity
- The *roof of the maxillary sinus* is formed by the floor of the orbit
- The *floor of the maxillary sinus* is formed by the alveolar part of the maxilla. The roots of the maxillary teeth, particularly the first two molars, often produce conical elevations in the floor of the sinus.

Each maxillary sinus drains by an opening—the **maxillary ostium**— into the middle meatus of the nasal cavity by way of the semilunar hiatus.

The **arterial supply** of the maxillary sinus is mainly from superior alveolar branches of the **maxillary artery**; however, branches of the **greater palatine artery** supply the floor of the sinus (Figs. 7.64*A* and 7.69). **Innervation** of the maxillary sinus is from the anterior, middle, and posterior **superior alveolar nerves**—branches of the maxillary nerve.

Infection of the Maxillary Sinuses

The maxillary sinuses are the most commonly infected, probably because their ostia are located high on their superomedial walls, a poor location for natural drainage ▶

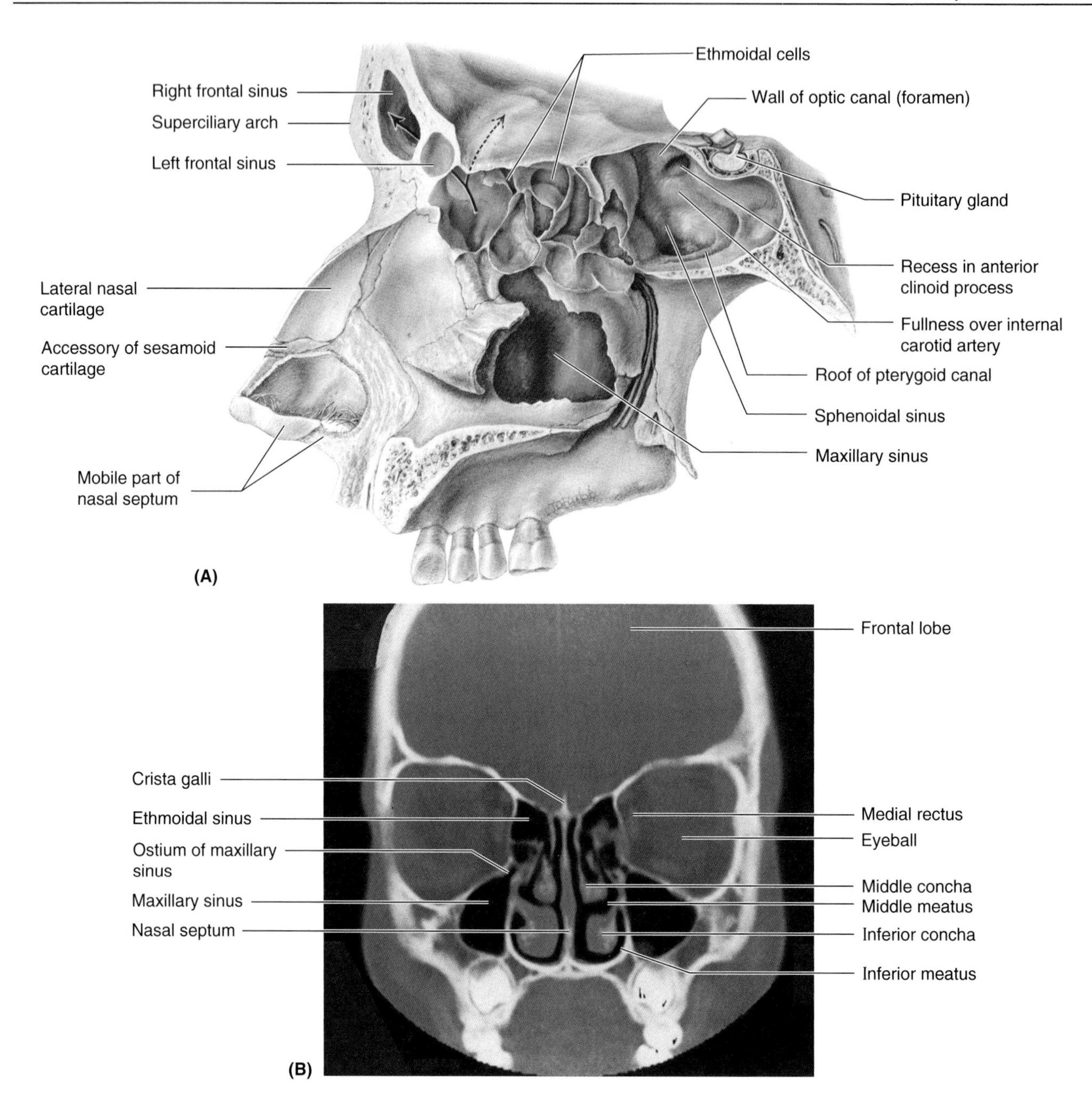

Figure 7.71. Paranasal sinuses. A. Sagittal section of the nasal cavity and palate, passing approximately a finger's breadth to the right of the midline (through the greater palatine foramen). The section also passes through the ethmoidal, sphenoidal, and frontal sinuses. The medial wall of the maxillary sinus has been chipped open. Collectively, the ethmoidal cells form the ethmoidal sinus. An anterior ethmoidal cell (*pink*) is invading the diploë of the frontal bone to become a frontal sinus. An offshoot (*broken arrow*) invades the orbital plate of the frontal bone. The sinuses continue to invade the surrounding bone, and marked extensions are common in the skulls of older individuals. The sphenoidal sinus in this specimen is extensive, extending (*a*) posteriorly, inferior to the pituitary gland, to the clivus; (*b*) laterally, inferior to the optic nerve (CN II), into the anterior clinoid process; and (*c*) inferior to the pterygoid process but leaving the pterygoid canal and rising as a ridge on the floor of the sinus. The maxillary sinus is pyramidal. **B.** Coronal CT scan. (Courtesy of Dr. D. Armstrong, Associate Professor of Medical Imaging, University of Toronto, Toronto, Ontario, Canada).

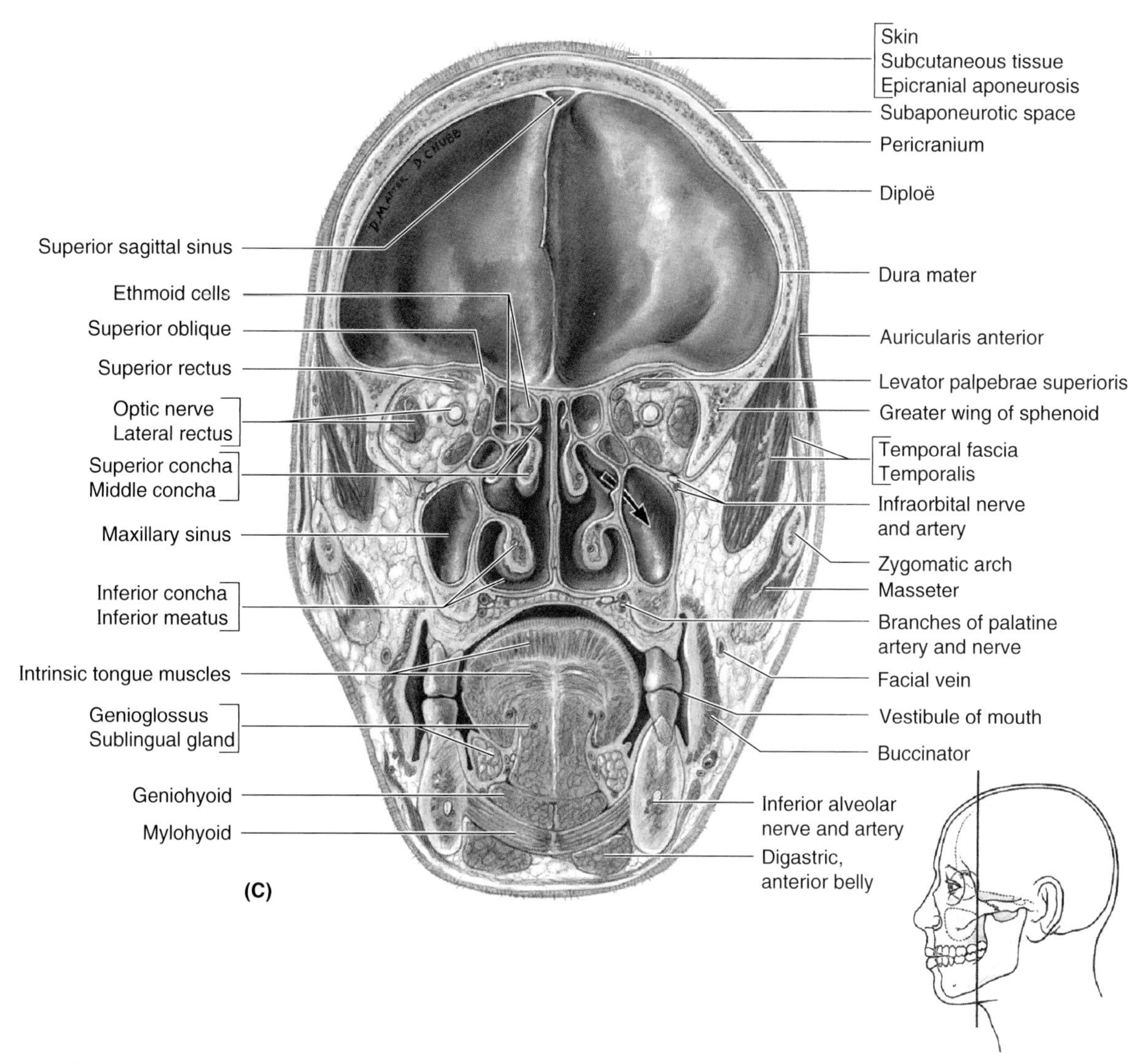

Figure 7.71. *(Continued)* **C.** Coronal section of the head. Observe the central position of the ethmoid bone, whose horizontal component forms the central part of the anterior cranial fossa superiorly and the roof of the nasal cavity inferiorly. The suspended ethmoidal cells give attachment to the superior and middle concha and form part of the medial wall of the orbit; the perpendicular plate of the ethmoid forms part of the nasal septum. Notice that the thin orbital part of the frontal bone forms a roof over the orbit and a floor for the anterior cranial fossa and that the palate forms the floor of the nasal cavity and the roof of the oral cavity. Note that the maxillary sinus forms the inferior part of the lateral wall of the nose; the middle concha shelters the hiatus semilunaris into which the maxillary ostium opens (*arrow*). The orientation drawing shows the plane of the section shown in (**B**) and (**C**).

▶ of the sinus. When the mucous membrane of the sinus is congested, the maxillary ostia are often obstructed. Because of the high location of the ostia, when the head is erect it is impossible for the sinuses to drain until they are full. Because the ostia of the right and left sinuses lie on the medial sides (i.e., are directed toward each other), only the upper ostium drains when lying on one's side. A cold or allergy involving both sinuses can result in nights of rolling from side-to-side in an attempt to keep the sinuses drained. The maxillary sinuses can be cannulated and drained by passing a cannula from the nostril through the maxillary ostium into the sinus.

Relationship of the Teeth to the Maxillary Sinus

The proximity of the maxillary molar teeth to the floor of the maxillary sinus poses potentially serious problems. During removal of a molar tooth, a fracture of a root may occur. If proper retrieval methods are not used, a piece ▶

▶ of the root may be driven superiorly into the maxillary sinus. A communication may be created between the oral cavity and the maxillary sinus as a result, and an infection may occur. Because the superior alveolar nerves—branches of the maxillary nerve—supply both the maxillary teeth and the mucous membrane of the maxillary sinus, inflammation of the mucosa of the sinus is frequently accompanied by a sensation of toothache in the molar teeth, especially when the maxilla is very thin in the floor of this sinus.

Transillumination of the Sinuses

Transillumination of the maxillary sinuses is performed in a darkened room. A bright light source is placed in the patient's mouth on one side of the hard palate. The light passes through the maxillary sinus and appears as a crescent-shaped, dull glow inferior to the orbit. If a sinus contains excess fluid, a mass, or a thickened mucosa, the glow is decreased. The frontal sinuses can also be transilluminated by directing the light superiorly under the medial aspect of the eyebrow, normally producing a glow superior to the orbit. Considerable variability of sinus illumination exists from patient to patient (Swartz, 1994). The ethmoidal and sphenoidal sinuses cannot be examined by transillumination. ○

Ear

The ear, or vestibulocochlear organ, is divided into external, middle, and internal parts (Fig. 7.72) and has two functions: equilibrium and hearing. The external and middle parts are mainly concerned with the transference of sound to the internal (inner) ear, which contains the organ for equilibrium—the condition of being evenly balanced—and hearing. The **tympanic membrane** separates the external ear from the middle ear or tympanic cavity. The *pharyngotympanic tube* joins the middle ear to the nasopharynx.

External Ear

The external ear comprises the *auricle*, which collects sound, and the *external acoustic meatus* (passage or canal), which conducts the sound to the tympanic membrane.

Auricle

Most of the auricle, consisting of several parts (Fig. 7.73), is composed of elastic cartilage covered with skin. The auricle has several depressions; the **concha** is the deepest one. The **lobule** (earlobe)—devoid of cartilage—consists of fibrous tissue, fat, and blood vessels. It is easily pierced for taking small blood samples and inserting earrings.

The **arterial supply** to the auricle is derived mainly from the *posterior auricular* and *superficial temporal arteries* (Fig. 7.74*A*). The **nerves to the skin** of the auricle are the great auricular and auriculotemporal nerves. The **great auricular nerve** supplies the superior surface and the lateral surface inferior to the external acoustic meatus. The **auriculotemporal nerve**, a branch of CN V_3, supplies the skin of the auricle superior to the external acoustic meatus. **Lymphatic drainage** of the lateral surface of the superior half of the auricle is to the **superficial parotid lymph nodes** (Fig. 7.74*B*). Lymph from the cranial (medial) surface of the superior half of the auricle drains to the **mastoid (retroauricular)** and deep cervical lymph nodes. Lymph from the remainder of the auricle, including the lobule, drains into the **superficial cervical lymph nodes**.

External Acoustic Meatus

The external acoustic meatus leads inward through the tympanic part of the temporal bone (Fig. 7.72). The canal extends from the deepest part of the concha to the tympanic membrane (Fig. 7.75), a distance of 2 to 3 cm in adults. The lateral third of this S-shaped canal is cartilaginous and is lined with skin, which is continuous with the skin of the auricle. Its medial two-thirds is bony and is lined with thin skin that is continuous with the external layer of the tympanic membrane. The ceruminous and sebaceous glands in the subcutaneous tissue of the cartilaginous part of the external acoustic meatus produce *cerumen* (earwax).

The **tympanic membrane**—approximately 1 cm in diameter—is a thin, oval semitransparent membrane at the medial end of the external acoustic meatus (Figs. 7.72 and 7.76, *A* and *B*). It forms a partition between the external acoustic meatus and the **tympanic cavity** of the middle ear. The tympanic cavity is an air chamber in the temporal bone containing the **auditory ossicles** (small ear bones)—**malleus**, **incus**, and **stapes**. The tympanic membrane is covered with very thin skin externally and mucous membrane of the middle ear internally. Viewed through an otoscope (p. 966), the tympanic membrane has a concavity toward the external acoustic meatus with a shallow, conelike central depression, the peak of which is the **umbo** (Fig. 7.75*A*). The central axis of the tympanic membrane, passing perpendicularly through the umbo like the handle of an umbrella, runs anteriorly and inferiorly as it runs laterally; thus, the tympanic membrane is oriented like a mini radar or satellite dish positioned to receive signals coming from the ground in front and to the side of the head. When being inspected with an *otoscope*, a bright reflection of the otoscope's illuminator—the **cone of light**—radiates anteroinferiorly from the umbo. Superior to the lateral process of the malleus (L. hammer), the membrane is thin and is called the **flaccid part** (L. pars flaccida); it lacks the radial and circular fibers present in the remainder of the membrane—the **tense part** (L. pars tensa). The flaccid part forms the lateral wall of the superior recess of the tympanic cavity.

The tympanic membrane moves in response to air vibrations that pass to it through the external acoustic meatus. Movements of the membrane are transmitted by the auditory ossicles through the middle ear to the internal ear (Fig. 7.72). The external surface of the tympanic membrane is supplied

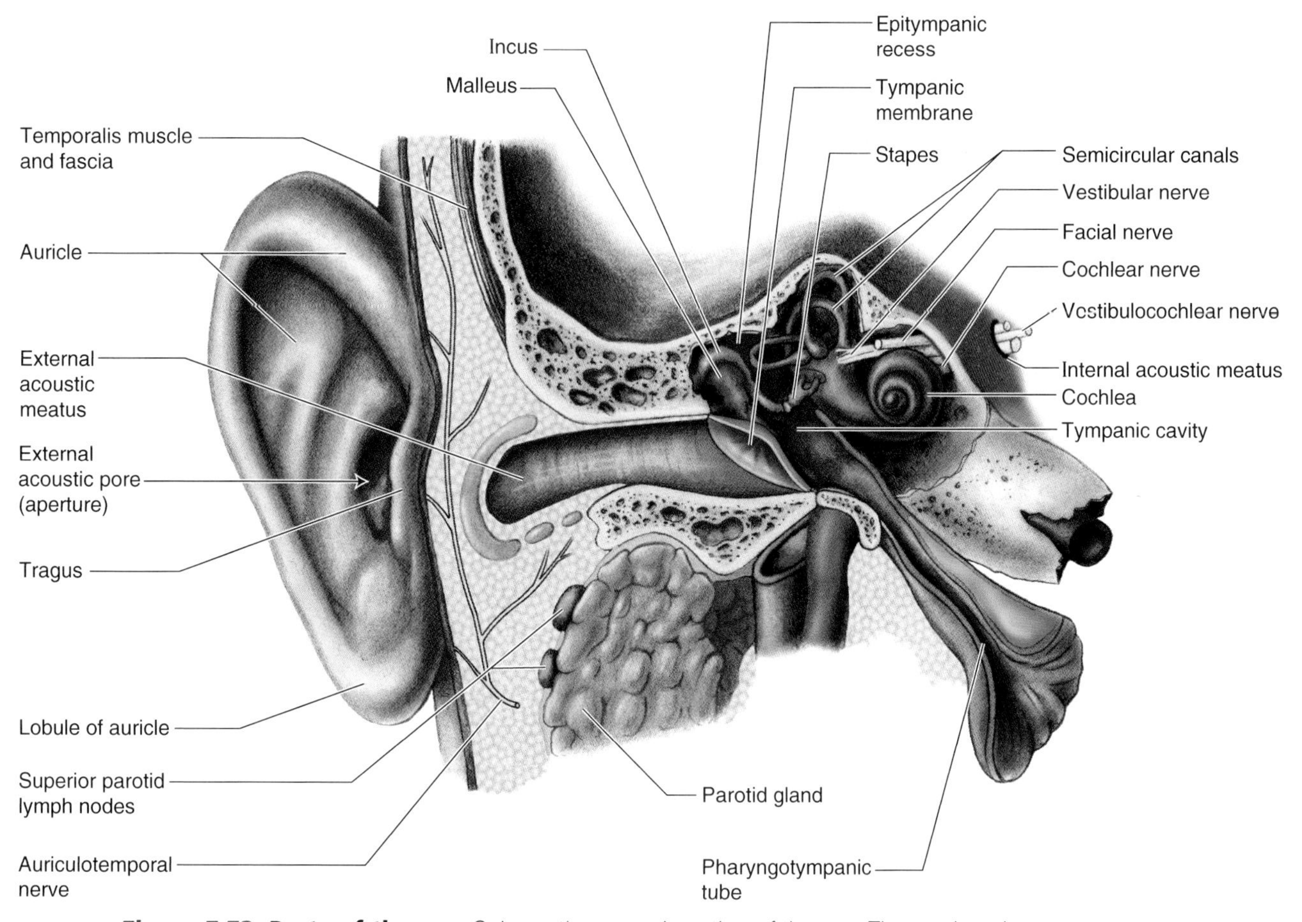

Figure 7.72. Parts of the ear. Schematic coronal section of the ear. The ear has three parts: external, middle, and internal. The external ear consists of the auricle (pinna) and external acoustic meatus (auditory canal). The middle ear, or tympanic cavity, is an air space in which the auditory ossicles are located. The internal (inner) ear contains the membranous labyrinth; its chief divisions are the cochlear labyrinth and the vestibular labyrinth.

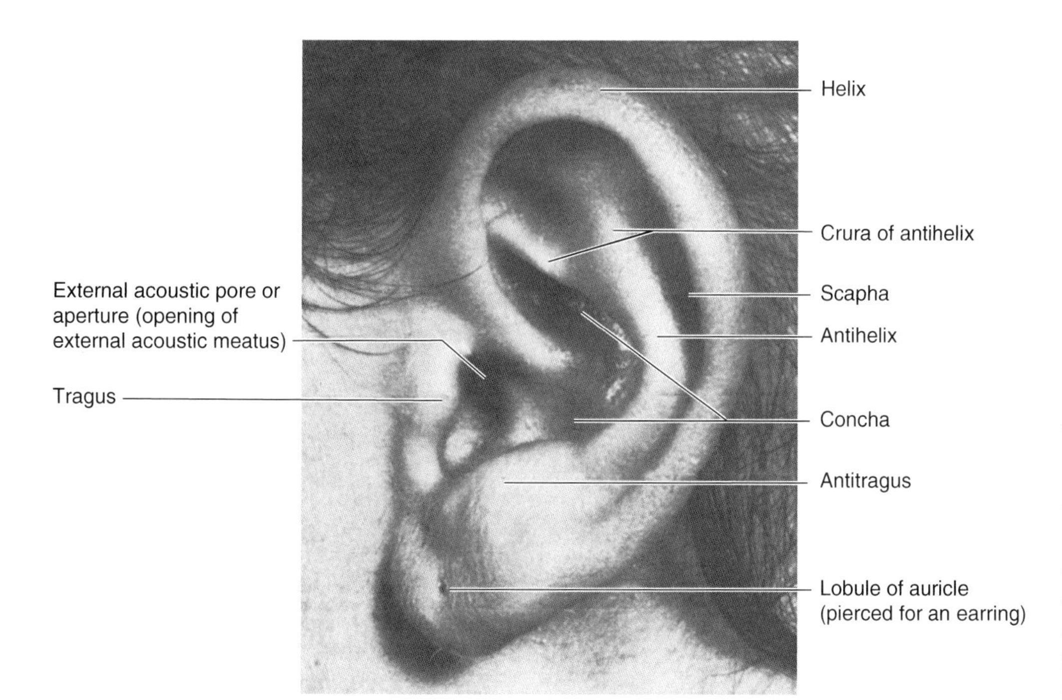

Figure 7.73. External ear of a 12-year-old girl. The names for the parts of the auricle (external ear, pinna) are those commonly used in clinical descriptions. The external acoustic meatus (auditory canal) extends from the concha of the auricle to the tympanic membrane (eardrum) (Fig. 7.72).

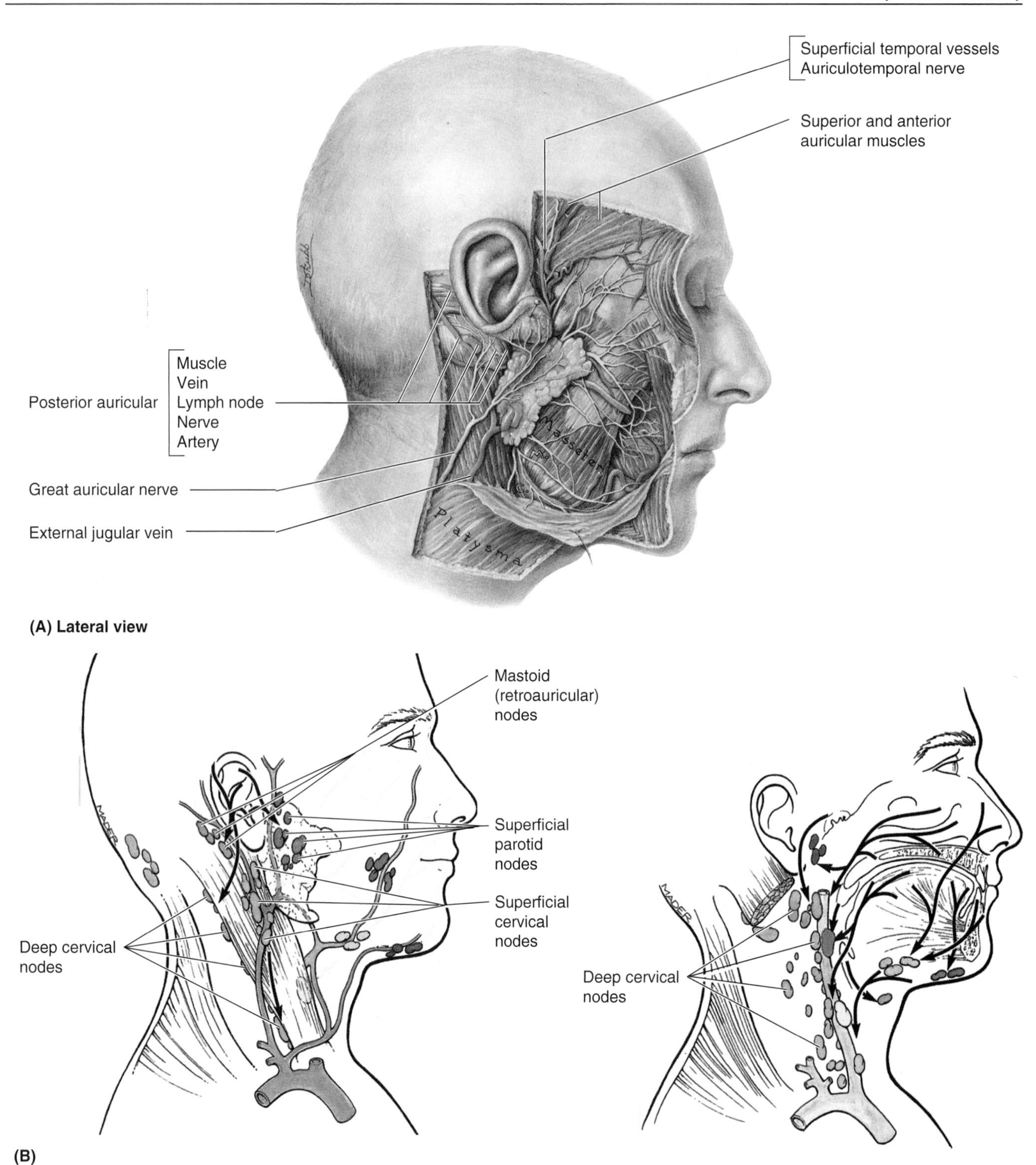

Figure 7.74. Dissection of the face and lymphatic drainage of the head. A. Lateral view. Note the posterior auricular and superficial temporal arteries and veins and the great auricular and auriculotemporal nerves, which provide the circulation and innervation of the external ear. One of the muscles associated with the external ear, the posterior auricular muscle, can also be seen. **B.** Lymphatic drainage. Drainage initially is to the parotid lymph nodes (especially a node directly anterior to the tragus) and the mastoid and superficial cervical nodes, all which drain to the deep cervical nodes.

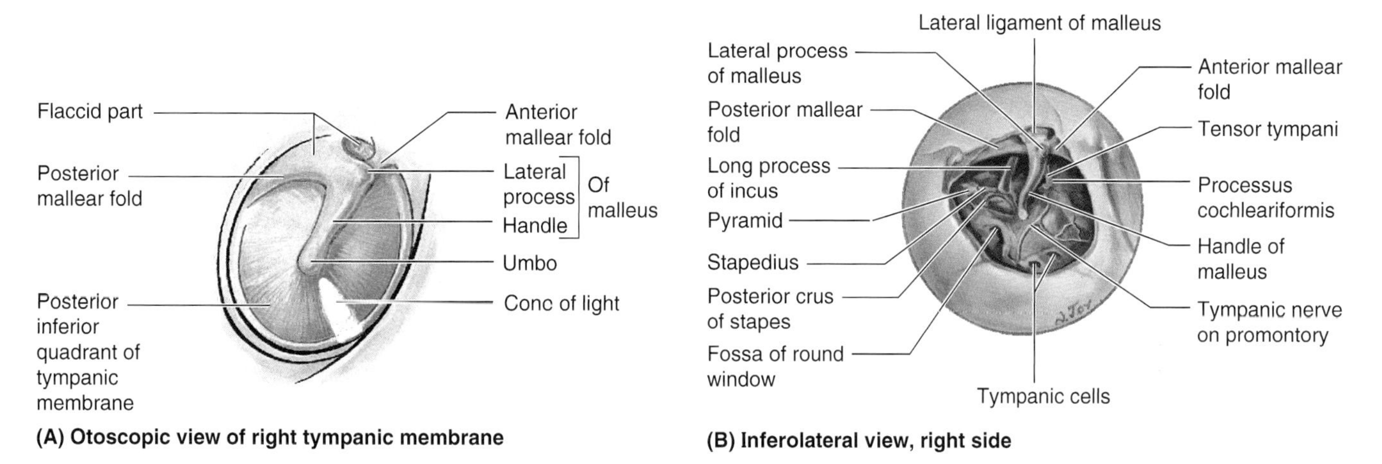

Figure 7.75. Tympanic membrane (eardrum) and tympanic cavity. A. Otoscopic view of the right tympanic membrane. The "cone of light" is a reflection of the light of the otoscope. **B.** The tympanic cavity after removal of the tympanic membrane. Observe the fullness of the promontory with grooves for the tympanic nerve (a branch of CN IX, the glossopharyngeal nerve) and its connections.

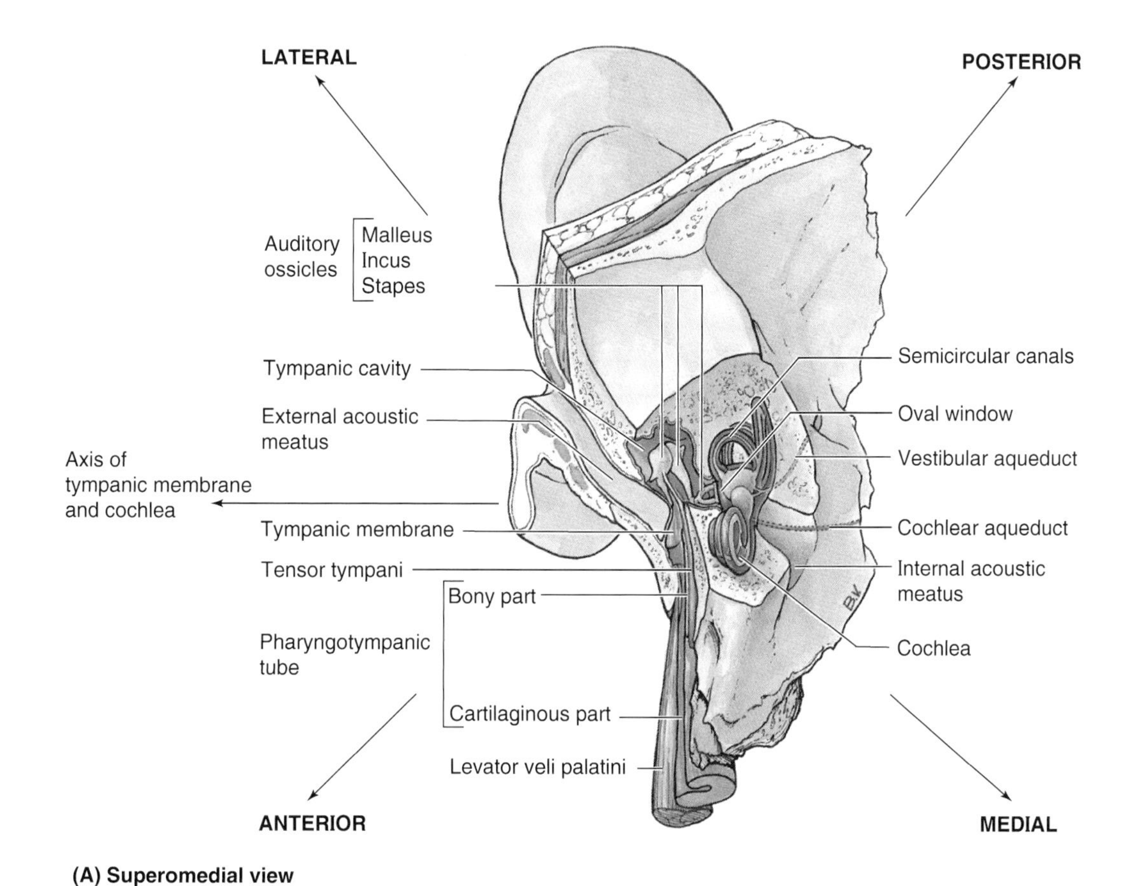

Figure 7.76. General scheme of the ear. A. The ear in situ. Notice the disposition of the various elements of the parts of the ear. The external acoustic meatus (auditory canal) runs lateral to medial; the axis of the tympanic membrane (eardrum) and the axis about which the cochlea (G. snail) winds runs inferiorly and anteriorly as it proceeds laterally. The long axes of the bony and membranous labyrinths and of the pharyngotympanic (auditory) tube and parallel tensor tympani and levator palati muscles lie perpendicular to those of the tympanic membrane and cochlea (i.e., they run inferiorly and anteriorly as they proceed medially).

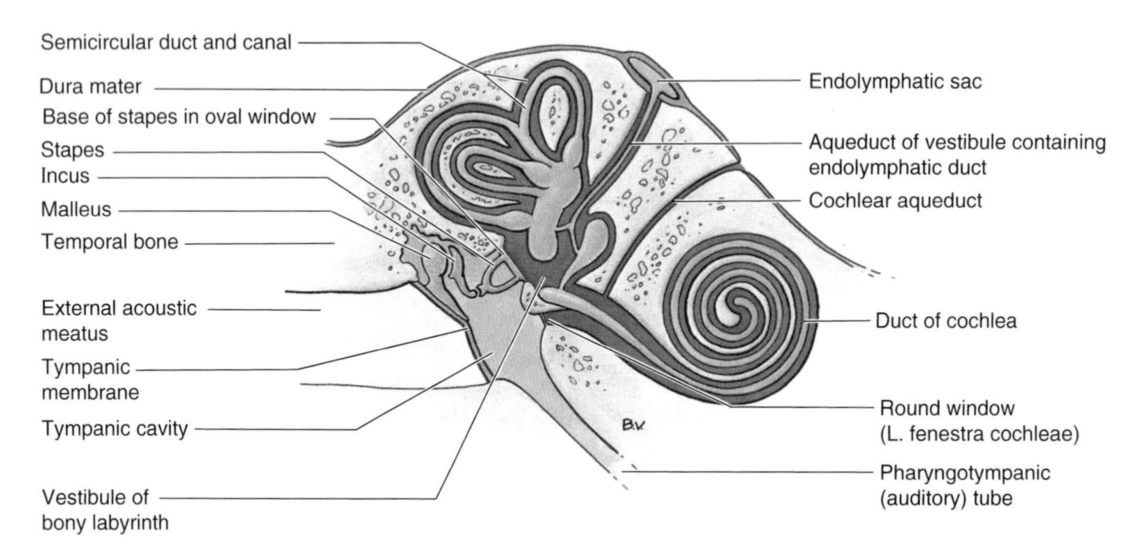

Figure 7.76. *(Continued)* **B.** The middle and internal parts of the ear. The *middle ear* (tympanic cavity) lies between the tympanic membrane and the internal ear. Three ossicles—the malleus (L. hammer), incus (L. anvil), and stapes (Mod. L. stirrup)—stretch from the lateral to the medial wall of the tympanic cavity. Of these, the malleus is attached to the tympanic membrane. The base (footplate) of the stapes is attached by an anular ligament to the oval window (L. fenestra vestibuli), and the incus connects these two ossicles. The pharyngotympanic tube is a communication between the anterior wall of the tympanic cavity and the lateral wall of the nasopharynx. The *internal ear* is composed of a closed system of membranous tubes and bulbs—the membranous labyrinth—that is filled with a fluid called endolymph (*orange*) and bathed in surrounding fluid called perilymph (*purple*).

External Ear Injury

Bleeding within the auricle resulting from trauma may produce an *auricular hematoma*. The localized collection of blood forms between the perichondrium and auricular cartilage. As the hematoma enlarges it compromises the blood supply to the cartilage. If untreated (e.g., by aspiration), fibrosis—formation of fibrous tissue—develops in the overlying skin, forming a malformed auricle (e.g., the "cauliflower ear" that occurs in some wrestlers).

Otoscopic Examination

Examination of the external acoustic meatus and tympanic membrane begins by straightening the meatus. In adults, the helix is grasped and pulled posterosuperiorly (up, out, and back). These movements reduce the curvature of the meatus, facilitating insertion of the *otoscope* (*A*). The external acoustic meatus is relatively short in infants; therefore, extra care must be exercised to prevent injury to the tympanic membrane. The meatus is straightened by ▶

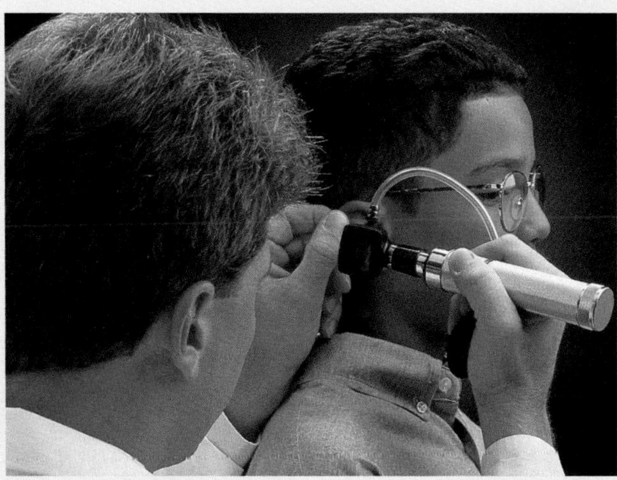

(A) Otoscopic examination

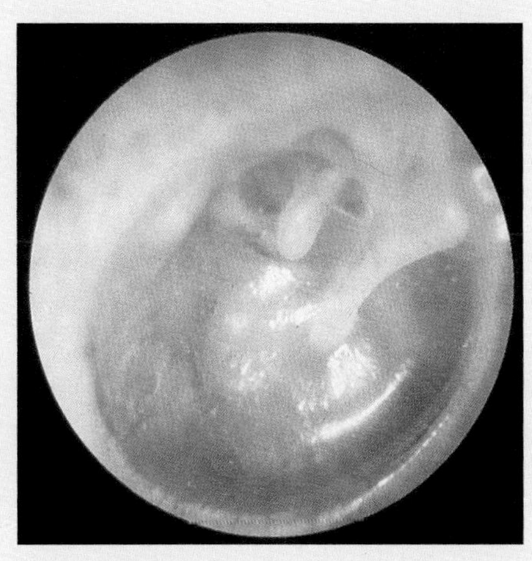

(B) Normal tympanic membrane

▶ pulling the auricle inferoposteriorly (down and back). The examination also provides a clue to tenderness, which can indicate inflammation of the auricle and/or the meatus—*otis externa*. The tympanic membrane is normally translucent and pearly gray (*B*). The handle of the malleus is usually visible near the center of the membrane. From the inferior end of the handle, a bright cone of light, reflected from the otoscope's illuminator—the *light reflex*—is often visible.

Otitis Externa

Otitis externa is a bacterial infection of the skin of the external acoustic meatus. The infection often develops in swimmers who do not dry their meatus after swimming and/or use ear drops. The affected individual complains of itching and pain in the external ear. Pulling the auricle increases the pain. Although the skin of the auricle does not contain organized nerve endings, stretching, pain, cold, and warmth can be felt.

Perforation of the Tympanic Membrane

Perforation of the tympanic membrane (a "ruptured eardrum") may result from otitis media and is one of several causes of middle ear deafness. Perforation of the membrane may also result from foreign bodies in the external acoustic meatus, trauma, or excessive pressure (e.g., during scuba diving). Minor ruptures of the tympanic membrane often heal spontaneously. Large ruptures usually require surgical repair. Because the superior half of the tympanic membrane is much more vascular than the inferior half, incisions to release pus from a middle ear abscess, for example, are made posteroinferiorly through the membrane. This incision also avoids injury to the chorda tympani nerve and auditory ossicles. ○

mainly by the **auriculotemporal nerve** (Fig. 7.74*A*), a branch of CN V_3. Some innervation is supplied by a small auricular branch of the vagus. The internal surface of the tympanic membrane is supplied by the glossopharyngeal nerve.

Middle Ear

The middle ear is in the petrous part of the temporal bone (Fig. 7.72 and 7.76). It includes the **tympanic cavity**, the space directly internal to the tympanic membrane, and the **epitympanic recess**, the space superior to the membrane. The middle ear is connected anteriorly with the nasopharynx by the **pharyngotympanic tube.** Posterosuperiorly, the tympanic cavity connects with the mastoid cells through the **mastoid antrum** (Fig. 7.77*A*). The tympanic cavity is lined with mucous membrane that is continuous with the lining of the pharyngotympanic tube, mastoid cells, and mastoid antrum. *The contents of the middle ear are the:*

- Auditory ossicles (small ear bones)—malleus, incus, and stapes (Mod. L. stirrup) (Fig. 7.76)
- Stapedius and tensor tympani muscles
- Chorda tympani nerve, a branch of CN VII (Fig. 7.77*B*)
- Tympanic plexus of nerves.

Walls of the Tympanic Cavity

The middle ear, shaped like a lozenge or narrow box with concave sides, has a roof, floor, and four walls (Fig. 7.77*B*).

- The *tegmental roof* (tegmental wall) is formed by a thin plate of bone, the *tegmen tympani*, which separates the tympanic cavity from the dura mater on the floor of the middle cranial fossa.
- The *floor* (jugular wall) is formed by a layer of bone that separates the tympanic cavity from the superior bulb of the IJV.
- The *lateral (membranous) wall* is formed almost entirely by the peaked convexity of the tympanic membrane; superiorly it is formed by the lateral bony wall of the *epitympanic recess*. The handle of the malleus is in the tympanic membrane and its head extends into the epitympanic recess.
- The *medial (labyrinthine) wall* separates the tympanic cavity from the internal ear, and features the promontory of the initial part (basal turn) of the cochlea (Fig. 7.76).
- The *anterior (carotid) wall* separates the tympanic cavity from the carotid canal; superiorly it has the opening of the auditory tube and the canal for the tensor tympani.
- The *posterior (mastoid) wall* features an opening in its superior part—the *aditus to the mastoid antrum*—connecting the tympanic cavity to the mastoid cells; the canal for the facial nerve descends between the posterior wall and the antrum, medial to the aditus.

The **mastoid antrum** is a cavity in the mastoid process of the temporal bone (Fig. 7.77*A*). The antrum—like the tympanic cavity—is separated from the middle cranial fossa by a thin bony roof, the **tegmen tympani**. The floor of the antrum has several apertures through which it communicates with the mastoid cells. The antrum and mastoid cells are lined by mucous membrane that is continuous with the lining of the middle ear. Anteroinferiorly, the mastoid antrum is related to the canal for the facial nerve.

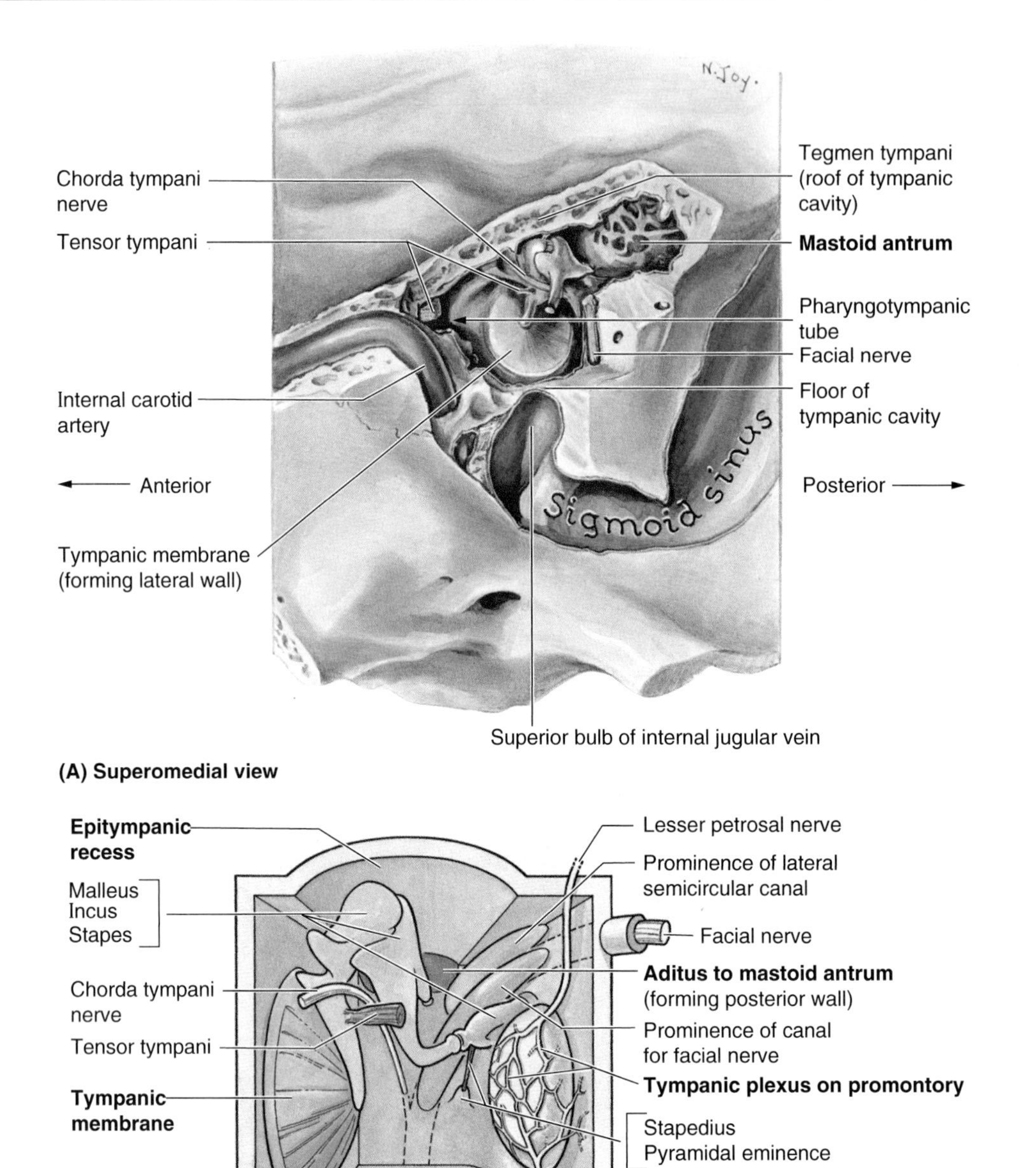

Figure 7.77. Walls of the tympanic cavity, or middle ear. A. Superomedial view. This specimen was dissected with a drill from the medial aspect. Observe the tegmen tympani forming the roof of the tympanic cavity and mastoid antrum, here fairly thick but most commonly it is extremely thin. Note that the internal carotid artery is the main relation of the anterior wall, the internal jugular vein (IJV) is the main relation of the floor, and the facial nerve (CN VII) is a main feature of the posterior wall. Note the *chorda tympani nerve* passing between the malleus (L. hammer) and incus (L. anvil). **B.** Schematic drawing of the middle ear. Anterior view. The anterior wall of the middle ear is removed. Observe the tympanic membrane (eardrum) forming much of the lateral wall; superior to it is the epitympanic recess in which are housed the larger parts of the malleus and incus. Branches of the tympanic plexus provide innervation to the mucosa of the middle ear and adjacent pharyngotympanic (auditory) tube, but one branch—the lesser petrosal nerve—is conveying presynaptic parasympathetic fibers to the otic ganglion for secretomotor innervation of the parotid gland.

Otitis Media

A bulging red tympanic membrane may indicate pus or fluid in the middle ear, a sign of otitis media. Infection of the middle ear is often secondary to upper respiratory infections. Inflammation and swelling of the mucous membrane lining the tympanic cavity may cause partial or complete *blockage of the pharyngotympanic tube.* The tympanic membrane becomes red and bulges, and the individual may complain of "ear popping." An amber-colored bloody fluid may be observed through the tympanic membrane. If untreated, otitis media may produce impaired hearing as the result of scarring of the auditory ossicles, limiting the ability of these bones to move in response to sound.

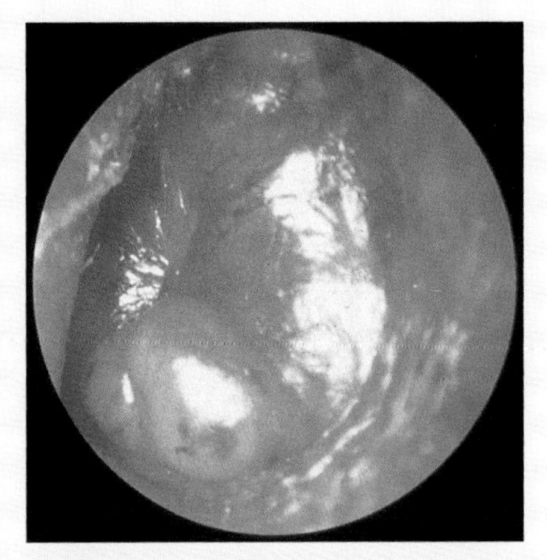

Otitis media

Mastoiditis

Infections of the mastoid antrum and mastoid cells (*mastoiditis*) result from a middle ear infection that causes inflammation of the mastoid process. Infections may spread superiorly into the middle cranial fossa through the petrosquamous fissure in children and cause *osteomyelitis* (bone infection) of the tegmen tympani. Since the advent of antibiotics, mastoiditis is uncommon. During operations for mastoiditis, surgeons are conscious of the course of the facial nerve so they will not injure it.

One point of access to the tympanic cavity is through the mastoid antrum. In children, only a thin plate of bone has to be removed from the lateral wall of the antrum to expose the tympanic cavity. In adults, bone must be penetrated for 15 mm or more. At present, most *mastoidectomies* are endaural (i.e., performed through the posterior wall of the external acoustic meatus).

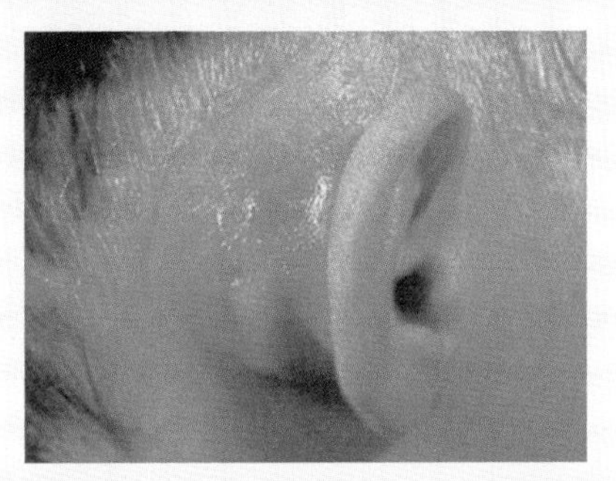

Mastoiditis (ruptured retroauricular abscess)

Earache

Earache is a common symptom that has multiple causes, two of which are otitis externa and otitis media. Earache may also be referred pain from distant lesions such as a dental abscess. ○

Pharyngotympanic Tube

The pharyngotympanic tube connects the tympanic cavity to the nasopharynx (Fig. 7.76, *A* and *B*), where it opens posterior to the inferior meatus of the nasal cavity. The posterolateral third of the tube is bony and the remainder is cartilaginous. The pharyngotympanic tube is lined by mucous membrane that is continuous posteriorly with that of the tympanic cavity and anteriorly with that of the nasopharynx.

The function of the pharyngotympanic tube is to equalize pressure in the middle ear with the atmospheric pressure, thereby allowing free movement of the tympanic membrane. By allowing air to enter and leave the tympanic cavity, this tube balances the pressure on both sides of the membrane. Because the walls of the cartilaginous part of the tube are normally in apposition, the tube must be actively opened. The tube is opened by the enlarged belly of the contracted levator veli palati pushing against one wall while the tensor veli palati pulls on the other. Because these are muscles of the soft palate, equalizing pressure ("popping the eardrums") is commonly associated with activities such as yawning and swallowing.

The **arteries** of the pharyngotympanic tube are derived from the **ascending pharyngeal artery** (Fig. 7.78*A*), a branch of the external carotid artery, and the **middle meningeal artery** and **artery of the pterygoid canal**—branches of the maxillary artery (Fig. 7.78*B*). The **veins** drain into the pterygoid venous plexus. The **nerves** of the pharyngotympanic tube arise from the *tympanic plexus* (Fig. 7.77*B*), which is formed by fibers of the facial and glossopharyngeal nerves. The tube also receives fibers from the *pterygopalatine ganglion* (Fig. 7.70).

Blockage of the Pharyngotympanic Tube

The pharyngotympanic tube forms a route for an infection to pass from the nasopharynx to the tympanic cavity. This tube is easily blocked by swelling of its mucous membrane, even as a result of mild infections (e.g., a ▶

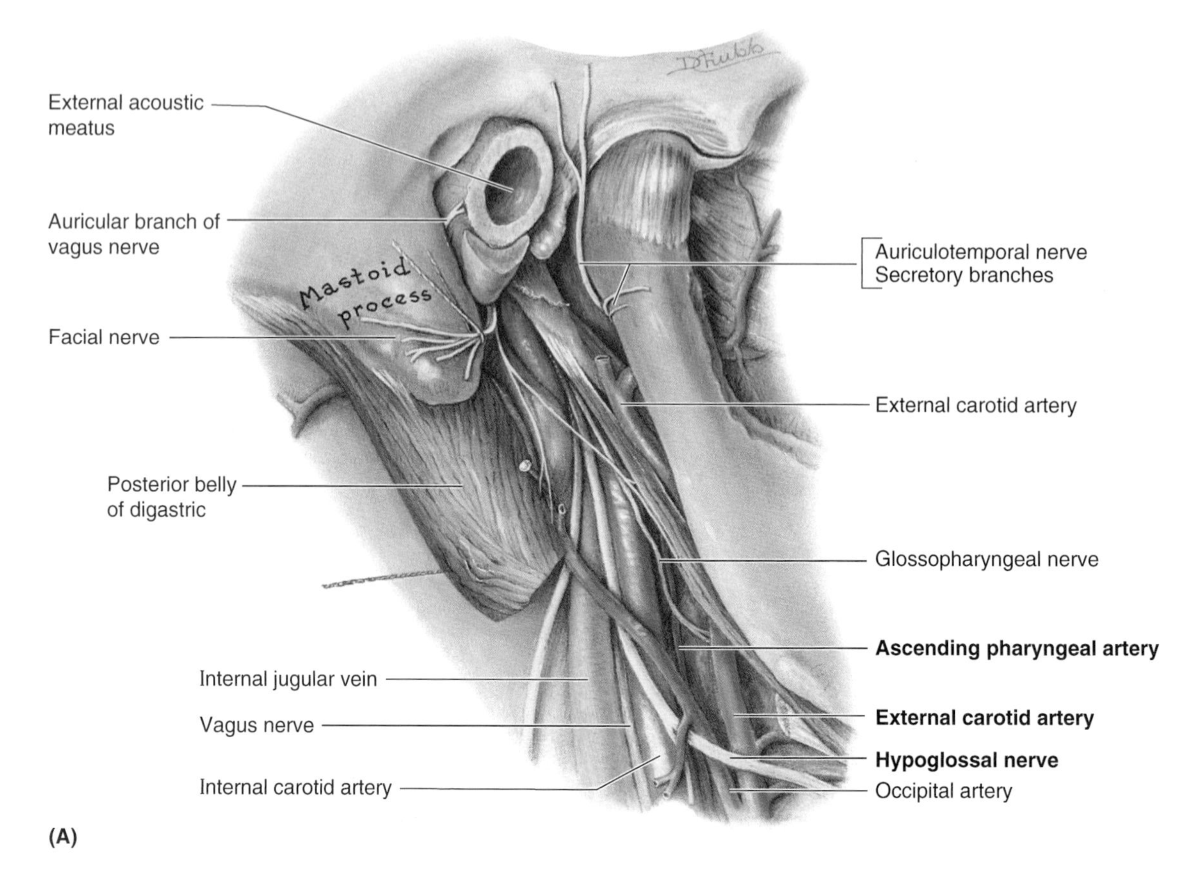

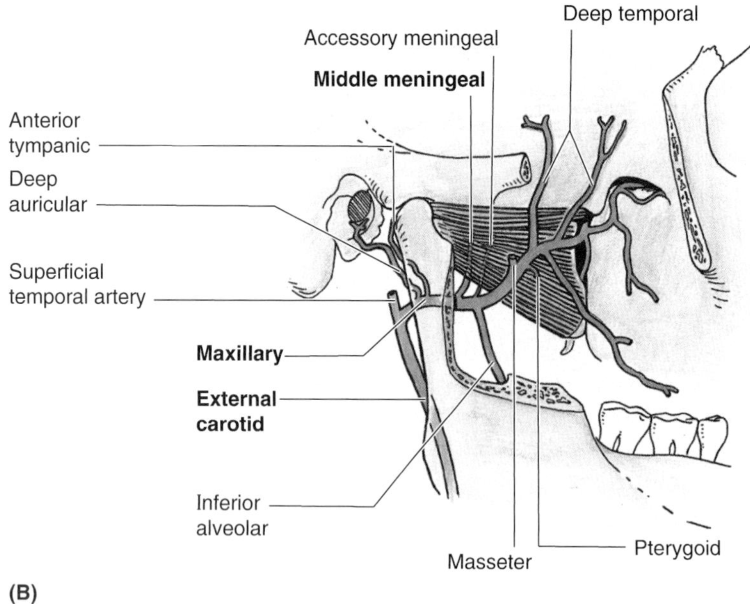

Figure 7.78. Dissection of structures deep to the parotid bed. A. The facial nerve (CN VII), the posterior belly of the digastric muscle, and the nerve to it are retracted. Note the deeply placed ascending pharyngeal artery, the only medial branch of the external carotid artery. It supplies the pharynx, palatine tonsil, pharyngotympanic (auditory) tube, and the medial wall of the tympanic cavity before it terminates by sending meningeal branches to the cranial cavity. **B.** Maxillary artery and its branches. The branches of the first (mandibular) part supply the external acoustic meatus (auditory canal) and tympanic membrane (eardrum). The middle meningeal artery sends branches to the pharyngotympanic tube before entering the skull through the foramen spinosum (L. spinous).

▶ head cold) given that the walls of its cartilaginous part are normally already in apposition. When the pharyngotympanic tube is occluded, residual air in the tympanic cavity is usually absorbed into the mucosal blood vessels, resulting in lower pressure in the tympanic cavity, retraction of the tympanic membrane, and interference with its free movement. Finally, hearing is affected. The more sudden, usually temporary pressure changes resulting from air flight can be equalized by swallowing (stimulated by gum chewing) or yawning; these movements open the pharyngotympanic tubes. ○

Auditory Ossicles

The auditory ossicles form a chain of small bones across the tympanic cavity from the tympanic membrane to the **oval window** (L. fenestra vestibuli)—an oval opening on the medial wall of the tympanic cavity leading to the vestibule of the internal ear (bony labyrinth). It is closed by the base of the stapes. The ossicles are the first bones to be fully ossified during development, being essentially mature at birth. The bone from which they are formed is exceptionally dense (hard). The **malleus** attaches to the tympanic membrane, and the **stapes** occupies the oval window (Figs. 7.76 and 7.79). The **incus** is located between these two bones and articulates with them. The ossicles are covered with the mucous membrane lining the tympanic cavity but, unlike other bones of the body, are not directly covered with a layer of periosteum.

Malleus. The rounded superior part, or *head of the malleus*, lies in the epitympanic recess (Fig. 7.79*A*). Its *neck* lies against the flaccid part of the tympanic membrane and its *handle* is embedded in the tympanic membrane—with its tip at the umbo—and moves with it. The head of the malleus articulates with the incus; the tendon of the tensor tympani inserts into its handle near the neck. The *chorda tympani nerve* crosses the medial surface of the neck of the malleus. The malleus functions as a lever, with the longer of its two processes and its handle attached to the tympanic membrane.

Incus. The large *body of the incus* lies in the epitympanic recess where it articulates with the head of the malleus. The *long limb process of the incus* lies parallel to the handle of the malleus and its interior end articulates with the stapes by way of a medially directed projection (lenticular process). Its *short limb* is connected by a ligament to the posterior wall of the tympanic cavity.

Stapes. The stapes—the smallest ossicle—has a head and a base that are united by two limbs (L. crura). Its **head**, directed laterally, articulates with the incus. The **base** (footplate) of the stapes fits into the oval window on the medial wall of the tympanic cavity. The base is considerably smaller than the tympanic membrane; as a result, the vibratory force of the stapes is increased approximately 10 times over that of the tympanic membrane. Consequently, the auditory ossicles increase the force but decrease the amplitude of the vibrations transmitted from the tympanic membrane.

Muscles Associated with the Ossicles. Two muscles dampen or resist movements of the auditory ossicles; one also dampens movements (vibration) of the tympanic membrane: tensor tympani and stapedius.

The **tensor tympani** (Figs. 7.75*B* and 7.77) is a short muscle that arises from the superior surface of the cartilaginous part of the pharyngotympanic tube, the greater wing of the sphenoid, and the petrous part of the temporal bone. The muscle inserts into the handle of the malleus. The tensor tympani pulls the handle medially, tensing the tympanic membrane and reducing the amplitude of its oscillations. This action tends to prevent damage to the internal ear when one is exposed to loud sounds. The tensor tympani is supplied by the mandibular nerve.

The **stapedius** (Fig. 7.75*B*) is a tiny muscle—the smallest voluntary (striated) muscle of the body—inside the *pyramidal eminence* (pyramid), a hollow, cone-shaped prominence on the posterior wall of the tympanic cavity. Its tendon enters the tympanic cavity by emerging from a pinpoint foramen in the apex of the eminence and inserts on the neck of the stapes. The nerve to the stapedius arises from the facial nerve. The stapedius pulls the stapes posteriorly and tilts its base in the oval window, thereby tightening the anular ligament and reducing the oscillatory range. It also prevents excessive movement of the stapes.

Paralysis of the Stapedius

The tympanic muscles have a protective action in that they dampen large vibrations of the tympanic membrane resulting from loud noises. Paralysis of the stapedius (e.g., resulting from a lesion of the facial nerve) is associated with excessive acuteness of hearing—*hyperacusis* or hyperacusia. This condition results from uninhibited movements of the stapes. ○

Internal Ear

The internal ear contains the *vestibulocochlear organ* concerned with the reception of sound and the maintenance of balance. Buried in the petrous part of the temporal bone (Figs. 7.76*A* and 7.80*A*), the internal ear consists of the sacs and ducts of the membranous labyrinth. The **membranous labyrinth** containing *endolymph* is suspended within the **bony labyrinth** by *perilymph*; both fluids carry sound waves to the end organs for hearing and balancing.

Bony Labyrinth

The bony labyrinth of the internal ear is a cave composed of three parts: cochlea, vestibule, and semicircular canals. It occupies much of the lateral part of the petrous part of the temporal bone (Figs. 7.76*A* and 7.80, *A* and *B*). Its walls are made of bone that is denser than the remainder of the petrous temporal bone and constitutes the bony **optic capsule**, which can be isolated (carved) from the surrounding matrix of bone using a dental drill.

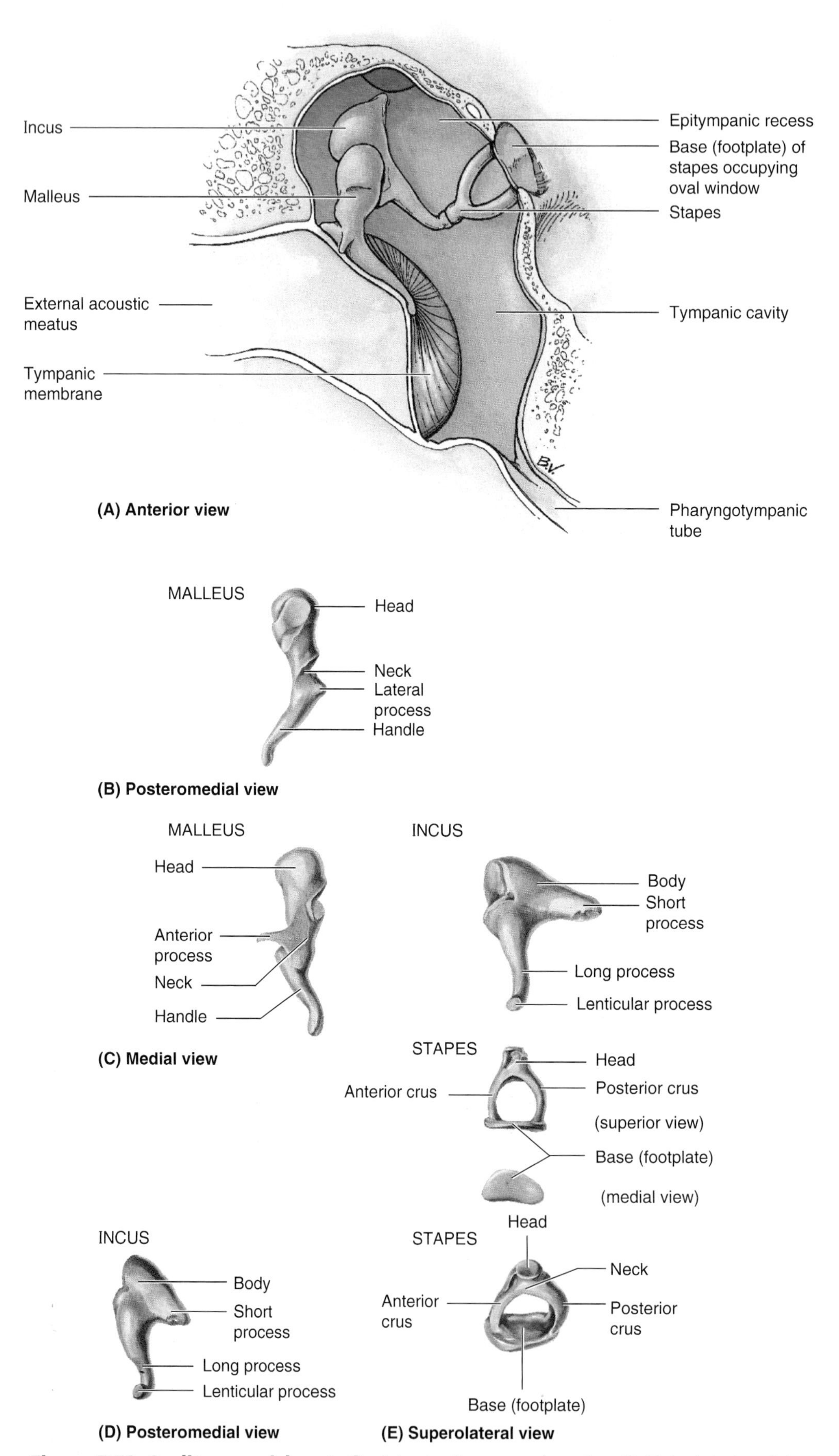

Figure 7.79. Auditory ossicles. A. Ossicles in situ, coronal section. **B–E.** Isolated ossicles of the middle ear.

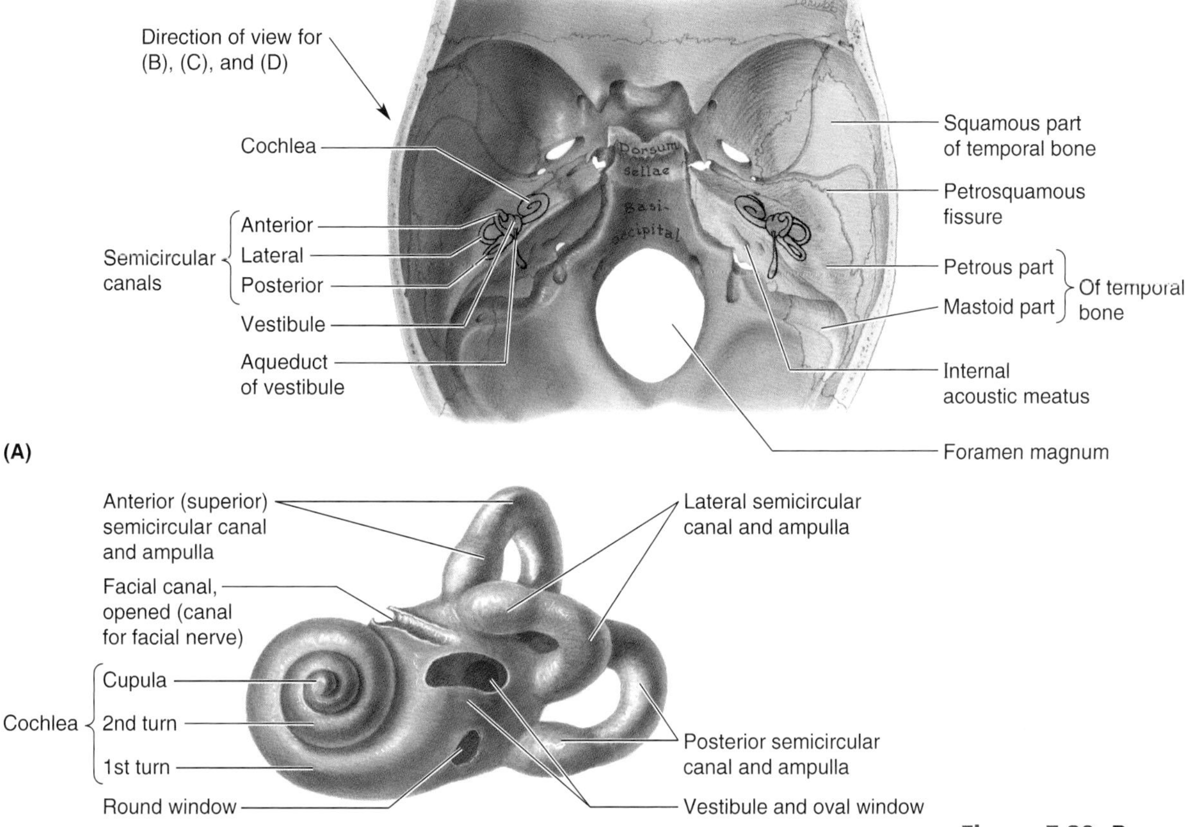

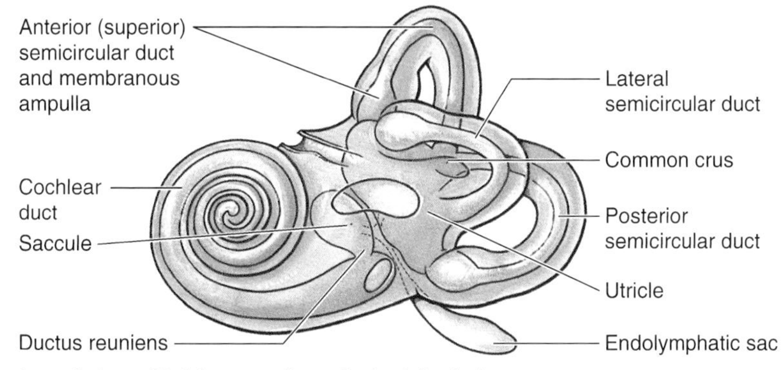

(B) Anterolateral view of left bony labyrinth (otic capsule)

(C) Anterolateral view of left bony and vestibular labyrinths

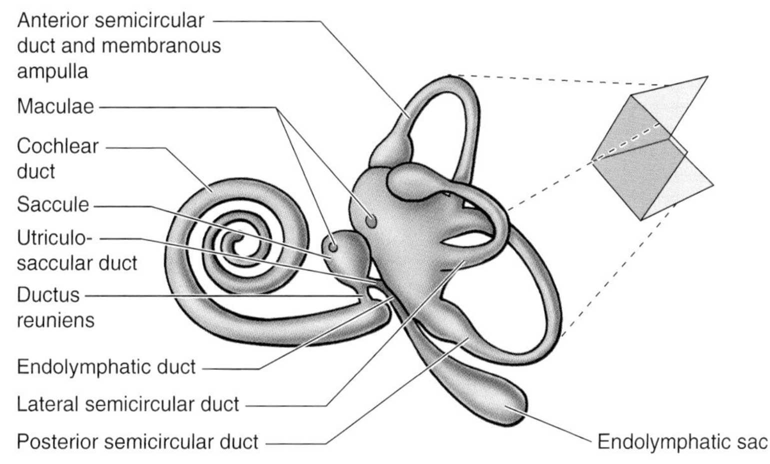

(D) Anterolateral view of left vestibular labyrinth

Figure 7.80. Bony and membranous labyrinths of internal ear. A. Superior view of the interior of the base of the skull, showing the temporal bone and the location of the bony labyrinth. **B.** Lateral view of the walls of the bony labyrinth (otic capsule) carved out of the petrous temporal bone. **C.** Similar view of the bony labyrinth (space within the otic capsule) occupied by perilymph and the membranous labyrinth. **D.** Lateral view of the membranous labyrinth after removal from the bony labyrinth. The membranous labyrinth is a closed system of ducts and chambers filled with endolymph and bathed by perilymph. Observe its three parts: the cochlear duct that occupies the cochlea (G. snail); the saccule and utricle that occupy the vestibule; and the three semicircular ducts that occupy the three semicircular canals. Note that the utricle communicates with the saccule through the utriculosaccular duct. The lateral semicircular duct lies in the horizontal plane and is more horizontal than it appears in this drawing.

The otic capsule is often illustrated and identified as being the bony labyrinth; however, the bony labyrinth is the *fluid-filled space*, which is surrounded by the otic capsule and is most accurately represented by a cast of the otic capsule following removal of the surrounding bone.

Cochlea. The cochlea is the shell-shaped part of the bony labyrinth that contains the **cochlear duct** (Fig. 7.80*C*), the part of the internal ear concerned with hearing. The **spiral canal of the cochlea** begins at the vestibule and makes 2.5 turns around a bony core, the **modiolus** (Fig. 7.81)—the cone-shaped core of spongy bone about which the spiral canal of the cochlea turns. The modiolus contains canals for blood vessels and for distribution of the cochlear nerve. The apex of the cone-shaped modiolus, like the axis of the tympanic membrane, is directed laterally, anteriorly, and inferiorly. The large basal turn of the cochlea produces the **promontory** on the medial wall of the tympanic cavity (Fig. 7.77*B*). At the basal turn, the bony labyrinth communicates with the subarachnoid space superior to the jugular foramen through the *cochlear aqueduct* (Fig. 7.76*B*); it also features the round window (L. fenestra cochleae), closed by the secondary tympanic membrane.

Vestibule. This small oval chamber (approximately 5 mm long) contains the **utricle** and **saccule** (Fig. 7.80*C*), parts of the balancing apparatus (membranous labyrinth). The vestibule features the oval window on its lateral wall, occupied by the base of the stapes. The vestibule is continuous with the bony cochlea anteriorly, the semicircular canals posteriorly, and the posterior cranial fossa by the **aqueduct of the vestibule** (Fig.7.80*A*). The aqueduct extends to the posterior surface of the petrous part of the temporal bone, where it opens posterolateral to the *internal acoustic meatus* (Fig. 7.80*A*). The aqueduct transmits the **endolymphatic duct** (Fig. 7.80*D*) and two small blood vessels.

Semicircular Canals. These canals (anterior, posterior, and lateral) communicate with the vestibule of the bony labyrinth. The canals lie posterosuperior to the vestibule into which they open; they are set at right angles to each other

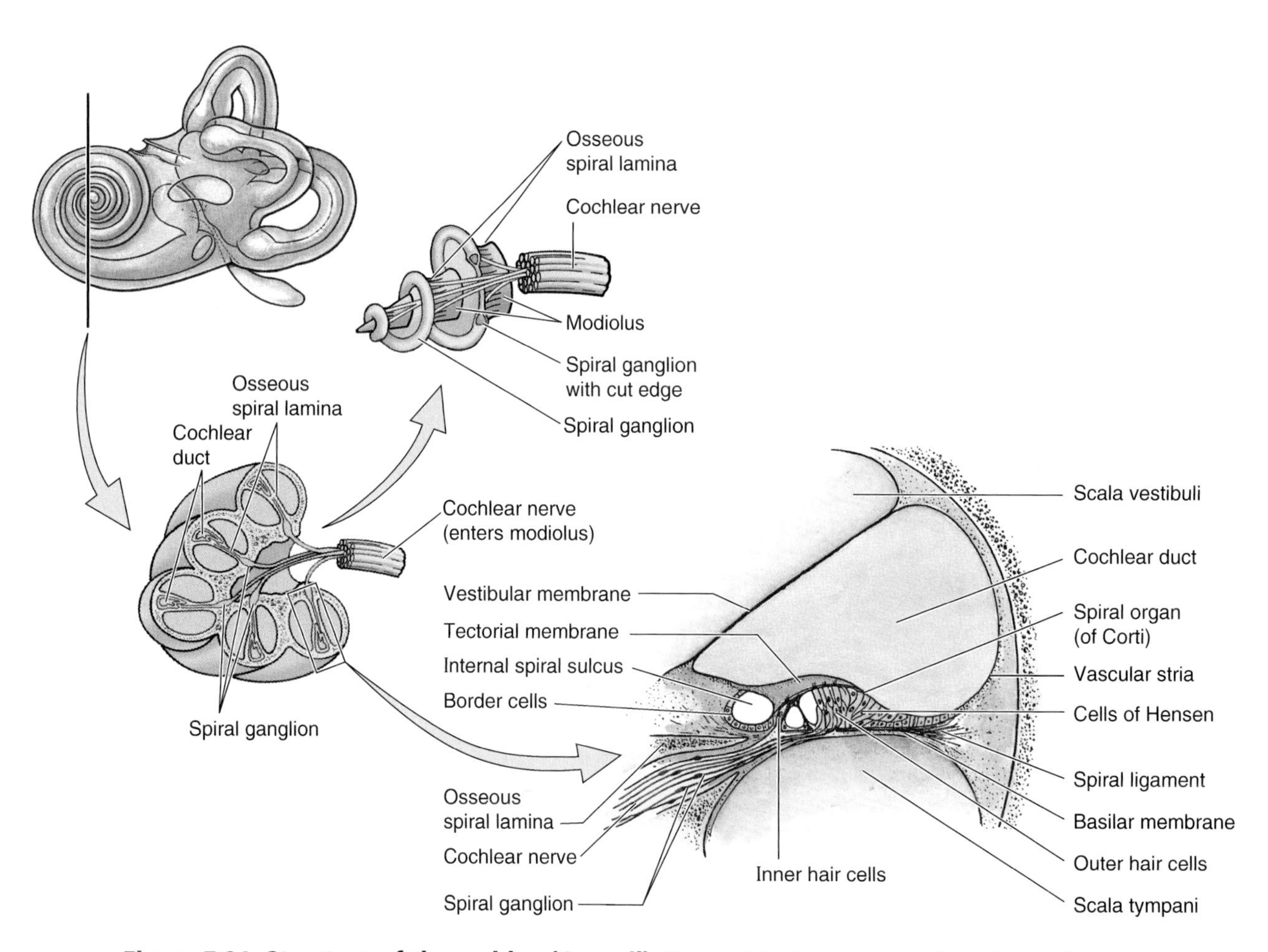

Figure 7.81. Structure of the cochlea (G. snail). The cochlea has been sectioned along the axis about which the cochlea winds (see the orientation figure in the upper left). An isolated, conelike, bony core of the cochlea—the modiolus—is shown after the turns of the cochlea are removed, leaving only the spiral lamina winding around it like the thread of a screw. The large drawing shows details of the area enclosed in the rectangle. Note the cochlear duct, the basilar membrane, the spiral organ (of Corti), and the tectorial membrane.

(Fig. 7.80*B*). They occupy three planes in space. Each semicircular canal forms approximately two-thirds of a circle and is approximately 1.5 mm in diameter, except at one end where there is a swelling—the **ampulla**. The canals have only five openings into the vestibule because the anterior and posterior canals have one limb common to both. Lodged within the canals are the **semicircular ducts** (Figs. 7.80, *C* and *D*).

Membranous Labyrinth

The membranous labyrinth consists of a series of communicating sacs and ducts that are suspended in the bony labyrinth (Fig. 7.80*C*). The membranous labyrinth contains *endolymph*, a watery fluid that differs in composition from the surrounding *perilymph* that fills the remainder of the bony labyrinth. *The membranous labyrinth*—composed of two divisions, the cochlear labyrinth and the vestibular labyrinth—*consists of two parts*:

- *Vestibular labyrinth*
  - Utricle and saccule, two small communicating sacs in the vestibule of the bony labyrinth
  - Three semicircular ducts in the semicircular canals
- *Cochlear labyrinth*
  - Cochlear duct in the cochlea.

The membranous labyrinth is suspended (does not "float") in the bony labyrinth; its chief divisions are the cochlear labyrinth and the vestibular labyrinth.

The **spiral ligament**, a spiral thickening of the periosteal lining of the cochlear canal, *secures the cochlear duct to the spiral canal of the cochlea* (Fig. 7.81). The various parts of the membranous labyrinth form a closed system of sacs and ducts that communicate with one another.

The **semicircular ducts** open into the utricle through five openings reflective of the way the surrounding semicircular canals open into the vestibule. The utricle communicates with the saccule through the **utriculosaccular duct** from which the **endolymphatic duct** arises (Fig. 7.81). The saccule is continuous with the cochlear duct through the **ductus reuniens**, a uniting duct.

The **utricle** and **saccule** have specialized areas of sensory epithelium—the maculae (Fig. 7.80). The **macula of the utricle** (L. macula utriculi) is in the floor of the utricle, parallel with the base of the skull, whereas the **macula of the saccule** (L. macula sacculi) is vertically placed on the medial wall of the saccule. The hair cells in the maculae are innervated by fibers of the vestibular division of the **vestibulocochlear nerve**. The primary sensory neurons are in the **vestibular ganglia** (Fig. 7.82), which are in the internal acoustic meatus. The **endolymphatic duct** traverses the *vestibular aqueduct of the bony labyrinth* (Fig. 7.76*B*) and emerges through the bone of the posterior cranial fossa, where it expands into a blind pouch—the **endolymphatic sac** (Figs. 7.80*C* and 7.82). It is located under the dura mater on the posterior surface of the petrous part of the temporal bone. The endolymphatic sac is a storage reservoir for excess endolymph formed by the blood capillaries in the membranous labyrinth.

Motion Sickness

The maculae of the membranous labyrinth are primarily static organs, which have small dense particles (otoliths) embedded among hair cells. Under the influence of gravity, the *otoliths* cause bending of the hair cells, which stimulate the vestibular nerve and provides awareness of the position of the head in space; the hairs also respond to quick tilting movements and to linear acceleration and deceleration. *Motion sickness* results mainly from fluctuating stimulation of the maculae. ○

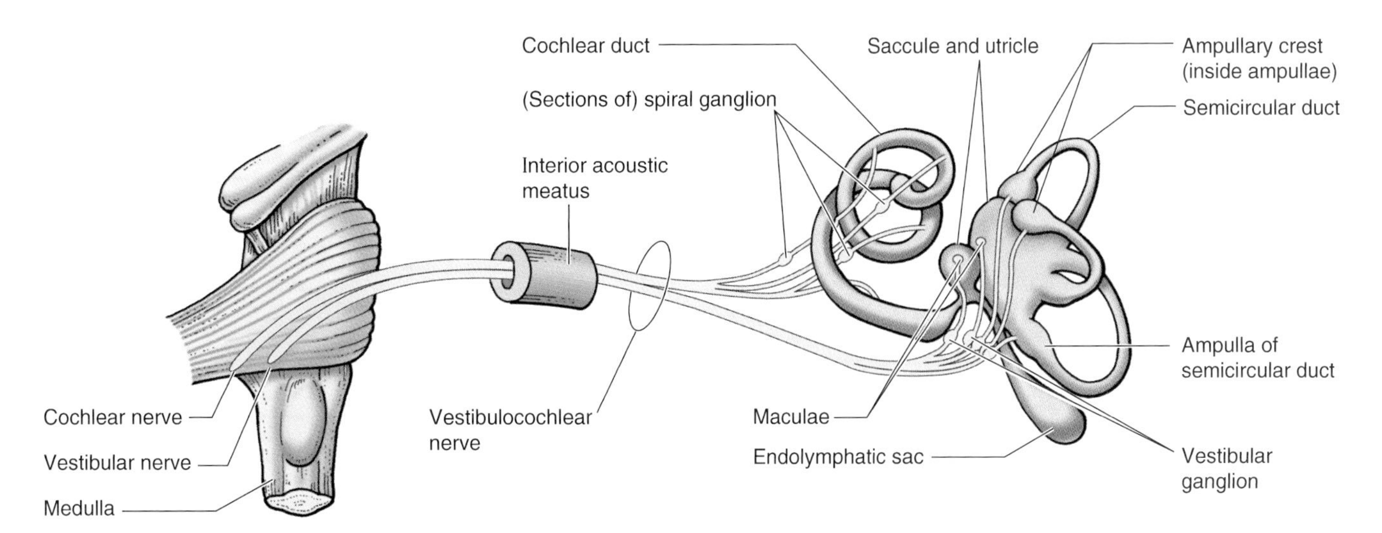

Figure 7.82. The vestibulocochlear nerve (CN VIII). Note that it has two parts: the cochlear nerve, or the nerve of hearing, and the vestibular nerve, or the nerve of balance. The cell bodies of the sensory fibers (only) that comprise the two parts of this nerve constitute the spiral and vestibular ganglia.

Semicircular Ducts. Each semicircular duct has an **ampulla** at one end (Fig. 7.82) containing a sensory area, the **ampullary crest** (L. crista ampullaris). The crests are sensors for recording movements of the endolymph in the ampulla resulting from rotation of the head in the plane of the duct. The hair cells of the crests, like those of the maculae, stimulate primary sensory neurons whose cell bodies are in the *vestibular ganglia.*

Cochlear Duct. The cochlear duct is a spiral, blind tube, triangular in cross-section, and firmly suspended across the cochlear canal between the **spiral ligament** on the external wall of the cochlear canal (Fig. 7.81) and the **osseous spiral lamina** of the modiolus. Spanning the spiral canal in this manner, the endolymph-filled cochlear duct divides the perilymph-filled spiral canal into two channels that communicate at the apex of the cochlea at the *helicotrema.* Waves of hydraulic pressure created in the perilymph of the vestibule by the vibrations of the base of the stapes ascend to the apex of the cochlea by one channel, the *scala vestibuli,* then the pressure waves pass through the helicotrema and then descend back to the basal turn by the other channel, the *scala tympani.* There the pressure waves again become vibrations, this time of the *secondary tympanic membrane,* which occupies the round window. Here the energy initially received by the (primary) tympanic membrane is finally dissipated into the air of the tympanic cavity. The roof of the cochlear duct is formed by the **vestibular membrane.** The floor of the duct is also formed by part of the duct, the **basilar membrane**, plus the outer edge of the osseous spiral lamina. The receptor of auditory stimuli is the **spiral organ** (of Corti), situated on the basilar membrane. It is overlaid by the gelatinous **tectorial membrane.** The spiral organ contains hair cells, the tips of which are embedded in the tectorial membrane. The spiral organ is stimulated to respond by deformation of the cochlear duct induced by the hydraulic pressure waves in the perilymph, which ascend and descend in the surrounding scalae vestibuli and tympani.

Dizziness and Hearing Loss

Injuries of the peripheral auditory system cause three major symptoms (Wazen, 1995): *hearing loss* (usually conductive hearing loss), *vertigo* (dizziness) when in the semicircular ducts, and *tinnitus* (buzzing or ringing) when localized in the cochlear duct. Tinnitus and hearing loss may result from lesions anywhere in the peripheral or central auditory pathways. The two types of hearing loss are:

- *Conductive hearing loss,* resulting from anything in the external or middle ear that interferes with movement of the oval or round windows. Patients often speak with a soft voice because, to them, their own voices sound louder than background sounds.
- *Sensorineural hearing loss,* resulting from defects in the cochlea, cochlear nerve, brainstem, or cortical connections.

Ménière Syndrome

This condition is related to *blockage of the cochlear aqueduct* and is characterized by recurrent attacks of tinnitus, hearing loss, and vertigo. These symptoms are accompanied by a sense of pressure in the ear, distortion of sounds, and sensitivity to noises (Wazen, 1995). Afflicted individuals have an increase in endolymphatic volume with ballooning of the cochlear duct, utricle, and saccule.

High Tone Deafness

Persistent exposure to excessively loud sounds causes degenerative changes in the spiral organ, resulting in *high tone deafness.* This type of hearing loss commonly occurs in workers who are exposed to loud noises and do not wear protective earmuffs (e.g., persons working for long periods around jet engines).

Otic Barotrauma

Injury caused to the ear by an imbalance in pressure between ambient (surrounding) air and the air in the middle ear is called *otic barotrauma.* This type of injury usually occurs in fliers and divers. ○

Internal Acoustic Meatus

The internal acoustic meatus is a narrow canal that runs laterally for approximately 1 cm within the petrous part of the temporal bone (Fig. 7.80*A*). The opening of the meatus is in the posteromedial part of this bone, in line with the external acoustic meatus. The internal acoustic meatus is closed laterally by a thin, perforated plate of bone that separates it from the internal ear. Through this plate pass the facial nerve, branches of the vestibulocochlear nerve, and blood vessels. The vestibulocochlear nerve divides near the lateral end of the internal acoustic meatus into two parts, a cochlear nerve and a vestibular nerve (Fig. 7.82).

Medical Imaging of the Head

Radiography

The common radiographic projections that are used to examine the skull are lateral, posteroanterior (PA), anteroposterior (AP), and axial. Skull radiographs are a reliable method for detecting skull fractures. Because skulls vary considerably in shape, one must examine radiographs carefully for abnormalities. Several pathological conditions cause the head to enlarge (*hydrocephaly*) or remain small (*microcephaly*). Lateral projections are often made to examine the calvaria, but they can also be useful for examining structures in the base of the skull (Fig. 7.83). The facial skeleton in adults forms approximately half the skull; in children, the facial area is proportionately smaller. The ▶

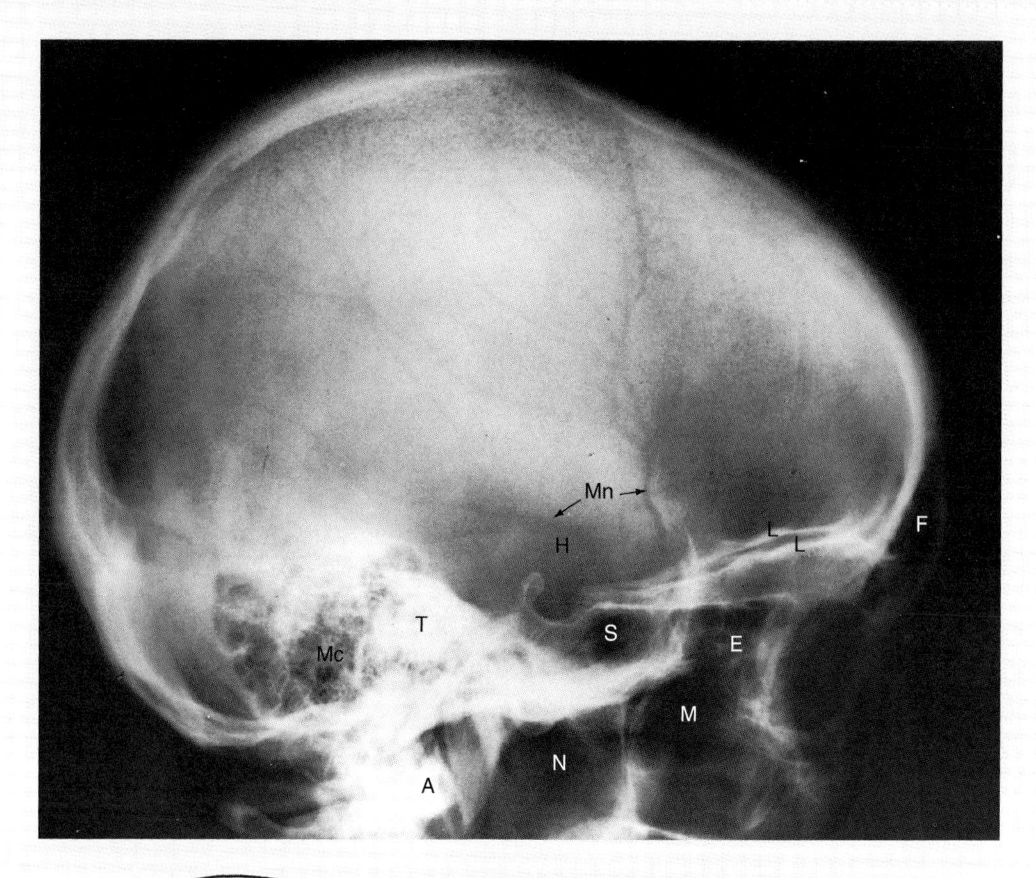

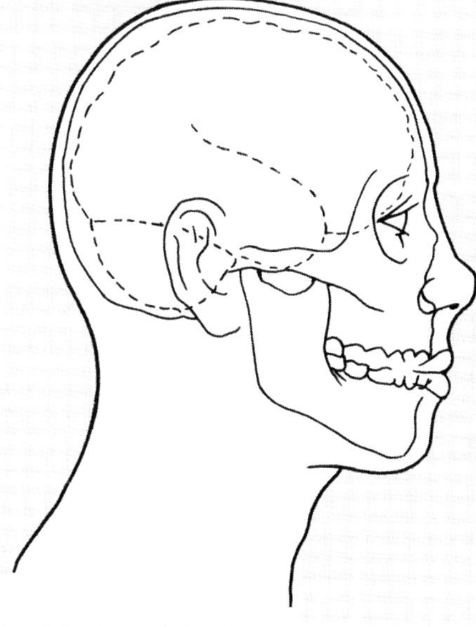

Figure 7.83. Lateral radiograph of the skull. Observe the paranasal sinuses: frontal (*F*), ethmoidal (*E*), sphenoidal (*S*), and maxillary (*M*). Also observe the hypophysial fossa (*H*) for the pituitary gland (L. hypophysis cerebri); the great density of the petrous (rocklike) part of the temporal bone (*T*); and the mastoid cells (*Mc*). Note that the right and left orbital parts, or plates, of the frontal bone are not superimposed and, thus, the floor of the anterior cranial fossa appears as two lines (*L*). Observe the bony grooves for the branches of the middle meningeal vessels (*Mn*); the arch of the atlas (*A*), and the nasopharynx (*N*). (Courtesy of Dr. E. Becker, Associate Professor of Diagnostic Imaging, University of Toronto, Toronto, Ontario, Canada)

▶ calvaria in adults varies from 3 to 8 mm in thickness, but the skull is much thicker around the external occipital protuberance. Observe the structures in Figure 7.83, especially the *hypophysial fossa* (*H*) for the pituitary gland and the grooves for the branches of the *middle meningeal vessels* (*Mn*). These vessels are often torn when someone is hit (e.g., with a stone) in the temporal fossa. Also observe the frontal (*F*), ethmoidal (*E*), sphenoidal (*S*), and maxillary (*M*) sinuses.

For examination of the orbits, nasal region, and certain paranasal sinuses, AP projections are also used (Fig. 7.84). In these views, the dens of the axis (C2 vertebra) and the lateral masses of the atlas (C1 vertebra) are superimposed on the facial skeleton. Observe that the right and left frontal sinuses are not the same size; this is normal. Also observe the lesser wing of the sphenoid (*S*), the superior orbital fissure (*Sr*), and the inferior and middle conchae (*I*) on the lateral wall of the nose.

To visualize the arteries of the brain, a radiopaque contrast medium is injected into the carotid artery and radiographs are taken, producing **carotid arteriograms** (Fig. 7.85). The four I's in the arteriogram indicate the parts of the internal carotid artery. This type of radiograph is useful for detecting cerebral aneurysms and arteriovenous malformations. ▶

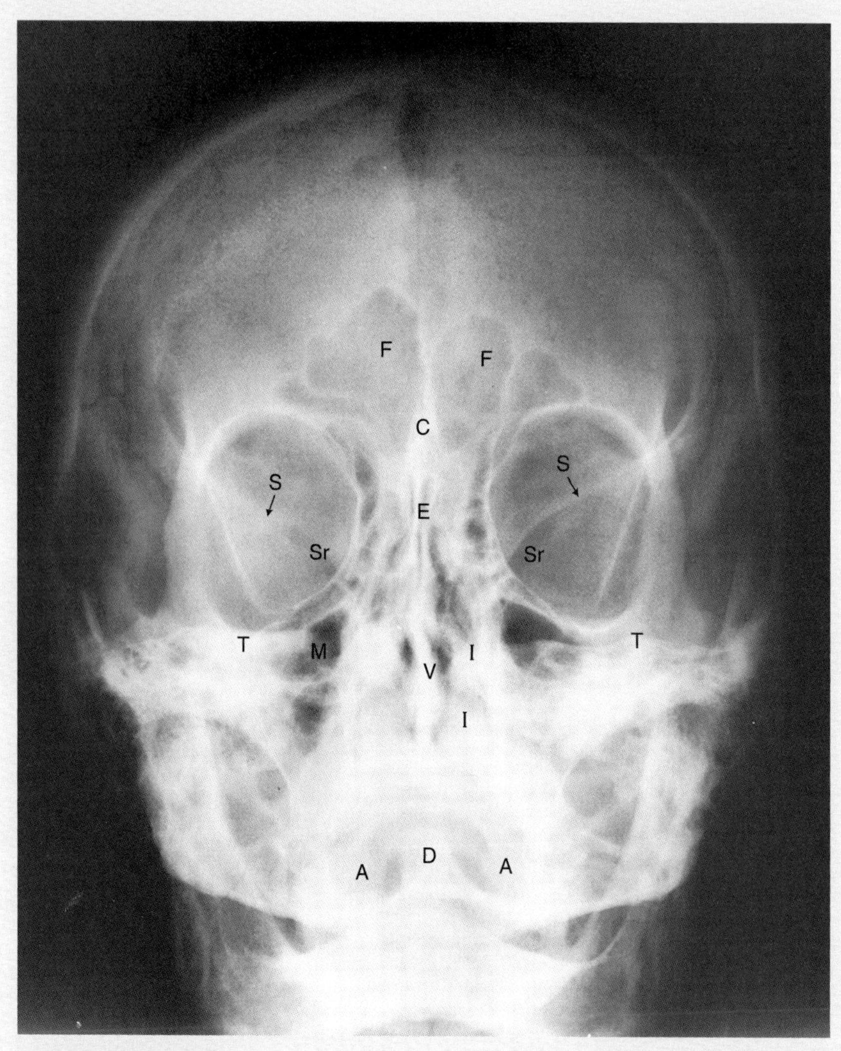

Anteroposterior view

Figure 7.84. Radiograph of the skull. Anteroposterior (AP) view. Observe the superior orbital fissure (*Sr*), the lesser wings of the sphenoid (*S*), and the superior surface of the petrous (rocklike) part of the temporal bone (*T*). Also observe that the nasal septum is formed by the perpendicular plate of the ethmoid (*E*) and the vomer (*V*). Examine the inferior and middle conchae (*I*) of the lateral wall of the nose; the crista galli (L. cock's comb) (*C*); the frontal sinus (*F*); and the maxillary sinus (*M*). Superimposed on the facial skeleton (viscerocranium or splanchnocranium) is the dens of the axis (*D*) and the lateral masses of the atlas (*A*). (Courtesy of Dr. E. Becker, Associate Professor of Medical Imaging, University of Toronto, Toronto, Ontario, Canada)

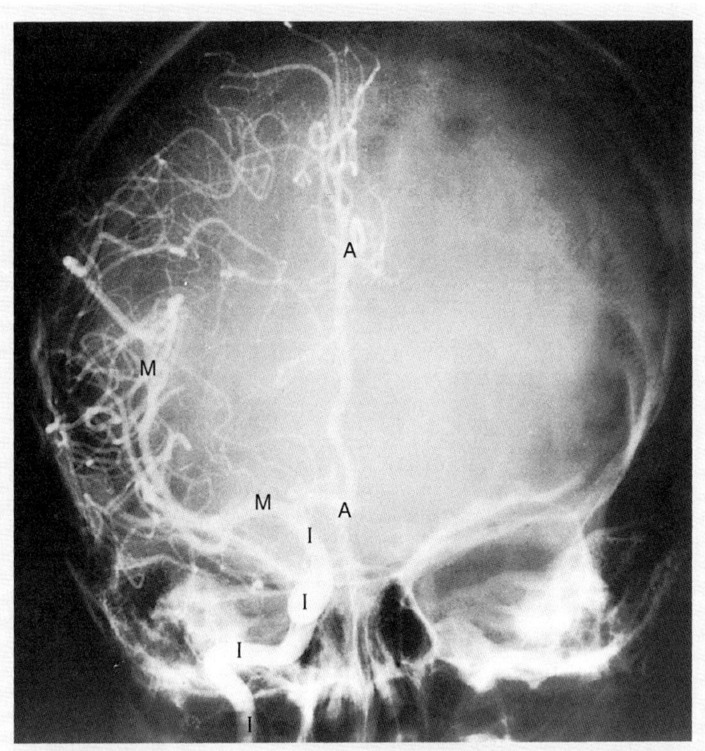

(A) Posteroanterior view

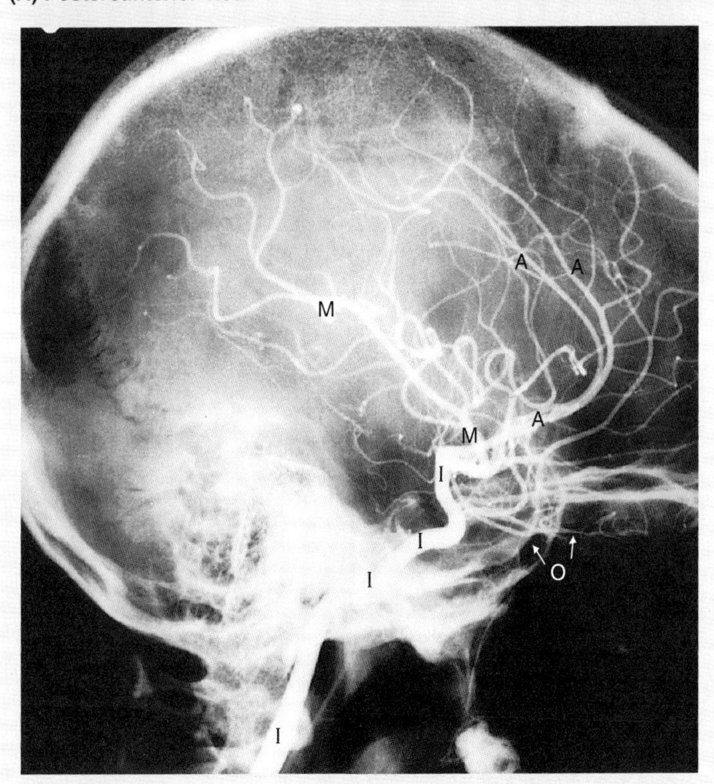

(B) Lateral view

Figure 7.85. Carotid arteriograms. A. Posteroanterior (PA) view. **B.** Lateral views. Observe the four letter *I's* indicating the parts of the internal carotid artery: cervical, before entering the skull; petrous (rock-like), within the temporal bone; cavernous, within the venous sinus; and cerebral, within the cranial subarachnoid (leptomeningeal) space. Also observe the anterior cerebral artery and its branches (*A*); the middle cerebral artery and its branches (*M*); and the ophthalmic artery (*O*). (Courtesy of Dr. D. Armstrong, Associate Professor of Medical Imaging, University of Toronto, Toronto, Ontario, Canada)

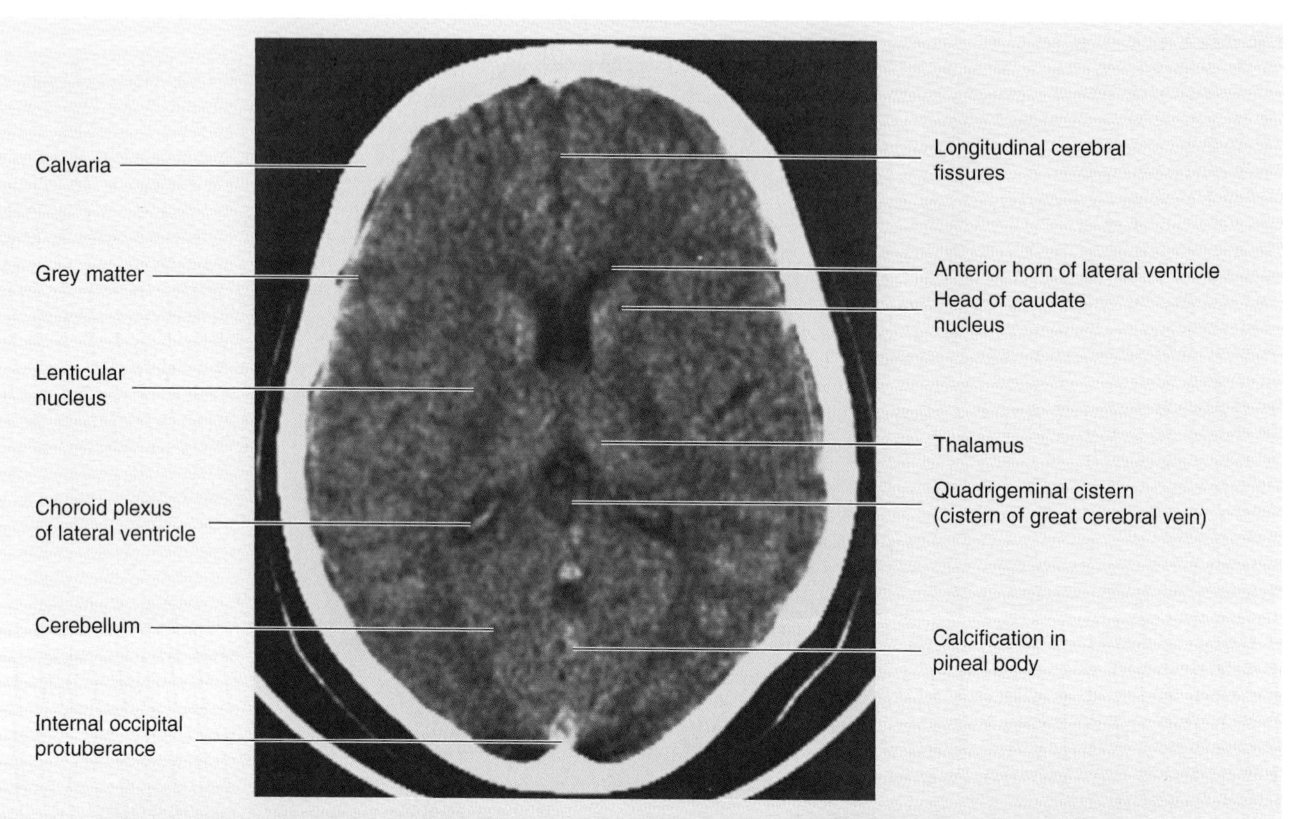

Figure 7.86. Transverse (axial) CT image of the brain. Observe the ventricles, various parts of the brain, and the choroid plexus of the lateral ventricle.

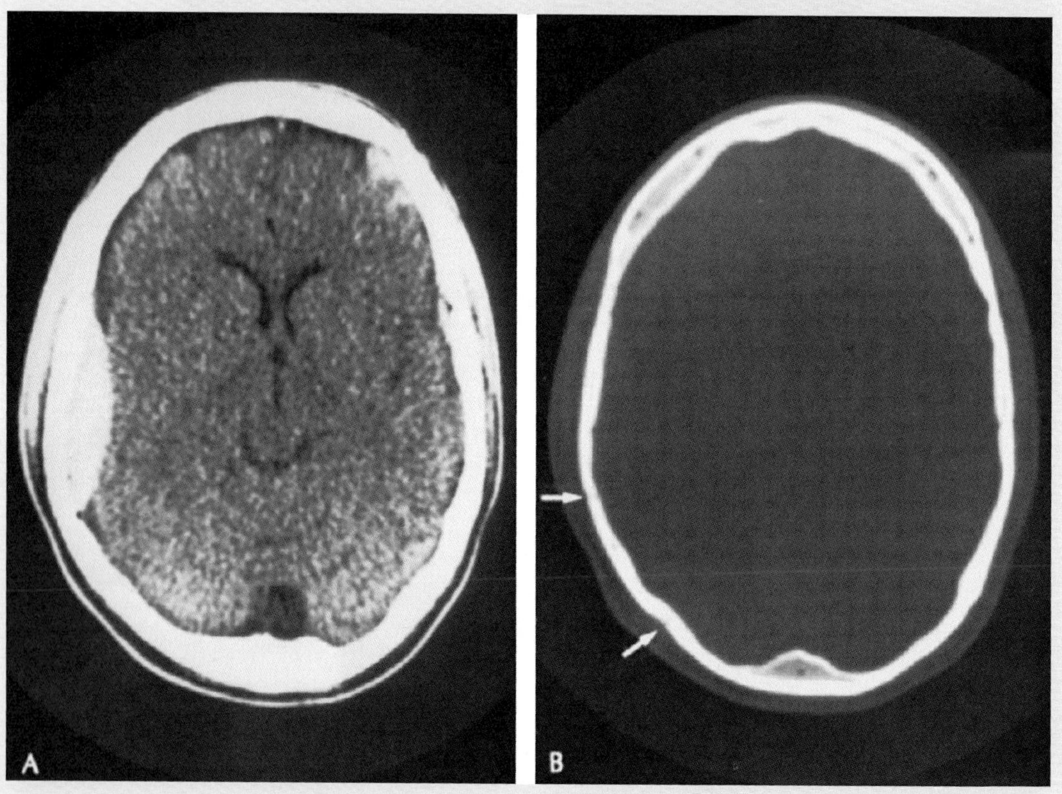

Figure 7.87. Extradural (epidural) hematoma. Transverse (axial) CT image of the calvaria and brain. **A.** The large white area on the right indicates the site of extradural hemorrhage. **B.** This CT scan with bone windows shows two adjacent fractures (*arrows*) of the calvaria; the anterior fracture is at the site of the groove for the middle meningeal artery.

Computed Tomography

CT is a premier imaging method in neurodiagnosis (Fig. 7.86); it is quicker and less expensive than magnetic resonance imaging (MRI) and is more informative than plain skull radiographs. "The major advantages of CT are the speed and ease with which each image is obtained, thereby obviating problems of patient discomfort and motion" (Chan et al., 1995). For reasons of cost, speed, and availability, CT is in wide use for evaluation of head injury. It is especially useful for patients who are neurologically or medically unstable, uncooperative, or claustrophobic, as well as for patients with pacemakers or other metallic implants (Chan et al., 1995). The diagnosis of an *extradural or epidural hematoma* can made by CT (Figs. 7.87 and 7.88). When the injuries are more severe, the meninges and cerebral cortex are torn. Traumatic hemorrhages often produce hematomas in the cerebral hemispheres, basal ganglia, and brainstem (Fig. 7.89).

Magnetic Resonance Imaging

MRIs show much more detail in the soft tissues than do CTs (Figs. 7.89 and 7.90). "MRI is the 'gold standard' for detecting and delineating intracranial and spinal lesions" (Chan et al., 1995). MRI provides good soft tissue contrast of normal and pathologic structures (Fig. 7.91). It also permits multiplanar capability, which provides three-dimensional information and relationships that are not so readily available with CT (Mohr and Prohovnik, 1995). MRI can also demonstrate blood and CSF flow. *Normal pressure hydrocephalus*—an occult form of hydrocephalus—is a potentially treatable cause of dementia (Prockop, 1995) that can be detected by a T_2-weighted MRI (Fig. 7.92). The syndrome often follows head trauma, subarachnoid hemorrhage, or meningitis. *Magnetic resonance angiography* (MRA) is useful for determining the patency of vessels of the cerebral arterial circle. Acute occlusions of major vessels in this circle and the basilar artery are detectable, but occlusion of small branches is not easily observed (Chan et al., 1995).

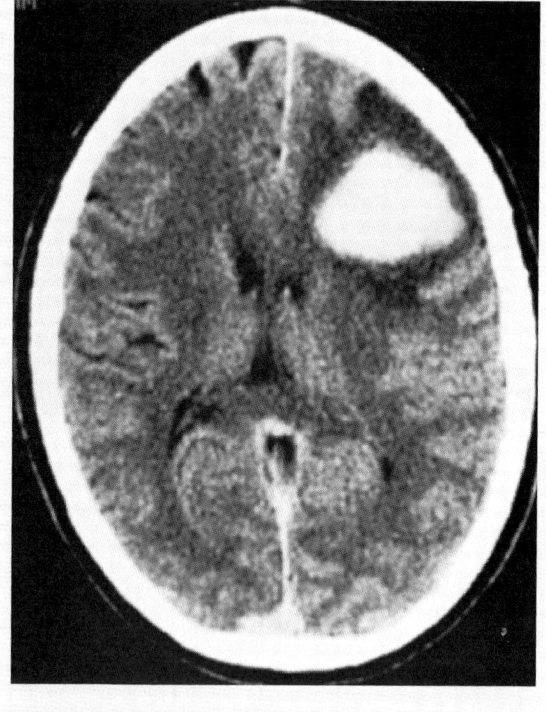

Figure 7.89. Traumatic hemorrhage, frontal lobe. This axial noncontrast CT scan demonstrates left frontal lobe density (hemorrhage), surrounding lucency (edema), and a mass effect (sulcal and ventricular effacement). (Courtesy of Drs. S.K. Hilal and J.A. Bello)

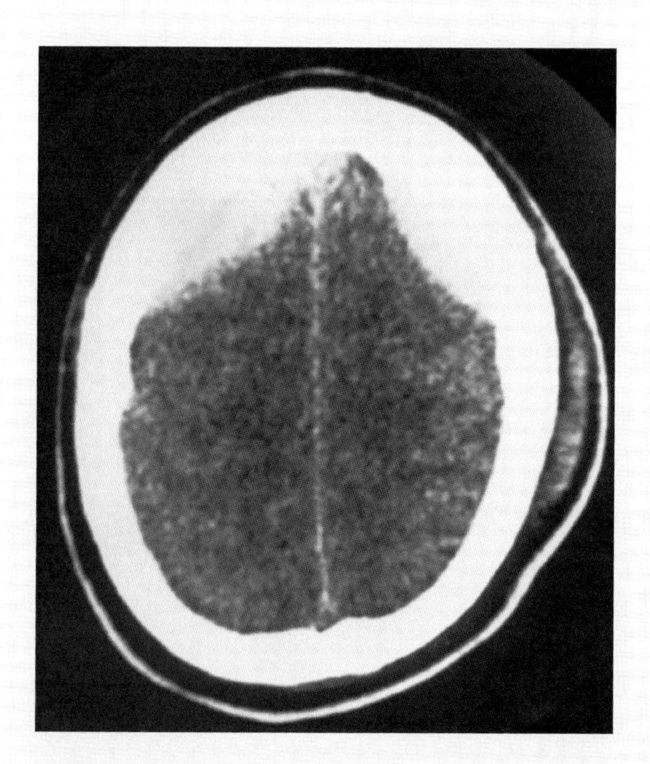

Figure 7.88. Cerebral trauma. The transverse (axial) CT scan shows bilateral acute extradural (epidural) hematomas. An extracranial soft tissue swelling is shown on the left.

Ultrasonography

Ultrasonography is useful in evaluating subependymal and intraventricular hemorrhage in high-risk premature infants and in following them for possible later development of hydrocephalus. Ultrasonography requires minimal manipulation of critically ill infants. ○

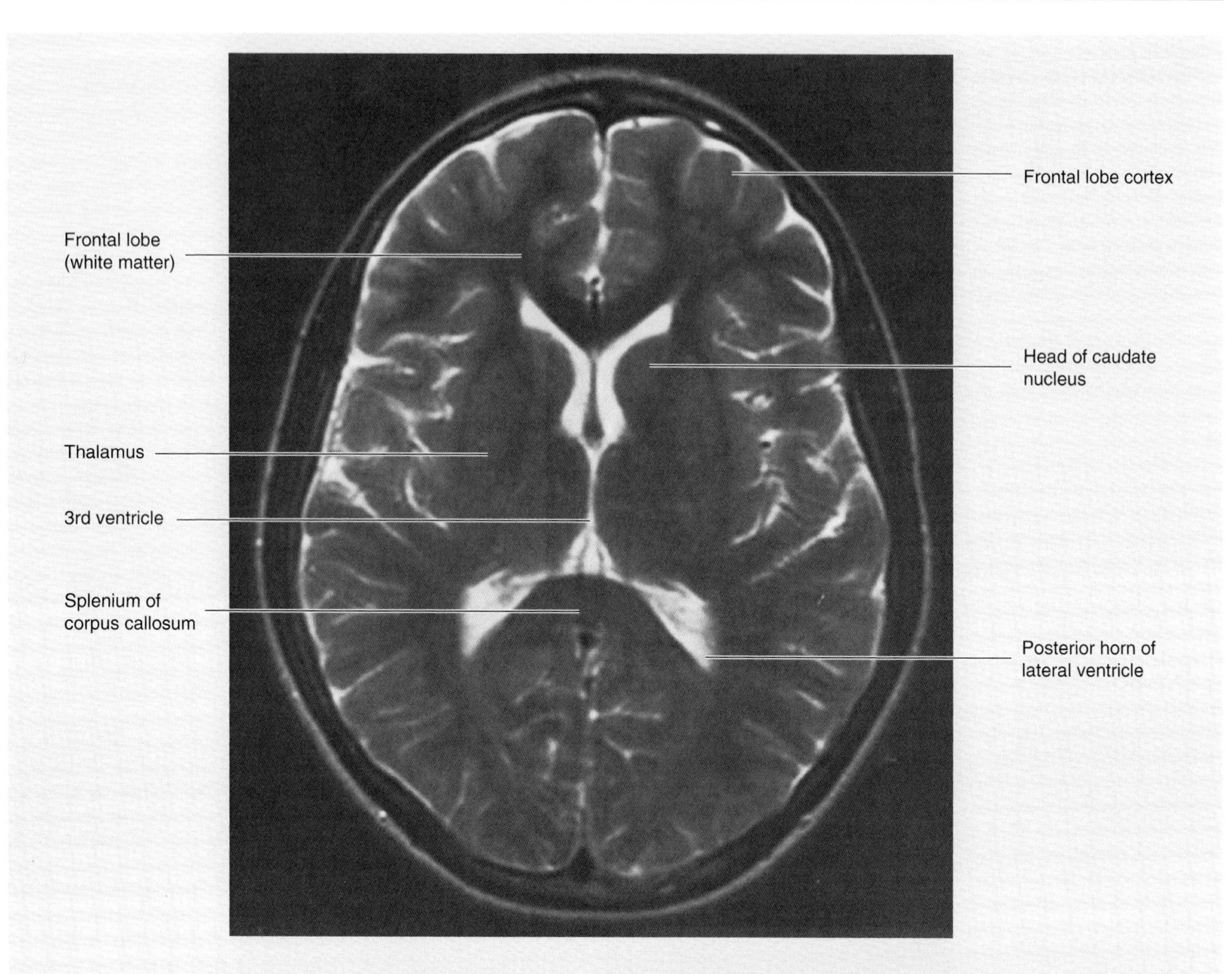

Figure 7.90. Transverse MRI of the brain. Examine the thalamus, caudate nucleus, and parts of the ventricular system—the 3rd ventricle, for example.

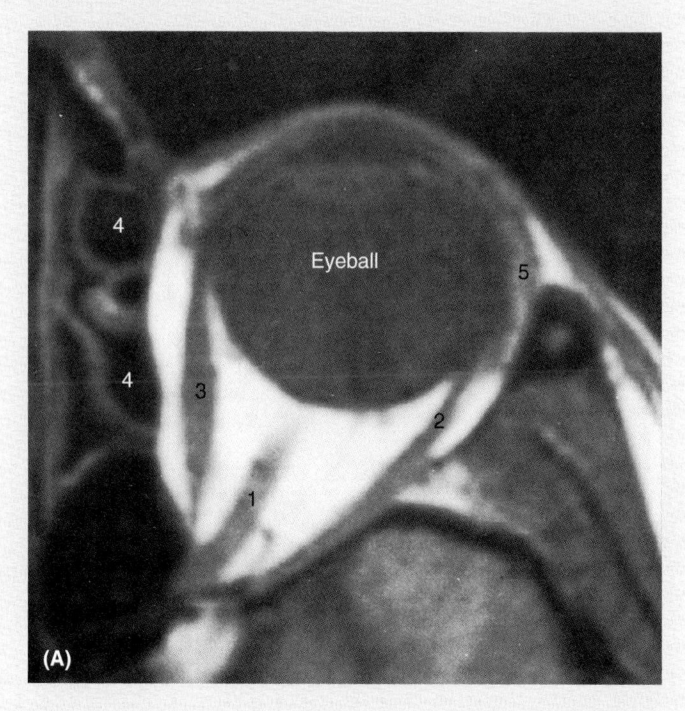

Figure 7.91. MRIs of the orbit and temporomandibular joint (TMJ). A. Transverse MRI of the orbit. *1*, optic nerve; *2*, lateral rectus muscle; *3*, medial rectus muscle; *4*, ethmoidal cells; *5*, lacrimal gland.

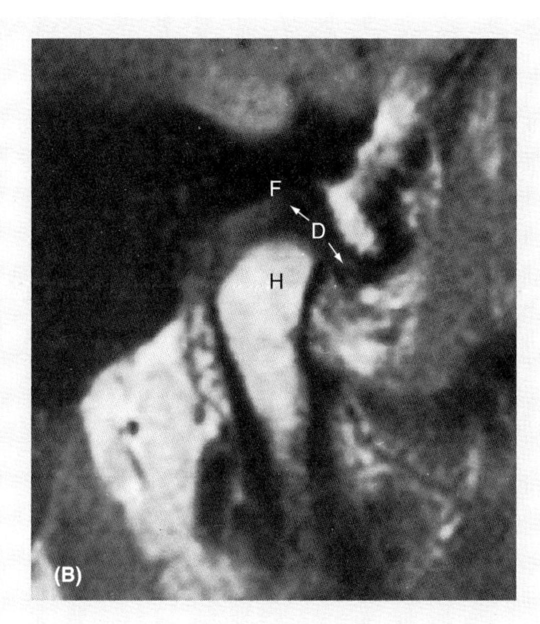

Figure 7.91. *(Continued)* **B.** Sagittal MRI of the TMJ with the mouth closed. Note the position of the articular disc (*D*) in relation to the head of the mandible (*H*) and mandibular fossa (*F*) of the temporal bone. (Courtesy of Dr. W. Kucharczyk, Chair of Diagnostic Imaging, University of Toronto, and Clinical Director of Tri-Hospital Resonance Centre, Toronto, Ontario, Canada)

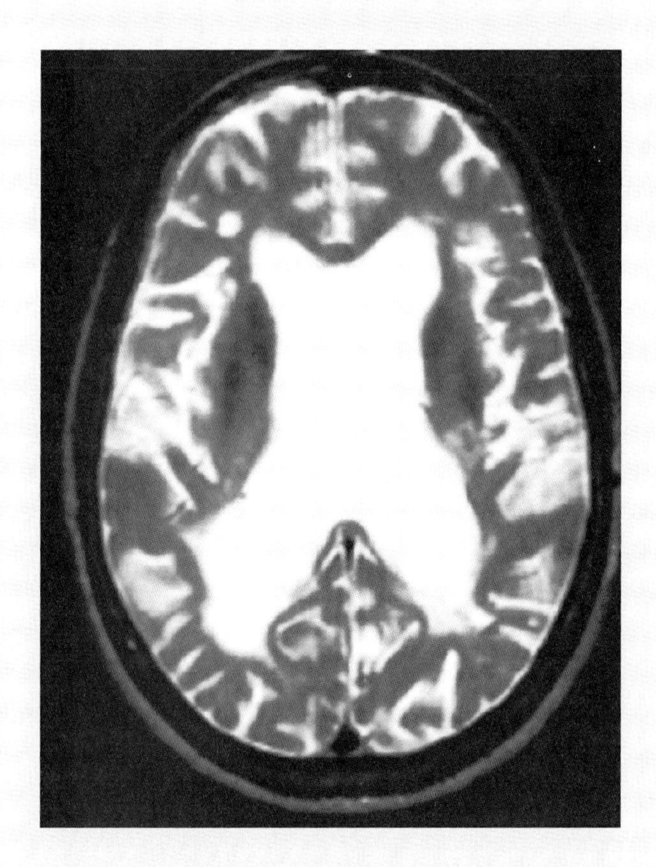

Figure 7.92. Normal pressure hydrocephalus. This T_2-weighted axial MRI scan shows dilated lateral ventricles.

CASE STUDIES

Case 7.1

During batting practice, a ball was foul-tipped and struck the left side of the head of a player standing nearby. He fell to the ground and was unconscious for more than 3 minutes. The initial examination by the trainer revealed that the skin was not broken, but there was a swelling in the temporal fossa. The player complained of an intense headache, disorientation, and blurred vision. His left pupil was moderately dilated and reacted sluggishly to light.

Clinicoanatomical Problems

- Which of the signs mentioned above indicate a possible cranial fracture and extradural (epidural) hematoma?
- What arterial branch was most likely torn? Where is it located? What type of cranial fracture is likely present? Where would the blood accumulate?
- If you were present at the time of injury and noted the above signs, what would you do?
- How do you think the neurosurgeon might remove the hematoma?

The problems are discussed on page 988.

Case 7.2

A young man was playing in a "pick-up" hockey game when he was knocked down. He was not wearing a helmet and hit his head hard on the ice. He was momentarily stunned and said that he "saw stars." The man's vision was blurred for approximately 20 seconds. He skated to the bench and showed no other signs of injury except that he complained of a lingering headache.

Clinicoanatomical Problems

- Do you think the person would have a fractured calvaria? Explain your answer.
- What may the lingering headache indicate?
- If you detected clear fluid dripping from the person's nose, what do you suspect might be the source of the fluid?

The problems are discussed on page 988.

Case 7.3

A hockey player was struck viciously on the lower face by an elbow during a scuffle in the corner of the rink. Profuse bleeding from the mouth was obvious and he was unable to close his jaws normally.

Clinicoanatomical Problems

- What bone may have been fractured?
- A loss of integrity of the mandible results in a change of bite. What is this condition called? What results from this deformity?
- What other structures may be fractured? Discuss these injuries.

The problems are discussed on page 989.

Case 7.4

A baseball batter was struck in the superolateral part of the right cheek by a fast ball. His cheek appeared flat and depressed. There was soon swelling and ecchymosis around the eye. The player complained of dizziness, double vision in the right eye, and numbness of the cheek.

Clinicoanatomical Problems

- What bone was most likely fractured?
- What other bones may have been fractured?
- What is the most common fracture of the upper cheek?
- What symptom suggests that the eye and orbit may be injured?

The problems are discussed on page 989.

Case 7.5

A shortstop fielding a ground ball was hit on the side of the nose when the ball bounced unexpectedly. The nose was deformed and the nasal bones displaced. Disruption of the nasal cartilages was also detected. *Epistaxis* was present, blood was spurting from his nose, and his nasal airway was obstructed.

Clinicoanatomical Problems

- Are nasal fractures common in body-contact sports?
- What is epistaxis?
- What causes spurting of blood from the nose?
- What vessels are usually torn?
- What causes obstruction to the nasal airway?
- If the nasal fracture extends into the cranium, what may be the result?

The problems are discussed on page 989.

Case 7.6

A 16-year-old youth was referred to a dermatologist for treatment of a severe case of acne (*acne vulgaris*). The physician observed an abscess (boil) on the ala of the youth's nose that had developed a yellow "head" at its apex. She treated the youth with antibiotics but warned him not to pick or squeeze the boil because it might cause the infection to spread to the meninges (meningitis) and brain (*encephalitis*).

Clinicoanatomical Problems

- Describe the danger triangle of the face.
- Explain anatomically how an infection could spread from the nose to the meninges and brain.
- Discuss the possible results of a meningeal infection.

The problems are discussed on page 989.

Case 7.7

A figure skater fell on the ice, hitting the back of her head hard on the ice. She did not lose consciousness but was slightly confused and complained of temporary dizziness. Her pupils appeared normal. The fist-sized swelling on the back of her head was tense on palpation. The physician who examined her said she probably had a hematoma of the scalp, but would monitor her condition for several hours.

Clinicoanatomical Problems

- Where in the scalp would the hematoma likely be located?
- What limits the spread of a superficial hematoma of the scalp?
- What initial treatment would likely be given?

The problems are discussed on page 989.

Case 7.8

A young woman hit her head on the dashboard of her car during a head-on collision. The frontal area of her scalp was lacerated and bleeding profusely. The wound was cleansed with a saline solution and covered with a sterile bandage. By the time the woman reached the hospital, she had two black eyes. Further examination revealed that there was no injury to her eyes.

Clinicoanatomical Problems

- How could the scalp wound bleeding be controlled?
- What is the anatomical basis for this procedure?
- How would the blood pass to both eyes when there was no injury to the orbital regions?

The problems are discussed on page 989.

Case 7.9

A 58-year-old man complained to his physician about a swelling of his cheek anterior to his ear lobule. He said that it had been growing rapidly for approximately 2 months and that this part of his face felt weak. He also experienced difficulty when he tried to whistle. The physical examination and subsequent pathological studies revealed a carcinoma of the parotid gland.

Clinicoanatomical Problems

- How could this tumor cause weakness of the face and make it difficult for the man to whistle?
- Where would tumor cells from this gland metastasize?
- Is the facial paralysis likely to be permanent?

The problems are discussed on page 989.

Case 7.10

An elderly man complained to his physician about a pea-sized swelling posterior to his auricle. Physical examination of the ear and scalp revealed an infected sebaceous cyst in the temporal region.

Clinicoanatomical Problems

- How could an infection in the temporal region cause a swelling posterior to the auricle?
- An infection of what other cells could produce a swelling posterior to the ear?

The problems are discussed on page 989.

Case 7.11

A young woman who was recovering from a severe cold and a "stuffed-up" nose complained to her physician about pain in her upper jaw. She said she had been to her dentist but he told her that her teeth were not infected.

Clinicoanatomical Problems

- Infection of what structures in her upper jaw could cause pain in this area of her cheek?
- Infection of what dental structures could cause this type of facial pain?

The problems are discussed on page 990.

Case 7.12

A 63-year-old man was hit by a motorcycle when he was crossing the street. His main complaint was his bleeding head. He was taken to the emergency department of a hospital where a deep wound of the scalp was observed and treated.

Clinicoanatomical Problems

- Why does the scalp bleed so profusely?
- Why does a deep wound of the scalp always require stitches?
- Why are deep scalp wounds potentially dangerous?

The problems are discussed on page 990.

Case 7.13

A 26-year-old man was seen in the emergency department after being hit in the eye with a pool cue during a brawl. Examination of the eye showed no serious injury except for bleeding into the anterior chamber. The physician was concerned about the condition of the eye and referred the patient to an ophthalmologist.

Clinicoanatomical Problems

- What type of orbital fracture may have occurred?
- How may a nonpenetrating blow to the eye cause serious eye problems?

The problems are discussed on page 990.

Case 7.14

A young boy was taken to a pediatrician because of a severe earache. An otoscopic examination revealed a bulging, inflamed tympanic membrane. His mother told the physician that the boy was recovering from a severe cold and throat infection.

Clinicoanatomical Problems

- Where do you think the ear infection was located?
- What is this type of inflammation called?
- If not adequately treated, where could the ear infection spread?
- How could an infection in the throat cause inflammation and bulging of the tympanic membrane?

The problems are discussed on page 990.

Case 7.15

A 27-year-old woman involved in a motorcycle accident was taken to the emergency department of a hospital. She had facial lacerations but no obvious fractures. An eye examination revealed a medial (internal) strabismus of her right eye.

Clinicoanatomical Problems

- What is the meaning of strabismus?
- Injury to which cranial nerve would cause this eye abnormality?
- Which muscle was paralyzed?

The problems are discussed on page 990.

Case 7.16

The wife of a 45-year-old commercial traveler was awakened by the unusual nature of her husband's snoring and was puzzled when she noticed that he was sleeping with his left eye open. In the morning she observed that the left side of his face was drooping. When he tried to examine his teeth, he found that his lips were also paralyzed on that side. He was unable to whistle or to puff out his cheek because the air blew out through his lips on the left side. He also found that he was unable to raise his eyebrow or to frown on that side. During breakfast he had trouble chewing his food because it dribbled out of the left side of his mouth. Fearing that he may have had a mild stroke during the night, he made an appointment to see his physician.

Physical Examination During her examination, the physician made the following observations: at rest, the left side of his face appeared flattened and expressionless; there were no lines on the left side of his forehead; there was sagging in the left lower half of his face, and saliva drooled from the left corner of his mouth. The patient also had a loss of taste sensation on the anterior two-thirds of the left side of his tongue and an absence of voluntary control of the left facial and platysma muscles. When he smiled, the lower portion of his face was pulled to the normal side and the right corner of his mouth was raised, but the left corner was not. During questioning the patient related how he had driven home late the night before and, because of drowsiness, had rolled the window down part way. He also recalled that he had had a severe head cold and an ear infection a few days previously. He recalled that the physician who treated him said that his illness resulted from a viral infection.

Diagnosis Bell's palsy.

Clinicoanatomical Problems

- Paralysis of what nerve would produce the signs exhibited by this patient?
- Why did his left eye remain open even when he was sleeping?
- Why was there loss of taste sensation on the anterior two-thirds of the left side of his tongue?
- Where would the lesion of the nerve probably be located?
- Is the paralysis permanent?

The problems are discussed on page 990.

Case 7.17

A 62-year-old man complained to his dentist about sudden short bouts of excruciating pain on the left side of his face. They were of approximately 2 months' duration and had been increasing in severity. Following examination, the dentist informed him that there was no dental cause for the pain. He stated that the disorder was probably neurological and that he should see a physician.

Physical Examination The man told the physician that the stabbing pains, lasting 15 to 20 seconds, occurred several times a day and were so severe that he had once contemplated suicide. He said that the onset of pain seemed to be triggered by chewing or a cold wind blowing on his upper lip. When the physician asked him to point out the area where the pains occurred, he pointed to his left upper lip and cheek. He said the pain also radiated to his lower eyelid, the lateral side of his nose, and the inside of his mouth. The physician applied firm, steady pressure over the patient's left cheek and over his infraorbital area. He detected no tenderness indicative of inflammation of the maxillary sinus. On further evaluation, the physician detected acute sensitivity to touch (*hyperesthesia*) on the left upper lip and to pin-pricking over the entire left maxillary region, but found no abnormality of sensation in the forehead or mandibular regions.

Diagnosis Tic douloureux.

Clinicoanatomical Problems

- Which branch of what major nerve supplies the area of skin and mucous membrane where the *paroxysms* (sudden recurring attacks) of stabbing pain were felt?
- Where does this nerve leave the skull?
- What are its branches and how are they distributed?

The problems are discussed on page 991.

Case 7.18

A 55-year-old farmer complained to his physician about a sore that had been on his lower lip for 6 months. He stated that he first thought it was a cold sore but he became worried because this one looked different and was getting larger.

Physical Examination On examination, the physician observed that the patient's ulcerated, indurated (L. indurare, to harden) lesion was on the central part of his lower lip. The man's face was darkly tanned. Systematic palpation of the patient's lymph nodes revealed enlarged, *hard submental lymph nodes*. None of the submandibular or deep cervical lymph nodes was enlarged. He took a biopsy of the labial lesion.

Diagnosis Examination of a small biopsy from the edge of the lesion revealed a *squamous cell carcinoma* (malignant tumor of epithelial origin).

Clinicoanatomical Problems

- Into which lymph nodes does lymph from the lower lip drain?
- Between the bellies of which muscles do these nodes lie?
- Into what structures, in addition to the central part of the lip, do afferent lymphatic vessels of these nodes drain?
- To which lymph nodes do lymph vessels from the lateral part of the lip pass?
- If the cancer had spread from the submental lymph nodes, where would you expect to find metastases?

The problems are discussed on page 991.

Case 7.19

A 22-year-old medical student was struck by a puck on the left temple during an interfaculty hockey game. He fell to the ice unconscious but regained consciousness in approximately 1 minute. A laceration approximately 3 cm superior to his left zygomatic arch was bleeding. The gash extended from the top of his auricle almost to his eyebrow. As you helped him to the bench, he said that he felt weak and unsteady. Realizing that he may have sustained a skull fracture, you asked a classmate to call a physician while you took him to the dressing room. During further examination, you found that the deep tendon reflexes in his upper and lower limbs were equal. His pupils were equal in size and contracted to light. In approximately half an hour, your friend said that he was sleepy and wanted to lie down. His left pupil was now moderately dilated and reacted sluggishly to light. By the time the physician arrived, he was unconscious.

Physical Examination The physician observed that the student's pupil on the left was widely dilated and did not respond to light, whereas the pupil on the right was slightly dilated but showed a normal reaction to light. The physician said, "We must get him to the hospital right away!"

Radiology Examination In the hospital, several skull radiographs and a CT scan of his head were taken. As the physician was almost certain of an intracranial hemorrhage, she called a neurosurgeon. When the specialist arrived, the radiologist reviewed the radiographs and CT images with him.

Diagnosis Compressed fracture of the temporal squama, posterior to the pterion, and an extradural hematoma (Fig. 7.93).

Clinicoanatomical Problems

- Define the region known as the temple.
- Delineate the cranial area known as the pterion.
- In what part of the temporal fossa is it located?
- Why is the pterion clinically important?
- What artery was most likely torn?
- What other vessel may have been torn?
- Where would the blood collect?

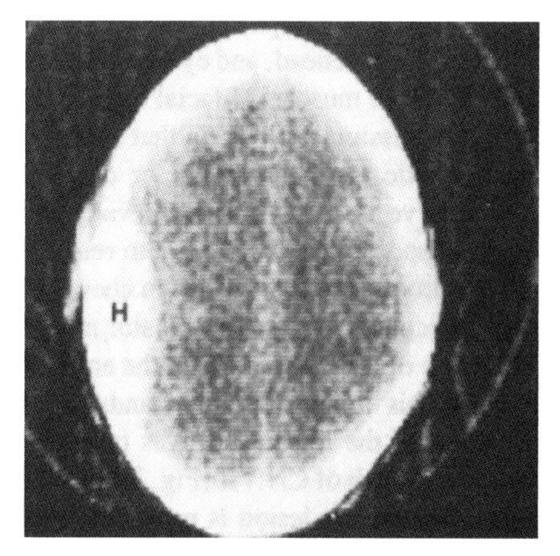

Figure 7.93. Transverse (axial) CT scan of the head. Observe the extradural (epidural) hematoma (*H*) in the left middle cranial fossa.

- Differentiate between an extradural and a subdural hemorrhage.
- What effect will an extradural hematoma likely have on the brain?

The problems are discussed on page 991.

Case 7.20

A 49-year-old woman developed a throbbing headache that lasted for approximately 30 minutes and then slowly faded away. Similar headaches occurred occasionally for the next week. One day as she was lifting a heavy chair, she experienced a sudden, severe headache that was accompanied by nausea, vomiting, and a general feeling of weakness. She decided to see her physician immediately.

Physical Examination The physician detected *nuchal rigidity* (neck stiffness) and an elevation in blood pressure. Visualization of the optic fundus through an ophthalmoscope showed *subhyaloid hemorrhages* (bleeding between the retina and vitreous body). Her deep tendon reflexes were symmetrical and all modalities of sensation were normal. On the basis of these distinct signs and symptoms, the physician made a tentative diagnosis of subarachnoid hemorrhage. He requested radiographic studies and lumbar puncture.

Radiology Report The arteriograms and CT scans show a large, saccular aneurysm of the anterior communicating artery. The CSF is grossly bloody; after centrifugation, the supernatant fluid was xanthochromatic (yellow-colored).

Diagnosis Ruptured aneurysm of the anterior communicating artery and subarachnoid hemorrhage.

Clinicoanatomical Problems

- Where would blood from the ruptured aneurysm most likely go?
- How do you explain anatomically the formation of subhyaloid hemorrhages?
- Why was the supernatant part of the CSF xanthochromatic?
- Where do most single aneurysms occur?

The problems are discussed on page 992.

Case 7.21.

A 23-year-old man went to a dentist to have a badly decayed mandibular 3rd molar (wisdom) tooth extracted. The dentist explained he would likely experience considerable pain with removal of the tooth and informed the patient that he was going to inject a "local" anesthetic to desensitize the tooth and associated soft tissues. When agreeing to the extraction, the patient requested that plenty of anesthetic be given because he was extremely sensitive to pain. The dentist inserted the needle through the mucous membrane on the inside of the patient's mouth, where the needle came to rest near the *lingula*, a bony projection on the medial surface of the ramus of the mandible. In a few minutes the patient stated that his gum, lip, chin, and tongue on the affected side were numb (anesthetized). During the extraction procedure the patient said he felt pain; the dentist injected more anesthetic. The tooth was removed without further incident.

As the patient was preparing to leave, he happened to look in the mirror. He was surprised to find that he was unable to close his eye and lips on the affected side and that his mouth sagged on this side, particularly when he attempted to expose his teeth. He also noted that his ear lobule was numb. When he reported these unusual symptoms, the dentist explained that because of the large amount of anesthetic injected, other nerves in addition to those supplying the teeth had been anesthetized. He assured the patient that these effects would disappear in 3 to 4 hours.

Clinicoanatomical Problems

- Name the nerve supplying the mandibular molar and premolar teeth.
- Why was the patient's chin, lower lip, and tongue on the injected side also anesthetized?
- When anesthetizing this nerve, what other nerves might be affected?
- What probably caused the patient's facial paralysis and loss of sensation in his ear lobule?

The problems are discussed on page 992.

DISCUSSION OF CASES

Case 7.1

The loss of consciousness for more than 3 minutes as the result of a blow to the side of the head indicates that a cranial fracture and extradural hematoma may be present. The gradual dilation of the left pupil and the mental confusion also suggests that a hematoma is present and enlarging. The anterior branch of the middle meningeal artery was most likely torn. This branch lies deep to the pterion and is frequently torn by fracture of the bones forming this bony landmark. The injury to the skull probably was a depressed fracture of the calvaria, in which the blood would accumulate between the dura and calvaria. If you were present and observed the signs described, you would have called an ambulance immediately. The neurosurgeon would likely operate immediately to decompress (surgically evacuate) the hematoma and control the bleeding from the meningeal vessels.

Case 7.2

The young man probably does not have a fractured calvaria because he did not lose consciousness. The fall on his head resulted in the slight alterations in neurological function: "seeing stars" and blurred vision. The lingering headache may indicate increasing intracranial pressure resulting from damage to the brain (e.g., contusion of the cerebral cortex). A fracture of the cribriform plate of the ethmoid may tear the meninges and result with loss of CSF through the nose (CSF rhinorrhea).

Case 7.3

Likely the player's lower jaw was fractured, which resulted in his teeth not fitting together (occluding) as they normally did. Fracture of the mandible is common in sports such as hockey, football, and rugby. Often two fractures, or a fracture-dislocation of the TMJ, occur in the jaw. The most common fracture site is near the angle of the mandible. When the integrity of the jaw is lost, the teeth do not fit together normally—a condition called *malocclusion*. This leads to changes in speech patterns because the articulation of words is difficult (dental sounds, such as the "S" sound, may not be producible in the normal way or may be inadvertently produced, and the required movements of the jaw may be difficult and painful). *Fractures of the teeth* may also occur when the jaws are hit by a hard blow. The fracture may occur through the enamel, pulp, or root of the tooth. Fractures exposing the pulp of the tooth cause severe pain and sensitivity to heat and cold. Root fractures cause mobility of the teeth.

Case 7.4

The zygomatic bone (or its attachment to the surrounding bones, or the surrounding bones themselves) was most likely fractured. The bones forming the orbit may also have fractured. Fractures of the zygomatic bone (zygoma) are the most common fractures of the upper cheek, the most serious of which is the "tripod" fracture of the zygomatic bone involving three separate breaks of the bones of the skull through the:

- Infraorbital foramen and canal to the infraorbital groove (a fracture of the maxilla)
- Zygomaticoparietal suture of the lateral margin of the orbit
- Zygomatic arch, usually at its narrowest point, where the suture between the zygomatic process of the temporal bone and the temporal process of the zygomatic bone occurs.

The *diplopia* suggests that the eyeball and orbit may have been injured. Diplopia indicates malalignment of the visual axes. Temporary blurred vision may be a complication of zygomatic fractures.

Case 7.5

Fractures of the nasal bones are the most common injuries in sports in which players do not wear protective face guards (e.g., baseball and boxing). *Epistaxis* refers to nasal hemorrhage (profuse bleeding from the nose). The bleeding results from the richness of the blood supply to the nasal mucosa. *Spurting of blood from the nose* results from ruptured arteries, particularly at the site of anastomosis of the sphenopalatine and greater palatine arteries on the nasal septum (Kiesselbach's area). The nasal airway is often obstructed with a nasal fracture because bone fragments can block the nose. if the fracture extends into the floor of the anterior cranial fossa, the cribriform plate may fracture and the meninges may tear, resulting in *CSF rhinorrhea*—discharge of CSF through the nose.

Case 7.6

The *danger triangle of the face* has its base at the vermillion border of the upper lip and its apex at the bridge of the nose. Inflammation of the facial vein associated with thrombus formation may result in the spread of infection from the facial vein through the superior and inferior ophthalmic veins to the meninges and the cavernous sinus (lateral sellar compartment). *Cavernous sinus thrombosis* may result in meningitis, thrombophlebitis of cortical veins, and *cerebral edema*, an increase in brain volume resulting from an increase in water and sodium content. When severe, brain edema can cause various forms of herniation of the brain and pressure on the brainstem, which may result in failure of respiration and circulation.

Case 7.7

The hematoma was probably located in the thick, richly vascularized subcutaneous layer of connective tissue (layer two of the scalp). A superficial hematoma of the scalp is limited by the fibrous tissue that binds the skin to the epicranial aponeurosis. The initial treatment of the hematoma was likely the application of an icepack to control the bleeding and swelling.

Case 7.8

The bleeding from a deep scalp wound can be stopped by direct pressure with sterile gauze. Once the bleeding has stopped, a circular dressing can be applied for transportation to the hospital for further treatment (e.g., suturing of the epicranial aponeurosis). Because the arteries enter the scalp inferiorly, this headband-type of bandage will compress the arteries. The skin of the scalp is continuous with that covering the forehead. Hence, bleeding in the loose connective layer can pass to the face, especially around the eyes, because this 4th layer of the scalp is somewhat like a sponge and may distend with blood from an injury. As the blood accumulates, it gravitates to the face and collects around the eyes.

Case 7.9

The facial nerve leaves the skull through the stylomastoid foramen and almost immediately enters the parotid gland. Its branches spread out like the abducted fingers of the hand (p. 863). Consequently, a malignant parotid tumor invades the facial nerve and interferes with its supply to the facial muscles, including those used to whistle. Tumor cells from the carcinoma would metastasize to the deep cervical lymph nodes. These nodes form a chain along the course of the IJV from the skull to the root of the neck. The facial paralysis is likely permanent because severely injured cranial nerves do not regenerate.

Case 7.10

An infected sebaceous cyst in the temporal region could spread through the lymphatics to the mastoid lymph nodes located posterior to the auricle. Inadequate treatment of a middle ear infection could spread through the mastoid antrum to the mastoid cells (*acute mastoiditis*) and produce a swelling posterior to the auricle.

Case 7.11

Infection of the maxillary sinuses—*maxillary sinusitis*—is a common complication of nasal infections. Infection of these sinuses is common because their ostia are near the roofs of the paranasal sinuses. Hence, the sinuses must be full before they will drain when the person is standing. An infection from an abscess in a maxillary molar could spread to the maxillary sinus because the apex of its root is in the thin floor of the maxillary sinus, and the infection could pass to the sinus.

Case 7.12

The arteries of the scalp are attached to the fibrous septa that bind the skin to the epicranial aponeurosis. A severed artery is therefore unable to contract to slow the circulation and allow clotting to occur. A deep wound of the scalp involving the epicranial aponeurosis gapes widely because of the pull of the anterior and posterior parts of the occipitofrontalis muscle. Therefore, stitches must be inserted in the aponeurosis to close the gap. A severe infection of the scalp may spread to the meninges and brain—because the diploic veins of the calvaria are connected with the dural venous sinuses—possibly causing infection of the bones of the calvaria (*osteomyelitis*) and thrombosis of the dural sinuses, which could in turn cause cerebral edema and death.

Case 7.13

A blow-out fracture of the orbit may have occurred as the result of the sudden pressure increase in the orbit caused by the blow from the pool cue. A nonpenetrating blow to the eye may produce herniation of the orbital contents inferiorly through a blow-out fracture in the thin floor of the orbit into the maxillary sinus.

Case 7.14

An infection in the middle ear was most likely the cause of the inflamed tympanic membrane. If not adequately treated, a middle ear infection could spread through the mastoid antrum to the mastoid cells and produce *mastoiditis*. Likely, the pathogenic organisms passed from the nasopharynx through the pharyngotympanic tube into the middle ear.

Case 7.15

Strabismus is a deviation of the eye that a person cannot overcome. In this case, the right abducent nerve was injured, causing paralysis of the right lateral rectus muscle. The patient cannot turn her right eye laterally because when the lateral rectus is paralyzed the medial rectus pulls the eyeball medially (medial strabismus).

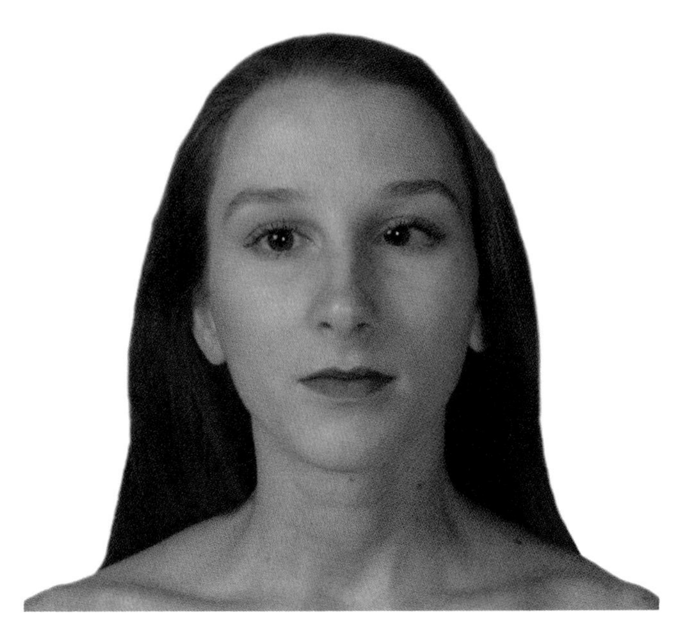

Abducent nerve injury

Case 7.16

Paralysis of the facial muscles may occur without any known cause. Bell's palsy (facial paralysis) often follows exposure to cold as occurs when riding in an open car or in one with the window open. Bell's palsy occurs at all ages but is slightly more common in the 3rd through 5th decades.

The characteristic facial appearance results from a *lesion of the facial nerve*. In this patient, the motor supply to the muscles of the left face, forehead, and eyelids were most severely affected. Paralysis of the muscles of facial expression on the left side explains the expressionless look on that side of his face and his inability to whistle, puff his cheek, or close his left eye. When the facial nerve is paralyzed, the levator palpebrae superioris (acting unopposed) causes the eye to remain open, even during sleep. The drooling and difficulty in chewing result from paralysis of the orbicularis oris and buccinator muscles. Loss of taste sensation on the anterior two-thirds of the left side of his tongue is understandable anatomically because this region of the tongue receives taste fibers through the chorda tympani branch of CN VII. This symptom also indicates that the nerve lesion is proximal to the origin of the chorda tympani nerve in the facial canal.

Because of paralysis of the orbicularis oculi, the lacrimal puncta and often the entire lower lid are no longer in contact with the cornea. As a result, tears tend to flow over the left lower eyelid onto the cheek. The cornea may dry out during sleep (if an ointment is not used) because the eyelids on the affected side remain open. Drying of the cornea can also occur during the day owing to the inability to blink; this dryness could result in *corneal ulceration*.

The site of the lesion was most likely in the facial canal in the petrous part of the temporal bone. The paralysis of the facial muscles is thought to be caused by inflammation of the facial nerve superior to the stylomastoid foramen. The cause of the inflammation is generally thought to be a viral infection, which causes edema of the facial nerve and compression of its fibers in the facial canal. If the lesion is complete, all facial muscles on that side are affected equally. Voluntary, emotional, and associated movements are all affected. In most cases the nerve fibers are not permanently damaged and nerve degeneration is incomplete. As a result, recovery is slow but generally good. Some facial asymmetry may persist (e.g., slight sagging of the corner of the mouth).

Case 7.17

The area of skin and mucosa where the stabbing pain was felt is supplied by the maxillary nerve, the 2nd division of the trigeminal nerve (CN V_2). This wholly sensory nerve leaves the skull through the foramen rotundum. At its termination as the infraorbital nerve, it gives rise to branches that supply the ala of the nose, the lower eyelid, and the skin and mucous membrane of the cheek and upper lip. Branches of the maxillary nerve also innervate the maxillary teeth and the mucous membranes of the nasal cavities, palate, mouth, and tongue.

The symptoms described by the patient are characteristic of *trigeminal neuralgia*. This disorder of the sensory division of the trigeminal nerve occurs most often in middle-aged and elderly persons. The pain may be so intense that the patient winces; hence the common term *tic* (twitch). In some cases the pain may be so severe that psychological changes occur leading to depression and even suicide attempts. The maxillary nerve distribution, as in this patient, is most frequently involved, then the mandibular and, least frequently, the ophthalmic.

The paroxysms of sudden stabbing pain are often set off by touching the face, brushing the teeth, drinking, or chewing. There may be an especially sensitive "trigger zone" (the left upper lip in the present case). The cause of trigeminal neuralgia is unknown. Some investigators believe that most patients with idiopathic trigeminal neuralgia have anomalous blood vessels that compress the trigeminal nerve. Other scientists believe the condition is caused by a pathological process affecting neurons in the trigeminal ganglion; still others believe that neurons in the nucleus of the spinal tract may be involved.

Medical or surgical treatment or both are used to alleviate the pain. Only anatomical aspects of these treatments are discussed here. Attempts have been made to block the nerve at the infraorbital foramen by using alcohol; this treatment usually relieves pain temporarily. The simplest surgical procedure is avulsion or cutting of the branches of the nerve at the infraorbital foramen.

Also in use is *radiofrequency selective ablation of parts of the trigeminal ganglion* by a needle electrode passing through the cheek and the foramen ovale. To prevent regeneration of nerve fibers, the sensory root of the trigeminal nerve may be partially cut between the ganglion and the brainstem (*rhizotomy*). Although the axons may regenerate, they do not do so within the brainstem. Surgeons attempt to differentiate and cut only the sensory fibers to the division of the trigeminal nerve involved. The same result may be achieved by sectioning the spinal tract of CN V (*tractotomy*). After this operation, the sensation of pain, temperature, and simple (light) touch is lost over the area of skin and mucous membrane supplied by the maxillary nerve. This may annoy the patient, who does not recognize the presence of food on the lip and cheek or feel it within the mouth on the side of the nerve section, but these disabilities are preferable to excruciating pain.

Case 7.18

Carcinomas of the lip usually involve the lower lip. Overexposure to sunshine over many years, as occurs in outdoor workers such as farmers, is a common feature of the history in these cases. Chronic irritation from pipe smoking appears to be a factor also, which may be related to long-term contact with tobacco tar. The submental lymph nodes lie on the fascia covering the mylohyoid muscle between the anterior bellies of the right and left digastric muscles. The central part of the lip, the floor of the mouth, and the apex of the tongue drain to these nodes, whereas the lateral part of the lip drains to the submandibular lymph nodes.

If cancer cells had spread further, metastases would have developed in the submandibular lymph nodes because efferents from the submental lymph nodes pass to them. In addition, lymphatic vessels from the submental lymph nodes pass directly to the jugulo-omohyoid node. As the submandibular nodes are situated beneath the deep cervical fascia in the submandibular triangle, the patient's chin may have to be lowered to slacken this fascia before the enlarged nodes can be palpated.

Because all parts of the head and neck ultimately drain into the deep cervical nodes, they might also be sites of metastases. As the jugulo-omohyoid node drains the submental and submandibular lymph nodes, it could be involved in the spread of tumor cells from a carcinoma of the lip. It is located where the omohyoid muscle crosses the IJV.

Case 7.19

The temple is the area of the temporal fossa between the temporal lines and the zygomatic arch, where the skull is thin and covered by the temporal muscle and fascia. *The pterion* is a somewhat variable H-shaped area that lies deep to the temporal muscle. Here, four bones (frontal, parietal, temporal, and sphenoid) approach each other or meet. The pterion is an important bony landmark because it indicates the location of the frontal branch of the *middle meningeal artery*. The center of the pterion is approximately 4 cm superior to the zygomatic arch and 3.5 cm posterior to the frontozygomatic suture. It lies in the anterior part of the temporal fossa.

The thin squamous part of the temporal bone is grooved on its internal aspect by the middle meningeal artery and its branches. The thin bones forming the temporal squama are easy to fracture, and the broken pieces may tear the artery and its branches as they pass superiorly on the external surface of the dura mater, especially when the artery is coursing in a bony groove. This situation results in a slow accumulation of blood in the extradural space, forming an *extradural hematoma*. The hematoma forms relatively slowly because the dura is firmly attached to the bones by fibers that, to a certain extent, resist stripping of the dura from the calvaria.

A *subdural hematoma* is a localized mass of extravasated blood close to the surface of the brain in the deep part of the dura. The "subdural space" is a potential space in the deep part of the dura. Injury to the dura as a result of a blow to the head produces a subdural space—as in this case—in which blood collects to form a subdural hematoma.

The middle meningeal artery, a branch of the first part of the maxillary artery, enters the skull through the foramen spinosum. It divides within the first 4 or 5 cm of its intracranial course. The frontal branch passes superiorly from the pterion, more or less parallel to the coronal suture of the skull. The parietal branch passes posterosuperiorly, with its exact site depending on its point of origin. In the present case, the frontal branch of the middle meningeal artery was almost certainly torn. This artery is usually accompanied by a meningeal vein that also may have been torn. The lucid interval that followed the patient's recovery from the brief loss of consciousness (resulting from cerebral concussion) occurs because of the slow formation of the *extradural hematoma*. In addition, this kind of a space-occupying intracranial lesion can be tolerated for a short time because some blood and CSF are squeezed out of the calvaria through the veins. However, because the cranium is nonexpansible, the intracranial pressure soon rises, producing drowsiness and then coma (G. koma, deep sleep).

The increased intracranial pressure forces the supratentorial part of the brain, usually the uncus, through the tentorial notch (incisure), squeezing the oculomotor nerve between the brain and the sharp, free edge of the tentorium. Compression of CN III causes *3rd nerve palsy*, which results in a dilated, nonreacting pupil on the side of the lesion. An extradural hemorrhage in the characteristic position, illustrated by the present case, primarily causes compression of the temporal lobe underlying the pterion. Immediate surgical intervention is necessary to relieve the intracranial pressure and avoid further compression of the brain, which could cause death by interfering with the cardiac and respiratory centers in the medulla.

Case 7.20

Unruptured saccular aneurysms are usually asymptomatic. In the present case, the initial headaches were probably caused by intermittent enlargement of the aneurysm or by slight bleeding from it into the subarachnoid space—the "warning leak." Her subsequent severe, almost unbearable headache was the result of gross bleeding from the aneurysm into the subarachnoid space.

Blood in the CSF causes meningeal irritation and a severe headache. As the anterior communicating artery is in the longitudinal fissure, rostral to the optic chiasm, blood escaping from the ruptured aneurysm would enter the chiasmatic cistern and other subarachnoid spaces around the brain and spinal cord. This explains why blood was found in the CSF obtained during the lumbar puncture.

Some authorities recommend against a lumbar puncture in an obvious case of subarachnoid hemorrhage, such as in this instance, because it may cause herniation of the brain. The lowering of CSF pressure in the spinal subarachnoid space by removing CSF might cause inferior movement of the brain, resulting in herniation (e.g., of part of the cerebellum).

Rupture of an aneurysm of the anterior communicating artery into the adjacent part of one frontal lobe may cause symptoms of a mass lesion in one hemisphere. In some cases, the *intracranial hematoma* may enter the ventricular system causing an acute expansion of the ventricle, which may cause death. Blockage of subarachnoid spaces by large amounts of blood in the CSF could impair circulation of this fluid, resulting in a further increase in intracranial pressure. This pressure could force the medial part of the temporal lobe (usually the uncus) through the tentorial notch and the cerebellar tonsils (parts of the posterior lobe of cerebellum) through the foramen magnum. *Herniation of the cerebellar tonsils* compresses the brainstem containing the vital respiratory and cardiovascular centers, producing a life-threatening situation.

The subhyaloid hemorrhages observed during *funduscopy* resulted from the abrupt rise in intracranial pressure transmitted to the subarachnoid space around the optic nerve, compressing and obstructing the central retinal vein where it crosses this space. This situation results in increased pressure in the retinal capillaries and hemorrhages between the retina and vitreous body. After centrifugation of the CSF, the supernatant fluid was xanthochromatic because it contained serum bilirubin and products of hemolyzed red blood cells. Most single aneurysms are on the anterior circulation, usually on the internal carotid artery (Brust, 1995). The most common site is its junction with the posterior communicating artery. Aneurysms of the anterior cerebral artery are usually on the anterior communicating artery, as in this case, or close to it.

Case 7.21

The inferior alveolar nerve supplies the mandibular molar and premolar teeth, and its branches supply the canine and incisor teeth. The mental nerve supplies the skin of the chin and the lower lip on that side. Hence, the *inferior alveolar nerve* supplies all teeth in one half of the mandible. Anesthetizing this nerve also anesthetizes the chin and lower lip because the mental nerve supplying these structures is a terminal branch of the inferior alveolar nerve. Because the *lingual nerve* descends just anterior to the inferior alveolar nerve near the mandibular foramen, it was also anesthetized. This is advantageous because the lingual nerve supplies sensory fibers to the mandibular gingiva as well as the tongue.

Because a relatively large amount of anesthetic was injected, it must have spread into the parotid gland. Paralysis of the muscles of facial expression resulted from anesthetized branches of the *facial nerve*. Because the parotid gland and these nerves occupy the space around the posterior margin of the ramus of the mandible, they could easily be affected as the anesthetic agent infiltrated the area. The injection was probably made posteriorly so that the anesthetic solution passed through the stylomandibular ligament, a sheet of fascia condensed between the parotid and submandibular glands that is continuous with the fascia covering the parotid gland. Like the anesthetic effects on the teeth and gingivae, the effects on the muscles of facial expression and on mastication disappear in a few hours.

The patient's ear lobule was numb because intermediate branches of the *great auricular nerve* were also anesthetized. The anterior branches of this nerve supply skin on the posteroinferior part of the face, and its intermediate branches supply the inferior part of the auricle on both surfaces.

References and Suggested Readings

Anderson MK, Hall SJ: *Sports Injury Management.* Baltimore, Williams & Wilkins, 1995.

Barr ML, Kiernan JA: *The Human Nervous System: An Anatomical Viewpoint,* 6th ed. Philadelphia, JB Lippincott, 1993.

Behrman RE, Kliegman RM, Arvin AM (eds): *Nelson Textbook of Pediatrics,* 15th ed. Philadelphia, WB Saunders, 1996.

Berg L, Klebanoff LM: Focal infections. *In* Rowland LP (ed): *Merritt's Textbook of Neurology,* 9th ed. Baltimore, Williams & Wilkins, 1995.

Brust JCM: Coma. *In* Rowland LP (ed): *Merritt's Textbook of Neurology,* 9th ed. Baltimore, Williams & Wilkins, 1995.

Chan S, Khandji AG, Hilal SK: CT and MRI. *In* Rowland LP (ed): *Merritt's Textbook of Neurology,* 9th ed. Baltimore, Williams & Wilkins, 1995.

Corbett JJ, Haines DE, Ard MD: The ventricles, choroid plexus, and cerebrospinal fluid. *In* Haines DE (ed): *Fundamental Neuroscience.* New York, Churchill Livingstone, 1997.

Fishman RA: Cerebral veins and sinuses. *In* Rowland LP (ed): *Merritt's Textbook of Neurology,* 9th ed. Baltimore, Williams & Wilkins, 1995a.

Fishman RA: Brain edema and disorders of intracranial pressure. *In* Rowland LP (ed): *Merritt's Textbook of Neurology,* 9th ed. Baltimore, Williams & Wilkins, 1995b.

Ger R, Abrahams P, Olson T: *Essentials of Clinical Anatomy,* 3rd ed. New York, Parthenon Publishing Group, 1996.

Haines DE: On the Question of a Subdural Space. *Anat Rec* 230:3–21, 1991.

Haines DE: *Neuroanatomy. An Atlas of Structures, Sections, and Systems,* 4th ed. Baltimore, Williams & Wilkins, 1995.

Haines DE (ed): *Fundamental Neuroscience.* New York, Churchill Livingstone, 1997.

Haines DE, Harkey HL, Al-Mefty O: The "subdural" space: A new look at an outdated concept. *Neurosurgery* 32:111, 1993.

Lange DJ, Trojaborg W, Rowland LP: Peripheral and cranial nerve lesions. *In* Rowland LP (ed): *Merritt's Textbook of Neurology,* 9th ed. Baltimore, Williams & Wilkins, 1995.

Liebgott B: *The Anatomical Basis of Dentistry.* St. Louis, Mosby, 1986.

Miller JR, Jubelt B: Bacterial infections. *In* Rowland LP (ed): *Merritt's Textbook of Neurology,* 9th ed. Baltimore, Williams & Wilkins, 1995.

Mohr JP, Prohovnik I: Neurovascular imaging. *In* Rowland LP (ed): *Merritt's Textbook of Neurology,* 9th ed. Baltimore, Williams & Wilkins, 1995.

Moore KL, Persaud TVN: *The Developing Human: Clinically Oriented Embryology,* 6th ed. Philadelphia, WB Saunders, 1998.

Mortimer CB, Kraft S: Ophthalmology. *In* Gross A, Gross P, Langer B (eds): *Surgery: A Complete Guide for Patients and Their Friends.* Toronto, Harper & Collins, 1989.

Prockop LD: Disorders of cerebrospinal and brain fluids. *In* Rowland LP (ed): *Merritt's Textbook of Neurology,* 9th ed. Baltimore, Williams & Wilkins, 1995.

Raskin NH: Headache. *In* Rowland LP (ed): *Merritt's Textbook of Neurology,* 9th ed. Baltimore, Williams & Wilkins, 1995.

Rowland LP (ed): *Merritt's Textbook of Neurology,* 9th ed. Baltimore, Williams & Wilkins, 1995.

Sacco RL: Pathogenesis, classification, and epidemiology of cerebrovascular disease. *In* Rowland LP (ed): *Merritt's Textbook of Neurology,* 9th ed. Baltimore, Williams & Wilkins, 1995.

Swartz MH: *Textbook of Physical Diagnosis. History and Examination,* 2nd ed. Philadelphia, WB Saunders, 1994.

Wazen JJ: Ménière syndrome. *In* Rowland LP (ed): *Merritt's Textbook of Neurology,* 9th ed. Baltimore, Williams & Wilkins, 1995.

Williams PL, Bannister LH, Berry MM, Collins P, Dyson M, Dussek JE, Fergusson MWJ (eds): *Gray's Anatomy,* 38th ed. Edinburgh, Churchill Livingstone, 1995.

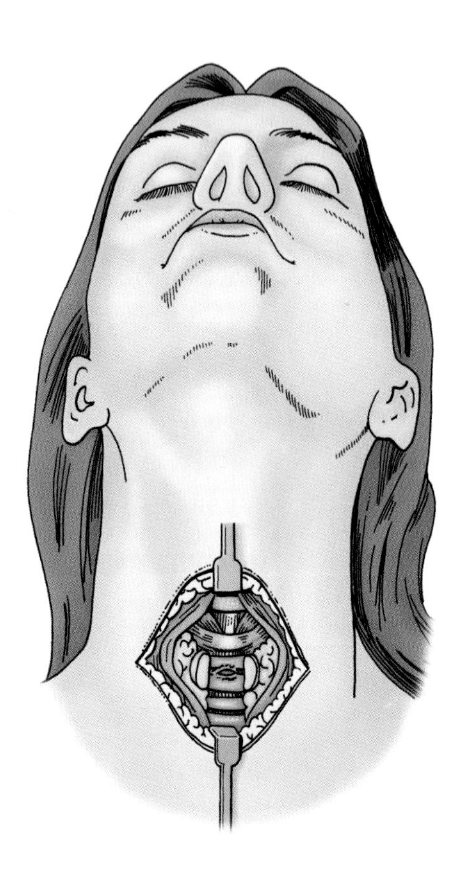

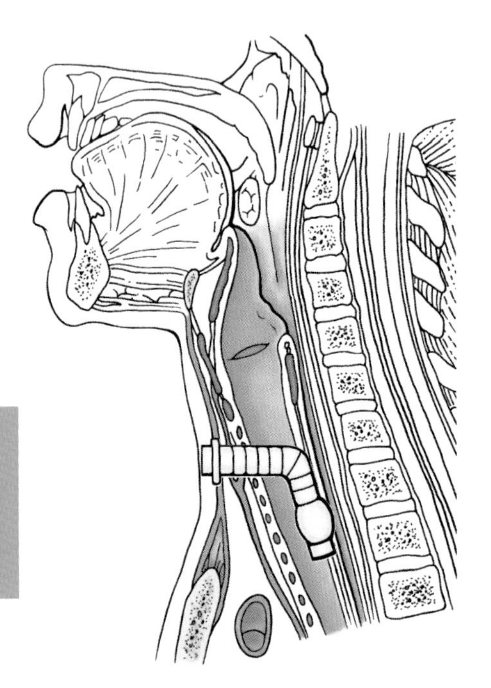

c h a p t e r

8 Neck

The neck is the major conduit between the head, trunk, and limbs. Many important structures are crowded together in the neck, such as muscles, glands, arteries, veins, nerves, lymphatics, trachea, esophagus, and vertebrae. Several structures that are important to life—such as the *thyroid gland, trachea, jugular veins*, and *carotid arteries*—lack the bony protection afforded to other parts of most of the systems to which these structures belong (Fig. 8.1). Blood vessels and nerves are the major structures commonly injured in penetrating wounds of the neck—such as those resulting from stab and gunshot injuries.

Derivatives of Latin terms for the neck, *cervix* and *collum*, are used in anatomical descriptions: the *cervical plexus*, an interjoining of nerves in the neck; *torticollis*, twisted neck; and *longus colli*, a long neck muscle, for example. The superior part of the **brachial plexus** of nerves supplying the upper limb is in the side of the neck. In the center of the neck is the **thyroid cartilage,** the largest of the cartilages of the larynx, and the **trachea**. Laterally is the main arterial bloodflow to the head and neck (e.g., the **common carotid arteries**) and the principal venous drainage (the **jugular veins**). Lymph from structures in the head and neck drain into *cervical lymph nodes*. Most lymph of the body enters the venous system through the **thoracic duct** in the root of the neck.

Cervical Pain

Cervical pain (neck pain) has several causes, including inflamed lymph nodes, muscle strain, and protrusion of intervertebral (IV) discs. As lymphatic vessels in the head drain into *cervical lymph nodes*, their enlargement may indicate a malignant tumor in the head; however, the primary cancer may be in the thorax or abdomen because the neck connects the trunk and head (e.g., bronchogenic [lung] cancer may metastasize spread] through the neck to the skull). "Most chronic neck pain is caused by bony abnormalities (cervical osteoarthritis or other forms of arthritis) or by local trauma" (Thompson and Rowland, 1995). Cervical pain may be affected by movement of the head and neck, and it may be exaggerated during coughing, sneezing, or straining during defecation (natural *Valsalva maneuvers*). ○

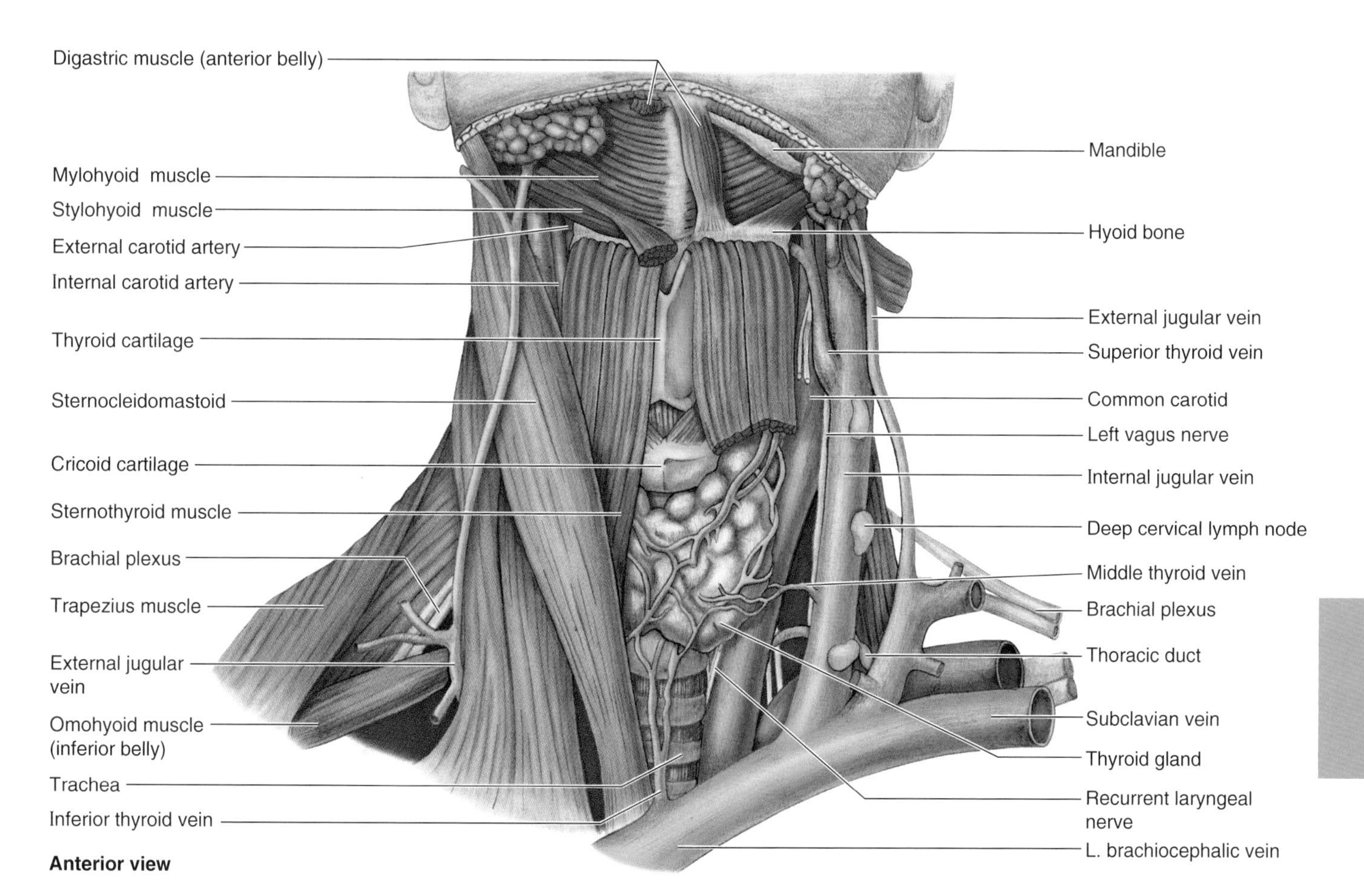

Figure 8.1. Dissection of the anterior neck. Anterior view. The fascia has been removed and the muscles on the left side have been reflected to show the hyoid bone, thyroid gland, and structures related to the carotid sheath (carotid artery, internal jugular vein [IJV], vagus nerve [CN X], and deep cervical lymph nodes).

Bones of the Neck

The *skeleton of the neck* is formed by the cervical vertebrae, hyoid bone, manubrium of the sternum (breast bone), and clavicles (collar bones) (Figs. 8.2 and 8.3). The cervical vertebrae, manubrium of the sternum, and U-shaped hyoid bone are the parts of the axial skeleton,and the clavicles are part of the superior appendicular skeleton.

Cervical Vertebrae

Seven cervical vertebrae form the cervical region of the vertebral column, which encloses the spinal cord and meninges (Fig. 8.2). Cervical vertebrae are described with the back (see Chapter 4); therefore, only a brief review follows.

Typical cervical vertebrae (3rd, 4th, 5th, and 6th) have the following *characteristics*:

- The vertebral body is small and longer from side to side than anteroposteriorly; the superior surface is concave (forming uncinate processes laterally), and the inferior surface is convex.
- The vertebral foramen is large and triangular.
- In the transverse processes of *all* cervical vertebrae (typical or atypical) is a foramen transversarium for the vertebral vessels (e.g., the vertebral vein and, except for C7, vertebral artery).

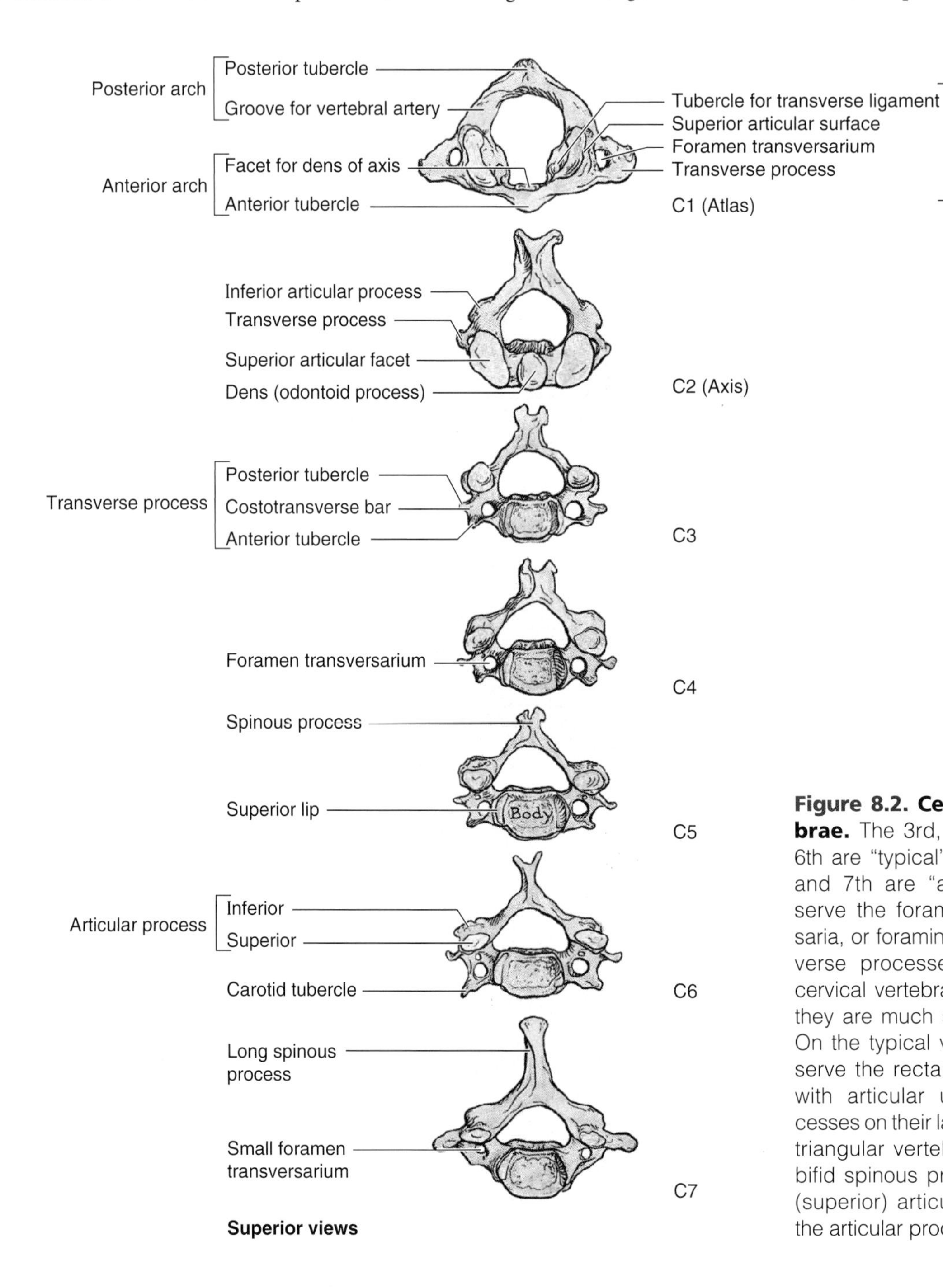

Figure 8.2. Cervical vertebrae. The 3rd, 4th, 5th, and 6th are "typical"; the lst, 2nd, and 7th are "atypical." Observe the foramina transversaria, or foramina of the transverse processes, of all the cervical vertebrae; notice that they are much smaller in C7. On the typical vertebrae, observe the rectangular bodies with articular uncinate processes on their lateral aspects, triangular vertebral foramina, bifid spinous processes, and (superior) articular facets on the articular processes.

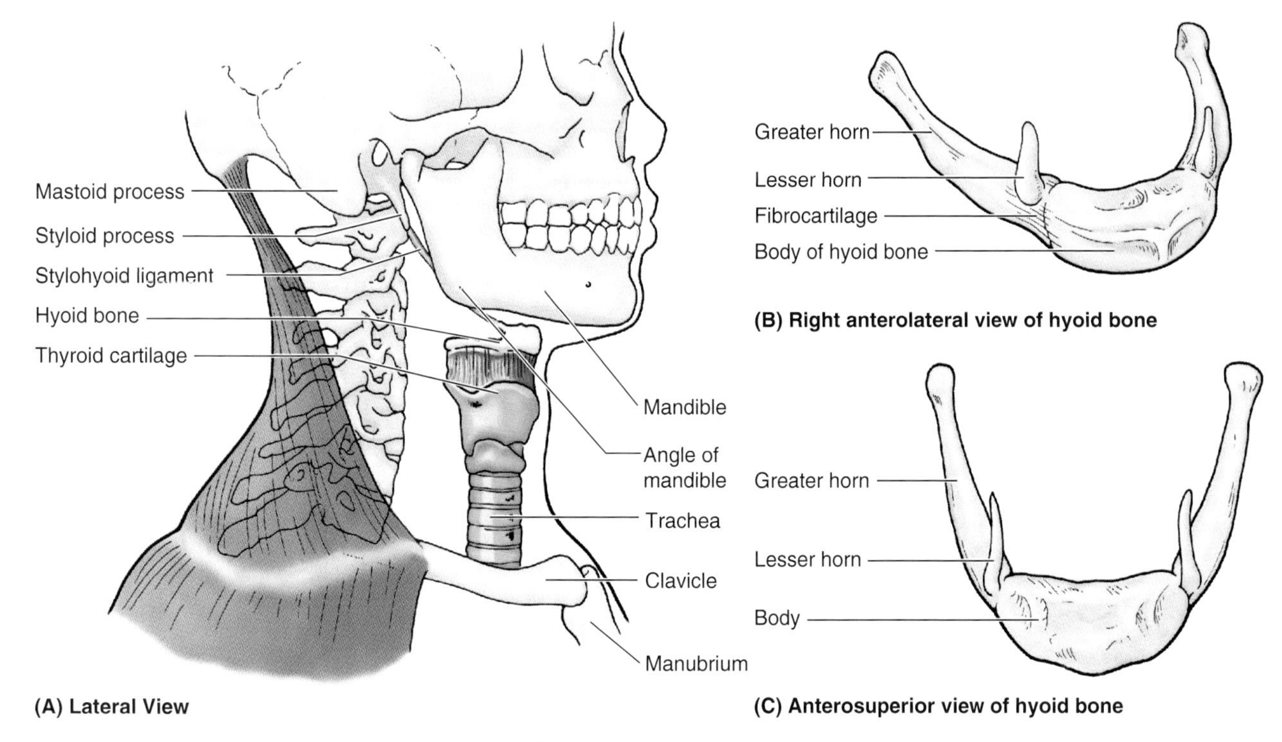

Figure 8.3. Bones and cartilages of the neck. A. Lateral view of the head and neck. Observe the bony and cartilaginous landmarks: vertebrae, mastoid and styloid processes, angle of the mandible (lower jaw), hyoid bone, thyroid cartilage, clavicle (collar bone), and manubrium of the sternum. **B.** Anterolateral and (**C**) anterosuperior views of the hyoid bone.

- The superior facets of the articular processes are directed superoposteriorly, and the inferior facets are directed inferoposteriorly.
- Short and bifid spinous processes.

Three cervical vertebrae (C1, C2, and C7) are *atypical*:

- C1 vertebra—the **atlas**—is a ringlike, kidney-shaped bone lacking a spinous process or body; it consists of two lateral masses connected by anterior and posterior arches. Its concave superior articular facets receive the occipital condyles. The atlas carries the skull.
- C2 vertebra—the **axis**—has a peglike **dens** (odontoid process), which projects superiorly from its body.
- C7 vertebra—the **vertebra prominens**—is so named because of its long spinous process, which is not bifid. Its transverse processes are large, but the foramina in them (L. foramina transversaria) are small and do not transmit the vertebral arteries.

Hyoid Bone

The mobile hyoid bone lies in the anterior part of the neck at the level of C3 vertebra in the angle between the mandible (lower jaw) and the thyroid cartilage (Fig. 8.3). The hyoid bone is suspended by muscles that connect it to the mandible, styloid processes, thyroid cartilage, manubrium of the sternum, and scapulae (shoulder blades). The hyoid bone is often fractured in people who are manually strangled by compression of the throat. The U-shaped hyoid bone derives its name from the Greek word *hyoeidés*, which means "shaped like the letter upsilon," the 20th letter in the Greek alphabet. The hyoid bone does not articulate with any other bone; it is suspended from the styloid processes of the temporal bones by the *stylohyoid ligaments* and is firmly bound to the thyroid cartilage. The hyoid bone consists of a body and greater and lesser horns (L. cornua). Functionally, the hyoid bone serves as an attachment for anterior neck muscles and a prop to keep the airway open.

The **body of the hyoid bone**, the middle part of the U-shaped bone, faces anteriorly and is approximately 2.5 cm wide and 1 cm thick (Fig. 8.3, *B* and *C*). Its anterior convex surface projects anterosuperiorly; its posterior concave surface projects posteroinferiorly. Each end of the body is united to a **greater horn** that projects posterosuperiorly and laterally from the body. Each greater horn in young people is united by fibrocartilage. In older people these horns are usually united by bone. Each **lesser horn** is a small bony projection from the superior part of the body of the bone near its union with the greater horn. It is connected to the body of the hyoid bone by fibrous tissue and sometimes to the greater horn by a synovial joint. The lesser horn projects superoposteriorly toward the styloid process; it may be partly or completely cartilaginous in some adults.

Fascia of the Neck

Structures in the neck are compartmentalized by layers of cervical fascia, the superficial and deep fascia. The fascial planes determine the direction in which an infection in the neck may be spread.

Superficial Cervical Fascia

The superficial cervical fascia is usually a thin layer of subcutaneous connective tissue that lies between the dermis of the skin and the investing layer of deep cervical fascia (Fig. 8.4*A*). It contains cutaneous nerves, blood and lymphatic vessels, superficial lymph nodes, and variable amounts of fat; anterolaterally it contains the *platysma muscle* (Fig. 8.4*B*).

Deep Cervical Fascia

The deep cervical fascia consists of three fascial layers (Fig. 8.4, A and *B*): *investing, pretracheal,* and *prevertebral.* These layers support the viscera (e.g., the thyroid gland), muscles, vessels, and deep lymph nodes. The deep cervical fascia also condenses around the common carotid arteries, internal jugular veins (IJVs), and vagus nerves to form the **carotid sheath** (Fig. 8.4, *B* and *C*). These fascial layers (sheaths) form natural cleavage planes through which tissues may be separated during surgery, and they limit the spread of abscesses (collections of pus) resulting from infections. The fascial layers sometimes deflect penetrating objects such as knives and low-velocity bullets away from vital structures. The deep cervical fascial layers also afford the slipperiness that allows structures in the neck to move and pass over one another without difficulty, such as when swallowing and turning the head and neck.

Investing Layer of Deep Cervical Fascia

The investing (superficial) layer of deep cervical fascia, the most superficial deep fascial layer, surrounds the entire neck deep to the skin and superficial cervical fascia. At the "four corners" of the neck, it splits into superficial and deep layers to enclose (invest) the trapezius and sternocleidomastoid (SCM) muscles. These muscles are derived from the same embryonic sheet of muscle, are innervated by the same nerve (cranial nerve XI), and have essentially continuous attachments to the skull superiorly and to the scapular spine and acromion and clavicle inferiorly (Fig. 8.4*C*). Superiorly, the investing layer of deep cervical fascia attaches to the:

- Superior nuchal line of the occipital bone
- Mastoid processes of the temporal bones
- Zygomatic arches
- Inferior border of the mandible
- Hyoid bone
- Spinous processes of cervical vertebrae.

Just inferior to its attachment to the mandible, the investing layer also splits to enclose the submandibular gland; posterior to the mandible, it splits to form the fibrous capsule of the parotid gland. The stylomandibular ligament is a thickened modification of this layer. Inferiorly, the investing layer of deep cervical fascia attaches to the:

- Manubrium of the sternum
- Clavicles
- Acromions and spines of the scapulae.

The investing layer of deep cervical fascia is continuous posteriorly with the periosteum covering the C7 spinous process and with the *nuchal ligament* (L. ligamentum nuchae), a triangular membrane that forms a median fibrous septum between the muscles of the two sides of the neck (Fig. 8.4*B*).

Inferiorly between the sternal heads of the SCM muscles and just superior to the manubrium, the investing layer of deep cervical fascia remains divided into the two layers that enclosed the muscle, with one layer attaching to the anterior and the other to the posterior surface of the manubrium. A **suprasternal space** lies between these layers (Fig. 8.4*A* and see Fig. 8.13) and encloses the inferior ends of the anterior jugular veins, the jugular venous arch, fat, and a few deep lymph nodes.

Pretracheal Layer of Deep Cervical Fascia

The thin pretracheal (visceral) layer of deep cervical fascia is limited to the anterior part of the neck (Fig. 8.4, *A–C*). It extends inferiorly from the hyoid bone into the thorax, where it blends with the fibrous pericardium covering the heart. The pretracheal layer of fascia includes a thin muscular layer, which encloses the infrahyoid muscles, and a visceral layer, which encloses the thyroid gland, trachea (see Fig. 8.13), and esophagus and is continuous posteriorly and superiorly with the *buccopharyngeal fascia* of the pharynx. The pretracheal fascia blends laterally with the *carotid sheaths*. In the region of the hyoid bone, a thickening of the pretracheal fascia forms a pulley or trochlea through which the intermediate tendon of the digastric muscle passes, suspending the hyoid bone. By wrapping around its lateral border, the pretracheal layer also tethers the two-bellied omohyoid muscle, redirecting the course of the muscle between the bellies (see Fig. 8.12).

Prevertebral Layer of Deep Cervical Fascia

The prevertebral layer of deep cervical fascia forms a tubular sheath for the vertebral column and the muscles associated with it, such as the longus colli and capitis anteriorly, the scalenes laterally, and the deep cervical muscles posteriorly (Fig. 8.4, *A–C*). The prevertebral layer of fascia extends from the base of the skull to T3 vertebra, where it fuses with the *anterior longitudinal ligament* (Fig. 8.4*A*). The prevertebral fascia extends laterally as the axillary sheath, which surrounds the axillary vessels and brachial

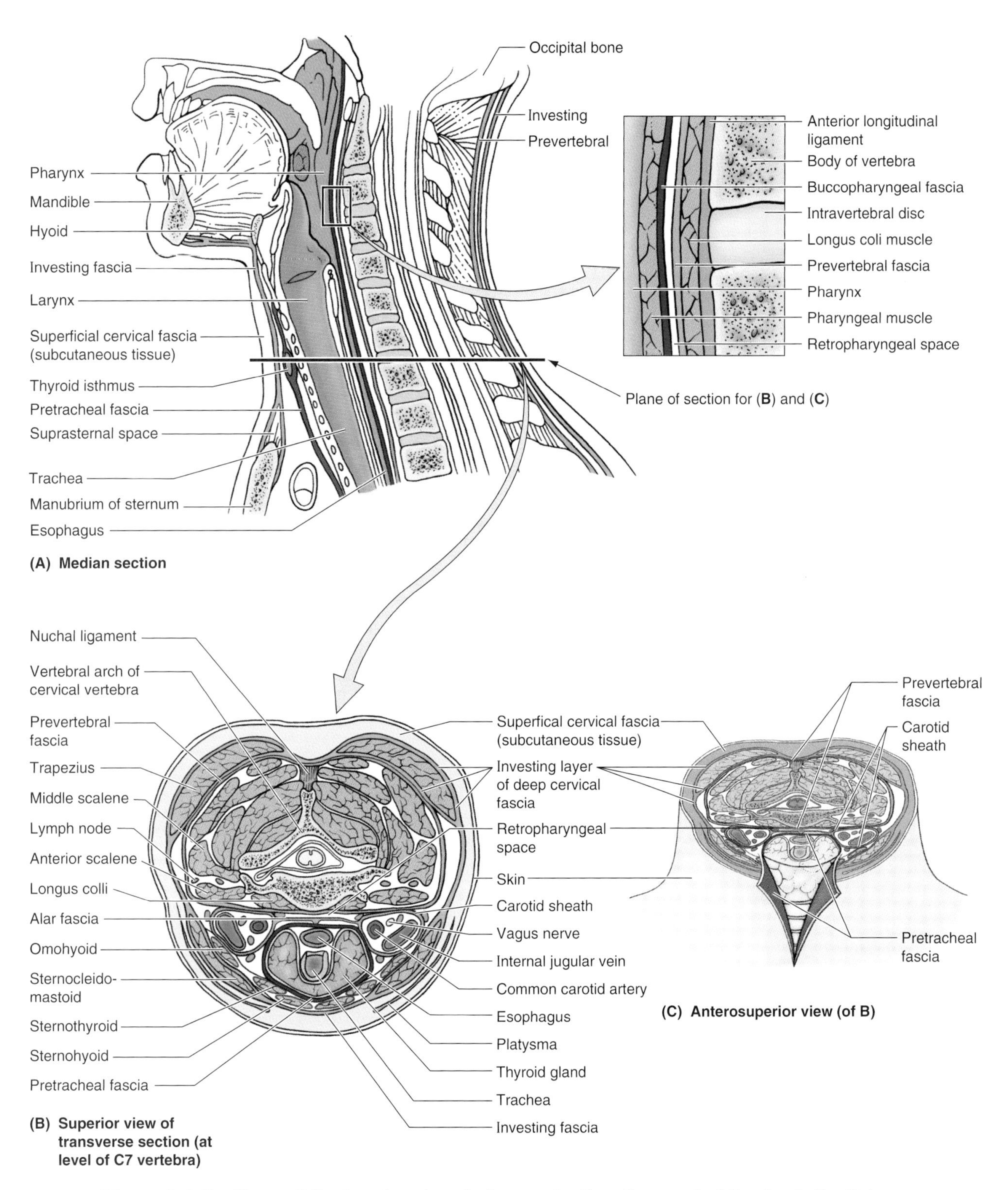

Figure 8.4. Sections of the head and neck demonstrating the cervical fascia. A. Sagittal section of the head and neck. The enlarged part on the right illustrates the fascia in the retropharyngeal region. **B.** Superior view of a transverse section of the neck passing through the isthmus of the thyroid gland at the level indicated in (**A**). Note the superficial cervical fascia (subcutaneous tissue) in which the platysma (G. flat plate) is embedded. Note also the way in which the outermost layer of deep cervical fascia, the investing (superficial) layer, splits to enclose the trapezius and the sternocleidomastoid (SCM) muscles at the "four corners" of the neck. The pretracheal (visceral) layer encloses muscles and viscera in the anterior neck. The prevertebral layer encircles the vertebral column and associated (especially the posterior deep cervical) muscles. Examine the carotid sheath and its contents. **C.** An illustration of the prevertebral and pretracheal fascia and the carotid sheath.

plexus (see Fig. 8.7*A*). The sympathetic trunks are embedded in this fascial layer.

The **carotid sheath** is a tubular, fascial investment that extends from the base of the skull to the root of the neck. This fascial sheath blends anteriorly with the investing and pretracheal layers of fascia and posteriorly with the prevertebral layer of deep cervical fascia (Fig. 8.4, *A–C*). *The carotid sheath contains:*

- The common and internal carotid arteries
- The IJV
- The vagus nerve (CN X)
- Some deep cervical lymph nodes
- The carotid sinus nerve
- Sympathetic nerve fibers (carotid periarterial plexuses).

The carotid sheath and pretracheal fascia communicate freely with the mediastinum of the thorax inferiorly and the cranial cavity superiorly. These communications represent potential pathways for the spread of infection and extravasated blood.

The **retropharyngeal space** is the largest and most important interfascial space in the neck (Fig. 8.4, *A* and *B*). It is a potential space consisting of loose connective tissue between the prevertebral layer of deep cervical fascia and the **buccopharyngeal fascia** surrounding the pharynx superficially (Fig. 8.4*A*). Inferiorly, the buccopharyngeal fascia is continuous with the pretracheal layer of deep cervical fascia. The *alar fascia* forms a further subdivision of the retropharyngeal space (Fig. 8.4*B*). This thin layer is attached along the midline of the buccopharyngeal fascia from the skull to the level of C7 vertebra. From this attachment it extends laterally and terminates in the carotid sheath. *The retropharyngeal space permits movement of the pharynx, esophagus, larynx, and trachea relative to the vertebral column during swallowing.* This space is closed superiorly by the base of the skull and on each side by the carotid sheath. It opens inferiorly into the superior mediastinum (see Chapter 1).

Spread of Infections in the Neck

The investing layer of deep cervical fascia helps prevent the spread of abscesses—circumscribed collections of purulent exudate (pus) caused by tissue destruction. If an infection occurs between the investing fascia and that surrounding the infrahyoid muscles, the infection will usually not spread beyond the superior edge of the manubrium. If, however, the infection occurs between the investing and pretracheal layers of fascia, it can spread into the thoracic cavity anterior to the pericardium. Pus from an abscess posterior to the prevertebral layer of deep cervical fascia may extend laterally in the neck and form a swelling posterior to the SCM muscle. The pus may perforate the prevertebral layer of deep cervical fascia and enter the retropharyngeal space, producing a bulge in the pharynx (*retropharyngeal abscess*). This swelling (edema) may cause difficulty in swallowing (dysphagia) and speaking (dysarthria). Infections in the head may also spread inferiorly posterior to the esophagus and enter the posterior mediastinum or anterior to the trachea and enter the anterior mediastinum (see Chapter 1). Infections in the retropharyngeal space may also extend inferiorly into the superior mediastinum. Similarly, air from a ruptured trachea, bronchus, or esophagus (*pneumomediastinum*) can pass superiorly in the neck. ⊙

Superficial and Lateral Muscles of the Neck

Three superficial and lateral muscles are in the neck: platysma, SCM (sternocleidomastoid), and trapezius. The platysma (G. flat plate), like the facial and scalp muscles, develops from a continuous sheet of musculature that derives from mesenchyme in the 2nd pharyngeal arch of the embryo (Moore and Persaud, 1998). The platysma and facial muscles are supplied by branches of the facial nerve (CN VII), which supplies the 2nd embryonic pharyngeal arch. Likewise, the trapezius and SCM muscles are derived from the same embryonic muscle mass and are both supplied by the accessory nerve (CN XI).

Platysma

The **platysma** is a broad, thin sheet of muscle in the subcutaneous tissue of the neck. The combined thickness of the skin and platysma is only a few millimeters (Figs. 8.4 and 8.5, Table 8.1). The **external jugular vein (EJV)**, descending from the angle of the mandible to the middle of the clavicle (Fig. 8.1), and the main cutaneous nerves of the neck are deep to the platysma. The platysma covers the anterolateral aspect of the neck. Its fibers arise in the fascia covering the superior parts of the deltoid and pectoralis major muscles and sweep superomedially over the clavicle to the inferior border of the mandible. The anterior borders of the two muscles decussate over the chin and blend with the facial muscles; inferiorly the fibers diverge, leaving a gap anterior to the larynx and trachea (Fig. 8.5). Much variation exists in terms of the continuity (completeness) of this muscle sheet, which often occurs as isolated slips. Acting from its superior attachment, the platysma tenses the skin, producing vertical skin ridges and releasing pressure on the superficial veins. Men commonly use this action of the platysma when shaving their neck and when easing a tight collar. Acting from its

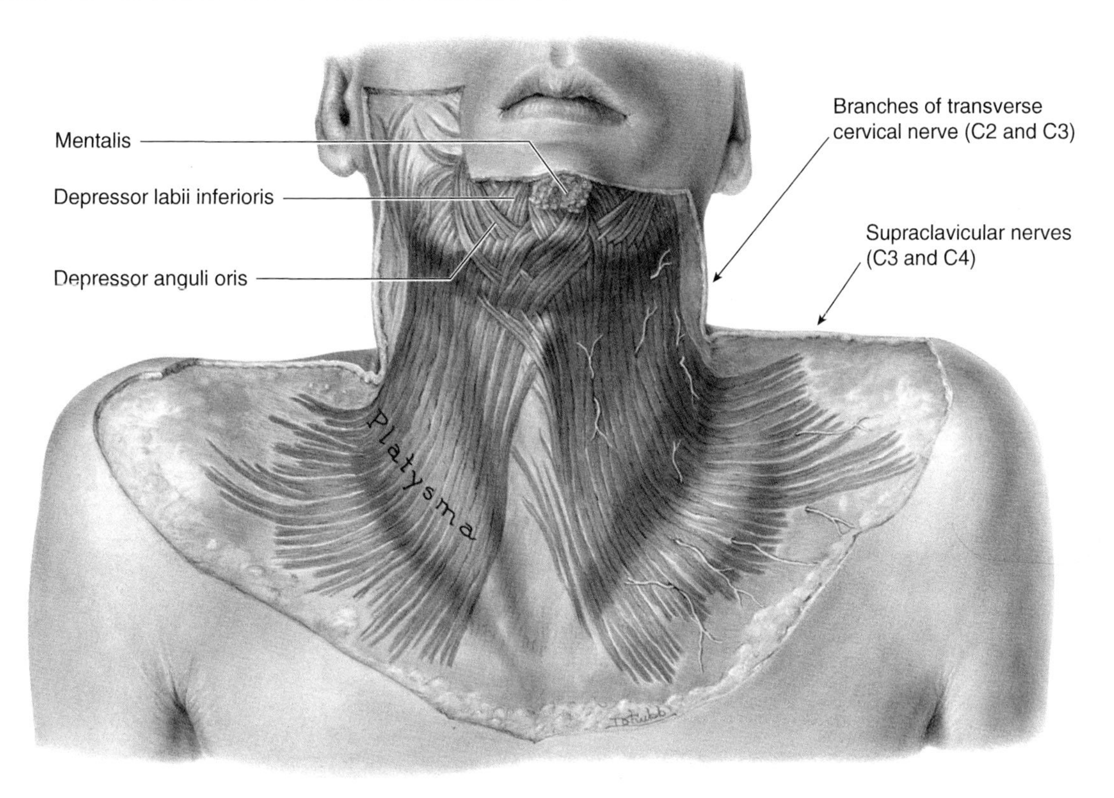

Figure 8.5. Platysma (G. flat plate) muscle. Observe that this thin muscle spreads subcutaneously like a sheet, passes over the clavicles, and is pierced by cutaneous nerves. Much variation is present in the continuity of this muscular sheet.

inferior attachment, the platysma helps depress the mandible and draw the corners of the mouth inferiorly, as in a grimace. As a muscle of facial expression, the platysma serves to convey tension or stress. The platysma is supplied by the cervical branch of the facial nerve.

Sternocleidomastoid

The SCM is the key muscular landmark in the neck because it divides each side of the neck into anterior and posterior triangles (Table 8.1 and 8.2). This broad, straplike muscle has two heads: the rounded tendon of the *sternal head* attaches to the manubrium of the sternum, and the thick fleshy *clavicular head* attaches to the superior surface of the medial third of the clavicle (L. cleido; refers to clavicle). The two heads of the SCM, separated by a space inferiorly, join as they pass obliquely upward toward the skull. The superior attachment of the SCM is the mastoid process of the temporal bone and the superior nuchal line of the occipital bone. The investing layer of deep cervical fascia splits to form a sheath for the SCM (Fig. 8.4*B*).

Acting bilaterally, the SCMs flex the neck; they can do this in two different ways: (a) acting alone, the SCMs "bend" the neck so that the chin approaches the manubrium, or (b) in conjunction with the extensors of the neck (deep cervical muscles), bilateral contraction of the SCMs can protrude the chin; this also occurs when lifting the head off the ground while lying supine (with gravity acting in place of the deep cervical muscles).

Unilaterally the SCM laterally flexes and rotates the head and neck so the ear approaches the shoulder of the same side. This turns the chin to the opposite side and directs it superiorly as the head rotates at the atlantoaxial joint (see Chapter 4). To test the SCM, turn the face to the opposite side against resistance (hand against chin). If it is acting normally, the muscle can be seen and palpated.

Trapezius

The trapezius is a large, flat, triangular muscle that covers the posterolateral aspect of the neck and thorax. It is a superficial muscle of the back (see Chapter 4), a muscle of the pectoral (shoulder) girdle (see Chapter 6), and a cervical muscle. The trapezius attaches the pectoral girdle to the skull and vertebral column and assists in suspending it. Its attachments, nerve supply, and main actions are described in Table 8.1. *To test the trapezius,* shrug the shoulder against resistance. If it is acting normally, its superior border can be seen and palpated. If the trapezius muscles are paralyzed, the shoulders droop; however, the combined actions of the levator scapulae and superior fibers of the serratus anterior (see Chapter 6) may compensate for their actions.

Table 8.1. **Superficial Muscles in the Lateral Aspect of the Neck**

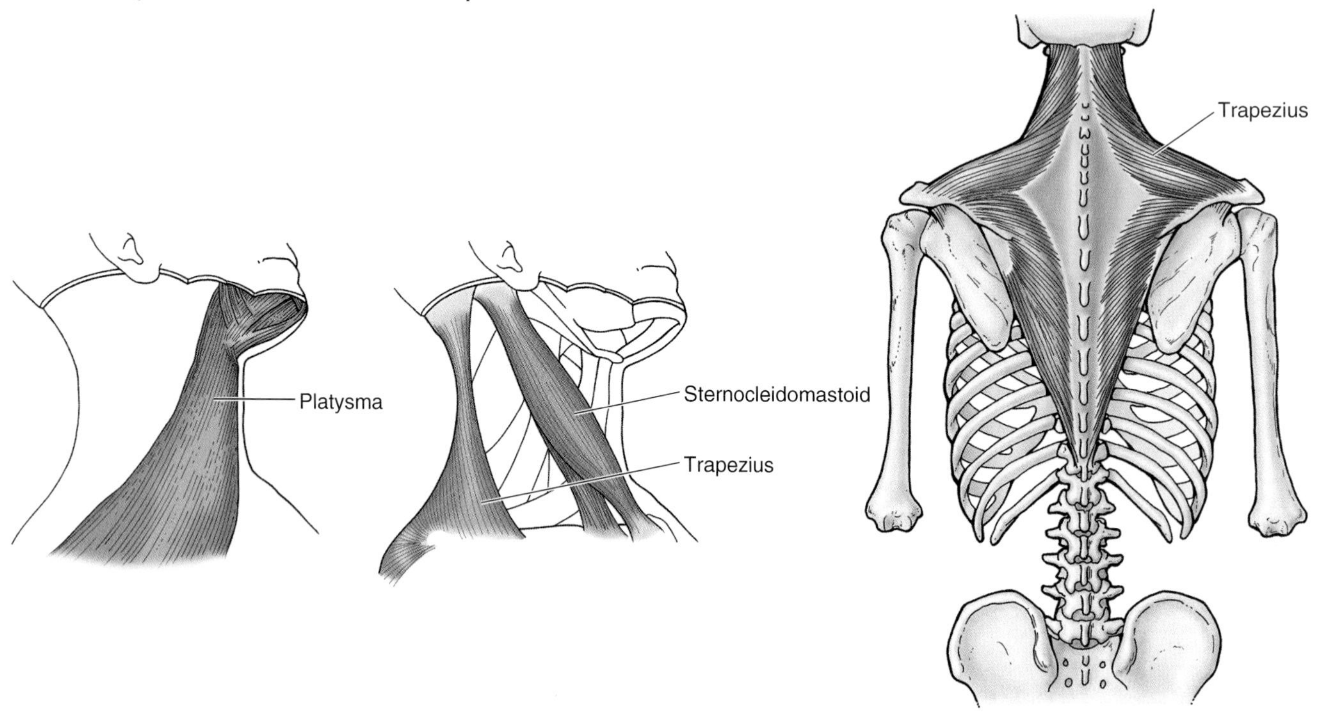

Muscle	Superior Attachment	Inferior Attachment	Innervation	Main Action
Platysma	Inferior border of mandible, skin, and subcutaneous tissues of lower face	Fascia covering superior parts of pectoralis major and deltoid muscles	Cervical branch of facial nerve	Draws corners of mouth inferiorly and widens it as in expressions of sadness and fright; draws the skin of neck superiorly when teeth are "clenched"
Sternocleidomastoid	Lateral surface of mastoid process of temporal bone and lateral half of superior nuchal line	Sternal head: anterior surface of manubrium of sternum Clavicular head: superior surface of medial third of clavicle	Spinal root of accessory nerve (motor) and C2 and C3 nerves (pain and proprioception)	Tilts head to one side, i.e., laterally; flexes neck and rotates it so face is turned superiorly toward opposite side; acting together, the two muscles flex the neck so chin is thrust forward
Trapezius	Medial third of superior nuchal line, external occipital protuberance, ligamentum nuchae, spinous processes of C7–T12 vertebrae, and lumbar and sacral spinous processes	Lateral third of clavicle, acromion, and spine of scapula	Spinal root of accessory nerve (motor) and C3 and C4 nerves (pain and proprioception)	Elevates, retracts, and rotates scapula; superior fibers elevate the scapula, middle fibers retract it, and inferior fibers depress it

Paralysis of the Platysma

Paralysis of the platysma resulting from injury to the cervical branch of the facial nerve causes the skin to fall away from the neck in slack folds. Consequently, during surgical dissections of the neck, extra care is necessary to preserve the cervical branch of the facial nerve. When suturing wounds of the neck, unless surgeons carefully suture the skin and edges of the platysma, the skin wound will be distracted (pulled into different directions) by the contracting muscle fibers, and a broad ugly scar may develop.

Congenital Torticollis

The most common type of torticollis (wry neck) results from a *fibrous tissue tumor* (fibromatosis colli) that develops in the SCM before birth (Raffensperger, 1990). The lesion causes the head to turn to the side and the face to turn away from the affected side. The position of the infant's head usually necessitates a breech delivery. Occasionally, the SCM is injured when an infant's head is pulled too much during a difficult birth, tearing its fibers (*muscular torticollis* [Behrman, et al., 1996]). A *hematoma* ▶

▶ (mass of extravasated blood) usually occurs that may develop into a fibrotic mass that entraps a branch of the accessory nerve; this denervates part of the SCM. If untreated, the lesion may result in *torticollis,* a flexion deformity of the neck. Stiffness of the neck results from fibrosis and shortening of the SCM. Surgical release of the SCM from its distal attachments, or division of the muscle inferior to the level of CN XI, may be necessary to enable the person to hold and rotate the head normally.

Spasmodic Torticollis

Cervical dystonia (abnormal tonicity), commonly known as *spasmodic torticollis* or wry neck, usually begins between ages 20 and 60 years. It may involve any bilateral combination of lateral neck muscles, especially the SCM and trapezius. Characteristics of this disorder are sustained turning, tilting, flexing, or extending of the neck. Shifting the head laterally or anteriorly can occur involuntarily (Fahn et al., 1995). The shoulder is usually elevated and anteriorly displaced on the side to which the chin turns. Neck pain occurs in most patients. ⭘

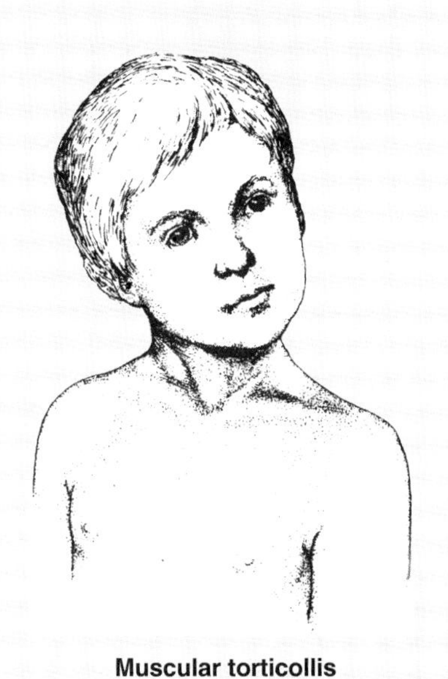

Muscular torticollis

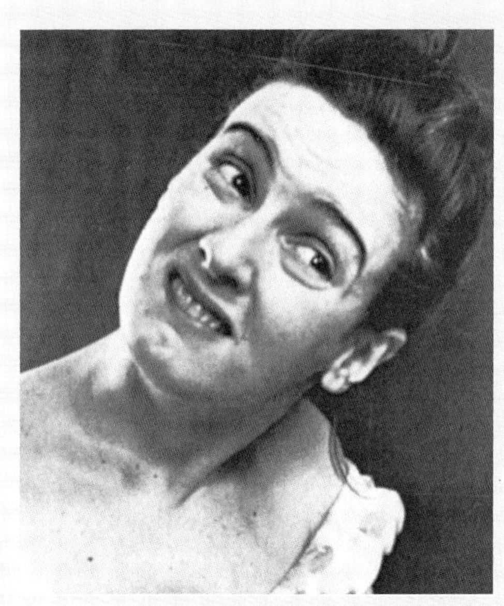

Spasmodic torticollis

Triangles of the Neck

To facilitate description of cervical anatomy, each side of the neck is divided into two triangles, anterior and posterior, by the obliquely placed SCM (Table 8.2). *The posterior triangle of the neck has:*

- An *anterior boundary,* formed by the posterior border of the SCM
- A *posterior boundary,* formed by the anterior border of the trapezius
- An *inferior boundary* (base), formed by the middle third of the clavicle between the trapezius and SCM
- Its *apex,* where the SCM and trapezius meet on the superior nuchal line of the occipital bone
- A *roof,* formed by the investing layer of deep cervical fascia
- A *floor,* formed by muscles covered by the prevertebral layer of deep cervical fascia.

For more precise localization of structures, the posterior triangle is divided into supraclavicular (subclavian) and occipital triangles by the inferior belly of the omohyoid muscle (Table 8.2).

The anterior triangle of the neck has:

- An *anterior boundary,* formed by the median line of the neck
- A *posterior boundary,* formed by the anterior border of the SCM
- A *superior boundary,* formed by the inferior border of the mandible
- Its *apex,* at the jugular (suprasternal) notch in the manubrium
- A *roof,* formed by subcutaneous tissue containing the platysma
- A *floor,* formed by the pharynx, larynx, and thyroid gland.

For more precise localization of structures, the anterior triangle is divided into the unpaired submental triangle and

Table 8.2. **Cervical Triangles and Contents**

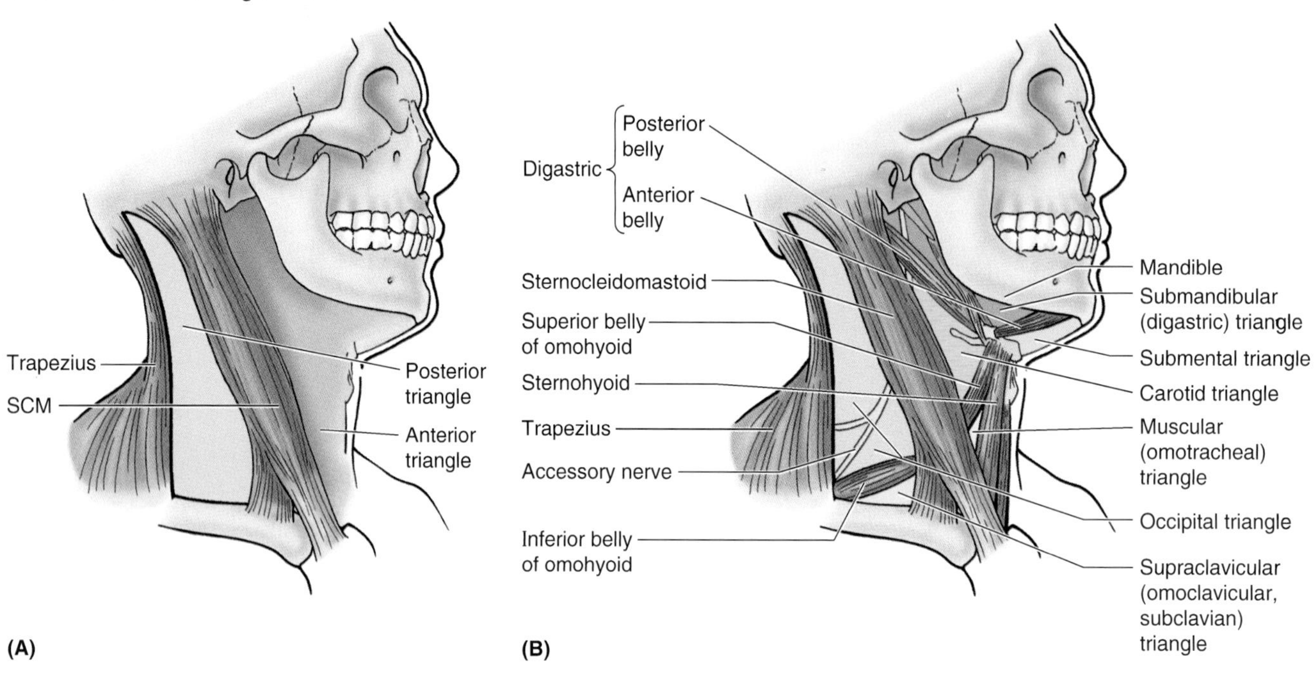

Posterior Triangle	Main Contents
Occipital triangle	Part of external jugular vein, posterior branches of cervical plexus of nerves, accessory nerve, trunks of brachial plexus, transverse cervical artery, cervical lymph nodes
Supraclavicular (omoclavicular, subclavian) triangle	Subclavian artery (3rd part), part of subclavian vein (sometimes), suprascapular artery, supraclavicular lymph nodes

Anterior Triangle	Main Contents
Submandibular (digastric) triangle	Submandibular gland almost fills triangle; submandibular lymph nodes, hypoglossal nerve, mylohyoid nerve, parts of facial artery and vein
Submental triangle	Submental lymph nodes, small veins that unite to form anterior jugular vein
Carotid triangle	Carotid sheath containing common carotid artery and its branches, internal jugular vein and its tributaries, and vagus nerve; external carotid artery and some of its branches; hypoglossal nerve and superior root of ansa cervicalis; accessory nerve; thyroid, larynx, and pharynx; deep cervical lymph nodes; branches of cervical plexus
Muscular (omotracheal) triangle	Sternothyroid and sternohyoid muscles, thyroid and parathyroid glands

three small paired triangles (submandibular, carotid, and muscular) by the digastric and omohyoid muscles.

Posterior Cervical Triangle

The posterior triangle is the area of the neck bounded by the SCM, trapezius, and clavicle (Fig. 8.6, Table 8.2). The posterior triangle wraps around the lateral surface of the neck like a spiral. The triangle is covered by skin and subcutaneous tissue containing the platysma.

Muscles in the Posterior Triangle

The floor of the posterior triangle is usually formed by the prevertebral fascia overlying four muscles (Fig. 8.6, see Table 8.4 [p. 1026]):

- Splenius capitis
- Levator scapulae
- Middle scalene (L. scalenus medius)
- Posterior scalene (L. scalenus posterior).

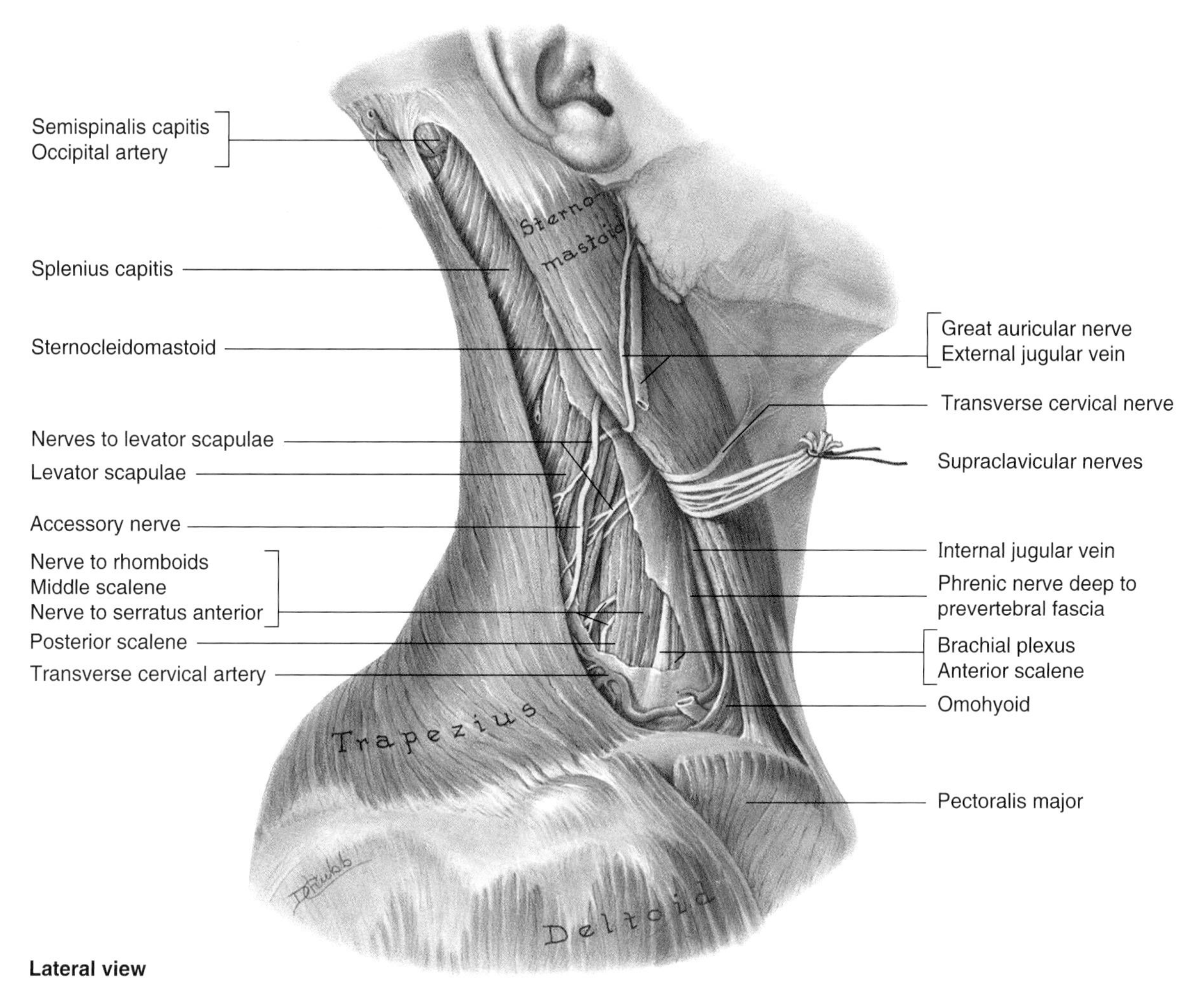

Figure 8.6. Deep dissection of the posterior triangle of the neck. Lateral view. The investing (superficial) layer of deep cervical fascia has been removed. Observe the brachial plexus and the motor nerves that run deep to the prevertebral layer of deep cervical fascia, covering the floor of the triangle.

Sometimes the inferior part of the anterior scalene (L. scalenus anterior) appears in the inferomedial angle of the posterior triangle, but it is usually hidden by the SCM. An occasional offshoot of the anterior scalene, the little scalene (L. scalenus minimus), passes posterior to the subclavian artery to attach to the 1st rib (Agur, 1991). The inferior belly of the omohyoid muscle divides the posterior cervical triangle into a large occipital triangle superiorly and a small supraclavicular triangle inferiorly (Table 8.2).

- The **occipital triangle** is so called because the *occipital artery* appears in its apex (Fig. 8.6). The most important nerve crossing the occipital triangle is the *accessory nerve.*
- The **supraclavicular triangle**, the smaller subdivision of the posterior triangle, is indicated on the surface of the neck by the supraclavicular fossa (p. 1023). The *EJV* and *suprascapular artery* cross this triangle superficially, and the *subclavian artery* lies deep in it (Figs. 8.6 and 8.7).

These vessels are covered by the investing layer of deep cervical fascia. Because of the presence of the subclavian artery in this region, the supraclavicular triangle is often called the subclavian triangle. The third part of the subclavian artery lies on the 1st rib, and its pulsations can be felt on deep pressure.

Vessels in the Posterior Triangle

Veins. The **EJV** drains most of the scalp and side of the face. The **EJV** begins near the angle of the mandible just inferior to the lobule of the auricle by the union of the posterior division of the *retromandibular vein* with the *posterior auricular vein* (Figs. 8.8 and 8.9). The EJV crosses the SCM obliquely, deep to the platysma, and enters the anteroinferior part of the posterior triangle. It then pierces the investing layer of deep cervical fascia forming the roof of the triangle at the posterior border of the SCM. The EJV descends to the inferior part of the posterior triangle and terminates in the *sub-*

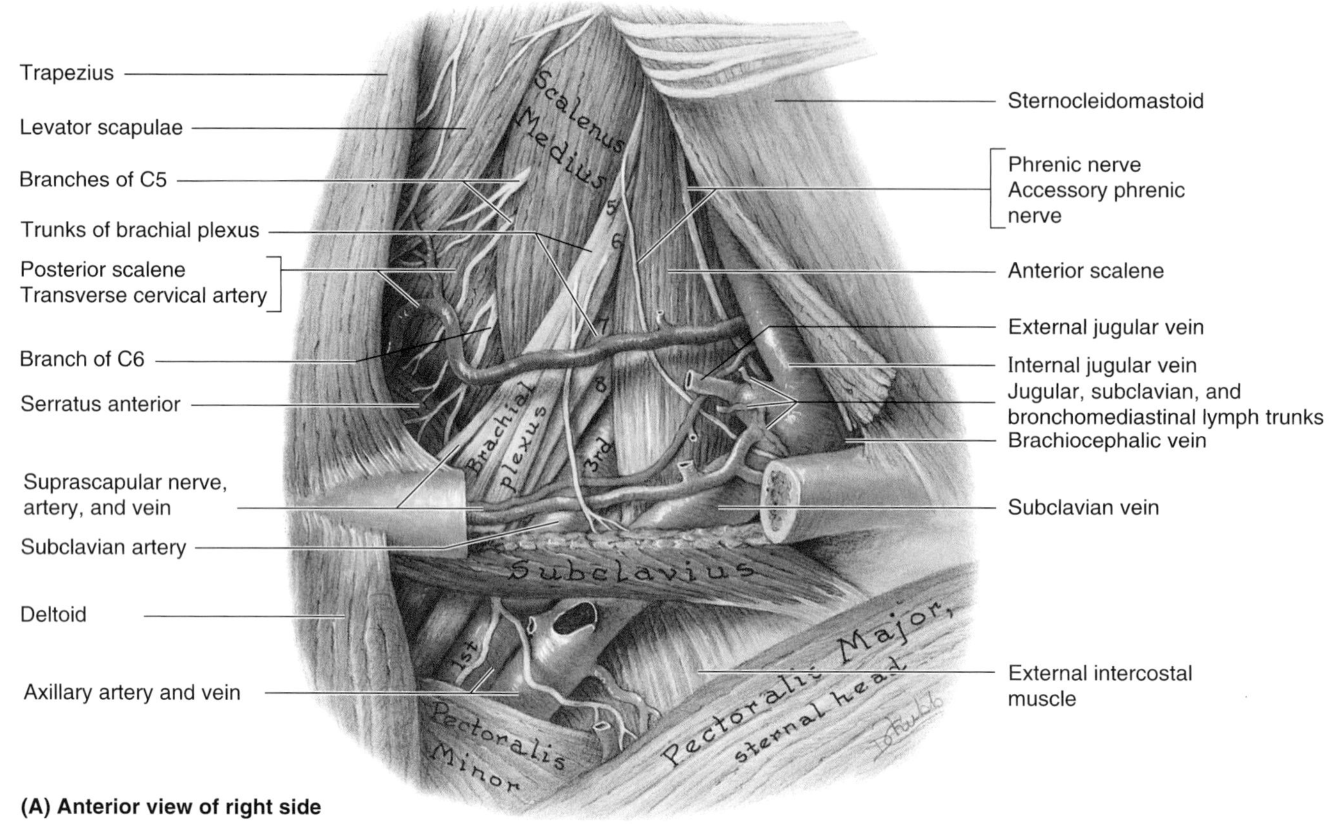

Figure 8.7. Deep dissection of the posterior triangle of the neck. Observe the brachial plexus of nerves passing to the upper limb and part of the subclavian vessels. All fascia, the omohyoid muscle, and the clavicular head of the pectoralis major have been removed to show the third part of the supraclavicular artery and the subclavian vein. Note that the internal jugular vein (IJV), deep to the sternocleidomastoid (SCM), is not in the posterior triangle but is close to it.

clavian vein. Just superior to the clavicle, the EJV receives the *transverse cervical, suprascapular*, and *anterior jugular veins*.

The **subclavian vein**, the major venous channel draining the upper limb, curves through the inferior part of the posterior triangle (Fig. 8.7*A*). It passes anterior to the anterior scalene muscle and phrenic nerve and unites at the medial border of the muscle with the **IJV** to form the **brachiocephalic vein** (Figs. 8.7*A* and 8.9), posterior to the medial end of the clavicle.

Subclavian Vein Puncture

To administer parenteral (nutritional) fluids and medications and to measure central venous pressure, the right subclavian vein is often the point of entry to the venous system for *central line placement* (Ger et al., 1996). When using the infraclavicular subclavian vein approach, the physician inserts the needle along the inferior surface of the middle part of the clavicle and moves it medially toward the jugular notch in the manubrium and along the posterior surface of the clavicle where the subclavian vein ascends. If the needle is not inserted carefully, it may tear the subclavian vein and parietal pleura, resulting in *hemothorax* (bleeding into the pleural cavity). Furthermore, if the needle goes too far posteriorly, it may enter the subclavian artery. When the needle has been inserted correctly, the physician introduces a soft, flexible catheter into the subclavian vein using the needle as a guide.

Puncture of the Internal Jugular Vein

For *right cardiac catheterization* (to take measurements of pressures within the right chambers of the heart), puncture of the IJV can be a way to introduce a catheter through the right brachiocephalic vein into the superior vena cava and the right side of the heart. Although the preferred route is ▶

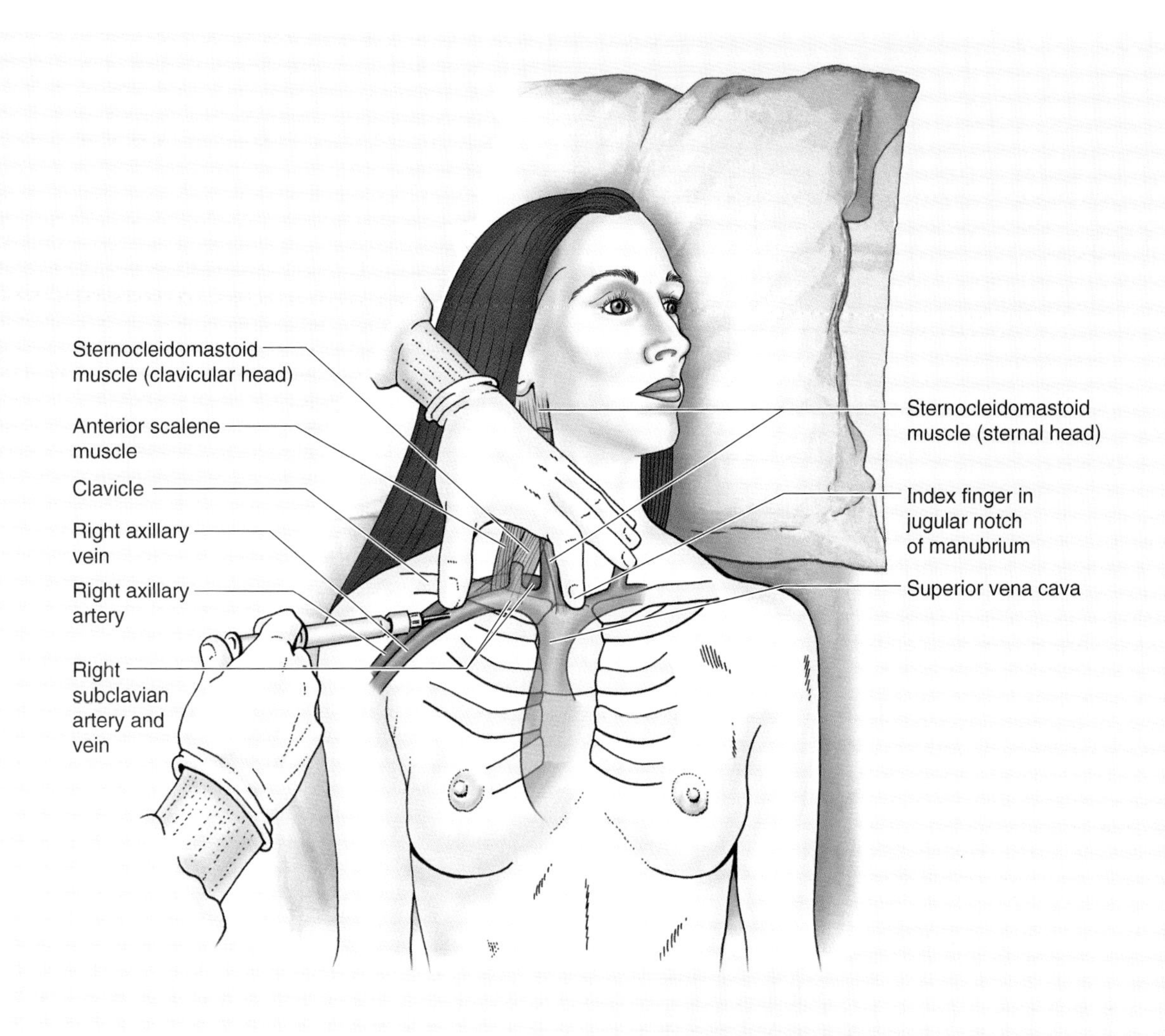

▶ through the IJV or the subclavian vein, it may be necessary in some patients to use the EJV. This vein is not ideal for catheterization because its angle of junction with the subclavian vein makes passage of the catheter difficult (Ger et al., 1996).

Severance of the External Jugular Vein

If the EJV is severed (e.g., by a knife slash) along the posterior border of the SCM where it pierces the roof of the posterior triangle, its lumen is held open by the tough investing layer of deep cervical fascia, and the negative intrathoracic pressure air will suck air into the vein. This action produces a churning noise in the thorax and *cyanosis*—a bluish discoloration of the skin and mucous membranes resulting from an excessive concentration of reduced hemoglobin in the blood. A *venous air embolism* produced in this way will fill the right side of the heart with froth, which nearly stops bloodflow through it, resulting in *dyspnea* (G. difficulty with breathing). The application of pressure to the severed jugular vein until it can be sutured will stop the bleeding and entry of air into the blood.

Prominence of the External Jugular Vein

The EJV may serve as an "internal barometer." When venous pressure is in the normal range, the EJV is usually visible above the clavicle for only a short distance. However, when venous pressure rises (e.g., as in heart failure), the vein becomes prominent throughout its course along the side of the neck. Consequently, routine observation of the EJV during physical examinations may give diagnostic signs of *heart failure*, obstruction of the superior vena cava (e.g., by tumor cells), enlarged supraclavicular lymph nodes, or *increased intrathoracic pressure*. ○

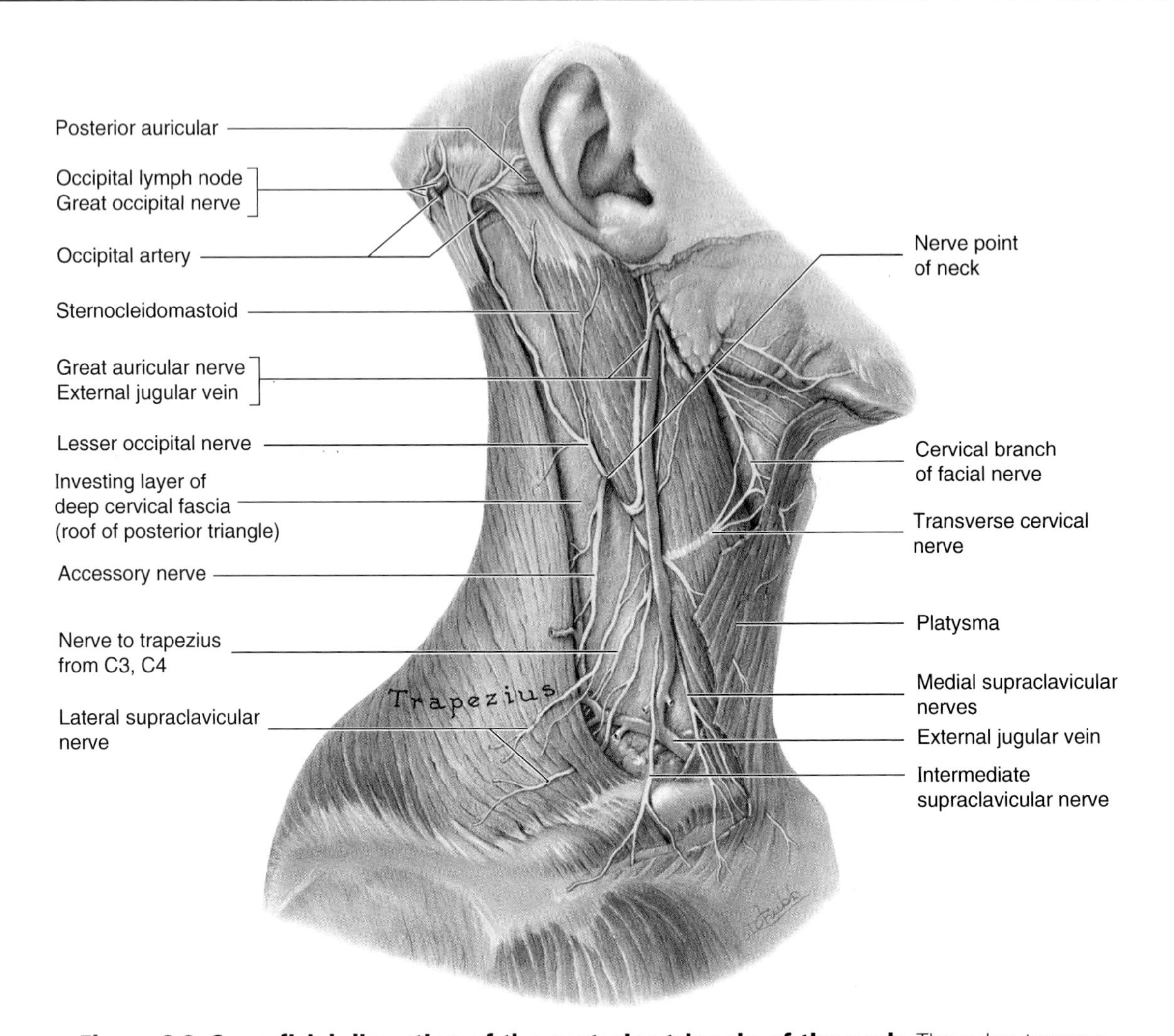

Figure 8.8. Superficial dissection of the posterior triangle of the neck. The subcutaneous tissue and the superficial layer of the investing (superficial) layer of the deep fascia have been removed from the superficial aspects of the trapezius and sternocleidomastoid (SCM) muscles. Between the muscles, the investing layer of deep cervical fascia forms a roof over the posterior triangle. Note the accessory nerve (CN XI), the only motor nerve superficial to (or embedded in) this fascia.

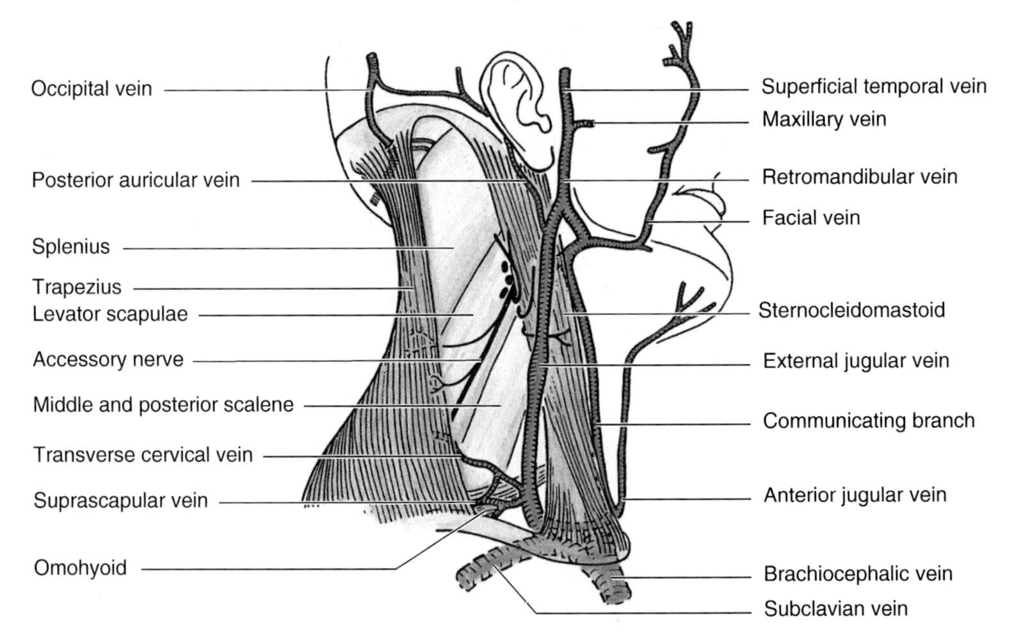

Arteries. The arteries in the posterior triangle are the transverse cervical artery, suprascapular artery, superficial cervical artery, and part of the occipital artery.

The **transverse cervical artery**, or transverse artery of the neck (Fig. 8.7), originates from the **thyrocervical trunk**, a branch of the subclavian artery. The transverse cervical artery runs superficially and laterally across the *phrenic nerve* and *anterior scalene muscle,* 2 to 3 cm superior to the clavicle. It then crosses (passes through) the **trunks of the brachial plexus** supplying branches to their vasa nervorum and passes deep to the trapezius. Its superficial branch accompanies the accessory nerve on the underside of the muscle, and its deep branch runs anterior to the rhomboid muscles as the dorsal scapular artery, accompanying the nerve of the same name.

The **suprascapular artery**, another branch of the thyrocervical trunk (Fig. 8.7, *A* and *B*), passes inferolaterally across the anterior scalene muscle and phrenic nerve. It then crosses the subclavian artery (3rd part) and the cords of the brachial plexus and passes posterior to the clavicle to supply muscles on the posterior aspect of the scapula.

The **occipital artery** (Fig. 8.8), a branch of the external carotid artery, enters the posterior triangle at its superior angle and ascends over the head to supply the posterior half of the scalp.

The **third part of the subclavian artery** (Fig. 8.7), its longest portion, supplies blood to the upper limb. It begins approximately a finger's breadth superior to the clavicle, opposite the lateral border of the anterior scalene muscle. It is hidden in the inferior part of the posterior triangle, posterosuperior to the subclavian vein. The third part is the most superficial portion of the subclavian artery, and its pulsations can be felt on deep pressure in the supraclavicular triangle of the neck. The artery is in contact with the 1st rib posterior to the anterior scalene muscle; consequently, compression of the subclavian artery against this rib can control bleeding in the upper limb. Posteriorly, the third part of the artery lies against the inferior trunk of the brachial plexus.

Nerves. The **accessory nerve** enters the posterior triangle at or inferior to the junction of the superior and middle thirds of the posterior border of the SCM (Fig. 8.8). It passes posteroinferiorly through the triangle, superficial to the *investing layer of deep cervical fascia.* CN XI then disappears deep to the anterior border of the trapezius at the junction of its superior two-thirds with its inferior one-third. CN XI has cranial and spinal roots (see Chapter 9). The *spinal root of the accessory nerve* separates immediately from the cranial root and passes posteroinferiorly. It supplies the SCM and then crosses the posterior triangle, superficial to the investing layer of deep cervical fascia covering its floor (Fig. 8.8). It passes deep to this muscle approximately 5 cm superior to the clavicle and supplies the trapezius.

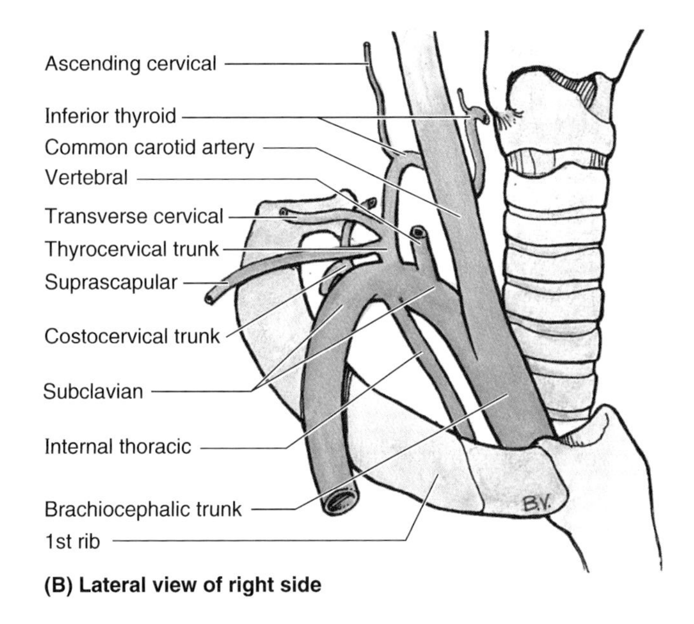

(B) Lateral view of right side

Figure 8.9. Superficial veins of the right side of the neck. Lateral view. Outside the skull, the superficial temporal and maxillary veins form the retromandibular vein, the posterior division of which unites with the posterior auricular vein to form the external jugular vein (EJV). The facial vein receives the anterior division of the retromandibular vein before emptying into the internal jugular vein (IJV) (posterior to the the sternocleidomastoid [SCM]; see Fig. 8.7*A*). The anterior jugular veins may lie superficial or deep to the investing (superficial) layer of the deep cervical fascia. **B.** (Above) The subclavian artery and its branches.

Lesions of the Spinal Root of CN XI

Lesions of the spinal root of the accessory nerve are uncommon. *CN XI may be damaged by:*

- Penetrating trauma such as a stab wound
- Surgical procedures in the posterior triangle
- Tumors at the base of the skull or cancerous cervical lymph nodes
- Fractures of the jugular foramen through which CN XI leaves the skull.

Although contraction of one SCM turns the head to one side, a unilateral lesion of CN XI usually does not produce an abnormal position of the head. However, patients with accessory nerve damage usually have weakness in turning the head to the opposite side against resistance.

Lesions of the spinal root of CN XI (e.g., resulting from a knife or bullet wound) produce weakness and atrophy of the trapezius, which impair neck movements. *Unilateral paralysis of the trapezius* is evident by the patient's inability to elevate and retract the shoulder and by difficulty in elevating the arm (L. brachium) superior to the horizontal level. The normal ridge in the neck produced by the trapezius is also depressed. *Drooping of the shoulder is an obvious sign of injury to the spinal root of CN XI.* During extensive surgical dissections in the posterior triangle—for example, during removal of malignant (cancerous) lymph nodes—the surgeon isolates the accessory nerve to preserve it, if possible. It is important to remember the superficial location of this nerve in the posterior triangle because *the accessory nerve is the most commonly iatrogenically injured nerve* (injury caused by the physician)—the injuries often occurring during superficial procedures. ⊙

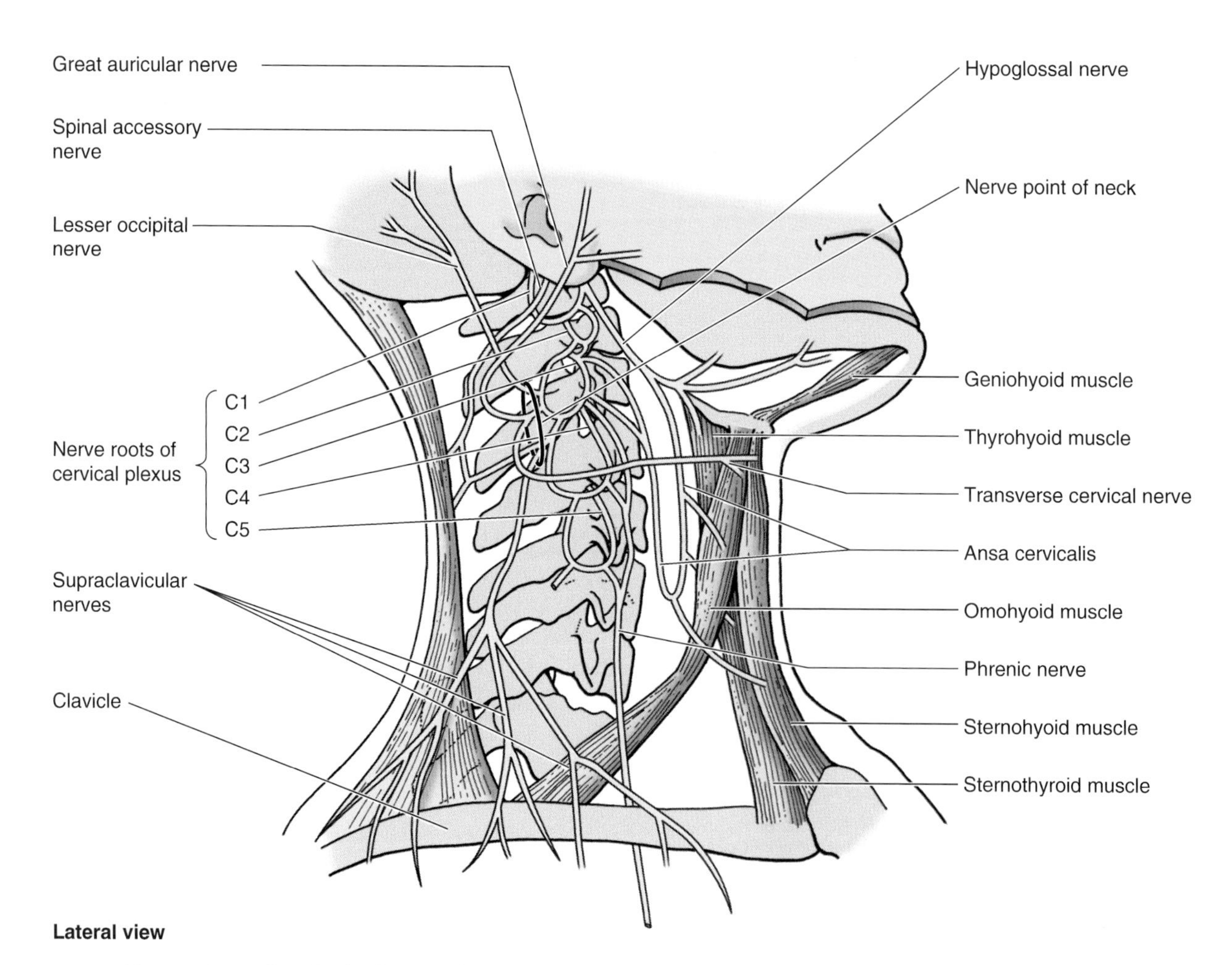

Figure 8.10. Cervical plexus of nerves. Lateral view. The plexus is formed by loops joining the adjacent ventral primary rami of the first four cervical nerves and receiving gray communicating rami from the superior cervical sympathetic ganglion (not shown here—see Fig. 8.18*B*). The branches of the plexus arise from the loops. The *ansa cervicalis* is a second-level loop, the superior limb of which arises from the loop between C1 (atlas) and C2 (axis) but travels initially with the hypoglossal nerve (CN XII)—not part of the cervical plexus.

The **ventral rami (roots) of the** brachial plexus appear in the neck between the anterior and middle scalene muscles (Fig. 8.7*A*). The five rami (C5 through C8 and T1) unite to form the *three trunks of the brachial plexus* that descend inferolaterally through the posterior triangle. The plexus then passes between the 1st rib and clavicle (*cervicoaxillary canal*) to enter the axilla (armpit), providing innervation for most of the upper limb (see Chapter 6). The ventral rami of the first four cervical nerves form the **cervical plexus** (Fig. 8.10); each nerve, except the first one, divides into ascending and descending branches that unite with the branches of the adjacent spinal nerve to form loops. These loops and the branches of these loops, which lie anterolateral to the levator scapulae and middle scalene muscles and deep to the SCM, compose the cervical plexus. Whereas its posterior branches provide cutaneous branches to the anterolateral neck and the superior part of the thorax (Figs. 8.10 and 8.11*A*), its anterior branches form the **ansa cervicalis**, a nerve loop that supplies the infrahyoid muscles in the anterior cervical triangle (Fig. 8.12).

Cutaneous branches from the cervical plexus emerge around the middle of the posterior border of the SCM, known clinically as *the nerve point of the neck*, and supply the skin of the neck, superolateral thoracic wall, and scalp between the auricle (pinna) and the external occipital protuberance. Close to their origin, the nerves of the cervical plexus receive gray *rami communicantes*, most of which descend from the large *superior cervical ganglion* in the superior part of the neck. The **branches of cervical plexus** (Figs. 8.10–8.12) arising from the loop between the ventral rami of C2 and C3 are the:

- *Lesser occipital nerve (C2)*, supplying skin of the neck and the scalp posterosuperior to the auricle
- *Great auricular nerve (C2 and C3)*, ascending diagonally across the SCM onto the parotid gland, where it divides and supplies the skin over the gland, the posterior aspect of the auricle, and an area of skin extending from the angle of the mandible to the mastoid process
- *Transverse cervical nerve (C2 and C3)*, supplying skin covering the anterior triangle; the nerve curves around the middle of the posterior border of the SCM and crosses this muscle deep to the platysma.

Branches of the cervical plexus arising from the loop formed between the ventral rami of C3 through C4 are the:

- *Supraclavicular nerves (C3 and C4)*, which emerge as a common trunk under cover of the SCM and send small branches to the skin of the neck; they then cross the clavicle and supply skin over the shoulder.

Branches of the ventral primary rami of cervical nerves supply motor branches to the rhomboids (dorsal scapular nerve [C4 and C5]), serratus anterior (long thoracic nerve [C5, **C6**, and **C7**]), and nearby prevertebral muscles.

The **suprascapular nerve**, which arises from the superior trunk of the **brachial plexus** (not cervical), runs laterally across the posterior triangle (Fig. 8.7) and supplies the supraspinatus and infraspinatus muscles; it also sends branches to the glenohumeral (shoulder) joint.

The **phrenic nerve** (Figs. 8.6 and 8.7) originates chiefly from the 4th cervical nerve (C4) but receives contributions from the 3rd and 5th cervical nerves (C3 and C5; Fig. 8.10). The phrenic nerves contain motor, sensory, and sympathetic nerve fibers. These nerves provide the sole motor supply to the diaphragm as well as sensation to its central part. In the thorax, each phrenic nerve supplies the mediastinal pleura and pericardium (see Chapter 1). Receiving variable communicating fibers in the neck from the cervical sympathetic ganglia or their branches, each phrenic nerve forms at the superior part of the lateral border of the anterior scalene muscle at the level of the superior border of the thyroid cartilage.

The phrenic nerve descends obliquely with the IJV across the anterior scalene, deep to the prevertebral layer of deep cervical fascia and the transverse cervical and suprascapular arteries. *On the left*, the phrenic nerve crosses anterior to the first part of the subclavian artery; *on the right*, it lies on the anterior scalene muscle and crosses anterior to the second part of the subclavian artery. On both sides, the phrenic nerve runs posterior to the subclavian vein and anterior to the internal thoracic artery as it enters the thorax.

The contribution of the 5th cervical nerve to the phrenic nerve may derive from an **accessory phrenic nerve** (Fig. 8.7). Frequently it is a branch of the nerve to the subclavius and may contain a large number of phrenic nerve fibers. If present, it lies lateral to the main nerve and descends posterior and sometimes inferior to the subclavian vein. The accessory phrenic nerve joins the phrenic nerve either in the root of the neck or in the thorax.

Severance of the Phrenic Nerve

Severance of a phrenic nerve results in paralysis of the corresponding half of the diaphragm, and a *phrenic nerve block* will produce a short period of paralysis of the diaphragm on one side (e.g., for a lung operation). The anesthetic is injected around the nerve where it lies on the anterior surface of the middle third of the anterior scalene. A *surgical phrenic nerve crush* will produce a longer period of paralysis (e.g., for weeks after surgical repair of a diaphragmatic hernia). If an accessory phrenic nerve is present, it must also be crushed to produce complete paralysis of the hemidiaphragm.

Nerve Blocks in the Posterior Triangle

For regional anesthesia before surgery, a nerve block of the nerves of the cervical and brachial plexuses inhibits nerve impulse conduction. A dentist blocks the nerves ▶

▶ supplying the teeth before performing dental work in a similar fashion. In a *cervical plexus block,* an anesthetic agent is injected at several points along the posterior border of the SCM, mainly at the junction of its superior and middle thirds—*the nerve point of the neck.* Because the phrenic nerve supplying half the diaphragm is usually paralyzed by a cervical nerve block, this procedure is not performed on patients with pulmonary or cardiac disease. For anesthesia of the upper limb, the anesthetic agent in a supraclavicular *brachial plexus block* goes around the supraclavicular part of the brachial plexus. The main injection site is superior to the midpoint of the clavicle.

Injury to the Suprascapular Nerve

The suprascapular nerve is vulnerable to injury in fractures of the middle third of the clavicle. A lesion of this nerve results in loss of lateral rotation of the humerus (L. arm bone) at the shoulder so that when relaxed the limb rotates medially in the *waiter's tip position* (p. 716). The ability to initiate abduction of the limb is also affected. ⊙

Lymph Nodes in the Posterior Triangle

The lymph from superficial tissues in the posterior triangle enters the **superficial cervical lymph nodes** that lie along the EJV superficial to the SCM. Efferent vessels from the superficial cervical nodes drain into the **deep cervical lymph nodes**, which form a chain along the course of the IJV and are embedded in the fascia of the carotid sheath.

Anterior Cervical Triangle

The anterior triangle of the neck is bounded by the anterior border of the SCM, the anterior midline of the neck, and the mandible. The anterior triangle is subdivided into four smaller triangles for descriptive purposes (Table 8.2):

- The **submandibular triangle** is a *glandular area* between the inferior border of the mandible and the anterior and posterior bellies of the digastric muscle (Fig. 8.11*A*). Some people refer to it as the "digastric triangle." The floor of the submandibular triangle is formed by the mylohyoid muscle, hyoglossus muscle, and middle constrictor of the pharynx. The **submandibular gland** nearly fills this triangle. Approximately half the size of the parotid gland, the submandibular gland is usually palpable as a soft mass between the body of the mandible and the mylohyoid muscle. **Submandibular lymph nodes** lie on each side of the submandibular gland and along the inferior border of the mandible. The **submandibular duct** (Fig. 8.11*B*), approximately 5 cm in length, passes from the deep process of the

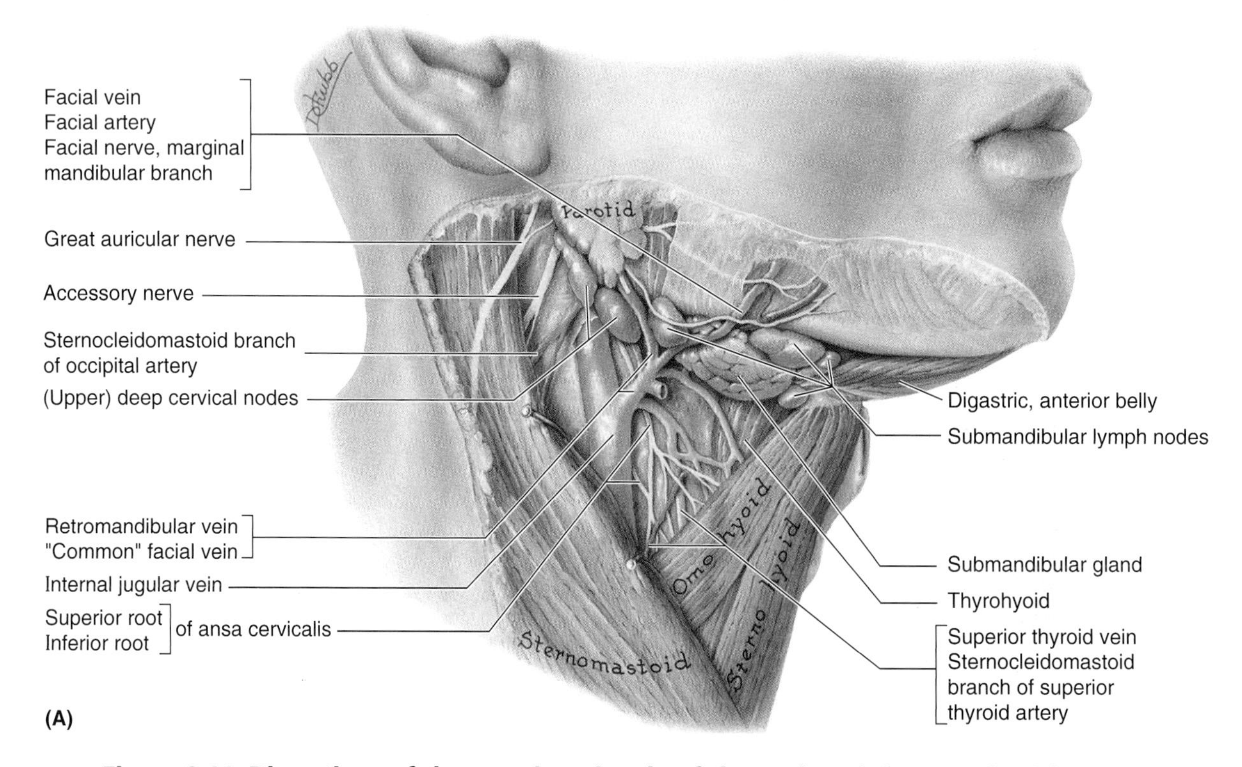

Figure 8.11. Dissections of the anterior triangle of the neck and the suprahyoid region. Lateral views. **A.** Superficial dissection of the neck. Observe the submandibular gland and lymph nodes. **B**. Suprahyoid region. The right half of the mandible (lower jaw) and the superior part of the mylohyoid muscle have been removed. Observe that the cut surface of the mylohyoid muscle becomes progressively thinner as it is traced anteriorly. Also observe the lingual nerve between the medial pterygoid muscle and the ramus of the mandible.

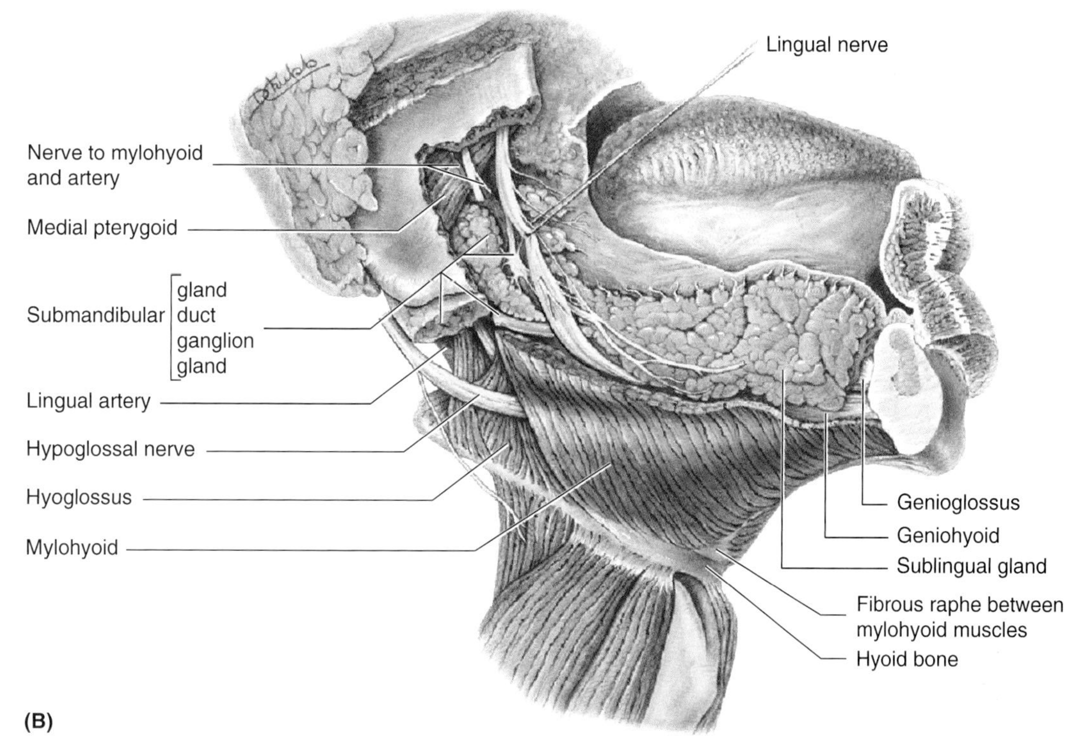

Figure 8.11. *(Continued)* **B.** Suprahyoid region. (See legend with (**A**) on previous page.)

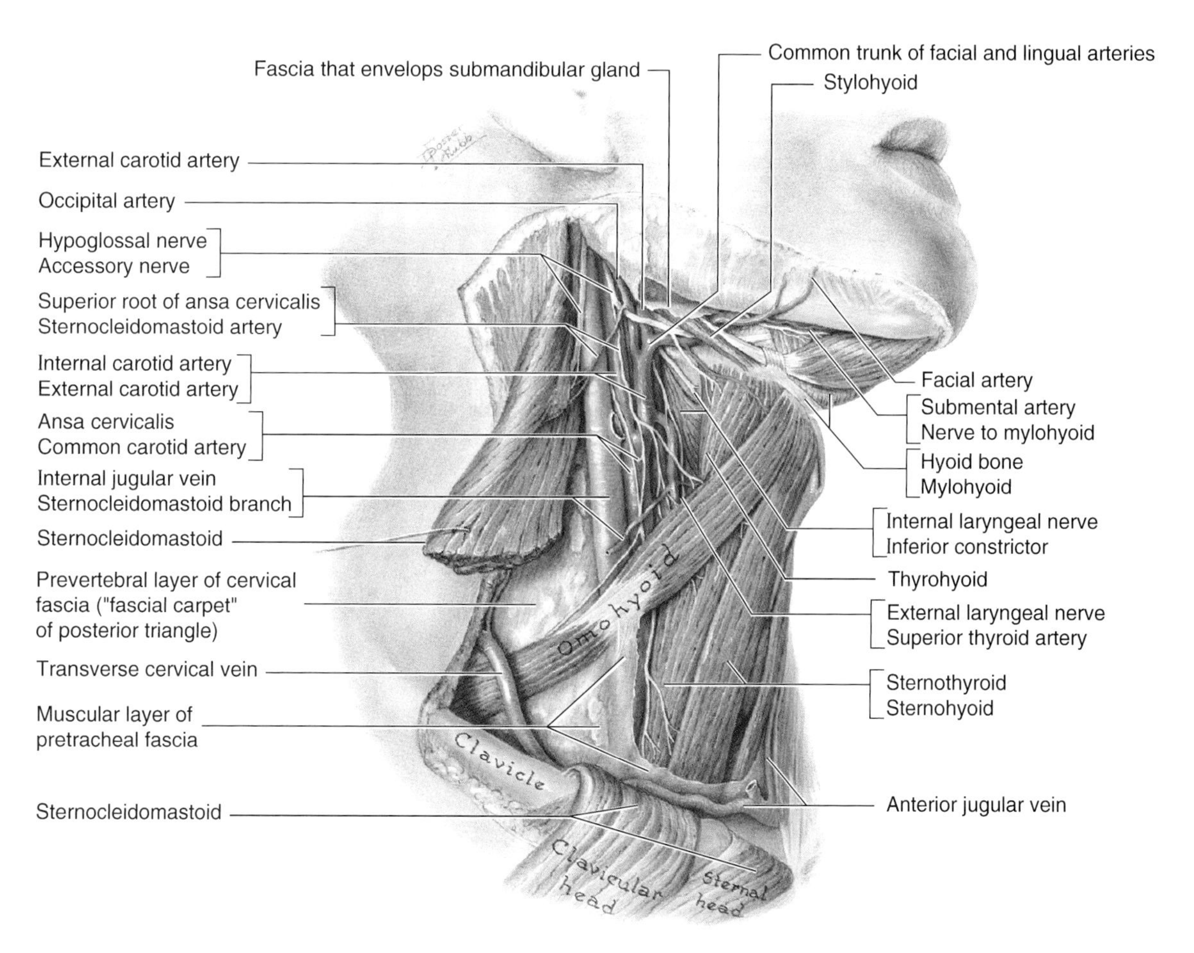

Figure 8.12. Deep dissection of the anterior triangle of the neck. Right side. Observe that the facial and lingual arteries in this person arise by a common trunk that passes deep to the stylohyoid and digastric muscles to enter the submandibular triangle.

submandibular gland, parallel to the tongue (L. lingua; G. glossa), to open by one to three orifices into the oral cavity. The openings are on an elevation, the *sublingual papilla*, which is produced at the side of the lingual frenulum by the sublingual gland. The *hypoglossal nerve* (CN XII), motor to the intrinsic and extrinsic muscles of the tongue, passes into the submandibular triangle, as do the *nerve to the mylohyoid muscle* (which also supplies the anterior belly of the digastric), parts of the *facial artery* and vein, and the *submental artery*, a branch of the facial artery (Figs. 8.11 and 8.12).

- The **submental triangle**—inferior to the chin—is an unpaired *suprahyoid area* bounded inferiorly by the body of the hyoid bone and laterally by the right and left anterior bellies of the digastric muscles (Table 8.2). The floor of the submental triangle is formed by the two mylohyoid muscles, which meet in a median **fibrous raphe** (Fig. 8.11*B*, Fig. 8.13). The apex of the submental triangle is at the mandibular symphysis (L. symphysis menti), the site of union of the mandibular halves during infancy, and its base is formed by the hyoid bone (Fig. 8.13). This triangle contains several small **submental lymph nodes.** The submental triangle also contains small veins that unite to form the *anterior jugular vein* (Fig. 8.12).
- The **carotid triangle** is a *vascular area* bounded by the superior belly of the omohyoid, the posterior belly of the digastric, and the anterior border of the SCM (Figs. 8.11*A* and 8.12). The carotid triangle is an important area because the **common carotid artery** ascends into and is where its pulse can be auscultated or palpated by compressing it lightly against the transverse processes of the cervical vertebrae. At the level of the superior border of the thyroid cartilage, the common carotid artery divides into the *internal* and *external carotid arteries* (Figs. 8.12, 8.14, and 8.15). Found at this site are:
    - The **carotid sinus**, a slight dilation of the proximal part of the internal carotid artery; the dilation may involve the common carotid (Figs. 8.14 and 8.15). Innervated principally by the glossopharyngeal nerve (CN IX)

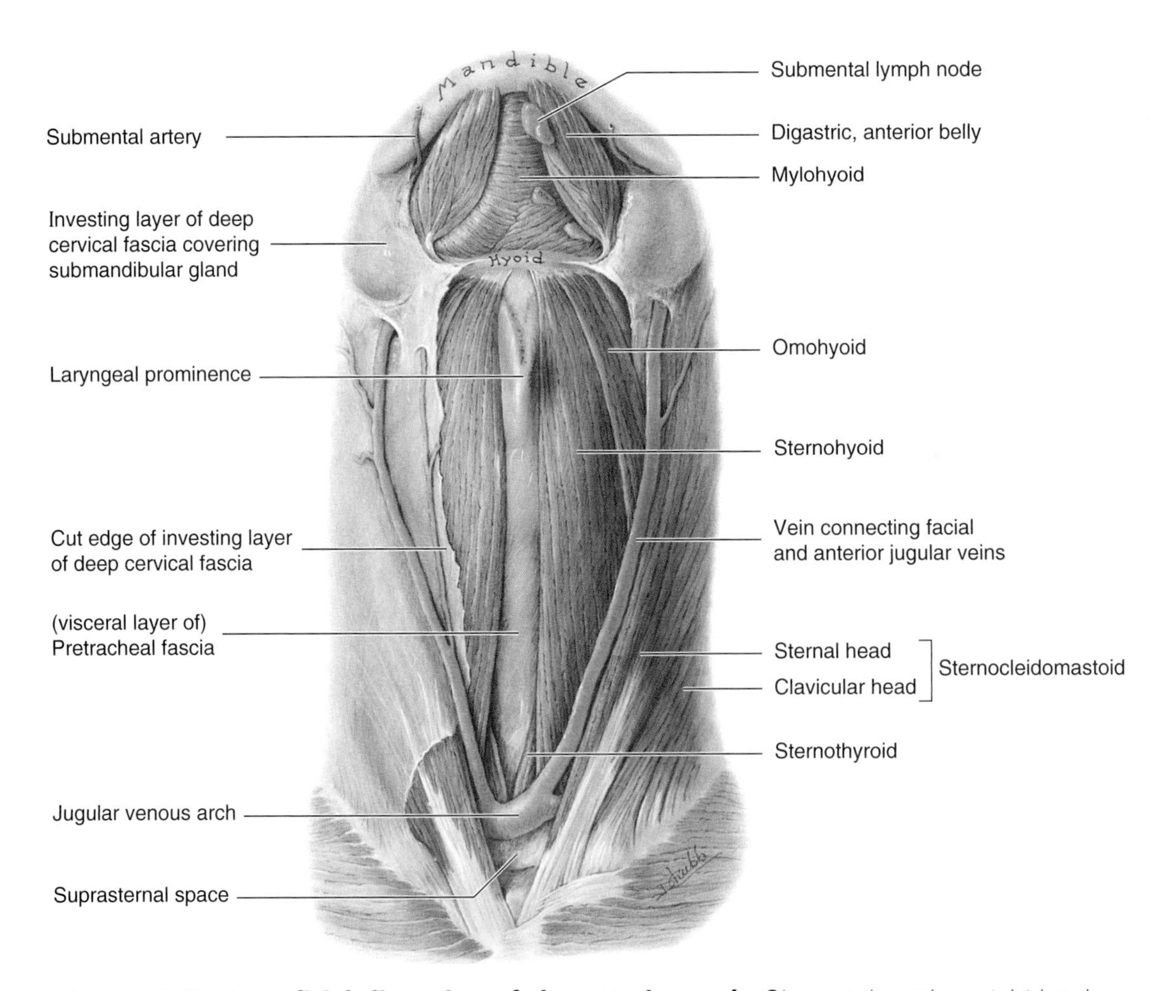

Figure 8.13. Superficial dissection of the anterior neck. Observe the submental triangle, bounded inferiorly by the body of the hyoid bone and laterally by the right and left anterior bellies of the digastric muscles. Note that the triangle contains submental lymph nodes and that its floor is formed by the two mylohyoid muscles (the raphe common to the right and left mylohyoids is not apparent here).

through the carotid sinus nerve, as well as by the vagus nerve, the *carotid sinus is a baroreceptor* (pressoreceptor) that reacts to changes in arterial blood pressure.

- The **carotid body**, a small, reddish-brown ovoid mass of tissue, lies on the medial (deep) side of the bifurcation of the common carotid artery in close relation to the carotid sinus (Fig. 8.15). Supplied mainly by the carotid sinus nerve (CN IX) and by CN X, *the carotid body is a chemoreceptor that monitors the level of oxygen in the blood.* It is stimulated by low levels of oxygen and initiates a reflex, which increases the rate and depth of respiration, cardiac rate, and blood pressure.

The **carotid sheath**, a tubular, thickly matted fascial condensation on each side of the neck, extends from the base of the skull to the root of the neck. It is formed by fascial extensions of all three layers of deep cervical fascia that fuse together. The inferior part of the carotid sheath contains the following clinically important structures (Fig. 8.12, see Fig. 8.17):

- The *common carotid artery* medially
- The *IJV* laterally
- The *vagus nerve* posteriorly.

Superiorly, the common carotid is replaced by the internal carotid artery. The *ansa cervicalis* usually lies on (or is embedded in) the anterolateral aspect of the sheath. Many *deep cervical lymph nodes* lie along the carotid sheath and the IJV.

- The **muscular triangle** is bounded by the superior belly of the omohyoid muscle, the anterior border of the SCM, and the median plane of the neck. This triangle contains the *infrahyoid muscles* and viscera of the neck, such as the *thyroid* and *parathyroid glands.*

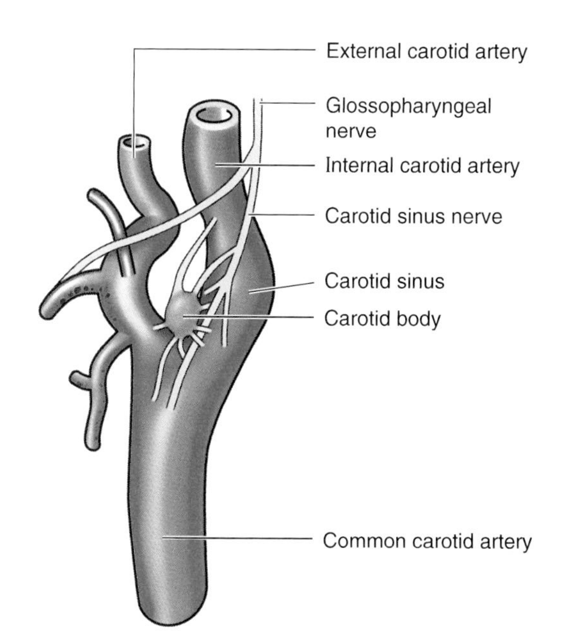

Figure 8.14. Carotid body. Observe that this small epithelioid body lies within the bifurcation of the common carotid artery. Also observe the carotid sinus and the associated network of sensory fibers of the glossopharyngeal nerve (CN IX).

Muscles in the Anterior Triangle

In the anterolateral part of the neck, the **hyoid bone** provides attachments for the suprahyoid muscles superior to it and the infrahyoid muscles inferior to it. These **hyoid muscles** steady or move the hyoid bone and larynx (Figs. 8.12 and 8.13). For descriptive purposes, they are divided into suprahyoid and infrahyoid muscles whose attachments, innervation, and main actions are presented in Table 8.3.

The **suprahyoid muscles** are superior to the hyoid bone and connect it to the skull (Figs. 8.11–8.13). The suprahyoid group includes the mylohyoid, geniohyoid, stylohyoid, and digastric muscles.

- *Mylohyoid muscles* form the mobile but stable floor of the mouth and a muscular sling inferior to the tongue. These muscles support the tongue and elevate it and the hyoid bone when swallowing or protruding the tongue.
- *Geniohyoid muscles* are superior to the mylohyoid muscles, where they reinforce the floor of the mouth.
- The *stylohyoid muscles* form a slip on each side, which is nearly parallel to the posterior belly of the digastric muscle (Table 8.3).
- *Digastric muscles,* each of which has two bellies that descend toward the hyoid bone, are joined by an *intermediate tendon.* A fibrous sling derived from the deep cervical fascia allows it to slide anteriorly and posteriorly as it connects this tendon to the body and greater horn of the hyoid bone. The difference in nerve supply between the two bellies of each digastric muscle results from the embryological origin of the anterior and posterior bellies from the 1st and 2nd pharyngeal arches, respectively (Moore and Persaud, 1998). CN V is the nerve to the 1st arch, and CN VII supplies the 2nd arch.

The **infrahyoid muscles**—often called strap muscles because of their ribbonlike appearance—are inferior to the hyoid bone (Figs. 8.12 and 8.13, Table 8.3). These four muscles anchor the hyoid bone, sternum, clavicle, and scapula and depress the hyoid bone and larynx during swallowing and speaking. They also work with the suprahyoid muscles to steady the hyoid bone, providing a firm base for the tongue. The infrahyoid group includes the sternohyoid, omohyoid, sternothyroid, and thyrohyoid muscles, which are arranged in two planes: a *superficial plane* comprising the sternohyoid and omohyoid and a *deep plane* comprising the sternothyroid and thyrohyoid.

- The *sternohyoid,* a thin narrow muscle, lies superficially parallel and adjacent to the anterior median line.
- The *omohyoid muscle,* lateral to the sternohyoid, has two bellies united by an intermediate tendon that connects to the clavicles by a fascial sling.

Table 8.3. Suprahyoid and Infrahyoid Muscles

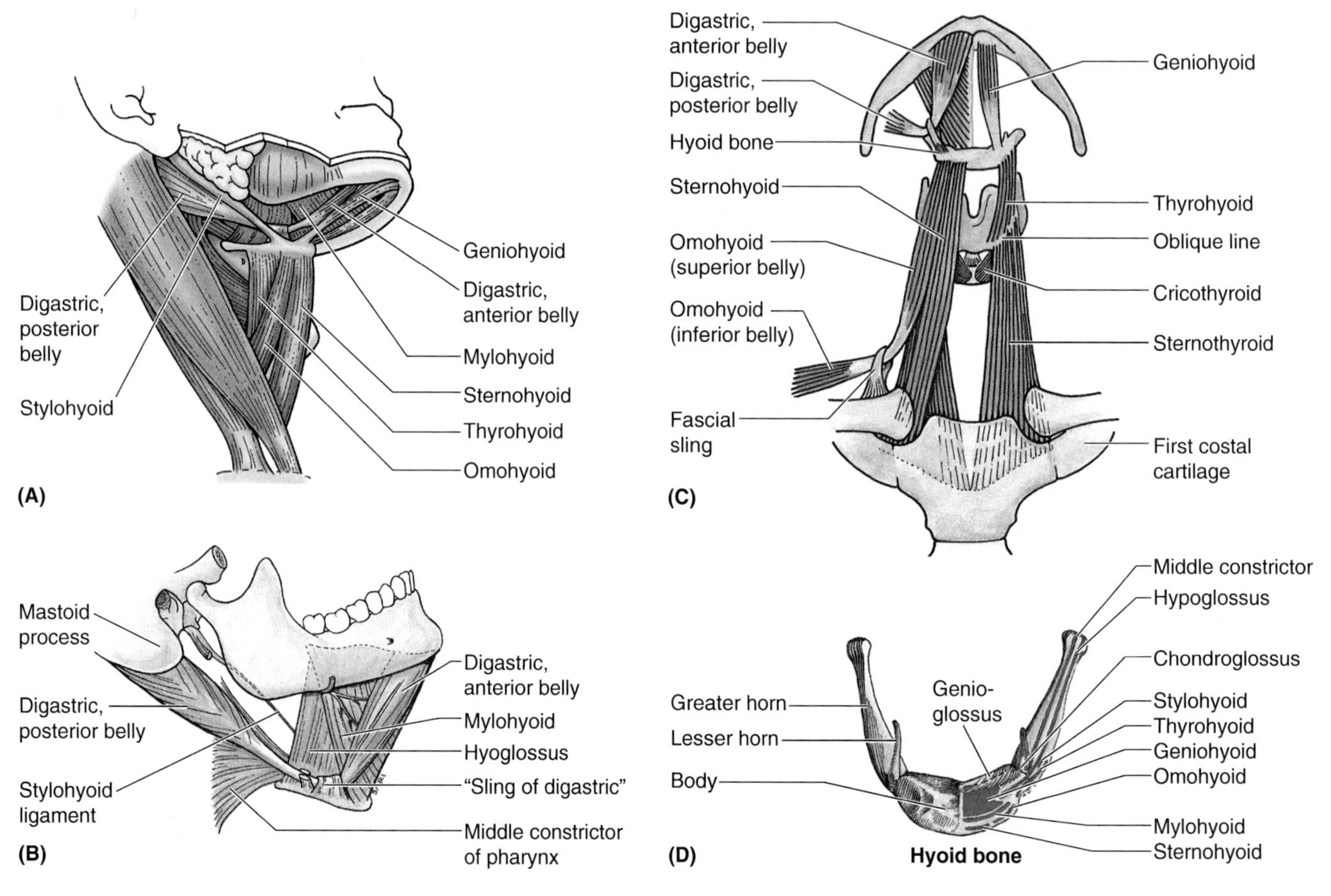

Muscle	Origin	Insertion	Innervation	Main Action
Suprahyoid				
Mylohyoid	Mylohyoid line of mandible	Raphe and body of hyoid bone	Mylohyoid nerve, a branch of inferior alveolar nerve of CN V_3	Elevates hyoid bone, floor of mouth, and tongue during swallowing and speaking
Geniohyoid	Inferior mental spine of mandible	Body of hyoid bone	C1 via the hypoglossal nerve	Pulls hyoid bone anterosuperiorly, shortens floor of mouth, and widens pharynx
Stylohyoid	Styloid process of temporal bone		Cervical branch of facial nerve	Elevates and retracts hyoid bone, thereby elongating floor of mouth
Digastric	Anterior belly: digastric fossa of mandible Posterior belly: mastoid notch of temporal bone	Intermediate tendon to body and greater horn of hyoid bone	Anterior belly: mylohyoid nerve, a branch of inferior alveolar nerve Posterior belly: facial nerve	Depresses mandible; raises hyoid bone and steadies it during swallowing and speaking
Infrahyoid				
Sternohyoid	Manubrium of sternum and medial end of clavicle	Body of hyoid bone	C1–C3 by a branch of ansa cervicalis	Depresses hyoid bone after it has been elevated during swallowing
Omohyoid	Superior border of scapula near suprascapular notch	Inferior border of hyoid bone		Depresses, retracts, and steadies hyoid bone
Sternothyroid	Posterior surface of manubrium of sternum	Oblique line of thyroid cartilage	C2 and C3 by a branch of ansa cervicalis	Depresses hyoid bone and larynx
Thyrohyoid	Oblique line of thyroid cartilage	Inferior border of body and greater horn of hyoid bone	C1 via hypoglossal nerve	Depresses hyoid bone and elevates larynx

- The *sternothyroid* is wider than the sternohyoid, under which it lies. The sternothyroid covers the lateral lobe of the thyroid gland, attaching to the oblique line of the lamina of the thyroid cartilage immediately above it. This muscle limits upward expansion of the thyroid gland; thus, tumors or goiters, which may enlarge it, cause it to expand anteriorly or inferiorly into the mediastinum.
- The *thyrohyoid* appears to be the continuation of the sternothyroid muscle, running superiorly from the oblique line of the thyroid cartilage to the hyoid bone.

Vessels in the Anterior Triangle

The anterior triangle of the neck contains the carotid system of arteries—the common carotid artery and its terminal branches, the internal and external carotid arteries. It also contains the IJV and its tributaries and the anterior jugular veins (Figs. 8.12–8.15).

Carotid Arteries. The *common carotid artery* and one of its terminal branches, the *external carotid artery*, are the main arterial vessels in the carotid triangle (Figs. 8.12 and 8.14). Branches of the external carotid (e.g., the superior thyroid artery) also originate in the carotid triangle. Each common carotid artery ascends within the **carotid sheath** with the IJV and vagus nerve to the level of the superior border of the thyroid cartilage, then terminates by dividing into the internal and external carotid arteries. Whereas the internal carotid has no branches in the neck, the external carotid has several.

The **right common carotid artery** begins at the bifurcation of the brachiocephalic trunk (Fig. 8.14). The right subclavian artery is the other branch of the brachiocephalic trunk. From the arch of the aorta, the **left common carotid artery** ascends into the neck. Consequently, the left common carotid has a course of approximately 2 cm in the superior mediastinum (see Chapter 1) before entering the neck.

The **internal carotid arteries**, the direct continuation of the common carotids, *have no branches in the neck*. They enter the skull through the carotid canals in the petrous parts of the temporal bones (see Chapter 7; p. 874) and become the main arteries of the brain and structures in the orbits. Each

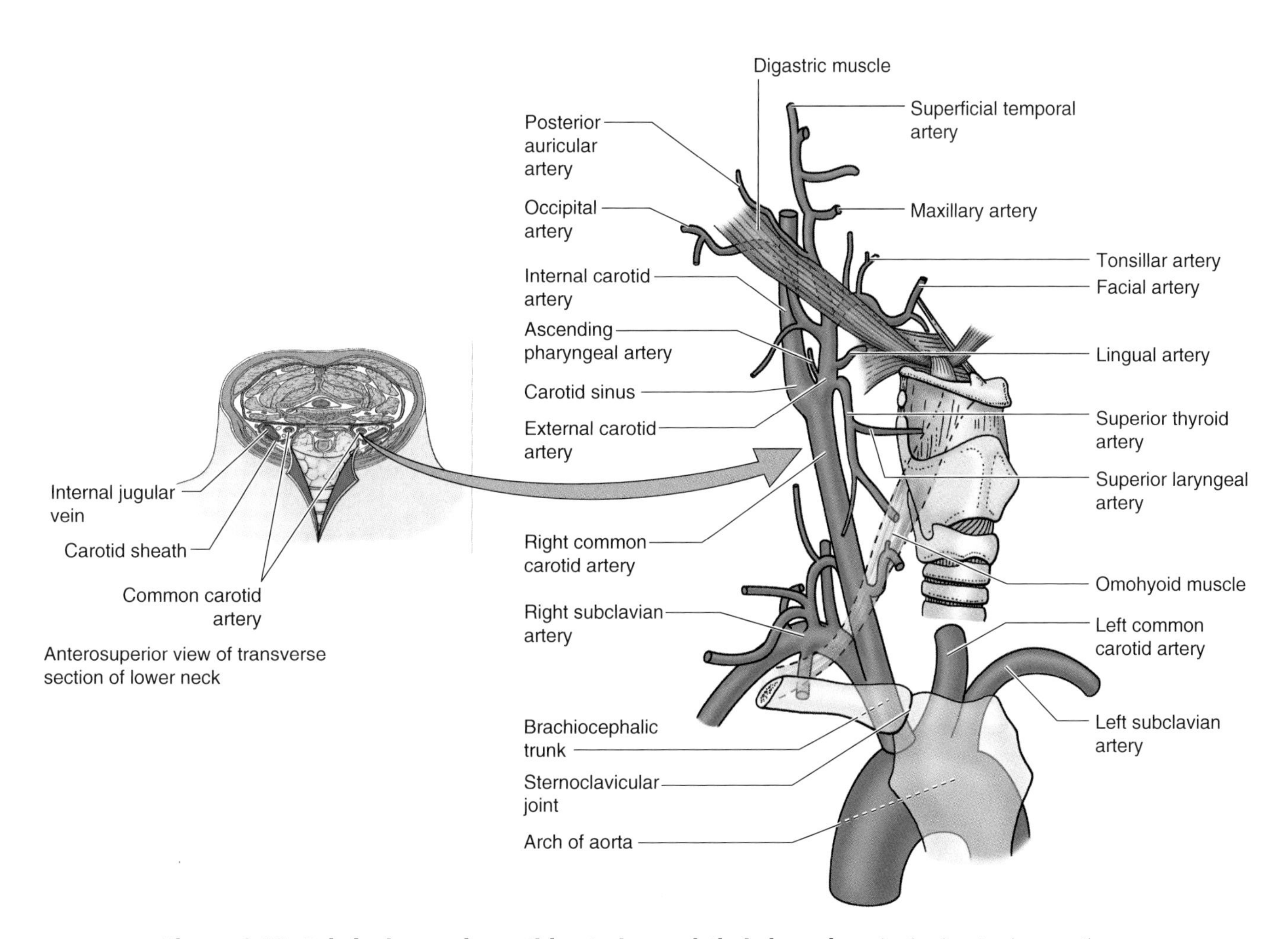

Figure 8.15. Subclavian and carotid arteries and their branches. In the inset, observe the carotid sheath, the dense fibrous investment of the common carotid artery, internal jugular vein (IJV), and vagus nerve (CN X). The carotid sheath is deep to the sternocleidomastoid (SCM).

internal carotid artery arises from the common carotid artery at the level of the superior border of the thyroid cartilage. The proximal part of each internal carotid artery is the site of the **carotid sinus**, a slight dilation that may involve the terminal part of the common carotid artery (Figs. 8.14 and 8.15). The wall of the sinus contains receptors that are sensitive to changes in blood pressure. An important regulator of bloodflow to the brain, the carotid sinus reacts to changes in head position and gravitational forces to keep the flow of blood nearly constant. The **carotid body** at the bifurcation of the common carotid artery is a *chemoreceptor* that responds to changes in the chemical composition of blood—specifically, the degree of oxygenation.

The **external carotid arteries** for the main part supply structures external to the skull (the middle meningeal artery is an obvious exception). Each artery runs posterosuperiorly to the region between the neck of the mandible and the lobule (earlobe) of the auricle, where it is embedded in the parotid gland, and terminates by dividing into two branches, the *maxillary artery* and *superficial temporal artery* (Fig. 8.14). The other six branches of the external carotid artery are the ascending pharyngeal, superior thyroid, lingual, facial, occipital, and posterior auricular arteries.

Ascending Pharyngeal Artery. The ascending pharyngeal artery, the 1st or 2nd branch of the external carotid artery, ascends on the pharynx deep (medial) to the internal carotid artery (see Fig. 7.78*A*) and sends branches to the pharynx, prevertebral muscles, middle ear, and cranial meninges.

Superior Thyroid Artery. The most inferior of the three anterior branches of the external carotid artery, the superior thyroid artery runs anteroinferiorly deep to the infrahyoid muscles to reach the thyroid gland. In addition to supplying this gland, it gives off branches to the infrahyoid muscles and SCM. It also gives rise to the *superior laryngeal artery* supplying the larynx (Fig. 8.15).

Lingual Artery. This artery arises from the external carotid where it lies on the middle constrictor muscle of the pharynx. It arches superoanteriorly and passes deep to the hypoglossal nerve (CN XII), the stylohyoid muscle, and the posterior belly of the digastric muscle (Fig. 8.17). It disappears deep to the hyoglossus and turns superiorly at the anterior border of this muscle to become the deep lingual and sublingual arteries.

Facial Artery. The facial artery arises from the external carotid artery, either in common with the lingual artery (Fig. 8.12) or immediately superior to it. The facial artery gives off a *tonsillar branch* (which, of the branches from five different arteries to the palatine tonsil, is the major one), as well as branches to the palate and submandibular gland. It then passes superiorly under cover of the digastric and stylohyoid muscles and the angle of the mandible. The facial artery loops anteriorly and enters a deep groove in the submandibular gland. It then hooks around the middle of the inferior border of the mandible (where its pulsations can be palpated) and enters the face.

Occipital Artery. The occipital artery arises from the posterior aspect of the external carotid artery, superior to the origin of the facial artery. It passes posteriorly, parallel and deep to the posterior belly of the digastric muscle, forming its own groove in the base of the skull medial to that of the origin of the muscle, and ends in the posterior part of the scalp. During its course, it passes superficial to the internal carotid artery and CN IX through CN XI.

Posterior Auricular Artery. This artery, a small posterior branch of the external carotid artery (Fig. 8.15; see also Table 7.5 Fig. (B), p. 866, and Fig. 7.74, p. 964), ascends posteriorly between the external acoustic meatus and the mastoid process to supply the adjacent muscles, parotid gland, facial nerve, structures in the temporal bone, auricle, and scalp.

Ligation of the External Carotid Artery

Sometimes ligation of an external carotid artery is necessary to control bleeding from one of its relatively inaccessible branches. This procedure decreases bloodflow through the artery and its branches but does not eliminate it. Blood flows retrogradely (passes backward) into the artery from the external carotid artery on the other side through communications between its branches (e.g., those in the face and scalp) or across the midline. When the external carotid or subclavian arteries are ligated, the descending branch of the occipital artery provides the main collateral circulation, anastomosing with the vertebral and deep cervical arteries.

Surgical Dissection of the Carotid Triangle

The carotid triangle provides an important surgical approach to the carotid system of arteries. It is also provides approaches to the IJV, the vagus and hypoglossal nerves, and the cervical sympathetic trunk. Damage or compression of the vagus and/or recurrent laryngeal nerves during surgical dissection of the carotid triangle may produce an alteration in the voice because these nerves supply laryngeal muscles.

Carotid Endarterectomy

Atherosclerotic thickening of the intima of the internal carotid artery, which obstructs bloodflow, can be observed in a *Doppler color study.* A doppler is a diagnostic instrument that emits an ultrasonic beam that reflects from moving structures. Partial occlusion of the internal carotid may also cause a *transient ischemic attack*—sudden focal loss of neurological function (e.g., dizziness and disorientation) that disappears within 24 hours. Arterial occlusion may also cause a *minor stroke*—loss of neurological function such as weakness or sensory loss on one side of the body that exceeds 24 hours but disappears within 3 weeks ▶

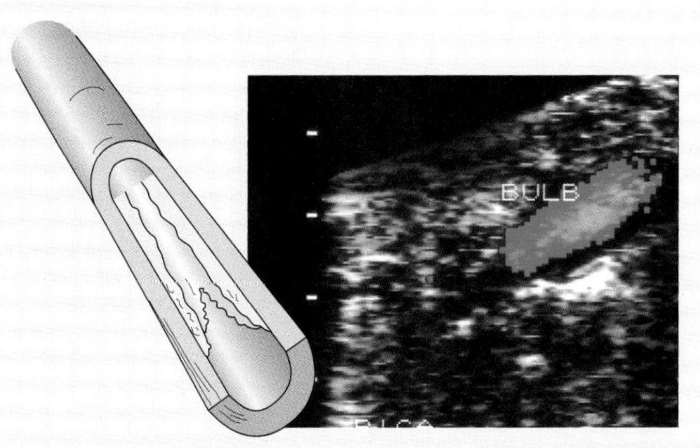

Occlusion of carotid artery (Doppler color flow study)

▶ (Sacco, 1995). The symptoms resulting from obstruction of bloodflow depend on the degree of obstruction and the amount of collateral bloodflow to the brain and structures in the orbit from other arteries.

Carotid stenosis (narrowing) in healthy patients can be relieved by opening the artery and stripping off the atherosclerotic plaque with the intima (Hallett, 1994). The common site for a carotid endarterectomy is the internal carotid artery, just superior to its origin. After the operation, administered drugs inhibit clot formation in the operated area until the endothelium has regrown. Because of the relations of the internal carotid artery, *cranial nerve injury* may occur during carotid endarterectomy and involve one or more of the following nerves: glossopharyngeal, vagus, accessory, hypoglossal, and superior laryngeal nerves. The superior laryngeal nerve is a branch of the vagus nerve at the inferior vagal ganglion.

Carotid Pulse

The carotid pulse (neck pulse) is easily felt by palpating the common carotid artery in the side of the neck, where it lies in a groove between the trachea and the infrahyoid (strap) muscles. It is usually easily palpated just deep to the anterior border of the SCM at the level of the superior border of the thyroid cartilage. It is routinely checked during *cardiopulmonary resuscitation*. Absence of a carotid pulse indicates cardiac arrest.

Carotid Artery Palpation

External pressure on the carotid artery in people with *carotid sinus hypersensitivity* may cause slowing of the heart rate, a fall in blood pressure, and cardiac ischemia with fainting (*syncope*). In all forms of syncope, symptoms result from a sudden and critical decrease in cerebral perfusion (Pedley and Ziegler, 1995). Consequently, this method of taking the pulse is not recommended for cardiac patients who are participating in cardiac rehabilitation programs. Because various types of vascular disease affect the sensitivity of the carotid sinus, the radial pulse at the wrist is most commonly checked.

Role of the Carotid Bodies

The carotid bodies are in an ideal position to monitor the oxygen content of the blood before it reaches the brain. A decrease in PO_2 (partial pressure tension] of oxygen), as occurs at high altitudes or in pulmonary disease, activates the aortic and carotid chemoreceptors, increasing alveolar ventilation. The carotid bodies respond either to increased carbon dioxide tension or to decreased oxygen tension in the blood. A fall in the oxygen content or an increase in carbon dioxide initiates reflexes through the glossopharyngeal and vagus nerves, which stimulate respiration, increasing the depth and rate of breathing; pulse rate and blood pressure also rise. Thus, more oxygen is taken in and more carbon dioxide is blown off. ⭘

Veins. Most veins in the anterior triangle are tributaries of the **IJV**, usually the largest vein in the neck (Figs. 8.12 and 8.16). The IJV drains blood from the brain, anterior face, cervical viscera, and deep muscles of the neck. It commences at the jugular foramen in the posterior cranial fossa as the direct continuation of the sigmoid sinus (see Chapter 7). From the dilation at its origin, the **superior bulb of the IJV** (Fig. 8.16), the vein runs inferiorly through the neck in the *carotid sheath* with the internal carotid, then the common carotid artery and vagus nerve. The artery is medial and the vein is lateral, and the nerve lies posteriorly between these vessels (Fig. 8.17). The *cervical sympathetic trunk lies posterior to the carotid sheath,* and, although closely related, the trunk does not lie within the sheath but instead is embedded in the prevertebral fascia (see Fig. 7.20). The IJV leaves the anterior triangle by passing deep to the SCM. The inferior end of the vein is deep to the gap between the sternal and clavicular heads of the SCM.

Posterior to the sternal end of the clavicle, the IJV unites with the subclavian vein to form the **brachiocephalic vein**

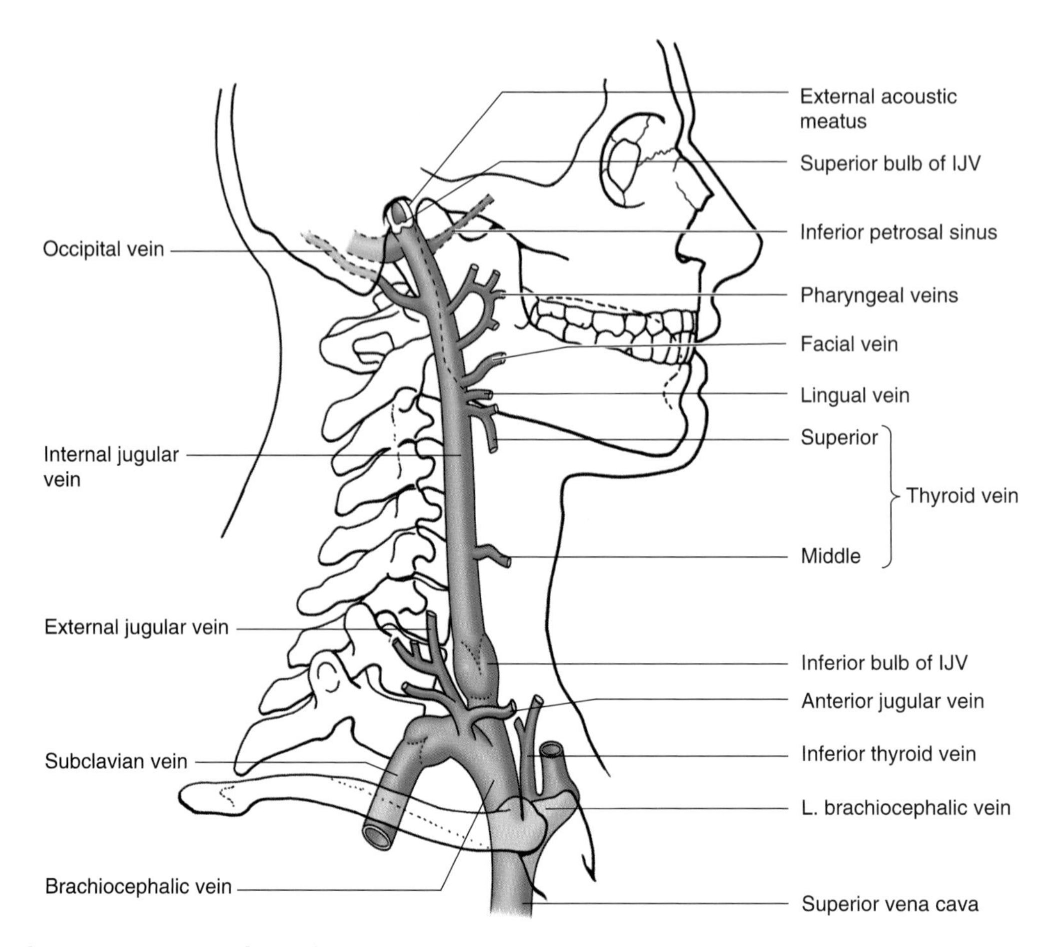

Figure 8.16. Internal jugular vein (IJV). This is the main venous structure of the neck. It originates as a continuation of the sigmoid (dural venous) sinus, is contained in the carotid sheath as it descends in the neck, and unites at the T1 vertebral level superior to the sternoclavicular (SC) joint with the subclavian vein to form the brachiocephalic vein. Notice the large valve near its termination that prevents reflux of blood into the vein.

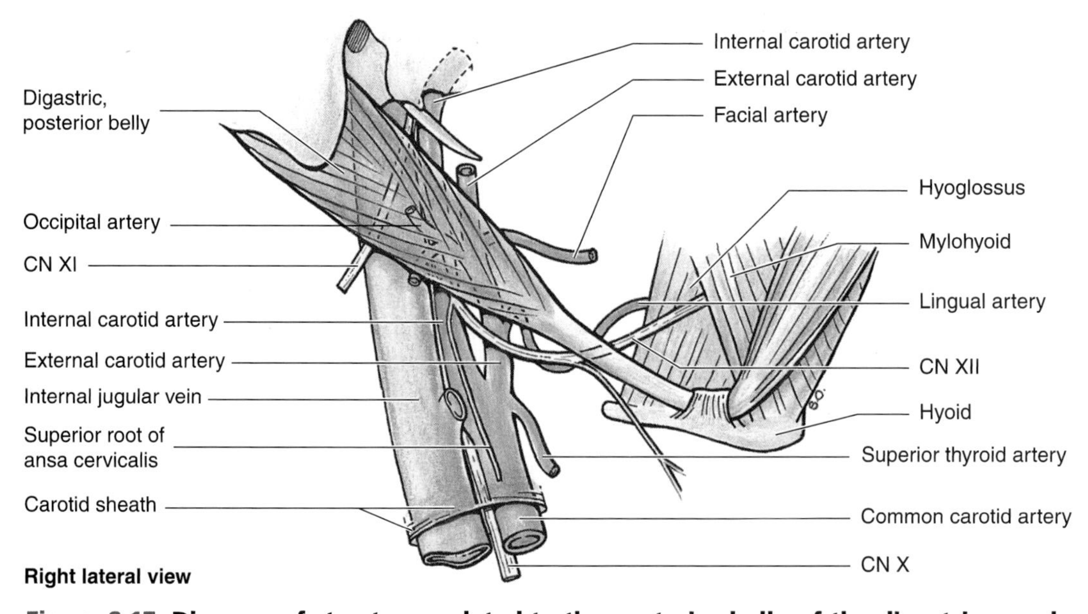

Figure 8.17. Diagram of structures related to the posterior belly of the digastric muscle. Observe the superficial and key position of the posterior belly of the digastric muscle that runs from the mastoid process to the hyoid bone. Note that all vessels and nerves cross deep to this belly except for the cervical branches of the facial nerve (CN VII), facial branches of the great auricular nerve, and the external jugular vein (EJV) and its connections (none of which are shown here).

(Fig. 8.16), and the inferior end of the IJV dilates to form the **inferior bulb of the IJV**. The inferior bulb contains a bicuspid valve that permits blood to flow toward the heart while preventing backflow into the vein, as might occur if inverted (standing on one's head or when intrathoracic pressure is increased). The *tributaries of the IJV* are the inferior petrosal sinus, facial and lingual (often by a common trunk), pharyngeal, and superior and middle thyroid veins. The **occipital vein** usually drains into the *suboccipital venous plexus,* drained by the deep cervical vein and the vertebral vein, but it may drain into the IJV.

The **inferior petrosal sinus** (Fig. 8.16) leaves the skull through the jugular foramen and enters the superior bulb of the IJV. The **facial vein** empties into the IJV opposite or just inferior to the level of the hyoid bone. The facial vein may receive the superior thyroid, lingual, or sublingual veins. The **lingual veins** form a single vein from the tongue, which empties into the IJV at the level of origin of the lingual artery. The **pharyngeal veins** arise from the venous plexus on the pharyngeal wall and empty into the IJV at the level of the angle of the mandible. The **superior** and **middle thyroid veins** leave the thyroid gland and drain into the IJV.

Internal Jugular Pulse

Pulsations of the IJV caused by contraction of the right ventricle of the heart may be palpable superior to the medial end of the clavicle. The pulsations are usually visible when the person's head is 10 to 25° lower than the feet. Because no valves are in the brachiocephalic vein or superior vena cava, a wave of contraction passes up these vessels to the inferior bulb of the IJV. The internal jugular pulse increases considerably in conditions such as mitral valve disease (see Chapter 1), which decreases pressure in the pulmonary circulation and the right side of the heart.

Internal Jugular Vein Puncture

A needle and catheter may be inserted into the IJV for diagnostic or therapeutic purposes. The right internal jugular is preferable because it is usually larger and straighter (Ger et al., 1996). During this procedure, the clinician palpates the common carotid artery and inserts the needle into the IJV just lateral to it at a 30° angle, aiming at the apex of the triangle between the sternal and clavicular heads of the SCM. The needle is then directed inferolaterally toward the ipsilateral nipple. For clinical details, see Ger et al. (1996). ○

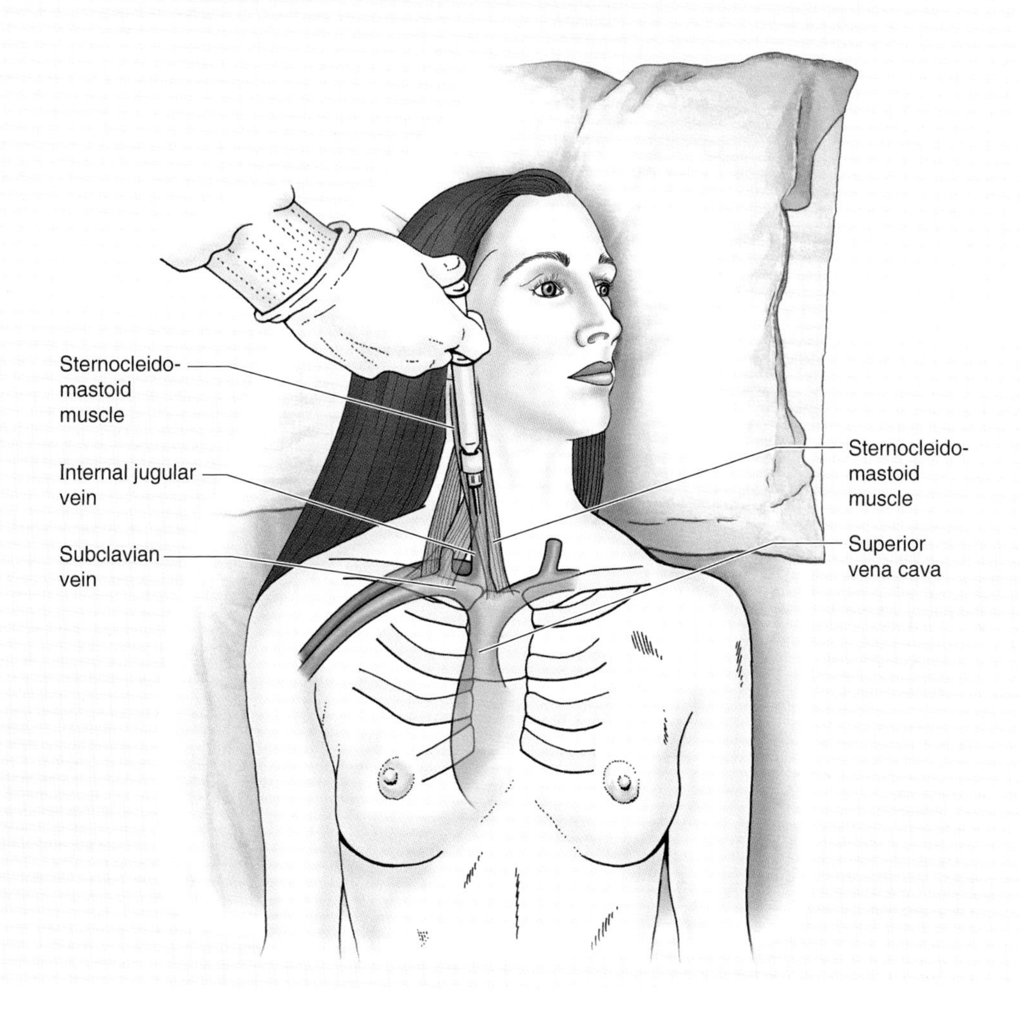

Nerves. Several nerves, including branches of cranial nerves, are in the anterior triangle (Fig. 8.17): the transverse cervical nerve, the hypoglossal nerve, and branches of the glossopharyngeal and vagus nerves.

The **transverse cervical nerve** (C2 and C3) supplies the skin covering the anterior triangle (Figs. 8.6 and 8.10). This nerve winds around the lateral border of the SCM just inferior to the great auricular nerve and crosses this muscle horizontally, deep to the platysma and EJV to reach the anterior triangle. Near the anterior border of the SCM, the transverse cervical nerve divides into superior and inferior branches that pass through the platysma and supply the skin covering the anterior triangle.

The **hypoglossal nerve** (CN XII): the motor nerve of the tongue, enters the submandibular triangle deep to the posterior belly of the digastric muscle to supply the intrinsic and four of the five extrinsic muscles of the tongue (Figs. 8.11, 8.12, and 8.17). CN XII passes between the external carotid and jugular vessels and gives off the superior root of the ansa cervicalis and then a branch to the geniohyoid muscle (Fig. 8.10). In both cases, the branch conveys only fibers from the C1 spinal nerve, which joined its proximal part—no hypoglossal fibers are conveyed in these branches. For details about CN XII, see Chapters 7 and 9.

Branches of the **glossopharyngeal** and **vagus nerves** are in the submandibular and carotid triangles (Figs. 8.17 and 8.18). CN IX is primarily related to the tongue and pharynx. In the neck, CN X gives rise to pharyngeal, laryngeal, and cardiac branches.

Surface Anatomy of the Triangles of the Neck

The skin of the neck is thin and pliable. The subcutaneous tissue contains the **platysma**, a thin sheet of striated muscle that ascends to the face (*arrows*). It can be observed by asking the person to contract it by pretending to ease a tight collar. Its fibers are apparent in most people, especially thin people. Cutaneous nerves, fat, and superficial blood and lymphatic vessels are also in this subcutaneous tissue. The EJV may be prominent, especially if the patient is asked to take a breath and hold it (*Valsalva maneuver*). It runs vertically across the SCM toward the angle of the mandible. The great auricular nerve parallels the vein, approximately a finger's breadth posterior to the vein.

The **SCM** is the key muscular landmark of the neck, which divides the neck into anterior and posterior triangles (Table 8.2). This broad, bulging muscle is easy to observe and palpate throughout its length as it passes superolaterally from the sternum and clavicle to the mastoid process. The SCM can be made to stand out by asking the person to move the chin to the shoulder on the opposite side. In this contracted state, the anterior and posterior borders of the muscle are clearly defined. The **jugular notch** in the manubrium of the sternum is in the fossa between the sternal heads of the SCMs. The suprasternal space and jugular venous arch are located superior to this notch (Figs. 8.4*A* and 8.13). A slight triangular depression lies between the sternal and clavicular heads of the SCM. The inferior end of the IJV lies deep to this depression, where it can be entered by a needle or catheter. Deep to the superior half of the SCM lies the *cervical plexus*, and deep to the inferior half of the SCM are the IJV, common carotid artery, and vagus nerve in the carotid sheath.

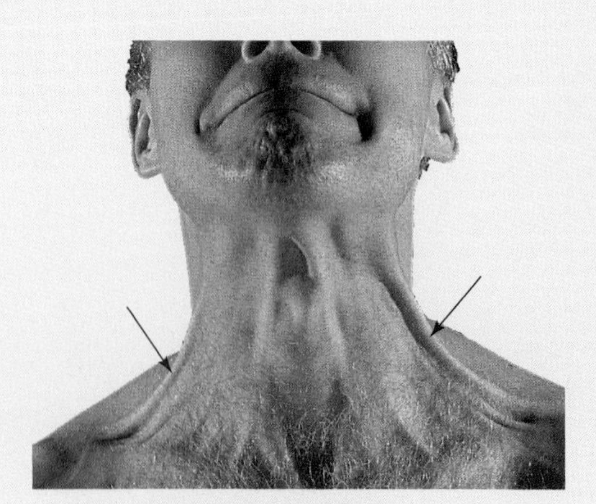

Contraction of platysma

Another large muscle of importance in the neck is the **trapezius**. It can be observed and palpated by asking the person to shrug (raise) his or her shoulder against resistance. The trapezius forms the posterior boundary of the posterior triangle of the neck. The *external occipital protuberance* and the *mastoid process* are important palpable bony landmarks of the neck. They are created by the downward pull of the trapezius and SCM during childhood. The inferior belly of the **omohyoid muscle** can just barely be seen and palpated as it passes superomedially across the inferior part of the posterior triangle. Easiest to observe in thin people, it can often be seen contracting when they are speaking. Just inferior to the inferior belly of the omohyoid is the **supraclavicular fossa,** the depression overlying the supraclavicular triangle. The third part of the subclavian artery passes through this triangle before coursing posterior to the clavicle and across the 1st rib. The supraclavicular fossa is clinically important because it is the pressure point for the subclavian artery, which can be occluded by pressing it against the 1st rib. The course of the **subclavian artery** in the neck is represented by a curved line from the sternoclavicular (SC) joint to the midpoint of the clavicle. The chief contents of the larger **occipital triangle**, superior to the omohyoid muscle, are the *accessory nerve*, cutaneous branches of *cervical nerves* C2, C3, and C4, and cervical lymph nodes. ▶

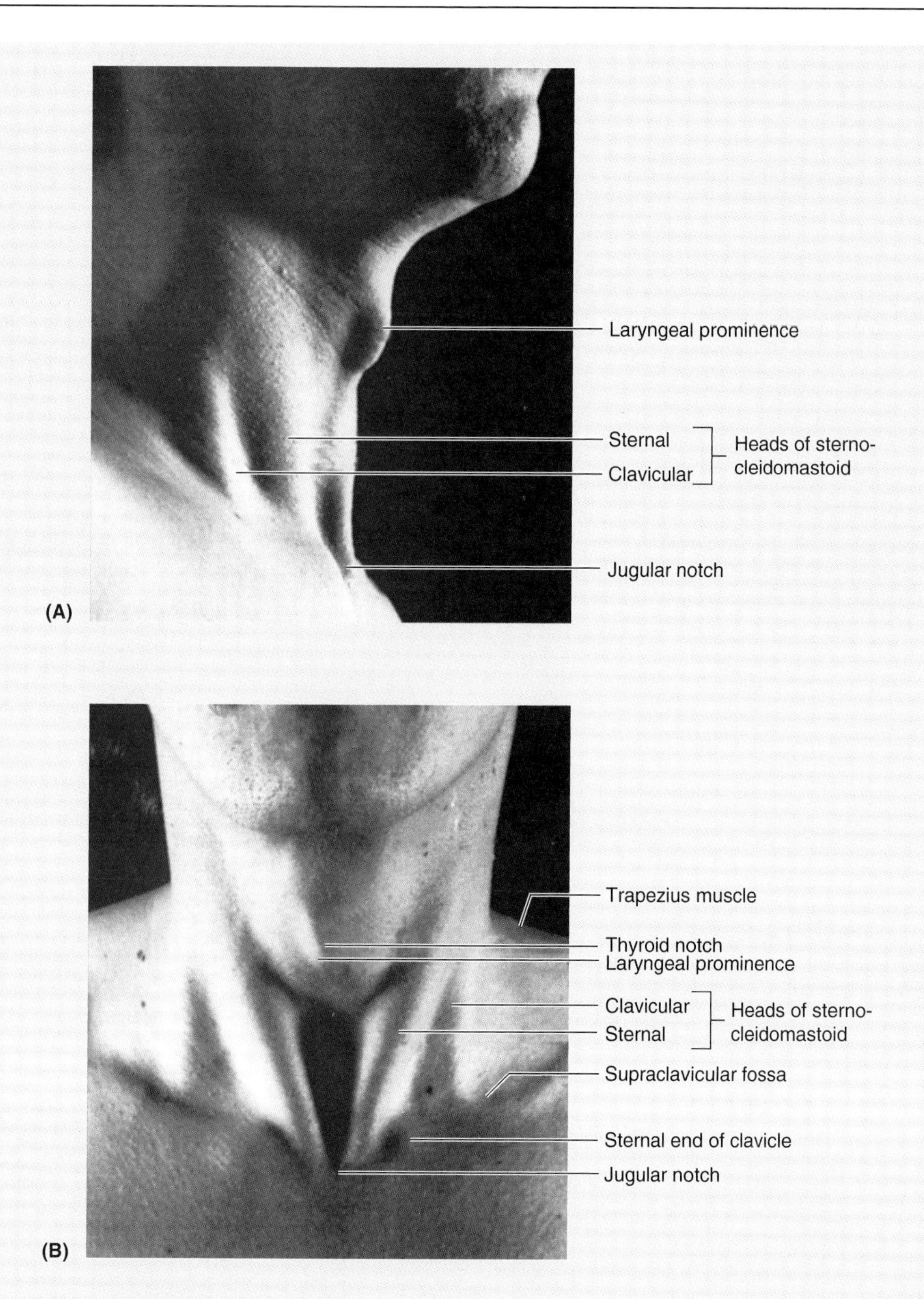

▶ The anterior triangle of the neck is the surgical approach to the cervical viscera and carotid arteries and their branches. It is subdivided by the bellies of the digastric and omohyoid muscles into four smaller triangles. The submandibular gland nearly fills the submandibular triangle; it is palpable as a soft mass inferior to the body of the mandible, especially when the apex of the tongue is forced against the maxillary incisor teeth. The submandibular lymph nodes lie superficial to the gland. If enlarged, these nodes can be palpated by moving the fingertips from the angle of the mandible along its inferior border. The *carotid arterial system* is located in the **carotid triangle**; hence, this area is important for surgical approaches to the carotid sheath containing the common carotid artery, IJV, and vagus nerve (Fig. 8.18). The carotid triangle also contains the hypoglossal nerve and cervical sympathetic trunk. The **carotid sheath** can be marked out by a line joining the SC joint to a point midway between the mastoid process and the angle of the mandible. ○

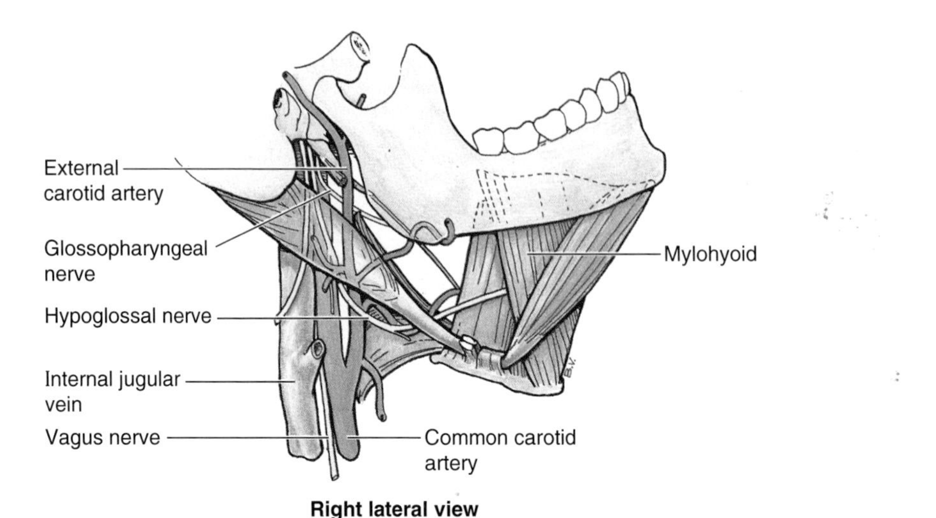

Figure 8.18. Diagram of the relationships of nerves and vessels to the suprahyoid muscles and cervical triangles.

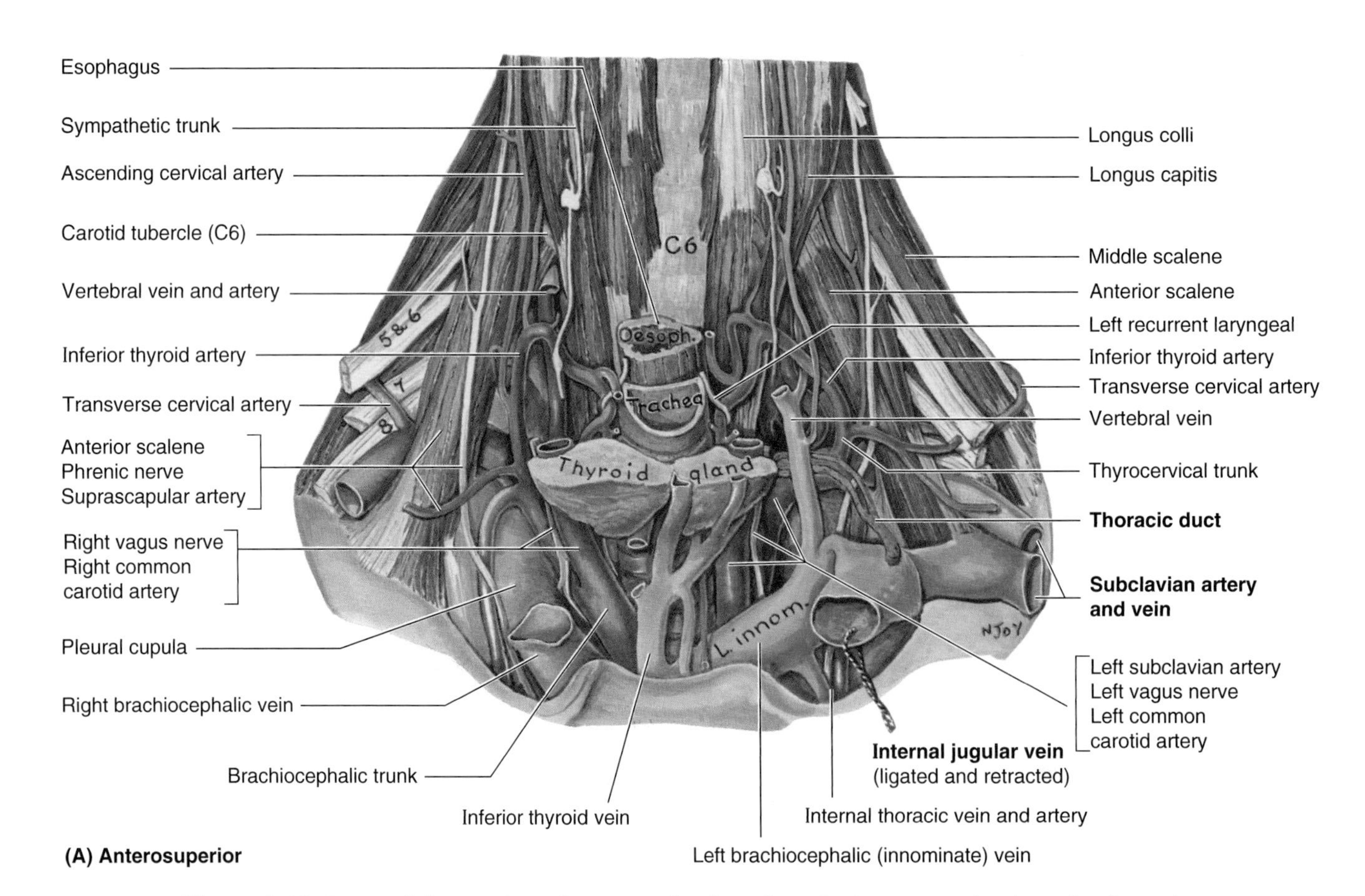

Figure 8.19. Root of the neck and prevertebral region. A. Anterosuperior view of a dissection of the root of the neck. Observe the *brachial plexus* and the third part of the subclavian artery emerging from between the anterior and middle scalene (L. scalenus medius) muscles. Note also the relationship of the first part of the subclavian arteries to the apex of the lung (pleural cupula), the branches of the subclavian arteries, and the thoracic duct entering the left venous angle.

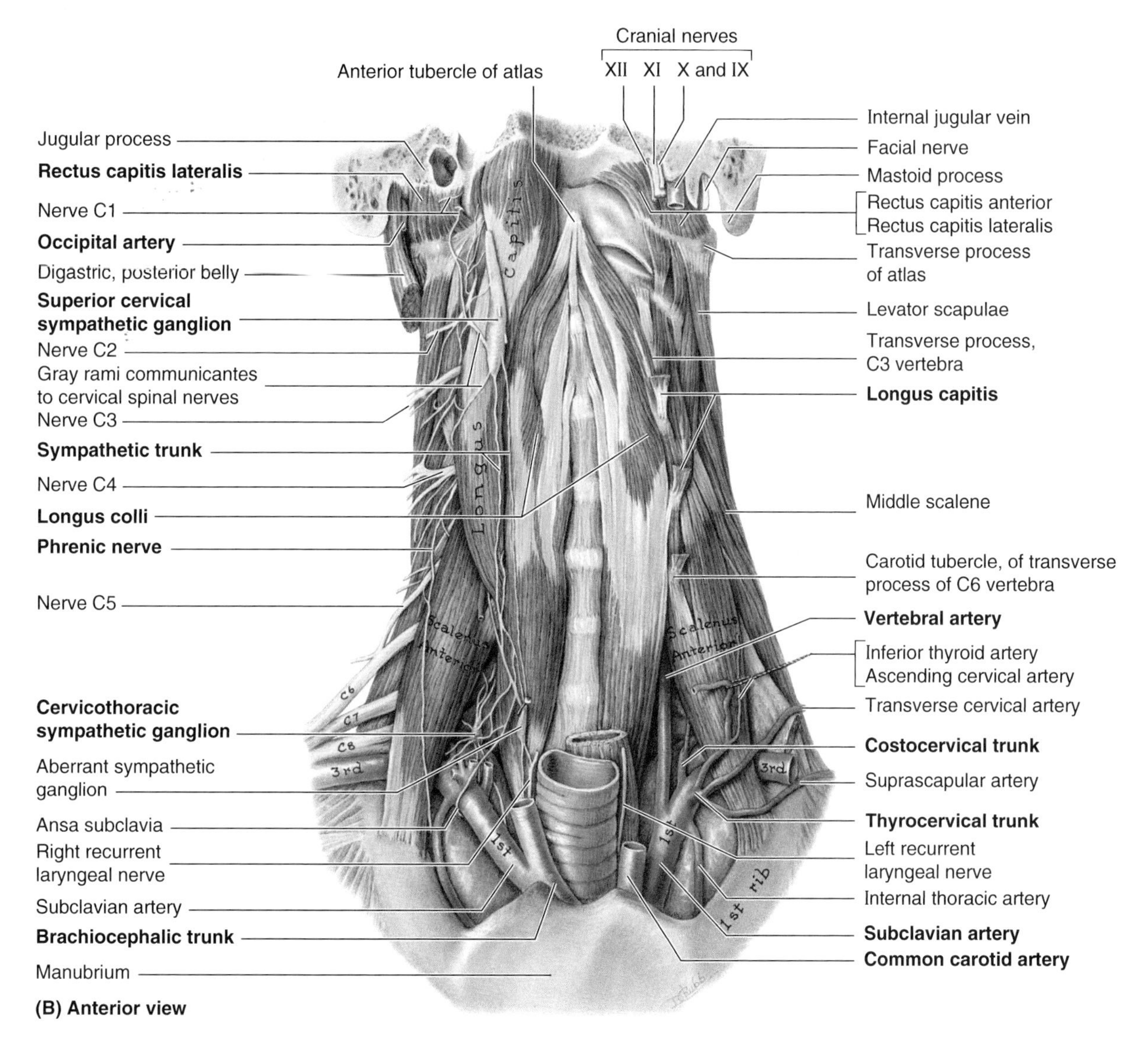

Figure 8.19. *(Continued)* **B.** Dissection of the prevertebral region and the root of the neck. The prevertebral fascia has been removed and the longus capitis muscle (long muscle of head) has been excised on the left side. On the right side, observe the cervical plexus of nerves arising from the ventral rami (roots) of C1 (atlas) through C4 and the brachial plexus of nerves arising from the ventral rami of C5 through C8 and T1. The sympathetic trunk lies on the prevertebral muscles, embedded in the prevertebral fascia (removed).

Deep Structures of the Neck

Notable deep structures of the neck are the brachial plexus, internal jugular and subclavian veins, muscles forming the floor of the posterior triangle, viscera of the neck (e.g., the thyroid gland), and hyoid and prevertebral muscles.

Prevertebral Muscles

The anterior and lateral vertebral or prevertebral muscles, posterior to the prevertebral layer of deep cervical fascia, lie in the floor of the anterior and posterior triangles of the neck (Figs. 8.4, 8.6, and 8.19, Table 8.4). The **anterior vertebral muscles,** consisting of the longus colli and capitis and rectus capitis anterior and lateralis, are related to the anterior triangle, and the **lateral vertebral muscles,** consisting of the splenius capitis, levator scapulae, and scalene muscles (anterior, middle, and posterior), are related to the posterior triangle.

- *Longus colli* (long muscle of neck) is applied to the anterior surface of the vertebral column, extending from the anterior tubercle of the atlas to the bodies of the 3rd thoracic vertebra and the transverse processes of C3 through C6 vertebrae.
- *Longus capitis* (long muscle of head), broad and thick superiorly and narrow inferiorly, passes superomedially from the transverse process of C3 through C6 vertebrae to the inferior surface of the basilar part of the occipital bone.

Table 8.4. **Prevertebral Muscles**

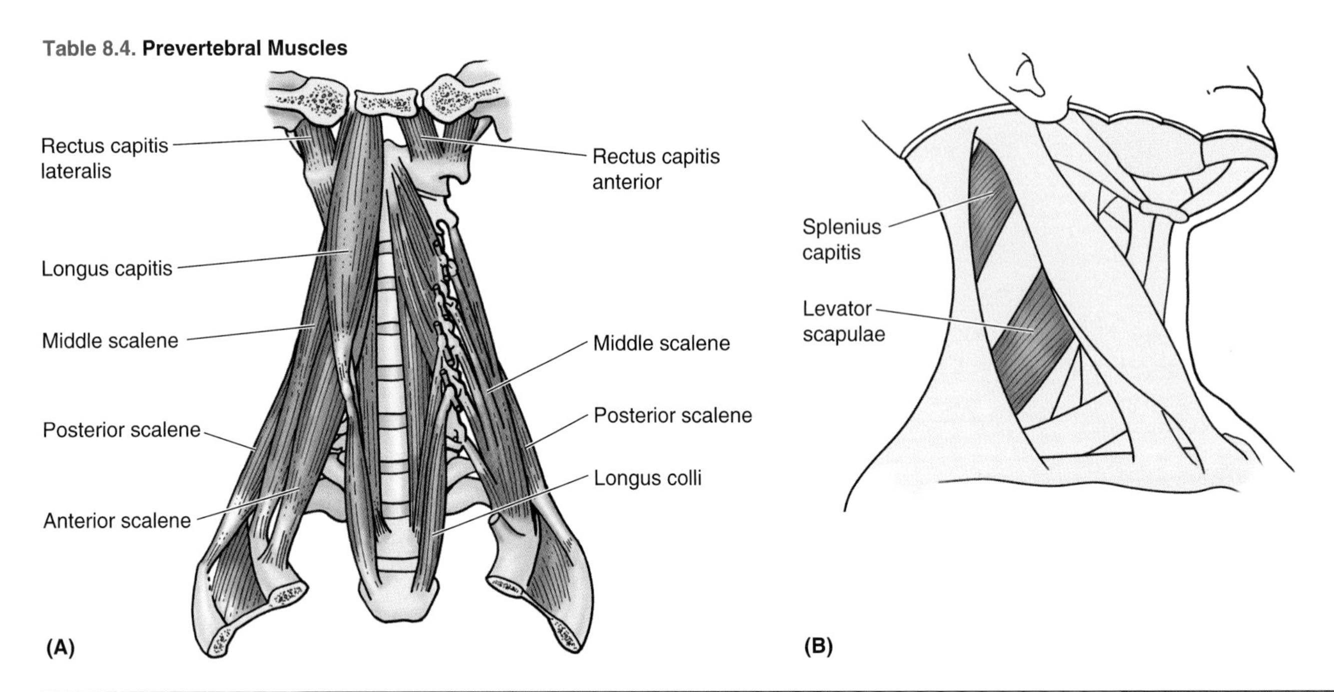

Muscle	Superior Attachment	Inferior Attachment	Innervation	Main Action
Anterior				
Longus colli	Anterior tubercle of C1 vertebra (atlas); bodies of C1–C3 and transverse processes of C3–C6 vertebrae	Bodies of C5–T3 vertebrae, transverse process of C3–C5 vertebrae	Ventral rami of C2–C6 spinal nerves	*Flexes neck with rotation (torsion) to opposite side if acting unilaterally
Longus capitis	Basilar part of occipital bone	Anterior tubercles of C3–C6 transverse processes	Ventral rami of C1–C3 spinal nerves	**Flexes head
Rectus capitis anterior	Base of skull, just anterior to occipital condyle	Anterior surface of lateral mass of C1 vertebra (atlas)	Branches from loop between C1 and C2 spinal nerves	
Rectus capitis lateralis	Jugular process of occipital bone	Transverse process of C1 vertebra (atlas)		Flexes head and helps to stabilize it
Lateral				
Splenius capitis	Inferior half of ligamentum nuchae and spinous processes of superior six thoracic vertebrae	Lateral aspect of mastoid process and lateral third of superior nuchal line	Dorsal rami of middle cervical spinal nerves	***Laterally flexes and rotates head and neck to same side; acting bilaterally, they extend head and neck
Levator scapulae	Posterior tubercles of transverse processes of C1–C4 vertebrae	Superior part of medial border of scapula	Dorsal scapular nerve C5 and cervical spinal nerves C3 and C4	Elevates scapula and tilts its glenoid cavity inferiorly by rotating scapula
Posterior scalene	Posterior tubercles of transverse processes of C4–C6 vertebrae	External border of 2nd rib	Ventral rami of cervical spinal nerves C7 and C8	Flexes neck laterally; elevates 2nd rib during forced inspiration
Middle scalene	Transverse process of axis (C2) and posterior tubercles of transverse processes of C3–C7 vertebrae	Superior surface of 1st rib, posterior to groove for subclavian artery	Ventral rami of cervical spinal nerves	Flexes neck laterally; elevates 1st rib during forced inspiration
Anterior scalene	Anterior tubercles of transverse processes of C3–C6 vertebrae	Scalene tubercle of 1st rib, anterior to groove for subclavian artery	Cervical spine nerves C4, C5, and C6	Elevates 1st rib; laterally flexes and rotates neck

* Flexion of neck = anterior (or lateral if so stated) bending of cervical vertebrae C2–C7

** Flexion of head = anterior (or lateral if so stated) bending of head relative to vertebral column of atlanto-occipital joints

*** Rotation of head occurs at atlantoaxial joints

- *Rectus capitis anterior* (anterior rectus muscle of the head) is a short, flat muscle posterior to the superior part of the longus capitis; it passes from the lateral mass of the atlas to the basilar part of the occipital bone.
- *Rectus capitis lateralis* (lateral rectus muscle of the head), also short and flat, passes from the transverse process of the atlas to the jugular process of the occipital bone.
- *Splenius capitis* (splenius muscle of head) is an inconstant extension of the splenius cervicus to the occipital bone, which forms part of the floor of the posterior triangle (Fig. 8.6, Table 8.4).
- *Anterior scalene* descends inferolaterally from the transverse processes of C3 through C6 vertebrae to the scalene tubercle of the 1st rib.
- *Middle scalene* descends inferolaterally from the transverse processes of C2 through C7 vertebrae to the 1st rib.
- *Posterior scalene*, incompletely separated from the middle scalene superiorly, passes from the transverse processes of C4 through C6 vertebrae to the 2nd rib.
- *Levator scapulae*, a thick, straplike muscle, passes from the transverse processes of the first three or four cervical vertebrae to the uppermost medial border (adjacent to the superior angle) of the scapula.

Root of the Neck

The root or base of the neck is the junctional area between the thorax and neck (Fig. 8.19). It opens into (and is the cervical side of) the superior thoracic aperture (root of the neck), through which pass all structures going from the head to the thorax and vice versa (see Chapter 1). *The root of the neck is bounded*:

- *Laterally* by the 1st pair of ribs and their costal cartilages
- *Anteriorly* by the manubrium of the sternum
- *Posteriorly* by the body of T1 vertebra.

Arteries in the Root of the Neck

The arteries in the root of the neck, the large *brachiocephalic trunk* on the right side and the *common carotid* and *subclavian arteries* on the left side (Figs. 8.14 and 8.19), originate from the arch of the aorta.

Brachiocephalic Trunk. Covered anteriorly by the sternohyoid and sternothyroid muscles, the brachiocephalic trunk is the largest branch of the arch of the aorta. From 4 to 5 cm long, it arises in the midline from the beginning of the aortic arch, posterior to the manubrium (Fig. 8.14), and passes superolaterally to the right where it divides into the right common carotid and right subclavian arteries posterior to the SC joint. The brachiocephalic trunk usually has no branches; occasionally (in approximately 10% of people) a thyroid ima artery (lowest thyroid artery) arises from it (p. 1032).

Subclavian Arteries. The subclavian arteries supply the upper limbs; however, they also supply the neck and brain (Fig. 8.19, *A* and *B*). Each subclavian artery arches superiorly, posteriorly, and laterally—grooving the pleura and lung. As the arteries ascend and reach their apex, they pass posterior to the anterior scalene muscles. When they descend, they pass inferiorly, posterior to the midpoints of the clavicles. Although the subclavian arteries of the two sides have different origins, the course for both in the neck begins posterior to the SC joints. From here the arteries arch superolaterally and then inferiorly to disappear posterior to the middle of the clavicles. For purposes of description, the anterior scalene divides each subclavian artery into three parts: the first part is medial to the muscle, the second is posterior to it, and the third is lateral to it (Fig. 8.19*B*).

The **right subclavian artery** arises from the brachiocephalic trunk, posterior to the right SC joint. The first part of the artery courses superolaterally, extending between its origin and the medial margin of the anterior scalene muscle. The cervical pleurae, apex of the lung, and sympathetic trunk lie posterior to this first part of the artery.

The **left subclavian artery** arises from the arch of the aorta, approximately 1 cm distal to the left common carotid artery. It ascends through the superior mediastinum (see Chapter 1) and enters the root of the neck posterior to the left SC joint. The left vagus nerve runs parallel to this part of the artery (Fig. 8.19*A*). The posterior relations of the artery are similar to those of the right subclavian artery.

The branches of subclavian arteries distribute widely to the brainstem, spinal cord, neck, upper limb, thoracic wall, and diaphragm. The **branches of the subclavian artery** are the:

- *Vertebral artery*, *internal thoracic artery*, and *thyrocervical trunk* from the first part of the subclavian artery
- *Costocervical trunk* from the second part of the subclavian artery
- *Dorsal scapular artery* from the third part of the subclavian artery.

The **cervical part of the vertebral artery** arises from the first part of the subclavian artery and ascends in the pyramidal space formed between the scalene and longus muscles (Figs. 8.19, *A* and *B*), passing deeply at its apex to course through the foramina of the transverse processes of C1 through C6 vertebrae (the **vertebral part of the vertebral artery**); however, it may enter a foramen more superior than C6 vertebra. In approximately 5% of people, the left vertebral artery arises from the arch of the aorta. The **suboccipital part of the vertebral artery** courses in a groove on the posterior arch of the atlas before it enters the cranial cavity through the foramen magnum (the beginning of the cranial part of the artery). Despite its name, the vertebral artery supplies mainly the posterior part of the brain. At the inferior border of the pons of the brainstem, the vertebral arteries join to form the basilar artery that participates in the formation of the cerebral arterial circle (of Willis) (see Chapter 7).

The **internal thoracic artery** arises from the anteroinferior aspect of the subclavian artery and passes inferomedially into the thorax (Fig. 8.19, *A* and *B*). The cervical part of the internal thoracic artery has no branches; its thoracic distribution is described in Chapter 1.

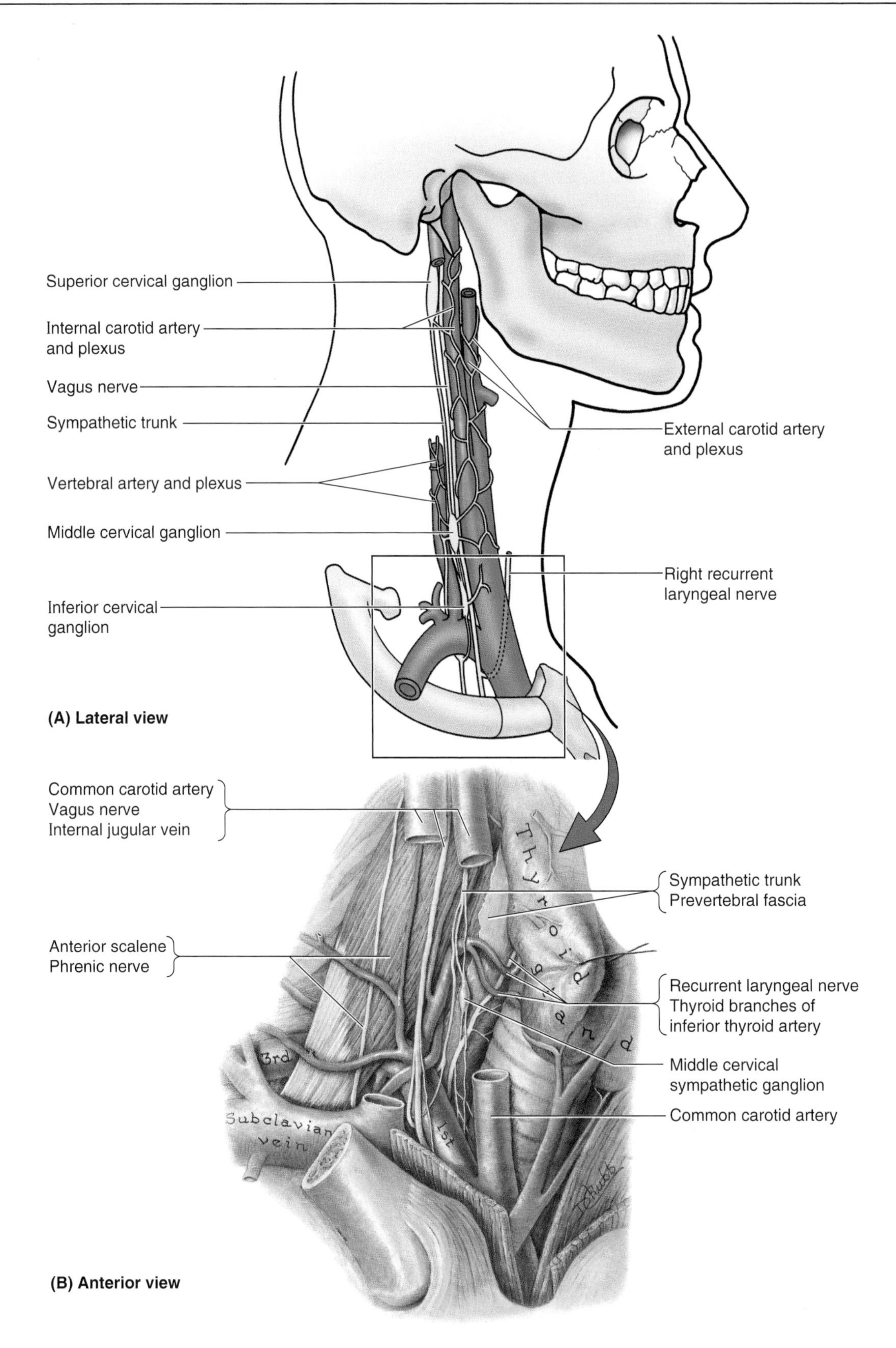

Figure 8.20. Nerves in the neck. A. Drawing of the head and neck showing the cervical sympathetic trunk and ganglia, the carotid arteries, and the sympathetic periarterial plexuses surrounding them. **B.** Root of the neck, right side. The clavicle is removed, sections are taken from the common carotid artery and internal jugular vein (IJV), and the right lobe of the thyroid gland is retracted to reveal the nerves and middle cervical ganglion in the root of the neck.

The **thyrocervical trunk** arises from the anterosuperior aspect of the first part of the subclavian artery, just medial to the anterior scalene muscle (Fig. 8.19, *A* and *B*). It has three branches, the largest and most important of which is the **inferior thyroid artery**, the primary visceral artery of the neck. Other branches of the thyrocervical trunk are the *suprascapular artery,* supplying muscles on the posterior scapula, and the *transverse cervical artery,* sending branches to muscles in the posterior cervical triangle, the trapezius and medial scapular muscles. These latter two arteries may arise from a common trunk or directly from the subclavian artery. The terminal branches of the thyrocervical trunk are the inferior thyroid and ascending cervical arteries. The latter artery supplies the lateral muscles of the upper neck.

The **costocervical trunk** arises from the posterior aspect of the second part of the subclavian artery (posterior to the anterior scalene muscle on the right side and usually just medial to it on the left side). The trunk passes posterosuperiorly and divides into the superior intercostal and deep cervical arteries, which supply the first two intercostal spaces and the posterior deep cervical muscles, respectively.

The **dorsal scapular artery** often arises as a branch of the transverse cervical artery, but it may be a branch of the second or third part of the subclavian artery (Fig. 8.19*A*). When it is a branch of the subclavian, the dorsal scapular artery passes laterally through the trunks of the brachial plexus, anterior to the middle scalene, and then runs deep to the levator scapulae to reach the scapula and supply the rhomboid muscles.

Veins in the Root of the Neck

Two large veins terminating in the root of the neck are the **EJV**, draining blood received mostly from the scalp and face (described on p. 1005), and the variable **anterior jugular vein,** usually the smallest of the jugular veins (Figs. 8.12 and 8.16). The anterior jugular vein typically arises near the hyoid bone from the confluence of superficial submandibular veins. It descends either in the subcutaneous tissue or deep to the investing layer of deep cervical fascia between the anterior median line and the anterior border of the SCM. At the root of the neck, the anterior jugular vein turns laterally, posterior to the SCM, and opens into the termination of the EJV or into the subclavian vein. Superior to the manubrium, the right and left anterior jugular veins commonly unite across the midline to form the **jugular venous arch** in the suprasternal space (Fig. 8.13). This common venous variation must be kept in mind when procedures such as tracheostomies (p. 1049) are performed in the midline of the neck.

The **subclavian vein**, the continuation of the axillary vein, begins at the lateral border of the 1st rib and ends when it unites with the **IJV**, posterior to the medial end of the clavicle, to form the **brachiocephalic vein** (Fig. 8.19*A*). The subclavian vein passes over the 1st rib anterior to the scalene tubercle parallel to the subclavian artery, but it is separated from it by the anterior scalene muscle. It usually has only one named tributary, the *EJV* (Fig. 8.16).

The **IJV** ends posterior to the medial end of the clavicle by uniting with the subclavian vein to form the brachiocephalic vein (Figs. 8.16 and 8.19*A*). This union is commonly referred to as the "venous angle" and is the site where the *thoracic duct* (left side) and the *right lymphatic trunk* (right side) drain lymph collected throughout the body into the venous circulation. Throughout its course, the IJV is enclosed by the carotid sheath (Fig. 8.17).

Nerves in the Root of the Neck

Vagus Nerves. Following their exit from the jugular foramen, each vagus nerve passes inferiorly in the neck within the posterior part of the carotid sheath in the angle between the IJV and common carotid artery (Figs. 8.18–8.20). The right vagus nerve passes anterior to the first part of the subclavian artery and posterior to the brachiocephalic vein and SC joint to enter the thorax. The left vagus nerve descends between the left common carotid and left subclavian arteries and posterior to the SC joint to enter the thorax. The **recurrent laryngeal nerves** arise from the vagus nerves in the inferior part of the neck (Fig. 8.20, A and B). The nerves of the two sides have essentially the same distribution; however, they recur (loop around) different structures and at different levels on the two sides. The right recurrent laryngeal nerve loops inferior to the right subclavian artery at approximately the T1/T2 vertebral level, and the left recurrent laryngeal nerve loops inferior to the arch of the aorta at approximately the T4/T5 vertebral level. After looping, both recurrent nerves ascend superiorly to the posteromedial aspect of the thyroid gland (Figs. 8.20, 8.21B, and 8.22), where they ascend in the tracheoesophageal groove to supply all the intrinsic muscles of the larynx except the cricothyroid. The cardiac branches of CN X originate in the neck as well as in the thorax and run along the arteries to the cardiac plexus of nerves (see Chapter 1).

Phrenic Nerves. The phrenic nerves are formed at the lateral borders of the anterior scalene muscles (Figs. 8.18 and 8.20)—mainly from C4 nerve with contributions from C3 and C5. The phrenic nerves descend anterior to the anterior scalene muscles under cover of the IJVs and the SCMs. They pass under the prevertebral layer of deep cervical fascia, between the subclavian arteries and veins, and proceed to the thorax and supply the diaphragm. The phrenic nerves are important mainly because—in addition to their sensory distribution—they provide the sole motor supply to their own half of the diaphragm (pp. 291–292).

Sympathetic Trunks. The sympathetic trunks are in the neck, anterolateral to the vertebral column, beginning at the level of C1 vertebra (Figs. 8.19 and 8.20). These trunks receive no white rami communicantes in the neck (recall that none are associated with cervical spinal nerves); however, they are associated with three **cervical sympathetic ganglia**: superior, middle, and inferior by way of gray rami communicantes. These ganglia receive presynaptic fibers from the superior thoracic spinal nerves and associated white rami communicantes through the sympathetic trunk. From the cervical sympathetic ganglia, postsynaptic fibers pass to splanchnic nerves to the cervical spinal nerves via gray rami communicantes or leave as direct visceral branches (splanchnic nerves). Branches to the head and viscera of the neck run with the arteries, especially the vertebral and internal and external carotid arteries (Fig. 8.20).

The **inferior cervical ganglion** in approximately 80% of people fuses with the 1st thoracic ganglion to form the large **cervicothoracic ganglion** (stellate ganglion). This star-shaped (L. stella, a star) ganglion lies anterior to the transverse process of C7 vertebra, just superior to the neck of the 1st rib on each side and posterior to the origin of the vertebral artery (Fig. 8.19*B*). Some postsynaptic fibers from the ganglion pass via gray rami communicantes to the ventral rami of the C7 and C8 spinal nerves (roots of the brachial plexus), and others pass to the heart via the inferior cervical cardiac nerve (a cardiopulmonary splanchnic nerve) that passes along the trachea to the deep *cardiac plexus.* Other fibers contribute to the periarterial sympathetic nerve plexus around the vertebral artery and pass into the cranial cavity (Fig. 8.20*A*).

The **middle cervical ganglion**, the smallest of the three ganglia, is occasionally absent. When present, it lies on the anterior aspect of the inferior thyroid artery at the level of the cricoid cartilage and the transverse process of C6 vertebra, just anterior to the vertebral artery (Figs. 8.20, *A* and *B*, and 8.22). Postsynaptic gray rami pass from the ganglion to the ventral rami of the C5 and C6 spinal nerves, via cardiopulmonary splanchnic nerves to the heart and via periarterial plexuses to the thyroid gland.

Cervicothoracic Ganglion Block

Anesthetic injected around the large cervicothoracic ganglion blocks transmission of stimuli through the cervical and superior thoracic ganglia. The cervicothoracic ganglion block may relieve vascular spasms involving the brain and upper limb. It is also useful when deciding if a surgical resection of the ganglion would be beneficial to a patient with excess vasoconstriction in the ipsilateral limb (Mathers et al., 1996).

Cervical Lesion of the Sympathetic Trunk

A lesion of a sympathetic trunk in the neck results in a sympathetic disturbance—the *Horner syndrome*—which is characterized by:

- Pupillary constriction—resulting from paralysis of the dilator pupillae muscle (see Chapter 7)
- Ptosis (drooping of the upper eyelid)—resulting from paralysis of the smooth (tarsal) muscle intermingled with the striated muscle of the levator palpebrae superioris
- Sinking in of the eye—possibly caused by paralysis of the smooth (orbitalis) muscle in the floor of the orbit
- Vasodilation and absence of sweating on the face and neck—caused by lack of a sympathetic (vasoconstrictive) nerve supply to the blood vessels and sweat glands. ⊙

The **superior cervical ganglion** is at the level of C1 and C2 vertebrae (Fig. 8.18 and 8.20*A*). Because of its large size it forms a good landmark for locating the sympathetic trunk, but it may need to be distinguished from a large sensory (nodose) ganglion of the vagus when present. Postsynaptic cephalic arterial rami pass from it to form the internal carotid sympathetic plexus along the internal carotid artery and enter the cranial cavity. This ganglion also sends arterial rami (branches) to the external carotid artery and gray rami to the ventral rami of the superior four cervical spinal nerves. Other postsynaptic fibers pass from it to the cardiac plexus of nerves via a cardiopulmonary splanchnic nerve (see Chapter 1).

Viscera of the Neck

The cervical viscera are disposed in three layers (Fig. 8.21). Superficial to deep, they are the:

- *Endocrine layer*—thyroid and parathyroid glands
- *Respiratory layer*—larynx and trachea
- *Alimentary layer*—pharynx and esophagus.

The names of these layers represent the functions of the viscera.

Endocrine Layer of the Cervical Viscera

The cervical organs in the endocrine layer are part of the body's endocrine system of ductless, hormone-secreting glands. The **thyroid gland**—prodigiously vascularized by the inferior and superior thyroid arteries—is the body's largest endocrine gland. It produces thyroid hormone, which controls the rate of metabolism and calcitonin, a hormone controlling calcium metabolism. The thyroid gland affects all areas of the body except itself and the adult brain, spleen, testes, and uterus. The hormone produced by the **parathyroid glands**—parathormone—controls the metabolism of phosphorus and calcium in the blood. The parathyroid glands target the skeleton, kidneys, and intestine.

Thyroid Gland

The thyroid gland lies deep to the sternothyroid and sternohyoid muscles from the level of C5 through T1 vertebrae (Fig. 8.21). It consists of two lobes, right and left, anterolateral to the larynx and trachea. An **isthmus** unites the lobes over the trachea, usually anterior to the 2nd and 3rd tracheal rings. The thyroid gland is surrounded by a thin fibrous capsule, which sends septa deeply into the gland. External to the capsule is a loose sheath formed by the visceral layer of the pretracheal deep cervical fascia. Dense connective tissue attaches the capsule of the thyroid gland to the *cricoid cartilage* and the *superior tracheal rings.*

Vessels of the Thyroid Gland. *Arteries.* The highly vascular thyroid gland is supplied by the superior and inferior **thyroid arteries** (Figs. 8.21*B* and 8.22). These vessels lie between the fibrous capsule and the pretracheal layer of deep cervical fascia. Usually the 1st branch of the external carotid, the **superior thyroid artery**,

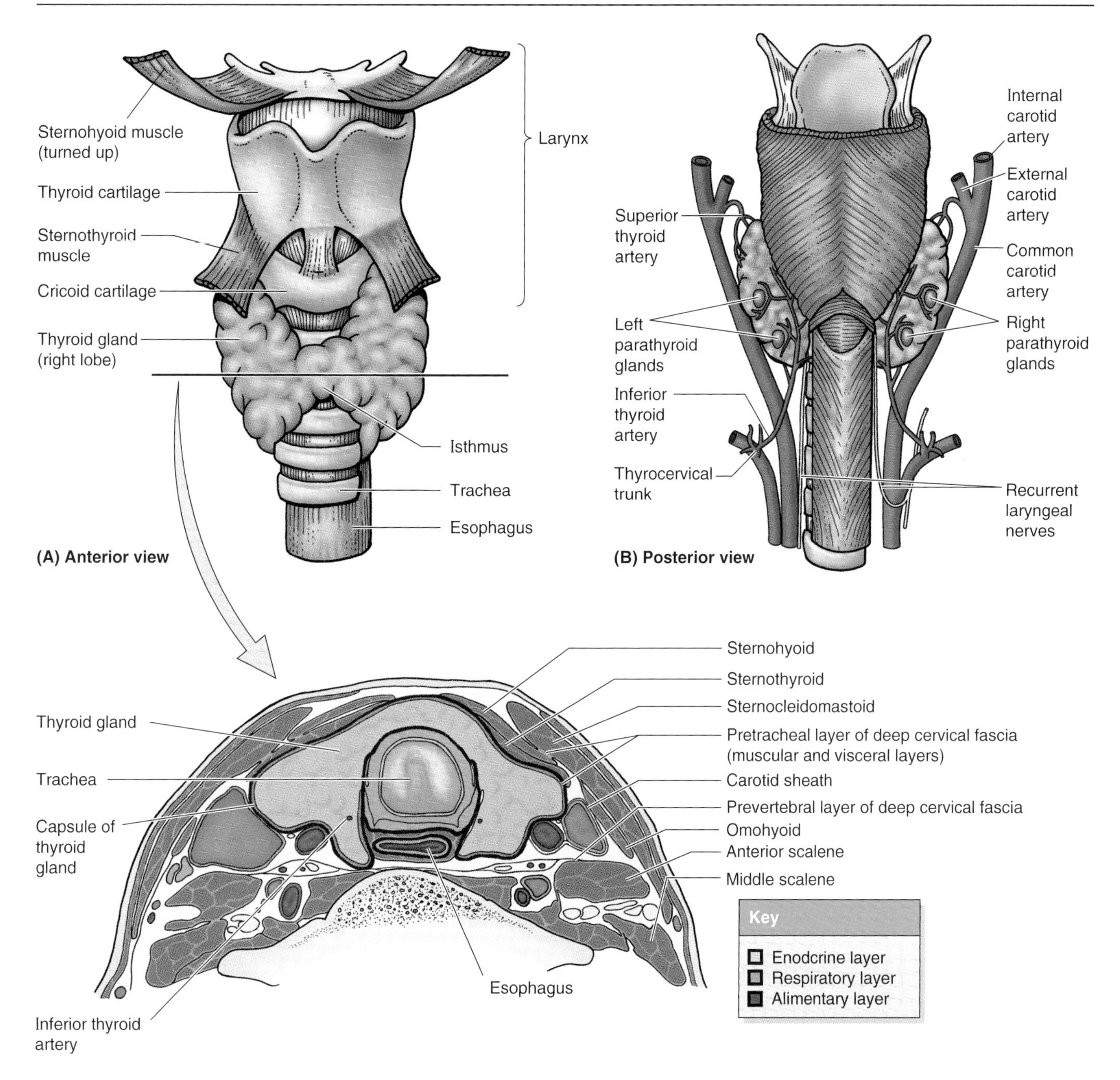

Figure 8.21. Relations of the thyroid gland. A. A normal thyroid gland showing its relationship to the trachea, esophagus, and cricoid cartilage. The sternothyroid muscles have been cut to expose the lobes of the gland. Note that the isthmus of the gland lies anterior to the 2nd and 3rd tracheal rings. **B.** Dissection of the posterior surface of the thyroid gland showing the parathyroid glands. Note the blood supply of the thyroid and parathyroid glands. **C.** Layers of the neck at the level shown in (**A**). Observe that the thyroid gland is asymmetrically enlarged in this specimen.

descends to the superior pole of each lobe of the gland, pierces the pretracheal layer of deep cervical fascia, and divides into anterior and posterior branches. The larger *anterior branch of the superior thyroid artery* descends along the anterior border of the thyroid gland and sends branches to its anterior surface. The anterior branches of the right and left sides anastomose across the midline. The *posterior branch of the superior thyroid artery* descends along the posterior surface of the thyroid gland and anastomoses with the inferior thyroid artery. The **inferior thyroid artery**, the largest branch of the thyrocervical trunk arising from the subclavian artery, runs superomedially posterior to the carotid sheath to reach the posterior aspect of the thyroid gland. It divides into several branches that pierce the pretracheal layer of the deep cervical fascia and supply the inferior pole of the gland.

Thyroid Ima Artery

In approximately 10% of people, a small, unpaired *thyroid ima artery* (L. thyroidea ima) usually arises from the brachiocephalic trunk; however, it may arise from the arch of the aorta or from the right common carotid, subclavian, or internal thoracic arteries. This small artery ascends on the anterior surface of the trachea, which it supplies, and continues to the isthmus of the thyroid gland, where it divides into branches that supply it. The possible presence of a thyroid ima artery must be considered when performing procedures in the midline of the neck inferior to the isthmus (e.g., for a tracheostomy [p. 1049]). As it runs anterior to the trachea, it is a potential source of bleeding.

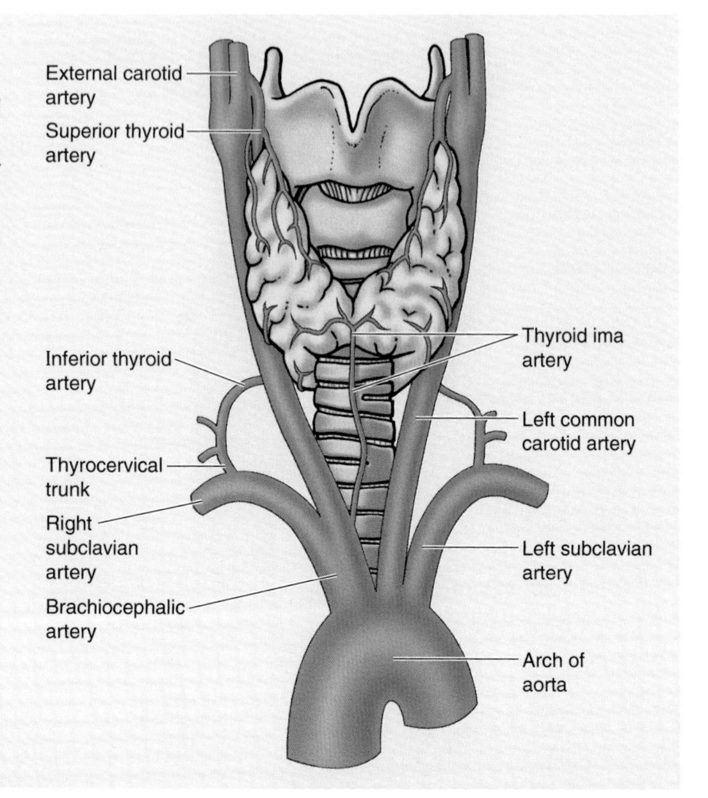

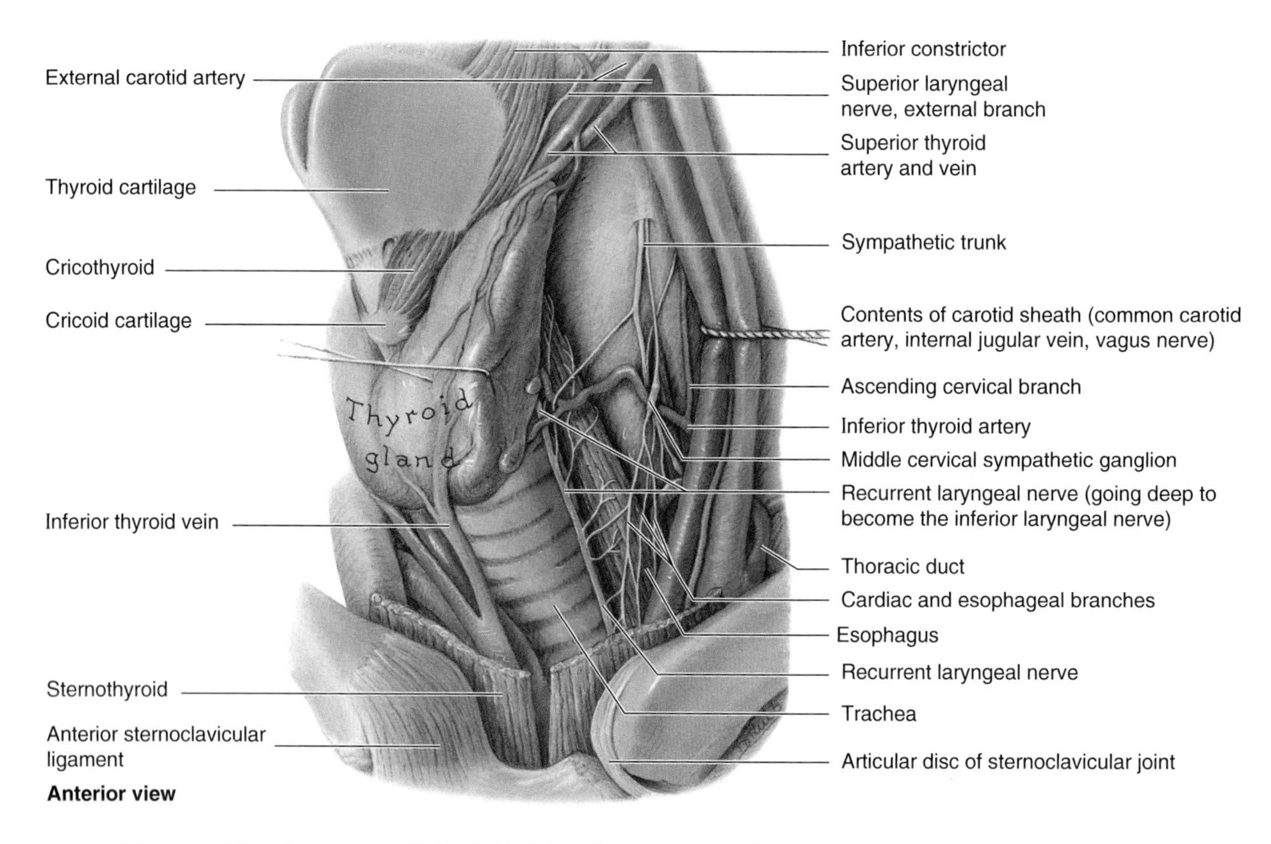

Figure 8.22. Dissection of the left side of the root of the neck. Anterior view. The viscera (thyroid gland, trachea, and esophagus) are retracted to the right, and the contents of the left carotid sheath are retracted to the left. The middle thyroid vein, severed to allow such retraction, is not apparent. Note the recurrent (inferior) laryngeal nerve ascending beside the trachea, just anterior to the angle between the trachea and esophagus. Note that the contents of the carotid sheath lie anterior to the thoracic duct and the thyrocervical trunk (not seen), which gives rise to the inferior thyroid artery.

Veins. Three pairs of thyroid veins usually drain the venous plexus on the anterior surface of the thyroid gland (Figs. 8.22 and 8.23). The **superior thyroid veins** drain the superior poles of the thyroid gland, the **middle thyroid veins** drain the middle of the lobes, and the **inferior thyroid veins** drain the inferior poles. The superior and middle thyroid veins drain into the IJVs and the inferior thyroid veins and drain into the brachiocephalic veins posterior to the manubrium of the sternum.

Lymphatic Drainage. Lymphatic vessels of the thyroid gland run in the interlobular connective tissue, often around the arteries, and communicate with a capsular network of lymphatic vessels. From here, the vessels pass to *prelaryngeal, pretracheal,* and **paratracheal lymph nodes** (Fig. 8.24). Laterally, lymphatic vessels located along the superior thyroid veins pass to the **inferior deep cervical lymph nodes**. Some lymphatic vessels may drain into the brachiocephalic lymph nodes or into the thoracic duct (Fig. 8.22).

Nerves. The nerves of the thyroid gland derive from the superior, middle, and inferior cervical sympathetic ganglia (Figs. 8.20 and 8.22). They reach the gland through the *cardiac* and *superior* and *inferior thyroid periarterial plexuses* that accompany the thyroid arteries. These fibers are vasomotor—they cause constriction of blood vessels—not secretomotor; the thyroid gland is hormonally regulated (by the pituitary gland).

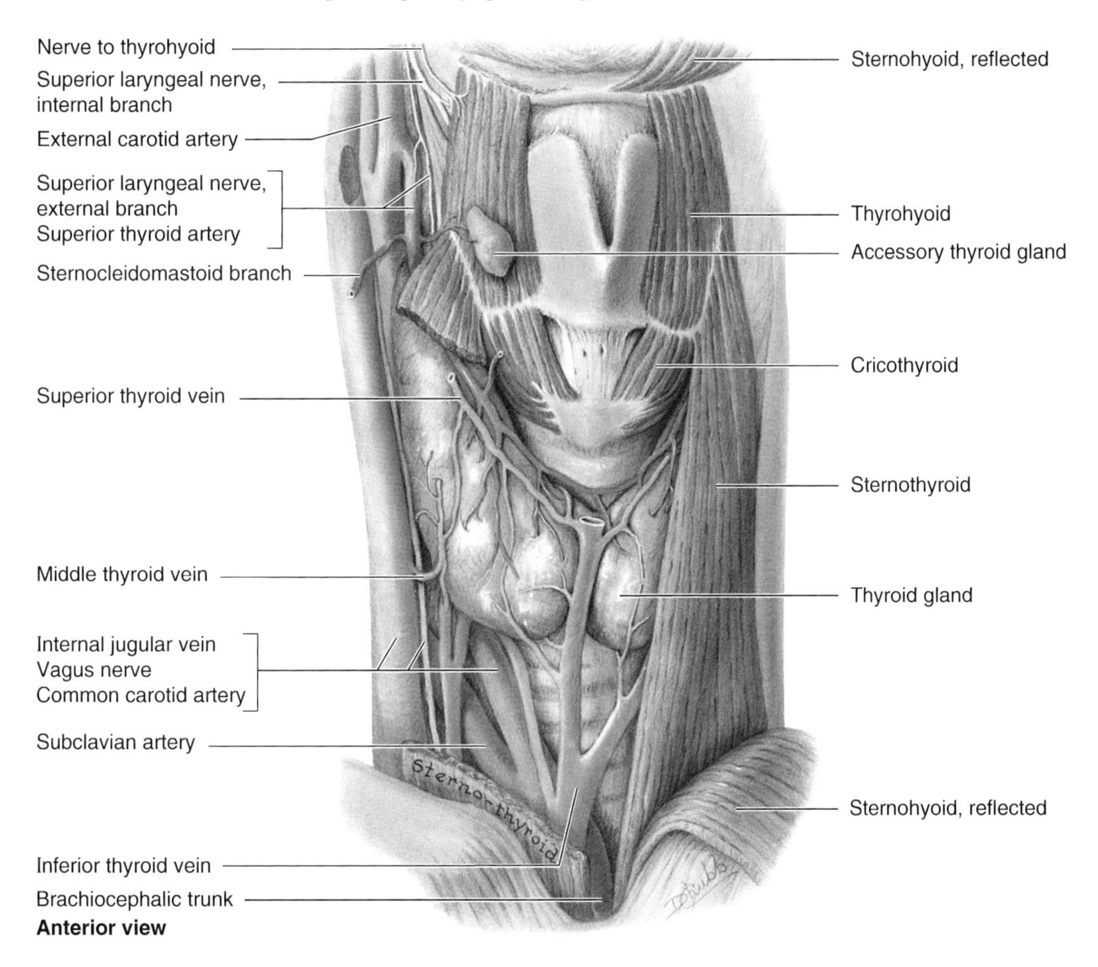

Figure 8.23. Thyroid gland. Anterior view. Dissection of the anterior aspect of the neck. In this specimen, an accessory thyroid gland is on the right, lying on the thyrohyoid muscle, lateral to the thyroid cartilage.

Thyroglossal Duct Cysts

The thyroid gland begins in the floor of the embryonic pharynx at the site indicated by the *foramen cecum* in the dorsum of the tongue. Subsequently, the gland descends through the tongue into the neck, passing anterior to the hyoid bone and thyroid cartilages to reach its final position anterolateral to the superior part of the trachea. During its migration, the thyroid gland is attached to the foramen cecum by a narrow tube, the **thyroglossal duct.** This duct normally disappears; however, remnants of it may remain and form thyroglossal duct cysts at any point along the path of descent of the developing thyroid gland. Cysts are usually near or within the body of the hyoid bone and form swellings in the anterior part of the neck. ▶

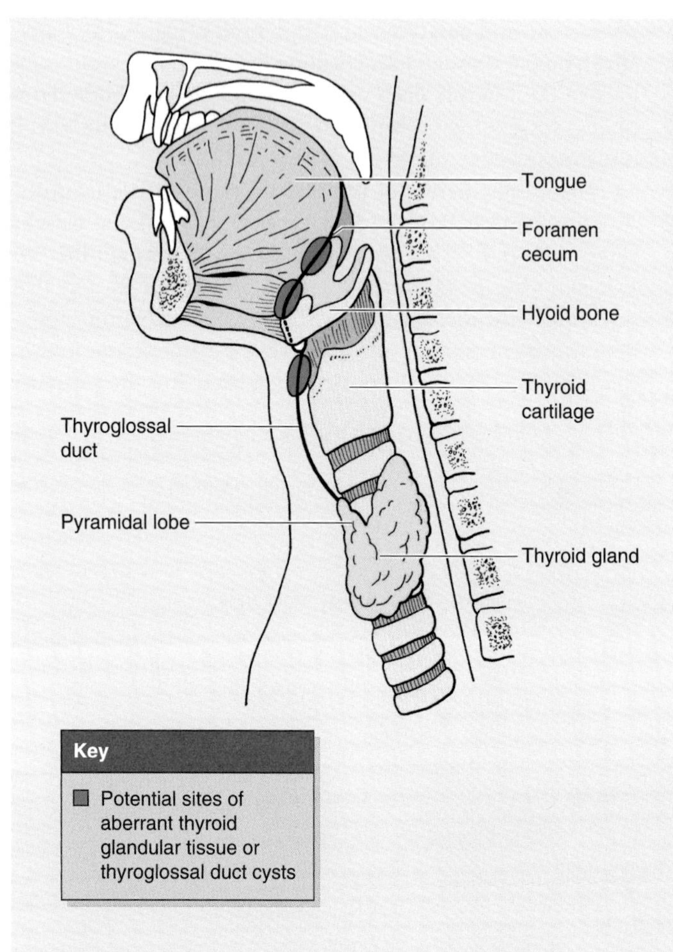

Ectopic Thyroid Gland

Uncommonly, the thyroid fails to descend from its embryonic origin in the base of the tongue (Moore and Persaud, 1998), resulting in a *lingual thyroid gland.* Incomplete descent results in the thyroid gland being high in the neck, at or just inferior to the hyoid bone. As a rule, an ectopic thyroid gland in the median plane of the neck is the only thyroid tissue present. Therefore, it is important to differentiate between an ectopic thyroid gland and a thyroglossal cyst; failure to do so may result in total thyroidectomy, leaving the person permanently dependent on thyroid medication (Leung et al., 1995). Occasionally thyroid glandular tissue is associated with a thyroglossal cyst.

Accessory Thyroid Glandular Tissue

Accessory thyroid gland tissue may appear in the thymus gland inferior to the thyroid gland. Although this tissue may be functional, it is often of insufficient size to maintain normal function if the thyroid gland is removed. An *accessory thyroid gland* may develop in the neck lateral to the thyroid cartilage; it usually lies on the thyrohyoid muscle (Fig. 8.23). Accessory thyroid glandular tissue originates from remnants of the thyroglossal duct.

Pyramidal Lobe of the Thyroid Gland

Approximately 50% of thyroid glands have a *pyramidal lobe* that varies in size (p. 1034). Frequently, the lobe extends superiorly from the isthmus of the thyroid, usually to the left of the median plane. The isthmus of the thyroid gland may be absent. A band of connective tissue may continue from the apex of the pyramidal lobe to the hyoid bone. The pyramidal lobe and its connective tissue continuation—often containing accessory thyroid tissue—develop from remnants of the thyroglossal duct.

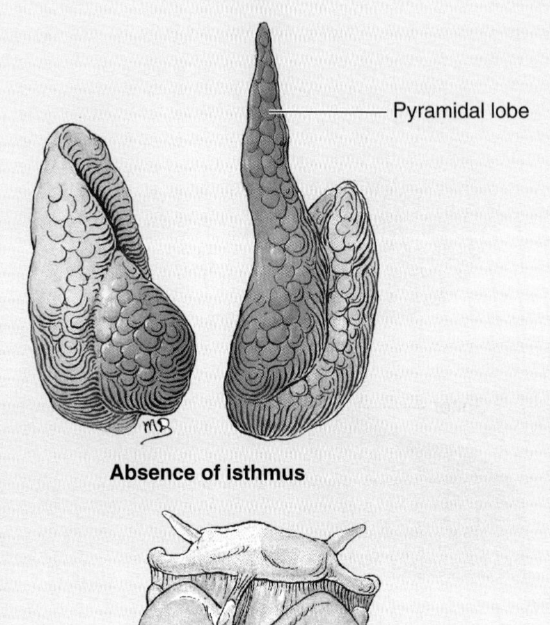

Absence of isthmus

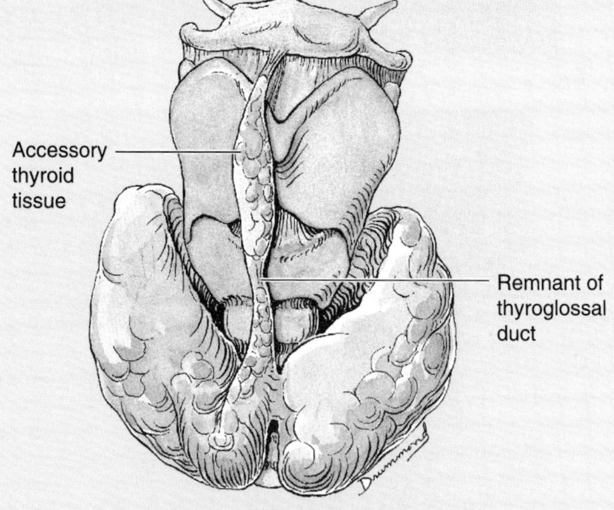

Enlargement of the Thyroid Gland

A nonneoplastic and noninflammatory enlargement of the thyroid gland, except for variable enlargement during menstruation and pregnancy, is called a goiter. *Goiter* is endemic—present in a community or among a group of people—in certain parts of the world where the soil and water are deficient in iodine. The enlarged thyroid gland causes a swelling in the neck that may compress the trachea, esophagus, and recurrent laryngeal nerves. Various ▶

▶ types of goiter exist. *Exophthalmic goiter,* for example, is a disorder caused by an excessive production of thyroid hormone. One sign of this disease, "bulging eyeballs," is *exophthalmos.* When the thyroid gland enlarges, it may do so anteriorly, posteriorly, inferiorly, or laterally. It cannot move superiorly because of the superior attachments of the sternothyroid and sternohyoid muscles (Table 8.3). Substernal extension of a goiter is also common.

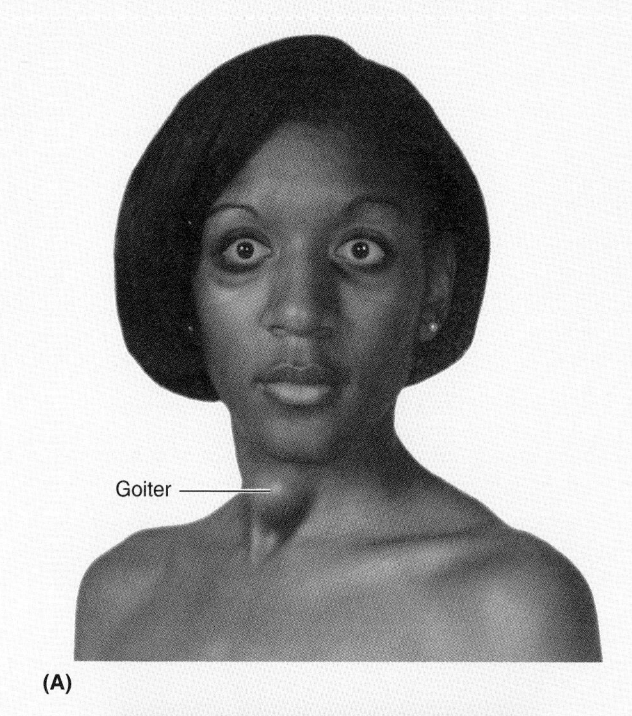

(A)

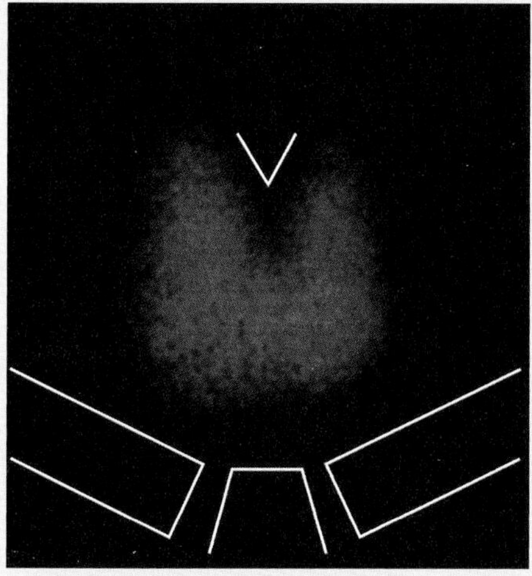

(B) Scintogram showing diffuse, enlarged thyroid gland

Thyroidectomy

Excision of a carcinoma (cancer) of the thyroid gland or other surgical procedures sometimes necessitate removal of the gland *(total thyroidectomy)*. In the surgical treatment of hyperthyroidism, the posterior part of each lobe of the enlarged thyroid is usually preserved—*subtotal (partial) thyroidectomy*—to protect the recurrent and superior laryngeal nerves and to spare the parathyroid glands. Postoperative hemorrhage after thyroid gland surgery may compress the trachea, making breathing difficult. The blood collects within the fibrous capsule of the thyroid gland, which is surrounded by the tough pretracheal layer of deep cervical fascia.

Injury to the Recurrent Laryngeal Nerves

The risk of injury to the recurrent laryngeal nerves is ever present during neck surgery. Near the inferior pole of the thyroid gland, *the right recurrent laryngeal nerve is intimately related to the inferior thyroid artery and its branches.* This nerve may cross anterior or posterior to branches of the artery, or it may pass between them. Because of this close relationship, the inferior thyroid artery is ligated some distance lateral to the thyroid gland, where it is not close to the nerve. Although the danger of injuring the left recurrent laryngeal nerve during surgery is not so great, the artery and nerve are also closely associated near the inferior pole of the thyroid gland (Fig. 8.22). *Hoarseness* is the

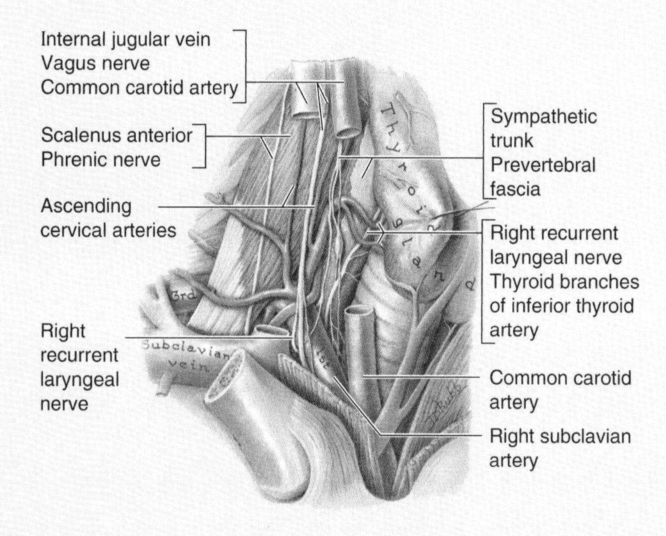

usual sign of unilateral recurrent nerve injury; however, *temporary aphonia* or disturbance of phonation (voice production) and *laryngeal spasm* may occur. These signs usually result from bruising the recurrent laryngeal nerves during surgery or from the pressure of accumulated blood, serous exudate, or both after the operation.

Injury to the External Laryngeal Nerve

Injury to the external laryngeal nerve (Fig. 8.22), a terminal branch of the superior laryngeal nerve (a branch of the vagus), results in a voice that is monotonous in character ▶

▶ because the *paralyzed cricothyroid muscle* supplied by it (see Table 8.5) is unable to vary the length and tension of the vocal fold (cord). To avoid injury to the external laryngeal nerve, the superior thyroid artery is ligated and sectioned more superior to the gland, where it is not as closely related to the nerve. Because an enlarged thyroid (goiter) may itself be the cause of impaired innervation of the larynx by compressing the laryngeal nerves, it is good practice to examine the vocal folds before an operation. In this way, damage to the larynx or its nerves resulting from a surgical mishap may be distinguished from a pre-existing injury resulting from nerve compression (e.g., by an enlarged thyroid gland). ⊙

Parathyroid Glands

The small, ovoid parathyroid glands usually lie external to the fibrous thyroid capsule on the medial half of the posterior surface of each lobe of the thyroid gland, but inside its sheath (Fig. 8.25*A*). The **superior parathyroid glands** usually lie slightly more than 1 cm superior to the point of entry of the inferior thyroid arteries into the thyroid, and the **inferior parathyroid glands** usually lie slightly more than 1 cm inferior to the arterial entry point (Skandalakis et al., 1995). Most people have four parathyroid glands, but approximately 5% of people have more than four glands. The *superior parathyroid glands*, more constant in position than the inferior ones, are usually at the level of the inferior border of the cricoid cartilage. The *inferior parathyroid glands* are usually near the inferior pole of the thyroid gland, but they may lie in various positions (Fig. 8.25*B*). In 1 to 5% of people, an inferior parathyroid gland is deep in the superior mediastinum (Norton and Wells, 1994).

Vessels of the Parathyroid Glands. The parathyroid glands are usually supplied by branches of the **inferior thyroid arteries** (Fig. 8.25*A*), but they may be supplied by the *superior thyroid arteries*; the *thyroid ima artery*; or the laryngeal, tracheal, and esophageal arteries. **Parathyroid veins** drain into the plexus of veins on the anterior surface of the thyroid gland and trachea (Fig. 8.23). **Lymphatic vessels** from the parathyroids drain with those from the thyroid gland into deep cervical lymph nodes and paratracheal lymph nodes (Fig. 8.24).

Nerves of the Parathyroid Glands. Nerves of the parathyroid glands derive from thyroid branches of the cervical sympathetic ganglia (Fig. 8.20).

Inadvertent Removal of the Parathyroid Glands

The variable position of the parathyroid glands, especially the inferior ones, puts them in danger of being damaged or removed during surgery on the thyroid gland. The superior parathyroid glands may be as far superior as the thyroid cartilage, and the inferior ones may be as far inferior as the superior mediastinum (see ▶

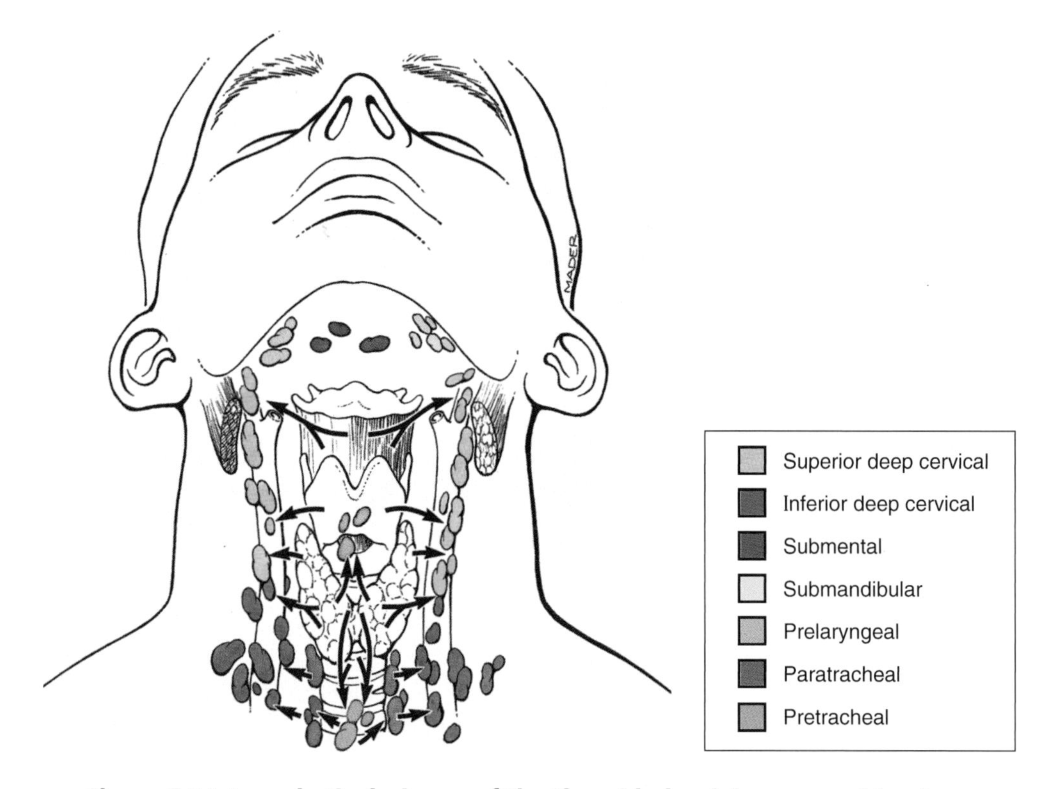

Figure 8.24. Lymphatic drainage of the thyroid gland, larynx, and trachea. The *arrows* indicate the direction of lymph flow.

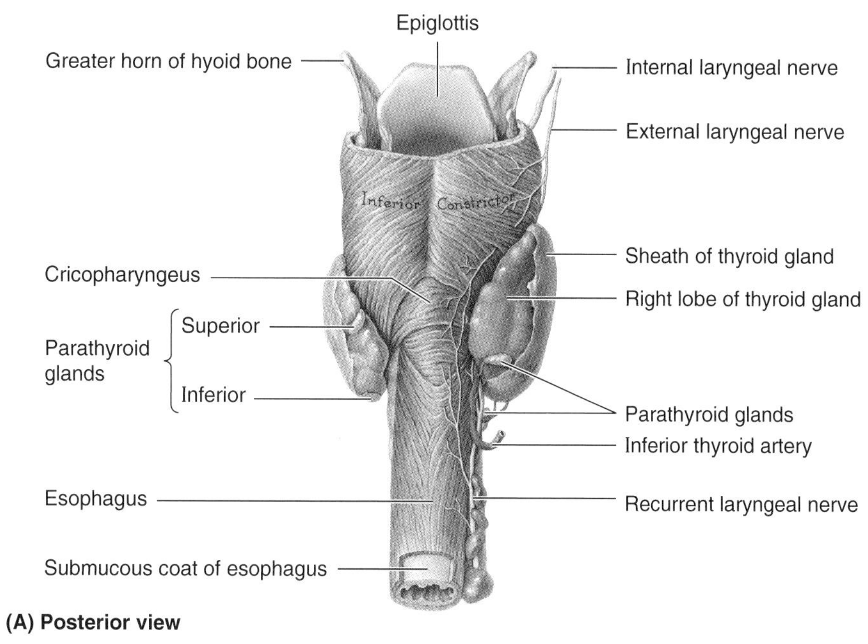

(A) Posterior view

0.3%
0.3%
11%
13%
3%
1%
1.6%
20%
38%
1%
6%
4%

(B) Anterior view

Figure 8.25. Thyroid and parathyroid glands. A. Posterior view. The sheath has been dissected from the posterior surface of the thyroid gland to reveal the three embedded parathyroid glands. Both parathyroid glands on the right side are rather low, with the inferior gland being inferior to the thyroid gland. **B.** Sites and frequency of aberrant parathyroid glandular tissue.

▶ Chapter 1). The aberrant sites of these glands are of concern when searching for abnormal parathyroid glands—for example, those with a *parathyroid adenoma*, an ordinarily benign tumor of epithelial tissue associated with hyperparathyroidism.

If the parathyroid glands atrophy or all of them are inadvertently removed during surgery, the patient suffers from *tetany*, a severe convulsive disorder. The generalized convulsive muscle spasms result from a *fall in serum calcium levels*. Because laryngeal and respiratory muscles are involved, failure to respond immediately with appropriate therapy can result in death. To safeguard these glands during thyroidectomy, surgeons usually preserve the posterior part of the lobe of the thyroid gland. In instances when it is necessary to do a total thyroidectomy (e.g., because of malignant disease), the parathyroid glands are carefully isolated with their blood vessels intact before removal of the thyroid gland. ⊙

Respiratory Layer of the Cervical Viscera

The viscera of the respiratory layer, the larynx and trachea, contribute to the respiratory function of the body. The **larynx** is the complex organ of voice production composed of nine cartilages connected by membranes and ligaments and containing the vocal folds. *The main functions of the respiratory layer are*:

- Routing air and food into the respiratory tract and esophagus, respectively
- Providing a patent airway, and a means of sealing it off temporarily
- Producing voice.

The **trachea**, extending from the larynx into the thorax, divides into right and left main bronchi. It transports air to and from the lungs, and its epithelium propels debris-laden mucus toward the pharynx for expulsion from the mouth.

Larynx

In the anterior neck at the level of the bodies of C3 through C6 vertebrae lies the **larynx**, *the phonating mechanism* designed for voice production (vocalization), which connects the inferior part of the pharynx (oropharynx) with the trachea (Fig. 8.26). It also guards the air passages, especially during swallowing, and maintains a patent airway.

Laryngeal Skeleton. The laryngeal skeleton consists of nine cartilages joined by ligaments and membranes. Three cartilages are single—thyroid, cricoid, and epiglottic—and three are paired—arytenoid, corniculate, and cuneiform (L. cuneus) (Fig. 8.27, *A* and *B*).

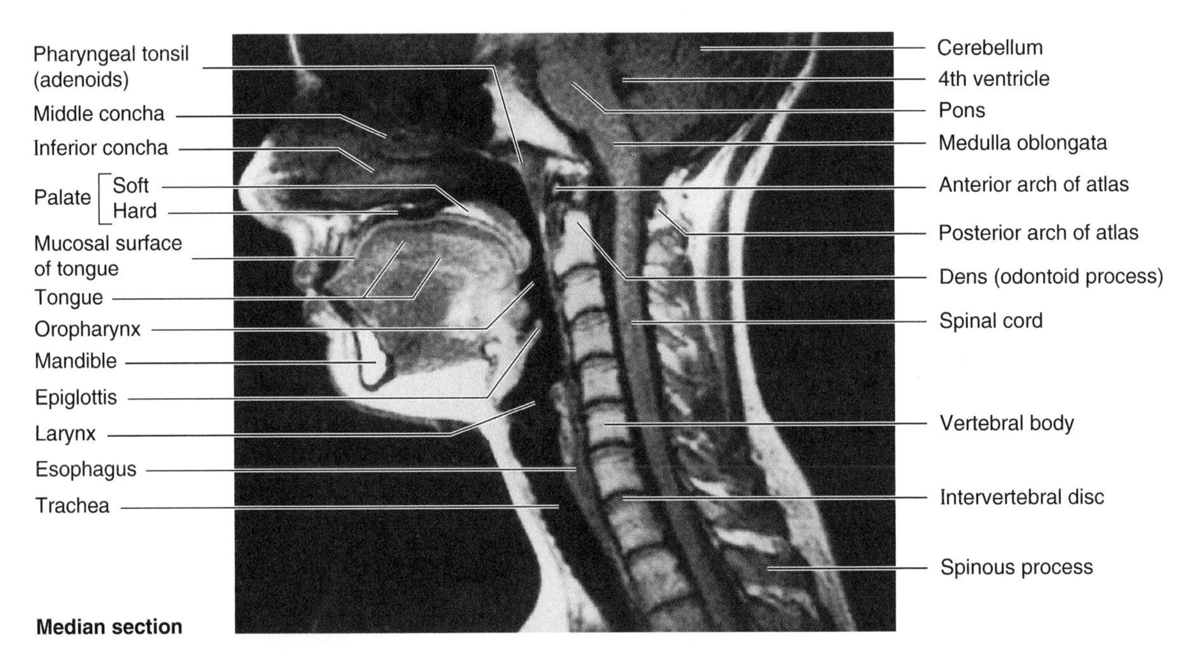

Figure 8.26. Median MRI through the head and neck. Note that the air and food passages share the oropharynx and so separation of food and air must occur to continue into the trachea (anterior) and esophagus (posterior). (Courtesy of Dr. W. Kucharczyk, Chair of Medical Imaging, University of Toronto, and Clinical Director, Tri-Hospital Resonance Centre, Toronto, Ontario, Canada)

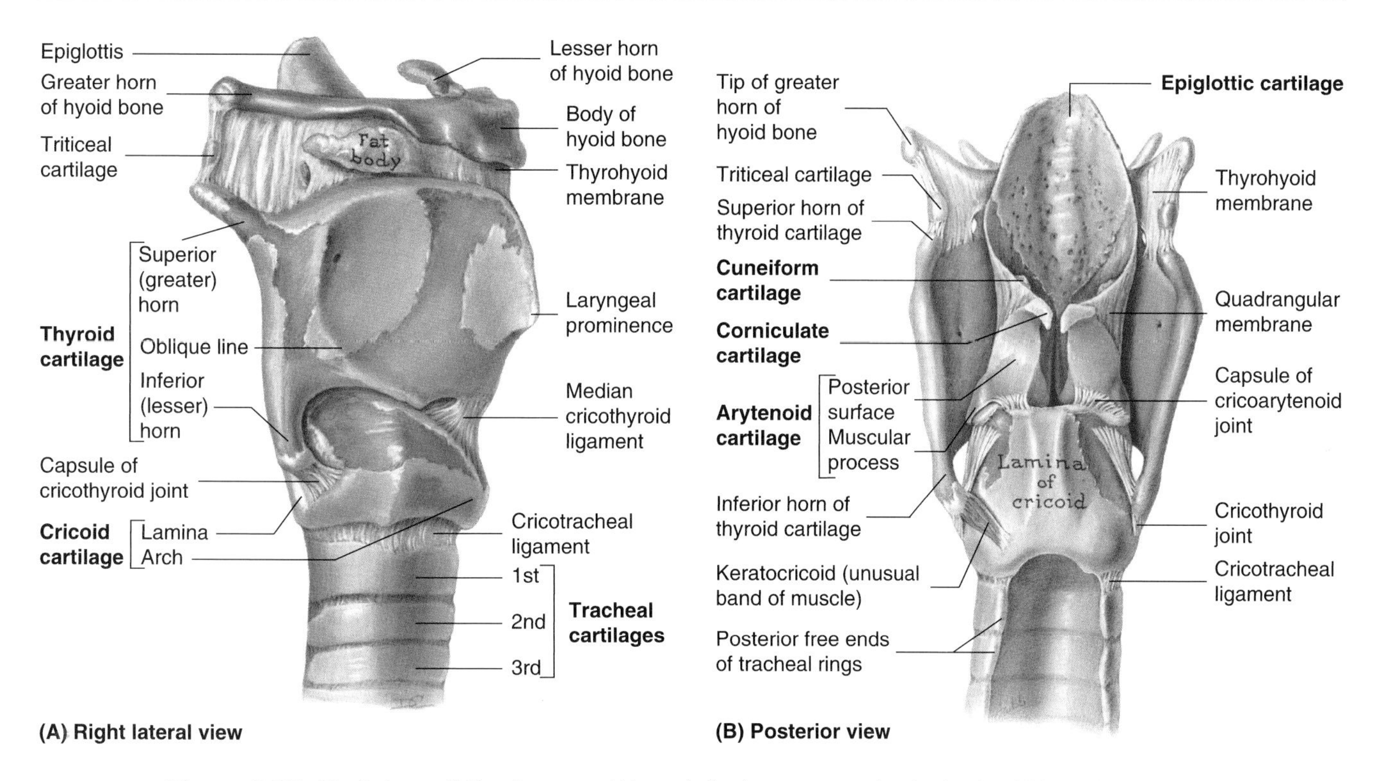

Figure 8.27. Skeleton of the larynx. Although firmly connected to it, the hyoid bone is not part of the larynx. **A.** Right lateral view. Notice that the larynx extends vertically from the tip of the heart-shaped epiglottis to the inferior border of the cricoid cartilage. **B.** Posterior view. Observe that the thyroid cartilage shields the smaller cartilages of the larynx. Although not part of the larynx, the hyoid bone shields the superior part of the epiglottic cartilage.

The **thyroid cartilage** is the largest of the cartilages. The inferior two-thirds of its two platelike **laminae** fuse anteriorly in the median plane to form the **laryngeal prominence** (Fig. 8.27, *A* and *C*). Superior to this prominence ("Adam's apple"), the laminae diverge to form a V-shaped **superior thyroid notch** (usually referred to as the thyroid notch); the inferior thyroid notch is a shallow indentation in the middle of the inferior border of the cartilage. The posterior border of each lamina projects superiorly as the **superior horn** and inferiorly as the **inferior horn**. The superior border and superior horns attach to the hyoid bone by the **thyrohyoid membrane** (Fig. 8.27, *A* and *B*). The thick median part of this membrane is the *median thyrohyoid ligament*; its lateral parts are the *lateral thyrohyoid ligaments* (Fig. 8.27*D*). The inferior horns articulate with the lateral surfaces of the cricoid cartilage at the **cricothyroid joints** (Fig. 8.27*B*). The main movements at these joints are rotation and gliding of the thyroid cartilage, which result in changes in the length of the vocal folds.

The **cricoid cartilage** is shaped like a signet ring with its band facing anteriorly. This ringlike opening of the cartilage fits an average finger. The posterior (signet) part of the cricoid is the *lamina*, and the anterior (band) part is the *arch* (Fig. 8.27*A*). Although much smaller than the thyroid cartilage, the cricoid cartilage is thicker and stronger and is the only complete ring of cartilage to encircle any part of the airway. It attaches to the inferior margin of the thyroid cartilage by the *median cricothyroid ligament* and to the 1st tracheal ring by the *cricotracheal ligament*. Where the larynx is closest to the skin and most accessible, the cricothyroid ligament may be felt as a soft spot during palpation inferior to the thyroid cartilage. The **arytenoid cartilages** are pairs of three-sided pyramids that articulate with lateral parts of the superior border of the cricoid cartilage lamina. Each cartilage has an apex superiorly, a vocal process anteriorly, and a large muscular process that projects laterally from its base. The apex bears the corniculate cartilage and attaches to the aryepiglottic fold, the vocal process provides the posterior attachment for the vocal ligament, and the muscular process serves as a lever to which the posterior and lateral cricoarytenoid muscles are attached.

The **cricoarytenoid joints**—between the bases of the arytenoid cartilages and the superolateral surfaces of the lamina of the cricoid cartilage (Fig. 8.27*B*)—permit the arytenoid cartilages to slide toward or away from one to another, to tilt anteriorly and posteriorly, and to rotate. These movements are important in approximating, tensing, and relaxing the vocal folds. The elastic **vocal ligament** extends from the junction of the laminae of the thyroid cartilage anteriorly to the vocal process of the arytenoid cartilage posteriorly (Fig. 8.27*D*). The vocal ligament forms the skeleton of the vocal

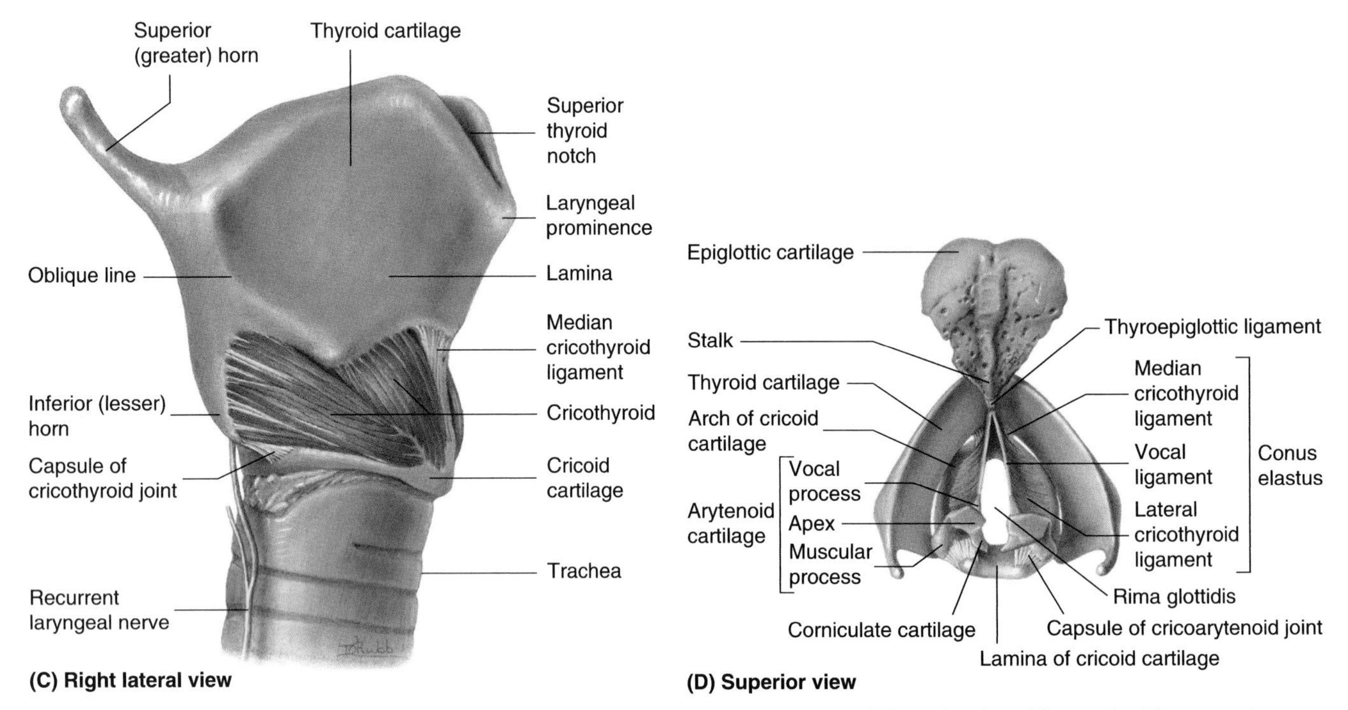

Figure 8.27. *(Continued)* **C.** Observe the thyroid cartilage and the cricothyroid muscle. The muscle produces movement at the cricothyroid joint. **D.** Observe that the epiglottic cartilage is pitted for mucous glands and that its stalk is attached by the thyroepiglottic ligament to the angle of the thyroid cartilage superior to the vocal ligaments. Note the vocal ligament, which forms the skeleton of the vocal fold (cord), extending from the vocal process to the "angle" of the thyroid cartilage, and there joining its fellow inferior to the thyroepiglottic ligament.

fold; it is the thickened, free, superior border of the **lateral cricothyroid ligament** (part of the *conus elasticus*). This ligament blends anteriorly with the **median cricothyroid ligament** (also part of the conus elasticus). The conus elasticus closes the laryngeal inlet except for rima glottidis (opening between the vocal ligaments).

The **epiglottic cartilage**, consisting of elastic cartilage, gives flexibility to the *epiglottis*—a heart-shaped cartilage covered with mucous membrane (Fig. 8.27*B*). Situated posterior to the root of the tongue and the hyoid bone and anterior to the laryngeal inlet, the epiglottic cartilage forms the superior part of the anterior wall and the superior margin of the laryngeal inlet. Its broad superior end is free, and its tapered inferior end (stalk) is attached to the angle formed by the thyroid laminae by the **thyroepiglottic ligament** (Fig. 8.27*D*). The **hyoepiglottic ligament** (Fig. 8.28) attaches the anterior surface of the epiglottic cartilage to the hyoid bone. The **quadrangular membrane** (Fig. 8.27*B*) is a thin, submucosal sheet of connective tissue that extends between the lateral aspects of the arytenoid and epiglottic cartilages. Its free inferior margin constitutes the *vestibular ligament,* which is covered loosely by the **vestibular fold** (Fig. 8.29). This fold lies superior to the vocal fold and extends from the thyroid cartilage to the arytenoid cartilage. The free superior margin of the quadrangular membrane forms the aryepiglottic ligament, which is covered with mucosa to form the aryepiglottic fold. The **corniculate** and **cuneiform cartilages** are small nodules in the posterior part of the aryepiglottic folds. The corniculate cartilages attach to the apices of the arytenoid cartilages; the cuneiforms do not directly attach to other cartilages.

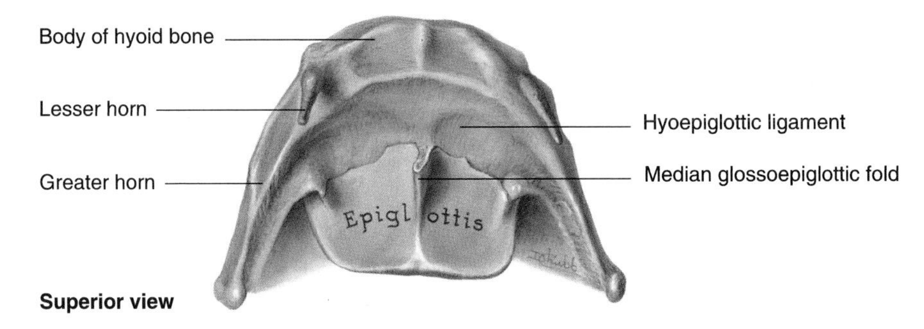

Figure 8.28. Epiglottis and hyoepiglottic ligament. Observe that the epiglottis is a spatula-like plate of elastic fibrocartilage, which is covered with mucous membrane (*pink*) and is attached anteriorly to the hyoid bone by the hyoepiglottic ligament (*blue*). It serves as a diverter valve over the superior aperture of the larynx during swallowing.

Fractures of the Laryngeal Skeleton

Fractures of the laryngeal skeleton often result from blows received in sports such as kick boxing, hockey, and karate or from compression by a shoulder strap during an automobile accident. Because of the frequency of this type of injury, most goalies in ice hockey and catchers in baseball have protective guards hanging from their masks that cover their larynges. *Laryngeal fractures* produce submucous hemorrhage and edema, respiratory obstruction, hoarseness, and sometimes a temporary inability to speak. ⊙

Interior of the Larynx. The laryngeal cavity extends from the laryngeal inlet, through which it communicates with the laryngopharynx, to the level of the inferior border of the cricoid cartilage. Here the **laryngeal cavity** is continuous with the cavity of the trachea (Figs. 8.29 and 8.30). *The laryngeal cavity is divided into three parts:*

- The *vestibule of the larynx*—superior to the vestibular folds
- The *ventricle of the larynx* (laryngeal sinus)—between the vestibular folds and superior to the vocal folds
- The *infraglottic* cavity—the inferior cavity of the larynx extending from the vocal folds to the inferior border of the cricoid cartilage, where it is continuous with the lumen of the trachea.

The **vocal folds** (true vocal cords) control sound production (Figs. 8.29–8.31). The apex of each wedge-shaped fold projects medially into the laryngeal cavity. *Each vocal fold has:*

- A *vocal ligament* consisting of thickened elastic tissue that is the medial free edge of the lateral cricothyroid ligament (conus elasticus)
- A *vocalis (vocal) muscle*, the exceptionally fine muscle fibers that form the most medial part of the thyroarytenoid muscle (Table 8.5).

The vocal folds are the source of the sounds that come from the larynx. These folds produce audible vibrations when their free margins are closely—but not tightly—opposed during phonation and air is forcibly expired intermittently (Fig. 8.31, *A* and *B*). The vocal folds also serve as the main inspi-

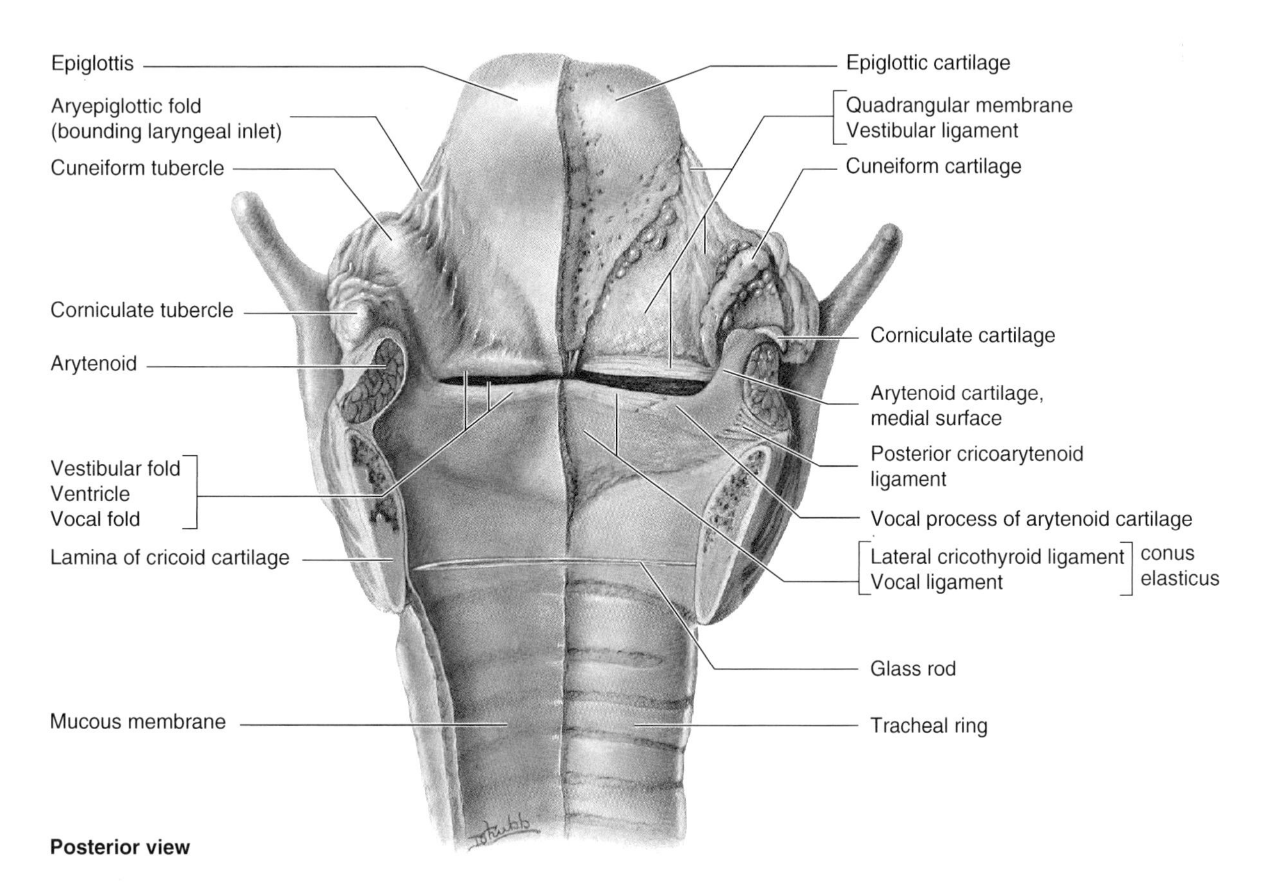

Figure 8.29. Interior of the larynx. Posterior view of a dissection of the interior of the larynx. The posterior wall of the larynx is split in the median plane, and the two sides are spread apart and held in place by a glass rod. On the left side the mucous membrane is intact; on the right side the mucous and submucous coats are peeled off and the skeletal coat, consisting of cartilages, ligaments, and the fibroelastic membrane, is laid bare.

ratory sphincter of the larynx when they are tightly closed. Complete adduction of the folds forms an effective sphincter that prevents entry of air (Fig. 8.31*C*).

The ***glottis*** (the vocal apparatus of the larynx) comprises the vocal folds and processes, together with the **rima glottidis**—the aperture between the vocal folds (Fig. 8.30*A*). The shape of the rima (L. slit) varies according to the position of the vocal folds (Fig. 8.31). During ordinary breathing, the rima is narrow and wedge shaped; during forced respiration it is wide and kite shaped. The rima glottidis is

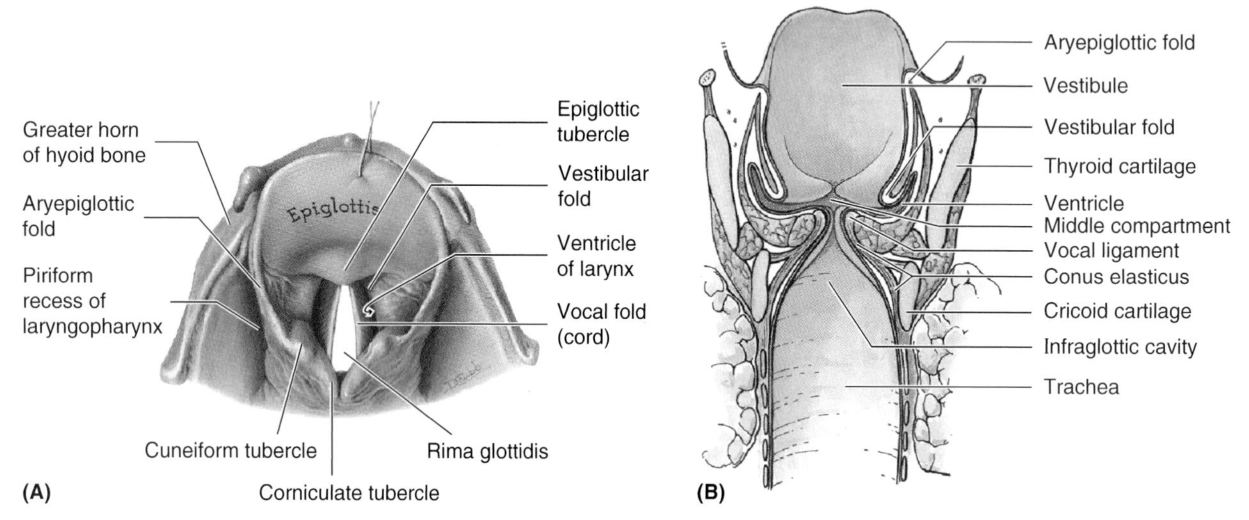

Figure 8.30. Larynx. A. Superior view. Observe the rima glottidis, the space between the vocal folds (cords). Also observe the inlet of the larynx (laryngeal aditus) and its boundaries: (*a*) anteriorly by the free curved edge of the epiglottis; (*b*) posteriorly by the arytenoid cartilages, the corniculate cartilages that cap them, and the interarytenoid fold that unites them; and (*c*) on each side, by the aryepiglottic fold that contains the superior end of the cuneiform (L. cuneus) cartilage. **B.** Coronal section showing the compartments of the larynx: the vestibule, middle compartment with left and right ventricles, and the infraglottic cavity.

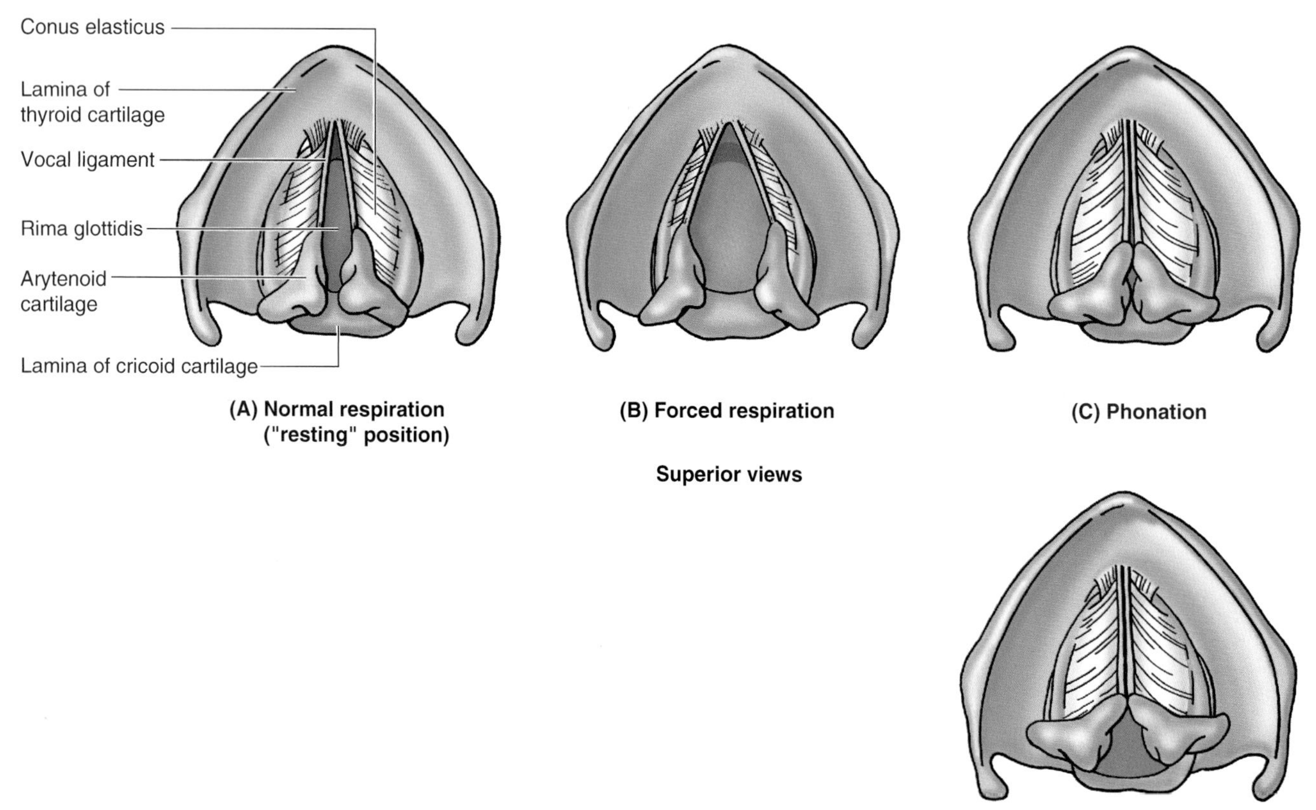

Figure 8.31. Variations in the rima glottidis. The rima glottidis is the aperture between the vocal folds (cords). Its shape varies according to the position of the vocal folds.

slitlike when the vocal folds are closely approximated during phonation. Variation in the tension and length of the vocal folds, in the width of the rima glottidis, and in the intensity of the expiratory effort produces changes in the pitch of the voice. The lower range of pitch of the voice of postpubertal males results from the greater length of the vocal folds.

The **vestibular folds** (false vocal cords), extending between the thyroid and arytenoid cartilages (Figs. 8.29 and 8.30), play little or no part in voice production; they are protective in function. They consist of two thick folds of mucous membrane enclosing the *vestibular ligaments*. The space between these ligaments is the rima vestibuli. The lateral indentations between the vocal and vestibular folds is the **ventricle of the larynx**.

Laryngoscopy

Laryngoscopy refers to any procedure used to examine the interior of the larynx. The larynx may be examined visually by *indirect laryngoscopy* using a laryngeal mirror. The anterior part of the tongue is gently pulled from the oral cavity to minimize the extent to which the posterior part of the tongue covers the epiglottis and laryngeal inlet. Because the rima vestibuli is larger than the rima glottidis during normal respiration, the vestibular folds and vocal folds are visible during a laryngoscopic examination. The larynx can also be viewed by *direct laryngoscopy* using a tubular, endoscopic instrument, a *laryngoscope*—any of several types of hollow tubes or flexible fiber optic endoscopes equipped with electrical lighting for examining or operating on the interior of the larynx through the mouth. The vestibular folds normally appear pink, whereas the vocal folds are usually pearly white.

Valsalva Maneuver

The sphincteric actions of the vestibular and vocal folds are important during the Valsalva maneuver—any forced expiratory effort against a closed airway—such as a cough, sneeze, or strain during a bowel movement or weight lifting.

- The lungs inflate by deep inspiration and the vestibular and vocal folds abduct widely.
- At the end of deep inspiration, both the vestibular and vocal folds are tightly adducted.
- The anterolateral abdominal muscles then contract strongly to increase the intrathoracic and intra-abdominal pressures.
- The relaxed diaphragm passively transmits the increased abdominopelvic pressure to the thoracic cavity.

Because high intrathoracic pressure impedes venous return to the right atrium, researchers use the Valsalva maneuver to study cardiovascular effects of raised peripheral venous pressure and decreased cardiac filling and cardiac output.

Aspiration of Foreign Bodies

A foreign object, such as a piece of steak, may accidentally aspirate through the laryngeal inlet into the vestibule of the larynx, where it becomes trapped superior to the vestibular folds. The resulting blockage may completely seal off the larynx and choke the person, leaving the individual speechless because the larynx is blocked. The person will die in approximately five minutes from lack of oxygen if the larynx is not opened. When a foreign object enters the vestibule of the larynx, the laryngeal muscles go into spasm, tensing the vocal folds. The rima glottidis closes and no air enters the trachea. Asphyxiation may occur if the foreign object is not dislodged.

Emergency therapy must be given to open the airway. The procedure to use depends on the condition of the patient, the facilities available, and the experience of the person giving first aid. The vestibular folds are part of the protective mechanism that closes the larynx. The mucosa of the ▶

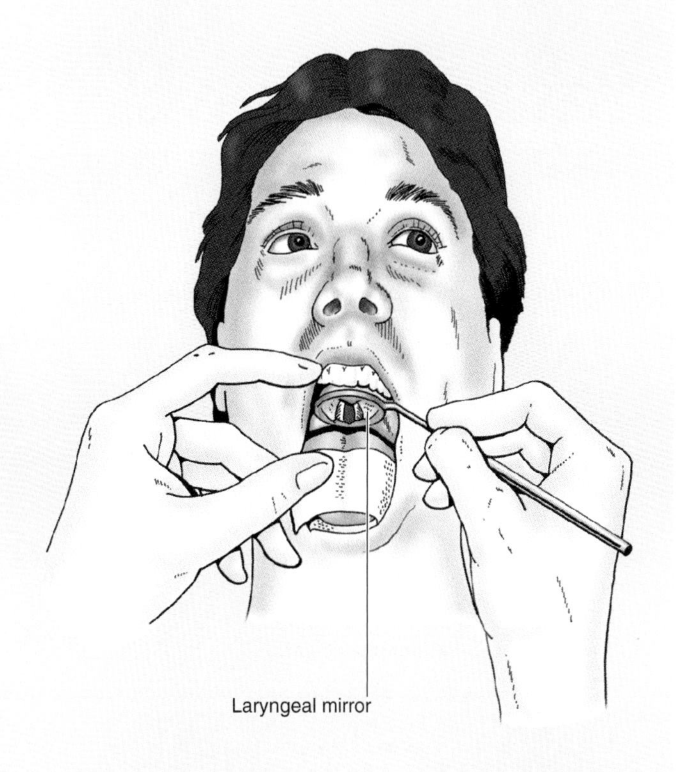

▶ vestibule is sensitive to foreign objects such as food. When an object passes through the laryngeal inlet and contacts the vestibular epithelium, violent coughing occurs in an attempt to expel the object. If this action fails, the aspirated food or other material may lodge in the rima glottidis, causing *laryngeal obstruction* (choking).

Because the lungs still contain air, sudden compression of the abdomen (*Heimlich maneuver*) causes the diaphragm to elevate and compress the lungs, expelling air from the trachea into the larynx. This maneuver usually dislodges the food or other material from the larynx. To perform the Heimlich maneuver, the person giving first aid uses subdiaphragmatic abdominal thrusts to expel the foreign object from the larynx. First, the closed fist, with the base of the palm facing inward, is placed on the victim's abdomen between the umbilicus and the xiphoid process of the sternum. The fist is grasped by the other hand and forcefully thrust inward and superiorly, forcing the diaphragm superiorly. This action forces air from the lungs and creates an artificial cough that usually expels the foreign object. Several abdominal thrusts may be necessary to remove the obstruction in the larynx.

In extreme cases, experienced persons (usually physicians) insert a large bore needle through the cricothyroid ligament (*needle cricothyrotomy* or "coniotomy") to permit fast entry of air. Later, a *surgical cricothyrotomy* may be performed, which involves an incision through the skin and cricothyroid ligament and insertion of a small *tracheostomy tube* into the trachea (p. 1048). ⊙

Laryngeal Muscles. The laryngeal muscles are divided into extrinsic and intrinsic groups.

- The **extrinsic laryngeal muscles** (discussed earlier with muscles of the anterior triangle [p. 1015]) move the larynx as a whole (Table 8.3). The infrahyoid muscles are depressors of the hyoid bone and larynx, whereas the suprahyoid and stylopharyngeus muscles are elevators of the hyoid bone and larynx.
- The **intrinsic laryngeal muscles** move the laryngeal parts, making alterations in the length and tension of the vocal folds and in the size and shape of the rima glottidis (Fig. 8.31). All but one of the intrinsic muscles of the larynx are supplied by the *recurrent laryngeal nerve* (Figs. 8.32–8.34), a branch of CN X; the cricothyroid is supplied by the external laryngeal nerve, one of the two terminal branches of the *superior laryngeal nerve.*

The actions of the intrinsic laryngeal muscles are easiest to understand when they are considered as functional groups: sphincters, adductors and abductors, and tensors and relaxers (Figs. 8.30–8.32, Table 8.5).

- **Adductors and abductors.** These muscles move the vocal folds to open and close the rima glottidis.

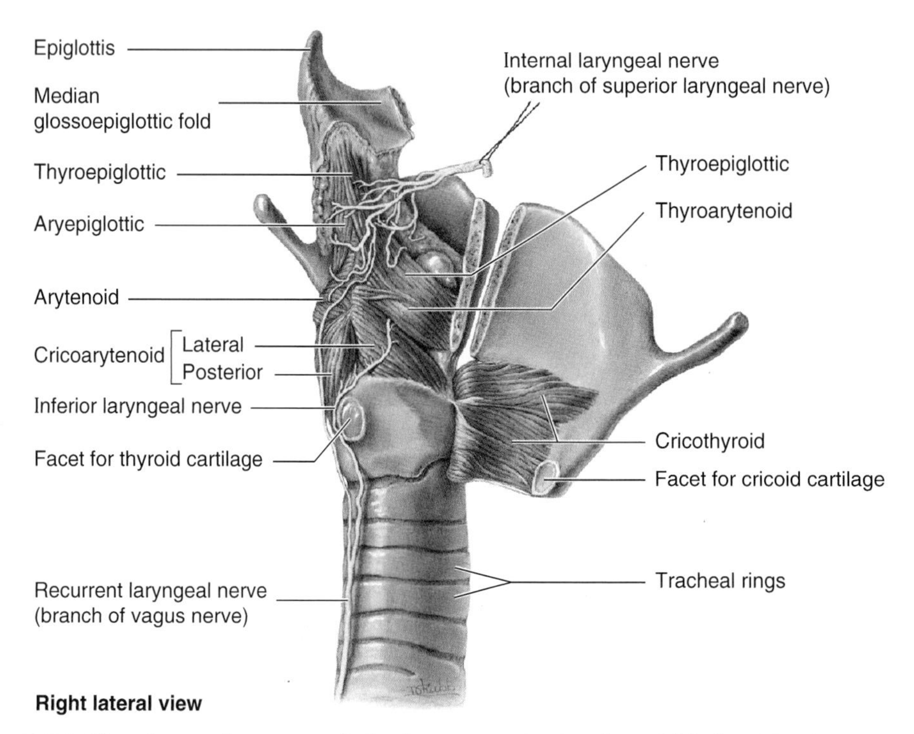

Figure 8.32. Muscles and nerves of the larynx and cricothyroid joint. The thyroid cartilage is sawn through to the right of the median plane. The cricothyroid joint is disarticulated and the right lamina of the thyroid cartilage is turned anteriorly (like opening a book), stripping the cricothyroid muscles off the arch of the cricoid cartilage.

Table 8.5. **Actions of Laryngeal Muscles**

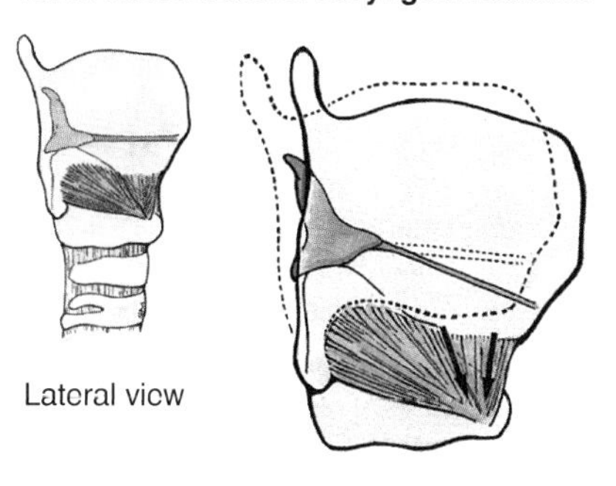
Lateral view

Lateral view
Cricothyroid

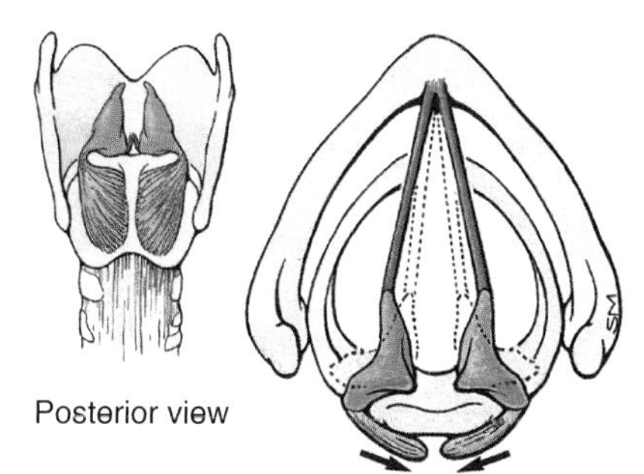
Posterior view

Superior view
Posterior cricoarytenoid

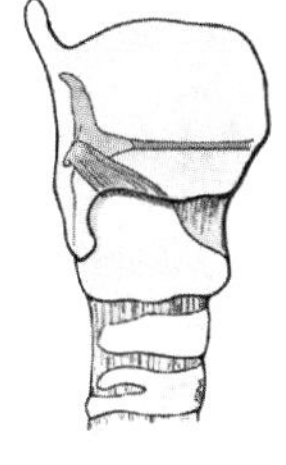
Lateral view

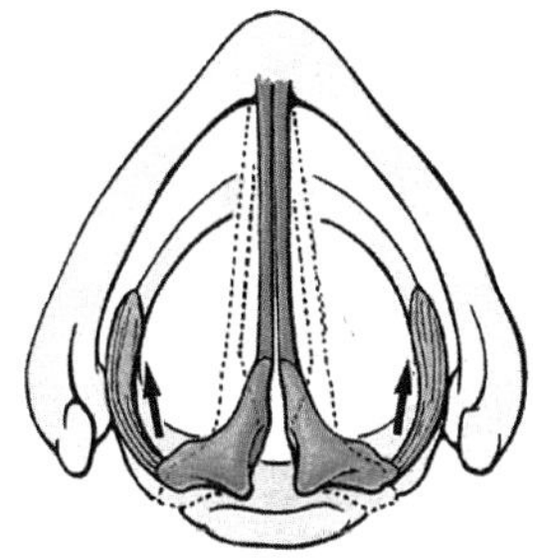
Superior view
Lateral cricoarytenoid

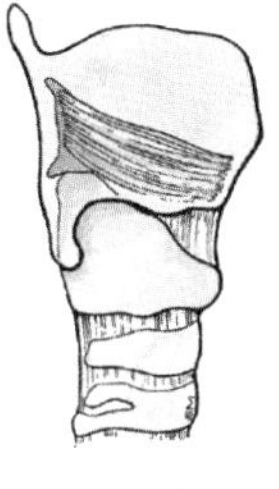
Lateral view

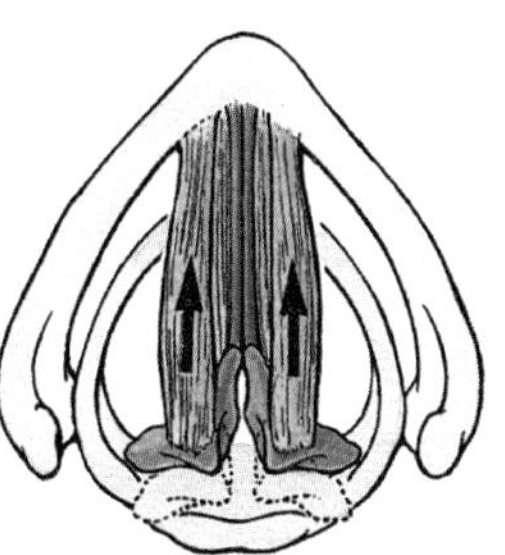
Superior view
Thyroarytenoid

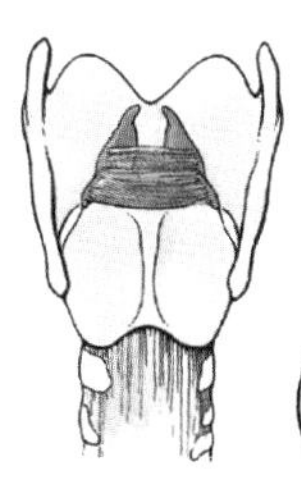
Transverse arytenoid (Posterior view)

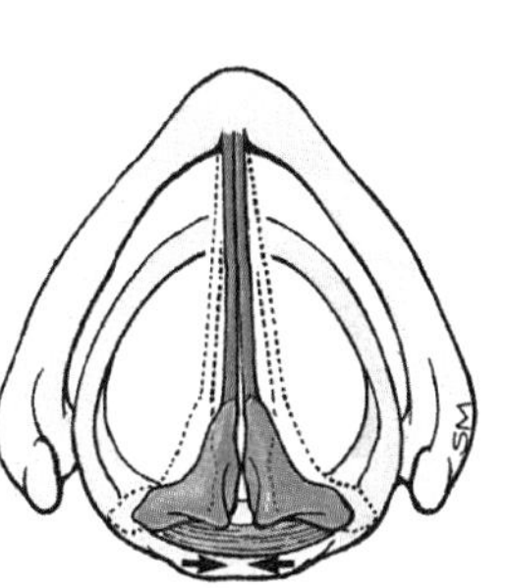
Superior view
Transverse & oblique arytenoids

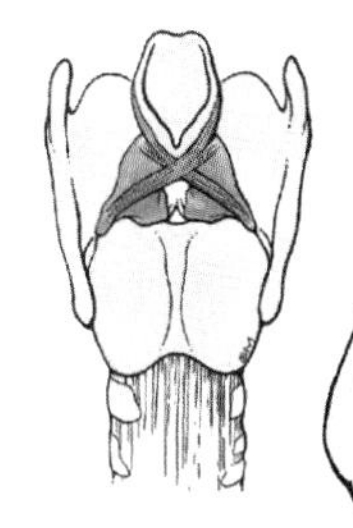
Oblique arytenoid (Posterior view)

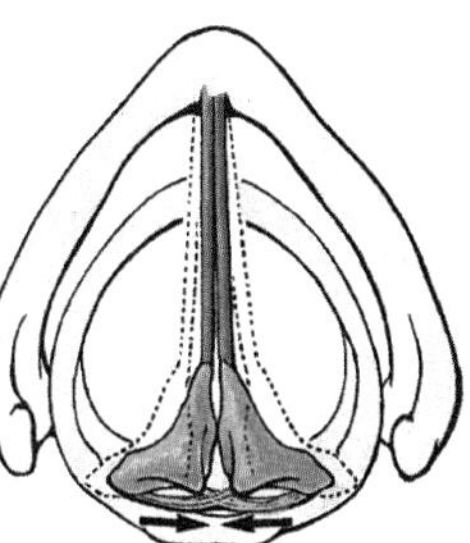
Superior view
Vocalis Muscle

Muscle	Origin	Insertion	Innervation	Main Action
Cricothyroid	Anterolateral part of cricoid cartilage	Inferior margin and inferior horn of thyroid cartilage	External laryngeal nerve	Stretches and tenses vocal fold
Posterior cricoarytenoid	Posterior surface of laminae of cricoid cartilage			Abducts vocal fold
Lateral cricoarytenoid	Arch of cricoid cartilage	Muscular process of arytenoid cartilage		Adducts vocal fold (inter-ligamentous portion)
Thyroarytenoid [a]	Posterior surface of thyroid cartilage		Recurrent laryngeal nerve	Relaxes vocal fold
Transverse and oblique arytenoids [b]	One arytenoid cartilage	Opposite arytenoid cartilage		Closes intercartilaginous portion of rima glottidis
Vocalis [c]	Vocal process of arytenoid cartilage	Vocal ligaments		Relaxes posterior vocal ligament while maintaining (or increasing) tension of anterior part

[a] Superior fibers of the thyroarytenoid muscle pass into the aryepiglottic fold, and some of them reach the epiglottic cartilage. These fibers constitue the thyroepiglottic muscle, which widens inlet of larynx.
[b] Some fibers of oblique arytenoid muscle continue as aryepiglottic muscle (Fig. 9.9A).
[c] This slender muscular slip is derived from inferior, deeper, and finer fibers of the thyroarytenoid muscle.

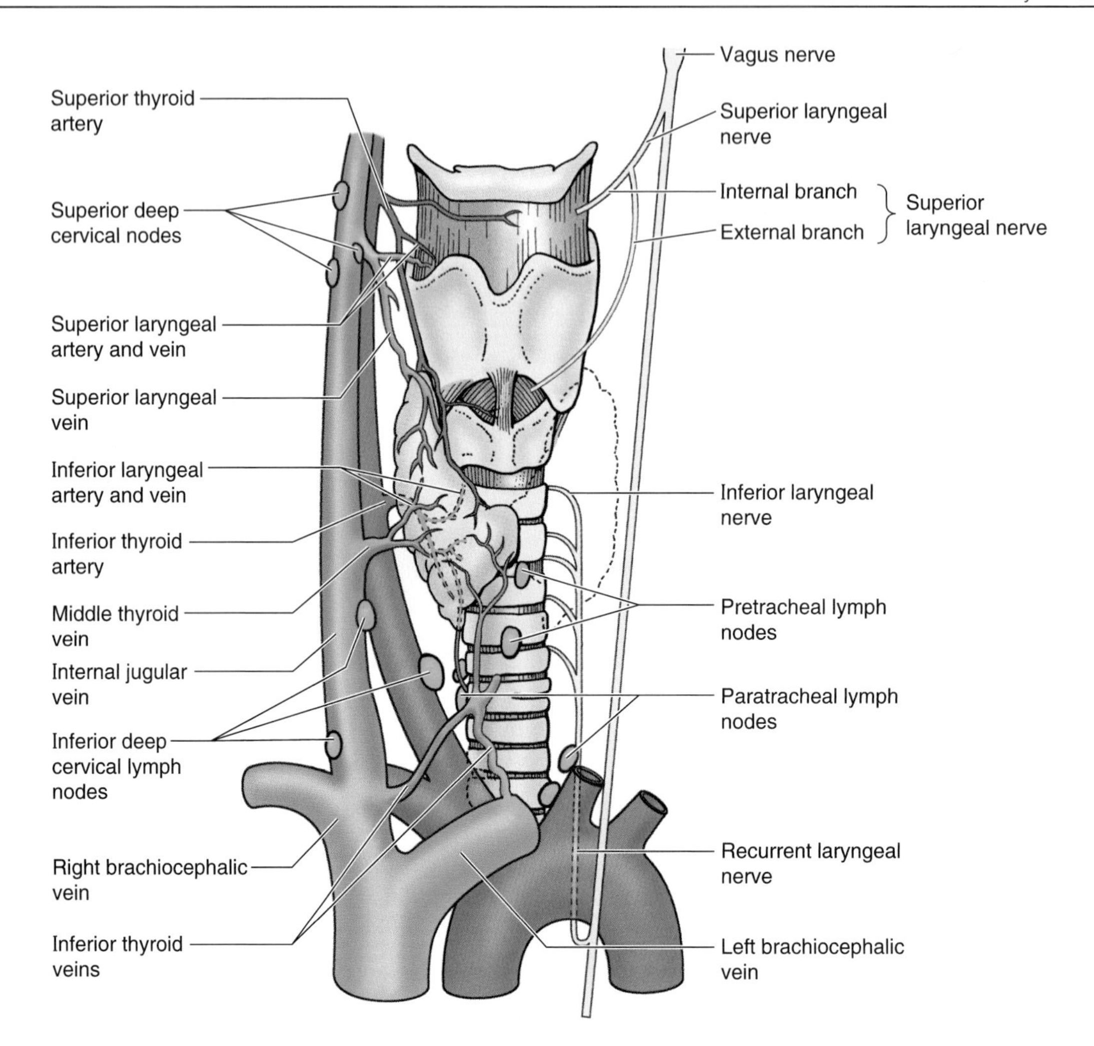

Figure 8.33. Vessels, nerves, and lymphatics of the larynx. Anterior view. Observe the anastomoses between the superior and inferior laryngeal arteries, which are branches of the superior and inferior thyroid arteries, respectively. The laryngeal nerves are derived from the vagus (CN X) through the internal and external branches of the superior laryngeal nerve, and the inferior laryngeal nerve from the recurrent (inferior) laryngeal nerve. Notice the left recurrent laryngeal nerve passing inferior to the arch of the aorta.

- *The principal adductors are the lateral cricoarytenoid muscles,* which pull the muscular processes anteriorly, rotating the arytenoids so that their vocal processes swing medially; when this action is combined with that of the *transverse arytenoid muscles* that pull the arytenoid cartilages together, air pushed through the rima glottidis causes vibrations of the vocal ligaments (phonation). When the vocal ligaments are adducted but the transverse arytenoids do not act, allowing the arytenoids to remain apart, air may bypass the ligaments; this is the position of whispering, when the breath is modified into voice in the absence of tone.
- *The sole abductors are the posterior cricoarytenoid muscles,* which pull the muscular processes posteriorly, rotating the vocal processes laterally, thus widening the rima glottidis.
- **Sphincters.** The combined actions of most of the muscles of the laryngeal inlet result in a sphincteric action that closes the laryngeal inlet as a protective mechanism during swallowing. Contraction of the lateral cricoarytenoids *transverse, and oblique arytenoid,* and *aryepiglottic* muscles brings the aryepiglottic folds together and pulls the arytenoid cartilages toward the epiglottis.
- **Tensors.** The principal tensors are the *cricothyroid muscles,* which tilt or pull the prominence or angle of the thyroid cartilage anteriorly and inferiorly toward the arch of the cricoid cartilage, increasing the distance between the thyroid prominence and arytenoid cartilages. Because the anterior ends of the vocal ligaments attach to the posterior aspect of the prominence, the vocal ligaments elongate and tighten, raising the pitch of the voice.

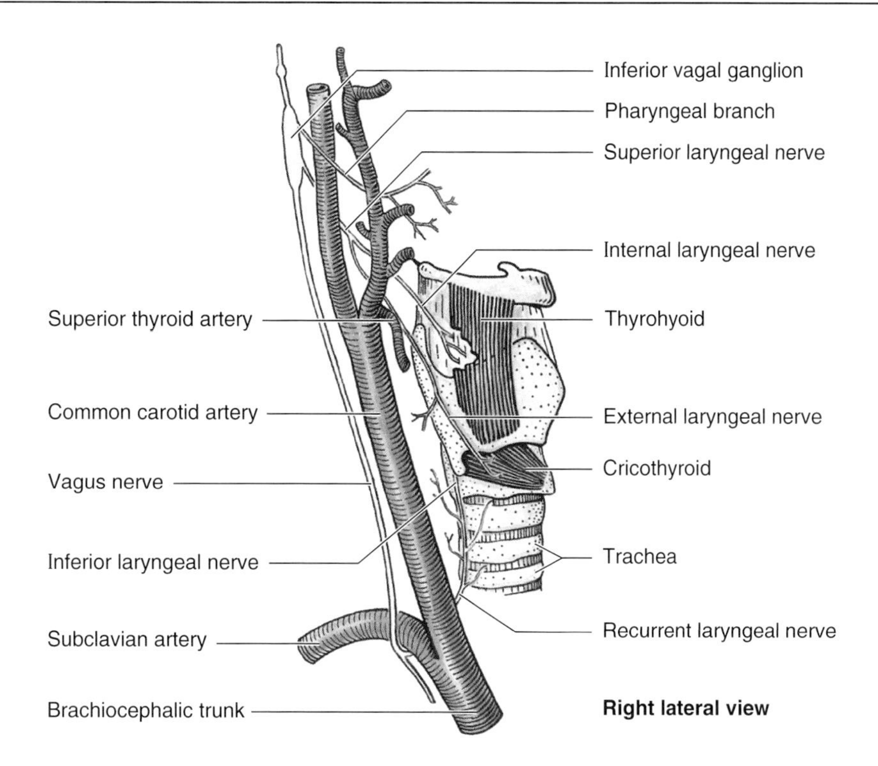

Figure 8.34. Laryngeal branches of the right vagus (CN X) nerve. Right lateral view. Observe that the nerves of the larynx are the internal and external branches of the superior laryngeal nerve and the inferior laryngeal nerve from the recurrent (inferior) laryngeal nerve. Notice the right recurrent laryngeal nerve passing inferior to the right subclavian artery.

- **Relaxers.** The principal relaxers are the *thyroarytenoid muscles*, which pull the arytenoid cartilages anteriorly, toward the thyroid angle (prominence), thereby relaxing the vocal ligaments. The *vocalis muscles* produce minute adjustments of the vocal ligaments, selectively tensing and relaxing parts of the vocal folds during animated speech and singing.

Vessels of the Larynx.

Arteries. The laryngeal arteries—branches of the superior and inferior thyroid arteries—supply the larynx (Fig. 8.33). The **superior laryngeal artery** accompanies the internal branch of the superior laryngeal nerve through the thyrohyoid membrane and branches to supply the internal surface of the larynx. The *cricothyroid artery*, a small branch of the superior thyroid artery, supplies the cricothyroid muscle. The **inferior laryngeal artery** accompanies the inferior laryngeal (terminal part of the recurrent laryngeal) nerve and supplies the mucous membrane and muscles in the inferior part of the larynx.

Veins. The laryngeal veins accompany the laryngeal arteries. The **superior laryngeal vein** usually joins the superior thyroid vein and through it drains into the IJV (Fig. 8.33). The **inferior laryngeal vein** joins the inferior thyroid vein or the venous plexus of thyroid veins on the anterior aspect of the trachea, which empties into the left brachiocephalic vein.

Lymphatics. The laryngeal lymphatic vessels superior to the vocal folds accompany the superior laryngeal artery through the thyrohyoid membrane and drain into the **superior deep cervical lymph nodes**. The lymphatic vessels inferior to the vocal folds drain into the **pretracheal** or **paratracheal lymph nodes**, which drain into the inferior deep cervical lymph nodes.

Laryngeal Nerves. The nerves of the larynx are the superior and inferior laryngeal branches of the vagus nerve.

The **superior laryngeal nerve** arises from the inferior vagal ganglion at the superior end of the carotid triangle (Fig. 8.34). It divides into two terminal branches within the carotid sheath: the internal laryngeal nerve (sensory and autonomic) and the external laryngeal nerve (motor). The **internal laryngeal nerve**, the larger of its terminal branches (Fig. 8.34), pierces the thyrohyoid membrane with the superior laryngeal artery and supplies sensory fibers to the laryngeal mucous membrane superior to the vocal folds, including the superior surface of these folds. The **external laryngeal nerve**, the smaller terminal branch of the superior laryngeal nerve, descends posterior to the sternothyroid muscle in company with the superior thyroid artery. At first the **external laryngeal nerve** lies on the inferior constrictor muscle of the pharynx, and then it pierces and supplies it and the cricothyroid muscle.

The **inferior laryngeal nerve,** the continuation of the recurrent laryngeal nerve—a branch of the vagus nerve—enters the larynx by passing deep to the inferior border of the inferior constrictor muscle of the pharynx (Fig. 8.33). It divides into anterior and posterior branches that accompany the inferior laryngeal artery into the larynx. The anterior branch supplies the lateral cricoartenoid, thyroarytenoid, vocalis, aryepiglottic, and thyroepiglottic muscles. The posterior branch supplies the posterior cricoartenoid and transverse and oblique arytenoid muscles. *Summary of the laryngeal nerves:*

- The internal laryngeal nerve is the sensory nerve of the larynx
- The recurrent (inferior) laryngeal nerve is the motor nerve of the larynx, supplying all muscles of the larynx, with one exception:
- The external laryngeal nerve supplies the cricothyroid muscle.

Injury to the Laryngeal Nerves

Because the inferior laryngeal nerve—the continuation of the recurrent laryngeal nerve—innervates the muscles moving the vocal fold, *paralysis of the vocal fold* results when the nerve is injured. The voice is poor because the paralyzed vocal fold cannot meet the normal vocal fold. When bilateral paralysis of the vocal folds occur, the voice is almost absent because the vocal folds cannot be adducted. *Hoarseness* is the most common symptom of serious disorders of the larynx such as carcinoma of the vocal folds. Paralysis of the superior laryngeal nerve causes anesthesia of the superior laryngeal mucosa. As a result, the protective mechanism designed to keep foreign bodies out of the larynx is inactive, and foreign bodies can easily enter the larynx.

Superior Laryngeal Nerve Block

A superior laryngeal block is often administered with *endotracheal intubation* in the awake patient. This technique is also in use for peroral endoscopy, transesophageal echocardiography, and laryngeal and esophageal instrumentation. The needle is inserted midway between the thyroid cartilage and the hyoid bone, 1 to 5 cm anterior to the greater horn of the hyoid bone. The needle passes through the thyrohyoid membrane, and the anesthetic agent bathes the internal laryngeal nerve, the larger terminal branch of the superior laryngeal nerve. Anesthesia of the laryngeal mucosa occurs superior to the vocal folds and includes the superior surface of these folds.

Cancer of the Larynx

The incidence of cancer of the larynx is high in individuals who smoke cigarettes or chew tobacco (Ruben and Farber, 1988). Most patients present with persistent hoarseness (Scher and Richtsmeier, 1994), often associated with earache (*otalgia*) and dysphagia. Enlarged pretracheal or paratracheal lymph nodes may indicate the presence of laryngeal cancer. *Laryngectomy* (removal of the larynx) is usually performed in severe cases of laryngeal malignancy. Vocal rehabilitation can be accomplished by the use of an electrolarynx, a tracheoesophageal prosthesis, or esophageal speech (regurgitation of ingested air).

Age Changes in the Larynx

The larynx grows steadily until approximately 3 years of age, after which time little growth occurs until approximately 12 years of age (onset of puberty). Before this time, no major laryngeal sex differences exist. Because of the presence of testosterone at puberty, the walls of the larynx strengthen, the laryngeal cavity enlarges, the vocal folds lengthen and thicken, and the laryngeal prominence becomes conspicuous in most males and some females. The length of the vocal folds increases gradually in both sexes up to puberty. During puberty, the increase in the length of the vocal folds is abrupt in males. The pitch of the voice lowers by an octave. The change in the length of the vocal folds is largely responsible for the voice changes that occur in males. The pitch of the voice of *eunuchs*—persons whose testes have been removed during childhood (*agonadal males*)—does not become lower without administration of male hormones. The thyroid, cricoid, and most of the arytenoid cartilages often ossify as age advances, commencing at approximately 25 years in the thyroid cartilage. By 65 years, the cartilages are frequently visible in radiographs. ⊙

Trachea

The trachea is a fibrocartilaginous tube supported by incomplete cartilaginous *tracheal rings* (Fig. 8.32). These rings, which keep the trachea patent, are deficient posteriorly where the trachea is adjacent to the esophagus. The posterior gap in the tracheal rings is spanned by the trachealis (smooth muscle). Hence, the posterior wall of the trachea is flat. In adults, the trachea is approximately 2.5 cm in diameter, whereas in an infant it has the diameter of a pencil. The trachea extends from the inferior end of the larynx at the level of the 6th cervical vertebra (Fig. 8.35). It ends at the level of the sternal angle or the T4/T5 IV disc, where it divides into the right and left main bronchi (see Chapter 1). Lateral to the trachea are the common carotid arteries (Fig. 8.23) and the thyroid lobes. Inferior to the isthmus of the thyroid gland are the jugular venous arch and the inferior thyroid veins. The brachiocephalic trunk is related to the right side of the trachea in the root of the neck.

Tracheostomy

A transverse incision through the skin of the neck and anterior wall of the trachea (*tracheostomy*) can establish an adequate airway in patients with upper airway obstruction or respiratory failure. The infrahyoid muscles are retracted laterally, and the isthmus of the thyroid gland is either divided or retracted superiorly. An opening is made in the trachea between the 1st and 2nd tracheal rings or through the 2nd through 4th rings. A *tracheostomy tube* is then inserted into the trachea and secured by neck straps. To avoid complications during a tracheostomy, the following anatomical relationships must be remembered:

- The *inferior thyroid veins* arise from a venous plexus on the thyroid and descend anterior to the trachea
- A small *thyroid ima artery* is present in approximately 10% of people and ascends to the isthmus of the thyroid
- The *left brachiocephalic vein*, jugular venous arch, and pleurae may be encountered, particularly in infants and children
- The *thymus* covers the inferior part of the trachea in infants and children
- The trachea is small, mobile, and soft in infants, making it easy to cut through its posterior wall and damage the esophagus. ✪

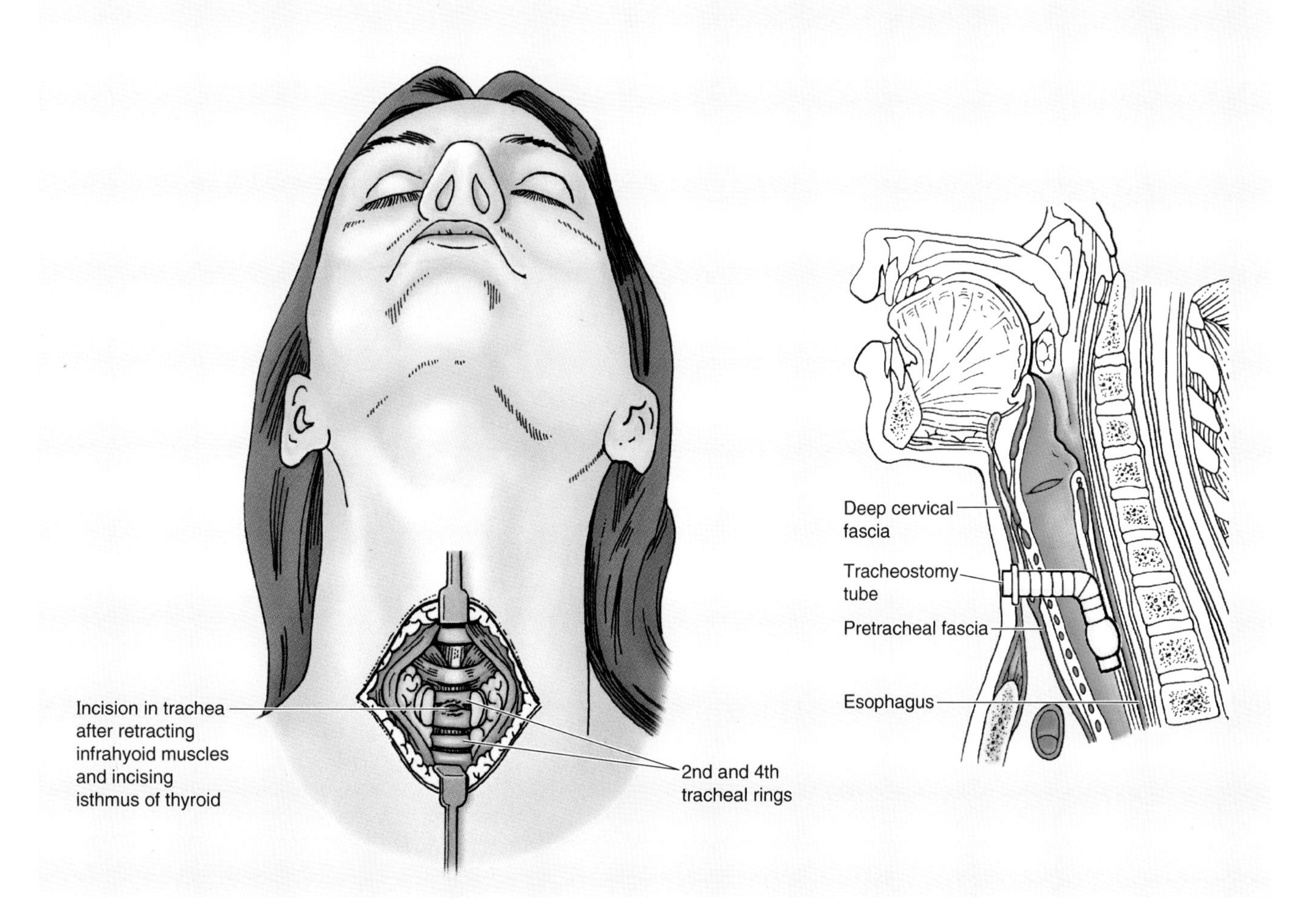

Alimentary Layer of the Cervical Viscera

In the alimentary (L. alimentum, nourishment) layer, cervical viscera take part in the digestive functions of the body. Although the **pharynx** conducts air to the larynx, trachea, and lungs, its constrictor muscles direct—and the epiglottis deflects—food to the esophagus. The **esophagus**, also involved in food propulsion, is the beginning of the alimentary canal (digestive tract).

Pharynx

The pharynx is the part of the digestive system posterior to the nasal and oral cavities, extending inferiorly past the larynx (Figs. 8.35 and 8.36). The pharynx extends from the base of

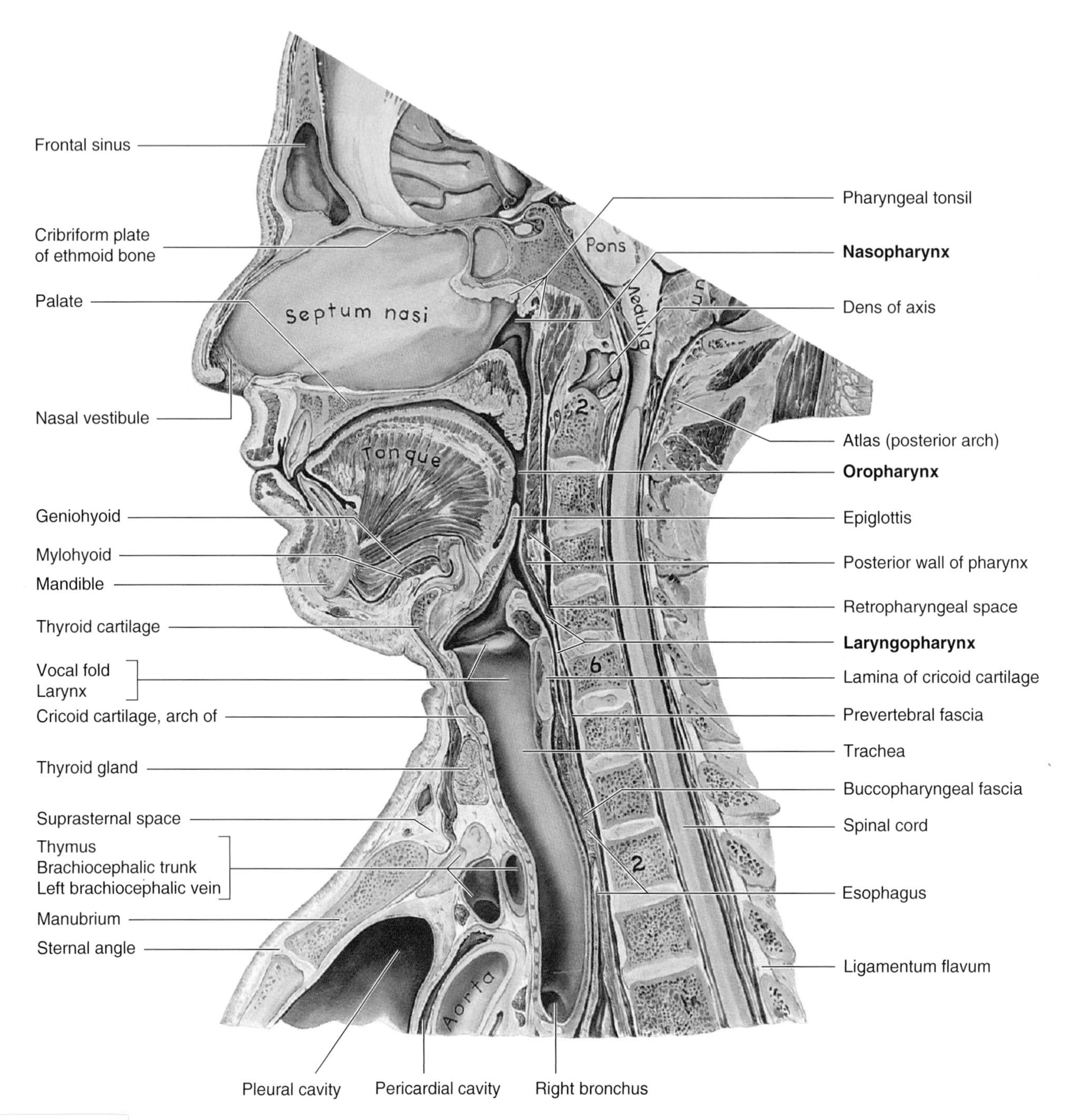

Figure 8.35. Median section of the head and neck. Observe that the pharynx extends from the base of the skull to below the level of the body of C6 vertebra (to C6/C7 IV discs here), where it is continuous with the esophagus.

the skull to the inferior border of the cricoid cartilage anteriorly and the inferior border of C6 vertebra posteriorly. It is widest (approximately 5 cm) opposite the hyoid bone and narrowest (approximately 1.5 cm) at its inferior end, where it is continuous with the esophagus. The posterior wall of the pharynx lies against the prevertebral layer of deep cervical fascia.

Interior of the Pharynx. The pharynx is divided into three parts (Figs. 8.35 and 8.36):

- *Nasopharynx,* posterior to the nose and superior to the soft palate
- *Oropharynx,* posterior to the mouth
- *Laryngopharynx,* posterior to the larynx.

The **nasopharynx** has a respiratory function. It lies superior to the *soft palate* and is the posterior extension of the nasal cavities (Fig. 8.35). The nose opens into the nasopharynx

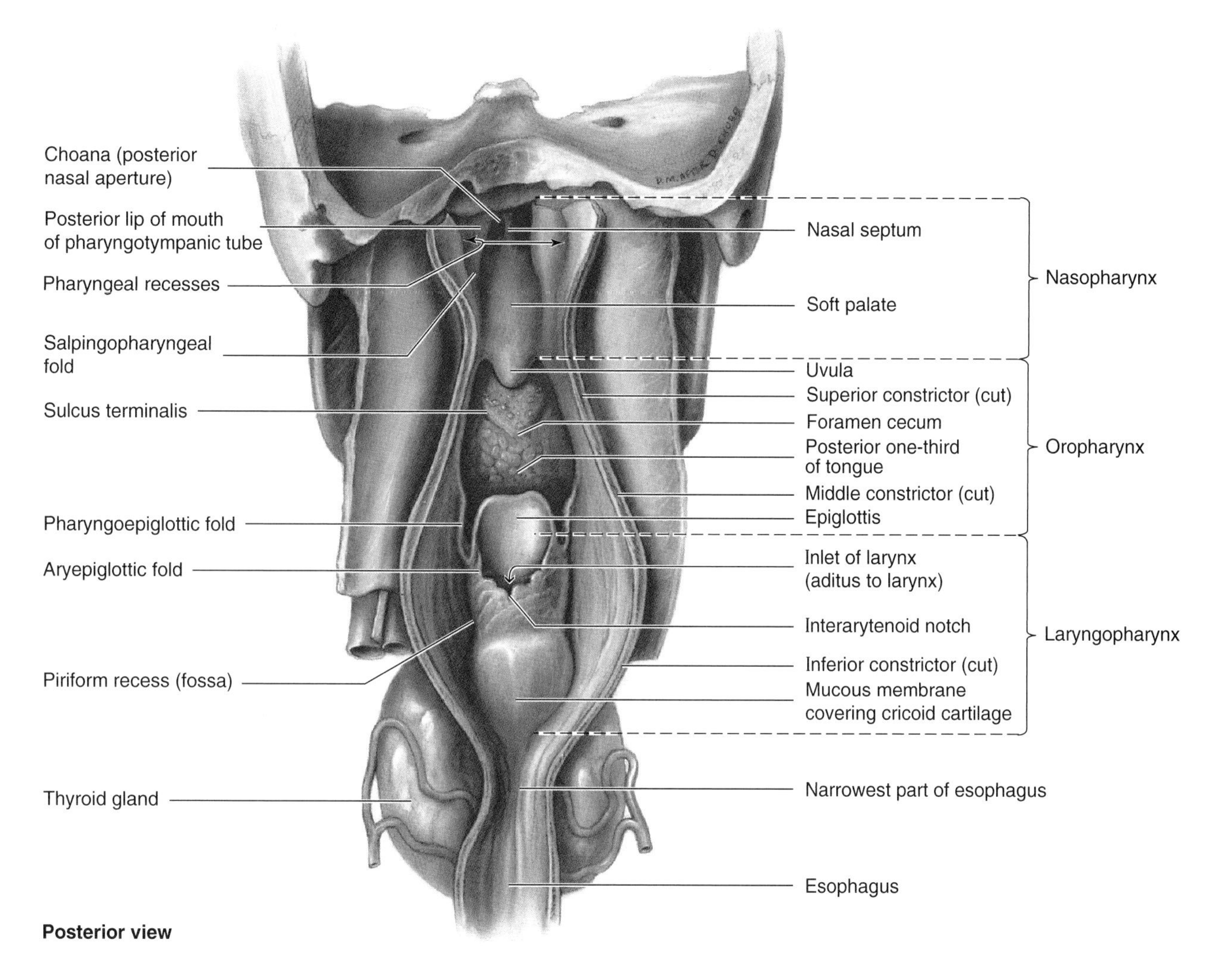

Figure 8.36. Interior of the pharynx. Posterior view of a dissection of the pharynx. The posterior wall has been incised along the midline and spread apart. On each side of the inlet of the larynx, separated from it by the aryepiglottic fold, observe a piriform recess formed by the invagination of the larynx into the anterior wall of the laryngopharynx (hypopharynx).

through two *choanae* (paired openings between the nasal cavity and nasopharynx). The roof and posterior wall of the nasopharynx form a continuous surface that lies inferior to the body of the sphenoid bone and the basilar part of the occipital bone (Fig. 8.36).

The abundant lymphoid tissue in the pharynx forms an incomplete tonsillar ring about the superior part of the pharynx—*Waldeyer's ring* (see Fig. 8.42). The lymphoid tissue is aggregated in certain regions to form masses called tonsils. The **pharyngeal tonsil** (commonly referred to as "adenoids"), is in the mucous membrane of the roof and posterior wall of the nasopharynx (Figs. 8.35 and 8.37). Extending inferiorly from the medial end of the pharyngotympanic (auditory) tube is a vertical fold of mucous membrane, the **salpingopharyngeal fold** (Figs. 8.36 and 8.37). It covers the salpingopharyngeus muscle, which opens the pharyngeal orifice of the pharyngotympanic tube during swallowing. The collection of lymphoid tissue in the submucosa of the pharynx near the pharyngeal orifice of the pharyngotympanic tube is the *tubal tonsil* (Fig. 8.37). Posterior to the *torus of the pharyngotympanic tube* and the salpingopharyngeal fold is a slitlike lateral projection of the pharynx, the **pharyngeal recess,** which extends laterally and posteriorly.

The **oropharynx** has a digestive function. It is bounded by the soft palate superiorly, the base of the tongue inferiorly, and the palatoglossal and palatopharyngeal arches laterally (Figs. 8.37 and 8.38*A*). It extends from the soft palate to the superior border of the epiglottis.

Deglutition (the act of swallowing) is the complex process that transfers a food bolus from the mouth through the phar-

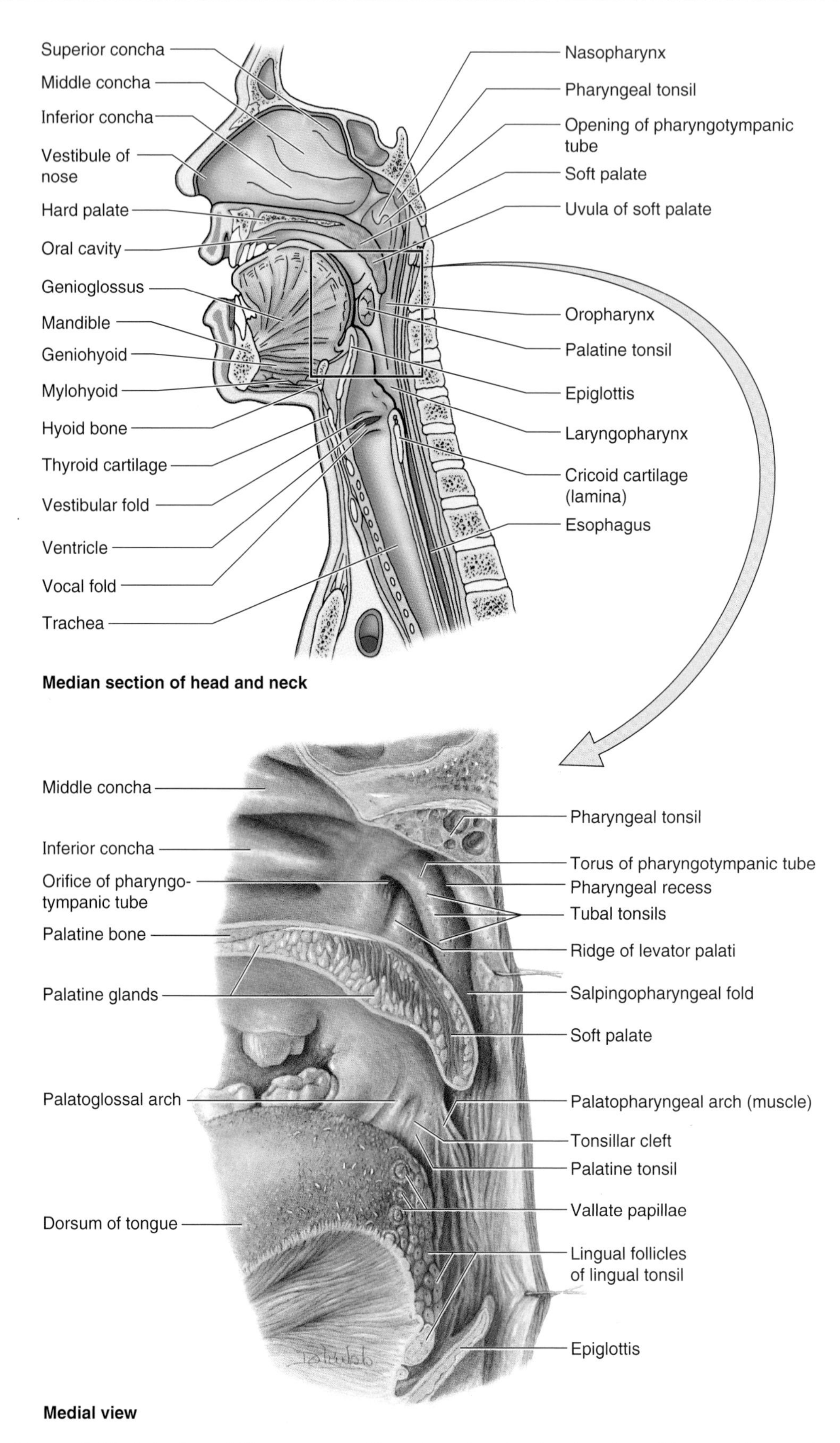

Figure 8.37. Interior of the pharynx. The orientation drawing (top figure) is a median section of the head and neck. The *blocked area* indicates the location of the medial view (lower figure).

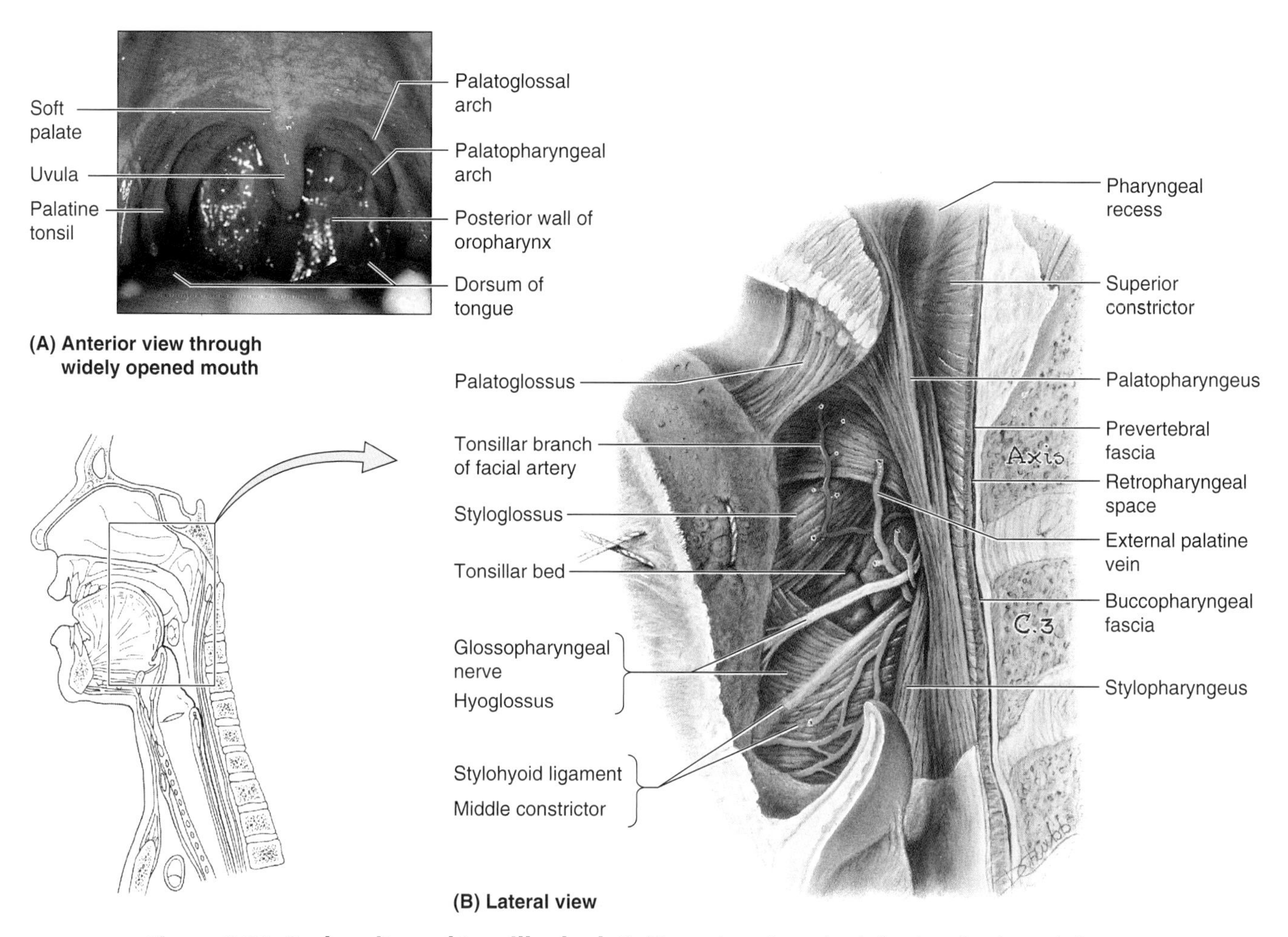

Figure 8.38. Oral cavity and tonsillar bed. A. The oral cavity and palatine tonsils of an adult man taken with the mouth wide open and the tongue protruding as far as possible. **B.** Deep dissection of the tonsillar bed after removal of the palatine tonsil. The tongue is pulled anteriorly and the inferior (lingual) attachment of the superior constrictor muscle is cut away.

ynx and esophagus into the stomach. Solid food is masticated (chewed) and mixed with saliva to form a soft bolus. *Deglutition occurs in three stages*:

- The 1st stage is voluntary; the bolus is compressed against the palate and pushed from the mouth into the oropharynx, mainly by movements of the muscles of the tongue and soft palate (Fig. 8.39, *A* and *B*)
- The 2nd stage is involuntary and rapid; the soft palate is elevated, sealing off the nasopharynx from the oropharynx and laryngopharynx (Fig. 8.39*C*), and the pharynx is wide and short to receive the bolus of food as the suprahyoid muscles and longitudinal pharyngeal muscles contract, elevating the larynx
- The 3rd stage is also involuntary; sequential contraction of all three constrictors forces the food bolus inferiorly into the esophagus (Fig. 8.39*D*).

The **palatine tonsils** are collections of lymphoid tissue on each side of the oropharynx in the interval between the palatine arches (Figs. 8.37 and 8.38*A*). The tonsil does not fill the *tonsillar cleft* (intratonsillar cleft) between the palatoglossal and palatopharyngeal arches in adults. The *tonsillar bed*, in which the palatine tonsil lies, is between these arches (Fig. 8.38*B*). The tonsillar bed is formed by the superior constrictor of the pharynx and the thin, fibrous sheet of **pharyngobasilar fascia** (Fig. 8.40). This fascia blends with the periosteum of the base of the skull and defines the limits of the pharyngeal wall in its superior part.

The **laryngopharynx** (hypopharynx) lies posterior to the larynx (Figs. 8.35 and 8.37), extending from the superior border of the epiglottis and the pharyngoepiglottic folds to the inferior border of the cricoid cartilage, where it narrows and becomes continuous with the esophagus. Posteriorly, the laryngopharynx is related to the bodies of C4 through C6 vertebrae. Its posterior and lateral walls are formed by the **middle** and **inferior constrictor muscles** (Fig. 8.40*A*), and internally the wall is formed by the *palatopharyngeus* and *stylopharyngeus muscles.* The laryngopharynx communicates with the larynx through the **inlet of the larynx**, or laryngeal inlet (Fig. 8.36), on its anterior wall.

The **piriform recess** (piriform fossa or sinus) is a small depression of the laryngopharyngeal cavity on either side of the

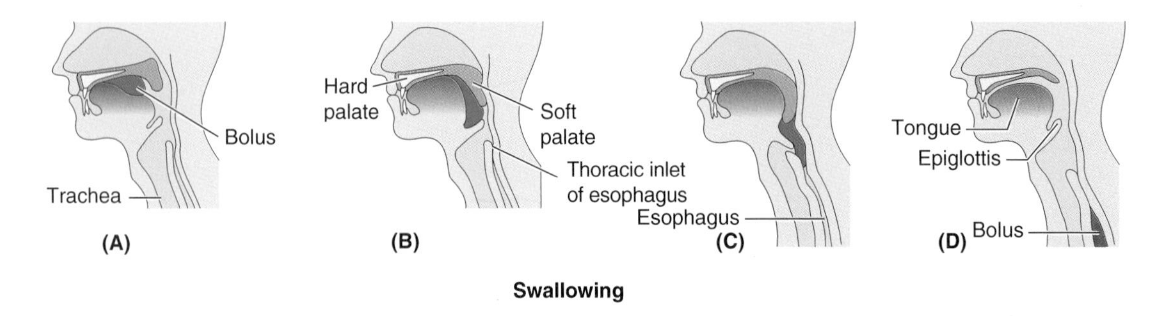

Figure 8.39. Swallowing. A. The bolus of food is pushed to the back of the mouth by pushing the tongue (L. lingua; G. glossa) against the palate. **B.** The nasopharynx is sealed off and the larynx is elevated, enlarging the pharynx to receive food. **C.** The pharyngeal sphincters contract sequentially, squeezing food into the esophagus. The epiglottis closes the trachea. **D.** The bolus of food moves down the esophagus by peristaltic contraction.

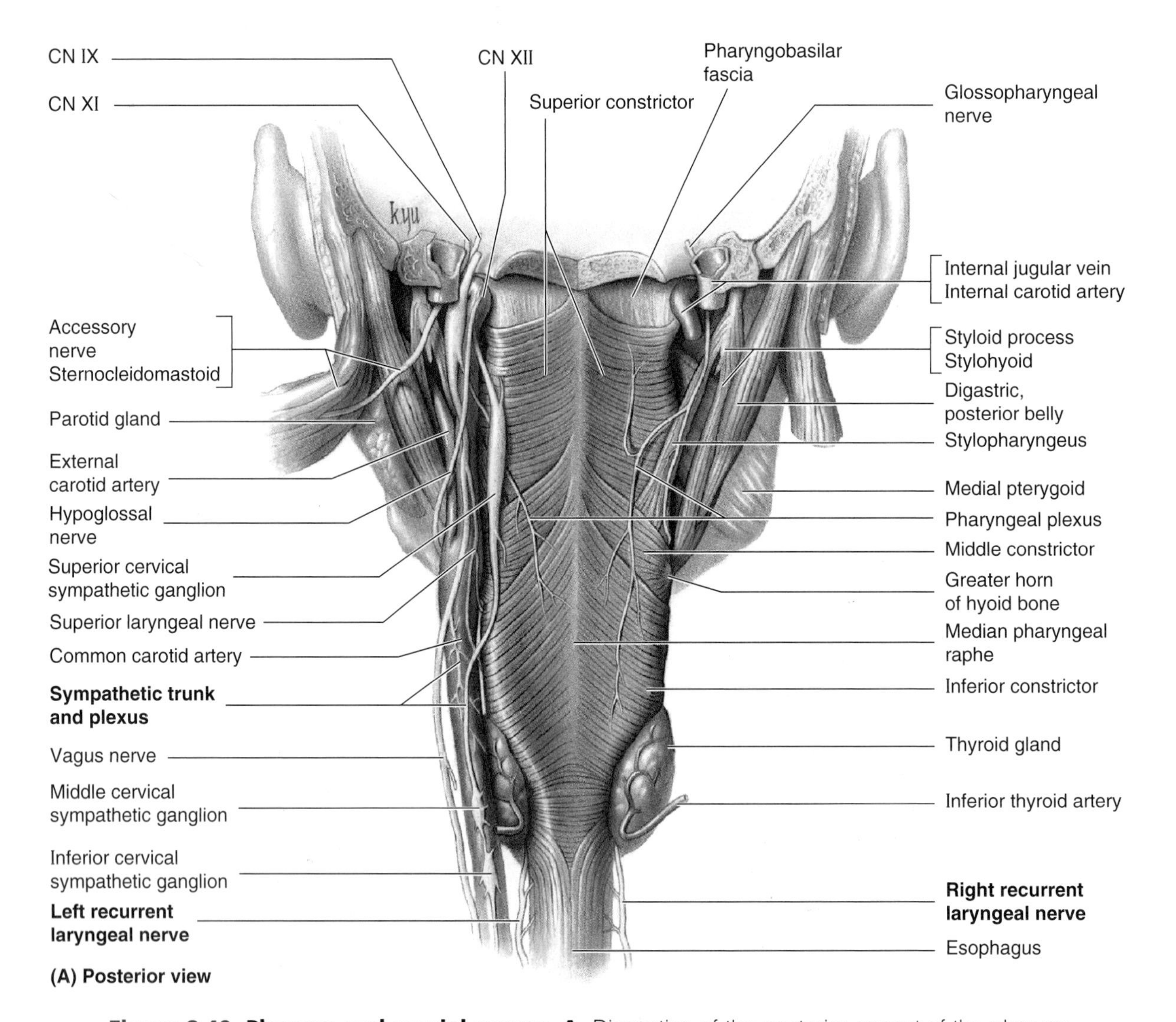

Figure 8.40. Pharynx and cranial nerves. A. Dissection of the posterior aspect of the pharynx and associated structures. The buccopharyngeal fascia has been removed. Examine the three pharyngeal constrictor muscles. Note that the inferior muscle overlaps the middle and the middle overlaps the superior, all three muscles forming a common raphe posteriorly.

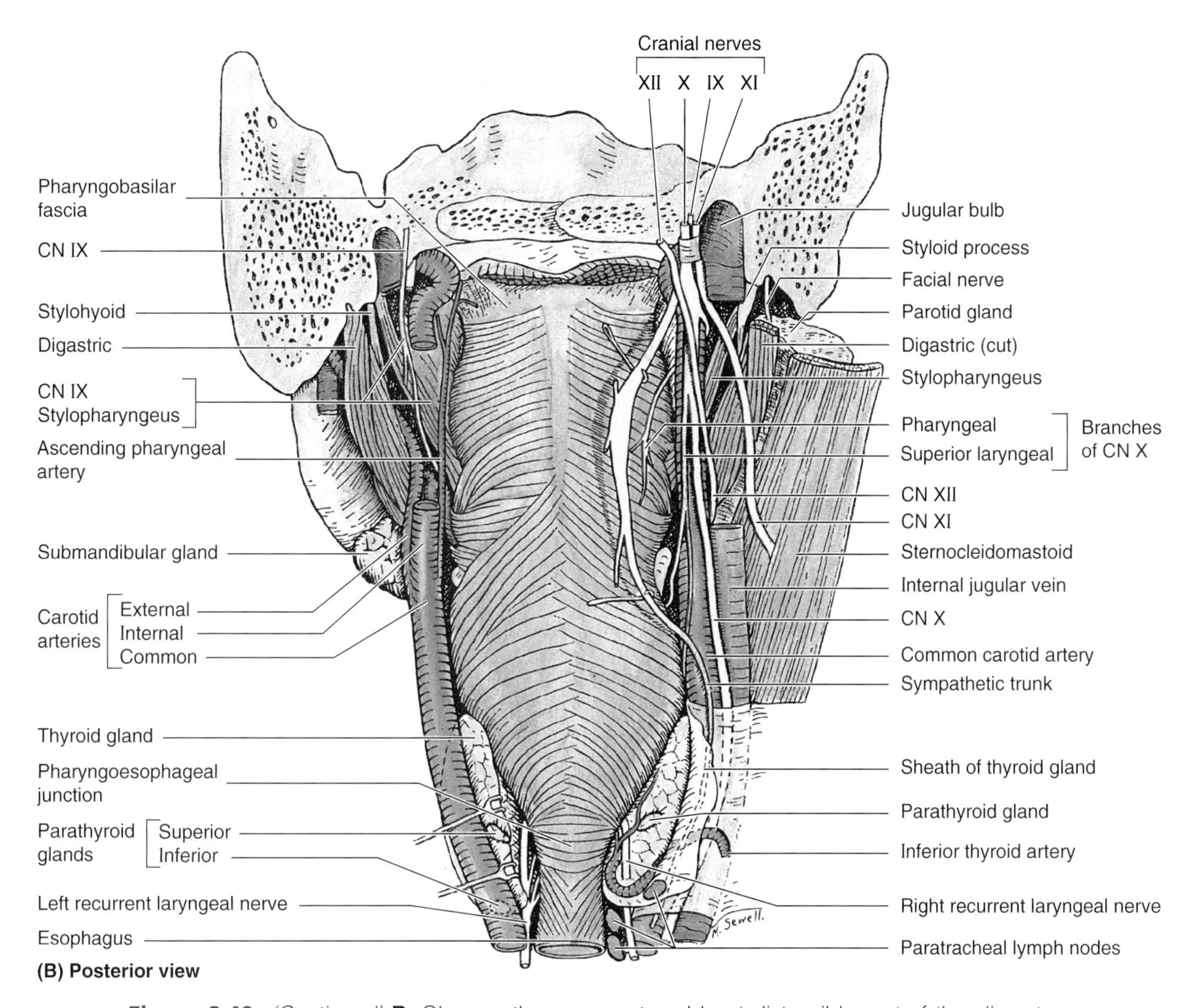

Figure 8.40. *(Continued)* **B.** Observe the narrowest and least distensible part of the alimentary tract, where the pharynx becomes the esophagus (pharyngoesophageal junction).

laryngeal inlet (Fig. 8.36). This mucosa-lined recess is separated from the laryngeal inlet by the *aryepiglottic fold*. Laterally, the piriform recess is bounded by the medial surfaces of the thyroid cartilage and the *thyrohyoid membrane*. Branches of the internal laryngeal and recurrent laryngeal nerves lie deep to the mucous membrane of the piriform recess and are vulnerable to injury when a foreign body lodges in the recess.

Pharyngeal Muscles. The wall of the pharynx is exceptional for the alimentary tract, being composed mainly of an external circular and an internal longitudinal layer of muscles; elsewhere, the arrangement is the opposite. The external circular layer of pharyngeal muscles consists of three constrictors (Figs. 8.38 and 8.40). The internal, mainly longitudinal layer of muscles consists of the **palatopharyngeus, stylopharyngeus**, and **salpingopharyngeus**. These muscles elevate the larynx and shorten the pharynx during swallowing and speaking. The attachments, nerve supply, and actions of the pharyngeal muscles are described in Table 8.6.

Pharyngeal Constrictors. The pharyngeal constrictors have a strong internal fascial lining, the **pharyngobasilar fascia** (Fig. 8.40*B*), and a thin external fascial lining, the *buccopharyngeal fascia*. Inferiorly, the buccopharyngeal fascia blends with the pretracheal layer of deep cervical fascia (p. 1000). The pharyngeal constrictors contract involuntarily so that contraction takes place sequentially from the superior to the inferior end of the pharynx, propelling food into the esophagus. *All three constrictors are supplied by the pharyngeal plexus of nerves* that is formed by pharyngeal branches of the vagus and glossopharyngeal nerves and by sympathetic branches from the superior cervical ganglion (Fig. 8.40*A*, Table 8.6). The pharyngeal plexus lies on the lateral wall of the pharynx, mainly on the middle constrictor.

The overlapping of the constrictor muscles leaves four gaps in the musculature for structures to enter or leave the pharynx (Fig. 8.40*A*):

- Superior to the superior constrictor, the levator veli palatini, pharyngotympanic tube, and ascending palatine artery pass through a gap between the superior constrictor and skull; it is here that the pharyngobasilar fascia blends

Table 8.6. **Muscles of the Pharynx**

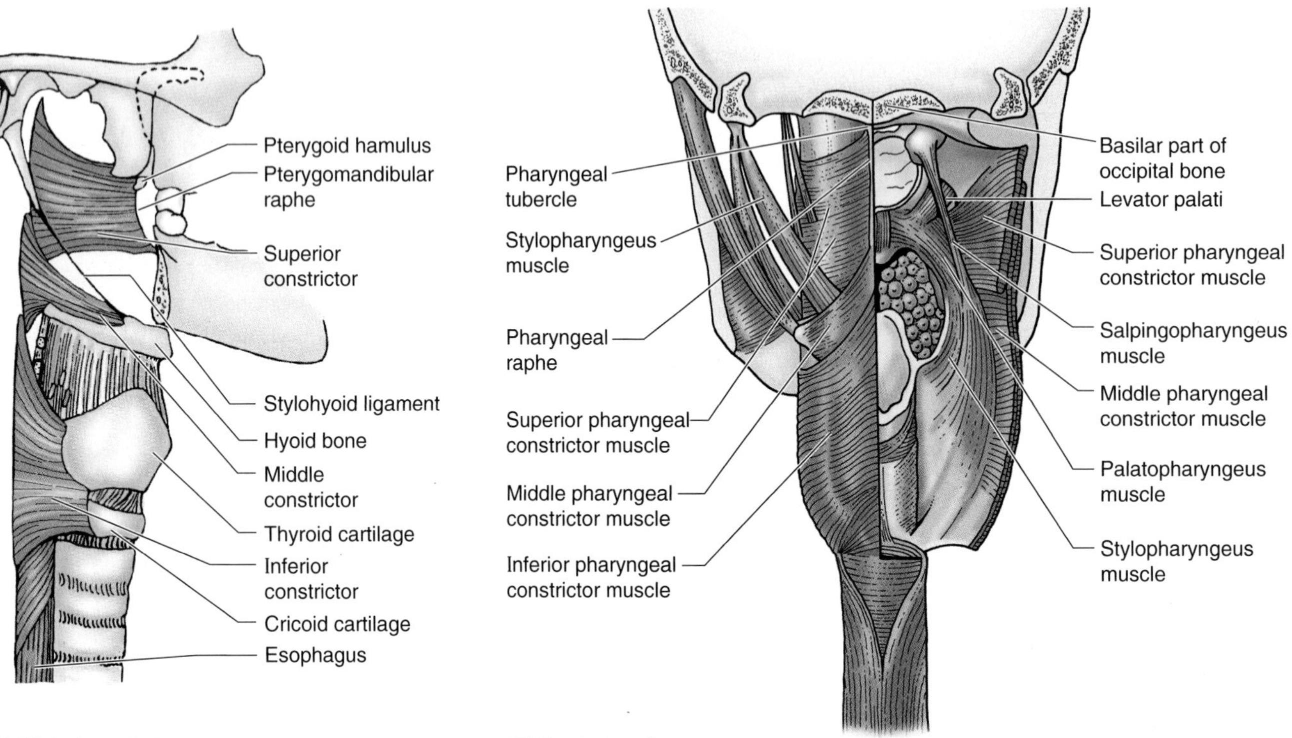

(A) Right lateral view **(B) Posterior view**

<table>
<tr><th>Muscle</th><th>Origin</th><th>Insertion</th><th>Innervation</th><th>Main Action</th></tr>
<tr><td colspan="5">External layer</td></tr>
<tr><td>Superior constrictor</td><td>Pterygoid hamulus, pterygomandibular raphe, posterior end of mylohyoid line of mandible, and side of tongue</td><td>Median raphe of pharynx and pharyngeal tubercle on basilar part of occipital bone</td><td>Cranial root of accessory nerve via pharyngeal branch of vagus and pharyngeal plexus</td><td rowspan="3">Constrict wall of pharynx during swallowing</td></tr>
<tr><td>Middle constrictor</td><td>Stylohyoid ligament and superior (greater) and inferior (lesser) horns of hyoid bone</td><td rowspan="2">Median raphe of pharynx</td><td rowspan="2">Cranial root of accessory nerve as above, plus branches of external and recurrent laryngeal nerves of vagus</td></tr>
<tr><td>Inferior constrictor</td><td>Oblique line of thyroid cartilage and side of cricoid cartilage</td></tr>
<tr><td colspan="5">Internal layer</td></tr>
<tr><td>Palatopharyngeus</td><td>Hard plate and palatine aponeurosis</td><td>Posterior border of lamina of thyroid cartilage and side of pharynx and esophagus</td><td rowspan="2">Cranial root of accessory nerve via pharyngeal branch of vagus and pharyngeal plexus</td><td rowspan="3">Elevate (shorten and widen) pharynx and larynx during swallowing and speaking</td></tr>
<tr><td>Salpingopharyngeus</td><td>Cartilaginous part of auditory tube</td><td>Blends with palatopharyngeus</td></tr>
<tr><td>Stylopharyngeus</td><td>Styloid process of temporal bone</td><td>Posterior and superior borders of thyroid cartilage with palatopharyngeus</td><td>Glossopharyngeal nerve</td></tr>
</table>

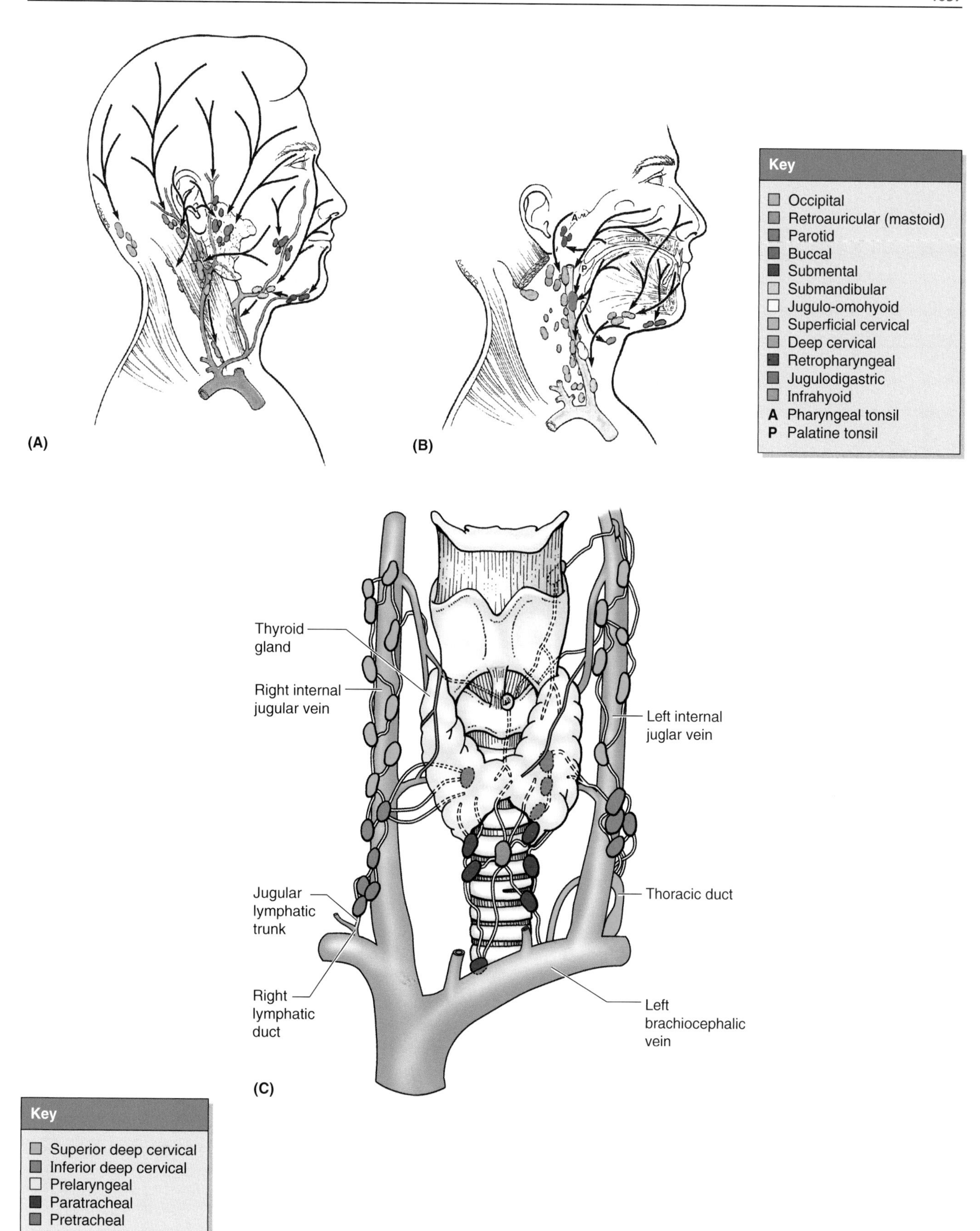

Figure 8.41. Lymphatic drainage of the head and neck. A. Superficial. **B.** Deep **C.** Lymphatic nodes, trunks, and thoracic duct.

with the buccopharyngeal fascia to form, with the mucous membrane, the thin wall of the **pharyngeal recess** (Fig. 8.36).

- Between the superior and middle constrictors is a gap that forms the gateway to the mouth, through which passes the stylopharyngeus, glossopharyngeal nerve, and stylohyoid ligament.
- Between the middle and inferior constrictors is a gap for the internal laryngeal nerve and superior laryngeal artery and vein to pass to the larynx.
- Inferior to the inferior constrictor is a gap for the recurrent laryngeal nerve and inferior laryngeal artery to pass superiorly into the larynx.

Vessels of the Pharynx. A branch of the facial artery, the **tonsillar artery** (Fig. 8.38), passes through the superior constrictor muscle and enters the inferior pole of the tonsil. The tonsil also receives arterial twigs from the ascending palatine, lingual, descending palatine, and ascending pharyngeal arteries. The large **external palatine vein** (paratonsillar vein) descends from the soft palate (Fig. 8.38) and passes close to the lateral surface of the tonsil before it enters the pharyngeal venous plexus.

The *tonsillar lymphatic vessels* pass laterally and inferiorly to the lymph nodes near the angle of the mandible and the **jugulodigastric node**, referred to as the *tonsillar node* because of its frequent enlargement when the tonsil is inflamed (*tonsillitis*) (Fig. 8.41). The palatine, lingual, and pharyngeal tonsils form the pharyngeal lymphoid (tonsillar) ring (Waldeyer's ring), an incomplete circular band of lymphoid tissue around the superior part of the pharynx (Fig. 8.42). The anteroinferior part of the ring is formed by the *lingual tonsil*, a collection of lymphoid nodules in the posterior part of the tongue. Lateral parts of the ring are formed by the palatine and tubal tonsils, and posterior and superior parts are formed by the pharyngeal tonsil.

Pharyngeal Nerves. The nerve supply to the pharynx (motor and most of sensory) derives from the **pharyngeal plexus of nerves** (Fig. 8.39). Motor fibers in the plexus derive from the cranial root of the accessory nerve and are carried by the vagus nerve—via its pharyngeal branch or branches—to all muscles of the pharynx and soft palate, except the stylopharyngeus (supplied by CN IX) and the tensor veli palatini (supplied by CN V_2). The inferior pharyngeal constrictor also receives some motor fibers from the external and recurrent laryngeal branches of the vagus. *Sensory fibers in the plexus* derive from the glossopharyngeal nerve. They supply most of the mucosa of all three parts of the pharynx. The sensory nerve supply of the mucous membrane of the anterior and superior nasopharynx is mainly from the maxillary nerve (CN V_2), a purely sensory nerve. The *tonsillar nerves* derive from the *tonsillar plexus of nerves* formed by branches of the glossopharyngeal and vagus nerves. Other branches are derived from the pharyngeal plexus of nerves (Fig. 8.39*B*).

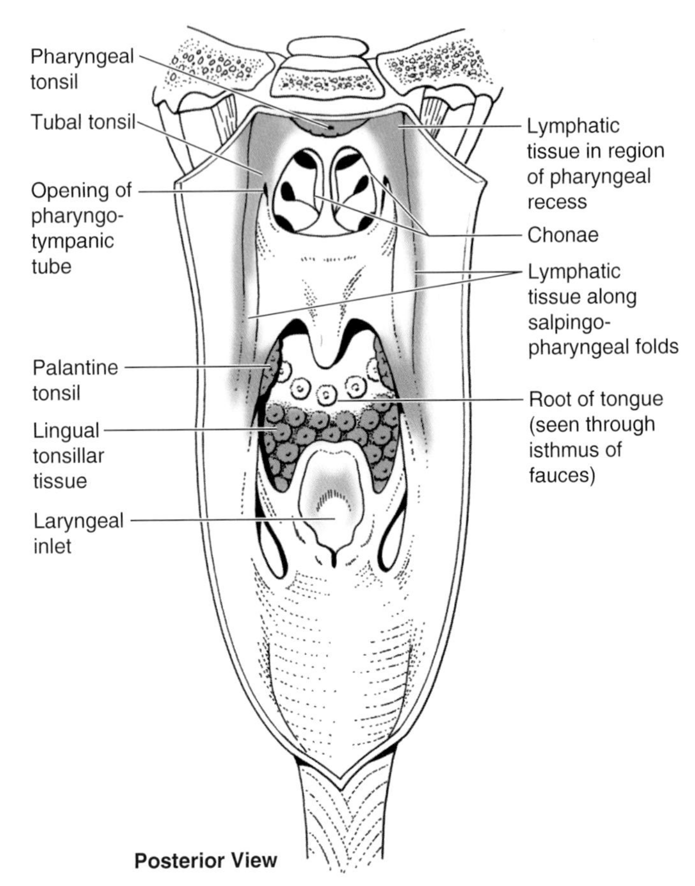

Figure 8.42. Lymphoid tissue in the tongue and pharynx. Observe the pharyngeal lymphoid (tonsillar) ring about the superior pharynx, formed of pharyngeal, tubal, palatine, and lingual tonsils.

Foreign Bodies in the Laryngopharynx

When food passes through the laryngopharynx during swallowing, some of it enters the piriform recesses. Foreign bodies (e.g., a chicken bone or fishbone) entering the pharynx may lodge in this recess. If the object is sharp, it may pierce the mucous membrane and injure the internal laryngeal nerve. The superior laryngeal nerve and its internal laryngeal branch are also vulnerable during removal of the object if the instrument used to remove the foreign body accidentally pierces the mucous membrane. Injury to these nerves may result in anesthesia of the laryngeal mucous membrane as far inferiorly as the vocal folds. Young children swallow a variety of objects, most of which reach the stomach and pass through the gastrointestinal tract without difficulty. In some cases, the foreign body stops at ▶

▶ the inferior end of the laryngopharynx, its narrowest part. A radiographic examination and/or a CT scan or magnetic resonance image will reveal the presence of a radiopaque foreign body (a chicken bone, for example). Foreign bodies in the pharynx are often removed under direct vision through a *pharyngoscope*.

Sinus Tract from the Piriform Recess

Although uncommon, a sinus tract may pass from the piriform recess to the thyroid gland and be a potential site for recurring *thyroiditis* (inflammation of the thyroid gland). This sinus tract apparently develops from a remnant of the thyroglossal duct that adheres to the developing laryngopharynx (Moore and Persaud, 1998). The thyroid gland develops from the inferior end of the thyroglossal duct. Removal of this sinus tract essentially involves a partial thyroidectomy because the piriform recess lies deep to the superior pole of the gland (Scher and Richtsmeier, 1994).

Tonsillectomy

Tonsillectomy is performed by dissecting the palatine tonsil from the tonsillar bed, or by a *guillotine or snare operation.* Each case involves removal of the tonsil and the fascial sheet covering the tonsillar bed. Because of the rich blood supply of the tonsil, bleeding may arise from the tonsillar artery or other arterial twigs; however, *bleeding commonly arises from the large external palatine vein* (Fig. 8.38*B*). The glossopharyngeal nerve accompanies the tonsillar artery on the lateral wall of the pharynx. Because this wall is thin, CN IX is vulnerable to injury. The internal carotid artery is especially vulnerable when it is tortuous and lies directly lateral to the tonsil.

Adenoiditis

Adenoiditis—inflammation of the pharyngeal tonsils *(adenoids)*—can obstruct the passage of air from the nasal cavities through the choanae into the nasopharynx, making mouth breathing necessary. Infection from the enlarged pharyngeal tonsils may spread to the tubal tonsils, causing swelling and closure of the pharyngotympanic tubes. Impairment of hearing may result from nasal obstruction and blockage of the pharyngotympanic tubes. Infection spreading from the nasopharynx to the middle ear causes *otitis media* (middle ear infection), which may produce temporary or permanent hearing loss (p. 969).

Branchial Fistula

A branchial fistula (abnormal canal) opens internally into the tonsillar cleft and externally on the side of the neck (*A*, p. 1060). This uncommon cervical canal results from persistence of remnants of the 2nd pharyngeal pouch and 2nd pharyngeal groove (cleft) (Moore and Persaud, 1998). The fistula ascends from its cervical opening—usually along the anterior border of the SCM in the inferior third of the neck—through the subcutaneous tissue, platysma, and fascia of the neck to enter the carotid sheath. It then passes between the internal and external carotid arteries on its way to its opening in the tonsillar cleft. Its course can be demonstrated by radiography (*B*, p. 1060). ▶

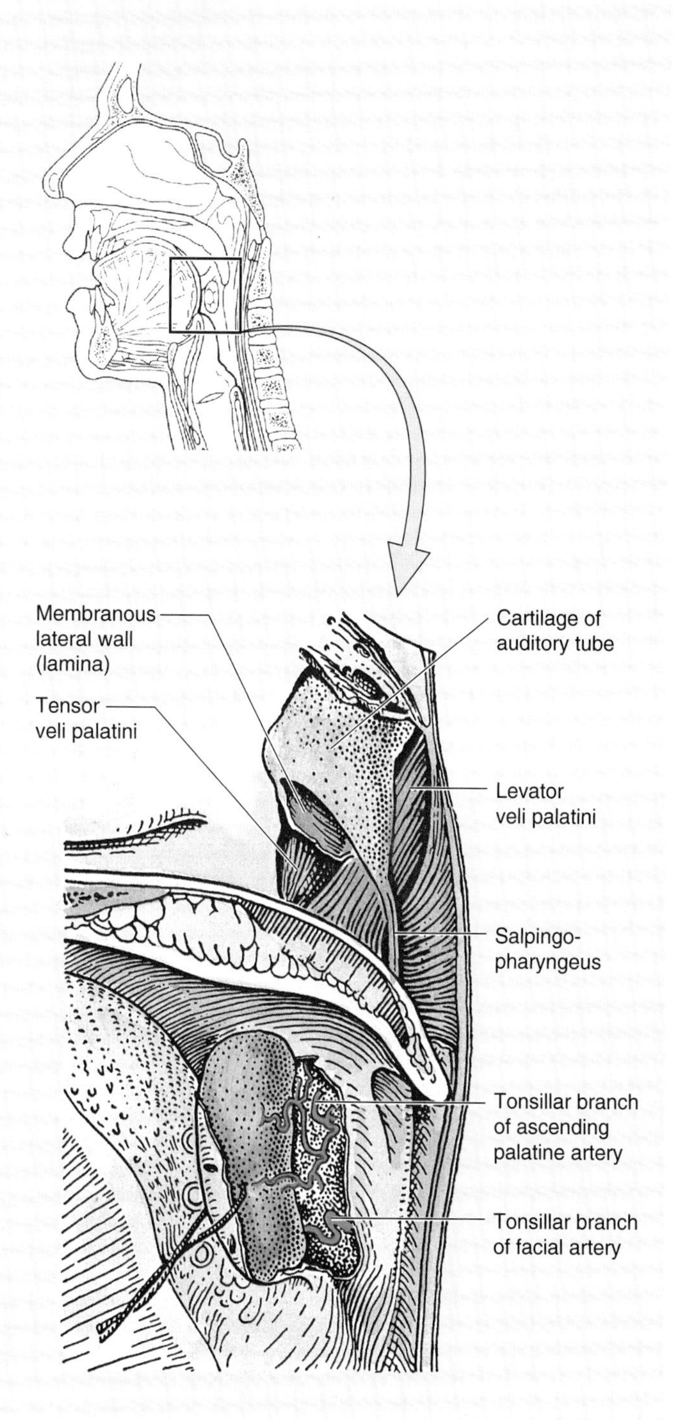

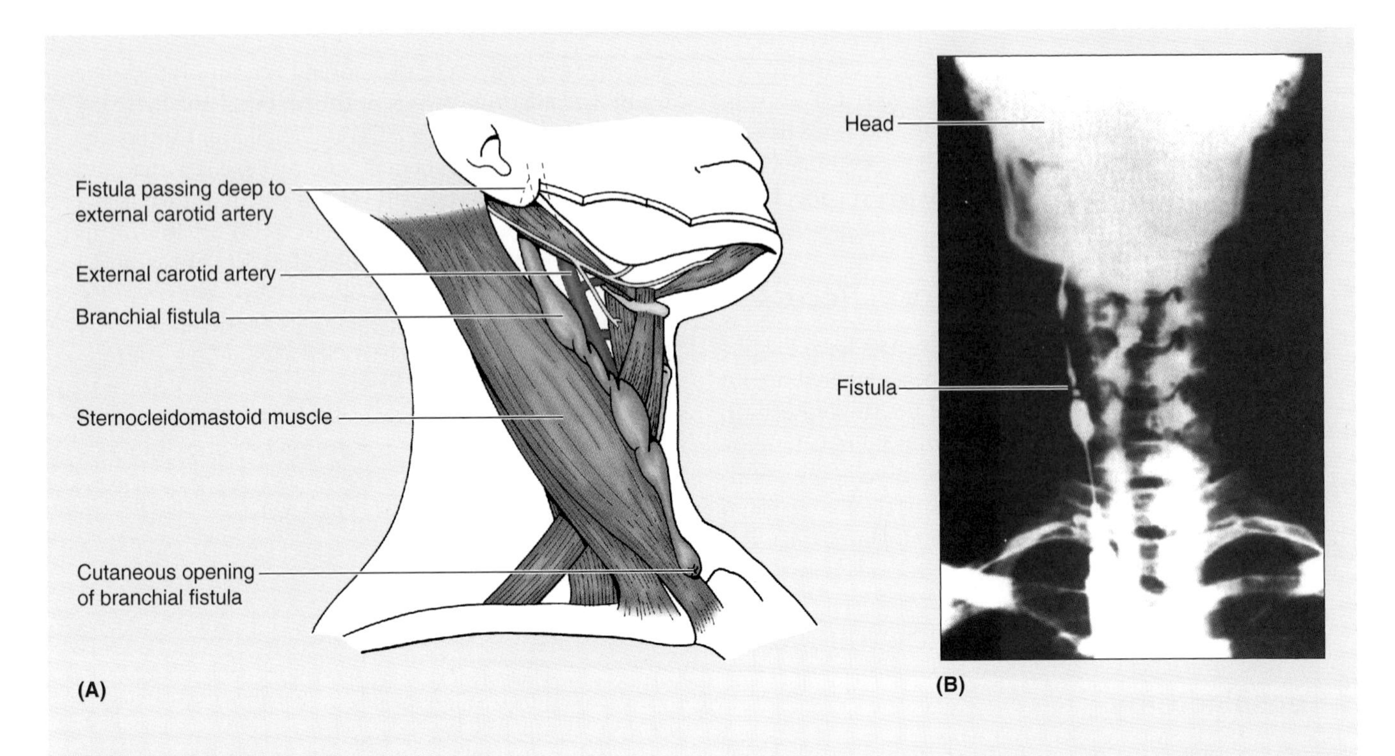

Branchial Sinuses and Cysts

When the embryonic *cervical sinus* fails to disappear, it may retain its connection with the lateral surface of the neck by *branchial sinus*, a narrow canal (Moore and Persaud, 1998). The opening of the sinus may be anywhere along the anterior border of the SCM. If the remnant of the cervical sinus is not connected with the surface, it may form a *branchial cyst* (lateral cervical cyst), usually located just inferior to the angle of the mandible. Although these cysts may be present in infants and children, they may not enlarge and become visible until early adulthood. The sinus and cyst are usually excised. ⊙

Esophagus

The esophagus is a muscular tube that is continuous with the laryngopharynx (Fig. 8.35). It consists of striated (voluntary) muscle in its upper third, smooth (involuntary) muscle in its lower third, and a mixture of striated and smooth muscle in between. Posterior to the cricoid cartilage at the level of C6 vertebra, the esophagus begins in the median plane at the inferior border of the cricoid cartilage and inclines slightly to the left as it descends. It continues through the superior mediastinum and then the posterior mediastinum (see Chapter 1), pierces the diaphragm, and enters the stomach at the cardial orifice (see Chapter 2). The esophagus lies between the trachea and cervical vertebral column (Fig. 8.37). On the right side, the esophagus is in contact with the cervical pleura in the root of the neck, whereas on the left side, the thoracic duct lies between the pleura and esophagus. When the esophagus is empty, it has a slitlike lumen. When a food bolus descends in it, the lumen expands, eliciting reflex peristalsis in the inferior two-thirds of the esophagus. A constriction at the pharyngoesophageal junction produced by the cricopharyngeal fibers of the inferior constrictor muscle is the narrowest part of the esophagus (Fig. 8.36).

The recurrent laryngeal nerve lies in the tracheoesophageal groove on each side of the esophagus (Fig. 8.40*A*). *On the right* of the esophagus is the right lobe of the thyroid gland and the right carotid sheath and its contents. The esophagus is in contact with the cervical pleura at the root of the neck. *On the left* is the left lobe of the thyroid gland and the left carotid sheath. The thoracic duct adheres to the left side of the esophagus and lies between the pleura and esophagus. For details concerning the thoracic and abdominal regions of the esophagus, see Chapters 1 and 2.

Vessels of the Esophagus. Arteries to the cervical esophagus are branches from the **inferior thyroid arteries**. Each artery gives off ascending and descending branches that anastomose with each other and across the midline. *Veins* from the cervical esophagus are tributaries of the **inferior thyroid veins.** *Lymphatic vessels* of the cervical part of the esophagus drain into the *paratracheal lymph nodes* and *inferior deep cervical lymph nodes* (Fig. 8.41).

Nerves of the Esophagus. The nerve supply is somatic motor and sensory to the upper half and parasympathetic (vagal) sympathetic and visceral sensory to the lower half. The cervical esophagus receives the somatic fibers by way of branches from the *recurrent laryngeal nerves* and vasomotor fibers from the *cervical sympathetic trunks* (Fig. 8.40) through the plexus around the inferior thyroid artery.

Esophageal Injuries

Esophageal injuries are the rarest kinds of penetrating neck trauma, but they cause the greatest *morbidity*—complications following a surgical procedure or other treatment. In the few patients who sustain esophageal injury, most have it in conjunction with an airway injury because the airway lies anterior to the esophagus and provides some protection to it. Esophageal injuries are often occult (hidden), which makes the injury difficult to detect, especially when it is isolated. Missed esophageal trauma causes death in nearly all patients who do not have surgery and in approximately 50% of those who do (Sinkinson, 1991).

Tracheoesophageal Fistula

The most common congenital anomaly of the esophagus is *tracheoesophageal fistula (TEF)*. Usually it is combined with some form of esophageal atresia. In the most common type of TEF (approximately 90% of cases), the proximal part of the esophagus ends in a blind pouch and the distal part communicates with the trachea (*A*). In these cases, the pouch fills with mucus, which the infant aspirates. In a few cases, the proximal esophagus communicates with the trachea and the distal esophagus joins the stomach (*C*). TEFs result from abnormalities in partitioning of the esophagus and trachea by the tracheoesophageal septum (Moore and Persaud, 1998).

Esophageal Cancer

The most common presenting complaint of cancer of the esophagus is *dysphagia*, which is not usually recognized until the lumen has reduced by 30 to 50%. *Esophagoscopy* is a common diagnostic tool for observing these tumors. Painful swallowing in some patients suggests extension of the tumor to periesophageal tissues. Enlargement of the inferior deep cervical lymph nodes also suggests esophageal cancer. *Compression of the recurrent laryngeal nerves by esophageal cancer produces hoarseness.*

Zones of Penetrating Neck Trauma

Three zones are common clinical guides to the seriousness of neck trauma. The zones give physicians an understanding of the structures that are at risk with penetrating neck injuries.

- **Zone I**: the root (base) of the neck extending from the clavicles and the manubrium of the sternum to the level of the inferior border of the cricoid cartilage. *Structures* ▶

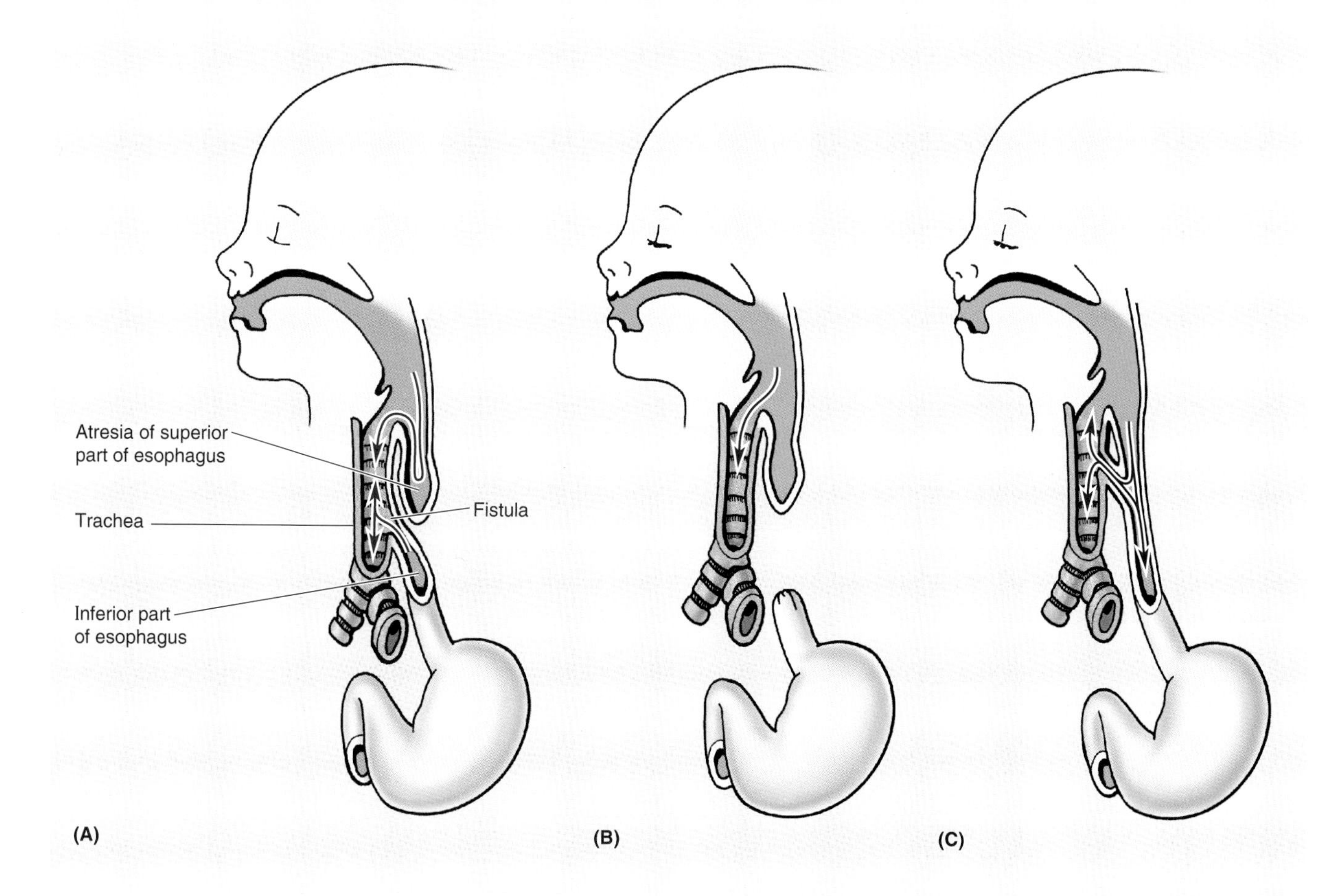

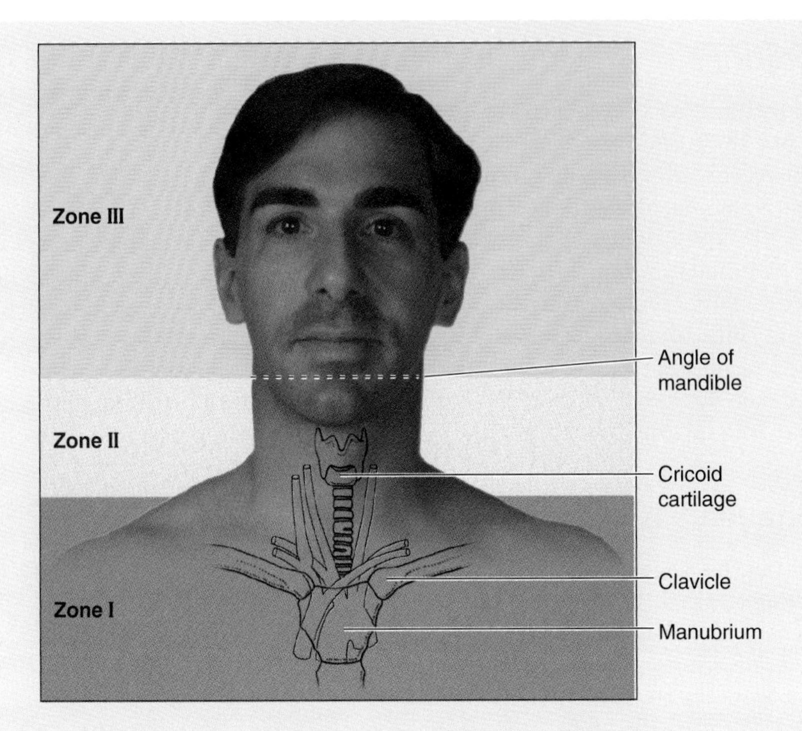

▶ *in jeopardy* are the cervical pleurae, apices of lungs, thyroid and parathyroid glands, trachea, esophagus, common carotid arteries, jugular veins, and cervical region of the vertebral column.

- **Zone II**: the cricoid cartilage to the level of the angles of the mandible. *Structures in jeopardy* are the superior poles of the thyroid gland, thyroid and cricoid cartilages, larynx, laryngopharynx, carotid arteries, jugular veins, esophagus, and cervical region of the vertebral column.
- **Zone III**: the angles of the mandibles superiorly. *Structures in jeopardy* are the salivary glands, oral and nasal cavities, oropharynx, and nasopharynx.

Injuries in Zones I and III obstruct the airway and have the greatest risk for morbidity (complications following surgical procedures and other treatments) and *mortality* (a fatal outcome) because the injured structures are difficult to visualize and repair and vascular damage is difficult to control. Injuries in Zone II are most common; however, morbidity and mortality are lower because physicians can control vascular damage by direct pressure and surgeons can visualize and treat injured structures more easily than they can in Zones I and III. ⊙

Lymphatics in the Neck

Most superficial tissues of the neck are drained by lymphatic vessels that enter **superficial cervical lymph nodes.** These nodes are located along the course of the EJV. Lymph from these nodes, like lymph from all of the head and neck, drains into **inferior deep cervical lymph nodes** (Figs. 8.41 and 8.42). The specific group of inferior deep cervical nodes involved here descends across the posterior triangle with the accessory nerve. Most lymph from the six to eight nodes then drains into the supraclavicular group of nodes that accompany the transverse cervical artery. The main group of deep cervical nodes forms a chain along the IJV, mostly under cover of the SCM. Other deep cervical nodes include the prelaryngeal, pretracheal, paratracheal, and retropharyngeal nodes. Efferent lymphatic vessels from the deep cervical nodes join to form the **jugular lymphatic trunks**, which usually join the *thoracic duct* on the left side and enter the junction of the internal jugular and subclavian veins (right venous angle) directly or via a short right lymphatic duct on the right.

The **thoracic duct**, a large lymphatic channel, begins at the chyle cistern (L. cisterna chyli) in the abdomen (see Chapter 2) and passes superiorly through the posterior mediastinum (see Chapter 1) and superior thoracic aperture along the left border of the esophagus. It arches laterally in the root of the neck, posterior to the carotid sheath and anterior to the sympathetic trunk and vertebral and subclavian arteries. The thoracic duct enters the left brachiocephalic vein at the junction of the subclavian and IJVs (*left venous angle*—Fig. 8.43). The thoracic duct drains lymph from the entire body, except the right side of the head and neck, the right upper limb, and the right side of the thorax, which drain through the *right lymphatic duct.*

Radical Neck Dissections

In radical neck dissections performed when cancer invades the lymphatics, the deep cervical lymph nodes and tissues around them are removed as completely as possible. The major arteries, brachial plexus, CN X, and phrenic nerve are preserved; however, most cutaneous branches of the cervical plexus are removed. The aim of the dissection is to remove all tissue that bears lymph nodes in one piece. The deep cervical lymph nodes, particularly those located along the transverse cervical artery, may be involved in the spread of cancer from the thorax and abdomen. As their enlargement may give the first clue to cancer in these regions, they are often referred to as the *cervical sentinel lymph nodes.* ⊕

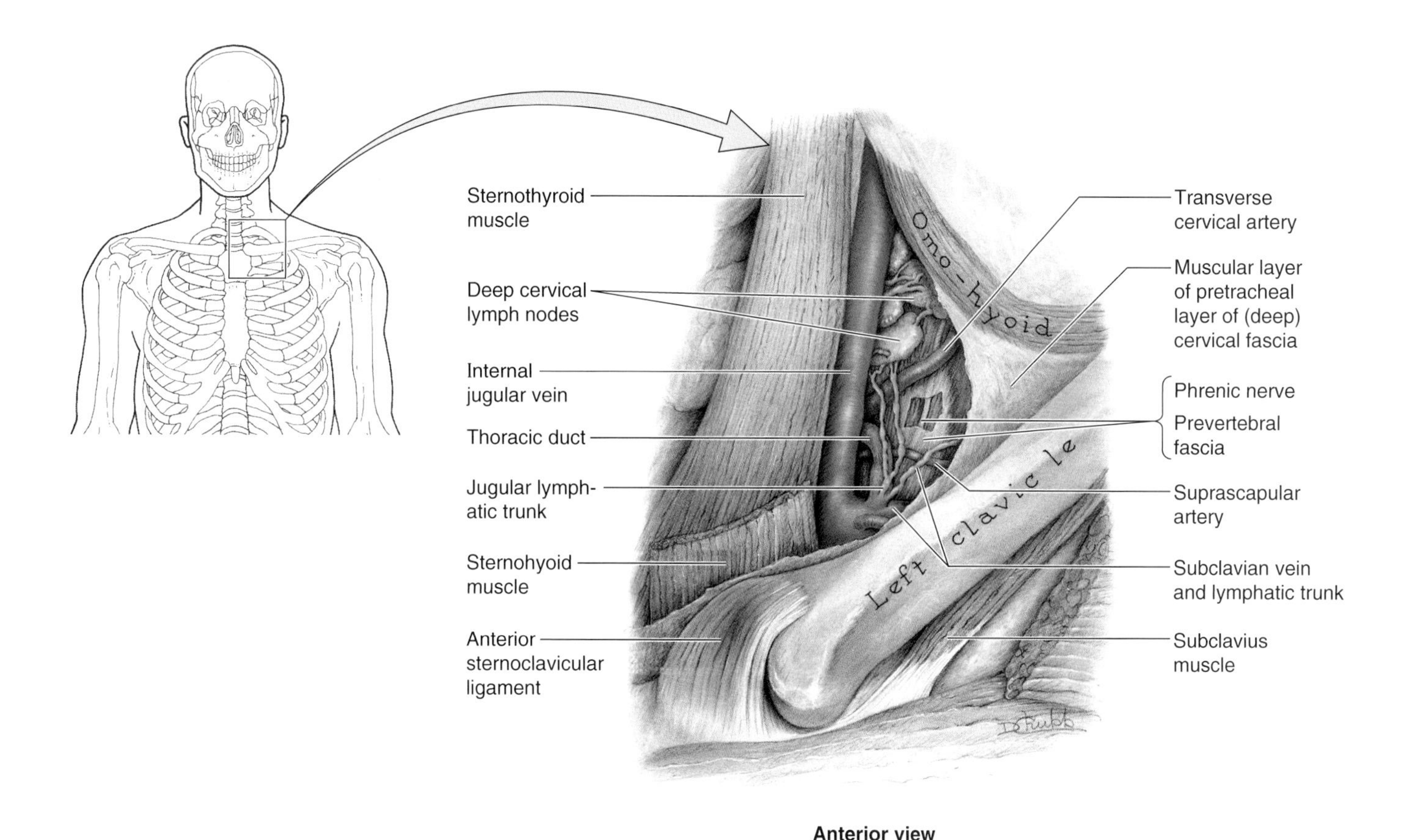

Figure 8.43. Lymphatics in the root of the neck. Dissection of the left side of the root of the neck, showing the deep cervical lymph nodes and the termination of the thoracic duct at the junction of the subclavian and internal jugular veins.

Surface Anatomy of the Neck

The neck of an infant is short; therefore, the cervical viscera are located more superiorly in infants than in adults. The viscera do not reach their final levels until after the 7th year. The elongation of the neck is accompanied by growth changes in the skin. Consequently, an incision in the lower neck of an infant results in a scar that lies over the upper sternum in a child.

Usually the EJV is visible as it crosses the neck obliquely, deep to the platysma and superficial to the SCM. The vein is less obvious in children and middle-aged women because their subcutaneous tissues tend to be thicker than they are in men. To assist visibility, the EJV can be distended by holding one's breath and expiring against resistance (e.g., with the mouth closed) or by gentle pressure on it in the supraclavicular triangle. These actions impede venous return to the right side of the heart. ▶

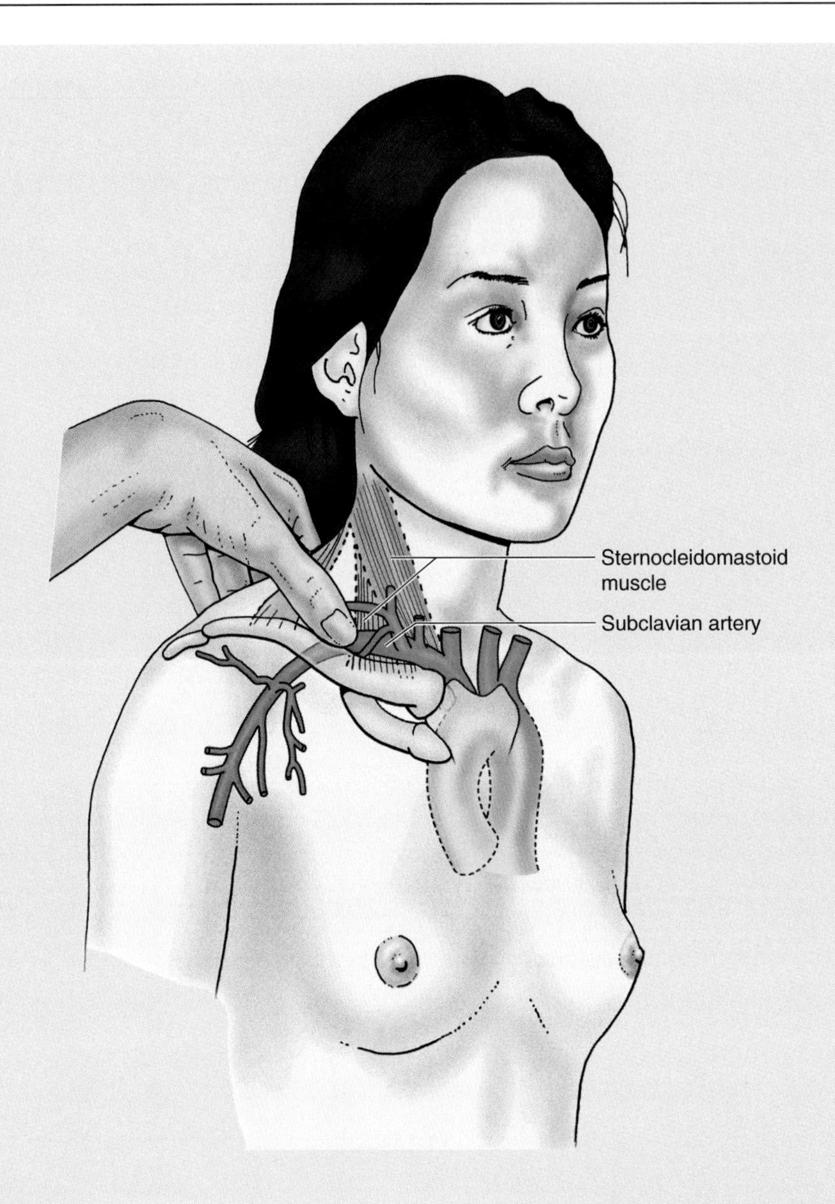

▶ A *subclavian arterial pulsation* can be felt in most people by pressing inferoposteriorly (down and back) from the posterior margin of the junction of the medial and middle thirds of the clavicle. The *submandibular lymph nodes* may be palpated between the mandible and the submandibular gland. They receive lymph from the face inferior to the eye and from the tongue. Move the fingers from the angles of the mandible along the inferior border of this bone until the fingers meet under the chin, where the *submental lymph nodes* may be felt.

The U-shaped **hyoid bone** lies in the anterior part of the neck in the deep angle between the mandible and thyroid cartilage at the level of C3 vertebra. Swallow, and it will move to your fingers. The greater horn of one side of the bone is palpable only when the greater horn of the opposite side is steadied. The **laryngeal prominence** is pro- ▶

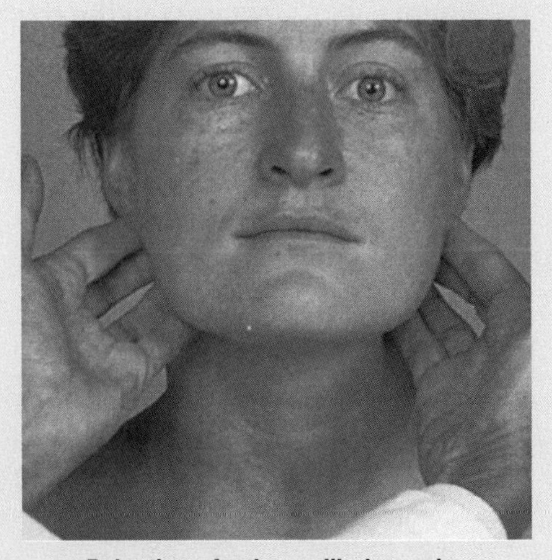

Palpation of submandibular nodes

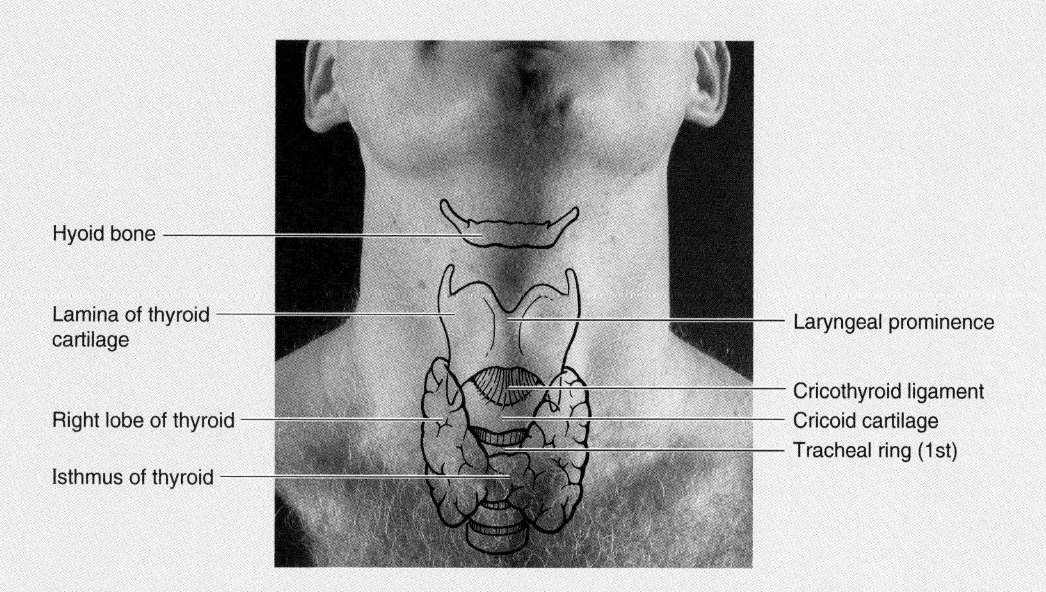

▶ duced by the meeting of the laminae of the *thyroid cartilage* at an acute angle in the anterior midline. This thyroid angle, most acute in postpubertal males, forms the laryngeal prominence ("Adam's apple"), which is palpable and frequently visible. The prominence can be felt to recede on swallowing. The vocal folds are at the level of the middle of the laryngeal prominence. The superior horn of the cartilage is palpable when the horn of the opposite side is steadied.

The *carotid pulse* can be palpated by placing the index and 3rd fingers on the thyroid cartilage and pointing them posterolaterally between the trachea and SCM. The pulse is palpable just medial to the SCM. The palpation is performed low in the neck to avoid pressure on the carotid sinus, which could cause a reflex drop in blood pressure and heart rate.

The **cricoid cartilage** can be felt inferior to the laryngeal prominence at the level of C6 vertebra. Extend your neck as far as possible and run your finger over the thyroid prominence. As your finger passes inferiorly from the prominence, feel the *cricothyroid ligament*, the site for a *needle cricothyrotomy* or coniotomy (p. 1044). After your finger passes over the arch of the cartilage, note that your fingertip sinks in because the arch of the cricoid cartilage projects further anteriorly than the rings of the trachea.

The cricoid cartilage, a key landmark in the neck, indicates the:

- Level of C6 vertebra
- Site where the carotid artery can be compressed against the transverse process of C6 vertebra
- Junction of the larynx and trachea
- Joining of the pharynx and esophagus
- Point where the recurrent laryngeal nerve enters the larynx
- Site that is approximately 3 cm superior to the isthmus of the thyroid gland.

The **tracheal rings** are palpable in the inferior part of the neck. The 2nd through 4th cartilaginous rings cannot be felt because the isthmus connecting the right and left thyroid lobes covers them. The 1st tracheal ring is just superior to the isthmus of the thyroid gland.

The **thyroid gland** may be palpated by anterior or posterior approaches (for details, see Willms et al., 1994). Although both approaches are usually performed, a perfectly normal thyroid gland may not be visible or distinctly palpable in some females except during menstruation or pregnancy. The normal gland has the consistency of muscle tissue (Swartz, 1994). The *isthmus of the thyroid gland* lies immediately inferior to the cricoid cartilage; it extends approximately 1.25 cm on either side of the midline. It can usually be felt by placing the fingertips of one hand on the midline below the cricoid arch and then having the person swallow. The isthmus will be felt moving up and then down. The apex of each *lateral lobe of the thyroid gland* is at the middle of the lamina of the thyroid cartilage.

The surface anatomy of the posterior aspect of the neck is described in Chapter 4. *Key points are*:

- Spinous processes of C6 and C7 vertebrae are palpable and visible, especially when the neck is flexed
- Transverse processes of C1, C6, and C7 vertebrae are palpable
- Tubercles of C1 can be palpated by deep pressure posteroinferior to the tips of the mastoid processes. ○

Medical Imaging of the Neck

Radiography

Standard radiographical examinations of the cervical region of the vertebral column include anteroposterior (AP), lateral, and oblique projections. In a routine AP projection of the cervical vertebral column, the mandible usually obscures the first two cervical vertebrae (Fig. 8.44). To visualize these bones, the neck is extended so that the occiput and maxillary teeth are in the same plane. The central x-ray beam aims through the open mouth, perpendicular to the x-ray film cassette (Fig. 8.45). This radiograph shows the relationship of the atlas to the axis. The ligaments of the atlantoaxial joint are radiolucent and therefore do not cast a shadow. The lateral masses of the atlas are triangular or wedge shaped and are easy to recognize. Laterally, the long transverse processes of the atlas are clearly visible, and between the lateral masses of the atlas, the dens of the axis is also readily discernible.

When taking a lateral projection of the cervical region of the vertebral column, the patient is usually sitting erect, with the neck slightly extended (Fig. 8.46). The central x-ray beam aims perpendicular to the x-ray film cassette at the level of the thyroid cartilage. Lateral projections are common for evaluating severe neck injuries. When a fracture is suspected, the lateral projection is examined before the person is moved for other projections. Observe the anterior and posterior margins of the vertebral bodies. Any deviation from the smooth curvature of these margins suggests a fracture and tearing of the associated ligaments. Observe that the IV disc spaces are wider anteriorly than posteriorly; this exists because the IV discs are wedge shaped. As the discs degenerate, the vertical height of the disc spaces decreases. Also observe the long, prominent spinous process of C7, the prominent vertebra (L. vertebra ▶

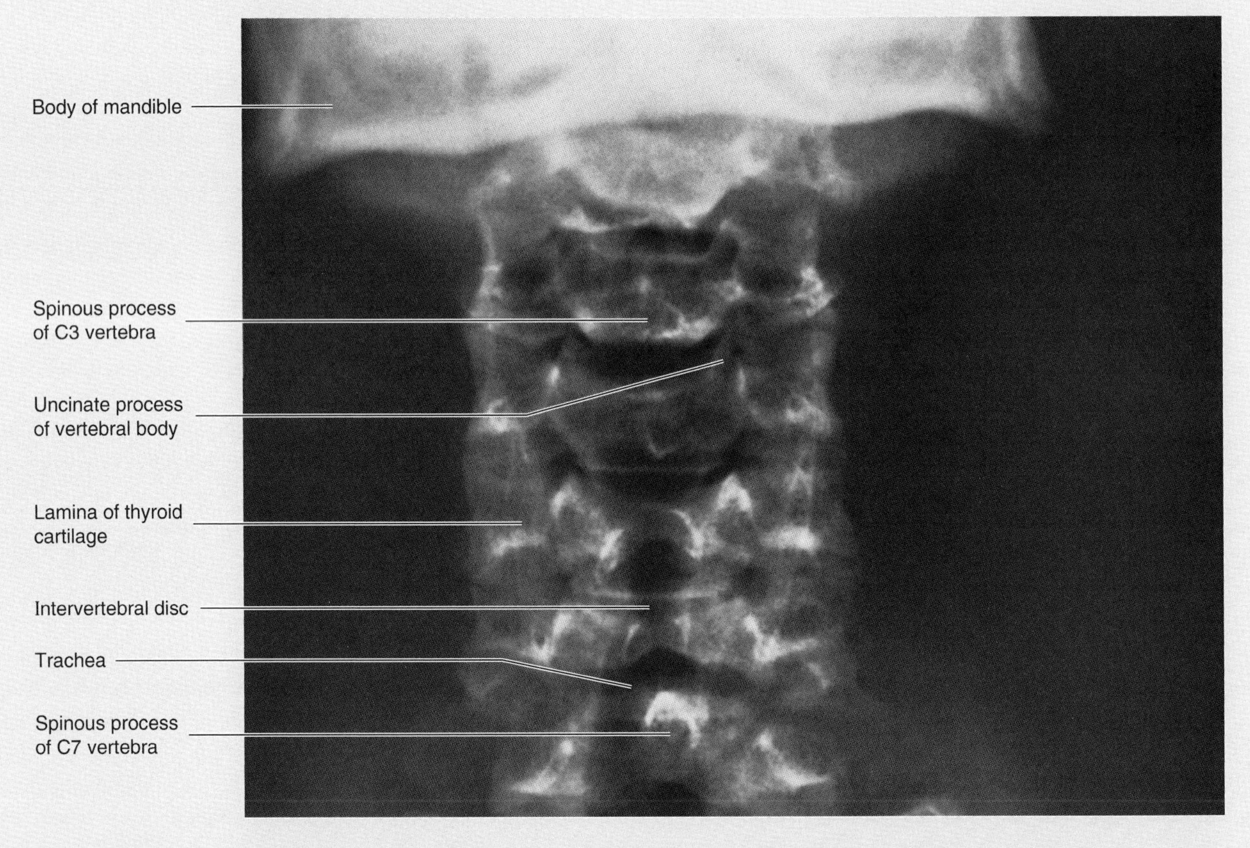

Figure 8.44. Radiograph of the cervical region of the vertebral column. Anteroposterior (AP) projection. Notice the superimposed shadow of the pharynx and trachea. Also notice the thoracic vertebrae and posterior parts of the uppermost ribs above the level of the clavicles.

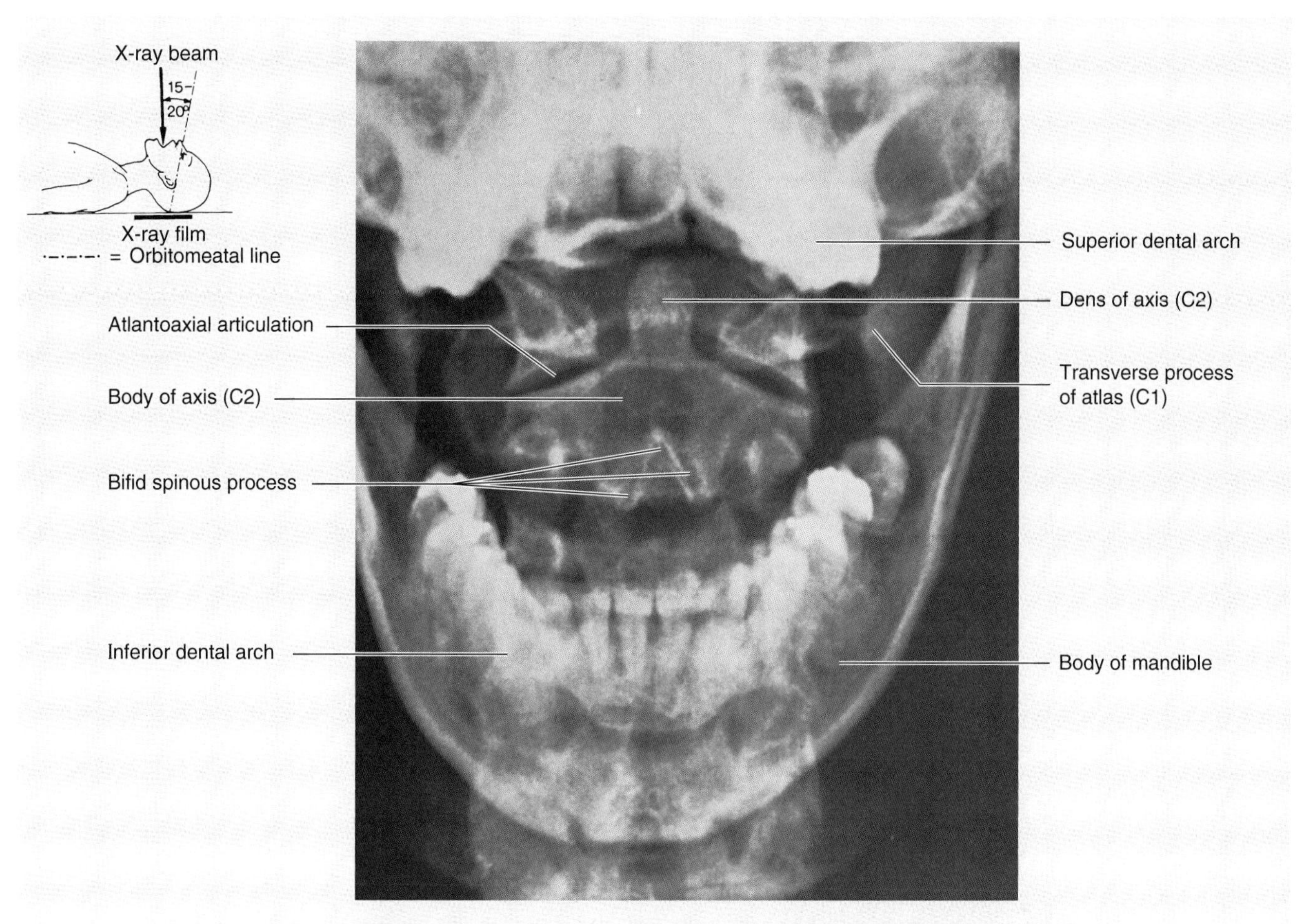

Figure 8.45. Radiograph of the dens through the open mouth. Anteroposterior (AP) projection. Observe the dens, a strong toothlike process projecting upward from the body of the axis (C2), around which the atlas (C1) rotates.

▶ prominens). Understand that the posterior margins of the bodies of cervical vertebra indicate the anterior aspect of the vertebral canal (spinal canal) containing the spinal cord.

Oblique radiographs of the neck are usually necessary for a complete study of the cervical vertebral column. The standard 45° oblique projection shows the IV foramina, as well as the uncovertebral joints (see Chapter 5). Bony outgrowths—such as *osteophytes* from the joints of the vertebral column—can be detected in oblique projections.

Contrast visualization of the esophagus is achieved by having the patient swallow a mixture of barium sulphate and water (Fig. 8.47). The first of four constrictions in the esophagus occurs at the pharyngoesophageal junction. The commencement of the esophagus is its narrowest part.

Computed Tomography

Transverse computed tomography (CT) scans through the thyroid gland provide sections of the neck (Fig. 8.48*A*). They are oriented to show how a horizontal section of the person's neck appears to the physician standing at the foot of the bed. The superior edge of the CT image therefore represents the anterior surface of the neck, and the right lateral edge of the image represents the left lateral surface. CT is used mainly as a diagnostic adjunct to conventional radiography. CT scans are superior to radiographs because they reveal radiodensity differences among and within soft tissues (e.g., the thyroid gland).

Magnetic Resonance Imaging

Magnetic resonance imaging (MRI) systems construct images of transverse, sagittal, and coronal sections of the neck and have the advantage of using no radiation. MRIs of the neck are superior to CTs for showing detail in the soft tissues, but they provide little information about bones (Figs. 8.26, 8.49*B*, 8.50, and 8.51).

Ultrasonography

Ultrasonography is also a useful diagnostic imaging technique for studying the soft tissues of the neck. Ultrasound provides images of many abnormal conditions of ▶

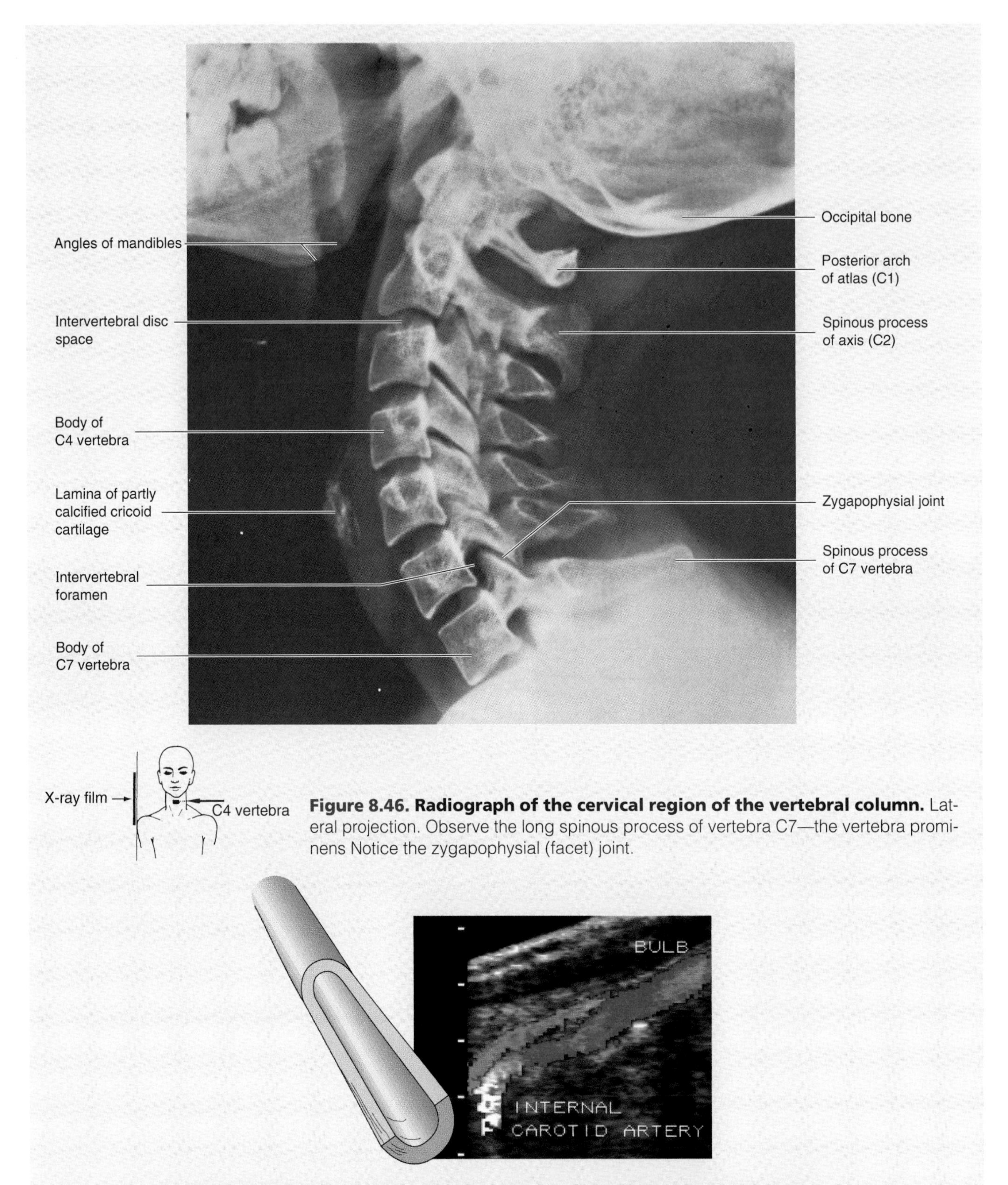

Figure 8.46. Radiograph of the cervical region of the vertebral column. Lateral projection. Observe the long spinous process of vertebra C7—the vertebra prominens Notice the zygapophysial (facet) joint.

Figure 8.47. Normal Doppler color flow study of the internal carotid artery.

▶ the neck noninvasively, at relatively low cost, and with minimal discomfort. Ultrasound is useful for distinguishing solid from cystic masses, for example, which may be difficult to determine during physical examinations. *Vascular imaging of arteries and veins* of the neck is possible using *intravascular ultrasonography*. The images are produced by placing a transducer within a blood vessel. *Doppler ultrasound techniques* help evaluate bloodflow through a vessel (Fig. 8.47) (e.g., for detecting stenosis of a carotid artery). ⊙

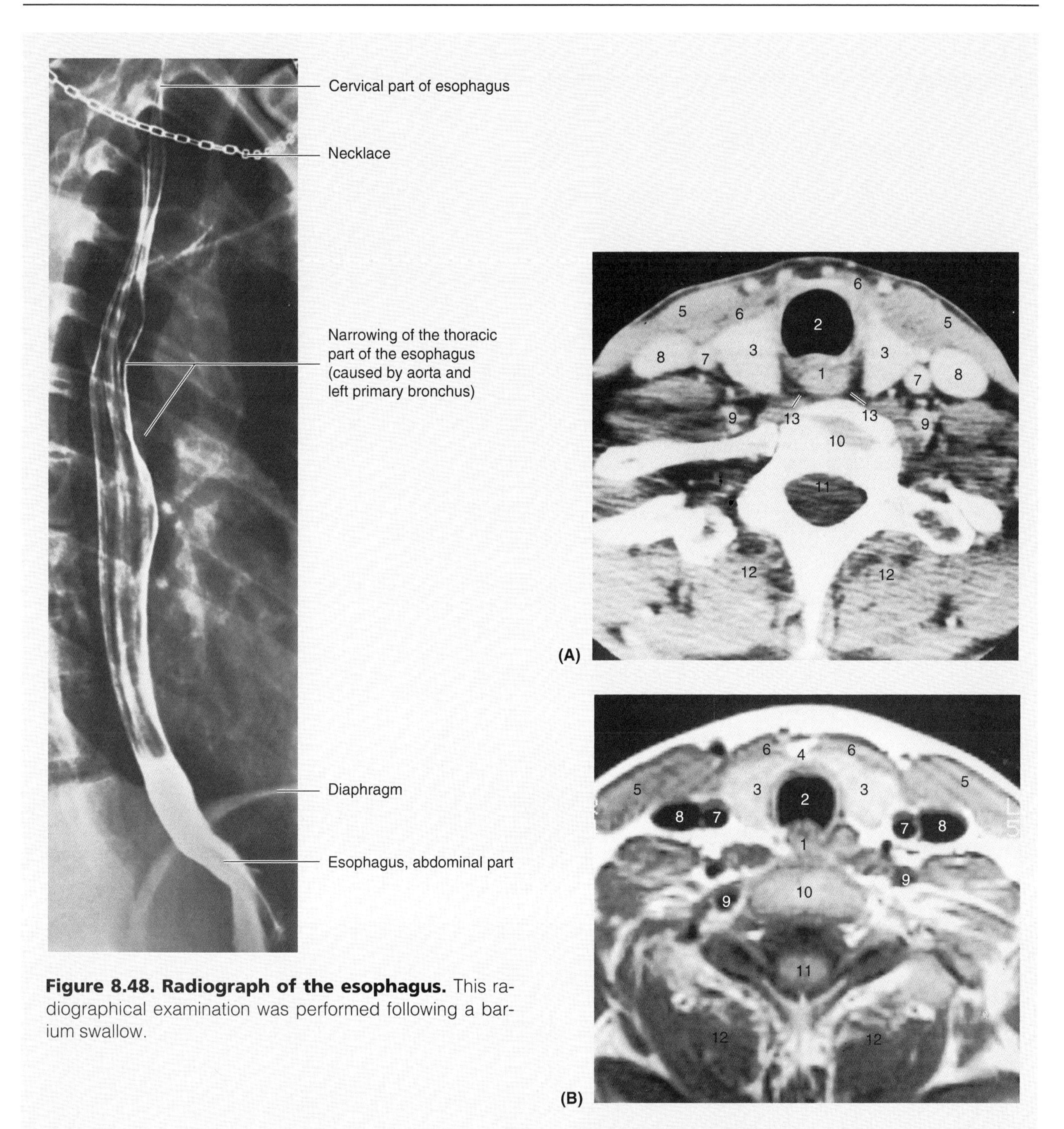

Figure 8.48. Radiograph of the esophagus. This radiographical examination was performed following a barium swallow.

Figure 8.49. Transverse scans of the neck through the thyroid gland. A. CT scan through the lobes of the thyroid gland. (Courtesy of Dr. M. Keller, Assistant Professor of Medical Imaging, University of Toronto, Toronto, Ontario, Canada) **B.** MRI scan through the isthmus of the thyroid gland. *1*, esophagus; *2*, trachea; *3*, lobes of thyroid gland; *4*, thyroid isthmus; *5*, SCM; *6*, sternohyoid muscles; *7*, common carotid artery; *8*, IJV; *9*, vertebral artery; *10*, vertebral body; *11*, spinal cord in cerebrospinal fluid; *12*, deep muscles of the back; *13*, retropharyngeal space. (Courtesy of Dr. W. Kucharczyk, Clinical Director, Tri-Hospital Resonance Centre, Toronto, Ontario, Canada)

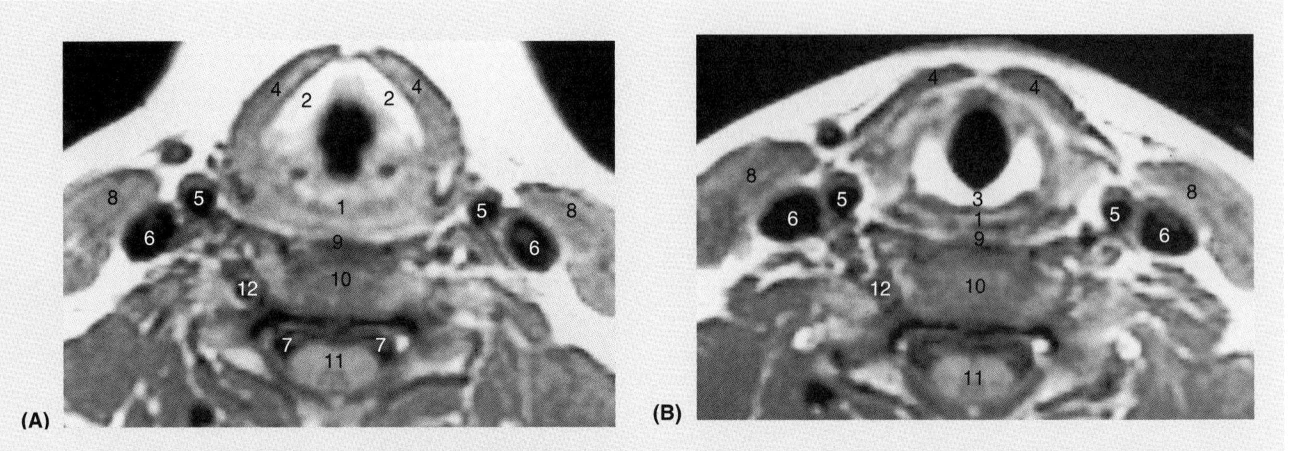

Figure 8.50. Transverse MRI of the larynx. A. Through the thyroid cartilage. **B.** Through the cricoid cartilage. *1*, esophagus; *2*, thyroid cartilage; *3*, lamina of cricoid cartilage; *4*, strap (sternothyroid and sternohyoid) muscles; *5*, common carotid artery; *6*, IJV; *7*, ventral root; *8*, SCM; *9*, inferior constrictor; *10*, vertebral body; *11*, spinal cord in cerebrospinal fluid; *12*, vertebral artery. (Courtesy of Dr. W. Kucharczyk, Chair of Medical Imaging, University of Toronto, and Clinical Director, Tri-Hospital Resonance Centre, Toronto, Ontario, Canada).

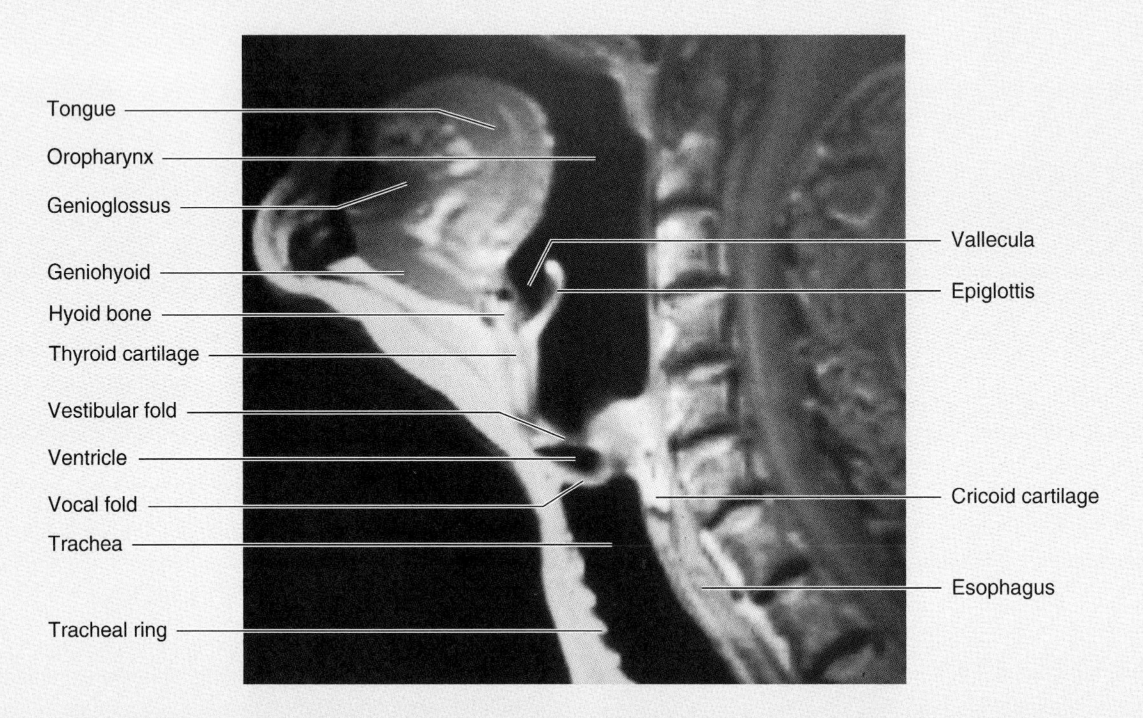

Figure 8.51. Median MRI through the head and neck. Observe the tongue, oropharynx, hyoid bone, thyroid and cricoid cartilages, vocal fold (cord), and trachea. (Courtesy of Dr. W. Kucharczyk, Chair of Medical Imaging, University of Toronto, and Clinical Director, Tri-Hospital Resonance Centre, Toronto, Ontario, Canada)

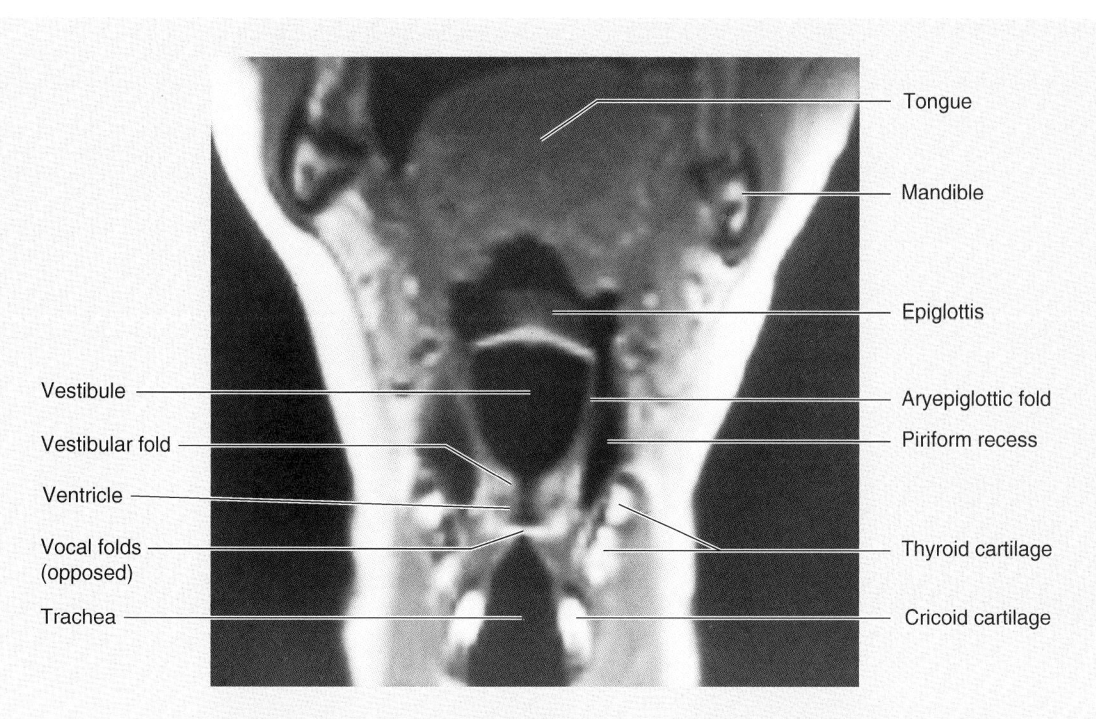

Figure 8.52. Coronal MRI through the oropharynx, larynx, and trachea. (Courtesy of Dr. W. Kucharczyk, Chair of Medical Imaging, University of Toronto, and Clinical Director, Tri-Hospital Resonance Centre, Toronto, Ontario, Canada)

CASE STUDIES

Case 8.1

A 4-year-old girl's grandmother noticed that she held her head to one side. The child's mother took the girl to the pediatrician, who confirmed the grandmother's observation. On examination he observed that the child's head was tilted to the right, her occiput was rotated toward her shoulder, and her chin was rotated to the left and elevated. The pediatrician also detected a palpable mass in the inferior part of the child's SCM. The remainder of the muscle was prominent throughout its course in the neck. He diagnosed the condition as *congenital muscular torticollis.*

Clinicoanatomical Problems

- What is the common name for this anomaly?
- What is the usual cause of this muscular abnormality?
- When does it usually occur?
- Why does it take so long for the torticollis to develop?
- Can the injury be diagnosed during infancy and treated before torticollis develops?
- If the muscular torticollis is not treated (e.g., by massaging and stretching the SCM and/or lengthening it by surgery), what further developmental abnormalities may occur?

The problems are discussed on page 1077.

Case 8.2

A 58-year-old woman consulted her physician about a slight swelling in her neck, inferior to her thyroid cartilage. Physical examination and ultrasound scanning revealed several thyroid nodules in the right lobe of her thyroid gland. Further examination of cells obtained by *fine-needle aspiration* revealed that the hypercellular aspirate was suspicious for malignancy. It was decided to perform a *hemithyroidectomy*. An endotracheal airway was inserted through the mouth before beginning the surgery. The patient's throat was sore for approximately 2 days, and her voice was hoarse.

Clinicoanatomical Problems

- Explain what is meant by a hemithyroidectomy?
- What do you think caused her sore throat?
- What was the likely cause of her hoarseness?

- During thyroid surgery, what nerves are vulnerable to injury?
- What structures would be affected by injury to these nerves?

The problems are discussed on page 1077.

Case 8.3

Following removal of a malignant tumor from the right posterosuperior region of the neck of a 52-year-old man, the surgeon decided to do a radical neck dissection of the region to remove enlarged lymph nodes. One enlarged lymph node was in the submandibular triangle, deep to the superior end of the SCM. After the operation, the man informed the surgeon that he had difficulty shrugging his right shoulder and turning his face to the left side against resistance.

Clinicoanatomical Problems

- What nerve was probably injured during the surgical removal of the enlarged lymph nodes?
- What is the relationship of this nerve to the superior end of the SCM?
- What lymph nodes do you think the surgeon removed?
- From what areas do these nodes receive lymph?
- If malignant cells from these nodes had metastasized, to which lymph nodes would they likely go?

The problems are discussed on page 1077.

Case 8.4

A 42-year-old woman told her family physician that she recently observed a swelling in the anterior part of her neck. She also said that her breathing seemed to be affected by this swelling. Physical examination revealed a firm swelling on the left side of her thyroid gland, which moved up and down during swallowing. An ultrasound scan revealed a solid nodule in the left lobe of her thyroid gland. A needle biopsy indicated that malignant changes had occurred in the cells.

Clinicoanatomical Problems

- Why did this nodular swelling move up and down when the woman swallowed?
- Why was her breathing affected?
- Based on your knowledge of the lymphatics of the thyroid gland, to which lymph nodes do you think the cancerous cells might metastasize?

The problems are discussed on page 1077.

Case 8.5

A 62-year-old man consulted his physician about his difficulty in swallowing and breathing. He said that his wife was also concerned about the swelling in his neck, which she thought was a *goiter*. Physical examination, an ultrasound scan, and a needle biopsy revealed that the man had *thyroid cancer*. It was decided to perform a thyroidectomy and a neck dissection to search for and remove enlarged lymph nodes.

Clinicoanatomical Problems

- What is a goiter?
- Why was the man having difficulty breathing and swallowing?
- Why is a total thyroidectomy usually not performed?
- How can one avoid damaging the recurrent laryngeal nerves during a thyroidectomy?

The problems are discussed on page 1077.

Case 8.6

A 65-year-old woman consulted her physician about muscle weakness, anorexia, nausea, constipation, and polyuria (passage of large volumes of urine). After taking a thorough history and performing an extensive physical examination, the physician ordered laboratory studies of her blood and urine. The laboratory reports revealed an elevated serum calcium concentration, an elevated serum level of parathyroid hormone, and an elevated urinary calcium excretion. A diagnosis of *parathyroid adenoma* was made. It was decided that resection of the enlarged gland(s) should be carried out.

The surgeon located the superior parathyroid glands without difficulty and found them to be normal in size. She could locate only one inferior parathyroid gland. Because it was enlarged and a frozen section suggested parathyroid hyperplasia and *parathyroid adenoma*, the gland was removed. She systematically searched the anterior part of the neck for the 4th parathyroid gland but was unable to find it. Continued searching resulted in detection of the gland.

Clinicoanatomical Problems

- How many parathyroid glands are usually present? How many may be present?
- Where would you expect to find a displaced (ectopic) parathyroid gland?
- If the ectopic parathyroid gland was not detected in the neck, where do you think the surgeon should search for it?
- Which parathyroid glands are most likely to be found in an ectopic position?
- How would you explain displacement of these glands?

The problems are discussed on page 1077.

Case 8.7

A newborn infant was referred by a family physician to a pediatrician because of respiratory stress and excessive salivation. The infant was coughing and choking during feeding. During the physical examination, the pediatrician observed an excessive amount of mucous secretion and saliva in the infant's mouth. She also observed that the infant experienced some difficulty in breathing and that gastric distension was present. She was unable to pass a nasogastric tube very far along the esophagus. She told the mother that the infant had an esophageal anomaly.

Clinicoanatomical Problems

- What congenital anomaly of the esophagus do you think was present?
- What other anomaly is usually associated with this type of esophageal anomaly? Describe this condition.
- What common association of congenital anomalies may occur with esophageal anomaly?
- What do you think might cause the gastric distension associated with the infant's esophageal and tracheal anomalies?

The problems are discussed on page 1078.

Case 8.8

A 3-year-old boy was taken to a physician because of an intermittent mucous discharge from a small opening in the side of his neck, around which was extensive redness and swelling. The physician informed the infant's mother that this discharge was probably from a remnant of the pharyngeal apparatus in the embryo.

Clinicoanatomical Problems

- What is the pharyngeal apparatus?
- What type of branchial or pharyngeal anomaly could result in the discharge of pus from the side of the boy's neck? Describe the origin of this anomaly.
- What is the common location of the abnormal external opening in the neck?
- If the cervical sinus opened internally, where would you expect the pus to be discharged?

The problems are discussed on page 1078.

Case 8.9

A 6-year-old child was taken to a family physician for treatment of a persistent sore throat. During physical examination the physician observed infection and hypertrophy of the tonsils and adenoids. He also detected an *enlarged tonsillar lymph node*. Although the boy's tonsils were the site of chronic infection, he was reluctant to recommend a T and A.

Clinicoanatomical Problems

- What is a T and A?
- Which lymph node is commonly referred to as the tonsillar lymph node?
- Where is this node located?
- Why do you think the physician was reluctant to recommend a T and A?
- Because of the vascularity of the palatine tonsils, tonsillectomy can be dangerous. Which vessels may bleed after tonsillectomy?

The problems are discussed on page 1078.

Case 8.10

A young man was slashed deeply with a knife in the middle of the right posterior triangle of the neck. The cut ended anterior to the middle of the SCM. The bleeding was arrested and the wound was sutured. The patient, who was right-handed, later complained that he had difficulty combing his hair and tilting his head to the right.

Clinicoanatomical Problems

- What blood vessel was likely severed?
- What large nerves would likely be injured? Describe the course of these nerves.
- Explain why the patient had difficulty combing his hair and laterally flexing his head.

The problems are discussed on page 1078.

Case 8.11

During an automobile accident, the neck of an 82-year-old man was injured by the safety belt as the vehicle stopped suddenly. He complained of difficulty breathing and a sore "Adam's apple." The physician who examined him realized that it was necessary to perform an emergency *cricothyroidotomy* to secure an adequate airway.

Clinicoanatomical Problems

- What structures in the anterior part of the neck were most likely injured by the safety belt during the automobile accident? How could this cause difficult breathing?
- The laryngeal skeleton is more easily fractured in elderly people. Why is this so?
- What structure is incised to enter the trachea during a cricothyroidectomy?
- What surgical procedure do you think would be performed to support an airway over an extended period of time?
- Based on your knowledge of the relations of the trachea, what structures may be injured during this procedure?

The problems are discussed on page 1078.

Case 8.12

A 58-year-old man consulted his physician about difficulty in swallowing. He said that he first had difficulty swallowing solid foods, but recently he has had difficulty swallowing soft foods and liquids. He also said that he has lost considerable weight in the last 2 months. Physical examination of the man's neck revealed a large, firm lump deep to the anterior border of the SCM. A biopsy of the tumor and surrounding tissues revealed a malignant tumor of the cervical esophagus that had begun to infiltrate the periesophageal tissues.

Clinicoanatomical Problems

- What caused the man's dysphagia?
- What do you think caused the large lump deep to the SCM?

- What is the lymphatic drainage of the cervical esophagus?
- Into which periesophageal tissues do you think the cancer would infiltrate?

The problems are discussed on page 1078.

Case 8.13

A 20-year-old man with moderate fever and a sore throat noticed a swelling on the side of his face anterior to his ear. He was also concerned when another swelling appeared below his jaw. He quickly arranged an appointment with his family physician. During the physical examination the physician observed enlargement of the right parotid and submandibular glands. Palpation of the parotid gland was painful. Examination of the oral cavity revealed redness of the openings of the ducts of these glands. The pain of the swollen parotid gland increased when he was asked to sip lemon juice. A diagnosis of *mumps* was made.

Clinicoanatomical Problems

- In which triangle of the neck is the submandibular gland located? Be specific.
- Where do the ducts of the parotid and submandibular glands open?
- Why is swelling of the parotid gland painful, especially during chewing?
- Why did sipping lemon juice elicit pain in the parotid gland?

The problems are discussed on page 1079.

Case 8.14

A 3-year-old boy was playing with some coins that he found on the floor. He put a five-cent piece into his mouth and accidentally swallowed it. He suddenly began to cough, drool, and choke. The boy was rushed to the Children's Hospital. A lateral radiograph of his neck revealed that the coin was lodged in his esophagus.

Clinicoanatomical Problems

- Where would the coin likely lodge in the cervical region of the esophagus?
- If the coin passes further down the esophagus into the thorax, where might it lodge?
- What would make the child choke?
- How do you think the coin might be removed?

The problems are discussed on page 1079.

Case 8.15

A 30-year-old man was eating a fish dinner when he suddenly started to choke. He told his wife that he believed he had a fishbone stuck in his throat. She drove him to the emergency department of a hospital. The attending physician examined the man's larynx with a laryngeal mirror; however, he was unable to see a fishbone. He then inserted an endoscope into the patient's laryngopharynx and was able to locate and remove the fishbone.

Clinicoanatomical Problems

- Where do you think the fishbone was lodged?
- What structure might be injured if the fishbone pierced the mucous membrane?
- What would be the result of injury to this structure?

The problems are discussed on page 1079.

Case 8.16

The mother of an 11-year-old child consulted her physician about a swelling in the anterior aspect of the child's neck. Although the swelling was painless, the mother was concerned because the lump seemed to be slowly getting larger. The physician explained to her that this type of neck swelling is common in children, adolescents, and young adults. He stated that these swellings often represent developmental anomalies that do not become apparent until childhood, adolescence, or early adulthood. He also stated that these midline masses tend to be benign (nonmalignant).

Physical Examination Physical examination revealed that none of the cervical lymph nodes was enlarged or tender. The physician noted that the swelling was located just inferior to the hyoid bone and that it was cystic and freely movable (Fig. 8.53). The physician held the swelling gently between his 1st and 2nd fingers and asked the patient to swallow and then to open her mouth and stick her tongue out. Feeling some movement of the mass, he asked the patient to stick her tongue out as far as possible and then retract it. The physician noted a definite superior tug on the mass as the patient's tongue protruded. The swelling

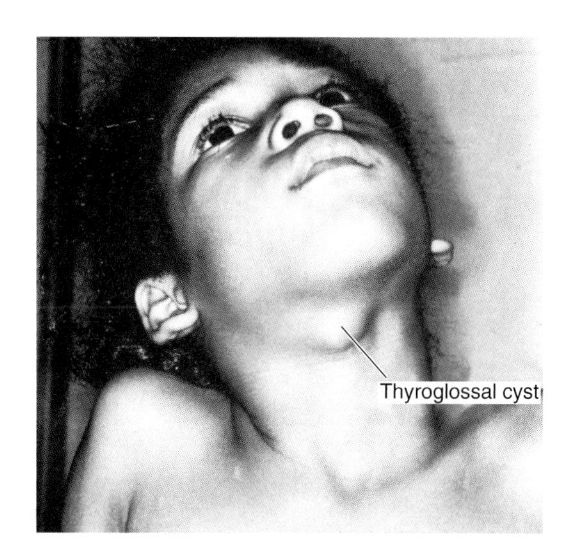

Figure 8.53. Thyroglossal duct cyst. This cyst formed from a remnant of the embryonic thyroglossal duct.

also moved superiorly during swallowing. The physician aspirated fluid from the swelling for laboratory investigation.

Laboratory Report The fluid consists of a thin, watery substance containing gelatinous material, suggesting the presence of a remnant of the embryonic thyroglossal duct.

Diagnosis Thyroglossal duct cyst.

Clinicoanatomical Problems

- Explain the embryological basis of this cyst.
- Where are these cysts likely to be found?
- What is the anatomical basis for movement of the cyst superiorly when the patient protruded her tongue and swallowed?
- What glandular tissue may be associated with this type of cyst?
- What would this condition be called if a midline cervical opening into the cyst had been present?

The problems are discussed on page 1079.

Case 8.17

A 27-year-old medical student consulted her clinical instructor about a painless, plum-shaped swelling in the anterior triangle of her neck, inferior to the angle of her mandible (Fig. 8.54). As her mandibular 3rd molar teeth (wisdom teeth) had not erupted, she thought the swelling might be caused by a dental abscess in the submandibular triangle because her 3rd molar on that side was producing some pain. She also feared that the swelling might be caused by a tumor of her submandibular gland or be an enlargement of the jugulodigastric lymph node.

History The physician asked the student several questions:

- *When did you first notice the lump?* I first saw it approximately 3 months ago.
- *Does it hurt?* No, but the swelling has been slowly getting larger.
- *Have you had any ear or throat infections recently?* No, but I had a cold approximately 2 weeks ago.
- *Have you detected any hoarseness since this swelling appeared?* No.

Physical Examination The physician palpated the neck lump and told her that it was a cystic swelling just anterior to the upper third of her left SCM. He aspirated fluid from the cyst for laboratory investigation. The physician examined her left 3rd molar tooth and said that it was impacted and that the gingiva was slightly inflamed; however, he assured her that the infection in her tooth was not related to the neck swelling.

Laboratory Report The cyst contains cloudy, cholesterol-crystal-laden fluid.

Diagnosis Branchial cyst (lateral cervical cyst, or branchial cleft cyst) (Figs. 8.54 and 8.55).

Surgical Treatment The physician arranged to have the cyst surgically removed (*cystectomy*). During excision of the cyst, a sinus tract was observed to pass superiorly from it. The sinus tract was carefully dissected and removed.

Clinicoanatomical Problems

- Explain the embryological basis of a branchial cyst.
- Where did the sinus tract probably terminate?

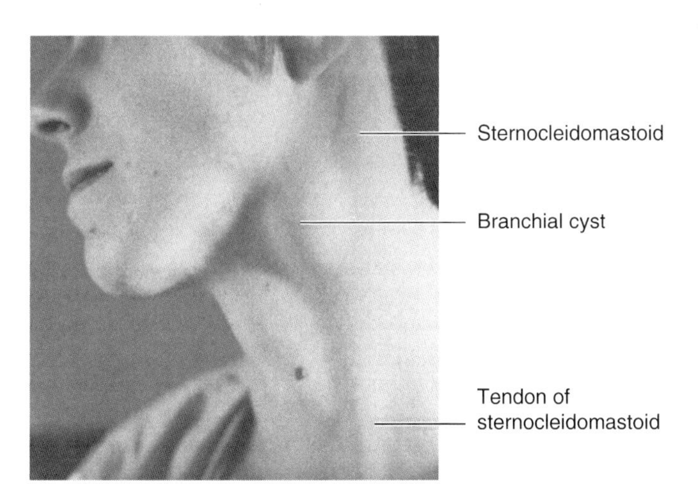

Figure 8.54. Branchial (cleft) cyst. A 27-year-old woman with a cervical swelling inferior to the angle of her left mandible (lower jaw) and anterior to her sternocleidomastoid (SCM) muscle.

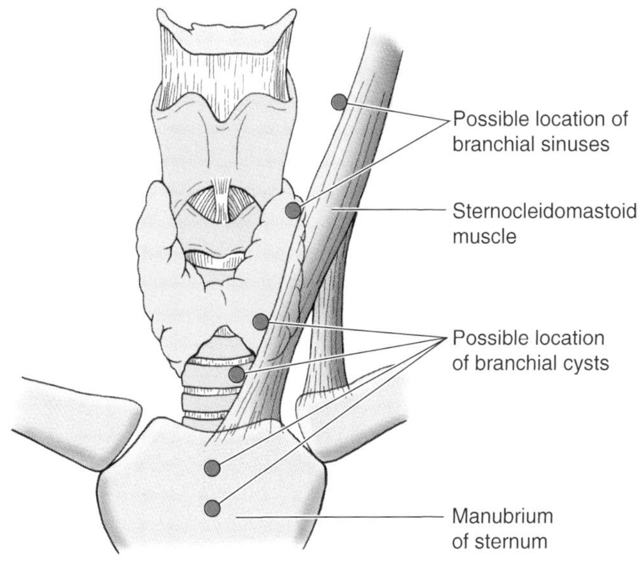

Figure 8.55. Possible locations of branchial sinuses and cysts. Branchial (cleft) cysts usually develop along the inferior part of the lateral border of the sternocleidomastoid (SCM).

- What nerve might be damaged during the cystectomy?
- What signs would be present if this nerve was injured?
- If the sinus tract had passed inferiorly, where would it probably open?
- If the cystic swelling was painful during palpation, do you think it could still be a branchial cyst?
- If the swelling had been firm and painful, what do you think might be the cause of the swelling?

The problems are discussed on page 1079.

Case 8.18

After completing your first anatomy examination, your father decided to celebrate and take you out for a steak dinner. After a few drinks you noted that his speech was slurred and that he was eating his steak rapidly. Later you noticed your father's face change suddenly. He had a terrified look and then collapsed on the floor. At first you suspected that he had passed out, but as you examined him more closely you thought that perhaps he was having a stroke, a heart attack, or some other seizure. Your examination also revealed that his pulse was strong, but his face began to turn blue. You then realized that your father was suffering from *asphyxia.* You opened his mouth widely and observed a large piece of steak caught in the posterior part of his throat. First you reached into his mouth with your index finger and tried to pull it out. On being unsuccessful, you rolled him into the prone position and performed the *Heimlich maneuver.* This increased his intra-abdominal pressure and moved his diaphragm superiorly, forcing the air out of his lungs and expelling the piece of steak.

Clinicoanatomical Problems

- Where was the piece of steak most likely lodged?
- If the Heimlich maneuver had not been successful and a physician had come to help you, what lifesaving measures do you think he or she might have taken?
- Discuss so-called "restaurant deaths."

The problems are discussed on page 1079.

Case 8.19

A 30-year-old woman was concerned about a swelling in the anterior part of her neck, nervousness, and loss of weight. She told her physician that her family complains that she is irritable, excitable, and cries easily.

Physical Examination During the examination the physician detected a swelling on each side of her neck, inferior to the larynx. During palpation of the patient's neck from a posterior position, the physician felt an enlarged thyroid gland and noted that it moved up and down during swallowing. The following signs were also detected: protrusion of the eyes, rapid pulse, tremor of the digits, moist palms, and loss of weight.

Diagnosis Hyperthyroidism (exophthalmic goiter, Graves' disease). When the patient did not respond to medical treatment, a subtotal (partial) thyroidectomy was performed. After the operation the patient complained of hoarseness.

Clinicoanatomical Problems

- What is the anatomical basis for the enlarged thyroid gland moving up and down during deglutition?
- Because the patient's thyroid gland was enlarged, what nerves might have been compressed or displaced?
- If a total thyroidectomy had been done, what other endocrine glands might inadvertently have been removed?
- What would result from this error?
- What was the probable cause of the patient's hoarseness?

The problems are discussed on page 1080.

Case 8.20

A 10-year-old boy was admitted to the hospital with a sore throat and earache. He had a high fever (temperature, 40.5°C), rapid pulse rate, and rapid respirations.

Physical Examination Examination of the boy's throat by the attending physician revealed diffuse redness and swelling of the pharynx, especially of the palatine tonsils. His left tympanic membrane was also bulging. The boy's medical history revealed that he had had chronic symptoms of inflammation of the nasal mucous membrane (*rhinitis*), including the pharyngeal tonsils (*adenoiditis*), resulting in persistent mouth breathing. On one occasion he had had a *peritonsillar abscess* (quinsy).

Diagnosis Tonsillitis—inflammation of the tonsils, especially the palatine tonsils.

Treatment Following antibiotic treatment, the infection cleared up. In view of the boy's history, it was decided to readmit him 3 or 4 months later for a *T and A* (tonsillectomy and adenoidectomy).

Clinicoanatomical Problems

- What is meant by the term tonsils?
- Explain the anatomical basis of the boy's earache.
- What lymph node in particular would likely be swollen and tender in this case?
- What is the probable source of hemorrhage during a tonsillectomy?
- Compression of what vessel would control severe arterial bleeding in the tonsillar bed?

The problems are discussed on page 1080.

DISCUSSION OF CASES

Case 8.1

This relatively common abnormality of the neck is usually called a twisted or wry neck by laypersons. Congenital muscular torticollis usually occurs at birth (L. congenitus, born with). Lesions of the SCM may result from fixed positioning of the head and neck in the uterus that causes tearing of the muscle fibers and *fibrosis*—formation of fibrous tissue as a reparative or reactive process. Stretching of the neck during a difficult delivery may also cause tearing of fibers and bleeding within the SCM. The resulting *hematoma* is contained within its own fascial compartment, which results in increased pressure on the muscle fibers. This damages the muscle and results in an area of *ischemia* (deficient blood supply). The damaged muscle fibers gradually undergo fibrosis.

The torticollis—twisting of the neck—develops slowly as fibrosis and contraction of the SCM occurs. It may not be noticed until the child is 5 or 6 years old. A thorough pediatric examination may reveal a hematoma in the inferior third of the SCM. If this swelling is massaged and daily passive neck stretching exercises are carried out, the hematoma may disappear and no fibrosis and shortening of the SCM may develop. Failure to correct muscular torticollis results in asymmetrical development of the facial bones. Wedge-shaped deformities of the cervical vertebrae may also develop.

Case 8.2

A hemithyroidectomy indicates that most but not all of one lobe is removed. The posterior part of the lobe is usually preserved to avoid removal of the parathyroid glands. It is common for patients to have a sore throat for 1 to 3 days after a thyroidectomy because of the insertion of an endotracheal airway. This procedure irritates the mucosal lining of the laryngopharynx, and some inflammation usually occurs. *Injury to the right recurrent laryngeal nerve during thyroid surgery could cause hoarseness*, but edema and infection developing after surgery may have caused compression of the nerve. The recurrent laryngeal nerves are vulnerable to injury during thyroid surgery because they are closely related to the thyroid gland and trachea. The left recurrent laryngeal nerve is more vulnerable to trauma and disease than the right nerve because of its longer course around the arch of the aorta. The recurrent laryngeal nerves supply sensation to the larynx inferior to the vocal folds and motor supply to all laryngeal muscles except the cricothyroids. Consequently, severance of these nerves would cause *aphonia* (loss of voice) because of paralysis of the vocal folds.

Case 8.3

Most likely the spinal accessory nerve was injured. This nerve pierces the deep surface of the superior part of the SCM and supplies it. CN XI then crosses the posterior triangle and supplies the trapezius. The SCM tilts the head laterally and rotates the neck, explaining why he had difficulty turning his head. The trapezius elevates, retracts, and rotates the scapula, explaining why he had difficulty shrugging his shoulder. The enlarged lymph nodes the surgeon removed would be *submandibular lymph nodes*, located beneath the investing layer of deep cervical fascia. The submandibular nodes receive lymph from a wide area, including lymph vessels from the submental, buccal, and lingual groups of lymph nodes. Efferent lymphatic vessels from the submandibular nodes pass to the superior and inferior deep cervical lymph nodes. Consequently, these nodes would also enlarge because of the presence of malignant cells from the primary tumor.

Case 8.4

The thyroid gland and the nodular swelling in it are invested by the visceral layer of pretracheal deep cervical fascia, which attaches the gland's capsule to the cricoid cartilage and tracheal rings. Consequently, when the woman swallows, her thyroid gland and the nodule within it move up and down as the larynx (and upper trachea) are elevated and then depressed by the suprahyoid and infrahyoid muscles, respectively. Her breathing was affected because each lobe of the thyroid gland is closely related to the trachea. Consequently, an enlarged lobe resulting from a large tumor is likely to compress the trachea and partially occlude its lumen, causing *dyspnea*. Because malignant cells were in the thyroid nodule, the cells would likely metastasize to the prelaryngeal, pretracheal, and paratracheal lymph nodes. From these nodes the malignant cells would likely pass to the inferior deep cervical lymph nodes.

Case 8.5

The term *goiter* refers to an enlargement of the thyroid gland that causes a swelling in the anterior part of the neck. The man had dyspnea and dysphagia because the enlarged thyroid gland could compress the trachea and esophagus and partially occlude their lumina. This would cause difficulty in breathing and swallowing. *Total thyroidectomy* is usually not performed because of the danger of injuring the recurrent laryngeal nerves and inadvertently removing all the parathyroid glands. Medially, the recurrent laryngeal nerves are closely related to the posterior part of the capsule of the thyroid gland. The parathyroid glands, when in their usual position, lie in close relationship to the posterior aspect of the thyroid gland, outside its fibrous capsule but inside its fascial sheath. Surgeons avoid injury to the recurrent laryngeal nerves during thyroidectomy by not cutting through the posterior aspect of the fibrous capsule of the thyroid, especially medially where these nerves lie in the tracheoesophageal grooves.

Case 8.6

Typically, a person has four parathyroid glands; however, more than four glands have been detected in approximately 5% of cases. An ectopic parathyroid gland is commonly found in association with the thymus or embedded in the inferior part of the thyroid gland. If the gland is not in the neck, the surgeon might explore the superior mediastinum (see Chapter 1). The superior parathyroid glands derive from the 4th pair of pharyngeal pouches and the inferior ones from the 3rd pair—as does the thymus. As the embryo develops, the thymus descends and usually separates from the inferior parathyroid glands. Usually the separation occurs when they lie posterior to the inferior lobe of the thyroid gland. This migration is extremely variable and, as a result, the inferior glands are more likely to be in an ectopic position.

Case 8.7

Inability to pass a catheter through the esophagus of a newborn infant into the stomach indicates *esophageal atresia.* Esophageal atresia in an infant is commonly associated with a *TEF.* Most often the esophagus forms a blind pouch and the distal esophagus is connected to the tracheobronchial tree just superior to the carina—the ridge separating the openings of the right and left main bronchi at their junction with the trachea. Sixty to seventy percent of infants with esophageal atresia have associated anomalies of the gastrointestinal, cardiovascular, genitourinary, musculoskeletal, and central nervous systems. *VACTERL association* is a common association of anomalies. The acronym indicates vertebral anomalies, anorectal anomalies, cardiac anomalies, TEF, esophageal atresia, renal anomalies, and limb anomalies (usually dysplasia of the radius). A TEF forces air through the fistula into the gastrointestinal tract, causing gastric distension, which precipitates passage of gastric contents into the trachea and bronchi.

Case 8.8

The pharyngeal (branchial) apparatus consists of pharyngeal arches, pharyngeal pouches, pharyngeal grooves, and pharyngeal membranes. These primitive embryonic structures contribute to the formation of the head and neck. Many congenital anomalies in the neck originate during transformation of the pharyngeal apparatus into its adult derivatives. An *infected branchial sinus* (lateral cervical sinus) could produce pus that could discharge from an opening in the neck. A branchial sinus occurs when the 2nd pharyngeal arch fails to grow caudally over the 3rd and 4th arches; hence, these sinuses are usually remnants of the embryonic cervical sinus. Usually the external opening of a branchial sinus is along the anterior border of the SCM in the inferior third of the neck. If the sinus opened internally, the pus would mostly be expelled into the tonsillar cleft of the palatine fossa.

Case 8.9

T and A refers to *tonsillectomy* and *adenoidectomy.* Lymph from the palatine tonsil drains to the superior deep cervical lymph nodes, especially the *jugulodigastric node*, often referred to as the tonsillar node. The jugulodigastric lymph node lies on the IJV where the posterior belly of the digastric muscle crosses the IJV.

Although both tonsils and adenoids are often removed in the same operation, separate tonsillectomy and adenoidectomy may be indicated, especially in children younger than 4 to 5 years (Behrman et al., 1996). The physician may be reluctant to recommend a tonsillectomy because tonsils are potentially important to the normal development of the immune system. He told the mother that a T and A does not decrease the incidence of recurrent throat infections. Sometimes the large *tonsillar artery*, a branch of the facial artery, may have to be ligated when bleeding occurs after a tonsillectomy. Bleeding often results from incising veins from the pharyngeal venous plexus, especially the large external palatine vein.

Case 8.10

The *EJV* was most likely severed. This large vessel passes inferolaterally across the SCM and pierces the investing layer of deep cervical fascia just superior to the clavicle. The *transverse cervical nerve* was likely cut because it turns around the middle of the posterior border of the SCM and crosses the muscle deep to the platysma. It then divides into branches that supply the skin on the side and anterior part of the neck. Superior to the middle of the posterior border of the SCM, the *accessory nerve* crosses the posterior triangle of the neck obliquely. It supplies the SCM and trapezius. The patient had difficulty combing his hair because the knife had cut his accessory nerve, paralyzing his trapezius. To raise his hand to his head, the trapezius, assisted by the serratus anterior, must rotate the scapula so the glenoid cavity faces superolaterally. He experienced some difficulty tilting his head to the right side because of paralysis of the SCM, which is also supplied by the accessory nerve.

Case 8.11

The laryngeal cartilages, particularly the protruding thyroid cartilage, were likely fractured. The displaced laminae of the thyroid cartilage probably blocked the airway. In addition, edema of the submucosa of the larynx may have obstructed the passage of air. The laryngeal skeleton—particularly the thyroid cartilage—is more easily fractured in elderly people because of ossification of the cartilages.

To facilitate breathing, the *cricothyroid membrane* was likely incised, and a small tracheostomy tube was inserted into the trachea. When an extended period of airway support is required and rapid entry into the trachea is not necessary, a *tracheostomy* is usually performed. This procedure may be performed either superior or inferior to the isthmus of the thyroid. Division of the isthmus allows exposure of the upper trachea (i.e., between the 1st and 2nd tracheal rings). Some surgeons prefer to make a vertical incision through rings 2 to 4. In a tracheal incision that extends too far posteriorly, the posterior tracheal wall and esophagus may be damaged, especially in children. In some cases, injury to the recurrent laryngeal nerves may occur during a tracheostomy.

Case 8.12

Dysphagia, present in approximately 80% of patients with malignant tumors of the esophagus, is caused by gradual pressure from the tumor and closure of the esophageal lumen. The dysphagia is usually progressive in nature. The hard lump deep to the SCM was likely a cancerous deep cervical lymph node located near the IJV. The malignancy had metastasized through the lymphatics to this node. The lymphatic drainage of the cervical esophagus is to the paratracheal and inferior deep cervical lymph nodes. Invasion of the trachea or main bronchi is likely to occur with advanced esophageal cancer.

Case 8.13

The submandibular gland is in the posterior part of the *submandibular triangle*, which is one of the four subdivisions of the anterior cervical triangle. The parotid gland opens into the oral cavity opposite the 2nd maxillary molar tooth. Inflammation of the duct results in redness of the papilla around its opening into the mouth. The submandibular duct opens by one to three orifices on the small sublingual papilla beside the lingual frenulum. Inflammation of the parotid gland (*parotiditis*) is painful because the gland is within a strong capsule that is continuous with the investing layer of deep cervical fascia and the capsule resists enlargement of the gland. Usually the pain is greater during chewing because the parotid gland is wrapped around the posterior border of the ramus of the mandible and compressed against the mastoid process when the mouth is open. Sipping lemon juice or eating a pickle causes pain in an inflamed parotid gland because the acid stimulates secretion of saliva, which causes pain in the inflamed parotid duct.

Case 8.14

The coin would most likely lodge where the pharynx joins the superior end of the esophagus, which is at the level of the inferior border of the cricoid cartilage. If the coin passed into the thoracic esophagus, it would likely lodge in the 2nd constriction at the level of the arch of the aorta. Choking of the infant would result from compression of the larynx or trachea by the localized enlargement of the esophagus. Removal of the coin would likely be performed under direct vision with an *esophagoscope*. Sometimes a Foley catheter with a retaining balloon is passed beyond the coin. The balloon is then inflated and the coin is removed with the catheter and balloon.

Case 8.15

The fishbone was likely lodged in the piriform recess between the aryepiglottic fold and the lateral wall of the laryngopharynx. Foreign bodies such as fishbones and chicken bones often pass into these pear-shaped recesses because when food passes through the laryngopharynx during swallowing, it is forced to flow through these recesses. Sharp objects may lodge in them and pierce the floors of the recesses. The internal laryngeal nerve may be injured because both nerves run immediately deep to the mucous membrane lining the recess. This nerve supplies sensory fibers to the laryngeal mucosa superior to the vocal folds, including the superior surface of these folds. The mucous membrane of the superior part of the larynx is sensitive, and contact by a foreign object causes immediate explosive coughing to expel it. Lack of sensation in this mucosa resulting from a nerve injury could allow food to enter the larynx, causing choking.

Case 8.16

A thyroglossal duct cyst develops from a remnant of the embryonic *thyroglossal duct*, which connects the thyroid gland with the base of the tongue in the embryo (Moore and Persaud, 1998). Normally the thyroglossal duct atrophies and degenerates as the thyroid descends to its final site in the neck. Remnants of this duct may persist anywhere along the anterior midline of the neck between the foramen cecum of the tongue and the thyroid gland. These remnants may give rise to cysts in the tongue or neck, but they usually lie just inferior to the hyoid. Often the cyst is in intimate contact with the anterior part of this bone. It may be connected superiorly by a duct with the foramen cecum, inferiorly with a pyramidal lobe or the isthmus of the thyroid gland, or both. These connections explain why thyroglossal cysts move up and down during deglutition and when the tongue is protruded. Sometimes thyroid tissue is associated with a thyroglossal duct cyst and, in unusual cases, the entire thyroid gland is attached to the cyst because it failed to descend to its normal position during the embryonic period. A thyroglossal duct cyst may develop an opening onto the surface of the neck (*thyroglossal sinus*). This results from erosion of cervical tissues following infection and rupture of the cyst.

Case 8.17

All the conditions that came to the student's mind could have caused the swelling in the side of her neck. *Branchial cysts* usually derive from remnants of the *cervical sinus*, the 2nd pharyngeal groove, or the 2nd pharyngeal pouch. Although they may be associated with branchial sinuses, as in the present case, and may drain through them, these cysts often lie in the neck just inferior to the angle of the mandible. They usually develop along the anterior border of the SCM. The cyst may extend deep to this muscle and involve other structures. In the present case, the cyst was probably derived from a remnant of the embryonic cervical sinus (Moore and Persaud, 1998). The sinus tract running superiorly from it was likely derived from the 2nd pharyngeal pouch. It probably passed between the internal and external carotid arteries. Likely it terminated in or close to the *tonsillar cleft*, the adult derivative of the cavity of the 2nd pharyngeal pouch.

The surgical excision of ascending sinuses associated with branchial cysts may bruise or injure the hypoglossal nerve, causing temporary or prolonged *unilateral lingual paralysis*. This condition would be indicated by hemiatrophy of the tongue and deviation of the tongue to the paralyzed side when it was protruded, a result of the unopposed action of the tongue muscles on the other side. If the sinus tract had passed inferiorly, it probably would have opened in the inferior third of the neck, along the anterior border of the SCM. Branchial sinuses that open externally are derived from remnants of the 2nd pharyngeal groove. If the cystic swelling had been painful during the physical examination, the cyst could be one that was secondarily infected (e.g., resulting from leakage of fluid from it into the surrounding tissues). If the swelling had been firm and painful, it could be neoplastic (i.e., it could be a metastatic lesion that metastasized, or spread, from a malignant tumor in the head).

Case 8.18

The piece of steak was probably lodged in the man's *laryngeal inlet*. Choking on food is a common cause of laryngeal obstruction, particularly in children, in persons who have consumed too much alcohol, and in persons with neurological impairment. Many "restaurant deaths," thought to be caused by heart attacks, have been shown to result from choking. Persons with dentures and/or who are impaired by alcohol intake are less able to chew their food properly and to detect a bite that is too large. The mucous membrane of the superior part of the larynx is sensitive, and contact by a foreign body such as a piece of steak causes immediate explosive coughing to expel it. However, neurological or alcohol impairment may reduce or eliminate this response.

Sometimes a foreign body (such as a fishbone) enters the *piriform recess* or passes through the larynx and becomes lodged in the trachea or a main bronchus. Usually, as in the present case, the piece of steak is only partly in the larynx, but entry of air into the trachea and lungs is largely prevented. The patient would likely have died within 5 minutes, almost certainly before there was time to get him to a hospital, if the piece of steak had not been dislodged using the *Heimlich maneuver*, which re-established adequate respiration.

Had your emergency procedure not been successful, the physician would likely have first tried to get the piece of steak out of the patient's larynx with a long spoon or a slender fork. If these procedures had failed, he would likely have performed a lifesaving *emergency inferior laryngotomy* (surgical incision of the larynx). If he happened to have a large bore needle with him, he would have inserted it through the *cricothyroid ligament*. If not, he probably would have used a penknife or a steak knife to make an incision through the midline of the neck into the cricothyroid ligament (*cricothyrotomy*). The physician would probably have inserted a large plastic straw or a tube of some sort (e.g., the empty barrel of a ballpoint pen) to enable the patient to breathe while he was being taken to a hospital for removal of the piece of steak from his larynx and repair of the cervical wound.

Case 8.19

The enlarged thyroid gland producing the swelling in the neck moves up and down during swallowing because the gland attaches to the larynx by the pretracheal layer of cervical fascia. The association of hyperthyroidism with *exophthalmos* was first described by an Irish physician, Dr. R.I. Graves. For many years his name has been associated with the disease. The cause of exophthalmos is not precisely known; however, a considerable increase in the size of the extrinsic eye muscles is certainly a factor.

The surgical treatment of hyperthyroidism is to remove part of each lobe of the thyroid (*subtotal thyroidectomy*), thereby leaving less glandular tissue to secrete hormones. As the *parathyroid glands* typically lie on the posterior surface of this gland, posterior parts of the lobes are left to prevent inadvertent removal of these glands. At least one of them is essential for secretion of parathyroid hormones, which maintain the normal level of calcium in the blood and body fluids. Removal of the parathyroid glands causes the patient to develop a convulsive disorder known as *tetany*. The signs are nervousness, twitching, and spasms in the facial and limb muscles.

Thyroidectomy may cause injury to the laryngeal nerves. Near the inferior pole of the thyroid gland, the *recurrent laryngeal nerves* are intimately related to the inferior thyroid arteries. The nerves may cross anterior or posterior to this artery or between its branches before ascending in or near the groove between the trachea and esophagus. Because of the close relation between the recurrent laryngeal nerves and the inferior thyroid arteries, a risk of injuring the nerves during surgery is present. These nerves supply all muscles of the larynx except the cricothyroids. Damage to or severance of one of the nerves could seriously affect speech (e.g., hoarseness, as in the present case, or a change in the quality of the voice to a brassy sound). Some patients also have difficulty clearing their throats.

Temporary malfunction of the recurrent laryngeal nerves may also result from postoperative edema. It must be remembered that a common cause of temporary hoarseness after surgery is trauma to the mucous membrane of the larynx by the endotracheal tube that the anesthetists insert as an airway. Injury to both recurrent laryngeal nerves—an unusual occurrence—may severely impair breathing and speech because the vocal folds remain partly abducted—the position of complete paralysis of the intrinsic muscles. Thus, the rima glottidis is not fully open. If the nerves are compressed as a result of inflammation or the accumulation of fluid, the breathing and speech defects will normally disappear following healing and drainage of the operative site.

Case 8.20

Numerous aggregates of lymphoid tissue are in the pharyngeal walls. Major aggregates of this tissue are called tonsils. Unless otherwise stated, reference to the tonsils refers to the *palatine tonsils*. The other tonsils are the lingual, pharyngeal, and tubal tonsils. Collectively, all the tonsils form a *tonsillar ring* around the pharyngeal isthmus, where the oropharynx communicates with the nasopharynx. Although it is often stated that the tonsillar ring acts as a barrier to infection, its function is not clearly understood. However, it is certain that this lymphatic tissue is important in the immune reaction to infection. In the present case, the infection had spread along the auditory tube into the middle ear, producing *otitis media* and bulging of the tympanic membrane. This bulging would likely be the chief cause of the boy's earache.

The tonsils are innervated by twigs from the glossopharyngeal nerve and, because the tympanic branch of this nerve supplies the mucous membrane of the tympanic cavity, some of the pain related to the tonsillitis may have also been referred to the ear. When the opening of the auditory tube is closed—as it probably was in the present case—pressure changes in the middle ear can also cause earache.

The numerous lymphatic vessels of the tonsils penetrate the pharyngeal wall and terminate principally in the *jugulodigastric node* of the deep chain of cervical lymph nodes. Enlargement of this *tonsillar node* is commonly associated with tonsillitis. The external palatine vein is usually the chief source of hemorrhage following tonsillectomy. This important and sometimes large vein descends from the soft palate and is immediately related to the lateral surface of the tonsil, before it pierces the superior constrictor muscle of the pharynx. In cases of severe and uncontrolled bleeding (e.g., from the tonsillar branch of the facial artery), hemorrhage may be controlled by compressing or clamping the external carotid artery at its origin. This artery supplies blood to the tonsillar arteries.

References and Suggested Readings

Agur AMR: *Grant's Atlas of Anatomy,* 9th ed. Baltimore, Williams & Wilkins, 1991.

Behrman RE, Kliegman RM, Arvin AM (eds): *Nelson Textbook of Pediatrics,* 15th ed. Philadelphia, WB Saunders, 1996.

Fahn S, Bressman SB, Brin MF: Dystonia. *In* Rowland LP (ed): *Merritt's Textbook of Neurology,* 9th ed. Baltimore, Williams & Wilkins, 1995.

Ger R, Abrahams P, Olson TR: *Essentials of Clinical Anatomy,* 2nd ed. New York, The Parthenon Publishing Group, 1996.

Hallett JW Jr: The arterial system. *In* Sabiston DC Jr, Lyerly HK (eds): *Sabiston Essentials of Surgery,* 2nd ed. Philadelphia, WB Saunders, 1994.

Johnson IJM, Smith I, Akintunde MO, et al: Assessment of pre-operative investigations of thyroglossal cysts. *J R Coll Surg Edinb* 41:48, 1996.

Lang DJ, Trojaberg W, Rowland LP: Peripheral and cranial nerve lesions. *In* Rowland LP (ed): *Merritt's Textbook of Neurology,* 9th ed. Baltimore, Williams & Wilkins, 1995.

Leung AKC, Wong AL, Robson WLLM: Ectopic thyroid gland simulating a thyroglossal duct cyst: A case report. *Can J Surg* 38:87, 1995.

Mathers LH, Chase RA, Dolph J, Glasgow EF, Gosling JA: *Clinical Anatomy Principles.* St. Louis, Mosby, 1996.

Moore KL, Persaud TVN: *The Developing Human: Clinically Oriented Embryology,* 6th ed. Philadelphia, WB Saunders, 1998.

Norton JA, Wells SA Jr: The parathyroid glands. *In* Sabiston DC Jr, Lyerly HK (eds): *Sabiston Essentials of Surgery,* 2nd ed. Philadelphia, WB Saunders, 1994.

Pedley TA, Ziegler DK: Syncope and seizure. *In* Rowland LP (ed): *Merritt's Textbook of Neurology,* 9th ed. Baltimore, Williams & Wilkins, 1995.

Raffensperger JG: Congenital cysts and sinuses of the neck. *In* Raffensperger JG (ed): *Swenson's Pediatric Surgery,* 5th ed. Norwalk, Appleton & Lange, 1990.

Ruben E, Farber JL (eds): *Pathology.* Philadelphia, JB Lippincott, 1988.

Sabiston DC Jr, Lyerly H: *Sabiston Essentials of Surgery,* 2nd ed. Philadelphia, WB Saunders, 1994.

Sacco RL: Pathogenesis, classification, and epidemiology of cerebrovascular disease. *In* Rowland LP (ed): *Merritt's Textbook of Neurology,* 9th ed. Baltimore, Williams & Wilkins, 1995.

Scher RL, Richtsmeier WJ: Otolaryngology: head and neck surgery. *In* Sabiston DC Jr, Lyerly HK (eds): *Sabiston Essentials of Surgery,* 2nd ed. Philadelphia, WB Saunders, 1994.

Sinkinson CA: The continuing saga of penetrating neck injuries. *Emerg Med* 12:135, 1991.

Skandalakis JE, Skandalakis PN, Skandalakis LJ: *Surgical Anatomy and Technique. A Pocket Manual.* New York, Springer-Verlag, 1995.

Swartz MH: *Textbook of Physical Diagnosis,* 2nd ed. Philadelphia, WB Saunders, 1994.

Thompson HG, Rowland LP: Diagnosis of pain and paresthesias. *In* Rowland LP (ed): *Merritt's Textbook of Neurology,* 9th ed. Baltimore, Williams & Wilkins, 1995.

Williams PH, Bannister LH, Berry MM, Collins P, Dyson M, Dussek JE, Ferguson MWJ: *Gray's Anatomy,* 38th ed. New York, Churchill Livingstone, 1995.

Willms JL, Schneiderman H, Algranati PS: *Physical Diagnosis: Beside Evaluation of Diagnosis and Function.* Baltimore, Williams & Wilkins, 1994.

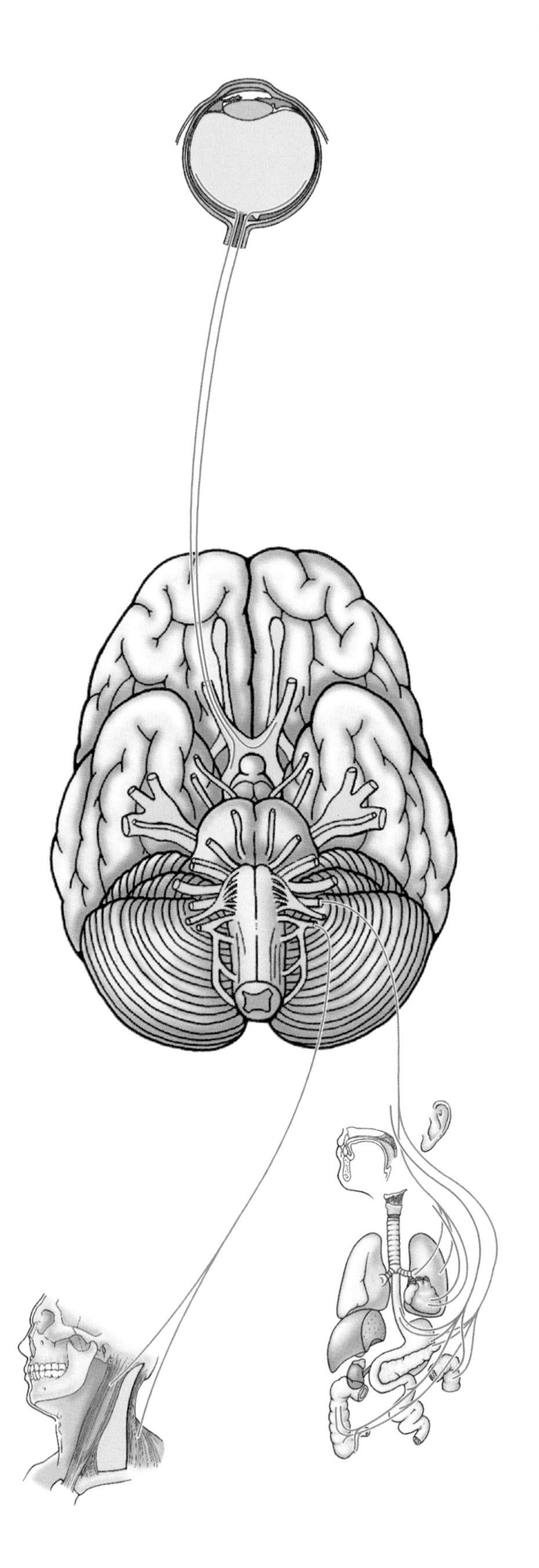

chapter

9 Summary of Cranial Nerves

The regional features of cranial nerves, especially those concerned with the head and neck, are described in preceding chapters. This chapter summarizes the cranial nerves in schematic and tabular forms. Cranial nerve lesions, illustrating important clinical features, are also described.

Overview of Cranial Nerves

Cranial nerves, like spinal nerves, contain sensory or motor fibers or a combination of these fiber types. Cranial nerves are bundles of processes from neurons that innervate muscles or glands or carry impulses from sensory areas. They were named cranial nerves because they emerge through foramina or fissures in the cranium (skull) and are covered by tubular sheaths derived from the cranial meninges. The *twelve pairs of cranial nerves* are numbered I through XII, from anterior to posterior, according to their attachments to the brain (Fig. 9.1, Table 9.1):

- The **olfactory nerve (CN I)** originates in the olfactory mucosa in the roof of the nasal cavity and along the nasal septum and medial wall of the superior concha (turbinated bone); it ends in the **olfactory bulb**, the rostral end of the **olfactory tract** that attaches to the base of the forebrain (prosencephalon).
- The **optic nerve (CN II)** originates from ganglion cells in the neural retina and extends from the posterior aspect of the eye to the **optic chiasm** (L. chiasma opticum). From the chiasm, the axons of retinal ganglion cells continue as the **optic tract,** which ends in the diencephalon—the part of the forebrain composed of the epithalamus, dorsal thalamus, subthalamus, and hypothalamus.
- The **oculomotor nerve (CN III)** and **trochlear nerve (CN IV)** attach to the midbrain (mesencephalon).
- The **trigeminal nerve (CN V)** attaches to the pons—the part of the brainstem between the medulla oblongata caudally and the midbrain rostrally. CN V emerges by two roots, sensory and motor.
- The **abducent nerve (CN VI)**, the **facial nerve (CN VII)**, and the **vestibulocochlear nerve (CN VIII)** attach to the junction between the pons and medulla.
- The **glossopharyngeal nerve (CN IX)**, the **vagus nerve (CN X)**, the cranial root of the **accessory nerve (CN XI)**, and the **hypoglossal nerve (CN XII)** attach to the medulla; however, the spinal root of CN XI arises from the superior part of the spinal cord.

Cranial nerves carry one or more of the following six functional components:

- *Somatic motor* (general somatic efferent) axons innervate the striated muscles in the orbit (e.g., ocular muscles) and tongue (L. lingua; G. glossa), which are not derived from the embryonic pharyngeal (branchial) arches.
- *Branchial motor* (special visceral efferent) axons innervate muscles (e.g., muscles of mastication, or chewing muscles), which are derived from the pharyngeal arches (face, larynx, and pharynx).
- *Visceral motor* (general visceral efferent) axons give rise to the cranial parasympathetic system that eventually innervates certain smooth muscles and glands (e.g., the sphincter pupillae and lacrimal gland).
- *Visceral sensory* (general visceral afferent) fibers convey visceral sensation from the parotid gland, carotid body and sinus, middle ear, pharynx, larynx, trachea, bronchi, lungs, heart, esophagus, stomach, and intestines as far as the left colic flexure; this sensory information does not normally reach consciousness.
- *General sensory* (general somatic afferent) fibers transmit general sensation (e.g., touch, pressure, heat, cold, etc.) from the skin and mucous membranes, mainly through CN V but also through CN VII, CN IX, and CN X; these sensations may or may not be experienced consciously.
- *Special sensory* (special visceral afferent) fibers transmit sensations of taste and smell, and special somatic afferent fibers serve the special senses of vision, hearing, and balance.

Some cranial nerves are wholly sensory, others are wholly motor, and several are mixed (Table 9.1). Four cranial nerves (CN III, CN VII, CN IX, and CN X) contain presynaptic parasympathetic axons as they emerge from the brainstem (Haines, 1997).

The fibers of cranial nerves connect centrally to **cranial nerve nuclei**—groups of neurons in which sensory or afferent fibers terminate and from which motor or efferent fibers originate (Fig. 9.2). Except for the olfactory areas of CN I, the nuclei of cranial nerves are located in the brainstem. Nuclei with general somatic and visceral motor components, and with general somatic and visceral sensory components correspond to functional columns of the spinal cord.

Cranial Nerve Injuries

Injury to the cranial nerves is a frequent complication of fracture in the base of the skull (Rowland, 1995). Furthermore, movement of the brain within the cranium may tear or bruise cranial nerve fibers, especially those of CN I. *Paralysis of cranial nerves* can usually be detected as soon as the patient's state of consciousness permits (Lange et al., 1995); however, in some patients the paralysis may not be evident for several days. Partial or complete recovery of function may occur following traumatic injuries to cranial nerves, except for CN I and CN II, which are really *tracts*—groups of nerve fibers that are extended portions of the brain. ⊙

Table 9.1. **Summary of Cranial Nerves**

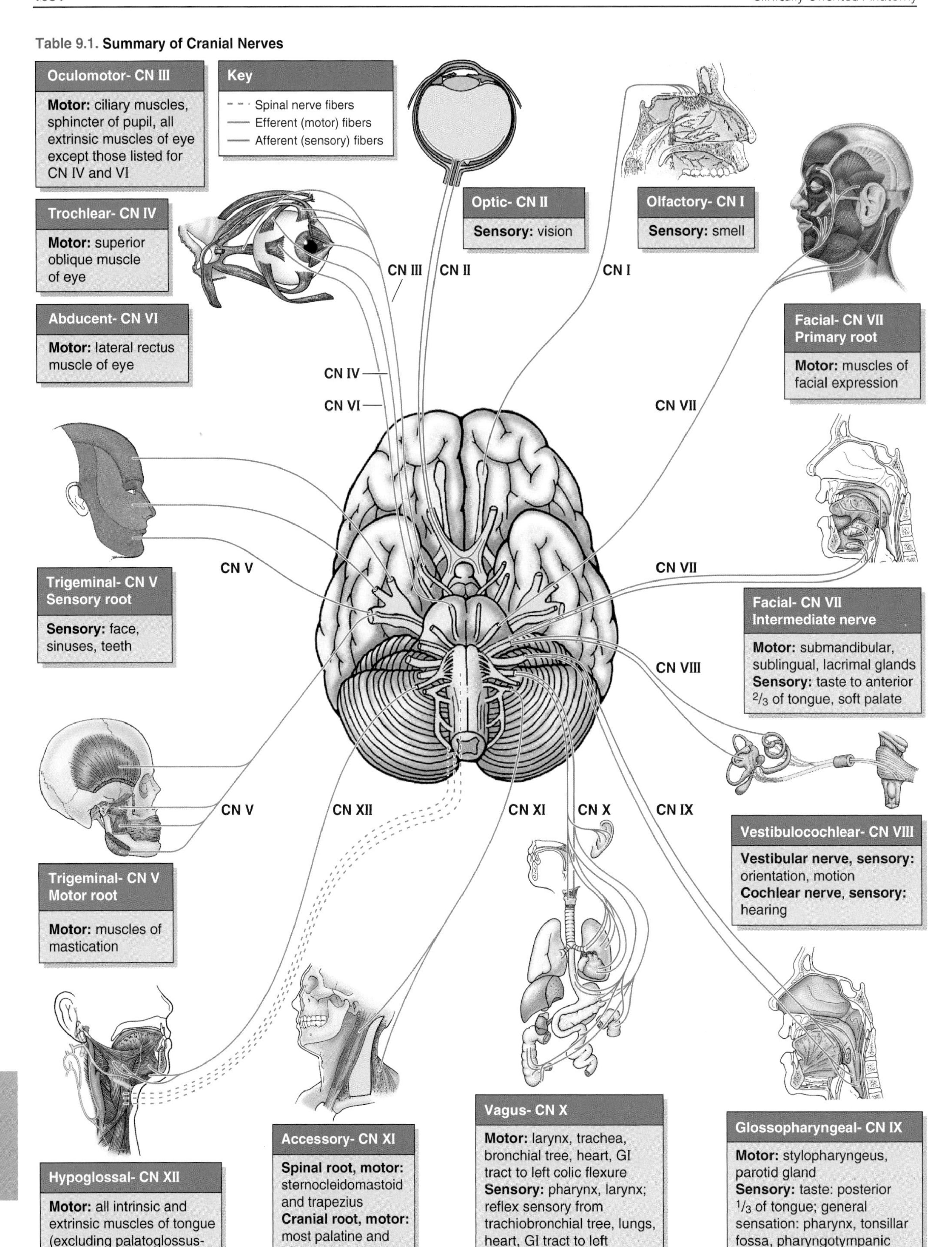

Nerve	Components	Location of Nerve Cell Bodies	Cranial Exit	Main Action
Olfactory (CN I)	Special sensory	Olfactory epithelium (olfactory cells)	Foramina in cribriform plate of ethmoid bone	Smell from nasal mucosa of roof of each nasal cavity and superior sides of nasal septum and superior concha
Optic (CN II)	Special sensory	Retina (ganglion cells)	Optic canal	Vision from retina
Oculomotor (CN III)	Somatic motor	Midbrain	Superior orbital fissure	Motor to superior, inferior, and medial rectus, inferior oblique, and levator palpebrae superioris muscles; raises upper eyelid; turns eyeball superiorly, inferiorly, and medially
	Visceral motor	Presynaptic: midbrain; postsynaptic: ciliary ganglion		Parasympathetic innervation to sphincter pupillae and ciliary muscle; constricts pupil and accommodates lens of eye
Trochlear (CN IV)	Somatic motor	Midbrain		Motor to superior oblique that assists in turning eye inferolaterally
Trigeminal (CN V)				
Ophthalmic division (CN V_1)	General sensory	Trigeminal ganglion	Superior orbital fissure	Sensation from cornea, skin of forehead, scalp, eyelids, nose, and mucosa of nasal cavity and paranasal sinuses
Maxillary division (CN V_2)	General sensory	Trigeminal ganglion	Foramen rotundum	Sensation from skin of face over maxilla including upper lip, maxillary teeth, mucosa of nose, maxillary sinuses, and palate
Mandibular division (CN V_3)	Branchial motor	Pons	Foramen ovale	Motor to muscles of mastication, mylohyoid anterior belly of digastric, tensor veli palatini, and tensor tympani
	General sensory	Trigeminal ganglion	Foramen ovale	Sensation from the skin over mandible, including lower lip and side of head, mandibular teeth, temporomandibular joint, and mucosa of mouth and anterior two-thirds of tongue
Abducent (CN VI)	Somatic motor	Pons	Superior orbital fissure	Motor to lateral rectus that turns eye laterally
Facial (CN VII)	Branchial motor	Pons	Internal acoustic meatus, facial canal, and stylomastoid foramen	Motor to muscles of facial expression and scalp; also supplies stapedius of middle ear, stylohyoid, and posterior belly of digastric
	Special sensory	Geniculate ganglion		Taste from anterior two-thirds of tongue, floor of mouth, and palate
	General sensory	Geniculate ganglion		Sensation from skin of external acoustic meatus
	Visceral motor	Presynaptic: pons; postsynaptic: pterygopalatine ganglion and submandibular ganglion		Parasympathetic innervation to submandibular and sublingual salivary glands, lacrimal gland, and glands of nose and palate
Vestibulocochlear (CN VIII)			Internal acoustic meatus	
Vestibular	Special sensory	Vestibular ganglion		Vestibular sensation from semicircular ducts, utricle, and saccule related to position and movement of head
Cochlear	Special sensory	Spiral ganglion		Hearing from spiral organ
Glossopharyngeal (CN IX)	Branchial motor	Medulla	Jugular foramen	Motor to stylopharyngeus that assists with swallowing
	Visceral motor	Presynaptic: medulla; postsynaptic: otic ganglion		Parasympathetic innervation to parotid gland
	Visceral sensory	Superior ganglion		Visceral sensation from parotoid gland, carotid body and sinus, pharynx, and middle ear
	Special sensory	Inferior ganglion		Taste from posterior third of tongue
	General sensory	Inferior ganglion		Cutaneous sensation from external ear
Vagus (CN X)	Branchial motor	Medulla		Motor to constrictor muscles of pharynx, intrinsic muscles of larynx, and muscles of palate except tensor veli palatini, and striated muscle in superior two-thirds of esophagus
	Visceral motor	Presynaptic: medulla; postsynaptic: neurons in, on, or near viscera		Parasympathetic innervation to smooth muscle of trachea, bronchi, digestive tract, and cardiac muscle of heart
	Visceral sensory	Superior ganglion		Visceral sensation from base of tongue, pharynx, larynx, trachea, bronchi, heart, esophagus, stomach, and intestine
	Special sensory	Inferior ganglion		Taste from epiglottis and palate
	General sensory	Superior ganglion		Sensation from auricle, external acoustic meatus, and dura mater of posterior cranial fossa
Accessory (CN XI)				
Cranial root	Somatic motor	Medulla		Motor to striated muscles of soft palate, pharynx via fibers that join CN X; larynx
Spinal root	Branchial motor	Spinal cord		Motor to sternocleidomastoid and trapezius
Hypoglossal (CN XII)	Somatic motor	Medulla	Hypoglossal canal	Motor to muscles of tongue (except palatoglossus)

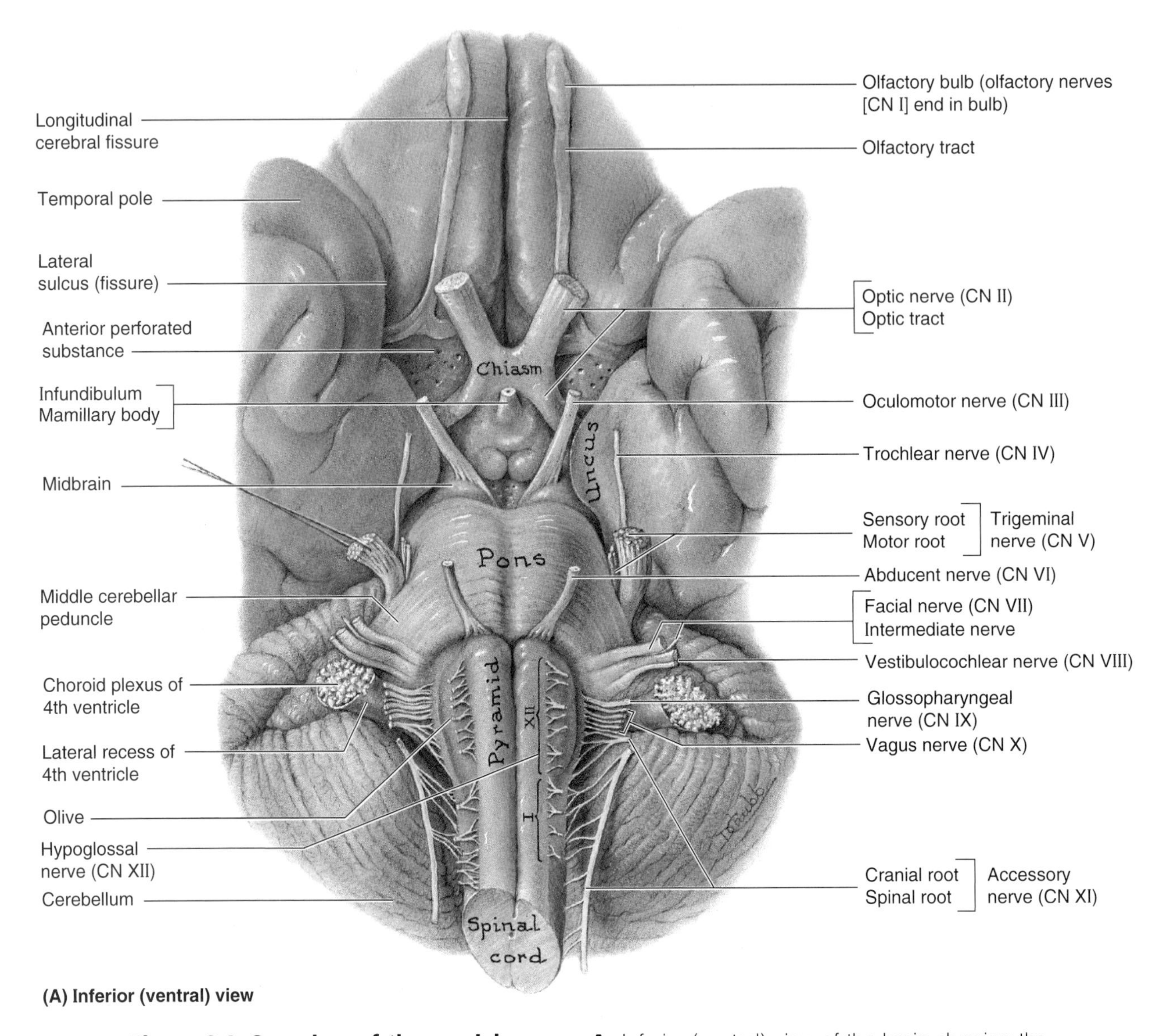

Figure 9.1. Overview of the cranial nerves. A. Inferior (ventral) view of the brain showing the cerebral hemispheres, diencephalon, and brainstem. The superficial origin of the cranial nerves, except for CN I and CN IV, is illustrated. CN I ends in the olfactory bulb, and CN IV arises from the dorsal aspect of the midbrain.

Olfactory Nerve (CN I)

Function: Special sensory (special visceral afferent)—*special sense of smell*

"*Olfaction* is the sensation of odors that results from the detection of odorous substances aerosolized in the environment" (Sweazy, 1997). The cell bodies of the *olfactory neurosensory cells* (primary olfactory neurons, or olfactory cells) are in the **olfactory epithelium** in the roof of the nasal cavity and along the nasal septum and medial wall of the superior concha (Figs. 9.1 and 9.3). The central processes of the bipolar **olfactory neurosensory cells** form approximately 20 bundles of olfactory nerve fibers that collectively form the olfactory nerve. The fibers pass through foramina in the **cribriform plate** of the ethmoid bone, pierce the dura and arachnoid of the brain, and enter the **olfactory bulb** in the anterior cranial fossa. The olfactory bulb lies in contact with the inferior or orbital surface of the frontal lobe of the cerebral hemisphere. The olfactory nerve fibers synapse with **mitral cells** in the olfactory bulb. The axons of these cells form the **olfactory tract**, which divides into lateral and medial olfactory striae (distinct fiber bands). The *lateral olfactory stria* terminates in the piriform cortex of the anterior part of the temporal lobe, and the *medial olfactory stria* projects through the anterior commissure to contralateral olfactory structures.

Olfactory stimuli arouse emotions and enable us to appreciate the aromas of food. They induce visceral responses (e.g., salivation resulting from the aroma of food) by modulating the activities of the autonomic nervous system.

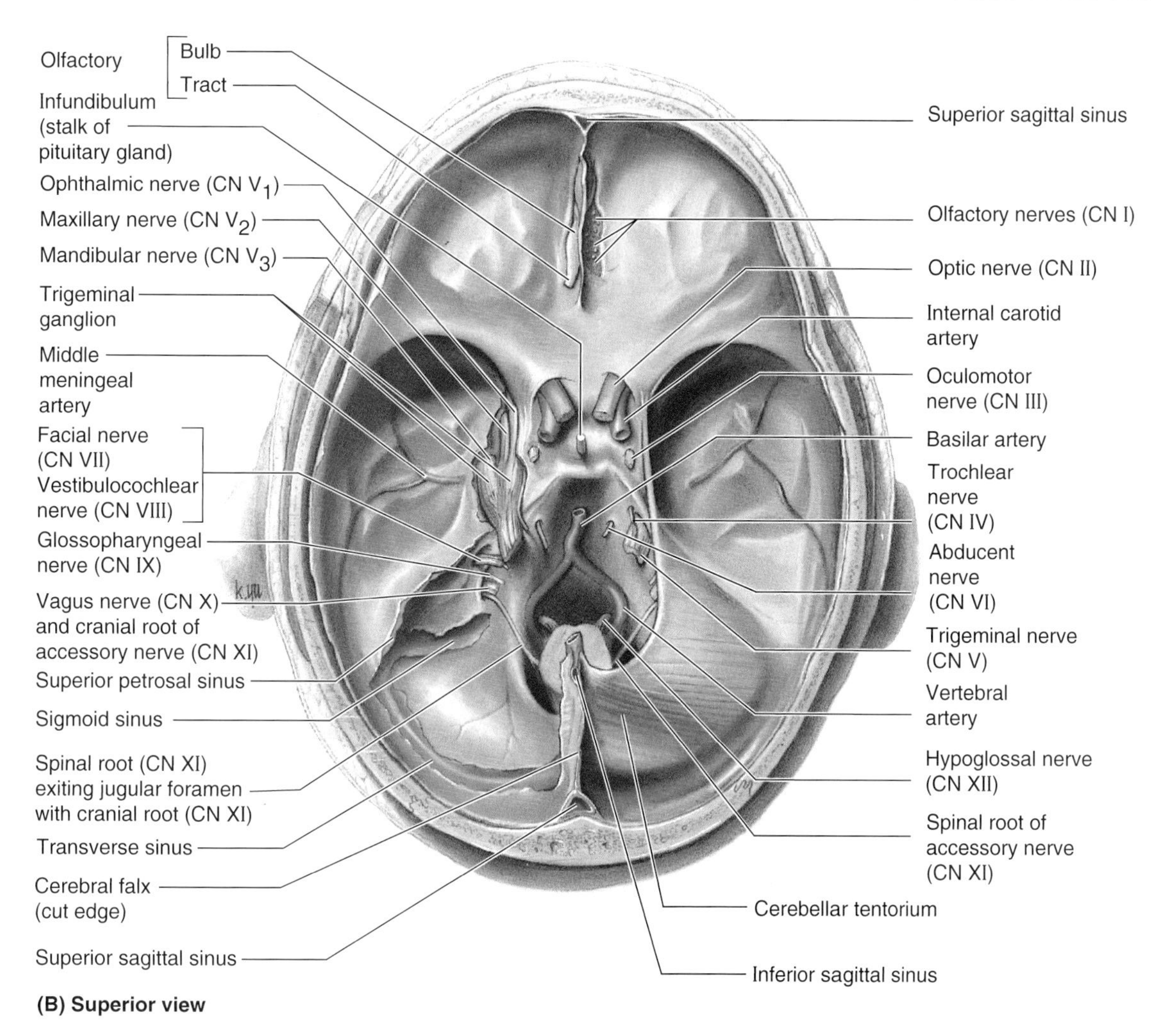

Figure 9.1. *(Continued)* **B.** Superior view of the interior of the base of the skull showing the cranial nerves, dura mater, and blood vessels. On the left side, the dura is cut away to expose the trigeminal cave (cavity) housing the trigeminal ganglion and its roots and the nerves arising from it: CN V_1, CN V_2, and CN V_3. The cerebellar tentorium (L. tentorium cerebelli) is also removed on the left side to show the nerves exiting the internal acoustic meatus (CN VII and CN VIII) and jugular foramen (CN IX, CN X, and CN XI), as well as the transverse and superior petrosal dural venous sinuses.

Anosmia—Loss of Sense of Smell

Loss of olfactory fibers usually occurs with aging. Consequently, elderly people often have a reduced acuity of the sensation of smell resulting from a progressive reduction in the number of olfactory neurosensory cells in the olfactory epithelium. The chief complaint of most people with anosmia is the loss or alteration of taste; however, clinical studies reveal that in all but a few people, the dysfunction is in the olfactory system (Sweazy, 1997). The reason for this is that most people confuse taste with flavor.

To test the sense of smell, the person is blindfolded and asked to identify common odors, such as freshly ground coffee, placed near the external nares (nostrils). One naris is occluded and the eyes are closed. Because anosmia is usually unilateral, each naris has to be tested separately. If the loss of smell is unilateral, the person may not be aware of it without clinical testing.

Transitory olfactory impairment occurs as a result of viral or allergic rhinitis—inflammation of the nasal mucous membrane. When a person complains of bilateral loss of smell, the complaint may result from loss of taste because the flavor of food depends on a normal olfactory system. Consequently, it is not uncommon for elderly people to complain that their food is tasteless. Injury to the nasal mucosa, olfactory nerve fibers, olfactory bulb, and olfactory tract may also impair smell.

In severe head injuries, the olfactory bulbs may be torn ▶

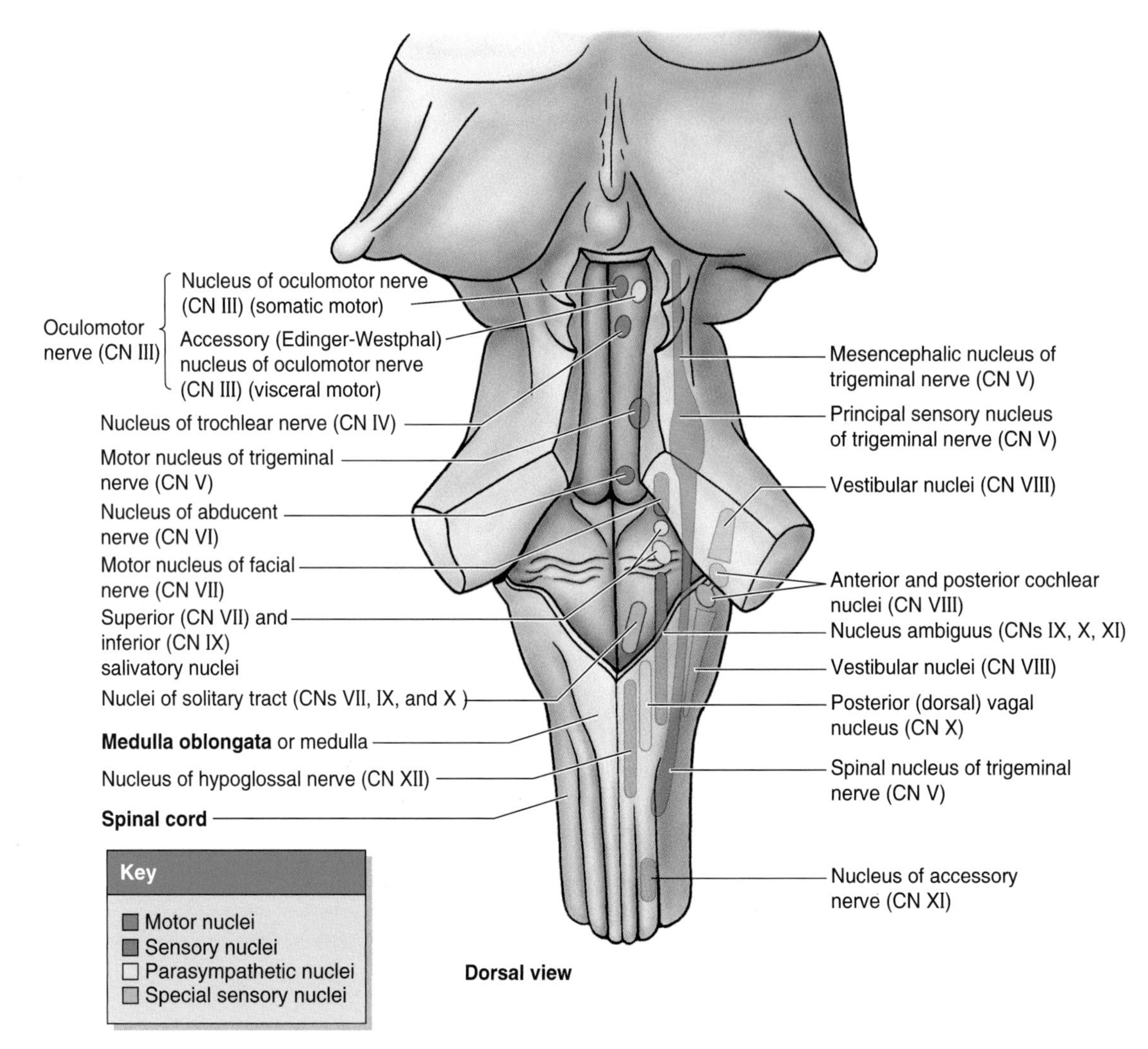

Figure 9.2. Cranial nerve nuclei (dorsal view of the brainstem). Observe the location of the motor, sensory, parasympathetic, and special sensory nuclei. Cranial nuclei are groups of nerve cell bodies in the brainstem and spinal cord that can be demarcated from neighboring cell groups on the basis of either differences in cell type or the presence of a surrounding zone of nerve fibers.

▶ away from the olfactory nerves, or some olfactory nerve fibers may be torn as they pass through a *fractured cribriform plate*. If all the nerve bundles on one side are torn, a complete loss of smell will occur on that side; consequently, *anosmia may be a clue to a fracture in the base of the cranium and cerebrospinal fluid (CSF) rhinorrhea*—leakage of CSF through the nose. A tumor and/or abscess (collection of pus) in the frontal lobe of the brain or a tumor of the meninges (*meningioma*) in the anterior cranial fossa may also cause anosmia by compressing the olfactory bulb and/or tract (Bruce and Fetell, 1995).

Olfactory Hallucinations and "Uncinate Fits"

Occasionally *olfactory hallucinations* (false perceptions of smell) may accompany lesions in the temporal lobe of the cerebral hemisphere. An irritating lesion that affects the lateral olfactory area (deep to the uncus, Fig. 9.1*A*) may cause *temporal lobe epilepsy or* "uncinate fits," which are characterized by imaginary disagreeable odors and involuntary movements of the lips and tongue. ○

Figure 9.3. Olfactory system showing olfactory structures. A. Sagittal section through the nasal cavity showing the relationship of the olfactory mucosa to the olfactory bulb. The bodies of the olfactory receptor cells are in the olfactory epithelium. **B.** The unmyelinated axons of these olfactory neurosensory cells form approximately 20 olfactory bundles on each side, which pass through openings in the cribriform plate of the ethmoid bone—collectively making up the olfactory nerve (CN I)—and end in the olfactory bulb. The axons of the principal neurons of the olfactory bulb form the olfactory tract. Most axons of the tract follow the lateral olfactory stria.

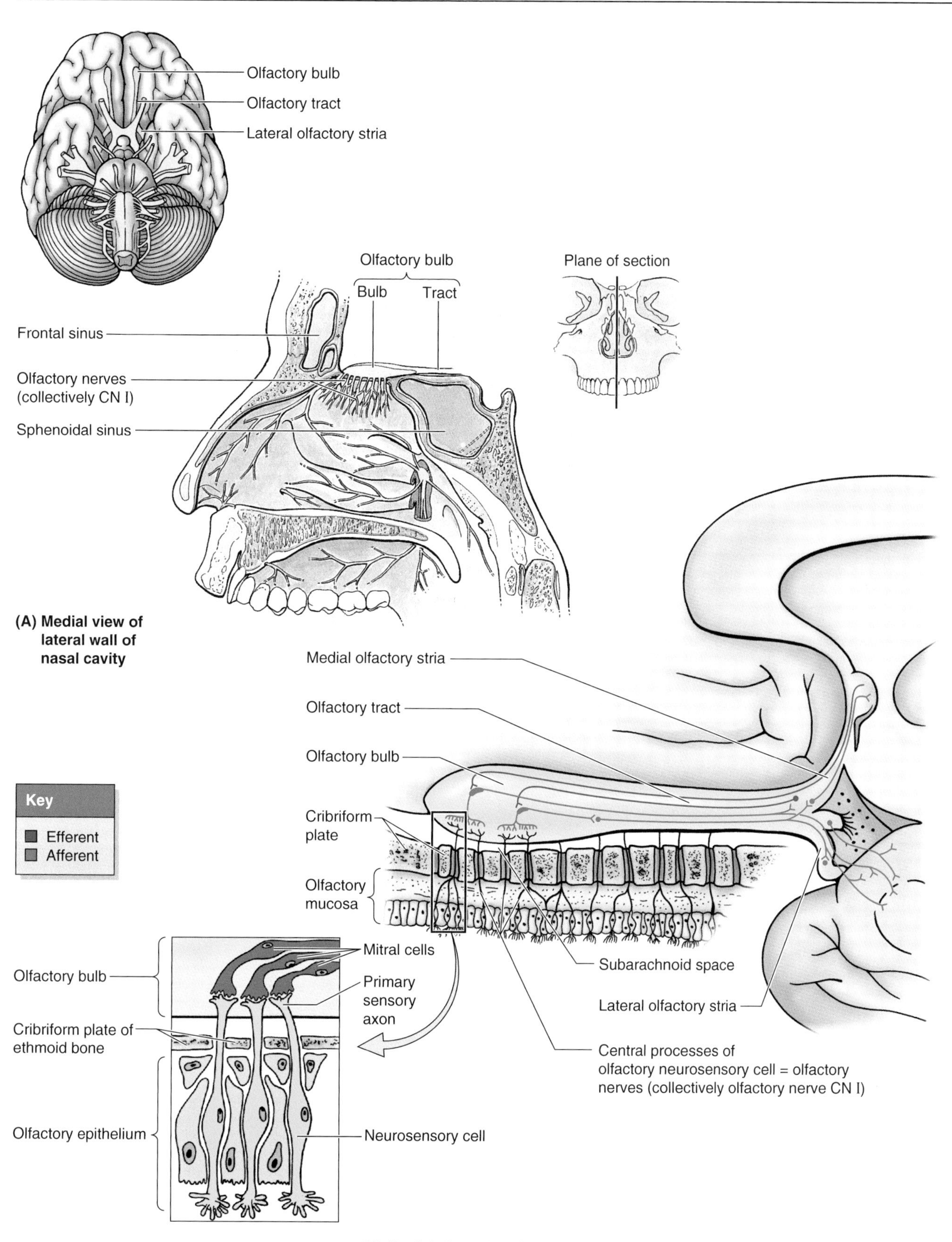

(A) Medial view of lateral wall of nasal cavity

(B) Medial view of sagittal section through cribriform plate of ethmoid bone

Optic Nerve (CN II)

Function: Special sensory (special somatic afferent)—*special sense of vision*

The optic nerve—*the nerve of sight* (Fig. 9.4)—is formed by axons of **retinal ganglion cells** (see Chapter 7). CN II is surrounded by extensions of the cranial meninges and subarachnoid (leptomeningeal) space filled with CSF. The central artery and vein of the retina traverse the meningeal layers and course in the anterior part of the optic nerve. *CN II begins where the axons of retinal ganglion cells pierce the sclera* (the opaque part of the external fibrous coat of the eyeball) deep to the optic disc. The nerve passes posteromedially in the orbit (eye socket) and exits through the *optic canal,* entering the middle cranial fossa where it forms the **optic chiasm.** Here, fibers from the nasal or medial half of each retina decussate in the chiasm and join uncrossed fibers from the temporal or lateral half of the retina to form the **optic tract.** The partial crossing of optic nerve fibers in the chiasm is a requirement for binocular vision, allowing depth of field (three-dimensional vision). Fibers from the nasal half of each retina cross to the opposite side, whereas those from the temporal half of each retina are uncrossed. Thus, fibers from the right halves of both retinas form the right optic tract, and those from the left halves form the left optic tract. The decussation of nerve fibers in the chiasm results in the right optic tract conveying impulses from the left visual field and vice versa. The **visual field** is what is seen by a person with both eyes wide open and looking straight ahead (Hutchins and Corbett, 1997). Most fibers in the optic tracts terminate in the **lateral geniculate bodies** of the thalamus. From these nuclei, axons are relayed to the *visual cortices of the occipital lobes* of the brain.

Papilledema

Papilledema is caused by a sustained increase in CSF pressure on the subarachnoid space surrounding the optic nerve, which compresses the central vein and impedes the return of venous blood from the retina. Papilledema is often the result of increased intracranial pressure. This causes a swelling of the tributaries of the retinal vein and *edema of the optic disc,* or papilla (papilledema). The characteristic swelling of the disc can be observed during ophthalmoscopy. *Papilledema is a valuable indicator of an increase in intracranial pressure* (Fetell, 1995). Some of the edema results from enlargement of axons in the optic disc, attributed to partial obstruction of axonal transport within the optic nerve fibers (Barr and Kiernan, 1993). Patients should be examined for papilledema before the performance of a puncture of the lumbar cistern; a sudden inferior release of intracranial pressure may result in herniation of brain tissue (cerebellum) through the foramen magnum into the spinal cord.

Optic Neuritis

Optic neuritis refers to lesions of the optic nerve that cause *diminution of visual acuity,* with or without changes in peripheral fields of vision (Lange et al., 1995). Optic neuritis may be caused by inflammatory, degenerative, demyelinating, or toxic disorders. The optic disc appears pale and smaller than usual on ophthalmoscopic examination. Many toxic substances (e.g., methyl and ethyl alcohol, tobacco, lead, or mercury) may also injure the optic nerve.

Visual Field Defects

In addition to undergoing routine tests for visual acuity and color perception, visual fields are tested for blindness (Hutchins and Corbett, 1997). *Visual field testing* detects lesions of the visual pathway. Patients may not be aware of changes in their visual fields until late in the course of disease because lesions in the visual pathway often develop insidiously. Visual field defects result from lesions that affect different parts of the visual pathway; the type of defect depends on where the pathway is interrupted:

- *Section of the right optic nerve* results in blindness in the temporal (*T*) and nasal (*N*) visual fields of the right eye (depicted in *black*)
- *Section of the optic chiasm* reduces peripheral vision ▶

Figure 9.4. Visual system. A. Orientation drawing showing the optic chiasm (L. chiasma opticum) in situ. The origin, course, and distribution of the visual pathway are shown. The axons of ganglion cells (retinal neurons) pass to the lateral geniculate body of the thalamus through the optic nerve (CN II) and optic tract. Fibers from the lateral geniculate body project to the visual cortices of the occipital lobes. The axons of ganglion cells of the nasal halves of the retinas cross in the optic chiasm; those from the temporal halves do not cross. **B.** The visual pathway begins with photoreceptor cells (rods and cones) in the retina. The responses of the photoreceptors are transmitted by bipolar cells to ganglion cells in the ganglion cell layer of the retina.

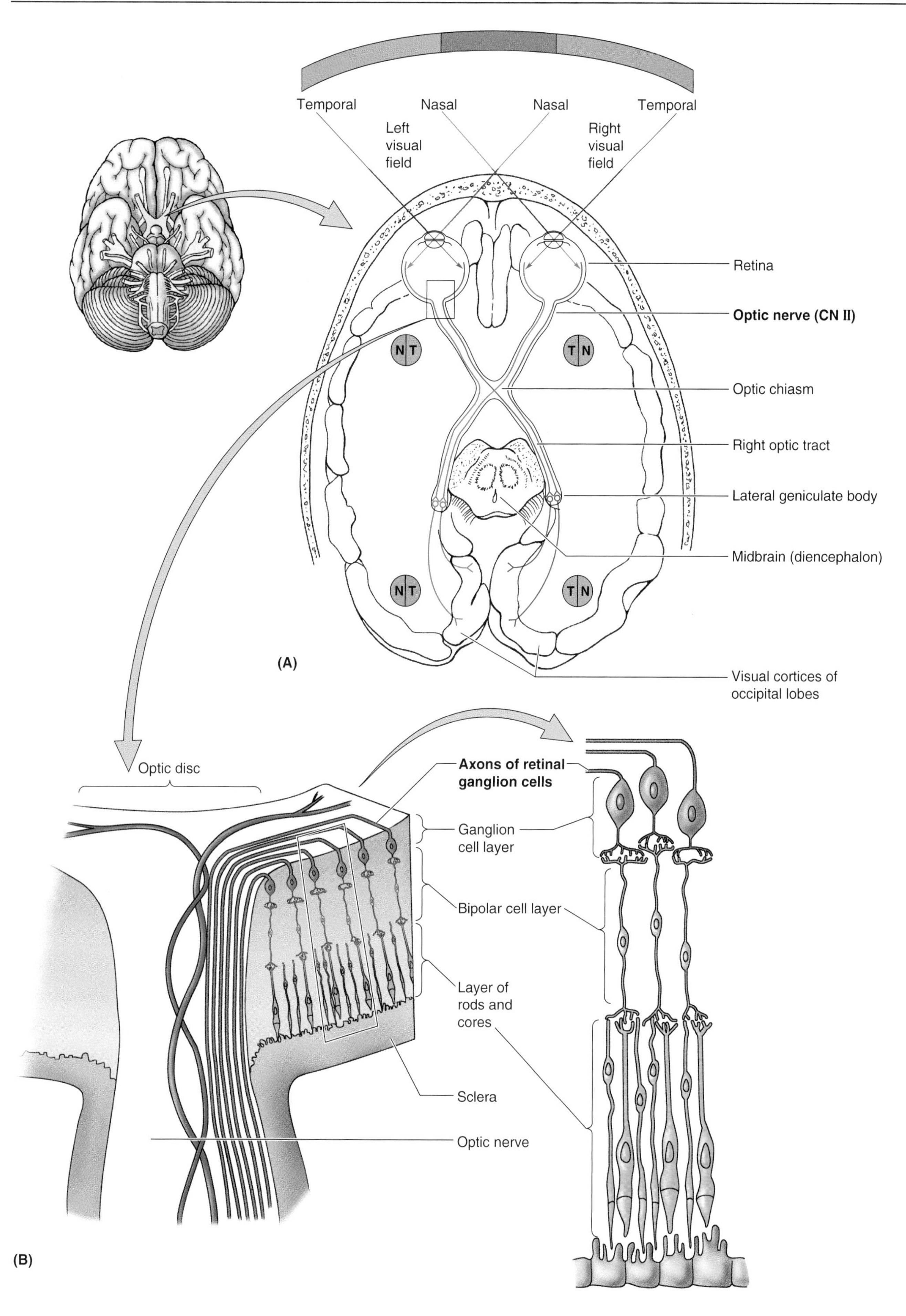
Temporal
Nasal
Nasal
Temporal
Left visual field
Right visual field
Retina
Optic nerve (CN II)
N T
T N
Optic chiasm
Right optic tract
Lateral geniculate body
Midbrain (diencephalon)
N T
T N
(A)
Visual cortices of occipital lobes
Optic disc
Axons of retinal ganglion cells
Ganglion cell layer
Bipolar cell layer
Layer of rods and cores
Sclera
Optic nerve
(B)

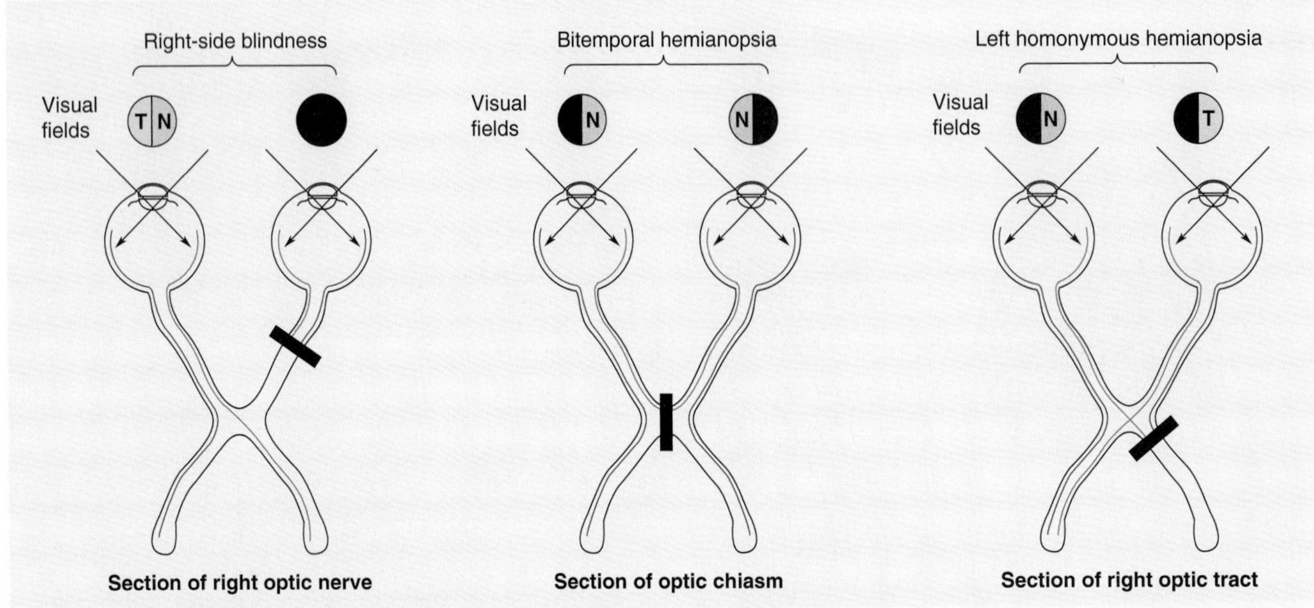

▶ that results in *bitemporal hemianopsia*—loss of vision of one-half of the visual field of both eyes

- *Section of the right optic tract* eliminates vision from the left temporal and right nasal visual fields. A lesion of the right or left optic tract causes a contralateral *homonymous hemianopsia*—indicating that the visual loss is in similar fields. This defect is the most common form of visual field loss and is often observed in patients with strokes (Swartz, 1994).

Defects of vision caused by compression of the optic chiasm may result from tumors of the pituitary gland and berry aneurysms of the internal carotid or the precommissural part of the anterior cerebral arteries (see Chapter 7). ⊙

Oculomotor Nerve (CN III)

Functions: Somatic motor (general somatic efferent) and visceral motor (general visceral efferent-parasympathetic)

Nuclei: There are two oculomotor nuclei. The *somatic motor nucleus* is in the periaqueductal gray matter of the midbrain, ventral to the cerebral aqueduct at the level of the superior colliculus (Fig. 9.2). The *visceral motor (parasympathetic) nucleus* lies dorsal to the rostral two-thirds of the somatic motor nucleus.

The oculomotor nerve is (Fig. 9.5):

- Motor to four of the six extraocular muscles (*superior, medial, and inferior rectus* and *inferior oblique*) and upper eyelid (*levator palpebrae superioris*)
- Proprioceptive to the above muscles
- Parasympathetic—through the ciliary ganglion—to the sphincter of the pupil (L. sphincter pupillae), which causes constriction of the pupil, and to the ciliary muscles of the lens, which produces accommodation of the lens.

CN III, the chief motor nerve to the ocular and extraocular muscles (Fig. 9.5), emerges from the midbrain, pierces the dura, and runs in the lateral wall of the *cavernous sinus.* CN III leaves the cranial cavity and enters the orbit through the *superior orbital fissure.* Within this fissure CN III divides into a *superior division* that supplies the superior rectus and levator palpebrae superioris, and an *inferior division* that supplies the inferior and medial rectus and inferior oblique. The inferior division also carries presynaptic autonomic fibers from the visceral nucleus (Edinger-Westphal nucleus) of CN III to the **ciliary ganglion,** where the parasympathetic fibers synapse. Postsynaptic fibers from this ganglion pass to the eyeball in the *short ciliary nerves* and supply the ciliary muscle (accommodation of lens) and sphincter pupillae (constriction of pupil).

Oculomotor Nerve Palsy

A lesion that interrupts CN III fibers causes paralysis of all extraocular muscles except the superior oblique and lateral rectus. The sphincter pupillae in the iris and the ciliary muscle in the ciliary body are also paralyzed.

Characteristic signs of a complete lesion of CN III are (Lange et al., 1995):

- *Ptosis* (drooping) of the upper eyelid, caused by paralysis of the levator palpebrae superioris
- *No pupillary (light) reflex* (constriction of pupil in response to bright light) in the affected eye
- *Dilation of the pupil,* resulting from interruption of parasympathetic fibers to the iris, leaving the dilator pupillae muscle unapposed ▶

- *Eyeball abducted* and directed slightly inferiorly ("down and out") because of unopposed actions of the lateral rectus and superior oblique
- *No accommodation of the lens* (adjustment to increased convexity for near vision), because of paralysis of the ciliary muscle.

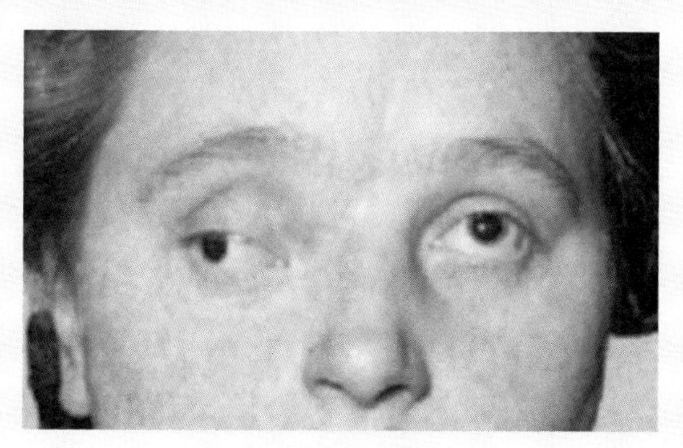

Oculomotor paralysis

Compression of CN III

Rapidly increasing intracranial pressure (e.g., resulting from an extradural hematoma) often compresses CN III against the crest of the petrous part of the temporal bone. Because autonomic fibers in CN III are superficial, they are affected first. As a result, the pupil dilates progressively on the injured side. Consequently, *the first sign of CN III compression is ipsilateral slowness of the pupillary response to light.*

Aneurysm of the Posterior Cerebral or Superior Cerebellar Artery

An aneurysm of a posterior cerebral or superior cerebellar artery may also exert pressure on CN III as it passes between these vessels. The effects this pressure produces depend on the extent of the exertion. Because CN III lies in the lateral wall of the cavernous sinus, injuries or infections may affect this sinus. ⊙

Trochlear Nerve (CN IV)

Functions: Somatic motor (general somatic efferent) to one extraocular muscle (superior oblique) and proprioceptive to it

Nucleus: The trochlear nucleus (Fig. 9.2) is located in the periaqueductal gray matter—surrounding the cerebral aqueduct—immediately caudal to the oculomotor nucleus at the level of the inferior colliculus.

CN IV emerges from the dorsal surface of the midbrain, winds around the brainstem—running the longest intracranial (subarachnoid) course of the cranial nerves—pierces the dura, and passes anteriorly in the lateral wall of the cavernous sinus (Fig. 9.5). CN IV continues past the sinus to pass through the superior orbital fissure into the orbit, where it supplies the *superior oblique.* **CN IV is the only cranial nerve to emerge dorsally from the brainstem.**

Trochlear Nerve Injury

CN IV is rarely paralyzed alone. Lesions of this nerve or its nucleus cause paralysis of the superior oblique and impairment of the ability to turn the affected eyeball inferomedially. CN IV may be torn in severe head injuries because of its long intracranial course. The characteristic sign of trochlear nerve injury is *diplopia* (double vision) when looking down (e.g., when going downstairs). Diplopia occurs because the inferior rectus normally assists the superior oblique muscle in moving the eyeball inferiorly, especially when it is in a medial position; thus, the direction of gaze is different for the two eyes when an attempt is made to look in this direction. The person can compensate for the diplopia by inclining the head anteriorly and to the side of the normal eye. ⊙

Trigeminal Nerve (CN V)

Functions: General sensory (general somatic afferent) and branchial motor (special visceral efferent)

Nuclei: There are four trigeminal nuclei: one motor and three sensory (Fig. 9.2).

- The *motor nucleus of CN V* is in the superior part of the pons, deep to the floor of the 4th ventricle.
- The *mesencephalic nucleus of CN V* is lateral to the cerebral aqueduct.
- The *principal sensory nucleus* is in the dorsolateral area of the pontine tegmentum at the level of entry of the sensory fibers.
- The *spinal nucleus of CN V* is in the inferior part of the pons and throughout the medulla.
- Central processes of neurons in the *trigeminal ganglion* (Fig. 9.6) enter the pons and terminate in the oval-shaped *principal (chief) sensory nucleus* (Fig. 9.2) and in the *spinal nucleus of CN V.*

CN V emerges from the pons by a *small motor root* and a *large sensory root* (Fig. 9.6). CN V is motor to the muscles of mastication, the mylohyoid, anterior belly of digastric, tensor veli palatini, and tensor tympani—all derivatives of the 1st pharyngeal (mandibular) arch (via the mandibular nerve [CN V_3] only)—and is the principal general sensory nerve for the head (face, teeth, mouth, nasal cavity, and dura). Fibers in the sensory root are mainly axons of neurons in the **trigeminal ganglion** (Fig. 9.6*B*). The peripheral processes of these neurons (Fig. 9.6*B*) form the ophthalmic nerve (CN V_1), the maxillary nerve (CN V_2), and the sensory component of the mandibular nerve. For a summary of CN V, see Table 9.2.

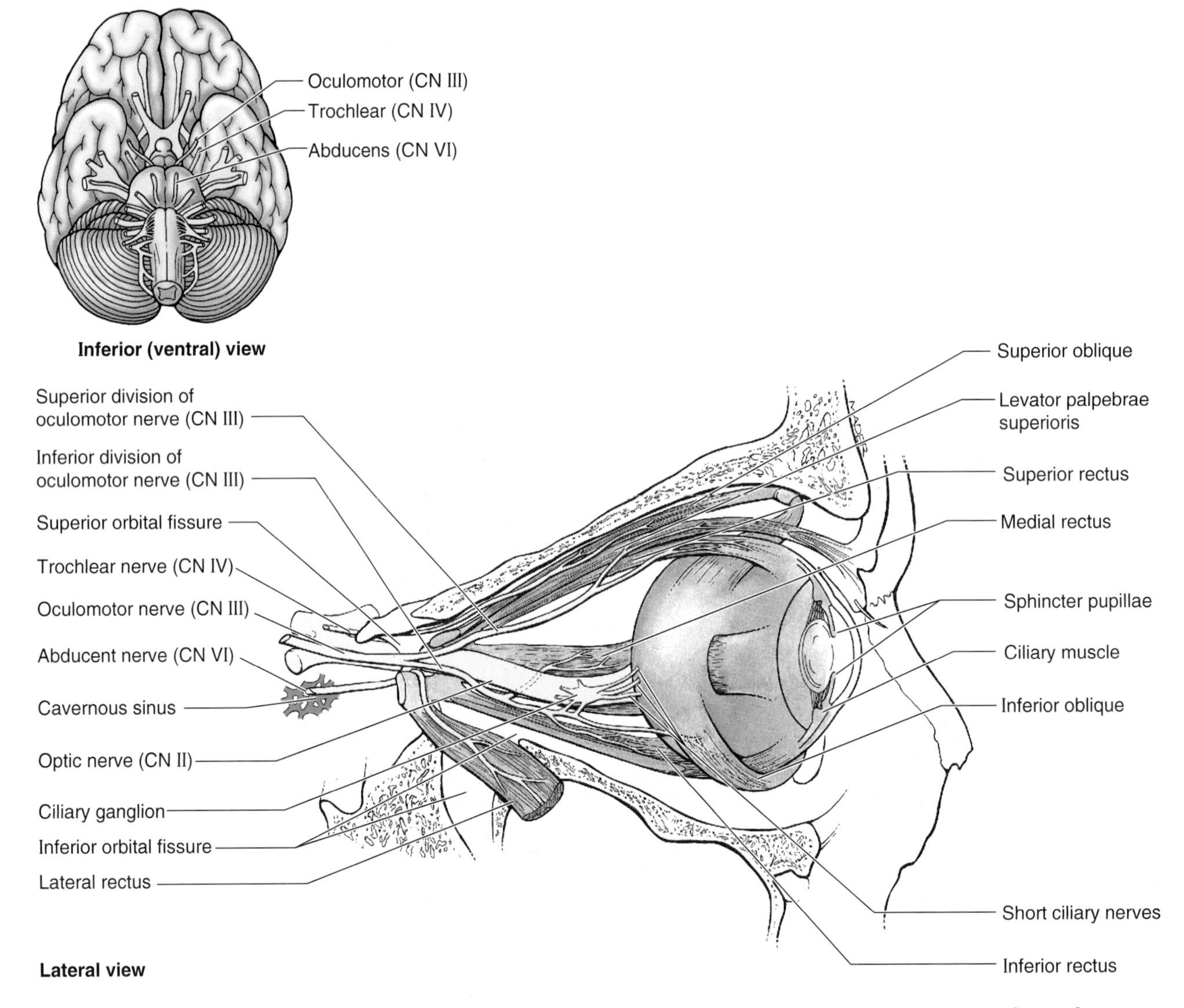

Figure 9.5. Distribution of the oculomotor (CN III), trochlear (CN IV), and abducent (CN VI) nerves. Innervation of the orbit (eye socket). CN IV supplies the superior oblique; CN VI supplies the lateral rectus; and CN III supplies the levator palpebrae superioris (superior eyelid), superior rectus, medial rectus, inferior rectus, and inferior oblique and extraocular muscles, the ciliary body, and the sphincter pupillae infraocular muscles. The orientation drawing of the inferior aspect of the brain shows the superficial origin of CN III and CN VI. CN IV arises from the dorsal aspect of the midbrain and curves around it to reach the ventral surface of the brainstem.

Figure 9.6. Distribution of the trigeminal nerve (CN V). The orientation drawing shows the trigeminal ganglion and its sensory and motor roots in situ. **A.** Distribution of the three divisions of the trigeminal nerve (CN V_1, the ophthalmic nerve; CN V_2, the maxillary nerve; and CN V_3, the mandibular nerve) to the skin of the face and scalp. All three divisions are sensory. **B.** Each division supplies the skin and subcutaneous tissue and sends a branch to the dura mater. Each division provides sensory fibers that pass through an autonomic ganglion and delivers the postsynaptic parasympathetic fibers from that ganglion: CN V_1 for the ciliary ganglion, CN V_2 for the pterygopalatine ganglion, and CN V_3 for the submandibular and otic ganglia. CN V_3 is also motor to four pairs of muscles: temporal, masseter, and the two pterygoid muscles. **C.** "Open book" illustration of the innervation of the anterior wall of the nasal cavity. Passing from the pterygopalatine ganglion, the nasopalatine nerve passes through the sphenopalatine foramen; the greater and lesser palatine nerves pass through the canals of the same name, and the pharyngeal nerve passes through the pharyngeal canal.

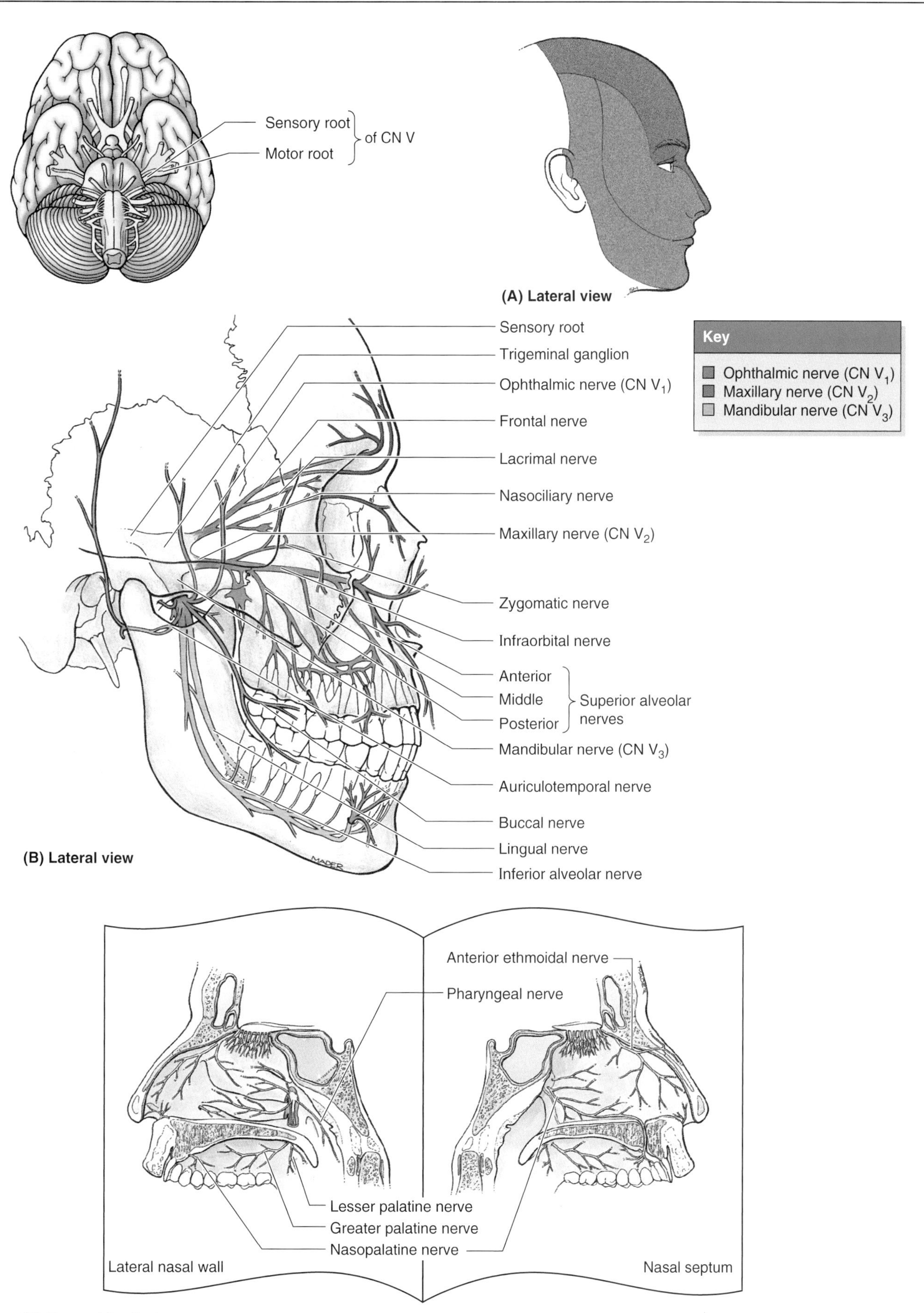
Sensory root
Motor root
of CN V
(A) Lateral view
Key
Ophthalmic nerve (CN V1)
Maxillary nerve (CN V2)
Mandibular nerve (CN V3)
Sensory root
Trigeminal ganglion
Ophthalmic nerve (CN V1)
Frontal nerve
Lacrimal nerve
Nasociliary nerve
Maxillary nerve (CN V2)
Zygomatic nerve
Infraorbital nerve
Anterior
Middle
Posterior
Superior alveolar nerves
Mandibular nerve (CN V3)
Auriculotemporal nerve
Buccal nerve
Lingual nerve
Inferior alveolar nerve
MADER
(B) Lateral view
Anterior ethmoidal nerve
Pharyngeal nerve
Lesser palatine nerve
Greater palatine nerve
Nasopalatine nerve
Lateral nasal wall
Nasal septum
(C) Opened book view

Table 9.2. Summary of Trigeminal Nerve (CN V)

Divisions	Branches
Ophthalmic nerve ($CN\ V_1$), a sensory nerve, passes through the superior orbital fissure and supplies the eyeball, conjuctiva, lacrimal gland and sac, nasal mucosa, frontal sinus, external nose, upper eyelid, forehead, and scalp	Tentorial nerve Lacrimal nerve Frontal nerve Supraorbital nerve Supratrochlear nerve Nasociliary nerve Short ciliary nerves Long ciliary nerves Infratrochlear nerve Anterior and posterior ethmoidal nerves
Maxillary nerve ($CN\ V_2$), a sensory nerve, passes through the foramen rotundum	Meningeal branch Zygomatic nerve Zygomaticofacial branch Zygomaticotemporal branch Posterior superior alveolar branches Infraorbital nerve Anterior and middle superior alveolar branches Superior labial branches Inferior palpebral branches External nasal branches Greater palatine nerves Posterior inferior lateral nasal nerves Lesser palatine nerves Posterior superior lateral nasal branches Nasopalatine nerve Pharyngeal nerve
Mandibular nerve ($CN\ V_3$), a motor and sensory nerve, passes through the foramen ovale	
General sensory branches	Meningeal branch (nervus spinosum) Buccal nerve Auriculotemporal nerve Lingual nerve Inferior alveolar nerve Nerve to mylohyoid Inferior dental plexus Mental nerve Incisive nerve
Branchial branches to muscles	Masseter Temporal Medial and lateral pterygoids Tensor veli palatini Mylohyoid Anterior belly of digastric Tensor tympani

Trigeminal Nerve Injury

CN V may be injured by trauma, tumors, aneurysms, or meningeal infections (Lange et al., 1995). Occasionally it may be involved in poliomyelitis and generalized *polyneuropathy*—a disease process involving several nerves. The sensory and motor nuclei in the pons and medulla may be destroyed by intramedullary tumors or vascular lesions. An isolated lesion of the spinal trigeminal tract also may occur with multiple sclerosis. Injury to the CN V causes:

- Paralysis of the muscles of mastication with deviation of the mandible (lower jaw) toward the side of the lesion (Table 9.3)
- Loss of the ability to appreciate soft tactile, thermal, or painful sensations in the face
- Loss of the corneal reflex (blinking in response to the cornea being touched) and the sneezing reflex.

Common causes of facial numbness are dental trauma, herpes zoster (infection caused by a herpes virus), cranial trauma, head and neck tumors, intracranial tumors, and idiopathic trigeminal neuropathy—a nerve disease of unknown cause.

Trigeminal neuralgia (tic douloureux), the principal disease affecting the sensory root of CN V, is characterized by attacks of excruciating pain in the area of distribution of the maxillary and/or mandibular divisions; the maxillary nerve is most frequently involved. The paroxysms of excruciating pain in the area of its distribution are often set off by touching an especially sensitive facial area. Usually the cause of the neuralgia—pain of a severe throbbing or stabbing character—is undetectable; however, inflammation of the petrous part of the temporal bone (osteitis) or an aberrant artery that lies close to the sensory root of CN V and compresses it is often present. ○

Abducent Nerve (CN VI)

Functions: Somatic motor (general somatic efferent) to one extraocular muscle (lateral rectus) and proprioceptive to this muscle

Nucleus: The abducent (abducens) nucleus is in the pons near the median plane (Fig. 9.2). It lies deep to the facial colliculus—a swelling in the floor of the 4th ventricle formed by fibers from the motor nucleus of CN VII looping around the abducent nucleus.

CN VI, the abducent nerve, emerges from the brainstem between the pons and medulla (Fig. 9.5) and enters the pontine cistern (see Chapter 7), where it runs alongside the basilar artery. It then pierces the dura and runs the longest course in the subarachnoid space of all the cranial nerves. It bends sharply over the crest of the petrous part of the temporal bone to enter the cavernous sinus, coursing through its venous blood-filled sinuses with the internal carotid artery. CN VI enters the orbit through the *superior orbital fissure* and runs anteriorly to supply the *lateral rectus,* which abducts the eye.

Abducent Nerve Injury

Because CN VI has a long intracranial course, it is often stretched when intracranial pressure rises, partly because ▶

▶ of the sharp bend it makes over the crest of the petrous part of the temporal bone after entering the dura. A space-occupying lesion such as a brain tumor may compress CN VI, causing paralysis. Complete paralysis of CN VI causes medial deviation of the affected eye; that is, it is fully adducted because of the unopposed action of the medial rectus, rendering the patient unable to abduct the eye. *Diplopia* is present in all ranges of movement of the eyeball, except on gazing to the side opposite the lesion. Paralysis of CN VI may also result from an *aneurysm of the cerebral arterial circle* (of Willis) at the base of the brain (see Chapter 7), from pressure from an atherosclerotic internal carotid artery in the cavernous sinus where CN VI is closely related to this artery, or from septic thrombosis of the sinus subsequent to suppuration (formation of pus) in the nasal cavities and/or paranasal sinuses. ○

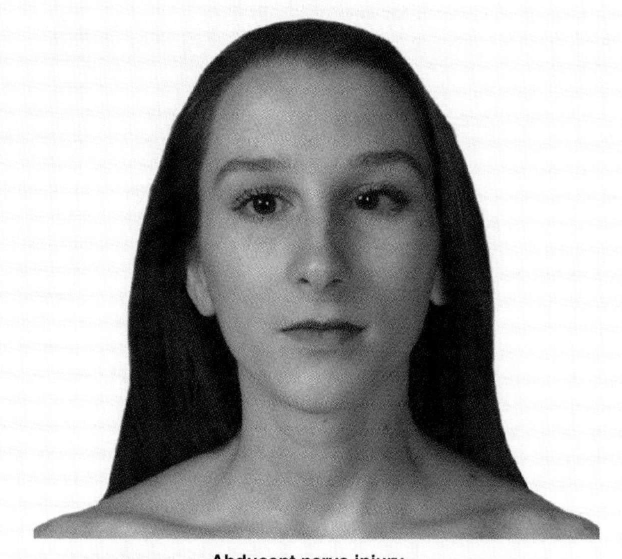

Abducent nerve injury

Facial Nerve (CN VII)

Functions: Sensory (general somatic afferent; special visceral afferent, and general visceral afferent), motor (special visceral efferent), and parasympathetic (general visceral efferent)

- Sense of taste from the anterior two-thirds of the tongue and soft palate
- Sensory from the external ear (auricular concha)
- Motor to the muscles of facial expression, fauces or throat (posterior belly of digastric, stylohyoid), and middle ear (stapedius)
- Proprioceptive to the above muscles
- Parasympathetic to the submandibular and sublingual salivary glands, lacrimal gland, and glands of the nasal cavity and palate.

Nuclei: The *motor nucleus of the facial nerve* is in the ventrolateral part of the tegmental area of the pons (Fig. 9.2). The *special sensory root* (taste) terminates in the rostral end of the *nucleus solitarius* (solitary nucleus) in the medulla. The cell bodies of primary sensory neurons are in the *geniculate ganglion* (Fig. 9.7), and their central processes end in the nucleus solitarius. General sensations (pain, touch, and thermal) from around the external ear end in the spinal nucleus of CN V.

CN VII emerges from the junction of the pons and medulla. CN VII has two divisions, the motor root and intermediate nerve (L. nervus intermedius). The larger *motor root* (facial nerve proper) innervates the muscles of facial expression, and the smaller root (*intermediate nerve*) carries taste, parasympathetic, and somatic sensory fibers (Fig. 9.7).

During its course, CN VII traverses the posterior cranial fossa, internal acoustic meatus, facial canal in the temporal bone, stylomastoid foramen, and parotid gland. At the medial wall of the tympanic cavity, the facial canal bends posteroinferiorly where the **geniculate ganglion** (sensory ganglion of CN VII) is located (Fig. 9.7). *Within the facial canal, CN VII gives rise to the*:

- Greater petrosal nerve
- Nerve to the stapedius
- Chorda tympani nerve.

After running the longest intraosseous course of any cranial nerve, CN VII emerges from the skull via the stylomastoid foramen and enters the parotid gland, forming the parotid plexus, which gives rise to the following six terminal branches:

- Posterior auricular
- Temporal
- Zygomatic
- Buccal
- Mandibular
- Cervical.

Branchial Motor

Terminal branches innervate the muscles of facial expression, occipitalis and auricular muscles, and posterior belly of digastric, stylohyoid, and stapedius muscles, all derivatives of the 2nd pharyngeal (branchial) arch.

Table 9.3. Summary of Cranial Nerve Lesions

Nerve	Type and/or Site of Lesion	Abnormal Findings
CN I	Fracture of cribiform plate	Anosmia (loss of smell); CSF rhinorrhea
CN II	Direct trauma to orbit or eyeball; fracture involving optic canal	Loss of pupillary constriction
	Pressure on optic pathway; laceration or intracerebral clot in the temporal, parietal, or occipital lobes of brain	Visual field defects
CN III	Pressure from herniating uncus on nerve; fracture involving cavernous sinus; aneurysms	Dilated pupil, ptosis, eye turns down and out; pupillary reflex on the side of the lesion will be lost
CN IV	Stretching of nerve during its course around brainstem; fracture of orbit	Inability to look down when the eye is adducted
CN V	Injury to terminal branches (particularly CN V_2) in roof of maxillary sinus; pathological processes affecting trigeminal ganglion	Loss of pain and touch sensations; paraesthesia; masseter and temporalis muscles do not contract; deviation of mandible to side of lesion when mouth is opened
CN VI	Base of brain or fracture involving cavernous sinus or orbit	Eye fails to move laterally; diplopia on lateral gaze
CN VII	Laceration or contusion in parotid region	Paralysis of facial muscles; eye remains open; angle of mouth droops; forehead does not wrinkle
	Fracture of temporal bone	As above, plus associated involvement of cochlear nerve and chorda tympani; dry cornea and loss of taste on anterior two-thirds of tongue
	Intracranial hematoma ("stroke")	Forehead wrinkles because of bilateral innervation of the frontalis muscle; otherwise paralysis of contralateral facial muscles
CN VIII	Tumor of nerve (acoustic neuroma)	Progressive unilateral hearing loss; tinnitus (noises in ear)
CN IX	Brainstem lesion or deep laceration of neck	Loss of taste on posterior one-third of tongue; loss of sensation on affected side of soft palate
CN X	Brainstem lesion or deep laceration of neck	Sagging of soft palate; deviation of uvula to normal side; hoarseness owing to paralysis of vocal fold
CN XI	Laceration of neck	Paralysis of sternocleidomastoid and superior fibers of trapezius; drooping of shoulder
CN XII	Neck laceration; basal skull fractures	Protruded tongue deviates toward affected side; moderate dysarthria (disturbance of articulation)

General Sensory

Some fibers from the geniculate ganglion supply a small area of skin around the external acoustic meatus. The parasympathetic distribution of the facial nerve is illustrated in Figure 9.8.

Postsynaptic fibers from the submandibular ganglion innervate the sublingual and submandibular salivary glands. The main features of parasympathetic ganglia associated with the facial nerve and other cranial nerves are summarized in Table 9.4. Parasympathetic fibers synapse in these ganglia, whereas sympathetic and other fibers pass through them.

Taste (Special Sensory)

Fibers of the chorda tympani join the lingual nerve to supply taste sensation from the anterior two-thirds of the tongue and soft palate.

Facial Nerve Injury

Among motor nerves, *CN VII is the most frequently paralyzed of all the cranial nerves.* Depending on the part of the nerve involved, injury to CN VII may cause paralysis of facial muscles without loss of taste on the anterior two-thirds of the tongue or altered secretion of the lacrimal and salivary glands.

A lesion of CN VII near its origin or near the geniculate ganglion is accompanied by loss of motor, gustatory ▶

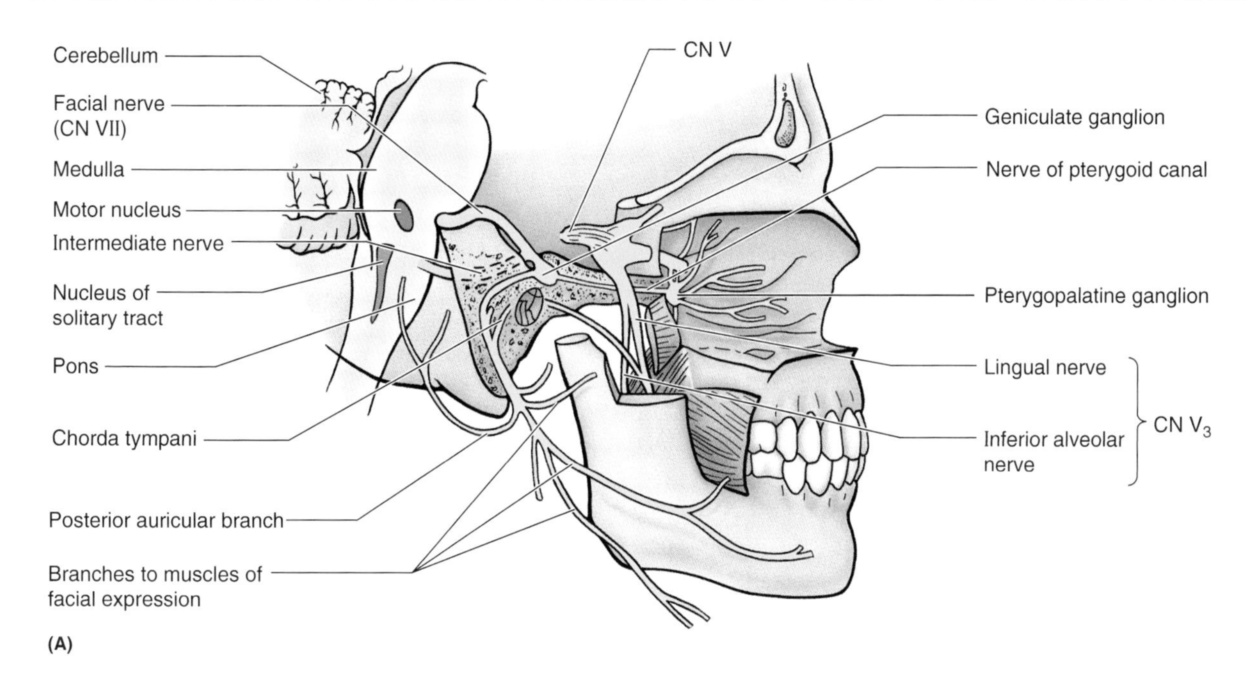

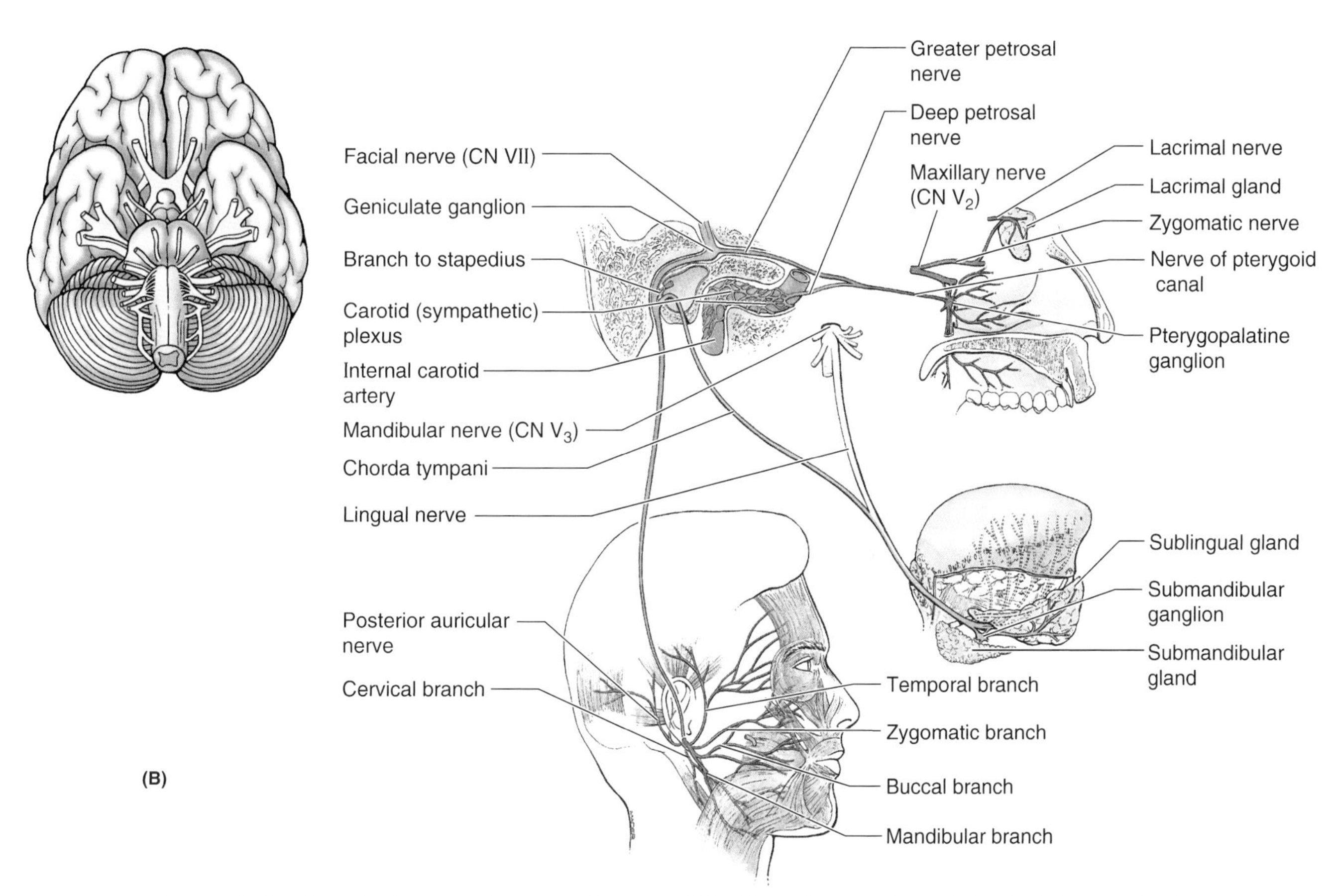

Figure 9.7. Distribution of the facial nerve (CN VII). A. Branches of the facial nerve in situ. **B.** Diagram showing the distribution of facial nerve fibers. Observe that CN VII supplies the muscles of facial expression, including the superficial muscles around the eye, nose, mouth, and ear. It also supplies the stylohyoid and the posterior belly of the digastric and the stapedius. Also observe that CN VII serves a special sense. Taste fibers with cell bodies in the geniculate ganglion pass (*a*) from the palate through the pterygopalatine ganglion—the nerve of the pterygoid canal—and the greater petrosal nerve to the geniculate ganglion, and (*b*) from the anterior two-thirds of the tongue through the chorda tympani to the facial nerve and through it to the geniculate ganglion.

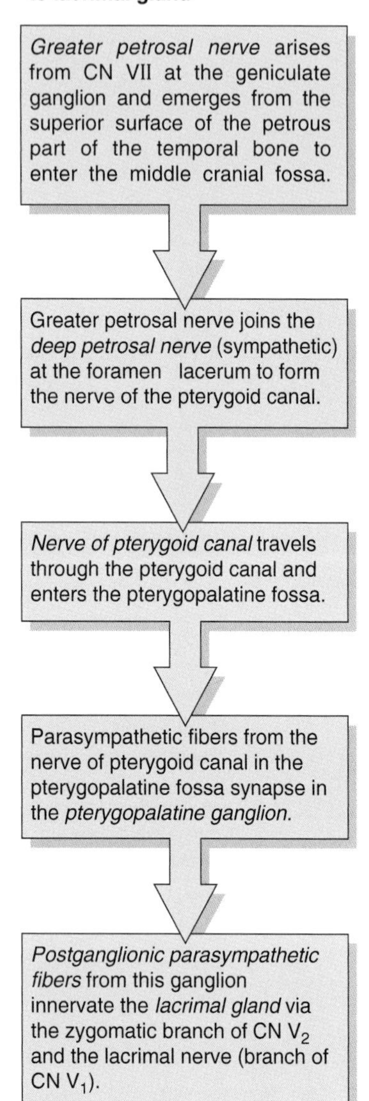

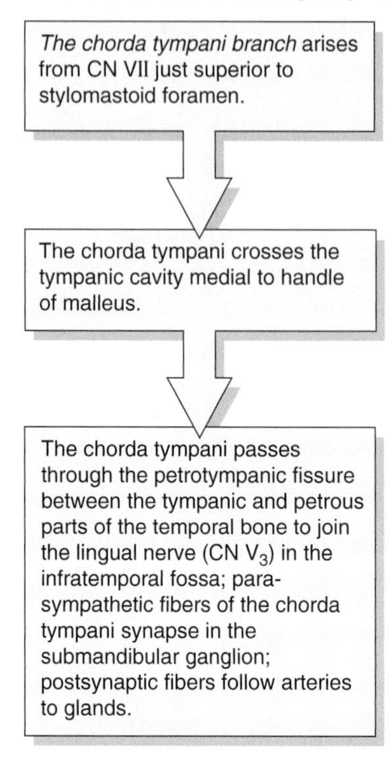

Figure 9.8. Flowchart showing the course of parasympathetic fibers in the facial nerve (CN VII). A. Parasympathetic (visceral motor) fibers to the lacrimal gland. **B.** Parasympathetic (visceral motor) fibers to the submandibular and sublingual glands.

▶ (taste), and autonomic functions. The motor paralysis of facial muscles involves upper and lower parts of the face on the ipsilateral (same) side. A *central lesion of CN VII* results in paralysis of muscles in the lower face on the contralateral (opposite) side; consequently, forehead wrinkling is not impaired. Lesions between the geniculate ganglion and the origin of the chorda tympani produce the same effects as that resulting from injury near the ganglion except that lacrimal secretion is not affected. Because it passes through the facial canal, CN VII is vulnerable to compression when a viral infection produces inflammation (*viral neuritis*) and swelling of the nerve just before it emerges from the stylomastoid foramen.

Bell's palsy (sometimes written Bell palsy) is the common disorder resulting from a CN VII lesion. The patient experiences a sudden loss of control of muscles of the entire left or right half of the face, usually without other neurological symptoms (Willms et al., 1994). This type of peripheral nerve paralysis, more fully illustrated and described on page 857, may occur without a known cause (Lange et al., 1995). Bell palsy often follows exposure to cold, as from an open window while sleeping. The peripheral part of CN VII may also be compressed by tumors of the parotid gland. The paralysis usually disappears in a few weeks; however, in severe cases, recovery (often incomplete) takes approximately 3 months.

Because the branches of CN VII are superficial, they are subject to injury from stab and gunshot wounds, cuts, and birth injury. Damage to CN VII is common with fracture of the temporal bone and is usually detectable immediately ▶

Table 9.4. Parasympathetic Ganglia Associated with CNs III, V, VII, and IX

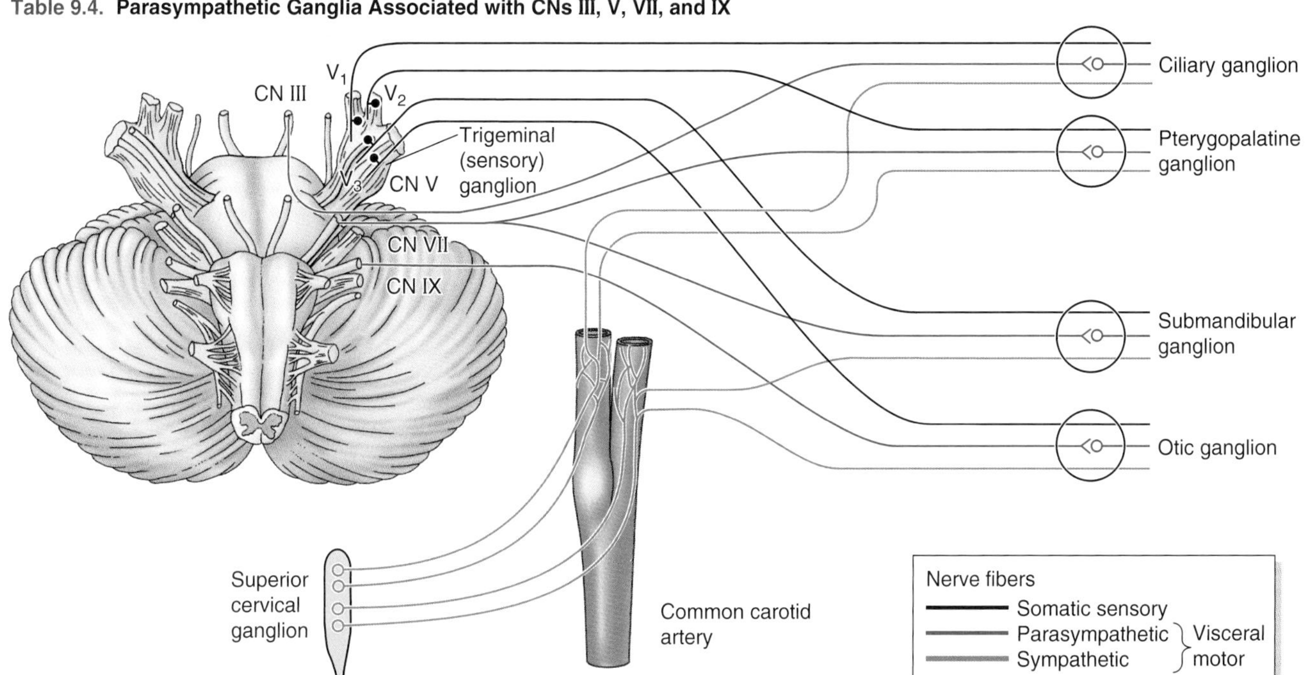

Ganglion	Location	Parasympathetic Root	Sympathetic Root	Main Distribution
Ciliary	Located between optic nerve and lateral rectus, close to apex of orbit	Inferior branch of oculomotor nerve (CN III)	Branches from internal carotid plexus in cavernous sinus	Parasympathetic postsynaptic fibers from ciliary ganglion pass to ciliary muscle and sphincter pupillae of iris; sympathetic post-ganglionic fibers from superior cervical ganglion pass to dilator pupillae and blood vessels of eye
Pterygo-palatine	Located in pterygopalatine fossa where it is suspended by pterygopalatine branches of maxillary nerve; located just anterior to opening of pterygoid canal and inferior to CN V_2	Greater petrosal nerve from facial nerve (CN VII)	Deep petrosal nerve, a branch of internal carotid plexus that is continuation of postsynaptic fibers of cervical sympathetic trunk; fibers from superior cervical ganglion pass through pterygopalatine ganglion and enter branches of CN V_2	Parasympathetic postganglionic fibers from pterygopalatine ganglion innervate lacrimal gland via zygomatic branch of CN V_2; sympathetic postsynaptic fibers from superior cervical ganglion accompany those branches of pterygopalatine nerve that are distributed to the blood vessels of the nasal cavity, palate, and superior parts of the pharynx
Otic	Located between tensor veli palatini and mandibular nerve (CN V_3); lies inferior to foramen ovale of sphenoid bone	Tympanic nerve from glosso-pharyngeal nerve (CN IX); from tympanic plexus tympanic nerve continues as lesser petrosal nerve	Fibers from superior cervical ganglion come from plexus on middle meningeal artery	Parasympathetic postsynaptic fibers from otic ganglion are distributed to parotid gland via auriculotemporal nerve (branch of CN V_3); sympathetic post-synaptic fibers from superior cervical ganglion pass to parotid gland and supply its blood vessels
Sub-mandibular	Suspended from lingual nerve by two short roots; lies on surface of hyo-glossus muscle inferior to submandibular duct	Parasympathetic fibers join facial nerve (CN VII) and leave it in its chorda tympani branch, which unites with lingual nerve	Sympathetic fibers from superior cervical ganglion come from the plexus on facial artery	Parasympathetic postsynaptic fibers from submandibular ganglion are distributed to the sub-lingual and submandibular glands; sympathetic fibers supply sub-lingual and submandibular glands and appear to be secretomotor

► after the injury. CN VII may also be affected by tumors of the brain and skull (Bruce and Fetell, 1995), aneurysms, meningeal infections, and herpesvirus. Although injuries to CN VII cause paralysis of facial muscles, sensory loss in the small area of skin on the posteromedial surface of the pinna (auricle) and around the opening of the external acoustic meatus is rare. Similarly, hearing usually is not impaired, but the ear may become more sensitive to low tones when the stapedius is paralyzed; this muscle dampens vibration of the stapes (see Chapter 7). ○

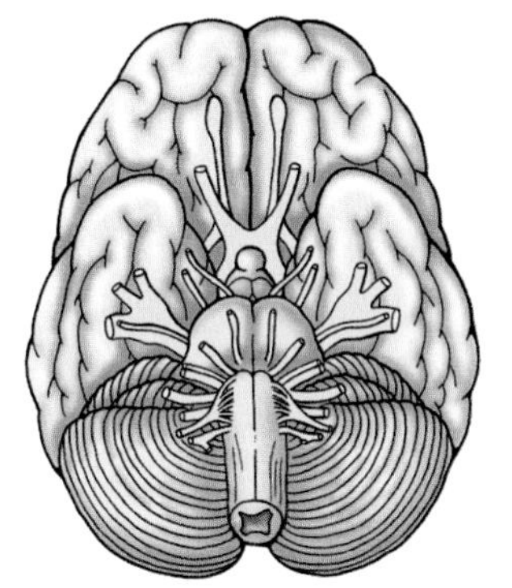

Figure 9.9. Vestibulocochlear nerve (CN VIII). A. Superior view of the interior of the base of the skull, showing the temporal bone and the location of the bony labyrinth of the internal (inner) ear. **B.** Schematic lateral view of the bony and membranous labyrinths showing that the membranous labyrinth is a closed system of ducts and chambers filled with endolymph and bathed by perilymph within the bony labyrinth. Observe the parts of the membranous labyrinth: the cochlear duct within the cochlea; the saccule and the utricle within the vestibule; and the semicircular ducts within the semicircular canals.

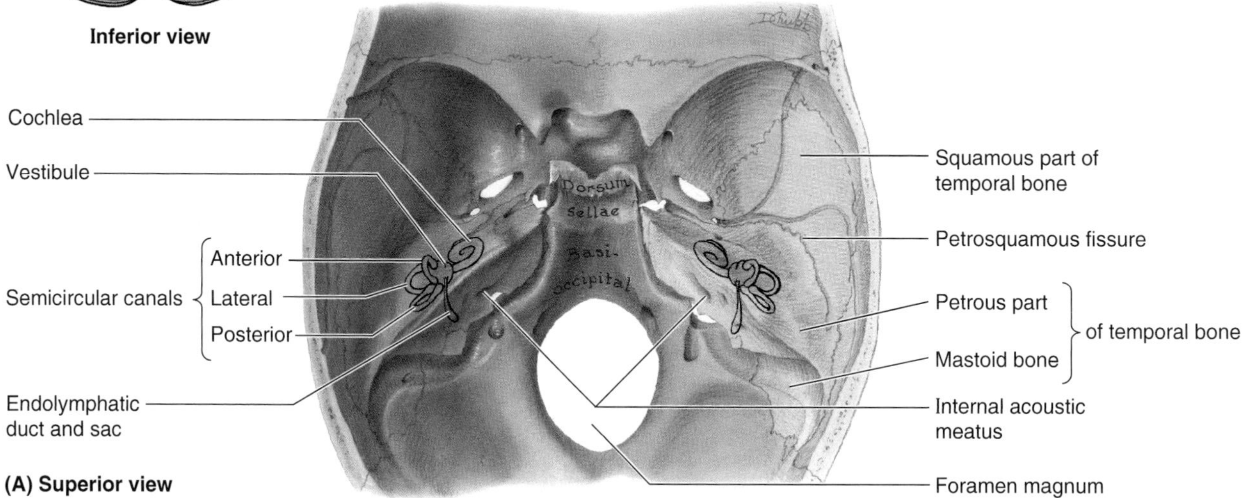

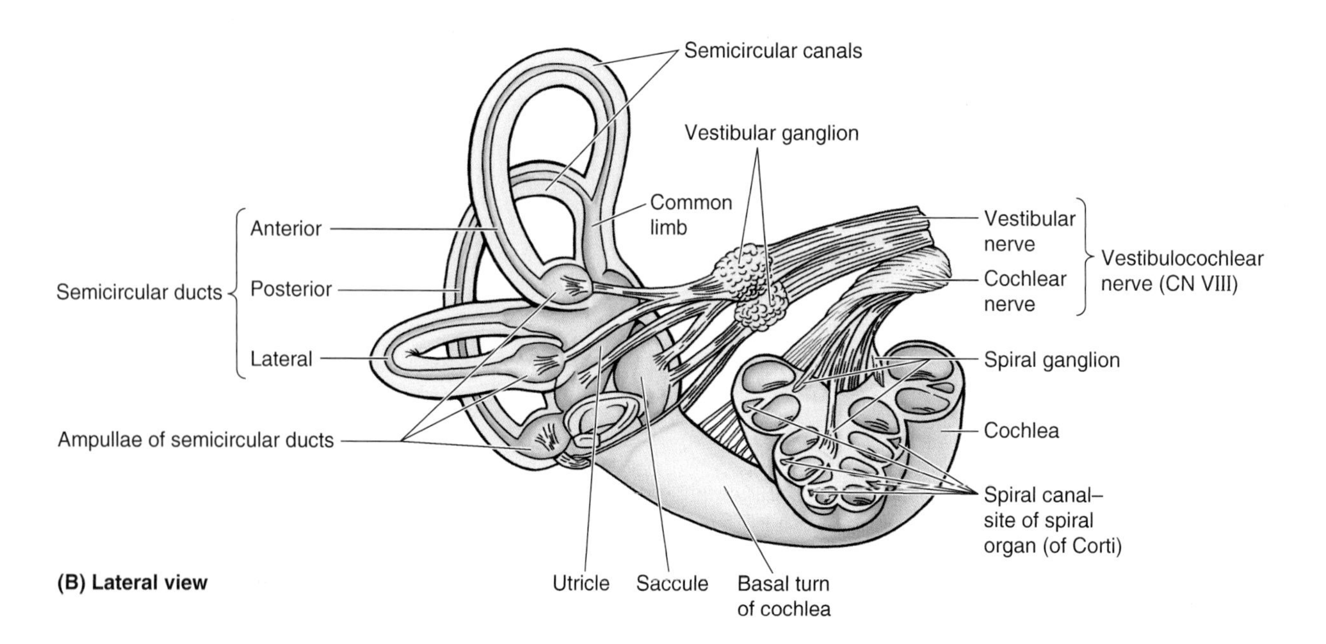

Vestibulocochlear Nerve (CN VIII)

Functions: Special sensory (special somatic afferent)—special sensations of hearing and equilibrium

Nuclei: *Four vestibular nuclei* are located at the junction of the pons and medulla in the lateral part of the floor of the 4th ventricle (Fig. 9.2); *two cochlear nuclei* are in the medulla. The dorsal and ventral nuclei are located superficially in the rostral end of the medulla, adjacent to the base of the inferior cerebellar peduncle.

CN VIII emerges from the junction of the pons and medulla and enters the **internal acoustic meatus** (Fig. 9.9). Here CN VIII separates into the vestibular and cochlear nerves.

- Vestibular fibers, concerned with equilibrium, are axons of neurons in the **vestibular ganglion**; the peripheral processes of the neurons enter the maculae of the utricle and saccule and the ampullae of the semicircular ducts.
- Cochlear fibers, concerned with hearing, are axons of neurons in the **spiral ganglion**; the peripheral processes of the neurons enter the spiral organ (of Corti). Within the internal acoustic meatus, CN VIII is accompanied by the two divisions of CN VII and the labyrinthine artery (see Chapter 7).

Vestibulocochlear Nerve Injuries

Although the vestibular and cochlear nerves are essentially independent, peripheral lesions often produce concurrent clinical effects because of their close relationship. Hence, lesions of CN VIII may cause tinnitus (ringing or buzzing in ears), vertigo (dizziness, loss of balance), and impairment or loss of hearing. Central lesions may involve either the cochlear or vestibular divisions of CN VIII.

Deafness

There are two kinds of deafness:

- Conductive deafness involving the external or middle ear (e.g., otitis media, inflammation in the middle ear)
- Sensorineural deafness is the result of disease in the cochlea or in the pathway from the cochlea to the brain.

Acoustic Neuroma

An acoustic neuroma (acoustic neurinoma, acoustic neurofibroma) is a slow-growing benign tumor of Schwann (neurolemma) cells. The tumor begins in the vestibular nerve while it is in the internal acoustic meatus, but the early symptom of an acoustic neuroma is usually loss of hearing. Dysequilibrium and tinnitus occur in approximately 70% of patients (Bruce and Fetell, 1995).

Trauma and Vertigo

Patients with head trauma often experience headache, dizziness, vertigo, and other features of postraumatic injury. *Vertigo is a hallucination of movement* involving the patient or the environment (Wazen, 1995). It often involves a spinning sensation but may be felt as a swaying back and forth or falling. These symptoms, often accompanied by nausea and vomiting, are usually related to a *peripheral vestibular nerve lesion.*

Ménière Syndrome

Ménière disease is characterized by recurrent attacks of tinnitus, hearing loss, and vertigo that are accompanied by a sense of pressure in the ear, distortion of sounds, and sensitivity to noises (Wazen, 1995). This syndrome affects people of all ages, especially those middle aged or older. Major attacks of vertigo can last for a few moments or many hours. Patients with this syndrome are usually asymptomatic between attacks. A consistent feature of the disease is *endolymphatic hydrops*—an increase in the volume of endolymph in the membranous labyrinth of the internal ear (often attributable to occlusion of the cochlear aqueduct), which results in ballooning of the cochlear duct, utricle, and saccule.

Vestibular Disease

Vestibular disease can result from the spread of an infection of the middle ear or be the result of *thrombosis in a labyrinthine artery.* A common cause of unilateral vestibular symptoms is a *transient ischemic attack* (TIA), resulting from temporary occlusion of part of the cerebral arterial circle (see Chapter 7). Usually the TIA lasts for approximately 15 minutes, but it may be followed weeks or months later by thrombosis of a cerebral artery and a stroke—an abrupt onset of focal or global neurologic symptoms caused by ischemia or hemorrhage within or around the brain resulting from diseases of the cerebral blood vessels (Sacco, 1995). ○

Glossopharyngeal Nerve (CN IX)

Functions: Sensory (general somatic afferent, special visceral afferent, general visceral afferent), motor (special visceral efferent), and parasympathetic (general visceral efferent)

- Sense of taste from the posterior third of the tongue
- Sensory from the mucosa of the pharynx, palatine tonsil, posterior third of tongue, pharyngotympanic (auditory) tube, and middle ear, and from the carotid sinus and carotid body

- Motor to the stylopharyngeus and proprioceptive to this muscle
- Parasympathetic to the parotid gland and glands in the posterior third of the tongue.

CN IX shares four nuclei in the medulla with CN X and CN XI: two motor and two sensory (Fig. 9.2).

- *Nucleus ambiguus,* deep in the superior part of the medulla
- *Inferior salivatory nucleus,* adjacent to the rostral part of the nucleus ambiguus in the inferior part of the pons
- *Nucleus of solitary tract* (L. tractus solitarius*),* lateral to the dorsal nucleus of the vagus in the superior part of the medulla
- *Spinal nucleus of the trigeminal nerve,* lateral to the nucleus ambiguus in the medulla.

CN IX emerges from the medulla and passes anterolaterally to leave the skull through the anterior aspect of the *jugular foramen* (see Figs. 9.11 and 9.12). At this foramen are superior and inferior ganglia, which contain the cell bodies for the afferent components of the nerve. CN IX follows the stylopharyngeus and passes between the superior and middle constrictor muscles of the pharynx to reach the oropharynx and tongue. It contributes to the *pharyngeal plexus* of nerves. CN IX is afferent from the tongue and pharynx (hence its name) and efferent to the stylopharyngeus and parotid gland.

Sensory (General Visceral)

The sensory branches of CN IX are (Fig. 9.11):

- *Tympanic nerve*
- *Carotid sinus nerve* to the carotid sinus and carotid body
- Nerves to the mucosa of tongue and oropharynx, palatine tonsil, soft palate, and posterior third of the tongue.

Taste (Special Sensory)

Taste fibers pass from the posterior third of the tongue.

Branchial Motor

Motor fibers pass to one muscle—the stylopharyngeus—derived from the 3rd pharyngeal (branchial) arch.

The parasympathetic distribution of CN IX is illustrated in Figure 9.10.

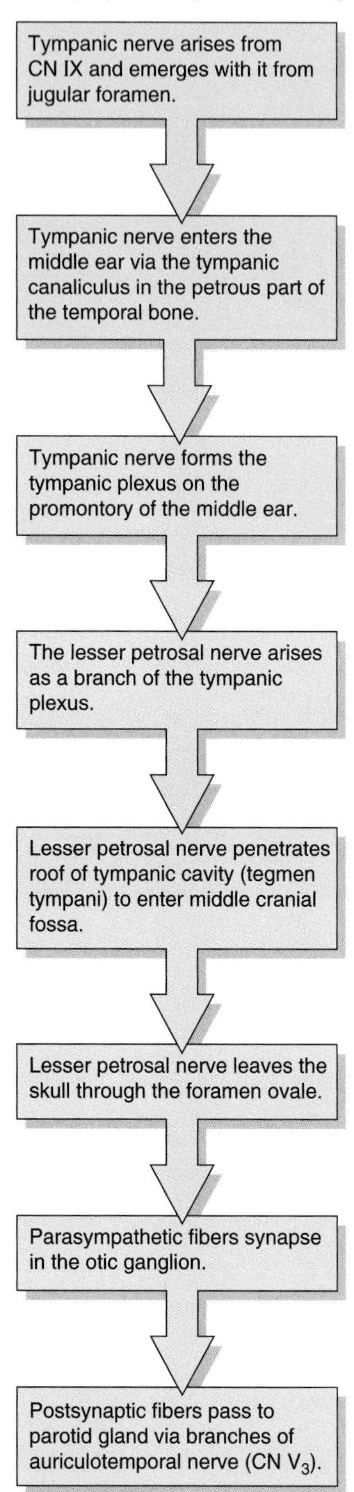

Figure 9.10. Flowchart showing the pathway followed by parasympathetic (visceral motor) fibers of the glossopharyngeal nerve (CN IX). Note that the parasympathetic component supplies presynaptic secretory fibers to the otic ganglion; postsynaptic fibers pass to the parotid gland via the auriculotemporal nerve (of CN V_3).

Lesions of the Glossopharyngeal Nerve

Isolated lesions of CN IX or its nuclei are uncommon and are not associated with perceptible disability (Lange et al., 1995). Taste is absent on the posterior third of ▶

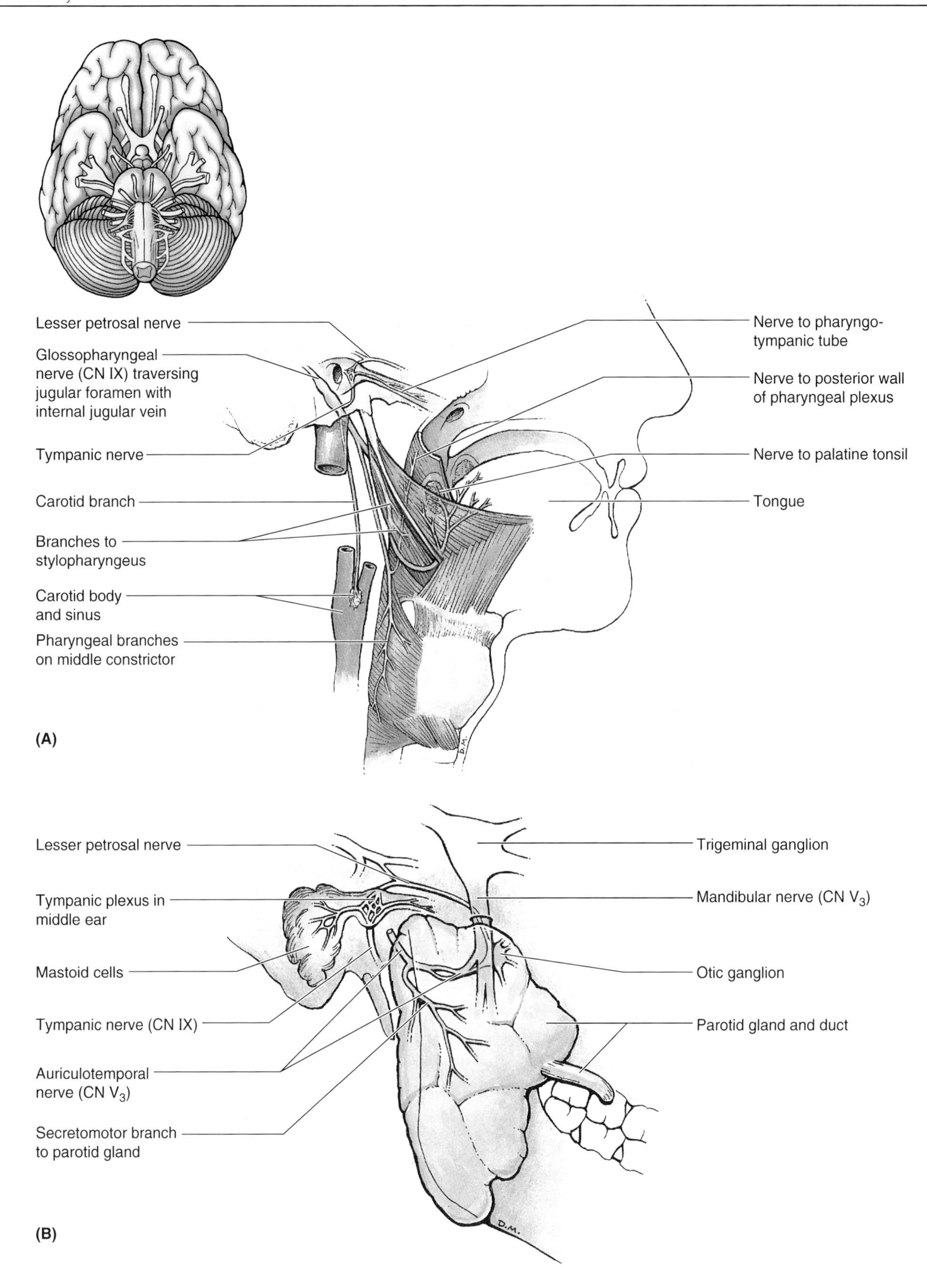

Figure 9.11. Distribution of the glossopharyngeal nerve (CN IX). A. The glossopharyngeal nerve is motor to one muscle, the stylopharyngeus. **B.** The parasympathetic component of CN IX supplies presynaptic secretory fibers to the otic ganglion; postsynaptic fibers pass to the parotid gland. The glossopharyngeal nerve also provides the special sense (taste) to the posterior third of the tongue.

▶ the tongue, and the gag reflex is absent on the side of the lesion. Injuries of CN IX resulting from infection or tumors are usually accompanied by signs of involvement of adjacent nerves. Because CN IX, CN X, and CN XI pass through the jugular foramen, tumors in this region produce multiple cranial nerve palsies—*jugular foramen syndrome*. A pain in the distribution of CN IX may be associated with injury to the nerve in the neck from a tumor.

Glossopharyngeal Neuralgia

Glossopharyngeal neuralgia (tic douloureux of CN IX) is uncommon and its cause unknown (Lange et al., 1995). The sudden intensification of pain is of a burning or stabbing nature. These paroxysms of pain are often initiated by swallowing, protruding the tongue, talking, or touching the palatine tonsil. Pain paroxysms occur during eating when trigger areas are stimulated. ⊙

Vagus Nerve (CN X)

Functions: Sensory (general somatic afferent, special visceral afferent, general visceral afferent), motor (special visceral efferent), and parasympathetic (general visceral efferent)

- Sensory from the inferior pharynx, larynx, and thoracic and abdominal organs
- Sense of taste from the root of the tongue and taste buds on the epiglottis
- Motor to the soft palate, pharynx, intrinsic laryngeal muscles (phonation), and a nominal extrinsic tongue muscle, the palatoglossus, which is actually a palatine muscle based on its derivation and innervation
- Proprioceptive to the above muscles
- Parasympathetic to thoracic and abdominal viscera.

Nuclei: CN X shares four nuclei (two motor and two sensory) with CN IX and CN XI (Fig. 9.2): *nucleus ambiguus* (motor fibers to muscles of the pharynx and larynx); *nucleus of solitary tract* (sensory fibers from thoracic and abdominal organs); *dorsal nucleus of vagus* (motor fibers that supply autonomic innervation to the heart, lungs, esophagus, and stomach); and *spinal nucleus of CN V* (sensory fibers from the oropharynx and upper gastrointestinal tract). The location of these nuclei is described with CN IX.

CN X arises by a series of rootlets from the medulla and leaves the skull through the *jugular foramen* in company with CN IX and CN XI (Figs. 9.12 and 9.13). CN X has a *superior ganglion* in the jugular foramen that is mainly concerned with the general sensory component of the nerve. Inferior to the foramen is an *inferior ganglion* concerned with the visceral sensory components of the nerve. In the region of the superior ganglion are connections with CN IX, CN XI, and the superior cervical ganglion. CN X continues inferiorly in the

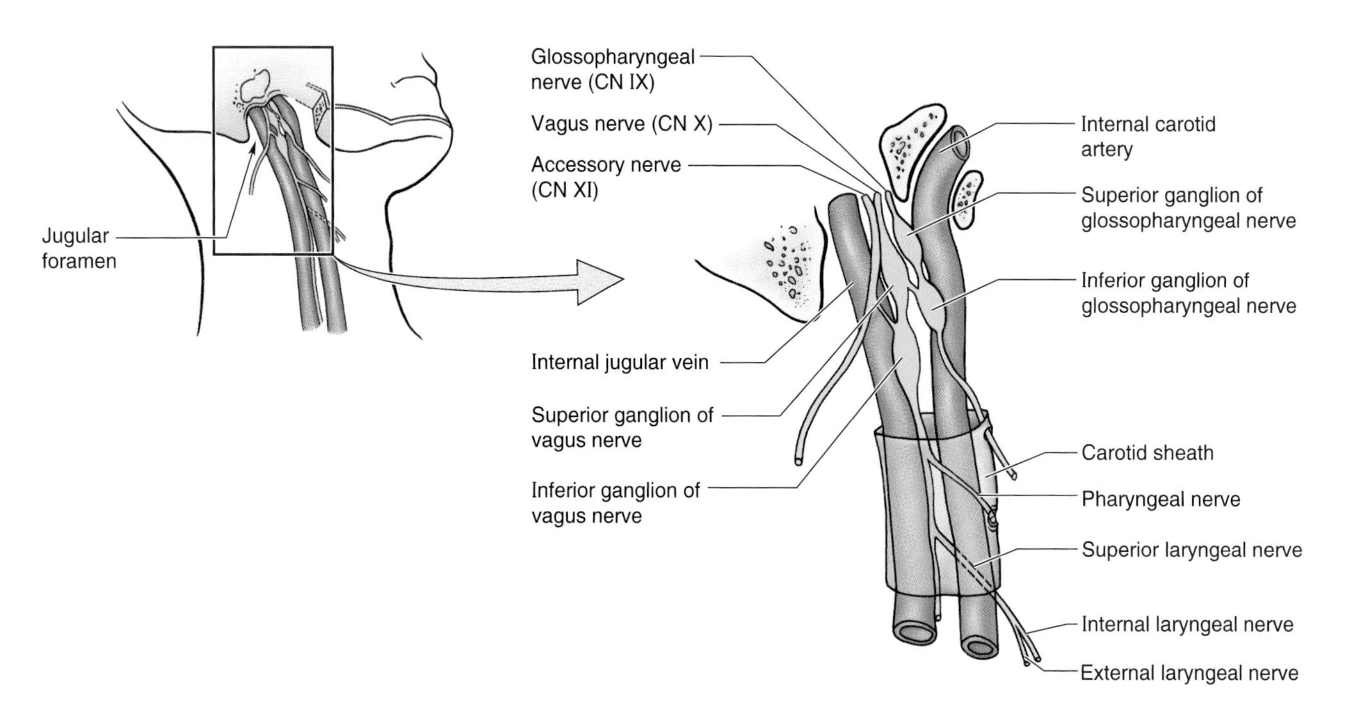

Figure 9.12. Relationship of the internal jugular vein and CN IX, CN X, and CN XI as they emerge from the jugular foramen. Observe also the relationship of these cranial nerves to the internal carotid artery and the carotid sheath.

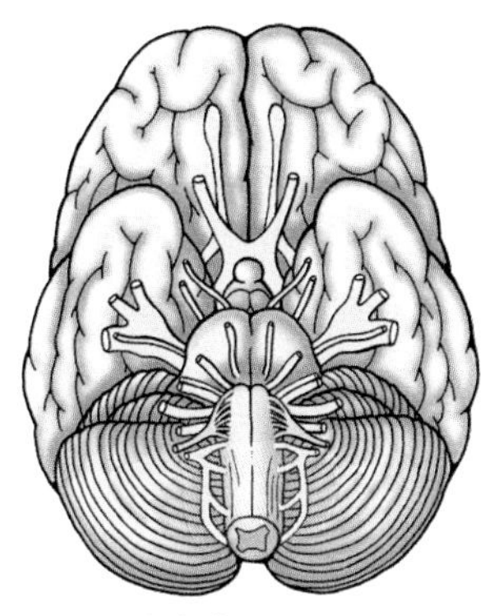

Figure 9.13. Distribution of the vagus nerve (CN X). Inferior view of the brain showing that CN X arises by a series of rootlets from the medulla. It leaves the skull through the jugular foramen in company with CN IX and CN XI (see Fig. 9.12). Observe the course of the vagus nerve in the thorax, supplying branches to the heart, bronchi, and lungs. Note that the left recurrent laryngeal nerve ascends to the larynx. In the abdomen, the anterior and posterior vagal trunks break up into branches that supply the esophagus, stomach, and intestinal tract as far as the left colic (splenic) flexure.

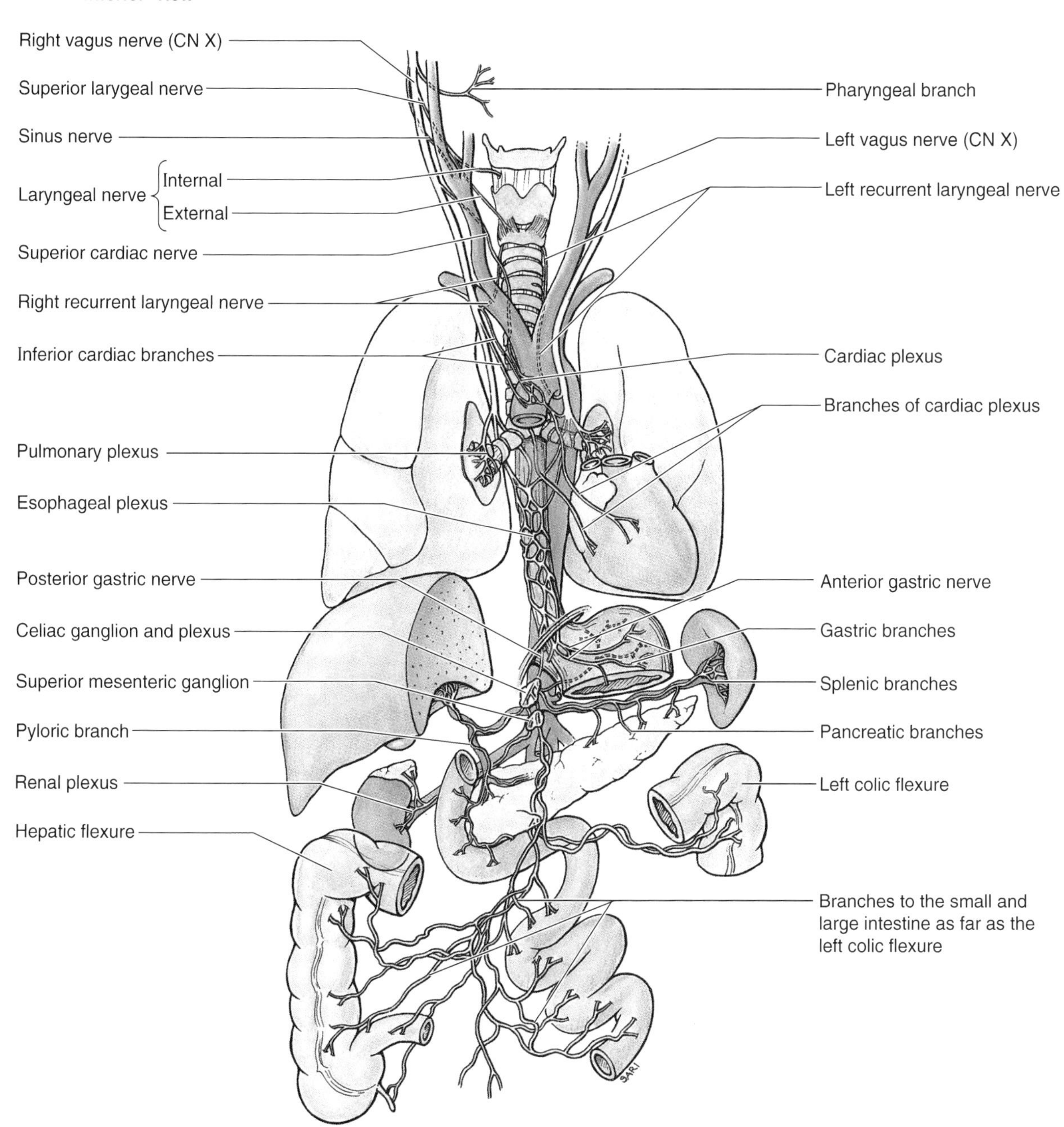

carotid sheath to the root of the neck (see Chapter 8). The course of CN X in the thorax differs on its two sides (p. 149, Fig. 1.58*B*). CN X supplies branches to the heart, bronchi, and lungs (Fig. 9.13, Table 9.5). The vagi join the *esophageal plexus* surrounding the esophagus, which is formed by branches of the vagi and sympathetic trunks. This plexus follows the esophagus through the diaphragm into the abdomen, where the anterior and posterior vagal trunks break up into branches that innervate the esophagus, stomach, and intestinal tract as far as the left colic flexure.

Table 9.5. Summary of Vagus Nerve

Divisions	Branches
Arises by a series of rootlets from medulla	
Leaves skull through jugular foramen	Receives cranial root of accessory nerve (CN XI) Meningeal branch to dura mater Auricular nerve
Enters carotid sheath and continues to root of neck	Pharyngeal nerves Superior laryngeal nerves Right recurrent laryngeal nerve Cardiac nerves
Passes through superior thoracic aperture into thorax	Left recurrent laryngeal nerve Cardiac nerves Pulmonary branches to bronchi and lungs Esophageal nerves
Passes through esophageal hiatus in diaphragm and enters abdomen	Esophageal branches Gastric branches Pancreatic branches Branches to gallbladder Branches to intestine as far as left colic flexure

Lesions of the Vagus Nerve

Isolated lesions of CN X are uncommon. *Injury to pharyngeal branches of CN X* results in *dysphagia* (difficulty in swallowing). *Lesions of the superior laryngeal nerve* produce anesthesia of the superior part of the larynx and paralysis of the cricothyroid muscle (see Chapter 8). The voice is weak and tires easily.

Injury of a recurrent laryngeal nerve may be caused by aneurysms of the arch of the aorta and may occur during neck operations. Recurrent nerve injury causes hoarseness and *dysphonia* (difficulty in speaking) because of paralysis of the vocal folds (cords). Paralysis of both recurrent laryngeal nerves causes *aphonia* (loss of voice) and *inspiratory stridor* (a harsh, high-pitched respiratory sound). Paralysis of recurrent laryngeal nerves usually results from cancer of the larynx and thyroid gland and/or from injury during surgery on the thyroid gland, neck, esophagus, heart, and lungs. Because of its longer course, lesions of the left recurrent laryngeal nerve are more common than those of the right. Proximal lesions of CN X also affect the pharyngeal and superior laryngeal nerves, causing difficulty in swallowing and speaking. ⊙

Accessory Nerve (CN XI)

Functions: Motor to the soft palate and pharynx (cranial root), and the sternocleidomastoid (SCM) and trapezius (spinal root).

Nuclei: Two motor nuclei are associated with the *accessory nerve*. The cranial root arises from neurons in the caudal part of the *nucleus ambiguus* in the medulla (Fig. 9.2), and the spinal root arises from the *spinal nucleus*, a column of anterior horn cells in the superior five or six cervical segments of the spinal cord.

CN XI has cranial and spinal roots; they are united for only a short distance (Fig. 9.14). The *cranial root* arises by a series of rootlets from the medulla, and the *spinal root* emerges as a series of rootlets from the first five cervical segments of the spinal cord. The cranial and spinal roots join as they pass through the *jugular foramen* and then separate. The *cranial root of CN XI* joins the vagus, and its fibers are distributed by vagal branches to striated muscle of the soft palate, pharynx, larynx, and esophagus. *The accessory nerve* descends along the internal carotid artery, penetrates and innervates the SCM, and emerges from it at its posterior border near its middle. It crosses the posterior triangle of the neck and innervates the superior part of the trapezius. The spinal root of the accessory nerve provides somatic motor fibers to the SCM and trapezius. Branches of the cervical plexus conveying sensory fibers from spinal nerves C2 through C4 join the accessory nerve in the posterior triangle of the neck, providing these muscles with pain and proprioceptive fibers.

Injury to the Spinal Root of CN XI

Because of its passage through the posterior cervical triangle, the spinal root of CN XI is susceptible to injury during surgical procedures such as lymph node biopsy, cannulation of the internal jugular vein, and carotid endarterectomy (see Chapter 8). Lesions of CN XI produce weakness and atrophy of the trapezius and impairment of rotary movements of the neck and chin to the opposite side as a result of weakness of the SCM. Weakness of shrugging movements of the shoulder and winging of the scapula (L. shoulder blade) while the upper limbs are at the side are the result of weakness of the trapezius. The scapular winging becomes worse on abduction of the shoulder (Lange et al., 1995). Conse- ▶

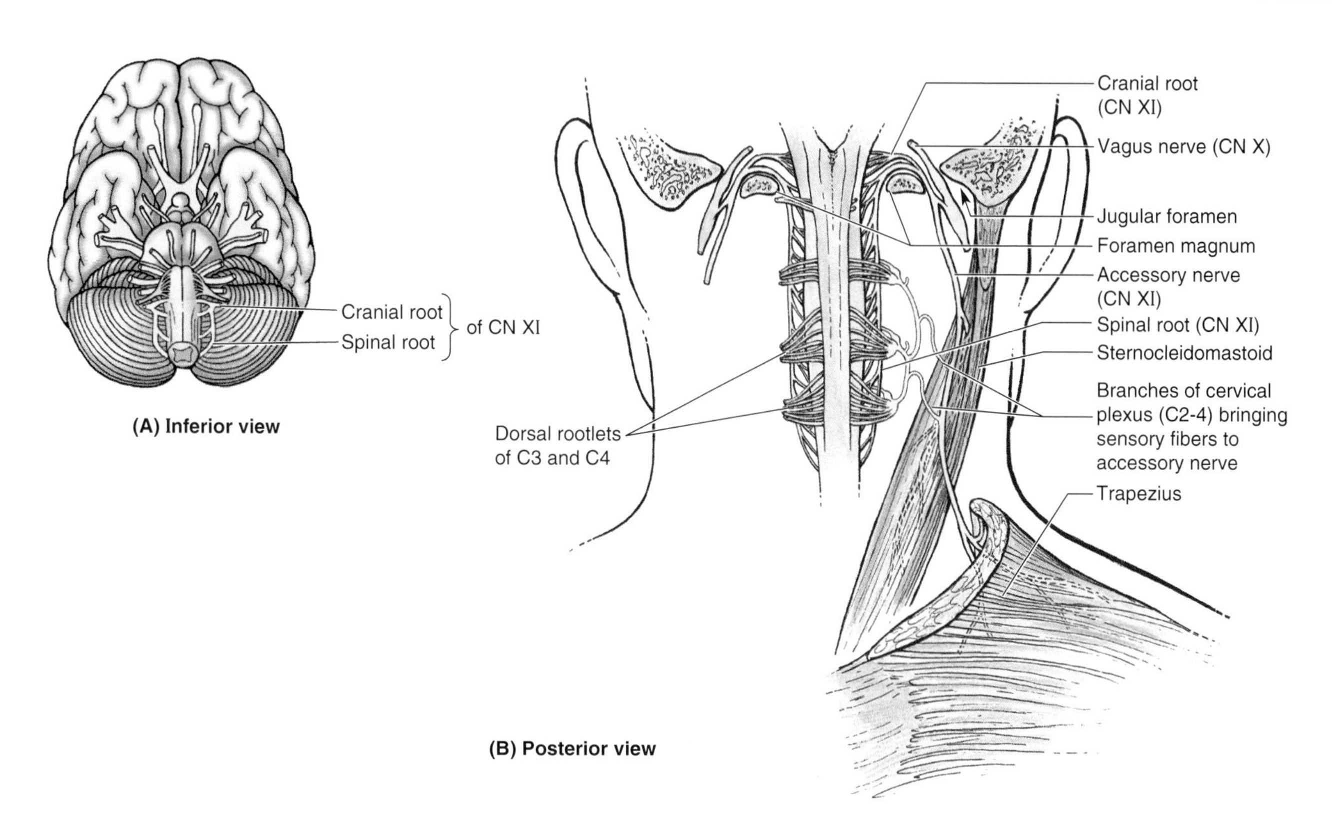

Figure 9.14. Distribution of the accessory nerve (CN XI). A. Inferior view of the brain and spinal cord showing the superficial attachment of CN XI. It has cranial and spinal roots; the cranial root arises by four or five rootlets from the lateral part of the medulla, and the spinal root arises from the cervical region of the spinal cord. **B.** The cranial and spinal roots join as they pass through the jugular foramen and then separate. The cranial root joins the vagus nerve (CN X) and its fibers are distributed by the vagal branches. The spinal root descends to supply the sternocleidomastoid (SCM) and trapezius. The spinal root is purely motor as it emerges from the jugular foramen. In the neck, it is joined by branches from the cervical plexus (spinal nerves C2 through C4), which provide afferent fibers for pain and proprioception.

▶ quently, scapular winging from paralysis of the trapezius can be distinguished from that resulting from injury to the long thoracic nerve and paralysis of the serratus anterior, in which the winging is negligible when the upper limbs are at one's side. ⊙

Hypoglossal Nerve (CN XII)

Functions: Motor (general somatic efferent) to the intrinsic and extrinsic muscles of the tongue (styloglossus, hyoglossus, genioglossus). It also conveys general somatic motor fibers from spinal nerves C1 and C2 to the hyoid muscles (thyrohyoid and geniohyoid), and general sensory (proprioceptive) fibers to these muscles and to the dura of the posterior cranial fossa.

CN XII arises as a purely motor nerve by several rootlets from the medulla and leaves the skull through the hypoglossal canal. After emerging from the canal, the nerve is joined by a branch at the cervical plexus conveying motor fibers from C1 and C2 spinal nerves, and by sensory fibers from the spinal ganglion of C2 spinal nerve. CN XII passes inferiorly medial to the angle of the mandible and then curves anteriorly to enter the tongue (Fig. 9.15). CN XII ends in many branches that supply all the extrinsic muscles of the tongue, except the palatoglossus (which is actually a palatine muscle). The nerve is joined in the hypoglossal canal by the superior division of C1 nerve. *CN XII has the following branches*:

- A meningeal branch returns to the skull through the hypoglossal canal and innervates the dura on the floor and posterior wall of the posterior cranial fossa. The nerve fibers conveyed are from the sensory spinal ganglion of spinal nerve C2, not from hypoglossal fibers.
- A descending branch joins the ansa cervicalis to supply the infrahyoid muscles (sternohyoid, sternothyroid, omohyoid, and thyrohyoid). This branch actually conveys fibers from the cervical plexus (the loop between the ventral rami of C1 and C2) to the muscles rather than to the hypoglossal fibers. The cervical spinal nerve fibers join the hypoglossal nerve proximally.
- Terminal branches to the styloglossus, hyoglossus, genioglossus muscles, and intrinsic muscles of the tongue.

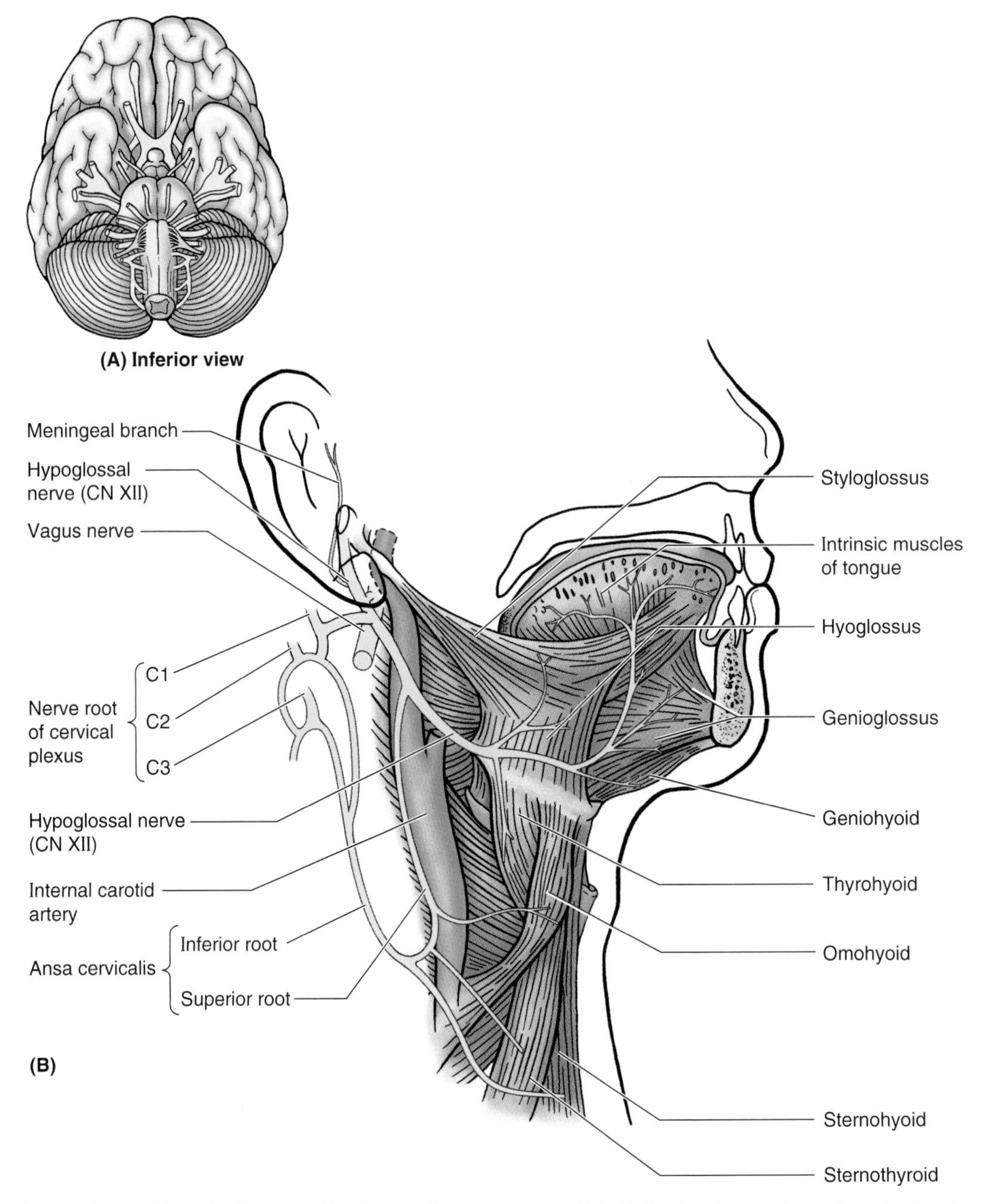

Figure 9.15. Distribution of the hypoglossal nerve (CN XII). A. Inferior view of the brain showing CN XII arising by several rootlets from the medulla. **B.** CN XII leaves the skull through the hypoglossal canal and passes inferolaterally to the angle of the mandible. From here it curves anteriorly to enter the tongue, where it supplies all intrinsic muscles except the palatoglossus. CN XII is joined immediately distal to the hypoglossal canal by a branch from the C1 and C2 loop of the cervical plexus. A descending branch from CN XII conveying these cervical spinal nerve fibers joins the ansa cervicalis and supplies the infrahyoid muscles.

Injury to the Hypoglossal Nerve

Injury to CN XII paralyzes the ipsilateral half of the tongue (Lange et al., 1995). After some time, the tongue atrophies, making it appear shrunken and wrinkled. When the tongue is protruded, its tip deviates toward the paralyzed side because of the unopposed action of the genioglossus muscle on the normal side of the tongue. ⊙

References and Suggested Readings

Barr ML, Kiernan JA: *The Human Nervous System: An Anatomical Viewpoint*, 6th ed. Philadelphia, JB Lippincott, 1993.

Bruce JN, Fetell MR: Tumors of the skull and cranial nerves. *In* Rowland LP (ed): *Merritt's Textbook of Neurology*, 9th ed. Baltimore, Williams & Wilkins, 1995.

Fetell MR: General considerations. *In* Rowland LP (ed): *Merritt's Textbook of Neurology*, 9th ed. Baltimore, Williams & Wilkins, 1995.

Haines DE: *Neuroanatomy: An Atlas of Structures, Sections, and Systems*, 4th ed. Baltimore, Williams & Wilkins, 1995.

Haines DE (ed): *Fundamental Neuroscience*. New York, Churchill Livingstone, 1997.

Haines DE, Mihailoff GA: An overview of the brainstem. *In* Haines DE (ed): *Fundamental Neuroscience*. New York, Churchill Livingstone, 1997.

Hutchins JB, Corbett JJ: The visual system. *In* Haines DE (ed): *Fundamental Neuroscience*. New York, Churchill Livingstone, 1997.

Lange DL, Trojaborg W, Rowland LP: Peripheral and cranial nerve lesions. *In* Rowland LP (ed): *Merritt's Textbook of Neurology*, 9th ed. Baltimore, Williams & Wilkins, 1995.

Rowland LP (ed): *Merritt's Textbook of Neurology*, 9th ed. Baltimore, Williams & Wilkins, 1995.

Sacco RL: Pathogenesis, classification, and epidemiology of cerebrovascular disease. *In* Rowland LP (ed): *Merritt's Textbook of Neurology*, 9th ed. Baltimore, Williams & Wilkins, 1995.

Swartz MH: *Textbook of Physical Diagnosis*, 2nd ed. Philadelphia, WB Saunders, 1994.

Sweazy RD: Olfaction and taste. *In* Haines DE (ed): *Fundamental Neuroscience*. New York, Churchill Livingstone, 1997.

Wazen JJ: Dizziness and hearing loss. *In* Rowland LP (ed): *Merritt's Textbook of Neurology*, 9th ed. Baltimore, Williams & Wilkins, 1995.

Willms JL, Schneiderman H, Algranati PS: *Physical Diagnosis: Bedside Evaluations of Diagnosis and Functions*. Baltimore, Williams & Wilkins, 1994.

Wilson-Pauwels L, Akesson EJ, Stewart PA: *Cranial Nerves: Anatomy and Clinical Comments*. Toronto, BC Decker, 1988.

Index

Note: Page numbers in italics indicate figures; page numbers followed by t indicate tables.

D

E

H

J

K

L

P